PHYSICAL CONSTANTS

Quantity	Symbol	Value
Universal gravitational constant	G	6.673×10^{-11} m^3/(kg$\cdot$s^2)
Speed of light in vacuum	c	2.998×10^8 m/s
Elementary charge	e	1.602×10^{-19} C
Planck's constant	h	6.626×10^{-34} J$\cdot$s
		4.136×10^{-15} eV$\cdot$s
	$\hbar = h/(2\pi)$	1.055×10^{-34} J$\cdot$s
		6.582×10^{-16} eV$\cdot$s
Universal gas constant	R	8.315 J/(mol$\cdot$K)
Avogadro's number	N_A	6.022×10^{23} mol^{-1}
Boltzmann constant	k_B	1.381×10^{-23} J/K
		8.617×10^{-5} eV/K
Coulomb force constant	$k = 1/(4\pi\epsilon_0)$	8.988×10^9 N$\cdot$m^2/C^2
Permittivity of free space (electric constant)	ϵ_0	8.854×10^{-12} C^2/(N$\cdot$m^2)
Permeability of free space (magnetic constant)	μ_0	$4\pi \times 10^{-7}$ T$\cdot$m/A
Electron mass	m_e	9.109×10^{-31} kg
		0.000 548 580 u
Electron rest energy	$m_e c^2$	0.5110 MeV
Proton mass	m_p	1.673×10^{-27} kg
		1.007 276 5 u
Proton rest energy	$m_p c^2$	938.272 MeV
Neutron mass	m_n	1.675×10^{-27} kg
		1.008 664 9 u
Neutron rest energy	$m_n c^2$	939.565 MeV
Compton wavelength	λ_C	2.426×10^{-12} m
Stefan-Boltzmann constant	σ	5.670×10^{-8} W/(m$^2\cdot$K^4)
Rydberg constant	R	1.097×10^7 m^{-1}
Bohr radius	a_0	5.292×10^{-11} m

College PHYSICS

Alan Giambattista
Cornell University

Betty McCarthy Richardson
Cornell University

Robert C. Richardson
Cornell University

McGraw Hill **Higher Education**

Boston   Burr Ridge, IL   Dubuque, IA   Madison, WI   New York   San Francisco   St. Louis
Bangkok   Bogotá   Caracas   Kuala Lumpur   Lisbon   London   Madrid   Mexico City
Milan   Montreal   New Delhi   Santiago   Seoul   Singapore   Sydney   Taipei   Toronto

The McGraw-Hill Companies

COLLEGE PHYSICS

Published by McGraw-Hill, a business unit of The McGraw-Hill Companies, Inc., 1221 Avenue
of the Americas, New York, NY 10020. Copyright © 2004 by The McGraw-Hill Companies, Inc.

Some ancillaries, including electronic and print components, may not be available to customers outside
the United States.

This book is printed on acid-free paper.

International   1 2 3 4 5 6 7 8 9 0 VNH/VNH 0 9 8 7 6 5 4 3
Domestic   1 2 3 4 5 6 7 8 9 0 VNH/VNH 0 9 8 7 6 5 4 3

ISBN 0–07–052407–6
ISBN 0–07–121462–3 (ISE)

Publisher: *Kent A. Peterson*
Sponsoring editor: *Daryl Bruflodt*
Developmental editor: *Mary E. Haas*
Marketing manager: *Debra B. Hash*
Lead project manager: *Jill R. Peter*
Senior production supervisor: *Laura Fuller*
Senior media project manager: *Stacy A. Patch*
Media technology producer: *Jeff Huettman*
Design manager: *Stuart Paterson*
Cover/interior designer: *Jamie E. O'Neal*
Cover image: *Stone*
Lead photo research coordinator: *Carrie K. Burger*
Photo research: *Karen Pugliano*
Supplement producer: *Brenda A. Ernzen*
Compositor: *Precision Graphics*
Typeface: *10/12 Times Roman*
Printer: *Von Hoffmann Corporation*

The credits section for this book begins on page 1137 and is considered an extension of the copyright page.

Library of Congress Cataloging-in-Publication Data

Giambattista, Alan.
  College physics / Alan Giambattista, Betty McCarthy Richardson, Robert C. Richardson. — 1st ed.
    p.   cm.
  Includes index.
  ISBN 0–07–052407–6 (set : acid-free paper) — ISBN 0–07–253724–8 (v. 1 : acid-free paper) —
ISBN 0–07–253725–6 (v. 2 : acid-free paper)
  1. Physics. I. Richardson, Betty McCarthy. II. Richardson, Robert C. (Robert Coleman), 1937–.
III. Title.

QC21.3 .G53   2004
530—dc21   2002038915
                                                           CIP

www.mhhe.com

ABOUT THE AUTHORS

Alan Giambattista grew up in Nutley, New Jersey. In his junior year at Brigham Young University he decided to pursue a physics major, after having explored math, music, and psychology. He did his graduate studies at Cornell University and has taught introductory college physics in one form or another for the past 20 years. When not found at the computer keyboard working on *College Physics,* he can often be found at the keyboard of a harpsichord or piano. He is a member of the Cayuga Chamber Orchestra and has given performances of the Bach harpsichord concerti at several regional Bach festivals. He met his wife Marion in a singing group and presently they live in an 1824 former parsonage surrounded by a dairy farm. Besides music and taking care of the house, gardens, and fruit trees, they love to travel together.

Betty McCarthy Richardson was born and grew up in Marblehead, Massachusetts, and tried to avoid taking any science classes after eighth grade but managed to avoid only ninth grade science. After discovering that physics tells how things work, she decided to become a physicist. She attended Wellesley College and did graduate work at Duke University. While at Duke, Betty met and married fellow graduate student Bob Richardson and had two daughters, Jennifer and Pamela. Betty began teaching physics at Cornell in 1977. Twenty-five years later, she is still teaching the same course, Physics 101/102, an algebra-based course with all teaching done one-on-one in a Learning Center. From her own early experience of math and science avoidance, Betty has empathy with students who are apprehensive about learning physics. Betty's hobbies include collecting old children's books and vintage toys, reading, enjoying music, travel, and dining with royalty. Betty's highlight during the Nobel Prize festivities in 1996 was being escorted to dinner at the Stockholm Royal Palace on the arm of King Carl XVI Gustav of Sweden and sitting between the king and the prime minister. Currently she is spending spare time pushing wooden trains around wooden tracks with grandson Jasper (the 1-m child in Chapter 1) and caring for the family cats.

Robert C. Richardson was born in Washington, D.C., attended Virginia Polytechnic Institute, spent time in the United States Army, and then returned to graduate school in physics at Duke University where his thesis work involved NMR studies of solid helium-3. In the fall of 1966 Bob began work at Cornell University in the laboratory of David M. Lee. Their research goal was to observe the nuclear magnetic phase transition in solid ^{3}He that could be predicted from Richardson's thesis work with Professor Horst Meyer at Duke. In collaboration with graduate student Douglas D. Osheroff, they worked on cooling techniques and NMR instrumentation for studying low-temperature helium liquids and solids. In the fall of 1971, they made the accidental discovery that liquid ^{3}He undergoes a pairing transition similar to that of superconductors. The three were awarded the Nobel Prize for that work in 1996. Bob is currently the F. R. Newman Professor of Physics and the Vice Provost for Research. He has been active in teaching various introductory physics courses throughout his time at Cornell. In his spare time he enjoys gardening and photography.

For Marion
 Alan

In memory of our daughter Pamela,
and for Jasper, Jennifer, and Jim Merlis
 Bob and Betty

CONTENTS IN BRIEF

CONTENTS

Part ONE

Mechanics

Part TWO

Thermal Physics

Part THREE

Electromagnetism

Part
FOUR

Electromagnetic Waves and Optics

Part
FIVE

Quantum and Particle Physics and Relativity

LIST OF SELECTED APPLICATIONS

Biology/Life Science

Forces supporting a water strider, Section 2.2
Forces on bird in air, Section 2.4
Function of tendons in the body, Section 2.7
Force exerted by deltoid muscle, Examples 4.3, 8.10
Traction applied to a leg, Example 4.5
Seagull dropping a clam, Section 4.5
Energy conversion in a jumping cricket, Section 6.6
Energy conversion in animal jumping, Section 6.6, Example 6.10
Flexor versus extensor muscles, Section 8.5
Forces on the human spine during heavy lifting, Section 8.5
Sphygmomanometer and blood pressure, Section 9.5
Floating and sinking of fish and animals, Example 9.8
Speed of blood flow, Example 9.9
Air circulation in underground burrows, Section 9.8
Arterial blockage, Example 9.12
Surface tension of alveoli in the lungs, Example 9.14
Tension and compression in bone, Example 10.2
Size limitations on organisms, Section 10.3
Comparison of walking speeds for various creatures, Example 10.10
Sound waves from a songbird, Example 12.2
Human ear, Section 12.5
Echolocation of bats and dolphins, Section 12.10
Medical applications of ultrasound, ultrasonic imaging, Section 12.10
Evolutionary advantages of warm-blooded versus cold-blooded animals, Section 13.7
Diffusion of oxygen into the bloodstream, Example 13.10
Using ice to protect buds from freezing, Section 14.5
Thermal radiation from the human body, Example 14.15
Electrolocation in fish, Section 16.4
Transmission of nerve impulses, Section 17.2, Example 17.11
Electrocardiogram (EKG), Section 17.2
Electroencephalogram (EEG), Section 17.2
Defibrillator, Example 17.12, Section 18.11
Magnetotactic bacteria, Section 19.1
Electromagnetic blood flow meter, Section 19.5
Magnetoencephalography, Section 20.3
Thermograms of the human body, Section 22.4
Fluorescence, Section 22.4
X-rays in medicine and dentistry, CAT scans, Section 22.4, Example 27.4
Navigation of bees, Section 22.8
Endoscope, Section 23.4
Human eye, Section 24.3

Correcting myopia/hyperopia, Section 24.3, Examples 24.4, 24.5
Iridescent colors in butterfly wings, Section 25.3
Resolution of the human eye, Section 25.8
Electron microscopes, Section 28.3
Laser surgery, Section 28.9
Radiocarbon dating, Section 29.4
Biological effects of radioactivity, Section 29.5
Radioactive tracers in medical diagnosis, Section 29.5
Positron emission tomography (PET) scans, Section 29.5
Radiation therapy, Section 29.5
Gamma knife radiosurgery, Section 29.5

Chemistry

Collision between krypton atom and water molecule, Example 7.10
Change of reaction rate with temperature, Example 13.8
Polarization of charge in water, Section 16.1
Current in electrolytes, Section 18.1
Spectroscopic analysis of elements, Section 27.6
Electron orbitals, electron configurations of atoms, Section 28.7
Periodic Table, Section 28.7

Geology/Earth Science

Forces on a submersible, Example 2.1
Angular speed of Earth, Example 5.1
Thunderclouds and lightning, Example 17.1, Section 17.6
Earth's magnetic field, Section 19.1, Examples 19.1, 19.2
Intensity of sunlight reaching the Earth, Example 22.6
Colors of the sky during the day and at sunset, Section 22.8
Rainbows, Section 23.3
Cosmic rays, Examples 26.2, 26.4
Radioactive dating of geologic formations, Section 29.4
Neutron activation analysis of geological objects, Section 29.5

Astronomy/Space Science

Spacecraft to Mars, Unit conversions, Section 1.5
Detection of extrasolar planets, Section 2.3
Speed of Hubble Telescope orbiting the Earth, Example 5.8
Kepler's laws of planetary motion, Sections 5.4, 8.8
Orbit of geostationary satellite, Examples 5.9, 6.7
Apparent weightlessness of orbiting astronauts, Section 5.7
Work on an orbiting satellite, Section 6.2
Escape speed from Earth, Example 6.9

Architecture

Technology/Machines

Nuclear fission reactors, Section 29.7
Nuclear fusion reactors, Section 29.8
High-energy particle accelerators, Section 30.4

Transportation

Motion of a train, Sections 3.1, 3.2
Velocity and acceleration of a motor scooter,
  Example 3.4
Acceleration of a sports car, Example 3.5
Coupling forces on freight cars, Example 3.6
Length of runway for airplane takeoff, Example 3.12
Airplane flight in a wind, Example 4.15
Rowing a boat in a current, Example 4.16
Angular speed of a motorcycle wheel, Example 5.3
Banked roadways, Section 5.3
Banking angle of an airplane, Section 5.3
Circular motion of stunt pilot, Example 5.14
Pulling a sled through snow, Example 6.2
Damage in a high-speed collision, Example 6.3
Power of a car climbing a hill, Example 6.12
Momentum of a moving car, Example 7.1
Force acting on a car passenger in a crash, Example 7.3
Analyzing speed of cars in a highway collision,
  Example 7.11
Torque on a spinning bicycle wheel, Example 8.3
Shock absorbers in a car, Section 10.9
Shock wave of a supersonic plane, Section 12.9
Air temperature changes in car tires, Example 13.5
Battery needed to start a car, Example 18.5
Maglev trains, Section 19.10
Bicycle generator, Example 20.2

Sports

Displacement in a relay race, Example 3.1
Terminal speed of skydivers, Example 3.15
Tension in a drawn bowstring, Example 4.4
Velocity and acceleration of an inline skater, Example 4.7
"Monkey and hunter" demonstration, Example 4.10
Bungee jumping, Example 6.4
Work done in drawing a bow, Sections 6.4,
  Example 6.5
Energy in a dart gun, Example 6.6
Speed of a downhill skier, Example 6.8
Rock climbers rappelling, Example 6.11
Elastic collision in a game of pool, Example 7.13
Muscle forces for the iron cross (gymnastics), Section 8.5
Rotational inertia of a figure skater, Section 8.8

Pressure on a diver, Example 9.3
Hot air balloons, Section 9.6
Compressed air tanks for a scuba diver, Example 13.6
Chill caused by perspiration, Section 14.5

Everyday Life

Buying clothes, Unit conversions, Example 1.5
Calibrating a scale, Example 2.2
Forces acting on a horse pulling a sleigh, Example 2.13
Forces acting on a mule, Example 3.2
Acceleration of a grocery cart, Example 3.8
Apparent weight in an elevator, Example 3.17
Pulling a sled to collect maple syrup, Example 4.1
Dragging a suitcase along the ground, Example 4.12
Whirling a stone in a sling, Section 5.1, Example 5.5
Circular motion of a CD, Example 5.4
Speed of roller-coaster car in vertical loop, Example 5.11
Circular motion of a potter's wheel, Examples 5.13, 8.5
Motion of a raft on a still lake, Example 7.5
Screen door closer, Example 8.4
Climbing a ladder on a slippery floor, Example 8.7
Pushing a file cabinet so it doesn't tip, Example 8.9
Torque on a grinding wheel, Example 8.11
Pressure exerted by high-heeled shoes, Example 9.1
Volume of an iceberg, Example 9.7
Household plumbing, Section 9.8
Stiffness of a wire, Example 10.1
Cutting action of a pair of scissors, Example 10.4
Sound of car horn in air and water, Example 11.5
Sound from a guitar, Section 12.1
Sound from a loudspeaker, Section 12.1
Sound intensity of a jackhammer, Example 12.3
Sources of musical sound, Section 12.4
Tuning a piano, Example 12.7
Heating a saucepan of water, Example 14.4
Offshore and onshore breezes, Section 14.7
Static charge from walking across a carpet, Example 16.1
Electrostatic charge of adhesive tape, Section 16.2
Colors from reflection and absorption of light,
  Section 23.1
Mirages, Section 23.3
Sparkle of a diamond, Section 23.4
Height needed for a full-length mirror, Example 23.5
Side-view mirrors on cars, Section 23.8, Example 23.7
Shaving or cosmetic mirrors, Section 23.8
Colors in soap films, oil slicks, Section 25.3
Neon signs, Section 27.6
Fluorescent dyes in laundry detergent, Section 27.7

PREFACE

College Physics is intended for a two-semester college course in introductory physics using algebra and trigonometry. Our main goals in writing this book are

- To present the basic concepts of physics that students need to know for later courses and future careers,
- To emphasize that physics is a tool for understanding the real world, and
- To teach transferable problem-solving skills that students can use throughout their lives.

We have kept these goals in mind while developing the main themes of the book.

COMPREHENSIVE COVERAGE

Students should be able to get the whole story from the book. The manuscript was tested for five semesters in a self-paced course, where students *must* rely on the text-book as their primary learning resource. Nonetheless, completeness and clarity are equally advantageous when the book is used in a more traditional classroom setting. Over a dozen other class tests have shown that *College Physics* frees the instructor from having to try to "cover" everything and from filling in the gaps left by the text. The instructor can then tailor class time to the students' needs, whether it be going over particularly difficult concepts, working through examples, engaging the students in cooperative learning activities, describing applications, or presenting demonstrations.

INTEGRATING CONCEPTUAL PHYSICS INTO A QUANTITATIVE COURSE

Some students approach introductory physics with the idea that physics is just the memorization of a long list of equations and the ability to plug numbers into those equations. We want to help students see that a relatively small number of basic physics concepts are applied to a wide variety of situations. In our presentation, based on years of teaching this course, we blend conceptual understanding with analytical skills, using plain language with simple mathematics throughout the book. In our experience, this approach anticipates many of the conceptual difficulties that students may have.

Physics education research has shown that students do not automatically acquire conceptual understanding; the concepts must be explained and the students given a chance to grapple with them. To that end, we include many **Conceptual Examples** and **Conceptual Practice Problems** in the text and a selection of **Conceptual Questions** and **Multiple Choice Questions** at the end of each chapter.

INTRODUCING CONCEPTS INTUITIVELY

We introduce key concepts and quantities in an informal way by establishing why the quantity is needed, why it is useful, and why it must be defined precisely. Then we make a transition from the informal, intuitive idea to a formal definition and name. We find that concepts motivated in this way are easier for students to grasp and remember than are concepts introduced by seemingly arbitrary, formal definitions.

For example, in Chapter 8, the idea of rotational inertia emerges in a natural way from the concept of rotational kinetic energy. Students can understand that a rotating rigid body has kinetic energy due to the motion of its particles. We discuss why it is useful to be able to write this kinetic energy in terms of a single quantity common to all the particles, rather than as a sum involving particles with many different speeds. Using this approach, the definition of rotational inertia is motivated; students understand why it is defined the way it is. Then, equipped with an understanding of rotational inertia, students are better prepared for the concept of torque.

We avoid as much as possible presenting definitions or formulas without any motivation. When an equation cannot be derived using algebra and trigonometry, we at least indicate where the equation comes from or give a plausibility argument. For example, Section 9.9 introduces Poiseuille's law with two identical pipes in series to show why the volume flow rate must be proportional to the pressure drop per unit length. Then we discuss why $\Delta V/\Delta t$ is proportional to the fourth power of the radius (rather than to R^2, as for an ideal fluid).

Similarly, we have found that the definitions of the displacement and velocity vectors seem arbitrary and counterintuitive to students if introduced without any motivation. Therefore, we precede any discussion of kinematic quantities with an introduction to Newton's laws, so students know that forces determine how the state of motion of an object changes. Then, when we define the kinematic quantities to give a precise definition of acceleration, we can apply Newton's second law quantitatively to see how forces affect the motion. We give particular attention to laying the groundwork for a concept when its name is a common English word such as *velocity* or *work*.

HELPING STUDENTS SEE THE RELEVANCE OF PHYSICS IN THEIR LIVES

Students in an introductory college physics course have a wide range of backgrounds and interests. We stimulate interest in physics by relating the principles to applications relevant to students' lives. We appeal to topics familiar to students and in line with their interests.

The text, examples, and end-of-chapter problems draw from the everyday world, from familiar technological applications, and from other fields such as biology, medicine, archaeology, astronomy, sports, environmental science, and geophysics. Applications in

the text are marked with an icon in the margin , **Making The Connection**.

In addition, the **Physics at Home** experiments give the students an opportunity to explore and see physics principles operate in their everyday lives. These activities are chosen for their simplicity and for the effective demonstration of physics principles. Each **Chapter Opener Vignette** is designed to capture student interest and maintain it through the chapter. The question asked in each opener is answered somewhere in the chapter.

WRITTEN IN CLEAR AND FRIENDLY STYLE

We have kept the writing down-to-earth and conversational in tone—the kind of language an experienced teacher uses when sitting at a table working one-on-one with a student. We believe students will find the book pleasant to read, informative, accurate without seeming threatening, and filled with analogies that make abstract concepts easier to grasp. We want students to feel confident that they can learn by studying the textbook.

While learning correct physics terminology is essential, we avoid all *unnecessary* jargon—terminology that just gets in the way of the student's understanding.

PROBLEM-SOLVING APPROACH

Problem-solving skills are central to an introductory physics course. We illustrate these skills in the example problems. Lists of problem-solving strategies are sometimes useful; we provide such strategies when appropriate. However, the most elusive skills—perhaps the most important ones—are subtle points that defy being put into a neat list. To develop real problem-solving expertise, students must learn how to think critically and analytically. Problem solving is a multidimensional, complex process; an algorithmic approach is not adequate to instill real problem-solving skills.

Strategy We begin each example with a discussion—in language that the students can understand—of the **Strategy** to be used in solving the problem. The strategy illustrates the kind of analytical thinking students must do when attacking a problem: How

"The major strength of this text is its approach, which makes students think out the problems, rather than always relying on a formula to get an answer. The way the authors encourage students to investigate whether the answer makes sense, and compare the magnitude of the answer with common sense is good also."—Dr. Jose D'Arruda, University of North Carolina, Pembroke

do I decide what approach to use? What laws of physics apply to the problem and which of them are *useful* in this solution? What clues are given in the statement of the question? What information is implied rather than stated outright? If there are several valid approaches, how do I determine which is the most efficient? What assumptions can I make? What kind of sketch or graph might help me solve the problem? Is a simplification or approximation called for? If so, how can I tell if the simplification is valid? Can I make a preliminary estimate of the answer? Only after considering these questions can the student effectively solve the problem.

Solution   Next comes the detailed **Solution** to the problem. Explanations are intermingled with equations and step-by-step calculations to help the student understand the approach used to solve the problem. We want the student to be able to follow the mathematics without wondering, "Where did that come from?"

Discussion   The numerical or algebraic answer is not the end of the problem; our examples end with a **Discussion**. Students must learn how to determine whether their answer is consistent and reasonable by checking the order of magnitude of the answer, comparing the answer to a preliminary estimate, verifying the units, and doing an independent calculation when more than one approach is feasible. When there are several different approaches, the Discussion looks at the advantages and disadvantages of each approach. We also discuss the implications of the answer—what can we learn from it? We look at special cases and look at "what if" scenarios. The Discussion sometimes generalizes the problem-solving techniques used in the solution.

Practice Problem   After each Example, a **Practice Problem** gives students a chance to gain experience using the same physics principles and problem-solving tools. By comparing their answers to those provided at the end of each chapter, they can gauge their understanding and decide whether to move on to the next section.

Our many years of experience in teaching college physics in a one-on-one setting has enabled us to anticipate where we can expect students to have difficulty. In addition to the consistent problem-solving approach, we offer several other means of assistance to the student throughout the text. A boxed problem-solving strategy gives detailed information on solving a particular type of problem, while an icon for problem-solving tips draws attention to techniques that can be used in a variety of contexts. A hint in a worked example or end-of-chapter problem provides a clue on what approach to use or what simplification to make. A warning icon emphasizes explanations that clarify possible points of confusion or common student misconceptions.

An important problem-solving skill that many students lack is the ability to extract information from a graph or to sketch a graph without plotting individual data points. Graphs often help students visualize physical relationships more clearly than they can do with algebra alone. We emphasize the use of graphs and sketches in the text, in worked examples, and in the problems.

USING APPROXIMATION, ESTIMATION, AND PROPORTIONAL REASONING

College Physics is up front about the constant use of simplified models and approximations in solving physics problems. One of the most difficult aspects of problem solving that students need to learn is that some kind of simplified model or approximation is usually required. We discuss how to know when it is reasonable to ignore friction or air resistance, treat *g* as constant, ignore viscosity, treat a charged object as a point charge, or ignore diffraction. A brief discussion of air resistance and terminal velocity in Chapter 3 enables us to discuss when it is reasonable to ignore air resistance—and also to show students that physics can account for these other effects.

"I understood the math, mostly because it was worked out step-by-step, which I like."—student, Bradley University

"The math was really clear. I was impressed with how easy the math and steps involved were to understand."—student, Bradley University

"The 'Strategy & Discussion' in each example were extremely helpful in understanding the ideas."—student, Houston Community College

Some Examples and Problems require the student to make an estimate—a useful skill both in physics problem solving and in many other fields. Similarly, we teach proportional reasoning as not only an elegant shortcut but also as a means to understanding patterns. We frequently use percentages and ratios to give students practice in using and understanding them.

SHOWCASING AN INNOVATIVE ART PROGRAM

Throughout the book we emphasize the value of drawing diagrams to help visualize and understand physical situations. We live up to our own advice by including many drawings, using color to distinguish different kinds of physical quantities so students can better see what concepts are important in each situation.

To help show that physics is more than a collection of principles that explain a set of contrived problems, in every chapter we have developed several innovative **Showcase Illustrations** to bring to life the connections between physics concepts and the complex ways in which they are applied. We believe these illustrations, with subjects ranging from three-dimensional views of electric field lines to the biomechanics of the human body and from representations of waves to the distribution of electricity in the home, will help students see the power and beauty of physics. We also provide many photographs to convey the application of physics to the real world and to help motivate many topics.

INNOVATIVE ORGANIZATION

There are a few places where, for pedagogical reasons, the organization of our text differs from that of most textbooks. The most significant reorganization is in the treatment of forces and motion. In *College Physics,* the central theme of Chapters 2–4 is *force.* Kinematics is woven into the fabric of the three force chapters as a tool to understand how forces affect motion. Overall, we spend less time on kinematics and more time on forces. This approach has these advantages:

- The first few chapters in any text set up student expectations that are hard to change later. If the course starts with a series of definitions of the kinematic quantities, with no explanation of *why* we are interested in those quantities, students may see physics as a series of equations to memorize and manipulate. Similarly, it is not a good precedent for students to learn all of the vector techniques at the outset when they are not yet needed.

- We explain to students that we develop the kinematic concepts so we can understand the effect of a net force on the motion of an object. Newton's second law is the key reason why we need a precise definition of acceleration; to define acceleration requires that we first define displacement and velocity. If the definitions of these quantities are imprecise, we cannot hope to understand how forces change the state of motion of an object. We have found that laying out this clear program at the start of the course helps students realize that in physics, physical concepts determine understanding, not mathematics.

- Learning kinematics first may suggest to students that physics is not connected to the real world. If they are told that objects all fall with the same acceleration—which they know from experience to be false—they learn not to trust the principles they're learning. With an understanding of forces and Newton's laws, *College Physics* enables the student to learn that constant acceleration is an approximation and to learn how to judge when that approximation is reasonable.

- A gradual introduction to vectors lets students acquire basic vector skills before moving on to more difficult ideas. We spread the vector material over three chapters, introducing vector techniques *as needed,* so students are not overloaded with formalism. We emphasize understanding how the physical vector quantities behave and can be analyzed, so that students are not required to learn the operations of an abstract mathematical entity before they have any inkling of how essential they are to problem solving in physics. It is our experience that by the time students reach Chapter 4, on forces and nonuniform motion in two dimensions, they have already attained a high level of proficiency with vector techniques (and with the conceptual framework of Newton's laws).

"I am thoroughly delighted to see a text which presents Newton's laws before kinematics. This in my opinion is an intuitively better approach for the students. I believe it is better to deal with why things move before we attempt to quantitatively describe the resulting motion. . . . Compared to the text we have been using, this approach is better."—Dr. Kelly Roos, Bradley University

"It was easier to understand these ideas since we came from reading about Newton's laws."—student, Bradley University

"I was concerned about having Newton's laws before kinematics until I read the chapters. After reading the chapters I think that this ordering is preferable to the more traditional order. . . . The approach that is taken with Newton's laws, introducing them at the beginning and spreading them out throughout the next three chapters is also very good. Understanding Newton's laws is one of the most difficult things that the students have to do in this class, and teaching them over and over again over a period of time is the best way to do it. I really liked the approach here."—Dr. Grant Hart, Brigham Young University

- We use correct vector terminology and methods from the very beginning. Even in one dimension, displacements, velocities, and accelerations are treated as vector quantities. Chapter 2 starts with the graphical addition of one-dimensional vectors. Chapter 3 introduces subtraction and component notation for one-dimensional vectors. For example, we write "$v_x = -5$ m/s" rather than "$v = -5$ m/s" when we are talking about a velocity component; we carefully distinguish components from magnitudes. Thus, the students carry over everything they learned about vectors in one dimension to two dimensions, without any change of notation. Some of our class testers report fewer students struggling with the concept of vector components using *College Physics* compared to their previous textbook.

- We begin in Chapter 2 with Newton's laws of motion so the students can build a solid conceptual framework in situations where the mathematics is simple. If forces were not introduced until Chapter 4, the students would have much less time to overcome conceptual difficulties associated with Newton's laws and would have much less practice applying them.

ACCURACY ASSURANCE

The authors and the publisher acknowledge the fact that inaccuracies can be a source of frustration for both the instructor and students. Therefore, throughout the writing and production of this textbook we have worked diligently to eliminate errors. We would like to describe the process used to assure the accuracy of this textbook.

Dr. Larry Rowan, of the University of North Carolina, checked the accuracy of all textual examples, practice problems and solutions, and end-of-chapter questions and problems in the second draft of the manuscript. Corrections were made to the final draft manuscript by the authors.

Dr. Richard Heinz, of Indiana University, and Dr. Marllin Simon, of Auburn University, then independently checked the accuracy of all textual examples and practice problems and solutions and worked all end-of-chapter questions and problems in the final draft of the manuscript.

Bill Fellers of Laurel Technology also conducted an independent accuracy check and worked all end-of-chapter questions and problems in the final draft of the manuscript. He then coordinated his efforts with Dr. Heinz, Dr. Simon, and the authors to resolve any discrepancies to ensure the accuracy of not only the text, but also the end-of-book answer section and the solutions manual. Corrections were then made to the manuscript before it was typeset.

The page proofs of the text were double-proofread against the manuscript to ensure the correction of any errors introduced when the manuscript was typeset. The textual examples, practice problems and solutions, end-of-chapter questions and problems, and problem answers were accuracy checked by Laurel Technology again at the page proof stage after the manuscript was typeset. This last round of corrections was then cross-checked against the solutions manual.

TO THE STUDENT: HOW TO USE THIS TEXTBOOK

Welcome! We hope that you enjoy your physics course. While studying physics does require hard work, in writing this book we have tried to remove the obstacles that sometimes make introductory physics unnecessarily difficult. We have also tried to show the beauty inherent in the principles of physics and how these principles are manifest all around you.

In our years of teaching experience, we have found that studying physics is a skill that must be learned. It's much more effective to *study* a physics textbook, which involves *active participation* on your part, rather than to read through passively. Even though active study takes more time initially, in the long run it will *save* you time; you learn more in one active study session than in three or four superficial readings. Here are some suggestions to help you get the most out of your study:

- Study the text *before* attending class on the same topic. Knowing what's in the book helps you take notes selectively rather than scrambling to write everything down. You'll be able to absorb more of the subtleties and difficult points from class. You'll also have some good questions to ask your instructor.

- Don't try to read an entire chapter in one sitting; study one or two sections at a time. It's difficult to maintain your concentration in a long session with so many new concepts and skills to learn.

- As you study, take particular note of these elements:

  Boxed laws, rules, and equations indicate the most important and central concepts.

   indicates a warning about a common student misconception or point of confusion.

   indicates a helpful problem-solving tip.

  Boxed problem-solving strategies give detailed information on solving a particular type of problem.

- When you come to an Example, pause after you've read the problem. Think about the strategy you would use to solve the problem. See if you can work through the problem on your own. Now study the Strategy, Solution, and Discussion in the textbook. Sometimes you will find that your own solution is right on the mark; if not, you can focus your attention on the place where you got stuck or any mistakes you may have made.

- Work the Practice Problem after each Example to practice applying the physics concepts and problem-solving skills you've just learned. Check your answer with the one given at the end of the chapter. If your answer isn't correct, review the previous section in the textbook to try to find your mistake.

- Take time to study the figures and graphs carefully.

- Find a study partner and get together regularly. Go over the difficulties either of you may be having. Take turns explaining things to each other—you learn a tremendous amount when you teach someone else. Compare your solutions to the Practice Problems. Discuss some of the Conceptual Questions from the end of the chapter.

- Try the Physics at Home experiments. They reinforce key physics concepts and help you see how these concepts operate in the world around you.

- Write your *own* chapter summary or outline and then compare it with the summary provided at the end of the chapter. This will help you identify the most important and fundamental concepts in each chapter.

We hope that these suggestions will help you get the most out of *College Physics*. We have spent many years working with students, both in the classroom and one-on-one in a self-paced course. We wrote this book so you could benefit from our experience as we carefully address the points that have caused difficulties for our students in the past. We also wish to share with you some of the pleasure and excitement we have found in learning about the physical laws that govern our world.

GUIDED TOUR

Forces and Motion Along a Line

A sailplane (or "glider") is a small, unpowered, high-performance aircraft. A sailplane must be initially towed a few thousand feet into the air by a small airplane, after which it relies on regions of upward-moving air such as thermals and ridge currents to ascend further. Suppose a small plane requires about 120 m of runway to take off by itself. When it is towing a sailplane, how much more runway does it need?

Chapter Opener

Each *Chapter Opener* includes a chapter opening photo and vignette designed to capture student interest and maintain it throughout the chapter.
The vignette describes the situation shown in the photo and asks students to consider the relevant physics. This question is then answered within the chapter.

Concepts & Skills to Review

- net force: vector addition (Section 2.4)
- freebody diagrams (Section 2.4)
- gravitational force (Section 2.5)
- internal and external forces (Section 2.4)

Concepts and Skills to Review

Presented on the first page of each chapter, this unique feature lists important material from previous chapters that students should understand before they start reading.

Illustrations and Page Layout to Facilitate and Reinforce Learning

Showcase Illustrations and Diagrammatic Illustrations

Showcase Illustrations and straightforward **Diagrammatic Illustrations** are used in combination to help students grasp the concepts. The Showcase Illustrations are designed to be very appealing in order to capture student interest and help them visualize difficult concepts. The Diagrammatic Illustrations are used to further explain and reinforce concepts.

Tendon

Biceps muscle (flexor)

Triceps muscle (extensor)

Insertion

$\mathbf{F_m}$

15°

CG

axis

12 cm

$\mathbf{F_s}$

$\mathbf{F_g}$

Reinforcement Notes

Various **Reinforcement Notes** appear in the margin to emphasize the important points in the text.

The symbol Δ stands for *the change in*. If the initial value of a quantity Q is Q_0 and the final value is Q_f, then $\Delta Q = Q_f - Q_0$. ΔQ is read "delta Q."

Equations

Important **equations** are numbered for ease in reference. Any equations that correspond to important laws are boxed for quick identification.

Displacement

$$\Delta \vec{\mathbf{r}} = \vec{\mathbf{r}}_f - \vec{\mathbf{r}}_0 \tag{3-1}$$

Rules or Laws

Statements of important physics **Rules** or **Laws** are highlighted within the text to help identify the most significant topics.

Newton's Laws of Motion (2-2)

1. If no forces act on an object, then its speed and direction of motion do not change. (If the object is at rest, it remains at rest with a speed of zero; if it is moving, it moves in a straight line with constant speed.)

2. A non-zero net force acting on an object causes its state of motion to change. Quantitatively,

$$\mathbf{F}_{net} = m\mathbf{a}$$

where $\mathbf{F}_{net}$ is the net force, m is the mass, and $\mathbf{a}$ is the acceleration (which measures the rate of change of the velocity.)

3. In an interaction between two objects, the forces that each exerts on the other are equal in magnitude and opposite in direction.

Consistent Problem-Solving Approach

Worked Example

A multipart *Quantitative* or *Conceptual Worked Example* appears when an important new concept or skill is introduced. A step-by-step approach is used for every worked example presented in the text.

Four-Step Problem-Solving Approach

1. Strategy
Each example begins with a discussion of the strategy to be used in solving the problem. The strategy illustrates the issues that should be considered when attacking a problem: How do we decide what approach to use? What laws of physics apply to the problem? What clues are given in the statement of the question? What assumptions can be made?

2. Solution
The strategy is followed by a numeric *Solution*. Explanations are intermingled with equations and step-by-step calculations to help the student understand the approach used to solve the problem.

Example 3.6

Coupling Force on First and Last Freight Cars

A train engine pulls out of a station along a straight track with five identical freight cars behind it, each of which weigh 90.0 kN. The train reaches a speed of 15.0 m/s within 5.00 min of starting out. Assuming the acceleration is constant, with what magnitude of force must the coupling between cars pull forward on the first and last of the freight cars? Ignore friction and air resistance. Assume $g = 9.80$ N/kg.

Strategy A sketch of the situation is shown in Fig. 3.15a. We can calculate the acceleration of the train from the initial and final velocities and the elapsed time. Then we can relate the acceleration to the net force using Newton's second law. To find the force of the first coupling, we can consider all five cars to be one system so that we do not have to worry about the force exerted on the first car by the second car. Once we identify a system, we draw a free-body diagram before applying Newton's second law.

Given: W = weight of each freight car = 90.0 kN = 9.00×10^4 N; v_x = 15.0 m/s at t = 5.00 min = 300 s; v_{0x} = 0 since the train starts from rest; a_x = constant

To find: tensions T_1 and T_5

Solution The acceleration of the train is
$$a_x = \frac{\Delta v_x}{\Delta t} = \frac{15.0 \text{ m/s}}{300 \text{ s}} = 0.0500 \text{ m/s}^2$$

First consider the last freight car (car 5). If we ignore friction and air resistance, the only forces acting are the force $\vec{T}_5$ due to the tension in the coupling, the normal force $\vec{N}$, and the car's weight $\vec{W}$; an FBD is shown in Fig. 3.15b. The normal force and the weight are vertical and in act opposite directions. They must be equal in magnitude; the vertical component of the net force is

zero since the vertical component of the acceleration is zero. Then the net force is equal to the tension in the coupling. The mass of the car is $m = W/g$, where $g = 9.80$ N/kg = 9.80 m/s². Then, from Newton's second law,
$$T_5 = \Sigma F_x = ma_x = \frac{W}{g} a_x$$
$$T_5 = 90.0 \times 10^3 \text{ N} \times \frac{0.0500 \text{ m/s}^2}{9.80 \text{ m/s}^2} = 459 \text{ N}$$

For the tension in the first coupling, consider the five cars as *one system*. Fig. 3.15c shows an FBD in which cars 1–5 are treated as a single object. Again, the vertical forces on the system add to zero. The only external horizontal force is the force $\vec{T}_1$ due to the tension in the first coupling. The mass of the system is five times the mass of one car. Therefore,
$$T_1 = \Sigma F_x = ma_x = (5 \times 90.0 \times 10^3 \text{ N}) \times \frac{0.0500 \text{ m/s}^2}{9.80 \text{ m/s}^2} = 2.30 \text{ kN}$$

Discussion The solution to this problem is much simpler when Newton's second law is applied to a system comprised of all five cars, rather than to each car individually. Although the problem can be solved by looking at individual cars, to find the tension in the first coupler you would have to draw five free-body diagrams (one for each car) and apply Newton's second law five times. That's because each car, except the fifth, is acted on by the unequal tensions in the couplers on either side. You'd have to first find the tension in the fifth coupler, then in the fourth, then the third, and so on.

Practice Problem 3.6 Coupling force between first and second freight cars
With what force does the coupling between the first and second cars pull forward on the second car? [*Hint:* Try two methods. One of them is to draw a FBD for the first car and apply Newton's *third* law as well as the second.]

Figure 3.15
(a) An engine pulling five identical freight cars. The entire train has a constant acceleration $\vec{a}$ to the right.
(b) FBD for car 5. (c) FBD for cars 1–5.

Hints

Hints appear within the text to help identify important information and to give clues on what approaches to use or laws to apply. These are plentiful in earlier chapters where students are learning to apply problem-solving techniques.

3. Discussion
The examples do not end with the solution. Students must also learn how to determine whether an answer is consistent and reasonable by checking order-of-magnitude, verifying the units, and doing an independent calculation when more than one approach is feasible. The *Discussion* also draws attention to the problem-solving techniques that were used in the solution.

4. Practice Problems
Every example is immediately followed by a *Practice Problem*. This allows students to further practice the problem-solving skills illustrated in the example and to reinforce what they have learned. The answer for each practice problem appears at the end of the chapter to allow students to check their work.

Warning Notes

An icon indicates a **Warning Note** that describes possible points of confusion or any common misconceptions that may apply to a particular concept.

Problem-Solving Tip

An icon points out **Problem-Solving Tips** that guide students in applying problem-solving techniques.

3.4 NEWTON'S SECOND LAW: FORCE AND ACCELERATION

According to Newton's second law, the acceleration of an object is proportional to the net force on it and is in the same direction. The larger the net force, the larger the acceleration. If the net force is zero, the acceleration is zero and the object moves with constant velocity—possibly zero velocity, but not necessarily. Newton's second law also says that the acceleration is inversely proportional to the object's mass. The same net force acting on two different objects causes a smaller acceleration on the object with greater mass. Mass is a measure of an object's inertia—the amount of resistance to *changes in velocity*.

 Mass and weight measure different physical properties. The mass of a body is a measure of its inertia, while weight is a measure of the gravitational force acting on it. Imagine taking a shuffleboard puck to the moon. Since the moon's gravitational field is weaker than the earth's, the puck's weight $\vec{W}$ would be smaller. A smaller normal force $\vec{N}$ would be required to hold it up. On the other hand, the puck's mass, an intrinsic property, is the same. Neglecting the effects of friction, an astronaut playing shuffleboard on the moon would have to exert the same horizontal force on the puck as on earth to give it the same acceleration (Fig. 3.14).

Newton's law relating net force and acceleration is

Newton's Second Law
$\vec{F}_{net} = m\vec{a}$ (3-8)

or

$$\Sigma\vec{F} = m\vec{a}$$

where Σ, the Greek capital letter sigma, stands for *the sum of*. $\Sigma\vec{F}$ means the sum of all the forces acting on a system. The order of the symbols in Eq. (3-8) does not reflect a cause-and-effect relationship; the net force causes the acceleration, not the other way around. The SI unit of force, the newton, is defined in terms of SI base units so that a 1-N net force acting on a 1-kg mass produces an acceleration of 1 m/s^2; therefore

$$1\,N = 1\,kg \cdot m/s^2$$

When calculating the net force on a system, only *external* forces need be considered. According to Newton's third law, internal forces always add to zero.

Problem-Solving Strategies

Problem-Solving Strategies give detailed information on solving a particular type of problem. These are supplied for the most fundamental physics rules and laws.

Problem-Solving Strategies for Newton's Second Law
•   Decide what objects will have Newton's second law applied to them.
•   Identify all the interactions affecting that object.
•   Draw a free-body diagram to show all the forces acting on the object.
•   Find the net force by adding the forces as vectors.
•   Use Newton's second law to relate the net force to the acceleration.

Applying Physics to the Real World

Physics at Home

Go to a balcony or climb up a ladder and drop a basket-style paper coffee filter (or a cupcake paper) and a penny simultaneously. Air resistance on the penny is negligible unless it is dropped from a very high balcony. At the other extreme, the effect of air resistance on the coffee filter is very noticeable; it reaches its terminal speed almost immediately. Stack several (two to four) coffee filters together and drop them simultaneously with a single coffee filter. Why is the terminal speed higher for the stack? Crumple a coffee filter into a ball and drop it simultaneously with the penny. Air resistance on the coffee filter is now reduced, but still noticeable.

Physics at Home

Throughout the chapters of this text students will find **Physics at Home** activities. These are designed to be performed individually at home to help students further understand and visualize certain physics concepts.

Making The Connection

Making The Connection: motion of a train

problem. The engineer wants to call the railroad office [...] he tell them where to find the train? He might say som[...] of the old trestle bridge." Notice that he uses a point of [...] Then he states how far the train is from that point and i[...] of the three pieces (the reference point, the distance, an[...] tion of the train's whereabouts is ambiguous.

The same thing is done in physics. First, we cho[...] **origin.** Then, to describe the location of something, we [...]

Making The Connection

The **Making The Connection** icon indicates areas in the text where physics can be applied to other areas in your students' lives. Familiar topics and interests are discussed in the text and examples, drawing from biology, archaeology, astronomy, sports, and the everyday world.

End-of-Chapter Material

Master the Concepts

The **Master the Concepts** section gives students a quick review of the chapter. A **Summary** lists important concepts and equations, while the

Highlighted Figures and Tables section identifies important ideas that are graphically represented within the chapter.

Multiple Choice Questions

The **Multiple Choice Questions** allow students to do a quick self-test and also provide practice for the type of questions found on the MCAT exam. Answers can be found at the end of the book.

Conceptual Questions

The **Conceptual Questions** are designed to test students' qualitative understanding of the key ideas within the chapter.

Problems

A large variety of **Problems** are given by chapter section to check the student's quantitative understanding of the chapter. **Comprehensive Problems** also allow students to test their overall understanding of the chapter.

Paired by Concept
Some problems are paired by concept, and their numbers are connected by a ruled box. Only one problem of the pair has an answer available at the end of the book and a solution available in the Student Solutions Manual.

Color Coded
Blue problem numbers indicate those with solutions available in the College Physics Student Solutions Manual.

Difficulty Level
Difficulty level is designated by ✦ for problems of intermediate difficulty and ✦✦ for more demanding problems.

Conceptual and Quantitative
A **ⓒ** indicates a problem that combines conceptual and quantitative problem solving.

---

acceleration when it falls at 75% of its

✦52. In free fall, we assume the accelerati
only is air resistance neglected, but
strength is assumed to be constant. (a)
an object fall to the earth's surface suc
field strength changes less than 1.0%
most cases, which do we have to worr
ance becoming significant or g changin

3.7 Apparent Weight

53. Refer to Example 3.17. What is the a
same passenger (weighing 598 N) in th
In each case, the magnitude of the ele
0.50 m/s^2. (a) After having stopped at
ger pushes 8th floor; the elevator is beg
ward. (b) Elevator is slowing down as i

ⓒ 54. You are standing on a bathroom scale i
weight is 140 lb, but the reading o
(a) What is the magnitude and directio

Supplements

Multimedia Supplements

Instructor's Presentation CD

The *Instructor's Presentation CD* is a collection of visual resources that allows instructors to utilize elements from the text such as:
- artwork
- worked examples
- tables
- interactive applets

With the help of these valuable resources, you can create:
- customized classroom presentations
- visually based tests and quizzes
- dynamic course website content
- attractive printed support materials

The digital assets on this cross-platform CD-ROM are grouped by chapter within easy-to-use folders.

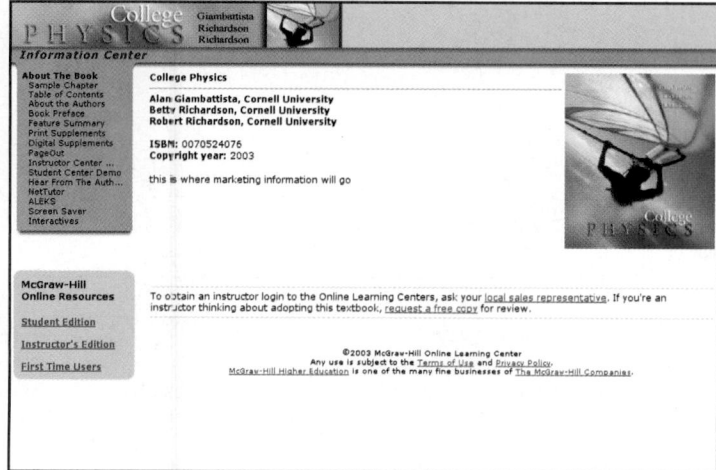

Online Learning Center (OLC)

The *Online Learning Center* is an online repository for teaching and learning aids. It houses downloadable and printable versions of traditional ancillaries plus a wealth of online content, some of which includes:

Instructor Center

Homework Systems
- CAPA
- Web Assign
- GradeSummit

Course Management Systems
- PageOut
- WebCT

Student Center
- Quizzes
- Interactive learning games
- Applets
- Power web

- Study tools that correlate to each chapter of the text

Supplements

Multimedia Supplements

Physics Interactive Applets

McGraw-Hill is proud to bring you an assortment of outstanding *Interactive Applets* like no other. These "Interactives" offer a fresh and dynamic method to teach the physics basics by providing students with applets that are *completely accurate and work with real data*. Interactives allow students to manipulate parameters and gain a better understanding of 20 of the more difficult physics topics by watching the effect of these manipulations. Each Interactive includes:

- Analysis Tool (interactive model)
- Tutorial describing its function
- Content describing its principle themes

Students will easily be able to jump between the Tutorial and the Analysis Tools with just the click of the mouse.

Instructor's Testing and Resource CD

The *Instructor's Testing and Resource CD* contains the *College Physics* Testbank within the Brownstone DIPLOMA© test generator as well as the Word and PDF files of the:

- worked examples and end-of-chapter problems from the text
- multiple-choice version of the end-of-chapter problems
- solutions to the end-of-chapter problems from the Instructor's Resource Guide
- the testbank

The Brownstone software includes a(n):

- test generator
- online testing program
- Internet testing site
- Grade Management System

This user-friendly software's testing capability is consistently ranked number one in evaluations over other products. The Word files for the end-of-chapter problems and the solutions can be used in combination with the Brownstone software or independently.

ALEKS®

Help students master the math skills needed to understand difficult physics problems.

ALEKS (Assessment and LEarning in Knowledge Spaces) is an artificial intelligence-based system for individualized math learning available via the World Wide Web.

ALEKS is . . .

- A robust course management system
  It tells you exactly what your students know and don't know.

- Focused and efficient
  Enables students to quickly master the math needed for college physics.

- Artificial Intelligence
  It totally individualizes assessment and learning.

- Customizable
  Click on or off each course topic.

- Web Based
  Use a standard browser for easy Internet access.

- Inexpensive
  There are no setup fees or site license fees.

ALEKS® is a registered trademark of ALEKS Corporation.

Printed Supplements

Instructor's Resource Guide

The *Instructor's Resource Guide* includes many unique assets for instructors, such as demonstrations, suggested reform ideas from physics education research, and ideas for incorporating Just-In-Time teaching techniques. It also includes answers to the end-of-chapter conceptual questions and complete, worked-out solutions for all the end-of-chapter problems from the text.

Transparencies

This boxed set of 300 full-color *Transparencies* feature images from the text. The images have been modified to ensure maximum readability in both small and large class-room settings.

Testbank

The *College Physics Testbank* is also available in a printed version. The testbank includes over 2000 test questions in multiple-choice format at a variety of difficulty levels.

Student Solutions Manual

The *Student Solutions Manual* contains complete, worked-out solutions to selected end-of-chapter problems from the text.

For more information, contact your McGraw-Hill sales representative at (800) 338-3987, or email www.mhhe.com.
To locate your sales representative, go to www.mhhe.com/catalogs/rep/.

ACKNOWLEDGMENTS

We would like to express our gratitude to the faculty, staff, and students at Cornell University who helped us in a myriad of ways. We especially thank our friend and colleague Bob Lieberman who shepherded us through the process as our literary agent and who inspired us as an exemplary physics teacher. Donald F. Holcomb, Persis Drell, Peter Lepage, and Phil Krasicky read portions of the manuscript and provided us with many helpful suggestions. Raphael Littauer contributed many innovative ideas and served as a model of a highly creative, energetic teacher.

We are indebted to all those who helped us class test the manuscript for five semesters at Cornell: David G. Cassel, Robert M. Cotts, Douglas B. Fitchen, Richard Galik, Robin Hughes, Jeevak Parpia, Joseph Rogers, Janet Scheel, many capable graduate teaching assistants, and, above all, the students in Physics 101 and 103. We also appreciate all the work contributed by our support staff of Bridget Meeds, Karen Fleckenstein, Rosemary French, and Margaret Slaght.

We acknowledge a great debt to David Chelton, who has been aptly called "the best physics developmental editor on the planet," for his expert guidance and sense of humor.

We are grateful for the guidance, patience, and enthusiasm of Daryl Bruflodt and Mary Haas at McGraw-Hill, whose tireless efforts were invaluable in bringing this project to fruition. We would like to thank the entire team of talented and hard-working professionals assembled by McGraw-Hill to publish this book, including Karen Allanson, Carrie Burger, David Dietz, Bill Fellers, Suzanne Guinn, Terri Hamer, Debra Hash, Kevin Kane, Mary Klein, Judith Kromm, Jenni Lang, John Paul Lenney, Richard Mickey, Jay Oberbroeckling, Stuart Paterson, Jill Peter, Kent Peterson, Karen Pugliano, Denise Schank, Pat Steele, and surely many others whose expertise and hard work have contributed to making the book a reality.

We are grateful to Bill Fellers, Richard Heinz, Larry Rowan, and Marllin Simon for accuracy checking the manuscript and for their many helpful suggestions.

Our thanks also go to Janet Scheel, Warren Zipfel, Rebecca Williams, and Mike Nichols for contributing some of the medical and biological applications, to Kyle Dawson and Tom Glickman for their work on the tutorials, and to Nick Taylor and Jason Marshall for writing solutions to the Conceptual Questions.

From Alan: I owe an immense debt of gratitude to Marion, Katie, Charlotte, Julia, and Denisha, without whose love, support, encouragement, and patience this book could never have been written.

From Bob and Betty: We thank our daughter Pamela's classmates and friends at Cornell and in the Vanderbilt Master's in Nursing program who were an early inspiration for the book, and we thank Dr. Philip Massey who was very special to Pamela and is dear to us. We thank our friends at *blur*, Alex, Damon, Dave, and Graham, who love physics and are inspiring young people of Europe to explore the wonders of physics through their work on the European Space Agency's *Beagle2* Mars Express mission. Finally we thank our daughter Jennifer, our grandson Jasper, and son-in-law Jim who endured our protracted hours of distraction while this book was being written.

REVIEWERS, CLASS TESTERS, AND ADVISORS

Valued Input Went into Striving to Meet Your Needs   The content of this first edition text reflects an extensive effort to evaluate the needs of college physics instructors and students, to learn how well we met those needs, and to make improvements where we fell short. The market feedback we received consisted of information gathered from numerous reviews, class tests, focus groups, and a board of advisors.

The primary stage of our market research began with commissioning reviews from instructors across the United States and Canada. We asked them to compare our manuscript to their current text and to submit suggestions for improvement on areas such as content, organization, illustrations, and ancillaries. These over 180 individual reviews constituted the major basis for the revision plan at each stage of the four drafts of manuscript.

The draft manuscript was not only class tested by the authors in their course at Cornell University, but also by thirteen other separate institutions. We solicited feedback from both professors and students, and the authors incorporated this information into their final revision of the manuscript.

We also organized focus groups in six U.S. cities in 2001 and 2002. Participants reviewed draft manuscript in comparison to the books they were using and supplied further input on improvements to the manuscript and ways in which we as publishers could help to improve the content of the college physics course.

Finally, we formed two boards of advisors consisting of professors from regions around the country. We relied on these seasoned instructors to supply opinions during the continued development of the text on discrepancies in reviewer comments, instructional design decisions, the quality of the photo and illustration program, and the content of the print and media ancillary packages.

Considering the total sum of these opinions, this text now embodies the collective knowledge, insight, and experience of more than 200 college physics instructors. Their influence can be seen in everything from the content, accuracy, and organization of the text to the quality of the illustrations.

We are grateful to these instructors for their thoughtful comments and advice.

Reviewers

Wathiq Abdul-Razzaq
West Virginia University

Maris Abolins
Michigan State University

B. N. Narahari Achar
University of Memphis

Brian Adrian
Bethany College

Zaven Altounian
McGill University

Farhang Amiri
Weber State University

Anjum Ansari
University of Illinois at Chicago

Sanjeev Arora
Fort Valley State University

Martina E. Bachlechner
West Virginia University

James R. Benbrook
University of Houston

Cornelius Bennhold
The George Washington University

Kyle A. Beran
Saint Mary College

Michael S. Berger
Indiana University

Nelson E. Bickers
University of Southern California

David L. Bixler
Angelo State University

Gary S. Blanpied
University of South Carolina

Donald E. Bowen
Stephen F. Austin
State University

Michael J. Bozack
Auburn University

Matthew E. Briggs
University of
Wisconsin–Madison

James J. Carroll
Youngstown State University

Gerald B. Cleaver
Baylor University

Sergio Conetti
University of Virginia

Ronald M. Cosby
Ball State University

Stephen R. Cotanch
North Carolina State
University

Michael Crescimanno
Youngstown State University

Pengcheng Dai
University of Tennessee, Knoxville

José D'Arruda
University of North
Carolina–Pembroke

Biman Das
State University of New York College
at Potsdam

Regina Demina
Kansas State University

Peter Douglas
University of Manitoba

Miles J. Dresser
Washington State University

Laurencin Dunbar
St. Louis Community College at
Florissant Valley

John Dykla
Loyola University–Chicago

Donald Elliott
Carroll College

Heidi Fencl
University of Wisconsin–Green Bay

Michael Fisher
Ohio Northern University

James Garner
University of North Florida

Lev Gasparov
University of North Florida

John Gault
Missouri Valley College

Simon George
California State University–Long
Beach

David Gerdes
University of Michigan

Michael Giangrande
Oakland Community College

Martin Guthold
Wake Forest University

Grant W. Hart
Brigham Young University

Richard Holman
Carnegie Mellon University

Keith R. Honey
West Virginia University Institute of Technology

Joey Huston
Michigan State University

Michael Jackson
University of Wisconsin–LaCrosse

Anne Johnson
Lorain County Community College

J. Bruce Johnson
Arkansas State University

Brian Jones
Colorado State University

Sanford Kern
Colorado State University

Richard Krantz
Metropolitan State College of Denver

Kwong H. Lau
University of Houston

Hans Laue
University of Calgary

David Lederman
West Virginia University

Paul L. Lee
California State University, Northridge

Bryan H. Long
Columbia State Community College

Rafael Lopez-Mobilia
University of Texas at San Antonio

Alfredo A. Louro
University of Calgary

Donald G. Luttermoser
East Tennessee State University

Igor Lyuksyutov
Texas A&M University

Dan MacIsaac
Northern Arizona University

Phil Matheson
Salt Lake Community College

Mark E. Mattson
James Madison University

Bill W. Mayes II
University of Houston

Marles L. "Mac" McCurdy
Tarrant County College, Northeast

Arthur R. McGurn
Western Michigan University

Charles R. Meitzler
Sam Houston State University

John Andrew Milsom
University of Arizona

Lisa Morris
Whitworth College

Richard Mowat
North Carolina State University

O. David Mylander
California State Polytechnic University, Pomona

Blaine E. Norum
University of Virginia

David P. Norwood
Southeastern Louisiana University

Ed Oberhofer
Lake Sumter Community College

Michael J. O'Shea
Kansas State University

Antonio Pagnamenta
University of Illinois at Chicago

James L. Pazun
Pfeiffer University

Ralph L. Place
Ball State University

Steven J. Pollock
University of Colorado, Boulder

E. W. Prohofsky
Purdue University

Oren Quist
South Dakota State University

Syed Rabbani
Central Texas College

George W. Rainey
California State Polytechnic University, Pomona

Michele Rallis
Ohio State University

Gordon P. Ramsey
Loyola University–Chicago

David D. Reid
Eastern Michigan University

William M. Robertson
Middle Tennessee State University

Ian Robinson
University of Illinois at Urbana–Champaign

Michael W. Roth
University of Northern Iowa

Ira Z. Rothstein
Carnegie Mellon University

John P. Rutherford
University of Arizona

Earl E. Scime
West Virginia University

Rob Sells
Alfred State College

Wesley Shanholtzer
Marshall University

Bartlett M. Sheinberg
Houston Community College System

William C. Shockley Jr.
University of Texas–Pan American

Michael A. Shupe
University of Arizona

Chandralekha Singh
University of Pittsburgh

David A. Slimmer
Lander University

Mark H. Slovak
Louisiana State University

Xiang-Ning Song
Richland College

Sudha Srinivas
Central Michigan University

Noel Stanton
Kansas State University

Jeffrey Sundquist
Palm Beach Community College

Wilfred R. Theisen
St. John's University

J. Marshall Thomsen
Eastern Michigan University

Lynn P. Thomson
Brigham Young University–Idaho

Stephen B. Turcotte
Brigham Young University–Idaho

Ashok Vaseashta
Marshall University

Michael F. Weber
Brigham Young University–Hawaii

Martha Riherd Weller
Middle Tennessee State University

John Wernegreen
Eastern Kentucky University

C. Steven Whisnant
James Madison University

Daniel Whitmire
University of Louisiana–Lafayette

Arthur W. Wiggins
Oakland Community College

Luc T. Wille
Florida Atlantic University

Suzanne Willis
Northern Illinois University

Robert J. Wilson
Colorado State University

Gary M. Wysin
Kansas State University

Sanichiro Yoshida
Southeastern Louisiana University

Michael K. Young
Santa Barbara City College

Gary R. Zauft
University of Wisconsin–Barron County

Darin T. Zimmerman
Penn State Altoona

Class Test Schools

Robert W. Arts
Pikeville College

Karla Bier
Stephens College

Regina Demina
Kansas State University

Sharmanthie Fernando
Northern Kentucky University

Esmail Hejazifar
Wilmington College

Susanne M. Lee
State University of New York, Albany

Ntungwa Maasha
Coastal Georgia Community College

Ponn Maheswaranathan
Winthrop University

Kelly R. Roos
Bradley University

Bartlett M. Sheinberg
Houston Community College System

David A. Slimmer
Lander University

Richard Summers
Reinhardt College

Richard J. Zabelka
Lake Superior State University

Focus Group Participants

Ricardo Alarcon
Arizona State University

Wolfgang Bauer
Michigan State University

Cornelius Bennhold
The George Washington University

Nancy Beverly
Mercy College

Donald E. Bowen
Stephen F. Austin State University

Matthew E. Briggs
University of Wisconsin–Madison

Myron Campbell
University of Michigan

John P. Cise
Austin Community College–Rio

Gerald B. Cleaver
Baylor University

J. M. Collins
Marquette University

Pengcheng Dai
University of Tennessee, Knoxville

Regina Demina
Kansas State University

Renee D. Diehl
Penn State University

Miles J. Dresser
Washington State University

John Dykla
Loyola University–Chicago

Robert Ehrlich
George Mason University

Jerry Feldman
The George Washington University

Nicholas Giordano
Purdue University

Gary Gladding
University of Illinois at Urbana–Champaign

Richard M. Heinz
Indiana University

Paul Holody
Henry Ford Community College

Kwong Lau
University of Houston

Susanne M. Lee
State University of New York, Albany

Rafael Lopez-Mobilia
University of Texas at San Antonio

Igor Lyuksyutov
Texas A&M University

Ponn Maheswaranathan
Winthrop University

David E. Meltzer
Iowa State University

David P. Norwood
Southeastern Louisiana University

Albert Osei
Oakwood College

Antonio Pagnamenta
University of Illinois at Chicago

Mark Plano Clark
Doane College

Gordon P. Ramsey
Loyola University–Chicago

David D. Reid
Eastern Michigan University

Carl Rotter
West Virginia University

Larry Rowan
University of North Carolina at Chapel Hill

Arthur G. Schmidt
Northwestern University–Illinois

Bartlett M. Sheinberg
Houston Community College System

Marllin Simon
Auburn University

Mark H. Slovak
Louisiana State University

Sudha Srinivas
Central Michigan University

Richard Summers
Reinhardt College

Linn Van Woerkom
Ohio State University

Gary Westfall
Michigan State University

Suzanne Willis
Northern Illinois University

Board of Advisors

A. Lewis Ford
Texas A&M University

Nicholas Giordano
Purdue University

Richard M. Heinz
Indiana University

Klaus Honscheid
Ohio State University

Susanne M. Lee
State University of New York, Albany

Alan Nathan
University of Illinois at Urbana–Champaign

Richard W. Robinett
Penn State University

Larry Rowan
University of North Carolina–Chapel Hill

Marllin Simon
Auburn University

Media Board of Advisors

Regina Demina
  Kansas State University

Grant Hart
  Brigham Young University

Brian Jones
  Colorado State University

Dale Sayers
  North Carolina State University

CONTRIBUTORS

Of course, we are deeply indebted to:

Professor Richard Heinz of Indiana University, Professor Larry Rowan of the University of North Carolina, and Professor Marllin Simon of Auburn University for their accuracy review of the *College Physics* manuscript;

Professor Suzanne Willis of Northern Illinois University and Professor Susanne M. Lee, Assistant Professor of Physics, University at Albany, SUNY, for their authoring of the instructor resources and demonstrations in the *College Physics* Instructors' Resource Guide;

Professor Larry Rowan of the University of North Carolina and Professor Edward Oberhofer of Lake Sumter Community College for their authoring of the *College Physics* Testbank;

Professor Marllin Simon of Auburn University for his contributions to Share Your Favorite Labs and for authoring the chapter practice quizzes on the *College Physics* Online Learning Center.

Introduction

Science fiction movies, like the *Star Wars* series, are a lot of fun to see. Some of the most exciting scenes involve battles taking place in outer space between the forces for good (Rebel alliance) and the forces for evil (Imperial storm troopers). When the rebels aboard the *Millennium Falcon* blow up a Death Star in *Return of the Jedi,* they see the bright flash and hear the loud roar of the explosion. Is that really what would happen?

Concepts & Skills to Review

• algebra, geometry, and trigonometry (Appendix A)

1.1   WHY STUDY PHYSICS?

Physics is the branch of science that describes matter, energy, space, and time at the most fundamental level. Whether you are planning to study biology, architecture, medicine, music, chemistry, or art, there are principles of physics that are relevant to your field.

Physicists look for patterns in the physical phenomena that occur in the universe. They try to explain what is happening and perform experiments to see if the proposed explanation is valid. The goal is to find the most basic laws that govern the universe and to formulate those laws in the most precise and simplified way possible.

The study of physics is valuable for several reasons:

• Since physics describes matter and its basic interactions, all natural sciences are built on a foundation of the laws of physics. A full understanding of chemistry requires a knowledge of the physics of atoms. A full understanding of biological processes in turn is based on the underlying principles of physics and chemistry. Centuries ago, the study of *natural philosophy* encompassed what later became the separate fields of biology, chemistry, geology, astronomy, and physics. Today there are scientists who call themselves biophysicists, chemical physicists, astrophysicists, and geophysicists, demonstrating how thoroughly the sciences are intertwined.

• In today's technological world, many important devices can be understood correctly only with a knowledge of the underlying physics. Just in the medical world, think of laser surgery, magnetic resonance imaging, instant-read thermometers, x-ray imaging, radioactive tracers, heart catheterizations, sonograms, pacemakers, microsurgery guided by optical fibers, ultrasonic dental drills, and radiation therapy.

• By studying physics, you acquire skills that are useful in other disciplines. These include thinking logically and analytically; solving problems; making simplifying assumptions; constructing mathematical models; using valid approximations; and making precise definitions.

• Society's resources are limited, so it is important to use them in beneficial ways and not squander them on scientifically impossible projects. Political leaders and the voting public are too often led astray by a lack of understanding of scientific principles. Can a nuclear power plant supply energy safely to a community? What is the truth about the greenhouse effect, the ozone hole, and the danger of radon in the home? By studying physics, you learn some of the basic scientific principles and acquire some of the intellectual skills necessary to ask probing questions and to formulate informed opinions on these important matters.

• Finally, by studying physics, we hope that you develop a sense of the beauty of the fundamental laws governing the universe. Physicists are driven to find explanations for the phenomena around them. Intellectual satisfaction accompanies a deeper understanding of some of these principles.

We asked if the bright flash and loud roar of the explosion could be seen and heard by the Rebel heroes on board their spaceship. A student of physics watching the film knows that light can easily travel through space—which is why sunlight, moonlight, and starlight can reach Earth. On the other hand, sound can only travel through a physical medium, such as the air in Earth's atmosphere. The near-vacuum of space is not an adequate medium to carry sound waves that can be heard by humans. The Rebel heroes would not hear the explosion.

The study of physics can deepen your understanding of what you observe in the real universe as well as in the imaginary universe of a science-fiction movie.

1.2   TALKING PHYSICS

Some of the words used in physics are familiar from everyday speech. This familiarity can be misleading, since the scientific definition of a word may differ considerably from its common meaning. In physics, words must be precisely defined so that anyone reading a scientific paper or listening to a science lecture understands exactly what is meant. Some of the basic defined quantities, whose names are also words used in everyday speech, include time, length, force, velocity, acceleration, mass, and temperature.

In everyday language, *speed* and *velocity* are synonyms. In physics, there is an important distinction between the two. As defined in Chapter 3, *velocity* includes the *direction* of travel as well as the distance traveled per unit time. *Acceleration* is another word that has a distinct meaning in physics; it can mean to speed up or to slow down or to maintain constant speed while changing direction. The scientific definition of acceleration includes an associated direction. In everyday use, acceleration means an increase in speed while another word, deceleration, is used for a reduction in speed. This is fine for conversational purposes, but the precise definition of acceleration, including the direction, is required for the term to be useful in physics.

There are two important reasons for the way in which we define physical quantities. First, physics is an experimental science. The results of an experiment must be stated unambiguously so that other scientists can perform similar experiments and compare their results. Quantities must be defined precisely to enable experimental measurements to be uniform no matter where they are made. Second, physics is a mathematical science. We use mathematics to quantify the relationships among physical quantities. These relationships can be expressed mathematically only if the quantities being investigated have precise definitions.

1.3   THE USE OF MATHEMATICS

A working knowledge of algebra, trigonometry, and geometry is essential to the study of introductory physics. Some of the more important mathematical tools are reviewed in Appendix A. If you know that your mathematics background is shaky, you might want to test your mastery by doing some problems from a math textbook.

Mathematical equations are shortcuts for expressing concisely in symbols relationships that are cumbersome to describe in words. Algebraic symbols in the equations stand for quantities that consist of numbers *and units*. The number represents a measurement and the measurement is made in terms of some standard; the unit indicates what standard is used. In physics, a number to specify a quantity is useless unless we know the unit attached to the number. When buying silk to make a sari, do we need a length of 5 millimeters, 5 meters, or 5 kilometers? Is the term paper due in 3 minutes, 3 days, or 3 weeks? Systems of units are discussed in Section 1.5.

In the language of physics, the word **factor** is used frequently, often in a rather idiosyncratic way. If the power emitted by a radio transmitter has doubled, we might say that the power has "increased by a factor of two." If the concentration of sodium ions in the bloodstream is half of what it was previously, we might say that the concentration has "decreased by a factor of two," or, in a blatantly inconsistent way, someone else might say that it has "decreased by a factor of one-half." The *factor* is the number by which a quantity is multiplied or divided when it is changed from one value to another. In other words, the factor is really a ratio. In the case of the radio transmitter, if P_0 represents the initial power and P represents the power after new equipment is installed, we write

$$\frac{P}{P_0} = 2$$

It is also common to talk about "increasing 5%" or "decreasing 20%." If a quantity increases $n\%$, that is the same as saying that it is multiplied by a factor of $1 + (n/100)$. If a quantity decreases $n\%$, then it is multiplied by a factor of $1 - (n/100)$. For example, an increase of 5% means something is 1.05 times its original value and a decrease of 4% means it is 0.96 times the original value.

Physicists talk about increasing "by some factor" because it often simplifies a problem to think in terms of **proportions**. When we say that A is proportional to B (written $A \propto B$), we mean that if B increases by some factor, then A must increase by the same factor. For instance, the circumference of a circle equals 2π times the radius: $C = 2\pi r$. Therefore $C \propto r$. If the radius doubles, the circumference also doubles. The area of a circle is proportional to the *square* of the radius ($A = \pi r^2$, so $A \propto r^2$). The area must increase by the same factor as the radius *squared*, so if the radius doubles, the area increases by a factor of $2^2 = 4$.

Example 1.1

Effect of Increasing Radius on the Volume of a Sphere

The volume of a sphere is given by the equation

$$V = \tfrac{4}{3}\pi r^3$$

where V is the volume and r is the radius of the sphere. If the radius of the sphere is increased by a factor of 3, by what factor does the volume of the sphere change?

Strategy The problem gives us the ratio of the new radius and the old radius:

$$\frac{r_2}{r_1} = 3$$

The subscripts help us keep track of which sphere's volume and radius we mean; the radius of the original sphere is r_1 and the radius of the new sphere is r_2. Since $\frac{4}{3}$ and π are constants, we can work in terms of proportions.

Solution The volume of a sphere is proportional to the cube of its radius:

$$V \propto r^3$$

Since the radius increased by a factor of three, and volume is proportional to the cube of the radius, the new volume should be bigger by a factor of $3^3 = 27$.

Discussion A slight variation on the solution is to write out the proportionality in terms of ratios of the corresponding sides of the two equations:

$$\frac{V_2}{V_1} = \frac{\tfrac{4}{3}\pi r_2^3}{\tfrac{4}{3}\pi r_1^3} = \left(\frac{r_2}{r_1}\right)^3$$

Substituting the ratio of r_2 to r_1 yields

$$\frac{V_2}{V_1} = 3^3 = 27$$

which says that V_2 is 27 times V_1.

Practice Problem 1.1 Power dissipated by a lightbulb

The electrical power P dissipated by a lightbulb of resistance R is $P = V^2/R$, where V represents the line voltage. During a brownout, the line voltage is 10.0% less than its normal value. How much power is drawn by a lightbulb during the brownout if it normally draws 100.0 watts? Assume that the resistance does not change.

 There are not enough letters in the alphabet to assign a unique letter to each quantity. The same letter V can represent volume in one context and voltage in another. Avoid attempting to solve problems by picking equations that seem to have the correct letters; a skilled problem-solver understands what quantity each symbol in a particular equation represents.

1.4   SCIENTIFIC NOTATION AND SIGNIFICANT FIGURES

In physics, we deal with some numbers that are very small and others that are very large. It can get cumbersome to write numbers in conventional decimal notation. In **scientific notation**, any number is written as the product of a number between 1 and 10 and a whole number power of ten. Thus the radius of Earth, approximately 6,380,000 m at the equator, can be written 6.38×10^6 m; the radius of a hydrogen atom, 0.000 000 000 053 m, can be written 5.3×10^{-11} m. Scientific notation eliminates the need to write zeros to locate the decimal point correctly.

In science, a measurement or the result of a calculation must indicate the **precision** to which the number is known. The precision of a device used to make a measurement is limited by the finest division on the scale. Using a meterstick with millimeter divisions as the smallest separations, we can measure a length to a precise number of millimeters and we can estimate a fraction of a millimeter between two divisions. If the meterstick has centimeter divisions as the smallest separations, we measure a precise number of centimeters and estimate the fraction of a centimeter that remains.

The most basic way to indicate the precision of a quantity is to write it with the correct number of **significant figures**. The significant figures are all the digits that are known accurately plus the one estimated digit. If we say that the distance from here to the state line is 12 km, that does not mean we know the distance to be *exactly* 12 kilometers. Rather, the distance is 12 km *to the nearest kilometer.* If instead we said that the distance is 12.0 km, that would indicate that we know the distance to the nearest *tenth* of a kilometer. More significant figures mean greater precision.

Rules for Identifying Significant Figures

1. Nonzero digits are always significant.
2. Final or ending zeros written to the right of the decimal point are significant.
3. Zeros written on either side of the decimal point for the purpose of spacing the decimal point are not significant.
4. Zeros written between significant figures are significant.

Example 1.2

Identifying the Number of Significant Figures

For each of these values, identify the number of significant figures and rewrite it in standard scientific notation.

(a) 409.8 s   (b) 0.058700 cm

(c) 9500 g   (d) 950.0×10^1 mL

Strategy We follow the rules for identifying significant figures as given. To rewrite a number in scientific notation, we move the decimal point so that the number to the left of the decimal point is between 1 and 10 and compensate by multiplying by the appropriate power of ten.

Solution (a) All four digits are significant. The zero is between two significant figures, so it is significant. To write the number in scientific notation, we move the decimal point two places to the left and compensate by multiplying by 10^2: 4.098×10^2 s.

(b) The first two zeros are not significant; they are used to place the decimal point. The digits 5, 8, and 7 are significant, as are the two final zeros. The answer has five significant figures: 5.8700×10^{-2} cm.

(c) This measurement has two significant figures (9 and 5). The two final zeros are used to space the decimal point: 9.5×10^3 g.

(d) The final zero is significant since it comes after the decimal point. The zero to its left is also significant since it comes between two other significant digits. The result has four significant figures. The number is not in *standard* scientific notation since 950.0 is not between 1 and 10; in scientific notation we write 9.500×10^3 mL.

Discussion Scientific notation clearly indicates the number of significant figures since all zeros are significant; none are used only to place the decimal point. In (c), there would be no way to know if the value had three or four significant figures instead of two. Are those final zeros just place holders, or are they significant? Most likely they are place holders. Writing 9.5×10^3 g makes it clear that the measurement has two significant figures; if there were four, we would write 9.500×10^3 g.

Practice Problem 1.2 Identifying significant figures

State the number of significant figures in each of these measurements and rewrite them in standard scientific notation.

(a) 0.00010544 kg   (b) 0.005800 cm   (c) 602000 s

When two or more quantities are added or subtracted, the result is as precise as the *least precise* of the quantities (Example 1.3). When quantities are multiplied or divided, the result has the same number of significant figures as the quantity with the *smallest number of significant figures* (Example 1.4).

Example 1.3

Significant Figures in Addition

Calculate the sum 44.56005 s + 0.0698 s + 1103.2 s.

Strategy   The sum cannot be more precise than the least precise of the three quantities. The quantity 44.56005 s is known to the nearest 0.00001 s, 0.0698 s is known to the nearest 0.0001 s, and 1103.2 s is known to the nearest 0.1 s. Therefore the least precise is 1103.2 s. The sum has the same precision; it is known to the nearest tenth of a second.

Solution   According to the calculator,

$$44.56005 + 0.0698 + 1103.2 = 1147.82985$$

We do *not* want to write all of those digits in the answer. That would imply greater precision than we actually have. Rounding to the nearest tenth of a second, the sum is written

$$= 1147.8 \text{ s}$$

and there are five significant figures in the result.

Discussion   Note that the least precise measurement is not necessarily the one with the fewest number of significant figures. The least precise is the one whose right-most significant figure represents the largest unit: the "2" in 1103.2 s represents 2 tenths of a second. In addition or subtraction, we are concerned with the precision rather than the number of significant figures. The three quantities to be added have seven, three, and five significant figures, respectively, while the sum has five significant figures.

Practice Problem 1.3   Significant figures in subtraction

Calculate the difference 568.42 m − 3.924 m and write the result in scientific notation. How many significant figures are in the result?

Example 1.4

Significant Figures in Division

Write the solution to 28.84 m divided by 6.2 s with the correct number of significant figures.

Strategy   The quotient should have the same number of significant figures as the factor with the least number of significant figures.

Solution   A calculator gives

$$\frac{28.84}{6.2} = 4.651612903$$

Since the answer should have only two significant figures, we round the answer to

$$\frac{28.84 \text{ m}}{6.2 \text{ s}} = 4.7 \text{ m/s}$$

Discussion   Writing the answer as 4.651612903 m/s would give the false impression that we know the answer to a precision of about 0.000000001 m/s, whereas we actually have a precision of only about 0.1 m/s.

Practice Problem 1.4   Significant figures in multiplication

Find the product of 45.26 m/s and 2.41 s. How many significant figures does the product have?

   When an integer, or a fraction of integers, is used in an equation, the precision of the result is not affected by the integer or the fraction; the number of significant figures is limited only by the measured values in the problem. The fraction 1/2 in an equation is *exact*; it does not reduce the number of significant figures to one. In an equation such as $C = 2\pi r$ for the circumference of a circle of radius r, the factors 2 and π are exact. We use as many digits for π as we need to maintain the precision of the other quantities.

Sometimes a problem may be too complicated to solve precisely, or information may be missing that would be necessary for a precise calculation. In such a case, an **order of magnitude** solution is the best we can do. By *order of magnitude,* we mean "roughly what power of ten?" An order of magnitude calculation is done to at most one significant figure. Even when a more precise solution is feasible, it is often a good idea to start with a quick, back-of-the-envelope estimate. Why? Because we can often make a good guess about the correct order of magnitude of the answer to a problem, even before we start solving the problem. If the answer comes out with a different order of magnitude, we go back and search out an error. Suppose a problem concerns a vase that is knocked off a fourth-story window ledge. We can guess by experience the order of magnitude of the time it takes the vase to hit the ground. It might be 1 s, or 2 s, but we are certain that it is *not* 1000 s or 0.00001 s.

1.5 UNITS

A **metric system** of units has been used for many years in scientific work and in European countries. The metric system is based on powers of ten (see Fig. 1.1). In 1960, the General Conference of Weights and Measures, an international authority on units, proposed a revised metric system called the *Système International d'Unités* in French (abbreviated **SI**), which uses the meter (m) for length, the kilogram (kg) for mass, and the second (s) for time. These are the three SI base units used in mechanics; there are four more SI base units (see Table 1.1). **Derived units** are constructed from combinations of the base units. For example, the SI unit of force is $kg \cdot m/s^2$; the combination of $kg \cdot m/s^2$ is given a special name, the newton (N), in honor of Isaac Newton. The newton is a derived unit because it is composed of a combination of base units. When units are named after famous scientists, the name of the unit is written with a lowercase letter, even though it is based on a proper name; the *abbreviation* for the unit is written with an uppercase letter. The inside front cover of the book has a complete listing of the derived SI units used in this book.

As an alternative to explicitly writing powers of ten, SI uses prefixes for units to indicate power of ten factors. Table 1.2 shows some of the powers of ten and the SI prefixes used for them. These are also listed on the inside front cover of the book.

SI units are preferred in physics and are emphasized in this book. Since other units are sometimes used, we must know how to convert units. Various scientific fields, even in physics, do use units other than SI units, whether for historical or practical reasons. For example, in atomic and nuclear physics, the SI unit of energy (the joule) is rarely used, for good reasons; instead the energy unit used is usually the electron-volt. Biologists and chemists use units that are not familiar to physicists. One reason that SI is preferred is that it provides a common denominator—all scientists are familiar with the SI units.

In most of the world, SI units are used in everyday life and in industry. In the United States, the U.S. Customary Units—sometimes called English units—are still used. The base units for this system are the foot (for length), the second (for time), and the pound (for weight—mass is a *derived* unit in this system).

In the autumn of 1999, to the chagrin of NASA, a \$125 million spacecraft designed to orbit Mars crashed right into the planet. The company building the booster rocket provided information about the rocket's thrust in U.S. Customary Units, but the NASA scientists who were controlling the rocket thought the figures provided were in metric units. Arthur Stephenson, chairman of the Mars Climate Orbiter Mission Failure Investigation Board, stated that, "The 'root cause' of the loss of the spacecraft was the failed translation of English units into metric units in a segment of ground-based, navigation-related mission software." A journey of 122 million miles ended up a few hundred miles off course thanks to the discrepancy in units, unfortunately resulting in a dramatic failure of the mission.

If the statement of a problem includes a mixture of different units, the units must be converted to a single, consistent set before the problem is solved. Usually the best way is to convert everything to SI units. Unit conversions are carried out by multiplying or dividing units algebraically (Example 1.5). Common conversion factors are listed on the inside front cover of this book.

Some conversions are exact by definition. One meter is defined to be *exactly* equal to 100 centimeters; all SI prefixes are exactly a power of ten. The use of an exact conversion factor such as 1 m = 100 cm, or 1 foot = 12 inches, does not affect the precision of the result: the number of significant figures is limited only by the other quantities in the problem.

Table 1.1

SI Base Units

Base Quantity	Unit Name	Symbol
Length	meter	m
Mass	kilogram	kg
Time	second	s
Electrical current	ampere	A
Temperature	kelvin	K
Amount of substance	mole	mol
Luminous intensity	candela	cd

Table 1.2

SI Prefixes

Prefix (abbreviation)	Power of Ten
peta- (P)	10^{15}
tera- (T)	10^{12}
giga- (G)	10^{9}
mega- (M)	10^{6}
kilo- (k)	10^{3}
deci- (d)	10^{-1}
centi- (c)	10^{-2}
milli- (m)	10^{-3}
micro- (μ)	10^{-6}
nano- (n)	10^{-9}
pico- (p)	10^{-12}
femto- (f)	10^{-15}

Making The Connection:
spacecraft to Mars

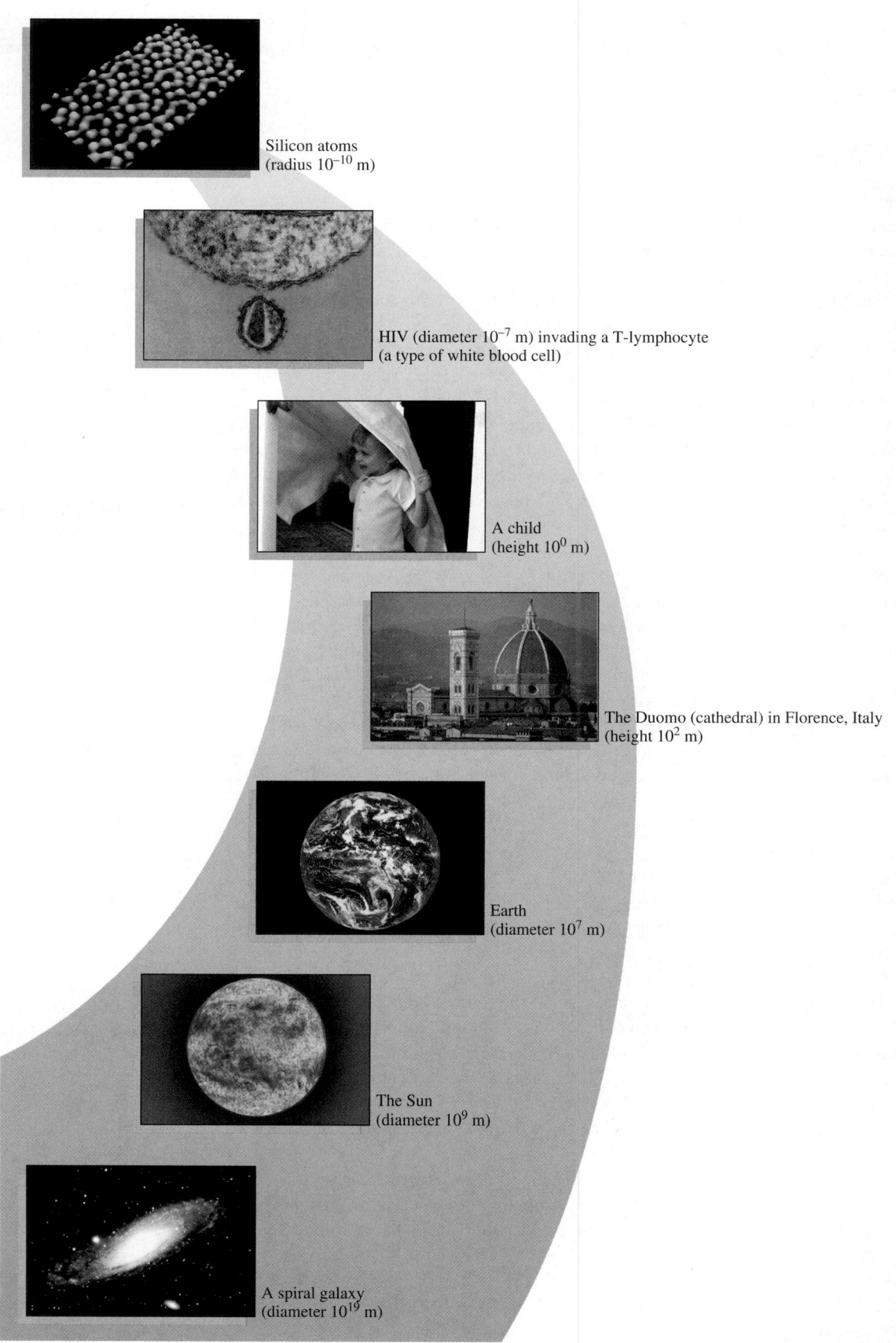

Silicon atoms
(radius 10^{-10} m)

HIV (diameter 10^{-7} m) invading a T-lymphocyte
(a type of white blood cell)

A child
(height 10^0 m)

The Duomo (cathedral) in Florence, Italy
(height 10^2 m)

Earth
(diameter 10^7 m)

The Sun
(diameter 10^9 m)

A spiral galaxy
(diameter 10^{19} m)

Figure 1.1   Scientific notation uses powers of ten to express quantities that have a wide range of values.

Example 1.5

Buying Clothes in a Foreign Country

 Michel, an exchange student from France, is studying in the United States. He wishes to buy a new pair of jeans, but the sizes are all in *inches*. He does remember that 1 m = 3.28 ft and that 1 ft = 12 in. If his waist size is 82 cm, what is his waist size in inches?

Strategy   Each conversion factor can be written as a fraction. If 1 m = 3.28 ft, then

$$\frac{3.28 \text{ ft}}{1 \text{ m}} = 1$$

We can multiply any quantity by 1 without changing its value. We arrange each conversion factor in a fraction and multiply one at a time to get from cm to in.

Solution   We first convert cm to meters.

$$82 \text{ cm} \times \frac{1 \text{ m}}{100 \text{ cm}}$$

Now, we convert meters to feet.

$$82 \text{ cm} \times \frac{1 \text{ m}}{100 \text{ cm}} \times \frac{3.28 \text{ ft}}{1 \text{ m}}$$

Finally, we convert feet to inches.

$$82 \text{ cm} \times \frac{1 \text{ m}}{100 \text{ cm}} \times \frac{3.28 \text{ ft}}{1 \text{ m}} \times \frac{12 \text{ in}}{1 \text{ ft}} = 32 \text{ in}$$

In each case, the fraction is written so that the unit we are converting *from* cancels out.

As a check:

$$\text{cm} \times \frac{\text{m}}{\text{cm}} \times \frac{\text{ft}}{\text{m}} \times \frac{\text{in}}{\text{ft}} = \text{in}$$

Discussion   This problem could have been done in one step using a direct conversion factor from inches to cm (1 in. = 2.54 cm). One of the great advantages of SI units is that all the conversion factors are powers of ten (Table 1.2); there is no need to remember that there are 12 inches in a foot, 4 quarts in a gallon, 16 ounces in a pound, 5280 feet in a mile, and so on.

Practice Problem 1.5   Driving on the autobahn

A BMW convertible travels on the German autobahn at a speed of 128 km/h. What is the speed of the car (a) in m/s? (b) in mi/h?

Example 1.6

Conversion of Volume

A beaker of water contains 255 mL of water. (1 mL = 1 milliliter; 1 liter = 1000 cm³.) What is the volume of the water in (a) cm³? (b) m³?

Strategy   First convert milliliters to liters; then convert liters to cm³. To convert cm³ to m³, use 100 cm = 1 m. Since there are *three* factors of cm to convert, we have to multiply by $\left(\frac{1 \text{ m}}{100 \text{ cm}}\right)$ *three times.*

Solution   (a) The prefix milli- means 10^{-3}, so 1 mL = 10^{-3} L. Then

$$255 \text{ mL} \times \frac{10^{-3} \text{ L}}{1 \text{ mL}} \times \frac{1000 \text{ cm}^3}{1 \text{ L}} = 255 \text{ cm}^3$$

(b) 1 m = 100 cm. Since we need to convert *cubic* centimeters to *cubic* meters, we must raise the conversion factor to the third power:

$$255 \text{ cm}^3 \times \left(\frac{1 \text{ m}}{100 \text{ cm}}\right)^3 = 255 \text{ cm}^3 \times \frac{(1 \text{ m})^3}{(100 \text{ cm})^3}$$

$$255 \text{ cm}^3 \times \frac{1 \text{ m}^3}{100^3 \text{ cm}^3} = 2.55 \times 10^{-4} \text{ m}^3$$

Discussion   Be careful when a unit is raised to a power other than one; the conversion factor must be raised to the same power. Writing out the units to make sure they cancel prevents mistakes. When a quantity is raised to a power, both the number and the unit must be raised to the same power. $(100 \text{ cm})^3$ is equal to $100^3 \text{ cm}^3 = 10^6 \text{ cm}^3$; it is *not* equal to 100 cm^3, nor is it equal to 10^6 cm.

Practice Problem 1.6   Surface area of Earth

If the radius of Earth is 6.4×10^3 km, find the surface area of Earth in square meters and in square miles. (Surface area of a sphere = $4\pi r^2$.)

Whenever a calculation is performed, always write out the units with each quantity. Combine the units algebraically to find the units of the result. This small effort has three important benefits:

1. It shows what the units of the result are. A common mistake is to get the correct numerical result of a calculation but to write it with the wrong units, making the answer wrong.

2. It shows where unit conversions must be done. If units that should have canceled do not, we go back and perform the necessary conversion. When a distance is calculated and the result comes out with units of $\frac{\text{m·s}}{\text{h}}$, we should convert hours (h) to seconds (s).

3. It helps locate mistakes. If a distance is calculated and the units come out as m/s, we know to look for an error.

1.6   DIMENSIONAL ANALYSIS

Dimensions are basic *types* of units, such as time, length, and mass. (Warning: the word *dimension* has several other meanings such as in "three-dimensional space" or "the dimensions of a soccer field.") Many different units of length exist: meters, inches, miles, nautical miles, fathoms, leagues, astronomical units, angstroms, and cubits, just to name a few. All have dimensions of length; each can be converted into any other.

We can add, subtract, or equate quantities only if they have the same dimensions (although they may not necessarily be given in the same units). It is possible to add 3 meters to 2 inches (after converting units), but it is not possible to add 3 meters to 2 kilograms. To analyze dimensions, treat them as algebraic quantities, just as we did with units in Section 1.5. Usually [M], [L], and [T] are used to stand for mass, length, and time dimensions. Equivalently, we can use the SI base units: kg for mass, m for length, and s for time.

Example 1.7

Dimensional Analysis for a Distance Equation

Analyze the dimensions of the equation $d = vt$, where d is distance traveled, v is speed, and t is elapsed time.

Strategy   Replace each quantity with its dimensions. Distance has dimensions [L]. Speed has dimensions of length per unit time [L]/[T]. The equation is dimensionally consistent if the dimensions are the same on both sides.

Solution   The right side has dimensions

$$\frac{[L]}{[T]} \times [T] = [L]$$

Since both sides of the equation have dimensions of length, the equation is dimensionally consistent.

Discussion   If, by mistake, we wrote $d = v/t$ for the relation between distance traveled and elapsed time, we could quickly

catch the mistake by looking at the dimensions. On the right side, v/t would have dimensions $[L]/[T]^2$, which is not the same as the dimensions of d on the left side.

A quick dimensional analysis of this sort is a good way to catch algebraic errors. Whenever we are unsure whether an equation is correct, we can check the dimensions.

Practice Problem 1.7   Testing dimensions of another equation

Test the dimensions of the following equation:

$$x = \tfrac{1}{2}at$$

where x is distance traveled, a is acceleration (which has SI units m/s^2), and t is the elapsed time. If incorrect, can you suggest what might have been omitted?

Dimensional analysis is good for more than just checking equations. In many cases, we can completely solve a problem—up to a dimensionless factor like $1/(2\pi)$ or $\sqrt{3}$—using dimensional analysis. To do this, first list all the relevant quantities upon which the answer might depend. Then determine what combinations of them have the same dimensions as the answer for which we are looking. If only one such combination exists, then we have the answer, except for a possible dimensionless multiplicative constant.

Example 1.8

Violin String Frequency

While it is being played, a violin string produces a tone with frequency f in s^{-1} (Fig. 1.2); the frequency is the number of vibrations *per second* of the string. The string has mass m, length L, and tension T. If the tension is increased 5.0%, how does the frequency change? Tension is a force with SI unit $\frac{kg \cdot m}{s^2}$.

Strategy We could make a study of violin strings, but let us see what we can find out by dimensional analysis. We want to find out how the frequency f can depend on m, L, and T. We won't know if there is a dimensionless constant involved, but we can work by proportions so any such constant will divide out.

Figure 1.2
A violinist

Solution The unit of tension T is $\frac{kg \cdot m}{s^2}$. The units of f do not contain kg or m; we can get rid of them from T by dividing the tension by the length and the mass:

$$\frac{T}{mL} \text{ has SI unit } s^{-2}$$

That is almost what we want; all we have to do is take the square root:

$$\sqrt{\frac{T}{mL}} \text{ has SI unit } s^{-1}$$

Therefore,

$$f = C\sqrt{\frac{T}{mL}}$$

where C is some dimensionless constant. To answer the question, let the original frequency and tension be f and T and the new frequency and tension be f' and T', where $T' = 1.050T$. Frequency is proportional to the square root of tension, so

$$\frac{f'}{f} = \sqrt{\frac{T'}{T}} = \sqrt{1.050} = 1.025$$

The frequency increases 2.5%.

Discussion The value of C is 1/2. That is the *only* thing we cannot get by dimensional analysis. There is *no* other way to combine T, m, and L to come up with a quantity that has the units of frequency.

Practice Problem 1.8   Increase in kinetic energy
When a body of mass m is moving with a speed v, it has kinetic energy associated with its motion. Energy is measured in $kg \cdot m^2 \cdot s^{-2}$. If the velocity of a moving body is increased by 25% while its mass remains constant, by what percentage does the kinetic energy increase?

1.7   PROBLEM-SOLVING TECHNIQUES

No single method exists that can be used to solve every physics problem. We demonstrate useful problem-solving techniques in the examples in every chapter of this text. Even for a particular problem, there may be more than one correct way to approach the solution. Problem-solving techniques are *skills* that must be *practiced* to be learned.

Think of the problem as a puzzle to be solved. Only in the easiest problems is the solution method immediately apparent. When you do not know the entire path to a solution, see where you can get by using the given information—find whatever you can.

Exploration of this sort may lead to a solution by suggesting a path that had not been considered. Be willing to take chances. You may even find the challenge enjoyable!

When having some difficulty, it helps to work with a classmate or two. One way to clarify your thoughts is to put them into words. After you have solved a problem, try to explain it to a friend. If you can explain the problem solution, you really do understand it. Both of you will benefit. But do not rely too much on help from others; the goal is for each of you to develop your own problem-solving skills.

General Guidelines for Problem Solving

1. Read the problem *carefully* and *all the way through.*

2. Reread the problem one sentence at a time and draw a sketch or diagram to help you visualize what is happening.

3. Write down and organize the given information. Some of the information can be written in labels on the diagram. Be sure that the labels are unambiguous. Identify in the diagram the object, the position, the instant of time, or the time interval to which the quantity applies. Sometimes information might be usefully written in a table beside the diagram. Look at the wording of the problem again for information that is implied or stated indirectly.

4. Identify the goal of the problem. What quantities need to be found?

5. If possible, make an estimate to determine the order of magnitude of the answer. This estimate is useful as a check on the final result to see if it is reasonable.

6. Think about how to get from the given information to the final desired information. Do not rush this step. Which principles of physics can be applied to the problem? Which will help get to the solution? How are the known and unknown quantities related? Which equations are relevant and may lead to the solution to the problem? This step requires skills developed only with much practice in problem solving.

7. Frequently, the solution involves more than one step. Intermediate quantities might have to be found first and then used to find the final answer. Try to map out a path from the given information to the solution. Whenever possible, a good strategy is to divide a complex problem into several simpler subproblems.

8. Perform algebraic manipulations with algebraic symbols (letters) as far as possible. Substituting the numbers in too early has a way of hiding mistakes.

9. Finally, if the problem requires a numerical answer, substitute the known numerical quantities, *with their units,* into the appropriate equation. Leaving out the units is a common source of error. Writing the units shows when a unit conversion needs to be done—and also may help identify an algebra mistake.

10. Once the solution is found, don't be in a hurry to move on. Check the answer—is it reasonable? Try to think of other ways to solve the same problem. Many problems can be solved in several different ways. Besides providing a check on the answer, finding more than one method of solution deepens our understanding of the principles of physics and develops problem-solving skills that will help solve other problems.

1.8   APPROXIMATIONS

Physics is about building models and comparing observations of the real world with the model. Simplified models help us to analyze complex situations. In various contexts we assume there is no friction, or no air resistance, no heat loss, or no wind blowing, and so forth. If we tried to take all these things into consideration with every problem, the problems would become vastly more complicated to solve. We never can take account of *every* possible influence. We freely make approximations whenever possible to turn a complex problem into an easier one, as long as the answer will be accurate enough for our purposes.

A valuable skill to develop is the ability to know when an assumption or approximation is reasonable. It might be permissible to neglect air resistance when dropping a stone, but not when dropping a beach ball. Why? We must always be prepared to justify any approximation we make by showing the answer is not changed very much by its use.

As well as making simplifying approximations in models, we also recognize that measurements are approximate. Every measured quantity has some uncertainty; it is impossible for a measurement to be exact to an arbitrarily large number of significant figures. Every measuring device has limits on the precision and accuracy of its measurements.

Sometimes it is difficult or impossible to measure precisely a quantity that is needed for a problem. Then we have to make a reasonable estimate. Suppose we need to know the surface area of a human being to determine the heat loss by radiation in a cold room. We can estimate the height of an average person. We can also estimate the average distance around the waist or hips. Approximating the shape of a human body as a cylinder, we can estimate the surface area by calculating the surface area of a cylinder with the same height and circumference (Fig. 1.3a).

If we need a better estimate, we use a slightly more refined model. For instance, we might approximate the arms, legs, trunk, and head and neck as cylinders of various sizes (Fig. 1.3b). How different is the sum of these areas from the original estimate? That gives an idea of how close the first estimate is.

(a)   (b)

Figure 1.3 Approximation of human body by one or more cylinders

Example 1.9

Number of Cells in the Human Body

Average-sized animal cells are about 20 μm in length; some human blood cells are shown in Fig. 1.4. How many cells are there in the human body? Make an order-of-magnitude estimate.

Strategy We divide this problem into three subproblems: estimating the volume of a human, estimating the volume of the average cell, and finally estimating the number of cells.

To find the volume of a human body, we approximate the body as a cylinder, as previously discussed. Next we assume the cells are cubical to find the volume of a cell. Third, the ratio of the two volumes (volume of the body to volume of the cell) shows how many cells are in the body.

Solution Model the body as a cylinder. A typical height is about 2 m. A typical *maximum* circumference (think waist or hip

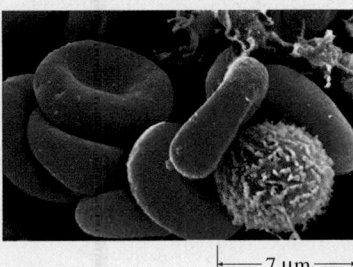

Figure 1.4
Red blood cells are about 7 μm in diameter, which is smaller than the average human cell length.

|← 7 μm →|

size) is about 1 m. The corresponding radius is $1/(2\pi)$ m or about 1/6 m. The *average* radius is somewhat smaller; say 0.1 m. The volume of a cylinder is the height times the cross-sectional area:

$$V = Ah = \pi r^2 h \approx 3 \times (0.1 \text{ m})^2 \times (2 \text{ m}) = 0.06 \text{ m}^3$$

The volume of a cube is $V = s^3$. Then the volume of an average cell is about

$$V_{cell} \approx (2 \times 10^{-5} \text{ m})^3 = 8 \times 10^{-15} \text{ m}^3$$

The number of cells is the ratio of the two volumes:

$$N = \frac{\text{volume of body}}{\text{average volume of cell}} \approx \frac{6 \times 10^{-2} \text{ m}^3}{8 \times 10^{-15} \text{ m}^3} \approx 8 \times 10^{12} \approx 10^{13}$$

Discussion This number is called "10 trillion" (in the United States) or "10 billion" (in Britain). Based on this rough estimate, we cannot rule out the possibility that a better estimate might be 3×10^{12}. On the other hand, we *can* rule out the possibility that the number of cells is, say, 100 million ($= 10^8$).

Practice Problem 1.9 Drinking water consumed in the United States

How many liters of water are swallowed by the people living in the United States in one year? This is a type of problem made famous by the physicist Enrico Fermi (1901–1954), who was a master at this sort of back-of-the-envelope calculation. Such problems are often called *Fermi problems* in his honor. (1 liter = 10^{-3} m^3 ≈ 1 quart.)

Besides estimating quantities that are not well known, another kind of approximation appears frequently in physics. In some cases, we can substitute a mathematical expression with a simpler one that is approximately true. For example, if you were offered a million dollars to calculate the square root of 1.0004 in your head, could you do it? Of course, it would be impossible to give the *exact* answer, which has an infinite number of digits. A calculator shows that the square root of 1.0004 is *very close to* 1.0002. That value can be found from the **binomial approximation** (Appendix A.5), derived in math classes and used in solving physics problems:

The symbol $|x|$ means the absolute value of x.
The symbol $\ll$ means *is much less than*.
The numeral "1" in the binomial approximation is exact.

$$(1 + x)^n \approx 1 + nx \quad \text{if } |x| \ll 1 \tag{1-1}$$

For finding a square root, n is replaced by $\frac{1}{2}$. Then we have

$$(1 + x)^{1/2} \approx 1 + \tfrac{1}{2}x \quad \text{for } |x| \ll 1$$

or $\quad (1 + 0.0004)^{1/2} \approx 1 + \tfrac{1}{2}(0.0004) = 1 + 0.0002 = 1.0002$

Any approximation is of limited validity. This one is valid as long as $|x| \ll 1$. The approximation becomes less useful as $|x|$ gets larger. See Problems 33 and 34 to test the limits of the approximation.

Other important mathematical approximations are found in Appendix A.5. One in particular that bears mentioning here is the **small angle approximation**:

$$\sin \theta \approx \tan \theta \approx \theta \quad \text{for small angles } \theta$$

where θ is measured in radians (rad) (see Appendix A.7). We discuss the radian more fully in Chapter 5. To convert radians to degrees, use $2\pi \text{ rad} = 360°$. As long as θ is small, $\sin \theta$, $\tan \theta$, and θ itself (measured in rad) all have very nearly the same value, so any one can be substituted for any of the others. The approximation is valid for $|\theta| \ll 1$ rad.

1.9   GRAPHS

Graphs are used to help us see a pattern in the relationship between two variables. It is much easier to see a pattern on a graph than to see it in a table of numerical values. When we do experiments in physics, we change one quantity (the **independent variable**) and see what happens to another (the **dependent variable**). We want to see how one variable *depends on* another. The value of the independent variable is usually plotted along the horizontal axis of the graph. In a plot of p versus q, normally p is the dependent variable and q is the independent variable.

Some general guidelines for recording data and making graphs are given next.

Recording Data and Making Data Tables

1. Label columns with the names of the data being measured and be sure to include the units for the measurements. Do not erase any data, but just draw a line through data that you think is erroneous. Sometimes you may decide later that the data were correct after all.

2. Try to make a realistic estimate of the precision of the data being taken when recording numbers. For example, if the timer says 2.3673 s, but you know your reaction time can vary by as much as 0.1 s, the time should be recorded as 2.4 s. When doing calculations using measured values, remember to round the final answer to the correct number of significant figures.

3. Do not wait until you have collected all of your data to start a graph. It is much better to graph each data point as it is measured. By doing so, you can often identify equipment malfunction or measurement errors that make your data unreliable. You can also spot where something interesting happens and take data points closer together there. Graphing as you go means that you need to find out the range of values for both the independent and dependent variables.

Graphing Data

1. Make *large, neat* graphs. A tiny graph is not very illuminating. Use at least half a page. A graph made carelessly obscures the pattern between the two variables.

2. Label axes with the name of the quantities graphed and their units. Write a meaningful title.

3. When a linear relation is expected, use a ruler or straightedge to draw the best-fit straight line. Do not *assume* that the line must go through the origin—make a measurement to find out, if possible. Some of the data points will probably fall above the line and some will fall below the line.

4. Determine the slope of a best-fit line by measuring the ratio $\Delta y/\Delta x$ using as large a range of the graph as possible. Do not choose two data points to calculate the slope; instead, read values from two points on the best-fit line. Show the calculations. Do not forget to write the units; slopes of graphs in physics have units, since the quantities graphed have units.

5. When a nonlinear relationship is expected between the two variables, the best way to test that relationship is to manipulate the data algebraically so that a linear graph is expected. The human eye is a good judge of whether a straight line fits a set of data points. It is not so good at deciding whether a curve is parabolic, cubic, or exponential. To test the relationship $x = \frac{1}{2}at^2$, where x and t are the quantities measured, graph x versus t^2 instead of x versus t.

6. If one data point does not lie near the line or smooth curve connecting the other data points, that data point should be investigated to see whether an error was made in the measurement or whether some interesting event is occurring at that point. If something unusual is happening there, obtain additional data points in the vicinity.

7. When the slope of a graph is used to calculate some quantity, pay attention to the equation of the line and the units along the axes. The quantity to be found may be the inverse of the slope or twice the slope or one-half the slope. For example, if you wish to find the value of a in the relationship $x = \frac{1}{2}at^2$, and you graph your measured values of x and t as x versus t^2, then the slope of the line is $\frac{1}{2}a$. The value of a you seek is twice the slope.

The equation of a straight line on a graph of y versus x can be written $y = mx + b$, where m is the slope and b is the y intercept (the value of y corresponding to $x = 0$).

The symbol Δ, the Greek letter delta, stands for the difference between two measurements. The notation Δy is read aloud as "delta y" and represents a change in the value of y.

Example 1.10

Length of a Spring

In an introductory physics laboratory experiment, students are investigating how the length of a spring varies with the weight hanging from it. Various weights (accurately calibrated to 0.01 N) ranging up to 6.00 N can be hung from the spring; then the length of the spring is measured with a meterstick (Fig. 1.5). The goal is to see if the weight F and length L are related by

$$F = kx$$

where $x = (L - L_0)$, L_0 is the length of the spring when no weight is hanging from it, and k is called the *spring constant* of the spring. Graph the data in the table and calculate k for this spring.

F (N):	0	0.50	1.00	2.50	3.00	3.50	4.00	5.00	6.00
L (cm):	9.4	10.2	12.5	17.9	19.7	22.5	23.0	28.8	29.5

Strategy Weight is the independent variable, so it is plotted on the horizontal axis. After plotting the data points, we draw the best-fit straight line. Then we calculate the slope of the line, using two points on the line that are widely separated and that cross gridlines of the graph (so the values are easy to read). The slope of the graph is not k; we must solve the equation for L, since length is plotted on the vertical axis.

Figure 1.5
A weight causes an extension in the length of a spring.

Solution Figure 1.6 shows a graph with data points and a best-fit straight line. There is some scatter in the data, but a linear relationship is plausible.

continued on next page

Example 1.10 *continued*

Two points where the line crosses gridlines of the graph are (0.80 N, 12.0 cm) and (4.40 N, 25.0 cm). From these, we calculate the slope:

$$\text{slope} = \frac{\Delta L}{\Delta F} = \frac{25.0\ \text{cm} - 12.0\ \text{cm}}{4.40\ \text{N} - 0.80\ \text{N}} = 3.60\ \frac{\text{cm}}{\text{N}}$$

By analyzing the units of the equation $F = k(L - L_0)$, it is clear that the slope cannot be the spring constant; k has the same units as weight divided by length (N/cm). Is the slope equal to $1/k$? The units would be correct for that case. To be sure, we solve the equation of the line for L:

$$L = \frac{F}{k} + L_0$$

We recognize the equation of a line with a slope of $1/k$. Therefore,

$$k = \frac{1}{3.60\ \text{cm/N}} = 0.278\ \frac{\text{N}}{\text{cm}}$$

Discussion As discussed in the graphing guidelines, the slope of the straight-line graph is calculated from two widely spaced values *along the best-fit line.* We do not subtract values of actual data points. We are looking for an average value from the data; using two data points to find the slope would defeat the purpose of plotting a graph or of taking more than two data measurements. The values read from the graph, including the units, are indicated in Fig. 1.6. The units for the slope are cm/N since we plotted cm versus N. For this particular problem the *inverse* of the slope is the quantity we seek, the spring constant in N/cm.

Practice Problem 1.10   Another weight on spring

What is the length of the spring of Example 1.10 when a weight of 8.00 N is suspended? Assume that the relationship found in Example 1.10 still holds for this weight.

Figure 1.6
Spring length versus weight hanging (Example 1.10)

MASTER THE CONCEPTS

Summary

- Terms used in physics must be precisely defined. The same term may have a different meaning in physics from the meaning of the same word in other contexts.

- A working knowledge of algebra, geometry, and trigonometry is essential in the study of physics.

- The *factor* by which a quantity is increased or decreased is the ratio of the new value to the original value.

- When we say that A is *proportional* to B (written $A \propto B$), we mean that if B increases by some factor, then A must increase by the same factor.

- In *scientific notation,* a number is written as the product of a number between 1 and 10 and a whole number power of ten.

- *Significant figures* are the basic *grammar* of precision. They enable us to communicate quantitative information and indicate the precision to which that information is known.

- When two or more quantities are added or subtracted, the result is as precise as the *least precise* of the quantities. When quantities are multiplied or divided, the result has the same number of significant figures as the quantity with the *smallest number of significant figures.*

- Order of magnitude estimates and calculations are made to be sure that the more precise calculations are realistic.

- The units used for scientific work are those from the *Système International* (*SI*). SI uses seven *base units,* which include the meter (m), the kilogram (kg), and the second (s) for length, mass, and time, respectively. Using combinations of the base units, we can construct other *derived units.*

MASTER THE CONCEPTS *continued*

- If the statement of a problem includes a mixture of different units, the units should be converted to a single, consistent set before the problem is solved. Usually the best way is to convert everything to SI units.

- Dimensional analysis is used as a quick check on the validity of equations. Whenever quantities are added, subtracted, or equated, they must have the same dimensions (although they may not necessarily be given in the same units).

- Mathematical approximations aid in simplifying complicated problems.

- Problem-solving techniques are *skills* that must be *practiced* to be learned.

- A graph is plotted to give a picture of the data and to show how one variable changes with respect to another. Graphs are used to help us see a pattern in the relationship between two variables.

- Whenever possible, a careful choice is made of the variables plotted so that the graph displays a linear relationship.

Highlighted Figures and Tables

F1.1 Powers of ten (p. 10)
T1.1 SI base units (p. 9)
T1.2 SI prefixes (p. 9)

CONCEPTUAL QUESTIONS

1. Give a few reasons for studying physics.
2. Why must words be carefully defined for scientific use?
3. Why are models used in scientific study if they do not exactly match real conditions?
4. By what factor does tripling the radius of a circle affect (a) the circumference of the circle? (b) the area of the circle?
5. What are some of the advantages of scientific notation?
6. After which numeral is the decimal point usually placed in scientific notation? What determines the number of numerical digits written in scientific notation?
7. Are all the digits listed as "significant figures" precisely known? Might any of the significant digits be less precisely known than others? Explain.
8. Why is it important to write quantities with the correct number of significant figures?
9. List three of the base units used in SI.
10. What are some of the differences between the SI and the U.S. Customary system of units? Why is SI preferred for scientific work?
11. Sort these units into three groups of dimensions and identify the dimensions: fathoms, grams, years, kilometers, miles, months, kilograms, inches, seconds.
12. What are the first two steps to be followed in solving almost any physics problem?
13. Why do scientists plot graphs of their data instead of just writing them in tables?
14. A student's lab report concludes, "The speed of sound in air is 327." What is wrong with that statement?
15. What are some of the reasons for making order-of-magnitude estimates?
16. Once the solution of a problem has been found, what should be done before moving on to solve another problem?

MULTIPLE CHOICE QUESTIONS

1. A kilometer is approximately
   (a) 2 miles   (b) 1/2 mile   (c) 1/10 mile   (d) 1/4 mile
2. 55 miles/hour is approximately
   (a) 90 km/h   (b) 30 km/h   (c) 10 km/h   (d) 2 km/h
3. By what factor does the volume of a cube increase if the length of the edges are doubled?
   (a) 16   (b) 8   (c) 4   (d) 2   (e) $\sqrt{2}$
4. If the length of a box is reduced to one-third of its original value and the width and height are doubled, by what factor has the volume changed?
   (a) 2/3   (b) 1   (c) 4/3   (d) 3/4
   (e) depends on relative proportion of length to height and width
5. If the area of a circle is found to be half of its original value after the radius is multiplied by a certain factor, what was the factor used?
   (a) $1/(2\pi)$ (b) 1/2   (c) $\sqrt{2}$ (d) $1/\sqrt{2}$ (e) 1/4

6. In terms of the original diameter d, what new diameter will result in a new spherical volume that is a factor of eight times the original volume?
   (a) $8d$ (b) $2d$ (c) $d/2$ (d) $d \times \sqrt[3]{2}$ (e) $d/8$
7. An equation for potential energy states $U = mgh$. If U is in $kg \cdot m^2 \cdot s^{-2}$, m is in kg, and g is in $m \cdot s^{-2}$, what are the units of h?
   (a) s   (b) s^2 (c) m^{-1} (d) m   (e) g^{-1}
8. The equation for the velocity of sound in a gas states that $v = \sqrt{\gamma k_B T/m}$. Velocity v is measured in m/s, γ is a dimensionless constant, T is temperature in kelvin (K), and m is mass in kg. What are the units for the Boltzmann constant, k_B?
   (a) $kg \cdot m^2 \cdot s^2$ (b) $kg \cdot m^2 \cdot s^{-2} \cdot K^{-1}$ (c) N/K
   (d) $kg \cdot m/s$ (e) $kg \cdot m^2 \cdot s^{-2}$

9. How many significant figures are in the sum 4.56 + 9.032 + 580.0078 + 540.439?

   (a) 3   (b) 4   (c) 5   (d) 6   (e) 7

10. How many significant figures are in the product of 0.0078406 × 9.45020?

    (a) 3   (b) 4   (c) 5   (d) 6   (e) 7

PROBLEMS

Note: **©** indicates a combination conceptual/quantitative problem. Gold diamonds ✦, ✦✦ are used to indicate the increasing level of difficulty of each problem. Problem numbers appearing in blue, 9., denote problems that have a detailed solution available in the Student Solutions Manual. Some problems are *paired* by concept; their numbers are connected by a ruled box.

1.3 The Use of Mathematics

1. A spherical balloon expands when it is taken from the cold outdoors to the inside of a warm house. If its surface area increases 16.0%, by what percentage does the radius of the balloon change?

✦ 2. The weight of an object at the surface of a planet is proportional to the planet's mass and inversely proportional to the square of the radius of the planet. Jupiter's radius is 11 times Earth's and its mass is 320 times Earth's. An apple weighs 1.0 N on Earth. How much would it weigh on Jupiter?

3. In cleaning out the artery of a patient, a doctor increases the radius of the opening by a factor of two. By what factor does the cross-sectional area of the artery change?

✦ 4. A scanning electron micrograph of xylem vessels in a corn root shows the vessels magnified by a factor of 600. In the micrograph the xylem vessel is 3.0 cm in diameter. (a) What is the diameter of the vessel itself? (b) By what factor has the cross-sectional area of the vessel been increased in the micrograph?

✦ 5. The average speed of a nitrogen molecule in air is proportional to the square root of the temperature in kelvins. If the average speed is 475 m/s on a warm summer day (temperature = 300.0 kelvins), what is the average speed on a cold winter day (250.0 kelvins)?

✦ 6. The electrical power P drawn from a generator by a lightbulb of resistance R is $P = \dfrac{V^2}{R}$, where V is the line voltage. The resistance of bulb B is 40% greater than the resistance of bulb A. What is the ratio $\dfrac{P_B}{P_A}$ of the power drawn by bulb B to the power drawn by bulb A, if the line voltages are the same?

1.4 Scientific Notation and Significant Figures

7. Given these measurements, identify the number of significant figures and rewrite in scientific notation.

   (a) 0.00574 kg   (b) 2 m   (c) 0.450×10^{-2} m
   (d) 45.0 kg   (e) 10.09×10^4 s   (f) 0.09500×10^5 mL

8. Write these numbers in scientific notation: (a) the U.S. population, 250 000 000; (b) the diameter of a Helium nucleus, 0.000 000 000 000 003 8 m.

9. Perform these operations with the appropriate number of significant figures.

   (a) $3.783 \times 10^6 + 1.25 \times 10^8$ (b) $(3.783 \times 10^6) \div (3.0 \times 10^{-2})$

10. Write your answers to these problems with the appropriate number of significant figures.

    (a) $6.85 \times 10^{-5} + 2.7 \times 10^{-7}$ (b) $702.35 + 1897.648$
    (c) 5.0×4.3 (d) $0.04/\pi$ (e) $0.040/\pi$

11. In these calculations, be sure to use an appropriate number of significant figures.

    (a) $4.759 \times 10^5 - 3.68 \times 10^7$ (b) $\dfrac{(6.497 \times 10^{-31})^5}{5.1037 \times 10^{26}}$

12. How many significant figures are in each of these measurements?

    (a) 7.68 g   (b) 0.420 kg   (c) 0.073 m   (d) 7.68×10^5 g
    (e) 4.20×10^3 kg   (f) 7.3×10^{-2} m   (g) 2.300×10^4 s

1.5 Units

13. Convert 1.00 km/h to m/s using 1 km = 1000 m and 1 h = 3600 s.

14. A sprinter can run at a top speed of 0.32 miles per minute. Express her speed in (a) m/s and (b) mi/h.

15. (a) How many center-stripe road reflectors, separated by 17.6 yards, are required along a 2.20-mile section of curving mountain roadway? (b) Solve the same problem for a road length of 3.54 km with the markers placed every 16.0 m. Would you prefer to be the highway engineer in a country with a metric system or a U.S. Common Unit system?

16. A furlong is 220 yards; a fortnight is 14 days. How fast is 1 furlong per fortnight (a) in μm/s? (b) in km/day?

17. The intensity of the Sun's radiation that reaches Earth's atmosphere is 1.4 kW/m² (kW = kilowatt; W = watt). Convert this to W/cm².

18. You are given these approximate measurements: (a) the radius of Earth is 6×10^6 m, (b) the length of a human body is 6 feet, (c) a cell's diameter is 10^{-6} m, (d) the width of the hemoglobin molecule is 3×10^{-9} m, and (e) the distance between two atoms (carbon and nitrogen) is 3×10^{-10} m. Write these measurements in the simplest possible metric prefix forms (in either nm, Mm, μm, or whatever works best).

19. The smallest "living" thing is probably a type of infectious agent know as a viroid. Viroids are plant pathogens that consist of a circular loop of single-stranded RNA, containing about 300 bases. (Think of the bases as beads strung on a circular RNA string.) The distance from one base to the next (measured along the circumference of the circular loop) is about 0.35 nm. What is the diameter of a viroid in (a) m, (b) μm, and (c) in.?

20. The largest living creature on Earth is the blue whale, which has an average length of 70 ft. The largest blue whale on record (and therefore the largest animal ever found) was 1.10×10^2 ft long. (a) Convert this length to meters. (b) If a double-decker London bus is 8.0 m long, how many double-decker-bus lengths is the record whale?

21. Density is the ratio of mass to volume. Mercury has a density of 1.36×10^4 kg/m^3. What is the density of mercury in units of g/cm^3?

22. An air molecule is moving at a speed of 459 m/s. How many meters would the molecule move during 7.00 ms (milliseconds) if it didn't collide with any other molecules?

23. A sheet of paper has length 27.95 cm, width 8.5 in., and thickness 0.10 mm. What is the volume of a sheet of paper in m^3? (Volume = length × width × thickness.)

24. Express this product in units of kg^3 with the appropriate number of significant figures: $(3.2 \text{ kg}) \times (4.0 \text{ g}) \times (13 \times 10^{-3} \text{ mg})$.

✦✦ 25. Three of the fundamental constants of physics are

the universal gravitational constant,

$$G = 6.7 \times 10^{-11} \frac{\text{m}^3}{\text{kg}\cdot\text{s}^2}$$

the speed of light, $\quad c = 3.0 \times 10^8$ m/s

and Planck's constant,

$$h = 6.6 \times 10^{-34} \text{ J}\cdot\text{s} \left(1 \text{ J} = \frac{1 \text{ kg}\cdot\text{m}^2}{\text{s}^2} \right)$$

(a) Find a combination of these three constants that has the dimensions of time. This time is called the *Planck time* and represents the age of the universe before which the laws of physics as presently understood cannot be applied.

(b) Using the formula for the Planck time derived in part (a), what is the time in seconds?

26. (a) How many square centimeters are there in 1 square foot? (1 in. = 2.54 cm.) (b) How many square centimeters are there in 1 square meter? (c) Using your answers to parts (a) and (b), but without using your calculator, roughly how many square feet are in one square meter?

27. The largest human chromosome (number 1) consists of a long chain of 2.5×10^8 base pairs of DNA. In the metaphase stage of cell division, the base pairs are all tightly wrapped up into a rod of about 9 μm in length. The distance between base pairs is about 3.5×10^{-10} m. If you could unwrap your chromosome 1, (a) how many meters long would it be? (b) how many inches?

28. The record blue whale in Problem 20 had a mass of 1.9×10^5 kg. Assuming that its average density was 0.85 g/cm^3, as has been measured for other blue whales, what was the volume of the whale in m^3? (Average density is the ratio of mass to volume.)

1.6 Dimensional Analysis

✦ 29. An object moving at speed v around a circle of radius r has an acceleration a directed toward the center of the circle. The SI unit of acceleration is m/s^2. (a) Use dimensional analysis to find a as a function of v and r. (b) If the speed is increased 10.0%, by what percentage does the centripetal acceleration increase?

✦ 30. The speed of ocean waves depends on their wavelength λ (measured in meters) and the gravitational field strength g (measured in m/s^2) in this way:

$$v = K\lambda^p g^q$$

where K is a dimensionless constant. Find the values of the exponents p and q.

✦✦ 31. The Space Shuttle astronauts use a *massing chair* to measure their mass. The chair is attached to a spring and is free to oscillate back and forth. The frequency of the oscillation is measured and that is used to calculate the total mass m attached to the spring. If the spring constant of the spring k is measured in kg/s^2 and the chair's frequency f is 0.50 s^{-1} for a 62-kg astronaut, what is the chair's frequency for a 75-kg astronaut? The chair itself has a mass of 10.0 kg. [*Hint:* use dimensional analysis to find out how f depends on m and k.]

32. Use dimensional analysis to determine how the period T of a pendulum (the elapsed time for a complete cycle of motion) depends on some, or all, of these properties: the length L of the pendulum, the mass m of the pendulum bob, and the gravitational field strength g (in m/s^2). Assume that the amplitude of the swing (the maximum angle that string makes with the vertical) has no effect on the period. (See Fig. 1.7.)

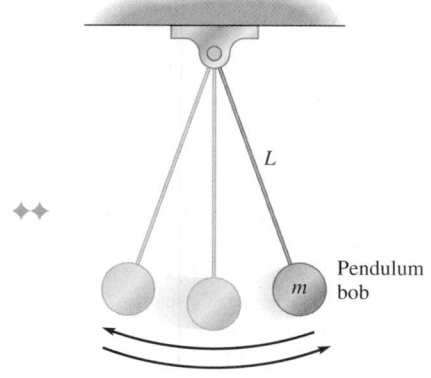

Figure 1.7
Swinging pendulum
(Problem 32)

1.8 Approximations

33. Use the binomial approximation (Eq. 1–1) to find the square root of (a) 1.008; (b) 1.04; (c) 1.4; (d) 1.66. Compare approximations with results from a calculator.

34. Use the binomial approximation (Eq. 1–1) to find the cube root of (a) 0.9994; (b) 0.994; (c) 0.92; (d) 1.3. Compare approximations with results from a calculator.

35. For each of the angles $\theta = 40°, 30°, 20°$, and $10°$, use a calculator to find the value of θ in *radians* and the value of tan θ. Organize the values in a table. At which of these angles in degrees does the difference between θ measured in radians and the value of tan θ become approximately: (a) 10% of θ in rad? (b) 1% of θ in rad?

36. For each of the angles $\theta = 35°, 25°, 18°$, and $8°$, make a table to compare values of tan θ and sin θ. At which of these angles in degrees does the difference between sin θ and tan θ become approximately: (a) 10% of sin θ? (b) 1% of sin θ?

37. What is the order of magnitude of the height (in meters) of a 40-story building?

38. What is the order of magnitude of the number of seconds in one year?

✦ 39. Without looking up any data, make an order-of-magnitude estimate of the annual consumption of gasoline (in gallons) by passenger cars in the United States. Make reasonable order-of-magnitude estimates for any quantities you need. Think in terms of average quantities. (1 gal ≈ 4 liters.)

✦ 40. How many cups of water are required to fill a bathtub?

1.9 Graphs

© 41. A nurse recorded the values shown in Table 1.3 for a patient's temperature. Plot a graph of temperature versus elapsed time and from the graph find (a) an estimate of the temperature at noon and (b) the slope of the graph. (c) Would you expect the graph to follow the same trend over the next 12 hours? Explain.

Table 1.3

Temperature Chart for Problem 41

Time	Temp (°F)
10:00 A.M.	100.00
10:30 A.M.	100.45
11:00 A.M.	100.90
11:30 A.M.	101.35
12:45 P.M.	102.48

© 42. The weight of a baby measured over an 11-month period is given in Table 1.4. (a) Plot the baby's weight versus age over the 11 months. (b) What was the average monthly weight gain for this baby over the period from birth to 5 months? How do you find this value from the graph? (c) What was the average monthly weight gain for the baby over the period from 5 to 10 months? (d) If a baby continued to grow at the same rate as in the first five months of life, what would the child weigh at age 12 years?

43. A physics student plots results of an experiment as v versus t. The equation that describes the line is given by $at = v - v_0$. (a) What is the slope of this line? (b) What is the "y" intercept of this line?

44. A linear plot of speed versus elapsed time has a slope of 6.0 m/s² and a "y" intercept of 3.0 m/s. (a) What is the change in speed in the time interval between 4.0 seconds and 6.0 seconds? (b) What is the speed when the elapsed time is equal to 5.0 seconds?

♦♦ 45. The population of a culture of yeast cells is studied in the laboratory to see the effects of limited resources (food, space) on population growth. At 2-hour intervals, the size of the population (measured as total mass of yeast cells) is recorded (Table 1.5). (a) Make a graph of the yeast population as a function of elapsed time. Draw a best-fit smooth curve. (b) Notice from the graph of part (a) that after a long time, the population approaches asymptotically a maximum known as the *carrying capacity*. From the graph, estimate the carrying capacity for this population. (c) When the population is much smaller than the carrying capacity, the growth is expected to be exponential: $m(t) = m_0 e^{rt}$, where m is the population at any time t, m_0 is the initial population, r is the *intrinsic growth rate* (that is, the growth rate in the absence of limits), and e is the base of natural logarithms (see Appendix A.3). To obtain a straight line graph from this exponential relationship, we can plot the natural logarithm of m/m_0:

$$\ln \frac{m}{m_0} = \ln e^{rt} = rt$$

Make a graph of $\ln \dfrac{m}{m_0}$ versus t from $t = 0$ to $t = 6.0$ h, and use it to estimate the intrinsic growth rate r for the yeast population. (The term ln stands for the natural logarithm; see Appendix A.3 if you need help with natural logs.)

Table 1.5

Yeast Population for Problem 45

Time (h)	Mass (g)
0.0	3.2
2.0	5.9
4.0	10.8
6.0	19.1
8.0	31.2
10.0	46.5
12.0	62.0
14.0	74.9
16.0	83.7
18.0	89.3
20.0	92.5
22.0	94.0
24.0	95.1

Table 1.4

Baby's Weight Chart for Problem 42

Weight (lb)	6.6	7.4	9.6	11.2	12.0	13.6	13.8	14.8	15.0	16.6	17.5	18.4
Age (months)	0 (birth)	1.0	2.0	3.0	4.0	5.0	6.0	7.0	8.0	9.0	10.0	11.0

COMPREHENSIVE PROBLEMS

46. A poster advertising a student election candidate is too large according to the election rules. The candidate is told she must reduce the length and width of the poster by 20.0%. By what percentage will the area of the poster be reduced?

47. If the radius of a circular garden plot is increased by 25%, by what percentage does the area of the garden increase?

48. A snail crawls at a pace of 5.0 cm/min. Express the snail's speed in (a) ft/s and (b) mi/h.

49. A marathon race is about 26 miles long. What is the length of the race in kilometers?

C 50. For what radius, *if any,* does the area of a circle equal the circumference of the circle? Explain.

51. The weight W of an object is given by $W = mg$, where m is the object's mass and g is the gravitational field strength. The SI unit of field strength g, expressed in SI base units, is m/s^2. What is the SI unit for weight, expressed in base units?

52. When a force acts over a distance, work is done (Chapter 6). If the force is parallel to the motion, the work W is given by $W = Fd$, where F is the magnitude of the force and d is the distance over which the force acts. (Note that in this context, W stands for work, not for weight.) Find the SI unit of work. Write it two ways: once in terms of base units only and again using the newton $(1\ N = 1\ kg{\cdot}m/s^2)$.

C 53. One morning you read in the *New York Times* that Bill Gates' net worth is $90,000,000,000. Later that day you see him on the street, and he gives you a $100 bill. What is his net worth now? (Think of significant figures.)

✦✦ 54. Make an order-of-magnitude estimate of the volume of water in the oceans. Do not look up any data in books. (Use your ingenuity to estimate the radius or circumference of Earth.)

✦✦ 55. According to the Stefan-Boltzmann law, an object radiates heat at a rate $\Delta Q/\Delta t = \epsilon\sigma AT^n$. In SI units, $\Delta Q/\Delta t$ is measured in watts (W), T is the temperature in kelvins (K), A is the object's surface area, $\sigma = 5.67 \times 10^{-8}\ W{\cdot}m^{-2}{\cdot}K^{-4}$, and ϵ is a dimensionless number called the emissivity. (a) What is the value of n? (b) If Earth's temperature in kelvins were to increase 5.0%, by how much would Earth's rate of radiation increase?

56. A typical virus is a packet of protein and DNA (or RNA) and can be spherical in shape. The influenza A virus is a spherical virus that has a diameter of 85 nm. If the volume of saliva coughed onto you by your friend with the flu is 0.010 cm^3 and $\dfrac{1}{10^9}$ of that volume consists of viral particles, how many influenza viruses have just landed on you?

57. A patient's temperature was 97.0°F at 8:05 A.M. and 101.0°F at 12:05 P.M. If the temperature change with respect to elapsed time was linear throughout the day, what would the patient's temperature be at 3:35 P.M.?

C ✦ 58. The data in Table 1.6 were obtained in a study of the effects of feeding two equal-sized flocks of hens with special diets. Initially, both flocks were producing an average of 10 eggs per day. The hens in Flock A had a common supplement added to their usual corn diet, while those in Flock B were fed the new NutruCorn. The number of eggs produced per day was then averaged over a week of egg laying for the next few weeks.

Plot the data showing average daily egg production versus time over the first 4 weeks for both groups of hens on one set of axes. Is the relationship linear? If not, can you think of a way to plot the data so that there will be a linear relationship? [*Hint:* You might need to make a new table. Consider the *increase* in average daily egg production per week.] After studying these results from the NutruCorn manufacturers, would you be persuaded to buy their product for your hens?

59. Suppose you have a pair of Seven League Boots. These are magic boots that enable you to stride along a distance of 7.0 leagues with each step. (a) If you march along at a military march pace of 120 paces/minute, what will be your speed in km/h? (b) Assuming you could march on top of the oceans when you step off the continents, how long (in minutes) will it take you to march around the Earth at the equator? (1 league = 3 miles = 4.8 km.)

Table 1.6

Average Daily Egg Production for Problem 58

Data Table	Average Daily # Eggs/Week	
Time (weeks)	Flock A	Flock B
0	10.0	10.0
1	11.2	12.1
2	13.6	16.4
3	17.2	22.8
4	22.0	31.3
5	22.1	31.0
6	22.0	31.2

ANSWERS TO PRACTICE PROBLEMS

1.1 81.0 W   **1.2** (a) five; 1.0544×10^{-4} kg; (b) four; 5.800×10^{-3} cm; (c) three; 6.02×10^5 s   **1.3** The least precise value is to hundredths of a meter; we round the result to the nearest hundredth of a meter: 564.50 m or, in scientific notation, 5.6450×10^2 m; five significant figures.   **1.4** 109   m;   three   significant   figures   **1.5** (a) 35.6 m/s; (b) 79.5 mi/h   **1.6** $5.1 \times 10^{14}\ m^2$; $2.0 \times 10^8\ mi^2$ **1.7** The equation is dimensionally inconsistent; the right side has dimensions [L]/[T]. To have matching dimensions we must multiply the right side by [T]; the equation must involve time squared: $x = \frac{1}{2}at^2$.   **1.8** kinetic energy = (constant) $\times\ mv^2$; kinetic energy increases by 56%.   **1.9** 10^{11} liters (Make a rough estimate of the population to be about 2.5×10^8 people; each drinking about 1.5 liters/day.)   **1.10** 37.9 cm

Forces and Introduction to Vectors

How is it possible for a water strider to stand on top of the surface of a pond without sinking into the water? What is holding it up?

Concepts & Skills to Review

- scientific notation and significant figures (Section 1.4)
- units (Section 1.5)
- problem-solving techniques (Section 1.7)

2.1   FORCES

This chapter begins our study of **classical mechanics**. Although it does not apply to the extremely small (atoms, nuclei, elementary particles), the cosmologically large (galaxies), or the incredibly fast (things moving near the speed of light), classical mechanics does apply to just about everything else. Classical mechanics enables us to predict the return of a comet, explain the operation of a centrifuge, control the motion of spacecraft and submarines, construct an artificial knee, design a longer-range golf ball, and understand how a water strider is able to stand on the surface of a pond.

The essence of classical mechanics is stated in three laws of motion, formulated by Isaac Newton (1642–1727), which tell us how the forces acting on an object determine the motion of that object. Together with his law of universal gravitation, Newton's laws showed for the first time that the heavenly bodies (the Sun, the planets and their satellites) obey the same physical laws of motion as do earthly bodies. To pre-Newtonian thinkers, the motion of the heavenly bodies was perfect and eternal, while earthly bodies always come to rest. For example, why does the Moon orbit the Earth month after month, without any engine to drive it, while an airplane that runs out of fuel falls to the ground? Pre-Newtonian thinkers naturally reasoned that there must be two different sets of physical laws.

To understand the causes of motion, we need to know how one body affects the motion of another. In other words, we need to study **interactions** between bodies. There are many sorts of interactions. Some are long-range, such as the gravitational interaction between the Earth and the Moon that keeps the Moon from wandering off. This interaction occurs even though the Earth and the Moon are far apart. Other long-range interactions are electrical or magnetic in nature.

Another type of interaction involves contact, as when a baseball bat contacts the pitcher's fastball and sends it over the fence for a home run. The bat has no noticeable effect on the ball's motion until the two come into contact. This brief interaction between the bat and ball causes a drastic change in the ball's motion. It also affects the bat, which is why the batter can feel the contact with the ball; the bat and ball are actually deformed slightly during the contact. The subsequent trajectory of the ball is affected by interactions with the bat, with the air, and with the Earth (gravity).

Physics at Home

On a dry day, stand in front of a mirror and run a comb vigorously through your hair until you hear some crackling. Now hold the comb a few centimeters from your hair. Observe the long-range electrical interaction between your hair and the comb.

Now take a refrigerator magnet. Hold it near but not touching the refrigerator door. You can feel the effect of a long-range magnetic interaction.

Just as human life would be dull without social interactions, the physical universe would be dull without physical interactions. Social interactions with friends and family change our behavior; physical interactions change the "behavior" (motion, temperature, etc.) of matter. Without interactions, that fastball could not be turned into a home run; there would be no gravity to make it fall to Earth and no air drag to slow it down. What *would* happen to the ball if it did not interact with anything else? According to Newton's first law (presented later in this chapter), the ball would continue on its course in a straight line with an undiminished speed.

Several different models are used in physics to describe the effect of an interaction. Some of the models that we study later in this book include potential energy functions (Chapter 6), momentum exchange (Chapter 7), and the exchange of messenger particles (Chapter 30). In this chapter we use *forces* to describe the interaction between two bodies. A **force** is a push or a pull that one body exerts on another. When you play handball, your hand exerts a force on the ball. The ball also exerts a force on your hand, the effect of

Force: a push or pull that one object exerts on another

which you can feel. A magnetic force pulls a magnet to the refrigerator door. A magnetic force also pulls the refrigerator door toward the magnet; this force is not as easy to observe because the door is so massive and is held in place by other forces. When two bodies interact, they exert forces on each other. We need to be able to analyze the forces acting on a body in order to relate them to changes in the body's state of motion.

Isolated forces do not exist; they always exist in pairs. Nothing can exert a force without having a force exerted on it. When you push open a door, the door pushes you. When two cars collide, each exerts a force on the other. Newton's third law (Section 2.3) specifies how the two forces that make up an interaction are related. Note that the two forces that make up an interaction *act on different objects*—the two objects that are interacting. For any force, we must be able to uniquely name both the object that exerts the force and the object on which the force is exerted. Doing so helps correctly identify the forces in a problem, which is a crucial first step in solving a mechanics problem.

Conceptual Example 2.1

Forces Acting on the *Alvin*

 The submersible *Alvin* was used in 1985 to find the remains of the *Titanic* on the ocean floor. Identify all the forces acting on the submersible while it sits at rest on the deck of the *Titanic*.

Strategy   There are two kinds of forces acting on the *Alvin:* long-range forces and contact forces. The possible long-range forces are gravity, electricity, and magnetism. Contact forces must involve contact with some other object, so we consider everything in contact with the *Alvin.* For each force, whether long-range or contact, we should be able to identify what object is exerting the force.

Solution   (1) The only long-range force on the *Alvin* is gravity, a downward pull due to the mass of the entire Earth. Long-range electrical and magnetic forces are unlikely, unless the *Alvin* is made of iron and happens to be sitting near some magnetized rocks.

(2) The *Alvin* is sitting on the deck without sinking into it. Therefore, the deck exerts a contact force upward on the *Alvin.*

(3) The only other thing in contact with the *Alvin* is seawater. There are contact forces exerted on the *Alvin* by the water.

Discussion   Did we omit the buoyant force—the force that makes some things float and that makes submerged objects *seem* to weigh less under water? The buoyant force is not one of the long-range forces, so it must be a contact force; we have already listed all possible contact forces. The buoyant force is a result of the water squeezing the *Alvin* inward on all sides. Similarly, if water currents exert a sideways force on the *Alvin,* the sideways force arises out of the contact force due to the water. At the microscopic level, the contact forces exerted on the *Alvin* by the water are due to vast numbers of water molecules colliding with the vessel's exterior. It is a great simplification to think of huge numbers of microscopic forces combined into one or two macroscopic forces.

 Correctly identifying all of the forces acting on an object is usually the first step to solving a mechanics problem. Overlooking one of the forces, or including a spurious one, makes it impossible to come up with the correct solution to the problem.

Conceptual Practice Problem 2.1   A sailboat on a lake

Identify the major forces acting on a sailboat sailing along the surface of a lake in a gentle breeze (Fig. 2.1).

Figure 2.1
A sailboat is acted upon by several forces.

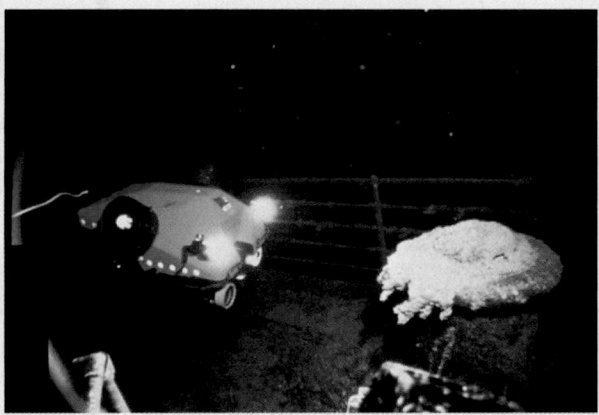

The submersible *Alvin* on the deck of the *Titanic*.

Measuring Forces

If the concept of force is to be useful in physics, there must be a way to measure forces. Consider a simple spring scale such as those used in a supermarket's produce section. As the scale's pan is pulled down, the spring above is stretched. The harder you pull, the more the spring stretches. As the spring stretches, an attached pointer moves. Then all we have to do to measure forces is to calibrate the scale so that the amount of stretch measures the strength (or *magnitude*) of the force.

As was observed by Robert Hooke (1635–1703), many springs have the property that the extension or compression—the increase or decrease in length from the unstretched length—is roughly proportional to the force exerted on the ends of the spring. **Hooke's law** says the deformation of an object tends to be proportional to the deforming force, as long as the force is not too great:

> **Hooke's law for an ideal spring**
>
> $$F = kx \qquad (2\text{-}1)$$

In Eq. (2-1), F is the *magnitude* of the force exerted on each end of the spring, x is the distance that the spring is stretched or compressed from its relaxed length, and k is called the **spring constant** for that particular spring. (Any *real* spring deviates from this simple linear behavior when stretched or compressed too much.) The SI unit for force is called the newton (N), as discussed in Section 1.5, and the SI unit of length is the meter (m), so the SI units for a spring constant are N/m.

The spring constant is a measure of how hard it is to stretch or compress a spring. A stiffer spring has a larger spring constant because larger forces must be exerted on the ends of the spring to stretch or compress it.

A supermarket scale is used to measure the weight of the produce. **Weight** is the force of the Earth's gravity pulling downward. The spring scale has no *direct* way to measure a gravitational force; what it *can* measure is the downward force that the produce exerts on the pan, which is a contact force.

In the United States, supermarket scales are generally calibrated to measure forces in pounds (lb). To convert pounds to newtons, use the approximate conversion factors

$$1.00 \text{ lb} = 4.448 \text{ N} \quad \text{or} \quad 1.00 \text{ N} = 0.2248 \text{ lb}$$

The newton is a derived unit in the SI system. It can be written in terms of three SI base units:

$$1 \text{ N} = 1 \, \frac{\text{kg} \cdot \text{m}}{\text{s}^2}$$

In most countries other than the United States, produce is sold in mass units (grams or kilograms) rather than in force units (pounds or newtons). The supermarket scale still measures a force, but the scale is calibrated to show the mass of the produce instead of the weight. A 1.0-kg mass has a weight at Earth's surface of approximately 9.8 N or 2.2 lb.

There are more sophisticated means for measuring forces than the supermarket scale. Even so, many operate on the same principle as the supermarket scale: a force is measured by the deformation—change of size or shape—it produces in some object. For the simple spring scale (Fig. 2.2), the larger the force pulling down on the scale, the more the spring stretches. We can measure a force by measuring the extension of the spring. For many springs, the extension is approximately proportional to the force [Eq. (2-1)], which makes calibration easy.

Hooke's law: the deformation is proportional to the deforming force.

The weight of an object near Earth's surface is the force of the Earth's gravity on the object.

Figure 2.2 A simple spring scale. Note that there is a pull on both ends of the scale; one pull is from the weight hanging below the scale and the other is from the hook attached to the ceiling and supporting the scale from above.

Example 2.2

Getting Down to Nuts and Bolts

 In many hardware stores, bulk nuts and bolts are sold by weight. A spring scale in the store stretches 4.8 cm when 24.0 N of bolts are weighed. On the scale, what is the distance between calibration marks, which are marked in increments of 1 N? Assume an ideal spring.

Strategy   For an ideal spring, Hooke's law holds. Rephrasing the question, we are asked how many centimeters the spring stretches for each newton of force, which is the *reciprocal* of the spring constant: k is a measure of how many newtons of force are required for each centimeter of stretch.

Solution   The reciprocal of the spring constant is

$$\frac{1}{k} = \frac{x}{F} = \frac{4.8 \text{ cm}}{24.0 \text{ N}} = 0.20 \text{ cm/N}$$

The spring stretches an additional 0.20 cm for each additional newton of force. Therefore the marks should be 0.20 cm apart.

Discussion   A variation on the solution is to find the spring constant and then use it to find the distance stretched for 1 N of applied force.

$$k = \frac{F}{x} = \frac{24.0 \text{ N}}{4.8 \text{ cm}} = 5.0 \text{ N/cm}$$

Now let $F = 1.00$ N.

$$x = \frac{F}{k} = \frac{1.00 \text{ N}}{5.0 \text{ N/cm}} = 0.20 \text{ cm}$$

The answer is reasonable: since it takes 5 N to make the spring stretch 1 cm, 1 N makes the spring stretch 1/5 cm.

Practice Problem 2.2   Stretching a spring

16.0 N of nuts are placed in the pan of the scale of Example 2.2. How far does the spring stretch?

Vector Quantities

The magnitude of a force is *not* a complete description of the force. The direction of the force is equally important. The direction of the force exerted by a baseball bat can make the difference between a home run and a foul ball; the direction of force exerted by a club face on a golf ball makes the difference as to whether the golf ball rolls into the hole or misses it.

Force is one of many quantities in physics that are called **vectors**. Because vectors are so important in physics, we introduce the mathematics of vectors gradually in Chapters 2–4 to give you time to learn about them and practice using them. Some vector quantities defined in the next few chapters include force, acceleration, velocity, and momentum. All vectors have a direction as well as a magnitude. Section 1.2 mentioned that velocity, as defined in physics, has magnitude and direction; the magnitude of the velocity vector is the speed and the direction is the direction of motion. The direction of a vector is always a physical direction in space such as up, down, north, or 35° south of west. If a homework or exam question has you calculate a vector quantity such as force or velocity, specify the direction as well as the magnitude in the answer. One without the other is incomplete. Furthermore, all vectors follow the same rules of addition—and these rules take into account the directions of the vectors being added.

Mass is not a vector, it is a **scalar**. A scalar quantity can have magnitude, algebraic sign, and units, but not a direction in space. Scalars add in the usual way: 3 kg of water plus 2 kg of water is always equal to 5 kg of water. Adding vectors is different. A 300-N force added to a 200-N force gives different results, depending on the directions of the two forces. If two friends are trying to push a car out of a snowbank, they help each other by applying forces *in the same direction*. If they push in *opposite* directions, the net effect of the two forces would be *much* smaller. In Section 2.4 we discuss how to add vectors in detail.

In this book, an arrow over a boldface symbol indicates a vector quantity ($\vec{\mathbf{F}}$). Some books use boldface without the arrow. When writing by hand, draw an arrow over a vector symbol to distinguish it from a scalar. When the symbol for a vector is written *without* the arrow and in italics rather than boldface (F), it stands for the *magnitude* of the vector (which is a scalar). Absolute value bars are also used to stand for the magnitude of a vector, so $F = |\vec{\mathbf{F}}|$.

Forces have both magnitude and direction.

Vector: a quantity with both magnitude and direction that is added according to special rules

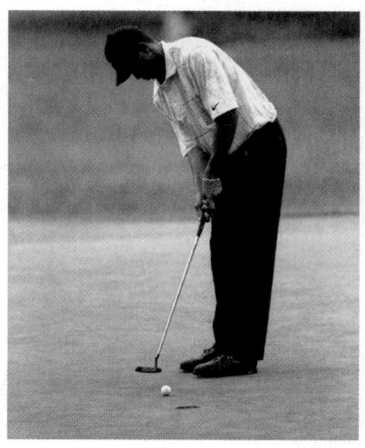

Conceptual Example 2.3

Vector or Scalar?

Can temperature be a vector quantity?

Strategy   If a quantity is a vector, it must have both a magnitude and a physical direction in space.

Solution and Discussion   Does temperature have a direction? A temperature in Fahrenheit or Celsius can be above or below zero—is that a direction? No. A vector must have a *physical direction* in space. It does not make sense to say that the temperature of your coffee is "85 degrees Celsius in the southwest direction." "The temperature is up 5 degrees today," means that it has increased, not that it is pointing vertically upward. Temperature cannot be a vector.

Conceptual Practice Problem 2.3   Bank balance

Is the balance of your checking account a vector quantity?

2.2   FUNDAMENTAL FORCES

One of the main goals of physics has been to understand the immense variety of forces in the universe in terms of the fewest number of fundamental laws. Physics has made great progress in this quest for *unification;* today all forces are understood in terms of just four fundamental interactions (Fig. 2.3). (At the high temperatures present in the early universe, two of these interactions—the electromagnetic and weak nuclear forces—can be understood as the effects of a single electroweak interaction.) The ultimate goal is to describe all forces in terms of a single interaction.

Gravity

Gravity keeps the planets in their orbits and interferes with any attempt of ours to get away from the Earth's surface, so you may be surprised to learn that gravity is the *weakest* of the fundamental forces. Gravity *seems* strong because in many cases it is the only appreciable force acting between two bodies. Any two objects exert gravitational forces on each other, but the force is tiny unless at least one of the masses is very large. We tend to notice the relatively large gravitational forces exerted by planets and stars, but not the feeble gravitational forces exerted by smaller objects.

Gravity is an unlimited-range force and is always attractive. The force gets weaker as the distance between two objects increases, but it never drops to zero, no matter how far apart the objects get.

Newton's law of gravity is an early example of unification. Before Newton, people did not understand that the same interaction that makes an apple fall from a tree also keeps the Moon in its orbit. Newton unified the Earth's gravitational force on objects with the gravitational force exerted by the Sun on the planets. A single law, Newton's law of universal gravitation, describes both.

The Strong and Weak Nuclear Forces

The strong nuclear force holds protons and neutrons together in the nucleus. The same force binds quarks (a family of elementary particles) in combinations so they can form protons and neutrons and many more exotic subatomic particles. The same force holds protons and neutrons together in the atomic nucleus. The strong nuclear force is by far the strongest of the fundamental forces—hence its name—but its range is short: its effect is negligible at distances much larger than the size of an atomic nucleus (about 10^{-15} m). We study the strong nuclear force in Chapters 29 and 30.

The range of the weak nuclear force is even shorter than that of the strong nuclear force (about 10^{-17} m). It is manifest in certain radioactive decay processes. In the Sun, the weak nuclear interaction allows thermonuclear reactions to occur, without which there would be no sunlight.

Figure 2.3 Schematic representation of the unification of forces

The Electromagnetic Force

The electromagnetic force is unlimited in range and acts between charged particles. The electrical and magnetic forces themselves were thought of separately until they were unified in the nineteenth century by James Clerk Maxwell (1831–1879). Place two magnets near each other and you will feel the macroscopic effect of the electromagnetic force. By reversing one magnet, you will see that this force can be attractive or repulsive. If you have ever experienced a shock upon touching a metal doorknob after walking across a carpet in dry weather, you have experienced the macroscopic effect of electrical forces.

What about the contact forces that are all around us? What *fundamental* interaction is responsible for the force of the wind on a kite, or the force of a muscle on a tendon? What fundamental interaction binds electrons to nuclei to form atoms, and binds atoms together in molecules? What interaction is responsible for the properties of solids, liquids, and gases, and forms the basis of the sciences of chemistry and biology? Gravity is much too weak to be responsible for these forces; the strong and weak nuclear forces are too short-ranged. The electromagnetic interaction between atoms is the cause of *all* contact forces.

The electromagnetic interactions behind these macroscopic forces can be quite complicated. Let's look again at the water strider. The force holding it up is ultimately electromagnetic in nature. How does it work? All liquids have surface tension; the molecules of the liquid pull on each other, causing the surface to be stretched like an elastic film.

Making The Connection:
forces supporting
a water strider

(a)   (b)

Figure 2.4 (a) The water molecule (H_2O) is arranged in a bent chain formation with positive and negative charge separated as shown, but held together by covalent bonds (bonds in which two electrons are shared by two atoms). (b) Hydrogen bonds, which are weaker than covalent bonds, result from an electrical attraction between a positively charged hydrogen atom in one water molecule and a negatively charged oxygen atom in another water molecule. Each oxygen atom in water is bonded to four hydrogen atoms, two by covalent bonds and two by the weaker hydrogen bonds.

Figure 2.5 Hydrogen bonds make the surface of the water like an elastic membrane and enable the water strider to stand on top of the water.

The surface tension of water is particularly strong because the water molecule is strongly *polar*: its net charge is zero, but the oxygen atom has a slight negative charge and the hydrogen atoms are slightly positive (Fig. 2.4). Since unlike charges attract each other, the hydrogen atom of one water molecule attracts the oxygen atom of another molecule, forming a hydrogen bond between the molecules.

At the beginning of this chapter we asked why a water strider can stand on the surface of a pond. For the water strider's leg to break through the surface, hydrogen bonds would have to be broken (Fig. 2.5). The gravitational force acting on the tiny mass of the water strider is too small to break the hydrogen bonds. (For more detail, see Problem 55.)

A better explanation of the hydrogen bond than this simplified model would entail quantum mechanics, but it is still the electromagnetic interaction that is responsible. The formation of hydrogen bonds gives water many of its unique properties. The hydrogen bond makes the boiling temperature of water higher than it would be otherwise. Without the hydrogen bond, there would be no oceans on Earth—just water vapor.

2.3   NEWTON'S LAWS OF MOTION

In 1687, Isaac Newton published his *Philosophiae Naturalis Principia Mathematica* (or *Principia* for short), a work whose Latin title translates as *The Mathematical Principles of Natural Philosophy*. The *Principia* contains three laws of motion that form the basis of classical mechanics.

Newton's Laws of Motion

1. If no forces act on an object, then its speed and direction of motion do not change. (If the object is at rest, it remains at rest with a speed of zero; if it is moving, it moves in a straight line with constant speed.)
2. A nonzero net force acting on an object causes its state of motion to change. Quantitatively,

$$\vec{\mathbf{F}}_{net} = m\vec{\mathbf{a}} \qquad (2\text{-}2)$$

    where $\vec{\mathbf{F}}_{net}$ is the net force, m is the mass, and $\vec{\mathbf{a}}$ is the acceleration (which measures the rate of change of the velocity).
3. In an interaction between two objects, the forces that each exerts on the other are equal in magnitude and opposite in direction.

Newton's First Law: The Law of Inertia

Newton's first law says that a body acted on by no forces either remains at rest or moves in a straight line with constant speed. Certainly it makes sense that a body at rest remains at rest unless some force acts upon it to make it move. On the other hand, it is not at all obvious that a body can continue to move with constant speed in a straight line without forces acting to keep it moving. In our experience, moving objects come to rest because of forces that oppose motion, like friction and air drag. A hockey puck can slide the entire length of a rink with very little change in speed or direction because the ice is slippery—frictional forces are small. If we could remove *all* the resistive forces, including friction and air drag, the puck would slide without changing its speed or direction at all. No force would be needed to keep it going. The brief but large contact force exerted by a hockey stick on the puck changes the puck's state of motion, but once the puck loses contact with the stick, it slides in a straight line at nearly constant velocity.

No force is required to keep a body in motion if there are no forces opposing its motion. A force is necessary to set a body into motion, if the body was previously at rest. If you throw a ball, the contact force due to your hand sets the ball into motion, but that force stops as soon as the ball leaves your hand.

Newton's first law is also called the **law of inertia**. In physics, *inertia* means resistance to *changes* in motion, not resistance to motion or "tendency to come to rest." No force needs to act on a body to keep it moving. On the contrary, in the absence of forces, a body continues moving at the same speed and in the same direction. A body at rest has a speed of zero.

Credit for the law of inertia really goes to Galileo Galilei (1564–1642), as Newton acknowledged in the *Principia*. In a series of clever experiments in which he rolled a ball up inclines of different angles, Galileo postulated that, if he could eliminate all resistive forces, a ball rolling on a horizontal surface would never stop (Fig. 2.6). Galileo made a brilliant conceptual leap from the real world with friction to an imagined, ideal world, free of friction. The law of inertia contradicted the view of the Greek philosopher Aristotle (384–322 B.C.). Almost 2000 years before Galileo, Aristotle had formulated his view that the natural state of a body is to be at rest; and, for a body to remain in motion, a force would have to act upon it continuously. Galileo conjectured that, in the absence of friction and other resistive forces, no continued force is needed to keep an object moving.

Galileo thought that the sustained motion of a body would be in a circle around the Earth—that the horizontal motion of the ball formed part of a great circle. Shortly after Galileo's death, René Descartes (1596–1650) argued that the motion of a body free of any forces should be in a straight line rather than a circle. Newton acknowledged his debt to Galileo, Descartes, and others when he wrote: "If I have seen farther, it is because I was standing on the shoulders of giants."

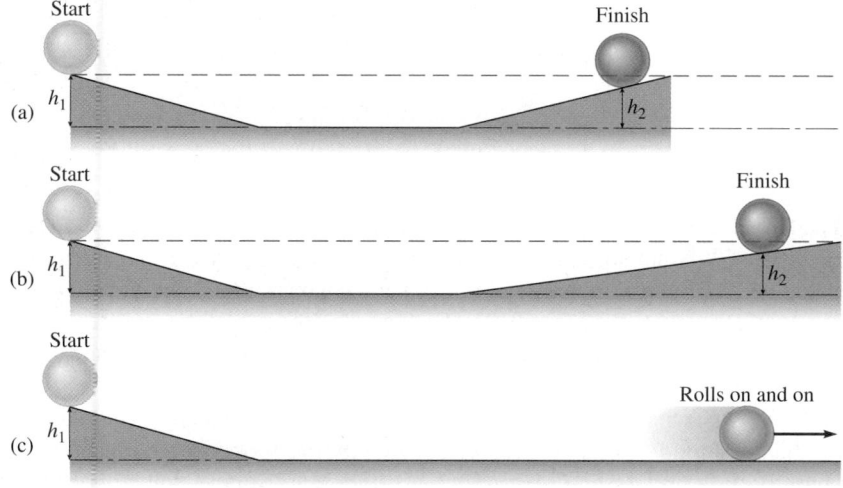

Figure 2.6 (a) Galileo found that a ball rolled down an incline stops when it reaches *almost* the same height on the second incline. He decided that it would reach the *same* height if resistive forces could be eliminated. (b) As the second incline is made less and less steep, the ball rolls farther and farther before stopping. (c) If the second incline is horizontal, and there are no resistive forces, the ball would never stop.

Conceptual Example 2.4

Snow Shoveling

The task of shoveling newly fallen snow from the driveway can be thought of as a struggle against the inertia of the snow. Without the application of a net external force, the snow remains at rest on the ground. However, there is an important way that the inertia of the snow makes it *easier* to shovel. Explain.

Strategy   Think about the physical motions used when shoveling snow. (If you live where there is no snow, think about shoveling gravel from a wheelbarrow to line a garden path.) To be "assisted by the law of inertia," there must be a time when the snow is moving on its own, without the shovel pushing it.

Solution and Discussion   Imagine scooping up a shovelful of snow and swinging the shovel forward toward the side of the driveway. The snow and the shovel are both in motion.

Then suddenly the forward motion of the shovel stops, but the snow continues to move forward because of its inertia; it slides forward off the shovel, to be pulled down to the ground by gravity. The snow does not stop moving forward when the forward force due to the shovel is removed.

This procedure works best with fairly dry snow. Wet sticky snow tends to cling to the shovel. The frictional force on the snow due to the shovel keeps it from moving forward and makes the job far more difficult.

Conceptual Practice Problem 2.4   Subway car inertia

Negar stands on a subway car, holding on to an overhead strap. As the train starts to pull out of the station, she feels thrust toward the rear of the car; as the train comes to a stop at the next station, she feels thrust forward. Explain the role played by inertia in this situation.

Physics at Home

The parlor trick in which a person pulls a tablecloth out from under dishes set for a formal dinner is a demonstration of inertia. Try a similar demonstration yourself. If you do not want to risk your dishes, place a quarter on top of an index card, or a credit card, balanced on top of a drinking glass (Fig. 2.7a). With your thumb and forefinger, flick the card so that it flies out horizontally from under the quarter. What happens to the quarter? The horizontal force on the coin due to friction is small. With a negligibly small horizontal force, the coin's inertia tends to keep it motionless while the card slides out from under it (Fig. 2.7b). Once the card is gone, gravity pulls the coin down into the glass (Fig. 2.7c).

Figure 2.7   A demonstration of inertia

(a)   (b)   (c)

The law of inertia is a postulate of classical mechanics—an assumption that is used as a starting point. It is not something we can prove experimentally. No object can ever be completely free of forces acting on it. Far from any galaxies, the gravitational force on an object would exist, even if it is small. The brilliant insights of Galileo and Descartes were to imagine what would happen in an *ideal* situation they could only approximate. Newton's second and third laws of motion are valid as long as an *inertial reference frame*—a reference frame in which the law of inertia holds—is used to observe the motion of objects.

At the beginning of this chapter we asked why the Moon orbits the Earth month after month, without any engine to drive it, while an airplane that runs out of fuel falls to the ground. A continuous force—such as that provided by an engine—is not necessary to keep an object moving if there are no resistive forces acting on it. The Moon is not slowed by friction or air resistance, but an airplane does encounter significant air resistance and must compensate using the forward force due to its engines.

The Moon does not move in a straight line because the force of Earth's gravity is pulling on it, changing its direction of motion. Imagine suddenly turning off the gravitational force on the Moon. It would then move in a straight line, getting farther and farther from Earth. The gravitational force on the Moon due to the Earth *does* cause the Moon to continually fall toward Earth—just the right amount so that it stays the same distance from Earth in its circular orbit.

Newton's Second Law

In Chapters 3 and 4, the state of motion of an object is characterized by its velocity—a vector quantity that conveys both the speed and direction of motion. The law of inertia says that if no forces act on an object, it moves with constant velocity; neither its speed nor its direction change.

Newton's second law describes what happens when forces do act. When more than one force acts on an object, an extremely important quantity is the **net force** (or **resultant force**) acting on the object. The net force is the vector sum of all the forces acting on an object. Newton's second law relates changes in an object's state of motion to the net force acting on that object. When the net force on a body is not zero, the body's velocity changes; a quantity called acceleration (which is also a vector quantity) measures the way velocity changes. Newton's second law states that the acceleration is proportional to the net force acting on the body and is in the same direction as the net force. Quantitative use of Newton's second law when the net force is nonzero depends upon careful definitions of velocity and acceleration, which we defer to Chapters 3 and 4.

For now we consider cases where the net force is zero. If the net force is zero, then the acceleration is zero and the velocity stays the same—the body moves as if there were no forces at all. Its direction and speed of motion do not change because the forces "cancel" each other. Such a body is said to be in **translational equilibrium**. *Equilibrium* conveys the idea that the forces are in balance; there is as much force upward as there is downward, as much to the right as to the left, and so forth.

An object in translational equilibrium is not necessarily at rest; it may be moving at constant speed in a straight line. If it *is* at rest, it is said to be in **static equilibrium**; otherwise, it moves at constant velocity in **dynamic equilibrium**.

The net force (or resultant force) is the vector sum of all the forces acting on an object.

An object in translational equilibrium has a net force of zero acting on it.

Newton's Third Law

In Section 2.1, we said that an interaction always consists of *two* forces; no object can exert a force without having a force exerted on it. Nature is quite democratic when it comes to interactions: not only are there always two forces, one on each object, but the two always have the *same magnitude*. The directions of the two forces are always opposite. This is Newton's third law: an interaction between two bodies consists of two forces, one acting on each body, that are equal in magnitude and opposite in direction.

When the Earth's gravity pulls down on a satellite, the satellite pulls upward just as hard on the Earth. Though the forces are the same strength, the effect they have on the motion of the two bodies is very different. The force on the satellite keeps it moving in a circular orbit around the Earth, while the force exerted by the satellite on the Earth has an immeasurably small effect on the Earth's motion. According to Newton's second law, the effect of a force acting upon an object depends on the object's mass, which is a measure of its inertia. The Earth, with vastly greater inertia—remember that means resistance to change in motion—reacts very little to the force. The satellite's mass is much less, so the same magnitude force has a much larger effect.

Figure 2.8 Two arctic terns fight over a fish. The forces exerted on the fish by the two birds, $\vec{\mathbf{F}}_{f1}$ and $\vec{\mathbf{F}}_{f2}$, *cannot* be third-law partners since they act on the same object.

Making The Connection:
detection of
extrasolar planets

Multiplying a vector by –1 reverses its direction. The vector $-\vec{\mathbf{A}}$ has the same magnitude as $\vec{\mathbf{A}}$ but its direction is opposite.

On the other hand, if a massive planet orbits a star in a relatively small orbit, the gravitational force that the planet exerts on the star can make the star wobble enough to be observed. The wobble enables astronomers to discover planets orbiting stars other than the Sun. The planets do not reflect enough light toward Earth to be seen, but their presence can be inferred from the effect they have on the star's motion.

Do not assume that Newton's third law is involved *every* time two forces *happen* to be equal and opposite—"it ain't necessarily so." Many situations involve two equal and opposite forces acting *on a single body*. Such forces cannot be *third-law partners* because they act on the same body. The third law deals with the two forces that comprise a *single interaction.* These two forces act on *different bodies*, one on each of the two bodies that are interacting.

Two children fighting over a toy *may* be pulling on it with equal and opposite forces, but they are not *required* to do so by Newton's laws. Suppose two arctic terns fight over a fish by pulling in opposite directions with equal magnitude forces. The two forces exerted on the fish, shown in Fig. 2.8, are *not* third-law partners. The third-law partner of the force exerted on the fish by one bird ($\vec{\mathbf{F}}_{f1}$) is the force exerted on that bird by the fish ($\vec{\mathbf{F}}_{1f}$).

To correctly identify two forces that are third-law partners, one of them must be a force exerted on object 1 by object 2 and the other must be a force exerted on object 2 by object 1. To keep track of what exerts a force on what, we can use subscripts. Thus, $\vec{\mathbf{F}}_{12}$ stands for the force exerted *on 1 by 2*, while $\vec{\mathbf{F}}_{21}$ stands for the force exerted on 2 by 1. These two forces must be equal in magnitude and opposite in direction: $\vec{\mathbf{F}}_{21} = -\vec{\mathbf{F}}_{12}$.

Conceptual Example 2.5

A Dangling Fish

A fish is suspended by a line from a fishing pole. Choose two forces acting on the fish and describe the third-law partner of each.

Strategy For a force exerted on the fish by object A, its third-law partner is the force exerted on A by the fish. The two forces must be equal in magnitude and opposite in direction, according to Newton's third law.

Solution and Discussion The weight of the fish is the downward gravitational force exerted on the fish by the Earth, $\vec{\mathbf{F}}_{fE}$. Its third-law partner must be the upward gravitational force on

the Earth by the fish, $\vec{\mathbf{F}}_{Ef}$. The fish is suspended from a line, so there is a downward force on the line due to the fish, $\vec{\mathbf{F}}_{lf}$; its third-law partner is the upward force on the fish due to the line, $\vec{\mathbf{F}}_{fl}$. Figure 2.9a shows these two pairs of forces.

 Since third-law partners act on different objects, the two forces of the pair can never appear in a diagram showing the forces on a single object. Figure 2.9b shows the fish in isolation with the forces acting on the fish; such forces acting on a single body *cannot* be third-law partners.

Conceptual Practice Problem 2.5 The fishing rod
Identify the Newton's third-law partner of each force acting on the fishing rod.

Figure 2.9 (a) Sketch to illustrate pairs of forces and (b) forces acting on the fish.

2.4   NET FORCE: VECTOR ADDITION

To find the net force acting on an object, we sum all of the forces acting on it. As previously mentioned, vector quantities have their own rules of addition. For now we add only *collinear*—parallel or antiparallel—vectors. Chapter 4 generalizes this method to include noncollinear vectors.

Adding Collinear Vectors

Suppose we want to add two force vectors, $\vec{\mathbf{F}}_1$ and $\vec{\mathbf{F}}_2$, to find the net force acting on an object. We can add them graphically. To draw a force vector, use an arrow pointing in the direction of the force. Make the length of the arrow proportional to the magnitude of the vector. To add two vectors, first draw one of them. Then draw the second one starting where the first left off. In other words, place the tail of the second vector at the tip of the first. Next, draw an arrow starting from the tail of the first and ending at the tip of the second. This arrow represents the **resultant vector**—the vector sum of the two. This graphical method shows what happens when two collinear vectors are added.

- If the two forces are in the same direction, the resultant force is in the same direction (Fig. 2.10a). Its magnitude is the sum of the magnitudes of the two. If you and your friend push a heavy trunk with forces of 200 N and 300 N in the same direction, the net effect is that of a 500-N force in that direction.

- If two forces are in opposite directions, or **antiparallel**, the magnitude of their sum is the *difference* between the magnitudes of the two vectors (Fig. 2.10b)—the larger minus the smaller, since vector magnitudes are never negative. The direction of the vector sum is the direction of the larger of the two. Pushing that trunk with a

Figure 2.10 Addition of collinear vectors that are (a) in the same direction and (b) in opposite directions. (c) The sum of two vectors with equal magnitudes and opposite directions is zero.

200 N   300 N

500 N

(a)

400 N

300 N

100 N

(b)

250 N

250 N

Sum is zero

(c)

400-N force to the right and a 300-N force to the left has the net effect of a 100-N force to the right.

- The only way that two vectors can add to zero is if they are opposites: they must have the same magnitude but opposite directions (Fig. 2.10c).

 Whenever quantities are added or subtracted, check whether the quantities are vectors. If so, be sure to add them correctly—do not just add their magnitudes.

Conceptual Example 2.6

Bird in Dynamic Equilibrium

A bird weighing 1 N is flying north in dynamic equilibrium. What is the force acting on the bird due to the air?

Strategy and Solution First, we identify all the forces acting on the bird. Since the bird is in dynamic equilibrium, the vector sum of the forces acting on it must be zero.

There are only two forces acting on the bird. One of them is long-range: the gravitational pull of the Earth, which is the bird's weight $\vec{\mathbf{W}}$. The other is a contact force due to the air ($\vec{\mathbf{F}}_{ba}$). Nothing else is in contact with the bird, so there are no other contact forces. Since the bird is in dynamic equilibrium, the net force acting on the bird must be zero:

$$\vec{\mathbf{W}} + \vec{\mathbf{F}}_{ba} = 0 \quad \text{or} \quad \vec{\mathbf{F}}_{ba} = -\vec{\mathbf{W}}$$

So the force on the bird due to the air is 1 N, directed upward.

Discussion The interaction between the bird and the air is complex at the microscopic level, as was the interaction between the *Alvin* and the seawater. The net effect of all the interactions between air molecules and the bird produces an upward force of 1 N.

With the framework of Newton's laws, we can now explain why a spring scale can measure weight (Section 2.1). If a melon is at rest in the pan of the scale, the net force on it must be zero. There are only two forces acting: gravity pulls down and the scale pulls up. These two forces must be equal in magnitude, so the reading of the scale is equal to the weight of the melon.

Conceptual Practice Problem 2.6   A crate of apples

An 80-N crate of apples sits at rest on a ramp that runs from the ground to the bed of a truck. The ramp is inclined at 20° to the ground. What is the force exerted on the crate by the ramp?

Making The Connection:
  forces on a
  bird in air

In Conceptual Example 2.6, we concluded that the air exerts an upward force on the bird to hold it up. How does the bird get the air to do that? By flapping its wings and pushing air downward, the bird takes advantage of Newton's third law—the air must push up on the bird. The shape of the wing is changed so that it pushes harder on the air while moving down than while moving up, so the average force on the air is downward. You might wonder whether there is a *drag* force—a force due to the air that resists the forward motion of the bird. There is, so the bird must flap its wings in such a way as to also push some air backward. The air then pushes forward on the bird. This push is called the *thrust*. If the bird is flying at constant velocity, the drag and the thrust are equal in magnitude. The total force on the bird due to the air—the sum of lift, drag, and thrust—is still 1 N upward.

Physics at Home

The next time you go swimming, you can feel the same effect that the flying bird does, except you will be in the water. Push down and backward on the water with your arms and legs. The water pushes up and forward on you. The various swimming strokes are devised so that you exert as large a force as possible backward on the water during the power part of the stroke, and then as small a force as possible forward on the water during the return part of the stroke.

Free-Body Diagrams

An essential tool used to find the net force acting on an object is a **free-body diagram** (FBD). In a free-body diagram, the body isn't free from interactions; the word *free* is taken to mean *isolated*. An FBD is a sketch of a single body with force vectors drawn to represent *every* force acting on that body. It does *not* include forces that act on other bodies. To draw an FBD:

- First draw the object. Feel free to draw it in a simplified way—you do not have to be Renoir to solve a mechanics problem. Almost any object can be represented as a box or a circle, or even a dot.

- Next, draw vector arrows representing every force acting on that object. Draw the arrows so that they correctly illustrate the directions of the forces. We usually draw the vectors as arrows that start on the object and point away from it, but remember that the vectors represent forces exerted *on* the object by something else. If you have enough information to do so, draw the lengths of the arrows so that they are proportional to the magnitudes of the forces. Figure 2.9b is an example of an FBD for the fish in Conceptual Example 2.5.

The most important thing is not to omit any forces and not to draw any extra ones. To make sure that you do not miss any, consider everything in contact with the object to see if there are contact forces; also include any long-range forces. Every force you draw must be a force exerted *on that object by something else*. Ask yourself: what is the other object that is exerting this force?

Internal and External Forces

When we say that a baseball has interactions with the Earth (gravity), with a baseball bat, and with the air, we are treating the baseball as a single entity. But the ball really consists of an enormous number of protons, neutrons, and electrons, all interacting with each other. The protons and neutrons interact with each other to form atomic nuclei; the nuclei interact with electrons to form atoms; interactions between atoms form molecules; and the molecules interact to form the structure of the thing we call a baseball. It would be difficult to have to deal with all of these interactions to predict the motion of a baseball.

Call the set of particles comprising the baseball a **system**. We can classify all the interactions as either **internal** or **external**. Internal interactions are those for which *both* interacting particles are part of the system. Suppose we want to add up all the forces acting on all of the particles comprising the baseball in order to find the net force. Every internal interaction consists of two forces, equal in magnitude and opposite in direction, so the sum of the two is zero. But for the external interactions, *only one of the two forces is exerted on the ball*. So to find the net force exerted on the ball, we only need to add external forces; the internal forces all add to zero.

The statement that internal forces always add to zero is particularly powerful because the choice of what comprises a system is completely arbitrary. We can choose *any* set of particles and think of it as a system. In one problem, it may be convenient to think of the baseball as a system; in another, we may choose a system comprising both the baseball and the bat together. The second choice might be useful if we do not have detailed information about the interaction between the bat and the ball.

Example 2.7

A Gold Crown

A gold crown of weight 17.5 N is immersed in a beaker of water while suspended from a spring scale (Fig. 2.11). When the crown is in static equilibrium, the reading on the spring scale is 9.0 N. (a) What is the force exerted on the crown by the water? (b) If the spring in the scale is 0.50 cm longer than when it is relaxed, what is its spring constant? (c) What is the

continued on next page

Example 2.7 *continued*

reading of the pan scale on which the beaker sits? The weight of the water and the beaker together is 23.5 N.

Strategy   (a) The question asks for one of the forces acting on the crown. First, we identify the forces acting on the crown. There are three: gravity, a force due to the spring scale, and a force due to the water. Since the crown is in static equilibrium, these three forces must sum to a net force of zero. A free-body diagram will help make sure we add them correctly.

(b) Assume the spring to be ideal and apply Hooke's law.

(c) The question asks for the reading of the pan scale, which is the magnitude of the force exerted on it by the beaker. Should we draw an FBD for the beaker?

Unfortunately, the weight of the beaker itself is not given, nor is the weight of the water. We do not have enough information to use an FBD for the beaker to solve for one of the forces; we would need to know the magnitude of all the other forces acting on the beaker. We do know the combined weight of the beaker including the water in it. Therefore we will consider the beaker and the water to be a single object or system throughout the problem. To solve part (c), we draw an FBD for the "water + beaker."

To keep the forces straight, use subscripts: c = crown; s = spring scale; w = water + beaker; p = pan scale; and e = Earth. The forces given in the problem are: (1) crown weight = $\vec{\mathbf{W}}_{ce}$ = 17.5 N, downward; (2) force on the spring scale due to the crown = $\vec{\mathbf{F}}_{sc}$ = 9.0 N, downward; and (3) weight of water + beaker = $\vec{\mathbf{W}}_{we}$ = 23.5 N, downward.

We need to find (a) the force on the crown by the water ($\vec{\mathbf{F}}_{cw}$), (b) the spring constant k, and (c) the pan scale reading = the magnitude of the force on the pan scale due to water + beaker = $|\vec{\mathbf{F}}_{pw}|$.

Solution   (a) Figure 2.12a shows an FBD for the crown. The forces due to the spring scale ($\vec{\mathbf{F}}_{cs}$) and the water ($\vec{\mathbf{F}}_{cw}$) are directed upward, while the force of gravity ($\vec{\mathbf{W}}_{ce}$) is downward.

The crown is in static equilibrium, so the sum of the forces acting on the crown must be zero.

$$\vec{\mathbf{F}}_{cs} + \vec{\mathbf{W}}_{ce} + \vec{\mathbf{F}}_{cw} = 0$$

This is *not* ordinary addition; it is vector addition. Vectors must be added according to the rules presented earlier.

The spring scale reads 9.0 N, which means that the force on the spring scale due to the crown ($\vec{\mathbf{F}}_{sc}$) has magnitude 9.0 N.

The force that the spring scale exerts on the crown must be equal in magnitude and opposite in direction. Therefore $\vec{\mathbf{F}}_{cs}$ has magnitude 9.0 N and is directed upward.

The two known forces are in opposite directions, so their sum is in the direction of the larger one and has a magnitude equal to the *difference* of the two magnitudes:

$$\vec{\mathbf{F}}_{cs} + \vec{\mathbf{W}}_{ce} = 9.0 \text{ N up} + 17.5 \text{ N down} = 8.5 \text{ N down}$$

To this sum, the force $\vec{\mathbf{F}}_{cw}$ must be added to give a net force of zero:

$$(\vec{\mathbf{F}}_{cs} + \vec{\mathbf{W}}_{ce}) + \vec{\mathbf{F}}_{cw} = 0$$

Two vectors that add to zero must be equal and opposite, so

$$\vec{\mathbf{F}}_{cw} = -(\vec{\mathbf{F}}_{cs} + \vec{\mathbf{W}}_{ce}) = 8.5 \text{ N up}$$

(b) For an ideal spring, Hooke's law says that

$$F = kx$$

where k is the spring constant, x is the extension of the spring from the relaxed length, and F is the force acting on the spring to cause the extension. The magnitude of the force acting on the spring due to the crown is 9.0 N. Solving for the spring constant,

$$k = \frac{F_{sc}}{x} = \frac{9.0 \text{ N}}{0.50 \text{ cm}} = 18 \text{ N/cm}$$

(c) Finally, what is the reading of the pan balance (the magnitude of $\vec{\mathbf{F}}_{pw}$)? It is not necessarily equal to the weight of the object sitting on it (the water + beaker) since other forces act on that system. Is it equal to the weight of water + beaker + crown? No, because there is an upward force on the crown due to the spring scale, which is partially supporting the crown. The reading of the pan scale should fall between the weight of the water + beaker (23.5 N) and the weight of water + beaker + crown (41.0 N).

The force on the crown due to the water is $\vec{\mathbf{F}}_{cw}$ = 8.5 N upward. The other force in the interaction between the crown and the water is the force exerted on the water by the crown, $\vec{\mathbf{F}}_{wc}$. It must act in the opposite direction and be of the same magnitude; therefore the crown is pushing down on the water + beaker system with a force of magnitude 8.5 N. The system

Figure 2.11
A gold crown is suspended from a spring scale into a beaker of water.

Figure 2.12
(a) A free-body diagram showing the forces acting on the crown and (b) a free-body diagram for the water + beaker system.

continued on next page

Example 2.7 *continued*

of water plus beaker pushes down on the pan balance with a force greater in magnitude than the weight of the water plus beaker system, $\vec{\mathbf{W}}_{we}$, as shown in the free-body diagram (Fig. 2.12b).

The system water + beaker is in static equilibrium:

$$\vec{\mathbf{F}}_{net} = \vec{\mathbf{W}}_{we} + \vec{\mathbf{F}}_{wc} + \vec{\mathbf{F}}_{wp} = 0$$

$$\vec{\mathbf{F}}_{wp} = -(\vec{\mathbf{W}}_{we} + \vec{\mathbf{F}}_{wc}) = -(23.5 \text{ N down} + 8.5 \text{ N down})$$

$$= -(32.0 \text{ N down}) = 32.0 \text{ N up}$$

The force that the water + beaker exerts on the pan scale is therefore 32.0 N down.

$$\vec{\mathbf{F}}_{pw} = -\vec{\mathbf{F}}_{wp} = 32.0 \text{ N down}$$

The pan scale reading, which is the magnitude of the force on the pan by the system of water + beaker, is 32.0 N, which, as expected, is between 23.5 N and 41.0 N.

Discussion Since the individual weights of the beaker and of the water are not known, it is not possible to draw an FBD for either one individually. Nor can we find out the forces that the water and beaker exert on each other. By considering the water and beaker as a single object or system, the forces that the water and beaker exert on each other become internal to the system. Therefore they are not included in an FBD for the composite object water + beaker.

Practice Problem 2.7 Force exerted by table

The weight of the pan scale in Example 2.7 is 13.0 N. What is the upward force exerted on it by the table on which it rests when it is arranged as shown in Fig. 2.11?

Example 2.8

Net Force on an Airplane

The external forces on an airplane in flight heading eastward are as follows: weight = 16.0 kN, downward; lift = 16.0 kN, upward; thrust = 1.8 kN, east; and drag = 0.8 kN, west. What is the net force on the plane?

Strategy After drawing an FBD, we must add the forces. They are not all along the same line, but we can first add those that *are* collinear and see what happens.

Figure 2.13
Free-body diagram for the airplane

Solution Figure 2.13 is a free-body diagram for the plane, using $\vec{\mathbf{W}}$, $\vec{\mathbf{L}}$, $\vec{\mathbf{T}}$, and $\vec{\mathbf{D}}$ for the weight, lift, thrust, and drag. There are two vertical vectors ($\vec{\mathbf{W}}$ and $\vec{\mathbf{L}}$). They are equal in magnitude and opposite in direction, so they add to zero:

$$\vec{\mathbf{W}} + \vec{\mathbf{L}} = 0$$

The two remaining vectors are horizontal. The thrust is larger, so their sum is in the direction of the thrust (east). The magnitude of their sum is the difference of their magnitudes:

$$T - D = 1.8 \text{ kN} - 0.8 \text{ kN} = 1.0 \text{ kN}$$

The net force is therefore

$$\vec{\mathbf{F}}_{net} = \vec{\mathbf{W}} + \vec{\mathbf{L}} + \vec{\mathbf{T}} + \vec{\mathbf{D}} = (\vec{\mathbf{W}} + \vec{\mathbf{L}}) + (\vec{\mathbf{T}} + \vec{\mathbf{D}})$$

$$= 0 + (1.0 \text{ kN east}) = 1.0 \text{ kN east}$$

Discussion When faced with vectors that are not *all* along the same line, usually the easiest thing is to first add groups of vectors that are collinear.

Practice Problem 2.8 New forces on the airplane

Find the net force on the airplane if the forces are weight = 16.0 kN, downward; lift = 15.5 kN, upward; thrust = 1.2 kN, north; drag = 1.2 kN, south.

2.5 GRAVITATIONAL FORCES

It is useful to summarize our knowledge about various forces, whether fundamental or not, in **force laws**. A force law states the magnitude and direction of a particular force.

Let us start with the gravitational force. According to Newton's law of universal gravitation, two bodies exert gravitational forces on each other that are proportional to

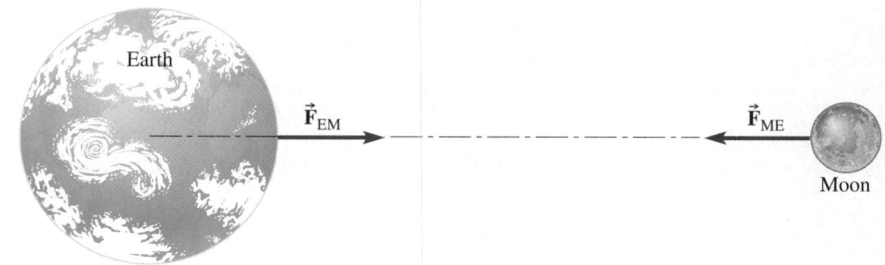

Figure 2.14 Gravity is always an attractive force. The force that each body exerts on the other is equal in magnitude, even though the masses may be very different. The force exerted on the Moon by the Earth ($\vec{\mathbf{F}}_{ME}$) is of the same magnitude as the force exerted on the Earth by the Moon ($\vec{\mathbf{F}}_{EM}$).

the masses (m_1 and m_2) of the two bodies and inversely proportional to the square of the distance (r) between their centers. Strictly speaking, the law of gravitation as presented here only applies to point particles and symmetric spheres. (The *point particle* is a common model in physics used when the size of an object is negligibly small and the internal structure is irrelevant.) But the law of gravitation is *approximately* true for any two bodies if the distance between their centers is large compared to their sizes.

In mathematical language, the law of universal gravitation is written:

$$F = \frac{Gm_1m_2}{r^2} \tag{2-3}$$

where the constant of proportionality ($G = 6.673 \times 10^{-11}$ N·m²/kg²) is called the universal gravitational constant. Gravity is an attractive force: each body is pulled toward the other's center (Fig. 2.14). The forces on the two bodies are equal in magnitude but the directions are opposite.

Example 2.9

Gravitational Force between Two Cars

Estimate the gravitational force exerted by one car on another when they are driving side-by-side in parallel lanes.

Strategy The cars are not spheres and are not far apart, but as long as we want only an order-of-magnitude estimate, we can apply the law of gravity. We need to make reasonable estimates for the masses of the cars and the distance between their centers. Then the universal law of gravitation will give an estimate of the force each exerts on the other. We expect the gravitational force to be quite small—otherwise, there would be many more auto accidents.

Solution A car might have a mass of around 1000 kg, so let $m_1 = m_2 = 1000$ kg. A reasonable approximation is that it is 4 m from the center of one car to the center of the other. Then

$$F = \frac{Gm_1m_2}{r^2} = \frac{6.673 \times 10^{-11} \text{ N·m}^2\text{·kg}^{-2} \times (1000 \text{ kg})^2}{(4 \text{ m})^2}$$

$$= 4 \times 10^{-6} \text{ N} = 4 \text{ μN}$$

which is about the same magnitude as the weight of a mosquito.

Discussion The estimate shows that gravitational forces exerted by ordinary objects are so small as to be negligible in most cases. The gravitational force on one car due to the other is tiny compared to other forces acting on the car, such as the gravitational force due to the Earth (about 10 kN) and the contact forces exerted by the road and by the air. The Earth exerts appreciable gravitational forces because its mass is so large.

Writing out the units of all quantities, including the constant G, prevents a mistake such as forgetting to square r. In the calculation in the solution section here, all the units other than newtons cancel, leaving a force in units of newtons:

$$\frac{\text{N·m}^2\text{·kg}^{-2} \times \text{kg}^2}{\text{m}^2} = \text{N}$$

Practice Problem 2.9 Gravitational force between two humans

Alex is on stage playing his bass guitar. Estimate the *gravitational* attraction between Alex and Frances, a fan who is standing 8 m from Alex. Alex has a mass of 55 kg and Frances has a mass of 40 kg.

For objects near Earth's surface, we often use a simpler expression for the *weight* of an object—which is the force of gravity on the object due to the mass of the Earth. At small distances above the Earth's surface, the distance between the object and the Earth's center is approximately equal to the Earth's mean radius, $R_E = 6.37 \times 10^6$ m. The mass of the Earth is $M_E = 5.98 \times 10^{24}$ kg, so the weight of an object of mass m near Earth's surface is

$$W = \frac{GM_E m}{R_E^2} = m \times \left(\frac{GM_E}{R_E^2}\right) = mg \qquad (2\text{-}4)$$

where g is called the **gravitational field strength** at the surface of the Earth. From Eq. (2-4),

$$g = \frac{GM_E}{R_E^2} = \frac{6.673 \times 10^{-11}\ \text{N·m}^2\text{·kg}^{-2} \times (5.98 \times 10^{24}\ \text{kg})}{(6.37 \times 10^6\ \text{m})^2} = 9.8\ \text{N/kg}$$

We have rounded to two significant figures because g varies from place to place. Since the newton is defined as $1\ \text{N} = 1\ \text{kg·m/s}^2$, g can be written in SI base units as:

$$g = 9.8\ \frac{\text{N}}{\text{kg}} \times \frac{1\ \text{kg·m/s}^2}{1\ \text{N}} = 9.8\ \text{m/s}^2$$

At most locations on the surface of the Earth, the gravitational field strength g is approximately 9.8 N/kg. The Earth is not a perfect sphere; it is slightly flattened at the poles. Since the distance from the surface to the center of the Earth is smaller there, the field strength is greatest at the poles: about 9.832 N/kg, in contrast to 9.814 N/kg at the equator (for sea level). Altitude also matters; as you climb above sea level, your distance from Earth's center increases and the field strength decreases. Tiny local variations in the field strength are also caused by geologic formations. On top of dense bedrock, g is a little greater than above less dense rock. Geologists and geophysicists measure these variations to study Earth's structure and also to locate deposits of various minerals, water, and oil. The device they use, a *gravimeter*, is essentially a mass hanging on a spring. As the gravimeter is carried from place to place, the extension of the spring increases where g is larger and decreases where g is smaller. The mass hanging from the spring does not change, but its weight does ($W = mg$).

Not only is the weight per unit mass at a given point the same for any object at that point, but so is the *direction* of the gravitational force. The **gravitational field** vector $\vec{g}$ is defined as the force per unit mass; its direction is that of the gravitational force exerted on an object at that point. Then $\vec{W} = m\vec{g}$, where $\vec{g}$ is the gravitational field at some point and $\vec{W}$ is the weight of an object at that point. Note that in this equation a vector ($\vec{g}$) is multiplied by a scalar (m). Since m is positive, the magnitude of the vector is multiplied by m but its direction is not changed. Multiplying a vector by a *negative* scalar reverses its direction.

Weight and mass are *not* the same thing. If you climb Mt. McKinley, the *weight* of your gear decreases as you get farther from the center of the Earth, but its *mass* does not change.

Gravitational field $\vec{g} = \dfrac{\vec{W}}{m}$

Gravitational force per unit mass

Example 2.10

Weight at High Altitude

When you are in a commercial airliner cruising at 6.40×10^3 m (21,000 ft), by what percentage has your weight (as well as the weight of the airplane) changed?

Strategy   This is a good problem to work by proportions. The masses and the universal gravitational constant are the same in both cases. Only the distance r changes.

Solution   Your weight is inversely proportional to the square of the distance r between you and Earth's center:

$$W = \frac{GM_E m}{r^2} \propto r^{-2}$$

At an altitude h above the Earth's surface, your weight is smaller by a factor

$$\frac{W}{W_{\text{surface}}} = \left(\frac{R_E + h}{R_E}\right)^{-2} = \left(1 + \frac{h}{R_E}\right)^{-2}$$

$$= \left(1 + \frac{6.40 \times 10^3\ \text{m}}{6.37 \times 10^6\ \text{m}}\right)^{-2} = 0.997994 \approx 1 - 0.0020$$

Your weight decreases by about 0.20%.

Discussion   Although 6400 m may seem like a significant altitude to us, it's a small fraction of the Earth's radius (0.100%), so the weight change is a small percentage. When

continued on next page

Example 2.10 *continued*

judging whether a quantity is small or large, always ask: small compared to what?

To find the weight of other objects on the plane, it would not be necessary to redo the whole calculation. *Every* object in the airplane—as well as the plane itself—changes weight by the same percentage. The weight of each object is still proportional to its mass, so we can still write $W = mg$, where g is the strength of the *local* gravitational field. Once g is known at a point, it can be used to find the weight of any object at that same point.

Practice Problem 2.10   A satellite put into orbit

The space shuttle carries a satellite in its cargo bay and places it into orbit around the Earth. Find the ratio of the Earth's gravitational force on the satellite when it is on a launch pad at the Kennedy Space Center to the gravitational force exerted when the satellite is orbiting 6.00×10^3 km above the launch pad.

Electrical Forces

There is another force law that is strikingly similar to Newton's law of universal gravitation. Just as all massive particles exert gravitational forces on each other, all *charged* particles exert *electrical* forces on each other. Coulomb's law says that two charged particles exert electrical forces on each other that are proportional to the magnitudes of the charges ($|q_1|$ and $|q_2|$) of the two particles and inversely proportional to the square of the distance (r) between them. In mathematical language, Coulomb's law is written

$$F = \frac{k|q_1||q_2|}{r^2} \tag{2-5}$$

where k is a constant of proportionality analogous to G.

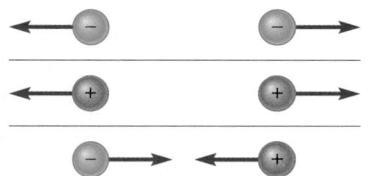

Figure 2.15 Like charges repel; opposite charges attract.

Unlike the gravitational force, which is always attractive, the electric force can be either attractive *or* repulsive (Fig. 2.15). The particles attract each other if their charges have opposite signs (like an electron and a proton); particles repel each other if their charges have the same sign (like two electrons). Just as for the law of universal gravitation, Coulomb's law explicitly obeys Newton's third law; the two forces are equal in magnitude and opposite in direction.

Physics at Home

Rub a balloon against your sweater and then place it against a wall. The rubbing transfers some charge between the balloon and the sweater, so they acquire net charges of opposite signs. When you place the balloon against the wall, it attracts opposite charges in the wall and repels like charges. The opposite charge in the wall is on average a little closer to the balloon than the like charge. As a result, the repulsive force due to the like charges is weaker than the attractive force due to the unlike charges. Thus there is a net attractive force on the balloon due to the wall. If there is a large enough charge on the balloon, this electrical attraction is capable of holding the balloon in place.

An **electric field** vector $\vec{\mathbf{E}}$ can be defined in a way analogous to the gravitational field. The gravitational field is the force per unit mass, but the electric force is proportional to charge, not mass. So the electric field is defined as the force per unit *charge*:

$$\vec{\mathbf{E}} = \frac{\vec{\mathbf{F}}_e}{q} \quad \left(\text{compare to } \vec{\mathbf{g}} = \frac{\vec{\mathbf{F}}_g}{m} \right)$$

We study electric fields in detail in Part 3 of this book.

2.6   CONTACT FORCES

Contact forces are the net effect of enormous numbers of electromagnetic interactions between atoms when two objects come into contact. Electromagnetic forces are fundamentally long-range, but due to the balance of positive and negative charge in macroscopic objects, the net effect is negligible unless atoms on the two surfaces come very close to each other—what we think of as *in contact*. On a microscopic level there is no fundamental difference between the contact forces and other electromagnetic forces.

Normal Forces

Consider a book resting on a horizontal desk surface. The state of motion of the book would be very different if the desk were removed. So there must be an upward force exerted by the desk on the book. The force acting on the book due to the desk is called the **normal force** because it is perpendicular to the contact surface between the book and the desk. (In geometry, the word *normal* means *perpendicular*.) If the surface of the desk is not horizontal, the normal force would not be vertical; by definition the normal force is perpendicular to the contact surface.

According to Newton's third law, there is also a normal force exerted by the book on the desk; this normal force acts downward and is of equal magnitude. In everyday language, we might say that the desk "feels the book's weight pushing down on it." That is not an accurate statement in the language of physics. The desk cannot "feel" the gravitational force on the book; it can only feel forces exerted *on it*. What the desk does "feel" is the normal force exerted on it by the book. The normal force always pushes one object *away* from the other.

Normal force: a contact force between two solid objects that is perpendicular to the contact surfaces. Each object pushes the other one away.

We can figure out what the magnitude of the normal force must be in various situations. In the case of the book, the normal force due to the desk must have just the right magnitude to keep the book from falling without making it shoot up into the air. If the book is in equilibrium and no other vertical forces act, the normal force on the book is equal in magnitude to the book's weight. However, if there are other vertical forces acting on the book, or if the book is not in equilibrium, then the normal force is *not* equal in magnitude to the book's weight.

How does the desk "know" how hard to push up on the book? First imagine putting the book on a bathroom scale instead of the desk. A spring inside the scale provides the upward force. The spring "knows" how hard to push because, as it is compressed, the force it exerts increases. When the book reaches equilibrium, the spring is exerting just the right amount of force, so there is no tendency to compress it further. The spring is compressed until it pushes up with a force equal to the book's weight. If the spring were stiffer, it would exert the same upward force but with less compression.

The forces that bind atoms together in a rigid solid, like the desk, act like extremely stiff springs that can provide large forces with little compression—so little that it's usually not noticed. The book makes a tiny indentation in the surface of the desk; a heavier book would make a slightly larger indentation. If the book were to be placed on a soft foam surface, the indentation would be much more noticeable.

Friction

A contact force *parallel* to the surfaces of contact is called **friction**. We distinguish two types: **static friction** and **kinetic friction**. When the two surfaces are sliding across one another, as when a loose shingle slides down a roof, the friction is kinetic. When no sliding occurs, such as between the tires of a car parked on a hill and the road surface, the friction is called static. Static friction acts to prevent objects from sliding; kinetic friction acts to make sliding objects slow down and stop moving relative to each other.

Frictional forces are enormously complicated on the microscopic level and the effects of friction are not very reproducible. For dry, solid surfaces, the amount of friction depends on how smooth the surfaces are and how many contaminants are present on the surface. Does polishing two steel surfaces decrease the frictional forces when they slide across each other? Yes, up to a point. If the surfaces are extremely smooth and all surface contaminants are removed, the steel surfaces form a "cold weld"—essentially, they become one piece of steel. If there are no oxides, oils, or other contaminants, the atoms at the surfaces have no way to "know" to which piece of steel they belong. They bond as strongly with their new neighbors as they do with the old. A bit of lubricant drastically decreases the frictional forces because the two surfaces will float past each other without much contact.

Despite the complexities, we can make some approximate statements about the frictional forces between dry, solid surfaces. In the simplest model, the *maximum* force of static friction is proportional to the normal force acting between the two surfaces.

$$(f_s)_{max} \propto N$$

where $(f_s)_{max}$ is the maximum magnitude of the force of static friction and N is the magnitude of the normal force at the same contact surfaces. If you want better traction between the tires of a rear-wheel-drive car and the road, it helps to put something heavy in the trunk to increase the normal force between the tires and the road.

The constant of proportionality is called the **coefficient of static friction** (symbol μ_s):

$$(f_s)_{max} = \mu_s N \tag{2-6a}$$

The magnitude of the static frictional force must be less than or equal to the maximum possible magnitude:

$$f_s \leq \mu_s N \tag{2-6b}$$

Since it is the ratio of the magnitudes of two forces, μ_s is dimensionless. The coefficient depends on the condition and nature of the surfaces. Only the *maximum* force of static friction can be found from Eqs. (2-6a) and (2-6b). The actual force of friction in a given situation is not necessarily the maximum possible.

For sliding or kinetic friction, the force of friction is only weakly dependent on the speed and is roughly proportional to the normal force. In the simplified model we will use, the force of kinetic friction is assumed to be proportional to the normal force and independent of speed:

$$f_k = \mu_k N \tag{2-6c}$$

where f_k is the magnitude of the force of kinetic friction and μ_k is called the **coefficient of kinetic friction**. The coefficient of static friction is always larger than the coefficient of kinetic friction for an object on a given surface. On a horizontal surface, a larger force is required to start the object moving than is required to keep it moving at a constant velocity.

Equations (2-6) are only *part* of the force law for friction; we must also specify the direction of the force:

- The static frictional force acts in the direction that tends to prevent the surfaces from beginning to slide.

- The direction of the kinetic frictional force is in the direction that would tend to make the sliding stop—the direction opposite to the object's motion.

According to Newton's third law, there must be a pair of frictional forces, one acting on each object. If a book slides to the left along a desk, the desk exerts a kinetic frictional force on the book to the right; the book exerts a kinetic frictional force on the desk of equal magnitude to the left.

> Equations (2-6) relate only the *magnitudes* of the frictional and normal forces. Their directions are perpendicular to each other.

Example 2.11

Pushing on a Small Chest

In order to slide a 750-N chest across the floor at constant speed, you must push horizontally with a force of 450 N. The chest is rather short and is in no danger of toppling over. (a) What is the coefficient of kinetic friction? (b) Suppose you push with a force of 110 N and the cabinet does not budge. Now what is the force of friction on the chest? Assume that $\mu_s = 1.2\mu_k$.

Strategy In both cases, the chest is in equilibrium— dynamic in (a) and static in (b). So in both cases, the net force acting on it is zero. In each case we draw a free-body diagram.

continued on next page

Example 2.11 *continued*

Solution   (a) Since you push at constant speed, the net force on the chest is zero (dynamic equilibrium). There are four forces acting on it (Fig. 2.16a): the weight $\vec{\mathbf{W}}$, the normal force $\vec{\mathbf{N}}$ exerted by the floor, kinetic friction $\vec{\mathbf{f}}_k$, and the force $\vec{\mathbf{F}}$ you apply—which is a normal force on another surface of the chest. Now we draw an FBD (Fig. 2.16b).

The sum of these four forces must give a net force of zero. There are two pairs of collinear vectors: one pair is vertical and the other is horizontal. Since we know how to add collinear vectors, let us group the vectors in collinear pairs:

$$\vec{\mathbf{F}}_{net} = (\vec{\mathbf{N}} + \vec{\mathbf{W}}) + (\vec{\mathbf{f}}_k + \vec{\mathbf{F}}) = \vec{\mathbf{0}}$$

The sum $\vec{\mathbf{N}} + \vec{\mathbf{W}}$ is a vector in the vertical direction and the sum $\vec{\mathbf{f}}_k + \vec{\mathbf{F}}$ is a vector in the horizontal direction. Imagine adding the two vectors $(\vec{\mathbf{N}} + \vec{\mathbf{W}})$ and $(\vec{\mathbf{f}}_k + \vec{\mathbf{F}})$ and setting the sum equal to zero. Two nonzero vectors can add to zero only if they are equal in magnitude and opposite in direction. These two

Figure 2.16
(a) Forces acting on the chest and (b) a free-body diagram for the chest.

vectors cannot be opposite in direction since one is vertical and the other is horizontal. Therefore, they must *each* be zero:

$$(\vec{\mathbf{N}} + \vec{\mathbf{W}}) = \vec{\mathbf{0}} \quad \text{and} \quad (\vec{\mathbf{f}}_k + \vec{\mathbf{F}}) = \vec{\mathbf{0}}$$

In each pair, the vectors are in opposite directions. Their magnitudes must be equal in order to add to zero:

$$N = W \quad \text{and} \quad f_k = F$$

Now we can calculate the coefficient of kinetic friction.

$$\mu_k = \frac{f_k}{N}$$

Substituting $N = W$ and $f_k = F$,

$$\mu_k = \frac{F}{W} = \frac{450}{750} = 0.60$$

(b) It might be tempting to write $f_s = \mu_s N$, but that's the *maximum possible* static frictional force for this contact surface. The actual f_s is whatever it takes to keep the cabinet in equilibrium, so long as it does not *exceed* $\mu_s N$. The free-body diagram is the same as before, except $\vec{\mathbf{f}}$ now represents static friction instead of kinetic friction. As before, the two horizontal forces must be equal and opposite, so $f_s = F = 110$ N.

Discussion   The static frictional force is like the normal force in one way: its magnitude and direction are whatever it takes to keep the surfaces from sliding across each other—up to a *maximum* of $\mu_s N$. Do not assume that its magnitude is *equal to $\mu_s N$.*

Practice Problem 2.11   Net force on a moving chest

What is the net force on the chest while you slide it across the floor by applying a horizontal force of 480 N?

Conceptual Example 2.12

Horse and Sleigh

To cause a sleigh to move through the snow (Fig. 2.17a), a horse pulls the sleigh with a forward horizontal force of magnitude T. (a) Draw free-body diagrams for the horse and for the sleigh when there is a net force on each in the forward direction. (According to Newton's second law, a net force in the forward direction is needed to start the sleigh moving forward.) (b) According to Newton's third law, the sleigh pulls back on the horse with a force of *the same* magnitude as the force with which the horse pulls the sleigh. Does this mean the horse can never budge the sleigh? Explain. (c) For each force acting on the sleigh, identify the other half of the interaction—the equal and opposite force exerted by the sleigh on another object.

Strategy   We can draw free-body diagrams for (1) the sleigh alone, (2) the horse alone, and (3) the system of horse plus

sleigh. In each, we include only the external forces acting *on that system*. Changes in the state of motion of a system are determined by the net force acting *on that system*.

Solution and Discussion   (a) There are four forces acting on the sleigh: the force exerted by the horse on the sleigh $\vec{\mathbf{T}}_{sh}$, the weight of the sleigh $\vec{\mathbf{W}}_{se}$, the normal force of the ground on the sleigh $\vec{\mathbf{N}}_{sg}$, and friction acting on the sleigh due to the ground $\vec{\mathbf{f}}_{sg}$.

Figure 2.17b shows an FBD for the sleigh. In order for the net force to be in the forward direction, the force with which the horse pulls ($\vec{\mathbf{T}}_{sh}$) must exceed the force of friction on the sleigh due to the ground ($\vec{\mathbf{f}}_{sg}$) and the normal force ($\vec{\mathbf{N}}_{sg}$) must be equal in magnitude to the weight of the sleigh ($\vec{\mathbf{W}}_{se}$).

$$T_{sh} > f_{sg}$$

$$N_{sg} = W_{se}$$

continued on next page

Conceptual Example 2.12 *continued*

There are four forces acting on the horse: the force exerted by the sleigh ($\vec{T}_{hs}$), the weight of the horse $\vec{W}_{he}$, the normal force of the ground on the horse $\vec{N}_{hg}$, and friction acting on the horse due to the ground $\vec{f}_{hg}$. Newton's third law says that $\vec{T}_{hs} = -\vec{T}_{sh}$; the sleigh pulls back on the horse with a force equal in magnitude to the forward pull of the horse on the sleigh. Therefore $\vec{T}_{hs}$ is to the left. In order for the net force on the horse to be directed to the right, the only other horizontal force, $\vec{f}_{hg}$, must be larger and directed *to the right*. How does the horse get friction to push it *forward*? By pushing *backward* on the ground with its feet. We all do the same thing when taking a step; by pushing backward on the ground, we get the ground to push forward on us.

Figure 2.17c shows an FBD for the horse. The normal force ($\vec{N}_{hg}$) must be equal in magnitude to the weight of the horse ($\vec{W}_{he}$).

$$f_{hg} > T_{hs}$$
$$N_{hg} = W_{he}$$

(b) The forces that the sleigh and horse exert on each other must be equal in magnitude and opposite in direction no matter what happens. What determines the motion of the sleigh is the net force acting on *it*. A force acting on the horse ($\vec{T}_{hs}$) is not included when finding the net force on the sleigh—the sum of all the forces *acting on the sleigh*.

The condition for the sleigh to start moving forward is that the net force on it be in the forward direction, which means that

$$T_{sh} > f_{sg}$$

For the horse to also move forward,

$$f_{hg} > T_{hs}$$

Since $T_{sh} = T_{hs}$, the horse can pull the sleigh forward if

$$f_{hg} > f_{sg}$$

The forward frictional force on the horse must exceed the backward frictional force on the sleigh. Imagine putting the horse but not the sleigh on slippery ice; it would be impossible for the horse to pull the sleigh forward.

(c)

Force Exerted on Sleigh	Third-Law Partner
Force exerted on the sleigh by the horse $\vec{T}_{sh}$	Force exerted on the horse by the sleigh $\vec{T}_{hs}$
The weight of the sleigh $\vec{W}_{se}$ (or $\vec{F}_{se}$)	Gravitational force exerted on Earth by the sleigh $\vec{F}_{es}$
The normal force on the sleigh by the ground $\vec{N}_{sg}$	The normal force on the ground by the sleigh $\vec{N}_{gs}$
Friction on sleigh due to the ground $\vec{f}_{sg}$	Friction on ground due to the sleigh $\vec{f}_{gs}$

Practice Problem 2.12   Find the missing force

A car is trying to get under way on a level north-south road. The contact forces exerted *by the tires on the road* are 11.0 kN downward and 3.3 kN southward.
(a) Draw an FBD for the car. (b) What is the weight of the car? (c) What is the net force acting on the car?

Figure 2.17   (a) Horse pulling sleigh, (b) free-body diagram for the sleigh, and (c) free-body diagram for the horse.

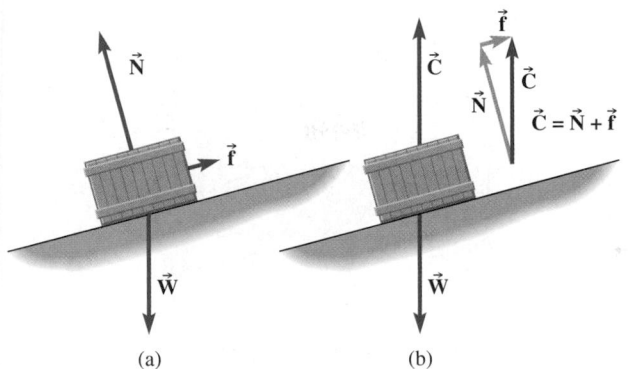

Figure 2.18 (a) Normal and frictional forces acting on a crate of weight $\vec{W}$ on an incline and (b) the total contact force $\vec{C}$ on the crate, which is directed straight up, is the vector sum of the normal and frictional forces. (In Chapter 4, we learn how to add vectors that are not collinear.)

Suppose that a crate of apples sits at rest on a ramp that is inclined at 20° to the horizontal. The normal force on the crate is perpendicular to the contact surfaces, or 20° from the vertical (Fig. 2.18a). The static frictional force points up the incline, 20° above the horizontal. In Practice Problem 2.6, we said that the contact force due to the ramp was straight up. Which is it?

Both are correct. Contact forces are the net macroscopic result of huge numbers of microscopic forces. Whether we think of a contact force as a single force or as a pair of perpendicular forces is entirely a matter of convenience. In Chapter 4, when the method of adding vectors is generalized to noncollinear vectors, we can show that the contact force $\vec{C}$ is the sum of the normal force and the frictional force (Fig. 2.18b).

2.7   TENSION

A flexible cord, rope, string, tendon, cable, or beam that is under tension exerts forces on the objects at each end, much as a spring does. There are two main differences. One is that the change in length is often negligibly small. The other is that a flexible rope or string can only pull; it cannot push. A rope may be under tension, but it cannot be compressed. A stiff cable or beam, on the other hand, can be compressed and therefore can push.

If the mass of the cord is small enough, then the forces exerted at the two ends are approximately equal in magnitude, assuming that nothing exerts forces on the cord except at the ends (see Conceptual Question 15). An *ideal cord* has zero mass and does not stretch at all for any amount of tension. The tension in an ideal cord is the same at the two ends (again assuming that nothing exerts forces on the cord except at the ends).

Pulleys

The pulley or a system of pulleys is sometimes called a *simple machine*. The name *machine* does not mean that a pulley can do work like an engine can, but a simple machine can change the direction or the magnitude of force required for performing some task. A single pulley (Fig. 2.19a) can change the direction of the force, exerted by a cord under tension, required to raise an object without changing the magnitude of the force required. If we are lifting a heavy weight it is easier to stand on the ground and pull down on the rope than to get above the weight on a platform and pull up on the rope.

Making The Connection:
advantages of a pulley

An ideal pulley has no mass and no friction. The tension of a cord that runs through an ideal pulley is the same on both sides of the pulley. The proof of this statement, using the concept of torque, is found in Chapter 8. An ideal pulley exerts no forces on the cord that are *tangent* to the cord, so it is not pulling in either direction along the cord. This property enables a pulley to change the direction of a force without changing its magnitude. As long as a real pulley has a small mass and negligible amount of friction, we can approximate it as an ideal pulley.

(a)

(b)

Figure 2.19 Two pulley systems: (a) a single pulley and (b) a two-pulley system

In Example 2.13, we will find that having more than one pulley enables us to pull with less force to raise an object. The pulley system of Fig. 2.19b requires only half as much force to lift the 45-kg mass as would be required to lift it directly since there are two ropes pulling up while only one is pulled down.

The pulley exerts forces on the cord and the cord exerts forces on the pulley. One way to think about the forces exerted on the pulley is to think of the part of the cord wrapped around the pulley as part of the pulley system; then there are two cords pulling on the pulley system, each with the same tension. A similar technique can be useful in other situations. Any cord can be thought of as two cords connected together at any arbitrary point; the two imagined cords pull on each other with forces whose magnitude is the tension in the cord.

Example 2.13

A Two-Pulley System

A 1804-N engine is hauled upward at constant speed (Fig. 2.20a). What are the tensions in the three ropes labeled A, B, and C? Assume ideal pulleys, L to the left and R to the right.

(a)   (b)   (c)

Figure 2.20
(a) A system of pulleys used to raise a heavy weight, (b) free-body diagram for the pulley on the left, and (c) free-body diagram for the pulley on the right.

Strategy Assuming that the pulleys are ideal means that the tension of a rope is the same on both sides of the pulley. Therefore the rope that is attached to the ceiling, passes through both pulleys, and is pulled downward at the other end has the same tension throughout. Call the tensions in the three ropes T_A, T_B, and T_C. The engine and the pulley from which it hangs move up at constant speed, so the net force on each of them is zero. The other pulley is at rest, so the net force on it is also zero.

Solution There are two forces acting on the engine: its weight (1804 N, downward) and the upward pull of rope A. These must be equal and opposite, since the net force is zero. Therefore $T_A = 1804$ N.

Figure 2.20b is an FBD for the pulley on the left. Rope A pulls down on it with a force of magnitude T_A. On *each side* of the pulley, rope C pulls upward. The rope has the same tension throughout so that all forces labeled T_C in Figs. 2.20b and 2.20c have the same magnitude. For the net force to equal zero,

$$2T_C = T_A$$

$$T_C = \tfrac{1}{2}T_A = 902.0 \text{ N}$$

Figure 2.20c is an FBD for the pulley on the right. Rope B pulls upward on it with a force of magnitude T_B. On *each side*

continued on next page

Example 2.13 *continued*

of the pulley, rope C pulls downward. For the net force to equal zero,

$$T_B = 2T_C = 1804 \text{ N}$$

Discussion   The engine is raised by pulling *down* on a rope—the pulleys change the direction of the applied force needed to lift the engine. In this case, they also change the *magnitude* of the required force. They do that by making the rope pull up on the engine twice, so the person pulling the rope only needs to

exert a force equal to half the engine's weight. Where does the "extra" force come from? It comes from the ceiling, which pulls up on the two ropes (B and C) with a total force of 2706 N.

Practice Problem 2.13   System of ropes, pulleys, engine

Consider the entire collection of ropes, pulleys, and the engine to be a single system. Draw an FBD for this system and show that the net force is zero.

Changing the direction of a tensile force is of interest in the study of animal motion, or biomechanics. You do not have pulleys inside your body; instead, tendons are often used to change the direction of the force exerted by a muscle. Tendons attach to bones and are lubricated by fluids to minimize friction (Fig. 2.21). For example, the quadriceps tendon connects to the kneecap (patella) and allows you to swing your lower leg back and forth or to kick a soccer ball. Muscles are usually connected by tendons, located on either end of the muscle, to two different bones, which in turn are linked at a joint. Usually one of the bones is more easily moved than the other. As an analogy, think of the muscle as a spring that can contract or relax. The tendons correspond to ropes tied to either end of the spring and then attached at their opposite ends to the movable bones.

Making The Connection:
functions of tendons
in the body

Figure 2.21   A muscle and two tendons attach two bones across a joint.

Physics at Home

Sit with your arm bent at the elbow with a heavy object on the palm of your hand. You can feel the contraction of the biceps muscle. With your other hand, feel the tendon that connects the biceps muscle to your forearm.

Now place your hand palm down on the desktop and push down. Now it is the triceps muscle that contracts, pulling up on the bone on the other side of the elbow joint. Muscles and tendons cannot push; they can only pull. The biceps muscle cannot push the forearm downward, but the triceps muscle can pull on the other side of the joint. In both cases, the arm acts as a lever.

MASTER THE CONCEPTS

Summary

Newton's laws of motion:

1. If no forces act on an object, then its state of motion does not change. If at rest, it remains at rest; if moving, it continues to move in a straight line with constant speed.

2. A nonzero net force acting on an object causes its state of motion to change.

3. In an interaction between two objects, the forces that each exerts on the other are equal in magnitude and opposite in direction.

- If the net external force acting on a body is zero, the body is in *equilibrium*. A body in equilibrium moves at a constant speed along a straight line. (The speed can be zero, in which case the body is at rest.)

- At the fundamental level, there are four interactions: gravitation, the strong and weak nuclear interactions, and the electromagnetic interaction. Contact forces are large-scale manifestations of many microscopic electromagnetic interactions.

- Gravity and electromagnetic forces are the only long-range forces. All other forces (in a macroscopic view) are contact forces.

- The SI unit of force is the newton. $1.00 \text{ N} = 0.2248 \text{ lb}$. $1 \text{ N} = 1 \text{ kg·m/s}^2$

- Force is a vector quantity—it has a direction as well as a magnitude. Vectors have their own rules for addition.

- The *net force* is the vector sum of all the forces acting on a body. Since all the internal forces come in third-law pairs, we need only sum the external forces.

- A *free-body diagram* (FBD) includes vector arrows representing every force acting on the chosen object, but no forces acting on other objects.

- The magnitude of the *gravitational force* between two objects is

$$F = \frac{Gm_1m_2}{r^2}$$

where r is the distance between their centers.

- The *weight* of an object near the Earth's surface is the gravitational force acting on the object due to the Earth. An object's weight is proportional to its mass: $\vec{\mathbf{W}} = m\vec{\mathbf{g}}$. Near Earth's surface, $\vec{\mathbf{g}} = 9.8 \text{ N/kg}$ downward.

- The magnitude of the *electric force* between two charged particles is

$$F = \frac{k|q_1||q_2|}{r^2}$$

where r is the distance between the particles.

- The *normal force* is a contact force perpendicular to the contact surfaces that pushes each object away from the other.

- *Friction* is a contact force parallel to the contact surfaces. In a simplified model, the kinetic frictional force and the maximum static frictional force are proportional to the normal force acting between the same contact surfaces.

$$f_s \leq \mu_s N$$
$$f_k = \mu_k N$$

The static frictional force acts in the direction that tends to keep the surfaces from beginning to slide. The direction of the kinetic frictional force is in the direction that would tend to make the sliding stop.

- Many springs obey *Hooke's law* as long as they are not stretched or compressed too far:

$$F = kx$$

- Cords and pulleys can be used to change the direction of a force without changing its magnitude.

Highlighted Figures and Tables

F2.2 Simple spring scale showing force measured by the deformation it produces in an object (p. 28)

F2.3 Schematic representation of the unification of forces (p. 31)

F2.6 Galileo postulated that if all resistive forces could be eliminated, a ball rolling on a horizontal surface would never stop. (p. 33)

F2.9 Illustration of pairs of forces (p. 37)

F2.10 Addition of collinear vectors that are in the same direction and in opposite directions (p. 38)

F2.12 Examples of free-body diagrams (p. 40)

F2.14 Gravity is always an attractive force. (p. 42)

F2.18 Contact force $\vec{\mathbf{C}}$ is the sum of the normal force and the frictional force. (p. 49)

F2.19 Single pulley and two-pulley systems (p. 50)

F2.20 Free-body diagrams of a pulley system (p. 50)

CONCEPTUAL QUESTIONS

1. Explain the need for automobile seat belts in terms of Newton's first law.

2. An American visitor to Finland is surprised to see heavy metal frames outside of all the apartment buildings. On Saturday morning the purpose of the frames becomes evident when several apartment dwellers appear, carrying rugs and carpet beaters to each frame. How does the law of inertia aid in the rug beating process? Do you see a similarity to the role inertia plays when you throw a baseball?

3. Explain why gravitation is the only appreciable interaction between the Moon and the Earth. Why are all of the other interactions negligible?

4. You are standing on a bathroom scale in an elevator. In which of these situations must the scale read the same as when the elevator is at rest? Explain. (a) Moving up at constant speed. (b) Moving up with increasing speed. (c) In free fall (after the elevator cable has snapped).

Figure 2.22 Question 5

5. A carpenter is working on the roof of a new house. The steel head of her hammer has become loose on the tapered upper end of its wooden handle. She holds the hammer vertically, raises it up, and then brings it down rapidly, hitting the bottom end of the wood handle on the sturdy ridge pole of the roof (Fig. 2.22). Explain how this tightens the head back onto the handle.

6. When a car begins to move forward, what force makes it do so? Remember that it has to be an *external* interaction; the internal forces all add to zero. How does the engine facilitate the propelling force?

7. Two cars are headed toward each other in opposite directions along a narrow country road. The cars collide head-on, crumpling up the hoods of both. Describe what happens to the car bodies in terms of inertia. Does the rear end of the car stop at the same time as the front end?

8. In some tables of unit conversions, you are told that 1 kg = 2.2 lb. What is wrong with that statement? On the surface of the Moon, does 1 kg "equal" 2.2 lb? Why or why not?

9. (a) What assumptions do you make when you call the reading of a bathroom scale your "weight"? What does the scale really tell you? (b) Under what circumstances might the reading of the scale *not* be equal to your weight?

10. A freight train consists of an engine and several identical cars on level ground. Determine whether each of these statements is correct or incorrect and explain why. (a) If the train is moving at constant speed, the engine must be pulling with a force greater than the train's weight. (b) If the train is moving at constant speed, the engine's pull on the first car must exceed that car's backward pull on the engine. (c) If the train is coasting, its inertia makes it slow down and eventually stop.

11. (a) Does a man weigh more at the North Pole or at the equator? (b) Does he weigh more at the top of Mt. Everest or at the base of the mountain?

12. What is the distinction between a vector and a scalar quantity? Give two examples of each.

13. Which of the fundamental forces governs the motion of planets in the solar system? Is this the strongest or the weakest of the fundamental forces? Explain.

14. An old steam locomotive uses coal as a fuel. A railroad worker, called a fireman, shovels the coal onto the fire in the firebox that powers the locomotive engine. Explain how he lets Newton's first law help him move the coal onto the fire.

15. Using Newton's laws, explain why a spring *must* exert forces of equal magnitude on the objects attached to each end, assuming that the spring's mass is negligibly small. The same line of reasoning applies to a rope or cord whose mass is negligible.

16. A boy has just won his first sailing race. After mooring the sailboat, he rows his dinghy up to the dock to greet his admiring mates. He ties the line attached to the bow of his dinghy to a cleat on the dock; initially he is only 2.0 m from the dock but the line is long enough so that the boat can float 5.0 m away from the dock. Then, feeling proud of himself, he decides to make a running jump from the dinghy to the dock. What happens next? Use Newton's third law to describe the action.

17. You are standing on one end of a light wooden raft that has floated 3 m away from the pier (Fig. 2.23). If the raft is 6 m long by 2.5 m wide and you are standing on the raft end nearest to the pier, can you propel the raft back toward the pier where a friend is standing with a pole and hook trying to reach you? You have no oars. Make suggestions of what to do without getting yourself wet.

18. Tell whether each of these objects is in equilibrium. If it is, is the equilibrium static or dynamic? (a) A car driving around a curve at constant speed on a flat road. (b) A car driving straight up a 6° incline at constant speed. (c) The Moon.

Figure 2.23 Question 17

19. A traffic light is hanging at an intersection. (a) If the air is still, is the light in equilibrium? (b) What about when a strong, steady wind blows so that the light makes an angle of 30° with its usual vertical position? Is the light in equilibrium?

20. Two boys are trying to break a cord. Gerardo says they should each pull in opposite directions on the two ends; Stefan says they should tie the cord to a pole and both pull together on the opposite end. Which plan is more likely to work?

MULTIPLE CHOICE QUESTIONS

1. A space probe leaves the solar system to explore interstellar space. Once it is far from any stars, when must it fire its rocket engines?

   (a) All the time, in order to keep moving.
   (b) Only when it wants to speed up.
   (c) When it wants to speed up or slow down.
   (d) Only when it wants to turn.
   (e) When it wants to speed up, slow down, or turn.

2. Your car won't start, so you are pushing it. You apply a horizontal force of 300 N to the car, but it doesn't budge. Which of these other forces *must* have a magnitude of 300 N?

   (a) the frictional force exerted by the road on the car
   (b) the force exerted by the car on you
   (c) the frictional force exerted on you by the road
   (d) (a), (b), and (c)
   (e) (a) and (c) but not (b)

3. The net force on a moving object suddenly becomes zero. The object then:

   (a) stops abruptly.
   (b) stops during a short time interval.
   (c) changes direction.
   (d) continues at constant velocity.
   (e) slows down gradually.

4. Which of these is *not* a long-range force?

   (a) the force that makes lightning flow between an electrically charged cloud and an oppositely charged area on the ground
   (b) the force that makes a compass point north
   (c) the force that a person exerts on a chair while sitting
   (d) the force that keeps the Moon in its orbital path around the Earth

5. The gravitational field strength on the surface of a distant planet is 1/5 that on the surface of Earth. An astronaut's mass and weight as measured on Earth are *m* and *W*, respectively. What are the astronaut's mass and weight as measured on the distant planet?

   (a) m, W (b) $\frac{1}{5}m, W$ (c) $m, \frac{1}{5}W$ (d) $\frac{1}{5}m, \frac{1}{5}W$

6. When an object is in translational equilibrium, which of these statements is *not* true?

   (a) The vector sum of the forces acting on the object is zero.
   (b) The object must be stationary.
   (c) The object has a constant velocity.
   (d) The speed of the object is constant.

7. A person is standing on a bathroom scale. Which of these is *not* a force exerted on the scale?

   (a) a contact force due to the floor
   (b) a contact force due to the person's feet
   (c) the weight of the person
   (d) the weight of the scale

8. Which of these forces bind electrons to nuclei to form atoms?

   (a) the strong nuclear force
   (b) the weak nuclear force
   (c) the electromagnetic force
   (d) the gravitational force

9. The Sun exerts a gravitational force on the Earth. Which statement is true concerning the gravitational force exerted by the Earth on the Sun?

   (a) It has the same magnitude and direction as the gravitational force exerted by the Sun on the Earth.
   (b) It has a much smaller magnitude than the gravitational force exerted by the Sun on the Earth and is opposite in direction.
   (c) It has a much smaller magnitude than the gravitational force exerted by the Sun on the Earth and is in the same direction.
   (d) It has the same magnitude as the gravitational force exerted by the Sun on the Earth and is opposite in direction.
   (e) The Earth exerts no gravitational force on the Sun; the Earth orbits the Sun but the Sun does not orbit the Earth.

10. To make an object start moving on a surface with friction requires

   (a) less force than to keep it moving on the surface.
   (b) the same force as to keep it moving on the surface.
   (c) more force than to keep it moving on the surface.
   (d) a force equal to the weight of the object.

PROBLEMS

Note: ⓒ indicates a combination conceptual/quantitative problem. Gold diamonds ✦, ✦✦ are used to indicate the increasing level of difficulty of each problem. Problem numbers appearing in blue, 9., denote problems that have a detailed solution available in the Student Solutions Manual. Some problems are *paired* by concept; their numbers are connected by a ruled box.

2.1 Forces

1. If the length of the Achilles tendon increases 0.50 cm when the force exerted on it by the muscle increases from 3200 N to 4800 N, what is the "spring constant" of the tendon?

2. If a force of 5.0 N causes a spring to stretch 3.5 cm from its relaxed length, how far does a force of 7.0 N cause the

same spring to stretch? What is the spring constant of this spring?

3. A block of wood is compressed 2.0 nm when inward forces of magnitude 120 N are applied to it on two opposite sides. (a) What is the effective spring constant of the block? (b) Assuming Hooke's law still holds, how much would the block be compressed by inward forces of magnitude 480 N?

4. A sack of flour has a weight of 19.8 N. What is its weight in pounds?

5. An astronaut weighs 175 lb. What is his weight in newtons?

2.3 Newton's Laws of Motion;
2.4 Net Force: Vector Addition

6. A person is standing on a bathroom scale. Identify the third-law partner of each of the forces exerted on the scale. In other words, for every interaction involving the scale, identify the force that the scale exerts on another object.

7. A skydiver is falling with her parachute open. Consider the apparatus that connects the parachute to the skydiver to be part of the parachute. (a) Identify the third-law partner of each of the forces exerted on the skydiver. (b) Identify the third-law partner of each of the forces exerted on the parachute. (c) If the skydiver is falling straight down at a constant speed, is she in equilibrium? If so, what kind of equilibrium?

8. A hummingbird is hovering motionless beside a flower. The blur of its wings shows that they are rapidly beating up and down. If the air pushes upward on the bird with a force of 0.30 N, what is the weight of the hummingbird?

9. A hanging potted plant is suspended by a cord from a hook in the ceiling. Draw a free-body diagram for each of these: (a) the system consisting of plant, soil, and pot; (b) the cord; (c) the hook; (d) the system consisting of plant, soil, pot, cord, and hook. Label each force arrow using subscripts (for example, $\vec{F}_{ch}$ would represent the force exerted on the cord by the hook).

10. A parked automobile slips out of gear, rolls unattended down a slight incline and then along a level road until it hits a stone wall. Draw a free-body diagram to show the forces acting on the car while it is in contact with the wall.

11. A sailboat, tied to a mooring with a line, weighs 820 N. The mooring line pulls horizontally toward the west on the sailboat with a force of 110 N. The sails are stowed away and the wind blows from the west. The boat is moored on a still lake—no water currents push on it. Draw a free-body diagram for the sailboat and indicate the magnitude of each force.

12. An airplane is cruising along in a horizontal level flight at a constant velocity, heading due west. (a) If the weight of the plane is 2.6×10^4 N, what is the net force on the plane? (b) With what force does the air push upward on the plane? (c) Is the plane in equilibrium? If so, what kind of equilibrium?

13. Find the magnitude and direction of the net force on the object in each of the free-body diagrams in Fig. 2.24.

14. A truck driving on a level highway is acted upon by the following forces: a downward gravitational force of 52 kN (kilonewtons); an upward contact force due to the road of 52 kN; another contact force due to the road of 7 kN, directed east; and a drag force due to air resistance of 5 kN, directed west. What is the net force acting on the truck?

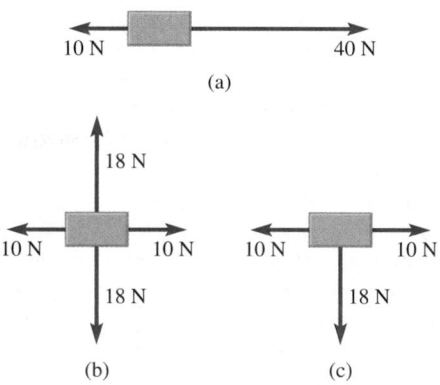

Figure 2.24 Problem 13

15. A woman who weighs 600.0 N sits on a chair with her feet on the floor and her arms resting on the chair's armrests. The chair weighs 100.0 N. Each armrest exerts an upward force of 25.0 N on her arms. The seat of the chair exerts an upward force of 500.0 N. (a) What force does the floor exert on her feet? (b) What force does the floor exert on the chair? (c) Consider the woman and the chair to be a single system. Draw a free-body diagram for this system that includes all of the *external* forces acting on it.

16. A man is lazily floating on an air mattress in a swimming pool. If the weight of the man and air mattress together is 806 N, what is the upward force of the water acting on the system of man and mattress?

17. Refer to Problem 7. Consider the skydiver and parachute to be a single system. What are the external forces acting on this system?

2.5 Gravitational Forces

18. (a) Calculate your weight in newtons. (b) What is the weight in newtons of 250 g of cheese? (c) Name a common object whose weight is about 1 N.

19. A young South African girl has a mass of 40.0 kg. Assume $g = 9.81$ N/kg. (a) What is her weight in newtons? (b) If she came to the United States, what would her weight be in pounds as measured on an American scale?

20. A man weighs 0.80 kN. What is his mass in kilograms?

21. The length of a spring increases by 7.2 cm from its relaxed length when a mass of 1.4 kg is suspended from the spring. (a) What is the spring constant? (b) A different mass is suspended and the spring length increases by 12.2 cm from its relaxed length. What is the second mass?

22. A spring scale in a French market is calibrated to show the *mass* of vegetables in grams and kilograms. (a) If the marks on the scale are 1.0 mm apart for every 25 grams, what maximum extension of the spring is required to measure up to 5.0 kg? (b) What is the spring constant of the spring? [Remember that the scale really measures *force*.]

23. Find the altitudes above the Earth's surface where Earth's gravitational field strength would be (a) two-thirds and (b) one-third of its value at the surface. [*Hint:* First find the radius for each situation; then recall that the altitude is the distance from the *surface* to a point above the surface. Use proportional reasoning.]

✦24. A spring hangs from a 2.44-m-high ceiling. The relaxed length of the spring is 0.305 m. A board, 1.98 m in length, hangs from the spring so that its lower end just reaches, but does not touch, the floor. The board has a mass of 10.4 kg and is perpendicular to the floor. Find the spring constant of the spring. Assume $g = 9.807$ N/kg.

✦25. A brick of mass 2.00 kg is "weighed under water" by hanging it from a spring scale (see Fig. 2.11). Let $g = 9.79$ N/kg. (a) If the spring scale reads 14.0 N, what is the upward force of water on the brick? (b) What is the reading of the pan scale on which the beaker sits? The weight of the water and the beaker together is 12.0 N. [Hint: refer to Example 2.7.]

✦26. The mass of the Moon is 0.0123 times that of the Earth. A spaceship is traveling along a line connecting the centers of the Earth and the Moon. At what distance from the Earth does the spaceship find the gravitational pull of the Earth equal in magnitude to that of the Moon? Express your answer as a percentage of the distance between the centers of the two bodies.

27. A 320-kg satellite is in orbit around the Earth 16,000 km above the Earth's surface. (a) What is the weight of the satellite when in orbit? (b) What was its weight when it was on the Earth's surface, before being launched? (c) While it orbits the Earth, what force does the satellite exert on the Earth?

28. At what altitude above the Earth's surface would your weight be half of what it is at the Earth's surface? [Hint: See Problem 23.]

29. (a) What is the magnitude of the gravitational force that the Earth exerts on the Moon? (b) What is the magnitude of the gravitational force that the Moon exerts on the Earth?

✦✦30. (a) If a spacecraft moves in a straight line between the Earth and the Sun, at what point would the force of gravity on the spacecraft due to the Sun be as large as that due to the Earth? (b) If the spacecraft is close to, but not at, this equilibrium point, does the net force on the spacecraft tend to push it toward or away from the equilibrium point? [Hint: imagine the spacecraft a small distance d closer to the Earth and find out which gravitational force is stronger.]

2.6 Contact Forces

31. Your textbook is sitting on your desk. (a) Draw a free-body diagram showing all the forces acting on the book. (b) Give the book a shove to the right. Draw a free-body diagram for the book while your hand was in contact with it. (c) Draw a free-body diagram for the book after your hand breaks contact with it, but while it is still sliding.

©✦32. A toy freight train consists of an engine and three identical cars. The train is moving to the right at constant speed along a straight, level track. Three spring scales are used to connect the cars as follows: spring scale A is located between the engine and the first car; scale B is between the first and second cars; scale C is between the second and third cars. (a) If air resistance and friction are negligible, what are the relative readings on the three spring scales A, B, and C? (b) Repeat part (a), taking air resistance and friction into consideration this time. [Hint: draw a free-body diagram for the car in the middle.] (c) If air resistance and friction together cause a force of magnitude 5.5 N on each car, directed toward the left, find the readings of scales A, B, and C.

33. The coefficient of static friction between a block and a horizontal floor is 0.40, while the coefficient of kinetic friction is 0.15. The mass of the block is 5.0 kg. If a horizontal force is slowly increased until it is barely enough to make the block start moving, what is the net force on the block the instant that it starts to slide?

34. A crate full of artichokes rests on a ramp that is inclined 10.0° above the horizontal. Give the direction of the normal force and the friction force acting on the crate in each of these situations. (a) The crate is at rest. (b) The crate is being pushed and is sliding up the ramp. (c) The crate is being pushed and is sliding down the ramp.

✦35. A 3.0-kg block is at rest on a horizontal floor. If you push horizontally on the 3.0-kg block with a force of 12.0 N, it just starts to move. (a) What is the coefficient of static friction? (b) A 7.0-kg block is stacked on top of the 3.0-kg block. What is the magnitude F of the force, acting horizontally on the 3.0-kg block as before, that is required to make the two blocks start to move?

36. A horse is trotting along pulling a sleigh through the snow. To move the sleigh, of mass m, straight ahead at a constant speed, the horse must pull with a force of magnitude T. (a) What is the net force acting on the sleigh? (b) What is the coefficient of kinetic friction between the sleigh and the snow?

✦37. A box full of books rests on a wooden floor. The normal force the floor exerts on the box is 250 N. (a) You push horizontally on the box with a force of 120 N, but it refuses to budge. What can you say about the coefficient of static friction between the box and the floor? (b) If you must push horizontally on the box with a force of at least 150 N to start it sliding, what is the coefficient of static friction? (c) Once the box is sliding, you only have to push with a force of 120 N to keep it sliding. What is the coefficient of kinetic friction?

38. Before hanging new William Morris wallpaper in her bedroom, Brenda sanded the walls lightly to smooth out some irregularities on the surface. Draw a free-body diagram for the sanding block as it moves straight up the wall.

39. Four separate blocks are placed side by side in a left-to-right row on a table. A horizontal force, acting toward the right, is applied to the block on the far left end of the row. Draw free-body diagrams for (a) the second block on the left and for (b) the system of four blocks.

2.7 Tension

40. A fly fisherman has hooked a Chinook salmon. The salmon is running out the line, moving at a constant velocity of 2 m/s through the water. Identify all the forces acting on the salmon.

41. A sailboat is tied to a mooring with a line. The wind is from the southwest. Identify all the forces acting on the sailboat.

42. A pulley is attached to the ceiling (Fig. 2.25a). Spring scale A is attached to the wall and a rope runs horizontally from it and over the pulley. The same rope is then attached to spring scale B. On the other side of scale B hangs a 120-N weight. What are the readings of the two scales A and B? The weights of the scales are negligible.

43. Spring scale A is attached to the floor and a rope runs vertically upward, loops over the pulley, and runs down on the other side to a 120-N weight (Fig. 2.25b). Scale B is attached

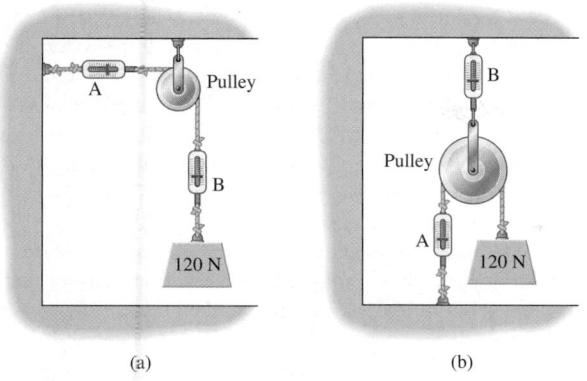

(a)   (b)

Figure 2.25 Problems 42 and 43

Figure 2.26 Problem 46

to the ceiling and the pulley is hung below it. What are the readings of the two spring scales, A and B? Neglect the weights of the pulley and scales.

✦ 44. The coefficient of static friction between block A and a horizontal floor is 0.45 and the coefficient of static friction between block B and the floor is 0.30. The mass of each block is 2.0 kg and they are connected together by a cord. (a) If a horizontal force $\vec{F}$ pulling on block B is slowly increased until it is barely enough to make the two blocks start moving, what is the magnitude of $\vec{F}$ at the instant that they start to slide? (b) What is the tension in the cord connecting blocks A and B at that same instant?

45. Two springs are connected in series so that spring scale A hangs from a hook on the ceiling and a second spring scale, B, hangs from the hook at the bottom of scale A (see springs A and B of Fig. 2.26). A 120-N weight hangs from the hook at the bottom of scale B. What are the readings on the upper scale A and the lower scale B? Neglect the weights of the scales.

✦ 46. Four *identical* spring scales, A, B, C, and D are used to hang a 220.0-N sack of potatoes (Fig. 2.26). (a) Assume the scales

have negligible weights and all four scales show the same reading. What is the reading of each scale? (b) Suppose that each scale has a weight of 5.0 N. If scales B and D show the same reading, what is the reading of each scale?

✦✦ 47. Two springs with equal spring constants k are connected first in series (one after the other) and then in parallel (side by side) with a weight hanging from the bottom of the combination. What is the effective spring constant of the two different arrangements? In other words, what would be the spring constant of a single spring that would behave exactly as (a) the series combination and (b) the parallel combination? Ignore the weight of the springs. [*Hint* for (a): *each spring stretches an amount* $x = F/k$, *but only one spring exerts a force on the hanging object. Hint for* (b): *each spring exerts a force* $F = kx$.]

COMPREHENSIVE PROBLEMS

© 48. When you hold up a 100-N weight in your hand, with your forearm horizontal and your palm up, the force exerted by your biceps is much larger than 100 N—perhaps as much as 1000 N. How can that be? What other forces are acting on your arm? (See Fig. 2.27.) Draw a free-body diagram for the

forearm, showing all of the forces. Assume that all the forces exerted on the forearm are purely vertical—either up or down.

49. In the sport of curling, popular in Canada and Ireland, a player slides a 20.0-kg granite stone down a 38-m-long ice rink. Draw free-body diagrams for the stone (a) while it sits at rest on the

Figure 2.27 Problem 48

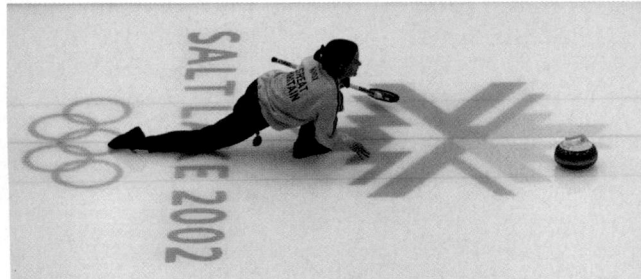

ice; (b) while it slides down the rink; (c) during a head-on collision with an opponent's stone that was at rest on the ice.

50. A car is driving on a straight, level road at constant speed. Draw a free-body diagram for the car, showing the significant forces that act upon it.

✦ 51. A refrigerator magnet weighing 0.14 N is used to hold up a photograph weighing 0.030 N. The magnet attracts the refrigerator door with a magnetic force of 2.10 N. (a) Identify the interactions between the magnet and other objects. (b) Draw a free-body diagram for the magnet, showing all the forces that act on it. (c) Which of these forces are long-range and which are contact forces? (d) Find the magnitudes of all the forces acting on the magnet.

52. A computer weighing 87 N sits on the surface of your desk. The coefficient of friction between the computer and the desk is 0.60. (a) Draw a free-body diagram for the computer. (b) What is the magnitude of the frictional force acting on the computer? (c) How hard would you have to push on it to get it to start to slide across the desk?

🄲 53. The readings of the two spring scales shown in Fig. 2.28 are the same. (a) Explain why they are the same. [*Hint:* draw free-body diagrams.] (b) What is the reading?

✦ 54. A 50.0-kg crate is suspended between the floor and the ceiling using two spring scales, one attached to the ceiling and one to the floor. If the lower scale reads 120 N, what is the reading of the upper scale? Ignore the weight of the scales.

Figure 2.28
Problem 53

✦✦ 55. The weight of a water strider is approximately 6×10^{-6} N. (a)
🄲 What is the force due to surface tension on the water strider when it is standing at rest on the water? (b) The *surface tension* γ of water is 0.07 N/m. Use dimensional analysis to determine whether the force of surface tension is proportional to the radius, the area, or the volume of the foot. In other words, to what power of r is the force proportional? (c) Explain why much larger animals cannot walk on water. Why can't they have feet big enough to make the surface tension of water support them? [*Hint:* Imagine a water strider that is twice as big in length, breadth, and height. Its volume—and therefore its weight—would be *eight* times as great.]

✦ 56. Spring scale A is attached to the ceiling. A 10.0-kg mass is suspended from the scale. A second spring scale, B, is hanging from a hook at the bottom of the 10.0-kg mass and a 4.0-kg mass hangs from the second spring scale. (a) What are the readings of the two scales if the masses of the scales are negligible? (b) What are the readings if each scale has a mass of 1.0 kg? Assume $g = 9.80$ N/kg.

✦ 57. A boy has stacked two blocks on the floor so that a 5.0-kg block is on top of a 2.0-kg block. (a) If the coefficient of static friction between the two blocks is 0.40 and the coefficient of static friction between the bottom block and the floor is 0.22, with what force should the boy push on the upper block to make both blocks in the system start to slide along the floor? (b) If he pushes too hard, the top block starts to slide off the lower block. What is the limit of the force with which he can push without that happening?

58. The forces required to extend a spring to various lengths are measured. The results are shown in the table below. Using the data in the table, plot a graph that helps you to answer the following two questions: (a) What is the spring constant? (b) What is the relaxed length of the spring?

Force (N)	1.00	2.00	3.00	4.00	5.00
Spring length (cm)	14.5	18.0	21.5	25.0	28.5

59. Plot a graph of this data for a spring resting horizontally on a table. Use your graph to find (a) the spring constant and (b) the relaxed length of the spring.

Force (N)	0.200	0.450	0.800	1.500
Spring length (cm)	13.3	15.0	17.3	22.0

✦ 60. A binary star consists of two stars of masses M_1 and $4.0M_1$ a distance d apart. Is there any point where the gravitational field due to the two stars is zero? If so, where is that point?

✦✦ 61. Two springs with spring constants k_1 and k_2 are connected in series. What is the effective spring constant of the combination? [See Problem 47(a).]

✦✦ 62. Two springs with spring constants k_1 and k_2 are connected in parallel. What is the effective spring constant of the combination? [See Problem 47(b).]

63. The tallest spot on Earth is Mt. Everest, which is 8850 m above sea level. If the radius of the Earth to sea level is 6370 km, how much does the magnitude of $\vec{g}$ change between sea level and the top of Mt. Everest?

64. In the fairy tale Rapunzel, the beautiful maiden let her long golden hair hang down from the tower in which she was held prisoner so that her prince could use her hair as a climbing rope to climb the tower and rescue her. Estimate how much force is required to pull a strand of hair out of your head. Next estimate how many hairs might be growing out of Rapunzel's head. If the prince has a mass of 60 kg, what is the force pulling on each strand of hair? Will Rapunzel be bald by the time the prince reaches the top of the 30-m tower?

65. A mass of 10 kg is hanging from a spring scale that is in turn hanging from a hook attached to the ceiling. The scale indicates a reading of 98 N for the force. Then two people hold the spring scale horizontally and pull on opposite ends until the scale again reads 98 N. With what force must each person pull to attain this result?

66. While trying to decide where to hang a framed picture, you press it against the wall to keep it from falling (Fig. 2.29). (a) What is the direction of the normal force exerted on the picture by your hand? (b) What is the direction of the normal force exerted on the picture by the wall? (c) The frictional force exerted on the picture by the wall can have two possible directions. Explain why.

Figure 2.29   Problem 66

ANSWERS TO PRACTICE PROBLEMS

2.1 The water pushes on the hull of the boat; the air pushes against the surface of the boat, including the sails; Earth's gravity pulls downward on the boat.

2.2 3.2 cm

2.3 No, the checkbook balance may be positive or negative, but there is no spatial direction associated with it. It is a scalar.

2.4 In the first case, inertia tends to keep Negar's body at rest with respect to the ground as the subway car begins to move forward. In the second case, inertia tends to keep Negar's body moving forward with respect to the ground, with constant speed as the subway car slows down.

2.5 downward force on the rod by the line and upward force on the line by the rod, downward gravitational force on the rod by the Earth and upward gravitational force on the Earth by the rod, upward force on the rod by the fisherman's hands and downward force on the hands by the rod

2.6 80 N upward

2.7 45.0 N

2.8 0.5 kN down

2.9 2 nN

2.10 3.77

2.11 30 N in the direction of the applied force

c = car
r = road
e = Earth

2.12 (a)   (b) weight of the car = 11.0 kN; (c) net force acting on the car = 3.3 kN northward

2.13

T_C = 902 N
T_B = 1804 N
W = 1804 N

Forces and Motion Along a Line

A sailplane (or "glider") is a small, unpowered, high-performance aircraft. A sailplane must be initially towed a few thousand feet into the air by a small airplane, after which it relies on regions of upward-moving air such as thermals and ridge currents to ascend farther. Suppose a small plane requires about 120 m of runway to take off by itself. When it is towing a sailplane, how much more runway does it need?

Concepts & Skills to Review

- net force: vector addition (Section 2.4)
- free-body diagrams (Section 2.4)
- gravitational force (Section 2.5)
- internal and external forces (Section 2.4)

Making The Connection:
motion of a train

3.1   POSITION AND DISPLACEMENT

A nonzero net force acting on an object changes the object's velocity. According to Newton's second law, the net force determines the *acceleration* of the object. In order to see how interactions affect motion, we first need to carefully define the quantities used to describe motion (position, displacement, velocity, and acceleration) and we must understand the relationships among them. The description of motion using these quantities is called **kinematics**. When we combine kinematics with Newton's laws, we are in the realm of **dynamics**. Chapter 3 considers motion only along a straight line.

To describe motion clearly, we first need a way to say *where* an object is located. Suppose that at 3:00 P.M. a train stops on an east-west track as a result of an engine problem. The engineer wants to call the railroad office to report the problem. How can he tell them where to find the train? He might say something like "three kilometers east of the old trestle bridge." Notice that he uses a point of reference: the old trestle bridge. Then he states how far the train is from that point and in what direction. If he omits any of the three pieces (the reference point, the distance, and the direction) then his description of the train's whereabouts is ambiguous.

The same thing is done in physics. First, we choose a reference point, called the **origin**. Then, to describe the location of something, we give its distance from the origin and the direction. These two quantities, direction and distance, together describe the **position** of the object. Position is a vector quantity; the direction is just as important as the distance. The position vector is written symbolically as $\vec{r}$. Graphically, the position vector can be drawn as an arrow starting at the origin and ending with the arrowhead on the object. When more than one position vector is to be drawn, a scale is chosen so that the length of the vector is proportional to the distance between the object and the origin.

Once the train's engine is repaired and it gets on its way, we might want to describe its motion. At 3:14 P.M. it leaves its initial position, 3 km east of the origin (Fig. 3.1). At 3:56 P.M. the train is 26 km west of the origin, which is 29 km to the west of its initial position. **Displacement** is defined as the change of position; it is written $\Delta \vec{r}$, where the

Figure 3.1   Initial ($\vec{r}_0$) and final ($\vec{r}_f$) position vectors for a train; the displacement vector $\Delta \vec{r}$ is found by subtracting the vector for the initial position from the vector for the final position.

symbol Δ means *the change in,* or the final value minus the initial value. If the initial and final positions are $\vec{\mathbf{r}}_0$ and $\vec{\mathbf{r}}_f$, then

---

Displacement

$$\Delta\vec{\mathbf{r}} = \vec{\mathbf{r}}_f - \vec{\mathbf{r}}_0 \qquad (3\text{-}1)$$

Since the positions are vector quantities, the operation indicated in Eq. (3-1) is a vector subtraction. To subtract a vector is to add its opposite, so $\vec{\mathbf{r}}_f - \vec{\mathbf{r}}_0 = \vec{\mathbf{r}}_f + (-\vec{\mathbf{r}}_0)$. Vectors representing the initial and final positions are shown as arrows in Fig. 3.1 and the vector subtraction is indicated. The displacement or change of position $\Delta\vec{\mathbf{r}}$ is shown; it is a vector of magnitude 29 km pointing west.

As a shortcut to subtract $\vec{\mathbf{r}}_0$ from $\vec{\mathbf{r}}_f$, instead of adding $-\vec{\mathbf{r}}_0$ to $\vec{\mathbf{r}}_f$, we can simply draw an arrow from the tip of $\vec{\mathbf{r}}_0$ to the tip of $\vec{\mathbf{r}}_f$ when the two vectors are drawn starting from the same point, as in the uppermost vector diagram of Fig. 3.1. This shortcut works because the displacement vector is the change of position; it takes us from the initial position to the final position. Since all vectors add and subtract according to the same rules, this same method can be used to subtract *any* kind of vector—as long as the vectors are drawn starting with their tails at the same point.

Notice that the magnitude of the displacement vector is not necessarily equal to the *distance traveled* by the train. Perhaps the train first travels 7 km to the east, putting it 10 km east of the origin, and then reverses direction and travels 36 km to the west. The total distance traveled in that case is 7 km + 36 km = 43 km, but the magnitude of the displacement—which is the distance between the initial and final positions—is 29 km. The displacement, since it is a vector quantity, must include the direction; it is 29 km to the west.

> The symbol Δ stands for *the change in*. If the initial value of a quantity Q is Q_0 and the final value is Q_f, then $\Delta Q = Q_f - Q_0$. ΔQ is read "delta Q."

> Displacement depends only on the starting and ending positions, not on the path taken.

Conceptual Example 3.1

A Relay Race

 A relay race is run along a straight line track of length L running south to north. The first runner starts at the south end of the track and passes the baton to a teammate at the north end of the track. The teammate races back to the start line and passes the baton to a third teammate who races 1/3 of the way northward to a finish line. What is the baton's displacement during the race?

Strategy The displacement of the baton is the vector from the starting point to the finish point; it is the *vector sum* of the *displacements* of each runner.

Solution The displacement is the vector sum of the three separate displacements that occurred during the race (Fig. 3.2). Since the first two displacements are equal in magnitude but in opposite directions, their vector sum is zero. The total displacement is then equal to the 3rd displacement, from the origin to the finish line, located 1/3 of the way along the track north of the starting point.

$$\Delta\vec{\mathbf{r}} = (\Delta\vec{\mathbf{r}}_1 + \Delta\vec{\mathbf{r}}_2) + \Delta\vec{\mathbf{r}}_3$$

$$= 0 + \Delta\vec{\mathbf{r}}_3 = \tfrac{1}{3}L, \text{ directed north}$$

Discussion Note that the displacement magnitude is less than the total distance traveled by the baton $(2\tfrac{1}{3}L)$. The distance

Figure 3.2
Baton displacements during race; vector sum of displacements

continued on next page

Conceptual Example 3.1 *continued*

traveled depends on the path followed, while the magnitude of the displacement is always the shortest distance between the two points of interest. The displacement vector shows the straight line path from starting point to finish point.

Conceptual Practice Problem 3.1   Around the bases

Casey hits a long fly ball over the heads of the outfielders. He runs from home plate to first base, to second base, to third base, and slides back to home plate safely (an inside-the-park home run). What is Casey's total displacement?

The sign of a vector component indicates the direction.

Vector Components

Since our universe has three spatial dimensions, positions in space are specified using a three-dimensional coordinate system, with x-, y-, and z-axes that are mutually perpendicular. Any vector can be specified either by its magnitude and direction or by its **components** along the x-, y-, and z-axes. The x-, y-, and z-components of vector $\vec{\mathbf{A}}$ are written with subscripts as follows: A_x, A_y, and A_z. One exception to this otherwise consistent notation is that the components of the position vector $\vec{\mathbf{r}}$ are usually written simply as x, y, and z rather than r_x, r_y, and r_z. The components themselves are scalars, which have a magnitude, units, and an algebraic sign. The sign indicates the direction; a positive x-component indicates the direction of the positive x-axis, while a negative x-component indicates the opposite direction (the negative x-axis). Specifying the components of a vector is as complete a description of the vector as is giving the magnitude and direction.

This chapter concentrates on motion along a straight line. If we choose one of the axes along that line, then the position, displacement, velocity, and acceleration vectors each have only one nonzero component. For horizontal linear motion, we usually choose the x-axis to lie along the direction of motion. For objects that move vertically, it is conventional to use the y-axis instead. One direction along the axis is chosen to be positive. Customarily the direction to the right of the origin is called the positive direction for a horizontal axis and upward is called the positive direction for a vertical axis. If vector $\vec{\mathbf{A}}$ points along the x-axis in the direction of $+x$ (to the right), then $A_x = +|\vec{\mathbf{A}}|$; if $\vec{\mathbf{A}}$ points in the opposite direction—in the direction of $-x$ (to the left)—then $A_x = -|\vec{\mathbf{A}}|$.

In Fig. 3.1 the compass arrows indicate that east is chosen as the direction for the positive x-axis. Using components, the train of Fig. 3.1 starts at $x = +3$ km and is later found at $x = -26$ km. The displacement (in component form) is $\Delta x = x_f - x_0 = -29$ km. The sign of the component indicates the direction. Since the positive x-axis is east, a negative x-component means that the displacement is to the west.

The rules for adding vectors translate into adding components with their algebraic signs. If $\vec{\mathbf{C}} = \vec{\mathbf{A}} + \vec{\mathbf{B}}$, then $C_x = A_x + B_x$, $C_y = A_y + B_y$, and $C_z = A_z + B_z$. Components become increasingly useful when adding vectors having more than one nonzero component. We can also subtract vectors by subtracting their components, since reversing the direction of a vector changes the sign of each of its components.

Example 3.2

A Stubborn Mule

In an attempt to get a mule moving in a horizontal direction, a farmer stands in front of the mule and pulls on the reins with a force of 320 N to the east while his son pushes on the mule's hindquarters with a force of 110 N to the east. In the meantime the mule pushes on the ground with a force of 430 N to the east (Fig. 3.3a). Use vector components to find the sum of the horizontal forces acting on the mule. Draw a vector diagram to verify the result.

Strategy We are interested in the forces acting *on the mule.* One force given in the problem statement is a force exerted *on the ground* (by the mule). From Newton's third law, the ground pushes back on the mule with a force of equal magnitude and opposite direction; therefore, we needed to know this force to determine one of the horizontal forces acting on the mule. Other forces acting on the mule that have not been explicitly stated are the mule's weight and the normal force on the mule due to the ground. These two are vertical forces: the weight is downward and the normal force is upward; these forces do not concern us since they have no effect on the horizontal motion of the mule. Therefore, we are left with three horizontal forces to sum. When finding vector components, we must first choose a direction for the positive x-axis. Here we choose east as the +x-direction.

Given: Force on mule by farmer: $\vec{F}_{mf}$ = 320 N to the east; force on mule by boy: $\vec{F}_{mb}$ = 110 N to the east; horizontal force on ground by mule: $\vec{F}_{gm}$ = 430 N to the east.

To find: Sum of horizontal forces on mule

Solution From Newton's third law, the ground pushes on the mule with a force equal and opposite to that of the mule on the ground; therefore, the ground pushes on the mule with a force $\vec{F}_{mg}$ of 430 N to the west. The other two horizontal forces acting

on the mule are directed to the east. Now we can drop the "m" subscripts since we deal only with the three forces acting on the mule ($\vec{F}_{mf}$ becomes $\vec{F}_f$, etc.). The x-components of forces directed to the east (F_{fx} and F_{bx}) are positive, since we chose east as the +x-direction. The x-component of the one force directed to the west (F_{gx}) is negative. Notice that the F's are not in bold-faced type because we are now talking about *components* of vectors; components are scalars. Then F_{fx} = +320 N, F_{bx} = +110 N, and F_{gx} = –430 N.

$$\Sigma F_x = F_{fx} + F_{bx} + F_{gx}$$
$$\Sigma F_x = 320 \text{ N} + 110 \text{ N} + (-430 \text{ N}) = 0$$

The sum of the horizontal forces is zero. Drawing the vectors to scale (Fig. 3.3b) verifies that the vector sum is zero: $\vec{F}_f + \vec{F}_b + \vec{F}_g = \vec{0}$.

Discussion Each force component has a sign that indicates the direction along the x-axis in which the force acts. Adding the three x-components with their algebraic signs gives the x-component of the sum. If the sum of the x-components had been positive, the sum of the vectors would be to the right (east); if it had been negative, the sum of the forces would be to the left (west).

Practice Problem 3.2   At last, the mule cooperates

The mule finally starts moving and hauls the farmer's wagon along a straight road for 4.3 km directly east to the neighboring farm, where a few bushels of corn are loaded onto the wagon. Then the farmer drives the mule back along the same straight road, heading west for 7.2 km to the market. Use vector components to find the displacement of the mule from the starting point.

Figure 3.3
(a) A farmer and his son encouraging their mule to move and (b) the vector sum of the horizontal forces on the mule

(a)   (b)

3.2   VELOCITY

A displacement vector indicates by how much and in what direction the position has changed, but implies nothing about how long it took to move from one point to the other. If we want to describe how fast something moves, we would give its speed. But we find that a more useful quantity than speed is **velocity**. Speed is a measure of "how fast": distance traveled divided by elapsed time (distance traveled *per unit time*). Velocity includes the direction of motion as well as the speed: it is defined as *displacement*

Velocity: a vector that measures how fast and in what direction something moves

Figure 3.4 Graph of position x versus time t for the train.

x (km)	+3	+3	+10	+10	0	−26
t (min)	0	14	23	28	40	56

divided by elapsed time (displacement per unit time). Speed, the magnitude of the velocity vector, is a scalar.

Average Velocity

Making The Connection:
    motion of a train

Figure 3.4 is a graph showing the position of the train considered in Section 3.1 as a function of time. The positions of the train at various times are marked with a dot. The position of the train would have to be measured at more frequent time intervals in order to accurately trace out the shape of the graph.

The graph of position versus time shows a curving line, but that does not mean the train travels along a curved path. The motion of the train is constrained along a straight line since the track runs in an east-west direction. The graph shows the x-component of the train's position as a function of time.

A horizontal portion of the graph (as from $t = 0$ to $t = 14$ min) indicates that the position is not changing during that time interval. Sloping portions of the graph indicate that the train is moving. The steeper the graph, the faster its position changes. The slope is a measure of the rate of change of position. A positive slope ($t = 14$ min to $t = 23$ min) indicates motion in the $+x$-direction while a negative slope ($t = 28$ min to $t = 56$ min) indicates motion in the $-x$-direction.

When a displacement $\Delta\vec{\mathbf{r}}$ occurs during a time interval Δt, the **average velocity** $\vec{\mathbf{v}}_{av}$ during that time interval is:

Average velocity is written $\vec{\mathbf{v}}_{av}$.

Average velocity
$$\vec{\mathbf{v}}_{av} = \frac{\Delta\vec{\mathbf{r}}}{\Delta t} \qquad (3\text{-}2)$$

Average velocity is a vector because it is the product of a vector, the displacement ($\Delta\vec{\mathbf{r}}$), and a scalar, the inverse of the time ($1/\Delta t$). Since Δt is always positive, the direction of the average velocity vector must be the same as the direction of the displacement vector. In terms of x-components,

$$v_{av,x} = \frac{\Delta x}{\Delta t} \qquad (3\text{-}3)$$

The symbol Δ does not stand alone and cannot be canceled in equations because it modifies the quantity that follows it; $\Delta x/\Delta t$ is *not* the same as x/t.

The *average* velocity does not convey detailed information about the motion during Δt, just the net effect. During the time interval in question, the object's motion could change direction and speed in many ways and still have the same *average* velocity. For any given displacement, the average velocity does indicate a constant speed and a straight line direction that would result in the same displacement during the same amount of time.

Suppose a butterfly flutters from a point O on one flower to a point D on another flower along the dashed line path shown in Fig. 3.5 (path $OABCD$). The displacement of the butterfly is the vector arrow $\Delta\vec{r}$ from O to D. The average velocity of the butterfly is the displacement per unit time, $\Delta\vec{r}/\Delta t$; the direction of the average velocity is the same as the direction of the displacement vector $\Delta\vec{r}$. The average velocity does not distinguish the actual path of the butterfly from any other path that begins at O and ends at D over the same time interval Δt.

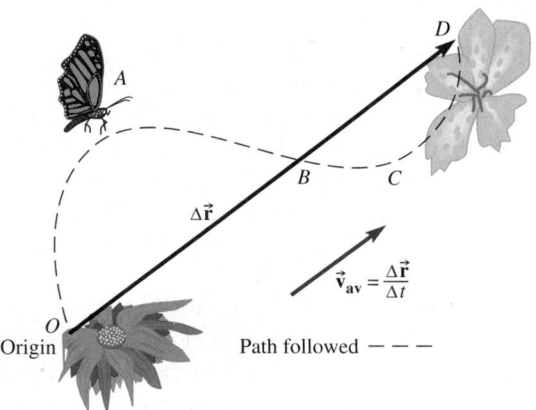

Origin   Path followed – – –

Figure 3.5 Path followed by a butterfly, fluttering from one flower to another.

Example 3.3

Average Velocity of Train

Find the average velocity of the train shown in Fig. 3.1 during the time interval between 3:14 P.M., when the train is 3 km east of the origin, and 3:56 P.M., when it is 26 km west of the origin.

Strategy We already know the displacement $\Delta\vec{r}$ from Fig. 3.1. The direction of the average velocity is the direction of the displacement.

Known: Displacement $\Delta\vec{r}$ = 29 km west; start time = 3:14 P.M.; finish time = 3:56 P.M.

To find: $\vec{v}_{av}$

Solution From Section 3.1, the displacement is 29 km to the west:

$$\Delta\vec{r} = 29 \text{ km west}$$

The time interval is

$$\Delta t = 56 \text{ min} - 14 \text{ min} = 42 \text{ min}$$

We convert the time interval to hours, so that we can use units of km/h.

$$\Delta t = 42 \text{ min} \times \frac{1 \text{ h}}{60 \text{ min}} = 0.70 \text{ h}$$

The average velocity is

$$\vec{v}_{av} = \frac{\text{displacement}}{\text{time interval}} = \frac{\Delta\vec{r}}{\Delta t}$$

$$\vec{v}_{av} = \frac{29 \text{ km west}}{0.70 \text{ h}} = 41 \text{ km/h to the west}$$

We can also use components to express the answer to this question. Assuming a positive x-axis pointing east, the x-component of the displacement vector is $\Delta x = -29$ km, where the negative sign indicates that the vector points to the west. Then

$$v_{av,x} = \frac{\Delta x}{\Delta t} = \frac{-29 \text{ km}}{0.70 \text{ h}} = -41 \text{ km/h}$$

The negative sign shows that the average velocity is directed along the negative x-axis, or to the west.

Discussion If the train had started at the same instant of time, 3:14 P.M., and had traveled directly west at a constant 41 km/h, it would have ended up in exactly the same place—26 km west of the trestle bridge—at 3:56 P.M.

Had we started measuring time from when we first spotted the motionless train at 3:00 P.M., instead of 3:14 P.M., we would have found the average velocity over a different time interval, changing the average velocity.

 The average velocity depends on the time interval considered.

Practice Problem 3.3 Average velocity of the baton from the relay race

Consider the baton from the relay race of Conceptual Example 3.1. For the winning team, the magnitudes of the average velocities of the first, second, and third runners are 7.30 m/s, 7.20 m/s, and 7.80 m/s, respectively. If the length of the track is 3.00×10^2 m, what is the average velocity of the baton for the entire race? [*Hint:* Find the time spent by each runner in completing her portion of the race.]

In Example 3.3, the magnitude of the train's average velocity is *not* equal to the total distance traveled divided by the time interval for the complete trip. That quantity is called the average speed:

$$\text{average speed} = \frac{\text{distance traveled}}{\text{total time}} = \frac{43 \text{ km}}{0.70 \text{ h}} = 61 \text{ km/h}$$

The distinction arises because the average velocity is the average of a vector quantity whereas the average speed is the average of a scalar.

Instantaneous Velocity

The speedometer of a car does not indicate the average speed, but shows how fast the car is going at any instant in time—the *instantaneous speed.* When a speedometer reads 55 mi/h, it does *not* necessarily mean that the car will travel 55 miles in the next hour; the car could change its speed or stop during that hour. The speedometer reading indicates how far the car will travel during a very short time interval—short enough that the speed does not change appreciably. For instance, at 55 mi/h (= 25 m/s), we can calculate that in 0.010 s the car moves 25 m/s × 0.010 s = 0.25 m—assuming the speed does not change significantly during that 0.010 s interval.

The **instantaneous velocity** is a *vector* quantity whose magnitude is the instantaneous speed and whose direction is the direction of motion. The direction of an object's instantaneous velocity at any point is tangent to the path of the object at that point. Repeating the word *instantaneous* can get cumbersome. When we refer simply to the *velocity,* we mean the *instantaneous* velocity.

The mathematical definition of instantaneous velocity starts with the average velocity during a short time interval. Then we consider shorter and shorter time intervals Δt; the corresponding displacements $\Delta\vec{\mathbf{r}}$ that take place during Δt get smaller and smaller. We let the time interval approach—but never reach—zero (Fig. 3.6). (This mathematical process is called finding the *limit* and is written $\lim_{\Delta t \to 0}$.) As Δt approaches zero, the average velocity during the increasingly short time interval approaches the instantaneous velocity.

> The *rate of change* of a quantity Q (whether scalar or vector) is
>
> $$\text{rate of change of } Q = \lim_{\Delta t \to 0} \frac{\Delta Q}{\Delta t}$$
>
> Velocity is the rate of change of position.

Definition of instantaneous velocity

$$\vec{\mathbf{v}} = \lim_{\Delta t \to 0} \frac{\Delta\vec{\mathbf{r}}}{\Delta t} \qquad (3\text{-}4)$$

Both numerator and denominator in Eq. (3-4) approach zero, but $\Delta\vec{\mathbf{r}}/\Delta t$ approaches a limiting value that is neither undefined nor infinite; it can be but is not necessarily zero. Suppose Δt is small enough that the velocity is approximately constant during the interval. If we now cut the time interval Δt in half, the displacement during the time interval is also cut in half (or nearly so), so $\Delta\vec{\mathbf{r}}/\Delta t$ (the average velocity during the interval) changes very little.

For motion along the x-axis, we can rewrite the definition of velocity in terms of x-components:

$$v_x = \lim_{\Delta t \to 0} \frac{\Delta x}{\Delta t} \qquad (3\text{-}5)$$

Figure 3.6 (a) A small displacement $\Delta\vec{\mathbf{r}}$ of a butterfly from point a to point b. If the time to travel from point a to point b is Δt, the average velocity during that time interval is $\vec{\mathbf{v}}_{av} = \dfrac{\Delta\vec{\mathbf{r}}}{\Delta t}$. (b) As points a and b get closer and closer to point A, the displacement $\Delta\vec{\mathbf{r}}$ becomes tangent to the path of travel at point A. The average velocity during the displacement from a to b approaches the instantaneous velocity at point A.

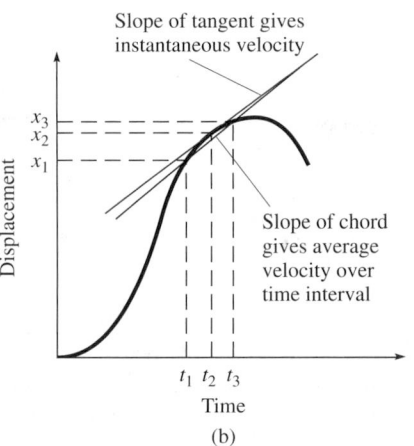

Figure 3.7 Instantaneous velocity at time t_2 is the slope of the tangent to the curve of displacement versus time at that time: (a) average velocity measured over a longer time interval and (b) average velocity measured over a shorter time interval

Table 3.1

Velocities Determined from the Graph of Fig. 3.4

t (min)	14	16	19	21	23	28	32	38	47	56
v_x (km/min)	0	0.5	1.8	0.7	0	0	−0.5	−1.3	−2.0	−1.3
v_x (m/s)	0	8.3	30	12	0	0	−8.3	−22	−33	−22

Graphical Relationship between Position and Velocity

How can we determine the instantaneous velocities from a graph of position versus time? Again, we look at the average velocity during a short time interval. For motion along the x-axis, the displacement (in component form) is Δx. Then the average velocity can be represented on the graph of x versus t as the slope of a line connecting two points (called a "chord"). In Fig. 3.7a, the displacement $\Delta x = x_3 - x_1$ is the "rise" of the graph and the time interval $\Delta t = t_3 - t_1$ is the "run" of the graph. The slope of the graph is the "rise over run," which is equal to the average velocity $\Delta x/\Delta t = v_{\mathrm{av},x}$ for that time interval.

To find the *instantaneous* velocity at some time $t = t_2$, we draw lines showing the average velocity for smaller and smaller time intervals. As the time interval is reduced (Fig. 3.7b), the average velocity changes. As $\Delta t \rightarrow 0$, the chord approaches a tangent to the graph. Therefore, v_x is the *slope of the line tangent to the graph of x versus t* at the chosen time.

Average and instantaneous velocities for the train can be found from the graph of x versus t (Fig. 3.4). Instantaneous velocities are found from the slopes of the tangents to the curve at various points. Table 3.1 shows some instantaneous velocities of the train in km/min along with equivalent velocities in m/s.

What about the other way around? Given a graph of the *velocity* as a function of time (v_x versus t), how can we determine displacements? For motion along a straight line it is most convenient to work in terms of the x-components of velocity and displacement (v_x and Δx). If v_x is constant during a time interval, then the average velocity is equal to the instantaneous velocity:

$$v_x = v_{\mathrm{av},x} = \frac{\Delta x}{\Delta t}$$

and therefore

$$\Delta x = v_x \, \Delta t \quad \text{(for constant } v_x\text{)}$$

The graph of Fig. 3.8 shows v_x versus t for an object moving along the x-axis with constant speed v_1 from time t_1 to t_2. The displacement Δx during the time interval $\Delta t = t_2 - t_1$ is $v_1 \Delta t$. The shaded rectangle has "height" v_1 and "width" Δt. Since the area of a rectangle is the product of the height and the width, the displacement Δx is represented by the area of the rectangle between the graph of $v_x(t)$ and the time axis for the time interval considered.

v_x is the slope of the graph of x versus t

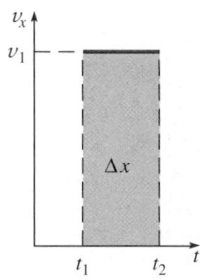

Figure 3.8 Displacement Δx between t_1 and t_2 is represented by the shaded area under the $v_x(t)$ graph.

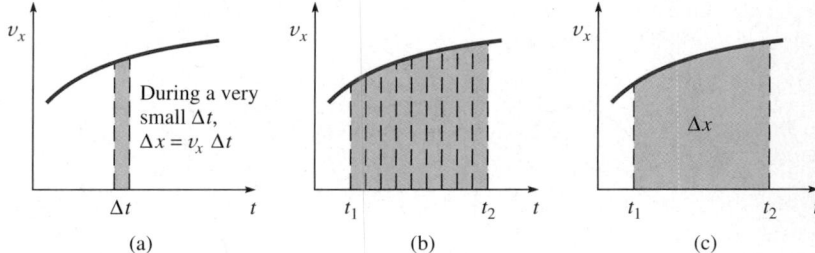

Figure 3.9 Displacement Δx is the area under the v_x versus time graph for the time interval considered.

When we speak of the area under a graph, we are not talking about the literal number of square centimeters of paper or computer screen. The figurative area under a graph usually does not have dimensions of an ordinary area ([L]^2). In a graph of v_x versus t, v_x has dimensions [L]/[T] and time has dimensions [T]; areas on such a graph have dimensions $\frac{[L]}{[T]} \times [T] = [L]$, which is correct for a displacement. The units of Δx are determined by the units used on the axes of the graph. If v_x is in meters per second and t is in seconds, then the displacement is in meters.

What if the velocity is not constant? The displacement Δx during a *very small* time interval Δt can be found in the same way as for constant velocity since, during a short enough time interval, the velocity does not change appreciably. Then v_x and Δt are the height and width of a narrow rectangle (Fig. 3.9a) and the displacement during that short time interval is the area of the rectangle. To find the total displacement during any time interval, the areas of all the narrow rectangles are added together.

In Fig. 3.9b the time from t_1 to t_2 is subdivided into many short time intervals so that many narrow rectangles of varying heights are formed. The width Δt is allowed to approach zero and the areas of the rectangles are added together. The total displacement Δx between t_1 and t_2 is the sum of the areas of the rectangles. Thus, *the displacement Δx during any time interval equals the area under the graph of $v_x(t)$* (Fig. 3.9c). If v_x is negative, x is decreasing and the displacement is negative, so we must count the area as negative when it is below the time axis.

The magnitude of the train's displacement from time $t = 14$ min to time $t = 23$ min is the area under the v_x versus t graph during that time interval. In Fig. 3.10 the area

Δx is the area under the graph of v_x versus t.
The area is negative when the graph is beneath the time axis ($v_x < 0$).

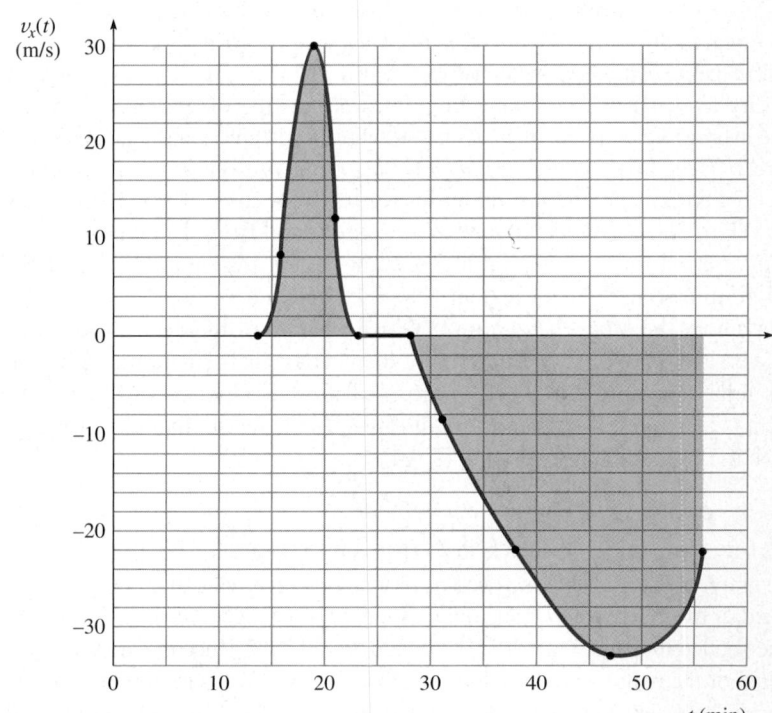

Figure 3.10 Train velocity versus time from $t = 14$ to 56 min using values from Table 3.1.

under the graph for that time interval is shaded. One way to estimate the area is to count the number of grid boxes under the curve. Each box is 2 m/s in height and 5 min (300 s) in width, so each box represents an "area" (displacement) of 2 m/s × 300 s = 600 m. When counting the number of boxes under the curve, we make our best estimate for the fraction of the boxes that are only partly below the curve.

The total number of shaded boxes for this time interval is about 12, so the displacement magnitude is 12×0.60 km ≈ 7 km, as expected—during this time interval the train went from +3 km to +10 km. To get more exact measurements of areas under a curve, we could divide the area into a finer grid.

The shaded area for the time interval $t = 28$ min to $t = 56$ min is below the x-axis, indicating a negative "area" or negative displacement. During this interval the train is headed west (in the $-x$-direction). The number of shaded grid boxes in this interval is approximately 60. The magnitude of the displacement is the product of the number of boxes times "area" of a single box and is negative since v_x is negative; $\Delta x \approx -(60) \times (0.60$ km$) \approx -36$ km. By adding this value to the displacement during the first 14 min, we have the total displacement from $t = 0$ to $t = 56$ min:

$$\Delta x = +7 \text{ km} + (-36 \text{ km}) = -29 \text{ km}$$

which agrees with Fig. 3.1.

3.3   ACCELERATION

One goal of this chapter is to quantify changes in velocity, so we can predict the effect on an object's motion of a nonzero net force. Just as we defined velocity in terms of displacements and time intervals, now we define **acceleration** in terms of changes in velocity and time intervals. The **average acceleration** during a time interval Δt is defined to be

$$\vec{\mathbf{a}}_{av} = \frac{\vec{\mathbf{v}}_f - \vec{\mathbf{v}}_0}{t_f - t_0} = \frac{\Delta \vec{\mathbf{v}}}{\Delta t} \tag{3-6}$$

The velocity vector can change magnitude, direction, or both. Acceleration is a vector quantity because it is the product of a scalar $t1/\Delta t$ and a vector $\Delta\vec{\mathbf{v}}$; *the direction of the average acceleration is the direction of the vector $\Delta\vec{\mathbf{v}}$*. The direction of the *change* in velocity $\Delta\vec{\mathbf{v}}$ is not necessarily the same as either the initial or the final velocity direction.

Just as instantaneous velocity is the limit of the average velocity as the time interval approaches zero, the **instantaneous acceleration** is defined as the limit of the average acceleration as the time interval approaches zero.

$$\vec{\mathbf{a}} = \lim_{\Delta t \to 0} \frac{\Delta \vec{\mathbf{v}}}{\Delta t} \tag{3-7}$$

The instantaneous acceleration is the rate of change of the velocity.

The SI units of acceleration are m/s², read as "meters per second squared." Just as with instantaneous velocity, the word *instantaneous* is not always repeated; acceleration without the adjective means *instantaneous* acceleration.

In physics, the word acceleration does not necessarily mean speeding up. A velocity can also change by decreasing speed or by changing direction. A car going around a curve at constant speed has a nonzero acceleration because its velocity is changing; the change is in the *direction* of the velocity rather than in the magnitude. Of course, both the magnitude and direction of the velocity can be changing simultaneously, as when a skateboarder goes up a curved ramp.

Acceleration is the rate of change of velocity.

Conceptual Example 3.4

Direction of Acceleration While Slowing Down

 When Damon approaches a stop sign on his motor scooter, he "decelerates" before coming to a full stop. If he moves in the negative x-direction while he slows down, is the scooter's acceleration component, a_x, positive or negative?

Strategy and Solution   The term "decelerate" is not a scientific term. In common usage it means the scooter is slowing: the scooter's velocity is decreasing in magnitude. The acceleration vector $\vec{\mathbf{a}}$ points in the direction of the *change* in the velocity vector $\Delta\vec{\mathbf{v}}$ during a short time interval. Since the velocity is decreasing in magnitude, the direction of $\Delta\vec{\mathbf{v}}$ is opposite to $\vec{\mathbf{v}}$ (Fig. 3.11). The acceleration vector is in the same direction as $\Delta\vec{\mathbf{v}}$—the positive x-direction. Thus a_x is positive (see Fig. 3.12).

Figure 3.11
The velocity vectors at two different times as the scooter moves to the left with decreasing speed. The change in velocity $\Delta\vec{\mathbf{v}} = \vec{\mathbf{v}}_2 - \vec{\mathbf{v}}_1$ is to the right.

Figure 3.12
In this graph of v_x versus t, as Damon is stopping, v_x is negative, while a_x (the slope) is positive. The value of v_x is increasing, but—since it is less than zero to begin with and is getting closer to zero as time goes on—the speed is *decreasing*.

Conceptual Practice Problem 3.4   Continuing on his way

As Damon pulls away from the stop sign, continuing in the $-x$-direction, his speed gradually increases. What is the sign of a_x?

---

a_x is the slope of a $v_x(t)$ graph.
Δv_x is the area under an $a_x(t)$ graph.

Both velocity and acceleration measure rates of change: velocity is the rate of change of position and acceleration is the rate of change of velocity. Therefore the graphical relationship of acceleration to velocity is the same as the graphical relationship of velocity to position: a_x is the slope of a $v_x(t)$ graph and Δv_x is the area under an $a_x(t)$ graph. On a graph of *any* quantity Q as a function of time, the slope of the graph represents the instantaneous rate of change of Q. On a graph of the *rate of change of Q* as a function of time, the area under the graph represents ΔQ.

Example 3.5

Acceleration of a Sports Car

 A sports car can accelerate from 0 to 30.0 m/s in 4.7 s according to the advertisements. Figure 3.13 shows data for the speed of the car as a function of time as the sports car starts from rest and accelerates to 60.0 m/s while traveling in a straight line headed east.

(a) What is the average acceleration of the sports car from 0 to 30.0 m/s?

(b) What is the maximum acceleration of the car?

(c) How far has the car traveled when it reaches 60.0 m/s?

(d) What is the car's average velocity during the entire 19.2-s interval?

Strategy   The velocity of the car is always in the same direction. If we choose that direction as the $+x$-axis, the x-component of the car's velocity is always positive. Therefore, v_x is equal to

the car's speed (which is the *magnitude* of the velocity). We can reinterpret the graph as a graph of v_x versus t.

(a) To find the average acceleration, the change in velocity for the time interval is divided by the time interval. (b) The instantaneous acceleration is the slope of the velocity graph, so it is maximum where the graph is steepest. At that point, the velocity is changing at a high rate. We expect the maximum acceleration to take place early on; the magnitude of acceleration must decrease as the velocity gets higher and higher—there is a maximum velocity for the car, after all. (c) The displacement is the area under the $v_x(t)$ graph. The graph is not a simple shape such as a triangle or rectangle, so an estimate of the area is made. (d) Once we have a value for the displacement, we can apply the definition of average velocity.

continued on next page

Example 3.5 continued

Given: Graph of $v(t)$ in Fig. 3.13

To find: (a) $a_{av,x}$ for $v_x = 0$ to 30.0 m/s; (b) a_{max}; (c) Δx from $v_x = 0$ to 60.0 m/s; (d) $v_{av,x}$ from $t = 0$ to 19.2 s

Solution   (a) During the first 4.7 s, the car is moving in a straight line to the east, so the average component of acceleration in the east direction is, by definition,

$$a_{av,x} = \frac{\Delta v_x}{\Delta t} = \frac{30.0 \text{ m/s}}{4.7 \text{ s}} = 6.4 \text{ m/s}^2$$

The positive sign indicates the acceleration is directed to the east. On average, then, the car's velocity increases 6.4 m/s each second during the first 4.7 s.

(b) The acceleration, at any instant of time, is the slope of the tangent line to the $v_x(t)$ graph, at that time. To find the maximum acceleration, notice where the graph is steepest. In this case, the largest slope occurs near $t = 0$, just as the car is starting out. In Fig. 3.13, a tangent line to the $v_x(t)$ graph at $t = 0$ passes through $t = 0$. Values for the change in velocity and change in time are read from the graph; the tangent line passes through the two points ($t = 0$, $v_x = 0$ and $t = 6.0$ s, $v_x = 55$ m/s) on the graph so that the change in velocity is 55 m/s for a change in time of 6.0 s. The slope of this line is

$$a_{max} = \frac{\text{rise}}{\text{run}} = \frac{55 \text{ m/s}}{6.0 \text{ s}} = +9.2 \text{ m/s}^2$$

Since the slope is positive, the direction of the acceleration is east.

v_x (m/s)

v_x (m/s)	0	15.0	20.0	25.0	30.0	35.0	40.0	45.0	50.0	55.0	60.0
t (s)	0	2.0	2.9	3.8	4.9	6.2	7.6	9.1	11.2	14.0	19.1

Figure 3.13
Speed as a function of time for a sports car.

(c) Displacement is the area under the $v_x(t)$ graph shown shaded in Fig. 3.13. The area can be estimated by counting the number of grid boxes under the curve. Each box is 5.0 m/s in height and 2.0 s in width, so each represents an "area" (displacement) of 10 m. When counting the number of boxes under the curve, a best estimate is made for the fraction of the boxes that are only partly below the curve. Approximately 75 boxes lie below the curve, so the displacement magnitude is $(75 \times 10 \text{ m}) = 750$ m.

(d) The average velocity for the 19.2-s interval is

$$v_{av,x} = \frac{\Delta x}{\Delta t}$$

$$= \frac{750 \text{ m}}{19.2 \text{ s}} = 39 \text{ m/s}$$

This result is reasonable; if the acceleration were constant, the average velocity would be $\frac{1}{2}(0 + 60.0 \text{ m/s}) = 30.0$ m/s. The actual average velocity is somewhat higher because the acceleration is greater at the start so less of the time interval is spent going (relatively) slowly and more is spent going fast. The speed is less than 30.0 m/s for only 4.7 s, but is greater than 30.0 m/s for 14.5 s.

 Discussion   The graph of velocity as a function of time is often the most helpful graph to have when solving a problem. If that graph is not given in the problem, it is useful to sketch one. The $v(t)$ graph shows all three kinematic quantities at once: the velocity is given by the points or the curve graphed, the displacement is the area under the curve, and the acceleration is the slope of the curve.

Practice Problem 3.5   Braking a car

An automobile is traveling along a straight road heading to the southeast at 24 m/s when the driver sees a deer begin to cross the road ahead of her. She steps on the brake and brings the car to a complete stop in an elapsed time of 8.0 s. A data recording device, triggered by the sudden braking action, records the following velocities and times as the car slows (Table 3.2). Let the positive x-axis be directed to the southeast. Plot a graph of v_x versus t and find (a) the average acceleration as the car comes to a stop and (b) the instantaneous acceleration at a time of 2.0 s after braking begins.

Table 3.2

Velocities and Times as Car Slows

v_x (m/s)	24	17.3	12.0	8.7	6.0	3.5	2.0	0.75	0
t (s)	0	1.0	2.0	3.0	4.0	5.0	6.0	7.0	8.0

When the acceleration and the velocity of an object are both pointing in the same direction, the object is speeding up. In terms of components, a_x and v_x must either both be positive or both be negative when speed is increasing. When they are both positive, the object is moving in the +x-direction and is speeding up. If they are both negative, the object is moving in the −x-direction and is speeding up.

When the velocity and acceleration point in opposite directions, so that their components have opposite signs, the object is slowing down. When v_x is positive and a_x is negative, the body is moving in the positive x-direction, but it is slowing down. When v_x is negative and a_x is positive, the body is moving in the negative x-direction and is slowing down (its speed is decreasing).

The acceleration does not contain any information about the initial velocity. Acceleration is the *rate of change* of velocity. Can an object have a velocity of zero and a nonzero acceleration at the same time? Yes, but the velocity does not *remain* zero. A ball thrown straight up into the air has a velocity of zero at its highest point. Its acceleration is *not* zero at that point; if it were, the ball would not fall back down. On the way up, the ball's velocity is upward and decreasing in magnitude. At the highest point, the velocity is zero but then the ball starts moving downward. On the way down, the velocity is downward and increasing in magnitude. For the entire flight of the ball, its acceleration is downward.

We can define the upward direction as positive or negative as we like; the choice is arbitrary, but we must remain consistent throughout a particular problem. For vertical motion, upward is usually defined as the positive y-direction. For a ball thrown into the air, v_y is positive on the way up and negative on the way down; a_y is negative and constant throughout the motion—on the way up, at the highest point, and on the way down. Since v_y and a_y have opposite signs on the way up, the ball is slowing down (the velocity is decreasing in magnitude). On the downward trip, v_y and a_y have the same sign so the ball's speed is increasing.

3.4 NEWTON'S SECOND LAW: FORCE AND ACCELERATION

According to Newton's second law, the acceleration of an object is proportional to the net force on it and is in the same direction. The larger the net force, the larger the acceleration.

Figure 3.14 An astronaut playing shuffleboard (a) on Earth and (c) on the Moon. Free-body diagrams for a puck being given the same acceleration on a *frictionless* court on (b) Earth and (d) on the Moon. The contact force on the puck due to the *pushing stick* ($\vec{F}_C$) must be the same since the mass of the puck is the same.

If the net force is zero, the acceleration is zero and the object moves with constant velocity—possibly zero velocity, but not necessarily. Newton's second law also says that the acceleration is inversely proportional to the object's mass. The same net force acting on two different objects causes a smaller acceleration on the object with greater mass. Mass is a measure of an object's inertia—the amount of resistance to *changes in velocity.*

Mass and weight measure different physical properties. The mass of a body is a measure of its inertia, while weight is a measure of the gravitational force acting on it. Imagine taking a shuffleboard puck to the Moon. Since the Moon's gravitational field is weaker than the Earth's, the puck's weight $\vec{W}$ would be smaller. A smaller normal force $\vec{N}$ would be required to hold it up. On the other hand, the puck's mass, an intrinsic property, is the same. Neglecting the effects of friction, an astronaut playing shuffleboard on the Moon would have to exert the same horizontal force on the puck as on Earth to give it the same acceleration (Fig. 3.14).

Newton's law relating net force and acceleration is

> **Newton's Second Law**
>
> $$\vec{F}_{net} = m\vec{a} \qquad (3\text{-}8)$$

or

$$\Sigma\vec{F} = m\vec{a}$$

where Σ, the Greek capital letter sigma, stands for *the sum of.* $\Sigma\vec{F}$ means the sum of all the forces acting on a system. The order of the symbols in Eq. (3-8) does not reflect a cause-and-effect relationship; the net force causes the acceleration, not the other way around. The SI unit of force, the newton, is defined in terms of SI base units so that a 1-N net force acting on a 1-kg mass produces an acceleration of 1 m/s^2; therefore,

$$1 \text{ N} = 1 \text{ kg·m/s}^2$$

When calculating the net force on a system, only *external* forces need be considered. According to Newton's third law, internal forces always add to zero.

Example 3.6

Coupling Force on First and Last Freight Cars

A train engine pulls out of a station along a straight track with five identical freight cars behind it, each of which weigh 90.0 kN. The train reaches a speed of 15.0 m/s within 5.00 min of starting out. Assuming the acceleration is constant, with what magnitude of force must

the coupling between cars pull forward on the first and last of the freight cars? Ignore friction and air resistance. Assume $g = 9.80$ N/kg.

Strategy A sketch of the situation is shown in Fig. 3.15a. We can calculate the acceleration of the train from the initial and final velocities and the elapsed time. Then we can relate the acceleration to the net force using Newton's second law.

(a)

Figure 3.15
(a) An engine pulling five identical freight cars. The entire train has a constant acceleration $\vec{a}$ to the right.
(b) FBD for car 5. (c) FBD for cars 1–5.

(b)   (c)

continued on next page

Example 3.6 *continued*

To find the force of the first coupling, we can consider all five cars to be one system so that we do not have to worry about the force exerted on the first car by the second car. Once we identify a system, we draw a free-body diagram before applying Newton's second law.

Given: W = weight of each freight car = 90.0 kN = 9.00×10^4 N; v_x = 15.0 m/s at t = 5.00 min = 300 s; v_{0x} = 0 since the train starts from rest; a_x = constant

To find: Tensions T_1 and T_5

Solution   The acceleration of the train is

$$a_x = \frac{\Delta v_x}{\Delta t} = \frac{15.0 \text{ m/s}}{300 \text{ s}} = 0.0500 \text{ m/s}^2$$

First consider the last freight car (car 5). If we ignore friction and air resistance, the only forces acting are the force $\vec{T}_5$ due to the tension in the coupling, the normal force $\vec{N}_5$, and the car's weight $\vec{W}_5$; an FBD is shown in Fig. 3.15b. The normal force and the weight are vertical and act in opposite directions. They must be equal in magnitude; the vertical component of the net force is zero since the vertical component of the acceleration is zero. Then the net force is equal to the tension in the coupling. The mass of the car is $m = W/g$, where g = 9.80 N/kg = 9.80 m/s². Then, from Newton's second law,

$$T_5 = \Sigma F_x = ma_x = \frac{W}{g} a_x$$

$$T_5 = 9.00 \times 10^4 \text{ N} \times \frac{0.0500 \text{ m/s}^2}{9.80 \text{ m/s}^2} = 459 \text{ N}$$

For the tension in the first coupling, consider the five cars as *one system*. Fig. 3.15c shows an FBD in which cars 1–5 are treated as a single object. Again, the vertical forces on the system add to zero. The only external horizontal force is the force $\vec{T}_1$ due to the tension in the first coupling. The mass of the system is five times the mass of one car. Therefore,

$$T_1 = \Sigma F_x = ma_x = (5 \times 9.00 \times 10^4 \text{ N}) \times \frac{0.0500 \text{ m/s}^2}{9.80 \text{ m/s}^2} = 2.30 \text{ kN}$$

Discussion   The solution to this problem is much simpler when Newton's second law is applied to a system comprised of all five cars, rather than to each car individually. Although the problem can be solved by looking at individual cars, to find the tension in the first coupler you would have to draw five free-body diagrams (one for each car) and apply Newton's second law five times. That's because each car, except the fifth, is acted on by the unequal tensions in the couplers on either side. You'd have to first find the tension in the fifth coupler, then in the fourth, then the third, and so on.

Practice Problem 3.6   Coupling force between first and second freight cars

With what force does the coupling between the first and second cars pull forward on the second car? [*Hint:* Try two methods. One of them is to draw an FBD for the first car and apply Newton's *third* law as well as the second.]

Example 3.7

Two Blocks Hanging on a Pulley

In Fig. 3.16a, two blocks are connected by a massless, flexible cord that does not stretch; the cord passes over a massless, frictionless pulley. If the masses are m_1 = 26.0 kg and m_2 = 42.0 kg,

(a)   (b)

Figure 3.16
(a) Two hanging blocks connected on either side of a frictionless pulley by a massless, flexible cord that does not stretch. (b) Free-body diagrams for the hanging blocks.

what are the accelerations of each block and the tension in the cord? The gravitational field strength is 9.80 N/kg.

Strategy   Since m_2 is greater than m_1, the downward pull of gravity is stronger on the right side than on the left. We expect m_2 to accelerate downward and m_1 to accelerate upward.

Since the cord does not stretch, the accelerations of the two blocks are equal in magnitude. If the accelerations had different magnitudes, then soon the two blocks would be moving with different speeds. That could only happen if the cord either stretches or contracts. The fixed length of the cord constrains the blocks to move with equal speeds (in opposite directions) at all times, so the magnitudes of their accelerations must be equal.

The tension in the cord must be the same everywhere along the cord since the masses of the cord and pulley are negligible and the pulley turns without friction.

We treat each block as a separate system, draw free-body diagrams for each, and then apply Newton's second law to each. It is convenient to choose the positive y-direction differently for the two blocks. For each, we choose the $+y$-axis in the direction of the acceleration. Doing so means that a_y has the same magnitude *and sign* for the two.

continued on next page

Example 3.7 *continued*

Given: $m_1 = 26.0$ kg and $m_2 = 42.0$ kg

To find: a_y; T

Solution   Figure 3.16b shows free-body diagrams for the two blocks. Two forces act on each: gravity and the pull of the cord. The acceleration vectors are drawn next to the free-body diagrams. Thus we know the direction of the net force: it is always the same as the direction of the acceleration. Then we know that the tension must be greater than $m_1 g$ to give block 1 an upward acceleration and less than $m_2 g$ to give block 2 a downward acceleration. The +y-axes are drawn for each block to be in the direction of the acceleration.

From the free-body diagram of block 1, the pull of the cord is in the +y-direction and the gravitational force is in the −y-direction. Then

$$\Sigma F_y = T - m_1 g = m_1 a_y$$

For block 2, the pull of the cord is in the −y-direction and the gravitational force is in the +y-direction. Therefore,

$$\Sigma F_y = m_2 g - T = m_2 a_y$$

Both T and a_y are identical in these two equations. We then have a system of two equations with two unknowns. Adding the equations, we obtain

$$m_2 g - m_1 g = m_2 a_y + m_1 a_y$$

Solving for a_y we find

$$a_y = \frac{(m_2 - m_1)g}{m_2 + m_1}$$

Substituting numerical values,

$$a_y = \frac{(42.0 \text{ kg} - 26.0 \text{ kg}) \times 9.80 \text{ N/kg}}{42.0 \text{ kg} + 26.0 \text{ kg}} = \frac{16.0 \text{ kg}}{68.0 \text{ kg}} \times 9.80 \text{ N/kg}$$

$$a_y = 2.31 \text{ m/s}^2$$

since

$$1 \frac{\text{N}}{\text{kg}} = 1 \frac{\text{kg·m/s}^2}{\text{kg}} = 1 \text{ m/s}^2$$

The blocks have the same magnitude acceleration. For block 1 the acceleration points upward and for block 2 it points downward.

To find T we can substitute the expression for a_y into either of the two original equations. Using the first equation,

$$T - m_1 g = m_1 \frac{(m_2 - m_1)g}{m_2 + m_1}$$

Solving for T yields

$$T = \frac{2m_1 m_2}{m_1 + m_2} g$$

Substituting,

$$T = \frac{2 \times 26.0 \text{ kg} \times 42.0 \text{ kg}}{68.0 \text{ kg}} \times 9.80 \text{ N/kg} = 315 \text{ N}$$

Discussion   A few quick checks:

- a_y is positive, which means that the accelerations are in the directions we expect.
- The tension (315 N) is between $m_1 g$ (255 N) and $m_2 g$ (412 N) as expected.
- The units and dimensions are correct for all equations.

It is also instructive to examine what happens to the expressions for a_y and T for special cases of hanging blocks with: equal masses, masses just slightly unequal, or one mass much greater than the other. We often have intuition about what should happen in such special cases. See Practice Problem 3.7 for some examples.

⚠ Note that we did *not* find out which way the blocks move. We found the directions of their *accelerations*. If the blocks start out at rest, then the block of mass m_2 moves downward and the block of mass m_1 moves upward. However, if initially m_2 is moving up and m_1 down, they continue to move in those directions, slowing down since their accelerations are opposite to their velocities. Eventually, they come to rest and then reverse directions.

Conceptual Practice Problem 3.7   Equal and slightly unequal masses

Suppose that the blocks attached to the pulley are of equal mass ($m_1 = m_2$). What do the expressions for a_y and T yield in that case? Can you explain why that must be correct? What if the blocks are only slightly unequal so that their masses $m_2 - m_1 \ll m_2 \approx m_1$? What is the tension in that case? What about the magnitude of the acceleration?

These steps are helpful in most problems that involve Newton's second law.

Problem-Solving Strategies for Newton's Second Law

- Decide what objects will have Newton's second law applied to them.
- Identify all the interactions affecting that object.
- Draw a free-body diagram to show all the forces acting on the object.
- Find the net force by adding the forces as vectors.
- Use Newton's second law to relate the net force to the acceleration.

Forces acting on an object determine the motion of the object. The sum of the forces is the net force, which causes an acceleration. The acceleration is a measure of the *rate of change* of the object's velocity. To describe the motion, one other piece of information is required: the object's initial velocity. From the initial velocity and the changes in velocity caused by the forces acting, the velocity at any later time can be calculated. The velocity is the *rate of change* of the position. This strategy for finding the velocity is a triumph of Newton's method of analyzing and then predicting the motion based on the initial conditions and the forces acting on the object.

Example 3.8

Forward Acceleration of a Grocery Cart

 Alfredo is shopping at the supermarket (Fig. 3.17a). Alfredo's mass is 72 kg and the mass of his shopping cart plus groceries is 46 kg. At some particular instant, Alfredo's foot pushes backward on the floor with a force of 147 N. What is the acceleration of the cart at that instant? The forces that oppose the forward motion of the cart—friction in the rotation of the cart wheels, air resistance, and so on—add to 5 N.

Strategy   To simplify this problem we may choose a system composed of Alfredo and the cart; they move together with the same acceleration. Two horizontal external forces act on the system: the forces opposing the motion of the cart, $\vec{\mathbf{f}}$, and a force of the floor pushing forward on Alfredo's foot. There is no motion in the vertical direction so the net vertical force is zero; the weight of the system is equal to the normal force with which the floor pushes up on the system. We apply both Newton's third law and Newton's second law to solve this problem.

Given: Alfredo's mass: $m_1 = 72$ kg
cart + groceries mass: $m_2 = 46$ kg
force on floor by Alfredo: $\vec{\mathbf{F}}_{fA} = 147$ N to the left
force opposing motion of cart: $\vec{\mathbf{f}} = 5$ N to the left

To find: $\vec{\mathbf{a}}$

Solution   From Newton's third law, we know that the floor pushes forward on the system with the same force that Alfredo pushes backward on the floor.

$$\vec{\mathbf{F}}_{Af} = -\vec{\mathbf{F}}_{fA}$$

Since $\vec{\mathbf{F}}_{fA}$ is to the left, $\vec{\mathbf{F}}_{Af}$ is to the right and of magnitude 147 N. We draw a free-body diagram (Fig. 3.17b), showing the external forces acting on the system. Since the vertical acceleration component is zero, the free-body diagram indicates that the weight of the system is balanced by an equal normal force. There may be a nonzero horizontal acceleration component. The +x-axis direction is chosen to the right, in the forward direction of the cart.

From Newton's second law for the vertical direction,

$$\Sigma F_y = N - (m_1 + m_2)g = ma_y = 0$$

and for the horizontal direction,

$$\Sigma F_x = F_{Af} - f = ma_x$$

Solving for the acceleration,

$$a_x = \frac{F_{Af} - f}{m_1 + m_2}$$

Substituting the given values,

$$a_x = \frac{147 \text{ N} - 5 \text{ N}}{72 \text{ kg} + 46 \text{ kg}} = \frac{142 \text{ N}}{118 \text{ kg}} = 1.20 \text{ m/s}^2$$

The direction of the acceleration is to the right.

Figure 3.17
(a) Alfredo pushes backward on the floor with force $\vec{\mathbf{F}}_{fA}$ and a force $\vec{\mathbf{f}}$ opposes the forward motion of the cart; (b) a free-body diagram for the system of Alfredo and cart + groceries.

(a)

(b)

continued on next page

Discussion By choosing the system to be composed of Alfredo and the cart and groceries, we did not have to worry about the forces internal to the system. Alfredo pushes on the cart handle while the cart handle pushes back on Alfredo's hands—these are two internal forces within the system that cancel each other. The only forces that are of concern for the motion of the system as a whole are the external forces, shown in Fig. 3.17b, that act on the system.

Practice Problem 3.8 Alfredo's pushing force

What is the force with which Alfredo's hands push on the grocery cart? [*Hint:* Choose a new system so that Alfredo is external to the system and consider forces acting external to the chosen system.]

3.5 MOTION WITH CONSTANT ACCELERATION

If the net force acting on an object is constant, then the acceleration of the object is also constant, both in magnitude and direction. *Uniform acceleration* is a synonym for constant acceleration.

Three essential relationships between position, velocity, and acceleration can be used to solve *any* constant acceleration problem. We write these relationships using these conventions:

- For motion along the x-axis, write position, velocity, and acceleration in terms of their x-components (x, v_x, a_x).
- At a time $t = 0$, we designate the initial position and velocity as x_0 and v_{0x}.
- At a later time $t > 0$, the position and velocity are x and v_x.

Just as the origin of a coordinate system can be chosen at any convenient point in space, the time at which $t = 0$ can be freely chosen to be any instant in time—whatever makes the problem easiest to solve.

The three essential relationships are: First, since the acceleration is constant, the change in velocity over a given time interval Δt is just the acceleration—the rate of change of velocity—times the elapsed time:

$$\Delta v_x = v_x - v_{0x} = a_x \Delta t \quad \text{for } a_x \text{ constant} \tag{3-9}$$

Second, the displacement is the average velocity times the time interval:

$$\Delta x = x - x_0 = v_{\text{av},x} \Delta t \tag{3-10}$$

Equation (3-10) is true whether the acceleration is constant or not; it comes directly from the definition of average velocity.

Third, since the velocity changes linearly with time, the average velocity is given by

$$v_{\text{av},x} = \frac{v_{0x} + v_x}{2} \quad \text{for } a_x \text{ constant} \tag{3-11}$$

Equation (3-11) is *not* true in *general*, but it is true for constant acceleration. To see why, refer to the velocity versus time graph in Fig. 3.18a. The graph is linear because the acceleration is constant. The displacement during any time interval is represented by the area under the $v(t)$ graph. The average velocity is found by forming a rectangle with an area equal to the area under the curve in Fig. 3.18a, because the average velocity should give the same displacement in the same time interval. Fig. 3.18b shows that, to make the area of one colored triangle equal to the area of the other, the average velocity must be exactly halfway between the initial and final velocities.

If the acceleration is *not* constant, there is no reason why the average velocity would have to be exactly halfway between the initial and the final velocity. As an illustration, imagine a trip where you drive along a straight highway at 80 km/h for 50 min and then at 60 km/h for 30 min. Your acceleration is zero for the entire trip *except* during the few seconds while you slowed from 80 km/h to 60 km/h. The magnitude of your average velocity would *not* be 70 km/h. You spent more time going 80 km/h than you did going

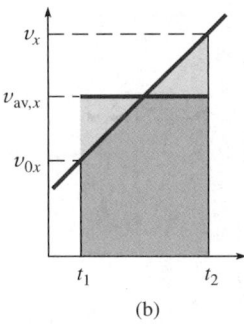

Figure 3.18 Finding average velocity with the aid of a graph

Tips

60 km/h, so the magnitude of your average velocity would be greater than 70 km/h (see Problem 16).

To summarize:

When the time interval is from $t = 0$ until a later time t, $\Delta t = t - 0 = t$

If a_x is constant during the entire time interval from $t = 0$ until a later time t,

$$\Delta v_x = v_x - v_{0x} = a_x \Delta t$$

$$\Delta x = x - x_0 = v_{\text{av},x} \Delta t$$

$$v_{\text{av},x} = \frac{v_{0x} + v_x}{2}$$

Other relationships can be formed between the various quantities (displacement, velocity, acceleration, and time interval), but it is usually better to start with the three basic relationships and work from there. Any constant acceleration problem can be solved using just these three, and they are not difficult to remember—two of them are really definitions.

Two such relationships that are often useful shortcuts can be derived from Eqs. (3-9) through (3-11) (see Problems 37 and 38). They are

$$\Delta x = x - x_0 = v_{0x}t + \tfrac{1}{2}a_x t^2 \tag{3-12}$$

$$v_x^2 - v_{0x}^2 = 2a_x \Delta x \tag{3-13}$$

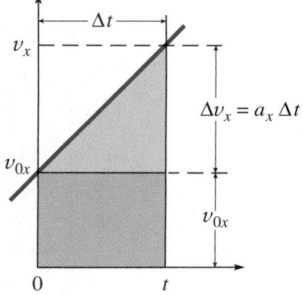

Figure 3.19 Graphical interpretation of Eq. (3-12)

Note that Δt is replaced by t because the time interval begins at $t = 0$ and $\Delta t = t$. Equation (3-12) is useful when the final velocity is not known, while Eq. (3-13) is useful when the elapsed time is not known.

We can interpret Eq. (3-12) graphically. Figure 3.19 shows a $v_x(t)$ graph for constant-acceleration motion. The displacement that occurs between $t = 0$ and a later time t is the area under the graph for that time interval. Partition this area into a rectangle plus a triangle. The area of the rectangle is

$$\text{height} \times \text{base} = v_{0x}t$$

The height of the triangle is the change in velocity, which is equal to $a_x t$. The area of the triangle is

$$\tfrac{1}{2} \times \text{height} \times \text{base} = \tfrac{1}{2} \times a_x t \times t = \tfrac{1}{2}a_x t^2$$

Adding these "areas" gives Eq. (3-12).

Equations (3-9) through (3-13) are called *kinematic* equations for motion with constant acceleration, since they deal with the relationships between position, velocity, acceleration, and time, but not with forces.

Example 3.9

Displacement of Motorboat

A motorboat accelerates from rest at a dock with a constant acceleration of magnitude 2.8 m/s². After traveling directly to the east for 140 m the motor is throttled down so that the boat slows down at 1.2 m/s² while still moving east until its speed is 16 m/s. Just as the boat attains the velocity of 16 m/s, it passes a buoy due east of the dock. What is the total displacement of the motorboat from the dock at that time?

Strategy   This problem involves two different values of acceleration, so it must be divided into two subproblems. The kinematic equations for constant acceleration cannot be applied to a time interval during which the acceleration changes. But for each of two time intervals, the acceleration of the boat is constant. The two subproblems are connected by the position and velocity of the boat at the instant that the acceleration changes.

For the first subproblem, the boat speeds up with a constant acceleration of 2.8 m/s² to the east. We know the acceler-

continued on next page

Example 3.9 continued

ation, the displacement (140 m east), and the initial velocity: the boat starts from rest, so the initial velocity is zero. We need to calculate the final velocity, which then becomes the initial velocity for the second subproblem. The boat is always headed to the east, so we let east be the positive x-direction.

> First subproblem: $v_{0x} = 0$; $a_x = +2.8$ m/s^2
>
> $\Delta x = 140$ m; find v_x

For the second subproblem, we know acceleration, final velocity, and we have just found initial velocity from the first subproblem. Since the boat is slowing down, its acceleration is in the direction opposite its velocity; therefore a_x is negative. From these three quantities we can find the displacement of the boat during the second time interval.

> Second subproblem: v_{0x} comes from first subproblem
>
> $a_x = -1.2$ m/s^2; $v_x = +16$ m/s; find Δx

Adding the displacements for the two time intervals gives the total displacement.

Solution   (1) To find v_x without knowing the time, the most convenient kinematic equation is

$$v_x^2 - v_{0x}^2 = 2a_x \Delta x$$

Solving for v_x,

$$v_x = \sqrt{v_{0x}^2 + 2a_x \Delta x}$$

$$v_x = \sqrt{0 + 2 \times 2.8 \text{ m/s}^2 \times 140 \text{ m}} = +28 \text{ m/s}$$

(2) For the second interval the *initial* velocity is the final velocity for the first interval: $v_{0x} = +28$ m/s. Then

$$v_x^2 - v_{0x}^2 = 2a_x \Delta x$$

This time we solve for the displacement.

$$\Delta x = \frac{v_x^2 - v_{0x}^2}{2a_x}$$

$$\Delta x = \frac{(16 \text{ m/s})^2 - (28 \text{ m/s})^2}{2 \times (-1.2 \text{ m/s}^2)} = +220 \text{ m}$$

And the total displacement is

$$\Delta x = \Delta x_1 + \Delta x_2 = 140 \text{ m} + 220 \text{ m} = +360 \text{ m}$$

The boat is 360 m east of the dock.

Discussion   This problem is solved by applying the same kinematic equation twice, once to find a velocity and once to find a displacement. The natural division of the problem into two parts occurs because the boat has two different constant accelerations during two different time periods.

In problems that can be subdivided in this way, the final velocity found in the first part becomes the initial velocity for the second part. The same is true for position.

Practice Problem 3.9   Time to reach buoy and average velocities

(a) What is the time required by the boat in the previous example to reach the buoy? (b) Find the average velocity for the entire trip from the dock to the buoy.

Visualizing Motion with Constant Acceleration

In Fig. 3.20 three carts move in the same direction with three different values of constant acceleration. The position of each cart is depicted as it would appear in a stroboscopic photograph with one picture taken every second.

The yellow cart has zero acceleration and therefore constant velocity. During each 1.0-s time interval its displacement is the same: 1.0 m/s × 1.0 s = 1.0 m to the right.

The red cart has a constant acceleration of 0.2 m/s^2 to the right. Although m/s^2 is normally read "meters per second squared," it can be useful to think of it as "m/s per second": the cart's velocity changes by 0.2 m/s during each 1.0-s time interval. In this case,

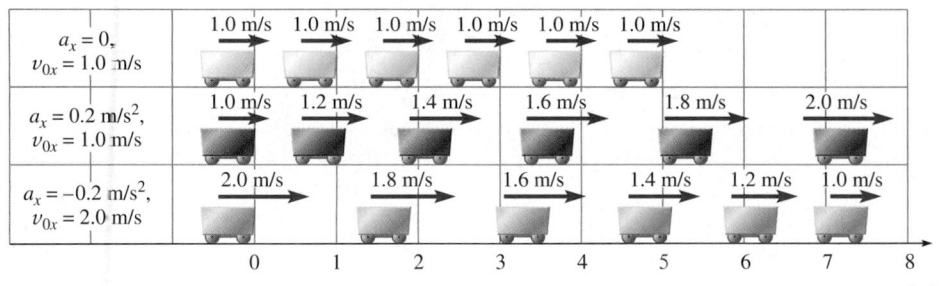

Figure 3.20   Each cart is shown as if stroboscopic photographs were taken with time intervals of 1.0 s between flashes. The arrows above each cart indicate velocity vectors as the strobe flashes occur.

Figure 3.21 (a) If the acceleration is parallel to the velocity, then the change in velocity ($\Delta \vec{v} = \vec{a}\,\Delta t$) is also parallel to the velocity. The result is an increase in the magnitude of the velocity: the object speeds up. (b) If the acceleration is antiparallel (parallel but pointed in the opposite direction) to the velocity, then the change in velocity ($\Delta \vec{v} = \vec{a}\,\Delta t$) is also antiparallel to the velocity. The result is a decrease in the magnitude of the velocity: the object slows down.

acceleration is in the same direction as the velocity, so the velocity increases (Fig. 3.21a). The displacement of the cart during successive 1.0-s time intervals gets larger and larger.

The blue cart experiences a constant acceleration of 0.2 m/s² in the –x-direction— the direction *opposite* to the velocity. The magnitude of the velocity then decreases (Fig. 3.21b); during each one-second interval the speed decreases by 0.2 m/s. Now the displacements during one-second intervals get smaller and smaller.

Figure 3.22 shows graphs of $x(t)$, $v_x(t)$, and $a_x(t)$ for each of the carts. The acceleration graphs are horizontal since each of the carts has a constant acceleration. All three v_x graphs are straight lines. Since a_x is the rate of change of v_x, the slope of the v_x graph at any value of t is a_x at that value of t. With constant acceleration, the slope is the same everywhere and the graph is linear. Note that a positive a_x does mean that v_x is increasing, but not necessarily that the *speed* is increasing; if v_x is negative then a positive a_x indicates a *decreasing* speed. (See Conceptual Example 3.4 and Fig. 3.12.) Speed is increasing when the acceleration and velocity are in the same direction (a_x and v_x both positive *or* both negative). Speed is decreasing when acceleration and velocity are in opposite directions—when a_x and v_x have opposite signs.

The position graph is linear for the yellow cart because it has constant velocity. For the red cart the slope of the $x(t)$ graph increases, showing that v_x is increasing; for the blue cart the slope of the $x(t)$ graph decreases, showing that v_x is decreasing.

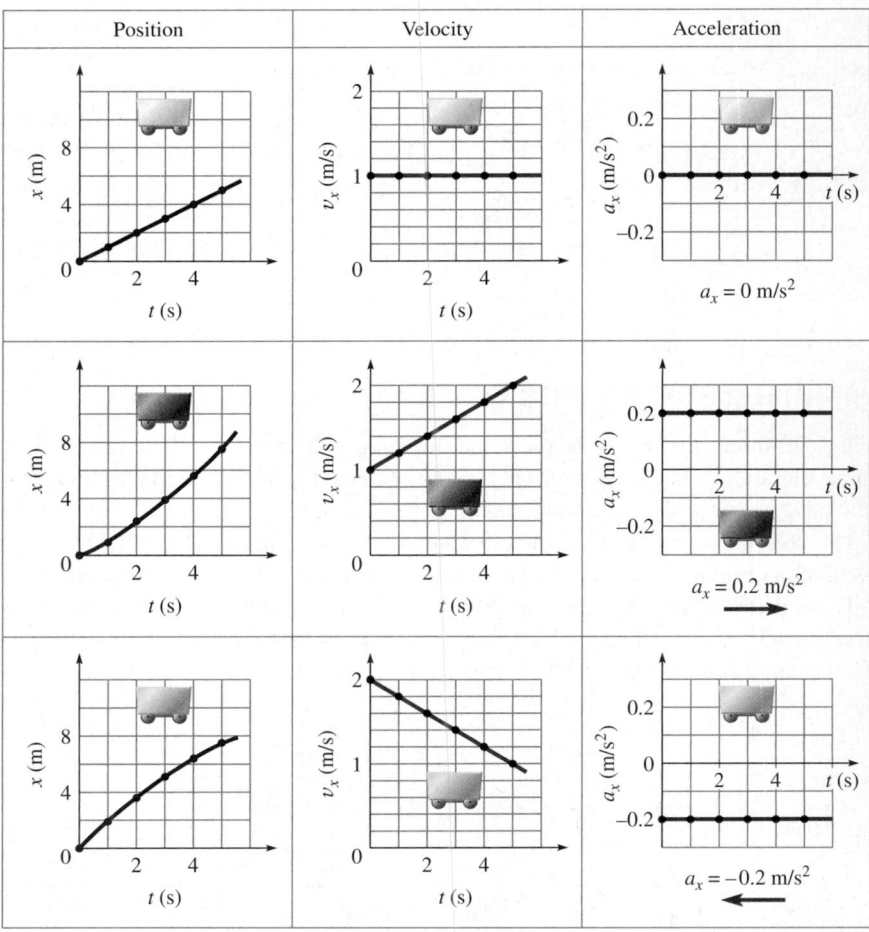

Figure 3.22 Plots of position, velocity, and acceleration along the x-axis for the carts of Fig. 3.20.

Example 3.10

Drag-Racing Spaceships

Two spaceships are moving from the same starting point in the +x-direction with constant accelerations. In component form, the silver spaceship starts with an initial velocity of +2.00 km/s and has an acceleration of +0.400 km/s^2. The black spaceship starts with a velocity of +6.00 km/s and has an acceleration of −0.400 km/s^2. Find the time at which the silver spaceship just overtakes the black spaceship.

Strategy   We can find the positions of the spaceships at later times from the initial velocities and the accelerations. At first, the black spaceship is moving faster, so it pulls out ahead. Later, the silver ship overtakes the black ship when their *positions are equal*.

Solution   The position of either spaceship at a later time is given by

$$x = x_0 + v_{0x}t + \tfrac{1}{2}at^2$$

We set the positions of the spaceships equal to each other ($x_{\text{silver}} = x_{\text{black}}$) and solve algebraically for the time at which this occurs. The initial positions are the same: $x_{0s} = x_{0b} = x_0$.

$$x_0 + v_{0sx}t + \tfrac{1}{2}a_{sx}t^2 = x_0 + v_{0bx}t + \tfrac{1}{2}a_{bx}t^2$$

Subtracting x_0 from each side, moving all terms to one side, and factoring out one power of t yields

$$t(v_{0sx} + \tfrac{1}{2}a_{sx}t - v_{0bx} - \tfrac{1}{2}a_{bx}t) = 0$$

One solution of the equation is $t = 0$ (the two spaceships at the same *initial* position). That is not the solution we seek, so the expression in the parentheses must be equal to zero. Solving for t,

$$t = \frac{2(v_{0sx} - v_{0bx})}{a_{bx} - a_{sx}} = \frac{2(2.00 \text{ km/s} - 6.00 \text{ km/s})}{-0.400 \text{ km/s}^2 - 0.400 \text{ km/s}^2} = 10.0 \text{ s}$$

The silver spaceship overtakes the black spaceship at $t = 10.0$ s.

Discussion   Quick check: the two ships must have the same displacement at $t = 10.0$ s.

$$\Delta x_s = v_{0sx}t + \tfrac{1}{2}a_{sx}t^2$$

$$= 2.00 \text{ km/s} \times 10.0 \text{ s} + \tfrac{1}{2} \times 0.400 \text{ km/s}^2 \times (10.0 \text{ s})^2 = 40.0 \text{ km}$$

$$\Delta x_b = v_{0bx}t + \tfrac{1}{2}a_{bx}t^2$$

$$= 6.00 \text{ km/s} \times 10.0 \text{ s} + \tfrac{1}{2} \times (-0.400 \text{ km/s}^2) \times (10.0 \text{ s})^2$$

$$= 40.0 \text{ km}$$

Practice Problem 3.10   Time to reach same velocity

Find the time at which the two spaceships have the same *velocity*.

Example 3.11

Two Blocks, One Sliding and One Hanging

A block of mass $m_1 = 3.0$ kg rests on a frictionless horizontal surface. A second block of mass $m_2 = 2.0$ kg hangs from a flexible cord of negligible mass that runs over an ideal pulley and then is connected to the first block (Fig. 3.23a). The blocks are released from rest. (a) Find the accelerations of the two blocks after they are released. (b) Find the tension in the cord connecting the blocks. (c) What is the velocity of the first block 1.2 s after the release of the blocks, assuming the first block does not run out of room on the table and the second block does not land on the floor? (d) What is the minimum distance from the pulley for block 1 to be located so that it can attain the velocity found in part (c)?

Strategy   We consider each block as a separate system and draw a free-body diagram for each. The tension in the cord is the same at both ends of the cord since the cord and pulley are ideal. We choose the +x-axis to the right and the +y-axis up. To find the accelerations of the blocks (which are equal in magnitude if the cord length is fixed) and the tension in the cord, we apply Newton's second law. Then we use kinematic equations for constant acceleration to answer (c) and (d).

Given: $m_1 = 3.0$ kg; $m_2 = 2.0$ kg; $\vec{v}_0 = 0$ for both; $\Delta t = 1.2$ s

Solution   (a) Figure 3.23b shows free-body diagrams for the two blocks. Block 1 slides along the table surface, so the vertical component of acceleration is zero; the normal force must

Figure 3.23
(a) Two blocks connected by a cord, one supported by a frictionless table and one hanging freely and
(b) free-body diagrams for blocks 1 and 2 with force magnitudes labeled.

(a)   (b)

continued on next page

Example 3.11 *continued*

be equal in magnitude to the weight. With the two vertical forces canceling, the only remaining force is horizontal: the pull of the cord.

$$\Sigma F_{1x} = T = m_1 a_{1x}$$

Only vertical forces act on block 2. From the free-body diagram for block 2, we write Newton's second law:

$$\Sigma F_{2y} = T - m_2 g = m_2 a_{2y}$$

The two accelerations have the same magnitude. The acceleration of block 1 is in the $+x$-direction while that of block 2 is in the $-y$-direction. Therefore we substitute

$$a_{1x} = a \quad \text{and} \quad a_{2y} = -a$$

where a is the magnitude of the acceleration.

Substituting $T = m_1 a$ and $a_{2y} = -a$ into the second equation,

$$m_1 a - m_2 g = -m_2 a$$

Now we solve for a:

$$a = \frac{m_2}{m_1 + m_2} g$$

Substituting the known quantities

$$a = \frac{2.0 \text{ kg}}{3.0 \text{ kg} + 2.0 \text{ kg}} \times 9.8 \text{ m/s}^2 = 3.9 \text{ m/s}^2$$

The acceleration of block 1 is 3.9 m/s^2 to the right and that of block 2 is 3.9 m/s^2 downward.

(b) The tension is now found from the acceleration:

$$T = m_1 a = 3.0 \text{ kg} \times 3.9 \text{ m/s}^2 = 12 \text{ N}$$

(c) Next we find the velocity of block 1 after 1.2 s. The problem gives the initial velocity, $v_{0x} = 0$ at $t = 0$, and the elapsed time, $\Delta t = t = 1.2$ s.

$$v_x = v_{0x} + a_x t$$

$$v_x = 0 + 3.9 \text{ m/s}^2 \times 1.2 \text{ s} = 4.7 \text{ m/s}$$

The positive sign indicates that block 1 moves to the right.

(d) To find the minimum distance from the pulley, we find the distance traveled during 1.2 s after starting from rest

$$\Delta x = v_{0x} t + \tfrac{1}{2} a_x t^2$$

$$\Delta x = 0 + \tfrac{1}{2} \left[3.9 \text{ m/s}^2 \times (1.2 \text{ s})^2 \right] = 2.8 \text{ m}$$

Discussion   An algebraic expression for the tension is

$$T = m_1 a = \frac{m_1 m_2}{m_1 + m_2} g$$

An algebraic expression can lead to insights that are lost when a numerical answer is calculated (see Practice Problem 3.11).

Conceptual Practice Problem 3.11   A quick check

Check the expressions for acceleration and tension in the special case $m_1 \gg m_2$. [*Hint:* What is the sum $m_1 + m_2$ approximately equal to if $m_1 \gg m_2$?]

Example 3.12

Towing a Glider

 A small plane of mass 760 kg requires 120 m of runway to take off by itself (120 m is the horizontal displacement of the plane just before it lifts off the runway, not the entire length of the runway). (a) When the plane is towing a 330-kg glider, how much runway does it need? (b) If the final speed of the plane just before it lifts off the runway is 28 m/s, what is the tension in the tow cable while the plane and glider are moving along the runway?

Strategy   The plane's engines produce thrust—the forward force on the plane due to the air. There is also a backward force on the plane due to the air: drag. The drag force increases as the plane's speed increases, but it is much less than the thrust. A small backward force is exerted on the rolling tires by the runway. As a simplified model we assume that the sum of the horizontal forces on the plane (thrust plus drag plus runway) is constant.

As the glider is towed along the runway, the tension in the cable pulls forward on the sailplane and backward on the plane. Ignoring the small mass of the cable, the tension is the same at both ends. Drag on the glider is negligible—it is designed to have very little drag.

Since the plane and glider are moving along the runway, the vertical component of acceleration is zero. We need not be concerned with gravity and with lift (the upward force on the aircraft's wings due to the air) since they add to zero to produce zero vertical acceleration.

Solution   (a) When the plane takes off by itself, we assume a constant horizontal net force. From Newton's second law,

$$\Sigma F_x = m_1 a_{1x}$$

where m_1 is the plane's mass and a_{1x} is its horizontal acceleration component.

When the glider is towed, we can consider the plane, glider, and cable to be a single system. The net horizontal force on the system is the same as ΣF_x above. The tension in the cable is an internal force; the glider produces no thrust; and we ignore drag on the glider. Therefore

$$m_1 a_{1x} = (m_1 + m_2) a_{2x}$$

where m_2 is the glider's mass and a_{2x} is the horizontal acceleration component of the plane and glider system. The same net force applied to a larger mass produces a smaller acceleration:

continued on next page

Example 3.12 continued

$a_{2x} < a_{1x}$. Rearranging the last equation shows that the acceleration is inversely proportional to the total mass:

$$\frac{a_{2x}}{a_{1x}} = \frac{m_1}{m_1 + m_2}$$

How is the acceleration related to the runway distance? The plane must get to the same final speed in order to lift off the runway. From

$$v_x^2 - v_{0x}^2 = 2a_x \Delta x$$

with the same values of v_x and v_{0x} in both cases, we see that Δx is inversely proportional to a_x; a smaller acceleration means a longer runway distance is required. Since the acceleration is inversely proportional to the total mass, and the runway distance is inversely proportional to the acceleration, the runway distance is directly proportional to the total mass:

$$\frac{\Delta x_2}{\Delta x_1} = \frac{a_{1x}}{a_{2x}} = \frac{m_1 + m_2}{m_1} = \frac{1090 \text{ kg}}{760 \text{ kg}} = 1.43$$

$$\Delta x_2 = 1.43 \times 120 \text{ m} = 172 \text{ m} \approx 170 \text{ m}$$

The plane uses 170 m of runway when towing the glider.

(b) The final speed given enables us to find the acceleration:

$$v_x^2 - v_{0x}^2 = 2a_x \Delta x$$

With $v_x = 28$ m/s, $v_{0x} = 0$, and $\Delta x = 172$ m,

$$a_x = \frac{v_x^2}{2\,\Delta x} = \frac{(28 \text{ m/s})^2}{2 \times 172 \text{ m}} = 2.28 \text{ m/s}^2$$

The tension in the cable is the only horizontal force acting on the glider. Therefore

$$\Sigma F_x = T = m_2 a_x = 330 \text{ kg} \times 2.28 \text{ m/s}^2 = 752 \text{ N}$$

The tension in the cable is approximately 750 N.

Discussion   This solution is based on a simplified model, so we can only regard the answers as approximate. Nevertheless, it illustrates Newton's second law. The same net force produces an acceleration inversely proportional to the mass of the object on which it acts. Here we have the same net force acting on two different objects: first the plane alone, then the plane and glider together.

Alternatively, we can look at forces acting only on the plane. When towing the glider, the cable pulls backward on the plane. The net force *on the plane* is smaller, so its acceleration is smaller. The smaller acceleration means that it takes more time to reach takeoff speed and travels a longer distance before lifting off the runway.

Practice Problem 3.12   Engine thrust

Neglecting air resistance, what is the thrust provided by the airplane's engines in the preceding example?

Example 3.13

Hauling a Crate Up to a Third-Floor Window

 A student is moving into a dorm room on the third floor and he decides to use a block and tackle arrangement (Fig. 3.24) to move a crate of mass 91 kg from the ground up to his window. If the breaking strength of the available cable is 550 N, what is the minimum time required to haul the crate to the level of the window, 30.0 m above the ground?

Strategy   The tension in the cable is T and is the same at both ends or anywhere along the cable, assuming the cable and pulleys are ideal. Two cable strands support the crate, each pulling upward with a force of magnitude T. The weight of the crate acts downward. We draw a free-body diagram and set the tension equal to the breaking force of the rope to find the maximum possible acceleration of the crate. Then from the maximum acceleration, we use kinematic relationships to find the minimum time to move the required distance to the third-floor window with that acceleration.

Given: $m = 91$ kg; $\Delta y = 30.0$ m; $g = 9.8$ m/s^2; $T_{max} = 550$ N; $v_{0y} = 0$

To find: Δt, the minimum time to raise the crate 30.0 m

continued on next page

Figure 3.24
(a) Block and tackle setup and (b) free-body diagram for the crate

Example 3.13 *continued*

Solution From the free-body diagram (Fig. 3.24b), if the forces acting up are greater than the force acting down, the net force is upward and the crate's acceleration is upward. In terms of components, with the +y-direction chosen to be upward,

$$\Sigma F_y = T + T - mg = ma_y$$

Solving for the acceleration,

$$a_y = \frac{T + T - mg}{m}$$

Setting $T = 550$ N, the maximum possible value before the cable breaks, and substituting the other known values,

$$a_y = \frac{550 \text{ N} + 550 \text{ N} - 91 \text{ kg} \times 9.8 \text{ m/s}^2}{91 \text{ kg}}$$

$$= 2.3 \text{ m/s}^2$$

The minimum time to move the crate up a distance Δy starting from rest can be found from

$$\Delta y = v_{0y}t + \tfrac{1}{2}a_y t^2$$

Setting $v_{0y} = 0$ and solving for t, we find

$$t = \pm\sqrt{\frac{2\,\Delta y}{a_y}}$$

Our equation applies only for $t \geq 0$ (the crate reaches the window *after* it leaves the ground). Taking the positive root and substituting numerical values,

$$t = \sqrt{\frac{2 \times 30.0 \text{ m}}{2.3 \text{ m/s}^2}} = 5.1 \text{ s}$$

This is the minimum time if a_y is the maximum acceleration.

Discussion If the crate is accelerated upward, the tension of the cable is greater than if the crate is moved at a constant velocity. For the crate to accelerate, there must be an upward net force. For motion with a constant velocity, the tension would be equal to half the weight of the crate, 450 N.

Practice Problem 3.13   Hauling the crate with a single pulley

If only a single pulley, attached to the pole above the fourth-floor, were available and if the student had a few friends to help him pull on the cable, could they haul the crate up to the third-floor window? If so, what is the minimum time required to do so?

3.6   FALLING OBJECTS

Suppose you are standing on a bridge over a deep gorge. If you drop a stone into the gorge, how fast does it fall? You know from experience that it does not fall at a constant velocity; the longer it falls, the faster it goes. A better question is: what is the stone's acceleration?

First, let us simplify the problem. If the stone were moving very fast, an appreciable force of air resistance would oppose its motion. When it is not falling so fast, air resistance is negligibly small. If air resistance is negligible, the only appreciable force is that of gravity. **Free fall** is a situation in which no forces act on an object other than the gravitational force that makes the object fall. On Earth, free fall is an idealization since there is always *some* air resistance.

What is the acceleration of an object in free fall? More massive objects are harder to accelerate: the acceleration of an object subjected to a given force is inversely proportional to its mass. However, the stronger gravitational force on a more massive object compensates for its greater inertia, giving it the same acceleration as a less massive object. The gravitational force on an object is

$$\vec{\mathbf{W}} = m\vec{\mathbf{g}}$$

From Newton's second law,

$$\vec{\mathbf{F}}_{\text{net}} = m\vec{\mathbf{g}} = m\vec{\mathbf{a}}$$

Dividing by the mass yields

$$\vec{\mathbf{a}} = \vec{\mathbf{g}} \tag{3-14a}$$

The acceleration of an object in free fall is $\vec{\mathbf{g}}$, regardless of the object's mass. Since $1 \text{ N} = 1 \text{ kg·m/s}^2$, an object in free fall near the Earth's surface has an acceleration of magnitude

$$a = g = 9.8 \frac{\text{N}}{\text{kg}} = 9.8 \frac{\cancel{\text{N}}}{\cancel{\text{kg}}} \times 1 \frac{\cancel{\text{kg}}\cdot\text{m/s}^2}{\cancel{\text{N}}} = 9.8 \text{ m/s}^2$$

Thus any object in free fall near the Earth's surface has a constant downward acceleration of magnitude 9.8 m/s². For this reason, $\vec{g}$ is sometimes called *the freefall acceleration*—the acceleration of an object near the surface of the Earth when the *only* force acting is gravity.

When dealing with vertical motion, the *y*-axis is usually chosen to be positive pointing upward. In two-dimensional motion, the *x*-axis is often used for the horizontal direction and the *y*-axis for the vertical direction. The direction of the acceleration is down, so in free fall

$$a_y = -g \qquad \text{(3-14b)}$$

The same techniques and equations used for other constant acceleration situations are used with free fall. The only change is that the constant acceleration in free fall is always directed toward the center of the Earth and has a known constant magnitude of approximately 9.8 m/s² for objects near the surface of the Earth.

In free fall near the Earth's surface, $a_y = -g$ (if the *y*-axis points up).

Earth's gravity always pulls downward, so the acceleration of an object in free fall is always downward, regardless of whether the object is moving up, down, or is at rest. If the object is moving downward, the downward acceleration makes it speed up; if it is moving upward, the downward acceleration makes it slow down; and if it is at rest, the downward acceleration makes it start moving downward.

If an object is thrown straight up, its velocity is zero at the highest point of its flight. Why? On the way up, the *y*-component of its velocity v_y is positive if the positive *y*-axis is pointing up. On the way down, v_y is negative. Since v_y changes continuously, at a rate of 9.8 m/s², it must pass through zero to change sign. At the one instant of time at its highest point, the object is neither moving up nor down. The object's acceleration is *not* zero at the top of flight. If the acceleration were to suddenly become zero at the top of flight, the velocity would no longer change; the object would get *stuck* at the top rather than fall back down! The velocity is zero at the top but it does not stay zero; it is still changing at the same rate.

Example 3.14

Throwing Stones

Standing on a bridge, you throw a stone straight upward. The stone hits a stream, 44.1 m below the point at which you release it, 4.00 s later. Assume g = 9.81 m/s². (a) What is the speed of the stone just after it leaves your hand? (b) What is the speed of the stone just before it hits the water?

Strategy   Ignoring air resistance, the acceleration is constant. Choose the positive *y*-axis pointing up. Let the stone be thrown at $t = 0$ and hit the stream at a later time t.

Known $a_y = -9.81$ m/s²; $\Delta y = -44.1$ m at $t = 4.00$ s
To find: $|v_{0y}|$ (speed at $t = 0$) and $|v_y|$ (speed at $t = 4.00$ s)

Solution   (a) Equation (3-12) can be used to solve for v_{0y}, since all the other quantities in it (Δy, t, and a_y) are known.

$$\Delta y = v_{0y}t + \tfrac{1}{2}a_y t^2$$

Solving for v_{0y},

$$v_{0y} = \frac{\Delta y}{t} - \frac{1}{2}a_y t = \frac{-44.1 \text{ m}}{4.00 \text{ s}} - \frac{1}{2}(-9.81 \text{ m/s}^2 \times 4.00 \text{ s}) \quad (1)$$

$$= -11.0 \text{ m/s} + 19.6 \text{ m/s} = 8.6 \text{ m/s}$$

The initial speed is 8.6 m/s.

(b) The change in v_y is $a_y t$ from Eq. (3-9):

$$v_y = v_{0y} + a_y t$$

Substituting the expression for v_{0y} found in Eq. (1),

$$v_y = \left(\frac{\Delta y}{t} - \frac{1}{2}a_y t\right) + a_y t \qquad (2)$$

$$= \frac{\Delta y}{t} + \frac{1}{2}a_y t$$

$$= \frac{-44.1 \text{ m}}{4.00 \text{ s}} + \frac{1}{2}(-9.81 \text{ m/s}^2 \times 4.00 \text{ s})$$

$$v_y = -11.0 \text{ m/s} - 19.6 \text{ m/s} = -30.6 \text{ m/s}$$

The final speed is 30.6 m/s.

Discussion   The final speed is greater than the initial speed, as expected.

Equations (1) and (2) have a direct interpretation, which is a good check on their validity. The first term, $\Delta y/t$, is the average velocity of the stone during the 4.00 s of free fall. The second term, $\frac{1}{2}a_y t$, is *half* the change in v_y since $(v_y - v_{0y}) = a_y t$. Because the acceleration is constant, the average velocity is halfway between the initial and final velocities. Therefore, the initial velocity is the average velocity minus half of the change, while the final velocity is the average velocity plus half of the change.

Practice Problem 3.14   Height attained by stone
(a) How high above the bridge does the stone go? [*Hint:* What is v_y at the highest point?] (b) If you dropped the stone instead of throwing it, how long would it take to hit the water?

Figure 3.25 A stroboscopic photo-graph shows two objects falling through the air with very different terminal speeds. The exposures are taken at fixed time intervals of 1/15 s.

Table 3.3

Some Typical Terminal Speeds

Object	Terminal Speed (m/s)
Feather	0.5
Snowflake	1
Raindrop	7
Skydiver (open parachute)	5–9
Basketball	20
Baseball	40
Skydiver (spread-eagle)	50–60
Skydiver (dive)	100
Bullet	100

Air Resistance—Falling with Varying Acceleration

A skydiver relies on a parachute to provide a large drag force of air resistance. Even with the parachute closed, drag is not negligible when the skydiver is falling rapidly. The drag force exerted on a body falling through air increases dramatically with speed; it is proportional to the *square* of the speed:

$$F_d = bv^2$$

where b is a constant that depends on the size and shape of the object. The direction of the drag force is opposite to the direction of motion.

Since the drag force increases as the speed increases, a falling object may eventually reach equilibrium when the drag force is equal in magnitude to the weight. The speed at which the drag force is equal in magnitude to the weight is called the object's **terminal speed**. As the speed gets near the terminal speed, the acceleration gets smaller and smaller. The acceleration is zero when the object falls at its terminal speed.

In Fig. 3.25, a baseball and a coffee filter are released from rest and fall through the air. The strobe photograph shows the positions of the two at equal time intervals. The baseball has a terminal velocity of about 40 m/s, so air resistance is negligible for the speeds shown in the photo. The displacement of the baseball in equal time intervals increases linearly, showing that its acceleration is constant. The coffee filter has a very large surface area for its small mass. As a result, its terminal speed is much smaller—about 1 m/s. The displacement of the filter barely changes from one strobe flash to the next, showing that it is falling at a small, nearly constant velocity.

At terminal speed v_t, the drag force is equal in magnitude to the weight. Therefore, $F_d = mg = bv_t^2$ and

$$b = \frac{mg}{v_t^2}$$

Therefore, at any speed v,

$$F_d = mg \frac{v^2}{v_t^2} \qquad (3\text{-}15)$$

The terminal speed of an object depends on its size, shape, and mass (see Table 3.3). A skydiver with the parachute closed will reach a terminal speed of about 50 m/s ($\approx$110 mi/h) in the spread-eagle position or as much as 100 m/s ($\approx$220 mi/h) in a dive. When the parachute is opened, the drag force increases dramatically—the larger surface area of the parachute means that more air has to be pushed out of the way. The terminal speed with the parachute open is typically about 9 m/s (20 mi/h). When the parachute is opened, the skydiver is initially moving *faster* than the new terminal speed. For $v > v_t$, the drag force is larger in magnitude than the weight and the acceleration is *upward*. The skydiver slows down, approaching the new terminal speed. Note that the terminal speed is not the maximum possible speed; it is the speed that the falling object approaches, regardless of initial conditions, when the only forces acting are drag and gravity.

Example 3.15

Skydivers Falling Freely

 Two skydivers have identical parachutes. Their masses (including parachutes) are 62.0 kg and 82.0 kg. Which of the skydivers has the larger terminal speed? What is the ratio of their terminal speeds?

Strategy With identical parachutes, we expect the same amount of drag at a given speed. The more massive skydiver must fall faster in order for the drag force to equal his weight, so the 82.0-kg skydiver should have a larger terminal speed. For the ratio of the terminal speeds, we first find how the terminal speed depends on mass, all other things being equal. Then we work by proportions.

Solution At terminal speed v_t, the drag force must be equal in magnitude to the weight.

continued on next page

Example 3.15 *continued*

$$mg = F_d = bv_t^2$$

Since the parachutes are identical, we expect the constant b to be the same for the two divers. Therefore,

$$v_t \propto \sqrt{m}$$

The more massive skydiver has a larger terminal speed—he must move faster in order for the drag force to equal his larger weight. The ratio of the terminal speeds is

$$\frac{v_{t2}}{v_{t1}} = \sqrt{\frac{m_2}{m_1}} = \sqrt{\frac{82.0}{62.0}} = 1.15$$

The terminal speed of the 82.0-kg diver is 1.15 times that of the less massive skydiver, or 15% faster.

Discussion   The 82.0-kg skydiver is 32% more massive:

$$\frac{82.0 \text{ kg}}{62.0 \text{ kg}} = 1.32$$

but his terminal speed is only 15% greater. That is because the drag force is proportional to the *square* of the speed. It only takes a 15% greater speed to make the drag force 32% greater:

$$(1.15)^2 = 1.32$$

Practice Problem 3.15   Air resistance at terminal speed

A pilot has bailed out of her airplane at a height of 2000 m above the surface of the Earth. The mass of pilot plus parachute is 112 kg. What is the force of air resistance when the pilot reaches terminal speed?

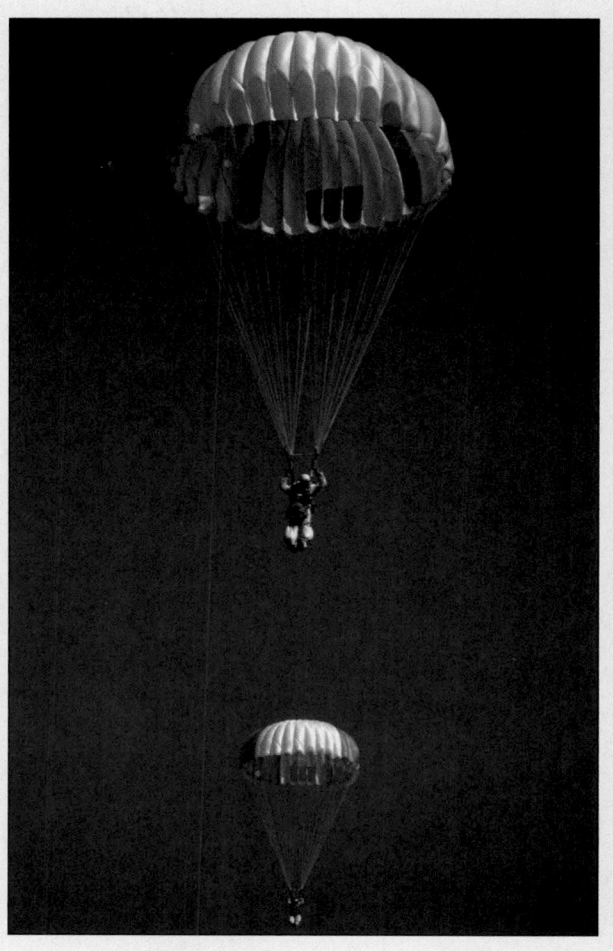

Example 3.16

Dropping the Ball

A basketball is dropped off a tall building. (a) What is the initial acceleration of the ball, just after it is released? (b) What is the acceleration of the ball when it is falling at its terminal speed? (c) What is the acceleration of the ball when falling at half its terminal speed?

Strategy   We choose the positive y-axis to point upward as usual. The ball is dropped from rest so *initially* the only force acting is gravity—the drag force is zero when the velocity is zero. Once the ball is moving, air drag contributes to the net force on the basketball.

Solution   (a) The initial acceleration is the free-fall acceleration ($\vec{a} = \vec{g}$) since the drag force is zero.

(b) Once the ball reaches terminal speed, the drag force is equal in magnitude to the weight of the ball but acts in the opposite direction. The net force on the ball is zero, so the acceleration is zero. At terminal speed, $\vec{a} = \vec{0}$.

(c) When the ball is falling at half its terminal speed, the drag force is significant but it is smaller than the weight. The net force is down and therefore the acceleration is still downward, though with a smaller magnitude. The drag force at any speed is given by

$$F_d = mg \frac{v^2}{v_t^2}$$

and this drag force acts in the opposite direction to the weight; it acts upward.

The net vertical force is

$$\Sigma F_y = F_d - mg = mg \frac{v^2}{v_t^2} - mg = mg \left(\frac{v^2}{v_t^2} - 1 \right)$$

Now we apply Newton's second law.

$$\Sigma F_y = ma_y$$

Solving for the acceleration yields

$$a_y = g \left(\frac{v^2}{v_t^2} - 1 \right)$$

continued on next page

Example 3.16 *continued*

At a time when the velocity is at half the terminal speed,

$$v = \frac{1}{2}v_t \quad \text{and} \quad \frac{v^2}{v_t^2} = \frac{1}{4}$$

$$a_y = g\left(\frac{1}{4} - 1\right) = -\frac{3}{4}g$$

so that the acceleration of the ball is

$$\vec{a} = \frac{3}{4}\vec{g}$$

where $\vec{a}$ and $\vec{g}$ both point downward.

Discussion   How do we know when air resistance is negligible? If we know the approximate terminal speed of an object, then air resistance is negligible as long as its speed moving through the air is small compared to the terminal speed.

Practice Problem 3.16   Acceleration graph sketch

Sketch a qualitative graph of $v_y(t)$ for the basketball using a y-axis that is positive pointing upward. [*Hint:* At first air resistance is negligible. After a long time the basketball is in equilibrium. Figure out what the beginning and end of the graph look like and then connect them smoothly.]

Physics at Home

Go to a balcony or climb up a ladder and drop a basket-style paper coffee filter (or a cupcake paper) and a penny simultaneously. Air resistance on the penny is negligible unless it is dropped from a very high balcony. At the other extreme, the effect of air resistance on the coffee filter is very noticeable; it reaches its terminal speed almost immediately. Stack several (two to four) coffee filters together and drop them simultaneously with a single coffee filter. Why is the terminal speed higher for the stack? Crumple a coffee filter into a ball and drop it simultaneously with the penny. Air resistance on the coffee filter is now reduced, but still noticeable.

3.7   APPARENT WEIGHT

Imagine being in an elevator when the cable snaps. Assume that some safety mechanism brings you to rest after you have been in free fall for a while. While you are in free fall, you *seem* to be "weightless," but your weight has not changed; the Earth still pulls downward with the same gravitational force. In free fall, gravity makes the elevator and everything in it accelerate downward at 9.8 m/s². The floor of the elevator stops pushing up on you, as it does when the elevator is at rest. If you jump up from the elevator floor, you seem to "float" up to the ceiling of the elevator. Your *weight* hasn't changed, but your **apparent weight** is zero while you are in free fall.

You don't need a disastrous elevator mishap to notice an apparent weight that differs from your true weight. A normally operating elevator will do quite nicely. Step in the elevator and push a button for a higher floor. When the elevator accelerates upward, you can feel your apparent weight increase. When the elevator slows down to stop, the elevator's acceleration is downward and your apparent weight is less than your true weight.

What is happening in the body while the elevator accelerates? Blood tends to collect in the lower extremities during acceleration upward and in the upper body during acceleration downward. The internal organs shift position within the body cavity resulting in a funny feeling in the gut as the elevator starts and stops. To avoid this problem, high-speed express elevators in skyscrapers keep the acceleration relatively small, but maintain that acceleration long enough to reach high speeds. That way, the elevator can travel quickly to the upper floors without making the passengers feel too uncomfortable.

Imagine an object resting on a bathroom scale. The scale measures the object's *apparent* weight W', which is equal to the true weight only if the object has zero acceleration. Newton's second law requires that

$$\vec{F}_{net} = \vec{N} + m\vec{g} = m\vec{a}$$

where $\vec{N}$ is the normal force of the scale pushing up. The apparent weight is the reading of the scale—the magnitude of $\vec{N}$:

$$W' = |\vec{N}| = N$$

(a)   (b)   (c)

Figure 3.26 (a) Apparent weight in an elevator with acceleration upward (b) Free-body diagram for the passenger (c) Normal force must be greater than the weight to have an upward net force.

In Fig. 3.26a, the acceleration of the elevator is upward. The normal force $\vec{\mathbf{N}}$ must be larger than the weight $m\vec{\mathbf{g}}$ in order for the net force to be upward (Fig. 3.26c). Writing the forces in component form where the $+y$-direction is upward,

$$\Sigma F_y = N - mg = ma_y$$

or

$$N = mg + ma_y$$

Therefore,

$$W' = N = m(g + a_y) \tag{3-16}$$

Since the elevator accelerates upward, $a_y > 0$; the apparent weight is greater than the true weight (Fig. 3.26c).

In Fig. 3.27a, the acceleration is downward. Then the net force must also point downward. The normal force is still upward, but it must be smaller than the weight in order to produce a downward net force (Fig. 3.27c). It is still true that

$$W' = m(g + a_y)$$

but now the acceleration is downward ($a_y < 0$). The apparent weight is less than the true weight.

(a)   (b)   (c)

Figure 3.27 (a) Apparent weight in an elevator with acceleration down (b) Free-body diagram for the passenger (c) Normal force must be less than the weight to have a downward net force.

In both these cases—and in general—the apparent weight is given by

$$W' = m|\vec{g} - \vec{a}| \tag{3-17}$$

where $\vec{g}$ and $\vec{a}$ are both vector quantities. When $\vec{g}$ and $\vec{a}$ point in opposite directions, the magnitude $|\vec{g} - \vec{a}|$ is greater than it is when they point in the same direction. If the acceleration of an elevator is upward, then $\vec{g}$ and $\vec{a}$ have opposite directions and the apparent weight is greater than the true weight. In an elevator with a downward acceleration, $\vec{g}$ and $\vec{a}$ have the same direction and the apparent weight is less than the true weight. If an elevator is in free fall, then $\vec{g}$ and $\vec{a}$ are in the same direction and are equal in magnitude; in free fall the apparent weight is zero. Draw the vectors and perform the vector subtraction to satisfy yourself that Eq. (3-17) is correct.

Physics at Home

Take an empty half-gallon paper milk carton or a plastic milk jug and poke one hole in the bottom and another in the side of the carton with a pencil. Fill the carton with water while sealing the holes with your fingers so the water does not pour out. Throw the carton straight up into the air and watch what happens at the holes. Does water start to pour out of the unsealed holes as the carton ascends? What about when the carton is falling downward? Can you explain your observations using the ideas of free fall and apparent weight?

Example 3.17

Apparent Weight in an Elevator

 A passenger weighing 598 N rides in an elevator. The gravitational field strength is 9.80 N/kg. What is the apparent weight of the passenger in each of the following situations? In each case, the magnitude of the elevator's acceleration is 0.500 m/s². (a) The passenger is on the 1st floor and has pushed the button for the 15th floor; the elevator is beginning to move upward. (b) The elevator is slowing down as it nears the 15th floor.

Strategy   Let the +y-axis be upward. The apparent weight is equal to the magnitude of the normal force exerted by the floor on the passenger. Newton's second law lets us find the normal force from the weight and the acceleration.

Given: $W = 598$ N; magnitude of the acceleration is $a = 0.500$ m/s²

Find: W'

Solution   (a) When the elevator starts up from the 1st floor it accelerates in the upward direction as its speed increases. Since the elevator accelerates upward, $a_y > 0$ (as in Fig. 3.26). We expect the apparent weight $W' = N$ to be greater than the true weight—the floor must push up with a force greater than W to cause an upward acceleration.

$$\Sigma F_y = N - W = ma_y$$

Since $m = W/g$,

$$W' = N = W + ma_y = W + \frac{W}{g}a_y = W\left(1 + \frac{a_y}{g}\right)$$

$$= 598\ \text{N} \times \left(1 + \frac{0.500\ \text{m/s}^2}{9.80\ \text{m/s}^2}\right) = 629\ \text{N}$$

(b) When the elevator approaches the 15th floor, it slows down while still moving upward; its acceleration is downward ($a_y < 0$) as in Fig. 3.27. The apparent weight is less than the true weight. Again, $\Sigma F_y = N - W = ma_y$, but this time $a_y = -0.500$ m/s².

$$N = W\left(1 + \frac{a_y}{g}\right)$$

$$= 598\ \text{N} \times \left(1 + \frac{-0.500\ \text{m/s}^2}{9.80\ \text{m/s}^2}\right) = 567\ \text{N}$$

Discussion   The apparent weight is greater when the direction of the elevator's acceleration is upward. That can happen in two cases: either the elevator is moving up with increasing speed, or it is moving down with decreasing speed.

Practice Problem 3.17   Elevator descending

What is the apparent weight of a passenger of mass 42.0 kg traveling in an elevator in each of the following situations? The gravitational field strength is 9.80 N/kg. In each case, the magnitude of the elevator's acceleration is 0.460 m/s². (a) The passenger is on the 15th floor and has pushed the button for the 1st floor; the elevator is beginning to move downward. (b) The elevator is slowing down as it nears the 1st floor.

MASTER THE CONCEPTS

Summary

- Position (symbol $\vec{r}$) is a vector from the origin to an object's location. Its magnitude is the distance from the origin and its direction points from the origin to the object.

- Displacement is the change in position: $\Delta\vec{r} = \vec{r}_f - \vec{r}_0$. The displacement depends only on the starting and ending positions, not on the path taken. The magnitude of the displacement vector is not necessarily equal to the total distance traveled; it is the straight line distance from the initial position to the final position.

- Vector components: If $\vec{A}$ points along the $+x$-axis, then $A_x = +A$; if $\vec{A}$ points in the opposite direction—along the negative x-axis—then $A_x = -A$. When adding or subtracting vectors, we can add or subtract their components.

- The average velocity states at what constant speed and in what direction to travel to cause that same displacement in the same amount of time.

$$\vec{v}_{av} = \frac{\Delta\vec{r}}{\Delta t}$$

- Velocity is a vector that states how fast and in what direction something moves. Its direction is the direction of the object's motion and its magnitude is the instantaneous speed.

$$\vec{v} = \lim_{\Delta t \to 0} \frac{\Delta\vec{r}}{\Delta t} \qquad (3\text{-}4)$$

- Average acceleration is the constant acceleration that would give the same velocity change in the same amount of time. In terms of changes in velocity and time,

$$\vec{a}_{av} = \frac{\Delta\vec{v}}{\Delta t} \qquad (3\text{-}6)$$

- Acceleration is the instantaneous rate of change of velocity:

$$\vec{a} = \lim_{\Delta t \to 0} \frac{\Delta\vec{v}}{\Delta t} \qquad (3\text{-}7)$$

Acceleration does not necessarily mean speeding up. A velocity can also change by decreasing speed or by changing directions.

- Interpreting graphs: On a graph of $x(t)$, the slope at any point is v_x. On a graph of $v_x(t)$, the slope at any point is a_x and the area under the graph during any time interval is the displacement during that time interval. If v_x is negative, the displacement is also negative, so we must count the area as negative when it is below the time axis. On a graph of $a_x(t)$, the area under the curve is the change in v_x during that time interval.

- Newton's second law:

$$\vec{F}_{net} = m\vec{a} \qquad (3\text{-}8)$$

- The SI unit of force is the newton; $1\ \text{N} = 1\ \text{kg}\cdot\text{m/s}^2$. One newton is the magnitude of the net force that gives a 1-kg object an acceleration of magnitude 1 m/s^2.

- Essential relationships for solving any constant acceleration problem: if a_x is constant during the entire time interval from $t = 0$ until a later time t,

$$\Delta v_x = v_x - v_{0x} = a_x t \qquad (3\text{-}9)$$

$$\Delta x = x - x_0 = v_{av,x} t \qquad (3\text{-}10)$$

$$v_{av,x} = \frac{v_{0x} + v_x}{2} \qquad (3\text{-}11)$$

$$\Delta x = x - x_0 = v_{0x}t + \tfrac{1}{2}a_x t^2 \qquad (3\text{-}12)$$

$$v_x^2 - v_{0x}^2 = 2a_x \Delta x \qquad (3\text{-}13)$$

Highlighted Figures and Tables

F3.1 Displacement is the change in position: $\Delta\vec{r} = \vec{r}_f - \vec{r}_0$ (p. 64)

F3.5 Displacement and average velocity (p. 67)

F3.6 As Δt approaches zero, the average velocity during the increasingly short time interval approaches the instantaneous velocity (p. 68)

F3.7 Graphical relationship between position and velocity (p. 69)

F3.9 Displacement Δx is the area under the v_x versus time graph for the time interval considered (p. 70)

F3.14 Mass versus weight (p. 76)

F3.18 Finding average velocity with the aid of a graph (p. 79)

F3.19 Graphical interpretation of Eq. (3-12) (p. 80)

F3.20 Visualizing motion with constant acceleration (p. 81)

T3.3 Some typical terminal speeds (p. 88)

CONCEPTUAL QUESTIONS

1. Explain the difference between distance traveled, displacement, and displacement magnitude.

2. Explain the difference between speed and velocity.

3. On a graph of v_x versus time, what quantity does the area under the graph represent?

4. On a graph of v_x versus time, what quantity does the slope of the graph represent?

5. On a graph of a_x versus time, what quantity does the area under the graph represent?

6. On a graph of x versus time, what quantity does the slope of the graph represent?

7. What is the relationship between average velocity and instantaneous velocity? An object can have different instantaneous velocities at different times. Can the same object have different average velocities? Explain.

8. If an object is traveling at a constant velocity, is it necessarily traveling in a straight line? Explain.

9. Can the average speed and the magnitude of the average velocity ever be equal? If so, under what circumstances?

10. If a feather and a lead brick are dropped simultaneously from the top of a ladder, the lead brick hits the ground first. What would happen if the experiment is repeated on the surface of the Moon?

11. Name a situation where the speed of an object is constant while the velocity is not.

12. Can the velocity of an object be zero and the acceleration be nonzero at the same time? Explain.

13. Why does a 1-kg sandbag fall with the same acceleration as a 5-kg sandbag? Explain in terms of Newton's second law and his law of gravitation.

14. If an object is acted on by a single constant force, is it possible for the object to remain at rest? Is it possible for the object to move with constant velocity? Is it possible for the object's speed to be decreasing? Is it possible for it to change direction?

15. If an object is acted on by two constant forces is it possible for the object to move at constant velocity? If so, what must be true about the two forces?

16. An object is placed on a scale. Under what conditions does the scale read something other than the object's weight, even though the scale is functioning properly and is calibrated correctly? Explain.

17. What is meant by the terminal speed of a falling object? Can an object ever move through air faster than the object's terminal speed? If so, give an example.

18. What force(s) act on a parachutist descending to Earth with a constant velocity? What is the acceleration of the parachutist?

19. A baseball is tossed straight up. Taking into consideration the force of air resistance, is the magnitude of the baseball's acceleration zero, less than g, equal to g, or greater than g on the way up? At the top of the flight? On the way down? Explain.

20. What is the acceleration of an object thrown straight up into the air at the highest point of its motion? Does the answer depend on whether air resistance is negligible or not? Explain.

21. You are bicycling along a straight north-south road. Let the x-axis point north. Describe your motion in each of the following cases. Example: $a_x > 0$ and $v_x > 0$ means you are moving north and speeding up. (a) $a_x > 0$ and $v_x < 0$. (b) $a_x = 0$ and $v_x < 0$. (c) $a_x < 0$ and $v_x = 0$. (d) $a_x < 0$ and $v_x < 0$. (e) Based on your answers, explain why it is not a good idea to use the expression "negative acceleration" to mean slowing down.

MULTIPLE CHOICE QUESTIONS

1. A go-kart travels around a circular track at a constant speed. Which of these is a true statement?

   (a) The go-kart has a constant velocity.
   (b) The go-kart has zero acceleration.
   (c) Both (a) and (b) are true.
   (d) Neither (a) nor (b) is true.

2. A ball is thrown straight up into the air. Neglect air resistance. While the ball is in the air its acceleration

   (a) increases   (b) is zero   (c) remains constant
   (d) decreases on the way up and increases on the way down
   (e) changes direction

3. A stone is thrown upward and reaches a height Δy. After an elapsed time Δt, measured from the time the stone was first thrown, the stone has fallen back down to the ground. The magnitude of the average velocity of the stone during this time is

   (a) zero   (b) $2\dfrac{\Delta y}{\Delta t}$ (c) $\dfrac{\Delta y}{\Delta t}$ (d) $\dfrac{1}{2}\dfrac{\Delta y}{\Delta t}$

4. A stone is thrown upward and reaches a height Δy. After an elapsed time Δt, measured from the time the stone was first thrown, the stone has fallen back down to the ground. The average speed of the stone during this time is

   (a) zero   (b) $2\dfrac{\Delta y}{\Delta t}$ (c) $\dfrac{\Delta y}{\Delta t}$ (d) $\dfrac{1}{2}\dfrac{\Delta y}{\Delta t}$

5. A ball is thrown straight up. At the top of its trajectory the ball is

   (a) instantaneously at rest.
   (b) instantaneously in equilibrium.
   (c) Both (a) and (b) are true.
   (d) Neither (a) nor (b) is true.

Multiple Choice Questions 6–15 refer to Fig. 3.28.

6. What distance does the jogger travel during the first 10.0 min ($t = 0$ to 10.0 min)?

   (a) 8.5 m   (b) 510 m   (c) 900 m   (d) 1020 m

7. What is the displacement of the jogger from $t = 18.0$ min to $t = 24.0$ min?

   (a) 720 m, south   (b) 720 m, north
   (c) 2160 m, south   (d) 3600 m, north

8. What is the displacement of the jogger for the entire 30.0 min?

   (a) 3120 m, south   (b) 2400 m, north
   (c) 2400 m, south   (d) 3840 m, north

9. What is the total distance traveled by the jogger in 30.0 min?

   (a) 3840 m   (b) 2340 m   (c) 2400 m   (d) 3600 m

10. What is the average velocity of the jogger during the 30.0 min?

    (a) 1.3 m/s, north   (b) 1.7 m/s, north
    (c) 2.1 m/s, north   (d) 2.9 m/s, north

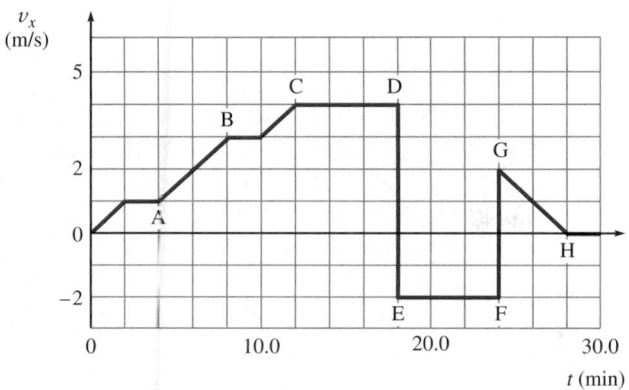

Figure 3.28 Multiple Choice Questions 6–15. A jogger is exercising along a long, straight road that runs north-south. She starts out heading north.

11. What is the average speed of the jogger for the 30 min?
    (a) 1.4 m/s   (b) 1.7 m/s   (c) 2.1 m/s   (d) 2.9 m/s
12. In what direction is she running at time $t = 20$ min?
    (a) south   (b) north   (c) not enough information
13. In which region of the graph is a_x positive?
    (a) A to B   (b) C to D   (c) E to F   (d) G to H
14. In which region is a_x negative?
    (a) A to B   (b) C to D   (c) E to F   (d) G to H
15. In which region is the velocity directed to the south?
    (a) A to B   (b) C to D   (c) E to F   (d) G to H

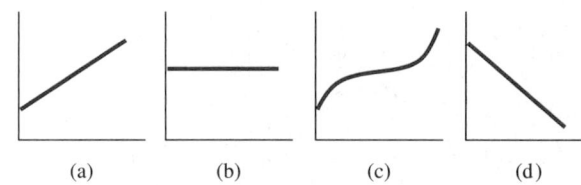

Figure 3.29 Multiple Choice Questions 16–20

Multiple Choice Questions 16–20 refer to Fig. 3.29.

16. If Fig. 3.29 shows four graphs of x versus time, which graph shows a constant, positive, nonzero velocity?
17. If Fig. 3.29 shows four graphs of v_x versus time, which graph shows a constant velocity?
18. If Fig. 3.29 shows four graphs of v_x versus time, which graph shows a_x constant and positive?
19. If Fig. 3.29 shows four graphs of v_x versus time, which graph shows a_x constant and negative?
20. If Fig. 3.29 shows four graphs of v_x versus time, which graph shows a changing a_x that is always positive?
21. Two blocks are connected by a light string passing over a pulley (see Fig. 3.23a). The block with mass m_1 slides on the frictionless horizontal surface, while the block with mass m_2 hangs vertically. ($m_1 > m_2$.) The tension in the string is:
    (a) zero   (b) less than m_2g
    (c) equal to m_2g (d) greater than m_2g, but less than m_1g
    (e) equal to m_1g (f) greater than m_1g

PROBLEMS

Note: **C** indicates a combination conceptual/quantitative problem. Gold diamonds ◆, ◆◆ are used to indicate the increasing level of difficulty of each problem. Problem numbers appearing in blue, 9., denote problems that have a detailed solution available in the Student Solutions Manual. Some problems are *paired* by concept; their numbers are connected by a ruled box.

3.1 Position and Displacement; 3.2 Velocity

1. Two cars, a Porsche and a Honda, are traveling in the same direction, although the Porsche is 186 m behind the Honda. The speed of the Porsche is 24.4 m/s and the speed of the Honda is 18.6 m/s. How much time does it take for the Porsche to catch the Honda? [*Hint:* What must be true about the displacements of the two cars when they meet?]
2. To get to a concert in time, a harpsichordist has to drive 121 mi in 2.00 h. If he drove at an average speed of 55 mi/h for the first 1.20 h, what must be his average speed for the remaining 0.80 h?
3. Figure 3.30 shows the vertical velocity component of an elevator versus time. How high is the elevator above the starting point ($t = 0$ s) after 20 s have elapsed?
4. Figure 3.30 shows the vertical velocity component of an elevator versus time. If the elevator starts to move at $t = 0$ s, when is the elevator at its highest location above the starting point?

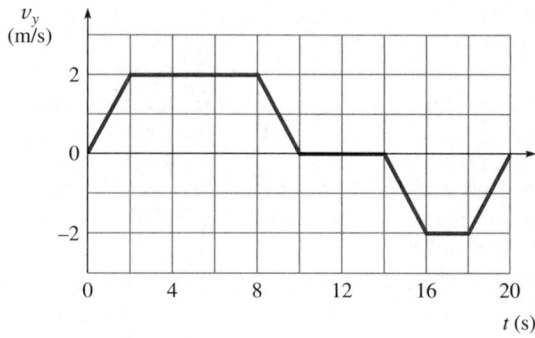

Figure 3.30 Problems 3 and 4

5. Figure 3.31 shows a graph of speedometer readings obtained as a car comes to a stop along a straight-line path. How far does the car move between $t = 0$ and $t = 16$ s?
6. Figure 3.32 shows a graph of speedometer readings, in meters per second (on the vertical axis), obtained as a car travels along a straight-line path. How far does the car move between $t = 3.00$ s and $t = 8.00$ s?

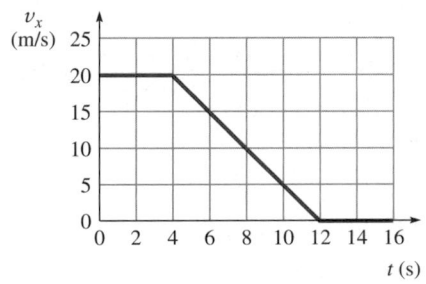

Figure 3.31 Problems 5, 17, and 79

7. Figure 3.32 shows a graph of $x(t)$ in meters, on the vertical axis, for an object traveling in a straight line. (a) What is $v_{av,x}$ for the interval from $t = 0$ to $t = 4.0$ s? (b) from $t = 0$ to $t = 5.0$ s?

8. Figure 3.32 shows a graph of $x(t)$ in meters for an object traveling in a straight line. What is v_x at $t = 2.0$ s?

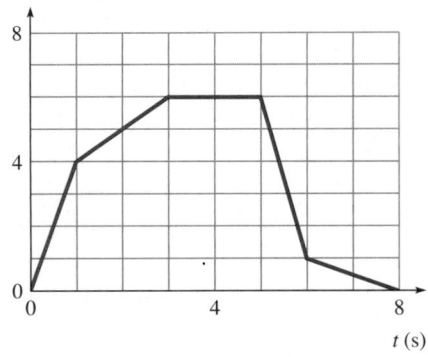

Figure 3.32 Problems 6, 7, and 8.

9. A rabbit, nervously trying to cross a road, first moves 80 cm to the right, then 30 cm to the left, then 90 cm to the right, and then 310 cm to the left. (a) What is the rabbit's total displacement? (b) If the elapsed time was 18 s, what was the rabbit's average speed? (c) What was its average velocity?

10. An object is moving along a straight line. The graph in Fig. 3.33 shows its position from the starting point as a function of time. (a) In which section(s) of the graph does the object have the highest speed? (b) At which time(s) does the object reverse its direction of motion? (c) How far does the object move from $t = 0$ to $t = 3$ s?

11. A motor scooter travels east at a speed of 12 m/s. The driver then reverses direction and heads west at 15 m/s. What was the change in velocity of the scooter? Give magnitude and direction.

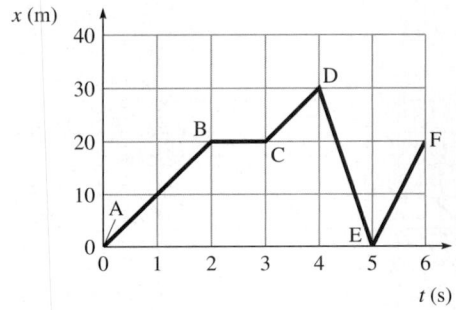

Figure 3.33 Problem 10

3.3 Acceleration

12. If a car traveling at 28 m/s is brought to a full stop in 4.0 s after the brakes are applied, find the average acceleration during braking.

13. If a pronghorn antelope accelerates from rest in a straight line with a constant acceleration of 1.7 m/s^2, how long does it take for the antelope to reach a speed of 22 m/s?

14. An airtrack glider, 8.0 cm long, blocks light as it goes through a photocell gate (Fig. 3.34). The glider is released from rest on a frictionless inclined track and the gate is positioned so that the glider has traveled 96 cm when it is in the middle of the gate. The timer gives a reading of 333 ms for the glider to pass through this gate. Friction is negligible. What is the acceleration (assumed constant) of the glider along the track?

15. Figure 3.35 shows a graph of v_x versus t for a body moving along a straight line. (a) What is a_x at $t = 11$ s? (b) What is a_x at $t = 3$ s? (c) How far does the body travel from $t = 12$ to $t = 14$ s?

16. Figure 3.36 shows a plot of $v_x(t)$ for a car. (a) What is $a_{av,x}$ between $t = 6$ s and $t = 11$ s? (b) What is $v_{av,x}$ for the same time interval? (c) What is $v_{av,x}$ for the interval $t = 0$ to $t = 20$ s? (d) What is the increase in the car's speed between 10 and 15 s? (e) How far does the car travel from time $t = 10$ s to time $t = 15$ s?

17. Figure 3.31 shows a graph of speedometer readings as a motorcycle comes to a stop. What is the magnitude of the acceleration at $t = 7.0$ s?

18. At 3:00 P.M., a bank robber is spotted driving north on I-15 at milepost 126. His speed is 112.0 mi/h. At 3:37 P.M. he is spotted at milepost 185 doing 105.0 mi/h. During this time interval, what are the bank robber's displacement, average velocity, and average acceleration? (Assume a straight highway.)

Figure 3.34 Problem 14

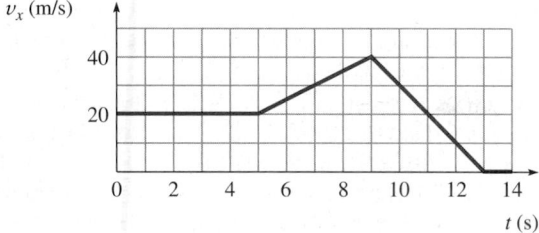

Figure 3.35 Problems 15, 33, and 34

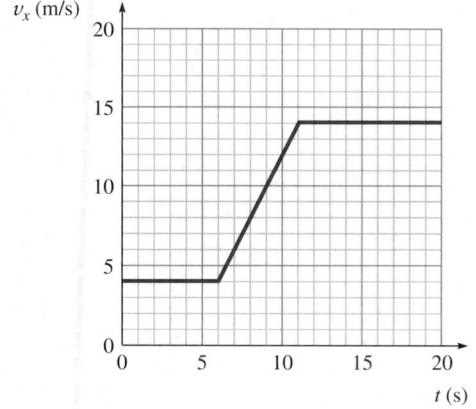

Figure 3.36 Problem 16

3.4 Newton's Second Law: Force and Acceleration

19. An engine accelerates a train of 20 freight cars, each having a mass of 5.0×10^4 kg, from rest to a speed of 4.0 m/s in 20.0 s on a straight track. Neglecting friction and assuming constant acceleration, what is the force with which the 10th car pulls the 11th one (at the middle of the train)?

20. In Fig. 3.16a, two blocks are connected by a lightweight, flexible cord that passes over a frictionless pulley. (a) If $m_1 = 3.0$ kg and $m_2 = 5.0$ kg, what are the accelerations of each block? (b) What is the tension in the cord?

21. A 2.0-kg toy locomotive is pulling a 1.0-kg caboose. The frictional force of the track on the caboose is 0.50 N backward along the track. If the train is accelerating forward at 3.0 m/s², what is the magnitude of the force exerted by the locomotive on the caboose?

22. A 2010-kg elevator accelerates upward at 1.5 m/s². What is the tension in the cable that supports the elevator?

23. A 2010-kg elevator accelerates downward at 1.5 m/s². What is the tension in the cable that supports the elevator?

24. A model sailboat is slowly sailing west across a pond. A gust of wind gives the sailboat a constant acceleration of magnitude 0.30 m/s² during a time interval of 2.0 s. If the net force on the sailboat during the 2.0-s interval has magnitude 0.375 N, what is the sailboat's mass?

25. The vertical component of the acceleration of a sailplane is zero when the air pushes up against its wings with a force of 3.0 kN. (a) Assuming that the only forces on the glider are that due to gravity and that due to the air pushing against its wings, what is the gravitational force on the Earth due to the glider? (b) If the wing stalls and the upward force decreases to 2.0 kN, what is the acceleration of the glider?

26. A man lifts a 2.0-kg stone vertically with his hand at a constant velocity of 1.5 m/s. What is the force exerted by his hand on the stone?

27. A man lifts a 2.0-kg stone vertically with his hand at a constant upward *acceleration* of 1.5 m/s². What is the magnitude of the total force of the stone on the man's hand?

✦ 28. A crate of oranges weighing 180 N rests on a flatbed truck 2.0 m from the back of the truck. The coefficients of friction between the crate and the bed are $\mu_s = 0.30$ and $\mu_k = 0.20$. The truck drives on a straight, level highway at a constant 8.0 m/s. (a) What is the force of friction acting on the crate? (b) If the truck speeds up with an acceleration of 1.0 m/s², what is the force of friction on the crate? (c) What is the maximum acceleration the truck can have without the crate starting to slide?

3.5 Motion with Constant Acceleration

29. A trolley car in New Orleans starts from rest at the St. Charles Street stop and accelerates uniformly at 1.20 m/s² for 12.0 s. (a) How far has the train traveled at the end of the 12.0 s? (b) What is the speed of the train at the end of the 12.0 s?

30. A train, traveling at a constant speed of 22 m/s, comes to an incline with a constant slope. While going up the incline, the train slows down with a constant acceleration of magnitude 1.4 m/s². (a) What is the speed of the train after 8.0 s on the incline? (b) How far has the train traveled up the incline after 8.0 s?

31. A car is speeding up and has an instantaneous speed of 10.0 m/s when a stopwatch reads 10.0 s. It has a constant acceleration of 2.0 m/s². (a) What change in speed occurs between $t = 10.0$ s and $t = 12.0$ s? (b) What is the speed when the stopwatch reads 12.0 s?

32. A train is traveling south at 24.0 m/s when the brakes are applied. It slows down at a constant rate to a speed of 6.00 m/s in a time of 9.00 s. (a) What is the acceleration of the train during the 9.00 s interval? (b) How far does the train travel during the 9.00 s?

33. Figure 3.35 shows the graph of v_x versus time for a body moving along the x-axis. How far does the body go between $t = 9.0$ s and $t = 13.0$ s? Solve using two methods: a graphical analysis and an algebraic solution using kinematic equations.

34. Figure 3.35 shows the graph of v_x versus time for a body moving along the x-axis. What is the average acceleration between $t = 5.0$ s and $t = 9.0$ s? Solve using two methods: a graphical analysis and an algebraic solution using kinematic equations.

35. An airplane lands and starts down the runway at a southwest velocity of 55 m/s. What constant acceleration allows it to come to a stop in 1.0 km?

✦ 36. The minimum stopping distance of a car moving at 30.0 mi/h is 12 m. Under the same conditions (so that the maximum braking force is the same), what is the minimum stopping distance for 60.0 mi/h? Work by proportions to avoid converting units.

✦37. Derive Eq. (3-12) using these steps. (a) Start with $\Delta x = v_{av,x}t$. Rewrite this equation with the average velocity expressed in terms of the initial and final velocities. (b) Replace the final velocity with the initial velocity plus the change in velocity $(v_{0x} + \Delta v_x)$. (c) Find the change in velocity in terms of a_x and t and substitute. Rearrange to form Eq. (3-12).

✦38. Derive Eq. (3-13) using these steps. (a) Start with $\Delta x = v_{av,x}t$. Rewrite this equation with the average velocity expressed in terms of the initial and final velocities. (b) Use $v_x - v_{0x} = a_xt$ to eliminate t. (c) Rearrange algebraically to form Eq. (3-13). [*Hint:* $A^2 - B^2 = (A + B)(A - B)$.]

39. A train of mass 55,200 kg is traveling along a straight, level track at 26.8 m/s (60.0 mi/h). Suddenly the engineer sees a truck stalled on the tracks 184 m ahead. If the maximum possible braking force has magnitude 84.0 kN, can the train be stopped in time?

40. In a television tube, electrons are accelerated from rest by a constant electric force of magnitude 6.4×10^{-17} N during the first 2.0 cm of the tube's length; then they move at essentially constant velocity another 45 cm before hitting the screen. (a) Find the speed of the electrons when they hit the screen. (b) How long does it take them to travel the length of the tube?

3.6 Falling Objects

In the problems, please assume $g = 9.8$ m/s^2 unless a more precise value is given in the problem.

41. A penny is dropped from the observation deck of the Empire State building (369 m above ground). With what velocity does it strike the ground? Ignore air resistance.

42. (a) How long does it take for a golf ball to fall from rest for a distance of 12.0 m? (b) How far would the ball fall in twice that time?

43. Grant Hill jumps 1.3 m straight up into the air to slam-dunk a basketball into the net. With what speed did he leave the floor?

✦44. A student, looking toward his fourth-floor dormitory window, sees a flowerpot with nasturtiums (originally on a window sill above) pass his 2.0-m-high window in 0.093 s. The distance between floors in the dormitory is 4.0 m. From a window on which floor did the flowerpot fall?

✦45. A balloonist, riding in the basket of a hot air balloon that is rising vertically with a constant velocity of 10.0 m/s, releases a sandbag when the balloon is 40.8 m above the ground. Neglecting air resistance, what is the bag's speed when it hits the ground? Assume $g = 9.80$ m/s^2.

46. A 55-kg lead ball is dropped from the leaning tower of Pisa. The tower is 55 m high. (a) How far does the ball fall in the first 3.0 s of flight? (b) What is the speed of the ball after it has traveled 2.5 m downward? (c) What is the speed of the ball 3.0 s after it is released?

47. During a walk on the Moon, an astronaut accidentally drops his camera over a 20.0-m cliff. It leaves his hands with zero speed, and after 2.0 s it has attained a velocity of 3.3 m/s downward. How far has the camera fallen after 4.0 s?

48. A stone is launched straight up by a slingshot. Its initial speed is 19.6 m/s and the stone is 1.50 m above the ground when launched. Assume $g = 9.80$ m/s^2. (a) How high above the ground does the stone rise? (b) How much time elapses before the stone hits the ground?

✦49. How far must something fall for its speed to get close to its terminal speed? Use dimensional analysis to come up with an estimate based on the object's terminal speed v_t and the gravitational field g. Estimate this distance for (a) a raindrop ($v_t \approx 7$ m/s) and (b) a skydiver in a dive ($v_t \approx 100$ m/s).

50. A paratrooper with a fully loaded pack has a mass of 120 kg. The force due to air resistance on him when falling with an unopened parachute has magnitude $F_d = bv^2$, where $b = 0.14 \dfrac{\text{N·s}^2}{\text{m}^2}$. (a) If he is falling with an unopened parachute at 64 m/s, what is the force of air resistance acting on him? (b) What is his acceleration? (c) What is his terminal speed?

51. A bobcat weighing 72 N jumps out of a tree. (a) What is the drag force on the bobcat when it falls at its terminal velocity? (b) What is the drag force on the bobcat when it falls at 75% of its terminal velocity? (c) What is the acceleration of the bobcat when it falls at its terminal velocity? (d) What is its acceleration when it falls at 75% of its terminal velocity?

✦52. In free fall, we assume the acceleration to be constant. Not only is air resistance neglected, but the gravitational field strength is assumed to be constant. (a) From what height can an object fall to the Earth's surface such that the gravitational field strength changes less than 1.000% during the fall? (b) In most cases, which do we have to worry about first: air resistance becoming significant or g changing?

3.7 Apparent Weight

In the problems, please assume $g = 9.8$ m/s^2 unless a more precise value is given in the problem.

53. Refer to Example 3.17. What is the apparent weight of the same passenger (weighing 598 N) in the following situations? In each case, the magnitude of the elevator's acceleration is 0.50 m/s^2. (a) After having stopped at the 15th floor, the passenger pushes the 8th floor button; the elevator is beginning to move downward. (b) Elevator is slowing down as it nears the 8th floor.

54. You are standing on a bathroom scale inside an elevator. Your weight is 140 lb, but the reading of the scale is 120 lb. (a) What is the magnitude and direction of the acceleration of the elevator? (b) Can you tell whether the elevator is speeding up or slowing down?

55. Felipe is going for a physical before joining the swim team. He is concerned about his weight, so he carries his scale into the elevator to check his weight while heading to the doctor's office on the 21st floor of the building. If his scale reads 750 N while the elevator is accelerating upward at 2.0 m/s^2, what does the nurse measure his weight to be?

✦56. Luke stands on a scale in an elevator that has a constant acceleration upward. The scale reads 0.960 kN. When Luke picks up a box of mass 20.0 kg, the scale reads 1.200 kN. (The acceleration remains the same.) Let $g = 9.80$ m/s^2. (a) Find the acceleration of the elevator. (b) Find Luke's weight.

COMPREHENSIVE PROBLEMS

In the problems, please assume $g = 9.8$ m/s^2 unless a more precise value is given in the problem statement.

57. The conduction electrons in a copper wire have instantaneous speeds of around 10^6 m/s. The electrons keep colliding with copper atoms and reversing direction, so that even when a current is flowing in the wire, they only drift along with average velocities on the order of centimeters per hour. If the drift velocity in a particular wire has magnitude 10 cm/h, calculate the distance an electron travels in order to move 1 m down the wire.

58. Imagine a trip where you drive along a straight east-west highway at 80.0 km/h for 50.0 min and then at 60.0 km/h for 30.0 min. What is your average velocity for the trip (a) if you drive east for both parts of the trip? (b) if you drive east for the first 50.0 min and then west for the last 30.0 min?

59. A rocket is launched from rest. After 8.0 min, it is 160 km above the Earth's surface and is moving at a speed of 7.6 km/s. Assuming the rocket moves up in a straight line, what are its (a) average velocity and (b) average acceleration?

C ◆ 60. Based on the information given in Problem 59, is it possible that the rocket moves with constant acceleration? Explain.

61. An unmarked police car starts from rest just as a speeding car passes at a speed of v_0. If the police car accelerates at a constant value of a, what is the speed of the police car when it catches up to the speeder, who does not realize she is being pursued and does not vary her speed?

◆◆ 62. In Fig. 3.23a, the block of mass m_1 slides to the right with coefficient of kinetic friction μ_k on a horizontal surface. The block is connected to a hanging block of mass m_2 by a light cord that passes over a light, frictionless pulley. (a) Find the acceleration of each of the blocks and the tension in the cord. (b) Check your answers in the special cases $m_1 \ll m_2$, $m_1 \gg m_2$, and $m_1 = m_2$. (c) For what value of m_2 (if any) do the two blocks slide at constant velocity? What is the tension in the cord in that case?

63. While passing a slower car on the highway, you accelerate uniformly from 39 mi/h to 61 mi/h in a time of 10.0 s. (a) How far do you travel during this time? (b) What is your acceleration in (mi/h)/s and m/s^2?

64. A cheetah can accelerate from rest to 24 m/s in 2.0 s. Assuming the acceleration is uniform, (a) what is the acceleration of the cheetah? (b) What is the distance traveled by the cheetah in these 2.0 s? (c) A runner can accelerate from rest to 6.0 m/s in the same time, 2.0 s. What is the acceleration of the runner? By what factor is the cheetah's average acceleration magnitude greater than that of the runner?

65. Neglecting air resistance, (a) from what height must a hockey puck drop if it is to attain a speed of 30.0 m/s (approximately 67 mi/h) before striking the ground? (b) If the puck comes to a full stop in a time of 1.00 s from initial impact, what acceleration (assumed constant) is experienced by the puck during the impact with the ground?

C ◆ 66. Locusts can jump to heights of 0.30 m. (a) Assuming the locust jumps straight up, and ignoring air resistance, what is the takeoff speed of the locust? (b) The locust actually jumps at an angle of about 55° to the horizontal, and air resistance is not negligible. The result is that the takeoff speed is about 40% higher than the value you calculated in part (a). If the

mass of the locust is 2.0 g and its body moves 4.0 cm in a straight line while accelerating from rest to the takeoff speed, calculate the acceleration of the locust (assumed constant). (c) Ignore the locust's weight and estimate the force exerted on the hind legs by the ground. Compare this force to the locust's weight. Was it OK to ignore the locust's weight?

67. In Fig. 3.16a, two blocks are connected by a lightweight, flexible cord that passes over a frictionless pulley. If m_1 is 3.6 kg and m_2 is 9.2 kg, and block 2 is initially at rest 140 cm above the floor, how long does it take block 2 to reach the floor?

68. A locomotive pulls a train of 10 identical cars with a force of 2.0×10^6 N directed east. What is the force with which the *last* car pulls on the rest of the train?

69. A woman of mass 51 kg is standing in an elevator. If the elevator floor pushes up on her feet with a force of 408 N, what is the acceleration of the elevator?

70. A drag racer crosses the finish line of a 400.0-m track with a final speed of 104 m/s. (a) Assuming constant acceleration during the race, find the racer's time and the minimum coefficient of static friction between the tires and the road. (b) If, because of bad tires or wet pavement, the acceleration were 30.0% smaller, how long would it take to finish the race?

◆◆ 71. The graph in Fig. 3.37 shows the position x of a switch engine in a rail yard as a function of time t. At which of the labeled times t_0 to t_7 is (a) $a_x < 0$, (b) $a_x = 0$, (c) $a_x > 0$, (d) $v_x = 0$, (e) the speed decreasing?

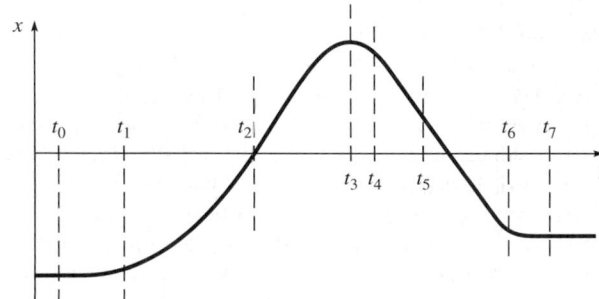

Figure 3.37 Problem 71

C ◆ 72. An elevator starts at rest on the ninth floor. At $t = 0$, a passenger pushes a button to go to another floor. The graph in Fig. 3.38 shows the acceleration a_y of the elevator as a function of time. Let the y-axis point upward. (a) Has the passenger gone to a higher or lower floor? (b) Sketch a graph of the velocity

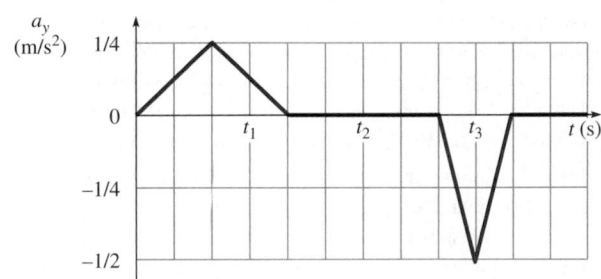

Figure 3.38 Problem 72

v_y of the elevator versus time. (c) Sketch a graph of the position y of the elevator versus time. (d) If a 1.40×10^2 lb person stands on a scale in the elevator, what does the scale read at times t_1, t_2, and t_3?

73. A streetcar named Desire travels between two stations 0.60 km apart. Leaving the first station, it accelerates for 10.0 s at 1.0 m/s^2 and then travels at a constant speed until it is near the second station, when it brakes at 2.0 m/s^2 in order to stop at the station. How long did this trip take? [*Hint:* What's the average velocity?]

© ✦ 74. The graph in Fig. 3.39 is the vertical velocity component v_y of a bouncing ball as a function of time. The y-axis points up. Answer these questions based on the data in the graph. (a) At what time does the ball reach its maximum height? (b) For how long is the ball in contact with the floor? (c) What is the maximum height of the ball? (d) What is the acceleration of the ball while in the air? (e) What is the average acceleration of the ball while in contact with the floor?

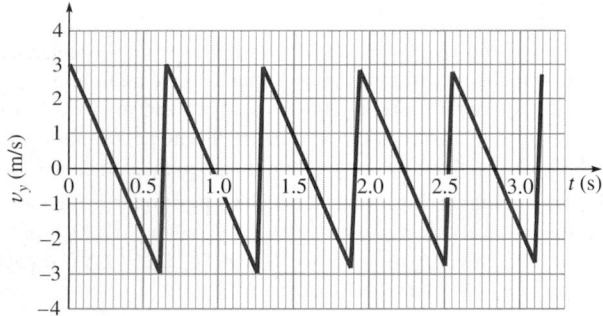

Figure 3.39 Problem 74

✦ 75. A rocket engine can accelerate a rocket launched from rest vertically up with an acceleration of 20.0 m/s^2. However, after 50.0 s of flight the engine fails. Neglect air resistance. (a) What is the rocket's altitude when the engine fails? (b) When does it reach its maximum height? (c) What is the maximum height reached? [*Hint:* a graphical solution may be easiest.] (d) What is the velocity of the rocket just before it hits the ground?

76. Two blocks lie side by side on a frictionless table. The block on the left is of mass m; the one on the right is of mass $2m$. The block on the right is pushed to the left with a force of magnitude F, pushing the other block in turn. What force does the block on the left exert on the block to its right?

✦ 77. A helicopter of mass M is lowering a truck of mass m onto the deck of a ship. (a) At first, the helicopter and the truck move downward together (the length of the cable doesn't change). If their downward speed is decreasing at a rate of $0.10g$, what is the tension in the cable? (b) As the truck gets close to the deck, the helicopter stops moving downward. While it hovers, it lets out the cable so that the truck is still moving downward. If the truck's downward speed is decreasing at a rate of $0.10g$, while the helicopter is at rest, what is the tension in the cable?

© ✦ 78. Fish don't move as fast as you might think. A small trout has a top swimming speed of only about 2 m/s, which is about the speed of a brisk walk (for a human, not a fish!). It may seem to move faster because it is capable of large *accelerations*—it

can dart about, changing its speed or direction very quickly. (a) If a trout starts from rest and accelerates to 2 m/s in 0.05 s, what is the trout's average acceleration? (b) During this acceleration, what is the average net force on the trout? Express your answer as a multiple of the trout's weight. (c) Explain how the trout gets the water to push it forward.

79. Figure 3.31 shows a graph of speedometer readings obtained as an automobile came to a stop. What is the displacement along a straight-line path in the first 16 s?

✦ 80. In Fig. 3.16a, two blocks are connected by a lightweight, flexible cord that passes over a frictionless pulley. If $m_1 \gg m_2$, find (a) the acceleration of each block and (b) the tension in the cord.

✦✦ 81. Find the point of no return for an airport runway of 1.50 mi in length if a jet plane can accelerate at 10.0 ft/s^2 and decelerate at 7.00 ft/s^2. The point of no return occurs when the pilot can no longer abort the takeoff without running out of runway. What length of time is available from the start of the motion in which to decide on a course of action?

82. Two blocks, masses m_1 and m_2, are connected by a massless cord (Fig. 3.40). If the two blocks are accelerated uniformly on a frictionless surface by applying a force of magnitude T_2 to a second cord connected to m_2, what is the ratio of the tensions in the two cords in terms of the masses? $T_1/T_2 = ?$

Figure 3.40 Problem 82

83. In the human nervous system, signals are transmitted along neurons as *action potentials* that travel at speeds of up to 100 m/s. (An action potential is a traveling influx of sodium ions through the membrane of a neuron.) The signal is passed from one neuron to another by the release of neurotransmitters in the synapse. Suppose someone steps on your toe. The pain signal travels along a 1.0-m-long sensory neuron to the spinal column, across a synapse to a second 1.0-m-long neuron, and across a second synapse to the brain. Suppose that the synapses are each 100 nm wide, that it takes 0.10 ms for the signal to cross each synapse, and that the action potentials travel at 100 m/s. (a) At what average speed does the signal cross a synapse? (b) How long does it take the signal to reach the brain? (c) What is the average speed of propagation of the signal?

84. A car traveling at 65 mi/h runs into a bridge abutment after the driver falls asleep at the wheel. (a) If the driver is wearing a seat belt and comes to rest within a 1.0-m distance, what is his acceleration (assumed constant)? (b) A passenger who isn't wearing a seat belt is thrown into the windshield and comes to a stop in a distance of 10.0 cm. What is the acceleration of the passenger? (c) Express these accelerations in terms of the free fall acceleration, $g = 9.8$ m/s^2, and find "how many g's" are felt in each case. (Test pilots can black out at accelerations of $4g$ or greater; pressure suits are designed to help force blood to the brain and prevent such a loss of consciousness.)

ANSWERS TO PRACTICE PROBLEMS

3.1 His displacement is zero since he ends up at the same place from which he started (home plate).

3.2 $\Delta x = -2.9$ km, so the displacement is 2.9 km to the west.

3.3 1.05 m/s to the north

3.4 The velocity is increasing in magnitude, so $\Delta \vec{v}$ and $\vec{a}$ are in the same direction as the velocity (the $-x$-direction). Thus a_x is negative.

3.5 (a) $a_{av,x} = -3.0$ m/s^2, where the negative sign means the acceleration is directed to the northwest; (b) $a_x = -4.2$ m/s^2 (northwest)

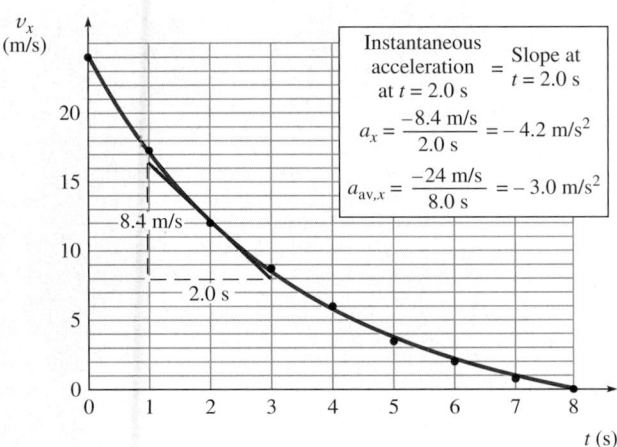

Instantaneous acceleration at $t = 2.0$ s = Slope at $t = 2.0$ s

$a_x = \dfrac{-8.4 \text{ m/s}}{2.0 \text{ s}} = -4.2$ m/s^2

$a_{av,x} = \dfrac{-24 \text{ m/s}}{8.0 \text{ s}} = -3.0$ m/s^2

3.6 1.84 kN

3.7 For equal masses, $a_y = 0$ and $T = mg$. The pull of gravity is equal on the two sides. For slightly unequal masses, $m_1 \approx m_2$, so again $T \approx mg$; the acceleration is very small since there is only a slight excess pull of gravity on the heavier side but plenty of inertia.

3.8 60 N

3.9 (a) 20 s; (b) 18 m/s east

3.10 5.00 s

3.11 If $m_1 \gg m_2$, a is very small and $T \approx m_2 g$. Block 1 is so massive that it barely accelerates; block 2 is essentially just hanging there, being supported by the cord.

3.12 2500 N

3.13 It is impossible to pull the crate up with a single pulley. The entire weight of the crate would be supported by a single strand of cable and that weight exceeds the breaking strength of the cable.

3.14 (a) 3.8 m; (b) 3.00 s

3.15 1.1 kN

3.16 At first, constant acceleration: straight line with slope $-g$. After a long time, constant velocity.

3.17 (a) 392 N; (b) 431 N

Forces and Motion
in Two Dimensions

A seagull scoops up a clam and takes it high above the ground. While flying parallel to the ground, the seagull lets go so that the clam lands on a rock below and cracks open. Then the seagull alights and enjoys lunch. A beachcomber on the beach sees the clam fall along a parabolic path, just as a projectile would. Why does the clam not drop straight down? What does the path of the falling clam look like to the bird?

The order does not matter in vector addition: $\vec{A} + \vec{B} = \vec{B} + \vec{A}$.

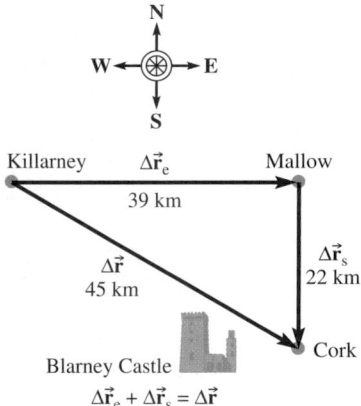

Figure 4.1 Displacement vectors for the trip from Killarney to Cork via Mallow

Compass headings are often used to specify vector directions. Any direction in the horizontal plane can be specified by giving an angle with respect to the directions of the compass. For example, the direction of the vector in Fig. 4.2 is "20° north of east," which means that the vector makes a 20° angle with the east direction and is on the north (rather than the south) side of east. The same direction could be described as "70° east of north," although it is customary to use the smaller angle. Northeast means "45° north of east" or, equivalently, "45° east of north."

Blarney Castle

4.1 ADDITION AND SUBTRACTION OF VECTORS IN TWO DIMENSIONS

This chapter is about analyzing motion in two dimensions. Before we delve into the topic, we must acquire the necessary mathematical tools. We start by extending the principles introduced in Chapter 3 to two-dimensional motion. Suppose we plan a trip in Ireland from Killarney to Cork in order to visit Blarney Castle. There are many possible routes. One route is to travel due east from Killarney for 39 km to Mallow and then south from Mallow for 22 km to Cork. Figure 4.1 shows the displacement vectors for the two legs of the trip. The total distance traveled is 39 km + 22 km = 61 km. (The direction doesn't matter; distance is a scalar quantity.) To find the *displacement*, a vector quantity, the displacement *vectors* are added. We have already learned the rules for adding vectors graphically. To add two or more vectors, place the tail of each successive vector at the tip of the preceding vector. The value of a vector is not changed by moving it as long as its direction and magnitude are not changed, so a vector can be drawn starting at any point. Moreover, it does not matter in what order the vectors are added since vector addition is commutative (that is, $\vec{A} + \vec{B} = \vec{B} + \vec{A}$; see Problem 1). The resultant vector—the sum of the two vectors—is then drawn from the tail of the first vector to the tip of the last vector. If the lengths and directions of the vectors are drawn accurately to scale, using a ruler and a protractor, then the length and direction of the resultant can be determined with the ruler and protractor.

The displacement vectors (Fig. 4.1) are labeled $\Delta\vec{r}_e$ for the trip from Killarney to Mallow and $\Delta\vec{r}_s$ for the trip from Mallow to Cork; the total displacement is labeled $\Delta\vec{r}$.

$$\Delta\vec{r} = \Delta\vec{r}_e + \Delta\vec{r}_s$$

The two separate displacements to the east and south are at right angles and the displacement $\Delta\vec{r}$ forms the hypotenuse of a right triangle so the magnitude can be found using the Pythagorean theorem. The magnitude of the displacement is

$$|\Delta\vec{r}| = \sqrt{(39\text{ km})^2 + (22\text{ km})^2} = 45\text{ km}$$

The vector arrow drawn from Killarney directly to Cork represents the total displacement for the trip; it indicates the route that an airplane would take in flying *directly* from one city to the other along a straight line. The magnitude of the displacement is the straight-line distance from Killarney to Cork.

Now suppose we take a different route from Killarney to Cork. We might drive at a compass heading of 27° west of south for 18 km to Kenmare, then directly south for 17 km to Glengariff, then at a compass heading of 13° north of east for 48 km to Cork, a total travel distance of 83 km. The total displacement (Fig. 4.3) is the same as for the previous route, since the trip begins and ends at the same two places.

In Fig. 4.4a, we select an origin and then draw position vectors from the origin to the starting and ending points of the trip. To subtract $\vec{r}_0$ from $\vec{r}_f$, the vectors are drawn with their tails at the same point. Then the vector drawn from the tip of $\vec{r}_0$ to the tip of $\vec{r}_f$ is the vector $\vec{r}_f - \vec{r}_0$ since it takes us from $\vec{r}_0$ to $\vec{r}_f$. To double-check, note that the initial position plus the change in position gives the final position:

$$\vec{r}_0 + \Delta\vec{r} = \vec{r}_f$$

We draw $\Delta\vec{r}$ so that, when added to $\vec{r}_0$, it gives $\vec{r}_f$. This procedure is equivalent to adding $-\vec{r}_0$ to $\vec{r}_f$ (Fig. 4.4b):

$$\Delta\vec{r} = \vec{r}_f - \vec{r}_0 = \vec{r}_f + (-\vec{r}_0)$$

The same methods of vector addition and subtraction are used for *any* vectors measuring like quantities, such as two force vectors or two displacement vectors. Example 4.1 considers the addition of two force vectors.

Figure 4.2   Measuring angles with respect to compass headings. The direction of this vector is 20° north of east (20° N of E).

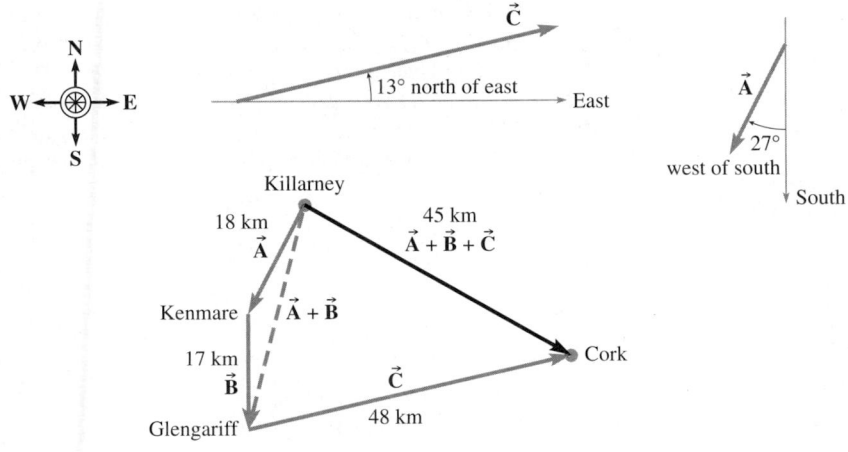

Figure 4.3   Graphical addition of the displacement vectors for the trip from Killarney to Cork via Kenmare and Glengariff.

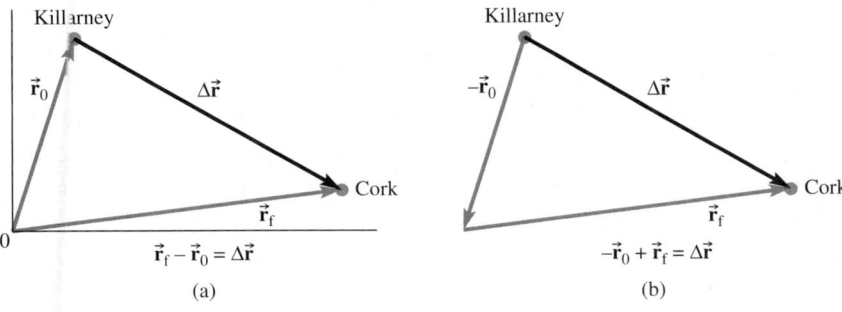

Figure 4.4   (a) Two position vectors, $\vec{r}_0$ and $\vec{r}_f$, drawn from an *arbitrary origin* to the starting point (Killarney) and to the ending point (Cork) of a trip. The final position vector minus the initial position vector is the displacement $\Delta\vec{r}$, drawn from the tip of $\vec{r}_0$ to the tip of $\vec{r}_f$. (b) Another way to find $\vec{r}_f - \vec{r}_0$ is to add $-\vec{r}_0$ to $\vec{r}_f$.

Example 4.1

Bringing Home the Maple Syrup

 Two draft horses, Sam and Bob, are dragging a sled loaded with jugs of maple syrup. They pull with forces of equal magnitude 1.50 kN on two ropes attached to the front of the sled. Each rope makes an angle of 15° with the forward direction (Fig. 4.5). Use the graphical method of vector addition to find the magnitude and direction of the sum of the forces exerted by the two horses.

Strategy   Forces are vector quantities, so they add just like displacement vectors. To add vectors graphically, we need to use a ruler and a protractor. The protractor is used to draw the vector arrows in the correct directions and the ruler is used to draw them with the correct lengths. We must choose a convenient scale for the lengths of the vector arrows. Here we choose to represent 200 N as a length of

continued on next page

Example 4.1 *continued*

1 cm, so that the 1.50 kN forces are drawn as vector arrows of length

$$1.50 \text{ kN} \times \frac{1 \text{ cm}}{200 \text{ N}} = 7.50 \text{ cm}$$

Once the resultant is drawn, its direction and magnitude are determined with the ruler and protractor.

Solution   We can redraw any vector starting at any point, as long as we do not change its direction or magnitude. Thus, we can either add vector $\vec{F}_B$ to vector $\vec{F}_S$ or add vector $\vec{F}_S$ to vector $\vec{F}_B$. Both possibilities are shown in Fig. 4.6.

The direction of the resultant vector is straight ahead since it lies on a horizontal ruling of the grid. Measurement of the length of the resultant vector arrow yields 14.5 cm. Therefore, the magnitude of the resultant is

$$14.5 \text{ cm} \times \frac{200 \text{ N}}{1 \text{ cm}} = 2.90 \text{ kN}$$

Discussion   That the resultant is straight ahead is not surprising given the symmetry of the situation: the two forces are equal in magnitude and pull at equal angles on either side of the forward direction.

A quick check on the magnitude of the resultant: if the two vectors being added were parallel, their sum would have magnitude 3.00 kN. Since they are not parallel, their sum must have a magnitude *smaller* than 3.00 kN. The magnitude is only slightly smaller because the angle between the two vectors is relatively small.

The accuracy of the graphical method depends on the accuracy with which the directions and magnitudes of the arrows are drawn. In Section 4.2 we introduce an algebraic method of vector addition that is both easier and more accurate.

Practice Problem 4.1   Pulling at other angles

(a) If Sam and Bob were to pull with forces of the same magnitude but angled at 10° to the forward direction, would the sum of the two be larger, smaller, or the same magnitude? Illustrate with a sketch. (b) If Sam pulls at 10° while Bob pulls at 15° to the forward direction, is it still possible for the sum of the two to be in the forward direction? If so, which force must have the larger magnitude? Illustrate with a sketch.

(a)

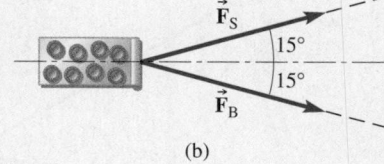

(b)

Figure 4.5
(a) Hauling the maple syrup (side view). (b) Forces exerted by Sam and Bob on the sled (top view)

Scale: 1 cm = 200 N   Each sq = 1 cm × 1 cm

Figure 4.6
Graphical addition of the two force vectors

4.2   COMPONENTS OF VECTORS IN TWO DIMENSIONS

It is generally easier and more accurate to add vectors algebraically rather than graphically. The algebraic method of vector addition relies on adding the components of the vectors. Vector components were introduced in Chapter 3, but only in cases where the vectors were parallel to one of the axes. If a vector lies in the *xy*-plane but *not* along one of the axes, then it has both *x*- and *y*-components.

A problem can be made easier to solve with a good choice of axes. We can choose any direction we want for the *x*- and *y*-axes, as long as they are perpendicular to each other. The two most common choices when near the surface of the Earth are

- *x*-axis horizontal and *y*-axis vertical, when the vectors all lie in a vertical plane; and

- *x*-axis east and *y*-axis north, when the vectors all lie in a horizontal plane.

Some problems are considerably easier to solve for a choice of axes other than the two common choices above.

Consider vector $\vec{A}$ in Fig. 4.7. Note that $\vec{A}$ is the sum of two vectors $\vec{A}_x$ and $\vec{A}_y$, the former along the *x*-axis and the latter along the *y*-axis:

$$\vec{A}_x + \vec{A}_y = \vec{A}$$

$\vec{A}_x$ and $\vec{A}_y$ are called the **vector components** of $\vec{A}$. Using these two vectors we can find the **scalar components** (A_x and A_y) of $\vec{A}$ exactly as in Chapter 3:

$$A_x = \pm \left| \vec{A}_x \right|$$

where the choice of sign reflects whether $\vec{A}_x$ points in the positive or negative *x*-direction.

In Fig. 4.7, A_x is positive and A_y is negative. Using the grid, we can measure the components of $\vec{A}$: $A_x = +5.0$ cm and $A_y = -8.0$ cm.

The process of finding the scalar components of a vector is called **resolving** or decomposing the vector into its components. From now on, when we refer simply to "components," we mean the *scalar* components.

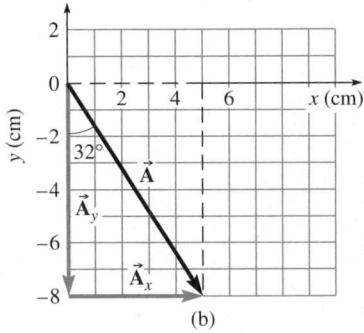

Figure 4.7 Resolving vector $\vec{A}$ into components along two axes in two equivalent ways

---

Finding the *x*- and *y*-Components of a Vector from Its Magnitude and Direction

1. Draw a right triangle with the vector as the hypotenuse and the other two sides parallel to the *x*- and *y*-axes.
2. Determine one of the angles in the triangle.
3. Use trigonometric functions to find the magnitudes of the components.
4. Supply the correct algebraic sign for each component.

Suppose that the magnitude of vector $\vec{A}$ is 9.4 cm and the direction is 58° below the horizontal. Choosing the *x*-axis to be horizontal and positive to the right and the *y*-axis to be vertical and positive upward, the vector is drawn as in Fig. 4.7 with the 58° angle labeled. Now, using trigonometry,

$$\cos 58° = \frac{\text{adjacent}}{\text{hypotenuse}} = \frac{|A_x|}{A}$$

where $A = |\vec{A}|$ is the magnitude of the vector $\vec{A}$. (See Appendix A.7 for a review of the trigonometric functions.) The *x*-component is positive so

$$A_x = +A \cos 58° = +(9.4 \text{ cm}) \times 0.530 = +5.0 \text{ cm}$$

For the *y*-component,

$$\sin 58° = \frac{\text{opposite}}{\text{hypotenuse}} = \frac{|A_y|}{A}$$

The *y*-component is *negative,* so

$$A_y = -A \sin 58° = -(9.4 \text{ cm}) \times 0.848 = -8.0 \text{ cm}$$

We must also know how to reverse the process:

Finding the Magnitude and Direction of a Vector from Its *x*- and *y*-Components

1. Sketch the vector on a set of *xy*-axes in the correct quadrant according to the signs of the components.
2. Draw a right triangle with the vector as the hypotenuse and the other two sides parallel to the *x*- and *y*-axes.
3. In the right triangle, choose which of the unknown angles you want to determine.
4. Use the inverse tangent function to find the angle. The lengths of the sides of the triangle represent $|A_x|$ and $|A_y|$. If θ is opposite the side parallel to the *x*-axis, then since $\tan \theta = $ opposite/adjacent,

$$\theta = \tan^{-1} \left| \frac{A_x}{A_y} \right|$$

If θ is opposite the side parallel to the *y*-axis, then

$$\theta = \tan^{-1} \left| \frac{A_y}{A_x} \right|$$

5. Interpret the angle: specify whether it is the angle below the horizontal, or the angle west of south, or the angle clockwise from the negative *y*-axis, and so on.
6. Use the Pythagorean theorem to find the magnitude:

$$A = \sqrt{A_x^2 + A_y^2}$$

For the vector $\vec{\mathbf{A}}$ in Fig. 4.7, let us choose to find the angle between $\vec{\mathbf{A}}$ and the +*x*-axis. Since $\tan \theta$ is the ratio of the opposite side to the adjacent side,

$$\theta = \tan^{-1} \frac{\text{opposite}}{\text{adjacent}} = \tan^{-1} \left| \frac{A_y}{A_x} \right| = \tan^{-1} \frac{8.0}{5.0} = 58°$$

The magnitude of $\vec{\mathbf{A}}$ is

$$A = \sqrt{A_x^2 + A_y^2} = \sqrt{(+5.0 \text{ cm})^2 + (-8.0 \text{ cm})^2} = 9.4 \text{ cm}$$

To add vectors algebraically, we work with the components of the vectors:

Adding Vectors Algebraically

1. Find the *x*- and *y*-components of each vector to be added.
2. Add the *x*-components (with their algebraic signs) of the vectors to find the *x*-component of the resultant.
3. Add the *y*-components (with their algebraic signs) of the vectors to find the *y*-component of the resultant.
4. Use the *x*- and *y*-components of the resultant to find the magnitude and direction of the resultant.

Example 4.2

Adding Displacements Algebraically

Use the algebraic method to add the displacement vectors for the trip from Killarney to Cork via Kenmare and Glengariff (Fig. 4.8a).

Strategy First we find the *x*- and *y*-components for the three displacements $\vec{\mathbf{A}}$, $\vec{\mathbf{B}}$, and $\vec{\mathbf{C}}$ using the *x*-axis to the east and the *y*-axis to the north. Then the *x*- and *y*-components Δx and Δy of the total displacement $\Delta \vec{\mathbf{r}}$ are

$$\Delta x = A_x + B_x + C_x$$

continued on next page

Example 4.2 *continued*

and

$$\Delta y = A_y + B_y + C_y$$

Finally, we use Δx and Δy to find the magnitude and direction of $\Delta \vec{r}$.

Solution The first displacement ($\vec{A}$) is directed 27° west of south. Both of its components are negative since west is the $-x$-direction and south is the $-y$-direction. Using the triangle in Fig. 4.8b, the side of the triangle opposite the 27° angle is parallel to the x-axis. The sine function relates the opposite side to the hypotenuse

$$A_x = -A \sin 27° = -18 \text{ km} \times 0.454 = -8.2 \text{ km}$$

while the cosine relates the adjacent side to the hypotenuse

$$A_y = -A \cos 27° = -18 \text{ km} \times 0.891 = -16 \text{ km}$$

Vector $\vec{B}$ has no x-component, since its direction is south. Therefore,

$$B_x = 0$$

$$B_y = -B = -17 \text{ km}$$

The direction of $\vec{C}$ is 13° north of east. Both its components are positive. From Fig. 4.8b, the side of the triangle opposite the 13° angle is parallel to the y-axis, so

$$C_x = +C \cos 13° = +48 \text{ km} \times 0.974 = +47 \text{ km}$$

$$C_y = +C \sin 13° = +48 \text{ km} \times 0.225 = +11 \text{ km}$$

Now we sum the x- and y-components separately to find the x- and y-components of the total displacement:

$$\Delta x = A_x + B_x + C_x = (-8.2 \text{ km}) + 0 + 47 \text{ km} = +39 \text{ km}$$

$$\Delta y = A_y + B_y + C_y = (-16 \text{ km}) + (-17 \text{ km}) + 11 \text{ km} = -22 \text{ km}$$

The magnitude and direction of $\Delta \vec{r}$ can be found from the triangle in Fig. 4.8c. The magnitude is represented by the hypotenuse:

$$\Delta r = \sqrt{(\Delta x)^2 + (\Delta y)^2} = \sqrt{(39 \text{ km})^2 + (-22 \text{ km})^2} = \sqrt{2005 \text{ km}^2} = 45 \text{ km}$$

The angle θ is

$$\theta = \tan^{-1} \frac{\text{opposite}}{\text{adjacent}} = \tan^{-1} \frac{22 \text{ km}}{39 \text{ km}} = 29°$$

Since $+x$ is east and $-y$ is south, the direction of the displacement is 29° south of east. The magnitude and direction of the resultant found algebraically agree with the resultant found graphically in Fig. 4.8a.

Discussion Note that A_x was found using the sine function, while C_x was found using the cosine. The x-component (or the y-component) of the vector can be related to *either* the sine or the cosine, depending on which angle in the triangle is used.

⚠ When finding the components of a vector, it is crucial to write them with the correct algebraic signs. Otherwise, when the components are added to find the components of the resultant, the resultant vector found will be wrong.

Practice Problem 4.2 Rotating the coordinate axes

Find the x- and y-components of $\vec{A}$, $\vec{B}$, and $\vec{C}$ if the x-axis points south and the y-axis points east.

(a)

(b)

(c)

Figure 4.8
Addition of displacement vectors using components

 Even though the algebraic method is the preferred way to add vectors, the graphical method is an important first step. Making a rough sketch of vector addition has two important benefits.

1. Drawing the vectors makes it much easier to get the angles and the signs of the components right.

2. The graphical addition serves as a check on the answer. Even a rough sketch of the vectors $\vec{A}$, $\vec{B}$, and $\vec{C}$ being added graphically, without carefully measuring their lengths or the angles, would give an approximate idea of the magnitude and direction of the resultant, which can be used to check the algebraic answer. Graphical addition gives you a mental picture of what is going on and an intuitive feel for the algebraic calculations.

Example 4.3

Exercise Is Good for You

 Suppose you are standing on the floor doing your daily exercises. For one exercise, you lift your arms up and out until they are horizontal. In this position, assume that the deltoid muscle exerts a force of 270 N at an angle of 15° above the horizontal on the humerus (Fig. 4.9a). What are the vertical and horizontal components of this force?

Strategy This problem gives the magnitude and direction of a force and asks for the components of the force. Let the x-axis be horizontal and the y-axis be vertical. To find the components, we draw a right triangle with the force vector as the hypotenuse and the sides parallel to the x- and y-axes.

Solution Figure 4.9b shows a triangle that can be used to find the components. With the choice of x- and y-axes shown, the x-component is negative; the +x-axis points to the right while the x-component of the force is to the left. The side of the triangle along the x-axis is adjacent to the 15° angle, so the cosine function is used for the x-component.

$$\cos 15° = \frac{\text{adjacent}}{\text{hypotenuse}} = \frac{|F_x|}{F}$$

$$F_x = -F \cos 15° = -270 \text{ N} \times 0.966 = -260 \text{ N}$$

The y-component is up, along the +y-axis. The sine function is used for the y-component, since the side representing F_y is opposite the 15° angle.

$$\sin 15° = \frac{\text{opposite}}{\text{hypotenuse}} = \frac{|F_y|}{F}$$

$$F_y = +F \sin 15° = 270 \text{ N} \times 0.259 = 70 \text{ N}$$

Discussion The vertical component of the muscle force supports the arm against the downward force of gravity. The horizontal component of the force stabilizes the joint by pulling the humerus in toward the scapula. The horizontal component is much larger than the vertical component because the force makes a small angle with the horizontal.

Practice Problem 4.3 Adding forces algebraically

In Practice Problem 4.1(b), suppose Sam pulls with a force of 1.50 kN at 10.0° to the forward direction. (a) What must be the magnitude of the force exerted by Bob at 15.0° on the other side of the forward direction so that the sum of the two forces is in the forward direction? (b) What is the magnitude of the sum? [*Hint*: Let the x-axis be in the forward direction. What must be true about the y-components of the two forces?]

(a)

(b)

Figure 4.9
(a) Force exerted by the deltoid muscle on the humerus; (b) drawing a triangle to find the force components.

4.3   EQUILIBRIUM

According to Newton's second law, the acceleration of an object is determined by the net force acting on it:

$$\Sigma \vec{F} = m\vec{a}$$

This is a vector equation that can be rewritten in terms of perpendicular components:

$$\Sigma F_x = ma_x \quad \text{and} \quad \Sigma F_y = ma_y$$

When an object is in equilibrium (either at rest or in motion at constant velocity), the net force acting on it is zero. Since the object's acceleration is zero, both of its components must be zero.

For an object in equilibrium,

$$\Sigma F_x = ma_x = 0 \quad \text{and} \quad \Sigma F_y = ma_y = 0 \qquad (4\text{-}1)$$

Example 4.4

Archery Practice

Figure 4.10a shows the bowstring of a bow and arrow just before it is released. The archer is pulling back on the midpoint of the bowstring with a horizontal force of 162 N. What is the tension in the bowstring?

Strategy The bowstring, arrow, bow, and the archer's hand are all in equilibrium. We must choose some object or system, draw a free-body diagram, and apply Newton's second law. The only object about which we have enough information is the bowstring. We know the force exerted on the string by the archer's fingers. The bow exerts two forces on the string—one at each end; the magnitude of each is equal to the tension in the string. (At either end, the magnitude of the force with which the string pulls on the bow is the tension in the string. From Newton's third law, the bow pulls on the string with a force equal in magnitude and opposite in direction.) We neglect the weight of the string, since it is small compared to the other forces on the bowstring.

Solution Figure 4.10b is a free-body diagram for the bowstring. The weight of the string is negligibly small compared to the other forces, so it is omitted. To find the components of the forces exerted by the bow, we draw a triangle as in Fig. 4.10c. From the measurements given, we can find the angle θ.

$$\sin \theta = \frac{\text{opposite}}{\text{hypotenuse}} = \frac{35 \text{ cm}}{72 \text{ cm}} = 0.486$$

$$\theta = \sin^{-1} 0.486 \approx 29°$$

The x-component of the force exerted on the upper end of the string is

$$F_x = -T \sin \theta$$

The x-component of the force exerted on the lower end of the string is the same. Therefore,

$$\Sigma F_x = -2T \sin \theta + F_a = 0$$

where F_a is the magnitude of the force applied by the archer. Solving for T,

$$T = \frac{F_a}{2 \sin \theta}$$

Substituting the value of $\sin \theta$ from the ratio of lengths (35 cm/72 cm = 0.486),

$$T = \frac{162 \text{ N}}{2 \times 0.486} = 170 \text{ N}$$

Discussion In this problem, only the x-components of the forces had to be used. The y-components also add to zero. At the upper end of the string, the y-component of the force exerted by the bow is $+T \cos \theta$ while at the lower end it is $-T \cos \theta$. Therefore, $\Sigma F_y = 0$.

The expression $T = F_a/(2 \sin \theta)$ can be evaluated for limiting values of θ to make sure that the expression is correct. As θ approaches 90°, the tension approaches $F_a/(2 \sin 90°) = F_a/2$. That is correct because the archer would be pulling to the right with a force F_a, while each side of the bowstring would pull to the left with a force of magnitude T. For equilibrium, $F_a = 2T$ or $T = F_a/2$.

As θ gets smaller, $\sin \theta$ decreases and the tension increases (for a fixed value of F_a). That agrees with our intuition. The larger the tension, the smaller the angle the string needs to make in order to supply the necessary horizontal force.

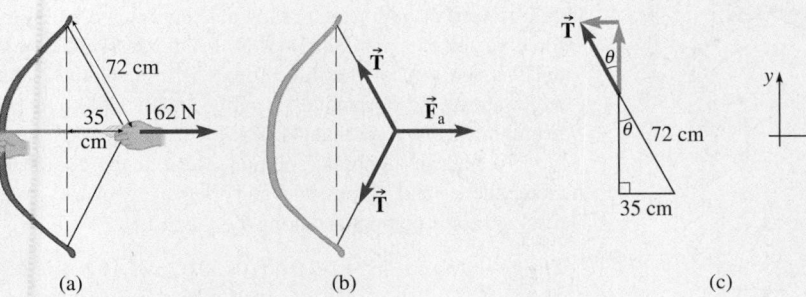

Figure 4.10
(a) The force applied to the bowstring by an archer. (b) Free-body diagram for a point on the bowstring with the forces labeled. The magnitudes of the two forces exerted by the bow are equal to the tension in the string T. (c) The angle θ is used to find the x- and y-components of the forces exerted at each end of the bowstring.

continued on next page

Example 4.4 *continued*

Practice Problem 4.4   Tightrope practice

Jorge decides to rig up a tightrope in the backyard so that his children can develop a good sense of balance (Fig. 4.11). For safety reasons he positions a horizontal cable only 0.60 m above the ground. If the 6.00 m long cable sags by 0.12 m from its taut horizontal position when Denisha (weight 250 N) is standing on the middle of it, what is the force on the eye-bolt due to the cable?

Figure 4.11
Tightrope for balancing practice

Example 4.5

Traction on a Foot

 Figure 4.12 shows a traction apparatus designed to apply a force to stretch the leg in a controlled fashion. What is the force exerted on the foot by this traction apparatus?

Strategy   Assuming ideal pulleys, the tension of the cord is the same everywhere along the cord ($T = 22.0$ N). The force exerted on the foot by the attached pulley is equal in magnitude and opposite in direction to the force that the foot exerts on the pulley according to Newton's third law. Since this pulley is in equilibrium, the net force acting on it is zero.

Figure 4.12
(a) A foot in traction; (b) forces acting on the pulley.

Solution   Four forces act on the pulley: three forces due to the cord and one due to the patient's foot. (The pulley's weight is negligible.) Then

$$\vec{F}_{PF} + \vec{F}_1 + \vec{F}_2 + \vec{F}_3 = 0$$

where $\vec{F}_{PF}$ is the force exerted on the pulley by the foot and $\vec{F}_1$, $\vec{F}_2$, and $\vec{F}_3$ are the forces on the pulley due to the three cords. Solving for $\vec{F}_{PF}$,

$$\vec{F}_{PF} = -(\vec{F}_1 + \vec{F}_2 + \vec{F}_3)$$

The problem asks for the force exerted on the foot by the pulley ($\vec{F}_{FP}$). From Newton's third law,

$$\vec{F}_{FP} = -\vec{F}_{PF} = \vec{F}_1 + \vec{F}_2 + \vec{F}_3$$

This equation confirms what intuition tells us: the resultant force exerted on the foot by the apparatus is the sum of the three forces due to the cord.

Figure 4.13a shows the graphical addition of the forces exerted on the pulley by the cord. The magnitudes of $\vec{F}_1$, $\vec{F}_2$, and $\vec{F}_3$ are equal to the tension in the cord (22.0 N). Two cords pull upward at a 45° angle; they are represented by a single force vector of magnitude 44.0 N. From this sketch, we can tell that the sum of the three forces is at a relatively small angle above the horizontal (roughly half of 45°) and has a magnitude a bit larger than 44.0 N.

To find an algebraic solution, we find the components along the x- and y-axes and add them (Fig. 4.13b-d). The resultant force component along the x-axis is

$$\Sigma F_x = (44.0 \text{ N}) \cos 45° + (22.0 \text{ N}) \cos 30.0° = 31.11 \text{ N} + 19.05 \text{ N}$$

$$= 50.16 \text{ N}$$

continued on next page

Example 4.5 *continued*

and along the y-axis,

$$\Sigma F_y = 44.0 \text{ N} \sin 45.0° + (-22.0 \text{ N} \sin 30.0°) = 31.11 \text{ N} - 11.0 \text{ N}$$

$$= 20.11 \text{ N}$$

The resultant force is found from the Pythagorean theorem (Fig. 4.13e):

$$F = \sqrt{(20.11 \text{ N})^2 + (50.16 \text{ N})^2} = 54.0 \text{ N}$$

and the direction of the force is found from

$$\tan \theta = \frac{\text{opposite}}{\text{adjacent}} = \left|\frac{F_y}{F_x}\right| = \frac{20.11 \text{ N}}{50.16 \text{ N}} = 0.4009$$

$$\theta = \tan^{-1} 0.4009 = 21.8°$$

The force exerted on the foot is 54.0 N at an angle 21.8° above the horizontal.

Discussion The result of the algebraic addition is what we expect from doing the graphical addition: the total force

(54.0 N) is somewhat larger than 44.0 N and is at an angle just about half of 45° above the horizontal.

A shortcut to the beginning of the solution would have been to consider the foot and the attached pulley as a single system. Then we immediately see that the total force acting on the foot–pulley system due to the three cords is the quantity to be found.

Practice Problem 4.5 Effect of changing the pulley angles

The pulleys are moved so that the two cords above the foot make an angle of 30.0° with the horizontal and the cord below the foot makes an angle of 60.0° with the horizontal. (a) What are the components of the traction force on the foot? (b) What is the magnitude of the traction force? (c) At what angle with the horizontal does the traction force act?

Figure 4.13
(a) Graphical sum of the forces on the pulley due to the cord, (b) finding the x- and y-components of the forces due to the cord, (c) adding the y-components, (d) adding the x-components, and (e) finding the resultant from its components.

Equilibrium on an Inclined Plane

Making The Connection:
an inclined plane

Suppose we wish to move a large box up to a loading dock platform. If we pick the box up, we have to support its entire weight. If we slide it along an incline, we can move it up to the same loading dock by exerting a force smaller than the box's weight. The inclined plane enables us to move an object up an incline by using less force than would be needed if we just picked up the object and lifted it. Part of the weight of the object is supported by the inclined plane itself. The inclined plane is one example of a *simple machine* that reduces the magnitude of the applied force needed for some task.

Figure 4.14a shows the three forces acting on a box being pushed up a frictionless incline. $\vec{F}_a$ represents the applied force. If we choose the x- and y-axes to be horizontal and vertical respectively, then two of the three forces have both x- and y-components. On the other hand, if we choose the x-axis parallel to the incline and the y-axis perpendicular to it, then only one of the three forces has both x- and y-components (the weight). In an equilibrium problem, choose x- and y-axes so that the fewest number of force vectors have both x- and y-components.

With axes chosen, the weight of the box is then resolved into two perpendicular components. In order to find the x- and y-components of $\vec{W}$, we must determine the

Figure 4.14 (a) A box of mass m pushed up an incline, (b) resolving the weight into components parallel to and perpendicular to the incline, and (c) free-body diagram for the box on the inclined plane

angle that $\vec{W}$ makes with one of the axes. In Fig. 4.14b, $\alpha = 90° - \phi$, since the three angles inside the large right triangle add up to 180°. The x- and y-axes are perpendicular, so $\alpha + \beta = 90°$ and, therefore, the angle between $\vec{W}$ and the $-y$-axis is $\beta = \phi$.

The y-component of the weight is perpendicular to the surface of the incline. From the triangle in Fig. 4.14b,

$$W_y = -mg \cos \phi$$

since the side parallel to the y-axis is adjacent to angle ϕ and $\cos \phi = \dfrac{\text{adjacent}}{\text{hypotenuse}}$. The x-component of the weight tends to make the box slide down the incline (in the $+x$-direction). Its value is

$$W_x = +mg \sin \phi$$

By pushing the box with a force equal in magnitude to W_x up the incline (in the negative x-direction), we can hold the box in place on the incline or move it with constant velocity. The component of the box's weight perpendicular to the incline is supported by the normal force, $\vec{N}$, that pushes the box away from the incline. Figure 4.14c is a free-body diagram in which the weight is separated into its x- and y-components.

If the box is in equilibrium, whether at rest or moving along the incline at constant velocity, the acceleration of the box is zero ($a_x = 0$ and $a_y = 0$). Then the force components along each axis sum to zero:

$$\Sigma F_x = (-F_a) + mg \sin \phi = 0$$

and

$$\Sigma F_y = N + (-mg \cos \phi) = 0$$

 The normal force is *not* equal in magnitude to the weight and it does not point straight up. If the applied force is of magnitude $mg \sin \phi$, we can push the box up the incline at constant velocity. If friction acts on the box, we must push with a force greater than $mg \sin \phi$ to move the box up the incline at constant velocity.

Example 4.6

Pushing a Safe Up an Incline

 A new safe is being delivered to the Corner Book Store. It is to be placed in the wall at a height of 1.5 m above the floor. The delivery people have a portable ramp that they plan to use to help them push the safe up and into position.

The mass of the safe is 510 kg, the coefficient of static friction along the incline is $\mu_s = 0.42$, and the coefficient of kinetic friction along the incline is $\mu_k = 0.33$. The ramp forms an angle $\theta = 15°$ above the horizontal. (a) Find the minimum applied force parallel to the incline needed to start the safe moving up

continued on next page

Example 4.6 *continued*

the incline. (b) If the movers continue to push with this same force, what is the safe's acceleration? (c) To move the safe up at a constant speed, with what force must the movers push?

Strategy   First we draw a free-body diagram for the safe. Before resolving the forces into components, we must choose *x*- and *y*-axes. The coefficient of friction relates the frictional force to the normal force. To use the coefficient of friction, we have to resolve the contact force on the safe due to the incline into components *parallel and perpendicular to the incline*— friction and the normal force, respectively—rather than into horizontal and vertical components. Therefore, we choose *x*- and *y*-axes parallel and perpendicular to the incline so that friction is along the *x*-axis and the normal force is along the *y*-axis. The applied force and the safe's acceleration are parallel to the incline as well. Using this choice of axes, the *only* vector that is not parallel to one of the axes is the weight. The problem would be much harder to solve using a horizontal *x*-axis and a vertical *y*-axis since only one force would be parallel to the axes; all the other forces and the acceleration would have to be resolved into horizontal and vertical components.

 When the direction of the acceleration is known in a nonequilibrium problem, it is usually easiest to choose axes parallel and perpendicular to the acceleration.

Solution   First we draw a diagram to show the forces acting (Fig. 4.15a) and sketch in the *x*- and *y*-axes. The weight $\vec{\mathbf{W}}$ must be resolved into two components: the component perpendicular to the incline, $W_y = -mg \cos \theta$, and the component along the incline, $W_x = -mg \sin \theta$ (Fig. 4.15b). Then

$\theta = 15°$

(a)

(b)   (c)

Figure 4.15
(a) Forces acting on the safe as it is moved up the incline, (b) resolving the weight into *x*- and *y*-components, and (c) a free-body diagram in which the weight is replaced with its *x*- and *y*-components.

we draw the FBD with the weight replaced by its components (Fig. 4.15c).

(a) Suppose that the safe is initially at rest. For the safe to *begin* moving up the ramp, a nonzero net force in the +*x*-direction must act on it. As the movers start to push, $F_{applied}$ gets larger and the force of static friction gets larger to "try" to keep the safe from sliding. Eventually, at some value of $F_{applied}$, static friction reaches its maximum possible value $\mu_s N$. If the movers push any harder than that, the safe starts to slide.

To *start* the box moving up the incline, it must have an acceleration directed up the incline:

$$\Sigma F_x = F_{applied} + (-mg \sin \theta) + (-f_s) = ma_x > 0$$

The maximum force of static friction is

$$(f_s)_{max} = \mu_s N$$

The normal force is *not* equal in magnitude to the weight of the safe. To find the normal force, sum the *y*-components of the forces:

$$\Sigma F_y = N + (-mg \cos \theta) = ma_y = 0$$

The safe slides along the surface of the incline; it never moves in the ±*y*-direction. Since v_y never changes (it is always zero), $a_y = 0$. Then

$$N = mg \cos \theta$$

The normal force is less than the weight since $\cos \theta < 1$.
In order for a_x to be positive,

$$F_{applied} > mg \sin \theta + f_s = mg \sin 15° + \mu_s mg \cos 15°$$

$$F_{applied} > (510 \text{ kg} \times 9.8 \text{ m/s}^2 \times 0.2588) + (0.42 \times 510 \text{ kg} \times 9.8 \text{ m/s}^2 \times 0.9659)$$

$$F_{applied} > 1293 \text{ N} + 2028 \text{ N} = 3321 \text{ N}$$

Rounding to two significant figures, we conclude that an applied force that exceeds 3300 N starts the box moving up the incline.

(b) Once the safe starts moving, the static friction force stops acting against the motion and the kinetic friction force replaces it. Since the kinetic friction force is less than the static friction force, the movers are pushing with more force than is necessary to maintain motion at constant speed. The net force along the incline, ΣF_x, determines the acceleration of the safe up the incline. If the movers continue to push with a force of 3321 N,

$$\Sigma F_x = F_{applied} + (-mg \sin 15°) + (-\mu_k mg \cos 15°) = ma_x$$

$$a_x = \frac{F_{applied} - mg \sin 15° - \mu_k mg \cos 15°}{m}$$

$$= \frac{3321 \text{ N} - (510 \text{ kg} \times 9.8 \text{ m/s}^2 \times 0.2588) - (0.33 \times 510 \text{ kg} \times 9.8 \text{ m/s}^2 \times 0.9659)}{510 \text{ kg}}$$

$$= \frac{3321 \text{ N} - 1293 \text{ N} - 1593 \text{ N}}{510 \text{ kg}}$$

$$= \frac{435 \text{ N}}{510 \text{ kg}} = 0.9 \text{ m/s}^2$$

continued on next page

Example 4.6 *continued*

(c) Once the safe starts moving, the movers need only push hard enough to make the net force on the safe equal to zero if they want the safe to slide at constant velocity.

$$\Sigma F_x = F_{\text{applied}} + (-mg \sin 15°) + (-\mu_k mg \cos 15°) = 0$$

$$F_{\text{applied}} = mg \sin 15° + \mu_k mg \cos 15°$$

$$= 1293 \text{ N} + 1593 \text{ N} = 2900 \text{ N}$$

Discussion   The force required to lift the safe straight up at constant speed would be equal to the weight.

$$F_{\text{lift}} = mg = (510 \text{ kg} \times 9.8 \text{ m/s}^2) = 5000 \text{ N}$$

The incline enables us to use a smaller force, but we must push over a longer distance.

Practice Problem 4.6   Smoothing the infield dirt

During the seventh-inning stretch of a baseball game, groundskeepers drag mats across the infield dirt to smooth it for the remainder of the game. A groundskeeper is pulling a mat at a constant velocity across the dirt surface of the infield at Fenway Park. If the force pulling the mat is 120 N at an angle of 22° above the ground, find the frictional force between the dirt and the mat and the weight of the mat assuming a coefficient of kinetic friction of 0.60.

Physics at Home

To estimate the coefficient of static friction between a penny and the cover of your physics book, place the penny on the book and slowly lift the cover. Note the angle of the cover when the penny starts to slide. Explain how you can use this angle to find the coefficient of static friction. Can you devise an experiment to find the coefficient of kinetic friction?

4.4   VELOCITY AND ACCELERATION

The definitions of average velocity, instantaneous velocity, average acceleration, and instantaneous acceleration in Chapter 3 still apply when the motion is not in a straight line. Suppose we want to know the instantaneous velocity of a race car at point P as it goes around a curved section of the Indy 500 (Fig. 4.16a). At a slightly later time the race car is at point Q. Let $\vec{\mathbf{r}}_0$ be the position of the car at P and $\vec{\mathbf{r}}_f$ be the position at point Q.

The displacement $\Delta\vec{\mathbf{r}}$ is $\vec{\mathbf{r}}_f - \vec{\mathbf{r}}_0$ (Fig. 4.16b). To subtract $\vec{\mathbf{r}}_0$ from $\vec{\mathbf{r}}_f$, the two vectors are drawn with their tails at the same point; the difference is the vector from the tip of $\vec{\mathbf{r}}_0$ to the tip of $\vec{\mathbf{r}}_f$. The average velocity for this time interval is the displacement $\Delta\vec{\mathbf{r}}$ divided by the time interval:

$$\vec{\mathbf{v}}_{\text{av}} = \frac{\vec{\mathbf{r}}_f - \vec{\mathbf{r}}_0}{t_f - t_0} = \frac{\Delta\vec{\mathbf{r}}}{\Delta t} \tag{3-2}$$

The direction of the average velocity is the direction of the displacement $\Delta\vec{\mathbf{r}}$.

The instantaneous velocity at P is the limit of the average velocity as Δt approaches zero. As we shorten the time interval between the initial and final positions by moving point Q closer and closer to P, the direction of the displacement vector $\Delta\vec{\mathbf{r}}$ gradually changes, approaching the tangent to the curved path at P (Fig. 4.16c). Expressed in mathematical terminology, the instantaneous velocity is the limit of $\Delta\vec{\mathbf{r}}/\Delta t$ as the time interval approaches zero:

$$\vec{\mathbf{v}} = \lim_{\Delta t \to 0} \frac{\Delta\vec{\mathbf{r}}}{\Delta t} \tag{3-4}$$

Figure 4.16 (a) Position vectors for two points on the curve. (b) The displacement $\Delta\vec{\mathbf{r}}$ from point P to point Q. (c) As time interval is decreased, the final point moves closer and closer to P; the direction of the displacement $\Delta\vec{\mathbf{r}}$ approaches the tangent to the curve at P. (d) Instantaneous velocity can be resolved into components along perpendicular axes.

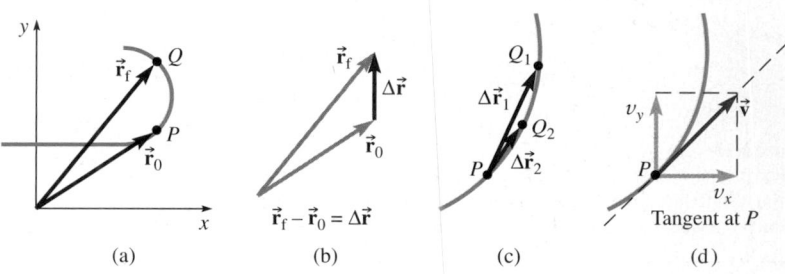

With this definition, the instantaneous velocity at P becomes tangent to the curve at P.

Since velocity is a vector quantity, it can be resolved into components (Fig. 4.16d). The vector equations defining average and instantaneous velocity can be rewritten in terms of x- and y-components:

$$v_{av,x} = \frac{\Delta x}{\Delta t} \quad \text{and} \quad v_{av,y} = \frac{\Delta y}{\Delta t}$$

$$v_x = \lim_{\Delta t \to 0} \frac{\Delta x}{\Delta t} \quad \text{and} \quad v_y = \lim_{\Delta t \to 0} \frac{\Delta y}{\Delta t}$$

Similarly, the average acceleration $\vec{\mathbf{a}}_{av}$ is the change in velocity divided by the elapsed time:

$$\vec{\mathbf{a}}_{av} = \frac{\vec{\mathbf{v}}_f - \vec{\mathbf{v}}_0}{t_f - t_0} = \frac{\Delta \vec{\mathbf{v}}}{\Delta t} \tag{3-6}$$

This vector equation is equivalent to two component equations:

$$a_{av,x} = \frac{\Delta v_x}{\Delta t} \quad \text{and} \quad a_{av,y} = \frac{\Delta v_y}{\Delta t}$$

Instantaneous acceleration is the limit of the average acceleration as the time interval approaches zero and point Q approaches point P:

$$\vec{\mathbf{a}} = \lim_{\Delta t \to 0} \frac{\Delta \vec{\mathbf{v}}}{\Delta t} \tag{3-7}$$

or

$$a_x = \lim_{\Delta t \to 0} \frac{\Delta v_x}{\Delta t} \quad \text{and} \quad a_y = \lim_{\Delta t \to 0} \frac{\Delta v_y}{\Delta t}$$

In straight-line motion the acceleration is always either parallel or antiparallel to the velocity. For more general motion in two dimensions, the acceleration can be in *any* direction, as illustrated in Example 4.7.

Example 4.7

Skating Uphill

An inline skater is traveling on a level road with a speed of 8.94 m/s. Then she comes to a long hill at a 15.0° angle of incline. 120.0 s later she is still climbing the hill while her speed has decreased to 7.15 m/s. (a) What is the change in her velocity? (b) What is her average acceleration during the 120.0-s time interval?

Strategy   The change in velocity is *not* 8.94 m/s – 7.15 m/s = 1.79 m/s. That is the change in *speed*. The change in velocity is found by subtracting the initial velocity *vector* from the final velocity *vector*. After first making a graphical sketch, we use the component method. The average acceleration is the change in velocity divided by the elapsed time.

Solution   (a) Figure 4.17a shows the initial and final velocity vectors and the slope of the hill. The initial velocity is horizontal as she skates on level ground. The final velocity is 15.0° above the horizontal. To subtract the two velocity vectors graphically, we place the tails of the vectors together. The change in velocity, $\Delta \vec{\mathbf{v}}$, is found by drawing a vector arrow from the tip of $\vec{\mathbf{v}}_0$ to the tip of $\vec{\mathbf{v}}_f$. Judging by the graphical subtraction in Fig. 4.17b, the change in velocity is roughly at a 45° angle above the –x-axis. Its magnitude is smaller than the magnitudes of the initial and final velocity vectors—something like 2 or 3 m/s.

The components v_{fx} and v_{fy} can be found from a right triangle (Fig. 4.17c):

$$v_{fx} = v_f \cos \theta = 7.15 \text{ m/s} \times 0.9659 = 6.91 \text{ m/s}$$

$$v_{fy} = v_f \sin \theta = 7.15 \text{ m/s} \times 0.2588 = 1.85 \text{ m/s}$$

continued on next page

Example 4.7 *continued*

Since v_0 has only an x-component,

$$v_{0y} = 0 \quad \text{and} \quad v_{0x} = v_0 = 8.94 \text{ m/s}$$

Now we subtract the components to find the components of $\Delta\vec{v}$:

$$\Delta v_x = v_{fx} - v_{0x} = (6.91 - 8.94) \text{ m/s} = -2.03 \text{ m/s}$$

and

$$\Delta v_y = v_{fy} - v_{0y} = (1.85 - 0) \text{ m/s} = +1.85 \text{ m/s}$$

To find the magnitude of $\Delta\vec{v}$, we apply the Pythagorean theorem (Fig. 4.17d):

$$(|\Delta\vec{v}|)^2 = (\Delta v_x)^2 + (\Delta v_y)^2 = (-2.03 \text{ m/s})^2 + (1.85 \text{ m/s})^2 = 7.54 \text{ (m/s)}^2$$

$$|\Delta\vec{v}| = 2.75 \text{ m/s}$$

The angle is found from

$$\tan\phi = \frac{\text{opposite}}{\text{adjacent}} = \left|\frac{\Delta v_y}{\Delta v_x}\right| = \frac{1.85 \text{ m/s}}{2.03 \text{ m/s}} = 0.911$$

$$\phi = \tan^{-1} 0.911 = 42.3°$$

The direction of the change in velocity $\Delta\vec{v}$ is 42.3° above the negative x-axis.

(b) The magnitude of the average acceleration is

$$|\vec{a}_{av}| = \frac{|\Delta\vec{v}|}{\Delta t} = \frac{2.75 \text{ m/s}}{120.0 \text{ s}} = 0.0229 \text{ m/s}^2$$

The direction of the average acceleration is the same as the direction of $\Delta\vec{v}$: 42.3° above the negative x-axis.

Discussion   Checking back with the graphical subtraction in Fig. 4.17b, the magnitude of $\Delta\vec{v}$ appears to be roughly $\frac{1}{4}$ to $\frac{1}{3}$ the magnitude of $\vec{v}_0$. Since $\frac{1}{4} \times 8.94$ m/s = 2.24 m/s and $\frac{1}{3} \times 8.94$ m/s = 2.98 m/s, the answer of 2.75 m/s is reasonable.

Figure 4.17b also shows the direction of $\Delta\vec{v}$ to be roughly midway between the $+y$- and $-x$-axes. We found the direction of $\Delta\vec{v}$ to be 42.3° above the $-x$-axis and therefore, 47.7° from the $+y$-axis. So the direction we calculated is also reasonable based on the graphical subtraction.

Practice Problem 4.7   Change in sailboat velocity

A C&C 30 sailboat is sailing at 12.0 knots heading directly east across the harbor. When a gust of wind comes up, the boat changes its heading to 11.0° north of east and its speed increases to 14.0 knots. What is the magnitude and direction of the change in velocity of the sailboat? [A boat's speed is customarily expressed in knots, which means nautical miles per hour. A nautical mile (6076 ft) is a little longer than a statute mile (5280 ft).]

(a)

(b)

(c)

(d)

Figure 4.17
(a) Change in velocity as the skater slows going uphill, (b) graphical subtraction of velocity vectors, (c) initial and final velocity vectors resolved into components, and (d) reconstruction of $\Delta\vec{v}$ from its components.

4.5   MOTION OF PROJECTILES

If an object moves in the xy-plane with constant acceleration, then both a_x and a_y are constant. By looking separately at the motion along two perpendicular axes, the y-direction and the x-direction, each component becomes a one-dimensional problem, which we already know how to solve. We can apply any of the constant-acceleration kinematic relations from Chapter 3 separately to the x-components and to the y-components. *It is generally easiest to choose the axes so that the acceleration has only one nonzero component.* Suppose we choose the axes so that the acceleration is in the positive or negative y-direction. Then $a_x = 0$ and v_x is constant.

As in Chapter 3, it is often convenient to write the kinematic relations for constant acceleration assuming that $t_0 = 0$.

$$\begin{array}{l}
\textbf{Kinematic Equations}\\
\text{for constant acceleration along the } y\text{-axis and } t_0 = 0
\end{array}$$

$a_x = 0$:	$v_x = v_{0x}$	(4-2)
	$x - x_0 = v_{0x}t$	(4-3)
$a_y = \text{constant}$:	$v_y = v_{0y} + a_y t$	(4-4)
	$\Delta y = v_{\text{av},y}t$	(4-5)
	$v_{\text{av},y} = \frac{1}{2}(v_{0y} + v_y)$	(4-6)
	$y - y_0 = v_{0y}t + \frac{1}{2}a_y t^2$	(4-7)
	$v_y^2 - v_{0y}^2 = 2a_y \Delta y$	(4-8)

Note that there is no mixing of components in Eqs. (4-2) through (4-8). Each equation pertains either to the x-components or to the y-components; none contains the x-component of one vector quantity and the y-component of another. The only quantity that appears in both x- and y-component equations is the time—a scalar.

An object in free fall near the Earth's surface has a constant acceleration. As long as air resistance is negligible, the constant downward pull of gravity gives the object a constant downward acceleration of 9.8 m/s^2. In Chapter 3 we considered objects in free fall, but only when they moved straight up or straight down. An object in free fall that has a nonzero horizontal velocity component is called a **projectile**; its motion takes place in a vertical plane. For such an object, Eqs. (4-2) through (4-8) describe projectile motion with

$$a_y = -g$$

assuming the y-axis points upward.

Figure 4.18   A medieval catapult

Suppose some medieval marauders are attacking a castle. They have a catapult that can propel large stones into the air to bombard the walls of the castle (Fig. 4.18). Picture a stone leaving the catapult with an initial velocity $\vec{v}_0$. The **angle of elevation** is the angle of the initial velocity above the horizontal. Once the stone is in the air, the only force acting on it is the downward gravitational force, provided that air resistance has a negligible effect on the motion. The **trajectory** (path) of the stone is shown in Fig. 4.19. The positive x-axis is chosen in the horizontal direction (to the right) and the positive y-axis is upward.

If the initial velocity is at an angle θ above the horizontal, then resolving it into components gives:

$$v_{0x} = v_0 \cos \theta$$
$$v_{0y} = v_0 \sin \theta$$

Since the gravitational force pulls downward, only the y-component of the velocity changes. Notice in Fig. 4.19 that the horizontal velocity component v_x never changes. With zero net force in the x-direction, $a_x = 0$; the stone moves along the horizontal axis at constant speed. The vertical velocity component v_y changes at a constant rate, exactly as if the stone were propelled straight up with an initial speed of v_{0y}. The initially positive v_y decreases until, at the top of flight, $v_y = 0$. Then the pull of gravity makes the projectile fall back downward. During the downward trip, velocity v_y is still changing at the same constant rate with which it changed on the way up and at the top of the path. The acceleration has the same constant value—magnitude *and* direction—for the entire path.

The displacement of the projectile at any instant is the vector sum of the displacements in the two mutually perpendicular directions. The motion of a projectile when air resistance is negligible is the superposition of horizontal motion with constant velocity and vertical

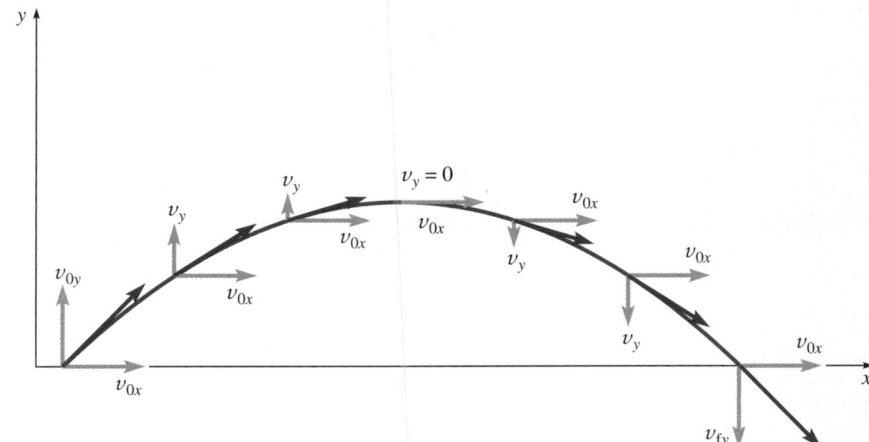

Figure 4.19 Trajectory of a projectile. The velocity vectors are drawn at equal time intervals along with the x- and y-components of $\vec{\mathbf{v}}$.

The horizontal and vertical motions of a projectile can be treated separately; they are independent of each other.

motion with constant acceleration. The vertical and horizontal motions each proceed independently, as if the other motion were not present. In the experiment of Fig. 4.20, one ball was dropped and, at the same instant, another was projected horizontally. The strobe photo shows snapshots of the two balls at closely spaced time intervals. The *vertical* motion of the two is identical; at every instant the two are at the same height. The fact that they have different horizontal motion does not affect their vertical motion. (This statement would *not* be true if air resistance were significant.)

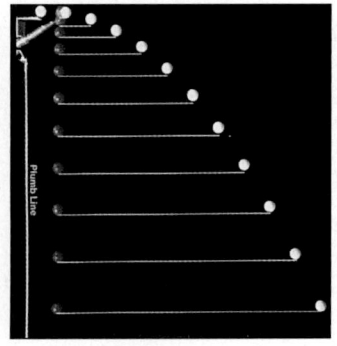

Figure 4.20 Independence of horizontal and vertical motion of a projectile in the absence of air resistance.

Physics at Home

Take two golf balls and a friend to an open window or balcony on the second or third floor of a building. After making sure that no one is below, you drop one ball while your friend, *at the same time*, throws the other ball horizontally. Which ball hits the ground first?

Conceptual Example 4.8

Trajectory of a Projectile

The graph of an equation of the form

$$y = kx^2, \quad k = \text{a nonzero constant}$$

is a parabola. Show that the trajectory of a projectile is a parabola. [*Hint:* Choose the origin at the highest point of the trajectory and let $t = 0$ at that instant.]

Strategy and Solution We start at the high point of the path and look at displacements from there. The horizontal displacement is proportional to the elapsed time, since the horizontal velocity is constant. The vertical displacement is the average vertical velocity component times the elapsed time. The average vertical velocity component is itself proportional to t since v_y changes at a constant rate. Therefore, the vertical displacement is proportional to t^2. Thus, the vertical displacement y is proportional to the square of the horizontal displacement x and $y = kx^2$, where k is a constant of proportionality. The path followed by a projectile in free fall is parabolic.

Discussion The same conclusion can be drawn algebraically. If we choose the origin and $t = 0$ at the top of flight, then x_0, y_0, and v_{0y} are all zero. We can write $y = v_{\text{av},y}t$, but y is *not* proportional to t. From Eq. (4-6)

$$v_{\text{av},y} = \tfrac{1}{2}(v_{0y} + v_y)$$

Substituting $v_{0y} = 0$ and $v_y = v_{0y} + a_y t = a_y t$ yields

$$v_{\text{av},y} = \tfrac{1}{2}a_y t$$

Then $y = v_{\text{av},y}t = \tfrac{1}{2}a_y t^2$. From $x = v_{0x}t$, we can substitute $t = x/v_{0x}$ into this expression for y:

$$y = \tfrac{1}{2}a_y\left(\frac{x}{v_{0x}}\right)^2$$

Thus, $y = kx^2$ where $k = a_y/(2v_{0x}^2)$ and $a_y = -g$.

Conceptual Practice Problem 4.8   Throwing stones

You stand at the edge of a cliff and throw stones horizontally into the river below. To double the horizontal displacement of a stone from the cliff to where it lands, by what factor must you increase the stone's initial speed? Neglect air resistance.

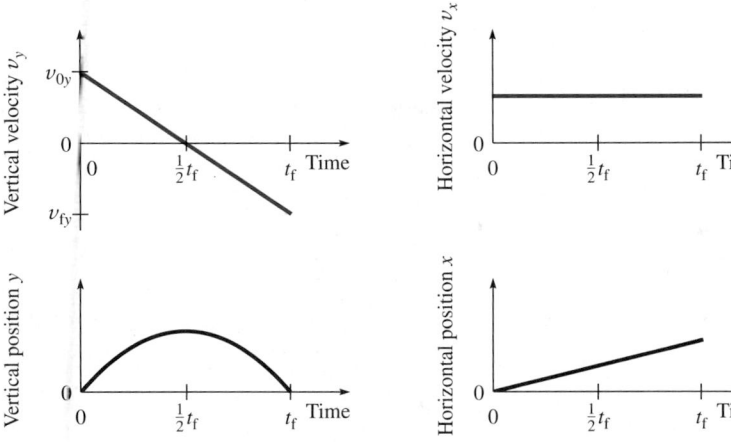

Figure 4.21 Projectile motion: separate vertical and horizontal quantities versus time.

Figure 4.21 shows graphs of the x- and y-components of the velocity and position of a projectile as functions of time. In this case, the projectile is launched above flat ground at $t = 0$ and returns to the same elevation at a later time t_f. The kinematic equations are valid only while the object is in free fall; once the projectile hits the ground, forces other than gravity act on it and its acceleration changes.

Note that the y-component graphs are symmetrical about the vertical line through the highest point in the trajectory. The y-component of velocity decreases linearly from its initial value; the slope of the line is $a_y = -g = -9.8$ m/s^2. When $v_y = 0$ the projectile is at the apex of its trajectory. Then v_y continues to decrease at the same rate and is now negative with its magnitude getting larger and larger. At t_f, when the projectile has returned to its original altitude, the y-component of the velocity has the same magnitude as at $t = 0$ but with the opposite sign ($v_y = -v_{0y}$).

The graph of y versus t indicates that the projectile moves upward, quickly at first and then gradually slowing, until it reaches the maximum height. The slope of the tangent to the y versus t graph at any particular moment of time is v_y at that instant. At the highest point of the y versus t graph, the tangent is horizontal and $v_y = 0$. After that gravity makes the projectile start to fall downward.

The horizontal velocity is constant because the projectile is not acted upon by any horizontal forces ($a_x = \Sigma F_x/m = 0$). Thus, the graph of v_x versus t is a horizontal line. The horizontal position x increases uniformly in time because the object is moving with a constant v_x.

Example 4.9

Attacking the Castle Walls

 The catapult used by the marauders hurls a stone of mass 32.0 kg with a velocity of 50.0 m/s at an angle of elevation of 30.0° (Fig. 4.22). Assume that $g = 9.80$ m/s^2. (a) What is the maximum height reached by the stone? (b) What is its range (defined as the horizontal distance traveled when the stone returns to its original height)? (c) How long has the stone been in the air when it returns to its original height?

Strategy The problem gives both the magnitude and direction of the initial velocity of the stone. Ignoring air resistance,

the stone has a constant downward acceleration once it has been launched—until it hits the ground or some obstacle. We choose the positive y-axis upward and the positive x-axis in the direction of horizontal motion of the stone (toward the castle). When the stone reaches its maximum height, the velocity component in the y-direction is zero, since the stone goes no higher. When the stone returns to its original height, $\Delta y = 0$ and $v_y = -v_{0y}$. The range can be found once the time of flight t_f is known—time is the quantity that connects the x-component equations to the y-component equations. Therefore, we solve part (c) before part (b). One way to find t_f is to find the time to

continued on next page

Example 4.9 *continued*

reach maximum height and then double it (see Fig. 4.21). (Other methods include finding the time when $\Delta y = 0$ (the stone returns to its original height) or the time when $v_y = -v_{0y}$.)

Solution   (a) First we find the x- and y-components of the initial velocity for an angle of elevation $\theta = 30.0°$.

$$v_{0y} = v_0 \sin \theta$$

$$v_{0x} = v_0 \cos \theta$$

The maximum height is the vertical displacement Δy when $v_y = 0$. From Eq. (4-8),

$$v_y^2 - v_{0y}^2 = 2a_y \Delta y$$

Solving for Δy,

$$\Delta y = \frac{v_y^2 - v_{0y}^2}{2a_y}$$

Substituting $v_y = 0$ and $v_{0y} = v_0 \sin \theta$ yields

$$\Delta y = \frac{-(v_0 \sin \theta)^2}{2a_y}$$

$$= \frac{-[(50.0 \text{ m/s}) \sin 30.0°]^2}{2 \times (-9.80 \text{ m/s}^2)} = 31.9 \text{ m}$$

The maximum height of the projectile is 31.9 m above its launch height.

(c) The time of flight (t_f) is *twice* the time it takes the projectile to reach its maximum height. The time to reach the maximum height can be found from

$$v_y = v_{0y} + a_y t$$

Solving for t,

$$t = \frac{v_y - v_{0y}}{a_y} = \frac{-v_{0y}}{a_y}$$

since $v_y = 0$. The time of flight is

$$t_f = 2 \times \frac{-v_{0y}}{a_y} = 2 \times \frac{-(50.0 \text{ m/s}) \sin 30.0°}{-9.80 \text{ m/s}^2} = 5.10 \text{ s}$$

(b) The range is

$$\Delta x = v_{0x} t_f = (50.0 \text{ m/s}) \cos 30.0° \times 5.10 \text{ s} = 221 \text{ m}$$

Discussion   Other kinematic relations can be used to check the answers. For example,

$$y - y_0 = v_{0y} t + \tfrac{1}{2} a_y t^2$$

can be used to check that $\Delta y = 31.9$ m at $t = \frac{1}{2} \times 5.10$ s and that $\Delta y = 0$ at $t = 5.10$ s. Here we check the first of these:

$$\Delta y = (50.0 \text{ m/s}) \sin 30.0° \times 2.55 \text{ s} + \tfrac{1}{2} \times (-9.80 \text{ m/s}^2) \times (2.55 \text{ s})^2$$

$$= 63.8 \text{ m} + (-31.9 \text{ m}) = 31.9 \text{ m}$$

which is correct. This is not an *independent* check, since this equation can be derived from the others, but it can reveal algebra or calculation errors.

Since we analyze the horizontal motion independently from the vertical motion, we start by resolving the initial velocity given into x- and y-components. Time is what connects the two.

Notice that we did not use the mass of the stone in the solution. The mass does not affect the maximum height, the time of flight, or the range, if we are given the initial velocity. Where mass does matter is in the ability of the catapult to accelerate the stone from rest to its launch velocity. The catapult (or, in other problems, a cannon, a human arm, or some other propelling machine) has an easier time imparting a particular velocity to a less massive projectile. A projectile that is too massive for the machine can only be given a small launch speed.

Practice Problem 4.9   Maximum height for arrows

Archers have joined in the attack on the castle and are shooting arrows over the walls. If the angle of elevation for an arrow is 45°, find an expression for the maximum height of the arrow in terms of v_0 and g. [*Hint:* Simplify the expression using $\sin 45° = \cos 45° = 1/\sqrt{2}$.]

Figure 4.22
A catapult projects a stone into the air in an attack on a castle wall.

Conceptual Example 4.10

Monkey and Hunter

 A thoughtless child aims and shoots an arrow, which fortunately has a rubber, suction-cup tip, straight at a monkey who is sitting in a tree and holding a coconut (Fig. 4.23). At the same instant the arrow leaves the bow, the monkey drops the coconut. Does the arrow hit the monkey?

Strategy and Solution   If there were no gravitational field, the arrow would fly straight to the monkey and coconut (along the dashed line from the bow to the monkey on the branch in Fig. 4.23). Since gravity gives the dropped coconut and the released arrow the same constant acceleration downward, they each fall the same vertical distance below the positions they would have had with no gravity. The coconut falls as shown by the dashed red line; the distance fallen at 0.25-s intervals is marked along a vertical axis. At the same time, the arrow drops below the gray dashed line by the amounts marked along its indicated trajectory at 0.25-s intervals.

The arrow ends up hitting the coconut no matter what the initial velocity of the arrow. The higher the velocity of the arrow, the sooner they meet and the shorter the vertical distance that the coconut falls before being hit.

Discussion   An experienced hunter would have aimed above the monkey on the branch to compensate for the amount his arrow would drop during the time of flight; he would have missed the dropping coconut but might have hit the monkey unless the monkey jumped down to retrieve the coconut.

Conceptual Practice Problem 4.10   Changes in position and velocity for consecutive arrows

An arrow is shot into the air. One second later, a second arrow is shot with the same initial velocity. While the two are both in the air, does the difference in their positions $(\vec{r}_2 - \vec{r}_1)$ stay constant or does it change with time? Does the difference in their velocities $(\vec{v}_2 - \vec{v}_1)$ stay constant or does it change with time?

Figure 4.23
A monkey drops a coconut at the very instant an arrow is shot toward the monkey. In each quarter second, the coconut and arrow have fallen the same distance below where their positions would be if there were no gravity.

Physics at Home

On a warm day, take a garden hose and aim the nozzle so that the water streams upward at an angle above the horizontal. Set the nozzle for a fast, narrow stream for best effect. Once the water leaves the nozzle, it becomes a projectile subject only to the force of gravity (neglecting the small effect of air resistance). The continuous stream of water lets us see the parabolic path easily. Stand in one place and try aiming the nozzle at different angles of elevation to find an angle that gives the maximum range. Aim for a particular spot on the ground (at a distance less than the maximum range) and see if you can find two different angles of elevated nozzle position that allow the stream to hit the target spot (see Fig. 4.24).

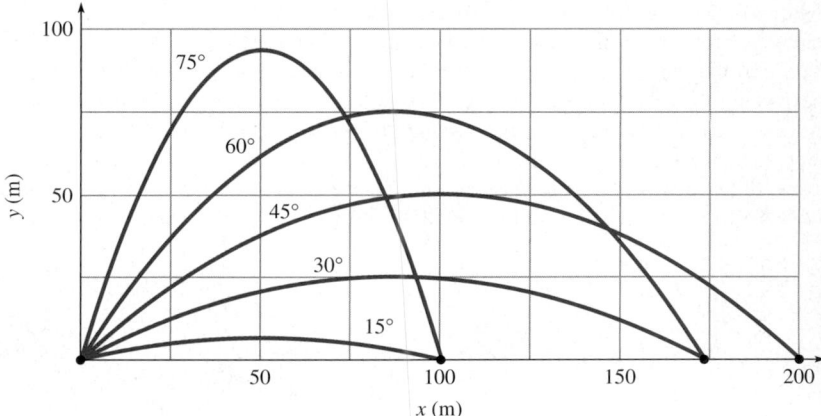

Figure 4.24 Maximum height versus range for various angles of elevation ($v_0 = 44.3$ m/s).

Example 4.11

A Bullet Fired Horizontally

A bullet is fired horizontally from the top of a cliff that is 20.0 m above a long lake. If the muzzle speed of the bullet is 500.0 m/s, how far from the bottom of the cliff does the bullet strike the surface of the lake? Ignore air resistance. Let $g = 9.80$ m/s^2.

Strategy   We need to find the total time of flight so that we can find the horizontal displacement. The bullet is starting from the high point of the parabolic path because $v_{0y} = 0$. As usual in projectile problems, we choose the y-axis to be the positive vertical direction.

Given: $\Delta y = -20.0$ m; $v_{0y} = 0$; $v_{0x} = 500.0$ m/s

To find: Δx

Solution   The vertical displacement through which the bullet falls is 20.0 m. The time to fall can be found from

$$\Delta y = v_{0y}t + \tfrac{1}{2}a_y t^2$$

Solving for t with $v_{0y} = 0$ yields

$$t = \sqrt{\frac{2\Delta y}{a_y}}$$

The horizontal displacement of the bullet is

$$\Delta x = v_{0x}t = v_{0x}\sqrt{\frac{2\Delta y}{a_y}}$$

$$= 500.0 \text{ m/s} \times \sqrt{\frac{2 \times (-20.0 \text{ m})}{-9.80 \text{ m/s}^2}} = 1.01 \text{ km}$$

Discussion   How did we know to start with the y-component equation when the question asks about the *horizontal* displacement? The question gives v_{0x} and asks for Δx. The missing information needed is the time during which the bullet is in the air; the time can be found from analysis of the *vertical* motion.

We neglected air resistance in this problem, which is not very realistic. The actual distance would be less than 1.01 km.

Practice Problem 4.11   Bullet velocity

Find the horizontal and vertical components of the bullet's velocity just before it hits the surface of the lake. At what angle does it strike the surface?

At the beginning of the chapter, we asked why the clam does not fall straight down when the seagull lets go. The seagull is flying horizontally with the clam, so the clam has the same horizontal velocity as the seagull. When the seagull lets go, the net force on the clam is downward due to gravity. The clam falls toward Earth, but since $a_x = 0$, the clam retains the same horizontal component of velocity as the seagull. Therefore, the clam is a projectile starting at the top of its parabolic trajectory.

Making The Connection:
seagull dropping
a clam

4.6   OTHER EXAMPLES OF CONSTANT ACCELERATION

Projectiles moving with negligible air resistance in a uniform gravitational field are not the only objects with constant acceleration. Any object moves with constant acceleration if the net force acting on it is constant. Once we have added all the forces and used Newton's second law to find the acceleration of the object, we can use the same kinematics relations as for projectiles. The only difference is that the acceleration is not necessarily 9.8 m/s² downward; the acceleration is determined by Newton's second law.

If the direction of the acceleration is known, it is easiest to choose x- and y-axes so that the acceleration has one component that is zero. Then we can view the motion as a superposition of constant velocity along one axis and constant acceleration along the other, as we do for projectiles.

Tips

Example 4.12

The Broken Suitcase

The wheels fell off of Beatrice's suitcase so she tied a rope to it and dragged it along the floor of the airport terminal (Fig. 4.25a). The rope makes an angle of 40.0° with the horizontal. The suitcase has a mass of 36.0 kg and Beatrice pulls on the rope with a force of 65.0 N. (a) What is the magnitude of the normal force acting on the suitcase due to the floor? (b) If the coefficient of kinetic friction between the suitcase and the marble floor is $\mu_k = 0.13$, find the frictional force acting on the suitcase. (c) What is the acceleration of the suitcase? (d) For how long a time must Beatrice pull with this force until the suitcase reaches a comfortable walking speed of 0.40 m/s?

Strategy   Since the suitcase is dragged along the floor, it can have no vertical acceleration component. (If it did have a vertical acceleration component, the suitcase would begin to move either down through the floor or up into the air.) Therefore, if we choose the +y-axis up and the +x-axis to be horizontal, $a_y = 0$. We resolve the forces acting on the suitcase into their components, draw a free-body diagram for the suitcase, and apply Newton's second law.

(a)   (b)   (c)

Figure 4.25
(a) Beatrice dragging her suitcase, (b) forces acting on a suitcase dragged along the floor, and (c) a free-body diagram.

continued on next page

Example 4.12 *continued*

Solution   (a) Figure 4.25b shows the forces acting on the suitcase, where $\vec{\mathbf{F}}$ is the force exerted by Beatrice. All the other forces are either parallel or perpendicular to the floor, so only $\vec{\mathbf{F}}$ needs to be resolved into x- and y-components.

$$F_x = F \cos 40.0° = 65.0 \text{ N} \times 0.766 = 49.8 \text{ N}$$
$$F_y = F \sin 40.0° = 65.0 \text{ N} \times 0.643 = 41.8 \text{ N}$$

We don't round to two significant figures yet because these intermediate results are used in subsequent calculations.

Figure 4.25c is a free-body diagram where $\vec{\mathbf{F}}$ is replaced by its two components. The vertical force components add to zero since $a_y = 0$.

$$\Sigma F_y = ma_y = 0$$
$$N + F \sin 40.0° - mg = 0$$

$N = mg - F \sin 40.0° = (36.0 \text{ kg} \times 9.8 \text{ m/s}^2) - (65.0 \text{ N} \times \sin 40.0°)$

$$= 353 \text{ N} - 41.8 \text{ N} = 311.2 \text{ N}$$

Now we round the final answer: the normal force is 310 N.

(b) The kinetic frictional force is

$$f_k = \mu_k N = 0.13 \times 311.2 \text{ N} = 40.46 \text{ N}$$

The frictional force is 40 N in the –x-direction (opposite the motion of the suitcase).

(c) The y-component of the acceleration is zero. To find the x-component, we apply Newton's second law to the x-components of the forces:

$$\Sigma F_x = ma_x$$
$$\Sigma F_x = +F \cos 40.0° + (-f_k)$$
$$= 49.8 \text{ N} - 40.46 \text{ N} = 9.34 \text{ N}$$
$$a_x = 9.34 \text{ N}/36.0 \text{ kg} = 0.259 \text{ m/s}^2$$

The acceleration is 0.26 m/s² in the +x-direction.

(d) With constant a_x,

$$v_x = v_{0x} + a_x t$$

The suitcase starts from rest so $v_{0x} = 0$. Then

$$t = \frac{v_x}{a_x} = \frac{0.40 \text{ m/s}}{0.259 \text{ m/s}^2} = 1.5 \text{ s}$$

Discussion   What Beatrice probably wants to do is to drag the suitcase along at constant velocity. To do that, she must first accelerate the suitcase from rest. Once the suitcase is moving, she pulls a little less hard so the net force is zero and the suitcase slides at constant speed. She would do so without thinking much about it, of course!

Practice Problem 4.12   The continuing suitcase story

How hard must Beatrice pull at a 40.0° angle so that the suitcase slides at constant velocity? [*Warning:* Do *not* assume that the normal force is the same as in the previous discussion.]

Example 4.13

A Sliding Brick

A brick of mass 1.0 kg slides down an icy roof inclined at 30.0° with respect to the horizontal. If the brick starts from rest 2.0 m from the edge of the roof, how fast is it moving at the edge? Ignore friction.

Strategy   First we draw a diagram and indicate the forces acting on the brick. Then we write down Newton's second law

and solve it by breaking it into components along two mutually perpendicular directions. The acceleration is directed along the roof, so we choose the x-axis parallel to the roof (toward the edge) and the y-axis perpendicular to the roof.

Solution   In Fig. 4.26a, $\vec{\mathbf{W}}$ is the weight of the brick and $\vec{\mathbf{N}}$ is the normal force exerted on the brick by the roof. The weight of the brick must be resolved into components in the x- and y-directions

Figure 4.26
(a) Forces acting on a brick sliding down an icy roof, (b) resolving the weight into components, and (c) a free-body diagram in which the weight is replaced by its components.

(a)   (b)   (c)

continued on next page

Example 4.13 *continued*

(Fig. 4.26b). Figure 4.26c is a free-body diagram in which the weight is separated into two perpendicular components.

Ignoring friction, the roof exerts no force in the *x*-direction on the brick. Then only the weight has an *x*-component. From Newton's second law,

$$\Sigma F_x = mg \sin 30.0° = ma_x$$

Solving for a_x,

$$a_x = g \sin 30.0°$$

Now that we have the acceleration, we use kinematic relationships to find the final velocity. We know that $\Delta x = 2.00$ m and $v_{0x} = 0$. Then,

$$v_x^2 - v_{0x}^2 = v_x^2 = 2a_x \Delta x$$

Solving for v_x,

$$v_x = \sqrt{2a_x \Delta x} = \sqrt{2g \sin 30.0° \times \Delta x}$$
$$= \sqrt{2 \times 9.8 \text{ m/s}^2 \times \tfrac{1}{2} \times 2.0 \text{ m}} = 4.4 \text{ m/s}$$

Discussion   In this problem it was not necessary to apply Newton's second law to the *y*-components of the forces. Doing so would enable us to find the normal force. If the roof were not frictionless, we would need to find the normal force in order to calculate the force of kinetic friction.

Practice Problem 4.13   Effect of friction on sliding brick

Repeat Example 4.13 assuming that the coefficient of kinetic friction is 0.20.

Example 4.14

A Pulley, an Incline, and Two Blocks

A block of mass $m_1 = 2.6$ kg rests upon an incline that is angled at 30.0° above the horizontal (Fig. 4.27a). A lightweight, flexible cord is connected from block 1 over an ideal, frictionless pulley to another block of mass $m_2 = 2.2$ kg that is hanging freely 2.00 m above the ground. The coefficient of kinetic friction between the incline and block 1 is 0.18. If the blocks are initially at rest, how long does it take for block 2 to reach the ground? Assume the cord is of negligible mass and does not stretch.

Strategy   It is not obvious which way the blocks start to move once they are released. Fortunately the problem tells us that the blocks start from rest and that block 2 hits the floor, so block 2 accelerates downward and block 1 accelerates up the incline. For block 1, we choose axes parallel and perpendicular to the incline so that its acceleration has only one nonzero component. The magnitudes of the accelerations of the two blocks are equal, since they are connected by a cord that does not stretch. The fixed length of the cord means that during any time interval, the blocks move the same distance. Since that is

true even for tiny time intervals, the speeds of the blocks are equal at any instant. Since the speeds change the same amount during any time interval, their accelerations have equal magnitudes. Since the cord's mass is neglected, the tension is the same at the two ends.

Solution   We start by drawing separate free-body diagrams for each block (Figs. 4.27b and 4.27c). Since block 1 slides up the incline, the frictional force f_k acts down the incline to oppose the sliding. The weight of block 1 is resolved into two components, one along the incline and one perpendicular to the incline.

Using the free-body diagrams we write Newton's second law in component form for each block. Block 1 has no acceleration in the *y*-direction. It does not sink into the incline or rise above it; it can only slide along the incline. Thus, the net force in the *y*-direction is zero.

$$\Sigma F_y = N - m_1 g \cos \theta = 0$$

or

$$N = m_1 g \cos \theta$$

Figure 4.27
(a) Block on an incline connected to a freely hanging block by a cord passing over a pulley, (b) free-body diagram for block 1, and (c) free-body diagram for block 2.

continued on next page

Example 4.14 *continued*

where $\theta = 30.0°$. Along the incline, in the x-direction, the acceleration is nonzero:

$$\Sigma F_x = T - m_1g \sin \theta - f_k = m_1a_x$$

The kinetic frictional force is related to the normal force:

$$f_k = \mu_k N = \mu_k m_1g \cos \theta$$

By substitution,

$$T - m_1g \sin \theta - \mu_k m_1g \cos \theta = m_1a_x \qquad (1)$$

For block 2, we choose an x-axis pointing downward. Doing so simplifies the solution since then the two blocks have the same a_x. Applying Newton's second law,

$$\Sigma F_x = m_2g - T = m_2a_x \qquad (2)$$

The tension in the cord, T, and the x-component of acceleration, a_x, are both unknown in Eqs. (1) and (2). We solve for T in Eq. (2) and substitute into Eq. (1),

$$T = m_2g - m_2a_x = m_2 (g - a_x)$$

$$m_2 (g - a_x) - m_1g \sin \theta - \mu_k m_1g \cos \theta = m_1a_x$$

Rearranging and solving for a_x yields

$$a_x = \frac{m_2g - m_1g (\sin \theta + \mu_k \cos \theta)}{m_1 + m_2} \qquad (3)$$

Substituting the known and given values,

$$a_x = \frac{2.2 \text{ kg} \times 9.8 \text{ m/s}^2 - 2.6 \text{ kg} \times 9.8 \text{ m/s}^2 \times (0.500 + 0.18 \times 0.866)}{2.6 \text{ kg} + 2.2 \text{ kg}}$$

$$= 1 \text{ m/s}^2$$

Block 2 has a distance of 2.00 m to travel starting from rest with a constant downward acceleration of 1.01 m/s^2.

$$\Delta x = \tfrac{1}{2}a_xt^2$$

The time to travel that distance is

$$t = \sqrt{\frac{2\Delta x}{a_x}} = \sqrt{\frac{2 \times 2.00 \text{ m}}{1 \text{ m/s}^2}} = 2 \text{ s}$$

Discussion There are several advantages to solving for a_x algebraically in Eq. (3) before substituting numerical values.

Dimensional analysis can easily be used to check for errors. In Eq. (3), the quantity in parentheses is dimensionless—the values of trigonometric functions are pure numbers as are coefficients of friction. Therefore the numerator is the sum of two quantities with dimensions of force, the denominator is the sum of two masses, and force divided by mass gives an acceleration.

Equation (3) also enables us to ask many "what if" questions. These both deepen our understanding of the physics behind the solution and serve as further checks on the validity of the solution. For one example, what if m_2 is much larger than m_1? In the numerator, the term m_2g would be much larger than m_1g ($\sin \theta + \mu_k \cos \theta$). In the denominator, m_2 would be much larger than m_1. Then Eq. (3) says that a_x would be approximately equal to g. If m_1 is negligibly small, then the tension in the cord would be small and block 2 would be almost in free fall.

Other such "what if" questions can be posed regarding the magnitudes of the masses, the angle of the incline, and the coefficient of friction.

What if the problem did not tell us the direction of the blocks' acceleration? We could figure it out by comparing the force with which gravity pulls down on block 2 (m_2g) with the component of the gravitational force pulling block 1 down the incline ($m_1g \sin \theta$). Whichever is greater "wins the tug-of-war." Once we know the direction of the acceleration, then we can determine the direction of the kinetic frictional force. If the blocks are not initially at rest, the kinetic frictional force opposes the direction of sliding, even though that may be opposite the direction of the acceleration.

Practice Problem 4.14   More fun with a pulley and an incline

Suppose that $m_1 = 3.8$ kg and $m_2 = 1.2$ kg and the coefficient of kinetic friction is 0.18. The blocks are released from rest and block 1 starts to slide. (a) Does block 1 slide up or down the incline? (b) In which direction does the kinetic frictional force act? (c) Find the acceleration of block 1.

4.7   RELATIVE VELOCITY

Suppose Wanda is walking down the aisle of a moving train (Fig. 4.28). Imagine asking, "How fast is she walking?" This question is not well defined. Do we mean her speed as measured by Tim, a passenger on the train, or her speed as measured by Greg, who is standing on the ground and looking into the train as it passes by? The answer to the question "How fast?" depends on the observer.

Figure 4.28b shows Wanda walking from one end of the car to the other during a time interval Δt. The displacement of Wanda as measured by Tim—her displacement relative to the train—is $\Delta \vec{\mathbf{r}}_{WT} = \vec{\mathbf{v}}_{WT} \Delta t$. During the same time interval, the *train's* displacement relative to the ground is $\Delta \vec{\mathbf{r}}_{TG} = \vec{\mathbf{v}}_{TG} \Delta t$. As measured by Greg, Wanda's displacement is partly due to her motion relative to the train and partly due to the motion of the train relative to the ground. Figure 4.28b shows that $\Delta \vec{\mathbf{r}}_{WT} + \Delta \vec{\mathbf{r}}_{TG} = \Delta \vec{\mathbf{r}}_{WG}$. Dividing by the time interval Δt gives the relationship between the three velocities:

$$\vec{\mathbf{v}}_{WT} + \vec{\mathbf{v}}_{TG} = \vec{\mathbf{v}}_{WG} \qquad (4-9)$$

(a)

$\Delta \vec{r}_{TG} = \vec{v}_{TG}\,\Delta t$ $\Delta \vec{r}_{WT} = \vec{v}_{WT}\,\Delta t$

$\Delta \vec{r}_{WG} = \vec{v}_{WG}\,\Delta t$

(b)

Figure 4.28 (a) Tim and Greg watch Wanda walk down the aisle of a train. Wanda's velocity with respect to Tim (or with respect to the train) is $\vec{v}_{WT}$; Tim's velocity with respect to Greg (or with respect to the ground) is $\vec{v}_{TG}$. (b) Wanda's displacement relative to the ground is the sum of her displacement relative to the train and the displacement of the train relative to the ground.

To be sure that you are adding the velocity vectors correctly, think of the subscripts as if they were fractions that get multiplied when the velocity vectors are added. In Eq. (4-9), $\dfrac{W}{T} \times \dfrac{T}{G} = \dfrac{W}{G}$, so the equation is correct.

Tips

Example 4.15

Flight from Denver to Chicago

 An airplane flies from Denver to Chicago (1770 km) in 4.4 h when no wind blows. On a day with a tail-wind, the plane makes the trip in 4.0 h. (a) What is the wind speed? (b) If a headwind blows with the same speed, how long does the trip take?

Strategy We assume the plane has the same *airspeed*—the same speed relative to the air—in both cases. Once the plane is up in the air, the behavior of the wings and the control surfaces depends on how fast the air is rushing by; the ground speed is irrelevant. But it is not irrelevant for the passengers, who are interested in a displacement relative to the ground.

Solution Let $\vec{v}_{PG}$ and $\vec{v}_{PA}$ represent the velocity of the plane relative to the ground and air, respectively. The wind velocity—the velocity of the air relative to the ground—can be written

$\vec{v}_{AG}$. Then $\vec{v}_{PA} + \vec{v}_{AG} = \vec{v}_{PG}$. The equation is correct since P/A × A/G = P/G. With no wind,

$$v_{PA} = v_{PG} = \frac{1770 \text{ km}}{4.4 \text{ h}} = 400 \text{ km/h}$$

(a) On the day with the tailwind,

$$v_{PG} = \frac{1770 \text{ km}}{4.0 \text{ h}} = 440 \text{ km/h}$$

We expect v_{PA} to be the same regardless of whether there is a wind or not. Since we are dealing with a tailwind, $\vec{v}_{PA}$ and $\vec{v}_{AG}$ are in the same direction, so their magnitudes add (Fig. 4.29a):

$$|\vec{v}_{PA}| + |\vec{v}_{AG}| = |\vec{v}_{PG}|$$

$$|\vec{v}_{AG}| = |\vec{v}_{PG}| - |\vec{v}_{PA}| = 440 \text{ km/h} - 400 \text{ km/h} = 40 \text{ km/h}$$

(b) With a 40 km/h headwind, $\vec{v}_{PA}$ and $\vec{v}_{AG}$ are in opposite directions (Fig. 4.29b); the ground speed of the plane is

$$|\vec{v}_{PG}| = |\vec{v}_{PA}| - |\vec{v}_{AG}| = 400 \text{ km/h} - 40 \text{ km/h} = 360 \text{ km/h}$$

continued on next page

Example 4.15 *continued*

Then the trip takes

$$\frac{1770 \text{ km}}{360 \text{ km/h}} = 4.9 \text{ h}$$

Discussion   If we choose the *x*-axis to the right, then the vector addition can be done by components: $v_{PAx} + v_{AGx} = v_{PGx}$. The vectors v_{PAx} and v_{PGx} are both positive. Vector v_{AGx} is positive in (a) and negative in (b).

Practice Problem 4.15   Rowing across the bay

Jamil, practicing to get on the crew team at school, rows a one-person racing shell to the north shore of the bay for a distance of 3.6 km to his friend's dock. On a day when the water is still (no current flowing), it takes him 20 min (1200 s) to reach his friend. On another day when a current flows southward, it takes him 30 min (1800 s) to row the same course. Ignore air resistance. (a) What is the speed of the current in m/s? (b) How long does it take Jamil to return home with that same current flowing?

Figure 4.29
Addition of velocity vectors in the case of (a) a tailwind and (b) a headwind. Lengths of vectors are not to scale.

Equation (4-9) applies to situations where the velocities are not all along the same line, as illustrated in Example 4.16.

Example 4.16

Rowing to the Dock

 Jack is enjoying a pleasant day at the ocean and is rowing his friend Kristina in a small boat in the harbor. A sudden squall comes up and the wind blows steadily along the beach, causing a current to travel from north to south at a uniform velocity of 0.61 m/s. Jack drops Kristina off at her sailboat and then proceeds 250 m directly west to the dock

(Fig. 4.30a). If Jack gets to the dock in a time of 4.2 min after leaving Kristina, in what direction did he head his rowboat? At what speed with respect to still water is Jack able to row?

Strategy   To keep the various velocities straight, we choose subscripts as follows: R = rowboat, W = water, and D = dock. The velocity of the current given is the velocity of the water relative to the dock ($\vec{v}_{WD}$). The velocity of the rowboat relative to the dock ($\vec{v}_{RD}$) is due west. The question asks for the magnitude and direction of the velocity of the rowboat relative to the water ($\vec{v}_{RW}$). The three velocities are related by

$$\vec{v}_{RW} + \vec{v}_{WD} = \vec{v}_{RD}$$

To compensate for the current's tendency to push the rowboat south of the dock, the rowboat is headed "against the current" at some angle to the north of west.

Given: $\vec{v}_{WD} = 0.61$ m/s; displacement from sailboat to dock = 250 m due west, time to travel from sailboat to dock = 4.2 min

To find: $\vec{v}_{RW}$

Solution   We start with a sketch of the vector addition (Fig. 4.30b). The velocity of the rowboat with respect to the water is at an angle θ north of west. With respect to the dock, Jack travels 250 m in 4.2 min, so his speed with respect to the dock is

$$v_{RD} = \frac{250 \text{ m}}{4.2 \text{ min} \times 60 \text{ s/min}} = 0.992 \text{ m/s}$$

(a)   (b)

Figure 4.30
(a) Seaside scene and (b) graphical addition of velocities.

continued on next page

Example 4.16 *continued*

We can find the angle at which the rowboat should be headed into the current by finding the tangent of the angle between $\vec{v}_{RW}$ and $\vec{v}_{RD}$.

$$\tan\theta = \frac{v_{WD}}{v_{RD}} = \frac{0.61 \text{ m/s}}{0.992 \text{ m/s}}$$

$$\theta = 32° \text{ N of W}$$

The speed at which Jack is able to row with respect to still water is the magnitude of $\vec{v}_{RW}$. Since $\vec{v}_{RD}$ and $\vec{v}_{WD}$ are perpendicular, the Pythagorean theorem yields

$$|\vec{v}_{RW}| = \sqrt{v_{WD}^2 + v_{RD}^2} = \sqrt{(0.61 \text{ m/s})^2 + (0.992 \text{ m/s})^2}$$

$$= 1.16 \text{ m/s}$$

Jack rows at a speed of 1.16 m/s with respect to the water.

Discussion   If Jack rowed directly west toward the dock, the ocean current would push him to the south, so he would end up traveling in a direction south of west. He has to compensate by heading against the ocean current by just the amount needed to have the velocity relative to the dock be directly west. By looking at the shore, Jack can see that he is headed straight west, but he knows that the prow of his boat is pointed to the north of west and that he is constantly fighting against the current.

Practice Problem 4.16   Paddling the canoe

A canoeist is attempting to paddle directly from east to west across a river that is flowing south at 5.0 km/h. If she can paddle the canoe so that it moves at 7.0 km/h with respect to the river, at what angle should she head the canoe to travel directly west as she intends?

At the beginning of this chapter, we asked what the path followed by the falling clam looks like as seen by the seagull flying through the air. With respect to a beachcomber on the ground, the clam has a constant horizontal velocity component given to it by the seagull and a changing vertical component of velocity due to the gravitational force (Fig. 4.31a); the clam moves in a parabolic path. If the gull continues to fly at the same horizontal velocity after dropping the clam, it is directly overhead when the clam hits the rock because they both have the same constant horizontal component of velocity with respect to Earth.

In its own reference frame—that is, using its own position as the origin of the coordinate axes—the seagull sees the clam drop straight down toward the ground while rocks and other objects on the beach are moving horizontally (Fig. 4.31b). The bird sees a collision between the horizontally moving rocks and the vertically falling clam. At any instant, if the velocity of the clam with respect to the seagull is $\vec{v}_{CS}$, the velocity of the clam with respect to the beach is $\vec{v}_{CB}$, and the velocity of the seagull with respect to the beach is $\vec{v}_{SB}$, then $\vec{v}_{CS} + \vec{v}_{SB} = \vec{v}_{CB}$.

(a)

(b)

Figure 4.31   (a) Beachcomber view: The bird flies along a horizontal line while the clam follows a parabolic path. (b) Bird's-eye view: The gull sees the ground and objects upon the ground moving while the clam drops straight down, landing upon the rocks just as the rocks move under the clam.

MASTER THE CONCEPTS

Summary

- Addition of vectors is commutative: $\vec{A} + \vec{B} = \vec{B} + \vec{A}$.

- Vectors are added graphically by moving each vector so that its tail is placed at the tip of the previous vector. The sum is drawn as a vector arrow from the tail of the first vector to the tip of the last.

- Vectors are subtracted graphically by drawing the vectors with their tails at the same point. Then the difference $\vec{A} - \vec{B}$ is a vector drawn from the tip of $\vec{B}$ to the tip of $\vec{A}$.

- Addition and subtraction of vectors algebraically using components is generally easier and more accurate than the graphical method. The graphical method is still a useful first step to get an approximate answer.

- To add vectors algebraically, add their components to find the components of the resultant: if $\vec{A} + \vec{B} = \vec{C}$, then $A_x + B_x = C_x$ and $A_y + B_y = C_y$.

- Components are found by drawing a right triangle with the vector as the hypotenuse and the other two sides parallel to the x- and y-axes. Trigonometric functions are used to find the magnitudes of the components. The correct algebraic sign must be supplied for each component. The same triangle can be used to find the magnitude and direction of a vector if its components are known.

- The x- and y-axes are chosen to make the problem easiest to solve. Any choice is valid as long as the two are perpendicular. In an equilibrium problem, choose x- and y-axes so that the fewest number of force vectors have to be resolved into both x- and y-components. In a nonequilibrium problem, if the direction of the acceleration is known, choose x- and y-axes so that the acceleration vector is parallel or antiparallel to one of the axes.

- The instantaneous velocity vector is tangent to the path of motion.

- The instantaneous acceleration vector does *not* have to be tangent to the path of motion, since velocities can change both in direction and in magnitude.

- The kinematic equations for an object moving in two dimensions with constant acceleration along the y-axis and $t_0 = 0$ are:

$$v_x = v_{0x} \tag{4-2}$$

$$x - x_0 = v_{0x}t \tag{4-3}$$

$$v_y = v_{0y} + a_y t \tag{4-4}$$

$$\Delta y = v_{av,y} t \tag{4-5}$$

$$v_{av,y} = \tfrac{1}{2}(v_{0y} + v_y) \tag{4-6}$$

$$y - y_0 = v_{0y}t + \tfrac{1}{2}a_y t^2 \tag{4-7}$$

$$v_y^2 - v_{0y}^2 = 2a_y \Delta y \tag{4-8}$$

- Problems involving Newton's second law—whether equilibrium or nonequilibrium—can be solved by treating the x- and y-components of the forces and the acceleration separately. The vector equation $\vec{F}_{net} = m\vec{a}$ is equivalent to the two scalar equations

$$\Sigma F_x = ma_x \quad \text{and} \quad \Sigma F_y = ma_y \tag{4-1}$$

- For a projectile or any object moving with constant acceleration in the $\pm y$-direction, the motion in the x- and y-directions can be treated separately. Since $a_x = 0$, v_x is constant. Thus, the motion is a superposition of constant velocity motion in the x-direction on constant acceleration motion in the y-direction.

- To relate the velocities of objects measured in different reference frames, use the equation

$$\vec{v}_{AC} = \vec{v}_{AB} + \vec{v}_{BC}$$

where $\vec{v}_{AC}$ represents the velocity of A relative to C, and so forth.

Highlighted Figures and Tables

F4.3 Graphical addition of displacement vectors (p. 105)

F4.4 The final position vector minus the initial position vector is the displacement $\Delta\vec{r}$ ($\vec{r}_f - \vec{r}_0 = \Delta\vec{r}$) (p. 105)

F4.7 Resolving vector $\vec{A}$ into components along two axes in two equivalent ways (p. 107)

F4.8 Addition of displacement vectors using components (p. 109)

F4.14 Forces acting on an object of mass m pushed up an inclined plane; resolving the weight into x- and y-components; and the free-body diagram (p. 114)

F4.16 Instantaneous velocity resolved into components along perpendicular axes (p. 116)

F4.19 Trajectory of a projectile (p. 120)

F4.21 Projectile motion: vertical and horizontal quantities versus time (p. 121)

F4.28 Relative velocity (p. 129)

CONCEPTUAL QUESTIONS

1. Why is the muzzle of a rifle not aimed directly at the center of the target?

2. Give an example of an object whose acceleration is (1) in the same direction as its velocity, (2) opposite its velocity, and (3) perpendicular to its velocity.

3. Does the monkey, coconut, and hunter demonstration still work if the gun is pointed *downward* at the monkey and coconut? Explain.

4. Can a body in free fall be in equilibrium? Explain.

5. Can the *x*-component of a vector ever be greater than the magnitude of the vector? Explain.

6. Explain how to combine two displacement vectors of magnitudes 3*L* and 4*L* so that the resultant vector has magnitude (a) *L*; (b) 7*L*; (c) 5*L*.

7. Compare the advantages and disadvantages of the two methods of vector addition (graphical and algebraic).

8. An object is subjected to two constant forces that are perpendicular to each other. Can a set of *x*- and *y*-axes be chosen so that the acceleration of the object has only one nonzero component? If so, how? Explain.

9. Is it possible for two identical projectiles with identical initial speeds, but with two different angles of elevation, to land in the same spot? Explain.

10. Pulleys and inclined planes are often called simple machines. Explain why in terms of the forces exerted with and without the use of the "machine."

11. Demonstrate with a vector diagram that a displacement is the same when measured in two different reference frames that are at rest with respect to each other.

12. If the trajectory is parabolic in one reference frame, is it always, never, or sometimes parabolic in another reference frame that moves at constant velocity with respect to the first reference frame? If the trajectory can be other than parabolic, what else can it be?

13. In free fall, neglecting air resistance, the *x*- and *y*-components of the motion are independent. Are the components independent when air resistance is significant? Why or why not? [*Hint:* The drag force has a magnitude determined by the *speed* of the projectile and a direction opposite to the velocity.]

MULTIPLE CHOICE QUESTIONS

1. Two balls, identical except for color, are projected horizontally from the roof of a tall building at the same instant. The initial speed of the red ball is twice the initial speed of the blue ball. Ignoring air resistance,

   (a) the red ball reaches the ground first.
   (b) the blue ball reaches the ground first.
   (c) both balls land at the same instant with different speeds.
   (d) both balls land at the same instant with the same speed.

2. A person stands on the roof garden of a tall building with one ball in each hand. If the red ball is thrown horizontally off the roof and the blue ball is simultaneously dropped over the edge, which statement is true?

   (a) Both balls hit the ground at the same time, but the red ball has a higher speed just before it strikes the ground.
   (b) The blue ball strikes the ground first, but with a lower speed than the red ball.
   (c) The red ball strikes the ground first with a higher speed than the blue ball.
   (d) Both balls hit the ground at the same time with the same speed.

3. A ball is thrown into the air and follows a parabolic trajectory. At the highest point in the trajectory,

   (a) the velocity is zero, but the acceleration is not zero.
   (b) both the velocity and the acceleration are zero.
   (c) the acceleration is zero, but the velocity is not zero.
   (d) neither the acceleration nor the velocity are zero.

4. A ball is thrown into the air and follows a parabolic trajectory. Point *A* is the highest point in the trajectory and point *B* is a point as the ball is falling back to the ground. Choose the correct relationship between the speeds and the magnitudes of the accelerations at the two points.

   (a) $v_A > v_B$ and $a_A = a_B$
   (b) $v_A < v_B$ and $a_A > a_B$
   (c) $v_A = v_B$ and $a_A \neq a_B$
   (d) $v_A < v_B$ and $a_A = a_B$

5. A small plane is flying directly west with an airspeed of 30 m/s. The plane flies into a region where the wind is blowing at 10 m/s directly to the southwest. If the pilot does not change the heading of the plane, the ground speed of the airplane

   (a) is greater than the airspeed.
   (b) is less than the airspeed.
   (c) is the same as the airspeed.
   (d) depends on the height of the plane.

6. A small plane is flying directly west with an airspeed of 30 m/s. The plane flies into a region where the wind is blowing at 10 m/s directly to the southwest. In that region, the pilot changes the directional heading to maintain the due west heading. The ground speed of the airplane

   (a) is greater than the airspeed.
   (b) is less than the airspeed.
   (c) is the same as the airspeed.
   (d) depends on the height of the plane.

7. A boy plans to cross a river in a rubber raft. The current flows from north to south at 1 m/s. In what direction should he head to get across the river to the east bank in the least amount of time if he is able to paddle the raft at 1.5 m/s in still water?

   (a) directly to the east
   (b) south of east
   (c) north of east
   (d) The three directions require the same time to cross the river.

8. A boy plans to paddle a rubber raft across a river to the east bank while the current flows downriver from north to south at 1 m/s. He is able to paddle the raft at 1.5 m/s in still water. In what direction should he head the raft to go straight east across the river to the opposite bank?

   (a) directly to the east   (b) south of east

   (c) north of east   (d) north

   (e) south

9. Vector $\vec{A}$ in Fig. 4.32 is equal to

   (a) $\vec{C} + \vec{D}$ (b) $\vec{C} + \vec{D} + \vec{E}$ (c) $\frac{1}{2}(\vec{D} - \vec{C})$

   (d) $\vec{B} + \vec{C}$ (e) $\vec{B} - \vec{D}$

10. Vector $\vec{E}$ in Fig. 4.32 is equal to

   (a) $-2(\vec{B} + \vec{C})$ (b) $-(\vec{C} + \vec{D})$ (c) $-2\vec{A}$

   (d) $-(\vec{A} + \vec{B} + \vec{C})$ (e) all of the above

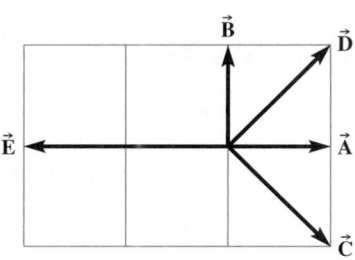

Figure 4.32   Multiple Choice Questions 9 and 10

PROBLEMS

Note: ⓒ indicates a combination conceptual/quantitative problem. Gold diamonds ✦, ✦✦ are used to indicate the increasing level of difficulty of each problem. Problem numbers appearing in blue, 9., denote problems that have a detailed solution available in the Student Solutions Manual. Some problems are *paired* by concept; their numbers are connected by a ruled box.

4.1 Addition and Subtraction of Vectors in Two Dimensions

1. Vectors $\vec{A}$ and $\vec{B}$ are drawn in Fig. 4.33. (a) Draw vectors $\vec{C}$ and $\vec{D}$, where $\vec{C} = \vec{A} + \vec{B}$ and $\vec{D} = \vec{A} - \vec{B}$. (b) Show that $\vec{A} + \vec{B} = \vec{B} + \vec{A}$ by graphical means.

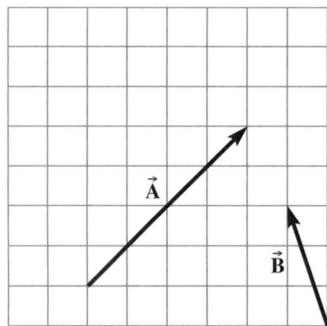

Figure 4.33   Problem 1

2. A scout troop is practicing its orienteering skills with map and compass. First they walk due east for 1.2 km. Next, they walk 45° west of north for 2.7 km. In what direction must they walk to go directly back to their starting point? How far will they have to walk? Use graph paper, ruler, and protractor to find a geometrical solution.

ⓒ 3. We have assumed that the displacement for a trip is equal to the sum of the displacements for each leg of the trip. Prove that this is true. [*Hint:* Imagine a trip that consists of n segments. The trip starts at position $\vec{r}_0$, proceeds to $\vec{r}_1$, then to $\vec{r}_2$, . . . , then to $\vec{r}_n - 1$, then finally to $\vec{r}_n$. Write an expression for each

displacement as the difference of two position vectors and then add them.]

4.2 Components of Vectors in Two Dimensions

4. The pilot of a small plane finds that the airport where he intended to land is fogged in. He flies 55 miles west to another airport to find that conditions there are too icy for him to land. He flies 25 miles south and is finally able to land at the third airport. (a) How far and in what direction must he fly the next day to go directly to his original destination? (b) How many extra miles beyond his original flight plan has he flown?

5. A vector is 20.0-m long and makes an angle of 60.0° counterclockwise from the y-axis (on the side of the $-x$-axis). What are the x- and y-components of this vector?

6. Vector $\vec{A}$ has magnitude 4.0 units; vector $\vec{B}$ has magnitude 6.0 units. The angle between $\vec{A}$ and $\vec{B}$ is 60.0°. What is the magnitude of $\vec{A} + \vec{B}$?

7. Vector $\vec{A}$ is directed along the positive y-axis and has magnitude $\sqrt{3.0}$ units. Vector $\vec{B}$ is directed along the negative x-axis and has magnitude 1.0 unit.

   (a) What are the magnitude and direction of $\vec{A} + \vec{B}$?

   (b) What are the magnitude and direction of $\vec{A} - \vec{B}$?

   (c) What are the x- and y-components of $\vec{B} - \vec{A}$?

8. Vector $\vec{a}$ has components $a_x = -3.0$ m/s² and $a_y = +4.0$ m/s². (a) What is the magnitude of $\vec{a}$? (b) What is the direction of $\vec{a}$? Give an angle with respect to one of the coordinate axes.

9. A roller coaster is towed up an incline at a steady speed of 0.50 m/s by a chain parallel to the surface of the incline. The slope is 3.0%, which means that the elevation increases by 3.0 m for every 100.0 m of distance along the surface. The mass of the roller coaster is 400.0 kg. Neglecting friction, what is the magnitude of the force exerted on the roller coaster by the chain?

10. Find the x- and y-components of the four vectors shown in Fig. 4.34.

11. In each of these, the x- and y-components of a vector are given. Find the magnitude and direction of the vector. (a) $x = -5.0$ cm,

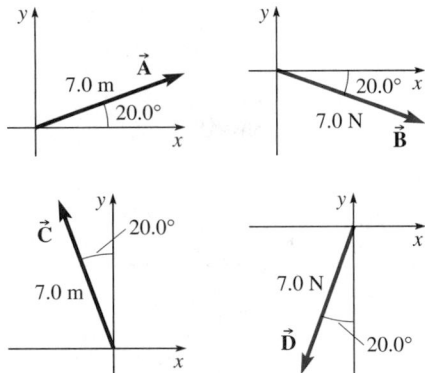

Figure 4.34 Problem 10

$y = +8.0$ cm. (b) $F_x = +120$ N, $F_y = -60.0$ N. (c) $v_x = -13.7$ m/s, $v_y = -8.8$ m/s. (d) $a_x = 2.3$ m/s^2, $a_y = 6.5$ cm/s^2.

12. Vector $\vec{b}$ has magnitude 7.1 and direction 14° below the +x-axis. Vector $\vec{c}$ has x-component $c_x = -1.8$ and y-component $c_y = -6.7$. Compute (a) the x- and y-components of $\vec{b}$; (b) the magnitude and direction of $\vec{c}$; (c) the magnitude and direction of $\vec{c} + \vec{b}$; (d) the x- and y-components of $\vec{c} - \vec{b}$.

13. Repeat Problem 2 using the component (algebraic) method.

14. A runner runs three-quarters of the way around a circular track, of radius 60.0 m, when she collides with another runner and trips. (a) How far had the runner traveled on the track before the collision? (b) What was the magnitude of the displacement of the runner from her starting position when the accident occurred?

15. When a particle undergoes diffusion, it follows a random path, bumping into obstacles here and there as it travels along. Suppose an electron is diffusing through a crystal. The black dots in Fig. 4.35 represent the regularly spaced atoms in the crystal. When the electron bumps into an atom, the direction of its motion is changed. If Fig. 4.35 shows a path taken by an electron, find the length of the path taken and the displacement of the electron.

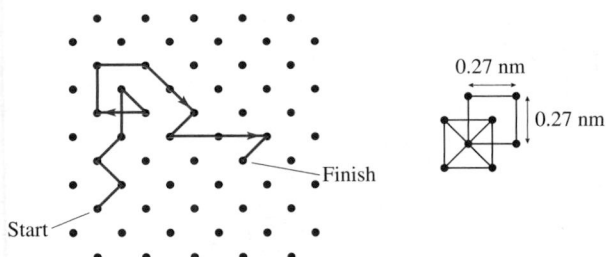

Figure 4.35 Problem 15: random path of diffusing electron

4.3 Equilibrium

16. (a) In Example 4.6, if the movers stop pushing on the safe, can the frictional force hold the safe in place without having it slide back down? (b) If not, what force needs to be applied to hold the safe in place?

17. Mechanical advantage is the ratio of the force required without the use of a simple machine to that needed when using the simple machine. Compare the force to lift an object to that needed to slide the same object up a frictionless incline and show that the mechanical advantage of the inclined plane is the length of the incline divided by the height of the incline (*d/h* in Fig. 4.14a).

18. A crow sits on a clothesline midway between two poles (Fig. 4.36). Each end of the rope makes an angle of θ below the horizontal where it connects to the pole. If the combined weight of the crow and the rope is W, what is the tension in the rope?

Figure 4.36 Problem 18

19. Tamar wants to cut down a large, dead poplar tree with her chain saw, but she does not want it to fall onto the nearby gazebo. Yoojin comes to help with a long rope. Yoojin, a physicist, suggests they tie the rope taut from the poplar to the oak tree and then pull *sideways* on the rope (Fig. 4.37). If the rope is 40.0 m long and Yoojin pulls sideways at the midpoint of the rope with a force of 360.0 N, causing a 2.00-m sideways displacement of the rope at its midpoint, what force will the rope exert on the poplar tree? Compare this with pulling the rope directly away from the poplar with a force of 360.0 N. [*Hint:* Until the poplar is cut through enough to start falling, the rope is in equilibrium.] Explain where the "extra" force comes from.

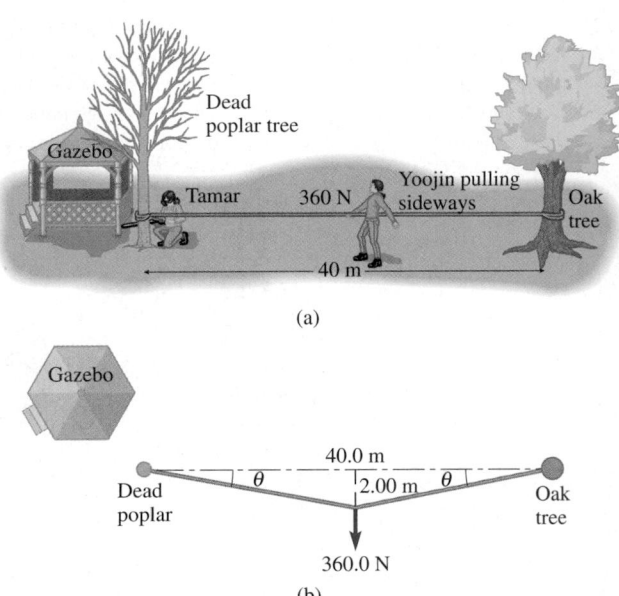

Figure 4.37 Problem 19: (a) side view of trees and rope and (b) overhead view of rope tied to trees.

20. A pulley is hung from the ceiling by a rope. A block of mass M is suspended by another rope that passes over the pulley and is attached to the wall (Fig. 4.38). The rope fastened to the wall makes a right angle with the wall. Neglect the masses of the rope and the pulley. Find (a) the tension in the rope from which the pulley hangs and (b) the angle θ that the rope makes with the ceiling.

Figure 4.38 Problem 20

21. A 2.0-kg ball tied to a string fixed to the ceiling is pulled to one side by a force $\vec{F}$ (Fig. 4.39). Just before the ball is released and allowed to swing back and forth, (a) how large is the force $\vec{F}$ that is holding the ball in position and (b) what is the tension in the string?

Figure 4.39 Problem 21

22. A young boy with a broken leg is undergoing traction (Fig. 4.40). (a) Find the magnitude of the total force of the traction apparatus applied to the leg, assuming the weight of the leg is 22 N and the weight hanging from the traction apparatus is also 22 N. (b) What is the horizontal component of the traction

force acting on the leg? (c) What is the magnitude of the force exerted on the femur by the lower leg?

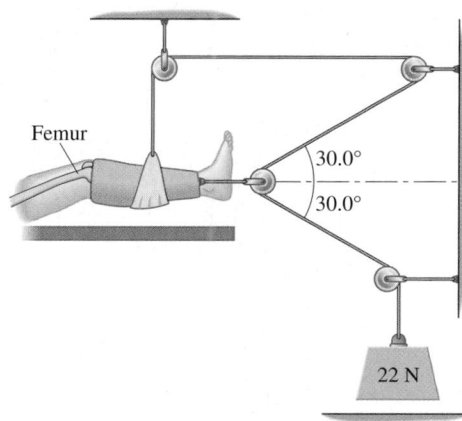

Figure 4.40 Problem 22

23. A large wrecking ball of mass m is resting against a wall. It hangs from the end of a cable that is attached at its upper end to a crane that is just touching the wall. The cable makes an angle of θ with the wall (Fig. 4.41). Neglecting friction between the ball and the wall, find the tension in the cable.

24. A crumb of bread is pulled upon by several ants from rival anthills. They exert the following forces: 0.06 N to the north, 0.08 N to the east, 0.02 N to the west, and $0.06\sqrt{2}$ N to the southeast (45° S of E). What additional force should be applied by a fifth ant to keep the crumb in equilibrium?

25. An 80.0-N crate of apples sits at rest on a ramp that runs from the ground to the bed of a truck. The ramp is inclined at 20.0° to the ground. (a) What is the normal force exerted on the crate by the ramp? (b) What is the static frictional force exerted on the crate by the ramp? (c) What is the minimum possible value of the coefficient of static friction? (d) The normal and frictional forces are perpendicular components of the contact force exerted on the crate by the ramp. Find the magnitude and direction of the contact force. Compare your answer to that of Practice Problem 2.6.

Figure 4.41 Problem 23

26. A 45-N lithograph is supported by two wires. One wire makes a 25° angle with the vertical and the other makes a 15° angle with the vertical. Find the tension in each wire.

27. A tire swing hangs at a 12° angle to the vertical when a stiff breeze is blowing. In terms of the tire's weight W, (a) what is the magnitude of the horizontal force exerted on the tire by the wind? (b) What is the tension in the rope supporting the tire? Ignore the weight of the rope.

4.4 Velocity and Acceleration

28. A car travels east at 96 km/h for 1 h. It then travels north at 128 km/h for 1 h. (a) What is the average speed for the trip? (b) What is the average velocity for the trip?

29. A car travels 3.2 km due east in 0.10 h, then 4.8 km due north in 0.15 h, and finally another 3.2 km due east in 0.10 h to reach its destination. The time lost in turning is negligible. What is the average velocity for the entire trip?

30. A car travels west at 108 km/h for 20.0 min. It then travels south at 90.0 km/h for 10.0 min. (a) What is the average speed for the trip? (b) What is the average velocity for the trip?

31. A car travels three-quarters of the way around a circle of radius 20.0 m in a time of 3.0 s at a constant speed. The initial velocity is west and the final velocity is south. (a) Find its average velocity for this trip. (b) What is the car's average acceleration during these 3.0 s? (c) Explain how a car moving at constant speed has a nonzero average acceleration.

32. At $t = 0$, an automobile traveling north begins to make a turn. It follows one-quarter of the arc of a circle of radius 10.0 m until, at $t = 1.60$ s, it is traveling east. The car does not alter its speed during the turn. Find (a) the car's speed, (b) the change in its velocity during the turn, and (c) its average acceleration during the turn.

33. Peggy drives from Cornwall to Atkins Glen in 45 min. Cornwall is 73.6 km from Illium in a direction 25° west of south. Atkins Glen is 27.2 km from Illium in a direction 15° south of west. Using Illium as your origin, (a) draw the initial and final position vectors, (b) find the displacement during the trip, and (c) find Peggy's average velocity for the trip.

34. A jetliner flies east for 600.0 km, then turns 30.0° toward the south and flies another 300.0 km. (a) How far is the plane from its starting point? (b) In what direction could the jetliner have flown directly to the same destination (in a straight-line path)? (c) If the jetliner flew at a constant speed of 400.0 km/h, how long did the trip take? (d) Moving at the same speed, how long would the direct flight have taken?

35. At the beginning of a 3.0-h plane trip, you are traveling due north at 192 km/h. At the end, you are traveling 240 km/h in the northwest direction (45° west of north). (a) Draw your initial and final velocity vectors. (b) Find the change in your velocity. (c) What is your average acceleration during the trip?

4.5 Motion of Projectiles

36. A ball is thrown from a point 1.0 m above the ground. The initial velocity is 19.6 m/s at an angle of 30.0° above the horizontal. (a) Find the maximum height of the ball above the ground. (b) Calculate the speed of the ball at the highest point in the trajectory.

37. The invention of the cannon in the fourteenth century made the catapult unnecessary and ended the safety of castle walls. Stone walls were no match for balls shot from cannons. Suppose a cannonball of mass 5.00 kg is launched from a height of 1.10 m, at an angle of elevation of 30.0° with an initial velocity of 50.0 m/s, toward a castle wall of height 30 m and located 215 m away from the cannon. (a) The range of a projectile is defined as the horizontal distance traveled when the projectile returns to its original height. Derive an equation for the range in terms of v_0, g, and angle of elevation θ. (b) What will be the range reached by the projectile, if it is not intercepted by the wall? (Let $g = 9.80$ m/s².) (c) If the cannonball travels far enough to hit the wall, find the height at which it strikes.

38. An arrow is shot into the air at an angle of 60.0° above the horizontal with a speed of 20.0 m/s. (a) What are the x- and y-components of the velocity of the arrow 3.0 s after it leaves the bowstring? (b) What are the x- and y-components of the displacement of the arrow during the 3.0-s interval?

39. Show that for a projectile launched at an angle of 45° the maximum height of the projectile is one quarter of the range (the distance traveled on flat ground).

40. Two ballplayers are practicing how to catch fly balls. One stands by home plate and hits balls into the air for the other, some distance away, to catch. A ball is hit with an initial velocity of 22.0 m/s at an angle of 60.0° above the horizontal. (a) How high will the ball rise? (b) How much time will elapse from the time the ball leaves the bat until it reaches the fielder? (c) At what distance from home plate will the fielder be when he catches the ball?

41. A ball is thrown horizontally off the edge of a cliff with an initial speed of 20.0 m/s. (a) How long does it take for the ball to fall to the ground 20.0 m below? (b) How long would it take for the ball to reach the ground if it were dropped from rest off the cliff edge? (c) How long would it take the ball to fall to the ground if it is thrown at an initial velocity of 20.0 m/s but 18° below the horizontal?

42. A marble is rolled so that it is projected horizontally off the top landing of a staircase. The initial speed of the marble is 3.0 m/s. Each step is 0.18 m high and 0.30 m wide. Which step does the marble strike first?

43. A suspension bridge is 60.0 m above the level base of a gorge. A stone is thrown or dropped from the bridge. Neglect air resistance. Let $g = 9.81$ m/s². (a) If you drop the stone, how long does it take for it to fall to the base of the gorge? (b) If you *throw* the stone straight down with a speed of 20.0 m/s, how long before it hits the ground? (c) If you throw the stone with a velocity of 20.0 m/s at 30.0° above the horizontal, how far from the point directly below the bridge will it hit the level ground?

44. A beanbag is thrown horizontally from a dorm room window a height h above the ground. It hits the ground a horizontal distance h (the *same* distance h) from the dorm directly below the window from which it was thrown. Neglecting air resistance, find the direction of the beanbag's velocity just before impact.

45. From the edge of the rooftop of a building, a boy throws a stone at an angle 25.0° above the horizontal. The stone hits the ground 4.20 s later, 105 m away from the base of the building. (Ignore air resistance.) (a) For the stone's path through the air, sketch graphs of x, y, v_x, and v_y as functions of time. These need to be only *qualitatively* correct—you

need not put numbers on the axes. (b) Find the initial velocity of the stone. (c) Find the initial height h from which the stone was thrown. (d) Find the maximum height H reached by the stone.

46. The citizens of Paris were terrified during World War I when they were suddenly bombarded with shells fired from a long-range gun known as Big Bertha. The barrel of the gun was 36.6 m long and it had a muzzle speed of 1.46 km/s. When the gun's angle of elevation was set to 55°, what would be the range? For the purposes of solving this problem, neglect air resistance. (The actual range at this elevation was 121 km; air resistance cannot be neglected for the high muzzle speed of the shells.)

✦ 47. A projectile is launched from the origin with initial speed v_0 at an angle θ above the horizontal. Show that the equation of the trajectory followed by the projectile is

$$y = \left(\frac{v_{0y}}{v_{0x}}\right)x + \left(\frac{-g}{2v_{0x}^2}\right)x^2$$

4.6 Other Examples of Constant Acceleration

48. A particle experiences a constant acceleration that is south at 2.50 m/s^2. At $t = 0$, its velocity is 40.0 m/s east. What is its velocity at $t = 8.00$ s?

49. A particle experiences a constant acceleration that is north at 100 m/s^2. At $t = 0$, its velocity vector is 60 m/s east. At what time will the magnitude of the velocity be 100 m/s?

50. In the physics laboratory, a glider is released from rest on a frictionless air track inclined at an angle similar to the one shown in Fig. 3.34. If the glider has gained a speed of 25.0 cm/s in traveling 50.0 cm from the starting point, what was the angle of inclination of the track?

51. A 10.0-kg block is released from rest on a frictionless track inclined at an angle of 45°. (a) What is the net force on the block after it is released? (b) What is the acceleration of the block? (c) If the block is released from rest, how long will it take for the block to attain a speed of 10.0 m/s?

52. A 6.0-kg block, starting from rest, slides down a frictionless incline of length 2.0 m. When it arrives at the bottom of the incline, its speed is v_f. At what distance from the top of the incline is the speed of the block $\frac{1}{2}v_f$?

53. A crate of books is to be put on a truck by rolling it up an incline of angle θ using a dolly. The total mass of the crate and the dolly is m. Assume that rolling the dolly up the incline is the same as sliding it up a frictionless surface. (a) What is the magnitude of the *horizontal* force that must be applied just to hold the crate in place on the incline? (b) What horizontal force must be applied to roll the crate up at constant speed? (c) In order to start the dolly moving, it must be accelerated from rest. What horizontal force must be applied to give the crate an acceleration up the incline of magnitude a?

✦ 54. A toy cart of mass m_1 moves on frictionless wheels as it is pulled by a string under tension T. A block of mass m_2 rests on top of the cart. The coefficient of friction between the cart and the block is μ. Find the maximum tension T that will not cause the block to slide on the cart if the cart rolls on (a) a horizontal surface; (b) up a ramp of angle θ above the horizontal. In both cases, the string is parallel to the surface on which the cart rolls.

55. A solar sailplane is going from Earth to Mars (Fig. 4.42). Its sail is oriented to give a solar radiation force of 8.00×10^2 N. The gravitational force due to the Sun is 173 N and the gravitational force due to the Earth and Moon is 1.00×10^2 N. All forces are in the plane formed by Earth, Sun, and sailplane. The mass of the sailplane is 14,500 kg. (a) What is the net force (magnitude and direction) acting on the sailplane? (b) What is the acceleration of the sailplane?

Figure 4.42 Problem 55

✦ 56. In a playground, two slides have different angles of incline θ_1 and θ_2 ($\theta_2 > \theta_1$). A child slides down the first at constant speed; on the second, his acceleration down the slide is a. Assume the coefficient of kinetic friction is the same for both slides. (a) Find a in terms of θ_1, θ_2, and g. (b) Find the numerical value of a for $\theta_1 = 45°$ and $\theta_2 = 61°$.

✦ 57. A 15-kg crate starts at rest at the top of a 60.0° incline. The coefficients of friction are $\mu_s = 0.40$ and $\mu_k = 0.30$. The crate is connected to a hanging 8.0-kg box by an ideal rope and pulley. (a) As the crate slides down the incline, what is the tension in the rope? (b) How long does it take the crate to slide 2.00 m down the incline? (c) To push the crate back up the incline at constant speed, with what force P should Pauline push on the crate (parallel to the incline)? (d) What is the smallest mass that you could substitute for the 8.0-kg box to keep the crate from sliding down the incline?

58. An accelerometer—a device to measure acceleration—can be as simple as a small pendulum hanging in the cockpit. Suppose you are flying a small plane in a straight horizontal line and your accelerometer hangs 12° behind the vertical (Fig. 4.43). What is your acceleration?

Figure 4.43 Problem 58

4.7 Relative Velocity

59. Two cars are driving toward each other on a straight, flat Kansas road. The Jeep Wrangler is traveling at 82 km/h north and the Ford Taurus is traveling at 48 km/h south, both measured relative to the road. What is the velocity of the Jeep relative to an observer in the Ford?

60. Two cars are driving toward each other on a straight and level road in Alaska. The BMW is traveling at 100.0 km/h north and the VW is traveling at 42 km/h south, both velocities measured relative to the road. At a certain instant, the distance between the cars is 10.0 km. Approximately how long will it take from that instant for the two cars to meet? [*Hint:* Consider a reference frame in which one of the cars is at rest.]

61. A Nile cruise ship takes 20.8 h to go upstream from Luxor to Aswan, a distance of 208 km, and 19.2 h to make the return trip downstream. Assuming the ship's speed relative to the water is the same in both cases, calculate the speed of the current in the Nile.

62. An airplane has a velocity relative to the ground of 210 m/s toward the east. The pilot measures his airspeed (the speed of the plane relative to the air) to be 160 m/s. What is the minimum wind velocity possible?

63. A boat that can travel at 4.0 km/h in still water crosses a river with a current of 1.8 km/h. At what angle must the boat be pointed upstream to travel straight across the river? In other words, in what direction is the velocity of the boat relative to the water?

64. A boy is attempting to swim directly across a river; he is able to swim at a speed of 0.500 m/s relative to the water. The river is 25.0 m wide and the boy ends up at 50.0 m downstream from his starting point. (a) How fast is the current flowing in the river? (b) What is the speed of the boy relative to a friend standing on the riverbank?

65. An aircraft has to fly between two cities, one of which is 600.0 km north of the other. The pilot starts from the southern city and encounters a steady 100.0 km/h wind that blows from the northeast. The plane has a cruising speed of 300.0 km/h in still air. (a) In what direction (relative to east) must the pilot head her plane? (b) How long does the flight take?

66. A Stanley Steamer automobile travels north at 40 km/h and a Pierce Arrow automobile travels east at 50 km/h. Relative to an observer riding in the Stanley Steamer, what are the *x*- and *y*-components of the velocity of the Pierce Arrow car?

67. A person climbs from a Paris metro station to the street level by walking up a stalled escalator in 94 s. It takes 66 s to ride the same distance when standing on the escalator when it is operating normally. How long would it take for him to climb from the station to the street by walking up the moving escalator?

68. Sheena can row a boat at 3.00 mi/h in still water. She needs to cross a river that is 1.20 mi wide with a current flowing at 1.60 mi/h. Not having her calculator ready, she guesses that to go straight across, she should head 60.0° upstream. (a) What is her speed with respect to the bank? (b) How long does it take her to cross the river? (c) How far upstream or downstream from her starting point will she reach the opposite bank? (d) In order to go straight across, what angle upstream should she have headed?

69. A dolphin wants to swim directly back to its home bay, which is 0.80 km due west. It can swim at a speed of 4.00 m/s relative to the water, but a uniform water current flows with speed 2.83 m/s in the southeast direction. (a) What direction should the dolphin head? (b) How long does it take the dolphin to swim the 0.80-km distance home?

70. A pilot starting from Athens, New York, wishes to fly to Sparta, New York, which is 320 km from Athens in the direction 20.0° N of E. The pilot heads directly for Sparta and flies at an airspeed of 160 km/h. After flying for 2.0 h, the pilot expects to be at Sparta, but instead he finds himself 20 km due west of Sparta. He has forgotten to correct for the wind. (a) What is the velocity of the plane relative to the air? (b) Find the velocity (magnitude and direction) of the plane relative to the ground. (c) Find the wind speed and direction.

COMPREHENSIVE PROBLEMS

71. A sailboat sails from Marblehead Harbor directly east for 45 nautical miles, then south for 20.0 nautical miles, returns to an easterly heading for 30.0 nautical miles, and sails north for 15.0 nautical miles, then west for 63 nautical miles. At that time the boat becomes becalmed and the auxiliary engine fails to start. The crew decides to notify the Coast Guard of their position. Using graph paper, ruler, and protractor, find the sailboat's displacement from the harbor. Then add the displacement vectors using the component method to obtain a more accurate answer.

72. A vector has magnitude 2.00 units and its *y*-component is +1.00 unit. What is the *x*-component of this vector? Is there more than one possibility? Explain.

73. A bike rider follows a trail 100.0 m south, then 200.0 m east, and finally 75.0 m northeast (45.0° north of east). (a) What total distance has he traveled from his starting point? (b) He decides to ride off across a field directly back to his starting point. How far must he travel to get back and in what direction?

74. A barge is hauled along a straight-line section of canal by two horses harnessed to tow ropes and walking along the tow paths on either side of the canal. Each horse pulls with a force of 560 N at an angle of 15° with the centerline of the canal. Find the resultant force on the barge.

75. A 200.0-N sign is suspended from a horizontal strut of negligible weight (Fig. 4.44). The force exerted on the strut by the

Figure 4.44 Problem 75

wall is horizontal. Draw a free-body diagram to show the forces acting on the strut. Find the tension T in the diagonal cable supporting the strut.

✦76. Peter is collecting paving stones from a quarry. He harnesses two dogs, Sandy and Rufus, in tandem to the loaded cart. Sandy pulls with force $\vec{F}$ at a 15° angle to the north of east; Rufus pulls with 1.5 times the force of Sandy and at an angle of 30.0° south of east. Use a ruler and protractor to draw the force vectors to scale (choose a simple scale, such as 2.0 cm ↔ F). Find the resultant net force vector graphically. Measure its length and find the magnitude of the net force from the scale used and the direction with the protractor. Will the cart stay on the road that runs directly west to east?

77. The seagull discussed in this chapter is flying horizontally 8.00 m above the ground at 6.00 m/s. Assume $g = 9.80$ m/s^2. Ignoring air resistance, (a) what is the horizontal distance to the rocks at the moment that the seagull should let go of the clam? (b) With what speed relative to the rocks does the clam smash into the rocks? (c) With what speed relative to the seagull does the clam smash into the rocks?

78. An airplane is traveling from New York to Paris, a distance of 5.80×10^3 km. Neglect the curvature of the Earth. (a) If the cruising speed of the airplane is 350.0 km/h, how much time will it take for the airplane to make the round-trip on a calm day? (b) If a steady wind blows from New York to Paris at 60.0 km/h, how much time will the round-trip take? (c) How much time will it take if there is a crosswind of 60.0 km/h?

79. A helicopter is flying horizontally at 8.0 m/s and an altitude of 18 m when a package of emergency medical supplies is ejected horizontally backward with a speed of 12 m/s *relative to the helicopter*. Ignoring air resistance, what is the horizontal distance between the package and the helicopter when the package hits the ground?

80. A particle has a constant acceleration of 5.0 m/s^2 to the east. At time $t = 0$, it is 2.0 m east of the origin and its velocity is 20 m/s north. What are the components of its position vector at $t = 2.0$ s?

🄒 81. A toboggan is sliding down a snowy slope. Table 4.1 shows the speed of the toboggan at various times during its trip. (a) Make a graph of the speed as a function of time. (b) Judging by the graph, is it plausible that the toboggan's acceleration is constant? If so, what is the acceleration? (c) Ignoring friction, what is the angle of incline of the slope? (d) If friction is significant, is the angle of incline larger or smaller than that found in (c)? Explain.

82. The range R of a projectile is defined as the magnitude of the horizontal displacement of the projectile *when it returns to its original altitude*. In other words, the range is the distance the projectile will travel on flat ground. A projectile is launched at $t = 0$ with initial speed v_0 at an angle θ above the horizontal. (a) Find the time t at which the projectile returns to its original altitude. (b) Show that the range is

$$R = \frac{v_0^2 \sin 2\theta}{g}$$

[*Hint:* Use the trigonometric identity $\sin 2\theta = 2 \sin \theta \cos \theta$.] (c) What value of θ gives the maximum range? What is this maximum range?

83. A projectile is launched at $t = 0$ with initial speed v_0 at an angle θ above the horizontal. (a) What are v_x and v_y at the projectile's highest point? (b) Find the time t at which the projectile reaches its maximum height. (c) Show that the maximum height H of the projectile is

$$H = \frac{(v_0 \sin \theta)^2}{2g}$$

✦✦84. An airplane of mass 2800 kg has just lifted off the runway. It is gaining altitude at a constant 2.3 m/s while the horizontal component of its velocity is increasing at a rate of 0.86 m/s^2. Assume $g = 9.81$ m/s^2. (a) Find the direction of the force exerted on the airplane by the air. (b) Find the horizontal and vertical components of the plane's acceleration if the force due to the air has the same magnitude but has a direction 2.0° closer to the vertical than its direction in part (a).

85. A locust jumps at an angle of 55.0° and lands 0.800 m from where it jumped. (a) What is the maximum height of the locust during its jump? Ignore air resistance and assume $g = 9.80$ m/s^2. (b) If it jumps with the same initial speed at an angle of 45.0°, would the maximum height be larger or smaller? (c) What about the range? (d) Calculate the maximum height and range for this angle.

✦86. A spotter plane sees a school of tuna swimming at a steady 5.00 km/h northwest. The pilot informs a fishing trawler; which is just then 100.0 km due south of the fish. The trawler sails along a straight-line course and intercepts the tuna after 4.0 h. How fast did the trawler move? [*Hint:* First find the velocity of the trawler relative to the tuna.]

Table 4.1	
Problem 81	
Time Elapsed, t (s)	**Speed of Toboggan, v (m/s)**
0	0
1.14	2.8
1.62	3.9
2.29	5.6
2.80	6.8

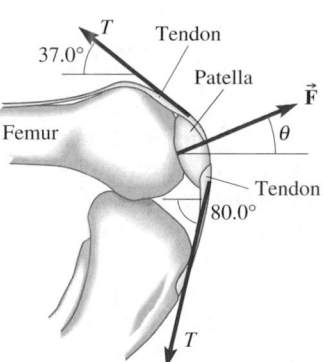

Figure 4.45 Problem 88

87. A baseball batter hits a long fly ball that rises to a height of 44 m. An outfielder on the opposing team can run at 7.6 m/s. What is the farthest the fielder can be from where the ball will land so that it is possible for him to catch the ball?

88. Figure 4.45 shows the quadriceps and the patellar tendons attached to the patella (the kneecap). If the tension T in each tendon is 1.30 kN, what is (a) the magnitude and (b) the direction of the contact force $\vec{F}$ exerted on the patella by the femur?

✦89. One of the tricky things about learning to sail is distinguishing the true wind from the apparent wind. When you are on a sailboat and you feel the wind on your face, you are experiencing the *apparent wind*—the motion of the air relative to you. The true wind is the speed and direction of the air relative to the water while the apparent wind is the speed and direction of the air relative to the *sailboat*. Fig. 4.46 shows three different directions for the true wind along with one possible sail orientation as indicated by the position of the boom attached to the mast. (a) In each case, draw a vector diagram to establish the magnitude and direction of the apparent wind. (b) In which of the three cases is the apparent wind speed greater than the true wind speed? (Assume that the speed of the boat relative to the water is less than the true wind speed.) (c) In which of the three cases is the direction of the apparent wind direction forward of the true wind? ("Forward" means coming from a direction more nearly straight ahead. For example, (1) is forward of (2) which is forward of (3).)

Figure 4.46 Problem 89

90. Figure 4.47 shows an elastic cord attached to two back teeth and stretched across a front tooth. The purpose of this arrangement is to apply a force $\vec{F}$ to the front tooth. (The figure has been simplified by running the cord straight from the front tooth to the back teeth.) If the tension in the cord is 1.2 N, what are the magnitude and direction of the force $\vec{F}$ applied to the front tooth?

Figure 4.47 Problem 90

91. Figure 4.48 shows a student's head bent over her physics book. The head weighs 50.0 N and is supported by the muscle force $\vec{F}_m$ exerted by the neck extensor muscles and by the contact force $\vec{F}_c$ exerted at the atlanto-occipital joint. Given that the magnitude of $\vec{F}_m$ is 60.0 N and is directed 35° below the horizontal, find (a) the magnitude and (b) the direction of $\vec{F}_c$.

Figure 4.48 Problem 91

✦ 92. When salmon head upstream to spawn, they often must make their way up a waterfall. If the water is not moving too fast, the salmon can swim right up through the falling water. If the water is falling with too great a speed, the salmon jump out of the water to get to a place in the waterfall where the water is not falling so fast. When humans build dams that interrupt the usual route followed by the salmon, artificial fish ladders must also be built to allow the salmon to get back uphill to the spawning area. These fish ladders consist of a series of small waterfalls with still pools of water in between them (Fig. 4.49). Assume that the water is at rest in the pools at the top and bottom of one "rung" of the fish ladder, that water falls straight down from one pool to the next, and that salmon can swim at 5.0 m/s with respect to the water. (a) What is the maximum height of a waterfall up which the salmon can swim without having to jump? (b) If a waterfall is 1.5 m high, how high must the salmon jump to get to water through which it can swim? Assume that they jump straight up. (c) What initial speed must a salmon have to jump the height found in part (b)? (d) For a 1.0-m-high waterfall, how fast will the salmon be swimming with respect to the ground when it starts swimming up the waterfall?

✦✦ 93. Jesse James and his gang are planning to rob the train carrying the payroll for the Lost Gulch mine. A Wells Fargo guard on the train spots Jesse astride his horse at a perpendicular distance of 0.300 km from the train tracks. The train is moving at 25.0 m/s on a straight track. The guard wants to frighten Jesse and the desperadoes away by shooting a hole in Jesse's ten-gallon hat, which happens to be at the same height as the guard's gun (Fig. 4.50). The muzzle velocity of the guard's gun is 0.350 km/s. Ignore air resistance in parts (a)–(c). (a) Should he aim the gun directly at Jesse's hat, in front of the hat, or behind the hat? Explain. (b) Unsure of the correct angle to aim the gun, the guard decides to aim his gun at a right angle to the track and fire the bullet a little before the time when he will be directly opposite to Jesse. At what distance before the point where the guard is directly opposite Jesse James should the guard fire? (c) Does the guard need to worry about the force of gravity on the bullet? If so, how far above the hat should he aim? (d) Given that the terminal speed of the bullet is about 120 m/s, should the guard fire earlier or later than the answer found in part (b)?

Figure 4.50 Problem 93 (figure not to scale)

Figure 4.49 Problem 92. A fish ladder

ANSWERS TO PRACTICE PROBLEMS

4.1 (a) larger; (b) yes, $F_S > F_B$

4.2 $A_x = +16$ km; $A_y = -8.2$ km; $B_x = +17$ km; $B_y = 0$; $C_x = -11$ km; $C_y = +47$ km

4.3 (a) 1.01 kN; (b) 2.45 kN

4.4 3.1 kN

4.5 (a) $F_x = 49.1$ N; $F_y = 2.9$ N; (b) $F = 49.2$ N; (c) 3.4° above the horizontal

4.6 $f_k = 110$ N; $W = 230$ N

4.7 3.19 knots directed 56.9° north of east

4.8 2

4.9 $y_{max} = \dfrac{v_0^2}{4g}$

4.10 $\vec{v}_2 - \vec{v}_1$ stays the same, since both velocities change at the same rate—both have the same acceleration. The difference in position ($\vec{r}_2 - \vec{r}_1$) changes, since their velocities have different y-components and, therefore, y_2 changes at a different rate from y_1. $x_2 - x_1$ stays constant, though, since both have the same constant horizontal velocity component.

4.11 $v_{fx} = 500.0$ m/s; $v_{fy} = -19.8$ m/s; the bullet enters the water at an angle of 2.27° below the horizontal.

4.12 54 N

4.13 $a_x = 3.2$ m/s^2 and $v_x = 3.6$ m/s

4.14 (a) down the incline; (b) up the incline; (c) 0.2 m/s^2 down the incline

4.15 (a) 1.0 m/s; (b) 15 min

4.16 46° north of west

Circular Motion

Alekos, a Greek goatherd, notices a wolf sneaking up toward his goats. After a moment of panic, he picks up a stone and puts it in his sling. He begins to whirl the stone faster and faster in a circle above his head, as the wolf inches closer and closer. To save the goats, Alekos must release the stone at precisely the right point in its circular path so it hits the wolf and drives it away. At what point should he release the stone?

Concepts & Skills to Review

- gravitational forces (Section 2.5)
- Newton's second law: force and acceleration (Section 3.4)
- velocity and acceleration (Section 4.4)

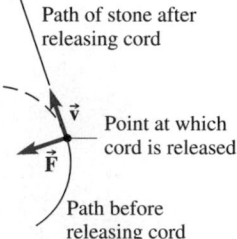

Making The Connection:
whirling a stone in a sling

Path of stone after
releasing cord

$\vec{v}$

Point at which
cord is released

$\vec{F}$

Path before
releasing cord

Figure 5.1   Path of a stone before and after a cord is released.

The abbreviation rpm means "revolutions per minute."

5.1   DESCRIPTION OF UNIFORM CIRCULAR MOTION

It might seem overly dramatic to introduce circular motion in life and death terms, but your life depends every day on *someone's* understanding of circular motion. Ask someone to name the most important machine ever invented by humans and you are likely to get *the wheel* as a response. Rotating objects are so essential to modern—and even not-so-modern—technology that we barely notice them. Examples include wheels on cars, bicycles, trains, and lawnmowers; propellers on airplanes and helicopters; CDs and DVDs; hard and floppy computer disks; the gears and hands of an analog watch; amusement park rides and centrifuges—the list is endless.

As Alekos whirls the stone, it travels in a circle. A nonzero net force must act on any object moving in a circle since its velocity is changing in direction (and, perhaps, also in magnitude). The stone rests in a leather pouch attached to two cords, which Alekos holds in his hand. The stone is held in its circular path by an inward force due to the pouch. If Alekos releases one of the cords, the pouch no longer holds the stone in its circular path; what is the subsequent path of the stone?

Initially, let us neglect the weight of the stone. Once the force due to the sling stops acting on the stone, the stone's velocity stops changing direction. The stone then continues to move in the direction of its velocity at the instant the force due to the sling stops acting. Newton's first law applies now: a body with no external forces acting on it moves at a constant speed along a straight line. Since the velocity is tangent to the stone's circular path, the stone initially heads off along a straight line tangent, to the circle (Fig. 5.1).

A satellite orbiting Earth may also travel in a circular path. Instead of being held in its path by the tension in a cord, it is held by the gravitational force on the satellite due to Earth's gravitational field. One of the main goals of this chapter is to find out what net force pulling inward is required to keep an object moving in a circle. The net force required turns out to depend both on the radius of the circular path and the speed at which the object moves.

To describe the circular motion of the stone or the satellite, we could use the familiar definitions of displacement, velocity, and acceleration. But much of the circular motion around us occurs in the rotation of a rigid object. A **rigid body** is one for which the distance between any two points of the body remains the same when the body is translated or rotated. When such an object rotates, every point on the object moves in a circular path. The radius of the path for any point is the distance between that point and the axis of rotation. When a compact disk spins inside a CD player, each point on the CD moves through a different distance and has a different velocity and acceleration. The velocity and acceleration of a given point keep changing direction as the CD spins. It would be clumsy to describe the rotation of the CD by talking about the motion of arbitrary points on it. However, some quantities are *the same* for every point on the CD. It is much simpler, for instance, to say "the CD spins at 210 rpm" instead of saying "a point 6.0 cm from the rotation axis of the CD is moving at 1.3 m/s."

To simplify the description of circular motion, we concentrate on angles instead of distances. If a CD spins through 1/4 of a turn, every point moves through the same angle (90°), but different points move different linear distances. On the CD shown in Fig. 5.2a, point 1 near the axis of rotation moves through a smaller distance than point 4 on the circumference. For this reason we define a set of variables that are analogous to displacement, velocity, and acceleration, but use angular measure instead of linear distance. Instead of displacement, we speak of **angular displacement** $\Delta\theta$, the angle through which the CD turns. A point on the CD moves along the circumference of a circle. As the point moves from the angular position θ_1 to angular position θ_2, a radial line drawn between the center of the circle and that point sweeps out an angle $\Delta\theta = \theta_2 - \theta_1$, which is the angular displacement of the CD during that time interval (Fig. 5.2b).

Definition of angular displacement

$$\Delta\theta = \theta_2 - \theta_1 \tag{5-1}$$

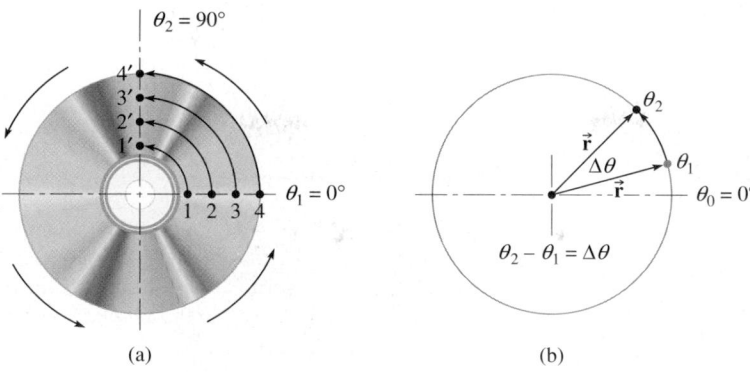

(a)   (b)

Figure 5.2 (a) A CD rotates through $\frac{1}{4}$ turn; points 1, 2, 3, and 4 travel through the same angle but different distances to reach their new positions, marked 1′, 2′, 3′, and 4′, respectively. (b) Angular positions such as θ_1 and θ_2 are measured from a reference axis (here marked $\theta_0 = 0°$).

Just as with displacement, angular displacement tells us in what direction the rotation occurred. The usual convention is that a positive angular displacement represents counterclockwise rotation and a negative angular displacement represents clockwise rotation. Counterclockwise and clockwise are only well defined for a particular viewing direction; counterclockwise rotation viewed from above is clockwise when viewed from below.

The **average angular velocity** ω_{av} is the average rate of change of the angular displacement.

Definition of average angular velocity

$$\omega_{av} = \frac{\Delta\theta}{\Delta t} \tag{5-2}$$

If we let the time interval Δt become shorter and shorter, we are averaging over smaller and smaller time intervals. In the limit $\Delta t \to 0$, ω_{av} becomes the **instantaneous angular velocity** ω.

Definition of instantaneous angular velocity

$$\omega = \lim_{\Delta t \to 0} \frac{\Delta\theta}{\Delta t} \tag{5-3}$$

The angular velocity also indicates—through its algebraic sign—in what direction the CD is spinning. Since angular displacements can be measured in degrees or radians, angular velocities have units such as degrees/second, radians/second, degrees/day, and the like. The SI unit is rad/s, which stands for radians per second.

You may be most familiar with measuring angles in degrees, but in many situations the most convenient measure is the *radian*. One such situation is when we relate the angular displacement or angular velocity of a rotating object with the distance traveled by, or the speed of, some point on the object.

In Fig. 5.3, an angle θ between two radii of a circle define an arc of length s. We say that θ is the angle *subtended* by the arc. The angle θ in radians is defined as

$$\theta \text{ (in radians)} = \frac{s}{r} \tag{5-4}$$

where r is the radius of the circle. Since an angle in radians is defined by the ratio of two lengths, it is dimensionless (a pure number). We use the term radians, abbreviated "rad" to keep track of the angular measure used. Since "rad" is not a physical unit like meters or kilograms, it does not have to balance in Eq. (5-4). For the same reason, we can drop "rad" whenever there is no chance of being misunderstood. We may write $\omega = 23 \text{ s}^{-1}$ as long as context makes it clear that we mean 23 radians per second.

Since the arc length for an angle of 360° is the circumference of the circle, the radian measure of an angle of 360° is

$$\theta = \frac{s}{r} = \frac{2\pi r}{r} = 2\pi \text{ rad}$$

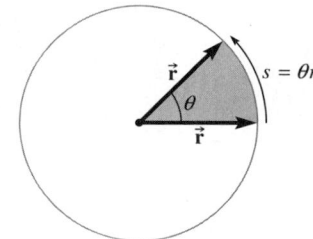

Figure 5.3 Definition of the radian: angle θ in radians is defined by the arc length s divided by the radius r.

Therefore, the conversion factor between degrees and radians is

$$360° = 2\pi \text{ rad}$$

Example 5.1

Angular Speed of Earth

Earth is rotating about its axis. What is its angular speed in rad/s? (The question asks for angular *speed*, so we do not have to worry about the direction of rotation.)

Strategy The Earth's angular velocity is constant, or nearly so. Therefore, we can calculate the average angular velocity for any convenient time interval and, in turn, the Earth's instantaneous angular speed $|\omega|$.

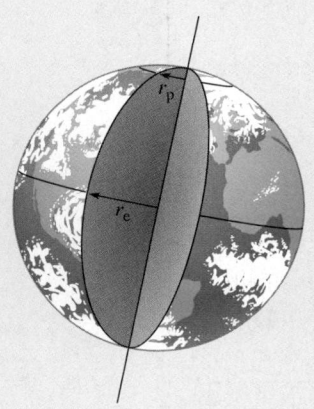

Solution It takes the Earth 1 day to complete one rotation, during which the angular displacement is 2π rad. More formally, during a time interval $\Delta t = 1$ day, the angular displacement of the Earth is $\Delta\theta = 2\pi$ rad. So the angular speed of the Earth is 2π rad/day and then convert days to seconds.

$$1 \text{ day} = 24 \text{ h} = 24 \text{ h} \times 3600 \text{ s/h} = 86{,}400 \text{ s}$$

$$|\omega| = \frac{2\pi \text{ rad}}{86{,}400 \text{ s}} = 7.27 \times 10^{-5} \text{ rad/s}$$

Discussion Notice that this problem is analogous to a problem in linear motion such as: "A car travels in a straight line at constant speed. In 3 h, it has traveled 192 mi. What is its velocity in m/s?" Just about everything in circular motion and rotation has this kind of analog—which means we can draw heavily on what we have already learned.

The *angular* velocity of a person near one of the poles due to Earth's rotation is the same as that of someone at the equator, although the linear speeds are very different. Both move through the same *angle* in the same time interval, but the person at the equator moves through a much larger distance.

Practice Problem 5.1   Angular speed of Venus

Given that a day on Venus is 243 Earth days, what is the angular speed of the rotation of Venus in rad/s?

Relation between Linear and Angular Speed

For a point moving in a circular path of radius r, the linear distance traveled along the circular path during an angular displacement of $\Delta\theta$ (in radians) is the arc length s where

$$s = r\Delta\theta = r(\theta_2 - \theta_1) \quad \text{(angles in radians)} \tag{5-5}$$

The point in question could be a point particle moving in a circular path, or it could be any point on a rotating rigid object. Since Eq. (5-5) comes directly from the definition of the radian, any equation derived from Eq. (5-5) is valid only when the angles are measured in radians.

What is the linear speed at which the point moves? The average linear speed is the distance traveled divided by the time interval,

$$v_{av} = \frac{s}{\Delta t} = \frac{r\Delta\theta}{\Delta t} \quad (\Delta\theta \text{ in radians})$$

We recognize $\Delta\theta/\Delta t$ as the average angular velocity ω_{av}. If we take the limit as Δt goes to zero, both average quantities (v_{av} and ω_{av}) become instantaneous quantities. Therefore, the relationship between linear speed and angular speed is

$$v = r\omega \quad (\omega \text{ in radians per unit time}) \tag{5-6}$$

For a rotating object, points farther from the axis move at higher linear speeds; they have a circle of bigger radius to travel and therefore cover more distance in the same time interval.

In uniform circular motion, speed is constant but velocity is *not* constant because the *direction* of the velocity is changing.

When the speed of a point moving in a circle is constant, its motion is called **uniform circular motion**. Even though the speed of the point is constant, the velocity is not: the direction of the velocity is changing. This distinction is important when we find the acceleration of an object in uniform circular motion (Section 5.2). The time for the point to travel completely around the circle is called the **period** of the motion, *T*. The **frequency** of the motion, which is the number of revolutions per unit time, is defined as

$$f = \frac{1}{T} \qquad (5\text{-}7)$$

since

$$\frac{\text{revolutions}}{\text{second}} = \frac{1}{\text{seconds/revolution}}$$

The speed is the total distance traveled divided by the time taken,

$$v = \frac{2\pi r}{T} = 2\pi r f \qquad (5\text{-}8)$$

Then, for uniform circular motion

$$\omega = \frac{v}{r} = 2\pi f \quad (\omega \text{ in radians per unit time}) \qquad (5\text{-}9)$$

where, in SI units, angular velocity ω is measured in rad/s and frequency f is measured in hertz (Hz). The hertz is a derived unit equal to 1 rev/s. The dimensions of Eq. (5-9) are correct since both revolutions and radians are pure numbers. The physical dimensions on both sides are a number per second (s^{-1}).

SI unit of frequency: 1 Hz = 1 rev/s

Example 5.2

Speed in a Centrifuge

A centrifuge is spinning at 5400 rpm. (a) Find the period (in s) and frequency (in Hz) of the motion. (b) If the radius of the centrifuge is 14 cm, how fast (in m/s) is an object at the outer edge moving?

Strategy Remember that rpm means *revolutions per minute*. 5400 rpm *is* the frequency, but in a unit other than Hz. After a unit conversion, the other quantities can be found using the relations already discussed.

Solution (a) First convert rpm to Hz:

$$f = 5400 \, \frac{\text{rev}}{\text{min}} \times \frac{1 \, \text{min}}{60 \, \text{s}} = 90 \, \text{rev/s}$$

so the frequency is $f = 90$ Hz $= 90 \, s^{-1}$. The period is

$$T = 1/f = 0.011 \, \text{s}$$

(b) To find the linear speed, we first find the angular speed in rad/s:

$$\omega = 90 \, \frac{\text{rev}}{\text{s}} \times 2\pi \frac{\text{rad}}{\text{rev}} = 180\pi \, \text{rad/s}$$

So $\omega = 2\pi f = 180\pi$ rad/s. The linear speed is

$$v = \omega r = 180\pi \, s^{-1} \times 0.14 \, \text{m} = 79 \, \text{m/s}$$

Discussion Notice that much of this problem was done with unit conversions. Instead of memorizing a formula such as $\omega = 2\pi f$, an understanding of where the formula came from (in this case, that 2π radians correspond to one revolution) is more useful and less prone to error.

Practice Problem 5.2   Clothing in the drier

An automatic clothing drier spins at 51.6 rpm. If the radius of the drier drum is 30.5 cm, how fast is the outer edge of the drum moving?

Rolling: Rotation and Translation Combined

When an object is rolling, it is both rotating and translating. The wheel rotates about an axle, but the axle is not at rest; it moves forward or backward. What is the relationship between the angular speed of the wheel and the linear speed of the axle? You might guess that $v = \omega r$ is the answer. You would be right, as long as the object rolls without slipping or skidding.

There is no fixed relationship between the linear and angular speeds of a wheel if it is allowed to skid or slip. When an impatient driver guns the engine the instant a traffic light turns green, the automobile wheels are likely to slip. The rubber sliding against the road surface makes the squealing sound and leaves tracks on the road. The driver could actually make the acceleration of the car greater by giving the engine *less* gas. When the wheels are skidding or slipping, *kinetic* friction propels the car forward instead of the potentially larger force of *static* friction.

For a wheel that rolls *without* slipping, as the wheel turns through one complete rotation, the axle moves a distance equal to the circumference of the wheel. Think of a paint roller leaving a line of paint as it rolls along a wall. After one complete rotation, the same point on the roller wheel is touching the wall as was initially touching it. The length of the line of paint is $2\pi r$. The elapsed time is T, so the axle's speed is

$$v_{\text{axle}} = \frac{2\pi r}{T}$$

while the angular velocity of the roller is

$$\omega = \frac{2\pi}{T}$$

Thus,

$$v_{\text{axle}} = \omega r \quad (\omega \text{ in radians per unit time})$$

Example 5.3

Angular Speed of a Rolling Wheel

 Kevin is riding his motorcycle at a speed of 13.0 m/s. If the diameter of the rear tire is 65.0 cm, what is the angular speed of the rear wheel?

Strategy   The given diameter of the tire enables us to find the circumference and thus the distance traveled in one revolution of the wheel. From the speed of the motorcycle we can find how many revolutions the tire must make per second.

Given: $v = 13.0$ m/s; $d_{\text{wheel}} = 65.0$ cm.

To find: angular velocity ω.

Solution   During one revolution of the wheel, the motorcycle travels a distance equal to the tire's circumference (Fig. 5.4). Then the time T to make one revolution is the time to travel a linear distance equal to $2\pi r$,

$$T = \frac{\text{distance}}{\text{speed}} = \frac{2\pi r}{v}$$

Since T is the time per revolution, the number of revolutions per second is $1/T$.

$$f = \text{number of revolutions/s} = \frac{v}{2\pi r}$$

For each revolution there is an angular displacement of 2π radians,

$$\omega = 2\pi \text{ radians} \times \text{number of revolutions/s} = 2\pi f = 2\pi \frac{v}{2\pi r}$$

so, as expected,

$$\omega = \frac{v}{r} = \frac{13.0 \text{ m/s}}{0.325 \text{ m}} = 40.0 \text{ rad/s}$$

Discussion   Check: time for one revolution is $\frac{2\pi \text{ rad}}{40.0 \text{ rad/s}} = 0.157$ s. Time to travel a distance $2\pi r = 2.04$ m is $\frac{2.04 \text{ m}}{13.0 \text{ m/s}} = 0.157$ s. Looks good.

You could have obtained this answer immediately by looking back through the text for the equation $\omega = v/r$ and plugging in numbers, but the solution here shows that you can re-create that equation. Here, and in many cases, there is no need to memorize a formula if you understand the concepts behind the formula. You are also less apt to make a mistake by forgetting a factor or constant in the equation, or by using an inappropriate formula. For another example, if an object moves along a straight line at a constant velocity, you know instantly that the displacement is the velocity times the time interval—

continued on next page

Figure 5.4 Position of a motorcycle wheel after one revolution.

Example 5.3 *continued*

not because you have memorized an equation ($\Delta \vec{\mathbf{r}} = \vec{\mathbf{v}} \Delta t$) but because you understand the concepts of displacement and velocity. This is the sort of internalization of scientific thinking that you will develop with more and more practice in problem solving.

Practice Problem 5.3   Rolling drum
A cylindrical steel drum is tipped over and rolled along the floor of a warehouse. If the drum has a radius of 0.40 m and makes one complete turn every 8.0 s, how long does it take to roll the drum 36 m?

5.2   CENTRIPETAL ACCELERATION

For a particle undergoing uniform circular motion, the magnitude of the velocity is constant, but the direction is continuously changing. At any instant of time, the direction of the instantaneous velocity is tangent to the path, as discussed in Section 4.4. Since the *direction* of the velocity continually changes, the particle has a nonzero acceleration.

In Fig. 5.5a, two velocity vectors of equal magnitude are drawn tangent to a circular path of radius r, representing the velocity at two different times of an object moving around a circular path with constant speed. At any instant, the velocity vector is perpendicular to a radius drawn from the center of the circle to the position of the object. As the time between velocity measurements approaches zero, the radii become closer together (Fig. 5.5b). To find the acceleration, $\vec{\mathbf{a}} = \lim\limits_{\Delta t \to 0} \dfrac{\Delta \vec{\mathbf{v}}}{\Delta t}$, we must first find the change in velocity $\Delta \vec{\mathbf{v}}$ for a very short time interval. Figure 5.5c shows that as the time interval Δt approaches zero, the angle between the two velocities also approaches zero and $\Delta \vec{\mathbf{v}}$ becomes perpendicular to the velocity.

Figure 5.5 Uniform circular motion at constant speed. (a) The velocity is always tangent to the circular path and perpendicular to the radius at that point. (b) As the time interval between two velocity measurements decreases, the angle between the velocity vectors decreases. (c) The change in velocity ($\Delta\vec{\mathbf{v}}$) is found by placing the tails of the two velocity vectors together. Then $\Delta\vec{\mathbf{v}}$ is drawn from the tip of the initial velocity ($\vec{\mathbf{v}}_1$) to the tip of the final velocity ($\vec{\mathbf{v}}_2$).

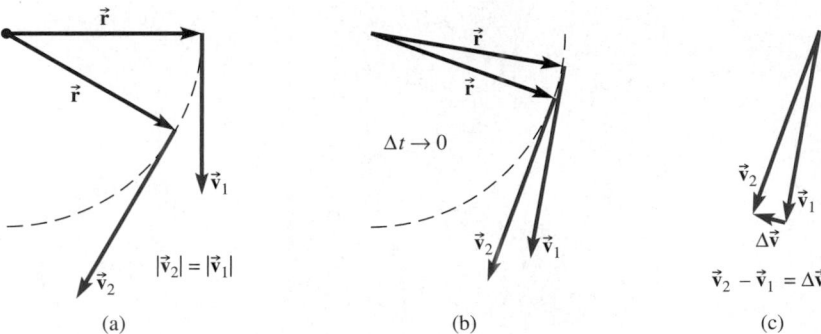

(a)   (b)   (c)

Since $\Delta\vec{\mathbf{v}}$ is perpendicular to the velocity, it is directed along a radius of the circle. Inspection of Figs. 5.5b and 5.5c shows that $\Delta\vec{\mathbf{v}}$ is radially *inward* (toward the center of the circle). Since the acceleration $\vec{\mathbf{a}}$ has the same direction as $\Delta\vec{\mathbf{v}}$ (in the limit $\Delta t \rightarrow 0$), the acceleration is also directed radially inward.

> In uniform circular motion, the direction of the acceleration is radially inward.

The word **centripetal** is a synonym for *radially inward*. Literally it means *toward the center*. The acceleration of an object undergoing *uniform* circular motion is often called the **centripetal acceleration, $\vec{\mathbf{a}}_c$.** The word "centripetal" here just reminds us of the direction of the acceleration.

Since the acceleration of an object in uniform circular motion is radially inward, the net force acting on that object must also be directed radially inward, perpendicular to the path of motion. For the stone Alekos whirls about in a sling, the tension in the cords pulls inward on the stone to hold it in its circular path. For a satellite orbiting Earth in a circular orbit, the gravitational force always pulls inward on the satellite toward the center of the Earth, which is the center of its circular orbit.

Magnitude of the Centripetal Acceleration

To find the magnitude of the centripetal acceleration for uniform circular motion, we must find the change in velocity $\Delta\vec{\mathbf{v}}$ for a time interval Δt in the limit $\Delta t \rightarrow 0$. The velocity keeps the same magnitude but changes direction at a steady rate, equal to the angular velocity ω. In a time interval Δt, the velocity $\vec{\mathbf{v}}$ rotates through an angle equal to the angular displacement $\Delta\theta = \omega\,\Delta t$. During this time interval, the velocity vector sweeps out an arc of a circle of "radius" v (Fig. 5.6). In the limit $\Delta t \rightarrow 0$, the magnitude of $\Delta\vec{\mathbf{v}}$ becomes equal to the arc length, since a very short arc approaches a straight line. Then

Figure 5.6 The velocity vector sweeps out an arc of a circle whose magnitude is equal to the magnitude of $\Delta\vec{\mathbf{v}}$ in the limit $\Delta t \rightarrow 0$.

$$|\Delta\vec{\mathbf{v}}| = \text{arc length} = \text{radius of circle} \times \text{angle subtended}$$

$$= v\,\Delta\theta = v\omega\,\Delta t$$

Acceleration is the rate of change of velocity, so the magnitude of the centripetal acceleration is

$$a_c = |\vec{\mathbf{a}}| = \frac{|\Delta\vec{\mathbf{v}}|}{\Delta t} = v\omega \quad (\omega \text{ in radians per unit time}) \qquad (5\text{-}10)$$

where absolute value symbols are used with the vector quantities to indicate their magnitudes. Velocity and angular velocity are not independent; $v = \omega r$. It is usually most convenient to write the magnitude of the centripetal acceleration in terms of one or the other of these two quantities. So we write the centripetal acceleration in two other equivalent ways using $v = \omega r$:

$$a_c = \frac{v^2}{r} \quad \text{or} \quad a_c = \omega^2 r \quad (\omega \text{ in radians per unit time}) \qquad (5\text{-}11)$$

Note that Eqs. (5-10) and (5-11) are valid only when ω is measured in *radians* per unit time (normally rad/s, but rad/min or rad/h would be correct).

Example 5.4

A Spinning CD

If a CD spins at 210 rpm, what is the centripetal acceleration of a point on the outer rim of the CD? The CD is 12 cm in diameter.

Strategy   From the number of revolutions per minute, we can find the frequency and the angular velocity. The angular velocity and the radius of the CD enable us to calculate the centripetal acceleration.

Solution   We convert 210 rpm into a frequency in revolutions per second (Hz).

$$f = 210 \frac{\text{rev}}{\text{min}} \times \frac{1}{60} \frac{\text{min}}{\text{s}} = 3.5 \frac{\text{rev}}{\text{s}} = 3.5 \text{ Hz}$$

For each revolution, the CD rotates through an angle of 2π radians. The angular velocity is

$$\omega = 2\pi f = 2\pi \frac{\text{radians}}{\text{rev}} \times 3.5 \frac{\text{rev}}{\text{s}} = 7.0\pi \text{ rad/s}$$

Then using Eq. (5-11), the centripetal acceleration is

$$a_c = \omega^2 r = (7.0\pi \text{ rad/s})^2 \times 0.060 \text{ m} = 29 \text{ m/s}^2$$

Discussion   When finding the centripetal acceleration, use whichever form of Eq. (5-11) is more convenient. For rotating objects such as the spinning CD, it's usually easiest to think in terms of the angular velocity. For an object moving around a circle, such as a satellite in orbit whose speed is known, it might be easier to use v^2/r. Since the two equations are equivalent, either can be used in any situation.

Practice Problem 5.4   Centripetal acceleration of a point on an old record

What is the centripetal acceleration of a point 25.4 cm from the center of a record that is rotating at 78 rpm on a turntable?

Now that we know the magnitude and direction of the acceleration of any object in uniform circular motion, we can use Newton's second law to relate the net force acting on the object to the speed and radius of its motion. The net force is found in the usual way: each of the individual forces acting on the object is identified and then the forces are added as vectors. Every force acting must be exerted *by some other object*. Resist the temptation to add in a new, separate force just because something moves in a circle. For an object to move in a circle at constant speed, real, physical forces such as gravity, tension, normal forces, and friction must act on it; these forces combine to produce a net force that has the correct magnitude and is always perpendicular to the velocity of the object.

Problem-Solving Strategy for an Object in Uniform Circular Motion

1.   Begin as for any Newton's second law problem: identify all the forces acting on the object and draw a free-body diagram.

2.   Choose perpendicular axes at the point of interest so that one is centripetal and the other is tangent to the circular path.

3.   Find the centripetal component of each force.

4.   Apply Newton's second law as follows:

$$\Sigma F_c = ma_c$$

where ΣF_c is the centripetal component of the net force and the centripetal component of the acceleration is

$$a_c = \frac{v^2}{r} = \omega^2 r$$

(For uniform circular motion, neither the net force nor the acceleration has a tangential component.)

Now let's consider Alekos and the whirling stone again. The path of the stone, once it is released, is tangent to the circular path it had been following. We had neglected the weight of the stone in comparison with the other forces acting. However, once the cord is released, the weight of the stone can no longer be neglected; the gravitational force pulls the stone down toward the ground. The stone then moves in a parabolic path with an initial velocity tangent to the original circular path and parallel to the ground (assuming the plane of the circular path is parallel to the plane of the ground). The stone is also at the high point of the parabolic path, on its journey to the ground, because it has no initial component of velocity in the vertical direction.

Example 5.5

Alekos and the Wolf

 Alekos, practicing his wolf deterrence strategies, is whirling a 0.50-kg stone about in a sling. The sling consists of two 80.0-cm-long cords tied to either side of a pouch that holds the stone. To release the stone, Alekos lets go of one cord. The sling makes 4.8 revolutions per second in a horizontal plane 2.0 m above the ground. (a) What is the speed of the stone when it leaves the sling? (b) What is the tension in each of the two cords (assumed equal) before one is released? (c) How far from the release point does the stone land? Ignore air resistance. (d) Sketch an overhead view of Alekos and the wolf and indicate where along the circular path he should release the cord.

Strategy We can find the linear speed of the stone at the time Alekos releases it from the radius and angular velocity. The tension in the cords can be found from the centripetal acceleration and the mass of the stone. Once the stone flies off, it follows the path of a projectile launched horizontally. We use the equations of motion for constant acceleration to find the landing distance of the projectile.

Solution (a) We find the angular velocity from the number of revolutions per second.

$$\omega \,(\text{rad/s}) = 4.8 \; \text{rev/s} \times 2\pi \; \text{rad/rev} = 30 \; \text{rad/s}$$

The tangential velocity component is

$$v_\text{t} = \omega r = 30 \; \text{rad/s} \times 0.800 \; \text{m} = 24 \; \text{m/s}$$

so that is the speed with which the stone leaves the sling.

(b) The net force on the stone and the leather pouch can be found from Newton's second law. We neglect the weight of the stone and the pouch, assuming that it is small compared to the tension in the cords. Then the tension in the cords is the only significant force acting on the pouch. Assuming that the tensions are equal and the cords are parallel,

$$F_\text{net} = 2T$$

For uniform circular motion, the net force must act radially inward. We write the centripetal acceleration component in terms of either ω or v_t. Choosing the former,

$$2T = F_\text{net} = ma_\text{c} = mr\omega^2$$

Substituting numerical values,

$$T = \tfrac{1}{2} \times 0.50 \; \text{kg} \times 0.800 \; \text{m} \times (30 \; \text{rad/s})^2 = 180 \; \text{N}$$

The tension is much larger than the weight of the stone (≈ 5 N), so the assumption that we could ignore the weight is justified.

(c) Once the cord is released, we treat the stone as a projectile with an initial horizontal velocity of 24 m/s. The time before the stone hits the ground is found by looking at the vertical motion. Initially the stone has no velocity component in the vertical direction, so it is acted on only by the downward force of gravity.

We can use a kinematic equation for constant acceleration to find the time it takes the stone to reach the ground:

$$\Delta y = v_{0y} t + \tfrac{1}{2} a_y t^2 \qquad (4\text{-}7)$$

Everything in this equation is given except for the time. The vertical displacement of the stone, starting 2.0 m above the ground and ending on the ground, is

$$\Delta y = -2.0 \; \text{m}$$

Since $v_{0y} = 0$ and $a_y = -g$,

$$\Delta y = \tfrac{1}{2} a_y t^2 = -\tfrac{1}{2} g t^2$$

so that the time before the stone lands is

$$t = \sqrt{\frac{2\,\Delta y}{-g}} = \sqrt{\frac{2 \times (-2.0 \; \text{m})}{-9.8 \; \text{m/s}^2}} = 0.64 \; \text{s}$$

The horizontal displacement of the stone during this time is

$$\Delta x = v_{0x} t = 24 \; \text{m/s} \times 0.64 \; \text{s} = 15 \; \text{m}$$

15 m is the horizontal distance between the release point of the stone and the point where it hits the ground—if it misses the wolf!

(d) The stone flies away at a tangent to the circle, so it must be let go at the position shown in Fig. 5.7 to be aimed directly at the wolf.

Discussion This example demonstrates the cumulative nature of physics concepts. The basic concepts keep reappearing, to be used over and over and to be extended for use in new contexts. Part of the problem involves new concepts (angular velocity and centripetal acceleration); the rest of the problem involves old material (Newton's second law, linear velocity and acceleration, projectile motion, and tension in a cord).

Practice Problem 5.5 Rotating carousel

A carousel is rotating at such a rate that a horse located 8.0 m from the central axis moves at a speed of 6.0 m/s. (a) What is the centripetal acceleration of the horse? (b) At what speed does a horse located 4.0 m from the center move?

Figure 5.7 The stone hits the wolf only if it is released at the correct point in its orbit.

Example 5.6

Conical Pendulum

Suppose you whirl a stone in a horizontal circle at a slower speed so that the weight of the stone is *not* negligible compared to the tension in the cord. Then the cord cannot be horizontal—the tension must have a vertical component to cancel the weight and leave a horizontal net force (Fig. 5.8). If the cord has length L, the stone has mass m, and the cord makes an angle ϕ with the vertical direction, what is the constant angular speed of the stone?

Strategy The net force must point toward the center of the circle, since the stone is in uniform circular motion. With the

Figure 5.8
(a) A stone is whirled in a horizontal circle of radius $r = L \sin \phi$. (b) A free-body diagram for the stone.

continued on next page

Example 5.6 *continued*

stone in the position depicted in Fig. 5.8a, the direction of the net force is along the +x-axis. This time the tension in the cord does not pull toward the center, but the *net* force does.

Solution   Start by drawing a free-body diagram (Fig. 5.8b). Now apply Newton's second law in component form. The acceleration has components $a_x = \omega^2 r$ and $a_y = 0$. For the x-components,

$$\Sigma F_x = T \sin \phi = ma_x = m\omega^2 r$$

Since the problem does not specify r, we must express r in terms of L and ϕ. In Fig. 5.8a, the radius forms a right triangle with the cord and the y-axis. Then

$$r = L \sin \phi$$

and

$$T \sin \phi = m\omega^2 L \sin \phi$$

$$\Sigma F_y = T \cos \phi - mg = ma_y = 0 \Rightarrow T \cos \phi = mg$$

Now we eliminate the tension:

$$(m\omega^2 L) \cos \phi = mg$$

Solving for ω,

$$\omega = \sqrt{\frac{g}{L \cos \phi}}$$

Discussion   We should check the dimensions of the final expression. Since cos ϕ is dimensionless,

$$\sqrt{\frac{[L]/[T]^2}{[L]}} = \frac{1}{[T]}$$

which is correct for ω (SI unit rad/s).

Another check is to ask how ω and ϕ are related for a given length cord. As ϕ increases toward 90°, the cord gets closer to horizontal and the radius increases. In our expression, as ϕ increases, cos ϕ decreases and therefore ω increases, in accordance with experience: the stone would have to be whirled faster and faster to make the cord more nearly horizontal.

Conceptual Practice Problem 5.6   Conical pendulum on the Moon

Examine the result of Example 5.6 to see how ω depends on g, all other things being equal. Where the gravitational field is weaker, do you have to whirl the stone faster or more slowly to keep the cord at the same angle ϕ? Is that in accord with your intuition?

5.3   BANKED CURVES

When you drive an automobile in a circular path along an unbanked roadway, friction acting on the tires due to the pavement acts to keep the automobile moving in a curved path. This frictional force acts *sideways*, toward the center of the car's circular path (Fig. 5.9) The frictional force might also have a tangential component; for example, if the car is braking, a component of the frictional force makes the car slow down by acting backward (opposite to the car's velocity). For now we assume that the car's speed is constant and that the tangential component of friction is negligible. As long as the tires roll without slipping, there is no relative motion between the bottom of the tire and the road, so it is the force of *static* friction that acts. If the car is in a skid, then it is the smaller force of kinetic friction that acts as the bottom portion of the tire slides along the pavement. As the speed of the car increases, or for slippery surfaces with low coefficients of friction, the static frictional force may not be enough to hold the car in its curved path.

Making The Connection:
banked roadways

To help prevent cars from going into a skid or losing control, the roadway is often banked (tilted at a slight angle) around curves so that the outer portion of the road—the part farthest from the center of curvature—is higher than the inner portion. Banking changes the angle and magnitude of the normal force, $\vec{N}$, so that it has a horizontal component N_x directed toward the center of curvature (in the centripetal direction—see Fig. 5.10). Then we need no longer rely solely on friction to keep the car moving in a circular path as it negotiates the curve; this component of the normal force acts to help the car remain on the curved path. Figure 5.10 shows an exaggerated illustration of a banked road with the normal force, the car's weight, and, in parts (c) and (d), the centripetal component of the normal force N_x. We choose the axes so that the x-axis is in the direction of the acceleration, which is to the left; the axes are *not* parallel and perpendicular to the incline.

Figure 5.9 A car negotiating a curve on an unbanked roadway: (a) top view and (b) head-on view showing the frictional force acting to the left. (c) A free-body diagram for the car, in which we choose the *x*-axis in the direction of the acceleration.

Figure 5.10 A car negotiating a curve on a banked roadway: (a) top view and (b) head-on view showing banked roadway. (c) Components of the normal force. (d) Free-body diagram for the car.

From Section 5.2, any object traveling at a constant speed along a circular path has centripetal acceleration component

$$a_c = \frac{v^2}{r} \tag{5-11}$$

To keep the car safely on the road and to prevent a skid, even under slippery conditions, we want the road to be banked so that the horizontal component of the normal force *alone* gives rise to the centripetal acceleration. With no friction,

$$N_x = \Sigma F_x = ma_x = m\frac{v^2}{r}$$

From Fig. 5.10c, the components of the normal force are

$$N_y = N \cos \theta$$

and

$$N_x = N \sin \theta$$

Since the y-component of the acceleration is zero,

$$\Sigma F_y = N_y - mg = ma_y = 0$$

Therefore, the y-component of the normal force is equal to mg.

Now we can determine the correct banking angle of the roadway. Note that if we divided N_x by N_y, the unknown magnitude of the normal force cancels out and leaves an expression for the banking angle:

$$\frac{N_x}{N_y} = \frac{N \sin \theta}{N \cos \theta} = \tan \theta$$

$$\frac{N_x}{N_y} = \frac{mv^2/r}{mg} = \frac{v^2}{rg}$$

Therefore,

$$\theta = \tan^{-1} \frac{v^2}{rg} \tag{5-12}$$

where v is the speed of the car, r is the radius of curvature of the circular path, and g is the gravitational field strength. Notice that the mass of the car does not appear in this equation; the same banking angle holds for a scooter, motorcycle, car, or tractor-trailer. Notice also that the banking angle depends on the square of the speed. Automobile racetracks and bicycle racetracks have highly banked road surfaces at hairpin curves to minimize skidding of the high-speed vehicles.

If there is no friction between the road and the tires, then there is only one speed—the speed that satisfies Eq. (5-12)—at which it is safe to drive around a given curve:

$$v = \sqrt{rg \tan \theta}$$

With friction, there is a *range* of safe speeds. The static frictional force can have any magnitude from 0 to μN and it can be directed either up or down the bank of the road.

Example 5.7

A Possible Skid

A car is going around an unbanked curve at the recommended speed of 11 m/s (25 mi/h). (a) If the radius of curvature of the path is 25 m and the coefficient of static friction between the rubber and the road is $\mu_s = 0.70$, does the car skid as it goes around the curve? (b) What happens if the driver ignores the highway speed limit sign and travels at 18 m/s (40 mi/h)? (c) What speed is safe for traveling around the curve if the road surface is wet from a recent rainstorm and the coefficient of static friction between the wet road and the rubber tires is $\mu_s = 0.50$? (d) For a car to safely negotiate the curve in icy conditions at a speed of 13 m/s (29 mi/h), what banking angle would be required?

Strategy The force of static friction is the only horizontal force acting on the car when the curve is not banked. The maximum force of static friction, which depends on road conditions, determines the maximum possible centripetal acceleration of the car. Therefore, we can compare the centripetal acceleration necessary to go around the curve at the specified speeds with the maximum possible centripetal acceleration determined by the coefficient of static friction. For part (d), in icy conditions we cannot rely much on friction, but the normal force has a horizontal component when the road is banked.

Solution (a) We find the centripetal acceleration required for a speed of 11 m/s:

$$a_c = \frac{v^2}{r} = \frac{(11 \text{ m/s})^2}{25 \text{ m}} = 4.8 \text{ m/s}^2$$

In order to have that acceleration, the component of the net force acting toward the center of curvature must be

$$\Sigma F_c = ma_c = m\frac{v^2}{r}$$

continued on next page

Example 5.7 *continued*

The only force with a horizontal component is the static frictional force acting on the tires due to the road (see the FBD in Fig. 5.10d). Therefore,

$$\Sigma F_c = f_s = m\frac{v^2}{r}$$

We must check to make sure that the maximum frictional force is not exceeded:

$$f_s \le \mu_s N$$

Since $N = mg$, the car can go around the curve without skidding as long as

$$m\frac{v^2}{r} \le \mu_s mg$$

Thus, the centripetal acceleration cannot exceed $\mu_s g$. That limits the car to speeds satisfying

$$v \le \sqrt{\mu_s gr}$$

Substituting numerical values,

$$v \le \sqrt{0.70 \times 9.8 \text{ m/s}^2 \times 25 \text{ m}} = 13 \text{ m/s}$$

Since 11 m/s is less than the maximum safe speed of 13 m/s, the car safely negotiates the curve.

(b) At 18 m/s, the car moves at a speed higher than the maximum safe speed of 13 m/s. The frictional force cannot supply the centripetal acceleration needed for the car to go around the curve—the car goes into a skid.

(c) The maximum safe speed occurs when the static frictional force has its maximum possible magnitude. For the wet road conditions, the static frictional force acting between the tires and the road is given by

$$f_s \le \mu_s N = \mu_s mg$$

where $\mu_s = 0.50$. The maximum possible acceleration is therefore

$$a_c = \mu_s g = 0.50 \times 9.8 \text{ m/s}^2 = 4.9 \text{ m/s}^2$$

Now we use the centripetal acceleration to find the maximum speed:

$$a_c = v^2/r$$

or

$$v = \sqrt{ra_c} = \sqrt{25 \text{ m} \times 4.9 \text{ m/s}^2} = 11 \text{ m/s}$$

which is the same speed recommended by the road sign. The highway engineer knew what he was doing when he had the sign placed along the road.

(d) Finally, we find the banking angle that would enable cars to travel around the curve at 13 m/s in icy conditions. Assuming that friction is negligible, the horizontal component of the normal force is the only horizontal force. With the x-axis pointing toward the center of curvature and the y-axis vertical,

$$\Sigma F_x = N \sin\theta = mv^2/r \qquad (1)$$

and

$$\Sigma F_y = N \cos\theta - mg = 0 \qquad (2)$$

Dividing Eq. (1) by Eq. (2) gives

$$\frac{N \sin\theta}{N \cos\theta} = \tan\theta = \frac{mv^2/r}{mg} = \frac{v^2}{rg}$$

$$\theta = \tan^{-1}\frac{v^2}{rg} = \tan^{-1}\frac{(13 \text{ m/s})^2}{25 \text{ m} \times 9.8 \text{ m/s}^2} = 35°$$

Discussion   A banking angle of 35° is far greater than those used in practice along roadways. Careful drivers would not try to drive around this curve in icy conditions at 13 m/s. What do you think might happen in icy conditions to a car that is traveling *very slowly* along a road banked at such a steep angle?

Highway curves are banked at slight angles to help drivers who are driving at reasonable speeds for the road conditions. They are not banked to save speed demons from their folly.

Practice Problem 5.7   A bobsled race

A bobsled races down an icy hill and then comes upon a horizontal curve, located 60.0 m from the bottom of the hill. The sled is traveling at 22.4 m/s (50 mph) as it approaches the curve that has a radius of curvature of 50.0 m. The curve is banked at an angle of 45° and the frictional force on the sled runners is negligible. Does the sled make it safely around the curve?

When an airplane pilot makes a turn in the air, the pilot makes use of a banking angle. The airplane itself is tilted as if it were traveling over an inclined surface. Because of the shape of the wings, an aerodynamic force called *lift* acts upward when the plane is in level flight. To go around a turn, the wings are tilted; the lift force stays perpendicular to the wings and, therefore, now has a horizontal component, just as the normal force has a horizontal component for a car on a banked curve. This component supplies the necessary centripetal acceleration, while the vertical component of the lift holds the plane up. Therefore,

$$L_x = ma_c = \frac{mv^2}{r} \quad \text{and} \quad L_y = mg$$

where the x-axis is horizontal and the y-axis is vertical. The lift force is different in its physical origin from the normal force, but its components split up the same way, so a plane in a turn banks its wings at the same angle that a road would be banked for the same speed and radius of curvature.

Making The Connection:
banking angle
of an airplane

5.4   CIRCULAR ORBITS

A satellite orbiting Earth travels in a circular path because of the long-range gravitational force between the satellite and the Earth. The same gravitational interaction that makes Earth-bound objects fall keeps satellites and planets from flying off into space. The law of universal gravitation is

$$F = \frac{Gm_1m_2}{r^2} \tag{2-3}$$

where the universal gravitational constant is

$$G = 6.67 \times 10^{-11}\ \text{N·m}^2/\text{kg}^2$$

We can use Newton's second law to find the speed of a satellite in circular orbit at constant speed. Let m be the mass of the satellite and M_E be the mass of the Earth. The direction of the gravitational force on the satellite is always toward the center of the Earth, which is the center of the orbit. Since gravity is the only force acting on the satellite,

$$\Sigma F_c = G\frac{mM_E}{r^2}$$

where r is the distance from the center of the Earth to the satellite. Then, from Newton's second law,

$$\Sigma F_c = ma_c = \frac{mv^2}{r}$$

Setting these equal,

$$G\frac{m M_E}{r^2} = \frac{m v^2}{r}$$

Solving for the speed yields

$$v = \sqrt{\frac{GM_E}{r}} \tag{5-13}$$

Notice that the mass of the satellite does not appear in the equation for speed; it has been algebraically canceled. The greater inertia of a more massive satellite is overcome by a proportionally greater gravitational force acting on it. Thus, the speed of a satellite in a circular orbit does not depend on the mass of the satellite. Equation (5-13) also shows that satellites in lower orbits (smaller radii) have greater speeds.

We have been discussing satellites orbiting Earth, but the same principles apply to the circular orbits of satellites around other planets and to the orbits of the planets around the Sun. For planetary orbits, the mass of the Sun would appear in Eq. (5-13) instead of the Earth's mass. Gravity is the force that keeps planets orbiting about the Sun. The paths of planetary orbits are actually ellipses instead of circles, although for most of the planets in the solar system the ellipses are nearly circular. Pluto and Mercury are the exceptions; their orbits are markedly different from circles.

Example 5.8

Speed of Satellite

The Hubble telescope is in a circular orbit 613 km above Earth's surface. The average radius of the Earth is 6.37×10^3 km and the mass of Earth is 5.98×10^{24} kg. What is the speed of the telescope in its orbit?

Strategy   We first need to find the orbital radius of the telescope. It is not 613 km; that is the distance from the *surface* of Earth to the telescope. We must add the radius of the Earth to

613 km to find the orbital radius, which is measured from the center of the Earth to the telescope. Then we use Newton's second law, along with what we know about centripetal acceleration.

Solution   The radius of the telescope's orbit is

$$r = 6.13 \times 10^2\ \text{km} + 6.37 \times 10^3\ \text{km} = (0.613 + 6.37) \times 10^3\ \text{km}$$

$$= 6.98 \times 10^3\ \text{km}$$

continued on next page

Example 5.8 *continued*

The net force on the telescope is equal to the gravitational force, given by Newton's law of gravity. Newton's second law relates the net force to the acceleration. Both are directed radially inward.

$$\Sigma F_c = \frac{GmM_E}{r^2} = \frac{mv^2}{r}$$

where m is the mass of the telescope. Solving for the speed, we find

$$v = \sqrt{\frac{GM_E}{r}}$$

$$v = \sqrt{\frac{6.67 \times 10^{-11}\ \text{N·m}^2/\text{kg}^2 \times 5.98 \times 10^{24}\ \text{kg}}{6.98 \times 10^6\ \text{m}}}$$

$$v = 7560\ \text{m/s} = 27{,}200\ \text{km/h}$$

Discussion   *Any* satellite orbiting Earth at an altitude of 613 km has this same speed, regardless of its mass.

Practice Problem 5.8   Speed of Earth in its orbit

What is the speed of Earth in its approximately circular orbit about the Sun? The average Earth–Sun distance is 1.496×10^{11} m and the mass of the Sun is 1.99×10^{30} kg. Once you find the speed, use it along with the distance traveled by the Earth during one revolution about the Sun to calculate the time in seconds for one orbit. Compare your answer with the number of seconds in one year.

Kepler's Laws of Planetary Motion

Making The Connection:
planetary motion

At the beginning of the seventeenth century, Johannes Kepler (1571–1630) proposed three laws to describe the motion of the planets. These laws predated Newton's laws of motion and his law of gravity. They offered a far simpler description of planetary motion than anything that had been proposed previously. We turn history on its head and look at one of Kepler's laws as a consequence of Newton's laws. The fact that Newton could derive Kepler's laws from his own work on gravity was seen as a confirmation of Newtonian mechanics.

Kepler's laws of planetary motion are

- The planets travel in elliptical orbits with the Sun at one focus of the ellipse.

- A line drawn from a planet to the Sun sweeps out equal areas in equal time intervals. (Fig. 5.11).

- The square of the orbital period is proportional to the cube of the average distance from the planet to the Sun.

Kepler's first law can be derived from the inverse square law of gravitational attraction. The derivation is a bit complicated, but for any two objects that have such an attraction, the orbit of one about the other is an ellipse, with the stationary object located at one focus. The circle is a special case of an ellipse where the two foci coincide.

We can derive Kepler's third law from Newton's law of universal gravitation for a circular orbit. The gravitational force gives rise to the centripetal acceleration:

$$\Sigma F_c = \frac{GmM_{\text{sun}}}{r^2} = \frac{mv^2}{r}$$

Solving for v yields

$$v = \sqrt{\frac{GM_{\text{sun}}}{r}}$$

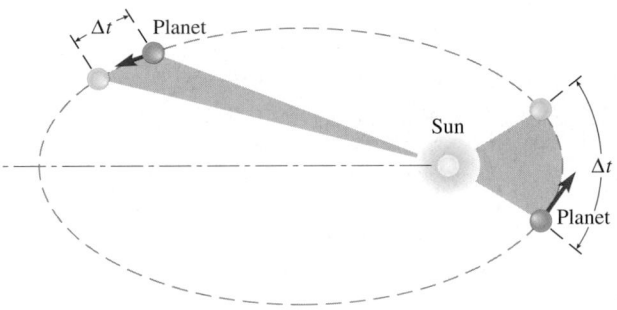

Figure 5.11   A planet orbiting the Sun. When the planet is closer to the Sun, it moves faster; when it is farther away, it moves more slowly. No matter where the planet is in its orbit, the radial line sweeps out the same area during the same time interval Δt. (The planetary orbits are nearly circular, unlike the elliptical orbit shown here, which is exaggerated for clarity.)

The distance traveled during one revolution is the circumference of the circle, which is equal to $2\pi r$. The speed is the distance traveled during one orbit divided by the period:

$$v = \sqrt{\frac{GM_{sun}}{r}} = \frac{2\pi r}{T}$$

Now we solve for T:

$$T = 2\pi \sqrt{\frac{r^3}{GM_{sun}}}$$

Squaring both sides yields

$$T^2 = \frac{4\pi^2}{GM_{sun}} r^3 = \text{constant} \times r^3 \tag{5-14}$$

Equation (5-14) is Kepler's third law: the square of the period of a planet is directly proportional to the cube of the average orbital radius.

Planetary orbits are affected by gravitational interactions with other planets; Kepler's laws ignore these small effects. Although Kepler's laws were derived for the motion of planets, they apply to satellites orbiting the Earth as well. Many satellites, such as those used for communications, are placed in a *geostationary* (or *geosynchronous*) orbit—a circular orbit in Earth's equatorial plane whose period is equal to Earth's rotational period (Fig. 5.12). A satellite in geostationary orbit remains directly above a particular point on the equator; to observers on the ground, it seems to hover above that point without moving. Due to their fixed positions with respect to Earth's surface, geostationary satellites are used as relay stations for communication signals. In Example 5.9, we find the speed of a geostationary satellite.

Example 5.9

Geostationary Satellite

 A 300.0-kg communications satellite is placed in a geostationary orbit 35,800 km above a relay station located in Kenya. What is the speed of the satellite in orbit?

Strategy   The period of the satellite is one day or approximately 24 h. To find the speed of the satellite in orbit we use Newton's law of gravity and his second law of motion along with what we know about centripetal acceleration.

Solution   Let m be the mass of the satellite and let M_E be the mass of the Earth. Gravity is the only force acting on the satellite in its orbit. From Newton's law of universal gravitation, Newton's second law, and the expression for centripetal acceleration,

$$\Sigma F_c = \frac{GmM_E}{r^2} = \frac{mv^2}{r}$$

Solving for the speed yields

$$v = \sqrt{\frac{GM_E}{r}}$$

We must add the mean radius of the Earth, $R_E = 6.37 \times 10^6$ m, to the height of the satellite above the Earth's surface to find the orbital radius.

$$r = h + R_E = 3.58 \times 10^7 \text{ m} + 0.637 \times 10^7 \text{ m}$$

$$r = 4.217 \times 10^7 \text{ m}$$

Substituting numerical values into the speed equation,

$$v = \sqrt{\frac{6.67 \times 10^{-11} \text{ N·m}^2/\text{kg}^2 \times 5.98 \times 10^{24} \text{ kg}}{4.217 \times 10^7 \text{ m}}}$$

$$= \sqrt{9.459 \times 10^6 \text{ m}^2/\text{s}^2}$$

$$v = 3.08 \times 10^3 \text{ m/s}$$

Figure 5.12
Geostationary satellite orbiting the Earth.

continued on next page

Example 5.9 *continued*

Discussion This result, an orbital speed of 3.08 km/s and a distance above Earth's surface of 35,800 km, applies to *all* geosynchronous satellites. The mass of the satellite does not matter; it cancels out of the equations for orbital radius and for speed.

If we were actually putting a satellite into orbit, we would use a more accurate value for the period. We should use a time of 23 h and 56 min, which is the length of a sidereal day—the time for Earth to complete one rotation about its axis relative to the fixed stars. The solar day, 24 h, is the period of time between the daily appearances of the Sun at its highest point in the sky. The fact that Earth moves around the Sun is what causes the difference between these two ways of measuring the length of a day. The error introduced by using the longer time is negligible in this problem.

We can use Kepler's third law to check the result. Examples 5.8 and 5.9 both concern circular orbits around the Earth. Is the square of the period proportional to the cube of the orbital radius? From Example 5.8, $r_1 = 6.98 \times 10^3$ km and

$$T_1 = \frac{2\pi r_1}{v} = \frac{2\pi \times 6.98 \times 10^3 \text{ km}}{7.56 \text{ km/s}} = 5800 \text{ s}$$

and, from the present example, $r_2 = 4.22 \times 10^7$ m and

$$T_2 = 24 \text{ h} \times \frac{3600 \text{ s}}{1 \text{ h}} = 86,400 \text{ s}$$

The ratio of the squares of the periods is

$$\left(\frac{T_2}{T_1}\right)^2 = \left(\frac{86,400 \text{ s}}{5800 \text{ s}}\right)^2 = 221$$

The ratio of the cubes of the radii is

$$\left(\frac{r_2}{r_1}\right)^3 = \left(\frac{4.22 \times 10^7 \text{ m}}{6.98 \times 10^6 \text{ m}}\right)^3 = 221$$

Practice Problem 5.9 Orbital radius of Venus

The period of the orbit of Venus around the Sun is 0.615 Earth years. Using this information, find the radius of its orbit in terms of R, the radius of Earth's orbit around the Sun.

Example 5.10

Orbiting Satellites

A satellite revolves about Earth with an orbital radius of r_1 and speed v_1. If an identical satellite were set into circular orbit with the same speed about a planet of mass three times that of Earth, what would its orbital radius be?

Strategy We can apply Newton's law of universal gravitation and set up a ratio to solve for the new orbital radius.

Solution From Newton's second law, the magnitude of the gravitational force on the satellite is equal to the satellite's mass times the magnitude of its centripetal acceleration:

$$\frac{Gm M_E}{r_1^2} = \frac{m v_1^2}{r_1}$$

where M_E and m are the masses of Earth and of the satellite, respectively. Solving for r_1 yields

$$r_1 = \frac{G M_E}{v_1^2}$$

Now we apply Newton's second law to the orbit of the second satellite about the planet of mass $3M_E$:

$$\frac{Gm \times 3M_E}{r_2^2} = \frac{m v_1^2}{r_2}$$

$$r_2 = \frac{G \times 3M_E}{v_1^2}$$

The ratio of r_2 to r_1 is

$$\frac{r_2}{r_1} = \frac{G \times 3M_E/v_1^2}{G M_E/v_1^2} = 3$$

Thus $r_2 = 3r_1$.

Discussion Notice that we did not rush to substitute numerical values for the constants G and M_E into the equations. We took the ratio r_2/r_1 so that these constants would cancel.

Practice Problem 5.10 Period of lunar lander

A lunar lander is orbiting about the Moon. If the radius of its orbit is $\frac{1}{3}$ the radius of Earth, what is the period of its orbit?

5.5 NONUNIFORM CIRCULAR MOTION

So far we have focused on *uniform* circular motion. Now we can extend the discussion to nonuniform circular motion, where the angular velocity changes with time.

Figure 5.13a shows the velocity vectors $\vec{v}_1$ and $\vec{v}_2$ at two different times for an object moving in a circle with changing speed. In this case the speed is increasing ($v_2 > v_1$). In Fig. 5.13b we subtract $\vec{v}_1$ from $\vec{v}_2$ to find the change in velocity. In the limit $\Delta t \to 0$, $\Delta\vec{v}$

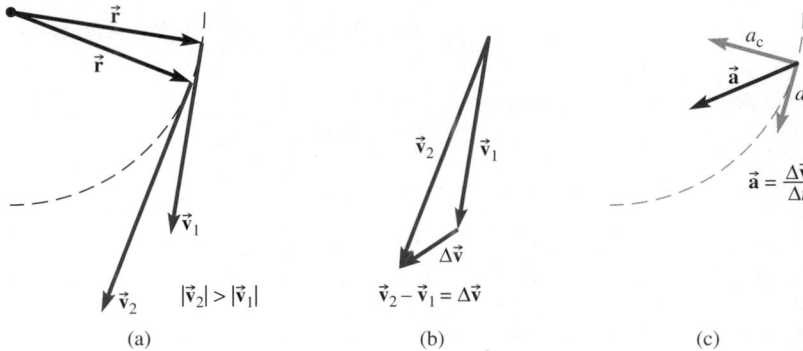

Figure 5.13 Motion along a curved path with a changing speed: (a) the magnitude of velocity $\vec{v}_2$ is greater than the magnitude of velocity $\vec{v}_1$, (b) the direction of $\Delta\vec{v}$ is not radial when the speed is changing, and (c) components of $\vec{a}$ can be taken along a tangent to the curved path (a_t) and along a radius (a_c).

does *not* become perpendicular to the velocity, as it did for uniform circular motion. Thus the direction of the acceleration is *not* centripetal if the speed is changing. However, we can resolve the acceleration into tangential and centripetal components (Fig. 5.13c). The centripetal component a_c changes the *direction* of the velocity, while the tangential component a_t changes the *magnitude* of the velocity. Since these are perpendicular components of the acceleration, the magnitude of the acceleration is

$$a = \sqrt{a_c^2 + a_t^2}$$

Using the same method as in Section 5.2 to find the centripetal acceleration, but working here with only the centripetal *component* of the acceleration, we find that

$$a_c = \frac{v^2}{r} = \omega^2 r \quad (\omega \text{ in radians per unit time}) \tag{5-11}$$

For circular motion, whether uniform or nonuniform, the centripetal component of the acceleration is given by Eq. (5-11). However, in *uniform* circular motion the centripetal component of the acceleration a_c is constant in magnitude, while for nonuniform circular motion a_c changes as the speed changes.

Also still true for nonuniform circular motion is the relationship between speed and angular speed:

$$v = r\omega \tag{5-6}$$

Many problems involving nonuniform circular motion are solved in the same way as for uniform circular motion. We find the *centripetal component of the net force* and then apply Newton's second law along the centripetal direction:

$$\Sigma F_c = ma_c$$

Problem-Solving Strategy for an Object in Nonuniform Circular Motion

1. Begin as for any Newton's second law problem: identify all the forces acting on the object and draw a free-body diagram.
2. Choose perpendicular axes at the point of interest so that one axis is centripetal and the other is tangent to the circular path.
3. Resolve each force into centripetal and tangential components.
4. Apply Newton's second law

$$\Sigma F_c = ma_c$$

where

$$a_c = \frac{v^2}{r} = \omega^2 r$$

and

$$\Sigma F_t = ma_t$$

The tangential acceleration component a_t determines how the speed of the body changes.

Example 5.11

Vertical Loop-the-Loop

 A roller coaster includes a vertical circular loop of radius 20.0 m (Fig. 5.14a). What is the minimum speed at which the car must move at the top of the loop so that it doesn't lose contact with the track?

Strategy A roller coaster car moving around a vertical loop is in nonuniform circular motion; its speed decreases on the way up and increases on the way back down. Nevertheless, it is moving in a circle and has a centripetal acceleration component as given in Eq. (5-11) as long as it moves in a circle. The only forces acting on the car are gravity and the normal force of the track pushing the car. Even if frictional or drag forces are present, at the top of the loop they act in the tangential direction and thus do not contribute to the centripetal component of the net force. At the top of the loop, the track exerts a normal force on the car as long as the car moves with a speed great enough to stay on the track. If the car moves too slowly, it loses contact with the track and the normal force is then zero.

Solution The normal force exerted by the track on the car at the top pushes the car *away* from the track (downward); the normal force cannot pull up on the car. Then, at the top of the loop, the gravitational force and the normal force both point straight down toward the center of the loop. Figure 5.14b is a free-body diagram for the car. From Newton's second law,

$$\Sigma F_c = N + mg = ma_c = \frac{mv_t^2}{r}$$

or

$$N = \frac{mv_t^2}{r} - mg$$

where v_t stands for the speed at the top. In this expression, N stands for the magnitude of the normal force. Since $N \geq 0$,

$$m\left(\frac{v_t^2}{r} - g\right) \geq 0$$

or

$$v_t \geq \sqrt{gr}$$

Imagine sending a roller coaster car around the loop many times with a slightly smaller speed at the top each time. As v_t approaches $\sqrt{gr}$, the normal force at the top gets smaller and smaller. When $v_t = \sqrt{gr}$, the normal force just becomes zero at the top of the loop. Any slower and the car loses contact with the track *before* getting to the highest point and would fall off the track unless prevented from falling by a backup safety mechanism. Therefore, the minimum speed at the top is

$$v_t = \sqrt{gr} = \sqrt{9.8 \text{ m/s}^2 \times 20.0 \text{ m}} = 14 \text{ m/s}$$

Discussion If the car is going faster than 14 m/s at the top, its centripetal acceleration is larger. The track pushing on the car provides the additional net force component that results in a larger centripetal acceleration. The minimum speed occurs when gravity alone provides the centripetal acceleration at the top of the loop. In other words, $a_c = g$ at the top of the loop for minimum speed.

Practice Problem 5.11 Normal force at bottom of track

If the speed of the roller coaster at the *bottom* of the loop is 25 m/s, what is the normal force exerted on the car by the track in terms of the car's weight mg? (See Fig. 5.14c.)

Physics at Home

Go outside on a warm day and fill a bucket with water. Swing the bucket around in a vertical circle over your head. What, if anything, keeps the water in the bucket when the bucket is upside down over your head? Why doesn't the water spill out? Do any upward forces act on the water at that point? [*Hint:* An FBD for the water when it is directly overhead is identical to the FBD for a roller coaster car at the top of a loop.] If it is warm enough and you don't mind getting wet, try swinging the bucket at lower and lower speeds to show that there is a minimum speed at the top if the water is to stay in the bucket.

Figure 5.14 (a) A roller coaster on a circular vertical loop, (b) free-body diagram for the car at the top of the loop, and (c) free-body diagram for the car at the bottom of the loop.

Conceptual Example 5.12

Acceleration of Pendulum Bob

A pendulum is released from rest at point A (Fig. 5.15). Sketch qualitatively the acceleration vector of the pendulum bob at points B and C.

Strategy The pendulum bob moves along the arc of a circle, but not at constant speed. To find the centripetal component, we use

$$a_c = \frac{v^2}{r}$$

Unless $v = 0$, the centripetal component is nonzero. As the pendulum bob swings toward the bottom, its speed is increasing; as it rises on the other side, its speed is decreasing. When the speed

is increasing, the tangential component of the acceleration a_t is in the same direction as the velocity; when the speed is decreasing, a_t is in the direction *opposite* to the velocity. At the lowest point, the speed is neither increasing nor decreasing and $a_t = 0$.

Solution and Discussion The bob's speed increases on the way down until it reaches point B. Therefore, between A and B, the tangential acceleration is parallel to the velocity. On the way back up, between points B and D, the speed is decreasing and the tangential acceleration is opposite to the velocity. At point B, where the bob has its maximum speed, the speed is neither increasing nor decreasing. At that point the tangential

continued on next page

Conceptual Example 5.12 *continued*

acceleration is zero. Therefore, at point *B*, the acceleration points in the centripetal direction: straight up (Fig. 5.15b).

The acceleration at point *C* has both tangential and centripetal components. The tangential acceleration is opposite to the velocity. Figure 5.15c shows the tangential and centripetal components added to form the acceleration vector $\vec{\mathbf{a}}$. We do not know the exact direction since we don't know the relative magnitudes of a_c and a_t.

Conceptual Practice Problem 5.12   Acceleration of bob at later time

Sketch the acceleration vector of the pendulum bob at point *D*, the highest point in its swing to the right.

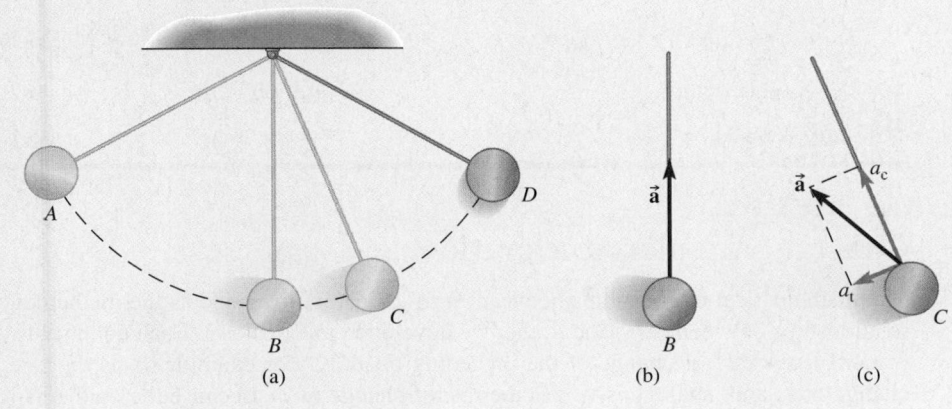

Figure 5.15
(a) A pendulum swings to the right, starting from rest at point *A*, (b) acceleration of the bob at point *B*, and (c) at point *C*, the bob has both tangential and centripetal acceleration components.

5.6   ANGULAR ACCELERATION

An object in nonuniform circular motion has a changing speed and a changing angular velocity. To describe how the angular velocity changes, we define an angular acceleration. If the angular velocity is ω_1 at time t_1 and is ω_2 at time t_2, the change in angular velocity is

$$\Delta\omega = \omega_2 - \omega_1$$

The time interval during which the angular velocity changes is $\Delta t = t_2 - t_1$. The average rate at which the angular velocity changes is called the **average angular acceleration, α_{av}.**

$$\alpha_{av} = \frac{\omega_2 - \omega_1}{t_2 - t_1} = \frac{\Delta\omega}{\Delta t} \tag{5-15}$$

As we let the time interval become shorter and shorter, α_{av} approaches the **instantaneous angular acceleration, α.**

$$\alpha = \lim_{\Delta t \to 0} \frac{\Delta\omega}{\Delta t} \tag{5-16}$$

If ω is in units of rad/s, α is in units of rad/s².

The angular acceleration is closely related to the tangential component of the acceleration. The tangential component of velocity is

$$v_t = r\omega \tag{5-6}$$

Equation (5-6) gives us a way to relate tangential acceleration to the angular acceleration. The tangential acceleration is the rate of change of the tangential velocity, so

$$a_t = \frac{\Delta v_t}{\Delta t} = r\frac{\Delta\omega}{\Delta t} \quad \text{(in the limit } \Delta t \to 0\text{)}$$

Therefore,

$$a_t = r\alpha \tag{5-17}$$

Table 5.1

Relationships between θ, ω, and α for Constant Angular Acceleration Are Analogous to the Relationships between x, v_x, and a_x for Constant Acceleration Along the x-Axis.

Linear Variables: x, v_x, a_x; Constant Linear Acceleration	Angular Variables: θ, ω, α; Constant Angular Acceleration	
$\Delta x = x - x_0 = v_{\text{av},x}\,\Delta t$	$\Delta\theta = \theta - \theta_0 = \omega_{\text{av}}\,\Delta t$	(5-18)
$v_x - v_{0x} = a_x t$	$\omega - \omega_0 = \alpha t$	(5-19)
$v_{\text{av},x} = \dfrac{v_{0x} + v_x}{2}$	$\omega_{\text{av}} = \dfrac{\omega_0 + \omega}{2}$	(5-20)
$x - x_0 = v_{0x} t + \frac{1}{2} a_x t^2$	$\theta - \theta_0 = \omega_0 t + \frac{1}{2}\alpha t^2$	(5-21)
$v_x^2 - v_{0x}^2 = 2 a_x \Delta x$	$\omega^2 - \omega_0^2 = 2\alpha\,\Delta\theta$	(5-22)

Constant Angular Acceleration

The mathematical relationships between θ, ω, and α are the same as the mathematical relationships between x, v_x, and a_x that we developed in Chapter 3. Each quantity is the instantaneous rate of change of the preceding quantity. For example, a_x is the rate of change of v_x and, analogously, α is the rate of change of ω. Because the mathematical relationships are the same, we can solve rotational kinematics problems with constant angular acceleration α in exactly the same way as we did for linear motion with constant acceleration a_x. All we have to do is take the kinematics relations and replace x with θ, v_x with ω, and a_x with α (see Table 5.1).

Equation (5-18) is the definition of average angular velocity. Equation (5-19) is the definition of average angular acceleration, with α_{av} replaced by α since the angular acceleration is constant. Constant α means that ω changes linearly with time; therefore, the average angular velocity is halfway between the initial and final angular velocities for any time interval [Eq. (5-20)]. Equations (5-21) and (5-22) can be derived from the preceding three relations. To interpret Eq. (5-21), note that the average angular velocity is halfway between the initial and final angular velocities, so the average angular velocity is ω_0 plus *half* the change in ω:

$$\omega_{\text{av}} = \omega_0 + \tfrac{1}{2}\alpha t$$

Now, using the average angular velocity to find $\Delta\theta$ yields Eq. (5-21).

Example 5.13

A Rotating Potter's Wheel

A potter's wheel rotates from rest to 210 rpm in a time of 0.75 s. (a) What is the angular acceleration of the wheel during this time, assuming constant angular acceleration? (b) How many revolutions does the wheel make during this time interval? (c) Find the tangential and centripetal components of the acceleration of a point 12 cm from the rotation axis when the wheel is spinning at 180 rpm.

Strategy We know the initial and final frequencies, so we can find the initial and final angular velocities. We also know the time it takes for the wheel to get to the final angular velocity.

That is all we need to find the average angular acceleration that, for constant angular acceleration, is equal to the instantaneous angular acceleration. To find the number of revolutions, we can find the angular displacement $\Delta\theta$ in radians and then divide by 2π rad/rev. We can find the angular velocity at $t = 0.75$ s and use it to find the centripetal acceleration component. The tangential acceleration is calculated from α.

Solution   (a) Initially the wheel is at rest, so the initial angular velocity is zero.

$$\omega_{\text{i}} = 0 \text{ rad/s}$$

continued on next page

Example 5.13 *continued*

Converting 210 rpm to rad/s gives the final angular velocity:

$$\omega = 210\,\frac{\text{rev}}{\text{min}} \times \frac{1}{60}\,\frac{\text{min}}{\text{s}} \times 2\pi\,\frac{\text{rad}}{\text{rev}} = 7.0\pi\,\text{rad/s}$$

The angular acceleration is the rate of change of the angular velocity. Since α is constant, we can calculate it by finding the *average* angular acceleration for the time interval:

$$\alpha = \frac{\omega_f - \omega_i}{t_f - t_i} = \frac{7.0\pi\,\text{rad/s} - 0}{0.75\,\text{s} - 0} = \frac{7.0\pi\,\text{rad/s}}{0.75\,\text{s}} = 29\,\text{rad/s}^2$$

(b) Since α is constant, the *average* angular velocity is

$$\omega_{av} = \tfrac{1}{2}(\omega_0 + \omega)$$

The angular displacement is

$$\Delta\theta = \omega_{av}\,\Delta t = \tfrac{1}{2}(7.0\pi\,\text{rad/s} + 0)(0.75\,\text{s}) = 8.25\,\text{rad}$$

Since 2π rad = one revolution, the number of revolutions is

$$\frac{8.25\,\text{rad}}{2\pi\,\text{rad/rev}} = 1.3\,\text{rev}$$

(c) At 180 rpm, the angular velocity is

$$\omega = 180\,\frac{\text{rev}}{\text{min}} \times \frac{1}{60}\,\frac{\text{min}}{\text{s}} \times 2\pi\,\frac{\text{rad}}{\text{rev}} = 6.0\pi\,\text{rad/s}$$

The centripetal acceleration component is

$$a_c = \omega^2 r = (6.0\pi\,\text{rad/s})^2 \times (0.12\,\text{m}) = 43\,\text{m/s}^2$$

and the tangential acceleration component is

$$a_t = \alpha r = 29\,\text{rad/s}^2 \times 0.12\,\text{m} = 3.5\,\text{m/s}^2$$

Discussion A quick check involves another of the equations for constant acceleration:

$$\omega^2 - \omega_0^2 = 2\alpha\,\Delta\theta$$

Since $\omega_0 = 0$, we can check

$$\omega = \sqrt{2\alpha\,\Delta\theta}$$

From the answers to (a) and (b),

$$\sqrt{2\alpha\,\Delta\theta} = \sqrt{2 \times 29\,\text{rad/s}^2 \times 8.25\,\text{rad}} = 22\,\text{rad/s}$$

The original value for ω in rad/s was 7.0π rad/s. Since $\pi \approx 22/7$, the check is successful.

Practice Problem 5.13 Rotation on a turntable

A Beatles phonograph record of diameter 0.305 m is revolving at 33.3 rpm on a turntable. (a) If the record comes to a stop in 2.0 s with a uniform angular acceleration when the turntable is switched off, what is that angular acceleration? (b) Through what angle does the record turn during this time interval?

5.7 ARTIFICIAL GRAVITY

Making The Connection:
apparent
weightlessness of
orbiting astronauts

You are no doubt familiar with pictures of astronauts "floating" while in orbit around the Earth. It seems as if the astronauts are weightless. To be truly weightless, the force of gravity acting on the astronauts due to Earth would have to be zero, or at least close to zero. Is it? We can calculate the weight of an astronaut in orbit. The orbital altitude for the space shuttle is typically about 600 km above the Earth. Then the orbital radius is 600 km + 6400 km = 7000 km. Comparing the astronaut's weight in orbit to his or her weight on Earth's surface,

$$\frac{W_{\text{orbit}}}{W_{\text{surface}}} = \frac{\dfrac{GM_E m}{(R_E + h)^2}}{\dfrac{GM_E m}{R_E^2}} = \frac{R_E^2}{(R_E + h)^2} = \frac{(6400\,\text{km})^2}{(7000\,\text{km})^2} = 0.84$$

The weight in orbit is 0.84 times the weight on the surface. The astronaut weighs less but certainly isn't *weightless!* Then why does the astronaut *seem* to be weightless?

Recall Section 3.7 on the apparent weightlessness of someone unfortunate enough to be in an elevator when the cable snaps. In that situation, the elevator and the passenger both have the same acceleration ($\vec{\mathbf{a}} = \vec{\mathbf{g}}$). Similarly, the astronaut has the same acceleration as the space shuttle, which is equal to the *local* gravitational field $\vec{\mathbf{g}}$. The magnitude of the apparent weight (the feeling of "weight" as experienced by the astronaut) is the magnitude of the normal force acting on the body:

$$W' = m|\vec{\mathbf{g}} - \vec{\mathbf{a}}| \qquad (3\text{-}17)$$

 Apparent weightlessness occurs when $\vec{\mathbf{a}} = \vec{\mathbf{g}}$, where $\vec{\mathbf{g}}$ is the *local* gravitational field.

Stunt pilots have to be careful about the accelerations to which they subject their bodies. An acceleration of about $3g$ can cause temporary blindness due to an inadequate supply of oxygen to the retina; the heart has difficulty pumping blood up to the head due to the blood's increased apparent weight. Larger accelerations can cause unconsciousness. Pressurized flight suits enable pilots to sustain accelerations up to about $5g$.

Example 5.14

Stunt Pilot

 Dave wants to practice vertical circles for a flying show exhibition (Fig. 5.16). (a) What must the minimum radius of the circle be to ensure that his acceleration at the bottom does not exceed $3.0g$? The speed of the plane is 78 m/s at the bottom of the circle. (b) What is Dave's apparent weight at the bottom of the circular path? Express your answer in terms of his true weight W.

Strategy For the *minimum* radius, we use the maximum possible centripetal acceleration since $a_c = v^2/r$. For the maximum centripetal acceleration, the *tangential* acceleration must be zero—the magnitude of the acceleration is $a = \sqrt{a_c^2 + a_t^2}$. Therefore, the centripetal acceleration component has magnitude $3.0g$ at the bottom. To find Dave's apparent weight, we do not need to use the numerical value of the radius found in part (a); we already know that his acceleration is upward and has magnitude $3.0g$.

Figure 5.16
Forces acting on the plane at the bottom of the circle.

Solution (a) The magnitude of the centripetal acceleration is

$$a_c = v^2/r$$

Solving for the radius,

$$r = \frac{v^2}{a_c} = \frac{v^2}{3.0g}$$

$$= \frac{(78 \text{ m/s})^2}{3.0 \times 9.8 \text{ m/s}^2} = 210 \text{ m}$$

(b) Dave's apparent weight is the magnitude of the normal force of the plane pushing up on him. Let the y-axis point upward. The normal force is up and the gravitational force is down. Then

$$\Sigma F_y = N - mg = ma_y$$

where $a_y = +3.0g$. Therefore,

$$W' = N = m(g + a_y) = 4.0mg$$

His apparent weight is 4.0 times his true weight.

Discussion It might have been tempting to jump to the conclusion that an acceleration of $3.0g$ means that his apparent weight is $3.0mg$. But is his apparent weight zero when his acceleration is zero? No.

According to Eq. (3-17), the apparent weight is $W' = m|\vec{\mathbf{g}} - \vec{\mathbf{a}}|$. In the solution we *added* $g + a_y$. Is there a contradiction? No; because $\vec{\mathbf{g}}$ is *down* while $\vec{\mathbf{a}}$ is *up*. Therefore,

$$\vec{\mathbf{g}} - \vec{\mathbf{a}} = \vec{\mathbf{g}} + (-\vec{\mathbf{a}}) = (g \text{ downward}) + (3.0g \text{ downward})$$

$$= (4.0g \text{ downward}).$$

Practice Problem 5.14   Astronaut's apparent weight

What is the apparent weight of a 730-N astronaut when her spaceship has an acceleration of magnitude $2.0g$ in the following two situations: (a) just above the surface of Earth, acceleration straight up; (b) far from any stars or planets?

In order for astronauts to spend long periods of time living in a space station without the deleterious effects of apparent weightlessness, *artificial gravity* must be created on the station. Many science fiction novels and movies feature ring-shaped space stations that rotate in order to create artificial gravity for the occupants.

The centrifuge is a device that creates artificial gravity on a smaller scale. Centrifuges are common not only in scientific and medical laboratories but also in everyday life. The first successful centrifuge was used to separate cream from milk in the 1880s. Water drips out of sopping wet clothes due to the pull of gravity when the clothes are hung on a clothesline, but the water is removed much faster by the artificial gravity created in the spin cycle of a washing machine.

If we attribute the apparent weight of an object to an apparent gravitational field $\vec{\mathbf{g}}'$, then

$$\vec{\mathbf{W}}' = m(\vec{\mathbf{g}} - \vec{\mathbf{a}}) = m\vec{\mathbf{g}}'$$

and

$$\vec{\mathbf{g}}' = \vec{\mathbf{g}} - \vec{\mathbf{a}} \tag{5-23}$$

In many cases, $g \ll a$ so that $\vec{\mathbf{g}}' \approx -\vec{\mathbf{a}}$. The apparent gravitational field is then equal in magnitude to the acceleration, but oppositely directed. In a rotating space station, the acceleration of an astronaut is inward (toward the rotation axis), but the apparent gravitational field is outward. Therefore, the ceiling of rooms on the station are closest to the rotation axis and the floor is farthest away (Fig. 5.17).

Figure 5.17 A rotating space station from the movie *2001: A Space Odyssey.*

MASTER THE CONCEPTS

Summary

- The angular displacement $\Delta\theta$ is the angle through which an object has turned. Positive and negative angular displacements indicate rotation in different directions. Conventionally, positive represents counterclockwise motion.

- Average angular velocity:

$$\omega_{av} = \frac{\theta_2 - \theta_1}{t_2 - t_1} = \frac{\Delta\theta}{\Delta t} \tag{5-2}$$

- Average angular acceleration:

$$\alpha_{av} = \frac{\omega_2 - \omega_1}{t_2 - t_1} = \frac{\Delta\omega}{\Delta t} \tag{5-15}$$

- The instantaneous angular velocity and acceleration are the limits of the average quantities as $\Delta t \to 0$.

- A useful measure of angle is the radian:

$$2\pi \text{ radians} = 360°$$

  Using radian measure for θ, the arc length s of a circle of radius r subtended by an angle θ is

$$s = \theta r \quad (\theta \text{ in radian measure}) \tag{5-5}$$

- Using radian measure for ω, the speed of an object in circular motion (including a point on a rotating object) is

$$v = r\omega \quad (\omega \text{ in radians per unit time}) \tag{5-6}$$

- Using radian measure for α, the tangential acceleration component is related to the angular acceleration by

$$a_t = r\alpha \quad (\alpha \text{ in radians per time}^2) \tag{5-17}$$

- An object moving in a circle has a centripetal acceleration component given by

$$a_c = \frac{v^2}{r} = \omega^2 r \quad (\omega \text{ in radians per unit time}) \tag{5-11}$$

  The tangential and centripetal accelerations are two perpendicular components of the acceleration. The centripetal acceleration component changes the direction of the velocity and the tangential acceleration component changes the speed.

- Uniform circular motion means that v and ω are constant. In uniform circular motion, the time to complete one revolution is constant and is called the period T. The frequency f is the number of revolutions completed per second.

$$f = 1/T \tag{5-7}$$
$$v = 2\pi r/T = 2\pi rf \tag{5-8}$$
$$\omega = v/r = 2\pi f \tag{5-9}$$

  where the SI unit of angular velocity is rad/s and that of frequency is rev/s = Hz.

- A rolling object is both rotating and translating. If the object rolls without skidding or slipping, then

$$v_{axle} = r\omega$$

- Kepler's third law says that the square of the period of a planetary orbit is proportional to the cube of the orbital radius:

$$T^2 = \text{constant} \times r^3 \tag{5-14}$$

MASTER THE CONCEPTS *continued*

- In the case of constant angular acceleration, we can solve rotational kinematics problems using relationships analogous to those we developed for linear motion:

$$\Delta\theta = \theta - \theta_0 = \omega_{av}\,\Delta t \qquad (5\text{-}18)$$

$$\omega - \omega_0 = \alpha t \qquad (5\text{-}19)$$

$$\omega_{av} = \frac{\omega_0 + \omega}{2} \qquad (5\text{-}20)$$

$$\theta - \theta_0 = \omega_0 t + \tfrac{1}{2}\alpha t^2 \qquad (5\text{-}21)$$

$$\omega^2 - \omega_0^2 = 2\alpha\,\Delta\theta \qquad (5\text{-}22)$$

- When an object is accelerating, its apparent weight differs from its true weight:

$$W' = m|\vec{\mathbf{g}} - \vec{\mathbf{a}}| \qquad (3\text{-}17)$$

where $\vec{\mathbf{g}}$ is the local gravitational field. The apparent gravitational field is

$$\vec{\mathbf{g}}' = \vec{\mathbf{g}} - \vec{\mathbf{a}} \qquad (5\text{-}23)$$

Highlighted Figures and Tables

F5.2 Angular displacement of a spinning CD (p. 147)

F5.3 Definition of the radian (p. 147)

F5.5 Uniform circular motion at a constant speed (p. 152)

F5.6 To find the magnitude of the centripetal acceleration for uniform circular motion, you must find the change in velocity $\Delta\vec{\mathbf{v}}$ for a time interval Δt in the limit $\Delta t \to 0$. (p. 152)

F5.9 Car negotiating a curve on an unbanked roadway (p. 157)

F5.10 Car negotiating a curve on a banked roadway (p. 157)

F5.13 Nonuniform circular motion (p. 164)

T5.1 Relationships between θ, ω, and α for constant angular acceleration are analogous to the relationships between x, v_x, and a_x for constant acceleration along the x-axis. (p. 168)

CONCEPTUAL QUESTIONS

1. Is depressing the "accelerator" (gas pedal) of a car the only way that the driver can make the car accelerate (in the physics sense of the word)? If not, what else can the driver do to give the car an acceleration?

2. Two children ride on a merry-go-round. One is 2 m from the axis of rotation and the other is 4 m from it. Which child has the larger (a) linear speed, (b) acceleration, (c) angular speed, and (d) angular displacement?

3. Explain why the orbital radius and the speed of a satellite in circular orbit are not independent.

4. In uniform circular motion, is the velocity constant? Is the acceleration constant? Explain.

5. In uniform circular motion, the net force is perpendicular to the velocity and changes the direction of the velocity but not the speed. If a projectile is launched horizontally, the net force (ignoring air resistance) is perpendicular to the initial velocity, and yet the projectile gains speed as it falls. What is the difference between the two situations?

6. The speed of a satellite in circular orbit around a planet does not depend on the mass of the satellite. Does it depend on the mass of the planet? Explain.

7. A flywheel (a massive disk) rotates with constant angular acceleration. For a point on the rim of the flywheel, is the tangential acceleration constant? Is the centripetal acceleration constant?

8. Explain why the force of gravity due to the Earth does not pull the Moon in closer and closer on an inward spiral until it hits Earth's surface.

9. When a roller coaster takes a sharp turn to the right, it feels as if you are pushed toward the left. Does a force push you to the left? If so, what is it? If not, why does there *seem* to be such a force?

10. Is there anywhere on Earth where a bathroom scale reads your true weight? If so, where? Where does your apparent weight due to Earth's rotation differ most from your true weight?

MULTIPLE CHOICE QUESTIONS

1. A spider sits on a turntable that is rotating at a constant 33 rpm. The acceleration $\vec{\mathbf{a}}$ of the spider is

    (a) greater the closer the spider is to the central axis.

    (b) greater the farther the spider is from the central axis.

    (c) nonzero and independent of the location of the spider on the turntable.

    (d) zero.

Questions 2–5. A satellite in orbit travels around the Earth in uniform circular motion. In Fig. 5.18, the satellite moves counterclockwise (*ABCDA*). Answer choices:

    (a) +*x*   (b) +*y*   (c) −*x*   (d) −*y*

    (e) 45° above +*x* (toward +*y*)   (f) 45° below +*x* (toward −*y*)

    (g) 45° above −*x* (toward +*y*)   (h) 45° below −*x* (toward −*y*)

2. What is the direction of the satellite's instantaneous velocity at point *D*?

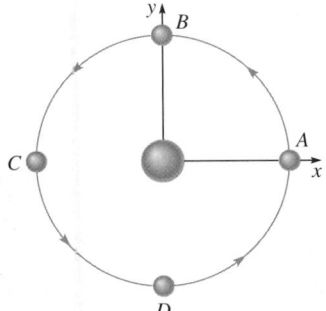

Figure 5.18 Multiple Choice Questions 2–5 and Problem 30

3. What is the direction of the satellite's average velocity for one quarter of an orbit, starting at C and ending at D?

4. What is the direction of the satellite's average acceleration for one half of an orbit, starting at C and ending at A?

5. What is the direction of the satellite's instantaneous acceleration at point C?

6. Two satellites are in orbit around Mars with the same orbital radius. Satellite 2 has twice the mass of satellite 1. The centripetal acceleration of satellite 2 has
   (a) twice the magnitude of the centripetal acceleration of satellite 1.
   (b) the same magnitude as the centripetal acceleration of satellite 1.

(c) half the magnitude of the centripetal acceleration of satellite 1.
(d) four times the magnitude of the centripetal acceleration of satellite 1.

Questions 7–8. A boy swings in a tire swing. Answer choices:
   (a) At the highest point of the motion
   (b) At the lowest point of the motion
   (c) At a point neither highest nor lowest
   (d) It is constant.

7. When is the tension in the rope the greatest?

8. When is the tangential acceleration the greatest?

Questions 9–10 concern these three statements:
   (1) Its acceleration is constant.
   (2) Its centripetal acceleration component is constant in magnitude.
   (3) Its tangential acceleration component is constant in magnitude.

9. An object is in uniform circular motion. Identify the correct statement(s).
   (a) 1 only   (b) 2 only   (c) 3 only
   (d) 1, 2, and 3   (e) 2 and 3   (f) 1 and 2
   (g) 1 and 3   (h) None of them

10. An object is in nonuniform circular motion with constant angular acceleration. Identify the correct statement(s). (Use the same answer choices as Question 9.)

PROBLEMS

Note: **C** indicates a combination conceptual/quantitative problem. Gold diamonds ✦, ✦✦ are used to indicate the increasing level of difficulty of each problem. Problem numbers appearing in blue, 9., denote problems that have a detailed solution available in the Student Solutions Manual. Some problems are *paired* by concept; their numbers are connected by a ruled box.

5.1 Description of Uniform Circular Motion

1. A carnival swing is fixed on the end of an 8.0-m-long beam. If the swing and beam sweep through an angle of 120°, what is the distance through which the riders move?

2. A soccer ball of diameter 31 cm rolls without slipping at a linear speed of 2.8 m/s. Through how many revolutions has the soccer ball turned as it moves a linear distance of 18 m?

3. Find the average angular speed of the second hand of a clock.

4. Convert these to radian measure: (a) 30.0°, (b) 135°, (c) 1/4 revolution, (d) 33.3 revolutions.

5. A bicycle is moving at 9.0 m/s. What is the angular speed of its tires if their radius is 35 cm?

6. An elevator cable winds on a drum of radius 90.0 cm that is connected to a motor. (a) If the elevator is moving down at 0.50 m/s, what is the angular speed of the drum? (b) If the elevator moves down 6.0 m, how many revolutions has the drum made?

7. Grace is playing with her dolls and decides to give them a ride on a merry-go-round. She places one of them on a record player and sets the angular speed at 33.3 rpm. (a) What is their angular speed in rad/s? (b) If the doll is 13 cm from the center of the spinning turntable platform, how fast (in m/s) is the doll moving?

8. A wheel is rotating at a rate of 2.0 revolutions every 3.0 s. Through what angle, in radians, does the wheel rotate in 1.0 s?

✦ 9. In the construction of railroads, it is important that curves be gentle, so as not to damage passengers or freight. Curvature is not measured by the radius of curvature, but in the following way. First a 100.0-ft-long chord is measured. Then the curvature is reported as the angle subtended by two radii at the endpoints of the chord (Fig. 5.19). (The angle is measured by determining the angle between two tangents 100 ft apart; since each tangent is perpendicular to a radius, the angles are the same.) In modern railroad construction, track curvature is kept below 1.5°. What is the radius of curvature of a "1.5° curve"? [*Hint:* Since the angle is small, the length of the chord is approximately equal to the arc length along the curve.]

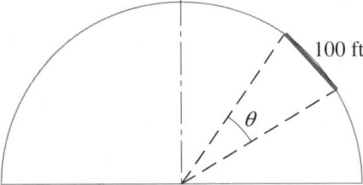

Figure 5.19 Problem 9

5.2 Centripetal Acceleration

10. Verify that all three expressions for centripetal acceleration ($v\omega$, v^2/r, and $\omega^2 r$) have the correct dimensions for an acceleration.

11. The apparatus of Fig. 5.20 is designed to study insects at an acceleration of magnitude 980 m/s² (= 100g). The apparatus consists of a 2.0-m rod with insect containers at either end.

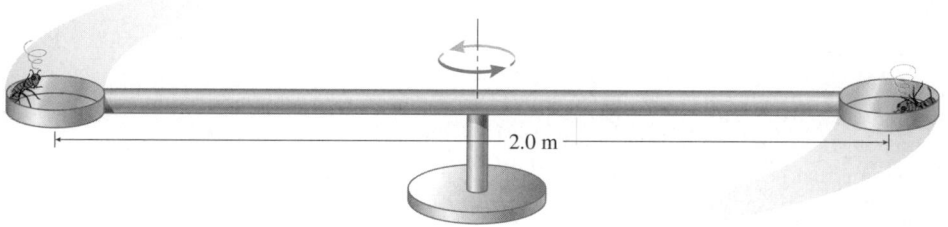

Figure 5.20 Problem 11

The rod rotates about an axis perpendicular to the rod and at its center. (a) How fast does an insect move when it experiences a centripetal acceleration of 980 m/s²? (b) What is the angular speed of the insect?

12. The rotor (Fig. 5.21) is an amusement park ride where people stand against the inside of a cylinder. Once the cylinder is spinning fast enough, the floor drops out. (a) What force keeps the people from falling out the bottom of the cylinder? (b) If the coefficient of friction is 0.40 and the cylinder has a radius of 2.5 m, what is the minimum angular speed of the cylinder so that the people don't fall out? (Normally the operator runs it considerably faster as a safety measure.)

Figure 5.21 Problem 12

13. Objects that are at rest relative to Earth's surface are in circular motion due to Earth's rotation. What is the centripetal acceleration of an African baobab tree located at the equator?

ⓒ 14. A curve in a stretch of highway has radius R. The road is unbanked. The coefficient of static friction between the tires and road is μ_s. (a) What is the fastest speed that a car can safely travel around the curve? (b) Explain what happens when a car enters the curve at a speed greater than the maximum safe speed. Illustrate with a free-body diagram.

ⓒ ✦ 15. Earth's orbit around the Sun is nearly circular. The period is 1 yr = 365.25 days. (a) In an elapsed time of 1 day, what is Earth's angular displacement? (b) What is the change in Earth's velocity, $\Delta\vec{\mathbf{v}}$? (c) What is Earth's average acceleration during 1 day? (d) Compare your answer for (c) to Earth's instantaneous centripetal acceleration. Explain.

✦ 16. A child swings a rock of mass m in a horizontal circle using a rope of length L. The rock moves at constant speed v. (a) Ignoring gravity, find the tension in the rope. (b) Now include gravity (the weight of the rock is no longer negligible, although the weight of the rope still is negligible). What is the tension in the rope? What angle does the rope make with the horizontal?

✦ 17. A *conical pendulum* consists of a bob (mass m) attached to a string (length L) swinging in a horizontal circle (Fig. 5.8). As the string moves, it sweeps out the area of a cone. The angle

that the string makes with the vertical is ϕ. (a) What is the tension in the string? (b) What is the period of the pendulum?

5.3 Banked Curves

18. A highway curve has a radius of 122 m. At what angle should the road be banked so that a car traveling at 26.8 m/s (60 mph) has no tendency to skid sideways on the road? [*Hint:* No tendency to skid means the frictional force is zero.]

19. A curve in a highway has radius of curvature 120 m and is banked at 3.0°. On a day when the road is icy, what is the safest speed to go around the curve?

20. A roller coaster car of mass 320 kg (including passengers) travels around a horizontal curve of radius 35 m. Its speed is 16 m/s. What is the magnitude and direction of the total force exerted on the car by the track?

21. A velodrome is built for use in the Olympics (Fig. 5.22). The radius of curvature of the surface is 20.0 m. At what angle should the surface be banked for cyclists moving at 18 m/s? (Choose an angle so that no frictional force is needed to keep the cyclists in their circular path. Large banking angles *are* used in velodromes.)

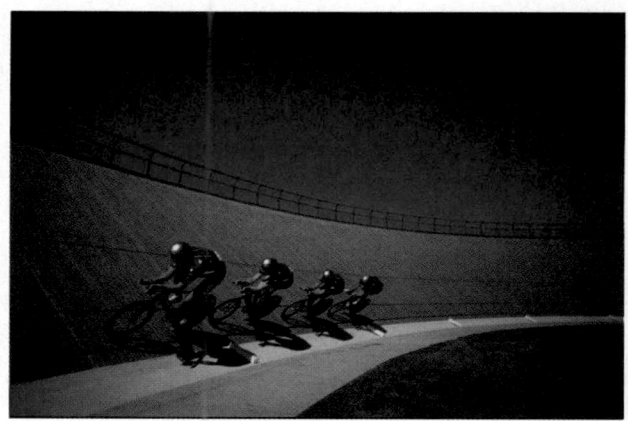

Figure 5.22 Problem 21

ⓒ 22. A car drives around a curve with radius 410 m at a speed of 32 m/s. The road is not banked. The mass of the car is 1400 kg. (a) What is the frictional force on the car? (b) Does the frictional force necessarily have magnitude $\mu_s N$? Explain.

✦ 23. A car drives around a curve with radius 410 m at a speed of 32 m/s. The road is banked at 5.0°. The mass of the car is 1400 kg. (a) What is the frictional force on the car? (b) At what speed could you drive around this curve so that the force of friction is zero?

✦✦ 24. A curve in a stretch of highway has radius R. The road is banked at angle θ to the horizontal. The coefficient of static friction between the tires and road is μ_s. What is the fastest speed that a car can travel through the curve?

25. An airplane is flying at constant speed v in a horizontal circle of radius r. The lift (aerodynamic force) on the wings is perpendicular to the wings. At what angle to the vertical must the wings be banked to fly in this circle?

5.4 Circular Orbits

26. A spy satellite is in circular orbit around Earth. It makes one revolution in 6.00 hours. (a) How high above Earth's surface is the satellite? (b) What is the satellite's acceleration?

✦27. Two satellites are in circular orbits around Jupiter. One, with orbital radius r, makes one revolution every 16 h. The other satellite has orbital radius $4.0r$. How long does the second satellite take to make one revolution around Jupiter?

28. The Hubble Space Telescope orbits Earth 613 km above Earth's surface. What is the period of the telescope's orbit?

29. The orbital speed of Earth about the Sun is 3.0×10^4 m/s and its distance from the Sun is 1.5×10^{11} m. The mass of Earth is approximately 6.0×10^{24} kg and that of the Sun is 2.0×10^{30} kg. What is the magnitude of the force exerted by the Sun on Earth? [*Hint:* Two different methods are possible. Try both.]

30. A satellite travels around Earth in uniform circular motion at altitude 35,800 km above Earth's surface. The satellite is in geosynchronous orbit (that is, the time for it to complete one orbit is exactly one day). In Fig. 5.18, the satellite moves counterclockwise (*ABCDA*). State directions in terms of the *x*- and *y*-axes. (a) What is the satellite's instantaneous velocity at point *C*? (b) What is the satellite's average velocity for one quarter of an orbit, starting at *A* and ending at *B*? (c) What is the satellite's average acceleration for one quarter of an orbit, starting at *A* and ending at *B*? (d) What is the satellite's instantaneous acceleration at point *D*?

✦✦31. A spacecraft is in orbit around Jupiter. The radius of the orbit is 3.0 times the radius of Jupiter (which is $R_J = 71{,}500$ km). The gravitational field at the surface of Jupiter is 23 N/kg. What is the period of the spacecraft's orbit? [*Hint:* You don't need to look up any more data about Jupiter to solve the problem.]

5.5 Nonuniform Circular Motion

32. A roller coaster has a vertical loop with radius 20.0 m. With what minimum speed should the roller coaster car be moving at the top of the loop so that the passengers do not lose contact with the seats?

Ⓒ 33. A pendulum is 0.80 m long and the bob has a mass of 1.0 kg. At the bottom of its swing, the bob's speed is 1.6 m/s. (a) What is the tension in the string at the bottom of the swing? (b) Explain why the tension is greater than the weight of the bob.

✦34. A pendulum (Fig. 5.23) is 0.800 m long and the bob has a mass of 1.00 kg. When the string makes an angle of $\theta = 15.0°$ with the vertical, the bob is moving at 1.40 m/s. Let $g = 9.80$ m/s². Find the tangential and centripetal acceleration components and the tension in the string. [*Hint:* Draw an FBD for the bob. Choose the *x*-axis to be tangential to the motion of the bob and the *y*-axis to be centripetal. Apply Newton's second law.]

✦35. Find the tangential acceleration of a freely swinging pendulum when it makes an angle θ with the vertical (Fig. 5.23).

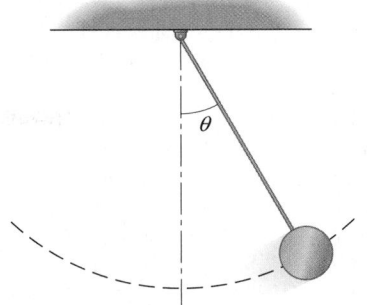

Figure 5.23
Problems 34, 35, and 48

5.6 Angular Acceleration

36. A wheel is subjected to uniform angular acceleration. Initially its angular velocity is zero. During the first 1.0-s time interval, it rotates through an angle of 90.0°. (a) Through what angle does it rotate during the next 1.0-s time interval? (b) Through what angle during the third 1.0-s time interval?

37. A cyclist starts from rest and pedals so that the wheels make 8.0 revolutions in the first 5.0 s. What is the angular acceleration of the wheels (assumed constant)?

38. During normal operation, a computer's hard disk spins at 7200 rpm. If it takes the hard disk 4.0 s to reach this angular velocity starting from rest, what is the average angular acceleration of the hard disk in rad/s²?

39. The turntable of a record player reaches its rated frequency of rotation, 33.3 rpm, in 2.0 s, starting from rest. (a) Assuming the angular acceleration is constant, what is its magnitude? (b) How many revolutions does the turntable make during this time interval?

40. In a Beams ultracentrifuge, the rotor is suspended magnetically in a vacuum. Since there is no mechanical connection to the rotor, the only friction is the air resistance due to the few air molecules in the vacuum. If the rotor is spinning with an angular speed of 5.0×10^5 rad/s and the driving force is turned off, its spinning slows down at an angular rate of 0.40 rad/s². (a) How long does the rotor spin before coming to rest? (b) During this time, through how many revolutions does the rotor spin?

41. The rotor of the Beams ultracentrifuge (see Problem 40) is 20.0 cm long. For a point at the end of the rotor, find the (a) initial speed, (b) tangential acceleration component, and (c) maximum centripetal acceleration component.

5.7 Artificial Gravity

Ⓒ✦42. Objects that are at rest relative to the Earth's surface are in circular motion due to Earth's rotation. (a) What is the centripetal acceleration of an object at the equator? (b) Is the object's apparent weight greater or less than its weight? Explain. (c) By what percentage does the apparent weight differ from the weight at the equator? (d) Does a bathroom scale measure your true weight or your apparent weight?

43. A space station is shaped like a ring and rotates to simulate gravity. If the radius of the space station is 120 m, at what frequency must it rotate so that it simulates Earth's gravity? [*Hint:* The apparent weight of the astronauts must be the same as their weight on Earth.]

44. A person of mass M stands on a bathroom scale inside a Ferris wheel compartment. The Ferris wheel has radius R and angular velocity ω. What is the apparent weight of the person (a) at the top and (b) at the bottom?

45. A person rides a Ferris wheel that turns with constant angular velocity. Her weight is 520.0 N. At the top of the ride her apparent weight is 1.5 N different from her true weight. (a) Is her apparent weight at the top 521.5 N or 518.5 N? Why? (b) What is her apparent weight at the bottom of the ride? (c) If the angular speed of the Ferris wheel is 0.025 rad/s, what is its radius?

The Millenium Wheel in London.

46. If a washing machine's drum has a radius of 25 cm and spins at 4.0 rev/s, what is the strength of the artificial gravity to which the clothes are subjected? Express your answer as a multiple of g.

47. Objects that are at rest relative to Earth's surface are in circular motion due to Earth's rotation. What is the centripetal acceleration of a painting hanging in the Prado Museum in Madrid, Spain, at a latitude of 40.2° North (Fig. 5.24)? (Note that the object's centripetal acceleration is not directed toward the center of the Earth.)

48. A little girl is having fun on a swing of length 8.0 m. If her weight is 180 N, find her apparent weight (a) at the lowest point, where her speed is 9.7 m/s; and (b) at the highest point, where the swing is at an angle $\theta = 75°$ to the vertical (Fig. 5.23).

(a) Three-dimensional view

(b) Cross-sectional view

Figure 5.24 Problem 47

COMPREHENSIVE PROBLEMS

49. The Earth rotates on its own axis once per day (24.0 h). What is the tangential speed of the summit of Mt. Kilimanjaro (elevation 5895 m above sea level), which is located approximately on the equator, due to the rotation of the Earth? The equatorial radius of Earth is 6378 km.

50. A trimmer for cutting weeds and grass near trees and borders has a nylon cord of 0.23-m length that whirls about an axle at 660 rad/s. What is the linear speed of the tip of the nylon cord?

51. A high-speed dental drill is rotating at 3.14×10^4 rad/s. Through how many degrees does the drill rotate in 1.00 s?

52. A jogger runs counterclockwise around a path of radius 90.0 m at constant speed. He makes 1.00 revolution in 188.4 s. At $t = 0$, he is heading due east. (a) What is the jogger's instantaneous velocity at $t = 376.8$ s? (b) What is his instantaneous velocity at $t = 94.2$ s?

53. The time to sunset can be estimated by holding out your arm, holding your fingers horizontally in front of your eyes, and counting the number of fingers that fit between the horizon and the setting Sun. (a) What is the angular speed, in rad/s, of

the Sun's apparent circular motion around the Earth? (b) Estimate the angle subtended by one finger held at arm's length. (c) How long in minutes does it take the Sun to "move" through this same angle?

54. In the professional videotape recording system known as quadriplex, four tape heads are mounted on the circumference of a drum of radius 2.5 cm that spins at 1500 rad/s. (a) At what speed are the tape heads moving? (b) Why are moving tape heads used instead of stationary ones, as in audio tape recorders? [*Hint:* How fast would the tape have to move if the heads were stationary?]

55. The Milky Way galaxy rotates about its center with a period of about 200 million years. The Sun is 2×10^{20} m from the center of the galaxy. How fast is the Sun moving with respect to the center of the galaxy?

56. A small body of mass 0.50 kg is attached by a 0.50-m-long cord to a pin set into the surface of a frictionless table top. The body moves in a circle on the horizontal surface with a speed of 2.0π m/s. (a) What is the magnitude of the centripetal acceleration of the body? (b) What is the tension in the cord?

57. What's the fastest way to make a U-turn at constant speed? Suppose that you need to make a 180° turn on a circular path. The minimum radius (due to the car's steering system) is 5.0 m, while the maximum (due to the width of the road) is 20.0 m. Your acceleration must never exceed 3.0 m/s² or else you will skid. Should you use the smallest possible radius, so the distance is small, or the largest, so you can go faster without skidding, or something in between? What is the minimum possible time for this U-turn?

58. The Milky Way galaxy rotates about its center with a period of about 200 million years. The Sun is 2×10^{20} m from the center of the galaxy. (a) What is the Sun's centripetal acceleration? (b) What is the net gravitational force on the Sun due to the other stars in the Milky Way?

59. A rotating flywheel slows down at a constant rate due to friction in its bearings. After 1 min its angular velocity has diminished to 0.80 of its initial value ω. At the end of the third minute what is the angular velocity in terms of the initial value?

60. A coin is placed on a record that is rotating at 33.3 rpm. If the coefficient of static friction between the coin and the record is 0.1, how far from the center of the record can the coin be placed without having it slip off?

61. Grace gives her dolls a merry-go-round ride on an old phonograph set at 33.3 rpm. The dolls are 5.0 in. from the central axis. She changes the setting to 45 rpm. (a) For this new setting, what is the linear speed of a point on the turntable at the location of the dolls? (b) If the coefficient of static friction between the dolls and the turntable is 0.13, do the dolls stay on the record?

62. Your car's wheels are 65 cm in diameter and the wheels are spinning at an angular velocity of 101 rad/s. How fast is your car moving in kilometers per hour (assume no slippage)?

63. In an amusement park rocket ride, cars are suspended from 4.25-m cables attached to rotating arms at a distance of 6.00 m from the axis of rotation (Fig. 5.25). The cables swing out at an angle of 45.0° when the ride is operating. What is the angular speed of rotation?

64. Centrifuges are commonly used in biological laboratories for the isolation and maintenance of cell preparations. For cell separation, the centrifugation conditions are typically 1.0×10^3 rpm using an 8.0-cm-radius rotor. (a) What is the acceleration of material in the centrifuge under these conditions? Express your answer as a multiple of g. (b) At 1.0×10^3 rpm (and with a 8.0-cm rotor) what is the net force on a red blood cell whose mass is 9.0×10^{-14} kg? (c) What is the net force on a virus particle of mass 5.0×10^{-21} kg under the same conditions? (d) To pellet out virus particles and even to separate large molecules such as proteins, super-high-speed centrifuges called ultracentrifuges are used in which the rotor spins in a vacuum to reduce heating due to friction. What is the centripetal acceleration inside an ultracentrifuge at 75,000 rpm with an 8.0-cm rotor? Express your answer as a multiple of g.

65. Bacteria swim using a corkscrew-like helical flagellum that rotates. For a bacterium with a flagellum that has a pitch of 1.0 μm that rotates at 110 rev/s, how fast could it swim if there were no "slippage" in the medium in which it is swimming? The pitch of a helix is the distance between "threads."

66. Massimo, a machinist, is cutting threads for a bolt on a lathe. He wants the bolt to have 18 threads per inch. If the cutting tool moves parallel to the axis of the would-be bolt at a linear velocity of 0.080 in./s, what must the rotational speed of the lathe chuck be to ensure the correct thread density? [*Hint:* One thread is formed for each complete revolution of the chuck.]

✦ 67. In Chapter 19 we will see that a charged particle can undergo uniform circular motion when acted on by a magnetic force and no other forces. (a) For that to be true, what must be the angle between the magnetic force and the particle's velocity? (b) The magnitude of the magnetic force on a charged particle is proportional to the particle's speed, $F = kv$. Show that two identical charged particles moving in circles at different speeds in the same magnetic field must have the same period. (c) Show that the radius of the particle's circular path is proportional to the speed.

✦✦ 68. Find the orbital radius of a geosynchronous satellite. Do not assume the speed found in Example 5.9. Start by writing an equation that relates the period, radius, and speed of the orbiting satellite. Then apply Newton's second law to the satellite. You will have two equations with two unknowns (the speed and radius). Eliminate the speed algebraically and solve for the radius.

Figure 5.25 Problem 63

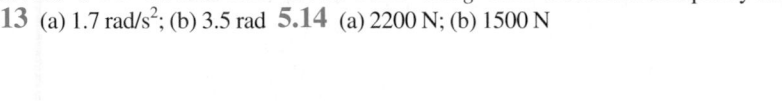

ANSWERS TO PRACTICE PROBLEMS

5.1 2.99×10^{-7} rad/s **5.2** 1.65 m/s **5.3** 1.9 min **5.4** 17 m/s² **5.5** (a) 4.5 m/s²; (b) 3.0 m/s **5.6** More slowly **5.7** No **5.8** 29.8 km/s; 3.16×10^7 s **5.9** 0.723R **5.10** 2.44 h **5.11** 4.2mg **5.12** Acceleration is purely tangential:
5.13 (a) 1.7 rad/s²; (b) 3.5 rad **5.14** (a) 2200 N; (b) 1500 N

Chapter 6

Energy

How high can a kangaroo jump? A kangaroo can jump about 0.70 m high when stretching its hind leg tendons approximately half of the maximum amount. How high can the same kangaroo jump when the tendons are stretched by an additional 10%?

Conservation law: a physical law phrased in terms of a quantity that does not change with time.

6.1   A CONSERVATION LAW

The hopping kangaroo is an example of the transformation of energy in a living thing. A kangaroo has long elastic tendons in its hind legs. As the kangaroo lands, these tendons store *elastic energy* to assist in making the next jump. Energy is then alternately expended and stored by the kangaroo as it propels itself forward in one jump after another. The storage of energy in the tendons reduces the amount of energy that the muscles must supply for subsequent jumps.

Energy comes in many different forms, including chemical energy, nuclear energy, mechanical energy, electrical energy, kinetic energy, and elastic energy. Energy can be changed from one form to another, or transferred from one place to another, but the total quantity of energy in any closed system cannot change. (A *closed system* is a set of interacting bodies upon which no net external force acts and that does not transfer energy to or from anything outside the system.) With energy we encounter our first **conservation law**: a physical law phrased in terms of a quantity that does not change with time. Conservation laws are often considered to be the most fundamental laws of physics.

The law of conservation of energy says that energy can neither be created nor destroyed. To apply this law, we need to learn how to calculate the amount of each kind of energy. There isn't one formula that applies to all.

Conservation of energy is a powerful tool in the search to understand nature. It is a great simplification to look at energy transformations in the complex interactions that occur, for instance, in the human body. The body converts chemical energy stored in food into various other forms of energy. Some of the energy is available to fuel physical activity (Fig. 6.1). Conservation of energy requires that when the amount of one form of energy decreases, the amount of other forms of energy increases such that the total energy remains the same. Over the years, scientists have faced situations where the conservation of energy seemed to be violated, only to find that energy had been converted into a previously unrecognized form.

6.2   WORK DONE BY CONSTANT FORCES

Since conservation of energy is an important law of nature, we need to define various forms of energy and learn how energy can be transformed from one form to another. We begin with an example. Suppose the trunk in Fig. 6.2a weighs 640 N and must be lifted a height $h = 1.5$ m. To lift it at constant speed, Rosie must exert a force of 640 N on the rope (assuming an ideal pulley and rope). As discussed in Example 2.13, the force required would be half as much (320 N) if she were to use the two-pulley system of Fig. 6.2b

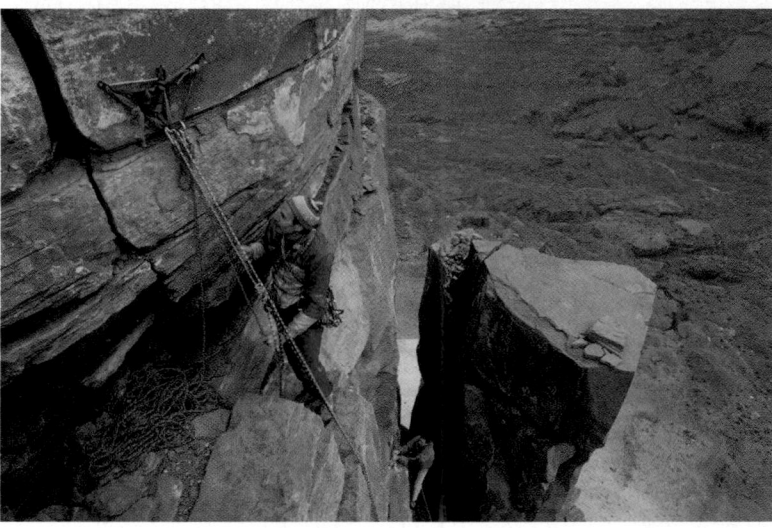

Figure 6.1 These rock climbers must convert chemical energy into other forms of energy.

(a) (b)

Figure 6.2 (a) Rosie is trying to move a trunk into her dorm room through the window. As long as she holds the trunk motionless in one position, she is doing no work on the trunk (in the physics sense of the word *work*). (b) The two-pulley system makes it easier for Rosie to lift the trunk: the force she must exert is halved. Is she getting something for nothing, or does she still have to do the same amount of work to lift the trunk?

instead. She doesn't get something for nothing, though. To lift the trunk 1.5 m, the sections of rope on *both* sides of pulley 2 must be shortened by 1.5 m, so Rosie must pull a 3.0-m length of rope. The two-pulley system enables her to pull with half the force, but now she must pull the rope twice as far.

Notice that the *product* of the magnitude of the force and the distance is the same for both cases:

$$640 \text{ N} \times 1.5 \text{ m} = 320 \text{ N} \times 3.0 \text{ m} = 960 \text{ N·m} = W$$

This product is called the **work** (W) done. Work is a scalar quantity; it does not have a direction. The same symbol W is also commonly used for the magnitude of the weight of an object. We can distinguish these two uses of the same symbol by context. Weight has dimensions of force, while work has dimensions of force times distance. On the other hand, the symbol $\vec{W}$ can never represent work, since work is not a vector quantity.

As with velocity, force, and acceleration, the word *work* has a precise definition in physics; don't be misled by the many different meanings it has in ordinary conversation. We talk about doing homework, or going to work, or having too much work to do to have time to go to the movies. Not everything we call "work" in conversation is *work* as defined in physics.

The SI unit of work and energy is the newton-meter (N·m), which is given the name joule (symbol: J). Using either method, Rosie must do 960 J of work to lift the trunk. When we say that Rosie does 960 J of work, we mean that Rosie supplies 960 J of energy—the amount of energy required to lift the trunk 1.5 m. *Work is an energy transfer due to the application of a force.* Simple machines such as levers, pulleys, and inclined planes can reduce the *force* needed to accomplish a task, but they cannot reduce the *work* that must be done.

According to the definition, Rosie does no work while she holds the trunk in one place. She can just as well fasten the rope to a tie-down and walk away (Fig. 6.3). Work is done only when she *moves* the trunk by moving the rope. *If there is no displacement, no work is done and no energy is transferred.* A more exact statement is that Rosie does no work *on the rope* or *on the trunk*. Work *is* being done inside her body by the muscle fibers, which continually contract and expand to hold a load in one place. Thus, Rosie does expend chemical energy to hold the trunk in place and she gets tired if she holds it there very long, but the energy expended is not transferred to the trunk.

SI unit of work and energy is the joule: 1 J = 1 N·m

Definition of work: an energy transfer due to the application of a force.

Figure. 6.3 While the trunk is held in place by tying the rope, no work is done and no energy transfers occur.

The force that Rosie exerts on the rope is in the same direction as the displacement of that end of the rope. How much work is done by a force that is at some other angle to the displacement? Only the component of the force in the direction of the displacement does work. So, in general, the work done by a constant force is defined as the product of the magnitude of the displacement and the *component* of the force *parallel to the displacement*. If the displacement is in the *x*-direction, then $W = F_x \Delta x$. If θ is the angle between the force and displacement vectors, then the force component in the direction of the displacement is $F \cos \theta$ (Fig. 6.4a). Therefore, work done by a constant force can be written $W = F \Delta r \cos \theta$, where F is the magnitude of the force and Δr is the magnitude of the displacement of the point of application of the force.

Work done by a constant force:
$$W = F \Delta r \cos \theta \tag{6-1}$$
For a displacement along the *x*-axis,
$$W = F_x \Delta x \tag{6-1a}$$

There is another way to calculate work. We identified $F \cos \theta$ as the component of the force along the direction of the displacement. We can just as well identify $\Delta r \cos \theta$ as the component of the *displacement* in the direction of the *force* (Fig. 6.4b). Therefore, the work done by a force is equal to the magnitude of the force times the parallel component of the displacement. For example, gravity pulls downward, so the work done by gravity is the weight times the *downward component* of the displacement. If the *y*-axis is taken to be upward, then Δy is the upward component of the displacement. The *downward* component of the displacement is $-\Delta y$.

$$W_g = F_g \times (-\Delta y) = -mg \, \Delta y \tag{6-2}$$
(in a uniform gravitational field with the *y*-axis pointing up)

Equation (6-2) holds even if the object does not move in a straight-line path.

The application of a force can decrease as well as increase the energy of the object to which the force is applied. If a force has a component parallel to the displacement of the object, then the work done by the force increases the object's energy. If a force has a component *antiparallel* to the displacement of the object, then the work done by the force decreases the object's energy. When the angle between $\vec{\mathbf{F}}$ and $\Delta \vec{\mathbf{r}}$ is less than 90°, $\cos \theta$ in Eq. (6-1) is positive. The force does positive work and the object's energy increases. If the angle between $\vec{\mathbf{F}}$ and $\Delta \vec{\mathbf{r}}$ is greater than 90°, $\cos \theta$ is negative. The force does negative work and the object's energy decreases. Pay careful attention to the algebraic sign when calculating work: negative work done on a system means that the energy of the system decreases, while positive work represents an energy increase.

If the force is perpendicular to the displacement, then there is no force component along the direction of the displacement and therefore no work is done. Equation (6-1) includes this case: if the angle between the force and the displacement is $\theta = 90°$, then the work done is zero since $\cos 90° = 0$. The normal force does no work on a sliding

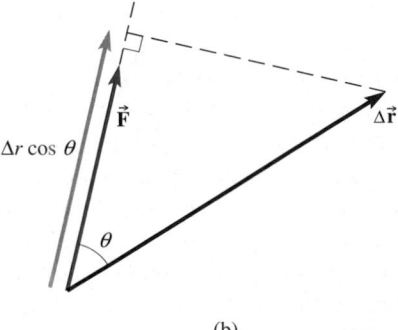

Figure 6.4 (a) The component of $\vec{\mathbf{F}}$ in the direction of $\Delta \vec{\mathbf{r}}$ is $F \cos \theta$. (b) The component of $\Delta \vec{\mathbf{r}}$ in the direction of $\vec{\mathbf{F}}$ is $\Delta r \cos \theta$.

(a)   (b)

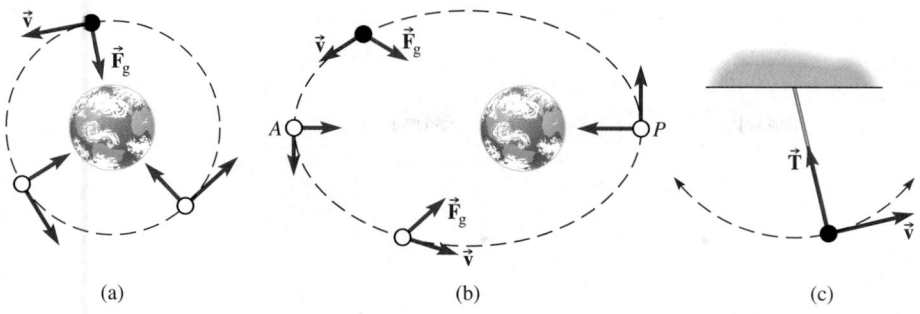

(a)   (b)   (c)

Figure 6.5 (a) No matter where the satellite is in its circular orbit, it experiences a gravitational force directed toward the center of the Earth. This force is always perpendicular to the satellite's velocity; thus, gravity does no work on the satellite. (b) In an elliptical orbit, the gravitational force is *not* always perpendicular to the velocity. As the satellite moves counterclockwise in its orbit from point *P* to point *A*, gravity does negative work; from *A* to *P*, gravity does positive work. (c) The tension in the string of a pendulum is always perpendicular to the velocity of the pendulum bob.

object since it is perpendicular to the displacement of the object, even if the surface is not horizontal. The direction of the normal force is always perpendicular to the velocity of an object sliding along the surface. During a short time interval, the normal force is perpendicular to the displacement $\vec{v}\,\Delta t$ of the object. Since the normal force and displacement are perpendicular for *every* small time interval, the normal force does no work.

The normal force is not the only instance of a force whose direction is always perpendicular to the velocity. No work is done by the Earth's gravitational force on a satellite in circular orbit (Fig. 6.5a). In a circular orbit, the gravitational force is always directed along a radius from the satellite to the center of the Earth. At every point in the orbit, the gravitational force is perpendicular to the velocity of the satellite (which is tangent to the circular orbit).

By contrast, gravity does work on a satellite in a noncircular orbit (Fig. 6.5b). Everywhere except at points *A* and *P*, there *is* a component of the gravitational force along the direction of the satellite's velocity. Wherever the gravitational force has a component parallel to the velocity, gravity does positive work; where the gravitational force has a component antiparallel to the velocity, gravity does negative work.

As a third example, no work is done by the tension in the string on a swinging pendulum bob because the tension is always perpendicular to the velocity (Fig. 6.5c).

Making The Connection:
work done on an orbiting satellite

No work is done on an object by a force that is perpendicular to the object's velocity.

Example 6.1

Antique Chest Delivery

A rare and valuable antique chest, made in 1907 by Gustav Stickley, is being moved into a truck using a 4.0-m-long ramp. The weight of the chest plus packing material is 1400 N. To get the chest from the ground onto the truck bed, which is 1.0 m higher, the movers must decide what to do. Should they lift it straight up, or should they push it up the ramp? Assume they push the chest on a dolly, which in a simplified model is equivalent to sliding it up a *frictionless* ramp.

(a) Find the work done by the movers on the chest if they lift it straight up 1.0 m at constant speed.

(b) Find the work done if they slide the chest up the 4.0-m-long *frictionless* ramp at constant speed by pushing parallel to the ramp.

(c) Find the work done by gravity on the chest in each case.

(d) Find the work done by the normal force of the ramp on the chest. Assume that all forces are constant.

Strategy In each case we must calculate the force required to move the chest, calculate the displacement, and then find the work done by multiplying the magnitude of one by the parallel component of the other. Drawing a free-body diagram helps us calculate the forces. The ramp is a simple machine—just as for Rosie's pulleys, the ramp cannot reduce the amount of work that must be done, so we expect the work done by the movers to be the same in both cases (neglecting friction). We expect the work done by gravity to be negative in both cases, since the chest is moving up rather than down. Gravity is *not* helping the movers lift the chest up into the truck. The normal force is perpendicular to the ramp so it does zero work on the chest.

Given: Weight of chest = 1400 N; length of ramp = 4.0 m; height of ramp = 1.0 m

To find: W_{movers} lifting straight up and W_{movers} using the ramp; W_{g}; W_{N}

Tips

When several forces act on a body, use subscripts to clarify which work is being calculated.

continued on next page

Figure 6.6 (a) An antique chest is pushed up a ramp into a truck, (b) free-body diagram for the chest, and (c) finding the angle of the incline.

Example 6.1 *continued*

Solution (a) The displacement is 1.0 m straight up. The movers must exert an upward force equal in magnitude to the weight of the chest to move it at constant speed. The work done to lift it 1.0 m is

$$W_{movers} = F\,\Delta r \cos\theta = 1400\ \text{N} \times 1.0\ \text{m} \times \cos 0 = +1400\ \text{J}$$

where $\theta = 0$ because the force and the displacement are in the same direction (upward).

(b) Figure 6.6a shows a sketch of the situation. We draw an FBD for the chest as it is pushed up the ramp (Fig. 6.6b). We take the *x*-axis along the inclined ramp and the *y*-axis perpendicular to the ramp and replace the gravitational force with its *x*- and *y*-components. Sliding along at constant speed, the chest's acceleration is zero, so the *x*-components of the forces add to zero.

continued on next page

Example 6.1 *continued*

The *x*-component of the gravitational force acts in the −*x*-direction and the force from the movers acts in the +*x*-direction.

$$\Sigma F_x = F_{movers} - mg \sin \phi = 0$$

From the right triangle formed by the ramp, the ground, and the truck bed in Fig. 6.6c:

$$\sin \phi = \frac{\text{height of truck bed}}{\text{distance along ramp}}$$

$$= \frac{h}{d} = \frac{1.0 \text{ m}}{4.0 \text{ m}} = 0.25$$

We can now solve for F_{movers}:

$$F_{movers} = mg \sin \phi = mg \times \frac{h}{d}$$

$$= 0.25mg = 0.25 \times 1400 \text{ N} = 350 \text{ N}$$

The force and displacement are parallel, so the work done by the movers is F_{movers} times the distance *d*.

Solving for the work done by the movers yields

$$W_{movers} = F_{movers} d$$

$$= 350 \text{ N} \times 4.0 \text{ m}$$

$$= +1400 \text{ J}$$

The work done is the same as in (a). The movers apply 1/4 the force over a displacement that is four times larger.

(c) In both cases, the force of gravity is the weight of magnitude *mg* acting downward. Also, in both cases, the upward component of the displacement in the direction of the gravitational force is $\Delta y = h = 1.0$ m. The work done by gravity is the same for the two cases.

$$W_g = -mg \, \Delta y$$

$$= -1400 \text{ N} \times 1.0 \text{ m} = -1400 \text{ J}$$

(d) The normal force of the ramp on the chest does no work because it acts in a direction perpendicular to the displacement of the chest.

$$W_N = N \Delta r \cos 90° = 0$$

Discussion   In (b), we could have worked with algebraic symbols as

$$W_{movers} = F_{movers} d = \frac{mgh}{d} \times d = mgh$$

Since *d*, the length of the ramp, cancels when multiplying the force times the distance, the work done by the movers is the same for *any* length ramp (as long as the height is the same). Since we found the work by multiplying 350 N by 4.0 m, we might not have noticed that *d* does not affect the final answer.

Using a frictionless ramp, the movers exert a smaller force than if they lift the chest straight up, but they exert it over a longer distance; the total work done is the same. With a *real* ramp, friction acts to oppose the motion of the chest, so the movers would have to do *more* than 1400 J of work to slide the chest up the ramp. There's no getting around it; if the movers want to get that chest into the truck, they're going to have to do *at least* 1400 J of work.

Practice Problem 6.1   Bicycling uphill

A bicyclist climbs a 2.0-km-long hill that makes an angle of 7.0° with the horizontal. The total weight of the bike and the rider is 750 N. How much work is done on the bike and rider by gravity?

Example 6.2

Fun on a Sled

 Diane pulls a sled along a snowy path on level ground with her little brother Jasper riding on the sled (Fig. 6.7a). The total mass of the sled and Jeremy is 26 kg. The cord makes a 20.0° angle with the ground. As a rough approximation, assume that the force of friction on the sled is determined by $\mu_k = 0.16$, even though the surfaces are not dry (some snow melts as the runners slide along it). Find the work done by Diane and the work done by the ground on the sled while the sled moves 120 m along the path at a constant 3 km/h.

Strategy   The sled's acceleration is zero, so the vector sum of all the external forces (gravity, friction, rope tension, and the normal force) is zero. We draw an FBD and use Newton's second law to find the tension in the rope.

Solution   The FBD is shown in Fig. 6.7b. The *x*- and *y*-axes are parallel and perpendicular to the ground, respectively. From Newton's second law with zero acceleration,

$$\Sigma F_x = +T \cos \theta - f_k = 0 \qquad (1)$$

$$\Sigma F_y = +T \sin \theta - mg + N = 0 \qquad (2)$$

where *T* is the tension and $\theta = 20.0°$. The force of kinetic friction is the coefficient of friction times the normal force,

$$f_k = \mu_k N$$

Substituting this into (1)

$$T \cos \theta - \mu_k N = 0 \qquad (3)$$

To find the tension, *T*, we need to eliminate the unknown normal force *N*. Equation (2) also involves the normal force *N*. We multiply Eq. (2) by μ_k,

$$\mu_k T \sin \theta - \mu_k mg + \mu_k N = 0 \qquad (4)$$

continued on next page

Example 6.2 *continued*

Adding Eqs. (3) and (4) to eliminate the normal force, N, and solving for T yields

$$T \cos \theta + \mu_k T \sin \theta - \mu_k mg = 0$$

$$T = \frac{\mu_k mg}{\mu_k \sin \theta + \cos \theta}$$

$$= \frac{0.16 \times 26 \text{ kg} \times 9.8 \text{ m/s}^2}{(0.16 \times \sin 20.0°) + \cos 20.0°} = 41.0 \text{ N}$$

Now the tension is used to find the work done by Diane. The component of the tension T acting parallel to the displacement is the force that does work. That component is $T \cos \theta$ and the displacement is $\Delta x = 120$ m. So the work done by Diane is

$$W_T = (T \cos \theta) \Delta x$$

$$= 41.0 \text{ N} \times \cos 20.0° \times 120 \text{ m} = +4600 \text{ J}$$

Displacement = 120 m

(a)

(b)

$\theta = 20.0°$

(c)

Figure 6.7
(a) Jasper being pulled on a sled; (b) free-body diagram; (c) resolving the tension into x- and y-components.

The force on the sled due to the ground has two components: N and f_k. The normal force does no work since it acts perpendicular to the displacement of the sled. Friction acts in a direction opposite to the displacement, so the work done by friction is

$$W_f = f_k \Delta x \cos 180° = -f_k \Delta x$$

From Eq. (1),

$$f_k = T \cos \theta$$

Therefore, the work done by the ground—the work done by the frictional force—is

$$W_f = -f_k \Delta x = -(T \cos \theta) \Delta x$$

Except for the minus sign, W_f is the same as W_T. Therefore, $W_f = -4600$ J.

Discussion   A force can be resolved into components parallel and perpendicular to the displacement. The perpendicular component does no work. Here, since the sled's displacement is in the x-direction, the work done is the x-component of the force times the displacement. The x-component of the tension is

$$T \cos \theta = 41.0 \text{ N} \cos 20.0° = 38.5 \text{ N}$$

Then the work done is

$$W = T_x \Delta x = 38.5 \text{ N} \times 120 \text{ m} = 4600 \text{ J}$$

Note that the speed (3 km/h) was not used in the solution. Assuming that the frictional force on the sled is independent of speed, Diane would exert the same force to pull the sled at *any* constant speed. Then the work she does is the same for a 120-m displacement. At a higher speed, though, she would have to do that amount of work in a shorter time interval.

Practice Problem 6.2   A different angle

Find the tension if Diane pulls at an angle $\theta = 30.0°$ instead of 20.0°, assuming the same coefficient of friction. What is the work done by Diane on the sled in this case for a 120-m displacement? Explain how the tension can be greater but the work done by Diane smaller.

Work Done by Dissipative Forces

The work done by kinetic friction was calculated in Example 6.2 according to a simplified model of friction. If friction truly does −4600 J of work on the sled, then 4600 J of energy would have to be transferred from the sled to the ground. What really happens is that 4600 J of energy is converted into thermal energy *shared* between the ground and the sled—both the ground and the sled warm up a little. So the 4600 J is not all transferred to the ground; some stays in the sled. A more accurate statement is that friction *dissipates* 4600 J of energy. **Dissipation** is the conversion of energy from an organized form to a disorganized form.

Kinetic friction is a complicated phenomenon. The friction force acts all along the bottom of the sled. The transfer of energy by friction is due to huge numbers of microscopic forces acting at many different points of contact. Each of these forces acts through some small displacement as the surfaces continually deform against one another. The work done is not the same as if a single force of kinetic friction were to act at a single point that moves through the same displacement as the sled.

As a practical matter, though, we usually are not concerned with *where* the thermal energy appears. In our simplified model of kinetic friction, we assume that there *is* a single force of kinetic friction moving through the same displacement as the object. Then we can calculate the work done by friction using Eq. (6-1) and get the correct amount of energy dissipated; we just won't know how much of it is transferred to the stationary surface and how much remains in the sliding object. This is how we apply the term *work* to kinetic friction or to other dissipative forces such as air resistance.

6.3   KINETIC ENERGY

When a single, constant force $\vec{F}$ directed along the x-axis acts on a point particle of mass m during a displacement $\Delta\vec{r}$, the work done is

$$W = F_x \Delta x$$

where Δx is the x-component of the displacement. If $\vec{F}$ is the only force acting, then Newton's second law tells us that $\vec{F} = \vec{F}_{net} = m\vec{a}$, so

$$W = F_x \Delta x = ma_x \Delta x \qquad (6\text{-}3)$$

Since the acceleration is constant, we can use any of the kinematic equations for constant acceleration from Chapter 3. The product of acceleration and displacement components occurs in Eq. (3-13):

$$a_x \Delta x = \tfrac{1}{2}(v_x^2 - v_{0x}^2)$$

Substituting Eq. (3-13) into Eq. (6-3) yields

$$W = m(a_x \Delta x) = \tfrac{1}{2}m(v_x^2 - v_{0x}^2)$$

Since the net force is along the x-axis, a_y and a_z are both zero. Then only the x-component of the velocity changes; v_y and v_z are constant. As a result,

$$v^2 - v_0^2 = (v_x^2 + \cancel{v_y^2} + \cancel{v_z^2}) - (v_{0x}^2 + \cancel{v_{0y}^2} + \cancel{v_{0z}^2}) = v_x^2 - v_{0x}^2$$

Therefore, the work done is

$$W = \tfrac{1}{2}m(v^2 - v_0^2)$$

The work done is equal to the change in the quantity $\tfrac{1}{2}mv^2$. This quantity is called the **kinetic energy** of the object (symbol K). Kinetic energy is a scalar quantity and is always nonnegative. The kinetic energy of an object moving with speed v is equal to the work that must be done on the object to accelerate it to that speed starting from rest. For a point particle, work done on the particle always appears as a change in the particle's kinetic energy. When a force does positive work on a point particle, it acts in a direction that tends to increase the particle's speed and thereby increase the object's kinetic energy. Negative work done on a point particle means that the force doing the work tends to slow the particle down and decrease its kinetic energy.

Kinetic energy:

$$K = \tfrac{1}{2}mv^2 \qquad (6\text{-}4)$$

$$W = \Delta K \quad \text{(point particle)} \qquad (6\text{-}5)$$

Equation (6-5) is a special case: it applies only to a **point particle**—an object with no internal structure. A point particle cannot be deformed, cannot change temperature, and cannot rotate. By assuming no internal structure, we can be sure that every part of the object goes through the same displacement while the force is applied and that the energy transferred all appears as kinetic energy. We must be careful when applying Eq. (6-5) to ordinary objects since some or all of the work done on the object may go into changing some other form of the object's energy, such as rotational kinetic energy or internal energy. Internal forces can do internal work, changing energy from one form to another within the object (as happens in our muscles when we hold something up without moving it). The internal energy of an object changes if it is deformed (its size or shape is altered), if its temperature changes, or if it undergoes a phase change such as melting. If we are sure that changes in these other forms of energy can be neglected, then we can apply Eq. (6-5) to ordinary objects.

Equation (6-4) defines the *translational* kinetic energy of an object—the kinetic energy due to the object's motion as a whole. Translational kinetic energy is the only kind of kinetic energy that a point particle can have. Extended objects have internal structure; they can have *rotational* and *vibrational* kinetic energy as well. For now, we consider only translational kinetic energy.

If more than one force acts on a point object, the change in kinetic energy is equal to the *total* work done by all the forces [Eq. (6-6)]. There are two ways to calculate the total work. The total work is the sum of the work done by each force individually, so we can calculate the work done by each force and add them [Eq. (6-7)]. An alternative is to calculate the work done by the *net* force [Eq. (6-8)]. Both methods give the same result as long as we consider only point particles, so that the point of application of every force moves through the same displacement.

$$W_{\text{total}} = \Delta K \quad \text{(point particle)} \tag{6-6}$$

$$W_{\text{total}} = \sum_i W_i \quad \text{(general)} \tag{6-7}$$

where W_i = work done by each individual force and Σ means *the sum of*.

$$W_{\text{total}} = F_{\text{net}} \, \Delta r \cos \theta \quad \text{(point particle)} \tag{6-8}$$

Conceptual Example 6.3

Collision Damage

Why is the damage caused by an automobile collision so much worse when the vehicles involved are moving at high speeds?

Strategy When a collision occurs, the kinetic energy of the automobiles gets converted into other forms of energy. Thus, we can use the kinetic energy as a rough measure of how much damage can be done in a collision.

Solution and Discussion Suppose we compare the kinetic energy of a car at two different speeds: 60.0 mph and 72.0 mph (which is 20.0% greater than 60.0 mph). If kinetic energy were proportional to speed, then a 20.0% increase in speed would mean a 20.0% increase in kinetic energy. However, since kinetic energy is proportional to the *square* of the speed, a 20.0% speed increase causes an increase in kinetic energy much greater than 20.0%. Working by proportions, we can find the percent increase in kinetic energy:

$$K \propto v^2$$

$$\frac{K'}{K} = \left(\frac{v'}{v}\right)^2 = (1.200)^2 = 1.440$$

Therefore, a 20.0% increase in speed causes a 44.0% increase in kinetic energy. What seems like a relatively modest difference in speed makes a lot of difference when a collision occurs.

Practice Problem 6.3   Two different cars collide with a stone wall

Suppose a sports utility vehicle and a small electric car both collide with a stone wall and come to a dead stop. If the SUV mass is 2.5 times that of the small car and the speed of the SUV is 60.0 mph while that of the other car is 40.0 mph, what is the ratio of the kinetic energy changes for the two cars (SUV to small car)? Does this ratio necessarily indicate the relative damage done to the two cars? Explain.

Example 6.4

Bungee Jumping

A bungee jumper makes a jump in the Gorge du Verdon in southern France. The jumping platform is 182 m above the bottom of the gorge. The jumper weighs 780 N. If the jumper falls to within 68 m of the bottom of the gorge, how much work is done by the bungee cord on the jumper during his descent? Ignore air resistance.

Strategy   Ignoring air resistance, only two forces act on the jumper during the descent: gravity and the tension in the cord. Since the jumper has zero kinetic energy at both the highest and lowest points of the jump, the change in kinetic energy for the descent is zero. Therefore, the total work done by the two forces on the jumper must equal zero.

Solution   Let W_g and W_c represent the work done on the jumper by gravity and by the cord. Then

$$W_{total} = W_g + W_c = \Delta K = 0$$

The work done by gravity is

$$W_g = -mg\,\Delta y$$

where the weight of the jumper is

$$mg = 780 \text{ N}$$

With $y = 0$ at the bottom of the gorge, the vertical component of the displacement is

$$\Delta y = y_f - y_i = 68 \text{ m} - 182 \text{ m} = -114 \text{ m}$$

Then the work done by gravity is

$$W_g = -(780 \text{ N}) \times (-114 \text{ m}) = +89 \text{ kJ}$$

The work done by the cord is therefore

$$W_c = -W_g = -89 \text{ kJ}$$

Discussion   The work done by gravity is positive, since the force and the displacement are in the same direction (downward). If not for the negative work done by the cord, the jumper would have a kinetic energy of 89 kJ after falling 114 m.

The length of the bungee cord is not given, but it does not affect the answer. At first the jumper is in free fall as the cord plays out to its full length; only then does the cord begin to stretch and exert a force on the jumper, ultimately bringing him to rest again. Regardless of the length of the cord, the total work done by gravity and by the cord must be zero since the change in the jumper's kinetic energy is zero.

Practice Problem 6.4   The bungee jumper's speed

Suppose that during the jumper's descent, at a height of 110 m above the bottom of the gorge, the cord has done −21.7 kJ of work on the jumper. What is the jumper's speed at that point?

6.4   WORK DONE BY VARIABLE FORCES

So far we have considered only constant forces when calculating work. How can we calculate the work done by a variable force? Consider an archer drawing back a compound bow (Fig. 6.8a). The compound bow is designed to make it easier to draw the string back and hold it back because, at a certain point, the force required to draw the string farther stops increasing. A convenient way to describe how the force varies with string position is to plot a graph. Figure 6.8b shows the force required to hold the string back as a function of distance. How can we calculate the work done by the archer as he draws the string back 40 cm?

We've asked analogous questions in previous chapters. Recall how we find the change in velocity Δv_x when the acceleration a_x is not constant (Section 3.3). During a short time interval Δt the acceleration change is negligible and $\Delta v_x = a_x \Delta t$. To find the velocity change for a longer time interval, we divide the time interval into a series of short time intervals and sum up all the velocity changes that occur during each one. On a graph of $a_x(t)$, each of these velocity changes is the area of a rectangle of height a_x and width Δt, so the area under a graph of a_x versus t is Δv_x.

To find the work done by a variable force F_x, we divide the displacement into a series of small displacements Δx. The displacements are small enough that the change

Making The Connection:
work done in drawing a bow

(a)

(b)

Figure 6.8 (a) A compound bow. (b) The force to draw back the compound bow depends on how far it is drawn. In this graph, the "area" represented by each rectangle is (0.050 m × 20.0 N) = 1.0 J.

in F_x during each one of them can be ignored. Then, during each small displacement, the work done is

$$\Delta W = F_x \, \Delta x \tag{6-9}$$

Each ΔW is the area of a rectangle of height F_x and width Δx on a graph of $F_x(x)$. The total work done is the sum of these areas. In the limit that the small displacements are infinitesimally small, *the total work done is the area under the graph of $F_x(x)$.*

In Fig. 6.8b, the "area" of each rectangle represents (0.050 m × 20.0 N) = 1.0 J of work. There are approximately 36 rectangles under the force curve between $x = 0$ and $x = 40$ cm, so the work done by the archer is +36 J.

Example 6.5

Archery Practice

To draw back a simple bow, the force the archer exerts on the string continues to increase as the displacement of the string increases and the bow bends slightly. The force-versus-position graph of Fig. 6.9 describes such a bow, which, like an ideal spring, requires a force proportional to the displacement. Calculate the work done to draw the string back 40.0 cm.

Strategy The work done is the area under the force-versus-position graph. This time, instead of counting rectangles, we can calculate the triangular area formed by the force-versus-position graph.

Solution We want to find the work done to draw the string back 40.0 cm, so the base of the triangle is 40.0 cm. The altitude of the triangle is the force at 40.0 cm: 160 N. Therefore, the work done is

$$W = \tfrac{1}{2}(0.400 \text{ m} \times 160 \text{ N}) = +32 \text{ J}$$

Discussion When the arrow is released from this displacement of 40.0 cm, the string and bow do +32 J of work on the arrow. The work done on the arrow is equal to the increase in the arrow's kinetic energy. Therefore, the arrow has a kinetic energy equal to +32 J as it leaves the string.

This simple bow is *initially* easier to pull back than the compound bow considered previously. The spring constant of

Figure 6.9
A simple bow requires a force proportional to the displacement of the string.

continued on next page

Example 6.5 *continued*

this bow is (200 N)/(50 cm) = 4 N/cm, while the initial slope of the $F(x)$ graph for the compound bow is (80 N)/(10 cm) = 8 N/cm. The compound bow is stiffer initially, but the force then levels out at 110 N, while the force required to pull the simple bow continues to increase to 160 N (at 40 cm). The work done to pull the compound bow back 40 cm is greater than the work done to pull the simple bow back 40 cm, but the work done to pull the compound bow back 50 cm is less than the work done to pull the simple bow back 50 cm.

Practice Problem 6.5 A gentle pull

How much work would you do to draw the string of each of the two bows back 10.0 cm instead of 40.0 cm?

We can generalize Example 6.5. For small amounts of stretch, springs tend to obey Hooke's law: $F = kx$, where F is the magnitude of the external force applied to the spring, x is the distance the spring is stretched or compressed from its equilibrium length, and k is the spring constant. The work done to stretch the spring from equilibrium a distance x is the area of a right triangle whose base is x and height is kx, or

$$W = \tfrac{1}{2}kx^2$$

Think of $\tfrac{1}{2}kx^2$ as the average force ($\tfrac{1}{2}kx$) times the displacement (x). The work done *by* the spring during the same displacement is $-\tfrac{1}{2}kx^2$, because the spring exerts the same magnitude force in the opposite direction.

More generally, if a spring starts at some displacement x_i, not necessarily at the equilibrium point, the work done to stretch or compress it more, to a new displacement x_f, is $\tfrac{1}{2}kx_f^2 - \tfrac{1}{2}kx_i^2$. Imagine the spring starting at equilibrium and ultimately ending up at a displacement x_f after passing through x_i. The total work done is $\tfrac{1}{2}kx_f^2$; then we subtract the work that was done to get the spring to position x_i.

Work done to compress or stretch an ideal spring:

$$W = \tfrac{1}{2}kx_f^2 - \tfrac{1}{2}kx_i^2$$

Example 6.6

The Dart Gun

In a dart gun (Fig. 6.10), a spring with $k = 400.0$ N/m is compressed 8.0 cm when the dart (mass 20.0 g) is loaded (Fig. 6.10a). What is the muzzle speed of the dart when the spring is released (Fig. 6.10b)?

Strategy As the spring expands, it does positive work on the dart because it exerts a force on the dart parallel to the dart's motion. Assuming that no other forces do work, the work done on the dart is equal to the change in the dart's kinetic energy. The dart loses contact with the spring when the spring is at its relaxed length, so $x_f = 0$ and $x_i = 8.0$ cm.

Solution Originally the dart was forced into the gun barrel, cocking the gun by compressing the spring 8.0 cm. Once the gun is fired, the compressed spring is released and the force exerted on the dart by the spring is equal in magnitude and opposite in direction to the force originally exerted on the spring by the dart during the compression. Both forces move through the same displacement. Therefore, the work done by the spring on the dart is the opposite of the work first done by the dart on the spring. As the spring does work it expands from its compressed position $x_i = 8.0$ cm to $x_f = 0$:

$$W_{\text{spring}} = -\tfrac{1}{2}k(x_f^2 - x_i^2) = \tfrac{1}{2}kx_i^2$$

The work done by the spring is positive because the spring pushes the dart in the direction of its motion and increases the

Figure 6.10
Dart gun (a) before and (b) after firing; note that the spring is compressed by a distance x when the gun is cocked before firing.

continued on next page

Example 6.6 *continued*

dart's kinetic energy. The work done equals the dart's gain in kinetic energy:

$$W_{\text{spring}} = \tfrac{1}{2}kx_i^2 = \tfrac{1}{2}mv^2$$

Solving for v,

$$v = \sqrt{\frac{kx_i^2}{m}} = \sqrt{\frac{400.0 \text{ N/m} \times (0.080 \text{ m})^2}{0.0200 \text{ kg}}} = 11 \text{ m/s}$$

Discussion   Notice that the muzzle speed is proportional to the distance the spring is compressed when the gun is cocked. If the spring is compressed halfway, it does only *one-quarter* as much work on the dart. The dart then acquires one-quarter

the kinetic energy, which means its speed is only half as much. A more massive dart fired from the same gun would have a smaller muzzle speed, but the *same* kinetic energy.

Practice Problem 6.6   A misfire

The same dart gun is cocked by compressing the spring the same distance (8.0 cm). This time the spring gets caught inside the gun, stopping at the point where it is still compressed by 4.0 cm. The dart is not caught inside the gun but is released. Find the muzzle speed of the dart. [*Hint:* Identify x_i and x_f and make use of both in finding the work.]

6.5   POTENTIAL ENERGY

An example of energy transfer occurs when the archer of Example 6.5 releases the string; the bent bow transfers energy to the arrow by doing work on it. The arrow acquires kinetic energy and flies off toward the target. Where did the bow get this energy?

The energy is transferred to the bow when the archer draws the string back. As the string is drawn back, the archer does positive work on it and therefore transfers energy to the string. What happens to this energy? It must be stored by the bow in a form other than kinetic energy, since nothing is moving. The farther back the string is drawn, the more the bow bends, and the more energy is stored in the bow. This kind of energy is called **potential energy** (symbol *U*). Kinetic energy is due to an object's motion, while potential energy is *stored* energy due to an object's position or configuration. The bow's potential energy depends on how far the string is drawn and thus how much the bow is bent.

The symbol for potential energy is *U*.

There are different kinds of potential energy. The kind stored in the bow is **elastic potential energy** since it is associated with the ability of the bent bow to spring back and straighten out. Other forms of potential energy include gravitational and electrical potential energy. Not every kind of interaction has an associated potential energy; for instance, there is no such thing as "frictional potential energy." Friction converts energy into a form that is not easily recoverable.

Let's calculate the potential energy for a spring that obeys Hooke's law (Fig. 6.11). In order to stretch or compress the spring a distance *x* starting from its relaxed position, we must do an amount of work equal to $+\tfrac{1}{2}kx^2$. In other words, the amount of energy we transfer to the spring is $+\tfrac{1}{2}kx^2$. Therefore, the potential energy of the spring increases by this same amount. The change in potential energy of the spring equals the work done by the external force to compress or stretch it. The change in *any* kind of potential energy is calculated in the same way: the potential energy change equals the work done by the external force in changing the object's position or configuration, assuming that all of the energy goes into potential energy and none into other forms.

Figure 6.11   An ideal spring is stretched a distance *x* beyond its relaxed length by the application of an external force $\vec{F}_{\text{ext}}$.

Potential energy:

$$\Delta U = W_{\text{ext}} \qquad\qquad (6\text{-}10)$$

(assuming that all of the work done by the external force goes into changing the potential energy and none goes into changing kinetic or internal energy)

If the spring starts with zero potential energy when it is relaxed ($U = 0$ at $x = 0$), then when it is stretched or compressed a distance *x*, its potential energy increases by the amount of work done to stretch or compress it—which was found in Section 6.4 to be $\tfrac{1}{2}kx^2$.

Is the potential energy really zero when the spring is relaxed? What about all the forces binding the atoms together? They are storing some potential energy even when the

---

Potential energy of an ideal spring:

$$U = +\tfrac{1}{2}kx^2 \qquad (6\text{-}11)$$

---

spring is relaxed. We really don't know what the absolute total amount of potential energy in the spring is. Luckily we don't have to; only *changes* in potential energy ever enter our calculations. The potential energy in Eq. (6-10) is the *additional* potential energy the spring acquires when it is stretched. As long as a problem concerns only stretching and compressing the spring—not burning it or putting it into the core of a nuclear reactor—we can use Eq. (6-10) to calculate the potential energy changes that occur.

Since our calculations only involve *changes* in potential energy, we can always choose the potential energy for any *one* configuration to be an arbitrary amount. Most often, we choose one configuration to have zero potential energy. Once that choice is made, the potential energy of every other configuration is determined by Eq. (6-11).

For example, consider the bow (Fig. 6.12). We can decide that the potential energy is zero before the string is pulled. Then the bow acquires *additional* potential energy when the string is pulled back and the bow is bent more. The potential energy is then greater than zero. When an arrow is shot, this additional potential energy is converted into other forms of energy, mostly the kinetic energy of the arrow. The potential energy of the bow is now zero again. It is possible for potential energy to be negative. If the bowstring is cut, the bow straightens out and its potential energy decreases from zero to negative values. We would have to do positive work to restring the bow and restore its original potential energy.

Gravitational Potential Energy

For another example, we calculate the potential energy of an object of mass m as it moves in the Earth's gravitational field near Earth's surface (Fig. 6.13). How much work is done to lift an object a height h? The force that must be applied to move the object up at constant speed—so that the kinetic energy stays constant—is mg upward; therefore, the work done by the external force is $+mgh$. Thus, the change in gravitational potential energy is $+mgh$. If we choose the potential energy to be zero initially, then $U = mgh$.

Potential energy for object of mass m in uniform gravitational field:

$$U = mgh \qquad (6\text{-}12)$$

where h is the height above a reference level at which $U = 0$

The potential energy depends only on changes in height; no work is done by gravity when an object moves horizontally since it then moves perpendicular to the gravitational force. Since only changes in potential energy are significant, we can choose $U = 0$ at any convenient place: on the floor, on a table, or at the top of a ladder. Potential energy is then positive above the chosen zero point and negative below it. If we put the origin at the point where $U = 0$, then h is the same as y.

The gravitational force is only *approximately* uniform near the surface of the Earth; the gravitational force between two bodies sufficiently far apart varies inversely as the square of the distance r between the centers of the two bodies:

$$F = \frac{Gm_1m_2}{r^2} \qquad (2\text{-}3)$$

The *general* expression for gravitational potential energy as a function of the distance between two bodies is

---

Gravitational potential energy:

$$U(r) = -\frac{Gm_1m_2}{r} \qquad (6\text{-}13)$$

where $U = 0$ when $r = \infty$

Only *changes* in potential energy are significant.

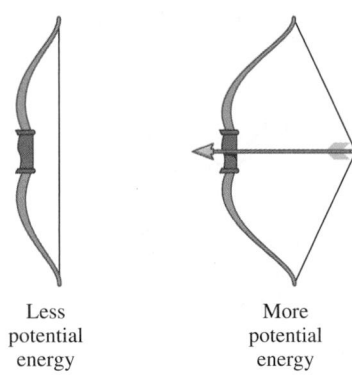

Figure 6.12 The elastic potential energy stored in a bow increases as the string is pulled back. We can define the potential energy of the bow on the left to be zero.

Less potential energy   More potential energy

Final position

$U = mgh$

$m\vec{g}$

h

Initial position

$U = 0$

Figure 6.13 An object of mass m moving in the Earth's gravitational field near the surface of the Earth.

Convention: The gravitational potential energy of two bodies infinitely far apart is taken to be zero in Eq. (6-13). For finite separation, gravitational potential energy is always negative.

Figure 6.14 Gravitational potential energy as a function of r, the distance between the centers of the two bodies.

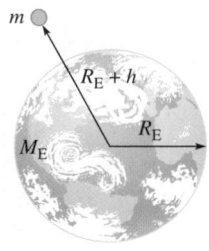

Figure 6.15 An object at a distance h above the Earth's surface is a distance $R_E + h$ from the Earth's center.

Equation (6-13) applies whenever the force law [Eq. (2-3)] does. A graph showing the variation of gravitational potential energy as a function of r is shown in Fig. 6.14.

For any finite separation r, the gravitational potential energy given by Eq. (6-13) is negative because $U = 0$ at infinite separation and the potential energy *decreases* as the bodies get closer together. But if we choose the Earth's *surface* to be the point where $U = 0$, as in Eq. (6-12), then the potential energy a little above the Earth's surface is *positive*. Which is it: positive or negative? That depends on the choice of zero; what is important is that potential energy *changes* are the same using either choice.

Consider an object of mass m at a height h above the surface of the Earth (Fig. 6.15). Let us write the potential energy at height h minus the potential energy at height 0:

$$\Delta U = U(R_E + h) - U(R_E) = \left(-\frac{GM_Em}{R_E + h}\right) - \left(-\frac{GM_Em}{R_E}\right)$$

Rearranging and factoring out the common factors GM_Em,

$$\Delta U = GM_Em\left(\frac{1}{R_E} - \frac{1}{R_E + h}\right)$$

Rewriting with a common denominator,

$$\Delta U = GM_Em\frac{R_E + h - R_E}{R_E(R_E + h)} \tag{1}$$

For values of h that are small compared to the radius of the Earth, $R_E + h \approx R_E$. Making that approximation in the denominator of Eq. (1),

$$\Delta U = \frac{GM_Emh}{R_E^2} \tag{2}$$

Recall that the gravitational field strength at the Earth's surface is

$$g = \frac{F_g}{m} = \frac{GM_E}{R_E^2}$$

Replacing GM_E/R_E^2 with g in Eq. (2) reduces that equation to

$$\Delta U = mgh$$

Therefore, Eqs. (6-12) and (6-13) *agree*: the difference between the potential energy for a mass m at a height h and the same mass on the Earth's surface is $\Delta U = mgh$ as long as $h \ll R_E$.

Example 6.7

Communications Satellite in Orbit

Calculate the change in potential energy of a 300.0-kg communications satellite when it is put into geosynchronous orbit from a launch pad on the surface of the Earth (see Example 5.9).

Strategy In Example 5.9 we found that a satellite in geosynchronous orbit at altitude $h = 35,800$ km above the Earth's surface moves at speed $v = 3.07$ km/s. We can calculate the initial

continued on next page

Example 6.7 *continued*

and final potential energies using the more general expression [Eq. (6-13)], where r stands for the distance between the satellite and the Earth's *center*. We cannot use Eq. (6-12) because the satellite's change in altitude is *not* small compared to the Earth's radius.

Solution   The initial potential energy is

$$U_i = -\frac{GM_E m}{R_E} = -\frac{6.67 \times 10^{-11}\ \text{N·m}^2\text{·kg}^{-2} \times 5.98 \times 10^{24}\ \text{kg} \times 300.0\ \text{kg}}{6.37 \times 10^6\ \text{m}}$$

$$= -1.88 \times 10^{10}\ \text{J}$$

The final potential energy is

$$U_f = -\frac{GM_E m}{R_E + h} = -\frac{6.67 \times 10^{-11}\ \text{N·m}^2\text{·kg}^{-2} \times 5.98 \times 10^{24}\ \text{kg} \times 300.0\ \text{kg}}{6.37 \times 10^6\ \text{m} + 3.58 \times 10^7\ \text{m}}$$

$$= -2.84 \times 10^9\ \text{J}$$

The change in potential energy is the final value minus the initial value:

$$\Delta U = U_f - U_i = (-0.284 \times 10^{10}\ \text{J}) - (-1.88 \times 10^{10}\ \text{J})$$

$$= +1.60 \times 10^{10}\ \text{J}$$

Discussion   What does this answer mean? It means that we would have to do a minimum of $+1.60 \times 10^{10}$ J of work to put the satellite 35,800 km above the Earth's surface *without changing its kinetic energy*. In order for the satellite to stay in a circular orbit, it must have a kinetic energy of

$$K = \tfrac{1}{2}mv^2 = \tfrac{1}{2}(300.0\ \text{kg}) \times (3070\ \text{m/s})^2 = +1.41 \times 10^9\ \text{J}$$

The increase in kinetic energy must also come from the work done on the satellite. To launch a satellite it is necessary to lift it up to the correct altitude *and* to get it *moving* at the correct speed and direction in its orbit.

Note the relationship between the satellite's potential and kinetic energies when it is in orbit:

$$U = -2.84 \times 10^9\ \text{J} \approx -2 \times (1.41 \times 10^9\ \text{J}) = -2K$$

It is *not* a coincidence that $U \approx -2K$. Problem 26 proves that $U = -2K$ for any gravitational circular orbit.

Practice Problem 6.7   Potential energy of the bungee jumper

What is the gravitational potential energy of the bungee jumper of Example 6.4 while he stands on the jumping platform? Take $U = 0$ at the bottom of the gorge.

Conservative Forces

Since potential energy is energy of configuration, potential energy can depend on only the positions of various objects, not on the *path* they took to get to those positions. The increase in the gravitational potential energy of an airplane that starts in New Orleans and ends in Denver is the same regardless of how many stops the plane makes on the way, or what its maximum cruising altitude is, or any other detail of the path it takes. Thus, the work done by the force associated with the potential energy *cannot depend on the path taken* by an object, but only on the initial and final positions of the object. Forces such as those we have considered in this section (ideal springs and gravity) that can have potential energies associated with them are called **conservative forces**.

The change in potential energy is equal in magnitude but opposite in sign to the work done by the conservative force associated with the potential energy:

$$\Delta U = -W_c \tag{6-14}$$

Why the opposite sign? If a conservative force does positive work W_c on an object, then the object's energy is increased by an amount W_c. Where does that energy come from? It is taken from stored potential energy. The conservative force dips into its "potential energy bank account" and gives the energy to the object. So the potential energy must decrease when the force does positive work. Contrast Eq. (6-14) with Eq. (6-10), which concerns the work done by an *external* force. For example, if we lift a bowling ball from the floor, we (the *external* force) do positive work while gravity does *negative* work; the change in potential energy is positive.

> Work done on an object by a conservative force depends only on the initial and final positions of the object, not on the path taken.

What *is* potential energy? Is it just a bookkeeping trick? No, *potential energy is field energy.* When an apple of weight *mg* hangs in a tree at a height *h* above the ground, the energy of the combined gravitational field of the Earth and the apple is greater by an amount *mgh* than when the apple is on the ground. Luckily we concern ourselves only with *changes* in potential energy, so we need not try to calculate the *total* energy of the Earth's gravitational field just to find the speed at which the apple hits the ground. Elastic potential energy is stored in the *electromagnetic* fields that bind atoms and molecules together.

There are two types of field energy. *Static* field energy—the potential energy that we discuss in this chapter—depends only on the configuration of the system. If the configuration is changed and then returns to its original state, the field energy returns to its original value. Field energy can also propagate as *waves*. For example, light is a wave that consists of oscillating electromagnetic fields. For now we are considering only static or nearly static fields, so we can ignore the possibility of energy being carried away by waves.

The work done by friction, air resistance, and other contact forces does depend on path, so these forces cannot have potential energies associated with them. We cannot use friction to store energy in a form that is completely recoverable. These forces are **nonconservative**. In a sense, though, it depends on your point of view. From a macroscopic point of view, these forces are certainly nonconservative. On the microscopic level, these contact forces are the result of large numbers of electromagnetic forces, which are *conservative*. On the microscopic level, every *one* of the electromagnetic interactions between two atoms is conservative. A nonconservative force like kinetic friction converts ordered energy (kinetic energy of a sliding object) into disordered energy (kinetic and potential energies of randomly vibrating atoms).

6.6   CONSERVATION OF MECHANICAL ENERGY

Making The Connection:
energy transformations in a jumping grasshopper

What is "conservative" about a conservative force? Imagine the jump of a grasshopper. Just before the jump (Fig. 6.16a), the grasshopper stores potential energy for the jump by stretching an elastic protein that acts like a spring. When the "spring" is released, the work done by the spring transforms the stored elastic energy into kinetic energy (Fig. 6.16b). Then, as the grasshopper moves upward, the gravitational force transforms kinetic energy into gravitational potential energy (Fig. 6.16c,d). As long as friction and air resistance are negligible, the sum of the kinetic and potential energies does not change during the jump.

We call the sum of the kinetic and potential energies the **mechanical energy**. Whenever nonconservative forces do no net work, mechanical energy is conserved (that is, the mechanical energy does not change). When a conservative force does positive

Figure 6.16 Energy transformations in the jump of a grasshopper.

Kinetic energy

Gravitational potential energy

Elastic potential energy

work, it transforms some of its potential energy into kinetic energy or other forms of potential energy; when it does negative work, the transformation goes the other way. A conservative force cannot change the mechanical energy because it never transforms energy into nonmechanical forms such as thermal energy. Now we see why forces associated with potential energies are called forces—they conserve mechanical energy.

Mechanical energy:

$$E = K + U \qquad \text{(6-15a)}$$

Conservation of mechanical energy:

If no nonconservative forces act, then the mechanical energy is constant.

$$E_i = E_f$$

$$\Delta E = \Delta K + \Delta U = 0 \qquad \text{(6-15b)}$$

Mechanical energy: the sum of the kinetic and potential energies

Example 6.8

A Quick Descent

 A ski trail makes a vertical descent of 78 m (Fig. 6.17). A novice skier, unable to control his speed, skis down this trail and is lucky enough not to hit any trees. What is his speed at the bottom of the trail?

Strategy   A skilled skier can control his speed by controlling the frictional force. For a worst-case scenario, we'll assume there's no friction at all; the only forces acting on the skier are the normal force and gravity. The normal force does no work, since it is always perpendicular to the skier's velocity, so mechanical energy is conserved.

Solution   As the skier descends, his change in potential energy is

$$\Delta U = -mgh$$

where $h = 78$ m. If he starts with zero kinetic energy, then

$$\Delta K = +\tfrac{1}{2}mv_f^2$$

Mechanical energy is conserved. Potential energy is converted to kinetic energy, but the total energy doesn't change:

$$\Delta K + \Delta U = \tfrac{1}{2}mv_f^2 - mgh = 0$$

We can solve for the final speed v_f:

$$v_f = \sqrt{2gh} = \sqrt{2 \times 9.8 \text{ m/s}^2 \times 78 \text{ m}} = 39 \text{ m/s}$$

Discussion   As an alternative solution, instead of looking at the energy changes directly, we could equate the initial and final mechanical energies:

$$E_i = E_f$$

which means

$$U_i + K_i = U_f + K_f$$

Initially, the kinetic energy is zero and the potential energy is mgh. At the bottom, the potential energy is zero and the kinetic energy is $\tfrac{1}{2}mv_f^2$. Substituting these energy values,

$$mgh + 0 = 0 + \tfrac{1}{2}mv_f^2$$

The two methods are equivalent, so we can use whichever is most convenient.

A speed of 39 m/s (87 mi/h) is dangerously fast. In reality, there would be some friction and air resistance, so the final speed would be smaller. How much smaller would depend on the length and contour of the trail, among other things. Without friction, the final speed doesn't depend on the detailed shape of the trail: it could have been a steep descent, or a long gradual one, or have a complicated profile with varying slope. It could even be a vertical descent—the final speed is the same for free-fall off a 78-m high building.

Practice Problem 6.8   Speeding roller coaster

A roller coaster is hauled to the top of the first hill of the ride by a motorized chain drive. After that, the train of cars is released and no more energy is supplied by an external motor. If the cars start from rest 35 m above the ground at the top of the first hill, how fast are they moving at the top of the second hill, which is 22 m above the ground? Ignore friction and air resistance.

78 m

Figure 6.17   The final speed of the skier depends only on the initial and final altitudes if no friction acts.

Example 6.9

Escape Speed

What is the minimum speed that a projectile must have at the Earth's surface so that the projectile can escape the Earth's gravitational attraction? Ignore air resistance.

Strategy   To just escape the Earth's gravity the projectile must have enough initial kinetic energy so that it can reach an infinite distance from the Earth with no leftover kinetic energy to spare. It is going to end up at the place where we have defined the potential energy to be zero. The projectile must be given enough kinetic energy at the beginning to overcome the initial negative value of potential energy it possesses by virtue of its position on the surface of the Earth—it has enough kinetic energy to climb the potential energy hill of Fig. 6.18.

Solution   By equating the initial mechanical energy at the Earth's surface to the final mechanical energy at the maximum value of r ($r = \infty$), we obtain an equation with initial speed v_i as the unknown variable.

Initial conditions: $K_i = \dfrac{1}{2}mv_i^2$

Since $r_i = R_E,\ U_i = -\dfrac{GM_Em}{R_E}$

Therefore, $E_i = \dfrac{1}{2}mv_i^2 + \left(-\dfrac{GM_Em}{R_E}\right)$

Final conditions:

$K_f = 0$ (the projectile is momentarily at rest at the maximum r)

At an infinite distance, $\dfrac{1}{r_f} = 0$, so $U_f = 0$

continued on next page

Example 6.9 *continued*

Figure 6.18
Gravitational potential energy for an object a distance r from the center of the Earth. The scale for the potential energy is in units of GM_Em/R_E.

Mechanical energy is conserved, so we equate the initial and final mechanical energies:

$$E_f = E_i$$

Substituting the kinetic and potential energies, we find

$$0 + 0 = \frac{1}{2}mv_i^2 + \left(-\frac{GM_Em}{R_E}\right)$$

Solving for v_i,

$$v_i = \sqrt{\frac{2GM_E}{R_E}}$$

This speed is called the **escape speed** of the Earth because it is the minimum initial speed for a projectile launched from the surface that can escape the influence of Earth's gravity.

$$v_{esc} = \sqrt{\frac{2GM_E}{R_E}} = 11.2 \text{ km/s (a little over 25,000 mi/hr)}$$

Discussion   Note that the answer is independent of the mass of the projectile because the mass cancels out in the solution for v_i. No matter what size projectile is launched from the

Earth, the same initial speed must be supplied to have the energy needed to escape the gravitational pull of the Earth.

We ignored air drag in the calculation. The friction of the Earth's atmosphere would actually cause so much heating that the projectile would burn up long before it got to a height of two Earth radii above the surface. This frictional heating caused by the dense lower Earth atmosphere makes it necessary for space scientists to gradually increase the speed of rockets launched from Earth by using more than one stage of rocket boosters. The rocket is eventually given enough kinetic energy to escape the Earth's gravity.

The concept of escape speed helps explain why there is little hydrogen or helium in Earth's atmosphere, even though these are by far the most common elements in the universe. We will see in Chapter 13 that the molecules in a gas have an average kinetic energy determined by the temperature of the gas. In a mixture of gases, the molecules with the smallest mass have the highest average speeds. The average speeds of hydrogen and helium in our atmosphere are large enough that a significant fraction of the molecules have speeds greater than the escape speed, and can therefore escape the atmosphere. A negligible fraction of the nitrogen, oxygen, or water molecules have speeds greater than the escape speed.

Practice Problem 6.9   Protons streaming away from the Sun

Particles such as protons and electrons are continually streaming away from the Sun in all directions. They carry off some of the energy released in the thermonuclear reactions occurring in the Sun. How fast must a proton be moving at a distance of 7.00×10^9 m from the center of the Sun in order for it to escape the Sun's gravitational pull and leave the solar system? The Sun's mass is 1.99×10^{30} kg.

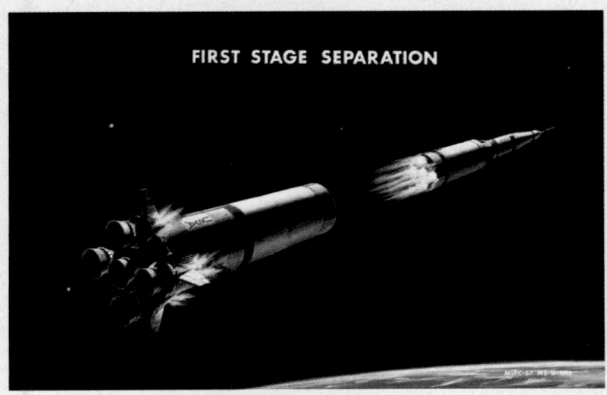

FIRST STAGE SEPARATION

When a human jumps (Fig. 6.19), the muscles provide the energy to propel the body upward. Try jumping as high as you can from a standing start. You no doubt start by crouching down. Then you accelerate upward, straightening your legs and your body; your muscles convert chemical energy into the mechanical energy of your jump. You might be able to jump about 1 m above the floor.

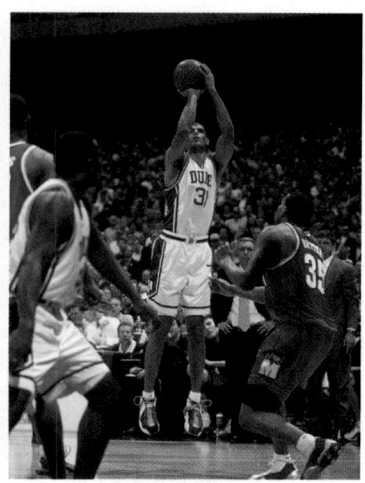

Figure 6.19 Shane Battier leaps to make a basket for Duke University.

The kangaroo uses a slightly different mechanism. It has long, elastic tendons and small muscles, in contrast to the relatively large muscles and short, stiffer tendons found in humans. The kangaroo slowly folds its legs before a jump, stretching the tendons to store elastic potential energy just as energy is stored in a stretched spring. The kangaroo then quickly extends its legs, relaxing the tendon like a released spring. The tendon supplies much of the energy needed for the jump; the rest is supplied by the kangaroo's leg muscles, which convert some chemical energy into mechanical energy. When the kangaroo lands on the ground, the tendons are stretched again as its legs bend, and it decelerates, storing up energy for the next jump. As the kangaroo takes off again and accelerates back up into the air, the stored energy from the previous jump's landing is used. The storage of elastic energy reduces the amount of chemical energy that must be supplied for each successive jump.

Insects use a catapult technique. The knee joint of a flea contains an elastic material called resilin (a rubber-like protein). The flea slowly bends its knee, stretching out the resilin and storing energy, and then locks its knee in place. When the flea is ready to jump, the knee is unlocked and the resilin quickly contracts with a sudden release of the stored elastic energy.

Example 6.10

The Hopping Kangaroo

Let's consider the hopping kangaroo (Fig. 6.20). If we assume that its tendons are like springs, a fair assumption in this case, then the potential energy stored is

$$U = \tfrac{1}{2}kx^2$$

If the kangaroo can jump 0.70 m straight up into the air when the tendons have been stretched by about half their maximum possible stretched length, how high can the kangaroo jump if the tendons are stretched 10% more?

Strategy By equating the initial mechanical energy before the jump to the final mechanical energy at the highest point of the jump, we can set up a ratio and find the relative change in height jumped. At first we imagine a kangaroo jumping straight up; then we try to generalize to more realistic hopping where there is forward motion as well.

Solution When the kangaroo is crouched, it is at rest momentarily and has zero kinetic energy. For convenience, we choose the initial gravitational potential energy to be zero. We must still account for the elastic potential energy stored in the tendon. The initial energies are

$$K_i = 0 \quad \text{and} \quad U_i = \tfrac{1}{2}kx_i^2$$

where x_i represents the initial stretch of the tendon. With the kangaroo at the high point of the jump, the kinetic energy is again zero if it jumped straight up. The tendon is no longer stretched, so the elastic potential energy is zero. But now there is gravitational potential energy. The final energies are

$$K_f = 0 \quad \text{and} \quad U_f = mgh$$

We equate initial and final mechanical energies to express the conservation of energy

$$E_i = E_f$$

$$\tfrac{1}{2}kx_i^2 = mgh$$

(a)	(b)

Figure 6.20
(a) Crouching and (b) leaping kangaroo.

continued on next page

Example 6.10 *continued*

We don't know all of the constants (mass, spring constant), so we set up a ratio,

$$\frac{E_{i1}}{E_{i2}} = \frac{E_{f1}}{E_{f2}} \quad \text{or} \quad \frac{x_1^2}{x_2^2} = \frac{h_1}{h_2}$$

For a 10% increase in stretch, $x_2 = 1.1x_1$ and

$$h_2 = (1.1)^2 \, h_1 = 1.21 \times 0.70 \text{ m} = 0.85 \text{ m}$$

Using a 10% increase in the stretch of the tendon, the kangaroo jumps about 20% higher.

When the kangaroo is hopping along, it does not jump straight up. Will the kangaroo's jump still be 20% higher when jumping at another angle? Imagine the kangaroo hopping along so that it leaves the ground at a 45° angle, which gives the maximum horizontal range per hop in the absence of air resistance. The elastic energy in the tendon is first converted to kinetic energy. This time, not all of the kinetic energy is converted to gravitational potential energy. The kinetic energy at the highest point of the jump is *not* zero because the kangaroo is still moving forward. The initial velocity can be resolved into components:

$$v^2 = v_x^2 + v_y^2 = 2v_x^2 \quad \text{(since } v_x = v_y \text{ for a 45° angle)}$$

At the highest point of the jump, the kinetic energy is $\frac{1}{2}mv_x^2$, which is half of the initial kinetic energy. Overall, *half* of the elastic energy of the tendon is converted to gravitational potential energy:

$$\frac{1}{2} \times \left(\frac{1}{2}kx_i^2\right) = mgh$$

Since h is still proportional to x_i^2, the height of the jump still increases by 20% if the stretch of the tendon is increased by 10%.

Discussion   The storage of elastic energy in the tendon is a clever way for the kangaroo to get more "miles per gallon." Without such an energy storage system, most of the kangaroo's kinetic energy would be "lost" at the end of each hop (converted to unrecoverable thermal energy). The tendons store some of the energy that would otherwise be lost and then release it to help the next jump. Since less energy is lost on each hop, the average power supplied by the kangaroo's body is less than it would be without the long, elastic tendons. Humans use a similar energy-saving mechanism when running (see Problem 81).

The analogy that the tendon behaves like an ideal spring fails for large changes in the amount of stretch because the tendon does not obey Hooke's law. For small changes in the amount of stretch, Hooke's law is a reasonable approximation.

Practice Problem 6.10   Jumping with joey

Suppose the kangaroo has a baby kangaroo (a *joey*) riding in her pouch. If the joey has grown to be one-sixth the mass of its mother, how high can the kangaroo jump with the additional load? Assume that, without the joey, she can jump 2.8 m.

Work Done by Nonconservative Forces

In general, a body can be acted on by both conservative and nonconservative forces, and both can do work. Then the total work done on the body can be written

$$W_{\text{total}} = W_c + W_{nc}$$

where W_c and W_{nc} represent the work done by conservative and nonconservative forces, respectively. For instance, a falling ball has work done on it by both gravity (conservative) and air resistance (nonconservative).

The work done by the conservative forces is accounted for by the change in potential energy [Eq. (6-14)]. This work doesn't change the mechanical energy. Work done by nonconservative forces does change the mechanical energy. If the work is done on a point particle, so that its internal energy cannot change, then the total work done is equal to the change in kinetic energy [see Eq. (6-5)]:

$$W_{\text{total}} = W_c + W_{nc} = \Delta K$$

Substituting $W_c = -\Delta U$,

$$W_{nc} = \Delta K - W_c = \Delta K + \Delta U = \Delta E \qquad \text{(6-16)}$$

The work done by the nonconservative forces is equal to the change in the mechanical energy. When nonconservative forces do no work, $W_{nc} = \Delta E = 0$ and mechanical energy is conserved.

When finding the change in mechanical energy, do not include the work done by conservative forces. Work done by conservative forces is already accounted for by the change in potential energy.

Example 6.11

Rock Climbing in Yosemite

 A team of climbers is rappelling down steep terrain in the Yosemite valley (Fig. 6.21). Mei-Ling (mass 60.0 kg) slides down a line starting from rest 12.0 m above a horizontal shelf. If she lands on the shelf below with a speed of 2.0 m/s, calculate the energy dissipated by the kinetic frictional forces acting between her and the line. The local value of g is 9.78 N/kg. Ignore air resistance.

Strategy   Since we know Mei-Ling's initial and final speeds as well as her mass, we can calculate the change in her kinetic energy. From the change in height, we can calculate the change in potential energy. Mechanical energy is *not* conserved here; mechanical energy decreases by an amount equal to the thermal energy generated by friction.

Given: mass of climber, $m = 60.0$ kg; $\Delta h = -12.0$ m;
$v_i = 0$ m/s and $v_f = 2.0$ m/s, just before stopping;
local field strength $g = 9.78$ N/kg

To find: change in mechanical energy, ΔE

Solution   Mei-Ling's kinetic energy is initially zero since she starts at rest. The change in her kinetic energy is

$$\Delta K = \tfrac{1}{2}mv_f^2 - \tfrac{1}{2}mv_i^2 = \tfrac{1}{2}mv_f^2 - 0 = \tfrac{1}{2}(60.0 \text{ kg}) \times (2.0 \text{ m/s})^2 = +120 \text{ J}$$

The change in her potential energy is

$$\Delta U = mgh_f - mgh_i$$
$$= (60.0 \text{ kg} \times 9.78 \text{ m/s}^2) \times (0 - 12.0 \text{ m}) = -7040 \text{ J}$$

Therefore, the change in her mechanical energy is

$$\Delta E = \Delta K + \Delta U = 120 \text{ J} + (-7040 \text{ J}) = -6920 \text{ J}$$

The energy dissipated by friction is 6920 J.

Discussion   One reason for wearing gloves while rappelling is to keep the hands from getting burned due to the thermal energy generated by kinetic friction.

If the line had broken when Mei-Ling was at the top, her final kinetic energy would have been +7040 J—disastrously large since it corresponds to a final speed of

$$v = \sqrt{\frac{K}{\tfrac{1}{2}m}} = \sqrt{\frac{7040 \text{ J}}{30.0 \text{ kg}}} = 15.3 \text{ m/s}$$

Instead, kinetic friction reduces her final kinetic energy to a manageable +120 J (which corresponds to a final speed of 2.0 m/s). Mei-Ling can absorb this much kinetic energy safely by landing on the shelf while bending her knees.

Practice Problem 6.11   Energy dissipated by air resistance

A ball thrown straight up at an initial speed of 14.0 m/s reaches a maximum height of 7.6 m. What fraction of the ball's initial kinetic energy is dissipated by air resistance as the ball moves upward?

6.7   GENERAL LAW OF ENERGY CONSERVATION

So far we have treated objects, no matter how large, as *point particles* that have all their mass located at a single point. We have ignored internal motion and rotation. A point particle can only have mechanical energy (translational kinetic energy plus the potential energy due to external interactions). A point particle has no rotational kinetic energy; nor does it have any way to change its internal energy. The principle of energy conservation, which we have applied only to point particles so far, is a fundamental pillar of physics that has much more general application.

In Chapter 8, we learn about the rotation of rigid objects and how to calculate the kinetic energy of rotation. Then we will be able to extend the conservation of energy to rotating bodies. After that, we will see how conservation of energy applies to fluid flow, heat flow, electromagnetic fields, and radioactivity. Many devices function by converting energy from one form to another. For instance, electric generators convert mechanical energy into electrical energy, while electric motors convert electrical energy into mechanical energy. All waves, whether sound, light, water waves, or earthquakes, transport energy from one place to another.

We can do little in the natural sciences—or for that matter, in life—without encountering the principle of energy conservation. Think about the energy conversions that make life possible. Green plants use photosynthesis to convert the energy they receive from the Sun into stored chemical energy. When animals eat the plants, that stored energy enables motion, growth, and maintenance of body temperature. Energy conservation governs every one of these processes.

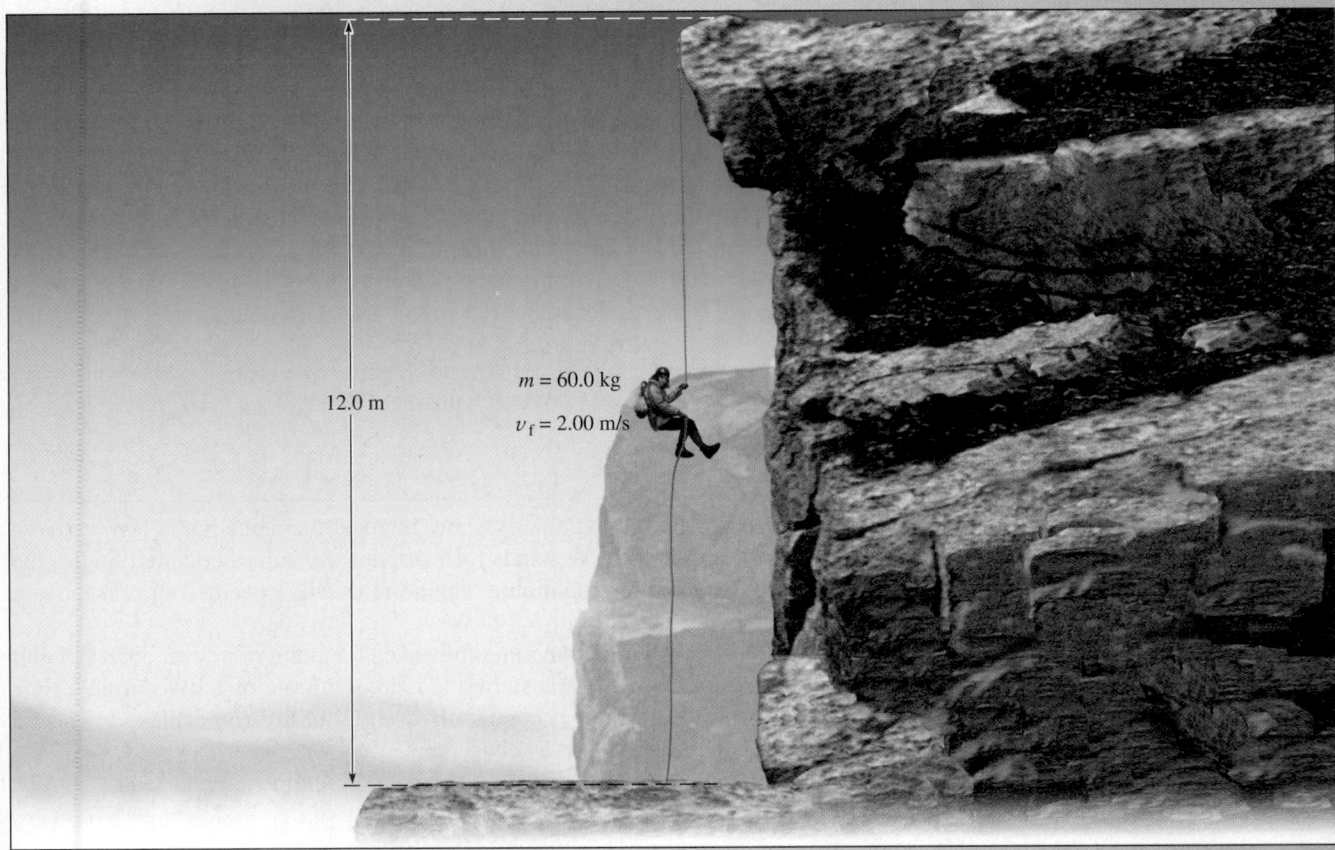

$m = 60.0$ kg

$v_f = 2.00$ m/s

12.0 m

Figure 6.21   Mei-Ling rappelling downward from a position 12.0 m above a shelf.

Since there are so many different forms of energy, we might wonder just what energy really *is*. Energy is sometimes defined as the ability to do work, but this definition is misleading. The availability of energy to do work depends on whether that energy exists in an ordered or a disordered form. Energy is a fundamental conserved quantity; work is just a way to *transfer* energy.

Another kind of energy transfer is heat flow. When two objects of different temperatures are put in contact, heat flows from the hotter object to the cooler one until the temperatures are equal. Before the nineteenth century, heat was thought to be the flow of an invisible fluid, called *caloric,* from one body to another. In the mid-nineteenth century, the English physicist James Prescott Joule (1818–1889) provided definitive evidence that heat flow is *energy* in transit; in other words, heat is energy. Before that time, heat and energy were measured in different units (the calorie was a unit of heat but *not* of energy). Joule determined the conversion factor, sometimes called "the mechanical equivalent of heat." More precisely, we might call this factor the mechanical energy equivalent of a given quantity of thermal energy. In recognition of Joule's experiments, the SI unit of energy was named for him.

Mechanical equivalent of heat

1 cal = 4.186 J

6.8 POWER

Sometimes the *rate* of energy transfer is important. When shopping for a sports car, you wouldn't ask the salesman how much work the engine can do. A tiny economy car like the Geo Metro does more work than a Ferrari if the Metro is used for daily commuting while the Ferrari sits in the garage most of the time. But the Ferrari can do work *at a much faster rate* than the Metro can. In other words, it can change chemical energy in the gasoline into mechanical energy of the car at a much faster rate—it has a larger maximum power output. We give the name **power** (symbol P) to the rate of energy transfer. The average power is the amount of energy transferred divided by the time the transfer takes.

Power is the rate of energy transfer.

Average power

$$P_{av} = \frac{\Delta E}{\Delta t} \qquad (6\text{-}17)$$

SI unit of power: watt
1 W = 1 J/s

The SI unit of power, the joule per second, is given the name watt (1 W = 1 J/s). Remember that the unit symbol W stands for *watt*, not *work*. In the United States, the maximum power output of an automobile engine is usually specified in horsepower, which is a non-SI unit of power (1 hp = 746 W).

The *kilowatt-hour* (kW·h) is sometimes mistaken for a unit of power, but it is a unit of *energy*—the amount of energy transferred at a constant rate of 1 kW during a time interval of 1 h. The kilowatt-hour is commonly used by utility companies to measure the amount of electrical energy used by consumers.

Since the work done by a constant force during a small time interval Δt is

$$\Delta W = F \, \Delta r \cos \theta$$

where the magnitude of the displacement is

$$\Delta r = v \, \Delta t$$

then power—the rate at which the force does work—can be found from the force and the velocity.

$$P = \frac{\Delta W}{\Delta t} = \frac{F \, \Delta r \cos \theta}{\Delta t} = F\frac{\Delta r}{\Delta t} \cos \theta = Fv \cos \theta$$

Instantaneous power

$$P = Fv \cos \theta \qquad (6\text{-}18)$$

Again, only the component of force in the direction of motion does work.

Example 6.12

Air Resistance on a Hill-Climbing Car

A 1000.0-kg car climbs a hill with a 4.0° incline at a constant 12.0 m/s (Fig. 6.22a). (a) At what rate is the gravitational potential energy increasing? (b) If the mechanical power output of the engine is 20.0 kW, find the force of air resistance on the car.

Strategy (a) We can find the rate of gravitational potential energy increase in two ways. One is to find the potential energy change during a time interval Δt and divide it by the time interval, which is equivalent to using the definition of average power [Eq. (6-17)]. The other possibility is to use Eq. (6-18) to find the rate at which the gravitational force does work.

(b) The car moves at constant speed, so its kinetic energy is not changing. Therefore, during any time interval, the work done by the engine plus the (negative) work done by air resistance is equal to the increase in the gravitational potential energy.

Given: car mass = 1000.0 kg; v = 12.0 m/s; 4.0° incline

To find: (a) rate of potential energy change, $\Delta U/\Delta t$; (b) force of air resistance, F_a

Solution (a) For a small change in elevation Δy, the change in potential energy is

$$\Delta U = mg \, \Delta y$$

continued on next page

Example 6.12 *continued*

Then the *rate* of potential energy change is

$$\frac{\Delta U}{\Delta t} = \frac{mg\,\Delta y}{\Delta t} = mg\,\frac{\Delta y}{\Delta t}$$

We recognize $\Delta y/\Delta t$ as the *y*-component of the velocity (v_y). Therefore,

$$\frac{\Delta U}{\Delta t} = mgv_y$$

From Fig. 6.22b, $v_y = v \sin \phi$, where $\phi = 4.0°$. Then

$$\frac{\Delta U}{\Delta t} = mgv \sin \phi$$

$$= 1000.0 \text{ kg} \times 9.8 \text{ m/s}^2 \times 12.0 \text{ m/s} \times \sin 4.0°$$

$$= 8200 \text{ W}$$

(b) During any time interval Δt, the work done by the engine must equal the increase in the gravitational potential energy plus the amount of energy dissipated by air resistance:

$$W_e + W_a = \Delta U$$

Dividing each term by Δt, we find

$$\frac{W_e}{\Delta t} + \frac{W_a}{\Delta t} = \frac{\Delta U}{\Delta t}$$

or

$$P_e + P_a = \frac{\Delta U}{\Delta t}$$

where P_e and P_a represent the power output of the engine and the rate at which air resistance dissipates energy, respectively. Then

$$P_a = \frac{\Delta U}{\Delta t} - P_e = 8.2 \text{ kW} - 20.0 \text{ kW} = -11.8 \text{ kW}$$

Figure 6.22
(a) Car climbing a hill at constant speed; (b) resolving the velocity into *x*- and *y*-components; and (c) the angle between the force and the velocity is 94°.

So, of the 20.0 kJ of mechanical work that the engine does each second, 8.2 kJ goes into gravitational potential energy and 11.8 kJ goes into stirring up the air.

The direction of the force of air resistance $\vec{F}_a$ on the car is opposite to the car's velocity, so

$$P_a = F_a v \cos 180° = -F_a v$$

Solving for F_a,

$$F_a = -\frac{P_a}{v} = -\frac{-11,800 \text{ W}}{12.0 \text{ m/s}} = 983 \text{ N}$$

Discussion   We can check (a) by using Eq. (6-18) to find the rate at which the gravitational force does work:

$$P = Fv \cos \theta$$

where $F = mg$. The angle θ is *not* the same as ϕ. In Eq. (6-18), θ is the angle between the force and velocity vectors, which is 94.0° (Fig. 6.22c). Then

$$P = mgv \cos 94.0°$$

$$= 1000.0 \text{ kg} \times 9.8 \text{ m/s}^2 \times 12.0 \text{ m/s} \times \cos 94.0°$$

$$= -8200 \text{ W}$$

Gravity does work on the car at a rate of –8200 W, which means the potential energy is *increasing* at a rate of +8200 W.

We can also figure out what mechanical power the engine must supply to go 12.0 m/s on level ground. With no change in potential energy, all of the mechanical power output of the engine goes into stirring up the air, so

$$P_e + P_a = 0$$

The magnitude of the force of air resistance is the same (983 N) since the speed is the same. Then air resistance dissipates energy at the same rate as before:

$$P_a = -Fv = -(983 \text{ N} \times 12.0 \text{ m/s}) = -11.8 \text{ kW}$$

Therefore, $P_e = 11.8$ kW. On level ground, the engine isn't increasing gravitational potential energy, so it only needs to do enough work to counteract the tendency of air resistance to slow down the car.

In this example we have assumed that all of the mechanical power output of the engine is delivered to the wheels to propel the car forward. In reality, some of the engine's power output is used to run auxiliary devices such as headlights, radios, and windshield wipers. Friction (in the moving parts of the engine, transmission, and drivetrain) also reduces the amount of power that is actually delivered to the wheels.

Practice Problem 6.12   Mechanical power output of an automobile engine

Assuming the force of air resistance on the car is proportional to the square of the car's speed, what mechanical power must the engine supply in order to drive on level ground at 15 m/s? What must the mechanical power output of the engine be to go up a 4.0° incline at 15 m/s?

When animals move, they depend on their muscles to supply the necessary power. We can think of muscles as engines for animal locomotion. The body supplies chemical energy to the muscle; the muscle requires oxygen to convert this chemical energy into mechanical energy. When an animal runs, jumps, or hops, it uses more oxygen than when at rest. A runner whose cardiovascular system is in excellent condition can run faster and farther than the rest of us partly because his body is much more efficient at delivering oxygen to the muscle tissues. The mechanical power output of an animal as it moves is limited by how fast it can deliver oxygen to the muscles.

The concept of energy can help us understand why the rat kangaroo (which is about the size of a rabbit) can jump just about as high as the larger wallaby and the much larger kangaroo. The work that can be done by the muscles of an animal is proportional to the animal's mass. If the work is used to propel the animal into the air in a jump, then the work done by the muscle (or some fraction of it) is equal to the increase in potential energy, mgh. Since both the work done by the muscles and the potential energy change are proportional to mass, we expect h to be *independent* of the animal mass. The larger animal can jump slightly higher mainly because it *starts out* higher above the ground. Similarly, the tiny flea and the much larger locust also have about the same jumping height (typically about 0.2 m).

MASTER THE CONCEPTS

Summary

- The principle of conservation of energy states that energy can be converted from one form to another, but the total energy in an isolated system never changes.

- The work done by a constant force is
$$W = F \, \Delta r \cos \theta \qquad (6\text{-}1)$$
where Δr is the displacement of the point of application of the force and θ is the angle between the force and displacement vectors.

- The kinetic energy of an object of mass m moving with speed v is
$$K = \tfrac{1}{2}mv^2 \qquad (6\text{-}4)$$

- Work done during a small displacement (variable forces) is
$$\Delta W = F_x \, \Delta x \qquad (6\text{-}9)$$

- The change in potential energy is equal to the negative of the work done by the conservative force that stores the potential energy
$$\Delta U = -W_c \qquad (6\text{-}14)$$
Why the minus sign? If the force does positive work, it takes energy *from* its stored potential energy and gives it away in some other form; therefore for a positive W_c, ΔU must be negative.

- The elastic potential energy of an ideal spring of spring constant k and stretched or compressed a distance x from its relaxed length is
$$U = +\tfrac{1}{2}kx^2 \qquad (6\text{-}11)$$

- The gravitational potential energy for an object of mass m raised a distance h above some reference level near Earth's surface, where h is small relative to the radius of the Earth, is
$$U = mgh \qquad (6\text{-}12)$$
if we choose $U = 0$, where $h = 0$.

- The gravitational potential energy as a function of the distance between the two bodies of masses m_1 and m_2 whose centers are separated by a distance r is
$$U(r) = -\frac{Gm_1 m_2}{r} \qquad (6\text{-}13)$$
where $U = 0$ at $r =$ infinity.

- The work done by a conservative force on an object does not depend on the path taken, but only on the initial and final positions of the object.

- The principle of conservation of mechanical energy states that for a system on which only conservative forces act, the sum of the kinetic and potential energies is constant.
$$E = K + U \qquad (6\text{-}15\text{a})$$
$$\Delta E = \Delta K + \Delta U = 0 \qquad (6\text{-}15\text{b})$$

- The work done on a point particle by nonconservative forces is equal to the change in the mechanical energy:
$$W_{nc} = \Delta E = \Delta K + \Delta U \qquad (6\text{-}16)$$

- Average power is the average rate of energy transfer.
$$P_{av} = \frac{\Delta E}{\Delta t} \qquad (6\text{-}17)$$
The instantaneous rate at which a force does work is the instantaneous power P:
$$P = Fv \cos \theta \qquad (6\text{-}18)$$

- The SI unit of work and energy is the joule. 1 J = 1 N·m. The SI unit of power is the watt. 1 W = 1 J/s.

Highlighted Figures and Tables

F6.4 Work done by a constant force and displacement in the direction of a force (p. 182)

F6.5 No work is done on an object by a force that is perpendicular to the object's velocity (p. 183)

F6.6 Work done on an antique chest (p. 185)

F6.13 An object of mass m moving in the Earth's gravitational field near the surface of the Earth (p. 193)

F6.14 The variation of gravitational potential energy as a function of r (p. 194)

F6.16 Conservation of mechanical energy (p. 196)

F6.17 Energy conservation applied to a skier (p. 198)

F6.21 Work done by friction on a rappeller (p. 203)

CONCEPTUAL QUESTIONS

1. An object moves in a circle. Is the total work done on the object by external forces necessarily zero? Explain.

2. You are walking to class with a backpack full of books. As you walk at constant speed on flat ground, does the force exerted on the backpack by your back and shoulders do any work? If so, is it positive or negative? Answer the same questions in two other situations: (1) you are walking down some steps at constant speed; (2) you start to run faster and faster on a level sidewalk to catch a bus.

3. Why do roads leading to the top of a mountain wind back and forth? [*Hint:* Think of the road as an inclined plane.]

4. A mango falls to the ground. During the fall, does the Earth's gravitational field do positive or negative work W_m on the mango? Does the mango's gravitational field do positive or negative work W_E on the Earth? Compare the signs and the magnitudes of W_m and W_E.

5. Can static friction do work? If so, give an example. [*Hint:* Static friction acts to prevent *relative* motion along the contact surface.]

6. In the design of a roller coaster, is it possible for any hill of the ride to be higher than the first one? If so, how?

7. When a ball is dropped to the floor from a height h, it strikes the ground and briefly undergoes a change of shape before rebounding to a maximum height less than h. Explain why it does not return to the same height h.

8. A gymnast is swinging in a vertical circle about a crossbar. In terms of energy conservation, explain why the speed of the gymnast's body is slowest at the top of the circle and fastest at the bottom.

9. A bicycle rider notices that he is approaching a steep hill. Explain, in terms of energy, why the bicyclist pedals hard to gain as much speed as possible on level road before reaching the hill.

10. You need to move a heavy crate by sliding it across a smooth floor. The coefficient of sliding friction is 0.2. You can either push the crate horizontally or pull the crate using an attached rope. When you pull on the rope, it makes a 30° angle with the floor. Which way should you choose to move the crate so that you will do the least amount of work? How can you answer this question without knowing the weight of the crate or the displacement of the crate?

11. The main effort of running is the work done by the muscles to accelerate and decelerate the legs. When a foot strikes the ground, it is momentarily brought to rest while the remainder of the animal's body continues to move forward. When the foot is picked up, it is accelerated forward by one set of muscles in order to move ahead of the rest of the body. Then the foot is slowed down by a second set of muscles until it is brought to rest on the ground again. The muscles expend energy both when accelerating and when decelerating the leg. How are thoroughbred horses, deer, and greyhounds adapted so that they can run at great speed?

MULTIPLE CHOICE QUESTIONS

1. After getting on the Santa Monica Freeway, a sports car accelerates from 30 mi/h to 90 mi/h. The ratio of its final to its initial kinetic energy is
   (a) $\sqrt{3}$ (b) $1/\sqrt{3}$ (c) 3
   (d) 1/3   (e) 9   (f) 1/9
   (g) Depends on the car's mass

2. If a kangaroo on Earth can jump from a standing start so that its feet reach a height h above the surface, approximately how high can the same kangaroo jump from a standing start on the Moon's surface? $g_{Moon} \approx \frac{1}{6} g_{Earth}$. (Assume the kangaroo has an oxygen tank and pressure suit with negligible mass.)
   (a) h (b) $6h$ (c) $\frac{1}{6}h$
   (d) $36h$ (e) $\frac{1}{36}h$ (f) $\sqrt{6}\,h$

Questions 3–6. The orbits of Mercury and Pluto are much more eccentric than the orbits of the other planets. That is, instead of being nearly circular, the orbits are noticeably elliptical. The point in the orbit nearest the Sun is called the *perihelion* and the point farthest from the Sun is called the *aphelion* (Fig. 6.23).
Answer choices for Questions 3–5:
   (a) its maximum value   (b) its minimum value
   (c) the same value as at every other point in the orbit

3. At perihelion, the gravitational potential energy of Pluto's orbit has

4. At perihelion, the kinetic energy of Pluto has

5. At perihelion, the mechanical energy of Pluto's orbit has

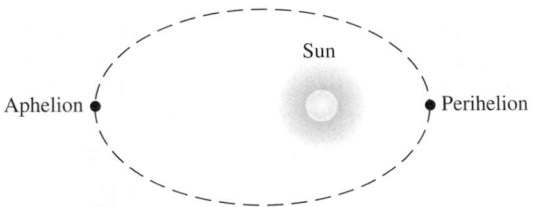

Figure 6.23   Multiple Choice Questions 3–6.

6. As Pluto moves from the perihelion to the aphelion, the work done by gravity on Pluto is
   (a) zero   (b) positive   (c) negative

7. Two balls are thrown from the roof of a building with the same initial speed. One is thrown horizontally while the other is thrown at an angle of 20° above the horizontal. Which hits the ground with the greater speed? Ignore air resistance.
   (a) the one thrown horizontally
   (b) the one thrown at 20°
   (c) they hit the ground with the same speed
   (d) the answer cannot be determined with the given information

8. A hiker descends from the South Rim of the Grand Canyon to the Colorado River. During this hike, the work done by gravity on the hiker is
   (a) positive and depends on the path taken
   (b) negative and depends on the path taken
   (c) positive and independent of the path taken
   (d) negative and independent of the path taken
   (e) zero

9. A stone is thrown straight down from the edge of a cliff. It hits the river below with a speed v_1. A second, identical stone is thrown straight up with the same initial speed from the same point at the edge of the cliff. When it hits the river, how does its speed v_2 compare to v_1? Do not ignore air resistance. [*Hint:* compare the work done by air resistance on the two stones.]
   (a) $v_2 > v_1$ (b) $v_2 < v_1$ (c) $v_2 = v_1$
   (d) The answer cannot be determined with the given information.

10. A simple catapult, consisting of a leather pouch attached to rubber bands tied to two forks of a wooden Y, has a spring constant k and is used to shoot a pebble horizontally. When the catapult is stretched by a distance d, it gives the pebble a speed v. What speed does it give the same pebble when it is stretched to a distance $3d$?
    (a) $\sqrt{3}v$ (b) $3v$ (c) $3\sqrt{3}v$ (d) $9v$ (e) $27v$

11. A projectile is launched at an angle θ above the horizontal. Ignoring air resistance, what fraction of its initial kinetic energy does the projectile have at the top of its trajectory?
    (a) $\cos\theta$ (b) $\sin\theta$ (c) $\tan\theta$ (d) $\dfrac{1}{\tan\theta}$ (e) $\tfrac{1}{2}$
    (f) $\cos^2\theta$ (g) $\sin^2\theta$ (h) 0 (i) 1

PROBLEMS

Note: **C** indicates a combination conceptual/quantitative problem. Gold diamonds ✦, ✦✦ are used to indicate the increasing level of difficulty of each problem. Problem numbers appearing in blue, 9., denote problems that have a detailed solution available in the Student Solutions Manual. Some problems are *paired* by concept; their numbers are connected by a ruled box.

Section 6.2 Work Done by Constant Forces

1. How much work must Denise do to drag her basket of laundry of mass 5.0 kg a distance of 5.0 m along a floor, if the force she exerts is 30.0 N at an angle of 60.0° with the horizontal?

2. A sled is dragged along a horizontal path at a constant speed of 1.5 m/s by a rope that is inclined at an angle of 30.0° with respect to the horizontal (Fig. 6.24). The total weight of the sled is 470 N. The tension in the rope is 240 N. How much work is done by the rope on the sled in a time interval of 10.0 s?

Figure 6.24 Problem 2

3. Hilda holds a gardening book of weight 10 N at a height of 1.0 m above her patio for 50 s. How much work does she do *on the book* during that 50 s?

4. The tension in the horizontal towrope pulling a water-skier is 240 N while the skier moves due west a distance of 54 m. How much work does the towrope do on the water-skier?

5. A barge of mass 5.0×10^4 kg is pulled along the Erie Canal by two mules, walking along towpaths parallel to the canal on either side of it. The ropes harnessed to the mules make angles of 45° to the canal. Each mule is pulling on its rope with a force of 1.0 kN. How much work is done on the barge by both of these mules together as they pull the barge 150 m along the canal?

6. A 402-kg pile driver (Fig. 6.25) is raised 12 m above ground. (a) How much work must be done to raise the pile driver? (b) How much work does gravity do on the driver as it is raised? (c) The driver is now dropped. How much work does gravity do on the driver as it falls?

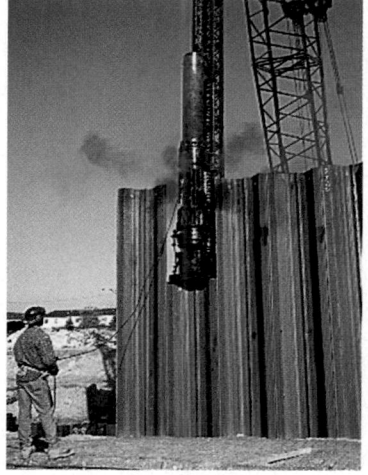

Figure 6.25
Problem 6

Section 6.3 Kinetic Energy

7. An automobile with a mass of 1600 kg has a speed of 30.0 m/s. What is its kinetic energy?

8. A record company executive is on his way to a TV interview and is carrying a promotional CD in his briefcase. The mass of the briefcase and its contents is 5.00 kg. The executive realizes that he is going to be late. Starting from rest, he starts to run, reaching a speed of 2.50 m/s. What is the work done by the executive on the briefcase during this time?

9. Sam pushes a 10.0-kg sack of bread flour on a frictionless horizontal surface with a constant horizontal force of 2.0 N starting from rest. (a) What is the kinetic energy of the sack after Sam has pushed it a distance of 35 cm? (b) What is the speed of the sack after Sam has pushed it a distance of 35 cm?

10. Josie and Charlotte push a 12-kg bag of playground sand for a sandbox on a frictionless, horizontal, wet polyvinyl surface with a constant, horizontal force for a distance of 8.0 m, starting from rest. If the final speed of the sand bag is 0.40 m/s, what is the magnitude of the force with which they pushed?

11. A ball of mass 0.10 kg moving with speed of 2.0 m/s hits a wall and bounces back with the same speed in the opposite direction. What is the change in the ball's kinetic energy?

12. A plane weighing 220 kN (25 tons) lands on an aircraft carrier. The plane is moving horizontally at 67 m/s (150 mi/h) when its tailhook grabs hold of the arresting cables. The cables bring the plane to a stop in a distance of 84 m. (a) How much work is done on the plane by the arresting cables? (b) What is the force (assumed constant) exerted on the plane by the cables? (Both answers will be *underestimates*, since the plane lands with the engines full throttle forward; in case the tailhook fails to grab hold of the cables, the pilot must be ready for immediate takeoff.)

13. A shooting star is a meteor that burns up when it reaches Earth's atmosphere. Many of these meteors are quite small. Calculate the kinetic energy of a meteor of mass 5.0 g moving at a speed of 48 km/s and compare it to the kinetic energy of a 1100-kg car moving at 29 m/s (65 mi/h).

Section 6.4 Work Done by Variable Forces

14. How much work is done on the bowstring of Example 6.5 to draw it back by 20.0 cm? [*Hint:* Rather than recalculate from scratch, use proportional reasoning.]

15. An ideal spring has a spring constant $k = 20.0$ N/m. What is the amount of work that must be done to stretch the spring 0.40 m from its relaxed length?

16. A spring is compressed from its relaxed position by a distance of 0.20 m. Figure 6.26 is a graph of the applied external force, F_x, versus the compression, x, of the spring. (a) Find the work done by the external force in compressing the spring 0.20 m starting from its relaxed position. (b) Find the work done by the external force to compress the spring from 0.10 m to 0.20 m.

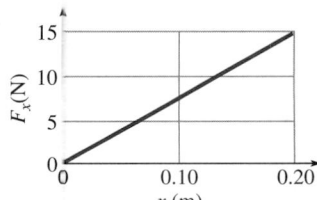

Figure 6.26 Problem 16

17. Rhonda keeps a 2.0-kg model airplane moving at constant speed in a horizontal circle at the end of a string of length 1.0 m. The tension in the string is 18 N. How much work does the string do on the plane during each revolution?

18. Figure 6.27 shows the force exerted on an object versus the position of that object along the x-axis. The force has no components other than along the x-axis. What is the work

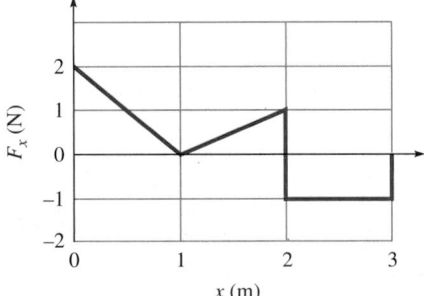

Figure 6.27
Problem 18

done by the force on the object as the object is displaced from 0 to 3.0 m?

19. The force that must be exerted to drive a nail into a wall is roughly as shown in the graph of Fig. 6.28. The first 1.2 cm are through soft drywall; then the nail enters the solid wooden stud. How much work must be done to hammer the nail a horizontal distance of 5.0 cm into the wall?

Figure 6.28 Problem 19

20. The bungee jumper of Example 6.4 made a jump into the Gorge du Verdon in southern France from a platform 182 m above the bottom of the gorge. The jumper weighed 780 N and came within 68 m of the bottom of the gorge. The cord's unstretched length is 30.0 m. (a) Assuming that the bungee cord follows Hooke's law when it stretches, find its spring constant. [*Hint:* The cord does not begin to stretch until the jumper has fallen 30.0 m.] (b) At what speed is the jumper falling when he reaches a height of 92 m above the bottom of the gorge?

Section 6.5 Potential Energy

21. Justin moves a desk 5.0 m across a level floor by pushing on it with a constant horizontal force of 340 N. (It slides for a negligibly small distance before coming to a stop when the force is removed.) Then, changing his mind, he moves it back to its starting point, again by pushing with a constant force of 340 N. (a) What is the change in the desk's gravitational potential energy during the round-trip? (b) How much work has Justin done on the desk? (c) If the work done by Justin is not equal to the change in gravitational potential energy of the desk, then where has the energy gone?

22. A spring used in an introductory physics laboratory stores 10.0 J of elastic potential energy when it is compressed 0.20 m. Suppose the spring is cut in half. When one of the halves is compressed by 0.20 m, how much potential energy is stored in it? [*Hint:* Does the half spring have the same k as the original uncut spring?]

23. Emil is learning to juggle with three oranges, each of mass 0.30 kg. (a) Emil throws one orange straight up and then catches it, throwing and catching it at the same point. How much work is done by gravity during the orange's free fall? (b) Emil throws another orange straight up, starting 1.0 m above the ground. He fails to catch it. How much work is done by gravity during this orange's free fall?

24. An airline executive decides to economize by reducing the energy, and thus the amount of fuel, required for long distance flights. He orders the ground crew to remove the paint from the outer surface of each plane. The paint removed from a single plane has a mass of approximately 100 kg. (a) If the airplane cruises at an altitude of 12,000 m, how much energy is saved in not having to lift the paint to that altitude? (b) How much energy is saved by not having to move that amount of paint from rest to a cruising speed of 250 m/s?

25. A satellite is placed in a noncircular orbit about the Earth. The farthest point of its orbit (*apogee*) is 4 Earth radii from the center of the Earth, while its nearest point (*perigee*) is 2 Earth radii from the Earth's center. If we define the gravitational potential energy U to be zero for an infinite separation of Earth and satellite, find the ratio $U_{perigee}/U_{apogee}$.

✦✦ 26. Prove that $U = -2K$ for any gravitational circular orbit. [*Hint:* Use Newton's second law to relate the gravitational force to the acceleration required to maintain uniform circular motion.]

Section 6.6 Conservation of Mechanical Energy; Section 6.7 General Law of Energy Conservation

27. How much energy is expended by an 80.0-kg person in climbing a vertical distance of 15 m? Assume that muscles have an efficiency of 22%; that is, the work done by the muscles to climb is 22% of the energy expended.

✦ 28. When a 0.20-kg mass is suspended from a vertically hanging spring, it stretches the spring from its original length of 5.0 cm to

a total length of 6.0 cm. The spring with the same mass attached is then placed on a horizontal frictionless surface. The mass is pulled so that the spring stretches to a *total* length of 10.0 cm; then the mass is released and it oscillates back and forth (Fig. 6.29). What is the maximum speed of the mass as it oscillates?

Figure 6.29 Problem 28

29. A cart starts from position 4 in Fig. 6.30 with a velocity of 15 m/s to the left. Find the speed with which the cart reaches positions 1, 2, and 3. Neglect friction.

30. A cart moving to the *right* passes point 1 in Fig. 6.30 at a speed of 20.0 m/s. Let $g = 9.81$ m/s². (a) What is the speed of the cart as it passes point 3? (b) Will the cart reach position 4? Neglect friction.

31. In an emergency, the passengers on an airplane slide down a curved escape ramp to safety (Fig. 6.31). The ramp has a 10.0-m radius of curvature and the passengers exit the plane 2.5 above the ground. (a) What is the change in the gravitational potential energy of a passenger of mass 72 kg as she slides down the ramp? (b) Neglecting friction, at what speed does the passenger reach the bottom of the ramp? (c) What is the apparent weight of the passenger near the bottom of the ramp? (d) If 33% of the gravitational potential energy change

Figure 6.30 Problems 29 and 30

Figure 6.31 Problem 31

is changed into thermal energy by friction, what is the passenger's speed at the bottom?

32. Bruce stands on a bank beside a pond, grasps the end of a 20.0-m long rope attached to a nearby tree and swings out to drop into the water. If the rope starts at an angle of 35.0° with the vertical, what is Bruce's speed at the bottom of the swing?

33. The maximum speed of a child on a swing is 4.9 m/s. The child's height above the ground is 0.70 m at the lowest point in his motion. How high above the ground is he at his highest point?

34. An object slides down an inclined plane of angle 30.0° and of incline length 2.0 m. If the initial speed of the object is 4.0 m/s directed down the incline, what is the speed at the bottom? Neglect friction.

Ⓒ 35. Rachel is on the roof of a building, h meters above ground. She throws a heavy ball into the air with a speed v, at an angle θ with respect to the horizontal. Ignore air resistance. (a) Find the speed of the ball when it hits the ground in terms of h, v, θ, and g. (b) For what value(s) of θ is the speed of the ball greatest when it hits the ground?

36. The escape speed from the surface of Planet Zoroaster is 12.0 km/s. The planet has no atmosphere. A meteor far away from the planet moves at speed 5.0 km/s on a collision course with Zoroaster. How fast is the meteor going when it hits the surface of the planet?

37. The escape speed from the surface of the Earth is 11.2 km/s. What would be the escape speed from another planet of the same density (mass per unit volume) as Earth but with a radius twice that of Earth?

◆ 38. The orbit of Halley's Comet around the Sun is a long thin ellipse. At its aphelion (point farthest from the Sun), the comet is 5.3×10^{12} m from the Sun and moves with a speed of 10.0 km/s. What is the comet's speed at its perihelion (closest approach to the Sun) where its distance from the Sun is 8.9×10^{10} m?

39. A spring with $k = 40.0$ N/m is at the base of a frictionless 30.0° inclined plane. A 0.50-kg object is pressed against the spring, compressing it 0.20 m from its equilibrium position. The object is then released. If the object is not attached to the spring, how far up the incline does it travel before coming to rest and then sliding back down? (See Fig. 6.32.)

Figure 6.32
Problems 39 and 78

40. A pendulum consists of a bob of mass m attached to the end of a cord of length L. The pendulum is released from a point at a height of $L/2$ above the lowest point of the swing. What is the tension in the cord as the bob passes the lowest point?

◆ 41. A pendulum, consisting of a bob of mass M on a cord of length L, is interrupted in its swing by a peg a distance d directly below its point of suspension (Fig. 6.33). (a) If the bob is to travel in a full circle of radius $(L - d)$ around the peg, what is the minimum possible speed it can have at the lowest point in its motion, just before it starts to go around? [*Hint:* Ignore any decrease in the length of the string due to the peg's circumference.] (b) From what minimum angle θ must the

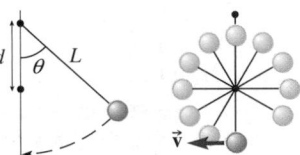

Figure 6.33   Problem 41

pendulum be released so that the bob attains the speed calculated in (a)?

42. A roller coaster car (mass = 988 kg including passengers) is about to roll down a track (Fig. 6.34). The diameter of the circular loop is 20.0 m and the car starts out from rest 40.0 m above the lowest point of the track. Ignore friction and air resistance and assume $g = 9.81$ m/s². (a) At what speed does the car reach the top of the loop? (b) What is the force exerted on the car by the track at the top of the loop? (c) From what minimum height above the bottom of the loop can the car be released so that it does not lose contact with the track at the top of the loop?

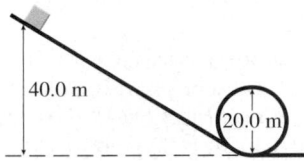

Figure 6.34   Problem 42

43. You shoot a 51-g pebble straight up with a catapult whose spring constant is 320 N/m. The catapult is initially stretched by 0.20 m. How high above the starting point does the pebble fly? Ignore air resistance.

44. A spring gun ($k = 28$ N/m) is used to shoot a 56-g ball horizontally. Initially the spring is compressed by 18 cm. The ball loses contact with the spring and leaves the gun when the spring is still compressed by 12 cm. What is the speed of the ball when it hits the ground, 1.4 m below the spring gun?

◆ 45. A block (mass m) hangs from a spring (spring constant k). The block is released from rest a distance d above its *equilibrium* position. (a) What is the speed of the block as it passes through the equilibrium point? (b) What is the maximum distance below the equilibrium point that the block will reach?

46. A hang glider moving at speed 9.5 m/s dives to an altitude 8.2 m lower. Neglecting drag, how fast is it then moving?

47. In Example 6.1, find the work done by the movers as they slide the chest up the ramp if the coefficient of friction between the chest and the ramp is 0.20.

48. A gymnast of mass 52 kg is jumping on a trampoline. She jumps so that her feet reach a maximum height of 2.5 m above the trampoline and, when she lands, her feet stretch the trampoline down 75 cm. How far does the trampoline stretch when she stands on it at rest? [*Hint:* Assume the trampoline obeys Hooke's law when it is stretched.]

Section 6.8 Power

49. Show that 1 kilowatt-hour (kW·h) is equal to 3.6 MJ.

50. If a man has an average useful power output of 40.0 W, what minimum time would it take him to lift fifty 10.0-kg boxes to a height of 2.00 m? The value of g is 9.80 m/s².

51. A bicycle and its rider together have a mass of 75 kg. What power output of the rider is required to maintain a constant

speed of 4.0 m/s (about 9 mph) up a 5.0% grade (a road that rises 5.0 m for every 100 m along the pavement)? Assume that frictional losses of energy are negligible.

✦52. A car with mass of 1000.0 kg accelerates from 0 m/s to 40.0 m/s in 10.0 s. Ignore air resistance. The engine has a 22% efficiency, which means that 22% of the thermal energy released by the burning gasoline is converted into mechanical energy. (a) What is the average mechanical power output of the engine? (b) What volume of gasoline is consumed? Assume that the burning of 1.0 L of gasoline releases 46 MJ of thermal energy.

53. A motorist driving a 1200-kg car on level ground accelerates from 20.0 m/s to 30.0 m/s in a time of 5.0 s. Neglecting friction and air resistance, determine the *average* mechanical power in watts the engine must supply during this time interval.

Ⓒ 54. A 62-kg woman takes 6.0 s to run up a flight of stairs. The landing at the top of the stairs is 5.0 m above her starting place. (a) What is the woman's average power output while she is running? (b) Would that be equal to her average power *input*—the rate at which chemical energy in food or stored fat—is used? Why or why not?

Ⓒ 55. How many grams of carbohydrate does a person of mass 74 kg need to metabolize to climb five flights of stairs (15 m height increase)? Each gram of carbohydrate provides 17.6 kJ of energy. Assume 10.0% efficiency—that is, 10.0% of the available chemical energy in the carbohydrate is converted to mechanical energy. What happens to the other 90% of the energy?

56. A race car with a mass of 500.0 kg completes a quarter-mile (402 m) drag race in a time of 4.2 s. The car's final speed is 125 m/s. What is the engine's average power output? Neglect friction and air resistance.

57. The power output of a cyclist moving at a constant speed of 6.0 m/s on a level road is 120 W. (a) What is the force exerted on the cyclist and the bicycle by the air? (b) By bending low over the handlebars, the cyclist reduces the air resistance to 18 N. If she maintains a power output of 120 W, what will her speed be?

✦58. Use this method to find how the speed with which animals of similar shape can run up a hill depends on the size of the animal. Let L represent some characteristic length such as the height or diameter of the animal. Assume that the maximum rate at which the animal can do work is proportional to the animal's surface area:

$$P_{max} \propto L^2$$

Set the maximum power output equal to the rate of increase of gravitational potential energy and determine how the speed v depends on L.

✦59. A 1500-kg car coasts in neutral down a 2.0° hill. The car attains a terminal speed of 20.0 m/s. (a) How much power must the engine deliver to drive the car on a *level* road at 20.0 m/s? (b) If the maximum useful power that can be delivered by the engine is 40.0 kW, what is the steepest hill the car can climb at 20.0 m/s?

COMPREHENSIVE PROBLEMS

60. Rosie lifts a trunk weighing 600 N up 1.5 m. If it takes her 15 s to lift the trunk, at what average rate does she do work?

61. If the skier of Example 6.8 is moving at 12 m/s at the bottom of the trail, calculate the total work done by friction and air resistance during the run. The skier's mass is 75 kg.

62. A child's playground swing is supported by chains that are 4.0 m long. If the swing is 0.50 m above the ground and moving at 6.0 m/s when the chains are vertical, what is the maximum height of the swing?

63. (a) Calculate the change in potential energy of 1 kg of water as it passes over Niagara Falls (a vertical descent of 50 m). (b) At what rate is gravitational potential energy lost by the water of the Niagara River? The rate of flow is 5.5×10^6 kg/s. (c) If 10% of this energy can be converted into electrical energy, how many households would the electricity supply? (An average household uses an average electrical power of about 1 kW.)

64. A 750-kg automobile is moving at 20.0 m/s at a height of 5.0 m above the bottom of a hill when it runs out of gasoline (Fig. 6.35). The car coasts down the hill and then continues coasting up the other side until it comes to rest. Ignoring frictional forces and air resistance, what is the value of h, the highest position the car reaches above the bottom of the hill?

65. A planet with a radius of 6.00×10^7 m has a gravitational field of magnitude 30.0 m/s^2 at the surface. What is the escape speed from the planet?

66. Suppose a satellite is in a circular orbit 3.0 Earth radii above the surface of the Earth (4.0 Earth radii from the center of the Earth). By how much does it have to increase its speed in order to be able to escape Earth? [*Hint:* You need to calculate the orbital speed and the escape speed.]

67. What is the minimum speed with which a meteor strikes the top of the Earth's stratosphere (about 40 km above Earth's surface), assuming that the meteor begins as a bit of interplanetary debris far from Earth?

68. A projectile with mass of 500 kg is launched straight up from the Earth's surface with an initial speed v_0. What magnitude of v_0 enables the projectile to just reach a maximum height of $5R_E$, measured from the *center* of the Earth? Ignore air friction as the projectile goes through the Earth's atmosphere.

✦69. The escape speed from Earth is 11.2 km/s, but that is only the minimum speed needed to escape *Earth's* gravitational pull; it does not give the object enough energy to leave the solar system. What is the minimum speed for an object near the Earth's surface so that the object escapes both the Earth's and the Sun's gravitational pulls? Ignore drag due to the atmosphere and the

Figure 6.35 Problem 64

gravitational forces due to the Moon and the other planets. Also ignore the rotation and the orbital motion of the Earth.

70. A car moving at 30 mi/h is stopped by jamming on the brakes and locking the wheels. The car skids 50 ft before coming to rest. How far would the car skid if it were initially moving at 60 mi/h? [*Hint:* You will not have to do any unit conversions if you set up the problem as a proportion.]

C ✦71. An elevator can carry a maximum load of 1202 kg (including the mass of the elevator car). The elevator has an 801-kg counterweight that always moves with the same speed but in the *opposite direction* to the car (Fig. 6.36). Let $g = 9.81$ m/s^2. (a) What is the average power that must be delivered by the motor to carry the maximum load up 40.0 m in 60.0 s? (b) How would your answer be different if there were no counterweight?

Motor

801 kg

1202 kg

Figure 6.36 Problem 71

72. If a high jumper needs to make his center of gravity rise 1.2 m, how fast must he be able to sprint? (For an extended object, the gravitational potential energy is $U = mgh$, where h is the height of the center of gravity.)

73. A pole-vaulter converts the kinetic energy of running to elastic potential energy in the pole, which is then converted to gravitational potential energy. If a pole-vaulter's center of gravity is 1.0 m above the ground while he sprints at 10.0 m/s, what is the maximum height of his center of gravity during the vault? For an extended object, the gravitational potential energy is $U = mgh$, where h is the height of the center of gravity. (In 1988, Sergei Bubka was the first pole-vaulter ever to clear 6 m.)

74. (a) How much work does a major-league pitcher do on the baseball when he throws a 90.0 mi/h (40.2 m/s) fastball? The mass of a baseball is 145 g. (b) How many fastballs would a pitcher have to throw to "burn off" a 1520 Calorie meal? (1 Calorie = 1000 cal = 1 kcal.) Assume that 80.0% of the chemical energy in the food is converted to thermal energy and only 20.0% becomes the kinetic energy of the fastballs.

75. The number of kilocalories per day required by a person resting under standard conditions is called the basal metabolic rate (BMR). (a) To generate 1 kcal, Jermaine's body needs approximately 0.010 mol of oxygen. If Jermaine's net intake of oxygen through breathing is 0.015 mol/min while he is resting, what is his BMR in kcal/day? (b) If Jermaine fasts for 24 h, how many pounds of fat does he lose? Assume that only fat is consumed. Each gram of fat consumed generates 9.3 kcal.

76. A kangaroo decides to see how high it can hop on *one leg*. Assuming the elastic energy stored in the tendon is the same as for Example 6.10, how high can it jump using a single leg?

C ✦77. A 4.0-kg block is released from rest at the top of a frictionless plane of length 8.0 m that is inclined at an angle of 15° to the horizontal. A cord is attached to the block and trails along behind it (Fig. 6.37). When the block reaches a point 5.0 m along the incline from the top, someone grasps the cord and exerts a constant tension parallel to the incline. The tension is such that the block just comes to rest when it reaches the bottom of the incline. (The person's force is a nonconservative force.) What is this constant tension? Solve the problem twice, once using work and energy and again using Newton's laws and constant-acceleration kinematics. Which method do you prefer?

8.0 m

5.0 m

Cord

15°

Figure 6.37 Problem 77

78. A 0.50-kg block, starting at rest, slides down a 30.0° incline with kinetic friction coefficient 0.25 (Fig. 6.32). After sliding 85 cm down the incline, it slides across a frictionless horizontal surface and encounters a spring ($k = 35$ N/m). (a) What is the maximum compression of the spring? (b) After the compression of part (a), the spring rebounds and shoots the block back up the incline. How far along the incline does the block travel before coming to rest?

79. The extinction of the dinosaurs and the majority of species on Earth in the Cretaceous period is thought to have been caused by an asteroid striking the Earth on the Yucatán peninsula about 65 million years ago (see Fig. 6.38). The collision put so much debris into the atmosphere that the amount of sunlight reaching Earth's surface was reduced enough to cause severe changes in climate. The asteroid was about 16 km (10 mi) in diameter, had a mass of about 10^{16} kg, and was moving at about 30 km/s (about 20 miles per *second*) when it struck the Earth. Calculate the energy released when the asteroid struck Earth and compare it to the energy released by a thermonuclear bomb (~10^{17} J).

Figure 6.38 The Barringer Meteor Crater in Arizona. The asteroid discussed in Problem 79 had a kinetic energy about 100 million times as large as that of the meteor that caused the Barringer crater.

✦80. The solar wind is a continuous stream of particles, mostly protons and electrons, moving away from the Sun. When the particles encounter Earth's upper atmosphere, they are moving at speeds of several hundred km/s. If such a particle reaches Earth's upper atmosphere with a speed of 92 km/s, how fast was it moving when it crossed Mercury's orbit at a distance of 5.8×10^7 km from the Sun? [*Hint:* Assume that the particles are slowed by the Sun's gravitational field as they move away from the Sun. Ignore the slight *increase* in speed that occurs close to Earth due to the Earth's gravitational field.]

81. Human feet and legs store elastic energy when walking or running. They are not nearly as efficient at doing so as kangaroo legs, but the effect is significant nonetheless. If not for the storage of elastic energy, a 70-kg man running at 4 m/s would lose about 100 J of mechanical energy each time he sets down a foot. Some of this energy is stored as elastic energy in the Achilles tendon and in the arch of the foot; the elastic energy is then converted back into the kinetic and gravitational potential energy of the leg, reducing the expenditure of metabolic energy. If the maximum tension in the Achilles tendon when the foot is set down is 4.7 kN and the tendon's spring constant is 350 kN/m, calculate how far the tendon stretches and how much elastic energy is stored in it.

Ⓒ ✦✦ 82. The graph in Fig. 6.39 shows the tension in a rubber band as it is first stretched and then allowed to contract. As you stretch a rubber band, the tension force at a particular length (on the way to a maximum stretch) is larger than the force at that same length as you let the rubber band contract. That is why Fig. 6.39 shows two separate lines, one for stretching and one for contracting; the lines are not superimposed as you might have thought they would be. (a) Make a rough estimate of the total work done by the external force applied to the rubber band for the entire process. (b) If the rubber band obeyed Hooke's law, what would the answer to (a) have to be? (c) While the rubber band is stretched, is all of the work done on it accounted for by the increase in elastic potential energy? If not, what happens to the rest of it? [*Hint:* Take a rubber band and stretch it rapidly several times. Then hold it against your wrist or your lip.]

Figure 6.39   Problem 82

Ⓒ ✦ 83. A wind turbine (Fig. 6.40) converts some of the kinetic energy of the wind into electrical energy. Suppose that the blades of a small wind turbine have length $L = 4.0$ m. (a) When a 10 m/s (22 mi/h) wind blows head-on, what volume of air (in m³) passes through the circular area swept out by the blades in 1.0 s? (b) What is the mass of this much air?

Each m³ of air has a mass of 1.2 kg. (c) What is the translational kinetic energy of this mass of air? (d) If the turbine can convert 40% of this kinetic energy into electrical energy, what is its electrical power output? (e) What happens to the power output if the wind speed decreases to $\frac{1}{2}$ of its initial value? What can you conclude about electrical power production by wind turbines?

✦ 84. Use dimensional analysis to show that the electrical power output of a wind turbine (Fig. 6.40) is proportional to the *cube* of the wind speed. The relevant quantities on which the power can depend are the length L of the rotor blades, the density ρ of air (SI units kg/m³), and the wind speed v.

Figure 6.40   Problems 83 and 84

✦ 85. Figure 6.41 is a graph of force versus extension for a wallaby tendon as it is stretched (upper curve) and then allowed to return back to its relaxed length (lower curve). The maximum extension of the tendon in the graph is similar to what occurs when a wallaby jumps. Estimate the percentage of the elastic energy stored in the tendon when the wallaby lands that is dissipated (and therefore cannot be recovered to assist the next jump).

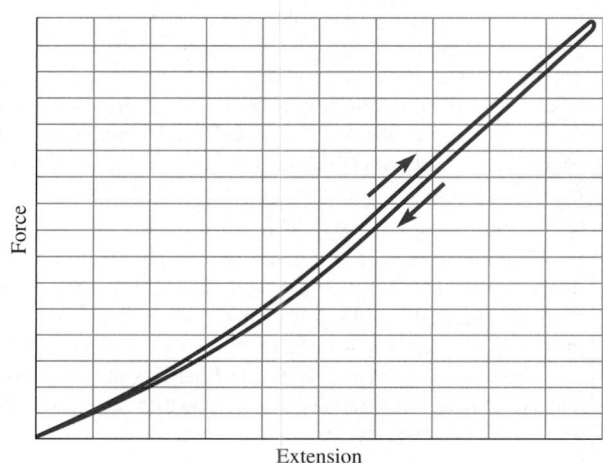

Figure 6.41   Problem 85

ANSWERS TO PRACTICE PROBLEMS

6.1 −180 kJ

6.2 43 N; 4500 J; she pulls with a greater force but its component in the direction of the displacement is smaller.

6.3 $\dfrac{(2.5m) \times (1.50v)^2}{mv^2} = 5.6$; it is difficult to judge the relative damage because the lighter car is made from lighter, less sturdy materials. It might be that both vehicles suffer the same amount of damage.

6.4 29 m/s

6.5 simple bow, 2.0 J; compound bow, 4.0 J

6.6 9.8 m/s

6.7 140 kJ

6.8 16 m/s

6.9 195 km/s

6.10 2.4 m

6.11 0.24

6.12 23 kW; 33 kW

Linear Momentum

Why were cannons on battleships mounted on wheels instead of being attached to the deck? According to a cautionary pirate tale, which may or may not be true, a ship called the *Ocean Queen* was captured in about 1820 by a French pirate, Captain la Bouche, and adapted for use as a pirate ship. The ship was relatively small, a little under 100 tons, and was equipped with 28 guns. As was the custom in the pirating trade, Captain la Bouche would sail alongside a merchant ship, run up the Jolly Roger pirate flag, and fire a broadside at the other ship if it did not surrender at once. (A "broadside" is the firing of all guns along one side of the ship at the same time.) When the *Ocean Queen* set off to hunt for unsuspecting ships, a young mate was put in charge because Captain la Bouche was sailing another ship. The ocean was rough, so the mate decided to bolt the guns firmly to the deck to prevent them from rolling about as the ship rocked from side to side. What happened when the *Ocean Queen* fired a broadside at another ship with the guns still bolted to the deck?

7.1   A VECTOR CONSERVATION LAW

In Chapter 3 we learned how to determine the acceleration of an object by finding the net force acting on it and applying Newton's second law of motion. If the forces happen to be constant, then the resulting constant acceleration enables us to calculate changes in velocity and position. Calculating velocity and position changes when the forces are not constant is much more difficult. In many cases the forces cannot even be easily determined. Conservation of energy is one tool that enables us to draw conclusions about motion without knowing all the details of the forces acting. Recall, for example, how easily we can calculate the escape speed of a projectile using conservation of energy, without even knowing the path the object takes. Now imagine how difficult the same calculation would be using Newton's second law, with a gravitational force that changes magnitude and direction depending on the path taken.

In this chapter we develop another conservation law. Conservation laws are powerful tools. If a quantity is conserved, then no matter how complicated the situation, we can set the value of the conserved quantity at one time equal to its value at a later time. The "before-and-after" aspect of a conservation law enables us to draw conclusions about the results of a complicated set of interactions without knowing all of the details.

The new conserved quantity, *momentum*, is a vector quantity, in contrast to energy, which is a scalar. When momentum is conserved, both the magnitude and the direction of the momentum must be constant. Equivalently, the *x*- and *y*-components of momentum are constant. When we find the total momentum of more than one object, we must add the momentum vectors according to the procedure by which vectors are always added.

7.2   MOMENTUM

The word *momentum* is often heard in broadcasts of sporting events. A sports broadcaster might say, "The home team has won five consecutive games; they have the momentum in their favor." The team with "momentum" is hard to stop; they are moving forward on a winning streak. A football player, running for the goal line with a football tucked under his arm, has momentum; he is hard to stop. This use of the word *momentum* is closer to the physics usage. In physics we would agree that the runner has momentum, but we have a precise definition in mind.

In everyday use, momentum has something to do with mass as well as with velocity. Would you rather have a running child bump into you, or a defensive lineman running with the same velocity? The child has much less momentum than the lineman, even though their velocities are the same.

Could a quantity combining mass and velocity be useful in physics? Imagine a collision between two spaceships (Fig. 7.1). Let the spaceships be so far from planets and stars that we can neglect gravitational interactions with celestial bodies. The spaceships exert forces on each other while they are in contact. According to Newton's third law, these forces are equal and opposite. The force on ship 2 exerted by ship 1 is equal and opposite to the force exerted on ship 1 by ship 2:

$$\vec{\mathbf{F}}_{21} = -\vec{\mathbf{F}}_{12}$$

$\vec{\mathbf{F}}_{21}$ is the force exerted *on* object 2 *by* object 1.

The changes in velocities of the two spaceships are *not* equal and opposite if the masses are different. Suppose a large spaceship (mass m_1) collides with a much smaller ship (mass $m_2 \ll m_1$). Assume for now that the forces are constant during the time interval Δt that the spaceships are in contact. Although the forces have the same magnitude, the magnitudes of the accelerations of the two ships are different because their masses are different. The ship with the larger mass has the smaller acceleration.

The acceleration of either spaceship causes its velocity to change by

$$\Delta \vec{\mathbf{v}} = \vec{\mathbf{a}} \, \Delta t = \frac{\vec{\mathbf{F}}}{m} \Delta t$$

Figure 7.1   Two spaceships about to collide.

The time interval Δt is the duration of the interaction between the two ships, so it must be the same for both ships.

Since the changes in velocity are inversely proportional to the masses, the changes in the *products* of mass and velocity are equal and opposite for the two bodies involved in the interaction:

$$m_1 \Delta \vec{\mathbf{v}}_1 = \vec{\mathbf{F}}_{12} \Delta t$$

$$m_2 \Delta \vec{\mathbf{v}}_2 = \vec{\mathbf{F}}_{21} \Delta t$$

Since $\vec{\mathbf{F}}_{21} = -\vec{\mathbf{F}}_{12}$,

$$m_2 \Delta \vec{\mathbf{v}}_2 = -m_1 \Delta \vec{\mathbf{v}}_1$$

This is a useful insight, so we give the product of mass and velocity a name and symbol: **linear momentum** (symbol $\vec{\mathbf{p}}$). Linear momentum (or just *momentum*) is a vector quantity having the same direction as the velocity.

Definition of linear momentum:

$$\vec{\mathbf{p}} = m\vec{\mathbf{v}} \qquad (7\text{-}1)$$

The collision of the two spaceships causes changes in their momenta that are equal in magnitude and opposite in direction:

$$\Delta \vec{\mathbf{p}}_1 = m_1 \vec{\mathbf{v}}_{1f} - m_1 \vec{\mathbf{v}}_{1i} = \vec{\mathbf{F}}_{12} \Delta t$$
$$\Delta \vec{\mathbf{p}}_2 = m_2 \vec{\mathbf{v}}_{2f} - m_2 \vec{\mathbf{v}}_{2i} = \vec{\mathbf{F}}_{21} \Delta t \qquad (7\text{-}2)$$

Since $\vec{\mathbf{F}}_{21} = -\vec{\mathbf{F}}_{12}$,

$$\Delta \vec{\mathbf{p}}_2 = -\Delta \vec{\mathbf{p}}_1$$

The momentum changes are equal and opposite.

Our new model of an interaction is this: in an interaction, an exchange of momentum takes place. Some momentum is transferred from one body to another. The momentum changes of the two bodies are equal and opposite, so the total momentum of the

During an interaction, momentum is transferred from one body to another.

isolated system composed of the two bodies is unchanged. The momentum of a system is the vector sum of the individual momenta of the bodies composing the system.

$$\text{change in momentum of system} = \Delta\vec{\mathbf{p}} = \Delta\vec{\mathbf{p}}_1 + \Delta\vec{\mathbf{p}}_2$$

Since $\Delta\vec{\mathbf{p}}_2 = -\Delta\vec{\mathbf{p}}_1$, the change in the total momentum of the system is

$$\Delta\vec{\mathbf{p}} = \Delta\vec{\mathbf{p}}_1 + \Delta\vec{\mathbf{p}}_2 = \Delta\vec{\mathbf{p}}_1 + (-\Delta\vec{\mathbf{p}}_1) = \vec{\mathbf{0}}$$

The momentum of either ship changes during the collision, but the total momentum of the two ships does *not* change.

Example 7.1

Change of Momentum of a Moving Car

A car weighing 12 kN is driving due north at 30.0 m/s. After driving around a sharp curve, the car is moving east at 13.6 m/s. What is the change in momentum of the car?

Strategy   The definition of momentum is $\vec{\mathbf{p}} = m\vec{\mathbf{v}}$. We can start by finding the car's mass. There are two potential pitfalls:

(1) momentum depends not on weight but on mass, and

(2) momentum is a vector, so we must take its direction into consideration as well as its magnitude. To find the change in momentum, we need to do a *vector* subtraction.

Solution   The car's mass is

$$m = \frac{W}{g} = \frac{1.2 \times 10^4 \text{ N}}{9.8 \text{ m/s}^2} = 1220 \text{ kg}$$

The car's initial velocity is

$$\vec{\mathbf{v}}_i = 30.0 \text{ m/s, north}$$

The car's initial momentum is then

$$\vec{\mathbf{p}}_i = m\vec{\mathbf{v}}_i = 1220 \text{ kg} \times 30.0 \text{ m/s north} = 3.66 \times 10^4 \text{ kg·m/s north}$$

After the curve, the final velocity is

$$\vec{\mathbf{v}}_f = 13.6 \text{ m/s, east}$$

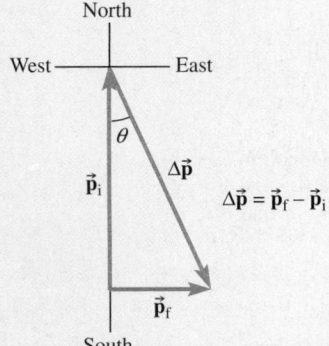

Figure 7.2
Vector subtraction to find the change in momentum.

The final momentum is

$$\vec{\mathbf{p}}_f = m\vec{\mathbf{v}}_f = 1220 \text{ kg} \times 13.6 \text{ m/s east} = 1.66 \times 10^4 \text{ kg·m/s east}$$

Momentum vectors are added and subtracted according to the same methods used for other vectors. To find the change in the momentum, we draw vector arrows representing the initial and final momenta with their tails at the same point (Fig. 7.2). Then the change in momentum $\Delta\vec{\mathbf{p}} = \vec{\mathbf{p}}_f - \vec{\mathbf{p}}_i$ is represented by the arrow drawn from the tip of $\vec{\mathbf{p}}_i$ to the tip of $\vec{\mathbf{p}}_f$. Since the three vectors in Fig. 7.2 form a right triangle, the magnitude of $\Delta\vec{\mathbf{p}}$ can be found from the Pythagorean theorem

$$|\Delta\vec{\mathbf{p}}| = \sqrt{p_i^2 + p_f^2} = \sqrt{(3.66 \times 10^4 \text{ kg·m/s})^2 + (1.66 \times 10^4 \text{ kg·m/s})^2}$$

$$= 4.02 \times 10^4 \text{ kg·m/s}$$

From the vector diagram, $\Delta\vec{\mathbf{p}}$ is directed at an angle θ east of south. Using trigonometry,

$$\tan\theta = \frac{\text{opposite}}{\text{adjacent}} = \frac{p_f}{p_i} = \frac{1.66 \times 10^4 \text{ kg·m/s}}{3.66 \times 10^4 \text{ kg·m/s}} = 0.454$$

$$\theta = \tan^{-1} 0.454 = 24.4°$$

Since the weight is given with two significant figures, we report the change in momentum of the car as 4.0×10^4 kg·m/s directed 24° east of south.

Discussion   As with displacements, velocities, accelerations, and forces, it is crucial to remember that momentum is a vector. When finding changes in momentum, we must find the difference between final and initial momentum *vectors*. If the initial and final momenta had not been perpendicular, we would have had to resolve the vectors into *x*- and *y*-components in order to subtract them.

Practice Problem 7.1   Falling apple

(a) What is the momentum of an apple weighing 1.0 N just before it hits the ground, if it falls out of a tree from a height of 3.0 m? (b) The apple falls because of the gravitational interaction between the apple and the Earth. How much does this interaction change the *Earth's* momentum? How much does it change the Earth's velocity?

7.3   THE IMPULSE-MOMENTUM THEOREM

Making The Connection:
safety features in a
modern car

We found that the change in momentum of an object due to an interaction is equal to the product of the force acting on the object and the time interval during which the force acts [Eq. (7-2)]:

$$\Delta \vec{\mathbf{p}} = \vec{\mathbf{F}} \, \Delta t$$

The product $\vec{\mathbf{F}} \Delta t$ is given the name **impulse**. Since the impulse is the product of a vector (the force) and a positive scalar (the time), impulse is a vector quantity having the same direction as that of the force. In words, $\Delta \vec{\mathbf{p}} = \vec{\mathbf{F}} \, \Delta t$ can be read as *"the change in momentum equals the impulse."* The SI units of impulse are newton-seconds (N·s) and those of momentum are kilogram-meters per second (kg·m/s). These are equivalent units, as can be demonstrated using the definition of the newton (Problem 16). The relation between impulse and momentum change is called the *impulse-momentum theorem* and is especially useful in solving problems that involve collisions and impacts.

Impulse = $\vec{\mathbf{F}} \Delta t$

Impulse-Momentum Theorem:

$$\Delta \vec{\mathbf{p}} = \vec{\mathbf{F}} \, \Delta t \qquad (7\text{-}2)$$

If an object is involved in more than one interaction, then its change in momentum during any time interval is equal to the *total* impulse during that time interval. The total impulse is the vector sum of the impulses due to each force. The total impulse is also equal to the net force times the time interval:

$$\text{total impulse} = \vec{\mathbf{F}}_1 \, \Delta t + \vec{\mathbf{F}}_2 \, \Delta t + \cdots$$
$$= (\vec{\mathbf{F}}_1 + \vec{\mathbf{F}}_2 + \cdots) \, \Delta t$$
$$= \vec{\mathbf{F}}_{\text{net}} \, \Delta t$$

Our discussion so far has assumed that the force acting during an interaction is constant. That is a rather unusual situation; the concept of momentum would be of limited use if it were applicable only when forces are constant. However, everything we have said still applies to situations where the forces are not constant, as long as we use the *average* force to calculate the impulse. The average force divided by the mass gives the average acceleration:

$$\vec{\mathbf{a}}_{\text{av}} = \frac{\vec{\mathbf{F}}_{\text{av}}}{m}$$

The change in velocity is always equal to the average acceleration times the time interval:

$$\Delta \vec{\mathbf{v}} = \vec{\mathbf{a}}_{\text{av}} \, \Delta t = \frac{\vec{\mathbf{F}}_{\text{av}} \, \Delta t}{m}$$

Then we conclude that impulse equals the change in momentum when impulse is the product of the average force and the time interval:

When the force is not constant, the impulse is found using the average force.

$$\Delta \vec{\mathbf{p}} = m \, \Delta \vec{\mathbf{v}} = \vec{\mathbf{F}}_{\text{av}} \, \Delta t \qquad (7\text{-}3)$$

Conceptual Example 7.2

Big Force–Short Time versus Small Force–Long Time

Which causes the larger change in momentum of an object, an average force of 5 N acting for 4 s or an average force of 2 N acting for 10 s? How might this principle be used when designing products to protect the human body from injury? Give an example.

Solution and Discussion   The change in momentum is found from the impulse-momentum theorem. The product of the force and the time interval gives the momentum change of the object. Over a period of 4 s, the 5-N force causes a momentum change of magnitude (5 N × 4 s) = 20 N·s, while the 2-N force acting for 10 s also causes a momentum change of magnitude (2 N × 10 s) = 20 N·s. The smaller force causes the same change in momentum because it acts for a longer time interval.

continued on next page

Conceptual Example 7.2 *continued*

Figure 7.3
A stuntman lands safely in an air bag to break his fall. The air bag reduces the risk of injury in two ways. It changes the stuntman's momentum more gradually, so that forces of smaller magnitude act on his body. It also spreads these forces over a larger area so they are less likely to cause serious injury.

When designing products to protect the human body, one goal is to lengthen the time period during which a velocity change occurs. For example, when a movie stuntman falls from a great height to simulate a victim being pushed off a balcony, he lands on a large air bag, which changes his momentum much more gradually than if he were to fall onto concrete (Fig. 7.3). The average force exerted by the air bag on the stuntman is much smaller than the average force exerted by concrete would be. Nets used under circus acrobats serve the same purpose. The net gives and dips downward when the acrobat falls into it, gradually reducing the speed of the fall over a longer time interval than if she fell directly onto the ground.

Practice Problem 7.2   Pole-vaulter landing on padded surface

A pole-vaulter vaults over the pole and falls onto thick padding. He lands with a speed of 9.8 m/s; the padding then brings him to a stop in a time of 0.40 s. What is the average force on his body *due to the padding* during that time interval? Express your answer as a fraction or multiple of his weight. [*Hint*: The force due to the padding is not the only force acting on the vaulter during the 0.40-s interval.]

Example 7.3

Collision between an Automobile and a Tree

A car moving at 20.0 m/s (44.7 mi/h) crashes into a tree. Find the magnitude of the average force acting on a passenger of mass 65 kg in each of the following cases. (a) The passenger is not wearing a seat belt. He is brought to rest by a collision with the windshield and dashboard that lasts 3 ms. (b) The passenger is wearing a seat belt. The force exerted by the seat belt for 0.3 s brings him to rest. (c) The car is equipped with a passenger-side air bag. The force due to the air bag acts for 1 s, bringing the passenger to rest.

Strategy   In each of the three cases, the passenger's momentum change is the same. What differs is the time interval during which the change occurs. It takes a larger force to change the momentum in a shorter time interval.

Solution   The magnitude of the passenger's initial momentum is

$$|\vec{\mathbf{p}}_i| = |m\vec{\mathbf{v}}_i| = (65\ \text{kg} \times 20.0\ \text{m/s}) = 1300\ \text{kg·m/s}$$

His final momentum is zero, so the magnitude of the momentum change is

$$|\Delta\vec{\mathbf{p}}| = 1300\ \text{kg·m/s}$$

This momentum change divided by the time interval gives the magnitude of the average force in each case.

(a) No seat belt: $|\vec{\mathbf{F}}_{av}| = \dfrac{|\Delta\vec{\mathbf{p}}|}{\Delta t} = \dfrac{1300\ \text{kg·m/s}}{0.003\ \text{s}} = 4 \times 10^5\ \text{N}$

(b) With seat belt: $|\vec{\mathbf{F}}_{av}| = \dfrac{|\Delta\vec{\mathbf{p}}|}{\Delta t} = \dfrac{1300\ \text{kg·m/s}}{0.3\ \text{s}} = 4 \times 10^3\ \text{N}$

(c) Air bag: $|\vec{\mathbf{F}}_{av}| = \dfrac{|\Delta\vec{\mathbf{p}}|}{\Delta t} = \dfrac{1300\ \text{kg·m/s}}{1\ \text{s}} = 1 \times 10^3\ \text{N}$

Discussion   Without a seat belt (or air bag), the average force is roughly 700 times the passenger's weight; with an air bag (and seat belt), the average force is only about 2 times the passenger's weight. The average forces required to bring the passenger to rest are inversely proportional to the time interval over which those forces act. It is a far happier situation to have the momentum change over as long a period as possible

continued on next page

Example 7.3 *continued*

to make the forces smaller. Automotive safety engineers design cars to minimize the average forces on the passengers during sudden stops and collisions.

the catcher moves his mitt backward toward his body by 8.0 cm as the ball lands in the glove, what is the magnitude of the average force acting on the catcher's mitt? Estimate the time interval required for the catcher to move his hands.

Practice Problem 7.3   *Catching a fastball*

A baseball catcher is catching a fastball that is thrown at 43 m/s (96 mi/h) by the pitcher. If the mass of the ball is 0.15 kg and if

One automotive design change implemented to minimize injury upon collision is the foam padding built into automobile dashboards (Fig. 7.4). Some steering wheels are collapsible in order to give way in a crash. Automobile bumpers have shock absorbers built in to lessen damage to the car body in small collisions. The structure of the car itself is often a single piece of metal with reinforced supports (*unibody* construction) so that the entire body can crumple and absorb the change in momentum more slowly than it would if it were made of separate sections of metal that would slide into or over each other or fall off the car. The safety glass in a windshield has two advantages. One is that it does not shatter and send sharp shards of glass into human tissue, but the other is that it distorts when struck by solid objects like human bones or a human head. The glass doesn't give much, but in a crash every little bit helps.

The use of seat belts plus the air bag is better than either alone. Without a seat belt, the body continues moving with the same speed the car had before the crash. The rapidly inflating air bag moves toward the body and the effective velocity is then the sum of the two velocities (air bag velocity + body velocity) when the two collide. The body flying

Figure 7.4   Some safety features of the modern automobile. Many of these features serve to lengthen the time interval during which a momentum change occurs in a crash, thereby lessening the forces acting on the passengers.

into the air bag can be injured more than a restrained body making more gradual contact with the air bag. An adult should sit at least 12 in. from the air bag container to avoid injury from the deploying air bag itself. Small children should always be placed in the back seat, in proper car seats for their size, to ensure their safety.

Physics at Home

Try playing catch with a friend (on the lawn) while using a raw egg or a water balloon as a ball. How do you move your hands to minimize the chance of breaking the egg or balloon when you catch it? What is likely to happen if you forget that the "ball" is an egg or balloon and catch it as you would a ball?

Impulse Due to a Changing Force

When a force is changing, how can we find the impulse? We've asked similar questions in previous chapters. For simplicity we consider components along the x-axis. Recall:

> A graph of $v_x(t)$ means the quantity v_x is plotted as a function of the variable t.

- displacement = $\Delta x = v_{av,x} \Delta t$ = area under $v_x(t)$ graph
- change in velocity = $\Delta v_x = a_{av,x} \Delta t$ = area under $a_x(t)$ graph

In both cases, the mathematical relationship is that of a rate of change. Velocity is the rate of change of position with time and acceleration is the rate of change of velocity with time. Now we have force as the rate of change of momentum with time. By analogy:

- impulse = $\vec{\mathbf{F}}_{av,x} \Delta t$ = area under $F_x(t)$ graph

So to find the impulse for a variable force, we find the area under the $F_x(t)$ graph. Then, if we wish to know the average force, we can divide the impulse by the time interval during which the force is applied.

The variable force of Fig. 7.5a increases linearly from 0 to 4 N in a time of 2 s; then it decreases from 4 N to 0 N in 2 s. The area under the $F(t)$ graph is found from the triangular area

$$\text{area} = \tfrac{1}{2}\,\text{base} \times \text{height} = 2\,\text{s} \times 4\,\text{N} = 8\,\text{N·s} = \text{impulse}$$

The average force during the 4-s time interval is

$$\text{average force} = \frac{\text{impulse}}{\text{time interval}} = \frac{8\,\text{N·s}}{4\,\text{s}} = 2\,\text{N}$$

Figure 7.5b shows the average force over the 4-s time interval; the area under the curve (the impulse) is the same as in Fig. 7.5a.

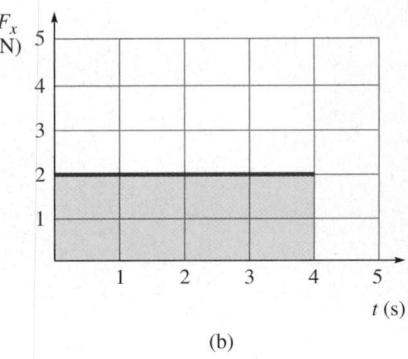

Figure 7.5   Areas under $F(t)$ graphs for (a) variable and (b) average force are a measure of the impulse.

Example 7.4

Hitting the Wall

A toy car of mass 10.2 kg moving at 1.2 m/s in the +x-direction crashes into a brick wall and rebounds. A force sensor on the car's bumper records the force that the wall exerts on the car as a function of time. These data are shown in graphical form in Fig. 7.6. (a) What is the maximum magnitude of the force exerted on the car? (b) What is the average force on the car during the collision? (c) At what speed does the car rebound from the wall?

Strategy   The maximum force can be read directly from the graph. To solve parts (b) and (c) of this problem, we must find the impulse exerted on the car. Since impulse is the area under the $F(t)$ curve, we'll make an estimate of the area. The impulse is then equal to the average force times the time interval and also to the car's change in momentum. Once we find the change in momentum, we use it to find the car's final speed.

Given: $m = 10.2$ kg; $v_{ix} = 1.2$ m/s; graph of $F_x(t)$
To find: (a) F_{max}; (b) $F_{av,x}$; (c) v_{fx}

Solution   (a) From Fig. 7.6, the maximum force is approximately 750 N in magnitude.

(b) Each division on the horizontal axis represents 0.01 s and each vertical division represents 200 N. Then the area of each grid box represents (200 N × 0.01 s) = 2 N·s. Counting the number of grid boxes between the $F(x)$ curve and the time axis, estimating fractions of boxes, yields about 10 boxes. Then the magnitude of the impulse is approximately

$$10 \text{ boxes} \times 2 \text{ N·s/box} = 20 \text{ N·s}$$

The collision is underway when the force is nonzero. So the collision begins at $t = 0.025$ s and ends at $t = 0.095$ s. The duration of the collision is

$$\Delta t = 0.07 \text{ s}$$

The magnitude of the average force is therefore approximately

$$|F_{av,x}| = \frac{\text{impulse}}{\Delta t} = \frac{20 \text{ N·s}}{0.07 \text{ s}} = 300 \text{ N}$$

(c) The impulse gives us the momentum change. The force exerted by the wall is in the $-x$-direction. Thus, the x-component of the impulse is negative. In the graph of F_x versus t, the area lies under the time axis and so is counted as negative. So, working with x-components,

$$\Delta p_x = mv_{fx} - mv_{ix} = F_{av,x} \, \Delta t = -20 \text{ N·s}$$

Solving for v_{fx},

$$v_{fx} = \frac{\Delta p_x + mv_{ix}}{m} = \frac{\Delta p_x}{m} + v_{ix}$$

Substituting numerical values in this expression yields

$$v_{fx} = \frac{-20 \text{ N·s}}{10.2 \text{ kg}} + 1.2 \text{ m/s} = -0.8 \text{ m/s}$$

The car rebounds at a speed of 0.8 m/s.

Discussion   As a check, we compare the average force to the maximum force. The average force is a bit less than half of the maximum force. If the force were a linear function of time, the average would be exactly half the maximum. Here, the average force is less than that because more time is spent at smaller values of force than at the larger values.

Practice Problem 7.4   Car-van collision

A car weighing 13.6 kN is moving at 10.0 m/s in the +x-direction when it collides head-on with a van weighing 33.0 kN. The horizontal force exerted on the car before, during, and after the collision is shown in Fig. 7.7. What is the car's velocity just after the collision?

Figure 7.6
Force versus time for toy car colliding with a wall.

Figure 7.7
Varying force on a car during a car-van collision.

Newton's Second Law Revisited

We can use the relationship between impulse and momentum ($\Delta \vec{\mathbf{p}} = \vec{\mathbf{F}}_{av} \Delta t$) to find a new way to understand Newton's second law. Let's rewrite the impulse-momentum theorem this way:

$$\vec{\mathbf{F}}_{av} = \frac{\Delta \vec{\mathbf{p}}}{\Delta t}$$

What happens if we let the time interval Δt get smaller and smaller, approaching zero? Then the average force is taken over a smaller and smaller time interval, approaching the instantaneous force:

$$\vec{\mathbf{F}} = \lim_{\Delta t \to 0} \frac{\Delta \vec{\mathbf{p}}}{\Delta t}$$

If more than one force acts, we must replace $\vec{\mathbf{F}}$ with $\vec{\mathbf{F}}_{net}$. Then our restatement of Newton's second law becomes

---

Newton's second law:

$$\vec{\mathbf{F}}_{net} = \lim_{\Delta t \to 0} \frac{\Delta \vec{\mathbf{p}}}{\Delta t} \qquad (7\text{-}4)$$

In words, *the net force is the rate of change of momentum.*

Equation (7-4) is *more general* than $\vec{\mathbf{F}}_{net} = m\vec{\mathbf{a}}$, the form of Newton's second law used in Chapters 2 through 6, which holds only when mass is constant. One situation in which mass is not constant is the rocket engine. In a rocket engine, fuel combustion produces hot gases that are then expelled at high speeds (Fig. 7.8). The rocket's mass decreases as the exhaust gases are expelled.

When mass is constant, then it can be factored out:

$$\vec{\mathbf{F}}_{net} = \lim_{\Delta t \to 0} \frac{\Delta \vec{\mathbf{p}}}{\Delta t} = \lim_{\Delta t \to 0} \frac{\Delta(m\vec{\mathbf{v}})}{\Delta t} = m \lim_{\Delta t \to 0} \frac{\Delta \vec{\mathbf{v}}}{\Delta t} = m\vec{\mathbf{a}}$$

Thus Eq. (7-4) reduces to the familiar form of Newton's second law from Chapters 2–6 when mass is constant.

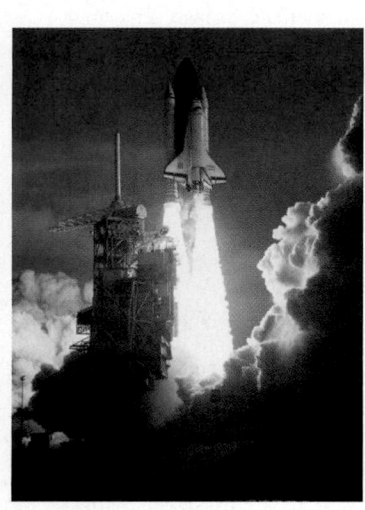

Figure 7.8 The space shuttle is propelled upward as hot gases are exhausted downward at high speeds.

7.4   CONSERVATION OF MOMENTUM

The idea of an interaction as a transfer of momentum is most useful when the forces acting are complicated and we are interested in knowing the state of motion after the interaction is finished. As an example, we might consider two pucks that bump into each other after sliding along on a frictionless table. Fig. 7.9 shows what happens to the two pucks before, during, and after their interaction. If we think of the two pucks as comprising a single system, then the gravitational interactions with the Earth and the contact interactions with the table are *external* interactions—they are interactions with objects external to the system. We know that the force of gravity on each object is balanced by the normal force on the same object and thus there is no net impulse up or down.

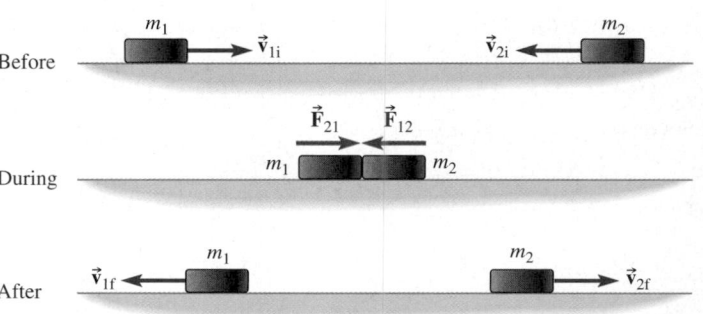

Figure 7.9 Two sliding pucks before, during, and after collision.

Together, these forces produce a net external force of zero, so they leave the system's momentum unchanged. Since these two always cancel, we can ignore these "external" interactions and just focus on the interaction between the pucks. Therefore, we omit the normal and gravitational forces in Fig. 7.9.

Until contact is made, there is no interaction between the pucks (ignoring the small gravitational interaction between the two). During the collision, the pucks exert forces on each other. Force $\vec{\mathbf{F}}_{12}$ is the contact force acting on mass m_1 and force $\vec{\mathbf{F}}_{21}$ is the contact force acting on mass m_2. If we continue to regard the two pucks as parts of a single interacting system, then those forces are *internal* forces of this system. When they collide, some momentum is transferred from one puck to the other. The changes in momentum of the two are equal and opposite:

$$\Delta\vec{\mathbf{p}}_1 = -\Delta\vec{\mathbf{p}}_2$$

Since the change in momentum is the final momentum minus the initial momentum, we write:

$$\vec{\mathbf{p}}_{1f} - \vec{\mathbf{p}}_{1i} = -(\vec{\mathbf{p}}_{2f} - \vec{\mathbf{p}}_{2i})$$

Moving the initial momenta to the right side and the final momenta to the left:

$$\vec{\mathbf{p}}_{1f} + \vec{\mathbf{p}}_{2f} = \vec{\mathbf{p}}_{1i} + \vec{\mathbf{p}}_{2i} \qquad (7\text{-}5)$$

The sum of the momenta of the pucks after the interaction is equal to the sum of the momenta before the interaction; or, more simply, the total momentum of the objects is unchanged by the collision. This isn't surprising since, if some momentum is just transferred from one to the other, the total hasn't changed. We say that momentum is *conserved* for this collision. The interaction between the pucks changes the momentum of each puck, but the total momentum of the system is unchanged.

To summarize:

- The total momentum of a system is the vector sum of the momenta of each object in the system.

- External interactions can change the total momentum of a system.

- Internal interactions do not change the total momentum of a system.

In the absence of external interactions, momentum is conserved:

Law of Conservation of Linear Momentum
If the net external force acting on a system is zero,
then the momentum of the system is conserved.

$$\text{if } \vec{\mathbf{F}}_{net} = 0, \quad \vec{\mathbf{p}}_i = \vec{\mathbf{p}}_f \qquad (7\text{-}6)$$

By definition, an isolated, or closed, system is subject to no external interactions; thus *linear momentum is always conserved for an isolated system.* Remember that momentum is a vector quantity, so both the magnitude and the direction of the momentum at the beginning and end of the interaction must be the same.

Example 7.5

Adrift on a Raft

Diana is standing on a raft of mass 100.0 kg that is floating on a still lake. She decides to walk the length of the raft (Fig. 7.10). If Diana's mass is 55 kg and she walks with a velocity of 0.91 m/s *relative to the water,* how fast and in what direction does the raft move while Diana is walking? Assume the raft is stationary with respect to the shore before Diana starts walking.

Strategy   Diana and the raft can be considered to be a single *isolated system:* as long as frictional forces on the raft due to the water and air are small enough to ignore, the net external force on the system is zero. Then the momentum of this system (raft + Diana) is conserved. We let the subscripts D stand for Diana and r for the raft and set the change in momentum of the system equal to zero.

continued on next page

Example 7.5 *continued*

Solution   To walk, Diana must exert a force on the raft: the static frictional force between her feet and the raft. This is an internal interaction within the isolated system, so it cannot change the total momentum of the system. Only something acting from outside the system could do that. As Diana walks in one direction, she acquires some momentum. The rest of the system (the raft) must acquire an equal and opposite momentum, because the momentum of the isolated system (Diana + raft) is conserved, which means that the change in momentum of the system is zero.

First we set the change in momentum of the system equal to zero:

$$\Delta \vec{p} = \vec{0} = \Delta \vec{p}_D + \Delta \vec{p}_r$$

or

$$\Delta \vec{p}_D = -\Delta \vec{p}_r$$

This means that the momentum changes of Diana and of the raft are equal and opposite. Since momentum is the product of mass and velocity and the masses of the raft and Diana do not change,

$$m_D \,\Delta \vec{v}_D = -m_r \,\Delta \vec{v}_r$$

Figure 7.10
Diana walking along a raft.

Solving for the change in velocity of the raft gives

$$\Delta \vec{v}_r = -\frac{m_D \,\Delta \vec{v}_D}{m_r}$$

Finally we substitute numerical values from the given information in the statement of the problem. Let Diana walk in the +*x*-direction.

$$\Delta \vec{v}_r = -\frac{55 \text{ kg} \times 0.91 \text{ m/s (in } +x\text{-direction)}}{100.0 \text{ kg}}$$

$$= 0.50 \text{ m/s in } -x\text{-direction}$$

The negative sign indicates that the raft moves in a direction opposite to Diana's movement to keep the momentum unchanged and thus conserved. Since the raft was originally stationary, this is the new velocity of the raft.

Discussion   In any momentum conservation problem there are two equivalent ways to proceed. In this example we set the momentum change of the system equal to zero. We could just as well write an equation that sets the initial total momentum equal to the final momentum of the system. The raft and Diana are initially at rest, so the initial momentum is zero:

$$\vec{0} = m_D \vec{v}_D + m_r \vec{v}_r$$

where $\vec{v}_D$ and $\vec{v}_r$ are the final velocities of Diana and the raft.

Practice Problem 7.5   Skaters pushing apart

Two skaters on in-line skates, Alexia and Bart, are initially at rest. They push apart and start moving in opposite directions. If Alexia's speed just after they push apart is 2.0 m/s and her mass is 85% of Bart's mass, how fast is Bart moving at that time?

When a bullet is fired from a rifle, the system of rifle plus bullet must conserve momentum. Suppose the rifle is at rest before the bullet is fired. The momentum of the system is zero. When the bullet is fired, part of the system's mass breaks away and travels in one direction with a certain momentum. The rifle, which is the remaining mass of the system, moves in the exact opposite direction such that the total momentum of the system is still zero. The rifle has a much larger mass than the bullet so it has a much smaller speed. The backward motion of the rifle is the *recoil* felt by anyone who has held a rifle against her shoulder and squeezed the trigger.

Jet engines and rockets operate by conservation of momentum. Hot combustion gases are forced out of nozzles at high speed by the engines. The increased backward momentum of the hot gases as they are expelled must be accompanied by an increased forward momentum of the engines.

Conceptual Example 7.6

Escape on Slippery Ice

A pilot parachutes from his disabled aircraft and lands on the frozen surface of a lake. There is no breeze blowing and the lake surface is too slippery to walk on. What can the pilot do to reach the shore?

Strategy and Solution   Since the person in jeopardy is a pilot, he begins to think about how hot gases forced backward from a jet engine cause the plane to move forward. That gives him an idea: he bundles the parachute into a package and pushes it as hard as possible in a direction away from the nearest point of the shore. If the net external force on the system of pilot plus parachute is zero, then the total momentum of the system cannot change. The momentum of the parachute plus the momentum of the pilot must still equal zero. By conservation of momentum, the pilot begins sliding in the opposite direction and glides toward the shore.

Discussion   If friction brings the pilot to rest before he reaches the shore, he can search his pockets and belt loops for other items to throw away. Once he reaches shore, he can tie one end of a rope to a tree and, holding onto the other end, venture back out onto the ice to retrieve any essential items. The rope provides him with an external force so he can get back to shore.

Conceptual Practice Problem 7.6   Bank robbery in the Old West

After tying the bank manager to his office chair, which has wheels, the James brothers begin emptying the bank safe. The manager faces away from his desk. His legs are securely tied, but he manages to free one arm, with which he retrieves a heavy gold nugget from the secret compartment below the armrest. How might the manager reach the telegraph on his desk to send a message for help? What might make this scheme fail?

Physics at Home

In case you and a friend ever end up stuck in the middle of the ice, practice the technique the pilot used to escape to the lakeshore. Bring a heavy medicine ball out to the middle of the ice rink and face each other with your skates aligned parallel. Toss the ball to your friend. What happens to you? What happens to your friend when he catches the ball? Can you both be "saved" by tossing the ball back and forth? (The same technique works using in-line skates.)

Example 7.7

A Broadside and Its Aftermath

The *Ocean Queen* has 14 guns mounted along the starboard side and the same number on the port side. The mass of the *Ocean Queen* is 8.5×10^4 kg. The cannonballs are each of mass 5.0 kg and leave the cannons with a horizontal velocity of 144 m/s. If all 14 cannons on the starboard side are fired in unison, with the firing taking place within a 0.50-s time interval, what is the average force exerted on the ship by the cannons during the firing?

Strategy   We can find the momentum change of the cannonballs. Then, using the impulse-momentum relation, we can find the average force on the ship.

Given: $M_{ship} = 8.5 \times 10^4$ kg; $m_{ball} = 5.0$ kg; $v_{ball} = 144$ m/s; $\Delta t = 0.50$ s

To find: F_{av}

Solution   The momentum of the cannonballs just after they are fired is

$$p = mv = (14 \text{ cannonballs} \times 5.0 \text{ kg/cannonball}) \times 144 \text{ m/s}$$

$$= 10{,}000 \text{ kg·m/s}$$

The cannonballs must exert an impulse of equal magnitude on the ship as they are fired:

$$F_{av} \times 0.50 \text{ s} = 1.0 \times 10^4 \text{ kg·m/s}$$

$$F_{av} = 2.0 \times 10^4 \text{ N}$$

The magnitude of the average force is 2.0×10^4 N—more than 2 tons. The direction is opposite to the momentum of the cannonballs as they leave the cannons.

Discussion   If the ship were free to recoil, it would do so at a speed given by momentum conservation:

$$M_{ship}v_{ship} = 1.0 \times 10^4 \text{ kg·m/s}$$

$$v_{ship} = \frac{1.0 \times 10^4 \text{ kg·m/s}}{8.5 \times 10^4 \text{ kg}} = 0.12 \text{ m/s}$$

The ship is prevented from moving sideways at 0.12 m/s by the resistance of the ocean water in which it is partially immersed, but it is relatively free to rotate and reacts to the impulse felt by rocking to one side in the water. Unfortunately for the young pirate, his ship tilted enough that water poured

continued on next page

Example 7.7 *continued*

through the open cannon portholes on the port side, scuttling the ship, and forcing the pirates to appeal to their intended victims for rescue.

One reason that cannons were mounted on wheels and held loosely in place by ropes is that the ropes allowed the cannons to roll backward as they were fired. The impulse exerted on the ship is thereby spread over a longer time interval, reducing the average force. The total impulse on the ship is *not* reduced; if the ship were free to recoil, it would do so at the same speed either way. However, the water exerts large forces on the ship, so it does not recoil as if it were free. The impor-

tant quantity is the *average force* exerted on the deck of the ship, which determines how much the ship tips in the water.

Practice Problem 7.7   Recoil of a rifle

During an afternoon of target practice, you fire a Winchester .308 rifle of mass 3.8 kg. The bullets have a mass of 9.72 g and leave the rifle at a muzzle velocity of 860 m/s. If you are sloppy and fire a round when the butt of the rifle is not firmly up against your shoulder, at what speed does the rifle butt smash into your shoulder? (Ouch!)

7.5   CENTER OF MASS

We have seen that the momentum of an isolated system is conserved even though parts of the system may interact with other parts; internal interactions transfer momentum between parts of the system but do not change the total momentum of the system. We can define a point called the **center of mass** (CM) that serves as an average location of the system. In Section 7.6 we prove that the center of mass of an isolated system must move with constant velocity, regardless of how complicated the motions of parts of the system may be. Then we can treat the mass of the system as if it were all concentrated at the center of mass, like a point particle. The center of mass of an object is not necessarily located within the object; for some objects, such as a boomerang, the center of mass is located outside of the object itself (Fig. 7.11a).

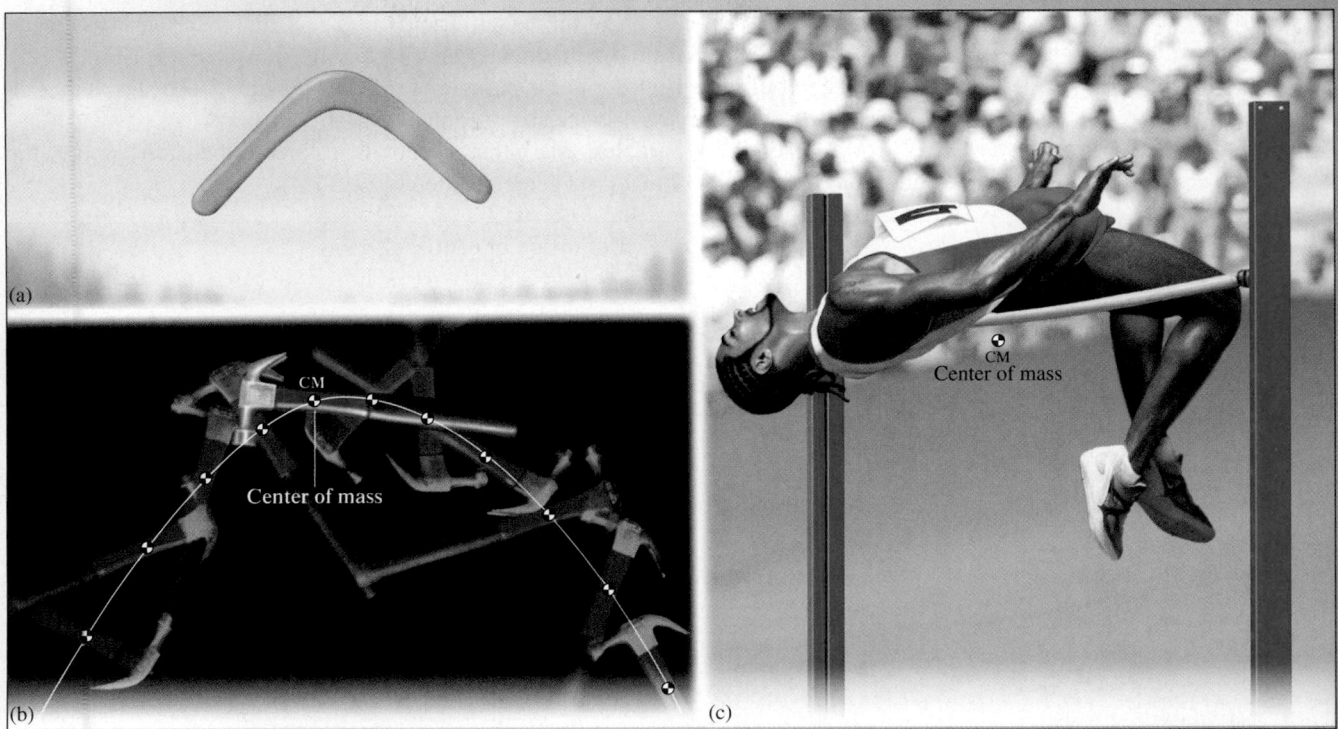

Figure 7.11 (a) The center of mass of a boomerang is a point outside of the boomerang. (b) The path followed by the center of mass when a hammer is tossed through the air. (c) Ben Challenger's center of mass actually passes *beneath* the bar as his body passes over the bar.

What if a system is not isolated, but has external interactions? Again imagine all of the mass of the system concentrated into a single point particle located at the center of mass. The motion of this fictitious point particle is determined by Newton's second law, where the net force is the sum of all of the external forces acting on *any part* of the system. In the case of a complex system composed of many parts interacting with each other, the motion of the center of mass is considerably simpler than the motion of an arbitrary particle of the system (Fig. 7.11b,c).

For a system composed of two particles, the center of mass lies somewhere on a line between the two particles. In Fig. 7.12, particles of masses m_1 and m_2 are located at positions x_1 and x_2, respectively. We define the location of the center of mass for these two particles as

$$x_{\text{CM}} = \frac{m_1 x_1 + m_2 x_2}{m_1 + m_2} \qquad (7\text{-}7)$$

The center of mass is a *weighted average* of the positions of the two particles. Here we use the word *weighted* in its statistical sense. The position of a particle with more mass counts more—carries more *statistical* weight—than does the position of a particle with a smaller mass. We can rewrite Eq. (7-7) as a weighted average:

$$x_{\text{CM}} = \frac{m_1}{M} x_1 + \frac{m_2}{M} x_2 \qquad (7\text{-}8)$$

Here $M = m_1 + m_2$ represents the total mass of the system. The statistical weight used for the location of each particle is the mass of that particle as a fraction of the total mass of the system.

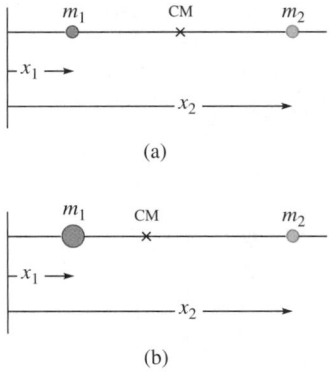

Figure 7.12 (a) Two particles of equal mass located at positions x_1 and x_2 from the origin. The center of mass is midway between the two. (b) Two particles of unequal mass. The center of mass is closer to the more massive particle.

Suppose masses m_1 and m_2 are equal. Then we expect the center of mass to be located midway between the two particles (Fig. 7.12a). If $m_1 = 2m_2$, as in Fig. 7.12b, then the center of mass is closer to the particle of mass m_1. Figure 7.12b shows that, in this case, the center of mass is twice as far from m_2 as from m_1.

For a system of N particles, at arbitrary locations in three-dimensional space, the definition of the center of mass is a generalization of Eq. (7-7).

Definition of center of mass

Vector form:
$$\vec{\mathbf{r}}_{CM} = \frac{\Sigma m_i \vec{\mathbf{r}}_i}{M} \qquad \text{(7-9)}$$

Component form:
$$x_{CM} = \frac{\Sigma m_i x_i}{M} \qquad y_{CM} = \frac{\Sigma m_i y_i}{M} \qquad z_{CM} = \frac{\Sigma m_i z_i}{M}$$

where $i = 1, 2, 3, \ldots, N$ and $M = \Sigma m_i$

The symbol Σ stands for *sum*. The shorthand notation $\Sigma m_i x_i$ is interpreted as

$$\Sigma m_i x_i = m_1 x_1 + m_2 x_2 + \cdots + m_N x_N$$

For particles in two-dimensional space, we use only two of these equations for the *x-y* plane and find the *x*- and *y*-components of the center of mass.

Example 7.8

Center of Mass of a Binary Star System

The two stars in a binary star system are positioned as shown in Fig. 7.13a. Due to the gravitational interaction between the two, the stars each move in a circular orbit around their center of mass. One star has a mass of 15.0×10^{30} kg; its center is located at $x = 1.0$ AU and $y = 5.0$ AU. The other has a mass of 3.0×10^{30} kg; its center is at $x = 4.0$ AU and $y = 2.0$ AU. Find the CM of the system composed of the two stars. (AU stands for *astronomical unit*. 1 AU = the average distance between the Earth and the Sun = 1.5×10^8 km.)

Strategy We treat the stars as point particles located at their centers. Since we are given *x*- and *y*-coordinates, the easiest way to proceed is to find the *x*- and *y*-coordinates of the CM. There is no particular advantage here in finding the position vector of the CM in terms of its length and direction.

Given: $m_1 = 15.0 \times 10^{30}$ kg $x_1 = 1.0$ AU $y_1 = 5.0$ AU
$m_2 = 3.0 \times 10^{30}$ kg $x_2 = 4.0$ AU $y_2 = 2.0$ AU

To find: x_{CM}; y_{CM}

Solution The total mass of the system is the sum of the individual masses:

$$M = m_1 + m_2 = 15.0 \times 10^{30} \text{ kg} + 3.0 \times 10^{30} \text{ kg} = 18.0 \times 10^{30} \text{ kg}$$

Figure 7.13
(a) Positions of two stars; (b) finding the CM for the system of two stars; and (c) three spheres located at *x, y* positions (1.0 cm, 1.0 cm), (2.0 cm, 3.0 cm), and (3.0 cm, 1.0 cm).

continued on next page

Example 7.8 *continued*

For the *x*-position, we find

$$x_{CM} = \frac{m_1}{M} x_1 + \frac{m_2}{M} x_2$$

$$= \frac{15.0 \times 10^{30} \text{ kg}}{18.0 \times 10^{30} \text{ kg}} \times 1.0 \text{ AU} + \frac{3.0 \times 10^{30} \text{ kg}}{18.0 \times 10^{30} \text{ kg}} \times 4.0 \text{ AU} = 1.5 \text{ AU}$$

and for the *y*-position, we find

$$y_{CM} = \frac{m_1}{M} y_1 + \frac{m_2}{M} y_2$$

$$= \frac{15.0}{18.0} \times 5.0 \text{ AU} + \frac{3.0}{18.0} \times 2.0 \text{ AU} = 4.5 \text{ AU}$$

Discussion   In Fig. 7.13b we mark the position of the center of mass. As we expect for the case of two particles, it is located closer to the larger mass and on a line connecting the two. Once the CM position is found in a problem, check to be sure its location is reasonable. Suppose we had made an error in this example and found the CM to be at $x = 1.5$ AU and $y = 1.7$ AU. This is not a reasonable location for the CM since it is not along the line connecting the two and is closer to the less massive star; we then would go back to look for the error.

Practice Problem 7.8   Three balls with unequal masses

Three spherical objects are shown in Fig. 7.13c. Their masses are $m_1 = m_3 = 1.0$ kg and $m_2 = 4.0$ kg. Find the location of the center of mass for the three objects.

Most objects we deal with in real life are not composed of a small set of point particles or spherically symmetric objects. In Example 7.8 we used the location of the center of each star to find the center of mass. Due to spherical symmetry, the center of mass of either star (by itself) is at its geometric center. The same technique can be applied to other shapes with symmetry. A standard 2 by 4, which is an 8-ft-long uniform piece of wood 1.5" deep × 3.5" high, has its center of mass at its geometric center: at half the length, half the depth, and half the height of the board. This assumes the object is uniform in construction. A "loaded" die does *not* have its CM at its geometric center, since a small metal plug has been inserted near one face. The definition of the CM [Eq. (7-9)] still holds as long as (x_i, y_i, z_i) are the coordinates of the center of mass of a part of the system with mass m_i.

7.6   MOTION OF THE CENTER OF MASS

Now that we know how to find the position of the CM of a system, we turn our attention to the motion of the CM. How is the velocity of the CM related to the velocities of the various parts of the system?

During a short time interval Δt, the displacement of the *i*th particle is

$$\Delta \vec{\mathbf{r}}_i = \vec{\mathbf{v}}_i \, \Delta t$$

and the displacement of the center of mass is

$$\Delta \vec{\mathbf{r}}_{CM} = \vec{\mathbf{v}}_{CM} \, \Delta t$$

From the definition of the center of mass [Eq. (7-9)], the displacements must be related as follows:

$$\Delta \vec{\mathbf{r}}_{CM} = \frac{\Sigma m_i \Delta \vec{\mathbf{r}}_i}{M}$$

or

$$M \, \Delta \vec{\mathbf{r}}_{CM} = m_1 \, \Delta \vec{\mathbf{r}}_1 + m_2 \, \Delta \vec{\mathbf{r}}_2 + \cdots + m_N \, \Delta \vec{\mathbf{r}}_N$$

We can replace each displacement by the product of a velocity and the time interval Δt.

$$M \vec{\mathbf{v}}_{CM} \, \Delta t = m_1 \vec{\mathbf{v}}_1 \, \Delta t + m_2 \vec{\mathbf{v}}_2 \, \Delta t + \cdots + m_N \vec{\mathbf{v}}_N \, \Delta t$$

The time interval Δt is the same in each term. Dividing both sides by Δt,

$$M \vec{\mathbf{v}}_{CM} = m_1 \vec{\mathbf{v}}_1 + m_2 \vec{\mathbf{v}}_2 + \cdots + m_N \vec{\mathbf{v}}_N \qquad (7\text{-}10)$$

The total momentum of a system is defined as the sum of the individual momenta of the particles comprising the system:

$$\vec{\mathbf{p}} = \Sigma \vec{\mathbf{p}}_i = \vec{\mathbf{p}}_1 + \vec{\mathbf{p}}_2 + \cdots + \vec{\mathbf{p}}_N$$

Equation (7-10) therefore says that the total momentum of a system is equal to the total mass of the system times the velocity of the center of mass:

$$\vec{\mathbf{p}} = M\vec{\mathbf{v}}_{CM} \tag{7-11}$$

For two-dimensional motion it is usually easiest to work with components of momenta in the x- and y-directions.

$$p_x = Mv_{CM,x} \quad \text{and} \quad p_y = Mv_{CM,y} \tag{7-12}$$

where p_x and p_y are the x- and y-components of the total momentum of the system.

In Section 7.4, we showed that, for an isolated system, the total linear momentum is conserved. In such a system, Eq. (7-11) implies that the center of mass must move with constant velocity regardless of the motions of the individual particles. On the other hand, what if the system is not isolated? If a net external force acts on a system, the center of mass does not move with constant velocity. Instead, it moves as if all the mass were concentrated there into a fictitious point particle with all the external forces acting on that point. The motion of the center of mass obeys the following statement of Newton's second law:

$$\Sigma\vec{\mathbf{F}}_{ext} = M\vec{\mathbf{a}}_{CM} \tag{7-13}$$

where M is the total mass of the system, $\Sigma\vec{\mathbf{F}}_{ext}$ is the net external force, and $\vec{\mathbf{a}}_{CM}$ is the acceleration of the center of mass. [Eq. (7-13) is proved in Problem 29.]

Example 7.9

An Exploding Rocket

A model rocket is fired from the ground in a parabolic trajectory. At the top of the trajectory, a horizontal distance of 260 m from the launch point, an explosion occurs within the rocket, breaking it into two fragments. One fragment, having one-third of the mass of the rocket, falls straight down to Earth as if it had been dropped from rest at that point. At what horizontal distance from the launch point does the other fragment land? Ignore air resistance.

Strategy   There are two different strategies that can be used to solve this problem.

Strategy 1: We apply conservation of momentum to the explosion. The momentum of the rocket *just before* the explosion is equal to the total momentum of the two fragments *just after* the explosion. Why can momentum conservation be assumed here? There is an external force—gravity—acting on the system. External forces change momentum. However, the explosion takes place in a *very short time interval*. From the impulse-momentum theorem [Eq. (7-2)], the momentum change of the system is the force of gravity multiplied by the time interval. As long as the time interval considered is sufficiently short, the momentum change of the system can be ignored.

Strategy 2: The explosion is caused by an *internal* interaction between two parts of the rocket. The motion of the center of mass of the system is unaffected by internal interactions, so it continues in the same parabolic path. The two pieces of the rocket land simultaneously; at that same instant, the center of mass also reaches the ground.

Solution 1   First we make a sketch of the situation (Fig. 7.14). At the top of the trajectory, where the explosion occurs, $v_y = 0$; the rocket is moving in the x-direction. The initial momentum just before the explosion is entirely in the x-direction. If M is the mass of the rocket, then

$$p_{ix} = Mv_{ix}$$

Just after the explosion, one-third of the mass of the rocket is at rest; it then drops straight down under the influence of the gravitational force. This piece has zero momentum just after the explosion. To conserve momentum, the other two-thirds of the rocket must have a momentum equal to the momentum just before the explosion.

$$p_{ix} = p_{1x} + p_{2x}$$
$$Mv_{ix} = 0 + (\tfrac{2}{3}M)v_{2x}$$

Solving for v_{2x}, we find

$$v_{2x} = \tfrac{3}{2}v_{ix}$$

The y-component of momentum must also be conserved:

$$p_{iy} = p_{1y} + p_{2y}$$

We know that both p_{iy} and p_{1y} are zero; therefore p_{2y} is zero as well. Just after the explosion both parts of the rocket have zero vertical components of velocity. Then both parts take the same time to fall to the ground as if the rocket had not exploded. With a horizontal velocity larger by a factor of 3/2, the second piece of the rocket travels a horizontal distance from the explosion a factor of 3/2 larger than 260 m (see Fig. 7.14). The distance from the launch point where this piece lands is

$$\Delta x = 260 \text{ m} + \tfrac{3}{2} \times 260 \text{ m} = 650 \text{ m}$$

continued on next page

Example 7.9 *continued*

Solution 2 The piece with mass $\frac{1}{3}M$ falls straight down and lands 260 m from the launch point. After the explosion, the CM continues to travel just as the rocket itself would have done if it had not broken apart. From the symmetry of the parabola, the CM touches the ground at a distance of 2×260 m $= 520$ m from the launch point. Since we know the location of the CM and that of one of the pieces, we can find where the second piece lands:

$$Mx_{CM} = \tfrac{1}{3}Mx_1 + \tfrac{2}{3}Mx_2$$

After canceling the common factor of M,

$$x_{CM} = \tfrac{1}{3}x_1 + \tfrac{2}{3}x_2$$

Solving for x_2 yields

$$x_2 = \frac{3x_{CM} - x_1}{2} = \frac{3 \times 520 \text{ m} - 260 \text{ m}}{2} = 650 \text{ m}$$

which is the same answer that we found in Solution 1.

Discussion The insight that the motion of the CM is unaffected by internal interactions can be of enormous help. Note, however, that Solution 2 would not be so simple if the two fragments did not land simultaneously. As soon as one fragment (fragment 1) hits the ground, the external force on the system is no longer due exclusively to gravity, so the CM doesn't continue to follow the same parabolic path. The normal and frictional forces acting on fragment 1 affect its subsequent motion and the subsequent motion of the CM even though the motion of fragment 2 is unaffected.

Practice Problem 7.9 Diana and the raft revisited

In Example 7.5, Diana (mass 55 kg) walks at 0.91 m/s (relative to the water) on a raft of mass 100.0 kg. The raft moves in the opposite direction at 0.50 m/s. Suppose it takes her 3.0 s to walk from one end of the raft to the other. (a) How far does Diana walk (relative to the water)? (b) How far does the raft move while Diana is walking? (c) How far does the center of mass of Diana and the raft move during the 3.0 s?

Figure 7.14
Rocket motion after explosion.

7.7 COLLISIONS IN ONE DIMENSION

What is a collision? In the macroscopic world, a moving body bumps into another body that may be at rest or in motion. The two bodies exert forces on each other while they are in contact; as a result, their velocities change. In the microscopic and submicroscopic world, our picture of a collision is different. When atoms collide, they don't "touch" each other: the atom doesn't have a definite spatial boundary, so there are no surfaces to make "contact." However, the collision model is still useful for atoms and subatomic particles whenever there is an interaction in which the forces are strong over a short time interval, so that there is a clear "before collision" and a clear "after collision."

The forces exerted during a collision between macroscopic objects generally vary with time in a way that is hard to predict. What we *are* able to do is look at the motion of the bodies before and after the interaction. Thanks to the law of conservation of linear momentum, we can make predictions of the motion after an interaction by knowing the conditions of motion before the interaction. The same techniques that are used for collisions in the macroscopic world (car crashes, billiard ball collisions, baseball bats hitting balls) are also used in collisions in the microscopic world (gas molecules colliding with each other and with surfaces, radioactive decays of nuclei). First, we study collisions limited to motion along a line; later, we consider collisions limited to motion in a plane (in two dimensions).

Figure 7.15 Colliding logs shown before and after a collision.

Suppose we observe a log traveling along a flume at speed v_0 toward a second log that is floating at rest in the water. The masses of the two logs are equal. When the first log hits the second, what happens?

Based on momentum considerations *alone*, there are many possible outcomes. One possibility is that the first log stops moving and the second log moves off with the same velocity that the first one had to begin with (Fig. 7.15). This possibility satisfies conservation of momentum because the total momentum is the same before and after.

Another possibility is that the two logs stick together, moving away together. With what speed do they move after the collision? If the momentum is to be the same with twice as much mass moving, the speed must be half the initial speed of the first object. There are many other possibilities. Conservation of momentum doesn't tell us which of these outcomes actually happens, but if we know one object's velocity after the collision, we can use momentum conservation to determine the other object's velocity.

Example 7.10

Collision in the Air

A krypton atom (mass 83.9 u) moving with a velocity of 0.80 km/s to the right and a water molecule (mass 18.0 u) moving with a velocity of 0.40 km/s to the left collide head-on. The water molecule has a velocity of 0.60 km/s to the right after the collision. What is the velocity of the krypton atom after the collision? (The symbol "u" stands for the atomic mass unit.)

Strategy Since we know both initial velocities and one of the final velocities, we can find the second final velocity by applying momentum conservation. Let the subscript "1" refer to the krypton atom and let the subscript "2" refer to the water

Before

After

Figure 7.16
Before and after snapshots of a collision.

molecule. Let the x-axis point to the right. Figure 7.16 shows before and after pictures of the collision.

Solution Momentum conservation requires that the final momentum be equal to the initial momentum:

$$\vec{p}_{1f} + \vec{p}_{2f} = \vec{p}_{1i} + \vec{p}_{2i}$$

Now we substitute $\vec{p} = m\vec{v}$ for each momentum. It is easiest to work in terms of components. For simplicity we drop the "x" subscripts, remembering that all quantities refer to x-components:

$$m_1 v_{1f} + m_2 v_{2f} = m_1 v_{1i} + m_2 v_{2i}$$

Since $m_1/m_2 = 83.9/18.0 = 4.661$, we can substitute $m_1 = 4.661 m_2$:

$$4.661 m_2 v_{1f} + m_2 v_{2f} = 4.661 m_2 v_{1i} + m_2 v_{2i}$$

The common factor m_2 cancels out. Solving for v_{1f},

$$v_{1f} = \frac{4.661 v_{1i} + v_{2i} - v_{2f}}{4.661}$$

$$= \frac{4.661 \times 0.80 \text{ km/s} + (-0.40 \text{ km/s}) - 0.60 \text{ km/s}}{4.661} = 0.59 \text{ Km/s}$$

After the collision, the krypton atom moves to the right with a speed of 0.59 Km/s.

Discussion To check this result, we calculate the total momentum (x-component) before and after the collision:

$$m_1 v_{1i} + m_2 v_{2i} = (83.9 \text{ u})(0.80 \text{ km/s}) + (18.0 \text{ u})(-0.40 \text{ km/s}) = 60 \text{ u·km/s}$$

$$m_1 v_{1f} + m_2 v_{2f} = (83.9 \text{ u})(0.59 \text{ km/s}) + (18.0 \text{ u})(0.60 \text{ km/s}) = 60 \text{ u·km/s}$$

continued on next page

Example 7.10 *continued*

Momentum is conserved. There is no need to convert u to kg since we only need to compare these two values.

⚠ If we made the mistake of thinking of momentum as a scalar, we would get the wrong answer. The sum of the *magnitudes* of the momenta before the collision is *not* equal to the sum of the *magnitudes* of the momenta after the collision. Conservation of energy is perhaps easier to understand intuitively since energy is a scalar quantity. Converting kinetic energy to potential energy is analogous to moving money from a checking account to a savings account; the total amount of money is the same before and after. This sort of analogy does *not* work with momentum!

Practice Problem 7.10   Head-on collision

A 5.0-kg ball is at rest when it is struck head-on by a 2.0-kg ball moving along a track at 10.0 m/s. If the 2.0-kg ball is at rest after the collision, what is the speed of the 5.0-kg ball after the collision?

Elastic and Inelastic Collisions

Collisions are often classified based on what happens to the kinetic energy of the colliding objects. A ball dropped from a height h does not rebound to the same height. The kinetic energy of the ball just after the collision with the floor or ground is less than it was just before the collision; the amount of the kinetic energy decrease depends on the makeup of the ball and the ground. A racquetball dropped onto a hard wooden floor may rebound nearly to its original height, but a watermelon rebounds very little or not at all. Why do some objects rebound much better than others?

Imagine a racquetball colliding with the floor (Fig. 7.17). The bottom of the ball is flattened. What makes the ball rebound from the floor? The forces holding the ball together are like springs; the kinetic energy of the ball has been transformed largely into potential energy in these springs. When the ball bounces back up, this energy is transformed back into kinetic energy. Then why does the watermelon not rebound? The watermelon, too, is deformed when it collides with the floor, but this deformation is not reversible. The kinetic energy of the watermelon is changed mostly into thermal energy rather than into potential energy.

A collision in which the total kinetic energy is the same before and after is called **elastic**. Not all collisions are elastic; when the final kinetic energy is less than the initial kinetic energy, the collision is said to be **inelastic**. A stick-together collision is a **perfectly inelastic collision**. The decrease of kinetic energy in a perfectly inelastic collision is as large as *possible* (consistent with the conservation of momentum). Now that we have defined elastic and inelastic collisions, we can put together a problem-solving strategy for collision problems.

Figure 7.17 Deformation of a racquetball during its collision with the floor.

Problem-Solving Strategy for Collisions Involving Two Objects

1. Draw before and after diagrams of the collision.
2. Collect and organize information on the masses and velocities of the two objects before and after the collision. Express the velocities in component form (with correct algebraic signs).
3. Set the sum of the momenta of the two before the collision equal to the sum of the momenta after the collision. Write one equation for each nonzero component:

$$m_1 v_{1ix} + m_2 v_{2ix} = m_1 v_{1fx} + m_2 v_{2fx}$$
$$m_1 v_{1iy} + m_2 v_{2iy} = m_1 v_{1fy} + m_2 v_{2fy}$$

4. If the collision is known to be perfectly inelastic, set the final velocities equal:

$$v_{1fx} = v_{2fx}$$
$$v_{1fy} = v_{2fy}$$

5. If the collision is known to be perfectly elastic, then set the final kinetic energy equal to the initial kinetic energy:

$$\tfrac{1}{2}m_1 v_{1i}^2 + \tfrac{1}{2}m_2 v_{2i}^2 = \tfrac{1}{2}m_1 v_{1f}^2 + \tfrac{1}{2}m_2 v_{2f}^2$$

6. Solve for the unknown quantities.

There is no *conservation law* for kinetic energy by itself. Total energy is always conserved, but that does not preclude some kinetic energy being transformed into another type of energy. The elastic collision is just a special kind of collision in which no kinetic energy is changed into other forms of energy.

In Practice Problem 7.11, the relative speed of the cars was the same before and after the collision. It can be proved (see Problem 43) that for *any* elastic collision between two objects, the relative speed is the same before and after the collision. This

Example 7.11

Collision at the Highway Entry Ramp

   At a Route 128 highway on-ramp, a car of mass 1.50×10^3 kg is stopped at a stop sign, waiting for a break in traffic before merging with the cars on the highway. Another car of mass 2.00×10^3 kg comes up from behind, hits the stopped car, and locks bumpers with it. How fast was the moving car going just before the collision if the two cars move together at 10.0 m/s just after the collision?

Strategy   The problem says that, after the collision, the two cars are locked together by their bumpers. The two separate cars then behave as one single, larger mass vehicle. We treat the cars as separate bodies before the collision and as a single object after the collision. The collision is perfectly inelastic since the cars "stick together." Let "1" refer to the car stopped at the stop sign and "2" refer to the other car. Fig. 7.18 shows a before and after diagram for the collision.

Given: $m_1 = 1.50 \times 10^3$ kg, $m_2 = 2.00 \times 10^3$ kg; before the collision, $v_1 = 0$; after the collision, $v_f = 10.0$ m/s (both cars have the same velocity)

To find: v_2 (speed of car 2 before the collision)

Solution   The momentum of the pair of locked cars after the collision is equal to the sum of the momenta of the two cars before the collision. All motions are in one direction so we need not include subscripts; let all p's and v's refer to x-components. From conservation of momentum,

$$p_f = p_i$$

The initial momentum is the sum of the momenta of the two cars.

$$p_f = p_1 + p_2$$

Since car 1 is initially at rest, $p_1 = 0$. Then

$$p_f = p_2$$

or

$$(m_1 + m_2)v_f = m_2 v_2$$

Solving for v_2,

$$v_2 = \frac{(m_1 + m_2)v_f}{m_2}$$

$$v_2 = \frac{(1500 \text{ kg} + 2000 \text{ kg}) \times 10.0 \text{ m/s}}{2000 \text{ kg}} = 17.5 \text{ m/s (or about 39 mph)}$$

Discussion   The entry ramp speed limit is 20 mph. By measuring the length of the skid marks from the stop sign, the investigating officer is able to determine that the stopped car was pushed onto the highway at a velocity of 10 m/s. Then the investigator calculates the speed of the other car before adding speeding to the charges against the offending driver.

Here the net external force acting on the cars wasn't zero because the friction from the road acts on the cars during the collision. We still use conservation of momentum because during the short time interval of the collision, friction doesn't have time to change the system's momentum significantly. The change of momentum of the system is equal to the external impulse, which is the product of the net external force with the time interval. For a short time interval, the impulse is small.

Practice Problem 7.11   Elastic collision between the cars

Instead of locking bumpers, suppose the two cars collided *elastically*. With the same initial conditions ($v_{1i} = 0$ and $v_{2i} = 17.5$ m/s), an elastic collision would push the stopped car onto the highway at 20.0 m/s (as you will confirm in this problem). (a) Given that $v_{1f} = 20.0$ m/s, calculate the velocity of car 2 just after the collision (v_{2f}) using momentum conservation. (b) Find the total kinetic energy of the two cars both before and after the collision and verify that the collision is elastic. (c) Find the relative speed of the two cars before and after the collision.

Figure 7.18
Before and after diagrams of the collision (side view).

fact is most useful in one-dimensional collisions; in two-dimensional collisions the *direction* of the relative velocity changes due to the collision. Since the relative velocity is in the opposite direction after the collision—first the objects move together, then they move apart—we can write:

$$v_{2ix} - v_{1ix} = -(v_{2fx} - v_{1fx}) \qquad (7\text{-}14)$$

for a one-dimensional elastic collision along the *x*-axis. For a one-dimensional elastic collision, Eq. (7-14) is a useful alternative to setting the final kinetic energy equal to the initial kinetic energy.

7.8   COLLISIONS IN TWO DIMENSIONS

Most collisions are not limited to motion in one dimension in the absence of a track or other device to constrain motion to a single line. In many collisions the initial velocities of the objects are neither parallel nor antiparallel (Fig. 7.19a). Even if the initial velocities are parallel or antiparallel, if the centers of mass do not move along a single straight line, then the final velocities do not have to be parallel or antiparallel to the initial velocities (Figs. 7.19b and c).

In a two-dimensional collision, we use the same techniques we used for one-dimensional collisions, as long as we remember that momentum is a vector. To apply conservation of momentum, it is usually easiest to work with *x*- and *y*-components.

See the Problem-Solving Strategy on p. 237.

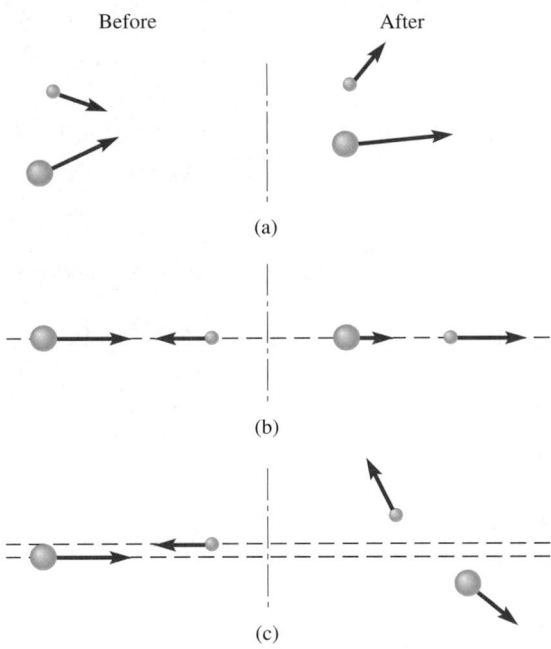

Figure 7.19 (a) A two-dimensional collision, (b) a head-on collision, and (c) a glancing collision; even though the initial velocities are in the +*x*- and –*x*-directions, the final velocities do not lie along the *x*-axis.

Example 7.12

Colliding Pucks on an Air Table

A small puck (mass $m_1 = 0.10$ kg) is sliding to the right with an initial speed of 8.0 m/s on an air table (Fig. 7.20a). An air table has many tiny holes through which air is blown; the resulting air

cushion allows objects to slide with very little friction. The puck collides with a larger puck (mass $m_2 = 0.40$ kg), which is initially at rest. Figure 7.20b shows the outcome of the collision: the pucks move off at angles $\phi_1 = 60.0°$ above and $\phi_2 = 30.0°$

continued on next page

Example 7.12 *continued*

below the initial direction of motion of the small puck. (a) What are the final speeds of the pucks? (b) Is this an elastic collision or an inelastic collision? (c) If inelastic, what fraction of the initial kinetic energy is converted to other forms of energy in the collision?

Strategy   The system of two pucks is an isolated system because the net external force is zero. Therefore we can apply conservation of momentum. Since motions in two dimensions are involved, we treat the horizontal and vertical components of momentum separately.

Figure 7.20 shows the pucks before and after the collision. Now we collect information on the known quantities, writing velocities in component form.

Masses: $m_1 = 0.10$ kg; $m_2 = 0.40$ kg

Before collision: $v_{1ix} = 8.0$ m/s; $v_{1iy} = v_{2ix} = v_{2iy} = 0$

After collision: $v_{1fx} = v_{1f} \cos \phi_1$; $v_{1fy} = v_{1f} \sin \phi_1$;

$v_{2fx} = v_{2f} \cos \phi_2$; $v_{2fy} = -v_{2f} \sin \phi_2$

($\phi_1 = 60.0°$ and $\phi_2 = 30.0°$)

To find: v_{1f} and v_{2f}; total kinetic energy before and after the collision

Solution   (a) Working with components means that we set the total x-component of momentum before the collision equal to the total x-component of momentum after the collision. We treat the y-components in the same way. The initial momentum is in the x-direction only. Thus, the total momentum y-component after the collision must be zero.

First we set the x-component of the total momentum after the collision equal to the x-component of the total momentum before the collision:

$$P_{fx} = P_{ix}$$

$$p_{1fx} + p_{2fx} = p_{1ix} + p_{2ix}$$

Each momentum component is now rewritten using $p_x = mv_x$:

$$m_1 v_{1f} \cos \phi_1 + m_2 v_{2f} \cos \phi_2 = m_1 v_{1i} + 0$$

Since $m_2 = 4m_1$,

$$m_1 v_{1f} \cos 60.0° + 4m_1 v_{2f} \cos 30.0° = m_1 v_{1ix}$$

After canceling the common factor m_1 and substituting numerical values for $\cos 60.0°$ and $\cos 30.0°$, this reduces to

$$0.500 v_{1f} + 3.46 v_{2f} = 8.0 \text{ m/s} \tag{1}$$

For conservation of the y-component of the momentum:

$$P_{fy} = P_{iy}$$

$$p_{1fy} + p_{2fy} = p_{1iy} + p_{2iy} = 0$$

The y-component of $\vec{p}_{2f}$ is negative because the y-component of $\vec{v}_{2f}$ is negative.

$$m_1 v_{1f} \sin \phi_1 + (-4m_1 v_{2f} \sin \phi_2) = 0$$

$$v_{1f} \sin 60.0° - 4v_{2f} \sin 30.0° = 0$$

We solve for v_{2f} in terms of v_{1f}:

$$v_{2f} = \frac{\sin 60.0°}{4 \sin 30.0°} v_{1f} = 0.433 v_{1f} \tag{2}$$

Equations (1) and (2) contain two unknowns. To eliminate one unknown, we substitute $0.433 v_{1f}$ for v_{2f} in Eq. (1):

$$0.500 v_{1f} + 3.46 (0.433 v_{1f}) = 2.00 v_{1f} = 8.0 \text{ m/s}$$

Solving this equation gives the value of v_{1f}:

$$v_{1f} = 4.0 \text{ m/s}$$

Then by substitution into Eq. (2), we find the value of v_{2f}:

$$v_{2f} = 0.433 v_{1f} = 1.7 \text{ m/s}$$

(b) Now that we have the final velocities, we can compare the initial and final kinetic energies.

$$K_i = \tfrac{1}{2} m_1 v_{1i}^2$$

$$K_i = \tfrac{1}{2}(0.10 \text{ kg}) \times (8.0 \text{ m/s})^2 = 3.2 \text{ J}$$

and

$$K_f = \tfrac{1}{2} m_1 v_{1f}^2 + \tfrac{1}{2} m_2 v_{2f}^2$$

$$= \tfrac{1}{2}(0.10 \text{ kg}) \times (4.0 \text{ m/s})^2 + \tfrac{1}{2}(0.40 \text{ kg}) \times (1.7 \text{ m/s})^2$$

$$K_f = 0.80 \text{ J} + 0.58 \text{ J} = 1.38 \text{ J}$$

The final kinetic energy is less than the initial kinetic energy so the collision is inelastic.

(c) The amount of kinetic energy converted to other forms of energy is

$$3.2 \text{ J} - 1.38 \text{ J} = 1.82 \text{ J}$$

We divide by the initial kinetic energy to find the fraction of the initial kinetic energy converted to other forms:

$$\frac{1.82 \text{ J}}{3.2 \text{ J}} = 0.57$$

continued on next page

(a)

(b)

Figure 7.20
Snapshots in time, (a) before and (b) after a collision.

Example 7.12 *continued*

Less than half of the kinetic energy of the incident puck therefore survives the collision as the kinetic energies of the two pucks.

Discussion   Although a two-dimensional collision problem tends to require more complicated algebra than a one-dimensional problem, the physical principles are the same. As long as the net external force on the system is zero (or negligibly small), the total vector momentum must be conserved.

Practice Problem 7.12   Colliding balls

A ball of mass m_1 moves at speed v_0 along the $+x$-axis toward a second ball of mass m_2, which is initially at rest. The second ball has five times the mass of the first ball. After the collision between these two objects, m_1 moves along the $+y$-axis at a speed v_1, and m_2 moves at a speed $v_2 = \frac{1}{4}v_0$ at an angle of 36.9° from the x-axis. Find v_1 in terms of v_0.

Example 7.13

Eric at the Pool Table

 Playing a game of pool, Eric is trying to decide whether to attempt a shot to sink the 4-ball in the pocket at corner B without *scratching* (sinking the cue ball "C" in corner A). The lines from the 4-ball to the two corner pockets make a right angle (Fig. 7.21a). The collision of the balls is nearly elastic. Assume Eric is an amateur player and does not know how to do fancy things, like putting sidespin on a ball. Should he attempt the shot?

Strategy   We assume a perfectly elastic collision between the balls. They have the same mass. The cue ball moves with an initial velocity $\vec{\mathbf{v}}_0$ and strikes the 4-ball, which is initially at rest. The 4-ball falls in pocket B if its velocity after the collision, $\vec{\mathbf{v}}_4$, points toward the pocket. Assuming it does, we use conservation of momentum and kinetic energy to find the direction of the cue ball velocity, $\vec{\mathbf{v}}_c$, after the collision.

continued on next page

Figure 7.21   (a) Should Eric try to sink the 4-ball? (b) Graphical addition of velocity vectors as required by the conservation of momentum. (c) Since $v_0^2 = v_c^2 + v_4^2$, the three velocities form a *right* triangle.

Example 7.13 *continued*

Solution   Conservation of momentum requires that

$$m\vec{\mathbf{v}}_0 = m\vec{\mathbf{v}}_c + m\vec{\mathbf{v}}_4$$

or

$$\vec{\mathbf{v}}_0 = \vec{\mathbf{v}}_c + \vec{\mathbf{v}}_4$$

This vector addition is shown graphically in Fig. 7.21b. Since the collision is elastic, the total kinetic energy doesn't change:

$$\tfrac{1}{2} m v_0^2 = \tfrac{1}{2} m v_c^2 + \tfrac{1}{2} m v_4^2$$

or

$$v_0^2 = v_c^2 + v_4^2$$

Since v_0, v_c, and v_4 are the sides of a triangle, this is a statement of the Pythagorean theorem—the triangle must be a *right* triangle with v_0 as the hypotenuse (Fig. 7.21c). Therefore, the velocities of the 4-ball and the cue ball after the collision are perpendicular to each other.

If Eric sinks the 4-ball, the cue ball falls into pocket A. He shouldn't attempt this shot until he learns how to put some spin on the ball.

Discussion   Note that we did not resolve the velocities into *x*- and *y*-components. Doing so would have made the solution longer in this case.

We found that the two balls must move at right angles after the collision if the collision is elastic. This is true in general only if the two objects colliding have equal masses and if one is initially at rest. In Example 7.12 the two pucks move at right angles after the collision, but the collision is inelastic—the masses are unequal.

Practice Problem 7.13   Finding the speed ratio

Suppose that the cue ball initially moves in the −*x*-direction. After the collision, the cue ball moves at 52.0° above the −*x*-axis and the 4-ball moves at 38.0° below the −*x*-axis. Find the ratio of the balls' speeds v_c/v_4 after the collision.

MASTER THE CONCEPTS

Summary

- Definition of linear momentum:

  $$\vec{\mathbf{p}} = m\vec{\mathbf{v}} \qquad (7\text{-}1)$$

- During an interaction, momentum is transferred from one body to another, but the total momentum of the two is unchanged.

  $$\Delta\vec{\mathbf{p}}_2 = -\Delta\vec{\mathbf{p}}_1$$

- Impulse is the average net force times the time interval. The impulse equals the change in momentum:

  $$\Delta\vec{\mathbf{p}} = \vec{\mathbf{F}}\,\Delta t \qquad (7\text{-}2)$$

- A conserved quantity is one that remains unchanged as time passes.

- Impulse is the area under a graph of force versus time.

- Force is the rate of change of momentum.

  $$\vec{\mathbf{F}} = \lim_{\Delta t \to 0} \frac{\Delta\vec{\mathbf{p}}}{\Delta t} \qquad (7\text{-}4)$$

- External interactions may change the total momentum of a system.

- Internal interactions do not change the total momentum of a system.

- Conservation of linear momentum: if the net external force acting on a system is zero, then the momentum of the system is conserved.

- The position of the center of mass of a system of *N* particles is

  $$x_{CM} = \frac{m_1 x_1 + m_2 x_2 + \cdots + m_N x_N}{M}$$

  and

  $$y_{CM} = \frac{m_1 y_1 + m_2 y_2 + \cdots + m_N y_N}{M} \qquad (7\text{-}9)$$

  where *M* is the total mass of the particles:

  $$M = m_1 + m_2 + \cdots + m_N$$

- The total momentum of a system is equal to the total mass times the velocity of the center of mass:

  $$\vec{\mathbf{p}} = \vec{\mathbf{p}}_1 + \vec{\mathbf{p}}_2 + \cdots + \vec{\mathbf{p}}_N = M\vec{\mathbf{v}}_{CM} \qquad (7\text{-}11)$$

- No matter how complicated a system is, the center of mass moves as if all the mass of the system were concentrated to a point particle with all the external forces acting on it:

  $$\Sigma\vec{\mathbf{F}}_{ext} = M\vec{\mathbf{a}}_{CM} \qquad (7\text{-}13)$$

- The center of mass of an isolated system moves at constant velocity.

- Conservation of momentum is used to solve problems involving collisions, explosions, and the like. Even when external forces are acting, the momentum of the system just before a collision is nearly equal to the momentum just after if the collision interaction is brief. The impulse, and therefore the change in momentum of the system, is small since the time interval is small.

MASTER THE CONCEPTS *continued*

Highlighted Figures and Tables

F7.2 Vector subtraction to find the change in momentum (p. 220)

F7.4 Safety features of the modern automobile that serve to lengthen the time interval of a momentum change during a crash (p. 223)

F7.5 Impulse on a graph of F_x versus t (p. 224)

F7.9 Conservation of momentum before, during, and after collision (p. 226)

F7.11 Center of mass (p. 231)

F7.14 Motion of the center of mass (p. 235)

F7.15 Collisions in one dimension (p. 236)

F7.19 Collisions in two dimensions (p. 239)

CONCEPTUAL QUESTIONS

1. You are trapped on the second floor of a burning building. The stairway is impassable, but there is a balcony outside your window. Describe what might happen in the following situations. (a) You jump from the second-story balcony to the pavement below, landing stiff-legged on your feet. (b) You jump into a privet hedge, landing on your back and rolling to your feet. (c) You jump into a firefighters' net, landing on your back. What happens to the net as you land in it? What do the firefighters do to cushion your fall even more?

2. A force of 30 N is applied for 5 s to each of two bodies of different masses. (a) Which one has the greatest momentum change? (b) The greatest velocity change? (c) The greatest acceleration?

3. If you take a rifle and saw off part of the barrel, the muzzle speed (the speed at which bullets emerge from the barrel) will be smaller. Why?

4. A firecracker at rest explodes, sending fragments off in all directions. Initially the firecracker has zero momentum, but after the explosion the fragments flying off each have quite a lot of momentum. Hasn't momentum been created? If not, explain why not.

5. An astronaut in deep space is taking a space walk when the tether connecting him to his spaceship breaks. How can he get back to the ship? He doesn't have a rocket propulsion backpack, unfortunately, but he is carrying a big wrench.

6. An astronaut hits a golf ball on the surface of the Moon. Is the momentum of the ball conserved while it is in flight? Is there a *component* of its momentum that is conserved?

7. Which would be more effective: a hammer that collides *elastically* with a nail, or one that collides perfectly *inelastically*? Assume that the mass of the hammer is much larger than that of the nail.

8. Squid are the fastest swimmers among invertebrates. A cavity within the squid is filled with water. The *mantle*, a powerful muscle, squeezes the cavity and expels the water through a narrow opening (the *siphon*)

at high speed. Using momentum conservation, explain how this propels the squid forward. How is the squid's swimming mechanism like a rocket engine?

9. In your own words, phrase each of Newton's three laws of motion as a statement about momentum.

10. Two objects with different masses have the same kinetic energy. Which has the larger magnitude of momentum?

11. A woman is 1.60 m tall. When standing straight, is her center of mass necessarily 0.80 m above the floor? Explain.

12. The momentum of a system can only be changed by an external force. What is the external force that changes the momentum of a bicycle (with its rider) as it speeds up, slows down, or changes direction? Is it true that changes in the bicycle's kinetic energy must come from an external force? Explain.

13. In an egg toss, two people try to toss a raw egg back and forth without breaking it as they move farther and farther apart. Discuss a strategy in terms of impulse and momentum for catching the egg without breaking it.

14. In the "executive toy" (Fig. 7.22), two balls are pulled back and then released. After the collision, two balls move away on the opposite side. Why do we never see three balls move

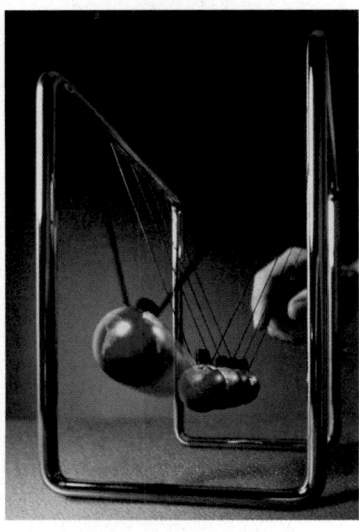

Figure 7.22
Conceptual Question 14

away following this action, although with a lower velocity so that linear momentum is still conserved?

15. A baseball batting coach emphasizes the importance of "follow-through" when a batter is trying for a home run. The coach explains that the follow-through allows the bat to remain in contact with the ball for a longer time so that the ball will travel a greater distance. Explain the reasoning behind this statement in terms of the impulse-momentum theorem.

MULTIPLE CHOICE QUESTIONS

1. A ball of mass m with initial speed v collides with another ball of mass M, initially at rest. After the collision the two balls stick together, moving with speed V. The ratio of the final speed V to the initial speed v is $V/v =$

   (a) $\dfrac{M}{M+m}$ (b) $\dfrac{M+m}{M}$

   (c) $\dfrac{m}{M+m}$ (d) $\dfrac{M+m}{m}$

   (e) $\sqrt{\dfrac{M}{M+m}}$ (f) $\sqrt{\dfrac{m}{M+m}}$

2. Two particles A and B of equal mass are located at some distance from each other. Particle A is at rest while B moves away from A at speed v. What happens to the center of mass of the system of two particles?

   (a) It does not move.
   (b) It moves with a speed v away from A.
   (c) It moves with a speed v toward A.
   (d) It moves with a speed $\frac{1}{2}v$ away from A.
   (e) It moves with a speed $\frac{1}{2}v$ toward A.

3. Two uniform spheres with equal mass per unit volume are in contact with one another. The mass of sphere A is five times that of sphere B. The center of mass of the system is

   (a) at the point where A and B touch.
   (b) inside sphere B somewhere on the line joining the centers of A and B.
   (c) inside sphere A somewhere on the line joining the centers.
   (d) at the center of sphere A.
   (e) outside of both spheres.

4. An object at rest suddenly explodes into three parts of equal mass (Fig. 7.23). Two of the parts move away at right angles to each other and with equal speeds v. What is the initial velocity of the third part?

   (a) Direction of vector 1 and magnitude $2v$

   (b) Direction of vector 2 and magnitude $\sqrt{2}v$

   (c) Direction of vector 3 and magnitude $\dfrac{1}{\sqrt{2}}v$

   (d) Direction of vector 2 and magnitude $\dfrac{1}{\sqrt{2}}v$

   (e) Direction of vector 1 and magnitude $\frac{1}{2}v$

5. A 3.0-kg object is initially at rest. It then receives an impulse of magnitude 15 N·s. After the impulse, the object has

   (a) a speed of 45 m/s.
   (b) a momentum of magnitude 5.0 kg·m/s.
   (c) a speed of 7.5 m/s.
   (d) a momentum of magnitude 15 kg·m/s.

6. An object of mass m drops from rest a little above the Earth's surface for a time t. Ignore air resistance. After time t the magnitude of its momentum is

   (a) mgt^2

   (b) mgt

   (c) $mg\sqrt{t}$

   (d) $\sqrt{mgt}$

   (e) $\dfrac{mgt^2}{2}$

Multiple Choice Questions 7–12 refer to a situation in which a golf ball is projected straight upward in the +y-direction. Ignore air resistance. The answer choices are found in Fig. 7.24.

7. Which graph shows the acceleration a_y of the ball as a function of time?

8. Which graph shows the vertical position y of the ball as a function of time?

9. Which graph shows the momentum p_y of the ball as a function of time?

10. Which graph shows the kinetic energy of the ball as a function of time?

11. Which graph shows the potential energy of the ball as a function of time?

12. Which graph shows the total energy of the ball as a function of time?

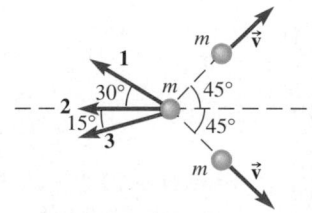

Figure 7.23 Multiple Choice Question 4

Figure 7.24 Multiple Choice Questions 7–12

PROBLEMS

Note: ⓒ indicates a combination conceptual/quantitative problem. Gold diamonds ✦, ✦✦ are used to indicate the increasing level of difficulty of each problem. Problem numbers appearing in blue, 9., denote problems that have a detailed solution available in the Student Solutions Manual. Some problems are *paired* by concept; their numbers are connected by a ruled box.

7.2 Momentum; 7.3 The Impulse-Momentum Theorem

1. A 2.0-kg block is moving to the right at 1.0 m/s just before it strikes and sticks to a 1.0-kg block initially at rest. What is the total momentum of the two blocks after the collision?

2. What is the momentum of an automobile (weight = 9800 N) when it is moving at 35 m/s to the south?

✦ 3. A pole-vaulter of mass 60.0 kg vaults to a height of 6.0 m before dropping to thick padding placed below to cushion her fall. (a) Find the speed with which she lands. (b) If the padding brings her to a stop in a time of 0.50 s, what is the average force on her body due to the padding during that time interval?

4. A cue stick hits a cue ball with an average force of 24 N for a duration of 0.028 s. If the mass of the ball is 0.16 kg, how fast is it moving after being struck?

5. A system consists of three particles with these masses and velocities: mass 3.0 kg, moving north at 3.0 m/s; mass 4.0 kg, moving south at 5.0 m/s; and mass 7.0 kg, moving north at 2.0 m/s. What is the total momentum of the system?

6. A sports car traveling along a straight line increases its speed from 20.0 mi/h to 60.0 mi/h. (a) What is the ratio of the final to the initial magnitude of its momentum? (b) What is the ratio of the final to the initial kinetic energy?

7. A ball of mass 5.0 kg moving with speed of 2.0 m/s in the +x-direction hits a wall and bounces back with the same speed in the −x-direction. What is the change of momentum of the ball?

8. An object of mass 3.0 kg is projected into the air at a 55° angle. It hits the ground 3.4 s later. What is its change in momentum while it is in the air? Ignore air resistance.

9. An object of mass 3.0 kg is allowed to fall from rest under the force of gravity for 3.4 s. What is the change in its momentum? Ignore air resistance.

10. What average force is necessary to bring a 50.0-kg sled from rest to a speed of 3.0 m/s in a period of 20.0 s? Assume frictionless ice.

11. For a safe re-entry into the Earth's atmosphere, the pilots of a space capsule must reduce their speed from 2.6×10^4 m/s to 1.1×10^4 m/s. The rocket engine produces a backward force on the capsule of 1.8×10^5 N. The mass of the capsule is 3800 kg. For how long must they fire their engine? [*Hint:* Ignore the change in mass of the capsule due to the expulsion of exhaust gases.]

✦ 12. Redo Problem 11, this time taking into account that 6.8 kg of exhaust gases are expelled per second. [*Hint:* The momentum change of the exhaust gases is already accounted for in the given force on the capsule; the force on the capsule is the rate of change of *its* momentum.]

13. An automobile traveling at a speed of 30.0 m/s applies its brakes and comes to a stop in 5.0 s. If the automobile has a mass of 1.0×10^3 kg, what is the average horizontal force exerted on it during braking? Assume the road is level.

14. A 3.0-kg body is initially moving northward at 15 m/s. Then a force of 15 N, toward the east, acts on it for 4.0 s. (a) At the end of the 4.0 s, what is the body's final velocity? (b) What is the change in momentum during the 4.0 s?

✦ 15. A boy of mass 60.0 kg is rescued from a hotel fire by leaping into a firefighters' net. The window from which he leapt was 8.0 m above the net. The firefighters lower their arms as he lands in the net so that he is brought to a complete stop in a time of 0.40 s. (a) What is his change in momentum during the 0.40-s interval? (b) What is the impulse on the net due to the boy during the interval? [*Hint:* Do not ignore gravity.] (c) What is the average force on the net due to the boy during the interval?

16. Verify that the SI unit of impulse is the same as the SI unit of momentum.

7.4 Conservation of Momentum

17. A rifle has a mass of 4.5 kg and it fires a bullet of mass 10.0 g at a muzzle speed of 820 m/s. What is the recoil speed of the rifle as the bullet leaves the gun barrel?

18. A submarine of mass 2.5×10^6 kg and initially at rest fires a torpedo of mass 250 kg. The torpedo has an initial speed of 100.0 m/s. What is the initial recoil speed of the submarine? Neglect the drag force of the water.

19. A uranium nucleus (mass 238 u), initially at rest, undergoes radioactive decay. After an alpha particle (mass 4.0 u) is emitted, the remaining nucleus is thorium (mass 234 u). If the alpha particle is moving at 0.050 times the speed of light, what is the recoil speed of the thorium nucleus? (Note: "u" is a unit of mass; it is *not* necessary to convert it to kg.)

✦✦ 20. A cannon on a railroad car is facing in a direction parallel to the tracks (Fig. 7.25). It fires a 98-kg shell at a speed of 105 m/s (relative to the ground) at an angle of 60.0° above the horizontal. If the cannon plus car have a mass of 5.0×10^4 kg, what is the recoil speed of the car if it was at rest before the cannon was fired? [*Hint:* A *component* of a system's momentum along an axis is conserved if the net external force acting on the system has no component along that axis.]

✦✦ 21. A marksman standing on a motionless railroad car fires a gun into the air at an angle of 30.0° from the horizontal (Fig. 7.26).

Figure 7.25 Problem 20

Figure 7.26 Problem 21

The bullet has a speed of 173 m/s (relative to the ground) and a mass of 0.010 kg. The man and car move to the left at a speed of 1.0×10^{-3} m/s after he shoots. What is the mass of the man and car? (See the hint in Problem 20.)

7.5 Center of Mass; 7.6 Motion of the Center of Mass

22. Particle A is at the origin and has a mass of 30.0 g. Particle B has a mass of 10.0 g. Where must particle B be located if the coordinates of the center of mass are $(x, y) = (2.0 \text{ cm}, 5.0 \text{ cm})$?

23. Particle A has a mass of 5.0 g, particle B has a mass of 1.0 g. Particle A is located at the origin, particle B at the point $(x, y) = (25 \text{ cm}, 0)$. What is the location of the center of mass?

24. The three bodies in Fig. 7.27 each have the same mass. If one of the bodies is moved 12 cm in the positive x-direction, by how much does the center of mass move?

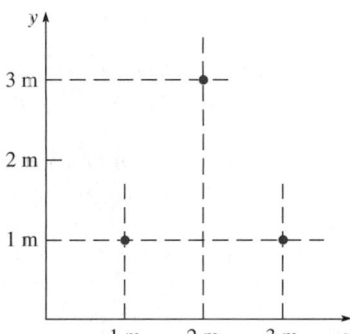

Figure 7.27 Problem 24

25. The positions of three particles, written as (x, y) coordinates, are: particle 1 (mass 4.0 kg) at (4.0 m, 0 m); particle 2 (mass 6.0 kg) at (2.0 m, 4.0 m); particle 3 (mass 3.0 kg) at (–1.0 m, –2.0 m). What is the location of the center of mass?

26. Consider two falling bodies. Their masses are 3.0 kg and 4.0 kg. At time $t = 0$, the two are released from rest. What is the velocity of their center of mass at $t = 10.0$ s? Ignore air resistance.

27. Body A of mass 3 kg is moving in the $+x$-direction with a speed of 14 m/s. Body B of mass 4 kg is moving in the $-y$-direction with a speed of 7 m/s. What are the x- and y-components of the velocity of the center of mass of the two bodies?

♦ 28. If a particle of mass 5.0 kg is moving east at 10 m/s and a particle of mass 15 kg is moving west at 10 m/s, what is the velocity of the center of mass of the pair?

29. Prove Eq. (7-13) $\Sigma \vec{F}_{\text{ext}} = M \vec{a}_{\text{CM}}$. [*Hint:* Start with $\Sigma \vec{F}_{\text{ext}} = \lim_{\Delta t \to 0} (\Delta \vec{p}/\Delta t)$, where $\Sigma \vec{F}_{\text{ext}}$ is the net external force acting on a system and $\vec{p}$ is the total momentum of the system.]

7.7 Collisions in One Dimension

30. In the railroad freight yard, an empty freight car of mass m rolls along a straight level track at 1.0 m/s and collides with an initially stationary, fully loaded boxcar of mass $4.0m$. The two cars couple together upon collision. (a) What is the speed of the two cars after the collision? (b) Suppose instead that the two cars are at rest after the collision. With what speed was the loaded boxcar moving before the collision if the empty one was moving at 1.0 m/s?

31. A 0.020-kg bullet traveling at 200.0 m/s east hits a motionless 2.0-kg block and bounces off it, retracing its original path with a velocity of 100.0 m/s west. What is the final velocity of the block? Assume the block rests on a perfectly frictionless horizontal surface.

32. A block of wood of mass 0.95 kg is initially at rest. A bullet of mass 0.050 kg traveling at 100.0 m/s strikes the block and becomes embedded in it. With what speed do the block of wood and the bullet move just after the collision?

33. A spring of negligible mass is compressed between two blocks, A and B, which are at rest on a frictionless horizontal surface at a distance of 1.0 m from a wall on the left and 3.0 m from a wall on the right. The sizes of the blocks and spring are small. When the spring is released, body A moves toward the left wall and strikes it at the same instant that body B strikes the right wall. The mass of A is 0.60 kg. What is the mass of B?

ⓒ ♦ 34. Two identical gliders on an air track are held together by a piece of string, compressing a spring between the gliders. While they are moving to the right at a common speed of 0.50 m/s, someone holds a match under the string and burns it, letting the spring force the gliders apart. One glider is then observed to be moving to the right at 1.30 m/s. (a) What velocity does the other glider have? (b) Is the total kinetic energy of the two gliders after the collision greater than, less than, or equal to the total kinetic energy before the collision? If greater, where did the extra energy come from? If less, where did the "lost" energy go?

35. A BMW of mass 2.0×10^3 kg is traveling at 42 m/s. It approaches a 1.0×10^3 kg Volkswagen going 25 m/s in the same direction and strikes it in the rear. Neither driver applies the brakes. Neglect the relatively small frictional forces on the cars due to the road and due to air resistance. If the collision slows the BMW down to 33 m/s, what is the speed of the VW after the collision?

36. A projectile of 1.0-kg mass approaches a stationary body of 5.0 kg at 10.0 m/s and, after colliding, rebounds in the reverse direction along the same line with a speed of 5.0 m/s. What is the speed of the 5.0-kg body after the collision?

37. A 2.0-kg object is at rest on a perfectly frictionless surface when it is hit by a 3.0-kg object moving at 8.0 m/s. If the two objects are stuck together after the collision, what is the speed of the combination?

38. Two cars, each of mass 1300 kg, are approaching each other on a head-on collision course. Each speedometer reads 19 m/s. What is the magnitude of the total momentum of the system?

39. A 75-kg man is at rest on ice skates. A 0.20-kg ball is thrown to him. The ball is moving horizontally at 25 m/s just before the man catches it. How fast is the man moving just after he catches the ball?

40. A 0.010-kg bullet traveling horizontally at 400.0 m/s strikes a 4.0-kg block of wood sitting at the edge of a table. The bullet is lodged into the wood. If the table height is 1.2 m, how far from the table does the block hit the floor?

41. Two objects with masses m_1 and m_2 approach each other with equal and opposite momenta so that the total momentum is zero. Show that, if the collision is elastic, the final *speed* of each object must be the same as its initial speed. (The final *velocity* of each object is *not* the same as its initial velocity, however.)

42. A helium atom (mass 4.0 u) moving at 600.0 m/s to the right collides with an oxygen molecule (mass 32 u) moving in the same direction at 400.0 m/s. After the collision, the oxygen molecule moves at 500.0 m/s to the right. What is the velocity of the helium atom after the collision?

43. Use the result of Problem 41 to show that in *any* elastic collision between two objects, the relative speed of the two is the same before and after the collision. [*Hints:* Look at the collision in its *center of mass frame*—the reference frame in which the center of mass is at rest. The *relative* speed of two objects is the same in any inertial reference frame.]

7.8 Collisions in Two Dimensions

44. Body A of mass M has an original velocity of 6.0 m/s in the +x-direction toward a stationary body (body B) of the same mass. After the collision, body A has velocity components of 1.0 m/s in the +x-direction and 2.0 m/s in the +y-direction. What is the magnitude of body B's velocity after the collision?

45. (a) With reference to Practice Problem 7.12, find the momentum change of the ball of mass m_1 during the collision. Give your answer in x- and y-component form; express the components in terms of m_1 and v_0. (b) Repeat for the ball of mass m_2. How are the momentum changes related?

46. A hockey puck moving at 0.45 m/s collides into another puck that was at rest. The pucks have equal mass. The first puck is deflected 37° to the right and moves off at 0.36 m/s. Find the speed and direction of the second puck after the collision.

47. Puck 1 sliding along the x-axis strikes stationary puck 2 of the same mass. After the collision, puck 1 moves off at speed v_{1f} in the direction 60.0° above the x-axis; puck 2 moves off at speed v_{2f} in the direction 30.0° below the x-axis. Find v_{2f} in terms of v_{1f}.

48. A projectile of mass 2.0 kg approaches a stationary target body at 5.0 m/s. The projectile is deflected through an angle of 60.0° and its speed after the collision is 3.0 m/s. What is the magnitude of the momentum of the target body after the collision?

49. In a nuclear reactor, a neutron moving at speed v_0 in the positive x-direction strikes a deuteron, which is at rest. The neutron is deflected by 90.0° and moves off with speed $v_0/\sqrt{3}$ in the positive y-direction. Find the x- and y-components of the deuteron's velocity after the collision. (The mass of the deuteron is twice the mass of the neutron.)

50. A firecracker is tossed straight up into the air. It explodes into three pieces of equal mass just as it reaches the highest point. Two pieces move off at 120 m/s at right angles to each other. How fast is the third piece moving?

COMPREHENSIVE PROBLEMS

51. An intergalactic spaceship is traveling through space far from any planets or stars, where no human has gone before. The ship carries a crew of 30 people (of total mass 2.0×10^3 kg). If the speed of the spaceship is 1.0×10^5 m/s and its mass (excluding the crew) is 4.8×10^4 kg, what is the magnitude of the total momentum of the ship and the crew?

52. A baseball player pitches a fastball toward home plate at a speed of 41 m/s. The batter swings, connects with the ball of mass 145 g, and hits it so that the ball leaves the bat with a speed of 37 m/s. Assume that the ball is moving horizontally just before and just after the collision with the bat. (a) What is the magnitude of the change in momentum of the ball? (b) What is the impulse delivered to the ball by the bat? (c) If the bat and ball are in contact for 3.0 ms, what is the magnitude of the average force exerted on the ball by the bat?

53. A tennis ball of mass 0.060 kg is served. It strikes the ground with a velocity of 54 m/s (120 mph) at an angle of 22° below the horizontal. Just after the bounce it is moving at 53 m/s at

an angle of 18° above the horizontal. If the interaction with the ground lasts 0.065 s, what average force did the ground exert on the ball?

54. A uniform rod of length 30.0 cm is bent into the shape of an inverted U (Fig. 7.28). Each of the 3 sides is of length 10.0 cm. Find the location, in x- and y-coordinates, of the center of mass as measured from the origin.

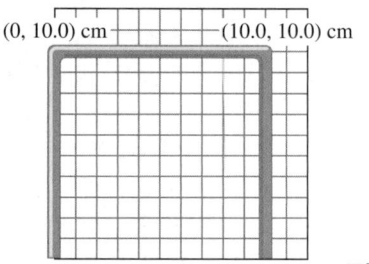

Figure 7.28 Problem 54

55. To contain some unruly demonstrators, the riot squad approaches with fire hoses. Suppose that the rate of flow of water through a fire hose is 24 kg/s and the stream of water from the hose moves at 17 m/s. What force is exerted by such a stream on a person in the crowd? Assume that the water comes to a dead stop against the demonstrator's chest.

56. An inexperienced catcher catches a 130 km/h fastball of mass 140 g within 1 ms, whereas an experienced catcher slightly retracts his hand during the catch, extending the stopping time to 10 ms. What are the average forces imparted to the two gloved hands during the catches?

◆57. A stationary 0.1-g fly encounters the windshield of a 1000-kg automobile traveling at 100 km/h. (a) What is the change in momentum of the car due to the fly? (b) What is the change of momentum of the fly due to the car? (c) Approximately how many flies does it take to reduce the car's speed by 1 km/h?

58. For a system of three particles moving along a line, an observer in a laboratory measures the following masses and velocities:

Mass (kg)	v_x (m/s)
3.0	+290
5.0	−120
2.0	+52

What is the velocity of the center of mass of the system?

◆59. A projectile of mass 2.0 kg approaches a stationary target body at 8.0 m/s. The projectile is deflected through an angle of 90.0° and its speed after the collision is 6.0 m/s. What is the speed of the target body after the collision if the collision is perfectly elastic?

60. An automobile weighing 13.6 kN is moving at 17.0 m/s when it collides with a stopped car weighing 9.0 kN. If they lock bumpers and move off together, what is their speed just after the collision?

61. A sled of mass 5.0 kg is coasting along on a frictionless ice-covered lake at a constant speed of 1.0 m/s. A 1.0-kg book is dropped vertically onto the sled. At what speed does the sled move once the book is on it?

62. A radioactive nucleus is at rest when it spontaneously decays by emitting an electron and neutrino (Fig. 7.29). The momentum of the electron is 8.20×10^{-19} kg·m/s and it is directed at right angles to that of the neutrino. The neutrino momentum is of magnitude 5.00×10^{-19} kg·m/s. (a) In what direction does the newly formed (daughter) nucleus recoil? Let the electron direction be along the positive x-axis and find direction of nucleus recoil with respect to the electron's direction. (b) What is its momentum?

◆63. A 60.0-kg woman stands at one end of a 120-kg raft that is 6.0 m long. The other end of the raft is 0.50 m from a pier. (a) The woman walks toward the pier until she gets to the other end of the raft and stops there. Now what is the distance between the raft and the pier? (b) In (a), how far did the woman walk (relative to the pier)?

◆◆64. A jet plane is flying at 130 m/s relative to the ground. There is no wind. The engines take in 81 kg of air per second. Hot gas (burned fuel and air) is expelled from the engines at high speed. The engines provide a forward force on the plane of magnitude 6.0×10^4 N. At what speed relative to the ground is the gas being expelled? [*Hint:* Look at the momentum change of the air taken in by the engines during a time interval Δt.] This calculation is approximate since we are ignoring the 3.0 kg of fuel consumed and expelled with the air each second.

◆65. Within cells, small organelles containing newly synthesized proteins are transported along microtubules by tiny molecular motors called kinesins. What force does a kinesin molecule need to deliver in order to accelerate an organelle with mass 0.01 pg (10^{-17} kg) from 0 to 1 μm/s within a time of 10 μs?

◆66. The pendulum bobs in Fig. 7.30 are made of soft clay so that they stick together after impact. The mass of bob A is half that of bob B. Bob B is initially at rest. What is the ratio of the kinetic energy of the combined bobs, just after impact, to the kinetic energy of bob A just before impact?

◆◆67. The pendulum bobs in Fig. 7.30 are made of soft clay so that they stick together after impact. The mass of bob A is half that of bob B. Bob B is initially at rest. If bob A is released from a height h above its lowest point, what is the maximum height attained by bobs A and B after the collision?

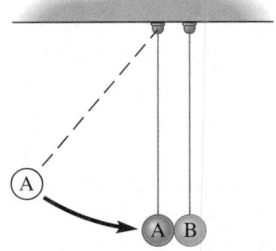

Figure 7.30 Problems 66 and 67

◆◆68. A flat, circular metal disk of uniform thickness has a radius of 3.0 cm. A hole is drilled in the disk that is 1.5 cm in radius. The hole is tangent to one side of the disk (Fig. 7.31). Where is the CM of the disk now that the hole has been drilled? [*Hint:* The original disk (before the hole is drilled) can be thought of as having two pieces—the disk with the hole plus the smaller disk of metal drilled out. Write an equation that expresses x_{CM} of the original disk in terms of the x_{CM}'s of the two pieces. Since the thickness is uniform, the mass of any piece is proportional to its area.]

Electron

Nucleus

Neutrino

Figure 7.29 Problem 62

Figure 7.31
Problem 68

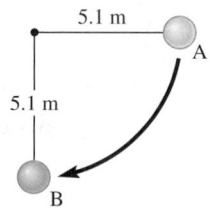

Figure 7.32   Problem 69

✦ 69. Two pendulum bobs have equal masses and lengths (5.1 m). Bob A is initially held horizontally while bob B hangs vertically at rest (Fig. 7.32). Bob A is released and collides elastically with bob B. How fast is bob B moving immediately after the collision?

70. Two identical gliders, each with elastic bumpers and mass 0.10 kg, are on a horizontal air track. Friction is negligible. Glider 2 is stationary. Glider 1 moves toward glider 2 from the left with a speed of 0.20 m/s. They collide. After the collision, what are the velocities of glider 1 and glider 2?

✦ 71. A radium nucleus (mass 226 u) at rest decays into a radon nucleus (symbol Rn, mass 222 u) and an alpha particle (symbol α, mass 4 u). (a) Find the ratio of the speeds v_α/v_{Rn} after the decay. (b) Find the ratio of the magnitudes of the momenta p_α/p_{Rn}. (c) Find the ratio of the kinetic energies K_α/K_{Rn}. (Note: "u" is a unit of mass; it is *not* necessary to convert it to kg.)

ANSWERS TO PRACTICE PROBLEMS

7.1 (a) 0.78 kg·m/s downward; (b) 0.78 kg·m/s toward the apple, 1.3×10^{-25} m/s

7.2 3.5 times his weight

7.3 1700 N; 0.0037 s

7.4 0.8 m/s in the –*x*-direction

7.5 1.7 m/s

7.6 The manager throws the nugget in a direction away from the desk. The scheme might fail if frictional forces bring the chair to rest before he reaches the telegraph.

7.7 2.2 m/s

7.8 (2.0 cm, 2.3 cm)

7.9 (a) 2.7 m; (b) 1.5 m in the other direction; (c) the center of mass does not move

7.10 4.0 m/s

7.11 (a) 2.5 m/s; (b) 306 kJ (same before and after, since the collision is elastic); (c) both are 17.5 m/s

7.12 $0.751v_0$

7.13 0.781

Torque and Angular Momentum

In gymnastics, the iron cross is a notoriously difficult feat that requires incredible strength. Why does it require such great strength? To perform the iron cross, the forces exerted by the gymnast's muscles must be much greater in magnitude than the gymnast's weight. How does the design of the human body make such large forces necessary?

Concepts & Skills to Review

- translational equilibrium (Sections 2.3, 4.3)
- uniform circular motion and circular orbits (Sections 5.1, 5.4)
- angular acceleration (Section 5.6)
- conservation of mechanical energy (Section 6.6)
- center of mass and its motion (Sections 7.5, 7.6)

The symbol Σ stands for a sum. $\sum\limits_{i=1}^{N}$ means the sum for $i = 1, 2, \ldots, N$.

8.1   ROTATIONAL KINETIC ENERGY AND ROTATIONAL INERTIA

When a rigid object is rotating in place, it has kinetic energy because each particle of the object is moving in a circle around the axis of rotation. In principle, we can calculate the kinetic energy of rotation by summing the kinetic energy of each particle. To say the least, that sounds like a laborious task. We need a simpler way to express the rotational kinetic energy of such an object so that we don't have to calculate this sum over and over. Our simpler expression exploits the fact that the speed of each particle is proportional to the angular speed of rotation ω.

If a rigid object consists of N particles, the sum of the kinetic energies of the particles can be written mathematically using a subscript to label the mass and speed of each particle:

$$K_{\mathrm{rot}} = \tfrac{1}{2}m_1 v_1^2 + \tfrac{1}{2}m_2 v_2^2 + \cdots + \tfrac{1}{2}m_N v_N^2 = \sum_{i=1}^{N} \tfrac{1}{2}m_i v_i^2$$

The speed of each particle is related to its distance from the axis of rotation. Particles that are farther from the axis move faster. In Section 5.1 we found that the speed of a particle moving in a circle is

$$v = r\omega \qquad\qquad (5\text{-}6)$$

where ω is the angular speed and r is the distance between the rotation axis and the particle. By substitution, the rotational kinetic energy can be written

$$K_{\mathrm{rot}} = \sum_{i=1}^{N} \tfrac{1}{2}m_i r_i^2 \omega^2$$

The entire object rotates at the same angular velocity ω, so the constants $\tfrac{1}{2}$ and ω^2 can be factored out of each term of the sum:

$$K_{\mathrm{rot}} = \tfrac{1}{2}\left(\sum_{i=1}^{N} m_i r_i^2\right) \omega^2$$

The quantity in the parentheses *cannot change* since the distance between each particle and the rotation axis stays the same if the object is rigid and doesn't change shape. However difficult it may be to compute the sum in the parentheses, we only need to do it *once* for any given mass distribution and axis of rotation.

Let's give the quantity in the brackets the symbol I. In Chapter 5, we found it useful to draw analogies between translational variables and their rotational equivalents. By using the symbol I, we can see that translational and rotational kinetic energy have similar forms: translational kinetic energy is

$$K_{\mathrm{tr}} = \tfrac{1}{2}mv^2$$

and **rotational kinetic energy** is

Rotational kinetic energy
$K_{\mathrm{rot}} = \tfrac{1}{2}I\omega^2$ $\qquad$ (8-1)

The quantity I is called the **rotational inertia**:

Rotational inertia
$I = \sum\limits_{i=1}^{N} m_i r_i^2$ $\qquad$ (8-2)
(SI unit: kg·m²)

Comparing the expressions for translational and rotational kinetic energy, we see that angular speed ω takes the place of speed v and rotational inertia I takes the place of

Tips   Since $v = r\omega$ was used to derive Eq. (8-1), ω must be expressed in radians per unit time (normally rad/s).

mass m. Mass is a measure of the inertia of an object, or in other words how difficult it is to change the object's velocity. Similarly, for a rigid rotating object, I is a measure of its rotational inertia—how hard it is to change its angular velocity. That is why the quantity I is called the rotational inertia; it is also sometimes called the **moment of inertia**.

When a problem requires you to find a rotational inertia, there are four principles to follow.

Finding the Rotational Inertia

1. If the object consists of a *small* number of particles, calculate the sum $I = \sum_{i=1}^{N} m_i r_i^2$ directly.

2. For symmetric objects with simple geometric shapes, advanced mathematical methods can be used to perform the sum in Eq. (8-2). For your convenience, Table 8.1 lists the results of these calculations for the shapes most commonly encountered.

3. Since the rotational inertia is a sum, you can always mentally decompose the object into several parts, find the rotational inertia of each part, and then add them. This is an example of the *divide-and-conquer* problem-solving technique.

4. Since the rotational inertia involves distances from the axis of rotation, you can move mass parallel to the rotation axis without changing I. For example, you might want to calculate the rotational inertia of a door about the axis through its hinges. Mentally "compress" the door vertically (parallel to the axis) into a horizontal rod with the same mass, as in Fig. 8.1. The rotational inertia is unchanged by the compression, since every particle maintains the same distance from the axis of rotation. Thus, the formula for the rotational inertia of a rod (listed in Table 8.1) can be used for the door.

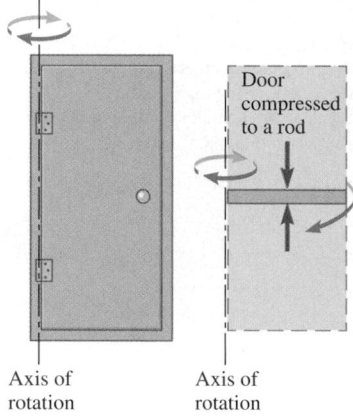

Axis of rotation   Axis of rotation

Figure 8.1 The rotational inertias of a door and a rod are similar.

Table 8.1

Rotational Inertia for Uniform Objects with Various Geometrical Shapes

Shape		Axis of Rotation	Rotational Inertia	Shape		Axis of Rotation	Rotational Inertia
Thin hollow cylindrical shell (or hoop)		Central axis of cylinder	MR^2	Solid sphere		Through center	$\frac{2}{5}MR^2$
Solid cylinder (or disk)		Central axis of cylinder	$\frac{1}{2}MR^2$	Thin hollow spherical shell		Through center	$\frac{2}{3}MR^2$
Hollow cylindrical shell or disk	Top view	Central axis of cylinder	$\frac{1}{2}M(a^2 + b^2)$	Thin rod		Perpendicular to rod through end	$\frac{1}{3}ML^2$
				Rectangular plate		Perpendicular to plate through center	$\frac{1}{12}M(a^2 + b^2)$

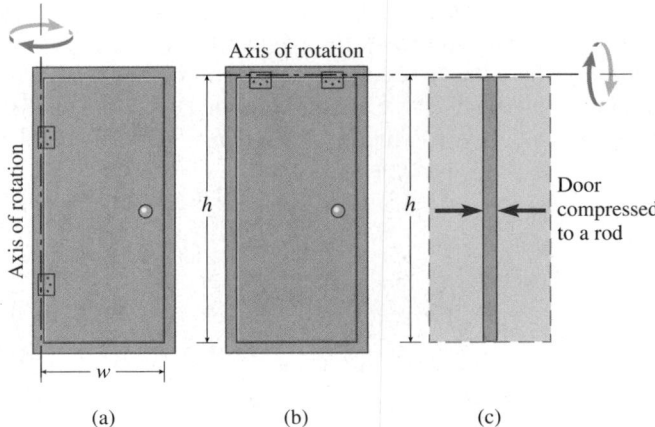

Figure 8.2 The rotational inertia of a door depends on the rotation axis. (a) The door with hinges at the side has a smaller rotational inertia, $I = \frac{1}{3}Mw^2$, than (b) the rotational inertia, $I = \frac{1}{3}Mh^2$, of the same door with hinges at the top. (c) The door can be mentally compressed horizontally into a rod.

Keep in mind that the rotational inertia of an object depends on the location of the rotation axis. For instance, imagine taking the hinges off the side of a door and putting them on the top, so that the door swings about a horizontal axis like a cat flap door (Fig. 8.2b). The door now has a considerably larger rotational inertia than before the hinges were moved. The door has the same mass as before, but its mass now lies on average much farther from the axis of rotation than that of the door in Fig. 8.2a. In applying Eq. (8-2) to find the rotational inertia of the door, the values of r_i range from 0 to the height of the door (h), whereas with the hinges in the normal position the values of r_i range from 0 only to the width of the door (w). If we think of compressing the door with hinges on top into a rod, the door would have to be compressed *horizontally* into a rod of length h (Fig. 8.2c).

Example 8.1

Rotational Inertia of a Rod About Its Midpoint

Table 8.1 gives the rotational inertia of a thin rod with the axis of rotation perpendicular to its length and passing through one end ($I = \frac{1}{3}ML^2$). Write an expression for the rotational inertia of a rod with mass M and length L that rotates about a perpendicular axis through its *midpoint* (Fig. 8.3).

Strategy In general, the same object rotating about a different axis has a different rotational inertia. With the axis at the midpoint, the rotational inertia is smaller than for the axis at the end, since the mass is closer to the axis. Imagine performing the

sum $\sum\limits_{i=1}^{N} m_i r_i^2$; for the axis at the end, the values of r_i range from 0 to L, whereas with the axis at the midpoint, r_i is never larger than $\frac{1}{2}L$.

Since the answer is not in Table 8.1, there are two options: either calculate the sum $\sum\limits_{i=1}^{N} m_i r_i^2$ directly, or decompose the object into parts. Imagine cutting the rod in half; then there are two rods, each with its axis of rotation at one of its *ends*. The rotational inertia of a rod about its end is listed in Table 8.1. Then, since rotational inertia is additive, the rotational inertia for two such rods is just twice the value for one rod.

Solution Each of the halves has mass $\frac{1}{2}M$ and length $\frac{1}{2}L$ and rotates about an axis at its endpoint. Table 8.1 gives $I = \frac{1}{3}ML^2$ for a rod with mass M and length L rotating about its end, so each of the halves has

$$I_{\text{half}} = \frac{1}{3} \times \text{mass of half} \times (\text{length of half})^2$$
$$= \frac{1}{3} \times (\frac{1}{2}M) \times (\frac{1}{2}L)^2 = \frac{1}{3} \times \frac{1}{2} \times \frac{1}{4} \times ML^2 = \frac{1}{24}ML^2$$

Since there are two such halves, the total rotational inertia is twice that:

$$I = I_{\text{half}} + I_{\text{half}} = \frac{1}{12}ML^2$$

Figure 8.3
(a) A rod rotating about a vertical axis through its center. (b) The same rod, viewed as two rods, each half as long, rotating about an axis through an end.

continued on next page

Example 8.1 *continued*

Discussion The rotational inertia is less than $\frac{1}{3}ML^2$, as expected. That it is $\frac{1}{4}$ as much is a result of the distances r_i being *squared* in the definition of rotational inertia. The various particles that compose the rod are at distances that range from 0 to $\frac{1}{2}L$ from the rotation axis, instead of from 0 to L. Think of it as if the rod were compressed to half its length, still pivoting about the endpoint. All the distances r_i are half as much as before; since each r_i is squared in the sum, the rotational inertia is $(\frac{1}{2})^2 = \frac{1}{4}$ its former value.

See the Physics at Home that follows the Practice Problem.

Practice Problem 8.1 Playground merry-go-round

A playground merry-go-round is essentially a uniform disk that rotates about a vertical axis through its center (Fig. 8.4).

Suppose the disk has a radius of 2.0 m and a mass of 160 kg; a child weighing 180 N sits at the edge of the merry-go-round. What is the merry-go-round's rotational inertia, including the contribution due to the child? [*Hint:* Treat the child as a point mass at the edge of the disk.]

2.0 m

180 N

Figure 8.4
Child on merry-go-round.

Physics at Home

The change in rotational inertia of a rod as the rotation axis changes can be easily felt. Hold a baseball bat in the usual way, with your hands gripping the bottom of the bat. Swing the bat a few times. Now "choke up" on the bat—move your hands up the bat—and swing a few times. The bat is easier to swing because it now has a smaller rotational inertia. Children often choke up on a bat that is too massive for them. Even major league baseball players occasionally choke up on the bat when they want more control over their swing to place a hit in a certain spot (Fig. 8.5). On the other hand, choking up on the bat makes it impossible to hit a home run. To hit a long fly ball, you want the pitched baseball to encounter a bat that is swinging with a lot of rotational inertia.

Figure 8.5 Hank Aaron choking up on the bat.

Example 8.2

Atwood's Machine

Atwood's machine consists of a cord around a pulley of rotational inertia I, radius R, and mass M, with two blocks (masses m_1 and m_2) hanging from the ends of the cord as in Fig. 8.6. (Note that in Example 3.7 we analyzed Atwood's machine for the special case of a massless pulley; for a massless pulley $I = 0$.) Assume that the pulley is free to turn without friction and that the cord does not slip. Ignore air resistance. If the masses are released from rest, find how fast they are moving after they have moved a distance h (one up, the other down).

Strategy Neglecting both air resistance and friction means that no nonconservative forces act on the system; therefore its mechanical energy is conserved. Gravitational potential energy is converted into the translational kinetic

Pulley mass M

R M

m_2

m_1

h

h Initial position

Figure 8.6
Atwood's machine

continued on next page

Example 8.2 *continued*

energies of the two blocks and the rotational kinetic energy of the pulley.

Solution For our convenience, we assume that $m_1 > m_2$. Mass m_1 therefore moves down and m_2 moves up. After the masses have each moved a distance h, the changes in gravitational potential energy are

$$\Delta U_1 = -m_1 g h$$

$$\Delta U_2 = +m_2 g h$$

Since mechanical energy is conserved,

$$\Delta U + \Delta K = 0$$

The mechanical energy of the system includes the kinetic energies of three objects: the two masses and the pulley. All start with zero kinetic energy, so

$$\Delta K = \tfrac{1}{2}(m_1 + m_2)v^2 + \tfrac{1}{2}I\omega^2$$

The speed v of the masses is the same since the cord's length is fixed. The speed v and the angular speed of the pulley ω are related if the cord does not slip: the tangential speed of the pulley must equal the speed at which the cord moves. The tangential speed of the pulley is its angular speed times its radius:

$$v = \omega R$$

After substituting v/R for ω, the energy conservation equation becomes

$$\Delta U + \Delta K = [-m_1 g h + m_2 g h] + \left[\tfrac{1}{2}(m_1 + m_2)v^2 + \tfrac{1}{2}I\left(\frac{v}{R}\right)^2\right] = 0$$

or

$$\left[\tfrac{1}{2}(m_1 + m_2) + \tfrac{1}{2}\frac{I}{R^2}\right]v^2 = (m_1 - m_2)g h$$

Solving this equation for v yields

$$v = \sqrt{\frac{2(m_1 - m_2)g h}{m_1 + m_2 + I/R^2}}$$

Discussion This answer is rich in information, in the sense that we can ask many "What if?" questions. Not only do these questions provide checks as to whether the answer is reasonable, they also enable us to perform thought experiments,

which could then be checked by constructing an Atwood's machine and comparing the results.

For instance: What if m_1 is only slightly greater than m_2? Then the final speed v is small—as m_2 approaches m_1, v approaches 0. This makes intuitive sense: a small imbalance in weights produces a small acceleration. You should practice this kind of reasoning by making other such checks.

It is also enlightening to look at terms in an algebraic solution and connect them with physical interpretations. The quantity $(m_1 - m_2)g$ is the imbalance in the gravitational forces pulling on the two sides. The denominator $(m_1 + m_2 + I/R^2)$ is a measure of the total inertia of the system—the sum of the two masses plus an inertial contribution due to the pulley. The pulley's contribution is *not* simply equal to its mass. If, for example, the pulley is a uniform disk with $I = \tfrac{1}{2}MR^2$, the term I/R^2 would be equal to *half* the mass of the pulley.

The same principles used to analyze Atwood's machine have many applications in the real world. One such application is in elevators, where one of the hanging masses is the elevator and the other is the counterweight. Of course, the elevator and counterweight are not allowed to hang freely from a pulley—we must also consider the energy supplied by the motor.

Practice Problem 8.2   Modified Atwood's machine

Figure 8.7 shows a modified form of Atwood's machine where one of the blocks slides on a table instead of hanging from the pulley. The coefficient of kinetic friction between the sliding mass and the table is μ_k. The blocks are released from rest. Find the speed of the blocks after they have moved a distance h in terms of μ_k, m_1, m_2, I, R, and h.

Figure 8.7
Modified Atwood's machine

8.2   TORQUE

Suppose you place a bicycle upside down, as if you're going to repair it. First, you give one of the wheels a spin. If everything is working as it should, the wheel spins for quite a while; its angular acceleration is small. If the wheel doesn't spin for very long, then its angular velocity changes rapidly and the angular acceleration is large in magnitude; there must be excessive friction somewhere. Perhaps the brakes are rubbing on the rim or the bearings need to be repacked.

If we could eliminate *all* the frictional forces acting on the wheel, including air resistance, then we would expect the wheel to keep spinning without diminishing angular speed. In that case, its angular acceleration would be zero. The situation is reminiscent of Newton's first law: a body with no external interactions, or at least no net force acting on it, moves with constant velocity. We can state a "Newton's first law for rotation": a rotating body with no external interactions, and whose rotational inertia doesn't change, keeps rotating at constant angular velocity.

Of course, the hypothetical frictionless bicycle wheel does have external interactions. The Earth's gravitational field exerts a downward force and the axle exerts an upward force to keep the wheel from falling. Then is it true that, as long as there is no net external force, the angular acceleration is zero? No; it is easy to give the wheel an angular acceleration while keeping the net force zero. Imagine bringing the wheel to rest by pressing two hands against the tire on opposite sides. On one side, the motion of the rim of the tire is downward and the kinetic frictional force is upward (see Fig. 8.8). On the other side, the tire moves upward and the frictional force is downward. In a similar way, we could apply equal and opposite forces to the opposite sides of a wheel at rest to make it start spinning. In either case, we exert equal magnitude forces, so that the net force is zero, and still give the wheel an angular acceleration.

A quantity related to force, called **torque**, plays the role in rotational dynamics that force itself played in translational dynamics. A torque is not separate from a force; it is impossible to exert a torque without exerting a force. Torque is a measure of how effective a given force is at twisting or turning something. For something rotating about a fixed axis such as the bicycle wheel, a torque can *change* the rotational motion either by making it rotate faster or by slowing it down.

When stopping the bicycle wheel with two equal and opposite forces, as in Fig. 8.8, the net applied force is zero and thus the wheel is in translational equilibrium; but the net torque is not zero so it is not in rotational equilibrium. Both forces tend to give the wheel the same sign of angular acceleration; they are both making the wheel slow down. The two torques are in fact equal, with the same sign.

What determines the torque produced by a particular force? Imagine trying to push open a massive bank vault door. Certainly you would push as hard as you can; the torque is proportional to the magnitude of the force. It also matters where and in what direction the force is applied. For maximum effectiveness, you would push perpendicular to the door (Fig. 8.9a). If you pushed radially, straight in toward the axis of rotation that passes through the hinges, the door wouldn't rotate, no matter how hard you push (Fig. 8.9b). A force acting in any other direction could be decomposed into radial and perpendicular components, with the radial component contributing nothing to the torque. Only the perpendicular component of the force produces a torque.

Furthermore, *where* you apply the force is critical (Fig. 8.10). Instinctively, you would push at the outer edge, as far from the rotation axis as possible. If you pushed close to the axis, it would be difficult to open the door. Torque is proportional to the distance between the rotation axis and the **point of application** of the force (the point at which the force is applied).

Figure 8.8 A spinning bicycle wheel slowed to a stop by friction. Each hand exerts a normal force $\vec{N}$ and a frictional force $\vec{f}$ on the tire.

The *radial* direction is directly toward or away from the axis of rotation. The *perpendicular* or *tangential* direction is perpendicular to both the radial direction and the axis of rotation; it is tangent to the circular path followed by a point as the object rotates.

(a) Maximum torque   (b) Zero torque

(a) Larger torque   (b) Smaller torque

Figure 8.9 Torque on a bank vault door depends on the direction of the applied force. (a) Pushing perpendicularly gives the maximum torque. (b) Pushing radially inward with the same magnitude force gives zero torque.

Figure 8.10 Torques; the same force at different distances from the axis.

The symbol ⊥ stands for *perpendicular*; ∥ stands for *parallel*.

To satisfy the requirements of the previous paragraphs we define the magnitude of the torque as the product of the distance between the rotation axis and the point of application of the force (r) with the perpendicular component of the force ($F_\perp$):

Definition of torque:

$$\tau = \pm r F_\perp \qquad (8\text{-}3)$$

---

where r is the shortest distance between the rotation axis and the point of application of the force and $F_\perp$ is the perpendicular component of the force.

The symbol for torque is τ, the Greek letter tau. The SI unit of torque is the N·m. The SI unit of *energy*, the joule, is equivalent to N·m, but we do not write torque in joules. Even though both energy and torque can be written using the same SI base units, the two quantities have different meanings; torque is not a form of energy. To help maintain the distinction, the joule is used for energy but *not* for torque.

The sign of the torque indicates the direction of the angular acceleration that torque would cause *by itself*. Recall from Section 5.1 that by convention a positive angular velocity ω means counterclockwise (CCW) rotation and negative angular velocity ω means clockwise (CW) rotation. A positive angular acceleration α either increases the rate of CCW rotation (increases the magnitude of a positive ω) or decreases the rate of CW rotation (decreases the magnitude of a negative ω).

Figure 8.11 (a) When the cyclist climbs a hill, the top half of the chain exerts a large force $\vec{F}$ on one of the gears attached to the rear wheel. As viewed here, the torque about the axis of rotation (the axle) due to this force is clockwise. By convention, we call this a negative torque. (b) When the brakes are applied, the brake pads are pressed onto the rim, giving rise to frictional forces on the rim. As viewed here, the frictional force $\vec{f}$ causes a counterclockwise (positive) torque on the wheel about the axle.

We use the same sign convention for torque. A force whose perpendicular component tends to cause rotation in the CCW direction gives rise to a positive torque; if it is the only torque acting, it would cause a positive angular acceleration α (see Fig. 8.11). A force whose perpendicular component tends to cause rotation in the CW direction gives rise to a negative torque. The symbol $\pm$ in Eq. (8-3) reminds us to assign the appropriate algebraic sign each time we calculate a torque.

The sign of the torque is *not* determined by the sign of the angular velocity (in other words, whether the wheel is spinning CCW or CW); rather, it is determined by the sign of the angular *acceleration* the torque would cause if acting alone. To determine the sign of a torque, imagine which way the torque would make the object begin to spin if it is initially not rotating.

Example 8.3

A Spinning Bicycle Wheel

 To stop a spinning bicycle wheel, suppose you push radially inward on opposite sides of the wheel, as shown in Fig. 8.8, with equal forces of magnitude 10.0 N. The radius of the wheel is 32 cm and the coefficient of kinetic friction between the tire and your hand is 0.75. The wheel is spinning in the clockwise sense. What is the net torque on the wheel?

Strategy The 10.0-N forces are directed radially toward the rotation axis, so they produce no torques themselves; only perpendicular components of forces give rise to torques. The forces of kinetic friction between the hands and the tire are tangent to the tire, so they do produce torques. The normal force applied to the tire is 10.0 N on each side; using the coefficient of friction, we can find the frictional forces.

Solution The frictional force exerted by each hand on the tire has magnitude

$$f = \mu_k N = (0.75) \times (10.0\ \text{N}) = 7.5\ \text{N}$$

The frictional force is tangent to the wheel, so $f_\perp = f$. Then the magnitude of each torque is

$$|\tau| = rf_\perp = (0.32\ \text{m}) \times (7.5\ \text{N}) = 2.4\ \text{N·m}$$

The two torques have the same sign, since they are both tending to slow down the rotation of the wheel. Is the torque positive or negative? The angular velocity of the wheel is negative since it rotates clockwise. The angular acceleration has the opposite sign because the angular speed is decreasing. Since $\alpha > 0$, the net torque is also positive. Therefore,

$$\tau_{\text{net}} = +4.8\ \text{N·m}$$

Discussion The trickiest part of calculating torques is determining the sign. To check, look at the frictional forces in Fig. 8.8. Imagine which way the forces would make the wheel begin to rotate if the wheel were not originally rotating. The frictional forces point in a direction that would tend to cause a counterclockwise rotation, so the torques are positive.

Practice Problem 8.3 Disk brakes

In the disk brakes that slow down a car, the brake pads squeeze a spinning rotor; friction between the pads and the rotor provides the torque that slows down the car. If the normal force that a pad exerts on a rotor is 85 N and the coefficient of friction is 0.62, what is the frictional force on the rotor? If this force acts 8.0 cm from the rotation axis, what is the magnitude of the torque on the rotor?

Lever Arms

There is another, completely equivalent, way to calculate torques that is often more convenient than finding the perpendicular component of the force. Figure 8.12 shows a force $\vec{F}$ acting at a distance r from an axis. The distance r is the length of a line perpendicular to the axis that runs from the axis to the force's point of application. The force makes an angle θ with that line. The torque is then

$$\tau = \pm rF_\perp = \pm r(F \sin \theta)$$

The factor $\sin \theta$ could be grouped with r instead of with F. Then $\tau = \pm(r \sin \theta)F$, or

$$\tau = \pm r_\perp F \tag{8-4}$$

The distance $r_\perp$ is called the **lever arm** (or **moment arm**). The magnitude of the torque is therefore the magnitude of the force times the lever arm.

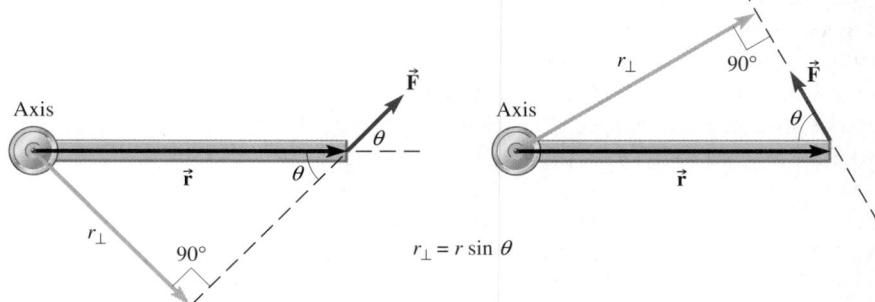

Figure 8.12 Finding the magnitude of a torque using lever arms.

$$r_\perp = r \sin \theta$$

Finding Torques Using the Lever Arm

1. Draw a line parallel to the force through the force's point of application; this line (dashed in Fig. 8.12) is called the force's **line of action**.

2. Draw a line from the rotation axis to the line of action. This line must be perpendicular to both the axis and the line of action. The distance from the axis to the line of action along this perpendicular line is the lever arm ($r_\perp$). If the line of action of the force goes through the rotation axis, the lever arm and the torque are both zero (Fig. 8.9b).

3. The magnitude of the torque is the magnitude of the force times the lever arm:

$$\tau = \pm r_\perp F$$

4. Determine the algebraic sign of the torque as before.

Example 8.4

Screen Door Closer

An automatic screen door closer attaches to a door 47 cm away from the hinges and pulls on the door with a force of 25 N, making an angle of 15° with the door (Fig. 8.13). Find the magnitude of the torque exerted on the door due to this force about the rotation axis through the hinges using (a) the perpendicular component of the force and (b) the lever arm. (c) What is the sign of this torque as viewed from above?

Strategy For method (a), we must find the component of the 25-N force perpendicular to the radial direction. Then this component is multiplied by the length of the radial line. For method (b), we draw in the line of action of the force. Then the lever arm is the perpendicular distance from the line of action to the rotation axis. The torque is the magnitude of the force times the lever arm. We must be careful not to combine the two methods: the torque is *not* equal to the perpendicular force component times the lever arm. For (c), we determine whether this torque would tend to make the door rotate CCW or CW.

Figure 8.13
Screen door with automatic closing mechanism.

continued on next page

Example 8.4 *continued*

Solution (a) As shown in Fig. 8.14a, the radial component of the force ($F_\parallel$) passes through the rotation axis. The angle labeled 15° would actually be a bit larger than 15°, but since the thickness of the door is much less than 47 cm, we approximate it as 15°. The perpendicular component is

$$F_\perp = F \sin 15°$$

The magnitude of the torque is

$$|\tau| = rF_\perp = 0.47 \text{ m} \times 25 \text{ N} \times \sin 15° = 3.0 \text{ N·m}$$

(b) Figure 8.14b shows the line of action of the force, drawn parallel to the force and passing through the point of application. The lever arm is the perpendicular distance between the rotation axis and the line of action. The distance r is approximately 47 cm (again neglecting the thickness of the door). Then the lever arm is

$$r_\perp = r \sin 15°$$

and the magnitude of the torque is

$$|\tau| = r_\perp F = 0.47 \text{ m} \times \sin 15° \times 25 \text{ N} = 3.0 \text{ N·m}$$

(c) Using the top view of Fig. 8.13, the torque tends to close the door by making it rotate counterclockwise (assuming the door is initially at rest and no other torques act). The torque is therefore positive as viewed from above.

Discussion The most common mistake to make in either solution method would be to use cosine instead of sine (or, equivalently, to use the complementary angle 75° instead of 15°). A check is a good idea. If the automatic closer were more nearly parallel to the door, the angle would be less than 15°. The torque would be smaller because the force is more nearly pulling straight in toward the axis. Since the sine function gets smaller for angles closer to zero, the expression checks out correctly.

It might seem silly for a door closer to pull at such an angle that the perpendicular component is relatively small. The reason it's done that way is so the door closer does not get in the way. A closer that pulled in a perpendicular direction would stick straight out from the door. As discussed in Section 8.5, the situation is much the same in our bodies. In order to not inhibit the motion of our limbs, our tendons and muscles are nearly parallel to the bones. As a result, the forces they exert must be much larger than we might expect.

Practice Problem 8.4 Exercise is good for you

A person is lying on an exercise mat and lifts one leg at an angle of 30.0° from the horizontal with an 89-N (20-lb) weight attached to the ankle (Fig. 8.15). The distance between the ankle weight and the hip joint (which is the rotation axis for the leg) is 84 cm. What is the torque due to the ankle weight on the leg?

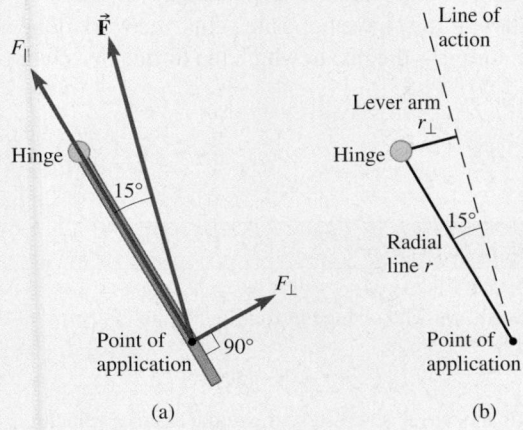

(a)

(b)

Figure 8.14
(a) Finding the perpendicular component of the force. (b) Finding the lever arm.

Figure 8.15
Exercise leg lifts.

Center of Gravity

We have seen that the torque produced by a force depends on the point of application of the force. What about gravity? The gravitational force on a body is not exerted at a single point, but is distributed throughout the volume of the body. When we talk of "the" force of gravity on something, we really mean the total force of gravity acting on each particle making up the system.

Fortunately, when we need to find the total torque due to the forces of gravity acting on an object, the total force of gravity can be considered to act at a single point. This point is called the **center of gravity**. The torque found this way is the same as finding all the torques due to the forces of gravity acting at every point in the body and adding

Tips When calculating the torque due to gravity, consider the entire gravitational force to act at the center of gravity.

them together. As you can verify in Problem 71, if the gravitational field is uniform in magnitude and direction, then the center of gravity of an object is located at the object's center of mass.

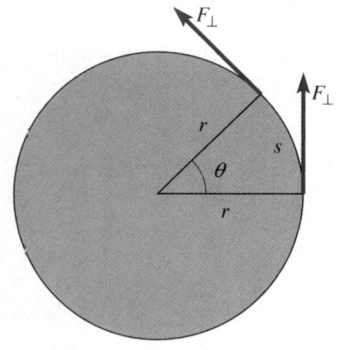

Figure 8.16 Work done by torque.

8.3　WORK DONE BY A TORQUE

Torques can do work, as anyone who has started a lawnmower with a pull cord can verify. Actually it is the force that does the work, but in rotational problems it is often simpler to calculate the work done from the torque. Just as the work done by a constant force is the product of force and the parallel component of displacement, work done by a constant torque can also be calculated as the torque times the *angular* displacement.

Imagine a torque acting on a wheel that spins through an angular displacement θ while the torque is applied. The work done by the force that gives rise to the torque is the product of the perpendicular component of the force ($F_\perp$) with the arc length s through which the point of application of the force moves (see Fig. 8.16). We use the perpendicular force component because that is the component parallel to the *displacement*, which is instantaneously tangent to the arc of the circle. Thus,

$$W = F_\perp s \qquad (8\text{-}5)$$

To write the work in terms of torque, note that $\tau = rF_\perp$ and $s = r\theta$; then

$$W = F_\perp s = \frac{\tau}{r} \times r\theta = \tau\theta$$

$$W = \tau\theta \quad (\theta \text{ in radians}) \qquad (8\text{-}6)$$

Work is indeed the product of torque and the angular displacement. If τ and θ have the same sign, the work done is positive; if they have opposite signs, the work done is negative. The *power* due to a constant torque—the rate at which the torque does work—is

$$P = \tau\omega \qquad (8\text{-}7)$$

Example 8.5

Work Done on Potter's Wheel

A potter's wheel is a heavy stone disk upon which the pottery is shaped. Potter's wheels were once driven by the potter pushing on a foot treadle; today most potter's wheels are driven by electric motors. (a) If the potter's wheel is a uniform disk of mass 40.0 kg and diameter 0.50 m, how much work must be done by the motor to bring the wheel from rest to 80.0 rpm? (b) If the motor delivers a constant torque of 8.2 N·m during this time, through how many revolutions does the wheel turn in coming up to speed?

Strategy　Work is an energy transfer. In this case, the motor is increasing the rotational kinetic energy of the potter's wheel. Thus, the work done by the motor is equal to the change in rotational kinetic energy of the wheel, ignoring frictional losses. In the expression for rotational kinetic energy, we must express ω in rad/s; we cannot substitute 80.0 rpm for ω. Once we know the work done, we use the torque to find the angular displacement.

Solution　(a) The change in rotational kinetic energy of the wheel is

$$\Delta K = \tfrac{1}{2}I\,(\omega_f^2 - \omega_i^2) = \tfrac{1}{2}I\,\omega_f^2$$

Initially the wheel is at rest, so the initial angular velocity ω_i is zero. From Table 8.1, the rotational inertia of a uniform disk is

$$I = \tfrac{1}{2}MR^2$$

Substituting this for I,

$$\Delta K = \tfrac{1}{4}MR^2\omega_f^2$$

Before substituting numerical values, we convert 80.0 rpm to rad/s:

$$\omega_f = 80.0\,\frac{\text{rev}}{\text{min}} \times 2\pi\,\frac{\text{rad}}{\text{rev}} \times \frac{1}{60}\,\frac{\text{min}}{\text{s}} = 8.38\ \text{rad/s}$$

Substituting the known values for mass and radius,

$$\Delta K = \tfrac{1}{4} \times 40.0\ \text{kg} \times (\tfrac{0.50}{2}\ \text{m})^2 \times (8.38\ \text{rad/s})^2 = 43.9\ \text{J}$$

Therefore, the work done by the motor, rounded to two significant figures, is 44 J.

continued on next page

Example 8.5 *continued*

(b) The work done by a constant torque is

$$W = \tau\theta$$

Solving for the angular displacement θ gives

$$\theta = \frac{W}{\tau} = \frac{43.9\,\text{J}}{8.2\,\text{N·m}} = 5.35\,\text{rad} \qquad (1)$$

Since 2π rad = 1 revolution,

$$\theta = 5.35\,\text{rad} \times \frac{1\,\text{rev}}{2\pi\,\text{rad}} = 0.85\,\text{rev}$$

Discussion As always, work is an energy transfer. In this problem, the work done by the motor is the means by which the potter's wheel acquires its rotational kinetic energy. But work done by a torque does not *always* appear as a change in rotational kinetic energy. For instance, when you wind up a mechanical clock or a windup toy, the work done by the torque you apply is stored as elastic potential energy in some sort of spring.

Practice Problem 8.5   Work done on air conditioner

A belt wraps around a pulley of radius 7.3 cm that drives the compressor of an automobile air conditioner. The tension in the belt on one side of the pulley is 45 N and on the other side of the pulley it is 27 N (Fig. 8.17). How much work is done by the belt on the compressor during one revolution of the pulley?

Figure 8.17
Practice Problem 8.5

8.4   EQUILIBRIUM REVISITED

In Chapter 2, we said that an object is in equilibrium when the net force acting on it is zero. That statement is true but incomplete. It is quite possible for the net force acting to be zero, while the net torque is nonzero; the object would then have a nonzero angular acceleration. When designing a bridge or a new house, it would be unacceptable for any of the parts to have nonzero angular acceleration! Zero net force is sufficient to ensure *translational* equilibrium; if an object is also in *rotational* equilibrium, then the net torque acting on it must also be zero.

Conditions for equilibrium:

$$\vec{\mathbf{F}}_{\text{net}} = 0 \ \text{ and } \ \tau_{\text{net}} = 0 \qquad (8\text{-}8)$$

Before tackling equilibrium problems, we must resolve a conundrum: if something is not rotating, then where is the axis of rotation? How can we calculate torques without knowing where the axis of rotation is? In some cases, perhaps involving axles or hinges, there may be a clear axis about which the object would rotate if the balance of forces and torques is disturbed. In many cases, though, it is not clear what the rotation axis would be, and in general it depends on exactly how the equilibrium is upset. Fortunately, the axis can be chosen *arbitrarily* when calculating torques *in equilibrium problems*.

In equilibrium, the net torque about *any* rotation axis must be zero. Does that mean that we have to write down an infinite number of torque equations, one for each possible axis of rotation? Fortunately, no. Although the proof is complicated, it can be shown that if the net force acting on an object is zero and the net torque about one rotation axis is zero, then the net torque about every other axis parallel to that axis must also be zero. Therefore, one torque equation is all we need.

Since the torque can be calculated about any desired axis, a judicious choice can greatly simplify the solution of the problem. The best place to choose the axis is usually at the point of application of an unknown force so that the unknown force does not appear in the torque equation.

Example 8.6

Carrying a 6 × 6 Beam

Two carpenters are carrying a uniform 6 × 6 beam. The beam is 2.44 m (8.00 ft) long and weighs 425 N (95.5 lb). One of the carpenters, being a bit stronger than the other, agrees to carry the beam 1.00 m in from the end; the other carries the beam at its opposite end. What is the upward force exerted on the beam by each carpenter?

Strategy The conditions for equilibrium are that the net external force equal zero and the net external torque equal zero. Should we start with forces or with torques? In this problem, it is easiest to start with torques. If we choose the axis of rotation where one of the unknown forces acts, then that force has a lever arm of zero and its torque is zero. The torque equation can be solved for the other unknown force. Then with only one force still unknown, we set the sum of the y-components of the forces equal to zero.

Solution The first step is to draw a force diagram (Fig. 8.18). Each force is drawn at the point where it acts. Known distances are labeled.

We choose a rotation axis perpendicular to the xy-plane and passing through the point of application of $\vec{\mathbf{F}}_2$. The simplest way to find the torques for this example is to multiply each force by its lever arm. The lever arm for $\vec{\mathbf{F}}_1$ is

$$2.44 \text{ m} - 1.00 \text{ m} = 1.44 \text{ m}$$

and the magnitude of the torque due to this force is

$$|\tau| = Fr_\perp = F_1 \times 1.44 \text{ m}$$

Since the beam is uniform, its center of gravity is at its midpoint. We imagine the entire gravitational force to act at this point. Then the lever arm for the gravitational force is

$$\tfrac{1}{2} \times 2.44 \text{ m} = 1.22 \text{ m}$$

and the torque due to gravity has magnitude

$$|\tau| = Fr_\perp = 425 \text{ N} \times 1.22 \text{ m} = 518.5 \text{ N·m}$$

The torque due to $\vec{\mathbf{F}}_1$ is negative since, if it were the only torque, it would make the beam start to rotate clockwise about our chosen axis of rotation. The torque due to gravity is positive since, if it were the only torque, it would make the beam start to rotate counterclockwise. Therefore,

$$\tau_{\text{net}} = -F_1 \times 1.44 \text{ m} + 518.5 \text{ N·m} = 0$$

Solving for F_1,

$$F_1 = \frac{518.5 \text{ N·m}}{1.44 \text{ m}} = 360 \text{ N}$$

Since another condition for equilibrium is that the net force be zero,

$$\Sigma F_y = F_1 + F_2 - mg = 0$$

Solving for F_2,

$$F_2 = 425 \text{ N} - 360 \text{ N} = 65 \text{ N}$$

Discussion A good way to check this result is to make sure that the net torque about a *different axis* is zero—for an object in equilibrium, the net torque about any axis must be zero. Suppose we choose an axis through the point of application of $\vec{\mathbf{F}}_1$. Then the lever arm for $m\vec{\mathbf{g}}$ is 1.22 m − 1.00 m = 0.22 m and the lever arm for $\vec{\mathbf{F}}_2$ is 2.44 m − 1.00 m = 1.44 m. Setting the net torque equal to zero:

$$\tau_{\text{net}} = -425 \text{ N} \times 0.22 \text{ m} + F_2 \times 1.44 \text{ m} = 0$$

Solving for F_2 gives

$$F_2 = \frac{425 \text{ N} \times 0.22 \text{ m}}{1.44 \text{ m}} = 65 \text{ N}$$

which agrees with the value calculated before. We could have used this second torque equation to find F_2 instead of setting ΣF_y equal to zero.

Figure 8.18
Diagram of the beam with rotation axis, forces, and distances shown.

continued on next page

Example 8.6 *continued*

Practice Problem 8.6 A diving board

A uniform diving board of length 5.0 m is supported at two points; one support is located 3.4 m from the end of the board and the second is at 4.6 m from the end (Fig. 8.19). The supports exert vertical forces on the diving board. A diver stands at the end of the board over the water. Determine the directions of the support forces. [*Hint:* In this problem, consider torques about different rotation axes.]

├── 1.2 m ──┤├────── 3.4 m ──────┤

Figure 8.19
Diving board

A diving board is an example of a cantilever—a beam or pole that extends beyond its support. The forces exerted by the supports on a diving board are considerably larger than if the same board were supported at both ends (see Problem 28). The advantage is that the far end of the board is left free to vibrate; as it does, the support forces adjust themselves to keep the board from tipping over. The architect Frank Lloyd Wright was fond of using cantilever construction to open up the sides and corners of a building, allowing corner windows that give buildings a lighter and more spacious feel (Fig. 8.20).

Making The Connection:
cantilever building
construction

Figure 8.20 The cantilevered master bedroom in the north wing of Wingspread by Frank Lloyd Wright juts well out over its brick foundation. The cypress trellis extending even farther beyond the bedroom balcony filters the natural light and serves to emphasize the free-floating nature of the structure with views of the landscape below.

Example 8.7

The Slipping Ladder

A 15.0-kg uniform ladder leans against a wall in the atrium of a large hotel (Fig. 8.21a). The ladder is 8.00 m long; it makes an angle $\theta = 60.0°$ with the floor. The coefficient of static friction between the floor and the ladder is $\mu_s = 0.45$. How far along the ladder can a

continued on next page

Example 8.7 *continued*

60.0-kg person climb before the ladder starts to slip? Assume that the wall is frictionless.

Strategy Normal forces act on the ladder due to the wall ($\vec{N}_w$) and the floor ($\vec{N}_f$). A frictional force acts on the base of the ladder due to the floor ($\vec{f}$), but no frictional force acts on the top of the ladder since the wall is frictionless. Gravitational forces act on the ladder and on the person climbing it. Consider the ladder and the climber as a single system. Until the ladder starts to slip, this system is in equilibrium. Therefore, the net external force and the net external torque acting on the system are both equal to zero. As the person ascends the ladder, the frictional force $\vec{f}$ has to increase to keep the ladder in equilibrium. The ladder begins to slip when the frictional force required to maintain equilibrium is larger than its maximum possible value $\mu_s N_f$. The ladder is about to slip when $f = \mu_s N_f$.

Solution The first step is to make a careful drawing of the ladder and label all distances and forces (Fig. 8.21b). Instead of cluttering the diagram with numerical values, we use L for the length of the ladder, d for the unknown distance from the bottom of the ladder to the point where the person stands, and M and m for the masses of the person and ladder, respectively. The weight of the ladder acts at the ladder's center of gravity, which is the ladder's midpoint since it is uniform.

The conditions for equilibrium are

$$\Sigma F_x = 0, \Sigma F_y = 0, \text{ and } \Sigma \tau = 0$$

Starting with $\Sigma F_x = 0$, we find

$$N_w - f = 0$$

where, if the climber is at the highest point possible, the frictional force must have its maximum possible magnitude:

$$f = \mu_s N_f$$

Combining these two equations, we obtain a relationship between the magnitudes of the two normal forces:

$$N_w = \mu_s N_f$$

Next we use the condition $\Sigma F_y = 0$, which gives

$$N_f - Mg - mg = 0$$

The only unknown quantity in this equation is N_f, so we can solve for it:

$$Mg = 60.0 \text{ kg} \times 9.8 \text{ m/s}^2 = 588 \text{ N}$$

$$mg = 15.0 \text{ kg} \times 9.8 \text{ m/s}^2 = 147 \text{ N}$$

$$N_f = Mg + mg = 588 \text{ N} + 147 \text{ N} = 735 \text{ N}$$

Now we can find the other normal force, N_w:

$$N_w = \mu_s N_f = 0.45 \times 735 \text{ N} = 331 \text{ N}$$

At this point, we know the magnitudes of all the forces. We do not know the distance d, which is the goal of the problem. To find d we must set the net torque equal to zero.

First we choose a rotation axis. The most convenient choice is an axis perpendicular to the plane of Fig. 8.21 and passing through the bottom of the ladder. Since two of the five forces ($\vec{N}_f$ and $\vec{f}$) act at the bottom of the ladder, these two forces have zero lever arms and thus produce zero torque. Another reason why this is a convenient choice of axis is that the distance d is measured from the bottom of the ladder.

continued on next page

Figure 8.21   (a) A ladder and (b) forces acting on the ladder.

Example 8.7 *continued*

To find torques we must either find the perpendicular component of each force or find the lever arm for each force. In this situation, with the forces either vertical or horizontal, it is probably easiest to find the lever arms. In three diagrams (Fig. 8.22), we first draw the line of action for each force; then the lever arm is the perpendicular distance between the axis and the line of action.

Using the usual convention that counterclockwise torques are positive, the torque due to $\vec{N}_w$ is negative while the torques due to gravity are positive. The magnitude of each torque is the magnitude of the force times its lever arm:

$$\tau = Fr_\perp$$

Setting the net torque equal to zero yields:

$$-N_w L \sin\theta + mg\left(\tfrac{1}{2}L\cos\theta\right) + Mgd\cos\theta = 0$$

Substituting known values,

$$-331\ \text{N} \times 8.00\ \text{m} \times \sin 60.0° + 147\ \text{N} \times 4.00\ \text{m} \times \cos 60.0° +$$
$$588\ \text{N} \times d\cos 60.0° = 0$$

$$-2293\ \text{N·m} + 294\ \text{N·m} + 294\ \text{N} \times d = 0$$

$$d = \frac{2293\ \text{N·m} - 294\ \text{N·m}}{294\ \text{N}} = 6.8\ \text{m}$$

The person can climb 6.8 m up the ladder without having it slip. This is the distance *along the ladder*, not the height above the ground, which is

$$h = 6.8\ \text{m} \times \sin 60.0° = 5.9\ \text{m}$$

Discussion   If the person goes any higher, then his weight produces a larger CCW torque about our chosen rotation axis. To stay in equilibrium, the total CW torque would have to get larger. The only force providing a CW torque is the normal force due to the wall, which pushes to the right. However, if this force were to get larger, the frictional force would have to get larger to keep the net horizontal force equal to zero. Since friction already has its maximum magnitude, there is no way for the ladder to be in equilibrium if the person climbs any higher.

Practice Problem 8.7   Another ladder leaning on a wall

A uniform ladder of mass 10.0 kg and length 3.2 m leans against a frictionless wall with its base located 1.5 m from the wall. If the ladder is not to slip, what must be the minimum coefficient of static friction between the bottom of the ladder and the ground? Assume the wall is frictionless.

Figure 8.22
Finding the lever arm for each force.

Physics at Home

Take a dumbbell and wrap some string around the center of its axle. (An alternative: slide two spools of thread onto a pencil near its center with a small gap between the spools. Wrap some thread around the pencil between the two spools.) Place the dumbbell on a table (or on the floor). Unwind a short length of string and try pulling perpendicular to the axle at different angles to the horizontal (see Fig. 8.23). Depending on the direction of your pull, the dumbbell can roll in either direction. Try to find the angle at which the rolling changes direction; at this angle the dumbbell does not roll at all. (If using the pencil and spools of thread, pull gently and try to find the angle at which the whole thing *slides* along the table without any rotation.)

What is special about this angle? Since the dumbbell is in equilibrium when pulling at this angle, we can analyze the torques using any rotation axis we choose. A convenient choice is the axis that passes through point P, the point of contact with the table. Then the contact force between the table and the dumbbell acts at the rotation axis and its torque is zero. The torque due to gravity is also zero since the line of action passes through point P. The dumbbell can only be in equilibrium if the torque due to the remaining force (the tension in the string) is zero. This torque is zero if the lever arm is zero, which means the line of action passes through point P.

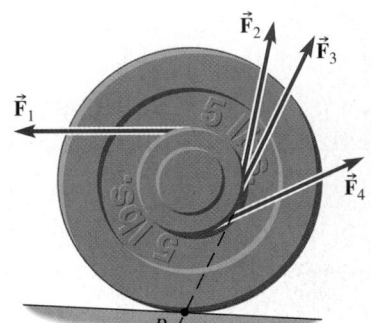

Figure 8.23   Forces $\vec{F}_1$ and $\vec{F}_2$ make the dumbbell roll to the left; $\vec{F}_4$ makes it roll to the right; $\vec{F}_3$ does not make it roll.

Example 8.8

The Sign and the Breaking Cord

A uniform beam of weight 196 N and of length 1.00 m is attached to a hinge on the outside wall of a restaurant. A cord is attached at the center of the beam and is attached to the wall, making an angle of 30.0° with the beam (Fig. 8.24a). The cord keeps the beam perpendicular to the wall. If the breaking tension of the cord is 620 N, how large can the mass of the sign be without breaking the cord?

Strategy The beam is in equilibrium; both the net force and the net torque acting on it must be zero. To find the maximum weight of the sign, we let the tension in the cord have its maximum value of 620 N. We do not know the force exerted by the hinge on the beam, so we choose an axis of rotation through the hinge. Then the force exerted by the hinge on the beam has a zero lever arm and does not enter the torque equation.

Before doing anything else, we draw a diagram showing each force acting on the beam and the chosen rotation axis. The free-body diagrams in previous chapters often placed all the force vectors starting from a single point. Now we draw each force vector starting at its point of application so that we can find the torque—either by finding the lever arm or by finding the perpendicular force component and the distance from the axis to the point of application.

Solution Figure 8.24b shows the forces acting on the beam; three of these contribute to the torque. The gravitational force on the beam can be taken to act at the midpoint of the beam

since it is uniform. The force due to the cord has a perpendicular component (Fig. 8.24c) of

$$F_\perp = 620 \text{ N} \sin 30.0° = 310 \text{ N}$$

The two gravitational forces tend to rotate the beam clockwise, while the tension in the cord tends to rotate it counterclockwise. As an alternative to setting the net torque equal to zero, we can set the total magnitude of the CW torques equal to the total magnitude of the CCW torques:

$$0.50 \text{ m} \times 196 \text{ N} + 1.00 \text{ m} \times Mg = 0.50 \text{ m} \times 310 \text{ N}$$

or

$$1.00 \text{ m} \times Mg = 0.50 \text{ m} \times (310 \text{ N} - 196 \text{ N})$$

Now we solve for the unknown mass M:

$$M = \frac{0.50 \text{ m} \times (310 \text{ N} - 196 \text{ N})}{1.00 \text{ m} \times 9.8 \text{ N/kg}} = 5.8 \text{ kg}$$

Discussion In this problem we did not have to set the net force equal to zero. By placing the axis of rotation at the hinge we eliminated two of the three unknowns from the torque equation: the horizontal and vertical components of the hinge force (or, equivalently, its magnitude and direction). If we wanted to find the hinge force as well, setting the net force equal to zero would be necessary.

Practice Problem 8.8   Hinge forces

Find the vertical component of the force exerted by the hinge in two different ways: (a) setting the net force equal to zero and (b) using a torque equation about a different axis.

(a)

620 N

30.0°

Axis

0.50 m

1.00 m

196 N $M\vec{g}$

(b)

620 N $F_\perp$

30.0°

$F_\parallel$

(c)

Figure 8.24
(a) A sign outside a restaurant.
(b) Forces acting on the beam.
(c) Finding the components of the tension in the cord.

Distributed Forces

Gravity is not the only force that is distributed rather than acting at a point. Contact forces, including both the normal component and friction, are spread over the contact surface. Just as for gravity, we can consider the contact force to act at a single point, but the location of that point is often not at all obvious. For a book sitting on a horizontal table, it seems reasonable that the normal force effectively acts at the geometric center of the book cover that touches the table. It is less clear where that effective point is if the book is on an incline or is sliding. As Example 8.9 shows, when something is about to topple over, contact is about to be lost everywhere except at the corner around which the toppling object is about to rotate. That corner then must be the location of the contact forces.

Example 8.9

The Toppling File Cabinet

A file cabinet of height a and width b is on a ramp at angle θ (Fig. 8.25a). The file cabinet's center of gravity is at its geometric center. Find the largest θ for which the file cabinet does not tip over. Assume the coefficient of static friction is large enough to prevent sliding.

Strategy Until the file cabinet begins to tip over, it is in equilibrium; the net force acting on it must be zero and the total torque about any axis must also be zero. We first draw a force diagram showing the three forces (gravity, normal, friction) acting on the file cabinet. The point of application of the two contact forces (normal, friction) must be at the lower edge of the file cabinet if it is on the steepest possible incline, just about to tip over. In that case, contact has been lost over the rest of the bottom surface of the file cabinet so that only the lower edge makes good contact with the ramp.

As in all equilibrium problems, a good choice of rotation axis makes the problem easier to solve. We know that, at the maximum angle, the contact forces act at the bottom edge of the file cabinet. A good choice of rotation axis is along the bottom edge of the file cabinet, because then the normal and frictional forces have zero lever arm.

Solution Figure 8.25b shows the forces acting on the file cabinet at the maximum angle θ. The gravitational force is drawn at the center of gravity. Instead of drawing a single vector arrow for the gravitational force, we represent the gravitational force by its components parallel and perpendicular to the ramp. Then we find the lever arm for each of the components. The lever arm for the parallel component of the weight ($mg \sin \theta$) is $\frac{1}{2}a$ and the lever arm for the perpendicular component ($mg \cos \theta$) is $\frac{1}{2}b$. Setting the net torque equal to zero:

$$\tau_{\text{net}} = -mg \cos \theta \times \tfrac{1}{2}b + mg \sin \theta \times \tfrac{1}{2}a = 0$$

After dividing out the common factors of $\frac{1}{2}mg$,

$$\cos \theta \times b = \sin \theta \times a$$

Solving for θ,

$$\theta = \tan^{-1} \frac{b}{a}$$

Discussion As a check, we can regard the normal and friction forces as two components of a single contact force. We can think of that contact force as acting at a single point—a "center of contact" analogous to the center of gravity. As the file cabinet is put on steeper and steeper surfaces, the effective point of application of the contact force moves toward the lower edge of the file cabinet (see Fig. 8.26). If we take the rotation axis through the center of gravity so there is no gravitational torque, then the torque due to the contact force must be zero. The only way that can happen is if its lever arm is zero, which means that the contact force must point directly toward the center of gravity. If the angle θ has its maximum

(a)   (b)

Figure 8.25
(a) File cabinet on an incline. (b) Forces acting on the file cabinet.

Point of application of contact force

Figure 8.26
Contact force for various incline angles.

continued on next page

Example 8.9 continued

value, the contact force acts at the lower edge and $\tan\theta = b/a$. The file cabinet is about to tip when its center of gravity is directly above the lower edge. Any object supported only by contact forces can be in equilibrium only if the point of application of the total contact force is directly below the object's center of gravity.

Conceptual Practice Problem 8.9   Gymnast holding a pike position

Figure 8.27 shows a gymnast holding a pike position. What can you say about the location of the gymnast's center of gravity?

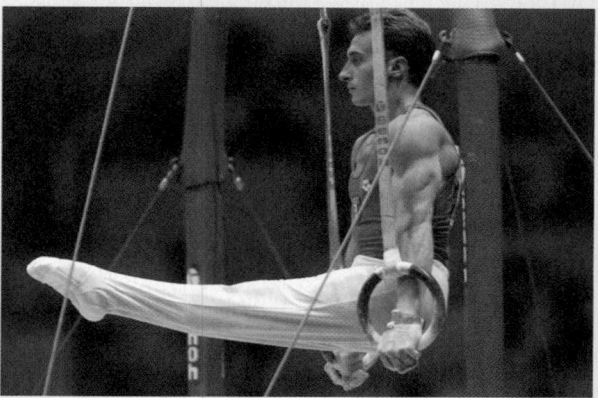

Figure 8.27
Yuri Chechi of Italy holds the pike position on the rings at the World Gymnastic Championships in Sabae, Japan.

(a)   (b)

Figure 8.28 Location of the center of gravity when (a) standing and (b) touching toes.

Physics at Home

When a person stands up straight, the body's center of gravity lies directly above a point between the feet, about 3 cm in front of the ankle joint (see Fig. 8.28a). When a person bends over to touch her toes, the center of gravity lies outside the body (Fig. 8.28b). Note that the lower half of the body must move backward to keep the center of gravity from moving out in front of the toes, which would cause the person to fall over.

An interesting experiment can be done that illustrates what happens to your balance when you shift your center of gravity. Stand against a wall with the heels of your feet touching the wall and your back pressed against the wall. Then carefully try to bend over as if to touch your toes, without bending your knees. Can you do this without falling over? Explain.

Problem-Solving Steps in Equilibrium Problems

- Identify an object or system in equilibrium. Draw a diagram showing all the forces acting on that object, each drawn at its point of application. Use the center of gravity as the point of application of any gravitational forces.

- To apply the force condition $\Sigma\vec{F} = 0$, choose a convenient coordinate system and resolve each force into its x- and y-components.

- To apply the torque condition $\Sigma\tau = 0$, choose a convenient rotation axis—generally one that passes through the point of application of an unknown force. Then find the torque due to each force. Use whichever method is easier: either the lever arm times the magnitude of the force or the distance times the perpendicular component of the force. Determine the direction of each torque; then either set the sum of all the torques (with their correct signs) equal to zero or set the magnitude of the CW torques equal to the magnitude of the CCW torques.

- Not all problems require all three equations (two force component equations and one torque equation). Sometimes it is easier to use more than one torque equation, with a different axis. Before diving in and writing down all the equations, think about which approach is the easiest and most direct.

Making The Connection:
flexor versus extensor muscles

8.5   EQUILIBRIUM IN THE HUMAN BODY

We can use the concepts of torque and equilibrium to understand some of how the musculoskeletal system of the human body works. A muscle has tendons at each end that connect it to two different bones across a joint (the flexible connection between the bones). When the muscle contracts, it pulls the tendons, which in turn pull on the bones. Thus, the muscle produces a pair of forces of equal magnitude, one acting on each of the two bones. The biceps muscle (Fig. 8.29) in the upper arm attaches the scapula to the forearm (radius) across the inside of the elbow joint. When the biceps contracts, the forearm is pulled toward the upper arm. The biceps is a *flexor* muscle; it moves one bone closer to another.

A muscle can pull but not push, so a flexor muscle such as the biceps cannot reverse its action to push the forearm away from the upper arm. The *extensor* muscles make bones move apart from each other. In the upper arm an extensor muscle, the triceps (Fig. 8.29), connects the humerus to the ulna (another bone in the forearm parallel to the radius) across the outside of the elbow. Since the biceps and triceps connect to the forearm on opposite sides of the elbow joint, they tend to cause rotation about the joint in opposite directions. When the triceps contracts it pulls the forearm away from the upper arm. Using flexor and extensor muscles on opposite sides of the joint, the body can produce both positive and negative torques, even though both muscles pull in the same direction.

Suppose the arm is held in a horizontal position. The deltoid muscle exerts a force $\vec{\mathbf{F}}$ on the humerus at an angle of about 15° above the horizontal. This force has to do two things. The vertical component (magnitude $F \sin 15° \approx 0.26F$) supports the weight of the arm, while the horizontal component (magnitude $F \cos 15° \approx 0.97F$) stabilizes the joint by pulling the humerus in against the shoulder (scapula). In Example 8.10 we estimate the magnitude of $\vec{\mathbf{F}}$.

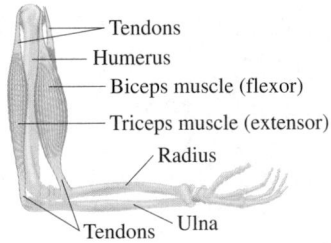

Figure 8.29 Muscles in the upper arm.

Example 8.10

Force to Hold Arm Horizontal

   A person is standing with his arm outstretched in a horizontal position. The weight of the arm is 30.0 N and its center of gravity is at the elbow joint, 27.5 cm from the shoulder joint (Fig. 8.30). The deltoid pulls on the upper arm at an angle of 15° above the horizontal and at a distance of 12 cm from the joint. What is the magnitude of the force exerted by the deltoid muscle on the arm?

Strategy   The arm is in equilibrium, so we can apply the conditions for equilibrium: $\vec{\mathbf{F}}_{net} = 0$ and $\tau_{net} = 0$. When calculating torques, we choose the rotation axis at the shoulder joint because then the unknown force $\vec{\mathbf{F}}_s$, which acts on the arm at the joint, has a zero lever arm and produces zero torque. With only one unknown in the torque equation, we can solve immediately for F_m. We do not need to apply the condition $\vec{\mathbf{F}}_{net} = 0$ unless we want to find $\vec{\mathbf{F}}_s$.

Solution   The gravitational force is perpendicular to the line between its point of application and the rotation axis. Gravity produces a clockwise torque of magnitude

$$|\tau| = Fr = 30.0\ \text{N} \times 0.275\ \text{m} = 8.25\ \text{N·m}$$

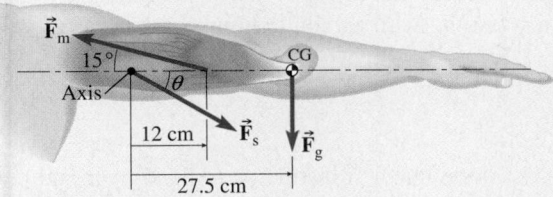

Figure 8.30
Forces exerted on an outstretched arm by the deltoid muscle ($\vec{\mathbf{F}}_m$), the scapula ($\vec{\mathbf{F}}_s$), and gravity ($\vec{\mathbf{F}}_g$)

For the torque due to $\vec{\mathbf{F}}_m$, we find the component of $\vec{\mathbf{F}}_m$ that is perpendicular to the line between its point of application and the rotation axis. Since this line is horizontal, we need the vertical component of $\vec{\mathbf{F}}_m$, which is $F_m \sin 15°$. Then the magnitude of the counterclockwise torque due to $\vec{\mathbf{F}}_m$ is

$$|\tau| = F_\perp r = F_m \sin 15° \times 0.12\ \text{m}$$

These torques must be equal in magnitude:

magnitude of CCW torque = magnitude of CW torque

$$F_m \sin 15° \times 0.12\ \text{m} = 8.25\ \text{N·m}$$

Solving for F_m,

$$F_m = \frac{8.25\ \text{N·m}}{\sin 15° \times 0.12\ \text{m}} = 270\ \text{N}$$

Discussion   The force exerted by the muscle is much larger than the 30.0-N weight of the arm. The muscle must exert a larger force because the lever arm is small; the point of application is less than half as far from the joint as the center of gravity (0.12 m/0.275 m ≈ 4/9). Also, the muscle cannot pull straight up on the arm; the vertical component of the muscle force is only about $\frac{1}{4}$ of the magnitude of the force. These two factors together make the weight supported (30.0 N) only $\frac{4}{9} \times \frac{1}{4} = \frac{1}{9}$ as large as the force exerted by the muscle.

Practice Problem 8.10   Holding a juice carton

Find the force exerted by the same person's deltoid muscle when holding a 1-L juice carton (weight 9.9 N) with the arm outstretched and parallel to the floor (as in Fig. 8.30). Assume that the juice carton is 60.0 cm from the shoulder.

Making The Connection:
muscle forces for
the iron cross
(gymnastics)

The Iron Cross

When a gymnast does the iron cross (Fig. 8.31a), the primary muscles involved are the latissimus dorsi ("lats") and pectoralis major ("pecs"). Since the rings are supporting the gymnast's weight, they exert an upward force on the gymnast's arms. Thus, the task for the muscles is not to hold the arm up, but to pull it down. The lats pull on the humerus about 3.5 cm from the shoulder joint (Fig. 8.31b). The pecs pull on the humerus about 5.5 cm from the joint (Fig. 8.31c). The other ends of these two muscles connect to bone in many places, widely distributed over the back (lats) and chest (pecs). As a reasonable simplification we can assume that these muscles pull at a 45° angle below the horizontal in the iron cross maneuver. We also assume that the two muscles exert equal forces, so we can replace the two with a single force acting at 4.5 cm from the joint.

To determine the force exerted, we look at the entire arm as a system in equilibrium. This time we can ignore the weight of the arm itself since the force exerted on the arm by the ring is much larger—half the gymnast's weight is supported by each ring. The ring exerts an upward force that acts on the hand about 60 cm from the shoulder joint (see Fig. 8.31d). Taking torques about the shoulder, in equilibrium we have

$$| \text{ CW torque } | = | \text{ CCW torque } |$$

$$F_m \times 0.045 \text{ m} \times \sin 45° = \tfrac{1}{2}W \times 0.60 \text{ m}$$

$$F_m = \frac{\tfrac{1}{2}W \times 0.60 \text{ m}}{0.045 \text{ m} \times \sin 45°}$$

$$= 9.4W$$

Thus, the force exerted by the lats and pecs *on one side* of the gymnast's body is more than nine times his weight.

The design of the human body makes large muscular forces necessary. Are there advantages to the design? Due to the small lever arms, the muscle forces are much larger than they would otherwise be, but the human body has traded this for a wide range of movement of the bones. The biceps and triceps muscles can move the lower arms through almost 180° while they change their lengths by only a few centimeters. The muscles also remain nearly parallel to the bones. If the biceps and triceps muscles were attached to the lower arm much farther from the elbow, there would have to be a large flap of skin to allow them to move so far away from the bones. The arrangement of our bones and muscles favors a wide range of movement.

Another advantage of the design is that it tends to minimize the rotational inertia of our limbs. For example, the muscles that control the motion of the lower arm are contained mostly within the *upper* arm. This keeps the rotational inertia of the lower arms about the elbow smaller. It also keeps the rotational inertia of the entire arm about the shoulder smaller. Smaller rotational inertia means that the energy we have to expend to move our limbs around is smaller.

The biceps muscle with its tendons is almost parallel to the humerus. One interesting observation is that the tendon connects to the radius at different points in different people. In one person this point may be 5.0 cm from the elbow joint, while in another person whose arm is the same length it may be 5.5 cm from the elbow. Thus, some people are naturally stronger than others because of their internal structure. Chimpanzees have an advantage over humans because their biceps muscle has a longer lever arm. Do not make the mistake of arm wrestling with an adult chimp; challenge the chimp to a game of chess instead.

Making The Connection:
forces on the
human spine during
heavy lifting

Heavy Lifting

When lifting an object from the floor, our first instinct is to bend over and pick it up. This is not a good way to lift something heavy. The spine is an ineffective lever and is susceptible to damage when a heavy object is lifted with bent waist. It is much better to squat down and use the powerful leg muscles to do the lifting instead of using our back muscles. Analyzing torques in a simplified model of the back can illustrate why.

Figure 8.31   (a) Gymnast doing the iron cross. The principal muscles involved are (b) the "lats" and (c) the "pecs." (d) Simplified model of the forces acting on the arm of the gymnast.

Figure 8.32 A simplified model of the human back when bent over.

The spine can be modeled as a rod with an axis at the tailbone (the sacrum). The sacrum exerts a force, marked $\vec{\mathbf{F}}_s$ in Fig. 8.32, when a person bends at the waist with the back horizontal. The forces due to the complicated set of back muscles can be replaced with a single equivalent force $\vec{\mathbf{F}}_b$ as shown. This equivalent force makes an angle of 12° with the spine and acts about 44 cm from the sacrum. The weight of the upper body, $m\vec{\mathbf{g}}$ in Fig. 8.32, is about 65% of total body weight; its center of gravity is about 38 cm from the sacrum. By placing an axis at the sacrum we can ignore the force $\vec{\mathbf{F}}_s$ in our torque equation. Since the vertical component of $\vec{\mathbf{F}}_b$ is $F_b \sin 12° \approx 0.21 F_b$, only about $\frac{1}{5}$ the magnitude of the forces exerted by the back muscles is supporting the body weight. The rest, the horizontal component, is pressing the rod representing the spine into the sacrum.

If we put some numbers into this example we can get an idea of the forces required for just supporting the upper body in this position. If the person's total weight is 710 N (160 lb), then the upper body weight is

$$mg = 0.65 \times 710 \text{ N}$$

Now we set the magnitude of the CCW torques about the axis equal to the magnitude of the CW torques:

$$F_b \times 0.44 \text{ m} \times \sin 12° = mg \times 0.38 \text{ m}$$

Substituting and solving,

$$F_b = \frac{0.65 \times 710 \text{ N} \times 0.38 \text{ m}}{0.44 \text{ m} \times \sin 12°} = 1920 \text{ N}$$

The muscular force that compresses the spine is the horizontal component of $\vec{\mathbf{F}}_b$:

$$F_b \cos 12° = 1900 \text{ N}$$

or about 430 lb. This is over four times the weight of the upper body.

Now if the person tries to lift something with his arms in this position, the lever arm for the weight of the load is even longer than for the weight of the upper body. The back muscles must supply a much larger force. The spine is now compressed with a dangerously large force. A cushioning disk called the lumbosacral disk, at the bottom of the spine, separates the last vertebra from the sacrum. This disk can be ruptured or deformed, causing great pain when the back is misused in such a fashion.

If, instead of bending over, we bend our knees and lower our body, keeping it vertically aligned as much as possible while lifting a load, the centers of gravity of the body and load are positioned more closely in a line above the sacrum, as in Fig. 8.33. Then the lever arms of these forces with respect to an axis through the sacrum are relatively small and the force on the lumbosacral disk is roughly equal to the upper body weight plus the weight being lifted.

8.6   ROTATIONAL FORM OF NEWTON'S SECOND LAW

The concepts of torque and rotational inertia can be used to formulate a "Newton's second law for rotation"—a law that fills the role of $\Sigma\vec{F} = m\vec{a}$ for rotation about a fixed axis. What is that role? Newton's second law determines the translational acceleration of an object if its mass and the net force acting on it are known. For rotation, we want to determine the angular acceleration, so α takes the place of $\vec{a}$. Net torque takes the place of net force. If the net torque is zero, then there is zero angular acceleration; the greater the net torque, the greater the angular acceleration. In place of mass, rotational inertia measures how difficult it is to change the angular velocity. The law that enables us to calculate the angular acceleration of a rigid object given the applied torques takes a form analogous to $\Sigma\vec{F} = m\vec{a}$:

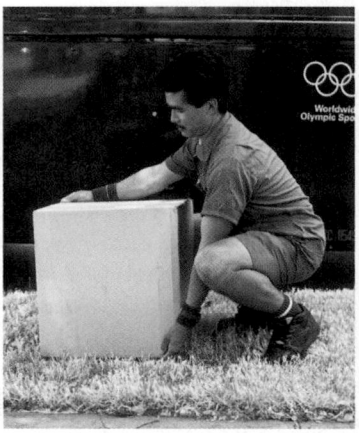

Figure 8.33   A safer way to lift a heavy object.

Rotational form of Newton's second law
$\Sigma\tau = I\alpha$ (8-9)

Thus, the angular acceleration of a rigid body is proportional to the net torque (more torque causes a larger α) and is inversely proportional to the rotational inertia (more inertia causes a smaller α). In equilibrium, the angular acceleration must be zero; Eq. (8-9) then requires that the net torque be zero. We used $\Sigma\tau = 0$ as one of the conditions of equilibrium in Sections 8.4 and 8.5.

Equation (8-9) is proved in Problem 48. It is subject to an important restriction. Just as $\Sigma\vec{F} = m\vec{a}$ is valid only if the mass of the object is constant, $\Sigma\tau = I\alpha$ is valid only if the rotational inertia of the object is constant. For a rigid object rotating about a fixed axis, I cannot change, so Eq. (8-9) is always applicable.

Tips   When calculating the net torque, remember to assign the correct algebraic sign to each torque before adding them.

Tips   The sum of the torques due to internal forces acting on a rigid object is always zero. Therefore, only *external* torques need be included in Eq. (8-9)

Example 8.11

The Grinding Wheel

A grinding wheel is a solid, uniform disk of mass 2.50 kg and radius 9.00 cm. Starting from rest, what constant torque must a motor supply so that the wheel attains a rotational speed of 126 rev/s in a time of 6.00 s?

Strategy   Since the grinding wheel is a uniform disk, we can find its rotational inertia using Table 8.1. After converting the revolutions per second to radians per second, we can find the angular acceleration from the change in angular velocity over the given time interval. Once we have I and α, we can find the net torque from Newton's second law for rotation.

Solution   The grinding wheel is a uniform disk, so its rotational inertia is

$$I = \tfrac{1}{2}mr^2$$

$$\tfrac{1}{2} \times 2.50 \text{ kg} \times (0.0900 \text{ m})^2 = 0.010125 \text{ kg·m}^2$$

A single rotation of the wheel is equivalent to 2π radians, so

$$\omega = 126\frac{\text{rev}}{\text{s}} \times 2\pi\frac{\text{rad}}{\text{rev}}$$

The angular acceleration is

$$\alpha = \frac{\Delta\omega}{\Delta t}$$

continued on next page

Example 8.11 continued

Then the torque required is

$$\tau_{net} = I\alpha = I\frac{\Delta\omega}{\Delta t} = (0.010125 \text{ kg·m}^2) \times \frac{(126 \text{ rev/s} \times 2\pi \text{ rad/rev})}{6.00 \text{ s}}$$

$$= 1.34 \text{ N·m}$$

If there are no other torques on the wheel, the motor must supply a constant torque of 1.34 N·m.

Discussion   We assumed that no other torques are exerted on the wheel. There is certain to be at least a small frictional

torque on the wheel with a sign opposite to the sign of the motor's torque. Then the motor would have to supply a torque larger than 1.34 N·m. The *net* torque would still be 1.34 N·m.

Practice Problem 8.11   Another approach

Verify the answer to Example 8.11 by: (a) finding the angular displacement of the wheel using kinematic equations for constant α; (b) finding the change in rotational kinetic energy of the wheel; and (c) finding the torque from $W = \tau\theta$.

Newton's second law for rotation explains why a tightrope walker carries a long pole to help maintain balance. If the acrobat is about to topple over sideways, the pole would have to go with him, rotating up and over the rope in a large arc. The pole has a large rotational inertia due to its length, so a large torque is required to make the pole start to rotate. The angular acceleration of the system (acrobat plus pole) due to a small gravitational torque is much smaller than it would be without the pole. The pole greatly increases the stability of the acrobat.

8.7   THE DYNAMICS OF ROLLING OBJECTS

A rolling object combines translational motion of the center of mass with rotation about an axis that passes through the center of mass (Section 5.1). For an object that is rolling without slipping, $v_{CM} = \omega R$. As a result, there is a specific relationship between the rolling object's translational and rotational kinetic energies. The total kinetic energy of a rolling object is the sum of its translational and rotational kinetic energies.

A wheel with mass M and radius R has a rotational inertia that is some pure number times MR^2; it couldn't be anything else and still have the right units. We can write the rotational inertia about an axis through the center of mass as $I_{CM} = \beta MR^2$ where β is a pure number that measures how far from the axis of rotation the mass is distributed. Larger β means the mass is, on average, farther from the axis. From Table 8.1, a hoop has $\beta = 1$; a disk, $\beta = \frac{1}{2}$; and a solid sphere, $\beta = \frac{2}{5}$.

Using $I_{CM} = \beta MR^2$ and $v_{CM} = \omega R$, the rotational kinetic energy for a rolling object can be written

$$K_{rot} = \frac{1}{2}I_{CM}\omega^2 = \frac{1}{2} \times \beta MR^2 \times \left(\frac{v_{CM}}{R}\right)^2 = \beta \times \frac{1}{2}Mv_{CM}^2$$

Since $\frac{1}{2}Mv_{CM}^2$ is the translational kinetic energy,

$$K_{rot} = \beta K_{tr} \tag{8-10}$$

This is convenient since β depends only on the shape, not on the mass or radius of the object. For a given shape rolling without slipping, the ratio of its rotational to translational kinetic energy is always the same (β).

The total kinetic energy can be written

$$K = K_{tr} + K_{rot}$$

$$K = \frac{1}{2}Mv_{CM}^2 + \frac{1}{2}I_{CM}\omega^2 \tag{8-11}$$

or in terms of β,

$$K = (1 + \beta) K_{tr}$$

$$K = (1 + \beta) \frac{1}{2}Mv_{CM}^2 \tag{8-12}$$

Thus, two objects of the same mass rolling at the same translational speed do *not* have the same kinetic energy. The object with the larger value of β has more rotational kinetic energy.

Conceptual Example 8.12

Hollow and Solid Rolling Balls

Starting from rest, two balls are rolled down a hill as in Fig. 8.34. One is solid, the other hollow. Which one is moving faster when it reaches the bottom of the hill?

Strategy and Solution   Energy conservation is the best way to approach this problem. As a ball rolls down the hill, its gravitational potential energy decreases as its kinetic energy increases by the same amount. The total kinetic energy is the sum of the translational and rotational contributions.

We do not know the mass or the radius of either ball and we cannot assume they are the same. Since both kinetic and potential energies are proportional to mass, mass does not affect the final speed. Also, the total kinetic energy does not depend on the radius of the ball [see Eq. (8-12)]. The final speeds of the two balls differ because a different *fraction* of their total kinetic energies is translational.

One ball is a solid sphere and the other is approximately a spherical shell. The mass of a spherical shell is all concentrated on the surface of a sphere, while a solid sphere has its mass distributed throughout the sphere's volume. Therefore, the shell has a larger β than the solid sphere. When the shell rolls, it converts a bigger fraction of the lost potential energy into rotational kinetic energy; therefore, a smaller fraction becomes translational kinetic energy. The final speed of the solid sphere is larger since it puts a larger fraction of its kinetic energy into translational motion.

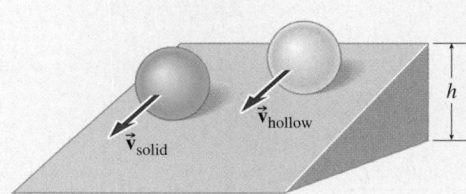

Figure 8.34
Rolling balls

Discussion   We can make this conceptual question into a quantitative one: what is the ratio of the speeds of the two balls at the bottom of the hill?

Let the height of the hill be h. Then for a ball of mass M, the loss of gravitational potential energy is Mgh. This amount of gravitational potential energy is converted into translational and rotational kinetic energy:

$$Mgh = K_{tr} + K_{rot} = (1 + \beta)K_{tr} = (1 + \beta)\frac{Mv^2}{2}$$

Mass cancels out, as expected. We can solve for the final speed in terms of g, h, and β. The final speed is independent of the ball's mass and radius.

$$v_{CM} = \sqrt{\frac{2gh}{1 + \beta}}$$

The ratio of the final speeds for two balls rolling down the same hill is therefore

$$\frac{v_1}{v_2} = \sqrt{\frac{1 + \beta_2}{1 + \beta_1}}$$

To evaluate the ratio, we look up the rotational inertias in Table 8.1. The solid sphere has $\beta = \frac{2}{5}$ and the spherical shell has $\beta = \frac{2}{3}$. Then

$$\frac{v_{solid}}{v_{hollow}} = \sqrt{\frac{1 + \frac{2}{3}}{1 + \frac{2}{5}}} \approx 1.091$$

The solid ball's final speed is therefore 9.1% faster than that of the hollow ball. This ratio depends neither on the masses of the balls, the radii of the balls, the height of the hill, nor the slope of the hill.

Practice Problem 8.12   Fraction of kinetic energy that is rotational energy

What fraction of a rolling ball's kinetic energy is rotational kinetic energy? Answer both for a solid ball and a hollow one.

What is the acceleration of a ball rolling down an incline? Figure 8.35 includes an FBD for such a ball. Static friction is the force that makes the ball rotate; if there were no friction, instead of rolling the ball would just *slide* down the incline. This is true because friction is the only force acting that yields a nonzero torque about the rotation axis, which is through the ball's center of mass. Gravity gives zero torque because it acts at the axis, so the lever arm is zero. The normal force points directly at the axis, so its lever arm is also zero.

The frictional force $\vec{\mathbf{f}}$ provides a torque

$$\tau = rf$$

where r is the ball's radius. An analysis of the forces and torques combined with Newton's second law in both forms enables us to calculate the acceleration of the ball in Example 8.13.

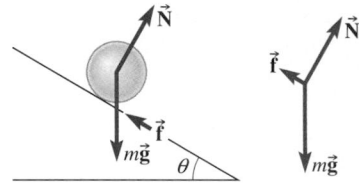

Figure 8.35   Forces acting on a ball rolling downhill.

Example 8.13

Acceleration of a Rolling Ball

Calculate the acceleration of a solid ball rolling down a slope inclined at an angle θ to the horizontal (Fig. 8.36a).

Strategy The axis of rotation is through the ball's center of mass. As already discussed, neither gravity nor the normal force produce a torque about this axis; the net torque is $\tau_{net} = rf$, where f is the magnitude of the frictional force. The net torque is related to the angular acceleration by $\tau_{net} = I\alpha$, Newton's second law for rotation. Similarly, the net force acting on the ball gives the acceleration of the center of mass: $\vec{F}_{net} = m\vec{a}_{CM}$. One problem is that the force of friction is unknown. We must resist the temptation to assume that $f = \mu_s N$; there is no reason to assume that static friction has its maximum possible magnitude. We do know that the two accelerations, translational and rotational, are related. We know that v_{CM} and ω are proportional since r is constant. To stay proportional they must change in lock step; their rates of change, a_{CM} and α, are proportional to each other by the same factor of r. Thus, $a_{CM} = \alpha r$. This connection should enable us to eliminate f and solve for the acceleration. Since the speed of a ball after rolling a certain distance was found to be independent of the mass and radius of the ball in Example 8.12, we expect the same to be true of the acceleration.

Solution Since the net torque is

$$\tau_{net} = rf$$

the angular acceleration is

$$\alpha = \frac{\tau}{I} = \frac{rf}{I} \quad (1)$$

where I is the ball's rotational inertia about its center of mass.

Figure 8.36b shows the forces along the incline acting on the ball. The acceleration of the center of mass is found from Newton's second law. The component of the net force acting along the incline (in the direction of the acceleration) is

$$F = mg \sin \theta - f = ma_{CM} \quad (2)$$

Because the ball is rolling without slipping, the acceleration of the center of mass and the angular acceleration are related by

$$a_{CM} = \alpha r$$

Now we try to eliminate the unknown frictional force f from the equations above. Solving Eq. (1) for f gives

$$f = \frac{I\alpha}{r}$$

Substituting this into Eq. (2), we get

$$mg \sin \theta - \frac{I\alpha}{r} = ma_{CM}$$

Now to eliminate α, we can substitute $\alpha = a_{CM}/r$:

$$mg \sin \theta - \frac{Ia_{CM}}{r^2} = ma_{CM}$$

Solving for a_{CM},

$$a_{CM} = \frac{g \sin \theta}{1 + (I/mr^2)}$$

For a solid sphere, $I = \frac{2}{5} mr^2$, so

$$a_{CM} = \frac{g \sin \theta}{1 + \frac{2}{5}} = \frac{5}{7}g \sin \theta$$

Discussion The acceleration of an object *sliding* down an incline without friction is $a = g \sin \theta$. The acceleration of the rolling ball is smaller than $g \sin \theta$ due to the frictional force directed up the incline.

We can check the answer using the result of Example 8.12. Since the ball's acceleration is constant, the kinematic quantities are related by familiar equations. If the ball starts from rest as in Fig. 8.36a, after it has rolled a distance d, its speed v is

$$v = \sqrt{2ad} = \sqrt{2\left(\frac{g \sin \theta}{1 + \beta}\right)d}$$

where $\beta = \frac{2}{5}$. The vertical drop during this time is $h = d \sin \theta$, so

$$v = \sqrt{\frac{2gh}{1 + \beta}}$$

Practice Problem 8.13   Acceleration of hollow cylinder

Calculate the acceleration of a hollow cylinder rolling down a slope inclined at an angle θ to the horizontal.

Force along incline = $mg \sin \theta - f$

(a)   (b)

Figure 8.36
(a) A ball accelerating downhill and (b) forces along the incline with the gravitational force resolved into components perpendicular and parallel to the incline.

8.8   ANGULAR MOMENTUM

Newton's second law for translational motion can be written in two ways:

$$\vec{\mathbf{F}}_{\text{net}} = \lim_{\Delta t \to 0} \frac{\Delta \vec{\mathbf{p}}}{\Delta t} \text{ (general form) } \quad \text{or} \quad \vec{\mathbf{F}}_{\text{net}} = m\vec{\mathbf{a}} \text{ (constant mass)}$$

In Eq. (8-9) we wrote Newton's second law for rotation as $\tau_{\text{net}} = I\alpha$, which applies only when I is constant—that is, for a rigid body rotating about a fixed axis. A more general form of Newton's second law for rotation uses the concept of **angular momentum** (symbol L).

> The net external torque acting on a system is equal to
> the rate of change of the angular momentum of the system.
>
> $$\tau_{\text{net}} = \lim_{\Delta t \to 0} \frac{\Delta L}{\Delta t} \qquad (8\text{-}13)$$

The angular momentum of a rigid body rotating about a fixed axis is the rotational inertia times the angular velocity, which is analogous to the definition of linear momentum (mass times velocity):

> **Angular momentum**
>
> $$L = I\omega \qquad (8\text{-}14)$$
> (rigid body, fixed axis)

Either Eq. (8-13) or Eq. (8-14) can be used to show that the SI units of angular momentum are kg·m²/s.

For a rigid body rotating around a fixed axis, angular momentum doesn't tell us anything new. The rotational inertia is constant for such a body since the distance r_i between every point on the object and the axis stays the same. Then any change in angular momentum must be due to a change in angular velocity ω:

$$\tau_{\text{net}} = \lim_{\Delta t \to 0} \frac{\Delta L}{\Delta t} = \lim_{\Delta t \to 0} \frac{I\Delta\omega}{\Delta t} = I \lim_{\Delta t \to 0} \frac{\Delta\omega}{\Delta t} = I\alpha$$

However, Eq. (8-13) is *not* restricted to rigid objects or to fixed rotation axes. In particular, if the net external torque acting on a system is zero, then the angular momentum of the system cannot change. This is the **law of conservation of angular momentum**:

> **Conservation of angular momentum**
>
> If $\tau_{\text{net}} = 0$,
>
> $$L_i = L_f \qquad (8\text{-}15)$$

Tips Conservation of angular momentum can be applied to any system if the net external torque on the system is zero (or negligibly small).

Here L_i and L_f represent the angular momentum of the system at two different times. Conservation of angular momentum is one of the most basic and fundamental laws of physics, along with the two other conservation laws we have studied so far (energy and linear momentum). For an isolated system, the total energy, total linear momentum, and total angular momentum of the system are each conserved. None of these quantities can change unless some external agent causes the change.

In this section, we restrict our consideration to cases where the axis of rotation is fixed but where the rotational inertia is not necessarily constant. One familiar example of a changing rotational inertia occurs when a figure skater spins (Fig. 8.37). To start the spin, the skater glides along with her arms outstretched and then begins to rotate her body about a vertical axis by pushing against the ice with a skate. The push of the ice against the skate provides the external torque that gives the skater her initial angular momentum. Initially the skater's arms and the leg not in contact with the ice are extended away from her body. The mass of the arms and leg when extended contribute

Making The Connection:
rotational inertia of
a figure skater

(a)

(b)

Figure 8.37 Figure skater Lucinda Ruh at the (a) beginning and (b) end of a spin. Her angular velocity is much higher in (b) than in (a).

more to her rotational inertia than they do when held close to the body. As the skater spins, she pulls her arms and leg close to her body to decrease her rotational inertia. As she does, her angular velocity increases dramatically. The concept of angular momentum helps explain why she spins faster as she pulls in her arms and leg.

As the skater spins, there is a negligibly small external torque due to friction between her skates and the ice. With no significant external torque acting, her angular momentum must be conserved:

$$L_i = L_f$$

Even though her rotational inertia changes, at the beginning and again at the end of the spin we can think of her as a rigid body. Then

$$I_i \omega_i = I_f \omega_f$$

where the subscripts "i" and "f" refer to before and after the skater pulls her arms and leg close to her body. The final rotational inertia is less than the initial rotational inertia because the arms and leg are much closer to the axis of rotation. The skater decreases her inertia by moving her arms and leg in toward her body:

$$I_f < I_i$$

In order for her angular momentum to stay the same,

$$\omega_f = \frac{I_i}{I_f} \omega_i > \omega_i$$

which means that she is spinning faster at the end than she was initially.

Many natural phenomena can be understood in terms of angular momentum. In a hurricane, circulating air is sucked inward by a low pressure region at the center of the storm (the *eye*). As the air moves closer and closer to the axis of rotation, it circulates faster and faster. An even more dramatic example is the formation of a pulsar. Under certain conditions, a star can implode under its own gravity, forming a neutron star (a collection of tightly packed neutrons). If the Sun were to collapse into a neutron star, its radius would be only about 13 km. If a star is rotating before its collapse, then as its rotational inertia decreases dramatically, its angular velocity must increase to keep its angular momentum constant. Such rapidly rotating neutron stars are called pulsars because they emit regular pulses of x-rays, at the same frequency as their rotation, that can be detected when they reach Earth. Some pulsars rotate in only a few thousandths of a second per revolution.

Example 8.14

Mouse on a Wheel

A 0.10-kg mouse is perched at point B on the rim of a 2.00-kg bicycle wheel that rotates freely in a horizontal plane at 1.00 rev/s (Fig. 8.38). The mouse crawls to point A at the center. Assume the mass of the wheel is concentrated at the rim. What is the frequency of rotation in rev/s when the mouse arrives at point A?

Strategy Assuming that frictional torques are negligibly small, there is no external torque acting on the mouse/wheel system. Then the angular momentum of the mouse/wheel system must be conserved; it takes an external torque to change angular momentum. The mouse and wheel exert torques on each other, but these *internal* torques only transfer some angular momentum between the wheel and the mouse without changing the total angular momentum. We can think of the system as initially being a rigid body with rotational inertia I_i. When the mouse reaches the center, we think of the system as a rigid body with a different rotational inertia I_f. The mouse changes the rotational inertia of the mouse/wheel system by moving from the outer rim, where its mass makes the maximum possible contribution to the rotational inertia, to the rotation axis, where its mass makes no contribution to the rotational inertia.

Solution Initially, all of the mass of the system is at a distance R from the rotation axis, where R is the radius of the wheel. Therefore,

$$I_i = (M + m)R^2$$

Figure 8.38
Mouse on a rotating wheel.

where M is the mass of the wheel and m is the mass of the mouse. After the mouse moves to the center of the wheel, its mass contributes nothing to the rotational inertia of the system:

$$I_f = MR^2$$

From conservation of angular momentum,

$$I_i \, \omega_i = I_f \, \omega_f$$

Substituting the rotational inertias and $\omega = 2\pi f$,

$$(M + m)R^2 \times 2\pi f_i = MR^2 \times 2\pi f_f$$

Factors of $2\pi R^2$ cancel from each side, leaving

$$(M + m)f_i = Mf_f$$

Solving for f_f,

$$f_f = \frac{M + m}{M} f_i = \frac{2.10 \text{ kg}}{2.00 \text{ kg}} (1.00 \text{ rev/s}) = 1.05 \text{ rev/s}$$

Discussion Conservation laws are powerful tools. We do not need to know the details of what happens as the mouse crawls along the spoke from the outer edge of the wheel; we need only look at the initial and final conditions.

A common mistake in this sort of problem is to say that the initial rotational kinetic energy is equal to the final rotational kinetic energy. This is not true because the mouse crawling in toward the center must expend energy to do so. In other words, the mouse does work, converting some internal energy into rotational kinetic energy.

Practice Problem 8.14 Change in rotational kinetic energy

What is the percent change in the rotational kinetic energy of the mouse/wheel system?

Angular Momentum in Planetary Orbits

Making The Connection:
Kepler's laws of planetary motion

Conservation of angular momentum applies to planets orbiting the Sun in elliptical orbits. Kepler's second law says that the orbital speed varies in such a way that the planet sweeps out area at a constant rate (Fig. 8.39a). In Problem 83 you can show that Kepler's second law is a direct result of conservation of angular momentum, where the angular momentum of the planet is calculated using an axis of rotation perpendicular to the plane of the orbit and passing through the Sun. When the planet is closer to the Sun, it moves faster; when it is farther away, it moves more slowly. Conservation of angular momentum can be used to relate the orbital speeds and radii at two different points in the orbit. The same applies to satellites and moons orbiting planets.

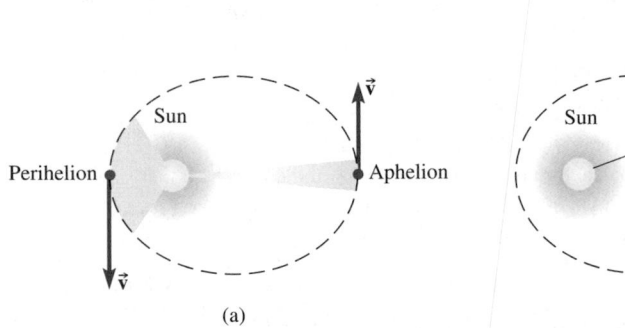

Figure 8.39 The planet's speed varies such that it sweeps out equal areas in equal time intervals. The eccentricity of the planetary orbit is exaggerated for clarity.

Example 8.15

Earth's Orbital Speed

 At perihelion (closest approach to the Sun), Earth is 1.47×10^8 km from the Sun and its orbital speed is 30.3 km/s. What is Earth's orbital speed at aphelion (greatest distance from the Sun), when it is 1.52×10^8 km from the Sun? Note that at these two points Earth's velocity is perpendicular to a radial line from the Sun (see Fig. 8.39a).

Strategy We take the axis of rotation through the Sun. Then the gravitational force on Earth points directly toward the axis; with zero lever arm, the torque is zero. With no other external forces acting on the Earth, the net external torque is zero. Earth's angular momentum about the rotation axis through the Sun must therefore be conserved. To find Earth's rotational inertia, we treat it as a point particle since its radius is much less than its distance from the axis of rotation.

Solution The rotational inertia of the Earth is

$$I = mr^2$$

where m is Earth's mass and r is its distance from the Sun. The angular velocity is

$$\omega = \frac{v_\perp}{r}$$

where $v_\perp$ is the component of the velocity perpendicular to a radial line from the Sun. At the two points under consideration, $v_\perp = v$. As the distance from the Sun r varies, its speed v must vary to conserve angular momentum:

$$I_i\,\omega_i = I_f\,\omega_f$$

By substitution,

$$mr_i^2 \times \frac{v_i}{r_i} = mr_f^2 \times \frac{v_f}{r_f}$$

or

$$r_i v_i = r_f v_f \qquad (1)$$

Solving for v_f,

$$v_f = \frac{r_i}{r_f}v_i = \frac{1.47 \times 10^8 \text{ km}}{1.52 \times 10^8 \text{ km}} \times 30.3 \text{ km/s} = 29.3 \text{ km/s}$$

Discussion Earth moves slower at a point farther from the Sun. This is what we expect from energy conservation. The potential energy is greater at aphelion than at perihelion. Since the mechanical energy of the orbit is constant, the kinetic energy must be smaller at aphelion.

Equation (1) implies that the orbital speed and orbital radius are inversely proportional, but strictly speaking this equation only applies to the perihelion and aphelion. At a general point in the orbit, the *perpendicular component $v_\perp$* is inversely proportional to r (see Fig. 8.39b). The orbits of Earth and most of the other planets are nearly circular so that $\theta \approx 0°$ and $v_\perp \approx v$.

Practice Problem 8.15   Puck on a string

A puck on a frictionless, horizontal air table is attached to a string that passes down through a hole in the table. Initially the puck moves at 12 cm/s in a circle of radius 24 cm. If the string is pulled through the hole, reducing the radius of the puck's circular motion to 18 cm, what is the new speed of the puck?

8.9   THE VECTOR NATURE OF ANGULAR MOMENTUM

Until now we have treated torque and angular momentum as scalar quantities. Such a treatment is adequate in the cases we have considered so far. However, the law of conservation of angular momentum applies to *all* systems, including rotating objects whose axis of rotation changes direction. Torque and angular momentum are actually vector quantities. Angular momentum is conserved in *both magnitude and direction* in the absence of external torques.

An important special case is that of a symmetric object rotating about an axis of symmetry, such as the spinning disk in Fig. 8.40. The magnitude of the angular

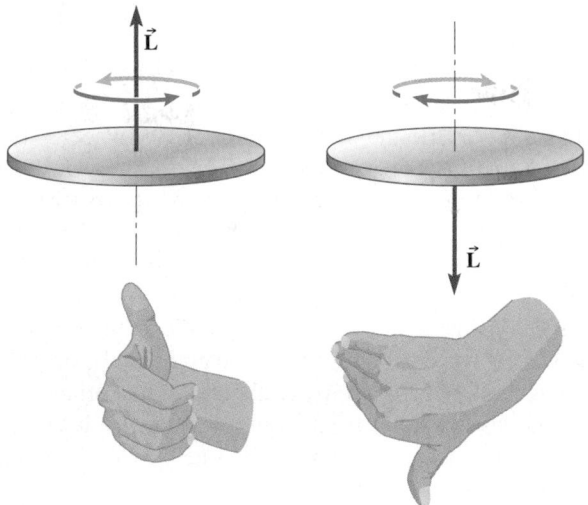

Figure 8.40 Right-hand rule for finding the direction of the angular momentum of a spinning disk.

momentum of such an object is $L = I\omega$. The direction of the angular momentum vector points along the axis of rotation. To find which of the two directions along the axis is correct, a **right-hand rule** is used. Align your right hand so that, as you curl your fingers in toward your palm, your fingertips follow the object's rotation; then your thumb points in the direction of $\vec{L}$.

A disk with a large rotational inertia can be used as a *gyroscope*. When the gyroscope spins at a large angular velocity, it has a large angular momentum. It is then difficult to change the orientation of the gyroscope's rotation axis, because to do so requires changing its angular momentum. To change the direction of a large angular momentum requires a correspondingly large torque. Thus a gyroscope can be used to maintain stability. Gyroscopes are used in guidance systems in airplanes, submarines, and space vehicles to maintain a constant direction in space.

The same principle explains the great stability of rifle bullets and spinning tops. A rifle bullet is made to spin as it passes through the rifle's barrel. The spinning bullet then keeps its correct orientation—nose first—as it travels through the air. Otherwise a small torque due to air resistance could make the bullet turn around randomly, greatly increasing air resistance and undermining accuracy. A properly thrown football is made to spin for the same reasons. A spinning top can stay balanced for a long time, while the same top falls over immediately when it is not spinning.

The Earth's rotation gives it a large angular momentum. As the Earth orbits the Sun, the axis of rotation stays in a fixed direction in space. The axis points nearly at Polaris (the North Star), so even as the Earth rotates around its axis, Polaris maintains its position in the northern sky. The fixed direction of the rotation axis also gives us the regular progression of the seasons (see Fig. 8.41).

Making The Connection:
angular momentum
of a gyroscope

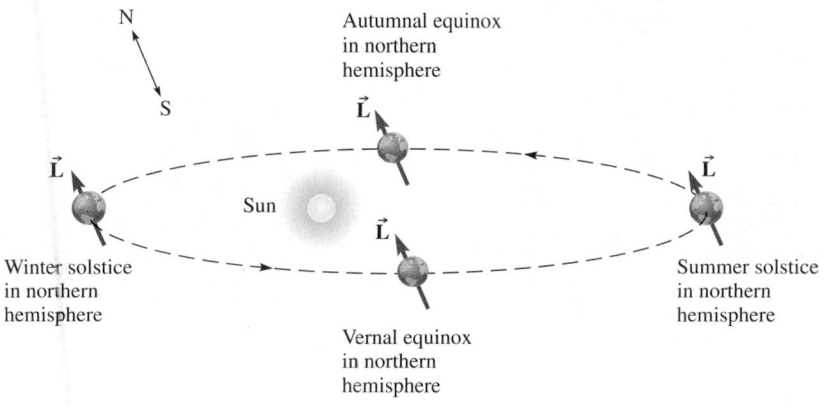

Figure 8.41 Spinning like a top, the Earth maintains the direction of its angular momentum due to rotation as it revolves around the Sun.

A Classic Demonstration

A demonstration often done in physics classes is for a student to hold a spinning bicycle wheel while standing on a platform that is free to rotate. The wheel's rotation axis is initially horizontal (Fig. 8.42a). Then the student repositions the wheel so that its axis of rotation is vertical (Fig. 8.42b). As he repositions the wheel, the platform begins to rotate opposite to the wheel's rotation. If we assume *no* friction acts to resist rotation of the platform, then the platform continues to rotate as long as the wheel is held with its axis vertical. If the student returns the wheel to its original orientation, the rotation of the platform stops.

The platform is free to rotate about a vertical axis. As a result, once the student steps onto the platform, *the vertical component L_y of the angular momentum of the system* (student + platform + wheel) is conserved. The horizontal components of $\vec{L}$ are *not* conserved. The platform is not free to rotate about any horizontal axis since the floor can exert external torques to keep it from doing so. In vector language, we would say that only the vertical component of the external torque is zero, so only the vertical component of angular momentum is conserved.

Initially $L_y = 0$ since the student and the platform have zero angular momentum and the wheel's angular momentum is horizontal. When the wheel is repositioned so that it spins with an upward angular momentum ($L_y > 0$), the rest of the system (the student and the platform) must acquire an equal magnitude of downward angular momentum ($L_y < 0$) so that the vertical component of the total angular momentum is still zero. Thus, the platform and student rotate in the opposite sense from the rotation of the wheel. Since the platform and student have more rotational inertia than the wheel, they do not spin as fast as the wheel, but their vertical angular momentum is just as large.

The student and the wheel apply torques to each other to transfer angular momentum from one part of the system to the other. These torques are equal and opposite and they have both vertical and horizontal components. As the student lifts the wheel, he feels a strange twisting force that tends to rotate him about a horizontal axis. The platform prevents the horizontal rotation by exerting unequal normal forces on the student's feet. The horizontal component of the torque is so counterintuitive that, if the student is not expecting it, he can easily be thrown from the platform!

Figure 8.42 A demonstration of angular momentum conservation.

(a)

(b)

MASTER THE CONCEPTS

Summary

- The rotational kinetic energy of a rigid object with rotational inertia I and angular velocity ω is

$$K_{rot} = \tfrac{1}{2}I\omega^2 \qquad (8\text{-}1)$$

In this expression, ω must be measured in *radians* per unit time.

- Rotational inertia is a measure of how difficult it is to change an object's angular velocity. It is defined as:

$$I = \sum_{i=1}^{N} m_i r_i^2 \qquad (8\text{-}2)$$

where r_i is the perpendicular distance between a particle of mass m_i and the rotation axis. The rotational inertia depends on the location of the rotation axis.

- Torque measures the effectiveness of a force for twisting or turning an object. It can be calculated in two equivalent ways: either as the product of the perpendicular component of the force with the shortest distance between the rotation axis and the point of application of the force

$$\tau = \pm r F_\perp \qquad (8\text{-}3)$$

or as the product of the magnitude of the force with its lever arm (the perpendicular distance between the line of action of the force and the axis of rotation)

$$\tau = \pm r_\perp F \qquad (8\text{-}4)$$

- A force whose perpendicular component tends to cause rotation in the CCW direction gives rise to a positive torque; a force whose perpendicular component tends to cause rotation in the CW direction gives rise to a negative torque.

- The work done by a constant torque is the product of the torque and the angular displacement:

$$W = \tau\theta \quad (\theta \text{ in radians}) \qquad (8\text{-}6)$$

- The conditions for equilibrium are

$$\vec{F}_{net} = 0 \text{ and } \tau_{net} = 0 \qquad (8\text{-}8)$$

The rotation axis can be chosen *arbitrarily* when calculating torques in equilibrium problems. Generally, the best place to choose the axis is at the point of application of an unknown force so that the unknown force does not appear in the torque equation.

- Newton's second law for rotation is

$$\Sigma\tau = I\alpha \qquad (8\text{-}9)$$

where radian measure must be used for α. A more general form is

$$\Sigma\tau = \lim_{\Delta t \to 0} \frac{\Delta L}{\Delta t} \qquad (8\text{-}13)$$

where L is the angular momentum of the system.

- The total kinetic energy of a body that is rolling without slipping is the sum of the rotational kinetic energy about an axis through the center of mass and the translational kinetic energy:

$$K = \tfrac{1}{2}Mv_{CM}^2 + \tfrac{1}{2}I_{CM}\omega^2 \qquad (8\text{-}11)$$

- The angular momentum of a rigid body rotating about a fixed axis is the rotational inertia times the angular velocity:

$$L = I\omega \qquad (8\text{-}14)$$

- The law of conservation of angular momentum: if the net external torque acting on a system is zero, then the angular momentum of the system cannot change.

$$\text{If } \tau_{net} = 0, L_i = L_f \qquad (8\text{-}15)$$

- This table summarizes the analogous quantities in translational and rotational motion.

Translation	Rotation
m	I
$\vec{F}$	τ
$\vec{a}$	α
$\Sigma\vec{F} = m\vec{a}$	$\Sigma\tau = I\alpha$
Δx	θ
$W = F_x \Delta x$	$W = \tau\theta$
$\vec{v}$	ω
$K = \tfrac{1}{2}mv^2$	$K = \tfrac{1}{2}I\omega^2$
$\vec{p} = m\vec{v}$	$L = I\omega$
$\Sigma\vec{F} = \lim_{\Delta t\to 0}\dfrac{\Delta\vec{p}}{\Delta t}$	$\Sigma\tau = \lim_{\Delta t\to 0}\dfrac{\Delta L}{\Delta t}$
If $\Sigma\vec{F} = 0$, $\vec{p}$ is conserved	If $\Sigma\tau = 0$, L is conserved

Highlighted Figures and Tables

F8.1 and **F8.2** Rotational inertia depends on the rotational axis (pp. 253, 254)

T8.1 Rotational inertia for uniform objects with various geometrical shapes (p. 253)

F8.8 Nonzero net torque with zero net force (p. 257)

F8.9 Torque as a product of the perpendicular component of the force (p. 257)

F8.11 Positive and negative torque (p. 258)

F8.12 Using lever arms to find torques (p. 260)

F8.16 Work done by torque (p. 262)

F8.18 Rotational axis, forces, and distances in equilibrium problems (p. 264)

F8.22 Finding the lever arm for each force (p. 267)

F8.26 Contact force for various incline angles (p. 269)

F8.35 Forces acting on an object rolling downhill (p. 277)

F8.39 Angular momentum in planetary orbits (p. 282)

F8.42 Angular momentum conservation on a platform that is free to rotate (p. 284)

CONCEPTUAL QUESTIONS

1. Explain why it is easier to drive a wood screw using a screwdriver with a large diameter handle rather than one with a thin handle.

2. One way to find the center of gravity of an irregular flat object is to suspend it from various points so that it is free to rotate (Fig. 8.43). When the object hangs in equilibrium, a vertical line is drawn downward from the support point. After drawing lines from several different support points, the center of gravity is the point where the lines all intersect. Explain how this works.

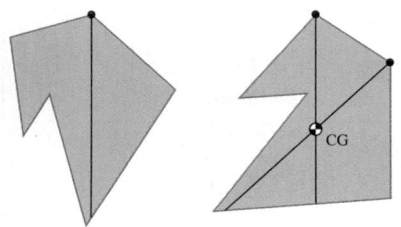

Figure 8.43
Conceptual Question 2

3. One of the effects of significant global warming would be the melting of part or all of the polar ice caps. This, in turn, would change the length of the day (the period of the Earth's rotation). Explain why. Would the day get longer or shorter?

4. A book measures 3 cm by 16 cm by 24 cm. About which of the axes shown in Fig. 8.44 is its rotational inertia smallest?

Figure 8.44
Conceptual Question 4

5. A body in static equilibrium has only two forces acting on it. We found in Section 2.4 that the forces must be equal in magnitude and opposite in direction in order to give a net force of zero. What else must be true of the two forces? [*Hint:* Consider the lines of action of the forces.]

6. Why do many helicopters have a small propeller attached to the tail that rotates in a vertical plane? Why is this attached at the tail rather than somewhere else? [*Hint:* Most of the helicopter mass is forward, in the cab.]

7. In the "Pinewood Derby," Cub Scouts construct cars and then race them down an incline. Some say that, everything else being equal (friction, drag coefficient, same wheels, etc.), a heavier car will win; others maintain that the weight of the car does not matter. Who is right? Explain. [*Hint:* Think about the fraction of the car's kinetic energy that is rotational.]

8. A large barrel lies on its side. In order to roll it across the floor, you apply a horizontal force, as shown in Fig. 8.45. If the applied force points toward the axis of rotation, which runs down the center of the barrel through the center of mass, it produces zero torque about that axis. How then can this applied force make the barrel start to roll?

Figure 8.45
Conceptual Question 8

9. Animals that can run fast always have thin legs. Their leg muscles are concentrated close to the hip joint; only tendons extend into the lower leg. Using the concept of rotational inertia, explain how this helps them run fast.

10. Figure 8.46a shows a simplified model of how the triceps muscle connects to the forearm. As the angle θ is changed, the tendon wraps around a nearly circular arc. Explain how this is much more effective than if the tendon is connected as in Fig. 8.46b. [*Hint:* Look at the lever arm as θ changes.]

Figure 8.46
Conceptual
Question 10

11. Figure 8.47a shows a simplified model of how the biceps muscle enables the forearm to support a load. What are the

Figure 8.47
Conceptual Question 11

advantages of this arrangement as opposed to the alternative shown in Fig. 8.47b, where the flexor muscle is in the forearm instead of in the upper arm? Are the two equally effective when the forearm is horizontal? What about for other angles between the upper arm and the forearm? Consider also the rotational inertia of the forearm about the elbow and of the entire arm about the shoulder.

12. In Section 8.6, it was asserted that the sum of all the internal torques (that is, the torques due to internal forces) acting on a rigid object is zero. Figure 8.48 shows two particles in a rigid object. The particles exert forces $\vec{F}_{12}$ and $\vec{F}_{21}$ on each other. These forces are directed along a line that joins the two particles. Explain why the torques due to these two forces must be equal and opposite even though the forces are applied at different points (and therefore possibly different distances from the axis).

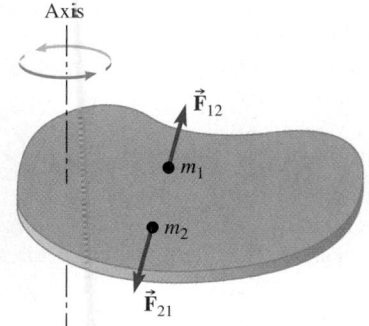

Figure 8.48
Conceptual Question 12

13. A playground merry-go-round (Fig. 8.4) spins with negligible friction. A child moves from the center out to the rim of the merry-go-round platform. Let the system be the merry-go-round plus the child. Which of these quantities change: angular velocity of the system, rotational kinetic energy of the system, angular momentum of the system? Explain your answer.

14. Figure 8.49 shows a common balancing toy with weights extending on either side. The toy is extremely stable. It can be pushed quite far off center one way or the other but it does not fall over. Explain why it is so stable.

Figure 8.49
Conceptual Question 14

15. Explain why the posture taken by defensive football linemen (Fig. 8.50) makes them more difficult to push out of the way. Consider both the height of the center of gravity and the size

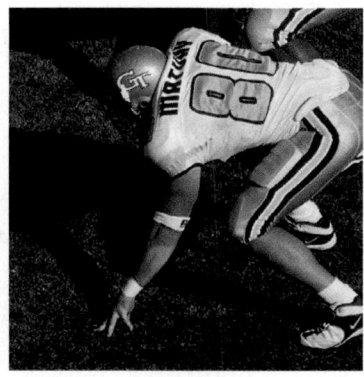

Figure 8.50
Conceptual Question 15

of the support base (the area on the ground bounded by the hands and feet touching the ground). In order to knock a person over, what has to happen to the center of gravity? Which do you think needs a more complex neurological system for maintaining balance: four legged animals or humans?

16. The CG of the upper body of a bird is located below the hips; in a human, the CG of the upper body is located well above the hips (see Fig. 8.51). Since the upper body is supported by the hips, are birds or humans more stable? Consider what happens if the upper body is displaced a little so that its CG is not directly above or below the hips. In what direction does the torque due to gravity tend to make the upper body rotate about an axis through the hips?

Figure 8.51 Conceptual Question 16

MULTIPLE CHOICE QUESTIONS

1. A heavy box is resting on the floor (Fig. 8.52). You would like to push the box to tip it over on its side, using the minimum force possible. Which of the force vectors in the diagram shows the correct location and direction of the force? Assume enough friction so that the box does not slide; instead it rotates about point P.

Figure 8.52 Multiple Choice Question 1

2. When both are expressed in terms of SI *base* units, torque has the same units as

   (a) angular acceleration   (b) angular momentum
   (c) force   (d) energy
   (e) rotational inertia   (f) angular velocity

Questions 3–4. A uniform solid cylinder rolls without slipping down an incline. At the bottom of the incline, the speed, v, of the cylinder is measured and the translational and rotational kinetic energies (K_{tr}, K_{rot}) are calculated. A hole is drilled through the cylinder along its axis and the experiment is repeated; at the bottom of the incline the cylinder now has speed v' and translational and rotational kinetic energies K'_{tr} and K'_{rot}.

3. How does the speed of the cylinder compare to its original value?

   (a) $v' < v$ (b) $v' = v$ (c) $v' > v$
   (d) answer depends on the radius of the hole drilled

4. How does the ratio of rotational to translational kinetic energy of the cylinder compare to its original value?

   (a) $\dfrac{K'_{rot}}{K'_{tr}} < \dfrac{K_{rot}}{K_{tr}}$ (b) $\dfrac{K'_{rot}}{K'_{tr}} = \dfrac{K_{rot}}{K_{tr}}$ (c) $\dfrac{K'_{rot}}{K'_{tr}} > \dfrac{K_{rot}}{K_{tr}}$
   (d) answer depends on the radius of the hole drilled

5. The SI units of angular momentum are

   (a) $\dfrac{\text{rad}}{\text{s}}$ (b) $\dfrac{\text{rad}}{\text{s}^2}$ (c) $\dfrac{\text{kg}\cdot\text{m}}{\text{s}^2}$

   (d) $\dfrac{\text{kg}\cdot\text{m}^2}{\text{s}^2}$ (e) $\dfrac{\text{kg}\cdot\text{m}^2}{\text{s}}$ (f) $\dfrac{\text{kg}\cdot\text{m}}{\text{s}}$

6. Which of the forces in Fig. 8.53 produces the largest magnitude torque about the rotation axis indicated?

   (a) 1   (b) 2   (c) 3   (d) 4

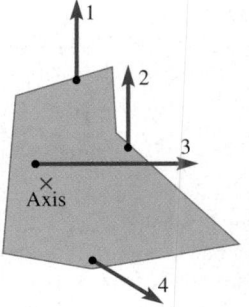

Figure 8.53 Multiple Choice Questions 6–8

7. Which of the forces in Fig. 8.53 produces a clockwise torque about the rotation axis indicated?

   (a) 3 only   (b) 4 only   (c) 1 and 2
   (d) 1, 2, and 3   (e) 1, 2, and 4

8. Which pair of forces in Fig. 8.53 might produce equal magnitude torques with opposite signs?

   (a) 2 and 3   (b) 2 and 4   (c) 1 and 2
   (d) 1 and 3   (e) 1 and 4   (f) 3 and 4

9. A high diver in midair pulls her legs inward toward her chest in order to rotate faster. Doing so changes which of these quantities: her angular momentum L, her rotational inertia I, and her rotational kinetic energy K_{rot}?

   (a) L only   (b) I only   (c) K_{rot} only
   (d) L and I only   (e) I and K_{rot} only   (f) all three

10. A student, moving with angular velocity ω on a platform that is free to rotate, is holding a 3-kg dumbbell in each hand at arm's length from his axis of rotation. Without changing the positions of his arms, he drops the two dumbbells to the ground. What effect does this have on the system of student and platform?

    (a) The angular velocity increases.
    (b) The angular velocity decreases.
    (c) The angular velocity is unchanged.
    (d) Both the angular velocity and the rotational kinetic energy increase.
    (e) Both the angular velocity and the rotational kinetic energy decrease.

PROBLEMS

Note: **C** indicates a combination conceptual/quantitative problem. Gold diamonds ✦, ✦✦ are used to indicate the increasing level of difficulty of each problem. Problem numbers appearing in blue, 9., denote problems that have a detailed solution available in the Student Solutions Manual. Some problems are *paired* by concept; their numbers are connected by a ruled box.

8.1 Rotational Kinetic Energy and Rotational Inertia

1. Verify that $\frac{1}{2}I\omega^2$ has dimensions of energy.
2. Four point masses of 3.0 kg each are arranged in a square on massless rods (Fig. 8.54). The length of a side of the square is

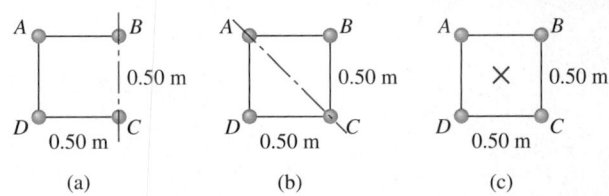

Figure 8.54 Problem 2

0.50 m. What is the rotational inertia for rotation about an axis (a) passing through masses *B* and *C*? (b) passing through masses *A* and *C*? (c) passing through the center of the square and perpendicular to the plane of the square?

3. What is the rotational inertia of a solid iron disk of mass 49 kg, with a thickness of 5.00 cm and radius of 20.0 cm, about an axis through its center and perpendicular to it?

4. A bowling ball made for a child has half the radius of an adult bowling ball. They are made of the same material (and therefore have the same mass *per unit volume*). By what factor is the (a) mass and (b) rotational inertia of the child's ball reduced compared to the adult ball?

5. How much work is done by the motor in a CD player to make a CD spin, starting from rest? The CD has a diameter of 12.0 cm and a mass of 15.8 g. The laser scans at a constant tangential velocity of 1.20 m/s. Assume that the music is first detected at a radius of 20.0 mm from the center of the disk. Ignore the small circular hole at the CD's center.

6. Find the ratio of the rotational inertia of the Earth for rotation about its own axis to its rotational inertia for rotation about the Sun.

7. A bicycle has wheels of radius 0.32 m. Each wheel has a rotational inertia of 0.080 kg·m² about its axle. The total mass of the bicycle including the wheels and the rider is 79 kg. When coasting at constant speed, what fraction of the total kinetic energy of the bicycle (including rider) is the rotational kinetic energy of the wheels?

© 8. In many problems in previous chapters (especially Chapter 6), cars and other objects that roll on wheels were considered to act as if they were sliding without friction. (a) Can the same assumption be made for a wheel rolling *by itself*? Explain your answer. (b) If a moving car of total mass 1300 kg has four wheels, each with rotational inertia of 0.705 kg·m² and radius of 35 cm, what fraction of the total kinetic energy is rotational?

9. A centrifuge has a rotational inertia of 6.5×10^{-3} kg·m². How much energy must be supplied to bring it from rest to 420 rad/s (4000 rpm)?

8.2 Torque

10. A mechanic turns a wrench using a force of 25 N at a distance of 16 cm from the rotation axis. The force is perpendicular to the wrench handle. What magnitude torque does she apply to the wrench?

11. The pull cord of a lawnmower engine is wound around a drum of radius 6.00 cm. While the cord is pulled with a force of 75 N to start the engine, what magnitude torque does the cord apply to the drum?

© ✦ 12. Any pair of equal and opposite forces acting on the same object is called a *couple*. Consider the couple in Fig. 8.55a. The rotation axis is perpendicular to the page and passes through point *P*. (a) Show that the net torque due to this couple is equal to *Fd*, where *d* is the distance between the lines of action of the two forces. Because the distance *d* is independent of the location of the rotation axis, this shows that the torque is the same for any rotation axis. (b) Repeat for the couple in Fig. 8.55b. Show that the torque is still *Fd* if *d* is the *perpendicular* distance between the lines of action of the forces.

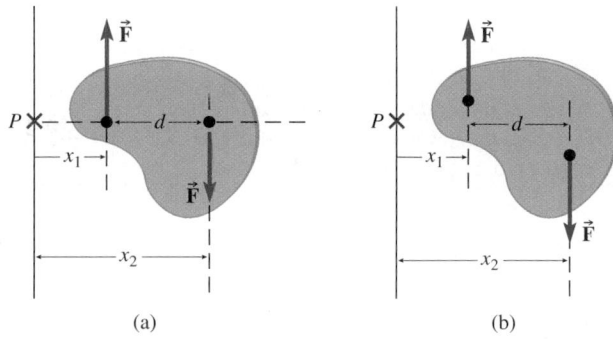

(a)   (b)

Figure 8.55 (a) A couple acting along a radius from the rotation axis located at *P* and perpendicular to the page; (b) a couple offset from the radial line containing the rotation axis. (Problem 12)

13. A uniform door weighs 50.0 N and is 1.0 m wide and 2.6 m high. What is the magnitude of the torque due to the door's own weight about a horizontal axis perpendicular to the door and passing through a corner?

14. A child of mass 40.0 kg is sitting on a horizontal seesaw at a distance of 2.0 m from the supporting axis. What is the magnitude of the torque about the axis due to the weight of the child?

15. A 124-g mass is placed on one pan of a balance, at a point 25 cm from the support of the balance. What is the magnitude of the torque about the support exerted by the mass?

16. A tower outside the Houses of Parliament in London has a famous clock commonly referred to as Big Ben, the name of its 13-ton chiming bell. The hour hand of each clock face is 2.7 m long and has a mass of 60.0 kg. Assume the hour hand to be a uniform rod attached at one end. (a) What is the torque on the clock mechanism due to the weight of one of the four hour hands when the clock strikes noon? The axis of rotation is perpendicular to a clock face and through the center of the clock. (b) What is the torque due to the weight of one hour hand about the same axis when the clock tolls 9:00 A.M.?

17. A weightless rod, 10.0 m long, supports three weights (Fig. 8.56). Where is its center of gravity?

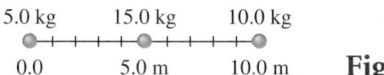

5.0 kg   15.0 kg   10.0 kg

0.0   5.0 m   10.0 m   **Figure 8.56** Problem 17

18. A door weighing 300.0 N measures 2.00 m × 3.00 m and is of uniform density; that is, the mass is uniformly distributed throughout the volume. A doorknob is attached to the door as

Figure 8.57
Problem 18

shown in Fig. 8.57. Where is the center of gravity if the door-knob weighs 5.0 N and is located 0.25 m from the edge?

✦ 19. A plate of uniform thickness is shaped as shown in Fig. 8.58. Where is the center of gravity? Assume the origin (0, 0) is located at the lower left corner of the plate; the upper left corner is at (0, s) and upper right corner is at (s, s).

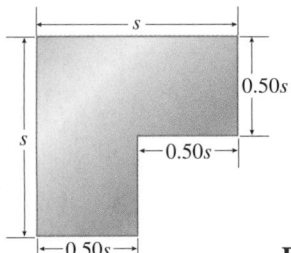

Figure 8.58 Problem 19

8.3 Work Done by a Torque

20. The radius of a wheel is 0.500 m. A rope is wound around the outer rim of the wheel. The rope is pulled with a force of magnitude 5.00 N, unwinding the rope and making the wheel spin counterclockwise about its central axis. Ignore the mass of the rope. (a) How much rope unwinds while the wheel makes 1.00 revolution? (b) How much work is done by the rope on the wheel during this time? (c) What is the torque on the wheel due to the rope? (d) What is the angular displacement θ, in radians, of the wheel during 1.00 revolution? (e) Show that the numerical value of the work done is equal to the product $\tau\theta$.

21. A stone used to grind wheat into flour is turned through 12 revolutions by a constant force of 20.0 N applied to the rim of a 10.0-cm-radius shaft connected to the wheel. How much work is done on the stone during the 12 revolutions?

✦ 22. A flywheel of mass 182 kg has an effective radius of 0.62 m (assume the mass is concentrated along a circumference located at the effective radius of the flywheel). (a) What torque is required to bring this wheel from rest to a speed of 120 rpm in a time interval of 30.0 s? (b) How much work is done during the 30.0 s?

✦ 23. A Ferris wheel rotates because a motor exerts a torque on the wheel. The radius of the London Eye, a huge observation wheel on the banks of the Thames, is 67.5 m and its mass is 1.90×10^6 kg. The cruising angular speed of the wheel is 3.50×10^{-3} rad/s. (a) How much work does the motor need to do to bring the stationary wheel up to cruising speed? [*Hint:* Treat the wheel as a hoop.] (b) What is the torque (assumed constant) the motor needs to provide to the wheel if it takes 20.0 seconds to reach the cruising angular speed?

8.4 Equilibrium Revisited

24. A sculpture is 4.00 m tall and has its CG located 1.80 m above the center of its base. The base is a square with a side of 1.10 m (Fig. 8.59). To what angle θ can the sculpture be tipped before it falls over?

25. A sign painter is standing on a uniform, horizontal platform that is held in equilibrium at the tenth-story level by two cables attached to supports on the roof of the building (Fig. 8.60). The painter has a mass of 75 kg and the mass of

Figure 8.59 Problem 24

Figure 8.60 Problem 25

the platform is 20.0 kg. The distance from the left end of the platform to where the painter is standing is $d = 2.0$ m and the total length of the platform is 5.0 m. (a) How large is the force exerted by the left-hand cable on the platform? (b) How large is the force exerted by the right-hand cable?

26. A rod being used as a lever is shown in Fig. 8.61. The fulcrum is 1.2 m from the load and 2.4 m from the applied force. If the load has a mass of 20.0 kg, what force must be applied to lift the load?

Figure 8.61 Problem 26

27. A weight of 1200 N rests on a lever at a point 0.50 m from a support (Fig. 8.62). On the same side of the support, at a distance of 3.0 m from it, an upward force with magnitude F is applied. Neglect the weight of the board itself. If the system is in equilibrium, what is F?

Figure 8.62 Problem 27

✦ 28. A uniform diving board, of length 5.0 m and mass 55 kg, is supported at two points; one support is located 3.4 m from the end of the board and the second is at 4.6 m from the end (see Fig. 8.19). What are the forces acting on the board due to the two supports when a diver of mass 65 kg stands at the end of the board over the water? Assume that these forces are vertical. [*Hint:* In this problem, consider using two different torque equations about different rotation axes. This may help you determine the directions of the two forces.]

✦ 29. A house painter stands 3.0 m above the ground on a 5.0-m-long ladder that leans against the wall at a point 4.7 m above the ground (Fig. 8.63.) The painter weighs 680 N and the ladder weighs 120 N. Assuming no friction between the house and the upper end of the ladder, find the force of friction that the driveway exerts on the bottom of the ladder.

Figure 8.63
Problem 29

✦ 30. A mountain climber is rappelling down a vertical wall (Fig. 8.64). The rope attaches to a buckle strapped to the climber's waist 15 cm to the right of his center of gravity. If the climber weighs 770 N, find (a) the tension in the rope and (b) the magnitude and direction of the contact force exerted by the wall on the climber's feet.

Figure 8.64
Problem 30

✦ 31. A boom of mass *m* supports a steel girder of weight *W* hanging from its end (Fig. 8.65). One end of the boom is hinged at the floor; a cable attaches to the other end of the boom and pulls horizontally on it. The boom makes an angle θ with the horizontal. Find the tension in the cable as a function of *m*, *W*, θ, and *g*. Comment on the tension at $\theta = 0$ and $\theta = 90°$.

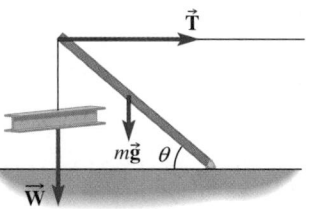

Figure 8.65
Problem 31

Ⓒ 32. A sign is supported by a uniform horizontal boom of length 3.00 m and weight 80.0 N. A cable, inclined at an angle of 35° with the boom, is attached at a distance of 2.38 m from the hinge at the wall (Fig. 8.66). The weight of the sign is 120 N. What is the tension in the cable and what are the horizontal and vertical forces F_x and F_y exerted on the boom by the hinge? Comment on the magnitude of F_y.

Figure 8.66
Problem 32

33. A man is doing push-ups (Fig. 8.67). He has a mass of 68 kg and his center of gravity is located at a horizontal distance of 0.70 m from his palms and 1.00 m from his feet. Find the forces exerted by the floor on his palms and feet.

Figure 8.67   Problem 33

8.5 Equilibrium in the Human Body

34. Find the tension in the Achilles tendon and the force that the tibia exerts on the ankle joint when a person who weighs 750 N supports himself on the ball of one foot (Fig. 8.68). The normal force $N = 750$ N pushes up on the ball of the foot on one side of the ankle joint, while the Achilles tendon pulls up on the foot on the other side of the joint.

Figure 8.68
Problem 34

Figure 8.69 Problem 36

35. In the movie *Terminator*, Arnold Schwarzenegger lifts someone up by the neck and, with both arms fully extended and horizontal, holds the person off the ground. If the person being held weighs 700 N, is 60 cm from the shoulder joint, and Arnold has an anatomy analogous to that in Fig. 8.30, what force must *each* of the deltoid muscles exert to perform this task?

✦ 36. A person is doing leg lifts with 3.0-kg ankle weights. She is sitting in a chair with her legs bent at a right angle initially. The quadriceps muscles are attached to the patella via a tendon; the patella is connected to the tibia by the patellar tendon, which attaches to bone 10.0 cm below the knee joint (Fig. 8.69). Assume that the tendon pulls at an angle of 20.0° *with respect to the lower leg*, regardless of the position of the lower leg. The lower leg has a mass of 5.0 kg and its center of gravity is 22 cm below the knee. The ankle weight is 41 cm from the knee. If the person lifts one leg, find the force exerted by the patellar tendon to hold the leg at an angle of (a) 30.0° and (b) 90.0° with respect to the vertical.

37. Find the force exerted by the biceps muscle in holding a 1-L milk carton (weight 9.9 N) with the forearm parallel to the floor (Fig. 8.70). Assume that the hand is 35.0 cm from the elbow and that the upper arm is 30.0 cm long. The elbow is bent at a right angle and one tendon of the biceps is attached to the forearm at a position 5.00 cm from the elbow, while the other tendon is attached at 30.0 cm from the elbow. The weight of the forearm and empty hand is 18.0 N and the center of gravity of the forearm is at a distance of 16.5 cm from the elbow.

✦ 38. A man is trying to lift 60.0 kg off the floor. (a) First he does so by bending at the waist (see Fig. 8.32). Assume that the man's upper body weighs 455 N and the upper body's center of gravity is 38 cm from the sacrum (tailbone). If, when bent over, the hands are a horizontal distance of 76 cm from the sacrum, what torque must be exerted by the erector spinae muscles to lift 60.0 kg off the floor? (The axis of rotation passes through the sacrum, as shown in Fig. 8.32.) (b) When bent over, the erector spinae muscles are a horizontal distance of 44 cm from the sacrum and act at a 12° angle above the horizontal. What force ($\vec{F}_b$ in Fig. 8.32) do the erector spinae

Figure 8.70 Problem 37

muscles need to exert to lift the weight? (c) What is the component of this force that compresses the spinal column?

8.6 Rotational Form of Newton's Second Law

39. Verify that the units of the rotational form of Newton's second law [Eq. (8-15)] are consistent. In other words, show that the product of a rotational inertia expressed in kg·m² and an angular acceleration expressed in rad/s² is a torque expressed in N·m.

40. A spinning flywheel has rotational inertia $I = 400.0$ kg·m². Its angular velocity decreases from 20.0 rad/s to zero in 300.0 s due to friction. What is the frictional torque acting?

41. An LP turntable must spin at 33.3 rpm (3.49 rad/s) to play a record. How much torque must the motor deliver if the turntable is to reach its final angular speed in 2.0 revolutions, starting from rest? The turntable is a uniform disk of diameter 30.5 cm and mass 0.22 kg.

© ✦ 42. Refer to Atwood's machine (Example 8.2). (a) Assuming that the cord does not slip as it passes around the pulley, what is the relationship between the angular acceleration of the pulley (α) and the magnitude of the linear acceleration of the blocks (a)? (b) What is the net torque on the pulley about its axis of rotation in terms of the tensions T_1 and T_2 in the left and right sides of the cord? (c) Explain why the tensions cannot be equal if $m_1 \neq m_2$. (d) Apply Newton's second law to each of the blocks and Newton's second law for rotation to the pulley. Use these three equations to solve for a, T_1, and T_2. (e) Since the blocks move with constant acceleration, use the result of Example 8.2 along with the constant acceleration equation $v_y^2 - v_{0y}^2 = 2a_y \Delta y$ to check your answer for a.

43. Four masses are arranged as shown in Fig. 8.71. They are connected by rigid, massless rods of lengths 0.75 m and 0.50 m. What torque must be applied to cause an angular acceleration of 0.75 rad/s^2 about the axis shown?

A 4.0 kg
B 3.0 kg
C 5.0 kg
D 2.0 kg

Figure 8.71 Problem 43

44. A grinding wheel, with a mass of 20.0 kg and a radius of 22.4 cm, is a uniform cylindrical disk. (a) Find the rotational inertia of the wheel about its central axis. (b) When the grinding wheel's motor is turned off, friction causes the wheel to slow from 1200 rpm to rest in 60.0 s. What torque must the motor provide to accelerate the wheel from rest to 1200 rpm in 4.00 s? Assume that the frictional torque is the same regardless of whether the motor is on or off.

45. A playground merry-go-round (Fig. 8.4), made in the shape of a solid disk, has a diameter of 2.50 m and a mass of 350.0 kg. Two children, each of mass 30.0 kg, sit on opposite sides at the edge of the platform. Approximate the children as point masses. (a) What torque is required to bring the merry-go-round from rest to 25 rpm in 20.0 s? (b) If two other bigger children are going to push on the merry-go-round rim to produce this acceleration, with what force magnitude must each child push?

46. Two children standing on opposite sides of a merry-go-round (Fig. 8.4) are trying to rotate it. They each push in opposite directions with forces of magnitude 10.0 N. (a) If the merry-go-round has a mass of 180 kg and a radius of 2.0 m, what is the angular acceleration of the merry-go-round? (Assume the merry-go-round is a uniform disk.) (b) How fast is the merry-go-round rotating after 4.0 s?

47. A bicycle wheel, of radius 0.30 m and mass 2 kg (concentrated on the rim), is rotating at 4.00 rev/s. After 50 s the wheel comes to a stop because of friction. What is the magnitude of the average torque due to frictional forces?

✦ 48. Derive the rotational form of Newton's second law as follows. Consider a rigid object that consists of a large number N of particles (Fig. 8.72). Let F_i, m_i, and r_i represent the tangential

component of the net force acting on the ith particle, the mass of that particle, and the particle's distance from the axis of rotation, respectively. (a) Use Newton's second law to find a_i, the particle's tangential acceleration. (b) Find the torque acting on this particle. (c) Replace a_i with an equivalent expression in terms of the angular acceleration α. (d) Sum the torques due to all the particles and show that $\sum\limits_{i=1}^{N} \tau_i = I\alpha$.

8.7 The Dynamics of Rolling Objects

49. A hollow cylinder, a uniform solid sphere, and a uniform solid cylinder all have the same mass m. The three objects are rolling on a horizontal surface with identical translational speeds v. Find their total kinetic energies in terms of m and v and order them from smallest to largest.

50. A solid sphere of mass 0.600 kg rolls without slipping along a horizontal surface with a translational speed of 5.00 m/s. It comes to an incline that makes an angle of 30° with the horizontal surface. Neglecting energy losses due to friction, (a) what is the total energy of the rolling sphere? and (b) to what vertical height above the horizontal surface does the sphere rise on the incline?

Ⓒ ✦51. A uniform solid cylinder rolls without slipping down an incline. A hole is drilled through the cylinder along its axis. The radius of the hole is 0.50 times the (outer) radius of the cylinder. (a) Does the cylinder take more or less time to roll down the incline now that the hole has been drilled? Explain. (b) By what percentage does drilling the hole change the time for the cylinder to roll down the incline?

✦✦52. A solid sphere of radius R and mass M *slides* without friction down a loop-the-loop track (Fig. 8.73). The sphere starts from rest at a height of h above the horizontal. Assume that the radius of the sphere is small compared to the radius r of the loop. (a) Find the minimum value of h in terms of r so that the sphere remains on the track all the way around the loop. (b) Find the minimum value of h if, instead, the sphere rolls without slipping on the track.

✦53. A hollow cylinder, of radius R and mass M, rolls without slipping down a loop-the-loop track of radius r (Fig. 8.73). The cylinder starts from rest at a height h above the horizontal section of track. What is the minimum value of h so that the cylinder remains on the track all the way around the loop?

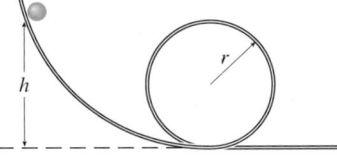

Figure 8.73
Problems 52 and 53

✦✦54. The string in a yo-yo is wound around an axle of radius 0.500 cm. The yo-yo has both rotational and translational motion, like a rolling object, and has mass 0.200 kg and outer radius 2.00 cm. Starting from rest, it rotates and falls a distance of 1.00 m (the length of the string). Assume for simplicity that the yo-yo is a uniform circular disk and that the string is thin compared to the radius of the axle. (a) What is the speed of the yo-yo when it reaches the distance of 1.00 m? (b) How long does it take to fall? [*Hint:* The translational and rotational kinetic energies are related, but the yo-yo is *not* rolling on its outer radius.]

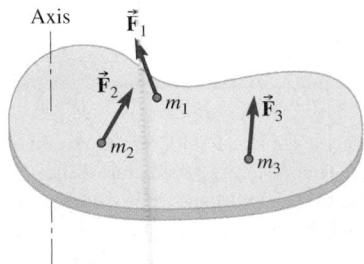

Figure 8.72 A rigid object, showing three of its constituent particles. (Problem 48)

8.8 Angular Momentum

55. A turntable of mass 5.00 kg has a radius of 0.100 m and spins with a frequency of 0.550 rev/s. What is its angular momentum? Assume the turntable is a uniform disk.

56. Assume the Earth is a uniform solid sphere with radius of 6.37×10^6 m and mass of 5.98×10^{24} kg. Find the magnitude of the angular momentum of the Earth due to rotation about its axis.

57. The mass of a flywheel is 5.6×10^4 kg. This particular flywheel has its mass concentrated at the rim of the wheel. If the radius of the wheel is 2.6 m and it is rotating at 350 rpm, what is the magnitude of its angular momentum?

58. The angular momentum of a spinning wheel is 240 kg·m²/s. After application of a constant braking torque for 2.5 s, it slows and has a new angular momentum of 115 kg·m²/s. What is the torque applied?

59. How long would a braking torque of 4.00 N·m have to act to just stop a spinning wheel that has an initial angular momentum of 6.40 kg·m²/s?

60. A rotating star collapses under the influence of gravitational forces to form a pulsar. The radius of the star after collapse is 1.0×10^{-4} times the radius before collapse. There is no change in mass. In both cases, the mass of the star is uniformly distributed in a spherical shape. Find the ratios of the (a) angular momentum, (b) angular velocity, and (c) rotational kinetic energy of the star after collapse to the value before collapse. (d) If the period of the star's rotation before collapse is 1.0×10^7 s, what is its period after collapse?

61. A figure skater is spinning at a rate of 1.0 rev/s with her arms outstretched. She then draws her arms in to her chest, reducing her rotational inertia to 67% of its original value. What is her new rate of rotation?

62. A skater is initially spinning at a rate of 10.0 rad/s with a rotational inertia of 2.50 kg·m² when her arms are extended. What is her angular velocity after she pulls her arms in and reduces her rotational inertia to 1.60 kg·m²?

63. A diver can change his rotational inertia by drawing his arms and legs close to his body in the tuck position. After he leaves the diving board (with some unknown angular velocity), he pulls himself into a ball as closely as possible and makes 2.00 complete rotations in 1.33 s. If his rotational inertia decreases by a factor of 3.00 when he goes from the straight to the tuck position, what was his angular velocity when he left the diving board?

✦64. The rotational inertia for a diver in a pike position is about 15.5 kg·m²; it is only 8.0 kg·m² in a tucked position (Fig. 8.74). (a) If the diver gives himself an initial angular momentum of 106 kg·m²/s as he jumps off the board, how many turns can he make when jumping off a 10.0-m platform in a

(a)

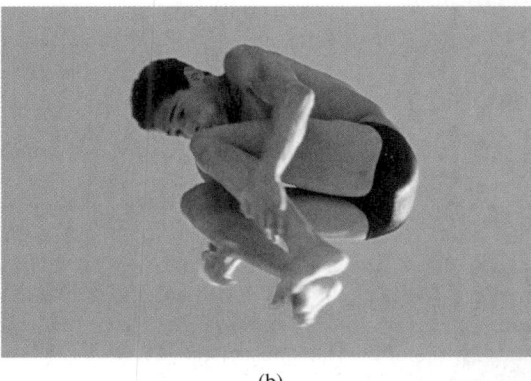

(b)

Figure 8.74 Problem 64. (a) Gregory Louganis in the pike position. (b) Mark Ruiz in the tuck position.

tuck position? (b) How many in a pike position? [*Hint:* Gravity exerts no torque on the person as he falls; assume he is rotating throughout the 10.0 m dive.]

8.9 The Vector Nature of Angular Momentum

Problems 65 and 66. A solid cylindrical disk is to be used as a stabilizer in a ship. By using a massive disk rotating in the hold of the ship, the captain knows that a large torque is required to tilt its angular momentum vector. The mass of the disk to be used is 1.00×10^5 kg and it has a radius of 2.00 m.

✦65. If the cylinder rotates at 300.0 rpm, what is the magnitude of the average torque required to tilt its axis by 60.0° in a time of 3.00 s? [*Hint:* Draw a vector diagram of the initial and final angular momenta.]

G 66. How should the disk be oriented to prevent rocking from side to side and from bow to stern? Does this orientation make it difficult to steer the ship? Explain.

COMPREHENSIVE PROBLEMS

67. The Moon's distance from Earth varies between 3.56×10^5 km at perigee and 4.07×10^5 km at apogee. What is the ratio of its orbital speed around Earth at perigee to that at apogee?

✦68. A painter (mass 61 kg) is walking along a trestle, consisting of a uniform plank (mass 20.0 kg, length 6.00 m) balanced on

two sawhorses. Each sawhorse is placed 1.40 m from an end of the plank. A paint bucket (mass 4.0 kg, diameter 28 cm) is placed as close as possible to the right-hand edge of the plank while still having the whole bucket in contact with the plank (Fig. 8.75). (a) How close to the right-hand edge of the

−1.40 m− ←−1.40 m→ ←0.28 m

←−−−−−− 6.00 m −−−−−−→

Figure 8.75
Problem 68

plank can the painter walk before tipping the plank and spilling the paint? (b) How close to the left-hand edge can the same painter walk before causing the plank to tip? [*Hint:* As the painter walks toward the right-hand edge of the plank and the plank starts to tip clockwise, what is the force acting upward on the plank from the left-hand sawhorse support?]

✦69. An experimental flywheel, used to store energy and replace an automobile engine, is a solid disk of mass 200.0 kg and radius 0.40 m. (a) What is its rotational inertia? (b) When driving at 22.4 m/s (50 mph), the fully energized flywheel is rotating at an angular speed of 3160 rad/s. What is the initial rotational kinetic energy of the flywheel? (c) If the total mass of the car is 1000.0 kg, find the ratio of the initial rotational kinetic energy of the flywheel to the translational kinetic energy of the car. (d) If the force of air resistance on the car is 670.0 N, how far can the car travel at a speed of 22.4 m/s (50 mph) with the initial stored energy? Ignore losses of mechanical energy due to means other than air resistance.

✦✦70. (a) Assume the Earth is a uniform solid sphere. Find the kinetic energy of the Earth due to its rotation about its axis. (b) Suppose we could somehow extract 1.0% of the Earth's rotational kinetic energy to use for other purposes. By how much would that change the length of the day? (c) For how many years would 1.0% of the Earth's rotational kinetic energy supply the world's energy usage (assume a constant 1.0×10^{21} J per year)?

✦71. A flat object in the xy-plane is free to rotate about the z-axis. The gravitational field is uniform in the $-y$-direction. Think of the object as a large number of particles with masses m_i located at coordinates (x_i, y_i), as in Fig. 8.76. (a) Show that the torques on the particles about the z-axis can be written $\tau_i = -x_i m_i g$. (b) Show that if the center of gravity is located at (x_{CG}, y_{CG}), the

total torque due to gravity on the object must be $\Sigma \tau_i = -x_{CG} M g$, where M is the total mass of the object. (c) Show that $x_{CG} = x_{CM}$. (This same line of reasoning can be applied to objects that are not flat and to other axes of rotation to show that $y_{CG} = y_{CM}$ and $z_{CG} = z_{CM}$.)

72. The operation of the Princeton Tokomak Fusion Test Reactor requires large bursts of energy. The power needed exceeds the amount that can be supplied by the utility company. Prior to pulsing the reactor, energy is stored in a giant flywheel of mass 7.27×10^5 kg and rotational inertia 4.55×10^6 kg·m². The flywheel rotates at a maximum angular speed of 386 rpm. When the stored energy is needed to operate the reactor, the flywheel is connected to an electrical generator, which converts some of the rotational kinetic energy into electrical energy. (a) If the flywheel is a uniform disk, what is its radius? (b) If the flywheel is a hollow cylinder with its mass concentrated at the rim, what is its radius? (c) If the flywheel slows to 252 rpm in 5.00 s, what is the average power supplied by the flywheel during that time?

73. The distance from the center of the breastbone to a man's hand, with the arm outstretched and horizontal to the floor, is 1.0 m. The man is holding a 10.0-kg dumbbell, oriented vertically, in his hand, with the arm horizontal. What is the torque due to this weight about a horizontal axis through the breastbone perpendicular to his chest?

74. A uniform rod of length L is free to pivot around an axis through its upper end. If it is released from rest when horizontal, at what speed is the lower end moving at its lowest point? [*Hint:* The gravitational potential energy change is determined by the change in height of the center of gravity.]

✦75. A gymnast is performing a giant swing on the high bar (Fig. 8.77). In a simplified model of the giant swing, assume that the gymnast keeps his arms and body straight as he swings all the way around the upper bar. Assume also that the gymnast does no work during the swing. With what angular speed should he be moving at the bottom of the giant swing in order to make it all the way around? The distance from the bar to his feet is 2.0 m and his center of gravity is 1.0 m from his feet.

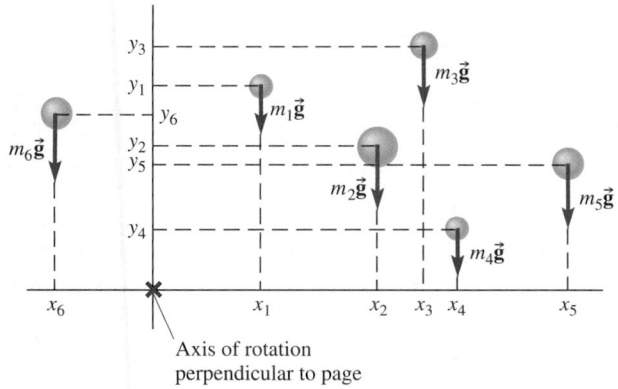

y_3 y_1 y_6 $m_3\vec{g}$ $m_1\vec{g}$

$m_6\vec{g}$ y_2 y_5

$m_2\vec{g}$ $m_5\vec{g}$

y_4 $m_4\vec{g}$

x_6 x_1 x_2 x_3 x_4 x_5

Axis of rotation
perpendicular to page

Figure 8.76 Problem 71

Figure 8.77
Problem 75

76. A box of mass 42 kg sits on top of a ladder (Fig. 8.78). Neglecting the weight of the ladder, find the tension in the brace. Assume that the brace exerts horizontal forces on the ladder at each end. [*Hint:* Use a symmetry argument; then analyze the forces and torques on one side of the ladder.]

Figure 8.78
Problem 76

77. A person is trying to lift a ladder of mass 15 kg and length 8.0 m. The person is exerting a vertical force on the ladder at a point of contact 2.0 m from the center of gravity. The opposite end of the ladder rests on the floor. (a) When the ladder makes an angle of 60.0° with the floor, what is this vertical force? (b) A person tries to help by lifting the ladder at the point of contact with the floor. Does this help the person trying to lift the ladder?

78. A crustacean (*Hemisquilla ensigera*) rotates its anterior limb to strike a mollusk, intending to break it open. The limb reaches an angular velocity of 175 rad/s in 1.50 ms. We can approximate the limb as a thin rod rotating about an axis perpendicular to one end (the joint where the limb attaches to the crustacean). (a) If the mass of the limb is 28.0 g and the length is 3.80 cm, what is the rotational inertia of the limb about that axis? (b) If the extensor muscle is 3.00 mm from the joint and acts perpendicular to the limb, what is the muscular force required to achieve the blow?

79. A block of mass m_2 hangs from a string. The string wraps around a pulley of rotational inertia I and then attaches to a second block of mass m_1, which sits on a frictionless table (Fig. 8.79). What is the acceleration of the blocks when they are released?

Figure 8.79 Problem 79
(see Fig. 8.7)

80. A 2.0-kg uniform flat disk is thrown into the air with a linear speed of 10.0 m/s. As it travels, the disk spins at 3.0 revolutions per second. If the radius of the disk is 10.0 cm, what is the magnitude of its angular momentum?

81. A hoop of 2.00-m circumference is rolling down an inclined plane of length 10.0 m in a time of 10.0 s. It started out from rest. (a) What is its angular velocity when it arrives at the bottom? (b) If the mass of the hoop, concentrated at the rim, is 1.50 kg, what is the angular momentum of the hoop when it reaches the bottom of the incline? (c) What force(s) supplied

the net torque to change the hoop's angular momentum? Explain. [*Hint:* Use a rotation axis through the hoop's center.] (d) What is the magnitude of this force?

82. A large clock has a second hand with a mass of 0.10 kg concentrated at the tip of the pointer. (a) If the length of the second hand is 30.0 cm, what is its angular momentum? (b) The same clock has an hour hand with a mass of 0.20 kg concentrated at the tip of the pointer. If the hour hand has a length of 20.0 cm, what is its angular momentum?

83. A planet moves around the Sun in an elliptical orbit (see Fig. 8.39). (a) Show that the external torque acting on the planet about an axis through the Sun is zero. (b) Since the torque is zero, the planet's angular momentum is constant. Write an expression for the planet's angular momentum in terms of its mass m, its distance r from the Sun, and its angular velocity ω. (c) Given r and ω, how much area is swept out during a short time Δt? [*Hint:* Think of the area as a fraction of the area of a *circle*, like a slice of pie; if Δt is short enough, the radius of the orbit during that time is nearly constant.] (d) Show that the area swept out per unit time is constant. You have just proved Kepler's second law!

84. A 68-kg woman stands straight with both feet flat on the floor. Her center of gravity is a horizontal distance of 3.0 cm in front of a line that connects her two ankle joints. The Achilles tendon attaches the calf muscle to the foot a distance of 4.4 cm behind the ankle joint. If the Achilles tendon is inclined at an angle of 81° with respect to the horizontal, find the force that each calf muscle needs to exert while she is standing. [*Hint:* Consider the equilibrium of the part of the body *above* the ankle joint.]

85. A merry-go-round (radius R, rotational inertia I_0) spins with negligible friction. Its initial angular velocity is ω_0. A child (mass m) on the merry-go-round moves from the center out to the rim. (a) Calculate the angular velocity after the child moves out to the rim. (b) Calculate the rotational kinetic energy and angular momentum of the system (merry-go-round + child) before and after.

86. In a motor, a flywheel (solid disk of radius R and mass M) is rotating with angular velocity ω_0. When the clutch is released, a second disk (radius r and mass m) initially not rotating is brought into frictional contact with the flywheel. The two disks spin around the same axle with frictionless bearings. After a short time, friction between the two disks brings them to a common angular velocity. (a) Ignoring external influences, what is the final angular velocity? (b) Does the total angular momentum of the two change? If so, account for the change. If not, explain why it does not. (c) Repeat (b) for the rotational kinetic energy.

87. Since humans are generally not symmetrically shaped, the height of our center of gravity is generally not half of our height. One way to determine the location of the center of gravity is shown in Fig. 8.80. A 2.2-m-long uniform plank is supported by two bathroom scales, one at either end. Initially the scales each read 100.0 N. A 1.60-m-tall student then lies on top of the plank, with the soles of his feet directly above scale B. Now scale A reads 394.0 N and scale B reads 541.0 N. (a) What is the student's weight? (b) How far is his center of gravity from the soles of his feet? (c) When standing, how far above the floor is his center of gravity, expressed as a fraction of his height?

Figure 8.80 Problem 87

88. A spool of thread of mass m rests on a plane inclined at angle θ. The end of the thread is tied as shown in Fig. 8.81. The outer radius of the spool is R and the inner radius (where the thread is wound) is r. The rotational inertia of the spool is I. Give all answers in terms of m, θ, R, r, I, and g. (a) If there is no friction between the spool and the incline, describe the motion of the spool and calculate its acceleration. (b) If the coefficient of friction is large enough to keep the spool from slipping, calculate the magnitude and direction of the frictional force. (c) What is the minimum possible coefficient of friction to keep the spool from slipping in part (b)?

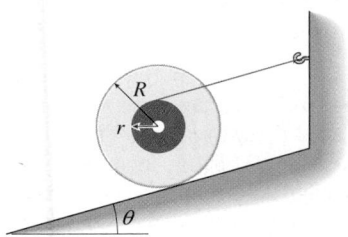

Figure 8.81
Problem 88

89. A bicycle travels up an incline at constant velocity (Fig. 8.82). The magnitude of the frictional force due to the road on the rear wheel is $f = 3.8$ N. The upper section of chain pulls on the sprocket wheel, which is attached to the rear wheel, with a force $\vec{\mathbf{F}}_C$. The lower section of chain is slack. If the radius of the rear wheel is 6.0 times the radius of the sprocket wheel, what is the magnitude of the force $\vec{\mathbf{F}}_C$ with which the chain pulls?

Figure 8.82
Problem 89

90. A circus roustabout is attaching the circus tent to the top of the main support post of length L when the post suddenly breaks at the base (Fig. 8.83). The worker's weight is negligible compared to that of the uniform post. What is the speed with which the roustabout reaches the ground if (a) he jumps at the instant he hears the post crack or (b) if he clings to the

post and rides to the ground with it. (c) Which is the safest course of action for the roustabout?

Figure 8.83 Problem 90

91. The 12.2-m crane (Fig. 8.84) weighs 18 kN and is lifting a 67-kN load. The hoisting cable (tension T_1) passes over a pulley at the top of the crane and attaches to an electric winch in the cab. The pendant cable (tension T_2), which supports the crane, is fixed to the top of the crane. Find the tensions in the two cables and the force $\vec{\mathbf{F}}_p$ at the pivot.

Figure 8.84 Problem 91

92. A collection of objects is set to rolling, without slipping, down a slope inclined at 30°. The objects are a solid sphere, a hollow sphere, a solid cylinder, and a hollow cylinder. A frictionless cube is also allowed to slide down the same incline. Which one gets to the bottom first? List the others in the order they arrive at the finish line.

○✦ 93. A student stands on a platform that is free to rotate and holds two dumbbells, each at a distance of 65 cm from his central axis (Fig. 8.85). Another student gives him a push and starts the system of student, dumbbells, and platform rotating at 0.50 rev/s. The student on the platform then pulls the dumbbells in close to his chest so that they are each 22 cm from his central axis. Each dumbbell has a mass of 1.00 kg and the rotational inertia of the student, platform, and dumbbells is initially 2.40 kg·m². Model each arm as a uniform rod of mass 3.00 kg with one end at the central axis; the length of the arm is initially 65 cm and then is reduced to 22 cm. (a) What is his new rate of rotation? (b) The student becomes nervous and drops the two dumbbells so that they fall away from the platform onto the floor. What happens to the student's rate of rotation? Explain.

| 65 cm | 65 cm |
| 22 cm |

Figure 8.85
Problem 93

94. A person places his hand palm downward on a scale and pushes down on the scale until it reads 96 N (Fig. 8.86). The triceps muscle is responsible for this arm extension force. Find the force exerted by the triceps muscle. The bottom of the triceps muscle is 2.5 cm to the left of the elbow joint and the palm is pushing at approximately 38 cm to the right of the elbow joint.

38 cm

2.5 cm

96 N

Figure 8.86   Problem 94

95. The posture of small animals may prevent them from being blown over by the wind. For example, with wind blowing from the side, a small insect stands with bent legs (Fig. 8.87); the more bent the legs, the lower the body, and thus the smaller the angle θ. The wind exerts a force on the insect, which causes a torque about the point where the downwind feet touch. The torque due to the weight of the insect must be equal and opposite to keep the insect from being blown over. For example, the drag force on a blowfly due to a sideways wind is $F_{wind} = cAv^2$, where v is the velocity of the wind, A is the cross-sectional area on which the wind is blowing, and $c \approx 1.3$ N·s²·m⁻⁴. (a) If the blowfly has a cross-sectional side area of 0.10 cm², a mass of 0.070 g, and crouches such that $\theta = 30.0°$, what is the maximum wind speed in which the blowfly can stand? (Assume that the drag force acts at the center of gravity.) (b) How about if it stands such that $\theta = 80.0°$? (c) Compare to the maximum wind velocity that a dog can withstand, if the dog stands such that $\theta = 80.0°$, has a cross-sectional area of 0.030 m², and weighs 10.0 kg. (Assume the same value of c.)

Leg   Body   Leg

$\vec{F}_{wind}$

θ

$m\vec{g}$

Figure 8.87
Problem 95

✦✦ 96. (a) Redo Example 8.7 to find an algebraic solution for d in terms of M, m, μ_s, L, and θ. (b) Use this to show that placing the ladder at a larger angle θ (that is, more nearly vertical) enables the person to climb farther up the ladder without having it slip, all other things being equal. (c) Using the numerical values from Example 8.7, find the minimum angle θ that enables the person to climb all the way to the top of the ladder.

ANSWERS TO PRACTICE PROBLEMS

8.1 $390\ \text{kg}\cdot\text{m}^2$

8.2 $v = \sqrt{\dfrac{2(m_2 - \mu_k m_1)gh}{m_1 + m_2 + I/R^2}}$

8.3 $53\ \text{N};\ 4.2\ \text{N}\cdot\text{m}$

8.4 $-65\ \text{N}\cdot\text{m}$

8.5 $8.3\ \text{J}$

8.6 left support, downward; right support, upward

8.7 0.27

8.8 $57\ \text{N}$, downward

8.9 It must lie in the same vertical plane as the two ropes holding up the rings. Otherwise the gravitational force would have a nonzero lever arm with respect to a horizontal axis that passes through the contact points between his hands and the rings; thus, gravity would cause a net torque about that axis.

8.10 $460\ \text{N}$

8.11 (a) $2380\ \text{rad}$; (b) $3.17\ \text{kJ}$; (c) $1.34\ \text{N}\cdot\text{m}$

8.12 solid ball, $\frac{2}{7}$; hollow ball, $\frac{2}{5}$.

8.13 $\frac{1}{2}g\sin\theta$

8.14 5% increase

8.15 $16\ \text{cm/s}$

Fluids

A hippopotamus in Kruger National Park, South Africa, wants to feed on the vegetation growing on the bottom of a pond. When the hippo wades into the pond, it floats. How does a hippopotamus get its floating body to sink to the bottom of a pond?

Fluids (liquids and gases) are materials that flow.

9.1   STATES OF MATTER

Macroscopic matter is usually classified into three states or phases: solids, liquids, and gases. Solids tend to hold their shapes. Many solids are quite rigid; they are not easily deformed by external forces because strong forces due to neighboring atoms hold each atom in a particular position. Although the atoms vibrate around fixed equilibrium positions, they do not have enough energy to break the bonds with their neighbors. To bend an iron bar, for example, the arrangement of the atoms must be altered, which is not easy to do. A blacksmith heats iron in a forge to loosen the bonds between atoms so that he can bend the metal into the required shape.

In contrast to solids, liquids and gases do not hold their shapes. A liquid flows and takes the shape of its container and a gas expands to fill its container. **Fluids**—both liquids and gases—are easily deformed by external forces. This chapter deals mainly with properties that are common to both liquids and gases.

The atoms or molecules in a fluid do not have fixed positions, so a fluid does not have a definite shape. An applied force can easily make a fluid flow; for instance, the squeezing of the heart muscle exerts an applied force that pumps blood through the blood vessels of the body. However, this squeezing does not change the *volume* of the blood by much. In many situations we can think of liquids as **incompressible**—that is, as having a fixed volume that is impossible to change. The shape of the liquid can be changed by pouring it from a container of one shape into a container of a different shape, but the volume of the liquid remains the same.

In most liquids, the atoms or molecules are almost as closely packed as those in the solid phase of the same material. The intermolecular forces in a liquid are almost as strong as those in solids, but the molecules are not locked in fixed positions as they are in solids. That is why the volume of the liquid can remain nearly constant while the shape is easily changed. Water is one of the exceptions: in cold water, the molecules in the liquid phase are actually *more* closely packed than those in the solid phase (ice).

Gases, on the other hand, cannot be characterized by a definite volume nor by a definite shape. A gas expands to fill its container and can easily be compressed. The molecules in a gas are very far apart compared to the molecules in liquids and solids. The molecules are almost free of interactions with each other except when they collide.

Gases are much more compressible than liquids.

9.2   PRESSURE

A **static** fluid does not flow; it is everywhere at rest. In the study of fluid statics (*hydrostatics*), we also assume that any solid object in contact with the fluid—whether a vessel containing the fluid or an object submerged in the fluid—is at rest. The atoms or molecules in a static fluid are not themselves static; they are continually moving. The motion of people bouncing up and down and bumping into each other in a mosh pit gives you a rough idea of the motion of the closely packed atoms or molecules in a liquid; in gases, the atoms or molecules are much farther apart than in liquids so they travel greater distances between collisions.

Fluid pressure is caused by collisions of the fast-moving atoms or molecules of a fluid. When a single molecule hits a container wall and rebounds, its momentum changes due to the force exerted on it by the wall. Figure 9.1a shows a molecule of a fluid within a container making an elastic collision with one of the container walls. In

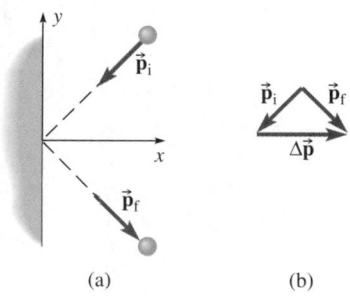

Figure 9.1 A single fluid molecule bouncing off a container wall.

Physics at Home

Drop a very tiny speck of dust or lint into a container of water and push the speck below the surface. The motion of the speck—called *Brownian motion*—is easily observed as it is pushed and bumped about randomly by collisions with water molecules. The water molecules themselves move about randomly, but at much higher speeds than the speck of dust due to their much smaller mass.

this case, the y-component of momentum is unchanged while the x-component reverses direction. The momentum change is in the $+x$-direction (Fig. 9.1b), which occurs because the wall exerts a force to the right on the molecule. By Newton's third law, the molecule exerts a force to the left on the wall during the collision. If we consider all the molecules colliding with this wall, *on average* they exert no force on the wall in the $\pm y$-direction, but all exert a force in the $-x$-direction. The frequent collisions of fluid molecules with the walls of the container cause a net force pushing outward on the walls.

A static fluid exerts a force on any surface with which it comes in contact; the direction of the force is perpendicular to the surface (Fig. 9.2). A static fluid *cannot* exert a force *parallel* to the surface. If it did, the surface would exert a force on the fluid parallel to the surface, by Newton's third law. This force would make the fluid flow along the surface, contradicting the premise that the fluid is static.

The force due to a static fluid on a surface is always perpendicular to the surface.

The *average pressure* of a fluid at points on a planar surface is

---

Definition of average pressure

$$P_{av} = \frac{F}{A} \qquad (9\text{-}1)$$

---

where F is the magnitude of the force acting perpendicular to the surface and A is the area of the surface. By imagining a tiny surface at various points within the fluid and measuring the force that acts on it, we can define the pressure at any point within the fluid. In the limit of a small area A, $P = F/A$ is the **pressure** P of the fluid.

Pressure is a scalar quantity; it does not have a direction. The force acting on an object submerged in a fluid—or on some portion of the fluid itself—is a vector quantity; its direction is perpendicular to the contact surface. Pressure is defined as a scalar because, at a given location in the fluid, the magnitude of the force per unit area is the same for any orientation of the surface. The molecules in a static fluid are moving in random directions; there can be no preferred direction since that would constitute fluid flow. There is no reason that a surface would have a greater number of collisions, or collisions with more energetic molecules, for one particular surface orientation compared with any other orientation.

The SI unit for pressure is the newton per square meter (N/m^2), which is named the *pascal* (symbol Pa) after the French scientist Blaise Pascal (1623–1662). Another commonly used unit of pressure is the *atmosphere* (atm). One atmosphere is the average air pressure at sea level. The conversion factor between atmospheres and pascals is

$$1 \text{ atm} = 101.3 \text{ kPa}$$

Other units of pressure in common use are introduced in Section 9.5.

Figure 9.2   Forces due to a static fluid acting on the walls of the container and on a submerged object. Here the weight of the fluid is neglected.

Example 9.1

The Danger of Stiletto-Heeled Shoes

A young woman weighing 534 N (120 lb) walks to her bedroom while wearing tennis shoes. She then gets dressed for her evening date, putting on her new "spike-heeled" dress shoes. The area of the heel section of her tennis shoe is 60.0 cm^2 and the area of the heel of her dress shoe is 1.00 cm^2. For each pair of shoes, find the average pressure caused by the heel making contact with the floor when her entire weight is supported by one heel.

Strategy   The average pressure is the force applied to the floor divided by the contact area. The force that the heel exerts on the floor is 534 N. To keep the units straight, we convert the areas from square centimeters to square meters.

Solution   To convert the area of the tennis shoe heel and the dress shoe heel from cm^2 to m^2, we multiply by the conversion factor $\left(\dfrac{1 \text{ m}}{10^2 \text{ cm}}\right)^2$.

For the tennis shoe

$$A = 60.0 \text{ cm}^2 \times \left(\frac{1 \text{ m}}{10^2 \text{ cm}}\right)^2 = 6.00 \times 10^{-3} \text{ m}^2$$

and for the dress shoe heel

$$A = 1.00 \text{ cm}^2 \times \left(\frac{1 \text{ m}}{10^2 \text{ cm}}\right)^2 = 1.00 \times 10^{-4} \text{ m}^2$$

continued on next page

Example 9.1 *continued*

The average pressure is the woman's weight divided by the area of the heel. For the tennis shoe:

$$P = \frac{F}{A} = \frac{534\,\text{N}}{6.00 \times 10^{-3}\,\text{m}^2} = 8.90 \times 10^4\,\text{N/m}^2 = 89.0\,\text{kPa}$$

For the "spike-heeled" shoe:

$$P = \frac{534\,\text{N}}{1.00 \times 10^{-4}\,\text{m}^2} = 5.34 \times 10^6\,\text{N/m}^2 = 5.34\,\text{MPa}$$

Discussion In atmospheres, these pressures are 0.879 atm and 52.7 atm, respectively. The pressure due to the dress shoe is 60 times the pressure due to the tennis shoe since the same force is spread over $\frac{1}{60}$ the area. This high pressure caused a problem for the manufacturers of vinyl floors when such shoes were in fashion. Permanent dents were made in floors due to the stress of tremendous pressure exerted on a tiny area

by such shoes. Finally the manufacturers sent out warnings that their floors were not guaranteed from damage caused by stiletto-heeled shoes. Women had to remove their shoes before stepping onto a vinyl kitchen floor.

Practice Problem 9.1 Pressure from ordinary dress shoe heel

Fortunately for floor manufacturers, and for women's feet, stiletto heels are out of fashion more often than they are in fashion. Suppose that a woman's dress shoes have heels that are each 4.0 cm² in area. Find the pressure on the floor, when the entire weight is on a single heel, for such a shoe worn by the same woman as in Example 9.1. Find the factor by which this pressure exceeds the pressure from the tennis shoe.

Atmospheric Pressure

On the surface of the Earth, we live at the bottom of an ocean of fluid called air. The forces exerted by air on our bodies and on surfaces of other objects may be surprisingly large: 1 atm is approximately 10 N/cm² of surface area, or nearly 15 lb/in². We are not crushed by this pressure because most of the fluids in our bodies are at approximately the same pressure as the air around us. As an analogy, consider a suitcase. Why is an empty suitcase not crushed by the air pushing in on all sides? Because the air inside the suitcase is at the same pressure and pushes out on the sides of the suitcase. The pressure of the fluids inside our cells matches the pressure of the surrounding fluids pushing in on the cell membranes, so the cells do not rupture.

By contrast, the blood pressure in the arteries is as much as 20 kPa higher than atmospheric pressure. The strong, elastic arterial walls are stretched by the pressure of the blood inside; the walls squeeze the arterial blood to keep its higher pressure from being transmitted to other fluids in the body.

Changing weather conditions cause variations of approximately 5% in the actual value of air pressure at sea level; 101.3 kPa is only the average value. Air pressure also decreases with increasing elevation. The average air pressure in Leadville, Colorado, the highest incorporated city in the United States (elevation 3100 m), is 70 kPa. Some Tibetans live at altitudes of over 5000 m, where the average air pressure is only half its value at sea level.

9.3 PASCAL'S PRINCIPLE

If the weight of a static fluid is negligible, then the pressure must be the same everywhere in the fluid. Why? In Fig. 9.3, imagine the submerged cube to be composed of the same fluid as its surroundings. Neglecting the fluid's weight, the only forces acting on the cubical piece of fluid are those due to the surrounding fluid pushing inward. The forces pushing on each pair of opposite sides of the cube must be equal in magnitude, since the fluid inside the cube is in equilibrium. Therefore, the pressure must be the same on both sides. Since we can extend this argument to any size and shape piece of fluid, *the fluid pressure must be the same everywhere*.

Even when the weight of the fluid is not negligible, the same reasoning can be used (see Conceptual Question 15) to establish **Pascal's principle**.

Figure 9.3 Forces acting on a cube of fluid

Pascal's principle

A *change in pressure* at any point in a confined fluid is transmitted everywhere throughout the fluid.

Figure 9.4 A hydraulic lift

For example, when a truck needs to have its muffler replaced, it is lifted into the air by a mechanism called a hydraulic lift (Fig. 9.4). A force is exerted on a liquid by a piston with a relatively small area; the resulting increase in pressure is transmitted everywhere throughout the liquid. Then the truck is lifted by the fluid pressure on a piston of much larger area. The upward force on the truck is much larger than the force applied to the small piston. Pascal's principle has many other applications, such as the hydraulic brakes in cars and trucks and the hydraulic controls in airplanes.

To analyze the forces in the hydraulic lift, let force F_1 be applied to the small piston of area A_1, causing a pressure increase:

$$\Delta P = \frac{F_1}{A_1}$$

A truck is supported by a piston of much larger area A_2 on the other side of the lift. The increase in pressure due to the small piston is transmitted everywhere in the liquid, so

$$\frac{F_1}{A_1} = \frac{F_2}{A_2}$$

Since A_2 is larger than A_1, the force exerted on the large piston (F_2) is larger than the force applied to the small piston (F_1). We are not getting something for nothing; just as for the pulley systems discussed in Chapter 6, the advantage of the smaller force applied to the small piston is balanced by a greater distance it must be moved. The small piston has to move a long distance d_1 while the large piston moves a short distance d_2. Assuming the liquid to be incompressible, the volume of fluid displaced by each piston is the same, so

$$A_1 d_1 = A_2 d_2$$

The displacements of the pistons are inversely proportional to their areas, while the forces are directly proportional to the areas; then the product of force and displacement is the same:

$$\frac{F_1}{A_1} \times A_1 d_1 = \frac{F_2}{A_2} \times A_2 d_2; \text{ therefore, } F_1 d_1 = F_2 d_2$$

The work (force times displacement) done by moving the small piston equals the work done by the large piston in raising the truck.

Example 9.2

The Hydraulic Lift

 In a hydraulic lift, if the diameter of the smaller piston is 4.0 cm and the diameter of the larger piston is 40.0 cm, what weight can the larger piston support when a force of 250 N is applied to the smaller piston?

Strategy According to Pascal's principle, the pressure increases the same amount at every point in the fluid. A natu-

ral way to work is in terms of proportions since the forces are proportional to the areas of the pistons.

Solution Since the pressure on the two pistons increases by the same amount,

$$\frac{F_1}{A_1} = \frac{F_2}{A_2}$$

continued on next page

Example 9.2 *continued*

Equivalently, the forces are proportional to the areas:

$$\frac{F_2}{F_1} = \frac{A_2}{A_1}$$

The ratio of the diameters is $d_2/d_1 = 10$, so the ratio of the areas is $A_2/A_1 = (d_2/d_1)^2 = 100$. Then the weight that can be supported is

$$F_2 = 100F_1 = 25{,}000 \text{ N} = 25 \text{ kN}$$

⚠ **Discussion**   One common error in this sort of problem is to think of the area and the force as a *trade-off*—in other words, that the piston with the *large* area has the *small* force and

vice versa. Since the pressures are the same, the force exerted by the fluid on either piston is proportional to the piston's area. We make the piston that lifts the truck large because we know the force on it will be large, *in direct proportion to* its area.

Practice Problem 9.2   Application of Pascal's principle

Consider the hydraulic lift of the preceding example. (a) What is the increase in pressure caused by the 250-N force on the small piston? (b) If the larger piston moves 5.0 cm, how far does the smaller piston move?

9.4   THE EFFECT OF GRAVITY ON FLUID PRESSURE

On a drive through the mountains or on a trip in a small plane, the feeling of our ears popping is evidence that pressure is not the same everywhere in a static fluid. Gravity makes fluid pressure increase as you move down and decrease as you move up. To understand more about this variation, we must first define the density of a fluid.

The density of a uniform substance is its mass divided by its volume.

The **density** of a substance is its mass per unit volume. The Greek letter ρ (rho) is used to represent density. The density of a uniform substance of mass m and volume V is

$$\rho = \frac{m}{V} \tag{9-2}$$

The SI units of density are kilograms per cubic meter: kg/m^3. For a nonuniform substance, Eq. (9-2) defines the **average density**.

Table 9.1 lists the densities of some common substances. Note that temperatures and pressures are specified in the table. For solids and liquids, density is only weakly dependent on temperature and pressure. On the other hand, gases are highly compressible, so even a relatively small change in temperature or pressure can change the density of a gas significantly.

Now, using the concept of density, we can find how pressure increases with depth due to gravity. Suppose we have a glass beaker containing a static liquid of uniform density ρ. Within this liquid, imagine a cylinder of liquid with cross-sectional area A and height d (Fig. 9.5a). The mass of the liquid in this cylinder is

$$m = \rho V$$

where the volume of the cylinder is

$$V = Ad$$

The weight of the cylinder of liquid is therefore

$$mg = (\rho Ad)g$$

The vertical forces acting on this column of liquid are shown in Fig. 9.5b. The pressure at the top of the cylinder is P_1 and the pressure at the bottom is P_2. Since the liquid in the column is in equilibrium, the net vertical force acting on it must be zero by Newton's second law:

$$\Sigma F_y = P_2 A - P_1 A - \rho A dg = 0$$

Dividing by the common factor A and rearranging yields:

Pressure variation with depth in a static fluid

$$P_2 = P_1 + \rho g d$$

where point 2 is a depth d below point 1   (9-3)

Table 9.1

Densities of Common Substances
(at 0°C and 1 atm unless otherwise indicated)

Gases	Density (kg/m³)	Liquids	Density (kg/m³)	Solids	Density (kg/m³)
Hydrogen	0.090	Gasoline	680	Polystyrene	100
Helium	0.18	Ethanol	790	Cork	240
Steam (100°C)	0.60	Oil	800–900	Wood (pine)	350–550
Nitrogen	1.25	Water (0°C)	999.87	Wood (oak)	600–900
Air (20°C)	1.20	Water (3.98°C)	1000.00	Ice	917
Air (0°C)	1.29	Water (20°C)	1001.80	Wood (ebony)	1000–1300
Oxygen	1.43	Seawater	1025	Bone	1500–2000
Carbon dioxide	1.98	Blood (37°C)	1060	Concrete	2000
		Mercury	13,600	Quartz, granite	2700
				Aluminum	2702
				Iron, steel	7860
				Copper	8920
				Lead	11,300
				Gold	19,300
				Platinum	21,500

(a) (b)

Figure 9.5 (a) A cylinder of liquid of height d and area A. (b) Vertical forces on the cylinder of liquid.

Since we can put the top and bottom of the cylinder anywhere we choose within the liquid, Eq. (9-3) relates the pressure at any two points in a static liquid where point 2 is a depth d below point 1. Note that in deriving this result, we assumed a constant density.

For gases, Eq. (9-3) can be applied as long as the depth d is small enough that changes in the density due to gravity are negligible. Since liquids are nearly incompressible, Eq. (9-3) holds to great depths in liquids.

For a liquid that is open to the atmosphere, suppose we take point 1 at the surface and point 2 a depth d below. Then $P_1 = P_{atm}$, so the pressure at a depth d below the surface is

---

Pressure at a depth d below the surface of a liquid open to the atmosphere

$$P = P_{atm} + \rho g d \qquad (9\text{-}4)$$

Example 9.3

A Diver

A diver swims to a depth of 3.2 m in a freshwater lake. What is the increase in the force pushing in on her eardrum, compared to what it was at the lake surface? The area of the eardrum is 0.60 cm².

Strategy We can find the increase in pressure at a depth of 3.2 m and then find the corresponding increase in force on the eardrum. If the force on the eardrum at the surface is P_1A and the force at a depth of 3.2 m is P_2A, then the increase in the force is $(P_2 - P_1)A$.

Solution The increase in pressure depends on the depth d and the density of water. From Table 9.1, the density of water is 1000 kg/m³ to two significant figures for any reasonable temperature.

$$P_2 - P_1 = \rho g d$$
$$\Delta P = 1000 \text{ kg/m}^3 \times 9.8 \text{ m/s}^2 \times 3.2 \text{ m}$$
$$= 31.4 \text{ kPa}$$

The increase in force on the eardrum is

$$\Delta F = \Delta P \times A$$

where $A = 0.60 \text{ cm}^2 = 6.0 \times 10^{-5} \text{ m}^2$. Then

$$\Delta F = (3.14 \times 10^4 \text{ Pa}) \times (6.0 \times 10^{-5} \text{ m}^2)$$
$$= 1.9 \text{ N}$$

Discussion A force also pushes *outward* on the eardrum due to the pressure inside the ear canal. If the diver descends rapidly so that the pressure inside the ear canal does not change, then a 1.9-N net force due to fluid pressure pushes inward on the eardrum. When the diver's ear "pops," the pressure inside the ear canal increases to equal the fluid pressure outside the eardrum, so that the net force due to fluid pressure on the eardrum is zero.

Practice Problem 9.3   Limits on submarine depth

A submarine is constructed so that it can safely withstand a pressure of 1.6×10^7 Pa. How deep may this submarine descend in the ocean if the average density of seawater is 1025 kg/m³?

Conceptual Example 9.4

The Hydrostatic Paradox

Three vessels have different shapes, but the same base area and the same weight when empty (Fig. 9.6). The vessels are filled with water to the same level and then the air is pumped out. The top surface of the water is then at a low pressure that, for simplicity, we assume to be zero. (a) Are the water pressures at the bottom of each vessel the same? If not, which is largest and which is smallest? (b) If the three vessels containing water are weighed on a scale, do they give the same reading? If not, which weighs the most and which weighs the least? (c) If the water exerts the same downward force on the bottom of each vessel, is that force equal to the weight of water in the vessel? Is there a paradox here? [*Hint:* Think about the forces due to fluid pressure on the *sides* of the containers; do they have vertical components?]

Solution and Discussion (a) The water at the bottom of each vessel is the same depth d below the surface. Water at the surface of each vessel is at a pressure $P_{\text{surface}} = 0$. Therefore, the pressures at the bottom must be equal:

$$P = P_{\text{surface}} + \rho g d = \rho g d$$

(b) The weight of each filled vessel is equal to the weight of the vessel itself plus the weight of the water inside. The vessels themselves have equal weights, but vessel A holds more water than C while vessel B holds less water than C. Vessel A weighs the most and vessel B weighs the least.

continued on next page

Figure 9.6
(a) Three differently shaped vessels filled with water to same level. (b) Forces exerted on the containers by the water. (c) Two different points on the bottom of an open vessel.

Conceptual Example 9.4 *continued*

(c) Each container supports the water inside by exerting an upward force equal in magnitude to the weight of the water. By Newton's third law, the water exerts a downward force on the container of the same magnitude. Figure 9.6b shows the forces acting on each container due to the water. In vessel C, the horizontal forces on any two diametrically opposite points on the walls of the container are equal and opposite; thus, the net force on the container walls is zero. The force on the bottom is

$$F = PA = \rho g d \times \pi r^2 = \rho g \times \text{volume of water} = \text{weight of water}$$

as expected. However, the force on the bottom of vessel A is less than the weight of the water in the container, while the force on the bottom of vessel B is greater than the weight of the water. Then how can the water be in equilibrium? In vessel A, the forces on the container walls have downward components as well as horizontal components. The horizontal components of the forces on any two diametrically opposite points are equal and opposite, so the horizontal components add to zero. The sum of the downward components of the forces on the walls and the downward force on the bottom of the container is equal to the weight of the water. Similarly, the forces on the walls of vessel B have upward components. In each case, the *total* force on the bottom *and sides* of the container due to the water is equal to the weight of the water.

Conceptual Practice Problem 9.4 Is pressure determined by column height?

Figure 9.6c shows a vessel with two points marked at the bottom of the water in the vessel. A narrow column of water is drawn above each point. (a) Is the pressure at point 2, P_2, the same as the pressure at point 1, P_1, even though the column of water above point 2 is not as tall? (b) Does $P = P_{atm} + \rho g d$ imply that $P_2 < P_1$? Explain.

"Water seeks its own level."

9.5 MEASURING PRESSURE

Many other units are used for pressure besides atmospheres and pascals. In the United States, the pressure in an automobile tire is measured in pounds per square inch (lb/in²). Weather bureaus record atmospheric pressure in bars or millibars. In the United States, television weather reports and home barometers measure pressure in inches of mercury. One atmosphere is equal to approximately 1 bar (1000 millibars), 76 cm of mercury, or 29.9 inches of mercury. Blood pressures are measured in millimeters of mercury (mm Hg), also called the torr. Inches or millimeters of mercury may seem like strange units for pressure: how can a force per unit area be equal to a *distance* (so many mm of Hg)? There is an assumption inherent in using these pressure units that we can understand by studying the mercury manometer.

Units of pressure: 1 atm = 101.3 kPa = 1.013 bar = 14.7 lb/in² = 760.0 mm Hg = 760.0 torr = 29.9 in Hg

Manometer

A mercury manometer consists of a vertical U-shaped tube, containing some mercury, with one side typically open to the atmosphere and the other connected to a vessel containing a gas whose pressure we want to measure. Figure 9.7a shows the manometer before it is connected to such a vessel. When both sides of the manometer are open to the atmosphere, the mercury levels are the same.

Now we connect a vessel containing a gas to the left side of the U-tube (Fig. 9.7b). If the gas is at a higher pressure than the atmosphere, the gas pushes the mercury down on the left side and forces it up on the right side. The density of a gas is small compared to the density of mercury, so every point within the gas is assumed to be at the same pressure. At point B, the mercury pushes on the gas with the same magnitude force with which the gas pushes on the mercury, so point B is at the same pressure as the gas. Since point B' is at the same height within the mercury as point B, the pressure at B' is the same as at B. Point C is at atmospheric pressure.

Open to the atmosphere

Open to the atmosphere

Gas

A'

A Starting level

C

d

B B'

Hg

Hg

(a)

(b)

Figure 9.7 (a) A mercury manometer open on both sides. Points A and A' are at atmospheric pressure. (b) The manometer connected on one side to a container of gas at a pressure greater than atmospheric pressure.

The pressure at B is

$$P_B = P_{B'} = P_C + \rho g d$$

where ρ is the density of mercury. The difference in the pressures on the two sides of the manometer is

$$\Delta P = P_B - P_C = \rho g d \qquad (9\text{-}5)$$

Thus, the difference in mercury levels d is a measure of the pressure *difference*—commonly reported in millimeters of mercury (mm Hg).

The pressure measured when one side of the manometer is open is the *difference* between atmospheric pressure and the gas pressure rather than the absolute pressure of the gas. This difference is called the **gauge pressure**, since it is what most gauges measure:

Gauge pressure

$$P_{\text{gauge}} = P_{\text{abs}} - P_{\text{atm}} \qquad (9\text{-}6)$$

Since the density of mercury is 13,600 kg/m³, 1.00 mm Hg can be converted to pascals by substituting $d = 1.00$ mm in Eq. (9-5):

$$1.00 \text{ mm Hg} = \rho g d = (13{,}600 \text{ kg/m}^3)(9.80 \text{ m/s}^2)(0.00100 \text{ m}) = 133 \text{ Pa}$$

The liquid in a manometer may be something other than mercury, such as water or oil. Equation (9-5) still applies, as long as we use the correct density ρ of the liquid in the manometer.

Example 9.5

The Mercury Manometer

A manometer is attached to a container of gas to determine its pressure. Before the container is attached, both sides of the manometer are open to the atmosphere. After the container is attached, the mercury on the side attached to the gas container rises 12 cm above its previous level. (a) What is the gauge pressure of the gas in Pa? (b) What is the absolute pressure of the gas in Pa?

Strategy The mercury column is higher on the side connected to the container of gas, so we know that the pressure of the enclosed gas is lower than atmospheric pressure. We need to find

the *difference* in levels of the mercury columns on the two sides.

⚠ Careful: it is *not* 12 cm! If one side went up by 12 cm, then the other side has gone down by 12 cm, since the same volume of mercury is contained in the manometer.

Solution (a) The difference d in the mercury levels is 24 cm. Since the mercury on the gas side went *up*, the absolute pressure of the gas is lower than atmospheric pressure. Therefore, the gauge pressure of the gas is *less than zero*. The gauge pressure in Pa is

$$P_{\text{gauge}} = \rho g d$$

continued on next page

Example 9.5 *continued*

where $d = -24$ cm (the mercury is 24 cm *lower* on the gas side). Then

$$P_{\text{gauge}} = 13{,}600 \text{ kg/m}^3 \times 9.8 \text{ m/s}^2 \times (-0.24 \text{ m}) = -32 \text{ kPa}$$

(b) The absolute pressure of the gas is

$$P = P_{\text{gauge}} + P_{\text{atm}}$$
$$= -32 \text{ kPa} + 101 \text{ kPa} = 69 \text{ kPa}$$

Discussion As a check, the manometer tells us directly that the gauge pressure of the gas is -240 mm Hg. Converting to pascals gives

$$(-240 \text{ mm Hg}) \times (133 \text{ Pa/mm Hg}) = -32 \text{ kPa}$$

Practice Problem 9.5 Column heights in manometer

A mercury manometer is connected to a container of gas. (a) The height of the mercury column on the side connected to the gas is 22.0 cm (measured from the bottom of the manometer). What is the height of the mercury column on the open side if the gauge pressure is measured to be 13.3 kPa? (b) If the gauge pressure of the gas doubles, what are the new heights of the two columns?

Barometer

A manometer can act as a **barometer**—a device to measure atmospheric pressure. Instead of attaching a container with a gas to one end of the manometer, attach a container and a vacuum pump. Pump the air out of the container to get as close to a vacuum—zero pressure—as possible. Then the atmosphere pushes down on one side and pushes the fluid up on the other side toward the evacuated container.

Figure 9.8 shows a barometer in which the vacuum is not created by a vacuum pump. A tube, of length greater than 76 cm and closed at one end, is filled with mercury. The tube is then inverted into an open container of mercury. Some mercury flows down from the tube into the bowl. The space left at the top of the tube is nearly a vacuum because nothing is left but a negligible amount of mercury vapor. Points A and B are at the same level in the mercury and therefore are both at atmospheric pressure since the bowl is open to the air. The distance d from A to the top of the mercury column in the closed tube is a measure of the atmospheric pressure (often called *barometric pressure* because it is measured with a barometer). The barometer was invented by Evangelista Torricelli, an assistant to Galileo, in the 1600s; in his honor, one millimeter of mercury is called one torr.

Figure 9.8 A barometer

Physics at Home

When you next have a drink with a straw, insert the straw into the drink and place your finger over the upper opening of the straw so that no more air can enter the straw. Raise the lower end of the straw up out of your drink. Does the liquid in the straw flow back down into your glass? What do you suppose is holding the liquid in place? Make a free-body diagram on your paper napkin.

Some air is trapped between your finger and the top of the liquid in the straw; that air exerts a downward force on the liquid of magnitude $P_1 A$ (Fig. 9.9). A downward gravitational force mg also acts on the liquid. The air at the bottom of the straw exerts an upward force on the liquid of magnitude $P_{\text{atm}} A$; this upward force is what holds the liquid in place. Because the liquid does not pour out of the straw, but instead is in equilibrium,

$$P_{\text{atm}} A = P_1 A + mg$$

Thus, the pressure P_1 of the air trapped above the liquid must be less than atmospheric pressure.

How did P_1 become less than atmospheric pressure? As you pulled the straw up and out, the liquid in the straw falls a bit, expanding the volume available to the air trapped above the liquid. As we learn in Chapter 15, when a gas expands under conditions like this, its pressure decreases.

When you remove your finger from the top of the straw, air can get in at the top of the straw. Then the pressures above and below the liquid are equal, so the gravitational force pulls the liquid down and out of the straw.

Figure 9.9 Force acting on the liquid inside a straw.

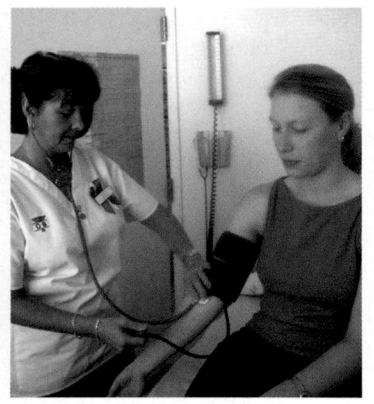

Figure 9.10 A sphygmomanometer

Sphygmomanometer

Blood pressure is measured with a sphygmomanometer (Fig. 9.10). The oldest kind of sphygmomanometer consists of a mercury manometer on one side attached to a closed bag—the cuff. The cuff is wrapped around the upper arm at the level of the heart and is then pumped up with air. The manometer measures the gauge pressure of the air in the cuff.

When the cuff pressure exceeds the highest pressure in the brachial artery, blood flow into the forearm is cut off. A valve on the cuff is then opened to allow the air to slowly escape. The sound of turbulent blood flow past the constriction is heard through a stethoscope when the pressure of the cuff begins to allow some blood to get through the brachial artery. At this point the pressure of the air in the cuff is equal to the maximum or *systolic* pressure—the pressure when the heart contracts.

As air continues to escape from the cuff, the sound of blood flowing through the constriction in the artery continues to be heard. When the pressure in the cuff reaches the *diastolic* pressure in the artery—the minimum pressure that occurs when the heart muscle is relaxed—there is no longer a constriction in the artery, so the pulsing sounds cease. The *gauge* pressures for a healthy heart are nominally around 120 mm Hg (systolic) and 80 mm Hg (diastolic).

9.6   ARCHIMEDES' PRINCIPLE

When an object is immersed in a fluid, the pressure on the lower surface of the object is higher than the pressure on the upper surface. The difference in pressures leads to an upward net force acting on the object due to the fluid pressure. If you try to push a beach ball underwater, you feel the effects of the buoyant force pushing the ball back up. It takes a rather large force to hold such an object completely underwater; the instant you let go, the object pops back up to the surface.

Consider a rectangular solid immersed in a fluid of density ρ (Fig. 9.11a). For each vertical face (left, right, front, and back), there is a face of equal area opposite it. The forces on these two faces due to the fluid are equal in magnitude since the areas and the average pressures are the same. The directions are opposite, so the forces acting on the vertical faces cancel in pairs.

Let the top and bottom surfaces each have area A. The force on the lower face of the block is $F_2 = P_2 A$; the force on the upper face is $F_1 = P_1 A$. The total force on the block due to the fluid, called the **buoyant force** F_B, is upward since $F_2 > F_1$ (Fig. 9.11b).

$$\vec{F}_B = \vec{F}_1 + \vec{F}_2$$
$$F_B = (P_2 - P_1)A$$

Figure 9.11 (a) Forces due to fluid pressure on the top and bottom of an immersed rectangular solid. (b) The buoyant force is the sum of $\vec{F}_1$ and $\vec{F}_2$.

(a)   (b)

Since $P_2 - P_1 = \rho g d$, the magnitude of the buoyant force can be written

> **Buoyant force**
>
> $$F_B = \rho g d A = \rho g V \tag{9-7}$$

where $V = Ad$ is the volume of the block.

Note that ρV is the mass of the volume V of the fluid that the block displaces. Thus, the buoyant force on the submerged block is equal to the weight of an equal volume of fluid, a result called **Archimedes' principle**.

> **Archimedes' principle**
>
> A fluid exerts an upward buoyant force on a submerged object equal in magnitude to the weight of the volume of fluid displaced by the object.

Archimedes' principle applies to a submerged object of any shape even though we derived it for a rectangular block. Why? Imagine replacing an irregular submerged object with enough fluid to fill the object's place. This "piece" of fluid is in equilibrium, so the buoyant force must be equal to its weight. The buoyant force is the net force exerted on the "piece" of fluid by the surrounding fluid, which is identical to the buoyant force on the irregular object since the two have the same shape and surface area.

The same argument can be used to show that if an object is only partly submerged, the buoyant force is still equal to the weight of fluid displaced. Equation (9-7) applies as long as V is the part of the object's volume below the fluid surface rather than the entire volume of the object.

The net force due to gravity and buoyancy acting on an object totally or partially immersed in a fluid (Fig. 9.12) is

$$\vec{F} = \vec{F}_B + \vec{W}$$

Figure 9.12   Forces acting on a floating ice cube

The force of gravity for an object of volume V_o and average density ρ_o is

$$W = mg = \rho_o g V_o$$

and the buoyant force is

$$F_B = \rho_f g V_f$$

where V_f and ρ_f are the volume of fluid displaced and the fluid density, respectively. Choosing up to be the $+y$-direction, the net force is

$$F_y = \rho_f g V_f - \rho_o g V_o \tag{9-8a}$$

Here F_y can be positive or negative, depending on which density is larger. Imagine releasing a pebble and an air bubble underwater. The pebble's average density is greater than the density of water, so the net force on it is downward; the pebble sinks. An air bubble's average density is less than the density of water, so the net force is upward, causing the bubble to rise toward the surface of the water.

As long as the object is completely submerged, the volumes of the object and the displaced fluid are the same and

$$F_y = (\rho_f - \rho_o)gV \tag{9-8b}$$

We might think of F_y as the *apparent weight* of the object when immersed in the fluid. If the object is suspended from a scale, the reading of the scale is the magnitude of F_y in Eq. (9-8b).

If $\rho_o < \rho_f$, the object floats with only part of its volume submerged. In equilibrium, the object displaces a volume of fluid whose weight is equal to the object's weight. At that point the gravitational force equals the buoyant force and the object floats. Setting $F_y = 0$ in Eq. (9-8a) yields

$$\rho_f g V_f = \rho_o g V_o$$

which can be rearranged as:

$$\frac{V_f}{V_o} = \frac{\rho_o}{\rho_f}$$

On the left side of this equation is the fraction of the object's volume that is submerged; it is equal to the ratio of the densities.

This ratio of densities is called the **specific gravity** of the material when ρ_f is the density of water at 4°C. Specific gravity is without units because it is a ratio of two densities. Water at 4°C is chosen as the reference material because at that temperature, the density of water is a maximum (at atmospheric pressure). The gram was originally defined as the mass of one cubic centimeter of water at 4°C. Thus, water at 4°C has a density of 1 g/cm³ (1000 kg/m³). The specific gravity of seawater is 1.025, which means that seawater has a density of 1.025 g/cm³ (1025 kg/m³).

Figure 9.13 The buoyant force due to the air keeps these balloons aloft.

Making The Connection:
hot air balloons

Specific gravity

$$\text{S.G.} = \frac{\rho}{\rho_{\text{water}}} = \frac{\rho}{1000 \text{ kg/m}^3} \qquad (9\text{-}9)$$

Gases such as air are fluids and exert buoyant forces just as liquids do. The buoyant force due to air is often negligible if an object's average density is much larger than the density of air. To see a significant buoyant force in air, we must use an object with a small average density. A hot air balloon has an opening at the bottom and a burner for heating the air within (Fig. 9.13). Many molecules of the heated air escape through the opening, decreasing the balloon's average density. When the balloon is less dense on average than the surrounding air, it rises; at higher altitudes, the surrounding air becomes less and less dense. At some particular altitude, the buoyant force is equal in magnitude to the weight of the balloon. Then, by Newton's second law, the net force on the balloon is zero. The balloon is in *stable* equilibrium at this altitude: if the balloon rises a bit, it experiences a net force downward, while if the balloon sinks down a bit, it is pushed back upward.

Example 9.6

The Golden (?) Falcon

A small statue in the shape of a falcon is painted black and has a weight of 24.1 N. The owner of the statue claims it is made of solid gold. When the statue is completely submerged in a container brimful of water, the weight of the water that spills over the top and into a bucket is 1.25 N. Find the density and specific gravity of the metal. Is the density consistent with the claim that the falcon is solid gold? Let g = 9.80 N/kg.

Strategy When the solid is totally submerged, it displaces a volume of water equal to its own volume. Thus, the volume of water spilled is equal to the volume of the statue. From the weight and the volume of the object, we can calculate its average density. Dividing by the density of water gives the specific gravity. We can compare the density with that of gold found in Table 9.1.

Solution The mass of 1.25 N of water is
$$m = \frac{W}{g}$$

$$= \frac{1.25 \text{ N}}{9.80 \text{ N/kg}} = 0.1276 \text{ kg}$$

The volume of the water is then
$$V = \frac{m}{\rho} = \frac{0.1276 \text{ kg}}{1000 \text{ kg/m}^3} = 1.276 \times 10^{-4} \text{ m}^3$$

This is also the volume of the statue. The mass of the metal is
$$\frac{24.1 \text{ N}}{9.80 \text{ N/kg}} = 2.459 \text{ kg}$$

and its average density is
$$\rho = \frac{m}{V} = \frac{2.459 \text{ kg}}{1.276 \times 10^{-4} \text{ m}^3} = 1.93 \times 10^4 \text{ kg/m}^3$$

Its specific gravity is therefore
$$\text{S.G.} = \frac{1.93 \times 10^4 \text{ kg/m}^3}{1000 \text{ kg/m}^3} = 19.3$$

From Table 9.1, we find that gold also has a density of 1.93×10^4 kg/m³. The metal has the correct density; it may *possibly* be gold.

Discussion An alternative solution recognizes that 24.1 N and 1.25 N are the weights of equal volumes of metal and water. Then the ratio of the weights gives us the ratio of the densities, which is the specific gravity of the statue:
$$\frac{W_{\text{metal}}}{W_{\text{water}}} = \frac{\rho_{\text{metal}} V g}{\rho_{\text{water}} V g} = \frac{\rho_{\text{metal}}}{\rho_{\text{water}}} = \text{S.G.}$$

$$\text{S.G.} = \frac{24.1 \text{ N}}{1.25 \text{ N}} = 19.3$$

This method shows that the value of g does not affect the result.

According to legend, this method to determine the specific gravity of a solid was discovered by Archimedes in the third century B.C.E. King Hieron II asked Archimedes to find a way to check whether his crown was made of pure gold—without melting down the crown, of course! Archimedes came up with his method while he was taking a bath; he noticed the water level rising as he got in and connected the rising water level

continued on next page

Example 9.6 *continued*

with the volume of water displaced by his body. In his excitement, he jumped out of the bath and ran naked through the streets of Siracusa (a city in Sicily) shouting "Eureka!"

Practice Problem 9.6   Identifying an unknown substance

An unknown solid substance has a weight of 142.0 N. The object is suspended from a scale and hung so that it is completely submerged in the water (but not touching bottom). The scale reads 129.4 N. Find the specific gravity of the object and determine whether the substance could be anything listed in Table 9.1.

Example 9.7

Hidden Depths of an Iceberg

 What percentage of a floating iceberg's volume is above water? The specific gravity of ice is 0.917 and the specific gravity of the surrounding seawater is 1.025.

Strategy   The ratio of the density of ice to the density of seawater tells us the ratio of the volume of ice that is submerged in the seawater to the total volume of the iceberg. The rest of the ice is above the water.

Solution   We could calculate the densities of seawater and of ice in SI units from their specific gravities, but that is unnecessary; the ratio of the specific gravities is equal to the ratio of the densities:

$$\frac{\text{S.G.}_{\text{ice}}}{\text{S.G.}_{\text{seawater}}} = \frac{\rho_{\text{ice}}/\rho_{\text{water}}}{\rho_{\text{seawater}}/\rho_{\text{water}}} = \frac{\rho_{\text{ice}}}{\rho_{\text{seawater}}}$$

The fraction of the iceberg's volume that is submerged is equal to the ratio of the densities of ice and seawater. Thus, the ratio of the volume submerged to the total volume of ice is

$$\frac{V_{\text{submerged}}}{V_{\text{ice}}} = \frac{\rho_{\text{ice}}}{\rho_{\text{seawater}}} = \frac{\text{S.G.}_{\text{ice}}}{\text{S.G.}_{\text{seawater}}}$$

$$= \frac{0.917}{1.025} = 0.895$$

89.5% of the ice is below the surface of the water, leaving only 10.5% above the surface.

Discussion   An alternative solution does not depend on remembering that the ratio of the volumes is equal to the ratio of the densities. The buoyant force is equal to the weight of a volume $V_{\text{submerged}}$ of water:

$$\text{buoyant force} = \rho_{\text{seawater}} V_{\text{submerged}} g$$

The weight of the iceberg is $mg = \rho_{\text{ice}} V_{\text{ice}} g$. From Newton's second law, the buoyant force must be equal in magnitude to the weight of the iceberg when it is floating in equilibrium:

$$\rho_{\text{seawater}} V_{\text{submerged}} g = \rho_{\text{ice}} V_{\text{ice}} g$$

or

$$\frac{V_{\text{submerged}}}{V_{\text{ice}}} = \frac{\rho_{\text{ice}}}{\rho_{\text{seawater}}}$$

The fact that ice floats is of great importance for the balance of nature. If ice were more dense than water, it would gradually fill up the ponds and lakes *from the bottom*. It would not form on top of lakes and remain there. The consequences for fish and other bottom dwellers of solidly frozen lakes would be catastrophic. The water below the surface layer of ice formed in winter remains just above freezing so that the fish are able to survive.

Practice Problem 9.7   Floating in freshwater versus seawater

If the average density of a human being is 980 kg/m³, what fraction of a human body floats above water in a freshwater pond and what fraction floats above seawater in the ocean? The specific gravity of seawater is 1.025.

Blood tests often include determination of the specific gravity of the blood—normally around 1.040 to 1.065. A reading that is too low may indicate anemia, since the presence of red blood cells increases the average density of the blood. Before taking blood from a donor, a drop of the blood is placed in a saline solution. If the drop does not sink, it is not safe for the donor to give blood because the concentration of red blood cells is too low. Urinalysis also includes a specific gravity measurement (normally 1.015 to 1.030); too high a value indicates an abnormally high concentration of dissolved salts, which can signal a medical problem.

Freighters, aircraft carriers, and cruise ships float, although they are made from steel and other materials that are more dense than seawater. When a ship floats, the buoyant force acting on the ship is equal to the ship's weight. A ship is constructed so that it displaces a volume of seawater larger than the volume of the steel and other construction materials. The *average* density of the ship is its weight divided by its total volume. A large part of a ship's interior is filled with air. All of the "empty" space contributes to the total volume; the resulting average density is less than that of seawater, allowing the ship to float.

Now we can understand how a hippopotamus can sink to the bottom of a pond: it can expel some of the air in its body by exhaling. Exhalation increases the average density of the hippopotamus so that it is just slightly above the density of the water; thus, it sinks. An armadillo does just the opposite: it swallows air, inflating its stomach and intestines, to increase the buoyant force for a swim across a large lake.

Conceptual Example 9.8

A Hovering Fish

 How is it that a fish is able to hover almost motionless in one spot—until some attractive food is spotted and, with a flip of the tail, off it swims after the food? Fish have a thin-walled bladder, called a swim bladder, located under the spinal column. The swim bladder contains a mixture of oxygen and nitrogen obtained from the blood of the fish. How does the swim bladder help the fish keep the buoyant and gravitational forces balanced so that it can hover?

Solution and Discussion If the fish's average density is greater than that of the surrounding water, it will sink; if its average density is smaller than that of the water, it will rise. By varying the volume of the swim bladder, the fish is able to vary its own volume and thus its average density. By adjusting its average density to match the density of the surrounding water, the fish can remain suspended in position. The fish can also adjust the volume of the bladder when it wants to rise or sink.

Conceptual Practice Problem 9.8   The diving beetle

A diving beetle traps a bubble of air under its wings. While under the water, the beetle uses the air in the bubble to breathe, gradually exchanging the oxygen for carbon dioxide. (a) What does the beetle do to the air bubble so that it can dive under the water? (b) Once under water, what does the beetle do so that it can rise to the surface? [*Hint:* Treat the beetle and the air bubble as a single system. How can the beetle change the buoyant force acting on the system?]

9.7   FLUID FLOW

Fluid dynamics (or *hydrodynamics*) is the study of *moving* fluids. The blood flowing in our bodies is a hydrodynamic system. The air all around us is another hydrodynamic system. We simplify this wonderfully complex subject—to illustrate some important ideas in less complex situations—by limiting our study at first to fluids flowing under special conditions.

One difference between moving fluids and static fluids is that a moving fluid *does* exert a force *parallel* to any surface over or past which it flows; a static fluid does not. Since the moving fluid exerts a force against a surface, the surface must also exert a force on the fluid. This **viscous force** opposes the flow of the fluid; it is the counterpart to the kinetic frictional force between solids. An external force must act on a viscous fluid (and thereby do work) to keep it flowing. Viscosity is considered in Section 9.9. Until then, we consider only nonviscous fluids—fluid flow where the viscous forces are negligibly small.

Fluid flow can be characterized as steady or unsteady. When the flow is **steady**, the velocity of the fluid *at any point* is constant in time. The velocity is not necessarily the same everywhere, but at any particular point, the velocity of the fluid passing that point remains constant in time. The density and pressure of a steadily flowing fluid are also constant in time.

Steady flow is **laminar**. The fluid flows in neat layers so that each small portion of fluid that passes a particular point follows the same path as every other portion of fluid that passes the same point. The path that the fluid follows, starting from any point, is called a **streamline** (see Fig. 9.14). The streamlines may curve and bend, but they cannot

Figure 9.14 A wind tunnel shows the streamlines in the flow of air past a car.

cross each other; if they did, the fluid would have to "decide" which way to go when it gets to such a point. The direction of the fluid velocity at any point must be tangent to the streamline passing through that point. Streamlines are a convenient way to depict fluid flow in a sketch.

The special case that we consider first is the flow of an **ideal fluid**. An ideal fluid is incompressible, undergoes laminar flow, and has no viscosity. Under some conditions real fluids can be modeled as (nearly) ideal, but not under all conditions.

The flow of an ideal fluid is governed by two principles: the continuity equation and Bernoulli's equation. The continuity equation is an expression of conservation of mass: since no fluid is created or destroyed, the total mass of the fluid must be constant. Bernoulli's equation, discussed in Section 9.8, is a form of the energy conservation law applied to fluid flow. Together, these two equations enable us to predict the flow of an ideal fluid.

The Continuity Equation

We start by deriving the continuity equation, which relates the speed of flow to the cross-sectional area of the fluid. Suppose an incompressible fluid flows into a pipe of nonuniform cross-sectional area under conditions of steady flow. In Fig. 9.15, the fluid on the left moves at speed v_1. During a time Δt, the fluid travels a distance

$$x_1 = v_1 \Delta t$$

If A_1 is the cross-sectional area of this section of pipe, then the mass of water moving past point 1 in time Δt is

$$\Delta m_1 = \rho V_1 = \rho A_1 x_1 = \rho A_1 v_1 \Delta t$$

During this same time interval, the mass of fluid moving past point 2 is

$$\Delta m_2 = \rho V_2 = \rho A_2 x_2 = \rho A_2 v_2 \Delta t$$

Figure 9.15 An incompressible fluid flowing horizontally through a nonuniform pipe

But, if the flow is steady, the mass passing through one section of pipe in time interval Δt must pass through any other section of the pipe in the same time interval. Therefore,

$$\Delta m_1 = \Delta m_2$$

or

$$\rho A_1 v_1 \Delta t = \rho A_2 v_2 \Delta t \tag{9-10}$$

The quantity $\rho A v$ is the *mass flow rate* of the fluid:

Mass flow rate

$$\frac{\Delta m}{\Delta t} = \rho A v \quad \text{(SI unit: kg/s)} \tag{9-11}$$

Since the time intervals Δt are the same, Eq. (9-11) says that the mass flow rate past any two points is the same. Since the density of an incompressible fluid is constant, the volume flow rate past any two points must also be the same:

Volume flow rate

$$\frac{\Delta V}{\Delta t} = A v \quad \text{(SI unit: m}^3\text{/s)} \tag{9-12}$$

The **continuity equation** for an ideal fluid equates the volume flow rates past two different points:

Continuity equation for ideal fluid

$$A_1 v_1 = A_2 v_2 \tag{9-13}$$

The same volume of fluid that enters the pipe in a given time interval exits the pipe in the same time interval. Where the diameter of the tube is large, the speed of the fluid is small; where the diameter is small, the fluid speed is large. A familiar example is what happens when you use your thumb to partially block the end of a garden hose to make a jet of water. The water moves past your thumb, where the cross-sectional area is smaller, at a greater speed than it moves in the hose. Similarly, water traveling along a river speeds up, forming rapids, when the riverbed narrows or is partially blocked by rocks and boulders.

Streamlines are closer together where the fluid flows faster and farther apart where it flows more slowly (Fig. 9.16). Thus streamlines help us visualize fluid flow. The fluid velocity at any point is tangent to a streamline through that point.

Figure 9.16 Streamlines in a pipe of varying diameter.

Figure 9.17 Demonstrating the continuity equation at the kitchen sink.

Physics at Home

The continuity equation applies to an ideal fluid even if it is not flowing through a pipe. Turn on a faucet so that the water flows out in a moderate stream [Fig. 9.17]. The falling water is in free fall, accelerated by gravity until it hits the sink below. As the water falls, its speed increases. The stream of water gradually narrows as it falls so that the product of speed and cross-sectional area is constant, as predicted by the continuity equation.

Example 9.9

Speed of Blood Flow

The heart pumps blood into the aorta, which has an inner radius of 1.0 cm. The aorta feeds 32 major arteries. If blood in the aorta travels at a speed of 28 cm/s, at approximately what average speed does it travel in the arteries? Assume that blood can be treated as an ideal fluid and that the arteries each have an inner radius of 0.21 cm.

Strategy   Since we have assumed blood to be an ideal fluid, we can apply the continuity equation. The main tube (the aorta) is connected to multiple tubes (the major arteries), so this problem seems to be more complicated than a single pipe with a constriction in it. What matters here is the total cross-sectional area into which the blood flows.

Solution   We start by finding the cross-sectional area of the aorta

$$A_1 = \pi r_{aorta}^2$$

and then the total cross-sectional area of the arteries

$$A_2 = 32\pi r_{artery}^2$$

Now we apply the continuity equation and solve for the unknown speed.

$$A_1 v_1 = A_2 v_2$$

$$v_2 = v_1 \frac{A_1}{A_2} = 0.28 \text{ m/s} \times \frac{\pi \times (0.010 \text{ m})^2}{32\pi \times (0.0021 \text{ m})^2} = 0.20 \text{ m/s}$$

Discussion   The blood flow slows in the arteries because the total cross-sectional area is greater than that of the aorta alone. From the arteries, the blood travels to the many capillaries of the body. Each capillary has a tiny cross-sectional area, but there are so many of them that the blood flow slows greatly once it reaches the capillaries—allowing time for the exchange of oxygen, carbon dioxide, and nutrients between the blood and the body tissues.

Practice Problem 9.9   Hosing down a wastebasket

A garden hose fills a 32-L wastebasket in 120 s. The opening at the end of the hose has a diameter of 1.95 cm. (a) How fast is the water traveling as it leaves the hose? (b) How fast does the water travel if half the exit area is obstructed by placing a finger over the opening?

9.8   BERNOULLI'S EQUATION

The continuity equation relates the flow velocities of an ideal fluid at two different points, based on the change in cross-sectional area of the pipe. However, its application requires that one of the two velocities be known. We haven't yet addressed the question of how the fluid is made to flow in the first place. Now we do so by analyzing the forces that act on the fluid.

According to the continuity equation, the fluid must speed up as it enters a constriction (Fig. 9.18) and then slow down to its original speed when it leaves the constriction. What forces cause the flow speed to change? Consider a small volume of fluid entering the constriction at position A. Its acceleration is to the right since it is speeding up; therefore, the net force must be to the right. The only horizontal forces acting on that volume of fluid are due to the pressure of the surrounding fluid. The pressure ahead of that volume (P_2) must be smaller than the pressure behind it (P_1). At position B, the volume of fluid has an acceleration to the left since it is slowing down. The net force is to the left, so the pressure ahead of it (P_1) is larger than the pressure behind it (P_2). For horizontal flow *the speed is higher where the pressure is lower*. This principle is often called the **Bernoulli effect**.

The Bernoulli effect: Fluid flows faster where the pressure is lower.

$$P_2 < P_1$$
$$v_2 > v_1$$
(a)

(b)

Figure 9.18   (a) A small volume of fluid speeds up as it moves into a constriction (position A) and then slows down as it moves out of the constriction (position B). (b) Unequal forces due to fluid pressure cause acceleration to the right at position A and to the left at position B.

The Bernoulli effect can seem counterintuitive at first; isn't rapidly moving fluid at *high* pressure? For instance, if you were hit with the fast-moving water out of a firehose, you would be knocked over easily. The force that knocks you over is indeed due to fluid pressure; you would justifiably conclude that the pressure was high. However, the pressure is not high *until you slow down the water* by getting in its way. The rapidly moving water in the jet is, in fact, approximately at atmospheric pressure (zero gauge pressure). So your conclusion is correct: when you stop the water, its pressure increases dramatically.

To find the quantitative relationship between pressure changes and flow speed changes for an ideal fluid, we apply conservation of energy. In Fig. 9.19, the shaded volume of fluid flows to the right. If the left end moves a distance Δx_1, then the right end moves a distance Δx_2. Since the fluid is incompressible,

$$A_1 \Delta x_1 = A_2 \Delta x_2 = V$$

Work is done by the neighboring fluid during this flow. Fluid behind (to the left) pushes forward, doing positive work, while fluid ahead pushes backward, doing negative work. The total work done on the shaded volume by neighboring fluid is

$$W = P_1 A_1 \Delta x_1 - P_2 A_2 \Delta x_2 = (P_1 - P_2)V$$

Since no dissipative forces act on an ideal fluid, the work done is equal to the total change in kinetic and gravitational potential energy. The net effect of the displacement is to move a volume V of fluid from height y_1 to height y_2 and to change its speed from v_1 to v_2. The energy change is therefore

$$\Delta E = \Delta K + \Delta U = \tfrac{1}{2}m(v_2^2 - v_1^2) + mg(y_2 - y_1)$$

where the +y-direction is up. Substituting $m = \rho V$ and equating the work done on the fluid to the change in its energy yields

$$(P_1 - P_2)V = \tfrac{1}{2}\rho V(v_2^2 - v_1^2) + \rho Vg(y_2 - y_1)$$

Dividing both sides by V and rearranging yields Bernoulli's equation, named after Swiss mathematician Daniel Bernoulli.

Bernoulli's equation

$$P_1 + \rho g y_1 + \tfrac{1}{2}\rho v_1^2 = P_2 + \rho g y_2 + \tfrac{1}{2}\rho v_2^2$$

$$\text{(or } P + \rho g y + \tfrac{1}{2}\rho v^2 = \text{constant)} \tag{9-14}$$

Bernoulli's equation relates the pressure, flow speed, and height at two points in an ideal fluid. Although we derived Bernoulli's equation in a relatively simple situation, it applies to the flow of any ideal fluid as long as points 1 and 2 are on the same streamline.

Each term in Bernoulli's equation has units of pressure, which in the SI system is Pa or N/m². Since a joule is a newton-meter, the pascal is also equal to a joule per cubic meter (J/m³). Each term represents energy per unit volume (energy density). The pressure is the work done by the fluid on the fluid ahead of it per unit volume of flow. The kinetic energy density (kinetic energy per unit volume) is $\tfrac{1}{2}\rho v^2$ and the gravitational potential energy density is $\rho g y$.

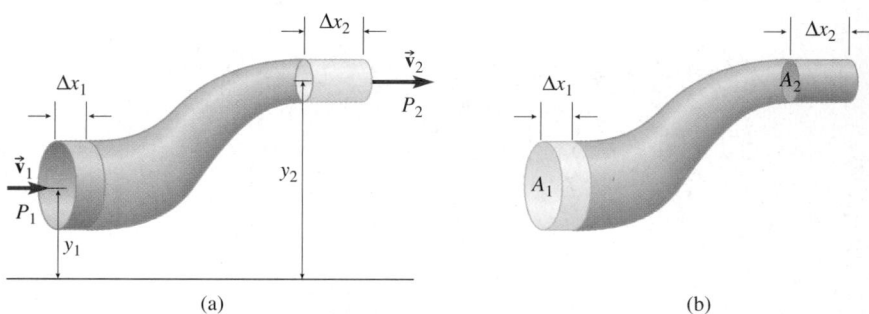

Figure 9.19 Applying conservation of energy to the flow of an ideal fluid. The shaded volume of fluid in (a) is flowing to the right; (b) shows the same volume of fluid a short time later.

We can check Bernoulli's equation in some special cases. For horizontal flow, $y_1 = y_2$. Then the pressure is higher where the speed is smaller, as expected. For a static fluid, $v_1 = v_2 = 0$. In this case Bernoulli's equation reduces to

$$P_1 + \rho g y_1 = P_2 + \rho g y_2$$

or, if $d = y_1 - y_2$,

$$P_2 - P_1 = -(\rho g y_2 - \rho g y_1) = \rho g d$$

which is the pressure dependence with depth for a static fluid that we found in Section 9.4.

Example 9.10

The Venturi Meter

A *Venturi meter* (Fig. 9.20) measures fluid speed in a pipe. A constriction (of cross-sectional area A_2) is put in a pipe of normal cross-sectional area A_1. Two vertical tubes, open to the atmosphere, rise from two points, one of which is in the constriction. The vertical tubes function like manometers, enabling the pressure to be determined. From this information the flow speed in the pipe can be determined.

Suppose that the pipe in question carries water, $A_1 = 2.0A_2$, and the fluid heights in the vertical tubes are $h_1 = 1.20$ m and $h_2 = 0.80$ m. Let $g = 9.80$ N/kg. (a) Find the ratio of the flow speeds v_2/v_1. (b) Find the gauge pressures P_1 and P_2. (c) Find the flow speed v_1 in the pipe.

Strategy Neither of the two flow speeds is given. We need more than Bernoulli's equation to solve this problem. Since we know the ratio of the areas, the continuity equation gives us the ratio of the speeds. The height of the water in the vertical tubes enables us to find the pressures at points 1 and 2. The fluid pressure at the bottom of each vertical tube is the same as the pressure of the moving fluid just beneath each tube—otherwise water would flow into or out of the vertical tubes until the pressure equalized. The water in the vertical tubes is static, so the gauge pressure at the bottom is $P = \rho g d$. Once we have the ratio of the speeds and the pressures, we apply Bernoulli's equation.

Solution (a) From the continuity equation, the product of flow speed and area must be the same at points 1 and 2. Therefore,

$$\frac{v_2}{v_1} = \frac{A_1}{A_2} = 2.0$$

The water flows twice as fast in the constriction as in the rest of the pipe.

(b) The gauge pressures are:

$$P_1 = \rho g d_1 = (1000 \text{ kg/m}^3) \times (9.80 \text{ N/kg}) \times (1.20 \text{ m}) = 11.8 \text{ kPa}$$

$$P_2 = \rho g d_2 = (1000 \text{ kg/m}^3) \times (9.80 \text{ N/kg}) \times (0.80 \text{ m}) = 7.8 \text{ kPa}$$

(c) Now we apply Bernoulli's equation. We can use gauge pressures as long as we do so on both sides—in effect we are just subtracting atmospheric pressure from both sides of the equation:

$$P_1 + \rho g y_1 + \tfrac{1}{2}\rho v_1^2 = P_2 + \rho g y_2 + \tfrac{1}{2}\rho v_2^2$$

Since the tube is horizontal, $y_1 \approx y_2$ and we can neglect the small change in gravitational potential energy density $\rho g y$. Then

$$P_1 + \tfrac{1}{2}\rho v_1^2 = P_2 + \tfrac{1}{2}\rho v_2^2$$

We are trying to find v_1, so we can eliminate v_2 by substituting $v_2 = 2.0v_1$:

$$P_1 + \tfrac{1}{2}\rho v_1^2 = P_2 + \tfrac{1}{2}\rho(2v_1)^2$$

Simplifying,

$$P_1 - P_2 = \tfrac{3}{2}\rho v_1^2$$

$$v_1 = \sqrt{\frac{2 \times (11{,}800 \text{ Pa} - 7800 \text{ Pa})}{3 \times 1000 \text{ kg/m}^3}} = 1.6 \text{ m/s}$$

Discussion The assumption that $y_1 \approx y_2$ is fine as long as the pipe radius is small compared to the difference between the static water heights (40 cm). Otherwise, we would have to account for the different y values in Bernoulli's equation.

One subtle point: recall that we assumed that the fluid pressure at the bottom of the vertical tubes was the same as the pressure of the moving fluid just beneath. Does that contradict Bernoulli's equation? Since there is an abrupt change in fluid speed, shouldn't there be a significant difference in the pressures? No, because these points are not on the same streamline.

Practice Problem 9.10   Garden hose

Water flows horizontally through a garden hose of radius 1.0 cm at a speed of 1.4 m/s. The water shoots horizontally out of a nozzle of radius 0.25 cm. What is the gauge pressure of the water inside the hose?

Streamlines

Figure 9.20
Venturi meter

Example 9.11

Torricelli's Theorem

A barrel full of rainwater has a spigot near the bottom, at a depth of 0.80 m beneath the water surface. (a) When the spigot is directed horizontally (Fig. 9.21a) and is opened, how fast does the water come out? (b) If the opening points upward (Fig. 9.21b), how high does the resulting "fountain" go?

Strategy The water at the surface is at atmospheric pressure. The water emerging from the spigot is *also* at atmospheric pressure since it is in contact with the air. If the pressure of the emerging water were different than that of the air, the stream would expand or contract until the pressures were equal. We apply Bernoulli's equation to two points: point 1 at the water surface and point 2 in the emerging stream of water.

Solution (a) Since $P_1 = P_2$, Bernoulli's equation is

$$\rho g y_1 + \tfrac{1}{2}\rho v_1^2 = \rho g y_2 + \tfrac{1}{2}\rho v_2^2$$

Point 1 is 0.80 m above point 2, so

$$y_1 - y_2 = 0.80 \text{ m}$$

Barrel filled to top   Barrel filled to top

0.80 m $d = ?$

$\vec{v}$ $\vec{v}$

(a)   (b)

Figure 9.21
Full barrel of rainwater with open spigot (a) horizontal and (b) upward.

The speed of the emerging water is v_2. What is v_1, the speed of the water at the surface? The water at the surface is moving slowly, since the barrel is draining. The continuity equation requires that

$$\frac{v_2}{v_1} = \frac{A_1}{A_2}$$

Since the cross-sectional area of the spigot A_2 is much smaller than the area of the top of the barrel A_1, the speed of the water at the surface v_1 is negligibly small compared with v_2. Setting $v_1 = 0$, Bernoulli's equation reduces to

$$\rho g y_1 = \rho g y_2 + \tfrac{1}{2}\rho v_2^2$$

After dividing through by ρ, we solve for v_2:

$$g(y_1 - y_2) = \tfrac{1}{2}v_2^2$$

$$v_2 = \sqrt{2g(y_1 - y_2)} = 4.0 \text{ m/s}$$

(b) Now take point 2 to be at the top of the fountain. Then $v_2 = 0$ and Bernoulli's equation reduces to

$$\rho g y_1 = \rho g y_2$$

The "fountain" goes right back up to the top of the water in the barrel!

Discussion The result of part (b) is called Torricelli's theorem. In reality, the fountain does not reach as high as the original water level; some energy is dissipated due to viscosity and air resistance.

Practice Problem 9.11   Fluid in free fall

Verify that the speed found in part (a) is the same as if the water just fell 0.80 m straight down. That shouldn't be too surprising since Bernoulli's equation is an expression of energy conservation.

Figure 9.22   A perfume atomizer

Making The Connection:
air circulation in
underground burrows

Making The Connection:
household plumbing

Bernoulli's equation applies in an approximate way to moving air. Even though air is not incompressible, in many instances the density changes are small enough to be negligible. A perfume atomizer (Fig. 9.22) is an application of the Bernoulli effect. A squeeze bulb forces air at a high speed over the top of an open tube with its lower end in the perfume liquid. The reduced pressure at the top of the tube then pulls the perfume up into the airflow so that it is ejected with the air in a spray of fine droplets.

Prairie dogs, woodchucks, and rabbits need air circulation in their underground burrows (Fig. 9.23). The burrows always have at least two entrances. Due to differences in the size, shape, and location of the holes, the wind blows across them with different speeds. Where the wind blows across an entrance with a higher speed, the pressure is lower, so air flows out of the burrow at that entrance and air is pulled in at the other entrance. The pressure difference is increased if one entrance is higher than another, since the wind speed is generally higher on higher ground.

Household plumbing also must take the Bernoulli effect into account. In the drain pipe under a sink you can see a U-shaped pipe fitting called a trap (Fig. 9.24). Some water remains in the trap at all times to prevent sewer gases from backing up into the house. The drain that leads to the sewer line is vented to the outside air. Without such a vent, a rush of water through the sewer line from a toilet or washing machine would reduce the pressure at one side of the trap. Atmospheric pressure on the other side of the trap could then push the water out of the trap. The vent to the outside air maintains atmospheric pressure at both sides of the trap.

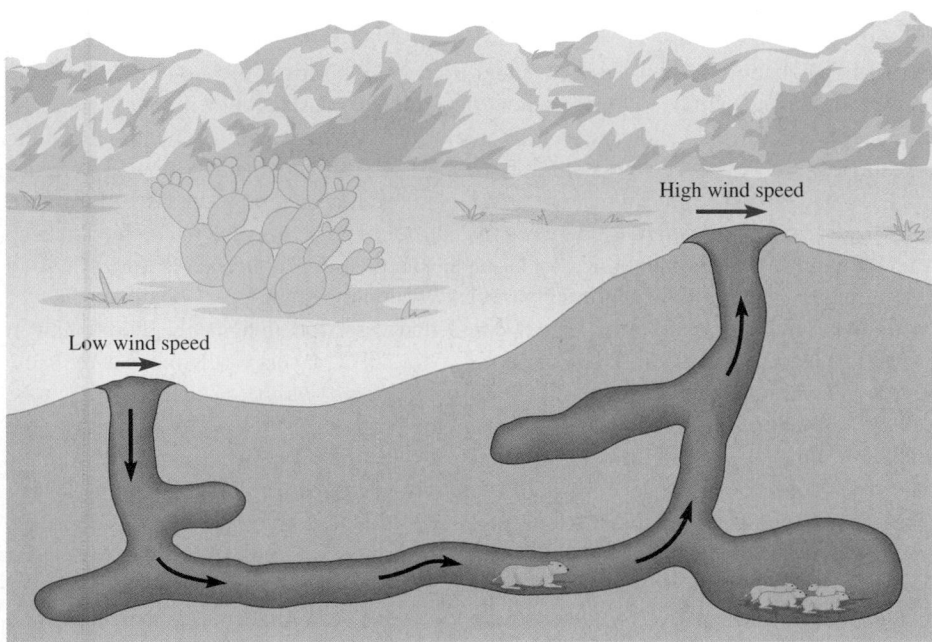

Figure 9.23 Air circulation in a prairie dog burrow.

Figure 9.24 Household plumbing must be vented due to the Bernoulli effect.

9.9   VISCOSITY

Bernoulli's equation ignores viscosity (fluid friction). According to Bernoulli's equation, an ideal fluid can continue to flow in a horizontal pipe at constant velocity on its own, just as a hockey puck would slide across frictionless ice at constant velocity without anything pushing it along. However, all real fluids have some viscosity; to maintain flow in a viscous fluid, we have to apply an external force since viscous forces oppose the flow of the fluid (Fig. 9.25). A *pressure difference* between the ends of the pipe must be maintained to keep a real liquid moving through a horizontal pipe. The pressure difference is important—in everything from blood flowing through arteries to oil pumped through a pipeline.

To visualize viscous flow in a tube of circular cross section, imagine the fluid to flow in cylindrical layers, or shells. If there were no viscosity, all the layers would move at the same speed (Fig. 9.26a). In viscous flow, the fluid speed depends on the distance from the tube walls (Figs. 9.26b and c). The fastest flow is at the center of the tube. Layers closer to the wall of the tube move more slowly. The outermost layer of fluid, which is in contact with the tube, does not move. Each layer of fluid exerts viscous forces on the neighboring layers; these forces oppose the relative motion of the layers. The outermost layer exerts a viscous force on the tube. In reality, layers of discrete thickness do not exist; rather, the fluid speed increases continuously from zero (at the wall of the tube) to the maximum (at the center).

A liquid is more viscous if the cohesive forces between molecules are stronger. The viscosity of a liquid decreases with increasing temperature because the molecules become less tightly bound. A decrease in the temperature of the human body is dangerous because the viscosity of the blood increases and the flow of blood through the body is hindered. Gases, on the other hand, have an increase in viscosity for an increase in temperature. At higher temperatures the gas molecules move faster and collide more often with each other.

The coefficient of viscosity (or simply *the viscosity*) of a fluid is written as the Greek letter eta (η) and has units of pascal-seconds (Pa·s) in SI. Other viscosity units in common use are the poise (pronounced *pwäz*, symbol P; 1 P = 0.1 Pa·s) and the centipoise (1 cP = 0.01 P = 0.001 Pa·s). Table 9.2 lists the viscosities of some common fluids.

Poiseuille's Law

The volume flow rate $\Delta V/\Delta t$ for laminar flow of a viscous fluid through a horizontal, cylindrical pipe depends on several factors. First of all, the volume flow rate is proportional to the *pressure drop per unit length* ($\Delta P/L$)—also called the pressure gradient. If a pressure drop ΔP maintains a certain flow rate in a pipe of length L, then a similar pipe of length $2L$ needs twice the pressure drop to maintain the same flow rate (ΔP across the first half and another ΔP across the second half). Thus, the flow rate must be proportional to the pressure drop per unit length.

Figure 9.25 (a) To maintain viscous flow, a net force due to fluid pressure $(P_1 - P_2)A$ must be applied in the direction of flow to balance the viscous force F_v due to the pipe, which opposes flow. (b) The pressure in the fluid decreases from P_1 at the left end to P_2 at the right end.

Figure 9.26 (a) Nonviscous flow. (b) Viscous flow. (c) Viscous flow of blood in an artery.

Next, the flow rate is inversely proportional to the viscosity of the fluid. The more viscous the fluid, the smaller the flow rate, if all other factors are equal.

The only other consideration is the radius of the pipe. In the nineteenth century, during a study of flow in blood vessels, French physician Jean-Louis Marie Poiseuille discovered that the flow rate is proportional to the *fourth power* of the pipe radius:

Poiseuille's law

$$\frac{\Delta V}{\Delta t} = \frac{\pi}{8} \frac{\Delta P / L}{\eta} r^4 \qquad (9\text{-}15)$$

---

where $\Delta V / \Delta t$ is the volume flow rate, ΔP is the pressure difference between the ends of the pipe, r and L are the inner radius and length of the pipe, and η is the viscosity of the fluid. Poiseuille's name is pronounced *pwahzoy*, in a rough English approximation.

It isn't often that we encounter a *fourth-power* dependence. Why such a strong dependence on radius? First of all, if fluids are flowing through two different pipes at the *same speed*, the volume flow rates are proportional to radius squared (flow rate = speed multiplied by cross-sectional area). But, in viscous flow, the average flow speed is larger for wider pipes; fluid farther away from the walls can flow faster. It turns out that the average flow speed for a given pressure gradient is also proportional to radius squared, giving the overall fourth power dependence on the pipe radius of Poiseuille's law.

The strong dependence of flow rate on radius is important in blood flow. A person with cardiovascular disease has arteries narrowed by plaque deposits. To maintain the necessary

Table 9.2

Viscosities of Some Fluids

Substance	Temperature	Viscosity (Pa·s)
Gases		
Water vapor	100°C	1.3×10^{-5}
Air	0°C	1.7×10^{-5}
	20°C	1.8×10^{-5}
	30°C	1.9×10^{-5}
	100°C	2.2×10^{-5}
Liquids		
Acetone	30°C	0.30×10^{-3}
Methanol	30°C	0.51×10^{-3}
Ethanol	30°C	1.0×10^{-3}
Water	0°C	1.8×10^{-3}
	20°C	1.0×10^{-3}
	30°C	0.80×10^{-3}
	40°C	0.66×10^{-3}
	60°C	0.47×10^{-3}
	80°C	0.36×10^{-3}
	100°C	0.28×10^{-3}
Blood plasma	37°C	1.3×10^{-3}
Blood, whole	20°C	3.0×10^{-3}
	37°C	2.1×10^{-3}
SAE #10 oil	30°C	0.20
Glycerin	20°C	0.83
	30°C	0.63

blood flow to keep the body functioning, the blood pressure increases. If the diameter of an artery narrows to $\frac{1}{2}$ of its original value due to plaque deposits, the blood flow rate decreases to $\frac{1}{16}$ of its original value should the pressure drop across it stay the same. To compensate for some of this decrease in blood flow, the heart pumps harder, increasing the blood pressure. High blood pressure is not good either; it introduces its own set of health problems, not least of which is the increased demands placed on the heart muscle.

Example 9.12

Arterial Blockage

A cardiologist reports to her patient that the radius of the left anterior descending artery of the heart has narrowed by 10.0%. What percent increase in the blood pressure drop across the artery is required to maintain the normal blood flow through this artery?

Strategy   We assume that the viscosity of the blood has not changed, nor has the length of the artery. To maintain normal blood flow, the volume flow rate $\Delta V/\Delta t$ must stay the same. From Poiseuille's law, then, the product of r^4 and ΔP must be the same before and after the narrowing of the artery.

Solution   If r_1 is the normal radius and r_2 is the actual radius, a 10.0% reduction in radius means $r_2 = 0.900 r_1$. Then, from Poiseuille's law,

$$r_1^4 \Delta P_1 = r_2^4 \Delta P_2$$

We solve for the ratio of the pressure drops:

$$\frac{\Delta P_2}{\Delta P_1} = \frac{r_1^4}{r_2^4} = \frac{1}{(0.900)^4} = 1.52$$

Discussion   A factor of 1.52 means there is a 52% increase in the blood pressure difference across that artery. The

continued on next page

Example 9.12 *continued*

increased pressure must be provided by the heart. If the normal pressure drop across the artery is 10 mm Hg, then it is now 15.2 mm Hg. The person's blood pressure either must increase by 5.2 mm Hg or there will be a reduction in blood flow through this artery. The heart is under greater strain as it works harder, attempting to maintain an adequate flow of blood.

Practice Problem 9.12   New water pipe

The town water supply is operating at nearly full capacity. The town board decides to replace the water main with a bigger one to increase capacity. If the maximum flow rate is to increase by a factor of 4.0, by what factor should they increase the radius of the water main?

Turbulence

When the fluid velocity at a given point changes, the flow is *unsteady*. **Turbulence** is an extreme example of unsteady flow (Fig. 9.27). In turbulent flow, swirling vortices—little whirlpools of fluid—appear. The vortices are not stationary; they move with the fluid. The flow velocity at any point changes erratically; prediction of the direction or speed of fluid flow under turbulent conditions is difficult.

9.10 VISCOUS DRAG

When an object moves through a fluid, the fluid exerts a drag force on it. When the relative velocity between the object and the fluid is low enough for the flow around the object to be laminar, the drag force derives from viscosity and is called **viscous drag**. The viscous drag force is proportional to the speed of the object. For larger relative speeds, the flow becomes turbulent and the drag force is proportional to the square of the object's speed. The air resistance discussed in Chapter 3 was due to the turbulent flow of air.

The viscous drag force depends also on the shape and size of the object. For a spherical object, the viscous drag force is given by Stokes's law:

Figure 9.27   Colored smoke rising from the ground shows the turbulent flow of air past the wingtip of an airplane.

Stokes's law (viscous drag on a sphere)	
$F_D = 6\pi \eta r v$	(9-16)

where r is the radius of the sphere, η is the viscosity of the fluid, and v is the speed of the object with respect to the fluid.

An object's **terminal velocity** is the velocity that produces just the right drag force so that the net force is zero. An object falling at its terminal velocity has zero acceleration, so it continues moving at that constant velocity. Using Stokes's law, we can find the terminal velocity of a spherical object falling through a fluid. When the object moves at terminal velocity, the net force acting on it is zero. If $\rho_o > \rho_f$, the object sinks; the terminal velocity is downward and the viscous drag force acts upward to oppose the motion. For an object, such as a helium balloon in air or an air bubble in oil, that rises rather than sinks ($\rho_o < \rho_f$), the terminal velocity is *upward* and the drag force is *downward*.

Example 9.13

Falling Droplet

In an experiment to measure the electric charge of the electron, a fine mist of oil droplets is sprayed into the air and observed through a telescope as they fall. These droplets are so tiny that they soon reach their terminal velocity. If the radius of the droplets is 2.40 μm and the average density of the oil is 862 kg/m³, find the terminal speed of the droplets. The density of air is 1.29 kg/m³ and the viscosity of air is 1.8×10^{-5} Pa·s.

continued on next page

Example 9.13 *continued*

Strategy   When the droplets fall at their terminal velocity, the net force on them is zero. We set the net force equal to zero and use Stokes' law for the drag force.

Solution   We set the sum of the forces equal to zero when $v = v_t$.

$$\Sigma F_y = +F_D + F_B - W = 0$$

If m_{air} is the mass of displaced air, then

$$6\pi\eta r v_t + m_{air}g - m_{oil}g = 0$$

Solving for v_t,

$$v_t = \frac{g(m_{oil} - m_{air})}{6\pi\eta r}$$

$$= \frac{\frac{4}{3}\pi r^3 g(\rho_{oil} - \rho_{air})}{6\pi\eta r}$$

After dividing the numerator and denominator by πr, we substitute numerical values:

$$v_t = \frac{\frac{4}{3} \times (2.40 \times 10^{-6}\ \text{m})^2 \times 9.8\ \text{N/kg} \times (862\ \text{kg/m}^3 - 1.29\ \text{kg/m}^3)}{6 \times 1.8 \times 10^{-5}\ \text{Pa·s}}$$

$$= 6.0 \times 10^{-4}\ \text{m/s} = 0.60\ \text{mm/s}$$

Discussion   We should check the units in the final expression:

$$\frac{\text{m}^2 \times \text{N/kg} \times \text{kg/m}^3}{\text{Pa·s}} = \frac{\text{N/m}}{\text{N/m}^2 \times \text{s}} = \text{m/s}$$

Stokes's law was applied in this way by Robert Millikan in his experiments in 1909–1913 to measure the charge of the electron. Using an atomizer, Millikan produced a fine spray of oil droplets. The droplets picked up electric charge as they were sprayed through the atomizer. Millikan kept a droplet suspended without falling by applying an upward electric force. After removing the electric force, he measured the terminal speed of the droplet as it fell through the air. He calculated the mass of the droplet from the terminal speed and the density of the oil using Stokes's law. By setting the magnitude of the electric force equal to the weight of a suspended droplet, Millikan calculated the electric charge of the droplet. He measured the charges of hundreds of different droplets and found that they were all multiples of the same quantity—the charge of an electron.

Practice Problem 9.13   Rising bubble

Find the terminal velocity of an air bubble of 1.00 mm diameter in a cup of vegetable oil. The specific gravity of the oil is 0.840 and the viscosity is 0.160 Pa·s. Assume the diameter of the bubble does not change as it rises.

Physics at Home

A demonstration of terminal velocity can be done at home. Climb up a small stepladder, or lean over an upstairs balcony, and drop two objects at the same time: a coin and two or three nested cone-shaped paper coffee filters. You will see the effects of viscous drag on the coffee filters as they fall with a constant terminal velocity. Enlist the help of a friend so you can get a side view of the two objects falling. (See also Fig. 3.25.) Why do the coffee filters work so well?

For small particles falling in a liquid, the terminal velocity is also called the *sedimentation velocity*. The sedimentation velocity is often small for two reasons. First, if the particle isn't much more dense than the fluid, then the apparent weight (vector sum of the gravitational and buoyant forces) is small. Second, notice that the terminal velocity is proportional to r^2; viscous drag is most important for small particles. Thus, it can take a long time for the particles to sediment out of solution. Because the sedimentation velocity is proportional to g, it can be increased by the use of a centrifuge, a rotating container that creates artificial gravity of magnitude $g_{eff} = \omega^2 r$ (see Eq. (5-11) and Section 5.7). Ultracentrifuges are capable of rotating at 100,000 rev/min and produce artificial gravity approaching a million times g.

9.11   SURFACE TENSION

The surface of a liquid has special properties not associated with the interior of the liquid. The surface acts like a stretched membrane under tension. The **surface tension** (symbol γ, the Greek letter gamma) of a liquid is the force per unit *length* with which the surface pulls *on its edge*. The direction of the force is tangent to the surface at its edge. Surface tension is caused by the cohesive forces that pull the molecules toward each other.

Physics at Home

Place a needle (or a flat plastic-coated paper clip) gently on the surface of a glass of water. It may take some practice, but you should be able to get it to "float" on top of the water (Fig. 9.28). Now add some detergent to the water and try again. The detergent reduces the surface tension of the water so it is unable to support the needle. Soaps and detergents are *surfactants*—substances that reduce the surface tension of a fluid. The reduced surface tension allows the water to spread out more, wetting more of a surface to be cleaned.

Figure 9.28   Surface tension supports the weight of a needle.

The high surface tension of water enables water striders and other small insects to walk on the surface of a pond. The foot of the insect makes a small indentation in the water surface (Fig. 2.5); the deformation of the surface enables it to push upward on the foot as if the water surface were a thin sheet of rubber. Visually it looks similar to a person walking across the mat of a trampoline. Other small water creatures, such as mosquito larvae and planaria, hang from the surface of water, using surface tension to hold themselves up. In plants, surface tension aids in the transport of water from the roots to the leaves.

The high surface tension of water is a hindrance in the lungs. The exchange of oxygen and carbon dioxide between inspired air and the blood takes place in tiny sacs called *alveoli*, 0.05 to 0.15 mm in radius, at the end of the bronchial tubes.

If the mucus coating the alveoli had the same surface tension as other body fluids, the pressure difference between the inside and outside of the alveoli would not be great enough for them to expand and fill with air. The alveoli secrete a surfactant that decreases the surface tension in their mucous coating.

Bubbles

In an underwater air bubble, the surface tension of the water surface tries to contract the bubble while the pressure of the enclosed air pushes outward on the surface. In equilibrium, the air pressure inside the bubble must be larger than the water pressure outside so that the net outward force due to pressure balances the inward force due to surface tension. The excess pressure $\Delta P = P_{in} - P_{out}$ depends both on the surface tension and the size of the bubble. In Problem 62, you can show that the excess pressure is

$$\Delta P = \frac{2\gamma}{r} \tag{9-17}$$

Look closely at a glass of champagne and you can see strings of bubbles rising, originating from the same points in the liquid. Why don't bubbles spring up from random locations? A very small bubble would require an insupportably large excess pressure. The bubbles need some sort of nucleus—a small dust particle, for instance—on which to form so they can start out larger, with excess pressures that aren't so large. The strings of bubbles in the glass of champagne are showing where suitable nuclei have been "found."

Example 9.14

Lung Pressure

During inhalation the gauge pressure in the alveoli is about −400 Pa to allow air to flow in through the bronchial tubes. Suppose the mucous coating on an alveolus of initial radius 0.050 mm had the same surface tension as water (0.070 N/m). What lung pressure outside the alveoli would be required to begin to inflate the alveolus?

Strategy   We model an alveolus as a sphere coated with mucus. Due to the surface tension of the mucus, the alveolus must have a lower pressure outside than inside, as for a bubble.

Solution   The excess pressure is

$$\Delta P = \frac{2\gamma}{r} = \frac{2 \times 0.070 \text{ N/m}}{0.050 \times 10^{-3} \text{ m}} = 2.8 \text{ kPa}$$

Thus, the pressure inside the alveolus would be 2.8 kPa higher than the pressure outside. The gauge pressure inside is −400 Pa, so the gauge pressure outside would be

$$P_{out} = -0.4 \text{ kPa} - 2.8 \text{ kPa} = -3.2 \text{ kPa}$$

Discussion   The *actual* gauge pressure outside the alveoli is about −0.5 kPa rather than −3.2 kPa; then $\Delta P = P_{in} - P_{out} =$

continued on next page

Example 9.14 *continued*

−0.4 kPa − (−0.5 kPa) = 0.1 kPa rather than 2.8 kPa. Here the surfactant comes to the rescue; by decreasing the surface tension in the mucus, it decreases ΔP to about 0.1 kPa and allows the expansion of the alveoli to take place. For a newborn baby, the alveoli are initially collapsed, making the required pressure difference about 4 kPa. That first breath is as difficult an event as it is significant.

Practice Problem 9.14   Champagne bubbles

A bubble in a glass of champagne is filled with CO_2. When it is 2.0 cm below the surface of the champagne, its radius is 0.50 mm. What is the gauge pressure inside the bubble? Assume that champagne has the same average density as water and a surface tension of 0.070 N/m.

MASTER THE CONCEPTS

Summary

- Fluids are materials that flow and include both liquids and gases. A liquid is nearly incompressible, whereas a gas expands to fill its container.

- Pressure is the perpendicular force per unit area that a fluid exerts on any surface with which it comes in contact. The SI unit of pressure is the pascal (1 Pa = 1 N/m²).

- The average air pressure at sea level is 1 atm = 101.3 kPa.

- Pascal's principle: A change in pressure at any point in a confined fluid is transmitted everywhere throughout the fluid.

- The average density of a substance is the ratio of its mass to its volume

$$\rho = \frac{m}{V} \qquad (9\text{-}2)$$

- The specific gravity of a material is the ratio of its density to that of water at 4°C.

- Pressure variation with depth in a static fluid:

$$P_2 = P_1 + \rho g d \qquad (9\text{-}3)$$

where point 2 is a depth d below point 1.

- Instruments to measure pressure include the manometer and the barometer. The barometer measures the pressure of the atmosphere. The manometer measures a pressure difference.

- Gauge pressure is the amount by which the absolute pressure exceeds atmospheric pressure:

$$P_{\text{gauge}} = P_{\text{abs}} - P_{\text{atm}} \qquad (9\text{-}6)$$

- Archimedes' principle: a fluid exerts an upward buoyant force on a completely or partially submerged object equal in magnitude to the weight of the volume of fluid displaced by the object.

- In steady flow, the velocity of the fluid *at any point* is constant in time. In laminar flow, the fluid flows in neat layers so that each small portion of fluid that passes a particular point follows the same path as every other portion of fluid that passes the same point. The path that the fluid follows, starting from any point, is called a streamline. Laminar flow is steady. Turbulent flow is chaotic and unsteady. The viscous force opposes the flow of the fluid; it is the counterpart to the frictional force for solids.

- An ideal fluid exhibits laminar flow, has no viscosity, and is incompressible. The flow of an ideal fluid is governed by two principles: the continuity equation and Bernoulli's equation.

- The continuity equation states that the volume flow rate for an ideal fluid is constant:

$$\frac{\Delta V}{\Delta t} = A_1 v_1 = A_2 v_2 \qquad (9\text{-}12, 9\text{-}13)$$

- Bernoulli's equation relates pressure changes to changes in flow speed and height:

$$P_1 + \rho g y_1 + \tfrac{1}{2}\rho v_1^2 = P_2 + \rho g y_2 + \tfrac{1}{2}\rho v_2^2 \qquad (9\text{-}14)$$

- Poiseuille's law gives the volume flow rate $\Delta V/\Delta t$ for viscous flow in a horizontal pipe:

$$\frac{\Delta V}{\Delta t} = \frac{\pi}{8}\frac{\Delta P/L}{\eta} r^4 \qquad (9\text{-}15)$$

where ΔP is the pressure difference between the ends of the pipe, r and L are the inner radius and length of the pipe, and η is the viscosity of the fluid.

- Stokes's law gives the viscous drag force on a spherical object moving in a fluid:

$$F_{\text{D}} = 6\pi\eta r v \qquad (9\text{-}16)$$

- The surface tension γ (the Greek letter gamma) of a liquid is the force *per unit length* with which the surface pulls on its edge.

Highlighted Figures and Tables

F9.1 Fluid pressure caused by collisions (p. 302)

F9.2 Forces due to a static fluid (p. 303)

F9.4 Pascal's principle and the hydraulic lift (p. 305)

T9.1 Densities of some common substances (p. 307)

F9.7 Mercury manometer and measuring pressure (p. 310)

F9.11 Archimedes' principle (p. 312)

F9.15 Continuity equation (p. 317)

F9.17 Streamlines of fluid (p. 318)

F9.18 Unequal forces due to fluid pressure cause accelerations (p. 319)

F9.19 Applying conservation of energy to the flow of an ideal fluid (p. 320)

CONCEPTUAL QUESTIONS

1. Does a manometer (with one side open) measure absolute pressure or gauge pressure? How about a barometer? A tire pressure gauge? A sphygmomanometer?

2. A volunteer firefighter holds the end of a firehose as a strong jet of water emerges. (a) The hose exerts a large backward force on the firefighter. Explain the origin of this force. (b) If another firefighter steps on the hose, forming a constriction (a place where the area of the hose is smaller), the hose begins to pulsate wildly. Explain.

3. The weight of a boat is listed on specification sheets as its "displacement." Explain.

4. In tall buildings, the water supply system uses multiple pumping stations on different floors. At each station, water pumped up from below collects in a storage tank held at atmospheric pressure before it enters the pump. The storage tank supplies water to the floors below it. What are some of the reasons why these multiple pumping stations are used?

5. Can an astronaut on the Moon use a straw to drink from a normal drinking glass? How about if he pokes a straw through an otherwise sealed juice box? Explain.

6. It is commonly said that wood floats because it is "lighter than water" or that a stone sinks because it is "heavier than water." Are these accurate statements? If not, correct them.

7. Why must a blood pressure cuff be wrapped around the arm at the same vertical level as the heart?

8. A hot air balloon is floating in equilibrium with the surrounding air. (a) How does the pressure inside the balloon compare to the pressure outside? (b) How does the density of the air inside compare to the density outside?

9. When helium weather balloons are released, they are purposely underinflated. Why? [*Hint:* The balloons go to very high altitudes.]

10. Bernoulli's equation applies only to *steady flow*. Yet Bernoulli's equation allows the fluid velocity at one point to be different than the velocity at another point. For the fluid velocity to change, the fluid must be accelerated as it moves from one point to another. In what way is the flow *steady*, then?

11. Before getting an oil change, it is a good idea to drive a few miles to warm up the engine. Why?

12. Your ears "pop" when you change altitude quickly—such as during takeoff or landing in an airplane, or during a drive in the mountains. Curiously, if you are a passenger in a high-speed train, your ears sometimes pop as the speed of the train increases rapidly—even though there is little or no change in altitude. Explain.

13. It is easier to get a good draft in a chimney on a windy day than when the outside air is still, all other things being equal. Why?

14. Two soap bubbles of *different radii* are formed at the ends of a tube with a closed valve in the middle. What happens to the bubbles when the valve is opened? (If the alveoli in the lung did not have a surfactant that reduces surface tension in the smaller alveoli, the same thing would happen in the lung, with disastrous results!)

15. *Pascal's principle: proof by contradiction.* Points A and B are near each other at the same height in a fluid. Suppose $P_A > P_B$. (a) Can both v_A and v_B be zero? Explain. (b) Point C is just above point D in a static fluid. Suppose the pressure at C increases by an amount ΔP. What would happen if the pressure at D did not increase by the same amount?

16. Two pilots are arguing over competing explanations of what holds an airplane up. One says that air flows over the top of the wing faster than the air beneath the wing; therefore, the pressure beneath is higher than the pressure above. The other says that the wing deflects air downward; by Newton's third law (or conservation of momentum), the air pushes up on the wing. Use Fig. 9.29, which shows streamlines of air flowing past an airplane wing, to answer these questions. (a) Does air above the wing move faster than air beneath the wing? (b) Is the pressure beneath greater than the pressure above? (Changes in the air's density are negligible at subsonic speeds.) (c) Does the wing deflect air downward? (d) Which pilot is right—or are they both right?

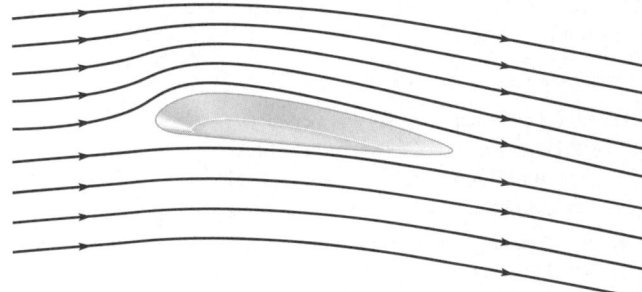

Figure 9.29 Streamlines about an airplane wing

17. What are the advantages of using hydraulic systems rather than mechanical systems to operate automobile brakes or the control surfaces of an airplane?

18. In any hydraulic system, it is important to "bleed" air out of the line. Why?

19. A large gravel truck is loosely covered with a tarpaulin. The edges of the tarp are fastened to the truck. If the truck moves fast enough, the tarp lifts up off the gravel. Why? [*Hint:* The air under the tarp is stationary with respect to the tarp, while the air above is moving with respect to the tarp.]

20. Is it possible for a skin diver to dive to any depth as long as his snorkel tube is sufficiently long? (A snorkel is a face mask with a breathing tube that sticks above the surface of the water.)

MULTIPLE CHOICE QUESTIONS

1. A glass of ice water is filled to the brim with water; the ice cubes stick up above the water surface. After the ice melts, which is true?

   (a) The water level is below the top of the glass.
   (b) The water level is at the top of the glass but no water has spilled.
   (c) Some water has spilled over the sides of the glass.
   (d) Impossible to say without knowing the initial densities of the water and the ice.

2. A dam holding back the water in a reservoir exerts a horizontal force on the water. The magnitude of this force depends on

   (a) The maximum depth of the reservoir.
   (b) The depth of the water at the location of the dam.
   (c) The surface area of the reservoir.
   (d) Both (a) and (b).
   (e) All three—(a), (b), and (c).

3. Bernoulli's equation applies to

   (a) Any fluid.
   (b) An incompressible fluid, whether viscous or not.
   (c) An incompressible, nonviscous fluid, whether the flow is turbulent or not.
   (d) An incompressible, nonviscous, nonturbulent fluid.
   (e) A static fluid only.

Questions 4–5. Two spheres, A and B, fall through the same viscous fluid.

Answer choices for Questions 4 and 5:

   (a) A has the larger terminal velocity
   (b) B has the larger terminal velocity
   (c) A and B have the same terminal velocity
   (d) Insufficient information is given to reach a conclusion

4. A and B have the same radius; A has the larger mass. Which has the larger terminal velocity?

5. A and B have the same density; A has the larger radius. Which has the larger terminal velocity?

6. Bernoulli's equation is an expression of

   (a) Conservation of mass.
   (b) Conservation of energy.
   (c) Conservation of momentum.
   (d) Conservation of angular momentum.

7. The continuity equation is an expression of

   (a) Conservation of mass.
   (b) Conservation of energy.
   (c) Conservation of momentum.
   (d) Conservation of angular momentum.

8. What is the gauge pressure of the gas in the closed tube in Fig. 9.30? (Take the atmospheric pressure to be 76 cm Hg.)

   (a) 20 cm Hg   (b) –20 cm Hg   (c) 96 cm Hg
   (d) 56 cm Hg   (e) –96 cm Hg   (f) –56 cm Hg

Figure 9.30 Multiple Choice Question 8

9. A manometer contains two different fluids of different densities (Fig. 9.31). Both sides are open to the atmosphere. Which pair(s) of points have equal pressure?

   (a) $P_1 = P_5$ (b) $P_2 = P_5$
   (c) $P_3 = P_4$ (d) Both (a) and (c)
   (e) Both (b) and (c)

Figure 9.31 Multiple Choice Question 9

10. A Venturi meter is used to measure the flow speed of a *viscous* fluid (Fig. 9.32). Which is true?

    (a) $h_3 = h_1$ (b) $h_3 > h_1$
    (c) $h_3 < h_1$ (d) Insufficient information to determine

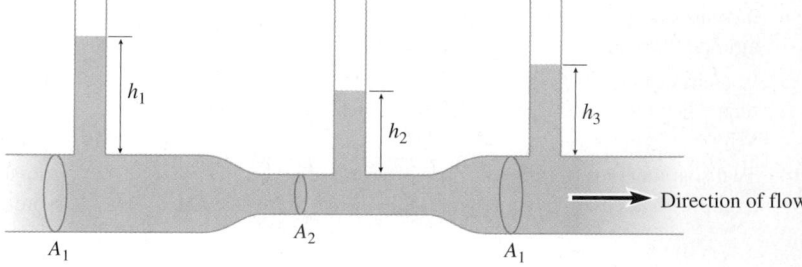

Figure 9.32 Multiple Choice Question 10

PROBLEMS

Note: **C** indicates a combination conceptual/quantitative problem. Gold diamonds ✦, ✦✦ are used to indicate the increasing level of difficulty of each problem. Problem numbers appearing in blue, 9., denote problems that have a detailed solution available in the Student Solutions Manual. Some problems are *paired* by concept; their numbers are connected by a ruled box.

9.2 Pressure

1. Someone steps on your toe, exerting a force of 500 N on an area of 1.0 cm² . What is the average pressure on that area in atm?

C 2. Atmospheric pressure is about 1.0×10^5 Pa on average. (a) What is the downward force of the air on a desktop with surface area 1.0 m²? (b) Convert this force to pounds so you really understand how large it is. (c) Why does this huge force not crush the desk?

3. What is the average pressure on the soles of the feet of a standing 90.0-kg person due to the contact force with the floor? Each foot has a surface area of 0.020 m².

4. The pressure inside a bottle of champagne is 4.5 atm higher than the air pressure outside. The neck of the bottle has an inner radius of 1.0 cm. What is the frictional force on the cork due to the neck of the bottle?

9.3 Pascal's Principle

5. A container is filled with gas at a pressure of 4.0×10^5 Pa. The container is a cube, 0.10 m on a side, with one side facing south. What is the magnitude and direction of the force on the south side of the container due to the gas inside?

6. A nurse applies a force of 4.40 N to the piston of a syringe. The piston has an area of 5.00×10^{-5} m². What is the pressure increase in the fluid within the syringe?

7. In a hydraulic lift, the diameters of the pistons are 5.00 cm and 20.0 cm. A car weighing $W = 10.0$ kN is to be lifted by the force of the large piston. (a) What force F_a must be applied to the small piston? (b) When the small piston is pushed in by 10.0 cm, how far is the car lifted? (c) Find the mechanical advantage of the lift, which is the ratio W/F_a.

8. A hydraulic lift is lifting a car that weighs 12 kN. The area of the piston supporting the car is A, the area of the other piston is a, and the ratio A/a is 100.0. How far must the small piston be pushed down to raise the car a distance of 1.0 cm? [*Hint:* Consider the work to be done.]

✦ 9. Depressing the brake pedal in a car pushes on a piston with cross-sectional area 3.0 cm². The piston applies pressure to the brake fluid, which is connected to two pistons, each with area 12.0 cm². Each of these pistons presses a brake pad against one side of a rotor attached to one of the rotating wheels (Fig. 9.33). (a) When the force applied by the brake pedal to the small piston is 7.5 N, what is the normal force applied to each side of the rotor? (b) If the coefficient of kinetic friction between a brake pad and the rotor is 0.80 and each pad is (on average) 12 cm from the rotation axis of the rotor, what is the torque on the rotor due to the two pads?

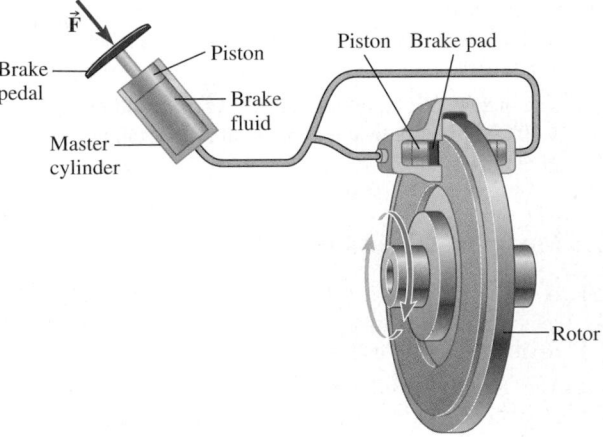

Figure 9.33 Problem 9

9.4 The Effect of Gravity on Fluid Pressure

10. At the surface of a freshwater lake the air pressure is 1.0 atm. At what depth under water in the lake is the water pressure 4.0 atm?

11. What is the pressure on a fish 10 m under the ocean surface?

12. How high can you suck water up a straw? The pressure in the lungs can be reduced to about 10 kPa below atmospheric pressure.

13. At the surface of a freshwater lake the pressure is 105 kPa. (a) What is the pressure increase in going 35.0 m below the surface? (b) What is the approximate pressure decrease in going 35 m above the surface? Air at 20°C has density of 1.20 kg/m³.

14. The maximum pressure most organisms can survive is about 1000 times atmospheric pressure. Only small, simple organisms such as tadpoles and bacteria can survive such high pressures. What then is the maximum depth at which these organisms can live under the sea (assuming that the density of seawater is 1025 kg/m³)?

15. In the Netherlands, a dike holds back the sea from a town below sea level. The dike springs a leak 3.0 m below the water surface. If the area of the hole in the dike is 1.0 cm², what force must the Dutch boy exert to save the town?

16. A container has a large cylindrical lower part with a long thin cylindrical neck (Fig. 9.34). The lower part of the container holds 12.5 m³ of water and the surface area of the

Figure 9.34 Problem 16

bottom of the container is 5.00 m². The height of the lower part of the container is 2.50 m and the neck contains a column of water 8.50 m high. The total volume of the column of water in the neck is 0.200 m³. What is the magnitude of the force exerted by the water on the bottom of the container?

17. The density of platinum is 21,500 kg/m³. Find the volume of 1.00 kg of platinum and compare it to the volume of 1.00 kg of aluminum.

9.5 Measuring Pressure

18. A woman's systolic blood pressure when resting is 160 mm Hg. What is this pressure in (a) Pa, (b) lb/in², (c) atm, (d) torr?

19. The gauge pressure of the air in an automobile tire is 32 lb/in². Convert this to (a) Pa, (b) torr, (c) atm.

20. An IV is connected to a patient's vein. The blood pressure in the vein has a gauge pressure of 12 mm Hg. At least how far above the vein must the IV bag be hung in order for fluid to flow into the vein? Assume the fluid in the IV has the same density as blood.

21. When a mercury manometer is connected to a gas main, the mercury stands 40.0 cm higher in the tube that is open to the air than in the tube connected to the gas main. A barometer at the same location reads 74.0 cm Hg. Determine the absolute pressure of the gas in cm Hg.

Figure 9.35   Problem 22

22. An experiment to determine the specific heat of a gas makes use of a water manometer attached to a flask (Fig. 9.35). Initially the two columns of water are even. Atmospheric pressure is 1.0×10^5 Pa. After heating the gas, the water levels change to those shown. Find the change in pressure of the gas in Pa.

23. A manometer using oil (density 0.90 g/cm³) as a fluid is connected to an air tank. Suddenly the pressure in the tank increases by 0.74 cm Hg. (a) By how much does the fluid level rise in the side of the manometer that is open to the atmosphere? (b) What would your answer be if the manometer used mercury instead?

24. Estimate the average blood pressure in a person's foot, if the foot is 1.37 m below the aorta, where the average blood pres-

sure is 104 mm Hg. For the purposes of this estimate, assume the blood isn't flowing.

9.6 Archimedes' Principle

25. A Canada goose floats with 25% of its volume below water. What is the average density of the goose?

26. A flat-bottomed barge, loaded with coal, has a mass of 3.0×10^5 kg. The barge is 20.0 m long and 10.0 m wide. It floats in freshwater. What is the depth of the barge below the waterline?

27. (a) What is the buoyant force on 0.90 kg of ice floating freely in liquid water? (b) What is the buoyant force on 0.90 kg of ice that is held completely submerged under water?

28. A block of wood floats in oil with 90.0% of its volume submerged. What is the density of the oil? The density of the block of wood is 0.67 g/cm³.

◆29. A piece of metal is released under water. The volume of the metal is 50.0 cm³ and its specific gravity is 5.0. What is its initial acceleration? [*Hint:* When $v = 0$, there is no drag force.]

◆30. (a) A piece of balsa wood with density 0.50 g/cm³ is released under water. What is its initial acceleration? [*Hint:* When $v = 0$, there is no drag force.] (b) Repeat for a piece of pine with density 0.750 g/cm³. (c) Repeat for a ping-pong ball with an average density of 0.125 g/cm³.

31. A cylindrical disk has volume 8.97×10^{-3} m³ and mass 8.16 kg. The disk is floating on the surface of some water with its flat surfaces horizontal. The area of each flat surface is 0.640 m². (a) What is the specific gravity of the disk? (b) How far below the water level is its bottom surface? (c) How far above the water level is its top surface?

32. An aluminum cylinder weighs 1.03 N. When this same cylinder is completely submerged in alcohol, the volume of the displaced alcohol is 3.90×10^{-5} m³. The apparent weight of the cylinder when completely submerged is 0.730 N. What is the specific gravity of the alcohol?

33. A fish uses a swim bladder to change its density so it is equal to that of water, enabling it to remain suspended under water. If a fish has an average density of 1080 kg/m³ and mass 10.0 g with the bladder completely deflated, to what volume must the fish inflate the swim bladder in order to remain suspended in seawater of density 1060 kg/m³?

34. The average density of a fish can be found by first weighing it in air and then finding the apparent weight of the fish as it hangs completely immersed in water. If a fish has weight 200.0 N and apparent weight 15.0 N in water, what is the average density of the fish?

9.7 Fluid Flow; 9.8 Bernoulli's Equation

35. A garden hose of inner diameter 2.0 cm carries water at 2.0 m/s. The nozzle at the end has diameter 0.40 cm. How fast does the water move through the nozzle?

36. If the average volume flow of blood through the aorta is 8.5×10^{-5} m³/s and the cross-sectional area of the aorta is

3.0×10^{-4} m^2, what is the average speed of blood in the aorta?

37. A nozzle of inner radius 1.00 mm is connected to a hose of inner radius 8.00 mm. The nozzle shoots out water moving at 25.0 m/s. (a) At what speed is the water in the hose moving? (b) What is the volume flow rate? (c) What is the mass flow rate?

38. Water entering a house flows with a speed of 0.20 m/s through a pipe of 2.0 cm inside diameter. What is the speed of the water at a point where the pipe tapers to a diameter of 5.0 mm?

39. Use Bernoulli's equation to estimate the upward force on an airplane's wing if the average flow speed of air is 190 m/s above the wing and 160 m/s below the wing. The density of the air is 1.3 kg/m^3 and the area of each wing surface is 28 m^2.

40. An airplane flies on a level path. There is a pressure difference of 500 Pa between the lower and upper surfaces of the wings. The area of each wing surface is about 100 m^2. The air moves below the wings at a speed of 80.5 m/s. Estimate (a) the weight of the plane and (b) the air speed above the wings.

✦41. The volume flow rate of the water supplied by a well is 2.0×10^{-4} m^3/s. The well is 40.0 m deep. (a) What is the power output of the pump—in other words, at what rate does the well do work on the water? (b) Find the pressure difference the pump must maintain. (c) Can the pump be at the top of the well or must it be at the bottom? Explain.

42. A nozzle is connected to a horizontal hose. The nozzle shoots out water moving at 25 m/s. What is the gauge pressure of the water in the hose? Neglect viscosity and assume that the diameter of the nozzle is much smaller than the inner diameter of the hose.

9.9 Viscosity

43. Using Poiseuille's law [Eq. (9-15)], show that viscosity has SI units of pascal-seconds.

44. A viscous liquid is flowing steadily through a pipe of diameter D. Suppose you replace it by two parallel pipes, each of diameter $D/2$, but the same length as the original pipe. If the pressure difference between the ends of these two pipes is the same as for the original pipe, what is the total rate of flow in the two pipes compared to the original flow rate?

45. A hypodermic syringe is attached to a needle that has an internal radius of 0.300 mm and a length of 3.00 cm. The needle is filled with a solution of viscosity 2.00×10^{-3} Pa·s; it is injected into a vein at a gauge pressure of 16.0 mm Hg. (a) What must the pressure of the fluid in the syringe be in order to inject the solution at a rate of 0.250 mL/s? (b) What force must be applied to the plunger, which has an area of 1.00 cm^2?

Problems 46–48. Four identical sections of pipe are connected in various ways to pumps that supply water at the pressures indicated in Fig. 9.36 (in units of 10^5 Pa). The water exits at the right at atmospheric pressure. Assume viscous flow.

Figure 9.36 Problems 46–48

46. If the *total* volume flow rates in systems A and C are the same and the flow speed in each of the pipes in C is 3.0 m/s, what is the flow speed in system A?

47. If the total volume flow rate in system B is 0.020 m^3/s, what is the total volume flow rate in system C?

48. If the total volume flow rates in systems A and B are the same, at what pressure does the pump supply water in system A?

49. (a) What is the pressure difference required to make blood flow through an artery of inner radius 2.0 mm and length 0.20 m at a speed of 6.0 cm/s? (b) What is the pressure difference required to make blood flow at 0.60 mm/s through a capillary of radius 3.0 μm and length 1.0 mm? (c) Compare both answers to your average blood pressure, about 100 torr.

50. (a) Since the flow rate is proportional to the pressure difference, show that Poiseuille's law can be written in the form $\Delta P = IR$, where I is the volume flow rate and R is a constant of proportionality called the fluid flow *resistance*. (Written this way, Poiseuille's law is analogous to *Ohm's law* for electrical current to be studied in Chapter 18: $\Delta V = IR$, where ΔV is the potential drop across a conductor, I is the electrical current flowing through the conductor, and R is the electrical resistance of the conductor.) (b) Find R in terms of the viscosity of the fluid and the length and radius of the pipe.

9.10 Viscous Drag

51. A dinoflagellate takes 5.0 s to travel 1.0 mm. Approximate a dinoflagellate as a sphere of diameter 70.0 μm (ignoring the flagellum). (a) What is the drag force on the dinoflagellate in seawater of viscosity 0.0010 Pa·s? (b) What is the power output of the flagellate?

52. An air bubble of 1.0-mm radius is rising in a container with vegetable oil of specific gravity 0.85 and viscosity 0.12 Pa·s. The container of oil and the air bubble are at 20°C. What is its terminal velocity?

53. Two identical spheres are dropped into two different columns: one column contains a liquid of viscosity 0.5 Pa·s, while the other contains a liquid of the same density but unknown viscosity. The sedimentation velocity in the second tube is 20% higher than the sedimentation velocity in the first tube. What is the viscosity of the second liquid?

54. A sphere of radius 1.0 cm is dropped into a glass cylinder filled with a viscous liquid. The mass of the sphere is 12.0 g

and the density of the liquid is 1200 kg/m³. The sphere reaches a terminal speed of 0.15 m/s. What is the viscosity of the liquid?

♦55. An aluminum sphere (specific gravity = 2.7) falling through water reaches a terminal speed of 5.0 cm/s. What is the terminal speed of an air bubble of the same radius rising through water? Assume viscous drag in both cases and ignore the possibility of changes in size or shape of the air bubble; the temperature is 20°C. [*Hint:* Compare the apparent weights of the two.]

© 56. This table gives the terminal speeds of various spheres falling through the same fluid. The spheres all have the same radius.

$m =$	5.0	11.3	20.0	31.3	45.0	80.0	(g)
$v_t =$	1.0	1.5	2.0	2.5	3.0	4.0	(cm/s)

Is the drag force primarily viscous or turbulent? Explain your reasoning.

© 57. This table gives the terminal speeds of various spheres falling through the same fluid. The spheres all have the same radius.

$m =$	8	12	16	20	24	28	(g)
$v_t =$	1.0	1.5	2.0	2.5	3.0	3.5	(cm/s)

Is the drag force primarily viscous or turbulent? Explain your reasoning.

58. *What keeps a cloud from falling?* A cumulus (fair-weather) cloud consists of tiny water droplets of average radius 5.0 μm. Find the terminal velocity for these droplets at 20°C, assuming viscous drag. (Besides the viscous drag force, there are also upward air currents called *thermals* that push the droplets upward.)

9.11 Surface Tension

59. An underwater air bubble has an excess inside pressure of 10 Pa. What is the excess pressure inside an air bubble with twice the radius?

60. Assume a water strider has a roughly circular foot of radius 0.02 mm. (a) What is the maximum possible upward force on the foot due to surface tension of the water? (b) What is the maximum mass of this water strider so that it can keep from breaking through the water surface? The strider has six legs.

♦♦61. There is potential energy associated with surface tension much like the elastic potential energy of a stretched spring or a balloon. Suppose we do work on a puddle of liquid, spreading it out through a distance of Δs along a line L perpendicular to the force (see Fig. 9.37). (a) What is the work done on the fluid surface in terms of γ, L, and Δs? (b) The work done is equal to the increase in surface energy of the fluid. Show that the increase in energy is proportional to the increase in area. (c) Show that we can think of γ as the surface energy per unit area. (d) Show that the SI units of surface tension can be expressed either as N/m (force per unit length) or J/m² (energy per unit area).

$$\Delta A = L \, \Delta s$$

Figure 9.37 Problem 61

♦♦62. A hollow hemispherical object is filled with air (Fig. 9.38a). (a) Show that the magnitude of the force due to fluid pressure on the curved surface of the hemisphere has magnitude $F = \pi r^2 P$, where r is the radius of the hemisphere and P is the pressure of the air. Neglect the weight of the air. [*Hint:* First find the force on the *flat* surface. What is the net force on the hemisphere due to the air?] (b) Consider an underwater air bubble to be divided into two hemispheres along the circumference (Fig. 9.38b). One hemisphere of the water surface exerts a force of magnitude $2\pi r\gamma$ (circumference times force per unit length) on the other due to surface tension. Show that the air pressure inside the bubble must exceed the water pressure outside by $\Delta P = (2\gamma)/r$.

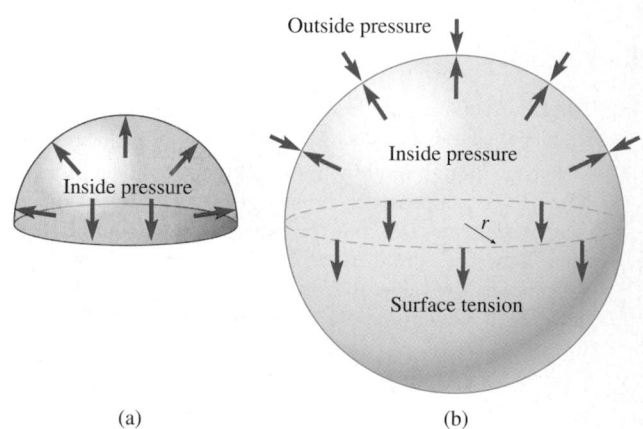

(a) (b)

Figure 9.38 Problem 62

COMPREHENSIVE PROBLEMS

63. The deepest place in the ocean is the Mariana Trench in the South Pacific, which is over 11.0 km deep. On January 23, 1960, the research sub *Trieste* went to a depth of 10.915 km, nearly to the bottom of the trench. This still is the deepest dive on record. The density of seawater is 1025 kg/m³. (a) What is the water pressure at that depth? (b) What was the force due to water pressure on a flat section of area 1.0 m² on the top of the sub's hull?

64. The pressure in a water pipe in the basement of an apartment house is 4.10×10^5 Pa, but on the seventh floor it is only 1.85 $\times 10^5$ Pa. What is the height between the basement and the seventh floor? Assume the water is not flowing; no faucets are opened.

65. The body of a 90.0-kg person contains 0.020 m³ of body fat. If the density of fat is 890 kg/m³, what percentage of the person's body weight is composed of fat?

66. Near sea level, how high a hill must you ascend for the reading of a barometer you are carrying to drop by 1.0 cm Hg? Assume the temperature remains at 20°C as you climb. The reading of a barometer on an average day at sea level is 76.0 cm Hg.

✦ 67. A plastic beach ball has radius 20.0 cm and mass 0.10 kg, not including the air inside. (a) What is the weight of the beach ball including the air inside? Assume the air density is 1.3 kg/m³ both inside and outside. (b) What is the buoyant force on the beach ball in air? The thickness of the plastic is about 2 mm—negligible compared to the radius of the ball. (c) The ball is thrown straight up in the air. At the top of its trajectory, what is its acceleration? [*Hint:* When $v = 0$, there is no drag force.]

68. A stone of weight W has specific gravity 2.50. (a) What is the apparent weight of the stone when under water? (b) What is its apparent weight in oil (specific gravity = 0.90)?

69. A simple hydrometer is an instrument for measuring the specific gravity of a liquid. Markings along a stem are calibrated to indicate the specific gravity for the level at which the hydrometer floats in a liquid (Fig. 9.39). The weighted base ensures that the hydrometer floats vertically. Suppose the hydrometer has a cylindrical stem of cross-sectional area 0.400 cm². The total volume of the bulb and stem is 8.80 cm³ and the mass of the hydrometer is 4.80 g. (a) How far from the top of the cylinder should a mark be placed to indicate a specific gravity of 1.00? (b) When the hydrometer is placed in alcohol, it floats with 7.25 cm of stem above the surface. What is the specific gravity of the alcohol? (c) What is the lowest specific gravity that can be measured with this hydrometer?

✦ 70. Are evenly spaced specific gravity markings on the cylinder of a hydrometer (Fig. 9.39) equal distances apart? In other words, is the depth d to which the cylinder is submerged linearly related to the density ρ of the fluid? To answer this question, assume that the cylinder has radius r and mass m. Find an expression for d in terms of ρ, r, and m and see if d is a linear function of ρ.

Hydrometer

Fluid to be tested

Figure 9.39
Simple hydrometer
(Problems 69 and 70)

71. The average speed of blood in the aorta is 0.3 m/s and the radius of the aorta is 1 cm. There are about 2×10^9 capillaries with an average radius of 6 μm. What is the approximate average speed of the blood flow in the capillaries?

72. If the cardiac output of a small dog is 4.1×10^{-3} m³/s, the diameter of its aorta is 1.0 cm, and the aorta length is 40.0 cm, determine the pressure drop across the aorta of the dog. Assume the viscosity of blood is 4.0×10^{-3} Pa·s.

✦ 73. To measure the airspeed of a plane, a device called a Pitot tube is used. A simplified model of the Pitot tube (Fig. 9.40) is a manometer with one side connected to a tube facing directly into the "wind" (stopping the air that hits it head-on) and the other side connected to a tube so that the "wind" blows across its openings. If the manometer uses mercury and the levels differ by 25 cm, what is the plane's airspeed? The density of air at the plane's altitude is 0.90 kg/m³.

25 cm

Figure 9.40   Problem 73

✦ 74. A house with its own well has a pump in the basement with an output pipe of inner radius 6.3 mm. Assume that the pump can maintain a gauge pressure of 410 kPa in the output pipe. A showerhead on the second floor (6.7 m above the pump's output pipe) has 36 holes, each of radius 0.33 mm. The shower is on "full blast" and no other faucet in the house is open. (a) Ignoring viscosity, with what speed does water leave the showerhead? (b) With what speed does water move through the output pipe of the pump?

75. In an aortic aneurysm, a bulge forms where the walls of the aorta are weakened. If blood flowing through the aorta (radius 1.0 cm) enters an aneurysm with a radius of 3.0 cm, how much on average is the blood pressure higher inside the aneurysm than the pressure in the unenlarged part of the aorta? The average flow rate through the aorta is 120 cm^3/s. Assume the blood is nonviscous and the patient is lying down so there is no change in height.

✦✦76. The diameter of a certain artery has decreased by 25% due to arteriosclerosis. (a) If the same amount of blood flows through it per unit time as when it was unobstructed, by what percentage has the blood pressure difference between its ends increased? (b) If, instead, the pressure drop across the artery stays the same, by what factor does the blood flow rate through it decrease? (In reality we are likely to see a combination of some pressure increase with some reduction in flow.)

77. Scuba divers are admonished not to rise faster than their air bubbles when rising to the surface. This rule helps them avoid the rapid pressure changes that cause the bends. Air bubbles of 2.0 mm diameter are rising from a scuba diver to the surface of the sea. Assume a water temperature of 20°C. (a) If the viscosity of the water is 1.0×10^{-3} Pa·s, what is the terminal velocity of the bubbles? (b) What is the largest rate of pressure change tolerable for the diver according to this rule?

✦78. A U-shaped tube is partly filled with water and partly filled with a liquid that does not mix with water (Fig. 9.41). Both sides of the tube are open to the atmosphere. What is the density of the liquid (in g/cm^3)?

0.50 m   0.30 m   0.45 m

Water

Figure 9.41
Problem 78

Ⓒ ✦79. Atmospheric pressure is equal to the weight of a vertical column of air, extending all the way up through the atmosphere, divided by the cross-sectional area of the column. (a) Explain why that must be true. [*Hint:* Apply Newton's second law to the column of air.] (b) If the air all the way up had a uniform density of 1.29 kg/m^3 (the density at sea level at 0°C), how high would the column of air be? (c) In reality, the density of air decreases with increasing altitude. Does that mean that the height found in (b) is a lower limit or an upper limit on the height of the atmosphere?

Ⓒ ✦80. On a nice day when the temperature outside is 20°C, you take the elevator to the top of the Sears Tower in Chicago, which is 440 m tall. (a) How much less is the air pressure at the top than the air pressure at the bottom? Express your answer both in pascals and atm. [*Hint:* The altitude change is small enough to treat the density of air as constant.] (b) How many pascals does the pressure decrease for every meter of altitude? (c) If the pressure gradient—the pressure decrease per meter of altitude—were uniform, at what altitude would the atmospheric pressure reach zero? (d) Atmospheric pressure does *not* decrease with a uniform gradient since the density of air decreases as you go up. Which is true: the pressure reaches zero at a *lower* altitude than your answer to (c), or the pressure is nonzero at that altitude and the atmosphere extends to a higher altitude? Explain.

Ⓒ 81. A shallow well usually has the pump at the top of the well. (a) What is the deepest possible well for which a surface pump will work? [*Hint:* A pump maintains a pressure difference, keeping the outflow pressure higher than the intake pressure.] (b) Why is there not the same depth restriction on wells with the pump at the bottom?

✦82. A bug from South America known as *Rhodnius prolixus* extracts the blood of animals. Suppose *Rhodnius prolixus* extracts 0.30 cm^3 of blood in 25 minutes from a human arm through its feeding tube of length 0.20 mm and radius 5.0 μm. What is the absolute pressure at the bug's end of the feeding tube if the absolute pressure at the other end (in the human arm) is 105 kPa? Assume the viscosity of blood is 0.0013 Pa·s. [*Note:* Negative absolute pressures are possible in liquids in very slender tubes.]

ANSWERS TO PRACTICE PROBLEMS

9.1 1.3×10^6 N/m^2 = 1.3 MPa; pressure is a factor of 15 greater than the pressure from the tennis shoe.

9.2 (a) 2.0×10^5 Pa; (b) 5.0 m

9.3 1.6 km

9.4 (a) Yes, $P_2 = P_1$. The column above point 2 is not as tall, but the pressure at the top of that column is *greater than* atmospheric pressure. (b) No, $P = P_{atm} + \rho g d$ gives the pressure at a depth d below a point where the pressure is P_{atm}.

9.5 (a) 32.0 cm; (b) 17.0 cm and 37.0 cm

9.6 S.G. = 11.3; could be lead

9.7 2% and 4%

9.8 (a) The beetle can squeeze the air bubble with its wings, compressing the air to reduce the bubble volume and decreasing the buoyant force. (b) When it is time to rise to the surface, the beetle relaxes the tension on the bubble, allowing it to expand again.

9.9 (a) 0.89 m/s; (b) 1.8 m/s

9.10 250 kPa

9.11 $\sqrt{2gh}$ = 4.0 m/s

9.12 1.4

9.13 2.9 mm/s upward

9.14 480 Pa

Elasticity and Oscillations

In the fairy tale, Jack's beanstalk grew taller and taller until it reached the sky, enabling Jack to climb up to the giant's kingdom. The giant is depicted as a greatly enlarged version of a man with his body having the same relative proportions as a human. If there were giants with bones composed of the same material as human bones, could their proportions be the same as those of humans? Is there a limit to how high a beanstalk might grow and still support its own weight?

Concepts & Skills to Review

- Hooke's law (Section 2.1)
- graphical relationship of position, velocity, and acceleration (Sections 3.2, 3.3)
- elastic potential energy (Section 6.5)

Figure 10.1   A tennis ball is flattened by contact with the strings of the tennis racquet.

10.1   ELASTIC DEFORMATIONS OF SOLIDS

If the net force and the net torque on an object are zero, the object is in equilibrium—but that does not mean that the forces and torques have no effect. An object is deformed when contact forces are applied to it (Fig. 10.1). A **deformation** is a change in the size or shape of the object. Many solids are stiff enough that the deformation cannot be seen with the human eye; a microscope or other sensitive device is required to detect the change in size or shape.

When the contact forces are removed, an **elastic** object returns to its original shape and size. Many objects are elastic as long as the deforming forces are not too large. On the other hand, any object may be permanently deformed or even broken if the forces acting are too large. An automobile that collides with a tree at a low speed may not be damaged; but at a higher speed the car suffers a permanent deformation of the bodywork and the driver may suffer a broken bone.

10.2   HOOKE'S LAW FOR TENSILE AND COMPRESSIVE FORCES

Suppose we stretch a wire by applying tensile forces of magnitude F to each end. The length of the wire increases from L to $L + \Delta L$. How does the elongation ΔL depend on the original length L? Conceptual Example 10.1 helps answer this question.

Conceptual Example 10.1

Stretching Wires

 If a given tensile force stretches a wire an amount ΔL, by how much would the same force stretch a wire twice as long but identical in thickness and composition?

Strategy and Solution   Think of the wire of length $2L$ as two wires of length L placed end-to-end (Fig. 10.2). Under the same tension, each of the two imagined wires stretches by an amount ΔL, so the total deformation of the long wire is $2\Delta L$.

Practice Problem 10.1   Cutting a spring in half

If a spring (spring constant k) is cut in half, what is the spring constant of each of the two newly formed springs?

Figure 10.2
Two identical wires are joined end-to-end and stretched by tensile forces. Each wire stretches an amount ΔL. The elongation of a wire is proportional to its original length L.

When stretched by the same tensile forces, the two wires in Conceptual Example 10.1 get longer by an amount proportional to their original lengths: $\Delta L \propto L$. In other words, the two wires have the same *fractional length change* $\Delta L/L$. The fractional length change is called the **strain**; it is a dimensionless measure of the degree of deformation.

$$\text{strain} = \frac{\Delta L}{L} \tag{10-1}$$

Suppose we had wires of the same composition and length but different thicknesses. It would require larger tensile forces to stretch the thicker wire the same amount as the thinner one; a thick steel cable is harder to stretch than the same length of a thin strand of steel. In Conceptual Question 13, we conclude that the tensile force required is proportional to the cross-sectional area of the wire ($F \propto A$). Thus, the same applied force *per unit area* produces the same deformation on wires of the same length and composition. The force per unit area is called the **stress**:

$$\text{stress} = \frac{F}{A} \qquad (10\text{-}2)$$

The SI units of stress are the same as those of pressure: N/m^2 or Pa.

Suppose that a solid object of initial length L is subjected to tensile or compressive forces of magnitude F. As a result of the forces, the length of the object is changed by magnitude ΔL. According to Hooke's law, the deformation is proportional to the deforming forces as long as they are not too large:

$$F = k\,\Delta L \qquad (10\text{-}3)$$

In Eq. (10-3), k is a constant analogous to the spring constant of a spring. This constant k depends on the length and cross-sectional area of the object. A larger cross-sectional area A makes k larger; a greater length L makes k smaller.

It is more convenient to write Hooke's law in terms of stress (F/A) and strain ($\Delta L/L$):

Hooke's law

stress $\propto$ strain

$$\frac{F}{A} = Y\frac{\Delta L}{L} \qquad (10\text{-}4)$$

Equation (10-4) still says that the length change (ΔL) is proportional to the magnitude of the deforming forces (F). Stress and strain account for the effects of length and cross-sectional area; the proportionality constant Y depends only on the inherent stiffness of the material from which the object is composed; it is independent of the length and cross-sectional area. Comparing Eqs. (10-3) and (10-4), the "spring constant" k for the object is

$$k = Y\frac{A}{L} \qquad (10\text{-}5)$$

The constant of proportionality Y in Eqs. (10-4) and (10-5) is called the **elastic modulus** or **Young's modulus**; Y has the same units as those of stress (Pa or N/m^2) since strain is dimensionless. Young's modulus can be thought of as the inherent stiffness of a material; it measures the resistance of the material to elongation or compression. Material that is flexible and stretches easily (for example, rubber) has a *low* Young's modulus. A stiff material (such as steel) has a high Young's modulus; it takes a larger stress to produce the same strain. Table 10.1 gives Young's modulus for a variety of common materials.

Table 10.1

Approximate Values of Young's Modulus for Various Substances

Substance	Young's Modulus $(10^9\,\text{Pa})$	Substance	Young's Modulus $(10^9\,\text{Pa})$
Rubber	0.002–0.008	Wood, along the grain	10–15
Human cartilage	0.024	Brick	14–20
Human vertebra	0.088 (compression); 0.17 (tension)	Concrete	20–30 (compression)
		Marble	50–60
Collagen, in bone	0.6	Aluminum	70
Human tendon	0.6	Cast iron	100–120
Wood, across the grain	1	Copper	120
Nylon	2–6	Wrought iron	190
Spider silk	4	Steel	200
Human femur	9.4 (compression); 16 (tension)	Diamond	1200

Hooke's law holds up to a maximum stress called the *proportional limit*. For many materials, Young's modulus has the same value for tension and compression. Some composite materials, such as bone and concrete, have significantly different Young's moduli for tension and compression. The components of bone include fibers of collagen (a protein found in all connective tissue) that give it strength under tension and hydroxyapatite crystals (composed of calcium and phosphate) that give it strength under compression. The different properties of these two substances lead to different values of Young's modulus for tension and compression.

Example 10.2

Compression of Femur

A man whose weight is 0.80 kN is standing upright (Fig. 10.3). By approximately how much is his femur (thighbone) shortened compared to when he is lying down? Assume that the compressive force on each femur is about half his weight. The average cross-sectional area of the femur is 8.0 cm^2 and the length of the femur when lying down is 43.0 cm.

Strategy   A change in length of the femur involves a strain. After finding the stress and looking up the Young's modulus, we can find the strain using Hooke's law. We assume that each femur supports *half* the man's weight.

Solution   The strain is proportional to the stress:

$$\frac{F}{A} = Y \frac{\Delta L}{L}$$

Solving this equation for ΔL gives

$$\Delta L = \frac{F/A}{Y} L$$

From Table 10.1, Young's modulus for a femur *in compression* is:

$$Y = 9.4 \times 10^9 \text{ Pa}$$

We need to convert the cross-sectional area to m^2 since 1 Pa = 1 N/m^2:

$$A = 8.0 \text{ cm}^2 \times \left(\frac{1 \text{ m}}{100 \text{ cm}}\right)^2 = 0.00080 \text{ m}^2$$

The force on each leg is 0.40 kN, or 4.0×10^2 N. The length change is then

$$\Delta L = \frac{F/A}{Y} L = \frac{(4.0 \times 10^2 \text{ N})/(0.00080 \text{ m}^2)}{9.4 \times 10^9 \text{ Pa}} \times 43.0 \text{ cm}$$

$$= 5.3 \times 10^{-5} \times 43.0 \text{ cm} = 0.0023 \text{ cm}$$

Discussion   The strain—or fractional length change—is 5.3×10^{-5}. Since the strain is much smaller than 1, we are justified in not worrying about whether the length is 43.0 cm with or without the compressive load; we would calculate the same value of ΔL (to two significant figures) either way.

Practice Problem 10.2   Fractional length change of a cable

A steel cable of diameter 3.0 cm supports a load of 2.0 kN. What is the fractional length increase of the cable compared to the length when there is no load if $Y = 2.0 \times 10^{11}$ Pa?

Figure 10.3
Compression of the femur

10.3   BEYOND HOOKE'S LAW

If the tensile or compressive stress exceeds the proportional limit, the strain is no longer proportional to the stress (Fig. 10.4). The solid still returns to its original length when the stress is removed as long as the stress does not exceed the *elastic limit*. If the stress exceeds the elastic limit, the material is permanently deformed. For still larger stresses, the solid fractures when the stress reaches the *breaking point*. The maximum stress that can be withstood without breaking is called the *ultimate strength*. The ultimate strength

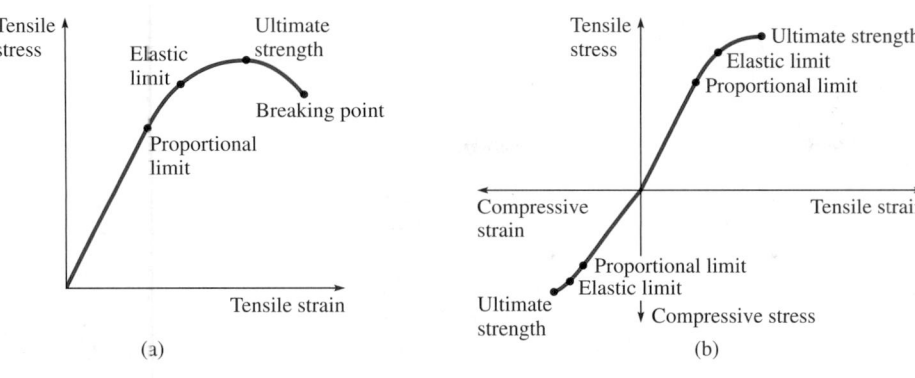

(a)

(b)

Figure 10.4   A stress-strain curve showing limits for (a) a ductile material, (b) a brittle material, and (c) compact bone. The elastic limit, ultimate strength, and breaking point are well separated for ductile materials, but close together for a brittle material. The breaking point is so close to the ultimate strength for the brittle material that it is not shown as a separate point on the stress-strain curve.

(c)

Figure 10.5   This machine, used to make thin wire, stretches metal beyond its elastic limit.

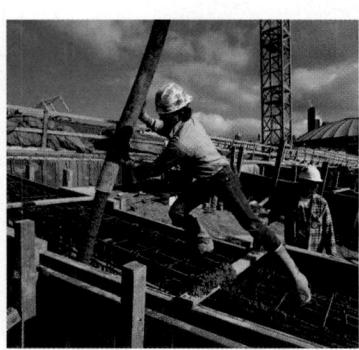

Figure 10.6   Pouring concrete over stretched steel rods

Making The Connection:
strength of
building materials

can be different for compression and tension; then we refer to the compressive strength or the tensile strength of the material.

A *ductile* material continues to stretch beyond its ultimate tensile strength without breaking; the stress then *decreases* from the ultimate strength (Fig. 10.4a). Examples of ductile solids are the relatively soft metals, such as gold, silver, copper, and lead. These metals can be pulled like taffy, becoming thinner and thinner until finally reaching the breaking point (Fig. 10.5).

For a *brittle* substance, the ultimate strength and the breaking point are close together (Fig. 10.4b). Bone is an example of a brittle material; it fractures abruptly if the stress becomes too large. Under either tension or compression, its elastic limit, breaking point, and ultimate strength are approximately the same. For compression, the stress and strain are roughly proportional right up to the breaking point, whereas for tension there is a marked deviation from Hooke's law. Babies have more flexible bones than adults because they have built up less of the calcium compound hydroxyapatite. As people age, their bones become more brittle as the collagen fibers lose flexibility and they also become weaker as calcium gets reabsorbed.

Like bone, concrete has one component for tensile strength and another for compressive strength. Reinforced concrete contains steel rods that provide tensile strength that concrete itself lacks. In prestressed concrete, the steel rods are stretched when the concrete is poured (Fig. 10.6). After the concrete hardens, the frame holding the rods under tension is removed. The rods contract, compressing the concrete. Then, when the prestressed concrete is subject to a tensile force, the compression of the concrete is lessened but not eliminated so that the concrete itself is never subjected to a tensile stress.

Example 10.3

Crane with Steel Cable

A crane is required to lift loads of up to 1.0×10^5 N (11 tons). (a) What is the minimum diameter of the steel cable that must be used? (b) If a cable of twice the minimum diameter is used and it is 8.0 m long when no load is present, how much longer is it when supporting 1.0×10^5 N? (Data for steel: $Y = 2.0 \times 10^{11}$ Pa; proportional limit $= 2.0 \times 10^8$ Pa; elastic limit $= 3.0 \times 10^8$ Pa; tensile strength $= 5.0 \times 10^8$ Pa.)

Strategy   The data given for steel consists of four quantities that all have the same units. It would be easy to mix them up if we didn't understand what each one means. Young's modulus is the proportionality constant between stress and strain. That will be useful in part (b) where we find the elongation of the cable; the elongation is the strain times the original length. However, we should first check that the stress is less than the proportional limit before using Young's modulus to find the strain.

The elastic limit is the maximum stress so that no permanent deformation occurs; the tensile strength is the maximum stress so that the cable does not break. We certainly don't want the cable to break, but it would be prudent to keep the stress under the elastic limit to give the cable a long useful life. Therefore, we choose a minimum diameter in (a) to keep the stress below the elastic limit.

Solution   (a) We choose the minimum diameter to keep the stress less than the elastic limit:

$$\frac{F}{A} < \text{elastic limit} = 3.0 \times 10^8 \text{ Pa}$$

for $F = 1.0 \times 10^5$ N. Then

$$A > \frac{F}{\text{elastic limit}} = \frac{1.0 \times 10^5 \text{ N}}{3.0 \times 10^8 \text{ Pa}} = 3.33 \times 10^{-4} \text{ m}^2$$

The minimum cross-sectional area corresponds to the minimum diameter. The cross-sectional area of the cable is πr^2 or $\pi d^2/4$, so

$$d = \sqrt{\frac{4A}{\pi}} = \sqrt{\frac{4 \times 3.33 \times 10^{-4} \text{ m}^2}{\pi}} = 2.1 \text{ cm}$$

The minimum *diameter* is therefore 2.1 cm.

(b) If we double the diameter and keep the same load, the stress is reduced by a factor of four since the cross-sectional area is proportional to the square of the diameter. Therefore, the stress is

$$\frac{F}{A} = \frac{3.0 \times 10^8 \text{ Pa}}{4} = 7.5 \times 10^7 \text{ Pa}$$

The strain is then

$$\frac{\Delta L}{L} = \frac{F/A}{Y} = \frac{7.5 \times 10^7 \text{ Pa}}{2.0 \times 10^{11} \text{ Pa}} = 0.000375$$

The strain is the fractional length change. Then the length change is

$$\Delta L = 0.000375L = 0.000375 \times 8.0 \text{ m} = 0.0030 \text{ m} = 3.0 \text{ mm}$$

Discussion   By using a cable twice as thick as the minimum, we build in a safety factor. We don't want to be right at the edge of disaster! Since doubling the diameter of the cable increases the cross-sectional area of the cable by a factor of four, the maximum stress on the cable is $\frac{1}{4}$ of the elastic limit.

Practice Problem 10.3   Tuning a harpsichord string

A harpsichord string is made of yellow brass (Young's modulus 9.0×10^{10} Pa, tensile strength 6.3×10^8 Pa). When tuned correctly, the tension in the string is 59.4 N, which is 93% of the maximum tension that the string can endure without breaking. What is the radius of the string?

Human anatomy has special features for adapting to the compressive stress associated with standing upright. For example, the vertebrae in the spinal column gradually increase in size from the neck to the tailbone. Such an arrangement places the stronger vertebrae in the lower positions, where they must support more weight. The vertebrae are separated by fluid-filled disks, which have a cushioning effect by spreading out the compressive forces.

Height Limits

What about the question of the height of Jack's beanstalk and the proportions of the giant? For simplicity, imagine the beanstalk to be a single cylindrical column. If the beanstalk grows too tall, it could be crushed under its own weight. The maximum height of a cylindrical column is limited since the compressive stress at the bottom cannot exceed the compressive strength of the material (see Problem 78). However, the beanstalk would *buckle* long before being crushed; the maximum height at which a vertical column buckles is generally less than the height at which it would be crushed.

The bones of our limbs are hollow; the inside of the structural material is filled with marrow, which is structurally weak. A hollow bone is better able to resist fracture from bending and twisting forces than a solid bone with the same amount of structural material, although the hollow bone would buckle more easily under a compressive force along the central axis.

Physics at Home

Challenge a friend to use a single sheet of 8.5" × 11" paper and two paper clips (or tape) to support a book at least 8 in. above a table. If your friend has no idea what to do, roll the sheet of paper into a narrow cylinder about 2.5 cm (an inch) in diameter; then fasten the cylinder at the top and bottom with paper clips (or with tape). Carefully place the book so that it is balanced on top of the cylinder. If you have difficulty, try using thicker paper or a lighter book.

Use the same "apparatus" to get some insight into the buckling of columns. Try making the diameter of the paper cylinder twice as large. The walls of this column are thinner because there are fewer layers of the paper in the cylinder wall, although the same cross-sectional area *of paper* supports the book. If nothing happens, try again with a heavier book. You will likely see the walls crumple in on themselves as the cylinder buckles and the book falls to the table.

Why would the design of a giant's bones have to be different from a human's? If the giant's average density is the same as a human's, then his weight is larger by the same factor that his *volume* is larger. If the giant is five times as tall as a human, for instance, and has the same relative proportions, then his volume is $5^3 = 125$ times as large, since each of the three dimensions of any body part has increased by a factor of five. On the other hand, the cross-sectional area of a bone is proportional to the *square* of its radius. So while the leg bones must support 125 times as much weight, the maximum compressive force they can withstand has only increased by a factor of 25. The giant would need much thicker legs (in relation to their length) to support his increased weight. Similar analysis can be applied to the twisting and bending forces that are more likely to break bones than are compressive forces. The result is the same: the bones of a giant could not have human proportions.

Some science fiction or horror movies portray giant insects as greatly magnified versions of a normal insect. Such a giant insect's legs would collapse under the weight of the insect.

Making The Connection:
size limitations
on organisms

10.4   SHEAR AND VOLUME DEFORMATIONS

In this section we consider two other kinds of deformation. In each case we define a stress (force per unit area), a strain (dimensionless), and a modulus (the constant of proportionality between stress and strain).

Shear Deformation

Unlike tensile and compressive forces, which act perpendicular to two opposite surfaces of an object, a shear deformation is the result of a pair of equal and opposite forces that act *parallel* to two opposite surfaces. Consider a book placed on a desk. If we push horizontally on the top cover of the book while pushing in the opposite direction on the bottom cover to hold it in place, the book is deformed as shown in Fig. 10.7. Such a deformation is called a **shear deformation**.

Shear forces produce the same kind of deformation in a solid block; the amount of the deformation is just smaller. The **shear stress** is the magnitude of the shear force divided by the area of the surface on which the force acts:

$$\text{shear stress} = \frac{\text{shear force}}{\text{area of surface}} = \frac{F}{A} \qquad (10\text{-}6)$$

Shear strain is the ratio of the relative displacement Δx to the separation L of the two surfaces:

$$\text{shear strain} = \frac{\text{displacement of surfaces}}{\text{separation of surfaces}} = \frac{\Delta x}{L} \qquad (10\text{-}7)$$

(a)

(b)

Figure 10.7 A book under shear stress

Figure 10.8 Scissors apply shear stress to a sheet of paper. The shear stress is the force exerted by a blade divided by the cross-sectional area of the paper—the thickness of the paper times the length of blade that is in contact with the paper.

The shear strain is proportional to the shear stress as long as the stress is not too large. The constant of proportionality is the **shear modulus S**.

Hooke's law for shear deformations:

shear stress ∝ shear strain

$$\frac{F}{A} = S\frac{\Delta x}{L} \qquad\qquad (10\text{-}8)$$

The units of shear stress and the shear modulus are the same as for tensile or compressive stress and Young's modulus: Pa or N/m^2. The strain is once again dimensionless. Table 10.2 lists shear moduli for various substances.

   An example of shear stress is the cutting action of a pair of scissors (or "shears") on a piece of paper. The forces acting on the paper from above and below are offset from each other and act parallel to the cross-sectional surfaces of the paper (see Fig. 10.8).

Example 10.4

Cutting Paper

A sheet of paper of thickness 0.20 mm is cut with scissors that have blades of length 10.0 cm and width 0.20 cm. While cutting, the scissors blades each exert a force of 3.0 N on the paper; the length of each blade

that makes contact with the paper is approximately 0.5 mm. What is the shear stress on the paper?

Strategy   Shear stress is a force divided by an area. In this problem, identifying the correct area is tricky. The blades

continued on next page

Example 10.4 *continued*

push two *cross-sectional* paper surfaces in opposite directions so they are displaced with respect to each other. The shear stress is the force exerted by each blade divided by this cross-sectional area—the thickness of the paper times the length of blade *in contact with the paper*. The total length and the width of the blades are irrelevant.

Solution The cross-sectional area is

$$A = \text{thickness} \times \text{contact length}$$
$$= 2.0 \times 10^{-4} \text{ m} \times 5 \times 10^{-4} \text{ m} = 1 \times 10^{-7} \text{ m}^2$$

The shear stress is

$$\frac{F}{A} = \frac{3.0 \text{ N}}{1 \times 10^{-7} \text{ m}^2} = 3 \times 10^7 \text{ N/m}^2$$

 Discussion To identify the correct area, remember that shear forces act *in the plane of* the surfaces that are displaced with respect to each other. By contrast, tensile and compressive forces act perpendicular to the area used to find tensile and compressive stresses.

Practice Problem 10.4 Shear stress due to hole punch

A hole punch has a diameter of 8.0 mm and presses onto 10 sheets of paper with a force of 6.7 kN. If each sheet of paper is of thickness 0.20 mm, find the shear stress. [*Hint:* Be careful in deciding what area to use. Remember that a shear force acts *parallel* to the surface whose area is relevant.]

Table 10.2

Shear and Bulk Moduli for Various Materials

Material	Shear Modulus S (10^9 Pa)	Bulk Modulus B (10^9 Pa)
Gases		
Air (1)		0.00010
Air (2)		0.00014
Liquids		
Ethanol		0.9
Water		2.2
Mercury		25
Solids		
Cast iron	40–50	60–90
Marble		70
Aluminum	25–30	70
Copper	40–50	120–140
Steel	80–90	140–160
Diamond		620

(1) At 0°C and 1 atm; constant temperature expansion or compression

(2) At 0°C and 1 atm; no heat flow during expansion or compression

When a bone is twisted, it is subjected to a shear stress. Shear stress is a more common cause of fracture than a compressive or tensile stress along the length of the bone. The twisting of a bone can result in a spiral fracture (Fig. 10.9).

Volume Deformation

As discussed in Chapter 9, a fluid exerts inward forces on an immersed solid object. These forces are perpendicular to the surfaces of the object. Since the fluid presses inward on all sides of the object (Fig. 10.10), the solid is compressed—its volume is reduced. The fluid pressure P is the force per unit surface area; it can be thought of as the **volume stress** on the solid object. Pressure has the same units as the other kinds of stress: N/m² or Pa.

$$\text{volume stress} = \text{pressure} = \frac{F}{A} = P$$

(a)

(b)

Figure 10.9 (a) An Olympic skier falls and his leg is subjected to a shear stress. (b) X-ray of a spiral fracture of the tibia

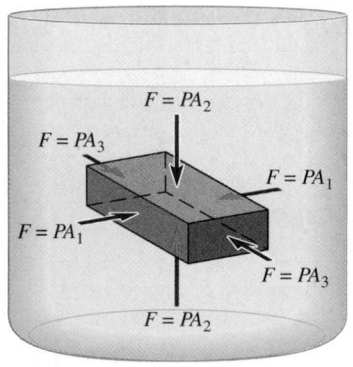

Figure 10.10 Forces on an object when submerged in a fluid

The resulting deformation of the object is characterized by the **volume strain**, which is the fractional change in volume:

$$\text{volume strain} = \frac{\text{change in volume}}{\text{original volume}} = \frac{\Delta V}{V} \qquad (10\text{-}9)$$

Unless the stress is too large, the stress and strain are proportional within a constant of proportionality called the **bulk modulus** B. A substance with a large bulk modulus is more difficult to compress than a substance with a small bulk modulus.

An object at atmospheric pressure is already under volume stress: the air pressure already compresses the object slightly compared to what its volume would be in vacuum. For solids and liquids the volume strain due to atmospheric pressure is, for most purposes, negligibly small (5×10^{-5} for water). Since we are usually concerned with the deformation due to a *change* in pressure from atmospheric pressure, we can write Hooke's law as:

Hooke's law for volume deformations

$$\Delta P = -B\frac{\Delta V}{V} \qquad (10\text{-}10)$$

---

where V is the volume at atmospheric pressure. The negative sign in Eq. (10-10) allows the bulk modulus to be positive—an increase in the volume stress causes a *decrease* in volume, so ΔV is negative. Table 10.2 lists bulk moduli for various substances.

Unlike the stresses and strains discussed previously, volume stress may be applied to fluids (liquids and gases) as well as solids. The bulk moduli of liquids are generally not much less than those of solids, since the atoms in liquids are nearly as close together as those in solids. In Chapter 9 we assume that liquids are incompressible, which is often a good approximation since the bulk moduli of liquids are generally large. In gases, the atoms are much farther apart on average than in solids or liquids. Gases are much easier to compress than solids or liquids, so their bulk moduli are much smaller.

Example 10.5

Marble Statue Under Water

A marble statue of volume 1.5 m³ is being transported by ship from Athens to Cyprus. The statue topples into the ocean when an earthquake-caused tidal wave sinks the ship; the statue ends up on the ocean floor, 1.0 km below the surface. Find the change in volume of the statue in cm³ due to the pressure of the water. The density of seawater is 1025 kg/m³.

Strategy The water pressure is the volume stress; it is the force per unit area pressing inward and perpendicular to all the surfaces of the statue. The water pressure at a depth d is greater than the pressure at the water surface; we can find the pressure using the given density of seawater. Then, using the bulk modulus of marble given in Table 10.2, we find the change in volume from Hooke's law.

Solution The pressure at a depth $d = 1.0$ km is larger than atmospheric pressure by

$$\Delta P = \rho g d$$
$$= 1025 \text{ kg/m}^3 \times 9.8 \text{ N/kg} \times 1000 \text{ m}$$
$$= 1.005 \times 10^7 \text{ Pa}$$

According to Table 10.2, the bulk modulus for marble is 70×10^9 Pa. This is the constant of proportionality between the volume stress (pressure increase) and the strain (fractional change in volume).

$$\Delta P = -B\frac{\Delta V}{V}$$

Solving for ΔV,

$$\Delta V = -\frac{\Delta P}{B}V = -\frac{1.005 \times 10^7 \text{ Pa}}{70 \times 10^9 \text{ Pa}} \times 1.5 \text{ m}^3$$
$$= -2.15 \times 10^{-4} \text{ m}^3 \times \left(\frac{100 \text{ cm}}{1 \text{ m}}\right)^3 = -200 \text{ cm}^3$$

The statue's volume decreases approximately 200 cm³.

Discussion The decrease in volume is 1.005×10^7 Pa/ 70×10^9 Pa $\approx 1/7000$ of the initial volume; a reduction of 0.014%.

In calculating the pressure increase, we assumed that the density of seawater is constant—the equation $\Delta P = \rho g d$ is derived for a constant fluid density ρ. Should we worry that

continued on next page

Example 10.5 *continued*

our calculation of ΔP is wrong? The result of Practice Problem 10.5 shows that density of seawater at a depth of 1.0 km is only about 0.43% greater than its density at the surface. The calculation of ΔP is inaccurate by less than 0.5%—negligible here since we only know the depth to two significant figures.

Practice Problem 10.5   Compression of water

Show that a pressure increase of 1.0×10^7 Pa (100 atm) on one cubic meter of seawater causes a 0.43% decrease in volume. The bulk modulus of seawater is 2.3×10^9 Pa.

10.5 SIMPLE HARMONIC MOTION

Vibration, one of the most common kinds of motion, is repeated motion back and forth along the same path. A common and important example of vibration is sound. At a rock concert, vibrations of the singer's vocal cords cause the air to vibrate. Vibrations in the air make the diaphragm of a microphone vibrate; the electrical signal generated by the microphone and amplifier makes the speaker cones vibrate. The speakers produce more vibrations in the air; we hear the music when the air makes our eardrums vibrate.

Vibrations occur in the vicinity of a point of **stable equilibrium**. An equilibrium point is *stable* if the net force on an object when it is displaced from equilibrium points back toward the equilibrium point (Fig. 10.11). Such a force is called a **restoring force** since it tends to restore equilibrium.

A special kind of vibratory motion—called **simple harmonic motion** (or **SHM**)—occurs whenever the restoring force is proportional to the displacement from equilibrium. The ideal spring is a favorite model of physicists because the restoring force it provides is proportional to the displacement from equilibrium ($F = kx$). Hooke's law applies to small deformations of many kinds of objects, not just springs. Thus, simple harmonic motion occurs in many situations as long as the vibrations are not too large.

Figure 10.12a shows an $F(x)$ curve for some restoring force. Since the curve is not linear, the resulting oscillations are not SHM—unless the amplitude is small. For small amplitudes, we can approximate the $F(x)$ curve near equilibrium by a straight line tangent to the curve at the equilibrium point (Fig. 10.12b). For small amplitude oscillations, the restoring force is approximately linear, so the resulting oscillations are (approximately) SHM.

Consider a relaxed ideal spring with spring constant k and zero mass. The spring is fixed at one end and attached at the other to an object of mass m (Fig. 10.13) that slides without friction. Since the normal force is equal and opposite to the weight of the object, the net force on the object is that due to the spring. When the spring is relaxed, the net force is zero; the object is in equilibrium.

If the object is now pulled to the right to the position $x = A$ and then released, the net force on the object is

$$F_x = -kx \tag{10-11}$$

where the negative sign tells us that the spring force is opposite in direction to the displacement from equilibrium. At first the object is to the right of the equilibrium position and the spring pulls to the left. Notice that the force exerted by the spring is in the correct direction to restore the object to the equilibrium position; it always pushes or pulls back toward the equilibrium point.

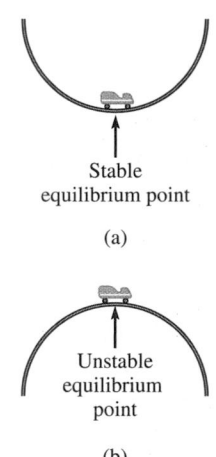

Stable equilibrium point

(a)

Unstable equilibrium point

(b)

Figure 10.11   (a) A point of *stable* equilibrium for a roller-coaster car. If the car is displaced slightly from its position at the bottom of the track, gravity pulls the car back toward the equilibrium point. (b) A point of *unstable* equilibrium for a roller-coaster car. If the car is displaced slightly from the very top of the track, gravity pulls the car *away from* the equilibrium point.

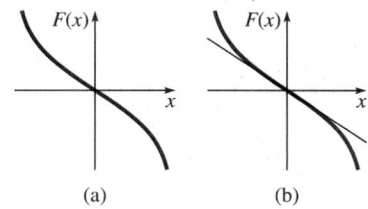

(a)   (b)

Figure 10.12   (a) A nonlinear restoring force. (b) For small vibrations we can approximate the restoring force as linear.

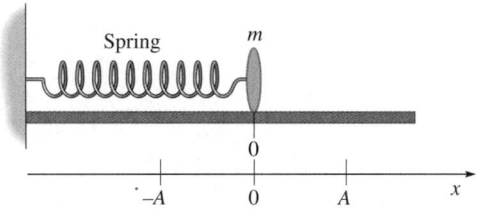

Figure 10.13   Spring in relaxed position

Figure 10.14 Positions of an oscillating body at equal time intervals over half a period. The spring is omitted for clarity.

Imagine taking a series of photos at equal time intervals as the object oscillates back and forth. In Fig. 10.14 the blue dots are the positions of the object at equal time intervals over one-half of a full cycle, from one endpoint to the other. (A full cycle would include the return trip.)

Figure 10.14 suggests that the speed is greatest as the object passes through the equilibrium position. The object slows as it approaches the endpoints and gains speed as it approaches the equilibrium point. At the endpoints ($x = \pm A$), the body is instantaneously at rest before heading back in the other direction. Conservation of energy supports these observations. The total mechanical energy of the mass and spring is constant.

$$E = K + U = \text{constant}$$

where K is the kinetic energy and U is the elastic potential energy stored in the spring. As the object oscillates back and forth, energy is converted from potential to kinetic and back to potential in the half-cycle shown in Fig. 10.14. From Chapter 6, the elastic potential energy of the spring is

$$U = \tfrac{1}{2}kx^2 \tag{6-11}$$

The speed at any point x can be found from the energy equation

$$E = \tfrac{1}{2}kx^2 + \tfrac{1}{2}mv_x^2 \tag{10-12}$$

The maximum displacement of the body is the **amplitude** A. At the maximum displacement, where the motion changes direction, the velocity is zero. Since the kinetic energy is zero at $x = \pm A$, all the energy is elastic potential energy at the endpoints. Therefore, the total energy E at the endpoints is

$$E_{\text{total}} = \tfrac{1}{2}kA^2 \tag{10-13}$$

and, since energy is conserved, this must be the total energy at any point in the object's motion. The maximum speed v_{m} occurs at $x = 0$ where all the energy is kinetic. Thus, at $x = 0$, the total energy equals the kinetic energy

$$E_{\text{total}} = \tfrac{1}{2}mv_{\text{m}}^2$$

and, from Eq. (10-13),

$$\tfrac{1}{2}mv_{\text{m}}^2 = \tfrac{1}{2}kA^2$$

Solving for v_{m} yields

$$v_{\text{m}} = \sqrt{\frac{k}{m}}\,A \tag{10-14}$$

The maximum speed is proportional to the amplitude; larger-amplitude oscillations have a larger maximum speed.

The force on the object at any point x is given by Hooke's law; Newton's second law then gives the acceleration:

$$F_x = -kx = ma_x$$

Solving for the acceleration,

$$a_x(t) = -\frac{k}{m}x(t) \tag{10-15}$$

Thus, the acceleration is a negative constant $(-k/m)$ times the displacement; the acceleration and displacement are always in opposite directions. Whenever the acceleration is a negative constant times the displacement, the motion is SHM.

The acceleration has its maximum magnitude a_m, where the force is largest, which is at the maximum displacement $x = \pm A$:

$$a_m = \frac{k}{m}A \qquad (10\text{-}16)$$

Example 10.6

Oscillating Model Rocket

A model rocket of 1.0-kg mass is attached to a horizontal spring with a spring constant of 6.0 N/cm. The spring is compressed by 18.0 cm and then released. The intent is to shoot the rocket horizontally, but the release mechanism fails to disengage, so the rocket starts to oscillate horizontally. Ignore friction and assume the spring to be ideal. (a) What is the amplitude of the oscillation? (b) What is the maximum speed? (c) What are the rocket's speed and acceleration when it is 12.0 cm from the equilibrium point?

Strategy   Initially all of the energy is elastic potential energy and the kinetic energy is zero. The initial displacement must be the maximum displacement—or amplitude—of the oscillations since to get farther from equilibrium would require more elastic energy than the total energy available. The speed at any position can be found using energy conservation $(\frac{1}{2}kx^2 + \frac{1}{2}mv_x^2 = \frac{1}{2}kA^2)$. The maximum speed occurs when all of the energy is kinetic. The acceleration can be found from Newton's second law.

Solution   (a) The amplitude of the oscillation is the maximum displacement, so $A = 18.0$ cm.

(b) From energy conservation, the maximum kinetic energy is equal to the maximum elastic potential energy:

$$K_m = \tfrac{1}{2}mv_m^2 = E = \tfrac{1}{2}kA^2$$

Solving for v_m,

$$v_m = \sqrt{\frac{k}{m}}A = \sqrt{\frac{6.0 \times 10^2 \text{ N/m}}{1.0 \text{ kg}}} \times 0.180 \text{ m} = 4.4 \text{ m/s}$$

(c) For the speed at a displacement of 0.120 m, we again use energy conservation.

$$\tfrac{1}{2}kx^2 + \tfrac{1}{2}mv^2 = \tfrac{1}{2}kA^2$$

Solving for v,

$$v = \sqrt{\frac{kA^2 - kx^2}{m}} = \sqrt{\frac{k}{m}(A^2 - x^2)}$$

$$= \sqrt{\frac{6.0 \times 10^2 \text{ N/m}}{1.0 \text{ kg}\,[(0.180 \text{ m})^2 - (0.120 \text{ m})^2]}} = 3.3 \text{ m/s}$$

From Newton's second law,

$$F_x = -kx = ma_x$$

At $x = \pm 0.120$ m,

$$a_x = -\frac{k}{m}x = \frac{6.0 \times 10^2 \text{ N/m}}{1.0 \text{ kg}} \times (\pm 0.120 \text{ m}) = \pm 72 \text{ m/s}^2$$

The magnitude of the acceleration is 72 m/s²; the direction is toward the equilibrium point.

Discussion   Note that at a given position (say $x = +0.120$ m), we can find the *speed* of the rocket, but the direction of the velocity can be either left or right; the rocket passes through each point (other than the endpoints) both on its way to the left and on its way to the right. By contrast, the *acceleration* at $x = +0.120$ m is always in the $-x$-direction, regardless of whether the rocket is moving to the left or to the right. If the rocket is moving to the right, then it is slowing down as it approaches $x = +A$; if it is moving to the left, then it is speeding up as it approaches $x = 0$.

Practice Problem 10.6   Maximum acceleration of rocket

What is the maximum acceleration of the rocket and at what position(s) does it occur?

10.6   THE PERIOD AND FREQUENCY FOR SHM

SHM is *periodic* motion because the same motion repeats over and over—a particle goes back and forth over the same path in exactly the same way. Each time the particle repeats its original motion, we say that it has completed another cycle. To complete one cycle of motion, the particle must be at the same point *and heading in the same direction* as it was at the start of the cycle. The period and frequency are defined exactly as for uniform circular motion, which is another kind of periodic motion. The **period** T is the time taken by one complete cycle. The **frequency** f is the number of cycles per unit time:

$$f = \frac{1}{T} \qquad \text{(SI unit: Hz = cycles per second)} \qquad (5\text{-}7)$$

Figure 10.15 An electrocardiogram

SHM is a special kind of periodic motion in which the restoring force is proportional to the displacement from equilibrium. Not all periodic vibrations are examples of simple harmonic motion since not all restoring forces are proportional to the displacement. Any restoring force can cause oscillatory motion. An electrocardiogram (Fig. 10.15) traces the periodic pattern of a beating heart, but the motion of the recorder needle is not simple harmonic motion. As we are about to show, in SHM the position is a sinusoidal function of time.

To learn more about SHM, imagine setting up an experiment (Fig. 10.16). We attach an object to an ideal spring, move the object away from the equilibrium position, and then release it. The object vibrates back and forth in simple harmonic motion with amplitude A. At the same time a horizontal circular disk, of radius $r = A$ and with a pin projecting vertically up from its outer edge, is set into rotation with uniform circular motion. Both the pin and the object attached to the spring are illuminated so that shadows of the vibrating object and of the pin on the rotating disk are seen on a screen. The speed of the disk is adjusted until the shadows oscillate with the same period. We will show that the motion of the two shadows is identical, so the mathematical description of one can be used for the other.

To find the mathematical description of SHM, we analyze the uniform circular motion of the pin as done in Chapter 5. Figure 10.16b shows the pin P moving counterclockwise around a circle of radius A at a constant angular velocity ω. For simplicity, let the pin start at $\theta = 0$ at time $t = 0$. The location of the pin at any time is then given by the angle θ:

$$\theta(t) = \omega t$$

The motion of the pin's shadow has the same x-component as the pin itself. Using a right triangle (Fig. 10.16c), we find that

$$x(t) = A \cos \theta = A \cos \omega t \tag{10-17}$$

Since the pin moves in uniform circular motion, its acceleration is constant in *magnitude* but not in direction; the acceleration is toward the center of the circle. In Chapter 5, the magnitude of the centripetal acceleration is shown to be

$$a = \omega^2 r = \omega^2 A \tag{5-11}$$

At any instant the direction of the acceleration vector is opposite to the direction of the displacement vector in Fig. 10.16b—that is, toward the center of the circle. Therefore,

$$a_x = -a \cos \theta = -\omega^2 A \cos \omega t \tag{10-18}$$

Comparing Eqs. (10-17) and (10-18), we see that, at any time t,

$$a_x(t) = -\omega^2 x(t) \tag{10-19}$$

Figure 10.16 (a) An experiment to show the relation between uniform circular motion and SHM. (b) A pin *P* moving counterclockwise around a circle as a disk rotates with constant angular velocity ω. (c) Finding the *x*-component of the displacement.

In Eq. (10-15) we showed that in SHM the acceleration is proportional to the displacement:

$$a_x = -\frac{k}{m}x \qquad (10\text{-}15)$$

The motions of the two shadows are identical as long as

$$\omega = \sqrt{\frac{k}{m}} \qquad (10\text{-}20a)$$

The position and acceleration of an object in SHM are sinusoidal functions of time [Eqs. (10-17) and (10-18)]. The sinusoidal functions are sine and cosine. In Problem 46, you can show that v_x is also a sinusoidal function of time.

Since the object in SHM and the pin in circular motion have the same frequency and period, the relationships between ω, f, and T still apply. Therefore, the frequency and period of a mass-spring system are

$$f = \frac{\omega}{2\pi} = \frac{1}{2\pi}\sqrt{\frac{k}{m}} \qquad (10\text{-}20b)$$

and

$$T = \frac{1}{f} = 2\pi\sqrt{\frac{m}{k}} \qquad (10\text{-}20c)$$

In the context of SHM the quantity ω is called the **angular frequency**. Note that the angular frequency is determined by the mass and the spring constant but is independent of the amplitude.

With the identification of ω for a mass-spring system, we can write the maximum speed and acceleration from Eqs. (10-14) and (10-16):

$$v_{\mathrm{m}} = \omega A \qquad (10\text{-}21)$$

$$a_{\mathrm{m}} = \omega^2 A \qquad (10\text{-}22)$$

These expressions are more general than Eqs. (10-14) and (10-16)—they apply to any system in SHM, not just a mass-spring system.

To Find the Angular Frequency for Any Object in SHM

- Write down the restoring force as a function of the displacement from equilibrium. Since the restoring force is linear, it always takes the form $F = -kx$, where k is a constant.
- Use Newton's second law to relate the restoring force to the acceleration.
- Solve for ω using $a_x = -\omega^2 x$ [Eq. (10-19)].

The term *harmonic* in *simple harmonic motion* refers to a sinusoidal vibration; this usage is related to similar usage in music and acoustics. The sinusoidal functions are also called harmonic functions. In Chapter 12 we show that a complex vibration can be formed by combining harmonic vibrations at different frequencies, which is why the study of SHM is the basis for understanding more complex vibrations. The term *simple* in SHM means that the amplitude of the vibration is constant; we assume there is no energy dissipation to cause the vibration to die out.

A Vertical Mass and Spring

The mass and spring systems discussed so far oscillate horizontally. An oscillating mass on a vertical spring also exhibits SHM; the difference is that the equilibrium point is shifted downward by gravity. In our discussions, we assume ideal springs that obey Hooke's law and have a negligibly small mass of their own.

Suppose that an object of weight mg is hung from an ideal spring of spring constant k (Fig. 10.17). In equilibrium, the spring is stretched downward a distance d from its relaxed length so that the spring pulls up with a force equal to mg. Taking the $+y$-axis in the upward direction, the condition for equilibrium is

$$\Sigma F_y = +kd - mg = 0 \qquad (10\text{-}23)$$

Let us take the origin ($y = 0$) at the equilibrium point. If the object is displaced vertically from the equilibrium point to a position y, the spring force becomes

$$F = k(d - y)$$

If y is positive, the object is displaced upward and the spring force is less than kd. The y-component of the net force is then

$$\Sigma F_y = k(d - y) - mg = kd - ky - mg$$

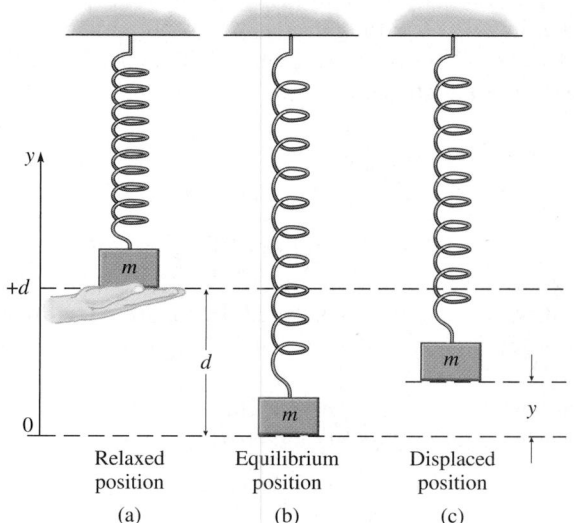

Figure 10.17 (a) A relaxed spring, of spring constant k, with mass m attached. (b) The same spring is extended to its equilibrium position, a distance d below the relaxed position, after mass m is allowed to hang freely. (c) The spring is displaced a distance y from the equilibrium position.

From Eq. (10-23), we know that $kd = mg$; therefore,

$$\Sigma F_y = -ky$$

The restoring force provided by the spring and gravity together is $-k$ times the displacement from equilibrium. Therefore, the vertical mass-spring exhibits SHM with the same period and frequency as if it were horizontal.

Example 10.7

A Vertical Spring

A spring with spring constant k is suspended vertically. A model goose of mass m is attached to the unstretched spring and then released so that the bird oscillates up and down. (Ignore friction and air resistance; assume an ideal massless spring.) Calculate the kinetic energy, the elastic potential energy, the gravitational potential energy, and the total mechanical energy at (a) the point of release and (b) the equilibrium point. Take the gravitational potential energy to be zero at the equilibrium point.

Strategy   The bird oscillates in SHM about its equilibrium point $y = 0$ between two extreme positions $y = +A$ and $y = -A$ (see Fig. 10.18). The amplitude A is equal to the distance the spring is stretched at the equilibrium point; it can be found by setting the net force on the bird equal to zero. The total mechanical energy is the sum of the kinetic energy, the elastic potential energy, and the gravitational potential energy. We expect the total energy to be the same at the two points; since no dissipative forces act, mechanical energy is conserved.

Solution   The equilibrium point is where the net force on the bird is zero:

$$\Sigma F_y = +kd - mg = 0 \qquad (10\text{-}23)$$

In this equation, d is the extension of the spring at equilibrium. Since the bird is released where the spring is relaxed, d is also the amplitude of the oscillations:

$$A = d = \frac{mg}{k}$$

(a) At the point of release, $v = 0$ and the kinetic energy is zero. The elastic energy is also zero—the spring is unstretched. The gravitational potential energy is

$$U_g = mgy = mgA = \frac{(mg)^2}{k}$$

The total mechanical energy is the sum of the kinetic and potential (elastic + gravitational) energies,

$$E = K + U_e + U_g = \frac{(mg)^2}{k}$$

(b) At the equilibrium point, the bird moves with its maximum speed $v_m = \omega A$. The angular frequency is the same as for a horizontal spring: $\omega = \sqrt{k/m}$. Then the kinetic energy is

$$K = \tfrac{1}{2}mv_m^2 = \tfrac{1}{2}m\omega^2 A^2$$

Substituting $A = mg/k$ and $\omega^2 = k/m$,

$$K = \frac{1}{2}m\frac{k}{m}\frac{(mg)^2}{k^2} = \frac{1}{2}\frac{(mg)^2}{k}$$

The spring is stretched a distance A, so the elastic energy is

$$U_e = \tfrac{1}{2}kA^2 = \frac{1}{2}k\frac{(mg)^2}{k^2} = \frac{1}{2}\frac{(mg)^2}{k}$$

The gravitational potential energy is zero at $y = 0$. Therefore, the total mechanical energy is

$$E = K + U_e + U_g = \frac{1}{2}\frac{(mg)^2}{k} + \frac{1}{2}\frac{(mg)^2}{k} + 0 = \frac{(mg)^2}{k}$$

which is the same as at $y = +A$.

continued on next page

Figure 10.18
(a) The spring is unstretched before the model bird is released at position $y = +A$; (b) the model bird passes through the equilibrium position $y = 0$ with maximum speed; (c) the spring's maximum extension occurs when the bird is at $y = -A$.

Example 10.7 *continued*

Discussion   As the bird moves down from the release point toward the equilibrium point, gravitational potential energy is converted into elastic energy and kinetic energy. After the bird passes the equilibrium point, both kinetic and gravitational energy are converted into elastic energy. At the lowest point in the motion, the gravitational potential energy has its lowest value while the elastic potential energy has its greatest value. The *total* potential energy (gravitational plus elastic) has its

minimum value at the equilibrium point since the kinetic energy is maximum there.

Practice Problem 10.7   Energy at maximum extension

Calculate the energies at the lowest point in the oscillations in Example 10.7.

10.7   GRAPHICAL ANALYSIS OF SHM

We have shown that the position of a particle vibrating along the *x*-axis is

$$x(t) = A \cos \omega t \tag{10-17}$$

Since the cosine function goes from −1 to +1, multiplying it by *A* gives us a displacement from −*A* to +*A*. Figure 10.19a is a graph of the position as a function of time.

The velocity at any time is the slope of the $x(t)$ graph. Note that the maximum slope in Fig. 10.19a occurs when $x = 0$, which confirms what we already know from energy conservation: the velocity is maximum at the equilibrium point. Note also that the velocity is zero when the displacement is a maximum (+*A* or −*A*). Figure 10.19b shows a graph of $v_x(t)$. The equation describing this graph is (see Problem 46):

$$v_x(t) = -v_m \sin \omega t = -\omega A \sin \omega t \tag{10-24}$$

The acceleration is the slope of the $v_x(t)$ graph. Figure 10.19c is a graph of $a_x(t)$, which is described by the equation

$$a_x(t) = -a_m \cos \omega t = -\omega^2 A \cos \omega t \tag{10-18}$$

Observe the interrelationships between the three graphs. The velocity graph is one quarter cycle ahead of the position graph; that is, $v_x(t)$ reaches its positive maximum one quarter period before $x(t)$ reaches its positive maximum. Likewise, the acceleration is one quarter cycle ahead of the velocity and one half cycle ahead of the position.

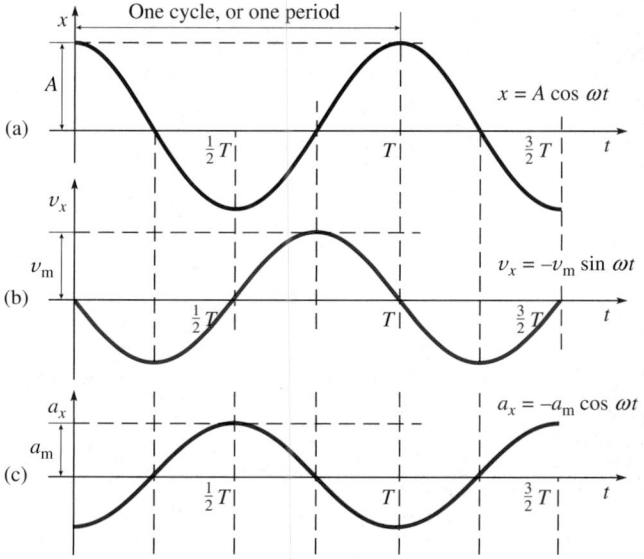

Figure 10.19   (a) Position of a particle in simple harmonic motion as a function of time, (b) velocity as a function of time, and (c) acceleration as a function of time

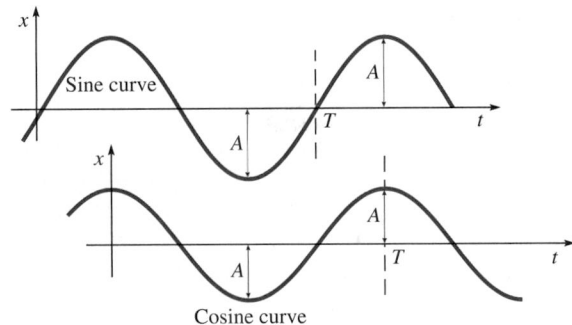

Figure 10.20 Sine and cosine curves of displacement versus time

We have written the position as a function of time in terms of the cosine function, but we can just as correctly use the sine function. The difference between the two is the initial position at time $t = 0$ (see Fig. 10.20). If the position is at a maximum ($x = A$) at $t = 0$, $x(t)$ is a cosine function. If the position is at the equilibrium point ($x = 0$) at $t = 0$, $x(t)$ is a sine function. By analyzing the slopes of the graphs, you can show (Problem 43) that if the position as a function of time is

$$x(t) = A \sin \omega t \qquad \text{(10-25a)}$$

then the velocity is

$$v_x(t) = v_m \cos \omega t \qquad \text{(10-25b)}$$

and the acceleration is

$$a_x(t) = -a_m \sin \omega t \qquad \text{(10-25c)}$$

Example 10.8

A Vibrating Loudspeaker Cone

A loudspeaker has a movable diaphragm (the *cone*) that vibrates back and forth to produce sound waves. The displacement of a loudspeaker cone playing a sinusoidal test tone is graphed in Fig. 10.21. Find (a) the amplitude of the motion, (b) the period of the motion, and (c) the frequency of the motion. (d) Write equations for $x(t)$ and $v_x(t)$.

Strategy   The amplitude and period can be read directly from the graph. The frequency is the inverse of the period. Since $x(t)$ begins at the maximum displacement, it is described by a cosine function. By looking at the slope of $x(t)$ we can tell whether the velocity is a positive or negative sine function.

Solution   (a) The amplitude is the maximum displacement shown on the graph: $A = 0.015$ m.

(b) The period is the time for one complete cycle. From the graph: $T = 0.040$ s.

(c) The frequency is the inverse of the period.

$$f = \frac{1}{T} = \frac{1}{0.040 \text{ s}} = 25 \text{ Hz}$$

(d) Since $x = +A$ at $t = 0$, we write $x(t)$ as a cosine function:

$$x(t) = A \cos \omega t$$

where $A = 0.015$ m and

$$\omega = 2\pi f = 160 \text{ rad/s}$$

The slope of $x(t)$ is initially zero and then goes negative. Therefore, $v_x(t)$ is a negative sine function:

$$v_x(t) = -v_m \sin \omega t$$

where $\omega = 2\pi f = 160$ rad/s and

$$v_m = \omega A = 160 \text{ rad/s} \times 0.015 \text{ m} = 2.4 \text{ m/s}$$

Discussion   As a check, the velocity should be $\frac{1}{4}$ of a cycle ahead of the position. If we imagine shifting the vertical axis to the right (ahead) by 0.01 s, the graph would have the shape of a negative sine function.

Practice Problem 10.8   Acceleration of speaker cone

Sketch a graph and write an equation for $a_x(t)$.

Figure 10.21
Horizontal displacement of a vibrating cone as a function of time

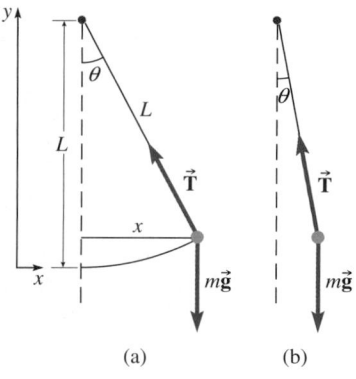

Figure 10.22 Forces on a pendulum bob (a) far from equilibrium and (b) near equilibrium.

10.8   THE PENDULUM

Simple Pendulum

When a pendulum swings back and forth, a string or thin rod constrains the bob to move along a circular arc (Fig. 10.22). However, *for oscillations with small amplitude*, we assume that the bob moves *back and forth along the x-axis*; the vertical motion of the bob is negligible.

Since the weight of the bob has no *x*-component, the restoring force is the *x*-component of the force due to the string. We expect the restoring force to be proportional to the displacement for small oscillations. From Fig. 10.22,

$$\Sigma F_x = -T \sin \theta \approx -\frac{Tx}{L}$$

where L is the length of the string and $\sin \theta = x/L$. The *y*-component of the acceleration is negligibly small, so

$$\Sigma F_y = T \cos \theta - mg = ma_y \approx 0$$

Since $\cos \theta \approx 1$ for small θ, $T \approx mg$. Then

$$\Sigma F_x \approx -\frac{mgx}{L} = ma_x$$

Solving for a_x:

$$a_x = -\frac{g}{L}x$$

To identify the angular frequency, we recall that

$$a_x = -\omega^2 x \tag{10-19}$$

Therefore, the angular frequency is

$$\omega = \sqrt{\frac{g}{L}} \tag{10-26a}$$

and the period is

$$T = \frac{2\pi}{\omega} = 2\pi\sqrt{\frac{L}{g}} \tag{10-26b}$$

 Note that the period depends on L and g but not on the mass of the pendulum.

Be careful not to confuse the *angular frequency* of the pendulum with its *angular velocity*. Even though the two have the same units (rad/s in SI) and are written with the same symbol (ω), for a pendulum they are *not* the same. When dealing with the pendulum, we use the symbol ω to stand for the *angular frequency* only. The angular frequency $\omega = 2\pi f$ of a given pendulum is constant, while the angular velocity (the rate of change of θ) changes with time between zero (at the extremes) and its maximum magnitude (at the equilibrium point).

Physics at Home

The relation between the period and the length of the pendulum is easily tested at home. Make a simple pendulum by tying a thin string to one end of a paper clip and sliding the clip over a coin. Some tape can be used to help hold the coin if it slips out of place. Holding the end of the string, let the coin swing through a small arc and note the time for the coin to make ten complete oscillations, starting from one extreme position and returning to the same position ten times. Divide the time by ten to get the period. (This gives a more accurate value than timing a single period.) Measure the length of the pendulum and test Eq. (10-26b).

Repeat the experiment by holding the string at a position closer to the coin, effectively shortening the length of the pendulum. What do you find? Is the period for the shorter pendulum longer, shorter, or the same as that measured for the longer pendulum?

The effect of a different mass on the period can also be tested by using two or three coins taped together, with the same length pendulum as used for the first measurement. Does a heavier coin affect the result?

Example 10.9

Grandfather Clock

A grandfather clock uses a pendulum with period 2.0 s to keep time. In one such clock, the pendulum bob has mass 150 g; the pendulum is set into oscillation by displacing it 33 mm to one side. (a) What is the length of the pendulum? (b) Does the initial displacement satisfy the small angle approximation?

Strategy   The period depends on the length of the pendulum and on the gravitational field strength g. It does not depend on the mass of the bob. It also does not depend on the initial displacement, as long as it is small compared to the length.

Solution   (a) Assuming small amplitudes, the period is

$$T = 2\pi \sqrt{\frac{L}{g}}$$

Solving for L,

$$L = \frac{T^2 g}{(2\pi)^2}$$

$$= \frac{(2.0 \text{ s})^2 \times 9.8 \text{ m/s}^2}{(2\pi)^2} = 0.99 \text{ m}$$

(b) The small angle approximation is valid if the maximum displacement is small compared to the length of the pendulum.

$$\frac{x}{L} = \frac{33 \text{ mm}}{990 \text{ mm}} = 0.033$$

Is that small enough? If $\sin\theta = x/L = 0.033$, then

$$\theta = \sin^{-1} 0.033 = 0.033006$$

Sin θ and θ differ by less than 0.02%. Since we only know T and g to two significant figures, the approximation is good.

Discussion   We should check that we didn't write the expression for the period "upside down," which is the most likely error we could make. Besides checking that the units work out, we know that a longer pendulum has a longer period, so L must go in the numerator. On the other hand, if g were larger, the restoring force would be larger and we would expect the period to shorten; thus, g belongs in the denominator.

Practice Problem 10.9   Pendulum on the Moon

A pendulum of length 0.99 m is taken to the Moon by an astronaut. The period of the pendulum is 4.9 s. What is the gravitational field strength on the surface of the Moon?

The period of a pendulum as just determined is valid only for small amplitudes. For larger amplitudes, the pendulum's motion is still periodic (though not SHM). Why would the period be any different for large amplitudes? Remember that we assumed the bob was moving horizontally back and forth along the x-axis. This simplification breaks down for large amplitudes. For instance, if we pull the pendulum out horizontally ($\theta = 90°$), the tangential component of the weight is mg, but $F_x = 0$! Since we have overestimated F_x, we have underestimated the time for the bob to return to $x = 0$; thus, the period for large amplitudes is greater than $2\pi\sqrt{L/g}$. Another way of looking at it is in terms of the tangential force. The expression for the tangential component of the weight is correct even for large amplitudes. However, the distance the bob must move to return to equilibrium is larger than x. For instance, starting at $\theta = 90°$, the bob must move one-quarter of the circumference, a distance $\frac{1}{4}(2\pi L) \approx 1.6L$, to return to equilibrium. Assuming linear motion along the x-axis would make the distance only L. With a longer distance to travel, the time is longer.

Physical Pendulum

Imagine that you have a simple pendulum of length L. Beside it you have a uniform metal bar of the same length, which is free to swing about an axis at one end (see Fig. 10.23). Would the two have the same period if they are set into oscillation?

For the simple pendulum, the bob is assumed to be a point mass; all the mass of the pendulum is at a distance L from the rotation axis. For the metal bar, however, the mass is uniformly distributed from the axis to a *maximum* distance L away from the axis. The center of mass is located at the midpoint, a distance $d = \frac{1}{2}L$ from the axis (Fig. 10.23). Since the mass is on average closer to the axis, the period is shorter than that of the simple pendulum.

Would this bar have a period equal to that of a simple pendulum of length $\frac{1}{2}L$? That is a good guess, since the center of mass of the bar is a distance $\frac{1}{2}L$ away from the rotation axis. Unfortunately, it isn't quite that easy. The gravitational force acts at the center of mass, but we *cannot* think of all the mass as being concentrated at that point—that would give the wrong rotational inertia. When set into oscillation, the bar, or any other rigid object free to rotate about a fixed axis, is called a **physical pendulum**.

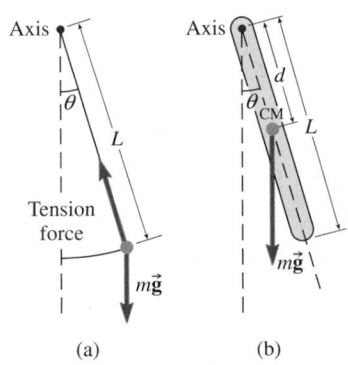

Figure 10.23   (a) A simple pendulum and (b) a physical pendulum

To find the period of a physical pendulum, we first find the net torque acting on the physical pendulum and then use the rotational form of Newton's second law. Taking torques about the rotation axis, only gravity gives a nonzero torque. If the pendulum has mass m and the distance from the axis to the center of mass is d, then the torque is

$$\tau = F_\perp r = -(mg \sin \theta)d$$

where θ is the angle indicated in Fig. 10.23. The other component of the gravitational force, $mg \cos \theta$, passes through the axis of rotation so it does not contribute to the torque. In the equation above, both τ and θ are positive if they are counterclockwise; the minus sign says that they always have opposite sign, since the torque always acts to bring θ closer to zero. Assuming small amplitudes, $\sin \theta \approx \theta$ (in radians) and the torque is

$$\tau = -mgd\theta$$

Thus, the restoring torque is proportional to the displacement angle θ, just as the restoring force was proportional to the displacement for the simple pendulum and the mass on a spring.

The net torque is equal to the rotational inertia times the angular acceleration:

$$\tau = -mgd\theta = I\alpha$$

and the angular acceleration is

$$\alpha = -\frac{mgd}{I}\theta \qquad (10\text{-}27)$$

Since the angular acceleration is a negative constant times the angular displacement from equilibrium, we indeed have SHM. Equation (10-27) is analogous to the equation for the linear acceleration of the oscillating spring

$$a_x = -\omega^2 x$$

where

$$\frac{mgd}{I} = \omega^2$$

Therefore, the angular frequency of the physical pendulum is

$$\omega = 2\pi f = \sqrt{\frac{mgd}{I}} \qquad (10\text{-}28\text{a})$$

and the period is

$$T = \frac{2\pi}{\omega} = 2\pi \sqrt{\frac{I}{mgd}} \qquad (10\text{-}28\text{b})$$

where d is the distance from the axis to the CM and I is the rotational inertia.

For a uniform bar of length L, the center of mass is halfway down the bar:

$$d = \tfrac{1}{2}L$$

From Table 8.1, the rotational inertia of a uniform bar rotating about an axis through an endpoint is $I = \tfrac{1}{3}mL^2$. The period of oscillation is

$$T = 2\pi \sqrt{\frac{I}{mgd}}$$

Substituting for I and d,

$$T = 2\pi \sqrt{\frac{\tfrac{1}{3}mL^2}{(mg)\tfrac{1}{2}L}} = 2\pi \sqrt{\frac{2L}{3g}}$$

The bar has the same period as a simple pendulum of length $\tfrac{2}{3}L$.

Example 10.10

Comparison of Walking Frequencies and Speeds for Various Creatures

 During a relaxed walking pace, an animal's leg can be thought of as a physical pendulum of length L that pivots about the hip. (a) What is the relaxed walking frequency for a cat ($L = 30$ cm), dog (60 cm), human (1 m), giraffe (2 m), and a mythological titan (10 m)? (b) Derive an equation that gives the walking speed (amount of ground covered per unit time) for a given walking frequency f.

continued on next page

Figure 10.24 The forward motion of a leg during walking is similar to the swing of a physical pendulum. From (a) to (b), the right leg swings forward like a pendulum. In (c), the right foot is on the ground and the left leg is about to swing forward.

Example 10.10 *continued*

[*Hint:* Start by drawing a picture of the leg position at the start of the swing (leg back) and the end of the swing (leg forward) and assume a comfortable angle of about 30° between these two positions. To how many steps does a complete period of the pendulum correspond?] (c) Find the walking speed for each of the animals listed in part (a).

Strategy We have to use an idealized model of the leg, since we don't know the exact location of the center of mass or the rotational inertia. The simple pendulum is not a good model, since it would assume all the mass of the leg at the foot! A much better model is to think of the leg as a uniform cylinder pivoting about one end.

Solution (a) For a uniform cylinder, the center of mass is a distance $d = \frac{1}{2}L$ from the pivot and the rotational inertia about an axis at one end is $I = \frac{1}{3}mL^2$. Then the period is

$$T = 2\pi \sqrt{\frac{I}{mgd}} = 2\pi \sqrt{\frac{\frac{1}{3}mL^2}{(mg)\frac{1}{2}L}} = 2\pi \sqrt{\frac{2L}{3g}}$$

and the frequency f is

$$f = \frac{1}{T} = \frac{1}{2\pi}\sqrt{\frac{3g}{2L}} \approx 0.2\sqrt{\frac{g}{L}}$$

Substituting the numerical values of L for each animal, we find the frequencies to be 1 Hz (cat), 0.8 Hz (dog), 0.6 Hz (human), 0.4 Hz (giraffe), and 0.2 Hz (titan).

(b) One period of the "pendulum" corresponds to two steps. In Fig. 10.24a, the right leg is about to step forward. The step occurs as the pendulum swings forward through half a cycle. In Fig. 10.24b, the right foot is about to touch the ground; in

Fig. 10.24c the right foot touches the ground and now the left leg is about to step forward. During this step, the right foot stays in place on the ground, but the right leg is swinging backward relative to the hip joint. During each step, the distance covered is approximately the length of a 30° arc of radius L, which is $\frac{1}{12}$ the circumference of a circle of radius L. So during one period, the distance walked is

$$D = 2 \times \frac{1}{12} \times 2\pi L = \frac{\pi}{3}L \approx L$$

and the walking speed is

$$v = \frac{D}{T} = Lf = 0.2\sqrt{gL}$$

(c) The speeds are 0.3 m/s (cat), 0.5 m/s (dog), 0.6 m/s (human), 0.9 m/s (giraffe), and 2 m/s (titan).

Discussion You may be more familiar with walking speeds in mi/h. Converting the units, 0.6 m/s ≈ 1.3 mi/h, which is just about right for a leisurely walk. A brisk walk is about 3 mi/h for most people; to go much faster than that, you need to jog or run.

The solution says that longer legs walk faster, but the frequency of the steps is lower. You can verify that by walking beside a friend who is much taller or much shorter than you, or by taking your dog for a walk.

Practice Problem 10.10 Walking speed for human

A more realistic model of a human leg of length 1.0 m has the center of mass 0.45 m from the hip and a rotational inertia of $\frac{1}{6}mL^2$. What is the walking speed predicted by this model?

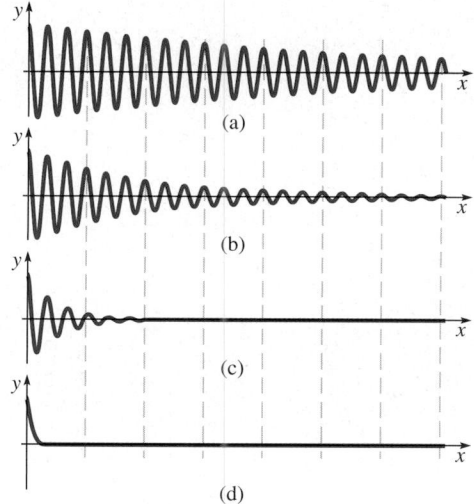

Figure 10.25 Graphs of $x(t)$ for a mass-spring system with increasing amounts of damping (a), (b), (c), (d). In (d) the damping is sufficient to prevent oscillations from occurring.

10.9 DAMPED OSCILLATIONS

In simple harmonic motion, we assume that no dissipative forces such as friction or viscous drag exist. Since the mechanical energy is constant, the oscillations continue forever with constant amplitude. SHM is a simplified model. The oscillations of a swinging pendulum or a vibrating tuning fork gradually die out as energy is dissipated. The amplitude of each cycle is a little smaller than that of the previous cycle (see Figs. 10.25a and b). This kind of motion is called **damped oscillation**, where the word *damped* is used in the sense of *extinguished* or *restrained*. For a small amount of damping, oscillations occur at approximately the same frequency as if there were no damping. A greater degree of damping lowers the frequency slightly (Fig. 10.25c). Even more damping prevents oscillations from occurring at all (Fig. 10.25d).

Damping is not always a disadvantage. The suspension system of a car includes shock absorbers that cause the vibration of the body—a mass connected to the chassis by springs—to be quickly damped. The shock absorbers reduce the discomfort that passengers would otherwise experience due to the bouncing of an automobile as it travels along a bumpy road. Figure 10.26 shows how a shock absorber works. In order to compress or expand the shock absorber, a viscous oil must flow through the holes in the piston. The viscous force dissipates energy regardless of which direction the piston moves. The shock absorber enables the spring to smoothly return to its equilibrium length without oscillating up and down (see Fig. 10.25d). When the oil leaks out of the shock absorber, the damping is insufficient to prevent oscillations. After hitting a bump, the body of the car oscillates up and down (Fig. 10.25c).

Making The Connection:
shock absorbers
in a car

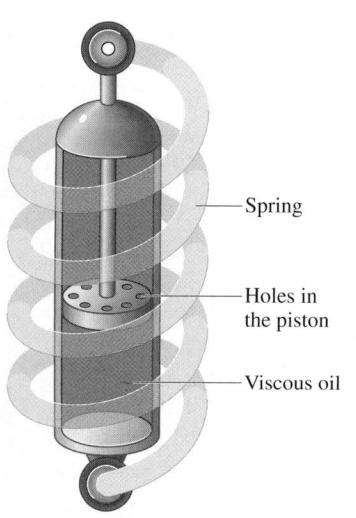

Spring

Holes in the piston

Viscous oil

Figure 10.26 A shock absorber

10.10 FORCED OSCILLATIONS AND RESONANCE

When damping forces are present, the only way to keep the amplitude of oscillations from diminishing is to replace the dissipated energy from some other source. When a child is being pushed on a swing, the parent replaces the energy dissipated with a small push. In order to keep the amplitude of the motion constant, the parent gives a little push once per cycle, adding just enough energy each time to compensate for the energy dissipated in one cycle. The frequency of the *driving force* (the parent's push) matches the *natural frequency* of the system (the frequency at which it would oscillate on its own).

Forced oscillations (or driven oscillations) occur when a periodic external driving force acts on a system that can oscillate. The frequency of the driving force does not have to match the natural frequency of the system. Ultimately, the system oscillates at the driving frequency, even if it is far from the natural frequency. However, the amplitude of the oscillations is generally quite small unless the driving frequency is close to the natural frequency (see Fig. 10.27). When the driving frequency is equal to the natural frequency of the system, the amplitude of the motion is a maximum. This condition is called **resonance**.

At resonance, the driving force is always in the same direction as the object's velocity. Since the driving force is always doing positive work, the energy of the oscillator builds up until the energy dissipated balances the energy added by the driving force. For an oscillator with little damping, this requires a large amplitude. When the driving and natural frequencies differ, the driving force and velocity are no longer synchronized; sometimes they are in the same direction and sometimes in opposite directions. The driving force is not at resonance, so it sometimes does negative work. The net work done by the driving force decreases as the driving frequency moves away from resonance. Therefore, the oscillator's energy and amplitude are smaller than at resonance.

Large-amplitude vibrations due to resonance can be dangerous in some situations. In 1940, a strong wind set the Tacoma Narrows Bridge in Washington state into vibration with increasing amplitude. Turbulence in the air as it flowed across the bridge caused the air pressure to fluctuate with a frequency matching one of the bridge's resonant frequencies. As the amplitude of the oscillations grew, the bridge was closed; soon after, the bridge collapsed (Fig. 10.28). Engineers now design bridges with much higher resonant frequencies so the wind cannot cause resonant vibrations.

In the nineteenth century, bridges were sometimes set into resonant vibration when the cadence of marching soldiers matched a resonant frequency of the bridge. After the collapse of several bridges due to resonance, soldiers were told to break step when crossing a bridge to eliminate the danger of their cadence setting the bridge into resonance.

Figure 10.27 Two resonance curves for an oscillator with natural frequency f_0. The amplitude of the driving force is constant. In the red graph, the oscillator has one-fourth as much damping as in the blue graph.

Making The Connection:
vibration of a bridge

Figure 10.28 The collapse of the Tacoma Narrows Bridge

MASTER THE CONCEPTS

Summary

- A deformation is a change in the size or shape of an object.

- When deforming forces are removed, an *elastic* object returns to its original shape and size.

- Hooke's law, in a generalized form, says that the deformation of a material (measured by the strain) is proportional to the magnitude of the forces causing the deformation (measured by the stress). The definitions of stress and strain are as given in the table.

	Type of Deformation		
	Tensile or Compressive	**Shear**	**Volume**
Stress	Force per unit cross-sectional area F/A	Shear force divided by the parallel area of the surface on which it acts F/A	Pressure P
Strain	Fractional length change $\Delta L/L$	Ratio of the relative displacement Δx to the separation L of the two parallel surfaces $\Delta x/L$	Fractional volume change $\Delta V/V$
Constant of Proportionality	Young's modulus Y	Shear modulus S	Bulk modulus B

- If the tensile or compressive stress exceeds the *proportional limit*, the strain is no longer proportional to the stress. The solid still returns to its original length when the stress is removed as long as the stress does not exceed the *elastic limit*. If the stress exceeds the elastic limit, the material is permanently deformed. For larger stresses yet, the solid fractures when the stress reaches the *breaking point*. The maximum stress that can be withstood without breaking is called the *ultimate strength*.

- Vibrations occur in the vicinity of a point of stable equilibrium. An equilibrium point is *stable* if the net force on an object when it is displaced from equilibrium points back toward the equilibrium point. Such a force is called a restoring force since it tends to restore equilibrium.

- Simple harmonic motion is periodic motion that occurs whenever the restoring force is proportional to the displacement from equilibrium. In SHM, the position, velocity, and acceleration as functions of time are sinusoidal (i.e., sine or cosine functions). Any oscillatory motion is approximately SHM if the amplitude is small, because for small oscillations the restoring force is approximately linear.

- The maximum velocity and acceleration in SHM are

$$v_m = \omega A \quad \text{and} \quad a_m = \omega^2 A \quad (10\text{-}21, 10\text{-}22)$$

where ω is the angular frequency. The acceleration is proportional to and in the opposite direction from the displacement:

$$a_x(t) = -\omega^2 x(t) \quad (10\text{-}19)$$

- The equations of motion for SHM are

If $x = A$ at $t = 0$,	If $x = 0$ at $t = 0$,
$x = A \cos \omega t$	$x = A \sin \omega t$
$v_x = -v_m \sin \omega t$	$v_x = v_m \cos \omega t$
$a_x = -a_m \cos \omega t$	$a_x = -a_m \sin \omega t$

In either case, the velocity is $\frac{1}{4}$ of a cycle ahead of the position and the acceleration is $\frac{1}{4}$ of a cycle ahead of the velocity.

- The angular frequency for a mass-spring system is

$$\omega = \sqrt{\frac{k}{m}} \quad (10\text{-}20a)$$

For a simple pendulum it is

$$\omega = \sqrt{\frac{g}{L}} \quad (10\text{-}26a)$$

and for a physical pendulum it is

$$\omega = \sqrt{\frac{mgd}{I}} \quad (10\text{-}28a)$$

- In the absence of dissipative forces, the total mechanical energy of a simple harmonic oscillator is constant and proportional to the square of the amplitude:

$$E = \tfrac{1}{2}kA^2 \quad (10\text{-}13)$$

where the potential energy has been chosen to be zero at the equilibrium point. At any point, the sum of the kinetic and potential energies is constant:

$$E = \tfrac{1}{2}kx^2 + \tfrac{1}{2}mv_x^2 = \tfrac{1}{2}kA^2 \quad (10\text{-}12)$$

Highlighted Figures and Tables

T10.1 Approximate values of Young's modulus for various substances (p. 343)

F10.4 Stress-strain curve showing limits for ductile material, brittle material, and compact bone (p. 345)

F10.7 Shear stress (p. 347)

T10.2 Shear and bulk moduli for various materials (p. 349)

F10.10 Forces on an object when submerged in fluid (p. 350)

F10.11 Points of stable and unstable equilibrium (p. 351)

F10.14 Positions of an oscillating body at equal time intervals (p. 352)

F10.16 Relation between uniform circular motion and SHM (p. 355)

F10.17 Relaxed, equilibrium, and displaced positions for a relaxed spring of constant k with mass m attached (p. 356)

F10.19 Position, velocity, and acceleration of a particle in SHM as a function of time (p. 358)

F10.22 Forces on a pendulum bob far from equilibrium and near equilibrium (p. 360)

F10.23 A simple and a physical pendulum (p. 361)

F10.25 Oscillations with increasing amounts of damping (p. 364)

CONCEPTUAL QUESTIONS

1. Young's modulus for diamond is about 20 times as large as that of glass. Does that tell you which is stronger? If not, what does it tell you?

2. A grandfather clock is running too fast. To fix it, should the pendulum be lengthened or shortened? Explain.

3. A karate student hits downward on a stack of concrete blocks supported at both ends (Fig. 10.29). A block breaks. Explain where it starts to break first, at the bottom or at the top. (The block experiences shear, compressive, and tensile stresses. Recall that concrete has much less tensile strength than compressive strength. Which part of the block is stretched and which is compressed when the block bends in the middle?)

Figure 10.29
Conceptual Question 3

4. A cylindrical steel bar is compressed by the application of forces of magnitude F at each end. What magnitude forces would be required to compress by the same amount (a) a steel bar of the same cross-sectional area but one half the length? (b) a steel bar of the same length but one half the radius?

5. The columns built by the ancient Greeks and Romans to support temples and other structures are tapered; they are thicker at the bottom than at the top (Fig. 10.30). This certainly has

Figure 10.30
Conceptual Question 5

an aesthetic purpose, but is there an engineering purpose as well? What might it be?

6. Explain how the period of a mass-spring system can be independent of amplitude, even though the distance traveled during each cycle is proportional to the amplitude.

7. In a saber saw, a *Scotch yoke* converts the rotation of the motor into the back-and-forth motion of the blade. The Scotch yoke is a mechanical device used to convert oscillatory motion to circular motion or *vice versa* (Fig. 10.31). A wheel with a fixed knob rotates at constant angular velocity; the knob is constrained within a vertical slot causing the saw blade to move left and right without moving up and down. Is the motion of the saw blade SHM? Explain.

Figure 10.31
Conceptual Question 7

8. A mass hanging vertically from a spring and a simple pendulum both have a period of oscillation of 1 s on Earth. An astronaut takes the two devices to another planet where the gravitational field is stronger than that of Earth. For each of the two systems, state whether the period is now longer than 1 s, shorter than 1 s, or equal to 1 s. Explain your reasoning.

9. A bungee jumper leaps from a bridge and comes to a stop a few centimeters above the surface of the water below. At that lowest point, is the tension in the bungee cord equal to the jumper's weight? Explain why or why not.

10. Does it take more force to break a longer rope or a shorter rope? Assume the ropes are identical except for their lengths and are ideal—there are no weak points. Does it take more *energy* to break the long rope or the short rope? Explain.

11. A pilot is performing vertical loop-the-loops over the ocean at noon. The plane speeds up as it approaches the bottom of the circular loop and slows as it approaches the top of the loop. An observer in a helicopter is watching the shadow of the plane on the surface of the water. Does the shadow exhibit SHM? Explain.

12. Are you more likely to find steel rods in a horizontal concrete beam or in a vertical concrete column? Is concrete more in need of reinforcement under tensile or compressive stress?

13. Suppose that it takes tensile forces of magnitude F to produce a given strain $\Delta L/L$ in a steel wire of cross-sectional area A. If you had two such wires side by side and stretched them simultaneously, what magnitude tensile forces would be required to produce the same strain? By thinking of a thick wire as two (or more) thinner wires side by side, explain why the force to produce a given strain must be proportional to the cross-sectional area. Thus, the strain depends on the stress—the force per unit area.

14. Think of a crystalline solid as a set of atoms connected by ideal springs (Fig. 10.32). When a wire is stretched, how is the elongation of the wire related to the elongation of each of the interatomic springs? Use your answer to explain why a given tensile stress produces an elongation of the wire proportional to the wire's initial length—or, equivalently, that a given stress produces the same strain in wires of different lengths.

Figure 10.32 Conceptual Question 14

15. What are the advantages of using the concepts of stress and strain to describe deformations?

16. An old highway is built out of concrete blocks of equal length. A car traveling on this highway feels a little bump at the joint between blocks. The passengers in the car feel that the ride is uncomfortable at a speed of 45 mi/h, but much smoother at speeds either lower or higher than that. Explain.

17. The period of oscillation of a simple pendulum does not depend on the mass of the bob. By contrast, the period of a mass-spring system does depend on mass. Explain the apparent contradiction. [*Hint:* What provides the restoring force in each case? How does the restoring force depend on mass?]

18. A mass connected to an ideal spring is oscillating without friction on a horizontal surface. Sketch graphs of the kinetic energy, potential energy, and total energy as functions of time for one complete cycle.

MULTIPLE CHOICE QUESTIONS

Questions 1–4. A body is suspended vertically from an ideal spring. The spring is initially in its relaxed position. The body is then released and oscillates about the equilibrium position. Answer choices for Questions 1–4:
  (a) The spring is relaxed.
  (b) The body is at the equilibrium point.
  (c) The spring is at its maximum extension.
  (d) The spring is somewhere between the equilibrium point and maximum extension.

1. The acceleration is greatest in magnitude and is directed upward when:

2. The speed of the body is greatest when:

3. The acceleration of the body is zero when:

4. The acceleration is greatest in magnitude and is directed downward when:

5. Two simple pendulums, A and B, have the same length, but the mass of A is twice the mass of B. Their vibrational amplitudes are equal. Their periods are T_A and T_B, respectively, and their energies are E_A and E_B. Choose the correct statement.
  (a) $T_A = T_B$ and $E_A > E_B$ (b) $T_A < T_B$ and $E_A > E_B$
  (c) $T_A > T_B$ and $E_A < E_B$ (d) $T_A = T_B$ and $E_A < E_B$

6. A force F applied to each end of a steel wire (length L, diameter d) stretches it by 1.0 mm. How much does F stretch another steel wire, of length $2L$ and diameter $2d$?
  (a) 0.50 mm   (b) 1.0 mm   (c) 2.0 mm
  (d) 4.0 mm   (e) 0.25 mm

7. A stiff material is characterized by
  (a) High ultimate strength.   (b) High breaking strength.
  (c) High Young's modulus.   (d) High proportional limit.

8. A brittle material is characterized by
  (a) High breaking strength and low Young's modulus.
  (b) Low breaking strength and high Young's modulus.
  (c) High breaking strength and high Young's modulus.
  (d) Low breaking strength and low Young's modulus.

9. Which pair of quantities can be expressed in the same units?
  (a) Stress and strain   (b) Young's modulus and strain
  (c) Young's modulus and stress
  (d) Ultimate strength and strain

10. Two wires have the same diameter and length. One is made of copper, the other brass. The wires are connected together end to end. When the free ends are pulled in opposite directions, the two wires *must* have the same
  (a) Stress.   (b) Strain.   (c) Ultimate strength.
  (d) Elongation.   (e) Young's modulus.

Figure 10.33 Multiple Choice Questions 11–20

Questions 11–20. Figure 10.33 is the graph of $v_x(t)$ for an object in SHM. Answer choices for each question:
  (a) 1 s, 2 s, 3 s   (b) 5 s, 6 s, 7 s   (c) 0 s, 1 s, 7 s, 8 s
  (d) 3 s, 4 s, 5 s   (e) 0 s, 4 s, 8 s   (f) 2 s, 6 s
  (g) 3 s, 5 s   (h) 1 s, 3 s   (i) 5 s, 7 s
  (j) 3 s, 7 s   (k) 1 s, 5 s

11. When is the kinetic energy maximum?

12. When is the kinetic energy zero?

13. When is the potential energy maximum?

14. When is the potential energy minimum?

15. When is the object at the equilibrium point?

16. When does the acceleration have its maximum magnitude?

17. Which answer specifies times when the net force is in the $+x$-direction?

18. Which answer specifies times when the object is on the $-x$-side of the equilibrium point ($x < 0$)?

19. Which answer specifies times when the object is moving away from the equilibrium point?

20. Which answer specifies times when the potential energy is decreasing?

PROBLEMS

10.2 Hooke's Law for Tensile and Compressive Forces

1. A steel beam is placed vertically in the basement of a building to keep the floor above from sagging. The load on the beam is 5.8×10^4 N, the length of the beam is 2.5 m, and the cross-sectional area of the beam is 7.5×10^{-3} m^2. Find the vertical compression of the beam.

2. A 91-kg man's thighbone has a relaxed length of 0.50 m, a cross-sectional area of 7.0×10^{-4} m^2, and a Young's modulus of 1.1×10^{10} N/m^2. By how much does the thighbone compress when the man is standing on both feet?

3. A brass wire with Young's modulus of 9.2×10^{10} Pa is 2.0 m long and has a cross-sectional area of 5.0 mm^2. If a weight of 5.0 kN is hung from the wire, by how much does it stretch?

4. A wire of length 5.00 m with a cross-sectional area of 0.100 cm^2 stretches by 6.50 mm when a load of 1.00 kN is hung from it. What is the Young's modulus for this wire?

◆ 5. It takes a flea 1.0×10^{-3} s to reach a peak velocity of 0.74 m/s. (a) If the mass of the flea is 0.45×10^{-6} kg, what is the average power required? (b) Insect muscle has a maximum output of 60 W/kg. If 20% of the flea's weight is muscle, can the muscle provide the power needed? (c) The flea has a resilin pad at the base of the hind leg that compresses when the flea bends its leg to jump. If we assume the pad is a cube with a side of 6.0×10^{-5} m. and the pad compresses fully, what is the energy stored in the compression of the pads of the two hind legs? Assume the Young's modulus for resilin is 1.7×10^6 N/m^2. (d) Does this provide enough power for the jump?

6. A 0.50-m-long guitar string, of cross-sectional area 1.0×10^{-6} m^2, has Young's modulus $Y = 2.0 \times 10^9$ N/m^2. By how much must you stretch the string to obtain a tension of 20 N?

7. Two steel wires (of the same length and different radii) are connected together, end to end, and tied to a wall (Fig. 10.34). An applied force stretches the combination by 1.0 mm. How far does the *midpoint* move?

Figure 10.34 Problem 7

8. Abductin is an elastic protein found in scallops, with a Young's modulus of 4.0×10^6 N/m^2. It is used as an inner hinge ligament, with a cross-sectional area of 0.78 mm^2 and a relaxed length of 1.0 mm. When the muscles in the shell relax, the shell opens. This increases efficiency as the muscles do not need to exert any force to open the shell, only to close it. If the muscles must exert a force of 1.5 N to keep the shell closed, by how much is the abductin ligament compressed?

10.3 Beyond Hooke's Law

C 9. Using the stress-strain graph for bone (Fig. 10.4c), calculate Young's moduli for tension and for compression. Consider only small stresses.

10. An acrobat of mass 55 kg is going to hang by her teeth from a steel wire and she does not want the wire to stretch beyond its elastic limit. The elastic limit for the wire is 2.5×10^8 Pa. What is the minimum diameter the wire should have to support her?

11. A hair breaks under a tension of 1.2 N. What is the diameter of the hair? The tensile strength is 2.0×10^8 Pa.

12. The ratio of the tensile (or compressive) strength to the density of a material is a measure of how strong the material is "pound for pound." (a) Compare tendon (tensile strength 80.0 MPa, density 1100 kg/m^3) to steel (tensile strength 0.50 GPa, density 7700 kg/m^3): which is stronger "pound for pound" under tension? (b) Compare bone (compressive strength 160 MPa, density 1600 kg/m^3) to concrete (compressive strength 0.40 GPa, density 2700 kg/m^3): which is stronger "pound for pound" under compression?

13. What is the maximum load that could be suspended from a copper wire of length 1.0 m and radius 1.0 mm without permanently deforming the wire? Copper has an elastic limit of 2.0×10^8 Pa and a tensile strength of 4.0×10^8 Pa.

14. What is the maximum load that could be suspended from a copper wire of length 1.0 m and radius 1.0 mm without breaking the wire? Copper has an elastic limit of 2.0×10^8 Pa and a tensile strength of 4.0×10^8 Pa.

15. A marble column with a cross-sectional area of 25 cm^2 supports a load of 7.0×10^4 N. (Marble: $Y = 6.0 \times 10^{10}$ Pa; compressive strength $= 2.0 \times 10^8$ Pa.) (a) What is the stress in the column? (b) What is the strain in the column? (c) If the column is 2.0 m high, how much is its length changed by supporting the load? (d) What is the maximum weight the column can support?

16. A copper wire of length 3.0 m is observed to stretch by 2.1 mm when a weight of 120 N is hung from one end. (a) What is the diameter of the wire and what is the tensile stress in the wire? (b) If the tensile strength of copper is 4.0×10^8 N/m^2, what is the maximum weight that may be hung from this wire?

17. The leg bone (femur) breaks under a compressive force of about 5×10^4 N for a human and 10×10^4 N for a horse. The human femur has a compressive strength of 1.6×10^8 Pa, while the horse femur has a compressive strength of 1.4×10^8 Pa. What is the effective cross-sectional area of the femur in a human and in a horse? (*Note:* Since the center of the femur contains bone marrow, which has essentially no compressive strength, the effective cross-sectional area is about 80% of the total cross-sectional area.)

18. The maximum strain of a steel wire ($Y = 2.0 \times 10^{11}$ N/m^2), just before breaking, is 0.20%. What is the stress at its breaking point, assuming that strain is proportional to stress up to the breaking point?

10.4 Shear and Volume Deformations

19. A sphere of copper is subjected to 1.0×10^8 N/m^2 of pressure. The copper has a bulk modulus of 130 GPa. By what fraction does the volume of the sphere change? By what fraction does the radius of the sphere change?

20. By what percentage does the density of water increase at a depth of 1.0 km below the surface?

21. Two steel plates are fastened together using four bolts (Fig. 10.35). The bolts each have a shear modulus of 8.0×10^{10} Pa and a shear strength of 6.0×10^8 Pa. The radius of each bolt is 1.0 cm. What is the maximum shearing force F on the plates that the four bolts can withstand?

Figure 10.35 Problem 21

22. An anchor, made of cast iron of bulk modulus 60.0×10^9 Pa and of volume 0.230 m^3, is lowered over the side of the ship to the bottom of the harbor where the pressure is greater than sea level pressure by 1.75×10^6 Pa. Find the change in the volume of the anchor.

23. The upper surface of a cube of gelatin, 5.0 cm on a side, is displaced 0.64 cm by a tangential force. If the shear modulus of the gelatin is 940 Pa, what is the magnitude of the tangential force?

24. A large sponge has forces of magnitude 12 N applied in opposite directions to two opposite faces of area 42 cm^2 (see Fig. 10.7). The thickness of the sponge (L) is 2.0 cm. The deformation angle (γ) is 8.0°. (a) What is Δx? (b) What is the shear modulus of the sponge?

10.5 Simple Harmonic Motion; 10.6 The Period and Frequency for SHM

25. The period of oscillation of an object in an ideal spring-and-mass system is 0.50 s and the amplitude is 5.0 cm. What is the speed at the equilibrium point?

26. The period of oscillation of a spring-and-mass system is 0.50 s and the amplitude is 5.0 cm. What is the magnitude of the acceleration at the point of maximum extension of the spring?

27. A sewing machine needle moves with a rapid vibratory motion, rather like SHM, as it sews a seam. Suppose the needle moves 8.4 mm from its highest to its lowest position and it makes 24 stitches in 9.0 s. What is the maximum needle speed?

28. The prong of a tuning fork moves back and forth when it is set into vibration. The distance the prong moves between its extreme positions is 2.24 mm. If the frequency of the tuning fork is 440.0 Hz, what are the maximum velocity and the maximum acceleration of the prong? Assume SHM.

29. Show that the equation $a = -\omega^2 x$ is consistent for units, and that $\sqrt{k/m}$ has the same units as ω.

30. A small bird's wings can undergo a maximum displacement amplitude of 5.0 cm (distance from the tip of the wing to the horizontal). If the maximum acceleration of the wings is 12 m/s^2, and we assume the wings are undergoing simple harmonic motion when beating, what is the oscillation frequency of the wing tips?

Figure 10.36 Problem 31

31. An empty cart, tied between two ideal springs, oscillates with $\omega = 10.0$ rad/s (Fig. 10.36). A load is placed in the cart, making the total mass 4.0 times what it was before. What is the new value of ω?

32. The air pressure variations in a sound wave cause the eardrum to vibrate. (a) For a given vibration amplitude, are the maximum velocity and acceleration of the eardrum greatest for high-frequency sounds or low-frequency sounds? (b) Find the maximum velocity and acceleration of the eardrum for vibrations of amplitude 1.0×10^{-8} m at a frequency of 20.0 Hz. (c) Repeat (b) for the same amplitude but a frequency of 20.0 kHz.

33. Show that, for SHM, the maximum displacement, velocity, and acceleration are related by $v_m^2 = a_m A$.

34. Equipment to be used in airplanes or spacecraft is often subjected to a shake test to be sure it can withstand the vibrations that may be encountered during flight. A radio receiver of mass 5.24 kg is set on a platform that vibrates in SHM at 120 Hz and with a maximum acceleration of 98 m/s^2 ($= 10g$). Find the radio's (a) maximum displacement, (b) maximum speed, and (c) the maximum net force exerted on it.

35. In an aviation test lab, pilots are subjected to vertical oscillations on a shaking rig to see how well they can recognize objects in times of severe airplane vibration. The frequency can be varied from 0.02 to 40.0 Hz and the amplitude can be set as high as 2 m for low frequencies. What are the maximum velocity and acceleration to which the pilot is subjected if the frequency is set at 25.0 Hz and the amplitude at 1.00 mm?

36. The diaphragm of a speaker has a mass of 50.0 g and responds to a signal of frequency 2.0 kHz by moving back and forth with an amplitude of 1.8×10^{-4} m at that frequency. (a) What is the maximum force acting on the diaphragm? (b) What is the mechanical energy of the diaphragm?

37. An ideal spring has a spring constant $k = 25$ N/m. The spring is suspended vertically. A 1.0-kg body is attached to the unstretched spring and released. It then performs oscillations. (a) What is the magnitude of the acceleration of the body when the extension of the spring is a maximum? (b) What is the maximum extension of the spring?

38. An ideal spring with a spring constant of 15 N/m is suspended vertically. A body of mass 0.60 kg is attached to the unstretched spring and released. (a) What is the extension of the spring when the speed is a maximum? (b) What is the maximum speed?

39. A 0.50-kg object, suspended from an ideal spring of spring constant 25 N/m, is oscillating vertically. How much change of kinetic energy occurs while the object moves from the equilibrium position to a point 5.0 cm lower?

40. A baby jumper consists of a cloth seat suspended by an elastic cord from the lintel of an open doorway. The unstretched length of the cord is 1.2 m and the cord stretches by 0.20 m

when a baby of mass 6.8 kg is placed into the seat. The mother then pulls the seat down by 8.0 cm and releases it. (a) What is the period of the motion? (b) What is the maximum speed of the baby?

10.7 Graphical Analysis of SHM

41. The displacement of an object in SHM is given by $y(t) = (8.0\ \text{cm}) \sin [(1.57\ \text{rad/s})t]$. What is the frequency of the oscillations?

42. A body is suspended vertically from an ideal spring of spring constant 2.5 N/m. The spring is initially in its relaxed position. The body is then released and oscillates about its equilibrium position. The motion is described by

$$y = (4.0\ \text{cm}) \sin [(0.70\ \text{rad/s})t]$$

What is the maximum kinetic energy of the body?

43. (a) Sketch a graph of $x(t) = A \sin \omega t$ (the position of an object in SHM that is at the equilibrium point at $t = 0$). (b) By analyzing the slope of the graph of $x(t)$, sketch a graph of $v_x(t)$. Is $v_x(t)$ a sine or cosine function? (c) By analyzing the slope of the graph of $v_x(t)$, sketch $a_x(t)$. (d) Verify that $v_x(t)$ is $\frac{1}{4}$ cycle ahead of $x(t)$ and that $a_x(t)$ is $\frac{1}{4}$ cycle ahead of $v_x(t)$.

44. A mass-and-spring system oscillates with amplitude A and angular frequency ω. (a) What is the *average* speed during one complete cycle of oscillation? (b) What is the maximum speed? (c) Find the ratio of the average speed to the maximum speed. (d) Sketch a graph of $v_x(t)$, and refer to it to explain why this ratio is greater than $\frac{1}{2}$.

45. A ball is dropped from a height h onto the floor and keeps bouncing. No energy is dissipated, so the ball regains the original height h after each bounce. Sketch the graph for $y(t)$ and list several features of the graph that indicate that this motion is *not* SHM.

46. (a) Given that $x(t) = A \cos \omega t$, show that $v_x(t) = -\omega A \sin \omega t$.[*Hint:* Draw the velocity vector for point P in Fig. 10.16b and then find its x-component.] (b) Verify that the expressions for $x(t)$ and $v_x(t)$ are consistent with energy conservation. [*Hint:* Use the trigonometric identity $\sin^2 \omega t + \cos^2 \omega t = 1$.]

10.8 The Pendulum

47. What is the period of a pendulum consisting of a 6.0-kg mass oscillating on a 4.0-m-long string?

48. A pendulum of length 75 cm and mass 2.5 kg swings with a mechanical energy of 0.015 J. What is the amplitude?

49. A 0.50-kg mass is suspended from a string, forming a pendulum. The period of this pendulum is 1.5 s when the amplitude is 1.0 cm. The mass of the pendulum is now reduced to 0.25 kg. What is the period of oscillation now, when the amplitude is 2.0 cm?

50. A bob of mass m is suspended from a string of length L, forming a pendulum. The period of this pendulum is 2.0 s. If the pendulum bob is replaced with one of mass $\frac{1}{3}m$ and the length of the pendulum is increased to $2L$, what is the period of oscillation?

51. A pendulum (mass m, unknown length) moves according to $x = A \sin \omega t$. (a) Write the equation for $v_x(t)$ and sketch one cycle of the $v_x(t)$ graph. (b) What is the maximum kinetic energy?

52. A clock has a pendulum that performs one full swing every 1.0 s (back *and* forth). The object at the end of the pendulum weighs 10.0 N. What is the length of the pendulum?

53. A pendulum of length L_1 has a period $T_1 = 0.950$ s. The length of the pendulum is adjusted to a new value L_2 such that $T_2 = 1.00$ s. What is the ratio L_2/L_1?

54. A pendulum of length 120 cm swings with an amplitude of 2.0 cm. Its mechanical energy is 5.0 mJ. What is the mechanical energy of the same pendulum when it swings with an amplitude of 3.0 cm?

55. A grandfather clock is constructed so that it has a simple pendulum that swings from one side to the other, a distance of 20.0 mm, in 1.00 s. What is the maximum speed of the pendulum bob? Use two different methods. First, assume SHM and use the relationship between amplitude and maximum speed. Second, use energy conservation. Assume $g = 9.81\ \text{m/s}^2$.

10.9 Damped Oscillations

56. (a) What is the energy of a pendulum ($L = 1.0$ m, $m = 0.50$ kg) oscillating with an amplitude of 5.0 cm? (b) The pendulum's energy loss (due to damping) is replaced in a clock by allowing a 2.0-kg mass to drop 1.0 m in 1 week. What percentage of the pendulum's energy is lost during one cycle?

57. Because of dissipative forces, the amplitude of an oscillator decreases 5.0% in 10 cycles. How many cycles does it take for the *energy* to decrease 5.0%? [*Hint:* Assume the same amount of loss per cycle.]

58. The amplitude of oscillation of a pendulum decays by a factor of 20.0 in 120 s. By what factor has its energy decayed in that time?

COMPREHENSIVE PROBLEMS

59. A pendulum passes $x = 0$ with a speed of 0.50 m/s; it swings out to $A = 0.20$ m. What is the period T of the pendulum? (Assume the amplitude is small.)

60. A person drops a cylindrical steel bar ($Y = 2.0 \times 10^{11}$ Pa) from a height of 1.0 m (distance between the floor and the bottom of the vertically oriented bar). The bar, of length 0.50 m, radius 0.75 cm, and mass 0.70 kg, hits the floor and bounces up, maintaining its vertical orientation. Assuming the collision with the floor is elastic, and that no rotation occurs, what is the maximum

compression of the bar? [*Hint:* At maximum compression, the kinetic energy is zero and the elastic potential energy in the bar is maximum. Start by finding the "spring constant" of the bar.]

61. What is the period of a pendulum formed by placing a horizontal axis (a) through the end of a meterstick (100-cm mark)? (b) through the 75-cm mark? (c) through the 60-cm mark? Assume $g = 9.80\ \text{m/s}^2$.

62. A naval aviator had to eject from her plane before it crashed at sea. She is rescued from the water by helicopter and dangles

from a cable that is 45 m long while being carried back to the aircraft carrier. What is the period of her vibration as she swings back and forth while the helicopter hovers over her ship?

63. A spider's web can undergo SHM when a fly lands on it and displaces the web. For simplicity, assume that a web obeys Hooke's law (which it does not really as it deforms permanently when displaced). If the web is initially horizontal, and a fly landing on the web is in equilibrium when it displaces the web by 0.030 mm, what is the frequency of oscillation when the fly lands?

✦64. Spider silk has a Young's modulus of 4.0×10^9 N/m² and can withstand stresses up to 1.4×10^9 N/m². A single web strand has a cross-sectional area of 1.0×10^{-11} m², and a web is made up of 50 radial strands. A bug lands in the center of a horizontal web so that the web stretches downward (Fig. 10.37). (a) If the maximum stress is exerted on each strand, what angle θ does the web make with the horizontal? (b) What does the mass of a bug have to be in order to exert this maximum stress on the web? (c) If the web is 0.10 m in radius, how far down does the web extend?

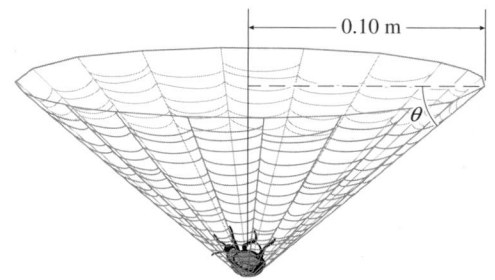

Figure 10.37 Problem 64

✦✦65. A pendulum is made from a uniform rod of mass m_1 and a small block of mass m_2 attached at the lower end. (a) If the length of the pendulum is L and the oscillations are small, find the period of the oscillations in terms of m_1, m_2, L, and g. (b) Check your answer to part (a) in the two special cases $m_1 \gg m_2$ and $m_1 \ll m_2$.

✦66. A thin circular hoop is suspended from a knife edge (Fig. 10.38). Its rotational inertia about the rotation axis (along the knife) is $I = 2mr^2$. Show that it oscillates with the same frequency as a simple pendulum of length equal to the diameter of the hoop.

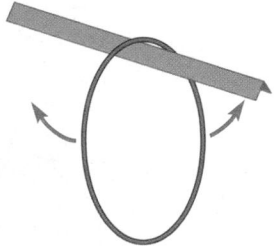

Figure 10.38 Problem 66

67. A hedge trimmer has a blade that moves back and forth with a frequency of 28 Hz. The blade motion is converted from the rotation provided by the electric motor to an oscillatory motion by means of a Scotch yoke (see Conceptual Question 7). The blade moves 2.4 cm during each stroke. Assuming that the blade moves with SHM, what are the maximum speed and maximum acceleration of the blade?

Ⓖ✦68. The motion of a simple pendulum is approximately SHM only if the amplitude is small. Consider a simple pendulum that is released from a horizontal position ($\theta_i = 90°$ in Fig. 10.22). (a) Using conservation of energy, find the speed of the pendulum bob at the bottom of its swing? Express your answer in terms of the mass m and the length L of the pendulum. Do *not* assume SHM. (b) Assuming (incorrectly, for such a large amplitude) that the motion *is* SHM, determine the maximum speed of the pendulum. Based on your answers, is the period of a pendulum for large amplitudes larger or smaller than that given by Eq. (10-26b)?

✦69. The gravitational potential energy of a pendulum is $U = mgy$. (a) Taking $y = 0$ at the lowest point, show that $y = L(1 - \cos \theta)$, where θ is the angle the string makes with the vertical. (b) If θ is small, $(1 - \cos \theta) \approx \frac{1}{2}\theta^2$ and $\theta \approx x/L$ (Appendix A.7). Show that the potential energy can be written $U \approx \frac{1}{2}kx^2$ and find the value of k (the equivalent of the spring constant for the pendulum).

70. The simple pendulum can be thought of as a special case of the physical pendulum where all of the mass is at a distance L from the rotation axis. For a simple pendulum of mass m and length L, show that the expression for the angular frequency of a physical pendulum (Eq. 10-28a) reduces to the expression for the angular frequency of a simple pendulum (Eq. 10-26a).

71. A 4.0-N body is suspended vertically from an ideal spring of spring constant 250 N/m. The spring is initially in its relaxed position. Write an equation to describe the motion of the body if it is released at $t = 0$. [*Hint:* Let $y = 0$ at the equilibrium point and take $+y =$ up.]

72. Show using dimensional analysis that the frequency f at which a mass-spring system oscillates is independent of the amplitude A and proportional to $\sqrt{k/m}$. [*Hint:* Start by assuming that f does depend on A (to some power).]

73. A steel piano wire ($Y = 2.0 \times 10^{11}$ Pa) has a diameter of 0.80 mm. At one end it is wrapped around a tuning pin of diameter 8.0 mm (Fig. 10.39). The length of the wire (not including the wire wrapped around the tuning pin) is 66 cm. Initially, the tension in the wire is 381 N. To tune the wire, the tension must be increased to 402 N. Through what angle must the tuning pin be turned?

Figure 10.39 Problem 73

74. When the tension is 402 N, what is the tensile stress in the piano wire in Problem 73? How does that compare to the elastic limit of steel piano wire (8.26×10^8 Pa)?

✦✦75. An ice cube slides back and forth without friction in a shallow bowl of radius R (Fig. 10.40). If the amplitude of oscillation is small, what is the period of oscillation?

Figure 10.40 Problem 75

76. A gibbon, hanging onto a horizontal tree branch with one arm, swings with a small amplitude. The gibbon's center of mass is

0.40 m from the branch and its rotational inertia divided by its mass is $I/m = 0.25$ m^2. Estimate the frequency of oscillation.

77. A tightrope walker who weighs 640 N walks along a steel cable. When he is halfway across, the cable makes an angle of 0.040 rad below the horizontal (Fig. 10.41). (a) What is the strain in the cable? Assume the cable is horizontal with a tension of 80 N before he steps onto it. Neglect the weight of the cable itself. (b) What is the tension in the cable when the tightrope walker is standing at the midpoint? (c) What is the cross-sectional area of the cable? (d) Has the cable been stretched beyond its elastic limit (2.5×10^8 Pa)?

Figure 10.41   Problem 77 (the 0.040-rad angles are greatly exaggerated)

G 78. The maximum height of a cylindrical column is limited by the compressive strength of the material; if the compressive stress at the bottom were to exceed the compressive strength of the material, the column would be crushed under its own weight. (a) For a cylindrical column of height h, radius r, made of material of density ρ, calculate the compressive stress at the bottom of the column. (b) Since the answer to part (a) is independent of the radius r, there is an absolute limit to the height of a cylindrical column, regardless of how wide it is. For marble, which has a density of 2.7×10^3 kg/m^3 and a compressive strength of 2.0×10^8 Pa, find the maximum height of a cylindrical column. (c) Is this limit a practical concern in the construction of marble columns? Might it limit the height of a beanstalk?

79. In Problem 8.34 we found that the force of the tibia (shinbone) on the ankle joint for a person (of weight 750 N) standing on the ball of one foot was 2800 N (Fig. 8.68). The ankle joint therefore pushes upward on the bottom of the tibia with a force of 2800 N, while the top end of the tibia must feel a net downward force of approximately 2800 N (ignoring the weight of the tibia itself). The tibia (Fig. 10.42) has a length of 0.40 m, an

Figure 10.42   Problem 79

average inner diameter of 1.3 cm, and an average outer diameter of 2.5 cm. (The central core of the bone contains marrow that has negligible compressive strength.) (a) Find the average cross-sectional area of the tibia. (b) Find the compressive stress in the tibia. (c) Find the change in length for the tibia due to the compressive forces.

G 80. A bungee jumper leaps from a bridge and undergoes a series of oscillations. Assume $g = 9.78$ m/s^2. (a) If a 60.0-kg jumper uses a bungee cord that has an unstretched length of 33.0 m and she jumps from a height of 50.0 m above a river, coming to rest just a few centimeters above the water surface on the first downward descent, what is the period of the oscillations? Assume the bungee cord follows Hooke's law. (b) The next jumper in line has a mass of 80.0 kg. Should he jump using the same cord? Explain.

✦81. Resilin is a rubber-like protein that helps insects to fly more efficiently. The resilin, attached from the wing to the body, is relaxed when the wing is down and is extended when the wing is up. As the wing is brought up, some elastic energy is stored in the resilin. The wing is then brought back down with little muscular energy, since the potential energy in the resilin is converted back into kinetic energy. Resilin has a Young's modulus of 1.7×10^6 N/m^2. (a) If an insect wing has resilin with a relaxed length of 1.0 cm and with a cross-sectional area of 1.0 mm^2, how much force must the wings exert to extend the resilin to 4.0 cm? (b) How much energy is stored in the resilin?

ANSWERS TO PRACTICE PROBLEMS

10.1 2k (When the original spring is stretched an amount L, each of the half-springs stretches only $\frac{1}{2}L$. Each of the newly formed springs stretches half as far as the original spring for a given applied force.) **10.2** 1.4×10^{-5} **10.3** 0.18 mm **10.4** 1.3×10^8 Pa

10.5 $-\dfrac{\Delta P}{B} = -\dfrac{1.0 \times 10^7 \text{ Pa}}{2.3 \times 10^9 \text{ Pa}} = -0.0043 = \dfrac{\Delta V}{V}$ and $\Delta V = -0.43\% \ V$

10.6 110 m/s^2 at $x = \pm A$ **10.7** $K = 0$, $U_e = 2(mg)^2/k$, $U_g = -(mg)^2/k$, $E = (mg)^2/k$

10.8

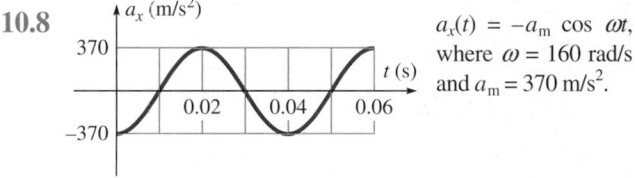

$a_x(t) = -a_m \cos \omega t$, where $\omega = 160$ rad/s and $a_m = 370$ m/s^2.

10.9 1.6 m/s^2 (about 1/6 that of the Earth) **10.10** 0.82 m/s or 1.8 mi/h

Waves

On January 17, 1995, a terrible earthquake struck the Hanshin region of Japan, killing over 6400 people and injuring about 40,000 others. Some 200,000 homes and buildings were damaged, causing the evacuation to shelters of 320,000 people. The heaviest damage occurred in the city of Kobe, including the buckling and collapse of an elevated highway. However, geologists found that the point of origin of the earthquake was 15–20 km below the northern tip of Awaji Island, about 20 km southwest of Kobe. How did the energy released by the earthquake travel that far with enough energy to cause such great devastation?

11.1   WAVES AND ENERGY TRANSPORT

Physicists use only a few basic models to describe the physical world. One such model is the particle: a point-like object with no inner structure and with certain characteristics such as mass and electrical charge. Another basic model is the **wave**. Water waves are familiar examples. When a pebble is dropped into a pond, it disturbs the surface of the water. Ripples on the surface of the pond travel away from the spot where the pebble landed. This traveling disturbance of the water from its equilibrium position is a wave on the surface of the water.

Any wave is characterized as some sort of "disturbance" that travels away from its source. In Chapters 11 and 12, we concentrate on mechanical waves traveling through a material medium, such as water waves, sound waves, and the seismic waves caused by earthquakes. Particles in the medium are disturbed from their equilibrium positions as the wave passes, returning to their equilibrium positions after the wave has passed. In Chapter 22, we discuss electromagnetic waves such as radio waves and light waves, in which the disturbance consists of oscillating electromagnetic fields. Two of our five human senses are wave detectors: the ear is sensitive to the tiny fluctuations in air pressure caused by compressional waves in air (audible sound) and the eye is sensitive to electromagnetic waves in a certain frequency range (visible light).

Energy Transport

Suppose we drop a pebble into a still pond. The kinetic energy of the pebble just before it hits the pond is partly converted into the energy carried off by the water wave. That waves carry energy is clear to anyone who has been surfing or swimming in the ocean. Speaking of surfing, information on the Internet is carried by waves of various sorts: electrical waves in wires, microwaves between Earth and communications satellites, light waves in optical fibers. Microwaves in ovens carry energy from their source to the food; the electromagnetic energy of the microwaves is absorbed by water molecules in the food and appears as thermal energy. Electromagnetic waves from the Sun bring the energy that makes life possible. Seismic waves carry energy released by an earthquake to other parts of the Earth, sometimes with devastating results.

When a baseball pitcher throws a ball to the catcher, the ball carries energy with it (Fig. 11.1a). The pitcher gives the ball kinetic energy; the catcher receives the energy when the ball hits his hand and his hand recoils. Suppose instead that they hold a rope stretched between them. If the pitcher suddenly moves his hand up and down quickly, a wave pulse travels along the rope until it reaches the catcher's hand (Fig. 11.1b). Once again, the

Figure 11.1 Two different ways to transfer energy. (a) The baseball carries energy from the pitcher to the catcher. (b) A wave pulse also carries energy from the pitcher to the catcher, but the piece of rope held by the pitcher's hand does not move to the catcher's hand.

pitcher sends the energy and the catcher receives it when the rope makes his hand recoil. However, in this case the pitcher is still holding his end of the rope; it never leaves his hand. Energy is transferred *without any matter moving from the pitcher to the catcher*.

Physics at Home

Stretch a heavy rope between yourself and a friend and test out the transfer of energy from one to the other. Does your hand recoil when the pulse arrives?

Similarly, seismic waves travel away from the *focus* of an earthquake (the point of origin) both through the Earth (*body waves*) and along the Earth's crust (*surface waves*), transporting vibrations and energy. However, the material through which the waves travel is *not* transported. Most earthquake damage is caused by seismic waves rather than caused by fault movement. In the Hanshin earthquake, damage to buildings was caused by seismic waves at distances over 100 km from the *epicenter* (the point on the surface directly above the focus) but the motion of the vibrating particles in the ground never moved more than about 1.5 m.

The sound of thunder travels for miles in all directions, but none of the air molecules zapped by lightning travel more than a meter or so during the few seconds that it takes the sound to reach our ears. A wave can transmit energy from one point to another without transporting any matter between the two points.

Physics at Home

Observe carefully what happens when you snap your fingers. You start by pressing your thumb against your fingers and then sideways, the thumb in one direction and the fingers in the opposite direction. Initially friction keeps them from moving sideways, but suddenly they slip, releasing the built-up energy partly as a sound wave.

Similarly, the rock on two sides of a fault line is pressed together and sideways. Friction keeps them from moving sideways as elastic (or strain) energy builds up. Then suddenly they slip, releasing a tremendous amount of energy largely in the form of seismic waves that carry vibrations far from the focus of the earthquake.

Intensity

For a wave that travels in a three-dimensional medium (such as sound waves or seismic waves traveling through the Earth), the **intensity** (symbol I, SI unit W/m^2) is a measure of the *average power per unit area* carried by the wave past a surface perpendicular to the wave's direction of propagation. For example, if a sound wave's intensity is a fairly loud 10^{-5} W/m^2 when it reaches the eardrum and the area of the eardrum is 10^{-4} m^2, then the power delivered to the eardrum is 10^{-9} W (assuming that all the energy incident on the eardrum is absorbed). The energy absorbed by the eardrum at this rate in *one hour* would be

$$10^{-9} \, W \times 3600 \, s \approx 4 \, \mu J$$

The human ear is a very sensitive detector indeed.

For most waves, the intensity decreases as the distance from the source increases. Some of the energy can be absorbed (dissipated) by the wave medium. The amount of energy absorbed depends on the medium. Air absorbs relatively little sound energy, which is why we can hear sounds generated far away.

Another reason that intensity decreases with distance is that, as the wave spreads out, the energy gets spread over a larger and larger area. Consider a point source emitting a wave uniformly in all directions—an *isotropic* source (Fig. 11.2). The average power (energy per unit time) emitted is constant. Imagine a sphere surrounding the source; the rate at which energy passes through the surface of the sphere is the same no matter what the radius. The surface area of a sphere is $4\pi r^2$, so as the wave moves farther from the source, the energy spreads out over a larger and larger area. Thus, the power per unit area (intensity) decreases with distance. Assuming that no energy is absorbed by the medium and there are no obstacles to reflect or absorb sound,

$$I = \frac{\text{power}}{\text{area}} = \frac{P}{4\pi r^2} \tag{11-1}$$

(point source emitting uniformly in all directions; no reflection or absorption)

Therefore, if energy absorption by the medium can be neglected, the intensity of the sound is inversely proportional to the square of the distance from the source. This "inverse square law" is the result of a conserved quantity (here, energy) radiating uniformly from a point source in three-dimensional space.

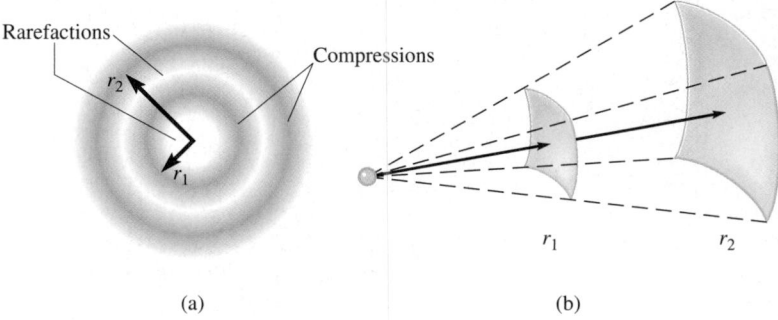

Figure 11.2 (a) A point source of sound radiating energy uniformly in all directions. (b) The intensity at a distance r_2 is smaller than the intensity at a distance r_1 since the same power is spread out over a greater area.

11.2   TRANSVERSE AND LONGITUDINAL WAVES

A Slinky toy can be used to demonstrate two different kinds of wave. In a **transverse** wave, the motion of particles in the medium is perpendicular to the direction of propagation of the wave. To send a transverse wave down a Slinky, wiggle the end of the Slinky back and forth in a direction perpendicular to the length of the Slinky (Fig. 11.3a). In a **longitudinal** wave, the motion of particles in the medium is along the same line as the direction of propagation of the wave. To send a longitudinal wave down the Slinky, jiggle the end in and out along its length to alternately stretch and compress the coils (Fig. 11.3b). A red dot painted on one coil of the Slinky helps illustrate the difference. In a transverse wave, the dot moves back and forth about a fixed position with its motion perpendicular to the direction of propagation of the wave; in a longitudinal wave, the dot also moves back and forth about a fixed position but along the direction of propagation of the wave. In both cases, the wave itself moves from one end of the Slinky to the other while the dot is moving about its fixed position.

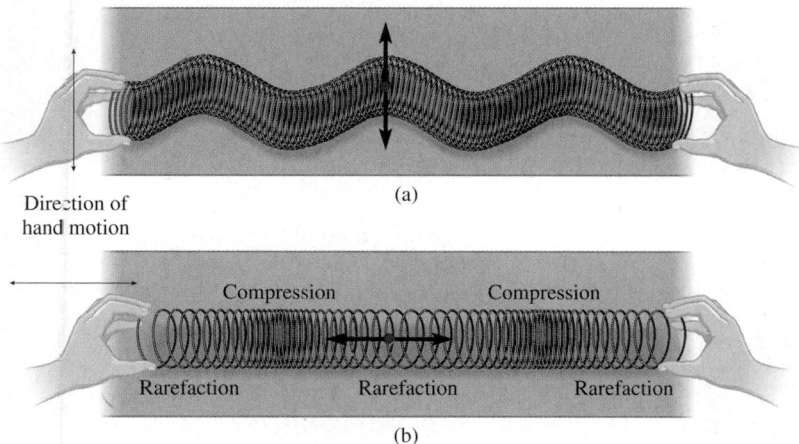

Direction of hand motion

(a)

Compression   Compression

Rarefaction   Rarefaction   Rarefaction

(b)

Figure 11.3 (a) Transverse and (b) longitudinal waves on a Slinky

Physics at Home

Ask a friend to sit at a table and hold one end of a long, loose spring (or a Slinky). With the spring supported by the table surface, grasp the other end and stretch the spring. Figure out how to move your hand to send transverse and longitudinal waves down the spring.

The Slinky—or any long spring—is a better approximation to solid materials than the stretched rope. In solids both types of waves can exist; a transverse wave results from a shear disturbance and a longitudinal wave from a compressional disturbance. An earthquake generates both longitudinal and transverse seismic waves. The two travel at different speeds so the time lapse between the arrival of the two wave types at a detector gives a clue to the distance to the earthquake's focus.

Fluids can be compressed but, because they flow, they do not sustain shear stresses. Therefore, longitudinal waves travel through fluids but transverse waves do not. However, gravity or surface tension can provide the transverse restoring force that allows a transverse wave to travel *along the surface* of a liquid.

A sound wave is longitudinal; each small volume of air vibrates back and forth along the direction of travel of the wave. Air molecules are compressed together in some places and more thinly spaced (*rarefied*) in others; the air has regions of higher and lower density called **compressions** and **rarefactions** (see Fig. 11.3b).

Seismic body waves can be either longitudinal or transverse. Longitudinal body waves (*P waves*) are the fastest seismic waves (typically 4–8 km/s). They are similar to sound waves in air: particles in the Earth's interior are pushed together and pulled apart in the same direction that the wave propagates (Fig. 11.4a). Transverse body waves

Figure 11.4 The motion of ground particles in (a) P waves, (b) S waves, and (c) one kind of surface wave. (d) Motion of a swimmer as a water wave passes by.

(*S waves*) travel more slowly—typically 2–5 km/s. In an S wave, particles in the Earth's interior vibrate at right angles to the direction that the wave travels (Fig. 11.4b). By measuring the time between the first arrivals of the two types of waves at different detection stations, geologists are able to determine the origin of the earthquake.

Waves That Combine Transverse and Longitudinal Motion

Not all seismic waves are purely transverse or purely longitudinal. In a surface wave, the ground near the surface rolls approximately in a circle. Thus, the motion of the ground has components both parallel and perpendicular to the direction of propagation. The transverse component can either be up and down (as shown in Fig. 11.4c) or side to side. The motion of the ground is greatest at the surface. Surface waves are generally slower than body waves.

Ocean waves are similar to the surface seismic wave shown in Fig. 11.4c. Deep underwater, the wave is mostly longitudinal (Fig. 11.4d); as the wave passes, water moves back and forth along the direction of propagation of the wave. Higher up, the wave has both transverse and longitudinal components; water moves in an oval as the wave passes. A swimmer floating at the surface moves approximately in a circle. The air above the surface presents little resistance, so water swells upward more easily there and then is pulled back downward by gravity (or, for small amplitudes, by surface tension). If the wave were purely transverse the swimmer would move only up and down, but the longitudinal component of the wave superimposed on the transverse component results in an approximately circular motion. When the wave gets close to shore, the crest often collapses or *breaks*; the motion of the water particles is then much more complex.

When a guitar string is plucked *gently*, the wave on the string is almost purely transverse; stretching of the string is negligible. When it is plucked more forcefully, the resulting wave is a combination of transverse and longitudinal waves. At any instant, the string is stretched more in some places than in others; a point on the string has longitudinal motion as well as transverse motion.

11.3   SPEED OF TRANSVERSE WAVES ON A STRING

The speed of a mechanical wave depends on properties of the wave medium. What properties of a string determine the speed of a transverse wave moving along it? Suppose that a string of length L and mass m is under tension F. In Problem 41, you can show that $\sqrt{FL/m}$ is the only combination of those three quantities with the correct units for speed. There could be a dimensionless constant multiplier, but a derivation using more advanced mathematics shows that the constant is 1; the speed of a transverse wave on a string is

$$v = \sqrt{\frac{FL}{m}} \tag{11-2}$$

We can rewrite Eq. (11-2) in another form. Length and mass are not independent; for a given string composition and diameter (say, a yellow brass string of 0.030 in. diameter), the mass of the string is proportional to its length. By defining the **linear mass density** (mass per unit length) of the string to be

$$\mu = m/L \tag{11-3}$$

the speed of a transverse wave on a string can be written

$$v = \sqrt{\frac{F}{\mu}} \tag{11-4}$$

An advantage of Eq. (11-4) is that it shows clearly that the wave speed depends on *local* properties of the medium; it does not depend on how much of the medium there is. The wave speed in the vicinity of some point P, for instance, does not depend on how long the string is; only properties of the string in the immediate vicinity of point P can determine how fast the wave travels past that point.

Note that as tension increases, wave speed increases; as mass density increases, wave speed decreases. A somewhat more general way to think about it, applicable to other waves as well, is *more restoring force makes faster waves; more inertia makes slower waves.*

The speed at which a wave propagates is not the same as the speed at which a particle in the medium moves. Suppose a horizontal string is stretched along the x-axis and a transverse pulse in the y-direction is sent down the string. The speed of propagation of the wave v is the speed at which the *pattern* or disturbance moves along the string (in the x-direction); for a uniform string, the wave speed is constant. A point on the string vibrates up and down in the $\pm y$-direction with a *different* speed that is *not* constant.

The wave speed is not the same as the speed of a particle in the wave medium.

Example 11.1

A Piñata

A string of length 0.50 m has a mass of 125 mg. The string is attached to the ceiling and a piñata of mass 0.500 kg hangs from the other end. A tall child grabs the piñata and shakes it to see if there is candy inside; as a result, a transverse pulse travels up the string toward the ceiling. At what speed does the pulse travel?

Strategy We start with a diagram of the situation (Fig. 11.5). The piñata puts the string under tension. The tension in the string is equal to the weight of the piñata. The mass and length of the string are given, so the linear mass density can be found. Then we can find the wave speed.

Figure 11.5
A transverse pulse travels up the string supporting a piñata.

0.500 kg

Solution The speed of a transverse wave on a string is given by Eq. (11-4):

$$v = \sqrt{F/\mu}$$

where F is the tension in the string and μ is the linear mass density of the string. The tension is equal to the weight hanging on the string:

$$F = Mg$$

The linear mass density of the string is mass per unit length ($\mu = m/L$). Substituting the tension and mass density, we have

$$v = \sqrt{\frac{F}{m/L}} = \sqrt{\frac{(Mg)L}{m}}$$

$$= \sqrt{\frac{0.500\ \text{kg} \times 9.8\ \text{m/s}^2 \times 0.50\ \text{m}}{125 \times 10^{-6}\ \text{kg}}} = 140\ \text{m/s}$$

Discussion The *weight* of the string (mg) is negligible in comparison with the weight hanging from the end of the string (Mg). That is not always the case, as can be seen in Practice Problem 11.1.

Practice Problem 11.1　Initial velocity of another wave pulse traveling on a string

A string of length 10.0 m has a linear mass density of 25 g/m. The string is fixed at the top and has an object of mass 0.200 kg hanging from the bottom. (a) What is the *initial* wave speed of a pulse sent up the string from the bottom? (b) What is the speed of the pulse as it approaches the top of the string? [*Hint:* Does the weight of the string itself affect the tension in either case?]

11.4　PERIODIC WAVES

A **periodic** wave repeats the same pattern over and over, each repeating section transporting the energy that was used to generate it. A periodic water wave can be produced by steadily dropping a series of pebbles into the water; a periodic wave on a cord can be produced by taking one end of the cord and moving it up and down, over and over, in a repeating pattern. As the wave propagates along the cord, every point on the cord oscillates with the same up and down pattern, though with a time delay that depends on the wave speed. While musical sounds are often periodic waves, noise is *aperiodic*. The human voice makes a periodic sound wave when a vowel is sung at a steady pitch (constant frequency); most of the consonant sounds are aperiodic (Fig. 11.6).

The terminology for periodic waves is similar to that used for uniform circular motion (Chapter 5) and for simple harmonic motion (SHM, Chapter 10). At any given point, the wave repeats itself after a time T called the **period**. The inverse of the period is the **frequency** f.

$$f = \frac{1}{T} \quad \text{(SI unit Hz = s}^{-1}\text{)} \tag{5-7}$$

The frequency tells how often the pattern of motion repeats itself at any single point. For instance, if the frequency is 20 Hz, then there are 20 repetitions, or cycles, per second. Each cycle takes a time $T = 1/f = 0.05$ s. The angular frequency is $\omega = 2\pi f$ and is measured in rad/s.

During one period T, a periodic wave traveling at speed v moves a distance vT. In Fig. 11.7, note that, at any instant, points separated by a distance vT along the direction of propagation of a wave move "in sync" with each other. Thus, vT is the *repetition distance* of the wave, just as the period is the *repetition time*. This distance is called the **wavelength** (symbol λ, the Greek letter lambda).

$$\lambda = vT \tag{11-5}$$

Combining this relation and the expression for frequency, we obtain

$$v = \frac{\lambda}{T} = f\lambda \tag{11-6}$$

Equations (11-5) and (11-6) are true for all periodic waves, no matter how the wave is produced or what the shape of the wave.

The maximum displacement of any particle from its equilibrium position is the **amplitude** A of the wave. For a sinusoidal wave traveling along a stretched string in the x-direction, the amplitude A is the maximum displacement in the positive or negative y-direction.

$$y_\mathrm{m} = \pm A$$

For surface water waves, the amplitude is the height of a crest (a high point) above or the depth of a trough (a low point) below the undisturbed water level.

Harmonic waves are a special kind of periodic wave in which the disturbance is sinusoidal (either a sine or cosine function). In a harmonic transverse wave on a string, for instance, every point on the string moves in SHM with the same amplitude and angular frequency, although different points reach their maximum displacements at different times. The maximum speed and maximum acceleration of a point on the string depend on both the angular frequency and the amplitude of the wave:

$$v_\mathrm{m} = \omega A \tag{10-21}$$

$$a_\mathrm{m} = \omega^2 A \tag{10-22}$$

Since the total energy of an object moving in SHM is proportional to the amplitude squared (Section 10.5), the total energy of a harmonic wave is proportional to the square of its amplitude. That turns out to be a general result not limited to harmonic waves:

> The intensity of a wave is proportional to the square of its amplitude.

11.5   MATHEMATICAL DESCRIPTION OF A WAVE

A wave is represented mathematically by a variation in some quantity (such as pressure or displacement) that is described as a function of both position and time. For a transverse wave on a guitar string, the function specifies the displacement of each point on the string from its equilibrium position. If the string is oriented along the x-axis and the displacement of any point on the string is in the $\pm y$-direction, then the wave is described by a function of two variables: $y(x, t)$.

Consider a long stretched string along the x-axis. One end of the string (at $x = 0$) is moved by an external agent according to some function $y = h(t)$; as a result, a transverse

(a)

(b)

Figure 11.6 Sound wave pattern produced by a singer (a) singing the vowel "ah" and (b) hissing the consonant "s."

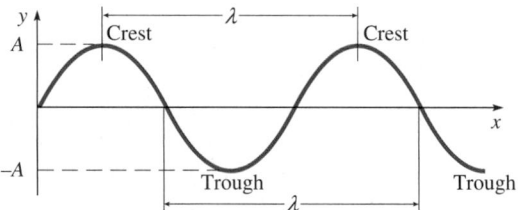

Figure 11.7 Sinusoidal wave moving with speed v in the x-direction. The amplitude A and the wavelength λ are shown.

wave is produced that travels in the +x-direction with wave speed v. If the wave retains the same shape as it moves down the string, then the motion of any point x on the string copies the motion of the left end after a time delay x/v (the time it takes the wave to travel a distance x at speed v—see Fig. 11.8). Thus, $y(x, t) = h(t - x/v)$. Even though the function that describes the wave has two variables (x and t), these variables must occur in the particular combination $(t - x/v)$ in order to describe a wave that retains its form as it propagates in the +x-direction. For a wave moving in the −x-direction, the variables would occur in the combination $(t + x/v)$. A wave that retains its shape as it moves in a single direction is called a *traveling wave*.

+x-direction: $y(x, t) = h(t - x/v)$
−x-direction: $y(x, t) = h(t + x/v)$

Tips

A Harmonic Traveling Wave

Suppose the motion of the left end of the string is described by $y = A \cos \omega t$. By substituting $(t - x/v)$ for t, we obtain the function that describes the motion of *any* point $x > 0$:

$$y(x, t) = A \cos \omega(t - x/v)$$

To simplify the writing, we introduce a constant called the **wavenumber** (symbol k, SI unit rad/m):

$$k = \frac{\omega}{v} = \frac{2\pi f}{v} = \frac{2\pi}{\lambda} \tag{11-7}$$

Then the equation for the harmonic wave can be written

$$y(x, t) = A \cos (\omega t - kx) \tag{11-8}$$

The argument of the sine or cosine function, $(\omega t \pm kx)$, is called the **phase** of the wave at x and t. Phase is measured in units of angle (usually radians). The phase of a wave at a given point and instant of time tells us how far along that point is in the repeating pattern of its motion. Since a sine or cosine function repeats every 2π radians, the motions of two different points x_1 and x_2 that differ in phase by an integer times 2π are exactly the same; the points move "in sync" or *in phase* with each other. The distance between the two points is an integral number of wavelengths:

If

$$k(x_2 - x_1) = 2\pi n \quad \text{(where } n \text{ is any integer)}$$

then

$$x_2 - x_1 = \frac{2\pi n}{k} = \frac{2\pi n}{2\pi/\lambda} = n\lambda$$

Figure 11.8 A wave pulse, with the same shape, at successive times. The motion of the point x repeats the motion of the point $x = 0$ with a time delay $\Delta t = x/v$.

Example 11.2

A Traveling Wave on a String

A wave on a string is described by $y(x, t) = a \sin (bt + cx)$, where a, b, and c are constants. (a) Does this wave retain its shape as it travels? (b) In what direction does the wave travel? (c) What is the wave speed?

Strategy   We try to manipulate the function to see if it can be written as a function of either $(t - x/v)$ or $(t + x/v)$ as in the general harmonic wave equation $y(x, t) = A \cos \omega(t - x/v)$. The wave speed v does not appear explicitly in the function as

continued on next page

Example 11.2 *continued*

written, but it may be some combination of the other constants in the function. We are starting from an equation similar to that of Eq. (11-8), $y(x, t) = A \cos(\omega t - kx)$.

Solution   The coefficient of t in our equation should be the constant that represents ω. Factoring out that constant, we have

$$y(x, t) = a \sin b\left(t + \frac{cx}{b}\right) = a \sin b\left(t + \frac{x}{b/c}\right)$$

Now we see that $y(x, t)$ is a function of $t + x/v$, where $v = b/c$,

$$y(x, t) = a \sin b\left(t + \frac{x}{v}\right) \text{ where } v = \frac{b}{c}$$

Therefore: (a) yes, the wave retains its shape since it is a function of $(t + x/v)$; (b) it travels in the $-x$-direction since the t and x/v terms have the same sign; and (c) the wave speed is b/c.

Discussion   Before being completely satisfied with this solution, it is a good idea to check that b/c has the right units for wave speed. The two terms bt and cx that are added together must have the same units. In SI, the argument of a sine function is measured in radians. Then b is measured in rad/s and c is measured in rad/m. Then the units of b/c are (rad/s)/(rad/m) = m/s, which is correct for wave speed.

Practice Problem 11.2   Another traveling wave on a string

A wave on a string is described by

$$y(x,t) = 0.0050 \sin(4.0t - 0.50x)$$

where x and y are in meters and t is in seconds. (a) Does this wave retain its shape as it travels? (b) In what direction does the wave travel? (c) What is the wave speed?

11.6   GRAPHING WAVES

To graph a one-dimensional wave $y(x, t)$, only one of the two independent variables (x, t) can be plotted. The other must be "frozen"; it is treated as a constant. If x is held constant, then one particular point (determined by the value of x) is singled out; the graph shows the motion of *that point* as a function of time (Fig. 11.9a). If instead t is held constant and y is plotted as a function of x, then the graph is like a snapshot—an instantaneous picture of what the wave looks like *at that particular instant* (Fig. 11.9b).

(a)

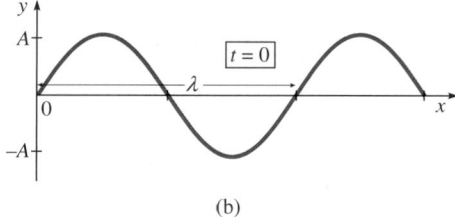

(b)

Figure 11.9   Two graphs of a harmonic wave $y(x, t) = A \sin(\omega t - kx)$ on a string. (a) The vertical displacement of a particular point on the string ($x = 0$) as a function of time. (b) The vertical displacement as a function of horizontal position at a single instant of time ($t = 0$).

Example 11.3

A Transverse Harmonic Wave

A transverse harmonic wave travels in the $+x$-direction on a string at a speed of 5.0 m/s. Figure 11.10 shows a graph of $y(t)$ for the point $x = 0$. (a) What is the period of the wave? (b) What is the wavelength? (c) What is the amplitude? (d) Write the

function $y(x, t)$ that describes the wave. (e) Sketch a graph of $y(x)$ at $t = 0$.

Strategy   Since the graph uses time as the independent variable, the period can be read from the graph as the time for one

continued on next page

Example 11.3 *continued*

cycle. The wavelength is the distance traveled by the wave during one period. The amplitude can be read from the graph as the maximum displacement. These are all the constants needed to write the function $y(x, t)$. We do have to think about the direction of travel and whether to write sine or cosine.

Figure 11.10
Graph of a transverse harmonic wave

Solution (a) The period T is the time for one cycle. From the graph, $T = 2.0$ s.

(b) The wavelength λ is the distance traveled by the wave at speed $v = 5.0$ m/s during one period:

$$\lambda = vT = 5.0 \text{ m/s} \times 2.0 \text{ s} = 1.0 \times 10^1 \text{ m}$$

(c) The amplitude A is the maximum displacement from equilibrium. From the graph, $A = 0.030$ m.

(d) Figure 11.10 is a sine function. The motion of the point $x = 0$ is

$$y(t) = A \sin 2\pi \frac{t}{T}$$

Since the wave moves in the $+x$-direction, a point at $x > 0$ duplicates the motion of $x = 0$ with a time delay of $\Delta t = x/v$ (the time for the wave to travel a distance x). Then

$$y(x, t) = A \sin 2\pi \frac{t - x/v}{T}$$

where $v = 5.0$ m/s and $T = 2.0$ s.

(e) Substituting $t = 0$,

$$y(x) = A \sin \left(-2\pi \frac{x}{vT} \right)$$

Substituting $vT = \lambda$ and using the identity $\sin(-\theta) = -\sin \theta$ (Appendix A.7), we have

$$y(x, t = 0) = -A \sin 2\pi \frac{x}{\lambda}$$

A graph of this function is an inverted sine function with amplitude $A = 0.030$ m and wavelength $\lambda = vT = 5.0$ m/s $\times$ 2.0 s $= 1.0 \times 10^1$ m.

Discussion Figure 11.10 shows that the point $x = 0$ is initially at $y = 0$ and then moves up (in the $+y$-direction) until it reaches the crest (maximum y) at $t = 0.50$ s. Imagine the graph in (e) to represent the first frame (at $t = 0$) of a movie of the wave. Since the wave moves to the right, the sinusoidal patterns shifts a little to the right in each successive frame. The point $x = 0$ moves up until it reaches the crest when the wave has traveled 2.5 m to the right. Since the wave speed is 5.0 m/s, the point $x = 0$ reaches the crest at $t = \dfrac{2.5 \text{ m}}{5.0 \text{ m/s}} = 0.50$ s.

Practice Problem 11.3 Another harmonic transverse wave

A wave is described by $y(x, t) = (1.2 \text{ cm}) \sin (10.0\pi t + 2.5\pi x)$, where x is in meters and t is in seconds. (a) Sketch a graph of $y(t)$ at $x = 0$. (b) Sketch a graph of $y(x)$ at $t = 0$. (c) What is the period of the wave? (d) What is the wavelength? (e) What is the amplitude? (f) What is the speed of the wave? (g) In what direction does the wave move?

11.7   PRINCIPLE OF SUPERPOSITION

Suppose two waves of the same type pass through the same region of space. Do the waves affect each other? If the amplitudes of the waves are large enough, then particles in the medium are displaced far enough from their equilibrium positions that Hooke's law (restoring force $\propto$ displacement) no longer holds; in that case the waves *do* affect each other. However, for small amplitudes, the waves can pass through each other and emerge *unchanged*. More generally, when the amplitudes are not too large, the principle of superposition applies:

Principle of superposition
When two or more waves overlap, the net disturbance at any point is the sum of the individual disturbances due to each wave.

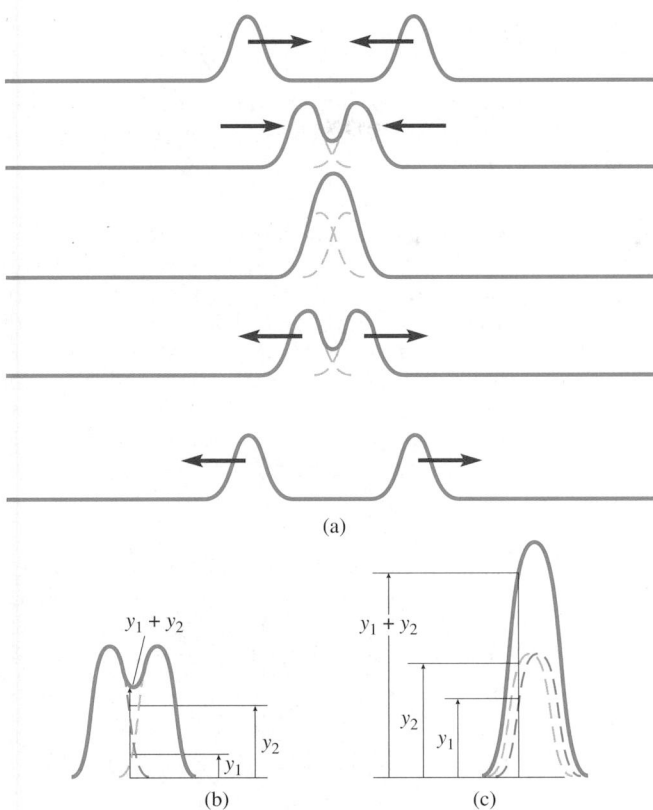

(a)

$y_1 + y_2$

y_2

y_1

(b)

$y_1 + y_2$

y_2

y_1

(c)

Figure 11.11   (a) Two identical wave pulses traveling toward and through each other. (b), (c) Details of the wave pulse summation; dashed lines are the separate wave pulses and solid line is the sum.

Suppose two wave pulses are traveling toward each other on a string (Fig. 11.11a). If one of the pulses (acting alone) would produce a displacement y_1 at a certain point and the other would produce a displacement y_2 at the same point, the result when the two overlap is a displacement of $y_1 + y_2$. Figures 11.11b, c show in greater detail how y_1 and y_2 add together to produce the net displacement when the pulses overlap. The dashed curves represent the individual pulses; the solid line represents the superposition of the pulses. In Fig. 11.11b the pulses are starting to overlap and in Fig. 11.11c they are just about to coincide.

The wave pulses pass right through each other without affecting each other; once they have separated, their shapes and heights are the same as before the overlap (Fig. 11.11a). The principle of superposition enables us to distinguish two voices speaking in the same room at the same time; the sound waves pass through each other unaffected.

Example 11.4

Superposition of Two Wave Pulses

Two identical wave pulses travel at 0.5 m/s toward each other on a long cord (Fig. 11.12). Sketch the shape of the cord at t = 1.0, 1.5, and 2.0 s.

Strategy   We start by sketching the two pulses in their new positions at each time given. Wherever they overlap, we

apply superposition by adding the individual displacements at each point to find the net displacement of the cord at that point.

Solution   Using graph paper, we draw the wave pulses at t = 0 (Fig 11.13a). At t = 1.0 s, each pulse has moved 0.5 m toward the other. The leading edges of the pulses are just starting to overlap (Fig. 11.13b). At t = 1.5 s, each pulse has moved another 0.25 m; the crests overlap exactly. By adding the displacements point by point, we see that the string has the shape of a single pulse twice as high as either of the individual pulses (Fig. 11.13c). At t = 2.0 s, the pulses have each moved another 0.25 m (Fig. 11.13d).

Discussion   When the two pulses exactly overlap, the displacement of points on the string is larger than for corresponding points on a single pulse because we add displacements *in the same direction* ($y > 0$ for both). However, superposition does not *always* produce larger displacements (see Practice Problem 11.4).

Figure 11.12
Two wave pulses at t = 0

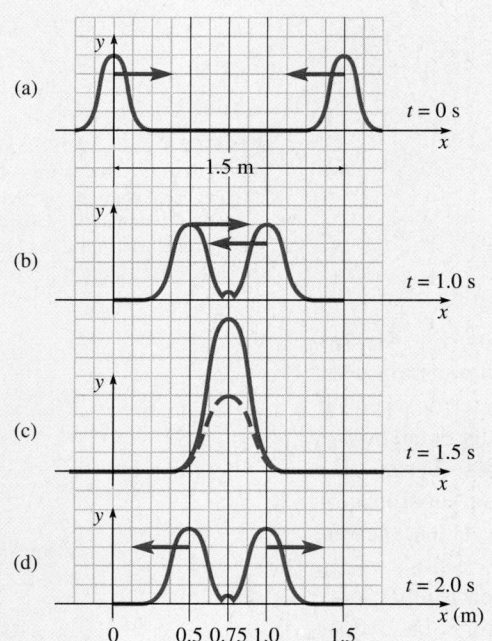

Figure 11.13
Wave positions at times t = 0, 1.0, 1.5, and 2.0 s

Practice Problem 11.4   Superposition of two opposite wave pulses

Repeat Example 11.4, except now let the pulse on the right be inverted (Fig. 11.14). [*Hint:* Points on the string below the x-axis have negative displacements ($y < 0$).]

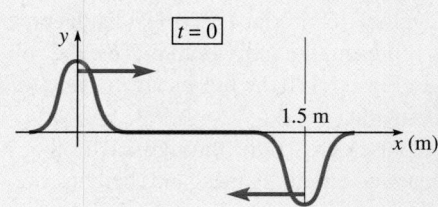

Figure 11.14
Wave pulses for Practice Problem 11.4

11.8   REFLECTION AND REFRACTION

Reflection

A traveling wave sends energy in the direction of propagation without sending any backward, as long as the wave medium is uniform. At an abrupt boundary between one medium and another, **reflection** occurs; a reflected wave carrying some of the energy of the incident wave travels backward from the boundary. A sound wave in air, for instance, does not produce a reflection as long as it travels through air; it *does* reflect when it reaches a wall. Reflection occurs whenever the wave speed changes abruptly. The amount of energy reflected depends on the ratio of the wave speeds in the two media; the farther the ratio is from 1, the more energy is reflected. There is no reason why sound waves cannot travel through a wall; in fact, they do. However, the speed of sound in plaster or wallboard is considerably faster than the speed in air; consequently, much of the sound wave's energy is reflected rather than transmitted through the wall.

The reflected wave can be inverted or not, depending on which side of the boundary has the faster wave speed. Let's look at an extreme example: a string tied to a wall. If you send a wave pulse down the string, the reflected pulse is inverted (Fig. 11.15). One way to understand why it is inverted is that, by the principle of superposition, the shape of the string at any point is the sum of the incident and reflected waves. This is true even at the fixed point at the end. The only way the end can stay in place is if the reflected wave is an upside down version of the incident wave. Another way to understand the inversion is by considering the force exerted on the string by the wall. When an upward pulse reaches the fixed end, the force exerted by the string on the wall has an upward component. By Newton's third law, the wall exerts a force on the string with a downward component. This downward force produces a downward reflected pulse.

Now, instead of tying the string to the wall, tie it to another string with an enormous linear mass density—so large that its motion is too small to measure. The original string doesn't know the difference; it just knows that one end is fixed in place. The second string with the huge density has a much slower wave speed than the first string. Now make the mass density of the second string not huge, but still greater than the first string. The greater inertia inhibits the motion of the boundary point and causes the reflected wave to be inverted. In general, when a wave reflects from a boundary with a region of slower wave speed, the reflected wave is inverted. On the other hand, when a wave reflects from a boundary with a region of *faster* wave speed, the reflected wave is *not* inverted.

Figure 11.15 Snapshots of the reflection of a wave pulse from a fixed end. The reflected pulse is upside down.

A wave that reflects from a boundary with a *slower* medium is inverted.

Refraction

When there is an abrupt change in wave speed, an incident wave splits up at the boundary; part is reflected and part is transmitted past the boundary into the other medium. The frequencies of both the reflected and transmitted waves are the same as the frequency of the incident wave. To understand why, think of a wave incident on the knot between two different strings. Both the reflected and the transmitted waves are generated by the up-and-down motion of the knot; the knot vibrates at the frequency dictated by the incident wave. However, since the wave speed changes at the boundary, *the wavelength of the transmitted wave is not the same* as the wavelength of the incident and reflected waves. Since $v = \lambda f$ and the frequencies are the same,

$$f = \frac{v_1}{\lambda_1} = \frac{v_2}{\lambda_2} \qquad (11\text{-}9)$$

When a wave passes from one medium into another, the frequency of the transmitted wave is the same as that of the incident wave.

Equation (11-9) applies to any kind of wave and is of particular importance in the study of optics.

Example 11.5

Wavelength in Air and Under Water

  A horn near the beach emits a 440-Hz sound wave. (a) What is the wavelength of the sound wave in air? The speed of sound in air is 340 m/s. (b) What is the wavelength of the sound wave in seawater? The speed of sound in seawater is 1520 m/s.

Strategy   The *frequency* of the sound wave in water is the same as in air. The wavelengths depend on both the frequency and the speed of sound in the medium. Sound travels faster in solids and liquids than in gases; during one period, the wave

travels farther in water than it does in air, so the wavelength is longer in water.

Solution   The wavelength in air is related to the speed of sound in air and the frequency:

$$\lambda_{\text{air}} = v_{\text{air}} T = \frac{v_{\text{air}}}{f}$$

Substituting numerical values,

$$\lambda_{\text{air}} = \frac{340 \text{ m/s}}{440 \text{ Hz}} = 0.77 \text{ m}$$

continued on next page

Example 11.5 *continued*

The wave in the water has the *same frequency*, but the speed of sound is different:

$$\lambda_{\text{water}} = \frac{v_{\text{water}}}{f} = \frac{1520 \text{ m/s}}{440 \text{ Hz}} = 3.5 \text{ m}$$

Discussion   The wavelength in water is longer, as expected. As a quick check, the ratio of the wavelengths should be equal to the ratio of the wave speeds:

$$\frac{0.77 \text{ m}}{3.5 \text{ m}} = 0.22; \quad \frac{340 \text{ m/s}}{1520 \text{ m/s}} = 0.22$$

Practice Problem 11.5   Working on the railroad

A railroad worker, driving in spikes, misses the spike and hits the iron rail; a sound wave travels through the air and through the rail. (We ignore the *transverse* wave that also travels in the rail.) The wavelength of the sound in air is 0.548 m. The speed of sound in air is 340 m/s; the speed of sound in iron is 5300 m/s. A coworker is standing 980 m away. (a) What is the frequency of the wave? (b) What is the wavelength of the sound wave in the rail?

A transmitted wave not only has a different wavelength than the incident wave; it also travels in a different direction unless the incident wave's direction of propagation is along the *normal* (the direction perpendicular to the boundary). This change in propagation direction is called **refraction**. If the change in wave speed is gradual, then the change in direction is gradual as well. The speed of ocean waves depends on the depth of the water; the waves are slower in shallower water. As waves approach the shore, they gradually slow down; as a result, they gradually bend until they reach shore nearly head-on. A sudden change in wave speed causes a sudden refraction. When a beam of light passes from air into water, in which the speed of light is slower, it bends toward the normal (Fig. 11.16a).

The law of refraction can be derived from the geometry of Fig. 11.16b. In a time interval Δt, the wave moving with speed v_1 through air travels a distance $v_1 \Delta t$. During that same time t, the wave in the water moving with speed v_2 travels a distance $v_2 \Delta t$. Note in Fig. 11.16b the two right triangles sharing a common side, s, along the boundary between air and water. From the definition of the sine function,

$$\sin \theta_1 = \frac{v_1 \Delta t}{s} \quad \text{and} \quad \sin \theta_2 = \frac{v_2 \Delta t}{s}$$

Dividing one equation by the other, we obtain the law of refraction.

Law of refraction

$$\frac{\sin \theta_1}{\sin \theta_2} = \frac{v_1}{v_2} \tag{11-10}$$

The angles in Eq. (11-10) are called the *angle of incidence* and the *angle of refraction*. Note that these angles are measured between the propagation direction of the wave and the *normal*.

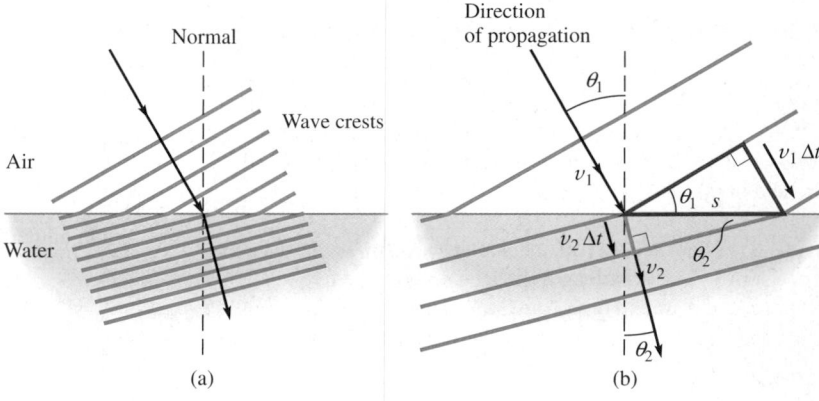

Figure 11.16 (a) A broad beam of light refracts when it passes from air into water. The reflected wave is omitted for clarity. The normal is the direction perpendicular to the boundary. The wave's propagation direction is closer to the normal in the slower medium (water, in this case). (b) Detailed view for derivation of the law of refraction

Although Fig. 11.16 shows a beam entering water from air, the same diagram holds for the reverse situation. If a flashlight under water is pointed toward the surface of the water at an angle of incidence θ_2 with the normal, the light beam is bent away from the normal as it enters the air, so that the angle of refraction becomes θ_1. Where the speed of the waves is greater, the angle with respect to the normal is greater.

Example 11.6

Refraction of an S Wave During an Earthquake

An earthquake emits seismic waves of two types, called P waves (longitudinal waves) and S waves (transverse waves), which travel at different speeds through the Earth. A typical P-wave speed is around 9.0 km/s and an S-wave speed is around 5.0 km/s. Suppose the S wave passes through a boundary in rock where its speed decreases from 5.0 km/s to 4.0 km/s. If the angle of incidence at the boundary is 40.0°, what is the angle of refraction?

Strategy   We know the speed of the S wave in the two materials and we know the angle of incidence. From the law of refraction, we can find the angle of refraction. We let the subscript 1 represent the first material and 2 the second. Then $\theta_1 = 40.0°$, $v_1 = 5.0$ km/s, and $v_2 = 4.0$ km/s; we want to find θ_2.

Solution   The law of refraction is

$$\frac{\sin \theta_1}{\sin \theta_2} = \frac{v_1}{v_2}$$

Solving for $\sin \theta_2$ yields

$$\sin \theta_2 = \frac{v_2}{v_1} \sin \theta_1$$

Now we substitute numerical values:

$$\sin \theta_2 = \frac{4.0 \text{ km/s}}{5.0 \text{ km/s}} \sin 40.0° = 0.514$$

Solving for θ_2,

$$\theta_2 = \sin^{-1} 0.514 = 31°$$

Discussion   The wave speed is slower in the second material so we expect the angle of refraction to be smaller than the angle of incidence; the wave is bent toward the normal.

Practice Problem 11.6   Refraction of a P wave

A P wave from the earthquake is traveling at 9.0 km/s. It is incident on a boundary between materials at an angle of incidence of 45°; the angle of refraction is 34°. What is the speed of the P wave in the other material?

Understanding the propagation of seismic waves, including reflection and refraction due to boundaries between geological features, is an essential part of the effort to reduce damage from future earthquakes. Scientists create small seismic waves with a large vibrator, then use seismographs to record ground vibrations at various locations. The goal is to produce a seismic hazard map so that preventative measures can be targeted to areas with the highest risk of earthquake damage.

11.9   INTERFERENCE AND DIFFRACTION

Interference

The principle of superposition leads to dramatic effects when applied to coherent waves. Two waves are **coherent** if they have the *same frequency* and they maintain a *fixed phase relationship* with each other. One way to obtain coherent waves is to get them from the same source. Such is the case, for example, if one *monophonic* amplifier sends the same signal to two speakers. Should some fluctuation occur in the amplifier driving the speakers, the same fluctuation occurs in both speakers at the same time and they maintain their coherence. Waves are **incoherent** if the phase relationship between them varies randomly. Whenever two waves come from two different sources they are incoherent.

Suppose coherent waves with amplitudes A_1 and A_2 pass through the same point in space. If the waves are *in phase* at that point—that is, the phase difference is any *even*

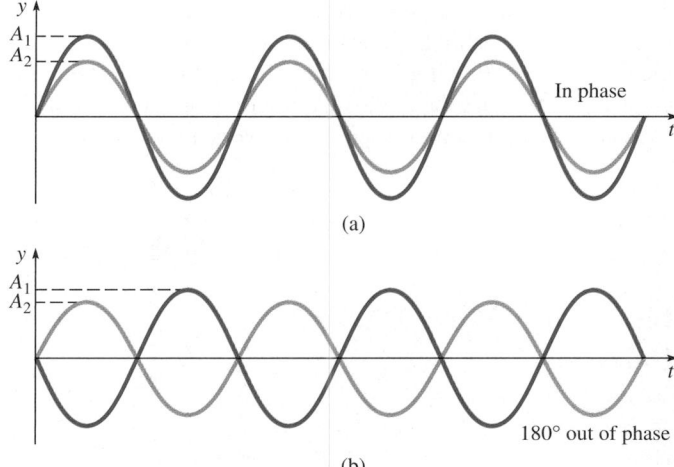

(a)

(b)

180° out of phase

In phase

Figure 11.17 Coherent waves (a) in phase and (b) 180° out of phase. (One wave is drawn with a lighter line to distinguish it from the other.)

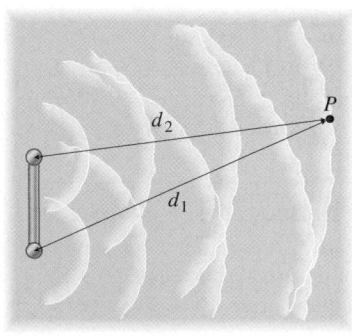

Figure 11.18 Overhead view of coherent surface water waves. The two waves travel different distances d_1 and d_2 to reach a point P. The phase difference between the waves at point P is $k(d_1 - d_2)$.

When two coherent waves are interfering and have a phase difference = $n\pi$, the interference is constructive for $n = 2, 4, 6, 8$, and so forth, and the interference is destructive for $n = 1, 3, 5, 7$, and so forth.

integral multiple of π rad—then the two waves consistently reach their maxima at exactly the same instants of time (Fig. 11.17a). The superposition of the waves that are in phase with each other is called **constructive interference**; the amplitude of the combined waves is the sum of the amplitudes of the two individual waves ($A_1 + A_2$).

Two waves that are *180° out of phase* at a given point have a phase difference of π rad, 3π rad, 5π rad, and so on. The waves are half a cycle apart; when one reaches its maximum, the other reaches its minimum (Fig. 11.17b). The superposition of waves that are 180° out of phase is called **destructive interference**—the amplitude of the combined waves is the *difference* of the amplitudes of the two individual waves ($|A_1 - A_2|$). For any other fixed phase relationship between the two waves, the superposition has an amplitude between $A_1 + A_2$ and $|A_1 - A_2|$.

Suppose two coherent waves start out in phase with each other. In Fig. 11.18, two rods vibrate up and down in step with each other to generate circular waves on the surface of the water. If the two waves travel the same distance to reach a point on the water surface, they arrive *in phase* and interfere constructively. At points where the distances are different, we calculate the phase difference as follows. One wavelength of path difference corresponds to a phase difference of 2π radians (one full cycle). Then, working by proportions,

$$\frac{d_1 - d_2}{\lambda} = \frac{\text{phase difference}}{2\pi \text{ rad}}$$

Thus, the phase difference is

$$\text{phase difference} = \frac{2\pi \text{ rad}}{\lambda} \times (d_1 - d_2) = k(d_1 - d_2) \tag{11-11}$$

If the phase difference is an even integral multiple of π rad, then constructive interference occurs at point P; if the phase difference is an odd integral multiple of π rad, then destructive interference occurs at point P.

Intensity Effects for Interfering Waves

When coherent waves interfere, the amplitudes add (for constructive interference) or subtract (for destructive interference). However, since intensity is proportional to amplitude *squared*, we cannot simply add or subtract the *intensities* of coherent waves when they interfere. *Incoherent* waves, on the other hand, have no fixed phase relationship; interference effects are averaged out due to the rapidly varying phase difference. In the superposition of incoherent waves, the total intensity is the sum of the individual intensities.

Why don't we see and hear interference effects all the time? Light from ordinary sources—incandescent bulbs, fluorescent bulbs, or the Sun—is incoherent because it is

Example 11.7

Intensity of Interfering Waves

Two coherent waves interfere. The intensity of one of them (alone) is 9.0 times the intensity of the other. What is the ratio of the maximum possible intensity to the minimum possible intensity of the resulting wave?

Strategy The intensity is *not* the sum or difference of the individual intensities because the waves are coherent. Since the waves maintain a fixed phase relationship, the principle of superposition tells us that the maximum and minimum *amplitudes* of the interfering waves are the sum and difference of the *individual amplitudes*. Intensity is proportional to amplitude squared, so we find the ratio of the amplitudes and then add or subtract them.

Solution The intensities of the two individual waves are related by $I_1 = 9.0I_2$ or $I_1/I_2 = 9.0$. Since intensity is proportional to amplitude squared,

$$\frac{A_1}{A_2} = \sqrt{\frac{I_1}{I_2}} = 3.0$$

Thus, $A_1 = 3.0A_2$. The maximum possible amplitude for the superposition occurs if the waves are in phase:

$$A_{max} = A_1 + A_2 = 4.0A_2$$

The minimum possible amplitude for the superposition occurs if the waves are 180° out of phase:

$$A_{min} = |A_1 - A_2| = 2.0A_2$$

The ratio of the maximum to minimum intensity is

$$\frac{I_{max}}{I_{min}} = \left(\frac{A_{max}}{A_{min}}\right)^2 = \left(\frac{4.0}{2.0}\right)^2 = 4.0$$

Discussion Had we added and subtracted the *intensities* instead of the amplitudes, we would have found a ratio of $10/8 = 1.25$ between the maximum and minimum intensities. We must be careful to add or subtract the *amplitudes* of the interfering waves instead of the intensities themselves when the waves are coherent.

Practice Problem 11.7 Two more coherent waves

Repeat Example 11.7, but change the ratio of the individual intensities to 4.0 (instead of 9.0).

---

generated by large numbers of independent atomic sources. A single source of sound normally contains many different frequencies, so a point of constructive interference for one frequency is not a point of constructive interference for other frequencies. Furthermore, in most situations there are many different sound waves that reach our ears after traveling different paths due to the reflection of sound from walls, ceilings, chairs, and so forth.

Diffraction

Diffraction is the bending of a wave around an obstacle in its path. The amount of diffraction depends on the relative size of the obstacle and the wavelength of the waves. Figure 11.19 shows water waves encountering a boat at anchor. If the distance

(a)

(b)

Direction of
wave motion

Direction of
wave motion

Figure 11.19 Diffraction of a wave when it encounters an obstacle. (a) Short wavelength waves are blocked by the boat without much bending. (b) Longer wavelength waves bend easily around the boat.

between crests (the wavelength) is small compared to the length of the boat, the waves do not bend significantly; the "shadow" of the boat is quite clear. If the distance between crests (the wavelength) is comparable to the size of the boat, the waves bend significantly so that the boat has a "shadow" up close but not farther away. Diffraction enables you to hear around a corner but not to see around a corner—the wavelengths of visible light are less than 1 mm while the wavelengths of sound waves in air are typically around 1 m. Higher frequencies have shorter wavelengths and do not diffract as much, so you hear mostly lower frequencies when a sound wave bends around a corner.

11.10 STANDING WAVES

Standing waves occur when a wave is reflected at a boundary and the reflected wave interferes with the incident wave so that the wave appears to stand still. Suppose that a harmonic wave on a string, coming from the right, hits a boundary where the string is fixed. The equation of the incident wave is

$$y(x, t) = A \sin(\omega t + kx)$$

The + sign is chosen in the phase because the wave travels to the left.

The reflected wave travels to the right, so $+kx$ is replaced with $-kx$; and the reflected wave is inverted, so $+A$ is replaced with $-A$. Then the reflected wave is described by

$$y(x, t) = -A \sin(\omega t - kx)$$

Applying the principle of superposition, the motion of the string is described by

$$y(x, t) = A[\sin(\omega t + kx) - \sin(\omega t - kx)]$$

This can be rewritten in a form that shows the motion of the string more clearly. Using the trigonometric identity (Appendix A.7),

$$\sin \alpha - \sin \beta = 2 \cos\left[\tfrac{1}{2}(\alpha + \beta)\right] \sin\left[\tfrac{1}{2}(\alpha - \beta)\right]$$

where

$$\alpha = \omega t + kx \quad \text{and} \quad \beta = \omega t - kx$$

we have

$$y(x, t) = 2A \cos \omega t \sin kx$$

Notice that t and x are separated. Every point moves in SHM with the same frequency. However, in contrast to a *traveling* harmonic wave, every point reaches its maximum distance from equilibrium simultaneously. In addition, different points move with different amplitudes; the amplitude at any point x is $2A \sin kx$.

Figure 11.20 shows the string at time intervals of $\frac{1}{8}T$, where T is the period. What you actually see when looking at a standing wave is a blur of moving string, with points that never move (**nodes**, labeled "N") halfway between points of maximum amplitude (**antinodes**, labeled "A"). The nodes are the points where $\sin kx = 0$. Since $\sin n\pi = 0$ ($n = 0, 1, 2, \ldots$), the nodes are located at $x = n\pi/k = n\lambda/2$. Thus, the distance between two adjacent nodes is $\frac{1}{2}\lambda$. The antinodes occur where $\sin kx = \pm 1$, which is exactly halfway between a pair of nodes. So the nodes and antinodes alternate, with one-quarter of a wavelength between a node and the neighboring antinode.

Node–node distance is $\frac{1}{2}\lambda$.
Node–antinode distance is $\frac{1}{4}\lambda$.

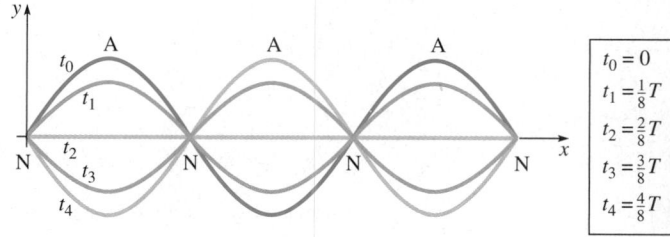

Figure 11.20 A standing wave at various times: $t = 0, \frac{1}{8}T, \frac{2}{8}T, \frac{3}{8}T,$ and $\frac{4}{8}T$, where T is the period.

$t_0 = 0$	
$t_1 = \frac{1}{8}T$	
$t_2 = \frac{2}{8}T$	
$t_3 = \frac{3}{8}T$	
$t_4 = \frac{4}{8}T$	

So far we have ignored what happens at the other end of the string. If the other end is fixed as well, then it too is a node. The string thus has two or more nodes, with one at each end. The distance between each pair of nodes is $\frac{1}{2}\lambda$, so

$$\frac{n\lambda}{2} = L$$

where L is the length of the string and $n = 1, 2, 3, \ldots$. The possible wavelengths for standing waves on a string are

$$\lambda_n = \frac{2L}{n} \quad (n = 1, 2, 3, \ldots) \tag{11-12}$$

The frequencies are

$$f_n = \frac{v}{\lambda_n} = \frac{nv}{2L} \quad (n = 1, 2, 3, \ldots) \tag{11-13}$$

The lowest frequency standing wave ($n = 1$) is called the **fundamental**. Notice that the higher frequency standing waves are all integral multiples of the fundamental; the set of standing wave frequencies makes an evenly spaced set:

$$f_1, 2f_1, 3f_1, 4f_1, \ldots, nf_1, \ldots$$

These frequencies are called the *natural frequencies* or *resonant frequencies* of the string. *Resonance* occurs when a system is driven at one of its natural frequencies; the resulting vibrations are large in amplitude compared to when the driving frequency is not close to any of the natural frequencies.

Resonance is responsible for much of the structural damage caused by seismic waves. If the frequency at which the ground vibrates is close to a resonant frequency of a structure, the vibration of the structure builds up to a large amplitude. Thus, to construct a building that can survive an earthquake, it is not enough to make it stronger. Either the building must be designed so it is isolated from ground vibrations, or a damping mechanism—something like a shock absorber—must be incorporated to dissipate energy and reduce the amplitude of the vibrations. Damping is becoming increasingly common in large buildings since it is just as effective and much less expensive than isolation.

Large sections of the Hanshin expressway vibrated in a twisting motion due to ground vibrations near a resonant frequency. The road now has rubber base isolators instead of steel bearings connecting the roadway to the concrete piers. Part of their function is to act like shock absorbers.

Figure 11.21 shows the first four standing wave patterns on a string. The two ends are always nodes since they are fixed in place. Notice that each successive pattern has one more node and one more antinode than the previous one. The fundamental has the fewest possible number of nodes (2) and antinodes (1).

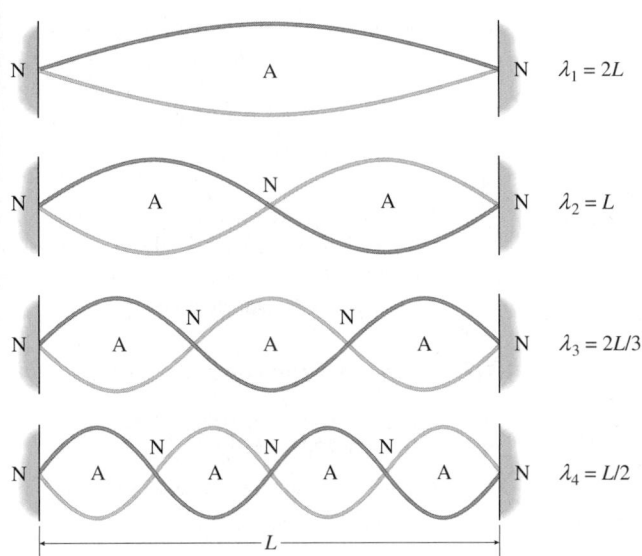

Figure 11.21 Standing waves on a string fixed at both ends. "N" marks the locations of the nodes and "A" marks the locations of the antinodes.

Example 11.8

Wavelength of a Standing Wave

A string is attached to a vibrator driven at 1.20×10^2 Hz. A weight hangs from the other end of the string; the weight is adjusted until a standing wave is formed (Fig. 11.22). What is the wavelength of the standing wave on the string?

Strategy The measured distance of 42 cm encompasses six "loops"—that is, six segments of string between one node and the next. Each of the loops represents a length of $\frac{1}{2}\lambda$.

Solution The length of one loop is

$$42 \text{ cm} \times \frac{1}{6} = 7.0 \text{ cm}$$

Since the length of one loop is $\frac{1}{2}\lambda$, the wavelength is 14 cm.

Discussion This string is *not* fixed at both ends. The left end is connected to a moving vibrator, so it is not a node. The right end wraps around a pulley; it may not be easy to determine precisely where the 'end' is. For this case, it is more accurate to measure the distance between two actual nodes rather than to assume that the ends are nodes.

Practice Problem 11.8   Standing wave with seven loops

The vibrator frequency is increased until there are seven loops within the 42-cm length. What is the new standing wave frequency for this string (assuming the same tension)?

$f = 120$ Hz

42 cm

Figure 11.22
Measuring distance between nodes for a standing wave

String vibrator

MASTER THE CONCEPTS

Summary

- An isotropic source radiates sound uniformly in all directions. Assuming that no energy is absorbed by the medium and there are no obstacles to reflect or absorb sound,

$$I = \frac{\text{power}}{\text{area}} = \frac{P}{4\pi r^2} \tag{11-1}$$

- In a transverse wave, the motion of particles in the medium is perpendicular to the direction of propagation of the wave. In a longitudinal wave, the motion of particles in the medium is along the same line as the direction of propagation of the wave.

- The speed of a transverse wave on a string is

$$v = \sqrt{\frac{F}{\mu}} \tag{11-4}$$

where

$$\mu = m/L \tag{11-3}$$

- A periodic wave repeats the same pattern over and over. Harmonic waves are a special kind of periodic wave characterized by a sinusoidal function (either a sine or cosine function).

- If a periodic wave has period T and travels at speed v, the repetition distance of the wave is the wavelength:

$$\lambda = vT \tag{11-5}$$

- The principle of superposition: When two or more waves overlap, the net disturbance at any point is the sum of the individual disturbances due to each wave.

- A harmonic traveling wave can be described by

$$y(x, t) = A \cos(\omega t - kx) \tag{11-8}$$

- The argument of the sinusoidal function, $(\omega t \pm kx)$, is called the phase of the wave at x and t. The constant k is the wave number

$$k = \frac{\omega}{v} = \frac{2\pi f}{v} = \frac{2\pi}{\lambda} \tag{11-7}$$

- Reflection occurs at a boundary between media with different wave speeds. Some energy may be transmitted into the new medium and the rest is reflected. The wave transmitted past the boundary is refracted:

$$\frac{\sin \theta_1}{\sin \theta_2} = \frac{v_1}{v_2} \tag{11-10}$$

- Coherent waves have the *same frequency* and maintain a *fixed phase relationship* with each other. Coherent waves that are in phase with each other interfere constructively; those that are 180° out of phase interfere destructively.

- Diffraction occurs when a wave bends around an obstacle in its path.

- In a standing wave on a string, every point moves in SHM with the same frequency. Nodes are points of zero amplitude; antinodes are points of maximum amplitude. The distance between two adjacent nodes is $\frac{1}{2}\lambda$.

Highlighted Figures and Tables

F11.2 A point source of sound radiating energy uniformly in all directions (p. 378)

F11.3 Transverse and longitudinal waves (p. 379)

F11.4 Motion of ground particles in P waves and S waves (p. 380)

MASTER THE CONCEPTS *continued*

F11.7 Sinusoidal wave moving with speed v in the x-direction (p. 383)

F11.8 A wave pulse at successive times (p. 384)

F11.9 Two graphs of a harmonic wave $y(x, t) = A \sin(\omega t - kx)$ on a string (p. 385)

F11.11 Two identical wave pulses traveling toward and through each other (p. 387)

F11.15 Reflection of a wave pulse from a fixed end (p. 389)

F11.16 Change in wave speed causing sudden refraction and geometric derivation (p. 390)

F11.17 Coherent waves in phase and 180° out of phase (p. 392)

F11.19 Diffraction of a wave when it encounters an obstacle (p. 393)

F11.21 Standing waves on a string fixed at both ends (p. 395)

CONCEPTUAL QUESTIONS

1. The piano strings that vibrate with the lowest frequencies consist of a steel wire around which a thick coil of copper wire is wrapped. Only the inner steel wire is under tension. What is the purpose of the copper coil?

2. Is the vibration of a string in a piano, guitar, or violin a *sound* wave? Explain.

3. The wavelength of the fundamental standing wave on a cello string depends on which of these quantities: length of the string, mass per unit length of the string, or tension? The wavelength of the *sound wave* resulting from the string's vibration depends on which of the same three quantities?

4. If the length of a guitar string is decreased while the tension remains constant, what happens to each of these quantities? (a) the wavelength of the fundamental, (b) the frequency of the fundamental, (c) the time for a pulse to travel the length of the string, (d) the maximum velocity of a point on the string (assuming the amplitude is the same both times), (e) the maximum acceleration of a point on the string (assuming the amplitude is the same both times).

5. Why is it possible to understand the words spoken by two people at the same time?

6. A cello player can change the frequency of the sound produced by her instrument by (a) increasing the tension in the string, (b) pressing her finger on the string at different places along the fingerboard, or (c) bowing a different string. Explain how each of these methods affects the frequency.

7. Why is a transverse wave sometimes called a shear wave?

8. Figure 11.23 shows a complex wave moving to the right along a cord. Draw the shape of the cord an instant later and determine which parts of the cord are moving upward and which are moving downward. Indicate the directions on your drawing with arrows.

9. When an earthquake occurs, the S waves (transverse waves) are not detected on the opposite side of the Earth while the P waves (longitudinal waves) are. How does this provide evidence that the Earth's solid core is surrounded by liquid?

10. Water waves can travel in any direction when far offshore, but they always arrive at the beach nearly head on. Explain.

11. Simple ear-protection devices use materials that reflect or absorb sound before it reaches the ears. A newer technology, sometimes called *noise cancellation*, uses a microphone to produce an electrical signal that mimics the noise. The signal is modified electronically, then fed to the speakers in a pair of headphones. The speakers emit sound waves that *cancel* the noise. On what principle is this technology based? What kind of modification is made to the electrical signal?

12. When connecting speakers to a stereo, it is important to connect them with the correct polarity so that, if the same electrical signal is sent, they both move in the same direction. If the wires going to one speaker are reversed, the listener hears a noticeably weaker bass (low frequencies). Explain what causes this and why low frequencies are affected more than high frequencies.

Figure 11.23 Conceptual Question 11.8

MULTIPLE CHOICE QUESTIONS

1. Standing waves are produced by the superposition of two waves with
   (a) the same amplitude, frequency, and direction of propagation.
   (b) the same amplitude and frequency, and opposite propagation directions.
   (c) the same amplitude and direction of propagation, but different frequencies.
   (d) the same amplitude, different frequencies, and opposite directions of propagation.

2. A transverse wave travels on a string of mass m, length L, and tension F. Which statement below is correct?
   (a) The energy of the wave is proportional to the square root of the wave amplitude.
   (b) The speed of a moving point on the string is the same as the wave speed.
   (c) The wave speed is determined by the values of m, L, and F.
   (d) The wavelength of the wave is proportional to L.

3. A transverse wave on a string is described by $y(x, t) = A \cos(\omega t + kx)$. It arrives at the point $x = 0$ where the string is fixed in place. Which function describes the reflected wave?

(a) $A \cos(\omega t + kx)$ (b) $A \cos(\omega t - kx)$ (c) $-A \sin(\omega t + kx)$
(d) $-A \cos(\omega t - kx)$ (e) $A \sin(\omega t + kx)$

4. A violin string of length L is fixed at both ends. Which one of these is *not* a wavelength of a standing wave on the string?

(a) L (b) $2L$ (c) $L/2$ (d) $L/3$ (e) $2L/3$ (f) $3L/2$

5. The speed of waves in a stretched string depends upon which one of the following?

(a) The tension in the string   (b) The amplitude of the waves
(c) The wavelength of the waves
(d) The gravitational field strength

6. The higher the frequency of a wave

(a) the smaller its speed.   (b) the shorter its wavelength.
(c) the greater its amplitude.   (d) the longer its period.

7. In a transverse wave, the individual particles of the medium

(a) move in circles.   (b) move in ellipses.
(c) move parallel to the direction of the wave's travel.
(d) move perpendicular to the direction of the wave's travel.

8. Of these properties of a wave, the one that is independent of the others is its

(a) amplitude   (b) wavelength
(c) speed   (d) frequency

9. Two successive transverse pulses, one caused by a brief displacement to the right and the other by a brief displacement to the left, are sent down a Slinky that is fastened at the far end. At the point where the first reflected pulse meets the second advancing pulse, the deflection (compared with that of a single pulse) is

(a) quadrupled.   (b) doubled.
(c) canceled.   (d) halved.

10. The intensity of an isotropic sound wave is

(a) directly proportional to the distance from the source.
(b) inversely proportional to the distance from the source.
(c) directly proportional to the square of the distance from the source.
(d) inversely proportional to the square of the distance from the source.
(e) none of the above.

PROBLEMS

Note: **ⓒ** indicates a combination conceptual/quantitative problem. Gold diamonds ◆, ◆◆ are used to indicate the increasing level of difficulty of each problem. Problem numbers appearing in blue, 9., denote problems that have a detailed solution available in the Student Solutions Manual. Some problems are *paired* by concept; their numbers are connected by a ruled box.

11.1 Waves and Energy Transport

1. The intensity of sunlight that reaches the Earth's atmosphere is 1400 W/m². What is the intensity of the sunlight that reaches Jupiter? Jupiter is 5.2 times as far from the sun as Earth. [*Hint:* Treat the Sun as an isotropic source of light waves.]

2. The intensity of the sound wave from a jet airplane is 9.2 W/m² at a distance of 120 m. What is the intensity of the sound wave that reaches the ground when the airplane flies overhead at an altitude of 8.6 km? Assume that the sound wave radiates from the airplane equally in all directions.

3. At what rate does the jet airplane in Problem 2 radiate energy in the form of sound waves?

4. The Sun emits electromagnetic waves (including light) equally in all directions. The intensity of the waves at the Earth's upper atmosphere is 1.4 kW/m². At what rate does the Sun emit electromagnetic waves? (In other words, what is the power output?)

11.3 Speed of Transverse Waves on a String

5. (a) What is the speed of propagation of the pulse shown in Fig. 11.24? (b) At what average speed does the point at $x = 2.0$ m move during this time interval?

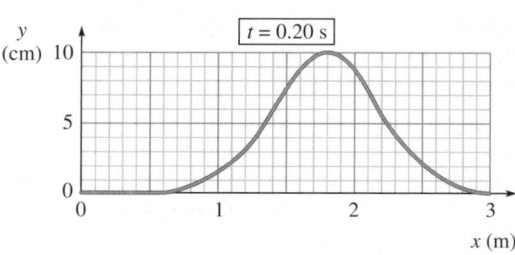

Figure 11.24 Problems 5, 6, 32, 33, 58, and 59. A pulse on a string at times $t = 0$ and $t = 0.20$ s. (Note that the horizontal and vertical scales are different.)

6. (a) What is the position of the peak of the pulse shown in Fig. 11.24 at $t = 3.00$ s? (b) When does the peak of the pulse arrive at $x = 4.00$ m?

7. A metal guitar string has a linear mass density of $\mu = 3.20$ g/m. What is the speed of transverse waves on this string when its tension is 90.0 N?

8. When the tension in a cord is 75 N, the wave speed is 140 m/s. What is the linear mass density of the cord?

9. A uniform string of length 10.0 m and weight 0.25 N is attached to the ceiling. A weight of 1.00 kN hangs from its lower end. The lower end of the string is suddenly displaced horizontally. How long does it take the resulting wave pulse to travel to the upper end? [*Hint*: Is the weight of the string negligible in comparison with that of the hanging mass?]

11.4 Periodic Waves

10. What is the speed of a wave whose frequency and wavelength are 500.0 Hz and 0.500 m, respectively?

11. What is the wavelength of a wave whose speed and period are 75.0 m/s and 5.00 ms, respectively?

12. What is the frequency of a wave whose speed and wavelength are 120 m/s and 30.0 cm, respectively?

13. The speed of sound in air at room temperature is 340 m/s. (a) What is the frequency of a sound wave in air with wavelength 1.0 m? (b) What is the frequency of a radio wave with the same wavelength? (Radio waves are electromagnetic waves that travel at 3.0×10^8 m/s in air or in vacuum.)

14. Light visible to humans consists of electromagnetic waves with wavelengths (in air) in the range 400–700 nm (4.0×10^{-7} m – 7.0×10^{-7} m). The speed of light in air is 3.0×10^8 m/s. What are the frequencies of electromagnetic waves that are visible?

15. A fisherman notices a buoy bobbing up and down in the water in ripples produced by waves from a passing speedboat. These waves travel at 2.5 m/s and have a wavelength of 7.5 m. At what frequency does the buoy bob up and down?

11.5 Mathematical Description of a Wave

16. You are swimming in the ocean as water waves with wavelength 9.6 m pass by. What is the closest distance that another swimmer could be so that his motion is exactly opposite yours (he goes up when you go down)?

17. The equation of a wave is

$$y(x, t) = 3.5 \sin \frac{\pi}{3.0}(x - 66t) \text{ cm}$$

where t is in seconds and x and y are both in cm. Find (a) the amplitude and (b) the wavelength of this wave.

18. Write an equation for a sine wave with amplitude 0.120 m, wavelength 0.300 m, and wave speed 6.40 m/s traveling in the $-x$-direction.

♦ 19. A wave on a string has equation $y(x, t) = (4.0 \text{ mm}) \sin (\omega t - kx)$, where $\omega = 6.0 \times 10^2$ rad/s and $k = 6.0$ rad/m. (a) What is the amplitude of the wave? (b) What is the wavelength? (c) What is the period? (d) What is the wave speed? (e) In which direction does the wave travel?

11.6 Graphing Waves

♦ 20. Figure 11.25 shows a snapshot of a transverse wave traveling along a string at 10.0 m/s. The equation for the wave is $y(x, t) = A \cos (\omega t + kx)$. (a) Is the wave moving to the right or to the left? (b) What are the numerical values of A, ω, and k? (c) At

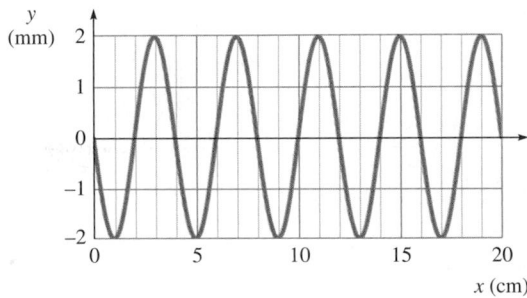

Figure 11.25 Problem 20

what times could this snapshot have been taken? (Give the three smallest nonnegative possibilities.)

21. (a) Plot a graph for $y(x, t) = 4.0 \sin (378t - 314x)$, where t is in s and x and y are in cm, versus x at $t = 0$ and at $t = \frac{1}{480}$ s. From the plots determine the amplitude, wavelength, and speed of the wave. (b) For the same function, plot a graph of $y(x, t)$ versus t at $x = 0$ and find the period of the vibration. Show that $\lambda = vT$.

22. Figure 11.26 shows a sine wave traveling to the right on a cord. The lighter line represents the shape of the cord at time $t = 0$; the darker line represents the shape of the cord at time $t = 0.10$ s. (Note that the horizontal and vertical scales are different.) What are (a) the amplitude and (b) the wavelength of the wave? (c) What is the speed of the wave? What are (d) the frequency and (e) the period of the wave?

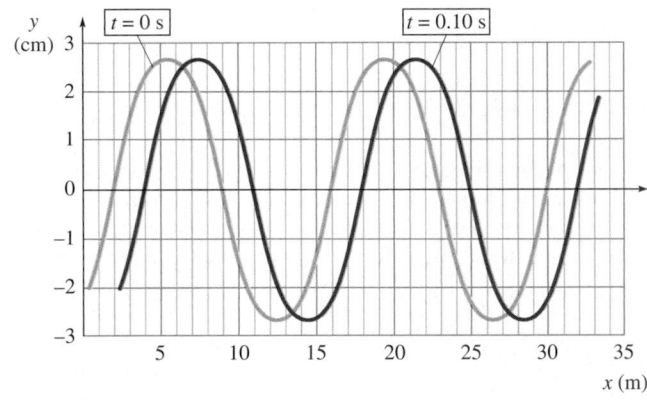

Figure 11.26 Problem 22

23. For $y(x, t) = 0.0050 \cos (4.0\pi t - 1.0\pi x)$ m, where all measurements are in meters and seconds, find the maximum speed and the maximum acceleration of a point on the string. Plot graphs for one cycle of displacement y versus t, velocity v_y versus t, and acceleration a_y versus t.

24. A transverse wave on a string is described by $y(x, t) = (1.2 \text{ mm}) \sin (2.0\pi t - 0.50\pi x)$, where x is in meters and t is in seconds. Plot the displacement y and the velocity v_y versus t for one complete cycle of the point $x = 0$ on the string.

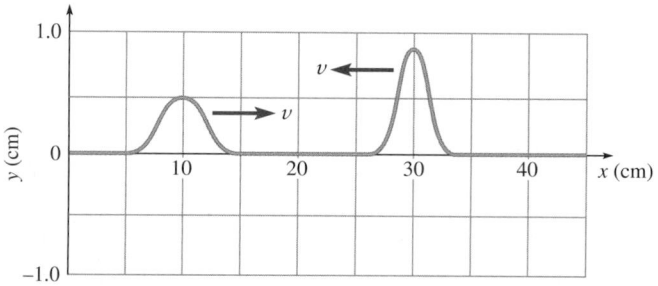

Figure 11.27 Problem 25

11.7 Principle of Superposition

25. Figure 11.27 shows two pulses on a cord at time $t = 0$. The pulses are moving toward each other; the speed of each pulse is 40 cm/s. Sketch the shape of the cord at 0.15, 0.25, and 0.30 s.

26. Figure 11.28 shows two pulses on a cord at time $t = 0$. The pulses are moving toward each other; the speed of each pulse is 2.5 m/s. Sketch the shape of the cord at 0.60, 0.80, and 0.90 s.

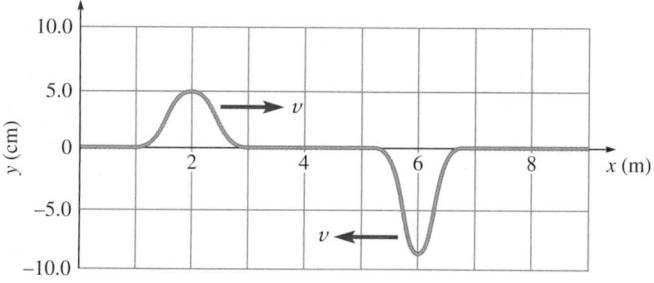

Figure 11.28 Problem 26

27. A traveling sine wave is the result of the superposition of two other sine waves with equal amplitudes, wavelengths, and frequencies. The two component waves each have amplitude 5.00 cm. If the resultant wave has amplitude 6.69 cm, what is the phase difference ϕ between the component waves? [*Hint:* Let $y_1 = A \sin(\omega t + kx)$ and $y_2 = A \sin(\omega t + kx - \phi)$. Make use of the trigonometric identity (Appendix A.7) for $\sin \alpha + \sin \beta$ when finding $y = y_1 + y_2$ and identify the new amplitude in terms of the original amplitude.]

28. Two traveling sine waves, identical except for a phase difference ϕ, add so that their superposition produces another traveling wave with the same amplitude as the two component waves. What is the phase difference between the two waves?

29. Using graph paper, sketch two identical sine waves of amplitude 4.0 cm that differ in phase by (a) 60.0° and (b) 90.0°. Use superposition to find the amplitude of the resultant wave.

11.8 Reflection and Refraction

30. Light of wavelength 0.500 μm (in air) enters the water in a swimming pool. The speed of light in water is 0.750 times the speed in air. What is the wavelength of the light in water?

31. An S wave crosses the boundary between two different types of rock. The angle of incidence is 22° and the angle of refraction is 34°. If the speed of the S wave before the boundary is 3.2 m/s, what is the speed past the boundary?

Problems 32–33: The pulse of Fig. 11.24 travels to the right on a string whose ends at $x = 0$ and $x = 4.0$ m are both fixed in place. Imagine a reflected pulse that begins to move onto the string at an endpoint at the same time the incident pulse reaches that endpoint. The superposition of the incident and reflected pulses gives the shape of the string.

32. When is the next time t that the string referred to in Fig. 11.24 looks exactly as it does at $t = 0$?

33. The pulse of Fig. 11.24 travels on a string whose ends at $x = 0$ and $x = 4.0$ m are both fixed in place. When does the string look completely flat? Find the first time for $t > 0$.

11.9 Interference and Diffraction

34. Two waves with identical frequency but different amplitudes $A_1 = 5.0$ cm and $A_2 = 3.0$ cm, occupy the same region of space (are superimposed). (a) At what phase difference does the resulting wave have the largest amplitude? What is the amplitude of the resulting wave in that case? (b) At what phase difference does the resulting wave have the smallest amplitude and what is its amplitude? (c) What is the ratio of the largest and smallest amplitudes?

35. Two waves with identical frequency but different amplitudes $A_1 = 6.0$ cm and $A_2 = 3.0$ cm, occupy the same region of space (i.e., are superimposed). (a) At what phase difference will the resulting wave have the highest intensity? What is the amplitude of the resulting wave in that case? (b) At what phase difference will the resulting wave have the lowest intensity and what will its amplitude be? (c) What is the ratio of the two intensities?

36. A sound wave with intensity 25 mW/m² interferes constructively with a sound wave with intensity 15 mW/m². What is the intensity of the superposition of the two?

37. A sound wave with intensity 25 mW/m² interferes destructively with a sound wave with intensity 28 mW/m². What is the intensity of the superposition of the two?

38. Two coherent sound waves have intensities of 0.040 and 0.090 W/m² where you are listening. (a) If the waves interfere constructively, what is the resultant intensity that you hear? (b) What if they interfere destructively? (c) If they were incoherent, what would be the resultant intensity? [*Hint:* If your answers are correct, then (c) is the average of (a) and (b).]

11.10 Standing Waves

39. In order to decrease the fundamental frequency of a guitar string by 4.0%, by what percentage should you reduce the tension?

40. The longest "string" (a thick metal wire) on a particular piano is 2.0 m long and has a tension of 300.0 N. It vibrates with a fundamental frequency of 27.5 Hz. What is the total mass of the wire?

41. Suppose that a string of length L and mass m is under tension F. (a) Show that $\sqrt{FL/m}$ has units of speed. (b) Show that there is no other combination of L, m, and F with units of speed. [*Hint:* Of the dimensions of the three quantities L, m, and F, only F includes time.] Thus, the speed of transverse waves on the string can only be some dimensionless constant times $\sqrt{FL/m}$.

42. A standing wave has wavenumber 2.0×10^2 rad/m. What is the distance between two adjacent nodes?

43. A piano string of length 1.50 m and mass density 25.0 mg/m vibrates at a (fundamental) frequency of 450.0 Hz. (a) What is the speed of the transverse string waves? (b) What is the tension? (c) What are the wavelength and frequency of the sound wave in air produced by vibration of the string? The speed of sound in air at room temperature is 340 m/s.

44. A cord of length 1.5 m is fixed at both ends. Its mass per unit length is 1.2 g/m and the tension is 12 N. (a) What is the frequency of the fundamental oscillation? (b) What tension is required if the $n = 3$ mode has a frequency of 0.50 kHz?

45. A guitar's E-string has length 65 cm and is stretched to a tension of 82 N. It vibrates at a fundamental frequency of 329.63 Hz. Determine the mass per unit length of the string.

46. A string 2.0 m long is held fixed at both ends. If a sharp blow is applied to the string at its center, it takes 0.050 s for the pulse to travel to the ends of the string and return to the middle. What is the fundamental frequency of oscillation for this string?

COMPREHENSIVE PROBLEMS

47. The speed of waves on a lake depends on frequency. For waves of frequency 1.0 Hz, the wave speed is 1.56 m/s; for 2.0 Hz waves, the speed is 0.78 m/s. The 2.0-Hz waves from a speedboat's wake reach you 120 s after the 1.0-Hz waves generated by the same boat. How far away is the boat?

48. The formula for the speed of transverse waves on a spring is the same as for a string. (a) A spring is stretched to a length much greater than its relaxed length. Explain why the tension in the spring is approximately proportional to the length. (b) A wave takes 4.00 s to travel from one end of such a spring to the other. Then the length is increased 10.0%. Now how long does a wave take to travel the length of the spring? [*Hint:* Is the mass per unit length constant?]

49. What is the wavelength of the radio waves transmitted by an FM station at 90 MHz? (Radio waves travel at 3.0×10^8 m/s.)

50. Figure 11.29 shows the graph of ground vibrations recorded by a seismograph 180 km from the focus of a small earthquake. It took the waves 30.0 s to travel from their source to the seismograph. Estimate the wavelength.

51. (a) Write an equation for a surface seismic wave moving along the $-x$-axis with amplitude 2.0 cm, period 4.0 s, and wavelength 4.0 km. Assume the wave is harmonic, x is measured in m, and t is measured in s. (b) What is the maximum speed of the ground as the wave moves by? (c) What is the wave speed?

52. A transverse wave on a string is described by $y(x, t) = (1.2 \text{ cm}) \sin (0.50\pi t - \pi x)$, where x is in meters and t is in seconds. Find the maximum velocity and the maximum acceleration of a point on the string. Plot graphs for displacement y versus t, velocity v_y versus t, and acceleration a_y versus t at $x = 0$.

53. Deep-water waves are *dispersive* (their wave speed depends on the wavelength). The restoring force is provided by gravity. Using dimensional analysis, find out how the speed of deep-water waves depends on wavelength λ, assuming that λ and g are the only relevant quantities. (For the curious: Mass density does not enter into the expression because the restoring force, arising from the weight of the water, is itself proportional to the mass density.)

54. In contrast to deep-water waves, shallow ripples on the surface of a pond are due to surface tension. The surface tension γ of water characterizes the restoring force; the mass density ρ of water characterizes the water's inertia. Use dimensional analysis to determine whether the surface waves are *dispersive* (the wave speed depends on the wavelength) or *nondispersive* (their wave speed is independent of wavelength). [*Hint:* Start by assuming that the wave speed is determined by γ, ρ, and the wavelength λ.]

55. An underground explosion sends out both transverse (S waves) and longitudinal (P waves) mechanical wave pulses (seismic waves) through the crust of the Earth. Suppose the speed of transverse waves is 8.0 km/s and that of longitudinal waves is 10.0 km/s. On one occasion, both waves follow the same path from a source to a detector (a seismograph); the longitudinal pulse arrives 2.0 s before the transverse pulse. What is the distance between the source and the detector?

Figure 11.29 Problem 50

Figure 11.30 Problem 56

© 56. Figure 11.30 shows a snapshot of a transverse wave moving to the left on a string. The wave speed is 10.0 m/s. At the instant the snapshot is taken, (a) in what direction is point A moving? (b) In what direction is point B moving? (c) At which of these points is the speed of the string segment (not the wave speed) larger? Explain.

57. When the string of a guitar is pressed against a fret, the shortened string vibrates at a frequency 5.95% higher than when the previous fret is pressed. If the length of the part of the string that is free to vibrate is 64.8 cm, how far from one end of the string are the first three frets located?

Problems 58–59: The pulse of Fig. 11.24 travels to the right on a string whose ends at $x = 0$ and $x = 4.0$ m are both fixed in place. Imagine a reflected pulse that begins to move onto the string at an endpoint at the same time the incident pulse reaches that endpoint. The superposition of the incident and reflected pulses gives the shape of the string.

◆◆58. The pulse of Fig. 11.24 travels on a string whose ends at $x = 0$ and $x = 4.0$ m are both fixed in place. Sketch the shape of the string at $t = 1.6$ s.

◆◆59. The pulse of Fig. 11.24 travels on a string whose ends at $x = 0$ and $x = 4.0$ m are both fixed in place. Sketch the shape of the string at $t = 2.2$ s.

60. A guitar string has a fundamental frequency of 300.0 Hz. (a) What are the next three lowest standing wave frequencies? (b) If you press a finger *lightly* against the string at its midpoint so that both sides of the string can still vibrate, you force there to be a node at the midpoint. What are the lowest four standing wave frequencies now? (c) If you press *hard* at the same point, only one side of the string can vibrate. What are the lowest four standing wave frequencies?

◆◆61. A harpsichord string is made of yellow brass (Young's modulus 9.0×10^{10} Pa, tensile strength 6.3×10^8 Pa, mass density 8500 kg/m³). When tuned correctly, the tension in the string is 59.4 N, which is 93% of the maximum tension that the string can endure without breaking. The length of the string that is free to vibrate is 9.4 cm. What is the fundamental frequency?

◆62. Two speakers spaced a distance 1.5 m apart emit coherent sound waves at a frequency of 680 Hz in all directions. The waves start out in phase with each other. A listener walks in a circle of radius greater than one meter centered on the midpoint of the two speakers. At how many points does the listener observe destructive interference? The listener and the speakers are all in the same horizontal plane and the speed of sound is 340 m/s. [*Hint:* Start with a diagram; then determine the *maximum* path difference between the two waves at points on the circle.] Experiments like this must be done in a special room so that reflections are negligible.

ANSWERS TO PRACTICE PROBLEMS

11.1 (a) 8.9 m/s; (b) 13 m/s

11.2 (a) Yes, the traveling wave retains its shape; (b) it travels in the +x-direction because the t and x/v terms have *opposite* signs; (c) the wave speed is 8.0 m/s.

11.3

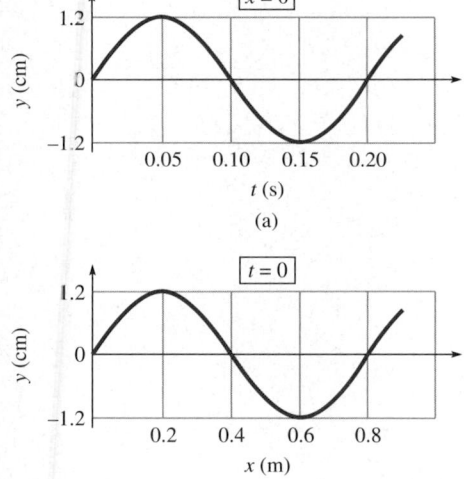

(a)

(b)

(c) $T = 0.20$ s; (d) $\lambda = 0.80$ m; (e) $A = 1.2$ cm; (f) $v = 4.0$ m/s; (g) the wave travels in the −x-direction because the signs of the terms containing x and t are the same.

11.4

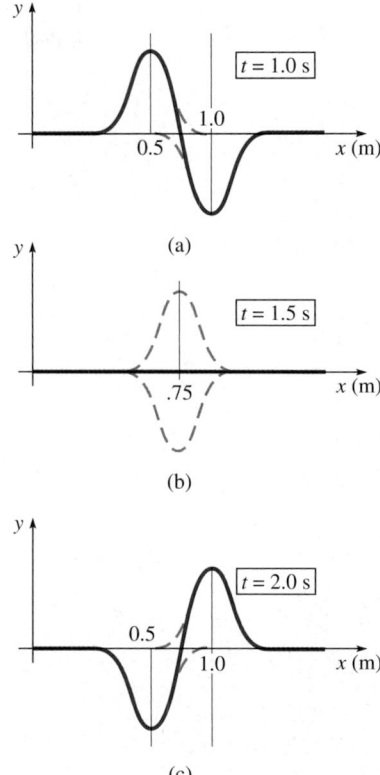

(a)

(b)

(c)

11.5 (a) 620 Hz; (b) 8.5 m

11.6 7.1 km/s

11.7 9.0

11.8 140 Hz

Sound

Ultrasonic imaging of the fetus is an increasingly important part of prenatal care. Could an image of the fetus be produced just as well using sound in the audible range rather than ultrasound? Why is ultrasound used rather than some other imaging technology, such as x-rays? Are there other medical applications of ultrasound?

Concepts & Skills to Review

- gauge pressure (Section 9.5)
- bulk modulus (Section 10.4)
- relation between energy and amplitude in SHM (Section 10.5)
- period and frequency in SHM (Section 10.6)
- longitudinal waves, intensity, standing waves, superposition principle (Chapter 11)

Making The Connection:
sound from a guitar

Making The Connection:
sound from a loudspeaker

12.1 SOUND WAVES

When a guitar string is plucked, a transverse wave travels along the string. The wave on the string is not what we hear, of course, since the string has no direct connection to our eardrums. The vibration of the string is transmitted through the bridge to the body of the guitar, which in turn transmits the vibration to the air—a sound wave. A transverse wave on a guitar string is not a sound wave, though it does *cause* a sound wave.

In the absence of a sound wave, air molecules dart around in random directions. On average, they are uniformly distributed and the pressure is the same everywhere (neglecting the insignificant variation of pressure due to small changes in altitude). In a sound wave, the uniform distribution of air molecules is disturbed. A loudspeaker produces pressure fluctuations that travel through the air in all directions (Fig. 12.1). In some regions (*compressions*), the air molecules are bunched together and the pressure is higher than the average pressure. In other regions (*rarefactions*), the molecules are spread out and the pressure is lower than average. The sound wave can be described mathematically by the gauge pressure p (the difference between the pressure at a given point and the average pressure in the surroundings) as a function of position and time (Fig. 12.2a).

The speaker cone produces these pressure variations by displacing air molecules from their uniform distribution (Fig. 12.2b). When the cone moves to the left of its equilibrium position, air molecules spread out into a region of lower pressure (rarefaction). When the cone moves to the right, air molecules are shoved together into a region of higher pressure (compression).

Thus, the regions of higher and lower pressure are formed when air molecules are displaced from a uniform distribution. A sound wave can be described equally well by the displacement s of an *element* of the air—a region of air that can be considered to move together as a unit (Fig. 12.2c). An element is much smaller than the wavelength of the wave but still large enough to contain many molecules. Near a compression, where the air molecules are bunched together, elements on the left have been displaced to the right, and those on the right have been displaced to the left, as shown by the displacement arrows in Fig. 12.2c. Elements at points of maximum or minimum pressure have zero displacement; they stay put while the neighboring elements move in toward them (a compression) or away from them (a rarefaction). Conversely, where the gauge pressure is zero, the displacement has its maximum magnitude. In a standing wave, a pressure node is a displacement antinode; a pressure antinode is a displacement node.

If the pressure is higher on one side than on the other, the net force pushes air toward the side with lower pressure. The uneven distribution of pressure results in air molecules being pushed toward rarefactions and away from compressions, as shown by the force arrows in Fig. 12.2b. Note that the directions of these force arrows, pointing

Figure 12.1   The vibrating speaker cone in this boombox creates alternating regions of high and low pressure in the air. Air nearby is affected by a net force due to the nonuniform air pressure; as a result, variations in pressure travel in all directions away from the speaker. This traveling disturbance is a sound wave.

opposite to the displacement arrows in a corresponding region, are such that where there is a compression at a given instant, there will later be a rarefaction, and *vice versa;* the pressure at a given point fluctuates above and below the average pressure.

What makes the sound wave propagate away from the speaker cone? At the instant shown in Fig. 12.2, the sound wave has propagated out a certain distance from the speaker. Beyond the leading edge of the wave, the air molecules are not yet disturbed from their uniform distribution. The higher pressure at the leading edge pushes molecules to the right, so that the compression moves outward and the leading edge of the wave moves away from the speaker. When the compressions and rarefactions reach our ear, the changing force due to the air pressure on the eardrum causes the eardrum to vibrate in and out.

Frequencies of Sound Waves

The human ear responds to sound waves within a limited range of frequencies. We generally consider the **audible range** to extend from 20 Hz to 20 kHz. Very few people can actually hear sounds over that entire range. Even for a person with excellent hearing, the sensitivity of the ear declines rapidly below 100 Hz and above 10 kHz. The terms **infrasound** and **ultrasound** are used to describe sound waves with frequencies below 20 Hz and above 20 kHz, respectively.

The audible ranges for animals can be quite different. Dogs can hear frequencies as high as 50 kHz, which is why we can make a dog whistle that is inaudible to humans.

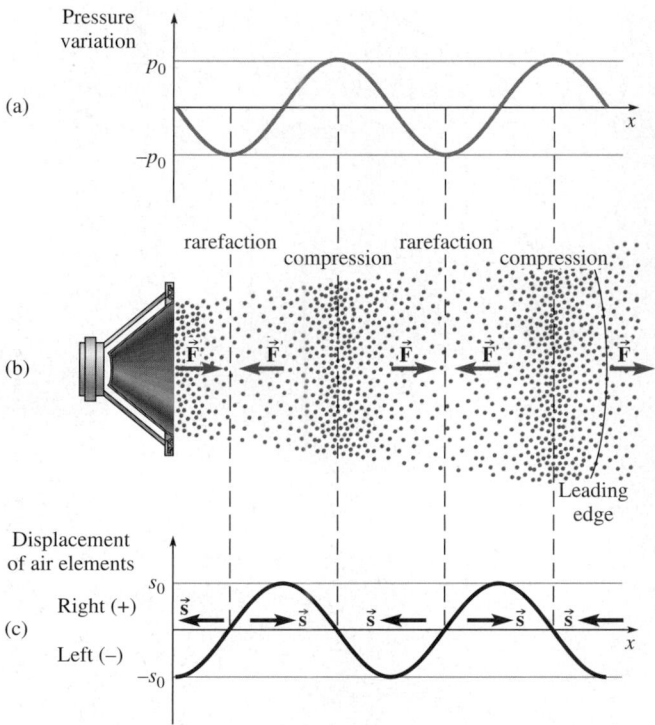

Figure 12.2 A sound wave generated by a loudspeaker. (a) Graph of the pressure variation of the air as a function of position. Pressure is high where air is squeezed together and low where it is more spread out. (b) Elements of the air are displaced from their equilibrium positions. Since the pressure is not uniform, air elements experience a net force due to air pressure; the force arrows indicate the direction of this net force. The force is always directed away from a compression (higher pressure) and toward a rarefaction. (c) Graph of the displacement of air elements from their equilibrium positions as a function of position; the arrows indicate the directions of the displacements in each region. Air elements are displaced towards compressions and away from rarefactions.

Dolphins make use of frequencies as high as 250 kHz. Elephants communicate over long distances (up to 4 km) using sounds with fundamental frequencies as low as 14 Hz. A rhinoceros uses frequencies down to 10 Hz. Such low-frequency sounds cannot be heard by humans, but the vibrations can be felt and the sounds can be recorded using special equipment.

12.2   THE SPEED OF SOUND WAVES

Just as for transverse waves on a string, the speed of sound waves is determined by a balance between two characteristics of the wave medium: the restoring force and the inertia. For string waves, the restoring force is characterized by the tension in the string F and the inertia is characterized by the linear mass density μ (mass per unit length). The speed of transverse waves on a string is

$$v = \sqrt{\frac{F}{\mu}} \tag{11-4}$$

For sound waves in a fluid, the restoring force is characterized by the bulk modulus B, defined in Section 10.4 as the increase in pressure needed to compress the fluid a certain fraction of its volume:

$$\Delta P = -B \frac{\Delta V}{V} \tag{10-10}$$

The inertia of the fluid is characterized by its mass density ρ. Following our dictum "more restoring force makes faster waves; more inertia makes slower waves," we expect the speed of sound to be faster in a medium with a larger bulk modulus (harder to compress means more restoring force) and slower in a medium with a larger density. By analogy with Eq. (11-4), we might *guess* that

$$v = \sqrt{\frac{\text{a measure of the restoring force}}{\text{a measure of the inertia}}} = \sqrt{\frac{B}{\rho}} \quad \text{(in fluids)} \tag{12-1}$$

In Problem 4, you can show that no other combination of B and ρ can have dimensions of speed, so that Eq. (12-1) *must* be correct except for the possibility of multiplication

More restoring force $\Rightarrow$ faster waves; more inertia $\Rightarrow$ slower waves.

by a dimensionless constant. The dimensionless constant turns out to be 1; Eq. (12-1) is the correct expression for the speed of sound in fluids.

The bulk modulus B of an ideal gas turns out to be directly proportional to the density ρ and to T, the *absolute temperature* ($B \propto \rho T$). As a result, the speed of sound in an ideal gas is proportional to the absolute temperature but is independent of pressure and density (at a fixed temperature):

$$v = \sqrt{\frac{B}{\rho}} \propto \sqrt{\frac{\rho T}{\rho}} \propto \sqrt{T} \quad \text{(ideal gas)}$$

The SI unit of absolute temperature is the kelvin (symbol K). To find absolute temperature in kelvins, add 273.15 to the temperature in degrees Celsius:

$$T \text{ (in K)} = T_C \text{ (in °C)} + 273.15 \tag{12-2}$$

Since $v \propto \sqrt{T}$, the speed of sound in an ideal gas at any absolute temperature T can be found if it is known at one temperature:

$$v = v_0 \sqrt{\frac{T}{T_0}} \tag{12-3}$$

where the speed of sound is v_0 at absolute temperature T_0. For example, the speed of sound in air at 0°C (or 273 K) is 331 m/s. At room temperature (20°C or 293 K), the speed of sound in air is

$$v = 331 \text{ m/s} \times \sqrt{\frac{293 \text{ K}}{273 \text{ K}}} = 343 \text{ m/s}$$

An *approximate* formula that can be used for the speed of sound in air is

$$v = (331 + 0.606 T_C) \text{ m/s} \tag{12-4}$$

where T_C is air temperature *in degrees Celsius* (see Problem 5 at the end of this chapter). The speed of sound in air increases 0.606 m/s for each degree Celsius increase in temperature. Equation (12-4) gives speeds accurate to better than 1% all the way from –66°C to +89°C.

The speed of sound in a *solid* depends on the Young's modulus Y and the shear modulus S. For sound waves traveling along the length of a thin solid rod, the speed is approximately

$$v = \sqrt{\frac{Y}{\rho}} \quad \text{(thin solid rod)} \tag{12-5}$$

Table 12.1 gives the speed of sound in various materials.

Table 12.1

Speed of Sound in Various Materials (at 0°C and 1 atm unless otherwise noted)

Medium	Speed (m/s)
Carbon dioxide	259
Air	331
Nitrogen	334
Air (20°C)	343
Helium	972
Hydrogen	1284
Mercury (25°C)	1450
Fat (37°C)	1450
Water (25°C)	1493
Seawater (25°C)	1533
Blood (37°C)	1570
Muscle (37°C)	1580
Lead	1322
Concrete	3100
Copper	3560
Bone (37°C)	4000
Pyrex glass	5640
Aluminum	5100
Steel	5790
Granite	6500

Conceptual Example 12.1

Speed of Sound in Hydrogen and Mercury

From Table 12.1, the speed of sound in hydrogen gas at 0°C is almost as large as the speed of sound in mercury, even though the density of mercury is 150,000 times larger than the density of hydrogen. How is that possible? Shouldn't the speed in mercury be much smaller, since it has so much more inertia?

Solution and Discussion   The speed of sound depends on *two* characteristics of the medium: the restoring force (measured by the bulk modulus) and the inertia (measured by the

density). The bulk modulus of mercury is much larger than the bulk modulus of hydrogen. The bulk modulus is a measure of how hard it is to compress a material. Liquids (such as mercury) are much more difficult to compress than are gases. Thus, the restoring forces in mercury are much larger than those in hydrogen; this allows sound to travel a bit faster in mercury than it does in hydrogen gas.

Conceptual Practice Problem 12.1   Speed of sound in solids versus liquids

Why does sound generally travel faster in a solid than in a liquid?

12.3   AMPLITUDE AND INTENSITY OF SOUND WAVES

Since there are two ways to describe a sound wave—pressure and displacement—the amplitude of a sound wave can take one of two forms: the pressure amplitude p_0 or the displacement amplitude s_0. The pressure amplitude p_0 is the maximum pressure fluctuation above or below the pressure in the absence of a sound wave; the displacement amplitude s_0 is the maximum displacement of an element of the medium from its equilibrium position. The pressure amplitude is proportional to the displacement amplitude. For a harmonic sound wave at angular frequency ω, an advanced analysis shows that

$$p_0 = \omega v \rho s_0 \qquad (12\text{-}6)$$

where v is the speed of sound and ρ is the mass density of the medium.

Is a larger amplitude sound wave perceived as *louder*? Yes, all other things being equal. However, the relationship between our perception of loudness and the amplitude of a sound wave is complex. Loudness is a subjective aspect of how sound is perceived; it has to do with how the ear responds to sound and how the brain interprets signals from the ear. Perceived loudness turns out to be *roughly* proportional to the logarithm of the amplitude. If the amplitude of a sound wave doubles repeatedly, the perceived loudness does not double; it increases by a series of roughly equal steps.

Discussions of loudness are more often phrased in terms of intensity rather than amplitude since we are interested in how much energy the sound wave carries. The intensity of a sound wave is

$$I = \frac{p_0^2}{2\rho v} \qquad (12\text{-}7)$$

Intensity $\propto$ (Amplitude)^2

where ρ is the mass density of the medium and v is the speed of sound in that medium. The most important thing to remember is that *intensity is proportional to amplitude squared*, which is true for all waves, not just sound. It is closely related to the fact that energy in SHM is proportional to amplitude squared [see Section 10.5, Eq. (10-13)].

Table 12.2 gives the pressure amplitudes and intensities for a wide range of sounds. Notice that the pressure amplitudes are given in *atmospheres;* thus, even for sounds that

Table 12.2

Pressure Amplitudes, Intensities, and Intensity Levels of a Wide Range of Sounds in Air at 20°C

Sound	Pressure Amplitude (atm)	Pressure Amplitude (Pa)	Intensity (W/m²)	Intensity Level (dB)
Threshold of hearing	3×10^{-10}	3×10^{-5}	10^{-12}	0
Leaves rustling	1×10^{-9}	1×10^{-4}	10^{-11}	10
Whisper (1 m away)	3×10^{-9}	3×10^{-4}	10^{-10}	20
Library background noise	1×10^{-8}	0.001	10^{-9}	30
Living room background noise	3×10^{-8}	0.003	10^{-8}	40
Office or classroom	1×10^{-7}	0.01	10^{-7}	50
Normal conversation at 1 m	3×10^{-7}	0.03	10^{-6}	60
Inside a moving car, light traffic	1×10^{-6}	0.1	10^{-5}	70
City street (heavy traffic)	3×10^{-6}	0.3	10^{-4}	80
Shout (at 1 m); or inside a subway train; risk of hearing damage if exposure lasts several hours	1×10^{-5}	1	10^{-3}	90
Car without muffler at 1 m	3×10^{-5}	3	10^{-2}	100
Construction site	1×10^{-4}	10	10^{-1}	110
Indoor rock concert; threshold of pain; hearing damage occurs rapidly	3×10^{-4}	30	1	120
Jet engine at 30 m	1×10^{-3}	100	10	130

are quite loud, the pressure fluctuations due to sound waves are small compared to the "background" atmospheric pressure.

Example 12.2

The Brown Creeper

The song of the Brown Creeper (*Certhia americana*) is very high in frequency—as high as 8 kHz. Many people who have lost some of their high-frequency hearing can't hear it at all. Suppose that you are out in the woods and hear the song. If the intensity of the song at your position is 1.4×10^{-8} W/m^2 and the frequency is 6.0 kHz, what are the pressure and displacement amplitudes? (Assume the temperature is 20°C.)

Strategy The displacement and pressure amplitudes are related through Eq. (12-6); the pressure amplitude is related to the intensity through Eq. (12-7). These relationships can be used to solve for both pressure amplitude, p_0, and displacement amplitude, s_0. The density of air at 20°C is $\rho = 1.20$ kg/m^3 (Table 9.1). The speed of sound in air at 20°C is $v = 343$ m/s. We need to multiply the frequency by 2π to get the angular frequency ω.

Solution Intensity and pressure amplitude are related by

$$I = \frac{p_0^2}{2\rho v} \tag{12-7}$$

Solving for p_0,

$$p_0 = \sqrt{2I\rho v} = \sqrt{2 \times 1.4 \times 10^{-8} \text{ W/m}^2 \times 1.20 \text{ kg/m}^3 \times 343 \text{ m/s}}$$
$$= 3.4 \times 10^{-3} \text{ Pa}$$

The pressure and displacement amplitudes are related by

$$p_0 = \omega v \rho s_0 \tag{12-6}$$

Substituting in Eq. (12-7) yields

$$I = \frac{(\omega v \rho s_0)^2}{2\rho v}$$

Solving for s_0,

$$s_0 = \sqrt{\frac{2I}{\rho \omega^2 v}} = \sqrt{\frac{2 \times 1.4 \times 10^{-8} \text{ W/m}^2}{1.20 \text{ kg/m}^3 \times (2\pi \times 6000 \text{ Hz})^2 \times 343 \text{ m/s}}}$$
$$= 2.2 \times 10^{-10} \text{ m}$$

Discussion This problem illustrates how sensitive the human ear is. The pressure amplitude is a fluctuation of one part in 30 million in the air pressure. Since the pressure amplitude is 3.4×10^{-3} Pa, the maximum force on the eardrum would be about

$$F_{\text{max}} = 3.4 \times 10^{-3} \text{ N/m}^2 \times 10^{-4} \text{ m}^2 \approx 3 \times 10^{-7} \text{ N}$$

which is about the weight of a large amoeba. The displacement amplitude is about the size of an atom.

Practice Problem 12.2   Pressure and intensity at an outdoor concert

At a distance of 5.0 m from the stage at an outdoor rock concert, the sound intensity is 1.0×10^{-4} W/m^2. Estimate the intensity and pressure amplitude at a distance of 25 m if there were no speakers other than those on stage. Explain the assumptions you make.

Decibels

Since the perception of loudness by the human ear is roughly proportional to the logarithm of the intensity, it is also roughly proportional to the logarithm of the amplitude (since $\log x^2 = 2 \log x$). An intensity of $I_0 = 10^{-12}$ W/m^2 is about the lowest intensity sound wave that can be heard under ideal conditions by a person with excellent hearing; it is therefore called the **threshold of hearing**. The threshold of hearing is used as a reference intensity in the definition of the intensity level.

A sound intensity I is compared to the reference level I_0 by taking the ratio of the two intensities. Suppose a sound has an intensity of 10^{-5} W/m^2; the ratio is

$$\frac{I}{I_0} = \frac{10^{-5} \text{ W/m}^2}{10^{-12} \text{ W/m}^2} = 10^7$$

so the intensity is 10^7 times that of the hearing threshold level. The power to which 10 is raised is the **sound intensity level** β in units of bels (after Alexander Graham Bell). A ratio of 10^7 indicates a sound intensity of 7 bels or, as it is more commonly stated, 70 decibels (dB). Since $\log_{10}(10^x) = x$, the sound intensity level in decibels is

$$\beta = (10 \text{ dB}) \log_{10} \frac{I}{I_0} \tag{12-8}$$

An intensity level of 0 dB corresponds to the threshold of hearing ($I = 10^{-12}$ W/m^2). Although the intensity level is really a pure number, the "units" (dB) remind us what the number means.

Example 12.3

Decibels from a Jackhammer

 The sound intensity 5 m from a jackhammer is 4.20×10^{-2} W/m^2. What is the sound intensity level in decibels? (Use a reference level of $I_0 = 1.00 \times 10^{-12}$ W/m^2.)

Strategy We are given the intensity in W/m^2 and asked for the intensity level in dB. First we find the ratio of the given intensity to the reference level. Then we take the logarithm of the result (to get the level in bels) and multiply by 10 (to convert from bels to dB).

Solution The ratio of the intensity to the reference level is

$$\frac{I}{I_0} = \frac{4.20 \times 10^{-2} \text{ W/m}^2}{10^{-12} \text{ W/m}^2} = 4.20 \times 10^{10}$$

The intensity level in bels is

$$\log_{10} \frac{I}{I_0} = \log_{10} 4.20 \times 10^{10} = 10.6 \text{ bels}$$

The intensity level in decibels is

$$\beta = 10.6 \text{ bels} \times (10 \text{ dB/bel}) = 106 \text{ dB}$$

Discussion As a quick check, 100 dB corresponds to $I = 10^{-2}$ W/m^2 and 110 dB corresponds to $I = 10^{-1}$ W/m^2; since the intensity is between 10^{-2} W/m^2 and 10^{-1} W/m^2, the intensity level must be between 100 dB and 110 dB.

Practice Problem 12.3 Consequences of a hole in the muffler

When rust creates a hole in the muffler of a car, the sound intensity level inside the car is 26 dB higher than when the muffler was intact. By what factor does the intensity increase?

As we saw in Section 11.9, when two sounds are coming from different sources, the waves are incoherent. If we know the intensity of each wave alone at a certain point, then the intensity due to the two waves together at that point is the sum of the two intensities:

$$I = I_1 + I_2 \quad \text{(incoherent waves)}$$

This is *not* true for two coherent waves, where the total intensity depends on the phase relationship between the waves. Since there is no fixed phase relationship between two incoherent waves, on average there is neither constructive nor destructive interference. The total power per unit area is the sum of the power per unit area of each wave.

Example 12.4

The Sound Intensity of Two Lathes Compared to That of One Lathe

A metal lathe in a workshop produces a 90.0-dB sound intensity level at a distance of 1 m. What is the intensity level when a second identical lathe starts operating? Assume the listener is at the same distance from both lathes.

Strategy The noise is coming from two different machines and thus they are incoherent sources. We *cannot* add 90.0 dB to 90.0 dB to get 180.0 dB, which would be a senseless result—two lathes are not going to drown out a jet

engine at close range (see Table 12.1). Instead, what doubles is the *intensity*. We must work in terms of intensity rather than intensity level.

Solution First find the intensity due to one lathe:

$$\beta = 90.0 \text{ dB} = (10 \text{ dB}) \log_{10} \frac{I}{I_0}$$

$$\log_{10} \frac{I}{I_0} = 9.00, \quad \text{so } \frac{I}{I_0} = 1.00 \times 10^9$$

continued on next page

Example 12.4 *continued*

We could solve for I numerically but it is not necessary. With two machines operating, the intensity doubles, so

$$\frac{I'}{I_0} = 2.00 \times 10^9$$

and the new intensity level is

$$\beta' = (10 \text{ dB}) \log_{10} \frac{I'}{I_0} = (10 \text{ dB}) \log_{10} (2.00 \times 10^9) = 93.0 \text{ dB}$$

 Discussion The new intensity level is just 3 dB higher than the original one, even though the intensity is twice as big. This turns out to be a general result: a 3-dB increase represents a doubling of the intensity.

Practice Problem 12.4 Intensity change for an increment of 5 dB

Maximum recommended exposure time to a sound level of 90 dB is 8 hours. For every increase of 5.0 dB in sound level up to 120 dB, the exposure time should be reduced by a factor of 2. (At 120 dB, damage occurs almost immediately; there is no safe exposure time.) What factor of intensity change does an intensity level increment of 5.0 dB represent?

Sound intensity level is useful because it roughly approximates the way we perceive loudness (since it is a logarithmic function of intensity). Equal increments in intensity level roughly correspond to equal increases in loudness. Two useful rules of thumb: every time the intensity increases by a *factor* of 10, the intensity level *adds* 10 dB; since $\log_{10} 2 = 0.30$, adding 3.0 dB to the intensity level *doubles* the intensity (see Problem 14). In Example 12.4, when both lathes are running at the same time, the intensity is twice as big as for one lathe, but the two do not sound twice as loud as one. Intensity *level* is a better guide to loudness; two lathes produce a level 3 dB higher than one lathe.

Decibels can also be used in a relative sense; instead of comparing an intensity to I_0, we can compare two intensities directly. Suppose we have two intensities I_1 and I_2 and two corresponding intensity levels β_1 and β_2. Then

$$\beta_2 - \beta_1 = 10 \text{ dB} \left(\log_{10} \frac{I_2}{I_0} - \log_{10} \frac{I_1}{I_0} \right)$$

Since $\log x - \log y = \log \frac{x}{y}$ [see Appendix A.3, Eq. (A-21)],

$$\beta_2 - \beta_1 = (10 \text{ dB}) \log_{10} \frac{I_2/I_0}{I_1/I_0} = (10 \text{ dB}) \log_{10} \frac{I_2}{I_1} \qquad (12\text{-}9)$$

Example 12.5

Variation of Intensity Level with Distance

At a distance of 30 m from a jet engine the sound intensity level is 130 dB. Serious, permanent hearing damage occurs rapidly at intensity levels this high, which is why you see air-

port personnel using hearing protection out on the runway. Assume the engine is an isotropic source of sound and ignore reflections and absorption. At what distance is the intensity level 110 dB—still quite loud but below the threshold of pain?

Strategy The intensity level drops 20 dB. According to the rule of thumb, each 10-dB change represents a factor of 10 in intensity. Therefore, we must find the distance at which the intensity is 2 factors of 10 smaller—that is, $\frac{1}{100}$ the original intensity. The intensity is proportional to $1/r^2$ since we assume an isotropic source.

Solution We set up a ratio between the intensities and the inverse square of the distances:

$$\frac{I_1}{I_2} = \left(\frac{r_2}{r_1} \right)^2$$

continued on next page

Example 12.5 *continued*

From the rule of thumb, we know that $I_2 = \frac{1}{100} I_1$. Then

$$\frac{r_2}{r_1} = \sqrt{\frac{I_1}{I_2}} = \sqrt{100} = 10$$

$$r_2 = 10r_1 = 300 \text{ m}$$

Discussion   It is not necessary to use the rule of thumb. Let $\beta_1 = 130$ dB and $\beta_2 = 110$ dB. Then

$$\beta_1 - \beta_2 = 20 \text{ dB} = (10 \text{ dB}) \log_{10} \frac{I_1}{I_2}$$

From this, we find that

$$\log_{10} \frac{I_1}{I_2} = 2 \quad \text{or} \quad \frac{I_1}{I_2} = 100$$

We can only consider 300 m an estimate. The jet engine may not radiate sound equally in all directions; it might be louder in front than on the side. Sound is partly absorbed and partly reflected by the runway, by the plane, and by any nearby objects. The air itself absorbs some of the sound energy—that is, some of the energy of the wave is dissipated.

Practice Problem 12.5   A plane as quiet as a library

At what distance from the jet engine would the intensity level be comparable to the background noise level of a library (30 dB)? Is your answer realistic?

Making The Connection:
sources of
musical sound

12.4   STANDING SOUND WAVES

Pipe Open at Both Ends

Recall (Section 11.8) that a transverse wave on a string is reflected from a fixed end. A string fixed at both ends reflects the wave at each end. Such a string supports standing waves, caused by the superposition of two waves traveling in opposite directions on the string, only at certain frequencies, since there must be a node at each end. Standing sound waves are also caused by reflections at boundaries. Standing wave patterns for sound waves are in general more complex, since sound is a three-dimensional wave. However, the air inside a pipe open at both ends gives rise to standing waves closely analogous to those on a string, as long as the pipe's diameter is small compared to its length. Such a pipe is an excellent model of some organ pipes and flutes.

If the pipe is open at both ends, then the pipe has the same boundary condition at each end. At each open end, the column of air inside the pipe communicates with the outside air, so the pressure at the ends can't deviate much from atmospheric pressure. The open ends are therefore *pressure nodes* (Fig. 12.3). They are also *displacement antinodes*—elements of air vibrate back and forth with maximum amplitude at the ends. Since nodes and antinodes alternate with equal spacing ($\lambda/4$), the wavelengths of standing sound waves in a pipe open at both ends are the same as for a string fixed at both ends (compare Fig. 12.3 with Fig. 11.21), regardless of whether you consider the pressure or the displacement description.

Standing sound waves (thin pipe open at both ends)

$$\lambda_n = \frac{2L}{n} \tag{11-12}$$

$$f_n = \frac{v}{\lambda_n} = n\frac{v}{2L} = nf_1 \tag{11-13}$$

where $n = 1, 2, 3, \ldots$

Pipe Closed at One End

Some organ pipes are *closed at one end* and open at the other (Fig. 12.4). The closed end is a pressure *antinode;* the air at the closed end meets a rigid surface, so there is no restriction on how far the pressure can deviate from atmospheric pressure. The closed end is also a *displacement node* since the air near it cannot move beyond that rigid surface. Some wind instruments are effectively pipes closed at one end. The reed of a clarinet admits only brief puffs of air into the instrument; the rest of the time the reed closes

Figure 12.3 Standing waves in a pipe open at both ends

off that end of the pipe. The pressure at the reed end fluctuates above and below atmospheric pressure. The reed end is a pressure antinode and a displacement node.

The wavelengths and frequencies of the standing waves can be found using either the pressure or displacement descriptions of the wave. Using displacement, the fundamental has a node at the closed end, an antinode at the open end, and no other nodes or antinodes (see Fig. 12.5). The distance from a node to the nearest antinode is always $\frac{1}{4}\lambda$, so for the fundamental

$$L = \tfrac{1}{4}\lambda \text{ or } \lambda = 4L$$

which is twice as large as the wavelength ($2L$) of the fundamental in a pipe of the same length open at both ends. Two thin organ pipes of the same length, one open at both

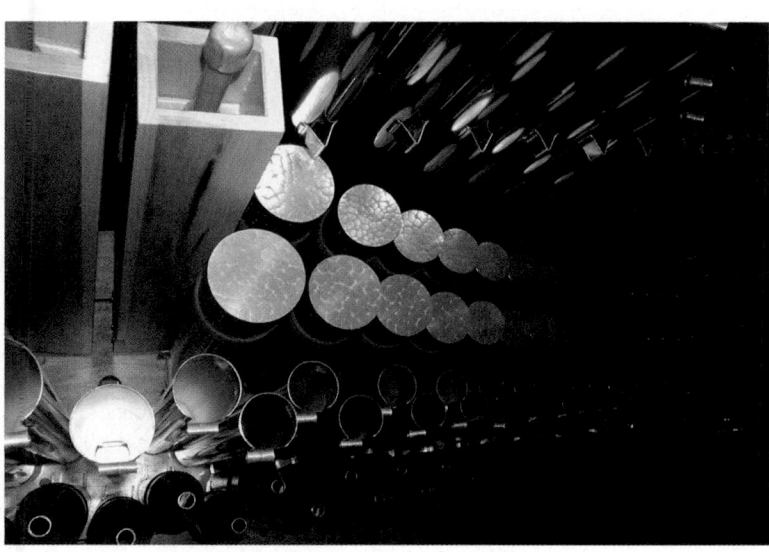

Figure 12.4 Some organ pipes are open at the top; others are closed.

Figure 12.5   Standing waves in a pipe closed at one end

ends and one closed at one end, do not have the same wavelength fundamental. The pipe closed at one end has a wavelength twice as large and therefore a frequency half as large, assuming the pipes are thin. (For musicians: the pitch of the pipe closed at one end sounds an octave lower than the other, since the interval of an octave corresponds to a factor of two in frequency.)

What are the other standing wave frequencies? The next standing wave mode is found by adding one node and one antinode. Then the length of the pipe is 3 quarter-cycles: $L = \frac{3}{4}\lambda$ or $\lambda = \frac{4}{3}L$. This is $\frac{1}{3}$ the wavelength of the fundamental and the frequency is 3 times that of the fundamental. Adding one more node and one more antinode, the wavelength is $\frac{4}{5}L$. Continuing the pattern, we find that the wavelengths and frequencies for standing waves are

---

Standing sound waves (thin pipe closed at one end)

$$\lambda_n = \frac{4L}{n} \tag{12-10a}$$

$$f_n = \frac{v}{\lambda_n} = n\frac{v}{4L} = nf_1 \tag{12-10b}$$

where $n = 1, 3, 5, 7, \ldots$

 Note that the standing wave frequencies for a pipe closed at one end are only *odd* multiples of the fundamental.

The "missing" standing wave patterns for even values of n require a clarinet to have many more keys and levers than a flute (Fig. 12.6). What the keys do is effectively shorten the length of the pipe, making the standing wave frequencies higher. If a flute's fundamental frequency is f_1 with no keys pressed, the next highest frequency possible without using any keys is $2f_1$—the flutist overblows, exciting the next highest standing wave frequency rather than the fundamental. The flute needs enough keys to fill in all the notes with frequencies between f_1 and $2f_1$. For a clarinet, if the fundamental frequency is

Figure 12.6 A clarinet (a pipe open at one end) needs many more keys than a flute (a pipe open at both ends).

f_1 with no keys pressed, the next highest frequency possible without using any keys is $3f_1$. The clarinet must have more keys because it has to accommodate all the notes with frequencies between f_1 and $3f_1$.

Problem-Solving Strategy for Standing Waves

There is no need to memorize equations for standing wave frequencies and wavelengths. Just sketch the standing wave patterns as in Figs. 12.3 and 12.5. Make sure that nodes and antinodes alternate and that the boundary conditions at the ends are correct. Then determine the wavelengths by setting the distance between a node and antinode equal to $\frac{1}{4}\lambda$. Once the wavelengths are known, the frequencies are found from $v = f\lambda$.

Example 12.6

A Demonstration of Resonance

A thin hollow tube of length 1.00 m is inserted vertically into a tall container of water (Fig. 12.7a). A tuning fork ($f = 520.0$ Hz) is struck and held near the top of the tube as the tube is slowly pulled up and out of the water. At certain distances (L) between the top of the tube and the water surface, the otherwise faint sound of the tuning fork is greatly amplified. At what values of L does this occur? The temperature of the air in the tube is 18°C.

Strategy Sound waves in the air inside the tube reflect from the water surface. Thus, we have an air column of variable length L, closed at one end by the water surface and open at the other end. The sound is amplified due to resonance; when the frequency of the tuning fork matches one of the natural frequencies of the air column, a large-amplitude standing wave builds up in the column. For standing waves in a column of air, the wavelength and frequency are related by the speed of sound in air. We start by finding the speed of sound in air from the temperature given. Then we can find the wavelength of the sound waves emanating from the tuning fork. Last, we find the column lengths that support standing waves of that wavelength.

Solution The speed of sound in air at 18°C is

$$v = (331 + 0.606 \times 18) \text{ m/s} = 342 \text{ m/s}$$

With the speed of sound and the frequency known, we can find the wavelength. The wavelength is the distance traveled by a wave during one period:

$$\lambda = vT = \frac{v}{f}$$

$$\lambda = \frac{342 \text{ m/s}}{520.0 \text{ Hz}} = 0.6577 \text{ m} = 65.77 \text{ cm}$$

The first possible resonance for a tube closed on one end occurs when there is a pressure node at the open end, a pressure antinode at the closed end, and no other pressure nodes or antinodes. Therefore,

$$L_1 = \frac{1}{4}\lambda = \frac{1}{4} \times 65.77 \text{ cm} = 16.4 \text{ cm}$$

continued on next page

Example 12.6 *continued*

To reach other resonances, the tube must be pulled out to accommodate additional pressure nodes and antinodes. To add one node and one antinode, the additional distance is $\frac{1}{2}\lambda = 32.9$ cm. The resonances occur at intervals of 32.9 cm:

$$L_2 = 16.4 \text{ cm} + 32.9 \text{ cm} = 49.3 \text{ cm}$$

$$L_3 = 49.3 \text{ cm} + 32.9 \text{ cm} = 82.2 \text{ cm}$$

The next one would require a tube longer than 1.00 m, so there are three values of L that produce resonance in this tube.

Discussion   As a check, we can sketch the standing wave pattern for the third resonance (Figs. 12.7b,c). There are 5 quarter-wavelengths in the length of the column so

$$L_3 = \frac{5}{4}\lambda = \frac{5}{4} \times 65.77 \text{ cm} = 82.2 \text{ cm}$$

At the open end of the tube, the node for pressure and the antinode for maximum displacement is actually a little *above* the opening. For this reason it is best to measure the distance between two successive resonances to find an accurate value for a half-wavelength rather than measuring the distance for the first possible resonance, the shortest distance between the opening and the water surface, and setting it equal to a quarter-wavelength.

Practice Problem 12.6   A roundabout way to measure temperature

A tuning fork of frequency 440.0 Hz is held above the hollow tube in Example 12.6. If the distance ΔL that the tube is moved between resonances is 39.3 cm, what is the temperature of the air inside the tube?

Displacement   Pressure

Figure 12.7
(a) Experimental setup for Example 12.6.
(b) Standing wave pattern, showing *pressure* nodes and antinodes, for the third resonance. (c) Standing wave pattern, showing *displacement* nodes and antinodes, for the third resonance.

(a)   (b)   (c)

Physics at Home

You can set up a resonance in an empty water bottle by blowing horizontally across the top of the bottle. Once you have heard one resonance, add varying amounts of water to raise the level within and listen for other resonances. The resonant sound is noticeably louder than the nonresonant sounds. Notice that the longer the air column within the bottle, the lower the pitch heard.

Making The Connection:
human ear

12.5   THE HUMAN EAR

Figure 12.8 shows the structure of the human ear. The human ear has an external part or *pinna* that acts something like a funnel, collecting sound waves and concentrating them at the opening of the auditory canal. The pinna is better at collecting sound coming from in front than from behind, which helps with localization. Resonance in the *auditory canal* (see Problem 44) boosts the ear's sensitivity in the 2- to 5-kHz frequency range— a crucial range for understanding speech.

At the end of the auditory canal, the eardrum (*tympanum*) vibrates in response to the incident sound wave. The region just beyond the eardrum is called the middle ear.

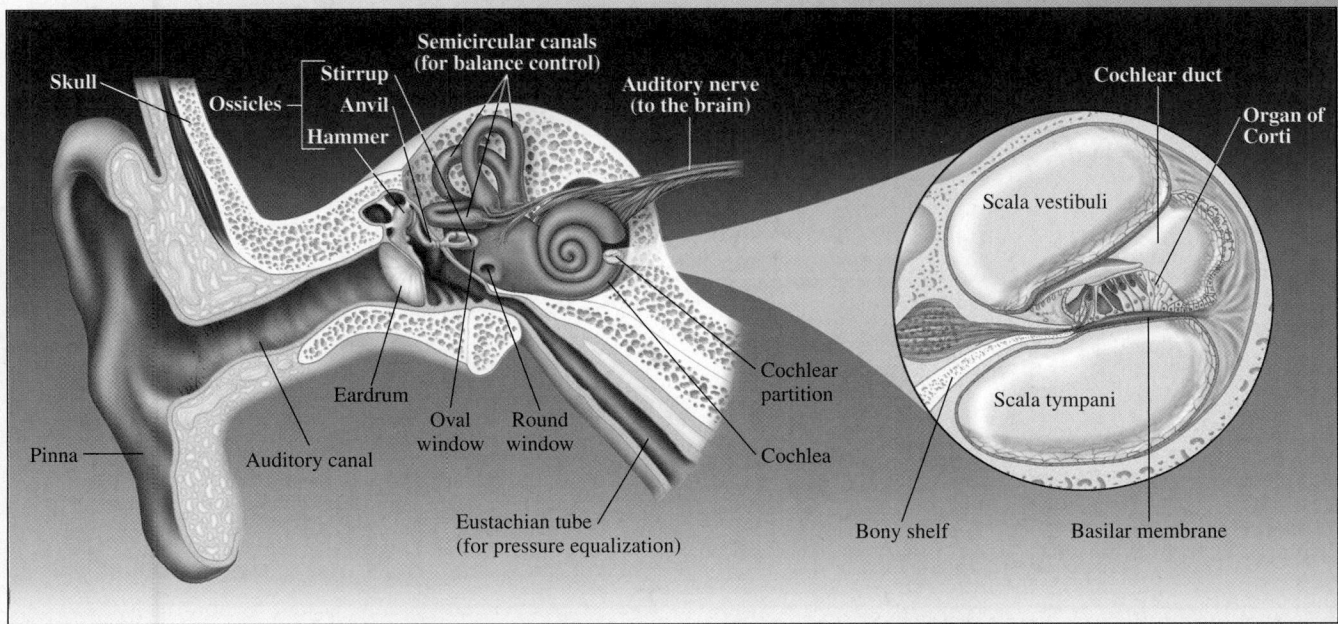

Figure 12.8   Structure of the human ear with a cross section of the cochlea

The vibrations of the eardrum are transmitted through three tiny bones of the middle ear (the *auditory ossicles*) to the *oval window* of the *cochlea*, a tapered spiral-shaped organ filled with fluid. The oval window is a membrane that is in contact with the fluid in the cochlea. The ossicles act as levers; the force exerted by the "stirrup" on the oval window is 1.5 to 2.0 times the force the eardrum exerts on the "hammer." The area of the oval window is one-twentieth that of the eardrum, so there is an overall amplification in pressure by a factor of 30 to 40. The ossicles protect the ear from damage: in response to a loud sound, a muscle pulls the stirrup away from the oval window. At the same time, another muscle increases the eardrum tension. These two changes make the ear temporarily less sensitive. It takes a few milliseconds for the muscles to respond in this way, so they provide no protection against *sudden* loud sounds.

The *cochlear partition* (Fig. 12.8) runs most of the length of the cochlea, separating it into two chambers (the *scala vestibuli* and the *scala tympani*). Vibration of the oval window sends a compressional wave down the fluid in the scala vestibuli, around the end of the partition, and back up the scala tympani to the *round window*. This wave sets the *basilar membrane*, located on the cochlear partition, into vibration. The basilar membrane is thinnest and under greatest tension near the oval and round windows; it gradually increases in thickness and decreases in tension toward its other end. High-frequency waves cause the membrane to vibrate with maximum amplitude near its thin, high-tension end; low-frequency waves cause maximum amplitude vibrations near its thicker, lower-tension end. The location of the maximum amplitude vibrations is one way the ear determines frequency; for low-frequency sounds (up to about 1 kHz), the ear sends periodic nerve signals to the brain at the frequency of the sound wave. For complex sounds, which consist of the superposition of many different frequencies (see Section 12.6), the ear performs a spectral analysis—it decomposes the complex sound into its constituent frequencies.

Located on the basilar membrane is the sensory organ (the *organ of Corti*). Rows of hair cells on the basilar membrane excite neurons when they bend in response to vibration. These neurons send electrical signals to the brain.

Figure 12.9 Curves of equal loudness

Loudness

While loudness is most closely correlated to intensity level, it also depends on frequency (as well as other factors). In other words, the sensitivity of the ear is frequency-dependent. Figure 12.9 shows a set of *curves of equal loudness* for a typical person. Each curve shows the intensity levels required so that sounds of different frequencies are equally loud. The curves show that the ear is most sensitive to frequencies between 3 kHz and 4 kHz, partly due to resonance in the auditory canal (see Problem 44). The ear's sensitivity falls off rapidly below 800 Hz and above 10 kHz. At any given frequency between 800 Hz and 10 kHz, the curves are approximately evenly spaced: equal steps in intensity level produce equal steps in loudness, which is why intensity level is often used as an approximate measure of loudness. In this frequency range, 1 dB is about the smallest change in intensity level that is perceptible as a change in loudness.

The threshold of hearing is shown by the lowest curve in the set; a person with excellent hearing cannot hear sounds with intensity levels below this curve. The threshold of hearing is at an intensity level of 0 dB only in the vicinity of 1 kHz.

Pitch

Pitch is the perception of frequency. If you sing or play up and down a scale, it is the pitch that is rising and falling. Although pitch is the aspect of sound perception most closely tied to a single physical quantity, frequency, our sense of pitch is affected to a small extent by other factors such as intensity and timbre (Section 12.6).

Our sense of pitch is a *logarithmic* function of frequency, just as loudness is approximately a logarithmic function of intensity. If you start at the lowest note on the piano (which has a fundamental frequency of 27.5 Hz) and play a chromatic scale—every white and black key in turn—all the way to the highest note (4190 Hz), you hear a series of equal steps in pitch. The frequencies do *not* increase in equal steps; the fundamental frequency of each note is 5.95% higher than the previous note. Under ideal conditions, most people can sense frequency changes as small as 0.3%. A trained musician can sense a frequency change of 0.1% or so.

Localization

How can you tell where a sound comes from? The ear has several different tools it uses to localize sounds:

- The principal method for high-frequency sounds (> 4 kHz) is the difference in intensity sensed by the two ears. The head casts a "sound shadow," so a sound coming from the right has a larger intensity at the right ear than at the left ear.

- The shape of the pinna makes it slightly preferential to sounds coming from the front. This helps with front-back localization for high-frequency sounds.

- For lower-frequency sounds, both the difference in arrival time and the phase difference between the waves arriving at the two ears is used for localization.

12.6   TIMBRE

The sound produced by the vibration of a tuning fork is nearly a pure sinusoid at a single frequency. In contrast, most musical instruments produce complex sounds that are the superposition of many different frequencies. The standing wave on a string or in a column of air is almost always the superposition of many standing wave patterns at different frequencies. The lowest frequency in a complex sound wave is called the fundamental; the rest of the frequencies are called **overtones**. All the overtones of a periodic sound wave have frequencies that are integral multiples of the fundamental; the fundamental and the overtones are then called **harmonics**.

Middle C played on an oboe does not sound the same as middle C played on a trumpet, even though the fundamental frequency is the same, largely because the two instruments produce overtones with different relative amplitudes. What is different about the two sounds is the **tone quality** or **timbre** (pronounced either "tamber" or "timber").

Any periodic wave, no matter how complicated, can be decomposed into a set of harmonics, each of which is a simple sinusoid. The characteristic wave form for a note played on a clarinet, for example, can be decomposed into its harmonic series (Fig. 12.10). This process is called harmonic analysis, or Fourier analysis, in honor of the French mathematician, Jean Baptiste Joseph Fourier (1768–1830), who developed mathematical methods for analyzing periodic functions.

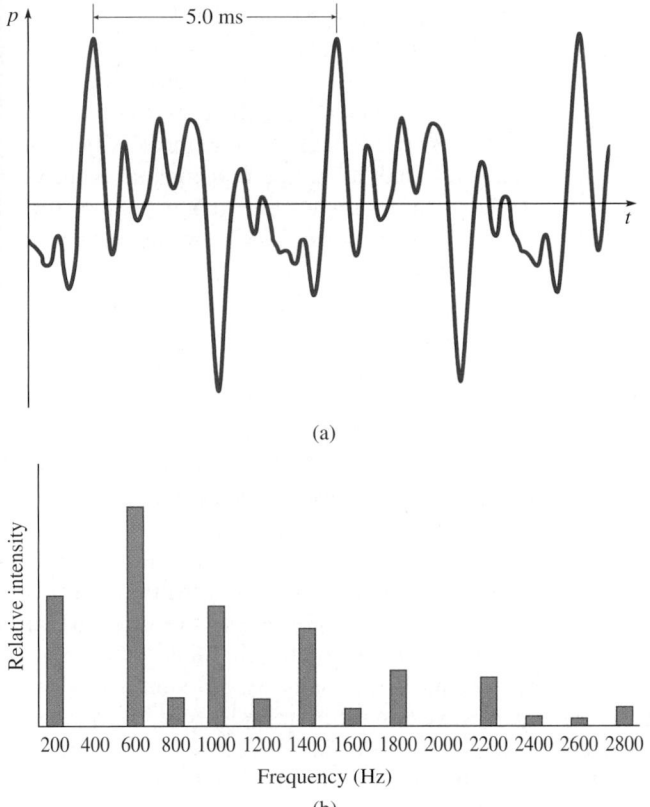

(a)

(b)

Figure 12.10   (a) A graph of the sound wave produced by a clarinet. (b) A bar graph showing the relative intensities of the harmonics, often called the *spectrum*. Notice that *odd* multiples of the fundamental dominate the spectrum. (Data courtesy of P. D. Krasicky, Cornell University.)

Figure 12.11 Complex wave form (bottom wave) composed by superposition of three sinusoidal waves (three upper waves). Note that the fundamental of the sum is not present in any of the constituent waves.

The opposite of harmonic analysis is harmonic synthesis: combining various harmonics to produce a complex wave. Electronic synthesizers can mimic the sounds of various instruments. Realistic-sounding synthesizers must also allow the adjustment of other parameters such as the attack and decay of the sound.

Although the spectrum of a periodic wave consists only of members of a harmonic sequence, not all members of the sequence need be present, not even the fundamental. A wave with three harmonic components having frequencies of 110, 165, and 220 Hz repeats at a frequency of 55 Hz because each of these three frequencies is an integral multiple of 55 Hz (Fig. 12.11). Even though the fundamental is missing—there is no harmonic component at 55 Hz—the ear is clever enough to "reconstruct" a 55-Hz tone. That's why you can listen to and recognize music on an inexpensive radio whose speaker may reproduce only a small range of frequencies.

12.7   BEATS

When two sound waves are close in frequency (within about 15 Hz of each other), the superposition of the two produces a pulsation that we call **beats**. Beats can be produced by any kind of wave; they are a general result of the principle of superposition when applied to two waves of nearly the same frequency.

Beats are caused by the slow change in the phase difference between the two waves. Suppose that at one instant ($t = 0$ in Fig. 12.12), the two waves are in phase with each other and interfere constructively. According to the superposition principle, the resultant amplitude is maximum—the sum of the amplitudes of the two waves shown in Fig. 12.12a, b. However, since the frequencies are different, the waves do not stay in phase. The higher-frequency wave has a shorter cycle, so it gets ahead of the other one. The phase difference between the two steadily increases; as it does, the resultant amplitude decreases. At a later time ($t = 5T_0$), the phase difference reaches 180°; now the waves are half a cycle out of phase and interfere destructively (Fig. 12.12c). Now the

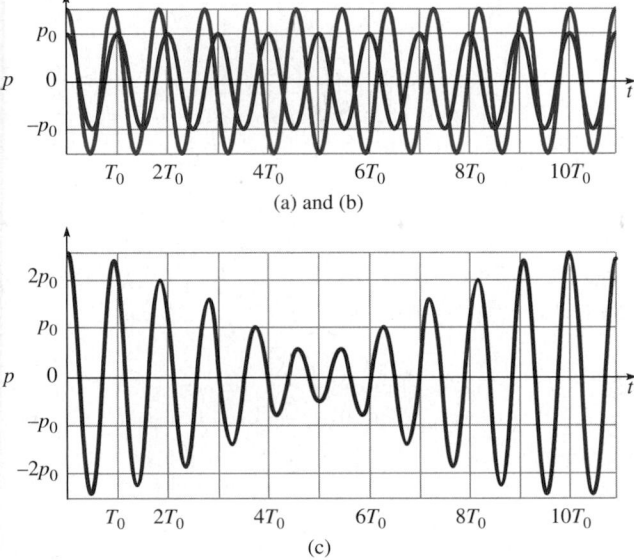

Figure 12.12 (a) Graph (red) of a
sound wave with frequency $f_1 = 1/T_0$ and
amplitude p_0. (b) Graph (blue) of a second sound wave with frequency $f_2 = 1.1f_1$
and amplitude $1.5p_0$. (c) The superposition of the two has maximum amplitude
$2.5p_0$ and minimum amplitude $0.5p_0$.

resultant amplitude is minimum—the difference between the amplitudes of the two waves. As the phase difference continues to increase, the amplitude increases until constructive interference occurs again ($t = 10T_0$). The ear perceives the amplitude (and intensity) cycling from large to small to large to small as a pulsation in loudness.

At what frequency do the beats occur? It depends on how far apart the frequencies of the two waves are. We can measure the time between beats T_{beat} as the time to go from constructive interference to the next occurrence of constructive interference. During that time, each wave must go through a whole number of cycles, with one of them going through one more cycle than the other. Since frequency (f) is the number of cycles per second, the number of cycles a wave goes through during a time T_{beat} is fT_{beat}. (To illustrate: in Fig. 12.12, $T_{beat} = 10T_0$. During that time, wave 1 goes through $f_1T_{beat} = 10$ cycles, while wave 2 goes through $f_2T_{beat} = 1.1/T_0 \times 10T_0 = 11$ cycles.) If $f_2 > f_1$, then wave 2 goes through one cycle more than wave 1:

$$f_2T_{beat} - f_1T_{beat} = (\Delta f)T_{beat} = 1$$

The beat frequency f_{beat} is $1/T_{beat}$:

$$f_{beat} = 1/T_{beat} = \Delta f \tag{12-11}$$

Thus, we obtain the remarkably simple result that the beat frequency is the difference between the frequencies of the two waves. If the difference in frequencies exceeds roughly 15 Hz, then the ear no longer perceives a pulsating intensity; instead, we hear two tones at different pitches.

Piano tuners listen for beats as they tune. The tuner sounds two strings and listens for the beats. The beat frequency indicates whether the interval is correct or not. If the two strings are played by the same key, they are tuned to the same fundamental frequency, so the beat frequency should be (nearly) zero. If the two strings belong to two different notes, the beat frequency is nonzero. Actually, in this case the tuner listens to beats between two *overtones* that are close in frequency, not the fundamentals. The fundamental frequencies are too far apart for the beats to be perceptible.

Example 12.7

The Piano Tuner

A piano tuner strikes his tuning fork ($f = 523.3$ Hz) and strikes a key on the piano at the same time. The two have nearly the same frequency; he hears 3.0 beats per second. As he tightens the piano string, he hears the beat frequency gradually decrease to 2.0 beats per second when the two sound together. (a) What was the frequency of the piano string before it was tightened? (b) By what percentage did the tension increase?

Strategy   The beat frequency is the difference between the two frequencies; we only have to determine which is higher. The wavelength of the string is determined by its length, which does not change. The increase in tension increases the speed of waves on the string, which in turn increases the frequency.

Solution   (a) Since the piano tuner heard 3.0 beats per second, the difference in the two frequencies was 3.0 Hz:

$$\Delta f = 3.0 \text{ Hz}$$

Is the piano string's frequency 3.0 Hz higher or 3.0 Hz lower than the tuning fork's frequency? As the tension increases gradually, the beat frequency decreases, which means that the frequency of the piano string is getting *closer* to the frequency of the tuning fork. Therefore, the string frequency must be 3.0 Hz *lower* than the tuning fork frequency:

$$f_{\text{string}} = 523.3 \text{ Hz} - 3.0 \text{ Hz} = 520.3 \text{ Hz}$$

(b) The tension (F) is related to the speed of the wave on the string (v) and the mass per unit length (μ) by

$$v = \sqrt{\frac{F}{\mu}} \qquad (11\text{-}4)$$

The mass per unit length does not change, so $v \propto \sqrt{F}$. The speed of the wave on the string is related to its wavelength and frequency by

$$v = \lambda f$$

⚠ The wavelength λ in this expression is the wavelength of the transverse wave on the string, *not* the wavelength of the sound wave in air. Since λ does not change, $v \propto f$. Therefore, $f \propto \sqrt{F}$ or

$$F \propto f^2$$

This means that the ratio of the tension F to the original tension F_0 is equal to the ratio of the frequencies squared:

$$\frac{F}{F_0} = \left(\frac{f}{f_0}\right)^2 = \left(\frac{521.3 \text{ Hz}}{520.3 \text{ Hz}}\right)^2 = 1.004$$

The tension was increased 0.4%.

Discussion   We needed to find whether the original frequency was too high or too low. As the beat frequency decreases, the frequency of the string is getting closer to the frequency of the tuning fork. Tightening the string makes the string's frequency increase; since increasing the string's frequency brings it closer to the tuning fork's frequency, we know that the original frequency of the string was lower than the frequency of the tuning fork. Had an increase in tension *increased* the beat frequency instead, we would know that the original frequency was already too high; the tension would have to be relaxed to tune the string.

Practice Problem 12.7   Tuning a violin

A tuning fork with a frequency of 440.0 Hz produces 4.0 beats per second when sounded together with a violin string of nearly the same frequency. What is the frequency of the string if a slight increase in tension increases the beat frequency?

12.8   THE DOPPLER EFFECT

A police car races by, its sirens screaming. As it passes, we hear the pitch change from higher to lower. The frequency change is called the **Doppler effect**, after the Austrian physicist Christian Andreas Doppler (1803–1853). The observed frequency is different from the frequency transmitted by the source when one or both are in motion relative to the wave medium.

We consider only the motion of the source and observer directly toward or away from each other in the reference frame in which the wave medium is at rest. Velocities of the source and observer are expressed as components along the direction of propagation of the sound wave (from source to observer). A positive component means the velocity is in the direction of propagation of the wave, while a negative component means the velocity is opposite the direction of propagation.

Moving Source

First we consider a moving source. A source emits a sound wave at frequency f_s, which means that wave crests (regions of maximum amplitude, indicated by circles in Fig. 12.13) leave the source spaced by a time interval $T_s = 1/f_s$. If the source is moving at velocity v_s

Figure 12.13 (a) A steamboat is moving to the right at speed v_s while it blows its whistle. The whistle emits wave crests at positions 1, 2, 3, 4, and 5; each wave crest moves outward in all directions, from the point at which it was emitted, at speed v. (b) The observed wavelength λ_o is the distance between wave crests.

toward a stationary observer on the right, Fig. 12.13a shows that the wavelength—the distance between crests—is smaller in front of the source and larger behind the source. In Fig. 12.13b, at the instant that crest 6 is emitted, crest 5 has traveled outward a distance vT_s from point 5, where v is the speed of sound. During the same time interval, the source has advanced a distance v_sT_s. The wavelength λ_o as measured by the observer on the right is the distance between crests 5 and 6:

$$\lambda_o = vT_s - v_sT_s$$

The frequency at which the crests arrive at the observer is the *observed* wave frequency f_o. The observed period T_o between the arrival of two crests is the time it takes sound to travel a distance $(v - v_s)T_s$:

$$T_o = \frac{(v - v_s)T_s}{v}$$

The observed frequency is

$$f_o = \frac{1}{T_o} = \frac{v}{(v - v_s)} \times \frac{1}{T_s}$$

Dividing numerator and denominator by v and substituting $f_s = 1/T_s$ yields

---

Doppler effect (moving source)

$$f_o = \left(\frac{1}{1 - v_s/v}\right)f_s \qquad (12\text{-}12)$$

Since the denominator $1 - v_s/v$ is less than 1, the observed frequency is higher than the source frequency when the source moves in the same direction as the wave (toward the observer). If the source instead moves *away* from the observer, the correct observed frequency is given by Eq. (12-12) as long as we make v_s negative (the source moves opposite the direction of the wave). With v_s negative, $1 - v_s/v$ is *greater* than 1, so the observed frequency is *less* than the source frequency.

Moving Observer

Now we consider motion of the observer. A stationary source emits a sound wave at frequency f_s and wavelength $\lambda_s = v/f_s$, where v is the speed of sound. A stationary observer would measure the arrival of wave crests spaced by a time interval $T_s = 1/f_s$. An observer moving away from the source at velocity v_o would observe a longer time interval between crests. Just as crest 1 reaches the observer, the next (crest 2) is a distance λ_s away. Crest 2 catches up with the observer at a time T_o later when the distance the wave crest travels toward the observer is equal to the distance the observer travels away from the wave crest plus the wavelength (Fig. 12.14):

$$vT_o = v_oT_o + \lambda_s \quad \text{or} \quad (v - v_o)T_o = \lambda_s = v/f_s$$

Solving for T_o,

$$T_o = \frac{v/f_s}{v - v_o}$$

The observed frequency is

$$f_o = \frac{1}{T_o} = \frac{v - v_o}{v}f_s$$

Dividing numerator and denominator by v yields

Doppler effect (moving observer)

$$f_o = \left(1 - \frac{v_o}{v}\right)f_s \qquad (12\text{-}13)$$

An observer moving away from the source measures a frequency lower than f_s. An observer moving *toward* the source moves opposite to the direction of the wave; in that case, v_o is negative and the observed frequency is *higher* than f_s.

Example 12.8

Train Whistle and Doppler Shift

A monorail train approaches a platform at a speed of 10.0 m/s while it blows its whistle. A musician with perfect pitch standing on the platform hears the whistle as "middle C," a frequency of 261 Hz. There is no wind and the temperature is a chilly 0°C. What is the observed frequency of the whistle when the train is at rest?

Strategy   In this case, the source—the whistle—is moving and the observer is stationary. The source is moving *toward* the observer, so v_s is *positive*. With the source approaching the observer, the observed frequency is higher than the source frequency. When the train is at rest, there is no Doppler shift; the observed frequency then is equal to the source frequency.

Solution   For a moving source, the source (f_s) and observed (f_o) frequencies are related by

$$f_o = \left(\frac{1}{1 - v_s/v}\right)f_s$$

where $v = 331$ m/s (the speed of sound in air at 0°C), $v_s = +10.0$ m/s, and $f_o = 261$ Hz. Solving for f_s,

$$f_s = (1 - v_s/v)f_o$$

$$= \left(1 - \frac{10.0 \text{ m/s}}{331 \text{ m/s}}\right) \times 261 \text{ Hz}$$

$$= 253 \text{ Hz}$$

The source frequency is less than the observed frequency, as expected. The observed frequency when the train is at rest is equal to the source frequency: 253 Hz.

continued on next page

Example 12.8 *continued*

Discussion When the train is moving toward the platform, the distance between source and observer is decreasing. Wave crests emitted later take *less time* to reach the observer than if the train were at rest, so the time between arrival of wave crests is smaller than if the train were stationary. When the distance between source and observer is decreasing, the observed frequency is higher than the source frequency; when the distance is increasing, the observed frequency is lower than the source frequency.

Practice Problem 12.8 A sports car racing by

Justine is gardening in her front yard when a Mazda Miata races by at 32.0 m/s (71.6 mi/h). If she hears the sound of the Miata's engine at 220.0 Hz as it approaches her, what frequency does she hear after it passes? Assume the temperature is 20°C and there is no wind.

Motion of Both Source and Observer

If both source and observer are moving, we combine the two Doppler shifts (see Conceptual Question 11) to obtain

$$f_o = \left(\frac{1 - v_o/v}{1 - v_s/v} \right) f_s \qquad (12\text{-}14)$$

Remember that the signs of v_o and v_s are positive for motion in the direction of propagation of the wave and negative for motion opposite the direction of propagation.

Figure 12.14 An observer moving away from a stationary source. The observed frequency is lower than the source frequency.

Example 12.9

Determining Speed from Horn Frequency

Two cars, with equal ground speeds, are moving in opposite directions away from each other on a straight highway. One driver blows a horn with a frequency of 111 Hz; the other measures the frequency as 105 Hz. If the speed of sound is 338 m/s and there is no wind, what is the ground speed of each car?

Strategy   A sound wave travels from source to observer. The source moves opposite the direction of the wave, so v_s is negative. The observer moves in the direction of the wave, so v_o is positive. The speeds are the same, so $v_s = -v_o$.

Solution   With both the source and observer moving, the frequencies are related by

$$f_o = \left(\frac{1 - v_o/v}{1 - v_s/v}\right) f_s$$

To simplify the algebra, we let $\alpha = v_o/v = -v_s/v$. Then

$$f_o = \left(\frac{1 - \alpha}{1 + \alpha}\right) f_s$$

Now we solve for α:

$$(1 + \alpha)\frac{f_o}{f_s} = 1 - \alpha$$

$$\alpha = \frac{1 - f_o/f_s}{1 + f_o/f_s} = \frac{1 - (105\ \text{Hz}/111\ \text{Hz})}{1 + (105\ \text{Hz}/111\ \text{Hz})} = 0.02777$$

Now we can find v_o:

$$v_o = \alpha v = 0.02777 \times 338\ \text{m/s} = 9.4\ \text{m/s}$$

The speed of each car is 9.4 m/s.

Discussion   Quick check on the algebra: substituting $v = 338$ m/s, $f_s = 111$ Hz, $v_o = 9.4$ m/s, and $v_s = -9.4$ m/s directly into Eq. (12-14),

$$f_o = \frac{1 - (9.4\ \text{m/s})/(338\ \text{m/s})}{1 - (-9.4\ \text{m/s})/(338\ \text{m/s})} \times 111\ \text{Hz} = 105\ \text{Hz}$$

Practice Problem 12.9   Finding speed from the Doppler shift

A car is driving due west at 15 m/s and sounds its horn with a frequency of 260.0 Hz. A passenger in a car heading east away from the first car hears the horn at a frequency of 230.0 Hz. How fast is the second car traveling? The speed of sound is 350 m/s.

12.9   SHOCK WAVES

Let's examine two interesting special cases of the Doppler formula [Eq. (12-14)]. First, what if the observer moves away from the source at the speed of sound ($v_o = v$)? The Doppler-shifted frequency would be zero according to Eq. (12-14). What does that mean? If the observer moves away from the source with a speed equal to (or greater than) the wave speed, the wave crests *never reach the observer*.

Second, what if the source moves toward the observer at a speed approaching the speed of sound ($v_s \rightarrow v$)? Then Eq. (12-14) gives an observed frequency that increases without bound ($f_o \rightarrow \infty$). Figure 12.15 helps us understand what that means. For a plane moving slower than sound, the wave crests in front of it are closer together due to the plane's motion (Fig. 12.15a). An observer to the right would measure a frequency higher than the source frequency. As the plane's speed increases, the wave crests in front of it get closer and closer together and the observed frequency increases. For a plane moving at the speed of sound (Fig. 12.15b), the wave crests pile up on top of each other; they move to the right at the same speed as the plane, so they can't get ahead of it. This wall of high-pressure air is called the *sound barrier*. An observer to the right

Making The Connection:
shock wave of a supersonic plane

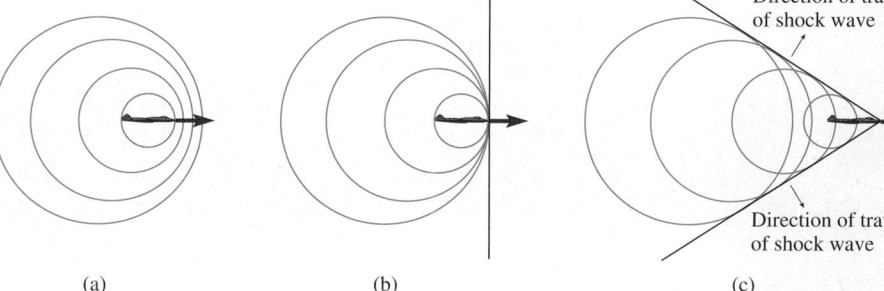

Figure 12.15   (a) Wave crests for a plane moving slower than sound. (b) A plane moving at the speed of sound; the wave crests pile up on each other since the plane moves to the right as fast as the wave crests. (c) Shock wave for a supersonic plane.

Direction of travel of shock wave

Direction of travel of shock wave

(a)   (b)   (c)

would measure a wavelength of zero—zero distance between wave crests—and therefore an infinite frequency.

What happens if the source moves at a speed *greater than* the speed of sound? Figure 12.15c shows that the wave crests pile up on top of one another to form cone-shaped *shock waves*, which travel outward in the direction indicated. There are two principal shock waves formed, one starting at the nose of the plane and one at the tail. The sound of a shock wave is referred to as a *sonic boom*.

The pressure variation shown in Fig. 12.16 is the sonic boom caused by an airplane flying at a constant velocity faster than sound. The variation is called an *N-wave* (because the pressure graph is shaped like the letter N). The initial, sudden rise in pressure is the shock wave formed at the nose of the plane. Then the pressure decreases linearly, becoming less than atmospheric, until it abruptly returns to atmospheric as the tail wave passes. The duration of the sonic boom ranges from about 0.1 s for small military planes up to about 0.5 s for larger craft such as the Concorde or the Space Shuttle.

The shock cone spreads out as it moves away from the plane; the higher the plane, the lower the intensity of the sonic boom. The largest intensity boom occurs directly under the flight path, but the boom is heard throughout the cone area. The sudden onset and release of pressure can even do physical damage to something under its path, breaking a window or shaking a building. It is not that the magnitude of the pressure change is so large, but the change happens in a very short time interval. The overpressure for supersonic planes flying in normal conditions ranges from 50 to 500 Pa.

The speed of a supersonic plane is often given as a *Mach number*, named for Austrian physicist Ernst Mach (1838–1916). The Mach number is the ratio of the speed of the plane to the speed of sound. A plane flying at Mach 3.3 where the air temperature is 11°C is moving at 3.3×338 m/s = 1100 m/s with respect to the air.

The pressure variation for the Concorde is 93 Pa above and below atmospheric pressure at a speed of Mach 2 and at a height of 15.8 km; the noise level heard from the Concorde's boom is usually about 120 dB directly under the flight path.

When the Space Shuttle lands, two "crack" sounds can easily be heard, due to the two shock waves created by the nose and tail. The pressure variation for the Space Shuttle is 60 Pa above and below atmospheric pressure at Mach 1.5 when it is making a landing approach at a height of 18 km. With small fighter planes, the two booms are usually heard as a single boom; there is not enough time between them for two separate sounds to be distinguished by the ear.

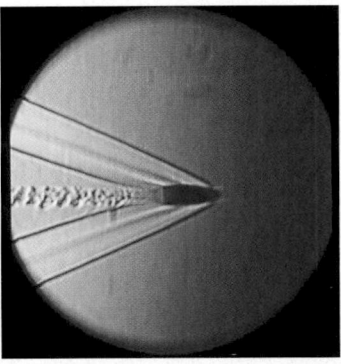

A shock wave is formed when a bullet moves through air faster than sound.

Breaking the sound barrier.

Physics at Home

You can make a visible shock wave by trailing your finger along the surface of the water in a sink or tub. If your finger pushes the water faster than water waves travel, water piles up in front of your finger and forms a V-shaped shock wave. See if you can approximate the case of a plane moving at the speed of sound with rounded waves moving outward from your finger (Fig. 12.15b) instead of a V-shaped wave. The next time you are in a boat, notice the V-shaped bow wave that extends from the prow of the boat when you are moving faster than the speed of water waves.

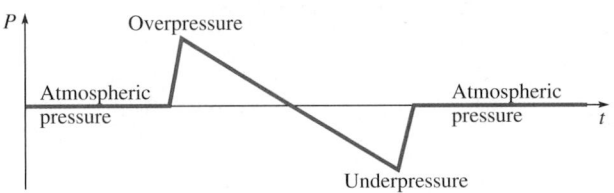

Figure 12.16 The variation in pressure at a point on the ground as a supersonic plane flies overhead.

Making The Connection:
echolocation of bats
and dolphins

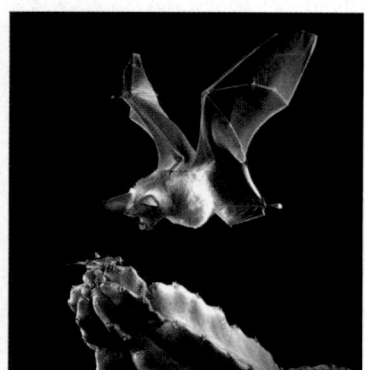

12.10 ECHOLOCATION AND MEDICAL IMAGING

Bats, dolphins, whales, and some birds use *echolocation* to locate prey and to "see" their environment. To find their way around in the darkness of caves, oilbirds of northern South America and cave swiftlets of Borneo and East Asia emit sound waves and listen for the echoes. The time it takes for the echoes to return tells them how far they are from an obstacle or cave wall. Differences between the echoes that reach the two sides of the head provide information on the direction from which the echo comes.

The sounds used by oilbirds and cave swiftlets for echolocation are audible to humans, but dolphins, whales, and most bats use ultrasound (20 to 200 kHz) instead. Bats and dolphins can also determine an object's velocity by sensing the Doppler shift between the emitted and reflected waves—a clear advantage in locating prey that are darting around to avoid being eaten. Some horseshoe bats can detect frequency differences as small as 0.1 Hz.

Prey are not completely helpless. Moths, lacewings, and praying mantises have primitive ears containing a few nerve cells to detect the ultrasound emitted by a nearby bat. A group of moths fluttering about at some distance from a cave may, for no apparent reason, fold their wings and drop suddenly to the ground. Folding their wings both reduces the amount of reflected sound and helps them drop quickly to the ground to evade the swooping bat. The moths' bodies are furry rather than smooth to help absorb some of the sound waves and thus reduce the intensity of reflected sound.

When the tiger moth detects the ultrasound from a bat, it emits its own ultrasound by flexing a part of its exoskeleton. The extra sounds mixed in with the echoes tend to confuse the bat, perhaps encouraging it to hunt elsewhere.

Echolocation is a useful navigational tool for seafarers. To find the depth of water below a ship, a *sonar* (**so**und **na**vigation and **r**anging) device sends out ultrasonic pulses (Fig. 12.17). The time delay Δt between an emitted ultrasonic pulse and the return of its reflection is used to determine the distance to the seafloor. Some autofocusing cameras use a similar system. The U.S. Navy has developed a low-frequency active (LFA) sonar for long-range detection of submarines using audible sound (100 to 500 Hz) rather than ultrasound. Seismic P waves—sound waves traveling through the Earth—generated by explosions or air guns are used to study the interior structure of Earth and to find oil beneath the surface.

Radar is a form of echolocation that uses electromagnetic waves instead of sound waves, but otherwise the concept is similar. Weather forecasting relies on *Doppler radar* to show not only the location of a storm, but also the wind velocity.

Making The Connection:
operation of sonar

Figure 12.17 A ship with a sonar device to locate the depth of the seafloor; an ultrasound pulse, sent out from the ship by a transmitter, is reflected from the sandy sea bottom and detected by a receiver on the ship.

Medical Applications of Ultrasound

 Making The Connection: medical applications of ultrasound, ultrasonic imaging

Millions of expectant parents see their unborn child for the first time when the mother has an ultrasonic examination. Ultrasonic imaging uses a pulse-echo technique similar to that used by bats and in sonar. Pulses of ultrasound are reflected at boundaries between different types of tissue.

In the early stages of pregnancy (10th to 14th weeks), the scan is used to verify that the fetus is alive and to check for twins. The length of the fetus is measured to help determine the due date more accurately. Some abnormalities can be discovered even at this early stage. For example, some chromosomal abnormalities can be detected by measuring the thickness of the skin at the back of the neck. After the 18th week, the fetus can be examined in even more detail. The major organs are examined to be sure they are developing normally. After the 30th week, the flow of blood in the umbilical cord is checked to ensure that oxygen and nutrients reach the fetus. The position of the placenta is also checked.

Why are sound waves used rather than, say, electromagnetic waves such as x-rays? X-ray radiation is damaging to tissue—especially to rapidly growing fetal tissue. After decades of use, ultrasound has no known adverse effects. In addition, ultrasound images are captured in real time, so they are available immediately and can show movement. A third reason is that regular x-rays detect the amount of radiation that passes through tissue, but cannot resolve details at different depths, and so cannot produce an image of a "slice" of the abdomen; a more complicated and expensive diagnostic tool such as a CAT scan (Computed Axial Tomography) would be required to resolve details at different depths. Fourth, some kinds of tissue are not detected well by x-rays but are clearly resolved in ultrasound.

Why is ultrasound used rather than sound waves of audible frequencies or lower? Sound waves with high frequencies have small wavelengths. Waves with small wavelengths diffract less around the same obstacle than do waves with larger wavelengths (see Section 11.9). Too much diffraction would obscure details in the image. As a rough rule of thumb, the wavelength is a lower limit on the smallest detail that can be resolved. The frequencies used in imaging are typically in the range 1 to 15 MHz, which means that the wavelengths in human tissue are in the range 0.1 to 1.5 mm. As a comparison, if sound waves at 15 kHz were used, the wavelength inside the body would be 10 cm. Higher frequencies give better resolution but at the expense of less penetration; sound waves are absorbed within a distance of about 500λ in tissue.

The medical applications of ultrasonic imaging are not limited to prenatal care. Ultrasound is also used to examine organs such as the heart, liver, gallbladder, kidneys, bladder, breasts, and eyes, and to locate tumors. It can be used to diagnose various heart conditions and to assess damage after a heart attack (Fig. 12.18). Ultrasound can show

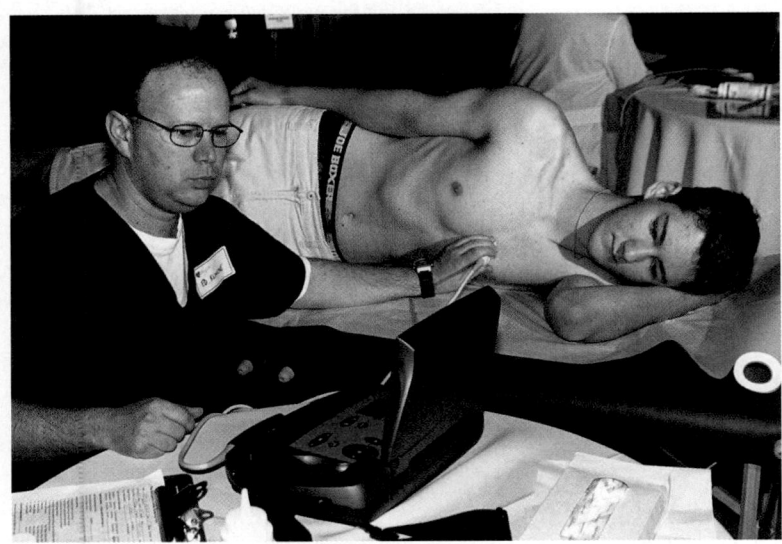

Figure 12.18 Ultrasonic imaging is used to diagnose heart disease.

movement, so it is used to assess heart valve function and to monitor blood flow in large blood vessels. Because ultrasound provides real-time images, it is sometimes used to guide procedures such as biopsies, in which a needle is used to take a sample from an organ or tumor for testing.

Doppler ultrasound is a newer technique that is used to examine blood flow. It can help reveal blockages to blood flow, show the formation of plaque in arteries, and provide detailed information on the heartbeat of the fetus during labor and delivery. The Doppler-shifted reflections interfere with the emitted ultrasound, producing beats. The beat frequency is proportional to the speed of the reflecting object (see Problem 41).

MASTER THE CONCEPTS

Summary

- A sound wave can be described either by the gauge pressure p, which measures the pressure fluctuations above and below the ambient atmospheric pressure, or by the displacement s of each point in the medium from its undisturbed position.

- Humans with excellent hearing can hear frequencies from 20 Hz to 20 kHz. The terms infrasound and ultrasound are used to describe sound waves with frequencies below 20 Hz and above 20 kHz, respectively.

- The speed of sound in a fluid is

$$v = \sqrt{\frac{B}{\rho}} \qquad (12\text{-}1)$$

- The speed of sound in an ideal gas at any absolute temperature T can be found if it is known at one temperature:

$$v = v_0 \sqrt{\frac{T}{T_0}} \qquad (12\text{-}3)$$

where the speed of sound at absolute temperature T_0 is v_0.

- The speed of sound in air at 0°C (or 273 K) is 331 m/s.

- For sound waves traveling along the length of a thin solid rod, the speed is approximately

$$v = \sqrt{\frac{Y}{\rho}} \quad \text{(thin solid rod)} \qquad (12\text{-}5)$$

- The pressure amplitude of a sound wave is proportional to the displacement amplitude. For a harmonic sound wave at angular frequency ω,

$$p_0 = \omega v \rho s_0 \qquad (12\text{-}6)$$

where v is the speed of sound and ρ is the mass density of the medium.

- The intensity of a sound wave is related to the pressure amplitude as follows:

$$I = \frac{p_0^2}{2\rho v} \qquad (12\text{-}7)$$

where ρ is the mass density of the medium and v is the speed of sound in that medium. The most important thing to remember is that *intensity is proportional to amplitude squared*, which is true for all waves, not just sound.

- Sound intensity level in decibels is

$$\beta = (10 \text{ dB}) \log_{10} \frac{I}{I_0} \qquad (12\text{-}8)$$

where $I_0 = 10^{-12} \text{ W/m}^2$. Sound intensity level is useful since it roughly corresponds to the way we perceive loudness.

Equal increments in intensity level roughly correspond to equal increases in loudness.

- In a standing sound wave in a thin pipe, an open end is a pressure node and a displacement antinode; a closed end is a pressure antinode and a displacement node.

For a pipe open at both ends,

$$\lambda_n = \frac{2L}{n} \qquad (11\text{-}12)$$

$$f_n = n \frac{v}{2L} = n f_1 \qquad (11\text{-}13)$$

where $n = 1, 2, 3, \ldots$.

For a pipe closed at one end,

$$\lambda_n = \frac{4L}{n} \qquad (12\text{-}10a)$$

$$f_n = n \frac{v}{4L} = n f_1 \qquad (12\text{-}10b)$$

where $n = 1, 3, 5, 7, \ldots$.

- When two sound waves are close in frequency, the superposition of the two produces a pulsation called *beats*.

$$f_{\text{beat}} = \Delta f \qquad (12\text{-}11)$$

- Doppler effect: if v_s and v_o are the velocities of the source and observer, the observed frequency is

$$f_o = \left(\frac{1 - v_o/v}{1 - v_s/v} \right) f_s \qquad (12\text{-}14)$$

where v_s and v_o are positive in the direction of propagation of the wave and the wave medium is at rest.

Highlighted Figures and Tables

F12.1 Sound waves (p. 407)

F12.2 Graphs of the pressure variation and the displacement of air elements for a sound wave generated by a loudspeaker (p. 408)

T12.1 Speed of sound in various materials (p. 409)

T12.2 Pressure amplitudes, intensities, and intensity levels of a range of sounds in air at 20°C (p. 410)

F12.3 Standing waves in a pipe open at both ends (p. 415)

F12.5 Standing waves in a pipe closed at one end (p. 416)

F12.7 Standing wave pattern showing pressure nodes and antinodes and displacement nodes and antinodes (p. 418)

MASTER THE CONCEPTS *continued*

F12.12 Beats are caused by the superposition of two waves with different frequencies (p. 423)

F12.13 Doppler effect for a moving source (p. 425)
F12.14 Doppler effect for a moving observer (p. 427)

CONCEPTUAL QUESTIONS

1. Explain why the pitch of a bassoon is more sensitive to a change in air temperature than the pitch of a cello. (That's why wind players keep blowing air through the instrument to keep it in tune.)

2. On a warm day, a piano is tuned to match an organ in an auditorium. Will the piano still be in tune with the organ the next morning, when the room is cold? If not, will the organ be higher or lower in pitch than the piano? (Assume that the piano's tuning doesn't change. Why is that a reasonable assumption?)

3. Many real estate agents have an ultrasonic rangefinder that enables them to quickly and easily measure the dimensions of a room. The device is held to one wall and reads the distance to the opposite wall. How does it work?

4. For high-frequency sounds, the ear's principal method of localization is the difference in intensity sensed by the two ears. Why can't the ear reliably use this method for low-frequency sounds? Doesn't the head cast a "sound shadow" regardless of the frequency? Explain. [*Hint:* Consider diffraction of sound waves around the head.]

5. For low-frequency sounds, the ear uses the phase difference between the sound waves arriving at the two ears to determine direction. Why can't the ear reliably use phase difference for high-frequency sounds? Explain.

6. What would happen if you tried to use an underwater autofocusing camera to take a picture in air? Would the picture be in focus? If not, would the camera focus too far away or too close?

7. Why do autofocusing cameras that use ultrasonic rangefinders focus incorrectly if a picture is taken through glass (such as through a window)?

8. Why is it that your own voice sounds strange to you when you hear it played back on a tape recorder, but your friends all agree that it is just what your voice sounds like? [*Hint:* Consider the media through which the sound wave travels when you usually hear your own voice.]

9. What is the purpose of the gel that is spread over the skin before an ultrasonic imaging procedure? [*Hint:* The speed of sound in the gel is similar to the speed in the body, while the speed in air is much slower. What happens to a wave at an abrupt change in wave speed?]

10. A stereo system whose amplifier can produce 60 watts per channel is replaced by one rated 120 watts per channel. Would you expect the new stereo to be able to play twice as loudly as the old one? Explain.

11. A moving source emits a sound wave that is heard by a moving observer. Imagine a thin wall at rest between the source and observer. The wall completely absorbs the sound and instantaneously emits an *identical* sound wave. Use this scenario to explain why we can combine the Doppler shifts due to motion of the source and observer as in Eq. (12-14). [*Hint:* What is the net effect of this imaginary wall?]

12. Explain why the displacement of air elements at condensations and rarefactions is zero.

13. Why is the speed of sound in solids generally much faster than the speed of sound in air?

14. If the pressure amplitude of a sound wave is doubled, what happens to the displacement amplitude, the intensity, and the intensity level?

15. The source and observer of a sound wave are both at rest with respect to the ground. The wind blows in the direction from source to observer. Is the observed frequency Doppler-shifted? Explain.

16. Many brass instruments have valves that increase the total length of the pipe from mouthpiece to bell. When a valve is depressed, is the fundamental frequency raised or lowered? What happens to the pitch?

17. When the viola section of an orchestra with six members plays together, is the sound 6 times as loud as when a single viola plays? Explain. Is the intensity 6 times what it would be for a single viola? [*Hint:* The six sound waves are not coherent.]

18. The fundamental frequency of the highest note on the piano is 4.186 kHz. Most musical instruments do not go that high; only a few singers can produce sounds with fundamental frequencies higher than around 1 kHz. Yet a good-quality stereo system must reproduce frequencies up to at least 16 to 18 kHz. Explain.

MULTIPLE CHOICE QUESTIONS

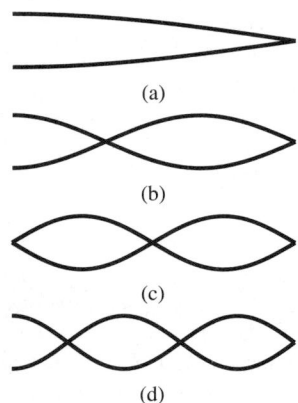

(a)

(b)

(c)

(d)

Figure 12.19 Multiple Choice Questions 1 and 2

1. An organ pipe is closed at one end. Several standing wave patterns are sketched in Fig. 12.19. Which one is *not* a possible standing wave pattern for this pipe?

2. Of the standing wave patterns sketched in Fig. 12.19, which shows the lowest frequency standing wave for an organ pipe closed at one end?

3. The speed of sound in water is 4.3 times the speed of sound in air. A whistle on land produces a sound wave with a frequency f_0. When this sound wave enters the water, its frequency becomes

   (a) $4.3f_0$ (b) f_0 (c) $\dfrac{f_0}{4.3}$

   (d) not enough information given

4. The intensity of a sound wave is directly proportional to

   (a) the frequency.   (b) the amplitude.
   (c) the square of the amplitude.
   (d) the square of the speed of sound.
   (e) none of the above.

5. The fundamental frequency of a pipe closed at one end is f_1. How many nodes are present in a standing wave of frequency $9f_1$?

   (a) 4   (b) 5   (c) 6   (d) 8   (e) 9   (f) 10

6. The length of a pipe closed at one end is L. In the standing wave whose frequency is 7 times the fundamental frequency, what is the closest distance between nodes?

   (a) $\frac{1}{14}L$ (b) $\frac{1}{7}L$ (c) $\frac{2}{7}L$ (d) $\frac{4}{7}L$ (e) $\frac{8}{7}L$
   (f) none of the above

7. The three lowest resonant frequencies of a system are 50 Hz, 150 Hz, and 250 Hz. The system could be

   (a) a tube of air closed at both ends.
   (b) a tube of air open at one end.
   (c) a tube of air open at both ends.
   (d) a vibrating string with fixed ends.

8. A source of sound with frequency 620 Hz is placed on a moving platform that approaches a physics student at speed v; the student hears sound with a frequency f_1. Then the source of sound is held stationary while the student approaches it at the same speed v; the student hears sound with a frequency f_2. Choose the correct statement.

   (a) $f_1 = f_2$; both are greater than 620 Hz.
   (b) $f_1 = f_2$; both are less than 620 Hz.
   (c) $f_1 > f_2 > 620$ Hz.
   (d) $f_2 > f_1 > 620$ Hz.

9. A moving van and a small car are traveling in the same direction on a two-lane road. The van is moving at twice the speed of the car and overtakes the car. The driver of the car sounds his horn, frequency = 440 Hz, to signal the van that it is safe to return to the lane. Which is the correct statement?

   (a) The car driver and van driver both hear the horn frequency as 440 Hz.
   (b) The car driver hears 440 Hz, but the van driver hears a lower frequency.
   (c) The car driver hears 440 Hz, but the van driver hears a higher frequency.
   (d) Both drivers hear the same frequency and it is lower than 440 Hz.

10. A trombone and a bassoon play notes of equal loudness with the same fundamental frequency. The two sounds differ primarily in

    (a) pitch.
    (b) intensity level.
    (c) amplitude.
    (d) timbre.
    (e) wavelength.

PROBLEMS

Note: **ⓒ** indicates a combination conceptual/quantitative problem. Gold diamonds ✦, ✦✦ are used to indicate the increasing level of difficulty of each problem. Problem numbers appearing in blue, 9., denote problems that have a detailed solution available in the Student Solutions Manual. Some problems are *paired* by concept; their numbers are connected by a ruled box.

Note: Assume a temperature of 20.0°C in all problems unless otherwise indicated.

12.2 The Speed of Sound Waves

1. Bats emit ultrasonic waves with a frequency as high as 1.0×10^5 Hz. What is the wavelength of such a wave in air of temperature 15°C?

2. Dolphins emit ultrasonic waves with a frequency as high as 2.5×10^5 Hz. What is the wavelength of such a wave in seawater at 25°C?

3. At a baseball game, a spectator is 60.0 m away from the batter. How long does it take the sound of the bat connecting with the ball to travel to the spectator's ears? The air temperature is 27.0°C.

✦ 4. (a) Show that since the bulk modulus has SI units N/m² and linear mass density has SI units kg/m³, Eq. (12-1) gives the speed of sound in m/s. Thus, the equation is dimensionally consistent. (b) Show that no other combination of B and ρ can give dimensions of speed. Thus, Eq. (12-1) *must* be correct except for the possibility of a dimensionless constant.

5. Derive Eq. (12-4) as: (a) Starting with Eq. (12-3), substitute $T = T_C + 273.15$. (b) Apply the binomial approximation to the square root (see Appendix A.5) and simplify.

6. Find the speed of sound in mercury, which has a bulk modulus of 2.8×10^{10} Pa and a density of 1.36×10^4 kg/m³.

7. A copper alloy has a Young's modulus of 1.1×10^{11} Pa and a density of 8.92×10^3 kg/m³. What is the speed of sound in a thin rod made from this alloy? Compare your result with that given in Table 12.1.

8. During a thunderstorm, you can easily estimate your distance from a lightning strike. Count the number of seconds that elapse from when you see the flash of lightning to when you hear the thunder. The rule of thumb is that 5 seconds elapse for each mile of distance. Verify that this rule of thumb is (approximately) correct. (One mile is 1.6 km and light travels at a speed of 3×10^8 m/s.)

9. A lightning flash is seen in the sky and 8.2 s later the boom of the thunder is heard. The temperature of the air is 12°C. (a) What is the speed of sound at that temperature? [*Hint:* Light is an electromagnetic wave that travels at a speed of 3.00×10^8 m/s.] (b) How far away is the lightning strike?

12.3 Amplitude and Intensity of Sound Waves

10. A sound wave with an intensity level of 80.0 dB is incident on an eardrum of area 0.600×10^{-4} m². How much energy is absorbed by the eardrum in 3.0 minutes?

11. The sound level 25 m from a loudspeaker is 71 dB. What is the rate at which sound energy is produced by the loudspeaker, assuming it to be an isotropic source?

12. (a) What is the pressure amplitude of a sound wave with an intensity level of 120.0 dB in air? (b) What force does this exert on an eardrum of area 0.550×10^{-4} m²?

13. An intensity level change of +1.00 dB corresponds to what percentage change in intensity?

14. (a) Show that if $I_2 = 10.0 I_1$, then $\beta_2 = \beta_1 + 10.0$ dB. (A factor of 10 increase in intensity corresponds to a 10.0-dB increase in intensity level.) (b) Show that if $I_2 = 2.0 I_1$, then $\beta_2 = \beta_1 + 3.0$ dB. (A factor of 2 increase in intensity corresponds to a 3.0-dB increase in intensity level.)

15. At a rock concert, the engineer decides that the music isn't loud enough. He turns up the amplifiers so that the amplitude of the sound, where you're sitting, increases by 50.0%. (a) By what percentage does the intensity increase? (b) How does the intensity level (in dB) change?

12.4 Standing Sound Waves

16. (a) What should be the length of an organ pipe, closed at one end, if the fundamental frequency is to be 261.5 Hz? (b) What is the fundamental frequency of the organ pipe of part (a) if the temperature drops to 0.0°C?

17. Repeat Problem 16 for an organ pipe that is open at both ends.

18. An organ pipe that is open at both ends has a fundamental frequency of 382 Hz at 0.0°C. What is the fundamental frequency for this pipe at 20.0°C?

19. What is the length of the organ pipe in Problem 18?

20. In an experiment to measure the speed of sound in air, standing waves are set up in a narrow pipe open at both ends, using a speaker driven at 702 Hz. The length of the pipe is 2.0 m. What is the air temperature inside the pipe (assumed reasonably near room temperature, 20°C to 35°C)? [*Hint:* The standing wave is not necessarily the fundamental.]

© ✦21. An aluminum rod, 1.0 m long, is held lightly in the middle. One end is struck head-on with a rubber mallet so that a longitudinal pulse—a sound wave—travels down the rod. The fundamental frequency of the longitudinal vibration is 2.55 kHz. (a) Describe the location of the node(s) and antinode(s) for the fundamental mode of vibration. Use either displacement or pressure nodes and antinodes. (b) Calculate the speed of sound in aluminum from the information given in the problem. (c) The vibration of the rod produces a sound wave in air that can be heard. What is the wavelength of the sound wave in the air? Take the speed of sound in air to be 334 m/s. (d) Do the two ends of the rod vibrate longitudinally in phase or out of phase with each other? That is, at any given instant, do they move in the same direction or in opposite directions?

22. How long a pipe is needed to make a tuba whose lowest note is low C (frequency 130.8 Hz)? Assume that a tuba is a long straight pipe open at both ends.

23. When a tuning fork is held over the open end of a very thin tube, as in Fig. 12.7a, the smallest value of L that produces resonance is found to be 30.0 cm. (a) What is the wavelength of the sound? [*Hint:* Assume that the displacement antinode is at the open end of the tube.] (b) What is the next larger value of L that will produce resonance with the same tuning fork? (c) If the frequency of the tuning fork is 282 Hz, what is the speed of sound in the tube?

12.7 Beats

24. A violin is tuned by adjusting the tension in the strings. Brian's A string is tuned to a slightly lower frequency than Jennifer's, which is correctly tuned to 440.0 Hz. (a) What is the frequency of Brian's string if beats of 2.0 Hz are heard when the two bow the strings together? (b) Does Brian need to tighten or loosen his A string to get in tune with Jennifer?

25. A piano tuner sounds two strings simultaneously. One has been previously tuned to vibrate at 293.0 Hz. The tuner hears 3.0 beats per second. The tuner increases the tension on the as-yet untuned string, and now when they are played together the beat frequency is 1.0 s⁻¹. (a) What was the original frequency of the untuned string? (b) By what percentage did the tuner increase the tension on that string?

✦26. A cello string has a fundamental frequency of 65.40 Hz. What beat frequency is heard when this cello string is bowed at the same time as a violin string with frequency of 196.0 Hz? [*Hint:* The beats occur between the third harmonic of the cello string and the fundamental of the violin.]

27. An auditorium has organ pipes at the front and at the rear of the hall. Two identical pipes, one at the front and one at the back, have fundamental frequencies of 264.0 Hz at 20.0°C. During a performance, the organ pipes at the back of the hall are at 25.0°C, while those at the front are still at 20.0°C. What is the beat frequency when the two pipes sound simultaneously?

12.8 The Doppler Effect

28. A source of sound waves of frequency 1.0 kHz is traveling through the air at 0.50 times the speed of sound. (a) Find the frequency of the sound received by a stationary observer if the source moves towards her. (b) Repeat if the source moves away from her instead.

29. A source of sound waves of frequency 1.0 kHz is stationary. An observer is traveling at 0.50 times the speed of sound. (a) What is the observed frequency if the observer moves toward the source? (b) Repeat if the observer moves away from the source instead.

30. A source and an observer are *each* traveling at 0.50 times the speed of sound. The source emits sound waves at 1.0 kHz. Find the observed frequency if (a) the source and observer are moving *toward* each other; (b) the source and observer are moving *away* from each other; (c) the source and observer are moving in the same direction.

31. Blood flow rates can be found by measuring the Doppler shift in frequency of ultrasound reflected by red blood cells (known as *angiodynography*). If the speed of the red blood cells is v, the velocity of sound in blood is c, the ultrasound source emits waves of frequency f, and we assume that the blood cells are moving directly toward the ultrasound source,

show that the frequency f_r of reflected waves detected by the apparatus is given by

$$f_r = f \frac{(1 + v/c)}{(1 - v/c)}$$

[*Hint:* There are *two* Doppler shifts. A red blood cell first acts as a moving observer; then it acts as a moving source when it reradiates the reflected sound at the same frequency that it received.]

✦32. Show that for a moving source, the fractional shift in observed frequency is equal to v_s/v, the source's speed as a fraction of the speed of sound. [*Hint:* Use the binomial approximation from Appendix A.5.]

✦33. The pitch of the sound from a racecar engine drops the musical interval of a fourth when it passes the spectators. This means the frequency of the sound after passing is 0.75 times what it was before. How fast is the racecar moving?

12.9 Shock Waves

34. A supersonic plane moves at speed v_{plane}; the speed of sound is v_{sound}. The conical shock wave makes an angle θ with the direction of motion of the plane (Fig. 12.20). Show that $\sin \theta = v_{sound}/v_{plane}$. [*Hint:* Consider how far and in what direction the plane and the shock wave move during a time interval Δt.]

35. A plane is flying at supersonic speed at an elevation where the speed of sound is 322 m/s. The shock wave cone forms with an angle $\theta = 22.0°$ with the direction of motion of the plane (see Problem 34 and Fig. 12.20). (a) What is the Mach number for this plane? (b) How fast is the plane traveling?

12.10 Echolocation and Medical Imaging

36. A boat is using sonar to detect the bottom of a freshwater lake. If the echo from a sonar signal is heard 0.540 s after it is emitted, how deep is the lake? Assume the temperature of the lake is uniform and at 25°C.

37. A geological survey ship mapping the floor of the ocean sends sound pulses down from the surface and measures the

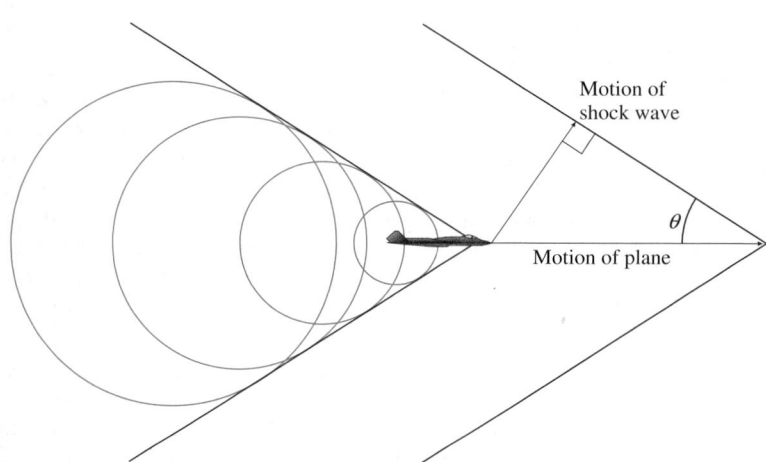

Motion of shock wave

θ

Motion of plane

Figure 12.20 Problems 34, 35, 52, and 53

time taken for the echo to return. How deep is the ocean at a point where the echo time (down and back) is 7.07 s? The temperature of the seawater is 25°C.

✦38. A bat emits chirping sounds of frequency 82.0 kHz while hunting for moths to eat. If the bat is flying toward the moth at a speed of 4.40 m/s and the moth is flying away from the bat at 1.20 m/s, what is the frequency of the sound wave reflected from the moth as observed by the bat? Assume it is a cool night with a temperature of 10.0°C. [*Hint:* There are two Doppler shifts. Think of the moth as a receiver, which then becomes a source as it "retransmits" the reflected wave.]

39. The bat of Problem 38 emits a chirp that lasts for 2.0 ms and then is silent while it listens for the echo. If the beginning of the echo returns just after the outgoing chirp is finished, how close to the moth is the bat? [*Hint:* Is the change in distance between the two significant during a 2.0-ms time interval?]

✦40. Doppler ultrasound is used to measure the speed of blood flow (see Problem 31). The reflected sound interferes with the emitted sound, producing beats. If the speed of red blood cells is 0.10 m/s, the ultrasound frequency used is 5.0 MHz, and the velocity of sound in blood is 1570 m/s, what is the beat frequency?

✦41. (a) In Problem 31, find the beat frequency between the outgoing and reflected sound waves. (b) Show that the beat frequency is proportional to the speed of the blood cell if $v \ll c$. [*Hint:* Use the binomial approximation from Appendix A.5.]

42. A ship is lost in a dense fog in a Norwegian fjord that is 1.80 km wide. The air temperature is 5.0°C. The captain fires a pistol and hears the first echo after 4.0 s. (a) How far from one side of the fjord is the ship? (b) How long after the first echo does the captain hear the second echo?

COMPREHENSIVE PROBLEMS

43. What are the four lowest standing-wave frequencies for an organ pipe that is 4.80 m long and closed at one end?

44. The length of the auditory canal in humans averages about 2.5 cm. What are the lowest three standing-wave frequencies for a pipe of this length open at one end? What effect might resonance have on the sensitivity of the ear at various frequencies? (Refer to Fig. 12.9. Note that frequencies critical to speech recognition are in the range 2 to 5 kHz.)

45. The Vespertilionidae family of bats detect the distance to an object by timing how long it takes for an emitted signal to reflect off the object and return. Typically they emit sound pulses 3 ms long and 70 ms apart while cruising. (a) If an echo is heard 60 ms later ($v_{sound} = 331$ m/s), how far away is the object? (b) When an object is only 30 cm away, how long will it be before the echo is heard? (c) Will the bat be able to detect this echo?

46. Some bats determine their distance to an object by detecting the difference in intensity between echoes. (a) If intensity falls off at a rate that is inversely proportional to the distance squared, show that the echo intensity is inversely proportional to the fourth power of distance. (b) The bat was originally 0.60 m from one object and 1.10 m from another. After flying closer, it is now 0.50 m from the first and at 1.00 m from the second object. What is the percentage increase in the intensity of the echo from each object?

47. Horseshoe bats use the Doppler effect to determine their location. A Horseshoe bat flies toward a wall at a speed of 15 m/s while emitting a sound of frequency 35 kHz. What is the beat frequency between the emission frequency and the echo?

48. At what frequency f does a sound wave in air have a wavelength of 15 cm, about half the diameter of the human head? Some methods of localization work well only for frequencies below f while others work well only above f. (See Conceptual Questions 4 and 5.)

✦✦49. A periodic wave is composed of the superposition of three sine waves whose frequencies are 36, 60, and 84 Hz. The speed of the wave is 180 m/s. What is the wavelength of the wave?

✦✦50. Analysis of the periodic sound wave produced by a violin's G string includes three frequencies: 392, 588, and 980 Hz. What is the fundamental frequency? [*Hint:* The wave on the string is the superposition of several different standing wave patterns.]

51. According to a treasure map, a treasure lies at a depth of 40.0 fathoms on the ocean floor due east from the lighthouse. The treasure hunters use sonar to find where the depth is 40.0 fathoms as they head east from the lighthouse. What is the elapsed time between an emitted pulse and the return of its echo at the correct depth if the water temperature is 25°C? [*Hint:* One fathom is 1.83 m.]

✦52. An airplane is flying 1 km directly over your position on the ground at Mach 2. How far from that overhead position will the airplane have moved along its horizontal flight path when you hear the sonic boom? [*Hint:* See Problem 34 and Fig. 12.20.]

53. A wind tunnel is used to simulate the flight of a plane. Air at 20°C is blown past the model plane at very high speeds. If a shock cone angle of $\theta = 40.0°$ develops, how fast is the air moving? [*Hint:* See Problem 34 and Fig. 12.20.]

⊙
✦✦54. In this problem, you will estimate the smallest kinetic energy of vibration that the human ear can detect. Suppose that a harmonic sound wave at the threshold of hearing ($I = 1.0 \times 10^{-12}$ W/m^2) is incident on the eardrum. The speed of sound is 340 m/s and the density of air is 1.3 kg/m^3. (a) What is the maximum speed of an element of air in the sound wave? [*Hint:* See Eq. (10-21).] (b) Assume the eardrum vibrates with displacement s_0 at angular frequency ω; its maximum speed is then equal to the maximum speed of an air element. The mass of the eardrum is approximately 0.1 g.

What is the *average* kinetic energy of the eardrum? (c) The average kinetic energy of the eardrum due to collisions with air molecules *in the absence of a sound wave* is about 10^{-20} J. Compare your answer to (b) and discuss.

✦ 55. When playing *fortissimo* (very loudly), a trumpet emits sound energy at a rate of 0.800 W out of a bell (opening) of diameter 12.7 cm. (a) What is the sound intensity level right in front of the trumpet? (b) If the trumpet radiates sound waves uniformly in all directions, what is the sound intensity level at a distance of 10.0 m?

✦✦ 56. During a rehearsal, all eight members of the first violin section of an orchestra play a very soft passage. The sound intensity level at a certain point in the concert hall is 38.0 dB. What is the sound intensity level at the same point if only one of the violinists plays the same passage? [*Hint:* When playing together, the violins are *incoherent* sources of sound.]

ANSWERS TO PRACTICE PROBLEMS

12.1 Although solids usually have somewhat higher densities than liquids, they have *much* higher bulk moduli—they are much stiffer. The greater restoring forces in solids cause sound waves to travel faster.

12.2 Assumptions: Treat the stage as a point source; neglect reflection and absorption of waves. 4.0×10^{-6} W/m^2, 0.057 Pa.

12.3 400

12.4 a factor of 3.2

12.5 3000 km. No; it is not realistic to ignore absorption and reflection over such a great distance.

12.6 24°C

12.7 444.0 Hz

12.8 182 Hz

12.9 27 m/s

Temperature and the Ideal Gas

In warm-blooded or homeothermic (constant temperature) animals, body temperature is carefully regulated. The hypothalamus, located in the brain, acts as the master thermostat to keep body temperature constant to within a fraction of a degree Celsius in a healthy animal. If the body temperature starts to deviate much from the desired constant level, the hypothalamus causes changes in blood flow and initiates other processes, such as shivering or perspiration, to bring the temperature back to normal. Why is this careful temperature regulation necessary? What would be the problem if body temperature fluctuated with environmental conditions? What evolutionary advantage does a constant body temperature give the warm-blooded animals (birds, mammals) over the cold-blooded (such as reptiles and insects)? What are the disadvantages?

Concepts & Skills to Review

- energy conservation (Sections 6.3, 6.6, 6.7)
- momentum conservation (Section 7.4)
- collisions (Sections 7.7, 7.8)

Heat is the flow of energy due to a temperature difference. Heat always flows from the hotter object to the colder object.

13.1　TEMPERATURE

The measurement of **temperature** is part of everyday life. We measure the temperature of the air outdoors to decide how to dress when going outside; a thermostat measures the air temperature indoors to control heating and cooling systems to keep our homes and offices comfortable. Regulation of oven temperature is important in baking. When we feel ill, we measure our body temperature to see if we have a fever. Despite how matter-of-fact it may seem, temperature is a subtle concept. Although our subjective sensations of hot and cold are related to temperature, they can easily mislead, as the next Physics at Home demonstrates.

The definition of temperature is based on the concept of **thermal equilibrium**. Suppose two objects or systems are allowed to exchange energy. The net flow of energy is always from the object at the higher temperature to the object at the lower temperature. As energy flows, the temperatures of the two objects approach each other. When the temperatures are the same, there is no longer any net flow of energy; the objects are now said to be in thermal equilibrium. Thus, *temperature is a quantity that determines when objects are in thermal equilibrium.* (The objects do *not* necessarily have the same *energy* when in thermal equilibrium.) The flow of energy that occurs between two objects or systems due to a temperature difference between them is called **heat flow**. In Chapter 14 we discuss heat flow in detail. If heat can flow between two objects or systems, the objects or systems are said to be in **thermal contact**.

Temperature measurement relies on the **zeroth law of thermodynamics**.

Zeroth law of thermodynamics:

If two objects are each in thermal equilibrium with a third object, then the two are in thermal equilibrium with each other.

The rather odd name *zeroth* law of thermodynamics came about because this law was formulated historically *after* the first, second, and third laws of thermodynamics and yet it is so fundamental that it should come *before* the others. **Thermodynamics**, the subject of Chapters 13 to 15, concerns temperature, heat flow, and the internal energy of systems.

Physics at Home

Try an experiment described by the philosopher John Locke in 1690. Fill one container with water that is hot (but not too hot to touch); fill a second container with lukewarm water; and fill a third container with cold water. Put one hand in the hot water and one in the cold water (Fig. 13.1) for about 10 to 20 seconds. Then plunge both hands into the container of lukewarm water. Although both hands are now immersed in water that is at a single temperature, the hand that had been in the hot water feels cool while the hand that had been in the cold water feels warm. This demonstration shows that we cannot trust our subjective senses to measure temperature.

Cold　　Lukewarm　　Hot

Figure 13.1　It is easy to trick our sense of temperature.

To measure the temperature of an object, we put a thermometer into thermal contact with the object. If the zeroth law were not true, it would be impossible to define temperature, since different thermometers could give different results. The zeroth law may *seem* obvious, but it is possible to imagine a universe in which there is no net flow of energy when A is in thermal contact with B, nor when B is in thermal contact with C, and yet energy does flow when A is in contact with C.

13.2 TEMPERATURE SCALES

Thermometers measure temperature by exploiting some property of matter that is temperature-dependent. The familiar liquid-in-glass thermometer relies on thermal expansion: the mercury or alcohol expands as its temperature rises (or contracts as its temperature drops) and we read the temperature on a calibrated scale. Since some materials expand more than others, these thermometers must be calibrated on a scale using some easily reproducible phenomenon, such as the melting point of ice or the boiling point of water. The assignment of temperatures to these phenomena is arbitrary.

The most commonly used temperature scale in the world is the Celsius scale. On the Celsius scale, 0°C is the freezing temperature of water at $P = 1$ atm (the *ice point*) and 100°C is the boiling temperature of water at $P = 1$ atm (the *steam point*).

In the United States, the Fahrenheit scale is still commonly used. The ice point is 32°F and the steam point is 212°F, so that the difference between the steam and ice points is 180°F. The size of the Fahrenheit degree interval is therefore smaller than the Celsius degree interval: a temperature difference of 1°C is equivalent to a difference of 1.8°F:

$$\Delta T_F = \Delta T_C \times 1.8 \, \frac{°F}{°C} \qquad (13\text{-}1)$$

Since the two scales also have an offset—0°C is not the same temperature as 0°F—conversion between the two is:

$$T_F = 1.8 T_C + 32°F \qquad (13\text{-}2a)$$

$$T_C = \frac{T_F - 32°F}{1.8} \qquad (13\text{-}2b)$$

The SI unit of temperature is the **kelvin** (symbol K, *without* a degree sign). The kelvin has the same degree size as the Celsius scale; that is, a temperature *difference* of 1°C is the same as a difference of 1 K. However, 0 K represents *absolute zero*—there are no temperatures below 0 K. The ice point is 273.15 K, so temperature in °C (T_C) and temperature in kelvins (T) are related.

$$T_C = T - 273.15 \qquad (13\text{-}3)$$

Equation (13-3) is the definition of the Celsius scale in terms of the kelvin. Table 13.1 shows some temperatures in kelvins, °C and °F.

Table 13.1

Some Reference Temperatures in K, °C, and °F

	K	°C	°F		K	°C	°F
Absolute zero	0	−273.15	−459.67	Human body temperature	310	37	98.6
Lowest transient temperature achieved (laser cooling)	10^{-9}			Water boils	373.15	100.00	212.0
				Campfire	1,000	700	1,300
Intergalactic space	3	−270	−454	Gold melts	1,337	1,064	1,947
Helium boils	4.2	−269	−452	Lightbulb filament	2,500	2,200	4,000
Nitrogen boils	77	−196	−321	Surface of Sun; iron welding arc	6,300	6,000	11,000
Carbon dioxide freezes ("dry ice")	195	−78	−108	Center of Earth	16,000	15,700	28,300
Mercury freezes	234	−39	−38	Center of Sun	10^7	10^7	10^7
Ice melts/water freezes	273.15	0	32.0	Interior of neutron star	10^9	10^9	10^9

Example 13.1

A Sick Friend

A friend suffering from the flu has a fever; her body temperature is 38.6°C. What is her temperature in (a) K and (b) °F?

Strategy   (a) Kelvins and °C differ only by a shift of the zero point. Converting from °C to K requires only the addition of 273.15 K since 0°C (the ice point) corresponds to 273.15 K. (b) The °F is a different size than the °C, as well as having a different zero. In the Celsius scale, the zero is at the ice point. First multiply by 1.8°F/°C to find how many °F above the ice point. Then add 32°F (the Fahrenheit temperature of the ice point).

Solution   (a) The temperature is 38.6 K *above* the ice point of 273.15 K. Therefore, the kelvin temperature is

$$T = 38.6 \text{ K} + 273.15 \text{ K} = 311.8 \text{ K}$$

(b) First find how many °F above the ice point:

$$\Delta T_F = 38.6°C \times (1.8°F/°C) = 69.5°F$$

The ice point is 32°F, so

$$T_F = 32.0°F + 69.5°F = 101.5°F$$

Discussion   The answer is reasonable since 98.6°F is normal body temperature.

Practice Problem 13.1   Normal body temperatures with two scales

Convert the normal human body temperature (98.6°F) to degrees Celsius and kelvins.

Figure 13.2   The blacksmith heats rods of iron in his hot forge and, before the iron cools, works quickly to shape it at his anvil into hoops for barrels, rims for wagon wheels, and shoes for horses.

13.3 THERMAL EXPANSION OF SOLIDS AND LIQUIDS

Most objects expand as their temperature increases. Long before the cause of thermal expansion was understood, the phenomenon was put to practical use. The cooper (barrel maker) was an important person in days gone by. Barrels made from wooden staves were often held together by bands of iron that fit tightly around the middle (Fig. 13.2). The iron band was heated red hot to make it expand. Then it was fit around the barrel and allowed to cool so that it would shrink and tighten in place. A circular iron rim was fit around a wooden wagon wheel in the same way.

Tensile stresses and strains were discussed in Chapter 10. An object subject to a tensile stress gets slightly longer. An object's length can also change due to a *change in temperature* rather than an applied force. For tensile stresses, Hooke's law says that the fractional change in length is proportional to the stress (up to some maximum stress). Similarly, as long as the temperature change is not too great, the fractional change in length of a solid is proportional to the temperature change. If the length of a wire, rod, or pipe is L_0 at temperature T_0 (Fig. 13.3), then

Figure 13.3 Expansion of a solid rod with increasing temperature.

$$\frac{\Delta L}{L_0} = \alpha \Delta T \qquad (13\text{-}4)$$

where $\Delta L = L - L_0$ and $\Delta T = T - T_0$. The length at temperature T is

$$L = L_0 + \Delta L = (1 + \alpha \Delta T) L_0 \qquad (13\text{-}5)$$

The constant of proportionality, α, is called the **coefficient of thermal expansion** of the substance. It plays a role in thermal expansion similar to that of the elastic modulus in tensile stress. If T is measured in kelvins or in degrees Celsius, then α has units of K^{-1} or $°C^{-1}$. Since only the *change* in temperature is involved in Eq. (13-4), either Celsius or Kelvin temperatures can be used to find ΔT; a temperature change of 1 K is the same as a temperature change of 1°C.

As is true for the elastic modulus, the coefficient of thermal expansion has different values for different solids and also depends to some extent on the starting temperature of the object. Table 13.2 lists the coefficients of thermal expansion for various solids at room temperature (20°C).

Figure 13.4 is a graph of the length of a steel girder as a function of temperature over a *wide* range of temperatures. The curvature of this graph shows that the thermal expansion of the girder is in general *not* proportional to the temperature change. However, over a *limited* temperature range, the curve can be approximated by a straight line; the slope of the tangent line is the coefficient α at the temperature T_0. For small temperature changes near T_0, the change in length of the girder can be treated as being proportional to the temperature change with only a small error.

Allowances must be made in building sidewalks, roads, bridges, and buildings to leave space for expansion in hot weather. Old subway tracks have small spaces left between rail sections to prevent the rails from pushing into each other and causing the

Making The Connection:
expansion joints in
bridges and buildings

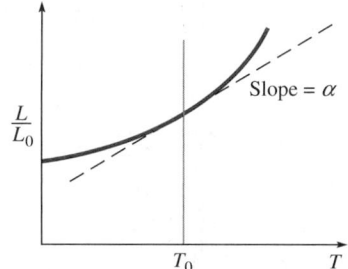

Figure 13.4 The relative length of a steel girder as a function of temperature. The dashed tangent line shows what Eq. (13-4) predicts for small temperature changes in the vicinity of T_0. The slope of this tangent line is the value of α at $T = T_0$.

Table 13.2

Coefficients of Linear Expansion α and Volume Expansion β (at $T_0 = 20°C$ except for ice)

Material	$\alpha(10^{-6}\ K^{-1})$	$\beta(10^{-6}\ K^{-1})$
Solids		
Glass (Vycor)	0.75	2.25
Brick	1.0	3.0
Glass (Pyrex)	3.25	9.75
Granite	8	24
Glass, most types	9.4	28.2
Cement or concrete	12	36
Iron or steel	12	36
Copper	16	48
Silver	18	54
Brass	19	57
Aluminum	22.5	69
Lead	29	87
Ice (at 0°C)	51	153
Liquids		
Mercury		182
Water		207
Gasoline		950
Ethyl alcohol		1120
Benzene		1240
Gases		
Air (and most other gases) at 1 atm		3340

Example 13.2

Expanding Rods

Two metal rods, one aluminum and one brass, are clamped down on one end (Fig. 13.5). At 0.0°C, the rods are each 50.0 cm long and are separated by 0.024 cm at their unfastened ends. At what temperature will the rods just come into contact? (Assume that the base to which the rods are clamped undergoes a negligibly small thermal expansion.)

Strategy Two rods of different materials expand by different amounts. The sum of the two expansions ($\Delta L_{br} + \Delta L_{Al}$) must equal the space between the rods. After finding ΔT, we add it to $T_0 = 0°C$ to obtain the temperature at which the two rods touch.

Known: $L_0 = 50.0$ cm, $T_0 = 0.0°C$ for both
Look up: $\alpha_{br} = 19 \times 10^{-6}\ K^{-1}$; $\alpha_{Al} = 22.5 \times 10^{-6}\ K^{-1}$
Requirement: $\Delta L_{br} + \Delta L_{Al} = 0.024$ cm
Find: $T_f = T_0 + \Delta T$

Solution The brass rod expands by

$$\Delta L_{br} = (\alpha_{br}\ \Delta T) L_0$$

and the aluminum rod by

$$\Delta L_{Al} = (\alpha_{Al}\ \Delta T) L_0$$

The sum of the two expansions is known:

$$\Delta L_{br} + \Delta L_{Al} = 0.024\ \text{cm}$$

Figure 13.5
Two clamped rods

Example 13.2 *continued*

Since both the initial lengths and the temperature changes are the same,

$$(\alpha_{br} + \alpha_{Al}) \Delta T \times L_0 = 0.024 \text{ cm}$$

Solving for ΔT,

$$\Delta T = \frac{0.024 \text{ cm}}{(\alpha_{br} + \alpha_{Al}) L_0}$$

$$= \frac{0.024 \text{ cm}}{(19 \times 10^{-6} \text{ K}^{-1} + 22.5 \times 10^{-6} \text{ K}^{-1}) \times 50.0 \text{ cm}}$$

$$= 11.6°C$$

The temperature at which the two touch is

$$T_f = T_0 + \Delta T = 0.0°C + 11.6°C = 11.6°C$$

Discussion As a check on the solution, we can find how much each individual rod expands and then add the two amounts.

$$\Delta L_{Al} = \alpha_{Al} \Delta T L_0$$

$$= 22.5 \times 10^{-6} \text{ K}^{-1} \times 11.6 \text{ K} \times 50.0 \text{ cm} = 0.013 \text{ cm}$$

$$\Delta L_{br} = \alpha_{br} \Delta T L_0$$

$$= 19 \times 10^{-6} \text{ K}^{-1} \times 11.6 \text{ K} \times 50.0 \text{ cm} = 0.011 \text{ cm}$$

$$\text{total expansion} = 0.013 \text{ cm} + 0.011 \text{ cm} = 0.024 \text{ cm}$$

which is correct.

Practice Problem 13.2   Expansion of a wall

The outer wall of a building is constructed from concrete blocks. If the wall is 5.00 m long at 20.0°C, how much longer is the wall on a hot day (30.0°C)? How much shorter is it on a cold day (−5.0°C)?

---

track to bow. A train riding on such tracks is subject to a noticeable amount of "clickety-clack" as it goes over these small expansion breaks in the tracks. Expansion joints are easily observed in bridges (Fig. 13.6). Concrete roads and sidewalks have tar joints between sections. In hot weather the tar softens, allowing the expanding concrete to push into the joint. Homeowners sometimes build their own sidewalks without realizing the necessity for such joints; these sidewalks begin to crack almost immediately!

Allowances must also be made for contraction in cold weather. If an object is not free to expand or contract, then as the temperature changes it is subjected to *thermal stress* as its environment exerts forces on it to prevent the thermal expansion or contraction that would otherwise occur.

Differential Expansion

When two strips made of different metals are joined together and then heated, one expands more than the other (unless they have the same coefficient of expansion). This differential expansion can be put to practical use: the joined strips bend into a curve, allowing one strip to expand more than the other.

The bimetallic strip (Fig. 13.7) is made by joining a material with a lower coefficient of expansion, such as steel, and one of a higher coefficient of expansion, such as brass. Unequal expansions or contractions of the two materials force the bimetallic strip

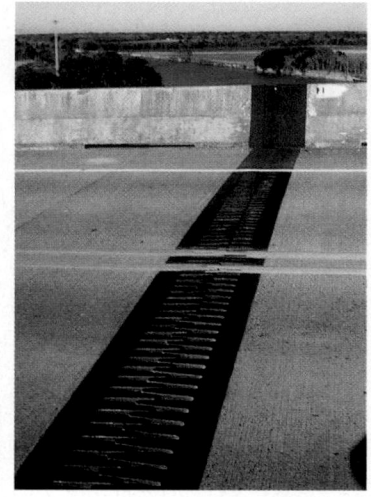

Figure 13.6 Expansion joints permit the roadbed of a bridge to expand and contract as the temperature changes.

Making The Connection:
bimetallic strip in a thermostat

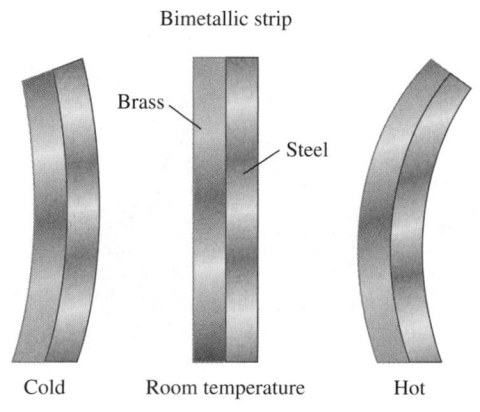

Bimetallic strip

Brass

Steel

Cold   Room temperature   Hot

Figure 13.7 A bimetallic strip bends when its temperature changes; brass expands and contracts more than steel for the same temperature change.

to bend. In Fig. 13.7 the brass expands more than the steel when the bimetallic strip is heated. As the strip is cooled, the brass contracts more than the steel.

The bimetallic strip is used in many wall thermostats. The bending of the bimetallic strip closes or opens an electrical switch in the thermostat that turns the furnace or air conditioner on or off. Inexpensive oven thermometers also use a bimetallic strip wound into a spiral coil; the coil winds tighter or unwinds as the temperature changes.

Area Expansion

As you might suspect, *each dimension* of an object expands when the object's temperature increases. For instance, a pipe expands not only in length, but also in radius. An isotropic substance expands uniformly in all directions; any two points move farther apart as the temperature rises. Thus, there are changes in area and volume caused by expansion in all directions. In Problem 14 you can show that, for small temperature changes, the area of any flat surface of a solid changes in proportion to the temperature change:

$$\frac{\Delta A}{A_0} = 2\alpha \, \Delta T \tag{13-6}$$

The factor of two in Eq. (13-6) arises because the surface expands in two perpendicular directions.

Volume Expansion

The fractional change in volume of a solid or liquid is also proportional to the temperature change as long as the temperature change is not too large:

$$\frac{\Delta V}{V_0} = \beta \, \Delta T \tag{13-7}$$

The coefficient of volume expansion, β, is the fractional change in volume per unit temperature change. For solids, the coefficient of volume expansion is three times the coefficient of linear expansion (as shown in Problem 15):

$$\beta = 3\alpha \tag{13-8}$$

The factor of three in Eq. (13-8) arises because the object expands in three-dimensional space. For liquids, the volume expansion coefficient is the only one given in tables. Since liquids do not necessarily retain the same shape as they expand, the quantity that is uniquely defined is the change in volume. Table 13.2 provides values of β for some common solids and liquids.

 A hollow cavity in a solid expands exactly as if it were filled—the interior of a steel gasoline can expands just as if it were a solid steel block. The steel wall of the can does *not* expand inward to make the cavity smaller; that would require pairs of points in the steel to get *closer together* instead of farther apart.

In an ordinary alcohol-in-glass or mercury-in-glass thermometer, it is not just the liquid that expands as temperature rises. The reading of the thermometer is determined by the difference in the volume expansion of the liquid and that of the interior of the glass. The calibration of an accurate thermometer must account for the expansion of the glass. A glance at Table 13.2 shows that, as is usually the case, the liquid expands much more than the glass for a given temperature change.

Example 13.3

Hollow Cylinder Full of Water

A hollow copper cylinder is filled to the brim with water at 20.0°C. If the water and the container are heated to a temperature of 91°C, what percentage of the water spills over the top of the container?

continued on next page

Example 13.3 continued

Strategy The volume expansion coefficient for water is greater than that for copper, so the water expands more than the interior of the cylinder. The cavity expands just as if it were solid copper. Since the problem does not specify the initial volume, we call it V_0. We need to find out how much a volume V_0 of water expands and how much a volume V_0 of copper expands; the difference is the water volume that spills over the top of the container.

Known: Initial copper cylinder interior volume = initial water
        volume = V_0
        Initial temperature = $T_0 = 20.0°C$
        Final temperature = $91°C$; $\Delta T = 71°C$

Look up: $\beta_{Cu} = 48 \times 10^{-6} \,°C^{-1}$; $\beta_{H_2O} = 207 \times 10^{-6} \,°C^{-1}$
Find: $\Delta V_{H_2O} - \Delta V_{Cu}$ as a percentage of V_0

Solution The volume expansion of the interior of the copper cylinder is

$$\Delta V_{Cu} = \beta_{Cu} \,\Delta T \, V_0$$

The volume expansion of the water is

$$\Delta V_{H_2O} = \beta_{H_2O} \,\Delta T \, V_0$$

The amount of water that spills is

$$\begin{aligned}
\Delta V_{H_2O} - \Delta V_{Cu} &= \beta_{H_2O} \,\Delta T \, V_0 - \beta_{Cu} \,\Delta T \, V_0 \\
&= (\beta_{H_2O} - \beta_{Cu}) \,\Delta T \, V_0 \\
&= (207 \times 10^{-6} \,°C^{-1} - 48 \times 10^{-6} \,°C^{-1}) \times 71 \,°C \times V_0 \\
&= 0.011 V_0
\end{aligned}$$

The percentage of water that spills is therefore 1.1%.

Discussion As a check we can find the change in volume of the copper container and of the water and find the difference.

$$\Delta V_{Cu} = \beta_{Cu} \,\Delta T \, V_0 = 48 \times 10^{-6} \,°C^{-1} \times 71 \,°C \times V_0 = 0.0034 V_0$$

$$\Delta V_{H_2O} = \beta_{H_2O} \,\Delta T \, V_0 = 207 \times 10^{-6} \,°C^{-1} \times 71 \,°C \times V_0 = 0.0147 V_0$$

volume of water that spills = $0.0147 V_0 - 0.0034 V_0 = 0.0113 V_0$

which again shows that 1.1% spills.

Practice Problem 13.3 Overflowing gas can

A driver fills an 18.9-L steel gasoline can with gasoline at 15.0°C right up to the top. He forgets to replace the cap and leaves the can in the back of his truck. The temperature climbs to 30.0°C by 1 P.M. How much gasoline spills out of the can?

13.4 MOLECULAR PICTURE OF A GAS

As we saw in Chapter 9, the densities of liquids are generally not much less than the densities of solids. Gases are *much* less dense than liquids and solids because the molecules are, on average, much farther apart. The mass density—mass per unit volume—of a substance depends on two microscopic quantities: the mass m of a single molecule and the number of molecules packed into a given volume of space (see Fig. 13.8). This second quantity, the number of molecules per unit volume, is called the **number density** to distinguish it from mass density. In SI units, number density is written as the number of molecules per cubic meter, usually written simply as m^{-3} (read "per cubic meter"). If a gas has a total mass M, occupies a volume V, and each molecule has a mass m, then the number of gas molecules is

$$N = \frac{M}{m} \tag{13-9}$$

(a)

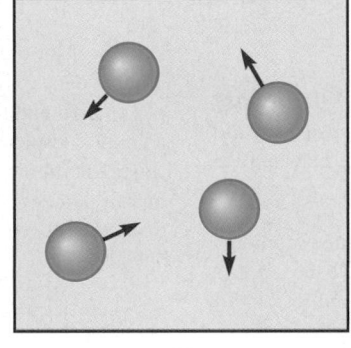

(b)

Figure 13.8 These two gases have the same mass per unit volume but different number densities. In (a), there are a larger number of molecules in a given volume, but the mass of each molecule in (b) is greater.

and the average number density is

$$\frac{N}{V} = \frac{M}{mV} = \frac{\rho}{m} \qquad (13\text{-}10)$$

where $\rho = M/V$ is the mass density.

It is common to express the amount of a substance in units of **moles** (abbreviated mol). The mole is an SI base unit and is defined as follows: one mole of anything contains the same number of units as there are atoms in 12 *grams* (not kilograms) of carbon-12. This number is called **Avogadro's number** and has the value

$$N_A = 6.022 \times 10^{23} \text{ mol}^{-1} \quad \text{(Avogadro's number)}$$

Avogadro's number is written with units, mol^{-1}, to show that this is the number *per mole*. The number of moles, n, is therefore given by

$$\text{number of moles} = \frac{\text{number of molecules}}{\text{number of molecules per mole}}$$

$$n = \frac{N}{N_A} \qquad (13\text{-}11)$$

The mass of a molecule is often expressed in units other than kg. The most common is the **atomic mass unit** (symbol u). By definition, one atom of carbon-12 has a mass of 12 u (exactly). Using Avogadro's number, the relationship between atomic mass units and kilograms can be calculated (see Problem 19):

$$1 \text{ u} = 1.66 \times 10^{-27} \text{ kg} \qquad (13\text{-}12)$$

The proton, neutron, and hydrogen atom all have masses within 1% of 1 u—which is why the atomic mass unit is so convenient. More precise values are 1.007 u for the proton, 1.009 u for the neutron, and 1.008 u for the hydrogen atom. The mass of an atom is *approximately* equal to the number of nucleons (neutrons plus protons)—the "atomic mass number"—times 1 u.

 Tables commonly express the mass of a substance as the **molar mass**—the mass of *one mole* of the substance *in grams*. For an element with several isotopes (such as carbon-12, carbon-13, and carbon-14), the molar mass is averaged according to the naturally occurring abundance of each isotope. The atomic mass unit is chosen so that the mass of a molecule in "u" is numerically the same as the molar mass in g/mol. For example, the molecular mass of O_2 is 32.0 g/mol and the mass of one molecule is 32.0 u. Probably the most important thing to remember from all of this is: *be careful* with units!

The mass of a molecule is the sum of the masses of its constituent atoms. The molecular mass is therefore equal to the sum of the atomic masses. For example, the atomic mass of carbon is 12.0 g/mol and the atomic mass of (atomic) oxygen is 16.0 g/mol; therefore, the molecular mass of carbon dioxide (CO_2) is $(12.0 + 2 \times 16.0)$ g/mol = 44.0 g/mol.

Example 13.4

A Helium Balloon

A helium balloon of volume 0.010 m^3 contains 0.40 mol of He gas. (a) Find the number of atoms, the number density, and the mass density. (b) Estimate the average distance between He atoms.

Strategy   The number of moles tells us the number of atoms as a fraction of Avogadro's number. Once we have the number of atoms, N, the next quantity we are asked to find is N/V. To find the mass density, we can look up the atomic mass of

helium in the Periodic Table. The mass per atom times the number density (atoms per m^3) equals the mass density (mass per m^3). To find the average distance between atoms, imagine a simplified picture in which each atom is at the center of a spherical volume equal to the total volume of the gas divided by the number of atoms. In this approximation, the average distance between atoms is equal to the diameter of each sphere.

continued on next page

Example 13.4 *continued*

Solution   (a) The number of atoms is

$$N = n N_A$$

$$= 0.40 \text{ mol} \times 6.022 \times 10^{23} \text{ atoms/mol} = 2.4 \times 10^{23} \text{ atoms}$$

The number density is

$$\frac{N}{V} = \frac{2.4 \times 10^{23} \text{ atoms}}{0.010 \text{ m}^3} = 2.4 \times 10^{25} \text{ atoms/m}^3$$

The atomic mass of helium is 4.00 u. Then the mass in kg of a helium atom is

$$m = 4.00 \text{ u} \times 1.66 \times 10^{-27} \text{ kg/u} = 6.64 \times 10^{-27} \text{ kg}$$

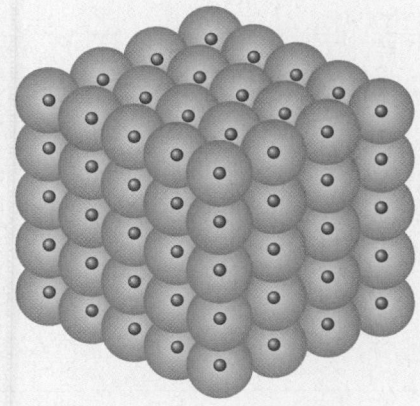

Figure 13.9
Simplified model in which equally spaced gas molecules sit at the center of a spherical volume of space

and the mass density of the gas is

$$\rho = \frac{M}{V} = m \times \frac{N}{V}$$

$$= 6.64 \times 10^{-27} \text{ kg} \times 2.4 \times 10^{25} \text{ m}^{-3} = 0.16 \text{ kg/m}^3$$

(b) We assume that each atom is at the center of a sphere of radius r (Fig. 13.9). The volume of the sphere is

$$\frac{V}{N} = \frac{1}{N/V} = \frac{1}{2.4 \times 10^{25} \text{ atoms/m}^3} = 4.2 \times 10^{-26} \text{ m}^3 \text{ per atom}$$

Then

$$\frac{V}{N} = \frac{4}{3}\pi r^3 \approx 4r^3 \quad (\text{since } \pi \approx 3)$$

Solving for r,

$$r \approx \left(\frac{V}{4N}\right)^{1/3} = 2.2 \times 10^{-9} \text{ m} = 2.2 \text{ nm}$$

The average distance between atoms is $d = 2r \approx 4$ nm (since this is an estimate).

Discussion   For comparison, in *liquid* helium the average distance between atoms is about 0.4 nm, so in the gas the average separation is about ten times larger.

Practice Problem 13.4   Number density for water
The mass density of liquid water is 1000.0 kg/m^3. Find the number density.

13.5   ABSOLUTE TEMPERATURE AND THE IDEAL GAS LAW

We have examined the thermal expansion of solids and liquids. What about gases? Is the volume expansion of a gas proportional to the temperature change? We must be careful; since gases are easily compressed, we must also specify what happens to the pressure. The French scientist Jacques Charles (1746–1823) found experimentally that, if the pressure of a gas is held constant, the change in temperature is indeed proportional to the change in volume (Fig. 13.10a).

Charles's law

$$\Delta V \propto \Delta T \quad (\text{for constant } P)$$

According to Charles's law, a graph of V versus T for a gas held at constant pressure is a straight line, but the line does not necessarily pass through the origin (Fig. 13.10b). Temperature scales such as the Celsius and Fahrenheit scales assign $T = 0$ to some arbitrary phenomenon. We would *not* expect a gas at 1 atm pressure and 0°C to occupy a volume of zero!

However, if we graph V versus T (at constant P) for various gases, something interesting happens. If we extrapolate the straight line to where it reaches $V = 0$, the temperature at that point is the *same* regardless of what gas we use, how many moles of gas are present, or what the pressure of the gas is (Fig. 13.10c). (The reason that we have to extrapolate is that all gases become liquids or solids before they reach $V = 0$.) This temperature, –273.15°C or –459.67°F, is called **absolute zero**—the lower limit of attainable

Absolute zero is the lower limit of attainable temperatures.

Figure 13.10 (a) Apparatus to verify Charles's law. The pressure of the enclosed gas is held constant by the fixed quantity of mercury resting on top of it. If the temperature of the gas is changed, it expands or contracts, moving the mercury column above it. (b) Charles's law: for a gas held at constant pressure, changes in temperature are proportional to changes in volume. (c) Volume versus temperature graphs for various gases, each at a constant pressure, are extrapolated to $V = 0$. The graphs intersect the temperature axis at the same temperature, T_{limit}. (d) An absolute temperature scale sets $T_{\text{limit}} = 0$.

temperatures. In kelvins—an *absolute* temperature scale—absolute zero is defined as 0 K (Fig. 13.10d). As long as it is understood that an absolute temperature scale is to be used, then Charles's law can be written

$$V \propto T \quad \text{(for constant } P)$$

Physics at Home

Take an empty 2-L soda bottle, cap it tightly, and put it in the freezer. Check it an hour later; what has happened? Estimate the percentage change in the volume of the air inside and compare to the percentage change in absolute temperature (if you don't have a thermometer handy, a typical freezer temperature is about −10°C).

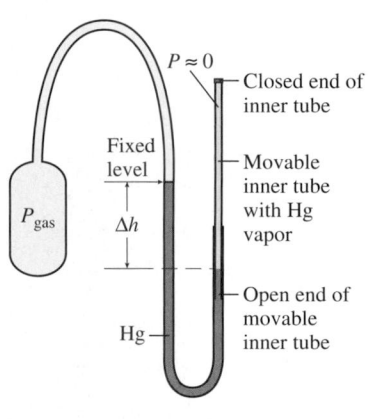

Figure 13.11 A constant volume gas thermometer. A dilute gas is contained in the vessel on the left, which is connected to a mercury manometer. The right side has a tube that can be moved up or down to keep the mercury level on the left at a fixed level, so that the volume of gas is kept constant. Then the manometer is used to measure the absolute pressure of the gas.

Thermal expansion of a gas can be used to measure temperature. Gas thermometers have advantages as well as disadvantages. One huge advantage is that a gas thermometer is universal: it does not matter what gas is used or how many moles of gas are present, as long as the number density is sufficiently low. Gas thermometers give absolute temperature in a natural way and they are extremely accurate and reproducible. The main disadvantage is that they are much less convenient to use than most other thermometers. Gas thermometers are mainly used to calibrate other, more convenient thermometers.

A thermometer based on Charles's law would be called a *constant pressure gas thermometer*. More common is the *constant volume gas thermometer* (Fig. 13.11), which is based on Gay-Lussac's law:

$$P \propto T \quad \text{(for constant } V)$$

Here we keep the volume of the gas constant, measure the pressure and use that to indicate the temperature. (It is much easier to keep the volume constant and measure the pressure than to do the opposite.)

Both Charles's law and Gay-Lussac's law are valid only for a **dilute** gas—a gas where the number density is low enough, or equivalently the average distance between gas molecules is large enough, that interactions between the molecules are negligible except when they collide. Two other experimentally discovered laws that apply to dilute gases are Boyle's law and Avogadro's law. Boyle's law states that the pressure of a gas is inversely proportional to its volume at constant temperature:

$$P \propto \frac{1}{V} \quad \text{(for constant } T)$$

Avogadro's law states that the volume occupied by a gas at a given temperature and pressure is proportional to the number of gas molecules N:

$$V \propto N \quad \text{(constant } P, T)$$

(A constant number of gas molecules was assumed in the statements of Boyle's, Gay-Lussac's, and Charles's laws.)

One equation combines all four of these gas laws—the **ideal gas law**:

Ideal gas law (microscopic form)

$$PV = NkT \quad (N = \text{number of molecules}) \tag{13-13}$$

The constant of proportionality is a universal quantity known as **Boltzmann's constant** (symbol k); its value is

$$k = 1.38 \times 10^{-23} \text{ J/K} \tag{13-14}$$

> In the ideal gas law, T stands for *absolute* temperature (in K).

The macroscopic form of the ideal gas law is written in terms of n, the number of *moles* of the gas, in place of N, the number of molecules. Substituting

$$N = nN_A$$

into the microscopic form yields

$$PV = nN_A kT$$

The product of N_A and k is called the **universal gas constant**:

$$R = N_A k = 8.31 \frac{\text{J/K}}{\text{mol}} \tag{13-15}$$

Then the ideal gas law in macroscopic form is written

Ideal gas law (macroscopic form)

$$PV = nRT \quad (n = \text{number of moles}) \tag{13-16}$$

Many problems deal with the changing pressure, volume, and temperature in a gas with a constant number of molecules (and a constant number of moles). In such problems, it is often easiest to write the ideal gas law as follows:

$$\frac{P_1 V_1}{T_1} = \frac{P_2 V_2}{T_2}$$

Example 13.5

Temperature of the Air in a Tire

Before starting out on a long drive, you check the air in your tires to make sure they are properly inflated. The pressure gauge reads 31.0 lb/in^2 (214 kPa), and the temperature is 15°C. After a few hours of highway driving, you stop and check the pressure again. Now the gauge reads 35.0 lb/in^2 (241 kPa). What is the temperature of the air in the tires now?

Strategy We treat the air in the tire as an ideal gas. We must work with absolute temperatures and absolute pressures when using the ideal gas law. The pressure gauge reads *gauge* pressure; to get absolute pressure we add 1 atm = 101 kPa. We don't know the number of molecules inside the tire or the volume, but we can reasonably assume that neither changes. The number is

constant as long as the tire does not leak. The volume may actually change a bit as the tire warms up and expands, but this change is small. Since N and V are constant, we can rewrite the ideal gas law as a proportionality between P and T.

Solution First convert the initial and final gauge pressures to absolute pressures:

$$P_i = 214 \text{ kPa} + 101 \text{ kPa} = 315 \text{ kPa}$$

$$P_f = 241 \text{ kPa} + 101 \text{ kPa} = 342 \text{ kPa}$$

Now convert the initial temperature to an absolute temperature:

$$T_i = 15°C + 273 \text{ K} = 288 \text{ K}$$

continued on next page

Example 13.5 *continued*

According to the ideal gas law, pressure is proportional to temperature, so

$$\frac{T_f}{T_i} = \frac{P_f}{P_i} = \frac{342 \text{ kPa}}{315 \text{ kPa}}$$

Then

$$T_f = \frac{P_f}{P_i} T_i = \frac{342}{315} \times 288 \text{ K} = 313 \text{ K}$$

Now convert back to °C:

$$313 \text{ K} - 273 \text{ K} = 40°C$$

Discussion The final answer of 40°C seems reasonable since, after a long drive, the tires are noticeably warm, but not hot enough to burn your hand.

 It is often most convenient to work with the ideal gas law by setting up a proportion. In this problem, we did not know the volume or the number of molecules, so

we had no choice. In essence, what we used was Gay-Lussac's law. Starting with the ideal gas law, we can "rederive" Gay-Lussac's law or Charles's law or any other proportionality inherent in the ideal gas law.

Practice Problem 13.5
Air pressure in tire after temperature decreases

Suppose you now (unwisely) decide to bleed air from the tires, since the manufacturer suggests keeping the pressure between 28 lb/in^2 and 32 lb/in^2. (The manufacturer's specification refers to when the tires are "cold.") If you let out enough air so that the pressure returns to 31 lb/in^2, what percentage of the air molecules did you let out of the tires? What is the gauge pressure after the tires cool back down to 15°C?

Physics at Home

The next time you take a car trip, check the tire pressure with a gauge just before the trip and then again after an hour or more of highway driving. Calculate the temperature of the air in the tires from the two pressure readings and the initial temperature. Feel the tire with your hand to see if your calculation is reasonable.

Example 13.6

Scuba Diver

 A scuba diver needs air delivered at a pressure equal to the pressure of the surrounding water—the pressure in the lungs must match the water pressure on the diver's body to prevent the lungs from collapsing. Since the pressure in the air tank is much higher, a regulator delivers air to the diver at the appropriate pressure. The compressed air in a diver's tank lasts 90 minutes at the water surface. About how long does the same tank last at a depth of 20 meters underwater? (Assume that the volume of air breathed per minute does not change.)

Strategy The compressed air in the *tank* is at a pressure much higher than the pressure at which the diver breathes, whether at the surface or at 20 m depth. The constant quantity is *N*, the number of gas molecules in the tank. We also assume that the temperature of the gas remains the same; it may change slightly but much less than the pressure or volume.

Solution Since *N* and *T* are constant,

$$PV = \text{constant}$$

or

$$P \propto 1/V$$

The pressure at the surface is (approximately) 1 atm, while the pressure at 20 m underwater is

$$P = 1 \text{ atm} + \rho gh$$

$$\rho gh = (1000 \text{ kg/m}^3) \times (9.8 \text{ m/s}^2) \times (20 \text{ m}) = 196 \text{ kPa} \approx 2 \text{ atm}$$

Therefore, at a depth of 20 m,

$$P = 3 \text{ atm}$$

To match the pressure of the surrounding water, the pressure of the compressed air is three times larger at a depth of 20 m; then the volume of air is one-third what it was at the surface. The diver breathes the same volume per minute, so the tank will last one-third as long—30 minutes.

continued on next page

Example 13.6 *continued*

Discussion   To do the same thing a bit more formally, we could write:

$$P_i V_i = P_f V_f$$

After setting $P_i = 1$ atm and $P_f = 3$ atm, we find that $V_f/V_i = \frac{1}{3}$.

In this problem, the only numerical values given (indirectly) were the initial and final pressures. Assuming that N and T remain constant, we then can find the ratio of the final and

 Tips   initial volumes. Whenever there *seems* to be insufficient numerical information given in a problem, think in terms of ratios and look for constants that cancel out.

Practice Problem 13.6   Pressure in air tank after temperature increase

A tank of compressed air is at an absolute pressure of 580 kPa at a temperature of 300.0 K. The temperature increases to 330.0 K. What is the pressure in the tank now?

Problem-Solving Tips for the Ideal Gas Law

- In most problems, some change occurs; decide which of the four quantities (P, V, N or n, and T) remain constant during the change.
- Use the microscopic form if the problem deals with the number of molecules and the macroscopic form if the problem deals with the number of moles.
- Use subscripts (i and f) to distinguish initial and final values.
- Work in terms of ratios so that constant factors cancel out.
- Write out the units when doing calculations.
- Remember that P stands for *absolute* pressure (not gauge pressure) and T stands for *absolute* temperature (in kelvins, not °C or °F).

13.6   KINETIC THEORY OF THE IDEAL GAS

In a gas, the interaction between two molecules gets weak rapidly as the distance between the molecules increases. In a dilute gas, the average distance between gas molecules is large enough that we can ignore interactions between the molecules except when they collide. In addition, the volume of space occupied by the molecules themselves is a small fraction of the total volume of the gas—the gas is mostly "empty space." The **ideal gas** is a simplified model of a dilute gas in which we think of the molecules as pointlike particles that move *independently* in free space with no interactions except for elastic collisions.

This simplified model is a good approximation for many gases under ordinary conditions. Many properties of gases can be understood from this model; the microscopic theory based on it is called the **kinetic theory** of the ideal gas.

Microscopic Basis of Pressure

The force that a gas exerts on a surface is due to collisions that the gas molecules make with that surface. For instance, think of the air inside an automobile tire. Whenever an air molecule collides with the inner tire surface, the tire exerts an inward force to turn the air molecule around and return it to the bulk of the gas. By Newton's third law, the gas molecule exerts an outward force on the tire surface. The net force per unit area on the inside of the tire due to all the collisions of the many air molecules is equal to the air pressure in the tire. The pressure depends on three things: how many molecules there are, how often each one collides with the wall, and the momentum transfer due to each collision.

We want to find out how the pressure of an ideal gas is determined by the motions of the gas molecules. To simplify the discussion, consider a gas contained in a box of length L and side area A (Fig. 13.12a)—the result does not depend on the shape of the container. Figure 13.12b shows a gas molecule about to collide with the rightmost wall of the container. For simplicity we assume that the collision is elastic; a more advanced analysis shows that the result is correct even though not all collisions are elastic.

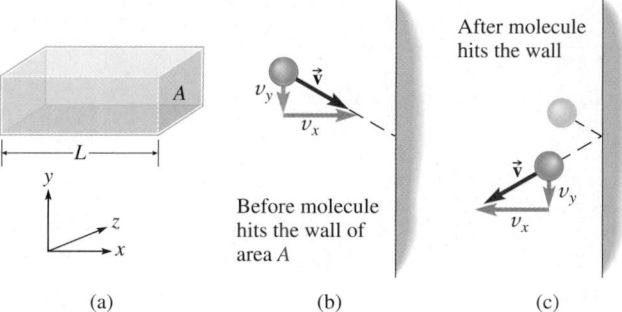

Figure 13.12 (a) Gas molecules confined to a container of length L and area A. (b) A molecule is about to collide with the wall of area A. (c) After an elastic collision, v_x has changed sign, while v_y and v_z are unchanged.

For an elastic collision, the x-component of the molecule's momentum is reversed in direction since the wall is much more massive than the molecule. Since the gas exerts only an outward force on the wall (a static fluid exerts no tangential force on a boundary), the y- and z-components of the molecule's momentum are unchanged. Thus, the molecule's momentum change is $\Delta p_x = 2m|v_x|$.

When does this molecule next collide with the same wall? Ignoring for now collisions with other molecules, its x-component of velocity never changes magnitude—only the sign of v_x changes when it reverses direction (Fig. 13.12c). The time it takes the molecule to travel the length L of the container and hit the other wall is $L/|v_x|$. Then the round-trip time is

$$\Delta t = 2\frac{L}{|v_x|}$$

The average force exerted by the molecule on the wall during one complete round-trip is

$$F_{av,x} = \frac{\Delta p_x}{\Delta t} = \frac{2m|v_x|}{2L/|v_x|} = \frac{m|v_x|^2}{L} = \frac{mv_x^2}{L}$$

The total force on the wall is the sum of the forces due to each molecule in the gas. If there are N molecules in the gas, we can simply multiply N by the *average* force due to one molecule to get the total force on the wall. To represent such an average, we use angle brackets $\langle\rangle$; the quantity inside the brackets is averaged over all the molecules in the gas.

$$F = N\langle F_{av}\rangle = \frac{Nm}{L}\langle v_x^2\rangle$$

The pressure is then

$$P = \frac{F}{A} = \frac{Nm}{AL}\langle v_x^2\rangle$$

The volume of the box is $V = AL$, so

$$P = \frac{Nm}{V}\langle v_x^2\rangle \tag{13-17}$$

which is true regardless of the shape of the container enclosing the gas. Since we end up averaging over all the molecules in the gas, the simplifying assumption about no collisions with other molecules does not affect the result.

The product $m\langle v_x^2\rangle$ suggests kinetic energy. It certainly makes sense that if the average kinetic energy of the gas molecules is larger, the pressure is higher. The average translational kinetic energy of a molecule in the gas is $\langle K_{tr}\rangle = \frac{1}{2}m\langle v^2\rangle$. For any gas molecule, $v^2 = v_x^2 + v_y^2 + v_z^2$, since velocity is a vector quantity. The gas as a whole is at rest, so there is no preferred direction of motion. Then the average value of v_x^2 must be the same as the averages of v_y^2 and v_z^2, so

$$\langle v_x^2\rangle = \tfrac{1}{3}\langle v^2\rangle$$

Therefore,

$$m\langle v_x^2\rangle = \tfrac{1}{3}m\langle v^2\rangle = \tfrac{2}{3}\langle K_{tr}\rangle$$

Substituting this into Eq. (13-17), the pressure is

$$P = \frac{2}{3}\frac{N\langle K_{tr}\rangle}{V} = \frac{2}{3}\frac{N}{V}\langle K_{tr}\rangle \qquad (13\text{-}18)$$

This expression is written with the variables grouped in two different ways to give two different insights. Since $N\langle K_{tr}\rangle$ is the total internal kinetic energy of the gas, the expression with the first grouping says that pressure is proportional to the kinetic energy density (the kinetic energy per unit volume). The second says that pressure is proportional to the product of the number density N/V and the average molecular kinetic energy. The pressure of a gas increases if either the gas molecules are packed closer together or if the molecules have more kinetic energy.

Note that $\langle K_{tr}\rangle$ is the average *translational* kinetic energy of a gas molecule and v is the center-of-mass speed of a molecule. If the gas molecule has more than one atom (such as in N_2), the atoms also have vibrational and rotational kinetic energies, but these are *not* included in $\langle K_{tr}\rangle$.

What about the assumption that the gas molecules never collide with each other? It certainly is *not* true that the same molecule returns to collide with the same wall at a fixed time interval and has the same v_x each time it returns! However, the derivation really only relies on average quantities. In a gas at equilibrium, an average quantity like $\langle v_x^2\rangle$ remains unchanged even though any one particular molecule changes its velocity components as a result of each collision.

Temperature and Translational Kinetic Energy

The temperature of an ideal gas has a direct physical interpretation that we can now bring to light. We found that in an ideal gas, the pressure, volume, and number of molecules are related to the average translational kinetic energy of the gas molecules:

$$P = \frac{2}{3}\frac{N}{V}\langle K_{tr}\rangle \qquad (13\text{-}18)$$

Solving for the average kinetic energy,

$$\langle K_{tr}\rangle = \frac{3}{2}\frac{PV}{N} \qquad (13\text{-}19)$$

The ideal gas law relates P, V, and N to the temperature:

$$PV = NkT \qquad (13\text{-}13)$$

By rearranging the ideal gas law, we find that P, V, and N occur in the same combination as in Eq. (13-19):

$$\frac{PV}{N} = kT$$

Then by substituting kT for $(PV)/N$ in Eq. (13-19), we find that

$$\langle K_{tr}\rangle = \tfrac{3}{2}kT \qquad (13\text{-}20)$$

Therefore, *the absolute temperature of an ideal gas is proportional to the average translational kinetic energy of the gas molecules.* Temperature then is a way to describe the average translational kinetic energy of the gas molecules. At higher temperatures, the gas molecules have (on average) greater kinetic energy.

The speed of a gas molecule that has the average kinetic energy is called the **rms** (root mean square) **speed**. The rms speed is *not* the same as the average speed. Instead, the rms speed is the square *root* of the *mean* (average) of the speed *squared*. Since

$$\langle K_{tr}\rangle = \tfrac{1}{2}m\langle v^2\rangle = \tfrac{1}{2}mv_{rms}^2 \qquad (13\text{-}21)$$

the rms speed is

$$v_{rms} = \sqrt{\langle v^2\rangle}$$

Squaring before averaging emphasizes the effect of the faster-moving molecules, so the rms speed is a bit higher than the average speed—about 9% higher as it turns out.

The pressure of an ideal gas is proportional to the average translational kinetic energy of its molecules and to the number of molecules per unit volume.

Since the average kinetic energy of molecules in an ideal gas depends only on temperature, Eq. (13-21) implies that more massive molecules move more slowly on average than lighter ones at the same temperature. If two different gases are placed in a single chamber so that they reach equilibrium and are at the same temperature, their molecules must have the same average translational kinetic energies. If one gas has molecules of larger mass, its molecules must move with a slower average velocity than those of the gas with the lighter mass molecules. In Problem 62 you can show that

$$v_{rms} = \sqrt{\frac{3kT}{m}} \qquad (13\text{-}22)$$

where k is Boltzmann's constant and m is the mass of a molecule. Therefore, at a given temperature, the rms speed is inversely proportional to the square root of the mass of the molecule.

Example 13.7

O$_2$ Molecules at Room Temperature

Find the average translational kinetic energy and the rms speed of the O$_2$ molecules in air at room temperature (20°C).

Strategy The average translational kinetic energy depends only on temperature. We must remember to use absolute temperature. The rms speed is the speed of a molecule that has the average kinetic energy.

Solution The absolute temperature is
$$20°C + 273 \text{ K} = 293 \text{ K}$$
Therefore, the average translational kinetic energy is

$$\langle K_{tr} \rangle = \tfrac{3}{2}kT = 1.50 \times 1.38 \times 10^{-23} \text{ J/K} \times 293 \text{ K} = 6.07 \times 10^{-21} \text{ J}$$

From the Periodic Table, we find the atomic mass of oxygen to be 16.0 u; the molecular mass of O$_2$ is twice that (32.0 u). First we convert that to kg:

$$32.0 \text{ u} \times 1.66 \times 10^{-27} \text{ kg/u} = 5.31 \times 10^{-26} \text{ kg}$$

The rms speed is the speed of a molecule with the average kinetic energy:

$$\langle K_{tr} \rangle = \tfrac{1}{2}mv_{rms}^2$$

$$v_{rms} = \sqrt{\frac{2\langle K_{tr} \rangle}{m}} = \sqrt{\frac{2 \times 6.07 \times 10^{-21} \text{ J}}{5.31 \times 10^{-26} \text{ kg}}} = 478 \text{ m/s}$$

Discussion How can we decide if the result is reasonable, since we have no first-hand experience watching molecules bounce around? Recall from Chapter 12 that the speed of sound in air at room temperature is 343 m/s. Since sound waves in air propagate by the collisions that occur between air molecules, the speed of sound must be of the same order of magnitude as the average speeds of the molecules.

Practice Problem 13.7   CO$_2$ molecules at room temperature

Find the average translational kinetic energy and the rms speed of the CO$_2$ molecules in air at room temperature (20°C).

Maxwell-Boltzmann Distribution

So far we have considered only the *average* kinetic energy and *rms* speed of a molecule. Sometimes we may want to know more: how many molecules have speeds in a certain range? The distribution of speeds is called the **Maxwell-Boltzmann distribution**. The distribution for oxygen at two different temperatures is shown in Fig. 13.13. The interpretation of the graphs is that the number of gas molecules having speeds between any two values v_1 and v_2 is proportional to the area under the curve between v_1 and v_2. In Fig. 13.13, the shaded areas represent the number of oxygen molecules having speeds above 800 m/s at the two selected temperatures. A relatively small temperature change has a significant effect on the number of gas molecules with high speeds.

Any given molecule changes its kinetic energy often—at each collision, which means billions of times per second. However, the total number of gas molecules in a given kinetic energy range in the gas stays the same, as long as the temperature is constant. In fact, it is the frequent collisions that maintain the stability of the Maxwell-Boltzmann distribution. The collisions keep the kinetic energy distributed among the gas molecules *in the most disordered way possible*, which is the Maxwell-Boltzmann distribution.

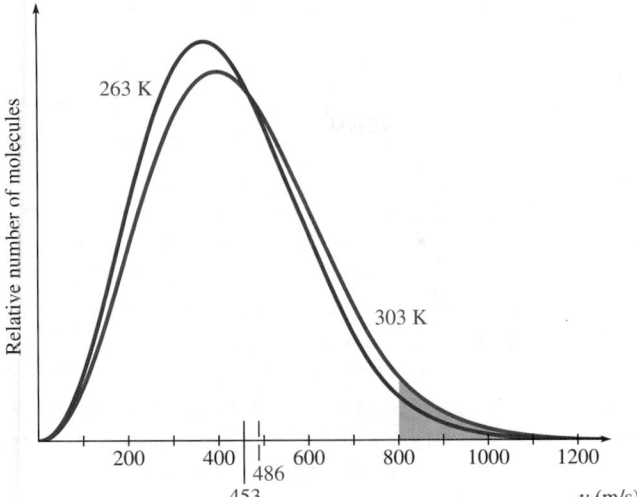

Figure 13.13 The probability distribution of kinetic energies in oxygen at two temperatures: −10°C (263 K) and +30°C (303 K). The area under either curve for any range of speeds is proportional to the number of molecules whose speeds lie in that range. Despite the relatively small difference in rms speeds (453 m/s at 263 K and 486 m/s at 303 K), the number of molecules in the high-speed tail is quite different.

Making The Connection:
composition of
planetary
atmospheres

The Maxwell-Boltzmann distribution helps us understand planetary atmospheres. Why does the Earth's atmosphere contain nitrogen, oxygen, and water vapor, among other gases, but not hydrogen or helium? Gas molecules in the upper atmosphere that are moving faster than the *escape speed* (Section 6.6) have enough kinetic energy to escape from the planetary atmosphere to outer space. Those that are heading away from the planet's surface will escape if they avoid colliding with another molecule. The high-energy tail of the Maxwell-Boltzmann distribution does not get depleted by molecules that escape. Other molecules will get boosted to those high kinetic energies as a result of collisions; these replacements will in turn also escape. Thus, the atmosphere gradually leaks away.

How fast the atmosphere leaks away depends on how far the rms speed is from the escape speed. If the rms speed is too small compared to the escape speed, the time for all the gas molecules to escape is so long that the gas is present in the atmosphere indefinitely. This is the case for nitrogen, oxygen, and water vapor in the Earth's atmosphere. On the other hand, since hydrogen and helium are much less massive, their rms speeds are higher. Though only a tiny fraction of the molecules are above the escape speed, the fraction is sufficient for these gases to escape quickly from the Earth's atmosphere (see Fig. 13.14). The Moon is often said to lack an atmosphere. The Moon's low escape speed (2400 m/s) allows most gases to escape, but it does have an atmosphere about 1 cm tall composed of krypton (a gas with molecular mass 83.8 u, about 2.6 times that of oxygen).

Figure 13.14 Maxwell-Boltzmann distributions for oxygen and hydrogen at $T = 300$ K. Escape speed from the Earth is 11,200 m/s (not shown on the graph).

13.7 TEMPERATURE AND REACTION RATES

What we have learned about the distribution of kinetic energies and its relationship to temperature has a great relevance to the dependence of chemical reaction rates on temperature. Imagine a mixture of two gases, N_2 and O_2, which can react to form nitric oxide (NO):

$$N_2 + O_2 \rightarrow 2NO$$

In order for the reaction to occur, a molecule of nitrogen must collide with a molecule of oxygen. But the reaction does not occur every time such a collision takes place. The reactant molecules must possess enough kinetic energy to initiate the reaction, because the reaction involves the rearrangement of chemical bonds between atoms. Therefore, some chemical bonds must be broken before new ones form; the energy to break these bonds must come from the energy of the reactants. The minimum kinetic energy of the reactant molecules that allows the reaction to proceed is called the **activation energy** (E_a).

If a molecule of N_2 collides with one of O_2, but their total kinetic energy is less than the activation energy, then the two just bounce off each other. Some energy may be transferred from one molecule to the other, or converted between translational, rotational, and vibrational energy, but we are still left with one molecule of N_2 and one of O_2. If the molecules do have enough energy to react, then the reaction may take place.

Now we begin to see why, with few exceptions, rates of reaction increase with temperature. At higher temperatures, the average kinetic energy of the reactants is higher and therefore a greater fraction of the collisions have total kinetic energies exceeding the activation energy. Thus, at higher temperatures the reaction rate is higher. In many situations, only a small fraction of the reactants have enough kinetic energy to react. This situation occurs when the activation energy is much greater than the average kinetic energy of the reactants:

$$E_a \gg \tfrac{3}{2}kT \tag{13-23}$$

Thus, the only candidates for reaction are molecules far off in the high-energy tail of the Maxwell-Boltzmann distribution. In this situation, a small increase in temperature can have an unexpectedly large effect on the reaction rate. Because of this dramatic effect, it turns out that the reaction rate depends *exponentially* on temperature:

$$\text{reaction rate} \propto e^{-E_a/kT} \tag{13-24}$$

Example 13.8

Increase in Reaction Rate with Temperature Increase

 The activation energy for the reaction $N_2O \rightarrow N_2 + O$ is 4.0×10^{-19} J. By what percentage does the reaction rate increase if the temperature is increased from 700.0 K to 707.0 K (a 1% increase in absolute temperature)?

Strategy We should first check that $E_a \gg \tfrac{3}{2}kT$; otherwise, Eq. (13-24) does not apply. Assuming that checks out, we can set up a ratio of the reaction rates at the two temperatures.

Solution Start by calculating $E_a/(kT_1)$, where $T_1 = 700.0$ K:

$$\frac{E_a}{kT_1} = \frac{4.0 \times 10^{-19} \text{ J}}{1.38 \times 10^{-23} \text{ J/K} \times 700.0 \text{ K}} = 41.41$$

So E_a is about 41 times kT, or about 28 times $\tfrac{3}{2}kT$. The activation energy is much greater than the average kinetic energy;

thus, only a small fraction of the collisions might cause a reaction to occur.

At $T_2 = 707.0$ K,

$$\frac{E_a}{kT_2} = \frac{4.0 \times 10^{-19} \text{ J}}{1.38 \times 10^{-23} \text{ J/K} \times 707.0 \text{ K}} = \frac{41.41}{1.01} = 41.00$$

The ratio of the reaction rates is

$$\frac{\text{new rate}}{\text{old rate}} = \frac{e^{-41.00}}{e^{-41.41}} = e^{-(41.00-41.41)} = e^{0.41} = 1.5$$

The reaction rate at 707.0 K is 1.5 times the rate at 700.0 K—a 50% increase in reaction rate for a 1% increase in temperature!

Discussion Normally we might suspect an error when a 1% change in one quantity causes a 50% change in another! However, this illustrates the dramatic effect of an *exponential* dependence. Reaction rates can be *extremely* sensitive to small temperature changes.

Practice Problem 13.8 Decrease in reaction rate for lower temperature

What is the percentage decrease in the rate of the same reaction if the temperature is lowered from 700.0 K to 699.0 K?

Figure 13.15 Warm-blooded animals use different strategies to maintain a constant body temperature. (a) The fur of an Arctic fox serves as a layer of insulation to help it stay warm. (b) Dogs pant and (c) people sweat when their bodies are in danger of overheating. In cases (b) and (c) the evaporation of water has a cooling effect on the body.

Although we have discussed reactions in terms of gases, the same principles apply to reactions in liquid solutions. In order for two molecules to react, they must collide with sufficient kinetic energy to overcome the "barrier" presented by the activation energy. The temperature in turn determines what fraction of the collisions have enough energy to react, and therefore the reaction rates show the same kind of temperature dependence whether the reaction occurs in a gas mixture or a liquid solution.

At the beginning of this chapter we asked about the necessity for temperature regulation in warm-blooded animals (birds and mammals) (see Fig. 13.15). The temperature dependence of chemical reaction rates has a profound effect on biological functions. In warm-blooded animals, vital internal organs and muscles must be kept at their optimal temperatures so they can perform their functions. If our internal temperatures varied we would have a varying metabolic rate, becoming sluggish in cold weather.

Making The Connection:
evolutionary
advantages of
warm-blooded versus
cold-blooded animals

By being able to regulate internal temperatures and maintain a constant body temperature higher than that of the environment, warm-blooded animals are able to tolerate a wider range of environmental temperatures than cold-blooded animals (such as reptiles and insects). Temperature fluctuation in the environment is much more severe on land than in water; thus, land animals are more likely to be homeothermic than aquatic animals. Keeping muscles at their optimal temperatures contributes to the much larger effort required to move around on land or in the air as opposed to moving through water. Keeping the muscles and vital organs warm allows the high level of aerobic metabolism needed to sustain intense physical activity.

Cold-blooded animals are dependent on the environment for temperature regulation; thus, we see a snake lying on a rock heated by the Sun in an attempt to keep warm. The temperature of the snake's blood depends on the temperature of the environment; as its blood temperature goes down in cold weather, the snake becomes inactive and lethargic. Most insects are inactive below 10°C and many cannot survive the cold of winter.

However, if environmental conditions become too extreme, it may be difficult to maintain ideal body temperature. Hypothermia occurs when the central core of the body becomes too cold; bodily processes slow and eventually cease. People caught outside in blizzards are urged to stay awake and to keep moving; the energy produced by exercise may be up to 20 times that produced by the resting body and can compensate for heat loss in extreme cold.

Warm-blooded animals must consume much more food than cold-blooded animals of a similar size; metabolic processes in warm-blooded animals act like a furnace to keep the body warm. A human must consume about 1500 kcal of food energy per day just to keep warm when resting at 20°C; an alligator of similar body weight needs only 60 kcal/day at rest at 20°C.

13.8   COLLISIONS BETWEEN GAS MOLECULES

How far does a gas molecule move, on average, between collisions? The mean (average) length of the path traveled by a gas molecule as a free particle (no interactions with other particles) is called the **mean free path** (Λ, the Greek capital lambda). The mean free path depends on two things: how large the molecules are and how many of them occupy a given volume. In a simplified model, imagine a molecule of diameter d. Its cross-sectional area is

$$A = \pi r^2 = \tfrac{1}{4}\pi d^2$$

As the molecule moves in a straight line, it sweeps out a cylindrical volume of space (Fig. 13.16). Once the volume it has swept out equals V/N, the total volume divided by the number of molecules, it has "used up" its own space and begins to infringe on the space "belonging" to another molecule—which means that a collision is imminent. Thus, a collision occurs when

$$\text{volume} = A\,\Lambda \approx V/N$$

or

$$\Lambda \approx \frac{1}{\tfrac{1}{4}\pi d^2\,(N/V)}$$

This simplified calculation is correct except for the dimensionless constant of proportionality; a detailed calculation yields

Figure 13.16 A molecule of diameter d moves an average distance Λ before colliding. The volume of space swept out while moving this distance is the volume of a cylinder of length Λ and cross-sectional area $A = \tfrac{1}{4}\pi d^2$.

$A = \tfrac{1}{4}\pi d^2$

Mean free path	
$$\Lambda = \frac{1}{\sqrt{2}\,\pi d^2\,(N/V)}$$	(13-25)

Example 13.9

Collisions per Second for N_2 at 20°C and 1 atm

Estimate the average number of collisions per second that each N_2 molecule undergoes in air at room temperature and atmospheric pressure. The diameter of an N_2 molecule is 0.3 nm.

Strategy   First find the mean free path. The average time between collisions is the time it takes to travel that distance at the average molecular speed. For the purposes of this estimate, we can use the rms speed instead of the average speed—the rms speed is only 9% higher.

Solution   Use the ideal gas law to find the number density:

$$PV = NkT$$

$$\frac{N}{V} = \frac{P}{kT} = \frac{1.01 \times 10^5 \text{ Pa}}{1.38 \times 10^{-23} \text{ J/K} \times 293 \text{ K}} = 2.50 \times 10^{25} \text{ m}^{-3}$$

The mean free path is then

$$\Lambda = \frac{1}{\sqrt{2} \pi d^2 (N/V)}$$

$$\Lambda = \frac{1}{\sqrt{2} \pi \times (3 \times 10^{-10} \text{ m})^2 \times 2.50 \times 10^{25} \text{ m}^{-3}}$$

$$\Lambda = 1 \times 10^{-7} \text{ m} = 0.1 \text{ μm}$$

An estimate of the average time between collisions is the time it takes to travel a distance Λ at speed v_{rms}. In Example 13.7, we found the rms speed of O_2 molecules at 293 K to be 478 m/s. The rms speed of N_2 molecules at that temperature is somewhat higher since the molecular mass of N_2 is smaller (28 u, as opposed to 32 u for O_2). Since average speed is inversely proportional to the square root of molecular mass, the difference is insignificant for our estimate. Therefore, taking $v_{rms} \approx$ 500 m/s, the average time between collisions is

$$\langle t \rangle = \frac{\Lambda}{v_{rms}} \approx \frac{1 \times 10^{-7} \text{ m}}{500 \text{ m/s}} = 2 \times 10^{-10} \text{ s} = 0.2 \text{ ns}$$

This is the average time per collision, so the number of collisions per second is

$$\frac{1}{\langle t \rangle} = 5 \times 10^9 \text{ s}^{-1}$$

Each molecule collides about 5×10^9 times per second.

Discussion   Note that the mean free path is larger than the average distance between a molecule and its nearest neighbor. At room temperature and atmospheric pressure, the latter distance is about 4 nm; the mean free path is 25 times larger. We should suspect an error if we found the mean free path to be comparable to or smaller than the distance between a molecule and its *nearest* neighbor—since only occasionally does a molecule collide with a nearest neighbor.

Practice Problem 13.9   Mean free path of a hydrogen atom in space

Intergalactic space is nearly a vacuum: there is on average approximately one hydrogen atom per cubic centimeter. The diameter of a hydrogen atom is about 0.1 nm. (a) Estimate the mean free path of a hydrogen atom under these conditions. (b) Find the rms speed of the hydrogen atoms at temperature 2.7 K. (c) Use (a) and (b) to estimate the average time between collisions in years.

Diffusion

A gas molecule moves in a straight line between collisions—the effect of gravity on the velocity of the molecule is negligible during a time interval of only 0.2 ns. At each collision, both the speed and direction of the molecule's motion change. The mean free path tells us the *average* length of the molecule's straight line paths between collisions. The result is that a given molecule follows a *random walk* trajectory (Fig. 13.17).

After an elapsed time t, how far on average has a molecule moved from its initial position? The answer to this question is relevant when we consider **diffusion**. Someone across the room opens a bottle of perfume: how long until the scent reaches you? As gas molecules diffuse into the air, the frequent collisions are what determine how long it takes the scent to travel across the room (assuming, as we do here, that there are no air currents). When there is a difference in concentrations between different points in a gas, the random thermal motion of the molecules tends to even out the concentrations (other things being equal). The net flow from regions of high concentration (near the perfume bottle) to regions of lower concentration (across the room) is diffusion (Fig. 13.18).

Consider a molecule of perfume in the air. It has a mean free path Λ. After a large number of collisions N, it has traveled a total *distance* $N\Lambda$. However, its displacement from its original position is much less than that, since at each collision it changes direction. It can be shown using statistical analysis of the random walk that the rms magnitude of its displacement after N collisions is proportional to $\sqrt{N}$. Since the number of collisions is proportional to the elapsed time, the rms displacement is proportional to $\sqrt{t}$.

Figure 13.17 Successive straight-line paths traveled by a molecule between collisions

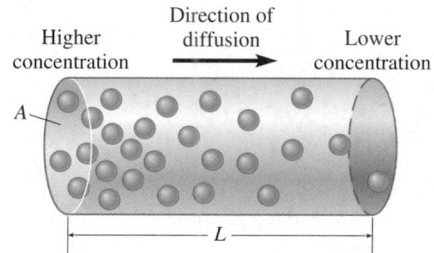

Figure 13.18 Molecules diffuse from regions of higher concentration to regions of lower concentration.

The root mean squared displacement in one direction is

$$x_{rms} = \sqrt{2Dt}$$ (13-26)

where D is a diffusion constant such as those given in Table 13.3. The diffusion constant D depends on the molecule or atom that is diffusing and the medium through which it is moving.

Diffusion is crucial in biological processes such as the transport of oxygen. Oxygen molecules diffuse from the air in the lungs through the walls of the alveoli and then through the walls of the capillaries to oxygenate the blood. The oxygen is then carried by hemoglobin in the blood to various parts of the body, where it again diffuses through capillary walls into intercellular fluids and then through cell membranes into cells. Diffusion is a slow process over long distances but can be quite effective over short distances—which is why cell membranes must be thin and capillaries must have small diameters. Evolution has seen to it that the capillaries of animals of widely different sizes are all about the same size—as small as possible while still allowing blood cells to flow through them.

Table 13.3

Diffusion Constants at 1 atm and 20°C

Diffusing Molecule	Medium	D (m²/s)
DNA	Water	1.3×10^{-12}
Oxygen	Tissue (cell membrane)	1.8×10^{-11}
Hemoglobin	Water	6.9×10^{-11}
Sucrose ($C_{12}H_{22}O_{11}$)	Water	5.0×10^{-10}
Glucose ($C_6H_{12}O_6$)	Water	6.7×10^{-10}
Oxygen	Water	1.0×10^{-9}
Oxygen	Air	1.8×10^{-5}
Hydrogen	Air	6.4×10^{-5}

Example 13.10

Diffusion Time for Oxygen into Capillaries

 How long on average does it take an oxygen molecule in an alveolus to diffuse into the blood? Assume for simplicity that the diffusion constant for oxygen passing through the two membranes (alveolus and capillary walls) is the same: 1.8×10^{-11} m²/s. The total thickness of the two membranes is 1.2×10^{-8} m.

Strategy Take the x-direction to be through the membranes. Then we want to know how much time elapses until $x_{rms} = 1.2 \times 10^{-8}$ m.

continued on next page

Example 13.10 *continued*

Solution Solving Eq. (13-26) for t yields

$$t = \frac{x_{rms}^2}{2D}$$

Now substitute $x_{rms} = 1.2 \times 10^{-8}$ m and $D = 1.8 \times 10^{-11}$ m^2/s:

$$t = \frac{(1.2 \times 10^{-8} \text{ m})^2}{2 \times 1.8 \times 10^{-11} \text{ m}^2/\text{s}} = 4.0 \times 10^{-6} \text{ s}$$

Discussion The time is proportional to the *square* of the membrane thickness. It would take four times as long for an oxygen molecule to diffuse through a membrane twice as thick. The rapid increase of diffusion time with distance is a principal reason why evolution has favored thin membranes over thicker ones.

Practice Problem 13.10 Time for oxygen to get halfway through the membrane

How long on average does it take an oxygen molecule to get *halfway* through the alveolus and capillary wall?

MASTER THE CONCEPTS

Summary

- Temperature is a quantity that determines when objects are in thermal equilibrium. The flow of energy that occurs between two objects or systems due to a temperature difference between them is called heat flow. If heat can flow between two objects or systems, the objects or systems are said to be in thermal contact. When two systems in thermal contact have the same temperature, there is no net flow of heat between them; the objects are said to be in thermal equilibrium.

- Zeroth law of thermodynamics: if two objects are each in thermal equilibrium with a third object, then the two are in thermal equilibrium with each other.

- The SI unit of temperature is the kelvin (symbol K, *without* a degree sign). The kelvin scale is an absolute temperature scale, which means that $T = 0$ is set to absolute zero.

- Temperature in °C (T_C) and temperature in kelvins (T) are related as

$$T_C = T - 273.15 \quad (13\text{-}3)$$

- As long as the temperature change is not too great, the fractional length change of a solid is proportional to the temperature change:

$$\frac{\Delta L}{L_0} = \alpha \Delta T \quad (13\text{-}4)$$

The constant of proportionality, α, is called the coefficient of thermal expansion of the substance.

- The fractional change in volume of a solid or liquid is also proportional to the temperature change as long as the temperature change is not too large:

$$\frac{\Delta V}{V_0} = \beta \Delta T \quad (13\text{-}7)$$

For solids, the coefficient of volume expansion is three times the coefficient of linear expansion: $\beta = 3\alpha$.

- The mole is an SI base unit and is defined as: one mole of anything contains the same number of units as there are atoms in 12 *grams* (not kilograms) of carbon-12. This number is called Avogadro's number and has the value

$$N_A = 6.02 \times 10^{23} \text{ mol}^{-1}$$

- The mass of a molecule is often expressed in the atomic mass unit (symbol u). By definition, one atom of carbon-12 has a mass of 12 u (exactly).

$$1 \text{ u} = 1.66 \times 10^{-27} \text{ kg} \quad (13\text{-}12)$$

The atomic mass unit is chosen so that the mass of a molecule in "u" is numerically the same as the molar mass in g/mol.

- In an ideal gas, the molecules move independently in free space with no interactions except when two molecules collide. The ideal gas is a useful model for many real gases, provided that the gas is sufficiently dilute. The ideal gas law:

$$\text{microscopic form: } PV = NkT \quad (13\text{-}13)$$

$$\text{macroscopic form: } PV = nRT \quad (13\text{-}16)$$

where Boltzmann's constant and the universal gas constant are

$$k = 1.38 \times 10^{-23} \text{ J/K} \quad (13\text{-}14)$$

$$R = N_A k = 8.31 \frac{\text{J/K}}{\text{mol}} \quad (13\text{-}15)$$

- According to kinetic theory, the pressure of a gas is proportional to the average translational kinetic energy of the molecules:

$$P = \frac{2}{3} \frac{N}{V} \langle K_{tr} \rangle \quad (13\text{-}18)$$

- The average translational kinetic energy of the molecules is proportional to the absolute temperature:

$$\langle K_{tr} \rangle = \frac{3}{2} kT \quad (13\text{-}20)$$

- The speed of a gas molecule that has the average kinetic energy is called the rms speed:

$$\langle K_{tr} \rangle = \frac{1}{2} m v_{rms}^2 \quad (13\text{-}21)$$

- The distribution of molecular speeds in an ideal gas is called the Maxwell-Boltzmann distribution.

- If the activation energy for a chemical reaction is much greater than the average kinetic energy of the reactants, the reaction rate depends *exponentially* on temperature:

$$\text{reaction rate} \propto e^{-E_a/kT} \quad (13\text{-}24)$$

MASTER THE CONCEPTS *continued*

- The mean free path (Λ) is the average length of the path traveled by a gas molecule as a free particle (no interactions with other particles) between collisions:

$$\Lambda = \frac{1}{\sqrt{2}\pi d^2\,(N/V)} \qquad (13\text{-}25)$$

- The root mean square displacement of a diffusing molecule along the *x*-axis is

$$x_{\text{rms}} = \sqrt{2Dt} \qquad (13\text{-}26)$$

where *D* is a diffusion constant.

Highlighted Figures and Tables

T13.1 Reference temperatures in K, °C, and °F (p. 443)

T13.2 Coefficients of linear expansion α and volume expansion β (p. 447)

F13.10 Charles's law: If the pressure of a gas is held constant, the change in temperature is proportional to the change in volume (p. 452)

F13.12 How the pressure of an ideal gas is determined by the motions of the gas molecules (p. 456)

F13.13 The probability distribution of kinetic energies in oxygen at two temperatures (p. 459)

F13.14 Maxwell-Boltzmann distributions for oxygen and hydrogen at $T = 300$ K (p. 459)

F13.15 Warm-blooded animals use different strategies to maintain a constant body temperature (p. 461)

CONCEPTUAL QUESTIONS

1. Explain why it would be impossible to uniquely define the temperature of an object if the zeroth law of thermodynamics were violated?

2. Why do we call the temperature 0 K "absolute zero"? How is 0 K fundamentally different from 0°C or 0°F?

3. Under what special circumstances can kelvins or Celsius degrees be used interchangeably?

4. What happens to a hole in a flat metal plate when the plate expands on being heated? Does the hole get larger or smaller?

5. Why would silver and brass probably not be a good choice of metals for a bimetallic strip (leaving aside the question of the cost of silver)? (See Table 13.2.)

6. One way to loosen the lid on a glass jar is to run it under hot water. How does that work?

7. Why must we use absolute temperature (temperature in kelvins) in the ideal gas law ($PV = NkT$)? Explain how using the Celsius scale would give nonsensical results.

8. Natural gas is sold by volume. In the United States, the price charged is usually per cubic foot. Given the price per cubic foot, what other information would you need in order to calculate the price per mole?

9. What are the SI units of mass density and number density? If two different gases have the same number density, do they have the same mass density?

10. Suppose we have two tanks, one containing helium gas and the other nitrogen gas. The two gases are at the same temperature and pressure. Which has the higher number density (or are they equal)? Which has the higher mass density (or are they equal)?

11. The mass of an aluminum atom is 27.0 u. What is the mass of *one mole* of aluminum atoms? (No calculation required!)

12. A ping-pong ball that has been dented during hard play can often be restored by placing it in hot water. Explain why this works.

13. Why does a helium weather balloon expand as it rises into the air? Assume the temperature remains constant.

14. Explain why there is almost no hydrogen (H_2) or helium (He) in the Earth's atmosphere, while both are present in Jupiter's atmosphere. [*Hint:* Escape velocity from the Earth is 11.2 km/s and escape velocity from Jupiter is 60 km/s.]

15. Explain how it is possible that more than half of the molecules in an ideal gas have kinetic energies less than the average kinetic energy. Shouldn't half have less and half have more?

16. In air under ordinary conditions (room temperature and atmospheric pressure), the average intermolecular distance is about 4 nm and the mean free path is about 0.1 μm. The diameter of a nitrogen molecule is about 0.3 nm. Explain how the mean free path can be so much larger than the average distance between molecules.

17. In air under ordinary conditions (room temperature and atmospheric pressure), the average intermolecular distance is about 4 nm and the mean free path is about 0.1 μm. The diameter of a nitrogen molecule is about 0.3 nm. Which two distances should we compare to decide that air is dilute and can be treated as an ideal gas? Explain.

18. In air under ordinary conditions (room temperature and atmospheric pressure), the average intermolecular distance is about 4 nm and the mean free path is about 0.1 μm. The diameter of a nitrogen molecule is about 0.3 nm. What would it mean if the intermolecular distance and the molecular diameter were about the same? In that case, would it make sense to speak of a mean free path? Explain.

19. Explain how an automobile airbag protects the passenger from injury. Why would the airbag be ineffective if the gas pressure inside is too low when the passenger comes into contact with it? What about if it is too high?

20. It takes longer to hard-boil an egg in Mexico City (2200 m above sea level) than it does in Amsterdam (parts of which are below sea level). Why? [*Hint:* At higher altitudes, water boils at less than 100°C.]

MULTIPLE CHOICE QUESTIONS

1. In a mixed gas such as air, the rms speeds of different molecules are
   (a) independent of molecular mass.
   (b) proportional to molecular mass.
   (c) inversely proportional to molecular mass.
   (d) proportional to $\sqrt{\text{molecular mass}}$.
   (e) inversely proportional to $\sqrt{\text{molecular mass}}$.

2. The average kinetic energy of the molecules in an ideal gas increases with the volume remaining constant. Which of these statements *must* be true?
   (a) The pressure increases and the temperature stays the same.
   (b) The number density decreases.
   (c) The temperature increases and the pressure stays the same.
   (d) Both the pressure and the temperature increase.

3. Which of these will increase the average kinetic energy of the molecules in an ideal gas?
   (a) reduce the volume, keeping P and N constant
   (b) increase the volume, keeping P and N constant
   (c) reduce the volume, keeping T and N constant
   (d) increase the pressure, keeping T and V constant
   (e) increase N, keeping V and T constant

4. The absolute temperature of an ideal gas is directly proportional to
   (a) the number of molecules in the sample.
   (b) the average momentum of a molecule of the gas.
   (c) the average translational kinetic energy of the gas.
   (d) the diffusion constant of the gas.

5. The rms speed is
   (a) the speed at which all the gas molecules move.
   (b) the speed of a molecule with the average kinetic energy.
   (c) the average speed of the gas molecules.
   (d) the maximum speed of the gas molecules.

6. What are the most favorable conditions for real gases to approach ideal behavior?
   (a) high temperature and high pressure
   (b) low temperature and high pressure
   (c) low temperature and low pressure
   (d) high temperature and low pressure

7. An ideal gas has the volume V_0. If the temperature and the pressure are each tripled during a process, the new volume is
   (a) V_0.
   (b) $9V_0$.
   (c) $3V_0$.
   (d) $0.33V_0$.

8. The average kinetic energy of a gas molecule can be found from which of these quantities?
   (a) pressure only
   (b) number of molecules only
   (c) temperature only
   (d) pressure and temperature are both required

9. If the temperature of an ideal gas is doubled and the pressure is held constant, the rms speed of the molecules
   (a) remains unchanged.
   (b) is 2 times the original speed.
   (c) is $\sqrt{2}$ times the original speed.
   (d) is 4 times the original speed.

10. A metal box is heated until each of its sides has expanded by 0.1%. By what percent has the *volume* of the box changed?
    (a) –0.3% (b) –0.2% (c) +0.1% (d) +0.2% (e) +0.3%

PROBLEMS

Note: **ⓒ** indicates a combination conceptual/quantitative problem. Gold diamonds ◆, ◆◆ are used to indicate the increasing level of difficulty of each problem. Problem numbers appearing in blue, 9., denote problems that have a detailed solution available in the Student Solutions Manual. Some problems are *paired* by concept; their numbers are connected by a ruled box.

13.2 Temperature Scales

1. On a warm summer day, the air temperature is 84°F. Express this temperature in (a) °C and (b) kelvins.

2. The temperature at which liquid nitrogen boils (at atmospheric pressure) is 77 K. Express this temperature in (a) °C and (b) °F.

3. A room air conditioner causes a temperature change of –6.0°C. (a) What is the temperature change in kelvins? (b) What is the temperature change in °F?

13.3 Thermal Expansion of Solids and Liquids

4. Steel railroad tracks of length 18.30 m are laid at 10.0°C (Fig. 13.19). How much space should be left between the track sections if they are to just touch when the temperature is 50.0°C?

5. A highway is made of concrete slabs that are 15 m long at 20.0°C. (a) If the temperature range at the location of the highway is from –20.0°C to +40.0°C, what size expansion gap should be left (at 20.0°C) to prevent buckling of the highway? (b) How large are the gaps at –20.0°C?

Figure 13.19 Problem 4

6. A lead rod and a common glass rod both have the same length when at 20.0°C. The lead rod is heated to 50.0°C. To what temperature must the glass rod be heated so that they are again at the same length?

7. The coefficient of linear expansion of brass is 1.9×10^{-5} °C^{-1}. At 20.0°C, a hole in a sheet of brass has an area of 1.00 mm^2. How much larger is the area of the hole at 30.0°C?

8. Aluminum rivets used in airplane construction are made slightly too large for the rivet holes to be sure of a tight fit. The rivets are cooled with dry ice (–78.5°C) before they are driven into the holes. If the holes have a diameter of 0.6350 cm at 20.5°C, what should be the diameter of the rivets at 20.5°C if they are to just fit when cooled to the temperature of dry ice?

9. A temperature change ΔT causes a volume change ΔV but has no effect on the mass of an object. (a) Show that the change in density $\Delta \rho$ is given by $\Delta \rho = -\beta \rho \, \Delta T$. (b) Find the fractional change in density ($\Delta \rho/\rho$) of a brass sphere when the temperature changes from 32°C to –10.0°C.

10. An ordinary drinking glass is filled to the brim with water (268.4 mL) at 2.0°C and placed on the sunny pool deck for a swimmer to enjoy. If the temperature of the water rises to 32.0°C before the swimmer reaches for the glass, how much water will have spilled over the top of the glass?

✦ 11. A steel ring of inner diameter 7.00000 cm at 20.0°C is to be heated and placed over a brass shaft of outer diameter 7.00200 cm at 20.0°C. (a) To what temperature must the ring be heated to fit over the shaft? The shaft remains at 20.0°C. (b) Once the ring is on the shaft and has cooled to 20.0°C, to what temperature must the ring plus shaft combination be cooled to allow the ring to slide off the shaft again?

12. A long, narrow steel rod of length 2.5000 m at 25°C is oscillating as a pendulum about a horizontal axis through one end. If the temperature changes to 0°C, what will be the fractional change in its period?

13. The George Washington Bridge crosses the Hudson River in New York. The span of the steel bridge is about 1.6 km. If the temperature in New York can vary from a low of –15°F in winter to a high of 105°F in summer, by how much might the length of the span change over an entire year?

14. A flat square of side s_0 at temperature T_0 expands by Δs in both length and width when the temperature increases by ΔT. The original area is $s_0^2 = A_0$ and the final area is $(s_0 + \Delta s)^2 = A$. Show that if $\Delta s \ll s_0$,

$$\frac{\Delta A}{A_0} = 2\alpha \, \Delta T \qquad (13\text{-}6)$$

(Although we derive this relation for a square plate, it applies to a flat area of any shape.)

15. The volume of a solid cube with side s_0 at temperature T_0 is $V_0 = s_0^3$. Show that if $\Delta s \ll s_0$, the change in volume ΔV due to a change in temperature ΔT is given by

$$\frac{\Delta V}{V_0} = 3\alpha \, \Delta T$$

and therefore that $\beta = 3\alpha$. (Although we derive this relation for a cube, it applies to a solid of any shape.)

16. A square brass plate, 8.00 cm on a side, has a hole cut into its center of area 4.90874 cm^2 (at 20.0°C). The hole in the plate is to slide over a cylindrical steel shaft of cross-sectional area 4.91000 cm^2 (also at 20.0°C) as shown in Fig. 13.20. To what

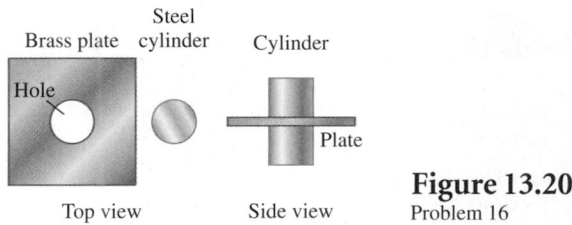

Figure 13.20
Problem 16

temperature must the brass plate be heated so that it can just slide over the steel cylinder (which remains at 20.0°C)? [*Hint:* The steel cylinder is not heated so it does not expand; only the brass plate is heated.]

17. A copper washer is to be fit in place over a steel bolt (Fig. 13.21). Both pieces of metal are at 20.0°C. If the diameter of the bolt is 1.0000 cm and the inner diameter of the washer is 0.9980 cm, at what temperature will the washer fit over the bolt?

Figure 13.21
Problem 17

✦ 18. A steel rule is calibrated for measuring lengths at 20.00°C. The rule is used to measure the length of a Vycor glass brick; when both are at 20.00°C, the brick is found to be 25.00 cm long. If the rule and the brick are both at 80.00°C, what would be the length of the brick as measured by the rule?

13.4 Molecular Picture of a Gas

19. Use the definition that one mol of ^{12}C (carbon-12) atoms has a mass of exactly 12 g, along with Avogadro's number, to derive the conversion between atomic mass units and kg.

20. Find the molecular mass of ammonia (NH_3).

21. Find the mass (in kg) of one molecule of CO_2.

22. The mass of one mole of ^{13}C (carbon-13) is 13.003 g. (a) What is the mass in u of one ^{13}C atom? (b) What is the mass in kg of one ^{13}C atom?

23. Estimate the number of H_2O molecules in a human body of mass 80.2 kg. Assume that, on average, water makes up about 62% of the mass of a human body.

24. At 0.0°C and 1.00 atm, 1.00 mol of a gas occupies a volume of 0.0224 m^3. (a) What is the number density? (b) Estimate the average distance between the gas molecules. (c) If the gas is nitrogen (N_2), the principal component of air, what is the total mass and mass density?

25. Sand is composed of SiO_2. Find the order of magnitude of the number of silicon (Si) atoms in a grain of sand. Approximate the sand grain as a sphere of diameter 0.5 mm and an SiO_2 molecule as a sphere of diameter 0.5 nm.

26. The mass density of diamond (a crystalline form of carbon) is 3500 kg/m^3. How many carbon atoms per cm^3 are there?

27. How many hydrogen atoms are present in 684.6 g of sucrose ($C_{12}H_{22}O_{11}$)?

28. How many moles of He are in 13 g of He?

29. The principal component of natural gas is methane (CH_4). How many moles of CH_4 are present in 144.36 g of methane?

30. What is the mass of one gold atom in kg?

31. Air at room temperature and atmospheric pressure has a mass density of 1.2 kg/m^3. The average molecular mass of air is 29.0 u. How many air molecules are there in 1.0 cm^3 of air?

13.5 Absolute Temperature and the Ideal Gas Law

32. Verify, using the ideal gas law, the assertion in Problem 24 that 1.00 mol of a gas at 0.0°C and 1.00 atm occupies a volume of 0.0224 m^3.

33. Verify that the SI units of PV (pressure times volume) are joules.

34. Incandescent lightbulbs are filled with an inert gas to lengthen the filament life. With the current off (at $T = 20.0$°C), the gas inside a lightbulb has a pressure of 115 kPa. When the bulb is burning, the temperature rises to 70.0°C. What is the pressure at the higher temperature?

35. What fraction of the air molecules in a house must be pushed outside while the furnace raises the inside temperature from 16.0°C to 20.0°C? The pressure does not change since the house is not 100% airtight.

36. What is the mass density of air at $P = 1.0$ atm and $T =$ (a) –10°C and (b) 30°C? The average molecular mass of air is approximately 29 u.

37. A constant volume gas thermometer containing helium is immersed in boiling ammonia (–33°C) and the pressure is read once equilibrium is reached. The thermometer is then moved to a bath of boiling water (100.0°C). After the manometer was adjusted to keep the volume of helium constant, by what factor was the pressure multiplied?

✦ 38. A bubble rises from the bottom of a lake of depth 80.0 m, where the temperature is 4°C. The water temperature at the surface is 18°C. If the bubble's initial diameter is 1.00 mm, what is its diameter when it reaches the surface? (Ignore the surface tension of water. Assume the bubble warms as it rises to the same temperature as the water and retains a spherical shape. Assume $P_{atm} = 1.0$ atm.)

✦ 39. A bubble with a volume of 1.00 cm^3 forms at the bottom of a lake that is 20.0 m deep. The temperature at the bottom of the lake is 10.0°C. The bubble rises to the surface where the water temperature is 25.0°C. Assume that the bubble is small enough that its temperature always matches that of its surroundings. What is the volume of the bubble just before it breaks the surface of the water? Ignore surface tension.

40. A hydrogen balloon at the Earth's surface has a volume of 5.0 m^3 on a day when the temperature is 27°C and the pressure is 1.00×10^5 N/m^2. The balloon rises and expands as the pressure drops. What would the volume of the same number of moles of hydrogen be at an altitude of 40 km where the pressure is 0.33×10^3 N/m^2 and the temperature is –13°C?

41. An ideal gas that occupies 1.2 m^3 at a pressure of 1.0×10^5 Pa and a temperature of 27°C is compressed to a volume of 0.60 m^3 and heated to a temperature of 227°C. What is the new pressure?

42. A diver rises quickly to the surface from a 5.0-m depth. If she did not exhale the gas from her lungs before rising, by what factor would her lungs expand? Assume the temperature to be constant

and the pressure in the lungs to match the pressure outside the diver's body. The density of seawater is 1.03×10^3 kg/m^3.

43. An emphysema patient is breathing pure O_2 through a face mask. The cylinder of O_2 contains 0.60 ft^3 of O_2 gas at a pressure of 2200 lb/in^2. (a) What volume would the oxygen occupy at atmospheric pressure (and the same temperature)? (b) If the patient takes in 8 L/min of O_2 at atmospheric pressure, how long will the cylinder last?

44. A scuba diver has an air tank with a volume of 0.010 m^3. The air in the tank is initially at a pressure of 1.0×10^7 Pa. Assuming that the diver breathes 0.500 L/s of air, find how long the tank will last at depths of (a) 2.0 m and (b) 20.0 m.

45. In intergalactic space there is an average of about one hydrogen atom per cm^3 and the temperature is 3 K. What is the absolute pressure?

46. A tank of compressed air of volume 1.0 m^3 is pressurized to 20.0 atm at $T = 273$ K. A valve is opened and air is released until the pressure in the tank is 15.0 atm. How many air molecules were released?

47. A mass of 0.532 kg of molecular oxygen is contained in a cylinder at a pressure of 1.0×10^5 Pa and a temperature of 0.0°C. What volume does the gas occupy?

48. Consider the expansion of an ideal gas at constant pressure. The initial temperature is T_0 and the initial volume is V_0. (a) Show that $\Delta V/V_0 = \beta \Delta T$, where $\beta = 1/T_0$. (b) Compare the coefficient of volume expansion β for an ideal gas at 20°C to the values for liquids and gases listed in Table 13.2.

13.6 Kinetic Theory of the Ideal Gas

49. What is the temperature of an ideal gas whose molecules have an average translational kinetic energy of 3.20×10^{-20} J?

50. What is the total translational kinetic energy of the gas molecules of 0.420 mol of air at atmospheric pressure that occupies a volume of 1.00 L (0.0010 m^3)?

51. What is the kinetic energy per unit volume in an ideal gas at (a) $P = 1.00$ atm and (b) $P = 300.0$ atm?

52. Show that, for an ideal gas,

$$P = \tfrac{1}{3}\rho v_{rms}^2$$

where P is the pressure, ρ is the mass density, and v_{rms} is the rms speed of the gas molecules.

53. Estimate the percentage of the O_2 molecules in air at 0.0°C and 1.00 atm that are moving faster than the speed of sound (see Fig. 13.13).

54. What is the total internal kinetic energy of 1.0 mol of an ideal gas at 0.0°C and 1.00 atm?

55. There are two identical containers of gas at the same temperature and pressure, one containing argon and the other neon. What is the ratio of the rms speed of the argon atoms to that of the neon atoms? The atomic mass of argon is twice that of neon.

56. A smoke particle has a mass of 1.38×10^{-17} kg and it is randomly moving about in thermal equilibrium with room temperature air at 27°C. What is the rms speed of the particle?

57. Find the rms speed in air at 0.0°C and 1.00 atm of (a) the N_2 molecules, (b) the O_2 molecules, and (c) the CO_2 molecules.

58. What are the rms speeds of helium atoms, and nitrogen, hydrogen, and oxygen molecules at 25°C?

59. If the upper atmosphere of Jupiter has a temperature of 160 K and the escape speed is 60 km/s, would an astronaut expect to find much hydrogen there?

60. A sealed cylinder contains a sample of ideal gas at a pressure of 2.0 atm. The rms speed of the molecules is v_0. If the rms speed is then reduced to $0.90v_0$, what is the pressure of the gas?

61. What is the temperature of an ideal gas whose molecules in random motion have an average translational kinetic energy of 3.20×10^{-20} J?

62. Show that the rms speed of a molecule in an ideal gas at absolute temperature T is given by

$$v_{\mathrm{rms}} = \sqrt{\frac{3kT}{m}}$$

where k is Boltzmann's constant and m is the mass of a molecule.

63. Show that the rms speed of a molecule in an ideal gas at absolute temperature T is given by

$$v_{\mathrm{rms}} = \sqrt{\frac{3RT}{M}}$$

where M is the *molar mass*—the mass of the gas per mole.

13.7 Temperature and Reaction Rates

♦64. The reaction rate for the hydrolysis of benzoyl-l-arginine amide by trypsin at 10.0°C is 1.878 times faster than that at 5.0°C. Assuming that the reaction rate is exponential as in Eq. (13-24), what is the activation energy?

♦65. The reaction rate for the prepupal development of male *Drosophila* is temperature-dependent. Assuming that the reaction rate is exponential as in Eq. (13-24), the activation energy for this development is then 2.81×10^{-19} J. A *Drosophila* is originally at 10.00°C and its temperature is increasing. If the rate of development has increased 3.5%, how much has its temperature increased?

♦66. At high altitudes, water boils at a temperature lower than 100.0°C due to the lower air pressure. A rule of thumb states that the time to hard-boil an egg doubles for every 10.0°C drop in temperature. What activation energy does this rule imply for the chemical reactions that occur when the egg is cooked?

13.8 Collisions between Gas Molecules

67. Estimate the mean free path of a N_2 molecule in air at (a) sea level ($P \approx 100$ kPa and $T \approx 290$ K), (b) the top of Mt. Everest (altitude = 8.8 km, $P \approx 50$ kPa, and $T \approx 230$ K), and (c) an altitude of 100 km ($P \approx 0.03$ Pa and $T \approx 1300$ K). For simplicity, assume that air is pure nitrogen gas. The diameter of a N_2 molecule is approximately 0.3 nm.

68. About how long will it take a perfume molecule to diffuse a distance of 5.00 m in one direction in a room if the diffusion constant is 1.00×10^{-5} m²/s? Assume that the air is perfectly still—there are no air currents.

69. Estimate the time it takes a sucrose molecule to move 5.00 mm in one direction by diffusion in water. Assume there are no currents in the water.

COMPREHENSIVE PROBLEMS

70. The automobile driver from Practice Problem 13.3 fills his 18.9-L steel gasoline tank in the cool of the morning when the temperature of the tank and the gasoline is 15.0°C. The temperature climbs to 30.0°C by 1 P.M., but the steel tank is reinforced so that it cannot expand and there is a tight-fitting gas cap with no allowance for an air space or pressure release. What is the pressure on the tank from the heated gasoline, which cannot expand in volume? The bulk modulus for gasoline is 1.00×10^9 N/m².

71. An iron bridge girder ($Y = 2.0 \times 10^{11}$ N/m²) is constrained between two rock faces whose spacing doesn't change. At 20.0°C the girder is relaxed. How large a stress develops in the iron if the sun heats the girder to 40.0°C?

♦72. A wine barrel has a diameter at its widest point of 134.460 cm at a temperature of 20.0°C. A circular iron band, of diameter 134.448 cm, is to be placed around the barrel at the widest spot. The iron band is 5.00 cm wide and 0.500 cm thick. (a) To what temperature must the band be heated to be able to fit it over the barrel? (b) Once the band is in place and cools to 20.0°C, what will be the tension in the band?

♦73. The inner tube of a Pyrex glass mercury thermometer has a diameter of 0.120 mm. The bulb at the bottom of the thermometer contains 0.200 cm³ of mercury. How far will the thread of mercury move for a change of 1.00°C? Remember to take into account the expansion of the glass.

74. An iron cannonball of radius 0.08 m has a cavity of radius 0.05 m that is to be filled with gunpowder. If the measurements were made at a temperature of 22°C, how much extra volume of gunpowder, if any, will be required to fill 500 cannonballs when the temperature is 30°C?

♦♦75. A bimetallic strip (Fig. 13.22) is made from metals with expansion coefficients α_1 and α_2 (with $\alpha_2 > \alpha_1$). The thickness of each layer is s. At some temperature T_0 the bimetallic strip is relaxed and straight. (a) Show that, at temperature $T_0 + \Delta T$, the radius of curvature of the strip is

$$R \approx \frac{s}{(\alpha_2 - \alpha_1)\Delta T}$$

[*Hint:* At T_0 the lengths of the two layers are the same. At temperature $T_0 + \Delta T$, the layers form circular arcs of radii R and $R + s$, which subtend the same angle θ. Assume a small ΔT so that $\alpha\Delta T \ll 1$ (for either α).] (b) If the layers are made of iron and brass, with $s = 0.1$ mm, what is R for $\Delta T = 20.0$°C?

Figure 13.22
Problem 75

76. Estimate the average distance between air molecules at 0.0°C and 1.00 atm.

77. If you wanted to make a scale model of air at 0.0°C and 1.00 atm, using ping-pong balls (diameter, 3.75 cm) to

represent the N_2 molecules (diameter, 0.30 nm), (a) how far apart on average should the ping-pong balls be at any instant? (b) How far would a ping-pong ball travel on average before colliding with another?

78. For divers going to great depths, the composition of the air in the tank must be modified. The ideal composition is to have approximately the same number of O_2 molecules per unit volume as in surface air (to avoid oxygen poisoning), and to use helium instead of nitrogen for the remainder of the gas (to avoid nitrogen narcosis, which results from nitrogen dissolving in the bloodstream). Of the molecules in dry surface air, 78% are N_2, 21% are O_2, and 1% are Ar. (a) How many O_2 molecules per m^3 are there in surface air at 20.0°C and 1.00 atm? (b) For a diver going to a depth of 100.0 m, what percentage of the gas molecules in the tank should be O_2? (Assume that the density of seawater is 1025 kg/m^3 and the temperature is 20.0°C.)

79. Show that, in two gases at the same temperature, the rms speeds are inversely proportional to the square root of the molecular masses:

$$\frac{(v_{rms})_1}{(v_{rms})_2} = \sqrt{\frac{m_2}{m_1}}$$

80. Agnes Pockels, a housewife, was able to determine Avogadro's number using only a few household chemicals, in particular oleic acid, whose formula is $C_{18}H_{34}O_2$. (a) What is the molecular mass of this acid? (b) The mass of one drop of oleic acid is 2.3×10^{-5} g and the volume is 2.6×10^{-5} cm^3. How many moles of oleic acid are there in one drop? (c) Now all Agnes Pockels needed was to find the number of molecules of oleic acid. Luckily, when oleic acid is spread out on water, it lines up in a layer one molecule thick. If the base of the molecule of oleic acid is a square of side d, the height of the molecule is known to be $7d$. Agnes Pockels spread out one drop of oleic acid on some water, and measured the area to be 70.0 cm^2. Using the volume and the area of oleic acid, what is d? (d) If we assume that this film is one molecule thick, how many molecules of oleic acid are there in the drop? (e) What value does this give you then for Avogadro's number?

81. A certain acid has a molecular mass of 63 u. By weight, it consists of 1.6% hydrogen, 22.2% nitrogen, and 76.2% oxygen. What is the chemical formula for this acid?

82. These data are from a constant-volume gas thermometer experiment. The volume of the gas was kept constant, while the temperature was changed. The resulting pressure was measured. Plot the data on a pressure versus temperature diagram. Based on these data, estimate the value of absolute zero in Celsius.

T (°C)	P (atm)
0	1.00
20	1.07
100	1.37
−33	0.88
−196	0.28

83. Given that our body temperature is 98.6°F, (a) what is the average kinetic energy of the air molecules in our lungs? (b) If our temperature has increased to 100.0°F, by what percentage has the kinetic energy of the molecules increased?

84. The volume of air taken in by a warm-blooded vertebrate in the Andes mountains is 210 L/day at standard temperature and pressure (i.e., 0°C and 1 atm). If the air in the lungs is at 39°C, under a pressure of 450 mm Hg, and we assume that the vertebrate takes in an average volume of 100 cm^3 per breath at the temperature and pressure of its lungs, how many breaths does this vertebrate take per day?

85. The alveoli (see Section 13.8) have an average radius of 0.125 mm and are approximately spherical. If the pressure in the sacs is 1.00×10^5 Pa, and the temperature is 310 K (average body temperature), how many air molecules are in an alveolus?

86. In plants, water diffuses out through small openings known as stomatal pores. If $D = 2.4 \times 10^{-5}$ m^2/s for water vapor in air, and the length of the pores is 2.5×10^{-5} m, how long does it take for a water molecule to diffuse out through the pore?

87. During hibernation, an animal's metabolism slows down, and its body temperature lowers. For example, a California ground squirrel's body temperature lowers from 40.0°C to 10.0°C during hibernation. If we assume that the air in the squirrel's lungs is 75.0% N_2 and 25.0% O_2, by how much will the rms speed of the air molecules in the lungs have decreased during hibernation?

88. A 10.0-L vessel contains 12 g of N_2 gas at 20°C. (a) Estimate the nearest-neighbor distance. (b) Is the gas dilute? [*Hint:* Compare the nearest-neighbor distance to the diameter of an N_2 molecule, about 0.3 nm.]

ANSWERS TO PRACTICE PROBLEMS

13.1 37.0°C; 310.2 K

13.2 0.60 mm longer; 1.5 mm shorter

13.3 0.26 L

13.4 3.34×10^{28} molecules/m^3

13.5 7.9% of the air molecules; 189 kPa (27 lb/in^2)

13.6 640 kPa

13.7 v_{rms} = 408 m/s—lower than that of oxygen since the molecule is more massive

13.8 6% decrease

13.9 (a) 2×10^{10} km ≈ 100 × the Earth-Sun distance!; (b) 260 m/s; (c) 2000 yr

13.10 7.0×10^{-6} s

Heat

The weather forecast predicts a late spring hard freeze one night; the temperature is to fall several degrees below 0°C and the apple crop is in danger of being ruined. To protect the tender buds, farmers rush out and spray the trees with water. How does that protect the buds?

14.1 INTERNAL ENERGY

From Section 13.6, the average translational kinetic energy $\langle K_{tr} \rangle$ of the molecules of an ideal gas is proportional to the absolute temperature of the gas:

$$\langle K_{tr} \rangle = \tfrac{3}{2}kT \tag{13-20}$$

The molecules move about in random directions even though, on a macroscopic scale, the gas is neither moving nor rotating. Equation (13-20) also gives the average translational kinetic energy of the random motion of molecules in liquids, solids, and nonideal gases except at very low temperatures. This random microscopic kinetic energy is *part* of what we call the *internal energy* of the system:

> The **internal energy** of a system is the total energy of all of the molecules
> in the system *except* for the macroscopic kinetic energy (kinetic energy
> associated with macroscopic translation or rotation) and the external
> potential energy (energy due to external interactions).

A **system** is whatever we define it to be: one object or a group of objects. Everything that is not part of the system is considered to be external to the system, or in other words, in the surroundings of the system.

Internal energy includes

- Translational and rotational kinetic energy of molecules *due to their individual random motions*.

- Vibrational energy—both kinetic and potential—of molecules and of atoms within molecules due to random vibrations about their equilibrium points.

- Potential energy due to interactions between the atoms and molecules of the system.

- Chemical and nuclear energy—the kinetic and potential energy associated with the binding of atoms to form molecules, the binding of electrons to nuclei to form atoms, and the binding of protons and neutrons to form nuclei.

Internal energy does *not* include

- The kinetic energy of the molecules due to translation, rotation, or vibration of the whole system or of a macroscopic part of the system.

- Potential energy due to interactions of the molecules of the system with something outside of the system (such as a gravitational field due to something outside of the system).

Example 14.1

Dissipation of Energy by Friction

A block of mass 10.0 kg starts at point A at a height of 2.0 m above the horizontal and slides down a frictionless incline (Fig. 14.1). It then continues sliding along the horizontal surface of a table that has friction. The block comes to rest at point C, a distance of 1.0 m along the table surface. How much has the internal energy of the system (block + table) increased?

Strategy Gravitational potential energy is converted to macroscopic translational kinetic energy as the block's speed increases. Friction then converts this macroscopic kinetic energy into internal energy—some of it in the block and some in the table. Since total energy is conserved, the increase in

continued on next page

Figure 14.1
An object sliding down a frictionless incline and then across a horizontal surface with friction.

Example 14.1 *continued*

internal energy is equal to the decrease in gravitational potential energy:

decrease in PE from A to B = increase in KE from A to B

= decrease in KE from B to C

= increase in internal energy from B to C

Solution   The initial potential energy (taking $U_g = 0$ at the horizontal surface) is

$$U_g = mgh = 10.0 \text{ kg} \times 9.8 \text{ m/s}^2 \times 2.0 \text{ m} = 200 \text{ J}$$

The final potential energy is zero. The initial and final translational kinetic energies of the block are both zero. Therefore, the increase in the internal energy of the block and table is 200 J.

Discussion   We do not know how much of this internal energy increase appears in the object and how much in the table; we can only find the total. We call friction a *nonconservative* force, but that only means that *macroscopic mechanical* energy is not conserved; total energy is always conserved. Friction merely converts some macroscopic mechanical energy into internal energy of the block and the table. This internal energy increase manifests itself as a slight temperature increase. We often say that mechanical energy is *dissipated* by friction or other nonconservative forces; in other words, energy in an ordered form (translational motion of the block) has been changed into disordered energy (random motion of molecules within the block and table).

Practice Problem 14.1   On the rebound

If a rubber ball of mass 1.0 kg is dropped from a height of 2.0 m and rebounds on the first bounce to 0.75 of the height from which it was dropped, how much energy is dissipated during the collision with the floor?

Physics at Home

Take an elastic band and hold a portion to your lip; it should feel cool to the touch. Then grasp a section of the elastic between your thumbs and forefingers and stretch it back and forth rapidly several times. Hold the stretched region to your lip. For contrast, hold an unstretched portion of the same elastic band to your lip. The unstretched portion feels relatively cool and the stretched portion feels warm. The work you did in stretching the elastic increased its internal energy.

A change in the internal energy of a system does not always cause a temperature change. As we explore further in Section 14.5, the internal energy of a system can change while the temperature of the system remains constant—for instance, when ice melts.

Conceptual Example 14.2

Internal Energy of a Bowling Ball

A bowling ball at rest has a temperature of 18°C. The ball is then rolled down a bowling alley. Neglecting the dissipation of energy by friction and drag forces, is the internal energy of the ball higher, lower, or the same as when the ball was at rest? Is the temperature of the ball higher, lower, or the same as when the ball was at rest?

Strategy, Solution, and Discussion   The only change is that the ball is now rolling—the ball has macroscopic translational and rotational kinetic energy. However, the definition of *internal* energy does *not* include the kinetic energy of the molecules due to translation, rotation, or vibration *of the system as a whole*. Therefore, the *internal* energy of the ball is the same. Temperature is associated with the average translational kinetic energy due to the *individual random* motions of molecules; the temperature is still 18°C.

Conceptual Practice Problem 14.2   Total translational KE

Is the *total* translational kinetic energy of the molecules in the ball higher, lower, or the same as when the ball was at rest?

14.2   HEAT

We defined heat in Chapter 13:

Definition of heat:
Heat is a *flow* of energy between two objects or systems due to a temperature difference between them.

Many eighteenth-century scientists thought that heat was a fluid called "caloric." The flow of heat into an object was thought to cause the object to expand in volume in order to accommodate the additional fluid; why no mass increase occurred was a mystery. Now we know that heat is not a substance but is a flow of energy. One experiment that led to this conclusion was carried out by Count Rumford (Benjamin Thompson, 1753–1814). While supervising the boring of cannon barrels, he noted that the drill doing the boring became quite hot. At the time it was thought that the grinding up of the cannon metal into little pieces caused caloric to be released because the tiny bits of metal could not hold as much caloric as the large piece from which they came. But Rumford noticed that the drill got hot even when it became so dull that metal was no longer being bored out of the cannon and that he could create a limitless amount of what we now call internal energy. He decided that "heat" must be a form of microscopic motion instead of a material substance.

It was not until later experiments were done by James Prescott Joule (1818–1889) that Rumford's ideas were finally accepted. In his most famous experiment (Fig. 14.2), Joule showed that a temperature increase can be caused by mechanical means. In a series of such experiments, Joule determined the "mechanical equivalent of heat," or the amount of mechanical work required to produce the same effect on a system as a given amount of heat. In those days heat was measured in calories, where one calorie was defined as the heat required to change the temperature of 1 g of water by 1°C (specifically from 14.5 to 15.5°C). Joule's experimental results were within 1% of the currently accepted value, which is

$$1 \text{ cal} = 4.186 \text{ J} \tag{14-1}$$

Equation (14-1) is now the *definition* of the calorie. The Calorie (with an uppercase letter C) used by dietitians and nutritionists is actually a kilocalorie:

$$1 \text{ Calorie} = 1 \text{ kcal} = 10^3 \text{ cal}$$

Although the calorie is still used, the SI unit for internal energy and for heat is the same as that used for all forms of energy and all forms of energy transfer: the joule.

Heat and work are similar in that both describe a particular kind of energy *transfer*. Work is an energy transfer due to a force acting through a displacement. Heat is a microscopic form of energy transfer involving large numbers of particles; the exchange of energy occurs due to the individual interactions of the particles. No macroscopic displacement occurs when heat flows and no macroscopic force is exerted by one object on the other.

It does not make sense to say that a system *has* 15 kJ of heat, just as it does not make sense to say that a system *has* 15 kJ of work. Similarly, we cannot say that the heat of a system has changed (nor that the work of a system has changed). A system can possess

Figure 14.2 Joule's experiment. As the two masses fall under the effects of gravity, they cause a paddle wheel to rotate within an insulated container filled with water. The paddles agitate the water and cause its temperature to rise. By measuring the distance through which the masses fell and the temperature change of the known quantity of water, Joule determined the mechanical work done and the internal energy increase of the water.

energy in various forms (including internal energy); heat and work are two ways that energy can be transferred from one system to another. Joule's experiments showed that a quantity of work done on a system or the same quantity of heat flowing into the system causes the same increase in the system's internal energy. If the internal energy increase comes from mechanical work, as from Joule's paddle wheel, no heat flow occurs.

Heat flows from a system at higher temperature to one at lower temperature. Temperature is associated with the microscopic translational kinetic energy of the molecules; thus the flow of heat tends to equalize the average microscopic translational kinetic energy of the molecules. When two systems are in thermal contact and no net heat flow occurs, the systems are in thermal equilibrium and have the same temperature. In Chapter 13, we saw that heat always flows in the direction from a hotter object to a colder one; we will see in Chapter 15 that this is one way of stating the second law of thermodynamics.

Example 14.3

A Joule Experiment

In an experiment similar to that done by Joule, an object of mass 12.0 kg descends a distance of 1.25 m at constant speed while causing the rotation of a paddle wheel in an insulated container of water. If the descent is repeated 20.0 times, what is the internal energy increase of the water in joules? How many calories of heat flowing into the water would cause the same change in internal energy?

Strategy   Each time the mass descends, it converts gravitational potential energy into kinetic energy of the paddle wheel, which in turn agitates the water and converts kinetic energy into internal energy.

Solution   The change in gravitational potential energy during 20.0 downward trips is

$\Delta U_g = mg\Delta h = 12.0\,\text{kg} \times 9.80\,\text{N/kg} \times 20.0\,\text{descents} \times 1.25\,\text{m/descent}$

$= 2.94\,\text{kJ}$

If all of this energy goes into the water, the internal energy increase of the water is 2.94 kJ. From the definition of the calorie (1 cal = 4.186 J), the number of calories of heat that would cause an internal energy change of 2.94 kJ is

$$\text{number of calories} = \frac{2.94 \times 10^3\,\text{J}}{4.186\,\text{J/cal}} = 702\,\text{cal}$$

Discussion   To perform an experiment like Joule's, we can vary the amount of energy delivered to the water. One way is to change the number of times the object is allowed to descend. Other possibilities include varying the mass of the descending object or raising the apparatus so that the object can descend a greater distance. All of these variations allow a change in the amount of gravitational potential energy converted into internal energy without requiring any changes in the complicated mechanism involving the paddle wheel.

Practice Problem 14.3   Temperature change of the water

If the water temperature in the insulated container is found to have increased 2.0°C after 20.0 descents of the falling object, what mass of water is in the container? Assume all of the internal energy increase appears in the water (neglect any internal energy change of the paddle wheel itself). [*Hint*: Recall that the calorie was defined as the heat required to change the temperature of 1 g of water by 1°C.]

The Cause of Thermal Expansion

If not to accommodate additional "caloric," then why do objects generally expand when their temperatures increase? (See Section 13.3.) An object expands when the *average* distance between the atoms (or molecules) increases. The atoms are not at rest; even in a solid, where each atom has a fixed equilibrium position, they *vibrate* to and fro about their equilibrium positions. The energy of vibration is part of the internal energy of the object. When heat flows into the object, raising its temperature, the internal energy increases. Some of the increase goes into vibration, so the average vibrational energy of an atom increases with increasing temperature.

The forces between atoms are highly asymmetric. Two atoms separated by *less* than their equilibrium distance repel each other *strongly*, while two atoms separated by *more* than their equilibrium distance attract each other much less strongly. Therefore, as vibrational energy increases, the maximum distance between the atoms increases more than the minimum distance decreases; the *average* distance between the atoms increases.

The coefficient of expansion varies from material to material because the strength of the interatomic (or intermolecular) bonds varies. As a general rule, the stronger the atomic bond, the smaller the coefficient of expansion. A stronger atomic bond has a potential energy curve with steeper sides; as temperature increases, the maximum displacements of the atoms from their equilibrium positions change relatively little. Liquids have much greater coefficients of volume expansion than do solids because the molecules are more loosely bound in a liquid than in a solid.

14.3   HEAT CAPACITY AND SPECIFIC HEAT

Suppose we have a system on which no mechanical work is done, but we allow heat to flow into the system by placing it in thermal contact with another system at higher temperature. As the internal energy of the system increases, its temperature increases (provided that no part of the system undergoes a change of phase, such as from solid to liquid). Thus, if heat flows into the system, without work being done or a phase change occurring, the temperature increase can be used to measure the heat—the more heat that flows, the larger the temperature increase. If Q represents the heat into the system and U represents the internal energy of the system, then the change in internal energy (ΔU) of the system is

$$\Delta U = Q \qquad (14\text{-}2)$$

(no work done; Q = heat *into* the system; U = internal energy of the system)

Remember that this is a *system on which no work is being done*; otherwise, conservation of energy would require that the change in internal energy be the sum of the heat into the system and the mechanical work done on the system. If heat flows *out* of the system rather than into the system, the internal energy of the system decreases ($\Delta U < 0$). We account for that possibility in Eq. (14-2) by making Q negative if heat flows out of the system; since Q is defined as the heat *into* the system, a negative heat represents heat flow *out of* the system.

Q is defined as the heat into the system.

Heat Capacity

For a large number of substances, under normal conditions, the temperature change ΔT is approximately proportional to the heat Q. The constant of proportionality is called the **heat capacity**:

$$\text{heat capacity} = \frac{Q}{\Delta T} \qquad (14\text{-}3)$$

The heat capacity depends both on the substance and on how much of it is present: 1 cal of heat into 1 g of water causes a temperature increase of 1°C, but 1 cal of heat into 2 g of water causes a temperature increase of 0.5°C. The SI unit of heat capacity is J/K. We can write J/K or J/°C interchangeably since only temperature *changes* are involved; a temperature change of 1 K is equivalent to a temperature change of 1°C.

 The term *heat capacity* is unfortunate since it has nothing to do with a capacity to *hold* heat, or a limited ability to absorb heat, as the name seems to imply. Instead, it relates the heat into a system to the temperature increase. It is better to think of heat capacity as a measure of how much heat must flow into or out of the system to produce a given temperature change.

Specific Heat

Since the heat capacity of a system is proportional to the mass of the system, the **specific heat capacity** (symbol c) of a substance is defined as the heat capacity per unit mass:

Specific heat capacity

$$c = \frac{Q}{m\,\Delta T} \qquad (14\text{-}4)$$

The SI units of specific heat capacity are J/(kg·K). In SI units, the specific heat is the number of joules of heat required to produce a 1 K temperature change in 1 kg of the substance. Again, since only temperature changes are involved, we can equivalently write J/(kg·°C). Specific heat capacity is often abbreviated to *specific heat*.

specific heat = heat capacity per unit mass

Table 14.1 lists specific heats for some common substances at 1 atm and 20°C (unless otherwise specified). For the range of temperatures in our examples and problems, assume these specific heat values to be valid. Note that water has a relatively large specific heat compared with most other substances. The relatively large specific heat of water causes the oceans to warm slowly in the spring and to cool slowly as winter approaches, moderating the temperature along the coast.

Rearrangement of Eq. (14-4) leads to an expression for the heat required to produce a known temperature change in a system:

$$Q = mc\,\Delta T \qquad (14\text{-}5)$$

Note that in Eqs. (14-4) and (14-5), the sign convention for Q is consistent: a temperature increase ($\Delta T > 0$) is caused by heat flowing *into* the system ($Q > 0$) while a temperature decrease ($\Delta T < 0$) is caused by heat flowing *out* of the system ($Q < 0$).

Equations (14-3) and (14-4) can be combined to represent the heat capacity:

$$\text{heat capacity} = \frac{Q}{\Delta T} = mc \qquad (14\text{-}6)$$

Equation (14-6) shows that the heat capacity of a system is proportional to its mass. The specific heat of the water in a drinking glass is the same as that for water in Lake Superior, but the glass of water and the lake have vastly different heat capacities.

Table 14.1

Specific Heats of Common Substances at 1 atm and 20°C

Substance	Specific Heat $\left(\dfrac{\text{kJ}}{\text{kg·K}}\right)$	Specific Heat $\left(\dfrac{\text{kcal}}{\text{kg·K}}\right)$
Gold	0.128	0.0306
Lead	0.13	0.031
Mercury	0.139	0.0332
Silver	0.235	0.0561
Brass	0.384	0.0917
Copper	0.385	0.0920
Steel	0.45	0.11
Iron	0.44	0.11
Granite	0.80	0.19
Marble	0.86	0.21
Aluminum	0.900	0.215
Air (50°C)	1.05	0.25
Wood (average)	1.68	0.40
Steam (110°C)	2.01	0.480
Ice (0°C)	2.1	0.50
Alcohol (ethyl)	2.4	0.57
Human tissue (average)	3.5	0.84
Water (15°C)	4.186	1.000

Example 14.4

Heating Water in a Saucepan

 A saucepan containing 5.00 kg of water initially at 20.0°C is heated over a gas burner for 10.0 min. The final temperature of the water is 30.0°C. (a) What is the internal energy increase of the water? (b) What is the expected final temperature if the water were heated for an additional 5.0 min? (c) Is it possible to estimate the flow of heat from the burner during the first 10.0 min?

Strategy   We are interested in the internal energy and the temperature *of the water,* so we define a system that consists of the water in the saucepan. Although the pan is also heated, it is not part of this system. The pan, the burner, and the room are all outside the system.

Since no work is done on the water, the internal energy increase is equal to the heat flowing into the water. The heat can be found from the mass of the water, the specific heat of water, and the temperature change. As long as the burner delivers heat at a constant rate, we can find the additional heat delivered in the additional time. Since the temperature change is proportional to the heat delivered, the temperature changes at a constant rate (a constant number of °C per minute). So, in half the time, half as much energy is delivered and the temperature change is half as much.

Solution   (a) First find the temperature change:

$$\Delta T = T_f - T_i = 30.0°C - 20.0°C = 10.0 \text{ K}$$

(A change of 10.0°C is equivalent to a change of 10.0 K.) The increase in the internal energy of the water is

$$\Delta U = Q = mc \, \Delta T$$
$$= 5.00 \text{ kg} \times 4.186 \text{ kJ/(kg·K)} \times 10.0 \text{ K} = 209 \text{ kJ}$$

(b) We assume that the heat delivered is proportional to the elapsed time. The temperature change is proportional to the energy delivered, so if the temperature changes 10.0°C in 10.0 min, it changes an additional 5.0°C in an additional 5.0 min. The final temperature is

$$T = 20.0°C + 15.0°C = 35.0°C$$

(c) Not all of the heat flows into the water. Heat also flows from the burner into the saucepan and into the room. All we can say is that *more than* 209 kJ of heat flows from the burner during the 10.0 min.

Discussion   As a check, the heat capacity of the water is 5.00 kg × 4.186 kJ/(kg·K) = 20.9 kJ/K; 20.9 kJ of heat must flow for each 1.0 K change in temperature. Since the temperature change is 10.0 K, the heat required is 20.9 kJ/K × 10.0 K = 209 kJ.

Practice Problem 14.4   Price of a bubble bath

If the cost of electricity is $0.080 per kW·h, what does it cost to heat 160 L of water for a bubble bath from 10.0°C (the temperature of the well water entering the house) to 70.0°C? [*Hint:* 1 L of water has a mass of 1 kg. 1 kW·h = 1000 J/s × 3600 s.]

What happens if more than two substances exchange heat? Suppose some water is heated in a large iron pot by dropping a hot piece of copper into the pot. We can define the system to be the water, the copper, and the iron pot; the environment is the room containing the system. Heat continues to flow between the three substances (iron pot, water, copper) until thermal equilibrium is reached—that is, until all three substances are at the same temperature. From energy conservation, if losses to the environment are negligible, all the heat that flows out of the copper flows into either the iron or the water:

$$Q_{Cu} + Q_{Fe} + Q_{H_2O} = 0$$

In this case Q_{Cu} is negative since heat flows *out* of the copper; Q_{Fe} and Q_{H_2O} are positive since heat flows into both the iron and the water.

Calorimetry

A calorimeter is an insulated container that enables the careful measurement of heat (Fig. 14.3). The calorimeter is designed to minimize the heat flow to or from the surroundings. A typical constant volume calorimeter, called a *bomb calorimeter*, consists of a hollow aluminum cylinder of known mass containing a known quantity of water; the cylinder is inside a larger aluminum cylinder with insulated walls. An insulating air space separates the two cylinders. An insulated lid fits over the opening of the cylinders; often there are two small holes in the lid, one for a thermometer to be inserted into the contents of the inner cylinder and one for a stirring device to help the contents reach equilibrium faster.

Suppose an object at one temperature is placed in a calorimeter with the water and aluminum cylinder at another temperature. By conservation of energy, all the heat that flows

Thermometer
Stirrer
Lid
Insulated jacket

Figure 14.3   A calorimeter

out of one substance ($Q < 0$) flows into some other substance ($Q > 0$). If no heat flows to or from the environment, the total heat into the object, water, and aluminum must equal zero:

$$Q_o + Q_w + Q_a = 0$$

Example 14.5 illustrates the use of a calorimeter to measure the specific heat of an unknown substance. The measured specific heat can be compared with a table of known values to help identify the substance.

Example 14.5

Specific Heat of an Unknown Metal

A sample of unknown metal of mass 0.550 kg is heated in a pan of hot water until it is in equilibrium with the water at a temperature of 75.0°C. The metal is then carefully removed from the heat bath and placed into the inner cylinder of an aluminum calorimeter that contains 0.500 kg of water at 15.5°C. The mass of the inner cylinder is 0.100 kg. When the contents of the calorimeter reach equilibrium, the temperature inside is 18.8°C. Find the specific heat of the metal sample and determine whether it could be any of the metals listed in Table 14.1.

Strategy   Heat flows from the sample to the water and to the aluminum until thermal equilibrium is reached, at which time all three have the same temperature. We use subscripts to keep track of the three heat flows and three temperature changes. Let T_f be the final temperature of all three. Initially, the water and aluminum are both at 15.5°C while the sample is at 75.0°C. When thermal equilibrium is reached, all three are at 18.8°C. We assume negligible heat flow to the environment—in other words, that no heat flows into or out of the system of aluminum + water + sample.

Solution   Heat flows out of the sample ($Q_s < 0$) and into the water and aluminum cylinder ($Q_w > 0$ and $Q_a > 0$). Assuming no heat into or out of the surroundings,

$$Q_s + Q_w + Q_a = 0$$

For each substance, the heat is related to the temperature change. Substituting $Q = mc\,\Delta T$ for each gives

$$m_s c_s \Delta T_s + m_w c_w \Delta T_w + m_a c_a \Delta T_a = 0 \qquad (1)$$

A table helps organize the given information:

Substituting known values into Eq. (1) yields

$$0.550 \text{ kg} \times c_s \times (-56.2°C) + (0.500 \text{ kcal/°C} + 0.0215 \text{ kcal/°C}) \times 3.3°C = 0$$

Now we solve for c_s.

$$0.550 \text{ kg} \times c_s \times 56.2°C = 1.721 \text{ kcal}$$

$$c_s = \frac{1.721 \text{ kcal}}{0.550 \text{ kg} \times 56.2°C} = 0.0557 \frac{\text{kcal}}{\text{kg·°C}}$$

By comparing this result with the values in Table 14.1, it appears that the unknown sample could be silver.

Discussion   As a quick check, the heat capacity of the sample is approximately $\frac{1}{17}$ that of the water since its temperature change is 56.2°C/3.3°C ≈ 17 times as much—ignoring the small heat capacity of the aluminum. Since the masses of the water and sample are about equal, the specific heat of the sample is roughly $\frac{1}{17}$ that of the water:

$$\frac{1}{17} \times 1.00 \frac{\text{kcal}}{\text{kg·°C}} = 0.059 \frac{\text{kcal}}{\text{kg·°C}}$$

That is quite close to our answer.

Practice Problem 14.5   Final temperature

If 0.25 kg of water at 90.0°C is added to 0.35 kg of water at 20.0°C in an aluminum calorimeter with an inner cylinder of mass 0.100 kg, find the final temperature of the mixture.

Substance	Mass (m)	Specific Heat (c)	Heat Capacity (mc)	ΔT
Sample	0.550 kg	c_s (unknown)	0.550 kg × c_s	18.8°C – 75.0°C = –56.2°C
H_2O	0.500 kg	$1.00 \dfrac{\text{kcal}}{\text{kg·K}}$	0.500 kcal/°C	18.8°C – 15.5°C = 3.3°C
Al	0.100 kg	$0.215 \dfrac{\text{kcal}}{\text{kg·K}}$	0.0215 kcal/°C	18.8°C – 15.5°C = 3.3°C

14.4   SPECIFIC HEAT OF IDEAL GASES

Since the average translational kinetic energy of a molecule in an ideal gas is

$$\langle K_{tr} \rangle = \frac{3}{2}kT \qquad\qquad \textbf{(13-20)}$$

the *total* translational kinetic energy of a gas containing N molecules (n moles) is

$$K_{tr} = \frac{3}{2}NkT = \frac{3}{2}nRT$$

Table 14.2

Molar Specific Heats at Constant Volume of Gases at 25°C

	Gas	$C_V \left(\dfrac{\text{J/K}}{\text{mol}} \right)$
Monatomic	He	12.5
	Ne	12.7
	Ar	12.5
Diatomic	H_2	20.4
	N_2	20.8
	O_2	21.0
Polyatomic	CO_2	28.2
	N_2O	28.4

Suppose we allow heat to flow into a *monatomic* ideal gas—one where the gas molecules consist of single atoms—while keeping the volume of the gas constant. Since the volume is constant, no work is done on the gas, so the change in the internal energy is equal to the heat. If we think of the atoms as point particles, the only way for the internal energy to change when heat flows into the gas is for the translational kinetic energy of the atoms to change. The rest of the internal energy is "locked up" in the atoms and does not change unless something else happens, such as a phase transition or a chemical reaction—neither of which can happen in an ideal gas. Then

$$Q = \Delta U = \Delta K_{tr} = \tfrac{3}{2} nR\, \Delta T \tag{14-7}$$

From Eq. (14-7) we can find the specific heat of the monatomic ideal gas. However, with gases it is more convenient to define the **molar specific heat** at constant volume (C_V) as

$$Q = nC_V\, \Delta T \tag{14-8}$$

The subscript "V" is a reminder that the volume of the gas is held constant during the heat flow. The molar specific heat is the heat capacity *per mole* rather than *per unit mass*. In essence, specific heat and molar specific heat can be thought of as the same quantity— heat capacity per amount of substance—expressed in different units. In one case we measure the amount of substance by the mass; in the other case by the number of moles.

From Eqs. (14-7) and (14-8) we can find the molar specific heat of a monatomic ideal gas:

$$Q = \tfrac{3}{2} nR\, \Delta T = nC_V\, \Delta T$$

$$C_V = \frac{3}{2}R = 12.5\ \frac{\text{J/K}}{\text{mol}} \quad \text{(\textit{monatomic} ideal gas)} \tag{14-9}$$

A glance at Table 14.2 shows that this calculation is remarkably accurate at room temperature for monatomic gases.

Diatomic gases have larger molar specific heats than monatomic gases. Why? We cannot model the diatomic molecule as a point mass; the two atoms in the molecule are separated, giving the molecule a much larger rotational inertia about two perpendicular axes (see Fig. 14.4). The molar specific heat is *larger* because not all of the internal energy increase goes into the translational kinetic energy of the molecules; some goes into rotational kinetic energy.

It turns out that the molar specific heat of a diatomic ideal gas at room temperature is approximately

$$C_V = \frac{5}{2}R = 20.8\ \frac{\text{J/K}}{\text{mol}} \quad \text{(\textit{diatomic} ideal gas)} \tag{14-10}$$

Why $\frac{5}{2}R$ instead of $\frac{3}{2}R$? The diatomic molecule has rotational kinetic energy about two perpendicular axes (Fig. 14.4b and c) in addition to translational kinetic energy associated with motion in three independent directions. Thus, the diatomic molecule has five ways to "store" internal energy while the monatomic molecule has only three. The theorem of **equipartition of energy**—which we cannot prove here—says that internal energy is

Figure 14.4 Rotation of a model diatomic molecule about three perpendicular axes. The rotational inertia about the *x*-axis (a) is negligible. The rotational inertias about the *y*- and *z*-axes (b) and (c) are much larger than for a single atom of the same mass because of the larger distance between the atoms and the axis of rotation.

distributed equally among all the possible ways in which it can be stored (as long as the temperature is sufficiently high). Each independent form of energy has an average of $\frac{1}{2}kT$ of energy per molecule and contributes $\frac{1}{2}R$ to the molar specific heat at constant volume.

Example 14.6

Heating Some Xenon Gas

A cylinder contains 250 L of xenon gas (Xe) at 20.0°C and a pressure of 5.0 atm. How much heat is required to raise the temperature of this gas to 50.0°C, holding the volume constant? Treat the xenon as an ideal gas.

Strategy   The molar heat capacity is the heat required per degree per mole. The number of moles of xenon (n) can be found from the ideal gas law, $PV = nRT$. Xenon is a monatomic gas, so we expect $C_V = \frac{3}{2}R$.

Solution   First we convert the known quantities into SI units.

$$P = 5.0 \text{ atm} = 5 \times 1.01 \times 10^5 \text{ Pa} = 5.05 \times 10^5 \text{ Pa}$$

$$V = 250 \text{ L} = 250 \times 10^{-3} \text{ m}^3$$

$$T = 20.0°C = 293.15 \text{ K}$$

From the ideal gas law, we find the number of moles,

$$n = \frac{PV}{RT} = \frac{5.05 \times 10^5 \text{ Pa} \times 250 \times 10^{-3} \text{ m}^3}{8.31 \text{ J/(mol·K)} \times 293.15 \text{ K}} = 51.8 \text{ mol}$$

We should check the units. Since Pa = N/m²,

$$\frac{\text{Pa} \times \text{m}^3}{\text{J/(mol·K)} \times \text{K}} = \frac{\text{N/m}^2 \times \text{m}^3}{\text{J/mol}} = \frac{\text{N·m}}{\text{J}} \times \text{mol} = \text{mol}$$

For a monatomic gas at constant volume, the energy all goes into increasing the translational kinetic energy of the gas molecules. The molar specific heat is defined by $Q = nC_V \Delta T$, where, for a monatomic gas, $C_V = \frac{3}{2}R$. Then,

$$Q = \frac{3}{2}nR\,\Delta T$$

where

$$\Delta T = 50.0°C - 20.0°C = 30.0°C$$

Substituting,

$$Q = \frac{3}{2} \times 51.8 \text{ mol} \times 8.31 \text{ J/(mol·°C)} \times 30.0°C = 19 \text{ kJ}$$

Discussion   Constant volume implies that all the heat is used to increase the internal energy of the gas; if the gas were to expand it could transfer energy by doing work. When we find the number of moles from the ideal gas law, we must remember to convert the Celsius temperature to kelvins. Only when an equation involves a *change* in temperature can we use kelvin or Celsius temperatures interchangeably.

Practice Problem 14.6   Heating some helium gas
A storage cylinder of 330 L of helium gas is at 21°C and is subjected to a pressure of 10.0 atm. How much energy must be added to raise the temperature of the helium in this container to 75°C?

You may wonder why we can ignore rotation for the monatomic molecule—which in reality is not a point particle—or why we can ignore rotation about one axis for the diatomic molecule. The answer comes from quantum mechanics. Energy cannot be added to a molecule in arbitrarily small amounts; energy can only be added in discrete amounts or "steps." At room temperature, there is not enough internal energy to excite the rotational modes with small rotational inertias, so they do not participate in the specific heat. We also ignored the possibility of vibration for the diatomic molecule. That is fine at room temperature, but at higher temperatures vibration becomes significant, adding two more energy modes (one kinetic and one potential). Thus, as temperature increases, the molar specific heat of a diatomic gas increases, approaching $\frac{7}{2}R$ at high temperatures.

14.5   PHASE TRANSITIONS

If heat continually flows into the water in a pot, the water eventually begins to boil; liquid water becomes steam. If heat flows into ice cubes, they eventually melt and turn into liquid water. A **phase transition** occurs whenever a material is changed from one phase, such as the solid phase, to another, such as the liquid phase.

When some ice cubes at 0°C are placed into a glass in a room at 20°C, the ice gradually melts. A thermometer in the water that forms as the ice melts reads 0°C until all the ice is melted. At atmospheric pressure, ice and water can only coexist in equilibrium at 0°C. Once all the ice is melted, the water gradually warms up to room temperature. Similarly, water boiling on a stove remains at 100°C until all the water has boiled away.

Figure 14.5 Temperature versus heat for 1 kg of ice that starts at a temperature below 0°C. (Horizontal axis not to scale.)

During a phase transition, the temperature of the mixture of two phases does not change.

Suppose we change 1.0 kg of ice at –25°C into steam at 125°C. A graph of the temperature versus heat is shown in Fig. 14.5. If the heat flows at a steady rate, a temperature versus *time* graph would have the same shape. During the two phase transitions, *heat flow continues, but the temperature of the mixture of two phases does not change.* Table 14.3 shows the heat during each step of the process.

The heat required *per unit mass* of substance to produce a phase change is called the **latent heat** (L). The word "latent" is related to the lack of temperature change during a phase transition.

Definition of latent heat
$$|Q| = mL \qquad (14\text{-}11)$$

The heat per unit mass for the solid-liquid phase transition (in either direction) is called the **latent heat of fusion** (L_f). From Table 14.3, it takes 333.7 kJ to change 1 kg of ice to water at 0°C, so for water $L_f = 333.7$ kJ/kg. For the liquid-gas phase transition (in either direction), the heat per unit mass is called the **latent heat of vaporization** (L_v). From Table 14.3, to change 1 kg of water to steam at 100°C takes 2256 kJ, so for water $L_v = 2256$ kJ/kg. Table 14.4 lists latent heats of fusion and vaporization for various materials.

All phase changes can go in either direction. Heat flowing into a substance can cause melting (solid to liquid) or boiling (liquid to gas). Heat flowing out of a substance can cause freezing (liquid to solid) or condensation (gas to liquid). If 2256 kJ must be supplied to turn 1 kg of water into steam, then 2256 kJ of heat is *released* from 1 kg of steam when it condenses to form water. A burn caused by 100°C steam is much more severe than a burn caused by 100°C water because the steam releases a large amount of heat as it condenses into water on the skin; much more energy is transferred to the skin than would be the case for the same amount of water at 100°C. The reverse process, evaporation, facilitates heat flow from a perspiring body.

The large latent heat of fusion of water is partly why spraying fruit trees with water can protect the buds from freezing. Before the buds can freeze, first the water must be cooled to 0°C and then it must freeze. In the process of freezing, the water gives up a large amount of heat and keeps the temperature of the buds from going below 0°C. Even if the water freezes, then the layer of ice over the buds acts like insulation since ice is not a particularly good conductor of heat.

Making The Connection:
using ice to protect buds from freezing

Table 14.3

Heat to Turn 1 kg of Ice at –25°C to Steam at 125°C

Phase Transition or Temperature Change	Q (kJ)
Ice: –25°C to 0°C	52.3
Melting: ice at 0°C to water at 0°C	333.7
Water: 0°C to 100°C	419
Boiling: water at 100°C to steam at 100°C	2256
Steam: 100°C to 125°C	50

Table 14.4

Latent Heats of Some Common Substances

Substance	Melting Point (°C)	Heat of Fusion		Boiling Point (°C)	Heat of Vaporization	
		(kJ/kg)	(kcal/kg)		(kJ/kg)	(kcal/kg)
Alcohol (ethyl)	−114	104	24.8	78	854	204
Aluminum	660	397	94.8	2,450	11,400	2,720
Copper	1083	205	49	2,340	5,070	1,210
Gold	1063	66.6	15.9	2,660	1,580	377
Lead	327	22.9	5.47	1,620	871	208
Silver	960.8	88.3	21.1	1,950	2,340	558
Water	0.0	333.7	79.7	100	2,256	539

To understand what is happening during a phase change, we must consider the substance on the molecular level. When a substance is in solid form, bonds between the atoms or molecules hold them near fixed equilibrium positions. Energy must be supplied to break the bonds and change the solid into a liquid. When the substance is changed from the liquid to the vapor phase, energy is used to separate the molecules from the loose bonds holding them together and to move the molecules apart as the gas expands. Temperature does not change during these phase transitions because the *kinetic energy* of the molecules is not changing. Instead, the *potential energy* of the molecules changes as work is done against the forces holding them together.

Example 14.7

Making Silver Charms

A jewelry designer plans to make some specially ordered silver charms for a commemorative bracelet. If the melting point of silver is 960.8°C, how much heat (in kcal and in kJ) must the jeweler add to 0.500 kg of silver at 20.0°C to be able to pour silver into her charm molds?

Strategy The solid silver first needs to be heated to its melting point; then more heat has to be added to melt the silver.

Solution The total heat flow into the silver is the sum of the heat to raise the temperature of the solid and the heat that causes the phase transition:

$$Q = mc\,\Delta T + mL_f$$

The temperature change of the solid is

$$\Delta T = 960.8°C - 20.0°C = 940.8°C$$

We look up the specific heat of solid silver and the latent heat of fusion of silver. Substituting numerical values into the equation for Q yields

$$Q = 0.500\,\text{kg} \times 0.0561\,\text{kcal/(kg·°C)} \times 940.8°C + 0.500\,\text{kg} \times 21.1\,\text{kcal/kg}$$

$$= 26.39\,\text{kcal} + 10.55\,\text{kcal} = 36.94\,\text{kcal}$$

Using values of c and L_f in kcal, we get an answer of 36.9 kcal. To convert to kJ, we use the conversion factor 1 kcal = 4.186 kJ:

$$Q = 36.94\,\text{kcal} \times 4.186\,\text{kJ/kcal} = 155\,\text{kJ}$$

⚠ **Discussion** An easy mistake to make is to use the wrong latent heat. Here we were dealing with melting, so we need the latent heat of fusion. Another possible error is to use the specific heat for the wrong phase: here we raised the temperature of *solid* silver, so we need the specific heat of *solid* silver. With water, we must always be careful to use the specific heat of the correct phase; the specific heats of ice, water, and steam have three different values.

Practice Problem 14.7 Making gold medals

Some gold medals are to be made from 750 g of solid gold at 24°C (Fig. 14.6). How much heat is required to melt the gold so that it can be poured into the molds for the medals?

Figure 14.6
A gold medal: the Nobel Prize for physics

Example 14.8

Turning Water into Ice

Ice cube trays are filled with 0.500 kg of water at 20.0°C and placed into the freezer compartment of a refrigerator. How much energy must be removed from the water to turn it into ice cubes at –5.0°C?

Strategy   We can think of this process as three consecutive steps. First, the liquid water is cooled to 0°C. Then the phase change occurs at constant temperature. Now the water is frozen; the ice continues to cool to –5.0°C. The energy that must be removed for the whole process is the sum of the energy removed in each of the three steps.

Solution   For liquid water going from 20.0°C to 0.0°C,

$$Q_1 = mc_{water} \Delta T_1$$

where

$$\Delta T_1 = 0.0°C - 20.0°C = -20.0°C$$

Since ΔT_1 is negative, Q_1 is negative: heat must flow *out of* the water in order for its temperature to decrease. Next the water freezes. The heat is found from the latent heat of fusion:

$$Q_2 = -mL_f$$

Again, heat flows *out* so Q_2 is negative. For phase transitions we supply the correct sign of Q according to the direction of the phase transition (minus sign for freezing, plus sign for melting). Finally, the ice is cooled to –5.0°C:

$$Q_3 = mc_{ice} \Delta T_2$$

where

$$\Delta T_2 = -5.0°C - 0.0°C = -5.0°C$$

We use subscripts on the specific heats to distinguish the specific heat of ice from that of water. The total heat is

$$Q = m \left(c_{water} \Delta T_1 - L_f + c_{ice} \Delta T_2 \right)$$

Now we look up c_{water}, L_f, and c_{ice} in Tables 14.1 and 14.4 and substitute:

$$Q = 0.500 \text{ kg} \left(1.00 \frac{\text{kcal}}{\text{kg·°C}} \times (-20.0°C) - 79.7 \frac{\text{kcal}}{\text{kg}} + 0.50 \frac{\text{kcal}}{\text{kg·°C}} \times (-5.0°C) \right)$$

$$= -0.500 \text{ kg} \times 102.2 \frac{\text{kcal}}{\text{kg}} = -51.1 \text{ kcal}$$

So 51.1 kcal of heat flows out of the water that becomes ice cubes.

 Discussion   We *cannot* consider the entire temperature change from +20°C to –5°C in one step. A phase change occurs, so we must include the flow of heat during the phase change. Also, the specific heat of ice is different from the specific heat of liquid water; we must find the heat to cool water 20°C and then the heat to cool ice 5°C.

Practice Problem 14.8   Frozen popsicles

Nigel pulls a tray of frozen popsicles out of the freezer to share with his friends. If the popsicles are at –4°C and go directly into hungry mouths at 37°C, how much energy is used to bring a popsicle of mass 0.080 kg to body temperature? Assume the frozen popsicles have the same specific heat as ice and the melted popsicle has the specific heat of water.

Example 14.9

Cooling a Drink

Two 50.0-g ice cubes are placed into 0.200 kg of water in a Styrofoam cup. The water is initially at a temperature of 25.0°C and the ice is initially at a temperature of –15.0°C. What is the final temperature of the drink? The average specific heat for ice between –15°C and 0°C is $0.49 \frac{\text{kcal}}{\text{kg·°C}}$.

Strategy   We need to raise the temperature of the ice from –15°C to 0°C before the ice can melt, so we first find how much heat this requires. Then we find how much heat is needed to melt all the ice. Once the ice is melted, the water from the melted ice can be raised to the final temperature of the drink. The heat for all three steps (raising temperature of ice, melting ice, raising temperature of water from melted ice) all comes from the water initially at 25°C. That water cools as heat flows out of it. Assuming no heat flow into or out of the room, the quantity of heat that flows out of the water initially at 25°C flows into the ice or melted ice (before, during, and after melting).

Given: $m_{ice} = 0.1000$ kg at –15.0°C; $m_w = 0.200$ kg at 25.0°C, $c_{ice} = 0.49$ kcal/(kg·°C)
Look up: L_f for water = 79.7 kcal/kg; $c_{water} = 1.00$ kcal/(kg·°C)
Find: T_f

Solution   Since heat flows out of the water and into ice, $Q_{water} < 0$ and $Q_{ice} > 0$. Their sum is zero:

$$Q_{ice} + Q_w = 0$$

The heat flow into the ice is the sum of three terms:

$$Q_{ice} = m_{ice}c_{ice} \Delta T_{ice} + m_{ice}L_f + m_{ice}c_{water} (T_f - 0.0°C)$$

The heat flow out of the water is

$$Q_w = m_w c_{water} (T_f - 25.0°C)$$

The heat required to bring the ice from –15.0°C to 0°C is

$$m_{ice}c_{ice} \Delta T_{ice} = 0.1000 \text{ kg} \times 0.49 \frac{\text{kcal}}{\text{kg·°C}} \times 15.0°C = 0.735 \text{ kcal}$$

The heat required to melt the ice at 0.0°C is

$$m_{ice}L_f = 0.1000 \text{ kg} \times 79.7 \text{ kcal/kg} = 7.97 \text{ kcal}$$

continued on next page

Example 14.9 *continued*

The heat to raise the temperature of the melted ice from 0.0°C to T_f is

$$m_{ice}c_{water}(T_f - 0.0°C) = 0.1000 \text{ kg} \times 1.00\frac{\text{kcal}}{\text{kg}\cdot°C} \times T_f$$

$$= 0.100\frac{\text{kcal}}{°C} \times T_f$$

The heat supplied by the water that was initially at 25.0°C is

$$m_w c_{water}(T_f - 25.0°C) = 0.200 \text{ kg} \times 1.00\frac{\text{kcal}}{\text{kg}\cdot°C} \times (T_f - 25.0°C)$$

$$= 0.200\frac{\text{kcal}}{°C} \times T_f - 5.00 \text{ kcal}$$

Now we substitute these values back into the original equation, $Q_{ice} + Q_w = 0$.

$$0.735 \text{ kcal} + 7.97 \text{ kcal} + \left(0.100\frac{\text{kcal}}{°C} \times T_f\right) + \left(0.200\frac{\text{kcal}}{°C} \times T_f - 5.00 \text{ kcal}\right) = 0$$

Simplifying yields

$$0.300\frac{\text{kcal}}{°C} \times T_f + 3.705 \text{ kcal} = 0$$

Solving for T_f, we find

$$T_f = -\frac{3.705 \text{ kcal}}{0.300 \text{ kcal/°C}} = -12.4°C$$

This result does not make sense: we assumed that all of the ice would melt and that the final mixture would be all liquid, but we cannot have liquid water at –12.4°C. Let's take another look at the solution.

What if the water initially at 25°C cools all the way to 0°C? From cooling the water, how much heat is available to warm the ice and melt it?

$$Q_w = m_w c_{water}(0°C - 25.0°C)$$

$$= 0.200 \text{ kg} \times 1.00\frac{\text{kcal}}{\text{kg}\cdot°C} \times (-25.0°C) = -5.00 \text{ kcal}$$

Thus, the water can supply 5.00 kcal when it cools to 0°C. But to warm the ice requires 0.735 kcal and to melt all of the ice requires another 7.97 kcal. The ice can be warmed to 0°C, but there is not enough heat available to melt all of the ice. Only some of the ice melts, so the drink ends up as a mixture of water and ice in equilibrium at 0°C.

 Discussion This example shows the value of checking a result to make sure it is reasonable. We started by assuming incorrectly that all of the ice would melt. When we obtained an answer that was impossible, we went back to see if there was enough heat available to melt all of the ice. Since there was not, the final temperature of the drink can only be 0°C—the only temperature at which ice and water can be in thermal equilibrium at atmospheric pressure.

Practice Problem 14.9 Melting ice

How much of the ice of Example 14.9 melts?

Evaporation

If you leave a cup of water out at room temperature, the water eventually evaporates. Recall that the temperature of the water reflects the average kinetic energy of the water molecules; some have higher than average energies and some have lower. The most energetic molecules have enough energy to break loose from the molecular bonds at the surface of the water. As these highest-energy molecules leave the water, the average energy of those left behind decreases—which is why evaporation is a cooling process. Approximately the same latent heat of vaporization applies to evaporation as to boiling, since the same molecular bonds are being broken. Perspiring basketball players cover up while sitting on the bench for a short time during a game to prevent getting a chill even though the air in the stadium may be warm.

When the humidity is high—meaning there is already a lot of water vapor in the air—evaporation proceeds more slowly. Water molecules in the air can also condense into water; the net evaporation rate is the difference in the rates of evaporation and condensation. A hot, humid day is uncomfortable because our bodies have trouble staying cool when perspiration evaporates slowly.

 Making The Connection:
chill caused by
perspiration

Physics at Home

The effects of evaporation can easily be felt. Rub some water on the inside of your forearm and then blow on your arm. The motion of the air over your arm removes the newly evaporated molecules from the vicinity of your arm and allows other molecules to evaporate more quickly. You can feel the cooling effect. If you have some rubbing alcohol, repeat the experiment. Since the alcohol evaporates faster, the cooling effect is noticeably greater.

Phase Diagrams

A useful tool in the study of phase transitions is the **phase diagram**—a diagram on which pressure is plotted on the vertical axis and temperature on the horizontal axis.

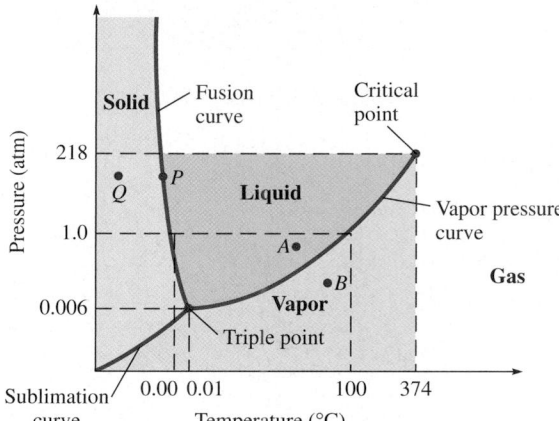

Figure 14.7 A phase diagram for water. The term vapor is often used to indicate a substance in the gaseous state below its critical temperature; above the critical temperature it is called gas. (Note that the axes do not use a linear scale.)

Figure 14.7 is a phase diagram for water. A point on the phase diagram represents water in a state determined by the pressure and the temperature at that point. The curves on the phase diagram are the demarcations between the solid, liquid, and vapor phases. For most temperatures, there is one pressure at which two particular phases can coexist in equilibrium. Since point P lies on the fusion curve, water can exist as liquid, or as solid, or as a mixture of the two at that temperature and pressure. At point Q, water can only be a solid. Similarly, at A water is a liquid; at B it is a vapor.

The one exception is at the **triple point**, where all three phases (solid, liquid, and vapor) can coexist in equilibrium. Triple points are used in precise calibrations of thermometers. The triple point of water is precisely 0.01°C at 0.006 atm.

From the vapor pressure curve, we see that as the pressure is lowered, the temperature at which water boils decreases. It takes longer to cook a hard-boiled egg in the high mountains because the temperature of the boiling water is less than 100°C; the chemical reactions that "cook" the egg proceed more slowly at a lower temperature. It might take as long as half an hour to hard-boil an egg on Pike's Peak, where the pressure is 0.6 atm.

If either the temperature or the pressure or both are changed, the point representing the state of the water moves along some path on the phase diagram. If the path crosses one of the curves, a phase transition occurs and the latent heat for that phase transition is either absorbed or released (depending on direction). Crossing the fusion curve represents freezing or melting; crossing the vapor pressure curve represents condensation or vaporization.

Notice that the vapor pressure curve ends at the **critical point**. Thus, if the path for changing a liquid to a gas goes around the critical point without crossing the vapor pressure curve, no phase transition occurs. At temperatures above the critical temperature or pressures above the critical pressure, it is *impossible* to make a clear distinction between the liquid and gas phases.

The sublimation curve represents another phase change, called **sublimation**, that occurs when a solid becomes vapor (or *vice versa*) without passing through the liquid phase. Sublimation occurs when ice on a car windshield becomes water vapor on a cold dry day. Mothballs and dry ice (solid carbon dioxide) also pass directly from solid to vapor. Sublimation has its own latent heat; the latent heat for sublimation is not the sum of the latent heats for fusion and vaporization.

The phase diagram of water has an unusual feature: the slope of the fusion curve is negative. The fusion curve has a negative slope only for substances (such as water, gallium, and bismuth) that expand on freezing. In these substances the molecules are *closer together* in the liquid than they are in the solid! As liquid water starting at room temperature is cooled, it contracts until it reaches 3.98°C. At this temperature water has its highest density (at a pressure of 1 atm); further cooling makes the water *expand*. When water freezes, it expands even more; ice is less dense than water.

One consequence of the expansion of water upon freezing is that cell walls might rupture when foods are frozen and thawed. The taste of frozen food suffers as a result. Another

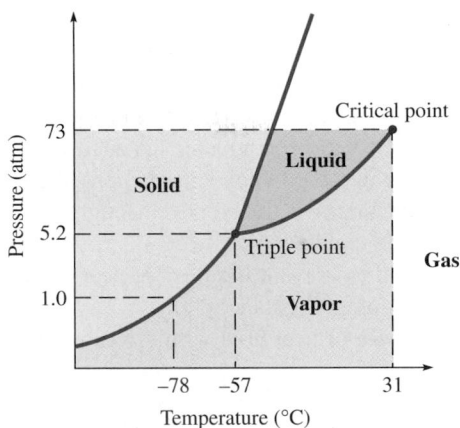

Figure 14.8 Phase diagram for carbon dioxide. (The axes do not use a linear scale.)

consequence is that lakes, rivers, and ponds do not freeze solid in the winter. A layer of ice forms on *top* since ice is less dense than water; underneath the ice, liquid water remains, which permits fish, turtles, and all manner of aquatic life to survive until spring.

The phase diagram for carbon dioxide (Fig. 14.8) is similar to that for water except that the fusion curve has a positive slope, which is the more common situation. At atmospheric pressure, only the solid and vapor phases of CO_2 exist. The liquid phase is not stable below 5.2 atm of pressure, so carbon dioxide does not melt at atmospheric pressure. Instead it sublimes; it goes from the solid directly to the vapor phase. Solid CO_2 is called *dry ice* because it is cold and looks like ice, but does not melt.

14.6   CONDUCTION

Until now we have considered the *effects* of heat flow, but not the mechanism of how heat flow occurs. We now turn our attention to three types of heat flow—*conduction, convection,* and *radiation.*

The **conduction** of heat can take place within solids, liquids, and gases. Conduction is due to collisions between atoms (or molecules) in which energy is exchanged. If the average energy is the same everywhere, there is no net flow of heat. If, on the other hand, the temperature is not uniform, then on average the atoms with more energy transfer some energy to those with less. The net result is that heat flows from the higher temperature region to the lower temperature region.

Conduction also occurs between objects that are in contact. A teakettle on an electric burner receives heat by conduction since the heating coil of the burner is in contact with the bottom of the kettle. The atoms that are vibrating in the object at higher temperature (the coil) collide with atoms in the object at lower temperature (the bottom of the kettle), resulting in a net transfer of energy to the colder object. If conduction is allowed to proceed, with no heat flow to or from the surroundings, then the objects in contact eventually reach thermal equilibrium when the average translational kinetic energies of the atoms are equal.

Suppose we consider a simple geometry such as an object with uniform cross section in which heat flows in a single direction. Examples are a plate of glass, with different temperatures on the inside and outside surfaces, or a cylindrical bar, with its ends at different temperatures (Fig. 14.9). The rate of heat conduction depends on the temperature difference $\Delta T = T_{hot} - T_{cold}$, the length (or thickness) d, the cross-sectional area A through which heat flows, and the nature of the material itself. The greater the temperature difference, the greater the heat flow. The thicker the material, the longer it takes for the heat to travel through—since the energy transfer has to be passed along a longer "chain" of atomic collisions—making the rate of heat flow smaller. A larger cross-sectional area allows more heat to flow.

The nature of the material is the final thing that affects the rate of energy transfer. In metals the electrons associated with the atom are free to move about and they carry the

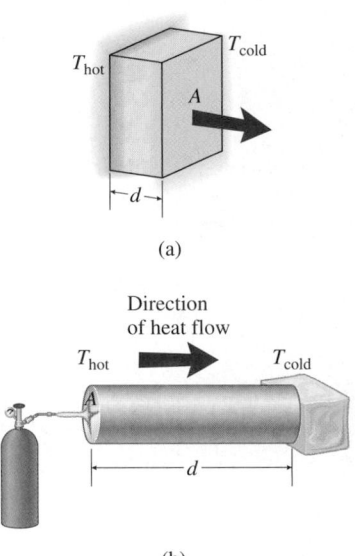

(a)

(b)

Figure 14.9 (a) Heat conduction through a slab of material of thickness d. (b) Heat conduction along a cylindrical bar of length d.

Table 14.5

Thermal Conductivities at 20°C

Material	$\kappa\left(\dfrac{W}{m \cdot K}\right)$
Air	0.023
Rock wool	0.038
Cork	0.046
Wood	0.13
Soil (dry)	0.14
Asbestos	0.17
Snow	0.25
Sand	0.39
Water	0.6
Glass	0.63
Concrete	1.7
Ice	1.7
Stainless steel	14
Lead	35
Steel	46
Nickel	60
Tin	66.8
Platinum	71.6
Iron	72.8
Brass	122
Zinc	116
Tungsten	173
Aluminum	237
Gold	318
Copper	401
Silver	429

Figure 14.10 Conduction of heat through two different layers ($T_1 > T_2 > T_3$).

heat. When a material has free electrons, the transfer rate is faster; if the electrons are all tightly bound, as in nonmetallic solids, the transfer is slower. Liquids, in turn, conduct heat less readily than solids, because the forces between atoms are weaker. A gas is even less efficient as a conductor of heat than solids or liquids since the atoms of a gas are so much farther apart and have to travel a greater distance before collisions occur. The **thermal conductivity** (symbol κ, the Greek letter kappa) of a substance is directly proportional to the rate at which energy is transferred through the substance. Higher values of κ are associated with good conductors of heat, smaller values with poor conductors or *insulators*. (Insulators tend to prevent the flow of heat.) Table 14.5 lists the thermal conductivities for several common substances.

The dependence of the rate of heat flow within a substance on all the factors mentioned is given by

---

Fourier's law of heat conduction

$$I_{cd} = \frac{Q}{t} = \kappa A \frac{\Delta T}{d} \tag{14-12}$$

---

where $I_{cd} = Q/t$ is the rate of heat flow (or power delivered), κ is the thermal conductivity of the material, A is the cross-sectional area, d is the thickness (or length) of the material, and ΔT is the temperature difference between one side and the other. The quantity $\Delta T/d$ is called the *temperature gradient*; it tells how many °C or K the temperature changes per unit of distance moved along the path of heat flow. Inspection of Eq. (14-12) shows that the SI units of κ are W/(m·K).

In Fig. 14.9a, a slab of material is shown that conducts heat because of a temperature difference between the two sides. By rearranging Eq. (14-12) and solving for ΔT,

$$\Delta T = I_{cd} \frac{d}{\kappa A} = I_{cd} R \tag{14-13}$$

The quantity $d/\kappa A$ is called the **thermal resistance R**.

$$R = \frac{d}{\kappa A} \tag{14-14}$$

Thermal resistance has SI units of K/W (kelvins per watt). Notice that the thermal resistance depends on the nature of the material (through the thermal conductivity κ) and the geometry of the object (d/A). Equation (14-13) is useful for solving problems when heat flows through one material after another.

Suppose we have two layers of material between two temperature extremes as in Fig. 14.10. These layers are in *series* because the heat flows through one and then through the other. Looking at one layer at a time,

$$T_1 - T_2 = I_{cd} R_1 \quad \text{and} \quad T_2 - T_3 = I_{cd} R_2$$

Then, adding the two together

$$(T_1 - T_2) + (T_2 - T_3) = I_{cd} R_1 + I_{cd} R_2$$

$$\Delta T = T_1 - T_3 = I_{cd} (R_1 + R_2)$$

The rate of heat flow through the first layer is the same as the rate through the second layer because otherwise the temperatures would be changing. For n layers,

$$\Delta T = I_{cd} \Sigma R_n \quad n = 1, 2, 3, \dots \tag{14-15}$$

Equation (14-15) shows that the effective thermal resistance for layers in series is the sum of each layer's thermal resistance.

Equation (14-15) has a familiar form. For a viscous fluid flowing through a pipe, Poiseuille's law can be written in the form $\Delta P = IR$, where I is the volume flow rate and R is called the fluid flow *resistance* (see Problem 9.50). For two or more pipes in series, where the volume flow rate is the same through each, the effective fluid flow resistance is the *sum of the resistances* of the individual pipes. Another equation of this form occurs in the study of electric currents, where the effective electrical resistance of two or more electrical conductors in series is the sum of the resistances of the individual conductors.

Example 14.10

The Rate of Heat Flow through Window Glass

 A windowpane that measures 20.0 cm by 15.0 cm is set into the front door of a house. The glass is 0.32 cm thick. The temperature outdoors is –15°C and inside is 22°C. At what rate does heat leave the house through that one small window?

Strategy   We assume one side of the glass to be at the temperature of the air inside the house and the other to be at the outdoor temperature.

Given: $\Delta T = 22°C - (-15°C) = 37°C$
thickness of windowpane $d = 0.32 \times 10^{-2}$ m
area of windowpane $A = 0.200$ m $\times 0.150$ m $= 0.0300$ m^2

Look up: thermal conductivity for glass = 0.63 W/(m·K)

Find: rate of heat flow, I_{cd}

Solution   The temperature gradient is

$$\frac{\Delta T}{d} = \frac{37°C}{0.32 \times 10^{-2}\,\text{m}} = 1.16 \times 10^4 \text{ K/m}$$

Now we have all the information we need to find the rate of conductive heat flow:

$$I_{cd} = \kappa A \frac{\Delta T}{d}$$
$$= 0.63 \text{ W/(m·K)} \times 0.0300 \text{ m}^2 \times 1.16 \times 10^4 \text{ K/m}$$
$$= 220 \text{ W}$$

Discussion   A loss of 220 W through one small window is significant. Our assumption about the temperatures of the two glass surfaces exaggerates the temperature difference across the glass. In reality, the inside surface of the glass is colder than the air inside the house, while the outside surface is warmer than the air outside.

Practice Problem 14.10   An igloo

A group of children build an igloo in their garden. The snow walls are 0.30 m thick. If the inside of the igloo is at 10.0°C and the outside is at –10.0°C, what is the rate of heat flow through the snow walls of area 14.0 m^2?

Air has a low thermal conductivity; it is an excellent insulator *when it is still*. An accurate calculation of the energy loss through a single-paned window *must* take into account the thin layer of stagnant air, due to viscosity, on each side of the glass. If the temperature is measured near a window, the temperature of the air just beside the window is intermediate in value between the temperatures of the room air and the outside air (see Fig. 14.11). Thus, the temperature gradient *across the glass* is considerably smaller than the difference between indoor and outdoor temperatures. In fact, much of the thermal resistance of a window is due to the stagnant air layers rather than to the glass.

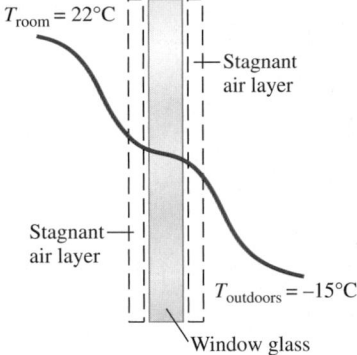

Figure 14.11   Temperature variation on either side of a windowpane. A plot of temperature versus position is superimposed on a cross section of the window glass and the air layers on either side.

Example 14.11

Heat Loss through a Double-Paned Window

 The single-paned window of Example 14.10 is replaced by a double-paned window with an air gap of 0.50 cm between the two panes. For the same temperature difference (22°C to –15°C), what is the new rate of heat loss through the double-paned window?

Strategy   Now there are three layers to consider: two layers of glass and one layer of air. We find the thermal resistance of

each layer and then add them together to find the total thermal resistance. Then we find the temperature difference between the inside of the house and the air outdoors and divide by the total thermal resistance to find the rate at which heat is lost through the replacement window.

Solution   For the first layer of glass,

$$R_1 = \frac{d}{\kappa A} = \frac{0.32 \times 10^{-2}\,\text{m}}{0.63 \text{ W/(m·K)} \times 0.0300 \text{ m}^2} = 0.169 \text{ K/W}$$

continued on next page

Example 14.11 *continued*

For the air gap,

$$R_2 = \frac{d}{\kappa A} = \frac{0.50 \times 10^{-2}\text{ m}}{0.023\text{ W}(\text{m·K}) \times 0.0300\text{ m}^2} = 7.246\text{ K/W}$$

The second layer of glass has the same thermal resistance as the first:

$$R_3 = R_1$$

The total thermal resistance is

$$\Sigma R_n = 0.169 + 7.246 + 0.169 = 7.584\text{ K/W}$$

and the rate of conductive heat flow is

$$I_{cd} = \frac{Q}{t} = \frac{\Delta T}{\Sigma R_n} = \frac{37\text{ K}}{7.584\text{ K/W}} = 4.9\text{ W}$$

Discussion   The reduction in the rate of heat loss by replacing a single-paned window with a double pane is significant. This example, however, overestimates the reduction since we assume that heat can only be conducted through the air layer. In reality, heat can also flow through air by convection and radiation. A more accurate calculation would have to account for the other methods of heat flow.

Practice Problem 14.11   Two panes of glass without the air gap

Repeat the problem of Example 14.11 if the two panes of glass are touching each other, without the intervening layer of air.

The United States building industry rates materials used in construction with *R-factors*. The R-factor is not quite the same as the thermal resistance; thermal resistance cannot be specified without knowing the cross-sectional area. The "R-factor" is the thickness divided by the thermal conductivity:

$$\text{R-factor} = \frac{d}{\kappa} = RA$$

$$\frac{I_{cd}}{A} = \frac{\Delta T}{\text{R-factor}}$$

Unfortunately, SI units are not used. The R-factors quoted in the United States are in units of °F·ft²/(Btu/h)! R-factors are added, just as thermal resistances are, when heat flows through several different layers.

14.7   CONVECTION

Convection involves *fluid currents* that carry heat from one place to another. In conduction, energy flows through a material but the material itself does not move. In convection, *the material itself moves* from one place to another. Thus, convection can occur only in fluids, not in solids. When a wood stove is burning, convection currents in the air carry heat upward to the ceiling. The heated air is less dense than cooler air, so the buoyant force causes it to rise, carrying heat with it. Meanwhile, cooler air that is more dense sinks toward the floor. An example of convection currents at the seashore is shown in Fig. 14.12. Air is a poor *conductor* of heat, but it can easily flow and carry heat by *convection*.

Making The Connection:
offshore and
onshore breezes

The use of sealed, double-paned windows replaces the large air gap of about 6 or 7 cm between a storm window and regular window with a much smaller gap. The smaller air gap minimizes circulating convection currents between the two panes. Down jackets and quilts are good insulators because air is trapped in many little spaces among the feathers, minimizing heat flow due to convection. Materials such as rock wool, glass wool, or fiberglass are used to insulate walls; much of their insulating value is due to the air trapped around and between the fibers.

Convection currents also occur in liquids. As water is heated in a pan on the stove burner, the heat travels from the burner through the bottom surface of the pot by conduction and heats the layer of water in contact with the pot bottom. The heated water then expands, becoming less dense, so buoyant forces help it rise into the cooler regions of water within the pot where it loses heat, coming to equilibrium with cooler water. The slightly cooled water then sinks down to become heated some more and repeat the process (Fig. 14.13).

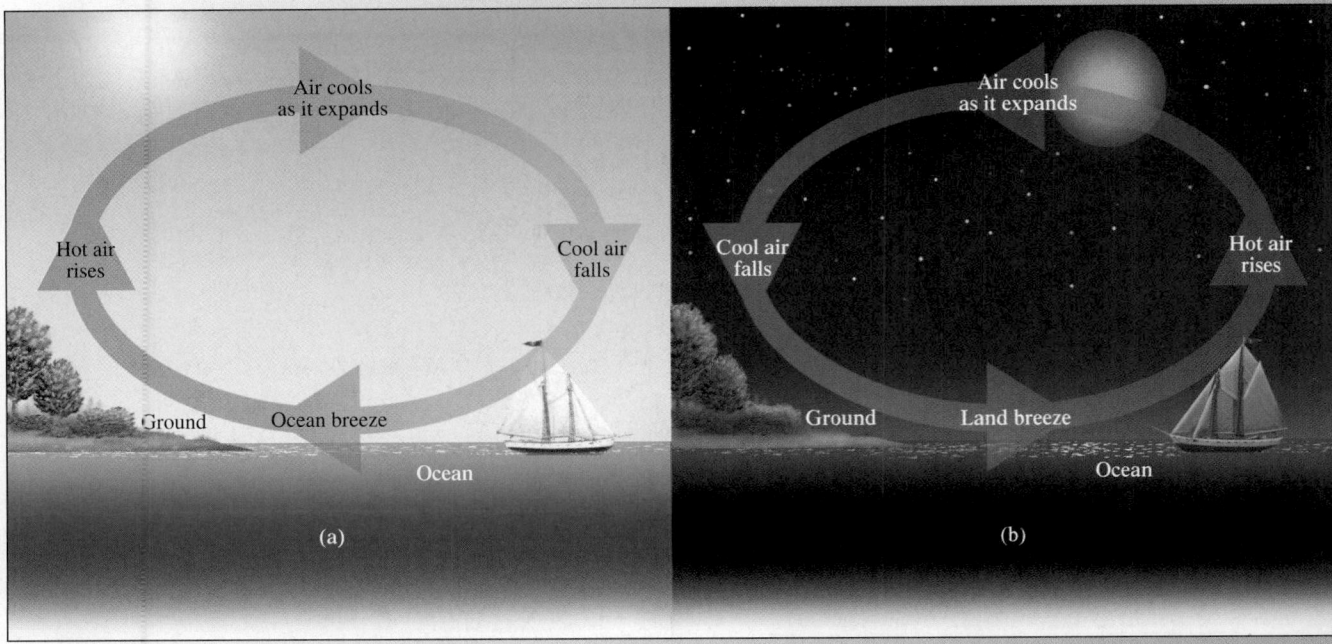

Figure 14.12 (a) During the day, air coming off the ocean is heated as it passes over the warm ground on shore; the heated air rises and expands. The expansion cools the air; it becomes more dense and falls back down. This cycle sets up a convection current that brings cool breezes from the sea to the shore. (b) The reverse circulation occurs at night when the land is cold and the sea is warmer, retaining heat absorbed during the day.

In *natural convection*, the currents are due to gravity. Fluid with a higher density sinks because the buoyant force is smaller than the weight; less dense fluid rises because the buoyant force exceeds the weight (Fig. 14.14). In *forced convection*, fluid is pushed around by mechanical means such as a fan or pump. In forced-hot-air heating, warm air is blown into rooms by a fan; in hot water baseboard heating, hot water is pumped through baseboard radiators (Fig. 14.15). Another example of forced convection is blood circulation in the body. The heart pumps blood around the body. When our body temperature is too high, the blood vessels near the skin dilate so that more blood can be pushed into them by the heart. The blood carries heat from the interior of the body to the skin; heat then flows from the skin into the cooler surroundings. If the surroundings are *hotter* than the skin, such as in a hot tub, this strategy backfires and can lead to dangerous overheating of the body. The hot water delivers heat to the dilated blood vessels; the blood carries the heat back to the central core of the body, *raising* the core temperature.

Figure 14.13 Convection currents in heated water

Physics at Home

In very hot climates, people drink *hot* drinks to cool off. Instead of reaching for a cold drink on a hot day, see if a hot cup of tea or cocoa makes you feel cooler. You are heating your central core so that your blood will carry heat to your skin; then the heat flows to the cooler air surrounding you and you feel cooler as a result.

A Global Warming Worry

One worry for scientists studying global warming is that northern Europe might be plunged into a deep freeze—a seeming contradiction that results from an interruption of the natural convection cycle. Earth's climate is influenced by convection currents

Making The Connection:
ocean currents and
global warming

Figure 14.14 Birds and people flying sailplanes take advantage of thermal updrafts.

Chimney

Chimney

Expansion tank

Baseboard radiator—Valve

Cold water return pipe

Baseboard radiator—Valve

Cold air

Hot air

Hot water

Exhaust Furnace

Exhaust Furnace

Figure 14.15 Household heating systems rely on forced convection.

Figure 14.16 Convection currents in the oceans. The Gulf Stream is a current of warm water flowing across the Atlantic.

caused by temperature differences between the poles and the tropics (Fig. 14.16). Massive sea currents travel through the Pacific and Atlantic oceans, carrying about half of the heat from the tropics to the poles, where it is dissipated. Storms moving north from the tropics carry much of the rest of the heat. If the polar regions warm at a faster rate than the tropics, the smaller temperature difference between them changes the patterns of the prevailing winds, the tracks followed by storms, the speed of ocean currents, and the amount of precipitation.

For example, the melting of the ice shelves combined with increased precipitation could lead to a layer of freshwater lying on top of the more dense salt water in the North Atlantic. Normally, the cold ocean water at the surface sinks and starts the process of convection. With the buoyancy of the less dense freshwater layer keeping it from sinking, the convection currents slow down or are stopped. Without the pull of the convection current, the usual northward movement of water from the warm Gulf Stream would slow or cease, causing *colder* temperatures in northern Europe.

Such an effect on climate is not without precedent. At the end of the last Ice Age, freshwater from melting glaciers flowed out the St. Lawrence River and into the North Atlantic. A freshwater layer, buoyed up by the more dense salt water, disrupted the usual ocean currents. The Gulf Stream was effectively shut down and Europe experienced a thousand years of deep freeze.

Mathematical analysis of convection is quite difficult, but a relatively simple approximation is useful in some cases. We can define a **coefficient of convection** (*h*) for various conditions, such as for dry air moving at various speeds across skin. The rate of

Table 14.6

Coefficients of Convection for Dry Air and Bare Skin

Wind Speed (m/s)	Convective Coefficient W/(m²·°C)
1	15
2	22
3	26
4	28.5
5	32

heat flow due to convection when a fluid moves along a surface is proportional to the surface area and to the temperature difference:

Rate of convective heat flow

$$I_{cv} = \frac{Q}{t} = hA\,\Delta T \tag{14-16}$$

Here A is the surface area over which the fluid moves, ΔT is the temperature difference between the surface and the fluid, and h is the coefficient of convection.

If the fluid is at a higher temperature than the surface, the heat flows from the fluid to the surface. If the surface is warmer, then the heat flows from the surface to the fluid. Equation (14-16) can be applied to air moving over skin with the help of Table 14.6, which lists convective coefficients for dry air moving over bare skin.

Example 14.12

Roller Blading in Still Air

A young woman is roller blading in still, dry air at a temperature of 30.0°C. She is moving along at 1.0 m/s and has a body surface area of 1.2 m², of which approximately 75% is exposed to the air. What is the rate of convective heat flow from her skin, at a temperature of 35.0°C, to the outside air?

Strategy The roller blader is moving at 1.0 m/s along the road, but we can consider her to be still and the air to be moving past her at 1.0 m/s; what matters is the airspeed with respect to the roller blader's skin. The area of skin exposed to the moving air is 75% of the total skin area.

Given: $T_{fluid} = 30.0°C$; $T_{surface} = 35.0°C$;
surface area of bare skin exposed is $A = 0.75 \times 1.2\ \text{m}^2$

Look up: coefficient of convection h for dry air and bare skin at wind speed of 1.0 m/s

Find: rate of heat flow, I_{cv}

Solution The surface area of exposed skin is

$$A = 0.75 \times 1.2\ \text{m}^2 = 0.90\ \text{m}^2$$

From Table 14.6, the coefficient of convection h for dry air and bare skin at a wind speed of 1.0 m/s is

$$h = 15\frac{\text{W}}{\text{m}^2 \cdot °\text{C}}$$

To find the rate of heat flow by convection, we use the equation

$$I_{cv} = hA\,\Delta T$$

Substituting values yields

$$I_{cv} = 15\frac{\text{W}}{\text{m}^2 \cdot °\text{C}} \times 0.90\ \text{m}^2 \times (30.0°\text{C} - 35.0°\text{C}) = -68\ \text{W}$$

Heat flows from her skin at a rate of 68 W.

Discussion The result can only be considered an estimate because the speed at which air moves past the skin is not a uniform 1.0 m/s. Due to the complicated pattern of airflow around the roller blader's body, the air moves faster past some surfaces than others.

In addition to convection, the body loses heat due to radiation (Section 14.8). This heat loss estimate represents only *part* of the heat lost by the body.

Practice Problem 14.12 A sailor standing on the bow of a ship

A sailor on an America's Cup racing yacht is wearing shorts and no shirt; the area of exposed skin is 1.3 m². The apparent wind (speed of the air relative to the sailor's body) is 5.0 m/s. What is the approximate rate of convective heat loss if his skin is at 35.0°C and the air temperature is 29.0°C? Assume that the air is dry.

14.8   RADIATION

Radiation is a third method of heat flow. All bodies emit energy through electromagnetic radiation—due to the oscillation of electric charges in the atoms. Radiation consists of electromagnetic waves that travel at the speed of light. Unlike conduction and convection, radiation need not travel through a material medium; the Sun radiates heat to Earth through the near vacuum of space.

When solar radiation reaches Earth, it is partially absorbed and partially reflected. The absorbed portion increases the Earth's internal energy. If absorption and reflection were the whole story, Earth's internal energy would continuously increase; but the Earth also emits radiation, which carries energy away. Since the temperature of the Earth remains relatively constant, it must emit energy at the same rate, on average, that it absorbs energy from the Sun. Thus, there is an equilibrium between absorption and emission.

Conceptual Example 14.13

An Alligator Lying in the Sun

An alligator crawls out into the Sun to get warm. Solar radiation is incident on the alligator at the rate of 300 W; 70 W of it is reflected. (a) What happens to the other 230 W? (b) If the alligator emits 100 W, does its body temperature rise, fall, or stay the same? Ignore heat flow by conduction and convection.

Solution and Discussion   (a) When radiation falls on an object, some can be absorbed, some can be reflected, and—for a transparent or translucent object—some can be transmitted through the object without being absorbed or reflected. Since the alligator is opaque, no radiation is transmitted through it. All the radiation is either absorbed or reflected, so the other 230 W is absorbed. (b) Since 230 W is absorbed while 100 W

is emitted, the alligator absorbs more radiation than it emits. Absorption increases internal energy while emission decreases it, so the alligator's internal energy is increasing at a rate of 130 W. Thus, we expect the alligator's body temperature to rise. (The actual rate of increase of internal energy would be smaller since conduction and convection carry heat away as well.)

Conceptual Practice Problem 14.13
Maintaining constant temperature

After some time elapses, the alligator's body temperature reaches a constant level. The rate of absorption is still 230 W. If the alligator loses heat by conduction and convection at a rate of 90 W, at what rate does it emit radiation?

Stefan's Radiation Law

Some bodies emit more radiation than others per unit surface area even though their temperatures are the same. An idealized body that absorbs all the radiation incident upon it is called a **blackbody**. A *blackbody* absorbs not only all visible light, but infrared, ultraviolet, and all other wavelengths of electromagnetic radiation. Black velvet and a substance called lampblack come close to behaving like blackbodies. (Objects appear black when they absorb most of the visible light incident on them, so that little is reflected. However, an ideal blackbody absorbs all the radiation incident on it, not just the visible radiation.)

What does this have to do with *emission*? It turns out (see Conceptual Question 23) that a good absorber is also a good emitter of radiation. A blackbody emits more radiant power per unit surface area than any real object at the same temperature. The rate at which a blackbody emits radiation per unit surface area is proportional to the fourth power of the *absolute* temperature:

> **Stefan's law of radiation (blackbody)**
>
> $$I_{em} = \frac{Q}{t} = \sigma A T^4 \qquad (14\text{-}17)$$

In Eq. (14-17), A is the surface area and T is the surface temperature of the blackbody *in kelvins*. Since Stefan's law involves the absolute temperature and not a temperature difference, °C *cannot* be substituted. The universal constant σ (Greek letter sigma) is called *Stefan's constant*:

$$\sigma = 5.670 \times 10^{-8} \text{ W/(m}^2 \cdot \text{K}^4) \qquad (14\text{-}18)$$

The fourth-power temperature dependence implies that the power emitted is extremely sensitive to temperature changes. If the absolute temperature of a body doubles, the energy emitted increases by a factor of $2^4 = 16$.

Since real bodies are not perfect absorbers and therefore emit less than a blackbody, we define the **emissivity** (e) as the ratio of the emitted power of the body to that of a blackbody at the same temperature. Then Stefan's law becomes

> **Stefan's law of radiation**
>
> $$I_{em} = \frac{Q}{t} = e\sigma A T^4 \qquad (14\text{-}19)$$

The emissivity ranges from 0 to 1; $e = 1$ for a perfect radiator and absorber (a blackbody) and $e = 0$ for a perfect reflector. The emissivity for polished aluminum, an excellent reflector, is about 0.05; for soot (carbon black) it is about 0.95. Human skin, no matter what the pigmentation, has an emissivity of about 0.97 in the infrared part of the spectrum. Many objects have high emissivities in the infrared even though they may reflect a fair amount of the visible light incident on them.

Radiation Spectrum

The electromagnetic radiation we are concerned with falls into three wavelength ranges. Infrared radiation includes wavelengths from about 100 μm down to 0.7 μm. The wavelengths of visible light range from about 0.7 μm to about 0.4 μm. Ultraviolet wavelengths are less than 0.4 μm.

The total power radiated is not the only thing that varies with temperature. Figure 14.17 shows the radiation spectrum—a graph of how much radiation occurs as a function of wavelength—for blackbodies at two different temperatures. The wavelength at which the maximum power is emitted decreases as temperature increases. Objects at ordinary temperatures emit primarily in the infrared—around 10 μm in wavelength for 300 K. The Sun, since it is much hotter, radiates primarily at shorter wavelengths. Its radiation peaks in the visible (no surprise there) but includes plenty of infrared and ultraviolet as well. The wavelength of maximum radiation is inversely proportional to the absolute temperature:

Wien's law

$$\lambda_{max}T = 2.898 \times 10^{-3}\ \text{m·K} \qquad (14\text{-}20)$$

---

where the temperature T is the temperature in kelvins and λ_{max} is the wavelength of maximum radiation in meters.

As the temperature of the blackbody rises to 1000 K and above, the peak intensity shifts toward shorter wavelengths until some of the emitted radiation falls in the visible. Since the longest visible wavelengths are for red light, the heated body glows dull red. As the temperature of the blackbody continues to increase, the red glow becomes brighter red, then orange, then yellow-white, and eventually blue-white as the blackbody emits more and more visible light. When the body is emitting all the colors of the visible spectrum, the glow appears white to the eye. When something is red-hot, it is not as hot as something that is white-hot.

Figure 14.17 Graphs of blackbody radiation as a function of wavelength at two temperatures. At the higher temperature, the wavelength of maximum radiation is shorter, as predicted by Wien's law.

Example 14.14

Temperature of the Sun

 The maximum rate of energy emission from the Sun occurs in the middle of the visible range—at about $\lambda = 0.5$ μm. Estimate the temperature of the Sun's surface.

Strategy We assume the Sun to be a blackbody. Then the wavelength of maximum emission and the surface temperature are related by Wien's law.

Solution Given: $\lambda_{max} = 0.5$ μm $= 5 \times 10^{-7}$ m. Then from Wien's law, we know that the product of the wavelength for maximum power emission and the corresponding temperature for the power emission is

$$\lambda_{max}T = 2.898 \times 10^{-3} \text{ m·K}$$

We can solve for the temperature since we know λ_{max}

$$T = \frac{2.898 \times 10^{-3} \text{ m·K}}{5 \times 10^{-7} \text{ m}}$$

$$= 6000 \text{ K}$$

Discussion Quick check: an object at 300 K has $\lambda_{max} \approx$ 10 μm, which is 20 times the λ_{max} in the radiation from the Sun (0.5 μm). Since λ_{max} and T are inversely proportional, the Sun's surface temperature is 20 times 300 K = 6000 K.

Practice Problem 14.14 Wavelengths of maximum power emission for skin

The temperature of skin varies from 30°C to 35°C depending on the blood flow near the skin surface. What is the range of wavelengths of maximum power emission from skin?

A body emits energy even if it is at the same temperature as its surroundings; it just emits at the same rate that it absorbs. If the surroundings are at a lower temperature than the body, the body emits more than it absorbs. The difference between the power emitted by the body and that absorbed by the body from its surroundings is

$$I_{em} - I_{ab} = e\sigma A(T^4 - T_0^4) \tag{14-21}$$

| Net rate of energy transfer due to emission and absorption of radiation

In Eq. (14-21), T_0 is the temperature of the surroundings. We have assumed that the emissivity is equal to the fraction of incident energy absorbed. If a body emits and absorbs at very different wavelengths—for example, if T and T_0 differ greatly—this assumption may not be true. *At any particular wavelength*, the fraction absorbed is always equal to the emissivity at that wavelength.

Do not substitute temperature in Celsius degrees into Eq. (14-21). The quantity inside the parentheses might look like a temperature difference, but it is not. The two kelvin temperatures are raised to the fourth power, *then* subtracted—which is not the same as the corresponding two Celsius temperatures subjected to the same mathematical operations. By the same token, do not subtract the temperatures in kelvins and then raise to the fourth power. The difference of the fourth powers is not equal to the difference raised to the fourth power, as can be readily demonstrated:

$$1 = (2-1)^4 \neq (2^4 - 1^4) = 15$$

Medical Applications

Thermal radiation from the body is used as a diagnostic tool in medicine. A thermogram shows whether one area is radiating more heat than it should, indicating a higher temperature due to abnormal cellular activity. For example, when a broken bone is healing, heat can be detected at the location of the break just by placing a hand lightly on the area of skin over the break. Infrared detectors, originally developed for military uses (nightscopes, for example), can be used to detect radiation from the skin. The radiation is absorbed and an electrical signal is produced that is then used to produce a visual display (Fig. 14.18).

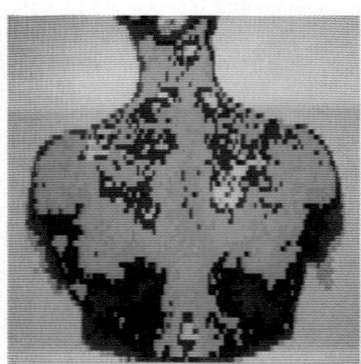

Figure 14.18 Thermography of a backache. The magenta areas are warmer than the surrounding tissue, revealing the location of the source of pain.

Example 14.15

Thermal Radiation from Human Body

 A person of body surface area 2.0 m^2 is sitting in a doctor's examining room with no clothing on. The temperature of the room is 22°C and the person's average skin temperature is 34°C. Skin emits about 97% as much as a blackbody at the same temperature for wavelengths in the infrared region, where most of the emission occurs. At what rate is energy radiated away from the body?

Strategy Both radiation and absorption occur in the infrared—the absolute temperatures of the skin and the room are not very different. Therefore, we can assume that 97% of the incident radiation from the room is absorbed. Equation (14-21) therefore applies. We must convert the Celsius temperatures to kelvin.

Given: surface area, $A = 2.0$ m^2; $T_{room} = 22$°C; skin temperature, $T = 34$°C;
    fraction of energy emitted, $e = 0.97$
To find: net rate of energy transfer, $I_{em} - I_{ab}$

Solution The temperature of the skin surface is
$$T = 273 + 34 = 307 \text{ K}$$
and of the room is
$$T_0 = 273 + 22 = 295 \text{ K}$$

The net rate of energy transfer between the room and the body is
$$I_{em} - I_{ab} = e\sigma A(T^4 - T_0^4)$$
Substituting,
$$I_{em} - I_{ab} = 0.97 \times 5.67 \times 10^{-8} \text{ W/(m}^2 \cdot \text{K}^4) \times 2.0 \text{ m}^2 \times (307^4 - 295^4) \text{ K}^4$$
$$= 140 \text{ W}$$

Discussion 140 W is a significant heat loss because the body also loses about 10 W by convection and conduction. To stay at a constant body temperature, an inactive person must give off heat at a rate of 90 W to account for basal metabolic activity; if the rate of heat loss exceeds that, the body temperature starts to drop. The patient had better wrap a blanket around his body or start running in place.

We need only the fraction of energy emitted and absorbed by the body; the emissivity of the walls of the room is irrelevant. If the walls are poor emitters, then they also absorb poorly, so they reflect radiation. The amount of radiation incident on the body is the same.

Practice Problem 14.15 The roller blader radiates

Find how much energy per unit time the roller blader in Example 14.12 loses by radiation from her body. Her skin temperature is 35°C and the air temperature is 30°C. Her surface area is 1.2 m^2 of which 75% is exposed to the air. Assume skin has $e = 0.97$.

Example 14.16

Radiative Equilibrium of Earth

Radiant energy from the Sun reaches Earth at a rate of 1.7×10^{17} W. An average of about 30% is reflected and the rest is absorbed. Energy is also radiated by the atmosphere. Assuming that the atmosphere emits as a blackbody in the infrared ($e = 1$), calculate the temperature of the atmosphere. (The Sun's radiation peaks in the visible part of the spectrum, but the Earth's radiation peaks in the infrared due to its much lower surface temperature.)

Strategy Earth must radiate the same power as it absorbs. We use Stefan's law to find the rate at which energy is radiated as a function of temperature and then equate that to the rate of energy absorption.

Solution Earth absorbs 70% of the incident solar radiation. To have a relatively constant temperature, it must emit radiation at the same rate:
$$I_{em} = 0.7 \times 1.7 \times 10^{17} \text{ W} = 1.2 \times 10^{17} \text{ W}$$
From Stefan's law,
$$I_{em} = e\sigma A T^4$$

where we take $e = 1$ since the atmosphere is assumed to emit as a blackbody. The surface area of the Earth is
$$A = 4\pi R_E^2$$
Solving Stefan's law for T yields
$$T = \left(\frac{I_{em}}{e\sigma A}\right)^{1/4}$$

Now we substitute numerical values:
$$T = \left(\frac{I_{em}}{e\sigma 4\pi R_E^2}\right)^{1/4} = \left(\frac{1.2 \times 10^{17} \text{ W}}{1 \times 5.67 \times 10^{-8} \text{ W/(m}^2 \cdot \text{K}^4) \times 4\pi(6.4 \times 10^6 \text{ m})^2}\right)^{1/4}$$
$$= 253 \text{ K} = -20°\text{C}$$

Discussion Remember that –20°C is supposed to be the average temperature of the *atmosphere*, not of the Earth's surface. This relatively simple calculation gives impressively accurate results. To find the temperature of the Earth's surface, we must take the greenhouse effect into account.

Practice Problem 14.16 Reflecting less incident radiation

If the Earth were to reflect 25% of the incident radiation instead of 30%, what would be the average temperature of the atmosphere?

The Greenhouse Effect

Making The Connection:
global warming and
the greenhouse effect

The Earth receives heat by radiation from the Sun. The atmosphere helps trap some of the radiation, acting rather like the glass in a greenhouse. When sunlight falls upon the glass of a greenhouse, most of the visible radiation and short-wavelength infrared (*near-infrared*) travel right on through; the glass is transparent to those wavelengths. The glass absorbs much of the incoming ultraviolet radiation. The radiation that gets through the glass is mostly absorbed inside the greenhouse. Since the inside of the greenhouse is much cooler than the Sun, it emits primarily infrared radiation (IR). The glass is not transparent to this longer-wavelength IR; much of it is absorbed by the glass. The glass itself also emits IR, but in both directions: half of it is emitted back inside the greenhouse. The absorption of IR by the glass keeps the greenhouse warmer than it would otherwise be. (The glass in a greenhouse has a second function not mirrored in the Earth's atmosphere—it prevents heat from being carried away by convection.)

The Earth is something like a greenhouse, where the atmosphere fulfills the role of the glass. Like glass, the atmosphere is largely transparent to visible and near-IR; the ozone layer in the upper atmosphere absorbs some of the ultraviolet. The atmosphere absorbs a great deal of the longer-wavelength IR emitted by Earth's surface. The atmosphere *radiates* IR in two directions: back toward the surface and out toward space (Fig. 14.19). "Greenhouse gases" such as CO_2 are particularly good absorbers of IR. The higher the concentration of greenhouse gases in the atmosphere, the more IR is absorbed and the warmer the Earth's surface becomes. Even small changes in the average surface temperature can have dramatic effects on climate.

In applying Stefan's radiation law to the Earth, there are some complications. One is the effect of the cloud cover. Clouds are quite reflective, but they are sometimes there and sometimes not. The heating of the lakes and oceans causes water to evaporate and form clouds. The clouds then serve as a screen and reflect sunlight away from the Earth, reducing the temperature again.

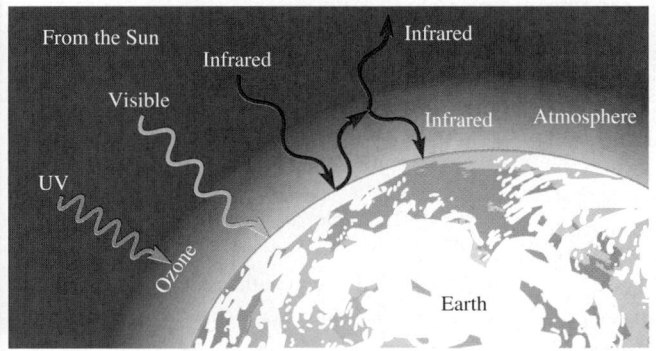

Figure 14.19 The global greenhouse effect. In this *simplified* diagram, all the UV from the Sun is absorbed by the atmosphere, while all the visible and IR from the Sun is transmitted. The Earth absorbs the visible and IR and radiates longer-wavelength IR. The longer-wavelength IR is absorbed by the atmosphere, which itself radiates IR both back toward the surface and out toward space.

MASTER THE CONCEPTS

Summary

- The internal energy of a system is the total energy of all of the molecules in the system except for the macroscopic kinetic energy (kinetic energy associated with macroscopic translation or rotation) and the external potential energy (energy due to external interactions).

- Heat is a *flow* of energy that occurs due to a temperature difference.

- The joule is the SI unit for all forms of energy, for heat, and for work. An alternate unit often used for heat and internal energy is the calorie:

$$1 \, cal = 4.186 \, J \qquad (14\text{-}1)$$

- The ratio of heat flow into a system to the temperature change of the system is the heat capacity of the system:

$$\text{heat capacity} = \frac{Q}{\Delta T} \qquad (14\text{-}3)$$

MASTER THE CONCEPTS *continued*

- The heat capacity per unit mass is the specific heat capacity (or specific heat) of a substance:

$$c = \frac{Q}{m\,\Delta T} \qquad (14\text{-}4)$$

- The *molar specific heat, C,* is the heat capacity per mole:

$$Q = nC_V\,\Delta T \qquad (14\text{-}8)$$

At room temperature, the molar heat capacity at constant volume for a monatomic ideal gas is approximately $C_V = \frac{3}{2}R$; for a diatomic ideal gas it is approximately $C_V = \frac{5}{2}R$.

- Phase transitions occur at constant temperature. The heat *per unit mass* that must flow to melt a solid or to freeze a liquid is the latent heat of fusion L_f. The latent heat of vaporization L_v is the heat *per unit mass* that must flow to change the phase from liquid to gas or from gas to liquid.

- *Sublimation* occurs when a solid changes directly to a gas without going into a liquid form.

- A phase diagram is a graph of pressure versus temperature that indicates solid, liquid, and vapor regions for a substance. The sublimation, fusion, and vapor pressure curves separate the three phases. Crossing one of these curves represents a phase transition.

- Heat flows by three processes: conduction, convection, and radiation.

- Conduction is due to atomic (or molecular) collisions within a substance or from one object to another when they are in contact. The rate of heat flow within a substance is:

$$I_{cd} = \frac{Q}{t} = \kappa A \frac{\Delta T}{d} \qquad (14\text{-}12)$$

where $I_{cd} = Q/t$ is the rate of heat flow (or power delivered), κ is the thermal conductivity of the material, A is the cross-sectional area, d is the thickness (or length) of the material, and ΔT is the temperature difference between one side and the other.

- Convection involves *fluid currents* that carry heat from one place to another. In convection, the material itself moves from one place to another.

- Radiation does not have to travel through a material medium. The energy is carried by electromagnetic waves that travel at the speed of light. All bodies emit energy through electromagnetic radiation. An idealized body that absorbs all the radiation incident on it is called a blackbody. A blackbody emits more radiant power per unit surface area than any real object at the same temperature. Stefan's law of radiation is

$$I_{em} = \frac{Q}{t} = e\sigma A T^4 \qquad (14\text{-}19)$$

where the emissivity e ranges from 0 to 1, A is the surface area, T is the surface temperature of the blackbody *in kelvins*, and Stefan's constant is $\sigma = 5.670 \times 10^{-8}$ W/(m²·K⁴). The wavelength of maximum power emission is inversely proportional to the absolute temperature:

$$\lambda_{max}T = 2.898 \times 10^{-3}\ \text{m·K} \qquad (14\text{-}20)$$

The difference between the power emitted by the body and that absorbed by the body from its surroundings is the net power emitted:

$$I_{em} - I_{ab} = e\sigma A(T^4 - T_0^4) \qquad (14\text{-}21)$$

Highlighted Figures and Tables

F14.2 Joule's experiment (p. 476)

T14.1 Specific heats of common substances at 1 atm and 20°C (p. 479)

F14.3 A calorimeter (p. 480)

T14.2 Molar specific heats at constant volume of gases at 25°C (p. 482)

F14.5 Temperature versus heat for 1 kg of ice that starts at a temperature below 0°C (p. 484)

T14.3 Heat to turn 1 kg of ice at –25°C to steam at 125°C (p. 484)

T14.4 Latent heats of some common substances (p. 485)

F14.7 Phase diagram for water (p. 488)

F14.8 Phase diagram for carbon dioxide (p. 489)

F14.9 Conduction through a slab and along a cylindrical bar (p. 489)

T14.5 Thermal conductivities at 20°C (p. 490)

F14.10 Conduction of heat through two different layers $(T_1 > T_2 > T_3)$ (p. 490)

F14.12 Example of convection currents at the seashore (p. 493)

F14.14 Thermal updrafts used in flight (p. 494)

F14.15 Household heating systems and forced convection (p. 494)

F14.16 Convection currents in the ocean (p. 495)

T14.6 Coefficients of convection for dry air and bare skin (p. 496)

F14.17 Blackbody radiation spectrum (p. 498)

CONCEPTUAL QUESTIONS

1. What determines the direction of heat flow when two objects at different temperatures are placed in thermal contact?

2. When an old movie has a scene of someone ironing, the person is often shown testing the heat of a hot flat iron with a moistened finger. Why is this safe to do?

3. Why do lakes and rivers freeze first at their surfaces?

4. Why is drinking water in a camp located near the equator often kept in porous jars?

5. Why are several layers of clothing warmer than one coat of equal weight?

6. Why are vineyards planted along lakeshores or riverbanks in cold climates?

7. A metal plant stand on a wooden deck feels colder than the wood around it. Is it necessarily colder? Explain.

8. Near a large lake, in what direction does a breeze passing over the land tend to blow at night?

9. What is the purpose of having fins on an automobile or motorcycle radiator?

10. Why do roadside signs warn that bridges ice before roadways? Explain.

11. Why do cooking directions on packages advise different timing to be followed for some locations?

12. Explain the theory behind the pressure cooker. How does it speed up cooking times?

13. When you eat a pizza that has just come from the oven, why is it that you are apt to burn the roof of your mouth with the first bite although the crust of the pizza feels only warm to your hand?

14. Explain why the molar specific heat of a diatomic gas such as O_2 is larger than that of a monatomic gas such as Ne.

15. At very low temperatures, the molar specific heat of hydrogen (H_2) is $C_V = 1.5R$. At room temperature, $C_V = 2.5R$. Explain.

16. When the temperature as measured in °C of a radiating body is doubled (such as a change from 20°C to 40°C), is the radiation rate necessarily increased by a factor of 16?

17. A cup of hot coffee has been poured, but the coffee drinker has a little more work to do at the computer before she picks up the cup. She intends to add some milk to the coffee. To keep the coffee hot as long as possible, should she add the milk at once, or wait until just before she takes her first sip?

18. Would heat loss be reduced or increased by increasing the usual air gap, 1 to 2 cm, between commercially made double-paned windows? Explain your reasoning. [Hint: Consider convection.]

19. A study of food preservation in Britain discovered that the temperature of meat that is kept in transparent plastic packages and stored in open and lighted freezers can be as much as 12°C above the temperature of the freezer. Why is this? How could this be prevented?

20. Which possesses more total internal energy, the water within a large, partially ice-covered lake in winter or a 6-cup teapot filled with hot tea? Explain.

21. A room in which the air temperature is held constant may feel warm in the summer but cool in the winter. Explain. [Hint: The walls are not necessarily at the same temperature as the air.]

22. Many homes are heated with "radiators," which are hollow metal devices filled with hot water or steam and located in each room of the house. They are sometimes painted with metallic, high-gloss silver paint so that they look well polished. Does this make them better radiators of heat? If not, what might be a more efficient finish to use?

23. Two objects with the same surface area are inside an evacuated container. The walls of the container are kept at a constant temperature. Suppose one object absorbs a larger fraction of incident radiation than the other. Explain why that object must emit a correspondingly greater amount of radiation than the other. Thus a good absorber must be a good emitter.

MULTIPLE CHOICE QUESTIONS

1. The main loss of heat from the Earth is by
(a) radiation.   (b) convection.   (c) conduction.
(d) All three processes are significant modes of heat loss from the Earth.

2. Assume the average temperature of the Earth's atmosphere to be 253 K. What would be the eventual average temperature of the Earth's atmosphere if the surface temperature of the Sun were to drop by a factor of 2?
(a) 253 K   (b) $\frac{253\text{ K}}{2} = 127$ K
(c) $\frac{253\text{ K}}{4} = 63$ K   (d) $\frac{253\text{ K}}{2^4} = 16$ K

3. In equilibrium, Mars emits as much radiation as it absorbs. If Mars orbits the Sun with an orbital radius that is 1.5 times the orbital radius of the Earth about the Sun, what is the approximate atmospheric temperature of Mars? Assume the atmospheric temperature of Earth to be 253 K.
(a) $\frac{253\text{ K}}{1.5} = 170$ K   (b) $\frac{253\text{ K}}{1.5^2} = 112$ K
(c) $\frac{253\text{ K}}{1.5^4} = 50$ K   (d) $\frac{253\text{ K}}{\sqrt{1.5}} = 207$ K

4. Which term best represents the relation between a blackbody and radiant energy? A blackbody is an ideal _____ of radiant energy.
(a) emitter   (b) absorber
(c) reflector   (d) emitter and absorber

5. A window conducts power P from a house to the cold outdoors. What power is conducted through a window of half the area and half the thickness?
(a) $4P$ (b) $2P$ (c) P (d) $P/2$ (e) $P/4$

6. Iron has a specific heat that is about four times that of gold. A cube of gold and a cube of iron, both of equal mass and at 20°C, are placed in two different Styrofoam cups, each filled with 100 g of water at 40°C. The Styrofoam cups have negligible heat capacities. After equilibrium has been attained,
(a) the temperature of the gold is lower than that of the iron.
(b) the temperature of the gold is higher than that of the iron.
(c) the temperatures of the water in the two cups are the same.
(d) (a) or (b) depending on how much mass is involved.

7. Sublimation is involved in which of these phase changes?
(a) liquid to gas   (b) solid to liquid
(c) solid to gas   (d) gas to liquid

8. When a vapor condenses to a liquid,
(a) its internal energy increases.
(b) its temperature rises.
(c) its temperature falls.
(d) it gives off internal energy.

9. When a substance is at its triple point, it
(a) is in its solid phase.   (b) is in its liquid phase.
(c) is in its vapor phase.
(d) may be in any or all of these phases.

10. The phase diagram for water is shown in Fig. 14.20. If the temperature of a certain amount of ice is increased by following the path represented by the dashed line from A to B in the phase diagram, which of the graphs of temperature as a function of heat added is correct?

11. Two thin rods are made from the same material and are of lengths L_1 and L_2. The two ends of the rods have the same temperature difference. What should the relation be between their diameters and lengths so that they conduct equal amounts of heat energy in a given time?

   (a) $\dfrac{L_1}{L_2} = \dfrac{d_1}{d_2}$ (b) $\dfrac{L_1}{L_2} = \dfrac{d_2}{d_1}$

   (c) $\dfrac{L_1}{L_2} = \dfrac{d_1^2}{d_2^2}$ (d) $\dfrac{L_1}{L_2} = \dfrac{d_2^2}{d_1^2}$

12. If you place your hand underneath, but not touching, a kettle of hot water, you *mainly* feel the presence of heat from

   (a) conduction.   (b) convection.   (c) radiation.

(a)

(b)

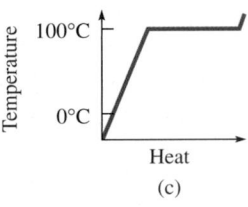

(c)

Figure 14.20 Multiple Choice Question 10

PROBLEMS

Note: **C** indicates a combination conceptual/quantitative problem. Gold diamonds ✦, ✦✦ are used to indicate the increasing level of difficulty of each problem. Problem numbers appearing in blue, 9., denote problems that have a detailed solution available in the Student Solutions Manual. Some problems are *paired* by concept; their numbers are connected by a ruled box.

14.1 Internal Energy

C 1. A mass of 1.4 kg of water at 22°C is poured from a height of 2.5 m into a vessel containing 5.0 kg of water at 22°C. (a) How much does the internal energy of the 6.4 kg of water increase? (b) Is it likely that the water temperature increases? Explain.

2. The water passing over Victoria Falls, located along the Zambezi River on the border of Zimbabwe and Zambia, drops about 105 m. How much internal energy is produced per kg as a result of the fall?

3. How much internal energy is generated when a 20.0-g lead bullet, traveling at 7.00×10^2 m/s, comes to a stop as it strikes a metal plate?

4. Nolan threw a baseball, of mass 147.5 g, at a speed of 162 km/h to a catcher. How much internal energy was generated when the ball struck the catcher's mitt?

C 5. A child of mass 15 kg climbs to the top of a slide that is 1.7 m above a horizontal run that extends for 0.50 m at the base of the slide. After sliding down, the child comes to rest just before reaching the very end of the horizontal portion of the slide. (a) How much internal energy was generated during this process? (b) Where did the generated energy go? (To the slide, to the child, to the air, or to all three?)

6. A 64-kg sky diver jumped out of an airplane at an altitude of 0.90 km. She opened her parachute after a while and eventually landed on the ground with a speed of 5.8 m/s. How much energy was dissipated by air resistance during the jump?

✦ 7. During basketball practice Shane made a jump shot, releasing a 0.60-kg basketball from his hands at a height of 2.0 m above the floor with a speed of 7.6 m/s. The ball swooshes through the net at a height of 3.0 m above the floor and with a speed of 4.5 m/s. How much energy was dissipated by air drag from the time the ball left Shane's hands until it went through the net?

14.2 Heat; 14.3 Heat Capacity and Specific Heat

8. An experiment is conducted with a basic Joule apparatus, where a mass is allowed to descend by 1.25 m and rotate a paddle wheel within an insulated container of water. There are several different sizes of descending masses to choose among. If the investigator wishes to deliver 1.00 kcal to the water within the insulated container after 30.0 descents, what descending mass value should be used? Let $g = 9.80$ m/s^2.

9. Convert 1.00 kcal to kilowatt-hours (kWh).

10. What is the heat capacity of 20.0 kg of silver?

11. What is the heat capacity of a gold ring that has a mass of 5.00 g?

12. If 30.0 kcal of heat are supplied to 5.00×10^2 g of water at 22°C, what is the final temperature of the water?

13. It is a damp, chilly day in a New England seacoast town suffering from a power failure. To warm up the cold, clammy

sheets, Jen decides to fill hot water bottles to tuck between the sheets at the foot of the beds. If she wishes to heat 2.0 L of water on the wood stove from 20.0°C to 80.0°C, how much heat must flow into the water?

14. An 83-kg man eats a banana of energy content 1.00×10^2 Calories (food calories). If all of the energy from the banana is converted into kinetic energy of the man, how fast is he moving, assuming he starts from rest?

15. A high jumper of mass 60.0 kg consumes a meal of 3.00×10^3 Calories (food calories) prior to a jump. If 3.3% of the energy from the food could be converted to gravitational potential energy in a single jump, how high could the athlete jump?

16. What is the heat capacity of a 30.0-kg block of ice?

17. What is the heat capacity of 1.00 m^3 of (a) aluminum? (b) iron? See Table 9.1 for density values.

18. What is the heat capacity of a system consisting of (a) a 0.450-kg brass cup filled with 0.050 kg of water? (b) 7.5 kg of water in a 0.75-kg aluminum bucket?

19. A 0.400-kg aluminum teakettle contains 2.00 kg of water at 15.0°C. How much heat is required to raise the temperature of the water (and kettle) to 100.0°C?

20. How much heat is required to raise the body temperature of a 50.0-kg woman from 37.0°C to 38.4°C?

21. It takes 210 cal to raise the temperature of 350 g of lead from 0 to 20.0°C. What is the specific heat of lead?

22. A mass of 1.00 kg of water at temperature T is poured from a height of 0.100 km into a vessel containing water of the same temperature T, and a temperature change of 0.100°C is measured. What mass of water was in the vessel? Neglect heat flow into the vessel, the thermometer, etc.

23. A thermometer containing 0.10 g of mercury is cooled from 15.0°C to 8.5°C. How much energy was lost by the mercury in this process?

24. A bit of space debris penetrates the hull of a spaceship traversing the asteroid belt and comes to rest in a container of water that was at 20.0°C before being hit. The mass of the space rock is 1.0 g and the mass of the water is 1.0 kg. If the space rock traveled at 8.4×10^3 m/s and if all of its kinetic energy is used to heat the water, what is the final temperature of the water?

25. A pot containing 2.00 kg of water is sitting on a hot stove and the water is stirred violently by a mixer that does 6.0 kJ of mechanical work on the water. The temperature of the water rises by 4.00°C. What quantity of heat flowed into the water from the stove during the process?

26. A heating coil inside an electric kettle delivers 2.1 kW of electrical power to the water in the kettle. How long will it take to raise the temperature of 0.50 kg of water from 20.0°C to 100.0°C?

27. A 10.0-g iron bullet with a speed of 4.00×10^2 m/s and a temperature of 20.0°C is stopped in a 0.500-kg block of wood, also at 20.0°C. (a) At first all of the bullet's kinetic energy goes into the internal energy of the bullet. Calculate the temperature increase of the bullet. (b) After a short time the bullet and the block come to the same temperature T. Calculate T, assuming no heat is lost to the environment.

14.4 Specific Heat of Ideal Gases

28. Imagine that 501 people are present in a movie theater of volume 8.00×10^3 m^3 that is sealed shut so no air can escape. Each person gives off heat at an average rate of 110 W. By how much will the temperature of the air have increased during a 2.0-h movie? The initial pressure is 1.01×10^5 Pa and the initial temperature is 20.0°C. Assume that all the heat output of the people goes into heating the air (a diatomic gas).

29. A chamber with a fixed volume of 1.0 m^3 contains a monatomic gas at 3.00×10^2 K. The chamber is heated to a temperature of 4.00×10^2 K. This operation requires 10.0 J of heat. (Assume all the energy is transferred to the gas.) How many gas molecules are in the chamber?

30. A cylinder contains 250 L of hydrogen gas (H_2) at 0.0°C and a pressure of 10.0 atm. How much energy is required to raise the temperature of this gas to 25.0°C?

14.5 Phase Transitions

♦ 31. You are given 2.5×10^2 g of coffee (same specific heat as water) at 80.0°C (too hot to drink). In order to cool this to 60.0°C, how much ice (at 0.0°C) must be added? Neglect heat content of the cup and heat exchanges with the surroundings.

32. Given these data, compute the heat of fusion of water.

Mass of calorimeter = 3.00×10^2 g	Specific heat of calorimeter = 0.090 cal/g·K
Mass of water = 2.00×10^2 g	Initial temperature of water and calorimeter = 20.0°C
Mass of ice = 11.6 g	Initial temperature of ice = −5.0°C
	Final temperature of calorimeter = 15.0°C

33. Given these data, compute the heat of vaporization of water.

Mass of calorimeter = 3.00×10^2 g	Specific heat of calorimeter = 0.090 cal/g·K
Mass of water = 2.00×10^2 g	Initial temperature of water and calorimeter = 15.0°C
Mass of condensed steam = 18.5 g	Initial temperature of steam = 100.0°C
	Final temperature of calorimeter = 62.0°C

34. What mass of water at 25.0°C added to a Styrofoam cup containing two 50.0-g ice cubes from a freezer at −15.0°C will result in a final temperature of 5.0°C for the drink?

35. How much heat in kcal is required to change 1.0 kg of ice, originally at −20.0°C, into steam at 110.0°C? Assume 1.0 atm of pressure.

36. Ice at 0.0°C is mixed with 5.00×10^2 mL of water at 25.0°C. How much ice must melt to lower the water temperature to 0.0°C?

ⓒ✦ 37. A phase diagram is shown in Fig. 14.21. Starting at point A, follow the dashed line to point E and consider what happens to the substance represented by this diagram as its pressure and temperature are changed. (a) Explain what happens for each line segment, AB, BC, CD, and DE. (b) What is the significance of point a and of point b?

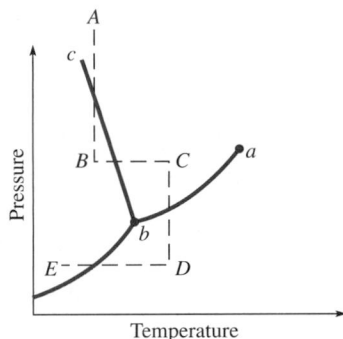

Figure 14.21
Problem 37

Temperature

✦ 38. Compute the heat of fusion of a substance from these data: 11.06 kcal will change 0.500 kg of the solid at 21°C to liquid at 401°C; melting point 325°C; specific heat of solid, 0.030 kcal/kg·K; and specific heat of liquid, 0.10 kcal/kg·K.

39. The graph in Fig. 14.22 shows the change in temperature as heat is supplied to a certain mass of ice at –80.0°C. What is the mass of the ice?

Figure 14.22
Problem 39

Heat (kcal)

40. A 20.0-g lead bullet leaves a rifle at a temperature of 87.0°C and hits a steel plate. If the bullet melts, what is the minimum speed it must have?

41. How many grams of aluminum at 80.0°C would have to be dropped into a hole in a block of ice at 0.0°C to melt 10.0 g of ice?

42. Is it possible to heat the aluminum of Problem 41 to a high enough temperature so that it melts an equal mass of ice? If so, what temperature must the aluminum have?

43. If a leaf is to maintain a temperature of 40°C (reasonable for a leaf), it must lose 250 W/m² by transpiration (evaporative heat loss). Note that the leaf also loses heat by radiation, but we will neglect this. How much water is lost after 1 h through transpiration only? The area of the leaf is 0.005 m².

44. A birch tree loses 618 mg of water per minute through transpiration (evaporation of water through stomatal pores). What is the rate of heat lost through transpiration?

✦ 45. A dog loses a lot of heat through panting. The air rushing over the upper respiratory tract causes evaporation and thus heat loss. A dog typically pants at a rate of 670 pants per minute. As a rough calculation, assume that one pant causes 0.010 g of water to be evaporated from the respiratory tract. What is the rate of heat loss for the dog through panting?

14.6 Conduction

46. A window whose glass has $\kappa = 1.0$ W/(m·K) is covered completely with a sheet of foam of the same thickness as the glass, but with $\kappa = 0.025$ W/(m·K). How is the rate at which heat is conducted through the window changed by the addition of the foam?

47. Given a slab of material that is 1.0 m² and 2.0×10^{-2} m thick, (a) what is the thermal resistance if the material is asbestos? (b) What is the thermal resistance if the material is iron? (c) What is the thermal resistance if the material is copper?

48. A copper rod of length 0.50 m and cross-sectional area 6.0×10^{-2} cm² is connected to an iron rod with the same cross section and length 0.25 m (Fig. 14.23). One end of the copper is immersed in boiling water and the other end is at the junction with the iron. If the far end of the iron rod is in an ice bath at 0°C, find the rate of heat transfer passing from the boiling water to the ice bath.

Figure 14.23
Problem 48

10.0 W/m² for $\Delta T = 20.0°C$ 20.0 W/m² for $\Delta T = 20.0°C$ $\Delta T_{tot} = 20.0°C$

Figure 14.24
Problem 49

49. For a temperature difference $\Delta T = 20.0°C$, one slab of material conducts 10.0 W/m²; another of the same shape conducts 20.0 W/m². What is the rate of heat flow per m² of surface area when the slabs are placed side by side, as shown in Fig. 14.24, with $\Delta T_{tot} = 20.0°C$?

ⓒ 50. A wall consists of a layer of wood and a layer of cork insulation of the same thickness. The temperature inside is 20.0°C and the temperature outside is 0.0°C. (a) What is the temperature at the interface between the wood and cork if the cork is on the inside and the wood on the outside? (b) What is the temperature at the interface if the wood is inside and the cork is outside? (c) Does it matter whether the cork is placed on the inside or the outside of the wooden wall? Explain. (See Fig. 14.25.)

$T = ?$ $T = ?$

Outside | Inside   Outside | Inside

$T = 0.0°C$ $T = 20.0°C$ $T = 0.0°C$ $T = 20.0°C$

Wood Cork   Cork Wood

Figure 14.25 Problem 50

✦ 51. A copper bar of thermal conductivity 401 W/(m·K) has one end at 104°C and the other end at 24°C. The length of the bar is

0.10 m and the cross-sectional area is 1.0×10^{-6} m². (a) What is the rate of heat conduction, I_{cd}, along the bar? (b) What is the temperature gradient in the bar? (c) If two such bars were placed in series (end to end) between the same temperature baths, what would I_{cd} be? (d) If two such bars were placed in parallel (side by side) with the ends in the same temperature baths, what would I_{cd} be? (e) In the series case, what is the temperature at the junction where the bars meet?

52. A hiker is wearing wool clothing of 0.50-cm thickness to keep warm. Her skin temperature is 35°C and the outside temperature is 4.0°C. Her body surface area is 1.2 m². (a) If the thermal conductivity of wool is 0.040 W/(m·K), what is the rate of heat conduction through her clothing? (b) If the hiker is caught in a rainstorm, the thermal conductivity of the soaked wool increases to 0.60 W/(m·K) (that of water). Now what is the rate of heat conduction?

53. The thermal conductivity of the fur (including the skin) of a male Husky dog is 0.026 W/(m·K). The dog's heat output is measured to be 51 W, its internal temperature is 38°C, its surface area is 1.31 m² and the thickness of the fur is 5.0 cm. How cold can the outside temperature be before the dog must increase its heat output?

54. The thermal resistance of a seal's fur and blubber combined is 0.33 K/W. If the seal's internal temperature is 37°C and the temperature of the sea is about 0°C, what must be the heat output of the seal in order for it to maintain its internal temperature?

✦55. One cross-country skier is wearing a down jacket that is 2.0 cm thick. The thermal conductivity of goose down is 0.025 W/(m·K). Her companion on the ski outing is wearing a wool jacket that is 0.50 cm thick. The thermal conductivity of wool is 0.040 W/(m·K). (a) If both jackets have the same surface area and the skiers both have the same body temperature, which one will stay warmer longer? (b) How much longer can the person with the warmer jacket stay outside for the same amount of heat loss?

14.7 Convection

56. A marathon runner is moving along the road at 2.0 m/s in still air that is at a temperature of 29.0°C. His surface area is 1.4 m², of which approximately 85% is exposed to the air. What is the rate of convective heat loss from his skin, at a temperature of 35.0°C, to the outside air?

57. A small child is being pulled in a cart behind a bicycle that is traveling at 1.80 m/s in still air at a temperature of 25.0°C. If the exposed skin area of the child is 0.500 m², what is the rate of convective heat loss from the child's skin? The convective coefficient for that wind speed is 21.0 W/(m²·°C) and the child's skin is at 35.0°C.

✦58. A leaf exposed to the noonday Sun needs to remove energy at the rate of 156 W/m² in order to remain at a temperature of 40.0°C. One method of heat loss is through convection. What is the rate of heat loss per unit area if the temperature difference between the leaf and the air is 5.0°C and the convective coefficient for dry air across a leaf is 84.8 W/(m²·°C) for a wind speed of 4.5 m/s?

14.8 Radiation

59. A black wood stove has a surface area of 1.20 m² and a surface temperature of 175°C. What is the net rate at which heat is radiated into the room? The room temperature is 20°C.

60. An incandescent lightbulb has a tungsten filament that is heated to a temperature of 3.00×10^3 K when an electric current passes through it. If the surface area of the filament is approximately 1.00×10^{-4} m² and it has an emissivity of 0.32, what is the power radiated by the bulb?

61. A tungsten filament in a lamp is heated to a temperature of 2.6×10^3 K by an electric current. The tungsten has an emissivity of 0.32. What is the surface area of the filament if the lamp delivers 40.0 W of power?

✦62. At a tea party, a coffeepot and a teapot are placed upon the serving table. The coffeepot is a shiny silver-plated pot with emissivity of 0.12; the teapot is ceramic and has an emissivity of 0.65. Both pots hold 1.00 L of liquid at 98°C when the party begins. If the room temperature is at 25°C, what is the rate of radiative heat loss from the two pots? [*Hint:* To find the surface area, approximate the pots with cubes of similar volume.]

✦63. A lizard of mass 3.0 g is warming itself in the bright sunlight. It casts a shadow of 1.6 cm² on a piece of paper held perpendicularly to the Sun's rays. The intensity of sunlight at the Earth is 1.4×10^3 W/m² but only half of this energy penetrates the atmosphere and is absorbed by the lizard. (a) If the lizard has a specific heat of 4.2 J/(g·°C), what is the rate of increase of the lizard's temperature? (b) Assuming that there is no heat loss by the lizard (to simplify), how long must the lizard lie in the Sun in order to raise its temperature by 5.0°C?

64. A person of surface area 1.80 m² is risking skin cancer by lying out in the sunlight to get a tan. If the intensity of the incident sunlight is 7.00×10^2 W/m², at what rate must heat be lost by the person in order to maintain a constant body temperature? (Assume the effective area of skin exposed to the Sun is 42% of the total surface area, 57% of the incident radiation is absorbed, and that internal metabolic processes contribute another 90 W for an inactive person.)

✦65. If the total power per unit area from the Sun incident on a horizontal leaf is 9.00×10^2 W/m², and we assume that 70.0% of this energy goes into heating the leaf, what would be the rate of temperature rise of the leaf? The specific heat of the leaf is 3.70 kJ/(kg·°C), the leaf area is 5.00×10^{-3} m², and its mass is 0.500 g.

✦✦66. Consider the leaf of Problem 65. Assume that the top surface of the leaf absorbs 70.0% of 9.00×10^2 W/m² of radiant energy, while the bottom surface absorbs all of the radiant energy incident on it due to its surroundings at 25.0°C. (a) If the only method of heat loss for the leaf were thermal radiation, what would be the temperature of the leaf? (Assume that the leaf radiates like a blackbody.) (b) If the leaf is to remain at a temperature of 25.0°C, how much power per unit area must be lost by other methods such as transpiration (evaporative heat loss)?

COMPREHENSIVE PROBLEMS

✦67. A stainless steel saucepan, with a base that is made of 0.350-cm-thick steel (κ = 46.0 W/(m·K)) fused to 0.150 cm thickness of copper (κ = 401 W/(m·K)), sits on a ceramic heating element at 104.00°C. The diameter of the pan is 18.0 cm and it contains boiling water at 100.00°C. (a) If the copper-clad bottom is touching the heat source, what is the temperature at the copper-steel interface? (b) At what rate will the water evaporate from the pan?

68. The inner vessel of a calorimeter contains 2.50×10^2 g of tetrachloromethane, CCl_4, at 40.00°C. The vessel is surrounded by 2.00 kg of water at 18.00°C. After a time, the CCl_4 and the water reach the equilibrium temperature of 18.54°C. What is the specific heat of CCl_4?

69. If the temperature surrounding the sunbather in Problem 64 is greater than normal body temperature of 37°C and the air is still, so that radiation, conduction, and convection play no part in cooling the body, how much water (in liters per hour) from perspiration must be given off to maintain the body temperature? The heat of vaporization of water is 580 cal/g at normal skin temperature.

70. Two 62-g ice cubes are dropped into 186 g of water in a glass. If the water is initially at a temperature of 24°C and the ice is at −15°C, what is the final temperature of the drink? **C** ✦

71. A 0.500-kg slab of granite is heated so that its temperature increases by 7.40°C. The amount of heat supplied to the granite is 0.700 kcal. Based on this information, what is the specific heat of granite?

72. A spring of force constant $k = 8.4 \times 10^3$ N/m is compressed by 0.10 m. It is placed into a vessel containing 1.0 kg of water and then released. Assuming all the energy from the spring goes into heating the water, find the change in temperature of the water.

✦73. A 75-kg block of ice at 0.0°C breaks off from a glacier, slides along the frictionless ice to the ground from a height of 2.43 m, and then slides along a horizontal surface consisting of gravel and dirt. Find how much of the mass of the ice is melted by the friction with the rough surface, assuming 75% of the internal energy generated is used to heat the ice.

74. A blacksmith heats a 0.38-kg piece of iron to 498°C in his forge. After shaping it into a decorative design, he places it into a bucket of water to cool. If the available water is at 20.0°C, what minimum amount of water must be in the bucket to cool the iron to 23.0°C? The water in the bucket should remain in the liquid phase.

75. If 4.0 g of steam at 100.0°C condenses to water on a burn victim's skin and cools to 45.0°C, (a) how much heat is given up by the steam? (b) If the skin was originally at 37.0°C, how much tissue mass was involved in cooling the steam to water? See Table 14.1 for the specific heat of human tissue.

76. If 4.0 g of boiling water at 100.0°C was splashed onto a burn **C** ✦ victim's skin, and if it cooled to 45.0°C on the 37.0°C skin, (a) how much heat is given up by the water? (b) How much tissue mass, originally at 37.0°C, was involved in cooling the water? See Table 14.1. Compare the result with that found in Problem 75.

77. The amount of heat generated during the contraction of muscle in an amphibian's leg is given by Q = 0.13 mcal + (0.35 mcal/cm) × Δx, where Δx is the length shortened in cm and Q is in mcal. If a muscle of length 3.0 cm and mass 0.10 g is shortened by 1.5 cm during a contraction, what is the temperature rise? Assume that the specific heat of muscle is 1.0 cal/(g·°C).

78. Many species cool themselves by sweating, because as the sweat evaporates, heat is given up to the surroundings. A human exercising strenuously has an evaporative heat loss rate of about 650 W. If a person exercises strenuously for 30.0 min, how much water must he drink to replenish his fluid loss? The heat of vaporization of water is 580 cal/g at normal skin temperature.

79. Your hot water tank is insulated, but not very well. To reduce heat loss, you wrap some old blankets around it. With the water at 81°C and the room at 21°C, a thermometer inserted between the outside of the original tank and your blanket reads 36°C. By what factor did the blanket reduce the heat loss?

80. A wall consists of a layer of wood outside and a layer of insulation inside. The temperatures inside and outside the wall are +22°C and −18°C; the temperature at the wood/insulation boundary is −8.0°C. By what factor would the heat loss through the wall increase if the insulation were not present?

C ✦81. Small animals eat much more food per kg of body mass than do larger animals. The basal metabolic rate (BMR) is the minimal energy intake necessary to sustain life in a state of complete inactivity. Table 14.7 lists the BMR, mass, and surface area for five animals. (a) Calculate the BMR/kg of body mass for each animal. Is it true that smaller animals must consume much more food per kg of body mass? (b) Calculate the BMR/m^2 of surface area. (c) Can you explain why the BMR/m^2 is approximately the same for animals of different sizes? Consider what happens to the food energy metabolized by an animal in a resting state.

Table 14.7

Basal Metabolic Rates for Several Animals

Animal	BMR (kcal/day)	Mass (kg)	Surface Area (m²)
Mouse	3.80	0.018	0.0032
Dog	770	15	0.74
Human	2050	64	2.0
Pig	2400	130	2.3
Horse	4900	440	5.1

C ✦82. Imagine a person standing naked in a room at 23.0°C. The walls are well insulated, so they also are at 23.0°C. The person's surface area is 2.20 m^2 and his basal metabolic rate is 2167 kcal/day. His emissivity is 0.97. (a) If the person's skin temperature were 37.0°C (the same as the internal body temperature), at what net rate would heat be lost through radiation? (Ignore losses by conduction and convection.) (b) Clearly the heat loss in (a) is not sustainable—but skin temperature is less than internal body temperature. Calculate the skin temperature such that the net heat loss due to radiation is equal to the basal metabolic rate. (c) Does wearing clothing slow the loss of heat by radiation, or does it only decrease losses by conduction and convection? Explain.

83. On a very hot summer day Daphne is off to the park for a picnic. She puts 0.10 kg of ice at 0°C in a thermos and then adds a grape-flavored drink, which she has mixed from a powder, using room temperature water (35°C). How much grape-flavored drink will just melt all the ice?

84. It requires 4.085 kcal to melt 1.00×10^2 g of urethane ($CO_2(NH_2)C_2H_5$) at 48.7°C. What is the latent heat of fusion of urethane in kcal/mol?

85. A 20.0-g lead bullet leaves a rifle at a temperature of 47.0°C and travels at a velocity of 5.00×10^2 m/s until it hits a large block of ice at 0°C and comes to rest within it. How much ice will melt?

✦86. Bare, dark-colored basalt has a thermal conductivity of 3.1 W/(m·K), whereas light-colored sandstone's thermal conductivity is only 2.4 W/(m·K). Even though the same amount of radiation is incident on both and their surface temperatures are the same, the temperature gradient within the two materials will differ. For the same patch of area, what is the ratio of the depth in basalt as compared to the depth in sandstone that gives the same temperature difference?

✦87. The power expended by a cheetah is 160 kW while running at 110 km/h, but its body temperature cannot exceed 41.0°C. If 70.0% of the energy expended is dissipated within its body, how far can it run before it overheats? Assume that the initial temperature of the cheetah is 38.0°C, its specific heat is 3.5 kJ/(kg·°C), and its mass is 50.0 kg.

88. A 2.0-kg block of copper at 100.0°C is placed into 1.0 kg of water in a 2.0-kg iron pot. The water and the iron pot are at 25.0°C just before the copper block is placed into the pot. What is the final temperature of the water, assuming negligible heat flow to the environment?

89. A piece of gold of mass 0.250 kg and at a temperature of 75.0°C is placed into a 1.500-kg copper pot containing 0.500 L of water. The pot and water are at 22.0°C before the gold is added. What is the final temperature of the water?

©✦✦90. A scientist working late at night in her low-temperature physics laboratory decides to have a cup of hot tea, but discovers the lab hot plate is broken. Not to be deterred, she puts about 8 oz of water, at 12°C, from the tap into a lab Dewar (essentially a large thermos bottle) and begins shaking it up and down. With each shake the water is thrown up and falls back down a distance of 33.3 cm. If she can complete 30 shakes per minute, how long will it take to heat the water to 87°C? Would this really work? If not, why not?

©✦91. For a cheetah, 70.0% of the energy expended during exertion is internal work done on the cheetah's system and is dissipated within his body; for a dog only 5.00% of the energy expended is dissipated within the dog's body. Assume that both animals expend the same total amount of energy during exertion, both have the same heat capacity, and the cheetah is 2.00 times as heavy as the dog. (a) How much higher is the temperature change of the cheetah compared to the temperature change of the dog? (b) If they both start out at an initial temperature of 35.0°C, and the cheetah has a temperature of 40.0°C after the exertion, what is the final temperature of the dog? Which animal probably has more endurance? Explain.

ANSWERS TO PRACTICE PROBLEMS

14.1 4.9 J

14.2 Higher. The molecules have the same amount of *random* translational kinetic energy plus the additional kinetic energy associated with the ball's translation and rotation.

14.3 350 g

14.4 at least $0.89

14.5 48°C

14.6 92 kJ

14.7 36 kcal (150 kJ)

14.8 9.5 kcal (40 kJ)

14.9 53.5 g

14.10 230 W

14.11 110 W

14.12 250 W

14.13 To maintain constant temperature, the net heat must be zero. The rate at which energy is emitted is 140 W.

14.14 9.4 μm (at 35°C) to 9.6 μm (at 30°C)

14.15 28 W

14.16 –16°C

Thermodynamics

The gasoline engines in cars are terribly inefficient. Of the chemical energy that is released in the burning of gasoline, typically only about 25% is converted into useful mechanical work done on the car to move it forward. Yet scientists and engineers have been working for decades to make a more efficient gasoline engine. Is there some fundamental limit to the efficiency of a gasoline engine? Is it possible to make an engine that converts all—or nearly all—of the chemical energy in the fuel into useful work?

The choice of a *system* is made in any way convenient for a given problem.

Tips

15.1   THE FIRST LAW OF THERMODYNAMICS

Both work and heat can change the internal energy of a system. Work can be done on a rubber ball by squeezing it, stretching it, or slamming it into a wall. Heat can be supplied to the ball by leaving it out in the sun or putting it into a hot oven. These two methods of increasing the internal energy of a system lead to the **first law of thermodynamics**:

First law of thermodynamics:
The change in internal energy of a system is equal to the heat flow into the system minus the work done *by* the system.

The statement of the first law is actually a statement of energy conservation. The change in internal energy of a system is equal to the energy transferred to it from its surroundings. If the internal energy increases, for example, then the energy in the surroundings decreases by the same amount and the total energy of the universe (system plus surroundings) is unchanged.

We must agree on sign conventions (Table 15.1) for heat and work in order to write the first law as an equation. Q is the heat flow *into* a system; positive Q causes an increase in internal energy ($\Delta U > 0$). If heat flows out of the system, then Q is negative. However, by convention W stands for the work done *by* a system. When a system does positive work, it transfers energy to its surroundings; positive W causes a *decrease* in internal energy ($\Delta U < 0$) while negative W causes an *increase* in internal energy. Therefore,

First law of thermodynamics	
$\Delta U = Q - W$	(15-1)

These sign conventions for work and heat came from the application of the first law to the steam engine. The usual state of affairs for a steam engine is to put heat into the engine (boiling the water to create steam) and to get work out of the engine. Thus heat flow into the engine and work done by the engine were chosen to be positive quantities. The sign conventions are easily remembered by thinking of the steam engine example.

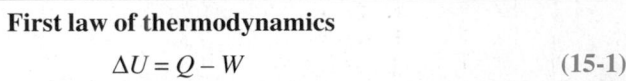
Table 15.1

Sign Conventions for the First Law of Thermodynamics

Quantity	Definition	Meaning of + Sign	Meaning of – Sign
Q	Heat flow into the system	Heat flow into the system	Heat flow *out of* the system
W	Work done by the system	Work done by the system	Work done *on* the system
ΔU	Internal energy change	Internal energy increase	Internal energy *decrease*

Example 15.1

A Hot Drink at Camp

A pot containing 6.00 kg of water is placed over a campfire (Fig. 15.1). An impatient camper, eager for a hot drink, does 9.2 kJ of mechanical work on the water by agitating it with a spoon. If the internal energy of the water increases by 45.2 kJ, how much heat is delivered to the water by the fire during that time?

Strategy   The first law of thermodynamics relates internal energy changes of a system—in this case, the water—to the

heat flow into the system and the work done by the system. The fire is at a higher temperature than the pot of water so heat flows from the fire through the pot and into the water. The spoon does work on the water; this also increases the internal energy of the water.

Solution   The increase in the internal energy of the water is

$$\Delta U = 45.2 \text{ kJ}$$

continued on next page

Example 15.1 *continued*

Figure 15.1
Energy transfers into a pot heated over a campfire and stirred by a camper.

Work done by camper

Heat flow from fire

Mechanical work is done *on* the water so W is negative, according to our sign convention:

$$W = -9.2 \text{ kJ}$$

From the first law,

$$\Delta U = Q - W$$

Solving for Q,

$$Q = \Delta U + W = 45.2 \text{ kJ} + (-9.2 \text{ kJ}) = 36.0 \text{ kJ}$$

Discussion   Fortunately we were not asked to find the *total* heat output of the fire; some went to heating the pot that contains the water, some went to heating the rack over the fire, and some heated the air and campers around the fire. Trying to heat water with a spoon is rather awkward and not very efficient; Joule's experiment requires a lot of patience.

Practice Problem 15.1   Changing internal energy of a gas

If the change in internal energy of a gas is +42 kJ due to a process involving a heat flow into the gas of 14 kJ, was the work done on the gas or by the gas? What is the value of the work, W, using the sign convention already discussed?

15.2   THERMODYNAMIC PROCESSES

A thermodynamic process is the method by which a system is changed from one *state* to another. The state of a system is described by a set of **state variables** such as pressure, temperature, volume, number of moles, and internal energy. State variables describe the state of a system at some instant of time but not how the system got to that state. Thus, heat and work are *not* state variables—they describe *how* a system gets from one state to another.

If a system is changed slowly so that it is always very near equilibrium, the changes in state can be represented by a curve on a plot of pressure versus volume (called a **PV diagram**). Each point on the curve represents a state of the system. The *PV* diagram is a useful tool for analyzing thermodynamic processes.

Figure 15.2a shows the expansion of a gas, starting with volume V_i and pressure P_i; Fig. 15.2b is the *PV* diagram for the process. As the gas slowly expands, it does work on the piston. The pressure of the gas pushes against the piston with a force $F = PA$, where A is the cross-sectional area of the piston. This force is not constant since the pressure decreases. As was shown in Section 6.4, the work done by a variable force is the area under a graph of $F(x)$. In thermodynamics, one of the great uses of a *PV* diagram is that the work done is the area under the *PV* curve.

To see how work is related to the area under the curve, first note that the units of $P \times V$ are those of work:

$$[\text{pressure} \times \text{volume}] = [\text{Pa}] \times [\text{m}^3] = \frac{[\text{N}]}{[\text{m}^2]} \times [\text{m}^3] = [\text{N}] \times [\text{m}] = [\text{J}]$$

So far, so good. Imagine that the piston is let out a *small* distance d—small enough that the pressure change is insignificant. The work done by the gas is

$$W = Fd = PAd$$

The volume increase of the gas is

$$\Delta V = Ad$$

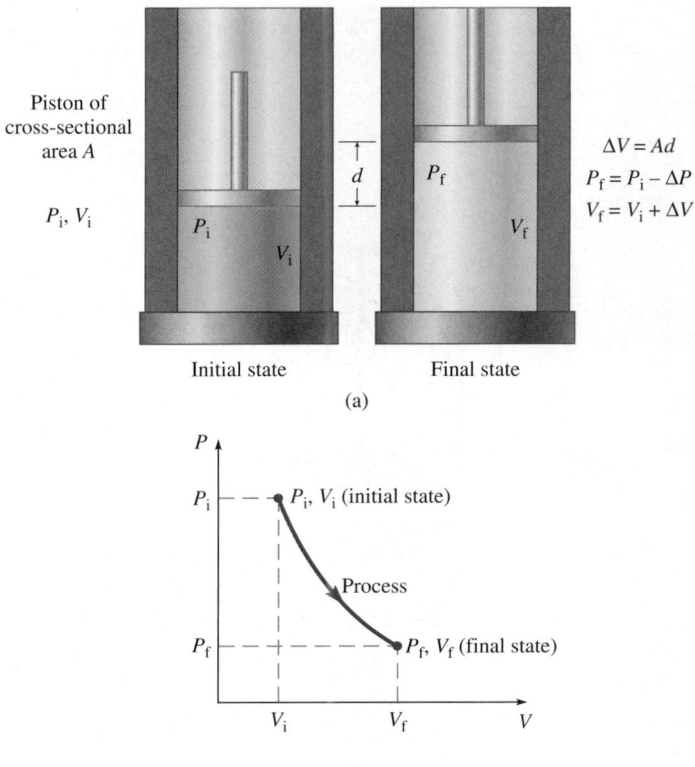

Figure 15.2 (a) Expansion of a gas from initial pressure P_i and volume V_i to final pressure P_f and volume V_f. Work is done by the gas as it pushes on the moving piston. (b) PV diagram for the expansion.

Figure 15.3 (a) The area under the PV curve is divided into many narrow rectangles of width ΔV and of varying heights P. The sum of the areas of the rectangles is the total area under the PV curve. (b) An enlarged view of one rectangle under the curve; its area is $P \Delta V$.

So the work done by the gas is

$$W = P \Delta V \tag{15-2}$$

To find the *total* work done, we add up the work done during each small volume change. During each small ΔV, the work done is the area of a rectangle of height P and width ΔV under the PV curve (Fig. 15.3). Therefore, the total work done is the total area under the PV curve.

 Note that the work done depends not only on the initial and final states of the system, but also on the intermediate states, as depicted by the path taken on the PV curve. Figure 15.4 shows two other possible paths between the same initial and final states as those of Fig. 15.3.

In Fig. 15.4a, the pressure is kept constant at the initial value P_i while the volume is increased from V_i to V_f. Then the volume is kept constant while the pressure is reduced from P_i to P_f. The work done is shown by the shaded area under the PV curve; it is greater than the work done in Fig. 15.3a. Alternatively, in Fig. 15.4b the pressure is first reduced from P_i to P_f while the volume is held fixed; then the volume is allowed to increase from V_i to V_f while the pressure is kept at P_f. We see by the shaded area that the

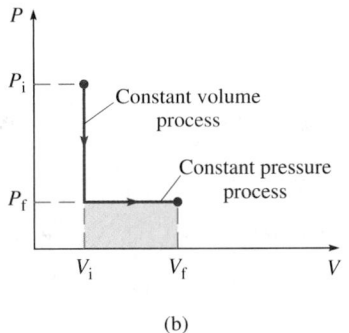

Figure 15.4 Work done (shaded areas) by following different processes between the same initial and final states. (a) Volume is increased from V_i to V_f while pressure is constant at P_i; then pressure is reduced to P_f. (b) Pressure is first decreased from P_i to P_f; then volume is increased from V_i to V_f.

work done this way is less than the work done in Fig. 15.3a. The work done differs from one process to another, even though the initial and final states are the same in each case.

Constant Pressure Processes

A process by which the state of a system is changed while the pressure is held constant is called an *isobaric* process. The word *isobaric* comes from the same Greek root as the word *barometer*. In Fig. 15.4a, the first change of state from V_i to V_f along the line from 1 to 2 occurs at the constant initial pressure P_i. A constant pressure process appears as a horizontal line on the PV diagram. The work done is the area under this horizontal line:

$$W = P_i (V_f - V_i) = P \, \Delta V \quad \text{(constant pressure)} \tag{15-3}$$

Constant Volume Processes

A process by which the state of a system is changed while the *volume* remains constant is called an *isochoric* process. Such a process is illustrated in Fig. 15.4a when the system moves along the line from 2 to 3 as the pressure changes from P_i to P_f at the constant volume V_f. No work is done during a constant volume process; without a displacement, work cannot be done. The area under the PV curve—a vertical line—is zero:

$$W = 0 \quad \text{(constant volume)} \tag{15-4}$$

If no work is done, then from the first law of thermodynamics, the change in internal energy is equal to the heat flow into the system:

$$\Delta U = Q \quad \text{(constant volume)} \tag{15-5}$$

In Section 14.4 we discussed the molar specific heat of an ideal gas *at constant volume*. The first law of thermodynamics enables us to calculate the internal energy change ΔU. Since no work is done during a constant volume process, $\Delta U = Q$. For a constant volume process, $Q = nC_v \Delta T$ and therefore

$$\Delta U = nC_v \Delta T \quad \text{(ideal gas)} \tag{15-6}$$

Internal energy is a state variable—its value depends only on the current state of the system, not on the path the system took to get there. Therefore, as long as the number of moles is constant, *the change in internal energy of an ideal gas is determined solely by the temperature change*. Equation (15-6) therefore gives the internal energy change of an ideal gas for *any* thermodynamic process, not just for constant volume processes.

Constant Temperature Processes

A process in which the temperature of the system remains constant is called an **isothermal** process. Such a process must proceed along a path on a PV diagram for which the temperature is constant. Such a path is called an **isotherm** (Fig. 15.5). All the points on an isotherm represent states of the system with the same temperature.

How can we keep the temperature of the system constant? One way is to put the system in thermal contact with a **reservoir**—a system with such a large heat capacity that it can exchange heat in either direction with a negligibly small temperature change.

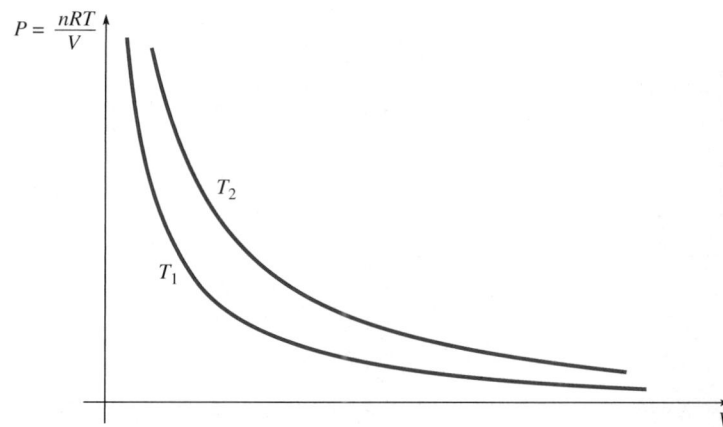

Figure 15.5 Two isotherms for an ideal gas at temperatures $T_1 < T_2$. Each isotherm is a graph of $P = nRT/V$ for a constant temperature.

Then as long as the state of the system does not change too rapidly, heat flows between the system and the reservoir to keep the system's temperature constant.

Isothermal Process for an Ideal Gas

For an ideal gas, we can plot isotherms using the ideal gas law $PV = nRT$. Since the change in internal energy of an ideal gas is determined solely by the temperature change,

$$\Delta U = 0 \quad \text{(ideal gas, isothermal)} \tag{15-7}$$

From the first law of thermodynamics, $\Delta U = 0$ means that $Q = W$; the heat flow into the gas is equal to the work done by the gas. Note that this is only true for an *ideal gas* at constant temperature. Other systems can change internal energy without changing temperature; one example is when the system undergoes a phase change.

It can be shown (using calculus to find the area under the *PV* curve) that the work done by an ideal gas during a constant temperature expansion (or contraction) from volume V_i to volume V_f is

$$W = nRT \ln\left(\frac{V_f}{V_i}\right) \quad \text{(ideal gas, isothermal)} \tag{15-8}$$

In Eq. (15-8), "ln" stands for the natural (or base-*e*) logarithm.

Example 15.2

Constant Temperature Compression of an Ideal Gas

An ideal gas is kept in thermal contact with a heat reservoir at 7°C (280 K) while it is compressed from a volume of 20.0 L to a volume of 10.0 L. During the compression an average force of 33.3 kN is used to move the piston a distance of 0.15 m. How much heat is exchanged between the gas and the reservoir? Does the heat flow into or out of the gas?

Strategy   We can find the work done on the piston from the average force applied and the distance moved. This is equal to the work done on the gas in compressing it. For isothermal compression of an ideal gas, $\Delta U = 0$. Then $W = Q$; the heat is equal to the work done.

Solution   The work done on the gas is

$$\text{work} = Fd = 33.3 \text{ kN} \times 0.15 \text{ m} = 5.0 \text{ kJ}$$

This work adds 5.0 kJ to the internal energy of the gas. Then 5.0 kJ of heat must flow out of the gas if its internal energy does not change. The work done on the gas is positive since the piston is pushed with an inward force as it moves inward. Since positive work is done *on* the gas, negative work is done *by* the gas. Using the sign conventions for the first law,

$$Q = W = -5.0 \text{ kJ}$$

Since Q represents heat flow *into* the gas, the negative sign tells us that heat flows out of the gas into the reservoir.

 Discussion   Although the temperature remains constant during the process, it does not mean that no heat

continued on next page

Example 15.2 *continued*

flows. To maintain a constant temperature when work is done on the gas, some heat must flow out of the gas. If the gas were thermally isolated so no heat could flow, then the work done on the gas would increase the internal energy, resulting in an increase in the temperature of the gas.

Practice Problem 15.2 Work done during constant temperature expansion of a gas

Suppose 2.0 mol of an ideal gas are kept in thermal contact with a heat reservoir at 57°C (330 K) while the gas expands from a volume of 20.0 L to a volume of 40.0 L (Fig. 15.6). Does heat flow into or out of the gas? How much heat flows? [*Hint*: Use $W = nRT \ln (V_f/V_i)$, which applies to an ideal gas at constant temperature.]

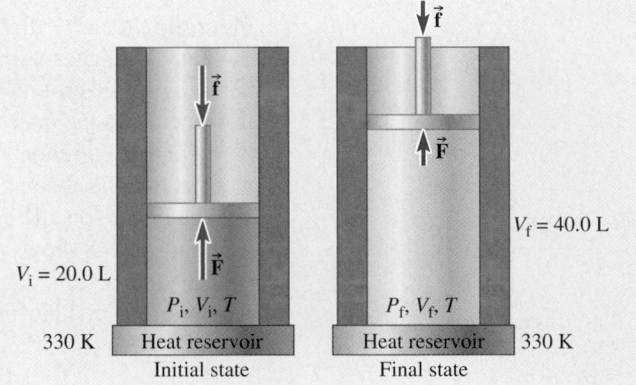

Figure 15.6
Isothermal expansion of an ideal gas

Adiabatic Processes

A process in which no heat is transferred into or out of the system is called an **adiabatic** process. An adiabatic process is *not* the same as a constant temperature (isothermal) process. Heat flows in or out as necessary to maintain a constant temperature in an isothermal process; *no* heat flow occurs in an adiabatic process, so if work is done, the temperature of the system may change. In the adiabatic compression of an ideal gas, work is done on the gas while no heat flows in or out. The internal energy increase due to the work done results in a temperature increase. One way to perform an adiabatic process is to completely insulate the system so that no heat can get in or out; another way is to perform the process so quickly that there is no time for heat to flow in or out.

From the first law of thermodynamics

$$\Delta U = Q - W \qquad (15\text{-}1)$$

With $Q = 0$,

$$\Delta U = -W \quad \text{(adiabatic)}$$

Table 15.2 summarizes all of the thermodynamic processes discussed.

Table 15.2

Summary of Thermodynamic Processes

Process	Name	Condition	Consequences
Constant temperature	Isothermal	T = constant; for an ideal gas, $\Delta U = 0$	For an ideal gas, $Q = W$
Constant pressure	Isobaric	P = constant	$W = P\,\Delta V$
Constant volume	Isochoric	V = constant	$W = 0$; $\Delta U = Q$
No heat flow	Adiabatic	$Q = 0$	$\Delta U = -W$

15.3   CONSTANT PRESSURE EXPANSION OF AN IDEAL GAS

Figure 15.7 is a *PV* diagram for heat flow into an ideal gas at constant volume. The initial and final states are shown as points on two different isotherms. The area under the vertical line is zero; no work is done when the volume is constant. With $W = 0$, the heat flow increases the internal energy of the gas, so the temperature increases.

Another common situation is when the *pressure* of the gas is constant. In this case, work is done as the gas expands. The first law of thermodynamics enables us to calculate the molar specific heat at constant pressure (C_p).

Figure 15.8 shows a *PV* diagram for the constant pressure expansion of an ideal gas starting and ending at the same temperatures as for the constant volume process. Applying the first law to the constant pressure process requires that

$$\Delta U = Q - W$$

where the work done is, from the ideal gas law,

$$W = P\,\Delta V = nR\,\Delta T$$

The definition of C_p is

$$Q = nC_p\,\Delta T \tag{15-9}$$

Figure 15.7   A *PV* diagram for constant volume process for ideal gas.

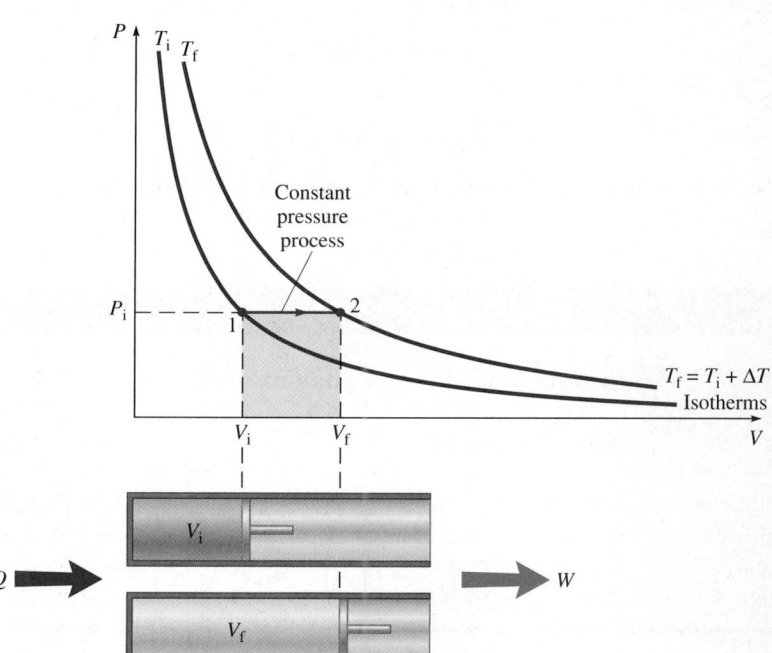

Figure 15.8   A *PV* diagram of a constant pressure expansion; heat is added to the ideal gas of volume V_i increasing the internal energy and doing work as the expanding gas moves the frictionless piston and expands to volume V_f. The work is equal to the shaded area under the constant pressure process path from point 1 to point 2.

Substituting into the first law, we obtain

$$\Delta U = nC_p \Delta T - nR \Delta T \qquad (15\text{-}10)$$

Since both of the processes discussed ended up at the same temperature and since the internal energy of an ideal gas depends only on the temperature, we can replace ΔU with its equivalent:

$$\Delta U = nC_v \Delta T \qquad (14\text{-}8)$$

Then

$$nC_v \Delta T = nC_p \Delta T - nR \Delta T$$

Canceling common factors of n and ΔT, this reduces to

$$C_v = C_p - R \qquad (15\text{-}11)$$

Since R is a positive constant, the molar specific heat of an ideal gas at constant pressure is larger than the molar specific heat at constant volume.

Is this result reasonable? When heat flows into the gas at constant pressure, the gas expands, doing work on the surroundings. Thus, not all of the heat goes into increasing the internal energy of the gas. More heat has to flow into the gas at constant pressure for a given temperature increase than at constant volume.

Example 15.3

Warming a Balloon at Constant Pressure

A weather balloon is filled with helium gas at 20.0°C and 1.0 atm of pressure. The volume of the balloon after filling is measured to be 8.50 m³. The helium is heated until its temperature is 55.0°C. During this process, the balloon expands at constant pressure (1.0 atm). What is the heat flow into the helium?

Strategy   We can find how many moles of gas n are present in the balloon by using the ideal gas law. For this problem, we consider the helium to be a system. Helium is a monatomic gas so its molar specific heat at constant *volume* is $C_v = \frac{3}{2}R$. The molar specific heat at constant *pressure* is then $C_p = C_v + R = \frac{5}{2}R$. Then the heat flow into the gas during its expansion is $Q = nC_p \Delta T$.

Solution   The ideal gas law is

$$PV = nRT$$

We know the pressure, volume, and temperature: $P = 1$ atm $= 1.01 \times 10^5$ Pa, $V = 8.50$ m³, and $T = 273$ K $+ 20.0°$C $= 293$ K. Solving for the number of moles yields

$$n = \frac{PV}{RT} = \frac{1.01 \times 10^5 \text{ Pa} \times 8.50 \text{ m}^3}{8.31 \text{ J/(mol·K)} \times 293 \text{ K}} = 352.6 \text{ mol}$$

For an ideal gas at constant pressure, the heat required to change the temperature is

$$Q = nC_p \Delta T$$

where $C_p = \frac{5}{2}R$. The temperature change is

$$\Delta T = 55.0°\text{C} - 20.0°\text{C} = 35.0 \text{ K}$$

Now we have everything we need to find Q:

$$Q = nC_p \Delta T = 352.6 \text{ mol} \times \tfrac{5}{2} \times 8.31 \text{ J/(mol·K)} \times 35.0 \text{ K}$$
$$= 256 \text{ kJ}$$

Discussion   We do not have to find the work $W = P\Delta V$ separately and then add it to the change in internal energy to find Q. The work done is *already* accounted for by the molar specific heat at constant pressure. This simplifies the problem since we use the same method for constant pressure as we use for constant volume; the only change is the choice of C_v or C_p.

Practice Problem 15.3   Air instead of helium

Suppose the balloon were filled with air instead of helium. Find Q for the same temperature change. (Air is mostly N_2 and O_2 so assume an ideal diatomic gas.)

15.4   REVERSIBLE AND IRREVERSIBLE PROCESSES

Have you ever wished you could make time go backward? Perhaps you accidentally broke an irreplaceable treasure in a friend's house, or missed a one-time opportunity to meet your favorite movie star, or said something unforgivable to someone close to you. Why can't the clock be turned around?

Imagine a perfectly elastic collision between two billiard balls. If you were to watch a movie of the collision, you would have a hard time telling whether the movie was being played forward or backward. The laws of physics for an elastic collision are valid even if the direction of time is reversed. Since the total momentum and the total kinetic energy are the same before and after the collision, the reversed collision is physically possible.

The perfectly elastic collision is one example of a **reversible** process. A reversible process is one that does not violate any laws of physics if "played in reverse." Most of the laws of physics do not distinguish forward from backward. A projectile moving in the absence of air resistance (on the Moon, say) is reversible: if we play the movie backward, the total mechanical energy is still conserved and Newton's second law ($\Sigma \vec{F} = m\vec{a}$) still holds at every instant in the projectile's trajectory.

Notice the caveats in the examples: "perfectly elastic" and "in the absence of air resistance." If friction or air resistance is present, then the process is **irreversible**. If you played *backward* a movie of a projectile with noticeable air resistance, it would be easy to tell that something is wrong. The force of air resistance on the projectile would act in the wrong direction—in the direction of the velocity instead of opposite it. The same would be true for sliding friction. Slide a book across the table; friction slows it down and brings it to rest. The macroscopic kinetic energy of the book—due to the orderly motion of the book in one direction—has been converted into disordered energy associated with the random motion of molecules; the table and book will be at slightly higher temperatures. The reversed process certainly would never occur, even though it does not violate the first law of thermodynamics (energy conservation). We would not expect a slightly warmed book placed on a slightly warmed table surface to spontaneously begin to slide across the table, gaining speed and cooling off as it goes, even if the total energy is the same before and after. It is easy to convert ordered energy into disordered energy, but not so easy to do the reverse. *The presence of energy dissipation (sliding friction, air resistance) always makes a process irreversible.*

Irreversible processes do *not* violate energy conservation.	

Heat flow is another example of an irreversible process. Put a container of warm lemonade into a cooler with some ice (Fig. 15.9). Some of the ice melts and the lemonade gets cold as heat flows out of the lemonade and into the ice. The reverse would never happen: putting cold lemonade into a cooler with some partially melted ice, we would not find that the lemonade gets warmer as the liquid water freezes. *Heat flow from a hotter body to a colder body is always irreversible.*

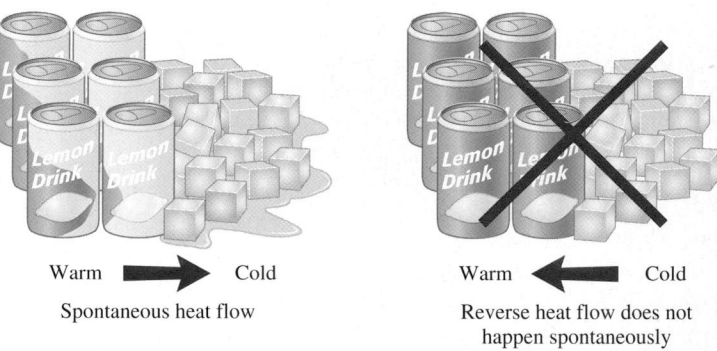

Figure 15.9 Spontaneous heat flow goes from warm to cool; the reverse does not happen spontaneously.

Warm ➡ Cold
Spontaneous heat flow

Warm ⬅ Cold
Reverse heat flow does not happen spontaneously

Conceptual Example 15.4

Eggs for Breakfast

Discuss the scrambling of an egg in terms of irreversible processes and in terms of disorder.

Solution and Discussion Scrambling the egg greatly increases the disorder, since now the yolk and egg-white molecules are mixed together randomly. Once the egg is scrambled, further stirring rearranges molecules but leaves the egg still scrambled—we would never expect the egg to unscramble as we stir. The scrambling is an irreversible process.

The unscrambled egg is in a highly ordered state, with the yolk and the white clearly separated, while the scrambled egg is in a highly disordered state. Scrambling changes the egg from an ordered state to a disordered state; once it is disordered, further stirring cannot return it to the ordered state.

Conceptual Practice Problem 15.4   A campfire

On a camping trip, you gather some twigs and logs and start a fire. Discuss the campfire in terms of irreversible processes and in terms of disorder.

Irreversibility, whether in the dissipation of energy or in heat flow or in scrambling of an egg, is based on statistics. If we looked *microscopically* at any of these processes, any one of the collisions between molecules would be reversible. As we will see in this chapter, the frictional dissipation of energy and the spontaneous flow of heat from a hotter to a colder body also can be thought of in terms of disorder. A system never goes from a disordered state to a more ordered state *spontaneously*. Reversible processes are those that do not change the total amount of disorder in the universe; irreversible processes increase the amount of disorder.

The second law of thermodynamics is the law of physics that determines the direction of time—all the other physical laws we have learned work equally well forward and backward. The second law can be stated in many equivalent forms. One statement is that the total amount of disorder in the universe never decreases. Since we have not yet learned how to measure disorder, we start with an equivalent statement involving heat flow:

> **Second law of thermodynamics** (Clausius statement):
> Heat never flows spontaneously from a colder body to a hotter body.

Spontaneous heat flow from a colder body to a hotter body would decrease the total disorder in the universe.

15.5   HEAT ENGINES

We said in Section 15.4 that it is far *easier* to convert ordered energy into disordered energy than to do the reverse. Converting ordered into disordered energy occurs spontaneously, but the reverse does not. A **heat engine** is a device designed to convert disordered energy into ordered energy. We will see that there is a fundamental limitation on how much mechanical work can be produced by a heat engine from a given amount of heat.

The invention of the steam engine—a heat engine that uses steam as its working substance—at the beginning of the eighteenth century was one of the crucial elements in the industrial revolution. The steam engine was the first practical machine with a sustained work output that used an energy source other than muscle, wind, or moving water. Efforts to improve the performance of the steam engine in the nineteenth century gave birth to the branch of physics we call *thermodynamics*. Steam engines are still used in many electrical power plants.

The source of energy in a heat engine is most often the burning of some fuel such as gasoline, coal, oil, natural gas, and the like. A nuclear power plant is a heat engine using energy released by a nuclear reaction instead of a chemical reaction (as in burning). A geothermal engine uses the high temperature found beneath the Earth's crust (which comes to the surface in places such as volcanoes and hot springs).

All practical engines operate in cycles—a "one-shot" engine would be of little use. Each cycle consists of several thermodynamic processes that are repeated the same way during each cycle. In order for these processes to repeat the same way, the engine must end the cycle in the same state in which it started. In particular, the internal energy of the engine must be the same at the end of a cycle as it was in the beginning. Then for one complete cycle,

$$\Delta U = 0$$

From the first law of thermodynamics (energy conservation),

$$Q = W$$

Therefore, for a cyclical heat engine,

> The work done by the engine is equal to the *net* heat input.

We stress that it is the *net* heat input since an engine not only takes in heat but exhausts some as well. Figure 15.10 shows the energy transfers involved.

Figure 15.10 A heat engine. The engine is represented by a circle and the arrows indicate the direction of the energy flow. The energy entering the engine equals the energy leaving the engine.

Making The Connection:
operation of
an internal
combustion engine

One familiar engine is the internal combustion engine found in automobiles. *Internal combustion* refers to the fact that gasoline is burned inside a cylinder; the resulting hot gases push against a piston and do work. A steam engine is an *external* combustion engine. The coal burned, for example, releases heat that is used to make steam; the steam is the working substance of the engine that drives the turbines.

Most automobile engines work in a cyclic thermodynamic process called the *Otto cycle*. The Otto cycle consists of six steps (Fig. 15.11):

1. Intake stroke: The piston is pulled out, drawing the fuel-air mixture into the cylinder at atmospheric pressure ($a \rightarrow b$).

2. Compression stroke: The piston moves back in, compressing the fuel-air mixture ($b \rightarrow c$). The gas in the cylinder does *negative* work—which means work is done *on* the gas.

3. Ignition: A spark ignites the gases, quickly and dramatically raising the temperature and pressure ($c \rightarrow d$).

4. Power stroke: The high pressure that results from ignition pushes the piston out ($d \rightarrow e$). As the piston is pushed out, the gases do positive work.

5. Cooling: Heat flows out of the cylinder ($e \rightarrow b$).

6. Exhaust stroke: A valve is opened and the exhaust gases are pushed out of the cylinder ($b \rightarrow a$).

An automobile engine is a *four-stroke* engine because each cycle has four strokes (steps 1, 2, 4, and 6) during which the piston moves.

Of the energy released by the burning of the gasoline, only about 25% is turned into mechanical work used to move the car forward and run other systems. The other 75% is discarded. The hot exhaust gases carry energy out of the engine, as does the liquid cooling system.

Making The Connection:
efficiency of a
heat engine

To measure how effectively an engine converts heat into mechanical work, we define the engine's **efficiency** e as what you get (work) divided by what you pay for (heat input):

Efficiency of an engine	
$e = \dfrac{\text{work output}}{\text{heat input}} = \dfrac{W}{Q_{\text{in}}}$	(15-12)

The efficiency is stated as either a fraction or a percentage. It gives the fraction of the heat input that is turned into useful work. Note that the heat input is *not* the same as the net heat Q; rather,

$$Q = Q_{\text{in}} + Q_{\text{out}} \qquad (15\text{-}13)$$

where Q_{in} is positive and Q_{out} is negative for a heat engine. The efficiency of an engine is less than 100% because some of the heat input, instead of being converted into useful work, is exhausted.

If an engine does work at a constant rate and its efficiency does not change, then it also takes in and exhausts heat at constant rates. The work done, heat input, and heat exhausted during any time interval are all proportional to the elapsed time. Therefore, all the same relationships that are true for the amounts of heat flow and work done apply to the *rates* at which heat flows and work is done. For example, the efficiency is

$$e = \frac{\text{work done}}{\text{heat input}} = \frac{\textit{rate} \text{ of doing work}}{\textit{rate} \text{ of taking in heat}} = \frac{W/\Delta t}{Q_{\text{in}}/\Delta t}$$

Figure 15.11 (a) The four-stroke Otto cycle. (b) *PV* diagram for the Otto cycle; the net work output is indicated by the area shaded in the diagram.

Example 15.5

Rate at Which Heat Is Exhausted from an Engine

An engine operating at 25% efficiency produces work at a rate of 0.10 MW. At what rate is heat exhausted into the environment?

Strategy   We are given that the engine does work at a constant *rate*. The efficiency is also constant.

Solution   The engine efficiency is the ratio of the rate at which work is done to the rate of heat input:

$$e = \frac{W/\Delta t}{Q_{in}/\Delta t} = \frac{W}{Q_{in}}$$

The *net* rate of heat input Q is

$$\frac{Q}{\Delta t} = \frac{Q_{in}}{\Delta t} + \frac{Q_{out}}{\Delta t}$$

where Q_{out} is negative. Since the internal energy of the engine does not change over a complete cycle, energy conservation (or the first law of thermodynamics) requires that

$$Q - W = 0 \quad \text{or} \quad Q_{in} + Q_{out} = W$$

In terms of the rate at which heat is delivered or exhausted and the rate at which work is done,

$$\frac{Q_{in}}{\Delta t} + \frac{Q_{out}}{\Delta t} = \frac{W}{\Delta t}$$

In other words, the rate at which the engine does work is equal to the *net* rate of heat input. We are asked to find the rate of heat exhausted $Q_{out}/\Delta t$:

$$\frac{Q_{out}}{\Delta t} = \frac{W}{\Delta t} - \frac{Q_{in}}{\Delta t} = \frac{W}{\Delta t} - \frac{W/\Delta t}{e}$$

$$\frac{Q_{out}}{\Delta t} = \frac{W}{\Delta t}\left(1 - \frac{1}{e}\right) = 0.10 \text{ MW} \times \left(1 - \frac{1}{0.25}\right) = -0.30 \text{ MW}$$

Discussion   The negative sign here means that exhausted heat flows *out* of the engine at a rate of 0.30 MW.

As a check: 25% efficiency means that $\frac{1}{4}$ of the heat input does work and $\frac{3}{4}$ of it is exhausted. Therefore, the ratio of work to exhaust is $\frac{1/4}{3/4} = \frac{1}{3} = \frac{0.10 \text{ MW}}{0.30 \text{ MW}}$.

For simplicity, we could have let W, Q_{in}, and Q_{out} refer to *rates* instead of to total amounts—to do this we just cancel common factors of Δt out of the equations for efficiency and energy conservation.

Practice Problem 15.5   Heat engine efficiency

An engine "wastes" 4.0 J of heat for every joule of work done. What is its efficiency?

According to the first law of thermodynamics, the efficiency of a heat engine cannot exceed 100%. An efficiency of 100% would mean that all of the heat input is turned into useful work and no "waste" heat is exhausted. It might seem theoretically possible to make a 100% efficient engine by eliminating all of the imperfections in design such as friction and lack of perfect insulation. However, it is not, as we see in Section 15.7.

15.6   REFRIGERATORS AND HEAT PUMPS

The second law of thermodynamics says that heat cannot *spontaneously* flow from a colder body to a hotter body; but machines such as refrigerators and heat pumps can make that happen. In a refrigerator, heat is pumped "uphill"—out of the food compartment into the warmer room. That doesn't happen by itself; it requires the *input* of work. The electricity used by a refrigerator turns the compressor motor, which does the work required to make the refrigerator function (Fig. 15.12). An air conditioner is essentially the same thing: it pumps heat out of the house into the hotter outdoors.

The only difference between a refrigerator (or an air conditioner) and a heat pump is which end is performing the useful task. Refrigerators and air conditioners pump heat out of a compartment that they are designed to keep cool. Heat pumps pump heat from the colder outdoors into the warmer house. The idea isn't to keep the outdoors cold; it is to heat the house.

Just as for the heat engine, we can apply the first law of thermodynamics to the heat pump. Once again, we assume that it operates in a cycle and $\Delta U = 0$. Therefore,

$$Q = W$$

This time both Q and W are negative. We do positive work on the heat pump, so the heat pump does *negative* work. Heat flows into the heat pump but more flows out since the work input is "converted" to heat.

Figure 15.12 In a refrigerator, a fluid is compressed, increasing its temperature. Heat is exhausted as the fluid passes through the condenser. Now the fluid is allowed to expand; its temperature falls. Heat flows from the food compartment into the cold fluid. The fluid returns to the compressor to begin the same cycle again.

Notice that the heat pump is a heat engine operating in reverse (Fig. 15.13). In the heat engine, heat flows from hot to cold, with work as the output. In a heat pump, heat flows from cold to hot, with work as the input. It will be most convenient to distinguish the heat transfers not by which is input and which output (since that switches in going from an engine to a heat pump), but rather by the temperature of the environment where the exchange is made, using subscripts "H" and "C" for hot and cold.

Thus, the efficiency of the heat engine is

$$e = \frac{\text{work output}}{\text{heat input}} = \frac{W}{Q_H} \tag{15-12}$$

The efficiency can also be expressed in terms of the heat flows. Since $W = Q_H + Q_C$,

$$e = \frac{Q_H + Q_C}{Q_H} = 1 + \frac{Q_C}{Q_H} \tag{15-14}$$

Since $Q_C < 0$, the efficiency is *less* than 1.

The same quantity e will be useful in our discussions of heat pumps (and refrigerators), but it does not measure *efficiency*. A more efficient heat pump uses a *smaller* work input for a given amount of heat output; therefore, we call e the *inefficiency* of the heat pump.

Figure 15.13 A heat engine and a heat pump (or refrigerator)

Inefficiency of a heat pump or refrigerator

$$e = \frac{\text{work input}}{\text{heat output}} = \frac{W}{Q_H} = 1 + \frac{Q_C}{Q_H} \tag{15-15}$$

In a heat pump, the directions of all the energy transfers are reversed from those for the heat engine; the signs of Q_H, Q_C, and W are opposite what they are for an engine. The heat input comes from the *colder* side; the heat exhausted goes to the *hotter* side. Work is not done *by* the pump, it is done *on* the pump. Since the signs are all reversed, e is still positive and less than 1.

Since manufacturers would be loathe to advertise the "inefficiency" of a heat pump or refrigerator, it's common practice to state a **coefficient of performance** K instead. Just as for the efficiency of an engine, the coefficients of performance are ratios of what you get divided by what you put in:

- for a heat pump:

$$K_p = \frac{\text{heat delivered}}{\text{work input}} = \frac{Q_H}{W} = \frac{1}{e} \tag{15-16}$$

- for a refrigerator or air conditioner:

$$K_r = \frac{\text{heat removed}}{\text{work input}} = \left| \frac{Q_C}{W} \right| = \frac{1}{e} - 1 \tag{15-17}$$

A higher coefficient of performance means a better heat pump or refrigerator.

The second law says that heat cannot flow spontaneously from cold to hot—we need to do some work to make that happen. That's equivalent to saying that the coefficient of performance can't be infinite and e can't be zero. So a restatement of the second law is:

Second law of thermodynamics

No refrigerator or heat pump can have $e = 0$.

Example 15.6

A Heat Pump

A heat pump has a performance coefficient of 2.5. (a) How much heat is delivered to the house for every joule of electrical energy consumed? (b) In an electric heater, for each joule of electrical energy consumed, one joule of heat is delivered to the house. Where does the "extra" heat delivered by the heat pump come from? (c) What would its coefficient of performance be when run as an air conditioner instead?

Strategy There are two slightly different meanings of coefficient of performance. For a heat pump, whose object is to deliver heat to the house, the coefficient of performance is the heat delivered (Q_H) per unit of work done to run the pump. With an air conditioner, the object is to *remove* heat from the house. The coefficient of performance is the heat *removed* (Q_C) per unit of work done.

Solution   (a) As a heat pump,

$$K_p = \frac{\text{heat output}}{\text{work input}} = \frac{Q_H}{W} = 2.5$$

$$Q_H = 2.5 \, W$$

For every joule of electrical energy (= work), 2.5 J are delivered to the house.

(b) The 2.5 J include the 1.0 J of work plus 1.5 J that were pumped in from the outside. The electric heater just transforms the joule of work into a joule of heat.

(c) From (b), the coefficient is 1.5:

$$K_r = \frac{\text{heat removed}}{\text{work done}} = \left| \frac{Q_C}{W} \right| = \left| \frac{1.5 \, J}{-1.0 \, J} \right| = 1.5$$

Discussion One thing that makes a heat pump economical in many situations is that the same machine can function as a heat pump in winter and as an air conditioner in summer. The heat pump delivers heat to the interior of the house, while the air conditioner pumps heat out.

Practice Problem 15.6   Heat exhausted by air conditioner

An air conditioner with a coefficient of performance $K_r = 3$ consumes electricity at an average rate of 1 kW. During 1 h of use, how much heat is exhausted to the outdoors?

15.7   REVERSIBLE ENGINES AND HEAT PUMPS

As we have already seen, a heat pump (or refrigerator) is essentially a reversed heat engine, since all of the energy transfers reverse direction. Many real engines actually can be run in reverse as heat pumps. However, in general the quantity $e = W/Q_H$ is *not* the same when reversed, due to irreversible processes. Recall that a real engine has an efficiency e less than that of an ideal engine operating in the same environment, while a real heat pump has an inefficiency e *greater* than that of an ideal heat pump. You might guess that a real engine's efficiency e_e would therefore be less than the inefficiency e_p of the reversed engine (= heat pump). That guess is correct, as we will see shortly.

An ideal engine or heat pump is one in which no irreversible processes occur: there is no friction and heat only flows across infinitesimal temperature differences. Since all the processes are reversible, we call an ideal engine or heat pump a **reversible engine** or a reversible heat pump.

In order to simplify the analysis of reversible engines, let us assume the engine takes heat input from a reservoir at temperature T_H and exhausts heat into another reservoir at temperature T_C (with $T_C < T_H$). Recall that a reservoir is a system with such a large heat capacity that it can exchange heat in either direction with a negligibly small temperature change. Therefore, the cold reservoir stays at temperature T_C and the hot reservoir stays at T_H.

Now imagine connecting an engine to a heat pump, both using the same two reservoirs. The engine has efficiency e_e and the pump has inefficiency e_p. The heat pump uses all the work output of the engine as its work input (Fig. 15.14) so $W_p = -W_e$. The engine takes heat $Q_{H,e}$ from the hot reservoir where $Q_{H,e} = W_e/e_e$. The pump delivers heat $Q_{H,p} = W_p/e_p$ to the hot reservoir. Is there a net flow of heat out of the hot reservoir or into it? If the engine and pump are reversible, then $e_e = e_p$ and $Q_{H,e} = -Q_{H,p}$. There is no net flow of heat into or out of the hot reservoir. The same is true for the cold reservoir. The net effect is: nothing.

What happens if $e_e \neq e_p$? If $e_p > e_e$, then $|Q_{H,p}| < |Q_{H,e}|$. There is a net flow of heat out of the hot reservoir and into the cold. The net effect is heat flowing from a hotter system to a colder system. Nothing unexpected here; that could happen spontaneously. But what if $e_e > e_p$? Then there is a net flow of heat into the hot reservoir and out of the cold reservoir. Without any input of work from the outside, heat is flowing from cold to hot. Since this violates the second law—heat cannot flow spontaneously from cold to hot—we can restate the second law as:

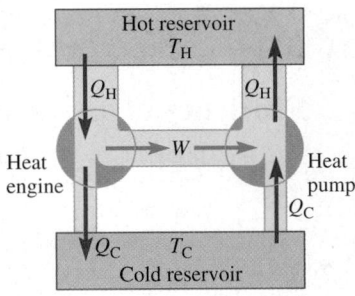

Figure 15.14 Heat engine connected to a heat pump

Second law of thermodynamics:
No engine can have an efficiency greater than the inefficiency of any pump operating between the same hot and cold reservoirs.

If e_r is the efficiency of a reversible engine or the inefficiency of a heat pump operating between the same two reservoirs, then equivalently:

Every engine has $e_e \leq e_r$ and every heat pump has $e_p \geq e_r$.

Since every reversible engine operating between the same temperatures, no matter what the details of its construction, has the same efficiency, the efficiency must depend only on the temperatures of the hot and cold reservoirs. It turns out that e_r is given by the remarkably simple expression:

$$e_r = 1 - \frac{T_C}{T_H} \qquad (15\text{-}18)$$

The temperatures in Eq. (15-18) must be *absolute* temperatures—in SI, kelvin temperatures. [Absolute temperature is also called *thermodynamic temperature* because you can use the efficiency of reversible engines to set a temperature scale. In fact, the definition of the kelvin is based on Eq. (15-18).]

Using Eq. (15-18), the ratio of the heat exhaust to the heat input *for a reversible engine* is

$$\frac{Q_C}{Q_H} = \frac{W - Q_H}{Q_H} = \frac{W}{Q_H} - 1 = e_r - 1 = -\frac{T_C}{T_H} \qquad (15\text{-}19)$$

Since Q_C and Q_H always have opposite signs, while the temperatures are always positive, we can make the following statement:

For a reversible engine, the ratio of the magnitudes of the heats is equal to the ratio of the temperatures.

The efficiency of a reversible engine is always less than 100%, assuming that the cold reservoir is not at absolute zero. Even an ideal, perfectly reversible engine must exhaust some heat, so the efficiency can never be 100% *even in principle*. Efficiencies of real engines are less than those of ideal engines. This provides a way to rephrase the second law of thermodynamics:

Second law of thermodynamics (Kelvin statement):
No engine can have 100% efficiency.

Example 15.7

Efficiency of an Automobile Engine

 In an automobile engine, the combustion of the fuel-air mixture can reach temperatures as high as 3000°C and the exhaust gases leave the cylinder at about 1000°C. (a) Find the efficiency of a *reversible* engine operating between reservoirs at those two temperatures. (b) Theoretically, we might be able to have the exhaust gases leave the engine at the temperature of the outside air (20°C). What would be the efficiency of the hypothetical reversible engine in this case?

Strategy   First we identify the temperatures of the hot and cold reservoirs in each case. We must convert the reservoir temperatures to kelvins in order to find the efficiency of a reversible engine.

Solution   (a) The reservoir temperatures in kelvins are found using

$$T = T_C + 273 \text{ K}$$

Therefore,

$$T_H \approx 3000°C \approx 3273 \text{ K}$$
$$T_C \approx 1000°C \approx 1273 \text{ K}$$

The efficiency of a reversible engine operating between these temperatures is

$$e_r = 1 - \frac{T_C}{T_H} = 1 - \frac{1273 \text{ K}}{3273 \text{ K}} = 0.61 \approx 60\%$$

(b) The high-temperature reservoir is still at 3273 K, while the low-temperature reservoir is now

$$T_C \approx 293 \text{ K}$$

This gives a higher efficiency:

$$e_r = 1 - \frac{T_C}{T_H} = 1 - \frac{293 \text{ K}}{3273 \text{ K}} = 0.91 \approx 90\%$$

Discussion   As mentioned in the chapter opener, real gasoline engines achieve efficiencies of only about 25%. While improvement is possible, the second law of thermodynamics limits the theoretical maximum efficiency to that of a reversible engine operating between the same temperatures. The *theoretical* maximum efficiency can only be increased by using a hotter hot reservoir or a colder cold reservoir. The higher the temperature achieved by the burning gasoline inside the cylinder, the higher the theoretical maximum efficiency.

Practice Problem 15.7   Temperature of hot gases

If the efficiency of a reversible engine is 75% and the temperature of the outdoor world into which the engine sends its exhaust is 27°C, what is the combustion temperature in the engine cylinder?

Example 15.8

Coal-Burning Power Plant

A coal-burning electrical power plant burns coal at 706°C. Heat is exhausted into a river near the power plant; the average river temperature is 19°C. What is the minimum possible rate of thermal pollution (heat exhausted into the river) if the station generates 125 MW of electricity?

Strategy   The minimum discharge of heat into the river would occur if the engine generating the electricity were reversible. As in Example 15.5, we can take all of the rates to be constant.

Solution   First find the absolute temperatures of the reservoirs:

$$T_H = 706°C = 979 \text{ K}$$
$$T_C = 19°C = 292 \text{ K}$$

The efficiency of a reversible engine operating between these temperatures is

$$e_r = 1 - \frac{T_C}{T_H} = 1 - \frac{292 \text{ K}}{979 \text{ K}} = 0.702$$

We want to find the rate at which heat is exhausted, which is $Q_C/\Delta t$. The efficiency is equal to the ratio of the work output to the heat input from the hot reservoir:

$$e = \frac{W}{Q_H}$$

Conservation of energy requires that

$$Q_H + Q_C = W$$

Solving for Q_C,

$$Q_C = W - Q_H = W - \frac{W}{e} = W \times \left(1 - \frac{1}{e}\right)$$

Assuming that all the rates are constant,

$$\frac{Q_C}{\Delta t} = 125 \text{ MW} \times \left(1 - \frac{1}{0.702}\right) = -53 \text{ MW}$$

The rate at which heat enters the river is 53 MW.

Discussion   We expect the actual rate of thermal pollution to be higher. A real, irreversible engine would have a lower efficiency, so more heat is dumped into the river.

Practice Problem 15.8   Generating electricity from coal

What is the minimum possible quantity of heat *input* (from the burning of coal) needed to generate 125 MW of electricity in this same plant?

15.8   CARNOT CYCLE

Sadi Carnot (1796–1832), a French engineer, published a treatise in 1824 that greatly expanded the understanding of how heat engines work. His treatment introduced a hypothetical, ideal engine that uses two heat reservoirs at different temperatures as the source and sink for heat and an ideal gas as the working substance of the engine. We now call this engine a **Carnot engine** and its cycle of operation the **Carnot cycle**. Remember that the Carnot engine is an *ideal* engine, not a real engine.

The Carnot engine is a particular kind of reversible engine. We must assume that all friction has somehow been eliminated—otherwise an irreversible process takes place. We also must avoid heat flow across a finite temperature difference, which would be irreversible. Therefore, whenever the ideal gas takes in or gives off heat, the gas must be at the same temperature as the reservoir with which it exchanges energy.

How can we get heat to flow without a temperature difference? Imagine putting the gas in good thermal contact with a reservoir at the same temperature. Now *slowly* pull a piston so that the gas expands. Since the gas does work, it must lose internal energy—and therefore its temperature drops, since in an ideal gas the internal energy is proportional to absolute temperature. But heat flows in to keep the temperature constant. As long as the expansion occurs slowly, the heat flows in fast enough to keep the temperature constant.

So, to keep everything reversible, we must exchange heat in *isothermal* processes. To take in heat, we expand the gas; to exhaust heat, we compress it. We also need processes to change the gas temperature from T_H to T_C and back to T_H. These processes must be *adiabatic* (no heat flow) since otherwise an irreversible heat flow would occur.

Now that the ground rules are set, here is the four-step Carnot cycle (Fig. 15.15):

1→2: Isothermal expansion. Take in heat Q_H from the hot reservoir, keeping the gas at constant temperature T_H.

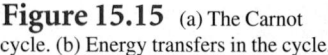

Figure 15.15 (a) The Carnot cycle. (b) Energy transfers in the cycle.

2→3: Adiabatic expansion. The gas does work without any heat flow in, so the temperature decreases. Continue until the gas temperature is T_C.

3→4: Isothermal compression. Heat Q_C is exhausted at constant temperature T_C.

4→1: Adiabatic compression until the temperature is back to T_H.

Carnot was able to calculate the efficiency of an engine operating in this cycle and obtained Eq. (15-18). Since *all* reversible engines operating between the same two reservoirs must have the same efficiency, deriving the efficiency for one particular kind of reversible engine is sufficient to derive it for all of them.

Example 15.9

Carnot Engine

A Carnot engine using 0.020 mol of an ideal gas operates between reservoirs at 1000.0 K and 300.0 K. The engine takes in 25 J of heat from the hot reservoir per cycle. Find the work done by the engine during the two isothermal steps in the cycle.

Strategy During the isothermal processes, the internal energy of the ideal gas stays the same, so the heat input is equal to the work done.

Solution 1→2: During the isothermal expansion, the work done is equal to the heat input—otherwise the temperature of the gas would change.

$$W_{1 \to 2} = +25 \text{ J (per cycle)}$$

3→4: During the isothermal compression, the work done is equal to the heat output. This time the work done is negative: the gas does negative work as it is compressed.

$$W_{3 \to 4} = Q_C < 0$$

The heats are proportional to the temperatures:

$$\frac{Q_C}{Q_H} = -\frac{T_C}{T_H}$$

Therefore,

$$W_{3 \to 4} = Q_C = -\frac{T_C}{T_H} Q_H = -\frac{300.0 \text{ K}}{1000.0 \text{ K}} \times 25 \text{ J} = -7.5 \text{ J (per cycle)}$$

Discussion We will not prove it here, but the total work done during the two adiabatic processes is zero. Then the total work done per cycle is

$$25 \text{ J} + (-7.5 \text{ J}) = 17.5 \text{ J}$$

The efficiency is then

$$e = \frac{17.5 \text{ J}}{25 \text{ J}} = 0.70$$

This should equal

$$e = 1 - \frac{T_C}{T_H} = 1 - \frac{300.0 \text{ K}}{1000.0 \text{ K}} = 0.7000$$

Conceptual Practice Problem 15.9   Adiabatic process in Carnot cycle

Since there is no heat flow during the adiabatic processes and the work done during them adds to zero, why do we need adiabatic processes in the Carnot cycle? Why not just eliminate them?

15.9 ENTROPY

When two systems of different temperatures are in thermal contact, heat flows out of the hotter system and into the colder system. There is no change in the total energy of the two systems; energy just flows out of one and into the other. Why then does heat flow? As we will see, heat flow into a system not only increases the system's internal energy, it also increases the *disorder* of the system. Heat flow out of a system decreases its internal energy and its disorder.

The **entropy** of a system (symbol S) is a quantitative measure of its disorder. Entropy is a state variable (like U, P, V, and T): a system in equilibrium has a unique entropy that does *not* depend on the past history of the system. (Recall that heat and work are *not* state variables.) The word *entropy* was coined by Rudolf Clausius (1822–1888) in 1865; its Greek root means *evolution* or *transformation*.

When heat flows from a hotter system to a colder one, the total energy change is zero, but the total entropy change is *not* zero. The heat entering the colder system increases its entropy *more* than the entropy of the hotter system decreases, so the total entropy of the two—and thus the total entropy of the universe—increases. This makes sense since we expect a spontaneous process to increase the amount of disorder. Heat flow in the opposite direction, from the colder system to the hotter system, would decrease the total entropy of the universe. This would be like the unscrambling of an egg: it *never* happens. The second law of thermodynamics can be stated in terms of entropy:

Second law of thermodynamics (entropy statement):
The entropy of the universe never decreases.

If an amount of heat Q flows into a system at constant absolute temperature T, the entropy change of the system is

$$\Delta S = \frac{Q}{T} \qquad (15\text{-}20)$$

From this, we can see that the SI unit for entropy is J/K. Heat flowing into a system increases the system's entropy (both ΔS and Q are positive); heat leaving a system decreases the system's entropy (both ΔS and Q are negative). Equation (15-20) is valid as long as the temperature of the system is constant, which is true if

- The heat capacity is large (as for a reservoir);
- The amount of heat flow Q is small; or
- The system does work $W = Q$ so that $\Delta U = 0$.

Note that Eq. (15-20) gives only the *change* in entropy, not the initial and final values of the entropy. As with potential energy, the *change* is what's important in most situations.

If a small amount of heat Q flows between two systems of different temperatures ($T_H > T_C$), the total entropy change is

$$\Delta S = \frac{-Q}{T_H} + \frac{Q}{T_C}$$

Since $T_H > T_C$,

$$\frac{Q}{T_H} < \frac{Q}{T_C}$$

so the increase in the colder system's entropy is larger than the decrease of the hotter system's entropy:

$$\Delta S > 0 \qquad (15\text{-}21)$$

A process that increases the entropy of the universe is irreversible, because the reverse process would decrease the universe's entropy. A reversible process must therefore be one that leaves the total entropy unchanged. For example, a reversible engine removes heat $-Q_H$ from a hot reservoir at temperature T_H and exhausts $-Q_C$ to a cold reservoir at T_C. (The minus signs are needed because we consider the heat flow in and out of the *reservoirs* rather than in and out of the engine.) The entropy of the engine itself is left unchanged since it operates in a cycle. The entropy of the hot reservoir decreases by an amount $-Q_H/T_H$ and that of the cold reservoir increases by $-Q_C/T_C$. Since the entropy of the universe must be unchanged by a reversible engine, it must be true that

$$\frac{-Q_H}{T_H} + \frac{-Q_C}{T_C} = 0 \quad \text{or} \quad \frac{Q_C}{Q_H} = -\frac{T_C}{T_H}$$

The efficiency of the engine is therefore

$$e = \frac{W}{Q_H} = \frac{Q_H + Q_C}{Q_H} = 1 + \frac{Q_C}{Q_H} = 1 - \frac{T_C}{T_H}$$

as stated in Section 15.7.

Entropy is *not* a conserved quantity like energy. The entropy of the universe is always increasing. It is possible to decrease the entropy of a *system*, but only at the expense of increasing the entropy of the surroundings by at least as much (usually more).

Example 15.10

Entropy Change of a Freely Expanding Gas

Suppose 1.0 mol of an ideal gas is allowed to freely expand into an evacuated container of equal volume so that the volume of the gas doubles (Fig. 15.16). No work is done by the gas as it expands, since there is nothing for it to push against. No heat flows into or out of the gas. What is the entropy change of the gas?

Strategy   The only way to calculate entropy changes that we've learned so far is for heat flow at a constant temperature. In free expansion, there is no heat flow—but that does not necessarily mean there is no entropy change. Since entropy is a state variable, ΔS depends only on the initial and final states of the gas, not the intermediate states. We can therefore find the entropy change using *any* thermodynamic process with the same initial and final states. The initial and final temperatures of the gas are identical since the internal energy does not change; therefore we find the entropy change for an *isothermal* expansion.

Valve closed

Figure 15.16
Two chambers connected by a valve. One chamber contains a gas and the other has been evacuated. When the valve is opened, the gas expands until it fills both chambers.

Solution   Imagine the gas confined to a cylinder with a moveable piston. In an isothermal expansion, heat flows into the gas at a constant temperature T. As the gas expands, it does work on the piston. If the temperature is to stay constant, the work done must equal the heat flow into the gas:

$$\Delta U = 0 \text{ implies } Q = W$$

In section 15.2 we found that the work done during the isothermal expansion of an ideal gas is

$$W = nRT \ln\left(\frac{V_f}{V_i}\right)$$

The volume of the gas doubles, so $V_f/V_i = 2$:

$$W = nRT \ln 2$$

Now we have an equal amount of heat flow into the gas. The entropy change is therefore

$$\Delta S = \frac{Q}{T} = nR \ln 2 = (1.0 \text{ mol}) \times \left(8.31 \frac{J}{mol \cdot K}\right) \times 0.693 = +5.8 \text{ J/K}$$

Discussion   The entropy change is positive, as expected. Free expansion is an irreversible process; the gas molecules do not spontaneously collect back in the original container. The reverse process would cause a decrease in entropy, without a larger increase elsewhere, and so violates the second law.

Practice Problem 15.10   Entropy change of the universe when lump of clay is dropped

A room-temperature lump of clay of mass 400 g is dropped from a height of 2 m and makes a totally inelastic collision with the floor. Approximately what is the entropy change of the universe due to this collision? [*Hint:* The temperature of the clay rises, but only slightly.]

What about the evolution of life on Earth? Some have said that the second law of thermodynamics cannot be true because life has developed from simple life-forms to complex organisms spontaneously. Is this not a spontaneous increase in order—a decrease in entropy?

The second law of thermodynamics does *not* say that the entropy of a system can never decrease, just that the *total* entropy of the universe cannot decrease. When heat flows from a hot body to a cold body, the entropy of the hot body decreases, but the increase in the cold body's entropy is greater, so the entropy of the universe increases. A living organism is not a closed system and neither is the Earth. An adult human, for instance, requires roughly 10 MJ of chemical energy from food per day. What happens to this energy? Some is turned into useful work by the muscles, some more is used to repair body tissues, but most of it is dissipated and leaves the body as heat. The human body therefore is constantly increasing the entropy of its environment. As evolution progresses from simpler to more complicated organisms, the increase in order within the organisms must be accompanied by a larger increase in disorder in the environment.

The Availability of Energy

When people speak of "conserving energy," they usually mean using fuel and electricity sparingly. In the physics sense of the word *conserve*, energy is *always* conserved. Burning natural gas to heat your house does not change the amount of energy around; it just changes it from one form to another.

What we need to be careful not to waste is *high-quality* energy. Our concern is not the total amount of energy but rather whether the energy is in a form that is useful and convenient. The chemical energy stored in fuel is relatively high-quality energy. In a fuel cell, this energy can be turned into useful work with a theoretical maximum efficiency of 100%. When fuel is burned, the energy is degraded into lower-quality energy. The burned fuel could supply the high-temperature reservoir for a heat engine—but now the maximum possible efficiency is no longer 100%. The energy is higher quality if it is burned at a higher temperature, because the theoretical efficiency of a heat engine converting the energy to mechanical work is higher.

A heat engine can only work if a temperature *difference* is available. The smaller the ratio of T_C/T_H, the more work can be extracted.

It may be that, as the entropy of the universe increases inexorably, the universe as a whole comes to thermal equilibrium—the same temperature everywhere. This is called the "heat death of the universe." There would still be just as much energy as ever, but with everything at the same temperature, nothing interesting would happen—there certainly couldn't be any life.

15.10 STATISTICAL INTERPRETATION OF ENTROPY

Suppose that we take four coins, number them, and toss them. We could report the outcome in two different ways: either by specifying the outcome of each coin toss individually (e.g., coin 1 is heads, coin 2 is tails, coin 3 is heads, and coin 4 is heads), or just by reporting the overall outcome as the number of heads. We can use this simple system as an analog of a thermodynamic system. We will find that the second law of thermodynamics is based on statistics, as is the outcome of the four-coin system.

Specifying the outcome of each coin toss individually is analogous to describing the **microstate** of a thermodynamic system. A microstate specifies the state of each constituent particle. For instance, in a monatomic ideal gas with N atoms, a microstate is specified by the position and velocity of each of the N atoms. As the atoms move about and collide, the system changes from one microstate to another. The total number of heads for coin tossing is analogous to a **macrostate** of a thermodynamic system. A macrostate of an ideal gas is determined by the values of the macroscopic state variables (the pressure, volume, and temperature).

In the four-coin model, each of the microstates is equally likely to occur on any toss. Each of the coins has equal probability of landing heads or tails. Since each of 4 coins has 2 possible outcomes, there are $2^4 = 16$ different but equally probable microstates. There are only five macrostates: the number of heads can range from zero to four. The macrostates are *not* equally likely. A good guess would be that 2 heads is much more likely than 4 heads. To find the probability of a macrostate, we count up the number of microstates corresponding to that macrostate and divide by the total number of microstates for all the possible macrostates. From Table 15.3, the probability of the most likely macrostate (2 heads) is $6/16 = 0.375$. The probability of 4 heads is only $1/16 = 0.0625$.

Table 15.3

Possible Results of Tossing Four Coins

Macrostate	Microstates	Number of Microstates
4H (4 heads)	HHHH	1
3H (3 heads)	HHHT HHTH HTHH THHH	4
2H (2 heads)	HHTT HTHT HTTH THHT THTH TTHH	6
1H (1 head)	HTTT THTT TTHT TTTH	4
0H (0 heads)	TTTT	1
	Total number of microstates = 16	

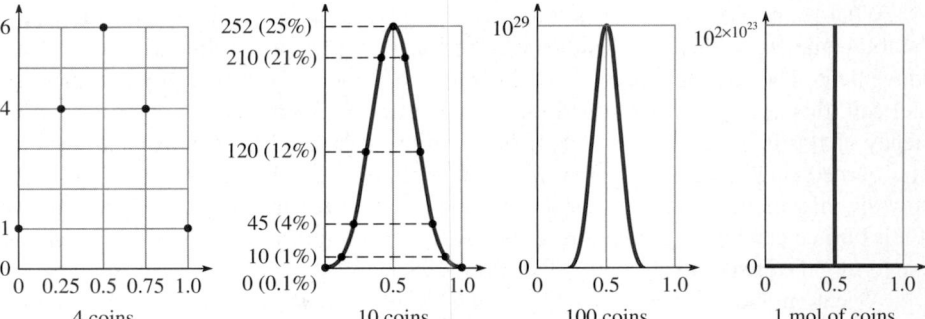

Figure 15.17 Graphs of the number of macrostates versus n/N ($n = 0, 1, \ldots, N$), where $n =$ number of heads for $N = 4$ coins, 10 coins, 100 coins, 1 mole of coins.

Physics at Home

Toss repeatedly a collection of 10 coins. After each toss, count and record the number of heads. After a large number of tosses, are your results similar to the results of a statistical analysis (see Fig. 15.17)? Why are your results not *exactly* the same?

We used 4 coins in our example for convenience, but thermodynamic systems have very large numbers of particles (for instance, there are 6×10^{23} particles in one mole). What happens to the coin-tossing problem if the number of coins gets large? In Fig. 15.17 we have graphed the number of microstates for the various macrostates for systems with $N = 4$ coins, 10 coins, 100 coins, and 1 mole of coins. The horizontal axes for the four graphs specify the macrostate as the fractional number of heads, or (number of heads)/N, which ranges from 0 to 1. The probability of obtaining any macrostate is proportional to the number of microstates since the microstates are equally likely. Notice what happens to the probability peak: as N gets large, the probability of obtaining a macrostate having a number of heads significantly different (say, more than 1%) from $\frac{1}{2}N$ gets smaller and smaller. What if there were 6×10^{23} coins? The probability of getting anything more than 1% away from 3×10^{23} heads is so small that we can call it zero—it is impossible.

This kind of statistical analysis is the basis for the second law of thermodynamics. It turns out, remarkably, that the number of microstates for a given macrostate is related to the entropy of that macrostate in a simple way. Letting W stand for the number of microstates, the relationship is

$$S = k \ln W \tag{15-22}$$

Equation (15-22) is inscribed on the tombstone of Ludwig Boltzmann (1844–1906), the Austrian physicist who made the connection between entropy and statistics in the late nineteenth century. The relationship between S and W has to be logarithmic because entropy is additive: if system 1 has entropy S_1 and system 2 has entropy S_2, then the total entropy is $S_1 + S_2$. However, the number of microstates is *multiplicative*. Think of dice: if die 1 has 6 microstates and die 2 also has 6, the total number of microstates is not 12, but $6 \times 6 = 36$. The entropy is additive since $\ln 6 + \ln 6 = \ln 36$.

Suppose we have a box with 100 coins arranged so that all are showing heads. There is only one way of obtaining this macrostate. This means $W_i = 1$ and

$$S_i = k \ln 1 = k \times 0 = 0$$

Next we shake the box for awhile and we end up with 50 coins showing heads and 50 showing tails.

Statistically, it can be shown that the number of ways to have N coins arranged so that N_1 are in one group (heads) and the remainder, $N - N_1 = N_2$, are in the other group (tails) is

$$\frac{N!}{N_1! \, N_2!} \tag{15-23}$$

where $N!$ is *N-factorial*:

$$N! = N \times (N-1) \times (N-2) \times (N-3) \times \cdots \times 3 \times 2 \times 1$$

Thus, for a collection of 100 coins, with 50 showing heads (and the other 50 showing tails), we have a number of possible arrangements (microstates) given by

$$W_f = \frac{100!}{50! \; 50!}$$

$$= \frac{9.33 \times 10^{157}}{(3.04 \times 10^{64}) \times (3.04 \times 10^{64})}$$

$$= 1.01 \times 10^{29}$$

There are about 10^{29} ways of having 100 coins showing 50 heads. The entropy for this outcome is

$$S_f \approx k \ln 10^{29} = 29k \ln 10 = 66.8k$$

The change in entropy, ΔS, is

$$\Delta S = S_f - S_i = 66.8k - 0 = +66.8k$$

The entropy has increased. No matter what the original arrangement of the coins in the box, we expect to end up with *approximately* 50 heads after we shake the box for awhile.

$$W_H < W_{50} \quad \text{(for H = initial number of heads < 50)}$$

$$\Delta S = k \ln W_{50} - k \ln W_H > 0$$

Thus, the entropy increases no matter what the starting condition.

Entropy never decreases because the macrostate with the highest entropy is the one with the greatest number of microstates, and thus the highest probability. (Recall that since the microstates are equally likely, the probability of a macrostate is proportional to W.) The probability peak is so sharp and narrow in thermodynamic systems that the probability is zero of finding a macrostate not in that peak. The equilibrium macrostate is the one with most microstates. The spontaneous evolution of a system always goes toward higher probability (higher entropy). A process that increases the probability is irreversible—because the reversed process would decrease the probability. Reversible processes leave the probability unchanged.

Example 15.11

Increased Number of Microstates in Free Expansion

Refer to the free expansion of an ideal gas (Example 15.10). How does the number of microstates change when the volume of the gas (containing N molecules) is doubled?

Strategy Since in Example 15.10 we found the entropy change for this process, we can now use the entropy change to find how the number of microstates changes. Since the relationship between S and W is logarithmic, an increase in S will tell us by what *factor* W increases.

Solution The entropy change for n moles was found to be

$$\Delta S = nR \ln 2$$

Since $nR = Nk$, the entropy increase can be written in terms of N:

$$\Delta S = Nk \ln 2$$

If W_i and W_f are the initial and final number of microstates, then

$$\Delta S = k \ln W_f - k \ln W_i = k (\ln W_f - \ln W_i) = k \ln \frac{W_f}{W_i}$$

Equating these last two expressions for ΔS, we find

$$N \ln 2 = \ln \frac{W_f}{W_i}$$

Since $\ln 2^N = N \ln 2$,

$$\frac{W_f}{W_i} = 2^N$$

continued on next page

Example 15.11 *continued*

Discussion   To get an idea of how large the increase in the number of microstates is, let $N = N_A$ (1 mol of gas). To write the number 2^N in ordinary base 10 notation, we would need 2×10^{23} *digits.*

The temperature is the same before and after, so the number of velocity states, rotational states, and vibrational states before and after is the same. But each molecule has twice as much volume in which it can be found, so *for each molecule* there are twice as many possible positions. Therefore, the number of microstates is multiplied by 2 for each molecule, or by 2^N overall.

Practice Problem 15.11   Change in entropy for 100 coins

What is the change in entropy (expressed as a multiple of the Boltzmann constant) if a box of 100 coins starts with 35 heads showing and then is shaken until 50 heads are showing? [*Hint*: Use $\ln A - \ln B = \ln (A/B)$ to simplify the math. $35! = 1.03 \times 10^{40}$ and $65! = 8.25 \times 10^{90}$.]

15.11 THE THIRD LAW OF THERMODYNAMICS

Like the second law, the third law of thermodynamics can be stated in several equivalent ways. We will state just one of them:

> **Third law of thermodynamics:**
> It is impossible to cool a system to absolute zero.

Why does the third law matter? According to one statement of the *second* law, heat can never be 100% converted into work. A reversible engine operating with a cold reservoir at $T = 0$ *could* do just that, since its efficiency would be 100%:

$$e = 1 - \frac{T_C}{T_H} = 1 - \frac{0}{T_H} = 1$$

Since, according to the third law, there can be no cold reservoir at absolute zero, no engine can operate with 100% efficiency.

Suppose we wished to use a refrigerator to cool some object down to absolute zero temperature. The first law tells us

$$W = Q_C + Q_H$$

where Q_H is the heat discharged to the hot reservoir. Assuming an ideal, reversible refrigerator,

$$\frac{Q_H}{T_H} = -\frac{Q_C}{T_C}$$

we can calculate the minimum amount of work that it takes to run the appliance through one cycle,

$$W = Q_H + Q_C = Q_C \times \left(\frac{Q_H}{Q_C} + 1 \right)$$

$$W = Q_C \left(1 - \frac{T_H}{T_C} \right) \tag{15-24}$$

Given a certain amount of heat Q_C to extract in each cycle and a temperature T_H of the hot reservoir, we are going to have to do an infinite amount of work as $T_C \to 0$; so we will find it impossible to reach absolute zero temperature.

While it is impossible to *reach* absolute zero, there is no limit on how *close* we can get. Scientists who study low-temperature physics have attained equilibrium temperatures as low as 1 μK and have sustained temperatures of 2 mK; transient temperatures in the nano- and picokelvin range have been observed.

MASTER THE CONCEPTS

Summary

- The first law of thermodynamics is a statement of energy conservation:

$$\Delta U = Q - W \qquad (15\text{-}1)$$

where Q is the heat flow *into* the system and W is the work done *by* the system.

- Pressure, temperature, volume, number of moles, internal energy, and entropy are state variables; they describe the state of a system at some instant of time but *not* how the system got to that state. Heat and work are *not* state variables—they describe *how* a system gets from one state to another.

- The work done by a system when the pressure is constant—or for a volume change small enough that the pressure change is insignificant—is

$$W = P \Delta V \qquad (15\text{-}2)$$

In general, the work done is the total area under the *PV* curve.

- The change in internal energy of an ideal gas is determined solely by the temperature change. Therefore,

$$\Delta U = 0 \quad \text{(ideal gas, isothermal)} \qquad (15\text{-}7)$$

- A process in which no heat is transferred into or out of the system is called an adiabatic process.

- The molar specific heats of an ideal gas at constant volume and constant pressure are related by

$$C_v = C_p - R \qquad (15\text{-}11)$$

- Heat flow from a hotter body to a colder body is always irreversible.

- For a cyclical engine, heat pump, and the like, since $\Delta U = 0$ for a cycle, conservation of energy requires

$$Q = Q_H + Q_C = W$$

- The efficiency of an engine (or the inefficiency of a heat pump, refrigerator, or air conditioner) is defined as

$$e = \frac{W}{Q_H} \qquad (15\text{-}12)$$

- The coefficient of performance for a heat pump is

$$K_p = \frac{\text{heat delivered}}{\text{work input}} = \frac{Q_H}{W} = \frac{1}{e} \qquad (15\text{-}16)$$

- The coefficient of performance for a refrigerator or air conditioner is

$$K_r = \frac{\text{heat removed}}{\text{work input}} = \left| \frac{Q_C}{W} \right| = \frac{1}{e} - 1 \qquad (15\text{-}17)$$

- A reservoir is a system with such a large heat capacity that it can exchange heat in either direction with a negligibly small temperature change.

- The second law of thermodynamics can be stated in various equivalent ways: (1) heat never flows spontaneously from a colder body to a hotter body, (2) no engine can have 100% efficiency, (3) no engine can have an efficiency greater than the inefficiency of any heat pump operating between the same hot and cold reservoirs, (4) no heat pump (or refrigerator) can have an inefficiency of zero, (5) every engine has $e_e \le e_r$ and every heat pump has $e_p \ge e_r$, (6) the entropy of the universe never decreases.

- The efficiency of a *reversible* engine is determined only by the *absolute* temperatures of the hot and cold reservoirs:

$$e_r = 1 - \frac{T_C}{T_H} \qquad (15\text{-}18)$$

- If an amount of heat Q flows into a system at constant absolute temperature T, the entropy change of the system is

$$\Delta S = \frac{Q}{T} \qquad (15\text{-}20)$$

- The number of microstates for a given macrostate is related to the entropy S of that macrostate

$$S = k \ln W \qquad (15\text{-}22)$$

- The third law of thermodynamics: it is impossible to cool a system to absolute zero.

Highlighted Figures and Tables

T15.1 Sign conventions for the first law of thermodynamics (p. 512)

F15.2 Expansion of a gas from initial pressure P_i and volume V_i to final pressure P_f and volume V_f and the *PV* diagram for expansion (p. 514)

F15.3 The *PV* curve shows the work done (p. 514)

F15.4 Work done by following two different processes between the same initial and final states (p. 515)

T15.2 Summary of thermodynamic processes (p. 517)

F15.9 Spontaneous heat flow goes from warm to cool (p. 520)

F15.10 The energy transfers of a heat engine (p. 521)

F15.11 The four-stroke Otto cycle and its *PV* diagram (p. 523)

F15.12 How a refrigerator works (p. 525)

F15.13 A heat engine and a heat pump (p. 525)

F15.15 The Carnot cycle (p. 529)

F15.17 Graphs of the number of macrostates versus n/N (p. 534)

CONCEPTUAL QUESTIONS

1. The entropy of a system increases by 10 J/K. Does this mean that the process is necessarily irreversible? Explain.

2. Is it possible to make a heat pump with an inefficiency equal to 1? Explain.

3. A perfectly elastic collision is reversible. What about an inelastic collision? Justify your answer.

4. An electric baseboard heater can convert 100% of the electrical energy used into heat that flows into the house. Since a gas furnace might be located in a basement and sends exhaust gases up the chimney, the heat flow into the living space is less than 100% of the chemical energy released by burning. Does this mean that electric heating is better? Which heating method consumes less fuel? In your answer, consider how the electricity might have been generated and the efficiency of that process.

5. One whimsical statement of the laws of thermodynamics—probably not one favored by gamblers—goes like this:
   I. You can never win; you can only lose or break even.
   II. You can only break even at absolute zero.
   III. You can never get to absolute zero.
   What do we mean by "win," "lose," and "break even?" [*Hint:* Think about a heat engine.]

6. Why must all reversible engines (operating between the same reservoirs) have the same efficiency? Try an argument by contradiction: imagine that two reversible engines exist with

   $e_1 > e_2$. Reverse one of them (into a heat pump) and use the work output from the engine to run the heat pump. What happens? (If it seems fine at first, switch the two.)

7. When supplies of fossil fuels such as petroleum and coal dwindle, people might call the situation an "energy crisis." From the standpoint of physics, why is that not an accurate name? Can you think of a better one?

8. If you leave the refrigerator door open and the refrigerator runs continuously, does the kitchen get colder or warmer? Explain.

9. Most heat pumps incorporate an electric heater as a "backup." For relatively mild outdoor temperatures, the electric heater is not used. However, if the outdoor temperature gets very low, the heat pump shuts off and the heater is used instead. Why?

10. Why are heat pumps more often used in mild climates than in areas with severely cold winters?

11. Are entropy changes always caused by the flow of heat? If not, give some other examples of processes that increase entropy.

12. Can a heat engine be made to operate without creating any "thermal pollution," that is, without making its cold reservoir get warmer in the long run? The net work output must be greater than zero.

13. A warm pitcher of lemonade is put into an ice chest. Describe what happens to the entropies of lemonade and ice as heat flows from the lemonade to the ice within the chest.

MULTIPLE CHOICE QUESTIONS

1. A real heat engine, when run between reservoirs at temperatures of 300°C and 30°C, has efficiencies e_e and e_p in the forward and reverse directions (i.e., as an engine or as a heat pump). Which of these pairs of values e_e, e_p are possible?
   (a) 50%, 56%　　(b) 56%, 50%　　(c) 82%, 95%
   (d) 42%, 57%　　(e) 57%, 42%

2. If two different systems are put in thermal contact, so that heat can flow from one to the other, then heat will flow until the systems have the same
   (a) energy.　　　　(b) heat capacity.
   (c) entropy.　　　　(d) temperature.

3. As moisture from the air condenses on the outside of a cold glass of water, the entropy of the condensing moisture
   (a) stays the same.　　(b) increases.
   (c) decreases.　　　　(d) not enough information.

4. As a system undergoes a constant volume process
   (a) the pressure does not change.
   (b) the internal energy does not change.
   (c) the work done on or by the system is zero.
   (d) the entropy stays the same.
   (e) the temperature of the system does not change.

5. Which of these statements are implied by the second law of thermodynamics?
   (a) The entropy of an engine (including its fuel and/or heat reservoirs) operating in a cycle never decreases.
   (b) The increase in internal energy of a system in any process is the sum of heat absorbed plus work done on the system.

   (c) A heat engine, operating in a cycle, that rejects no heat to the high temperature reservoir is impossible.
   (d) Both (a) and (c).　　　　(e) All three [(a), (b), and (c)].

6. On a summer day, you keep the air conditioner in your room running. From the list numbered 1 to 4 choose the hot reservoir and the cold reservoir.
   1. the air outside
   2. the compartment inside the air conditioner where the air is compressed
   3. the freon gas that is the working substance (expands and compresses in each cycle)
   4. the air in the room
   (a) 1 is hot, 2 is cold reservoir.　　(b) 1 is hot, 3 is cold reservoir.
   (c) 1 is hot, 4 is cold reservoir.　　(d) 2 is hot, 3 is cold reservoir.
   (e) 2 is hot, 4 is cold reservoir.　　(f) 3 is hot, 4 is cold reservoir.

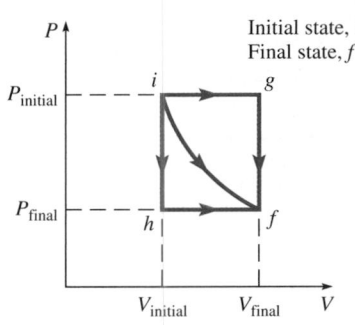

Figure 15.18
Multiple Choice Question 7

7. Figure 15.18 shows a *PV* diagram with several paths to get from an initial to a final state. For which path does the system do the most work?

   (a) path *igf*   (b) path *if*
   (c) path *ihf*   (d) All paths represent equal work.

8. An ideal gas is confined to the left chamber of an insulated container. The right chamber is evacuated. A valve is opened between the chambers, allowing gas to flow into the right chamber. After equilibrium is established, the temperature of the gas ____. [*Hint:* What happens to the internal energy?]

   (a) is lower than the initial temperature
   (b) is higher than the initial temperature
   (c) is the same as the initial temperature
   (d) could be higher than, the same as, or lower than the initial temperature

9. In the first law of thermodynamics ($\Delta U = Q - W$), the variables Q and W stand for

   (a) the heat flow *out of* the system and the work done *on* the system.
   (b) the heat flow *out of* the system and the work done *by* the system.
   (c) the heat flow *into* the system and the work done *by* the system.
   (d) the heat flow *into* the system and the work done *on* the system.

10. As an ideal gas is adiabatically expanding,

    (a) the temperature of the gas does not change.
    (b) the internal energy of the gas does not change.

    (c) work is not done on or by the gas.
    (d) no heat is given off or taken in by the gas.
    (e) both (a) and (d).
    (f) both (a) and (b).

11. As an ideal gas is compressed at constant temperature,

    (a) heat flows out of the gas.
    (b) the internal energy of the gas does not change.
    (c) the work done on the gas is zero.
    (d) none of the above.
    (e) both (a) and (b).
    (f) both (a) and (c).

12. Given 1 mole of an ideal gas, in a state characterized by P_A, V_A, a change occurs so that the final pressure and volume are equal to P_B, V_B, where $V_B > V_A$. Which of these is true?

    (a) The heat supplied to the gas during the process is completely determined by the values P_A, V_A, P_B, and V_B.
    (b) The change in the internal energy of the gas during the process is completely determined by the values P_A, V_A, P_B, and V_B.
    (c) The work done by the gas during the process is completely determined by the values P_A, V_A, P_B, and V_B.
    (d) All three are true.
    (e) None of these is true.

13. Which choice correctly identifies the three processes shown in Fig. 15.19?

    (a) 1 = isobaric; 2 = isochoric; 3 = adiabatic
    (b) 1 = isothermal; 2 = isothermal; 3 = isobaric
    (c) 1 = isochoric; 2 = adiabatic; 3 = isobaric
    (d) 1 = isobaric; 2 = isothermal; 3 = isochoric

(1)

(2)

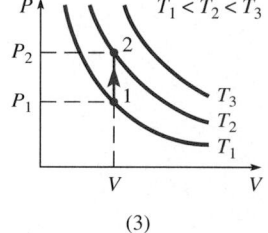
(3)

Figure 15.19 Three processes illustrated on *PV* diagrams. (Multiple Choice Question 13)

PROBLEMS

Note: 🄒 indicates a combination conceptual/quantitative problem. Gold diamonds ✦, ✦✦ are used to indicate the increasing level of difficulty of each problem. Problem numbers appearing in blue, 9., denote problems that have a detailed solution available in the Student Solutions Manual. Some problems are *paired* by concept; their numbers are connected by a ruled box.

15.1 The First Law of Thermodynamics;
15.2 Thermodynamic Processes;
15.3 Constant Pressure Expansion of an Ideal Gas

1. A system takes in 550 J of heat while performing 840 J of work. What is the change in internal energy of the system?

2. The internal energy of a system increases by 400 J while 500 J of work are performed on it. What was the heat flow into or out of the system?

3. A model steam engine of 1.00 kg mass pulls eight cars of 1.00-kg mass each. The cars start at rest and reach a velocity of 3.00 m/s in a time of 3.00 s while moving a distance of 4.50 m. During that time the engine takes in 135 J of heat. What is the change in the internal energy of the engine?

4. A monatomic ideal gas at 27°C undergoes a constant pressure process from *A* to *B* and a constant volume process from *B* to *C*, as shown in Fig. 15.20a. Find the total work done during these two processes.

5. A monatomic ideal gas at 27°C undergoes a constant volume process from *A* to *B* and a constant pressure process from *B* to *C*, as shown in Fig. 15.20b. Find the total work done during these two processes.

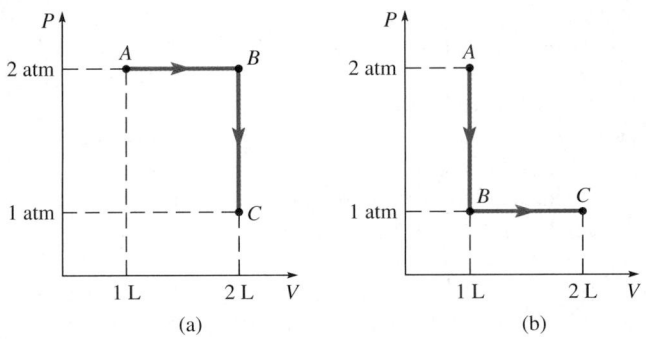

Figure 15.20   Problems 4 and 5

6. Suppose a monatomic ideal gas is changed from state A to state D by one of the processes shown on the PV diagram of Fig. 15.21. (a) Find the total work done by the gas if it follows the constant volume path A–B followed by the constant pressure path B–C–D. (b) Calculate the total change in internal energy of the gas during the entire process and the total heat flow into the gas.

7. Repeat Problem 6 for the case when the gas follows the constant temperature path A–C followed by the constant pressure path C–D.

8. Repeat Problem 6 for the case when the gas follows the constant pressure path A–E followed by the constant temperature path E–D.

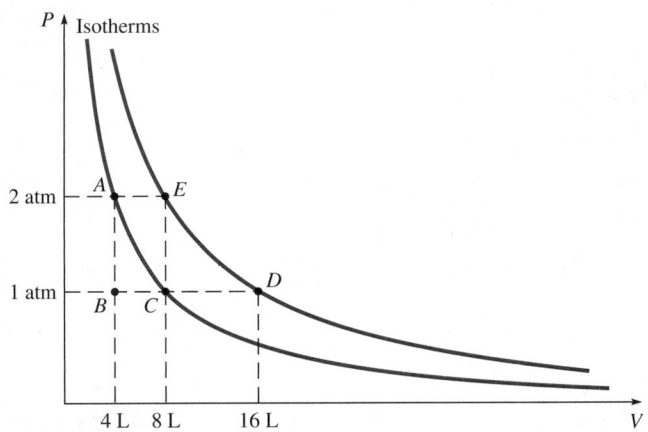

Figure 15.21   Problems 6–8

9. An ideal gas is in contact with a heat reservoir so that it remains at a constant temperature of 300.0 K. The gas is compressed from a volume of 24.0 L to a volume of 14.0 L. During the process the mechanical device pushing the piston to compress the gas is found to expend 5.00 kJ of energy. How much heat flows between the heat reservoir and the gas and in what direction does the heat flow occur?

10. Suppose 1.00 mol of oxygen is heated at constant pressure of 1.00 atm from 10.0°C to 25.0°C. (a) How much heat is absorbed by the gas? (b) Using the ideal gas law, calculate the change of volume of the gas in this process. (c) What is the work done by the gas during this expansion? (d) From the first law, calculate the change of internal energy of the gas in this process.

15.5 Heat Engines; 15.6 Refrigerators and Heat Pumps

11. (a) How much heat does an engine with an efficiency of 33.3% absorb in order to deliver 1.00 kJ of work? (b) How much heat is exhausted by the engine?

12. The efficiency of an engine is 0.21. For every 1.00 kJ of heat absorbed by the engine, how much (a) net work is done by it and (b) heat is released by it?

13. The United States generates about 5.0×10^{16} J of electric energy a day. This energy is equivalent to work, since it can be converted into work with almost 100% efficiency by an electric motor. (a) If this energy is generated by power plants with an average efficiency of 0.30, how much heat is dumped into the environment each day? (b) How much water would be required to absorb this heat if the water temperature is not to increase more than 2.0°C?

14. The intensity (power per unit area) of the sunlight incident on the Earth's surface, averaged over a 24-h period, is about 0.20 kW/m^2. If a solar power plant is to be built with an output capacity of 1.0×10^9 W, how big must the area of the solar energy collectors be for photocells operating at 20.0% efficiency?

15. An engine releases 0.450 kJ of heat for every 0.100 kJ of work it does. What is the efficiency of the engine?

16. An engine works at 30.0% efficiency. The engine raises a 5.00-kg crate from rest to a vertical height of 10.0 m, at which point the crate has a speed of 4.00 m/s. How much heat input is required for this engine?

17. How much heat does a heat pump with an inefficiency of 33.3% deliver when supplied with 1.00 kJ of electricity?

18. An air conditioner whose coefficient of performance is 2.00 removes 1.73×10^8 J of heat from a room per day. How much does it cost to run the air conditioning unit per day if electricity costs $0.10 per kilowatt-hour? (Note that 1 kilowatt-hour = 3.6×10^6 J.)

15.7 Reversible Engines and Heat Pumps; 15.8 Carnot Cycle

19. In a certain steam engine, the boiler temperature is 127°C and the cold reservoir temperature is 27°C. While this engine does 2.0 kcal of work, what minimum amount of heat must be discharged into the cold reservoir?

20. Calculate the maximum possible efficiency of a heat engine that uses surface lake water at 18.0°C as a source of heat and rejects waste heat to the water 0.100 km below the surface where the temperature is 4.0°C.

21. An ideal refrigerator removes heat at a rate of 0.10 kW from its interior (+2.0°C) and exhausts heat at 40.0°C. How much electrical power is used?

22. A heat pump is used to heat a house with an interior temperature of 20.0°C. On a chilly day with an outdoor temperature of −10.0°C, what is the minimum work that the pump requires in order to deliver 1.0 kJ of heat to the house?

23. A coal-fired electrical generating station can use a higher T_H than a nuclear plant; for safety reasons the core of a nuclear reactor is not allowed to get as hot as coal. Suppose that $T_H = 727$°C for a coal station but $T_H = 527$°C for a nuclear station. Both power plants exhaust waste heat into a lake at $T_C = 27$°C. How much waste heat does each plant

exhaust into the lake to produce 1.00 MJ of electricity? Assume both operate as reversible engines.

24. (a) Calculate the efficiency of a reversible engine that operates between the temperatures 600.0°C and 300.0°C. (b) If the engine absorbs 100.0 kcal of heat from the hot reservoir, how much does it exhaust to the cold reservoir?

25. A reversible engine with a 30.0% efficiency has $T_C = 310.0$ K. (a) What is T_H? (b) How much heat is exhausted for every 0.100 kJ of work done?

26. Show that in a reversible engine the amount of heat Q_C is related to the work done W by

$$Q_C = -\frac{T_C}{T_H - T_C} W$$

27. An electrical power station generates steam at 500.0°C and condenses it with river water at 27°C. By how much would its theoretical maximum efficiency decrease if it had to switch to cooling towers that condense the steam at 47°C?

28. An ideal refrigerator keeps its contents at 0.0°C and exhausts heat into the kitchen at 40.0°C. For every 1.0 kJ of work done, (a) how much heat is exhausted? (b) How much heat is removed from the contents?

29. The outdoor temperature on a winter's day is –4°C. If you use 1.0 kJ of electrical energy to run a heat pump, how much heat does that put into your house at 21°C? Assume that the heat pump is reversible.

30. An oil-burning electric power plant uses steam at 773 K to drive a turbine, after which the steam is expelled at 373 K. The engine has an efficiency of 0.40. What is the theoretical maximum efficiency possible at those temperatures?

31. An inventor proposes a heat engine to propel a ship, using the temperature difference between the water at the surface and the water 10 m below the surface as the two reservoirs. If these temperatures are 15.0°C and 10.0°C, respectively, what is the maximum possible efficiency of the engine?

32. A heat engine uses the warm air at the ground as the hot reservoir and the cooler air at an altitude of several thousand meters as the cold reservoir. If the warm air is at 37°C and the cold air is at 25°C, what is the maximum possible efficiency for the engine?

33. A reversible refrigerator has an inefficiency of 0.25. How much work must be done to freeze 1.0 kg of liquid water initially at 0°C?

✦ 34. On a hot day, you are in a sealed, insulated room. The room contains a refrigerator, operated by an electric motor. The motor does work at the rate of 250 W when it is running. Assume the motor is ideal (no friction or electrical resistance) and that the refrigerator operates on a reversible cycle. In an effort to cool the room, you turn on the refrigerator and open its door. Let the temperature in the room be 320 K when this process starts, and the temperature in the cold compartment of the refrigerator be 256 K. At what net rate is heat added to (+) or subtracted from (–) the room and all of its contents?

✦✦ 35. Imagine that a car engine could be replaced by a Carnot engine with an ideal gas as the working substance. When the car is traveling at 65 miles per hour, the Carnot engine goes through its cycle 900.0 times per minute. The engine's hot reservoir is at 1000.0°C (the temperature of the exploding gas in a real car engine) and the cold reservoir is at 20.0°C (the outside temperature). During the isothermal expansion

part of each cycle, the volume of the ideal gas increases by a factor of 10.0. The cylinders contain 0.223 mol of gas. What is the power output of the engine?

15.9 Entropy

36. List these in order of increasing entropy: (a) 0.5 kg of ice and 0.5 kg of (liquid) water at 0°C; (b) 1 kg of ice at 0°C; (c) 1 kg of (liquid) water at 0°C; (d) 1 kg of water at 20°C.

🄖 37. List these in order of increasing entropy: (a) 0.01 mol of N_2 gas in a 1-L container at 0°C; (b) 0.01 mol of N_2 gas in a 2-L container at 0°C; (c) 0.01 mol of liquid N_2.

38. An ice cube at 0.0°C is slowly melting. What is the change in the ice cube's entropy for each 1.00 g of ice that melts?

39. From Table 14.4 we know that approximately 540 kcal are needed to transform 1 kg of water at 100°C to steam at 100°C. What is the change in entropy of 1.0 kg of water evaporating at 100.0°C? (Specify whether the change in entropy is an increase, +, or a decrease, –.)

40. A large block of copper initially at 20.0°C is placed in a vat of hot water (80.0°C). For the first 1.0 J of heat that flows from the water into the block, find (a) the entropy change of the block, (b) the entropy change of the water, and (c) the entropy change of the universe. Note that the temperatures of the block and water are essentially unchanged by the flow of only 1.0 J of heat.

41. A large, cold (0.0°C) block of iron is immersed in a tub of hot (100.0°C) water. In the first 10.0 s, 10.0 kcal of heat are transferred, although the temperatures of the water and the iron do not change much in this time. Calculate the change in entropy of the system in this time.

42. On a cold winter day, the outside temperature is –15.0°C. Inside the house the temperature is +20.0°C. Heat flows out of the house through a window at a rate of 220.0 W. At what rate is the entropy of the universe changing due to this heat conduction through the window?

43. Within an insulated system, 100.0 kcal of heat is conducted through a copper rod from a hot reservoir at +200.0°C to a cold reservoir at +100.0°C. (The reservoirs are so big that this heat exchange does not change their temperatures appreciably.) What is the net change in entropy of the system, in kcal/K?

44. Plot the temperature versus entropy for the four stages of the Carnot engine discussed in Example 15.9. [*Hint*: First plot the constant temperature stages and then fill in the adiabatic stages.]

45. A student eats 2000 kcal per day. (a) Assuming that all of the food energy is released as heat, what is the rate of heat released (in watts)? (b) What is the rate of change of entropy of the surroundings if all of the heat is released into air at room temperature (20°C)?

15.10 Statistical Interpretation of Entropy

46. Suppose the macrostate of a system of 100 coins is specified by the number of heads. What is the entropy of the state with one head (in terms of Boltzmann's constant, k_B)?

47. For a system composed of two dice, let the macrostate be defined by the sum of the numbers showing on the top faces. What is the maximum entropy of this system in units of Boltzmann's constant, k_B?

◆48. Two dice are thrown. A macrostate is specified by the sum of the two numbers that come up on the dice. (a) How is a microstate specified for this system? (b) How many different microstates are there? (c) How many different macrostates are there? (d) What is the most probable macrostate? (e) What is the probability of getting this result? (f) What is the probability of rolling "snake eyes" (two 1s)?

◆49. Six coins are tossed simultaneously. The macrostate is specified by the number of "heads." (a) What is/are the most probable macrostate(s)? (b) What is/are the least probable macrostate(s)? (c) What is the probability of obtaining the most probable macrostate?

50. If 1.0-g of ice at 0.0°C melts into liquid water at 0.0°C, by what factor has the number of microstates increased?

51. If the number of microstates for a thermodynamic system doubles, how much has the system's entropy increased?

◆52. Four indistinguishable marbles must be placed into two distinguishable boxes. (a) How many microstates are there? (b) How many macrostates? (c) What is the most probable macrostate? (d) What is the entropy of that macrostate? (e) What is the least probable macrostate? (f) What is the entropy of the least probable macrostate?

COMPREHENSIVE PROBLEMS

53. The motor that drives a reversible refrigerator produces 148 W of useful power. The hot and cold temperatures of the heat reservoirs are 20.0°C and –5.0°C. What is the maximum amount of ice it can produce in 2.0 h from water that is initially at 8.0°C?

54. An engineer designs a ship that gets its power in the following way. The engine draws in warm water from the ocean, and after extracting some of the water's internal energy, returns the water to the ocean at a temperature 14.5°C lower than the ocean temperature. If the ocean is at a uniform temperature of 17°C, is this an efficient engine? Will the engineers design work?

55. A balloon contains 200.0 L of nitrogen gas at 20.0°C and at atmospheric pressure. How much energy must be added to raise the temperature of the nitrogen to 40.0°C while allowing the balloon to expand at atmospheric pressure?

56. An ideal gas is heated at a constant pressure of 2.0×10^5 Pa from a temperature of –73°C to a temperature of +27°C. The initial volume of the gas is 0.10 m^3. The heat energy supplied to the gas in this process is 25 kJ. What is the increase in internal energy of the gas?

57. For a reversible engine, will you obtain a better efficiency by increasing the high-temperature reservoir by an amount ΔT or decreasing the low-temperature reservoir by the same amount ΔT?

58. A 0.50-kg block of iron [$c = 0.44$ kJ/(kg·K)] at 20.0°C is in contact with a 0.50-kg block of aluminum [$c = 0.900$ kJ/(kg·K)] at a temperature of 20.0°C. The system is completely isolated from the rest of the universe. Suppose heat flows from the iron into the aluminum until the temperature of the aluminum is 22.0°C. (a) From the first law, calculate the final temperature of the iron. (b) Estimate the entropy change of the system. (c) Explain how the result of part (b) shows that this process is impossible. [*Hint*: Since the system is isolated, $\Delta S_{\text{System}} = \Delta S_{\text{Universe}}$.]

59. List these in order of increasing entropy: (a) 1 mol of water at 20°C and 1 mol of ethanol at 20°C in separate containers; (b) a mixture of 1 mol of water at 20°C and 1 mol of ethanol at

20°C; (c) 0.5 mol of water at 20°C and 0.5 mol of ethanol at 20°C in separate containers; (d) a mixture of 1 mol of water at 30°C and 1 mol of ethanol at 30°C.

60. List these in order of increasing entropy: (a) 1000 He atoms moving at random velocities with an average speed of 400 m/s; (b) 1000 He atoms all moving at 400 m/s in the same direction; (c) 1000 He atoms all moving at 400 m/s in random directions.

61. If the pressure on a fish increases from 1.1 to 1.2 atm, its swim bladder decreases in volume from 8.16 mL to 7.48 mL while the temperature of the air inside remains constant. How much work is done on the air in the bladder?

◆62. A fish at a pressure of 1.1 atm has its swim bladder inflated to an initial volume of 8.16 mL. If the fish starts swimming horizontally, its temperature increases from 20.0°C to 22.0°C as a result of the exertion. (a) Since the fish is still at the same pressure, how much work is done by the air in the swim bladder? [*Hint*: First find the new volume from the temperature change.] (b) How much heat is gained by the air in the swim bladder? Assume air to be a diatomic ideal gas. (c) If this quantity of heat is lost by the fish, by how much will its temperature decrease? The fish has a mass of 5.00 g and its specific heat is about 3.5 J/(g·°C).

63. Suppose you mix 4.0 mol of a monatomic gas at 20.0°C and 3.0 moles of another monatomic gas at 30.0°C. If the mixture is allowed to reach equilibrium, what is the final temperature of the mixture? [*Hint*: Use energy conservation.]

64. A balloon contains 160 L of nitrogen gas at 25°C and 1.0 atm. How much energy must be added to raise the temperature of the nitrogen to 45°C while allowing the balloon to expand at atmospheric pressure?

65. The efficiency of a muscle during weight lifting is equal to the work done in lifting the weight divided by the total energy output of the muscle (work plus internal energy dissipated in the muscle). Let $g = 9.80 \text{ m/s}^2$. Fill in this table to determine the efficiency of the muscle:

Table for Problem 65				
Mass Lifted (g)	**Height Raised (cm)**	**Internal Energy Dissipated (J)**	**Work Done (J)**	**Efficiency (%)**
16.4	0.577	1.39×10^{-3}		
8.09	0.745	1.64×10^{-3}		
1.05	0.880	1.85×10^{-3}		

66. (a) What is the entropy change of 1.00 mol of H_2O when it changes from ice to water at 0.0°C? (b) If the ice is in contact with an environment at a temperature of 10.0°C, what is the entropy change of the universe when the ice melts?

67. Estimate the entropy change of 850 g of water when it is heated from 20.0°C to 50.0°C. [*Hint*: Assume that the heat flows into the water at an average temperature.]

68. For a more realistic estimate of the maximum coefficient of performance of a heat pump, assume that a heat pump takes in heat from outdoors at *10°C below* the ambient outdoor temperature, to account for the temperature difference across its heat exchanger. Similarly, assume that the output must be *10°C hotter* than the house (which itself might be kept at 20°C), to make the heat flow into the house. Make a graph of the coefficient of performance of a reversible heat pump under these conditions as a function of outdoor temperature (from −15°C to +15°C in 5°C increments).

69. A 0.500-kg block of iron at 60.0°C is placed in contact with a 0.500-kg block of iron at 20.0°C. (a) The blocks soon come to a common temperature of 40.0°C. *Estimate* the entropy change of the universe when this occurs. [*Hint*: Assume that all the heat flow occurs at an average temperature for each block.] (b) Estimate the entropy change of the universe if, instead, the temperature of the hotter block increased to 80.0°C while the temperature of the colder block decreased to 0.0°C. [*Hint*: The answer is negative, indicating that the process is impossible.]

70. A container holding 1.20 kg of water at 20.0°C is placed in a freezer that is kept at −20.0°C. The water freezes and comes to thermal equilibrium with the interior of the freezer. What is the minimum amount of electrical energy required by the freezer to do this if it operates between reservoirs at temperatures of 20.0°C and −20.0°C?

71. A reversible heat engine has an efficiency of 33.3%, removing heat from a hot reservoir and rejecting heat to a cold reservoir at 0°C. If the engine now operates in reverse, how long would it take to freeze 1.0 kg of water at 0°C, if it operates on a power of 186 W?

✦72. Consider a heat engine that is *not* reversible. It has the *PV* cycle shown in Fig. 15.22. In the first stage (A) there is a constant temperature expansion while in contact with a warm reservoir at 373 K from $P_1 = 1.55 \times 10^5$ Pa and $V_1 = 2.00 \times 10^{-2}$ m^3 to $P_2 = 1.24 \times 10^5$ Pa and $V_2 = 2.50 \times 10^{-2}$ m^3. Then (B) a heat reservoir at the cooler temperature of 273 K is used to cool the gas at constant volume to 273 K from P_2 to P_3 = 0.91 × 10^5 Pa. This is followed by (C) a constant temperature compression while still in contact with the cold reservoir at 273 K from P_3, V_2 to $P_4 = 1.01 \times 10^5$ Pa, V_1. The final stage (D)

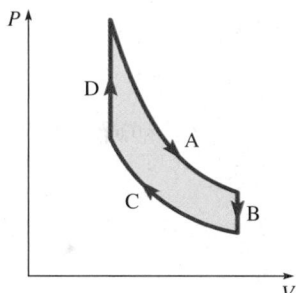

Figure 15.22 Problem 72

is heating the gas at constant volume from 273 K to 373 K by being in contact with the warm reservoir again, to return from P_4, V_1 to P_1, V_1. The engine uses 1.000 mol of a diatomic gas. (a) Fill out this chart, showing for each step in the cycle the work done, the heat input, and the change in internal energy.

Stage	W (Work Done)	ΔU (Change in Internal Energy)	Q (Net Heat Input)
A			
B			
C			
D			
Cycle ABCD			

(b) Find the efficiency of this engine. (c) Compare to the efficiency of a reversible engine.

73. (a) For the gas cycle in Problem 72 find the change in entropy of the cold reservoir in step B. Remember that the gas is always in contact with the cold reservoir. (b) What is the change in entropy of the hot reservoir in step D? (c) Using this information, find the change in entropy of the total system of gas plus reservoirs during the whole cycle.

✦74. A town is considering using its lake as a source of power. The average temperature difference from the top to the bottom is 15°C, and the average surface temperature is 22°C. (a) Assuming that the town can set up a reversible engine using the surface and bottom of the lake as heat reservoirs, what would be its efficiency? (b) If the town needs about 1.0×10^8 W of power to be supplied by the lake, how many m^3 of water does the heat engine use per second? (c) The surface area of the lake is 8.0×10^7 m^2 and the average incident intensity (over 24 h) of the sunlight is 200 W/m^2. Can the lake supply enough heat to meet the town's energy needs with this method?

ANSWERS TO PRACTICE PROBLEMS

15.1 Work was done on the gas; −28 kJ. **15.2** Heat flows into the gas; Q = 3.8 kJ **15.3** 359 kJ **15.4** The fire is irreversible: smoke and carbon dioxide and ash will not come together to form logs and twigs. The wood is in a highly ordered state, whereas the products of the fire are much less ordered. **15.5** 20% **15.6** 4 kW

15.7 1200 K **15.8** 178 MW **15.9** The adiabatic processes are needed to change the temperature of the working substance in the engine back and forth between T_C and T_H. **15.10** 0.03 J/K **15.11** +4.52k

Electric Forces and Fields

The elegant fish in the photograph is the *Gymnarchus niloticus*, a native of Africa found in the Nile River. *Gymnarchus* has some interesting traits. It swims gracefully with equal facility either forward or backward. Instead of propelling itself by lashing its tail sideways, as most fish do, it keeps its spine straight—not only when swimming straight ahead, but even when turning. Its propulsion is accomplished by means of the undulations of the fin along its back.

Gymnarchus navigates with great precision, darting after its prey and evading obstacles in its path. What is surprising is that it does so just as precisely when swimming backward. Furthermore, *Gymnarchus* is nearly blind; its eyes respond only to extremely bright light. How then, is it able to locate its prey in the dim light of a muddy river?

PART THREE   Electromagnetism

Figure 16.1 Jewelry made from amber. Note the insect trapped inside the amber.

Net charge = the algebraic sum of all the charges in a system

16.1 ELECTRIC CHARGE

In Part Three of this book, we study electric and magnetic fields in detail. Recall from Chapter 2 that all interactions in the universe fall into one of four categories: gravitational, electromagnetic, strong, and weak. All of the familiar, everyday forces other than gravity—contact forces, tension in cables, and the like—are fundamentally electromagnetic. What we think of as a single interaction is really the net effect of huge numbers of microscopic interactions between electrons and atoms. Electromagnetic forces bind electrons to nuclei to form atoms and molecules. They hold atoms together to form liquids and solids, from skyscrapers to trees to human bodies. Technological applications of electromagnetism abound, especially once we realize that radio waves, microwaves, light, and other forms of electromagnetic radiation consist of oscillating electric and magnetic fields.

Many everyday manifestations of electromagnetism are complex; hence we study simpler situations in order to gain some insight into how electromagnetism works. The hybrid word *electromagnetism* itself shows that electricity and magnetism, which were once thought to be completely separate forces, are really aspects of the same fundamental interaction. This unification of the studies of electricity and magnetism occurred in the late nineteenth century. However, understanding comes more easily if we first tackle electricity (Chapters 16–18), then magnetism (Chapter 19), and finally see how they are closely related (Chapters 20–22).

The existence of electrical forces has been familiar to humans for at least 3000 years. The ancient Greeks used pieces of amber—a hard, fossilized form of the sap from pine trees—to make jewelry (Fig. 16.1). When a piece of amber was polished by rubbing it with a piece of fabric, it was observed that the amber would subsequently attract small objects, such as bits of string or hair. Using modern knowledge, we say that the amber is *charged* by rubbing: some electric charge is transferred between the amber and the cloth. Our word *electric* comes from the Greek word for amber (*elektron*).

A similar phenomenon occurs on a dry day when you walk across a carpeted room wearing rubber-soled shoes. Charge is transferred between the carpet and your shoes and between your shoes and your body. Some of the charge you have accumulated may be unintentionally transferred from your fingertips to a doorknob or to a friend—accompanied by the sensation of a shock.

Types of Charge

Electric charge is not *created* by these processes; it is just transferred from one object to another. The law of **conservation of charge** is one of the fundamental laws of physics; no exceptions to it have ever been found.

> **Conservation of charge**:
> The net charge of a closed system never changes.

Experiments with amber and other materials that can be charged show that there are two types of charge; one type is called *positive* (+) and the other *negative* (−). The **net charge** of a system is the algebraic sum—taking care to include the positive and negative signs—of the charges of the constituent particles in the system. When a piece of glass is rubbed by silk, the glass acquires a positive charge and the silk a negative charge; the net charge of the system of glass and silk does not change. An object that is **electrically neutral** has equal amounts of positive and negative charge and thus a net charge of zero. The symbols used for quantity of charge are q or Q.

Ordinary matter consists of atoms, which in turn consist of electrons, protons, and neutrons. The protons and neutrons are called *nucleons* because they are found in the nucleus. The neutron is electrically neutral (thus the name *neutron*). The charges on the proton and the electron are of *equal magnitude* but of opposite sign. The charge on the proton is arbitrarily chosen to be positive; that on the electron is therefore negative. A neutral atom has equal numbers of protons and electrons, a balance of positive and negative charge. If the number of electrons and protons is not equal, then the atom is called an *ion* and has a nonzero net charge. If the ion has more electrons than protons, its net charge is negative; if the atom has fewer electrons than protons, its net charge is positive.

Elementary Charge

The *magnitude* of charge on the proton and electron is the same (see Table 16.1). That amount of charge is called the **elementary charge** (symbol e). In terms of the SI unit of charge, the coulomb (C), the value of e is

$$e = 1.602 \times 10^{-19} \text{ C} \tag{16-1}$$

Since ordinary objects have only slight imbalances between positive and negative charge, the coulomb is often an inconveniently large unit. One would have to remove 6.25×10^{18} electrons from an object to give it a net charge of +1 C. For this reason charges are often given in millicoulombs (mC), microcoulombs (μC), nanocoulombs (nC), or picocoulombs (pC). The coulomb is named after the French physicist Charles Coulomb (1736–1806), who developed the expression for the electric force between two charged particles.

The net charge of any object is an integral multiple of the elementary charge. Even in the extraordinary matter found in exotic places like the interior of stars, the upper atmosphere, or in particle accelerators, the observable charge is always in units of e.

Table 16.1

Masses and Electrical Charges of the Proton, Electron, and Neutron

Particle	Mass	Electrical Charge
Proton	$m_p = 1.673 \times 10^{-27}$ kg	$q_p = +e = +1.602 \times 10^{-19}$ C
Electron	$m_e = 9.109 \times 10^{-31}$ kg	$q_e = -e = -1.602 \times 10^{-19}$ C
Neutron	$m_n = 1.675 \times 10^{-27}$ kg	$q_n = 0$

Example 16.1

An Unintentional Shock

 The magnitude of charge transferred when you walk across a carpet, reach out to shake hands, and unintentionally give a shock to a friend might be typically about 1 nC. (a) If the charge is transferred by electrons only, how many electrons are transferred? (b) If your body has a net charge of –1 nC, estimate the percentage of excess electrons. [*Hint:* The mass of the electron is only about 1/2000 that of a nucleon, so most of the mass of the body is in the nucleons. Many of the atoms present in the body (such as C, N, O) have roughly equal numbers of protons and neutrons. An important exception is hydrogen, which has one proton and no neutrons in its most common form. Since the problem involves an order-of-magnitude calculation, there is no need for great precision; we can just assume that $\frac{1}{2}$ of the nucleons are protons.]

Strategy Since the coulomb (C) is the SI unit of charge, the "n" must be the prefix "nano-" (= 10^{-9}). We know the size of the elementary charge in coulombs. For part (b), we first make an order-of-magnitude estimate of the number of electrons in the human body.

Solution (a) The number of electrons transferred is the quantity of charge transferred divided by the charge of each electron:

$$\frac{-1 \times 10^{-9} \text{ C}}{-1.6 \times 10^{-19} \text{ C per electron}} = 6 \times 10^9 \text{ electrons}$$

Notice that the magnitude of the charge transferred is 1 nC, but since it is transferred by electrons, the sign of the charge transferred is negative.

(b) We estimate a typical body mass of around 70 kg. Most of the mass of the body is in the nucleons, so

$$\text{number of nucleons} = \frac{\text{mass of body}}{\text{mass per nucleon}} = \frac{70 \text{ kg}}{1.7 \times 10^{-27} \text{ kg}}$$

$$= 4 \times 10^{28} \text{ nucleons}$$

continued on next page

Example 16.1 *continued*

Assuming that roughly $\frac{1}{2}$ of the nucleons are protons,

$$\text{number of protons} = \tfrac{1}{2} \times 4 \times 10^{28} = 2 \times 10^{28} \text{ protons}$$

In an electrically neutral object, the number of electrons is equal to the number of protons. With a net charge of −1 nC, the body has 6×10^9 extra electrons. The percentage of excess electrons is then

$$\frac{6 \times 10^9}{2 \times 10^{28}} \times 100\% = (3 \times 10^{-17})\%$$

Discussion   As shown in this example, charged macroscopic objects have *tiny* differences between the magnitude of the positive charge and the magnitude of the negative charge. For this reason, electrical forces between macroscopic bodies are often negligible.

Practice Problem 16.1   Excess electrons on a balloon

How many excess electrons are found on a balloon with a net charge of −12 nC?

In Section 2.5 we pointed out the close similarities between the gravitational force and the electrical force. One of the important *differences* comes about because charge has either a positive or a negative sign, while mass is always a positive quantity. The gravitational force between two massive bodies is always an attractive force, while the electrical force between two charged particles can be attractive or repulsive depending on the signs of the charges. Two particles with charges of the same sign repel one another, while two particles with charges of opposite sign attract one another. More briefly,

> *Like charges repel one another; unlike charges attract one another.*

Polarization

Polarization = charge separation within an object

An electrically neutral object may have regions of positive and negative charge within it, separated from one another. Such an object is **polarized**. A polarized object can experience an electric force even though its net charge is zero. A rubber rod charged negatively after being rubbed with fur attracts small bits of paper. So does a glass rod that is *positively* charged after being rubbed with silk (Fig. 16.2). The bits of paper are electrically neutral, but a charged rod polarizes the paper—it attracts the unlike charge in the paper a bit closer and pushes the like charge in the paper a bit farther away (Fig. 16.2c). The attraction between the rod and the unlike charge then becomes a little stronger than the repulsion between the rod and the like charge, since the electrical force gets weaker as the separation increases and the like charge is farther away. (Recall from Chapter 2 that electrical force between two charged particles is inversely proportional to the square of the distance between them.) Thus, the net force on the paper is always attractive, regardless of the sign of charge on the rod.

In this case, we say that the paper is *polarized by induction;* the polarization of the paper is induced by the charge on the nearby rod. When the rod is moved away, the paper is no longer polarized. Some objects, including some molecules, are intrinsically polarized. An electrically neutral water molecule, for example, has equal amounts of

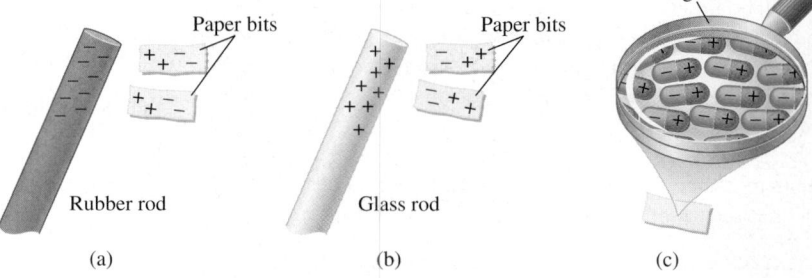

Figure 16.2 (a) Negatively charged rubber rod attracting bits of paper. (b) Positively charged glass rod attracting bits of paper. (c) Magnified view of polarized molecules within a bit of paper.

positive and negative charge (10 protons and 10 electrons), but the center of positive charge and the center of negative charge do not coincide. The electrons in the molecule are shared in such a way that the oxygen end of the molecule has a negative charge, while the hydrogen atoms are positive.

Making The Connection:
polarization of
charge in water

Physics at Home

On a dry day, run a comb through your hair (this works best if your hair is clean and dry and you have not used conditioner) or rub the comb on a wool sweater. When you are sure the comb is charged (by observing the behavior of your hair, listening for crackling sounds, etc.), go to a sink and turn the water on so that a *thin* stream of water comes out. It does not matter if the stream breaks up into droplets near the bottom. Hold the charged comb near the stream of water. You should see that the water experiences a force due to the charge on the comb (Fig. 16.3). Is the force attractive or repulsive? Does this mean that the water coming from the tap has a net charge? Explain your observations.

Figure 16.3   A stream of water is deflected by a charged comb.

16.2   CONDUCTORS AND INSULATORS

Ordinary matter consists of atoms containing electrons and nuclei. The electrons differ greatly in how tightly they are bound to the nucleus. In atoms with many electrons, most of the electrons are tightly bound—under ordinary circumstances nothing can tear them away from the nucleus. Some of the electrons are much more weakly bound and can be removed from the nucleus in one way or another.

Materials vary dramatically in how easy or difficult it is for charge to move within them. Materials in which some charge can move easily are called electrical **conductors**, while materials in which charge does not move easily are called **insulators**.

Metals are materials in which *some* of the electrons are so weakly bound that they are not tied to any one particular nucleus; they are free to wander about within the metal. The *free electrons* in metals make them good conductors. Some metals are better conductors than others, with copper being one of the best. Glass, plastics, rubber, wood, paper, and many other familiar materials are insulators. Insulators do not have free electrons; each electron is bound to a particular nucleus.

The terms *conductor* and *insulator* are applied frequently to electrical wires, which are omnipresent in today's society (Fig. 16.4). The copper wires allow free electrons to flow, conducting an electrical current. The plastic or rubber insulator surrounding the wires keeps the electrical current from leaving the wires (and entering your hand, for instance).

Water is usually thought of as a conductor. It is wise to assume so and take precautions such as not handling electrical devices with wet hands. Actually, *pure* water is an insulator. Pure water consists mostly of complete water molecules (H_2O), which carry no net charge as they move about; there is only a tiny concentration of ions (H^+ and OH^-). But tap water is by no means pure—it contains dissolved minerals. These minerals can

Figure 16.4   Some electrical wires. The metallic conductor is surrounded by insulating material. The insulation must be stripped away where the wire makes an electrical connection with something else.

greatly increase the concentration of ions, making tap water a conductor. The human body contains many ions and therefore is a conductor.

Similarly, air is a good insulator, because most of the molecules in air are electrically neutral, carrying no charge as they move about. However, air does contain some ions; air molecules are ionized by radioactive decays or by cosmic rays.

Intermediate between conductors and insulators are the **semiconductors**. The part of the computer industry clustered in northern California is referred to as "Silicon Valley" because silicon is a common semiconductor used in making computer chips and other electronic devices. *Pure* semiconductors are good insulators, but by *doping* them—adding tiny amounts of impurities in a controlled way—their electrical properties can be fine-tuned.

Charging by Rubbing

When objects are given net charges by rubbing them against one another, both electrons and ions (charged atoms) can be transferred from one object to the other. Charging works best in dry air. When the humidity is high, a film of moisture condenses on the surfaces of objects; charge can then leak off more easily, so it is difficult to build up charge.

Notice that we rub two *insulators* together to separate charge. A piece of metal can be rubbed all day with fur or silk without charging the metal; it is too easy for the charge in the metal to move around and avoid getting scraped off. Once an insulator is charged, the charge remains where it is.

How can a conductor be charged? First rub two insulators together to separate charge; then touch one of the charged insulators to the conductor (Fig. 16.5). Since the charge transferred to the conductor spreads out, the process can be repeated to build up more and more charge on the conductor.

Grounding

How can a conductor be discharged? One way is to *ground* it. The Earth is a conductor because of the presence of metal ions and moisture. The Earth is large enough that for many purposes it may be thought of as a limitless reservoir of charge. The word *reservoir*

(a)

Figure 16.5 Charging a conductor. (a) After rubbing a glass rod with a silk cloth, the glass rod is left with a net positive charge and the silk is left with a net negative charge. (b) Touching the glass rod to a metal sphere. The positively charged glass attracts some of the free electrons from the metal onto the glass. (c) The glass rod is removed. The metal sphere now has fewer electrons than protons, so it has a net positive charge. Even though negative charge is actually transferred (electrons), it is often said that "positive charge is transferred to the metal" since the net effect is the same.

(b)

(c)

is used deliberately to call to mind heat reservoirs. A heat reservoir has such a large heat capacity that it is possible to exchange heat with it without changing its temperature appreciably. To *ground* a conductor means to provide a conducting path between it and the ground (or to another charge reservoir). A charged conductor that is grounded discharges because the charge spreads out by moving off the conductor and onto the Earth.

A buildup of even a relatively small amount of charge on a truck that delivers gasoline could be dangerous—a spark could trigger an explosion. To prevent such a charge buildup, the truck grounds its tank before starting to deliver gasoline to the service station.

The third opening of modern electrical outlets is called *ground*. It is literally connected by a conducting wire to the ground, either through a metal rod driven into the Earth or through underground metal water pipes. The purpose of the ground connection is more fully discussed in the next chapter, but you can understand one purpose already: it prevents static charges from building up on the conductor that is grounded.

Charging by Induction

A conductor is not necessarily discharged when it is grounded if there are other charges nearby. It is even possible to charge an initially neutral conductor by grounding it. In the process shown in Fig. 16.6, the charged insulator never touches the conducting sphere. The positively charged rod first polarizes the sphere, attracting the negative charges on the sphere while repelling the positive charges. The resulting separation of charge on the conducting sphere causes negative charges from the Earth to be attracted along the grounding wire and onto the sphere by the nearby positive charges.

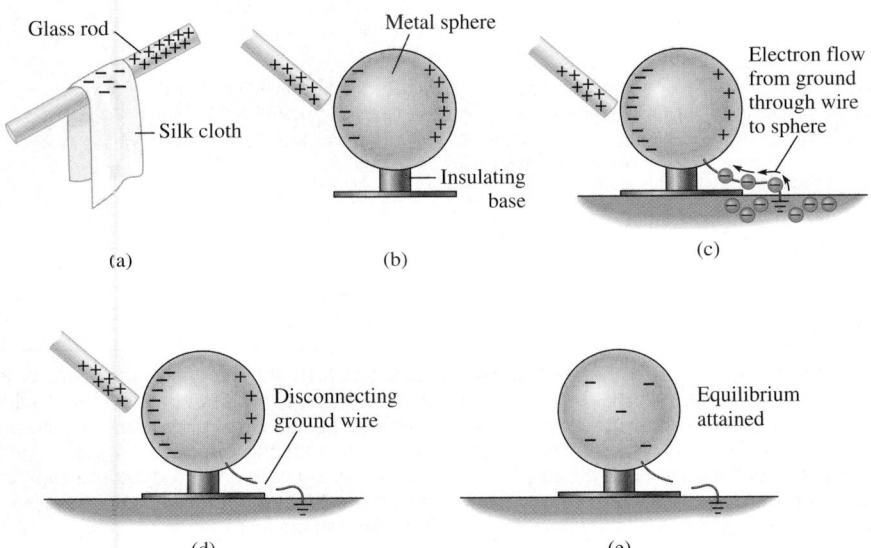

(a)   (b)   (c)

(d)   (e)

The symbol ⏚ represents a connection to ground.

Figure 16.6   Charging by induction. (a) A glass rod is charged by rubbing it with silk. (b) The positively charged glass rod is held near a metal sphere, but does not touch it. The sphere is polarized as free electrons within the sphere are attracted toward the glass rod. (c) When the sphere is grounded, electrons from the ground move onto the sphere, attracted there by positive charges on the sphere. (d) The ground connection is broken without moving the glass rod. (e) Now the glass rod is removed. Charge spreads over the metal surface as the like charges repel each other. The sphere is left with a net negative charge because of the excess electrons.

Conceptual Example 16.2

The Electroscope

An electroscope is charged negatively and the gold foil leaves hang apart as in Fig. 16.7a. What happens to the leaves as the following operations are carried out in the order listed? Explain what you see after each step. (a) You touch the metal bulb at the top of the electroscope with your hand. (b) You bring a glass rod that has been rubbed with silk *near* the bulb without touching it. [*Hint:* A glass rod rubbed with silk is positively charged.] (c) The glass rod touches the metal bulb.

Solution and Discussion   (a) By touching the electroscope bulb with your hand, you ground it. Charge is transferred between your hand and the bulb until the bulb's net charge is zero. Since the electroscope is now discharged, the foil leaves hang down as in Fig. 16.7b. (b) When the positively charged rod is held near the bulb, the electroscope becomes polarized by induction. Negatively charged free electrons are drawn toward the bulb, leaving the foil leaves with a positive net

continued on next page

Conceptual Example 16.2 *continued*

charge (Fig. 16.7c). The leaves hang apart due to the mutual repulsion of the positive charges on them. (c) When the positively charged rod touches the bulb, some negative charge is transferred from the bulb to the rod. The electroscope now has a positive net charge. The glass rod still has a positive net charge that repels the positive charge on the electroscope, pushing it as far away as possible—toward the foil leaves.

The leaves hang farther apart, since they now have *more* positive charge on them than before.

Conceptual Practice Problem 16.2   Removing the glass rod

What happens to the leaves as the glass rod is moved away?

Figure 16.7
An electroscope is a device used to demonstrate the presence of charge. A conducting pole has a metallic bulb at the top and a pair of flexible leaves of gold foil at the bottom. (a) The leaves are pushed apart due to the repulsion of the negative charges. (b) With no net charge, the leaves hang straight down. (c) With a positively charged rod near the bulb, the electroscope has no net charge but it is polarized: the bulb is negative and the leaves are positive. Repulsion between the positive charges on the leaves pushes them apart.

Making The Connection:
electrostatic charge
of adhesive tape

Physics at Home

Ordinary transparent tape has an adhesive that allows it to stick to paper and many other materials. Since the sticking force is electrical in nature, it is not too surprising that adhesive can be used to separate charge. If you have ever peeled a roll of tape too quickly and noticed that the strip of tape curls around and behaves strangely, you have seen effects of this charge separation—the strip of tape has a net charge (and so does the tape left behind, but of opposite sign). Tape pulled *slowly* off a surface does not tend to have a net charge. There are some instructive experiments you can perform:

*   Pull a strip of tape quickly from the roll. How can you tell if the tape has a net charge?
*   Take the roll of tape into a dark closet. What do you see when you pull a strip quickly from the roll?
*   See if the strip is attracted or repelled when you hold it near a paper clip. Explain what you see.
*   Rub the tape on both sides between your thumb and forefinger. Now try the paper clip again. What has happened? Explain.
*   Pull a second strip of tape *slowly* from the roll. Is the force between the two strips attractive or repulsive? What does that tell you?
*   Hold the second strip near the paper clip. Is there a net force? What can you conclude?
*   Can you think of a way of reliably making two strips of tape with like charges? With unlike charges?
*   Enough suggestions—have some fun and see what you can discover!

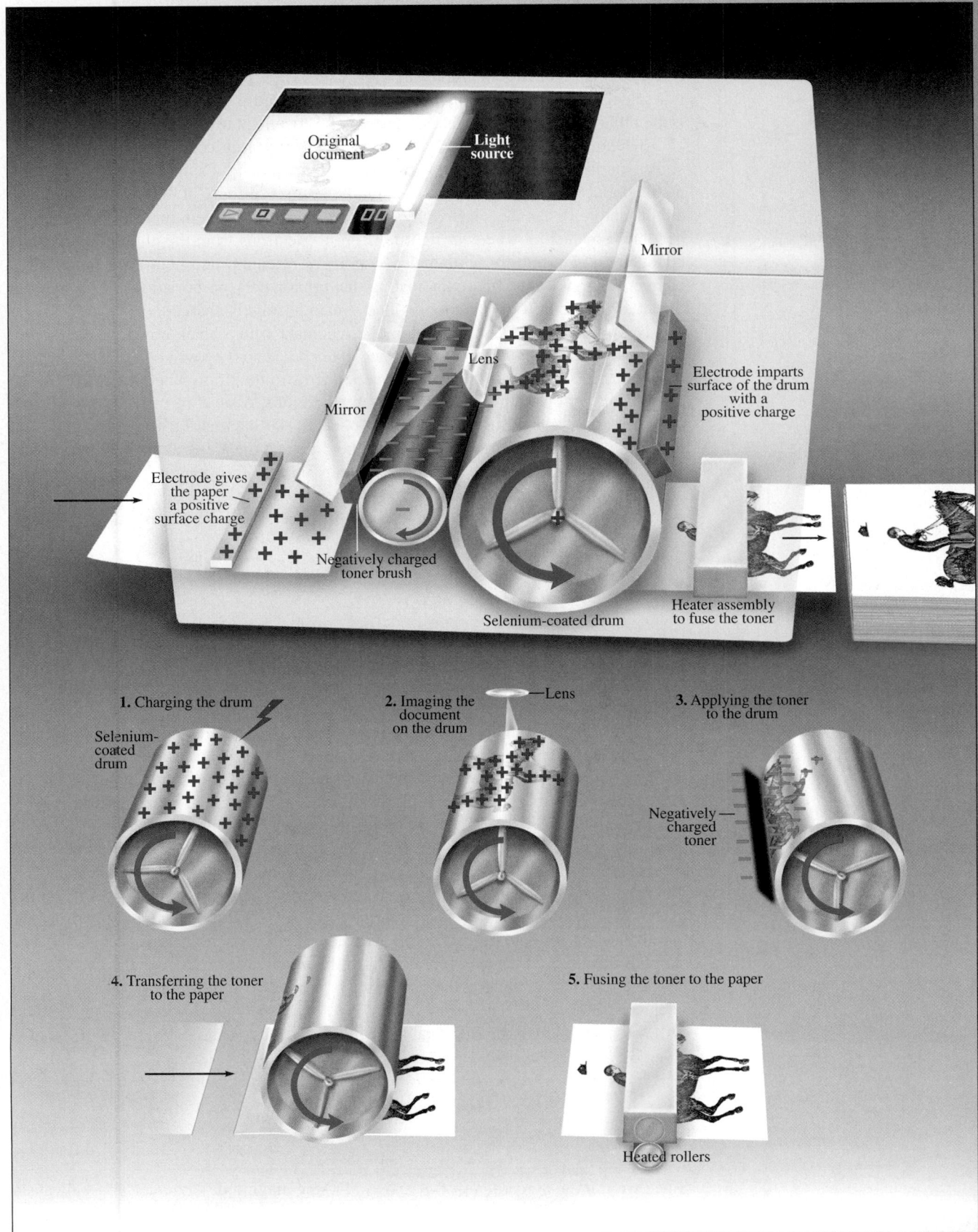

Figure 16.8 The operation of a photocopier is based on the attraction of negatively charged toner particles to regions on the drum that are positively charged.

The Photocopier

The operation of photocopiers (and laser printers) is based on the separation of charge and the attraction between unlike electric charges (Fig. 16.8). Positive charge is applied to a selenium-coated aluminum drum by rotating the drum under an electrode. The drum is then illuminated with a projected image of the document to be copied (or by a laser).

Selenium is a *photoconductor*—a light-sensitive semiconductor. When no light shines on the selenium, it is a good insulator; but when light shines on it, it becomes a good conductor. The selenium coating on the drum is initially in the dark. Behaving as an insulator, it can be electrically charged. When the selenium is illuminated, it becomes conducting wherever light falls on it. Electrons from the aluminum—a good conductor—pass into the illuminated regions of selenium and neutralize the positive charge. Regions of the selenium coating that remain dark do not allow electrons from the aluminum to flow in, so those regions remain positively charged.

Next, the drum is allowed to come into contact with a black powder called *toner*. The toner particles have been given a negative charge so they will be attracted to positively charged regions of the drum. Toner adheres to the drum where there is positive charge, but no toner adheres to the uncharged regions. A sheet of paper is now rotated onto the drum and positive charge is applied to the back surface of the paper. The charge on the paper is larger than that on the drum, so the paper attracts the negatively charged toner away from the drum, forming an image of the original document on the paper. The final step is to fuse the toner to the paper by passing the paper between hot rollers. With the ink sealed into the fibers of the paper, the copy is finished.

16.3   COULOMB'S LAW

Let's now begin a quantitative treatment of electrical forces among charged objects. We first encountered Coulomb's law in Chapter 2; it gives the electric force acting between two *point charges*. A **point charge** is a point-like object with a nonzero electric charge. Recall that a point-like object is small enough that its internal structure is of no importance. The electron can be treated as a point charge, since there is no experimental evidence for any internal structure. The proton *does* have internal structure—it contains three particles called *quarks* bound together—but, since its size is only about 10^{-15} m, it too can be treated as a point charge for most purposes. A charged metal sphere of radius 10 cm can be treated as a point charge if it interacts with another such sphere 100 m away, but not if the two spheres are only a few centimeters apart. Context is everything! A common shorthand is to say "a charge" instead of saying "a particle with charge."

Like gravity, the electric force is an *inverse square law* force. That is, the strength of the force decreases as the separation increases such that the force is proportional to the inverse square of the separation r between the two point charges ($F \propto 1/r^2$). The strength of the force is also proportional to the *magnitude* of each of the two charges ($|q_1|$ and $|q_2|$) just as the gravitational force is proportional to the *mass* of each of two interacting objects.

The magnitude of the electrical force that each of two charges exerts on the other is given by

Coulomb's law	
$$F = \dfrac{k|q_1|\,|q_2|}{r^2}$$	(2-5)

Since we use the *magnitudes* of q_1 and q_2, F—the magnitude of a vector—is always a positive quantity. The proportionality constant k is experimentally found to have the value

$$k = 8.99 \times 10^9 \ \frac{\text{N·m}^2}{\text{C}^2} \qquad (16\text{-}2\text{a})$$

The constant k, which we call the *Coulomb constant*, can be written in terms of another constant ϵ_0, the *permittivity of free space*:

$$\epsilon_0 = \frac{1}{4\pi k} = 8.85 \times 10^{-12} \ \frac{C^2}{N \cdot m^2} \qquad (16\text{-}2b)$$

Using ϵ_0, Coulomb's law is

$$F = \frac{|q_1| \, |q_2|}{4\pi\epsilon_0 r^2}$$

Coulomb's law gives us the *magnitude* of the force. The direction of the force is always along the line that joins the two point charges. We must remember that, unlike the gravitational force, the electric force can either be attractive or repulsive depending on the signs of the charges. Coulomb's law is in agreement with Newton's third law: the forces on the two charges are equal in magnitude and opposite in direction (Fig. 16.9).

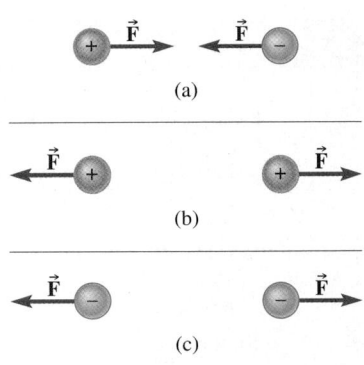

Figure 16.9 The electric force on (a) two opposite charges; (b) and (c) two like charges. Vectors are drawn showing the force on each of the two interacting charges.

Problem-Solving Tips for Coulomb's Law

1. Use consistent units; since we know k in standard SI units ($N \cdot m^2/C^2$), distances should be in meters and charges in coulombs. When the charge is given in μC or nC, be sure to change the units to coulombs: $1 \ \mu C = 10^{-6} \ C$ and $1 \ nC = 10^{-9} \ C$.

2. When finding the electric force on a single charge due to two or more other charges, find the force due to each of the other charges separately. The net force on a particular charge is the vector sum of the forces acting on that charge due to each of the other charges. Often it helps to separate the forces into x- and y-components, add the components separately, then find the resultant force from the net x- and net y-components.

3. If several charges lie along the same line, we do not have to worry about an intermediate charge "shielding" the charge located on one side from the charge on the other side. The electric force is long-range just as is gravity; the gravitational force on the Earth due to the Sun does not stop when the Moon passes between the two.

Example 16.3

Electric Force on a Point Charge

Suppose three point charges are arranged as shown in Fig. 16.10. A charge $q_1 = +1.2 \ \mu C$ is located at the origin of an x-, y-coordinate system; a second charge $q_2 = -0.60 \ \mu C$ is located at (1.20 m, 0.50 m) and the third charge $q_3 = +0.20 \ \mu C$ is located at (1.20 m, 0). What is the force on q_3 due to the other two charges?

Figure 16.10
Location of point charges in Example 16.3

Strategy The force on q_3 due to q_1 and the force on q_3 due to q_2 are determined separately. The sum of these two forces is the resultant force on q_3. After sketching a free-body diagram, we add the two forces. Adding the repulsive force due to q_1 and the attractive force due to q_2 gives a resultant force with components upward and to the right. Charge q_1 is more than twice as far away as q_2; even though $|q_1| = 2|q_2|$, the electric force is inversely proportional to distance *squared*, so the force due to q_2 is stronger than the force due to q_1. Let the distance between charges 1 and 3 be $r_{13} = 1.20$ m and the distance between charges 2 and 3 be $r_{23} = 0.50$ m.

Solution Charges 1 and 3 are both positive. The force $\vec{F}_1$ on q_3 due to q_1 is repulsive; it can be represented by a vector pointing in the positive x-direction (Fig. 16.11a). Charge q_2 is negative so it attracts q_3 along the line joining those two charges; $\vec{F}_2$ points in the positive y-direction.

The magnitudes of the charges and the separations between the charges were given in the statement of the problem. We first find the magnitude of force $\vec{F}_1$ on q_3 due to q_1 from Coulomb's law and then repeat the same process to find the magnitude of force $\vec{F}_2$ on q_3 due to q_2.

continued on next page

Example 16.3 *continued*

From Coulomb's law and the given information

$$F_1 = \frac{k|q_1||q_3|}{r_{13}^2}$$

$$= 8.99 \times 10^9 \frac{\text{N·m}^2}{\text{C}^2} \times \frac{(1.2 \times 10^{-6}\,\text{C}) \times (0.20 \times 10^{-6}\,\text{C})}{(1.20\,\text{m})^2}$$

$$= 1.50 \times 10^{-3}\,\text{N}$$

Now for the force due to charge 2.

$$F_2 = \frac{k|q_2||q_3|}{r_{23}^2}$$

$$= 8.99 \times 10^9 \frac{\text{N·m}^2}{\text{C}^2} \times \frac{(0.60 \times 10^{-6}\,\text{C}) \times (0.20 \times 10^{-6}\,\text{C})}{(0.50\,\text{m})^2}$$

$$= 4.32 \times 10^{-3}\,\text{N}$$

As expected, $F_2 > F_1$.

We draw a free-body diagram for q_3 to show the directions of the forces $\vec{F}_1$ and $\vec{F}_2$ (Fig. 16.11a). Adding the two forces gives the resultant force $\vec{F}$. Since the vectors are perpendicular,

we can use the Pythagorean theorem to find the magnitude of the resultant (Fig. 16.11b).

The magnitude of the resultant force F is

$$F = \sqrt{F_1^2 + F_2^2}$$

$$= \sqrt{(1.50 \times 10^{-3}\,\text{N})^2 + (4.32 \times 10^{-3}\,\text{N})^2}$$

$$= 4.6 \times 10^{-3}\,\text{N}$$

where we have rounded to two significant figures. With the aid of Fig. 16.11b, we can find the direction of the force.

$$\tan \theta = \frac{F_2}{F_1} = \frac{4.32 \times 10^{-3}\,\text{N}}{1.50 \times 10^{-3}\,\text{N}} = 2.88$$

$$\theta = 71° \text{ above the } x\text{-axis}$$

Discussion   The net force has a direction compatible with our expectation—it has components in the $+x$- and $+y$-directions. The force due to q_2 is stronger, so the y-component of the net force is larger than the x-component; the force makes a smaller angle with the y-axis (19°) than with the x-axis (71°).

Practice Problem 16.3   Electric force at a vertex of an equilateral triangle

Find the magnitude and direction of the electric force on charge 3 due to charges 1 and 2 in Fig. 16.12. The charges are $q_1 = q_2 = +0.40\ \mu\text{C}$ and $q_3 = +0.20\ \mu\text{C}$.

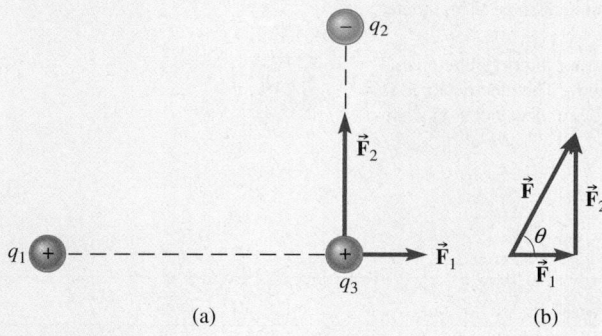

(a)

Figure 16.11
(a) The directions of forces $\vec{F}_1$ and $\vec{F}_2$. (b) Vectors $\vec{F}_1$ and $\vec{F}_2$ and their sum $\vec{F}$.

(b)

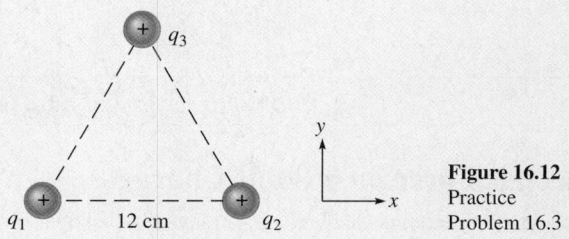

Figure 16.12
Practice
Problem 16.3

If we consider the forces acting on the microscopic building blocks of matter (such as atoms, molecules, ions, and electrons), we find that the electrical forces are much stronger than the gravitational forces between them. Only when we put a large number of atoms and molecules together to make a massive object can the gravitational force dominate. This domination occurs only because there is an almost perfect balance between positive and negative charges in a large object, leading to nearly zero net charge.

Example 16.4

Two Charged Balls, Hanging in Equilibrium

Two Styrofoam balls of mass 10.0 g are suspended by threads of length 25 cm. The balls are charged, after which they hang apart, each at $\theta = 15.0°$ to the vertical (Fig. 16.13a). (a) Are the signs of the charges the same or opposite? (b) Are the magnitudes of

the charges necessarily the same? Explain. (c) Find the net charge on each ball, *assuming* that the charges are equal.

Strategy   Each ball exerts an electric force on the other since both are charged. The *gravitational* forces that the balls

continued on next page

Example 16.4 *continued*

exert on one another are negligibly small, but the gravitational forces that the Earth exerts on the balls are not negligible. The third force acting on each of the balls is due to the tension in a thread. We choose to analyze the forces acting on the ball using a free-body diagram. The sum of the three forces must be zero since the ball is in equilibrium.

Solution Each ball experiences three forces: the electrical force, the gravitational force, and the pull of the thread, which is under tension. Figure 16.13b shows a free-body diagram for one of the balls.

(a) The electric force is clearly repulsive—the balls are pushed apart—so the charges must have the same sign. There is no way to tell whether they are both positive or both negative.

(b) At first glance it *might* appear that the charges must be the same; the balls are hanging at the same angle, so there is no clue as to which charge is larger. But look again at Coulomb's law: the force on either of the balls is proportional to the product of the two charges; $F \propto q_1 q_2$. In accordance with Newton's third law, Coulomb's law says that the two forces that make up the interaction are equal in magnitude and opposite in direction. The charges are not necessarily equal.

(c) Let us choose the x- and y-axes in the horizontal and vertical directions, respectively. Of the three forces acting on a ball, only one, that due to the tension in the thread, has both x-

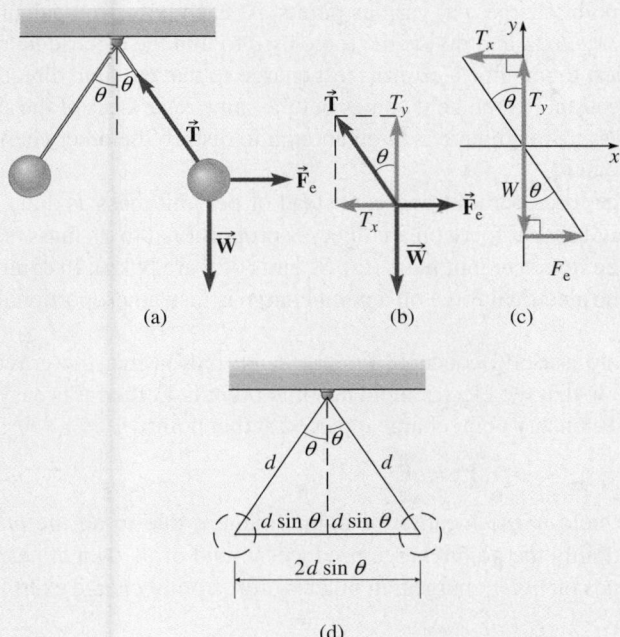

(a) (b) (c)

(d)

Figure 16.13
(a) Sketch of the situation; (b) a free-body diagram for one ball; (c) similar triangles used in the solution; (d) finding separation between the two balls.

and y-components. From Fig. 16.13b, the tension in the thread has a y-component equal in magnitude to the weight of the ball, and an x-component equal in magnitude to the electric force on the ball. By similar triangles (Fig. 16.13c),

$$\frac{F_e}{W} = \tan \theta \quad \text{or} \quad F_e = W \tan \theta$$

From Coulomb's law,

$$F_e = \frac{k|q|^2}{r^2}$$

where $|q|$ is the magnitude of the charge on each of the two balls (now assumed to be equal). The separation of the balls (Fig. 16.13d) is

$$r = 2(d \sin \theta)$$

where $d = 25$ cm is the length of the thread.

Some algebra now enables us to solve for $|q|$. From Coulomb's law,

$$|q|^2 = \frac{F_e r^2}{k}$$

We can substitute these expressions for F_e and r:

$$|q|^2 = \frac{(W \tan \theta)(2d \sin \theta)^2}{k}$$

$$= \frac{4d^2 mg \tan \theta \sin^2 \theta}{k}$$

$$|q| = \sqrt{\frac{4 \times (0.25 \text{ m})^2 \times 0.0100 \text{ kg} \times 9.8 \text{ N/kg} \times \tan 15.0° \times \sin^2 15.0°}{8.99 \times 10^9 \text{ N·m}^2/\text{C}^2}}$$

$$= 0.22 \text{ μC}$$

The charges can either be both positive or both negative, so the charges are either both +0.22 μC or both –0.22 μC.

Discussion We can check the units in the final expression for q:

$$\sqrt{\frac{\text{m}^2 \times \text{kg} \times \text{N/kg}}{\text{N·m}^2/\text{C}^2}} = \sqrt{\frac{\text{N·m}^2}{\text{N·m}^2/\text{C}^2}} = \sqrt{\text{C}^2} = \text{C} \quad \text{(OK!)}$$

Another check: if the balls were uncharged, they would hang straight down ($\theta = 0$). Substituting $\theta = 0$ into the final algebraic expression does give $q = 0$.

How large a charge would make the threads horizontal? As the charge on the balls is increased, the angle of the thread *approaches* 90° but can never reach 90° because the tension in the thread must always have an upward component to balance gravity. In the algebraic answer, as $\theta \to 90°$, $\tan \theta \to \infty$ and $\sin \theta \to 1$, which would yield a charge q approaching ∞. The threads cannot be horizontal for any *finite* amount of charge.

Practice Problem 16.4   Three point charges

Three identical point charges $q = -2.0$ nC are at the vertices of an equilateral triangle with sides of length 1.0 cm. What is the magnitude of the electrical force acting on any one of them?

16.4   THE ELECTRIC FIELD

In Section 2.5, we learned that the gravitational field at a point is defined to be the gravitational force per unit mass on an object placed at that point. If the gravitational force on an apple of mass m due to the Earth is $\vec{\mathbf{F}}_g$, then the Earth's gravitational field $\vec{\mathbf{g}}$ at the location of the apple is given by

$$\vec{\mathbf{g}} = \frac{\vec{\mathbf{F}}_g}{m}$$

The directions of $\vec{\mathbf{F}}_g$ and $\vec{\mathbf{g}}$ are the same since m is positive. The gravitational field we encounter most often is that due to the Earth, but the gravitational field could be due to any astronomical body, or to more than one body. For instance, an astronaut may be concerned with the gravitational field at the location of her spacecraft due to the Sun, the Earth, and the Moon combined. Since gravitational forces add as vectors—as do all forces—the gravitational field at the location of the spacecraft is the vector sum of the separate gravitational fields due to the Sun, the Earth, and the Moon.

Similarly, if a point charge q is in the vicinity of other charges, it experiences an electric force. The **electric field** (symbol $\vec{\mathbf{E}}$) at any point is defined to be the electric force per unit *charge* at that point:

$$\vec{\mathbf{E}} = \frac{\vec{\mathbf{F}}_e}{q} \qquad\qquad \text{(16-3a)}$$

The SI units of the electric field are N/C.

In contrast to the gravitational force, which is always parallel to the gravitational field, the electrical force can either be parallel or antiparallel to the electric field depending on the sign of the charge q that is sampling the field. *If q is positive*, the direction of the electric force $\vec{\mathbf{F}}_e$ is the same as the direction of the electric field $\vec{\mathbf{E}}$; if q is negative, the two vectors have opposite directions. To probe the electric field in some region, imagine placing a point charge q at various points. At each point you calculate the electric force on this *test charge* and divide the force by q to find the electric field at that point. It is usually easiest to imagine a *positive* test charge so that the field direction is the same as the force direction, but the field comes out the same regardless of the sign or magnitude of the q, unless its magnitude is large enough to disturb the other charges and thus change the electric field.

Why is $\vec{\mathbf{E}}$ defined as the force per unit *charge* instead of per unit mass as done for gravitational field? The gravitational force on an object is proportional to its mass, so it makes sense to talk about the force per unit mass (the SI units of $\vec{\mathbf{g}}$ are N/kg). In contrast to the gravitational force, the electrical force on a point charge is instead proportional to its *charge*.

Why is the electric field a useful concept? For the same reason that the gravitational field is. Once we know that the electric field at some point is $\vec{\mathbf{E}}$, then it is easy to calculate the electric force $\vec{\mathbf{F}}_e$ on any point charge q placed at that point:

The electric force $\vec{\mathbf{F}}_e$ on a charge q at a point where the electric field is $\vec{\mathbf{E}}$ is $q\vec{\mathbf{E}}$.

$$\vec{\mathbf{F}}_e = q\vec{\mathbf{E}} \qquad\qquad \text{(16-3b)}$$

Note that $\vec{\mathbf{E}}$ is the electric field at the location of point charge q due to all the *other* charges in the vicinity. Certainly the point charge produces a field of its own at nearby points; this field causes forces on *other* charges. In other words, a point charge exerts no force on itself.

Electric Field due to a Point Charge

The electric field due to a single point charge Q can be found using Coulomb's law. Imagine a positive test charge q placed at various locations. Coulomb's law says that the force acting on the test charge is

$$F = \frac{k|q|\,|Q|}{r^2}$$

The electric field strength is then

$$E = \frac{F}{|q|} = \frac{k|Q|}{r^2} \qquad (16\text{-}4)$$

The field falls off as $1/r^2$, following the same inverse square law as the electrical force.

What is the direction of the field? If Q is positive, then a positive test charge would be repelled, so the field would point away from Q (or *radially outward*). If Q is negative, then the field points toward Q (*radially inward*).

The electric field due to more than one point charge can be found using the **principle of superposition**:

> The electric field at any point is the vector sum of the fields
> at that point caused by each charge separately.

Example 16.5

Electric Field at a Point in Space

Two point charges are located on the x-axis (Fig. 16.14). Charge $q_1 = +0.60\ \mu C$ is located at $x = 0$; charge $q_2 = -0.50\ \mu C$ is located at $x = 0.40$ m. Point P is located at $x = 1.20$ m. What is the magnitude and direction of the electric field at point P due to the two charges?

Strategy   We can determine the field at P due to q_1 and the field at P due to q_2 separately using Coulomb's law and the definition of the electric field. In each case, the electric field points in the direction of the electric force on a *positive* test charge at point P. The sum of these two fields is the resultant field at P. We sketch a vector diagram to help add the fields correctly. Since there are two different distances in the problem, subscripts help to distinguish them. Let the distance between charge 1 and point P be $r_1 = 1.20$ m and the distance between charge 2 and point P be $r_2 = 0.80$ m.

Solution   Charge 1 is positive. We imagine a tiny positive test charge, q_0, located at point P. Since charge 1 repels the positive test charge, the force $\vec{F}_1$ on the test charge due to q_1 is in the positive x-direction (Fig. 16.15a). The direction of the

electric field due to charge 1 is also in the $+x$-direction since $\vec{E}_1 = \vec{F}_1/q_0$ and $q_0 > 0$. Charge q_2 is negative so it attracts the imaginary test charge along the line joining the two charges; the force $\vec{F}_2$ on the test charge due to q_2 is in the negative x-direction. Therefore $\vec{E}_2 = \vec{F}_2/q_0$ is in the $-x$-direction.

We first find the magnitude of the field $\vec{E}_1$ at P due to q_1 and then repeat the same process to find the magnitude of field $\vec{E}_2$ at P due to q_2. From the given information,

$$E_1 = \frac{k|q_1|}{r_1^2}$$

$$= 8.99 \times 10^9\ \frac{\text{N} \cdot \text{m}^2}{\text{C}^2} \times \frac{(0.60 \times 10^{-6}\ \text{C})}{(1.20\ \text{m})^2}$$

$$= 3.75 \times 10^3\ \text{N/C}$$

Now for the magnitude of field $\vec{E}_2$ at P due to charge 2.

$$E_2 = \frac{k|q_2|}{r_2^2}$$

$$= 8.99 \times 10^9\ \frac{\text{N} \cdot \text{m}^2}{\text{C}^2} \times \frac{(0.50 \times 10^{-6}\ \text{C})}{(0.80\ \text{m})^2}$$

$$= 7.02 \times 10^3\ \text{N/C}$$

A vector diagram (Fig. 16.15a) indicates the directions of the fields $\vec{E}_1$ and $\vec{E}_2$. Figure 16.15b shows the vector addition $\vec{E}_1 + \vec{E}_2$ and the resultant field $\vec{E}$, which points in the $-x$-direction since $E_2 > E_1$. The magnitude of E at point P is

$$E = 7.02 \times 10^3\ \text{N/C} - 3.75 \times 10^3\ \text{N/C} = 3.3 \times 10^3\ \text{N/C}$$

The resultant electric field at P is 3.3×10^3 N/C in the $-x$-direction.

Figure 16.14
Two point charges on the x-axis

Figure 16.15
(a) Directions of electric field vectors at point P due to charges q_1 and q_2. (b) Vector addition of $\vec{E}_1$ and $\vec{E}_2$.

continued on next page

Example 16.5 *continued*

Discussion This same method is used to find the electric field at a point due to *any* number of point charges. The direction of the electric field due to each charge alone is the direction of the electric force on an imaginary positive test charge at that point. The magnitude of each electric field is found from Eq. (16-4). Then the electric field vectors are added. If the charges and the point do not all lie on the same line, then the fields can be added by resolving them into x- and y-components and summing the components.

Even when electric fields are not due to a small number of point charges, the principle of superposition still applies: the electric field at any point is the vector sum of the fields at that point caused by each charge or set of charges separately.

Practice Problem 16.5   Electric field at point P due to two charges

Find the magnitude and direction of the electric field at point P due to charges 1 and 2 located along the x-axis in Fig. 16.16. The charges are $q_1 = +0.040\ \mu C$ and $q_2 = +0.010\ \mu C$. Charge q_1 is at the origin, charge q_2 is at $x = 0.30$ m and point P is at $x = 1.50$ m.

Figure 16.16
Practice Problem 16.5

Electric Field Lines

It is not easy to make a visual representation of an electric field using vector arrows; it is hard to draw vectors at enough different points to be useful without all the vector arrows overlapping and becoming impossible to distinguish. A more useful representation of the electric field is a sketch of the **electric field lines**, a set of continuous lines that represent both the magnitude and the direction of the electric field as follows.

Interpretation of Electric Field Lines

- The direction of the electric field at any point is *tangent to the field line* passing through that point and in the direction indicated by arrows on the field line (Fig. 16.17a).

- The electric field is strong where field lines are close together and weak where they are far apart. More specifically, if you imagine a small surface perpendicular to the field lines, the magnitude of the field is proportional to the number of lines that cross the surface divided by the area (Fig. 16.17b).

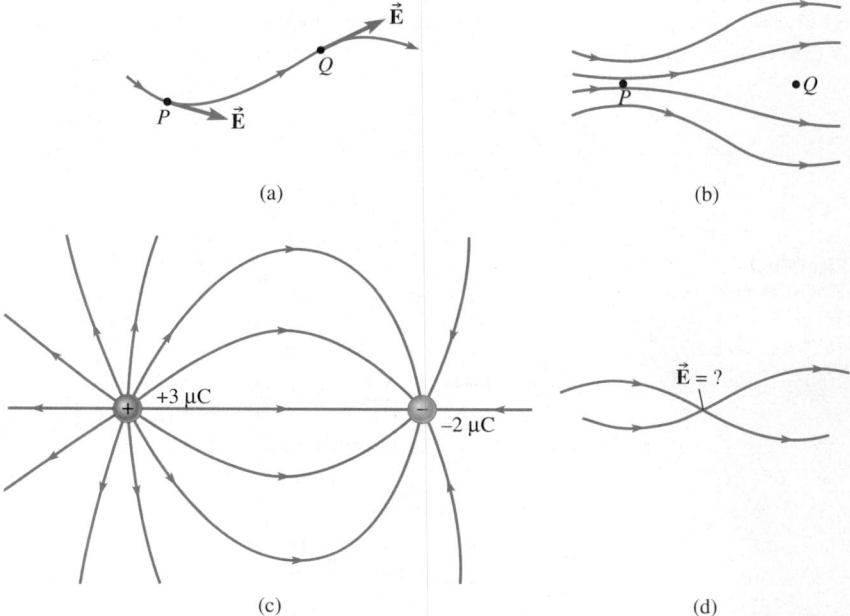

Figure 16.17 Field line rules illustrated. (a) The electric field direction at points P and Q. (b) The magnitude of the electric field at point P is larger than the magnitude at Q. (c) If 12 lines are drawn starting on a point charge $+3\ \mu C$, then 8 lines must be drawn ending on a $-2\ \mu C$ point charge. (d) If field lines were to cross, the direction of $\vec{E}$ at the intersection would be undetermined.

Once field lines are drawn, the rules given on page 560 are used to interpret them. To help sketch the field lines, these three additional rules are useful.

Rules for Sketching Field Lines

- Field lines start only on positive charges and end only on negative charges.
- The number of lines starting on a positive charge (or ending on a negative charge) is proportional to the magnitude of the charge (Fig. 16.17c). (The total number of lines you draw is arbitrary; the more lines you draw, the better the representation of the field.)
- Field lines never cross. The electric field at any point has a unique direction; if field lines crossed, the field would have two directions at the same point (Fig. 16.17d).

Field Lines for a Point Charge

Figure 16.18 shows sketches of the field lines due to single point charges. The field lines show clearly that the direction of the field is radial (away from a positive charge or toward a negative charge). The lines are close together near the point charge, where the field is strong, and are more spread out farther from the point charge, showing that the field strength diminishes with distance. No other nearby charges are shown in these sketches, so the lines go out to infinity as if the point charge were the only thing in the universe. If the field of view is enlarged, so that other charges are shown, the lines starting on the positive point charge would end on some faraway negative charges, and those that end on the negative charge would start on some faraway positive charges.

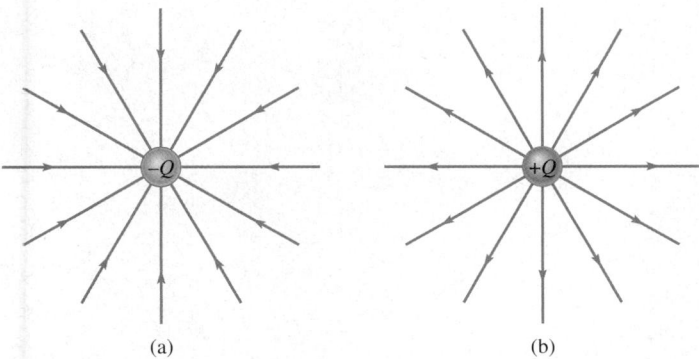

(a)   (b)

Figure 16.18 Electric field lines of isolated point charges. (a) Field of negative point charge; (b) field of positive point charge.

Electric Field due to a Dipole

What if there are two point charges with equal and opposite charges? This configuration is called a **dipole** (literally *two poles*). To find the electric field for the dipole at various points by using Coulomb's law would be extremely tedious, but sketching some field lines immediately gives an approximate idea of the electric field (Fig. 16.19).

Dipole = two equal and opposite point charges

Because the charges in the dipole have equal charge magnitudes, the same number of lines that start on the positive charge end on the negative charge. Close to either of the charges, the field lines are evenly spaced in all directions, just as if the other charge were not present. As we approach one of the charges, the field due to that charge gets so large ($F \propto 1/r^2$, $r \to 0$) that the field due to the other charge is negligible in comparison and we are left with the spherically symmetric field due to a single point charge.

The field at other points has contributions from both charges. Figure 16.19 shows, for one point P, how the field vectors ($\vec{E}_-$ and $\vec{E}_+$) due to the two separate charges add, following vector addition rules, to give the total field $\vec{E}$ at point P. Note that the total field $\vec{E}$ is tangent to the field line through point P.

The principles of superposition and symmetry are two powerful tools for determining electric fields. The use of symmetry is illustrated in Conceptual Example 16.6.

Figure 16.19   Electric field lines for a dipole. The electric field vector $\vec{\mathbf{E}}$ at a point P is tangent to the field line through that point and is the sum of the fields ($\vec{\mathbf{E}}_-$ and $\vec{\mathbf{E}}_+$) due to each of the two point charges.

Conceptual Example 16.6

Field Lines for a Thin Spherical Shell

A thin metallic spherical shell of radius R carries a total charge Q, which is positive. The charge is spread out evenly over the shell's outside surface. Sketch the electric field lines in two different views of the situation: (a) The spherical shell is tiny and you are looking at it from distant points; (b) you are looking at the field inside the shell's cavity.

Strategy   Since the charge on the shell is positive, field lines begin on the shell. A sphere is a highly symmetric shape: standing at the center, it looks the same in any chosen direction. This symmetry helps in sketching the field lines.

Solution   (a) A tiny spherical shell located far away cannot be distinguished from a point charge. The sphere looks like a point when seen from a great distance and the field lines look just like those emanating from a positive point charge (Fig. 16.20a).

(b) Field lines begin on the positive charges on the shell surface. Some go outward, representing the electric field outside the shell, while others may *perhaps* go inward, representing the field inside the shell. Any field lines inside must start evenly spaced on the shell and point directly toward the center of the shell (Fig. 16.20b); the lines cannot deviate from the radial direction due to the symmetry of the sphere. But what would happen to the field lines when they reach the center?

The lines can only end at the center if a negative point charge is found there—but there is no point charge. If the lines do not end, they would cross at the center. That cannot be right since the field must have a *unique* direction at every point—field lines never cross. The inescapable conclusion: *there are no field lines inside the shell* (Fig. 16.20c).

Discussion   We conclude that the electric field *inside* a spherical shell of charge is zero. This conclusion, which we reached using field lines and symmetry considerations, can also be proved using Coulomb's law, the principle of superposition, and some calculus—a much more difficult method!

The field line picture also shows that *the electric field pattern outside a spherical shell is the same as if the charge were all condensed into a point charge at the center of the sphere.*

Conceptual Practice Problem 16.6   Field lines after a negative point charge is inserted

Suppose the spherical shell of evenly distributed positive charge Q has a point charge $-Q$ placed at its center. (a) Sketch the field lines. [*Hint:* Since the charges are equal in magnitude, the number of lines starting on the shell is equal to the number ending on the point charge.] (b) Defend your sketch using the principle of superposition (total field = field due to shell + field due to point charge).

(a)   (b)   (c)

Figure 16.20
(a) Field lines outside the shell are directed radially outward. (b) If there are field lines inside the shell, they must start on the shell and point radially inward. Then what? (c) There can be no field lines—and therefore no electric field—inside the shell.

Example 16.7

Charged Sphere Hanging in a Uniform $\vec{\mathbf{E}}$ Field

A small sphere of mass 5.10 g is hanging vertically from an insulating thread that is 12.0 cm long (Fig. 16.21a). By charging some nearby flat metal plates, the sphere is subjected to a horizontal electric field of magnitude 7.20×10^5 N/C. As a result, the sphere is displaced 6.00 cm horizontally in the direction of the electric field. (a) What is the angle θ that the thread makes with the vertical? (b) What is the tension in the thread? (c) What is the charge on the sphere?

Strategy   We assume that the sphere is small enough to be treated as a point charge. Then the electric force on the sphere is given by $\vec{\mathbf{F}}_e = q\vec{\mathbf{E}}$. Figure 16.21a shows that the sphere is pushed to the right by the field; therefore, $\vec{\mathbf{F}}_e$ is to the right. Since $\vec{\mathbf{F}}_e$ and $\vec{\mathbf{E}}$ have the same direction, the charge on the sphere is positive. After drawing a free-body diagram showing all the forces acting on the sphere, we set the net force on the sphere equal to zero since it hangs in equilibrium.

continued on next page

Example 16.7 *continued*

Solution (a) The angle θ can be found from the geometry of Fig. 16.21a. The thread's length (12.0 cm) is the hypotenuse of a right triangle. The side of the triangle opposite angle θ is the horizontal displacement (6.00 cm). Thus,

$$\sin \theta = \frac{6.00 \text{ cm}}{12.0 \text{ cm}} = 0.500 \quad \text{and} \quad \theta = 30.0°$$

(b) We start by drawing a free-body diagram (Fig. 16.21b). The gravitational force must balance the vertical component of the thread's pull on the sphere ($\vec{F}_T$). The electrical force must balance the horizontal component of the same force. In Fig. 16.21c we show the components of $\vec{F}_T$. The magnitude of $\vec{F}_T$ is the tension in the thread T.

Looking at the free-body diagram and reviewing what we have in the given information, we cannot determine $\vec{F}_e$ yet because the value of the charge q is unknown. There is also no information on the tension in the thread. The weight, however, can be determined since the mass is known:

$$F_g = mg$$

The weight must be equal to the upward component of $\vec{F}_T$:

$$mg = T \cos \theta$$

Substituting known values,

$$(5.10 \times 10^{-3} \text{ kg}) \times (9.80 \text{ N/kg}) = 0.0500 \text{ N} = T \cos 30.0°$$

$$T = \frac{0.0500 \text{ N}}{\cos 30.0°} = 5.77 \times 10^{-2} \text{ N}$$

This result gives the magnitude of $\vec{F}_T$. The direction is along the thread toward the support point, at an angle of 30.0° from the vertical.

(c) We set the horizontal component of the thread force equal to the magnitude of the electrical force.

$$T \sin \theta = F_e = |q|E$$

We can now solve for $|q|$.

$$|q| = \frac{T \sin \theta}{E} = \frac{(5.77 \times 10^{-2} \text{ N}) \sin 30.0°}{7.20 \times 10^5 \text{ N/C}} = 40.1 \text{ nC}$$

We have determined the magnitude of the charge. The sign of the charge is positive because the electric force on the sphere is in the direction of the electric field. Therefore,

$$q = 40.1 \text{ nC}$$

Discussion This problem has many steps, but, taken one by one, each step helps to solve for one of the unknowns and leads the way to find the next unknown. At first glance it may appear that not enough information is given, but after a figure is drawn to aid in the visualization of the forces and their components, the steps to follow are more easily determined.

Practice Problem 16.7 Effect of doubling the charge on the hanging mass

If the charge on the sphere were doubled in Example 16.7, what angle would the thread make with the vertical?

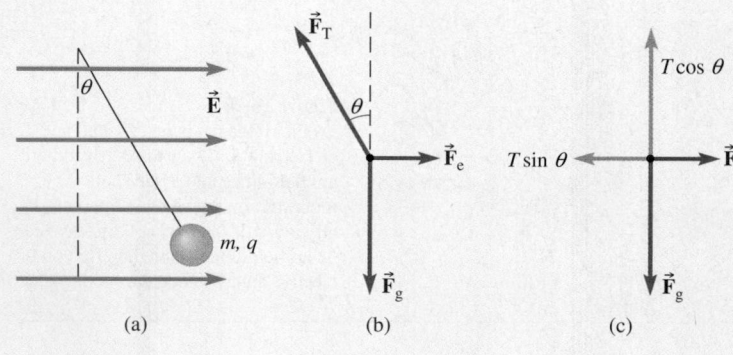

(a)   (b)   (c)

Figure 16.21
(a) A charged sphere hanging in a uniform $\vec{E}$ field.
(b) A free-body diagram showing forces acting on the sphere. (c) Free-body diagram in which the force due to the cord is replaced by its vertical and horizontal components.

Making The Connection:
electrolocation in fish

Electrolocation

Long before scientists learned how to detect and measure electric fields, certain animals and fish evolved organs to produce and detect electric fields. *Gymnarchus niloticus* (see the Chapter Opener) has electrical organs running along the length of its body; these organs set up an electric field around the fish (Fig. 16.22). When a nearby object distorts the field lines, *Gymnarchus* detects the change through sensory receptors, mostly near the head, and responds accordingly. This extra sense enables the fish to detect prey or predators in muddy streams where eyes are less useful.

Since *Gymnarchus* relies primarily on electrolocation, where slight changes in the electric field are interpreted as the presence of nearby objects, it is important that it be

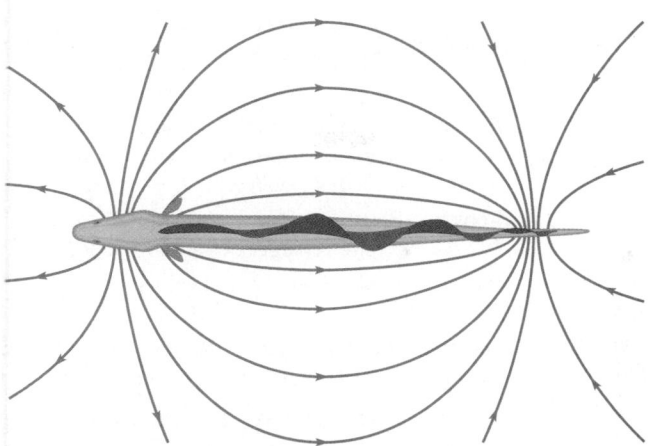

Figure 16.22 The electric field generated by *Gymnarchus*. The field is approximately that of a dipole. The head of the fish is positively charged and the tail is negatively charged.

able to create the same electric field over and over. For this reason *Gymnarchus* swims by undulating its long dorsal fin while holding its body rigid. Keeping the backbone straight keeps the negative and positive charge centers aligned and at a fixed distance apart. A swishing tail would cause variation in the electric field and that would make electrolocation much less accurate.

16.5   MOTION OF A POINT CHARGE IN A UNIFORM ELECTRIC FIELD

The simplest example of how a charged object responds to an electric field is when the electric field (due to other charges) is **uniform**—that is, has the same magnitude and direction at every point. The field due to single point charge is *not* uniform; it is radi-ally directed and its magnitude follows the inverse square law. To create a uniform field requires a large number of charges. The most common way to create a (nearly) uniform electric field is to put equal and opposite charges on two parallel metal plates (Fig. 16.23). If the charges are $\pm Q$ and the plates have area A, the magnitude of the field between the plates is

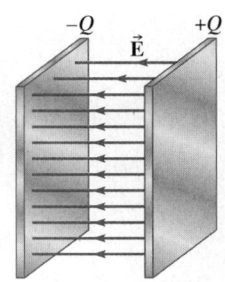

Figure 16.23 Uniform electric field between two parallel metal plates

$$E = \frac{Q}{\epsilon_0 A} \tag{16-5}$$

The direction of the field is perpendicular to the plates, from the positively charged plate toward the negatively charged plate.

Assuming the uniform field $\vec{\mathbf{E}}$ is known, a point charge q experiences an electric force

$$\vec{\mathbf{F}} = q\vec{\mathbf{E}}$$

If this is the only force acting on the point charge, then the net force is constant and therefore so is the acceleration:

$$\vec{\mathbf{a}} = \frac{\vec{\mathbf{F}}}{m} = \frac{q\vec{\mathbf{E}}}{m} \tag{16-6}$$

With a constant acceleration, the motion can take one of two forms. If the initial veloc-ity of the point charge is zero or is parallel or antiparallel to the field, then the motion is along a straight line. If the point charge has an initial velocity component perpendicular to the field, then the trajectory is parabolic (just like a projectile in a uniform gravita-tional field if other forces are negligible). All the tools developed in Chapters 2–4 to analyze motion with constant acceleration can be used here. The direction of the accel-eration is either parallel to $\vec{\mathbf{E}}$ (for a positive charge) or antiparallel to $\vec{\mathbf{E}}$ (for a negative charge).

Example 16.8

Electron Beam

 A cathode ray tube (CRT) is used to accelerate electrons in televisions, computer monitors, oscilloscopes, and x-ray tubes. Electrons from a heated filament pass through a hole in the cathode; they are then accelerated by an electric field between the cathode and the anode (Fig. 16.24). Suppose electrons pass through the hole in the cathode at a velocity of 10.0 m/s toward the anode. The electric field is uniform between the anode and cathode and has a magnitude of 1.0×10^4 N/C. (a) What is the acceleration of the electrons? (b) If the anode and cathode are separated by 2.0 cm, what is the final velocity of the electrons?

Strategy   Because the field is uniform, the acceleration of the electrons is constant. Then we can apply Newton's second law and use any of the methods we previously developed for motion with constant acceleration.

Given: Initial velocity $v_0 = 10.0$ m/s;
separation between plates $d = 0.020$ m;
Electric field magnitude $E = 1.0 \times 10^4$ N/C

Look up: Electron mass $m_e = 9.109 \times 10^{-31}$ kg;
electron charge $q = -e = -1.602 \times 10^{-19}$ C

Find: (a) acceleration; (b) final velocity

Solution   (a) First, check that gravity is negligible. The weight of the electron is

$$F_g = mg = 9.109 \times 10^{-31} \text{ kg} \times 9.8 \text{ m/s}^2 = 8.9 \times 10^{-30} \text{ N}$$

The magnitude of the electric force is

$$F_e = eE = (1.602 \times 10^{-19} \text{ C}) \times (1.0 \times 10^4 \text{ N/C}) = 1.6 \times 10^{-15} \text{ N}$$

continued on next page

Electron gun

(A)
Plates for
horizontal deflection

(B)
Plates for
vertical deflection

Heated filament
(source of electrons)

Cathode
(−)

Anode
(+)

Electron beam

Uniform $\vec{E}$ field
seen from side

Cathode —— Anode

Electron beam

Side view

Conductive coating

Fluorescent screen

Figure 16.24   In a cathode ray tube (CRT), electrons are accelerated to high speeds by an electric field.

Example 16.8 *continued*

which is about 14 orders of magnitude larger. Gravity is completely negligible. While between the plates, the electrons' acceleration is therefore

$$a = \frac{F}{m_e} = \frac{eE}{m_e} = \frac{(1.602 \times 10^{-19}\,\text{C}) \times (1.0 \times 10^4\,\text{N/C})}{9.109 \times 10^{-31}\,\text{kg}} = 1.76 \times 10^{15}\,\text{m/s}^2$$

To two significant figures, $a = 1.8 \times 10^{15}\,\text{m/s}^2$. Since the charge on the electron is negative, the direction of the acceleration is opposite to the electric field, or to the right in the figure.

(b) The initial velocity of the electrons is also to the right. We have a one-dimensional constant acceleration problem since the initial velocity and the acceleration are collinear. From Eq. (3-13), the final velocity is

$$v_f = \sqrt{v_0^2 + 2ad}$$
$$= \sqrt{10.0^2 + 2 \times 1.76 \times 10^{15} \times 0.020}\,\text{m/s} = 8.4 \times 10^6\,\text{m/s to the right}$$

Discussion   The initial speed of 10.0 m/s is completely negligible compared to the final speed. Had we known this at the beginning, we could just as well have taken the electrons to start from rest; the final velocity would have been the same to within two significant figures.

The acceleration of the electrons seems large. This large value might cause some concern, but there is no law of physics against such large accelerations. Note that the final *speed* is less than the speed of light (3×10^8 m/s), the universe's ultimate speed limit.

You may suspect that this problem can also be solved using energy methods. We could indeed find the work done by the electric force and use the work done to find the change in kinetic energy. The energy approach for electric fields is developed in the next chapter.

Practice Problem 16.8   Slowing some protons

If a beam of *protons* were projected horizontally to the right through the hole in the cathode (Fig. 16.24) with an initial velocity of $v_0 = 3.0 \times 10^5$ m/s, with what velocity and in what direction would the protons reach the anode (if they do reach it)?

The electric field is used to speed up the electron beam in a CRT. In an oscilloscope—a device used to measure time-dependent quantities in circuits—it is also used to deflect the beam sideways. An electric field is *not* used to deflect the electron beam in a TV tube or computer monitor; that function is performed by a magnetic field.

Example 16.9

Deflection of an Electron Projected into a Uniform $\vec{\mathbf{E}}$ Field

An electron is projected horizontally into the uniform electric field directed vertically downward between two parallel plates (Fig. 16.25). The plates are 2.00 cm apart and are of length 4.00 cm. The initial speed of the electron is $v_0 = 8.00 \times 10^6$ m/s. As it enters the region between the plates, the electron is midway between the two plates; as it leaves, the electron just misses the upper plate. What is the magnitude of the electric field?

Strategy   Using the x- and y-axes in the figure, the electric field is in the −y-direction and initial velocity of the electron is in the +x-direction. The electric force on the electron is *upward* (in the +y-direction) since it has a negative charge. Thus the acceleration of the electron is upward. Since the

acceleration is in the +y-direction, the x-component of the velocity is constant. If the electron just misses the upper plate, its displacement is +1.00 cm in the y-direction and +4.00 cm in the x-direction. From v_x and Δx, we can find the time the electron spends between the plates. From Δy and the time, we can find a_y. From the acceleration we find the electric field using Newton's second law, $\vec{\mathbf{F}}_{\text{net}} = m\vec{\mathbf{a}}$.

We ignore the gravitational force on the electron because we assume it to be negligible. We can test this assumption later.

Given: $\Delta x = 4.00$ cm; $\Delta y = 1.00$ cm; $v_x = 8.00 \times 10^6$ m/s

Find: electric field strength, E

Solution   We start by finding the time the electron spends between the plates from Δx and v_x.

$$t = \frac{\Delta x}{v_x} = \frac{4.00 \times 10^{-2}\,\text{m}}{8.00 \times 10^6\,\text{m/s}} = 5.00 \times 10^{-9}\,\text{s}$$

From the time spent between the plates and Δy, we find the component of the acceleration in the y-direction.

$$\Delta y = \tfrac{1}{2} a_y t^2$$

$$a_y = \frac{2\,\Delta y}{t^2} = \frac{2 \times 1.00 \times 10^{-2}\,\text{m}}{(5.00 \times 10^{-9}\,\text{s})^2} = 8.00 \times 10^{14}\,\text{m/s}^2$$

Figure 16.25
An electron deflected by an electric field

continued on next page

4.00 cm

2.00 cm

Example 16.9 *continued*

This acceleration is produced by the electric force acting on the electron since we assume that no other forces act. From Newton's second law,

$$F_y = qE_y = m_e a_y$$

Solving for E_y, we have

$$E_y = \frac{m_e a_y}{q}$$

$$= \frac{(9.109 \times 10^{-31} \text{ kg}) \times (8.00 \times 10^{14} \text{ m/s}^2)}{-1.602 \times 10^{-19} \text{ C}} = -4.55 \times 10^3 \text{ N/C}$$

Since the field has no x-component, its magnitude is 4.55×10^3 N/C.

Discussion   We have ignored the gravitational force on the electron because we suspect that it is negligible in comparison

with the electric force. This should be checked to be sure it is a valid assumption.

$$\vec{F} = m_e \vec{g} = (9.109 \times 10^{-31} \text{ kg}) \times (9.80 \text{ N/kg downward})$$

$$= 8.93 \times 10^{-30} \text{ N downward}$$

$$\vec{F}_e = q\vec{E} = (-1.602 \times 10^{-19} \text{ C}) \times (4.55 \times 10^3 \text{ N/C downward})$$

$$= 7.29 \times 10^{-16} \text{ N upward}$$

The electric force is stronger than the gravitational force by a factor of approximately 10^{14}, so the assumption is valid.

Practice Problem 16.9   Deflection of a proton projected into a uniform $\vec{E}$ field

If the electron is replaced by a proton projected with the same initial velocity, will the proton exit the region between the plates or will it hit one of the plates? If it does not strike one of the plates, by how much is it deflected by the time it leaves the region of electric field?

16.6   CONDUCTORS IN ELECTROSTATIC EQUILIBRIUM

In Section 16.1 we described how a piece of paper can be polarized by nearby charges. The polarization is the paper's response to an applied electric field. By *applied* we mean a field due to charges *outside the paper*. The separation of charge in the paper produces an electric field of its own. The net electric field at any point—whether inside or outside the paper—is the sum of the applied field and the field due to the separated charges in the paper.

How much charge separation occurs depends on both the strength of the applied field and properties of the atoms and molecules that make up the paper. Some materials are more easily polarized than others. The *most* easily polarized materials are conductors because they contain highly mobile charges that can move freely through the entire volume of the material.

It is useful to examine the distribution of charge in a conductor, whether the conductor has a net charge or lies in an externally applied field, or both. We restrict our attention to a conductor in which the mobile charges are at rest in equilibrium, a situation called **electrostatic equilibrium**. If charge is put on a conductor, mobile charges move about until a stable distribution is attained. The same thing happens when an external field is applied or changed—charges move in response to the external field, but they soon reach an equilibrium distribution.

If the electric field inside a conductor is nonzero, it exerts a force on each of the mobile charges (usually electrons) and makes them move preferentially in a certain direction. With mobile charge in motion, the conductor cannot be in electrostatic equilibrium. Therefore, we can draw this conclusion:

In electrostatic equilibrium, there is no net motion of charge.

> *The electric field is zero at any point inside a conductor in electrostatic equilibrium.*

Electronic circuits and cables are often shielded from stray electric fields produced by other devices by placing them inside metal enclosures (see Conceptual Question 6). Free charges in the metal enclosure rearrange themselves as the external electric field changes. As long as the charges in the enclosure can keep up with changes in the external field, the external field is canceled inside the enclosure.

The electric field is zero *inside*, but is not necessarily zero *outside*. If there are field lines outside but none inside, field lines must either start or end at charges on the surface of the conductor. Field lines start or end on charges, so

> *When a conductor is in electrostatic equilibrium, only points on its surface(s) can have net charge.*

At any point inside the conductor, there are equal amounts of positive and negative charge. Imbalance between positive and negative charge can occur only on the surface(s) of the conductor.

It is also true that, in electrostatic equilibrium,

> *The electric field at the surface of the conductor is perpendicular to the surface.*

How do we know that? If the field had a component parallel to the surface, any free charges at the surface would feel a force parallel to the surface and would move in response. Thus, if there is a parallel component at the surface, the conductor cannot be in electrostatic equilibrium.

If the conductor has an irregular shape, the excess charge is concentrated more where the surface has its greatest curvature. We know that the excess charge must be located on the surface(s) of the conductor. On flat surfaces, repulsive forces between the charges push parallel to the surface, making the charges spread apart evenly. On a curved surface, the forces between charges are no longer parallel to the surface (Fig. 16.26). The greater the curvature, the smaller the component of repulsive force parallel to the surface. For this reason,

> *The excess charge on a conductor in electrostatic equilibrium is more concentrated at regions of greater curvature.*

Sharp points on a conducting surface have a large curvature, so charge tends to collect at sharp points.

The electric field lines just outside the conducting surface are more densely packed where the curvature is greatest because each line starts or ends on a surface charge. Since the density of field lines reflects the magnitude of the electric field, the electric field outside the conductor is largest near the sharpest points of the conducting surface.

Lightning rods are often found on the roofs of tall buildings and old farmhouses (Fig. 16.27). The rod always comes to a sharp point at the top. When a passing thunderstorm attracts charge to the top of the rod, the strong electric field at the point ionizes nearby air molecules. Neutral air molecules do not transfer net charge when they move, but ionized molecules do, so ionization allows charge to leak gently off the building through the air instead of building up to a dangerously large value. If the rod did not come to a sharp point, the electric field might not be large enough to ionize the air.

The conclusions we have reached about conductors in electrostatic equilibrium can be restated in terms of field line rules:

For a conductor in electrostatic equilibrium,
- There are no field lines inside a conductor;
- Field lines that start or stop on the surface of a conductor are perpendicular to the surface where they intersect it; and
- The density of field lines at the surface of the conductor is greatest at regions of greatest curvature.

Figure 16.26 On a curved conducting surface, net charge is concentrated more densely at the regions of greatest curvature. The force that is effective at pushing the surface charges apart is the force parallel to the surface. Where the curvature is greatest, the component of the repulsive force parallel to the surface, F_s, is small compared to the repulsive force $\vec{F}$ that exists parallel to flat and gently curved surfaces. Because there is less repulsive force at a curved point, more charge is concentrated there.

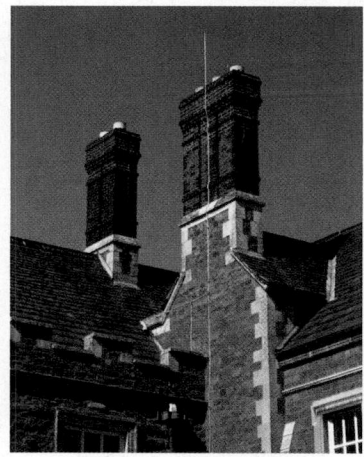

Figure 16.27 Lightning rods protect this college dormitory.

Making The Connection: lightning rods

Conceptual Example 16.10

Spherical Conductor in a Uniform Applied Field

Two oppositely charged parallel plates produce a uniform electric field between them (Fig. 16.28a). An uncharged metal sphere is placed between the plates. Assume that the sphere is small enough that it does not affect the charge distribution on the plates. Sketch the electric field lines between the plates once electrostatic equilibrium is reached.

Strategy We expect electrons in the metal sphere to be attracted to the positive plate, leaving the surface near the positive plate with a negative surface charge. The other side will have a positive surface charge. The electric field is changed by these surface charges, so that it is no longer uniform.

Solution and Discussion There are no field lines inside the metal sphere. The field lines cannot "go around" the sphere, since then there would be a field component parallel to the sphere's surface. Furthermore, since we already know that there is charge on the sphere's surface, some field lines must start on the positive side and others end on the negative side. The field lines must intersect the sphere perpendicular to the surface. Figure 16.28b shows a field line diagram for the sphere.

Conceptual Practice Problem 16.10   Oppositely charged spheres

Two metal spheres of the same radius R are given charges of equal magnitude and opposite sign. No other charges are nearby. Sketch the electric field lines when the center-to-center distance between the spheres is approximately $3R$.

(a)

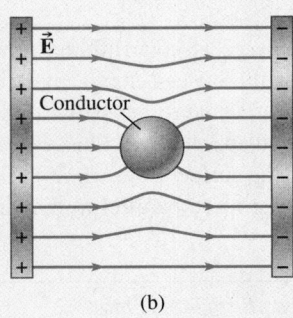

(b)

Figure 16.28
(a) Uniform field between two plates; (b) field lines when a metal sphere is placed between the plates.

Making The Connection:
electrostatic
precipitator

Electrostatic Precipitator

One direct application of electric fields is the *electrostatic precipitator*—a device that reduces the air pollution emitted from industrial smokestacks (Fig. 16.29). Many industrial processes, such as the burning of fossil fuels in electrical generating plants, release flue gases containing particulates into the air. To reduce the quantity of particulates released, the gases are sent through a precipitator chamber before leaving the smokestack.

Inside the precipitator chamber is a set of oppositely charged metal plates. The positively charged plates are fitted with needle-like wire projections that serve as discharge points. The electric field is strong enough at these points to ionize air molecules. The particulates are positively charged by contact with the ions. The electric field between

Figure 16.29   An electrostatic precipitator

the plates then attracts the particulates to the negatively charged collection plates. After enough particulate matter has built up on these plates, it falls to the bottom of the precipitator chamber from where it is easily removed.

16.7 GAUSS'S LAW FOR ELECTRIC FIELDS

Gauss's law is a powerful statement of properties of the electric field. It relates the electric field on a closed surface—*any* closed surface—to the net charge inside the surface. A **closed surface** encloses a volume of space, so that there is an inside and an outside. The surface of a sphere, for instance, is a closed surface, whereas the interior of a circle is not. Gauss's law says: I can tell you how much charge you have inside that "box" without looking inside; I'll just look at the field lines that enter or exit the box.

If a box has no charge inside of it, then the same number of field lines that go into the box must come back out; there is nowhere for field lines to end or to begin. Even if there is charge inside, but the *net* charge is zero, the same number of field lines that start on the positive charge must end on the negative charge, so again the same number of field lines that go in must come out. If there is net positive charge inside, then there will be field lines starting on the positive charge that leave the box; then more field lines come out than go in. If there is net negative charge inside, some field lines that go in end on the negative charge; more field lines go in than come out.

Field lines are a useful device for visualization but they are not quantifiable in any standard way. In order for Gauss's law to be useful, we formulate it mathematically so that numbers of field lines are not involved. To reformulate the law, there are two conditions to satisfy. First, a mathematical quantity must be found that is proportional to the number of field lines leaving a closed surface. Second, a proportionality must be turned into an equation by solving for the constant of proportionality.

Recall from the previous section that the magnitude of the electric field is proportional to the number of field lines *per unit cross-sectional area*:

$$E \propto \frac{\text{number of lines}}{\text{area}}$$

If a surface of area A is everywhere perpendicular to an electric field of uniform magnitude E, then the number of field lines that cross the surface is proportional to EA, since

$$\text{number of lines} = \frac{\text{number of lines}}{\text{area}} \times \text{area} \propto EA$$

This is only true if the surface is perpendicular to the electric field everywhere. As an analogy, think of rain falling straight down into a bucket. Less rainwater enters the bucket when it is tilted to one side than if the bucket rests with its opening perpendicular to the direction of rainfall. In general, the number of field lines crossing a surface is proportional to the *perpendicular component* of the field times the area:

$$\text{number of lines} \propto E_\perp A = EA \cos \theta$$

where θ is the angle that the field lines make with the normal (perpendicular) to the surface (Fig. 16.30a). The component of the electric field parallel to the surface, $E_\parallel$, does not contribute to the number of lines crossing the surface; only $E_\perp$ does (Fig. 16.30b). Equivalently, Fig. 16.30c shows that the number of lines crossing the surface is the same as the number crossing a surface of area $A \cos \theta$, which is the area perpendicular to the field.

The mathematical quantity that is proportional to the number of field lines crossing a surface is called the **flux of the electric field** (symbol Φ_E; Φ is the Greek capital phi).

Definition of flux:

$$\Phi_E = E_\perp A = EA_\perp = EA \cos \theta \qquad (16\text{-}7)$$

For a closed surface, flux is defined to be positive if more field lines leave the surface than enter, or negative if more lines enter than leave. Flux is then positive if the net enclosed charge is positive and negative if the net enclosed charge is negative.

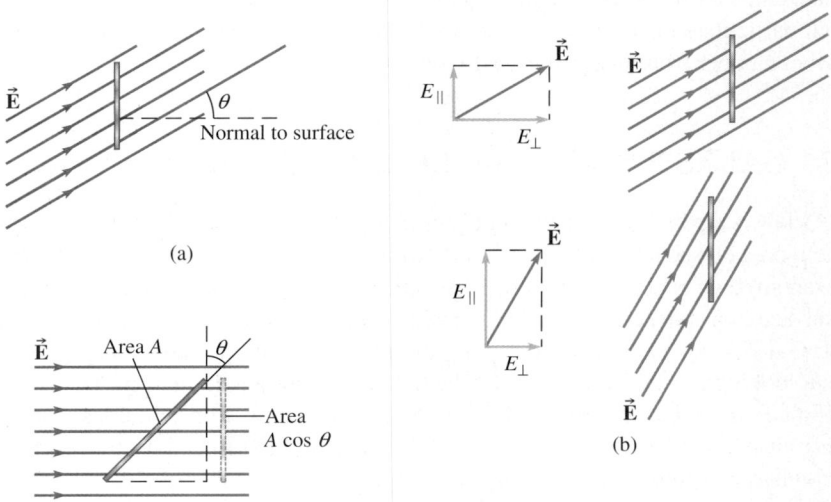

Figure 16.30 (a) Electric field lines crossing through a rectangular surface (side view). The angle between the field lines and the line *perpendicular* to the surface is θ. (b) In these two examples, the magnitude of the field is the same. Fewer field lines cross the surface when the perpendicular component of $\vec{E}$ is smaller. (c) The number of field lines that cross the surface of area A is the same as the number that cross the perpendicular surface of area $A \cos \theta$.

Since the net number of field lines is proportional to the net charge inside a closed surface, Gauss's law takes the form

$$\Phi_E = \text{constant} \times q$$

where q stands for the *net charge enclosed by the surface*. In Problem 43, you can show that the constant of proportionality is $4\pi k = 1/\epsilon_0$. Therefore,

---

Gauss's law
$$\Phi_E = 4\pi k q = q/\epsilon_0 \qquad (16\text{-}8)$$

Example 16.11

Flux through a Sphere

What is the flux through a sphere of radius $r = 5.0$ cm that has a point charge $q = -2.0$ μC at its center?

Strategy In this case there are two ways to find the flux. The electric field is known from Coulomb's law and it can be used to find the flux; or we can use Gauss's law.

Solution The electric field at a separation r from a point charge is

$$E = \frac{kq}{r^2}$$

For a negative point charge, the field is radially inward. The field has the same strength everywhere on the sphere, since the separation from the point charge is constant. Also, the field is always perpendicular to the surface of the sphere ($\theta = 0$ everywhere). Therefore,

$$\Phi_E = EA = \frac{kq}{r^2} \times 4\pi r^2 = 4\pi k q$$

This is exactly what Gauss's law tells us. The flux is independent of the radius of the sphere, since all the field lines cross the sphere regardless of its radius. A negative value of q gives a negative flux, which is correct since the field lines go inward. Then

$$\Phi_E = 4\pi k q$$

$$= 4\pi \times \left(9.0 \times 10^9 \, \frac{\text{N} \cdot \text{m}^2}{\text{C}^2}\right) \times (-2.0 \times 10^{-6} \, \text{C}) = -2.3 \times 10^5 \, \frac{\text{N} \cdot \text{m}^2}{\text{C}}$$

Discussion In this case, we can find the flux directly because the field at every point on the sphere is constant in magnitude and perpendicular to the sphere. However, Gauss's law tells us that the flux through *any* surface that encloses this charge, no matter what shape or size, must be the same.

Practice Problem 16.11 Flux through a side of a cube

What is the flux through *one side* of a cube that has a point charge −2.0 μC at its center? [*Hint:* Of the total number of field lines, what fraction passes through one side of the cube?]

As presented so far, Gauss's law is a way to determine how much charge is inside a closed surface given the electric field on the surface. Sometimes it can be turned around and used to *find the electric field* due to a distribution of charges. Why not just use

Coulomb's law? In many cases there are such a large number of charges that the charge can be viewed as being continuously spread along a line, or over a surface, or throughout a volume. Microscopically, charge is still limited to multiples of the electronic charge, but when there are large numbers of charges, it is simpler to view the charge as a continuous distribution.

For a continuous distribution, the **charge density** is usually the most convenient way to describe how much charge is present. There are three kinds of charge densities:

- If the charge is spread throughout a volume, the relevant charge density is the charge per unit *volume* (symbol ρ).

- If the charge is spread over a two-dimensional surface, then the charge density is the charge per unit *area* (symbol σ).

- If the charge is spread over a one-dimensional line or curve, the appropriate charge density is the charge per unit *length* (symbol λ).

Gauss's law alone is not sufficient to calculate the electric field. It can only be used in cases where there is enough *symmetry* to tell us something about the field lines. Example 16.12 illustrates this technique.

Example 16.12

Electric Field at a Distance from a Long Thin Wire

Charge is spread *uniformly* along a long thin wire. The charge per unit length on the wire is λ and is constant. Find the electric field at a distance r from the wire, far from either end of the wire.

Strategy The electric field at any point is the sum of the electric field contributions from the charge all along the wire. Coulomb's law tells us that the strongest contributions come from the charge on nearby parts of the wire, with contributions falling off as $1/r^2$ for faraway points. When concerned only with points near the wire, and far from either end, an approximately correct answer is obtained by assuming the wire is *infinitely long*.

How is it a simplification to add more charges? It would not be if we had to use Coulomb's law. When using Gauss's law, a symmetric situation is far simpler than a situation that lacks symmetry. An infinitely long wire with a uniform linear charge density has *axial symmetry*. Sketching the field lines first helps show what symmetry tells us about the electric field.

Solution We start by sketching field lines for an infinitely long wire. The field lines either start or stop on the wire (depending on whether the charge is positive or negative). Then what do the field lines do? The only possibility is that they move radially outward (or inward) from the wire. Figure 16.31a shows sketches of the field lines for positive and negative charges, respectively. The wire looks the same from all sides, so a field line could not start to curl around as in Fig. 16.31b: how would it determine which way to go? Also, the field lines cannot go along the wire as in Fig. 16.31c: again, how could the lines decide whether to go right or left? The wire looks exactly the same in both directions.

Once we recognize that the field lines are radial, the next step is to choose a surface. Gauss's law is easiest to handle if the electric field is constant in magnitude and perpendicular to the surface. A cylinder with a radius r with the wire as its axis

has the field perpendicular to the surface everywhere, since the lines are radial (Fig. 16.32a). The magnitude of the field must also be constant on the surface of the cylinder because every point on the cylinder is located an equal distance from the wire. Since a *closed* surface is necessary, the two circular ends of the cylinder are included. The flux through the ends is zero since no field lines pass through; equivalently, the *perpendicular component* of the field is zero (Fig. 16.32b).

Since the field is constant in magnitude and perpendicular to the surface, the flux is

$$\Phi_E = E_r A$$

where E_r is the radial component of the field. E_r is positive if the field is radially outward and negative if the field is radially inward. A is the area of a cylinder of radius r and . . . what length? Since the cylinder is imaginary, we can consider an arbitrary length denoted by L. The area of the cylinder is (Appendix A.6)

$$A = 2\pi r L$$

(a)

(b)   (c)

Figure 16.31
(a) Electric field lines emanating from a long wire, radially outward and radially inward; (b) lines circling a wire; (c) lines parallel to the wire.

continued on next page

Example 16.12 *continued*

How much charge is enclosed by this cylinder? The charge per unit length is λ and a length L of the wire is inside the cylinder, so the enclosed charge is

$$q = \lambda L$$

which can be either positive or negative. Gauss's law and the definition of flux yield

$$4\pi k q = \Phi_E = E_r A$$

Substituting the expressions for A and q into Gauss's law yields

$$E_r \times (2\pi r L) = 4\pi k \lambda L$$

Solving for E_r,

$$E_r = \frac{2k\lambda}{r}$$

The field direction is radially outward for $\lambda > 0$ and radially inward for $\lambda < 0$.

Discussion   The final expression for the electric field does not depend on the arbitrary length L of the cylinder. If L appeared in the answer, we would know to look for a mistake.

We should check the units of the answer: λ is the charge per unit length so it has SI units

$$[\lambda] = \frac{C}{m}$$

The constant k has SI units

$$[k] = \frac{N \cdot m^2}{C^2}$$

The factor of 2π is dimensionless and r is a distance. Then

$$\left[\frac{2k\lambda}{r}\right] = \frac{C}{m} \times \frac{N \cdot m^2}{C^2} \times \frac{1}{m} = \frac{N}{C}$$

which is the SI unit of electric field.

The electric field falls off as the inverse of the separation ($E \propto 1/r$). Wait a minute—does this violate Coulomb's law, which says $E \propto 1/r^2$? No, because that is the field at a separation r from a *point charge*. Here the charge is spread out in a line. The different geometry changes the field lines (they come radially outward from a line rather than from a point) and this changes how the field depends on distance.

Conceptual Practice Problem 16.12   Which area to use?

In Example 16.12, we wrote the area of a cylinder as $A = 2\pi r L$, which is only the area of the curved surface of the cylinder. The total area of a cylinder includes the area of the circles on each end (top and base): $A_{total} = 2\pi r L + 2\pi r^2$. Why did we not include the area of the ends of the cylinder when calculating flux?

Figure 16.32
(a) Electric field lines from a wire located along the axis of a cylinder are perpendicular to the surrounding imaginary cylindrical surface. (b) Top, bottom, and (c) side views of the cylinder and the field lines; the field lines are perpendicular to the cylindrical surface area but parallel to the plane of the top and bottom circular areas.

MASTER THE CONCEPTS

Summary

- Coulomb's law gives the electric force exerted on one point charge due to another. The magnitude of the force is

$$F = \frac{k|q_1||q_2|}{r^2} \qquad (2\text{-}5)$$

where the Coulomb constant is

$$k = 8.99 \times 10^9 \, \frac{\text{N} \cdot \text{m}^2}{\text{C}^2} \qquad (16\text{-}2a)$$

- The direction of the force on one point charge due to another is either directly toward the other charge (if the charges have opposite signs) or directly away (if the charges have the same sign).
- The electric field (symbol $\vec{\mathbf{E}}$) is the electrical force per unit *charge*. It is a vector quantity.
- If a point charge q is located where the electric field due to all other charges is $\vec{\mathbf{E}}$, then the electric force on the point charge is

$$\vec{\mathbf{F}}_e = q\vec{\mathbf{E}} \qquad (16\text{-}3b)$$

- The SI units of the electric field are N/C.
- Electric field lines are useful for representing an electric field.
- The direction of the electric field at any point is tangent to the field line passing through that point and in the direction indicated by the arrows on the field line.
- The electric field is strong where field lines are close together and weak where they are far apart.
- Field lines never cross.
- Lines start on positive charges and end on negative charges.
- The number of lines starting on a positive charge (or ending on a negative charge) is proportional to the magnitude of the charge.

- The principle of superposition says that the electric field due to a collection of charges at any point is the vector sum of the electric fields caused by each charge separately.
- Electric flux:

$$\Phi_E = E_\perp A = E A_\perp = E A \cos \theta \qquad (16\text{-}7)$$

- Gauss's law:

$$\Phi_E = 4\pi k q = q/\epsilon_0 \qquad (16\text{-}8)$$

Highlighted Figures and Tables

T16.1 Masses and electrical charges of the proton, electron, and neutron (p. 547)

F16.2 Polarized molecules within a bit of paper (p. 548)

F16.5 Charging a conductor (p. 550)

F16.6 Charging by induction (p. 551)

F16.9 The electric force on two opposite charges and two like charges (p. 555)

F16.17 Field line rules illustrated (p. 560)

F16.18 Electric field lines of isolated point charges (p. 561)

F16.19 Electric field lines for a dipole (p. 562)

F16.22 Electric field generated by *Gymnarchus* (p. 565)

F16.23 Uniform electric field between two parallel metal plates (p. 565)

F16.25 An electron deflected by an electric field (p. 567)

F16.26 Net charge is concentrated more densely at the regions of greatest curvature on a curved conducting surface (p. 569)

F16.29 An electrostatic precipitator (p. 570)

F16.30 Electric field lines crossing through a rectangular surface (p. 572)

CONCEPTUAL QUESTIONS

1. (a) List three similarities between gravity and the electric force. (b) List two differences.

2. Due to the similarity between Newton's law of gravity and Coulomb's law, a friend proposes these hypotheses: perhaps there is no gravitational interaction at all. Instead, what we call gravity might be *electric* forces acting between objects that are almost, but not quite, electrically neutral. Think up as many counterarguments as you can.

3. What makes clothes cling together—or to your body—after they've been through the dryer? Why do they not cling as much if they are taken out of the dryer while slightly damp? In which case would you expect your clothes to cling more, all other things being equal: when the clothes in the dryer are all made of the same material, or when they are made of several different materials?

4. Explain why any net charge on a solid metal conductor is found on the outside surface of the conductor instead of being distributed uniformly throughout the solid.

5. Explain why electric field lines begin on positive charges and end on negative charges. [*Hint:* What is the direction of the electric field near positive and negative charges?]

6. Electronic devices are usually enclosed in metal boxes. One function of the box is to shield the inside components from external electric fields. (a) How does this shielding work? (b) Why is the degree of shielding better for constant or slowly varying fields than for rapidly varying fields? (c) Explain the reasons why it is not possible to shield something from gravitational fields in a similar way.

7. A metal sphere is initially uncharged. After being touched by a charged rod, the metal sphere is positively charged. (a) Is

the mass of the sphere larger, smaller, or the same as before it was charged? Explain. (b) What sign of charge is on the rod?

8. Your laboratory partner hands you a glass rod and asks if it has negative charge on it. There is an electroscope in the laboratory. How can you tell if the rod is charged? Can you determine the sign of the charge? If the rod is charged to begin with, will its charge be the same after you have made your determination? Explain.

9. A lightweight plastic rod is rubbed with a piece of fur. A second plastic rod, hanging from a string, is attracted to the first rod and swings toward it. When the second rod touches the first, it is suddenly repelled and swings away. Explain what has happened.

10. The following *hypothetical* reaction shows a neutron (n) decaying into a proton (p^+) and an electron (e^-):

$$n \rightarrow p^+ + e^-$$

At first there is no charge, but then charge seems to be "created." Does this reaction violate the law of charge conservation? Explain. (In Section 29.3 it is shown that the neutron does not decay into just a proton and an electron; the decay products include a third, electrically neutral particle.)

11. A fellow student says that there is *never* an electric field inside a conductor. Do you agree? Explain.

12. Explain why electric field lines never cross.

13. A truck carrying explosive gases either has chains or straps that drag along the ground, or else it has special tires that conduct electricity (ordinary tires are good insulators). Explain why the chains, straps, or conducting tires are necessary.

14. An electroscope consists of a conducting sphere, conducting pole, and two metal foils (Fig. 16.7a). The electroscope is initially uncharged. (a) A positively charged rod is allowed to touch the conducting sphere and then is removed. What happens to the foils and what is their charge? (b) Next, another positively charged rod is brought near to the conducting sphere without touching it. What happens? (c) The positively charged rod is removed and a negatively charged rod is brought near the sphere. What happens?

15. A rod has been negatively charged by rubbing it with fur. It is brought near another rod of unknown composition and charge. There is a repulsive force between the two. (a) Is the first rod an insulator or a conductor? Explain. (b) What can you tell about the charge of the second rod?

16. A negatively charged rod is brought near a grounded conductor. After the ground connection is broken, the rod is removed. Is the charge on the conductor positive, negative, or zero? Explain.

17. In some textbooks, the electric field is called the *flux density*. Explain the meaning of this term. Does flux density mean the flux per unit volume? If not, then what does it mean?

18. The word *flux* comes from the Latin "to flow." What does the quantity $\Phi_E = E_\perp A$ have to do with flow? Figure 16.33 shows some streamlines for the flow of water in a pipe. The streamlines are actually field lines for the *velocity field*. What is the physical significance of the quantity $v_\perp A$? Sometimes physicists call positive charges *sources* of the electric field and negative charges *sinks*. Why?

Figure 16.33   Conceptual Question 18

19. The flux through a closed surface is zero. Is the electric field necessarily zero? Is the net charge inside the surface necessarily zero? Explain your answers.

20. Consider a closed surface (Fig. 16.34) that surrounds Q_1 and Q_2 but not Q_3 or Q_4. (a) Which charges contribute to the electric field at point P? (b) Would the value obtained for the flux through the surface, calculated using only the electric field due to Q_1 and Q_2, be greater than, less than, or equal to that obtained using the total field?

Figure 16.34
Conceptual Question 20

MULTIPLE CHOICE QUESTIONS

1. An α particle (charge $+2e$ and mass approximately $4m_p$) is on a collision course with a proton (charge $+e$ and mass m_p). Assume that no forces act other than the electrical repulsion. Which one of these statements about the accelerations of the two particles is true?

   (a) $\vec{\mathbf{a}}_\alpha = \vec{\mathbf{a}}_p$ (b) $\vec{\mathbf{a}}_\alpha = 2\vec{\mathbf{a}}_p$ (c) $\vec{\mathbf{a}}_\alpha = 4\vec{\mathbf{a}}_p$
   (d) $2\vec{\mathbf{a}}_\alpha = \vec{\mathbf{a}}_p$ (e) $4\vec{\mathbf{a}}_\alpha = \vec{\mathbf{a}}_p$ (f) $\vec{\mathbf{a}}_\alpha = -\vec{\mathbf{a}}_p$
   (g) $\vec{\mathbf{a}}_\alpha = -2\vec{\mathbf{a}}_p$ (h) $\vec{\mathbf{a}}_\alpha = -4\vec{\mathbf{a}}_p$ (i) $-2\vec{\mathbf{a}}_\alpha = \vec{\mathbf{a}}_p$
   (j) $-4\vec{\mathbf{a}}_\alpha = \vec{\mathbf{a}}_p$

2. The electric charge on a conductor is
   (a) uniformly distributed throughout the volume.
   (b) confined to the surface and is uniformly distributed.
   (c) mostly on the outer surface, but is not uniformly distributed.
   (d) entirely on the surface and is distributed according to the shape of the object.
   (e) dispersed throughout the volume of the object and distributed according to the object's shape.

3. The electric field at a point in space is a measure of
   (a) the total charge on an object at that point.
   (b) the electric force on any charged object at that point.
   (c) the charge-to-mass ratio of an object at that point.
   (d) the electric force per unit mass on a point charge at that point.
   (e) the electric force per unit charge on a point charge at that point.

4. Two charged particles attract each other with a force of magnitude F acting on each. If the charge of one is doubled and the distance separating the particles is also doubled, the force acting on each of the two particles has magnitude
   (a) $F/2$ (b) $F/4$ (c) F (d) $2F$ (e) $4F$
   (f) None of the above.

5. A charged insulator and an uncharged metal
   (a) exert no electric force on each other.
   (b) repel each other electrically.
   (c) attract each other electrically.
   (d) attract or repel, depending on whether the charge is positive or negative.

6. A tiny charged pellet of mass m is suspended at rest by the electric field between two horizontal, charged metallic plates (Fig. 16.35). The lower plate has a positive charge and the upper plate has a negative charge. Which statement in the answers below is *not* true?
   (a) The electric field between the plates points vertically upward.
   (b) The pellet is negatively charged.
   (c) The magnitude of the electric force on the pellet is equal to mg.
   (d) If the magnitude of charge on the plates is increased, the pellet begins to move upward.

Pellet

Figure 16.35 Multiple Choice Question 6

7. Which of these statements comparing electrical and gravitational forces is correct?
   1. The direction of the electric force exerted by one point particle on another is always the same as the direction of the gravitational force exerted by that particle on the other.
   2. The electric and gravitational forces exerted by two particles on one another are inversely proportional to the separation of the particles.
   3. The electric force exerted by one planet on another is typically stronger than the gravitational force exerted by that same planet on the other.
   (a) 1 only   (b) 2 only   (c) 3 only
   (d) none of them

8. In Fig. 16.36, which best represents the field lines due to two spheres with equal and opposite charges?

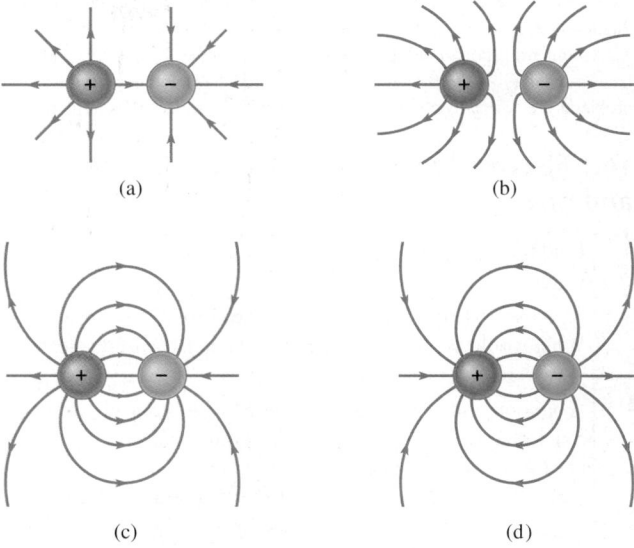

Figure 16.36 Multiple Choice Question 8

9. In Fig. 16.37 put points 1–4 in order of increasing field strength.
   (a) 2, 3, 4, 1   (b) 2, 1, 3, 4   (c) 1, 4, 3, 2
   (d) 4, 3, 1, 2   (e) 2, 4, 1, 3

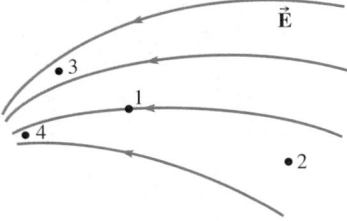

Figure 16.37
Multiple Choice Question 9

10. Two point charges q and $2q$ lie on the x-axis (Fig. 16.38). Which region(s) on the x-axis include a point where the electric field due to the two point charges is zero?
    (a) to the right of $2q$ (b) between $2q$ and point P
    (c) between point P and q (d) to the left of q
    (e) both (a) and (c)   (f) both (b) and (d)

Figure 16.38 Multiple Choice Question 10; Problems 15 and 27 $(q > 0)$

PROBLEMS

Note: **C** indicates a combination conceptual/quantitative problem. Gold diamonds ✦, ✦✦ are used to indicate the increasing level of difficulty of each problem. Problem numbers appearing in blue, 9., denote problems that have a detailed solution available in the Student Solutions Manual. Some problems are *paired* by concept; their numbers are connected by a ruled box.

16.1 Electric Charge; 16.2 Conductors and Insulators

1. Find the total positive charge of all the protons in 1.0 mol of water.

2. Suppose a 1.0-g nugget of pure gold has zero net charge. What would be its net charge after it has 1.0% of its electrons removed?

C 3. A balloon, initially neutral, is rubbed with fur until it acquires a net charge of –0.60 nC. (a) Assuming that only electrons are transferred, were electrons removed from the balloon or added to it? (b) How many electrons were transferred?

C 4. A metallic sphere has a charge of +4.0 nC. A negatively charged rod has a charge of –6.0 nC. When the rod touches the sphere, 8.2×10^9 electrons are transferred. What are the charges of the sphere and the rod now?

C 5. A positively charged rod is brought near two uncharged conducting spheres of the same size that are initially touching each other (Fig. 16.39a). The spheres are moved apart and then the charged rod is removed (Fig. 16.39b). (a) What is the sign of the net charge on sphere a in Fig. 16.39b? (b) In comparison with the charge on sphere a, how much and what sign of charge is on sphere b?

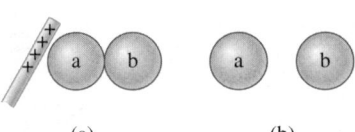

(a)   (b)

Figure 16.39
Problem 5

16.3 Coulomb's Law

6. If the electrical force of repulsion between two 1 C charges is 10 N, how far apart are they?

7. Two small metal spheres are 25.0 cm apart. The spheres have equal amounts of negative charge and repel each other with a force of 0.036 N. What is the charge on each sphere?

8. How many electrons must be removed from each of two 5.0-kg copper spheres to make the electric force of repulsion between them equal in magnitude to the gravitational attraction between them?

9. What is the ratio of the electric force to the gravitational force between a proton and an electron separated by 5.3×10^{-11} m (the radius of a hydrogen atom)?

10. Three point charges are fixed in place in a right triangle (Fig. 16.40). What is the electric force on the –0.60-μC charge due to the other two charges?

11. Three point charges are fixed in place in a right triangle (Fig. 16.40). What is the electric force on the +1.0-μC charge due to the other two charges?

Figure 16.40
Problems 10 and 11

12. A +2.0-nC point charge is 3.0 cm away from a –3.0-nC point charge. (a) What are the magnitude and direction of the electric force acting on the +2.0-nC charge? (b) What are the magnitude and direction of the electric force acting on the –3.0-nC charge?

13. Two Styrofoam balls with the same mass $m = 9.0 \times 10^{-8}$ kg and the same positive charge Q are suspended from the same point by insulating threads of length $L = 0.98$ m (Fig. 16.41). The separation of the balls is $d = 0.020$ m. What is the charge Q?

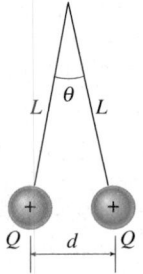

Figure 16.41 Problem 13

14. Two metal spheres separated by a distance much greater than either sphere's radius have equal mass m and equal electric charge q. What is the ratio of charge to mass q/m in C/kg if the electrical and gravitational forces balance?

15. In Fig. 16.38, a third point charge $-q$ is placed at point P. What is the electric force on $-q$ due to the other two point charges?

16. A K$^+$ ion and a Cl$^-$ ion are directly across from each other on opposite sides of a membrane 9.0 nm thick. What is the electric force on the K$^+$ ion due to the Cl$^-$ ion? Ignore the presence of other charges.

17. Two point charges are separated by a distance r and repel each other with a force F. If their separation is reduced to 0.25 times the original value, what is the magnitude of the force of repulsion between them?

✦ 18. Using the three point charges of Example 16.3, find the magnitude of the force on q_2 due to the other two charges, q_1 and q_3. [*Hint:* After finding the force on q_2 due to q_1, separate that force into x- and y-components.]

✦ 19. An equilateral triangle (Fig. 16.42) has a point charge $+q$ at each of the three vertices (A, B, C). Another point charge Q is placed at D, the midpoint of the side BC. Solve for Q if the total electric force on the charge at A due to the charges at B, C, and D is zero.

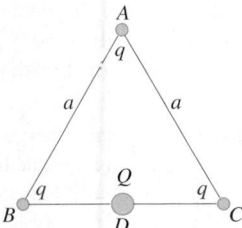

Figure 16.42 Problem 19

16.4 The Electric Field

20. A small sphere with a charge of -6.0×10^{-7} C is placed in a uniform electric field of magnitude 1.2×10^6 N/C pointing to the west. What is the magnitude and direction of the force on the sphere due to the electric field?

21. The electric field across a cellular membrane is 1.0×10^7 N/C directed into the cell. (a) If a pore opens, which way do sodium ions (Na^+) flow—into the cell or out of the cell? (b) What is the magnitude of the electric force on the sodium ion? The charge on the sodium ion is $+e$.

22. Sketch the electric field lines near two isolated and equal (a) positive point charges and (b) negative point charges. Include arrowheads to show the field directions.

23. Sketch the electric field lines in the plane of the page due to the charges shown in Fig. 16.43.

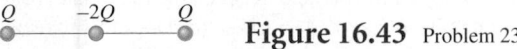

Figure 16.43 Problem 23

24. What are the magnitude and direction of the acceleration of an electron at a point where the electric field has magnitude 6100 N/C and is directed due north?

25. What are the magnitude and direction of the acceleration of a proton at a point where the electric field has magnitude 33 kN/C and is directed straight up?

26. What are the magnitude and direction of the electric field midway between two point charges, -15 μC and $+12$ μC, which are 8.0 cm apart?

27. In Fig. 16.38, what is the electric field at point P?

28. Two point charges, $q_1 = +20.0$ nC and $q_2 = +10.0$ nC, are located on the x-axis at $x = 0$ and $x = 1.00$ m, respectively. Where on the x-axis is the electric field equal to zero?

29. Two electric charges, $q_1 = +20.0$ nC and $q_2 = +10.0$ nC, are located on the x-axis at $x = 0$ m and $x = 1.00$ m, respectively. What is the magnitude of the electric field at the point $x = 0.50$ m, $y = 0.50$ m?

30. Two equal charges ($Q = +1.00$ nC) are situated at the diagonal corners A and B of a square of side 1.0 m (Fig. 16.44). What is the magnitude of the electric field at point D?

Figure 16.44 Problem 30

31. Suppose a charge q is placed at point $x = 0$, $y = 0$. A second charge q is placed at point $x = 8.0$ m, $y = 0$. What charge must be placed at the point $x = 4.0$ m, $y = 0$ in order that the field at the point $x = 4.0$ m, $y = 3.0$ m be zero?

32. An electron traveling horizontally from west to east enters a region where a uniform electric field is directed upward. What is the direction of the electric force exerted on the electron once it has entered the field?

33. A negative point charge $-Q$ is situated near a large metal plate that has a total charge of $+Q$ (Fig. 16.45). Sketch the electric field lines.

Figure 16.45 Problem 33

16.5 Motion of a Point Charge in a Uniform Electric Field

34. An electron ($q = -1.602 \times 10^{-19}$ C) is projected horizontally into the space between two oppositely charged metal plates. The electric field between the plates is 500.0 N/C, directed up. (a) While in the field, what is the force on the electron? (b) If the vertical deflection of the electron as it leaves the plates is 3.00 mm, how much has its kinetic energy increased due to the electric field?

35. A horizontal beam of electrons initially moving at 4.0×10^7 m/s is deflected vertically by the vertical electric field between oppositely charged parallel plates (Fig. 16.46). The magnitude of the field is 2.00×10^4 N/C. (a) What is the direction of the field between the plates? (b) What is the charge per unit area on the plates? (c) What is the vertical deflection d of the electrons as they leave the plates?

Figure 16.46
Problem 35

36. After the electrons in Example 16.8 pass through the anode, they are moving at a speed of 8.4×10^6 m/s. They next pass between a pair of parallel plates ((A) in Fig. 16.24). The plates each have an area of 2.50 cm by 2.50 cm and they are separated by a distance of 0.50 cm. The uniform electric field between them is 1.0×10^3 N/C and the plates are charged as shown. (a) In what direction are the electrons deflected? (b) By how much are the electrons deflected after passing through these plates?

37. After the electrons pass through the parallel plates in Problem 36, they pass between another set of parallel plates ((B) in Fig. 16.24). These plates also have an area of 2.50 cm by 2.50 cm and are separated by a distance of 0.50 cm. (a) In what direction must the field be oriented so that the electrons are deflected vertically upward? (b) If we neglect the gravitational force, how strong must the field be between these plates in order for the electrons to be deflected by 2.0 mm? (c) How much less will the electrons be deflected if we *do* include the gravitational force?

16.6 Conductors in Electrostatic Equilibrium

38. A hollow conducting sphere of radius R carries a negative charge $-q$. Write expressions for the electric field $\vec{E}$ inside $(r < R)$ and outside $(r > R)$ the sphere. Also indicate the direction of the field. (b) Sketch a graph of the field strength as a function of r. [*Hint:* See Conceptual Example 16.6.]

C 39. A conducting sphere that carries a total charge of -6 μC is placed at the center of a conducting spherical shell that carries a total charge of $+1$ μC (Fig. 16.47). Determine the charge on the *outer surface* of the shell. [*Hint:* Sketch a field line diagram.]

C 40. A conducting sphere that carries a total charge of $+6$ μC is placed at the center of a conducting spherical shell that also carries a total charge of $+6$ μC (Fig. 16.47). (a) Determine the charge on the inner surface of the shell. (b) Determine the total charge on the outer surface of the shell.

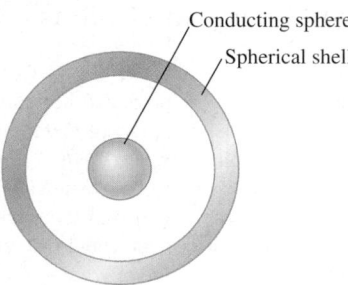

Figure 16.47
Problems 39 and 40

◆ 41. A conductor contains a cavity in which there are two point charges: $q_1 = +5$ μC and $q_2 = -12$ μC. The conductor itself carries a net charge -4 μC. How much charge is on (a) the inner surface of the conductor? (b) the outer surface of the conductor?

◆ 42. In fair weather, over flat ground, there is a downward electric field of about 150 N/C. (a) Assume that the Earth is a conducting sphere with charge on its surface. If the electric field just outside is 150 N/C radially inward, calculate the total charge on the Earth and the charge per unit area. (b) At an altitude of 250 m above the Earth's surface, the field is only 120 N/C. Calculate the charge density (charge per unit volume) of the air (assumed constant). [*Hint:* See Conceptual Example 16.6.]

16.7 Gauss's Law for Electric Fields

43. In this problem, you can show from Coulomb's law that the constant of proportionality in Gauss's law must be $1/\epsilon_0$. Imagine a sphere with its center at a point charge q. (a) Write an expression for the electric flux in terms of the field strength E and the radius r of the sphere. [*Hint:* The field strength E is the same everywhere on the sphere and the field lines cross the sphere perpendicular to its surface.] (b) Use Gauss's law in the form $\Phi_E = cq$ (where c is the constant of proportionality) and the electric field strength given by Coulomb's law to show that $c = 1/\epsilon_0$.

44. (a) Find the electric flux through each side of a cube of edge length a in a uniform electric field of magnitude E. The field direction is perpendicular to two of the faces. (b) What is the total flux through the cube?

45. In a uniform electric field of magnitude E, the field lines cross through a rectangle of area A at an angle of $60.0°$ with respect to the plane of the rectangle. What is the flux through the rectangle?

◆ 46. A thin, flat sheet of charge has a uniform surface charge density σ ($\sigma/2$ on each side). (a) Sketch the field lines due to the sheet. (b) Sketch the field lines for an infinitely large sheet with the same charge density. (c) For the infinite sheet, how does the field strength depend on the distance from the sheet? [*Hint:* Refer to your field line sketch.] (d) For points close to the finite sheet and far from its edges, can the sheet be approximated by an infinitely large sheet? [*Hint:* Again, refer to the field line sketches.] (e) Use Gauss's law to show that the magnitude of the electric field near a sheet of uniform charge density σ is $E = \sigma/(2\epsilon_0)$.

◆ 47. A flat *conducting* sheet of area A has a charge q on each surface. (a) What is the electric field inside the sheet? (b) Use Gauss's law to show that the electric field just outside the sheet is $E = q/(\epsilon_0 A) = \sigma/\epsilon_0$. (c) Does this contradict the result of Problem 46? Compare the field line diagrams for the two situations.

◆ 48. A *parallel-plate capacitor* consists of two flat metal plates of area A separated by a small distance d. The plates are given equal and opposite net charges $\pm q$. (a) Sketch the field lines and use your sketch to explain why almost all of the charge is on the inner surfaces of the plates. (b) Use Gauss's law to show that the electric field between the plates and away from the edges is $E = q/(\epsilon_0 A) = \sigma/\epsilon_0$. (c) Does this agree with or contradict the result of Problem 47? Explain. (d) Use the principle of superposition and the result of Problem 46 to arrive at this same answer. [*Hint:* The inner surfaces of the two plates are thin, flat sheets of charge.]

◆ 49. (a) Use Gauss's law to prove that the electric field *outside* any spherically symmetric charge distribution is the same as if all of the charge were concentrated into a point charge. (b) Now use Gauss's law to prove that the electric field *inside* a spherically symmetric charge distribution is zero if none of the charge is at a distance from the center less than that of the point where we determine the field.

50. Using the results of Problem 49, we can find the electric field at any radius for any spherically symmetric charge distribution. A solid sphere of charge of radius R has a total charge of q uniformly spread throughout the sphere. (a) Find the magnitude of the electric field for $r \geq R$. (b) Find the magnitude of the electric field for $r \leq R$. (c) Sketch a graph of $E(r)$ for $0 \leq r \leq 3R$.

◆ 51. A coaxial cable consists of a wire of radius a surrounded by a thin metal cylindrical shell of radius b (Fig. 16.48). The wire has a uniform linear charge density $\lambda > 0$ and the outer shell has a uniform linear charge density $-\lambda$. (a) Sketch the field

Figure 16.48
Problem 51

lines for this cable. (b) Find expressions for the magnitude of the electric field in the regions $r \leq a$, $a < r < b$, and $b \leq r$.

52. Power lines have a limit on the maximum size of the electric field they produce. In most states a maximum of 5 kN/C at about 20 m from the wires is allowed. This is quite large compared to the Earth's fair-weather electric field of about 100 N/C. Assume that the charge on the wire is static (not true, but a simplification here) and use the formula for the electric field for a wire derived in Example 16.12 to determine how much charge per unit length is on the wire.

✦ 53. Two concentric, infinitely long cylinders have radii r_1 and r_2 ($r_2 > r_1$), and corresponding surface charge densities σ_1 and σ_2. (a) Use Gauss's law to determine the electric field strength as a function of r between the two cylinders. (b) Sketch a graph of the electric field strength between the two cylinders as a function of r. Assume $\sigma_2 = 2\sigma_1 > 0$.

COMPREHENSIVE PROBLEMS

54. Consider two protons (charge $+e$), separated by a distance of 2.0×10^{-15} m (as in a typical atomic nucleus). The electric force between these protons is equal in magnitude to the gravitational force on an object of what mass near the Earth's surface?

55. In lab tests it was found that rats can detect electric fields of about 5.0 kN/C or more. If a point charge of 1.0 μC is sitting in a maze, how close must the rat come to the charge in order to detect it?

56. A raindrop inside a thundercloud has charge $-8e$. What is the electric force on the raindrop if the electric field at its location (due to other charges in the cloud) has magnitude 2.0×10^6 N/C and is directed up?

© ✦ 57. A thin wire with positive charge evenly spread along its length is shaped into a semicircle. What is the direction of the electric field at the center of curvature of the semicircle? Explain.

58. In a television tube, electrons initially at rest are accelerated by a uniform electric field of magnitude 4.0×10^5 N/C during the first 5.0 cm of the tube's length; then they move at essentially constant velocity another 45 cm before hitting the screen. (a) Find the speed of the electrons when they hit the screen. (b) How long does it take them to travel the length of the tube?

59. An electron beam in an oscilloscope is deflected by the electric field produced by oppositely charged metal plates. If the electric field between the plates is 2.00×10^5 N/C directed downward, what is the force on each electron when it passes between the plates?

60. A point charge $q_1 = +5.0$ μC is fixed in place at $x = 0$ and a point charge $q_2 = -3.0$ μC is fixed at $x = -20.0$ cm. Where can we place a point charge $q_3 = -8.0$ μC so that the net electric force on q_1 due to q_2 and q_3 is zero?

61. The Bohr model of the hydrogen atom proposes that the electron orbits around the proton in a circle of radius 5.3×10^{-11} m. The electric force is responsible for the centripetal acceleration of the electron. What is the speed of the electron in this model?

✦ 62. What is the electric force on the chloride ion in the lower right-hand corner in Fig. 16.49? Since the ions are in water, the "effective charge" on the chloride ions (Cl⁻) is -2×10^{-21} C and that of the sodium ions (Na⁺) is $+2 \times 10^{-21}$ C. The effective charge is a way to account for the partial shielding due to nearby water molecules. Assume that all four ions are coplanar.

© ✦ 63. (a) What would the net charges on the Sun and Earth have to be if the electric force instead of the gravitational force were responsible for keeping the Earth in its orbit? There are many possible answers, so restrict yourself to the case where the magnitude of the charges is proportional to the masses. (b) If the magnitude of the charges of the proton and electron were not exactly equal, astronomical bodies would have net charges that are approximately proportional to their masses. Could this possibly be an explanation for the Earth's orbit?

✦ 64. A dipole consists of two equal and opposite point charges ($\pm q$) separated by a distance d. (a) Write an expression for the electric field at a point $(0, y)$ on the dipole axis (see Fig. 16.50). Specify the direction of the field in all four regions: $y > \frac{1}{2}d$, $0 < y < \frac{1}{2}d$, $-\frac{1}{2}d < y < 0$, and $y < -\frac{1}{2}d$. (b) At distant points ($|y| \gg d$), write a simpler, approximate expression for the field. To what power of y is the field proportional? Does this conflict with Coulomb's law? [Hint: Use the binomial approximation $(1 \pm x)^n \approx 1 \pm nx$ for $x \ll 1$.]

✦ 65. A dipole consists of two equal and opposite point charges ($\pm q$) separated by a distance d. (a) Write an expression for the magnitude of the electric field at a point $(x, 0)$ a large distance ($x \gg d$) from the midpoint of the charges on a line perpendicular to the dipole axis (see Fig. 16.50). [Hint: Use small angle approximations.] (b) Give the direction of the field for $x > 0$ and for $x < 0$.

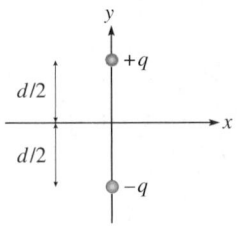

Figure 16.50 Problems 64 and 65

66. In a thunderstorm, charge is separated through a complicated mechanism that is ultimately powered by the Sun. A simplified model of the charge in a thundercloud represents the positive charge accumulated at the top and the negative charge at the bottom as a pair of point charges (Fig. 16.51). (a) What is the magnitude and direction of the electric field produced by the two point charges at point P, which is just above the surface

Figure 16.49
Problem 62

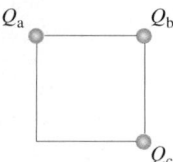

Figure 16.51
Problem 66

of the Earth? (b) Thinking of the Earth as a conductor, what sign of charge would accumulate on the surface near point P? (This accumulated charge increases the magnitude of the electric field near point P.)

67. Two point charges are located on the x-axis: a charge of +6.0 nC at $x = 0$ and an unknown charge q at $x = 0.50$ m. No other charges are nearby. If the electric field is zero at the point $x = 1.0$ m, what is q?

68. Three equal charges are placed on three corners of a square (Fig. 16.52). If the force that Q_a exerts on Q_b has magnitude F_{ab} and the force that Q_a exerts on Q_c has magnitude F_{ac}, what is the ratio of F_{ac} to F_{ab}?

Figure 16.52 Problem 68

69. Two identical conducting spheres carry charges of +5.0 μC and −1.0 μC respectively. They are initially a large distance L apart. The spheres are brought together, touched together, and then returned to their original separation L. What is the ratio of the magnitude of the force on either sphere after they are touched to that before they were touched?

70. In Fig. 16.53, regions A and C extend far to the left and right, respectively. The electric field due to the two point charges is zero at some point in which region or regions? Explain.

Figure 16.53
Problem 70

71. Two metal spheres of radius 5.0 cm carry net charges of +1.0 μC and +0.2 μC, respectively. (a) What (approximately) is the magnitude of the electrical repulsion on either sphere when their centers are 1.00 m apart? (b) Why cannot Coulomb's law be used to find the force of repulsion when their centers are 12 cm apart? (c) Would the actual force be larger or smaller than the result of using Coulomb's law with $r = 12$ cm? Explain.

72. A dipole consists of two opposite charges (q and $-q$) separated by a fixed distance d. The dipole is placed in an electric field in the $+x$-direction of magnitude E. The dipole axis makes an angle θ with the electric field as shown in Fig. 16.54. (a) Calculate the net electric force acting on the dipole. (b) Calculate the net torque acting on the dipole due to the electric forces as a function of θ. Let clockwise torque be positive. (c) Evaluate the net torque for $\theta = 0°$, $\theta = 36.9°$, and $\theta = 90.0°$. Let $q = \pm 3.0$ μC, $d = 7.0$ cm, and $E = 2.0 \times 10^4$ N/C.

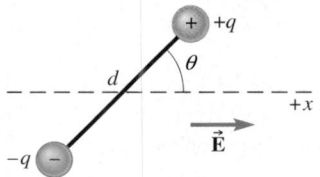

Figure 16.54 Problem 72

73. This problem illustrates the ideas behind the Millikan oil drop experiment—the first measurement of the electron charge. Millikan examined a fine spray of spherical oil droplets falling through air; the drops had picked up an electric charge as they were sprayed through an atomizer. He measured the terminal speed of a drop when there was no electric field and the electric field that kept the drop motionless between the plates of a capacitor. (a) With no electric field, the forces acting on the oil droplet were the gravitational force, the buoyant force, and the viscous drag. The droplets used were so tiny (a radius of about one thousandth of a millimeter) that they rapidly reached terminal velocity. Find the radius R of a drop in terms of the terminal velocity (v_t), the densities of the oil and of air (ρ_{oil} and ρ_{air}), the viscosity η, and the gravitational field strength (g). [*Hint:* At terminal velocity, the net force on the drop is zero. Use Stokes's law and don't forget buoyancy.] (b) Find the charge q of a drop in terms of the electric field strength between the plates (E), the distance between the plates (d), the radius of the drop (R), the densities of the oil and of air (ρ_{oil} and ρ_{air}), and the gravitational field strength (g). [*Hint:* The net force is again zero. There is an electric force but no viscous force (since the drop is at rest).]

ANSWERS TO PRACTICE PROBLEMS

16.1 7.5×10^{10} electrons

16.2 As the positively charged rod is moved away, the free electrons of the electroscope spread out more evenly. Since there is less net positive charge on the leaves, they do not hang as far apart.

16.3 8.7×10^{-2} N in the +y-direction

16.4 6.2×10^{-4} N

16.5 220 N/C to the right

16.6

(a) Inside the shell, field lines run from the positive charge spread about the surface to the negative charge located at the center of the shell.

(b) Outside the shell, we can imagine the charge +Q all concentrated at the center of the sphere where it cancels the −Q of the point charge. Therefore, E = 0 outside. Inside, the shell produces no electric field (as we found in the Example) so the field is just that due to the point charge −Q.

16.7 $\theta = 49.1°$

16.8 2.3×10^{5} m/s to the right

16.9 The proton is deflected downward, but it has a much smaller acceleration because it has a much larger mass than the electron ($m_p = 1.673 \times 10^{-27}$ kg). The proton's acceleration vertically downward is 4.36×10^{11} m/s². The y-displacement, after spending 5.00×10^{-9} s between the plates, is 5.44×10^{-6} m, or 5.44×10^{-4} cm. The proton is barely deflected at all before leaving the region between the plates.

16.10

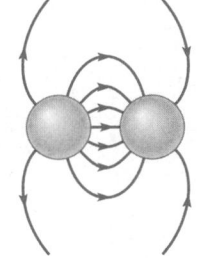

16.11 -3.8×10^{4} N·m²/C

16.12 On the ends, $\vec{E}$ is *parallel* to the surface, so the component of $\vec{E}$ perpendicular to the ends is zero and the flux through the ends is zero. No field lines *pass through* the ends of the cylinder.

Electric Potential

A tool widely used in medicine to diagnose the condition of the heart is the electrocardiogram (EKG). The EKG data is displayed on a graph that shows a pattern repeated with each beat of the heart. What physical quantity is measured in an EKG?

Concepts & Skills to Review

- potential energy (Section 6.5)
- gravitational forces (Section 2.5)
- gravitational potential energy (Section 6.5)
- Coulomb's law (Section 16.3)
- electric field inside a conductor (Section 16.4)

Potential energy is energy stored in a field.

17.1   ELECTRIC POTENTIAL ENERGY

In Chapter 6, we learned about gravitational potential energy—energy stored in a gravitational field. **Electric potential energy** is the energy stored in an *electric* field (Fig. 17.1). For both gravitational and electric potential energy, the *change* in potential energy when objects move around is equal in magnitude but opposite in sign to the work done by the field:

$$\Delta U = -W_{field} \qquad (6\text{-}14)$$

The minus sign indicates that, when the field does positive work W_{field} on an object, the *object's* energy is increased by an amount W_{field}. That amount of energy is *taken from* stored potential energy. The field dips into its "potential energy bank account" and gives the energy to the object, so the potential energy decreases when the force does positive work.

Some of the many similarities between gravitational and electric potential energy include:

- In both cases, the potential energy depends on only the *positions* of various objects, not on the *path* they took to get to those positions.

- Only *changes* in potential energy are physically significant, so we are free to assign the potential energy to be zero at any *one* convenient point. The potential energy in a given situation depends on the choice of the point where $U = 0$, but *changes* in potential energy are *not* affected by this choice.

- For two point particles, we usually choose $U = 0$ when the particles are infinitely far apart.

- Both the gravitational and electrical forces exerted by one point particle on another are inversely proportional to the square of the distance between them ($F \propto 1/r^2$). As a result, the gravitational and electric potential energies have the *same distance dependence* ($U \propto 1/r$, with $U = 0$ at $r = \infty$).

- The gravitational force and the gravitational potential energy for a pair of point particles are proportional to the product of the masses of the particles:

$$F = G\frac{m_1 m_2}{r^2} \qquad (2\text{-}3)$$

$$U_g = -G\frac{m_1 m_2}{r} \quad (U_g = 0 \text{ at } r = \infty) \qquad (6\text{-}13)$$

The electric force and the electric potential energy for a pair of point particles are proportional to the product of the *charges* of the particles:

$$F = k\frac{|q_1 q_2|}{r^2} \qquad (2\text{-}5)$$

Electric potential energy for a pair of point charges a distance r apart

$$U_E = k\frac{q_1 q_2}{r} \quad (U_E = 0 \text{ at } r = \infty) \qquad (17\text{-}1)$$

The minus sign in Eq. (6-13) indicates that gravity is always an attractive force: if two particles move closer together, gravity does positive work and thereby changes

Figure 17.1   (a) An object moving through a gravitational field; the gravitational potential energy changes. (b) A charged particle moving through an *electric* field; the *electric* potential energy changes.

(a)

(b)

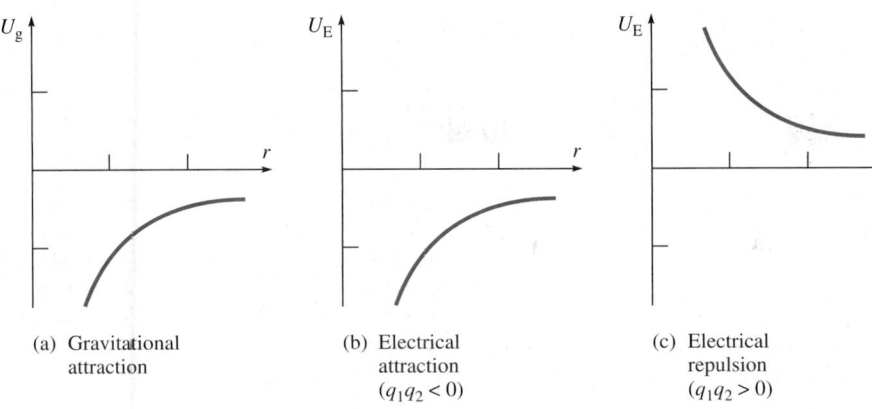

(a) Gravitational attraction

(b) Electrical attraction ($q_1q_2 < 0$)

(c) Electrical repulsion ($q_1q_2 > 0$)

Figure 17.2 Potential energies for pairs of point particles as a function of separation distance r. In each case, we choose $U = 0$ at $r = \infty$. For an attractive force, (a) and (b), the potential energy is negative. If two particles start far apart where $U = 0$, they "fall" spontaneously toward each other as the potential energy *decreases*. For a repulsive force (c), the potential energy is positive. If two particles start far apart, they have to be pushed together; some external agent has to do work to increase the potential energy.

some gravitational potential energy into kinetic energy or some other form of energy. If the two particles move farther apart, the gravitational potential energy increases.

Why is there no minus sign in Eq. (17-1)? If the two charges have opposite signs, the force is an attractive one. The potential energy should be negative, as it is for the attractive force of gravity. With opposite signs, the product q_1q_2 is negative and the potential energy has the correct sign (Fig. 17.2). If the two charges instead have the same sign—both positive or both negative—the product q_1q_2 is positive. The electric force between two like charges is *repulsive*; the potential energy *increases* as they move closer together. Thus, Eq. (17-1) automatically gives the correct sign in every case.

Coulomb's law is written in terms of the *magnitudes* of the charges ($|q_1q_2|$) since it gives the *magnitude* of a vector quantity—the force. In the potential energy expression [Eq. (17-1)], we do *not* write the absolute value bars. The signs of the two charges determine the sign of the potential energy, a scalar quantity that can be positive, negative, or zero.

 If potential energy is negative, it *increases* when its absolute value gets smaller (just as –6 is *greater than* –8).

Example 17.1

Electric Potential Energy in a Thundercloud

  In a thunderstorm, charge is separated through a complicated mechanism that is ultimately powered by the Sun. A simplified model of the charge in a thundercloud represents the positive charge accumulated at the top and the negative charge at the bottom as a pair of point charges (Fig. 17.3). (a) What is the electric potential energy of the pair of point charges, assuming that $U = 0$ when the two charges are infinitely far apart? (b) Explain the sign of the potential energy in light of the fact that *positive* work must be done by external forces in the thundercloud to *separate* the charges.

Strategy (a) The electric potential energy for a pair of point charges is given by Eq. (17-1), where $U = 0$ at infinite separation is assumed. The algebraic signs of the charges are included when finding the potential energy. (b) The work done by an external force to separate the charges is equal to the *change* in the electric potential energy as the charges are *moved apart* by forces acting within the thundercloud.

Solution and Discussion (a) The general expression for electric potential energy for two point charges is

$$U_E = k\frac{q_1q_2}{r}$$

Coulomb's constant is

$$k = 8.99 \times 10^9 \ \frac{\text{N·m}^2}{\text{C}^2}$$

We substitute the known values into the equation for electric potential energy.

$$U_E = 8.99 \times 10^9 \ \frac{\text{N·m}^2}{\text{C}^2} \times \frac{(+50 \ \text{C}) \times (-20 \ \text{C})}{8000 \ \text{m}}$$

$$= -1 \times 10^9 \ \text{J}$$

 +50 C

8 km

–20 C

Figure 17.3
Charge separation in a thundercloud

continued on next page

Example 17.1 *continued*

(b) Recall that we chose $U = 0$ at infinite separation. Negative potential energy therefore means that, *if the two point charges started infinitely far apart,* their electric potential energy would decrease as they are brought together—in the absence of other forces they would "fall" spontaneously toward one another. However, in the thundercloud, the unlike charges *start close together* and are moved *farther apart* by an external force; the external agent must do *positive* work to increase the potential energy and move the charges *apart*. Initially, when the charges are close together, the potential energy is *less than* -1×10^9 J; the *change* in potential energy as the charges are moved apart is *positive*.

Practice Problem 17.1   Two point charges with like signs

Two point charges, $Q = +6.0\ \mu$C and $q = +5.0\ \mu$C, are separated by 15.0 m. (a) What is the electric potential energy? (b) Charge q is free to move—no other forces act on it—while Q is fixed in place. Both are initially at rest. Does q move toward or away from charge Q? (c) How does the motion of q affect the electric potential energy? Explain how energy is conserved.

Potential Energy due to Several Point Charges

To find the potential energy due to more than two point charges, we add the potential energies of each *pair* of charges. For three point charges, there are three pairs, so the potential energy is

Electric potential energy due to three point charges ($U_E = 0$ when all three are infinitely separated)

$$U_E = k\left(\frac{q_1 q_2}{r_{12}} + \frac{q_1 q_3}{r_{13}} + \frac{q_2 q_3}{r_{23}}\right) \tag{17-2}$$

where, for instance, r_{12} is the distance between q_1 and q_2. The potential energy in Eq. (17-2) is the negative of the work done by the electric field as the three charges are put into their positions, *starting from infinite separation*. If there are more than three charges, the potential energy is a sum just like Eq. (17-2), which includes *one* term for each *pair* of charges. Be sure not to count the potential energy of the same pair twice. If the potential energy expression has a term $(q_1 q_2)/r_{12}$, it must not also have a term $(q_2 q_1)/r_{21}$.

Example 17.2

Electric Potential Energy due to Three Point Charges

Find the electric potential energy for the array of charges shown in Fig. 17.4. Charge $q_1 = +4.0\ \mu$C is located at (0.0, 0.0) m; charge $q_2 = +2.0\ \mu$C is located at (3.0, 4.0) m; and charge $q_3 = -3.0\ \mu$C is located at (3.0, 0.0) m.

Strategy   With three charges, there are three pairs to include in the potential energy sum [Eq. (17-2)]. The charges are given; we need only find the distance between each pair. Subscripts are useful to identify the three distances; r_{12}, for example, means the distance between q_1 and q_2.

Solution   From Fig. 17.4, $r_{13} = 3.0$ m and $r_{23} = 4.0$ m. The Pythagorean theorem enables us to find r_{12}:

$$r_{12} = \sqrt{3.0^2 + 4.0^2}\ \text{m} = \sqrt{25}\ \text{m} = 5.0\ \text{m}$$

The potential energy has one term for each pair:

$$U_E = k\left(\frac{q_1 q_2}{r_{12}} + \frac{q_1 q_3}{r_{13}} + \frac{q_2 q_3}{r_{23}}\right)$$

Substituting numerical values,

$$U_E = 8.99 \times 10^9\ \frac{\text{N} \cdot \text{m}^2}{\text{C}^2} \times$$

$$\left[\frac{(+4.0)(+2.0)}{5.0} + \frac{(+4.0)(-3.0)}{3.0} + \frac{(+2.0)(-3.0)}{4.0}\right] \times 10^{-12}\ \frac{\text{C}^2}{\text{m}}$$

$$= -0.035\ \text{J}$$

Discussion   To interpret the answer, assume that the three charges start far apart from each other. As the charges are

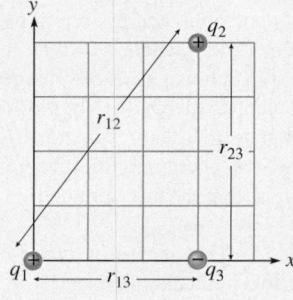

Figure 17.4
Three point charges

continued on next page

Example 17.2 *continued*

brought together and put into place, the electric fields do a total work of +0.035 J. Once the charges are in place, an external agent would have to supply 0.035 J of energy to separate them again.

Conceptual Practice Problem 17.2   Four charges
When finding the potential energy due to four point charges, how many *pairs* of charges are there?

17.2   ELECTRIC POTENTIAL

Imagine that a collection of point charges is somehow fixed in place while another charge q can move. Moving q may involve changes in electric potential energy since the distance between it and the fixed charges may change. Just as the electric field is defined as the electric force per unit charge, the **electric potential** V is defined as the electric potential energy *per unit charge* (Fig. 17.5).

$$V = \frac{U_E}{q}$$

(assuming $U_E = 0$ when q is far away)   (17-3)

Definition of potential: electric potential energy per unit charge

In Eq. (17-3) we assume $U_E = 0$ when q is infinitely far away but all other charges remain where they are. In other words, U_E in Eq. (17-3) includes the potential energy due to the interactions between q and each of the fixed charges, but does not include the potential energy due to the interactions between pairs of fixed charges.

The SI unit of electric potential is the joule per coulomb, which is named the volt (symbol V) after the Italian scientist Alessandro Volta (1745–1827). Volta invented the voltaic pile, an early form of battery. *Electric potential* is often shortened to *potential*. It is also informally called "voltage," especially in connection with electric circuits, just as weight is sometimes called "tonnage." Be careful to distinguish *electric potential* from *electric potential energy*. It is all too easy to confuse the two, but they are not interchangeable.

$$1 \text{ V} = 1 \text{ J/C} \qquad (17\text{-}4)$$

Definition of the volt

Since potential energy and charge are scalars, potential is also a scalar. The principle of superposition is easier to apply to potentials than to fields since fields must be added as vectors. Given the potential at various points, it is easy to calculate the potential energy change when a charge moves from one point to another. Potentials do not

Figure 17.5 Relationships between force, field, potential energy, and potential. (The meaning of *gradient* is discussed in Section 17.3.)

have direction in space; they are added just as any other scalar. Potentials can be either positive or negative and so must be added with their algebraic signs.

Since only changes in potential energy are significant, only changes in potential are significant. We are free to choose the potential arbitrarily at any one point. Equation (17-3) assumes that the potential is zero infinitely far away from the collection of fixed charges. We could add a constant to Eq. (17-3) without affecting the potential *change* that occurs when charge q moves from one point to another.

If the potential at a point due to a collection of fixed charges is V, then when a charge q is placed at that point, the electric potential energy is

$$U_E = qV \qquad (17\text{-}5)$$

In Eq. (17-5), the potential energy is zero when the moveable charge q is infinitely far away from the fixed charges. This is not the same as the potential energy when *all* of the charges are infinitely far apart. Equation (17-5) omits the potential energy change that took place when the fixed charges were assembled. We can ignore that energy since it remains constant as long as those charges remain fixed in place. Equation (17-5) takes advantage of the property that only *changes* in potential energy are significant; we don't have to worry about the absolute total amount of electric potential energy any more than we do with gravity, springs, or any other form of potential energy.

Potential Difference

When a point charge q moves from point A to point B, it moves through a *potential difference*

$$\Delta V = V_f - V_i = V_B - V_A \qquad (17\text{-}6)$$

The potential difference is the change in electric potential energy per unit charge:

$$\boxed{\Delta U_E = q\,\Delta V \qquad (17\text{-}7)}$$

The electric force on a charge is always directed toward regions of lower electric potential energy, just as the gravitational force on an object is directed toward regions of lower gravitational potential energy (that is, downward). For a positive charge, lower potential energy means lower potential, but for a negative charge, lower potential energy means *higher* potential. This shouldn't be surprising, since the force on a negative charge is opposite to the direction of $\vec{E}$, while the force on a positive charge is in the direction of $\vec{E}$. Since the electric field points toward lower potential energy for positive charges,

$$\boxed{\vec{E} \text{ points in the direction of decreasing } V.}$$

Example 17.3

A Battery-Powered Lantern

A battery-powered lantern is switched on for 5.0 min. During this time, electrons with total charge -8.0×10^2 C flow through the lamp; 9600 J of electric potential energy is converted to light and heat. Through what potential difference do the electrons move?

Strategy Equation (17-7) relates the change in electric potential energy to the potential difference. We could apply Eq. (17-7) to a single electron, but since all of the electrons move through the same potential difference, we can let q be

the total charge of the electrons and ΔU_E be the total change in electric potential energy.

Solution The total charge moving through the lamp is $q = -800$ C. The change in electric potential energy is *negative* since it is converted into other forms of energy. Therefore,

$$\Delta V = \frac{\Delta U_E}{q} = \frac{-9600 \text{ J}}{-8.0 \times 10^2 \text{ C}} = +12 \text{ V}$$

continued on next page

Example 17.3 *continued*

Discussion   The sign of the potential difference is positive: negative charges decrease the electric potential energy when they move through a potential increase.

Conceptual Practice Problem 17.3   An electron beam

A beam of electrons is deflected as it moves between oppositely charged parallel plates (Fig. 17.6). Which plate is at the higher potential?

Figure 17.6
An electron beam deflected by a pair of oppositely charged plates.

Potential due to a Point Charge

If q is in the vicinity of one other point charge Q, the potential energy is

$$U = \frac{kQq}{r} \qquad (17\text{-}1)$$

when Q and q are separated by a distance r. Therefore, the electric potential at a distance r from a point charge Q is

$$V = \frac{kQ}{r} \qquad (17\text{-}8)$$

The potential at a point P due to N point charges is the sum of the potentials due to each charge:

$$V = \sum V_i = \sum \frac{kQ_i}{r_i} \quad \text{for} \quad i = 1, 2, 3, \ldots, N \qquad (17\text{-}9)$$

where r_i is the distance from the ith point charge Q_i to point P. Potential has a unique value at each point in space.

Example 17.4

Potential due to Three Point Charges

Charge $Q_1 = +4.0 \ \mu C$ is located at $(0.0, 3.0)$ cm; charge $Q_2 = +2.0 \ \mu C$ is located at $(1.0, 0.0)$ cm; and charge $Q_3 = -3.0 \ \mu C$ is located at $(2.0, 2.0)$ cm (Fig. 17.7). (a) Find the electric potential at point A $(x = 0.0, y = 1.0$ cm$)$ due to the three charges. (b) A point charge $q = -5.0$ nC moves from a great distance to point A. What is the change in electric potential energy?

Strategy   The potential at A is the sum of the potentials due to each point charge. The first step is to find the distance from each charge to point A. We call these distances r_1, r_2, and r_3 to avoid using the wrong one by mistake. Then we add the potentials due to each of the three charges at A.

Solution   (a) From the grid, $r_1 = 2.0$ cm. The distance from Q_2 to point A is the diagonal of a square that is 1.0 cm on a side. Thus, $r_2 = \sqrt{2.0}$ cm $= 1.414$ cm. The third charge is located at a distance equal to the hypotenuse of a right triangle with sides of 2.0 cm and 1.0 cm. From the Pythagorean theorem,

$$r_3 = \sqrt{1.0^2 + 2.0^2} \ \text{cm} = \sqrt{5.0} \ \text{cm} = 2.236 \ \text{cm}$$

The potential at A is the sum of the potentials due to each point charge:

$$V = k \sum \frac{Q_i}{r_i}$$

Substituting numerical values:

$$V_A = 8.99 \times 10^9 \ \frac{\text{N} \cdot \text{m}^2}{\text{C}^2} \times$$

$$\left(\frac{+4.0 \times 10^{-6} \ \text{C}}{0.020 \ \text{m}} + \frac{+2.0 \times 10^{-6} \ \text{C}}{0.01414 \ \text{m}} + \frac{-3.0 \times 10^{-6} \ \text{C}}{0.02236 \ \text{m}} \right)$$

$$= +1.863 \times 10^6 \ \text{V}$$

To two significant figures, the potential at point A is $+1.9 \times 10^6$ V.

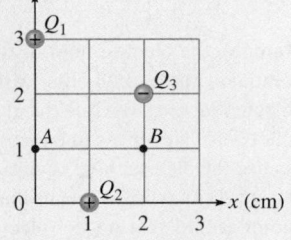

Figure 17.7
An array of three point charges

continued on next page

Example 17.4 *continued*

(b) The change in potential energy is

$$\Delta U_E = q\,\Delta V$$

Here ΔV is the potential difference through which charge q moves. If we assume that q starts from an infinite distance, then $V_i = 0$. Therefore,

$$\Delta U_E = q(V_A - 0) = (-5.0 \times 10^{-9}\,\text{C}) \times (+1.863 \times 10^6\,\text{J/C} - 0)$$

$$= -9.3 \times 10^{-3}\,\text{J}$$

Discussion   The positive sign of the potential indicates that a positive charge at point A would have positive potential energy. To bring in a positive charge from far away, the potential energy must be increased and therefore positive work must be done by the agent bringing in the charge. A negative charge at that point, on the other hand, has negative potential energy. When q moves from a potential of zero to a

positive potential, the potential increase causes a potential energy decrease ($q < 0$).

In Practice Problem 17.4, you are asked to find the work done by the field as q moves from A to B. One approach would be to find the component of the electric force in the direction of motion and then multiply that by the distance moved. The force is not constant in magnitude or direction, though, so we cannot just multiply force component times distance. In principle, the problem could be solved this way using calculus; but using the potential difference gives the same result without vector components or calculus.

Practice Problem 17.4   Potential at point B

Find the potential due to the same array of charges at point B ($x = 2.0$ m, $y = 1.0$ m) and the work done by the electric field if $q = -5.0$ nC moves from A to B.

Conceptual Example 17.5

Field and Potential at the Center of a Square

Four equal positive point charges q are fixed at the corners of a square of side s (Fig. 17.8a). (a) Is the electric field zero at the center of the square? (b) Is the potential zero at the center of the square?

Strategy and Solution   (a) The electric field at the center is the *vector* sum of the fields due to each of the point charges. Figure 17.8b shows the field vectors at the center of the square due to each charge. Each of these vectors has the same magnitude since the center is equidistant from each corner and the four charges are the same. From symmetry, the vector sum of the electric fields is zero.

(b) Since potential is a scalar rather than a vector, the potential at the center of the square is the *scalar* sum of the potentials due to each charge. These potentials are all equal since the distances and charges are the same. Each is positive since $q > 0$. The total potential at the center of the square is

$$V = 4\frac{kq}{r}$$

where $r = s/\sqrt{2}$ is the distance from a corner of the square to the center.

Discussion   In this example, the electric field is zero at a point where the potential is not zero. In other cases, there may be points where the potential is zero while the electric field at the same points is not zero. Never assume that the potential at a point is zero because the electric field is zero or *vice versa*. If the electric field is zero at a point, it means that a point charge placed at that point would feel no net electric

force. If the potential is zero at a point, it means zero total work would be done by the electric field as a point charge moves from infinity to that point.

Practice Problem 17.5   Field and potential for a different set of charges

Find the electric field and the potential at the center of a square of side 2.0 cm with a charge of $+9.0\,\mu\text{C}$ at one corner and with charges of $-3.0\,\mu\text{C}$ at the other three corners (Fig. 17.8c).

Figure 17.8

(a) Four equal point charges at the corners of a square. (b) Electric field vectors due to each of the point charges at the center of the square. (c) Charges for Practice Problem 17.5.

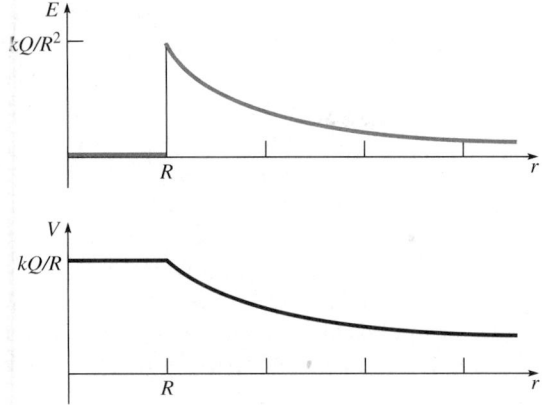

Figure 17.9 The electric field and the potential due to a hollow conducting sphere of radius R and charge Q as a function of r, the distance from the center.

Potential due to Spherical Conductors

In Chapter 16 we saw that the field outside a charged conducting sphere is the same as if all of the charge were concentrated into a point charge located at the center of the sphere. As a result, the electric potential due to a conducting sphere is similar to the potential due to a point charge.

Figure 17.9 shows graphs of the electric field strength and the potential as functions of the distance r from the center of a hollow conducting sphere of radius R and charge Q. The electric field inside the conducting sphere (from $r = 0$ to $r = R$) is zero. The magnitude of the electric field is greatest at the surface of the conductor and then drops off as $1/r^2$. Outside the sphere, the electric field is the same as for a charge Q located at $r = 0$.

The potential is chosen to be zero for $r = \infty$. The electric field outside the sphere ($r > R$) is the same as the field at a distance r from a *point charge Q*. Therefore, for any point at a distance $r > R$ from the center of the sphere, the potential is the same as the potential at a distance r from a *point charge Q*:

$$V = \frac{kQ}{r} \qquad (17\text{-}8)$$

For a positive charge Q, the potential is positive; and it is negative for a negative charge. At the surface of the sphere, the potential is

$$V = \frac{kQ}{R}$$

Since the field inside the conductor is zero, the field does no work as a point charge q moves within the sphere. Thus, the potential *anywhere inside* the sphere is the same as the potential at the surface of the sphere ($V = kQ/R$). (We assume that the charge q is so tiny that it does not affect the charge distribution on the surface of the conducting sphere.) Thus, for $r < R$, the potential is *not* the same as for a point charge. The magnitude of the potential due to a point charge continues to increase as $r \rightarrow 0$.

Van de Graaff Generator

An apparatus designed to charge a conductor to a high potential difference is the van de Graaff generator (Fig. 17.10). A large conducting sphere is supported on an insulating cylinder. In the cylinder, a motor-driven conveyor belt collects negative charge either by rubbing or from some other source of charge at the base of the cylinder. The charge is carried by the conveyor belt to the top of the cylinder, where it is collected by small metal rods and transferred to the conducting sphere. As more and more charge is

Figure 17.10 The van de Graaff generator

Making The Connection:

van de Graaff generator

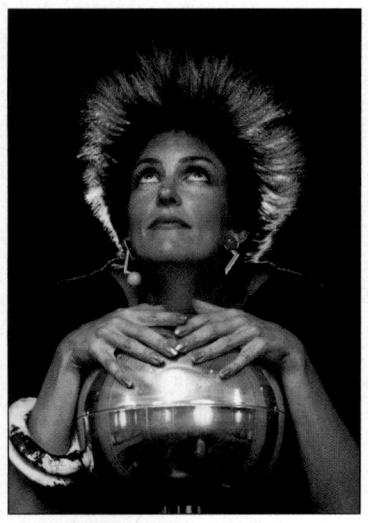

deposited onto the conducting sphere, the charges repel each other and move as far away from each other as possible, ending up on the outer surface of the conducting sphere.

Inside the conducting sphere, the electric field is zero, so no repulsion from charges already on the sphere is felt by the charge on the conveyor belt. Thus, a large quantity of charge can build up on the conducting sphere so that an extremely high potential difference can be established. Potential differences of millions of volts can be attained with a large sphere (Fig. 17.11). Commercial van de Graaff generators supply the large potential differences required to produce intense beams of high energy x-rays. The x-rays are used in medicine for cancer therapy; industrial uses include radiography (to detect tiny defects in machine parts) and the polymerization of plastics. Old science fiction movies often show sparks jumping from generators of this sort.

Figure 17.11 A hair-raising experience. A person touching the dome of a van de Graaff while electrically isolated from ground reaches the same potential as the dome. Although the effects are quite noticeable, there is no danger to the person since the whole body is at the same potential. A large potential *difference* between two parts of the person's body would be dangerous or lethal.

Example 17.6

Minimum Radius Required for a van de Graaff

You wish to charge a van de Graaff to a potential of 240 kV. On a day with average humidity, an electric field of 8.0×10^5 N/C or greater ionizes air molecules, allowing charge to leak off the van de Graaff. Find the minimum radius of the conducting sphere under these conditions.

Strategy We set the potential of a conducting sphere equal to $V_{max} = 240$ kV and require the electric field strength just outside the sphere to be less than $E_{max} = 8.0 \times 10^5$ N/C. Since both $\vec{E}$ and V depend on the charge on the sphere and its radius, we should be able to eliminate the charge and solve for the radius.

Solution The potential of a conducting sphere with charge Q and radius R is

$$V = \frac{kQ}{R}$$

The electric field strength just outside the sphere is

$$E = \frac{kQ}{R^2}$$

Comparing the two expressions, we see that $E = V/R$. Now let $V = V_{max}$ and require $E < E_{max}$:

$$E = \frac{V_{max}}{R} < E_{max}$$

Solving for R,

$$R > \frac{V_{max}}{E_{max}} = \frac{2.4 \times 10^5 \text{ V}}{8.0 \times 10^5 \text{ N/C}}$$

$$R > 0.30 \text{ m}$$

The minimum radius is 30 cm.

Discussion To achieve a large potential difference, a large conducting sphere is required. A small sphere—or a conductor with a sharp point, which is like part of a sphere with a small radius of curvature—cannot be charged to a high potential. Even a relatively small potential on a conductor with a sharp point, such as a lightning rod, enables charge to leak off into the air since the strong electric field ionizes the nearby air.

The equation $E = V/R$ derived in this example is *not* a general relationship between field and potential. The general relationship is discussed in Section 17.3.

Practice Problem 17.6 A small conducting sphere

What is the largest potential that can be achieved on a conducting sphere of radius 0.5 cm? Assume $E_{max} = 8.0 \times 10^5$ N/C.

Figure 17.12   (a) The structure of a neuron. (b) The action potential. The graph shows the potential difference between the inside and outside of the cell as a function of time.

Potential Differences in Biological Systems

In general, the inside and outside of a biological cell are *not* at the same potential. The potential difference across a cell membrane is due to different concentrations of ions in the fluids inside and outside the cell. These potential differences are particularly note-worthy in nerve and muscle cells.

Making The Connection:
transmission of
nerve impulses

A nerve cell or *neuron* consists of a cell body and a long extension, called an *axon* (Fig. 17.12a). Human axons are 10–20 μm in diameter. When the axon is in its resting state, negative ions on the inner surface of the membrane and positive ions on the outer surface cause the fluid inside to be at a potential of about –85 mV relative to the fluid outside.

A nerve impulse is a change in the potential difference across the membrane that gets propagated along the axon. The cell membrane at the end stimulated suddenly becomes permeable to positive sodium ions for about 0.2 ms. Sodium ions flow into the cell, changing the polarity of the charge on the inner surface of the membrane. The potential difference across the cell membrane changes from about –85 mV to +60 mV. The reversal of polarity of the potential difference across the membrane is called the *action potential* (Fig. 17.12b). The action potential propagates down the axon at a speed of about 30 m/s.

Restoration of the resting potential involves both the diffusion of potassium and the pumping of sodium ions out of the cell—called *active transport*. As much as 20% of the resting energy requirements of the body are used for the active transport of sodium ions.

Potential Differences in Medicine

Similar polarity changes occur across the membranes of muscle cells. When a nerve impulse reaches a muscle fiber, it causes a change in potential, which propagates along the muscle fiber and signals the muscle to contract.

Muscle cells, including those in the heart, have a layer of negative ions on the inside of the membrane and positive ions on the outside. Just before each heartbeat, positive ions are pumped into the cells, neutralizing the potential difference. Just as for the action potential in neurons, the *depolarization* of muscle cells begins at one end of the cell and proceeds toward the other end. Depolarization of various cells occurs at different times. When the heart relaxes, the cells are polarized again.

An electrocardiogram (EKG) measures the potential difference between points on the chest as a function of time (Fig. 17.13a). The depolarization and polarization of the cells in the heart causes potential differences that can be measured using electrodes connected to the skin. The potential difference measured by the electrodes is amplified and recorded on a chart recorder or a computer (Fig. 17.13b).

Potential differences other than those due to the heart are used for diagnostic purposes. In an electroencephalogram (EEG), the electrodes are placed on the head. The EEG measures potential differences caused by electrical activity in the brain (Fig. 17.14). In an electroretinogram (ERG), the electrodes are placed near the eye to measure the potential differences due to electrical activity in the retina when stimulated by a flash of light.

Making The Connection:
electrocardiogram
(EKG)

Making The Connection:
electroencephalogram
(EEG)

17.3   THE RELATIONSHIP BETWEEN ELECTRIC FIELD AND POTENTIAL

Figure 17.5 suggests a close relationship between electric field and electric potential. In this section we explore the relationship in detail, starting with visual representations of each.

Equipotential Surfaces

A field line sketch is a useful visual representation of the electric field. Is there a similar way to represent the electric potential? There cannot be "potential lines" since potential doesn't have direction. Instead, we can create something analogous to a contour map. An **equipotential surface** has the same potential at every point on the surface. The idea is similar to that of isobars on a weather map, which show places where the air pressure is the same, or lines of constant elevation on a topographic map, which show where the

Figure 17.13 (a) A stress test. The EKG is a graph of the potential difference measured between the electrodes as a function of time. These potential differences reveal whether the heart functions normally during exercise. (b) A normal EKG indicates that the heart is healthy. (c) An abnormal or irregular EKG indicates a problem. This EKG indicates ventricular fibrillation.

(a)

(b)

(c)

(a)

(b)

Figure 17.14 (a) A doctor explains to a patient receiving an EEG how the results are used to study brain function and diagnose illness. (b) The patterns of the potential difference are sometimes called brain waves.

elevation is the same (Fig. 17.15). Since the potential difference between any two points on such an equipotential surface is zero, no work is done by the field when a charge moves from one point on the surface to another.

Equipotential surfaces and field lines are closely related. Suppose you want to move a charge in a direction so that the potential stays constant. In order for the field to do no work on the charge, the displacement must be perpendicular to the electric force (and therefore perpendicular to the field). As long as you always move the charge in a direction perpendicular to the field, the work done by the field is zero and the potential stays the same.

> An equipotential surface is perpendicular to the electric field lines at all points.

Conversely, if you want to move a charge in a direction that *maximizes* the change in potential, you would move parallel or antiparallel to the electric field. Only the component of displacement perpendicular to an equipotential surface changes the potential. Think of a contour map: the steepest slope—the quickest change of elevation—is

(a)

(b)

Figure 17.15 (a) A weather map. The air pressure at any point on an isobar is the same. (b) A topographic map, showing the lines of constant elevation.

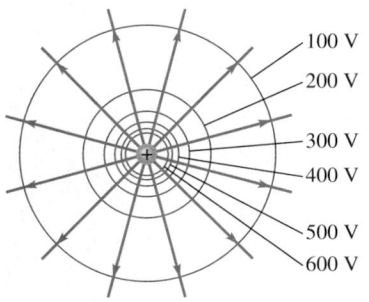

Figure 17.16 Equipotential surfaces near a positive point charge. The circles represent the intersection of the spherical surfaces with the plane of the page. The potential decreases as we move away from a positive charge. The electric field lines are perpendicular to the spherical surfaces and point toward lower potentials. The spacing between equipotential surfaces increases with increasing distance since the electric field gets weaker.

perpendicular to the lines of constant elevation. The electric field is sometimes called the **potential gradient**, where the word *gradient* suggests *grade* or *slope* (see Fig. 17.5). On a contour map a hill is steepest where the lines of constant elevation are close together; a diagram of equipotential surfaces is similar.

> If equipotential surfaces are drawn such that the potential difference between adjacent surfaces is constant, then the surfaces are closer together where the field is stronger.

Since $\vec{\mathbf{E}}$ points in the direction of *decreasing V*,

> The electric field always points in the direction of maximum potential decrease.

The simplest equipotential surfaces are those for a single point charge. The potential due to a point charge depends only on the distance from the charge, so the equipotential surfaces are spheres with the charge at the center (Fig. 17.16). There are an infinite number of equipotential surfaces, so we customarily draw a few surfaces equally spaced in potential—just like a contour map that shows places of equal elevation in 5-m increments.

Conceptual Example 17.7

Equipotential Surfaces for Two Point Charges

Sketch some equipotential surfaces for two point charges $+Q$ and $-Q$.

Strategy and Solution   One way to draw a set of equipotential surfaces is to first draw the field lines. Then we construct the equipotential surfaces by sketching lines that are perpendicular to the field lines at all points. Close to either point charge, the field is primarily due to the nearby charge, so the surfaces are nearly spherical.

Figure 17.17a shows a sketch of the field lines and equipotential surfaces for the two charges.

Discussion   This two-dimensional sketch shows only the intersection of the equipotential surfaces with the plane of the page. Figure 17.17b shows a three-dimensional representation of the equipotential surfaces.

Conceptual Practice Problem 17.7   Equipotential surfaces for two positive charges

Sketch some equipotential surfaces for two equal positive point charges.

Potential in a Uniform Electric Field

In a uniform electric field, the field lines are equally spaced parallel lines. Since equipotential surfaces are perpendicular to field lines, the equipotential surfaces are a set of parallel planes (Fig. 17.18). The potential decreases from one plane to the next in the direction of $\vec{\mathbf{E}}$. Since the spacing of equipotential planes depends on the magnitude of $\vec{\mathbf{E}}$, in a uniform field planes at equal potential increments are equal distances apart.

To find a quantitative relationship between the field strength and the spacing of the equipotential planes, imagine moving a point charge $+q$ a distance d in the direction of an electric field of magnitude E. The work done by the electric field is

$$W_E = F_E d = qEd$$

The change in electric potential energy is

$$\Delta U_E = -W_E = -qEd$$

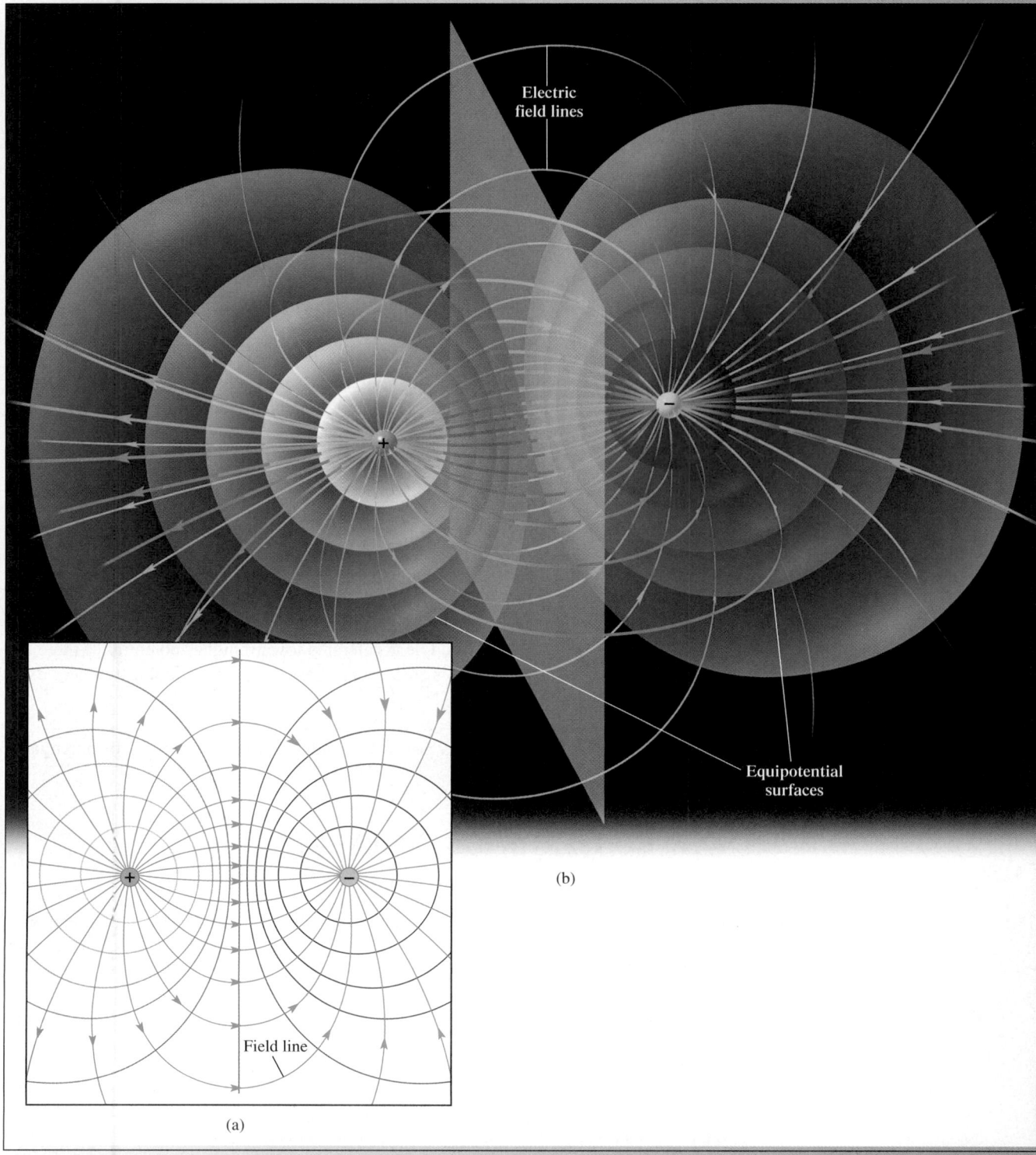

Figure 17.17   (a) A sketch of the equipotential surfaces and electric field lines for two point charges of the same magnitude but opposite in sign. (b) A more three-dimensional view of the equipotential surfaces.

From the definition of potential, the potential change is

$$\Delta V = \frac{\Delta U_E}{q} = -Ed \qquad (17\text{-}10)$$

The minus sign in Eq. (17-10) is correct because potential *decreases* in the direction of the electric field.

Equation (17-10) implies (and Problem 18 verifies) that the SI unit of electric field (N/C) can also be written *volts per meter* (V/m):

$$1\ \text{N/C} = 1\ \text{V/m} \qquad (17\text{-}11)$$

In Fig. 17.18, the equipotential planes differ in potential by 1 V. If the electric field magnitude is 25 N/C = 25 V/m, then the distance between the planes is

$$\frac{1\ \text{V}}{25\ \text{V/m}} = 0.04\ \text{m}$$

Where the field is strong, the equipotential surfaces are close together: with a large number of volts per meter, it doesn't take many meters to change the potential a given number of volts.

Potential Inside a Conductor

In Chapter 16 we learned that $E = 0$ at every point inside a conductor in electrostatic equilibrium (when no charges are moving). If the potential gradient (the field) is zero at every point, then the potential does not change as we move from one point to another. If there were potential differences within the conductor, then charges would move in response. Positive charge would be accelerated by the field toward regions of lower potential and negative charge would be accelerated toward higher potential. If there are no moving charges, then the field is zero everywhere and no potential differences exist within the conductor. Therefore:

> In electrostatic equilibrium, every point in a conductor must be at the same potential.

17.4   CONSERVATION OF ENERGY FOR MOVING CHARGES

When a charge moves from one position to another in an electric field, the change in electric potential energy must be accompanied by a change in other forms of energy so that the total energy is the same. Energy conservation simplifies problem solving just as it does with gravitational or elastic potential energy.

If no other forces act on a point charge, then as it moves in an electric field, the sum of the kinetic and electric potential energy is constant:

$$K_i + U_i = K_f + U_f = \text{constant}$$

Changes in gravitational potential energy are negligible compared with changes in electric potential energy when the gravitational force is much weaker than the electric force.

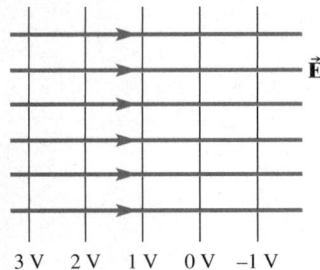

Figure 17.18   Field lines and equipotential surfaces (at 1-V intervals) in a uniform field. The equipotential surfaces are equally spaced *planes* perpendicular to the field lines.

Relationship between electric field and potential difference

3 V   2 V   1 V   0 V   −1 V

$\vec{E}$

Example 17.8

Electron Gun

Figure 17.19 shows a simplified model of the *electron gun*, the first stage of the cathode ray tube (CRT) found in computer monitors, TVs, oscillo-scopes, and x-ray tubes. Electrons are first "boiled" off a heated wire—the cathode. The electrons are then accelerated toward the anode, which is at a potential 8 kV to 20 kV higher

continued on next page

Example 17.8 *continued*

than the cathode. The electrons pass at high speed through a small hole in the center of the anode and toward the screen. If the potential difference between the cathode and anode is 12 kV, at what speed do the electrons move as they pass through the hole in the anode? Assume that the initial kinetic energy of the electrons as they leave the cathode is negligible.

Strategy Using energy conservation, we set the sum of the initial kinetic and potential energies equal to the sum of the final kinetic and potential energies. The initial kinetic energy is taken to be zero. Once we find the final kinetic energy, we can solve for the speed.

Known: $K_i = 0$; $\Delta V = +12$ kV

Find: v

Solution The change in electric potential energy is
$$\Delta U = U_f - U_i = q\,\Delta V$$
From conservation of energy,
$$K_i + U_i = K_f + U_f$$
Solving for the final kinetic energy,
$$K_f = K_i + (U_i - U_f) = K_i - \Delta U$$
$$= 0 - q\,\Delta V$$
To find the speed, we set $K_f = \frac{1}{2}mv^2$.
$$\tfrac{1}{2}mv^2 = -q\,\Delta V$$
Solving for the speed,
$$v = \sqrt{\frac{-2q\,\Delta V}{m}}$$
For an electron,
$$q = -e = -1.602 \times 10^{-19}\ \text{C}$$
$$m = 9.109 \times 10^{-31}\ \text{kg}$$

Substituting numerical values,
$$v = \sqrt{\frac{-2 \times (-1.602 \times 10^{-19}\ \text{C}) \times (12{,}000\ \text{V})}{9.109 \times 10^{-31}\ \text{kg}}} = 6.5 \times 10^7\ \text{m/s}$$

Discussion The answer is more than 20% of the speed of light (3×10^8 m/s). A more accurate calculation of the speed would have to account for Einstein's theory of relativity.

Using conservation of energy to solve this problem makes it clear that the final speed depends only on the potential difference between the cathode and anode, not on the distance between them. To solve the problem using Newton's second law, even if the electric field is uniform, we have to assume some distance d between the cathode and anode. Using d, we can find the magnitude of the electric field
$$E = \frac{\Delta V}{d}$$
The acceleration of the electron is
$$a = \frac{F_E}{m} = \frac{eE}{m} = \frac{e\,\Delta V}{md}$$
Now we can find the final speed. Since the acceleration is constant,
$$v = \sqrt{v_0^2 + 2ad} = \sqrt{0 + 2 \times \frac{e\,\Delta V}{md} \times d}$$
The distance d cancels and gives the same result as the energy calculation.

Practice Problem 17.8 Proton accelerated

A proton is accelerated from rest through a potential difference. Its final speed is 2.00×10^6 m/s. What is the potential difference? The mass of the proton is 1.673×10^{-27} kg.

Figure 17.19
An electron gun accelerates electrons to high speeds by means of a large potential difference between the cathode and anode.

17.5 CAPACITORS

Can a useful device be built to store electric potential energy? Yes. Many such devices, called *capacitors*, are found in every piece of electronic equipment (Fig. 17.20).

A **capacitor** is a device that stores electric potential energy by storing separated positive and negative charges. It consists of two conductors separated by either vacuum or an insulating material. Charge is separated, with positive charge put on one of the conductors and an equal amount of negative charge on the other conductor. Work must be done to separate positive charge from negative charge, since there is an attractive force between the two.

Figure 17.20 The arrows indicate a few of the many capacitors on a circuit board from the inside of an amplifier.

Figure 17.21 Side view of two parallel metal plates with charges of equal magnitude and opposite sign. There is a potential difference between the two plates; the positive plate is at the higher potential.

The work done to separate the charge ends up as electric potential energy. An electric field arises between the two conductors, with field lines beginning on the conductor with positive charge and ending on the conductor with negative charge (Fig. 17.21). The stored potential energy is associated with this electric field. We can recover the stored energy—that is, convert it into some other form of energy—by letting the charges come together again.

The simplest form of capacitor is a **parallel plate capacitor**, consisting of two parallel metal plates, each of the same area A, separated by a distance d (Fig. 17.22). A charge $+Q$ is put on one plate and a charge $-Q$ on the other. For now, assume there is air between the plates. One way to charge the plates is to connect the positive terminal of a battery to one and the negative terminal to the other. The battery removes electrons from one plate, leaving it positively charged, and puts them on the other plate, leaving it with an equal magnitude of negative charge. In order to do this, the battery has to do work (Fig. 17.23)—some of the battery's chemical energy is converted into electric potential energy.

In general, the field between two such plates does not have to be uniform (Fig. 17.21). However, if the plates are close together, then a good approximation is to say that the charge is evenly spread on the inner surfaces of the plates and none is found on the outer surfaces. The plates in a real capacitor are almost always close enough that this approximation is valid.

With charge evenly spread on the inner surfaces, a uniform electric field exists between the two plates (Fig. 17.22). We can neglect the nonuniformity of the field near the edges as long as the plates are close together. The electric field lines start on positive charges and end on negative charges. If charge of magnitude Q is evenly spread over each plate with surface of area A, then the *surface charge density* (the charge per unit area) is denoted by σ, the Greek letter sigma:

$$\sigma = Q/A \tag{17-12}$$

In Problem 40 you can show that the magnitude of the electric field just outside a conductor is

---

Electric field just outside a conductor

$$E = 4\pi k\sigma = \sigma/\epsilon_0 \tag{17-13}$$

Recall that the constant $\epsilon_0 = 1/(4\pi k) = 8.854 \times 10^{-12} \text{ C}^2/(\text{N·m}^2)$ is called the *permittivity of free space* [Eq. (16-2b)]. Since the field between the plates of the capacitor is uniform, Eq. (17-13) gives the magnitude of the field *everywhere* between the plates.

100 V 0 V
75 V 50 V 25 V

Figure 17.22 A parallel plate capacitor with a potential difference of 100 V between the plates. We arbitrarily assign the negative plate a potential of zero (remember that only potential differences are significant). The field is uniform except near the edges of the plates, so the equipotential surfaces are parallel planes that start to bend near the edges.

Note that the electric field between the plates is caused by the charges on *both* plates. If the plate on the right in Fig. 17.22 were removed, the charge on the other plate would spread over *both* its left and right surfaces. With only half as much charge per unit area, the electric field on the right side of the plate would be half as strong.

What is the potential difference between the plates? Since the field is uniform, the *magnitude* of the potential difference is

$$\Delta V = Ed \qquad (17\text{-}10)$$

The field is proportional to the charge and the potential difference is proportional to the field; therefore, *the charge is proportional to the potential difference*. That turns out to be true for any capacitor, not just a parallel plate capacitor. The constant of proportionality between charge and potential difference depends only on geometric factors (size and shapes of the plates) and the material between the plates. Conventionally, this proportionality is written

---
Definition of capacitance
$$Q = C\,\Delta V \qquad (17\text{-}14)$$
---

where Q is the magnitude of the charge on each plate and ΔV is the magnitude of the potential *difference* between the plates. The constant of proportionality C is called the **capacitance**. Think of capacitance as the capacity to hold charge for a given potential difference. The SI units of capacitance are coulombs per volt, which is called the *farad* (symbol F). Capacitances are commonly measured in μF (microfarads), nF (nanofarads), or pF (picofarads) because the farad is a rather large unit; a pair of plates with area 1 m^2 spaced 1 mm apart has a capacitance of only about 10^{-8} F = 10 nF.

We can now find the capacitance of a parallel plate capacitor. The electric field is

$$E = \frac{\sigma}{\epsilon_0} = \frac{Q}{\epsilon_0 A}$$

where A is the inner surface area of each plate. If the plates are a distance d apart, then the *magnitude* of the potential difference is

$$\Delta V = Ed = \frac{Qd}{\epsilon_0 A}$$

Figure 17.23 A charge $+q$ is moved from the negatively charged plate to the positively charged plate. Since the electric field opposes the motion of the charge, work must be done by some external agent such as a battery.

Figure 17.24 A disassembled capacitor, showing the foil conducting plates and the thin sheet of insulating material.

By rearranging, this can be rewritten in the form $Q = \text{constant} \times \Delta V$:

$$Q = \frac{\epsilon_0 A}{d} \Delta V$$

Comparing with the definition of capacitance, the capacitance of a parallel plate capacitor is

---

Capacitance of parallel plate capacitor

$$C = \frac{\epsilon_0 A}{d} = \frac{A}{4\pi k d} \qquad (17\text{-}15)$$

To produce a large capacitance, we make the plate area large and the plate spacing small. To get large areas while still keeping the physical size of the capacitor reasonable, the plates are often made of thin conducting foil that is rolled, with the insulating material sandwiched in between, into a cylinder (Fig. 17.24). The effect of using an insulator other than air or vacuum is discussed in Section 17.6.

Example 17.9

Computer Keyboard

In one kind of computer keyboard, each key is attached to one plate of a parallel plate capacitor; the other plate is fixed in position (Fig. 17.25a). The capacitor is maintained at a constant potential difference of 5.0 V by an external circuit. When the key is pressed down, the top plate moves closer to the bottom plate, changing the capacitance and causing charge to flow through the circuit. If each plate is a square of side 6.0 mm and the plate separation changes from 4.0 mm to 1.2 mm when a key is pressed, how much charge flows through the circuit? Does the charge on the capacitor increase or decrease? Assume that there is air between the plates instead of a flexible insulator.

Strategy Since we are given the area and separation of the plates, we can find the capacitance from Eq. (17-15). The charge is then found from the product of the capacitance and the potential difference across the plates: $Q = C \Delta V$.

Solution The capacitance of a parallel plate capacitor is given by Eq. (17-15):

$$C = \frac{A}{4\pi k d}$$

The area is $A = (6.0 \text{ mm})^2$. Since the potential difference ΔV is kept constant, the change in the magnitude of the charge on the plates is

$$Q_f - Q_i = C_f \Delta V - C_i \Delta V$$

$$= \left(\frac{A}{4\pi k d_f} - \frac{A}{4\pi k d_i} \right) \Delta V = \frac{A \Delta V}{4\pi k} \left(\frac{1}{d_f} - \frac{1}{d_i} \right)$$

Substituting numerical values,

$$Q_f - Q_i = \frac{(0.0060 \text{ m})^2 \times 5.0 \text{ V}}{4\pi \times 8.99 \times 10^9 \text{ N·m}^2/\text{C}^2} \times \left(\frac{1}{0.0012 \text{ m}} - \frac{1}{0.0040 \text{ m}} \right)$$

$$= +9.3 \times 10^{-13} \text{ C} = +0.93 \text{ pC}$$

continued on next page

Figure 17.25 (a) Basically, this kind of computer key is merely a capacitor with a variable plate spacing. A circuit detects the change in the plate spacing as charge flows between the plates. (b) This microphone uses a capacitor with one moving plate to create an electrical signal.

Example 17.9 *continued*

Since ΔQ is positive, the magnitude of charge on the plates increases.

Discussion   If the plates move closer together, the capacitance increases. A greater capacitance means that more charge can be stored for a given potential difference. Therefore, the magnitude of the charge increases.

Practice Problem 17.9   Capacitance and the charge stored

A parallel plate capacitor has plates of area 1.0 m² and a separation of 1.0 mm. The potential difference between the plates is 2.0 kV. Find the capacitance and the magnitude of the charge on each plate. Which of these quantities depends on the potential difference?

Other devices are based on a capacitor with one moveable plate. In a *condenser microphone* (Fig. 17.25b), one plate moves in and out in response to a sound wave. (*Condenser* is a synonym for capacitor.) The capacitor is maintained at a constant potential difference; as the plate spacing changes, charge flows onto and off the plates. The moving charge—an electric current—is amplified to generate an electrical signal. The design of many *tweeters* (speakers for high-frequency sounds) is just the reverse; in response to an electrical signal, one plate moves in and out, generating a sound wave.

Making The Connection:
condenser
microphone

Capacitors have many other uses. Each RAM (random-access memory) chip in a computer contains millions of microscopic capacitors. Each of the capacitors stores one bit (binary digit). To store a 1, the capacitor is charged; to store a 0, it is discharged. The insulation of the capacitors from their surroundings is not perfect, so charge would leak

Making The Connection:
random-access memory (RAM) chips

off if it were not periodically refreshed—which is why the contents of RAM are lost when the computer's power is turned off.

Besides storing charge and electrical energy, capacitors are also useful for the uniform electric field between the plates. This field can be used to accelerate or deflect charges in a controlled way. The oscilloscope—a device used to display time-dependent potential differences in electric circuits—is a cathode ray tube that sends electrons between the plates of two capacitors (see Fig. 16.24). One of the capacitors deflects the electrons vertically; the other deflects them horizontally.

Discharging a Capacitor

If we connect one plate of a charged capacitor to the other with a conducting wire, charge moves along the wire until there is no longer a difference in potential between the plates. Figure 17.26 shows a charged capacitor on the left and the same capacitor after its plates are connected with a conducting wire.

Making The Connection:
**camera flash
attachments**

Physics at Home

The next time you are taking flash pictures with a camera, try to take two pictures one right after the other. Unless you have a professional-quality camera, the flash does not work the second time. There is a minimum time interval of a few seconds between successive flashes. Many cameras have an indicator light to show when the flash is ready.

Did you ever wonder how the small battery in a camera produces such a bright flash? Compare the brightness of a flashlight with the same type of battery. By itself, a small battery cannot pump charge fast enough to produce the bright flash needed. During the time when the flash is inoperative, the battery pumps charge onto a capacitor. Once the capacitor is fully charged, the flash is ready. When the picture is taken, the capacitor is discharged through the bulb, producing a bright flash of light.

17.6   DIELECTRICS

There is a problem inherent in trying to store a large charge in a capacitor. To store a large charge without making the potential difference excessively large, we need a large capacitance. Capacitance is inversely proportional to the spacing d between the plates. One problem with making the spacing small is that the air between the plates of the capacitor breaks down at an electric field of about 3000 V/mm with dry air (less for humid air). The breakdown allows a spark to jump across the gap so the stored charge is lost.

One way to overcome this difficulty is to put a better insulator than air between the plates. Some insulating materials, which are also called **dielectrics**, can withstand electric fields larger than those that cause air to break down. Another advantage of placing a dielectric between the plates is that the capacitance itself is increased.

For a parallel plate capacitor in which a dielectric fills the entire space between the plates, the capacitance is

Capacitance of parallel plate capacitor with dielectric
$$C = \kappa \frac{\epsilon_0 A}{d} = \kappa \frac{A}{4\pi k d} \qquad (17\text{-}16)$$

Figure 17.26 (a) A charged capacitor. (b) When a conducting wire connects the two plates, the excess electrons on the lower plate (total charge $-Q$) flow to the upper plate, leaving both plates uncharged.

The effect of the dielectric is to increase the capacitance by a factor κ (Greek letter kappa), which is called the **dielectric constant**. The dielectric constant is a dimensionless number: the ratio of the capacitance with the dielectric to the capacitance without the dielectric. The value of κ varies from one dielectric material to another. Equation (17-16) is more general than Eq. (17-15), which applies only when $\kappa = 1$. When there is vacuum between the plates, $\kappa = 1$ by definition. Air has a dielectric constant that is only slightly larger than 1; for most practical purposes we can take $\kappa = 1$ for air also. The flexible insulator in a computer key (Example 17.9) increases the capacitance by a factor of κ. Thus, the amount of charge that flows when the key is pressed is larger than the value we calculated.

The dielectric constant depends on the insulating material used. Table 17.1 gives dielectric constants and the breakdown limit, or **dielectric strength**, for several materials. The dielectric strength is the electric field strength at which dielectric breakdown occurs and the material becomes a conductor. Since $\Delta V = Ed$ for a uniform field, the dielectric strength determines the maximum potential difference that can be applied across a capacitor per meter of plate spacing.

Do not confuse dielectric constant and dielectric strength; they are not related. The dielectric constant determines how much charge can be stored for a given potential difference, while dielectric strength determines how large a potential difference can be applied to a capacitor before dielectric breakdown occurs.

Polarization in a Dielectric

What is happening microscopically to a dielectric between the plates of a capacitor? Recall that polarization is a separation of the charge in an atom or molecule. The atom or molecule remains neutral, but the center of positive charge no longer coincides with the center of negative charge.

Figure 17.27 is a simplified diagram to indicate polarization of an atom. The unpolarized atom with a central positive charge is encircled by a cloud of electrons, so that the center of the negative charge coincides with the center of the positive charge. When a positively charged rod is brought near the atom, it repels the positive charge in the atoms and

Figure 17.27 A positively charged rod induces polarization in a nearby atom.

Table 17.1		

Dielectric Constants and Dielectric Strengths for Materials at 20°C (in order of increasing dielectric constant)

Material	Dielectric Constant κ	Dielectric Strength (kV/mm)
Vacuum	1 (exact)	—
Air (dry, 1 atm)	1.00054	3
Paraffined paper	2.0–3.5	40–60
Teflon	2.1	60
Rubber (vulcanized)	3.0–4.0	16–50
Paper (bond)	3.0	8
Mica	4.5–8.0	150–220
Bakelite	4.4–5.8	12
Glass	5–10	8–13
Diamond	5.7	100
Porcelain	5.1–7.5	10
Rubber (neoprene)	6.7	12
Titanium dioxide ceramic	70–90	4
Water	80	—
Strontium titanate	310	8
Nylon 11	410	27
Barium titanate	6000	—

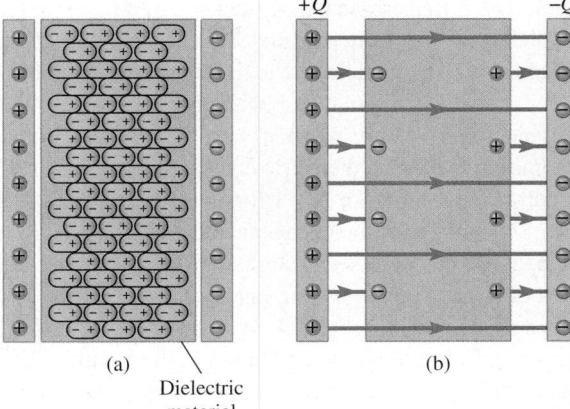

Figure 17.28 (a) Polarization of molecules in a dielectric material. (b) A dielectric with $\kappa = 2$ between the plates of a parallel plate capacitor.

attracts the negative. This separation of the charges means the centers of positive and negative charge no longer coincide; they are distorted by the influence of the charged rod.

In Fig. 17.28a, a slab of dielectric material has been placed between the plates of a capacitor. The charges on the capacitor plates induce a polarization of the dielectric. The polarization occurs throughout the material, so the positive charge is slightly displaced relative to the negative charge.

Throughout the bulk of the dielectric, there are still equal amounts of positive and negative charge. The net effect of the polarization of the dielectric is a layer of positive charge on one face and negative charge on the other (Fig. 17.28b). Each conducting plate faces a layer of opposing charge.

The layer of opposing charge induced on the surface of the dielectric helps attract more charge to the conducting plate, for the same potential difference, than would be there without the dielectric. Since capacitance is charge per unit potential difference, the capacitance must have increased. The dielectric constant of a material is a measure of the ease with which the insulating material can be polarized. A larger dielectric constant indicates a more easily polarized material. Thus, neoprene rubber ($\kappa = 6.7$) is more easily polarized than Teflon ($\kappa = 2.1$).

The induced charge on the faces of the dielectric reduces the strength of the electric field in the dielectric compared to the field outside. Some of the electric field lines end on the surface of the insulating dielectric material; fewer lines penetrate the dielectric and thus the field is weaker. With a weaker field, there is a smaller potential difference between the plates (recall that for a uniform field, $\Delta V = Ed$). A smaller potential difference makes it easier to put more charge on the capacitor. We have succeeded in having the capacitor store more charge with a smaller potential difference. Since there is a limiting potential difference before breakdown occurs, this is an important factor for reaching maximum charge storage capability.

Suppose a dielectric is immersed in an external electric field E_0. The *definition* of the **dielectric constant** is the ratio of the electric field in vacuum E_0 to the electric field E inside the dielectric material:

Definition of dielectric constant

$$\kappa = \frac{E_0}{E} \tag{17-17}$$

The electric field inside the dielectric (E) is thus *smaller* than the external applied field E_0:

$$E = E_0 / \kappa$$

In a capacitor, the dielectric is immersed in an applied field E_0 due to the charges on the plates. By reducing the field between the plates to E_0 / κ, the dielectric reduces the potential difference between the plates by the same factor $1/\kappa$. Since $Q = C\,\Delta V$, multiplying ΔV by $1/\kappa$ for a given charge Q means the capacitance is multiplied by a factor of κ due to the dielectric [see Eq. (17-16)].

Example 17.10

Parallel Plate Capacitor with Dielectric

A parallel plate capacitor has plates of area 1.00 m² and spacing of 0.500 mm. The insulator has dielectric constant 4.9 and dielectric strength 18 kV/mm. (a) What is the capacitance? (b) What is the maximum charge that can be stored on this capacitor?

Strategy   Finding the capacitance is a straightforward application of Eq. (17-16). The dielectric strength and the plate spacing determine the maximum potential difference; using the capacitance we can find the maximum charge.

Solution   (a) The capacitance is

$$C = \kappa \frac{A}{4\pi kd} = 4.9 \times \frac{1.00 \text{ m}^2}{4\pi \times 8.99 \times 10^9 \text{ N·m}^2/\text{C}^2 \times 5.00 \times 10^{-4} \text{ m}}$$

$$= 8.67 \times 10^{-8} \text{ F} = 86.7 \text{ nF}$$

(b) The maximum potential difference is

$$\Delta V = 18 \text{ kV/mm} \times 0.500 \text{ mm} = 9.0 \text{ kV}$$

Using the definition of capacitance, the maximum charge is

$$Q = C \Delta V = 8.67 \times 10^{-8} \text{ F} \times 9.0 \times 10^3 \text{ V} = 7.8 \times 10^{-4} \text{ C}$$

Discussion   Check: Each plate has a surface charge density of magnitude $\sigma = Q/A$ [Eq. (17-12)]. If the capacitor plates had this same charge density with no dielectric between them, the electric field between the plates would be [Eq. (17-13)]:

$$E_0 = 4\pi k\sigma = \frac{4\pi kQ}{A} = 8.8 \times 10^7 \text{ V/m}$$

From Eq. (17-17), the dielectric reduces the field strength by a factor of 4.9:

$$E = \frac{E_0}{\kappa} = \frac{8.8 \times 10^7 \text{ V/m}}{4.9} = 1.8 \times 10^7 \text{ V/m} = 18 \text{ kV/mm}$$

Thus, with the charge found in (b), the electric field has its maximum possible value.

Practice Problem 17.10   Changing the dielectric

If the dielectric were replaced with one having twice the dielectric constant and half the dielectric strength, what would happen to the capacitance and the maximum charge?

Example 17.11

Neuron Capacitance

 A neuron can be modeled as a parallel plate capacitor, where the membrane serves as the dielectric and the oppositely charged ions are the charges on the "plates" (Fig. 17.29). Find the capacitance of a neuron and the number of ions (assumed to be singly charged) required to establish a potential difference of 85 mV. Assume that the membrane has a dielectric constant of $\kappa = 3.0$, a thickness of 10.0 nm, and an area of $1.0 \times 10^{-10} \text{ m}^2$.

Strategy   Since we know κ, A, and d, we can find the capacitance. Then, from the potential difference and the capacitance, we can find the magnitude of charge Q on each side of the membrane. A singly charged ion has a charge of magnitude e, so Q/e is the number of ions on each side.

Solution   From Eq. (17-16),

$$C = \kappa \frac{A}{4\pi kd}$$

Substituting numerical values,

$$C = 3.0 \times \frac{1.0 \times 10^{-10} \text{ m}^2}{4\pi \times 8.99 \times 10^9 \text{ N·m}^2/\text{C}^2 \times 10.0 \times 10^{-9} \text{ m}}$$

$$= 2.66 \times 10^{-13} \text{ F} = 0.27 \text{ pF}$$

From the definition of capacitance,

$$Q = C \Delta V = 2.66 \times 10^{-13} \text{ F} \times 0.085 \text{ V}$$

$$= 2.26 \times 10^{-14} \text{ C} = 0.023 \text{ pC}$$

Each ion has a charge of magnitude $e = +1.602 \times 10^{-19}$ C. The number of ions on each side is therefore,

$$\text{number of ions} = \frac{2.26 \times 10^{-14} \text{ C}}{1.602 \times 10^{-19} \text{ C/ion}} = 1.4 \times 10^5 \text{ ions}$$

Discussion   To see if the answer is reasonable, we can estimate the average distance between the ions. If 10^5 ions are evenly spread over a surface of area 10^{-10} m^2, then the area per ion is 10^{-15} m^2. Assuming each ion to occupy a square of area 10^{-15} m^2, the distance from one ion to its nearest neighbor is the side of the square $s = \sqrt{10^{-15} \text{ m}^2} = 30$ nm. The size of a typical atom or ion is 0.2 nm. Since the distance between ions is much larger than the size of an ion, the answer is plausible; if the distance between ions came out to be less than the size of an ion, the answer would not be plausible.

Practice Problem 17.11   Action potential

How many ions must cross the membrane to change the potential difference from −0.085 V (with negative charge inside and positive outside) to +0.060 V (with negative charge outside and positive charge inside)?

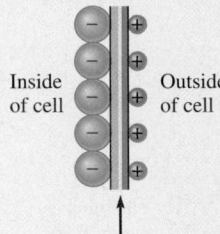

Inside of cell

Outside of cell

Figure 17.29
Cell membrane as a dielectric   Cell membrane

Making The Connection:
thunderclouds and
lightning

Lightning

Lightning (Fig. 17.30) involves the dielectric breakdown of air. Charge separation occurs within a thundercloud; the top of the cloud becomes positive and the lower part becomes negative (Fig. 17.31a). How this charge separation occurs is not completely understood, but one leading hypothesis is that collisions between ice particles or between an ice particle and a droplet of water tend to transfer electrons from the smaller particle to the larger. Updrafts in the thundercloud lift the smaller, positively charged particles to the top of the cloud, while the larger, negatively charged particles settle nearer the bottom of the cloud.

The negative charge at the bottom of the thundercloud induces positive charge on the Earth just underneath the cloud. When the electric field between the cloud and Earth reaches the breakdown limit for moist air (about 3.3×10^5 V/m), negative charge jumps from the cloud, moving in branching steps of about 50 m each. This stepwise progression of negative charges from the cloud is called a *stepped leader* (Fig. 17.31b).

Since the average electric field strength is $\Delta V/d$, the largest field occurs where d is the smallest—between tall objects and the stepped leader. *Positive streamers*—stepwise progressions of positive charge from the surface—reach up into the air from the tallest objects. If a positive streamer connects with one of the stepped leaders, a lightning channel is completed; electrons rush to the ground, lighting up the bottom of the channel. The rest of the channel then glows as more electrons rush down. The other stepped leaders also glow, but less brightly than the main channel because they contain fewer electrons. The flash of light starts at the ground and moves upward so it is called a *return stroke* (Fig. 17.31c). A total of about –20 to –25 C of charge is transferred from the thundercloud to the surface.

How can you protect yourself during a thunderstorm? Stay indoors or in an automobile if possible. If you are caught out in the open, keep low to prevent yourself from being the source of positive streamers. Do not stand under a tall tree; if lightning strikes the tree, charge traveling down the tree and then along the surface puts you in grave danger. Do not lie flat on the ground, or you risk the possibility of a large potential difference developing between your feet and head when a lightning strike travels through the ground. Go to a nearby ditch or low spot if there is one. Crouch with your head low and your feet as close together as possible to minimize the potential difference between your feet.

Making The Connection:
lightning rods

Lightning rods are used to protect barns and homes from lightning. A conducting rod higher than the roof might seem likely to attract lightning; in fact, lightning rods *prevent* lightning strikes. Charge is attracted from the ground along a cable to the sharp

Figure 17.30 Lightning illuminates the sky near the West Virginia state capitol building.

Figure 17.31 (a) Charge separation in a thundercloud. A thunderstorm acts as a giant heat engine; work is done by the engine to separate positive charge from negative charge. (b) A stepped leader extends from the bottom of the cloud toward the surface. (c) When a positive streamer from the surface connects to a stepped leader, a complete path is formed for charge to move between the cloud and the surface.

point at the top of the rod. Due to the sharp point, a strong electric field is produced, which ionizes nearby air molecules, allowing charge to leak from the rod into the surrounding air. A gradual leakage of charge keeps the potential difference between the rod and the cloud from building up enough to cause a lightning strike. Should there be a lightning strike anyway, it is more likely to strike the lightning rod than the building. Charge flows along the cable to the ground, bypassing the building and preventing a fire.

17.7 ENERGY STORED IN A CAPACITOR

A capacitor not only stores charge; it also stores energy. Figure 17.32 shows what happens when a battery is connected to an initially uncharged capacitor. Electrons are pumped off the upper plate and onto the lower until the potential difference between the capacitor plates is equal to the potential difference ΔV maintained by the battery.

The energy stored in the capacitor can be found by summing the work done by the battery to separate the charge. As the amount of charge on the plates increases, the potential difference ΔV through which charge must be moved also increases. Suppose we look at this process at some instant of time when one plate has charge $+q_i$, the other has charge $-q_i$, and the potential difference between the plates is ΔV_i.

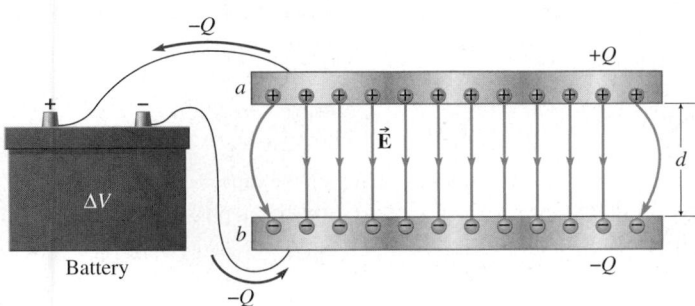

Figure 17.32 A parallel plate capacitor charged by a battery. Electrons with total charge $-Q$ are moved from the upper plate to the lower, leaving the plates with charges of equal magnitude and opposite sign.

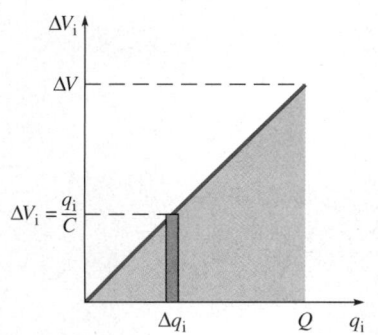

Figure 17.33 The total energy transferred is the area under the curve $\Delta V_i = q_i/C$.

To avoid writing a collection of minus signs, we imagine transferring positive charge instead of the negative charge; the result is the same whether we move negative or positive charges. From the definition of capacitance,

$$\Delta V_i = \frac{q_i}{C}$$

Now the battery transfers a little more charge Δq_i from one plate to the other, increasing the electric potential energy. If Δq_i is small, the potential difference is approximately constant during the transfer. The increase in energy is

$$\Delta U_i = \Delta q_i \times \Delta V_i$$

The total energy U stored in the capacitor is the sum of all the electric potential energy increases, ΔU_i:

$$U = \Sigma\, \Delta U_i = \Sigma\, \Delta q_i \times \Delta V_i$$

We can find this sum using a graph of the potential difference ΔV_i as a function of the charge q_i (Fig. 17.33). The graph is a straight line since $\Delta V_i = q_i/C$. The energy increase $\Delta U_i = \Delta q_i \times \Delta V_i$ when a small amount of charge is transferred is represented on the graph by the area of a rectangle of height ΔV_i and width Δq_i.

Summing the energy increases means summing the areas of a series of rectangles of increasing height. Thus, the total energy stored in the capacitor is represented by the triangular area under the graph. If the final values of the charge and potential difference are Q and ΔV, then

---

Energy stored in a capacitor

$$U = \text{area of triangle} = \tfrac{1}{2}\,(\text{base} \times \text{height})$$

$$U = \tfrac{1}{2}Q\,\Delta V \qquad\qquad (17\text{-}18a)$$

The factor of $\tfrac{1}{2}$ reflects the fact that the potential difference through which the charge was moved increases from zero to ΔV; the *average* potential difference through which the charge was moved is $\Delta V/2$. To move charge Q through an average potential difference of $\Delta V/2$ requires $Q\,\Delta V/2$ of work.

Equation (17-18a) can be written in other useful forms, using the definition of capacitance to eliminate either Q or ΔV.

$$U = \frac{1}{2}Q\,\Delta V = \frac{1}{2}(C\,\Delta V) \times \Delta V = \frac{1}{2}C\,(\Delta V)^2 \qquad\qquad (17\text{-}18b)$$

$$U = \frac{1}{2}Q\,\Delta V = \frac{1}{2}Q \times \frac{Q}{C} = \frac{Q^2}{2C} \qquad\qquad (17\text{-}18c)$$

Example 17.12

A Defibrillator

Fibrillation is a chaotic pattern of heart activity that is ineffective at pumping blood and is therefore life threatening. A device know as a *defibrillator* is used to shock the heart back to a normal beat pattern. The defibrillator discharges a capacitor through paddles on the skin, so that some of the charge flows through the heart (Fig. 17.34). (a) If an 11.0-μF capacitor is charged to 6.00 kV and then discharged through paddles into a patient's body, how much energy is stored in the capacitor? (b) How much charge flows through the patient's body if the capacitor discharges completely?

Strategy There are three equivalent expressions for energy stored in a capacitor. Since the capacitance and the potential difference are given, Eq. (17-18b) is the most direct. Since the capacitor is completely discharged, all of the charge initially on the capacitor flows through the patient's body.

Solution (a) The energy stored in the capacitor is

$$U = \tfrac{1}{2}C(\Delta V)^2 = \tfrac{1}{2} \times 11.0 \times 10^{-6}\,\text{F} \times (6.00 \times 10^3\,\text{V})^2 = 198\,\text{J}$$

(b) The charge initially on the capacitor is

$$Q = C\,\Delta V = 11.0 \times 10^{-6}\,\text{F} \times 6.00 \times 10^3\,\text{V} = 0.0660\,\text{C}$$

continued on next page

Example 17.12 *continued*

Discussion   To test our result, we make a quick check:

$$U = \frac{Q^2}{2C} = \frac{(0.0660 \text{ C})^2}{2 \times 11.0 \times 10^{-6} \text{ F}} = 198 \text{ J}$$

Practice Problem 17.12   Charge and stored energy for a parallel plate capacitor

A parallel plate capacitor of area 0.24 m² has a plate separation, in air, of 8.00 mm. The potential difference between the plates is 0.800 kV. Find (a) the charge on the plates and (b) the stored energy.

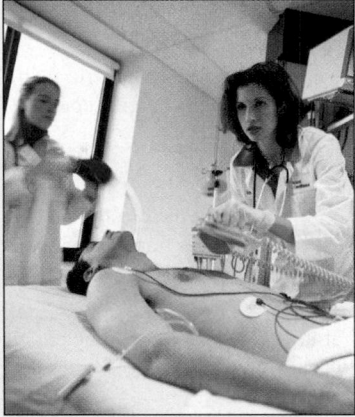

Figure 17.34
An emergency room doctor using a defibrillator to resuscitate a patient.

Energy Stored in an Electric Field

Potential energy is energy of interaction or field energy. The energy stored in a capacitor is stored in the electric field between the plates. We can use the energy stored in a capacitor to calculate how much energy *per unit volume* is stored in an electric field E. Why energy per unit volume? Two capacitors can have the same electric field but store different amounts of energy. The larger capacitor stores more energy, proportional to the volume of space between the plates.

In a parallel plate capacitor, the energy stored is

$$U = \frac{1}{2}C(\Delta V)^2 = \frac{1}{2}\kappa\frac{A}{4\pi kd}(\Delta V)^2$$

Since the field is uniform, the potential difference is

$$\Delta V = Ed$$

Substituting Ed for ΔV,

$$U = \frac{1}{2}\kappa\frac{A}{4\pi kd}(Ed)^2 = \frac{1}{2}\kappa\frac{Ad}{4\pi k}E^2$$

We recognize Ad as the volume of space between the plates of the capacitor. Then the **energy density** u—the electric potential energy per unit volume—is

$$u = \frac{U}{Ad} = \frac{\kappa}{8\pi k}E^2 = \frac{1}{2}\kappa\epsilon_0 E^2 \qquad (17\text{-}19)$$

The energy density is proportional to the square of the field strength. This is true in general, not just for a capacitor; there is energy associated with any electric field.

MASTER THE CONCEPTS

Summary

- Electric potential energy can be stored in an electric field. The electric potential energy of two point charges separated by a distance r is

$$U_E = k\frac{q_1 q_2}{r} \qquad (17\text{-}1)$$

- The signs of q and Q determine whether the electric potential energy is positive or negative. For more than two charges, the electric potential energy is the scalar sum of the individual potential energies for each *pair* of charges.

- The electric potential V at a point is the electric potential energy per unit charge when a point charge q is present, assuming $U = 0$ when q is not present:

$$V = \frac{U_E}{q} \qquad (17\text{-}3)$$

- Electric potential, like electric potential energy, is a scalar quantity. Potential has a unique value at each point in space. The SI unit for potential is the volt (1 V = 1 J/C).

- If a point charge q moves through a potential difference ΔV, then the change in electric potential energy is

$$\Delta U_E = q\,\Delta V \qquad (17\text{-}7)$$

- The electric potential at a distance r from a point charge Q is

$$V = \frac{kQ}{r} \qquad (17\text{-}8)$$

The potential at a point P due to N point charges is the sum of the potentials due to each charge.

- An equipotential surface has the same potential at every point on the surface. An equipotential surface is perpendicular to the electric field at all points. No change in electric potential energy occurs when a charge moves from one position to another on an equipotential surface. If equipotential surfaces are drawn such that the potential difference between adjacent surfaces is constant, then the surfaces are closer together where the field is stronger.

- The electric field always points in the direction of maximum potential decrease.

- The potential difference that occurs when you move a distance d in the direction of a uniform electric field of magnitude E is

$$\Delta V = -Ed \qquad (17\text{-}10)$$

- The electric field has units of

$$\text{N/C} = \text{V/m} \qquad (17\text{-}11)$$

- In electrostatic equilibrium, every point in a conductor must be at the same potential.

- A capacitor consists of two conductors (the *plates*) that are given opposite charges. A capacitor stores charge and electric potential energy. Capacitance is the ratio of the magnitude of charge on each plate (Q) to the electric potential difference between the plates (ΔV). Capacitance is measured in farads (F).

$$Q = C\,\Delta V \qquad (17\text{-}14)$$
$$1\ \text{F} = 1\ \text{C/V}$$

- The capacitance of a parallel plate capacitor is

$$C = \kappa\frac{A}{4\pi kd} = \kappa\frac{\epsilon_0 A}{d} \qquad (17\text{-}16)$$

where A is the area of the plates, d is their separation, and ϵ_0 is the permittivity of free space [$\epsilon_0 = 1/(4\pi k) = 8.854 \times 10^{-12}\ \text{C}^2/(\text{N·m}^2)$]. If vacuum separates the plates, $\kappa = 1$; otherwise $\kappa > 1$ is the dielectric constant of the dielectric (the insulating material). If a dielectric is immersed in an external electric field, the dielectric constant is the ratio of the external electric field E_0 to the electric field E in the dielectric.

$$\kappa = \frac{E_0}{E} \qquad (17\text{-}17)$$

- The dielectric constant is a measure of the ease with which the insulating material can be polarized.

- The dielectric strength is the electric field strength at which dielectric breakdown occurs and the material becomes a conductor.

- The energy stored in a capacitor is

$$U = \frac{1}{2}Q\,\Delta V = \frac{1}{2}C\,(\Delta V)^2 = \frac{1}{2}\frac{Q^2}{C} \qquad (17\text{-}18)$$

- The energy density u—the electric potential energy per unit volume—associated with an electric field is

$$u = \frac{\kappa}{8\pi k}E^2 = \frac{1}{2}\kappa\epsilon_0 E^2 \qquad (17\text{-}19)$$

Highlighted Figures and Tables

F17.1 Electrical potential energy (p. 586)

F17.2 Potential energies for pairs of point particles as a function of separation distance r (p. 587)

F17.5 Relationships between force, field, potential energy, and potential (p. 589)

F17.9 The electric field and the potential due to a hollow conducting sphere of radius R and charge q as a function of r, the distance from the center (p. 593)

F17.12 The structure and action potential of a neuron (p. 595)

F17.16 Equipotential surfaces near a positive point charge (p. 598)

F17.17 Equipotential surfaces near a dipole (p. 599)

F17.18 Field lines and equipotential surfaces in a uniform field (p. 600)

F17.21 Field lines for a charged capacitor (p. 602)

F17.22 Equipotential surfaces for a parallel plate capacitor (p. 603)

T17.1 Dielectric constants and dielectric strengths for materials at 20°C (p. 607)

F17.28 Polarization of molecules in a dielectric material (p. 608)

F17.31 Lightning involves the dielectric breakdown of air (p. 611)

CONCEPTUAL QUESTIONS

1. A negatively charged particle with charge $-q$ is far away from a positive charge $+Q$ that is fixed in place. As $-q$ moves closer to $+Q$, (a) does the electric field do positive or negative work? (b) Does $-q$ move through a potential increase or a potential decrease? (c) Does the electric potential energy increase or decrease? (d) Repeat questions (a)–(c) if the fixed charge is instead negative ($-Q$).

2. Air breaks down for a voltage of about 3000 V/mm. Is it possible to build a parallel plate capacitor with a plate spacing of 1 mm that can be charged to a potential difference greater than 3000 V? If so, explain how.

3. A bird is perched on a high-voltage power line whose potential varies between –100 kV and +100 kV. Why is the bird not electrocuted?

4. A positive charge is initially at rest in an electric field and is free to move. Does the charge start to move toward a position of higher or lower potential? What happens to a negative charge in the same situation?

5. Points A and B are at the same potential. What is the total work done by an external agent to move a charge from A to B? Does your answer mean that no external force need be applied? Explain.

6. A point charge moves to a region of higher potential and yet the electric potential energy *decreases*. How is this possible?

7. Why are all parts of a conductor at the same potential in electrostatic equilibrium?

8. If $E = 0$ at a single point, then a point charge placed at that point will feel no electric force. What does it mean if the *potential* is zero at a point? Are there any assumptions behind your answer?

9. If $E = 0$ everywhere throughout a region of space, what do we know is true about the potential at points in that region?

10. If the potential increases as you move from point P in the +x-direction, but the potential does not change as you move from P in the y- or z-directions, what is the direction of the electric field at P?

11. If the potential is the same at every point throughout a region of space, is the electric field the same at every point in that region? What can you say about the magnitude of $\vec{E}$ in the region? Explain.

12. If a uniform electric field exists in a region of space, is the potential the same at all points in the region? Explain.

13. When we talk about the potential difference between the plates of a capacitor, shouldn't we really specify two *points*, one on each plate, and talk about the potential difference between those points? Or doesn't it matter which points we choose? Explain.

14. A swimming pool is filled with water (total mass M) to a height h. Explain why the gravitational potential energy of the water (taking $U = 0$ at ground level) is $\frac{1}{2}Mgh$. Where does the factor of $\frac{1}{2}$ come from? How much work must be done to fill the pool, if there is a ready supply of water at ground level? What does this have to do with capacitors? [*Hint*: Make an analogy between the capacitor and the pool. What is the

analogous to the water? What quantity is analogous to M? What quantity is analogous to gh?]

15. The charge on a capacitor doubles. What happens to its capacitance?

16. During a thunderstorm, some cows gather under a large tree (Fig. 17.35). One cow stands facing directly toward the tree. Another cow stands at about the same distance from the tree, but it faces sideways (tangent to a circle centered on the tree). Which cow do you think is more likely to be killed if lightning strikes the tree? [*Hint*: Think about the potential difference between the cows' front and hind legs in the two positions.]

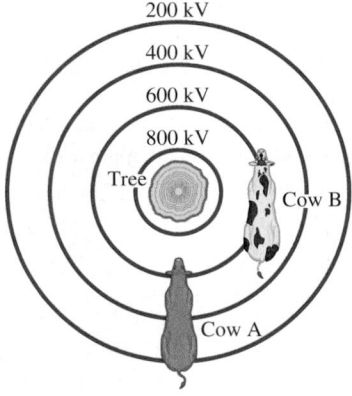

Figure 17.35 Conceptual Question 16 and Problem 42

17. If we know the potential at a single point, what (if anything) can we say about the magnitude of the electric field at that same point?

18. In Fig. 17.11, why is the person touching the dome of the van de Graaff generator not electrocuted even though there may be a potential difference of hundreds of thousands of volts between her and the ground?

19. The electric field just above Earth's surface on a clear day in an open field is about 150 V/m downward. Which is at a higher potential: the Earth or the upper atmosphere?

20. A parallel plate capacitor has the space between the plates filled with a slab of dielectric with $\kappa = 3$. While the capacitor is connected to a battery, the dielectric slab is removed. Describe *quantitatively* what happens to the capacitance, the potential difference, the charge on the plates, the electric field, and the energy stored in the capacitor as the slab is removed. [*Hint*: First figure out which quantities remain constant.]

21. Repeat Question 20 if the capacitor is charged and then disconnected from the battery before removing the dielectric slab.

22. A charged parallel plate capacitor has the space between the plates filled with air. The capacitor has been disconnected from the battery that charged it. Describe *quantitatively* what happens to the capacitance, the potential difference, the charge on the plates, the electric field, and the energy stored in the capacitor as the plates are moved closer together. [*Hint*: First figure out which quantities remain constant.]

Figure 17.36
Conceptual Question 23

23. A positive charge $+2\ \mu C$ and a negative charge $-5\ \mu C$ lie on a line (Fig. 17.36). In which region or regions (A, B, C) is there a point on the line a finite distance away where the potential is zero? Explain your reasoning. Are there any points where both the electric field and the potential are zero?

MULTIPLE CHOICE QUESTIONS

In all problems, we assign the potential due to a point charge to be zero at an infinite distance from the charge.

1. Two charges are located at opposite corners (A and C) of a square (Fig. 17.37). We do not know the magnitude or sign of these charges. What can be said about the potential at corner B relative to the potential at corner D?
   (a) It is the same as that at D.
   (b) It is different from that at D.
   (c) It is the same as that at D only if the charges at A and C are equal.
   (d) It is the same as that at D only if the charges at A and C are equal in magnitude and opposite in sign.

Figure 17.37   Multiple Choice Question 1

2. Among these choices, which is/are correct units for electric field?
   (a) N/kg only   (b) N/C only   (c) N only
   (d) N·m/C only   (e) V/m only   (f) Both N/C and V/m

3. In Fig. 17.38, the potential is zero at which of the points A–E?
   (a) B, D, and E (b) B only   (c) A, B, and C
   (d) All five points   (e) All except B

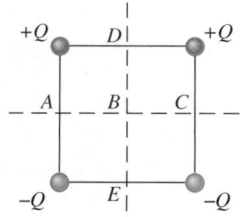

Figure 17.38   Multiple Choice Question 3

4. Which of these units can be used to measure electric potential?
   (a) N/C   (b) J   (c) V·m   (d) V/m   (e) $\dfrac{N \cdot m}{C}$

5. A parallel plate capacitor is attached to a battery that supplies a constant potential difference. While the battery is still attached, the parallel plates are separated a little more. Which statement describes what happens?
   (a) The electric field increases and the charge on the plates decreases.
   (b) The electric field remains constant and the charge on the plates increases.
   (c) The electric field remains constant and the charge on the plates decreases.
   (d) Both the electric field and the charge on the plates decrease.

6. A capacitor has been charged with $+Q$ on one plate and $-Q$ on the other plate. Which of these statements is true?
   (a) The potential difference between the plates is QC.
   (b) The energy stored is $\frac{1}{2}Q\,\Delta V$.
   (c) The energy stored is $\frac{1}{2}Q^2C$.
   (d) The potential difference across the plates is $Q^2/(2C)$.
   (e) None of the statements above is true.

7. Two solid metal spheres of different radii are far apart. The spheres are connected by a fine metal wire. Some charge is placed on one of the spheres. After electrostatic equilibrium is reached, the wire is removed. Which of these quantities will be the *same* for the two spheres?
   (a) The charge on each sphere
   (b) The electric field inside each sphere, at the same distance from the center of the spheres
   (c) The electric field just outside the surface of each sphere
   (d) The electric potential at the surface of each sphere
   (e) Both b and c   (f) Both b and d   (g) Both a and c

8. A large negative charge $-Q$ is located in the vicinity of points A and B. Suppose a positive charge $+q$ is moved at constant speed from A to B by an external agent. Along which of the paths shown in Fig. 17.39 will the work done by the field be the greatest?
   (a) path 1   (b) path 2   (c) path 3   (d) path 4
   (e) work is the same along all four paths

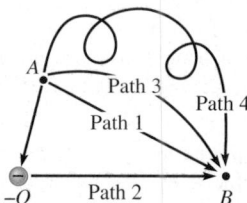

Figure 17.39   Multiple Choice Question 8

9. A tiny charged pellet of mass m is suspended at rest between two horizontal, charged metallic plates (Fig. 17.40). The lower plate has a positive charge and the upper plate has a negative charge. Which statement in the answers here is *not* true?
   (a) The electric field between the plates points vertically upward.
   (b) The pellet is negatively charged.
   (c) The magnitude of the electric force on the pellet is equal to mg.
   (d) The plates are at different potentials.

Figure 17.40   Multiple Choice Question 9

10. Two positive 2.0-μC point charges are placed as shown in Fig. 17.41a. The distance from each charge to the point P is 0.040 m. Then the charges are rearranged as shown in Fig. 17.41b. Which statement is now true concerning $\vec{E}$ and V at point P?

(a) The electric field and the electric potential are both zero.

(b) $\vec{E} = 0$, but V is the same as before the charges were moved.

(c) $V = 0$, but $\vec{E}$ is the same as before the charges were moved.

(d) $\vec{E}$ is the same as before the charges were moved, but V is less than before.

(e) Both $\vec{E}$ and V have changed and neither is zero.

(a)   (b)

Figure 17.41 Multiple Choice Question 10

11. In Fig. 17.42, which two points are closest to being at the same potential?

(a) A and D (b) B and C   (c) B and D (d) A and C

12. In Fig. 17.42, which point is at the lowest potential?

(a) A (b) B (c) C (d) D

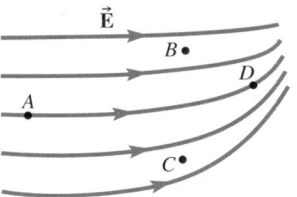

Figure 17.42 Multiple Choice Questions 11 and 12

PROBLEMS

Note: **C** indicates a combination conceptual/quantitative problem. Gold diamonds ✦, ✦✦ are used to indicate the increasing level of difficulty of each problem. Problem numbers appearing in blue, 9., denote problems that have a detailed solution available in the Student Solutions Manual. Some problems are *paired* by concept; their numbers are connected by a ruled box.

17.1 Electric Potential Energy

1. Two point charges, +5.0 μC and –2.0 μC, are separated by 5.0 m (Fig. 17.43). What is the electric potential energy?

C 2. A hydrogen atom has a single proton at its center and a single electron at a distance of approximately 0.0529 nm from the proton. (a) What is the electric potential energy in joules? (b) What is the significance of the sign of the answer?

$Q = +5.0$ μC $q = -2.0$ μC

$\leftarrow r = 5.0$ m $\rightarrow$

Figure 17.43 Problem 1

3. The nucleus of a helium atom contains two protons that are approximately 1 fm apart. How much work must be done by an external agent to bring the two protons from an infinite separation to a separation of 1.0 fm?

4. Two point charges (+10.0 nC and –10.0 nC) are located 8.00 cm apart (Fig. 17.44). What is the *total* electric potential energy of the three charges when a third point charge of –4.2 nC is placed at points a, b, and c in turn? (Let $U = 0$ when *all three* charges are separated by an infinite distance.)

5. Find the electric potential energy for the following array of charges: charge $q_1 = +4.0$ μC is located at $(x, y) = (0.0, 0.0)$ m; charge $q_2 = +3.0$ μC is located at $(4.0, 3.0)$ m; and charge $q_3 = -1.0$ μC is located at $(0.0, 3.0)$ m.

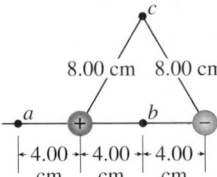

Figure 17.44 Problems 4 and 59

✦ 6. (a) In Fig. 17.45, how much work is done *by the electric field* as a charge of +2.00 nC moves from infinity to each of the three points a, b, and c? (b) How much work is done *by the field* as the charge (+2.00 nC) moves from a to b and from b to c in turn?

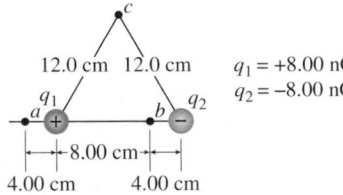

$q_1 = +8.00$ nC
$q_2 = -8.00$ nC

Figure 17.45 Problems 6 and 17

17.2 Electric Potential

7. Find the electric field and the potential at the center of a square of side 2.0 cm with charges of +9.0 μC at each corner (Fig. 17.46).

8. Find the electric field and the potential at the center of a square of side 2.0 cm with a charge of +9.0 μC at one corner of the square and with charges of –3.0 μC at the remaining three corners (Fig. 17.47).

Figure 17.46 Problem 7

Figure 17.47 Problem 8

9. A charge $Q = -50.0$ nC is located 0.30 m from point A and 0.50 m from point B (Fig. 17.48). (a) What is the potential at A? (b) What is the potential at B? (c) If a point charge q is moved from A to B while Q is fixed in place, through what potential difference does it move? Does its potential increase or decrease? (d) If $q = -1.0$ nC, what is the change in electric potential energy as it moves from A to B? Does the potential energy increase or decrease? (e) How much work is done by the electric field due to charge Q as q moves from A to B?

Figure 17.48 Problem 9

10. A charge of $+2.0$ mC is located at $x = 0$, $y = 0$ and a charge of -4.0 mC is located at $x = 0$, $y = 3.0$ m. What is the electric potential due to these charges at a point with coordinates $x = 4.0$ m, $y = 0$?

11. The electric potential at a distance of 20.0 cm from a point charge is $+1.0$ kV (assuming $V = 0$ at infinity). (a) Is the point charge positive or negative? (b) At what distance is the potential $+2.0$ kV?

12. An array of four charges, arranged along the x-axis at intervals of 1.0 m, is shown in Fig.17.49. (a) If two of the charges are $+1.0$ μC and two are -1.0 μC, draw a configuration of these charges that minimizes the potential at $x = 0$. (b) If three of the charges are the same, $q = +1.0$ μC, and the charge at the far right is -1.0 μC, what is the potential at the origin?

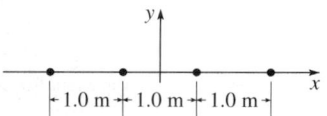

Figure 17.49 Problem 12

13. At a point P, a distance R_0 from a positive charge Q_0, the electric field has a magnitude $E_0 = 100$ N/C and the electric potential is $V_0 = 10$ V (Fig. 17.50). The charge is now increased by a factor of three, becoming $3Q_0$. (a) At what distance, R_E, from the charge $3Q_0$ will the electric field have the same value, $E = E_0$; and (b) at what distance, R_V, from the charge $3Q_0$ will the electric potential have the same value, $V = V_0$?

Figure 17.50 Problem 13

14. Charges of $+2.0$ nC and -1.0 nC are located at opposite corners, A and C, respectively, of a square which is 1.0 m on a side (Fig. 17.51). What is the electric potential at a third corner, B, of the square (where there is no charge)?

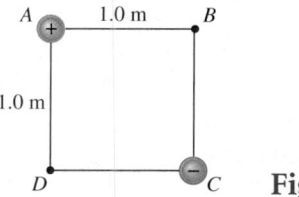

Figure 17.51 Problem 14

15. (a) Find the electric potential at points a and b for charges of $+4.2$ nC and -6.4 nC located as shown in Fig. 17.52. (b) What is the potential difference ΔV for a trip from a to b? (c) How much work must be done by an external agent to move a point charge of $+1.50$ nC from a to b?

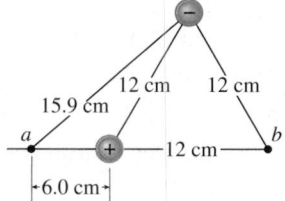

Figure 17.52 Problem 15

16. (a) Find the potential at points a and b in Fig. 17.53 for charges $Q_1 = +2.50$ nC and $Q_2 = -2.50$ nC. (b) How much work must be done by an external agent to bring a point charge q from infinity to point b?

Figure 17.53 Problem 16

◆17. (a) Find the potentials at points a, b, and c for the arrangement of charges in Fig. 17.45. (b) How much work must be done by an external agent to move the charge ($+2.00$ nC) from a to b and from b to c in turn? (If you have done Problem 6, compare your answers.)

17.3 The Relationship between Electric Field and Potential

18. By rewriting each unit in terms of kg, m, s, and C, show that 1 N/C = 1 V/m.

19. A uniform electric field has magnitude 240 N/C and is directed to the right (Fig. 17.54). A particle with charge +4.2 nC moves along the straight line from *a* to *b*. (a) What is the electric force that acts on the particle? (b) What is the work done on the particle by the electric field? (c) What is the potential difference $V_a - V_b$ between points *a* and *b*?

Figure 17.54 Problem 19

20. In a region where there is an electric field, the electric forces do $+8.0 \times 10^{-19}$ J of work on an electron as it moves from point *X* to point *Y*. (a) Which point, *X* or *Y*, is at a higher potential? (b) What is the potential difference, $V_Y - V_X$, between point *Y* and point *X*?

21. Suppose a uniform electric field of magnitude 100.0 N/C exists in a region of space. How far apart are a pair of equipotential surfaces whose potentials differ by 1.0 V?

22. Draw some electric field lines and a few equipotential surfaces outside a negatively charged hollow conducting sphere. What shape are the equipotential surfaces?

23. Draw some electric field lines and a few equipotential surfaces outside a positively charged conducting cylinder. What shape are the equipotential surfaces?

24. It is believed that a large electric fish known as *Torpedo occidentalis* uses electricity to shock its victims. A typical fish can deliver a potential difference of 0.20 kV for a duration of 1.5 ms. This pulse delivers charge at a rate of 18 C/s. (a) What is the rate at which this work is done by the electric organs during a pulse? (b) What is the total amount of work done during one pulse?

25. A positive point charge is located at the center of a hollow spherical metal shell with zero net charge (Fig. 17.55). (a) Draw some electric field lines and sketch some equipotential surfaces for this arrangement. (b) Sketch graphs of the electric field magnitude and the potential as functions of *r*.

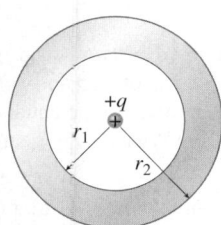

Figure 17.55 Problem 25

26. A positively charged oil drop is injected into a region of uniform electric field between two oppositely charged, horizontally oriented plates. If the electric force on the drop is found to be 9.6×10^{-16} N and the electric field magnitude is 3000 V/m, what is the magnitude of the charge on the drop in terms of the elementary charge *e*?

27. A positively charged oil drop of mass 1.0×10^{-15} kg is placed in the region of a uniform electric field between two oppositely charged, horizontal plates. The drop is found to remain stationary under the influence of the Earth's gravitational field and the uniform electric field of 6.1×10^4 N/C. What is the magnitude of the charge on the drop?

17.4 Conservation of Energy for Moving Charges

28. Point *P* is at a potential of 500.0 kV and point *S* is at a potential of 200.0 kV. The space between these points is evacuated. When a charge of +2e moves from *P* to *S*, by how much does its kinetic energy change?

29. An electron is accelerated from rest through a potential difference ΔV. If the electron reaches a speed of 7.26×10^6 m/s, what is the potential difference? Be sure to include the correct sign (does the electron move through an increase or a decrease in potential?).

30. Figure 17.56 shows a graph of electric potential versus position along the *x*-axis. An electron is originally at point *A*, moving in the positive *x*-direction. How much kinetic energy does it need to have at point *A* in order to be able to reach point *E* (with no forces acting on the electron other than those due to the indicated potential)? Points *B*, *C*, and *D* have to be passed on the way.

31. Repeat Problem 30 for a proton rather than an electron.

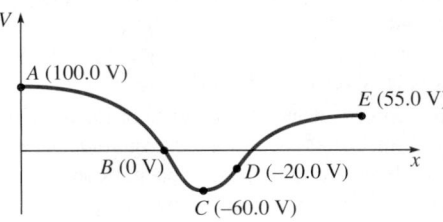

Figure 17.56 Problems 30 and 31

32. A helium nucleus (charge +2e) moves through a potential difference $\Delta V = -0.50$ kV. Its initial kinetic energy is 1.20×10^{-16} J. What is its final kinetic energy?

33. A beam of electrons of mass m_e is deflected vertically by the uniform electric field between two oppositely charged, parallel metal plates (Fig. 17.57). The plates are a distance *d* apart and the potential difference between the plates is ΔV. (a) What is the direction of the electric field between the plates? (b) If the *y*-component of the electrons' velocity as they leave the region between the plates is v_y, derive an expression for the time it takes each electron to travel through the region between the plates in terms of ΔV, v_y, m_e, *d*, and *e*. (c) Does the electric potential energy of an electron increase, decrease, or stay constant while it moves between the plates? Explain.

34. An electron (charge –e) is projected horizontally into the space between two oppositely charged parallel plates. The electric field between the plates is 500.0 N/C upward. If the vertical deflection of the electron as it leaves the plates has magnitude

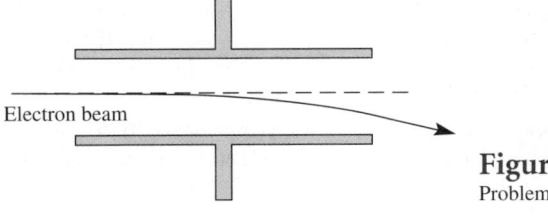

Figure 17.57
Problem 33

3.0 mm, how much has its kinetic energy increased due to the electric field? [*Hint*: First find the potential difference through which the electron moves.]

17.5 Capacitors

35. If a capacitor has a capacitance of 10.2 μF and we wish to lower the potential difference across the plates by 60.0 V, how much charge will we have to remove?

36. A parallel plate capacitor has a capacitance of 2.0 μF and plate separation of 1.0 mm. (a) How much potential difference can be placed across the capacitor before dielectric breakdown of air occurs ($E_{max} = 3 \times 10^6$ V/m)? (b) What is the magnitude of the greatest charge the capacitor can store before breakdown?

37. A parallel plate capacitor is charged by connecting it to a 12-V battery. The battery is then disconnected from the capacitor, leaving the plates with a 12-V potential difference. The parallel plates are then pulled so that the air spacing between the plates is enlarged slightly. What is the effect (a) on the electric field between the plates? (b) on the potential difference between the plates?

38. A parallel plate capacitor has a capacitance of 1.20 nF. There is a charge of magnitude 0.800 μC on each plate. (a) What is the potential difference between the plates? (b) If the plate separation is doubled, while the charge is kept constant, what will happen to the potential difference?

39. A shark is able to detect the presence of electric fields as small as 1.0 μV/m. To get an idea of the magnitude of this field, suppose you have a parallel plate capacitor connected to a 1.5-V battery. How far apart must the parallel plates be to have an electric field of 1.0 μV/m between the plates?

40. A spherical conductor of radius R carries a total charge Q. (a) Show that the magnitude of the electric field just outside the sphere is $E = \sigma/\epsilon_0$, where σ is the charge per unit area on the conductor's surface. (b) Construct an argument to show why the electric field at a point P just outside *any* conductor in electrostatic equilibrium has magnitude $E = \sigma/\epsilon_0$, where σ is the local surface charge density. [*Hint*: Consider a tiny area of an arbitrary conductor and compare it to an area of the same size on a spherical conductor with the same charge density. Think about the number of field lines starting or ending on the two areas.]

17.6 Dielectrics

41. Before a lightning strike can occur, the breakdown limit for *damp* air must be reached. If this occurs for an electric field of 3.33×10^5 V/m, what is the maximum possible height above the Earth for the bottom of a thundercloud, which is at a potential 1.00×10^8 V below Earth's surface potential, if there is to be a lightning strike?

42. Two cows, with approximately 1.8 m between their front and hind legs, are standing under a tree during a thunderstorm (Fig. 17.35). (a) If the equipotential surfaces about the tree just after a lightning strike are as shown, what is the average electric field between Cow A's front and hind legs? (b) Which cow is more likely to be killed? Explain.

43. A parallel plate capacitor has a charge of 0.020 μC on each plate with a potential difference of 240 V. The parallel plates are separated by 0.40 mm of bakelite. What is the capacitance of this capacitor?

44. To make a parallel plate capacitor, you have available two flat plates of aluminum (area 120 cm²), a sheet of paper (thickness = 0.10 mm, $\kappa = 3.5$), a sheet of glass (thickness = 2.0 mm, $\kappa = 7.0$), and a slab of paraffin (thickness = 10.0 mm, $\kappa = 2.0$). (a) What is the largest capacitance possible using one of these dielectrics? (b) What is the smallest?

45. A capacitor can be made from two sheets of aluminum foil separated by a sheet of waxed paper. If the sheets of aluminum are 0.30 m by 0.40 m and the waxed paper, of slightly larger dimensions, is of thickness 0.030 mm and dielectric constant $\kappa = 2.5$, what is the capacitance of this capacitor?

46. In capacitive electrostimulation, electrodes are placed on opposite sides of a limb. A potential difference is applied to the electrodes, which is believed to be beneficial in treating bone defects and breaks. If the capacitance is measured to be 0.59 pF, the electrodes are 4.0 cm² in area, and the limb is 3.0 cm in diameter, what is the (average) dielectric constant of the tissue in the limb?

17.7 Energy Stored in a Capacitor

47. Figure 17.31b shows a thundercloud before a lightning strike has occurred. The bottom of the thundercloud and the Earth's surface might be modeled as a charged parallel plate capacitor. The base of the cloud, which is roughly parallel to the Earth's surface, serves as the negative plate and the region of Earth's surface under the cloud serves as the positive plate. The separation between the cloud base and the Earth's surface is small compared to the length of the cloud. (a) Find the capacitance for a thundercloud of base dimensions 4.5 km by 2.5 km located 550 m above the Earth's surface. (b) Find the energy stored in this capacitor if the charge magnitude is 18 C.

48. A parallel plate capacitor of capacitance 6.0 μF has the space between the plates filled with a slab of glass with $\kappa = 3.0$. The capacitor is charged by attaching it to a 1.5-V battery. After the capacitor is disconnected from the battery, the dielectric slab is removed. Find (a) the capacitance, (b) the potential difference, (c) the charge on the plates, and (d) the energy stored in the capacitor after the glass is removed.

49. A parallel plate capacitor is composed of two square plates, 10.0 cm on a side, separated by an air gap of 0.75 mm. (a) What is the charge on this capacitor when there is a potential difference of 150 V between the plates? (b) What energy is stored in this capacitor?

50. The capacitor of Problem 49 is initially charged to a 150-V potential difference. The plates are then physically separated by another 0.750 mm in such a way that none of the charge can leak off the plates. Find (a) the new capacitance and (b) the new energy stored in the capacitor. Explain the result using conservation of energy.

51. Capacitors are used in many applications where you need to supply a short burst of energy. A 100.0-μF capacitor in an electronic flash lamp supplies an average power of 10.0 kW to the lamp for 2.0 ms. (a) To what potential difference must the capacitor initially be charged? (b) What is its initial charge?

52. A parallel plate capacitor has a charge of 0.020 µC on each plate with a potential difference of 240 V. The parallel plates are separated by 0.40 mm of air. What energy is stored in this capacitor?

53. A parallel plate capacitor has a capacitance of 1.20 nF. There is a charge of 0.80 µC on each plate. How much work must be done by an external agent to double the plate separation while keeping the charge constant?

54. A defibrillator is used to restart a person's heart after it stops beating. Energy is delivered to the heart by discharging a capacitor through the body tissues near the heart. If the capacitance of the defibrillator is 9 µF and the energy delivered is to be 300 J, to what potential difference must the capacitor be charged?

55. A defibrillator consists of a 15-µF capacitor that is charged to 9.0 kV. (a) If the capacitor is discharged in 2.0 ms, how much charge passes through the body tissues? (b) What is the average power delivered to the tissues?

56. A thundercloud is at a potential of -1.00×10^8 V with respect to Earth's surface. If a charge of 25.0 C is transferred to the Earth during a lightning strike, completely discharging the cloud, find the energy delivered to Earth.

✦ 57. (a) If the bottom of a thundercloud has a potential of -1.00×10^9 V with respect to the Earth and a charge of -20.0 C is discharged from the cloud to the Earth during a lightning strike, how much energy is released? (Assume that the system acts like a capacitor—as charge flows, the potential difference decreases to zero.) (b) If the energy is absorbed by a tree struck by the lightning bolt and 10.0% of the energy vaporizes sap in the tree, about how much sap is vaporized? (Assume the sap to have the same latent heat as water.) (c) If 10.0% of the energy released from the lightning strike could be stored and used by a homeowner who uses 400.0 kW·hr of electricity per month, for how long could the lightning bolt supply electricity to the home?

COMPREHENSIVE PROBLEMS

58. Charges of -12.0 nC and -22.0 nC are separated by 0.700 m. What is the potential midway between the two charges?

59. Two point charges ($+10.0$ nC and -10.0 nC) are located 8.00 cm apart (Fig. 17.44). (a) What is the potential energy of a point charge of -4.2 nC when it is placed at points a, b, and c in turn? Let $U = 0$ when the -4.2 nC charge is far away (but the other two are still in place). (b) How much work would an external force have to do to move the point charge from b to a?

60. If an electron moves from one point at a potential of -100.0 V to another point at a potential of $+100.0$ V, how much work is done by the electric field?

61. A van de Graaff generator has a metal sphere of radius 15 cm. To what potential can it be charged before the electric field at its surface exceeds 3.0×10^6 N/C (which is sufficient to break down dry air and initiate a spark)?

62. Find the potential at the sodium ion, Na⁺, which is surrounded by two chloride ions, Cl⁻, and a calcium ion, Ca²⁺, in water as shown in Fig. 17.58. The effective charge of the positive sodium ion in water is 2.0×10^{-21} C, of the negative chlorine ion is -2.0×10^{-21} C, and of the positive calcium ion is 4.0×10^{-21} C.

Figure 17.58 Problem 62

63. An infinitely long conducting cylinder sits near an infinite conducting sheet (side view in Fig. 17.59). The cylinder and sheet have equal and opposite charges; the cylinder is positive. (a) Sketch some electric field lines. (b) Sketch some equipotential surfaces.

Figure 17.59 Problem 63

64. Two parallel plates are 4.0 cm apart. The bottom plate is charged positively and the top plate is charged negatively, producing a uniform electric field of 5.0×10^4 N/C in the region between the plates. What is the time required for an electron, which starts at rest at the upper plate, to reach the lower plate? (Assume a vacuum exists between the plates.)

✦ 65. Draw some electric field lines and a few equipotential surfaces outside a positively charged metal cube. [*Hint:* What shape are the equipotential surfaces close to the cube? What shape are they far away?]

66. The potential difference across a cell membrane is -90 mV. If the membrane's thickness is 10 nm, what is the magnitude of the electric field in the membrane? Assume the field is uniform.

67. A beam of electrons traveling with a speed of 3.0×10^7 m/s enters a uniform, downward electric field of magnitude 2.0×10^4 N/C between the deflection plates of an oscilloscope (Fig. 17.60). The initial velocity of the electrons is perpendicular to the field. The plates are 6.0 cm long. (a) What is the direction and magnitude of the change in velocity of the electrons while they are between the plates? (b) How far are the electrons deflected in the ±y-direction while between the plates?

68. A parallel plate capacitor has a charge of 0.020 µC on each plate with a potential difference of 240 V. The parallel plates are separated by 0.40 mm of air. (a) What is the capacitance for this capacitor? (b) What is the area of a single plate?

Figure 17.60 Problem 67

Figure 17.61 Problem 74

(c) At what voltage will the air between the plates become ionized? Assume a dielectric strength of 3.0 kV/mm for air.

✦ 69. The potential difference across a cell membrane from outside to inside is initially at −90 mV (when in its resting phase). When a stimulus is applied, Na^+ ions are allowed to move into the cell such that the potential changes to +20 mV for a short amount of time. (a) If the membrane capacitance per unit area is 1 μF/cm^2, how much charge moves through a membrane of area 0.05 cm^2? (b) The charge on Na^+ is +e. How many ions move through the membrane?

70. An axon has the outer part of its membrane positively charged and the inner part negatively charged. The membrane has a thickness of 4.4 nm and a dielectric constant κ = 5. If we model the axon as a parallel plate capacitor whose area is 5 μm^2, what is its capacitance?

✦ 71. It has only been fairly recently that 1.0-F capacitors have been readily available. A typical 1.0-F capacitor can withstand up to 5.0 V. To get an idea why it isn't easy to make a 1.0-F capacitor, imagine making a 1.0-F parallel plate capacitor using titanium dioxide (κ = 90.0, breakdown strength 4.00 kV/mm) as the dielectric. (a) Find the minimum thickness of the titanium dioxide such that the capacitor can withstand 5.00 V. (b) Find the area of the plates so that the capacitance is 1.0 F.

72. A cell membrane has a surface area of 1.0×10^{-7} m^2, a dielectric constant of 5.2 and a thickness of 7.5 nm. The membrane acts like the dielectric in a parallel plate capacitor; a layer of positive ions on the outer surface and a layer of negative ions on the inner surface act as the capacitor plates. The potential difference between the "plates" is 90.0 mV. (a) How much energy is stored in this capacitor? (b) How many positive ions are there on the outside of the membrane? Assume that all the ions are singly charged (charge +e).

✦ 73. A parallel plate capacitor is connected to a battery. The space between the plates is filled with air. The electric field strength between the plates is 20.0 V/m. Then, *with the battery still connected*, a slab of dielectric (κ = 4.0) is inserted between the plates. The thickness of the dielectric is half the distance between the plates. Find the electric field inside the dielectric.

74. An electron beam is deflected upward through 3.0 mm while traveling in a vacuum between two deflection plates 12.0 mm apart (Fig. 17.61). The potential difference between the deflecting plates is 100.0 kV and the kinetic energy of each electron as it enters the space between the plates is 2.0 × 10^{-15} J. What is the kinetic energy of each electron when it leaves the space between the plates?

75. A point charge q = −2.5 nC is initially at rest adjacent to the negative plate of a capacitor. The charge per unit area on the plates is 4.0 μC/m^2 and the space between the plates is 6.0 mm. (a) What is the potential difference between the plates? (b) What is the kinetic energy of the point charge just before it hits the positive plate, assuming no other forces act on it?

76. An alpha particle (helium nucleus, charge +2e) starts from rest and travels a distance of 1.0 cm under the influence of a uniform electric field of magnitude 10.0 kV/m. What is the final kinetic energy of the alpha particle?

✦ 77. A parallel plate capacitor is attached to a battery that supplies a constant voltage. While the battery remains attached to the capacitor, the distance between the parallel plates increases by 25%. What happens to the energy stored in the capacitor?

✦ 78. A parallel plate capacitor is attached to a battery that supplies a constant voltage. While the battery is still attached, a dielectric of dielectric constant κ = 3.0 is inserted so that it just fits between the plates. What is the energy stored in the capacitor after the dielectric is inserted in terms of the energy U_0 before the dielectric was inserted?

79. The inside of a cell membrane is at a potential of 90.0 mV lower than the outside. How much work does the electric field do when a sodium ion (Na^+) with a charge of +e moves through the membrane from outside to inside?

✦ 80. (a) Calculate the capacitance per unit length of an axon of radius 5.0 μm (see Fig. 17.12). The membrane acts as an insulator between the conducting fluids inside and outside the neuron. The membrane is 6.0 nm thick and has a dielectric constant of 7.0. (*Note*: The membrane is thin compared to the radius of the axon, so the axon can be treated as a parallel plate capacitor.) (b) In its resting state (no signal being transmitted), the potential of the fluid inside is about 85 mV lower than the outside. Therefore, there must be small net charges ±Q on either side of the membrane. Which side has positive charge? What is the magnitude of the charge density on the surfaces of the membrane?

ⓒ ✦ 81. A 4.00-μF air gap capacitor is connected to a 100.0-V battery until the capacitor is fully charged. The battery is removed and then a dielectric of dielectric constant 6.0 is inserted between the plates without allowing any charge to leak off the plates. (a) Find the energy stored in the capacitor before and after the dielectric is inserted. [*Hint*: First find the new capacitance and potential difference.] (b) Does an external agent have to do positive work to insert the dielectric or to remove the dielectric? Explain.

ANSWERS TO PRACTICE PROBLEMS

17.1 (a) +0.018 J; (b) away from Q; (c) U decreases as the separation increases. The potential energy decrease accompanies an increase in kinetic energy as q moves faster and faster.

17.2 six pairs (with subscripts 12, 13, 14, 23, 24, 34)

17.3 the lower plate

17.4 $V_B = -1.5 \times 10^5$ V; work (done by $\vec{E}$) $= -\Delta U_E = -0.010$ J.

17.5 $\vec{E} = 5.4 \times 10^8$ N/C away from the +9.0-μC charge; $V = 0$.

17.6 4 kV

17.7

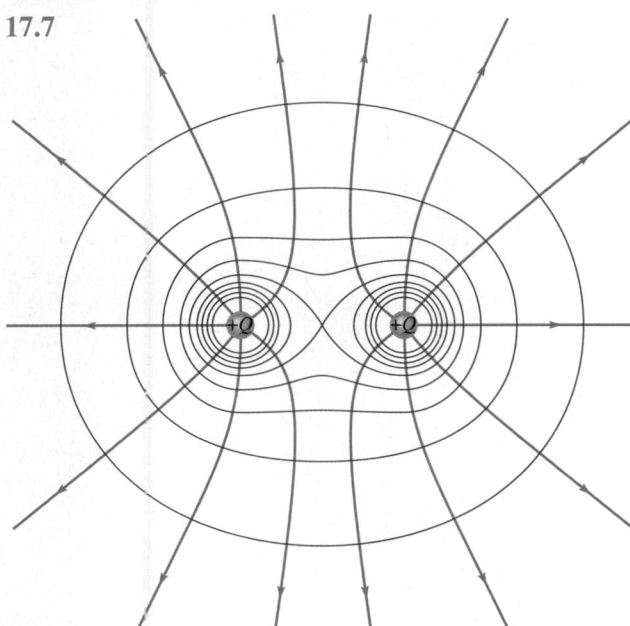

17.8 –20.9 kV (Note that a positive charge gains kinetic energy when it moves through a potential decrease; a negative charge gains kinetic energy when it moves through a potential increase.)

17.9 8.9 nF; 18 μC; charge (capacitance is independent of potential difference)

17.10 C doubles, maximum charge unchanged

17.11 2.4×10^5 ions

17.12 0.21 μC; 85 μJ

Chapter 18

Electric Current and Circuits

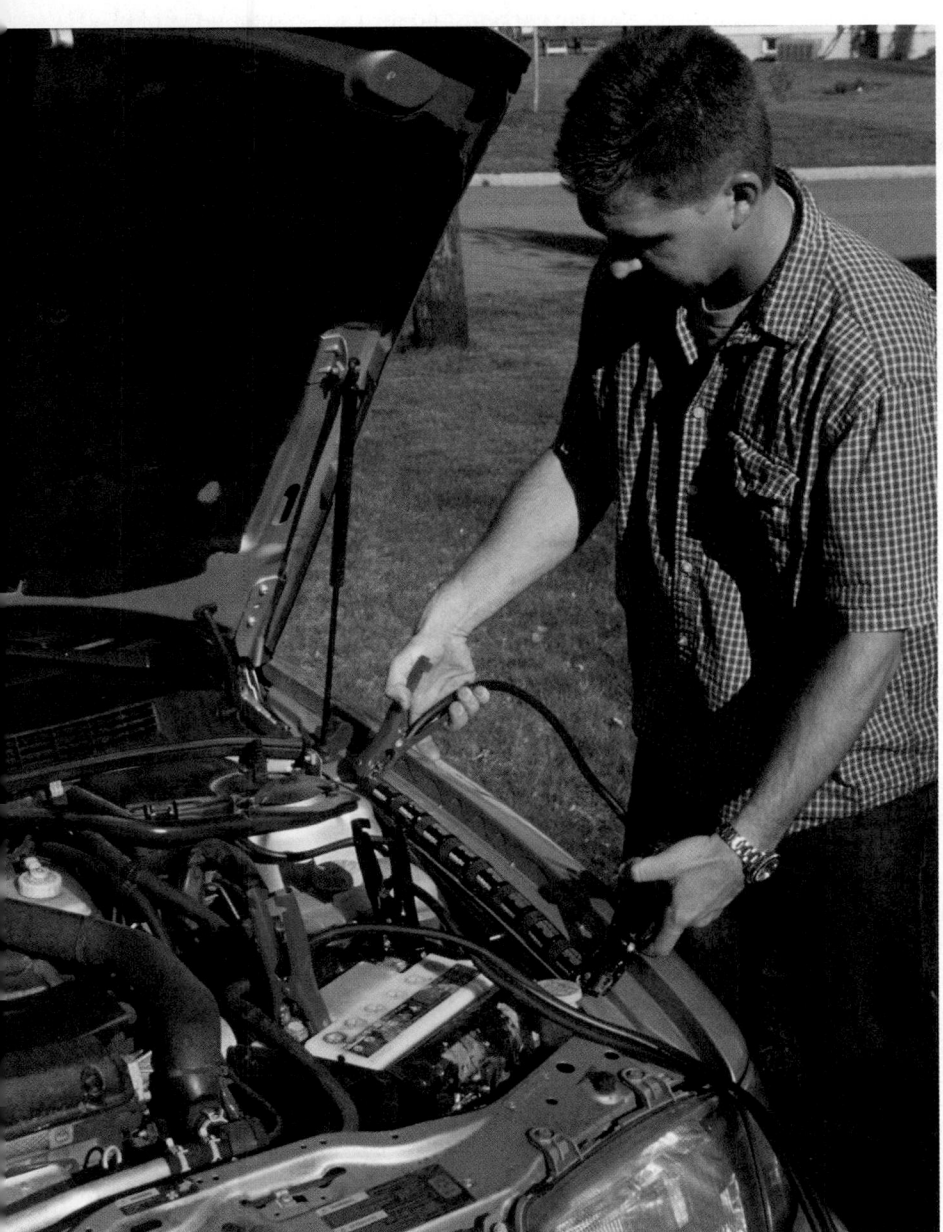

Graham's car won't start; the battery is dead after he left the headlights on overnight. In a kitchen drawer are several 1.5-V flashlight batteries. Graham decides to connect eight of them together, being careful to connect the positive terminal of one to the negative terminal of the next, just the way two 1.5-V batteries are connected inside a flashlight to provide 3.0 V. Eight 1.5-V batteries should provide 12 V, he reasons. Will this scheme work?

18.1   ELECTRIC CURRENT

Figure 18.1 shows two oppositely charged metal spheres on insulated stands. The two spheres are connected by a metal wire. The free electrons in the metal (also called **conduction electrons**) are free to move through the bulk of the spheres and the wire; they are not bound to any particular nucleus. When the switch is closed, an electric field exerts forces on the conduction electrons, causing them to move from the negative sphere through the wire to the positive sphere. The flow of electrons stops when the two spheres are at the same potential and the electric field everywhere inside the metal is zero.

A net flow of charge is called an **electric current**. In Fig. 18.1, current flows only briefly. If we can somehow *maintain* a potential difference between the spheres, by pumping charge from one sphere to the other, then the current in the wire would continue to flow. Under the influence of an electric field, all charges *of like sign* that are free to move head in *one direction*. In this case the negatively charged electrons all move in the direction opposite to the electric field. If there were mobile positive charges, they would move in the direction of the electric field.

The *current* (symbol I) is defined as the net amount of charge passing per unit time through an area perpendicular to the flow direction (Fig. 18.2). The magnitude of the current tells us the rate of the net flow of charge. If Δq is the net charge that passes through the shaded surface in Fig. 18.2 during a time interval Δt, then the current in the wire is defined as

Definition of current
$$I = \frac{\Delta q}{\Delta t} \qquad (18\text{-}1)$$

Currents are not necessarily steady. In order for Eq. (18-1) to define the instantaneous current, we must use a sufficiently small time interval Δt.

The SI unit of current, equal to one coulomb per second, is the ampere (A), named for the French scientist André-Marie Ampère (1775–1836). The ampere is one of the SI base units; the coulomb is a derived unit defined as one ampere-second:

$$1\,C = 1\,A\cdot s$$

Small currents are more conveniently measured in milliamperes (mA = 10^{-3} A) or in microamperes (μA = 10^{-6} A). The word amperes is often shortened to *amps*; for smaller currents, we speak of *milliamps* or *microamps*.

According to convention, the direction of an electric current is defined as the direction in which *positive* charge is transported. Benjamin Franklin established this convention (and decided which kind of charge would be called positive) long before scientists understood that the mobile charges in metals are electrons. If electrons move to the left

Direction of current = direction of flow of positive charge

The symbol —⌒○— represents an open switch (no electrical connection). The symbol —●○— represents a closed switch.

(a)   (b)

Figure 18.1   (a) Initially the spheres carry charges of opposite sign. (b) When the switch is closed, electrons flow through the wire until the two spheres are at the same potential.

in a metal wire, the direction of the current is to the *right*; negative charge moving to the left has the same effect on the net distribution of charge as positive charge moving to the right. Since the direction of the electric field is the direction of the force on a positive charge, current flows in the direction of the electric field regardless of whether positive or negative charge is moving.

In most situations, the motion of positive charge in one direction causes the same macroscopic effects as the motion of negative charge in the opposite direction. In circuit analysis, we always draw currents in the conventional direction regardless of the sign of the charge carriers.

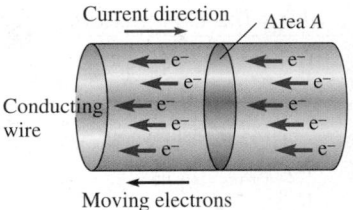

Figure 18.2 Close-up picture of the wire that connects the two spheres. The current is the rate of flow of charge through an area perpendicular to the direction of flow.

Example 18.1

Current in a Clock

Two wires of cross-sectional area 1.6 mm^2 connect the terminals of a battery to the circuitry in a clock. During a time interval of 0.040 s, 5.0×10^{14} electrons move to the right through a cross section of one of the wires. (Actually, electrons pass through the cross section in both directions; the number that cross to the right is 5.0×10^{14} more than the number that cross to the left.) What is the magnitude and direction of the current in the wire?

Strategy Current is the rate of flow of charge. We are given the number N of electrons; multiplying by the elementary charge e gives the magnitude of moving charge Δq.

Solution The magnitude of the charge of 5.0×10^{14} electrons is

$$\Delta q = Ne = 5.0 \times 10^{14} \times 1.60 \times 10^{-19}\,\text{C} = 8.00 \times 10^{-5}\,\text{C}$$

The magnitude of the current is therefore,

$$I = \frac{\Delta q}{\Delta t} = \frac{8.00 \times 10^{-5}\,\text{C}}{0.040\,\text{s}} = 0.0020\,\text{A} = 2.0\,\text{mA}$$

Negatively charged electrons moving to the right means that the direction of conventional current—the direction in which positive charge is effectively being transported—is to the *left*.

 Discussion To find the magnitude of the current, we use the *magnitude* of the charge on the electron. We *do* treat current as a signed quantity when analyzing circuits. We arbitrarily choose a direction for current when the actual direction is not known. If the calculations result in a negative current, the minus sign reveals that the actual direction of the current is opposite the chosen direction. The minus sign merely means the current flows in the direction opposite to the one we assumed.

In this problem, the cross-sectional area of the wire was extraneous information. To find the current, we need only the quantity of charge and the time for the charge to pass.

Practice Problem 18.1 Current in a calculator

(a) If 0.320 mA of current flow through a calculator, how many electrons pass through per second? (b) How long does it take for 1.0 C of charge to pass through the calculator?

Electric Current in Liquids and Gases

Electric currents can exist in liquids and gases as well as in solid conductors. In an ionic solution, both positive and negative charges contribute to the current by moving in opposite directions (Fig. 18.3). The electric field is to the right, away from the positive electrode and toward the negative electrode. In response, positive ions move in the direction of the electric field (to the right) and negative ions move in the opposite direction (to the left). Since positive and negative charges are moving in opposite directions, they both contribute to current in the *same* direction. Thus, we can find the magnitudes of the currents separately due to the motion of the negative charges and the positive charges and *add* them to find the total current. The direction of the current in Fig. 18.3 is to the right.

If positive and negative charges were moving in the *same* direction, they would represent currents in *opposite* directions and the individual currents would be *subtracted* to find the net current. Just remember this example: In a water pipe, there is an enormous amount of moving charge (the protons and electrons in the neutral water molecules), but equal quantities of positive and negative charges are moving in the same direction. As a result, the net electric current in the pipe is zero; there is no net transport of charge.

Figure 18.3 A current in a solution of potassium chloride consists of positive ions (K$^+$) and negative ions (Cl$^-$) moving in opposite directions. The direction of the current is the direction in which the positive ions move.

The circuit symbol for a battery is
—+|⊢— . The long line represents the terminal at higher potential and the short line represents the terminal at lower potential. Since many batteries consist of more than one chemical cell, an alternate form is
—|||⊢— .

Currents also exist in gases. Figure 18.4 shows a neon sign. A large potential difference is applied to the metal electrodes inside a glass container of neon gas. There are always some positive ions present in a gas due to bombardment by cosmic rays and natural radioactivity. The positive ions are accelerated by the electric field toward the cathode; if they have sufficient energy, they can knock electrons loose when they collide with the cathode. These electrons are accelerated toward the anode; they ionize more gas molecules as they pass through the container. Collisions between electrons and ions produce the characteristic red light of a neon sign. Fluorescent lights are similar, but the collisions produce ultraviolet radiation; a coating on the inside of the glass absorbs the ultraviolet and emits visible light.

18.2 EMF AND CIRCUITS

To maintain a current in a conducting wire, we need to maintain a potential difference between the ends of the wire. One way to do that is to connect the ends of the wire to the terminals of a battery (one end to each of the two terminals). An *ideal* battery maintains a constant potential difference between its terminals, regardless of how fast it must pump charge to do so. An ideal battery is analogous to an ideal water pump that maintains a constant pressure difference between intake and output regardless of the volume flow rate.

The potential difference maintained by an ideal battery is called the battery's **emf** (symbol $\mathscr{E}$). Emf originally stood for *electromotive force*, but emf is *not* a measure of the force applied to a charge or to a collection of charges; emf cannot be expressed in newtons. Rather, emf is measured in units of potential (volts) and is a measure of the work done by the battery per unit charge. To avoid this confusion, we just write "emf" (pronounced *ee-em-eff*). If the amount of charge pumped by an ideal battery of emf $\mathscr{E}$ is q, then the work done by the battery is

---

Work done by an ideal battery

$$W = \mathscr{E}q \qquad (18\text{-}2)$$

Any device that pumps charge is called a *source of emf* (or just an *emf*). Generators, solar cells, and fuel cells are other sources of emf. Fuel cells, used in the space shuttle and perhaps someday in cars and homes, are similar to batteries, but their reactants are supplied externally. Many living organisms also contain sources of emf (Fig. 18.5). The signals transmitted by the human nervous system are electrical in nature, so our bodies contain sources of emf. The same circuit symbol is used for *any* source of constant emf (—+|⊢—). All emfs are energy conversion devices; they convert some other form of energy into electrical energy. The energy sources used by emfs include chemical energy (batteries, fuel cells, biological sources of emf), sunlight (solar cells), and mechanical energy (generators).

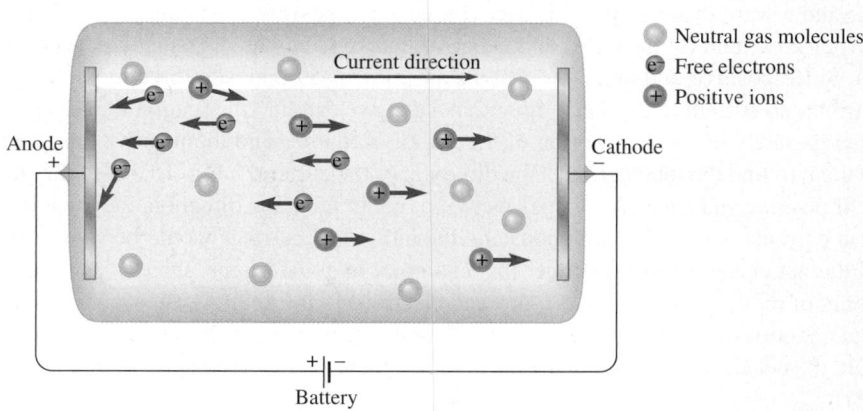

Figure 18.4 Simplified diagram of a neon sign. The neon gas inside the glass tube is ionized by a large potential difference between the electrodes.

Figure 18.5   The South American electric eel (*Electrophorous electricus*) has hundreds of thousands of cells (called *electroplaques*) that supply emf. The current supplied by the electroplaques is used to stun its enemies and to kill its prey.

In Fig. 18.6, imagine that the flow of water represents electric current (the flow of charge) in a circuit. The people take water from the place where its potential energy is lowest and do the work necessary to carry it uphill to the place where its potential energy is highest. The water then runs downhill, encountering resistance to its flow (the

people = battery

An electric circuit

+

−

Figure 18.6   Using the flow of water as an analogy to what happens in an electric circuit sluice gate) along the way before returning to the pool at the bottom. A battery (or other source of emf) plays a role something like that of the people who carry buckets of water. Thinking of current as the movement of positive charge, a battery takes positive charge from the place where its *electric potential* is lowest (the negative terminal of the battery) and does the work necessary to move it to the place where the electric potential is highest (the positive terminal). Then the charge flows through some device that offers resistance to the flow of current—perhaps a lightbulb or a CD player—before returning to the negative terminal of the battery.

A 9-V battery (such as the kind used in a smoke detector) maintains its positive terminal 9 V higher than its negative terminal—as long as conditions permit the battery to be treated as ideal. Since a volt is a joule per coulomb, the battery does 9 J of work for every coulomb of charge that it pumps. The battery does work by converting some of its stored chemical energy into electrical energy. When a battery is dead, its supply of chemical energy has been depleted and it can no longer pump charge. Some batteries can be recharged by forcing charge to flow through them in the opposite direction, reversing the direction of the electrochemical reaction and converting electrical energy into chemical energy.

Batteries come with various emfs (12 V, 9 V, 1.5 V, etc.) as well as in various sizes (Fig. 18.7). The size of a battery does *not* determine its emf. Common battery sizes AAA, AA, A, C, D all provide the same emf (1.5 V). However, the larger batteries have a larger quantity of the chemicals and thus store more chemical energy. A larger battery can supply more energy by pumping more charge than a smaller one, even though the two do the same amount of work *per unit charge*. The amount of charge that a battery can pump is often measured in A·h (ampere-hours). Another difference is that larger batteries can generally pump charge *faster*—in other words, they can supply larger currents.

Figure 18.7 Batteries come in many sizes and shapes. In the back is a lead-acid automobile battery. In front, from left to right are three types of rechargeable nickel-cadmium batteries, seven batteries commonly used in flashlights, cameras, and watches, and a zinc graphite dry cell.

Circuits

For currents to continue to flow, a complete circuit is required. That is, there must be a continuous conducting path from one terminal of the emf to one or more devices and then back to the other terminal. In Fig. 18.8a,b there is one complete circuit for the current to travel from the positive terminal of the battery, through a wire, through the lightbulb filament, through another wire, into the battery at the negative terminal, and

(a)

(b)

(c)

Figure 18.8 (a) Connecting a battery to a lightbulb. The bulb only lights up when current flows through its filament. (b) To maintain current flow, a complete circuit must exist. Note the use of the arrows to indicate the direction of current flow in the wires, lightbulb, and battery. (c) An analogous circuit dealing with the flow of water rather than of charge.

through the battery to return to the positive terminal. Since this circuit has only a single loop for current to flow, the current must be the same everywhere. Think of the battery as a water pump, the wires as hoses, and the lightbulb as the engine block and radiator of an automobile (Fig. 18.8c). Water must flow from the pump, through a hose, through the engine and radiator, through another hose, and back to the pump. The volume flow rate in this single-loop "water circuit" is the same everywhere. Current does not get "used up" in the lightbulb any more than water gets used up in the radiator.

In this chapter, we consider only circuits in which the current in any branch always moves in the same direction—a **direct current** (DC) circuit. In Chapter 21 we study **alternating current** (AC) circuits, in which the currents periodically reverse direction.

18.3   MICROSCOPIC VIEW OF CURRENT IN A METAL

Figure 18.2 showed a simplified picture of the conduction electrons in a metal, all moving with the same constant velocity due to an electric field. Why do the electrons not move with a constant *acceleration* due to a constant electric force? To answer this question and to understand the relationship between electric field and current in a metal, we need a more accurate picture of the motion of the electrons.

In the absence of an applied electric field, the conduction electrons in a metal are in constant random motion at high speed—about 10^6 m/s in copper. The electrons suffer frequent collisions with each other and with ions (the atomic nuclei with their bound electrons). In copper, a given conduction electron collides 4×10^{13} times per second, traveling on average about 40 nm between collisions. A collision can change the direction of the electron's motion, so each electron moves in a random path similar to that of a gas molecule (Fig. 18.9a). The average *velocity* of the conduction electrons in a metal is zero in the absence of an electric field, so there is no net transport of charge.

If a uniform electric field exists within the metal, the electric force on the conduction electrons gives them a uniform acceleration between collisions (when the net force due to nearby ions and other conduction electrons is small). The electrons still move about in random directions like gas molecules, but the electric force makes them move on average a little faster in the direction of the force than in the opposite direction—much like air molecules in a gentle breeze. As a result, the electrons slowly drift in the direction of the electric force (Fig. 18.9b). The electrons now have a nonzero average velocity called the **drift velocity** $\vec{v}_D$ (which corresponds to the wind velocity for air molecules). The magnitude of the drift velocity (the *drift speed*) is much smaller than the instantaneous speeds of the electrons—typically less than 1 mm/s—but since it is nonzero there is a net transport of charge. We show below that the current is proportional to the drift speed.

It might seem that a uniform acceleration should make the electrons move faster and faster. If there were no collisions, they would. An electron has a uniform acceleration *between collisions*, but every collision sends it off in some new direction with a different speed. Each collision between an electron and an ion is an opportunity for the electron to transfer some of its kinetic energy to the ion. The net result is that the drift velocity is constant and energy is transferred from the electrons to the ions at a constant rate.

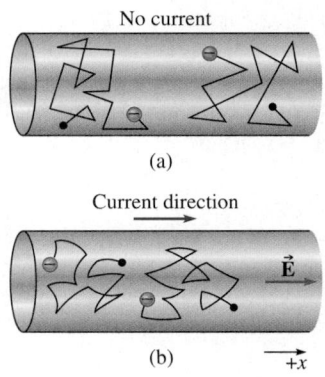

Figure 18.9 (a) Random paths followed by two conduction electrons in a metal wire in the absence of an electric field. (b) An electric field in the $+x$-direction gives the electrons a constant acceleration in the $-x$-direction between collisions. *On average*, the electrons drift in the $-x$-direction. The current in the wire is in the $+x$-direction.

Relationship between Current and Drift Velocity

To find out how current depends on drift velocity, we use a simplified model in which all the electrons move at a constant velocity $\vec{v}_D$ (Fig. 18.10). The number of conduction electrons per unit volume (n) is a characteristic of a particular metal. Suppose we calculate the current by finding how much charge moves through the

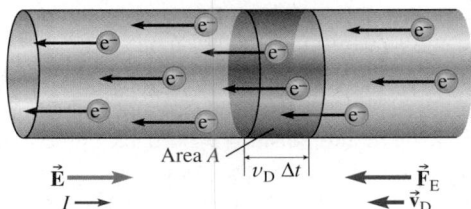

Figure 18.10 Simplified picture of the conduction electrons moving at a uniform velocity $\vec{v}_D$. In a time Δt, each electron moves a distance $v_D \Delta t$. The vector arrows show the displacement of each electron during Δt. All of the conduction electrons within a distance $v_D \Delta t$ pass through the shaded cross-sectional area in a time Δt.

shaded area in a time Δt. During that time, every electron moves a distance $v_D \Delta t$ to the left. Thus, every conduction electron in a volume $A v_D \Delta t$ moves through the shaded area. The number of electrons in this volume is $N = n A v_D \Delta t$; the magnitude of the charge is

$$\Delta Q = Ne = neA v_D \Delta t$$

Therefore, the magnitude of the current in the wire is

$$I = \frac{\Delta Q}{\Delta t} = neA v_D \qquad (18\text{-}3)$$

Remember that, since electrons carry negative charge, the direction of current flow is opposite the direction of motion of the electrons. The electric force on the electrons is opposite the electric field, so the current is in the direction of the electric field in the wire.

Equation (18-3) can be generalized to systems where the current carriers are not necessarily electrons, simply by replacing e with the charge of the carriers. In materials called semiconductors, there may be both positive and negative carriers. The negative carriers are electrons; the positive carriers are "missing" electrons (called *holes*) that act as particles with charge $+e$. The electrons and holes drift in opposite directions; both contribute to the current. Since the concentrations of electrons and holes may be different and they may have different drift speeds, the current is

$$I = n_+ eA v_+ + n_- eA v_- \qquad (18\text{-}4)$$

In Eq. (18-4), v_+ and v_- are drift *speeds*—both are positive.

When we turn on a light by flipping a wall switch, current flows through the lightbulb almost instantaneously. We do *not* have to wait for electrons to move from the switch to the lightbulb—which is a good thing, since it would be a long wait (see Example 18.2). Conduction electrons are present all along the wires that form the circuit. When the switch is closed, the *electric field* extends into the entire circuit almost instantaneously. The electrons start to drift as soon as the electric field is nonzero.

Example 18.2

Drift Speed in Household Wiring

 A #12 gauge copper wire, commonly used in household wiring, has a diameter of 2.053 mm. There are 8.00×10^{28} conduction electrons per cubic meter in copper. (a) If the wire carries a constant DC current of 5.00 A, what is the drift speed of the electrons?

(b) Explain why a thinner copper wire carrying a current of 5.00 A has a *larger* drift speed.

Strategy   From the diameter, we can find the cross-sectional area A of the wire. The number of conduction electrons per

continued on next page

Example 18.2 *continued*

cubic meter is n in Eq. (18-3). Then Eq. (18-3) enables us to solve for the drift speed. In (b), we consider how to get the same number of electrons per second flowing through a smaller cross-sectional area.

Solution   (a) The cross-sectional area of the wire is

$$A = \pi r^2 = \tfrac{1}{4}\pi d^2$$

The drift speed is given by

$$v_D = \frac{I}{neA} =$$

$$\frac{5.00 \text{ A}}{8.00 \times 10^{28} \text{ m}^{-3} \times 1.602 \times 10^{-19} \text{ C} \times \tfrac{1}{4}\pi \times (2.053 \times 10^{-3} \text{ m})^2}$$

$$= 1.18 \times 10^{-4} \text{ m·s}^{-1} = 0.118 \text{ mm/s}$$

(b) A thinner wire has fewer conduction electrons in a given length—the number per unit *volume* is the same but now the cross-sectional area is smaller. To produce the same current using fewer electrons, the electrons must move faster. This reasoning is confirmed by Eq. (18-3). The wire is still copper, so n is the same; the magnitude of the charge on the electron is also the same. If I is the same but A is smaller, v_D must be larger.

Discussion   The drift speed may seem surprisingly small: at an average speed of 0.118 mm/s, it takes an electron over 2 h

to move one meter along the wire! How can 5 C/s—a respectable amount of current—be carried by electrons with such small average velocities? Because there are so many of them. As a check: the number of conduction electrons per unit length of wire is

$$nA = 8.00 \times 10^{28} \text{ m}^{-3} \times \tfrac{1}{4}\pi \times (2.053 \times 10^{-3} \text{ m})^2$$

$$= 2.65 \times 10^{23} \text{ electrons/m}$$

Then the number of conduction electrons in a 0.118 mm length of wire is

$$2.65 \times 10^{23} \text{ electrons/m} \times 0.118 \times 10^{-3} \text{ m} = 3.12 \times 10^{19} \text{ electrons}$$

The magnitude of the total charge of these electrons is

$$3.12 \times 10^{19} \text{ electrons} \times 1.602 \times 10^{-19} \text{ C/electron} = 5.00 \text{ C}$$

Practice Problem 18.2   Current and drift speed in a silver wire

A silver wire has a diameter of 2.588 mm and contains 5.80×10^{28} conduction electrons per cubic meter. A battery of 1.50 V pushes 880 C through the wire in 45 min. Find (a) the current and (b) the drift speed in the wire.

18.4   RESISTANCE AND RESISTIVITY

Resistance

Suppose we maintain a potential difference across the ends of a conductor. How does the current I that flows through the conductor depend on the potential difference ΔV? For many conductors, the current is proportional to the potential difference. Georg Ohm (1789–1854) first observed this relationship, which is now called **Ohm's law**:

Ohm's law	
$I \propto \Delta V$	(18-5)

Ohm's law is not a universal law of physics like the conservation laws. It does not apply at all to some materials, whereas even materials that obey Ohm's law for a wide range of potential differences fail to do so when ΔV becomes too large. Hooke's law ($F \propto \Delta x$ or stress $\propto$ strain) is similar; it applies to many materials under many circumstances but is not a fundamental law of physics. Any *homogeneous* material follows Ohm's law for *some* range of potential differences; metals that are good conductors follow Ohm's law over a *wide* range of potential differences.

Ohm was inspired to look at the relationship between current and potential difference by Fourier's observation that the rate of heat flow through a conductor of heat is proportional to the temperature difference across it. Another analogous situation is the flow of oil (or any viscous fluid) through a pipe. Poiseuille's law says that the rate of flow of the fluid is proportional to the pressure difference between the ends of the pipe.

The electrical **resistance** R is defined to be the ratio of the potential difference (or *voltage*) ΔV across a conductor to the current I through the material:

Definition of resistance

$$R = \frac{\Delta V}{I} \qquad (18\text{-}6)$$

In SI units, electrical resistance is measured in ohms (symbol Ω, the Greek capital omega), defined as

$$1\ \Omega = 1\ \text{V/A} \qquad (18\text{-}7)$$

For a given potential difference, a large current flows through a conductor with a small resistance, while a small current flows through a conductor with a large resistance.

 An *ohmic* conductor—one that follows Ohm's law—has a resistance that is constant, regardless of the potential difference applied. Equation (18-6) is *not* a statement of Ohm's law, since it does not require that the resistance be constant; it is the *definition of resistance* for nonohmic conductors as well as for ohmic conductors. For an ohmic conductor, a graph of current versus potential difference is a straight line through the origin with slope $1/R$ (Fig. 18.11a). For some nonohmic systems, the graph of I versus ΔV is dramatically nonlinear (Fig. 18.11b,c).

Resistivity

Resistance depends on size and shape. Returning to the analogy with fluid flow: a longer pipe offers more resistance to fluid flow than does a short pipe and a wider pipe offers less resistance than a narrow one. By analogy, we expect a long wire to have higher resistance than a short one (everything else being the same) and a thicker wire to have a lower resistance than a thin one. The electrical resistance of a conductor of length L and cross-sectional area A can be written:

$$R = \rho \frac{L}{A} \qquad (18\text{-}8)$$

The constant of proportionality ρ (Greek letter rho), which is an intrinsic characteristic of a particular material at a particular temperature, is called the **resistivity** of the material. The SI unit for resistivity is $\Omega \cdot \text{m}$. Table 18.1 lists resistivities for various substances at 20°C. The resistivities of good conductors are small. The resistivities of pure semiconductors are significantly larger. By doping semiconductors (introducing controlled amounts of impurities), their resistivities can be changed dramatically, which is one reason that semiconductors are used to make computer chips and other electronic devices

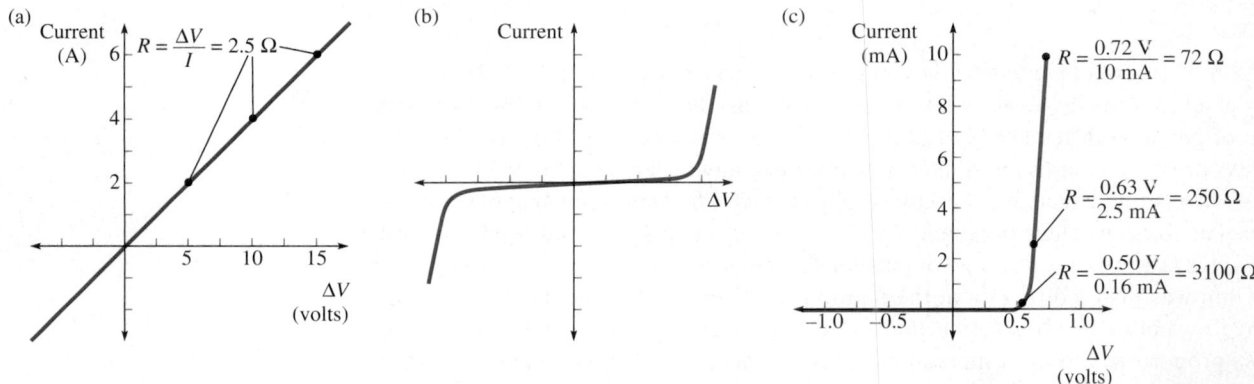

Figure 18.11 (a) Current as a function of potential difference for a tungsten wire. The resistance is the same for any value of ΔV on the graph, so the wire is an ohmic conductor. Similar graphs for (b) a diode and (c) the gas in a fluorescent light are far from linear; these systems are nonohmic.

Table 18.1

Resistivities and Temperature Coefficients at 20°C

	$\rho\,(\Omega\cdot m)$	$\alpha\,(°C^{-1})$		$\rho\,(\Omega\cdot m)$	$\alpha\,(°C^{-1})$
Conductors			**Semiconductors (pure)**		
Silver	1.59×10^{-8}	3.8×10^{-3}	Carbon	3.5×10^{-5}	-0.5×10^{-3}
Copper	1.67×10^{-8}	4.05×10^{-3}	Germanium	0.6	-50×10^{-3}
Gold	2.35×10^{-8}	3.4×10^{-3}	Silicon	2300	-70×10^{-3}
Aluminum	2.65×10^{-8}	3.9×10^{-3}			
Tungsten	5.40×10^{-8}	4.50×10^{-3}	**Insulators**		
Iron	9.71×10^{-8}	5.0×10^{-3}			
Lead	21×10^{-8}	3.9×10^{-3}	Glass	$10^{10} - 10^{14}$	
Platinum	10.6×10^{-8}	3.64×10^{-3}	Lucite	$> 10^{13}$	
Manganin	44×10^{-8}	0.002×10^{-3}	Quartz (fused)	$> 10^{16}$	
Constantan	49×10^{-8}	0.002×10^{-3}	Rubber (hard)	$10^{13} - 10^{16}$	
Mercury	96×10^{-8}	0.89×10^{-3}	Teflon	$> 10^{13}$	
Nichrome	108×10^{-8}	0.4×10^{-3}	Wood	$10^{8} - 10^{11}$	

Figure 18.12 A scanning electron microscope view of a microprocessor chip. Much of the chip is made of silicon. By introducing impurities into the silicon in a controlled way, some regions act as insulating material, others as conducting wires, and others as the transistors—circuit elements that act as switches. SOI stands for silicon on insulator, a new technology that reduces the heat generated within a chip.

(Fig. 18.12). Insulators have very large resistivities (about a factor of 10^{20} larger than for conductors). The inverse of resistivity is called *conductivity* [SI units $(\Omega \cdot m)^{-1}$].

Why is resistance proportional to length? Suppose we have two otherwise identical wires with different lengths. If the wires carry the same current, they must have the same drift speed; to have the same drift speed, the electric field must be the same. Since for a uniform field $\Delta V = EL$, the potential differences across the wires are proportional to their lengths. Therefore, $R = \Delta V/I$ is proportional to length.

Why is resistance inversely proportional to area? Suppose we have two otherwise identical wires with different areas. Applying the same potential difference produces the same drift speed, but the thicker wire has more conduction electrons per unit length. Since $I = neAv_{D}$ [Eq. (18-3)], the current is proportional to the area and $R = \Delta V/I$ is inversely proportional to area.

Example 18.3

Resistance of an Extension Cord

(a) A 30.0-m-long extension cord is made from two #19 gauge copper wires. (One wire carries current *to* an appliance, while the other wire carries current *from* it.) What is the resistance of each wire at 20.0°C? The diameter of #19 gauge wire is 0.912 mm. (b) If the copper wire is to be replaced by an aluminum wire of the same length, what is the minimum diameter so that the new wire has a resistance no greater than the old?

Strategy   After calculating the cross-sectional area of the copper wire from its diameter, we find the resistance of the copper wire from Eq. (18-8). The resistivities of copper and aluminum are found in Table 18.1.

Solution   (a) From Table 18.1, the resistivity of copper is

$$\rho = 1.67 \times 10^{-8}\ \Omega\cdot m$$

The wire's cross-sectional area is

$$A = \tfrac{1}{4}\pi d^2 = \tfrac{1}{4}\pi\ (9.12 \times 10^{-4}\ m)^2 = 6.533 \times 10^{-7}\ m^2$$

Resistance is resistivity times length over area:

$$R = \rho \frac{L}{A}$$

$$= \frac{1.67 \times 10^{-8}\ \Omega\cdot m \times 30.0\ m}{6.533 \times 10^{-7}\ m^2}$$

$$= 0.767\ \Omega$$

(b) We want the resistance of the aluminum wire to be less than or equal to the resistance of the copper wire ($R_a \le R_c$):

$$\frac{\rho_a L}{\tfrac{1}{4}\pi d_a^2} \le \frac{\rho_c L}{\tfrac{1}{4}\pi d_c^2}$$

which simplifies to $\rho_a d_c^2 \le \rho_c d_a^2$. Solving for d_a yields

$$d_a \ge d_c \sqrt{\frac{\rho_a}{\rho_c}} = 0.912\ mm \times \sqrt{\frac{2.65 \times 10^{-8}\ \Omega\cdot m}{1.67 \times 10^{-8}\ \Omega\cdot m}} = 1.15\ mm$$

Discussion   Check: the resistance of an aluminum wire of diameter 1.149 mm is

$$R = \frac{\rho L}{A} = \frac{2.65 \times 10^{-8}\ \Omega\cdot m \times 30.0\ m}{\tfrac{1}{4}\pi\ (1.149 \times 10^{-3}\ m)^2} = 0.767\ \Omega$$

Aluminum has a higher resistivity, so the wire must be thicker to have the same resistance.

Extension cords are rated according to the maximum safe current they can carry. For an appliance that draws a larger current, a thicker extension cord must be used; otherwise the potential difference across the wires would be too large ($\Delta V = IR$).

Practice Problem 18.3   Resistance of a lightbulb filament

Find the resistance at 20°C of a tungsten lightbulb filament of length 4.0 cm and diameter 0.020 mm.

Resistivity Depends on Temperature

Resistivity does not depend on the size or shape of the material, but it does depend on temperature. Two factors primarily determine the resistivity of a metal: the number of conduction electrons per unit volume and the rate of collisions between an electron and an ion. The second of these factors is sensitive to changes in temperature. At a higher temperature, the internal energy is greater; the ions vibrate with larger amplitudes. As a result, the electrons collide more frequently with the ions. With less time to accelerate between collisions, they acquire a smaller drift speed; thus, the current is smaller for a given electric field. Therefore, as the temperature of a metal is raised, its resistivity increases. The metal filament in a glowing incandescent lightbulb reaches a temperature of about 3000 K; its resistance is significantly higher than at room temperature.

For many materials, the relation between resistivity and temperature is linear over a fairly wide range of temperatures (about 500°C):

$$\rho = \rho_0(1 + \alpha\ \Delta T) \qquad (18\text{-}9)$$

Here ρ_0 is the resistivity at temperature T_0 and ρ is the resistivity at temperature $T = T_0 + \Delta T$. The quantity α is called the **temperature coefficient of resistivity** and has SI units °C^{-1} or K^{-1}. Temperature coefficients for some materials are listed in Table 18.1.

The relationship between resistivity and temperature is the basis of the *resistance thermometer*. The resistance of a conductor is measured at a reference temperature and at the temperature to be measured; the change in the resistance is then used to calculate the unknown temperature. For measurements over limited temperature ranges, the linear relationship of Eq. (18-9) can be used in the calculation; over larger temperature ranges, the resistance thermometer must be calibrated to account for the nonlinear variation of resistivity with temperature. Materials with high melting points (such as tungsten) can be used to measure high temperatures.

Temperature dependence of resistivity

Making The Connection:
resistance thermometer

Note that for semiconductors, $\alpha < 0$. A negative temperature coefficient means that the resistivity *decreases* with increasing temperature. It is still true, as for metals that are good conductors, that the collision rate increases with temperature. However, in semiconductors the number of carriers (conduction electrons and/or holes) per unit volume increases dramatically with increasing temperature; with more carriers, the resistivity is smaller.

Some materials become *superconductors* ($\rho = 0$) at low temperatures. Once a current is started in a superconducting loop, it continues to flow indefinitely *without* a source of emf. Experiments with superconducting currents have lasted more than two years without any measurable change in the current. Mercury was the first superconductor discovered (by Dutch scientist Kammerlingh Onnes in 1911). As the temperature of mercury is decreased, its resistivity gradually decreases—as for any metal—but at mercury's critical temperature ($T_C = 4.15$ K) its resistivity suddenly becomes zero. Many other superconductors have since been discovered. In the past two decades, scientists have created ceramic materials with much higher critical temperatures than those previously known. Above their critical temperatures, the ceramics are insulators.

Example 18.4

Change in Resistance with Temperature

The nichrome heating element of a toaster has a resistance of 12.0 Ω when it is red-hot (1200°C). What is the resistance of the element at room temperature (20°C)? Ignore changes in the length or diameter of the element due to temperature.

Strategy   Since we assume the length and cross-sectional area to be the same, the resistances at the two temperatures are proportional to the resistivities at those temperatures:

$$\frac{R}{R_0} = \frac{\rho L/A}{\rho_0 L/A} = \frac{\rho}{\rho_0}$$

Thus, we do not need the length or cross-sectional area of the heating element.

Given: $T_0 = 20°C$; $R = 12\ \Omega$ at $T = 1200°C$.

To find: R_0

Solution   From Eq. (18-9),

$$\frac{R}{R_0} = \frac{\rho L/A}{\rho_0 L/A} = \frac{\rho}{\rho_0} = 1 + \alpha\,\Delta T$$

The change in temperature is

$$\Delta T = T - T_0 = 1200°C - 20°C = 1180°C$$

For nichrome, Table 18.1 gives

$$\alpha = 0.4 \times 10^{-3}\ °C^{-1}$$

Solving for R_0 yields

$$R_0 = \frac{R}{1 + \alpha\,\Delta T} = \frac{12\ \Omega}{1 + 0.4 \times 10^{-3}\ °C^{-1} \times 1180°C} = 8\ \Omega$$

Discussion   Why do we write only one significant figure? Since the temperature change is so large (1180°C), the result must be considered an estimate. The relationship between resistivity and temperature may not be linear over such a large temperature range.

Practice Problem 18.4   Using a resistance thermometer

A platinum resistance thermometer has a resistance of 225 Ω at 20.0°C. When the thermometer is placed in a furnace, its resistance rises to 448 Ω. What is the temperature of the furnace? Assume the temperature coefficient of resistivity is constant over the temperature range in this problem.

Resistors

A **resistor** is a circuit element designed to have a known resistance. Resistors are found in virtually all electronic devices (Fig. 18.13). In circuit analysis, it is customary to write the relationship between voltage and current for a resistor as $V = IR$. Remember that V actually stands for the potential *difference* between the ends of the resistor even though the symbol Δ is omitted. Sometimes V is called the *voltage drop*. Current flows in the direction of the electric field, which points from higher to lower potential. Therefore, if you move across a resistor in the direction of current flow, the voltage *drops* by an amount IR. Remember a useful analogy: water flows downhill (toward lower potential energy); electric current flows toward lower potential.

Figure 18.13 The little cylinders with four colored bands on this computer circuit board are resistors. The colored bands specify the resistance of the resistor according to the resistance color code.

In a circuit diagram, the symbol —⩗⩗— represents a resistor or any other device in a circuit that dissipates electrical energy.
A straight line ——— represents a conducting wire with negligible resistance. (If a wire's resistance is appreciable, then we draw it as a resistor.)

(a)   (b)

Figure 18.14 (a) A lightbulb connected to a battery by conducting wires. (b) A circuit diagram for the same circuit. The emf and the internal resistance of the battery are enclosed by a dashed line as a reminder that in reality the two are not separate; we can't make a connection to the "wire" between the two!

Internal Resistance of a Battery

Figure 18.14a shows a circuit we've seen before. Figure 18.14b is a *circuit diagram* of the circuit. The lightbulb is represented by the symbol for a resistor (R). The battery is represented by two symbols surrounded by a dashed line. The battery symbol represents an *ideal* emf and the resistor (r) represents the *internal resistance* of the battery. If the internal resistance of a source of emf is negligible, then we just draw the symbol for an ideal emf.

When the current through a source of emf is zero, the **terminal voltage**—the potential difference between its terminals—is equal to the emf. When the source supplies current to a *load* (a lightbulb, a toaster, or any other device that uses electrical energy), its terminal voltage is less than the emf; there is a voltage drop due to the internal resistance of the source. If the current is I and the internal resistance is r, then the voltage drop across the internal resistance is Ir and the terminal voltage is

$$V = \mathscr{E} - Ir \qquad (18\text{-}10)$$

When the current is small enough, the voltage drop Ir due to the internal resistance is negligible compared to $\mathscr{E}$; then we can treat the emf as ideal ($V \approx \mathscr{E}$). A flashlight that is left on for a long time gradually dims because, as the chemicals in a battery are depleted, the internal resistance increases. As the internal resistance increases, the terminal voltage $V = \mathscr{E} - Ir$ decreases; thus, the voltage across the lightbulb decreases and the light gets dimmer.

Conceptual Example 18.5

Starting a Car Using Flashlight Batteries

Discuss the merits of Graham's scheme to start his car using eight D-cell flashlight batteries, each of which provides an emf of 1.50 V and has an internal resistance of 0.10 Ω. (A current of over 100 A is required to turn the starter motor in a car, while the current through the bulb in a flashlight is typically less than 1 A.)

Strategy We consider not only the values of the emfs, but also whether the batteries can supply the required *current*.

Solution and Discussion Connecting eight 1.5-V batteries as in a flashlight—with the positive terminal of one connected to the negative terminal of the next—does provide an emf of 12 V. Each battery does 1.5 J of work per coulomb of charge; if the charge must pass through all eight batteries in turn, the total work done is 12 J per coulomb of charge.

When the batteries are used to power a device that draws a *small* current (because the resistance of the load R is large compared to the internal resistance r of each battery), the terminal voltage of each battery is nearly 1.5 V and the terminal voltage of the combination is nearly 12 V. For instance, in a flashlight that draws 0.50 A of current, the terminal voltage of a D-cell is

$$V = \mathscr{E} - Ir = 1.50 \text{ V} - 0.50 \text{ A} \times 0.10 \text{ Ω} = 1.45 \text{ V}$$

continued on next page

Conceptual Example 18.5 *continued*

However, the current required to start the car is over 100 A. As the current increases, the terminal voltage decreases. We can estimate the *maximum* current that a battery can supply by setting its terminal voltage to zero (the smallest possible value):

$$V = \mathscr{E} - I_{max}r = 0$$

$$I_{max} = \mathscr{E}/r = (1.5 \text{ V})/(0.10 \text{ }\Omega) = 15 \text{ A}$$

(This estimate is optimistic since the battery's chemical energy would be rapidly depleted and the internal resistance

would increase dramatically.) The flashlight batteries cannot supply a current large enough to start the car.

***

Practice Problem 18.5   Terminal voltage of a battery in a clock

The current supplied by an alkaline D cell (1.500 V emf, 0.100 Ω internal resistance) in a clock is 50.0 mA. What is the terminal voltage of the battery?

Physics at Home

Turn on the headlights of a car and then start the car. Notice that the headlights dim considerably. If the car battery were an *ideal* emf of 12 V, it would continue to supply 12 V to the headlights regardless of how much current is drawn from it. The internal resistance of the battery is small—only a few milliohms—but while the battery supplies a few hundred amps of current to the starter, the terminal voltage of the battery is significantly less than 12 V.

18.5   KIRCHHOFF'S RULES

Two rules, developed by Gustav Kirchhoff (1824–1887), are essential in circuit analysis. **Kirchhoff's junction rule** states that the current that flows into a junction—any electrical connection—must equal the current that flows out of the same junction. The junction rule is a consequence of the law of conservation of charge. Since charge does not continually build up at a junction, the net rate of flow of charge into the junction must be zero. When applying the junction rule, let current flowing *into* a junction be positive and current flowing *out* of the junction be negative; then we can say that the net current flowing into a junction is zero.

Kirchhoff's Junction Rule

$$\Sigma I = 0 \qquad\qquad (18\text{-}11)$$

(Current into junction is positive; current out of junction is negative)

Figure 18.15a shows two streams joining to form a larger stream. Figure 18.15b shows an analogous junction (point *A*) in an electric circuit. Applying the junction rule to point *A* results in the equation $I_1 + I_2 - I_3 = 0$.

(a)   (b)

Figure 18.15 (a) The rate at which water flows into the junction from the two streams is equal to the rate at which water flows out of the junction into the larger stream. Equivalently, we can say that the net rate of flow of water into the junction is zero. (b) An analogous junction in an electric circuit. When applying the junction rule, I_1 and I_2 are positive because they flow into the junction; I_3 is negative because it flows out of the junction.

Kirchhoff's loop rule is an expression of energy conservation applied to changes in potential in a circuit. Recall that the electric potential must have a unique value at any point; the potential at a point cannot depend on the path one takes to arrive at that point. Therefore, if a closed path is followed in a circuit, beginning and ending at the same point, the algebraic sum of the potential changes must be zero. Think of taking a hike in the mountains, starting and returning at the same spot. No matter what path you take, the algebraic sum of all your elevation changes must equal zero.

Kirchhoff's Loop Rule

$$\Sigma \Delta V = 0 \qquad (18\text{-}12)$$

for any path in a circuit that starts and ends at the same point.
(Potential rises are positive; potential drops are negative.)

Be careful to get the signs right when applying the loop rule. If you follow a path through a resistor going in the same direction as the current, the potential drops ($\Delta V = -IR$). If your path takes you through a resistor in a direction opposite to the current ("upstream"), the potential rises ($\Delta V = +IR$). For an emf, the potential drops if you move from the positive terminal to the negative ($\Delta V = -\mathscr{E}$); it rises if you move from the negative to the positive ($\Delta V = +\mathscr{E}$).

18.6 SERIES AND PARALLEL CIRCUITS

Resistors in Series

Series: same current

When one or more electrical devices are wired so that the *same current* flows through each one, the devices are said to be wired in **series** (Figs. 18.16 and 18.17). The circuit of Fig. 18.17a shows two resistors in series. The straight lines represent wires, which we assume to have negligible resistance. Negligible resistance means negligible voltage drop ($V = IR$), so *points connected by wires of negligible resistance are at the same potential.* The junction rule, applied to any of the points A–D, tells us that the same current flows through the emf and the two resistors.

Let's apply the loop rule to a clockwise loop $DABCD$. From D to A we move from the negative terminal to the positive terminal of the emf, so $\Delta V = +1.5$ V. Since we move around the loop *with* the current, the potential *drops* as we move across each resistor. Therefore,

$$1.5 \text{ V} - IR_1 - IR_2 = 0$$

The same current I flows through the two resistors in series. Factoring out the common current I,

$$I(R_1 + R_2) = 1.5 \text{ V}$$

The current I would be the same if a single equivalent resistor $R_{eq} = R_1 + R_2$ replaced the two resistors in series:

$$IR_{eq} = I(R_1 + R_2) = 1.5 \text{ V}$$

Figure 18.17b shows how the circuit diagram can be redrawn to indicate the simplified, equivalent circuit.

We can generalize this result to any number of resistors in series:

For any number N of resistors connected in series,

$$R_{eq} = \Sigma R_i = R_1 + R_2 + \cdots + R_N \qquad (18\text{-}13)$$

Note that the equivalent resistance for two or more resistors in series is *larger* than *any* of the resistances.

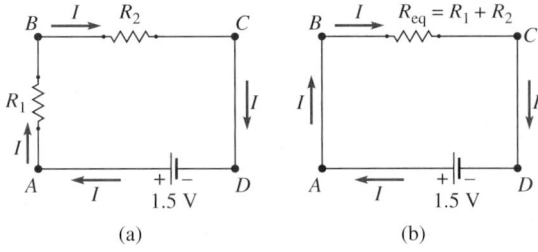

Figure 18.16 Just as water flows at the same mass flow rate through each of the two sluice gates, the same current flows through two resistors in series. Just as $\Delta y_1 + \Delta y_2 = \Delta y$, the potential difference ΔV across a series pair is the sum of the two potential differences.

Figure 18.17 (a) A circuit with two resistors in series. (b) Replacing the two resistors with an equivalent resistor.

Emfs in Series

In many devices, batteries are connected in series with the positive terminal of one connected to the negative terminal of the next. This provides a larger emf than a single battery can (Fig. 18.18). The emfs of batteries connected in this way are added just as series resistances are added. However, there is a disadvantage in connecting batteries in series: the internal resistance is larger because the internal resistances are in series as well.

Making The Connection:
battery connection in
a flashlight

(a)

(b)   (c)

Figure 18.18 (a) Two 1.5-V batteries connected in series in a flashlight to supply 3.0 V. (b) Circuit diagram, including the internal resistances of the batteries. (c) Simplified circuit diagram, where the two batteries are combined into a single source of emf $2\mathscr{E}$ with internal resistance $2r$.

Sources can be connected in series with the emfs in opposition. A common use for such a circuit is in a battery charger. In Fig. 18.19, as we move from point C to B to A, the potential decreases by $\mathscr{E}_2$ and then increases by $\mathscr{E}_1$, so the net emf is $\mathscr{E}_1 - \mathscr{E}_2$.

Capacitors in Series

Figure 18.20a shows three capacitors connected in series. We want to find the equivalent capacitance C_{eq} that would store the same amount of charge (Fig. 18.20b). When the switch is closed, the emf pumps charge until the potential difference between points A and D is equal to the emf. Since there are no voltage drops across the connecting wires, points B and b must be at the same potential. Likewise, points C and c are at the same potential. The potential difference between A and D is

$$\mathscr{E} = V_{AD} = V_{AB} + V_{BC} + V_{CD} \qquad (18\text{-}14)$$

Figure 18.19 Circuit for charging a rechargeable battery (shown as emf $\mathscr{E}_2$). The source supplying the energy to charge the battery must have a larger emf ($\mathscr{E}_1 > \mathscr{E}_2$). The net emf in the circuit is $\mathscr{E}_1 - \mathscr{E}_2$; the current is $I = (\mathscr{E}_1 - \mathscr{E}_2)/R$ (where R includes the internal resistances of the sources).

The magnitude of the charge on the three capacitors is the same. Why? Suppose that the battery puts charge $+Q$ on plate A. Then, since capacitor plates store opposite charges of equal magnitude, plate B has charge $-Q$. Plates B and b and their connecting wire are not connected to anything else, so their net charge must remain zero. Charge just shifts between plates B and b; therefore, plate b has charge $+Q$. Continuing the same line of reasoning leads to the conclusion that *capacitors in series have the same charge*. Since $Q = C\,\Delta V$,

$$V_{AB} = \frac{Q}{C_1}, \quad V_{BC} = \frac{Q}{C_2}, \quad \text{and } V_{CD} = \frac{Q}{C_3}$$

Substituting into Eq. (18-14) yields

$$\mathscr{E} = \frac{Q}{C_1} + \frac{Q}{C_2} + \frac{Q}{C_3}$$

Since

$$\mathscr{E} = \frac{Q}{C_{eq}}$$

the equivalent capacitance is given by

$$\frac{1}{C_{eq}} = \frac{1}{C_1} + \frac{1}{C_2} + \frac{1}{C_3}$$

This can be extended to the general case for any number of capacitors connected in series.

(a)

(b)

Figure 18.20 (a) Three capacitors connected in series. (b) Equivalent circuit.

For N capacitors connected in series,
$$\frac{1}{C_{eq}} = \sum \frac{1}{C_i} = \frac{1}{C_1} + \frac{1}{C_2} + \cdots + \frac{1}{C_N} \qquad (18\text{-}15)$$

Figure 18.21   Some water flows through one branch and some through the other. The mass flow rate before the water channels divide and after they come back together is equal to the sum of the flow rates in the two branches. The elevation change Δy for the two branches is equal since they start and end at the same elevations. For two resistors in parallel, the currents *add*; the potential differences are *equal*.

Note that the equivalent capacitor stores the same magnitude of charge as *each* of the capacitors it replaces.

Resistors in Parallel

When one or more electrical devices are wired so that the *potential difference across them is the same*, the devices are said to be wired in **parallel** (Fig. 18.21). In Fig. 18.22, an emf is connected to three resistors in parallel with one another. The left side of each resistor is at the same potential since they are all connected by wires of negligible resistance. Likewise, the right side of each resistor is at the same potential. Thus, there is a common potential difference across the three resistors. Applying the junction rule to point A yields

| Parallel: same potential difference

$$+I - I_1 - I_2 - I_3 = 0 \quad \text{or} \quad I = I_1 + I_2 + I_3 \qquad (18\text{-}16)$$

How much of the current I from the emf flows through each resistor? The current divides such that the potential difference $V_A - V_B$ must be the same along each of the three paths—and it must equal the emf $\mathscr{E}$. From the definition of resistance,

$$\mathscr{E} = I_1 R_1 = I_2 R_2 = I_3 R_3$$

Therefore, the currents are

$$I_1 = \frac{\mathscr{E}}{R_1}, \quad I_2 = \frac{\mathscr{E}}{R_2}, \quad I_3 = \frac{\mathscr{E}}{R_3}$$

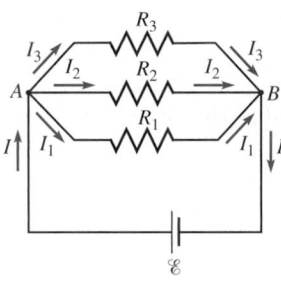

Figure 18.22   Three resistors connected in parallel

Substituting the currents into Eq. (18-16) yields

$$I = \frac{\mathscr{E}}{R_1} + \frac{\mathscr{E}}{R_2} + \frac{\mathscr{E}}{R_3}$$

Dividing by $\mathscr{E}$ yields

$$\frac{I}{\mathscr{E}} = \frac{1}{R_1} + \frac{1}{R_2} + \frac{1}{R_3}$$

The three parallel resistors can be replaced by a single equivalent resistor R_{eq}. In order for the same current to flow, R_{eq} must be chosen so that $\mathscr{E} = IR_{eq}$. Then $I/\mathscr{E} = 1/R_{eq}$ and

$$\frac{1}{R_{eq}} = \frac{1}{R_1} + \frac{1}{R_2} + \frac{1}{R_3}$$

Although we examined three resistors in parallel, the result applies to any number of resistors in parallel:

For N resistors connected in parallel,

$$\frac{1}{R_{eq}} = \sum \frac{1}{R_i} = \frac{1}{R_1} + \frac{1}{R_2} + \cdots + \frac{1}{R_N} \qquad (18\text{-}17)$$

Note that the equivalent resistance for two or more resistors in parallel is *smaller* than *any* of the resistances ($1/R_{eq} > 1/R_i$, so $R_{eq} < R_i$). Note also that the equivalent resistance for resistors in *parallel* is found in the same way as the equivalent capacitance for capacitors in *series*. The reason is that resistance is defined as $R = \Delta V/I$ and capacitance as $C = Q/\Delta V$. One has ΔV in the numerator, the other in the denominator.

Example 18.6

Current for Two Parallel Resistors

(a) Find the equivalent resistance for the two resistors in Fig. 18.23 if $R_1 = 20.0\ \Omega$ and $R_2 = 40.0\ \Omega$. (b) What is the ratio of the current through R_1 to the current through R_2?

Strategy   Points A and B are at the same potential; points C and D are at the same potential. Therefore, the voltage drops across the two resistors are equal; the two resistors are in parallel. The ratio of the currents can be found by equating the potential differences in the two branches in terms of the current and resistance.

Solution   (a) The equivalent resistance for two parallel resistors is

$$\frac{1}{R_{eq}} = \frac{1}{R_1} + \frac{1}{R_2} = \frac{1}{20.0\ \Omega} + \frac{1}{40.0\ \Omega} = 0.0750\ \Omega^{-1}$$

$$R_{eq} = \frac{1}{0.0750\ \Omega^{-1}} = 13.3\ \Omega$$

(b) The potential differences across the resistors are equal

$$I_1 R_1 = I_2 R_2$$

Therefore,

$$\frac{I_1}{I_2} = \frac{R_2}{R_1} = \frac{40.0\ \Omega}{20.0\ \Omega} = 2.00$$

Discussion   Note that the current in each branch of the circuit is inversely proportional to the resistance of that branch.

⚠ Since R_2 is twice R_1, it has half as much current flowing through it. At the junction of two or more parallel branches, the current does not all flow through the "path of least resistance," but *more* current flows through the branch of least resistance than through the branches with larger resistances.

Practice Problem 18.6   Three resistors in parallel

Find the equivalent resistance from point A to point B for the three resistors in Fig. 18.24.

Figure 18.23
Circuit with parallel resistors for Example 18.6

Figure 18.24
Three parallel resistors

Example 18.7

Equivalent Resistance for Network in Series and Parallel

(a) Find the equivalent resistance for the network of resistors in Fig. 18.25. (b) Find the current through the 1.0-Ω resistor connected to points A and B if $\mathscr{E} = 0.60$ V.

Figure 18.25
Network of resistors for Example 18.7

Strategy We simplify the network of resistors in a series of steps. At first, the only series or parallel combination is the two resistors in parallel between points B and C. No other pair of resistors has either the same current (for series) or the same voltage drop (for parallel). We replace those two with an equivalent resistor, redraw the circuit, and look for new series or parallel combinations, continuing until the entire network reduces to a single resistor.

Solution (a) For the two 2.0-Ω resistors in parallel between points B and C,

$$R_{eq} = \left(\frac{1}{R_1} + \frac{1}{R_2}\right)^{-1} = \left(\frac{1}{2.0\ \Omega} + \frac{1}{2.0\ \Omega}\right)^{-1} = 1.0\ \Omega$$

We redraw the circuit, replacing the two parallel resistors with an equivalent 1.0-Ω resistor.

(1)

The two 1-Ω resistors are in series since the same current must flow through them. They can be replaced with a single resistor,

$$R_{eq} = 1.0\ \Omega + 1.0\ \Omega = 2.0\ \Omega$$

The network of resistors now becomes

(2)

As before, the equivalent resistance for two 2.0-Ω resistors in parallel is 1.0 Ω. The network of resistors reduces to a single equivalent 1.0-Ω resistor.

(3)

(b) The current through the 1.0-Ω resistor is I_2 (Fig. 18.25). From circuit diagram (2), when I_2 flows through an equivalent resistance of 2.0 Ω, the voltage drop is 0.60 V. Therefore,

$$I_2 = \frac{0.60\ V}{2.0\ \Omega} = 0.30\ A$$

Discussion To reduce complicated arrangements of resistors to an equivalent resistance, look for resistors in parallel (resistors connected so that they must have the same potential difference) and resistors in series (connected so that they must have the same current). Replace all parallel and series combinations of resistors with equivalents. Then look for new parallel and series combinations in the simplified circuit. Repeat until there is only one resistor remaining.

Practice Problem 18.7 Three resistors connected

Find the equivalent resistance that can be placed between points A to B to replace the three equal resistors shown in Fig. 18.26. First try to decide whether these resistors are in series or parallel. Label the black dots with A or B by tracing the straight lines from A or B to their connections at one side or another of the resistors. Redraw the diagram if that helps you decide.

Figure 18.26
Three connected resistors

Emfs in Parallel

Two or more sources of *equal* emf are often connected in parallel with all the positive terminals connected together and all the negative terminals connected together (Fig. 18.27a). The equivalent emf for any number of equal sources in parallel is the same as the emf of each source. The advantage of connecting sources in this way is not to achieve a larger emf, but rather to lower the internal resistance and thus supply more current. In Fig. 18.27a, the two internal resistances (r) are equal. Since they are in parallel—note that points A and B are at the same potential—the equivalent internal resistance for the parallel combination is $\frac{1}{2}r$. To jump-start a car, one connects the two batteries in parallel, positive to positive and negative to negative.

We need not concern ourselves with unequal emfs in parallel or with arrangements like that shown in Fig. 18.27b. In such cases the two batteries drain each other and supply little or no current to the rest of the circuit.

Figure 18.27 (a) Two identical batteries (with internal resistances r) in parallel. The combination provides an emf $\mathscr{E}$ and can supply twice as much current as one battery since the equivalent internal resistance is $\frac{1}{2}r$. (b) Two identical batteries connected backwards. Points C and D are at the same potential, so the batteries supply no emf to the rest of the circuit; they just drain each other. If two car batteries are connected in this way, the large current that would flow would be dangerous.

(a)   (b)

Capacitors in Parallel

Capacitors in series have the same charge but may have different potential differences. Capacitors in parallel share a common potential difference but may have different charges. Suppose three capacitors are in parallel (Fig. 18.28). After the switch is closed, the source of emf pumps charge onto the plates of the capacitors until the potential difference across each capacitor is equal to the emf $\mathscr{E}$. Suppose that the total magnitude of charge pumped by the battery is Q. If the magnitude of charge on the three capacitors is q_1, q_2, and q_3, respectively, conservation of charge requires that

$$Q = q_1 + q_2 + q_3$$

The relation between the potential difference across a capacitor and the charge on either plate of the capacitor is $q = C\,\Delta V$. For each capacitor, $\Delta V = \mathscr{E}$. Therefore,

$$Q = q_1 + q_2 + q_3 = C_1\mathscr{E} + C_2\mathscr{E} + C_3\mathscr{E} = (C_1 + C_2 + C_3)\mathscr{E}$$

We can replace the three capacitors with a single equivalent capacitor. In order for it to store charge of magnitude Q for a potential difference $\mathscr{E}$, $Q = C_{eq}\mathscr{E}$. Therefore, $C_{eq} = C_1 + C_2 + C_3$. Once again, this result can be extended to the general case for any number of capacitors connected in parallel.

For N capacitors connected in parallel,

$$C_{eq} = \Sigma\, C_i = C_1 + C_2 + \cdots + C_N \tag{18-18}$$

Figure 18.28 (a) Three capacitors in parallel. (b) When the switch is closed, each capacitor is charged until the potential difference between its plates is equal to $\mathscr{E}$. If the capacitances are unequal, the charges on the capacitors are unequal.

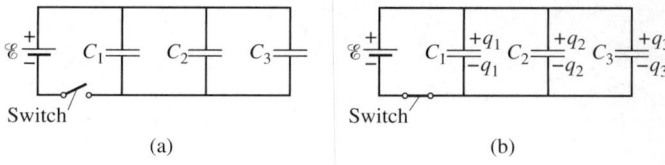

(a)   (b)

18.7 CIRCUIT ANALYSIS USING KIRCHHOFF'S RULES

Sometimes a circuit cannot be simplified by replacing parallel and series combinations alone. In such cases, we apply Kirchhoff's rules directly and solve the resulting equations simultaneously.

Problem-Solving Strategy: Using Kirchhoff's Rules to Analyze a Circuit

1. Assign variables to the currents in each branch of the circuit ($I_1, I_2, \ldots$) and choose directions for each current. Draw the circuit with the current directions indicated by arrows. It does not matter whether or not you choose the correct direction.

2. Apply Kirchhoff's junction rule to *all but one* of the junctions in the circuit. (Applying it to every junction produces one redundant equation.) Remember that current into a junction is positive; current out of a junction is negative.

3. Apply Kirchhoff's loop rule to enough loops so that every branch of the circuit is traversed. For each loop, choose a starting point and a direction to go around the loop. Be careful with signs. For a resistor, if your path through a resistor goes *with* the current ("downstream"), there is a potential drop; if your path goes against the current ("upstream"), the potential rises. For an emf, the potential drops or rises depending on whether you move from the positive terminal to the negative or *vice versa*; the direction of the current is irrelevant. A helpful method is to write "+" and "−" signs on the ends of each resistor and emf to indicate which end is at the higher potential and which is at the lower potential.

4. Solve the loop and junction equations simultaneously. If a current comes out negative, the direction of the current is opposite to the direction you chose.

5. Check your result using one or more loops or junctions. A good choice is a loop that you did not use in the solution.

Example 18.8

A Two-Loop Circuit

Find the currents through each branch of the circuit of Fig. 18.29.

$R_1 = 4.0 \ \Omega$
$R_2 = 6.0 \ \Omega$
$R_3 = 3.0 \ \Omega$

Figure 18.29
Circuit to be analyzed using Kirchhoff's rules

Strategy First we look for series and parallel combinations. R_1 and $\mathscr{E}_1$ are in series, but since one is a resistor and one an emf we cannot replace them with a single equivalent circuit element. No pair of resistors is either in series or in parallel. R_1 and R_2 might look like they're in parallel, but the emf $\mathscr{E}_1$ keeps points A and F at different potentials, so they are not. The two emfs might look like they're in series, but the junction at point F means that the current through the two is not the same. Since there are no series or parallel combinations to simplify, we proceed to apply Kirchhoff's rules directly.

Solution First we assign the currents variable names and directions on the circuit diagram:

Points C and F are junctions between the three branches of the circuit. We choose current I_1 for branch *FABC*, current I_2 for branch *FEDC*, and current I_3 for branch *CF*.

Now we can apply the junction rule. There are two junctions; we can choose either one. For point C, I_1 and I_2 flow into the junction and so are positive when using the junction rule; I_3 flows out of the junction and is negative. The resulting equation is

$$I_1 + I_2 - I_3 = 0 \qquad (1)$$

Before applying the loop rule, we write "+" and "−" signs on each resistor and emf to show which side is at the higher potential and which at the lower, given the directions assumed for the currents. In a resistor, current flows from higher to lower potential. The emf symbol uses the longer line for the positive terminal and the shorter line for the negative terminal.

continued on next page

Example 18.8 *continued*

Now we choose a closed loop and add up the potential rises and drops as we travel around the loop. Suppose we start at point A and travel around loop $ABCFA$. The starting point and direction to go around the loop are arbitrary choices, but once made, we stick with it regardless of the directions of the currents. From A to B, we move in the same direction as the current I_1. The current through a resistor travels from higher to lower potential, so going from A to B is a potential drop: $\Delta V_{A \to B} = -I_1 R_1$.

From B to C, since the wire is assumed to have negligible resistance, there is no potential rise or drop. From C to F, we move with current I_3, so there is another potential drop: $V_{C \to F} = -I_3 R_2$.

Finally, from F to A, we move from the negative terminal of an emf to the positive terminal. The potential *rises*: $V_{F \to A} = +\mathscr{E}_1$. A was the starting point, so the loop is complete. The loop rule says that the sum of the potential changes is equal to zero:

$$-I_1 R_1 - I_3 R_2 + \mathscr{E}_1 = 0 \qquad (2)$$

We must choose another loop since we have not yet gone through resistor R_3 or emf $\mathscr{E}_2$. There are two choices possible: the right-hand loop (such as $FCDEF$) or the outer loop ($ABCDEFA$). Let's choose $FCDEF$.

From F to C, we move *against* the current I_3 ("upstream"). The potential rises: $\Delta V_{F \to C} = +I_3 R_2$. From C to D, the potential does not change. From D to E, we again move upstream so $\Delta V_{D \to E} = +I_2 R_3$. From E to F, we move through a source of emf from the negative to the positive terminal. The potential increases: $V_{E \to F} = +\mathscr{E}_2$. Then the loop rule gives

$$+I_3 R_2 + I_2 R_3 + \mathscr{E}_2 = 0 \qquad (3)$$

Now we have three equations and three unknowns (the three currents). To solve them simultaneously, we first substitute known numerical values:

$I_1 + I_2 - I_3 = 0$	(1)
$-(4.0 \ \Omega)I_1 - (6.0 \ \Omega)I_3 + 1.5 \text{ V} = 0$	(2)
$(6.0 \ \Omega)I_3 + (3.0 \ \Omega)I_2 + 3.0 \text{ V} = 0$	(3)

To solve simultaneous equations, we can solve one equation for one variable and substitute into the other equations, thus eliminating one variable. Solving Eq. (1) for I_1 yields $I_1 = -I_2 + I_3$. Substituting in Eq. (2):

$$-(4.0 \ \Omega)(-I_2 + I_3) - (6.0 \ \Omega)I_3 + 1.5 \text{ V} = 0$$

Simplifying,

$$4.0 I_2 - 10.0 I_3 = -1.5 \text{ V}/\Omega = -1.5 \text{ A} \qquad (2a)$$

Eqs. (2a) and (3) now have only two unknowns. We can eliminate I_2 if we multiply Eq. (2a) by 3 and Eq. (3) by 4 so that I_2 has the same coefficient.

$$12.0 I_2 - 30.0 I_3 = -4.5 \text{ A} \quad 3 \times \text{Eq. (2a)}$$
$$12.0 I_2 + 24.0 I_3 = -12.0 \text{ A} \quad 4 \times \text{Eq. (3)}$$

Subtracting one from the other,

$$54.0 I_3 = -7.5 \text{ A}$$

Now we can solve for I_3:

$$I_3 = -\frac{7.5}{54.0} \text{ A} = -0.139 \text{ A}$$

Substituting the value of I_3 into Eq. (2a) enables us to solve for I_2:

$$4 I_2 + 10 \times 0.139 \text{ A} = -1.5 \text{ A}$$

$$I_2 = \frac{-1.5 - 1.39}{4} \text{ A} = -0.723 \text{ A}$$

Equation (1) now gives I_1:

$$I_1 = -I_2 + I_3 = +0.723 \text{ A} - 0.139 \text{ A} = +0.584 \text{ A}$$

Rounding to two significant figures, the currents are $I_1 = +0.58 \text{ A}$, $I_2 = -0.72 \text{ A}$, and $I_3 = -0.14 \text{ A}$. Since I_2 and I_3 came out negative, the actual directions of the currents in those branches are opposite to the ones we arbitrarily chose.

Discussion   Note that it did not matter that we chose some of the current directions wrong. It also doesn't matter which loops we choose (as long as we cover every branch of the circuit), which starting point we use for a loop, or which direction we go around a loop.

The hardest thing about applying Kirchhoff's rules is getting the signs correct. It is also easy to make an algebraic mistake when solving simultaneous equations. Therefore, it is a good idea to check the answer. A good way to check is to write down a loop equation for a loop that was not used in the solution (see Practice Problem 18.8).

Practice Problem 18.8   Verifying solution with loop rule

Apply Kirchhoff's loop rule to loop $CBAFEDC$ to verify the solution of Example 18.8.

18.8   POWER AND ENERGY IN CIRCUITS

From the definition of electric potential, if a charge q moves through a potential difference ΔV, the change in electric potential energy is

$$\Delta U_E = q \Delta V \qquad (17\text{-}7)$$

From energy conservation, a change in electric potential energy means that conversion between two forms of energy takes place. For example, a battery converts stored chemical energy into electric potential energy. A resistor converts electric potential energy into internal energy. The *rate* at which the energy conversion takes place is the *power P*. Since current is the rate of flow of charge, $I = q/\Delta t$ and

---

Power

$$P = \frac{\Delta U_E}{\Delta t} = \frac{q}{\Delta t}\Delta V = I\,\Delta V \qquad (18\text{-}19)$$

Thus, the power for *any circuit element* is the product of current and potential difference. We can verify that current times voltage comes out in the correct units for power by substituting coulombs per second for amperes and joules per coulomb for volts:

$$A \times V = \frac{C}{s} \times \frac{J}{C} = \frac{J}{s} = W$$

According to the definition of emf, if the amount of charge pumped by an ideal source of constant emf $\mathcal{E}$ is q, then the work done by the battery is

$$W = \mathcal{E}q \qquad (18\text{-}2)$$

The power supplied by the emf is the rate at which it does work:

$$P = \frac{\Delta W}{\Delta t} = \mathcal{E}\frac{q}{\Delta t} = \mathcal{E}I \qquad (18\text{-}20)$$

Since $\Delta V = \mathcal{E}$ for an ideal emf, Eqs. (18-20) and (18-19) are equivalent.

Power Dissipated by a Resistor

If an emf causes current to flow through a resistor, what happens to the energy supplied by the emf? Why must the emf continue supplying energy to maintain the current?

Current flows in a metal wire when an emf gives rise to a potential difference between one end and the other. The electric field makes the conduction electrons drift in the direction of lower electric potential energy (higher potential). If there were no collisions between electrons and atoms in the metal, the average kinetic energy of the electrons would continually increase. However, the electrons frequently collide with atoms; each such collision is an opportunity for an electron to give away some of its kinetic energy. For a steady current, the average kinetic energy of the conduction electrons does not increase; the rate at which the electrons gain kinetic energy (due to the electric field) is equal to the rate at which they lose kinetic energy (due to collisions). The net effect is that the energy supplied by the emf increases the vibrational energy of the atoms. The vibrational energy of the atoms is part of the internal energy of the metal, so the temperature of the metal rises.

From the definition of resistance, the potential drop across a resistor is

$$V = IR$$

Then the rate at which energy is dissipated in a resistor can be written

$$P = I \times IR = I^2R \qquad (18\text{-}21a)$$

or

$$P = \frac{V}{R} \times V = \frac{V^2}{R} \qquad (18\text{-}21b)$$

Is the power dissipated in a resistor directly proportional to the resistance [Eq. (18-21a)] or inversely proportional to the resistance [Eq. (18-21b)]? It depends on the situation. For two resistors with the *same current* (such as two resistors in series), the power is directly proportional to resistance—the voltage drops are not the same. For two resistors with the same voltage drop (such as two resistors in parallel), the power is inversely proportional to resistance; this time the currents are not the same.

Dissipation in a resistor is not necessarily undesirable. In any kind of electric heater—in portable or baseboard heaters, electric stoves and ovens, toasters, hair dryers, and electric clothes dryers—and in incandescent lights, the dissipation of energy and the resulting temperature increase of a resistor are put to good use.

Power Supplied by an Emf with Internal Resistance

If the source has internal resistance, then the net power supplied is *less* than $\mathcal{E}I$. Some of the energy supplied by the emf is dissipated by the internal resistance. The net power supplied to the rest of the circuit is

$$P = \mathcal{E}I - I^2 r \qquad (18\text{-}22)$$

where r is the internal resistance of the source. Equation (18-22) agrees with Eq. (18-19); remember that the potential difference is *not* equal to the emf when there is internal resistance (see Problem 59).

Example 18.9

Two Flashlights

A flashlight is powered by two batteries in series. Each has an emf of 1.50 V and an internal resistance of 0.10 Ω. The batteries are connected to the lightbulb by wires of total resistance 0.40 Ω. At normal operating temperature, the resistance of the filament is 9.70 Ω. (a) Calculate the power dissipated by the bulb—that is, the rate at which energy in the form of heat and light flows away from it. (b) Calculate the power dissipated by the wires and the net power supplied by the batteries. (c) A second flashlight uses *four* such batteries in series and the same resistance wires. A bulb of resistance 42.1 Ω (at operating temperature) dissipates approximately the same power as the bulb in the first flashlight. Verify that the power dissipated is nearly the same and calculate the power dissipated by the wires and the net power supplied by the batteries.

Strategy All the circuit elements are in series. We can simplify the circuit by replacing all the resistors (including the internal resistances of the batteries) with one series equivalent and the two emfs with one equivalent emf. Doing so enables us to find the current. Then we can use Eq. (18-21a) to find the power in the wires and in the filament. Equation (18-21b) could be used, but would require an extra step: finding the voltage drops across the resistors. Equation (18-22) gives the net power supplied by the batteries.

Solution (a) Figure 18.30 is a sketch of the circuit for the first flashlight. To find the power dissipated in the lightbulb, we need either the current through it or the voltage drop across it. We can find the current in this single loop circuit by replacing the two ideal emfs with a series equivalent emf of $\mathcal{E}_{eq} = 3.00$ V and all the resistors by a series equivalent resistance of

$$R_{eq} = 9.70\ \Omega + 0.40\ \Omega + 2 \times 0.10\ \Omega = 10.30\ \Omega$$

Then the current is

$$I = \frac{\mathcal{E}_{eq}}{R_{eq}} = \frac{3.00\ \text{V}}{10.30\ \Omega} = 0.2913\ \text{A}$$

The power dissipated by the filament is

$$P_f = I^2 R = (0.2913\ \text{A})^2 \times 9.70\ \Omega = 0.823\ \text{W}$$

(b) The power dissipated by the wires is

$$P_w = I^2 R = (0.2913\ \text{A})^2 \times 0.40\ \Omega = 0.034\ \text{W}$$

The net power supplied by the batteries is

$$P_b = \mathcal{E}_{eq} I - I^2 r_{eq}$$

where $r_{eq} = 0.20\ \Omega$ is the series equivalent for the two internal resistances. Then

$$P_b = 3.00\ \text{V} \times 0.2913\ \text{A} - (0.2913\ \text{A})^2 \times 0.20\ \Omega = 0.857\ \text{W}$$

(c) In the second circuit, $\mathcal{E}_{eq} = 6.00$ V and

$$R_{eq} = 42.1\ \Omega + 0.40\ \Omega + 4 \times 0.10\ \Omega = 42.90\ \Omega$$

The current is

$$I = \frac{\mathcal{E}_{eq}}{R_{eq}} = \frac{6.00\ \text{V}}{42.90\ \Omega} = 0.1399\ \text{A}$$

The power dissipated by the filament is

$$P_f = I^2 R = (0.1399\ \text{A})^2 \times 42.1\ \Omega = 0.824\ \text{W}$$

which is only 0.1% more than the filament in the first flashlight. The power dissipated by the wires is

$$P_w = I^2 R = (0.1399\ \text{A})^2 \times 0.40\ \Omega = 0.0078\ \text{W}$$

The series equivalent for the four internal resistances is $r_{eq} = 0.40\ \Omega$, so the net power supplied by the batteries is

$$P_b = \mathcal{E}_{eq} I - I^2 r_{eq}$$
$$= 6.00\ \text{V} \times 0.1399\ \text{A} - 0.0078\ \text{W} = 0.832\ \text{W}$$

Discussion Note that in each case, the net power supplied by the batteries is equal to the total power dissipated in the wires and the filament. Since there is nowhere else for the energy to go, the wires and filament must dissipate energy—convert electrical energy to light and heat—at the same rate that the battery supplies electrical energy.

Figure 18.30
Circuit for the first flashlight

continued on next page

Example 18.9 *continued*

The power supplied to the two filaments is about the same in the two cases. However, the power dissipated by the wires in the second flashlight is a bit less than one-fourth as much as in the first. By using a larger emf, the current required to supply a given amount of power is smaller. The current is smaller because the load resistance (the resistance of the filament) is larger. A smaller current means the power dissipated in the wires is smaller. Utility companies distribute power over long distances using high-voltage wires for exactly this reason: the smaller the current, the smaller the power dissipated in the wires.

Practice Problem 18.9 A simplified flashlight circuit

A flashlight takes two 1.5-V batteries connected in series. If the current that flows to the bulb in the flashlight is 0.35 A, find the power delivered to the lightbulb and the amount of energy dissipated after the light has been in the "on position" for three minutes. Treat the batteries as ideal and ignore the resistance of the wires. [*Hint*: It is not necessary to calculate the resistance of the filament since in this case the voltage drop across it is equal to the emf.]

18.9 MEASURING CURRENTS AND VOLTAGES

Current and potential difference in a circuit can be measured with instruments called **ammeters** and **voltmeters**, respectively. A multimeter functions as an ammeter or a voltmeter, depending on the setting of a switch. Meters can be either digital or analog; the latter uses a rotating pointer to indicate the value of current or voltage on a calibrated scale. At the heart of an analog voltmeter or analog ammeter is a **galvanometer**, a sensitive detector of current whose operation is based on magnetic forces.

Suppose a particular galvanometer has a resistance of 100.0 Ω and deflects full scale for a current of 100 μA. We want to build an ammeter to measure currents from 0 to 10 A—when a current of 10 A passes through the meter, the needle should deflect full scale. Therefore, when a current of 10 A goes through the ammeter, 100 μA should go through the galvanometer; the other 9.9999 A must bypass the galvanometer. We put a resistor in parallel with the galvanometer so that the 10 A current branches, sending 100 μA to deflect the needle and 9.9999 A through the *shunt resistor* (Fig. 18.31a).

(a)   (b)

Figure 18.31 (a) An ammeter constructed from a galvanometer. (b) The circuit diagram for the ammeter. The galvanometer is represented as a 100.0-Ω resistor.

Example 18.10

Constructing an Ammeter from a Galvanometer

If the internal resistance of a galvanometer (Fig. 18.31a) is 100.0 Ω and it deflects full scale for a current of 100.0 μA, what resistance should the shunt resistor have to make an ammeter for measuring currents up to 10.00 A?

Strategy When 10.00 A flows into the ammeter, 100.0 μA should go through the galvanometer and 9.9999 A should go

through the shunt resistor (Fig. 18.31b). Since the two are in parallel, the potential difference across the galvanometer is equal to the potential difference across the shunt resistor.

Solution The voltage drop across the galvanometer when it deflects full scale is

$$V = IR = 100.0 \ \mu A \times 100.0 \ \Omega$$

continued on next page

Example 18.10 continued

The voltage drop across the shunt resistor must be the same, so

$$V = 100.0\ \mu\text{A} \times 100.0\ \Omega = 9.9999\ \text{A} \times R_\text{S}$$

$$R_\text{S} = \frac{100.0\ \mu\text{A} \times 100.0\ \Omega}{9.9999\ \text{A}} = 0.001000\ \Omega = 1.000\ \text{m}\Omega$$

Discussion   The resistance of the ammeter is

$$\left(\frac{1}{0.001000\ \Omega} + \frac{1}{100.0\ \Omega}\right)^{-1} = 1.000\ \text{m}\Omega$$

A good ammeter should have a small resistance. When an ammeter is used to measure the current in a branch of a cir-cuit, it must be inserted *in series* in that branch—the amme-ter can only measure whatever current passes through it. Adding a small resistance in series has only a slight effect on the circuit.

Practice Problem 18.10   Changing the ammeter scale

What if the ammeter is to measure currents from 0 to 1.00 A? What shunt resistance should be used? What is the resistance of the ammeter? Use the same galvanometer as in Example 18.10.

An ammeter must have a small resistance.

In order to give accurate measurements, an ammeter must have a small resistance so its presence in the circuit does not change the current significantly from its value in the absence of the ammeter. An ideal ammeter has zero resistance.

It is also possible to construct a voltmeter by connecting a resistor (R_S) *in series* with the galvanometer (R_S, Fig. 18.32). The series resistor R_S is chosen so that the current through the galvanometer makes it deflect full scale when the desired full-scale voltage appears across the voltmeter. A voltmeter measures the potential difference between its terminals; to measure the potential difference across a resistor, for example, the voltmeter is connected in parallel with the resistor, with one terminal connected to

A voltmeter must have a large resistance.

each side of the resistor. So as not to affect the circuit too much, a good voltmeter must have a large resistance; then when measurements are taken, the current through the voltmeter (I_m) is small compared to I and the potential difference across the parallel combination is nearly the same as when the voltmeter is disconnected. An ideal volt-meter has infinite resistance.

To measure a resistance in a circuit, we can use a voltmeter to measure the potential difference across the resistor and an ammeter to measure the current through the resistor (Fig. 18.33). By definition, the ratio of the voltage to the current is the resistance.

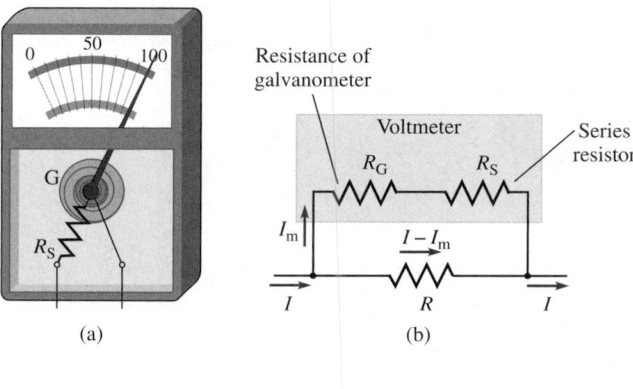

Figure 18.32 (a) A voltmeter con-structed from a galvanometer. (b) Circuit diagram of the voltmeter measuring the voltage across the resistor R.

Figure 18.33 Two ways to arrange meters to measure a resistance R. If the meters were ideal (an ammeter with zero resistance and a voltmeter with infinite resistance), the two arrangements would give exactly the same measurement. Note the symbols used for the meters.

18.10 *RC* CIRCUITS

Circuits containing both resistors and capacitors have many important applications. *RC circuits* are commonly used to control timing. When windshield wipers are set to operate intermittently, the charging of a capacitor to a certain voltage is the trigger that turns them on. The time delay between wipes is determined by the resistance and capacitance in the circuit; adjusting a variable resistor changes the length of the time delay. Similarly, an *RC* circuit controls the time delay in strobe lights and in some pacemakers. We can also use the *RC* circuit as a simplified model of the transmission of nerve impulses.

Charging *RC* Circuit

In Fig. 18.34, switch *S* is initially open and the capacitor is uncharged. When the switch is closed, current begins to flow and charge starts to build up on the plates of the capacitor. At any instant, Kirchhoff's loop law requires that

$$\mathscr{E} - V_R - V_C = 0$$

Figure 18.34   An *RC* circuit

where V_R and V_C are the voltage drops across the resistor and capacitor, respectively. As charge accumulates on the capacitor plates, it becomes increasingly difficult to push more charge onto them.

Just after the switch is closed, the potential difference across the resistor is equal to the emf since the capacitor is uncharged. Initially, a relatively large current $I_0 = \mathscr{E}/R$ flows. As the voltage drop across the capacitor increases, the voltage drop across the resistor decreases, and thus the current decreases. Long after the switch is closed, the potential difference across the capacitor is nearly equal to the emf and the current is small.

Using calculus, it can be shown that the voltage across the capacitor involves an exponential function (see Fig. 18.35):

$$V_C(t) = \mathscr{E}(1 - e^{-t/\tau}) \qquad (18\text{-}23)$$

where $e \approx 2.718$ is the base of the natural logarithm and the quantity $\tau = RC$ is called the **time constant** for the *RC* circuit.

Time constant

$$\tau = RC \qquad (18\text{-}24)$$

The product *RC* has time units:

$$[R] = \frac{\text{volts}}{\text{amps}} \quad \text{and} \quad [C] = \frac{\text{coulombs}}{\text{volts}} \quad \text{so } [RC] = \frac{\text{C}}{\text{A}} = \text{s}$$

The time constant is a measure of how fast the capacitor charges. At $t = \tau$, the voltage across the capacitor is

$$V_C(t = \tau) = \mathscr{E}(1 - e^{-1}) \approx 0.632\mathscr{E}$$

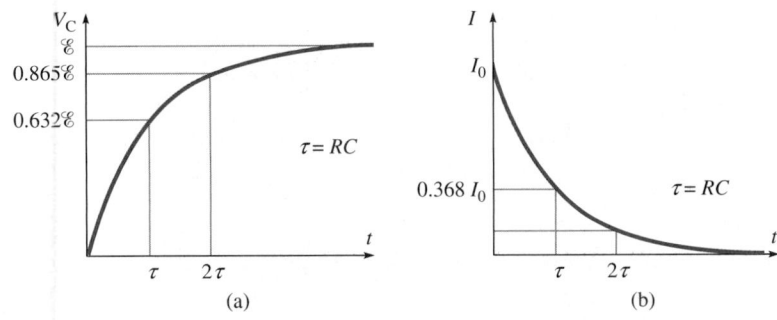

(a)   (b)

Figure 18.35   (a) The potential difference across the capacitor as a function of time as the capacitor is charged. (b) The current through the resistor as a function of time.

Since $Q = CV_C$, when one time constant has elapsed the capacitor has 63.2% of its final charge.

From Eq. (18-23), we can use the loop rule to find the current.

$$\mathcal{E} - IR - \mathcal{E}(1 - e^{-t/\tau}) = 0$$

Solving for I,

$$I(t) = \frac{\mathcal{E}}{R}e^{-t/\tau} \tag{18-25}$$

At $t = \tau$, the current is

$$I(t = \tau) = \frac{\mathcal{E}}{R}e^{-1} \approx 0.368\frac{\mathcal{E}}{R}$$

When one time constant has elapsed, the current is reduced to 36.8% of its initial value. The voltage drop across the resistor as a function of time can be found from $V_R = IR$.

Example 18.11

An *RC* Circuit with Two Capacitors in Series

Two 0.500-μF capacitors in series are connected to a 50.0-V battery through a 4.00-MΩ resistor at $t = 0$ (Fig. 18.36). The capacitors are initially uncharged. (a) Find the charge on the capacitors at $t = 1.00$ s and $t = 3.00$ s. (b) Find the current in the circuit at the same two times.

Figure 18.36
The circuit for Example 18.11

Strategy First we find the equivalent capacitance of two 0.500-μF capacitors in series. Then we can find the time constant using the equivalent capacitance. Equation (18-23) gives the voltage across the equivalent capacitor at any time t; once we know the voltage, we can find the charge from $Q = CV_C$. The charge on each of the two capacitors is equal to the charge on the equivalent capacitor. The current decreases exponentially according to Eq. (18-25).

Solution (a) For two equal capacitors C in series,

$$\frac{1}{C_{eq}} = \frac{1}{C} + \frac{1}{C} = \frac{2}{C}$$

Then $C_{eq} = \frac{1}{2}C = 0.250$ μF. The time constant is

$$\tau = RC_{eq} = 4.00 \times 10^6\ \Omega \times 0.250 \times 10^{-6}\ \text{F} = 1.00\ \text{s}$$

The final charge on the capacitor is

$$Q_f = C_{eq}\mathcal{E} = 0.250 \times 10^{-6}\ \text{F} \times 50.0\ \text{V} = 12.5 \times 10^{-6}\ \text{C} = 12.5\ \mu\text{C}$$

At any time t, the charge on each capacitor is

$$Q(t) = C_{eq}V_C(t) = C_{eq}\mathcal{E}(1 - e^{-t/\tau}) = Q_f(1 - e^{-t/\tau})$$

At $t = 1.00$ s, $t/\tau = 1.00$; the charge on each capacitor is

$$Q = Q_f(1 - e^{-1}) = \left(1 - \frac{1}{2.718}\right) \times 12.5\ \mu\text{C} = 7.90\ \mu\text{C}$$

At $t = 3.00$ s, $t/\tau = 3.00$; the charge on each capacitor is

$$Q = Q_f(1 - e^{-3}) = \left(1 - \frac{1}{2.718^3}\right) \times 12.5\ \mu\text{C} = 11.9\ \mu\text{C}$$

(b) The initial current is

$$I_0 = \frac{\mathcal{E}}{R} = \frac{50.0\ \text{V}}{4.00 \times 10^6\ \Omega} = 12.5\ \mu\text{A}$$

At a time t,

$$I = I_0 e^{-t/\tau}$$

At $t = 1.00$ s,

$$I = I_0 e^{-1} = 12.5\ \mu\text{A} \times \frac{1}{2.718} = 4.60\ \mu\text{A}$$

At $t = 3.00$ s,

$$I = I_0 e^{-3} = 12.5\ \mu\text{A} \times \frac{1}{2.718^3} = 0.623\ \mu\text{A}$$

Discussion The solution can be checked using the loop rule. At $t = \tau$, we found that $Q = 7.90$ μC and $I = 4.60$ μA. Then at $t = \tau$,

$$V_C = \frac{Q}{C_{eq}} = \frac{7.90\ \mu\text{C}}{0.250\ \mu\text{F}} = 31.6\ \text{V}$$

and

$$V_R = IR = 4.60\ \mu\text{A} \times 4.00\ \text{M}\Omega = 18.4\ \text{V}$$

Since 31.6 V + 18.4 V = 50.0 V = $\mathcal{E}$, the loop rule is satisfied.

Notice the pattern: the current is multiplied by $1/e$ during a time interval equal to the time constant. Thus, for a current of 4.60 μA at $t = \tau$, we expect a current of 4.60 μA × $1/e$ = 1.69 μA at $t = 2\tau$ and a current of 1.69 μA × $1/e$ = 0.623 μA at $t = 3\tau$.

Practice Problem 18.11   Another *RC* circuit

At $t = 0$ a capacitor of 0.050 μF is connected through a 5.0-MΩ resistor to a 12-V battery. Initially the capacitor is uncharged. Find the initial current, the charge on the capacitor at $t = 0.25$ s, the current at $t = 1.00$ s, and the final charge on the capacitor.

Discharging *RC* Circuit

In Fig. 18.37, the capacitor is first charged to a voltage $\mathcal{E}$ by closing switch S_1 with switch S_2 open. Once the capacitor is fully charged, S_1 is opened and then S_2 is closed at $t = 0$. Now the capacitor acts like a battery in the sense that it supplies energy in the circuit, though not at a constant potential difference. As the potential difference between the plates causes current to flow, the capacitor discharges.

The loop rule requires that the voltages across the capacitor and resistor be equal in magnitude. As the capacitor discharges, the voltage across it decreases. A decreasing voltage across the *resistor* means that the current must be decreasing. The current as a function of time is the same as for the charging circuit [Eq. (18-25)] with time constant $t = RC$. The voltage across the capacitor begins at its maximum value $\mathcal{E}$ and decreases exponentially (Fig. 18.38):

$$V_C(t) = \mathcal{E}e^{-t/\tau} \qquad (18\text{-}26)$$

The bulb in a camera flash needs a burst of current much larger than a small battery can supply (due to the battery's internal resistance). Therefore, the battery charges a capacitor (Fig. 18.39). When the capacitor is fully charged, the flash is ready; when the picture is taken, the capacitor is discharged quickly through the bulb. The resistance of the bulb is small so that the capacitor discharges quickly. After taking a picture, there is a delay of a second or two while the capacitor recharges. The time constant is longer for the charging circuit due to the internal resistance of the battery.

Figure 18.37 A capacitor is discharged through a resistor *R*.

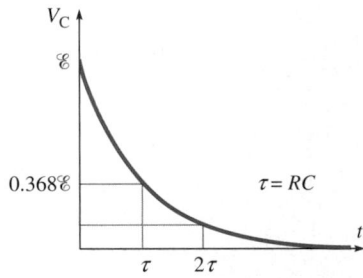

Figure 18.38 Decreasing capacitor voltage as it discharges through a resistor.

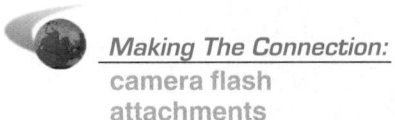

Making The Connection:
camera flash
attachments

Figure 18.39 A flash attachment for a camera. The large gray cylinder is the capacitor.

RC Circuits in Neurons

An *RC* time constant also determines the speed at which nerve impulses travel. Fig. 18.40a is a simplified model of a myelinated axon. Inside the axon is a fluid called the *axoplasm*, which is a conductor due to the presence of ions. Outside is the *interstitial fluid*, a conducting fluid with a much lower resistivity. Between the *nodes of Ranvier*, the cell membrane is covered with a *myelin sheath*—an insulator that reduces the capacitance of the section of axon (by increasing the distance between the conducting fluids) and reduces the leakage current that flows through the membrane.

A section of axon between nodes is modeled as an *RC* circuit in Fig. 18.40b. The interstitial fluid has little resistance, so it is modeled as a conducting wire. Current *I* travels inside the axon through the axoplasm (resistor *R*). The capacitor consists of the two conducting fluids as the plates, with the membrane and myelin sheath acting as insulator. For a section of axon 1 mm long with radius 5 μm, the resistance and

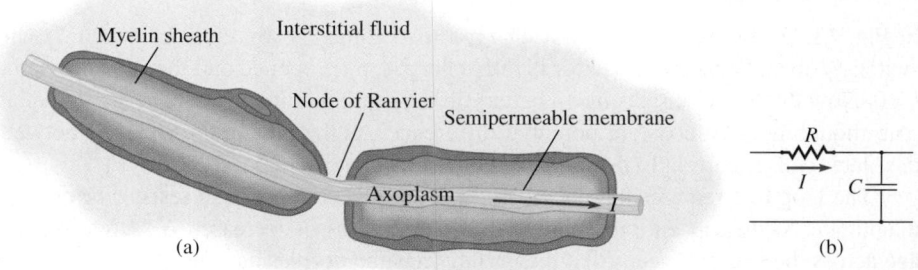

Figure 18.40 (a) A simplified picture of two sections of myelinated axon. (b) A simplified *RC* circuit model of a section of axon between nodes of Ranvier.

capacitance are approximately $R = 13$ MΩ and $C = 1.6$ pF. The time constant is therefore,

$$\tau = RC = 13 \text{ M}\Omega \times 1.6 \text{ pF} \approx 20 \text{ }\mu\text{s}$$

An estimate of how fast the electrical impulse travels is

$$v \approx \frac{\text{length of section}}{\tau} = \frac{1 \text{ mm}}{20 \text{ }\mu\text{s}} = 50 \text{ m/s}$$

This simple estimate is remarkably accurate; nerve impulses in a human myelinated axon of radius 5 μm travel at speeds ranging from 60 to 90 m/s.

Both *R* and *C* depend on the radius *r* of the axon. In humans, *r* ranges from under 2 μm to over 10 μm. The capacitance is proportional to *r* due to the larger plate area, but the resistance is inversely proportional to r^2 due to the larger cross-sectional area of the "wire." Thus, $RC \propto 1/r$ and $v \propto r$. The largest radius axons—those with the largest signal speeds—are those that must carry signals over relatively long distances.

18.11 ELECTRICAL SAFETY

Effects of Current on the Human Body

Electric currents passing through the body interfere with the operation of the muscles and the nervous system. Large currents also cause burns due to the energy dissipated in the tissues. A current of around 1 mA or less causes an unpleasant sensation but usually no other effect. The maximum current that can pass through the body without causing harm is about 5 mA. A current of 10 to 20 mA results in muscle contractions or paralysis; paralysis may prevent the person from letting go of the source of the current.

Making The Connection:
defibrillator

Currents of 100 to 300 mA may cause ventricular fibrillation (uncontrolled, arrhythmic contractions of the heart) if they pass through or near the heart. Once the heart begins to beat erratically, the person will die unless treated with a defibrillator to shock the heart back into a normal rhythm. Through the defibrillator paddles, a brief spurt of current of several amps is sent into the body near the heart (see Fig. 17.34). The shocked heart suffers a sudden muscular contraction, after which it may return to a normal state with regular contractions.

Most of the electrical resistance of the body is due to the skin. The fluids inside the body are good conductors due to the presence of ions. The total resistance of the body between distant points *when the skin is dry* ranges from around 10 kΩ to 1 MΩ. The resistance is much lower when the skin is wet—around 1 kΩ or even less.

A *short circuit* (a low resistance path) may occur between the circuitry inside an appliance to metal on the outside of the appliance. A person touching the appliance would then have one hand at 120 V with respect to ground. (To simplify the discussion we treat the emf as if it were DC rather than AC.) If his feet are in a wet tub, which makes good electrical contact to the grounded water pipes, he might have a resistance as low as 500 Ω. Then a current of magnitude 120 V/500 Ω = 0.24 A = 240 mA flows

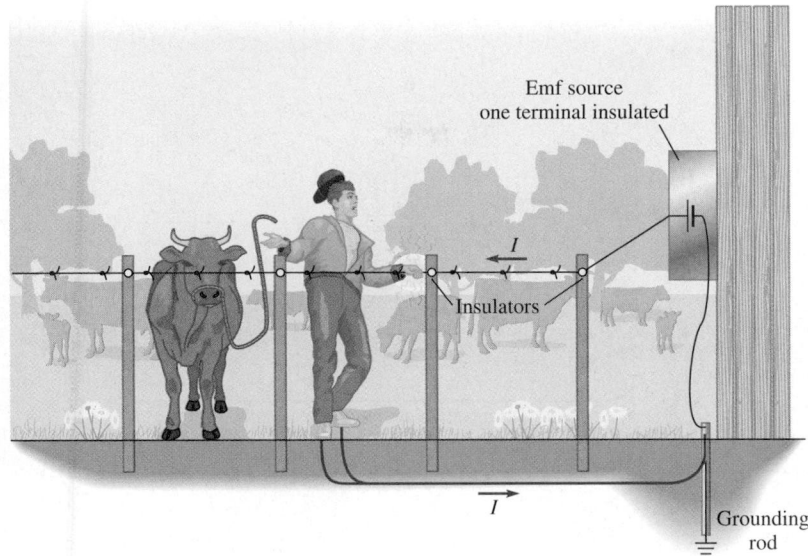

Remember that the symbol ⏚ represents a connection to ground.

Figure 18.41 An electric fence. The circuit is completed when a person touches the wire.

through the body past the heart. Ventricular fibrillation is likely to occur. If the person were not standing in the tub, but had one hand on the hair dryer and another hand on a metal faucet, which is also grounded through the household plumbing, he is still in trouble. The electrical resistance of a person from one damp hand to the other might be around 1600 Ω, resulting in a current of 75 mA, which could still be lethal.

An electrified fence (Fig. 18.41) keeps farm animals in a pasture or wild animals out of a garden. One terminal of an emf is connected to the wire; the wire is insulated from the fence posts by ceramic insulators. The other terminal of the emf is connected to ground by a metal rod driven into the ground. When an animal or person touches the metal wire, the circuit is completed from the wire through the body and back to the ground (Fig. 18.41). The current flowing through the body is limited so that it produces an unpleasant sensation without being dangerous.

Grounding of Appliances

A two-pronged plug does not protect against a short circuit. The case of the appliance is supposed to be insulated from the wiring inside. If, by accident, a wire breaks loose or its insulation becomes frayed, a short circuit might occur, providing a low-resistance path directly to the metal case of the appliance. If a person touches the case, which is now at a high potential, the current travels through the person and back to the ground (Fig. 18.42a).

With a three-pronged plug, the case of the appliance is connected directly to ground through the third prong (Fig. 18.42b). Then, if a short circuit occurs, the person touching the case does not complete the circuit to the ground. Instead the current travels from the case directly to the ground through low-resistance wiring via the third prong in the wall outlet. For safety reasons, the metal cases of many electrical appliances are grounded.

Hospitals must take care that patients, connected to various monitors and IVs, are protected from a possible short circuit. For this reason the patient's bed, as well as anything else that the patient might touch, is insulated from the ground. Then if the patient touches something at a high potential, there is no ground connection to complete the circuit through the patient's body.

Fuses and Circuit Breakers

A simple fuse is made from an alloy of lead and tin that melts at a low temperature. The fuse is put in series with the circuit and is designed to melt—due to I^2R heating—if the

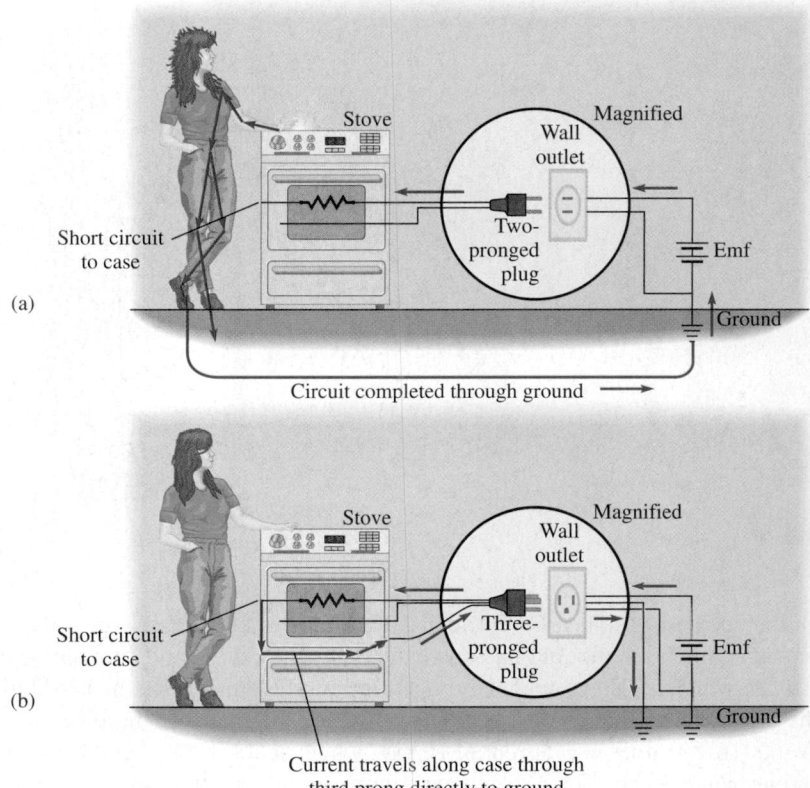

Figure 18.42 (a) If an electric stove were connected with a two-pronged plug to a wall outlet, a short circuit to the case of the stove allows the circuit to be completed through the body of a person touching the stove. (b) If a short circuit occurs with a three-pronged plug, the person is safe.

current to the circuit exceeds a given value. The melted fuse is an open switch, interrupting the circuit and stopping the current. Many appliances are protected by fuses. Replacing a fuse with one of a higher current rating is dangerous because too much current may go through the appliance, damaging it or causing a fire.

Most household circuits are now protected by circuit breakers instead of fuses. When too much current flows, perhaps because too many appliances are connected to the same circuit, a bimetallic strip or an electromagnet "trips" the circuit breaker, making it an open switch. After the problem causing the overload is corrected, the circuit breaker can be reset by flipping it back into the closed position.

Making The Connection:
household wiring

Household wiring (Fig. 18.43) is arranged so that several appliances can be connected in parallel to a single circuit with one side of the circuit grounded and the other side—the *hot* side—at a higher potential (120 V in our simplified DC model). Within one house or apartment, there are many such circuits; each one is protected by a circuit breaker (or fuse) placed in the hot side of the circuit. If a short circuit occurs, the large current that results trips the circuit breaker. If the breaker were placed on the grounded side, a blown circuit breaker would leave the hot side hot, possibly allowing a hazardous condition to continue. For the same reason, wall switches for overhead lights and for wall outlets are placed on the hot side.

Figure 18.43 Household wiring

MASTER THE CONCEPTS

Summary

- Electric current is the rate of net flow of charge:

$$I = \frac{\Delta q}{\Delta t} \qquad (18\text{-}1)$$

The SI unit of current is the ampere (1 A = 1 C/s), one of the base units of the SI. By convention, the direction of current is the direction of flow of positive charge. If the carriers are negative, the direction of the current is opposite the direction of motion of the carriers.

- A complete circuit is required for a continuous flow of charge.

- The current in a metal is proportional to the drift speed (v_D) of the conduction electrons, the number of electrons per unit volume (n), and the cross-sectional area of the metal (A):

$$I = \frac{\Delta Q}{\Delta t} = neAv_D \qquad (18\text{-}3)$$

- Electrical resistance is the ratio of the potential difference across a conducting material to the current through the material. It is measured in ohms: 1 Ω = 1 V/A.

$$R = \frac{\Delta V}{I} \qquad (18\text{-}6)$$

For an ohmic conductor, R is independent of ΔV and I; then ΔV is proportional to I.

- The electrical resistance of a wire is directly proportional to its length and inversely proportional to its cross-sectional area:

$$R = \rho \frac{L}{A} \qquad (18\text{-}8)$$

- The resistivity ρ is an intrinsic characteristic of a particular material at a particular temperature and is measured in $\Omega \cdot$m. For many materials resistivity varies linearly with temperature:

$$\rho = \rho_0 (1 + \alpha \Delta T) \qquad (18\text{-}9)$$

- A device that pumps charge is called a source of emf. The emf $\mathscr{E}$ is work done per unit charge [$W = \mathscr{E}q$, Eq. (18-2)]. The terminal voltage may differ from the emf due to the internal resistance r of the source:

$$V = \mathscr{E} - Ir \qquad (18\text{-}10)$$

- Kirchhoff's junction rule: $\Sigma I = 0$ at any junction [Eq. (18-11)]. Current into the junction is positive; current out of the junction is negative. Kirchhoff's loop rule: $\Sigma \Delta V = 0$ for any

path in a circuit that starts and ends at the same point [Eq. (18-12)]. Potential rises are positive; potential drops are negative.

- Circuit elements wired in series have the same current through them. Circuit elements wired in parallel have the same potential difference across them.

- The power—the rate of conversion between electrical energy and another form of energy—for any circuit element is

$$P = I \Delta V \qquad (18\text{-}19)$$

The SI unit for power is the watt (W). Electrical energy is dissipated (transformed into internal energy) in a resistor.

- The quantity $\tau = RC$ is called the time constant for an RC circuit. The currents and voltages are

$$V_C(t) = \mathscr{E}(1 - e^{-t/\tau}) \quad \text{(charging)} \qquad (18\text{-}23)$$

$$V_C(t) = \mathscr{E}e^{-t/\tau} \quad \text{(discharging)} \qquad (18\text{-}26)$$

$$I(t) = \frac{\mathscr{E}}{R}e^{-t/\tau} \quad \text{(both)} \qquad (18\text{-}25)$$

Highlighted Figures and Tables

F18.2 The current is the rate of flow of charge through an area perpendicular to the direction of flow (p. 627)

F18.3 The direction of the current is the direction in which the positive ions move (p. 628)

F18.6 Using the flow of water as an analogy to what happens in an electric circuit (p. 629)

F18.8 Because this circuit has only a single loop for current to flow, the current must be the same everywhere (p. 630)

F18.9 Electron drift velocity (p. 631)

F18.11 Graphs of I versus ΔV for ohmic and nonohmic systems (p. 634)

T18.1 Resistivities and temperature coefficients at 20°C (p. 635)

F18.14 A circuit and its circuit diagram (p. 638)

F18.15 Junction rule (p. 639)

F18.16 Current flow through two resistors in series (p. 641)

F18.35 The potential difference across a capacitor as a function of time as it is charged, and the current through a resistor as a function of time (p. 653)

F18.38 Decreasing capacitor voltage as it discharges through a resistor (p. 655)

CONCEPTUAL QUESTIONS

1. Draw a circuit diagram for automobile headlights, connecting two separate bulbs and a switch to a single battery so that: (1) one switch turns both bulbs on and off and (2) one bulb still lights up even if the other bulb burns out.

2. Ammeters often contain fuses that protect them from large currents, while voltmeters seldom do. Explain.

3. Why do lightbulbs usually burn out just after they are switched on and not when they have been on for a while?

4. Jeff needs a 100-Ω resistor for a circuit, but he only has a box of 300-Ω resistors. What can he do?

5. A friend says that electric current "follows the path of least resistance." Is that true? Explain.

6. Compare the resistance of an ideal ammeter with that of an ideal voltmeter. Which has the larger resistance? Why?

7. Why does the resistivity of a metallic conductor increase with increasing temperature?

8. Suppose a battery is connected to a network of resistors and capacitors. What happens to the energy supplied by the battery?

9. Why are electric stoves and clothes dryers supplied with 240 V, but lights, radios, and clocks are supplied with 120 V?

10. Why are ammeters connected in series with a circuit element in which the current is to be measured and voltmeters connected in parallel across the element for which the potential difference is to be measured?

11. Is it more dangerous to touch a "live" electric wire when your hands are dry or wet, everything else being equal? Explain.

12. Is the electric field inside a conductor always zero? If not, when is it not zero? Explain.

13. Some batteries can be "recharged." Does that mean that the battery has a supply of charge that is depleted as the battery is used? If "recharging" does not literally mean to put charge back into the battery, what *does* it mean?

14. A battery is connected to a clock by copper wires as shown in Fig. 18.44. What is the direction of current through the clock (*B* to *C* or *C* to *B*)? What is the direction of current through the battery (*D* to *A* or *A* to *D*)? Which terminal of the battery is at the higher potential (*A* or *D*)? Which side of the clock is

Figure 18.44 Conceptual Question 14

at the higher potential (*B* or *C*)? Does current *always* flow from higher to lower potential? Explain.

15. Think of a wire of length *L* as two wires of length *L*/2 in series. Construct an argument for why the resistance of a wire must be proportional to its length.

16. Think of a wire of cross-sectional area *A* as two wires of area *A*/2 in parallel. Construct an argument for why the resistance of a wire must be inversely proportional to its cross-sectional area.

17. An electrician working on "live" circuits wears insulated shoes and keeps one hand behind his or her back. Why?

18. A 15-A circuit breaker trips repeatedly. Explain why it would be dangerous to replace it with a 20-A circuit breaker.

19. A bird perched on a power line is not harmed, but if you are pruning a tree and your metal pole saw comes in contact with the same wire, you risk being electrocuted. Explain.

20. When batteries are connected in parallel, they should have the same emf. However, batteries connected in series need not have the same emf. Explain.

MULTIPLE CHOICE QUESTIONS

1. In an ionic solution, sodium ions (Na^+) are moving to the right and chlorine ions (Cl^-) are moving to the left. In which direction is the current due to the motion of (1) the sodium ions and (2) the chlorine ions?

   (a) Both are to the right.
   (b) Current due to Na^+ is to the left; current due to Cl^- is to the right.
   (c) Current due to Na^+ is to the right; current due to Cl^- is to the left.
   (d) Both are to the left.

2. A capacitor and a resistor are connected through a switch to an emf (Fig. 18.45). At the instant just after the switch is closed

   (a) The current in the circuit is zero.
   (b) The voltage across the capacitor is $\mathcal{E}$.
   (c) The voltage across the resistor is zero.
   (d) The voltage across the resistor is $\mathcal{E}$.
   (e) Both (a) and (c) are true.

3. Which is a unit of energy?
   (a) $A^2 \cdot \Omega$ (b) $V \cdot A$ (c) $\Omega \cdot m$
   (d) $\dfrac{N \cdot m}{V}$ (e) $\dfrac{A}{C}$ (f) $V \cdot C$

4. How does the resistance of a piece of conducting wire change if both its length and diameter are doubled?
   (a) Remains the same (b) 2 times as much
   (c) 4 times as much (d) 1/2 as much
   (e) 1/4 as much

Each of the graphs in Fig. 18.46 shows a relation between the potential drop across (*V*) and the current through (*I*) a circuit element.

5. Which depicts a circuit element whose resistance increases with increasing current?

6. Which depicts an ohmic circuit element?

Figure 18.45 Multiple Choice Question 2

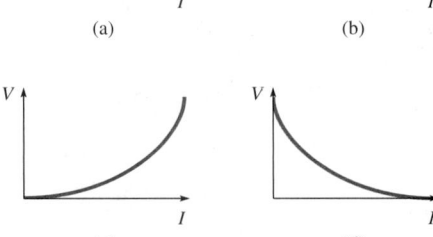

Figure 18.46 Multiple Choice Questions 5 and 6

7. The electrical properties of copper and rubber are different because

   (a) The positive charges are free to move in copper and stationary in rubber.
   (b) Many electrons are free to move in copper but nearly all are bound to molecules in rubber.
   (c) The positive charges are free to move in rubber but stationary in copper.
   (d) Many electrons are free to move in rubber but nearly all are bound to molecules in copper.

8. Consider these four statements. Choose true or false for each one in turn and then find the answer that matches your choices for all four together.

   (1) An ammeter should draw very little current compared with that in the rest of the circuit.
   (2) An ammeter should have a high resistance compared with the resistances of the other elements in the circuit.
   (3) To measure the current in a circuit element, the ammeter should be connected in series with that element.
   (4) Connecting the ammeter in series with a circuit element causes at least a small reduction of the current in that element.

   (a) (1) true,   (2) true,   (3) false,   (4) false
   (b) (1) true,   (2) false,   (3) true,   (4) true
   (c) (1) false,   (2) false,   (3) true,   (4) false
   (d) (1) false,   (2) false,   (3) true,   (4) true
   (e) (1) false,   (2) true,   (3) true,   (4) true
   (f) (1) false,   (2) false,   (3) false,   (4) true

9. A 12-V battery with internal resistance 0.5 Ω has initially no load connected across its terminals (Fig. 18.47). Then the switches S_1 and S_2 are closed successively. The voltmeter (assumed ideal) has which set of successive readings?

   (a) 12 V, 11 V, 10 V   (b) 12 V, 12 V, 12 V
   (c) 12 V, 9.6 V, 7.2 V   (d) 12 V, 9.6 V, 8 V
   (e) 12 V, 8 V, 4 V   (f) 12 V, zero, zero

10. Which of these is equal to the emf of a battery?

    (a) the chemical energy stored in the battery
    (b) the terminal voltage of the battery when no current flows
    (c) the maximum current that the battery can supply
    (d) the amount of charge the battery can pump
    (e) the chemical energy stored in the battery divided by the net charge of the battery

Figure 18.47 Multiple Choice Question 9

PROBLEMS

Note: **C** indicates a combination conceptual/quantitative problem. Gold diamonds ✦, ✦✦ are used to indicate the increasing level of difficulty of each problem. Problem numbers appearing in blue, 9., denote problems that have a detailed solution available in the Student Solutions Manual. Some problems are *paired* by concept; their numbers are connected by a ruled box.

18.1 Electric Current

1. The current in a wire is 0.500 A. (a) How much charge flows past a point in the wire in 10.0 s? (b) How many electrons pass the same point in 10.0 s?

2. (a) What is the direction of the current in the vacuum tube shown in Fig. 18.48? (b) Electrons hit the anode at a rate of 6.0×10^{12} per second. What is the current in the tube?

Figure 18.48
Problem 2

3. In an ion accelerator, 3.0×10^{13} helium-4 nuclei (charge $+2e$) per second strike a target. What is the beam current?

4. The current in the electron beam of a computer monitor is 320 μA. How many electrons per second hit the screen?

5. A potential difference is applied between the electrodes in a gas discharge tube. In 1.0 s, 3.8×10^{16} electrons and 1.2×10^{16} singly charged positive ions move in opposite directions through a surface perpendicular to the length of the tube. What is the current in the tube?

6. Two electrodes are placed in a calcium chloride solution and a potential difference is maintained between them. If 3.8×10^{16} Ca^{2+} ions and 6.2×10^{16} Cl^- ions per second move in opposite directions through an imaginary area between the electrodes, what is the current in the solution?

18.2 Emf and Circuits

7. A Vespa scooter and a Toyota automobile might both use a 12-V battery, but the two batteries are of different sizes and can pump different amounts of charge. Suppose the scooter battery can pump 4.0 kC of charge and the automobile battery can pump 30.0 kC of charge. How much energy can each battery deliver, assuming the batteries are ideal?

8. What is the energy stored in a small battery if it can move 675 C through a potential difference of 1.20 V?

9. The label on a 12.0-V truck battery states that it is rated at 180.0 A·h (ampere-hours). Treat the battery as ideal. (a) How much charge in coulombs can be pumped by the battery? [*Hint:* Convert A·h to A·s.] (b) How much electrical energy can the battery supply? (c) Suppose the radio in the truck is left on when the engine is not running. The radio draws a current of 3.30 A. How long does it take to drain the battery if it starts out fully charged?

10. The starter motor in a car draws 220.0 A of current from the 12.0-V battery for 1.20 s. (a) How much charge is pumped by the battery? (b) How much electrical energy is supplied by the battery?

11. A solar cell provides an emf of 0.45 V. (a) If the cell supplies a constant current of 18.0 mA for 9.0 h, how much electrical energy does it supply? (b) What is the power—the rate at which it supplies electrical energy?

18.3 Microscopic View of Current in a Metal

12. A silver wire of diameter 1.0 mm carries a current of 150 mA. The density of conduction electrons in silver is 5.8×10^{28} m^{-3}. How long (on average) does it take for a conduction electron to move 1.0 cm along the wire?

13. A strip of doped silicon 260 μm wide contains 8.8×10^{22} conduction electrons per cubic meter and an insignificant number of holes. When the strip carries a current of 130 μA, the drift speed of the electrons is 44 cm/s. What is the thickness of the strip?

✦ 14. A copper wire of cross-sectional area 1.00 mm^2 has a current of 2.0 A flowing along its length. What is the drift speed of the conduction electrons? Assume 1.3 conduction electrons per copper atom. The mass density of copper is 9.0 g/cm^3 and its atomic mass is 64 g/mol.

✦ 15. An aluminum wire of diameter 2.6 mm carries a current of 12 A. How long on average does it take an electron to move 12 m along the wire? Assume 3.5 conduction electrons per aluminum atom. The mass density of aluminum is 2.7 g/cm^3 and its atomic mass is 27 g/mol.

18.4 Resistance and Resistivity

16. 83 mA of current flow through the resistor of Fig. 18.49. (a) What is the resistance of the resistor? (b) In what direction does the current flow through the resistor?

4.50 V   **Figure 18.49**   Problem 16

17. A bird sits on a high-voltage power line with its feet 2.0 cm apart. The wire is made from aluminum, is 2.0 cm in diameter, and carries a current of 150 A. What is the potential difference between the bird's feet?

Ⓒ 18. A person can be killed if a current as small as 50 mA passes near the heart. An electrician is working on a humid day with hands damp from perspiration. Suppose his resistance from

one hand to the other is 1 kΩ and he is touching two wires, one with each hand. (a) What potential difference between the two wires would cause a 50-mA current from one hand to the other? (b) An electrician working on a "live" circuit keeps one hand behind his or her back. Why?

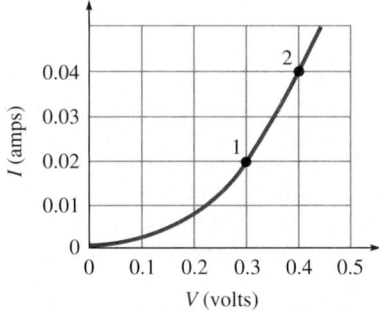

Figure 18.50
Problems 19 and 89

19. An electrical device has the current-voltage (*I-V*) graph in Fig. 18.50. What is its resistance at (a) point 1 and (b) point 2?

20. A diode is a semiconductor device that allows current to flow easily in one direction but not in the other. When 60.0 mA of current flows in the preferred ("forward biased") direction, the potential difference across the diode is 1.0 V. If a 1.0-V potential difference is applied to the diode in the opposite direction ("reverse biased"), the current is 0.020 mA. Calculate the resistance of the diode in each case.

21. If 46 m of nichrome wire is to have a resistance of 10.0 Ω at 20°C, what diameter wire should be used?

22. An aluminum bus bar has the dimensions 0.50 cm × 2.0 cm × 20.0 m. What is its resistance at 20°C? [*Hint:* The resistance is measured along the 20.0 m length.]

23. The resistance of a conductor is 19.8 Ω at 15.0°C and 25.0 Ω at 85.0°C. What is the temperature coefficient of resistance of the material?

24. A common flashlight bulb is rated at 0.300 A and 2.90 V (the values of current and voltage under operating conditions). If the resistance of the bulb's tungsten filament at room temperature (20.0°C) is 1.10 Ω, estimate the temperature of the tungsten filament when the bulb is turned on.

25. Find the maximum current that a fully charged D-cell can supply—if only briefly—such that its terminal voltage is at least 1.0 V. Assume an emf of 1.5 V and an internal resistance of 0.10 Ω.

26. (a) What are the ratios of the resistances of (a) silver and (b) aluminum wire to the resistance of copper wire (R_{Ag}/R_{Cu} and R_{Al}/R_{Cu}) for wires of the same length and the same diameter? (c) Which material is the best conductor, for wires of equal length and diameter?

✦ 27. What are the ratios of the resistances of (a) silver and (b) aluminum wire to the resistance of copper wire (R_{Ag}/R_{Cu} and R_{Al}/R_{Cu}) for wires of the same length and the same *mass* (not the same diameter? (c) Which material is the best conductor, for wires of equal length and equal mass? The densities are: silver 10.1×10^3 kg/m^3; copper 8.9×10^3 kg/m^3; aluminum 2.7×10^3 kg/m^3.

✦ 28. A wire with cross-sectional area A carries a current I. Show that the electric field strength E in the wire is proportional to the current per unit area (I/A) and identify the constant of proportionality. [*Hint:* Assume a length L of wire. How is the potential difference across the wire related to the electric field in the wire? (Which is uniform?) Use $V = IR$ and the connection between resistance and resistivity.]

© ✦ 29. A copper wire has a resistance of 24 Ω at 20°C. An aluminum wire has three times the length and twice the radius of the copper wire. The resistivity of copper is 0.6 times that of aluminum. Both Al and Cu have temperature coefficients of resistivity of 0.004°C⁻¹. (a) What is the resistance of the aluminum wire at 20°C? (b) Figure 18.51 shows an *I-V* plot for the copper wire. What is the resistance of the wire when operating steadily at a current of 10 A? (c) What must the temperature of the copper wire have been when operating at 10 A? Ignore changes in the wire's dimensions. (d) With the copper wire connected to an ideal battery, the current increases greatly when the wire is immersed in liquid nitrogen. Ignoring changes in the wire's dimensions, state whether each of the following quantities increases, decreases, or stays the same as the wire is cooled: the electric field in the wire, the resistivity, and the drift speed. Explain your answers.

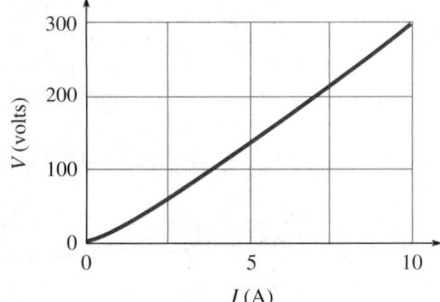

Figure 18.51
Problem 29

18.6 Series and Parallel Circuits

30. Suppose a collection of five batteries is connected as shown in Fig. 18.52. What is the equivalent emf of the collection? Treat them as ideal sources of emf.

31. Suppose four batteries are connected in series as shown in Fig. 18.53. What is the equivalent emf of the set of four batteries? Treat them as ideal sources of emf.

Figure 18.52
Problem 30

Figure 18.53 Problem 31

32. Find the equivalent capacitance between points *A* and *B* for the three capacitors in parallel in Fig. 18.54.

33. Find the equivalent capacitance between points *A* and *B* for the five capacitors in parallel in Fig. 18.55.

Figure 18.54 Problem 32

Figure 18.55
Problem 33

34. What is the resistance between points *A* and *B* in Fig. 18.56?
35. What is the resistance between points *A* and *B* in Fig. 18.57?

Figure 18.56 Problem 34   **Figure 18.57** Problem 35

36. Find the equivalent resistance between points *A* and *B* for the combination of resistors shown in Fig. 18.58.

Figure 18.58 Problem 36

37. Find the equivalent resistance between points *A* and *B* for the combination of resistors shown in Fig. 18.59.

Figure 18.59 Problem 37

✦ 38. (a) What is the resistance between points *A* and *B* in Fig. 18.60? Each resistor has the same resistance *R*. [*Hint:* Redraw the circuit.] (b) What is the resistance between points *B* and *C*?

Figure 18.60 Problem 38

39. Find the equivalent resistance between points A and B for the combination of resistors shown in Fig. 18.61.

Figure 18.61
Problem 39

40. A 6.0-pF capacitor is needed to construct a circuit. The only capacitors available are rated as 9.0 pF. How can a combination of 9.0-pF capacitors be assembled so that the equivalent capacitance of the combination is 6.0 pF? (There are many possible solutions. Just find *one* of them—preferably one that uses a small number of 9.0-pF capacitors.)

41. Find the value of a single capacitor to replace the three capacitors in Fig. 18.62.

Figure 18.62 Problem 41

42. (a) Find the equivalent resistance for all of the resistors in Fig. 18.63. (b) What current flows through the emf? (c) What is the current through the 4.00-Ω resistor at the bottom?

Figure 18.63 Problem 42

18.7 Circuit Analysis Using Kirchhoff's Rules

43. Find the current in each branch of the circuit of Fig. 18.64. Specify the direction of each.

44. Find the current in each branch of the circuit of Fig. 18.65. Specify the direction of each.

Figure 18.64 Problem 43

Figure 18.65 Problem 44

45. Find the unknown emf and the unknown currents for the circuit of Fig. 18.66.

46. Find the unknown emf and the unknown resistor in the circuit of Fig. 18.67.

Figure 18.66
Problem 45

Figure 18.67 Problem 46

47. Figure 18.68 shows a simplified circuit diagram for an automobile. The equivalent resistor R represents the total electrical load due to spark plugs, lights, radio, fans, starter, rear window defroster, and the like in parallel. (a) If $R = 0.850\ \Omega$, find the current in each branch. What is the terminal voltage of the battery? Is the battery charging or discharging? (b) For what range of values of R is the battery discharging?

Figure 18.68 Problem 47

18.8 Power and Energy in Circuits

48. What is the power dissipated by the resistor in the circuit of Fig. 18.69?

49. What is the power dissipated by the resistor in the circuit of Fig. 18.70?

Figure 18.69 Problem 48   **Figure 18.70** Problem 49

50. What is the current in a 60.0-W bulb when connected to a 120-V emf?

51. What is the resistance of a 40.0-W, 120-V lightbulb?

52. If a chandelier has a label stating 120 V, 5.0 A, can its power rating be determined? If so, what is it?

53. A portable CD player does not have a power rating listed, but it has a label stating that it draws a maximum current of

250.0 mA. The player uses three 1.50-V batteries connected in series. What is the maximum power consumed?

54. How much work are the batteries in the circuit of Fig. 18.71 doing in every 10.0-s time interval?

Figure 18.71   Problem 54

55. Show that $A^2 \times \Omega = W$ (amperes squared times ohms = watts).

56. Two immersion heaters, A and B, are both connected to a 120-V supply. Heater A can raise the temperature of 1.0 L of water from 20.0°C to 90.0°C in 2.0 min, while heater B can raise the temperature of 5.0 L of water from 20.0°C to 90.0°C in 5.0 min. What is the ratio of the resistance of heater A to the resistance of heater B?

57. At what rate is electrical energy converted to internal energy in the 4.00-Ω and 5.00-Ω resistors in Fig. 18.72?

Figure 18.72   Problem 57

58. During a "brownout," which occurs when the power companies cannot keep up with high demand, the voltage of the household circuits drops below its normal 120 V. (a) If the voltage drops to 108 V, what would be the power consumed by a "100-W" lightbulb (that is, a lightbulb that consumes 100.0 W when connected to 120 V)? Ignore (for now) changes in the resistance of the lightbulb filament. (b) More realistically, the lightbulb filament will not be as hot as usual during the brownout. Does this make the power drop more or less than that you calculated in part (a)? Explain.

✦59. A source of emf $\mathcal{E}$ has internal resistance r. (a) What is the terminal voltage when the source supplies a current I? (b) The net power supplied is the terminal voltage times the current. Starting with $P = I\Delta V$, derive Eq. (18-22) for the net power supplied by the source. Interpret each of the two terms. (c) Suppose that a battery of emf $\mathcal{E}$ and internal resistance r is being recharged: another emf sends a current I through the battery in the reverse direction (from positive terminal to negative). At what rate is electrical energy converted to chemical energy in the recharging battery? (d) What is the power supplied by the recharging circuit to the battery?

18.9 Measuring Currents and Voltages

60. A galvanometer has a coil resistance of 50.0 Ω. It is to be made into an ammeter with a full-scale deflection equal to 10.0 A. If the galvanometer deflects full scale for a current of 0.250 mA, what size shunt resistor should be used?

61. A galvanometer has a coil resistance of 34.0 Ω. It is to be made into a voltmeter with a full-scale deflection equal to

100.0 V. If the galvanometer deflects full scale for a current of 0.120 mA, what size resistor should be placed in series with the galvanometer?

62. A galvanometer is to be turned into a voltmeter that deflects full scale for a potential difference of 100.0 V. What size resistor should be placed in series with the galvanometer if it has an internal resistance of 75 Ω and deflects full scale for a current of 2.0 mA?

63. Many voltmeters have a switch by which one of several series resistors can be selected. Thus, the same meter can be used with different full-scale voltages. What size series resistors should be used in the voltmeter of Problem 62 to give it full-scale voltages of (a) 50.0 V and (b) 500.0 V?

64. Redraw the circuit of Fig. 18.73 (a) to show how an ammeter would be added to the circuit to measure the current through the 1.40-kΩ resistor; (b) to show how a voltmeter would be added to measure the voltage drop across the 83.0-kΩ resistor.

65. (a) In Fig. 18.73, find the current through the 1.40-kΩ resistor. (b) An ammeter with a resistance of 240 Ω is used to measure the current through the 1.40-kΩ resistor. What is the ammeter reading?

Figure 18.73   Problems 64, 65, and 96

✦66. A voltmeter has a switch that enables voltages to be measured with a maximum of 25.0 V or 10.0 V. For a range of voltages to 25.0 V, the switch connects a resistor of magnitude 9850 Ω in series with the galvanometer; for a range of voltages to 10.0 V, the switch connects a resistor of magnitude 3850 Ω in series with the galvanometer. Find the coil resistance of the galvanometer and the galvanometer current that causes a full-scale deflection. [*Hint:* There are two unknowns, so you will need to solve two equations simultaneously.]

18.10 *RC* Circuits

67. A capacitor is charged to an initial voltage $V_0 = 9.0$ V. The capacitor is then discharged by connecting its terminals through a resistor. The current $I(t)$ through this resistor, determined by measuring the voltage $V_R(t) = I(t)R$ with an oscilloscope, is shown in Fig. 18.74. (a) Find the capacitance C, the resistance R, and the total energy dissipated in the resistor. (b) At what time is the energy in the capacitor half its initial value? (c) Graph the voltage across the capacitor, $V_C(t)$, as a function of time.

68. A charging *RC* circuit controls the intermittent windshield wipers in a car. The emf is 12.0 V. The wipers are triggered when the voltage across the 125-μF capacitor reaches 10.0 V; then the capacitor is quickly discharged (through a much smaller resistor) and the cycle repeats. What resistance

should be used in the charging circuit if the wipers are to operate once every 1.80 s?

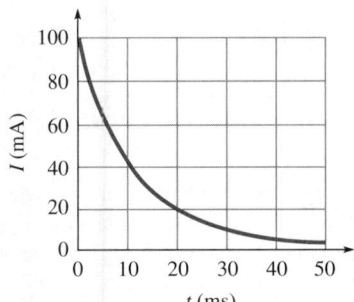

Figure 18.74
Problem 67

69. Capacitors are used in many applications where one needs to supply a short burst of relatively large current. A 100.0-μF capacitor in an electronic flash lamp supplies a burst of current that dissipates 20.0 J of energy (as light and heat) in the lamp. (a) To what potential difference must the capacitor initially be charged? (b) What is its initial charge? (c) Approximately what is the resistance of the lamp if the current reaches 5.0% of its original value in 2.0 ms?

© 70. A defibrillator passes a brief burst of current through the heart to restore normal beating. In one such defibrillator, a 50.0-μF capacitor is charged to 6.0 kV. Paddles are used to make an electrical connection to the patient's chest. A pulse of current lasting 1.0 ms partially discharges the capacitor through the patient. The electrical resistance of the patient (from paddle to paddle) is 240 Ω. (a) What is the initial energy stored in the capacitor? (b) What is the initial current through the patient? (c) How much energy is dissipated in the patient during the 1.0 ms? (d) If it takes 2.0 s to recharge the capacitor, compare the average power supplied by the power source to the average power delivered to the patient. (e) Referring to your answer to part (d), explain one reason a capacitor is used in a defibrillator.

✦✦ 71. A charged capacitor is discharged through a resistor. The current $I(t)$ through this resistor, determined by measuring the voltage $V_R(t) = I(t)R$ with an oscilloscope, is shown in Fig. 18.75. The total energy dissipated in the resistor is 2.0×10^{-4} J. (a) Find the capacitance C, the resistance R, and the initial charge on the capacitor. [*Hint:* You will need to solve three equations simultaneously for the three unknowns. You can find both the initial current and the time constant from the graph.] (b) At what time is the stored energy in the capacitor 5.0×10^{-5} J?

Figure 18.75
Problem 71

72. In the circuit of Fig. 18.76, the capacitor is initially uncharged. At $t = 0$ switch S is closed. Find the currents I_1 and I_2 and voltages V_1 and V_2 (assuming $V_3 = 0$) at points 1 and 2 at the following times: (a) $t = 0$ (i.e., just after the switch is closed), (b) $t = 1.0$ ms, and (c) $t = 5.0$ ms.

Figure 18.76   Problem 72

73. In the circuit of Fig. 18.77, the initial energy stored in the capacitor is 25 J. At $t = 0$ the switch is closed. (a) Sketch a graph of the voltage across the resistor (V_R) as a function of t. Label the vertical axis with key numerical value(s) and units. (b) At what time is the energy stored in the capacitor 1.25 J?

Figure 18.77   Problem 73

74. A 20-μF capacitor is discharged through a 5-kΩ resistor. The initial charge on the capacitor is 200 μC. (a) Sketch a graph of the current through the resistor as a function of time. Label both axes with numbers and units. (b) What is the initial power dissipated in the resistor? (c) What is the total energy dissipated?

75. (a) In a charging RC circuit, how many time constants have elapsed when the capacitor has 99.0% of its final charge? (b) How many time constants have elapsed when the capacitor has 99.90% of its final charge? (c) How many time constants have elapsed when the current has 1.0% of its initial value?

✦✦ 76. A capacitor is charged by a 9.0-V battery. The charging current $I(t)$ is shown in Fig. 18.78. (a) What, approximately, is the total charge on the capacitor in the end? [*Hint:* During a short time interval Δt, the amount of charge that flows in the circuit is $I\Delta t$.] (b) Using your answer to (a), find the capacitance C of the capacitor. (c) Find the total resistance R in the circuit. (d) At what time is the stored energy in the capacitor half of its maximum value?

Figure 18.78
Problem 76

18.11 Electrical Safety

77. In the physics laboratory, Oscar measured the resistance between his hands to be 2.0 kΩ. Being curious by nature, he then took hold of two conducting wires that were connected to the terminals of an emf with a terminal voltage of 100.0 V. (a) What current passes through Oscar? (b) If one of the conducting wires is grounded and the other has an alternate path to ground through a 15-Ω resistor (so that Oscar and the resistor are in parallel), how much current would pass through Oscar if the maximum current that can be drawn from the emf is 1.00 A?

78. A person bumps into a set of batteries with an emf of 100.0 V that can supply a maximum power of 5.0 W. If the man's resistance between the points where he contacts the batteries is 1.0 kΩ, how much current passes through him?

○ 79. The wiring circuit for a typical room is shown schematically in Fig. 18.79. (a) Of the six locations for a circuit breaker indicated by A, B, C, D, E, and F, which one would best protect the household against a short circuit in any one of the three appliances? Explain. (b) The room circuit is supplied with 120 V. Suppose the heater draws 1500 W, the lamp draws 300 W, and the microwave draws 1200 W. The circuit breaker is rated at 20.0 A. Can all three devices be operated without tripping the breaker? Explain.

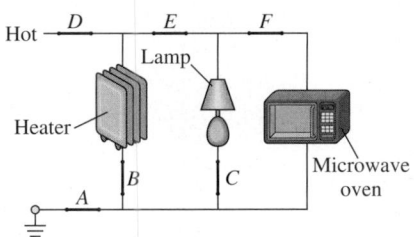

Figure 18.79
Problem 79

○ 80. Several possibilities are listed for what might or might not happen if the insulation in the current-carrying wires of Fig. 18.80 breaks down and point *b* makes electrical contact with point *c*. Discuss each possibility. (a) The person touching the microwave oven gets a shock; (b) the cord begins to smoke; (c) a fuse blows out; (d) an electrical fire breaks out inside the kitchen wall.

Figure 18.80
Problem 80

COMPREHENSIVE PROBLEMS

81. A 1.5-V flashlight battery can maintain a current of 0.30 A for 4.0 h before it is exhausted. How much chemical energy is converted to electrical energy in this process? (Assume zero internal resistance of the battery.)

82. In Fig. 18.81, the positive terminal of the 12-V battery is grounded—it is at zero potential. At what potential is point *X*?

Figure 18.81   Problem 82

83. In the circuit shown in Fig. 18.82, an emf of 150 V is connected across a resistance network. What is the current through R_2? Each of the resistors has a value of 10 Ω.

Figure 18.82   Problem 83

84. In Fig. 18.83, A_1 and A_2 represent ammeters with negligible resistance. What are the values of the currents (a) in A_1 and (b) in A_2?

85. Repeat Problem 84 if each of the ammeters has resistance 0.200 Ω.

Figure 18.83   Problems 84 and 85

♦ 86. About 5.0×10^4 m above Earth's surface, the atmosphere is sufficiently ionized that it behaves as a conductor. The Earth and the ionosphere form a giant spherical capacitor, with the lower atmosphere acting as a leaky dielectric. (a) Find the capacitance C of the Earth-ionosphere system by treating it as a *parallel plate* capacitor. Why is it OK to do that? [*Hint:* Compare the Earth's radius to the distance between the "plates."] (b) The fair-weather electric field is about 150 V/m, downward. How much energy is stored in this capacitor? (c) Due to radioactivity and cosmic rays, some air molecules are ionized even in fair weather. The resistivity of air is roughly 3.0×10^{14} Ω·m. Find the resistance of the lower atmosphere and the total current that flows between the Earth's surface and the ionosphere. [*Hint:* Since we treat the system as a parallel plate capacitor, treat the atmosphere as a dielectric of *uniform thickness* between the plates.] (d) If there were no lightning, the capacitor would discharge. In this model, how much time would elapse before the Earth's charge were reduced to 1% of its normal value? (Thunderstorms are the sources of emf that maintain the charge on this leaky capacitor.)

87. A 2.00-μF capacitor is charged using a 5.00-V battery and a 3.00-μF capacitor is charged using a 10.0-V battery. (a) What is the total energy stored in the two capacitors? (b) The batteries are disconnected and the two capacitors are connected together (+ to + and – to –). Find the charge on each capacitor and the total energy in the two capacitors after they are connected. (c) Explain what happened to the "missing" energy. [*Hint:* The wires that connect the two have some resistance.]

88. In a pacemaker used by a heart patient, a capacitor with a capacitance of 25 μF is charged to 1.0 V and then discharged through the heart every 0.80 s. What is the average discharge current?

89. A certain electrical device has the current-voltage (*I-V*) graph in Fig. 18.50. What is the power dissipated at points 1 and 2?

90. A 1.5-horsepower motor operates on 120 V. Ignoring I^2R losses, how much current does it draw?

91. Three identical lightbulbs are connected with wires to an ideal battery, as shown in Fig. 18.84. The two terminals on each socket connect to the two terminals of its lightbulb. Wires do *not* connect with one another where they appear to cross in the picture. Neglect the change of the resistances of the filaments due to temperature changes. (a) Which of the schematic circuit diagrams correctly represent(s) the circuit? (List more than one choice if more than one diagram is correct.) (b) Which bulb(s) is/are the brightest? Which is/are the dimmest? Or are they all the same? Explain. (c) Find the current through each bulb if the filament resistances are each 24.0 Ω and the emf is 6.0 V.

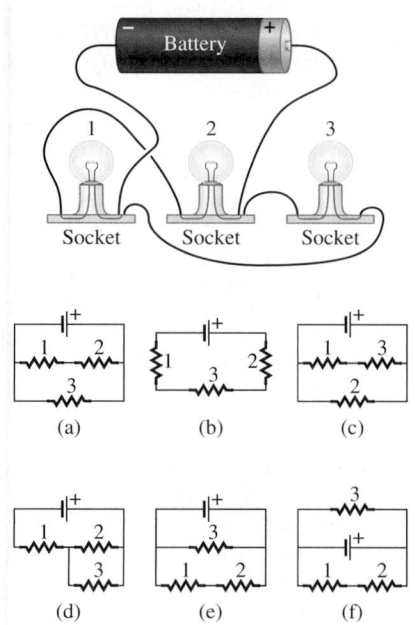

Figure 18.84   Problem 91

92. Near Earth's surface the air contains both negative and positive ions, due to radioactivity in the soil and cosmic rays from space. As a simplified model, assume there are 600.0 singly-charged positive ions per cm^3 and 500.0 singly charged negative ions per cm^3; ignore the presence of multiply-charged ions. The electric field is 100.0 V/m, directed downward. (a) In which direction do the positive ions move? The negative ions? (b) What is the direction of the current due to these

ions? (c) The measured resistivity of the air in the region is 4.0×10^{13} Ω·m. Calculate the drift speed of the ions, assuming it to be the same for positive and negative ions. [*Hint:* Consider a vertical tube of air of length *L* and cross-sectional area *A*. How is the potential difference across the tube related to the electric field strength?] (d) If these conditions existed over the entire surface of the Earth, what is the total current due to the movement of ions in the air?

93. A portable radio requires an emf of 4.5 V. Oliver has only two non-rechargeable 1.5-V batteries, but he finds a larger 6.0-V battery. (a) How can he arrange the batteries to produce an emf of 4.5 V? Draw a circuit diagram. (b) Is it advisable to use this combination with his radio? Explain.

94. We can model some of the electrical properties of an unmyelinated axon as an electrical cable covered with defective insulation so that current leaks out of the axon to the surrounding fluid. We assume the axon consists of a cylindrical membrane filled with conducting fluid. A current of ions can travel along the axon in this fluid and can also leak out through the membrane. The inner radius of the cylinder is 5.0 μm; the membrane thickness is 8.0 nm. (a) If the resistivity of the axon fluid is 2.0 Ω·m, calculate the resistance of a 1.0-cm length of axon to current flow along its length. (b) If the resistivity of the porous membrane is 2.5×10^7 Ω·m, calculate the resistance of the wall of a 1.0-cm length of axon to current flow across the membrane. (c) Find the length of axon for which the two resistances are equal. This length is a rough measure of the distance a signal can travel without amplification.

95. A piece of gold wire of length *L* has a resistance R_0. Suppose the wire is drawn out so that its length increases by a factor of three. What is the new resistance *R* in terms of the original resistance?

96. In Fig. 18.73, a voltmeter with a resistance of 670 kΩ is used to measure the voltage across the 83.0-kΩ resistor. What is the voltmeter reading?

97. A gold wire and an aluminum wire have the same dimensions and carry the same current. The electron density (in electrons/cm^3) in aluminum is three times larger than the density in gold. How do the drift speeds of the electrons in the two wires, v_{Au} and v_{Al}, compare?

98. (a) Given two identical, ideal batteries (emf = $\mathscr{E}$) and two identical lightbulbs (resistance = *R* assumed constant), design a circuit to make both bulbs glow as brightly as possible. (b) What is the power dissipated by each bulb? (c) Design a circuit to make both bulbs glow, but one more brightly than the other. Identify the brighter bulb.

99. Two circuits (Fig. 18.85) are constructed using identical, ideal batteries (emf = $\mathscr{E}$) and identical lightbulbs (resistance = *R*). If each bulb in circuit 1 dissipates 5.0 W of power, how much power does each bulb in circuit 2 dissipate? Ignore changes in the resistance of the bulbs due to temperature changes.

Figure 18.85
Problem 99

100. Given two identical, ideal batteries of emf $\mathcal{E}$ and two identical lightbulbs of resistance R (assumed constant), find the total power dissipated in the circuit of Fig. 18.86 in terms of $\mathcal{E}$ and R.

Figure 18.86 Problem 100

101. Consider a 60.0-W lightbulb and a 100.0-W lightbulb designed for use in a household lamp socket at 120 V. (a) What are the resistances of these two bulbs? (b) If they are wired together in a series circuit, which bulb shines brighter (dissipates more power)? Explain. (c) If they are connected in parallel in a circuit, which bulb shines brighter? Explain.

102. A 500-W electric heater unit is designed to operate with an applied potential difference of 120 V. (a) If the local power company imposes a voltage reduction to lighten its load, dropping the voltage to 110 V, by what percentage does the heat output of the heater drop? (Assume the resistance does not change.) (b) If you took the variation of resistance with temperature into account, would the actual drop in heat output be larger or smaller than calculated in part (a)?

103. Copper and aluminum are being considered for the cables in a high-voltage transmission line where each must carry a current of 50 A. The resistance of each cable is to be 0.15 Ω per kilometer. (a) If this line carries power from Niagara Falls to New York City (approximately 500 km), how much power is lost along the way in the cable? Compute for each choice of cable material (b) the necessary cable diameter and (c) the mass per meter of the cable. The electrical resistivities for copper and aluminum are given in Table 18.1; the mass density of copper is 8920 kg/m³ and that of aluminum is 2700 kg/m³.

104. The circuit shown in Fig. 18.87 is used to study the charging of a capacitor. (a) At $t = 0$, the switch is closed. What initial charging current is measured by the ammeter? (b) After the current has decayed to zero, what are the voltages at points A, B, and C?

Figure 18.87
Problem 104

105. A battery with an emf of 1.0 V is connected to a 1.0-kΩ resistor and a diode (a nonohmic device), as shown in Fig. 18.88a. The current that flows through the diode for a given voltage drop is shown in Fig. 18.88b. (a) What is the current through the diode? (b) What is the current through the battery? (c) What is the total power dissipated in the diode and resistor? (d) Suppose the battery emf were increased so that the power dissipated in the 1.0-kΩ resistor doubled. Would you expect the power dissipated in the diode to double? If not, would it increase by a factor greater than 2 or less than 2? Explain briefly.

(a)   (b)

Figure 18.88 Problem 105

106. A parallel plate capacitor is constructed from two square conducting plates of length $L = 0.10$ m on a side. There is air between the plates, which are separated by a distance $d = 89$ μm. The capacitor is connected to a 10.0-V battery (Fig. 18.89). (a) After the capacitor is fully charged, what is the charge on the upper plate? (b) The battery is disconnected from the plates and the capacitor is discharged through a resistor $R = 0.100$ MΩ. Sketch the current through the resistor as a function of time t ($t = 0$ corresponds to the time when R is connected to the capacitor). (c) How much energy is dissipated in R over the whole discharging process? (d) The capacitor is again charged up to 10 V using the same battery. The battery is disconnected. Then a dielectric slab (dielectric constant $\kappa = 2.0$) is inserted into the space between the plates, completely filling the space. The capacitor is then discharged through the same resistor as before. On the graph you made in part (c), sketch the current through the resistor as a function of time and label it as part (d). Explain the differences in the two graphs.

Figure 18.89 Problem 106

107. Poiseuille's law [Eq. (9-15)] gives the volume flow rate of a viscous fluid through a pipe. (a) Show that Poiseuille's law can be written in the form $\Delta P = IR$, where $I = \Delta V/\Delta t$ represents the volume flow rate and R is a constant of proportionality called the fluid flow *resistance*. (b) Find R in terms of the viscosity of the fluid and the length and radius of the pipe. (c) If two or more pipes are connected in series, so that the volume flow rate through them is the same, do the resistances of the pipes add as for electrical resistors ($R_{eq} = R_1 + R_2 + \cdots$)? Explain. (d) If two or more pipes are connected in parallel, so the pressure drop across them is the same, do the reciprocals of the resistances add as for electrical resistors ($1/R_{eq} = 1/R_1 + 1/R_2 + \cdots$)? Explain.

108. The *Wheatstone bridge* is a circuit used to measure unknown resistances. The bridge in Fig. 18.90 is balanced—no current flows through the galvanometer. What is

the unknown resistance R_x? [*Hint:* What is the potential difference between points A and B?]

Figure 18.90 Problem 108

C 109. A *potentiometer* is a circuit to measure emfs. In Fig. 18.91, with switch S_1 closed and S_2 open, there is no current through the galvanometer G for $R_1 = 20.0\ \Omega$ with a standard cell $\mathcal{E}_s$ of 2.00 V. With switch S_2 closed and S_1 open, there is no current through the galvanometer G for $R_2 = 80.0\ \Omega$. (a) What is the unknown emf $\mathcal{E}_x$? (b) Explain why the potentiometer accurately measures the emf even for a source with substantial internal resistance.

Figure 18.91 Problem 109

110. In Fig. 18.92, $\mathcal{E} = 45.0$ V and $R = 100.0\ \Omega$. If a voltage $V_x = 30.0$ V is needed for a circuit, what should resistance R_x be?

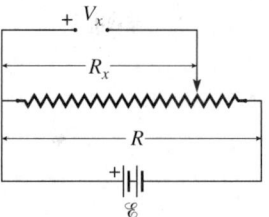

Figure 18.92 Problem 110

C ◆111. An ideal emf $\mathcal{E}$ is used to charge a capacitor C through a resistor R. (a) What is the final charge on the capacitor? (b) What is the final energy stored in the capacitor? (c) What is the total electrical energy supplied by the emf? [*Hint:* Use your answer to part (a).] (d) Explain why the final energy stored in the capacitor is less than the total electrical energy supplied by the emf.

ANSWERS TO PRACTICE PROBLEMS

18.1 (a) 2.00×10^{15} electrons; (b) 52 min

18.2 (a) 0.33 A; (b) 6.7 μm/s

18.3 6.9 Ω

18.4 292°C

18.5 1.495 V

18.6 1.0 Ω

18.7 $\frac{1}{3}R$ (the resistors are in parallel)

18.8 $+(0.58\ \text{A})(4.0\ \Omega) - 1.5\ \text{V} - 3.0\ \text{V} + (0.72\ \text{A})(3.0\ \Omega) = 0.0$

18.9 1.1 W, 190 J

18.10 10.0 mΩ; 10.0 mΩ

18.11 2.4 μA; 0.38 μC; 44 nA; 0.60 μC

Magnetic Forces and Fields

Some bacteria live in the mud at the bottom of the sea. As long as they are in the mud, all is well. Suppose that the mud gets stirred up, perhaps by a crustacean walking by. Now things are not so rosy. The bacteria cannot survive for long in the water, so it is imperative that they swim back down to the mud as soon as possible. The problem is that knowing which direction is down is not so easy. The mass density of the bacteria is almost identical to that of water, so the buoyant force prevents them from "feeling" the downward pull of gravity. Nevertheless, the bacteria are somehow able to swim in the correct direction to get back to the mud. How do they do it?

Electron micrograph of magnetotactic bacteria

Symbol for magnetic field: $\vec{\mathbf{B}}$

Making The Connection:
magnetic compass

19.1   MAGNETIC FIELDS

Permanent Magnets

Permanent magnets have been known at least since the time of the ancient Greeks, about 2500 years ago. A naturally occurring iron ore called lodestone (now called magnetite) was mined in various places, including the region of modern-day Turkey called Magnesia. Some of the chunks of lodestone were permanent magnets; they exerted magnetic forces on each other and on iron and could be used to turn a piece of iron into a permanent magnet. In China, the magnetic compass was used as a navigational aid at least a thousand years ago—possibly much earlier. Not until 1820 was a connection between electricity and magnetism established, when Danish scientist Hans Christian Oersted (1777–1851) discovered that a compass needle is deflected by a nearby electric current.

Figure 19.1a shows a plate of glass lying on top of a bar magnet. Iron filings have been sprinkled on the glass and then the glass has been tapped to shake the filings a bit and allow them to move around. The filings have lined up with the **magnetic field** (symbol: $\vec{\mathbf{B}}$) due to the bar magnet. Figure 19.1b shows a sketch of the magnetic field lines representing this magnetic field. As is true for electric field lines, the magnetic field lines represent both the magnitude and direction of the magnetic field. The magnetic field at any point is tangent to the field line and the magnitude of the field is proportional to the number of lines per unit area perpendicular to the lines.

Figure 19.1b may strike you as being similar to a sketch of the electric field lines for an electric dipole (Fig. 16.19). The similarity is not a coincidence; the bar magnet is one instance of a **magnetic dipole**. By *dipole* we mean *two opposite poles*. In an electric dipole, the electric poles are positive and negative electric charges. A magnetic dipole consists of two opposite magnetic poles. The end of the bar magnet where the field lines emerge is called the **north pole** and the end where the lines go back in is called the **south pole**. If two magnets are near each other, opposite poles (the north pole of one magnet and the south pole of the other) exert attractive forces on one another; like poles (two north poles or two south poles) repel each other.

The names *north pole* and *south pole* are derived from magnetic compasses. A compass is simply a small bar magnet that is free to rotate. Any magnetic dipole, including a compass needle, feels a torque that tends to line it up with an external magnetic

(a)

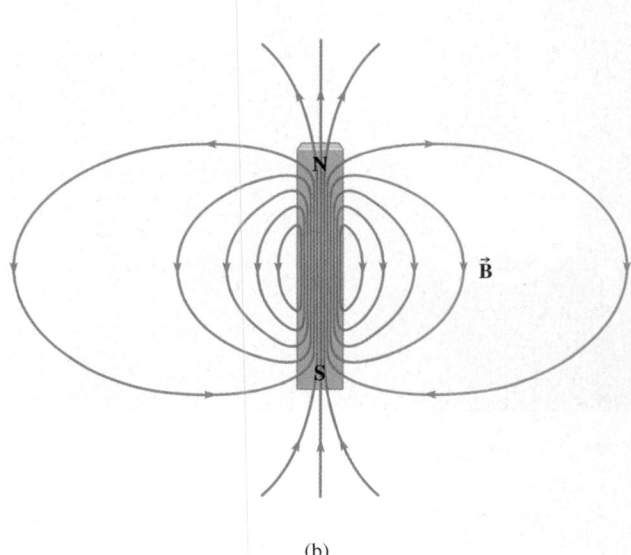

(b)

Figure 19.1   (a) Photo of a bar magnet. Nearby iron filings line up with the magnetic field. (b) Sketch of the magnetic field lines due to the bar magnet.

Physics at Home

Obtain a flexible magnetic card. With scissors, cut a thin strip (about 2 mm wide) from one edge. Rub the back of the strip across the back of the remaining piece. Try both orientations (both parallel and perpendicular to the side from which the strip was cut). Repeat with a strip cut from an adjacent side of the card. Estimate the orientation and the size of the magnetized strips. (See Fig. 19.3b.)

field (Fig. 19.2). The north pole of the compass needle is the end that points in the direction of the magnetic field. In a compass, the bar magnet needle is mounted to minimize frictional and other torques so it can swing freely in response to a magnetic field.

Permanent magnets come in many shapes other than the bar magnet. Figure 19.3 shows some others, with the magnetic field lines sketched. Notice in Fig. 19.3a that if the pole faces are parallel and close together, the magnetic field between them is nearly uniform. A magnet need not have only two poles; it must have *at least* one north pole and *at least* one south pole. Some magnets are designed to have a large number of north and south poles. The flexible magnetic card (Fig. 19.3b), commonly found on refrigerator doors, is designed to have many poles, both north and south, on one side and no poles on the other. The magnetic field is strong near the side with the poles and weak near the other side; the card sticks to an iron surface (such as a refrigerator door) on one side but not on the other.

Just as like charges repel and unlike charges attract, like magnetic poles repel and unlike poles attract. However, there is an important difference between electricity and magnetism. Coulomb's law for electrical forces gives the force acting between two point charges—two electric *monopoles*. However, as far as we know, there are no *magnetic* monopoles—that is, there is no such thing as an isolated north pole or an isolated south pole. If you take a bar magnet and cut it in half, you do not obtain one piece with a north pole and another piece with a south pole. Both pieces are magnetic dipoles (Fig. 19.4). There have been theoretical predictions of the existence of magnetic monopoles, but years of experiments have yet to turn up a single one. If magnetic monopoles do exist in our universe, they must be extremely rare.

Magnetic Field Lines

Figure 19.1b may suggest that magnetic field lines begin on north poles and end on south poles, but they do not. A careful look at Fig. 19.1b reveals that *magnetic field lines are all closed loops*. If there are no magnetic monopoles, there is no place for the field lines to begin or end, so they *must* be closed loops. Contrast Fig. 19.1b with Fig. 16.19—the field

Figure 19.2 Each compass needle is aligned with the magnetic field due to the bar magnet. The north end of the needle points in the direction of the magnetic field.

Magnetic field lines are always closed loops.

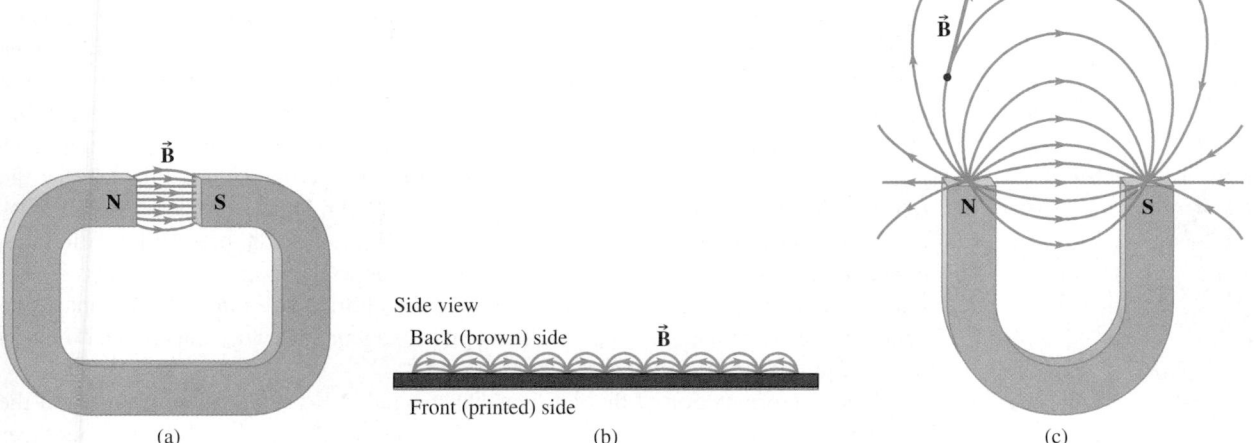

(a)   (b)   (c)

Figure 19.3 Other permanent magnets with their magnetic field lines. Only the field lines *outside* the magnets are sketched.

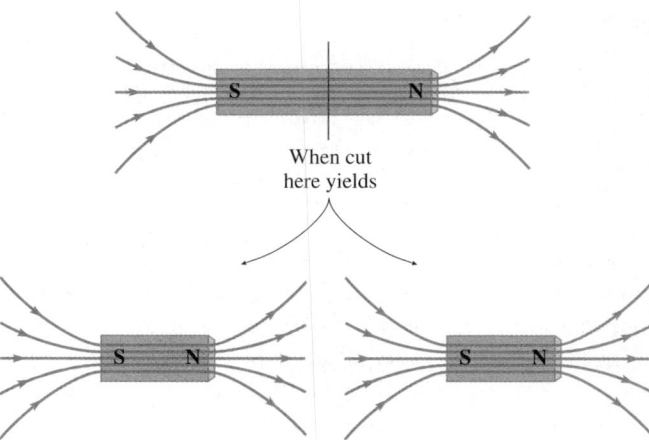

Figure 19.4 Sketch of a bar magnet that is subsequently cut in half. We have labeled the N and S poles and sketched field lines.

lines for an electric dipole. The field line patterns are similar away from the dipole, but nearby and between the poles they are quite different. The electric field lines are not closed loops; they start on the positive charge and end on the negative charge.

Despite these differences between electric and magnetic field lines, the *interpretation* of magnetic field lines is exactly the same as for electric field lines:

Interpretation of Magnetic Field Lines

- The direction of the magnetic field at any point is *tangent to the field line* passing through that point and is in the direction indicated by arrows on the field line (as in Fig. 19.3c).
- The magnetic field is strong where field lines are close together and weak where they are far apart. More specifically, if you imagine a small surface perpendicular to the field lines, the magnitude of the field is proportional to the number of lines that cross the surface, divided by the area.

Making The Connection:
Earth's magnetic field

The Earth's Magnetic Field

Figure 19.5 shows field lines for the Earth's magnetic field. Near Earth's surface, the magnetic field is approximately that of a dipole, as if a bar magnet were buried at the center of the Earth. Farther away from Earth's surface, the dipole field is distorted by the solar wind—charged particles streaming from the Sun toward the Earth. As discussed in Section 19.8, moving charged particles create their own magnetic fields, so the solar wind has a magnetic field associated with it.

In most places on the surface, the Earth's magnetic field is not horizontal; it has a significant vertical component. The vertical component can be measured directly using a *dip meter*, which is just a compass mounted so that it can rotate in a vertical plane (Fig. 19.6). In the northern hemisphere, the vertical component is downward, while in the southern hemisphere it is upward. In other words, magnetic field lines emerge from Earth's surface in the southern hemisphere and reenter in the northern hemisphere. A magnetic dipole that is free to rotate aligns itself with the magnetic field such that the north end of the dipole points in the direction of the field. Figure 19.2 shows a bar magnet with several compasses in the vicinity. Each compass needle points in the direction of the local magnetic field, which in this case is due to the magnet. A compass is normally used to detect the Earth's magnetic field. In a horizontally mounted compass, the needle is free to rotate only in a horizontal plane, so its north end points in the direction of the *horizontal component* of the Earth's field.

Note the orientation of the fictitious bar magnet in Fig. 19.5: the south pole of the magnet faces roughly toward geographic *north* and the north pole of the magnet faces roughly toward geographic *south*. The field lines emerge from Earth's surface in the southern hemisphere and return in the northern hemisphere.

Figure 19.5 The magnetic field due to the Earth. The field lines are approximately those of a magnetic dipole inside the Earth. The bar magnet sketched shows the orientation of the dipole.

The origin of the Earth's magnetic field is still under investigation. According to a leading theory, the field is created by the flow of molten iron and nickel in Earth's outer core more than 3000 km below the surface. The outer core has a net charge and rotates with a higher angular speed than Earth's surface. The Earth's magnetic field is slowly changing. In 1948, Canadian scientists discovered that the location of Earth's magnetic pole in the Arctic was about 250 km away from where it was found in 1831 by a British explorer. The magnetic poles move on average between 10 and 40 km per year. The magnetic poles have undergone a complete reversal in polarity (north becomes south and south becomes north) roughly 100 times in the past 5 million years.

South end — **S**

North end

Figure 19.6 A compass in a vertical plane indicating the direction of the Earth's magnetic field.

Magnetotactic Bacteria

In the electron micrograph of the bacterium discussed in the chapter opener, a line of crystals stands out. They are crystals of magnetite, the same iron oxide (Fe_3O_4) that was known to the ancient Greeks. The crystals are tiny permanent magnets that function essentially as compass needles. When the bacteria get stirred up into the water, their compass needles automatically rotate to line up with the magnetic field (Fig. 19.7). As the bacteria swim along, they follow a magnetic field line. In the northern hemisphere, the north end of the "compass needle" faces forward. The bacteria swim in the direction of the magnetic field, which has a downward component, so they return to their home in the mud. Bacteria in the southern hemisphere have the south pole forward; they must swim opposite to the magnetic field since the field has an *upward* component. If some of these *magnetotactic* (*-tactic* = feeling or sensing) bacteria are brought from the southern hemisphere to the northern, or *vice versa*, they swim up instead of down!

There is evidence of magnetic navigation in several species of bacteria and also in some higher organisms. Experiments with homing pigeons, robins, and bees have shown that these organisms have some magnetic sense. On sunny days, they primarily use the Sun's location for navigation, but on overcast days they use the Earth's magnetic field. Permanently magnetized crystals, similar to those found in the mud bacteria, have been found in the brains of these organisms, but the mechanism by which they can sense the Earth's field and use it to navigate is not understood. Some experiments have shown that even humans may have some sense of the Earth's magnetic field, which is not out of the realm of possibility since tiny magnetite crystals have been found in the brain.

19.2 MAGNETIC FORCE ON A POINT CHARGE

Before we go into more detail on the magnetic forces and torques on a magnetic dipole, we need to start with the simpler case of the magnetic force on a point charge. In Chapter 16 we defined the electric field as the electric force per unit charge. A point charge q located where the electric field is $\vec{E}$ experiences an electric force

$$\vec{F}_E = q\vec{E} \tag{16-3b}$$

The electric force is either in the same direction as $\vec{E}$ or in the opposite direction, depending on the sign of the point charge.

The magnetic force on a point charge is more complicated—it is *not* the charge times the magnetic field. The magnetic force *depends on the point charge's velocity* as well as on the magnetic field. If the point charge is at rest, there is no magnetic force. The magnitude and direction of the magnetic force depend on the direction and speed of the charge's motion. We have learned about other velocity-dependent forces, such as the drag force on an object moving through a fluid. Like drag forces, the magnetic force increases in magnitude with increasing velocity. However, the direction of the drag

⚠ The magnetic force is velocity-dependent.

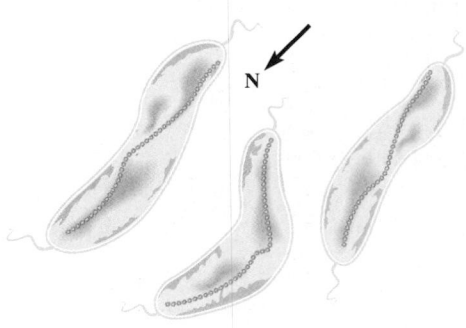

Figure 19.7 The magnetite crystals inside the bacteria line up with the Earth's magnetic field.

force is always opposite to the object's velocity, while the direction of the magnetic force on a charged particle is *perpendicular* to the velocity of the particle.

Imagine that a positive point charge q moves at velocity $\vec{v}$ at a point where the magnetic field is $\vec{B}$. The angle between $\vec{v}$ and $\vec{B}$ is θ (Fig. 19.8a). The magnitude of the magnetic force acting on the point charge is the product of

- The magnitude of the charge $|q|$,
- The magnitude of the field B, and
- The component of the velocity perpendicular to the field (Fig. 19.8b).

Magnetic force on a moving point charge

$$F_B = |q|v_\perp B = |q|(v \sin \theta)B$$

$$(\text{since } v_\perp = v \sin \theta) \tag{19-1}$$

Note that if the point charge is at rest ($v = 0$) or if its motion is parallel or antiparallel to the magnetic field ($v_\perp = 0$), then the magnetic force is zero.

In some cases it is convenient to look at the factor $\sin \theta$ from a different point of view. If we associate the factor $\sin \theta$ with the magnetic field instead of with the velocity, then $B \sin \theta$ is the component of the magnetic field perpendicular to the velocity of the charged particle (Fig. 19.8c):

$$F_B = |q|v(B \sin \theta) = |q|vB_\perp$$

From Eq. (19-1), the SI unit of magnetic field is

$$\frac{\text{force}}{\text{charge} \times \text{velocity}} = \frac{\text{N}}{\text{C} \cdot \text{m/s}} = \frac{\text{N}}{\text{A} \cdot \text{m}}$$

This combination of units is given the name tesla (symbol T) after Nikola Tesla (1856–1943), an American engineer who was born in Croatia.

$$1 \text{ T} = 1 \frac{\text{N}}{\text{A} \cdot \text{m}} \tag{19-2}$$

Cross Product

The direction and magnitude of the magnetic force depend on the vectors $\vec{v}$ and $\vec{B}$ in a special way that occurs often in physics and mathematics. The magnetic force can be written in terms of the **cross product** (or *vector product*) of $\vec{v}$ and $\vec{B}$. The cross product of two vectors $\vec{a}$ and $\vec{b}$ is written $\vec{a} \times \vec{b}$. The magnitude of the cross product is the magnitude of one vector times the perpendicular component of the other; it doesn't matter which is which.

$$|\vec{a} \times \vec{b}| = |\vec{b} \times \vec{a}| = a_\perp b = ab_\perp = ab \sin \theta \tag{19-3}$$

However, the order of the vectors *does* matter in determining the *direction* of the result. Switching the order reverses the direction of the product:

$$\vec{b} \times \vec{a} = -\vec{a} \times \vec{b} \tag{19-4}$$

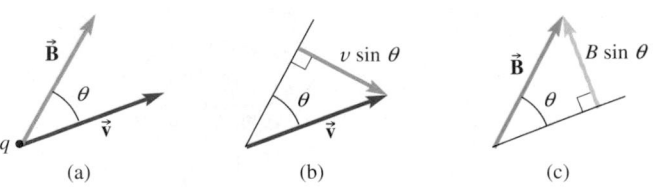
(a)   (b)   (c)

Figure 19.8 A positive charge moving in a magnetic field. (a) The particle's velocity vector $\vec{v}$ and the magnetic field vector $\vec{B}$ are drawn starting at the same point. θ is the angle between them. (b) The component of $\vec{v}$ perpendicular to $\vec{B}$ is $v \sin \theta$. (c) The component of $\vec{B}$ perpendicular to $\vec{v}$ is $B \sin \theta$.

The cross product of two vectors is perpendicular to both vectors.

The cross product of two vectors $\vec{a}$ and $\vec{b}$ is a vector that is perpendicular to both $\vec{a}$ and $\vec{b}$. Note that $\vec{a}$ and $\vec{b}$ do not have to be perpendicular to each other. For any two vectors that are neither parallel nor antiparallel, there are *two* directions perpendicular to both vectors. To choose between the two, we use a **right-hand rule**. Suppose we want to know the direction of $\vec{a} \times \vec{b}$ in Fig. 19.9a. The right-hand rule is used as described next.

Using Right-Hand Rule 1 (to Find the Direction of a Cross Product)

1. Determine the two directions that are perpendicular to both $\vec{a}$ and $\vec{b}$. (In Fig. 19.9a, since both $\vec{a}$ and $\vec{b}$ lie in the plane of the page, the cross product is perpendicular to the page: either into the page or out of the page.)

2. Choose one of the two directions to test. Point the thumb of your *right* hand in that direction and the outstretched four fingers of your right hand along the first vector in the product with the hand in the shape of a karate chop (Fig. 19.9b).

3. Keeping the thumb and palm stationary, curl your outstretched fingers inward toward your palm until your fingertips point in the direction of the second vector (Fig. 19.9c). If you can't do it—because your fingers would have to go through your palm to reach the second vector as they sweep through an angle greater than 180°—then the direction of your thumb is *opposite* to the direction of the cross product. If you *can* do it, sweeping your fingers through an angle ≤ 180°, then your thumb points in the direction of the cross product.

4. Verify your conclusion by testing the other of the two possible directions (Fig. 19.9d).

Vector symbols: • or ⊙ = out of the page; × or ⊗ = into the page

Since magnetism is inherently three-dimensional, we often need to draw vectors that are perpendicular to the page. The symbol • (or ⊙) represents a vector arrow pointing out of the page; think of the tip of an arrow coming toward you. The symbol × (or ⊗) represents a vector pointing into the page; it suggests the tail feathers of an arrow moving away from you.

Direction of the Magnetic Force

The magnetic force on a charged particle can be written as the charge times the cross product of $\vec{v}$ and $\vec{B}$:

Magnetic force on a moving point charge
$$\vec{F}_B = q\vec{v} \times \vec{B} \qquad (19\text{-}5)$$

Figure 19.9 Using the right-hand rule to find the direction of the cross product. (a) Two vectors, $\vec{a}$ and $\vec{b}$. We want to find the cross product $\vec{a} \times \vec{b}$. (b) Initial hand position to test the direction out of the page. The thumb points out of the page and the fingers point along $\vec{a}$. (c) The fingers are curled in through an angle < 180° until they point along $\vec{b}$. Therefore, $\vec{a} \times \vec{b}$ is out of the page. (d) Initial hand position to test the direction into the page. To curl from $\vec{a}$ to $\vec{b}$, the fingers would have to sweep through an angle greater than 180°, so the thumb points in the direction opposite to $\vec{a} \times \vec{b}$.

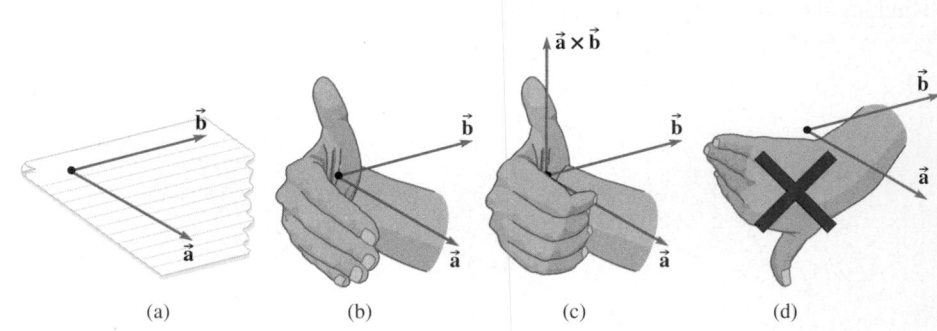

(a)   (b)   (c)   (d)

The direction of the magnetic force is not along the same line as the field (as is the case for the electric field); instead it is *perpendicular*. The force is also perpendicular to the charged particle's velocity. Therefore, if $\vec{v}$ and $\vec{B}$ lie in a plane, the magnetic force is always perpendicular to that plane; magnetism is inherently three-dimensional. A negatively charged particle feels a magnetic force in the direction *opposite* to $\vec{v} \times \vec{B}$; multiplying a *negative* scalar (q) by $\vec{v} \times \vec{B}$ reverses the direction of the magnetic force.

> The magnetic force on a point charge is *perpendicular* to the magnetic field.

Problem-Solving Technique: Finding the Magnetic Force on a Point Charge

1. The magnetic force is zero if (a) the particle is not moving ($\vec{v} = \vec{0}$), (b) its velocity has no component perpendicular to the magnetic field ($v_\perp = 0$), or (c) the magnetic field is zero.

2. Otherwise, determine the angle θ between the velocity and magnetic field vectors when the two are drawn tail-to-tail.

3. Find the magnitude of the force from Eq. (19-1), using the *magnitude* of the charge (since magnitudes of vectors are nonnegative).

4. Determine the two directions that are perpendicular to a plane that contains both $\vec{v}$ and $\vec{B}$ (and therefore are perpendicular to both $\vec{v}$ and $\vec{B}$).

5. Determine the direction of $\vec{v} \times \vec{B}$ by using the right-hand rule to choose between the two possibilities. The magnetic force is in the direction of $\vec{v} \times \vec{B}$ if the charge is positive. If the charge is negative, the force is in the direction opposite to $\vec{v} \times \vec{B}$.

Conceptual Example 19.1

Deflection of Cosmic Rays

 Cosmic rays are charged particles moving toward Earth at high speeds. The origin of the particles is not fully understood, but explosions of supernovae may produce a significant fraction of them. About $\frac{7}{8}$ of the particles are protons that move toward Earth with an average speed of about $\frac{2}{3}$ the speed of light. Suppose that a proton is moving straight down, directly toward the equator. (a) What is the direction of the magnetic force on the proton due to Earth's magnetic field? (b) Explain how the Earth's magnetic field shields us from bombardment by cosmic rays. (c) Where on Earth's surface is this shielding least effective?

Strategy and Solution   (a) First we sketch the Earth's magnetic field lines and the velocity vector for the proton (Fig. 19.10a). The field lines run from southern hemisphere to northern; high above the equator, the field is approximately horizontal (due north). To find the direction of the magnetic

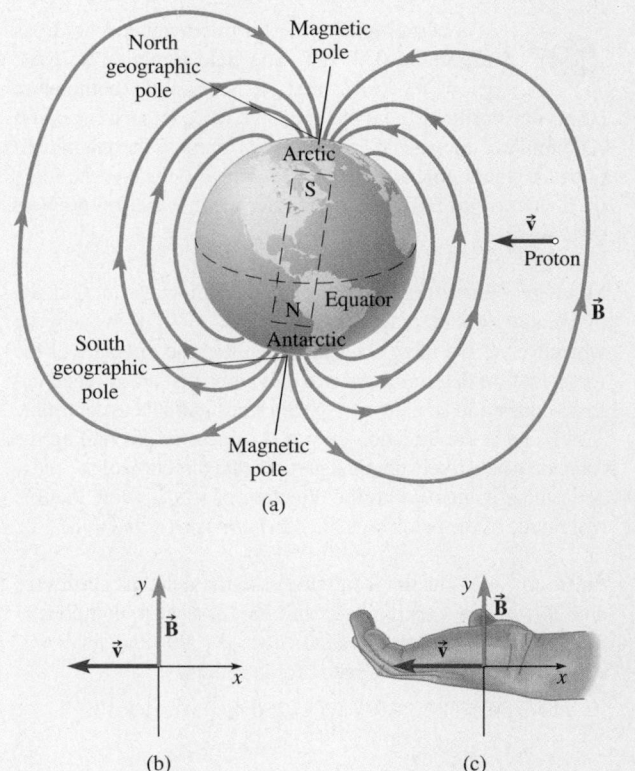

Figure 19.10
(a) A sketch of the Earth, its magnetic field lines, and the velocity vector $\vec{v}$ of the proton. (b) The vectors $\vec{v}$ and $\vec{B}$. The y-axis points north; the x-axis points away from the equator. (c) The right-hand rule shows that $\vec{v} \times \vec{B}$ is into the page. With the thumb pointing into the page, the fingers sweep from $\vec{v}$ to $\vec{B}$ through an angle of 90°.

continued on next page

Conceptual Example 19.1 *continued*

force, first we determine the two directions that are perpendicular to both $\vec{v}$ and $\vec{B}$; then we use the right-hand rule to determine which is the direction of $\vec{v} \times \vec{B}$. Figure 19.10b is a sketch of $\vec{v}$ and $\vec{B}$ in the xy-plane. The x-axis points away from the equator and the y-axis points north. The two directions that are perpendicular to both vectors are perpendicular to the xy-plane: into the page and out of the page. Using the right-hand rule, if the thumb points out of the page, the fingers of the right hand would have to curl from $\vec{v}$ to $\vec{B}$ through an angle of 270°. Therefore, $\vec{v} \times \vec{B}$ is into the page (Fig. 19.10c). Since $\vec{F}_B = q\vec{v} \times \vec{B}$ and q is positive, the magnetic force is into the page or east.

(b) Without the Earth's magnetic field, the proton would move straight down toward Earth's surface. The magnetic field deflects the particle sideways and keeps it from reaching the surface. Many fewer cosmic ray particles reach the surface than would do so if there were no magnetic field.

(c) Near the poles, particles headed toward Earth have velocities that are nearly parallel or antiparallel to the magnetic field lines; the component of $\vec{v}$ perpendicular to the field ($v_\perp$) is a small fraction of v. Since the magnetic force is proportional to $v_\perp$, the deflecting force is much less effective near the poles.

 Discussion When finding the direction of the magnetic force (or any cross product), a good sketch is essential. Since all three dimensions come into play, we must choose the two axes that lie in the plane of the sketch. In this example, both vectors $\vec{v}$ and $\vec{B}$ lie in the plane of the sketch, so we know that $\vec{F}_B$ is perpendicular to that plane.

Practice Problem 19.1   Acceleration of cosmic ray particle

If $v = 6.0 \times 10^7$ m/s and $B = 6.0$ μT, what is the magnitude of the magnetic force on the proton and the magnitude of the proton's acceleration?

Example 19.2

Magnetic Force on an Ion in the Air

 At a certain place, the Earth's magnetic field has magnitude 0.50 mT. The field direction is 70.0° below the horizontal; its horizontal component points due north. (a) Find the magnetic force on an oxygen ion (O_2^-) moving due east at 250 m/s. (b) Compare the magnitude of the magnetic force to the ion's weight and the electric force on it due to the Earth's fair-weather electric field (150 V/m downward).

Strategy Since there are two equivalent ways to find the magnitude of the magnetic force [Eq. (19-1)], we choose whichever seems most convenient. To find the direction of the force, first we determine the two directions that are perpendicular to both $\vec{v}$ and $\vec{B}$; then we use the right-hand rule to determine which one is the direction of $\vec{v} \times \vec{B}$. Since we are finding the force on a negatively charged particle, the direction of the magnetic force is *opposite* to the direction of $\vec{v} \times \vec{B}$. Note that the magnitude of the field is specified in *milli*teslas (1 mT = 10^{-3} T).

Solution (a) The ion is moving east; the field has northward and downward components, but no east-west component. Therefore, $\vec{v}$ and $\vec{B}$ are perpendicular; $\theta = 90°$ and $\sin \theta = 1$. The magnitude of the magnetic force is then

$$F = |q|vB = (1.6 \times 10^{-19}\,\text{C}) \times (250\,\text{m/s}) \times (5.0 \times 10^{-4}\,\text{T})$$

$$= 2.0 \times 10^{-20}\,\text{N}$$

Since $\vec{v}$ is east and the force must be perpendicular to $\vec{v}$, the force must lie in a plane perpendicular to the east/west axis. We draw the velocity and magnetic field vectors in this plane, using axes that run north/south and up/down (Fig. 19.11a, where east is out of the page). Since north is to the right in this

sketch, the viewer looks westward; west is into the page and east is out of the page. The force $\vec{F}$ must lie in this plane and be perpendicular to $\vec{B}$. There are two possible directions, shown with a dashed line in Fig. 19.11b. Now we try these

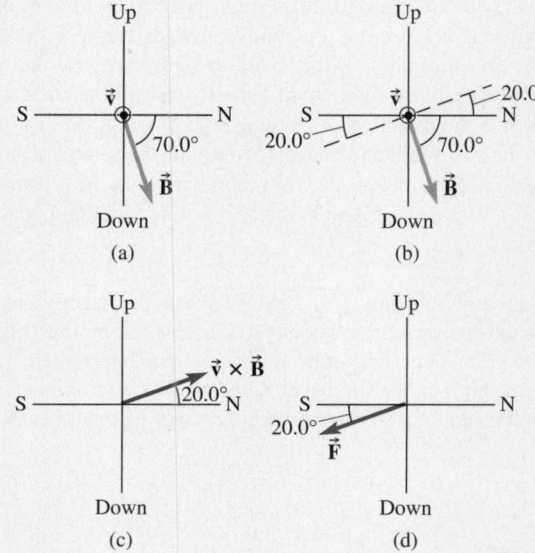

Figure 19.11
(a) The vectors $\vec{v}$ and $\vec{B}$, with $\vec{v}$ out of the page. West is into the page and east is out of the page. (b) Since $\vec{F}$ is perpendicular to both $\vec{v}$ and $\vec{B}$, it must lie along the dashed line. (c) The direction for $\vec{v} \times \vec{B}$ given by the right-hand rule. (d) Since the ion is negatively charged, the magnetic force direction is opposite $\vec{v} \times \vec{B}$.

continued on next page

Example 19.2 *continued*

two directions with the right-hand rule; the correct direction for $\vec{v} \times \vec{B}$ is shown in Fig. 19.11c. Since the ion is negatively charged, the magnetic force is in the direction opposite to $\vec{v} \times \vec{B}$; it is 20.0° below the horizontal, with its horizontal component pointing south.

(b) From Appendix B, the mass of an oxygen atom is 16 u. Therefore, the weight of an O_2^- ion is

$$W = mg = (2 \times 16 \text{ u} \times 1.66 \times 10^{-27} \text{ kg/u}) \times 9.8 \text{ N/kg}$$

$$= 5.2 \times 10^{-25} \text{ N}$$

The electric force has magnitude

$$F_E = |q|E = (1.6 \times 10^{-19} \text{ C}) \times 150 \text{ N/C} = 2.4 \times 10^{-17} \text{ N}$$

The magnetic force on the ion is much stronger than the gravitational force and much weaker than the electric force.

 Discussion Again, a key to solving this sort of problem is drawing a convenient set of axes. If one of the two vectors $\vec{v}$ and $\vec{B}$ lies along a reference direction—a point of the compass, up or down, or along one of the *xyz*-axes—and the other does not, a good choice is to sketch axes in a plane *perpendicular* to that reference direction. In this case, $\vec{v}$ is in a reference direction (east) but $\vec{B}$ is not, so we sketch axes in a plane perpendicular to east.

Practice Problem 19.2 Magnetic force on an electron

Find the magnetic force on an *electron* moving straight up at 3.0×10^6 m/s in the same magnetic field. [*Hint:* The angle between $\vec{v}$ and $\vec{B}$ is *not* 90°.]

Example 19.3

Electron in a Magnetic Field

An electron moves with speed 2.0×10^6 m/s in a uniform magnetic field of 1.4 T directed due north. At one instant, the electron experiences an upward magnetic force of 1.6×10^{-13} N. In what direction is the electron moving at that instant? [*Hint:* If there is more than one possible answer, find all the possibilities.]

Strategy This example is more complicated than the previous two. We need to apply the magnetic force law again, but this time we must deduce the direction of the velocity from the directions of the force and field.

Solution The magnetic force is always perpendicular to both the magnetic field and the particle's velocity. The force is upward, therefore the velocity must lie in a horizontal plane.

Figure 19.12 shows the magnetic field pointing north and a variety of possibilities for the velocity (all in the horizontal plane). The direction of the magnetic force is up, so the direction of $\vec{v} \times \vec{B}$ must be down since the charge is negative. Pointing the thumb of the right hand downward, the fingers curl in the clockwise sense. Since we curl from $\vec{v}$ to $\vec{B}$, the velocity must be somewhere in the left half of the plane; in other words, it must have a west component in addition to a north or south component.

The westward component is the component of $\vec{v}$ that is perpendicular to the field. Using the magnitude of the force, we can find the perpendicular component of the velocity:

$$F_B = |q|v_\perp B$$

$$v_\perp = \frac{F_B}{|q|B} = \frac{1.6 \times 10^{-13} \text{ N}}{1.6 \times 10^{-19} \text{ C} \times 1.4 \text{ T}} = 7.14 \times 10^5 \text{ m/s}$$

The velocity also has a component parallel or antiparallel to the field that can be found using the Pythagorean theorem:

$$v^2 = v_\perp^2 + v_\parallel^2$$

$$v_\parallel = \pm\sqrt{v^2 - v_\perp^2} = \pm 1.87 \times 10^6 \text{ m/s}$$

The ± sign would seem to imply that $v_\parallel$ could either be a north or a south component. The two possibilities are shown in Fig. 19.13. Use of the right-hand rule confirms that *either* gives $\vec{v} \times \vec{B}$ in the correct direction.

Now we need to find the direction of $\vec{v}$ given its components. From Fig. 19.12,

$$\sin \theta = \frac{v_\perp}{v} = \frac{7.14 \times 10^5 \text{ m/s}}{2.0 \times 10^6 \text{ m/s}} = 0.357$$

$$\theta = 21° \text{ W of N or } 159° \text{ W of N}$$

Since 159° W of N is the same as 21° W of S, the direction of the velocity is either 21° W of N or 21° W of S.

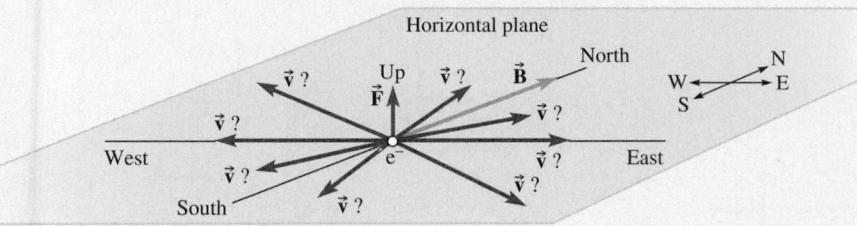

Figure 19.12
The velocity must be perpendicular to the force and thus in the plane shown. Various possibilities for the direction of $\vec{v}$ are considered. Only those in the west half of the plane give the correct direction for $\vec{v} \times \vec{B}$.

continued on next page

Example 19.3 *continued*

⚠️ **Discussion**   We *cannot* assume that $\vec{v}$ is perpendicular to $\vec{B}$. The magnetic force is always perpendicular to both $\vec{v}$ and $\vec{B}$, but there can be any angle between $\vec{v}$ and $\vec{B}$.

Practice Problem 19.3   Velocity component parallel to field

Suppose the electron moves with the same speed in the same magnetic field. If the magnetic force on the electron has magnitude 2.0×10^{-13} N, what is the component of the electron's velocity parallel to the magnetic field?

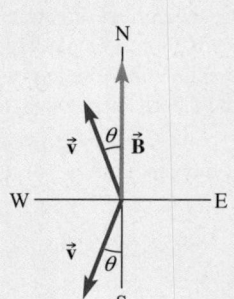

Figure 19.13
Two possibilities for the direction of $\vec{v}$.

19.3   CHARGED PARTICLE MOVING PERPENDICULAR TO A UNIFORM MAGNETIC FIELD

Using the magnetic force law and Newton's second law of motion, we can deduce the trajectory of a charged particle moving in a uniform magnetic field with no other forces acting. In this section, we discuss a case of particular interest: when the particle is initially moving perpendicular to the magnetic field.

Figure 19.14a shows the magnetic force on a positively charged particle moving perpendicular to a magnetic field. Since $v_\perp = v$, the magnitude of the force is

$$F = |q|vB \tag{19-6}$$

Since the force is perpendicular to the velocity, the particle changes direction but not speed. The force is also perpendicular to the field, so there is no acceleration component in the direction of $\vec{B}$. Thus, the particle's velocity remains perpendicular to $\vec{B}$. As the velocity changes direction, the magnetic force changes direction to stay perpendicular to both $\vec{v}$ and $\vec{B}$. The magnetic force acts as a steering force, curving the particle around in a trajectory of radius r at constant speed. The particle undergoes uniform circular motion, so its acceleration is centripetal and has magnitude v^2/r. From Newton's second law,

$$a_\perp = \frac{v^2}{r} = \frac{F}{m} = \frac{|q|vB}{m} \tag{19-7}$$

Magnetic fields can cause charges to move along circular paths.

where m is the mass of the particle. Since the radius of the trajectory is constant—r depends only on q, v, B, and m, which are all constant—the particle moves in a circle at constant speed (Fig. 19.14b). Negative charges move in the opposite sense from positive charges in the same field (Fig. 19.14c).

Making The Connection:
bubble chamber

Bubble Chamber

The circular motion of charged particles in uniform magnetic fields has many applications. The *bubble chamber*, invented by American physicist Donald Glaser (1926–), is

Figure 19.14   (a) Force on a positive charge moving to the right in a magnetic field that is into the page. (b) As the velocity changes direction, the magnetic force changes direction to stay perpendicular to both $\vec{v}$ and $\vec{B}$. The force is constant in magnitude, so the particle moves along the arc of a circle. (c) Motion of a negative charge in the same magnetic field.

(a)

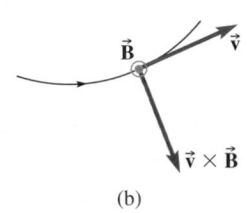

(b)

Figure 19.15 (a) Artistically enhanced tracks left by charged particles moving through the BEBC (Big European Bubble Chamber). The tracks are curved due to the presence of a magnetic field. The direction of curvature reveals the sign of the charge. (b) Analysis of the magnetic force on one particular particle. This particle must have a negative charge since the force is opposite in direction to $\vec{\mathbf{v}} \times \vec{\mathbf{B}}$.

a particle detector that was used in high-energy physics experiments from the 1950s into the 1970s. The chamber is filled with liquid hydrogen and is immersed in a magnetic field. When a charged particle moves through the liquid it leaves a trail of bubbles. Figure 19.15a shows tracks made by particles in a bubble chamber. The magnetic field is out of the page. The magnetic force on any particle points toward the center of curvature of the particle's trajectory. Figure 19.15b shows the directions of $\vec{\mathbf{v}}$ and $\vec{\mathbf{B}}$ for one particle. Using the right-hand rule, $\vec{\mathbf{v}} \times \vec{\mathbf{B}}$ is in the direction shown in Fig. 19.15b. Since $\vec{\mathbf{v}} \times \vec{\mathbf{B}}$ points away from the center of curvature, which is the direction of $\vec{\mathbf{F}}$, the particle must have a negative charge. The magnetic force law lets us determine the sign of the charge on the particle.

Mass Spectrometer

Making The Connection:
mass spectrometer

The basic purpose of a *mass spectrometer* is to separate ions (charged atoms or molecules) by mass and measure the mass of each type of ion. Although originally devised to measure the masses of the products of nuclear reactions, mass spectrometers are now used by researchers in many different scientific fields and in medicine to identify what atoms or molecules are present in a sample and in what concentrations. Even ions present in minute concentrations can be isolated, making the mass spectrometer an essential tool in toxicology and in monitoring the environment for trace pollutants. Mass spectrometers are used in food production, petrochemical production, the electronics industry, and in the international monitoring of nuclear facilities. They are also an important tool for investigations of crime scenes, as several popular TV shows demonstrate weekly.

Today, many different types of mass spectrometer are in use. The oldest type, now called a magnetic-sector mass spectrometer, is based on the circular motion of a charged particle in a magnetic field. The atoms or molecules are first ionized so that they have a known electric charge. They are then accelerated by an electric field that can be varied to adjust their speeds. The particles then enter a region of uniform magnetic field $\vec{\mathbf{B}}$ oriented perpendicular to their velocities $\vec{\mathbf{v}}$ so that they move in circular arcs. From the charge, speed, magnetic field strength, and radius of the circular arc, we can determine the mass of the particle.

In some magnetic-sector spectrometers, the ions start at rest or at low speed and are accelerated through a fixed potential difference, so that they all have the *same kinetic energy* when they enter the magnetic field. Then, if there is a mixture of ions with different masses, the speeds are not all the same. Another possibility is to use a *velocity*

Figure 19.16 (a) A simplified diagram of a magnetic-sector mass spectrometer that accelerates ions through a fixed potential difference so that they all enter the magnetic field with the same kinetic energy. (b) A mass spectrometer in which ions travel around a path of fixed radius.

selector (Section 19.5) to make sure that all ions, regardless of mass, have the same *speed* when they enter the magnetic field. In the spectrometer of Example 19.4, ions of different masses travel in circular paths of different radii (Fig. 19.16a). In other spectrometers, only ions that travel along a path of *fixed radius* reach the detector; either the speed of the ions or the magnetic field is varied to select which ions move with the correct radius (Fig. 19.16b).

Example 19.4

Separation of Lithium Ions in a Mass Spectrometer

In a mass spectrometer, a beam of ^{6}Li and ^{7}Li ions passes through a velocity selector so that the ions all have the same velocity. The beam then enters a region of uniform magnetic field. If the radius of the orbit of the ^{6}Li ions is 8.4 cm, what is the radius of the orbit of the ^{7}Li ions?

Strategy Much of the information in this problem is implicit. The charge of the ^{6}Li ions is the same as the charge of the ^{7}Li ions. The ions enter the magnetic field with the same speed. We do not know the magnitudes of the charge, velocity, or magnetic field, but they are the same for the two types of ion. With so many common quantities, a good strategy is to try to find the *ratio* between the radii for the two types of ion so that the common quantities cancel out.

Solution From Appendix B we find the masses of ^{6}Li and ^{7}Li:

$$m_6 = 6.015 \text{ u}$$

$$m_7 = 7.016 \text{ u}$$

We now apply Newton's second law to an ion moving in a circle. The acceleration is that of uniform circular motion:

$$a_\perp = \frac{v^2}{r} = \frac{F}{m} = \frac{|q|vB}{m} \qquad (1)$$

Since the charge q, the speed v, and the field B are the same for both types of ion, the radius must be directly proportional to the mass.

$$r \propto m$$

$$\frac{r_7}{r_6} = \frac{m_7}{m_6} = \frac{7.016 \text{ u}}{6.015 \text{ u}} = 1.166$$

$$r_7 = 8.4 \text{ cm} \times 1.166 = 9.8 \text{ cm}$$

Discussion To solve this sort of problem, there aren't any new formulas to learn. We apply Newton's second law with the net force given by the magnetic force law ($\vec{\mathbf{F}}_B = q\vec{\mathbf{v}} \times \vec{\mathbf{B}}$) and the magnitude of the centripetal acceleration being what it always is for uniform circular motion (v^2/r).

If the direct proportion between r and m is not apparent, we could proceed by solving (1) for the radius:

$$r = \frac{mv^2}{|q|vB}$$

Now, if we set up a ratio between r_7 and r_6, all quantities except the masses cancel, yielding

$$\frac{r_7}{r_6} = \frac{m_7}{m_6}$$

Practice Problem 19.4   Ion speed

The magnetic field strength used in the mass spectrometer of Example 19.4 is 0.50 T. At what speed do the Li$^+$ ions move through the magnetic field? (Each ion has charge $q = +e$ and moves perpendicular to the field.)

Cyclotrons

Making The Connection:
cyclotron

Another device that was originally used in experimental physics but is now used frequently in the life sciences and medicine is the *cyclotron*, invented in 1929 by American physicist E. O. Lawrence (1901–1958). Figure 19.17 shows a schematic diagram of a proton cyclotron. The two hollow metal shells are called *dees* after their shape (like the letter "D"). An alternating potential difference is maintained between the dees to accelerate the protons. When the protons are inside one of the dees, there is no electric field acting on them; inside the conductor they are all at the same potential. However, the uniform magnetic field causes the protons to travel in a circular arc at constant speed. The potential difference alternates so that, whenever a proton reaches the gap between the dees, the dee toward which it moves is at a lower potential. Thus, the electric field between the dees gives the proton a little kick every time it crosses the gap. As the proton speed increases, the radius of its path increases. When protons reach the maximum radius of the dees, they are taken out of the cyclotron and the high-energy proton beam is used to bombard some target.

The fortunate coincidence that makes the cyclotron work is that, as the protons increase their speed and kinetic energy, the time it takes them to move around one complete circle stays constant (see Problems 24 and 25). It turns out that, as the speed increases, so does the distance they travel (the circumference of the circular path), so that the time for one revolution stays the same. The potential difference between the dees can then be made to alternate at this same frequency (the *cyclotron frequency*) so that the protons gain rather than lose kinetic energy each time they cross the gap.

In hospitals, cyclotrons produce some of the radioisotopes used in nuclear medicine (Fig. 19.18). While nuclear reactors also produce medical radioisotopes, cyclotrons offer certain advantages. For one thing, a cyclotron is much easier to operate and is much smaller—typically 1 m or less in radius. A cyclotron can be located in or adjacent to a hospital so that short-lived radioisotopes can be produced as they are needed. It would be difficult to try to produce short-lived isotopes in a nuclear reactor and transport them to the hospital fast enough for them to be useful. Cyclotrons also tend to produce different kinds of isotopes than do nuclear reactors.

Another medical use of the cyclotron is *proton beam radiosurgery*, where the cyclotron's proton beam is used as a surgical tool. Proton beam radiosurgery offers advantages over surgical and other radiological methods in the treatment of unusually shaped brain tumors. For one thing, doses to the surrounding tissue are much lower than with other forms of radiosurgery.

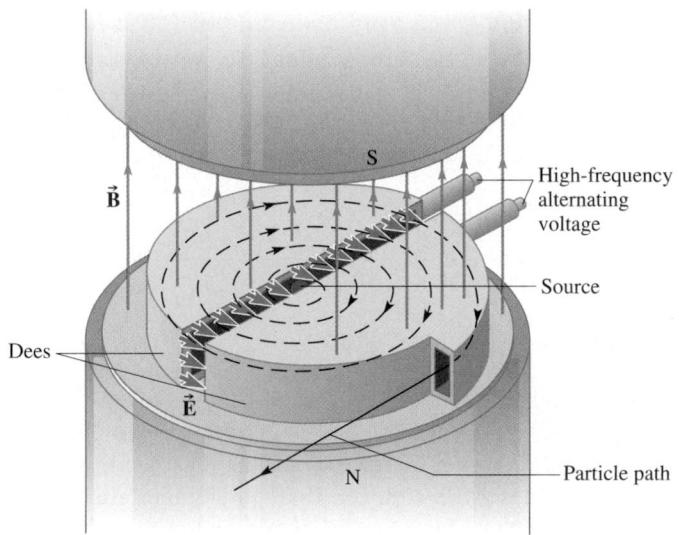

Figure 19.17 Schematic view of a cyclotron

Figure 19.18 A patient is prepared for surgery at the Northeast Proton Therapy Center of Massachusetts General Hospital. The protons are accelerated by a cyclotron.

Example 19.5

Maximum Kinetic Energy in a Proton Cyclotron

 A proton cyclotron uses a magnet that produces a 0.60-T field between its poles. The radius of the dees is 24 cm. What is the maximum possible kinetic energy of the protons accelerated by this cyclotron?

Strategy As a proton's kinetic energy increases, so does the radius of its path in the dees. The maximum kinetic energy is therefore determined by the maximum radius.

Solution While in the dees, the only force acting on the proton is magnetic. First we apply Newton's second law to a circular path.

$$F = |q|vB = \frac{mv^2}{r}$$

We can solve for v:

$$v = \frac{|q|Br}{m}$$

From v, we calculate the kinetic energy:

$$K = \frac{1}{2}mv^2 = \frac{1}{2}m\left(\frac{|q|Br}{m}\right)^2$$

For a proton, $q = +e$. The magnetic field strength is $B = 0.60$ T. For the maximum kinetic energy, we set the radius to its maximum value $r = 0.24$ m.

$$K = \frac{(qBr)^2}{2m} = \frac{(1.6 \times 10^{-19}\,\text{C} \times 0.60\,\text{T} \times 0.24\,\text{m})^2}{2 \times 1.67 \times 10^{-27}\,\text{kg}} = 1.6 \times 10^{-13}\,\text{J}$$

Discussion Just as in Example 19.4 (the mass spectrometer), this cyclotron problem is solved using Newton's second law. Once again the net force on the moving charge is given by the magnetic force law and the centripetal acceleration has magnitude v^2/r for motion at constant speed along the arc of a circle.

Practice Problem 19.5 Increasing kinetic energy in a proton cyclotron

Using the same magnetic field, what would the radius of the dees have to be to accelerate the protons to a kinetic energy of 1.6×10^{-12} J (ten times the previous value)?

Making The Connection:
synchrotron

Synchrotrons and Storage Rings

A cyclotron cannot accelerate particles to energies high enough for today's particle physics experiments. One reason is that, to achieve a higher particle energy, either the radius of the dees and the magnet must be larger (as illustrated by Practice Problem 19.5) or the magnetic field must be stronger. Both possibilities are difficult and expensive. A second reason is that, for particle speeds approaching the speed of light, Newton's second law no longer holds. When the motion of the charged particle is analyzed using Einstein's theory of relativity, the frequency of revolution *does* depend on speed at high speeds.

Most high-energy experiments today use particles accelerated by *synchrotrons* (or by storage rings, which are similar). In a synchrotron the particles move in a circular path of *fixed radius*. The magnetic field to bend the particles around in a circular path is adjusted according to the speed of the particles in a way that keeps the radius fixed. Synchrotrons have radii that are typically 1 km or more, as compared to the approximately 1 m typical of a cyclotron. Since the radius of the trajectory is fixed, magnets can be spaced at intervals around the ring of the synchrotron (Fig. 19.19). Accelerating stations are also spaced around the ring; as particles move through an accelerating station, they are accelerated by an electric field.

19.4   MOTION OF A CHARGED PARTICLE IN A UNIFORM MAGNETIC FIELD: GENERAL

What is the trajectory of a charged particle moving in a uniform magnetic field with no other forces acting? In Section 19.3, we saw that the trajectory is a circle *if* the velocity is perpendicular to the magnetic field. If $\vec{\mathbf{v}}$ is either parallel or antiparallel to $\vec{\mathbf{B}}$, then there is no perpendicular component. With $v_\perp = 0$, the magnetic force is zero and the particle moves at constant velocity.

In general, the velocity may have components both perpendicular to and parallel to the magnetic field. The component parallel to the field is constant, since the magnetic force is always perpendicular to the field. The particle therefore moves along a *helical* path. The helix is formed by circular motion of the charge in a plane perpendicular to the field superimposed onto motion of the charge at constant speed along a field line (Fig. 19.20a).

Even in nonuniform fields, charged particles tend to spiral around magnetic field lines. Above Earth's surface, charged particles from cosmic rays and the solar wind (charged particles streaming toward Earth from the Sun) are trapped by Earth's magnetic field. The particles spiral back and forth along magnetic field lines (Fig. 19.20b). Near the poles, the field lines are closer together, so the field is stronger. As the field strength increases, the radius of a spiraling particle's path gets smaller and smaller. As a result,

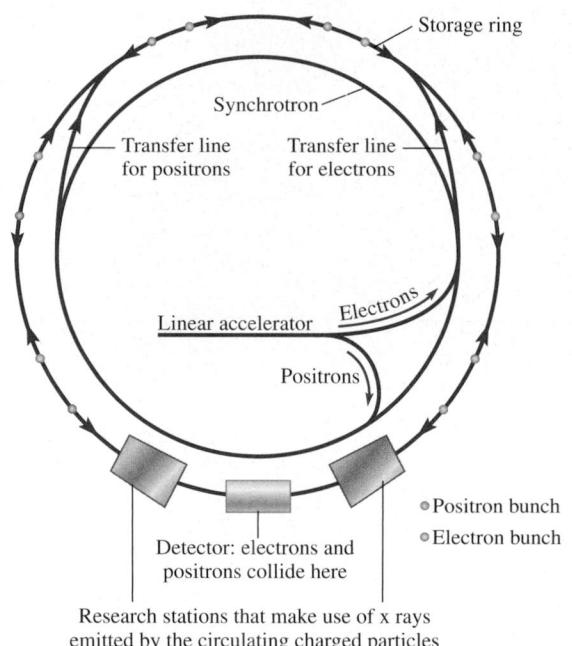

Storage ring

Synchrotron

Transfer line for positrons

Transfer line for electrons

Linear accelerator

Electrons

Positrons

Positron bunch

Electron bunch

Detector: electrons and positrons collide here

Research stations that make use of x rays emitted by the circulating charged particles

Figure 19.19 Schematic of a synchrotron

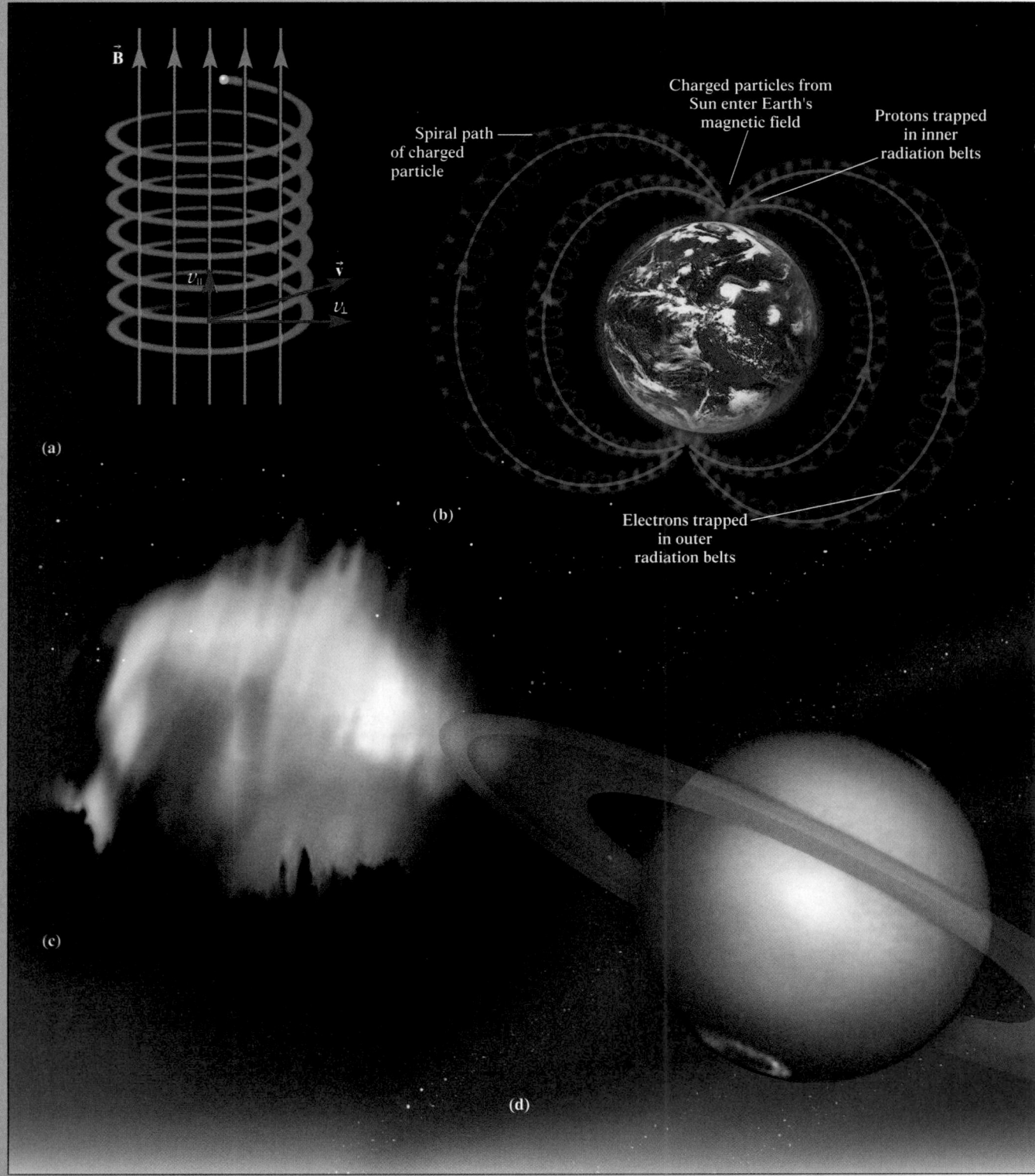

Figure 19.20 (a) Helical motion of a charged particle in a uniform magnetic field. (b) Charged particles spiral back and forth along field lines high in the atmosphere. (c) The *aurora borealis* over Lake Kenny (Alaska). (d) An ultraviolet image reveals aurorae at both poles of Saturn.

there is a concentration of these particles near the poles. The particles collide with and ionize air molecules. When the ions recombine with electrons to form neutral atoms, visible light is emitted—the *aurora borealis* in the northern hemisphere and the *aurora australis* in the southern hemisphere (Fig. 19.20c). Aurorae also occur on Jupiter and Saturn, which have much stronger magnetic fields than does the Earth (Fig. 19.20d).

Making The Connection:
aurorae on Earth,
Jupiter, and Saturn

19.5   A CHARGED PARTICLE IN CROSSED $\vec{\mathbf{E}}$ AND $\vec{\mathbf{B}}$ FIELDS

If a charged particle moves in a region of space where both electric and magnetic fields are present, then the electromagnetic force on the particle is the vector sum of the electric and magnetic forces:

$$\vec{\mathbf{F}} = \vec{\mathbf{F}}_E + \vec{\mathbf{F}}_B \tag{19-8}$$

A particularly important and useful case is when the electric and magnetic fields are perpendicular to each other and the velocity of a charged particle is perpendicular to both fields. Since the magnetic force is always perpendicular to both $\vec{\mathbf{v}}$ and $\vec{\mathbf{B}}$, it must be either parallel or antiparallel to the electric field and thus also to the electric force. If the magnitudes of the two forces are the same and the directions are opposite, then there is zero net force on the charged particle. For any particular combination of electric and magnetic fields, this balance of forces occurs only for one particular particle speed, since the magnetic force is velocity-dependent, while the electric force is not. The velocity that gives zero net force can be found from

The magnetic force is velocity-dependent, while the electric force is not.

$$\vec{\mathbf{F}} = \vec{\mathbf{F}}_E + \vec{\mathbf{F}}_B = 0$$

$$q\vec{\mathbf{E}} + q\vec{\mathbf{v}} \times \vec{\mathbf{B}} = 0$$

Dividing out the common factor of q,

$$\vec{\mathbf{E}} + \vec{\mathbf{v}} \times \vec{\mathbf{B}} = 0 \tag{19-9}$$

There is zero net force on the particle only if

$$v = \frac{E}{B} \tag{19-10}$$

and if the direction of $\vec{\mathbf{v}}$ is correct. Since $\vec{\mathbf{E}} = -\vec{\mathbf{v}} \times \vec{\mathbf{B}}$, it can be shown (see Conceptual Question 7) that the correct direction of $\vec{\mathbf{v}}$ is the direction of $\vec{\mathbf{E}} \times \vec{\mathbf{B}}$.

Velocity Selector

A **velocity selector** uses crossed electric and magnetic fields to select a single velocity out of a beam of charged particles. Suppose a beam of ions is produced in the first stage of a mass spectrometer. The beam may contain ions moving at a range of different speeds. If the second stage of the mass spectrometer is a velocity selector (Fig. 19.21),

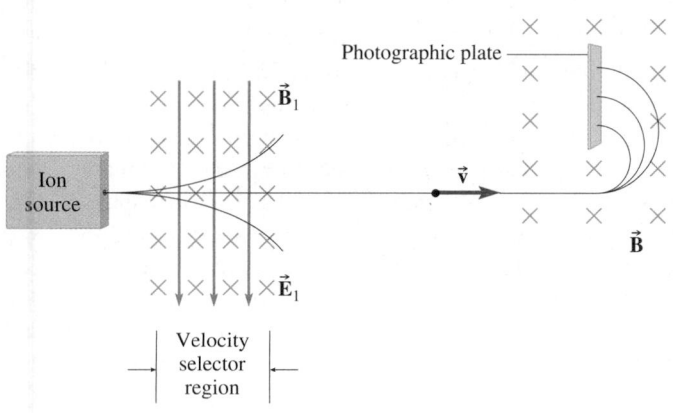

Figure 19.21 A mass spectrometer that uses a velocity selector to ensure that all ions enter the magnetic field with the same speed

only ions moving at a single speed $v = E/B$ pass through the velocity selector and into the third stage. The speed can be selected by adjusting the magnitudes of the electric and magnetic fields. For particles moving *faster* than the selected speed, the magnetic force is stronger than the electric force; fast particles curve out of the beam in the direction of the magnetic force. For particles moving *slower* than the selected speed, the magnetic force is weaker than the electric force; slow particles curve out of the beam in the direction of the electric force. The velocity selector ensures that only ions with speeds very near $v = E/B$ enter the magnetic sector of the mass spectrometer.

Example 19.6

Velocity Selector

A velocity selector is to be constructed to select ions moving to the right at 6.0 km/s. The electric field is 300.0 V/m into the page. What should be the magnitude and direction of the magnetic field?

Strategy First, in a velocity selector, $\vec{E}$, $\vec{B}$, and $\vec{v}$ are mutually perpendicular. That allows only two possibilities for the direction of $\vec{B}$. Setting the magnetic force equal and opposite to the electric force determines which of the two directions is correct and gives the magnitude of $\vec{B}$. The magnitude of the magnetic field is chosen so that the electric and magnetic forces on a particle moving at the given speed are exactly opposite.

Solution Since $\vec{v}$ is to the right and $\vec{E}$ is into the page, the magnetic field must either be up or down. The sign of the ions' charge is irrelevant—changing the charge from positive to negative would change the directions of *both* forces, leaving them still opposite to each other. For simplicity, then, we assume the charge to be positive.

The direction of the electric force on a positive charge is the same as the direction of the field, which here is into the page. Then we need a magnetic force that is out of the page. Using the right-hand rule to evaluate both possibilities for $\vec{B}$ (up and down), we find that $\vec{v} \times \vec{B}$ is out of the page if $\vec{B}$ is up.

The magnitudes of the forces must also be equal:

$$qE = qvB$$

$$B = \frac{E}{v} = \frac{300.0 \text{ V/m}}{6000 \text{ m/s}} = 0.050 \text{ T}$$

Discussion Let's check the units; is a tesla really equal to $\frac{\text{V/m}}{\text{m/s}}$? From $\vec{F} = q\vec{v} \times \vec{B}$, we can reconstruct the tesla:

$$[B] = \text{T} = \left[\frac{F}{qv}\right] = \frac{\text{N}}{\text{C·m/s}}$$

Recall that two equivalent units for electric field are N/C = V/m. By substitution,

$$\text{T} = \frac{\text{V}}{\text{m}^2/\text{s}} = \frac{\text{V/m}}{\text{m/s}}$$

so the units check out.

Another check: for a velocity selector the correct direction of $\vec{v}$ is the direction of $\vec{E} \times \vec{B}$. The velocity is to the right. Using the right-hand rule, $\vec{E} \times \vec{B}$ is to the right if $\vec{B}$ is up.

Practice Problem 19.6 Deflection of a particle moving too fast

If a particle enters this velocity selector with a speed greater than 6.0 km/s, in what direction is it deflected out of the beam?

The velocity selector can be used to determine the charge to mass ratio q/m of a charged particle. First, the particle is accelerated from rest through a potential difference ΔV, converting electric potential energy into kinetic energy. The change in its electric potential energy is $\Delta U = q\,\Delta V$, so the charge acquires a kinetic energy

$$K = \tfrac{1}{2}mv^2 = -q\,\Delta V$$

(K is positive regardless of the sign of q: a positive charge is accelerated by decreasing its potential, while a negative charge is accelerated by increasing its potential.) Now a velocity selector is used to determine the speed $v = E/B$, by adjusting the electric and magnetic fields until the particles pass straight through. The charge-to-mass ratio q/m can now be determined (see Problem 27). In 1897, British physicist J. J. Thomson (1856–1940) used this technique to show that "cathode rays" are charged particles. In a vacuum tube, he maintained two electrodes at a potential difference of a few thousand volts (Fig. 19.22) so that cathode rays were emitted by the negative electrode (the cathode). By measuring the charge-to-mass ratio, Thomson established that cathode rays are streams of negatively charged particles that all have the same charge-to-mass ratio—particles we now call *electrons*.

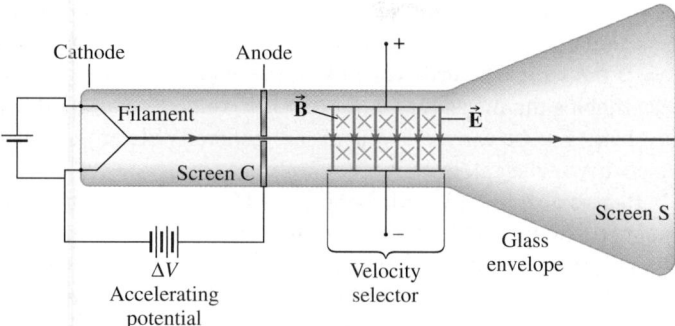

Figure 19.22 Modern apparatus, similar in principle to the one used by Thomson, to find the charge-to-mass ratio of the electron. Electrons emitted from the cathode are accelerated toward the anode by the electric field between the two. Some of the electrons pass through the anode and then enter a velocity selector. The deflection of the electrons is viewed on the screen. The electric and magnetic fields in the velocity selector are adjusted until the electrons are not deflected.

Electromagnetic Flowmeter

The principle of the velocity selector finds another application in the electromagnetic flowmeters used to measure the speed of blood flow through a major artery during cardiovascular surgery. Blood contains ions; the motion of the ions can be affected by a magnetic field. In an electromagnetic flowmeter, a magnetic field is applied perpendicular to the flow direction. The magnetic force on positive ions is toward one side of the artery, while the magnetic force on negative ions is toward the opposite side (Fig. 19.23a). This separation of charge, with positive charge on one side and negative charge on the other, produces an electric field across the artery (Fig. 19.23b). As the electric field builds up, it exerts a force on moving ions in a direction opposite to that of the magnetic field. In equilibrium, the two forces are equal in magnitude:

$$F_E = F_B$$
$$qE = qvB$$
$$E = vB$$

where v is the average speed of an ion, equal to the average speed of the blood flow. Thus, the flowmeter is just like a velocity selector, except that the ion speed determines the electric field instead of the other way around.

A voltmeter is attached to opposite sides of the artery to measure the potential difference. From the potential difference, we can calculate the electric field; from the electric field and magnetic field magnitudes, we can determine the speed of blood flow. A great advantage of the electromagnetic flowmeter is that it does not involve anything being inserted into the artery.

The Hall Effect

The **Hall effect** is similar to the electromagnetic flowmeter, but pertains to the moving charges in a current-carrying wire or other solid instead of to the moving ions in blood. A magnetic field perpendicular to the wire causes the moving charges to be deflected to

Making The Connection:
electromagnetic blood flowmeter

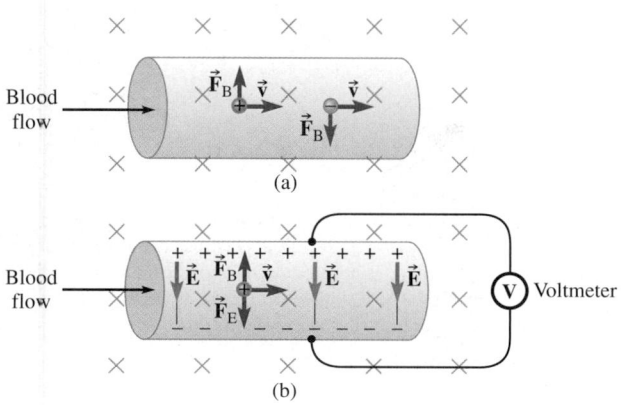

Figure 19.23 Principles behind the electromagnetic blood flowmeter. (a) When a magnetic field is applied perpendicular to the direction of blood flow, positive and negative ions are deflected toward opposite sides of the artery. (b) As the ions are deflected, an electric field develops across the artery. In equilibrium, the electric force on an ion due to this field is equal and opposite to the magnetic force; the ions move straight down the artery with an average velocity of magnitude $v = E/B$.

one side. An electric field then exists across the wire. The potential difference (or **Hall voltage**) *across* the wire is measured and then used to calculate the electric field (or **Hall field**) across the wire. The drift velocity of the charges is then given by $v_D = E/B$. The Hall effect enables the measurement of the drift velocity and the determination of the sign of the charges. (The carriers in metals are generally electrons, but semiconductors may have positive or negative carriers or both.)

The Hall effect is also the principle behind the **Hall probe**, a common device used to measure magnetic fields. As shown in Example 19.7, the Hall voltage across a conducting strip is proportional to the magnetic field strength. A circuit causes a fixed current flow through the strip. The probe is then calibrated by measuring the Hall voltage caused by magnetic fields of known strength. Once calibrated, measurement of the Hall voltage enables a quick and accurate determination of magnetic field strengths.

Example 19.7

Hall Effect

A flat slab of semiconductor has thickness $t = 0.50$ mm, width $w = 1.0$ cm, and length $L = 30.0$ cm. A current $I = 2.0$ A flows along its length to the right (Fig. 19.24a). A magnetic field $B = 0.25$ T is directed into the page, perpendicular to the flat surface of the slab. Assume that the carriers are electrons. There are 7.0×10^{24} mobile electrons per m^3. (a) What is the magnitude of the Hall voltage across the slab? (b) Which edge (top or bottom) is at the higher potential?

Strategy   We need to find the drift velocity of the electrons from the relation between current and drift velocity. Since the Hall field is uniform, the Hall voltage is the Hall field times the width of the slab.

Given: current $I = 2.0$ A, magnetic field $B = 0.25$ T, thickness $t = 0.50 \times 10^{-3}$ m, width $w = 0.010$ m, $n = 7.0 \times 10^{24}$ electrons/m^3

Solution   (a) The drift velocity is related to the current:

$$I = neAv_D \qquad (18\text{-}3)$$

The area is the width times the thickness of the slab:

$$A = wt$$

Solving for the drift velocity,

$$v_D = \frac{I}{newt}$$

We find the Hall field by setting the magnitude of the magnetic force equal to the magnitude of the electric force caused by the Hall field across the slab:

$$F_E = eE_H = F_B = ev_DB$$

$$E_H = v_DB$$

The Hall voltage is

$$V_H = E_Hw = Bv_Dw$$

Substituting the expression for drift velocity,

$$V_H = \frac{BIw}{newt} = \frac{BI}{net}$$

$$= \frac{0.25 \text{ T} \times 2.0 \text{ A}}{7.0 \times 10^{24} \text{ m}^{-3} \times 1.6 \times 10^{-19} \text{ C} \times 0.50 \times 10^{-3} \text{ m}}$$

$$= 0.89 \text{ mV}$$

(b) Since the current flows to the right, the electrons actually move to the left. Figure 19.24b shows that the magnetic force on an electron moving to the left is upward. The magnetic force deflects electrons toward the top of the slab, leaving the bottom with a positive charge. An upward electric field is set up across the slab (Fig. 19.24c). Therefore, the bottom edge is at the higher potential.

Figure 19.24
(a) Measuring the Hall voltage. (b) Magnetic force on an electron moving to the left. (c) With electrons deflected toward the top of the slab, the top is negatively charged and the bottom is positively charged. The Hall field in this case is directed upward, from the positive charges to the negative charges.

continued on next page

Example 19.7 *continued*

Discussion The width of the slab w does not appear in the final expression for the Hall voltage $V_H = BI/(net)$. Is it possible that the Hall voltage is independent of the width? If the slab were twice as wide, for instance, the same current means half the drift velocity v_D since the number of carriers per unit volume n and their charge magnitude e cannot change. With the carriers moving half as fast on average, the average magnetic force is half. Then in equilibrium, the electric force is half, which means the field is half. An electric field half as strong times a width twice as wide gives the same Hall voltage.

Practice Problem 19.7   Holes as carriers

If the carriers had been particles with charge $+e$ instead of electrons, with everything else the same, would the Hall voltage have been any different? Explain.

19.6 MAGNETIC FORCE ON A CURRENT-CARRYING WIRE

A wire carrying electric current has many moving charges in it. For a current-carrying wire in a magnetic field, the magnetic forces on the individual moving charges add up to produce a net magnetic force on the wire. Although the average force on one of the charges may be small, there are so many charges that the net magnetic force on the wire can be appreciable.

Say a straight wire segment of length L in a uniform magnetic field $\vec{B}$ carries a current I. The mobile carriers have charge q. The magnetic force on any one charge is

$$\vec{F} = q\vec{v} \times \vec{B}$$

where $\vec{v}$ is the instantaneous velocity of that charge. The net magnetic force on the wire is the vector sum of these forces. The sum isn't easy to carry out, since we don't know the instantaneous velocity of each of the charges. The charges move about in random directions at high speeds; their velocities suffer large changes when they collide with other particles. Instead of summing the instantaneous magnetic force on each charge, we can instead multiply the *average* magnetic force on each charge by the number of charges. Since each charge has the same average velocity—the drift velocity—each experiences the same average magnetic force $\vec{F}_{av}$.

$$\vec{F}_{av} = q\vec{v}_D \times \vec{B}$$

Then, if N is the total number of carriers in the wire, the total magnetic force on the wire is

$$\vec{F} = Nq\vec{v}_D \times \vec{B} \tag{19-11}$$

Equation (19-11) can be rewritten in a more convenient way. Instead of having to figure out the number of carriers and the drift velocity, it is more convenient to have an expression that gives the magnetic force in terms of the current I. The current I is related to the drift velocity:

$$I = nqAv_D \tag{18-3}$$

Here n is the number of carriers *per unit volume*. If the length of the wire is L and the cross-sectional area is A, then

$$N = \text{number per unit volume} \times \text{volume} = nLA$$

By substitution, the magnetic force on the wire can be written

$$\vec{F} = Nq\vec{v}_D \times \vec{B} = nqAL\vec{v}_D \times \vec{B}$$

Almost there! Since current is not a vector, we cannot substitute $\vec{I} = nqA\vec{v}_D$. Therefore, we define a *length vector* $\vec{L}$ to be a vector in the direction of the current with magnitude equal to the length of the wire. Then $nqAL\vec{v}_D = I\vec{L}$ and

---

Magnetic force on a straight segment of current-carrying wire

$$\vec{F} = I\vec{L} \times \vec{B} \tag{19-12a}$$

The current I times the cross product $\vec{L} \times \vec{B}$ gives the magnitude and direction of the force. The magnitude of the force is

$$F = IL_\perp B = ILB_\perp = ILB \sin \theta \qquad \text{(19-12b)}$$

The direction of the force is perpendicular to both $\vec{L}$ and $\vec{B}$. The same right-hand rule used for any cross product is used to choose between the two possibilities.

Problem-Solving Technique: Finding the Magnetic Force on a Straight Segment of Current-Carrying Wire

1. The magnetic force is zero if (a) the current in the wire is zero, (b) the wire is parallel to the magnetic field, or (c) the magnetic field is zero.
2. Otherwise, determine the angle θ between $\vec{L}$ and $\vec{B}$ when the two are drawn tail-to-tail.
3. Find the magnitude of the force from Eq. (19-12b).
4. Determine the two directions that are perpendicular to a plane that contains both $\vec{L}$ and $\vec{B}$.
5. Determine the direction of $\vec{L} \times \vec{B}$ by using the right-hand rule to choose between the two possibilities.

Example 19.8

Magnetic Force on Power Line

A 125-m-long power line is horizontal and carries a current of 2500 A toward the south. The Earth's magnetic field at that location is 0.52 mT, toward the north and inclined 62° below the horizontal (Fig. 19.25a). What is the magnetic force on the power line? (Ignore any drooping of the wire; assume it's straight.)

Strategy We are given all the quantities necessary to calculate the force:
$I = 2500$ A;
$\vec{L}$ has magnitude 125 m and direction south;
$\vec{B}$ has magnitude 0.52 mT. It has a downward component and a northward component.
We find the cross product $\vec{L} \times \vec{B}$ and then multiply by I.

Solution The magnitude of the force is given by
$$F = IL_\perp B = ILB_\perp$$

The second form is more convenient here, since $\vec{L}$ is southward. The perpendicular component of $\vec{B}$ is the vertical component, which is $B \sin 62°$ (see Fig. 19.25b). Then

$F = ILB \sin 62° = 2500\,\text{A} \times 125\,\text{m} \times 5.2 \times 10^{-4}\,\text{T} \times \sin 62°$

$\quad = 140\,\text{N}$

Figure 19.25b shows the vectors $\vec{L}$ and $\vec{B}$ sketched in the north/south–up/down plane. Since north is to the right, this is a view looking toward the west. The cross product $\vec{L} \times \vec{B}$ is out of the page by the right-hand rule. Therefore, the direction of the force is east.

Discussion The hardest thing in this sort of problem is choosing a plane in which to sketch the vectors. Here we chose a plane in which we could draw both $\vec{L}$ and $\vec{B}$; then the cross product has to be perpendicular to this plane.

Practice Problem 19.8   Magnetic force on a current-carrying wire

A vertical wire carries 10.0 A of current upward. What is the direction of the magnetic force on the wire if the magnetic field is the same as in Example 19.8?

(a)

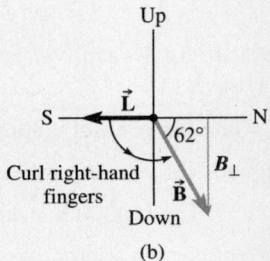

(b)

Figure 19.25
(a) The wire and the magnetic field vector. (b) The vectors $\vec{L}$ and $\vec{B}$ sketched in a vertical plane. The cross product of the two must then be perpendicular to this plane—either east (out of the page) or west (into the page). The right-hand rule enables us to choose between the two possibilities.

19.7   TORQUE ON A CURRENT LOOP

Consider a rectangular loop of wire carrying current I in a uniform magnetic field $\vec{B}$. In Fig. 19.26a, the field is parallel to sides 1 and 3 of the loop. There is no magnetic force on sides 1 and 3 since $\vec{B}$ is either parallel or antiparallel to $\vec{L}$ for each. The forces on sides 2 and 4 are equal in magnitude and opposite in direction. There is no net magnetic force on the loop, but the lines of action of the two forces are offset by a distance b, so there is a nonzero net torque. The torque tends to make the loop rotate about a central axis in the direction indicated in Fig. 19.26a. The magnitude of the magnetic force on sides 2 and 4 is

$$F = ILB = IaB$$

The lever arm for each of the two forces is $\frac{1}{2}b$, so the torque due to each is

$$\text{magnitude of force} \times \text{lever arm} = F \times \tfrac{1}{2}b = \tfrac{1}{2}IabB$$

Then the total torque on the loop is $\tau = IabB$. The area of the rectangular loop is $A = ab$, so

$$\tau = IAB$$

If, instead of a single turn, there are N turns forming a coil, then the magnetic torque on the coil is

$$\tau = NIAB \tag{19-13a}$$

Equation (19-13a) holds for a planar loop or coil of *any* shape (see Problem 47).

What if the field is not parallel to the plane of the coil? In Fig. 19.26b, the same loop has been rotated about the axis shown. The angle θ is the angle between the magnetic field and a line *perpendicular* to the current loop. Which perpendicular direction is

Figure 19.26 (a) A rectangular coil of wire in a uniform magnetic field. The current in the coil (counterclockwise as viewed from the top) causes a magnetic torque, which is clockwise as viewed from the front. (b) Side view of the same coil after it has been rotated in the field. The current in side 4 comes out of the page, along side 1 (diagonally down the page), and back into the page in side 2. The lever arms of the forces on sides 2 and 4 are now smaller: $\frac{1}{2}b \sin \theta$ instead of $\frac{1}{2}b$. The torque is then smaller by the same factor ($\sin \theta$). (c) Using the right-hand rule to choose the perpendicular direction from which θ is measured.

determined by the right-hand rule: take the cross product of $\vec{L}$ for any side with $\vec{L}$ for the next side, going around in the same direction as the current (Fig. 19.26c). Before, when the field was in the plane of the loop, θ was 90°. For $\theta \neq 90°$, the magnetic forces on sides 1 and 3 are no longer zero, but they are equal and opposite and act along the same line of action, so they contribute neither to the net force nor to the net torque. The magnetic forces on sides 2 and 4 are the same as before, but now the lever arms are smaller by a factor of sin θ: instead of $\frac{1}{2}b$, the lever arms are now $\frac{1}{2}b$ sin θ. Therefore,

---

Torque on a current loop

$$\tau = NIAB \sin \theta \qquad (19\text{-}13b)$$

The torque has maximum magnitude if the field is in the plane of the coil ($\theta = 90°$ or 270°). If $\theta = 0°$ or 180°, the field is perpendicular to the plane of the loop and the torque is zero. There are *two* positions of rotational equilibrium, but they are not equivalent. The position at $\theta = 180°$ is an *unstable* equilibrium, because at angles *near* 180° the torque tends to rotate the coil *away* from 180°. The position at $\theta = 0°$ is a *stable* equilibrium; the torque for angles *near* 0° makes the coil rotate back toward $\theta = 0°$ and thus tends to restore the equilibrium.

The torque on a current loop in a uniform magnetic field is analogous to the torque on an electric dipole in a uniform electric field (see Problem 46). This similarity is our first clue that

---

a current loop is a magnetic dipole.

The direction perpendicular to the loop chosen by the right-hand rule is the direction of the **magnetic dipole moment vector**. The dipole moment vector points from the dipole's south pole toward its north pole. (By comparison, the *electric* dipole moment vector points from the electric dipole's negative charge toward its positive charge.)

Making The Connection:
electric motor (DC)

Electric Motor

In a simple DC motor, a coil of wire is free to rotate between the poles of a permanent magnet (Fig. 19.27). When current flows through the loop, the magnetic field exerts a torque on the loop. If the direction of the current in the coil doesn't change, then the coil just oscillates about the stable equilibrium orientation ($\theta = 0°$). To make a motor we need the coil to keep turning in the same direction. The trick used to make a DC motor is to automatically reverse the direction of the current as soon as the coil passes $\theta = 0°$. In effect, just as the coil goes through the stable equilibrium orientation, we reverse the current to make the coil's orientation an *unstable* equilibrium. Then, instead of pulling

Figure 19.27 Simple DC motor. The two sides of the rotating coil are labeled *A* and *B*. In position (a) the coil rotates away from unstable equilibrium. The direction of the current in the coil causes side *A* to be a south magnetic pole and *B* to be a north magnetic pole. In position (b) brushes pass over the split in the commutator, interrupting the flow of current. If the current in the coil still flowed in the same direction as in (a), this would be stable equilibrium. When the coil turns a little farther, in position (c), the brushes reverse the direction of the current. Now side *A* is a north pole and side *B* is a south pole. The coil is pushed away from *unstable* equilibrium rather than pulled back toward stable equilibrium.

(a)   (b)   (c)

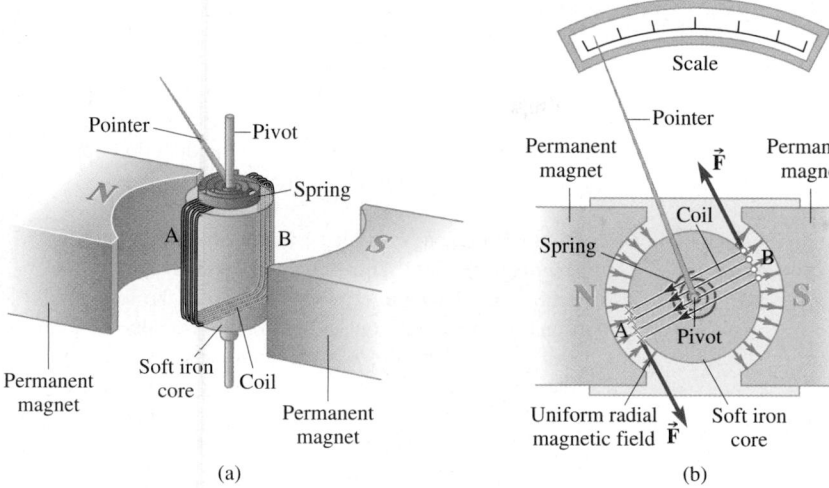

(a)

(b)

Figure 19.28   Two views of a galvanometer. (a) The galvanometer coil in a magnetic field. (b) Top view of coil with current directions shown.

the coil backward toward the (stable) equilibrium, the torque keeps turning the coil in the same direction by pushing it *away from* (unstable) equilibrium.

To reverse the current, the source of current is connected to the coil of the motor by means of two brushes (Fig. 19.27b). The brushes make electric contact with the *commutator*, which rotates with the coil. The commutator is a split ring with each side connected to one end of the coil. Every time the brushes pass over the split, the current to the coil is reversed.

Galvanometer

The magnetic torque on a current loop is also the principle behind the operation of a galvanometer—a sensitive device used to measure current. A rectangular coil of wire is placed between the poles of a magnet (Fig. 19.28a). The shape of the magnet's pole faces keeps the field acting on sides A and B of the coil perpendicular to the wires and constant in magnitude regardless of the angle of the coil, so that the torque does not depend on the angle of the coil. A hairspring provides a restoring torque that is proportional to the angular displacement of the coil. When a current passes through the coil, the magnetic torque is proportional to the current [Eqs. (19-13)]. The coil rotates until the restoring torque due to the spring is equal in magnitude to the magnetic torque. Thus, the angular displacement of the coil is proportional to the current in the coil. A pointer attached to the coil indicates the current on a calibrated scale. An analog ammeter is constructed by wiring a resistor in parallel with a galvanometer; an analog voltmeter is constructed by wiring a resistor in series with a galvanometer (see Section 18.9).

Making The Connection:
galvanometer

Conceptual Example 19.9

Force and Torque on Galvanometer Coil

Show that (a) there is zero net magnetic force on the pivoted coil in the galvanometer of Fig. 19.28; (b) there is a net torque; and (c) the torque is in the correct direction to swing the pointer in the plane of the page. (d) Determine which direction the current in the coil must flow to swing the pointer to the right. Assume that the magnetic field is radial and has uniform *magnitude* in the space between the magnet pole faces and the iron core and that the field is zero in the vicinity of the two sides of the coil that cross over the iron core.

Strategy   Since we do not know the direction of the current, we pick one arbitrarily; in part (d) we will find out whether the choice was correct. Only the two sides of the coil near the magnet pole faces experience magnetic forces, since the other two sides are in zero field.

Solution   We choose the current in the side near the north pole to flow into the page. The current must then flow out of

continued on next page

Conceptual Example 19.9 continued

the page in the side of the coil near the south pole. In Fig. 19.28b, the current directions are marked with symbols • and ×, which also represent the directions of the $\vec{L}$ vectors used to find the magnetic force. The magnetic field vectors are also shown. Note that, since the direction of the field is radial, the two magnetic vectors are the same (same direction *and* magnitude). The direction of the magnetic force on either side is given by

$$\vec{F} = N I \vec{L} \times \vec{B}$$

where N stands for the number of turns of wire in the coil. The force vectors are shown on Fig. 19.28b.

(a) Since the $\vec{B}$ vectors are the same and the $\vec{L}$ vectors are equal and opposite (same length but opposite direction), the forces are equal and opposite. Then the net magnetic force on the coil is zero. (b) The net torque is not zero because the lines of action of the forces are separated. (c) The forces make the pointer rotate counterclockwise in the plane of the page. (d) Since the meter shows positive current by rotating clockwise,

we have chosen the wrong direction for the current. The leads of the galvanometer should be attached so that positive current makes the current in the coil flow in the direction opposite to the one we chose initially.

Discussion   Equations (19-13a,b) were derived for a coil in a *uniform* magnetic field. They do not necessarily apply here. In particular, the galvanometer works because the torque is proportional to the current but independent of the orientation of the coil. In Eq. (19-13b), θ is the angle between the magnetic field and a line perpendicular to the coil. In the galvanometer, the magnetic field acting on the coil is always in the plane of the coil; in essence θ is a constant 90° even while the coil swings about the pivot.

Practice Problem 19.9   Torque on a coil

Starting with the magnetic forces on the sides of the coil, show that the torque on the coil *is* given by Eq. (19-13a).

Making The Connection:
audio speakers

Audio Speakers

In contrast to a coil in a uniform field, a coil of wire in a *radial* magnetic field may experience a nonzero net magnetic *force*. A coil in a radial field is the principle behind the operation of many audio speakers (Fig. 19.29a). An electric current passes through a coil of wire. The coil sits between the poles of a magnet shaped so that the magnetic field is radial (Fig. 19.29b). Even though the coil is not a straight wire, the field direction is such that the force on every part of the coil is in the same direction. Since the field is everywhere perpendicular to the wire, the magnetic force is

$$F = ILB$$

where L is the total length of the coil of wire. If the coil has N turns of radius r, then $L = 2\pi r N$ and

$$F = 2\pi r N I B$$

A spring-like mechanism exerts a linear restoring force on the coil so that when a magnetic force acts, the displacement of the coil is proportional to the magnetic force, which in turn is proportional to the current in the coil. Thus, the motion of the coil—and the motion of the attached cone—mirrors the current sent through the speaker by the amplifier.

19.8   MAGNETIC FIELD DUE TO AN ELECTRIC CURRENT

So far we have explored the magnetic forces acting on charged particles and current-carrying wires. We have not yet looked at *sources* of magnetic fields other than permanent magnets. It turns out that *any moving charged particle* creates a magnetic field. There is a certain symmetry about the situation:

- Moving charges experience magnetic forces and moving charges create magnetic fields;
- Charges at rest feel no magnetic forces and create no magnetic fields;
- Charges feel electric forces and create electric fields, whether moving or not.

Figure 19.29 (a) Simplified sketch of a loudspeaker. (b) The loudspeaker coil between the poles of the magnet. Compare to Fig. 19.28 to see how the radial magnetic fields and the coil orientations differ. In the case of the loudspeaker there is a net force on the coil; in the galvanometer there is zero net force, but a nonzero net torque on the coil.

Today we know that electricity and magnetism are closely intertwined. It may be surprising to learn that they were not known to be related until the nineteenth century. Hans Christian Oersted discovered in 1820 by happy accident that electric currents flowing in wires made nearby compass needles swing around. Oersted's discovery was the first evidence of a connection between electricity and magnetism.

The magnetic field due to a single moving charged particle is negligibly small in most situations. However, when an electric current flows in a wire, there are enormous numbers of moving charges. The magnetic field due to the wire is the sum of the magnetic fields due to each charge; the principle of superposition applies to magnetic fields just as it does to electric fields.

Magnetic Field due to a Long Straight Wire

Let us first consider the magnetic field due to a long, straight wire carrying a current I. What is the magnetic field at point P, a distance r from the wire? Point P is near the wire and far from its ends. Figure 19.30a is a photo of such a wire, passing through a glass plate on which iron filings have been sprinkled. The iron bits line up with the magnetic field due to the current in the wire. The photo suggests that the magnetic field lines are circles centered on the wire. Circular field lines are indeed the only possibility, given the symmetry of the situation. If the lines were any other shape, they would be farther from the wire in some directions than in others (Fig. 19.30b). However, the wire has axial or cylindrical symmetry, so the lines have to be the same distance from the wire on all sides.

(a)

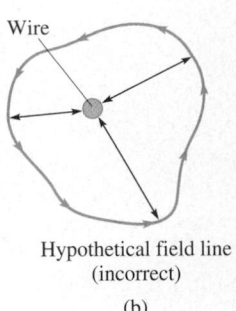

Wire

Hypothetical field line
(incorrect)

(b)

Figure 19.30 Magnetic field due to a long straight wire. (a) Photo of a long wire, with iron filings lining up with the magnetic field. (b) Any field line shape other than circular wouldn't make sense; we expect the magnitude of the field to be the same at equal distances in any direction. (c) Compasses show the direction of the field. (d) Sketch illustrating how to use the right-hand rule to determine the direction of the field lines.

(c)

(d)

The iron filings do not tell us the direction of the field. By using compasses instead of iron filings (Fig. 19.30c), the direction of the field is revealed—it is the direction indicated by the north end of each compass. The field lines due to the wire are shown in Fig. 19.30d, where the current in the wire flows upward. A second right-hand rule relates the current direction in the wire to the direction of the field around the wire:

Using Right-Hand Rule 2 (to Find the Direction of the Field due to a Current-Carrying Wire)

1. Point the thumb of the right hand in the direction of the current in the wire.
2. Curl the fingers inward toward the palm; the direction that the fingers curl is the direction of the magnetic field lines around the wire (Fig. 19.30d).

The magnitude of the magnetic field at a distance r from the wire is given by

Magnetic field due to a long straight wire
$$B = \frac{\mu_0 I}{2\pi r} \qquad\qquad (19\text{-}14)$$

where I is the current in the wire and μ_0 is a universal constant known as the **permeability of free space**. The permeability plays a role in magnetism similar to the role of the permittivity (ϵ_0) in electricity. In SI units, the value of μ_0 is

$$\mu_0 = 4\pi \times 10^{-7} \, \frac{\text{T·m}}{\text{A}} \quad \text{(exact, by definition)} \qquad (19\text{-}15)$$

Two parallel current-carrying wires that are close together exert magnetic forces on one another. The magnetic field of wire 1 causes a magnetic force on wire 2; the magnetic field of wire 2 causes a magnetic force on wire 1. From Newton's third law, we expect the forces on the wires to be equal and opposite. If the currents flow in the same direction, the force is attractive; if they flow in opposite directions, the force is repulsive (see Problem 56). Note that for current-carrying wires, "likes" (currents in the same direction) *attract* each other and "unlikes" (currents in opposite directions) *repel* each other.

The constant μ_0 can be assigned an exact value because the magnetic forces on two parallel wires are used to *define* the ampere, which is an SI base unit. One ampere is the current in each of two long parallel wires 1 m apart such that each exerts a magnetic force on the other of exactly 2×10^{-7} N per meter of length. The ampere, not the coulomb, is chosen to be an SI base unit because it can be defined in terms of forces and lengths that can be measured accurately. The coulomb is then defined as 1 ampere-second.

Example 19.10

Magnetic Field due to Household Wiring

In household wiring, two long parallel wires are separated and surrounded by an insulator. The wires are a distance d apart and carry currents of magnitude I in opposite directions. (a) Find the magnetic field at a distance $r \gg d$ from the center of the wires (point P in Fig. 19.31a). (b) Find the numerical

value of B if $I = 5$ A, $d = 5$ mm, and $r = 1$ m and compare to the Earth's magnetic field strength at the surface ($\sim 5 \times 10^{-4}$ T).

Strategy   The magnetic field is the vector sum of the fields due to each of the wires. The fields due to the wires at P are

continued on next page

Example 19.10 *continued*

equal in magnitude (since the currents and distances are the same), but the directions are not the same. Equation (19-14) gives the magnitude of the field due to either wire. Since the field lines due to a single long wire are circular, the direction of the field is tangent to a circle that passes through P and whose center is on the wire. The right-hand rule determines which of the two tangent directions is correct.

Solution (a) Since $r \gg d$, the distance from either wire to P is approximately r (see Fig. 19.31b). Then the magnitude of the field at P due to either wire is

$$B \approx \frac{\mu_0 I}{2\pi r}$$

In Fig. 19.31b, we draw radial lines from each wire to point P. The direction of the magnetic field due to a long wire is tangent to a circle and therefore perpendicular to a radius. Using the right-hand rule, the field directions are as shown in Fig. 19.31b. The y-components of the two field vectors add to zero; the x-components are the same:

$$B_x = \frac{\mu_0 I}{2\pi r} \sin \theta$$

Since $r \gg d$,

$$\sin \theta = \frac{\text{opposite}}{\text{hypotenuse}} \approx \frac{\frac{1}{2}d}{r}$$

The total magnetic field due to the two wires is in the +x-direction and has magnitude

$$B_{\text{net}} = 2B_x = \frac{\mu_0 I d}{2\pi r^2}$$

(b) By substitution,

$$B = \frac{\mu_0}{2\pi} \times \frac{Id}{r^2} = 2 \times 10^{-7} \frac{\text{T·m}}{\text{A}} \times \frac{5\,\text{A} \times 0.005\,\text{m}}{(1\,\text{m})^2} = 5 \times 10^{-9}\,\text{T}$$

The field due to the wires is 10^{-5} times the Earth's field.

Discussion The field due to the two wires decreases with distance proportional to $1/r^2$. It falls off much faster with distance than does the field due to a single wire, which is proportional to $1/r$. With equal currents flowing in opposite directions, we have a net current of zero. The only reason the field isn't zero is the small distance between the two wires.

Since the current in household wiring actually alternates at 60 Hz, so does the field. If 5.0 A is the maximum current, then 5×10^{-9} T is the maximum field strength.

Practice Problem 19.10   Field midway between two wires

Find the magnetic field at a point halfway between the two wires in terms of I and d.

(a)

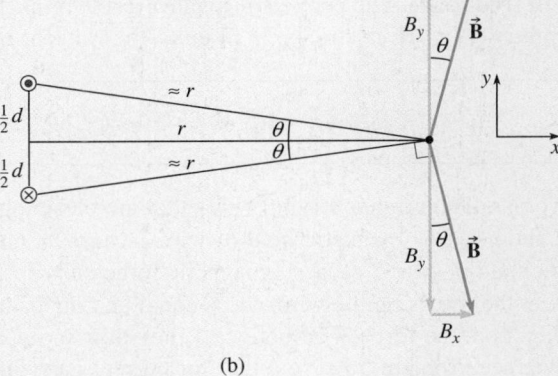

(b)

Figure 19.31
(a) The two wires are perpendicular to the plane of the page. They are marked to show that the current in the upper wire flows out of the page and the current in the lower wire flows into the page. (b) Field vectors due to each wire.

Magnetic Field due to a Current Loop

In Section 19.7 we saw the first clue that a loop of wire that carries current around in a complete circuit is a magnetic dipole. A second clue comes from the magnetic field produced by a current loop. As for a straight wire, the magnetic field lines circulate around the wire, but for a circular current loop, the field lines are not circular. The field lines are more concentrated inside the current loop and less concentrated outside (Fig. 19.32a). The field lines emerge from one side of the current loop (the north pole) and reenter the other side (the south pole). Thus, the field due to a current loop is similar to the field of a short bar magnet.

The direction of the field lines is given by right-hand rule 2. Point the thumb in the direction of the current in the loop at any point. Then curl the fingers of the right hand to find the direction of the field near that point (Fig. 19.32b).

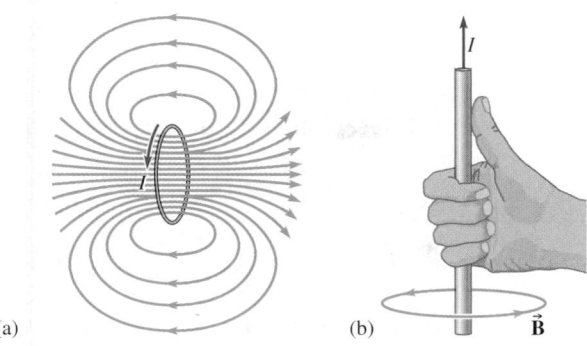

(a)   (b) $\vec{B}$

Figure 19.32 (a) Magnetic field lines due to a circular current loop. (b) Using right-hand rule 2 to determine the direction of the field inside the loop.

The magnetic fields due to coils of current-carrying wire are used in televisions and computer monitors to deflect the electron beam so that it lands on the screen in the desired spot.

Magnetic Field due to a Solenoid

An important source of magnetic field is that due to a **solenoid** because the field inside a solenoid is nearly uniform. In magnetic resonance imaging (MRI), the patient is immersed in a strong magnetic field inside a solenoid.

To construct a solenoid with a circular cross section, wire is tightly wrapped in a cylindrical shape, forming a helix (Fig. 19.33a). We can think of the field as the superposition of the fields due to a large number of circular loops. If the loops are sufficiently close together, then the field lines go straight through one loop to the next, all the way down the solenoid. Having a large number of loops, one next to the other, straightens out the field lines. Figure 19.33b shows the magnetic field lines due to a solenoid. Inside the solenoid and away from the ends, the field is nearly uniform and parallel to the solenoid's axis as long as the solenoid is long compared to its radius. To find in which direction the field points along the axis, use right-hand rule 2 to find the field due to one circular loop of wire.

(a)

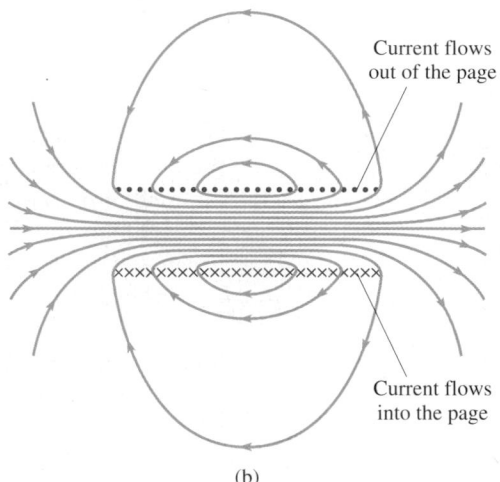

Current flows out of the page

Current flows into the page

(b)

Figure 19.33 (a) A solenoid. (b) Magnetic field lines due to a solenoid. Each dot represents the wire crossing the plane of the page with current out of the page; each cross represents the wire crossing the plane of the page with current into the page.

If a long solenoid has N turns of wire and length L, then the magnetic field strength inside is given by

---

Magnetic field strength inside an ideal solenoid

$$B = \frac{\mu_0 NI}{L} = \mu_0 nI \qquad \text{(19-16)}$$

In Eq. (19-16), I is the current in the wire and $n = N/L$ is the number of turns per unit length. Note that the field does *not* depend on the radius of the solenoid. The magnetic field near the ends is weaker and starts to bend outward; the field outside the solenoid is quite small—look how spread out the field lines are outside. A solenoid is therefore one way to produce a uniform magnetic field.

The similarity in the magnetic field lines due to a solenoid compared to those due to a bar magnet (Fig. 19.1b) suggested to André-Marie Ampère that the magnetic field of a permanent magnet might also be due to electric currents. The nature of these currents is explored in Section 19.10.

Magnetic Resonance Imaging

In magnetic resonance imaging (Fig. 19.34), the main solenoid is usually made with superconducting wire, which must be kept at low temperature (see Section 18.4). The main solenoid produces a strong, uniform magnetic field (typically 0.5–2 T). The nuclei of hydrogen atoms (protons) in the body act like tiny permanent magnets; a magnetic torque tends to make them line up with the magnetic field. A radio-frequency coil emits pulses of radio waves (rapidly varying electric and magnetic fields). If the radio wave has just the right frequency (the resonant frequency), the protons can absorb energy from the wave, which disturbs their magnetic alignment. When the protons flip back into alignment with the field, they emit radio wave signals of their own that can be detected by the radio-frequency coil.

The resonant frequency of the pulse that makes the protons flip depends on the total magnetic field due to the MRI machine and due to the neighboring atoms. Protons in different chemical environments have slightly different resonant frequencies. In order to image a slice of the body, three other coils create small (15 to 30-mT) magnetic fields that vary in the x-, y-, and z-directions. The magnetic fields of these coils are adjusted so that the protons are in resonance with the radio-frequency signal only in a single slice, a few mm thick, in any desired direction through the body.

19.9 AMPÈRE'S LAW

Ampère's law plays a role in magnetism similar to that of Gauss's law in electricity. Both relate the field to the source of the field. For the electric field, the source is charge. Gauss's law relates the net charge inside a closed surface to the flux of the electric field through that surface. The source of magnetic fields is current. Ampère's law must take a different form from Gauss's law: since magnetic field lines are always closed loops, the magnetic flux through a *closed surface* is always zero. (This fact is called *Gauss's law for magnetism* and is itself a fundamental law of electromagnetism.)

Instead of a closed surface, Ampère's law concerns any closed *path* or *loop*. For Gauss's law we would find the flux: the perpendicular component of the electric field times the surface area. If $E_\perp$ is not the same everywhere, then we break the surface into pieces and sum up $E_\perp \Delta A$. For Ampère's law, we multiply the component of the magnetic field *parallel* to the path (or the tangential component at points along a closed curve) times the *length* of the path. Just as for flux, if the magnetic field component is not constant then we take parts of the path (each of length Δl) and sum up the product. This quantity is called the **circulation**.

$$\text{circulation} = \Sigma B_{\parallel} \Delta l \qquad \text{(19-17)}$$

Figure 19.34 MRI apparatus

Ampère's law relates the circulation of the field to the *net* current *I* that crosses the interior of the path.

Ampère's law
$\Sigma B_{\parallel} \Delta l = \mu_0 I$ (19-18)

There is a symmetry between Gauss's law and Ampère's law. Table 19.1 summarizes the similarities and differences between the two.

Table 19.1

Comparison of Gauss's and Ampère's Laws

Gauss's Law	Ampère's Law
Electric field	Magnetic field (static only)
Applies to any closed *surface*	Applies to any closed *path*
Relates the electric field on the surface to the net *electric charge* inside the surface	Relates the magnetic field on the path to the net *current* cutting through interior of the path
Component of the electric field *perpendicular* to the surface ($E_\perp$)	Component of the magnetic field *parallel* to the path ($B_\parallel$)
Flux = perpendicular field component × *area* of surface	Circulation = parallel field component × *length* of path
$\Sigma E_\perp \Delta A$	$\Sigma B_\parallel \Delta l$
Flux = $1/\epsilon_0$ × net charge	Circulation = μ_0 × net current
$\Sigma E_\perp \Delta A = \dfrac{1}{\epsilon_0} q$	$\Sigma B_\parallel \Delta l = \mu_0 I$

Example 19.11

Magnetic Field due to a Long Straight Wire

Use Ampère's law to show that the magnetic field due to a long straight wire is $B = \mu_0 I/(2\pi r)$.

Strategy As with Gauss's law, the key is to exploit the symmetry of the situation. The field lines have to be circles around the wire, assuming the ends are far away. Choose a closed path around a circular field line (Fig. 19.35). The field is everywhere tangent to the field line and therefore tangent to the path; there is no perpendicular component. The field must also have the same magnitude at a uniform distance r from the wire.

Solution Since the field has no component perpendicular to the path, $B_\parallel = B$. Going around the circular path, B is constant, so

$$\text{circulation} = B_\parallel s = B \times 2\pi r = \mu_0 I$$

where I is the current in the wire. Solving for B,

$$B = \frac{\mu_0 I}{2\pi r}$$

Discussion Ampère's law shows why the magnetic field of a long wire varies inversely as the distance from the wire. A circle of any radius r around the wire has a length that is proportional to r, while the current that cuts through the interior of the circle is always the same (I). So the field must be proportional to $1/r$.

Practice Problem 19.11   Circulation due to three wires

What is the circulation of the magnetic field for the path in Fig. 19.36?

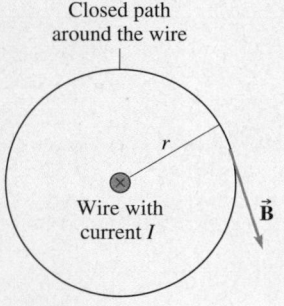

Closed path around the wire

r

Wire with current I

$\vec{B}$

Figure 19.35
Applying Ampère's law to a long straight wire. A closed path is chosen to follow a circular magnetic field line; the magnetic field is then calculated from Ampère's law.

$+2I$ ⊙

$\otimes$ $-2I$

$+3I$ ⊙

$-I$ $\otimes$

$-5I$ $\otimes$

$+7I$ ⊙

Figure 19.36
Six wires perpendicular to the page carry currents as indicated. A path is chosen to enclose three of the wires.

19.10 MAGNETIC MATERIALS

All materials are magnetic in the sense that they have magnetic properties. The magnetic properties of most substances are quite unremarkable, though. If a bar magnet is held near a piece of wood or aluminum or plastic, there is no noticeable interaction between the two. In common parlance, these substances might be called nonmagnetic.

In reality, all substances experience *some* force when near a bar magnet. For most substances, the magnetic force is so weak that it is not noticed.

Substances that experience a noticeable force due to a nearby magnet are called **ferromagnetic** (*ferro-* in Latin refers to iron). Iron is a well-known ferromagnet; others include nickel, cobalt, and chromium dioxide (used to make chrome audiotapes). Ferromagnetic materials experience a magnetic force toward a region of stronger magnetic field. Refrigerator magnets stick because the refrigerator door is made of a ferromagnetic metal. When a permanent magnet is brought near, there is an attractive force on the door, and from Newton's third law there must also be an attractive force on the magnet. The surfaces of the magnet and the door are pulled together by magnetic forces. As a result, each exerts a surface contact force on the other; the component of the contact force parallel to the contact surface—the frictional force—holds the magnet up.

The so-called nonmagnetic substances can be divided into two groups. **Paramagnetic substances** are like the ferromagnets in that they are attracted toward regions of stronger magnetic field, though the force is *much* weaker. **Diamagnetic** substances are weakly repelled by a region of stronger field. All substances, including liquids and gases, are either ferromagnetic, paramagnetic, or diamagnetic.

Any substance, whether ferromagnetic, diamagnetic, or paramagnetic, contains a large number of tiny magnets: the electrons. The electrons are like little magnets in two ways. First, an electron's orbital motion around the nucleus makes it a tiny current loop and thus is a magnetic dipole. Second, an electron has an *intrinsic* magnetic dipole moment *independent of its motion*. The intrinsic magnetism of the electron is one of its fundamental properties, just like its electric charge and mass. (Other particles, such as protons and neutrons, also have intrinsic magnetic dipole moments.) The net magnetic dipole moment of an atom or molecule is the vector sum of the dipole moments of its constituent particles.

In most materials—paramagnets and diamagnets—the atomic dipoles are randomly oriented. Even when the material is immersed in a strong external magnetic field, the dipoles only have a slight tendency to line up with it. The torque that tends to make dipoles line up with the external magnetic field is overwhelmed by the thermal tendency for the dipoles to be *randomly* aligned, so there is only a slight degree of large-scale alignment. The magnetic field inside the material is nearly the same as the applied field; the dipoles have little effect in paramagnets and diamagnets.

Ferromagnetic materials have much stronger magnetic properties because there is an interaction—the explanation of which requires quantum physics—that keeps the magnetic dipoles aligned, even in the *absence* of an external magnetic field. A ferromagnetic material is divided up into regions called **domains** in which the atomic or molecular dipoles line up with each other. Even though each atom is a weak magnet by itself, when all of them have their dipoles aligned in the same direction within a domain, the domain can have a significant dipole moment.

The moments of different domains are not necessarily aligned with each other, however. Some may point one way and some another (Fig. 19.37a). When the net dipole moment of all the domains is zero, the material is unmagnetized. If the material is placed in an external magnetic field, two things happen. Atomic dipoles at domain boundaries can "defect" from one domain to an adjacent one by flipping their dipole moments. Thus, domains with their dipole moments aligned or nearly aligned with the external field grow in size and the others shrink. The other thing that happens is that domains can change their direction of orientation, with all the atomic dipoles flipping to a new direction. When the net dipole moment of all the domains is nonzero, the material is magnetized (Fig. 19.37b).

(a)

(b)

Figure 19.37 Domains within a ferromagnetic material are indicated by arrows indicating the direction of each domain's magnetic field. In (a), the domains are randomly oriented; the material is unmagnetized. In (b), the material is magnetized; the domains show a high degree of alignment to the right.

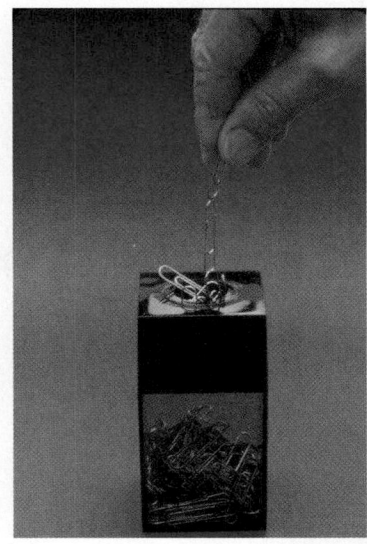

Figure 19.38 Each magnetized paper-clip is capable of magnetizing another paper-clip.

Physics at Home

If a needle is placed in contact with a strong magnet, the needle becomes magnetized. Then the magnetized needle attracts other needles. This phenomenon is easily observed in paper-clip containers with magnets that hold the paper-clips upright for ease in pulling one out. The magnetized paper-clips often drag out other paper-clips as well (Fig. 19.38). Try it.

Once a ferromagnet is magnetized, it does not necessarily lose its magnetization when the external field is removed. It takes some energy to align the domains with the field; there is a kind of internal friction that must be overcome. If there is a lot of this internal friction, then the domains stay aligned even after the external field is removed. The material is then a permanent magnet. If there is relatively little of this internal friction, then there is little energy required to reorient the domains. This kind of ferromagnet does not make a good permanent magnet; when the external field is removed, it retains only a small fraction of its maximum magnetization.

At high temperature, the interaction that keeps the dipoles aligned within a domain is no longer able to do so. Without the alignment of dipoles, there are no longer any domains; the material becomes paramagnetic. The temperature at which this occurs for a particular ferromagnetic material is called the *Curie temperature* of that material [after Pierre Curie (1859–1906), French physicist famous for studies of radioactive materials done with his wife, Marie Curie]. For iron, the Curie temperature is about 770°C.

Making The Connection:
electromagnets

Electromagnets

An *electromagnet* is made by inserting a *soft iron* core into the interior of a solenoid. Soft iron does not retain a significant permanent magnetization when the solenoid's field is turned off—soft iron does not make a good permanent magnet. When current flows in the solenoid, magnetic dipoles in the iron tend to line up with the field due to the solenoid. The net effect is that the field inside the iron is intensified by a factor known as the **relative permeability** κ_B. The relative permeability is analogous in magnetism to the dielectric constant in electricity. However, the dielectric constant is the factor by which the electric field is *weakened*, while the relative permeability is the factor by which the magnetic field is *strengthened*. The relative permeability of a ferromagnet can be in the hundreds or even thousands—the intensification of the magnetic field is significant. Not only that, but in an electromagnet the strength and even direction of the magnetic field can be changed by changing the current in the solenoid. Figure 19.39 shows the field lines in an electromagnet. Notice that the iron core channels the field lines; the windings of the solenoid need not be at the business end of the electromagnet.

Figure 19.39 An electromagnet with field lines sketched

Making The Connection:
computer hard disks, magnetic tape

Magnetic Storage

The basic principles of magnetic storage are the same, whether applied to computer hard disks, removable disks, or magnetic tape used to store audio, video, or computer data. To record or write, an electromagnet called a *head* is used to magnetize ferromagnetic particles in a coating on the disk or tape surface (Fig. 19.40). The ferromagnetic particles retain their magnetization even after the head has moved away, so the data persists until it is erased or written over. Data can be erased if a tape or disk is brought close to a strong magnet.

Making The Connection:
maglev trains

Maglev Trains

Maglev (<u>mag</u>netic <u>lev</u>itation) trains promise to be faster, quieter, smoother, and more energy-efficient than conventional trains. Two types are being developed, one in Japan and one in Germany. In the Japanese version, a repulsive force is exerted on large magnets on the undercarriage of the train by superconducting electromagnets running along the guideway track. The repulsive force levitates the train above the track by about 1 to 10 cm. In the German system, the undercarriage of the train wraps around a steel guideway; electromagnets on the undercarriage are attracted upward toward the guideway to levitate the train about 1 cm above the rail whether the train is moving or not.

Once levitated, both versions of maglev trains are then propelled forward by electromagnets along the track walls and embedded along the sides of the train; these magnets also stabilize and guide the train during travel. The electric current supplied to the

Computer hard drive

Read-write head
There is one head on each side of each platter

Platter
has a magnetizable coating on each side

Spindle motor
turns the platters at several thousand rpm

Audiotape recorder

To amplifier

Magnetic tape

To amplifier

Coil

$\vec{B}$ field line

Iron core

Unmagnetized tape

Gap

Record head

Record/playback head

S N

Magnetized tape
Signal recorded

Magnetic coating

Mylar substrate

Tape motion

Figure 19.40   The ferromagnetic particles in the tape or disk are magnetized in a direction determined by the current in the record head.

coils in the guideway walls can change the strength and the polarity of the magnetic coils that propel the train forward or slow it down. To propel the train, the polarity of the magnetic fields in the guideway are periodically reversed to pull the magnetic coils of the train forward (Fig. 19.41). Since there is no ground contact, bumps, noise, and energy losses are greatly reduced. Friction is eliminated, although there is still air

Figure 19.41 At a given instant of time, the polarities of the magnetic coils in the guideway and along the train are oriented to pull the train forward.

resistance, but with their aerodynamic design, these trains can reach speeds of 500 kph (310 mph). A commercial maglev train, built by the German company Transrapid International, is expected to operate between the city of Shanghai and its airport, 19 mi away; the trip will take less than 10 min.

MASTER THE CONCEPTS

Summary

- Magnetic field lines are interpreted just like electric field lines. The magnetic field at any point is tangent to the field line; the magnitude of the field is proportional to the number of lines per unit area perpendicular to the lines.

- Magnetic field lines are always closed loops because there are no magnetic monopoles.

- The smallest unit of magnetism is the magnetic dipole. Field lines emerge from the north pole and reenter at the south pole. A magnet can have more than two poles, but it must have at least one north pole and at least one south pole.

- The magnetic force on a charged particle is

$$\vec{F}_B = q\vec{v} \times \vec{B} \qquad (19\text{-}5)$$

If the charge is at rest ($v = 0$) or if its motion is parallel or antiparallel to the magnetic field ($v_\perp = 0$), then the magnetic force is zero. The force is always perpendicular to the magnetic field and to the velocity of the particle.

- The magnitude of the cross product of two vectors is the magnitude of one vector times the perpendicular component of the other:

$$|\vec{a} \times \vec{b}| = |\vec{b} \times \vec{a}| = a_\perp b = ab_\perp = ab \sin \theta \qquad (19\text{-}3)$$

- The direction of the cross product is the direction perpendicular to both vectors that is chosen using right-hand rule 1:

    Right-hand rule 1 for cross products (see Fig. 19.9)

Point the thumb of your right hand in a direction perpendicular to both vectors and the outstretched four fingers of your right hand along the first vector. Curl your outstretched fingers inward toward your palm until your fingertips point in the direction of the second vector. If the fingers sweep through an angle less than 180°, the thumb points in the correct direction of the cross product.

- The SI unit of magnetic field is the tesla:

$$1 \text{ T} = 1 \frac{\text{N}}{\text{A·m}} \qquad (19\text{-}2)$$

- If a charged particle moves at right angles to a uniform magnetic field, then its trajectory is a circle. If the velocity has a component parallel to the field as well as a component perpendicular to the field, then its trajectory is a helix.

- The magnetic force on a straight wire carrying current I is

$$\vec{F} = I\vec{L} \times \vec{B} \qquad (19\text{-}12a)$$

where $\vec{L}$ is a vector whose magnitude is the length of the wire and whose direction is along the wire in the direction of the current.

- The magnetic torque on a planar current loop is

$$\tau = NIAB \sin \theta \qquad (19\text{-}13b)$$

where θ is the angle between the magnetic field and the dipole moment vector of the loop. The direction of the dipole moment is perpendicular to the loop as chosen using right-hand rule 1 (take the cross product of $\vec{L}$ for any side with $\vec{L}$ for the *next* side, going around in the same direction as the current).

- The magnetic field at a distance r from a long straight wire has magnitude

$$B = \frac{\mu_0 I}{2\pi r} \qquad (19\text{-}14)$$

The field lines are circles around the wire with the direction given by right-hand rule 2:

    Right-hand rule 2 for the field due to a current
    (see Fig. 19.30d)

Point the thumb of the right hand in the direction of the current in the wire. Then curl the fingers inward toward the

MASTER THE CONCEPTS *continued*

palm; the direction that the fingers curl is the direction of the magnetic field lines around the wire.

- The permeability of free space is

$$\mu_0 = 4\pi \times 10^{-7} \frac{\text{T·m}}{\text{A}} \qquad (19\text{-}15)$$

- The magnetic field inside a long tightly wound solenoid is uniform:

$$B = \frac{\mu_0 N I}{L} = \mu_0 n I \qquad (19\text{-}16)$$

Its direction is along the axis of the solenoid, as given by right-hand rule 2.

- Ampère's law relates the circulation of the magnetic field around a closed path to the *net* current I that crosses the interior of the path.

$$\Sigma B_{\parallel} \Delta l = \mu_0 I \qquad (19\text{-}18)$$

- The magnetic properties of ferromagnetic materials are due to an interaction that keeps the magnetic dipoles aligned within regions called domains, even in the *absence* of an external magnetic field.

Highlighted Figures and Tables

F19.1 Magnetic field lines of a bar magnet (p. 674)

F19.5 Magnetic field due to the Earth (p. 677)

F19.8 Finding the magnetic force on a moving point charge (p. 679)

F19.9 Using the right-hand rule to find the direction of the cross product (p. 680)

F19.14 Circular motion of charges in a magnetic field (p. 684)

F19.26 A rectangular coil of wire in a uniform magnetic field (p. 697)

F19.29 Magnetic force on coil in a speaker (p. 701)

F19.30 Magnetic field due to a long straight wire (p. 702)

F19.32 Magnetic field lines due to a circular current loop (p. 705)

F19.33 Magnetic field lines due to a solenoid (p. 705)

F19.34 MRI apparatus (p. 707)

T19.1 Comparison of Gauss's and Ampère's laws (p. 708)

F19.40 Magnetic recording devices (p. 711)

F19.41 Maglev trains (p. 712)

CONCEPTUAL QUESTIONS

1. The electric field is defined as the electric force per unit charge. Explain why the magnetic field *cannot* be defined as the magnetic force per unit charge.

2. A charged particle moves through a region of space at constant velocity. Ignore gravity. In the region, is it possible that there is (a) a magnetic field but no electric field? (b) an electric field but no magnetic field? (c) a magnetic field and an electric field? For each possibility, what must be true about the direction(s) of the field(s)?

3. Suppose that a horizontal electron beam is deflected to the right by a uniform magnetic field. What is the direction of the magnetic field? If there is more than one possibility, what can you say about the direction of the field?

4. A circular metal loop carries a current I as shown in Fig. 19.42. The points are all in the plane of the page and the loop is perpendicular to the page. Sketch the loop, and draw vector arrows at the points A, B, C, D, and E to show the direction of the magnetic field at those points.

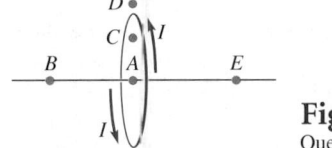

Figure 19.42 Conceptual Question 4

5. In a TV or computer monitor, a constant electric field accelerates the electrons to high speed; then a magnetic field is

used to deflect the electrons to the side. Why can't a constant magnetic field be used to speed up the electrons?

6. A uniform magnetic field directed upward exists in some region of space. In what direction(s) could an electron be moving if its trajectory is (a) a straight line? (b) a circle? Assume that the electron is subject only to magnetic forces.

7. In a velocity selector, the electric and magnetic forces cancel if $\vec{E} + \vec{v} \times \vec{B} = 0$. Show that $\vec{v}$ must be in the same direction as $\vec{E} \times \vec{B}$. (*Hint:* Since $\vec{v}$ is perpendicular to both $\vec{E}$ and $\vec{B}$ in a velocity selector, there are only two possibilities for the direction of $\vec{v}$: the direction of $\vec{E} \times \vec{B}$ or the direction of $-\vec{E} \times \vec{B}$.)

8. Two ions with the same velocity and mass but different charges enter the magnetic field of a mass spectrometer. One is singly charged ($q = +e$) and the other is doubly charged ($q = +2e$). Is the radius of their circular paths the same? If not, which is larger? By what factor?

9. The mayor of a city proposes a new law to require that magnetic fields generated by the power lines running through the city be zero outside of the electric company's right of way. What would you say at a public discussion of the proposed law?

10. A horizontal wire that runs east-west carries a steady current to the east. A C-shaped magnet (see Fig. 19.3a) is placed so that the wire runs between the poles, with the north pole above the wire and the south pole below. What is the direction of the magnetic force on the wire between the poles?

11. The magnetic field due to a long straight wire carrying steady current is measured at two points, P and Q (Fig. 19.43). Where is the wire and in what direction does the current flow?

Figure 19.43 Conceptual Question 11

12. A circular loop of current carries a steady current. (a) Sketch the magnetic field lines in a plane perpendicular to the plane of the loop. (b) Which side of the loop is the north pole of the magnetic dipole and which is the south pole?

13. Computer speakers that are intended to be placed near a computer monitor are magnetically shielded—either they don't use magnets or they are designed so that their magnets produce only a small magnetic field nearby. Why is the shielding important? What might happen if an ordinary speaker (*not* intended for use near a monitor) is placed next to a computer monitor?

14. One iron nail does not necessarily attract another iron nail, although both are attracted by a magnet. Explain.

15. Two wires at right angles in a plane carry equal currents (Fig. 19.44). At what points in the plane is the magnetic field zero?

Figure 19.44 Conceptual Question 15

16. Explain why a constant magnetic field does no work on a point charge moving through the field. Since the field does no work, what can we say about the speed of a point charge acted on only by a magnetic field?

17. In Fig. 19.45, a metal bar is shown at two different times. The arrows represent the alignment of the dipoles within each magnetic domain. (a) What happened between t_1 and t_2 to cause the change? (b) Is the metal a paramagnet, diamagnet, or ferromagnet? Explain.

Time t_1

Time $t_2 > t_1$

Figure 19.45 Conceptual Question 17

18. If a magnet is held near the screen of a TV, computer monitor, or oscilloscope, the picture is distorted. [*Don't try this*—see part (b).] (a) Why is the picture distorted? (b) With a color TV or monitor, the distortion remains even after the magnet is removed. Explain. (A color CRT has a metal mask just behind the screen with holes to line up the electrons from different guns with the red, green, and blue phosphors. What kind of metal is the mask made of?)

19. Refer to the bubble chamber tracks in Fig. 19.15a. Suppose that particle 2 moves in a smaller circle than particle 1. Can we conclude that $|q_2| > |q_1|$? Explain.

20. The trajectory of a charged particle in a uniform magnetic field is a helix if $\vec{v}$ has components both parallel to and perpendicular to the field. Explain how the two other cases (circular motion for $v_\parallel = 0$ and straight line motion for $v_\perp = 0$) can each be considered to be special cases of helical motion.

MULTIPLE CHOICE QUESTIONS

Multiple Choice Questions 1–4. In Fig. 19.46, four point charges move in the directions indicated in the vicinity of a bar magnet. The magnet, charge positions, and velocity vectors all lie in the plane of this page. Answer choices:

(a) ↑   (b) ↓   (c) ←   (d) →
(e) ✕ (into page) (f) • (out of page) (g) the force is zero

1. What is the direction of the magnetic force on charge 1 if $q_1 < 0$?

2. What is the direction of the magnetic force on charge 2 if $q_2 > 0$?

3. What is the direction of the magnetic force on charge 3 if $q_3 < 0$?

4. What is the direction of the magnetic force on charge 4 if $q_4 < 0$?

5. The magnetic force on a point charge in a magnetic field $\vec{B}$ is largest, for a given speed, when it

(a) moves parallel to the magnetic field.
(b) moves antiparallel to the magnetic field.
(c) moves perpendicular to the magnetic field.
(d) has velocity components both parallel to and perpendicular to the field.

Figure 19.46 Multiple Choice Questions 1–4

Multiple Choice Questions 6–9. A wire carries current as shown in Fig. 19.47. Charged particles 1, 2, 3, and 4 move in the directions indicated. Answer choices for Questions 6–8:

(a) ↑   (b) ↓   (c) ←   (d) →
(e) ✕ (into page) (f) • (out of page) (g) the force is zero

6. What is the direction of the magnetic force on charge 1 if $q_1 < 0$?

7. What is the direction of the magnetic force on charge 2 if $q_2 > 0$?

8. What is the direction of the magnetic force on charge 3 if $q_3 < 0$?

9. If the magnetic forces on charges 1 and 4 are equal and their velocities are equal,

   (a) the charges have the same sign and $|q_1| > |q_4|$.
   (b) the charges have opposite signs and $|q_1| > |q_4|$.
   (c) the charges have the same sign and $|q_1| < |q_4|$.
   (d) the charges have opposite signs and $|q_1| < |q_4|$.
   (e) $q_1 = q_4$.   (f) $q_1 = -q_4$.

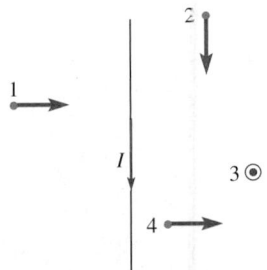

Figure 19.47 Multiple Choice Questions 6–9

10. The magnetic field lines *inside* a bar magnet run in what direction?

   (a) from north pole to south pole
   (b) from south pole to north pole
   (c) from side to side
   (d) None of the above—there are no magnetic field lines *inside* a bar magnet

11. The magnetic forces that two parallel wires with unequal currents flowing in opposite directions exert on each other are

   (a) attractive and unequal in magnitude.
   (b) repulsive and unequal in magnitude.
   (c) attractive and equal in magnitude.
   (d) repulsive and equal in magnitude.
   (e) both zero.
   (f) in the same direction and unequal in magnitude.
   (g) in the same direction and equal in magnitude.

12. What is the direction of the magnetic field at point P in Fig. 19.48? (P is on the axis of the coil.)

   (a) ↑   (b) ↓   (c) ←   (d) →
   (e) × (into page)   (f) • (out of page)

Figure 19.48 Multiple Choice Question 12

PROBLEMS

Note: ⓒ indicates a combination conceptual/quantitative problem. Gold diamonds ◆, ◆◆ are used to indicate the increasing level of difficulty of each problem. Problem numbers appearing in blue, 9., denote problems that have a detailed solution available in the Student Solutions Manual. Some problems are *paired* by concept; their numbers are connected by a ruled box.

19.1 Magnetic Fields

ⓒ 1. At which point in Fig. 19.49 is the magnetic field strength (a) the smallest and (b) the largest? Explain.

2. Draw vector arrows to indicate the direction and relative magnitude of the magnetic field at each of the points A–F in Fig. 19.49.

Figure 19.49 Problems 1 and 2

3. Two identical bar magnets lie next to one another on a table. Sketch the magnetic field lines if the north poles are at the same end (Fig. 19.50a).

4. Two identical bar magnets lie next to one another on a table. Sketch the magnetic field lines if the north poles are at opposite ends (Fig. 19.50b).

5. Two identical bar magnets lie on a table along a straight line with their north poles facing each other (Fig. 19.50c). Sketch the magnetic field lines.

6. Two identical bar magnets lie on a table along a straight line with opposite poles facing each other (Fig. 19.50d). Sketch the magnetic field lines.

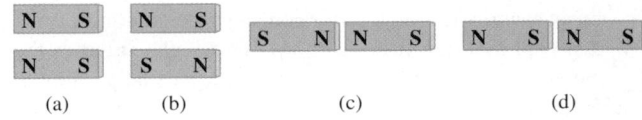

Figure 19.50 Problems 3–6

7. The magnetic forces on a magnetic dipole result in a torque that tends to make the dipole line up with the magnetic field. In this problem we show that the *electric forces* on an *electric dipole* result in a torque that tends to make the electric dipole line up with the electric field. (a) For each orientation of the dipole shown in Fig. 19.51, sketch the electric forces and determine the direction of the torque—clockwise or counterclockwise—about an axis perpendicular to the page through the center of the dipole. (b) The torque always tends to make the dipole rotate toward what orientation?

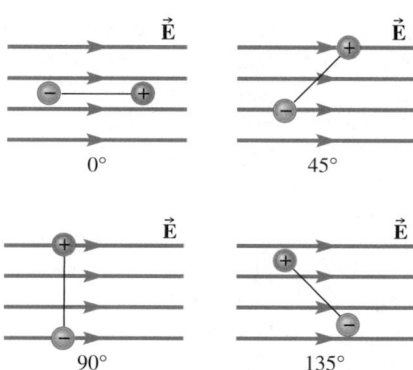

Figure 19.51 Problem 7

19.2 Magnetic Force on a Point Charge

8. Find the magnetic force exerted on an electron moving vertically upward at a speed of 2.0×10^7 m/s by a horizontal magnetic field of 0.50 T directed north.

9. A uniform magnetic field points north; its magnitude is 1.5 T. A proton with kinetic energy 8.0×10^{-13} J is moving vertically downward in this field. What is the magnetic force acting on it?

10. Find the magnetic force (magnitude and direction) on an electron moving at speed 8.0×10^5 m/s for each of the directions shown in Fig. 19.52. The magnetic field has magnitude $B = 0.40$ T.

Figure 19.52 Problem 10

11. Electrons in a TV picture tube are accelerated from rest by an electric field through a potential difference of 2.5 kV. In contrast to an oscilloscope, where the electron beam is deflected by an electric field, in a TV or computer monitor the beam is deflected by a magnetic field. (a) What is the speed of the electrons? (b) The beam is deflected by a perpendicular magnetic field of magnitude 0.80 T. What is the acceleration of the electrons while in the field? (c) What is the speed of the electrons after they travel 4.0 mm through the magnetic field? (d) What strength electric field would give the electrons the same magnitude acceleration as in (b)? (e) Why do we have to use an electric field in the first place to get the electrons up to speed? Why not use the large acceleration due to a magnetic field for that purpose?

12. A magnet produces a 0.30 T field between its poles, directed to the east. A dust particle with charge $q = -8.0 \times 10^{-18}$ C is moving straight down at 0.30 cm/s in this field. What is the magnitude and direction of the magnetic force on the dust particle?

13. An electron beam in vacuum moving at 1.8×10^7 m/s passes between the poles of an electromagnet (Fig. 19.53). The diameter of the magnet pole faces is 2.4 cm and the field between them is 0.20×10^{-2} T. How far and in what direction is the beam deflected when it hits the screen, which is 25 cm past the magnet? [*Hint:* The electron velocity changes relatively little while they are in the field, so you can approximate the magnetic force as a constant force.]

Figure 19.53 Problem 13

14. A positron ($q = +e$) moves at 5.0×10^7 m/s in a magnetic field of magnitude 0.47 T. The magnetic force on the positron has magnitude 2.3×10^{-12} N. (a) What is the component of the positron's velocity perpendicular to the magnetic field? (b) What is the component of the positron's velocity parallel to the magnetic field? (c) What is the angle between the velocity and the field?

15. An electron moves with speed 2.0×10^5 m/s in a 1.2-T uniform magnetic field. At one instant, the electron is moving due west and experiences an upward magnetic force of 3.2×10^{-14} N. What is the direction of the magnetic field? Be specific: give the angle(s) with respect to N, S, E, W, up, down. (If there is more than one possible answer, find all the possibilities.)

16. An electron moves with speed 2.0×10^5 m/s in a uniform magnetic field of 1.4 T, pointing south. At one instant, the electron experiences an upward magnetic force of 1.6×10^{-14} N. In what direction is the electron moving at that instant? Be specific: give the angle(s) with respect to N, S, E, W, up, down. (If there is more than one possible answer, find all the possibilities.)

19.3 Charged Particle Moving Perpendicular to a Uniform Magnetic Field

17. The magnetic field in a cyclotron is 0.50 T. Find the magnitude of the magnetic force on a proton with velocity of 1.0×10^7 m/s in a plane perpendicular to the field.

18. When two particles travel through a region of uniform magnetic field pointing out of the plane of the paper, they follow the trajectories shown in Fig. 19.54. What are the signs of the charges of each particle?

Figure 19.54 Problem 18

19. The magnetic field in a cyclotron is 0.50 T. What must be the radius of the vacuum chamber if the maximum proton velocity desired is 1.0×10^7 m/s?

20. A singly charged ion of unknown mass moves in a circle of radius 12.5 cm in a magnetic field of 1.2 T. The ion was accelerated through a potential difference of 7.0 kV before it entered the magnetic field. What is the mass of the ion?

21. Natural carbon consists of two different isotopes (excluding ^{14}C, which is present in only trace amounts). The isotopes have different masses, which is due to different numbers of neutrons in the nucleus; however, the number of protons is the same, and subsequently the chemical properties are the same. The most abundant isotope has an atomic mass of 12.0000 u. When natural carbon is placed in a mass spectrometer, two lines are formed on the photographic plate. The lines show that the more abundant isotope moved in a circle of radius 15.0 cm, while the rarer isotope moved in a circle of radius 15.6 cm. What is the atomic mass of the rarer isotope? (The ions are accelerated through the same potential difference before entering the magnetic field.)

22. After being accelerated through a potential difference of 5.0 kV, a singly charged carbon ion (^{12}C) moves in a circle of radius 21 cm in the magnetic field of a mass spectrometer. What is the magnitude of the field?

✦ 23. A sample containing carbon (atomic mass 12 u), oxygen (16 u), and an unknown element is placed in a mass spectrometer. The ions all have the same charge and are accelerated through the same potential difference before entering the magnetic field. The carbon and oxygen lines are separated by 2.250 cm on the photographic plate, and the unknown element makes a line between them that is 1.160 cm from the carbon line. (a) What is the mass of the unknown element? (b) Identify the element.

24. A sample containing sulfur (atomic mass 32 u), manganese (55 u), and an unknown element is placed in a mass spectrometer. The ions are accelerated through the same potential difference before entering the magnetic field. The sulfur and manganese lines are separated by 3.20 cm, and the unknown element makes a line between them that is 1.07 cm from the sulfur line. (a) What is the mass of the unknown element? (b) Identify the element.

✦ 25. In one type of mass spectrometer, ions having the *same velocity* move through a uniform magnetic field. The spectrometer is being used to distinguish ^{12}C and ^{14}C ions that have the same charge. The ^{12}C ions move in a circle of diameter 25 cm. (a) What is the diameter of the orbit of ^{14}C ions? (b) What is the ratio of the frequencies of revolution for the two types of ion? (c) Repeat parts (a) and (b) if the ions are given the same *kinetic energy* rather than the same velocity.

19.5 A Charged Particle in Crossed $\vec{E}$ and $\vec{B}$ Fields

26. Crossed electric and magnetic fields are established over a certain region. The magnetic field is 0.50 T vertically downward. The electric field is 2.0×10^6 V/m horizontally east. An electron, traveling horizontally northward, experiences zero resultant force from these fields and so continues in a straight line. What is the electron's speed?

✦ 27. A charged particle is accelerated from rest through a potential difference ΔV. The particle then passes straight through a velocity selector (field magnitudes E and B). Derive an expression for the charge-to-mass ratio (q/m) of the particle in terms of ΔV, E, and B.

🅒 28. A current $I = 40.0$ A flows through a strip of metal (Fig. 19.55). An electromagnet is switched on so that there is a uniform magnetic field of magnitude 0.30 T directed into the page. (a) How would you hook up a voltmeter to measure the Hall voltage? Show how the voltmeter is connected on a sketch of the strip. (b) Assuming the carriers are electrons, which lead of your voltmeter is at the higher potential? Mark it with a "+" sign in your sketch. Explain briefly.

29. In Problem 28, if the width of the strip is 3.5 cm, the magnetic field is 0.43 T, and the Hall voltage is measured to be 7.2 μV, what is the drift velocity of the carriers in the strip?

30. In Problem 28, the width of the strip is 3.5 cm, the magnetic field is 0.43 T, the Hall voltage is measured to be 7.2 μV, the thickness of the strip is 0.24 mm, and the current in the wire is 54 A. What is the density of carriers (number of carriers per unit volume) in the strip?

🅒 31. The strip in Fig. 19.55 is used as a Hall probe to measure magnetic fields. (a) What happens if the strip is not perpendicular to the field? Does the Hall probe still read the correct field strength? Explain. (b) What happens if the field is in the plane of the strip?

32. A strip of copper 2.0 cm wide carries a current $I = 30.0$ A to the right (Fig. 19.55). The strip is in a magnetic field $B = 5.0$ T into the page. (a) What is the direction of the average magnetic force on the conduction electrons? (b) The Hall voltage is 20.0 μV. What is the drift velocity?

Figure 19.55 Problems 28–32

33. An electromagnetic flowmeter is used to measure blood flow rates during surgery. Blood containing Na$^+$ ions flows due south through an artery with a diameter of 0.40 cm. The artery is in a downward magnetic field of 0.25 T and develops a Hall voltage of 0.35 mV across its diameter. (a) What is the blood speed (in m/s)? (b) What is the flow rate (in m^3/s)? (c) The leads of a voltmeter are attached to diametrically opposed points on the artery to measure the Hall voltage. Which of the two leads is at the higher potential?

34. An electromagnetic flowmeter is used to measure blood flow rates during surgery. Blood containing ions (primarily Na$^+$) flows through an artery with a diameter of 0.50 cm. The artery is in a magnetic field of 0.35 T and develops a Hall voltage of 0.60 mV across its diameter. (a) What is the blood speed (in m/s)? (b) What is the flow rate (in m^3/s)? (c) If the magnetic field points west and the blood flow is north, is the top or bottom of the artery at the higher potential?

19.6 Magnetic Force on a Current-Carrying Wire

35. A straight wire segment of length 0.60 m carries a current of 18.0 A and is immersed in a uniform external magnetic field of magnitude 0.20 T. (a) What is the magnitude of the maximum possible magnetic force on the wire segment? (b) Explain why the given information enables you to calculate only the *maximum possible* force.

36. A straight wire segment of length 25 cm carries a current of 33.0 A and is immersed in a uniform external magnetic field. The magnetic force on the wire segment has magnitude 4.12 N. (a) What is the minimum possible magnitude of the magnetic field? (b) Explain why the given information enables you to calculate only the *minimum possible* field strength.

37. Parallel conducting tracks, separated by 2.0 cm, run north and south (Fig. 19.56). There is a uniform magnetic field of 1.2 T pointing upward (out of the page). A 0.040-kg cylindrical metal rod is placed across the tracks and a battery is connected between the tracks, with its positive terminal connected to the east track. If the current through the rod is 3.0 A, find the magnitude and direction of the magnetic force on the rod.

Figure 19.56
Problem 37

38. A straight, stiff wire of length 1.00 m and mass 25 g is suspended in a magnetic field $B = 0.75$ T (Fig. 19.57). The wire is connected to an emf. How much current must flow in the wire and in what direction so that the wire is suspended and the tension in the supporting wires is zero?

Figure 19.57
Problem 38

39. A 20.0 cm × 30.0 cm rectangular loop of wire carries 1.0 A of current clockwise around the loop (Fig. 19.58). (a) Find the magnetic force on each side of the loop if the magnetic field is 2.5 T out of the page. (b) What is the net magnetic force on the loop?

40. Repeat Problem 39 if the magnetic field is 2.5 T to the left (in the −x-direction).

Figure 19.58 Problems 39, 40, and 75

41. A straight wire is aligned east-west in a region where the Earth's magnetic field has magnitude 0.48 mT and direction 72° below the horizontal, with the horizontal component directed due north. The wire carries a current I toward the west. The magnetic force on the wire per unit length of wire has magnitude 0.020 N/m. (a) What is the direction of the magnetic force on the wire? (b) What is the current I?

19.7 Torque on a Current Loop

42. In an electric motor, a circular coil with 100 turns of radius 2.0 cm can rotate between the poles of a magnet. When the current through the coil is 75 mA, the maximum torque that the motor can deliver is 0.0020 N·m. (a) What is the strength of the magnetic field? (b) Is the torque clockwise or counterclockwise as viewed in Fig. 19.59?

43. In an electric motor, a coil with 100 turns of radius 2.0 cm can rotate between the poles of a magnet. The magnetic field strength is 0.20 T. When the current through the coil is 50.0 mA, what is the maximum torque that the motor can deliver?

Figure 19.59 Problem 42

44. A certain fixed length L of wire carries a current I. (a) Show that if the wire is formed into a square coil, then the maximum torque in a given magnetic field B is developed when the coil has just one turn. (b) Show that the magnitude of this torque is $\tau = \frac{1}{16} L^2 IB$.

45. A square loop of wire of side 3.0 cm carries 3.0 A of current. A uniform magnetic field of magnitude 0.67 T makes an angle of 37° with the plane of the loop. (a) What is the magnitude of the torque on the loop? (b) What is the net magnetic force on the loop?

46. The torque on a loop of wire (a magnetic dipole) in a uniform magnetic field is $\tau = NIAB \sin\theta$, where θ is the angle between $\vec{B}$ and a line perpendicular to the loop of wire. Suppose an *electric* dipole, consisting of two charges ±q a fixed distance d apart is in a uniform electric field $\vec{E}$. (a) Show that the net electric force on the dipole is zero. (b) Let θ be the angle between $\vec{E}$ and a line running from the negative to the positive charge. Show that the torque on the electric dipole is $\tau = qdE \sin\theta$ for all angles $-180° \leq \theta \leq 180°$. (Thus, for both electric and magnetic dipoles, the torque is the product of the *dipole moment* times the field strength times $\sin\theta$. The quantity qd is the electric dipole moment; the quantity NIA is the magnetic dipole moment.)

47. Use the following method to show that the torque on an irregularly shaped planar loop is given by Eq. (19-13a). The irregular loop of current in Fig. 19.60a carries current I. There is a perpendicular magnetic field B. To find the torque on the irregular loop, sum up the torques on each of the smaller loops shown in Fig. 19.60b. The pairs of imaginary currents flowing across carry equal currents in opposite directions, so the magnetic forces on them would be equal and opposite; they would therefore contribute nothing to the net torque. Now generalize this argument to a loop of *any* shape. [*Hint:* Think of a curved loop as a series of tiny, straight, perpendicular segments.]

(a)   (b)

Figure 19.60 Problem 47

19.8 Magnetic Field due to an Electric Current

48. Two parallel long straight wires are suspended by strings of length $L = 1.2$ m (Fig. 19.61). Each wire has mass per unit length 0.050 kg/m. When the wires each carry 50.0 A of current, the wires swing apart. (a) How far apart are the wires in equilibrium? (Assume that this distance is small compared to L.) [*Hint:* Use a small angle approximation.] (b) Are the wires carrying current in the same or opposite directions?

Figure 19.61 Problem 48

49. Imagine a long straight wire perpendicular to the page and carrying a current I into the page. Sketch some $\vec{B}$ field lines with arrowheads to indicate directions.

50. Two conducting wires perpendicular to the page are shown in cross section as gray dots in Fig. 19.62. They each carry 10.0 A out of the page. What is the magnetic field at point P?

51. Two wires each carry 10.0 A of current (in *opposite directions*) and are 3.0 mm apart (Fig. 19.63). (a) Calculate the magnetic field 25 cm away at point P, in the plane of the wires. (b) What is the magnetic field at the same point if the currents instead both run to the left?

52. Point P is midway between two long, straight, parallel wires that run north-south in a horizontal plane. The distance between the wires is 1.0 cm. Each wire carries a current of

Figure 19.62 Problem 50

Figure 19.63
Problem 51

1.0 A toward the north. (a) Find the magnitude and direction of the magnetic field at point P. (b) Repeat the question if the current in the wire on the east side runs toward the south instead.

53. A long straight wire carries a current of 50.0 A. An electron, traveling at 1.0×10^7 m/s, is 5.0 cm from the wire. What force (magnitude and direction) acts on the electron if the electron's velocity is directed toward the wire?

54. Two long straight wires carry the same amount of current in the directions indicated in Fig. 19.64. The wires cross each other in the plane of the paper. Rank points A, B, C, and D in order of decreasing field strength.

Figure 19.64 Problem 54

55. Two long straight parallel wires separated by 8.0 cm carry currents of equal magnitude but heading in opposite directions. The wires are shown perpendicular to the plane of this page in Fig. 19.65. Point P is 2.0 cm from wire 1, and the magnetic field at point P is 1.0×10^{-2} T directed in the $-y$-direction. Calculate the current in wire 1 and its direction.

Figure 19.65
Problem 55

56. Two parallel wires in a horizontal plane carry currents I_1 and I_2 to the right. The wires each have length L and are separated by a distance d. (a) What are the magnitude and direction of the field due to wire 1 at the location of wire 2? (b) What are the magnitude and direction of the magnetic force on wire 2

due to this field? (c) What are the magnitude and direction of the field due to wire 2 at the location of wire 1? (d) What are the magnitude and direction of the magnetic force on wire 1 due to this field? (e) Do parallel currents attract or repel? (f) What about antiparallel currents?

19.9 Ampère's Law

57. In Fig. 19.66, a number of wires carry currents into or out of the page as indicated. (a) Using loop 1 for Ampère's law, what is the net current through the interior of the loop? (b) Repeat for loop 2.

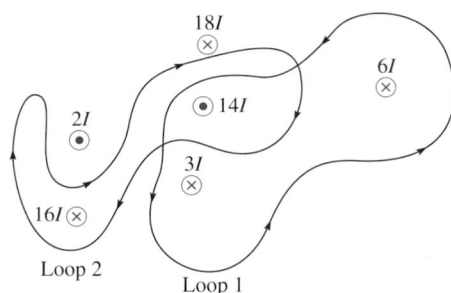

Figure 19.66 Problem 57

58. An infinitely long, thick cylindrical shell of inner radius a and outer radius b carries a current I uniformly distributed across a cross section of the shell (Fig. 19.67). (a) On a sketch of a cross section of the shell (such as Fig. 19.67b), draw some magnetic field lines. The current flows out of the page. Consider all regions ($r \le a$, $a \le r \le b$, $b \le r$). (b) Find the magnetic field for $r > b$.

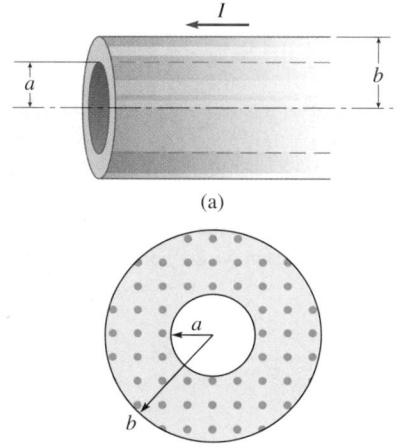

Figure 19.67
Problem 58

59. In this problem, use Ampère's law to show that the magnetic field inside a long solenoid is $B = \mu_0 nI$. Assume that the field inside the solenoid is uniform and parallel to the axis and that the field outside is zero. Choose a rectangular path for Ampère's law (Fig. 19.68). (a) Write down $B_{\parallel}s$ for each of the four sides of the path, in terms of B, a, and b. (b) Sum these to form the circulation. (c) Now, to find the current cutting

through the path: each loop carries the same current I, and some number N of loops cut through the path, so the total current is NI. Rewrite N in terms of the number of turns per unit length (n) and the physical dimensions of the path. (d) Solve for B.

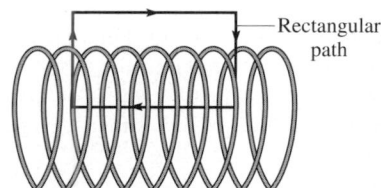

Figure 19.68
Problem 59

60. A toroid is like a solenoid that has been bent around in a circle until its ends meet. The field lines are circular as shown in Fig. 19.69. What is the magnitude of the magnetic field inside a toroid of N turns carrying current I? Apply Ampère's law, following a field line at a distance r from the center of the toroid. Work in terms of the total number of turns N, rather than the number of turns per unit length (why?). Is the field uniform, as it is for a long solenoid? Explain.

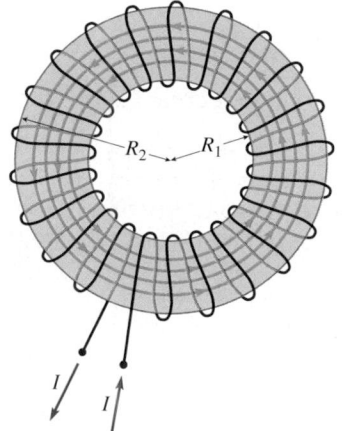

Figure 19.69 Problem 60

19.10 Magnetic Materials

61. A bar magnet (Fig. 19.70a) is broken into two parts (Fig. 19.70b). If care is taken not to disturb the magnetic domains, what are the polarities of the new ends (c) and (d) respectively?

Figure 19.70
Problem 61

62. The intrinsic magnetic dipole moment of the electron has magnitude 9.3×10^{-24} A·m². In other words, the electron acts as though it were a tiny current loop with $NIA = 9.3 \times 10^{-24}$ A·m². What is the maximum torque on an electron due to its intrinsic dipole moment in a 1.0-T magnetic field?

63. In a simple model, the electron in a hydrogen atom orbits the proton at a radius of 53 pm and at a constant speed of 2.2×10^6 m/s. The orbital motion of the electron gives it an orbital magnetic dipole moment. (a) What is the current I in this cur-

rent loop? [*Hint:* How long does it take the electron to make one revolution?] (b) What is the orbital dipole moment IA? (c) Compare the orbital dipole moment to the intrinsic magnetic dipole moment of the electron (9.3×10^{-24} A·m^2).

64. An electromagnet is made by inserting a soft iron core into a solenoid. The solenoid has 1800 turns, radius 2.0 cm, and length 15 cm. When 2.0 A of current flows through the solenoid, the magnetic field inside the iron core has magnitude 0.42 T. What is the relative permeability of the iron core?

65. Figure 19.71 shows *hysteresis curves* for three different materials. A hysteresis curve is a plot of the magnetic field strength inside the material (B) as a function of the externally applied field (B_0). (a) Which material would make the best

permanent magnet? Explain. (b) Which would make the best core for an electromagnet? Explain.

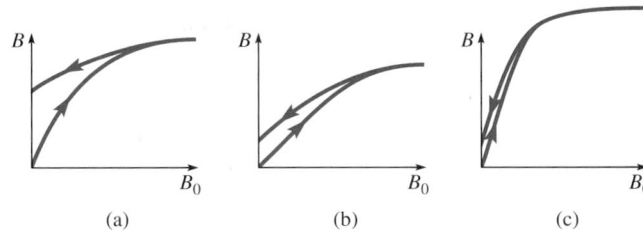

Figure 19.71   Problem 65

COMPREHENSIVE PROBLEMS

66. A compass is placed directly on top of a wire (Fig. 19.72, needle not shown). The current in the wire flows to the right. Which way does the north end of the needle point? Explain. (Neglect the Earth's magnetic field.)

Figure 19.72   Problem 66

67. You want to build a cyclotron to accelerate protons to a speed of 3.0×10^7 m/s. The largest magnetic field strength you can attain is 1.5 T. What must be the minimum radius of the dees in your cyclotron? Show how your answer comes from Newton's second law.

68. An electron moves in a circle of radius R in a uniform magnetic field $\vec{B}$ (Fig. 19.73). The field is into the page. (a) Does the electron move clockwise or counterclockwise? (b) How much time does the electron take to make one complete revolution? *Derive* an expression for the time, starting with the magnetic force on the electron. Your answer may include R, B, and any fundamental constants.

Figure 19.73   Problem 68

69. Prove that the time for one revolution of a charged particle moving perpendicular to a uniform magnetic field is independent of its speed. (This is the principle on which the cyclotron operates.) In doing so, write an expression that gives the period T (the time for one revolution) in terms of the mass of the particle, the charge of the particle, and the magnetic field strength.

70. (a) A proton moves with uniform circular motion in a magnetic field of magnitude 0.80 T. At what frequency f does it circulate? (b) Repeat for an electron.

71. The concentration of free electrons in silver is 5.85×10^{28} per m^3. A strip of silver of thickness 0.050 mm and width 20.0 mm is placed in a magnetic field of 0.80 T (Fig. 19.74). A current of 10.0 A is sent down the strip. (a) What is the drift velocity of the electrons? (b) What is the Hall voltage measured by the meter? (c) Which side of the voltmeter is at the higher potential?

Figure 19.74
Problem 71

72. An electromagnetic flowmeter is to be used to measure blood velocity. A magnetic field of 0.115 T is applied across an artery of inner diameter 3.80 mm. The Hall voltage is measured to be 88.0 μV. What is the average velocity of the blood flowing in the artery?

73. In a carbon-dating experiment, a particular type of mass spectrometer is used to separate ^{14}C from ^{12}C. Carbon ions from a sample are first accelerated through a potential difference ΔV_1 between the charged accelerating plates (Fig. 19.75). Then the ions enter a region of uniform vertical magnetic field $B = 0.20$ T. The ions pass between deflection plates spaced 1.0 cm apart. By adjusting the potential difference ΔV_2 between these plates, only one of the two isotopes (^{12}C or ^{14}C) is allowed to pass through to the next stage of the mass spectrometer. The distance from the entrance to the ion detector is a fixed 0.20 m. By suitably adjusting ΔV_1 and ΔV_2, the detector counts only one type of ion, so the relative abundances can be determined. (a) Are the ions positively or negatively charged? (b) Which of the accelerating plates (east or west) is positively charged? (c) Which of the deflection plates (north or south) is positively charged? (d) Find the correct values of ΔV_1 and ΔV_2 in order to count ^{12}C$^+$ ions. (e) Find the correct values of ΔV_1 and ΔV_2 in order to count ^{14}C$^+$ ions.

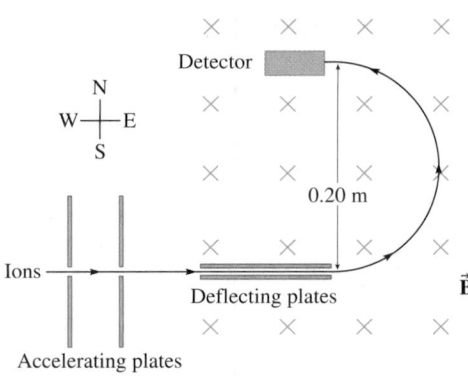

Figure 19.75   Problem 73

74. Sketch the magnetic field as it would appear inside the coil of wire in Fig. 19.76, looking down into the coil from the position shown.

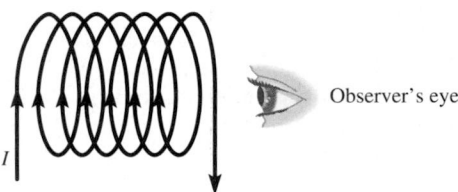

Observer's eye

Figure 19.76   Problem 74

✦ 75. Repeat Problem 39 if the magnetic field is 2.5 T in the plane of the loop, 60.0° below the +x-axis.

76. Four long parallel wires pass through the four corners of a square (Fig. 19.77). All four wires carry the same amount of current and the current directions are as indicated. (a) What is the direction of the magnetic field at the center of the square? (b) The current is 10.0 A in each wire and the square is 0.10 m on a side. What is the magnitude of the magnetic field?

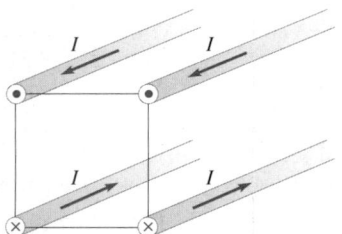

Figure 19.77
Problem 76

✦ 77. A *current balance* is a device to measure magnetic forces. It is constructed from two parallel coils, each with an average radius of 12.5 cm (Fig. 19.78). The lower coil rests on a balance; it has 20 turns and carries a constant current of 4.0 A. The upper coil, suspended 0.314 cm above the lower coil, has 50 turns and a current that can be varied. The reading of the balance changes as the magnetic force on the lower coil changes. What current is needed in the upper coil to exert a force of 1.0 N on the bottom coil? [*Hint:* Since the distance between the coils is small compared to the radius of the coils, approximate the setup as two long parallel straight wires.]

78. A rectangular loop of wire, carrying current $I_1 = 2.0$ mA, is next to a very long wire carrying a current $I_2 = 8.0$ A

Figure 19.78   Problem 77

(Fig. 19.79). (a) What is the direction of the magnetic force on each of the four sides of the rectangle due to the long wire's magnetic field? (b) Calculate the net magnetic force on the rectangular loop due to the long wire's magnetic field. [*Hint:* The long wire does *not* produce a uniform magnetic field.]

Figure 19.79   Problem 78

79. A strip of copper carries current in the +x-direction. There is an external magnetic field directed out of the page. What is the direction of the Hall electric field?

80. A bar magnet is held near the electron beam in an oscilloscope (Fig. 19.80). The beam passes directly below the south pole of the magnet. In what direction will the beam move on the screen? (*Don't try this* with a color TV tube. There is a metal mask just behind the screen that separates the pixels for red, green, and blue. If you succeed in magnetizing the mask, the picture will be permanently distorted.)

Figure 19.80   Problem 80

✦ 81. In a certain region of space, there is a uniform electric field $\vec{E} = 3.0 \times 10^4$ V/m directed due east and a uniform magnetic field $\vec{B} = 0.080$ T *also directed due east.* What is the electromagnetic force on an electron moving *due south* at 5.0×10^6 m/s?

♦ 82. An early cyclotron at Cornell University was used from the 1930s to the 1950s to accelerate protons, which would then bombard various nuclei. The cyclotron used a large electromagnet with an iron yoke to produce a uniform magnetic field of 1.3 T over a region in the shape of a flat cylinder. Two hollow copper dees of inside radius 16 cm were located in a vacuum chamber in this region. (a) What is the frequency of oscillation necessary for the alternating voltage difference between the dees? (b) What is the kinetic energy of a proton by the time it reaches the outside of the dees? (c) What would be the equivalent voltage necessary to accelerate protons to this energy from rest in one step (say between parallel plates)? (d) If the potential difference between the dees has a magnitude of 10.0 kV each time the protons cross the gap, what is the minimum number of revolutions each proton has to make in the cyclotron?

♦ 83. We wish to use a mass spectrometer to measure the mass m of the $^{238}\text{U}^+$ ion. Assume a source of $^{238}\text{U}^+$ ions exists and that the ions initially move slowly. We want the ions to move at a high speed v in the mass spectrometer. (a) To accelerate the ions to speed v, we use the electric field between the plates of a capacitor (Fig. 19.81). If the plates have area A and are a distance d apart, what should the charge on the plates be? Indicate which plate is positive and which negative. (Answer in terms of v, A, d, m, and universal constants.) (b) The ions now have speeds roughly equal to v, but because their initial speeds varied somewhat, there is a spread. We want to select ions with speeds very close to v and reject the rest. To do so, we pass the beam of ions through a velocity selector. In the

selector, there is a uniform magnetic field out of the page, with magnitude B. What are the magnitude and direction of the electric field in the selector? (Answer in terms of v, m, B, and universal constants.) (c) Sketch the trajectory of ions that enter the velocity selector with speeds less than v. (d) Some of the ions now enter the mass spectrometer. The magnetic field is the same as in part (b), but there is no electric field. The ions are detected when the detector is a distance D from the entry point of the ions. Find the mass of the ions. (Answer in terms of v, B, D, and universal constants.)

✪ ♦ 84. In a certain region of space, there is a uniform electric field $\vec{E} = 2.0 \times 10^4$ V/m to the east and a uniform magnetic field $\vec{B} = 0.0050$ T to the west. (a) What is the electromagnetic force on an electron moving north at 1.0×10^7 m/s? (b) With the electric and magnetic fields as specified, is there some velocity such that the net electromagnetic force on the electron would be zero? If so, give the magnitude and direction of that velocity. If not, explain briefly why not.

85. In the mass spectrometer of Fig. 19.82, neon ions ($q = +e$) come from the ion source and are accelerated through a potential difference V. The ions then pass through an aperture in a metal plate into a uniform magnetic field where they travel in semicircular paths until exiting into the detector. Neon ions having a mass of 20.0 u leave the field at a distance of 50.0 cm from the aperture. At what distance from the aperture do neon ions having a mass of 22.0 u leave the field? (1 u = 1.67×10^{-27} kg.)

Figure 19.82   Problem 85

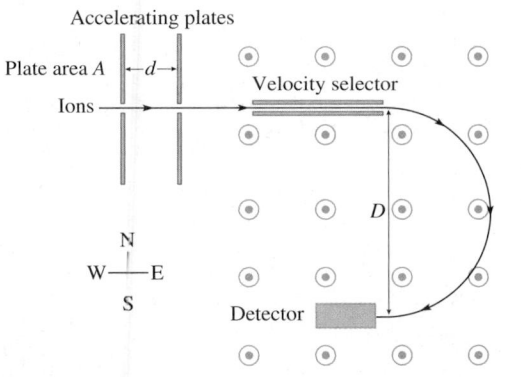

Figure 19.81   Problem 83

♦♦ 86. A proton moves in a helical path at speed $v = 4.0 \times 10^7$ m/s high in the atmosphere, where the Earth's magnetic field has magnitude $B = 1.0 \times 10^{-6}$ T. The proton's velocity makes an angle of 25° with the magnetic field. (a) Find the radius of the helix. [*Hint:* Use the perpendicular component of the velocity.] (b) Find the *pitch* of the helix—the distance between adjacent "coils." [*Hint:* Find the time for one revolution; then find how far the proton moves along a field line during that time interval.]

ANSWERS TO PRACTICE PROBLEMS

19.1 5.8×10^{-17} N, 3.4×10^{10} m/s^2

19.2 magnitude $= 8.2 \times 10^{-17}$ N, direction = east

19.3 $\pm 1.8 \times 10^6$ m/s

19.4 6.7×10^5 m/s

19.5 76 cm

19.6 out of the page (if the speed is too great, the magnetic force is larger than the electric force)

19.7 same magnitude Hall voltage, but opposite polarity: the top edge would be at the higher potential

19.8 west

19.9 (proof)

19.10 $\vec{B} = \dfrac{2\mu_0 I}{\pi d}$ in the $+x$-direction

19.11 $+4\mu_0 I$

Electromagnetic Induction

In many electric stoves, an electric current passes through the coiled heating elements. As electric energy is dissipated in the elements, they get hot. A newer kind of electric stove—the induction stove—is becoming more popular. It has two advantages over the older kind of electric stove. One is that there are no heating elements; the surface is smooth and flat with no place for food to fall, so cleaning is much easier. The other advantage is safety. If you touch the heating element of a traditional stove, you will burn your hand; a potholder carelessly left in contact with a hot coil may catch on fire. With an induction stove, the surface does not feel hot to the touch and the potholder does not catch fire. How can heat get to the food in a pot or pan if the stove surface is not hot? [*Hint:* Pots and pans made of Pyrex glass do not work on an induction stove; they must be made of metal.]

20.1 MOTIONAL EMF

The only sources of electric energy (emf) we've discussed so far are batteries. The amount of electric energy that can be supplied by a battery before it needs to be recharged or replaced is limited. Batteries also have internal resistances that impose limits on the current they can supply and thus on the *rate* at which they can supply energy.

Most of the world's electric energy is produced by generators. In this section we study **motional emf**—the emf produced when a conductor is moved in a magnetic field. Motional emf is the principle behind the electric generator.

Imagine a metal rod of length L in a uniform magnetic field $\vec{\mathbf{B}}$. When the rod is at rest, the conduction electrons move in random directions at high speed, but their average velocity is zero. Since their average velocity is zero, the average magnetic force on the electrons is zero; therefore, the total magnetic force on the rod is zero. The magnetic field affects the motion of individual electrons, but the rod as a whole feels no net magnetic force.

Now imagine that the rod is moving instead of being at rest. Figure 20.1a shows a uniform magnetic field into the page, the velocity $\vec{\mathbf{v}}$ of the rod is to the right, and the rod is oriented up and down—the field, velocity, and axis of the rod are mutually perpendicular. Now the electrons have a nonzero average velocity: it is $\vec{\mathbf{v}}$, since the electrons are being carried to the right along with the rod. Then the average magnetic force on each conduction electron is

$$\vec{\mathbf{F}}_{av} = -e\vec{\mathbf{v}} \times \vec{\mathbf{B}}$$

By the right-hand rule, the direction of this force is down (toward the lower end of the rod). The magnetic force causes electrons to accumulate at the lower end, giving it a negative charge and leaving positive charge at the upper end. This separation of charge by the magnetic field is similar to the Hall effect, but here the charges are moving due to the motion of the rod itself rather than due to a current flowing in a stationary rod.

As charge accumulates at the ends, an electric field develops in the rod, with field lines running from the positive to the negative charge. Eventually an equilibrium is reached: the electric field builds up until it causes a force equal and opposite to the magnetic force on electrons in the middle of the rod. Then there is no further accumulation of charge at the ends. Thus, in equilibrium,

$$\vec{\mathbf{F}}_E = q\vec{\mathbf{E}} = -\vec{\mathbf{F}}_B = -(q\vec{\mathbf{v}} \times \vec{\mathbf{B}})$$

or

$$\vec{\mathbf{E}} = -\vec{\mathbf{v}} \times \vec{\mathbf{B}}$$

just as for the Hall effect. Since $\vec{\mathbf{v}}$ and $\vec{\mathbf{B}}$ are perpendicular, $E = vB$. The potential difference between the ends is

$$\Delta V = EL = vBL \tag{20-1a}$$

In this case the direction of $\vec{\mathbf{E}}$ is parallel to the rod. If it were not, then the potential difference between the ends is found using only the *component* of $\vec{\mathbf{E}}$ parallel ($\parallel$) to the rod:

$$\Delta V = E_{\parallel}L = (\vec{\mathbf{v}} \times \vec{\mathbf{B}})_{\parallel}L \tag{20-1b}$$

As long as the rod keeps moving at constant speed, the separation of charge is maintained. The moving rod acts like a battery that is not connected to a circuit; positive

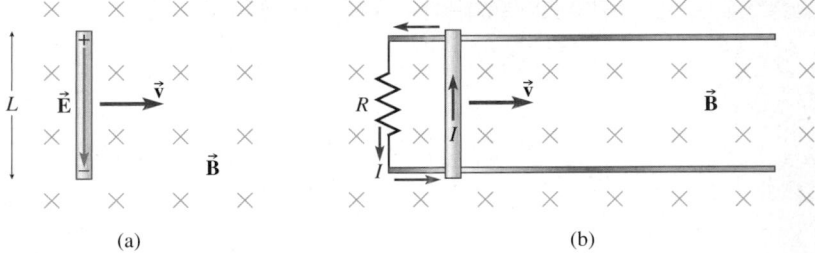

Figure 20.1 (a) A metal rod moving at velocity $\vec{\mathbf{v}}$ in uniform magnetic field $\vec{\mathbf{B}}$. (b) The rod is connected to a circuit with resistance R.

(a)   (b)

charge accumulates at one terminal and negative charge at the other, maintaining a constant potential difference. Now the important question: if we connect this rod to a circuit, does it act like a battery and cause current to flow?

Figure 20.1b shows the rod connected to a circuit. The rod slides on metal rails so that the circuit stays complete even as the rod continues to move. We assume the resistance R is large compared to the resistances of the rod and rails—in other words, the internal resistance of our source of emf (the moving rod) is negligibly small. The resistor R sees a potential difference ΔV across it, so current flows. The current tends to deplete the accumulated charge at the ends of the rod, but the magnetic force pumps more charge to maintain a constant potential difference. So the moving rod *does* act like a battery with an emf given by:

Motional emf

$$\mathcal{E} = vBL \qquad (20\text{-}2a)$$

More generally, if $\vec{\mathbf{E}}$ is not parallel to the rod, then

$$\mathcal{E} = (\vec{\mathbf{v}} \times \vec{\mathbf{B}})_{\parallel} L \qquad (20\text{-}2b)$$

A sliding rod would be a clumsy way to make a generator. No matter how long the rails are, the rod will eventually reach the end. In Section 20.2 we see that the principle of the motional emf can be applied to a *rotating coil* of wire instead of a sliding rod.

Where does the electric energy come from? The rod is acting like a battery, supplying electric energy that is dissipated in the resistor. How can energy be conserved? The key is to recognize that as soon as current flows through the rod, a magnetic force acts on the rod in the direction opposite to the velocity (see Problem 1). Left on its own, the rod would slow down as its kinetic energy gets transformed into electric energy. To maintain a constant emf, the rod must maintain a constant velocity, which can only happen if some other force pulls the rod. The work done by the force pulling the rod is the source of the electric energy (Problem 2).

Example 20.1

Loop Moving through a Magnetic Field

A square metal loop made of four rods of length L moves at constant velocity $\vec{\mathbf{v}}$ (Fig. 20.2). The magnetic field in the central region has magnitude B; elsewhere the magnetic field is zero. The loop has resistance R. At each position 1–5, state the direction (CW or CCW) and the magnitude of the current in the loop.

Strategy If current flows in the loop, it is due to the motional emf that pumps charge around. The vertical sides (a, c) have motional emfs as they move through the magnetic field, just as in Fig. 20.1b. We need to look at the horizontal sides (b, d) to see whether they also give rise to motional emfs. Once we figure out

the emf in each side, then we can determine whether they cooperate with each other—pumping charge around in the same direction—or tend to cancel each other.

Solution The vertical sides (a, c) have motional emfs as they move through the region of magnetic field. The emf acts to pump current upward (toward the top end). The magnitude of the emf is

$$\mathcal{E} = vBL$$

For the horizontal sides (b, d), the average magnetic force on a current-carrying electron is $\vec{\mathbf{F}}_{av} = -e\vec{\mathbf{v}} \times \vec{\mathbf{B}}$. Since the velocity is to the right and the field is into the page, the

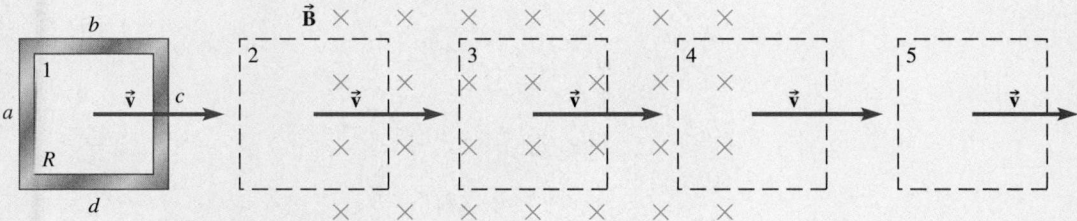

Figure 20.2
Loop moving through a magnetic field

continued on next page

Example 20.1 *continued*

right-hand rule shows that the direction of the force is down, just as in sides *a* and *c*. However, now the magnetic force does not move charge along the length of the rod; the magnetic force instead moves charge across the diameter of the rod. An electric field then develops *across* the rod. In equilibrium, the magnetic and electric forces cancel, exactly as in the Hall effect. The magnetic force does not push charge along the length of the rod, so there is no motional emf in sides *b* and *d*.

In positions 1 and 5, the loop is completely out of the region of magnetic field. There is no motional emf in any of the sides; no current flows.

In position 2, there is a motional emf in side *c* only; side *a* is still outside the region of $\vec{B}$ field. The emf makes current flow upward in side *c*, and therefore counterclockwise in the loop. The magnitude of the current is

$$I = \frac{\mathscr{E}}{R} = \frac{vBL}{R}$$

In position 3, there are motional emfs in both sides *a* and *c*. Since the emfs in both sides push current toward the top of the loop, the net emf around the loop is zero—as if two identical batteries were connected as in Fig. 20.3. No current flows around the loop.

$I = 0$

$\mathscr{E} = vBL$ ⊣⊢   ⊣⊢ $\mathscr{E} = vBL$

Figure 20.3
A loop of wire with two identical batteries

In position 4, there is a motional emf only in side *a*, since side *c* has left the region of the $\vec{B}$ field. The emf makes current flow upward in side *a* and therefore *clockwise* in the loop. The magnitude of the current is again

$$I = \frac{\mathscr{E}}{R} = \frac{vBL}{R}$$

 Discussion Figure 20.3 illustrates a useful technique: it often helps to draw battery symbols to represent the directions of the motional emfs and other induced emfs.

⚠ Note that if the loop were *at rest* instead of moving to the right at constant velocity, there would be no motional emf at any of the positions 1–5. The motional emf does not arise simply because one of the vertical sides of the loop is immersed in magnetic field while the other is not; it arises because one side *moves through* a magnetic field while the other does not.

Conceptual Practice Problem 20.1   Loop of different metal

Suppose a loop made of a different metal but with identical size, shape, and velocity moved through the same magnetic field. Of these quantities, which would be different: the magnitudes of the emfs, the directions of the emfs, the magnitudes of the currents, or the directions of the currents?

Making The Connection:
   electric generators

20.2   ELECTRIC GENERATORS

For practical reasons, electric generators use coils of wire that rotate in a magnetic field rather than rods that slide on rails. The rotating coil is called an *armature*. A simple electric generator is shown in Fig. 20.4. The rectangular coil is mounted on a shaft that is turned by some external power source such as a hand crank or the turbine of a steam engine.

Generators at Little Goose Dam in the state of Washington

Figure 20.4 An AC generator produces an emf that reverses polarity twice per complete turn of the coil. (a) Perspective view. (b) Side view, where the velocity vectors of sides 2 and 4 make an angle θ with the magnetic field. (c) Battery symbols indicate the direction of the emfs in the rotating loop of wire.

Let us begin with a single turn of wire—a rectangular loop—that rotates at a constant angular speed ω. The loop rotates in the space between the poles of a permanent magnet or an electromagnet that produces a nearly uniform field of magnitude B. Sides 2 and 4 are each of length L and are a distance r from the axis of rotation; the length of sides 1 and 3 is therefore $2r$.

None of the four sides of the loop moves perpendicularly to the magnetic field, so we must generalize the results of Section 20.1. In Problem 6, you can verify that there is zero induced emf in sides 1 and 3, so we concentrate on sides 2 and 4. Since these two sides do not, in general, move perpendicularly to $\vec{\mathbf{B}}$, the magnitude of the average magnetic force on the electrons is reduced by a factor of sin θ, where θ is the angle between the velocity of the wire and the magnetic field (Fig. 20.4b):

$$F_{av} = evB \sin \theta$$

The induced emf is then reduced by the same factor:

$$\mathscr{E} = vBL \sin \theta$$

Note that the induced emf is proportional to the component of the velocity perpendicular to $\vec{\mathbf{B}}$ ($v_\perp = v \sin \theta$). For a visual image, think of the induced emf as proportional to the rate at which the wire *cuts through magnetic field lines*. The component of the velocity *parallel* to $\vec{\mathbf{B}}$ moves the wire along the magnetic field lines, so it does not contribute to the rate at which the wire cuts through the field lines.

The loop turns at constant angular speed ω, so the speed of sides 2 and 4 is

$$v = \omega r$$

The angle θ changes at a constant rate ω. For simplicity, we choose $\theta = 0$ at $t = 0$, so that $\theta = \omega t$ and the emf $\mathscr{E}$ as a function of time t in each of sides 2 and 4 is

$$\mathscr{E}(t) = vBL \sin \theta = (\omega r)BL \sin \omega t$$

Sides 2 and 4 move in opposite directions, so electrons are pushed in opposite directions: in side 2, the electrons are pushed back, while in side 4 they are pushed forward. *Both* sides tend to send electrons around the loop in the same direction (clockwise as viewed in Fig. 20.4c). Therefore, the two emfs are connected with the negative "terminal" of one connected to the positive "terminal" of the other, and the *total* emf in the loop is the *sum* of the two:

$$\mathscr{E}(t) = 2\omega rBL \sin \omega t$$

The rectangular loop has sides L and $2r$, so the area of the loop is $A = 2rL$. Therefore, the total emf $\mathscr{E}$ as a function of time t is

$$\mathscr{E}(t) = \omega BA \sin \omega t \tag{20-3a}$$

When written in terms of the area of the loop, Eq. (20-3a) is true for a planar loop of *any* shape. If the coil consists of N turns of wire (N identical rectangular loops), the emf is N times as great:

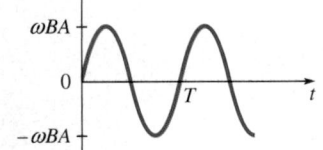

Emf produced by AC generator
$\mathscr{E}(t) = \omega NBA \sin \omega t$ (20-3b)

Figure 20.5 Generator-produced emf is a sinusoidal function of time.

The emf produced by a generator is not constant; it is a sinusoidal function of time (Fig. 20.5). The maximum emf (= ωBA) is called the *amplitude* of the emf (just as in simple harmonic motion, where the maximum displacement is called the amplitude). Sinusoidal emfs are used in AC (alternating current) circuits. Household electric outlets in the United States and Canada provide an emf with an amplitude of approximately 170 V and a frequency $f = \omega/2\pi = 60$ Hz. In much of the rest of the world, the amplitude is about 310–340 V and the frequency is 50 Hz.

This energy does not come for free; work must be done to turn the generator shaft. As current flows in the coil, the magnetic force on sides 2 and 4 cause a torque in the direction opposing the coil's rotation (Problem 50). To keep the coil rotating at constant angular speed, an equal and oppositely directed torque must be applied to the shaft. In an ideal generator, this external torque would do work at exactly the same rate as electric energy is generated. In reality, some energy is dissipated by friction and by the electrical resistance of the coil, among other things. Then the external torque does more work than the amount of electric energy generated. Since the rate at which electric energy is generated is

$$P = \mathscr{E}I$$

the external torque required to keep the generator rotating depends not only on the emf but also on the current it supplies. The current supplied depends on the *load*—the external circuit through which the current must flow.

In most power stations that supply our electricity, the work to turn the generator shaft is supplied by a steam engine. The steam engine is powered by burning coal, natural gas, or oil, or by a nuclear reactor. In a hydroelectric power plant, the gravitational potential energy of water is the energy source used to turn the generator shaft.

In electric and hybrid gas-electric cars, the drive train of the vehicle is connected to an electric generator when brakes are applied, which charges the batteries. Thus, instead of the kinetic energy of the vehicle being completely dissipated, much of it is stored in the batteries. This energy is used to propel the car after braking is finished.

The DC Generator

Note that the induced emf produced in an AC generator reverses direction twice per period. Mathematically, the sine functions in Eqs. (20-3) are positive half the time and negative half the time. When the generator is connected to a load, the current also reverses direction twice per period—which is why we call it alternating current.

What if the load requires a direct current (DC) instead? Then we need a DC generator, one in which the emf does *not* reverse direction. One way to make a DC generator is to equip the AC generator with a split-ring commutator and brushes, exactly as for the DC motor (Section 19.7). Just as the emf is about to change direction, the connections to the rotating loop are switched as the brushes pass over the gap in the split-ring. The commutator effectively reverses the connections to the outside load so that the emf and current supplied maintain the same direction. The emf and current are *not* constant, though. The emf is described by

$$\mathcal{E}(t) = \omega NBA \left| \sin \omega t \right| \tag{20-3c}$$

which is graphed in Fig. 20.6.

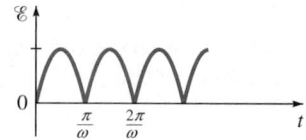

Figure 20.6 The emf as a function of time

A simple DC *motor* can be used as a DC *generator*, and *vice versa*. When configured as a motor, an external source of electric energy such as a battery causes current to flow through the loop. The magnetic torque makes the motor rotate. In other words, the current is the input and the torque is the output. When configured as a generator, an external torque makes the loop rotate, the magnetic field induces an emf in the loop, and the emf makes current flow. Now the torque is the input and the current is the output. The conversion between mechanical energy and electric energy can proceed in either direction.

More sophisticated DC generators have many coils, wired in series, distributed evenly around the axis of rotation. The emf *in each coil* still varies sinusoidally, but each emf has a slightly different phase. When they are added together, the output emf has only small fluctuations, which can be smoothed out by a circuit called a voltage regulator if necessary.

Example 20.2

A Bicycle Generator

A simple DC generator can be used to supply power for a bicycle's headlight. The generator has 150 turns of wire in a circular coil of radius 1.8 cm. The magnetic field strength in the region of the coil is 0.20 T. When the generator supplies an emf of amplitude 4.2 V to the lightbulb, the lightbulb consumes an average power of 6.0 W and a maximum instantaneous power of 12.0 W. (a) What is the rotational speed in rpm of the armature of the generator? (b) What is the average torque and maximum instantaneous torque that must be applied by the bicycle tire to the generator, assuming the generator to be ideal? (c) The radius of the tire is 32 cm and the radius of the shaft of the generator where it contacts the tire is 1.0 cm.

At what linear speed must the bicycle move to supply an emf of amplitude 4.2 V?

Strategy   The amplitude is the maximum value of the time-dependent emf [Eq. (20-3c)]. To find the torques, two methods are possible. One is to find the current in the coil, then the torque on the coil due to the magnetic field. To keep the armature moving at a constant angular velocity, an equal magnitude but oppositely directed torque must be applied to it. Another method is to analyze the energy transfers. The external torque applied to the armature must do work at the same rate that electric energy is dissipated in the lightbulb. The second

continued on next page

Example 20.2 *continued*

approach is easiest, especially since the problem states the power in the lightbulb. To find the linear speed of the bicycle, we set the tangential speeds of the tire and shaft equal (the shaft is "rolling" on the tire).

Solution (a) The emf as a function of time is

$$\mathcal{E}(t) = \omega NBA \, |\sin \omega t| \qquad (20\text{-}3c)$$

The emf has its maximum value when $\sin \omega t = \pm 1$. Thus, the amplitude of the emf is

$$\mathcal{E}_m = N\omega AB$$

where $N = 150$, $A = \pi r^2$, and $B = 0.20$ T. Solving for the angular frequency,

$$\omega = \frac{\mathcal{E}_m}{NAB} = \frac{4.2 \text{ V}}{150 \times \pi \times (0.018 \text{ m})^2 \times 0.20 \text{ T}} = 137.5 \text{ rad/s}$$

A check of the units verifies that $1\dfrac{\text{V}}{\text{T} \cdot \text{m}^2} = 1 \text{ s}^{-1}$. The question asks for the number of rpm, so we convert the angular frequency to rev/min:

$$\omega = 137.5 \, \frac{\text{rad}}{\text{s}} \times \frac{1 \text{ rev}}{2\pi \text{ rad}} \times \frac{60 \text{ s}}{1 \text{ min}} = 1300 \text{ rpm}$$

(b) Assuming the generator to be ideal, the torque applied to the crank must do work at the same rate that electric energy is generated:

$$P = \frac{\Delta W}{\Delta t}$$

Since for a small angular displacement $\Delta\theta$ the work done is $W = \tau \Delta\theta$,

$$P = \tau \frac{\Delta\theta}{\Delta t} = \tau\omega$$

The average torque is then

$$\tau_{av} = \frac{6.0 \text{ W}}{137.5 \text{ rad/s}} = 0.044 \text{ N·m}$$

and the maximum torque is

$$\tau_m = \frac{12 \text{ W}}{137.5 \text{ rad/s}} = 0.087 \text{ N·m}$$

(c) The tangential speed of the generator shaft is

$$v_{tan} = \omega r = 137.5 \text{ rad/s} \times 0.018 \text{ m} = 2.5 \text{ m/s}$$

The tangential speed of the tire where it touches the generator shaft is the same, since the shaft rolls without slipping on the tire. Since the generator is almost at the outside edge of the tire, the tangential speed at the outer radius of the tire is approximately the same. Assuming that the bicycle rolls without slipping on the road, its linear speed is approximately 2.5 m/s.

Discussion To check the result, we can find the maximum current in the coil and use it to find the maximum torque. The maximum current occurs when the power dissipated is maximum:

$$P_m = \mathcal{E}_m I_m$$

$$I_m = \frac{12 \text{ W}}{4.2 \text{ V}} = 2.86 \text{ A}$$

The magnetic torque on a current loop is

$$\tau = NIAB \sin \theta$$

where $\theta = \omega t$ is the angle between the magnetic field and the *normal* to the loop. At the position where the emf is maximum, $|\sin \theta| = 1$. Then

$$\tau_m = NI_m AB = 150 \times 2.86 \text{ A} \times \pi \times (0.018 \text{ m})^2 \times 0.20 \text{ T} = 0.087 \text{ N·m}$$

Practice Problem 20.2   Riding more slowly

What would the maximum power be if the bicycle moves half as fast? Assume that the resistance of the lightbulb does not change. Remember that the angular velocity affects the emf, which in turn affects the current. How does the power in the lightbulb depend on the bicycle's speed?

The adjective *normal* means *perpendicular* in geometry. *The normal to the loop* means *the direction perpendicular to the plane of the loop.*

20.3 FARADAY'S LAW

In 1820, Hans Christian Oersted accidentally discovered that an electric current produces a magnetic field (Section 19.1). Soon after hearing the news of that discovery, the English scientist Michael Faraday (1791–1867) started experimenting with magnets and electric circuits in an attempt to do the reverse—use a magnetic field to produce an electric current. Faraday's brilliant experiments led to the development of the electric motor, the generator, and the transformer.

In 1831, Faraday discovered two ways to produce an induced emf. One is to move a conductor in a magnetic field. The other does *not* involve movement of the conductor. Instead, Faraday found that a changing magnetic field induces an emf in a conductor even if the conductor is stationary. The induced emf due to a changing $\vec{\mathbf{B}}$ field cannot be understood in terms of the magnetic force on the conduction electrons: if the conductor is stationary, the average velocity of the electrons is zero, and the average magnetic force is zero.

Consider a circular loop of wire between the poles of an electromagnet (Fig. 20.7). The loop is perpendicular to the magnetic field; field lines cross the interior of the loop. Since the strength of the magnetic field is related to the spacing of the field lines, if the

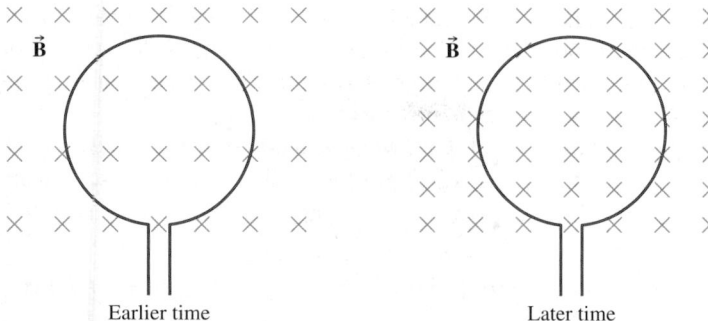

Earlier time   Later time

Figure 20.7 Circular loop in a magnetic field of increasing magnitude

strength of the field varies (by changing the current in the electromagnet), the number of field lines passing through the conducting loop changes. Faraday found that the emf induced in the loop is proportional to the *rate of change* of the number of field lines that cut through the interior of the loop.

We can formulate *Faraday's law* mathematically so that numbers of field lines are not involved. The magnitude of the magnetic field is proportional to the number of field lines *per unit cross-sectional area*:

$$B \propto \frac{\text{number of lines}}{\text{area}}$$

If a flat, open surface of area A is perpendicular to a uniform magnetic field of magnitude B, then the number of field lines that cross the surface is proportional to BA, since

$$\text{number of lines} = \frac{\text{number of lines}}{\text{area}} \times \text{area} \propto BA \qquad (20\text{-}4)$$

Equation (20-4) is correct only if the surface is perpendicular to the field. In general, the number of field lines crossing a surface is proportional to the *perpendicular component* of the field times the area:

$$\text{number of lines} \propto B_\perp A = BA \cos \theta$$

where θ is the angle between the field lines and the direction normal to the surface. The component of the magnetic field parallel to the surface $B_\parallel$ doesn't contribute to the number of lines crossing the surface; only $B_\perp$ does (see Fig. 20.8a). Equivalently, Fig. 20.8b shows that the number of lines crossing the surface area A is the same as the number crossing a surface of area $A \cos \theta$, which is perpendicular to the field.

The mathematical quantity that is proportional to the number of field lines cutting through a surface is called the **magnetic flux**. The symbol Φ (Greek capital phi) is used for flux; in Φ_B the subscript B indicates *magnetic* flux.

(a)

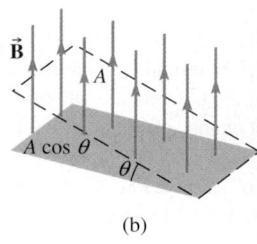

(b)

Figure 20.8 (a) The component of $\vec{B}$ perpendicular to the surface of area A is $B \cos \theta$. (b) The projection of the area A onto a plane perpendicular to $\vec{B}$ is $A \cos \theta$.

Magnetic flux through a flat surface

$$\Phi_B = B_\perp A = BA_\perp = BA \cos \theta \qquad (20\text{-}5)$$

The SI unit of magnetic flux is the weber (1 Wb = 1 T·m²).

Faraday's law says that the magnitude of the induced emf around a loop is equal to the rate of change of the magnetic flux through the loop.

Faraday's law

$$\mathscr{E} = -\frac{\Delta \Phi_B}{\Delta t} \qquad (20\text{-}6)$$

Problem 12 asks you to verify the units of Eq. (20-6). Faraday's law, if it is to give the *instantaneous* emf, must be taken in the limit of a very small time interval Δt. However, Faraday's law can be applied just as well to longer time intervals; then $\Delta \Phi_B / \Delta t$ represents the *average* rate of change of the flux, and $\mathscr{E}$ represents the *average* emf during that time interval.

The minus sign in Eq. (20-6) concerns the sense of the induced emf around the loop (clockwise or counterclockwise). The interpretation of the sign depends on a formal definition of the emf direction that we do not use. Instead, in Section 20.4, we introduce *Lenz's law*, which gives the direction of the induced emf.

If, instead of a single loop of wire, we have a coil of N turns, then Eq. (20-6) gives the emf induced in each turn; the total emf in the coil is then N times as great:

$$\mathcal{E} = -N\frac{\Delta\Phi_B}{\Delta t}$$

Alternatively, we can think of Φ_B in Eq. (20-6) as representing the *total* flux through the coil ($\Phi_{B,\text{total}} = N\Phi_{B,1} = NBA \cos\theta$).

Example 20.3

Induced Emf due to Changing Magnetic Field

A 40.0-turn coil of wire of radius 3.0 cm is placed between the poles of an electromagnet. The field increases from 0 to 0.75 T at a constant rate in a time interval of 225 s. What is the magnitude of the induced emf in the coil if (a) the field is perpendicular to the plane of the coil? (b) the field makes an angle of 30.0° with the plane of the coil?

Strategy First we write an expression for the flux through the coil in terms of the field. The only thing changing is the strength of the field, so the rate of flux change is proportional to the rate of change of the field. Faraday's law gives the induced emf.

Solution (a) The magnetic field is perpendicular to the coil, so the total flux is

$$\Phi_B = NBA$$

where $N = 40$, B is the field strength, and A is the area of the loop. Since the field increases at a constant rate, so does the flux. The rate of change of flux is then equal to the change in flux divided by the time interval. The flux changes at a constant rate, so the emf induced in the loop is constant.

By Faraday's law,

$$\mathcal{E} = -\frac{\Delta\Phi_B}{\Delta t} = -\frac{NB_fA - 0}{\Delta t}$$

$$|\mathcal{E}| = \frac{40 \times 0.75 \text{ T} \times \pi \times (0.030 \text{ m})^2}{225 \text{ s}} = 3.77 \times 10^{-4} \text{ V} = 0.38 \text{ mV}$$

(b) In Eq. (20-5), θ is the angle between $\vec{B}$ and the direction *normal* to the coil. If the field makes an angle of 30.0° with the plane of the coil, then it makes an angle

$$\theta = 90.0° - 30.0° = 60.0°$$

with the normal to the coil. The magnetic flux is

$$\Phi_B = NBA \cos\theta$$

The induced emf is therefore,

$$|\mathcal{E}| = \frac{\Delta\Phi_B}{\Delta t} = \frac{NB_fA \cos\theta - 0}{\Delta t} = 3.77 \times 10^{-4} \text{ V} \times \cos 60.0° = 0.19 \text{ mV}$$

Discussion If the rate of change of the field were not constant, then 0.38 mV would be the *average* emf during that time interval. The instantaneous emf would be sometimes higher and sometimes lower.

Practice Problem 20.3 Using the perpendicular component of $\vec{B}$

Draw a sketch that shows the coil, the direction normal to the coil, and the magnetic field lines. Find the component of $\vec{B}$ in the normal direction. Now use $\Phi_B = B_\perp A$ to verify the answer to part (b).

Sinusoidal Emfs

Emfs that are sinusoidal (sine or cosine) functions of time, such as in Example 20.2, are common in AC generators, motors, and circuits. A sinusoidal emf is generated whenever the flux is a sinusoidal function of time. It can be shown (see Fig. 20.9 and Problem 16) that:

$$\text{If } \Phi(t) = \Phi_0 \sin\omega t, \text{ then } \frac{\Delta\Phi}{\Delta t} = \omega\Phi_0 \cos\omega t \quad \text{(for small } \Delta t\text{)}; \tag{20-7a}$$

$$\text{if } \Phi(t) = \Phi_0 \cos\omega t, \text{ then } \frac{\Delta\Phi}{\Delta t} = -\omega\Phi_0 \sin\omega t \quad \text{(for small } \Delta t\text{)}. \tag{20-7b}$$

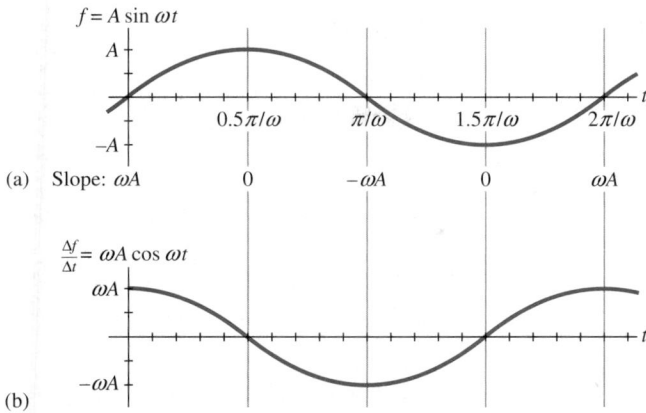

(a)   Slope: ωA 0 $-\omega A$ 0 ωA

(b)

Figure 20.9 (a) A graph of the function $f(t) = A \sin \omega t$ where A is constant. The slope of the graph is 0 at the maxima and minima and $\pm \omega A$ at the zero crossings. (b) A graph of the slope of $f(t)$—which represents the rate of change of $f(t)$. If $f(t) = A \sin \omega t$, then $\Delta f / \Delta t = \omega A \cos \omega t$.

Example 20.4

Rotating Field Generator

The generator discussed in Section 20.2 is a *rotating coil generator*, in which a coil of wire is rotated between the poles of a fixed magnet. Another type of generator, more frequently used in large applications, is the *rotating field generator*. In it, the coil of wire does not move; rather, an electromagnet rotates about the coil (Fig. 20.10). The magnetic field between the poles of the electromagnet has magnitude B, the electromagnet rotates with angular velocity ω, and the coil has N turns of area A. Find the emf induced in the coil.

Strategy The magnetic field strength does not vary, but its direction does. The number of field lines crossing through the coil depends on the angle that the field makes with the normal (the direction perpendicular to the coil). The changing magnetic flux induces an emf in the coil, according to Faraday's law.

Solution Let us choose $t = 0$ to be an instant when the field is perpendicular to the coil. At this instant, $\vec{B}$ is parallel to the normal, so $\theta = 0$. At a later time $t > 0$, the coil has rotated

through an angle $\Delta \theta = \omega t$. Thus, the angle that the field makes with the normal can be written

$$\theta = \omega t$$

just as for the rotating coil generator (Section 20.2). The only difference is that the angle θ changes due to the rotation of the field direction rather than to rotation of the coil. The flux through the coil is given by

$$\Phi = NBA \cos \theta = NBA \cos \omega t$$

To find the instantaneous emf, we need to know the instantaneous rate of change of the flux. Using Eq. (20-7b), where $\Phi_0 = NBA$,

$$\frac{\Delta \Phi}{\Delta t} = -\omega NBA \sin \omega t$$

From Faraday's law,

$$\mathcal{E} = -\frac{\Delta \Phi_B}{\Delta t} = \omega NBA \sin \omega t$$

exactly the same as for the rotating coil generator [Eq. (20-3b)].

Coils of wire in which current is induced

Induced current

Current to electro-magnet

Coil

N

S

Rotating electromagnet

Coil

(a)

Current to electro-magnet

S

N

S

N

Rotor (rotating electromagnet)

(b)

Stator

Figure 20.10
(a) Simplified model of a rotating field generator. The electromagnet rotates, while the two coils are stationary. The changing magnetic flux through the coils induces an electric current in the coils. (b) The *alternator* in an automobile is a rotating field generator.

continued on next page

Example 20.4 *continued*

Discussion Here we see that it does not matter whether a conducting coil rotates in a magnetic field or whether the magnetic field rotates about a stationary conducting coil. In either case there is a changing flux through the coil, which causes an emf to be induced. Since the flux as a function of time for the rotating field generator is identical to the flux for the rotating coil generator, the emfs generated are identical.

Practice Problem 20.4   Rotating coil generator

In a rotating coil generator, the magnetic field between the poles of an electromagnet has magnitude 0.40 T. A circular coil between the poles has 120 turns and radius 4.0 cm. The coil rotates with frequency 5.0 Hz. Find the maximum emf induced in the coil.

Earlier in this section, we wrote Faraday's law to give the magnitude of the induced emf due to a changing magnetic field. But that's only part of the story. Faraday's law gives the induced emf due to a changing magnetic flux, *no matter what the reason for the flux change*. The flux change can occur for reasons other than a changing magnetic field. A conducting loop might be moving through regions where the field is not constant, or it can be rotating, or changing size or shape; in all of these cases, Faraday's law as already stated gives the correct emf. The induced emf is equal to minus the rate of change of the flux, regardless of why the flux is changing. Recall that flux can be written

$$\Phi_B = BA \cos \theta \qquad (20\text{-}5)$$

Then the flux changes if the magnetic field strength (B) changes, or if the area of the loop (A) changes, or if the angle between the field and the normal changes.

Faraday's law says that, no matter what the reason for the change in flux, the induced emf is

$$\mathscr{E} = -\frac{\Delta \Phi_B}{\Delta t} \qquad (20\text{-}6)$$

The mobile charges in a moving conductor are pumped around due to the magnetic force on the charges. Since the conductor as a whole is moving, the mobile charges have a nonzero average velocity and therefore a nonzero average magnetic force. In the case of a changing magnetic field and a stationary conductor, the mobile charges aren't set into motion by the magnetic force—they have zero average velocity before current starts to flow. Exactly what *does* make current flow is considered in Section 20.8.

The Generation and Transmission of Electricity

An enormous amount of our technology depends on magnetic induction. There are so many applications of Faraday's law that it's hard to even begin a list. Certainly first on the list has to be the electric generator. Almost all of the electricity we use is produced by generators—either moving coil or moving field—that operate according to Faraday's law. Our entire system for distributing electricity is based on *transformers*, devices that use magnetic induction to change AC voltages (Section 20.6). Transformers raise voltages for transmission over long distances across power lines; transformers then reduce the voltages for safe use in homes and businesses. So our entire system for generating *and* distributing electricity depends on Faraday's law of induction.

Many other applications of Faraday's law are considered in this chapter, including in the questions and problems at the end of the chapter.

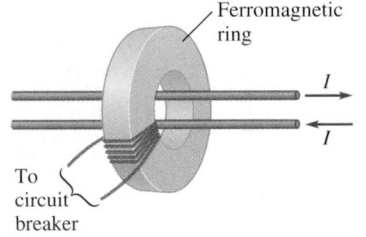

Figure 20.11   A ground fault interrupter

Making The Connection:
    **ground fault
    interrupter**

Ground Fault Interrupter

A *ground fault interrupter* (GFI) is a device commonly used in AC electric outlets in bathrooms and other places where the risk of electric shock is great. In Fig. 20.11, the two wires that supply the outlet normally carry equal currents in opposite directions at all times. These AC currents reverse direction 120 times per second. If a person with wet hands accidentally comes into contact with part of the circuit, a current may flow to ground through the person instead of through the return wiring. Then the currents in the two wires are unequal. The magnetic field lines due to the unequal currents are channeled by a ferromagnetic ring through a coil. The flux through the coil reverses directions 120 times per second, so there is an induced emf in the coil, which trips a circuit breaker that disconnects the circuit from the power lines. GFIs are sensitive and fast, so they are a significant safety improvement over a simple circuit breaker.

Magnetic Data Storage

When a recording is made on audiotape, ferromagnetic particles on the tape surface are magnetized by an electromagnet (the record head). The tape is played by passing it at constant speed beneath the gap in the playback head (Fig. 20.12). Magnetic field lines are channeled through the head, which is made of a ferromagnetic material. The field lines pass through a coil wrapped around the head. As the magnetization of the tape surface changes, the flux through this coil changes. The induced emf in the coil is amplified and then sent to a loudspeaker. In low-cost tape players, recording and playback are accomplished by the same head; higher-quality tape decks have separate record and playback heads.

Making The Connection:
computer hard disks, magnetic tape

Magnetic induction is at the heart of how other magnetic media are read. In a videocassette recorder, both the tape and the head are in motion to achieve a higher relative speed between the two. In a computer disk drive, the disk surfaces spin just below the read heads.

Figure 20.12 The playback head of an audiotape player

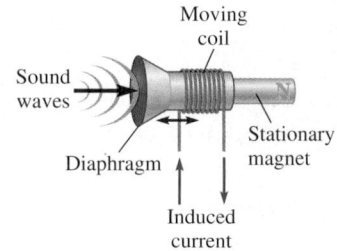

Figure 20.13 A moving coil microphone

Microphones

Figure 20.13 is a simplified sketch of a moving coil microphone. The coil of wire is attached to a diaphragm that moves back and forth in response to sound waves in the air. The magnet is fixed in place. An induced emf appears in the coil due to the changing magnetic flux. In another common type of microphone, the magnet is attached to the diaphragm and the coil is fixed in place.

Making The Connection:
moving coil microphone

Magnetoencephalography

Faraday's law provides another way to detect currents that flow in the human body. In addition to measuring potential differences between points on the skin, we can measure the magnetic fields generated by these currents. Since the currents are small, the magnetic fields are weak, so sensitive detectors called SQUIDs are used. When the currents change, changes in the magnetic field induce emfs in the SQUIDs. In a magnetoencephalogram, the induced emfs are measured at many points just outside the cranium (Fig. 20.14); then a computer calculates the location, magnitude, and direction of the currents in the brain that produce the field. Similarly, a magnetocardiogram detects the electric currents in the heart and surrounding nerves.

Making The Connection:
magnetoencephalography

20.4 LENZ'S LAW

The directions of the induced emfs and currents caused by a changing magnetic flux can be determined using **Lenz's law**:

Lenz's law
The direction of the induced emfs and currents always *opposes* the *change* that produces them.

Note that induced emfs and currents do not necessarily oppose the magnetic field or the magnetic flux; they oppose the *change* in the magnetic flux.

One way to apply Lenz's law is to look at the direction of the magnetic field produced by the induced current. The induced current around a loop produces its own magnetic field. This field may be weak compared to the external magnetic field. It cannot prevent the magnetic flux through the loop from changing, but its direction is always such that it "tries" to minimize the *change* in the flux. The magnetic field direction is related to the direction of the current by right-hand rule 2 (Section 19.8). Point the thumb in the direction of the current in the loop at any point. Then curl the fingers of the right hand to show the direction of the field lines around the wire near that point. As the fingers pass through the loop, they give the direction of the magnetic field inside.

Lenz's law is really an expression of energy conservation. That connection is not easy to make in general, but in specific cases the connection can often be quite apparent (see Conceptual Example 20.5).

Figure 20.14 In magneto-encephalography, brain function can be observed in real time through non-invasive means. The two white cryostats seen here contain sensitive magnetic field detectors cooled by liquid helium.

Conceptual Example 20.5

Faraday's and Lenz's Laws for the Moving Loop

Verify the emfs and currents calculated in Example 20.1 using Faraday's and Lenz's laws—that is, find the direction and magnitudes of the emfs and currents by looking at the changing magnetic flux through the loop.

 Strategy To apply Faraday's law, look for the reason why the flux is changing. In Example 20.1, a loop moves to the right at constant velocity into, through, and then out of a region of magnetic field. The magnitude and direction of the magnetic field within the region are not changing, nor is the area of the loop. What does change is the *portion* of that area that is immersed in the region of magnetic field.

Solution At positions 1, 3, and 5, the flux is *not* changing even though the loop is moving. In each case, a small displacement of the loop causes no flux change. The flux is zero at positions 1

and 5, and nonzero but constant at position 3. For these three positions, the induced emf is zero and so is the current.

If the loop were *at rest* at position 2, the magnetic flux would be constant. However, since the loop is moving into the region of field, the area of the loop through which magnetic field lines cross is increasing. Thus, the flux is increasing. According to Lenz's law, the direction of the induced current opposes the change in flux. Since the field is into the page, and the flux is increasing, the induced current flows in the direction that produces a magnetic field *out of* the page. By the right-hand rule, the current is counterclockwise.

At position 2, a length x of the loop is in the region of magnetic field. The area of the loop that is immersed in the field is Lx. The flux is then

$$\Phi_B = BA = BLx$$

Only x is changing. The rate of change of flux is

$$\frac{\Delta \Phi_B}{\Delta t} = BL\frac{\Delta x}{\Delta t} = BLv$$

continued on next page

Conceptual Example 20.5 *continued*

Therefore,

$$|\mathscr{E}| = BLv$$

and

$$I = \frac{|\mathscr{E}|}{R} = \frac{BLv}{R}$$

At position 4, the flux is decreasing as the loop leaves the region of magnetic field. Once again, let a length x of the loop be immersed in the field. Just as at position 2,

$$\Phi_B = BLx$$

$$|\mathscr{E}| = \left|\frac{\Delta\Phi_B}{\Delta t}\right| = BL\left|\frac{\Delta x}{\Delta t}\right| = BLv$$

and

$$I = \frac{|\mathscr{E}|}{R} = \frac{BLv}{R}$$

This time the flux is *decreasing*. To oppose a *decrease*, the induced current makes a magnetic field in the *same* direction as the external field—into the page. Then the current must be clockwise.

The magnitudes and directions of the emfs and currents are exactly as found in Example 20.1.

Discussion Another way to use Lenz's law to find the direction of the current is by looking at the magnetic force on the loop. The changing flux is due to the motion of the loop to the right. In order to oppose the change in flux, current flows in

the loop in whatever direction gives a magnetic force to the *left*, to try to bring the loop to rest and stop the flux from changing. At position 2, the magnetic forces on sides b and d are equal and opposite; there is no magnetic force on side a since $B = 0$ there. Then there must be a magnetic force on side c to the left. From $\vec{F} = I\vec{L} \times \vec{B}$, the current in side c is up and thus flows counterclockwise in the loop. Similarly, at position 4, the current in side a is upward to give a magnetic force to the left.

The connection between Lenz's law and energy conservation is more apparent when looking at the force on the loop. When current flows in the loop, electric energy is dissipated at a rate $P = I^2R$. Where does this energy come from? If there is no external force pulling the loop to the right, the magnetic force slows down the loop; the dissipated energy comes from the kinetic energy of the loop. To keep the loop moving to the right at constant velocity while current is flowing, an external force must pull it to the right. The work done by the external force replenishes the loop's kinetic energy.

Practice Problem 20.5 The magnetic force on the loop

(a) Find the magnetic force on the loop at positions 2 and 4 in terms of B, L, v, and R. (b) Verify that the rate at which an external force does work ($P = Fv$) to keep the loop moving at constant velocity is equal to the rate at which energy is dissipated in the loop ($P = I^2R$).

Conceptual Example 20.6

Lenz's Law for a Conducting Loop in a Changing Magnetic Field

A circular loop of wire moves toward a bar magnet at constant velocity (Fig. 20.15). The loop passes around the magnet and continues away from it on the other side. Use Lenz's law to find the direction of the current in the loop at positions 1 and 2.

Strategy The magnetic flux through the loop is changing because the loop moves from weaker to stronger field (at

position 1), and *vice versa* (at position 2). We can specify current directions as counterclockwise or clockwise as viewed from the left (with the loop moving away).

Solution At position 1, the magnetic field lines enter the magnet at the south pole, so the field lines cross the loop from left to right (Fig. 20.16a). Since the loop is moving closer to the magnet, the field is getting stronger; the number of field lines crossing the loop increases (Fig. 20.16b). The flux is therefore increasing. To oppose the increase, the current

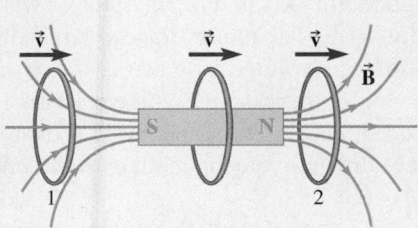

Figure 20.15
Conducting loop passing over a bar magnet

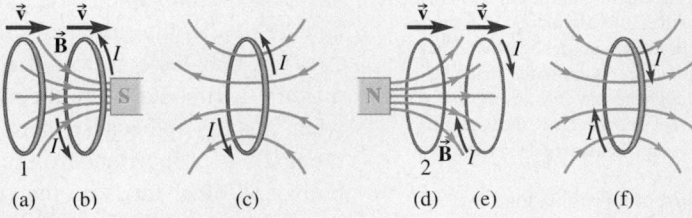

Figure 20.16
Loop moving toward magnet from position (a) to (b); (c) current induced in loop to produce a $\vec{B}$ field opposing the increasing strength of the nearing bar magnet; loop moving away from magnet from position (d) to (e); (f) current induced in loop to produce a $\vec{B}$ field opposing the decreasing strength of the retreating bar magnet.

continued on next page

Conceptual Example 20.6 *continued*

makes a magnetic field to the left (Fig. 20.16c). The right-hand rule gives the current direction to be counterclockwise as viewed from the left.

At position 2, the field lines still cross the loop from left to right (Fig. 20.16d), but now the field is getting weaker (Fig. 20.16e). The current must flow in the opposite direction—clockwise as viewed from the left (Fig. 20.16f).

 Discussion There's almost always more than one way to apply Lenz's law. An alternative way to think about the situation is to remember the current loop is a magnetic dipole and we can think of it as a little bar magnet. At position 1, the current loop is repelled by the (real) bar magnet. The flux change is due to the motion of the loop toward the magnet; to oppose the change there should be a force pushing away. Then the poles of the current loop must

be as in Fig. 20.17a; like poles repel. Point the thumb of the right hand in the direction of the north pole, and curl the fingers to find the current direction.

The same procedure can be used at position 2. Now the flux change is due to the loop moving away from the magnet, so to oppose the change in flux there must be a force attracting the loop toward the magnet (Fig. 20.17b).

Conceptual Practice Problem 20.6   Direction of induced emf in coil

(a) In Fig. 20.18, just after the switch is closed, what is the direction of the magnetic field in the iron core? (b) In what direction does current flow through the resistor connected to coil 2? (c) If the switch remains closed, does current continue to flow in coil 2? Why or why not? (d) Make a drawing in which coils 1 and 2, just after the switch is closed, are replaced by equivalent little bar magnets.

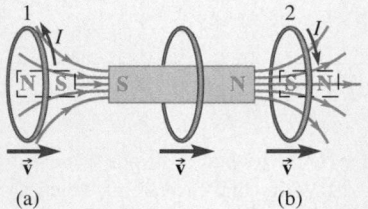

Figure 20.17
Current loops can be represented by small bar magnets.

Figure 20.18
Two coils wrapped about a common soft-iron core

Making The Connection:
back emf in a motor

Figure 20.19 An external emf ($\mathcal{E}_{ext}$) is connected to a DC motor. The resistance R represents the electrical resistance of the motor's windings plus the effects of friction inside the motor. The back emf ($\mathcal{E}_{back}$) is due to the changing flux through the windings. As the motor's rotational speed increases, the back emf increases and the current decreases.

The current through a load that is connected to an emf is sometimes called the current *drawn* by the load.

20.5   BACK EMF IN A MOTOR

If a generator and a motor are essentially the same device, is there an induced emf in the coil (or windings) of a motor? There must be, according to Faraday's law, since the magnetic flux through the coil changes as the coil rotates. By Lenz's law, this induced emf—called a **back emf**—opposes the flow of current in the coil, since it is the current that makes the coil rotate and thus causes the flux change. The magnitude of the back emf depends on the rate of change of the flux, so the back emf increases as the rotational speed of the coil increases.

Figure 20.19 shows a simple circuit model of the back emf in a DC motor. We assume that this motor has many coils (also called windings) at all different angles so that the torques, emfs, and currents are all constant. When the external emf is first applied, there is no back emf because the windings are not rotating. Then the current has a maximum value $I = \mathcal{E}_{ext}/R$. The faster the motor turns, the greater the back emf, and the smaller the current: $I = (\mathcal{E}_{ext} - \mathcal{E}_{back})/R$.

You may have noticed that when a large motor—as in a refrigerator or washing machine—first starts up, the room lights dim a bit. The motor draws a large current when it starts up because there is no back emf. The voltage drop across the wiring in the walls is proportional to the current flowing in them, so the voltage across light-bulbs and other loads on the circuit is reduced, causing a momentary "brown-out." As the motor comes up to speed, the current drawn is much smaller, so the brown-out ends.

If a motor is overloaded, so that it turns slowly or not at all, the current through the windings is large. Motors are designed to withstand such a large current only momentarily, as they start up; if the current is sustained at too high a level the motor "burns out"—the windings heat up enough to do damage to the motor.

20.6 TRANSFORMERS

Making The Connection:
transformers

In the late nineteenth century, there were ferocious battles over what form of current should be used to supply electric power to homes and businesses. Thomas Edison was a proponent of direct current, while George Westinghouse, who owned the patents for the AC motor and generator invented by Nikola Tesla, was in favor of alternating current. Westinghouse won mainly because AC permits the use of transformers to change voltages and to transmit over long distances with less power loss than DC, as we see in this section.

Figure 20.20 shows two simple transformers. In each, two separate strands of insulated wire are wound around a soft-iron core. Note that the two coils enclose the same magnetic field lines. An alternating voltage is applied to the *primary* coil; the AC current in the primary causes a changing magnetic flux through the *secondary* coil.

If the primary coil has N_1 turns, an emf $\mathscr{E}_1$ is induced in the primary coil according to Faraday's law:

$$\mathscr{E}_1 = -N_1 \frac{\Delta\Phi_B}{\Delta t} \qquad (20\text{-}8a)$$

Here $\Delta\Phi_B/\Delta t$ is the rate of change of the flux through *each turn* of the primary. Ignoring resistance in the coil and other energy losses, the induced emf is equal to the AC voltage applied to the primary.

If the secondary coil has N_2 turns, then the emf induced in the secondary coil is

$$\mathscr{E}_2 = -N_2 \frac{\Delta\Phi_B}{\Delta t} \qquad (20\text{-}8b)$$

At any instant, the flux through each turn of the secondary is equal to the flux through each turn of the primary, so $\Delta\Phi_B/\Delta t$ is the same quantity in Eqs. (20-8a) and (20-8b). Eliminating $\Delta\Phi_B/\Delta t$ from the two equations, we find the ratio of the two emfs to be

$$\frac{\mathscr{E}_2}{\mathscr{E}_1} = \frac{N_2}{N_1} \qquad (20\text{-}9)$$

The output—the emf in the secondary—is N_2/N_1 times the input emf applied to the primary. The ratio N_2/N_1 is called the **turns ratio**. A transformer is often called a *step-up* or a *step-down* transformer, depending on whether the secondary emf is larger or smaller than the emf applied to the primary. The same transformer may often be used as a step-up or step-down transformer depending on which coil is used as the primary.

Current Ratio

In an *ideal transformer*, power losses in the transformer itself are negligible. Most transformers are very efficient, so neglecting power loss is usually reasonable. Then the rate at which energy is supplied to the primary is equal to the rate at which energy is supplied by the secondary ($P_1 = P_2$). Since power equals voltage times current, the ratio of the currents is the inverse of the ratio of the emfs:

$$\frac{I_2}{I_1} = \frac{\mathscr{E}_1}{\mathscr{E}_2} = \frac{N_1}{N_2} \qquad (20\text{-}10)$$

Circuit symbol for a transformer

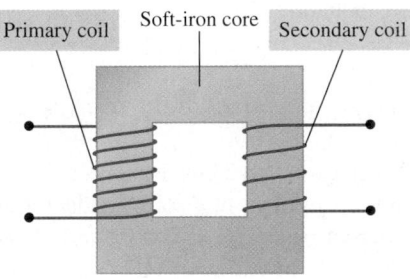

Figure 20.20 Two simple transformers. Each consists of two coils wound on a common soft-iron core so that nearly all the magnetic field lines produced by the primary coil pass through each turn of the secondary.

Example 20.7

A CD Player's Transformer

A transformer inside the power supply for a portable CD player has 500 turns in the primary coil. It supplies an emf of amplitude 6.8 V when plugged into the usual sinusoidal household emf of amplitude 170 V. (a) How many turns does the secondary coil have? (b) If the current drawn by the CD player has amplitude 1.50 A, what is the amplitude of the current in the primary?

Strategy   The ratio of the emfs is the same as the turns ratio. We know the two emfs and the number of turns in the primary, so we can find the number of turns in the secondary. To find the current in the primary, we assume an ideal transformer. Then the currents in the two are inversely proportional to the emfs.

Solution   (a) The turns ratio is equal to the emf ratio:

$$\frac{\mathscr{E}_2}{\mathscr{E}_1} = \frac{N_2}{N_1}$$

Solving for N_2 yields

$$N_2 = \frac{\mathscr{E}_2}{\mathscr{E}_1} N_1 = \frac{6.8 \text{ V}}{170 \text{ V}} \times 500 = 20 \text{ turns}$$

(b) The currents are inversely proportional to the emfs:

$$\frac{I_1}{I_2} = \frac{\mathscr{E}_2}{\mathscr{E}_1} = \frac{N_2}{N_1}$$

$$I_1 = \frac{\mathscr{E}_2}{\mathscr{E}_1} I_2 = \frac{6.8 \text{ V}}{170 \text{ V}} \times 1.50 \text{ A} = 0.060 \text{ A}$$

Discussion   The most likely error would be to get the turns ratio upside-down. Here we need a step-down transformer, so N_2 must be smaller than N_1. If the same transformer were hooked up backward, interchanging the primary and the secondary, then it would act as a step-up transformer. Instead of supplying 6.8 V to the CD player, it would supply

$$170 \text{ V} \times \frac{500}{20} = 4250 \text{ V}$$

We can check that the power input and the power output are equal:

$$P_1 = \mathscr{E}_1 I_1 = 170 \text{ V} \times 0.060 \text{ A} = 10.2 \text{ W}$$
$$P_2 = \mathscr{E}_2 I_2 = 6.8 \text{ V} \times 1.50 \text{ A} = 10.2 \text{ W}$$

(Since emfs and currents are sinusoidal, the instantaneous power is not constant. By multiplying the amplitudes of the current and emf, we calculate the *maximum* power.)

Practice Problem 20.7   An ideal transformer

An ideal transformer has five turns in the primary and two turns in the secondary. If the average power input to the primary is 10.0 W, what is the average power output of the secondary?

Transformers in the Distribution of Electricity

Making The Connection:
electric power distribution

Why is it so important to be able to transform voltages? The main reason is to minimize energy dissipation in power lines. Suppose that a power plant supplies a power P to a distant city. Since the power supplied is $P_s = I_s V_s$, where I_s and V_s are the current and voltage supplied to the load (the city), the plant can either supply a high voltage and a small current, or a low voltage and a large current. If the power lines have total resistance R, the rate of energy dissipation in the power lines is $I_s^2 R$. Thus, to minimize energy dissipation in the power lines, we want as small a current as possible flowing through them, which means the potential differences must be large—hundreds of kilovolts in some cases. (With a small current, the voltage across the power lines $V_{\text{power lines}} = I_s R$ is small compared to the voltage across the load V_s.) Transformers are used to raise the output emf of a generator to high voltages (Fig. 20.21). It would be unsafe to have such high voltages on household wiring, so the voltages are transformed back down before reaching the house.

20.7   EDDY CURRENTS

Whenever a conductor is subjected to a changing magnetic flux, the induced emf causes currents to flow. In a solid conductor, induced currents flow simultaneously along many different paths. These **eddy currents** are so named due to their resemblance to swirling eddies of current in air or in the rapids of a river. Though the pattern of current flow is complicated, we can still use Lenz's law to get a general idea of the direction of the current

Figure 20.21   Voltages are transformed in several stages. This step-up transformer raises the voltage from a generating station to 345 kV for transmission over long distances. Voltages are transformed back down in several stages. The last transformer in the series reduces the 3.4 kV on the local power lines to the 170 V used in the house.

flow (clockwise or counterclockwise). We can also determine the qualitative effects of eddy current flow using energy conservation. Since they flow in a resistive medium, the eddy currents dissipate electric energy.

Example 20.8

Eddy-Current Damping

A balance must have some damping mechanism. Without one, the balance arm would tend to oscillate for a long time before it settles down; determining the mass of an object would be a long, tedious process. A typical device used to damp out the oscillations is shown in Fig. 20.22.

A metal plate attached to the balance arm passes between the poles of a permanent magnet. (a) Explain the damping effect in terms of energy conservation. (b) When the plate moves upward through the magnetic field, make a sketch showing the general sense (CW or CCW) of the eddy currents flowing in it. The magnetic field is out of the page. (c) Use the direction of the eddy currents in the part of the plate in the magnetic field to find the magnetic force on the plate as it moves upward. (d) Does the damping force depend on the speed of the plate?

Figure 20.22
A balance. The damping mechanism is at the far right; as the balance arm oscillates, the metal plate moves between the poles of a magnet.

Strategy   As portions of the metal plate move into or out of the magnetic field, the changing magnetic flux induces emfs. These induced emfs cause the flow of eddy currents. Lenz's law determines the direction of the eddy currents.

Solution   (a) As the plate moves between the magnet poles, parts of it move into the magnetic field while other parts move out of the field. Due to the changing magnetic flux, induced emfs cause eddy currents to flow. The eddy currents dissipate energy; the energy must come from the kinetic energy of the balance arm, pan, and object on the pan. As the currents flow, the kinetic energy of the balance decreases and it comes to rest much sooner than it would otherwise.

(b) At the top edge of the magnet, the plate is moving out of the magnetic field. The flux decreases, so Lenz's law says that induced currents there produce a magnetic field out of the page to oppose the decrease. From the right-hand rule, the currents are counterclockwise.

At the bottom edge of the magnet, however, the plate is moving *into* the field. To oppose the flux increase, the currents there must be clockwise instead. Figure 20.23 shows the sense of the eddy currents in the plate.

(c) From Fig. 20.23, the currents in the region of magnetic field flow to the right. Using $\vec{F} = I\vec{L} \times \vec{B}$, the magnetic force is downward—as expected since the plate is moving up and the magnetic force must slow down the plate.

(d) If the plate is moving faster, the flux is changing faster. Faraday's law says that the induced emfs are proportional to the rate of change of the flux. Larger induced emfs cause larger currents to flow; therefore, the damping force is larger.

continued on next page

Example 20.8 *continued*

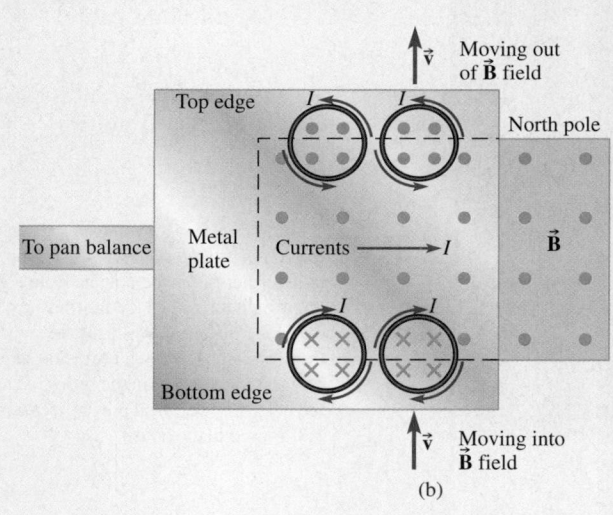

Figure 20.23
Metal plate serving as damping device

Discussion   Another way to approach part (a) is to use Lenz's law. The magnetic force acting on the eddy currents must oppose the flux change, so it must oppose the motion of the plate through the magnet. Slowing down the plate lessens the rate of flux change, while speeding up the plate would increase the rate of flux change—and increase the balance's kinetic energy, violating energy conservation.

Conceptual Practice Problem 20.8   Choosing a core for a transformer

In some transformers, the core around which wire is wrapped consists of parallel, insulated iron wires instead of solid iron (Fig. 20.24). Explain the advantage of using the insulated wires instead of the solid core. [*Hint:* Think about eddy currents. Why are eddy currents a disadvantage here?]

(a) Bundles of   (b) Solid soft-   **Figure 20.24**
   iron wires   iron core   Transformer cores

Physics at Home

If either the magnets or the metal plate are removed from a balance, it takes much longer for the oscillations of the balance arm to die out. If your instructor consents, test this on a laboratory balance. Usually a few screws need to be removed.

Eddy-Current Braking

Making The Connection:
eddy-current braking

The phenomenon described in Example 20.8 is called *eddy-current braking*. In the brakes on bicycles and most cars, the braking force is due to friction between two surfaces in contact—the brake pad and the metal rotor, drum, or rim. The constant scraping of brake pad and metal means that the brakes wear out and must periodically be replaced. There is no wear problem with eddy-current brakes, though there still is the issue of what to do with the heat generated. The eddy-current brake is ideal for a sensitive instrument such as a balance. The tiny amounts of heat generated are not a problem, the damping mechanism never wears out or needs adjustment, and we are guaranteed that it exerts no force when the balance arm is not moving. Eddy-current brakes are also used with modern rail vehicles such as the maglev monorail, tramways, locomotives, passenger coaches, freight cars, and the latest high-speed maglev trains.

The damping force due to eddy currents automatically acts opposite to the motion; its magnitude is also larger when the speed is larger. The damping force is much like the viscous force on an object moving through a fluid (see Problem 26).

The Induction Stove

Making The Connection:
induction stove

The induction stove discussed in the opening of this chapter operates via eddy currents. Under the cooking surface is an electromagnet that generates an oscillating magnetic field. When a metal pan is put on the stove, the emf causes currents to flow, and the energy dissipated by these currents is what heats the pan (Fig. 20.25). The pan must be made of metal; if a pan made of Pyrex glass is used, no currents flow and no heating occurs. For the same reason, there is no risk of starting a fire if a pot holder or sheet of paper is accidentally put on the induction stove. The cooking surface itself is a nonconductor; its temperature only rises to the extent that heat is conducted to it from the pan. The cooking surface therefore gets no hotter than the bottom of the pan.

20.8   INDUCED ELECTRIC FIELDS

When a conductor moves in a magnetic field, a motional emf arises due to the magnetic force on the mobile charges. Since the charges move along with the conductor, they have a nonzero average velocity. The magnetic force on these charges pushes them around the circuit if a complete circuit exists.

Figure 20.25 The eddy currents induced in a metal pan on an induction stove

What causes the induced emf in a stationary conductor in a changing magnetic field? Now the conductor is at rest and the mobile charges have an average velocity of zero. The average magnetic force on them is then zero, so it cannot be the magnetic force that pushes the charges around the circuit. An **induced electric field**, created by the changing magnetic field, acts on the mobile charge in the conductor, pushing it around the circuit. The same force law ($\vec{\mathbf{F}} = q\vec{\mathbf{E}}$) applies to induced electric fields as to any other electric field.

The induced emf around a loop is the work done per unit charge on a charged particle that moves around the loop. Thus, an induced electric field does nonzero work on a charge that moves around a closed path, starting and ending at the same point. In other words, the induced electric field is nonconservative. The work done by the induced $\vec{\mathbf{E}}$ field *cannot* be described as the charge times the potential difference. The concept of potential depends on the electric field doing zero work on a charge moving around a closed path—only then can the potential have a unique value at each point in space. Table 20.1 summarizes the differences between conservative and nonconservative $\vec{\mathbf{E}}$ fields.

In general, the problem of finding the induced electric field due to a changing magnetic field is beyond the scope of this book. However, there is one case in which we can calculate the induced electric field. Suppose an electromagnet with flat, circular pole faces produces a uniform magnetic field that is increasing at a constant rate $\Delta B/\Delta t$. A circular loop of thin wire of radius r is centered between the pole faces. An induced electric field causes current to flow in the loop.

The induced electric field must have field lines that are closed loops. Since the field is uniform within a cylindrical volume of space, the $\vec{\mathbf{E}}$ field lines must reflect that cylindrical symmetry: they must be concentric circles (Fig. 20.26). The loop coincides with one of the circular electric field lines, so the induced electric field is everywhere tangent to the loop and has the same magnitude at every point on the loop.

Table 20.1

Comparison of Conservative and Nonconservative $\vec{\mathbf{E}}$ Fields

	Conservative $\vec{\mathbf{E}}$ Fields	Nonconservative (Induced) $\vec{\mathbf{E}}$ Fields
Source	Charges	Changing $\vec{\mathbf{B}}$ fields
Field lines	Start on positive charges and end on negative charges	Closed loops
Can be described by an electric potential?	Yes	No
Work done over a closed path	Always zero	Can be nonzero work over a closed path
Circulation	Always zero	Can be nonzero; the emf around a closed path is equal to the circulation of the field around that path

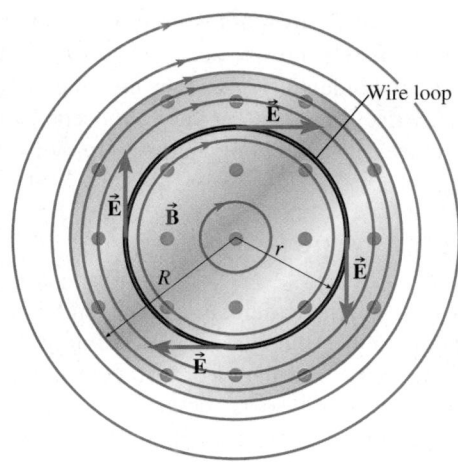

Figure 20.26 The magnetic field of an electromagnet with circular pole faces is out of the page and increasing in magnitude. The induced electric field lines are circular. The electric field is tangent to the circular wire loop.

Since the electric force is always parallel to the motion of a charge that goes around the loop, the work done on a charge q is

$$W = \Sigma F \, \Delta s$$

Here Δs is a small displacement of the charge. Both q and E are constant as we move around the loop. Since the sum of the displacements during one trip around the circle ($\Sigma \, \Delta s$) is the circumference ($2\pi r$), the work done on a charge during one revolution is

$$W = \Sigma F \, \Delta s = \Sigma qE \, \Delta s = qE \, \Sigma \, \Delta s = qE \times 2\pi r$$

Faraday's law gives the induced emf. For $r \leq R$,

$$|\mathscr{E}| = \frac{\Delta \Phi_B}{\Delta t} = A \frac{\Delta B}{\Delta t} = \pi r^2 \frac{\Delta B}{\Delta t}$$

The emf is the work done per unit charge:

$$|\mathscr{E}| = \frac{W}{q}$$

Equating the two expressions for emf yields

$$\pi r^2 \frac{\Delta B}{\Delta t} = 2\pi r E$$

Now we can solve for E:

$$E = \frac{1}{2} r \frac{\Delta B}{\Delta t} \quad (r \leq R) \tag{20-11}$$

 Note that the induced electric field would be present *whether the wire loop is present or not*. Without the loop no current would flow, of course, but the $\vec{\mathbf{E}}$ field would be the same. Equation (20-11) gives the magnitude of the electric field at all points between the electromagnet's pole faces.

Betatron

Making The Connection:
betatron

The *betatron* is a kind of electron accelerator. (Historically, electrons were called *beta rays* or *beta particles*.) In contrast to the cyclotron, in which the magnetic field is constant, the betatron features a magnetic field that increases in magnitude so that the electrons move in a circle of fixed radius R. The increasing magnetic field also gives rise to an induced electric field. This induced electric field does work on the electrons, increasing their kinetic energy. One use of the betatron is to accelerate electrons, then smash them into a target, which produces high-energy x-rays that can be used in cancer treatment.

Electromagnetic Fields

How can Faraday's law give the induced emf regardless of why the flux is changing—whether because of a changing magnetic field or because of a conductor moving in a magnetic field? A conductor that is moving in one frame of reference is at rest in another frame of reference. As we will see in Chapter 26, Einstein's theory of special relativity says that either reference frame is equally valid. In one frame, the induced emf is due to the motion of the conductor; in the other, the induced emf is due to a changing magnetic field.

The electric and magnetic fields are not really separate entities. They are intimately connected. Though it is advantageous in many circumstances to think of them as distinct fields, a more accurate view is to think of them as two aspects of the **electromagnetic field**. To use a loose analogy: a vector has different *x*- and *y*-components in different coordinate systems, but these components represent the same vector quantity. In the same way, the electromagnetic field has electric and magnetic parts (analogous to vector components) that depend on the frame of reference. A purely electric field in one frame of reference has both electric and magnetic "components" in another reference frame.

You may notice a missing symmetry. If a changing $\vec{\mathbf{B}}$ field is always accompanied by an induced $\vec{\mathbf{E}}$ field, what about the other way around? Does a changing electric field make an induced magnetic field? The answer to this important question—central to our understanding of light as an electromagnetic wave—is yes (Chapter 22).

20.9   MUTUAL- AND SELF-INDUCTANCE

Mutual-Inductance

Figure 20.27 shows two coils of wire. A power supply with variable emf causes current I_1 to flow in coil 1; the current produces magnetic field lines as shown. Some of these field lines cross through the turns of coil 2. If we adjust the power supply so that I_1 changes, the flux through coil 2 changes and an induced emf appears in coil 2. **Mutual-inductance**—when a changing current in one device causes an induced emf in another device—can occur between two circuit elements in the same circuit as well as between circuit elements in two different circuits. In either case, a changing current through one element induces an emf in the other. The effect is truly mutual: a changing current in coil 2 induces an emf in coil 1 as well.

At any point, the magnetic field due to coil 1 is proportional to I_1. For instance, if we double I_1, the magnetic field everywhere would be twice as large. The total flux through coil 2 is proportional to the magnetic field, and therefore to the current I_1:

$$\Phi_{21} \propto I_1$$

where the subscripts "21" remind us that Φ_{21} stands for the total flux through coil 2 due to the field produced by coil 1. The constant of proportionality is called the *mutual-inductance* (*M*):

$$\Phi_{21} = MI_1$$

The mutual-inductance depends on the shape and size of the two circuit elements, their separation, and their relative orientation. It is exceedingly difficult to calculate mutual-inductances from the geometry of the two elements. In every case, the mutual-inductance *M* turns out to be the same regardless of which way we look at it:

$$M = \frac{\Phi_{21}}{I_1} = \frac{\Phi_{12}}{I_2} \tag{20-12}$$

From Faraday's law, the induced emf in coil 2 is

$$\mathscr{E}_{21} = -\frac{\Delta\Phi_{21}}{\Delta t} = -M\frac{\Delta I_1}{\Delta t} \tag{20-13}$$

Similarly, the induced emf in coil 1 is

$$\mathscr{E}_{12} = -\frac{\Delta\Phi_{12}}{\Delta t} = -M\frac{\Delta I_2}{\Delta t}$$

Figure 20.27   An induced emf appears in coil 2 due to the changing current in coil 1.

Recall that, to give the *instantaneous* emf, Faraday's law must be applied to a very short time interval. If Δt represents a longer time interval, then Faraday's law gives the *average* emf over that time interval.

From Eq. (20-13), we can find the SI units of M:

$$[M] = \frac{[\mathscr{E}]}{[\Delta I/\Delta t]} = \frac{\text{V}}{\text{A/s}} = \frac{\text{V·s}}{\text{A}}$$

This combination of units is given the name henry (symbol: H) after Joseph Henry (1797–1878), the American scientist who was the first to wrap insulated wires around an iron core to make an electromagnet. Henry actually discovered induced emfs before Faraday, but Faraday published first.

Example 20.9

Mutual-Inducance

A circular loop of wire is placed near a solenoid (Fig. 20.28). When the current in the solenoid is 550 mA, the flux through the circular loop is 2.7×10^{-5} Wb. When the current in the solenoid changes at 6.0 A/s, the induced current in the circular loop is 0.36 mA. What is the resistance of the circular loop?

Strategy   The mutual-inductance is the proportionality constant between the current in the solenoid and the flux through the loop. M is also the proportionality constant between the rate of change of current in the solenoid and the *emf* in the loop. From the induced emf and the induced current, we can find the resistance of the loop.

Solution   The mutual-inductance is

$$M = \frac{\Phi_{\text{ls}}}{I_{\text{s}}} = \frac{2.7 \times 10^{-5}\ \text{Wb}}{0.550\ \text{A}} = 4.91 \times 10^{-5}\ \text{H}$$

Solenoid

Loop

Figure 20.28
A changing current in the solenoid induces an emf in the loop; a changing current in the loop also induces an emf in the solenoid.

The subscript "l" stands for "loop" and "s" stands for "solenoid." Then when the current in the solenoid changes at the rate $\Delta I_{\text{s}}/\Delta t = 6.0$ A/s, the induced emf in the loop is

$$|\mathscr{E}_{\text{ls}}| = M\left|\frac{\Delta I_{\text{s}}}{\Delta t}\right| = 4.91 \times 10^{-5}\ \frac{\text{V·s}}{\text{A}} \times 6.0\ \frac{\text{A}}{\text{s}} = 2.95 \times 10^{-4}\ \text{V}$$

The resistance of the loop is

$$R = \frac{|\mathscr{E}_{\text{ls}}|}{I} = \frac{0.295\ \text{mV}}{0.36\ \text{mA}} = 0.82\ \Omega$$

Discussion   The mutual-inductance determines the induced emf in the loop for a given rate of change of current in the solenoid. How much current flows in the loop in response to the induced emf depends on the loop's electrical resistance. A loop with a much higher resistance would have the *same* emf, but a much smaller current would flow.

Practice Problem 20.9   Flux through the solenoid

If a 1.5-V power supply is connected to the loop, what would be the total magnetic flux through the solenoid due to the loop's magnetic field?

Self-Inductance

Henry was the first to suggest that a changing current in a coil induces an emf in the *same* coil as well as in other coils—an effect called **self-inductance**. When the current through the coil is changing, the changing magnetic flux inside the coil produces an induced electric field that gives rise to an induced emf. When a coil, solenoid, toroid, or other circuit element is used in a circuit primarily for its self-inductance effects, it is often referred to as an **inductor**. *Self-inductance* is often shortened to *inductance*. A few inductors are shown in Fig. 20.29.

To calculate the self-inductance of coil 1 in Fig. 20.27, we follow the same steps that we used to find mutual-inductance. If a current I_1 flows in coil 1, then the total flux through coil 1 is proportional to I_1:

$$\Phi_{11} \propto I_1$$

The subscripts "11" remind us that Φ stands for the total flux through coil 1 due to the current in coil 1. When the context is clearly one of self-inductance, we write simply $\Phi \propto I$.

The circuit symbol for an inductor is

Figure 20.29 Inductors come in many sizes and shapes.

The self-inductance L of the coil is defined as the constant of proportionality between self-flux and current:

Definition of self-inductance:
$$\Phi = LI \qquad (20\text{-}14)$$

The most common form of inductor is the solenoid. In Problem 35, the self-inductance L of a long air-core solenoid of n turns per unit length, length ℓ, and radius r is found to be

$$L = \mu_0 n^2 \pi r^2 \ell \qquad (20\text{-}15a)$$

For a solenoid with $N = nl$ turns,

$$L = \frac{\mu_0 N^2 \pi r^2}{\ell} \qquad (20\text{-}15b)$$

According to Faraday's law, the induced emf in coil 1 is then

---

$$\mathcal{E} = -\frac{\Delta \Phi}{\Delta t} = -L\frac{\Delta I}{\Delta t} \qquad (20\text{-}16)$$

The SI unit of self-inductance is the same as that of mutual-inductance: the henry.

The behavior of an inductor in a circuit can be summarized as current *stabilizer*. The inductor "likes" the current to be constant—it "tries" to maintain the status quo. If the current is constant, there is no induced emf; to the extent that we can neglect the resistance of its windings, the inductor acts like a short circuit. When the current is changing, the induced emf is proportional to the rate of change. According to Lenz's law, the direction of the emf opposes the change that produces it. If the current is increasing, the direction of the emf in the inductor pushes back as if to make it harder for the current to increase (Fig. 20.30a). If the current is decreasing, the direction of the emf in the inductor is forward, as if to help the current keep flowing (Fig. 20.30b).

An inductor stores energy in a magnetic field, just as a capacitor stores energy in an electric field. Suppose the current in an inductor increases at a constant rate from 0 to I in a time T. We let lowercase i stand for the instantaneous current at some time t between 0 and T, and let uppercase I stand for the *final* current. The instantaneous rate at which energy accumulates in the inductor is

$$P = \mathcal{E}i$$

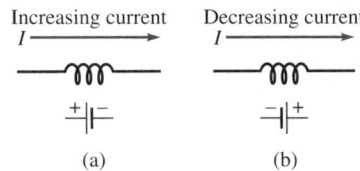

Increasing current   Decreasing current

(a)   (b)

Figure 20.30 The current through both these inductors flows to the right. In (a), the current is increasing; the induced emf in the inductor "tries" to prevent the increase. In (b), the current is decreasing; the induced emf in the inductor "tries" to prevent the decrease.

Since current increases at a constant rate, the magnetic flux increases at a constant rate, so the induced emf is constant. Also, since the current increases at a constant rate, the average current is $I_{av} = I/2$. Then the *average* rate at which energy accumulates is

$$P_{av} = \mathcal{E}I_{av} = \tfrac{1}{2}\mathcal{E}I$$

Using Eq. (20-16) for the emf, the average power is

$$P_{av} = \frac{1}{2}L\frac{\Delta i}{\Delta t}I$$

and the total energy stored in the inductor is

$$U = P_{av}T = \frac{1}{2}\left(L\frac{\Delta i}{\Delta t}\right)IT$$

Since the current changes at a constant rate, $\Delta i/\Delta t = I/T$, the total energy stored in the inductor is

---

Magnetic energy stored in an inductor

$$U = \tfrac{1}{2}LI^2 \qquad\qquad (20\text{-}17)$$

Although to simplify the calculation we assumed that the current was increased from zero at a constant rate, Eq. (20-17) for the energy stored in an inductor depends only on the current I and not on how the current reached that value (see Problem 37).

Compare the energy stored in an inductor and the energy stored in a capacitor:

$$U_C = \tfrac{1}{2}C^{-1}Q^2 \qquad\qquad (17\text{-}18c)$$

The energy in the inductor is proportional to the square of the current, just as the energy in the capacitor is proportional to the square of the charge. We can use the inductor to find the magnetic energy density in a magnetic field, just as we found the energy density in an electric field using a capacitor. In a long solenoid, the inductance is given by

$$L = \mu_0 n^2 \pi r^2 \ell$$

where n is the number of turns per unit length, ℓ is the length of the solenoid, and r is its radius. The energy stored in the inductor when a current I flows is

$$U = \tfrac{1}{2}LI^2 = \tfrac{1}{2}\mu_0 n^2 \pi r^2 \ell I^2$$

The volume of space inside the solenoid is the length times the cross-sectional area:

$$\text{volume} = \pi r^2 \ell$$

Then the magnetic energy density—energy per unit volume—is

$$u_B = \frac{U}{\pi r^2 \ell} = \frac{1}{2}\mu_0 n^2 I^2$$

To express the energy density in terms of the magnetic field strength, recall that $B = \mu_0 nI$ [Eq. (19-16)] inside a long solenoid. Therefore,

---

Magnetic energy density

$$u_B = \frac{1}{2\mu_0}B^2 \qquad\qquad (20\text{-}18)$$

Equation (20-18) is valid for more than the interior of an air-core solenoid; it gives the energy density for *any* magnetic field except for the field inside a ferromagnet. Both the magnetic energy density and the electric energy density are proportional to the square of the field strength: recall that the electric energy density is

$$u_E = \tfrac{1}{2}\kappa\epsilon_0 E^2 \qquad\qquad (17\text{-}19)$$

Example 20.10

Energy Stored in a Solenoid

An ideal air-core solenoid has radius of 2.0 cm, length 12 cm, and 9000.0 turns. The solenoid carries a current of 2.0 A. (a) Find the magnetic field inside the solenoid. (b) How much energy is stored in the solenoid?

Strategy   Since the solenoid is ideal, we ignore the nonuniformity in the magnetic field near the ends. We consider the magnetic field to be uniform in the entire volume inside. There are two ways to find the energy. We can either find the self-inductance and then use Eq. (20-17) for the energy stored in an inductor, or we can find the energy density [Eq. (20-18)] and multiply by the volume.

Given: $N = 9000$, $r = 0.020$ m, $\ell = 0.12$ m
Find B, U_B

Solution   (a) The magnetic field inside an ideal solenoid is

$$B = \mu_0 nI = \frac{\mu_0 NI}{\ell}$$

$$= \frac{4\pi \times 10^{-7} \text{ H/m} \times 9000.0 \times 2.0 \text{ A}}{0.12 \text{ m}} = 0.19 \text{ T}$$

(b) The magnetic energy density is

$$u_B = \frac{1}{2\mu_0} B^2 \qquad (20\text{-}18)$$

The total energy stored in the solenoid is

$$U_B = \frac{1}{2\mu_0} B^2 \times \pi r^2 \ell$$

since $\pi r^2 \ell$ is the volume of the interior of the solenoid. Substituting the expression for B yields

$$U_B = \frac{1}{2\mu_0}\left(\frac{\mu_0 NI}{\ell}\right)^2 \times \pi r^2 \ell$$

$$= \frac{\mu_0 N^2 I^2 \pi r^2}{2\ell}$$

Now we can substitute numerical values.

$$U_B = \frac{4\pi \times 10^{-7} \text{ H/m} \times 9000.0^2 \times (2.0 \text{ A})^2 \times \pi (0.020 \text{ m})^2}{2 \times 0.12 \text{ m}} = 2.1 \text{ J}$$

Discussion   Let's use the alternative method as a check. For a solenoid with N turns,

$$L = \frac{\mu_0 N^2 \pi r^2}{\ell} \qquad (20\text{-}15\text{b})$$

The magnetic energy stored in an inductor is

$$U_B = \tfrac{1}{2} LI^2 \qquad (20\text{-}17)$$

By substitution,

$$U_B = \frac{1}{2}\frac{\mu_0 N^2 \pi r^2}{\ell} I^2$$

which agrees with the expression found previously.

We should also verify the units. Since $1 \text{ H} = \dfrac{1 \text{ V}}{\text{A/s}}$,

$$\frac{\text{H/m} \times \text{A}^2 \times \text{m}^2}{\text{m}} = \text{H} \times \text{A}^2 = \frac{\text{V} \times \text{A}^2}{\text{A/s}} = \text{V} \times \text{A} \times \text{s} = \text{V} \times \text{C} = \text{J}$$

Practice Problem 20.10   Power in an inductor
Suppose the current in the inductor of Example 20.10 increases from 0 to 2.0 A during a time interval of 4.0 s. Calculate the average rate at which energy is stored in the inductor during this time interval. [*Hint:* Use one method to calculate the answer and another as a check.]

20.10 *LR* CIRCUITS

To get an idea of how inductors behave in circuits, let's first study them in DC circuits—that is, in circuits with batteries or other constant-voltage power supplies. Consider the *LR* **circuit** in Fig. 20.31a. The inductor is assumed to be ideal: its windings have negligible resistance. At $t = 0$, the switch S is closed. What is the subsequent current in the circuit?

The current through the inductor just before the switch is closed is zero. As the switch is closed, the current is initially zero. A sudden change in current through an inductor means a sudden change in its stored energy, since $U \propto I^2$. A sudden change in energy means that energy is supplied in zero time. Since nothing can provide infinite power,

> Current through an inductor must always change *continuously*, never instantaneously.

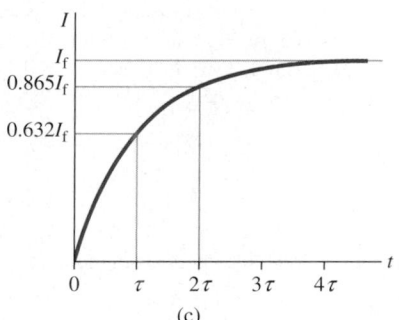

Figure 20.31 (a) A DC circuit with an inductor L, a resistor R, and a switch S. When the current is changing, an emf is induced in the inductor (represented by a battery symbol above the inductor). (b) The voltage drop across the inductor as the current builds up. (c) The current in the circuit as a function of time.

The initial current is zero, so there is no voltage drop across the resistor. The magnitude of the induced emf in the inductor ($\mathscr{E}_L$) is *initially* equal to the battery's emf ($\mathscr{E}_b$). Therefore, the current is rising at an initial rate given by

$$\frac{\Delta I}{\Delta t} = \frac{\mathscr{E}_b}{L}$$

As current builds up, the voltage drop across the resistor increases (Fig. 20.31b). Then the induced emf in the inductor ($\mathscr{E}_L$) gets smaller so that

$$(\mathscr{E}_b - \mathscr{E}_L) - IR = 0 \tag{20-19a}$$

or

$$\mathscr{E}_b = \mathscr{E}_L + IR \tag{20-19b}$$

Since the voltage across an *ideal* inductor is the induced emf, we can substitute $\mathscr{E}_L = L(\Delta I/\Delta t)$: [The minus sign has already been written explicitly in Eq. (20-19); $\mathscr{E}_L$ here stands for the *magnitude* of the emf.]

$$\mathscr{E}_b = L\frac{\Delta I}{\Delta t} + IR \tag{20-20}$$

The battery emf is constant. Thus, as the current increases, the voltage drop across the resistor gets larger and the induced emf in the inductor gets smaller. The only way that can happen is if the *rate* at which the current increases gets smaller (Fig. 20.31c).

After a very long time, the current reaches a stable value. Since the current is no longer changing, there is no voltage drop across the inductor, so $\mathscr{E}_b = I_f R$ or

$$I_f = \frac{\mathscr{E}_b}{R}$$

The equation for the current as a function of time $I(t)$ is an exponential function, similar to the charge on a charging capacitor:

$$I(t) = I_f(1 - e^{-t/\tau}) \tag{20-21}$$

The time constant τ for this circuit must be some combination of L, R, and $\mathscr{E}$. Dimensional analysis (Problem 49) shows that τ must be some dimensionless constant times L/R. It can be proved with calculus that the dimensionless constant is 1:

Time constant, *LR* circuit

$$\tau = \frac{L}{R} \tag{20-22}$$

The induced emf as a function of time is

$$\mathscr{E}_{L}(t) = \mathscr{E}_{b} - IR = \mathscr{E}_{b} - \frac{\mathscr{E}_{b}}{R}(1 - e^{-t/\tau})R = \mathscr{E}_{b}e^{-t/\tau} \tag{20-23}$$

The *LR* circuit in which the current is initially zero is analogous to the charging *RC* circuit. In both cases, the device starts with no stored energy and gains energy after the switch is closed. In charging a capacitor, the *voltage* eventually reaches a nonzero equilibrium value, while for the inductor the *current* reaches a nonzero equilibrium value. Compare the graphs of Fig. 18.35 with those of Fig. 20.31; current and voltage have switched places.

What about an *LR* circuit analogous to the discharging *RC* circuit? That is, once a steady current is flowing through an inductor, and energy is stored in the inductor, how can we stop the current and reclaim the stored energy? Simply opening the switch in Fig. 20.31a would *not* be a good way to do it. The attempt to suddenly stop the current would induce a *huge* emf in the inductor. Most likely, sparks would complete the circuit across the open switch, completing the circuit once again and allowing the current to die out more gradually. Sparking generally isn't good for the health of the switch, though.

A better way to stop the current is shown in Fig. 20.32. Initially switch S_1 is closed and a current $I_0 = \mathscr{E}_b/R_1$ is flowing through the inductor (Fig. 20.32a). Switch S_2 is closed and then S_1 is immediately opened at $t = 0$. Since the current through the inductor can only change continuously, the current flows as shown in Fig. 20.32b. At $t = 0$, the current is $I_0 = \mathscr{E}_b/R_1$. The current gradually dies out as the energy stored in the inductor is dissipated in resistor R_2. The current as a function of time is a decaying exponential:

$$I(t) = I_0 e^{-t/\tau} \tag{20-24}$$

where

$$\tau = \frac{L}{R_2}$$

The voltages across the inductor and resistor can be found from the loop rule and Ohm's law. Table 20.2 organizes what we know about *RC* and *LR* circuits.

(a)

(b)

Figure 20.32 A circuit that allows the current in the inductor circuit to be safely stopped. (a) Initially switch 1 is closed and switch 2 is open. (b) At $t = 0$, switch 2 is closed and then switch 1 immediately opened.

Table 20.2

Comparison of *RC* and *LR* Circuits

	Capacitor	Inductor
Voltage is proportional to	Charge	Rate of change of current
Can change discontinuously	Current	Voltage
Cannot change discontinuously	Voltage	Current
Energy stored (U) is proportional to	V^2	I^2
When $V = 0$ and $I \neq 0$	$U = 0$	$U = $ maximum
When $I = 0$ and $V \neq 0$	$U = $ maximum	$U = 0$
Energy stored (U) is proportional to	E^2	B^2
Time constant =	RC	L/R
"Charging" circuit	$I(t) \propto e^{-t/\tau}$	$I(t) \propto (1 - e^{-t/\tau})$
	$V_C(t) \propto (1 - e^{-t/\tau})$	$V_L(t) = \mathscr{E}_L(t) \propto e^{-t/\tau}$
"Discharging" circuit	$I(t) \propto e^{-t/\tau}$	$I(t) \propto e^{-t/\tau}$
	$V_C(t) \propto e^{-t/\tau}$	$V_L(t) = \mathscr{E}_L(t) \propto e^{-t/\tau}$

Example 20.11

Switching on a Large Electromagnet

A large electromagnet has an inductance $L = 15$ H. The resistance of the windings is $R = 8.2\ \Omega$. Treat the electromagnet as an ideal inductor in series with a resistor (as in Fig. 20.31a). When a switch is closed, a 24-V DC power supply is connected to the electromagnet. (a) What is the ultimate current through the windings of the electromagnet? (b) How long after closing the switch does it take for the current to reach 99.0% of its final value?

Strategy   When the current reaches its final value, there is no induced emf. The ideal inductor in Fig. 20.31a therefore has no potential difference across it. Then the entire voltage of the power source is across the resistor. The current follows an exponential curve as it builds to its final value. When it is at 99.0% of its final value, it has 1.0% left to go.

Solution   (a) After the switch has been closed for many time constants, the current reaches a steady value. When the current is no longer changing, there is no induced emf. Therefore, the entire 24 V of the power supply is dropped across the resistor:

$$\mathscr{E}_b = \mathscr{E}_L + IR$$

when $\mathscr{E}_L = 0, \quad I_f = \dfrac{\mathscr{E}_b}{R} = \dfrac{24\ \text{V}}{8.2\ \Omega} = 2.9\ \text{A}$

(b) The factor $e^{-t/\tau}$ represents the fraction of the current yet to build up. When the current reaches 99% of its final value,

$$1 - e^{-t/\tau} = 0.99$$

or

$$e^{-t/\tau} = 0.01$$

There is 1% yet to go. To solve for t, first take the natural log (ln) of both sides to get t out of the exponent:

$$\ln(e^{-t/\tau}) = -t/\tau = \ln 0.01 = -4.61$$

Now solve for t:

$$t = -\tau \ln 0.010 = -\dfrac{L}{R}\ln 0.010 = -\dfrac{15\ \text{H}}{8.2\ \Omega} \times (-4.61) = 8.4\ \text{s}$$

It takes 8.4 s for the current to build up to 99% of its final value.

Discussion   A slightly different approach is to write the current as a function of time:

$$I(t) = \dfrac{\mathscr{E}_b}{R}(1 - e^{-t/\tau}) = I_f(1 - e^{-t/\tau})$$

We are looking for the time t at which $I = 99\%$ of 2.9 A or $I/I_f = 0.99$. Then

$$0.99 = (1 - e^{-t/\tau}) \quad \text{or} \quad e^{-t/\tau} = 0.01$$

as before.

Practice Problem 20.11   Switching off the electromagnet

When the electromagnet is to be turned off, it is connected to a 50.0-Ω resistor, as in Fig. 20.33 to allow the current to decrease gradually. How long after the switch is opened does it take for the current to decrease to 0.1 A?

24 V

50.0 Ω

15 H

8.2 Ω

Electromagnet

Figure 20.33
Practice Problem 20.11

MASTER THE CONCEPTS

Summary

- A conductor moving through a magnetic field develops a motional emf given by

$$\mathcal{E} = vBL \qquad (20\text{-}2a)$$

  if both $\vec{v}$ and $\vec{B}$ are perpendicular to the rod.

- The emf due to an AC generator with one planar coil of wire turning in a uniform magnetic field is sinusoidal and has amplitude ωNBA:

$$\mathcal{E}(t) = \omega NBA \sin \omega t \qquad (20\text{-}3b)$$

  Here ω is the angular speed of the coil, A is its area, and N is the number of turns.

- Magnetic flux through a planar surface:

$$\Phi_B = B_\perp A = BA_\perp = BA \cos \theta \qquad (20\text{-}5)$$

  The magnetic flux is proportional to the number of magnetic field lines that cut through a surface. The SI unit of magnetic flux is the weber (1 Wb = 1 T·m^2).

- Faraday's law gives the induced emf whenever there is a changing magnetic flux, regardless of the reason the flux is changing:

$$\mathcal{E} = -\frac{\Delta \Phi_B}{\Delta t} \qquad (20\text{-}6)$$

- Lenz's law: the direction of an induced emf or an induced current *opposes* the *change* that caused it.

- The back emf in a motor increases as the rotational speed increases.

- For an ideal transformer,

$$\frac{\mathcal{E}_2}{\mathcal{E}_1} = \frac{N_2}{N_1} = \frac{I_1}{I_2} \qquad (20\text{-}9, 10)$$

  The ratio N_2/N_1 is called the turns ratio. There is no energy loss in an ideal transformer, so the power input is equal to the power output.

- Whenever a solid conductor is subjected to a changing magnetic flux, the induced emf causes eddy currents to flow simultaneously along many different paths. Eddy currents dissipate energy.

- A changing magnetic field gives rise to an induced electric field. The induced emf is the circulation of the induced electric field.

- A changing current in one circuit element induces an emf in another circuit element. The mutual-inductance is the constant of proportionality between the rate of change of the current and the induced emf.

$$M = \frac{\Phi_{21}}{I_1} = \frac{\Phi_{12}}{I_2} \qquad (20\text{-}12)$$

$$\mathcal{E}_{21} = -\frac{\Delta \Phi_{21}}{\Delta t} = -M \frac{\Delta I_1}{\Delta t} \qquad (20\text{-}13)$$

- Self-inductance is when a changing current induces an emf in the same device:

$$\Phi = LI \qquad (20\text{-}14)$$

$$\mathcal{E} = -L \frac{\Delta I}{\Delta t} \qquad (20\text{-}16)$$

- The energy stored in an inductor is

$$U = \tfrac{1}{2} LI^2 \qquad (20\text{-}17)$$

- The energy density (energy per unit volume) in a magnetic field is

$$u_B = \frac{1}{2\mu_0} B^2 \qquad (20\text{-}18)$$

- Current through an inductor must always change *continuously*, never instantaneously. In an *LR* circuit, the time constant is

$$\tau = \frac{L}{R} \qquad (20\text{-}22)$$

  The current in an *LR* circuit is

$$\text{If } I_0 = 0, \quad I(t) = I_f(1 - e^{-t/\tau}) \qquad (20\text{-}21)$$

$$\text{If } I_f = 0, \quad I(t) = I_0 e^{-t/\tau} \qquad (20\text{-}24)$$

Highlighted Figures and Tables

F20.1 Motional emf (p. 726)

F20.4 An AC generator (p. 729)

F20.5 Generator-produced emf is a sinusoidal function of time (p. 730)

F20.8 Finding the magnetic flux when $\vec{B}$ is not perpendicular to the surface (p. 733)

F20.17 Currents induced by a moving magnet (p. 740)

F20.20 Two simple transformers (p. 741)

T20.1 Comparison of conservative and nonconservative $\vec{E}$ fields (p. 745)

F20.26 Induced electric field (p. 746)

F20.30 Emf in an inductor (p. 749)

F20.31 *LR* circuit (p. 752)

F20.32 A circuit that allows the current in the inductor circuit to be safely stopped (p. 753)

T20.2 Comparison of *RC* and *LR* circuits (p. 753)

CONCEPTUAL QUESTIONS

1. In a metal detector, an alternating current passes through a coil, producing an oscillating magnetic field (Fig. 20.34). A second coil is mounted perpendicular to the first. An induced emf in the second coil signals the possible presence of a metal object near the detector. (a) What happens if a metal object is located in the oscillating magnetic field? (b) Why is an emf induced in the second coil when a metal object is nearby? (c) Does the detector work only for ferromagnetic metals (such as iron), or does it also work for nonmagnetic metals (such as copper, silver, gold)?

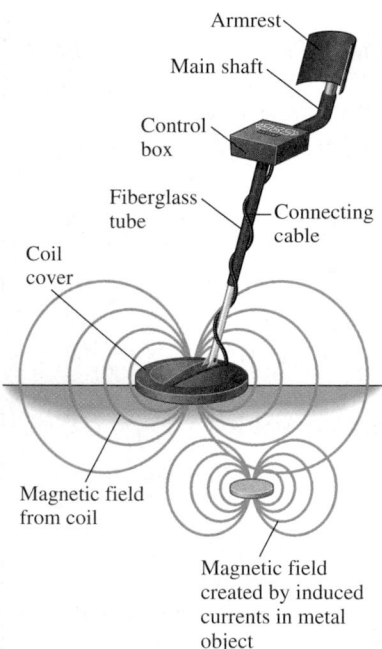

Figure 20.34
Conceptual Question 1

2. A transformer is essentially a mutual-inductance device. Two coils are wound around an iron core; an alternating current in one coil induces an emf in the second. The core is normally made of either laminated iron—thin sheets of iron with an insulating material between them—or of a bundle of parallel insulated iron wires. Why not just make it of solid iron?

3. A certain amount of energy must be supplied to increase the current through an inductor from 0 mA to 10 mA. Does it take the same amount of energy, more, or less to increase the current from 10 mA to 20 mA?

4. Figure 20.12 shows the playback head of an audiotape deck. Explain how an electrical signal that replicates changes in the magnetization of the tape is produced in the coil.

5. Computer disks and tapes operate on a principle similar to audiotapes. The bit 0 is represented by a magnetization that does not reverse direction ($\rightarrow\rightarrow$ or $\leftarrow\leftarrow$), while a 1 is represented by a change in the magnetization direction ($\rightarrow\leftarrow$ or $\leftarrow\rightarrow$). The magnetization direction is always changed at a bit boundary. For example, the sequence shown in Fig. 20.35 represents 10011. Why this complicated system? Why not

Figure 20.35
Conceptual Question 5

just use the two directions of magnetization, say 0 represented by $\rightarrow$ and 1 represented by $\leftarrow$?

C 6. Magnetic induction is the principle behind the operation of mechanical speedometers used in automobiles and bicycles. In Fig. 20.36, a simplified version of the speedometer, a metal disk is free to spin about the vertical axis passing through its center. Suspended above the disk is a horseshoe magnet. (a) If the horseshoe magnet is connected to the drive shaft of the vehicle so that it rotates about a vertical axis, what happens to the disk? [*Hint:* Think about eddy currents and Lenz's law.] (b) Instead of being free to rotate, the disk is restrained by a hairspring. The hairspring exerts a restoring torque on the disk proportional to its angular displacement from equilibrium. When the horseshoe magnet rotates, what happens to the disk? A pointer attached to the disk indicates the speed of the vehicle. How does the angular *position* of the pointer depend on the angular *speed* of the magnet?

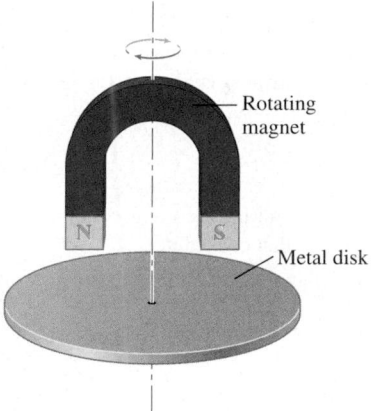

Figure 20.36
Conceptual Question 6

7. Wires that carry telephone signals are twisted as in Fig. 20.37. The twisting reduces the noise on the line from nearby electric devices that produce changing currents. How does the twisting reduce noise pickup?

Figure 20.37   Conceptual Question 7

8. In an MRI the patient must be immersed in a strong magnetic field. Why is the magnetic field turned on gradually rather than suddenly?

9. The magnetic flux through a flat surface is known. The area of the surface is also known. Is that information enough to calculate the average magnetic field on the surface? Explain.

10. Would a ground fault interrupter work if the circuit used DC current instead of AC? Explain.

11. In the study of thermodynamics, we thought of a refrigerator as a reversed heat engine. (a) Explain how a generator is a reversed electric motor. (b) What kind of device is a reversed loudspeaker? (c) What kind of device is a reversed playback head on a tape player?

12. Two identical circular coils of wire are separated by a fixed center-to-center distance. Describe the orientation of the coils that would (a) maximize or (b) minimize their mutual-inductance.

13. (a) Explain why a transformer works for AC but not for DC. (b) Explain why a transformer designed to be connected to an emf of amplitude 170 V would be damaged if connected to a DC emf of 170 V.

14. Credit cards have a magnetic strip that encodes information about the credit card account. Why do devices that read the magnetic strip often include the instruction to swipe the card rapidly? Why can't the magnetic strip be read if the card is swiped too slowly?

15. Think of an example that illustrates why an "anti-Lenz" law would violate the conservation of energy. (The "anti-Lenz" law is: The direction of induced emfs and currents always *reinforces* the *change* that produces them.)

16. A 2-m-long copper pipe is held vertically. When a marble is dropped down the pipe, it falls through in about 0.7 s. A magnet of similar size and shape dropped down the pipe takes *much* longer. Why?

17. An electric mixer is being used to mix up some cake batter. What happens to the motor if the batter is too thick, so the beaters are turning slowly?

18. A circular loop of wire can be used as an antenna to sense the changing magnetic fields in an electromagnetic wave (such as a radio transmission). What is the advantage of using a coil with many turns rather than a single loop?

19. Some low-cost tape recorders do not have a separate microphone. Instead, the speaker is used as a microphone when recording. Explain how this works.

20. High-voltage power lines run along the edge of a farmer's field. Describe how the farmer might be able to steal electric power without making any electrical connection to the power line. (Yes, it works. Yes, it has been done. Yes, it is illegal.)

MULTIPLE CHOICE QUESTIONS

1. An electric current is induced in a conducting loop by all but one of these processes. Which one does *not* produce an induced current?
   (a) rotating the loop so that it cuts across magnetic field lines
   (b) placing the loop so that its area is perpendicular to a changing magnetic field
   (c) moving the loop parallel to uniform magnetic field lines
   (d) expanding the area of the loop while it is perpendicular to a uniform magnetic field

2. A split-ring commutator is used in a DC generator to
   (a) rotate a loop so that it cuts through magnetic flux.
   (b) reverse the connections to an armature so that the current periodically reverses direction.
   (c) reverse the connections to an armature so that the current does not reverse direction.
   (d) prevent a coil from rotating when the magnetic field is changing.

3. Suppose the switch in Fig. 20.18 has been closed for a long time but is suddenly opened at $t = t_0$. Which of these graphs best represents the current in coil 2 as a function of time? I_2 is positive if it flows from A to B through the resistor.

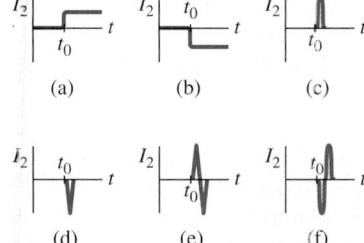

4. The current in the long wire in Fig. 20.38 is decreasing. What is the direction of the current induced in the conducting loop below the wire?
   (a) counterclockwise   (b) clockwise
   (c) CCW or CW depending on the shape of the loop
   (d) No current is induced.

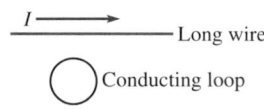

Figure 20.38 Multiple Choice Question 4 and Problems 10 and 11

5. In a bicycle speedometer, a bar magnet is attached to the spokes of the wheel and a coil is attached to the frame so that the north pole of the magnet moves past it once for every revolution of the wheel. As the magnet moves past the coil, a pulse of current is induced in the coil. A computer then measures the time between pulses and computes the bicycle's speed. Figure 20.39 shows the magnet about to move past the coil. Which of the graphs shows the resulting current pulse? Take current counterclockwise in Fig. 20.39a to be positive.

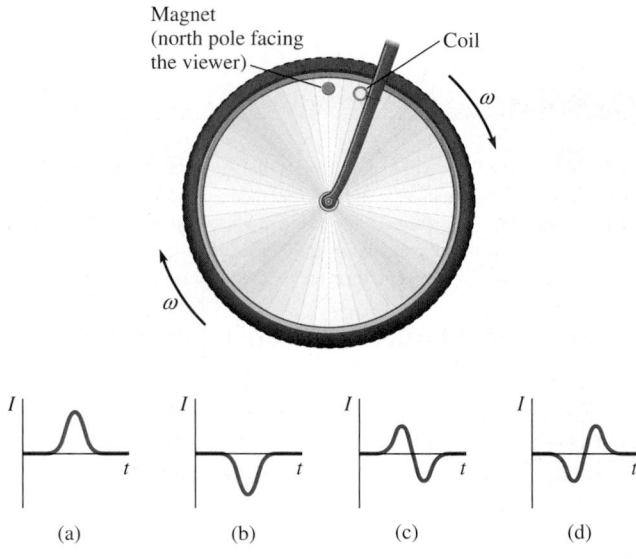

Figure 20.39 Multiple Choice Question 5

6. For each of the experiments (1, 2, 3, 4) shown in Fig. 20.40, in what direction does current flow *through the resistor*? Note that the wires are not always wrapped around the plastic tube in the same way.

	(1)	(2)	(3)	(4)
(a)	P to Q	P to Q	P to Q	P to Q
(b)	P to Q	Q to P	P to Q	Q to P
(c)	Q to P	P to Q	Q to P	P to Q
(d)	Q to P	P to Q	P to Q	Q to P
(e)	Q to P	Q to P	Q to P	Q to P
(f)	Q to P	Q to P	P to Q	P to Q

(1) S to be closed

(2) S to be opened

(3) Coil moves to right

(4) Coil moves left

Figure 20.40
Multiple Choice Question 6

7. Figure 20.41 shows a region of uniform magnetic field out of the page. Outside the region, the magnetic field is zero. Some rectangular wire loops move as indicated. Which of the loops would feel a magnetic force directed to the right?

(a) 1   (b) 2   (c) 3   (d) 4
(e) 1 and 2   (f) 2 and 4   (g) 3 and 4   (h) none of them

Figure 20.41
Multiple Choice Question 7

8. In a moving coil microphone, the induced emf in the coil at any instant depends mainly on

(a) the displacement of the coil.
(b) the velocity of the coil.
(c) the acceleration of the coil.

9. A moving magnet microphone is similar to a moving coil microphone (Fig. 20.13) except that the coil is stationary and the magnet is attached to the diaphragm, which moves in response to sound waves in the air. If, in response to a sound wave, the magnet moves according to $x(t) = A \sin \omega t$, the induced emf in the coil would be (approximately) proportional to which of these?

(a) $\sin \omega t$ (b) $\cos \omega t$
(c) $\sin 2\omega t$ (d) $\cos 2\omega t$

10. An airplane is flying due east. The Earth's magnetic field has a downward vertical component and a horizontal component due north. Which point on the plane's exterior is at the highest potential due to the motional emf?

(a) the nose (the point farthest east)
(b) the tail (the point farthest west)
(c) the tip of the left wing (the point farthest north)
(d) the tip of the right wing (the point farthest south)

PROBLEMS

Note: **C** indicates a combination conceptual/quantitative problem. Gold diamonds ◆, ◆◆ are used to indicate the increasing level of difficulty of each problem. Problem numbers appearing in blue, 9., denote problems that have a detailed solution available in the Student Solutions Manual. Some problems are *paired* by concept; their numbers are connected by a ruled box.

20.1 Motional Emf; 20.2 Electric Generators

1. In Fig. 20.1b, a metal rod moves to the right at speed v. (a) What is the current in the rod, in terms of v, B, L, and R? (b) In what direction does the current flow? (c) What is the direction of the magnetic force on the rod? (d) What is the magnitude of the magnetic force on the rod (in terms of v, B, L, and R)?

C 2. Suppose that the current were to flow in the direction *opposite* to that found in Problem 1. (a) In what direction would the magnetic force on the rod be? (b) In the absence of an external force, what would happen to the rod's kinetic energy? (c) Why is this not possible? Returning to the correct direction of the current, sketch a rough graph of the kinetic energy of the rod as a function of time.

C 3. To maintain a constant emf, the moving rod of Fig. 20.1b must maintain a constant velocity. In order to maintain a constant velocity, some external force must pull it to the right. (a) What is the magnitude of the external force required, in terms of v, B, L, and R? (See Problem 1.) (b) At what rate does this force do work on the rod? (c) What is the power dissipated in the resistor? (d) Overall, is energy conserved? Explain.

4. In Fig. 20.1b, what would the magnitude (in terms of v, L, R, and B) and direction (CW or CCW) of the current be if the direction of the magnetic field were: (a) into the page; (b) to the right (in the plane of the page); (c) up (in the plane of the

page); (d) such that it has components both out of the page and to the right, with a 20.0° angle between the field and the plane of the page?

◆◆ 5. A solid metal cylinder of mass m rolls down parallel metal rails spaced a distance L apart with a constant acceleration of magnitude a_0 (Fig. 20.42a). The rails are inclined at an angle θ to the horizontal. Now the rails are connected electrically at the top and immersed in a magnetic field of magnitude B that is perpendicular to the plane of the rails (Fig. 20.42b). (a) As it rolls down the rails, in what direction does current flow in the cylinder? (b) What direction is the magnetic force on the cylinder? (c) Instead of rolling at constant acceleration, the cylinder now approaches a terminal speed v_t. What is v_t in terms of L, m, R, a_0, θ, and B? R is the total electrical resistance of the circuit consisting of the cylinder, rails, and wire; assume R is constant (that is, the resistances of the rails themselves are negligible).

(a)   (b)

Figure 20.42
Problem 5

6. In Fig. 20.4, side 3 of the rectangular coil in the electric generator rotates about the axis at constant angular speed ω. Figure 20.43 shows side 3 by itself. (a) First consider the right half of side 3. Although the speed of the wire differs depending on the distance from the axis, the direction is the same for the entire right half. Use the magnetic force law to find the direction of the force on electrons in the right half of the wire. (b) Does the magnetic force tend to push electrons along the wire, either toward or away from the axis? (c) Is there an induced emf along the *length* of this half of the wire? (d) Generalize your answers to the left side of wire 3 and the two sides of wire 1. What is the net emf due to these two sides of the coil?

Figure 20.43
Problem 6

◆ 7. A square loop of wire of side 2.3 cm and electrical resistance 79 Ω is near a long straight wire that carries a current of 6.8 A in the direction indicated in Fig. 20.44. The long wire and loop both lie in the plane of the page. The left side of the loop is 9.0 cm from the wire. (a) If the loop is at rest, what is the induced emf in the loop? What are the magnitude and direction of the induced current in the loop? What are the magnitude and direction of the magnetic force on the loop? (b) Repeat if the loop is moving to the right at a constant speed of 45 cm/s. (c) In (b), find the electric power dissipated in the loop and show that it is equal to the rate at which an external force, pulling the loop to keep its speed constant, does work.

Figure 20.44
Problem 7

© 8. A solid copper disk of radius R rotates at angular velocity ω in a perpendicular magnetic field B. Figure 20.45 shows the disk rotating clockwise and the magnetic field into the page. (a) Is the charge that accumulates on the edge of the disk positive or negative? Explain. (b) What is the potential difference between the center of the disk and the edge? [*Hint:* Think of the disk as a large number of thin wedge-shaped rods. The center of such a rod is at rest, and the outer edge moves at speed $v = \omega R$. The rod moves through a perpendicular magnetic field at an *average* speed of $\frac{1}{2}\omega R$.]

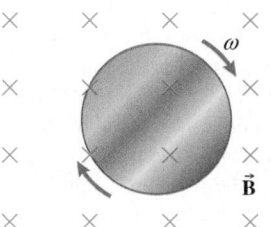

Figure 20.45 Problem 8

20.3 Faraday's Law; 20.4 Lenz's Law

9. A horizontal desk surface measures 1.3 m × 1.0 m. If the Earth's magnetic field has magnitude 0.44 mT and is directed 65° below the horizontal, what is the magnetic flux through the desk surface?

10. A long straight wire carrying a steady current is in the plane of a circular loop of wire (Fig. 20.38). (a) If the loop is moved closer to the wire, what direction does the induced current in the loop flow? (b) At one instant, the induced emf in the loop is 3.5 mV. What is the rate of change of the magnetic flux through the loop at that instant? Express your answer in T·m²/s.

11. A long straight wire carrying a current I is in the plane of a circular loop of wire (Fig. 20.38). The current I is decreasing. Both the loop and the wire are held in place by external forces. The loop has resistance 24 Ω. (a) In what direction does the induced current in the loop flow? (b) In what direction is the external force holding the loop in place? (c) At one instant, the induced current in the loop is 84 mA. What is the rate of change of the magnetic flux through the loop at that instant in Wb/s?

12. Verify that, in SI units, $\Delta\Phi_B/\Delta t$ can be measured in volts—in other words, that 1 Wb/s = 1 V.

13. The perpendicular component of the external magnetic field through a 50-turn coil of radius 5.0 cm increases from 0 to 1.8 T in 3.6 s. (a) If the resistance of the coil is 2.8 Ω, what is the magnitude of the induced current in the coil? (b) What is the direction of the current if the perpendicular component of the field is away from the viewer?

Figure 20.46 Problem 14

14. In Fig. 20.46, switch S is initially open. It is closed, and then opened again a few seconds later. (a) In what direction does current flow through the ammeter when switch S is closed? (b) In what direction does current flow when switch S is then opened? (c) Sketch a qualitative graph of the current through the ammeter as a function of time. Take the current to be positive to the right.

15. A circular metal ring is suspended above a solenoid. The magnetic field due to the solenoid is shown in Fig. 20.47. The current in the solenoid is increasing. (a) What is the direction of the current in the ring? (b) The flux through the ring is proportional to the current in the solenoid. When the current in the solenoid is 12.0 A, the magnetic flux through the ring is 0.40 Wb. When the current increases at a rate of 240 A/s, what is the induced emf in the ring? (c) Is there a net magnetic force on the ring? If so, in what direction? (d) If the ring is cooled by immersing it in liquid nitrogen, what happens to its electrical resistance, the induced current, and the magnetic force? The change in size of the ring is negligible. (With a sufficiently strong magnetic field, the ring can be made to shoot up high into the air.)

Metal ring

$\vec{B}$

Solenoid

Figure 20.47 Problem 15

16. (a) For a particle moving in simple harmonic motion, the position can be written $x(t) = x_m \cos \omega t$. What is the velocity $v_x(t)$ as a function of time for this particle? (b) Using the small-angle approximation for the sine function, find the slope of the graph of $\Phi(t) = \Phi_0 \sin \omega t$ at $t = 0$. Does your result agree with the value of $\Delta\Phi/\Delta t = \omega\Phi_0 \cos \omega t$ at $t = 0$?

17. The strings of an electric guitar are made of ferromagnetic metal. The pickup consists of two components. A magnet causes the part of the string near it to be magnetized. The vibrations of the string near the pickup coil produce an induced emf in the coil. The electrical signal in the coil is then amplified and used to drive the speakers. In Fig. 20.48, the string is moving away from the coil. What is the direction of the induced current in the coil?

Metal guitar string

Pickup coil

$\vec{v}$

N S
N S
N S
N S

Permanent magnet

To amplifier

Figure 20.48 Problem 17

18. Two loops of wire are next to each other in the same plane (Fig. 20.49). (a) If the switch S is closed, does current flow in loop 2? If so, in what direction? (b) Does the current flow for only a brief moment, or does it continue? (c) Is there a magnetic force on loop 2? If so, in what direction? (d) Is there a magnetic force on loop 1? If so, in what direction?

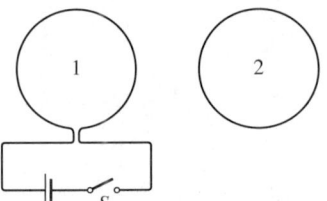

1   2

S

Figure 20.49 Problem 18

20.6 Transformers

19. A step-down transformer has 4000 turns on the primary and 200 turns on the secondary. If the primary voltage is 2.2 kV, what is the secondary voltage?

20. A step-down transformer has a turns ratio of 1/100. An AC voltage of amplitude 170 V is applied to the primary. If the primary current is 1.0 mA, what is the secondary current?

21. A doorbell uses a transformer to deliver 8.5 V when it is connected to a 170-V line. If there are 50 turns on the secondary, (a) what is the turns ratio? (b) How many turns does the primary have?

22. The primary coil of a transformer has 250 turns; the secondary coil has 1000 turns. An alternating current is sent through the primary coil. The emf in the primary is 16 V. What is the emf in the secondary?

23. When the emf for the primary of a transformer is 5.00 V, the secondary emf is 10.0 V. What is the transformer turns ratio (N_2/N_1)?

24. A transformer with a primary coil of 1000 turns is used to step up the standard 170-V amplitude line voltage to a 220-V amplitude. How many turns are required in the secondary coil?

25. A transformer for an answering machine takes an AC voltage of amplitude 170 V as its input and supplies a 7.8-V amplitude to the answering machine. The primary has 300 turns. (a) How many turns does the secondary have? (b) When idle, the answering machine uses a maximum power of 5.0 W. What is the amplitude of the current drawn from the 170-V line?

20.7 Eddy Currents

26. A 2-m-long copper pipe is held vertically. When a marble is dropped down the pipe, it falls through in about 0.7 s. A magnet of similar size and shape takes *much* longer to fall through the pipe. (a) As the magnet is falling through the pipe with its north pole below its south pole, what direction do currents flow around the pipe above the magnet? Below the magnet? (CW or CCW as viewed from the top?) (b) Sketch a graph of the speed of the magnet as a function of time. [*Hint:* What would the graph look like for a marble falling through honey?]

27. In Problem 26, the pipe is suspended from a spring scale. The weight of the pipe is 12.0 N; the weight of the marble and magnet are each 0.3 N. Sketch graphs to show the reading of the spring scale as a function of time for the fall of the marble and again for the fall of the magnet. Label the vertical axis with numerical values.

20.8 Induced Electric Fields

28. In the solenoid shown in Fig. 20.50, the current in the solenoid flows clockwise (as viewed from the left) and is increasing in magnitude. (a) What is the direction of the induced electric field at points A, B, and C? [*Hint:* The solenoid is long enough that cylindrical symmetry can be assumed.] (b) What is the magnitude of the induced electric field at point A if the current in the solenoid changes at a constant rate $\Delta I/\Delta t$? The solenoid has n turns per unit length and radius r.

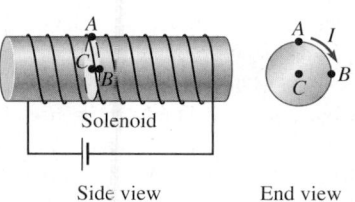

Side view   End view

Figure 20.50
Problem 28

29. An electromagnet with flat, circular, closely spaced pole faces produces a uniform magnetic field that is increasing at a constant rate of 0.30 T/s. A circular loop of thin wire is centered between the pole faces. The resistance of the loop is 7.2 Ω and its radius is 5.4 cm. What is the induced electric field that causes current to flow in the loop?

30. An electromagnet with flat, circular, closely spaced pole faces of radius 8.0 cm produces a uniform magnetic field that is increasing at a constant rate of 2.2 T/s. (a) What is the induced electric field at the center of the space between the pole faces ($r = 0$)? (b) What is the induced electric field at the outside edge of the space between the pole faces?

20.9 Mutual- and Self-Inductance

31. A solenoid is made of 300.0 turns of wire, wrapped around a hollow cylinder of radius 1.2 cm and length 6.0 cm. What is the self-inductance of the solenoid?

32. A solenoid of length 2.8 cm and diameter 0.75 cm is wound with 160 turns per cm. When the current through the solenoid is 0.20 A, what is the magnetic flux through *one* of the windings of the solenoid?

33. If the current in the solenoid in Problem 32 is decreasing at a rate of 35.0 A/s, what is the induced emf (a) in one of the windings? (b) in the entire solenoid?

34. An ideal solenoid has length ℓ. If the windings are compressed so that the length of the solenoid is reduced to $0.50\,\ell$, what happens to the inductance of the solenoid?

35. In this problem, you derive the expression for the self-inductance of a long solenoid [Eq. (20-15a)]. The solenoid has n turns per unit length, length ℓ, and radius r. Assume that the current flowing in the solenoid is I. (a) Write an expression for the magnetic field inside the solenoid in terms of n, ℓ, r, I, and universal constants. (b) Assume that all of the field lines cut through each turn of the solenoid. In other words, assume the field is uniform right out to the ends of the solenoid—a good approximation if the solenoid is tightly wound and sufficiently long. Write an expression for the magnetic flux through one turn. (c) What is the total flux through all turns of the solenoid? (d) Use the definition of self-inductance [Eq. (20-14)] to find the self-inductance of the solenoid.

36. Compare the electric energy that can be stored in a capacitor to the magnetic energy that can be stored in an inductor of the same size (that is, the same volume). For the capacitor, assume that air is between the plates; the maximum electric field is then the breakdown strength of air, about 3 MV/m. The maximum magnetic field attainable in an ordinary solenoid with an air core is on the order of 10 T.

37. In Section 20.9, in order to find the energy stored in an inductor, we assumed that the current was increased from zero at a constant rate. In this problem, you will prove that the energy stored in an inductor is $U_L = \frac{1}{2}LI^2$—that is, it only depends on the current I and not on the previous time-dependence of the current. (a) If the current in the inductor increases from i to $i + \Delta i$ in a very short time Δt, show that the energy added to the inductor is $\Delta U = Li\,\Delta i$. [*Hint:* Start with $\Delta U = P\,\Delta t$.] (b) Show that, on a graph of Li versus i, for any small current interval Δi, the energy added to the inductor can be interpreted as the area under the graph for that interval. (c) Now show that the energy stored in the inductor when a current I flows is $U = \frac{1}{2}LI^2$.

38. The current in a 0.080-H solenoid increases from 20.0 mA to 160.0 mA in 7.0 s. Find the average emf in the solenoid during that time interval.

39. Calculate the equivalent inductance L_{eq} of two ideal inductors, L_1 and L_2, connected in series in a circuit. Assume that their mutual-inductance is negligible. [*Hint:* Imagine replacing the two inductors with a single equivalent inductor L_{eq}. How is the emf in the series equivalent related to the emfs in the two inductors? What about the currents?]

40. Calculate the equivalent inductance L_{eq} of two ideal inductors, L_1 and L_2, connected in parallel in a circuit. Assume that their mutual-inductance is negligible. [*Hint:* Imagine replacing the two inductors with a single equivalent inductor L_{eq}. How is the emf in the parallel equivalent related to the emfs in the two inductors? What about the currents?]

20.10 *LR* Circuits

41. In the circuit of Fig. 20.51, switch S is opened at $t = 0$ after having been closed for a long time. (a) How much energy is stored in the inductor at $t = 0$? (b) What is the instantaneous rate of change of the inductor's energy at $t = 0$? (c) What is the *average* rate of change of the inductor's energy between $t = 0.0$ and $t = 1.0$ s? (d) How long does it take for the current in the inductor to reach 0.0010 times its initial value?

Figure 20.51
Problem 41

◆42. In Fig. 20.52, after the switch has been closed for a long time, it is opened. How long does it take for the energy stored in the inductor to decrease to 0.10 times its initial value?

Figure 20.52
Problems 42 and 43

43. No currents flow in the circuit of Fig. 20.52 before the switch is closed. Consider all circuit elements to be ideal. (a) At the instant the switch is closed, what are the values of the currents I_1 and I_2, the potential differences across the resistors, the power supplied by the battery, and the induced emf in the inductor? (b) After the switch has been closed for a long time, what are the values of the currents I_1 and I_2, the potential differences across the resistors, the power supplied by the battery, and the induced emf in the inductor?

◆44. A 0.30-H inductor and a 200.0-Ω resistor are connected in series to a 9.0-V battery. (a) What is the maximum current that flows in the circuit? (b) How long after connecting the battery does the current reach half its maximum value? (c) When the current is half its maximum value, find the energy stored in the inductor, the rate at which energy is being stored in the inductor, and the rate at which energy is dissipated in the resistor. (d) Redo parts (a) and (b) if, instead of being negligibly small, the internal resistances of the inductor and battery are 75 Ω and 20.0 Ω, respectively.

45. A coil has an inductance of 0.15 H and a resistance of 33 Ω. The coil is connected to a 6.0-V battery. After a long time elapses, the current in the coil is no longer changing. (a) What is the current in the coil? (b) What is the energy stored in the coil? (c) What is the rate of energy dissipation in the coil? (d) What is the induced emf in the coil?

46. A coil of wire is connected to an ideal 6.00-V battery at $t = 0$. At $t = 10.0$ ms, the current in the coil is 204 mA. One minute later, the current is 273 mA. Find the resistance and inductance of the coil. [*Hint:* Sketch $I(t)$.]

© 47. The windings of an electromagnet have inductance $L = 8.0$ H and resistance $R = 2.0\ \Omega$. A 100.0-V DC power supply is connected to the windings. (a) What is the current in the windings? (b) The electromagnet is to be shut off. Before disconnecting the power supply, a shunt resistor with resistance 20.0 Ω is connected in parallel across the windings. Why is the shunt resistor needed? Why must it be connected *before* the power supply is disconnected? (c) What is the maximum power dissipated in the shunt resistor? The shunt resistor must be chosen so that it can handle at least this much power without damage. (d) When the power supply is disconnected, how long does it take for the current in the windings to drop to 0.10 A? (e) Would a larger shunt resistor dissipate the energy stored in the electromagnet faster? Explain.

48. A coil has an inductance of 0.15 H and a resistance of 33 Ω. The coil is connected to a 6.0-V ideal battery. When the current reaches half its maximum value: (a) at what *rate* is magnetic energy being stored in the inductor? (b) At what rate is energy being dissipated? (c) What is the total power that the battery supplies?

49. The time constant τ for an *LR* circuit must be some combination of L, R, and $\mathcal{E}$. (a) Write the units of each of these three quantities in terms of V, A, and s. (b) Show that the only combination that has units of seconds is L/R.

COMPREHENSIVE PROBLEMS

© 50. In the AC generator of Fig. 20.4, the emf produced is $\mathcal{E}(t) = \omega NBA \sin \omega t$. If the generator is connected to a load of resistance R, then the current that flows is

$$I(t) = \frac{\omega NBA}{R} \sin \omega t$$

(a) Find the magnetic forces on sides 2 and 4 at the instant shown in Fig. 20.4b. (Remember that $\theta = \omega t$.) (b) Why do the magnetic forces on sides 1 and 3 not cause a torque about the axis of rotation? (c) From the magnetic forces found in (a), calculate the torque on the loop about its axis of rotation at the instant shown in Fig. 20.4b. (d) In the absence of other torques, would the magnetic torque make the loop increase or decrease its angular velocity? Explain.

◆51. A *flip coil* is a device used to measure a magnetic field. A coil of radius r, N turns, and electrical resistance R is initially perpendicular to a magnetic field of magnitude B. The coil is connected to a special kind of galvanometer that measures the total charge Q that flows through it. To measure the field, the flip coil is rapidly flipped upside down. (a) What is the change in magnetic flux through the coil in one flip? (b) If the time interval during which the coil is flipped is Δt, what is the average induced emf in the coil? (c) What is the average current that flows through the galvanometer? (d) What is the total charge Q in terms of r, N, R, and B?

✦ 52. A DC motor is connected to a constant emf of 12.0 V. The resistance of its windings is 2.0 Ω. At normal operating speed, the motor delivers 6.0 W of mechanical power. (a) What is the initial current drawn by the motor when it is first started up? (b) What current does it draw at normal operating speed? (c) What is the back emf induced in the windings at normal speed?

✦ 53. In Fig. 20.53, a bar magnet is initially far from a circular loop of wire. The magnet is moved at constant speed along the axis of the loop. It moves toward the loop, proceeds to pass through it, and then continues until it is far away on the right side of the loop. Sketch a qualitative graph of the current in the loop as a function of the position of the bar magnet. Take the current to be positive when it is counterclockwise as viewed from the left.

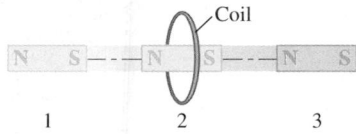

Figure 20.53
Problem 53

54. A bar magnet approaches a coil (Fig. 20.54a). (a) In which direction does current flow through the galvanometer as the magnet approaches? (b) In Fig. 20.54b, the magnet is initially at rest inside the coil. It is then pulled out from the left side. In which direction does current flow through the galvanometer as the magnet is pulled away? (c) In both situations, how does the magnitude of the current depend on the number of turns in the coil? (The resistance of the coil is negligible compared to the resistance of the galvanometer.) (d) How does the current depend on the speed of the magnet? (e) How does the magnitude of the current change if two such magnets were used, held together with the north poles together and the south poles together? (f) How does the magnitude of the current change if two such magnets were used, held together with the *opposite* poles together? (g) Would the experiment give similar results if the magnet remains stationary and the coil moves instead?

Figure 20.54
Problem 54

(a)   (b)

55. A toroid has a square cross section of side a. The toroid has N turns and radius R (Fig. 20.55). The toroid is narrow ($a \ll R$) so that the magnetic field inside the toroid can be considered to be uniform in magnitude. What is the self-inductance of the toroid?

56. An ideal toroid has N turns and self-inductance L. A single turn of wire is wrapped around the toroid (Fig. 20.56a).

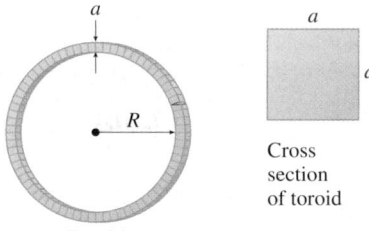

Figure 20.55
Problem 55

(a) What is the mutual-inductance between the toroid and the single turn of wire? (b) What would the mutual-inductance be if the turn of wire had an area twice as large as the cross-sectional area of the toroid (Fig. 20.56b)?

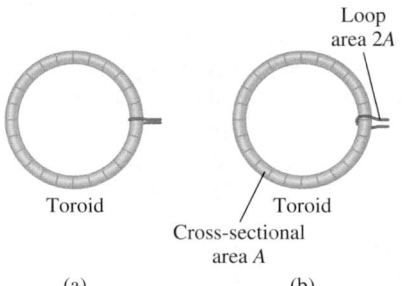

Figure 20.56
Problem 56

57. Two solenoids, of N_1 and N_2 turns respectively, are wound on the same form. They have the same length L and radius r. (a) What is the mutual-inductance of these two solenoids? (b) If an AC current

$$I_1(t) = I_m \sin \omega t$$

flows in solenoid 1 (N_1 turns), write an expression for the total flux through solenoid 2. (c) What is the maximum induced emf in solenoid 2? [*Hint:* Refer to Eq. (20-7).]

✦ 58. An ideal inductor of inductance L is connected to an AC power supply, which provides an emf $\mathcal{E}(t) = \mathcal{E}_m \sin \omega t$. (a) Write an expression for the current in the inductor as a function of time. [*Hint:* See Eq. (20-7).] (b) What is the ratio of the maximum emf to the maximum current? This ratio is called the *reactance*. (c) Do the maximum emf and maximum current occur at the same time? If not, how much time separates them?

59. Suppose you wanted to use the Earth's magnetic field to make an AC generator at a location where the magnitude of the field is 0.50 mT. Your coil has 1000.0 turns and a radius of 5.0 cm. At what angular velocity would you have to rotate it in order to generate an emf of amplitude 1.0 V?

60. A uniform magnetic field of magnitude 0.29 T makes an angle of 13° with the plane of a circular loop of wire. The loop has radius 1.85 cm. What is the magnetic flux through the loop?

61. A solenoid is 8.5 cm long, 1.6 cm in diameter, and has 350 turns. When the current through the solenoid is 65 mA, what is the magnetic flux through one turn of the solenoid?

62. An airplane is flying due north at 180 m/s. The Earth's magnetic field has a northward component of 0.30 mT and an upward component of 0.38 mT. (a) If the wingspan (distance between the wingtips) is 46 m, what is the potential difference between the wingtips? (b) Which wing is at the higher potential?

63. Repeat Problem 62 if the plane flies 30.0° west of south at 180 m/s instead.

64. How much energy due to the Earth's magnetic field is present in 1.0 m^3 of space near Earth's surface, at a place where the field has magnitude 0.45 mT?

65. The largest constant magnetic field achieved in the laboratory is about 40 T. (a) What is the magnetic energy density due to this field? (b) What magnitude electric field would have an equal energy density?

66. A TV tube requires a 20.0-kV-amplitude power supply. (a) What is the turns ratio of the transformer that raises the 170-V-amplitude household voltage to 20.0 kV? (b) If the tube draws 82 W of power, find the currents in the primary and secondary windings. Assume an ideal transformer.

67. The magnetic field between the poles of an electromagnet is 2.6 T. A coil of wire is placed in this region so that the field is parallel to the axis of the coil. The coil has electrical resistance 25 Ω, radius 1.8 cm, and length 12.0 cm. When the current supply to the electromagnet is shut off, the total charge that flows through the coil is 9.0 mC. How many turns are there in the coil?

68. The alternator in an automobile generates an emf of amplitude 12.6 V when the engine idles at 1200 rpm. What is the amplitude of the emf when the car is being driven on the highway with the engine at 2800 rpm?

69. The outside of an ideal solenoid (N_1 turns, length L, radius r) is wound with a coil of wire with N_2 turns. (a) What is the mutual-inductance? (b) If the current in the solenoid is changing at a rate $\Delta I_1/\Delta t$, what is the induced emf in the coil?

70. An ideal solenoid (N_1 turns, length L_1, radius r_1) is placed inside another ideal solenoid (N_2 turns, length $L_2 > L_1$, radius $r_2 > r_1$) such that the axes of the two coincide. (a) What is the mutual-inductance? (b) If the current in the outer solenoid is changing at a rate $\Delta I_2/\Delta t$, what is the induced emf in the inner solenoid?

71. A standard ammeter must be inserted in series into the circuit (Section 18.9). An *induction ammeter* has the great advantage of being able to measure currents without making any electrical connection to the circuit. An iron ring is hinged so that it can be snapped around a wire (Fig. 20.57). A coil is wrapped around the iron ring; the ammeter uses the induced emf in the coil to determine the current flowing in the wire. (a) Does the induction ammeter work equally well for AC and DC currents? Explain. (b) Can the induction ammeter be placed around both wires connected to an appliance to measure the current drawn by the appliance? Explain.

Figure 20.57
Problem 71

ANSWERS TO PRACTICE PROBLEMS

20.1 only the magnitudes of the currents

20.2 3.0 W. The power is proportional to the bicycle's speed *squared*.

20.3 $B_\perp = B \cos 60.0°$

20.4 7.6 V

20.5 (a) $F = B^2L^2v/R$ (to the left at position 2 and to the right at position 4); (b) $P = B^2L^2v^2/R$

20.6 (a) to the left; (b) from A to B through the resistor; (c) no; current only flows in coil 2 while the flux is *changing*. When the magnetic field due to coil 1 is constant, no current flows in coil 2.

(d)

N	S	S	N

Coil 1   Coil 2

20.7 10.0 W

20.8 In a solid core, eddy currents would flow around the axis of the core. The insulation between wires prevents these eddy currents from flowing. Since energy is dissipated by eddy currents, their existence reduces the efficiency of the transformer.

20.9 9.0×10^{-5} Wb

20.10 0.53 W

20.11 0.87 s

Alternating Current

Look closely at the overhead power lines that supply electricity to a house. There are three cables: two are insulated and one is uninsulated. Why are there three—aren't two sufficient to make a complete circuit? Do the three cables correspond to the three prongs of an electrical outlet?

21.1   SINUSOIDAL CURRENTS AND VOLTAGES; RESISTORS IN AC CIRCUITS

In a DC circuit, current flows in one direction only. Emfs in a DC circuit may be constant, as for an ideal battery, or not, but they never change polarity. In an alternating current (AC) circuit, currents and emfs periodically change direction. An AC power supply—like a battery that keeps changing polarity—periodically reverses the polarity of its emf.

The sinusoidally varying emf due to an AC generator (also called an AC source) can be written (Fig. 21.1a)

$$\mathscr{E}(t) = \mathscr{E} \sin \omega t$$

The emf varies continuously between $+\mathscr{E}$ and $-\mathscr{E}$; $\mathscr{E}$ is called the **amplitude** (or **peak** value) of the emf. In a circuit with a sinusoidal emf connected to a resistor (Fig. 21.1b), the potential difference across the resistor is equal to $\mathscr{E}(t)$, by Kirchoff's loop rule. Then the current $i(t)$ varies sinusoidally with amplitude $I = \mathscr{E}/R$:

$$i(t) = \frac{\mathscr{E}(t)}{R} = \frac{\mathscr{E}}{R} \sin \omega t = I \sin \omega t$$

It is important to distinguish the time-dependent quantities from their amplitudes. Note that we use lowercase letters ($\mathscr{E}$ and i) to indicate instantaneous values; uppercase letters ($\mathscr{E}, I$) represent amplitudes. We follow this convention throughout the chapter.

As simple as it may appear, the circuit of Fig. 21.1b has many applications. Electric heating elements found in toasters, hair dryers, electric baseboard heaters, electric stoves, and electric ovens are just resistors connected to an AC source. So is an incandescent lightbulb: the filament is a resistor whose temperature rises due to energy dissipation until it is hot enough to radiate a significant amount of visible light (Fig. 21.2).

The definitions of period, frequency, and angular frequency used in AC circuits are the same as for simple harmonic motion. The time T for one complete cycle is the period. The frequency f is the inverse of the period:

$$\text{cycles per second} = \frac{1}{\text{seconds per cycle}}$$

$$f = \frac{1}{T}$$

Since there are 2π radians in one complete cycle, the angular frequency in radians is

$$\omega = 2\pi f$$

In SI units the period is measured in seconds, the frequency is measured in hertz (Hz), and the angular frequency is measured in rad/s. The usual voltage at a wall outlet in a home in the United States has an amplitude of about 170 V and a frequency of 60 Hz.

Power Dissipated in a Resistor

The instantaneous power dissipated by a resistor in an AC circuit is

$$p(t) = i(t)v(t) = I \sin \omega t \times V \sin \omega t = IV \sin^2 \omega t \tag{21-1}$$

Circuit symbol for an AC generator (source of sinusoidal emf)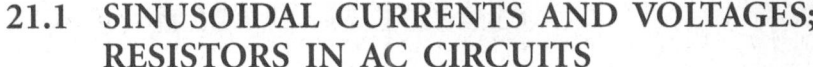

Figure 21.1   (a) A sinusoidal emf as a function of time. For each half period, the generator symbol indicates the polarity of the emf (+ = higher potential, – = lower potential). (b) An AC source connected to a resistor.

(a)   (b)

where $i(t)$ and $v(t)$ represent the current through and potential difference across the resistor. Since $V = IR$, the power can also be written as

$$p = I^2 R \sin^2 \omega t = \frac{V^2}{R} \sin^2 \omega t$$

Figure 21.3 shows the instantaneous power delivered to a resistor in an AC circuit; it varies from 0 to a maximum of IV. Since the sine function *squared* is always nonnegative, the power is always nonnegative. The direction of energy flow is always the same—energy is dissipated in the resistor—no matter what the direction of the current flow.

The maximum power is given by the product of the peak current and the peak voltage (IV). We are usually more concerned with average power than with instantaneous power, since the instantaneous power varies rapidly. In a toaster or lightbulb, the fluctuations in instantaneous power are too fast to be noticeable. The average power is IV times the average value of $\sin^2 \omega t$, which is 1/2 (see Problem 10). On a power versus time graph, the area under the curve represents the energy dissipated. In Fig. 21.3, the extra energy dissipated during the peaks (the half of the cycle when the power is greater than $IV/2$) just fills in the valleys (the half when the power is less than $IV/2$), so the average power is $IV/2$. Replacing $\sin^2 \omega t$ in Eq. (21-1) by its average value (1/2), the average power is shown to be one half of the maximum power:

Average power dissipated by a resistor

$$P_{av} = \tfrac{1}{2}IV = \tfrac{1}{2}I^2R \tag{21-2}$$

What DC current I_{dc} would dissipate energy at the same average rate as an AC current of amplitude I? Clearly $I_{dc} < I$ since I is the maximum current. To find I_{dc}, we set the average powers (through the same resistance) equal:

$$P_{av} = I_{dc}^2 R = \tfrac{1}{2}I^2R$$

Solving for I_{dc} yields

$$I_{dc} = \sqrt{\tfrac{1}{2}I^2}$$

This effective DC current is called the **root mean square (rms)** current because it is the square *root* of the *mean* (average) of the *square* of the AC current:

$$i^2(t) = I^2 \sin^2 \omega t$$

$$\text{average of } i^2 = \text{average of } (I^2 \sin^2 \omega t) = I^2 \times \tfrac{1}{2}$$

$$I_{rms} = \sqrt{\text{average of } i^2} = \frac{1}{\sqrt{2}} I$$

Thus, the rms current is equal to the peak current divided by $\sqrt{2}$. Similarly, the rms values of sinusoidal emfs and potential differences are also equal to the peak values divided by $\sqrt{2}$.

$$\text{rms} = \frac{1}{\sqrt{2}} \times \text{amplitude} \tag{21-3}$$

Figure 21.2 Inside an incandescent lightbulb is a tungsten filament. Tungsten is used because it is strong and has a high melting point. The inside of the bulb must be free of oxygen; otherwise the filament would quickly burn up.

Figure 21.3 Power delivered to a resistor in an AC circuit. The shaded areas of the peak and valley are equal. Since the area under $p(t)$ represents the energy dissipated, the additional energy dissipated during a peak exactly compensates for the smaller amount of energy dissipated during a valley. The average power is $IV/2$.

Rms values have the advantage that they can be treated like DC values for finding the average power dissipated in a resistor:

$$P_{av} = I_{rms} V_{rms} = I_{rms}^2 R = \frac{V_{rms}^2}{R} \qquad (21\text{-}4)$$

Meters designed to measure AC voltages and currents are usually calibrated to read rms values instead of peak values. In the United States, most electrical outlets supply an AC voltage of approximately 120 V rms; the peak voltage is 120 V × √2 = 170 V. Electrical devices are usually labeled with rms values. For instance, if a hair dryer is labeled "120 V, 10 A," both quantities are rms values; the hair dryer then consumes an average power of 120 V × 10 A = 1200 W.

Example 21.1

Resistance of a 100.0-W Lightbulb

A 100.0-W lightbulb is designed to be connected to an AC voltage of 120.0 V (rms). (a) What is the resistance of the lightbulb filament at normal operating temperature? (b) Find the rms and peak currents through the filament. (c) When the cold filament is initially connected to the circuit by flipping a switch, is the average power larger or smaller than 100.0 W?

Strategy The *average* power dissipated by the filament is 100.0 W. Since the rms voltage across the bulb is 120.0 V, if we connected the bulb to a *DC* power supply of 120.0 V, it would dissipate a constant 100.0 W.

Solution (a) Average power and rms voltage are related by

$$P_{av} = \frac{V_{rms}^2}{R} \qquad (21\text{-}4)$$

We solve for R:

$$R = \frac{V_{rms}^2}{P_{av}} = \frac{(120.0\ \text{V})^2}{100.0\ \text{W}} = 144.0\ \Omega$$

(b) Average power is rms voltage times rms current:

$$P_{av} = I_{rms} V_{rms}$$

We can solve for the rms current:

$$I_{rms} = \frac{P_{av}}{V_{rms}} = \frac{100.0\ \text{W}}{120.0\ \text{V}} = 0.8333\ \text{A}$$

The amplitude of the current is a factor of √2 larger.

$$I = \sqrt{2}\, I_{rms} = 1.179\ \text{A}$$

(c) For metals, resistance increases with increasing temperature. When the filament is cold, its resistance is smaller. Since it is connected to the same voltage, the current is larger and the average power dissipated is larger.

Discussion Check: The power dissipated can also be found from peak values:

$$P_{av} = \tfrac{1}{2}IV = \tfrac{1}{2}(1.179\ \text{A} \times 169.7\ \text{V}) = 100.0\ \text{W}$$

Another check: the amplitudes should be related by Ohm's law.

$$V = IR = 1.179\ \text{A} \times 144.0\ \Omega = 169.8\ \text{V}$$

Practice Problem 21.1   European wall outlet

The rms voltage at a wall outlet in Europe is 220 V. Suppose a space heater draws an rms current of 12.0 A. What are the amplitudes of the voltage and current? What are the peak power and the average power dissipated in the heating element? What is the resistance of the heating element?

Making The Connection:
household wiring

21.2   ELECTRICITY IN THE HOME

In a North American home, most electrical outlets supply an rms voltage of 110–120 V at a frequency of 60 Hz. However, some appliances with heavy demands—such as electric heaters, water heaters, stoves, and large air conditioners—are supplied with 220–240 V rms. At twice the voltage amplitude, they only need to draw half as much current for the same power to be delivered, reducing energy dissipation in the wiring (and the need for extra thick wires).

Local power lines are at voltages of several kilovolts. Step-down transformers reduce the voltage to 120/240 V rms. You can see these transformers wherever the power lines run above ground from pole to pole; they are the metal cans mounted to the poles (Fig. 21.4). The transformer has a center tap—a connection to the middle of the secondary coil; the voltage across the entire secondary coil is 240 V rms, but the voltage between the center tap and either end is only 120 V rms. The center tap is grounded at the transformer and runs to a building by an uninsulated cable. There it is connected to the *neutral* wire (which usually has white insulation) in every 120-V circuit in the building.

Figure 21.4 Electrical wiring in a North American home

The other two connections from the transformer run to the building by insulated cables and are called *hot*. The hot wires in an outlet box usually have either black or red insulation. Relative to the neutral wire, each of the hot wires is at 120 V rms, but the two are 180° out of phase with each other. Half of the 120-V circuits in the building are connected to one of the hot cables and half to the other. Appliances needing to be supplied with 240 V are connected to both hot cables; they have no connection to the neutral cable.

Older 120-V outlets have only two prongs: hot and neutral. The slot for the neutral prong is slightly larger than the hot; a *polarized* plug can only be connected one way, preventing the hot and neutral connections from being interchanged. This safety feature is now superseded in devices that use the third prong on modern outlets. The third prong is connected directly to ground through its own set of wires (usually uninsulated or with green insulation)—it is not connected to the neutral wires. The metal case of most electrical appliances is connected to ground as a safety measure. If something goes wrong with the wiring inside the appliance so that the case becomes electrically connected to the hot wire, the third prong provides a low-resistance path for the current to flow to ground; the large current trips a circuit breaker or fuse. Without the ground connection, the case of the appliance would be at 120 V rms with respect to ground; someone touching the case could get a shock by providing a conducting path to ground.

21.3 CAPACITORS IN AC CIRCUITS

Figure 21.5a shows a capacitor connected to an AC source. The AC source pumps charge as needed to keep the voltage across the capacitor equal to the voltage of the source. Since the charge on the capacitor is proportional to the potential difference,

$$q(t) = Cv(t)$$

the current is proportional to the *rate of change* of the voltage,

$$i(t) = \frac{\Delta q}{\Delta t} = C\frac{\Delta v}{\Delta t} \tag{21-5}$$

The time interval Δt must be small for i to represent the instantaneous current.

Figure 21.5b shows the voltage $v(t)$ and current $i(t)$ as functions of time for the capacitor. Note some important points:

- The current is maximum when the voltage is zero.
- The voltage is maximum when the current is zero.
- The capacitor repeatedly charges and discharges.

The voltage and the current are both sinusoidal functions of time with the same frequency, but they are out of phase: the current starts at its maximum positive value but

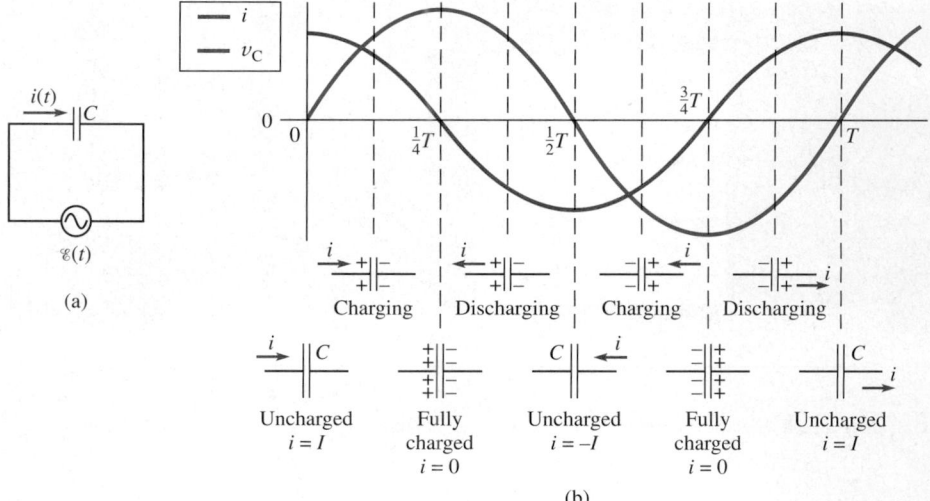

Figure 21.5 (a) An AC generator connected to a capacitor. (b) One complete cycle of the charge on a capacitor connected to an AC source as a function of time. Signs are chosen so that positive current (to the right) gives the capacitor a positive charge (left plate positive).

the voltage reaches its maximum positive value $\frac{1}{4}$ cycle later. The voltage stays a quarter cycle behind the current at all times. The period T is the time for one complete cycle of a sinusoidal function; one cycle corresponds to 360° since

$$\omega T = 2\pi \,\mathrm{rad} = 360°$$

For $\frac{1}{4}$ cycle, $\frac{1}{4}\omega T = \pi/2$ rad = 90°. Thus, we say that the voltage and current are $\frac{1}{4}$ cycle out of phase or 90° out of phase. The current *leads* the capacitor voltage by a phase constant of 90°; equivalently, the voltage *lags* the current by the same phase angle.

Current leads voltage by 90° in a capacitor in an AC circuit.

If the voltage across the capacitor is given by

$$v(t) = V \sin \omega t$$

then the current varies in time as

$$i(t) = I \sin (\omega t + \pi/2)$$

We add the $\pi/2$ radians to the argument of the sine function to give the current a head start of $\pi/2$ rad. (We use radians rather than degrees since angular frequency ω is generally expressed in rad/s.)

In the general expression

$$i = I \sin (\omega t + \phi)$$

the angle ϕ is called the **phase constant**, which, for the case of the current in the capacitive circuit, is $\phi = \pi/2$. A sine function shifted $\pi/2$ radians ahead is a cosine function, as can be seen in Fig. 21.5; that is,

$$\sin (\omega t + \pi/2) = \cos \omega t$$

so

$$i(t) = I \cos \omega t$$

The amplitude of the current I is proportional to the voltage amplitude V. A larger voltage means that more charge needs to be pumped onto the capacitor; to pump more charge in the same amount of time requires a larger current. We write the proportionality as

$$V_C = I X_C \qquad\qquad \text{(21-6)}$$

where the quantity X_C is called the **reactance** of the capacitor. Compare Eq. (21-6) to Ohm's law for a resistor ($v = iR$); reactance must have the same SI unit as resistance (ohms). We have written Eq. (21-6) in terms of the amplitudes (V, I), but it applies equally well to the rms values (since both are smaller by the same factor, $\sqrt{2}$).

By analogy with Ohm's law, we can think of the reactance as the "effective resistance" of the capacitor. The reactance determines how much current flows; the capacitor reacts in a way to impede the flow of current. A larger reactance means a smaller current, just as a larger resistance means a smaller current.

There are, however, important differences between reactance and resistance. A resistor dissipates energy, but an ideal capacitor does *not*; the average power dissipated by an ideal capacitor is zero, not $I_{\mathrm{rms}}^2 X_C$. Note also that Eq. (21-6) relates only the *amplitudes* of the current and voltage. Since the current and voltage in a capacitor are 90° out of phase, it does *not* apply to the instantaneous values:

$$v(t) \neq i(t) X_C$$

For a resistor, on the other hand, the current and voltage are *in phase* (phase difference of zero); it *is* true for a resistor that $v(t) = i(t)R$.

Another difference is that reactance depends on frequency. Recall from Chapter 20 that

$$\text{If } \Phi(t) = \Phi_0 \sin \omega t, \text{ then } \frac{\Delta\Phi}{\Delta t} = \omega\Phi_0 \cos \omega t \quad \text{(for small } \Delta t\text{);} \qquad \text{(20-7a)}$$

$$\text{if } \Phi(t) = \Phi_0 \cos \omega t, \text{ then } \frac{\Delta\Phi}{\Delta t} = -\omega\Phi_0 \sin \omega t \quad \text{(for small } \Delta t\text{).} \qquad \text{(20-7b)}$$

These are general mathematical relationships giving the rates of change of sinusoidal functions. We have seen the same relationships in simple harmonic motion: if the position of a particle is

$$x(t) = A \sin \omega t$$

then its velocity, the rate of change of position, is

$$v(t) = \frac{\Delta x}{\Delta t} = \omega A \cos \omega t \qquad (10\text{-}25b)$$

For a capacitor in an AC circuit, if the charge as a function of time is

$$q(t) = Q \sin \omega t$$

then the current (the rate of change of the charge on the capacitor) must be

$$i(t) = \frac{\Delta q}{\Delta t} = \omega Q \cos \omega t$$

Therefore, the peak current is

$$I = \omega Q$$

Since $Q = CV$, we can find the reactance:

$$X_C = \frac{V}{I} = \frac{V}{\omega Q} = \frac{V}{\omega CV}$$

$$X_C = \frac{1}{\omega C} \qquad (21\text{-}7)$$

The reactance is inversely proportional to the capacitance and to the angular frequency. To understand why, let us focus on the first quarter of a cycle ($0 \le t \le T/4$) in Fig. 21.5b. During this quarter cycle, a total charge $Q = CV$ flows onto the capacitor plates since the capacitor goes from being uncharged to fully charged. For a larger value of C, a proportionately larger charge must be put on the capacitor to reach a potential difference of V; to put more charge on in the same amount of time ($T/4$), the current must be larger. Thus, when the capacitance is larger, the reactance must be lower because more current flows for a given AC voltage amplitude.

The reactance is also inversely proportional to the frequency. For a higher frequency, the time available to charge the capacitor ($T/4$) is shorter. For a given voltage amplitude, a larger current must flow to achieve the same maximum voltage in a shorter amount of time. Thus, the reactance is smaller for a higher frequency.

At very high frequencies, the reactance approaches zero. The capacitor no longer impedes the flow of current; AC current flows in the circuit as if there were a conducting wire short-circuiting the capacitor. For the other limiting case, very low frequencies, the reactance approaches infinity. At a very low frequency, the applied voltage changes slowly; the flow of current stops as soon as the capacitor is charged to a voltage equal to the applied voltage.

Example 21.2

Capacitive Reactance for Two Frequencies

(a) Find the capacitive reactance and the rms current for a 4.00-μF capacitor when it is connected to an AC source of 12.0 V rms at 60.0 Hz. (b) Find the reactance and current when the frequency is changed to 15.0 Hz while the rms voltage remains at 12.0 V.

Strategy The reactance is the proportionality constant between the rms values of the voltage across and current through the capacitor. The capacitive reactance is given by Eq. (21-7). Frequencies in Hz are given; we need *angular* frequencies to calculate the reactance.

Solution   (a) Angular frequency is

$$\omega = 2\pi f$$

Then the reactance is

$$X_C = \frac{1}{2\pi f C}$$

$$= \frac{1}{2\pi \times 60.0\ \text{Hz} \times 4.00 \times 10^{-6}\ \text{F}} = 663.1\ \Omega \approx 663\ \Omega$$

The rms current is

$$I_{\text{rms}} = \frac{V_{\text{rms}}}{X_C} = \frac{12.0\ \text{V}}{663\ \Omega} = 18.096\ \text{mA} \approx 18.1\ \text{mA}$$

continued on next page

Example 21.2 continued

(b) We could redo the calculation in the same way. An alternative is to note that the frequency is multiplied by a factor $\frac{15}{60} = \frac{1}{4}$. Since reactance is *inversely* proportional to frequency,

$$X_C = 4 \times 663 \ \Omega = 2650 \ \Omega$$

A larger reactance means a smaller current:

$$I_{rms} = \tfrac{1}{4} \times 18.096 \ \text{mA} = 4.52 \ \text{mA}$$

Discussion   When the frequency is increased, the reactance decreases and the current increases. As we see in Section 21.7, capacitors can be used in circuits to filter out low frequencies

because at lower frequency, less current flows. When a PA system makes a humming sound (60-Hz hum), a capacitor can be inserted between the amplifier and the speaker to block much of the 60-Hz noise while letting the higher frequencies pass through.

Practice Problem 21.2   Capacitive reactance and rms current for a new frequency

Find the capacitive reactance and the rms current for a 4.00-μF capacitor when it is connected to an AC source of 220.0 V rms and 4.00 Hz.

Power

Figure 21.6 shows graphs for the instantaneous power $p(t) = v(t)i(t)$ for a capacitor superimposed on graphs of the current and voltage. The 90° phase difference between current and voltage has implications for the power in the circuit. During the first quarter cycle ($0 \leq t \leq T/4$), both the voltage and the current are positive. The power is positive: the generator is delivering energy to the capacitor to charge it. During the second quarter cycle ($T/4 \leq t \leq T/2$), the current is negative while the voltage remains positive. The power is negative; as the capacitor discharges, energy is returned to the generator from the capacitor.

The power continues to alternate between positive and negative as the capacitor stores and then returns electrical energy. The average power is zero since all the energy stored is given back and none of it is dissipated.

> The average power is zero for an ideal capacitor in an AC circuit.

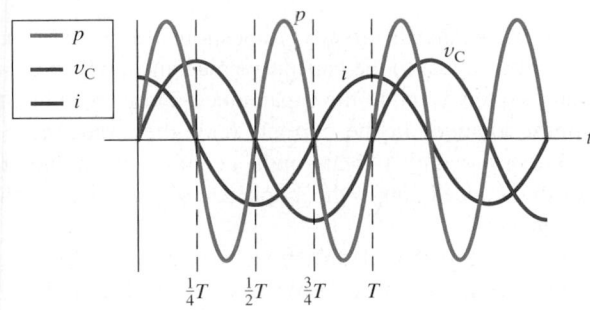

Figure 21.6 Current, voltage, and power for a capacitor in an AC circuit

21.4   INDUCTORS IN AC CIRCUITS

An inductor in an AC circuit develops an induced emf that opposes changes in the current, according to Faraday's law [Eq. (20-6)]. We use the same sign convention as for the capacitor: the current i through the inductor in Fig. 21.7a is positive when it flows to the right and the voltage across the inductor v_L is positive if the left side is at a higher potential than the right side. If current flows in the positive direction and is *increasing*, the induced emf *opposes the increase* (Fig. 21.7b) and v_L is positive. If current flows in the positive direction and is *decreasing*, the induced emf *opposes the decrease* (Fig. 21.7c) and v_L is negative. Since in the first case $\Delta i/\Delta t$ is positive and in the second case $\Delta i/\Delta t$ is negative, the voltage has the correct sign if we write

$$v_L = L\frac{\Delta i}{\Delta t} \tag{21-8}$$

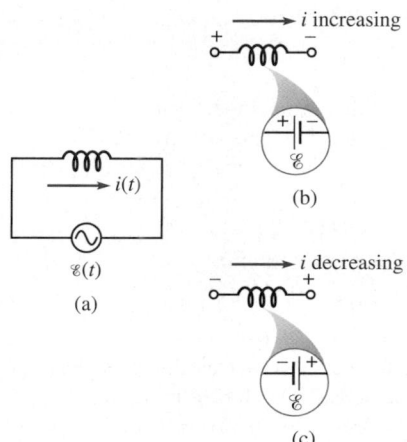

Figure 21.7 (a) An inductor connected to an AC source. (b) and (c) The potential difference across the inductor for current flowing to the right depends on whether the current is increasing or decreasing.

In Problem 27 you can verify that Eq. (21-8) also gives the correct sign when current flows to the left.

The voltage amplitude across the inductor is proportional to the amplitude of the current. The constant of proportionality is called the **reactance** of the inductor (X_L):

$$V_L = IX_L \tag{21-9}$$

As for the capacitive reactance, the inductive reactance X_L has units of ohms. As in Eq. (21-6), V and I in Eq. (21-9) can represent *either* the two amplitudes *or* the two rms values, but be careful not to mix them.

In Problem 25 you can show, using reasoning similar to that used for the capacitor, that the reactance of an inductor is

$$X_L = \omega L \tag{21-10}$$

Note that the inductive reactance is directly proportional to the inductance L and to the angular frequency ω, in contrast to the capacitive reactance, which is *inversely* proportional to the angular frequency and to the capacitance. The induced emf in the inductor always acts to oppose changes in the current. At higher frequency, the more rapid changes in current are opposed by a greater induced emf in the inductor. Thus, the ratio of the amplitude of the induced emf to the amplitude of the current—the reactance—is greater at higher frequency.

Figure 21.8 shows the potential difference across the inductor and the current through the inductor as functions of time. We assume an ideal inductor—one with no resistance in its windings. Since $v_L = L \Delta i/\Delta t$, the graph of $v_L(t)$ is proportional to the *slope* of the graph of $i(t)$ at any time t. The voltage and current are out of phase by $\frac{1}{4}$ cycle, but this time the current *lags* the voltage by 90°; current reaches its maximum $\frac{1}{4}$ cycle *after* the voltage reaches a maximum. A mnemonic device for remembering what leads and what lags is that the letter *c* (for *current*) appears in the second half of the

Current lags voltage across an inductor in an AC circuit

Figure 21.8 Current and potential difference across an inductor in an AC circuit. Note that when the current is maximum or minimum, its instantaneous rate of change—represented by its slope—is zero, so $v_L = 0$. On the other hand, when the current is zero, it is changing the fastest, so v_L has its maximum magnitude.

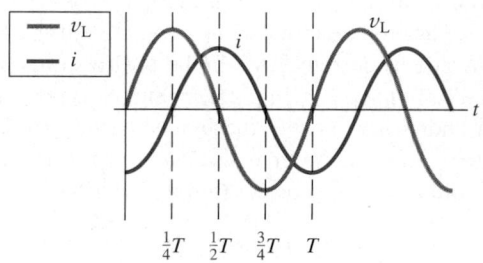

word *inductor* (current lags inductor voltage) and at the *beginning* of the word *capacitor* (current leads capacitor voltage).

In Fig. 21.8, the voltage across the inductor can be written

$$v_L(t) = V \sin \omega t$$

The current is

$$i(t) = -I \cos \omega t = I \sin (\omega t - \pi/2)$$

where we have used the trigonometric identity $-\cos \omega t = \sin (\omega t - \pi/2)$. We see explicitly that the current lags behind the voltage from the phase constant $\phi = -\pi/2$.

Power

As for the capacitor, the 90° phase difference between current and voltage means that the average power is zero. No energy is dissipated in an *ideal* inductor (one with no resistance). The generator alternately sends energy to the inductor and receives energy back from the inductor.

> The average power is zero for an ideal inductor in an AC circuit.

Example 21.3

Inductor in a Radio Tuning Circuit

 A 0.56-μH inductor is used as part of the tuning circuit in a radio. Assume the inductor is ideal. (a) Find the reactance of the inductor at a frequency of 90.9 MHz. (b) Find the amplitude of the current through the inductor if the voltage amplitude is 0.27 V. (b) Find the capacitance of a capacitor that has the same reactance at 90.9 MHz.

Strategy   The reactance of an inductor is the product of angular frequency and inductance. The reactance in ohms is the ratio of the voltage amplitude to the amplitude of the current. For the capacitor, the reactance is $1/\omega C$.

Solution   (a) The reactance of the inductor is
$$X_L = \omega L = 2\pi f L$$
$$= 2\pi \times 90.9 \text{ MHz} \times 0.56 \text{ μH} = 320 \text{ Ω}$$

(b) The amplitude of the current is
$$I = \frac{V}{X_L}$$
$$= \frac{0.27 \text{ V}}{320 \text{ Ω}} = 0.84 \text{ mA}$$

(c) We set the two reactances equal ($X_L = X_C$) and solve for C:

$$\omega L = \frac{1}{\omega C}$$

$$C = \frac{1}{\omega^2 L} = \frac{1}{4\pi^2 \times (90.9 \times 10^6 \text{ Hz})^2 \times 0.56 \times 10^{-6} \text{ H}}$$
$$= 5.5 \text{ pF}$$

Discussion   We can check by calculating the reactance of the capacitor:

$$X_C = \frac{1}{\omega C} = \frac{1}{2\pi \times 90.9 \times 10^6 \text{ Hz} \times 5.5 \times 10^{-12} \text{ F}} = 320 \text{ Ω}$$

In Section 21.6 we study tuning circuits in more detail.

Practice Problem 21.3   Reactance and rms current

Find the inductive reactance and the rms current for a 3.00-mH inductor when it is connected to an AC source of 10.0 mV (rms) at a frequency of 60.0 kHz.

21.5   *RLC* SERIES CIRCUITS

Figure 21.9a shows an *RLC* series circuit. Kirchoff's junction rule tells us that the instantaneous current through each element is the same, since there are no junctions. The loop rule requires the sum of the instantaneous voltage drops across the three elements to equal the applied AC voltage:

$$\mathscr{E}(t) = v_L(t) + v_R(t) + v_C(t) \qquad (21\text{-}11)$$

The three voltages are sinusoidal functions of time with the same frequency but different phase constants.

Suppose that we choose to write the current with a phase constant of zero. The voltage across the resistor is in phase with the current, so it also has a phase constant of zero

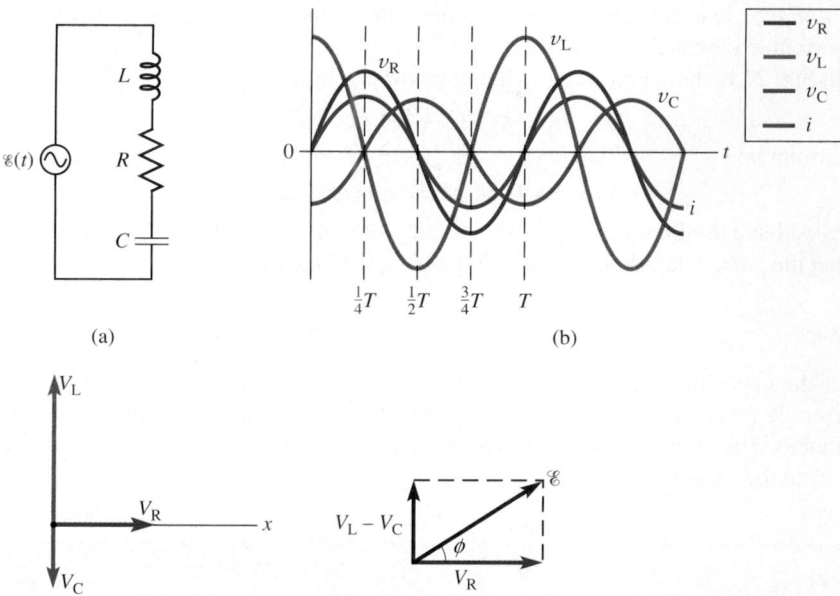

Figure 21.9 (a) An *RLC* series circuit. (b) The voltages across the circuit elements and the current as functions of time. The current is in phase with v_R, leads v_C by 90°, and lags v_L by 90°. (c) Phasor representation of the voltages. (d) The phase angle ϕ between the source emf and the voltage across the resistor (which is in phase with the current).

(see Fig. 21.9b). The voltage across the inductor leads the current by 90°, so it has a phase constant of $+\pi/2$. The voltage across the capacitor lags the current by 90°, so it has a phase constant of $-\pi/2$.

$$\mathscr{E}(t) = \mathscr{E} \sin(\omega t + \phi) = V_L \sin\left(\omega t + \frac{\pi}{2}\right) + V_R \sin \omega t + V_C \sin\left(\omega t - \frac{\pi}{2}\right) \quad (21\text{-}12)$$

We could simplify this sum using trigonometric identities, but there is an easier method. We can represent each sinusoidal voltage by a vector-like object called a **phasor**. The magnitude of the phasor represents the amplitude of the voltage; the angle of the phasor represents the phase constant of the voltage. We can then add phasors the same way we add vectors. (See Problem 41 for insight into why the phasor method works.) Although we draw them like vectors and *add like vectors*, they are not vectors in the usual sense. A phasor is not a quantity with a direction in space, like real vectors such as acceleration, momentum, or magnetic field.

Figure 21.9c shows three phasors representing the voltages $v_L(t)$, $v_R(t)$, and $v_C(t)$. An angle counterclockwise from the +x-axis represents a positive phase constant. First we add the phasors representing $v_L(t)$ and $v_C(t)$, which are in opposite directions. Then we add the sum of these two to the phasor that represents $v_R(t)$ (Fig. 21.9d). The resultant represents $\mathscr{E}(t)$. The amplitude of $\mathscr{E}(t)$ is the length of the resultant phasor; from the Pythagorean theorem,

$$\mathscr{E} = \sqrt{V_R^2 + (V_L - V_C)^2} \quad (21\text{-}13)$$

Each of the voltage amplitudes on the side of Eq. (21-13) can be rewritten as the amplitude of the current times a reactance or resistance:

$$\mathscr{E} = \sqrt{(IR)^2 + (IX_L - IX_C)^2}$$

Factoring out the current yields

$$\mathscr{E} = I\sqrt{R^2 + (X_L - X_C)^2}$$

Thus, the amplitude of the AC source voltage is proportional to the amplitude of the current. The constant of proportionality is called the **impedance** Z of the circuit.

$$\mathscr{E} = IZ \quad (21\text{-}14a)$$

$$Z = \sqrt{R^2 + (X_L - X_C)^2} \quad (21\text{-}14b)$$

Impedance is measured in ohms.

From Fig. 21.9d, the source voltage $\mathcal{E}(t)$ leads $v_R(t)$—and the current $i(t)$—by a phase angle ϕ where

$$\tan \phi = \frac{V_L - V_C}{V_R} = \frac{IX_L - IX_C}{IR} = \frac{X_L - X_C}{R} \qquad (21\text{-}15)$$

We have assumed $X_L > X_C$ in Fig. 21.9. If $X_L < X_C$, the phase angle ϕ is negative, which means that the source voltage *lags* the current. Figure 21.9d also implies that

$$\cos \phi = \frac{V_R}{\mathcal{E}} = \frac{IR}{IZ} = \frac{R}{Z} \qquad (21\text{-}16)$$

If one or two of the elements R, L, and C are not present in a circuit, the foregoing analysis is still valid. Since there is no potential difference across a missing element, we simply set the resistance or reactance of the missing element(s) to zero. For instance, since an inductor is made by coiling a long length of wire, it usually has an appreciable resistance. We can model a real inductor as an ideal inductor in series with a resistor. The impedance of the inductor is found by setting $X_C = 0$ in Eq. (21-14b).

Example 21.4

An *RLC* Series Circuit

In an *RLC* circuit, the following three elements are connected in series: a resistor of 40.0 Ω, a 22.0-mH inductor, and a 0.400-μF capacitor. The AC source has a peak voltage of 0.100 V and an angular frequency of 1.00×10^4 rad/s. (a) Find the amplitude of the current. (b) Find the phase angle between the current and the AC source. Which leads? (c) Find the peak voltages across each of the circuit elements.

Strategy   The impedance is the ratio of the source voltage amplitude to the amplitude of the current. By finding the reactances of the inductor and capacitor, we can find the impedance and then solve for the amplitude of the current. The reactances also enable us to calculate the phase constant ϕ. If ϕ is positive, the source voltage leads the current; if ϕ is negative the source voltage lags the current. The peak voltage across any element is equal to the peak current times the reactance or resistance of that element.

Solution   (a)   The inductive reactance is

$$X_L = \omega L = 1.00 \times 10^4 \text{ rad/s} \times 22.0 \times 10^{-3} \text{ H} = 220 \ \Omega$$

The capacitive reactance is

$$X_C = \frac{1}{\omega C} = \frac{1}{1.00 \times 10^4 \text{ rad/s} \times 0.400 \times 10^{-6} \text{ F}} = 250 \ \Omega$$

Then the impedance of the circuit is

$$Z = \sqrt{R^2 + X^2} = \sqrt{(40.0 \ \Omega)^2 + (-30 \ \Omega)^2} = 50 \ \Omega$$

For a source voltage amplitude $V = 0.100$ V, the amplitude of the current is

$$I = \frac{V}{Z} = \frac{0.100 \text{ V}}{50 \ \Omega} = 2.0 \text{ mA}$$

(b)   The phase angle ϕ is

$$\phi = \tan^{-1} \frac{X_L - X_C}{R} = \tan^{-1} \frac{-30 \ \Omega}{40 \ \Omega} = -0.64 \text{ rad} = -37°$$

Since $X_L < X_C$, the phase angle ϕ is negative, which means that the source voltage *lags* the current.

(c)   The voltage amplitude across the inductor is

$$V_L = IX_L = 2.0 \text{ mA} \times 220 \ \Omega = 440 \text{ mV}$$

For the capacitor and resistor,

$$V_C = IX_C = 2.0 \text{ mA} \times 250 \ \Omega = 500 \text{ mV}$$

and

$$V_R = IR = 2.0 \text{ mA} \times 40.0 \ \Omega = 80 \text{ mV}$$

Discussion   Since the voltage phasors in Fig. 21.9 are each proportional to I, we can divide each by I to form a phasor diagram where the phasors represent reactances or resistances (Fig. 21.10). Such a phasor diagram can be used to find the impedance of the circuit and the phase constant, instead of using Eqs. (21-14b) and (21-15).

⚠ Note that the sum of the voltage amplitudes across the three circuit elements is not the same as the source voltage amplitude:

$$100 \text{ mV} \neq 440 \text{ mV} + 80 \text{ mV} + 500 \text{ mV}$$

Figure 21.10
A phasor diagram used to find impedance and phase angle. (The lengths of the phasors are not to scale.)

continued on next page

Example 21.4 *continued*

The voltage amplitudes across the inductor and capacitor are each *larger* than the source voltage amplitude. The voltage amplitudes are *maximum* values; since the voltages are not in phase with each other, they do not attain their maximum values at the same instant of time. What is true is that the sum of the *instantaneous* potential differences across the three elements at any given time is equal to the instantaneous source voltage at the same time [Eq. (21-12)].

Practice Problem 21.4   Instantaneous voltages

If the current in this same circuit is written as $i(t) = I \sin \omega t$, what would be the corresponding expressions for $v_C(t)$, $v_L(t)$, $v_R(t)$, and $\mathcal{E}(t)$? (The main task is to get the phase constants correct.) Using these expressions, show that at $t = 80.0$ μs, $v_C(t) + v_L(t) + v_R(t) = \mathcal{E}(t)$. (The loop rule is true at *any* time t; we just verify it at one particular time.)

Physics at Home

Find an electrical device that has a label with two numerical ratings, one in V·A and one in W. The windings of a transformer have significant inductance, so try something with an external transformer (inside the power supply) or an internal transformer (such as a desktop computer). The windings of motors also have inductance, so something with a motor is also a good choice. Calculate the power factor for the device. Now find a device that has little reactance compared to its resistance, such as a heater or a lightbulb. Why is there no numerical rating in V·A?

Power Factor

No power is dissipated in an ideal capacitor or an ideal inductor; the power is dissipated only in the resistance of the circuit (including the resistances of the wires of the circuit and the windings of the inductor):

$$P_{av} = I_{rms} V_{R,rms} \tag{21-4}$$

We want to rewrite the average power in terms of the rms source voltage.

$$\frac{V_{R,rms}}{\mathcal{E}_{rms}} = \frac{I_{rms}R}{I_{rms}Z} = \frac{R}{Z}$$

From Eq. (21-16), $R/Z = \cos \phi$. Therefore,

$$V_{R,rms} = \mathcal{E}_{rms} \cos \phi$$

and

$$P_{av} = I_{rms} \mathcal{E}_{rms} \cos \phi \tag{21-17}$$

The factor $\cos \phi$ in Eq. (21.28) is called the **power factor**. When there is only resistance and no reactance in the circuit, $\phi = 0$ and $\cos \phi = 1$; then $P_{av} = I_{rms} \mathcal{E}_{rms}$. When there is only capacitance or inductance in the circuit, $\phi = \pm 90°$ and $\cos \phi = 0$, so that $P_{av} = 0$. Many electrical devices contain appreciable inductance or capacitance; the load they present to the source voltage is not purely a resistance. In particular, any device with a transformer has some inductance due to the windings. The label on an electrical device sometimes includes a quantity with units of V·A and a smaller quantity with units of W. The former is the product $I_{rms} \mathcal{E}_{rms}$; the latter is the average power consumed.

Example 21.5

Laptop Power Supply

A power supply for a laptop computer is labeled as follows: "45 W AC Adapter. AC input: 1.0 A max, 120 V, 60.0 Hz." A simplified circuit model for the power supply is a resistor R and an ideal inductor L in series with an ideal AC emf. The inductor represents primarily the

inductance of the windings of the transformer; the resistor represents primarily the load presented by the laptop computer. Find the values of L and R when the power supply draws the maximum rms current of 1.0 A.

continued on next page

Example 21.5 *continued*

Strategy   First we sketch the circuit (Fig. 21.11a). The next step is to identify the quantities given in the problem, taking care to distinguish rms quantities from amplitudes and average power from $I_{rms}\mathscr{E}_{rms}$. Since power is dissipated in the resistor but not in the inductor, we can find the resistance from the average power. Then we can use the power factor to find L. We assume no capacitance in the circuit, which means we can set $X_C = 0$.

Solution   The problem tells us that the maximum *rms* current is $I_{rms} = 1.0$ A. The rms source voltage is $\mathscr{E}_{rms} = 120$ V. The frequency is $f = 60.0$ Hz. The average power is 45 W when the power supply draws 1.0 A rms; the average power is smaller when the current drawn is smaller. Then

$$\mathscr{E}_{rms}I_{rms} = 120\text{ V} \times 1.0\text{ A} = 120\text{ V·A}$$

Note that the average power is less than $I_{rms}\mathscr{E}_{rms}$; it can never be greater than $I_{rms}\mathscr{E}_{rms}$ since $\cos\phi \leq 1$.

Since power is dissipated only in the resistor,

$$P_{av} = I_{rms}^2 R$$

The resistance is therefore

$$R = \frac{P_{av}}{I_{rms}^2} = \frac{45\text{ W}}{(1.0\text{ A})^2} = 45\text{ }\Omega$$

The ratio of the average power to $I_{rms}\mathscr{E}_{rms}$ gives the power factor:

$$\frac{\mathscr{E}_{rms}I_{rms}\cos\phi}{\mathscr{E}_{rms}I_{rms}} = \cos\phi = \frac{45\text{ W}}{120\text{ V·A}} = 0.375$$

The phase angle is $\phi = \cos^{-1} 0.375 = 68.0°$. From the phasor diagram of Fig. 21.11b,

$$\tan\phi = \frac{V_L}{V_R} = \frac{IX_L}{IR} = \frac{X_L}{R}$$

Solving for X_L,

$$X_L = R\tan\phi = (45\text{ }\Omega)\tan 68.0° = 111.4\text{ }\Omega = \omega L$$

Now we can solve for L:

$$L = \frac{X_L}{\omega} = \frac{111.4\text{ }\Omega}{2\pi \times 60.0\text{ Hz}} = 0.30\text{ H}$$

Discussion   Check: $\cos\phi$ should be equal to R/Z.

$$\frac{R}{Z} = \frac{R}{\sqrt{R^2 + X_L^2}} = \frac{45\text{ }\Omega}{\sqrt{(45\text{ }\Omega)^2 + (111.4\text{ }\Omega)^2}} = 0.375$$

which agrees with $\cos\phi = 0.375$.

Practice Problem 21.5   A more typical current draw

The adapter rarely draws the maximum rms current of 1.0 A. Suppose that, more typically, the adapter draws an rms current of 0.25 A. What is the average power? Use the same simplified circuit model with the same value of L but a *different* value of R. [*Hint:* Begin by finding the impedance $Z = \sqrt{R^2 + X_L^2}$.]

Figure 21.11
(a) A circuit diagram for the power supply. (b) Phasor addition of the voltages across the inductor and resistor.

21.6   RESONANCE IN AN *RLC* CIRCUIT

Suppose an *RLC* circuit is connected to an AC source with a fixed amplitude but variable frequency. The impedance depends on frequency, so the amplitude of the current depends on frequency. Figure 21.12 shows three graphs (called **resonance curves**) of the amplitude

Figure 21.12   The amplitude of the current *I* as a function of angular frequency ω for three different resistances in a series *RLC* circuit. The widths of each peak at half-maximum current are indicated. The horizontal scale is logarithmic.

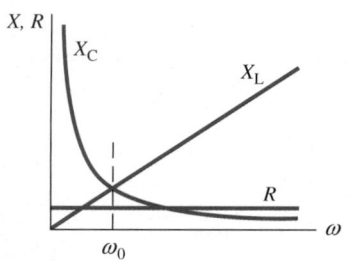

Figure 21.13 Frequency dependence of the inductive and capacitive reactances and of the resistance as a function of frequency.

of the current $I = \mathcal{E}/Z$ as a function of angular frequency for a circuit with $L = 1.0$ H, $C = 1.0$ μF, and $\mathcal{E} = 100$ V. Three different resistors were used: 200 Ω, 500 Ω, and 1000 Ω.

The shape of these graphs is determined by the frequency dependence of the inductive and capacitive reactances (Fig. 21.13). At low frequencies, the reactance of the capacitor $X_C = 1/(\omega C)$ is much greater than either R or X_L, so $Z \approx X_C$. At high frequencies, the reactance of the inductor $X_L = \omega L$ is much greater than either R or X_C, so $Z \approx X_L$. At extreme frequencies, either high or low, the impedance is larger and the amplitude of the current is therefore small.

The impedance of the circuit is

$$Z = \sqrt{R^2 + (X_L - X_C)^2} \qquad \text{(21-14b)}$$

Since R is constant, the minimum impedance $Z = R$ occurs at an angular frequency ω_0—called the **resonant** angular frequency—for which the reactances of the inductor and capacitor are equal so that $X_L - X_C = 0$.

$$X_L = X_C$$

$$\omega_0 L = \frac{1}{\omega_0 C}$$

Solving for ω_0,

$$\omega_0 = \frac{1}{\sqrt{LC}} \qquad \text{(21-18)}$$

Note that the resonant frequency of a circuit depends only on the values of the inductance and the capacitance, not on the resistance. In Fig. 21.12, the maximum current occurs at the resonant frequency for any value of R. However, the value of the maximum current depends on R since $Z = R$ at resonance. The resonance peak is higher for a smaller resistance. If we measure the width of a resonance peak where the amplitude of the current has half its maximum value, we see that the resonance peaks get narrower with decreasing resistance.

Resonance in an *RLC* circuit is analogous to resonance in mechanical oscillations (see Section 10.10). Just as a mass-spring system has a single resonant frequency, determined by the spring constant and the mass, the *RLC* circuit has a single resonant frequency, determined by the capacitance and the inductance. When either system is driven externally—by a sinusoidal applied force for the mass-spring or by a sinusoidal applied emf for the circuit—the amplitude of the system's response is greatest when driven at the resonant frequency. In both systems, energy is being converted back and forth between two forms. For the mass-spring, the two forms are kinetic and elastic potential energy; for the *RLC* circuit, the two forms are electric energy stored in the capacitor and magnetic energy stored in the inductor. The resistor in the *RLC* circuit fills the role of friction in a mass-spring system: dissipating energy.

A sharp resonance peak enables the tuning circuit in a TV or radio to select one out of many different frequencies being broadcast. With one type of tuner, common in old radios, the tuning knob adjusts the capacitance by rotating one set of parallel plates relative to a fixed set so that the area of overlap is varied (Fig. 21.14). By changing the capacitance, the resonant frequency can be varied. The tuning circuit is driven by a mixture of many different frequencies coming from the antenna, but only frequencies very near the resonance frequency produce a significant response in the tuning circuit.

Example 21.6

A Tuner for a Radio

A radio tuner has a 400.0-Ω resistor, a 0.50-mH inductor, and a variable capacitor connected in series. Suppose the capacitor is adjusted to 72.0 pF. (a) Find the resonant frequency for the circuit. (b) Find the reactances of the inductor and capacitor at the resonant fre-

quency. (c) The applied emf at the resonant frequency coming in from the antenna is 20.0 mV (rms). Find the rms current in the tuning circuit. (d) Find the rms voltages across each of the circuit elements.

continued on next page

Example 21.6 *continued*

Strategy The resonant frequency can be found from the values of the capacitance and the inductance. The reactances at the resonant frequency must be equal. To find the current in the circuit, we note that the impedance is equal to the resistance since the circuit is in resonance. The rms current is the ratio of the rms voltage to the impedance. The rms voltage across a circuit element is the rms current times the element's reactance or resistance.

Solution (a) The resonant angular frequency is given by

$$\omega_0 = \frac{1}{\sqrt{LC}}$$

$$= \frac{1}{\sqrt{0.50 \times 10^{-3}\,\text{H} \times 72.0 \times 10^{-12}\,\text{F}}}$$

$$= 5.27 \times 10^6 \, \text{rad/s}$$

The resonant frequency in Hz is

$$f_0 = \frac{\omega_0}{2\pi} = 840 \, \text{kHz}$$

(b) The reactances are

$$X_L = \omega L = 5.27 \times 10^6\,\text{rad/s} \times 0.50 \times 10^{-3}\,\text{H} = 2.6\,\text{k}\Omega$$

and

$$X_C = \frac{1}{\omega C} = \frac{1}{5.27 \times 10^6\,\text{rad/s} \times 72.0 \times 10^{-12}\,\text{F}} = 2.6\,\text{k}\Omega$$

They are equal, as expected.

(c) At the resonant frequency, the impedance is equal to the resistance.

$$Z = R = 400.0 \, \Omega$$

The rms current is

$$I_{\text{rms}} = \frac{\mathscr{E}_{\text{rms}}}{Z} = \frac{20.0\,\text{mV}}{400.0\,\Omega} = 0.0500 \, \text{mA}$$

(d) The rms voltages are

$$V_{\text{L-rms}} = I_{\text{rms}} X_L = 0.0500\,\text{mA} \times 2.6 \times 10^3\,\Omega = 130 \, \text{mV}$$

$$V_{\text{C-rms}} = I_{\text{rms}} X_C = 0.0500\,\text{mA} \times 2.6 \times 10^3\,\Omega = 130 \, \text{mV}$$

$$V_{\text{R-rms}} = I_{\text{rms}} R = 0.0500\,\text{mA} \times 400.0\,\Omega = 20.0 \, \text{mV}$$

Discussion The resonant frequency of 840 kHz is a reasonable result since it lies in the AM radio band (530 kHz–1700 kHz).

The rms voltages across the inductor and across the capacitor are equal at resonance, but the instantaneous voltages are opposite in phase (a phase difference of π rad or 180°), so the sum of the potential difference across the two is always zero. In a phasor diagram, the phasors for v_L and v_C are opposite in direction and equal in length, so they add to zero. Then the voltage across the resistor is equal to the applied emf in both amplitude and phase.

Practice Problem 21.6   Tuning the radio to a different station

The tuner of this radio works by varying the capacitance. Find the capacitance required to tune to a station broadcasting at 1420 kHz.

Figure 21.14 The variable capacitor inside an old radio. The radio is tuned to a particular resonant frequency by adjusting the capacitance.

21.7 CONVERTING AC TO DC; FILTERS

Diodes

A *diode* is a circuit component that allows current to flow much more easily in one direction than in the other. An *ideal* diode has zero resistance for current in one direction, so that the current flows without any voltage drop across the diode, and infinite

Making The Connection:
diodes and rectifiers

The circuit symbol for a diode is . The arrow indicates the direction of current flow.

resistance for current in the other direction, so that no current flows. The circuit symbol for a diode has an arrowhead to indicate the direction of allowed current.

The circuit in Fig. 21.15a is called a *half-wave rectifier*. If the input is a sinusoidal emf, the output (the voltage across the resistor) is as shown in Fig. 21.15b. The output signal can be smoothed out by a capacitor (Fig. 21.15c). The capacitor charges up when current flows through the diode; when the source voltage starts to drop and then changes polarity, the capacitor discharges through the resistor. (The capacitor cannot discharge through the diode because that would send current the wrong way through the diode.) The discharge keeps the voltage v_R up. By making the RC time constant ($\tau = RC$) long enough, the discharge through the resistor can be made to continue until the source voltage turns positive again (Fig. 21.15d).

Circuits involving more than one diode can be arranged to make a *full-wave rectifier*. The output of a full-wave rectifier (without a capacitor to smooth it) is shown in Fig. 21.15e. Circuits like these are found inside the AC adapter used with devices such as portable CD players, radios, and laptop computers. Many other devices have circuits to do AC-to-DC conversion inside of them.

This AC adapter from a portable CD player contains a transformer (labeled "CK-62") to reduce the amplitude of AC source voltage of 120 V rms to 15 V rms. The two red diodes serve as a full-wave rectifier circuit and the capacitor (labeled "470 μF") smooths out the ripples. The output is a nearly constant DC voltage of 12 V.

Filters

The capacitor in Fig. 21.15c serves as a *filter*. Figure 21.16 shows two *RC filters* commonly used in circuits. Figure 21.16a is a *low-pass filter*. For a high-frequency AC signal, the capacitor serves as a low reactance path to ground ($X_C \ll R$); the voltage across the resistor is much larger than the voltage across the capacitor, so the voltage across the output terminals is a small fraction of the input voltage. For a low-frequency signal, $X_C \gg R$, so the output voltage is nearly as great as the input voltage. For a signal consisting of a mixture of frequencies, the high frequencies are "filtered out" while the low frequencies "pass through."

The *high-pass filter* of Fig. 21.16b does just the opposite. Suppose a circuit connected to the input terminals supplies a mixture of a DC potential difference plus AC voltages at a range of frequencies. The reactance of the capacitor is large at low frequencies, so most of the voltage drop for low frequencies occurs across the capacitor; most of the high-frequency voltage drop occurs across the resistor and thus across the output terminals.

Combinations of capacitors and inductors are also used as filters. Figure 21.17 shows two *LC filters*, in which a combination of a capacitor and an inductor are placed between an AC input voltage and the output terminals. Figure 21.17a is a low-pass filter: a high-frequency input voltage sees a low reactance path to ground through the capacitor, so

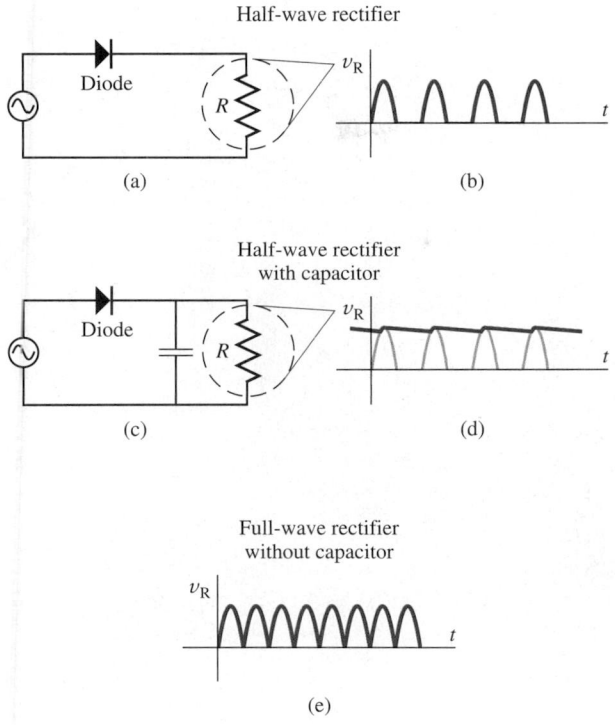

Figure 21.15 (a) A half-wave rectifier. (b) The voltage across the resistor. When the input voltage is negative, the output voltage v_R is zero, so the negative half of the "wave" has been cut off. (c) A capacitor inserted to smooth the output voltage. (d) The dark graph line shows the voltage across the resistor, assuming the RC time constant is much larger than the period of the sinusoidal input voltage. The light graph line shows what the output would have been without the capacitor. (e) Output of a full-wave rectifier.

little high-frequency voltage appears at the output terminals. Fig. 21.17b is a high-pass filter: a low-frequency input voltage sees a low reactance path to ground through the *inductor*.

For both RC and LC filters, there is a gradual transition between frequencies that are blocked and frequencies that pass through. The frequency range where the transition occurs can be selected by choosing the values of R and C (or L and C).

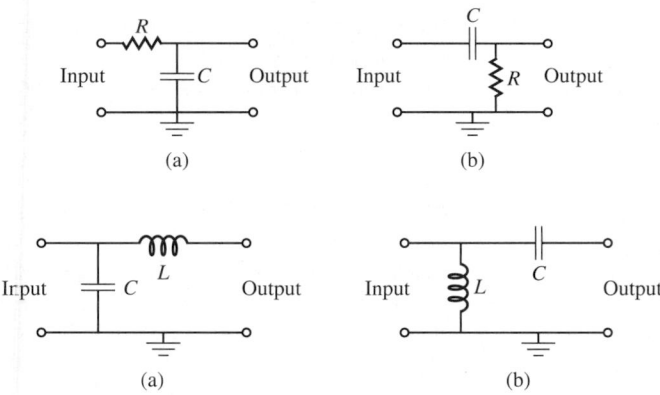

Figure 21.16 Two RC filters: (a) low-pass and (b) high-pass.

Figure 21.17 Two LC filters: (a) low-pass and (b) high-pass.

Crossover Networks

In a speaker used with an audio system, there are often two vibrating cones (the *drivers*) producing the sounds. The *woofer* produces the low-frequency sounds, while the *tweeter* produces the high-frequency sounds. A *crossover network* (Fig. 21.18a) separates the signal from the amplifier, sending the low frequencies to the woofer and the high frequencies to the tweeter. Figure 21.18b shows the relative amplitude of each current as a function of frequency. The crossover point is the frequency at which the current is evenly divided, half going to the woofer and half to the tweeter.

Making The Connection:
filters for audio tweeters and woofers

Figure 21.18 (a) Two speaker cones are connected to an amplifier by a crossover network. (b) The amplitude of the current I going to each of the drivers expressed as a fraction of the input amplitude I_0, graphed as a function of frequency.

(a)   (b)

MASTER THE CONCEPTS

Summary

- In the expression

$$v = V \sin (\omega t + \phi)$$

the lowercase letter (v) represents the instantaneous quantity while the uppercase letter (V) represents the *amplitude* (peak value). The quantity ϕ is called the *phase constant*.

- The *rms value* of a sinusoidal quantity is $1/\sqrt{2}$ times the amplitude.

- *Reactances* (X_C, X_L) and *impedance* (Z) are a generalization of the concept of resistance and are measured in ohms. The amplitude of the voltage across a circuit element or combination of elements is equal to the amplitude of the current through the element(s) times the reactance or impedance of the element(s). Except for a resistor, there is a phase difference between the voltage and current:

	Amplitude	Phase
Resistor	$V_R = IR$	v_R, i are in phase
Capacitor	$V_C = IX_C$ (21-6)	i leads v_C by 90° ($\pi/2$ rad)
	$X_C = \dfrac{1}{\omega C}$ (21-7)	
Inductor	$V_L = IX_L$ (21-9)	v_L leads i by 90° ($\pi/2$ rad)
	$X_L = \omega L$ (21-10)	
RLC series circuit	$\mathscr{E} = IZ$ (21-14a)	$\mathscr{E}$ leads i by
	$Z = \sqrt{R^2 + (X_L - X_C)^2}$ (21-14b)	$\phi = \tan^{-1}\dfrac{X_L - X_C}{R}$ (21-15)

- The average power dissipated in a resistor is

$$P_{av} = I_{rms} V_{rms} = I_{rms}^2 R = \frac{V_{rms}^2}{R} \qquad (21-4)$$

The average power dissipated in an ideal capacitor or ideal inductor is zero.

- The average power dissipated in a series *RLC* circuit can be written

$$P_{av} = I_{rms} \mathscr{E}_{rms} \cos \phi \qquad (21-17)$$

where ϕ is the phase difference between $i(t)$ and $\mathscr{E}(t)$. The *power factor* $\cos \phi$ is equal to R/Z.

- To add sinusoidal voltages, we can represent each voltage by a vector-like object called a *phasor*. The magnitude of the phasor represents the amplitude of the voltage; the angle of the phasor represents the phase constant of the voltage. We can then add phasors the same way we add vectors.

- The angular frequency at which *resonance* occurs in a series *RLC* circuit is

$$\omega_0 = \frac{1}{\sqrt{LC}} \qquad (21-18)$$

At resonance, the current amplitude has its maximum value, the capacitive reactance is equal to the inductive reactance, and the impedance is equal to the resistance. If the resistance in the circuit is small, the resonance curve (the graph of current amplitude as a function of frequency) has a sharp peak.

MASTER THE CONCEPTS *continued*

By adjusting the resonant frequency, such a circuit can be used to select a narrow range of frequencies from a signal consisting of a broad range of frequencies, as in radio or TV broadcasting.

- An *ideal* diode has zero resistance for current in one direction, so that the current flows without any voltage drop across the diode, and infinite resistance for current in the other direction, so that no current flows. Diodes can be used to convert AC to DC.

- Capacitors and inductors can be used to make filters to selectively remove unwanted high or low frequencies from an electrical signal.

Highlighted Figures and Tables

F21.1 A sinusoidal emf (p. 768)

F21.3 Power delivered to a resistor in an AC circuit (p. 769)

F21.4 Electrical wiring in a North American home (p. 771)

F21.5 The charge on a capacitor connected to an AC source as a function of time (p. 772)

F21.7 An inductor connected to an AC source (p. 776)

F21.8 Current and potential difference across an inductor in an AC circuit (p. 776)

F21.9 An *RLC* series circuit (p. 778)

F21.12 The amplitude of the current *I* as a function of angular frequency ω for three different resistances in a series *RLC* circuit (p. 781)

F21.13 Frequency dependence of the inductive and capacitive reactances and of the resistance as a function of frequency (p. 782)

F21.16 Two *RC* filters (low- and high-pass) (p. 785)

F21.17 Two *LC* filters (low- and high-pass) (p. 785)

CONCEPTUAL QUESTIONS

1. Explain why there is a phase difference between the current in an AC circuit and the potential difference across a capacitor in the same circuit.

2. Electric power is distributed long distances over transmission lines by using high AC voltages and therefore small AC currents. What is the advantage of using high voltages instead of safer low voltages?

3. Explain the differences between average current, rms current, and peak current in an AC circuit.

4. The United States and Canada use 120 V rms as the standard household voltage while most of the rest of the world uses 240 V rms for the household standard. What are the advantages and disadvantages of the two systems?

5. Some electrical appliances are able to operate equally well with either DC or AC voltage sources, while other appliances require one type of source or the other and cannot run on both. Explain and give a few examples of each type of appliance.

6. For an ideal inductor in an AC circuit, explain why the voltage across the inductor must be zero when the current is maximum.

7. For a capacitor in an AC circuit, explain why the current must be zero when the voltage across the capacitor is maximum.

8. An electric heater is plugged into an AC outlet. Since the AC current changes polarity, there is no net movement of electrons through the heating element; the electrons just tend to oscillate back and forth. How, then, does the heating element

heat up? Don't we need to send electrons *through* the element? Explain.

9. An electrical appliance is rated 120 V, 5 A, 500 W. The first two are rms values; the third is the average power consumption. Why is the power not 600 W (= 120 V × 5 A)?

10. How does adjusting the tuning knob on a radio tune in different stations?

11. Give a qualitative explanation for why the reactance of a capacitor increases with decreasing frequency.

12. What happens if a 40-W lightbulb, designed to be connected to an AC voltage with amplitude 170 V and frequency 60 Hz, is instead connected to a 170-V DC power supply? Explain. What DC voltage would make the lightbulb burn with the same brightness as the 170-V peak 60-Hz AC?

13. How can the lights in a home be dimmed using a coil of wire and a soft-iron core?

14. Explain what is meant by a *phase difference*. Sketch graphs of $i(t)$ and $v_C(t)$, given that the current leads the voltage by $\pi/2$ radians.

15. What does it mean if the power factor is 1? What does it mean if it is zero?

16. Give a qualitative explanation for why the reactance of an inductor increases with increasing frequency.

17. Suppose you buy a 120-W lightbulb in Europe (where the rms voltage is 240 V). What happens if you bring it back to the United States (where the rms voltage is 120 V) and plug it in?

MULTIPLE CHOICE QUESTIONS

1. Graphs (1, 2) in Fig. 21.19 could represent:

   (a) the (1-voltage, 2-current) for a capacitor in an AC circuit.
   (b) the (1-current, 2-voltage) for a capacitor in an AC circuit.
   (c) the (1-voltage, 2-current) for a resistor in an AC circuit.
   (d) the (1-current, 2-voltage) for a resistor in an AC circuit.
   (e) the (1-voltage, 2-current) for an inductor in an AC circuit.
   (f) the (1-current, 2-voltage) for an inductor in an AC circuit.
   (g) either (a) or (e).   (h) either (a) or (f).
   (i) either (b) or (e).   (j) either (b) or (f).

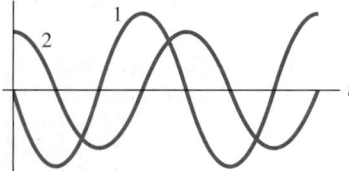

Figure 21.19 Multiple Choice Question 1

2. For a capacitor in an AC circuit, how much energy is stored in the capacitor at the instant when current is zero?

   (a) zero   (b) maximum
   (c) half of the maximum amount
   (d) $1/\sqrt{2} \times$ the maximum amount
   (e) impossible to answer without being given the phase angle

3. For an ideal inductor in an AC circuit, how much energy is stored in the inductor at the instant when current is zero?

   (a) zero   (b) maximum
   (c) half of the maximum amount
   (d) $1/\sqrt{2} \times$ the maximum amount
   (e) impossible to tell without being given the phase angle

4. For an ideal inductor in an AC circuit, the current through the inductor

   (a) is in phase with the induced emf.
   (b) leads the induced emf by 90°.
   (c) leads the induced emf by an angle less than 90°.
   (d) lags the induced emf by 90°.
   (e) lags the induced emf by an angle less than 90°.

5. A capacitor is connected to the terminals of a variable frequency oscillator. The peak voltage of the source is kept fixed while the frequency is increased. Which statement is true?

   (a) The rms current through the capacitor increases.
   (b) The rms current through the capacitor decreases.

   (c) The phase relation between the current and source voltage changes.
   (d) The current stops flowing when the frequency change is large enough.

6. A voltage of $v(t) = 120 \sin 302t$ volts, where the time t is in seconds, is produced by an AC generator. What is the rms voltage and the frequency of the source?

   (a) 170 V and 213 Hz   (b) 20 V and 427 Hz
   (c) 60 V and 150 Hz   (d) 85 V and 48 Hz

7. An AC source is connected to a series combination of a resistor, capacitor, and an inductor. Which statement is correct?

   (a) The current in the capacitor leads the current in the inductor by 180°.
   (b) The current in the inductor leads the current in the capacitor by 180°.
   (c) The current in the capacitor and the current in the resistor are in phase.
   (d) The voltage across the capacitor and the voltage across the resistor are in phase.

8. A series *RLC* circuit is connected to an AC generator. When the generator frequency varies (but the peak emf is constant), the average power is:

   (a) a minimum when $|X_L - X_C| = R$.
   (b) a minimum when $X_C = X_L$.
   (c) equal to $I_{rms}^2 R$ only at the resonant frequency.
   (d) equal to $I_{rms}^2 R$ at all frequencies.

The graphs in Fig. 21.20 show the peak current as a function of frequency for various circuit elements placed in the diagrammed circuit. The amplitude of the generator emf is constant, independent of the frequency.

9. Which graph is correct if the circuit element is a capacitor?
10. Which graph is correct if the circuit element is a resistor?

 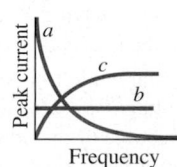

Figure 21.20
Multiple Choice Questions 9 and 10

PROBLEMS

Note: **C** indicates a combination conceptual/quantitative problem. Gold diamonds ✦, ✦✦ are used to indicate the increasing level of difficulty of each problem. Problem numbers appearing in blue, 9., denote problems that have a detailed solution available in the Student Solutions Manual. Some problems are *paired* by concept; their numbers are connected by a ruled box.

Unless the problem says otherwise, assume voltages and currents are rms values.

21.1 Sinusoidal Currents and Voltages; Resistors in AC Circuits; 21.2 Electricity in the Home

1. A lightbulb is connected to a 120-V (rms), 60-Hz source. How many times per second does the current reverse direction?

2. A European outlet supplies 220 V (rms) at 50 Hz. How many times per second is the magnitude of the voltage equal to 220 V?

3. A 1500-W heater runs on 120 V rms. What is the peak current through the heater?

4. A circuit breaker trips when the rms current exceeds 20.0 A. How many 100.0-W lightbulbs can run on this circuit without tripping the breaker? (The voltage is 120 V rms.)

5. A 4.0-kW heater is designed to be connected to a 120-V rms source. What is the power dissipated by the heater if it is instead connected to a 120-V DC source?

6. (a) What rms current is drawn by a 4200-W electric room heater when running on 120 V rms? (b) What is the power dissipation by the heater if the voltage drops to 105 V rms during a brown-out? Assume the resistance stays the same.

7. A television set draws an rms current of 2.50 A from a 60-Hz power line. Find (a) the average current, (b) the average of the square of the current, and (c) the amplitude of the current.

8. The instantaneous sinusoidal emf from an AC generator with an rms emf of 4.0 V oscillates between what values?

9. A hair dryer has a power rating of 1200 W at 120 V rms. Assume the hair dryer circuit contains only resistance. (a) What is the resistance of the heating element? (b) What is the rms current drawn by the hair dryer? (c) What is the maximum instantaneous power that the resistance must withstand?

✦ 10. Show that over one complete cycle, the average value of a sine function squared is $\frac{1}{2}$. [*Hint:* Use the following trigonometric identities: $\sin^2 a + \cos^2 a = 1$; $\cos 2a = \cos^2 a - \sin^2 a$.]

C ✦ 11. Figure 21.21 is a simplified household circuit. Resistor $R_1 = 240.0 \ \Omega$ represents a lightbulb; resistor $R_2 = 12.0 \ \Omega$ represents a hair dryer. The resistors $r = 0.50 \ \Omega$ (each) represent the resistance of the wiring in the walls. Assume that the generator

Figure 21.21
Problem 11

supplies a constant 120.0 V rms. (a) If the lightbulb is on and the hair dryer is off, find the rms voltage across the lightbulb and the power dissipated by the lightbulb. (b) If both the lightbulb and the hair dryer are on, find the rms voltage across the lightbulb, the power dissipated by the lightbulb, and the rms voltage between point A and ground. (c) Explain why lights sometimes dim when an appliance is turned on. (d) Explain why the neutral and ground wires in a junction box are not at the same potential even though they are both grounded.

21.3 Capacitors in AC Circuits

12. At what frequency is the reactance of a 6.0-μF capacitor equal to 1.0 kΩ?

13. A 0.400-μF capacitor is connected across the terminals of a variable frequency oscillator. (a) What is the frequency when the reactance is 6.63 kΩ? (b) Find the reactance for half of that same frequency.

14. A 0.250-μF capacitor is connected to a 220-V rms AC source at 50.0 Hz. (a) Find the reactance of the capacitor. (b) What is the rms current through the capacitor?

15. A capacitor is connected across the terminals of a 115-V rms, 60.0-Hz generator. For what capacitance is the rms current 2.3 mA?

16. Show, from $X_C = 1/\omega C$, that the units of capacitive reactance are ohms.

C 17. A capacitor (capacitance = C) is connected to an AC power supply with peak voltage V and angular frequency ω. (a) During a quarter-cycle when the capacitor goes from being uncharged to fully charged, what is the *average* current (in terms of C, V, and ω)? [*Hint:* $i_{av} = \Delta Q/\Delta t$.] (b) What is the rms current? (c) Explain why the average and rms currents are not the same.

18. Three capacitors (2.0 μF, 3.0 μF, 6.0 μF) are connected in series to an AC voltage source with amplitude 12.0 V and frequency 6.3 kHz. (a) What are the peak voltages across each capacitor? (b) What is the peak current that flows in the circuit?

19. A capacitor and a resistor are connected in parallel across an AC source. The reactance of the capacitor is equal to the resistance of the resistor. Assuming that $i_C(t) = I \sin \omega t$, sketch graphs of $i_C(t)$ and $i_R(t)$ on the same axes.

21.4 Inductors in AC Circuits

20. At what frequency is the reactance of a 20.0-mH inductor equal to 18.8 Ω?

21. What is the reactance of an air-core solenoid of length 8.0 cm, radius 1.0 cm, and 240 turns at a frequency of 15.0 kHz?

22. A 4.00-mH inductor is connected to an AC voltage source of 151.0 V rms. If the rms current in the circuit is 0.820 A, what is the frequency of the source?

23. Two ideal inductors (0.10 H, 0.50 H) are connected in series to an AC voltage source with amplitude 5.0 V and frequency 126 Hz. (a) What are the peak voltages across each inductor? (b) What is the peak current that flows in the circuit?

C ✦ 24. Suppose that an ideal capacitor and an ideal inductor are connected in series in an AC circuit. (a) What is the phase difference between $v_C(t)$ and $v_L(t)$? [*Hint:* Since they are in series, the same current $i(t)$ flows through both.] (b) If the rms voltages across the capacitor and inductor are 5.0 V and 1.0 V, respectively, what would an AC voltmeter (which reads rms voltages) connected across the series combination read?

✦25. The voltage across an inductor and the current through the inductor are related by $v_L = L\,\Delta i/\Delta t$. Suppose that $i(t) = I \sin \omega t$. (a) Write an expression for $v_L(t)$. [*Hint:* Use one of the relationships of Eq. (20-7).] (b) From your expression for $v_L(t)$, show that the reactance of the inductor is $X_L = \omega L$. (c) Sketch graphs of $i(t)$ and $v_L(t)$ on the same axes. What is the phase difference? Which one leads?

© ✦26. Make a figure analogous to Fig. 21.5 for an ideal *inductor* in an AC circuit. Start by assuming that the voltage across an ideal inductor is $v_L(t) = V_L \sin \omega t$. Make a graph showing one cycle of $v_L(t)$ and $i(t)$ on the same axes. Then, at each of the times $t = 0, \frac{1}{8}T, \frac{2}{8}T, \ldots, T$, indicate the direction of the current (or that it is zero), whether the current is increasing, decreasing, or (instantaneously) not changing, and the direction of the induced emf in the inductor (or that it is zero).

© 27. Suppose that current flows to the *left* through the inductor in Fig. 21.7a so that i is negative. (a) If the current is increasing in magnitude, what is the sign of $\Delta i/\Delta t$? (b) In what direction is the induced emf that opposes the increase in current? (c) Show that Eq. (21-8) gives the correct sign for v_L. [*Hint:* v_L is positive if the left side of the inductor is at a higher potential than the right side.] (d) Repeat these three questions if the current flows to the left through the inductor and is *decreasing* in magnitude.

21.5 *RLC* Series Circuits

28. A 25.0-mH inductor, with internal resistance of 25.0 Ω, is connected to a 110-V rms source. If the average power dissipated in the circuit is 50.0 W, what is the frequency? (Model the inductor as an ideal inductor in series with a resistor.)

29. An inductor has an impedance of 30.0 Ω and a resistance of 20.0 Ω at a frequency of 50.0 Hz. What is the inductance? (Model the inductor as an ideal inductor in series with a resistor.)

© ✦30. A 0.48-μF capacitor is connected in series to a 5.00-kΩ resistor and an AC source of voltage amplitude 2.0 V. (a) At $f =$ 120 Hz, what are the voltage amplitudes across the capacitor and across the resistor? (b) Do the voltage amplitudes add to give the amplitude of the source voltage (i.e., does $V_R + V_C = 2.0$ V)? Explain. (c) Draw a phasor diagram to show the addition of the voltages.

© ✦31. A series combination of a 22.0-mH inductor and a 145.0-Ω resistor are connected across the output terminals of an AC generator with peak voltage 1.20 kV. (a) At $f = 1250$ Hz, what are the voltage amplitudes across the inductor and across the resistor? (b) Do the voltage amplitudes add to give the source voltage (i.e., does $V_R + V_L = 1.20$ kV)? Explain. (c) Draw a phasor diagram to show the addition of the voltages.

32. A series combination of a resistor and a capacitor are connected to a 110-V rms, 60.0-Hz AC source. If the capacitance is 0.80 μF and the rms current in the circuit is 28.4 mA, what is the resistance?

33. A 300.0-Ω resistor and a 2.5-μF capacitor are connected in series across the terminals of a sinusoidal emf with a frequency

of 159 Hz. The inductance of the circuit is negligible. What is the impedance of the circuit?

© ✦34. A 3.3-kΩ resistor is in series with a 2.0-μF capacitor in an AC circuit. The rms voltages across the two are the same. (a) What is the frequency? (b) Would each of the rms voltages be half of the rms voltage of the source? If not, what fraction of the source voltage are they? (In other words, $V_R/\mathscr{E} = V_C/\mathscr{E} = ?$) [*Hint:* Draw a phasor diagram.] (c) What is the phase angle between the source voltage and the current? Which leads? (d) What is the impedance of the circuit?

© ✦35. A 150-Ω resistor is in series with a 0.75-H inductor in an AC circuit. The rms voltages across the two are the same. (a) What is the frequency? (b) Would each of the rms voltages be half of the rms voltage of the source? If not, what fraction of the source voltage are they? (In other words, $V_R/\mathscr{E} = V_L/\mathscr{E} = ?$) (c) What is the phase angle between the source voltage and the current? Which leads? (d) What is the impedance of the circuit?

36. (a) Find the power factor for the *RLC* series circuit of Example 21.4. (b) What is the average power delivered to each element (R, L, C)?

37. A computer draws an rms current of 2.80 A at an rms voltage of 120 V. The average power consumption is 240 W. (a) What is the power factor? (b) What is the phase difference between the voltage and current?

38. An *RLC* series circuit is connected to an AC power supply with a 12-V amplitude and a frequency of 2.5 kHz. If $R = 220$ Ω, $C = 8.0$ μF, and $L = 0.15$ mH, what is the average power dissipated?

✦39. A series circuit with a resistor and a capacitor has a time constant of 0.25 ms. The circuit has an impedance of 350 Ω at a frequency of 1250 Hz. What are the capacitance and the resistance?

✦40. (a) What is the reactance of a 10.0-mH inductor at the frequency $f = 250.0$ Hz? (b) What is the impedance of a series combination of the 10.0-mH inductor and a 10.0-Ω resistor at 250.0 Hz? (c) What is the maximum current through the same circuit when the AC voltage source has a peak value of 1.00 V? (d) By what angle does the current lag the voltage in the circuit?

41. Suppose that two sinusoidal voltages at the same frequency are added:

$$V_1 \sin \omega t + V_2 \sin (\omega t + \phi_2) = V \sin (\omega t + \phi)$$

A phasor representation is shown in Fig. 21.22. (a) Substitute $t = 0$ into the equation. Interpret the result by referring to the phasor diagram. (b) Substitute $t = \pi/(2\omega)$ and simplify using the trigonometric identity $\sin (\theta + \pi/2) = \cos \theta$. Interpret the result by referring to the phasor diagram.

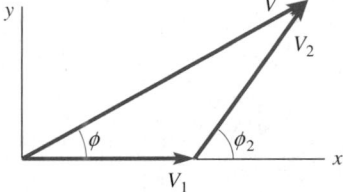

Figure 21.22
Problem 41

42. An AC circuit contains a 12.5-Ω resistor, a 5.00-μF capacitor, and a 3.60-mH inductor connected in series to an AC generator with an output voltage of 50.0 V (peak) and frequency of 1.59 kHz. Find the impedance, the power factor, and the phase difference between the source voltage and current for this circuit.

21.6 Resonance in an *RLC* Circuit

43. An *RLC* series circuit has a resistance of $R = 325$ Ω, an inductance $L = 0.300$ mH, and a capacitance $C = 33.0$ nF. (a) What is the resonant frequency? (b) If the capacitor breaks down for peak voltages in excess of 7.0×10^2 V, what is the maximum source voltage amplitude when the circuit is operated at the resonant frequency?

44. An *RLC* series circuit has $L = 0.300$ H and $C = 6.00$ μF. The source has a peak voltage of 440 V. (a) What is the angular resonant frequency? (b) When the source is set at the resonant frequency, the peak current in the circuit is 0.560 A. What is the resistance in the circuit? (c) What are the peak voltages across the resistor, the inductor, and the capacitor at the resonant frequency?

45. A series *RLC* circuit has $R = 500.0$ Ω, $L = 35.0$ mH, and $C = 87.0$ pF. What is the impedance of the circuit at resonance? Explain.

46. In an *RLC* series circuit, these three elements are connected in series: a resistor of 60.0 Ω, a 40.0-mH inductor, and a 0.0500 F capacitor. The series elements are connected across the terminals of an AC oscillator with an rms voltage of 10.0 V. Find the resonant frequency for the circuit.

21.7 Converting AC to DC; Filters

47. An *RC* filter is shown in Fig. 21.23. The filter resistance R is variable between 180 Ω and 2200 Ω and the filter capacitance is $C = 0.086$ μF. At what frequency is the output amplitude equal to $1/\sqrt{2}$ times the input amplitude if $R = $ (a) 180 Ω; (b) 2200 Ω? (c) Is this a low-pass or high-pass filter? Explain.

Figure 21.23
Problem 47

48. In the crossover network of Fig. 21.24, the crossover frequency is found to be 252 Hz. The capacitance is $C = 560$ μF. Assume the inductor to be ideal. (a) What is the impedance of the tweeter branch (the capacitor in series with the 8.0-Ω resistance of the tweeter) at the crossover frequency? (b) What is the impedance of the woofer branch at the crossover frequency? [*Hint:* The current amplitudes in the two branches are the same.] (c) Find L. (d) Derive an equation for the crossover frequency f_{co} in terms of L and C.

49. In the crossover network of Fig. 21.24, the inductance L is 1.20 mH. The capacitor is variable; its capacitance can be adjusted to set the crossover point according to the frequency response of the woofer and tweeter. What should the capacitance be set to for a crossover point of 180 Hz? [*Hint:* At the crossover point, the currents are equal in amplitude.]

Figure 21.24
Problems 48 and 49

50. The circuit of Fig. 21.25 has a source voltage of 440 V rms, resistance $R = 250$ Ω, inductance $L = 0.800$ H, and capacitance $C = 2.22$ μF. (a) Find the angular frequency ω_0 for resonance in this circuit. (b) Draw a phasor diagram for the circuit at resonance. (c) Find these rms voltages measured between various points in the circuit: V_{ab}, V_{bc}, V_{cd}, V_{bd}, and V_{ad}. (d) The resistor is replaced with one of $R = 125$ Ω. Now what is the angular frequency for resonance? (e) What is the rms current in the circuit operated at resonance with the new resistor?

Figure 21.25 Problem 50

COMPREHENSIVE PROBLEMS

51. A portable heater is connected to a 60-Hz AC outlet. How many times per second is the instantaneous power a maximum?

52. What is the rms voltage of the oscilloscope trace of Fig. 21.26, assuming that the signal is sinusoidal? The central horizontal line represents zero volts. The oscilloscope voltage knob has been clicked into its calibrated position.

53. A 22-kV power line that is 10.0 km long supplies the electrical energy to a small town at an average rate of 6.0 MW. (a) If a pair of aluminum cables of diameter 9.2 cm are used,

Figure 21.26 Problem 52

what is the average power dissipated in the transmission line? (b) Why is aluminum used rather than a better conductor such as copper or silver?

54. An x-ray machine uses 240 kV rms at 60.0 mA rms when it is operating. If the power source is a 420-V rms line, (a) what must be the turns ratio of the transformer? (b) What is the rms current in the primary? (c) What is the average power used by the x-ray tube?

55. A coil with an internal resistance of 120 Ω and inductance of 12.0 H is connected to a 60.0-Hz, 110-V rms line. (a) What is the impedance of the coil? (b) Calculate the current in the coil.

56. The field coils used in an AC motor are designed to have a resistance of 0.45 Ω and an impedance of 35.0 Ω. What inductance is required if the frequency of the AC source is (a) 60.0 Hz? and (b) 0.20 kHz?

57. A capacitor is rated at 0.025 μF. How much current flows when the capacitor is connected to a 110-V rms, 60.0-Hz line?

58. A capacitor to be used in a radio is to have a reactance of 6.20 Ω at a frequency of 520 Hz. What is the capacitance?

✦59. (a) What is the reactance of a 5.00-μF capacitor at the frequencies f = 12.0 Hz and 1.50 kHz? (b) What is the impedance of a series combination of the 5.00-μF capacitor and a 2.00-kΩ resistor at the same two frequencies? (c) What is the maximum current through the circuit of part (b) when the AC source has a peak voltage of 2.00 V? (d) For each of the two frequencies, does the current lead or lag the voltage? By what angle?

60. An alternator supplies a peak current of 4.68 A to a coil with a negligibly small internal resistance. The voltage of the alternator is 420-V peak at 60.0 Hz. When a capacitor of 38.0 μF is placed in series with the coil, the power factor is found to be 1.00. Find (a) the inductive reactance of the coil and (b) the inductance of the coil.

61. At what frequency does the maximum current flow through a series RLC circuit containing a resistance of 4.50 Ω, an inductance of 440 mH, and a capacitance of 520 pF?

62. What is the rms current flowing in a 4.50-kW motor connected to a 220-V rms line when (a) the power factor is 1.00 and (b) when it is 0.80?

✦63. A 40.0-mH inductor, with internal resistance of 30.0 Ω, is connected to an AC source $\mathscr{E}(t) = 286 \sin (390t)$ volts where t is in seconds. (a) What is the impedance of the inductor in the circuit? (b) What are the peak and rms voltages across the inductor (including the internal resistance)? (c) What is the peak current in the circuit? (d) What is the average power dissipated in the circuit? (e) Write an expression for the current through the inductor as a function of time.

✦64. In an RLC circuit, these three elements are connected in series: a resistor of 20.0 Ω, a 35.0-mH inductor, and a 50.0-μF capacitor. The AC source of the circuit has an rms voltage of 100.0 V and an angular frequency of 1.0×10^3 rad/s. Find (a) the reactances of the capacitor and inductor, (b) the impedance, (c) the rms current, (d) the current amplitude, (e) the phase angle, and (f) the rms voltages across each of the circuit

elements. (g) Does the current lead or lag the voltage? (h) Draw a phasor diagram.

65. A variable capacitor is connected in series to an inductor with negligible internal resistance and of inductance 2.4×10^{-4} H. The combination is used as a tuner for a radio. If the lowest frequency to be tuned in is 0.52 MHz, what is the maximum capacitance required?

✦66. An RLC series circuit is connected to a 240-V rms power supply at a frequency of 2.50 kHz. The elements in the circuit have the following values: R = 12.0 Ω, C = 0.26 μF, and L = 15.2 mH. (a) What is the impedance of the circuit? (b) What is the rms current? (c) What is the phase angle? (d) Does the current lead or lag the voltage? (e) What are the rms voltages across each circuit element? Draw a phasor diagram.

67. A large coil used as an electromagnet has a resistance of R = 450 Ω and an inductance of L = 2.47 H. The coil is connected to an AC source with a voltage amplitude of 2.0 kV and a frequency of 9.55 Hz. (a) What is the power factor? (b) What is the impedance of the circuit? (c) What is the peak current in the circuit? (d) What is the average power delivered to the electromagnet by the source?

68. An AC series circuit containing a capacitor, inductor, and resistance is found to have a current of amplitude 0.50 A for a source voltage of amplitude 10.0 V at an angular frequency of 200.0 rad/s. The total resistance in the circuit is 15.0 Ω. (a) What are the power factor and the phase angle for the circuit? (b) Can you determine whether the current leads or lags the source voltage? Explain.

69. A generator supplies an average power of 12 MW through a transmission line that has a resistance of 10.0 Ω (Fig. 21.27). What is the power loss in the transmission line if the rms line voltage $\mathscr{E}_{rms}$ is (a) 15 kV and (b) 110 kV? What percentage of the total power supplied by the generator is lost in the transmission line in each case?

70. (a) Calculate the rms current drawn by the load in Fig. 21.27 if $\mathscr{E}_{rms}$ = 250 kV and the average power supplied by the generator is 12 MW. (b) Suppose that the average power supplied by the generator is still 12 MW, but the load is not purely resistive; rather, the load has a power factor of 0.86. What is the rms current drawn? (c) Why would the power company want to charge more in the second case, even though the average power is the same?

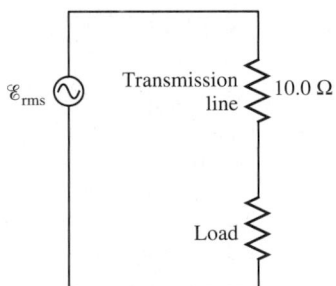

Figure 21.27 Problems 69 and 70

71. Transformers are often rated in terms of kilovolt-amps. A pole on a residential street has a transformer rated at 35 kV·A to serve four homes on the street. (a) If each home has a fuse that limits the incoming current to 60 A rms at 220 V rms, find the maximum load in kV·A on the transformer. (b) Is the rating of the transformer adequate? (c) Explain why the transformer rating is given in kV·A rather than in kW.

72. A variable inductor can be placed in series with a lightbulb to act as a dimmer. (a) What inductance would reduce the current through a 100-W lightbulb to 75% of its maximum value? Assume a 120-V rms, 60-Hz source. (b) Could a variable resistor be used in place of the variable inductor to reduce the current? Why is the inductor a much better choice for a dimmer?

ANSWERS TO PRACTICE PROBLEMS

21.1 $V = 310$ V, $I = 17.0$ A, $P_{max} = 5300$ W, $P_{av} = 2600$ W, $R = 18\ \Omega$

21.2 $9950\ \Omega$; 22.1 mA

21.3 1.13 kΩ, 8.84 μA

21.4 $v_C(t) = (500\ \text{mV}) \sin(\omega t - \pi/2)$, $v_L(t) = (440\ \text{mV}) \sin(\omega t + \pi/2)$, $v_R(t) = (80\ \text{mV}) \sin \omega t$, and $\mathscr{E}(t) = (100\ \text{mV}) \sin(\omega t - 0.64)$.

At $t = 80.0\ \mu$s, $\omega t = 0.800$ rad.
$v_C = (500\ \text{mV}) \sin(-0.771\ \text{rad}) = -350$ mV,
$v_L = (440\ \text{mV}) \sin(2.371\ \text{rad}) = +310$ mV,
$v_R(t) = (80\ \text{mV}) \sin(0.80\ \text{rad}) = +57$ mV,
and $\mathscr{E}(t) = (100\ \text{mV}) \sin(0.16\ \text{rad}) = +16$ mV.
$v_C + v_L + v_R = +17$ mV (discrepancy comes from roundoff error)

21.5 29 W

21.6 25 pF

Chapter 22

Electromagnetic Waves

Bees use the position of the Sun in the sky to navigate and find their way back to their hives. This is remarkable in itself—since the Sun moves across the sky during the day, the bees navigate with respect to a moving reference point rather than a fixed reference point. Even if the bees are kept in the dark for part of the day, they still navigate with reference to the Sun; they compensate for the motion of the Sun during the time they were in the dark. They must have some sort of internal clock that enables them to keep track of the Sun's motion.

What do they do when the Sun's position is obscured by clouds? Experiments have shown that the bees can still navigate as long as there is a patch of blue sky. To humans, a patch of blue sky doesn't reveal the position of the Sun behind the clouds, but to bees it does. How is this possible?

EM waves are produced only by
accelerating charges.

22.1 ACCELERATING CHARGES PRODUCE ELECTROMAGNETIC WAVES

In our study of electromagnetism so far, we have considered the electric and magnetic fields due to charges that are either at rest or moving slowly and whose acceleration is small. A point charge at rest gives rise to an electric field only. A charge moving at constant velocity gives rise to both electric and magnetic fields. Charges at rest or moving at constant velocity do not generate **electromagnetic waves**—waves that consist of oscillating electric and magnetic fields. Electromagnetic (EM) waves are produced only by charges that *accelerate*. EM waves, also called **electromagnetic radiation**, consist of oscillating electric and magnetic fields that travel away from the accelerating charges.

A brief acceleration of a charged particle causes a brief EM wave pulse. To create an EM wave that lasts longer than a pulse, the charge must continue to accelerate. Charged particles such as electrons that move around a circular path in a synchrotron or storage ring produce *synchrotron radiation*, a beacon of high-energy x-rays that is useful in its own right. (Synchrotron radiation is used in x-ray diffraction studies that reveal the molecular structures of complex biological molecules and viruses.) Even though a particle in a synchrotron may be moving at constant speed, it moves in a circle so its velocity is changing direction; thus, it has a nonzero acceleration. As the particle radiates, the energy in the radiation comes from its kinetic energy. To keep it moving at constant speed, its kinetic energy must be replenished by an applied electric field.

A charge that oscillates back and forth with frequency f produces oscillating electric and magnetic fields at that same frequency f. The oscillating fields travel away from the charge as EM waves. In practice, more common than a single oscillating charge is an oscillating electric dipole—two point charges $\pm q$ moving along a line that joins them in simple harmonic motion with the same amplitude and a phase difference of 180°. What does the electric field due to an oscillating electric dipole look like? Figure 22.1a shows the electric field due to a *static* electric dipole. The static dipole produces no magnetic field because the charges are at rest.

The oscillating electric dipole consists of *moving* charges and therefore generates a magnetic field as well as an electric field. Since the charges change directions every half period, the current oscillates; when the positive charge moves downward and the negative charge moves upward, the current is down. When the charges reverse direction, so does the current.

The oscillating fields affect each other. The magnetic field is not constant, since the current is changing. According to Faraday's law of induction, a changing magnetic field induces a changing electric field. The electric field of the oscillating dipole at any instant is therefore different from the electric field of a static dipole. Faraday's law liberates the electric field lines: they do not have to start and end on the source charges. Instead, they can circulate around magnetic field lines far from the oscillating dipole.

According to Ampère's law as we have stated it, the magnetic field lines must enclose the current that is their source. However, Scottish physicist James Clerk Maxwell (1831–1879) was puzzled by a lack of symmetry in the laws of electromagnetism. If a changing magnetic field gives rise to an electric field, might not a changing electric field give rise to a magnetic field? The answer turns out to be yes. A changing electric field does give rise to a magnetic field. The magnetic field lines need not enclose a current; they can circulate around electric field lines, which extend far from the oscillating dipole.

Figure 22.1b shows the electric and magnetic field lines due to an oscillating dipole. With changing electric fields as a source of magnetic fields, the field lines (both electric and magnetic) can break free of the dipole, form closed loops, and travel away from the dipole as an electromagnetic wave. The electric and magnetic fields sustain each other as the wave travels outward, so the fields diminish in strength, but much less rapidly than if the field lines were tied to the dipole. Since changing electric fields are a source of magnetic fields, a wave consisting of just an oscillating electric field without

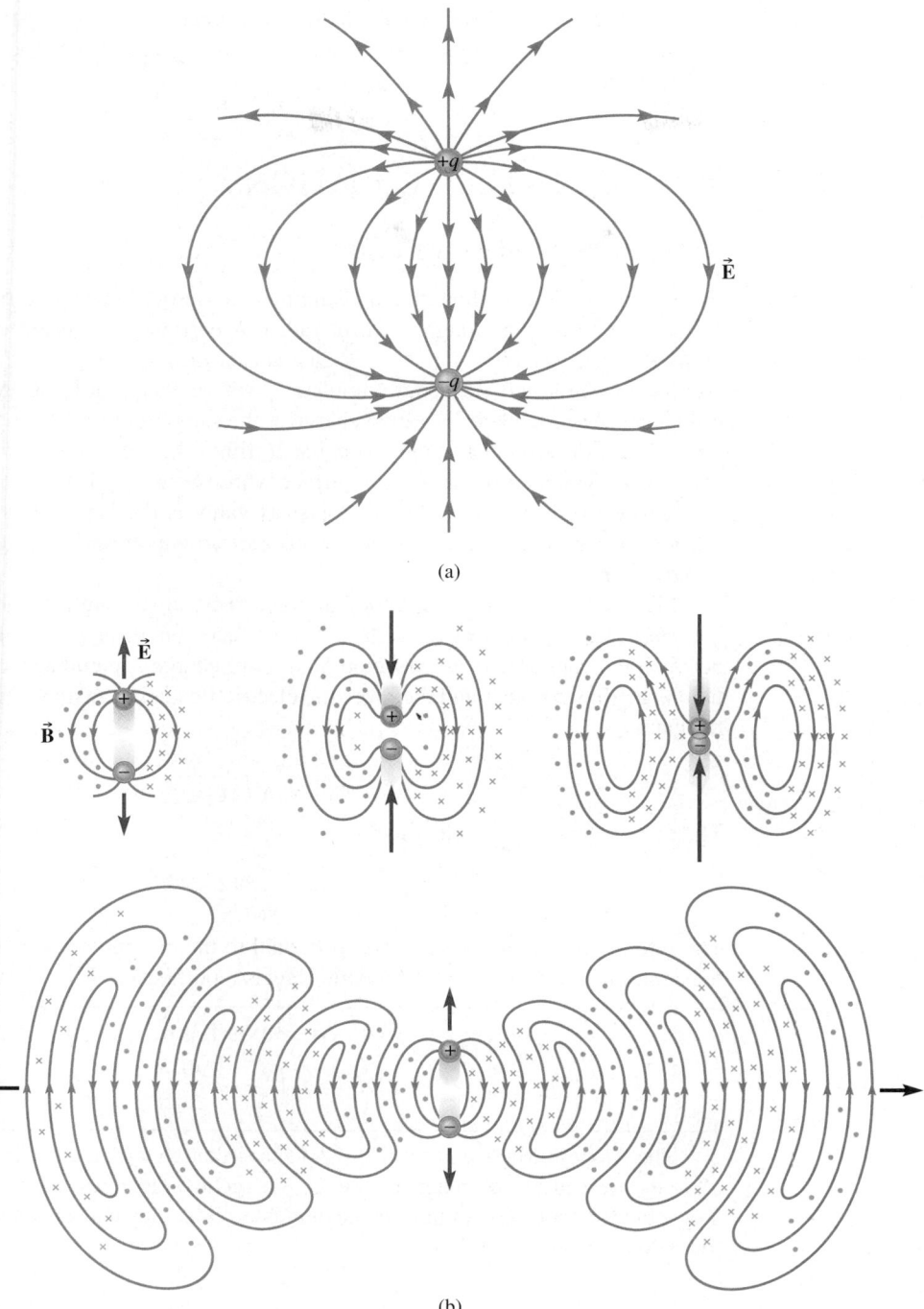

(a)

(b)

Figure 22.1   (a) Electric field lines due to an electric dipole at rest. (b) Electric and magnetic field lines due to an oscillating dipole. The loops are electric field lines in the plane of the page. The dots and crosses are magnetic field lines crossing the plane of the page; the magnetic field lines are closed loops. The field lines break free of the dipole and travel away from it as an electromagnetic wave.

an oscillating magnetic field is impossible. Since changing magnetic fields are a source of electric fields, a wave consisting of just an oscillating magnetic field without an oscillating electric field is also impossible.

There are no electric waves or magnetic waves; there are only *electromagnetic* waves.

Figure 22.1b illustrates that the fields far from an oscillating dipole are strongest—field lines are closest together—in directions perpendicular to the dipole axis and weakest along the axis.

22.2   MAXWELL'S EQUATIONS

The Ampère-Maxwell Law

Here is an example to show that a changing electric field *must* give rise to a magnetic field. Imagine a long straight wire of radius R carrying a constant current I. At one point, the wire has a tiny gap (Fig. 22.2a). The surfaces of the gap act as a capacitor; as current flows, the left surface accumulates positive charge at a rate $\Delta q/\Delta t = I$ and the right surface accumulates negative charge at the same rate. Ampère's law says that the circulation of $\vec{B}$ around a loop must equal μ_0 times the current that crosses the interior of the loop. The interior can be *any* surface whose boundary is the loop. The "butterfly net" surface (green in Fig. 22.2b) is just as valid as the circle (blue). The current I crosses through the blue circle, but *no current* crosses through the green surface—a contradiction.

Maxwell recognized that, although no current cuts through the blue surface, there is a *changing electric flux* through it. The surface cuts through the electric field lines between the capacitor plates; as more and more charge accumulates on the plates, the field is getting stronger, and therefore the electric flux is increasing. The electric field in the gap is

$$E = \frac{\sigma}{\epsilon_0} = \frac{q}{\epsilon_0 A} = \frac{q}{\epsilon_0 \pi R^2} \tag{17-13}$$

The rate of change of the electric flux is

$$\frac{\Delta \Phi_E}{\Delta t} = \frac{\Delta E \, \pi R^2}{\Delta t} = \frac{\Delta q}{\epsilon_0 \pi R^2} \times \frac{\pi R^2}{\Delta t} = \frac{I}{\epsilon_0}$$

The rate of change of the flux is proportional to the current! Maxwell recognized that the contradiction is resolved if Ampère's law is modified as:

Ampère-Maxwell law

$$\Sigma B_{\parallel} l = \mu_0 \left(I + \epsilon_0 \frac{\Delta \Phi_E}{\Delta t} \right) \tag{22-1}$$

The Ampère-Maxwell law (Eq. 22-1) says that changing electric fields as well as currents are sources of magnetic fields. Magnetic field lines are still always closed loops, but the loops do not have to surround currents; they can surround changing electric fields as well.

(a)   (b)

Figure 22.2   (a) A long cylindrical wire carrying a constant current I. Charge accumulates on the faces of the gap in the wire. (b) A circular loop around the wire, with two possible surfaces to use in Ampère's law. Both of the surfaces are bounded by the circular loop. The green surface does *not* cross the wire; it passes through the gap. However, electric field lines in the gap cut through the green surface.

Maxwell's Equations

Maxwell modified Ampère's law and then used it with the three other basic laws of electromagnetism to show that electromagnetic waves exist and to derive their properties. In honor of this achievement, the four laws are collectively called **Maxwell's equations**. They are:

1. **Gauss's law**: If an electric field line is not a closed loop, it can only start and stop on electric charges. The electric flux *through a closed surface* (or the net number of field lines leaving the surface) is proportional to the net charge inside the surface.

$$\Phi_E = \Sigma E_\perp A = q/\epsilon_0 \qquad \text{(16-7, 16-8)}$$

2. **Gauss's law for magnetism**: Magnetic field lines are always closed loops since there are no magnetic charges (*monopoles*). The magnetic flux *through a closed surface* (or the net number of field lines leaving the surface) is zero.

$$\Phi_B = \Sigma B_\perp A = 0 \qquad \text{(22-2)}$$

3. **Faraday's law**: Changing magnetic fields are another source of electric fields. The circulation of the electric field around a closed path is equal to the negative of the rate of change of magnetic flux through the interior of the path.

$$\mathcal{E} = \Sigma E_\parallel l = -\frac{\Delta\Phi_B}{\Delta t} \qquad \text{(20-6)}$$

Changing electric fields generate magnetic fields; changing magnetic fields generate electric fields.

4. The **Ampère-Maxwell law**: Both currents and changing electric fields are sources of magnetic fields. The circulation of the magnetic field around a closed path is related to the current and the rate of change of magnetic flux through the interior of the path. Magnetic field lines must surround either currents or changing electric fields (or both).

$$\Sigma B_\parallel l = \mu_0\left(I + \epsilon_0\frac{\Delta\Phi_E}{\Delta t}\right) \qquad \text{(22-1)}$$

22.3   ANTENNAS

As we saw in Section 22.1, an electric dipole that oscillates back and forth produces oscillating electric and magnetic fields that then travel away from the charge as EM waves. The **electric dipole antenna** generates EM waves in this way. It consists of two metal rods lined up as if they were a single long rod (Fig. 22.3). The rods are fed from the center with an oscillating current. For half of a cycle, the current flows toward the left side; that side of the antenna acquires a positive charge and the right side acquires an equal negative charge. Thus, an electric dipole is produced. When the current reverses direction, these accumulated charges diminish and then reverse direction so that the left side of the antenna becomes negatively charged and the right side becomes positively charged. The result of feeding an alternating current to the antenna is an oscillating electric dipole.

Figure 22.3   Current in an electric dipole antenna

The field lines for the EM wave emitted by an electric dipole antenna are similar to the field lines for an oscillating electric dipole (Fig. 22.1b). From the field lines, some of the properties of EM waves can be observed:

- For equal distances from the antenna, the amplitudes of the fields are smallest along the antenna's axis (in the $\pm x$-direction in Fig. 22.3) and largest in directions perpendicular to the antenna (in any direction perpendicular to the x-axis in Fig. 22.3).

- In directions perpendicular to the antenna, the electric field is lined up with the antenna's axis. In other directions, $\vec{\mathbf{E}}$ is *not* parallel to the antenna's axis, but is perpendicular to the *direction of propagation* of the wave—that is, perpendicular to the direction that energy travels from the antenna to the observation point.

- The magnetic field is perpendicular to both the electric field and to the direction of propagation.

Making The Connection:
radio/TV antennas

An electric dipole antenna used as a receiver should be aligned with the electric field of the wave.

An electric dipole antenna can be used as a receiver or detector of EM waves as well. In Fig. 22.4a, an EM wave travels past an electric dipole antenna. The electric field of the wave acts on free electrons in the antenna, causing an oscillating current. This current can then be amplified and the signal processed to decode the radio or TV transmission. The antenna is most effective if it is aligned with the electric field of the wave. If it is not, then only the component of $\vec{\mathbf{E}}$ parallel to the antenna acts to cause the oscillating current. The emf and the oscillating current are reduced by a factor of cos θ, where θ is the angle between $\vec{\mathbf{E}}$ and the antenna (Fig. 22.4b). If the antenna is perpendicular to the $\vec{\mathbf{E}}$ field, no oscillating current results.

Figure 22.4 (a) EM wave (drawn as a sine curve with vector arrows representing the $\vec{\mathbf{E}}$-field at various points) passing an electric dipole antenna. An oscillating current flows in the antenna. (b) The current in the antenna is smaller when it is not aligned with the electric field. Only the component of $\vec{\mathbf{E}}$ parallel to the antenna accelerates electrons along the antenna's length. (For clarity, only the $\vec{\mathbf{E}}$-field is shown, but a magnetic field is also present.)

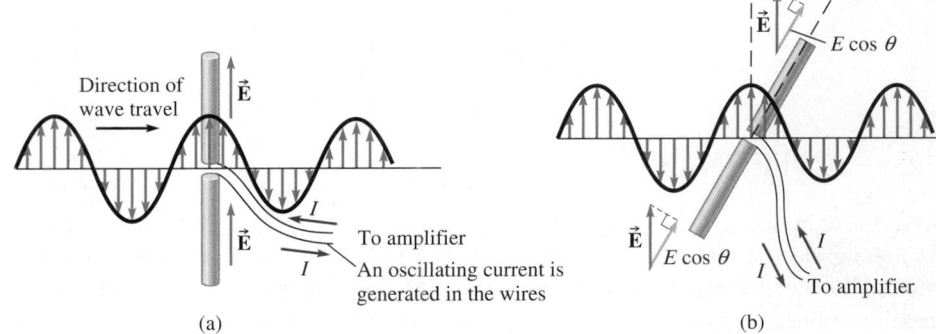

Example 22.1

Electric Dipole Antenna

An electric dipole antenna at the origin has length 84 cm. It is used as a receiver for an EM wave traveling in the $+z$-direction. The electric field of the wave is always in the $\pm y$-direction and varies sinusoidally with time. The electric field at the location of the antenna is

$$E_y(t) = E_m \cos \omega t; \quad E_x = E_z = 0$$

where the amplitude—the **m**aximum magnitude—of the electric field is $E_m = 3.2$ V/m. (a) How should the antenna be oriented for best reception? (b) What is the emf in the antenna if it is oriented properly?

Strategy For maximum amplitude the antenna must be oriented so that the full electric field can drive current along the

length of the antenna. The emf is defined as the work done by the electric field per unit charge.

Solution (a) We want the electric field of the wave to push free electrons along the antenna's length with a force directed along the length of the antenna. The electric field is always in the $\pm y$-direction, so the antenna should be oriented along the y-axis.

(b) The work done by the electric field E as it moves a charge q along the length of the antenna is

$$W = F_y \, \Delta y = qEL$$

continued on next page

Example 22.1 *continued*

The emf is the work per unit charge:

$$\mathcal{E} = \frac{W}{q} = EL$$

The emf varies with time because the electric field oscillates. The emf as a function of time is

$$\mathcal{E}(t) = EL = E_m L \cos \omega t$$

Therefore, it is a sinusoidally varying emf with the same frequency as the wave. The amplitude of the emf is

$$\mathcal{E}_{r1} = E_m L = 3.2 \text{ V/m} \times 0.84 \text{ m} = 2.7 \text{ V}$$

Discussion According to the problem statement, the oscillating electric field has the same amplitude and phase at every point on the antenna. As a result, the emf is proportional to the length of the antenna. If the antenna is so long that the phase of the electric field varies with position along the antenna, then the emf is no longer proportional to the length of the antenna and may even start to decrease with additional length. In practice, the ideal length of an antenna is $\lambda/2$.

Practice Problem 22.1 Location of transmitting antenna

(a) If the wave in Example 22.1 is transmitted from a distant electric dipole antenna, where is the transmitting antenna located relative to the receiving antenna? (Answer in terms of *xyz* coordinates.) (b) Write an equation for the electric field components as a function of position and time.

TV broadcasts in the United States and some other countries are transmitted from electric dipole antennas oriented horizontally. Thus, for best reception, a rooftop antenna should be oriented horizontally and with the antenna perpendicular to the direction to the transmitter. (Most rooftop antennas consist of several parallel pairs of metal rods; only one of these is actually the antenna, connected with wires to the TV set. The other rods help amplify the signal and make the antenna more directional.) In other countries, both the transmitting and receiving antennas are oriented vertically. Should you live where it is necessary to put up a rooftop antenna, it pays to know which convention is used in your area of the world.

Another kind of antenna is the **magnetic dipole antenna**. Recall that a loop of current is a magnetic dipole. The right-hand rule establishes the direction of the north pole of the dipole: if the fingers of the right hand are curled around the loop in the direction of the current, the thumb points "north." To make an oscillating magnetic dipole, we feed an alternating current into a loop or coil of wire. When the current reverses directions, the north and south poles of the magnetic dipole are interchanged.

If we consider the antenna axis to be the direction perpendicular to the coil, then the three observations made for the electric dipole antenna still hold, if we just substitute *magnetic* for *electric* and *vice versa*.

The magnetic dipole antenna works as a receiver as well (Fig. 22.5). The oscillating magnetic field of the wave causes a changing magnetic flux through the antenna. According to Faraday's law, an induced emf then makes an alternating current flow in the antenna. To maximize the rate of change of flux, the magnetic field should be perpendicular to the plane of the antenna.

Antennas can generate only EM waves with the longest wavelengths and lowest frequencies. It isn't practical to use an antenna to generate EM waves with short wavelengths and high frequencies such as visible light; the frequency at which the current would have to alternate to generate such waves is far too high to be achieved in an antenna, while the antenna itself cannot be made short enough. (To be most effective, the length of an antenna should not be larger than half the wavelength.)

A magnetic dipole antenna used as a receiver should be aligned so the magnetic field of the wave is perpendicular to the plane of the antenna.

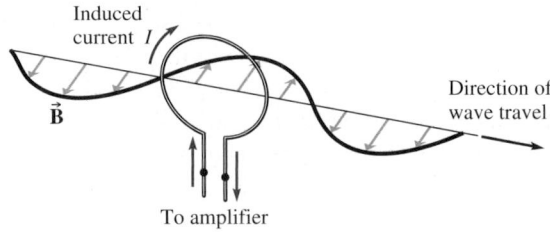

To amplifier

Figure 22.5 A loop of wire serves as a magnetic dipole antenna. As the EM wave travels to the right, the magnetic flux through the loop changes, causing an induced current in the loop.

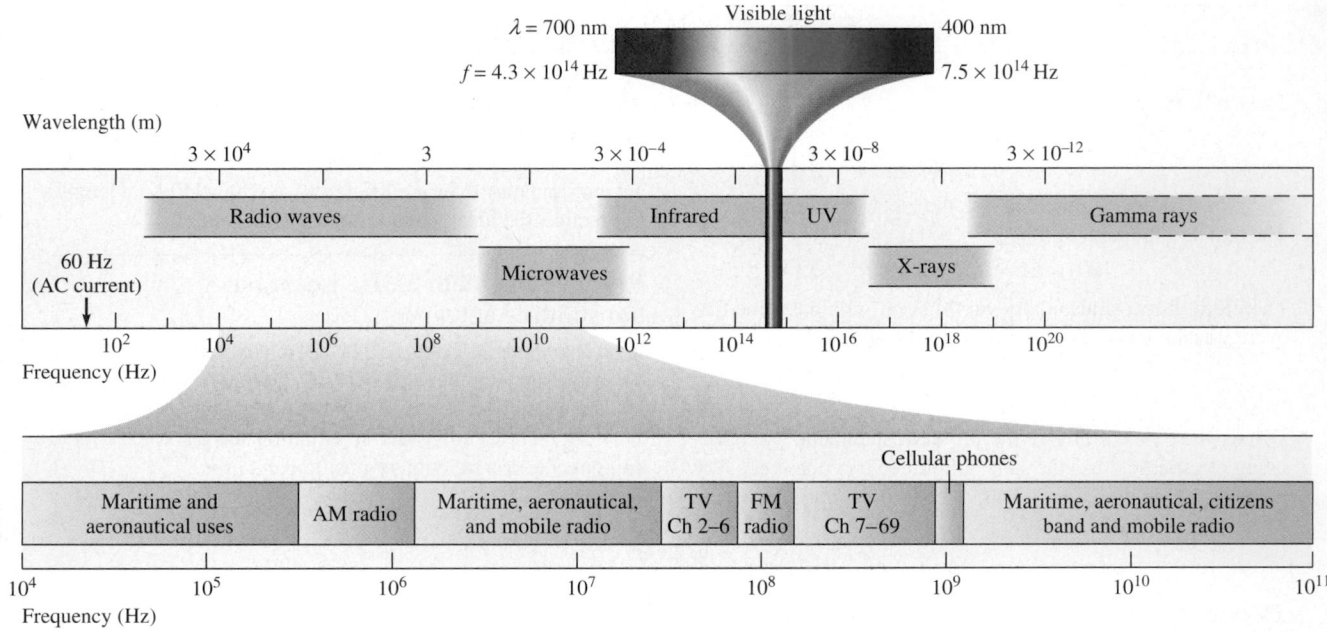

Figure 22.6 Regions of the EM spectrum

22.4   THE ELECTROMAGNETIC SPECTRUM

EM waves can exist at every frequency, without restriction. The properties of EM waves and their interactions with matter depend on the frequency of the wave. The **electromagnetic spectrum**—the range of frequencies (and wavelengths)—is traditionally divided into six or seven named regions (Fig. 22.6). The names persist partly for historical reasons—the regions were discovered at different times—and partly because the EM radiation of different regions interacts with matter in different ways. The boundaries between the regions are fuzzy and somewhat arbitrary. Throughout this section, the wavelengths given are those *in vacuum*; EM waves in vacuum or in air travel at a speed of 3.00×10^8 m/s.

Visible Light

Visible light is the part of the spectrum that can be detected by the human eye. This seems like a pretty cut-and-dried definition, but actually the sensitivity of the eye falls off gradually at both ends of the visible spectrum. Just as the range of frequencies of sound that can be heard varies from person to person, so does the range of frequencies of light that can be seen. For an average range we take frequencies of 430 THz (1 THz = 10^{12} Hz) to 750 THz, corresponding to wavelengths in vacuum of 700–400 nm. Light containing a mixture of all the wavelengths in the visible range appears white. White light can be separated by a prism into the colors red (700–620 nm), orange (620–600 nm), yellow (600–580 nm), green (580–490 nm), blue (490–450 nm), and violet (450–400 nm). Red has the lowest frequency (longest wavelength) and violet has the highest frequency (shortest wavelength).

It is not a coincidence that the human eye evolved to be most sensitive to the range of EM waves that are most intense in sunlight (see Fig. 22.7). However, other animals have visible ranges that differ from that of humans; the range is often well suited to the particular needs of the animal.

Lightbulbs, fire, the Sun, and fireflies are some *sources* of visible light. Most of the things we see are *not* sources of light; we see them by the light they *reflect*. When light strikes an object, some may be absorbed, some may be transmitted through the object,

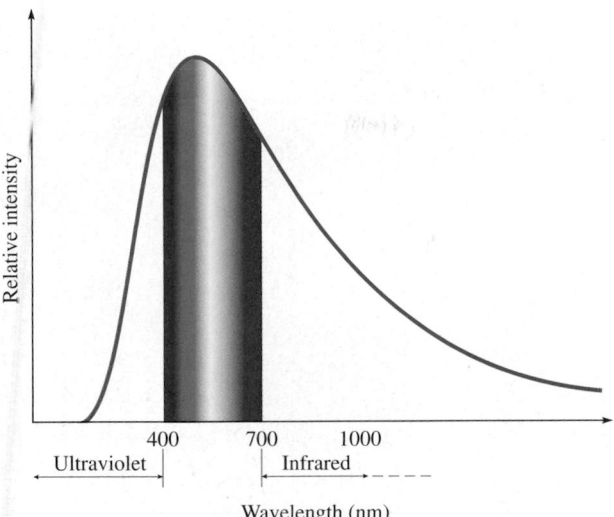

Figure 22.7 Graph of relative intensity (average power per unit area) of sunlight above Earth's atmosphere as a function of wavelength

and some may be reflected. The relative amounts of absorption, transmission, and reflection usually differ for different wavelengths. A lemon appears yellow because it reflects much of the incident yellow light and absorbs most of the other spectral colors.

The wavelengths of visible light are small on an everyday scale but large compared to atoms. The diameter of an average-sized atom—and the distance between atoms in solids and liquids—is about 0.2 nm. Thus, the wavelengths of visible light are 2000–4000 times larger than the size of an atom.

Infrared

After visible light, the first parts of the EM spectrum to be discovered were those on either side of the visible: infrared and ultraviolet (discovered in 1800 and 1801, respectively). The prefix *infra-* means *below;* **infrared** radiation (IR) is lower in frequency than visible light. IR extends from the low-frequency (red) edge of the visible to a frequency of about 300 GHz ($\lambda = 1$ mm). The astronomer William Herschel (1738–1822) discovered IR in 1800 while studying the temperature rise caused by the light emerging from a prism. He discovered that the thermometer reading was highest for levels just *outside* the illuminated region, adjacent to the red end of the spectrum. Since the radiation was not *visible*, Herschel deduced that there must be some invisible radiation beyond the red.

The thermal radiation given off by objects near room temperature is primarily infrared, with the peak of the radiated IR at a wavelength of about 0.01 mm = 10 μm (Fig. 22.8). At higher temperatures, the power radiated increases as the wavelength of peak radiation decreases. A roaring wood stove with a surface temperature of 500°F has an absolute temperature about 1.8 times room temperature (530 K); it radiates about 11 times more power than when at room temperature since $P \propto T^4$ [Stefan's law, Eq. (14-19)]. Nevertheless, the peak is still in the infrared. The wavelength of peak radiation is about 5.5 μm since $\lambda_{max} \propto 1/T$ [Wien's law, Eq. (14-20)]. If the stove gets even hotter, its radiation is still mostly IR but glows red as it starts to radiate significantly in the red part of the visible spectrum. (Call the fire department!) Even the filament of a lightbulb ($T \approx 2000$ K) radiates more IR than it does visible. The *peak* of the Sun's thermal radiation is in the visible; nevertheless about half the energy reaching us from the Sun is IR.

Remote controls used to control TVs, VCRs, and stereos communicate via a beam of IR. Reconnaissance satellites can record images of the IR radiated by buildings, planes, tanks, automobiles, and people (Fig. 22.8a, b).

(a)

(b)

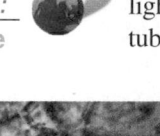

Figure 22.8   (a) A thermogram of a man's head. The red areas show regions of pain from a headache; these areas are warmer, so they give off more infrared radiation. (b) Thermogram of a house in winter, showing that most of the heat escapes through the roof. The scale shows that the blue areas are the coolest, while the white areas are the warmest. Note that some heat escapes around the window frame, while the window itself is cool due to double-pane glass.

Ultraviolet

The prefix *ultra-* means *above*; **ultraviolet** (UV) radiation is higher in frequency than visible light. UV ranges in wavelength from the shortest visible wavelength (about 380 nm) down to about 10 nm. There is plenty of UV in the Sun's radiation; its effects on human skin include tanning, sunburn, formation of vitamin D, and melanoma. Water vapor transmits much of the Sun's UV, so tanning and sunburn can occur on overcast days. On the other hand, glass absorbs most UV, so you can't get a tan through a window. Black lights emit UV; certain *fluorescent* materials—such as the coating on the inside of the glass tube in a fluorescent light—can absorb UV and then emit visible light (Fig. 22.9).

(a)

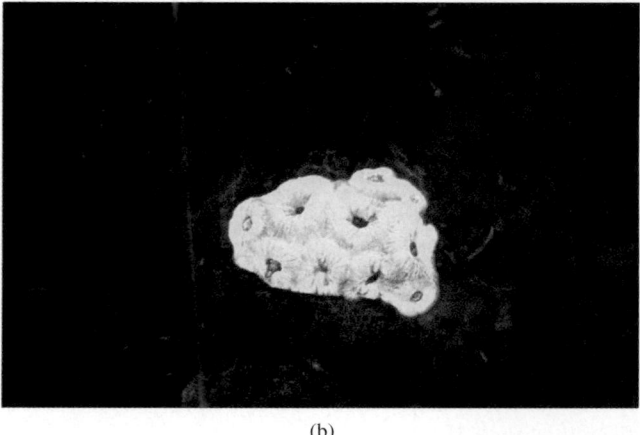

(b)

Figure 22.9   (a) The large star coral (*Montastraea cavernosa*) is dull brown when illuminated by white light. (b) When illuminated with an ultraviolet source, the coral absorbs UV and emits visible light that appears bright yellow. The red fluorescence around the coral comes from chlorophyll in algae that cover the surface. A small sponge (bottom right corner) looks bright red in white light due to selective reflection. It appears black when illuminated with UV because it does not fluoresce.

Radio Waves

After IR and UV were identified, most of the nineteenth century passed before any of the outlying regions of the EM spectrum were discovered. The lowest frequencies (up to about 1 GHz) and longest wavelengths (down to about 0.3 m) are called **radio waves**. Radio waves were discovered in 1888 by Heinrich Hertz. AM and FM radio, VHF and UHF TV broadcasts, and ham radio operators occupy assigned frequency bands within the radio wave part of the spectrum.

Microwaves

Microwaves are the part of the EM spectrum lying between radio waves and IR, with vacuum wavelengths roughly from 1 mm to 30 cm. Microwaves are used in communications (cell phones and satellite TV) and in radar. After the development of radar in World War II, the search for peacetime uses of microwaves resulted in the development of the microwave oven.

A microwave oven (Fig 22.10) immerses food in microwaves with a wavelength in vacuum of about 12 cm. Water is a good absorber of microwaves because the water molecule is polar. An electric dipole in an electric field feels a torque that tends to align the dipole with the field, since the positive and negative charges are pulled in opposite directions. As a result of the rapidly oscillating electric field of the microwaves ($f = 2.5$ GHz), the water molecules rotate back and forth; the energy of this rotation then spreads throughout the food. The frequency of the microwaves is chosen to be near, but

Making The Connection:
microwave ovens

Figure 22.10 A microwave oven. The microwaves are produced in a *magnetron*, a resonant cavity that produces the oscillating currents that give rise to microwaves at the desired frequency. Since metals reflect microwaves well, a metal waveguide directs the microwaves toward the rotating metal stirrer, which reflects the microwaves in many different directions to distribute them throughout the oven. (This reflective property is one reason why metal containers and aluminum foil should generally not be used in a microwave oven; no microwaves could reach the food inside the container or foil.) The oven cavity is enclosed by metal to reflect microwaves back in and minimize the amount leaking out of the oven. The sheet of metal in the door has small holes so we can see inside, but since the holes are much smaller than the wavelength of the microwaves, the sheet still reflects microwaves.

not exactly at, the resonant frequency of the molecule's rotation. We don't want the water to absorb the microwave energy *too* effectively; if it did, the surface of the food would absorb too much energy and not enough would penetrate to the interior.

In the early 1960s, Arno Penzias and Robert Wilson were having trouble with their radio telescope; they were plagued by noise in the microwave part of the spectrum. Subsequent investigation led them to discover that the entire universe is bathed in microwaves that correspond to blackbody radiation at a temperature of 2.7 K (peak wavelength about 1 mm). This *cosmic microwave background radiation* is left over from the origin of the universe—a huge explosion called the *Big Bang*.

Making The Connection:
cosmic microwave
background radiation

X-Rays and Gamma Rays

Higher in frequency and shorter in wavelength than UV are **x-rays** and **gamma rays**, which were discovered in 1895 and 1900, respectively. The two names are still used, based on the source of the waves, mostly for historical reasons. There is considerable overlap in the frequencies of the EM waves generated by these two methods, so today the distinction is somewhat arbitrary.

X-rays were unexpectedly discovered by Wilhelm Konrad Röntgen (1845–1923) when he accelerated electrons to high energies and smashed them into a target. The large deceleration of the electrons as they come to rest in the target produces the x-rays. Röntgen received the first Nobel Prize in physics for the discovery of x-rays.

Unfortunately the danger that x-ray radiation poses for human tissues was not recognized for another half century. In the 1940s, children getting new shoes put their feet into an x-ray machine so that parents could see a child's wiggling toe bones inside the shoe outline to make sure that the shoes allowed room to grow. In the 1950s, x-rays were used to treat acne and other skin conditions. The fear of tuberculosis in the 40s and 50s was such that children who had a reaction to a skin patch test were routinely sent to get chest x-rays, although they had no other symptoms.

Making The Connection:
x-rays in medicine
and dentistry,
CAT scans

Most diagnostic x-rays used in medicine and dentistry have wavelengths between 10 and 60 pm (1 pm = 10^{-12} m). In a conventional x-ray, film records the amount of x-ray radiation that passes through the tissue. Computer-assisted tomography (CAT or CT scan) allows a cross-sectional image of the body. An x-ray source is rotated around the body in a plane and a computer measures the x-ray transmission at many different angles. Using this data, the computer constructs an image of that slice of the body (Fig. 22.11).

Gamma rays were first observed in the decay of radioactive nuclei on Earth. Pulsars, neutron stars, black holes, and explosions of supernovae are sources of gamma rays that travel toward Earth, but—fortunately for us—gamma rays are absorbed by the atmosphere. Only when detectors were placed high in the atmosphere and above it by using balloons and satellites did the science of gamma-ray astronomy develop. In the late 1960s, scientists first observed bursts of gamma rays from deep space that last for times ranging from a fraction of a second to a few minutes; these bursts occur about once a day. A gamma-ray burst can emit more energy in 10 s than the Sun will emit in its entire lifetime. The source of the gamma-ray bursts is still under investigation.

22.5 SPEED OF EM WAVES IN VACUUM AND IN MATTER

Light travels so fast that it is not obvious that it takes any time at all to go from one place to another. Since high-precision electronic instruments were not available, early measurements of the speed of light had to be cleverly designed. In 1849, French scientist Armand Hippolyte Louis Fizeau (1819–1896) measured the speed of visible light to be approximately 3×10^8 m/s. Fizeau's experiment used a notched wheel (Fig. 22.12). When the apparatus is correctly aligned, a beam of light passes through one of the notches in the wheel, travels a long distance (over 8 km) to a mirror, reflects, and passes back through the same notch to the observer. When the wheel is made to rotate, the

Figure 22.11   Apparatus used for a CAT scan

notch moves out of position and the reflected beam is interrupted by the wheel. As the angular velocity of the wheel is increased, it reaches a value ω where the next notch moves into position just in time to allow the reflected beam to pass through. The observer can see the reflected beam for an integral multiple of ω, since any of the equally spaced notches allow the reflected beam to pass through. The speed of light can be determined from a measurement of the angular velocities at which the observer sees the reflected beam.

Figure 22.12 The apparatus used by Fizeau in 1849 to measure the speed of light

In Chapters 11 and 12 we saw that the speed of a mechanical wave depends on properties of the wave medium. Sound travels faster through steel than it does through water and faster through water than through air. In every case, the wave speed depended on two characteristics of the wave medium: one that characterizes the restoring force and another that characterizes the inertia.

Unlike mechanical waves, electromagnetic waves can travel through vacuum; they do not require a material medium. Light reaches Earth from galaxies billions of light years away, traveling the vast distances between galaxies without problem; but a sound wave can't even travel a few meters between two astronauts on a space walk, since there is no air or other medium to sustain a sound wave's pressure variations. What, then, determines the speed of light in vacuum?

Looking back at the laws that describe electric and magnetic fields, we find two universal constants. One of them is the permittivity of free space ϵ_0, found in Coulomb's law and Gauss's law; it is associated with the electric field. The second is the permeability of free space μ_0, found in Ampère's law; it is associated with the magnetic field. Since these are the only two quantities that can determine the speed of light in vacuum, there must be a combination of them that has the dimensions of speed.

The values of these constants in SI units are

$$\epsilon_0 = 8.85 \times 10^{-12} \frac{C^2}{N \cdot m^2}$$

and

$$\mu_0 = 4\pi \times 10^{-7} \frac{T \cdot m}{A}$$

The tesla can be written in terms of other SI units. Using $\vec{F} = q\vec{v} \times \vec{B}$ as a guide,

$$1 \, T = 1 \frac{N}{C \cdot m/s}$$

Multiplying $\epsilon_0 \times \mu_0$ gives

$$\epsilon_0 \mu_0 = 8.85 \times 10^{-12} \frac{\cancel{C}^2}{N \cdot m^2} \times 4\pi \times 10^{-7} \frac{N \cdot m}{\cancel{C} \cdot (m/s) \cdot (\cancel{C}/s)}$$

$$= 1.11 \times 10^{-17} \frac{s^2}{m^2}$$

To end up with m/s, we need to take the reciprocal of the square root:

$$\frac{1}{\sqrt{\epsilon_0 \mu_0}} = 3.00 \times 10^8 \, m/s$$

The dimensional analysis done here leaves the possibility of a multiplying factor such as $\frac{1}{2}$ or $\sqrt{\pi}$. In the mid-nineteenth century, Scottish physicist James Clerk Maxwell proved mathematically that an electromagnetic wave—a wave consisting of oscillating electric and magnetic fields propagating through space—could exist in vacuum. Starting from Maxwell's equations (Section 22.2), he derived the *wave equation*, an equation of a special mathematical form that describes wave propagation for *any* kind of wave. In the place of the wave speed appeared $(\epsilon_0\mu_0)^{-1/2}$. Using the values of ϵ_0 and μ_0 that had been measured in 1856, Maxwell proved that electromagnetic waves in vacuum travel at 3.00×10^8 m/s—very close to what Fizeau measured. Maxwell's derivation was the first evidence that light is an electromagnetic wave.

The speed of electromagnetic waves in vacuum is represented by the symbol c (for the Latin *celeritas*, "speed").

Speed of electromagnetic waves in vacuum

$$c = \frac{1}{\sqrt{\epsilon_0\mu_0}} = 3.00 \times 10^8 \text{ m/s} \qquad (22\text{-}3)$$

While c is usually called *the speed of light*, it is the speed of *any* electromagnetic wave in vacuum, regardless of frequency or wavelength, not just the speed for frequencies visible to humans.

Example 22.2

Light Travel Time from a "Nearby" Supernova

 A supernova is an exploding star; a supernova is billions of times brighter than an ordinary star. Most supernovae occur in distant galaxies and cannot be observed with the naked eye. The last two supernovae visible to the naked eye occurred in 1604 and 1987. Supernova SN1987a (Fig. 22.13) occurred 1.6×10^{21} m from Earth. *When* did the explosion occur?

Strategy   The light from the supernova travels at speed c. The time that it takes light to travel a distance 1.6×10^{21} m tells us how long ago the explosion occurred.

Solution   The time for light to travel a distance d at speed c is

$$\Delta t = \frac{d}{c} = \frac{1.6 \times 10^{21} \text{ m}}{3.00 \times 10^8 \text{ m/s}} = 5.33 \times 10^{12} \text{ s}$$

To get a better idea how long that is, we convert seconds to years:

$$5.33 \times 10^{12} \text{ s} \times \frac{1 \text{ h}}{3600 \text{ s}} \times \frac{1 \text{ d}}{24 \text{ h}} \times \frac{1 \text{ yr}}{365 \text{ d}} = 170{,}000 \text{ yr}$$

Discussion   When we look at the stars, the light we see was radiated by the stars long ago. By looking at distant galaxies, astronomers get a glimpse of the universe in the past. Beyond the Sun, the closest star to Earth is about 4 ly (light-years) away, which means that it takes light 4 yr to reach us from that star. The most distant galaxies observed are at a distance of about 10^{10} ly; looking at them, we see 10 billion yr into the past.

Practice Problem 22.2   A light-year

A light-year is the distance traveled by light (in vacuum) in one Earth year. Find the conversion factor from light-years to meters.

(a)

(b)

Figure 22.13
Photos of the sky (a) before and (b) after light from Supernova SN1987a reached Earth.

Speed of Light in Matter

The speed of an EM wave through matter is less than c.

When an EM wave travels through a material medium, it travels at a speed v that is less than c. For example, visible light travels through glass at speeds between about 1.6×10^8 m/s and 2.0×10^8 m/s, depending on the type of glass and the frequency of the light. Instead of specifying the speed, it is common to specify the **index of refraction** n:

Index of refraction
$$n = \frac{c}{v} \qquad (22\text{-}4)$$

Refraction refers to the bending of a wave as it passes from one medium to another; we study refraction in detail in Section 23.3. Since the index of refraction is a ratio of two speeds, it is a dimensionless number. For glass in which light travels at 2.0×10^8 m/s, the index of refraction is

$$n = \frac{3.0 \times 10^8 \text{ m/s}}{2.0 \times 10^8 \text{ m/s}} = 1.5$$

The speed of light in air (at 1 atm) is only slightly less than c; the index of refraction of air is 1.0003. Most of the time this 0.03% difference is not important so we can use c as the speed of light in air. The speed of visible light in an optically transparent medium is less than c, so the index of refraction is greater than 1.

When an EM wave passes from one medium to another, the frequency and wavelength cannot both remain unchanged since

$$v = f\lambda$$

A wave passing from one medium into another changes wavelength but retains the same frequency.

As is the case with mechanical waves, it is the wavelength that changes (Fig. 22.14); the frequency remains the same. The incoming wave (with frequency f) causes charges in the atoms at the boundary to oscillate with the same frequency f, just as for the charges in an antenna. The oscillating charges at the boundary radiate an EM wave at that same frequency into the second medium. Therefore, the electric and magnetic fields in the second medium *must* oscillate at the same frequency as the fields in the first medium. In just the same way, if a transverse wave of frequency f traveling down a string reaches a point at which an abrupt change in wave speed occurs (Fig. 22.14), the incident wave makes that point oscillate up and down at the same frequency f as any other point on the string. The oscillation of that point sends a wave of the same frequency to the other side of the string. Since the wave speed has changed but the frequency is the same, the wavelength has changed as well.

We sometimes need to find the wavelength λ of an EM wave in a medium of index n, given its wavelength λ_0 in vacuum. Since the frequencies are equal,

$$f = \frac{c}{\lambda_0} = \frac{v}{\lambda}$$

Solving for λ gives

$$\lambda = \frac{v}{c}\lambda_0 = \frac{\lambda_0}{n} \qquad (22\text{-}5)$$

Figure 22.14 A sinusoidal wave changes wavelength at a boundary where the wave speed changes.

Since $n > 1$, the wavelength is shorter than the wavelength in vacuum. The wave travels more slowly in the medium than in vacuum; since the wavelength is the distance traveled by the wave in one period $T = 1/f$, the wavelength in the medium is shorter.

If blue light of wavelength $\lambda_0 = 480$ nm enters glass that has an index of refraction of 1.5, it is still visible light, even though its wavelength in glass is 320 nm; it has not been transformed into UV radiation. When the light enters the eye, it has the same frequency and wavelength in the fluid in the eye regardless of how many material media it has passed through, since the frequency remains the same at each boundary.

Example 22.3

Wavelength Change from Glass to Water

The index of refraction of glass is 1.50 and that of water is 1.33. If light of wavelength 285 nm *in glass* passes into water, what is the wavelength in the water?

Strategy   The key is to remember that the frequency is the same as the wave passes from one medium to another.

Solution   Frequency, wavelength, and speed are related by

$$v = \lambda f$$

Solving for frequency, $f = v / \lambda$. Since the frequencies are equal,

$$\frac{v_\mathrm{w}}{\lambda_\mathrm{w}} = \frac{v_\mathrm{g}}{\lambda_\mathrm{g}}$$

The speed of light in a material is $v = c/n$. Solving for λ_w and substituting $v = c/n$ gives

$$\lambda_\mathrm{w} = v_\mathrm{w} \frac{\lambda_\mathrm{g}}{v_\mathrm{g}} = \frac{c}{n_\mathrm{w}} \times \frac{n_\mathrm{g} \lambda_\mathrm{g}}{c} = \frac{1.50 \times 285 \text{ nm}}{1.33} = 321 \text{ nm}$$

Discussion   Water has a smaller index of refraction so the speed of light in water is greater than in glass. Since wavelength is the distance traveled in one period, the wavelength in water is longer (321 nm > 285 nm).

Practice Problem 22.3   Wavelength change from air to water

The speed of visible light in water is 2.25×10^8 m/s. When light of wavelength 592 nm in air passes into water, what is its wavelength in water?

Dispersion

Although EM waves of every frequency travel through vacuum at the same speed c, the speed of EM waves in a material medium *does* depend on frequency. Therefore, the index of refraction is not a constant for a given material; it is a function of frequency. Variation of the speed of a wave with frequency is called **dispersion**. Dispersion causes white light to separate into colors when it passes through a glass prism (Fig. 22.15). The dispersion of the light into different colors arises because each color travels at a slightly different speed in the same medium (whether water or glass).

A **nondispersive** medium is one for which the variation in the index of refraction is negligibly small for the range of frequencies of interest. No medium (apart from vacuum) is truly nondispersive; but many can be treated as nondispersive for a restricted range of frequencies. For most optically transparent materials, the index of refraction rises as the frequency is increased; blue light travels more slowly through glass than does red light. In other parts of the EM spectrum, or even for visible light in unusual materials, n can decrease with increasing frequency instead.

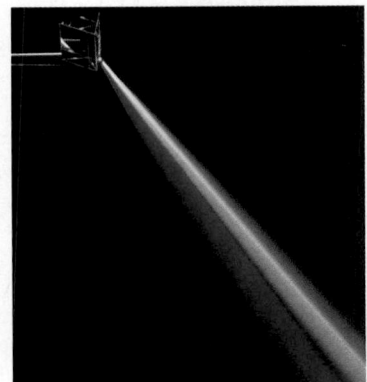

Figure 22.15   A prism separates white light into the colors of the spectrum.

22.6   CHARACTERISTICS OF ELECTROMAGNETIC WAVES IN VACUUM

The various characteristics of EM waves in vacuum can be derived from the basic laws of electromagnetism (Maxwell's equations, Section 22.2). Such a derivation requires higher level mathematics, so we state the characteristics without proof.

- EM waves in vacuum travel at speed $c = 3.00 \times 10^8$ m/s, independent of frequency. The speed is also independent of amplitude.

- The electric and magnetic fields oscillate at the *same frequency*. Thus, a single frequency f and a single wavelength $\lambda = c/f$ pertain to both the electric and magnetic fields of the wave.

- The electric and magnetic fields oscillate *in phase* with each other. That is, at a given instant, the electric and magnetic fields are at their maximum magnitudes at a common set of points. Similarly, the fields are both zero at a common set of points at any instant.

- The amplitudes of the electric and magnetic fields are proportional to each other. The ratio is c:

$$E_m = cB_m \tag{22-6}$$

- Since the fields are in phase and the amplitudes are proportional, the instantaneous magnitudes of the fields are proportional at any point:

$$E(x, y, z, t) = cB(x, y, z, t) \tag{22-7}$$

- The EM wave is *transverse*; that is, the electric and magnetic fields are each perpendicular to the direction of propagation of the wave.

- The fields are also perpendicular to *each other*. Therefore, $\vec{E}$, $\vec{B}$, and the velocity of propagation are three mutually perpendicular vectors.

- At any point, $\vec{E} \times \vec{B}$ is always in the direction of propagation.

- The electric energy density is equal to the magnetic energy density at any point. The wave carries exactly half its energy in the electric field and half in the magnetic field.

Example 22.4

Traveling EM Wave

The x-, y-, and z-components of the electric field of an EM wave in vacuum are

$$E_y(x, y, z, t) = -60.0\,\frac{V}{m} \times \cos\left[(4.0\,\text{m}^{-1})x + \omega t\right], \quad E_x = E_z = 0$$

(a) In what direction does the wave travel? (b) Find the value of ω. (c) Write an expression for the components of the magnetic field of the wave.

Strategy   Parts (a) and (b) require some general knowledge about waves, but nothing specific to EM waves. Turning back to Chapter 11 may help refresh your memory. Part (c) involves the relationship between the electric and magnetic fields, which *is* particular to EM waves. The instantaneous magnitude of the magnetic field is given by $B(x, y, z, t) = E(x, y, z, t)/c$. We must also determine the direction of the magnetic field: $\vec{E}$, $\vec{B}$, and the velocity of propagation are three mutually perpendicular vectors and $\vec{E} \times \vec{B}$ must be in the direction of propagation.

Solution   (a) Since the electric field depends on the value of x but not on the values of y or z, the wave moves parallel to the x-axis. Imagine riding along a crest of the wave—a point where

$$\cos\left[(4.0\,\text{m}^{-1})x + \omega t\right] = 1$$

Then

$$(4.0\,\text{m}^{-1})x + \omega t = 2\pi n$$

where n is some integer. A short time later, t is a little bigger, so x must be a little smaller so that $(4.0\,\text{m}^{-1})x + \omega t$ is still equal to $2\pi n$. Since the x-coordinate of a crest gets smaller as time passes, the wave is moving in the $-x$-direction.

(b) The constant multiplying x, 4.0 m^{-1}, is related to the wavelength. Since the wave repeats in a distance λ and the cosine function repeats every 2π radians, $k(x + \lambda)$ must be 2π radians greater than kx:

$$k(x + \lambda) = kx + 2\pi$$

or

$$k = \frac{2\pi}{\lambda}$$

Therefore, the wavenumber is $k = 4.0$ m^{-1}. The speed of the wave is c. Since any periodic wave travels a distance λ in a time T,

$$T = \frac{\lambda}{c}$$

$$\omega = \frac{2\pi}{T} = \frac{2\pi c}{\lambda} = kc = 4.0\,\text{m}^{-1} \times 3.00 \times 10^8\,\text{m/s} = 1.2 \times 10^9\,\text{rad/s}$$

(c) Since the wave moves in the $-x$-direction and the electric field is in the $\pm y$-direction, the magnetic field must be in the $\pm z$-direction to make three perpendicular directions. Since the

continued on next page

Example 22.4 *continued*

magnetic field is in phase with the electric field, with the same wavelength and frequency, it must take the form

$$B_z(x, y, z, t) = \pm B_m \cos [(4.0 \text{ m}^{-1})x + (1.2 \times 10^9 \text{ s}^{-1})t], \quad B_x = B_y = 0$$

The amplitudes are proportional:

$$B_m = \frac{E_m}{c} = \frac{60.0 \text{ V/m}}{3.00 \times 10^8 \text{ m/s}} = 2.00 \times 10^{-7} \text{ T}$$

The last step is to decide which sign is correct. At $x = t = 0$, the electric field is in the $-y$-direction. $\vec{\mathbf{E}} \times \vec{\mathbf{B}}$ must be in the $-x$-direction (the direction of propagation). Then

$$(-y) \times (\text{direction of } \vec{\mathbf{B}}) = (-x)$$

Trying both possibilities with the right-hand rule (Fig. 22.16), we find that $\vec{\mathbf{B}}$ is in the $+z$-direction at $x = t = 0$. Then the magnetic field is written

$$B_z(x,y,z,t) = (2.00 \times 10^{-7} \text{ T}) \cos [(4.0 \text{ m}^{-1})x + (1.2 \times 10^9 \text{ s}^{-1})t],$$
$$B_x = B_y = 0$$

Discussion   When $\cos [(4.0 \text{ m}^{-1})x + (1.2 \times 10^9 \text{ s}^{-1})t]$ is negative, then $\vec{\mathbf{E}}$ is in the $+y$-direction and $\vec{\mathbf{B}}$ is in the $-z$-direction. Since both fields reverse direction, it is still true that $\vec{\mathbf{E}} \times \vec{\mathbf{B}}$ is in the direction of propagation.

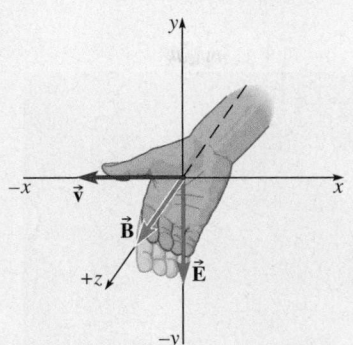

Figure 22.16
Using the right-hand rule to find the direction of $\vec{\mathbf{B}}$.

Practice Problem 22.4   Another traveling wave

The x-, y-, and z-components of the electric field of an EM wave in vacuum are

$$E_x(x, y, z, t) = 32 \frac{\text{V}}{\text{m}} \times \cos [ky - (6.0 \times 10^{11} \text{ s}^{-1})t], \quad E_y = E_z = 0$$

where k is positive. (a) In what direction does the wave travel? (b) Find the value of k. (c) Write an expression for the components of the magnetic field of the wave.

22.7   ENERGY TRANSPORT BY EM WAVES

Electromagnetic waves carry energy, as do all waves. Life on Earth exists only because the energy of EM radiation from the Sun can be harnessed by green plants, which through photosynthesis convert some of the energy in light to chemical energy. Photosynthesis sustains not only the plants themselves, but also animals that eat plants and fungi that derive their energy from decaying plants and animals—the entire food chain can be traced back to the Sun as energy source. Only a few exceptions exist, such as the bacteria that live in geothermal vents on the ocean floor. The heat flow from the interior of the Earth does not originate with the Sun; it comes from radioactive decay.

Industrial sources of energy are almost entirely derived from electromagnetic energy from the Sun. Fossil fuels—petroleum, coal, and natural gas—come from the remains of plants and animals. Solar cells convert the incident sunlight's energy directly into electricity; the Sun is also used to heat water and homes directly (Fig. 22.17). Hydroelectric power plants rely on the Sun to evaporate water, in a sense pumping it back uphill so that it can once again flow down rivers and turn turbines. Wind can be harnessed to generate electricity, but the winds are driven by uneven heating of Earth's surface by the Sun. The only energy sources we have that do not come from the Sun's EM radiation are nuclear fission and geothermal energy.

Energy Density

The energy in light is stored in the oscillating electric and magnetic fields in the wave. For an EM wave in vacuum, the energy densities are

$$u_E = \tfrac{1}{2}\epsilon_0 E^2 \tag{17-19}$$

and

$$u_B = \frac{1}{2\mu_0}B^2 \tag{20-18}$$

(a)   (b)

Figure 22.17   (a) A solar panel farm in the Sierra Nevada Mountains. (b) Solar water heating system

It can be proved (Problem 31) that the two energy densities are equal for an EM wave in vacuum, using the relationship between the magnitudes of the fields [Eq. (22-7)]. Thus, for the total energy density, we can write

$$u = \epsilon_0 E^2 = \frac{1}{\mu_0} B^2 \tag{22-8}$$

Since the fields vary from point to point and also change with time, so do the energy densities. Since the fields oscillate rapidly, in most cases we are concerned with the *average* energy densities—the average of the squares of the fields. Recall that an rms value is defined as the square root of the average of the square:

$$E_{rms} = \sqrt{\langle E^2 \rangle} \quad \text{and} \quad B_{rms} = \sqrt{\langle B^2 \rangle} \tag{22-9}$$

The angle brackets around a quantity denote the average value of that quantity. Squaring both sides, we have

$$E_{rms}^2 = \langle E^2 \rangle \quad \text{and} \quad B_{rms}^2 = \langle B^2 \rangle$$

Then the average energy density can be written in terms of the rms values of the fields:

$$\langle u \rangle = \epsilon_0 \langle E^2 \rangle = \epsilon_0 E_{rms}^2 \tag{22-10}$$

$$\langle u \rangle = \frac{1}{\mu_0} \langle B^2 \rangle = \frac{1}{\mu_0} B_{rms}^2 \tag{22-11}$$

Intensity

The energy density tells us how much energy is stored in the wave per unit volume; this energy is being carried with the wave at speed c. Suppose light falls at normal incidence on a surface (such as a photographic film or a leaf) and we want to know how much energy hits the surface. For one thing, the energy arriving at the surface depends on how long it is exposed—the reason exposure time is a critical parameter in photography. Also important is the surface area; a large leaf receives more energy than a small one, everything else being equal. Thus, the most useful quantity to know is how much energy arrives at a surface per unit time per unit area—or the average power per unit area. If light hits a surface of area A at normal incidence, the **intensity** (I) is

$$I = \frac{\langle P \rangle}{A} \qquad (22\text{-}12)$$

The SI units of I are

$$\frac{\text{energy}}{\text{time·area}} = \frac{\text{J}}{\text{s·m}^2} = \frac{\text{W}}{\text{m}^2}$$

The intensity depends on how much energy is in the wave (measured by u) and the speed at which the energy moves (which is c). If a surface of area A is illuminated by light at normal incidence, how much energy falls on it in a time Δt? The wave moves a distance $c\,\Delta t$ in that time, so all the energy in a volume $Ac\,\Delta t$ hits the surface during that time (Fig. 22.18). (We are not concerned with what happens to the energy—whether it is absorbed, reflected, or transmitted.) The intensity is then

$$I = \frac{\langle u \rangle V}{A\,\Delta t} = \frac{\langle u \rangle Ac\,\Delta t}{A\,\Delta t} = \langle u \rangle c \qquad (22\text{-}13)$$

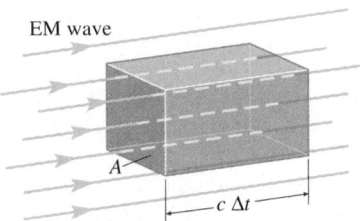

EM wave

A

$c\,\Delta t$

Figure 22.18 Geometry for finding the relationship between energy density and intensity

Example 22.5

EM Fields of a Lightbulb

At a distance of 4.00 m from a 100.0-W lightbulb, what are the intensity and the rms values of the electric and magnetic fields? Assume that all of the electric power consumed goes into EM radiation and that the radiation is isotropic (equal in all directions).

Strategy Since the radiation is isotropic, the intensity depends only on the distance from the lightbulb. Imagine a sphere surrounding the lightbulb at a distance of 4.00 m. Radiant energy must pass through the surface of the sphere at a rate of 100.0 W. We can figure out the intensity (average power per unit area) and from it the rms values of the fields.

Solution All of the energy radiated by the lightbulb crosses the surface of a sphere of radius 4.00 m. Therefore, the intensity at that distance is the power radiated divided by the surface area of the sphere:

$$I = \frac{\langle P \rangle}{A} = \frac{\langle P \rangle}{4\pi r^2} = \frac{100.0 \text{ W}}{4\pi \times 16.0 \text{ m}^2} = 0.497 \text{ W/m}^2$$

To solve for E_{rms}, we relate the intensity to the average energy density and then the energy density to the field:

$$\langle u \rangle = \frac{I}{c} = \epsilon_0 E_{\text{rms}}^2$$

$$E_{\text{rms}} = \sqrt{\frac{I}{\epsilon_0 c}} = \sqrt{\frac{0.497 \text{ W/m}^2}{8.85 \times 10^{-12} \dfrac{\text{C}^2}{\text{N·m}^2} \times 3.00 \times 10^8 \text{ m/s}}}$$

$$= 13.7 \text{ V/m}$$

Similarly, for B_{rms},

$$B_{\text{rms}} = \sqrt{\frac{\mu_0 I}{c}} = \sqrt{\frac{4\pi \times 10^{-7} \dfrac{\text{T·m}}{\text{A}} \times 0.497 \text{ W/m}^2}{3.00 \times 10^8 \text{ m/s}}}$$

$$= 4.56 \times 10^{-8} \text{ T}$$

Discussion A good check would be to calculate the ratio of the two rms fields:

$$\frac{E_{\text{rms}}}{B_{\text{rms}}} = \frac{13.7 \text{ V/m}}{4.56 \times 10^{-8} \text{ T}} = 3.00 \times 10^8 \text{ m/s} = c$$

as expected.

Practice Problem 22.5 Field of lightbulb at greater distance

What are the rms fields 8.00 m away from the lightbulb? [*Hint:* Look for a shortcut rather than redoing the whole calculation.]

Figure 22.19 The surface of area $A \cos \theta$, which is perpendicular to the incoming wave, intercepts the same light energy as would a surface of area A for which the incoming wave is incident at an angle θ from the normal.

If a surface is illuminated by light of intensity I, but the surface is not perpendicular to the incident light, the rate at which energy hits the surface is less than IA. As Fig. 22.19 shows, a perpendicular surface of area $A \cos \theta$ casts a shadow over the surface of area A and thus intercepts all the energy. The angle θ is measured between the direction of the incident light and the normal (a direction *perpendicular* to the surface). Thus, a surface that is not perpendicular to the incident wave receives energy at a rate

$$\langle P \rangle = IA \cos \theta \tag{22-14}$$

If Eq. (22-14) reminds you of flux, then congratulations on your alertness! The intensity is often called the *flux density*. Electric and magnetic fields are sometimes called *electric flux density* and *magnetic flux density*. However, the flux involved with intensity is not the same as the electric or magnetic fluxes that we defined in Eqs. (16-7) and (20-5). The intensity is the *power* flux density.

Example 22.6

Power per Unit Area from the Sun in Summer and Winter

The intensity of sunlight reaching Earth's surface on a clear day is about 1.0 kW/m². At a latitude of 40.0° north, find the average power per unit area reaching Earth at noon (a) on the summer solstice (Fig. 22.20a) and (b) on the winter solstice (Fig. 22.20b). The difference is due to the 23.5° inclination of Earth's rotation axis. In summer, the axis is inclined toward the Sun, while in winter it is inclined away from the Sun.

Strategy Because Earth's surface is not perpendicular to the Sun's rays, the power per unit area falling on Earth is less than 1.0 kW/m². We must find the angle that the Sun's rays make with the normal to the surface in each case.

Solution (a) A radius going from Earth's center to the surface is normal to the surface at that point, assuming the Earth to be a sphere. We need to find the angle between the normal and an incoming ray. At a latitude of 40.0°, the angle between the radius and Earth's axis of rotation is 90.0° − 40.0° = 50.0° (Fig. 22.20c). From the figure, θ + 50.0° + 23.5° = 90.0° and therefore θ = 16.5°. The average power per unit area is then

$$\frac{\langle P \rangle}{A} = I \cos \theta = 1.0 \times 10^3 \text{ W/m}^2 \times \cos 16.5° = 960 \text{ W/m}^2$$

(b) At the winter solstice, the rotation axis is inclined *away* from the Sun (Fig. 22.20d). Now θ = 90.0° − 26.5° = 63.5°. The average power per unit area is

$$\frac{\langle P \rangle}{A} = I \cos \theta = 1.0 \times 10^3 \text{ W/m}^2 \times \cos 63.5° = 450 \text{ W/m}^2$$

Discussion As expected, the power is greater in summer than in winter. The power per unit area at the winter solstice is less than half that at the summer solstice. The intensity of sunlight hasn't changed; what changes is how the energy is spread out on the surface. Fewer of the Sun's rays hit a given surface area when the surface is tilted more.

The Earth is actually a bit *closer* to the Sun in the northern hemisphere's winter than in summer. The angle at which the Sun's radiation hits the surface and the length of the day are much more important in determining the incident power than is the small difference in distance from the Sun.

___

Practice Problem 22.6   Average power at equinox

What is the average power per unit area at a latitude of 40.0° north at noon on the spring or autumn equinox (Fig. 22.20e)?

continued on next page

Example 22.6 *continued*

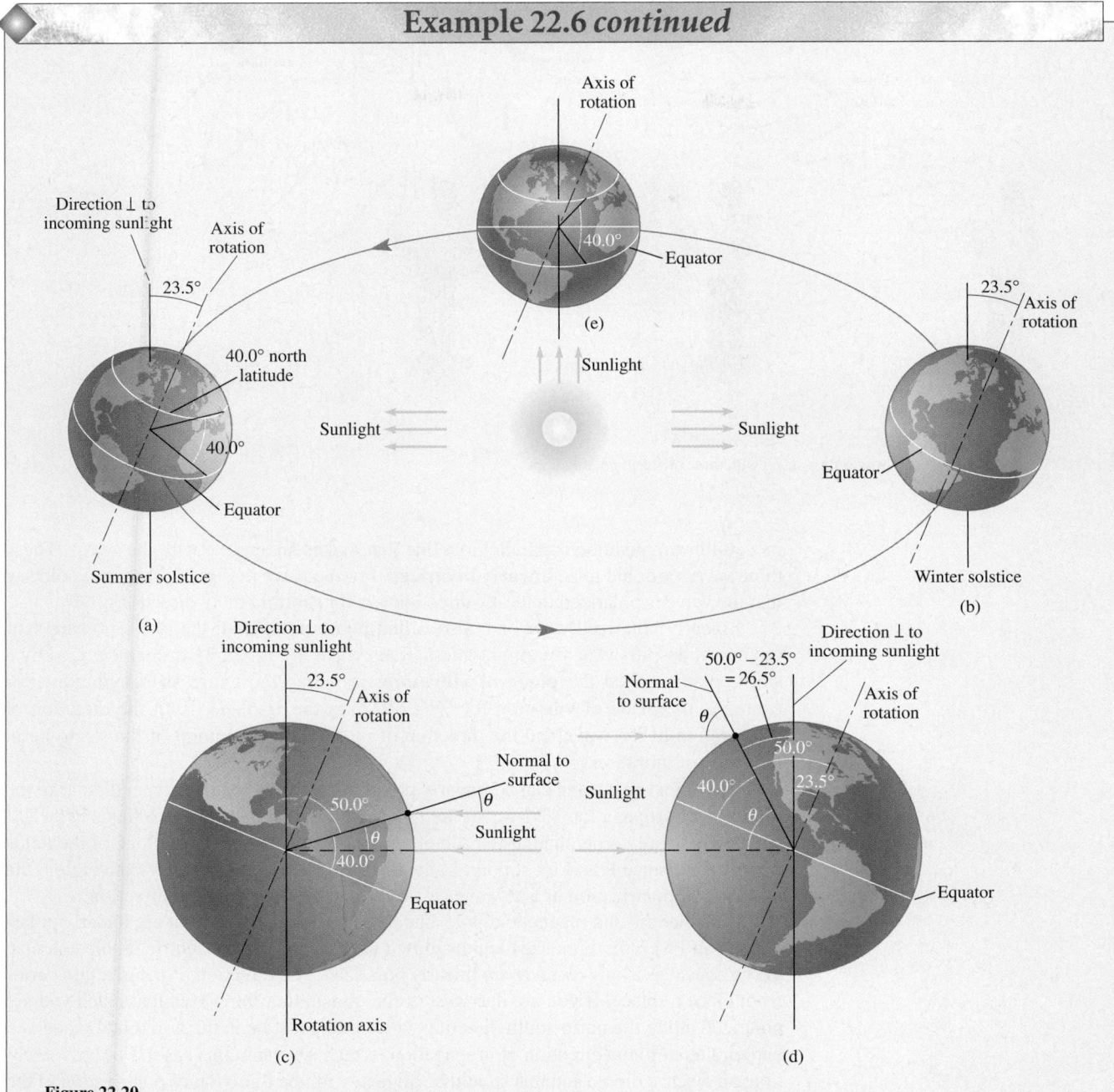

Figure 22.20
(a) At noon on the summer solstice, the rotation axis is inclined 23.5° toward the Sun. At a latitude of 40.0° north, the incoming sunlight is nearly normal to the surface of the Earth. (b) At noon on the winter solstice, the rotation axis is inclined 23.5° *away from* the Sun. At a latitude of 40.0° north, the incoming sunlight makes a large angle with the normal to the surface. (c) Geometry for the summer solstice. (d) Geometry for the winter solstice. (e) Spring equinox.

22.8   POLARIZATION

Linear Polarization

Imagine a transverse wave traveling in the $-z$-direction. Since this discussion applies to any transverse wave, let us use a transverse wave on a string as an example. The string is stretched along the z-axis. In what directions can the string be displaced to produce transverse waves on this string? The displacement could be in the $\pm x$-direction, as in Fig. 22.21a. Or it could be in the $\pm y$-direction, as in Fig. 22.21b. Or it could be in any direction in the xy-plane. In Fig. 22.21c, the displacement of any point on the string from

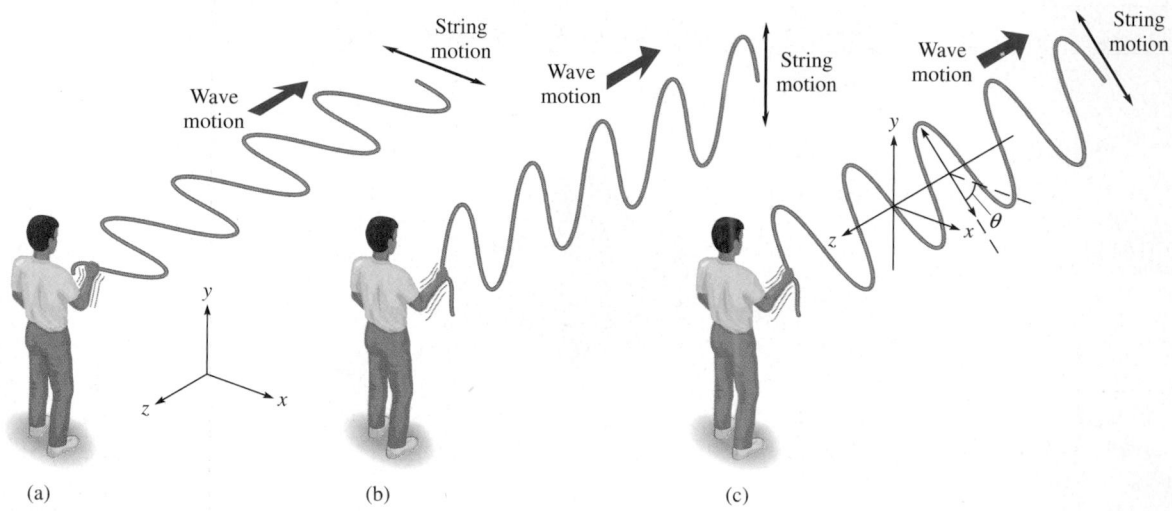

Figure 22.21 Transverse waves on a string with three different polarizations

its equilibrium position is parallel to a line that makes an angle θ with the x-axis. These three waves are said to be **linearly polarized**. For the wave in Fig. 22.21a, we would say that the wave is polarized in the $\pm x$-direction (or, for short, in the x-direction).

Linearly polarized waves are also called **plane polarized**; the two terms are synonymous, despite what you might guess. Each wave in Fig. 22.21 is characterized by a single plane, called the **plane of vibration**, in which the entire string vibrates. For example, the plane of vibration for Fig. 22.21a is the xz-plane. Both the direction of propagation of the wave and the direction of motion of every point of the string lie in the plane of vibration.

Any transverse wave can be linearly polarized in any direction perpendicular to the direction of propagation. EM waves are no exception. But there are two fields in an EM wave, which are perpendicular to each other. Knowing the direction of one of the fields is sufficient, since $\vec{\mathbf{E}} \times \vec{\mathbf{B}}$ must point in the direction of propagation. By convention, the direction of polarization of EM waves is taken to be the *electric* field direction.

Both electric and magnetic dipole antennas emit radio waves that are linearly polarized. If an FM radio broadcast is transmitted using a horizontal electric dipole antenna, the radio waves at any receiver are linearly polarized. The direction of polarization varies from place to place. If you are due west of the transmitter, the waves that reach you are polarized along the north-south direction, since they must be in the horizontal plane and perpendicular to the direction of propagation (which is west in this case). For best reception, an electric dipole antenna should be aligned with the direction of polarization of the radio waves, since it is the electric field that drives current in the antenna.

In Section 22.3, we said that if an electric dipole antenna is not lined up with the electric field of the wave, then the emf is reduced by a factor of cos θ, where θ is the angle between $\vec{\mathbf{E}}$ and the antenna. Think about this in terms of polarization. Any linearly polarized wave can be thought of as the superposition of two perpendicular linearly polarized waves along any axes we choose. Displacements are vectors and vectors can always be written as the sum of perpendicular components; therefore, the transverse wave on the string in Fig. 22.21c can be thought of as the superposition of two waves, one polarized in the x-direction and the other in the y-direction. If the amplitude of the wave is A, the amplitude of the "x-component wave" is $A \cos \theta$ and the amplitude of the "y-component wave" is $A \sin \theta$ (see Fig. 22.22).

The same is true for EM waves, since the electric and magnetic fields are vectors. Any linearly polarized EM wave can be regarded as the sum of two waves polarized along perpendicular axes. If an electric dipole antenna makes an angle θ with the electric field of a wave, only the component of $\vec{\mathbf{E}}$ along the antenna makes electrons move

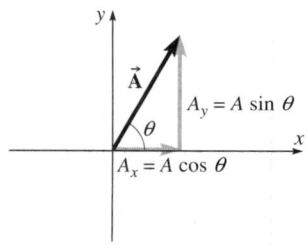

Figure 22.22 Any linearly polarized wave can be thought of as a superposition of two perpendicular polarizations, since displacements—as well as electric and magnetic fields—are vectors.

back and forth along the antenna. If we think of the wave as two perpendicular polarizations, the antenna responds to the polarization parallel to it while the perpendicular polarization has no effect.

Random Polarization

The light coming from an incandescent lightbulb is **unpolarized** or **randomly polarized**. The direction of the electric field changes rapidly and in a random way. Antennas emit linearly polarized waves because the motion of the electrons up and down the antenna is orderly and always along the same line. Thermal radiation (which is mostly IR, but also includes visible light) from a lightbulb is caused by the thermal vibrations of huge numbers of atoms. The atoms are essentially independent of each other; nothing makes them vibrate in step or in the same direction. The wave is therefore made up of the superposition of a huge number of waves whose electric fields are in random, uncorrelated directions. Thermal radiation is always unpolarized, whether it comes from a lightbulb, from a wood stove (mostly IR), or from the Sun.

Polarizers

Devices called *polarizers* transmit linearly polarized waves in a fixed direction regardless of the polarization state of the incident waves. A polarizer for transverse waves on a string is shown in Fig. 22.23. The vertical slot enables the string to slide vertically without friction, but prevents any horizontal motion. When a vertically polarized wave is sent down the string toward the slot, it passes through (Fig. 22.23a). A horizontally polarized wave does not pass through; it is reflected since the slot acts like a fixed end for horizontal motion. The direction of the slot is called the *transmission axis* since the polarizer transmits waves polarized in that direction.

What if a linearly polarized wave is sent toward the polarizer, as in Fig. 22.23c, where the incident wave is polarized at an angle θ to the transmission axis? The incident wave can be decomposed into components parallel and perpendicular to the transmission axis; the parallel wave passes through. If the incident wave has amplitude A, then the transmitted wave has amplitude $A \cos \theta$ (Fig. 22.23d).

A polarizer for microwaves consists of many parallel strips of metal (Fig. 22.24). The spacing of the strips must be significantly less than the wavelength of the microwaves. The strips act as little antennas. The parallel component of the electric

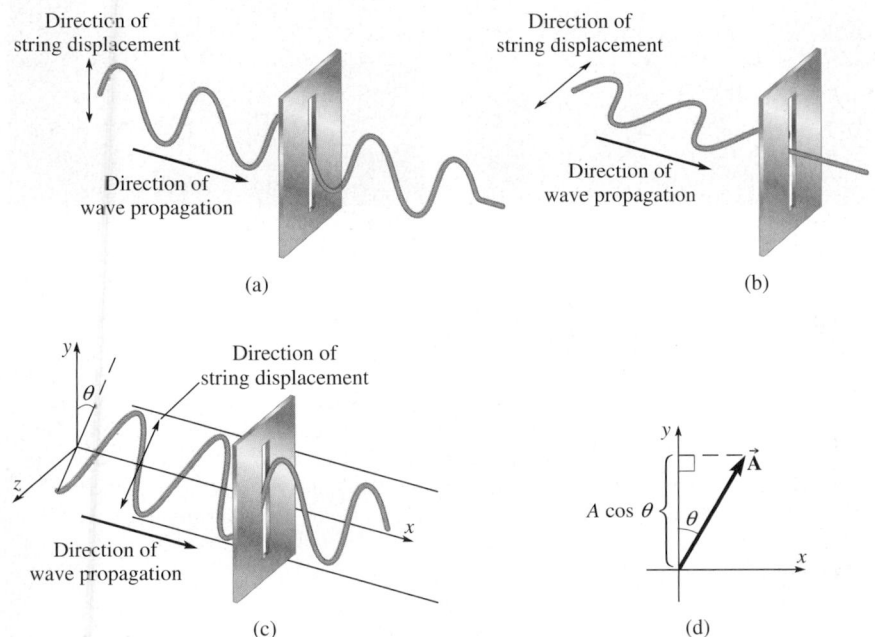

(a)

(b)

(c)

(d)

Figure 22.23   A vertical slot allows vertically polarized waves to pass through (a), but not horizontally polarized waves (b). If the incident wave is polarized at an angle θ to the vertical, a vertically polarized wave of amplitude $A \cos \theta$ is transmitted (c) and (d).

field of the incident wave makes currents flow up and down the metal strips. These currents dissipate energy, so some of the wave is absorbed. The antennas also produce a wave of their own; it is out of phase with the incident wave, so it cancels the forward-going wave and sends a reflected wave back. Between absorption and reflection, none of the electric field parallel to the metal strips gets through the polarizer. The microwaves that are transmitted are linearly polarized *perpendicular to the strips.* Although the microwave polarizer *looks* similar to the polarizer for waves on a string, the electric field does not pass through the "slots" between the metal strips! The transmission axis of the polarizer is *perpendicular* to the strips.

Sheet polarizers for visible light operate on a principle similar to that of the wire grid polarizer. A sheet polarizer contains many long hydrocarbon chains with iodine atoms attached. In production, the sheet is stretched so that these long molecules are all aligned in the same direction. The iodine atoms allow electrons to move easily along the chain, so the aligned polymers behave as parallel conducting wires, and their spacing is close enough that it does to visible light what a wire grid polarizer does to microwaves. The sheet polarizer has a transmission axis perpendicular to the aligned polymers.

If randomly polarized light is incident on a polarizer, the transmitted intensity is half the incident intensity, regardless of the orientation of the transmission axis (Fig. 22.25a). The randomly polarized wave can be thought of as two perpendicular polarized waves that are *uncorrelated*—the relative phase of the two varies rapidly with time. Half of the energy of the wave is associated with each of the two perpendicular polarizations.

$$I = \tfrac{1}{2}I_0 \quad \text{(incident wave unpolarized)} \tag{22-15}$$

If, instead, the incident wave is linearly polarized, then the component of $\vec{E}$ parallel to the transmission axis gets through (Fig. 22.25b). If θ is the angle between the incident polarization and the transmission axis, then

$$E = E_0 \cos \theta \tag{22-16a}$$

Since intensity is proportional to the square of the amplitude, the transmitted intensity is

$$I = I_0 \cos^2 \theta \tag{22-16b}$$

Equation (22-16b) is called **Malus's law** after its discoverer Étienne-Louis Malus (1775–1812), an engineer and one of Napoleon's captains.

Figure 22.24 A polarizer for microwaves. A horizontally polarized microwave beam passes through the polarizer if its strips are oriented vertically.

(a)   (b)

Figure 22.25 (a) Unpolarized light is incident on three polarizers oriented in different directions. The transmitted intensity is the same for all three. (b) Linearly polarized light is incident on the same three polarizers.

Example 22.7

Unpolarized Light Incident on Two Polarizers

Randomly polarized light of intensity I_0 is incident on two sheet polarizers (Fig. 22.26). The transmission axis of the first polarizer is vertical; that of the second makes a 30.0° angle with the vertical. What is the intensity and polarization state of the light after passing through the two?

Strategy We treat each polarizer separately. First we find the intensity of light transmitted by the first polarizer. The light transmitted by a polarizer is always linearly polarized parallel to the transmission axis of the polarizer, since only the component of $\vec{E}$ parallel to the transmission axis gets

continued on next page

Example 22.7 *continued*

through. Then we know the intensity and polarization state of the light that is incident on the second polarizer.

Solution When randomly polarized light passes through a polarizer, the transmitted intensity is half the incident intensity [Eq. (22-15)] since the wave has equal amounts of energy associated with its two perpendicular (but uncorrelated) components.

$$I_1 = \tfrac{1}{2}I_0$$

The light is now linearly polarized parallel to the transmission axis of the first polarizer, which is vertical.

The component of the electric field parallel to the transmission axis of the second polarizer passes through. The amplitude is thus reduced by a factor cos 30.0° and, since intensity is proportional to amplitude squared, the intensity is

reduced by a factor cos² 30.0° (Malus's law). The intensity transmitted through the second polarizer is

$$I_2 = I_1 \cos^2 30.0° = \tfrac{1}{2}I_0 \cos^2 30.0° = 0.375\, I_0$$

The light is now linearly polarized 30.0° from the vertical.

 Discussion For problems involving two or more polarizers in series, treat each polarizer in turn. Use the intensity and polarization state of the light that emerges from one polarizer as the incident intensity and polarization for the next polarizer.

Practice Problem 22.7 Minimum and maximum intensities

If randomly polarized light of intensity I_0 is incident on two polarizers, what is the maximum and minimum possible intensity of the transmitted light as the angle between the two transmission axes is varied?

Figure 22.26
The circular disks are polarizing sheets with their transmission axes marked.

Liquid Crystal Displays

Liquid crystal displays (LCDs) are commonly found in flat-panel computer screens, calculators, digital watches, and digital meters. In each segment of the display, a liquid crystal layer is sandwiched between two finely grooved surfaces with their grooves perpendicular (Fig. 22.27a). As a result the molecules twist 90° between the two surfaces. When a voltage is applied across the liquid crystal layer, the molecules line up in the direction of the electric field (Fig. 22.27b).

Unpolarized light from a small fluorescent bulb is polarized by one polarizing sheet. The light then passes through the liquid crystal and then through a second polarizing sheet with its transmission axis perpendicular to the first. When no voltage is applied to the liquid crystal, it rotates the polarization of the light by 90° and the light can pass through the second polarizer (Fig. 22.27a). When a voltage is applied, the liquid crystal transmits light without changing its polarization; the second polarizer blocks transmission of the light (Fig. 22.27b). When you look at an LCD display, you see the light transmitted by the second sheet. If a segment of liquid crystal has no voltage applied to it, no light is transmitted; we see a black segment. If a segment does have an applied voltage, it transmits light of the same gray color as the background.

Making The Connection:
liquid crystal displays

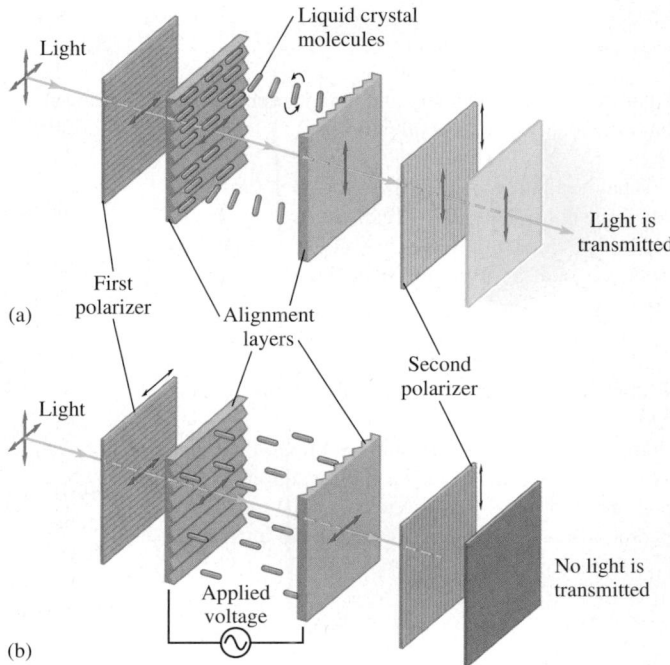

Figure 22.27 (a) When no voltage is applied to the liquid crystal, it rotates the polarization of the light so it can pass through the second polarizing sheet. (b) When a voltage is applied to the liquid crystal, no light is transmitted through the second polarizing sheet.

Polarization by Scattering and by Reflection

While the radiation emitted by the Sun is unpolarized, much of the sunlight that we see is **partially polarized**. Partially polarized light is a mixture of unpolarized and linearly polarized light. A sheet polarizer can be used to distinguish linearly polarized, partially polarized, and unpolarized light. The polarizer is rotated and the transmitted intensity at different angles is noted. If the incident light is unpolarized, the intensity stays constant as the polarizer is rotated. If the incident light is linearly polarized, the intensity is zero in one orientation and maximum at a perpendicular orientation. If partially polarized light is analyzed in this way, the transmitted intensity varies as the polarizer is rotated, but it is not zero for *any* orientation; it is maximum in one orientation and minimum (but nonzero) in a perpendicular orientation. A polarizer used to analyze the polarization state of light is often called an *analyzer*.

Natural, unpolarized light becomes partially polarized when it scatters or is reflected. So, unless you look straight at the Sun (which can cause severe eye damage— do not do this!), the sunlight that reaches you has been scattered and/or reflected and thus is partially polarized. Common Polaroid sunglasses consist of a sheet polarizer, oriented to absorb the preferential direction of polarization of light reflected from horizontal surfaces, such as a road or the water on a lake, and to reduce the glare of scattered light in the air. Polaroid sunglasses are often used in boating and aviation because they preferentially cut down on glare rather than indiscriminately reducing the intensity for all polarization states (Fig. 22.28).

How do scattering and reflection make the light partially polarized? Let's look at scattering; polarization by reflection is similar. The blue sky we see on sunny days is sunlight that is scattered by molecules in the air. On the Moon, there is no blue sky

Physics at Home

Take a pair of inexpensive Polaroid sunglasses outside on a sunny day and analyze the polarization state of light reflecting from various surfaces. Also analyze the polarization of the sky in various directions (but do not look directly at the Sun, even through sunglasses!). Get a second pair of sunglasses so you can put two polarizers in series. Rotate the one closest to you while holding the other in the same orientation. When is the transmitted intensity maximum? When is it minimum?

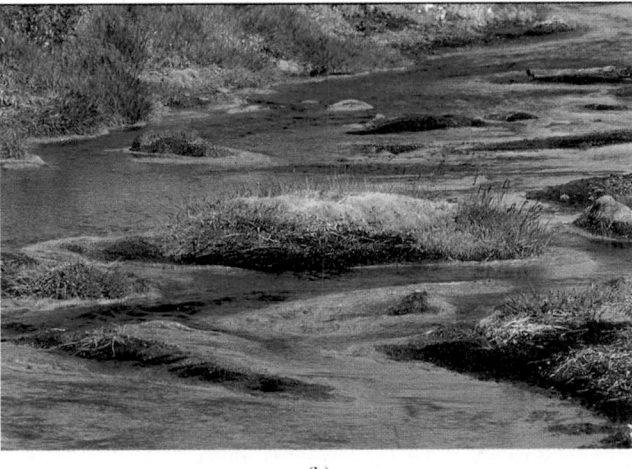

(a)   (b)

Figure 22.28 Photo of a stream in Iceland taken (a) without and (b) with a polarizing filter in front of the camera's lens. The filter reduces the amount of reflected glare from the surface of the stream. Polarizing sunglasses also reduce reflected glare.

because there is no atmosphere. Even during the day, the sky is as black as at night, although the Sun and the Earth may be brightly shining above (Fig. 22.29). Earth's atmosphere scatters blue light, with its shorter wavelengths, more than light with longer wavelengths. At sunrise and sunset, we see the light left over after much of the blue is scattered out—primarily red and orange.

Figure 22.30 shows an unpolarized sunlight wave being scattered by a single molecule. In response to the electric field of the wave, the positive and negative charges in the molecule oscillate, but *in opposite directions*. Thus, the molecule responds to the wave by becoming an oscillating dipole. Since the incoming wave is unpolarized, the dipole does not oscillate along a single axis, but does so in random directions perpendicular to the incident wave (since light is a transverse wave). As an oscillating dipole, the molecule radiates EM waves. An oscillating dipole radiates most strongly in directions perpendicular to its axis; it does not radiate at all in directions parallel to its axis.

Horizontal oscillation of the molecular dipole radiates in the three directions (A, B, C) equally, since those directions are all perpendicular to the horizontal axis of the dipole. Vertical oscillation of the molecular dipole radiates primarily in a horizontal plane (including A); it radiates more weakly as the angle from the horizontal increases (B). Directly below the molecule (C), the light scattered through 90° is linearly polarized.

Making The Connection:
colors of the sky
during the day
and at sunset

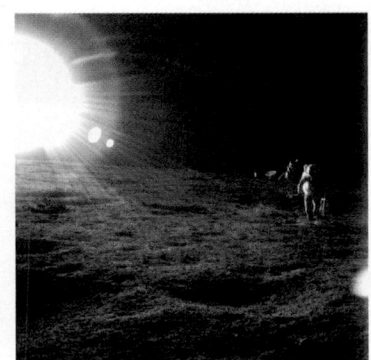

Figure 22.29 An astronaut walks away from the lunar module Intrepid while a brilliant Sun shines above the Apollo 12 base. Notice that the sky is dark even though the Sun is above the horizon; the Moon lacks an atmosphere to scatter sunlight and form a blue sky.

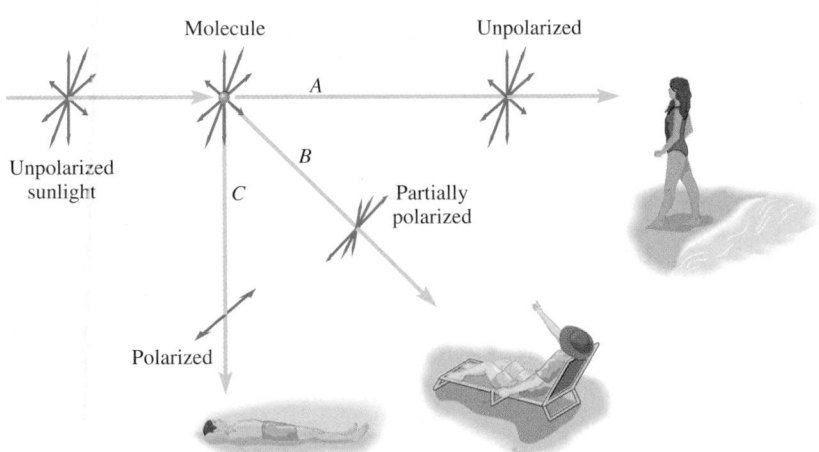

Figure 22.30 Sunlight scattered by a molecule in air

Conceptual Example 22.8

Light Polarized by Atmospheric Scattering

Just before sunset, if you look straight up at the sky, in what direction is the light polarized?

Strategy The light from the Sun just before sunset travels eastward (approximately). Some of the light is scattered by molecules in the air through roughly 90° and then travels downward toward the observer. We consider the unpolarized light from the Sun to be a mixture of two perpendicular polarizations. Looking at each polarization by itself, we determine how effectively a molecule can scatter the light downward.

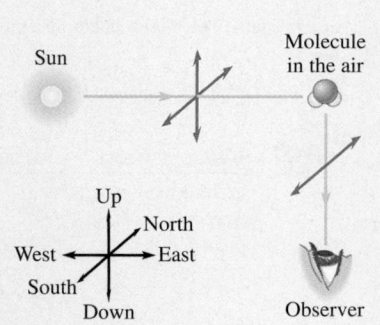

Figure 22.31
Light traveling eastward from the Sun is an uncorrelated mixture of both vertical and north-south polarizations. The two polarizations are represented by double-headed arrows. The light scattered downward is partially polarized along the north-south direction.

Solution and Discussion Figure 22.31 shows light traveling eastward from the Sun as a mixture of vertical and north-south polarizations. We choose these directions rather than two other perpendicular directions in the same plane because we are concerned with light scattered *downward*. Now we treat the two polarizations one at a time.

The north-south electric fields cause charges in the molecule to oscillate along a north-south axis. An oscillating dipole radiates most strongly in directions perpendicular to the dipole axis, so the molecule scatters light downward with a north-south polarization.

The vertical electric fields produce an oscillating dipole with a vertical axis. An oscillating dipole radiates only weakly in directions nearly parallel to its axis. Therefore, the light scattered downward is polarized along the north-south direction. Since not all of the light is scattered through exactly 90°, the light coming from directly overhead is *partially* polarized along the north-south axis.

Conceptual Practice Problem 22.8   Looking north

Just before sunset, if you look north at the sky just over the horizon, in what direction is the light partially polarized?

How Bees Navigate on Cloudy Days

Making The Connection:
navigation of bees

Human eyes are not sensitive to the polarization of light, but the eyes of bees are. A bee has a compound eye consisting of a large number of transparent fibers called the ommatidia. Each ommatidium has one end on the hemispherical surface of the compound eye (Fig. 22.32a, b) and is sensitive to light coming from the direction along which the fiber is aligned.

Each ommatidium is made up of eight cells. Each of the cells responds to a different polarization of the incident light, as shown in Fig. 22.32c. The bee can therefore detect the polarization state of light coming from various directions. When the Sun is not visible, the bee can infer the position of the Sun from the polarization of scattered light, as was established by a series of ingenious experiments by Karl von Frisch and others in the 1960s. Using polarizing sheets, von Frisch et al. could change the apparent polarization state of the scattered sunlight and watch the effects on the flight of the bees.

22.9   THE DOPPLER EFFECT FOR EM WAVES

In Section 12.8 we saw that the observed frequency of sound waves is affected by the motion of the source and/or the observer with respect to the wave medium. This frequency shift is called the *Doppler effect*. When the source and observer are approaching (getting closer together), the observed frequency is *higher* than the frequency of the source. Each successive wavefront generated by the source takes less and less time to reach the observer, since the distance the wave must travel is getting shorter and shorter. The time between the arrival of successive wavefronts at the observer (T_o) is less than the time interval between the origination of the wavefronts at the source (T_s); therefore,

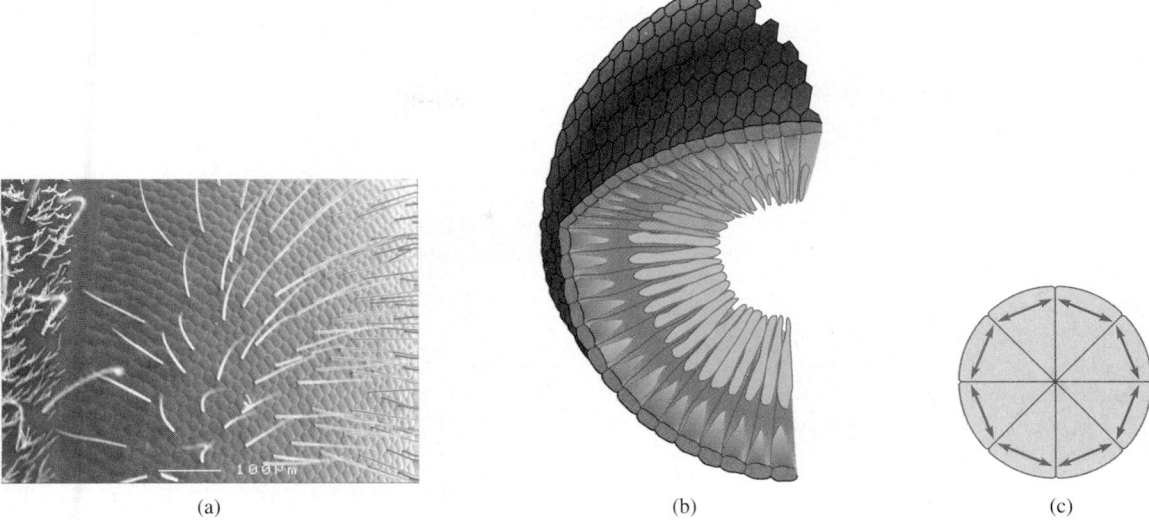

Figure 22.32   (a) Electron micrograph of the compound eye of a bee. The "bumps" are the outside surfaces of the ommatidia. (b) Cutaway drawing of the compound eye, showing the ommatidia, each facing a different direction. (c) Each of the eight cells in an ommatidium in a bee's eye is sensitive to light polarized in a different direction.

the observed frequency ($f_o = 1/T_o$) is greater than the source frequency ($f_s = 1/T_s$). If the source and observer are receding (getting farther apart), the observed frequency is less than the source frequency.

The Doppler effect exists for all kinds of waves, including EM waves. However, the Doppler formula [Eq. (12-14)] derived for sound cannot be correct for EM waves. Those equations involve the velocity of the source and the observer *relative to the medium through which the sound travels*. For sound waves in air, v_s and v_o are measured *relative to the air*. Since EM waves are not vibrations in a mechanical medium, the Doppler shift for light can only involve the *relative* velocity of the observer and the source.

Using Einstein's relativity, the Doppler shift formula for light can be derived:

$$f_o = f_s \sqrt{\frac{1 + v_{rel}/c}{1 - v_{rel}/c}} \qquad (22\text{-}17)$$

In Eq. (22-17), v_{rel} is positive if the source and observer are approaching and negative if receding. If the relative speed of source and observer is much less than c, a simpler expression can be found using the binomial approximations found in Appendix A.5:

$$\left(1 + \frac{v_{rel}}{c}\right)^{1/2} \approx 1 + \frac{v_{rel}}{2c}$$

and

$$\left(1 - \frac{v_{rel}}{c}\right)^{-1/2} \approx 1 + \frac{v_{rel}}{2c}$$

Substituting these approximations into Eq. (22-17),

$$f_o \approx f_s \left(1 + \frac{v_{rel}}{2c}\right)^2$$

$$f_o \approx f_s \left(1 + \frac{v_{rel}}{c}\right) \qquad (22\text{-}18)$$

where in the last step we used the binomial approximation once more.

Example 22.9

A Speeder Caught by Radar

A police car is moving at 38.0 m/s (85.0 mi/h) to catch up with a speeder directly ahead. The speed limit is 29.1 m/s (65.0 mi/h). A police car radar "clocks" the speed of the other car by emitting microwaves with frequency 3.0×10^{10} Hz and observing the frequency of the reflected wave. The reflected wave, when combined with the outgoing wave, produces beats at a rate of 1400 s^{-1}. How fast is the speeder going? [*Hint:* First find the frequency "observed" by the speeder. The electrons in the metal car body oscillate and emit the reflected wave with this same frequency. For the reflected wave, the speeder is the source and the police car is the observer.]

Strategy   There are *two* Doppler shifts, since the EM wave is reflected off the car. We can first think of the car as the observer, receiving a Doppler-shifted radar wave from the police car (Fig. 22.33a). Then the car "rebroadcasts" this wave back to the police car (Fig. 22.33b). This time the speeder's car is the source and the police car is the observer. The relative speed of the two cars is *much* less than the speed of light, so we use the approximate formula [Eq. (22-18)].

There are three different frequencies in the problem. Let's call the frequency emitted by the police car $f_1 = 3.0 \times 10^{10}$ Hz, the frequency received by the speeder f_2, and the frequency of the reflected wave as observed by the police car f_3. The police car is catching up to the speeder, so the source and observer are approaching; therefore, v_{rel} is positive and the Doppler shift is toward higher frequencies.

Solution   The beat frequency is

$$f_{beat} = f_3 - f_1 \qquad (12\text{-}11)$$

The frequency observed by the speeder is

$$f_2 = f_1\left(1 + \frac{v_{rel}}{c}\right)$$

Now the speeder's car emits a microwave of frequency f_2. The frequency observed by the police car is

$$f_3 = f_2\left(1 + \frac{v_{rel}}{c}\right) = f_1\left(1 + \frac{v_{rel}}{c}\right)^2$$

We need to solve for v_{rel}. We can avoid solving a quadratic equation by using the binomial approximation:

$$f_3 = f_1\left(1 + \frac{v_{rel}}{c}\right)^2 \approx f_1\left(1 + 2\frac{v_{rel}}{c}\right)$$

Solving for v_{rel},

$$v_{rel} = \frac{1}{2}c\left(\frac{f_3}{f_1} - 1\right) = \frac{1}{2}c\left(\frac{f_3 - f_1}{f_1}\right) = \frac{1}{2}c\left(\frac{f_{beat}}{f_1}\right)$$

$$= \frac{1}{2} \times 3.00 \times 10^8 \text{ m/s} \times \frac{1400 \text{ Hz}}{3.0 \times 10^{10} \text{ Hz}} = 7.0 \text{ m/s}$$

Since the two are approaching, the speeder is moving at less than 38.0 m/s. Relative to the road, the speeder is moving at

$$38.0 \text{ m/s} - 7.0 \text{ m/s} = 31.0 \text{ m/s} (= 69.3 \text{ mi/h})$$

Perhaps the police officer will be kind enough to give only a warning this time.

Discussion   Using the approximate form for the Doppler shift greatly simplifies the algebra. Using the exact form would be much more difficult and in the end would give the same answer. The speeds involved are so much less than c that the error is truly negligible.

Practice Problem 22.9   Reflection from stationary objects

Suppose the police car is moving at 23 m/s. What beat frequency results when the radar is reflected from stationary objects?

38.0 m/s

Emitted microwave travels toward speeder

$v = ?$

Police   Emitted at frequency f_1 Received at frequency f_2 Speeder

(a)

38.0 m/s

Microwave reflected back toward police

$v = ?$

Police   Received at frequency f_3 Emitted at frequency f_2 Speeder

(b)

Figure 22.33
(a) The police car emits microwaves at frequency f_1. The speeder receives them at a Doppler-shifted frequency f_2. (b) The wave is reflected at frequency f_2; the police car receives the reflected wave at frequency f_3.

Radar used by meteorologists can provide information about the position of storm systems. Now they use *Doppler radar*, which also provides information about the velocity of storm systems. Another important application of the Doppler shift of visible light is the evidence it gives for the expansion of the universe. Light reaching Earth from distant stars is *red-shifted*. That is, the spectrum of visible light is shifted downward in frequency toward the red. *Hubble's law* postulates that the speed at which a galaxy moves away from ours is proportional to how far from us the galaxy is. Assuming Hubble's law, the Doppler shift can be used to determine a star or galaxy's distance from Earth.

Looking out at the universe, the red shift tells us that stars are moving away from us in all directions; the farther away the star, the faster it is receding from us and the greater the Doppler shift of the light that reaches Earth. This doesn't mean that Earth is at the center of the universe; in a uniformly expanding universe, observers on a planet *anywhere* in the universe would see stars moving away from it in all directions. Ever since the Big Bang, the universe has been expanding. Whether it continues to expand forever, or whether the expansion will stop and the universe collapse into another big bang, is a central question studied by cosmologists and astrophysicists.

Making The Connection:
doppler shift for
distant stars
and galaxies

MASTER THE CONCEPTS

Summary

- Only accelerating charges produce electromagnetic (EM) waves (also called electromagnetic radiation). EM waves consist of oscillating electric and magnetic fields that propagate away from their source. EM waves always have both electric and magnetic fields. Just as changing magnetic fields give rise to electric fields, changing electric fields give rise to magnetic fields.

- The Ampère-Maxwell law is Ampère's law modified by Maxwell so that a changing electric field generates a magnetic field:

$$\Sigma B_{\parallel} l = \mu_0 \left(I + \epsilon_0 \frac{\Delta\Phi_E}{\Delta t} \right) \qquad (22\text{-}1)$$

- The Ampère-Maxwell law, along with Gauss's law, Gauss's law for magnetism, and Faraday's law, are called Maxwell's equations. They describe completely the electric and magnetic fields. Maxwell's equations say that $\vec{E}$- and $\vec{B}$-field lines do not have to be tied to matter. Instead, they can break free and electromagnetic waves can travel far from their sources.

- Radiation from a dipole antenna is weakest along the antenna's axis and strongest in directions perpendicular to the axis. Electric dipole antennas and magnetic dipole antennas can be used either as sources of EM waves or as receivers of EM waves.

- The electromagnetic spectrum—the range of frequencies and wavelengths of EM waves—is traditionally divided into named regions. From lowest to highest frequency, they are: radio waves, microwaves, infrared, visible, ultraviolet, x-rays, and gamma rays.

- EM waves of any frequency travel through vacuum at a speed

$$c = \frac{1}{\sqrt{\epsilon_0 \mu_0}} = 3.00 \times 10^8 \text{ m/s} \qquad (22\text{-}3)$$

- EM waves can travel through matter, but they do so at speeds less than c. The index of refraction for a material is defined as

$$n = \frac{c}{v} \qquad (22\text{-}4)$$

where v is the speed of EM waves through the material.

- The speed of EM waves (and therefore also the index of refraction) in matter depends on the frequency of the wave.

- When an EM wave passes from one medium to another, the wavelength changes; the frequency remains the same. The wave in the second medium is created by the oscillating charges at the boundary, so the fields in the second medium must oscillate at the same frequency as the fields in the first.

- Properties of EM waves in vacuum:

  The electric and magnetic fields oscillate at the *same frequency* and are *in phase*.

$$E(x, y, z, t) = cB(x, y, z, t) \qquad (22\text{-}7)$$

  $\vec{E}$, $\vec{B}$, and the direction of propagation are three mutually perpendicular directions.

  $\vec{E} \times \vec{B}$ is always in the direction of propagation.

  The electric energy density is equal to the magnetic energy density.

- Energy density of an EM wave in vacuum:

$$\langle u \rangle = \epsilon_0 \langle E^2 \rangle = \epsilon_0 E_{rms}^2 = \frac{1}{\mu_0} \langle B^2 \rangle = \frac{1}{\mu_0} B_{rms}^2 \qquad (22\text{-}10, 22\text{-}11)$$

- The intensity is

$$I = \langle u \rangle c \qquad (22\text{-}13)$$

- The average power incident on a surface of area A is

$$\langle P \rangle = IA \cos \theta \qquad (22\text{-}14)$$

where θ is 0° for normal incidence and 90° for grazing incidence.

MASTER THE CONCEPTS *continued*

- If unpolarized waves pass through a polarizer, the transmitted intensity is half the incident intensity:

$$I = \tfrac{1}{2}I_0 \qquad (22\text{-}15)$$

- If a linearly polarized wave is incident on a polarizer, the component of $\vec{E}$ parallel to the transmission axis gets through. If θ is the angle between the incident polarization and the transmission axis, then

$$E = E_0 \cos\theta \qquad (22\text{-}16\text{a})$$

Since intensity is proportional to the square of the amplitude, the transmitted intensity is

$$I = I_0 \cos^2\theta \qquad (22\text{-}16\text{b})$$

- The Doppler effect for EM waves:

$$f_o = f_s \sqrt{\frac{1 + v_{rel}/c}{1 - v_{rel}/c}} \qquad (22\text{-}17)$$

where v_{rel} is positive if the source and observer are approaching, and negative if receding. If the relative speed of source and observer is much less than c,

$$f_o \approx f_s \left(1 + \frac{v_{rel}}{c}\right) \qquad (22\text{-}18)$$

Highlighted Figures and Tables

F22.1 Electric field lines due to an electric dipole at rest and electric and magnetic field lines due to an oscillating dipole (p. 797)

F22.6 Regions of the EM spectrum (p. 802)

F22.14 A sinusoidal wave changes wavelength at a boundary where the wave speed changes (p. 810)

F22.16 Using the right-hand rule to find the direction of $\vec{B}$ (p. 813)

F22.19 Incident power depends on the angle of the surface (p. 816)

F22.21 Transverse waves on a string with three different polarizations (p. 818)

F22.22 Any linearly polarized wave can be thought of as a superposition of two perpendicular polarizations (p. 818)

F22.30 Polarization by scattering (p. 823)

CONCEPTUAL QUESTIONS

1. In Section 22.3, we stated that an electric dipole antenna should be aligned with the electric field of an EM wave for best reception. If a magnetic dipole antenna is used instead, should its axis be aligned with the magnetic field of the wave? Explain.

2. A magnetic dipole antenna has its axis aligned with the vertical. The antenna sends out radio waves. If you are due south of the antenna, what is the polarization state of the radio waves that reach you?

3. Linearly polarized light of intensity I_0 shines through two polarizing sheets. The second of the sheets has its transmission axis perpendicular to the polarization of the light before it passes through the first sheet. Must the intensity transmitted through the second sheet be zero, or is it possible that some light gets through? Explain.

4. Using Faraday's law, explain why it is impossible to have a magnetic wave without any electric component.

5. According to Maxwell, why it is impossible to have an electric wave without any magnetic component?

6. Polaroid sunglasses are useful for cutting out reflected glare due to reflection from horizontal surfaces. In which direction should the transmission axis of Polaroid sunglasses be oriented: vertically or horizontally? Explain.

7. Suppose you are facing due north at sunrise. Sunlight is reflected by a store's display windows as shown in Fig. 22.34. Is the reflected light partially polarized? If so, in what direction?

8. Describe the polarization of radio waves transmitted from a horizontal electric dipole antenna that travel parallel to the Earth's surface.

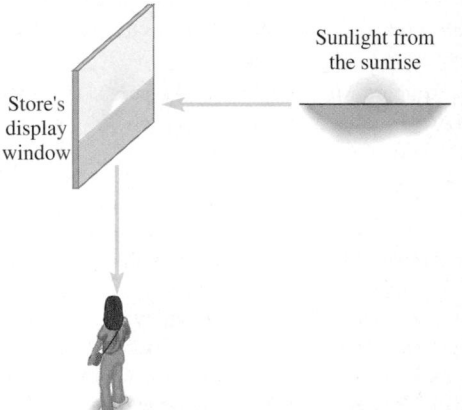

Figure 22.34   Question 7

9. Figure 22.35 shows a magnetic dipole antenna transmitting an electromagnetic wave. At a point P far from the antenna, what are the directions of the electric and magnetic fields of the wave?

Figure 22.35   Conceptual Question 9, Problem 26

10. In everyday experience, visible light seems to travel in straight lines while radio waves do not. Explain.

11. A light wave passes through a hazy region in the sky. If the electric field vector of the emerging wave is $\frac{1}{4}$ that of the incident wave, what is the ratio of the transmitted intensity to the incident intensity?

12. Can sound waves be polarized? Explain.

13. Until the Supreme Court ruled it to be unconstitutional, drug enforcement officers examined buildings at night with a camera sensitive to infrared. How did this help them identify marijuana growers?

14. What is the distinction between normal and anomalous dispersion?

15. The amplitudes of an EM wave are related by $E_m = cB_m$. Since $c = 3.00 \times 10^8$ m/s, a classmate says that the electric field in an EM wave is much larger than the magnetic field. How would you reply?

16. Why is it warmer in summer than in winter?

17. Why is the antenna on a cell phone shorter than the radio antenna on a car?

MULTIPLE CHOICE QUESTIONS

1. The radio station that broadcasts your favorite music is located exactly north of your home; it uses a horizontal electric dipole antenna directed north-south. In order to receive this broadcast, you need to

   (a) orient the receiving antenna horizontally, north-south.
   (b) orient the receiving antenna horizontally, east-west.
   (c) use a vertical receiving antenna.
   (d) move to a town farther to the east or to the west.
   (e) use a magnetic dipole antenna instead of an electric dipole antenna.

2. Which of these statements correctly describes the orientation of the electric field ($\vec{E}$), the magnetic field ($\vec{B}$), and the velocity of propagation ($\vec{v}$) of an electromagnetic wave?

   (a) $\vec{E}$ is perpendicular to $\vec{B}$; $\vec{v}$ may have any orientation relative to $\vec{E}$.
   (b) $\vec{E}$ is perpendicular to $\vec{B}$; $\vec{v}$ may have any orientation perpendicular to $\vec{E}$.
   (c) $\vec{E}$ is perpendicular to $\vec{B}$; $\vec{B}$ is parallel to $\vec{v}$.
   (d) $\vec{E}$ is perpendicular to $\vec{B}$; $\vec{E}$ is parallel to $\vec{v}$.
   (e) $\vec{E}$ is parallel to $\vec{B}$; $\vec{v}$ is perpendicular to both $\vec{E}$ and $\vec{B}$.
   (f) Each of the three vectors is perpendicular to the other two.

3. An electromagnetic wave is created by

   (a) all electric charges.
   (b) an accelerating electric charge.
   (c) an electric charge moving at constant velocity.
   (d) a stationary electric charge.
   (e) a stationary bar magnet.
   (f) a moving electric charge, whether accelerating or not.

4. The speed of an electromagnetic wave in vacuum depends on

   (a) the amplitude of the electric field but not on the amplitude of the magnetic field.
   (b) the amplitude of the magnetic field but not on the amplitude of the electric field.
   (c) the amplitude of both fields.
   (d) the angle between the electric and magnetic fields.
   (e) the frequency and wavelength.
   (f) none of the above.

5. If the wavelength of an electromagnetic wave is about the diameter of an apple, what type of radiation is it?

   (a) X-ray   (b) UV   (c) Infrared
   (d) Microwave   (e) Visible light (f) Radio wave

6. The Sun is directly overhead and you are facing toward the north. Light coming to your eyes from the sky just above the horizon is

   (a) partially polarized north-south.
   (b) partially polarized east-west.
   (c) partially polarized up-down.
   (d) randomly polarized.
   (e) linearly polarized up-down.

7. Light reflecting from the surfaces of lakes, roads, and automobile hoods is

   (a) partially polarized in the horizontal direction.
   (b) partially polarized in the vertical direction.
   (c) polarized only if the Sun is directly overhead.
   (d) polarized only if it is a clear day.
   (e) randomly polarized.

8. A vertical electric dipole antenna

   (a) radiates uniformly in all directions.
   (b) radiates uniformly in all horizontal directions, but more strongly in the vertical direction.
   (c) radiates most strongly and uniformly in the horizontal directions.
   (d) does not radiate in the horizontal directions.

9. A beam of light is linearly polarized. You wish to rotate its direction of polarization by 90° using one or more *ideal* polarizing sheets. To get maximum transmitted intensity, you should use how many sheets?

   (a) 1   (b) 2   (c) 3
   (d) As many as possible
   (e) There is no way to rotate the direction of polarization 90° using polarizing sheets.

10. Light passes from one medium (in which the speed of light is v_1) into another (in which the speed of light is v_2). If $v_1 < v_2$, as the light crosses the boundary,

    (a) both f and λ decrease.
    (b) neither f nor λ change.
    (c) f increases, λ decreases.
    (d) f does not change, λ increases.
    (e) both f and λ increase.
    (f) f does not change, λ decreases.
    (g) f decreases, λ increases.

PROBLEMS

Note: **C** indicates a combination conceptual/quantitative problem. Gold diamonds ✦, ✦✦ are used to indicate the increasing level of difficulty of each problem. Problem numbers appearing in blue, 9., denote problems that have a detailed solution available in the Student Solutions Manual. Some problems are *paired* by concept; their numbers are connected by a ruled box.

22.2 Maxwell's Equations

Problems 1–4. Apply the Ampère-Maxwell law to a circular path with its center on the central axis of the wire of Fig. 22.2a to find the magnitude of the magnetic field at the locations specified.

✦ 1. Find B outside the wire at a distance $r \geq R$ from the central axis. [*Hint:* The electric field inside the wire is constant, so there is no changing electric flux.]

✦✦ 2. Find B outside the gap in the wire at a distance $r \geq R$ from the central axis. [*Hint:* What is the rate of change of electric flux through the circle in terms of the current I?]

✦✦ 3. Find B inside the gap in the wire at a distance $r \leq R$ from the central axis. [*Hint:* Only the rate of change of electric flux $\Delta\Phi_E/\Delta t$ *through the interior of the circular path* goes into the Ampère-Maxwell law.]

✦✦ 4. Find B inside the wire at a distance $r \leq R$ from the central axis. [*Hint:* Only the current through the interior of the circular path goes into the Ampère-Maxwell law.]

22.3 Antennas

5. A dipole radio transmitter has its rod-shaped antenna oriented vertically. At a point due south of the transmitter, what is the orientation of the magnetic field of the radio waves?

6. A vertical antenna in Syracuse, New York, emits radio waves. This radiation is received in Cortland, New York, with an antenna in the form of a circular coil of wire. If Cortland is directly south of Syracuse, how should the coil be oriented for best reception?

7. Using Faraday's law, show that if a magnetic dipole antenna's axis makes an angle θ with the magnetic field of an EM wave, the induced emf in the antenna is reduced from its maximum possible value by a factor of cos θ. [*Hint:* Assume that, at any instant, the magnetic field everywhere inside the loop is uniform.]

✦ 8. A magnetic dipole antenna is used to detect an electromagnetic wave. The antenna is a coil of 50 turns with radius 5.0 cm. The EM wave has frequency 870 kHz, electric field amplitude 0.50 V/m, and magnetic field amplitude 1.7×10^{-9} T. (a) For best results, should the axis of the coil be aligned with the electric field of the wave, or with the magnetic field, or with the direction of propagation of the wave? (b) Assuming it is aligned correctly, what is the amplitude of the induced emf in the coil? (Since the wavelength of this wave is *much* larger than 5.0 cm, it can be assumed that at any instant the fields are uniform within the coil.) (c) What is the amplitude of the emf induced in an electric dipole antenna of length 5.0 cm aligned with the electric field of the wave?

22.4 The Electromagnetic Spectrum; 22.5 Speed of EM Waves in Vacuum and in Matter

9. What is the wavelength of the radio waves broadcast by an FM radio station with a frequency of 90.9 MHz?

10. What is the frequency of the microwaves in a microwave oven? The wavelength is 12 cm.

C 11. The currents in household wiring and power lines alternate at a frequency of 60.0 Hz. (a) What is the wavelength of the EM waves emitted by the wiring? (b) Compare this wavelength to the Earth's radius. (c) In what part of the EM spectrum are these waves?

12. In order to study the structure of a crystalline solid, you want to illuminate it with EM radiation whose wavelength is the same as the spacing of the atoms in the crystal (0.20 nm). (a) What is the frequency of the EM radiation? (b) In what part of the EM spectrum (radio, visible, etc.) does it lie?

13. In musical acoustics, a frequency ratio of 2:1 is called an octave. Humans with extremely good hearing can hear sounds ranging from 20 Hz to 20 kHz, which is approximately 10 octaves (since $2^{10} = 1024 \approx 1000$). (a) Approximately how many octaves of visible light are humans able to perceive? (b) Approximately how many octaves wide is the microwave region?

14. What is the speed of light in a diamond that has an index of refraction of 2.4168?

15. The speed of light in topaz is 1.85×10^8 m/s. What is the index of refraction of topaz?

16. How long does it take sunlight to travel from the Sun to the Earth?

17. How long does it take light to travel from this text to your eyes? Assume a distance of 50.0 cm.

18. The index of refraction of water is 1.33. (a) What is the speed of light in water? (b) What is the wavelength in water of a light wave with a vacuum wavelength of 515 nm?

19. Light of wavelength 692 nm in air passes into window glass with an index of refraction of 1.52. (a) What is the wavelength of the light inside the glass? (b) What is the frequency of the light inside the glass?

20. How far does a beam of light travel in 1 ns?

21. By expressing ϵ_0 and μ_0 in base SI units (kg, m, s, A), prove that the *only* combination of the two with dimensions of speed is $(\epsilon_0\mu_0)^{-1/2}$.

22.6 Characteristics of Electromagnetic Waves in Vacuum

22. The electric field in a microwave traveling through air has amplitude 0.60 mV/m and frequency 30 GHz. Find the amplitude and frequency of the magnetic field.

23. The magnetic field in a radio wave traveling through air has amplitude 2.5×10^{-11} T and frequency 3.0 MHz. (a) Find the

amplitude and frequency of the electric field. (b) The wave is traveling in the −y-direction. At $y = 0$ and $t = 0$, the magnetic field is 1.5×10^{-11} T in the +z-direction. What are the magnitude and direction of the electric field at $y = 0$ and $t = 0$?

◆ 24. The magnetic field of an EM wave is given by $B_y = B_m \sin (kz + \omega t)$, $B_x = 0$, and $B_z = 0$. (a) In what direction is this wave traveling? (b) Write expressions for the components of the electric field of this wave.

◆ 25. The electric field of an EM wave is given by $E_z = E_m \sin (ky - \omega t + \pi/6)$, $E_x = 0$, and $E_z = 0$. (a) In what direction is this wave traveling? (b) Write expressions for the components of the magnetic field of this wave.

◆ 26. An EM wave is generated by a magnetic dipole antenna as shown in Fig. 22.35. The current in the antenna is produced by an LC resonant circuit. The wave is detected at a distant point P. Using the coordinate system in the figure, write equations for the x-, y-, and z-components of the EM fields at a distant point P. (If there is more than one possibility, just give one consistent set of answers.) Define all quantities in your equations in terms of L, C, E_m (the electric field amplitude at point P), and universal constants.

22.7 Energy Transport by EM Waves

27. The intensity of the sunlight that reaches Earth's upper atmosphere is approximately 1400 W/m². (a) What is the average energy density? (b) Find the rms values of the electric and magnetic fields.

28. The cylindrical beam of a 10.0-mW laser is 0.85 cm in diameter. What is the rms value of the electric field?

29. A certain star is 14 million light-years from Earth. The intensity of the light that reaches Earth from the star is 4×10^{-21} W/m². At what rate does the star radiate EM energy?

30. The intensity of the sunlight that reaches Earth's upper atmosphere is approximately 1400 W/m². (a) What is the total average power output of the Sun, assuming it to be an isotropic source? (b) What is the intensity of sunlight incident on Mercury, which is 5.8×10^{10} m from the Sun?

31. Prove that, in an EM wave traveling in vacuum, the electric and magnetic energy densities are equal; that is, prove that

$$\frac{1}{2}\epsilon_0 E^2 = \frac{1}{2\mu_0}B^2$$

at any point and at any instant of time.

32. Verify that the equation $I = \langle u \rangle c$ is dimensionally consistent (that is, check the units).

❸ 33. The solar panels on the roof of a house measure 2.0 m by 6.0 m. Assume they convert 35% of the incident EM wave's energy to electrical energy. (a) What average power do the panels supply when the incident intensity is 1.0 kW/m² and the panels are perpendicular to the incident light? (b) What average power do the panels supply when the incident intensity is 0.40 kW/m² and the light is incident at an angle of 60.0° from the normal? (c) Take the average daytime power requirement of a house to be about 2 kW. How do your answers to (a) and (b) compare? What are the implications for the use of solar panels?

34. The radio telescope in Arecibo, Puerto Rico, has a diameter of 305 m. It can detect radio waves from space with intensities as small as 10^{-26} W/m². (a) What is the average power incident on the telescope due to a wave at normal incidence with intensity 1.0×10^{-26} W/m²? (b) What is the average power incident on the Earth's surface? (c) What are the rms electric and magnetic fields?

22.8 Polarization

35. Unpolarized light passes through two polarizers in turn with polarization axes at 45° to each other. What is the fraction of the incident light intensity that is transmitted?

36. Light polarized in the x-direction shines through two polarizing sheets. The first sheet's transmission axis makes an angle θ with the x-axis and the transmission axis of the second is parallel to the y-axis. (a) If the incident light has intensity I_0, what is the intensity of the light transmitted through the second sheet? (b) At what angle θ is the transmitted intensity maximum?

❸ 37. A polarized beam of light has intensity I_0. We want to rotate the direction of polarization by 90.0° using polarizing sheets. (a) Explain why we must use at least two sheets. (b) What is the transmitted intensity if we use two sheets, each of which rotates the direction of polarization by 45.0°? (c) What is the transmitted intensity if we use four sheets, each of which rotates the direction of polarization by 22.5°?

❸ 38. Vertically polarized microwaves traveling into the page are directed at each of three metal plates (a, b, c) that have parallel slots cut in them (Fig. 22.36).

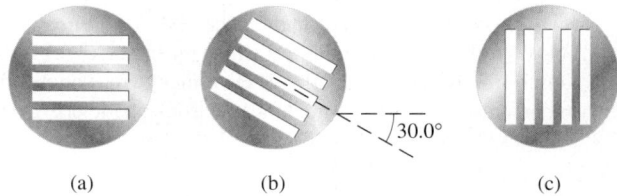

(a)   (b)   (c)

Figure 22.36 Problem 38

(a) Which plate transmits microwaves best? (b) Which plate reflects microwaves best? (c) If the intensity *transmitted through* the *best* transmitter is I_1, what is the intensity transmitted through the second-best transmitter?

39. Two sheets of polarizing material are placed with their transmission axes at right angles to each other. A third polarizing sheet is placed between them with its transmission axis at 45° to the axes of the other two. (a) If unpolarized light of intensity I_0 is incident on the system, what is the intensity of the transmitted light? (b) What is the intensity of the transmitted light when the middle sheet is removed?

22.9 The Doppler Effect for EM Waves

40. If the speeder in Example 22.9 were going *faster* than the police car, how fast would it have to go so that the reflected microwaves produce the same number of beats per second?

41. Light of wavelength 659.6 nm is emitted by a star. The wavelength of this light as measured on Earth is 661.1 nm. How fast is the star moving with respect to Earth? Is it moving toward Earth or away from it?

42. What must be the relative speed between source and receiver if the wavelength of an EM wave as measured by the receiver is twice the wavelength as measured by the source? Are source and observer moving closer together or farther apart?

43. How fast would you have to drive in order to see a red light as green? Take $\lambda = 630$ nm for red and $\lambda = 530$ nm for green.

COMPREHENSIVE PROBLEMS

44. Astronauts on the Moon communicated with mission control in Houston via EM waves. There was a noticeable time delay in the conversation due to the round-trip transit time for the EM waves between the Moon and the Earth. How long was the time delay?

45. The antenna on a cordless phone radiates microwaves at a frequency of 2.0 GHz. What is the maximum length of the antenna if it is not to exceed half of a wavelength?

46. An AM radio station broadcasts at 570 kHz. (a) What is the wavelength of the radio wave in air? (b) If a radio is tuned to this station and the inductance in the tuning circuit is 0.20 mH, what is the capacitance in the tuning circuit? (c) In the vicinity of the radio, the amplitude of the electric field is 0.80 V/m. The radio uses a coil antenna of radius 1.6 cm with 50 turns. What is the maximum emf induced in the antenna, assuming it is oriented for best reception? Assume that the fields are sinusoidal functions of time. [*Hint:* See Eq. (20-7).]

47. A 60.0-mW pulsed laser produces a pulse of EM radiation with wavelength 1060 nm (in air) that lasts for 20.0 ps (picoseconds). (a) In what part of the EM spectrum is this pulse? (b) How long (in cm) is a single pulse in air? (c) How long is it in water ($n = 1.33$)? (d) How many wavelengths fit in one pulse? (e) What is the total electromagnetic energy in one pulse?

48. The range of wavelengths allotted to the radio broadcast band is from about 190 m to 550 m. If each station needs exclusive use of a frequency band 10 kHz wide, how many stations can operate in the broadcast band?

49. Polarized light of intensity I_0 is incident on a pair of polarizing sheets. Let θ_1 and θ_2 be the angles between the direction of polarization of the incident light and the transmission axes of the first and second sheets, respectively. Show that the intensity of the transmitted light is $I = I_0 \cos^2 \theta_1 \cos^2 (\theta_1 - \theta_2)$.

50. An unpolarized beam of light (intensity I_0) is moving in the *x*-direction. The light passes through three ideal polarizers whose transmission axes are (in order) at angles 0.0°, 45.0°, and 30.0° counterclockwise from the *y*-axis in the *yz*-plane. (a) What is the intensity and polarization of the light that is transmitted by the last polarizer? (b) If the polarizer in the middle is removed, what is the intensity and polarization of the light transmitted by the last polarizer?

51. What are the three lowest angular speeds for which the wheel in Fizeau's apparatus (Fig. 22.12) allows the reflected light to pass through to the observer? Assume the distance between the notched wheel and the mirror is 8.6 km and that there are 5 notches in the wheel.

52. A sinusoidal EM wave has an electric field amplitude $E_m = 32.0$ mV/m. What are the intensity and average energy density? [*Hint:* Recall the relationship between amplitude and rms value for a quantity that varies sinusoidally.]

53. A 10.0-W laser emits a beam of light 4.0 mm in diameter. The laser is aimed at the Moon. By the time it reaches the Moon, the beam has spread out to a diameter of 85 km. Ignoring absorption by the atmosphere, what is the intensity of the light (a) just outside the laser and (b) where it hits the surface of the Moon?

ANSWERS TO PRACTICE PROBLEMS

22.1 (a) EM waves from the transmitting antenna travel outward in all directions. Since the wave travels from the transmitter to the receiver in the $+z$-direction (the direction of propagation), the direction from the receiver to the transmitter is the $-z$-direction. (b) $E_y(t) = E_m \cos(kz - \omega t)$ where $k = 2\pi/\lambda$ is the wavenumber; $E_x = E_z = 0$.

22.2 $1 \text{ ly} = 9.5 \times 10^{15} \text{ m}$

22.3 444 nm

22.4 (a) $+y$-direction; (b) $2.0 \times 10^3 \text{ m}^{-1}$; (c) $B_z(x,y,z,t) = (-1.1 \times 10^{-7} \text{ T}) \cos[(2.0 \times 10^3 \text{ m}^{-1})y - (6.0 \times 10^{11} \text{ s}^{-1})t]$, $B_x = B_y = 0$

22.5 The rms fields are proportional to $\sqrt{I}$ and I is proportional to $1/r^2$, so the rms fields are proportional to $1/r$. $E_{rms} = 6.84$ V/m; $B_{rms} = 2.28 \times 10^{-8}$ T

22.6 770 W/m^2

22.7 minimum zero (when transmission axes are perpendicular); maximum is $\frac{1}{2}I_0$ (when transmission axes are parallel)

22.8 vertically

22.9 4.6 kHz

Reflection and Refraction of Light

Alexander Graham Bell is famous today for the invention of the telephone in the 1870s. However, Bell believed his most important invention was the *Photophone*. Instead of sending electrical signals over metal wires, the Photophone sent light signals through the air, relying on focused beams of sunlight and reflections from mirrors. What prevented Bell's Photophone from becoming as commonplace as the telephone many years ago?

Making The Connection:
colors from reflection and absorption of light

Figure 23.1   The light flash of a firefly is caused by a chemical reaction between oxygen and the substance luciferin. The reaction is catalyzed by the enzyme luciferase.

23.1   WAVEFRONTS, RAYS, AND HUYGENS'S PRINCIPLE

Sources of Light

When we speak of *light*, we mean electromagnetic radiation that we can see with the unaided eye. Light is produced in many different ways. The filament of an incandescent lightbulb emits light due to its high surface temperature; at $T \approx 2000$ K, a significant fraction of the thermal radiation occurs in the visible range. The light emitted by a firefly is the result of a chemical reaction, not of a high surface temperature (Fig. 23.1). A fluorescent substance—such as the one painted on the inside of a fluorescent lightbulb—emits visible light after absorbing ultraviolet radiation.

Most objects we see are not sources of light; we see them by the light they reflect or transmit. Some fraction of the light incident on an object is absorbed, some fraction is transmitted through the object, and the rest is reflected. The nature of the material and its surface determine the relative amounts of absorption, transmission, and reflection at a given wavelength. Grass appears green because it reflects wavelengths that the brain interprets as green. Terra-cotta roof tiles reflect wavelengths that the brain interprets as red/orange (see Fig. 23.2).

Wavefronts and Rays

Since EM waves share many properties in common with all waves, we can use other waves (such as water waves) to aid visualization. A pebble dropped into a pond starts a disturbance that propagates radially outward in all directions on the surface of the water (Fig. 23.3). A **wavefront** is a set of points of equal phase. Each of the circular wave crests in Fig. 23.3 can be considered a wavefront. A water wave with straight line wavefronts can be created by repeatedly dipping a long bar into water.

A **ray** points in the direction of propagation of a wave and is perpendicular to the wavefronts. For a circular wave, the rays are radii pointing outward from the point of

Figure 23.2   Reflectivity—fraction of incident light that is reflected—as a function of wavelength for (a) grass and (b) some terra-cotta roof tiles.

Figure 23.3 Concentric circular ripples travel on the surface of a pond outward from the point where a fish broke the water surface to catch a bug.

origin of the wave (Fig. 23.4a); for a linear wave, the rays are a set of lines parallel to each other, perpendicular to the wavefronts (Fig. 23.4b).

While a surface water wave can have wavefronts that are circles or lines, a wave traveling in three dimensions, such as light, has wavefronts that are spheres, planes, or other *surfaces*. If a small source emits light equally in all directions, the wavefronts are spherical and the rays point radially outward (Fig. 23.4c). Far away from such a point source, the rays are nearly parallel to each other and the wavefronts nearly planar, so the wave can be represented as a plane wave (Fig. 23.4d). The Sun can be considered a point source when viewed from across the galaxy; even on Earth we can treat the sunlight falling upon a small lens as a collection of nearly parallel rays.

Huygens's Principle

Long before the development of electromagnetic theory, the Dutch scientist Christiaan Huygens (1629–1695) developed a geometric method for explaining the behavior of light when it travels through a medium, when it passes from one medium to another, or when it runs into a barrier from which it is reflected.

Huygens's Principle
At some time t, consider every point on a wavefront as a source of a new spherical wave. These *wavelets* move outward at the same speed as the original wave. At a later time $t + \Delta t$, each wavelet has a radius $v\, \Delta t$, where v is the speed of propagation of the wave. The wavefront at $t + \Delta t$ is a surface tangent to the wavelets. (In situations where no reflection occurs, we ignore the backward-moving wavefront.)

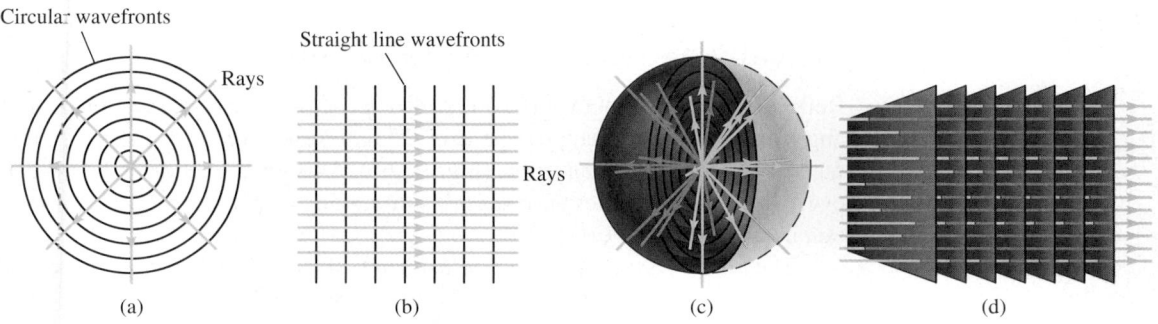

Figure 23.4 Wavefronts and rays for waves with (a) circular wavefronts, (b) straight line wavefronts, (c) spherical wavefronts, and (d) planar wavefronts.

Conceptual Example 23.1

Wavefronts from a Plane Wave

Apply Huygens's principle to a plane wave. In other words, draw the wavelets from points on a planar wavefront and use them to sketch the wavefront at a later time.

Strategy Since we are limited to a two-dimensional sketch, we draw a wavefront of a plane wave as a straight line. We choose a few points on the wavefront as sources of wavelets. Since there is no backward-moving wave, the wavelets are hemispheres; we draw them as semicircles. Then we draw a line tangent to the wavelets to represent the surface tangent to the wavefronts; this surface is the new wavefront.

Solution and Discussion In Fig. 23.5a we first draw a wavefront and four points. We imagine each point as a source of wavelets, so we draw four semicircles of equal radius, one centered on each of the four points. Finally, we draw a line tangent to the four semicircles; this line represents the wavefront at a later time.

Why draw a straight line instead of a wavy line that follows the semicircles along their edges as in Fig. 23.5b? Remember that Huygens's principle says that *every* point on the wavefront is a source of wavelets. We only draw wavelets from a few points, but we must remember that wavelets come from every point on the wavefront. Imagine drawing in more and more wavelets; the surface tangent to them would get less and less wavy, ultimately becoming a plane.

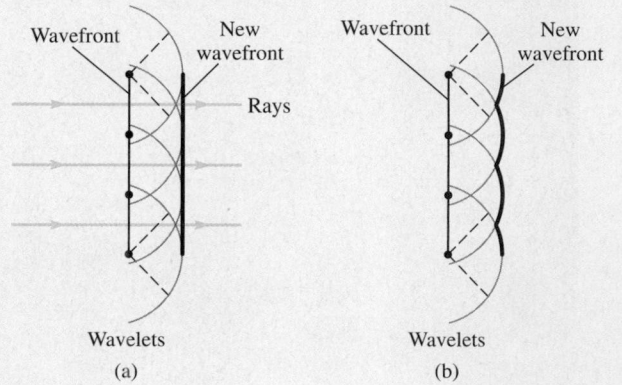

Figure 23.5
(a) Application of Huygens's principle to a plane wave. (b) This construction is not complete because it does not show wavelets coming from *every point* on the wavefront.

At the edges, the new wavefront is curved. This distortion of the wavefront at the edges is an example of diffraction. If a plane wavefront is large, then the wavefront at a later time is a plane with only a bit of curvature at the edges; for many purposes, the diffraction at the edges is negligible.

Conceptual Practice Problem 23.1   A spherical wave

Repeat Example 23.1 for the spherical light wave due to a point source.

Geometric Optics

Geometric optics is an *approximation* to the behavior of light that applies only when interference and diffraction are negligible. In order for diffraction to be negligible, the sizes of objects and apertures must be *large* compared to the wavelength of the light. In the realm of geometric optics, the propagation of light can be analyzed using rays alone. In a homogeneous material, the rays are straight lines. At a boundary between two different materials, both reflection and transmission may occur. Huygens's Principle enables us to derive the laws that determine the directions of the reflected and transmitted rays.

23.2   THE REFLECTION OF LIGHT

Specular and Diffuse Reflection

Reflection from a smooth surface is called *specular reflection*; rays incident at a given angle all reflect at the same angle (Fig. 23.6a). Reflection from a rough, irregular surface is called *diffuse reflection* (Fig. 23.6b). Diffuse reflection is more common in everyday life and enables us to see our surroundings. Specular reflection is more important in optical instruments.

The roughness of a surface is a matter of degree; what appears smooth to the unaided eye can be quite rough on the atomic scale. Thus, there is not a sharp distinction between diffuse and specular reflection. If the sizes of the pits and holes in the rough surface of Fig. 23.6b were small compared to the wavelengths of visible light, the

Incident
laser beam

Reflected rays

Smooth surface
(a)

Incident
laser beam

Reflected rays

Rough surface
(b)

Figure 23.6 (a) A laser beam reflecting from a mirror illustrates specular reflection. (b) Diffuse reflection occurs when the same laser reflects from a rough surface.

reflection would be specular. When the sizes of the pits are much larger than the wavelengths of visible light, the reflection is diffuse. A polished glass surface looks smooth to visible light, because the wavelengths of visible light are thousands of times larger than the spacing between atoms in the glass. The same surface looks rough to x-rays with wavelengths smaller than the atomic spacing. The metal mesh in the door of a microwave oven reflects microwaves well because the size of the holes is small compared to the 12-cm wavelength of the microwaves.

The Laws of Reflection

Huygens's principle illustrates how specular reflection occurs. In Fig. 23.7, plane wavefronts travel toward a polished metal surface. Every point on an incident wavefront serves as a source of secondary wavelets. Points on an incident wavefront just make the wavefront advance toward the surface. When a point on an incident wavefront contacts the metal, the wavelet propagates *away* from the surface—forming the reflected wavefront—since light cannot penetrate the metal. Wavelets emitted from these points all travel at the same speed, but they are emitted at different times. At any given instant, a wavelet's radius is proportional to the time interval since it was emitted.

Although Huygens's principle is a geometric construction, the construction is validated by modern wave theory. We now know that the reflected wave is generated by charges at the surface that oscillate in response to the incoming electromagnetic wave; the oscillating charges emit EM waves, which add up to form the reflected wave.

The laws of reflection summarize the relationship between the directions of the incident and reflected rays. The laws are formulated in terms of the angles between a ray and a *normal*—a line perpendicular to the surface where the ray touches the surface. The **angle of incidence** (θ_i) is the angle between an incident ray and the normal (Fig. 23.8); the **angle of reflection** (θ_r) is the angle between the reflected ray and the normal. In Problem 9 you can go on to prove that

$$\theta_i = \theta_r \qquad (23\text{-}1)$$

The other law of reflection says that the incident ray, the reflected ray, and the normal all lie in the same plane (the **plane of incidence**).

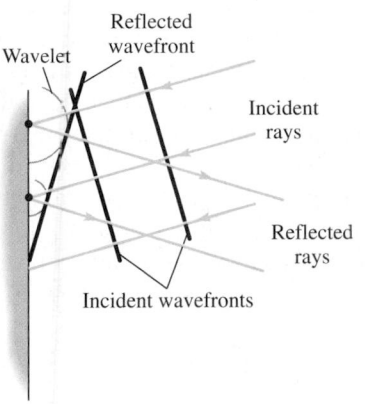

Wavelet

Reflected
wavefront

Incident
rays

Reflected
rays

Incident wavefronts

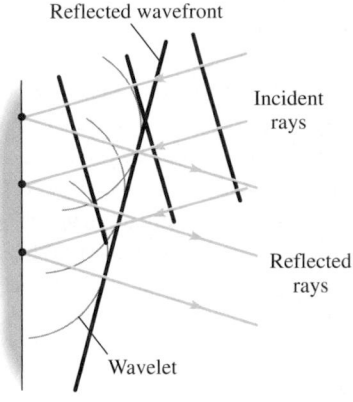

Reflected wavefront

Incident
rays

Reflected
rays

Wavelet

Figure 23.7 A plane wave strikes a metal surface. The wavelets emitted by each point on an incident wavefront when it reaches the surface form the reflected wave.

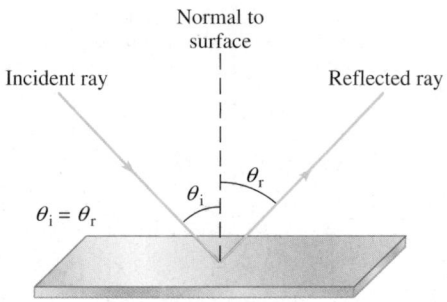

Normal to surface

Incident ray

Reflected ray

$\theta_i = \theta_r$

θ_i θ_r

Figure 23.8 The angles of incidence and of reflection are measured between the ray and the *normal* to the surface (not between the ray and the surface). The incident ray, the reflected ray, and the normal all lie in the same plane.

Laws of reflection
1. The angle of incidence equals the angle of reflection.
2. The reflected ray lies in the same plane as the incident ray and the normal to the surface at the point of incidence.

For diffuse reflection from rough surfaces, the angles of reflection for the incoming rays are still equal to their respective angles of incidence. However, the normals to the rough surface are at random angles with respect to each other, so the reflected rays travel in many directions (Fig. 23.6b).

Reflection and Transmission

So far we have considered only specular reflection from a totally reflecting surface such as polished metal. When light reaches a boundary between two *transparent* media, such as from air to glass, some of the light is reflected and some is transmitted into the new medium. The reflected light still follows the same laws of reflection (as long as the surface is smooth so that the reflection is specular). Generally, much more of the light is transmitted than is reflected. For normal incidence on an air-glass surface, only 4% of the incident intensity is reflected; 96% is transmitted.

23.3   THE REFRACTION OF LIGHT: SNELL'S LAW

In Section 22.5 we showed that when light passes from one transparent medium to another, the wavelength changes (unless the speeds of light in the two media are the same) while the frequency stays the same. In addition, Huygens's principle helps us understand why *light rays change direction* as they cross the boundary between the two media—a phenomenon known as **refraction**.

Refraction is the changing of direction of a light ray as it passes from one medium into another.

We can use Huygens's principle to understand how refraction occurs. Figure 23.9a shows a plane wave incident on a planar boundary between air and glass. In the air, a series of planar wavefronts moves toward the glass. The distance between the wavefronts is equal to one wavelength. Once the wavefront reaches the glass boundary and enters the new material, the wave slows down—light moves more slowly through glass than through air. Since the wavefront approaches the boundary at an angle to the normal, the portion of the wavefront that is still in air continues at the same merry pace while the part that has entered the glass moves more slowly. Figure 23.9b shows a Huygens's construction for a wavefront that is partly in glass. The wavelets have smaller radii in glass since the speed of light is smaller in glass than in air.

Figure 23.9c shows two right triangles that are used to relate the angle of incidence θ_1 to the angle of refraction θ_2. The two triangles share the same hypotenuse (h). Using some trigonometry, we find that

$$\sin \theta_1 = \frac{\lambda_1}{h} \quad \text{and} \quad \sin \theta_2 = \frac{\lambda_2}{h}$$

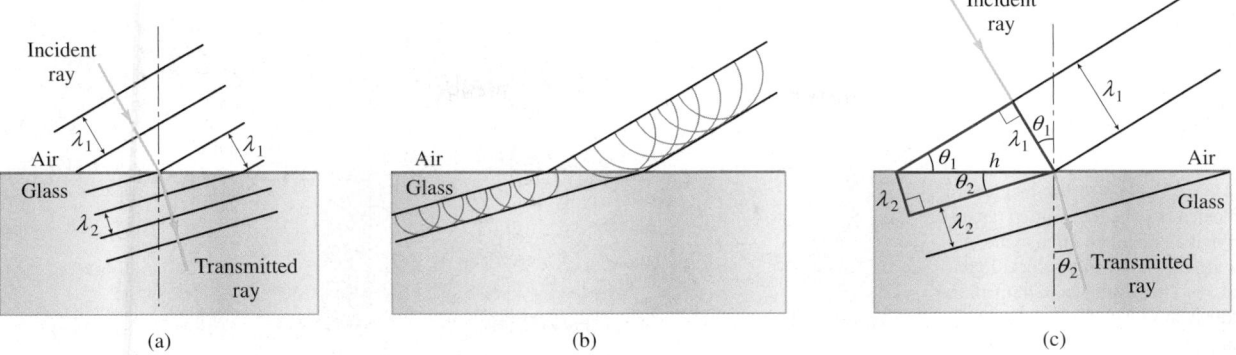

Figure 23.9 (a) Wavefronts and rays at a glass-air boundary. The reflected wavefronts are omitted. Note that the wavefronts are closer together in glass because the wavelength is smaller. (b) Huygens's construction for a wavefront partly in air and partly in glass. (c) Geometry for finding the angle of the transmitted ray.

Eliminating h yields

$$\frac{\sin \theta_1}{\sin \theta_2} = \frac{\lambda_1}{\lambda_2} \qquad (23\text{-}2)$$

It is more convenient to rewrite this relationship in terms of the indices of refraction. Recall that when light passes from one transparent medium to another, the *frequency f does not change* (see Section 22.5). Since $v = f\lambda$, λ is directly proportional to v. By definition [$n = c/v$, (Eq. 22-4)], the index of refraction n is *inversely* proportional to v. Therefore, λ is inversely proportional to n:

$$\frac{\lambda_1}{\lambda_2} = \frac{v_1/f}{v_2/f} = \frac{v_1}{v_2} = \frac{c/n_1}{c/n_2} = \frac{n_2}{n_1} \qquad (23\text{-}3)$$

By replacing λ_1/λ_2 with n_2/n_1 in Eq. (23-2) and cross multiplying, we obtain

---

Snell's law

$$n_1 \sin \theta_1 = n_2 \sin \theta_2 \qquad (23\text{-}4)$$

This law of refraction was discovered experimentally by Dutch professor Willebrord Snell (1580–1626). To determine the direction of the transmitted ray *uniquely*, two additional statements are needed:

Laws of Refraction

1. $n_1 \sin \theta_1 = n_2 \sin \theta_2$, where the angles are measured from the normal.
2. The incident ray, the transmitted ray, and the normal all lie in the same plane—the plane of incidence (see Fig. 23.10a).
3. The incident and transmitted rays are on *opposite sides* of the normal (see Fig. 23.10b).

The laws of refraction are reversible, as are the laws of reflection. Mathematically, Snell's law treats the two media as interchangeable, so the path of a light ray transmitted from one medium to another is correct if the direction of the ray is reversed.

The index of refraction of a material depends on the temperature and on the frequency of the light. Table 23.1 lists indices of refraction for several materials for yellow light with a *wavelength in vacuum* of 589.3 nm. (It is customary to specify the vacuum wavelength instead of the frequency.) In many circumstances the slight variation of n over the visible range of wavelengths can be ignored.

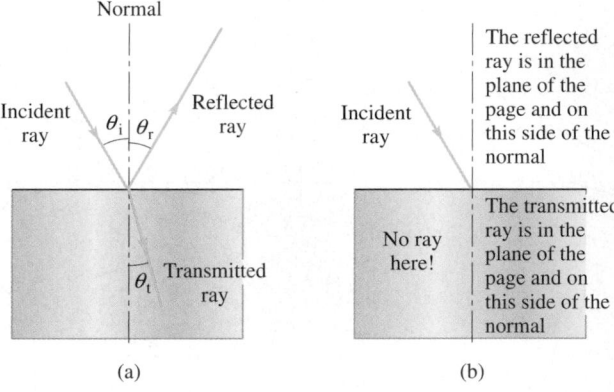

Figure 23.10 (a) The incident ray, the reflected ray, the transmitted ray, and the normal all lie in the same plane. All angles are measured with respect to the normal. (b) The incident and transmitted rays are always on opposite sides of the normal, no matter what the indices of refraction of the two media.

Table 23.1

Indices of Refraction for $\lambda = 589.3$ nm in Vacuum (at 20°C unless otherwise noted)

Material	Index	Material	Index
Solids		**Liquids**	
Ice (at 0°C)	1.309	Water	1.333
Fluorite	1.434	Acetone	1.36
Fused quartz	1.458	Ethyl alcohol	1.361
Polystyrene	1.49	Carbon tetrachloride	1.461
Lucite	1.5	Glycerine	1.473
Plexiglas	1.51	Sugar solution (80%)	1.49
Crown glass	1.517	Benzene	1.501
Plate glass	1.523	Carbon disulfide	1.628
Sodium chloride	1.544	Methylene iodide	1.74
Light flint glass	1.58		
Dense flint glass	1.655	**Gases at 0°C, 1 atm**	
Sapphire	1.77		
Zircon	1.923	Helium	1.000 036
Diamond	2.419	Ethyl ether	1.000 152
Titanium dioxide	2.9	Water vapor	1.000 250
Gallium phosphide	3.5	Dry air	1.000 293
		Carbon dioxide	1.000 449

Physics at Home

Fill a clear drinking glass with water and then put a pencil in the glass. Look at the pencil from many different angles. Why does the pencil look as if it is bent?

Example 23.2

Ray Traveling through a Window Pane

A beam of light strikes one face of a window pane with an angle of incidence of 30.0°. The index of refraction of the glass is 1.52. The beam travels through the glass and emerges from a parallel face on the opposite side. Ignore reflections.

(a) Find the angle of refraction for the ray inside the glass.
(b) Show that the rays in air on either side of the glass (the incident and emerging rays) are parallel to each other.

continued on next page

Example 23.2 *continued*

Strategy First we draw a ray diagram. We are only concerned with the rays transmitted at each boundary, so we omit reflected rays from the diagram. At each boundary we draw a normal, label the angles of incidence and refraction, and apply Snell's law. When the ray passes from air ($n = 1$) to glass ($n = 1.52$), it bends *closer to* the normal: since $n_1 \sin \theta_1 = n_2 \sin \theta_2$, a larger n means a smaller θ. Likewise, when the ray passes from glass to air, it bends *away from* the normal.

Solution (a) Figure 23.11 is a ray diagram. At the first air-glass boundary, Snell's law yields

$$n_1 \sin \theta_1 = n_2 \sin \theta_2$$

$$\sin \theta_2 = \frac{n_1}{n_2} \sin \theta_1 = \frac{1.00}{1.52} \sin 30.0° = 0.3289$$

The angle of refraction is

$$\theta_2 = \sin^{-1} 0.3289 = 19.2°$$

(b) At the second boundary, from glass to air, we apply Snell's law again. Since the surfaces are parallel, the two normals are parallel. The angle of refraction at the first boundary and the angle of incidence at the second are alternate interior angles, so the angle of incidence at the second boundary must be θ_2.

$$n_2 \sin \theta_2 = n_3 \sin \theta_3$$

We do not need to solve for θ_3 numerically. From the first boundary we know that $n_1 \sin \theta_1 = n_2 \sin \theta_2$; therefore, $n_1 \sin \theta_1 = n_3 \sin \theta_3$. Since $n_1 = n_3$, $\theta_3 = \theta_1$. The two rays—emerging and incident—are parallel to each other.

Discussion Note that the emerging ray is parallel to the incident ray, but it is *displaced* (see the dashed line in Fig. 23.11). If the two glass surfaces were not parallel, then the two normals would not be parallel. Then the angle of incidence at the second boundary would not be equal to the angle of refraction at the first; the emerging ray would *not* be parallel to the incident ray.

Practice Problem 23.2 Fish eye view

A fish hovers beneath the still surface of a pond. If the Sun is 33° above the horizon, at what angle above the horizontal does the fish see the Sun? [*Hint*: Draw a diagram that includes the normal to the surface; be careful to correctly identify the angles of incidence and refraction.]

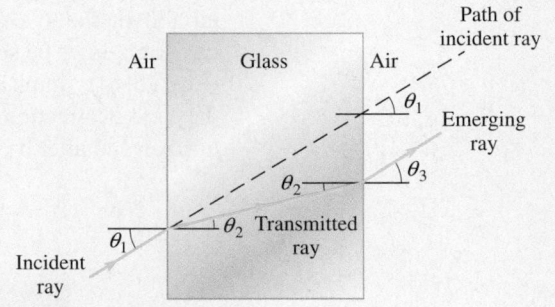

Figure 23.11
A ray of light travels through a window pane.

Physics at Home

Place a coin at the far edge of the bottom of an empty mug. Sit in a position so that you are just unable to see the coin. Then, without moving your head, utter the magic word *REFRACTION* as you pour water carefully into the mug on the near side; pour slowly so that the coin does not move. The coin becomes visible when the mug is filled with water (Fig. 23.12). Why?

(a)

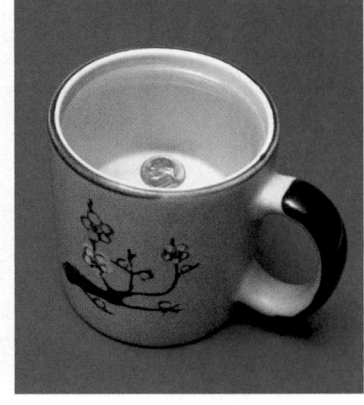

(b)

Figure 23.12 (a) The coin at the bottom of the mug is not visible. (b) After the mug is filled with water, the coin is visible.

Mirages

Refraction of light in the air causes the *mirages* seen in the desert or on a hot road in summer (Fig. 23.13a). The hot ground warms the air near it, so light rays from the sky travel through warmer and warmer air as they approach the ground. Since the speed of light in air increases with increasing temperature, light travels faster in the hot air near the ground than in the cooler air above. The temperature change is gradual, so there is no *abrupt* change in the index of refraction; instead of being bent abruptly, rays gradually curve upward (Fig. 23.13b).

    The wavelets from points on a wavefront travel at different speeds; the radius of a wavelet closer to the ground is larger than that of a wavelet higher up (Fig. 23.13c). The brain interprets the rays coming upward into the eye as coming from the ground even though they really come from the sky. What may look like a body of water on the ground is actually an image of the blue sky overhead.

Superior Mirage

A superior mirage occurs when the layer of air near Earth's surface is *colder* than the air above, due to a snowy field or to the ocean. A ship located just *beyond* the horizon can sometimes be seen because light rays from the ship are gradually bent downward (Fig. 23.14). Ships and lighthouses seem to float in the sky or appear much taller than they are. Refraction also allows the Sun to be seen before it actually rises above the horizon and after it is already below the horizon at sunrise and sunset.

(a)

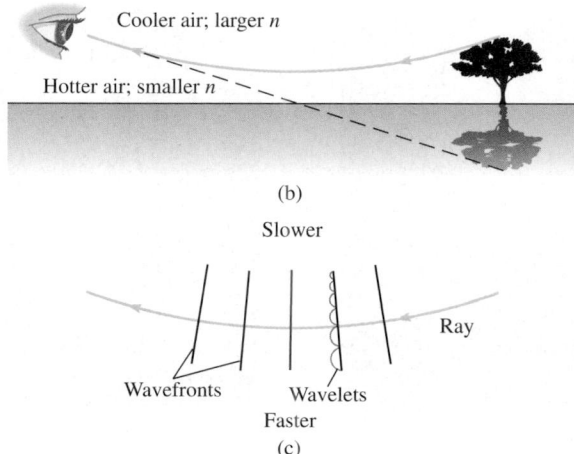

(b)

(c)

Figure 23.13 (a) Mirage seen in the desert in Namibia. (b) A ray from the Sun bends upward into the eye of the observer. (c) The bottom of the wavefront moves faster than the top.

(a)

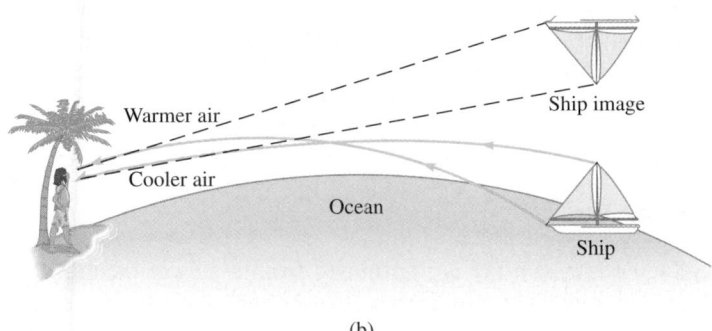

(b)

Figure 23.14 (a) Superior mirage of a house seen in Finland's southwestern archipelago; (b) a sketch of the light rays that form a superior mirage of a ship.

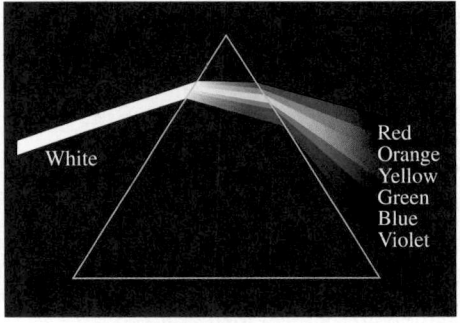

Figure 23.15 Dispersion of white light by a prism. (See also the photo in Fig. 22.15.)

Dispersion

When natural white light enters a triangular prism, the light emerging from the far side of the prism is separated into a continuous spectrum of colors from red to violet (Fig. 23.15). The separation occurs because the prism is dispersive—that is, the speed of light in the prism depends on the frequency of the light.

Natural white light is a mixture of light at all the frequencies in the visible range. At the front surface of the prism, each frequency refracts at an angle determined by the index of refraction of the prism at that frequency. The index of refraction increases with increasing frequency, so it is smallest for red and increases gradually until it is largest for violet. As a result, violet bends the most and red the least. Refraction occurs again as light leaves the prism. The geometry of the prism is such that the different colors are spread apart even more at the back surface.

Rainbows

Rainbows are formed by the dispersion of light in water. A ray of sunlight that enters a raindrop is separated into the colors of the spectrum. At each air-water boundary there may be both reflection and refraction. The rays that contribute to a *primary rainbow*—the brightest and often the only one seen—pass into the raindrop, reflect off the back of the raindrop, and then are transmitted back into the air (Fig. 23.16a). Refraction occurs both where the ray enters the drop (air-water) and again when it leaves (water-air), just as for a prism. Since the index of refraction varies with frequency, sunlight is separated into the spectral colors. For relatively large water droplets, as occur in a gentle summer shower, the rays emerge with an angular separation between red and violet of about 2° (Fig. 23.16b).

A person looking into the sky away from the Sun sees red light coming from raindrops higher in the sky and violet light coming from lower droplets (Fig. 23.16c). The rainbow is a circular arc that subtends an angle of 42° for red and 40° for violet, with the other colors falling in between.

In good conditions, a double rainbow can be seen. The secondary rainbow has a larger radius, is less intense, and has its colors reversed (Fig. 23.16d). It arises from rays that undergo *two reflections* inside the raindrop before emerging. The angles subtended by a secondary rainbow are 50.5° for red and 54° for violet.

23.4   TOTAL INTERNAL REFLECTION

According to Snell's law, if a ray is transmitted from a slower medium into a faster medium (from a higher index of refraction to a lower one), the refracted ray bends *away* from the normal (Fig. 23.17, ray *b*). That is, the angle of refraction is greater than the angle of incidence. As the angle of incidence is increased, the angle of refraction eventually reaches 90° (Fig. 23.17, ray *c*). At 90°, the refracted ray is parallel to the surface. It isn't transmitted into the faster medium; it just moves along the surface. The angle of incidence for which the angle of refraction is 90° is called the **critical angle** θ_c for the boundary between the two media. From Snell's law,

$$n_i \sin \theta_c = n_t \sin 90°$$

critical angle: $\theta_c = \sin^{-1} \dfrac{n_t}{n_i}$	(23-5a)

where the subscripts "i" and "t" refer to the media in which the incident and transmitted rays travel (sometimes called the *internal* and *external* media, respectively). Since we are discussing an incident ray in a slower medium, $n_i > n_t$.

> The critical angle is the minimum angle of incidence for which no light is transmitted past the boundary.

For an angle of incidence greater than θ_c, the refracted ray can't bend away from the normal *more* than 90°; to do so would be reflection rather than refraction, and a different law governs the angle of reflection. Mathematically, there is no angle whose sine is greater than 1 (= sin 90°), so it is impossible to satisfy Snell's law if $n_i \sin \theta_i > n_t$ (which is equivalent to saying $\theta_i > \theta_c$). If the angle of incidence is greater than θ_c, there cannot be a transmitted ray; if there is no ray transmitted into the faster medium, all the light must be reflected from the boundary (Fig. 23.17, ray *d*). This is called **total internal reflection**.

no transmitted ray for $\theta_i \geq \theta_c$	(23-5b)

Total reflection cannot occur when a ray in a faster medium hits a boundary with a slower medium. In that case the refracted ray bends *toward* the normal, so the angle of refraction is always less than the angle of incidence. Even at the largest possible angle of incidence, 90°, the angle of refraction is less than 90°. Total internal reflection can only occur when the incident ray is in the slower medium.

Figure 23.16 (a) Reflection and refraction of light rays falling on the upper half of a raindrop. While the incident rays are parallel, the emerging rays are not. The pair of rays along the bottom edge shows where the emerging light has the highest intensity. (b) Dispersion of light in a raindrop. At each boundary, both reflection and transmission occur; reflected or transmitted rays that do not contribute to the primary rainbow are omitted. (c) Light from many different raindrops contributes to the appearance of a rainbow. Angles are exaggerated for clarity. (d) Light rays that reflect twice inside the rain-drop form the secondary rainbow. Note that the order of the colors is reversed: now violet is highest and red is lowest.

Figure 23.17 Partial reflection and total internal reflection at upper surface of a rectangular glass prism. The intensities of the reflected and transmitted light are given as percentages of the incident intensity. The angles of incidence of rays a and b are less than the critical angle, ray c is incident at the critical angle θ_c, and ray d is incident at an angle greater than θ_c.

Example 23.3

Total Internal Reflection in a Triangular Glass Prism

A beam of light is incident on the triangular glass prism in air. What is the largest angle of incidence θ_i below the normal (as shown in Fig. 23.18a) so that the beam undergoes total reflection from the back of the prism (the hypotenuse)? The prism has an index of refraction $n = 1.50$.

Strategy In this problem it is easiest to work backward. Total internal reflection occurs if the angle of incidence at the back of the prism is greater than or equal to the critical angle. We start by finding the critical angle and then work backward using geometry and Snell's law to find the corresponding angle of incidence at the front of the prism.

Solution To find the critical angle from Snell's law, we set the angle of refraction equal to 90°.

$$n_i \sin \theta_c = n_a \sin 90°$$

The incident ray is in the internal medium (glass). Therefore, $n_i = 1.50$ and $n_a = 1.00$.

$$\sin \theta_c = \frac{n_a}{n_i} \sin 90° = \frac{1.00}{1.50} \times 1.00 = 0.667$$

$$\theta_c = \sin^{-1} 0.667 = 41.8°$$

In Fig. 23.18b, we draw an enlarged ray diagram and label the angle of incidence at the back of the prism as θ_c. The angles of incidence and refraction at the front are labeled θ_i and θ_t; they are related through Snell's law:

$$1.00 \sin \theta_i = 1.50 \sin \theta_t$$

What remains is to find the relationship between θ_t and θ_c. By drawing a line at the second boundary that is parallel to the normal at the first boundary, we can use alternate interior angles to label θ_t (see Fig. 23.18b). The angle between the two normals is 45.0°, so

$$\theta_t = 45.0° - \theta_c = 45.0° - 41.8° = 3.2°$$

Then

$$\sin \theta_i = 1.50 \sin \theta_t = 1.50 \times 0.05582 = 0.0837$$

$$\theta_i = \sin^{-1} 0.0837 = 4.8°$$

Discussion For a beam incident below the normal at angles from 0 to 4.8°, total internal reflection occurs at the back. If a beam is incident at an angle greater than 4.8°, then the angle of incidence at the back is less than the critical angle, so transmission into the air occurs there. Conceptual Practice Problem 23.3 considers what happens to a beam incident above the normal.

If we had mixed up the two indices of refraction, we would have wound up trying to take the inverse sine of 1.5. That would be a clue that we made a mistake.

Conceptual Practice Problem 23.3 Ray incident from above the normal

Draw a ray diagram for a beam of light incident from *above* the normal (Fig. 23.18). Show that at *any* angle of incidence, the beam undergoes total internal reflection at the back of the prism.

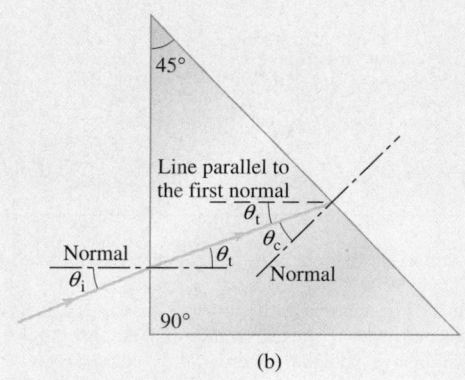

Figure 23.18
Example 23.3

(a)   (b)

Total Internal Reflection in Prisms

Optical instruments such as periscopes, single-lens reflex (SLR) cameras, binoculars, and telescopes often use prisms to reflect a beam of light. Fig. 23.19a shows a simple periscope. Light is reflected through a 90° angle by each of two prisms; the net result is a displacement of the beam. A similar scheme is used in binoculars (Fig. 23.19b). In an SLR camera, one of the prisms is replaced by a movable mirror. When the mirror is in place, the light through the camera lens is diverted up to the viewfinder, so you can see exactly what will appear on film. Depressing the shutter moves the mirror out of the way so the light falls onto the film instead. In binoculars and telescopes, *erecting prisms* are often used to turn an upside-down image right-side-up.

An advantage of using prisms instead of mirrors in these applications is that 100% of the light is reflected. A typical mirror reflects only about 90%—remember that the oscillating electrons that produce the reflected wave are moving in a metal with some electrical resistance, so energy loss occurs.

The brilliant sparkle of a diamond is due to total internal reflection. The cuts are made so that most of the light incident on the front faces is totally reflected several times inside the diamond and then re-emerges toward the viewer. A poorly cut diamond allows too much light to emerge away from the viewer (see Fig. 23.20).

Fiber Optics

Total internal reflection is the principle behind fiber optics, a technology that has revolutionized both communications and medicine. At the center of an optical fiber is a transparent cylindrical core made of glass or plastic with a relatively high index of refraction (Fig. 23.21). The core may be as thin as a few μm in diameter—quite a bit thinner than a human hair. Surrounding the core is a coating called the cladding, which is also transparent but has a lower index of refraction than the core. The "mismatch" in the indices of refraction is maximized so that the critical angle at the core-cladding boundary is as small as possible.

Light signals travel nearly parallel to the axis of the core. It is impossible to have light rays enter the fiber *perfectly* parallel to the axis of the fiber, so the rays eventually hit the cladding *at a large angle of incidence*. As long as the angle of incidence is greater than the critical angle, the ray is totally reflected back into the core; no light leaks out into the cladding. A ray may typically reflect from the cladding thousands of times per meter of fiber, but since the ray is totally reflected each time, the signal can travel long distances—kilometers in some cases—before any appreciable signal loss occurs.

The fibers are flexible so they can be bent as necessary. The smaller the critical angle, the more tightly a fiber can be bent. If the fiber is kinked (bent too tightly), rays strike the boundary at less than the critical angle, resulting in dramatic signal loss as light passes into the cladding.

Making The Connection:
periscope

Making The Connection:
**sparkle of a
diamond**

Making The Connection:
fiber optics

(a)

(b)

Figure 23.19 (a) A periscope uses two reflecting prisms to shift the beam of light. (b) In binoculars, the light undergoes total internal reflection twice in each prism.

Figure 23.20 (a) This ray undergoes total internal reflection twice before re-emerging from a front face of the diamond. (b) Due to a poor cut, a similar ray in this diamond would be incident on one of the back faces at less than the critical angle. The ray is mostly transmitted out the back of the diamond.

(a)   (b)

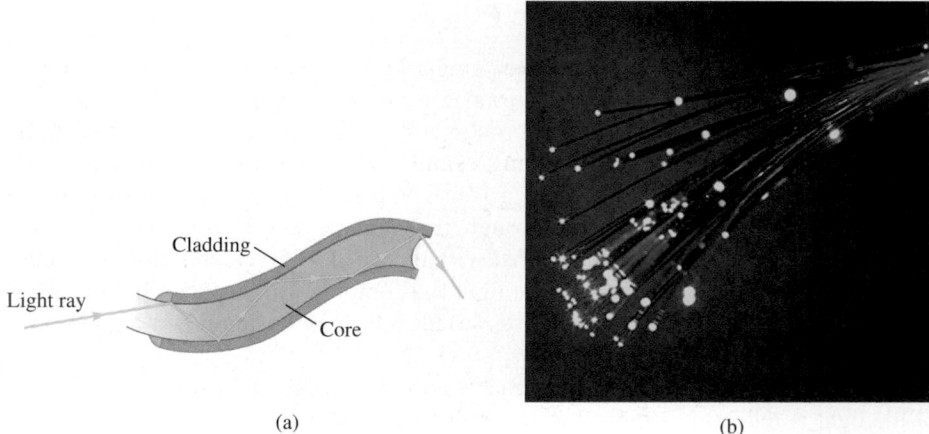

Cladding

Light ray

Core

Figure 23.21 (a) An optical fiber
(b) A bundle of optical fibers

(a)

(b)

Optical fiber is far superior to copper wire in its capacity to carry information. A single optical fiber can carry tens of thousands of phone conversations, while a pair of copper wires can only carry about 20 at the most. Electrical signals in copper wires also lose strength much more rapidly (due in part to the electrical resistance of the wires) and are susceptible to electrical interference. Over 80% of the long-distance phone calls in the world are carried by fiber optics; computer networks and video increasingly use fiber optics as well.

In medicine, bundles of optical fibers are at the heart of the endoscope (Fig. 23.22), which is fed through the nose, mouth, or rectum, or through a small incision, into the body. One bundle of fibers carries light into a body cavity or an organ and illuminates it; another bundle transmits an image back to the doctor for viewing.

Making The Connection:
endoscope

Eyepiece

Light, water,
and suction cable

Up/down control

Left/right control

Suction control

Air/water control

Air pipe
Control wires
Light guide

Image guide

Instrument channel
Light guide
Water pipe

Endoscope tube

(a)

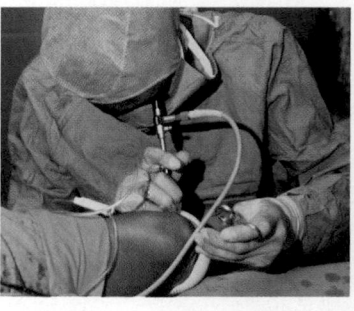

(b)

Figure 23.22 (a) An endoscope. (b) Arthroscopic knee surgery. An arthroscope is similar to an endoscope, but is used in the diagnosis and treatment of injuries to the joints.

The endoscope is not limited to diagnosis; it can be fitted with instruments enabling a physician to take tissue samples, perform surgery, cauterize blood vessels, or suction out debris. Surgery performed using an endoscope uses much smaller incisions than traditional surgery; as a result, recovery is much faster. A gallbladder operation that used to require an extended hospital stay can now be done on an outpatient basis in many cases.

Bell's Photophone

Almost a century before the invention of fiber optics, Bell's Photophone used light to carry a telephone signal. The Photophone projected the voice toward a mirror, which vibrated in response (Fig. 23.23). A focused beam of sunlight reflecting from the mirror captured the vibrations. Other mirrors were used to reflect the signal as necessary until it was transformed back into sound at the receiving end. The light traveled in straight line paths through air between the mirrors.

Bell's Photophone worked only intermittently. Many things could interfere with a transmission, including cloudy weather. With nothing to keep the beam from spreading out, it worked only over short distances. Not until the invention of fiber optics in the 1970s could light signals travel reliably over long distances without significant loss or interference.

23.5   BREWSTER'S ANGLE

In Section 22.8 we mentioned that light is partially or totally polarized by reflection. Using Snell's law, we can find the angle of incidence—called **Brewster's angle** θ_B—for which the reflected light is totally polarized. The reflected light is totally polarized *when the reflected and transmitted rays are perpendicular to each other* (Fig. 23.24). These rays are perpendicular if $\theta_B + \theta_t = 90°$. Since the two angles are complementary, $\sin \theta_t = \cos \theta_B$. Then

Brewster's angle is the angle of incidence for which the reflected light is totally polarized.

$$n_i \sin \theta_B = n_t \sin \theta_t = n_t \cos \theta_B$$

Brewster's angle: $\quad \theta_B = \tan^{-1} \dfrac{n_t}{n_i}$	(23-6)

The value of Brewster's angle depends on the indices of refraction of the two media. Unlike the critical angle for total internal reflection, Brewster's angle exists regardless of which index of refraction is larger.

Why is the reflected light totally polarized when the reflected and transmitted rays are perpendicular? In Fig. 23.24, the unpolarized incident light is represented as a mixture of two perpendicular polarization components: one perpendicular to the plane of incidence and one in the plane of incidence. Note that the polarization components in the plane of incidence, represented by red and blue arrows, are not in the same direction; polarization components must be perpendicular to the ray since light is a transverse wave.

Figure 23.23 One of Bell's sketches of the Photophone

Figure 23.24 When light is incident at Brewster's angle, the reflected and transmitted rays are perpendicular. As a result, the reflected light is totally polarized perpendicular to the plane of the page. (The three polarization directions are shown in different colors merely to help distinguish them; these colors have nothing to do with the color of the light.)

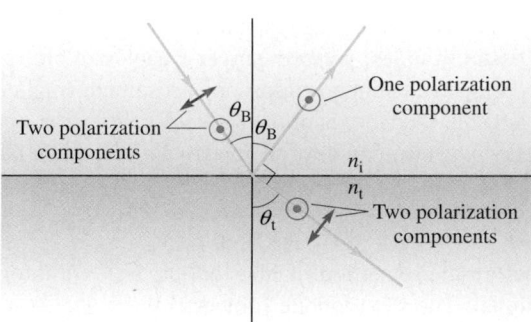

The same oscillating charges at the surface of the second medium radiate both the reflected light and the transmitted light. Oscillations in the plane of incidence are along the blue direction. This direction of oscillation contributes nothing to the reflected ray since an oscillating charge does not radiate along its axis of oscillation. Thus, when the light is incident at Brewster's angle, *the reflected light is totally polarized perpendicular to the plane of incidence*. At other angles of incidence, the reflected light is *partially* polarized perpendicular to the plane of incidence. If light is incident at Brewster's angle and is polarized in the plane of incidence (that is, it has no polarization component perpendicular to the plane of incidence), no light is reflected.

23.6 THE FORMATION OF IMAGES THROUGH REFLECTION OR REFRACTION

When you look into a mirror, you see an image of yourself. What do we mean by an *image*? It *appears* as if your twin were standing behind the mirror. If you were looking at an actual twin, each point on your twin would reflect light in many different directions. Some of that light enters your eye. In essence, what your eye does is take the rays diverging from a given point and trace them backward to figure out where they come from. Your eye interprets light reflected from the mirror in the same way: all the light rays from any point on you (the object whose image is being formed) must reflect from the mirror *as if they came from a single point behind the mirror*.

Ideally, in the formation of an image, there is a one-to-one correspondence of points on the object and points on the image. If rays from one point on the object seem to come from many different points, the overlap of light from different points would look blurry. (A real lens or mirror may deviate somewhat from ideal behavior, causing some degree of blurring in the image.)

There are two kinds of images. For the plane mirror, the light rays *seem* to come from a point behind the mirror, but we know there aren't actually any light rays back there. In a **virtual image**, we trace light rays from a point on the object back to a point from which they *appear* to diverge, even though the rays do not actually come from that point. In a **real image**, the rays actually *do* pass through the image point. A camera lens forms a real image of the object being photographed on the film. The light rays have to actually be there to expose the film! The rays from a point on the object must all reach the same point on the film or else the picture will come out blurry. If the film and the back of the camera were not there to interrupt the light rays, they would diverge from the image point (Fig. 23.25). An image must be real if it is projected onto a surface such as film, a viewing screen, or a detector.

 Projecting a real image onto a screen is only one way to view it. Real images can also be viewed directly (as virtual images are viewed) by looking into the lens or mirror. However, to view a real image, the viewer must be located *beyond the image* so that the rays from a point on the object all diverge from a point on the image. In Fig. 23.25, if the film is removed, the image can be viewed by looking into the lens from points beyond the image (that is, to the right of where the film is placed).

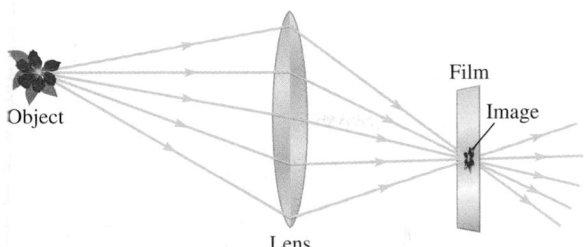

Figure 23.25 Formation of a real image by a camera lens. If the film and the back of the camera were not there, the rays would continue on, diverging from the image point.

Finding an Image Using a Ray Diagram

- Draw two (or more) rays coming from a single point on the object toward whatever forms the image (usually a lens or mirror). (Only two rays are necessary since they *all* map to the same image point.)
- Trace the rays, applying the laws of reflection and refraction as needed, until they reach the observer.
- Extrapolate the rays backward along straight line paths until they intersect at the image point.
- If light rays actually go through the image point, the image is real. If they do not, the image is virtual.
- To find the image of an extended object, find the images of two or more points on the object.

Example 23.4

A Kingfisher Looking for Prey

A small fish is at a depth d below the surface of a still pond. What is the *apparent* depth of the fish as viewed by a belted kingfisher—a bird that dives underwater to catch fish? Assume the kingfisher is directly above the fish. Use $n = \frac{4}{3}$ for water.

Strategy   The apparent depth is the depth of the *image* of the fish. Light rays coming from the fish toward the surface are refracted as they pass into the air. We choose a point on the fish and trace the rays from that point into the air; then we trace the refracted rays backward along straight lines until they meet at the image point. The kingfisher directly above sees not only a ray coming straight up ($\theta_i = 0$); it also sees rays at small but nonzero angles of incidence. We may be able to use small-angle approximations for these angles. However, for clarity in the ray diagram, we exaggerate the angles of incidence.

Solution   In Fig. 23.26a we sketch a fish under water at a depth d. From a point on the fish, rays diverge toward the surface. At the surface they are bent away from the normal (since air has a lower index of refraction). The image point is found by tracing the refracted rays straight backward (dashed lines) to where they meet. We label the image depth d'. From the ray diagram, we see that $d' < d$; the apparent depth is less than the actual depth.

Only two rays need be used to locate the image. To simplify the math, one of them can be the ray normal to the surface. The other ray is incident on the water surface at angle θ_i. This ray leaves the water surface at angle θ_t, where

$$n_w \sin \theta_i = n_a \sin \theta_t \qquad (1)$$

Since the angles are small, we set $\sin \theta = \theta$:

$$n_w \theta_i = n_a \theta_t \quad \text{or} \quad \frac{\theta_i}{\theta_t} = \frac{n_a}{n_w}$$

To find d', we use two right triangles (Fig. 23.26b) that share the same side s—the distance between the points at which the two chosen rays intersect the water surface. The angles θ_i and θ_t are known since they are alternate interior angles with the angles at the surface. From these triangles,

$$\tan \theta_i = \frac{s}{d} \quad \text{and} \quad \tan \theta_t = \frac{s}{d'}$$

For small angles, we can set $\tan \theta \approx \sin \theta$. Then Eq. (1) becomes

$$n_w \frac{s}{d} = n_a \frac{s}{d'}$$

After eliminating s, we solve for the ratio d'/d:

$$\frac{\text{apparent depth}}{\text{actual depth}} = \frac{d'}{d} = \frac{n_a}{n_w} = \frac{3}{4}$$

The apparent depth of the fish is $\frac{3}{4}$ of the actual depth.

Discussion   The result is valid only for small angles of incidence—that is, for a viewer directly above the fish. The apparent depth depends on the angle at which the fish is viewed.

The image of the fish is virtual. The light rays seen by the kingfisher *seem* to come from the location of the image, but they do not.

Practice Problem 23.4   Evading the predator

Suppose the fish looks upward and sees the kingfisher. If the kingfisher is a height h above the surface of the pond, what is its apparent height h' as viewed by the fish?

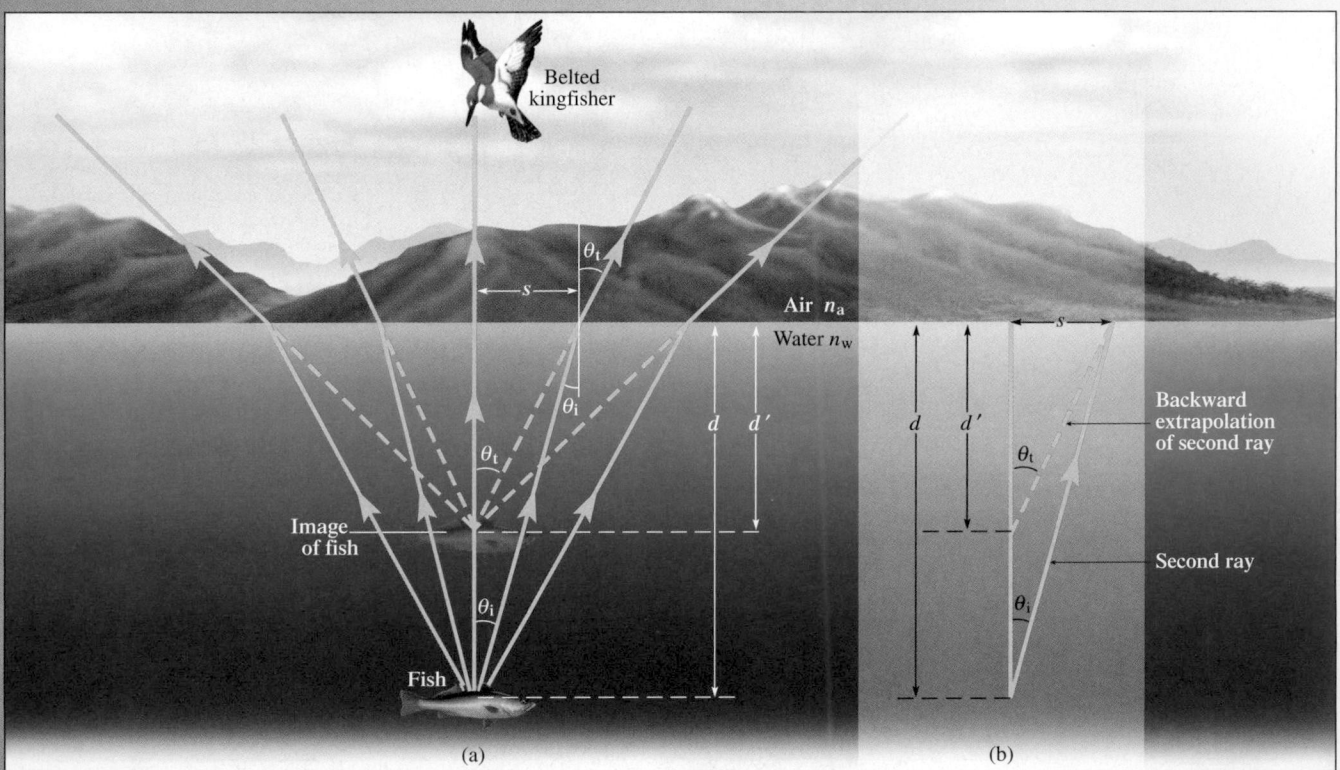

Figure 23.26 (a) Formation of the image of the fish. (b) Two triangles that share side *s* enable us to solve for the image depth *d′* in terms of *d*.

23.7 PLANE MIRRORS

A shiny metal surface is a good reflector of light. An ordinary mirror is *back-silvered*; that is, a thin layer of shiny metal is applied to the *back* of a flat piece of glass. A back-silvered mirror actually produces two reflections: a faint one, seldom even noticed, from the front surface of the glass and a strong one from the metal. *Front-silvered* mirrors are used in precision work, since they produce only one reflection; they are not practical for everyday use because the metal coating is easily scratched. If we ignore the faint reflection from the glass, then back-silvered mirrors are treated the same as front-silvered mirrors.

Light reflected from a mirror follows the laws of reflection discussed in Section 23.2. Figure 23.27 shows a point source of light located in front of a plane mirror; an observer looks into the mirror. If the reflected rays are extrapolated backward through the mirror, they all intersect at one point, which is the image of the point source. Using any two rays and some geometry, you can show (Problem 39) that

> For a plane mirror, a point source and its image are at the same distance from the mirror (on opposite sides); both lie on the same normal line.

The rays only *appear* to originate at the image behind the mirror; no rays travel through the mirror.

We treat an extended object in front of a plane mirror as a set of point sources (the points on the surface of the object). In Fig. 23.28, a pencil is in front of a mirror. To sketch the image, we first construct normals to the mirror from several points on the pencil. Then each image point is placed a distance behind the mirror equal to the distance from the mirror to the corresponding point on the object.

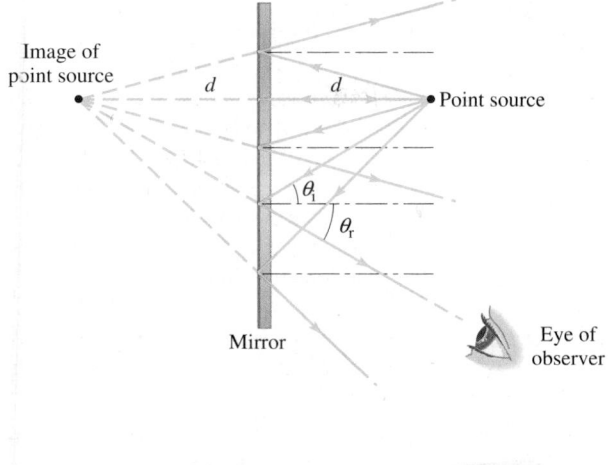

Figure 23.27   A plane mirror forms an image of a point source. The source and image are equidistant from the mirror and lie on the same normal line.

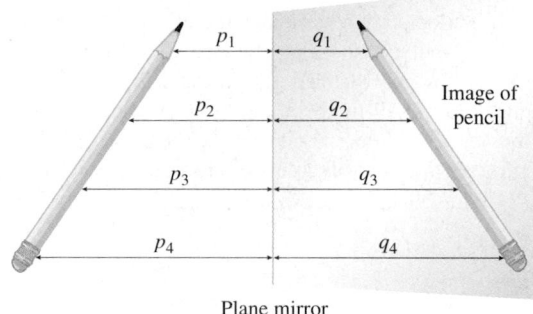

Figure 23.28   Sketching the image of a pencil in front of a plane mirror

Conceptual Example 23.5

Mirror Length for a Full-Length Image

 Grant is carrying his niece Dana on his shoulders (Fig. 23.29). What is the minimum vertical length of a plane mirror in which Grant can see a full image (from his toes to the top of Dana's head)? How should this minimum-length mirror be placed on the wall?

 Strategy   Ray diagrams are *essential* in geometric optics. A ray diagram is most helpful if we carefully decide which rays are most important to the solution. Here, we want to make sure Grant can see the images of two particular points: his toes and the top of Dana's head. If he can see those two points, he can see everything between them. In order for Grant to see the image of a point, a ray of light from that point must reflect from the mirror and enter Grant's eye.

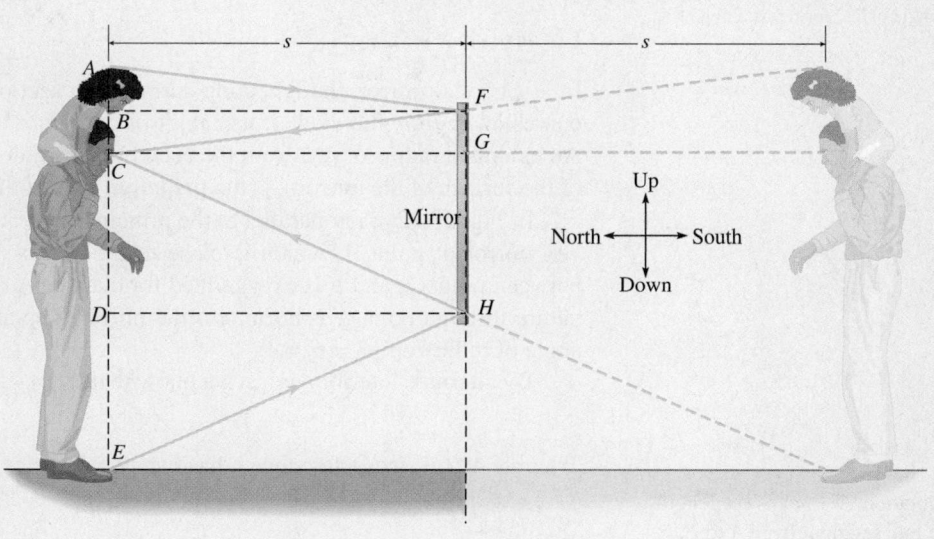

Figure 23.29
Conceptual Example 23.5

continued on next page

Conceptual Example 23.5 *continued*

Solution and Discussion After drawing Grant, Dana, and the mirror (Fig. 23.29), we want to draw a ray from Grant's toes that strikes the mirror and is reflected to his eye. The line *DH* is a normal to the mirror surface. Since the angle of incidence is equal to the angle of reflection, the triangles *CHD* and *EHD* are congruent and *CD = DE = GH*. Therefore,

$$GH = \tfrac{1}{2}CE$$

Similarly, we draw a ray from the top of Dana's head to the mirror that is reflected into Grant's eye and find that

$$FG = \tfrac{1}{2}AC$$

The length of the mirror is

$$FH = FG + GH = \tfrac{1}{2}(AC + CE) = \tfrac{1}{2}AE$$

Therefore, the length of the mirror must be *one-half* the distance from Grant's toes to Dana's head.

The minimum-length mirror only allows a full-length view if it is hung properly. The top of the mirror (*F*) must be a distance *AB* below the top of Dana's head. A full-length mirror is *not* necessary to get a full-length view. Extending the mirror all the way to the floor is of no use; the bottom of the mirror only needs to be halfway between the floor and the eyes of the shortest person who uses the mirror. Note that the distance *s* between Grant and the mirror has no effect on the result. That is, you need the same height mirror whether you're up close to it or farther back.

Practice Problem 23.5   Two sisters with one mirror

Sarah's eyes are 1.72 m above the floor when she is wearing her dress shoes, and the top of her head is 1.84 m above the floor. Sarah has a mirror that is 0.92 m in length, hung on the wall so she can just see a full-length image of herself. Suppose Sarah's sister Michaela is 1.62 m tall and her eyes are 1.52 m above the floor. If Michaela uses Sarah's mirror without moving it, can she see a full-length image of herself? Draw a ray diagram to illustrate.

Figure 23.30 Two plane mirrors at an angle of 72° form four images.

Physics at Home

You can easily demonstrate *multiple* images using two plane mirrors. Set up two plane mirrors at a 90° angle on a table and place an object with lettering on it between them. You should see three images. The image straight back is due to rays that reflect twice— once from each mirror. Draw a ray diagram to illustrate the formation of each image. In which of the images is the lettering reversed? (See Conceptual Question 4 for some insight into the apparent left-right reversal.)

One way to think about multiple images is that each mirror forms an image of the other mirror; then the image-mirrors produce images of images. To explore further, gradually reduce the angle between the mirrors (see Fig. 23.30). At what angles do you see a whole number of images? Can you figure out the pattern?

23.8 SPHERICAL MIRRORS

Convex Mirror

In a spherical mirror, the reflecting surface is a section of a sphere. A **convex mirror** curves *away from* the viewer; its *center of curvature* is *behind* the mirror (Fig. 23.31). An extended radius drawn from the center of curvature through the **vertex**—the center of the surface of the mirror—is the **principal axis** of the mirror.

In Fig. 23.32, a ray parallel to the principal axis is incident on the surface of a convex mirror at point *A*, which is close to the vertex *V*. (In the diagram, the distance between points *A* and *V* is exaggerated for clarity.) A radial line from the center of curvature through point *A* is normal to the mirror. The angle of incidence is equal to the angle of reflection: $\theta_i = \theta_r = \theta$.

By alternate interior angles, we know that

$$\angle ACF = \theta$$

Triangle *AFC* is isosceles since it has two equal angles; therefore,

$$\overline{AF} = \overline{FC}$$

The notation $\overline{AF}$ means the length of the line segment from *A* to *F*.

Since the incident ray is close to the principal axis, θ is small. As a result,

$$\overline{AF} + \overline{FC} \approx R \quad \text{and} \quad \overline{VF} \approx \overline{AF} \approx \tfrac{1}{2}R$$

(a)

(b)

Figure 23.31 (a) A convex mirror's center of curvature is behind the mirror. (b) This convex mirror gives a driver a wide viewing angle around the corner of the driveway.

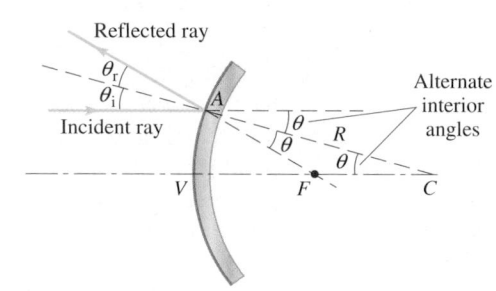

Figure 23.32 Location of the focal point (F) of a convex mirror

where $\overline{AC} = \overline{VC} = R$ is the radius of curvature of the mirror. Note that this derivation is true for *any* angle θ *as long as it is sufficiently small*. Thus, all rays parallel to the axis *that are incident near the vertex* are reflected by the convex mirror so that they *appear* to originate from point F, which is called the **focal point** of the mirror (Fig. 23.33). A convex mirror is also called a **diverging mirror** since the reflection of a set of parallel rays is a set of diverging rays.

> The focal point of a convex mirror is on the axis a distance $\frac{1}{2}R$ behind the mirror.

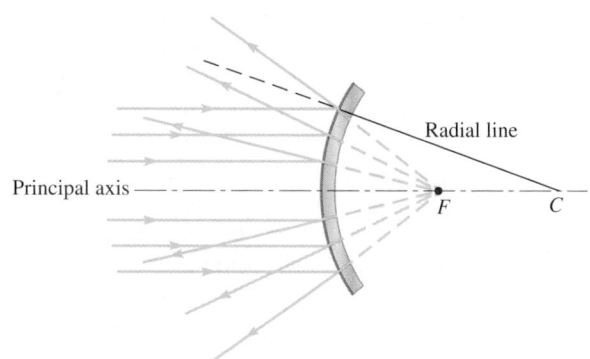

Figure 23.33 Parallel rays reflected from a convex mirror *appear to be* coming from the focal point.

Figure 23.34 (a) Using the three principal rays to locate the image formed by a convex mirror. The three rays are shown in different colors merely to help distinguish them; the actual color of the light along each ray is the same—whatever the color of the top of the object is. (b) Ray diagram showing many rays from the top of the object. When extrapolated backward along straight lines, they all meet at the same point—as long as the angles of incidence are sufficiently small.

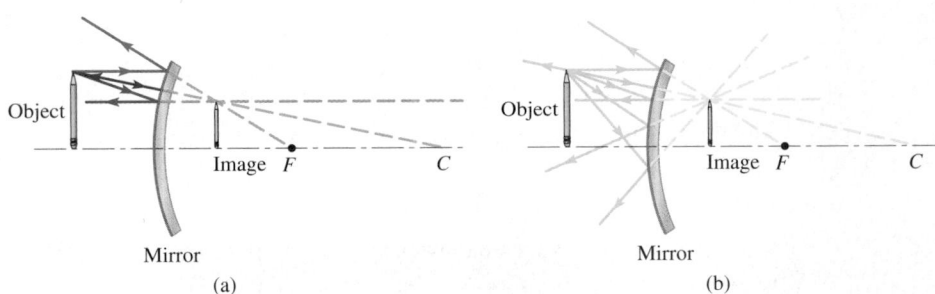

(a)   (b)

To find the image of an object placed in front of the mirror, we draw a few rays. Figure 23.34a shows an object in front of a convex mirror. Three rays are drawn from the point at the top of the object to the mirror surface. One ray (shown in green) is parallel to the principal axis; it is reflected as if it were coming from the focal point. Another ray (red) is headed along a radius toward the center of curvature C; it reflects back upon itself since the angle of incidence is zero. A third ray (blue) heads directly toward the focal point F. Since a ray parallel to the axis is reflected as if it came from F, a ray going toward F is reflected parallel to the axis. Why? Because the law of reflection is reversible; we can reverse the direction of a ray and the law of reflection still holds.

These three reflected rays—as well as other reflected rays from the top of the object (Fig. 23.34b)—meet at one point when extended behind the mirror. That is the location of the top of the image. The bottom of the image lies on the principal axis because the bottom of the object is on the principal axis; rays along the principal axis are radial rays, so they reflect back upon themselves at the surface of the mirror. From the ray diagram, we can conclude that the image is upright, virtual, smaller than the object, and closer to the mirror than the object. Note that the image is *not* at the focal point; the rays coming from a point on the object are *not* parallel to the principal axis. If the object were far from the mirror, then the rays from any point would be nearly parallel to each other. Rays from a point on the principal axis would meet at the focal point; rays from a point not on the axis would meet at a point in the **focal plane**—the plane perpendicular to the axis passing through the focal point.

⚠ **Principal rays** are easier to draw, but they are not more important than other rays in forming an image.

The three rays we chose to draw are called the **principal rays** only because they are easier to draw than other rays. Any two of them can be drawn to locate an image.

Principal rays for convex mirrors

1. A ray parallel to the principal axis is reflected as if it came from the focal point.
2. A ray along a radius is reflected back upon itself.
3. A ray directed toward the focal point is reflected parallel to the principal axis.

Making The Connection:
side-view mirrors
on cars

A convex mirror enables one to see a larger area than the same size plane mirror. The outward curvature of the convex mirror enables the viewer to see light rays coming from larger angles. Convex mirrors are commonly used in stores to enable clerks to watch for shoplifting. The passenger's side mirror in most cars is convex to enable the driver to see farther out to the side. Objects viewed in the mirror are "closer than they appear" because the convex mirror forms a smaller image than a plane mirror would.

Concave Spherical Mirror

A **concave mirror** curves *toward* the viewer; its center of curvature is *in front* of the mirror. A concave mirror is also called a **converging mirror** since it makes parallel rays converge to a point (Fig. 23.35). In Problem 47 you can show that rays parallel to the mirror's principal axis pass through the focal point F at a distance $R/2$ from the vertex, assuming the angles of incidence are small.

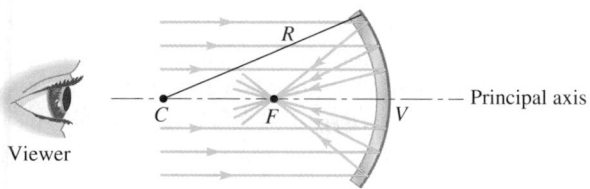

The location of the image of an object placed in front of a concave mirror can be found by drawing two or more rays. As for the convex mirror, there are three principal rays—rays that are easiest to draw. The principal rays are similar to those for a convex mirror, the difference being that the focal point is in front of a concave mirror.

Principal rays for concave mirrors
1. A ray parallel to the principal axis is reflected through the focal point.
2. A ray along a radius is reflected back on itself.
3. A ray along the direction from the focal point to the mirror is reflected parallel to the principal axis.

Figure 23.36 illustrates the use of two principal rays to find an image. In this case, the object is between the focal point and the center of curvature. The image is real because it is in front of the mirror; the principal rays actually do pass through the image point. Depending on the location of the object, a concave mirror can form either real or virtual images. The images can be larger or smaller than the object.

Mirrors designed for shaving or for applying cosmetics are often concave in order to form a magnified image (Fig. 23.37a). Dentists use concave mirrors for the same reason. Whenever an object is within the focal point of a concave mirror, the image is virtual, upright, and larger than the object (Fig. 23.37b).

Making The Connection:
shaving or
cosmetic mirrors

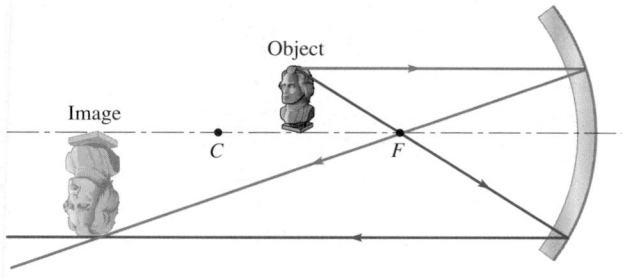

Figure 23.36 An object between the focal point and the center of curvature of a concave mirror forms a real image that is inverted and larger than the object. (The angles and the curvature of the mirror are exaggerated for clarity.)

(a)

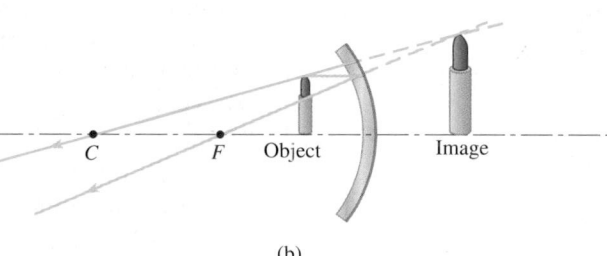

(b)

Figure 23.37 (a) Putting on makeup is made easier because the image is enlarged. (b) Formation of an image when the object is within the focal point of a concave mirror.

In automobile headlights, the lightbulb filament is placed at the focal point of a concave mirror. Light rays coming from the filament are reflected out in a beam of parallel rays. (Sometimes the shape of the mirror is parabolic rather than spherical; a parabolic mirror reflects *all* the rays from the focal point into a parallel beam, not just those close to the principal axis.)

Example 23.6

Scale Diagram for a Concave Mirror

Make a scale diagram showing a 1.5-cm-tall object located 10.0 cm in front of a concave mirror with a radius of curvature of 8.0 cm. Locate the image graphically and estimate its position and height.

Strategy For a scale diagram, we should use a piece of graph paper and choose a scale that fits on the paper— although sometimes it is helpful to make a rough sketch first to get some idea of where the image is. Drawing two principal rays enables us to find the image. Using the third principal ray is a good check. Since the mirror is concave, the center of curvature and the focal point are both in front of the mirror.

Solution To start, we draw the mirror and the principal axis; then we mark the focal point and center of curvature at the correct distances from the vertex (Fig. 23.38). Ray 1 goes from the top of the object to the mirror parallel to the principal axis. It is reflected by the mirror so that it passes through the focal point. Ray 2 travels from the tip of the object through the focal point F. This ray is reflected from the mirror along a line parallel to the principal axis. The intersection of the two rays determines the location of the tip of the image. By measuring the image on the graph paper, we find that the image is 6.7 cm from the mirror and is 1.0 cm high.

Discussion As a check, ray 3 travels through the center of curvature along a radius. Assuming the mirror extends far enough to reflect this ray, it strikes the mirror perpendicular to the surface since it is on a radial line. The reflected ray travels back along the same radial line. Ray 3 would intersect the other two rays at the tip of the image, verifying our result.

Practice Problem 23.6 Another graphical solution

Draw a scale diagram to locate the image of an object 1.5 cm tall and 6.0 cm in front of the same mirror. Estimate the position and height of the image. Is it real or virtual? [*Hint*: Draw a rough sketch first.]

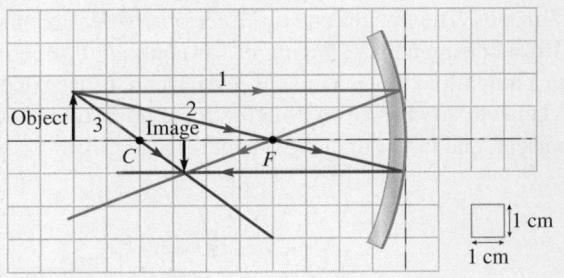

Figure 23.38
Example 23.6

Transverse Magnification

The image formed by a mirror or a lens is, in general, not the same size as the object. It may also be inverted (upside down). The **transverse magnification** m is a ratio that gives both the relative size of the image—in any direction perpendicular to the principal axis— and its orientation. The magnitude of m is the ratio of the image size to the object size:

$$|m| = \frac{\text{image size}}{\text{object size}} \tag{23-7}$$

The sign of m is determined by the orientation of the image. For an inverted (upside-down) image, $m < 0$; for an upright (right-side-up) image, $m > 0$.

Let h (> 0) be the height of the object (really the displacement of the top of the object from the axis) and h' be the height of the image. If the image is inverted, h' and h have opposite signs. Then the definition of the transverse magnification is

$$h' = mh \tag{23-8}$$

Using Fig. 23.39, we can find a relationship between the magnification and p and q, the **object distance** and the **image distance**. Note that p and q are measured along the principal axis to the vertex of the mirror. The two right triangles $\triangle OAV$ and $\triangle IJV$ are similar, so

$$\frac{h}{p} = \frac{-h'}{q}$$

Why the minus sign? In this case, if h is positive, then h' is negative, since the image is on the opposite side of the axis from the object. The magnification is then

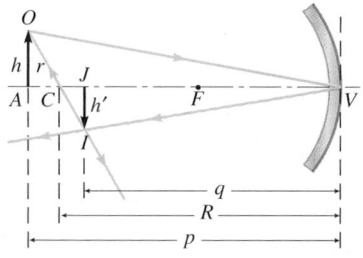

Magnification equation

$$m = \frac{h'}{h} = -\frac{q}{p} \qquad (23\text{-}9)$$

Although in Fig 23.39 the object is beyond the center of curvature, Eq. (23-9) is true regardless of where the object is placed. It applies to any spherical mirror, concave or convex (see Problem 45), as well as to plane mirrors.

The Mirror Equation

From Fig. 23.39 we can derive an equation relating the object distance p, the image distance q, and the **focal length** $f = \frac{1}{2}R$ (the distance from the focal point to the vertex). Note that p, q, and f are all measured along the principal axis to the vertex of the mirror. There are two other similar triangles, $\triangle OAC$ and $\triangle IJC$. Note that $\overline{AC} = p - R$ and $\overline{JC} = R - q$, where R is the radius of curvature. Then

$$\frac{\overline{OA}}{\overline{AC}} = \frac{\overline{IJ}}{\overline{JC}} \quad \text{or} \quad \frac{h}{p-R} = \frac{-h'}{R-q}$$

Rearranging yields

$$\frac{h'}{h} = -\frac{R-q}{p-R}$$

Since h'/h is the magnification,

$$\frac{h'}{h} = -\frac{q}{p} = -\frac{R-q}{p-R} \qquad (23\text{-}9)$$

Substituting $f = R/2$, cross multiplying, and dividing by p, q, and f (Problem 48), we obtain the **mirror equation**.

Mirror equation

$$\frac{1}{p} + \frac{1}{q} = \frac{1}{f} \qquad (23\text{-}10)$$

We derived the magnification and mirror equations for a concave mirror forming a real image, but the equations apply as well to convex mirrors and to virtual images if we use the sign conventions for q and f listed in Table 23.2. Note that q is negative when the image is behind the mirror and f is negative when the focal point is behind the mirror.

The magnification equation and the sign convention for q imply that *real images of real objects are always inverted* (if both p and q are positive, m is negative); *virtual images of real objects are always upright* (if p is positive and q is negative, m is positive). The same rule can be established by drawing ray diagrams. A real image is always in front of the mirror (where the light rays are); a virtual image is behind the mirror.

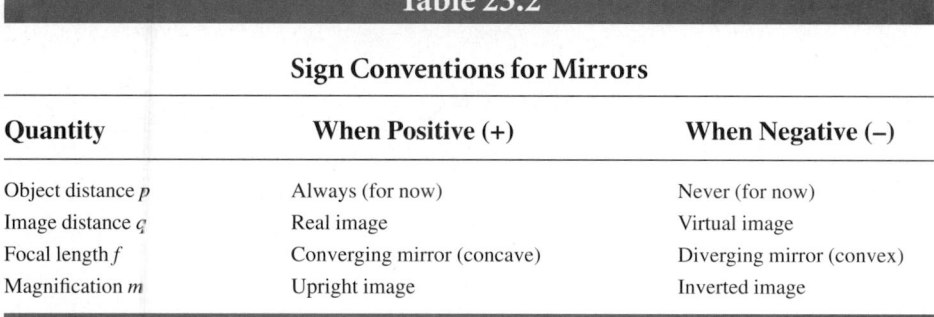

Figure 23.39 Right triangles $\triangle OAV$ and $\triangle IJV$ are similar because the angle of incidence for ray OVI equals the angle of reflection.

Table 23.2

Sign Conventions for Mirrors

Quantity	When Positive (+)	When Negative (–)
Object distance p	Always (for now)	Never (for now)
Image distance q	Real image	Virtual image
Focal length f	Converging mirror (concave)	Diverging mirror (convex)
Magnification m	Upright image	Inverted image

If an object is far from the mirror ($p = \infty$), the mirror equation gives $q = f$. Rays coming from a faraway object are nearly parallel to each other; parallel rays are focused to a point a distance $\frac{1}{2}R$ from the mirror. If the faraway object is not on the principal axis, the image is formed above or below the focal point (Fig. 23.40).

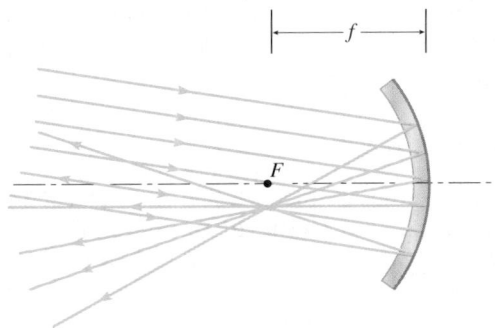

Figure 23.40 A faraway object above the principal axis forms an image at $q = f$.

Example 23.7

Passenger's Side Mirror

 An object is located 30.0 cm from a passenger's side mirror. The image formed is upright and $\frac{1}{3}$ the size of the object. (a) Is the image real or virtual? (b) What is the focal length of the mirror? (c) Is the mirror concave or convex?

Strategy   The magnitude of the magnification is the ratio of the image size to the object size, so $|m| = \frac{1}{3}$. The sign of the magnification is positive for an upright image and negative for an inverted image. Therefore, we know that $m = +\frac{1}{3}$. The object distance is $p = 30.0$ cm. The magnification is also related to the object and image distances, so we can find q. The sign of q indicates whether the image is real or virtual. Then the mirror equation can be used to find the focal length of the mirror. The sign of the focal length tells us whether the mirror is concave or convex.

Solution   (a) The magnification is related to the image and object distances:

$$m = -\frac{q}{p} \qquad (23\text{-}9)$$

Solving for the image distance,

$$q = -mp = -\frac{1}{3} \times 30.0 \text{ cm} = -10.0 \text{ cm}$$

Since q is negative, the image is virtual.

(b) Now we can use the mirror equation to find the focal length:

$$\frac{1}{f} = \frac{1}{p} + \frac{1}{q} = \frac{q + p}{pq}$$

$$f = \frac{pq}{q + p}$$

$$= \frac{30.0 \text{ cm} \times (-10.0 \text{ cm})}{-10.0 \text{ cm} + 30.0 \text{ cm}}$$

$$= -15.0 \text{ cm}$$

(c) Since the focal length is negative, the mirror is convex.

Discussion   As expected, the passenger's side mirror is convex. With all the distances known, we can sketch a ray diagram (Fig. 23.41) to check the result.

Practice Problem 23.7   A spherical mirror of unknown type

An object is in front of a spherical mirror; the image of the object is upright and twice the size of the object and it appears to be 12.0 cm behind the mirror. What is the object distance, what is the focal length of the mirror, and what type of mirror is it (convex or concave)?

continued on next page

Example 23.7 *continued*

Figure 23.41
Ray diagram for convex mirror (Example 23.7)

Physics at Home

Look at each side of a *shiny* metal spoon. (Stainless steel gets dull with use; the newer the spoon the better. A polished silver spoon would be ideal.) One side acts as a convex mirror; the other acts as a concave mirror. For each, notice whether your image is upright or inverted and enlarged or diminished in size. Next, decide whether each image is real or virtual. Which side gives you a larger field of view (in other words, enables you to see a bigger part of the room)? Try holding the spoon at different distances to see what changes. (Keep in mind that the focal length of the spoon is small. If you hold the spoon less than a focal length from your eye, you won't be able to see clearly—your eye cannot focus at such a small distance. Thus, it is not possible to get close enough to the concave side to see a virtual image.)

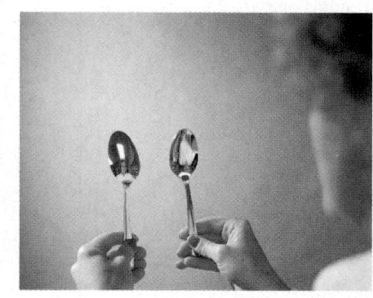

23.9   THIN LENSES

Whereas spherical mirrors form images through reflection, lenses form images through refraction. Most lenses are spherical: each of the two surfaces is a section of a sphere. The **principal axis** of a lens passes through the centers of curvature of the lens surfaces. The **optical center** of a lens is a point on the principal axis through which rays pass without deviation.

We can understand the behavior of a lens by regarding it as an assembly of prisms (Fig. 23.42a, b). The angle of deviation of the ray—the angle that the ray emerging from the prism makes with the incident ray—is proportional to the angle between the two faces of the prism (see Fig. 23.42c and Problem 17). The two faces of a lens are parallel where they intersect the principal axis. A ray striking the lens at the center emerges in the same direction as the incident ray since the refraction of an entering ray is undone as the ray emerges. However, the ray is *displaced*; it is not along the same line as the incident ray. As long as we consider only *thin lenses*—lenses whose thickness is small compared to the focal length—the displacement is negligible; the ray passes straight through the lens.

The curved surfaces of a lens mean that the angle β between the two faces gradually increases as we move away from the center. Thus, the angle of deviation of a ray increases as the point where it strikes the lens moves away from the center. We restrict our consideration to **paraxial rays**: rays that strike the lens close to the principal axis (so that β is small) and do so at a small angle of incidence. For paraxial rays and thin lenses, a ray incident on the lens at a distance d from the center has an angle of deviation δ that is proportional to d (see Fig. 23.42d and Problem 77).

Lenses are classified as **diverging** or **converging**, depending on what happens to the rays as they pass through the lens. A diverging lens bends light rays outward, away from the principal axis. A converging lens bends light rays inward, toward the principal axis. If the incident rays are already diverging, a converging lens may not be able to make them converge; it may only make them diverge less (Fig. 23.43a). Lenses take

Figure 23.42 (a) and (b) Lenses made by combining prism sections. (c) The angle of deviation δ increases as the angle β between the two faces increases. For small β, δ is proportional to β. (d) The angle of deviation of a paraxial ray striking the lens a distance d from the principal axis is proportional to d. To simplify ray diagrams, we draw rays as if they bend at a vertical line through the optical center rather than bending at each of the two lens surfaces.

many possible shapes (Fig. 23.43b); the two surfaces may have different radii of curvature. Note that converging lenses are thickest at the center; diverging lenses are thinnest at the center.

Focal Points and Principal Rays

Any lens has two focal points. The distance between each focal point and the optical center is the magnitude of the **focal length** of the lens. For a diverging lens, rays parallel to the principal axis appear to diverge from the **principal focal point**, which is *before* the lens (Fig. 23.42a); for a converging lens they converge to the principal focal point *past* the lens (Fig. 23.42b).

Two rays suffice to locate the image formed by a thin lens, but a third ray is useful as a check. The three rays that are generally the easiest to draw are called the **principal rays** (Table 23.3). The third principal ray makes use of the **secondary focal point**, which is on the opposite side of the lens from the principal focal point. The behavior of ray 3 can be understood by reversing the direction of all the rays, which also interchanges the two focal points. Figure 23.44 illustrates how to draw the principal rays.

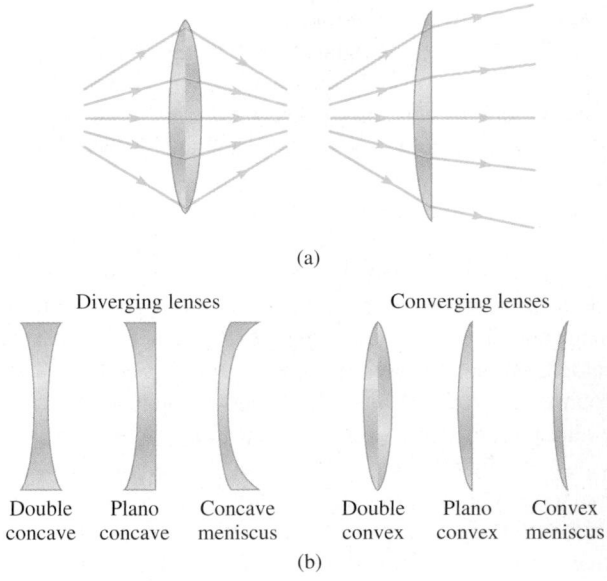

Figure 23.43 (a) When diverging rays strike a converging lens, the lens bends them inward. If they are diverging too rapidly, the lens may not be able to bend them enough to make them converge. In that case, the rays diverge less rapidly after they leave the lens. (b) Shapes of some diverging and converging lenses.

Table 23.3

Principal Rays and Principal Focal Points for Thin Lenses

	Converging Lens	Diverging Lens
Ray 1. An incident ray parallel to the principal axis	Passes through the principal focal point	Appears to come from the principal focal point
Ray 2. A ray incident at the optical center	Passes straight through the lens	Passes straight through the lens
Ray 3. A ray that *emerges* parallel to the principal axis	Appears to come from the secondary focal point	Appears to have been heading for the secondary focal point
Location of the principal focal point	Past the lens	Before the lens

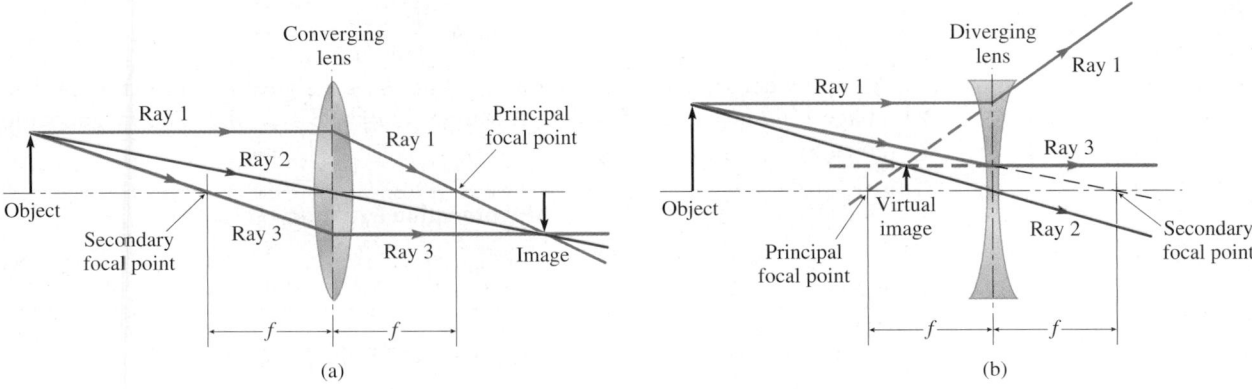

(a)   (b)

Figure 23.44 (a) The three principal rays for a converging lens forming a real image. (b) The three principal rays for a diverging lens forming a virtual image.

Conceptual Example 23.8

Orientation of Virtual Images

A lens forms an image of an object placed before the lens. Using a ray diagram, show that if the image is virtual, then it must also be upright, regardless of whether the lens is converging or diverging.

Strategy The principal rays are usually the easiest to draw. Principal rays 1 and 3 behave differently for converging and diverging lenses. They also deal with focal points, whereas the problem implies that the location of the object with respect to the focal points is irrelevant (except that we know a virtual image is formed). Ray 2 passes undeviated through the center of the lens. It behaves the same way for both types of lens and does not depend on the location of the focal points.

Solution and Discussion Figure 23.45 shows an object in front of a lens (which could be either converging or diverging). Principal ray 2 from the top of the object passes straight through the center of the lens. We extrapolate the refracted ray backward

and sketch a few possibilities for the location of the image—with only one ray we do not know the actual location. We do know that a point on a virtual image is located not where the

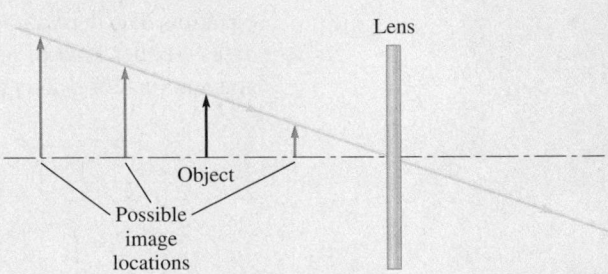

Figure 23.45
The principal ray passing undeviated through the center of a lens shows that virtual images of real objects are upright.

continued on next page

Conceptual Example 23.8 *continued*

rays emerging from the lens meet, but rather where the *backward extrapolation of those rays* meet. In other words, the position of a virtual image is always before the lens (on the same side as the *incident* rays). Therefore, the image is on the same side of the lens as the object. From Fig. 23.45, we see that, just as for mirrors, the virtual image is upright—the image of the point at the top of the object is always above the principal axis.

Conceptual Practice Problem 23.8 Orientation of real images

A converging lens forms a real image of an object placed before the lens. Using a ray diagram, show that the image is inverted.

The Magnification and Thin Lens Equations

We can derive the thin lens equation and the magnification equation from the geometry of Fig. 23.46. From the similar right triangles $\triangle EGC$ and $\triangle DBC$, we write

$$\tan \alpha = \frac{h}{p} = \frac{-h'}{q}$$

As in the derivation of the mirror equation, h' is a signed quantity. For an inverted image h' is negative; $-h'$ is the (positive) length of side BD. The magnification is given by

Magnification equation

$$m = \frac{h'}{h} = -\frac{q}{p} \qquad (23\text{-}9)$$

From two other similar right triangles $\triangle ACF$ and $\triangle DBF$,

$$\tan \beta = \frac{h}{f} = \frac{-h'}{q-f}$$

or

$$\frac{q-f}{f} = \frac{-h'}{h} = \frac{q}{p}$$

After dividing through by q and rearranging, we obtain the **thin lens equation**.

Thin lens equation

$$\frac{1}{p} + \frac{1}{q} = \frac{1}{f} \qquad (23\text{-}10)$$

The magnification and thin lens equations have exactly the same form as the mirror equation. The derivations used a converging lens and a real image, but they apply to all cases—either kind of lens and either kind of image—as long as we use the same sign conventions for q and f as for spherical mirrors (Table 23.4).

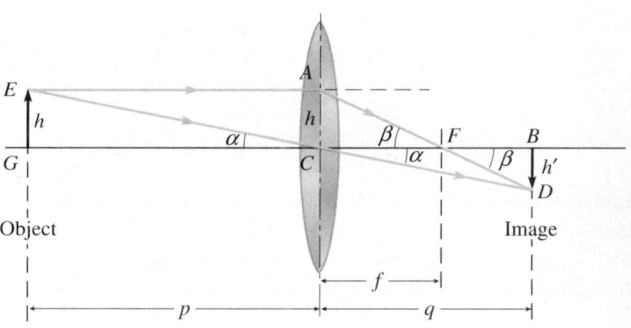

Figure 23.46 Ray diagram for derivation of thin lens equation and magnification

Table 23.4

Sign Conventions for Mirrors and Lenses

Quantity	When Positive (+)	When Negative (–)
Object distance p	Always (for now)	Never (for now)
Image distance q	Real image	Virtual image
Focal length f	Converging lens or mirror	Diverging lens or mirror
Magnification m	Upright image	Inverted image

Example 23.9

Zoom Lens

A wild daisy 1.2 cm in diameter is 90.0 cm from a camera's zoom lens. The focal length of the lens has magnitude 150.0 mm. (a) Find the distance between the lens and the film. (b) How large is the image of the daisy?

Strategy The problem can be solved using the lens and magnification equations. The lens must be converging to form a real image, so $f = +150.0$ mm. The image must be formed on the film, so the distance from the lens to the film is q. After finishing the algebraic solution, we sketch a ray diagram as a check.

Given: $h = 1.2$ cm; $p = 90.0$ cm; $f = +15.00$ cm
Find: q, h'

Solution (a) Since p and f are known, we find q from the lens equation

$$\frac{1}{p} + \frac{1}{q} = \frac{1}{f}$$

Let us solve for q.

$$q = \left(\frac{1}{f} - \frac{1}{p}\right)^{-1}$$

Substituting numerical values,

$$q = \left(\frac{1}{15.00 \text{ cm}} - \frac{1}{90.0 \text{ cm}}\right)^{-1} = +18.0 \text{ cm}$$

The film is 18.0 cm from the lens.

(b) From the magnification equation,

$$m = \frac{h'}{h} = -\frac{q}{p} = -\frac{18.0 \text{ cm}}{90.0 \text{ cm}} = -0.200$$

$$h' = mh = -0.200 \times 1.2 \text{ cm}$$

$$= -0.24 \text{ cm}$$

The image of the daisy is 0.24 cm in diameter.

Discussion Figure 23.47 shows a ray diagram using the three principal rays that confirms the algebraic solution.

Practice Problem 23.9   Finding the focal length of a lens

A 3.00-cm-tall object is placed 60.0 cm in front of a lens. The virtual image formed is 0.50 cm tall. What is the focal length of the lens? Is it converging or diverging?

Figure 23.47
Ray diagram for Example 23.9

Physics at Home

If you or a friend are farsighted and have eyeglasses, put the glasses on a table so that you can look through the lenses. Increase your distance from the lenses until you see a clear image of distant objects. Why is the image inverted? Is the image real or virtual? The eyeglasses certainly form an upright image when they are worn as intended. Are the lenses converging or diverging?

Objects and Images at Infinity

Suppose an object is a large distance from a lens ("at infinity"). Substituting $p = \infty$ in the lens equation yields $q = f$. The rays from a faraway object are nearly parallel to each other when they strike the lens, so the image is formed in the principal **focal plane** (the plane perpendicular to the axis passing through the principal focal point). Similarly, if an object is placed in the principal focal plane of a converging lens, then $p = f$ and $q = \infty$. The image is at infinity—that is, the rays emerging from the lens are parallel, so they appear to be coming from an object at infinity.

MASTER THE CONCEPTS

Summary

- A wavefront is a set of points of equal phase. A ray points in the direction of propagation of a wave and is perpendicular to the wavefronts. Huygens's principle is a geometric construction used to analyze the propagation of a wave. Every point on a wavefront is considered a source of spherical wavelets. A surface tangent to the wavelets at a later time is the wavefront at that time.

- Geometric optics deals with the propagation of light when interference and diffraction are negligible. The chief tool used in geometric optics is the ray diagram. At a boundary between two different media, light can be reflected as well as transmitted. The laws of reflection and refraction give the directions of the reflected and transmitted rays. In the laws of reflection and refraction, angles are measured between rays and a normal to the boundary.

- Laws of reflection:
  1. The angle of incidence equals the angle of reflection.
  2. The reflected ray lies in the same plane as the incident ray and the normal to the surface at the point of incidence.

- Laws of refraction:
  1. $n_1 \sin \theta_1 = n_2 \sin \theta_2$ (Snell's law).
  2. The incident ray, the transmitted ray, and the normal all lie in the same plane—the plane of incidence; the incident and transmitted rays are on *opposite sides* of the normal.

- When a ray is incident on a boundary from a material with a higher index of refraction to one with a lower index of refraction, total internal reflection occurs (there is no transmitted ray) if the angle of incidence exceeds the critical angle

$$\theta_c = \sin^{-1} \frac{n_t}{n_i} \qquad (23\text{-}5a)$$

- When a ray is incident on a boundary, the reflected ray is totally polarized perpendicular to the plane of incidence if the angle of incidence is equal to Brewster's angle

$$\theta_B = \tan^{-1} \frac{n_t}{n_i} \qquad (23\text{-}6)$$

- In the formation of an image, there is a one-to-one correspondence of points on the object and points on the image. In a virtual image, light rays *appear* to diverge from the image point, but they really don't. In a real image, the rays actually *do* pass through the image point.

- Finding an image using a ray diagram:
  1. Draw two (or more) rays coming from a single point on the object toward the lens or mirror.
  2. Trace the rays, applying the laws of reflection and refraction as needed, until they reach the observer.
  3. Extrapolate the rays backward along straight line paths until they intersect at the image point.

- The easiest rays to trace for a mirror or lens are called the principal rays.

- A plane mirror forms an unmagnified, upright, virtual image of an object that is located at the same distance behind the mirror as the object is in front of the mirror. The object and image are both located on a normal from the object to the mirror surface.

- The magnitude of the transverse magnification m is the ratio of the image size to the object size; the sign of m is determined by the orientation of the image. For an inverted (upside-down) image, $m < 0$; for an upright (right-side-up) image, $m > 0$. For either lenses or mirrors,

$$m = \frac{h'}{h} = -\frac{q}{p} \qquad (23\text{-}9)$$

MASTER THE CONCEPTS *continued*

- The mirror/thin lens equation relates the object and image distances to the focal length:

$$\frac{1}{p} + \frac{1}{q} = \frac{1}{f} \qquad (23\text{-}10)$$

- These sign conventions enable the magnification and mirror/thin lens equations to apply to all kinds of mirror and lenses and both kinds of image:

Quantity	When Positive (+)	When Negative (−)
Object distance p	Always (for now)	Never (for now)
Image distance q	Real image	Virtual image
Focal length f	Converging lens or mirror	Diverging lens or mirror
Magnification m	Upright image	Inverted image

Highlighted Figures and Tables

F23.4 Wavefronts and rays for waves with circular wavefronts, straight line wavefronts, spherical wavefronts, and planar wavefronts (p. 837)

F23.5 Application of Huygens's principle to a plane wave, and incomplete construction that does not show wavelets coming from *every point* on the wavefront (p. 838)

F23.6 A laser beam reflecting from a mirror illustrates specular reflection; diffuse reflection occurs when the same laser reflects from a rough surface (p. 839)

F23.7 A plane wave strikes a metal surface (p. 839)

F23.8 The angles of incidence and of reflection are measured between the ray and the *normal* to the surface (not between the ray and the surface) (p. 840)

F23.9 Wavefronts and rays at a glass-air boundary; Huygens's construction for a wavefront partly in air and partly in glass; geometry for finding the angle of the transmitted ray (p. 841)

F23.10 The incident ray, the reflected ray, the transmitted ray, and the normal all lie in the same plane; the incident and transmitted rays are always on opposite sides of the normal, no matter what the indices of refraction of the two media (p. 842)

T23.1 Indices of refraction for $\lambda = 589.3$ nm in vacuum (p. 842)

F23.16 Formation of rainbow by water droplets (p. 847)

F23.17 Partial reflection and total internal reflection at upper surface of a rectangular glass prism (p. 848)

F23.24 When light is incident at Brewster's angle, the reflected and transmitted rays are perpendicular (p. 852)

F23.25 Formation of a real image by a camera lens (p. 853)

F23.27 A plane mirror forms an image of a point source (p. 855)

F23.28 Sketching the image of a pencil in front of a plane mirror (p. 855)

F23.39 Right triangles $\triangle OAV$ and $\triangle IJV$ are similar because the angle of incidence for ray OVI equals the angle of reflection (p. 861)

T23.2 Sign conventions for mirrors (p. 861)

F23.40 A faraway object above the principal axis forms an image at $q = f$ (p. 862)

T23.4 Sign conventions for mirrors and lenses (p. 867)

CONCEPTUAL QUESTIONS

1. Describe the difference between specular and diffuse reflection. Give some examples of each.

2. What is the difference between a virtual and a real image? State a method for demonstrating the presence of a real image.

3. Water droplets in air create rainbows. Describe the physical situation that causes a rainbow. Should you look toward or away from the Sun to see a rainbow? Why is the secondary rainbow fainter than the primary rainbow?

4. Why does a mirror hanging in a vertical plane seem to interchange left and right but not up and down? [*Hint*: Refer to Fig. 23.29. Instead of calling Grant's hands left and right, call them east and west. In Grant's image, are the east and west hands reversed? Note that Grant faces south while his image faces north.]

5. A framed poster is covered with "nonglare glass." The surface of the glass is rougher than regular glass. How does a rough surface prevent glare?

6. Explain how a plane mirror can be thought of as a special case of a spherical mirror. What is the focal length of a plane mirror? Does the spherical mirror equation work for plane mirrors with this choice of focal length? What is the transverse magnification for any image produced by a plane mirror?

7. A ray of light passes from air into water, striking the surface of the water with an angle of incidence of 45°. Which of these quantities change as the light enters the water: wavelength, frequency, speed of propagation, direction of propagation?

8. If the angle of incidence is greater than the angle of refraction for a light beam passing an interface, what can be said about the relative values of the indices of refraction and the speed of light in the first and second media?

9. A concave mirror has focal length f. (a) If you look into the mirror from a distance less than f, is the image you see upright or inverted? (b) If you stand at a distance greater than $2f$, is the image upright or inverted? (c) If you stand at a distance

between f and $2f$, an image is formed but you cannot see it. Why not? Sketch a ray diagram and compare the locations of the object and image.

10. The focal length of a concave mirror is 4.00 m and an object is placed 3.00 m in front of the mirror. Describe the image in terms of real, virtual, upright, and inverted.

11. Why is the passenger's side mirror in many cars convex rather than plane or concave?

12. When a virtual image is formed by a mirror, is it in front of the mirror or behind it? What about a real image?

13. Light rays travel from left to right through a lens. If a virtual image is formed, on which side of the lens is it? On which side would a real image be found?

14. Why is the brilliance of an artificial diamond made of cubic zirconium ($n = 1.9$) distinctly inferior to the real thing ($n = 2.4$) even if the two are cut exactly the same way? How would an artificial diamond made of glass compare?

15. The surface of the water in a swimming pool is completely still. Describe what you would see looking straight up toward the surface from underwater. [*Hint*: Sketch some rays. Consider both reflected and transmitted rays at the water surface.]

16. A ray reflects from a spherical mirror at point P. Explain why a radial line from the center of curvature through point P always bisects the angle between the incident and reflected rays.

17. Why must projectors and cameras form real images? Does the lens in the eye form real or virtual images on the retina?

18. Is it possible for a plane mirror to produce a real image of an object in front of the mirror? Explain. If it is possible, sketch a ray diagram to demonstrate. If it is not possible, sketch a ray diagram to show which way a curved mirror must curve (concave or convex) to produce a real image.

19. A slide projector forms a real image of the slide on a screen using a converging lens. If the bottom half of the lens is blocked by covering it with opaque tape, does the bottom half of the image disappear, or does the top half disappear, or is the entire image still visible on the screen? If the entire image is visible, is anything different about it? [*Hint*: It may help to sketch a ray diagram.]

20. A lens is placed at the end of a bundle of optical fibers in an endoscope. The purpose of the lens is to make the light rays parallel before they enter the fibers (in other words, to put the image at infinity). What is the advantage of using a lens with the same index of refraction as the core of the fibers?

21. A converging lens made from dense flint glass is placed into a container of transparent glycerine. Describe what happens to the focal length.

MULTIPLE CHOICE QUESTIONS

1. The image of an object in a plane mirror
   (a) is always smaller than the object.
   (b) is always the same size as the object.
   (c) is always larger than the object.
   (d) can be larger, smaller, or the same size as the object, depending on the distance between the object and the mirror.

2. Which statements are true? The rays in a plane wave are
   1. parallel to the wavefronts.
   2. perpendicular to the wavefronts.
   3. directed radially outward from a central point.
   4. parallel to each other.
   (a) 1, 2, 3, 4   (b) 1, 4   (c) 2, 3   (d) 2, 4

3. The image of a slide formed by a slide projector is correctly described by which of the listed terms?
   (a) real, upright, enlarged
   (b) real, inverted, diminished in size
   (c) virtual, inverted, enlarged   (d) virtual, upright, diminished
   (e) real, upright, diminished   (f) real, inverted, enlarged
   (g) virtual, inverted, enlarged

4. During a laboratory experiment with an object placed in front of a concave mirror, the image distance q is determined for several different values of object distance p. How might the focal length f of the mirror be determined from a graph of the data?
   (a) Plot q versus p; slope $= f$
   (b) Plot q versus p; slope $= 1/f$
   (c) Plot $1/p$ versus $1/q$, y-intercept $= 1/f$
   (d) Plot q versus p; y-intercept $= 1/f$
   (e) Plot q versus p; y-intercept $= f$
   (f) Plot $1/p$ versus $1/q$, slope $= 1/f$

5. A man runs toward a plane mirror at 5 m/s and the mirror, on rollers, simultaneously approaches him at 2 m/s. The speed at which his image moves (relative to the ground) is
   (a) 14 m/s   (b) 7 m/s   (c) 3 m/s
   (d) 9 m/s   (e) 12 m/s

6. Two converging lenses, of exactly the same size and shape, are held in sunlight, the same distance above a sheet of paper. Fig. 23.48 shows the paths of some rays through the two lenses. Which lens is made of material with a higher index of refraction? How do you know?
   (a) Lens 1, because its focal length is smaller
   (b) Lens 1, because its focal length is greater
   (c) Lens 2, because its focal length is smaller
   (d) Lens 2, because its focal length is greater
   (e) Impossible to answer with the information given

Figure 23.48 Multiple Choice Question 6

7. Which of these statements correctly describe the images formed by an object placed before a single thin lens?

   1. Real images are always enlarged.
   2. Real images are always inverted.
   3. Virtual images are always upright.
   4. Convex lenses never produce virtual images.
   (a) 1 and 3   (b) 2 and 3
   (c) 2 and 4   (d) 2, 3, and 4
   (e) 1, 2, and 3   (f) 4 only

8. Figure 23.49 shows some rays passing through a diverging lens. Which of the rays is definitely *not* drawn correctly?

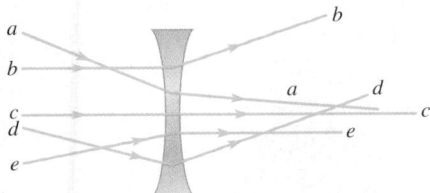

Figure 23.49 Multiple Choice Question 8

9. A point source of light is placed at the focal point of a converging lens; the rays of light coming out of the lens are parallel to the principal axis. Now suppose the source is moved closer to the lens but still on the axis. Which statement is true about the light rays coming out of the lens?

   (a) They diverge from each other.
   (b) They converge toward each other.
   (c) They still emerge parallel to the principal axis.
   (d) They emerge parallel to each other but not parallel to the axis.
   (e) No rays emerge because a virtual image is formed.

10. A light ray inside a glass prism is incident at Brewster's angle on a surface of the prism with air outside. Which of these is true?

   (a) There is no transmitted ray; the reflected ray is plane polarized.
   (b) The transmitted ray is plane polarized; the reflected ray is partially polarized.
   (c) There is no transmitted ray; the reflected ray is partially polarized.
   (d) The transmitted ray is partially polarized; the reflected ray is plane polarized.
   (e) The transmitted ray is plane polarized; there is no reflected ray.

PROBLEMS

Note: ⒸⒸ indicates a combination conceptual/quantitative problem. Gold diamonds ✦, ✦✦ are used to indicate the increasing level of difficulty of each problem. Problem numbers appearing in blue, 9., denote problems that have a detailed solution available in the Student Solutions Manual. Some problems are *paired* by concept; their numbers are connected by a ruled box.

23.1 Wavefronts, Rays, and Huygens's Principle

1. Sketch the wavefronts and rays for the light emitted by an isotropic point source (*isotropic* = same in all directions). Use Huygens's principle to illustrate the propagation of one of the wavefronts.

Ⓒ ✦ 2. Apply Huygens's principle to a 5-cm-long planar wavefront approaching a reflecting wall at normal incidence. The wavelength is 1 cm and the wall has a wide opening (width = 4 cm). The center of the incoming wavefront approaches the center of the opening. Repeat the procedure until you have wavefronts on both sides of the wall. Without worrying about the details of edge effects, what are the general shapes of the wavefronts on each side of the reflecting wall?

Ⓒ ✦ 3. Repeat Problem 2 for an opening of width 0.5 cm.

23.2 The Reflection of Light

4. A plane wave reflects from a spherical surface. Draw a ray diagram and sketch some wavefronts for the reflected wave.

5. A spherical wave (from a point source) reflects from a planar surface. Draw a ray diagram and sketch some wavefronts for the reflected wave.

6. Light rays from the Sun, which is at an angle of 35° above the western horizon, strike the still surface of a pond. (a) What is the angle of incidence of the Sun's rays on the pond? (b) What is the angle of reflection of the rays that leave the pond surface? (c) In what direction and at what angle from the pond surface are the reflected rays traveling?

7. A light ray reflects from a plane mirror as shown in Fig. 23.50. What is the angle of deviation δ?

Figure 23.50 Problem 7

8. Two plane mirrors form a 70.0° angle (Fig. 23.51). For what angle θ is the final ray horizontal?

Figure 23.51 Problem 8

✦ 9. Choose two rays in Fig. 23.8 and use them to prove that the angle of incidence is equal to the angle of reflection. [*Hint*: Choose a wavefront at two different times, one before reflection and one after. The time for light to travel from one wavefront to the other is the same for the two rays.]

23.3 The Refraction of Light; Snell's Law

10. Sunlight strikes the surface of a lake at an angle of incidence of 30.0°. At what angle with respect to the normal would a fish see the Sun?

11. Sunlight strikes the surface of a lake. A diver sees the Sun at an angle of 42.0° with respect to the vertical. What angle do the Sun's rays in air make with the vertical?

12. A light ray in the core ($n = 1.40$) of a cylindrical optical fiber (Fig. 23.52) travels at an angle $\theta_1 = 49.0°$ with respect to the *axis of the fiber*. A ray is transmitted through the cladding ($n = 1.20$) and into the air. What angle θ_2 does the exiting ray make with the outside surface of the cladding?

13. A light ray in the core ($n = 1.40$) of a cylindrical optical fiber (Fig. 23.52) is incident on the cladding. A ray is transmitted through the cladding ($n = 1.20$) and into the air. The emerging ray makes an angle $\theta_2 = 5.00°$ with the outside surface of the cladding. What angle θ_1 did the ray in the core make with the axis?

Air

θ_2

Cladding

θ_1 — Axis — Core

Figure 23.52
Problems 12 and 13

14. A glass lens has a scratch-resistant plastic coating on it. The speed of light in the glass is $0.67c$ and the speed of light in the coating is $0.80c$. A ray of light in the coating is incident on the plastic-glass boundary at an angle of 12.0° with respect to the normal. At what angle with respect to the normal is the ray transmitted into the glass?

15. A ray of light that contributes to a mirage (see Fig. 23.13) is traveling at an angle of 2.00° below the horizontal. At what angle to the horizontal does the ray travel when its elevation is lower and the speed of light is 0.010% faster?

✦ 16. A horizontal light ray is incident on a crown glass prism as shown in Fig. 23.53 where $\beta = 30.0°$. Find the angle of deviation δ of the ray—the angle that the ray emerging from the prism makes with the incident ray.

✦ 17. A horizontal light ray is incident on a prism as shown in Fig. 23.53 where β is a *small angle* (exaggerated in the figure). Find the angle of deviation δ of the ray—the angle that the ray emerging from the prism makes with the incident ray—as a function of β and n, the index of refraction of the prism and show that δ is proportional to β.

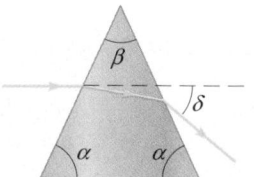

Figure 23.53   Problems 16 and 17

© 18. A diamond in air is illuminated with white light. On one particular facet, the angle of incidence is 26.00°. Inside the diamond, red light ($\lambda = 660.0$ nm in vacuum) is refracted at 10.48° with respect to the normal; blue light ($\lambda = 470.0$ nm in vacuum) is refracted at 10.33°. (a) What are the indices of refraction for red and blue light in diamond? (b) What is the ratio of the speed of red light to the speed of blue light in diamond? (c) How would a diamond look if there were no dispersion?

✦ 19. The prism in Fig. 23.54 is made of crown glass. Its index of refraction ranges from 1.517 for the longest visible wavelengths to 1.538 for the shortest. Find the range of refraction angles for the light transmitted into air through the right side of the prism.

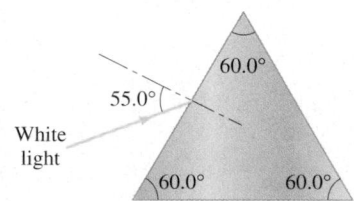

60.0°

55.0°

White light

60.0°   60.0°

Figure 23.54   Problem 19

23.4 Total Internal Reflection

20. Calculate the critical angle for a sapphire surrounded by air.

© 21. (a) Calculate the critical angle for a diamond surrounded by air. (b) Calculate the critical angle for a diamond under water. (c) Explain why a diamond sparkles less under water than in air.

© 22. Figure 23.55 shows some light rays reflected from a small defect in the glass toward the surface of the glass. (a) If $\theta_c = 40.00°$, what is the index of refraction of the glass? (b) Is there any point above the glass at which a viewer would not be able to see the defect? Explain.

© 23. A 45° prism (Fig. 23.56) has an index of refraction of 1.6. Light is normally incident on the left side of the prism. Does light exit the back of the prism (for example, at point P)? If so, what is the angle of refraction with respect to the normal at point P? If not, what happens to the light?

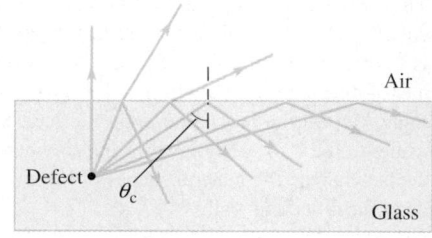

Air

Defect • θ_c

Glass

Figure 23.55   Problem 22

45°

P

45°

Figure 23.56   Problems 23 and 82

© 24. The angle of incidence θ of a ray of light in air is adjusted gradually as it enters a shallow tank made of Plexiglas and filled with carbon disulfide. Is there an angle of incidence for which light is transmitted into the carbon disulfide but not into the Plexiglas at the bottom of the tank? If so, find the angle. If not, explain why not.

© 25. Repeat Problem 24 for a Plexiglas tank filled with carbon tetrachloride instead of carbon disulfide.

26. What is the index of refraction of the core of an optical fiber if the cladding has $n = 1.20$ and the critical angle at the core-cladding boundary is 45.0°?

23.5 Brewster's Angle

© 27. (a) Sunlight reflected from the still surface of a lake is totally polarized when the incident light is at what angle with respect to the horizontal? (b) In what direction is the reflected light polarized? (c) Is any light incident at this angle transmitted into the water? If so, at what angle below the horizontal does the transmitted light travel?

© 28. (a) Sunlight reflected from the smooth ice surface of a frozen lake is totally polarized when the incident light is at what angle with respect to the horizontal? (b) In what direction is the reflected light polarized? (c) Is any light incident at this angle transmitted into the ice? If so, at what angle below the horizontal does the transmitted light travel?

© ✦ 29. Light travels in a medium with index n_1 toward a boundary with another material of index $n_2 < n_1$. (a) Which is larger, the critical angle or Brewster's angle? Does the answer depend on the values of n_1 and n_2 (other than assuming $n_2 < n_1$)? (b) What can you say about the critical angle and Brewster's angle for light coming the other way (from the medium with index n_2 toward the medium with n_1)?

23.6 The Formation of Images through Reflection or Refraction

30. A defect in a diamond appears to be 2.0 mm below the surface when viewed from directly above that surface. How far beneath the surface is the defect?

31. An insect is trapped inside a piece of amber ($n = 1.546$). Looking at the insect from directly above, it appears to be 7.00 mm below a smooth surface of the amber. How far below the surface is the insect?

✦ 32. A penny is at the bottom of a bowl full of water. When you look at the water surface from the side, with your eyes at the water level, the penny appears to be just barely under the surface and a horizontal distance of 3.0 cm from the edge of the bowl. If the penny is actually 8.0 cm below the water surface, what is the horizontal distance between the penny and the edge of the bowl? [*Hint*: The rays you see pass from water to air with refraction angles close to 90°.]

23.7 Plane Mirrors

33. Norah wants to buy a mirror so that she can check on her appearance from top to toe before she goes off to work. If Norah is 1.64 m tall, how tall a mirror does she need?

34. Daniel's eyes are 1.82 m from the floor when he is wearing his dress shoes, and the top of his head is 1.96 m from the floor. Daniel has a mirror that is 0.98 m in length. How high from the floor should the bottom edge of the mirror be located if Daniel is to see a full-length image of himself? Draw a ray diagram to illustrate your answer.

35. A rose in a vase is placed 0.250 m in front of a plane mirror. Nagar looks into the mirror from 2.00 m in front of it. How far away from Nagar is the image of the rose?

36. Entering a darkened room, Gustav strikes a match in an attempt to see his surroundings. At once he sees what looks like another match about 4 m away from him. As it turns out, a mirror hangs on one wall of the room. How far is Gustav from the wall with the mirror?

✦ 37. Maurizio is standing in a rectangular room with two adjacent walls and the ceiling all covered by plane mirrors. How many images of himself can Maurizio see?

38. Hannah is standing in the middle of a room with two opposite walls that are separated by 10.0 m and covered by plane mirrors (Fig. 23.57). There is a candle in the room 1.50 m from one mirrored wall. Hannah is facing the opposite mirrored wall and sees many images of the candle. How far from Hannah are the closest four images of the candle that she can see?

✦ 39. A point source of light is in front of a plane mirror. (a) Prove that all the reflected rays, when extended back behind the mirror, intersect in a single point. [*Hint*: See Fig. 23.27 and use similar triangles.] (b) Show that the image point lies on a line through the object and perpendicular to the mirror, and that the object and image distances are equal. [*Hint*: Use any pair of rays in Fig. 23.27.]

Figure 23.57   Problem 38

23.8 Spherical Mirrors

40. An object 2.00 cm high is placed 12.0 cm in front of a convex mirror with radius of curvature of 8.00 cm. Where is the image formed? Draw a ray diagram to illustrate.

41. A 1.80-cm-high object is placed 20.0 cm in front of a concave mirror with a 5.00-cm focal length. What is the position of the image? Draw a ray diagram to illustrate.

42. An object is placed in front of a concave mirror with a 25.0-cm radius of curvature. A real image twice the size of the object is formed. At what distance is the object from the mirror? Draw a ray diagram to illustrate.

43. An object is placed in front of a convex mirror with a 25.0-cm radius of curvature. A virtual image half the size of the object is formed. At what distance is the object from the mirror? Draw a ray diagram to illustrate.

C 44. A concave mirror has a radius of curvature of 5.0 m. An object, initially 2.0 m in front of the mirror, is moved back until it is 6.0 m from the mirror. Describe how the image location changes.

◆ 45. Derive the magnification equation, $m = h'/h = -q/p$, for a *convex* mirror. Draw a ray diagram as part of the solution. [*Hint*: Draw a ray that is not one of the three principal rays, as was done in the derivation for a concave mirror.]

C 46. In a subway station, a convex mirror allows the attendant to view activity on the platform. A woman 1.64 m tall is standing 4.5 m from the mirror. The image formed of the woman is 0.500 m tall. (a) What is the radius of curvature of the mirror? (b) The mirror is 0.500 m in diameter. If the woman's shoes appear at the bottom of the mirror, does her head appear at the top—in other words, does the image of the woman fill the mirror from top to bottom? Explain.

◆ 47. Show that when rays parallel to the principal axis reflect from a concave mirror, the reflected rays all pass through the focal point at a distance $R/2$ from the vertex. Assume that the angles of incidence are small. [*Hint*: Follow the similar derivation for a *convex* mirror in the text.]

48. Starting with Fig. 23.39, perform all the algebraic steps to obtain the mirror equation in the form of Eq. (23-10).

23.9 Thin Lenses

49. Sketch a ray diagram to show that when an object is placed more than twice the focal length away from a converging lens, the image formed is inverted, real, and diminished in size.

50. Sketch a ray diagram to show that when an object is placed at twice the focal length from a converging lens, the image formed is inverted, real, and the same size as the object.

51. Sketch a ray diagram to show that when an object is placed between twice the focal length and the focal length from a converging lens, the image formed is inverted, real, and enlarged in size.

52. Sketch a ray diagram to show that when an object is a distance equal to the focal length from a converging lens, the

emerging rays from the lens are parallel to each other, so the image is at infinity.

53. When an object is placed 6.0 cm in front of a converging lens, a virtual image is formed 9.0 cm from the lens. What is the focal length of the lens?

54. An object of height 3.00 cm is placed 12.0 cm from a diverging lens of focal length –12.0 cm. Draw a ray diagram to find the height and position of the image.

55. A diverging lens has a focal length of –8.00 cm. (a) What are the image distances for objects placed at these distances from the lens: 5.00 cm, 8.00 cm, 14.0 cm, 16.0 cm, 20.0 cm? In each case, describe the image as real or virtual, upright or inverted, and enlarged or diminished in size. (b) If the object is 4.00 cm high, what is the height of the image for the object distances of 5.00 cm and 20.0 cm?

56. A converging lens has a focal length of 8.00 cm. (a) What are the image distances for objects placed at these distances from the thin lens: 5.00 cm, 14.0 cm, 16.0 cm, 20.0 cm? In each case, describe the image as real or virtual, upright or inverted, and enlarged or diminished in size. (b) If the object is 4.00 cm high, what is the height of the image for the object distances of 5.00 cm and 20.0 cm?

57. Sketch a ray diagram to show that if an object is placed less than the focal length from a converging lens, the image is virtual and upright.

58. For each of the lenses in Fig. 23.58, list whether the lens is converging or diverging.

(a)   (b)   (c)   (d)   **Figure 23.58** Problem 58

C ◆ 59. A standard "35-mm" slide measures 24.0 mm by 36.0 mm. Suppose a slide projector produces a 60.0-cm by 90.0-cm image of the slide on a screen. The focal length of the lens is 12.0 cm. (a) What is the distance between the slide and the screen? (b) If the screen is moved farther from the projector, should the lens be moved closer to the slide or farther away?

60. An object that is 6.00 cm tall is placed 40.0 cm in front of a diverging lens. The magnitude of the focal length of the lens is 20.0 cm. Find the image position and size. Is the image real or virtual? Upright or inverted?

COMPREHENSIVE PROBLEMS

61. A converging lens made of glass ($n = 1.5$) is placed under water ($n = 1.33$). How does the focal length of the lens under water compare to the focal length in air? Draw a diagram to illustrate your answer.

62. A point source of light is placed 10 cm in front of a concave mirror; the reflected rays are parallel. What is the radius of curvature of the mirror?

✦63. A ray of light in air is incident at an angle of 60.0° with the surface of some benzene contained in a shallow tank made of crown glass (Fig. 23.59). What is the angle of refraction of the light ray when it enters the glass at the bottom of the tank?

Glass

Figure 23.59 Problem 63

✦64. A ray of light passes from air through dense flint glass and then back into air. The angle of incidence on the first glass surface is 60.0°. The thickness of the glass is 5.00 mm; its front and back surfaces are parallel. How far is the ray displaced as a result of traveling through the glass?

65. A glass prism bends a ray of blue light more than a ray of red light since its index of refraction is slightly higher for blue than for red. Does a diverging glass lens have the same focal point for blue light and for red light? If not, for which color is the focal point closer to the lens?

© 66. In many cars the passenger's side mirror says: "Objects in the mirror are closer than they appear." (a) Does this mirror form real or virtual images? (b) Since the image is diminished in size, is the mirror concave or convex? Why? (c) Show that the image must actually be *closer* to the mirror than is the object. (d) How then can the image seem to be farther away?

67. A scuba diver in a lake aims her underwater spotlight at the lake surface. (a) If the beam makes a 75° angle of incidence with respect to a normal to the water surface, is it reflected, transmitted, or both? Find the angles of the reflected and transmitted beams (if they exist). (b) Repeat for a 25° angle of incidence.

68. Laura is walking directly toward a plane mirror at a speed of 0.8 m/s relative to the mirror. At what speed is her image approaching the mirror?

69. Xi Yang is practicing for his driver's license test. He notices in the rearview mirror that a tree, located directly behind the automobile, is approaching his car as he is backing up. If the car is moving at 8.0 km/h in reverse, how fast *relative to the car* does the image of the tree appear to be approaching?

70. A plane mirror reflects a beam of light. Show that the rotation of the mirror by an angle α causes the beam to rotate through an angle 2α.

71. A 3.00-cm-high pin, when placed at a certain distance in front of a concave mirror, produces an upright image 9.00 cm high, 30.0 cm from the mirror. Find the position of the pin relative to the mirror and the image. Draw a ray diagram to illustrate.

72. A dentist holds a small mirror 1.9 cm from a surface of a patient's tooth. The image formed is upright and 5.0 times as large as the object. (a) Is the image real or virtual? (b) What is the focal length of the mirror? Is it concave or convex? (c) If the mirror is moved closer to the tooth, does the image get larger or smaller? (d) For what range of object distances does the mirror produce an upright image?

73. An object of height 5.00 cm is placed 20.0 cm from a converging lens of focal length 15.0 cm. Draw a ray diagram to find the height and position of the image.

74. A letter on a page of the compact edition of the *Oxford English Dictionary* is 0.60 mm tall. A magnifying glass (a single thin lens) held 4.5 cm above the page forms an image of the letter that is 2.4 cm tall. (a) Is the image real or virtual? (b) Where is the image? (c) What is the focal length of the lens? Is it converging or diverging?

75. An object is placed 10.0 cm in front of a lens. An upright, virtual image is formed 30.0 cm away from the lens. What is the focal length of the lens? Is the lens converging or diverging?

© 76. A manufacturer is designing a shaving mirror, which is intended to be held close to the face. If the manufacturer wants the image formed to be upright and as large as possible, what characteristics should he choose? (type of mirror?, long or short focal length relative to the object distance of face to mirror?)

✦77. Show that the deviation angle δ for a ray striking a thin converging lens at a distance d from the principal axis is given by $\delta = d/f$. Therefore, a ray is bent through an angle δ that is proportional to d and does *not* depend on the angle of the incident ray (as long as it is paraxial). [*Hint*: Look at Fig. 23.60 and use the small-angle approximation $\sin \theta \approx \tan \theta \approx \theta$ (in radians)].

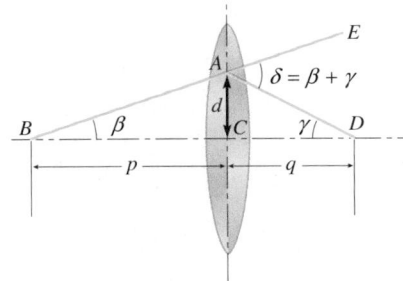

Figure 23.60
Problem 77 (Angles are greatly exaggerated for ease in labeling.)

78. The focal length of a thin lens is –20.0 cm. A screen is placed 160 cm from the lens. What is the *y*-coordinate of the point where the light ray shown hits the screen (Fig. 23.61)? The incident ray is parallel to the central axis and is 1.0 cm from that axis.

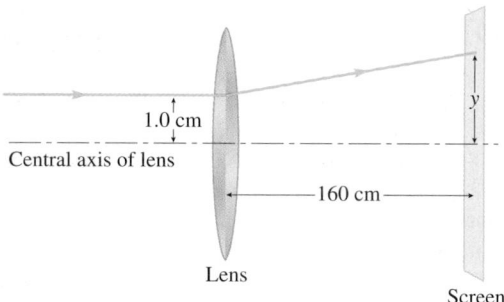

Figure 23.61   Problem 78

◆◆ 79. The angle of deviation through a triangular prism is defined as the angle between the incident ray and the emerging ray (angle δ). It can be shown that when the angle of incidence i is equal to the angle of refraction r' for the emerging ray, the angle of deviation is at a minimum. Show that the minimum deviation angle ($\delta_{min} = D$) is related to the prism angle A and the index of refraction n, by

$$n = \frac{\sin \frac{1}{2}(A + D)}{\sin \frac{1}{2}A}$$

[*Hint*: For an isosceles triangular prism, the minimum angle of deviation occurs when the ray inside the prism is parallel to the base, as shown in Fig. 23.62.]

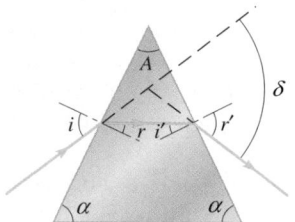

Figure 23.62   Problem 79

80. A ray of light is reflected from two mirrored surfaces as shown in Fig. 23.63. If the initial angle of incidence is 34°, what are the values of angles α and β? (The figure is not to scale.)

Figure 23.63   Problem 80

81. A beam of light consisting of a mixture of red, yellow, and blue light originates from a source submerged in some carbon disulfide. The light beam strikes an interface between the carbon disulfide and air at an angle of incidence of 37.5° as shown in Fig. 23.64. The carbon disulfide has the following indices of refraction for the wavelengths present: red (656.3 nm), $n = 1.6182$; yellow (589.3 nm), 1.6276; blue (486.1 nm), 1.6523. Which color(s) is/are recorded by the detector located above the surface of the carbon disulfide?

Figure 23.64   Problem 81

82. A ray of light is incident normally from air onto a glass ($n = 1.50$) prism as shown in Fig. 23.56. (a) Draw all of the rays that emerge from the prism and give angles to represent their directions. (b) Repeat part (a) with the prism immersed in water ($n = 1.33$). (c) Repeat part (a) with the prism immersed in a sugar solution ($n = 1.50$).

83. A concave mirror has a radius of curvature of 14 cm. If a pointlike object is placed 9.0 cm away from the mirror on its principal axis, where is the image?

84. A glass block ($n = 1.7$) is submerged in an unknown liquid (Fig. 23.65). A ray of light inside the block undergoes total internal reflection. What can you conclude concerning the index of refraction of the liquid?

Figure 23.65   Problem 84

85. Ray diagrams often show objects that conveniently have one end on the principal axis. Draw a ray diagram and locate the image for the object shown in Fig. 23.66 that passes through the principal axis.

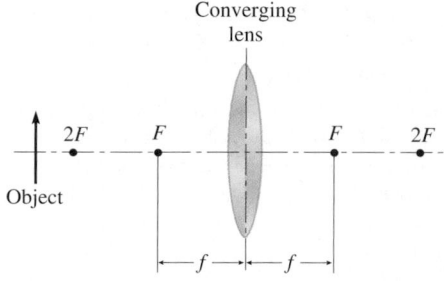

Figure 23.66
Problem 85

86. A 5.0-cm-tall object is placed 50.0 cm from a lens with focal length −20.0 cm. (a) How tall is the image? (b) Is the image upright or inverted?

◆ 87. The vertical displacement d of light rays parallel to the axis of a lens is measured as a function of the vertical displacement h of the incident ray from the principal axis (Fig. 23.67a). The data are graphed in Fig. 23.67b. The distance D from the lens to the screen is 1.0 m. What is the focal length of the lens for paraxial rays?

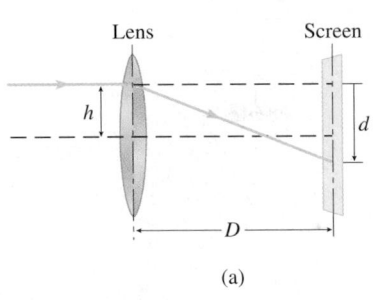

(a)   (b)

Figure 23.67 Problem 87

88. A rectangular block of glass ($n_g = 1.7$) is suspended in air. A ray of light enters the block at an angle α as shown in Fig. 23.68. The transmitted ray makes an angle of 30.0° with the normal. (a) What is α? (b) Sketch a ray diagram to show what happens when the ray strikes the bottom surface of the glass. Find the angle(s) that each ray makes with the normal at the glass surface. (c) The glass is now partially immersed in a liquid (index n_l). What value(s) can n_l have to ensure that there will be a ray leaving the glass at the bottom surface and entering the liquid?

Figure 23.68
Problem 88

ANSWERS TO PRACTICE PROBLEMS

23.1

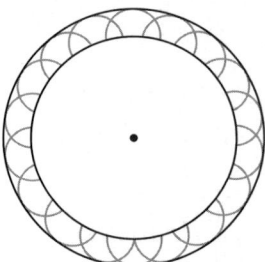

23.2 51°

23.3 If $\theta_i = 0$, then $\theta_t = 0$ and the angle of incidence at the back of the prism is 45°, which is larger than the critical angle (41.8°). If $\theta_i > 0$, then $\theta_t > 0$ and the angle of incidence at the back is greater than 45°.

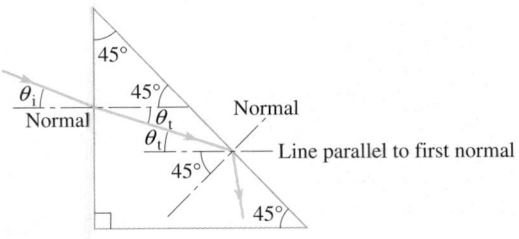

23.4 $\frac{4}{3}h$

23.5 No, she can't see her feet; the bottom of the mirror is 10 cm too high.

23.6 12 cm from the mirror, 3.0 cm tall, real

23.7 $p = 6.00$ cm, $f = +12$ cm, concave

23.8

23.9 –12.0 cm (diverging)

Chapter 24

Optical Instruments

The Hubble Space Telescope, orbiting the Earth at an altitude of about 600 km, was launched in 1990 by the crew of the space shuttle *Discovery*. What is the advantage of having a telescope in space when there are telescopes on Earth with larger light-gathering capabilities? What justifies the cost of two billion dollars to place this 12.5-ton instrument into orbit?

Concepts & Skills to Review

- distinction between real and virtual images (Section 23.6)
- magnification (Section 23.8)
- refraction (Section 23.3)
- thin lenses (Section 23.9)
- finding images with ray diagrams (Section 23.6)

In a series of lenses, the image formed by one lens serves as the object for the next lens.

Rays from a real object are diverging as they enter a lens; rays from a virtual object are *converging* as they enter a lens.

24.1   LENSES IN COMBINATION

Optical instruments generally involve two or more lenses in combination. Let's start this chapter by considering what happens when light rays emerging from a lens pass through another lens so that the image formed by the first lens serves as the object for the second lens.

Suppose that light rays diverge from a point on the image formed by the first lens. These rays are refracted by the second lens the same way as if they were coming from a point on an object. Therefore, the location and size of the image formed by the second lens can be found by applying the lens equation, where the object distance p is the distance from the image formed by the first lens to the second lens. For lenses in combination, we apply the lens equation to each lens in turn, where the object for a given lens is the image formed by the previous lens. Remember that for any application of the lens equation, the object and image distances p and q are measured from the center of the same lens. This same procedure holds true for combinations of lenses and mirrors.

In Chapter 23, all objects were real; p was always positive. With more than one lens, it is possible to have a **virtual object** for which p is *negative*. If one lens produces a real image that would have formed *past* the second lens—so that the rays are converging to a point past the second lens—that image becomes a virtual object for the second lens. Before the real image could form from the first lens, the presence of the second lens intervenes; the rays striking the second lens are converging to a point rather than diverging from a point. This seemingly complicated situation is treated simply by using a negative object distance for a virtual object.

When a lens forms a real image, *its position with respect to the second lens* determines whether it is a real or a virtual object for the second lens. If the first lens would have formed a real image past the second lens, the image becomes a virtual object for the second lens. If the first lens forms a real or virtual image before the second lens, the image is a real object for the second lens.

For a system of two thin lenses separated by a distance s, we can apply the thin lens equation separately to each lens:

$$\frac{1}{p_1} + \frac{1}{q_1} = \frac{1}{f_1}$$

$$\frac{1}{p_2} + \frac{1}{q_2} = \frac{1}{f_2}$$

The object distance p_2 for the second lens is

$$p_2 = s - q_1 \tag{24-1}$$

Equation (24-1) gives the correct sign for p_2 in every case. If $q_1 < s$, then the image formed by the first lens is on the incident side of the second lens and, thus, is a real object for the second lens ($p_2 > 0$). If $q_1 > s$, then the second lens interrupts the light rays before they form an image. The image that would have been formed by the first lens is beyond the second lens, so the image becomes a virtual object for the second lens ($p_2 < 0$).

Conceptual Example 24.1

Virtual Image as Object

Two lenses are used in combination. Suppose the first lens forms a virtual image. Does that image serve as a virtual object for the second lens?

Strategy   The distinction between a real and virtual object depends on whether the rays incident on the second lens are converging or diverging.

Solution and Discussion   If the first lens forms a virtual image, then the rays from any point on the object *diverge* as they emerge from the first lens. To find the image point, we trace those rays backward to find the point from which they seem to originate. Since the rays incident on the second lens

continued on next page

Conceptual Example 24.1 *continued*

are diverging, the image must become a *real* object for the second lens.

Another approach: the image formed by the first lens is located *before* the second lens (that is, on the same side as the incident light rays). Thus, the rays behave as if they diverge from an actual object at the same location—as a real object.

Conceptual Practice Problem 24.1   Real image as object

Two lenses are used in combination. Suppose the first lens forms a *real* image. Does that image serve as a real object or as a virtual object for the second lens? If either is possible, what determines whether the object is real or virtual?

Ray Diagrams for Two Lenses

In a ray diagram for a two-lens system, *only one of the principal rays for the first lens is a principal ray for the second lens*. Figure 24.1 shows a system where lens 1 is converging and lens 2 is diverging. Ray 1 comes from the object through focal point F_1' and emerges from lens 1 parallel to the principal axis. Ray 1 is a principal ray for lens 2, emerging as if it came directly from F_2. In the absence of lens 2, ray 1 would have continued parallel to the axis. To locate the image formed by lens 1, we choose another principal ray (ray 2) and trace it, ignoring lens 2. These two rays locate the image formed by lens 1. Since it lies beyond lens 2, it becomes a virtual object. We do not yet know what happens to ray 2 when it strikes lens 2.

To find the final image, we need another principal ray for lens 2. Ray 3 passes undeflected through the center of lens 2; we extrapolate it back through lens 1 to the object. The intersection of rays 1 and 3 locates the final image, which is virtual. Now we can finish ray 2; it must emerge from lens 2 as if coming from the image point.

Magnification

Suppose N lenses are used in combination. Let h be the size of the object, h_1 the size of the image formed by the first lens, and so forth. Since

$$\frac{h_N}{h} = \frac{h_1}{h} \times \frac{h_2}{h_1} \times \frac{h_3}{h_2} \times \cdots \times \frac{h_N}{h_{N-1}}$$

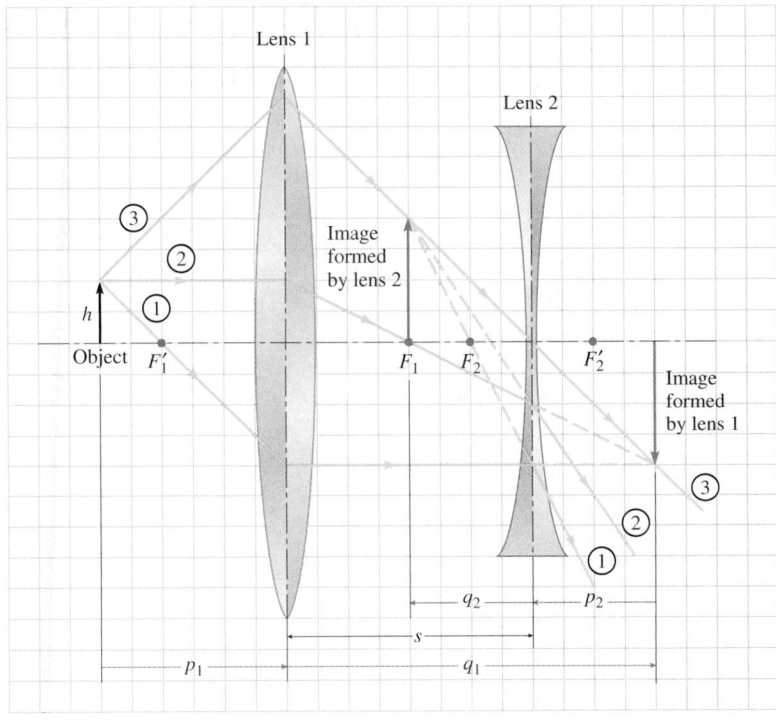

Figure 24.1 Ray diagram for a two-lens combination

the total transverse magnification due to the N lenses is the product of the magnifications due to the individual lenses:

Total transverse magnification

$$m_{\text{total}} = m_1 \times m_2 \times \cdots \times m_N \qquad (24\text{-}2)$$

Example 24.2

Two Converging Lenses

Two converging lenses, separated by a distance of 40.0 cm, are used in combination. The focal lengths are $f_1 = +10.0$ cm and $f_2 = +12.0$ cm. An object, 4.00 cm high, is placed 15.0 cm in front of the first lens. Find the intermediate and final image distances, the total transverse magnification, and the height of the final image.

Strategy   We draw a diagram to help visualize what is happening and then apply the lens equation to each lens in turn. The total magnification is the product of the separate magnifications due to the two lenses.

Given: $p_1 = +15.0$ cm; $f_1 = +10.0$ cm; $f_2 = +12.0$ cm; separation $s = 40.0$ cm

To find: q_1; q_2; m; h'

Solution   Figure 24.2 is a ray diagram that uses two principal rays for each lens to find the intermediate and final images. From the ray diagram, we expect that the intermediate image is real and to the left of lens 2; the final image is virtual, inverted, to the left of lens 1, and greatly enlarged.

The thin lens equation, applied to lens 1, enables us to solve for q_1.

$$\frac{1}{p_1} + \frac{1}{q_1} = \frac{1}{f_1} \qquad (23\text{-}10)$$

Rearranging the equation and substituting values, we have

$$\frac{1}{q_1} = \frac{1}{f_1} - \frac{1}{p_1} = \frac{1}{10.0 \text{ cm}} - \frac{1}{15.0 \text{ cm}} = \frac{1}{30.0 \text{ cm}}$$

Therefore, $q_1 = +30.0$ cm.

From Fig. 24.2, the object distance for lens 2 (p_2) is the separation of the two lenses (s) minus the image distance for the image formed by lens 1 (q_1).

$$p_2 = s - q_1 = 40.0 \text{ cm} - 30.0 \text{ cm} = 10.0 \text{ cm}$$

The object distance is positive because the object is real: it is on the left of lens 2 and the rays from the object are diverging as they enter lens 2. We apply the thin lens equation to the second lens to find q_2.

$$\frac{1}{q_2} = \frac{1}{f_2} - \frac{1}{p_2} = \frac{1}{12.0 \text{ cm}} - \frac{1}{10.0 \text{ cm}} = -\frac{1}{60.0 \text{ cm}}$$

$$q_2 = -60.0 \text{ cm}$$

The image is 60.0 cm to the left of lens 2 or, equivalently, 20.0 cm to the left of lens 1. The image distance is negative, so the image is virtual.

For a single lens the magnification is

$$m = -\frac{q}{p} \qquad (23\text{-}9)$$

For a combination of two lenses the total magnification is

$$m = m_1 \times m_2 = -\frac{q_1}{p_1} \times \left(-\frac{q_2}{p_2}\right)$$

$$= \left(-\frac{30.0 \text{ cm}}{15.0 \text{ cm}}\right) \times \left(-\frac{-60.0 \text{ cm}}{10.0 \text{ cm}}\right) = -12.0$$

The final image is inverted, as indicated by the negative value of m, and its height is

$$4.00 \text{ cm} \times 12.0 = 48.0 \text{ cm}$$

Discussion   Now we compare the numerical results with the ray diagram. As expected, the intermediate image is real and to the left of lens 2 ($q_1 = 30.0$ cm $< s = 40.0$ cm). The final image is virtual ($q_2 < 0$), inverted ($m < 0$), and enlarged ($|m| > 1$).

Practice Problem 24.2   Object located at more than twice the focal length

Repeat Example 24.2 if the same object is placed 25.0 cm before the first lens and the second lens is moved so it is only 10.0 cm from the first lens. Are you able to predict anything about the final image by sketching a ray diagram?

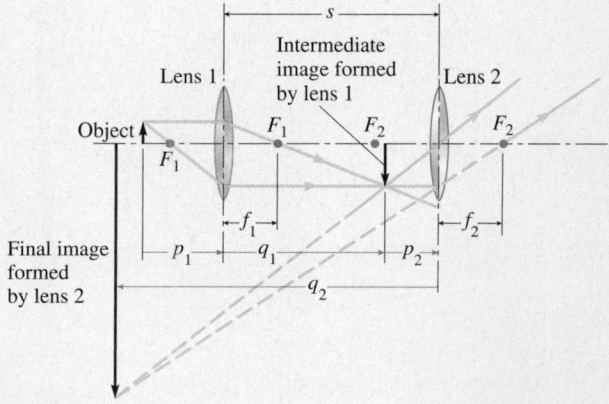

Figure 24.2
Tracing the rays for two converging lenses of Example 24.2

(a)

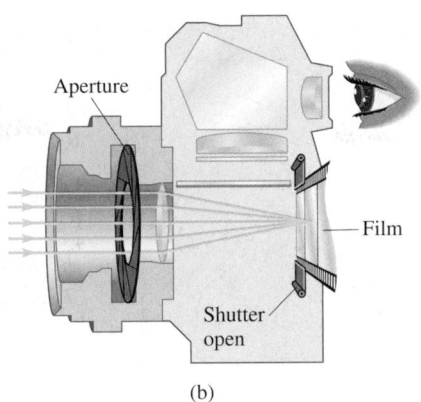
(b)

Figure 24.3 This 35-mm camera uses a single converging lens to form real images on the film. 35 mm is not the focal length of the lens; it is the width of the film. The camera is focused on objects at different distances by moving the lens closer to or farther away from the film. (a) The shutter is closed, preventing exposure of the film. (b) The mirror swings out of the way and the shutter opens for a short time to expose the film.

24.2 CAMERAS

One of the simplest optical instruments is the camera, which often has only one lens to produce an image, or even—in a pinhole camera—no lens. Figure 24.3 shows a simple 35-mm camera. The camera uses a converging lens to form a real image. The image must be real in order to *expose* the film (that is, cause a chemical reaction). Light rays from a point on an object being photographed must converge to a corresponding point on the film. In good-quality cameras the lens is movable so that objects at different distances can be focused onto the film. The lens is attached to a rotating barrel; turning the barrel changes the distance between lens and film. The distance between film and lens must be adjusted in accordance with the lens equation so that a sharp image forms on the film.

For distant objects, the lens must be one focal length from the film. For closer objects, the lens must be a little farther than that, since the image forms past the focal point. Simple fixed focus cameras have a lens that cannot be moved. Such cameras may give good results for faraway objects, but for closer objects it is more important that the lens position be adjustable.

A digital camera does away with film, replacing it with a CCD (charge-coupled device) array that receives the image in much the same way that film does. The digital image is stored by the CCD array until it is transferred to magnetic or some other type of storage. After the transfer, the registers of the CCD array are cleared in preparation for receiving a new image. Eventually the stored images are processed by a computer for viewing on the computer screen or for printing.

A slide or movie projector is the inverse of a camera. A light source is placed at the focal point of a converging lens so that nearly parallel light rays exit the lens and illuminate the slide. Another converging lens then forms an inverted, real image on a distant screen.

Making The Connection:
cameras

Cameras, slide projectors, and movie projectors form real images.

Example 24.3

Fixed Focus Camera

A camera lens has a focal length of 50.0 mm. Photographs are taken of objects located at various positions, from an infinite distance away to as close as 6.00 m from the lens. (a) For an object at infinity, at what distance from the lens is the image formed? (b) For an object at a distance of 6.00 m, at what distance from the lens is the image formed?

Strategy We apply the thin lens equation for the two object distances and find the two image distances.

Solution (a) The thin lens equation is

$$\frac{1}{p} + \frac{1}{q} = \frac{1}{f}$$

For an object at infinity, $1/p = 1/\infty = 0$. Then

$$\frac{1}{\infty} + \frac{1}{q} = \frac{1}{f}$$

Therefore, $q = f$. The image distance is equal to the focal length; the image is 50.0 mm from the lens.

continued on next page

Example 24.3 *continued*

(b) This time $p = 6.00$ m from the camera:

$$\frac{1}{6.00 \text{ m}} + \frac{1}{q} = \frac{1}{50.0 \times 10^{-3} \text{ m}}$$

Solving for q yields

$$\frac{1}{q} = \frac{1}{50.0 \times 10^{-3} \text{ m}} - \frac{1}{6.00 \text{ m}}$$

or

$$q = 50.4 \text{ mm}$$

Discussion   The images are formed within 0.4 mm of each other, so the camera can form reasonably well focused images for objects from 6 m to infinity with a fixed distance between lens and film.

Practice Problem 24.3   Close-up photograph

Suppose the same lens is used with an adjustable camera to take a photograph of an object at a distance of 1.50 m. To what distance from the film should the lens be moved?

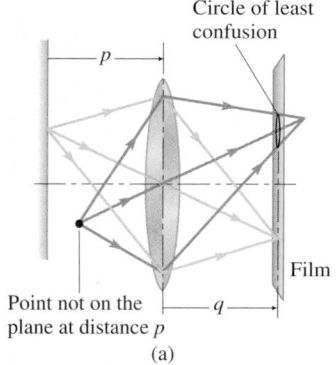

Circle of least confusion

Point not on the plane at distance p

Film

(a)

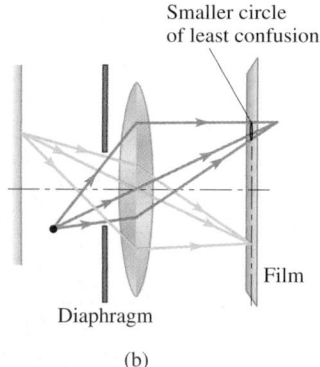

Smaller circle of least confusion

Film

Diaphragm

(b)

Figure 24.4 (a) The circle of least confusion for a point not on the plane in focus. (b) Reduction of the aperture size reduces the circle of least confusion and thereby increases the depth of field.

Regulating Exposure of the Film

A diaphragm made of overlapping metal blades acts like the iris of the eye; it regulates the size of the *aperture*—the opening through which light is allowed into the camera (see Fig. 24.3). The *shutter* is the mechanism that regulates the *exposure time*—the time interval during which light is allowed through the aperture. The aperture size and exposure time are selected so that the correct amount of light energy reaches the film. If they are chosen incorrectly, the film is over- or underexposed.

Depth of Field

Once a lens is focused by adjusting its distance q from the film, only objects in a plane at a particular distance p from the lens form sharp images on the film. Rays from a point on an object not in this plane expose a *circle* on the film (the *circle of least confusion*) instead of a single point (Fig. 24.4a). For some range of distances from the plane, the circle of least confusion is small enough to form an acceptably clear image on the film. This range of distances is called the *depth of field*.

A diaphragm can be placed before the lens to reduce the aperture size, reducing the size of the circle of least confusion (Fig. 24.4b). Thus, reducing the aperture size causes an increase in the depth of field. The trade-off is that, with a smaller aperture, a longer exposure time is necessary to correctly expose the film, which can be problematic if the subject is in motion or if the camera is not held steady by a tripod. Some compromise must be made between using a small aperture—so that more of the surroundings are focused—and using a short exposure time so that motion of the subject and/or camera does not blur the image.

Pinhole Camera

Even simpler than a camera with one lens is a **pinhole camera**, or *camera obscura* ("dark room"). To make a pinhole camera, a tiny pinhole is made in one side of a box (Fig. 24.5a). An inverted, real "image" is formed on the opposite side of the box. A small hole in one side enables the image to be viewed. A photographic plate (a glass plate coated with a photosensitive emulsion) or film placed on the back wall can record the image.

Artists made use of the camera obscura by working in a chamber with a small opening that admitted light rays from a scene outside the chamber (Fig. 24.5b). Using two mirrors or prisms, the image could be projected right side up onto a canvas. Then the artist could trace the outline of the scene on the canvas. Jan van Eyck, Titian, Caravaggio, Vermeer, and Canaletto are just a few of the artists known or believed to have used a camera obscura to achieve realistic naturalism (Fig. 24.5c). In the eighteenth and nineteenth centuries, the camera obscura was commonly used to copy paintings and prints.

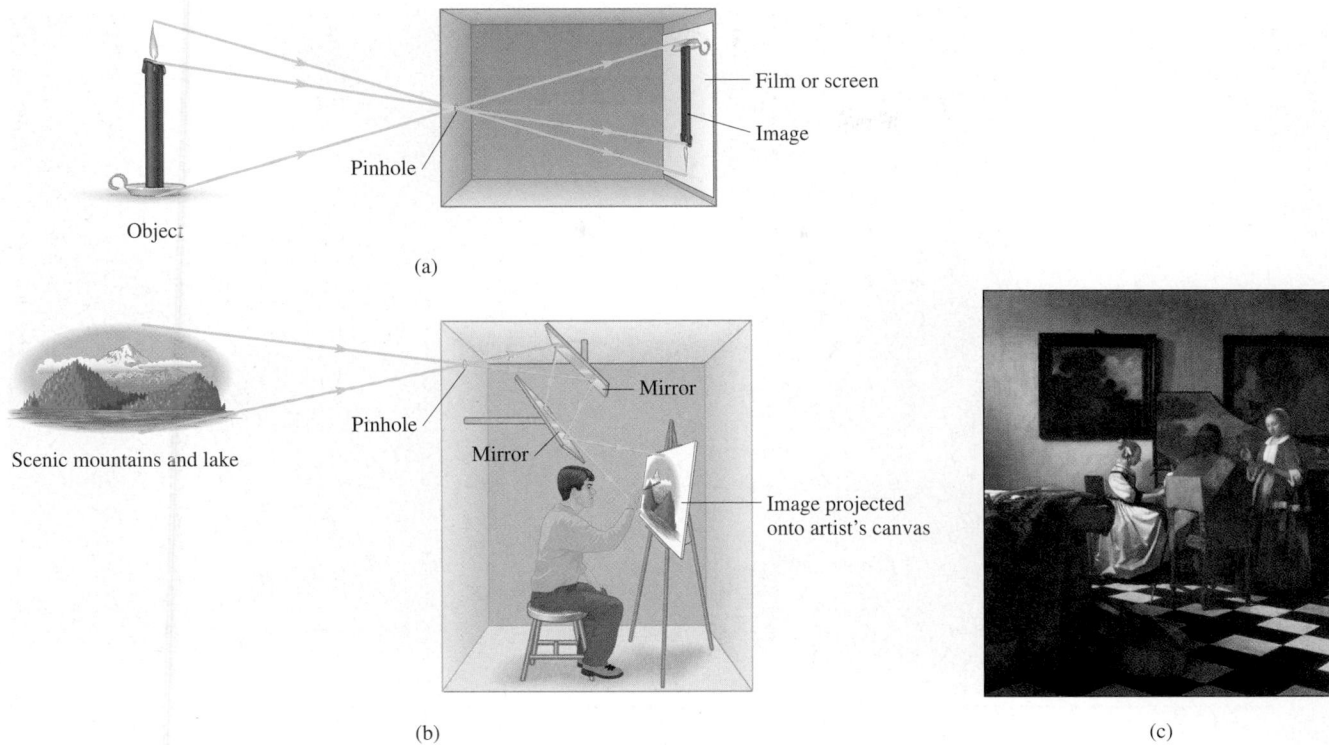

(a)

(b)

(c)

Figure 24.5   (a) A small pinhole camera. (b) A camera obscura used by an artist. (c) *The Concert* was painted by Jan Vermeer around 1666. A camera obscura probably contributed to the accuracy of the perspective and the near-photographic detail in Vermeer's paintings.

Physics at Home

A safe way to view the Sun is through a pinhole camera arrangement (Fig. 24.6). (This is a good way to view a solar eclipse.) Poke a pinhole in a piece of cardboard, a paper plate, or an aluminum pie pan. Then hold a white sheet of cardboard below the pinhole and view the image of the Sun on it. (Remember not to look directly at the Sun, even during an eclipse; severe damage to your eyes can occur.)

The pinhole camera does not form a *true* image—rays from a point on an object do not converge to a single point on the wall. The pinhole admits a narrow cone of rays diverging from each point on the object; the cone of rays makes a small circular spot on the wall. If this spot is small enough, the image appears clear to the eye. A smaller pinhole results in a sharper "image" unless the hole is so small that diffraction spreads the spots out significantly.

24.3   THE EYE

The human eye is similar to a digital camera. The camera forms a real image on a CCD array; the eye forms a real image on the *retina*, a membrane with approximately 125 million photoreceptor cells (the *rods* and *cones*). The focusing mechanism is different, though. In the camera, the lens moves toward or away from the film to keep the image on the film as the object distance changes. In the eye, the lens is at a fixed distance from the retina, but it has a variable focal length; the focal length is adjusted to keep the image distance constant as the object distance varies.

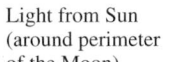

Light from Sun
(around perimeter
of the Moon)

Pinhole

Image of eclipse

Figure 24.6   A pinhole camera arrangement for viewing an eclipse of the Sun

Making The Connection:
human eye

Figure 24.7 Anatomy of the eyeball

Simplified model of the human eye: a single converging lens of variable focal length at a fixed distance from the retina

Figure 24.7 shows the anatomy of the eyeball. It is approximately spherical, with an average diameter of 2.5 cm. A bulge in front is filled with the *aqueous fluid* (or aqueous "humor") and covered on the outside by a transparent membrane called the *cornea*. The aqueous fluid is kept at an overpressure to maintain the slight outward bulge. The curved surface of the cornea does most of the refraction of light rays entering the eye. The adjustable *lens* does the fine tuning. For most purposes, we can consider the cornea and the lens to act like a single lens, about 2.0 cm from the retina, with adjustable focal length (see Problem 6). In order to see objects at distances of 25 cm or greater from the eye, which is considered normal vision, the focal length of the eye must vary between 1.85 cm and 2.00 cm if the retina is 2.00 cm from the eye (see Problem 13).

The spherical volume of the eyeball behind the lens is filled with a jelly-like material called the *vitreous fluid*. The indices of refraction of the aqueous fluid and the vitreous fluid are approximately the same as that of water (1.333). The index of the lens, made of a fibrous, jelly-like material, is a bit higher (1.437). The cornea has an index of refraction of 1.351.

The eye has an adjustable aperture (the *pupil*) that functions like the diaphragm in a camera to control the amount of light that enters. The size of the pupil is adjusted by the *iris*, a ring of muscular tissue (the colored portion of the eye). In bright light, the iris expands to reduce the size of the pupil and limit the amount of light entering the eye. In dim light, the iris contracts to allow more light to enter through the dilated pupil. The expansion and contraction of the iris is a *reflex* action in response to changing light conditions. In ordinary light the diameter of the pupil is about 2 mm; in dim light it is about 8 mm.

On the retina, the photoreceptor cells (mostly cones) are densely concentrated in a small region called the *macula lutea*. The cones contain visual pigments; there are three different types that respond to different wavelengths of light (Fig. 24.8). Thus, the cones are responsible for color vision. Centered within the macula lutea is the *fovea*

Figure 24.8 Sensitivity of the rods and three types of cones to different wavelengths

Physics at Home

Each retina has a *blind spot* with no rods or cones, located where the optic nerve leaves the retina. The blind spot is not usually noticed because the brain fills in the missing information. To observe the blind spot, draw a cross and a dot, about 10 cm apart, on a sheet of white paper. Cover your left eye and hold the paper far from your eyes with the dot on the right. Keep your eye focused on the cross as you slowly move the paper toward your face. The dot disappears when the image falls on the blind spot. Continue to move the paper even closer to your eye; you will see the spot again when its image moves off the blind spot.

centralis, of diameter 0.25 mm, where the cones are tightly packed together and where the most acute vision occurs in bright light. The muscles that control eye movement ensure that the image of an object being examined is centered on the fovea centralis.

The rods are more sensitive to dim light than the cones but do not have different types sensitive to different wavelengths, so we cannot distinguish colors in dim light. Outside the macula the photoreceptor cells are much less densely packed and they are all rods. However, the rods outside the macula are more densely packed than the rods inside the macula. If you are trying to see a dim star in the sky, it helps to look a little to the side of the star so the image of the star falls outside the macula where there are more rods.

Accommodation

Variation in the focal length of the flexible lens is called **accommodation**; it is the result of an actual change in the shape of the lens of the eye through the action of the *ciliary muscles*. The adjustable shape of the lens allows for accommodation for various object distances, while still forming an image at the fixed image distance determined by the separation of lens and retina. When the object being viewed is far away, the ciliary muscles relax; the lens is relatively flat and thin, giving it a longer focal length lens (Fig. 24.9a). For closer objects, the ciliary muscles squeeze the lens into a thicker, more rounded shape (Fig. 24.9b), giving the lens a shorter focal length.

Accommodation enables an eye to form a sharp image on the retina of objects at a range of distances from the **near point** to the **far point**. A young adult with good vision has a near point at 25 cm or less and a far point at infinity. A child can have a near point as small as 10 cm. Corrective lenses (eyeglasses or contact lenses) or surgery can compensate for an eye with a near point greater than 25 cm and/or a far point less than infinity.

Figure 24.9 The lens of the eye has (a) a longer focal length when viewing distant objects and (b) a shorter focal length when viewing nearby objects.

Viewing distant object, longer focal length
(a)

Viewing nearby object, shorter focal length
(b)

Optometrists write prescriptions in terms of the **refractive power** (P) of a lens rather than the focal length. The refractive power is simply the reciprocal of the focal length:

$$P = \frac{1}{f} \tag{24-3}$$

Refractive power is usually measured in **diopters** (symbol D), which is an inverse meter ($1\ \mathrm{D} = 1\ \mathrm{m}^{-1}$). The shorter the focal length, the more "powerful" the lens because the rays are bent more. Converging lenses have positive refractive powers and diverging lenses have negative refractive powers.

Why use refractive power instead of focal length? When two or more thin lenses with refractive powers $P_1, P_2, \ldots$ are sufficiently close together, they act as a single thin lens with refractive power

$$P = P_1 + P_2 + \cdots \tag{24-4}$$

as can be shown in Problem 6 by substituting P for $1/f$.

Making The Connection:
correcting myopia

Myopia

A myopic eye can see nearby objects clearly but not distant objects. Myopia (nearsightedness) occurs when the shape of the eyeball is elongated or when the curvature of the cornea is excessive. A myopic eye forms the image of a distant object *before* the retina (Fig. 24.10a). The refractive power of the lens is too large; the eye makes the rays converge too soon. A diverging corrective lens (with negative refractive power) can compensate for nearsightedness by bending the rays outward (Fig. 24.10b).

For objects at any distance from the eye, the diverging corrective lens forms a virtual image closer to the eye than is the object. For an object at infinity, the corrective lens forms an image *at the far point* of the eye (Fig. 24.10c). For less distant objects, the virtual image is closer than the far point. The eye is able to focus rays from this image onto the retina since it is never past the far point.

Example 24.4

Correction for Nearsighted Eye

Without her contact lenses, Dana cannot see clearly an object more than 40.0 cm away. What refractive power should her contact lenses have to give her normal vision?

Strategy The far point for Dana's eyes is 40.0 cm. For an object at infinity, the corrective lens must form a virtual image 40.0 cm from the eye. We use the lens equation with $p = \infty$ and $q = -40.0$ cm to find the focal length or refractive power of the corrective lens. The image distance is negative because the image is virtual—it is formed on the same side of the lens as the object.

Solution The thin lens equation is

$$\frac{1}{p} + \frac{1}{q} = \frac{1}{f} = P$$

Since $p = \infty$, $1/p = 0$. Then

$$0 + \frac{1}{-40.0\ \mathrm{cm}} = \frac{1}{f}$$

continued on next page

Figure 24.10 (a) In a nearsighted eye, parallel rays from a point on a distant object converge before they reach the retina. (b) A diverging lens corrects for the nearsighted eye by bending the rays outward just enough that the eye brings them back together at the retina. (c) The diverging lens forms a virtual image closer to the eye than the object; the eye can make the rays from this image converge into a real image on the retina.

Example 24.4 *continued*

Solving for the focal length,
$$f = -40.0 \text{ cm}$$
The refractive power of the lens in diopters is the inverse of the focal length in meters.

$$P = \frac{1}{f} = \frac{1}{-40.0 \times 10^{-2} \text{ m}} = -2.50 \text{ D}$$

Discussion   The focal length and refractive power are negative, as expected for a diverging lens. We might have anticipated that $f = -40.0$ cm without using the thin lens equation. Rays coming from a distant source are nearly parallel. Parallel

rays incident on a diverging lens emerge such that they appear to come from the focal point before the lens. Thus, the image is at the focal point on the incident side of the lens.

Practice Problem 24.4   What happens to the near point?

Suppose Dana's *near* point (without her contact lenses) is 10.0 cm. What is the closest object she can see clearly with her contact lenses on? [*Hint:* For what object distance do the contact lenses form a virtual image 10.0 cm before the lenses?]

Hyperopia

A hyperopic (farsighted) eye can see distant objects clearly but not nearby objects; the near point is too large. The refractive power of the eye is too small; the cornea and lens do not refract the rays enough to make them converge on the retina (Fig. 24.11a). A converging lens can correct for hyperopia by bending the rays inward so they converge sooner (Fig. 24.11b). In order to have normal vision, the near point should be 25 cm (or less). Thus, for an object at 25 cm from the eye, the corrective lens forms a virtual image at the eye's near point.

Making The Connection:
correcting hyperopia

Example 24.5

Correction for Farsighted Eye

Winifred is unable to focus on objects closer than 2.50 m from her eyes. What refractive power should her corrective lenses have?

Strategy   For an object 25 cm from Winifred's eye, the corrective lens must form a virtual image at the near point of Winifred's eye (2.50 m from the eye). We use the thin lens equation with

continued on next page

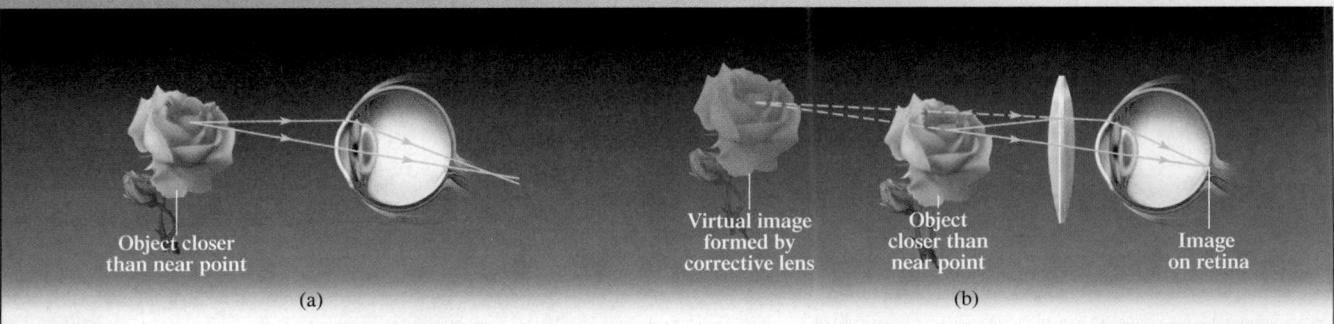

Object closer
than near point

(a)

Virtual image
formed by
corrective lens

Object
closer than
near point

Image
on retina

(b)

Figure 24.11   (a) A farsighted eye forms an image of a nearby object past the retina. (b) A converging corrective lens forms a virtual image farther away from the eye than the object. Rays from this virtual image can be brought together by the eye to form a real image on the retina.

Example 24.5 *continued*

$p = 25$ cm and $q = -2.50$ m to find the focal length. As in the last example, the image distance is negative because it is a virtual image formed on the same side of the lens as the object.

Solution   From the thin lens equation

$$\frac{1}{p} + \frac{1}{q} = \frac{1}{f}$$

Substituting $p = 0.25$ m and $q = -2.50$ m

$$\frac{1}{0.25 \text{ m}} + \frac{1}{-2.50 \text{ m}} = \frac{1}{f}$$

Solving for the focal length

$$f = 0.28 \text{ m}$$

The refractive power is

$$P = \frac{1}{f} = +3.6 \text{ D}$$

Discussion   This solution assumes that the corrective lens is very close to the eye, as for a contact lens. If Winifred wears eyeglasses that are 2.0 cm away from her eyes, then the object and image distances we should use—since they are measured from the *lens*—are $p = 23$ cm and $q = -2.48$ m. The thin lens equation then gives $f = 0.25$ m and $P = +4.0$ D.

Practice Problem 24.5   Using eyeglasses

A man can clearly see an object that is 2.00 m away (or more) without using his eyeglasses. If the eyeglasses have a refractive power of +1.50 D, how close can an object be to the eyeglasses and still be clearly seen by the man? Assume the eyeglasses are 2.0 cm from the eye.

Presbyopia

As a person ages, the lens of the eye becomes less flexible and the eye's ability to accommodate decreases, a phenomenon known as presbyopia. Older people have difficulty focusing on objects held close to the eyes; from the age of about 40 years many people need eyeglasses for reading. At age 60, a near point of 50 cm is typical; in some people it may be 1 m or even more. Reading glasses for a person suffering from presbyopia are similar to those used by a farsighted person.

24.4   THE SIMPLE MAGNIFIER

We use magnifiers and microscopes to enlarge objects too small to see with the naked eye. But what do we mean by *enlarged* in this context? The apparent size of an object depends on the size of the image *formed on the retina* of the eye. For the unaided eye, the retinal image size is proportional to the angle subtended by the object. Figure 24.12 shows two identical objects being viewed from different distances. Imagine rays from

Figure 24.12 Identical objects viewed from different distances. Rays drawn from the top and bottom of the nearer object illustrate the angle θ subtended by the object.

the top and bottom of each object that are incident on the center of the lens of the eye. The angle θ is called the **angular size** of the object. The image on the retina subtends the same angle θ; the angular size of the image is the same as that of the object. Rays from the object at a greater distance subtend a smaller angle; the angular size depends on distance from the eye.

A magnifying glass, microscope, or telescope serves to make the image on the retina larger *than it would be if viewed with the unaided eye*. Since the size of the image on the retina is proportional to the angular size, we measure the usefulness of an optical instrument by its **angular magnification**—the ratio of the angular size using the instrument to the angular size with the unaided eye.

Definition of angular magnification

$$M = \frac{\theta_{\text{aided}}}{\theta_{\text{unaided}}} \qquad (24\text{-}5)$$

The transverse magnification (the ratio of the image size to the object size) isn't as useful here. The transverse magnification of a telescope-eye combination is minute: the Moon is *much* larger than the image of the Moon on the retina, even using a telescope. The telescope makes the image of the Moon larger than it would be in unaided viewing.

Physics at Home

On a clear night with the Moon visible, go outside, shut one eye, and hold a penny or a quarter at arm's length between your open eye and the Moon so that the coin blocks your view of the Moon (Fig. 24.13). Notice that the angular size of the Moon is similar to the angular size of the coin. Estimate the distance from your eye to the coin and the coin's diameter. Use this information and the Earth-Moon distance (4×10^5 km) to estimate the diameter of the Moon. Compare your estimate to the actual diameter of the Moon (3.5×10^3 km).

Figure 24.13 Comparing the angular sizes of the Moon and a coin

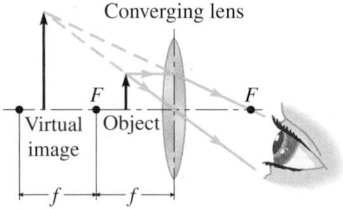

Converging lens

F   *F*

Virtual   Object
image

f   *f*

Figure 24.14 A converging lens used as a magnifying glass forms an enlarged virtual image. The object distance is less than the focal length.

When you want to see something in greater detail, you naturally move your eye closer to the object to increase the angular size of the object. But the eye's ability to accommodate for nearby objects is limited; anything closer than the near point cannot be seen clearly. Thus, the maximum angle subtended at the unaided eye by an object occurs when the object is located at the near point.

A **simple magnifier** is a converging lens placed so that the object distance is less than the focal length. The virtual image formed is enlarged, upright, and farther away from the lens than the object (Fig. 24.14). Using the magnifier, the object can be placed closer to the eye than the near point and the position of the magnifier adjusted so that an enlarged virtual image is formed no closer than the near point. Typically, the image is put well beyond the near point so that it is viewed by a more relaxed eye at the expense of a small reduction in angular magnification. The object being viewed is held closer to the eye than would be possible without the magnifying glass while the image seen by the eye is at a comfortable viewing distance. The angle subtended by the enlarged virtual image seen by the eye is much larger than the angle subtended by the object when placed at the near point.

If a small object of height h is viewed with the unaided eye (Fig. 24.15a), the angular size when it is placed at the near point (a distance N from the eye) is

$$\theta \approx \frac{h}{N}$$

where we assume $h \ll N$ and, thus, θ is small. If the object is now placed at the focal point of a converging lens, the image is formed at infinity so that it can be viewed with a relaxed eye (Fig. 24.15b). The angular size of the image is

$$\theta_\infty \approx \frac{h}{f}$$

Then the angular magnification M is

$$M = \frac{\theta_\infty}{\theta} = \frac{h/f}{h/N} = \frac{N}{f} \tag{24-6}$$

> When calculating the angular magnification of an optical instrument, it is customary to assume a typical near point of $N = 25$ cm.

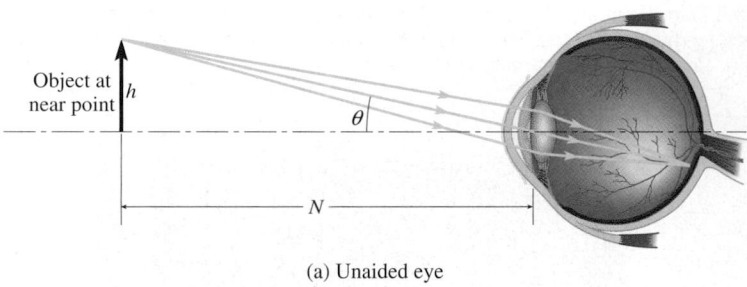

Object at near point h

θ

N

(a) Unaided eye

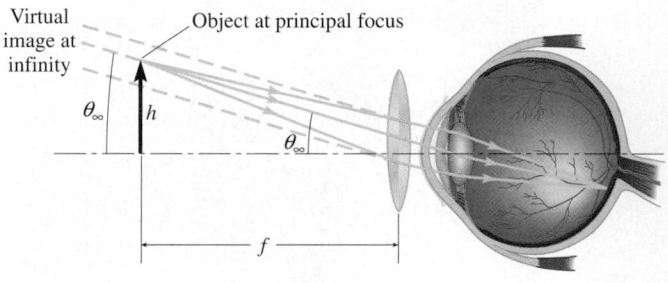

Virtual image at infinity

Object at principal focus

θ_∞ h

θ_∞

f

(b) Eye aided by converging lens

Figure 24.15 (a) The angle θ subtended at the eye by an object placed at the near point. (b) The magnifier forms a virtual image of the object at infinity. The angle θ_∞ subtended by the virtual image is larger than θ.

Equation (24-6) gives the angular magnification when the object is placed at the focal point of the magnifier. If the object is placed closer to the magnifier, the angular magnification is somewhat larger. In many cases the small increase in angular magnification is not worth the added eyestrain of viewing an image closer to the eye (see Problem 23).

Example 24.6

A Magnifying Glass

A converging lens with a focal length of 4.00 cm is used as a simple magnifier. The lens forms a virtual image at your near point, 25.0 cm from your eye. Where should the object be placed and what is the angular magnification? Assume that the magnifier is held close to your eye.

Strategy   We can use 25.0 cm as the image distance from the *lens*; if the magnifier is near the eye, distances from the lens are approximately the same as distances from the eye. We apply the thin lens equation to find the object distance with the focal length and image distance known.

Solution   By rearranging the thin lens equation to solve for the object distance, we obtain

$$p = \frac{fq}{q-f}$$

We now substitute $q = -25.0$ cm (negative for a virtual image) and $f = +4.00$ cm.

$$p = \frac{4.00 \text{ cm} \times (-25.0 \text{ cm})}{-25.0 \text{ cm} - 4.00 \text{ cm}}$$

$$= 3.45 \text{ cm}$$

The object is placed 3.45 cm from the lens. The angular size (in radians) of the image formed is

$$\theta = \frac{h}{p}$$

where h is the size of the object. The object is *not* at the focal point of the lens, so the angular size is not h/f as it is in Fig. 24.15b. If the object were to be viewed without the magnifier, while placed at the near point of $N = 25$ cm, the angular size would be

$$\theta_0 = \frac{h}{N}$$

The angular magnification is

$$M = \frac{h/p}{h/N} = \frac{N}{p} = \frac{25.0 \text{ cm}}{3.45 \text{ cm}} = 7.25$$

Discussion   If the object had been placed at the principal focus, 4.00 cm from the lens, to form a final image at infinity, the angular magnification would have been

$$M = \frac{N}{f} = \frac{25.0 \text{ cm}}{4.00 \text{ cm}} = 6.25$$

Practice Problem 24.6   Where to place an object with a magnifier

The focal length of a simple magnifier is 12.0 cm. Assume the viewer's eye is held close to the lens. (a) What is the angular magnification of an object if the magnifier forms a final image at the viewer's near point (25.0 cm)? (b) What is the angular magnification if the final image is at infinity?

24.5   COMPOUND MICROSCOPES

Making The Connection:
microscopes

The simple magnifier is limited to angular magnifications of 15–20 at most. By contrast, the **compound microscope**, which uses two converging lenses, enables angular magnifications of 2000 or more. Invention of the compound microscope is credited to Antonie van Leeuwenhoek (1632–1723) in the late seventeenth century; just a few years after its invention, it made the world of microorganisms available to study.

A small object to be viewed under the microscope is placed *just beyond* the focal point of a converging lens called the **objective**. The function of the objective is to form an enlarged real image. A second converging lens, called the **ocular** or **eyepiece**, is used to view the real image formed by the objective lens (Fig. 24.16). The eyepiece acts as a simple magnifier; it forms an enlarged virtual image. The position of the final image can be anywhere between the near point of the observer and infinity. Usually it is placed at infinity, since that enables viewing with a relaxed eye and doesn't decrease the angular magnification very much. To form a final image at infinity, the image formed by the objective is located at the focal point of the eyepiece. Inside the barrel of the microscope, the positions of the two lenses are adjusted so that the image formed by the objective falls at or within the focal point of the eyepiece.

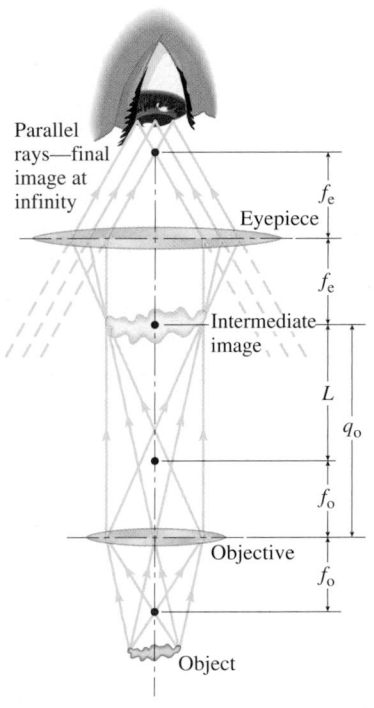

Figure 24.16 A compound microscope. To form a final image at infinity, the intermediate image must be located at the focal point of the eyepiece.

If we used just the eyepiece as a simple magnifier to view the object, the angular magnification would be

$$M_e = \frac{N}{f_e} \quad \text{(due to eyepiece)}$$

where f_e is the focal length of the eyepiece and the virtual image is at infinity for ease of viewing. Customarily we assume $N = 25$ cm. The objective forms an image that is larger than the object; as shown in Problem 30, the transverse magnification due to the objective is

$$m_o = -\frac{L}{f_o} \quad \text{(due to objective)}$$

where L (the **tube length**) is the distance between the focal points of the two lenses. Since the image of the objective is placed at the focal point of the eyepiece, as in Fig. 24.16, the tube length is

$$L = q_o - f_o \qquad (24\text{-}7)$$

Many microscopes are designed with a tube length of 16 cm.

When we view the image with the eyepiece, the eyepiece provides the same angular magnification as before (M_e), but it magnifies an image already m_o times as large as the object. The total angular magnification is the product of m_o and M_e:

Angular magnification due to a microscope
$$M_{total} = m_o M_e = -\frac{L}{f_o} \times \frac{N}{f_e} \qquad (24\text{-}8)$$

The minus sign in Eq. (24-8) means that the final image is inverted. Sometimes with microscopes and telescopes the sign is ignored and the angular magnification (also called the *magnifying power*) is reported as a positive number.

Equation (24-8) shows that, for large magnification, both focal lengths should be small. Microscopes are often made so that any one of several different objective lenses can be swung into position, depending on the magnification desired. The manufacturer usually provides the values of the magnification ($|m_o|$ and M_e) instead of the focal lengths of the lenses. For example, if an eyepiece is labeled "5×," then $M_e = 5$.

Example 24.7

Magnification by a Microscope

A compound microscope has an objective lens of focal length 1.40 cm and an eyepiece with a focal length of 2.20 cm. The objective and the eyepiece are separated by 19.6 cm. The final image is at infinity. (a) What is the angular magnification? (b) How far from the objective should the object be placed?

Strategy We first find the tube length of the microscope. From Fig. 24.16, the distance between the lenses is the sum of the two focal lengths plus the tube length. We assume the typical near point of 25 cm. Since the final image is at infinity, Eq. (24-8) can be used to find the angular magnification. To find where the object should be placed, we apply the thin lens equation to the objective. The image formed by the objective is at the focal point of the eyepiece since the final image is at infinity.

Given: $f_o = 1.40$ cm, $f_e = 2.20$ cm, lens separation = 19.6 cm

To find: (a) total angular magnification M; (b) object distance p_o

Solution   (a) The tube length is
$$L = \text{distance between lenses} - f_o - f_e$$
$$= 19.6 \text{ cm} - 1.40 \text{ cm} - 2.20 \text{ cm} = 16.0 \text{ cm}$$

Then the angular magnification is

$$M = -\frac{L}{f_o} \times \frac{N}{f_e}$$
$$= -\frac{16.0 \text{ cm}}{1.40 \text{ cm}} \times \frac{25 \text{ cm}}{2.20 \text{ cm}} = -130$$

The negative magnification indicates that the final image is inverted.

(b) To have the final image at infinity, the image formed by the objective lens must be located at the focal point of the eyepiece. From Fig. 24.16, the intermediate image distance is

$$q_o = L + f_o = 16.0 \text{ cm} + 1.40 \text{ cm} = 17.4 \text{ cm}$$

Then the object distance is found using the thin lens equation

$$\frac{1}{p_o} + \frac{1}{q_o} = \frac{1}{f_o}$$

continued on next page

Example 24.7 *continued*

Solving for the object distance, p

$$p_o = \frac{f_o q_o}{q_o - f_o}$$

$$= \frac{1.40 \text{ cm} \times 17.4 \text{ cm}}{17.4 \text{ cm} - 1.40 \text{ cm}}$$

$$= 1.52 \text{ cm}$$

Discussion We can check the result for part (b) to see if the object is just past the focal point of the objective. The object is 1.52 cm from the objective while the focal point is 1.40 cm, so the object is just 1.2 mm past the focal point.

Practice Problem 24.7 Object distance for good focus

An observer with a near point of 25 cm looks through a microscope with an objective lens of focal length $f_o = 1.20$ cm. When an object is placed 1.28 cm from the objective, the angular magnification is –198 and the final image is formed at infinity. (a) What is the tube length L for this microscope? (b) What is the focal length of the eyepiece?

The Transmission Electron Microscope

Many other kinds of microscope, both optical and nonoptical are in use. The one most similar to the optical compound microscope is the *transmission electron microscope* (TEM). In the 1920s, Ernst Ruska (1906–1988) found that a magnetic field due to a coil could act as a lens for electrons. An optical lens functions by changing the directions of the light rays; the magnetic coil changes the directions of the electrons' trajectories. Ruska was able to use the lens to form an image of an object irradiated with electrons. Eventually he coupled two such lenses together to form a microscope. By 1933 he had produced the first electron microscope, using an electron beam to form images of tiny objects with far greater clarity than the conventional optical microscope. Ruska's microscope is called a *transmission* microscope because the electron beam passes right through the thin slice of a sample being studied, just as light passes through the object in a light microscope.

Resolution

A large magnification is of little use if the image is blurry. *Resolution* is the ability to form clear and distinct images of points very close to each other on an object. High resolution is a desirable quality in a microscope. The ultimate limit on the resolution of an optical instrument is limited by diffraction—the spreading out of light rays. Diffraction limits the size of an object, or the separation of two objects, that can be distinguished to approximately the wavelength of the light used. Thus, we cannot expect to see anything smaller than about 400 nm using a compound optical microscope. An atom, roughly 0.2–0.5 nm in diameter, is much smaller than the wavelength of light, so a microscope cannot resolve details on the atomic scale. Ultraviolet microscopes can do a little better (about 100 nm) due to the shorter wavelength. Transmission electron microscopes can resolve details down to about 0.5 nm.

24.6 TELESCOPES

Making The Connection:
telescopes

Refracting Telescopes

The most common type of telescope for nonscientific work is the **refracting telescope**, which has two converging lenses that function just as those in a compound microscope. The refracting telescope has an objective lens that forms a real image of the object; the eyepiece (ocular) is used to view this real image. The microscope is used to view *tiny* objects placed close to the objective lens; the purpose of the objective is to form an *enlarged* image. The telescope is used to view objects whose *angular* sizes are small because they are far away; the objective forms an image that is tiny compared to the object, but the image is available for closeup viewing through the eyepiece.

In an *astronomical* refracting telescope, the object is so far away that the rays from a point on the object can be assumed to be parallel; the object distance is taken as infinity (Fig. 24.17). The objective forms a real, diminished image at its principal focus. By placing this image at the secondary focal point of the eyepiece, the final image is at infinity for ease of viewing. Thus, the principal focal point of the objective must coincide with the secondary focal point of the eyepiece, in contrast to the microscope in which the two are separated by a distance L (the tube length). Astronomical telescopes are often connected to a camera to record the images, in which case the camera lens replaces the eyepiece. The image formed by the objective is *not* placed at the focal point of the camera lens; the camera lens must form a real image on the film.

The objective is located at one end of the telescope barrel and the eyepiece is at the other end. Then the *barrel length* of the telescope is the sum of the focal lengths of the objective and the eyepiece.

$$\text{barrel length} = f_o + f_e \tag{24-9}$$

In Fig. 24.17, a highlighted ray passing through the secondary focal point of the objective leaves the lens parallel to the principal axis, then continues to the eyepiece and is refracted so that it goes through the principal focal point of the eyepiece. Two small right triangles from the diagram are redrawn below the diagram for clarity. The hypotenuse (AC, FD) of each triangle is along the highlighted ray. The leg (BC, EF) of each triangle from the principal axis to the hypotenuse is of length h because the line connecting C to F is parallel to the principal axis and passes through the tip of the image.

The angle that would be subtended if viewed by the unaided eye is the same as the angle subtended at the objective (θ_o). The angle subtended at the observer's eye looking through the eyepiece at the final image formed at infinity is θ_e. From the two small right triangles and the small angle approximation, the angular size of the object for the unaided eye is

$$\theta_o \approx \tan \theta_o = \frac{h}{AB} = \frac{h}{f_o}$$

The angular size of the final image is

$$\theta_e \approx \tan \theta_e = -\frac{h}{DE} = -\frac{h}{f_e}$$

The final image is inverted, so its angular size is negative. With a telescope the magnification that is of interest is again the *angular* magnification: the ratio of the angle subtended at the eye by the final magnified image to the angle subtended for the unaided eye. Then the angular magnification is

---

Angular magnification due to an astronomical telescope

$$M = \frac{\theta_e}{\theta_o} = -\frac{f_o}{f_e} \tag{24-10}$$

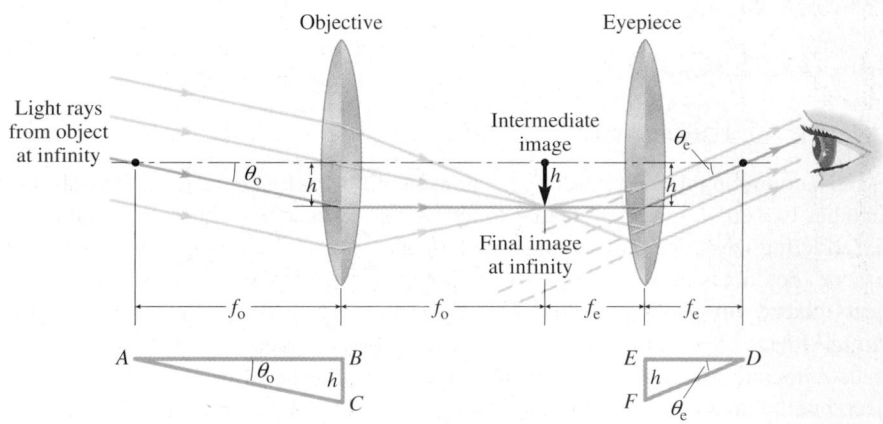

Figure 24.17 An astronomical refracting telescope

where the minus sign indicates an inverted image. As for microscopes, the angular magnification is usually reported as a positive number. For the greatest magnification, the objective lens has as long a focal length as possible, while the eyepiece has as short a focal length as possible.

Tips

Example 24.8

Yerkes Refracting Telescope

The Yerkes telescope in southern Wisconsin (Fig. 24.18) is the largest refracting telescope in the world. Its objective lens is 1.016 m (40 in.) in diameter and has a focal length of 19.8 m (65 ft). If the magnifying power is 508, what is the focal length of the eyepiece?

Strategy The magnifying power is the magnitude of the angular magnification. For an astronomical refracting telescope, the angular magnification is negative.

Solution From Eq. (24.10), the angular magnification is

$$M = \frac{\theta_e}{\theta_o} = -\frac{f_o}{f_e}$$

Solving for f_e yields

$$f_e = -\frac{f_o}{M}$$

Now we substitute $M = -508$ and $f_o = 19.8$ m:

$$f_e = -\frac{19.8 \text{ m}}{-508} = 3.90 \text{ cm}$$

Discussion The focal length of the eyepiece is positive, which is correct. The eyepiece serves as a simple magnifier used to view the image formed by the objective. The simple magnifier is a converging lens—that is, a lens with positive focal length.

Practice Problem 24.8 Replacing the eyepiece

If the eyepiece used with the Yerkes telescope in Example 24.8 is changed to one with focal length 2.54 cm that produces a final image at infinity, what is the new angular magnification?

Figure 24.18
The Yerkes refracting telescope was the world's largest telescope when it was first operated in 1897 and is still the largest *refracting* telescope in the world. It is part of the Department of Astronomy and Astrophysics at the University of Chicago.

An inverted image is no problem when the telescope is used as an astronomical telescope. When a telescope is used to view terrestrial objects, such as a bird perched high on a tree limb or a rock singer on stage at an outdoor concert, the final image must be upright. Binoculars are essentially a pair of telescopes. The reflecting prisms not only allow the eyepieces to come together so that they can be viewed simultaneously; they also invert the image so that the final image is upright.

Another way to make a terrestrial telescope is to add a third lens between the objective and the eyepiece to invert the image again so that the final image is upright. The Galilean telescope, invented by Galileo in 1609, produces an upright image without using a third lens. The upright image is obtained by using a *diverging* lens as the eyepiece (see Problem 34). The eyepiece is located so that the image formed by the objective becomes a *virtual* object for the eyepiece, which then forms an upright virtual image. The barrel length for a Galilean telescope is shorter than for telescopes with only converging lenses.

Reflecting Telescopes

Reflecting telescopes use one or more mirrors in place of lenses. There are several advantages to mirrors over lenses; these advantages become overwhelming in the large telescopes that must be used to gather enough light rays to be able to see distant, faint

Making The Connection:
reflecting
telescopes

stars. (Large telescopes also minimize the loss of resolution due to diffraction.) Since the index of refraction varies with wavelength, a lens has slightly different focal lengths for different wavelengths; thus, dispersion distorts the image. A mirror works by reflection rather than refraction, so it has the same focal length for all wavelengths. Large mirrors are much easier to build than large lenses. When making a large glass lens, the glass becomes so heavy that it deforms due to its own weight. It also suffers from stresses and strains as it cools from a molten state; such stresses reduce the optical quality of the lens. A large mirror need not be so heavy, since only the *surface* is important; it can be supported everywhere under its surface, while a lens can only be supported at the edge. Another advantage of the reflecting telescope is that the heaviest part—the large concave mirror—is located at the base of the telescope, making the instrument stable. The largest lens used with a refracting telescope—a little over 1 m (3.3 ft) in diameter—is in the Yerkes telescope. By comparison, the mirror in each of the twin Keck telescopes in Hawaii is ten times as large—10 m (33 ft) in diameter—making them the world's largest optical (and infrared) telescopes.

Figure 24.19 shows one kind of reflecting telescope, known as the Cassegrain arrangement. Parallel light rays from a distant star are reflected from a concave mirror toward its focal point F. Before the rays can reach the focal point, they are intercepted in their path by a smaller convex mirror. The convex mirror directs the rays through a hole in the center of the large concave mirror so that they come to a focus at a point P. A camera or an electronic recording instrument can be placed at point P, or a lens can be used to direct the rays to a viewer's eye.

Making The Connection:
Hubble Space
Telescope

Hubble Space Telescope

A famous telescope using the Cassegrain arrangement is the Hubble Space Telescope (HST). The HST orbits Earth at an altitude of over 600 km; its mirror is 2.4 m in diameter (Fig. 24.20). Why put a telescope in orbit? The atmosphere limits the amount of detail that is seen by any telescope on Earth. The density of the air in the atmosphere at any location is continually fluctuating; as a result, light rays from distant stars are bent by different amounts, making it impossible to bring the rays to a sharp focus. There are systems that correct for atmospheric fluctuations, but since the HST is above the atmosphere, it avoids the whole problem.

The HST had troubles at the beginning; the first images were blurry because of a mistake made in the lens specifications. Ever since astronauts fitted the telescope with corrective lenses in 1993, the images have been spectacular. NASA reports that 13,670 objects were found during the first decade of use. Accomplishments of the HST include clear images of quasars, the most energetic objects of the universe; the first surface map of Pluto; the discovery of intergalactic helium left over from the big bang (the birth of the universe); and clear evidence for the existence of black holes (objects so dense that nothing, not even light, can escape their gravitational pull). The HST has demonstrated the existence of gravitational lenses, in which the gravity from massive galaxies bends light rays inward like a lens to form images of even more distant objects behind them.

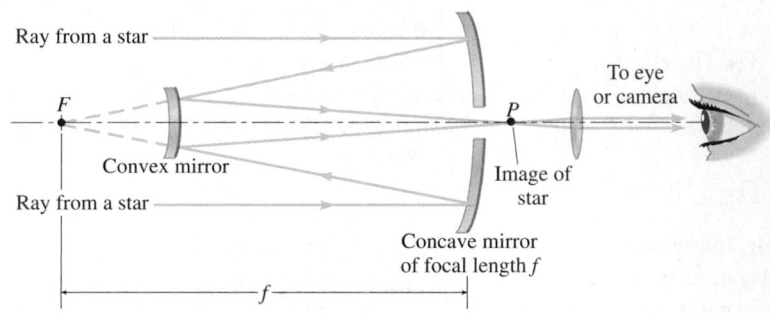

Figure 24.19   A reflecting telescope directing rays from a distant star to an eye or a camera.

Figure 24.20 Four stunning images captured by the Advanced Camera for Surveys aboard the Hubble Space Telescope. Clockwise from the upper left: (1) A galaxy known as the "Tadpole" collides with a much smaller galaxy, producing a long tail of stars and gas. (2) The Cone Nebula, a pillar of cold gas and dust. Hydrogen atoms absorb ultraviolet radiation and emit light, causing the red "halo" around the pillar. (3) Collision of two spiral galaxies known as the "Mice." A similar fate may await our galaxy a few billion years from now. (4) The center of the Omega Nebula, a region of flowing gas and newly formed stars surrounded by a cloud of hydrogen. Light emitted by excited atoms of nitrogen and sulfur produces the rose-colored region right of center. Other colors are produced by excited atoms of hydrogen and oxygen.

The HST has provided a deeper look back in time than any other optical telescope, providing views of galaxies at an early stage of the universe and evidence for the age of the universe. In 2010 NASA plans to replace the HST with the Next-Generation Telescope (NGT) with a mirror several times larger than the HST. It will be placed a million miles from Earth on the side away from the Sun.

Radio Telescopes

Making The Connection:
radio telescopes

The EM radiation traveling to Earth from celestial bodies is not limited to the visible part of the spectrum. Radio telescopes detect radio waves from space. The radio telescope at Arecibo, Puerto Rico (Fig. 24.21), is the most sensitive radio telescope in the world. The bowl of the telescope, 305 m (0.19 mi) in diameter and 51 m (167 ft) deep, is made from metallic mesh panels instead of solid metal; it reflects just as well as a solid metal surface because the holes are much smaller than the wavelengths of the radio waves. A detector is suspended in midair at the focal point, 137 m above the bowl. Arecibo takes only a few minutes to gather information from a radio source that would require several hours of observation with a smaller radio telescope.

A home satellite dish is a small version of a radio telescope. It is directed toward a satellite and forms a real image of the microwaves beamed down to Earth from the satellite. When the dish is properly aimed to receive the signal sent by the satellite from a TV station, the microwaves of that station are focused on the antenna of the receiver.

24.7   ABERRATIONS OF LENSES AND MIRRORS

Real lenses and mirrors deviate from the behavior of an ideal lens or mirror. There are many kinds of **aberrations**; in this section we consider only two of them. In Chapter 25 we consider in more detail the limits on the resolution of optical instruments due to diffraction.

Figure 24.21 The radio telescope at Arecibo, Puerto Rico, occupies nearly twenty acres of a remote hilltop region.

Spherical Aberration

The derivations of the lens and mirror equations assume that light rays are paraxial—that they are nearly parallel to the principal axis and not too far away from the axis. That assumption enabled us to use small-angle approximations. Rays for which the angle of incidence on the lens or mirror is *not* small do not all converge to the same focal point (or appear to diverge from the same focal point). Then rays from a point on the object do not correspond to a single image point; the image is blurred due to **spherical aberration**. Spherical aberration can be minimized by placing a diaphragm before the lens or mirror so that only rays traveling rather close to the principal axis can reach the lens and pass through it. The image formed is in better focus (sharper) but is less bright.

For mirrors, spherical aberration can be avoided by using a *parabolic* mirror instead of a spherical mirror. A parabolic mirror focuses parallel rays to a point even if they are not paraxial. Large astronomical reflecting telescopes use parabolic mirrors. Since light rays are reversible, if a point light source is placed at the focal point of a parabolic mirror, the reflected rays form a parallel beam. Searchlights and automobile headlights use parabolic reflectors to send out fairly parallel rays in a well-defined beam of light.

Chromatic Aberration

Another aberration of lenses (but not of mirrors) is due to dispersion. When light composed of several wavelengths passes through a lens, the various wavelengths are bent by differing amounts due to dispersion; they do not focus at the same point. This lens defect is called **chromatic aberration**. One way to correct for chromatic aberration is to use a combination of lenses consisting of different materials so that the aberrations from one lens cancel those from another.

Physics at Home

Look at a TV or computer monitor with your unaided eye (or through your usual eyeglasses). Then make a fist with one hand, leaving an opening like a small tunnel to peer through with one eye. Look at the same screen through your hand; you will see a much sharper image. Slowly expand the opening made by your fist and watch the image become fuzzy. People with poor or changing eyesight sometimes squint in order to see more clearly.

MASTER THE CONCEPTS

Summary

- If one lens produces a real image that would have formed *past* a second lens—so that the rays are converging to a point past the second lens—that image becomes a *virtual object* for the second lens. In the thin lens equation, p is *negative* for a virtual image.

- When the image formed by one lens serves as the object for a second lens a distance s away, the object distance p_2 for the second lens is

$$p_2 = s - q_1 \qquad (24\text{-}1)$$

- The total transverse magnification of an image formed by two or more lenses is the product of the magnifications due to the individual lenses.

$$m_{\text{total}} = m_1 \times m_2 \times \cdots \times m_N \qquad (24\text{-}2)$$

- A typical camera has a single converging lens. To focus on an object, the distance between the lens and the film is adjusted so that a real image is formed on the film.

- The aperture size and the exposure time must be chosen to allow just enough light to expose the film. The *depth of field* is the range of distances from the plane of sharp focus for which the lens forms an acceptably clear image on the film. Greater depth of field is possible with a smaller aperture.

- In the human eye, the cornea and the lens refract light rays to form a real image on the photoreceptor cells in the retina. For most purposes, we can consider the cornea and the lens to act like a single lens with an adjustable focal length. The adjustable shape of the lens allows for accommodation for various object distances, while still forming an image at the fixed image distance determined by the separation of lens and retina. The nearest and farthest object distances that the eye can accommodate are called the near point and far point. A young adult with good vision has a near point at 25 cm or less and a far point at infinity.

- The refractive power of a lens is the reciprocal of the focal length:

$$P = \frac{1}{f} \qquad (24\text{-}3)$$

Refractive power is measured in diopters ($1\ \text{D} = 1\ \text{m}^{-1}$).

- A myopic (nearsighted) eye has a far point less than infinity; for objects past the far point, it forms an image before the retina. A diverging corrective lens (with negative refractive power) can compensate for nearsightedness by bending light rays outward.

- A hyperopic (farsighted) eye has too large a near point; the refractive power of the eye is too small. For objects closer than the near point, the eye forms an image past the retina. A converging lens can correct for hyperopia by bending the rays inward so they converge sooner.

- As a person ages, the lens of the eye becomes less flexible and the eye's ability to accommodate decreases, a phenomenon known as presbyopia.

- *Angular* magnification is the ratio of the angular size using the instrument to the angular size as viewed by the unaided eye.

$$M = \frac{\theta_{\text{aided}}}{\theta_{\text{unaided}}} \qquad (24\text{-}5)$$

- The simple magnifier is a converging lens placed so that the object distance is less than the focal length. The virtual image formed is enlarged and upright. If the image is formed at infinity for ease of viewing, the angular magnification M is

$$M = \frac{N}{f} \qquad (24\text{-}6)$$

where N, the near point, is usually taken to be 25 cm.

- The compound microscope consists of two converging lenses. A small object to be viewed is placed *just beyond* the focal point of the objective, which forms an enlarged real image. The eyepiece (ocular) acts as a simple magnifier to view the image formed by the objective. If the final image is at infinity, the angular magnification due to the microscope is

$$M_{\text{total}} = m_{\text{o}} M_{\text{e}} = -\frac{L}{f_{\text{o}}} \times \frac{N}{f_{\text{e}}} \qquad (24\text{-}8)$$

where N is the near point (usually 25 cm) and L (the tube length) is the distance between the focal points of the two lenses.

- An astronomical refracting telescope uses two converging lenses. As in the microscope, the objective forms a real image and the eyepiece functions as a magnifier for viewing the real image. The angular magnification is

$$M = -\frac{f_{\text{o}}}{f_{\text{e}}} \qquad (24\text{-}10)$$

- In a reflecting telescope, a concave mirror takes the place of the objective lens.

- Spherical aberration occurs because rays that are not paraxial are brought to a focus at a different spot than are paraxial rays.

- Chromatic aberration is caused by dispersion in the lens.

Highlighted Figures and Tables

F24.1 Ray diagram for a two-lens combination (p. 881)

F24.3 This 35-mm camera uses a single converging lens to form real images on the film (p. 883)

F24.7 Anatomy of the eyeball (p. 886)

F24.8 Sensitivity of the rods and cones to different wavelengths (p. 887)

F24.9 The lens of the eye has a longer focal length when viewing distant objects and a shorter focal length when viewing nearby objects (p. 888)

F24.10 A nearsighted eye (p. 889)

F24.11 A farsighted eye (p. 890)

F24.14 A converging lens used as a magnifying glass forms an enlarged virtual image (p. 892)

F24.16 A compound microscope (p. 894)

F24.17 An astronomical refracting telescope (p. 896)

F24.19 A reflecting telescope (p. 898)

CONCEPTUAL QUESTIONS

1. Why must a camera or a slide projector use a converging lens? Why must the objective of a microscope or telescope be a converging lens (or a converging mirror)? Why can the eyepiece of a telescope be either converging or diverging?

2. A magnifying glass can be held over a piece of white paper and its position adjusted until the image of an overhead light is formed on the paper. Explain.

3. If a piece of white cardboard is placed at the location of a virtual object, what (if anything) would be seen on the cardboard?

4. Why is a refracting telescope with a large angular magnification longer than one with a smaller magnification?

5. Why are astronomical observatories often located on mountaintops?

6. Why do some telescopes produce an inverted image?

7. Why is the receiving antenna of a satellite dish placed at a set distance from the dish?

8. Two magnifying glasses are labeled with their angular magnifications. Glass A has a magnification of "2×" ($M = 2$) and glass B has a magnification of 4×. Which has the longer focal length? Explain.

9. What causes chromatic aberration? What can be done to compensate for chromatic aberration?

10. For human eyes about 70% of the refraction occurs at the cornea; less than 25% occurs at the two surfaces of the lens. Why? [*Hint:* Consider the indices of refraction.] Is the same thing true for fish eyes?

11. If rays from points on an object are converging as they enter a lens, is the object real or virtual?

12. What are some of the advantages of using mirrors rather than lenses for astronomical telescopes?

13. When snorkeling, you wear goggles in order to see clearly. Why is your vision blurry without the goggles? A nearsighted person notices that he is able to see nearby objects *more* clearly when she is underwater (without goggles or corrective lenses) than in air (without corrective lenses). Why might this be true?

14. When the muscles of the eye remain tensed for a significant period of time, eyestrain results. How much is this a concern for a person using (a) a microscope, (b) a telescope, and (c) a simple magnifier?

15. Both a microscope and a telescope can be constructed from two converging lenses. What are the differences? Why can't a telescope be used as a microscope? Why can't a microscope be used as a telescope?

16. In her bag, a photographer is carrying three exchangeable camera lenses with focal lengths of 400.0-mm, 50.0-mm, and 28.0-mm. Which lens should she use for (a) wide angle shots (a cathedral, taken from the square in front), (b) everyday use (children at play), and (c) telephoto work (lions in Africa taken from across a river)?

17. Figure 24.22 shows a schematic diagram of a defective eyeball. What is this defect called?

Figure 24.22 Conceptual Question 17

18. Draw a simple eye diagram labeling the cornea, the lens, the iris, the retina, and the aqueous and vitreous fluids.

19. On a camping trip, you discover that no one has brought matches. A friend suggests using his eyeglasses to focus sunlight onto some dry grass and shredded bark to get a fire started. Could this scheme work if your friend is nearsighted? What about if he is farsighted? Explain.

MULTIPLE CHOICE QUESTIONS

1. The compound microscope is made from two lenses. Which statement is true concerning the operation of the compound microscope?
   (a) Both lenses form real images.
   (b) Both lenses form virtual images.
   (c) The lens closest to the object forms a virtual image; the other lens forms a real image.
   (d) The lens closest to the object forms a real image; the other lens forms a virtual image.

2. Which of these statements best explains why a telescope enables us to see details of a distant object such as the Moon or a planet more clearly?
   (a) The image formed by the telescope is larger than the object.
   (b) The image formed by the telescope subtends a larger angle at the eye than the object does.
   (c) The telescope can also collect radio waves that sharpen the visual image.

3. Siu-Ling has a far point of 25 cm. Which statement here is true?
   (a) She may have normal vision.
   (b) She is myopic and requires diverging lenses to correct her vision.
   (c) She is myopic and requires converging lenses to correct her vision.
   (d) She is hyperopic and requires diverging lenses to correct her vision.
   (e) She is hyperopic and requires converging lenses to correct her vision.

4. Figure 24.23 shows a schematic diagram of a defective eyeball and some lenses. Which of the lenses shown can correct for this defect?

5. A nearsighted person wears corrective lenses. One of the focal points of the corrective lenses should be
   (a) at the cornea.   (b) at the retina.   (c) at infinity.
   (d) past the retina.   (e) at the near point.   (f) at the far point.

Figure 24.23 Multiple
Choice Question 4

(a) (b) (c) (d) (e)

6. An astronomical telescope has an angular magnification of 10. The length of the barrel is 33 cm. What are the focal lengths of the objective and the eyepiece, in that order respectively, from the choices listed?

(a) 3 cm, 30 cm   (b) 30 cm, 3 cm

(c) 20 cm, 13 cm   (d) 0.3 m, 3 m

7. A compound microscope has three possible objective lenses (focal lengths f_o) and two eyepiece lenses (focal lengths f_e). For maximum angular magnification, the objective and eyepiece should be chosen such that

(a) f_o and f_e are both the largest available.

(b) f_o and f_e are both the smallest available.

(c) f_o is the largest available, f_e is the smallest available.

(d) f_e is the largest available, f_o is the smallest available.

(e) f_e and f_o are nearly the same.

8. What causes chromatic aberration?

(a) Light is an electromagnetic wave and has intrinsic diffraction properties.

(b) Different wavelengths of light give different angles of refraction at the lens-air interface.

(c) The coefficient of reflection is different for light of different wavelengths.

(d) The outer edges of the lens produce a focus at a different point from that formed by the central portion of the lens.

(e) The absorption of light in the glass varies with wavelength.

9. What causes spherical aberration?

(a) Light is an electromagnetic wave and has intrinsic diffraction properties.

(b) Different wavelengths of light give different angles of refraction at the lens-air interface.

(c) The lens surface is not perfectly smooth.

(d) The outer edges of the lens produce a focus at a different point from that formed by the central portion of the lens.

10. Reducing the aperture on a camera

(a) reduces the depth of field and requires a longer exposure time.

(b) reduces the depth of field and requires a shorter exposure time.

(c) increases the depth of field and requires a longer exposure time.

(d) increases the depth of field and requires a shorter exposure time.

(e) does not change the depth of field and requires a longer exposure time.

(f) does not change the depth of field and requires a shorter exposure time.

PROBLEMS

Note: **Ⓒ** indicates a combination conceptual/quantitative problem. Gold diamonds ◆, ◆◆ are used to indicate the increasing level of difficulty of each problem. Problem numbers appearing in blue, 9., denote problems that have a detailed solution available in the Student Solutions Manual. Some problems are *paired* by concept; their numbers are connected by a ruled box.

24.1 Lenses in Combination

1. An object is placed 12.0 cm in front of a lens of focal length 5.0 cm. Another lens of focal length 4.0 cm is placed 2.0 cm past the first lens. (a) Where is the final image? Is it real or virtual? (b) What is the overall magnification?

2. A converging lens and a diverging lens, separated by a distance of 30.0 cm, are used in combination. The converging lens has a focal length of 15.0 cm. The diverging lens is of unknown focal length. An object is placed 20.0 cm in front of the converging lens; the final image is virtual and is formed 12.0 cm *before* the diverging lens. What is the focal length of the diverging lens?

3. An object is located 16.0 cm in front of a converging lens with focal length 12.0 cm. To the right of the converging lens, separated by a distance of 20.0 cm, is a diverging lens of focal length –10.0 cm. Find the location of the final image by ray tracing and verify using the lens equations.

4. An object is located 10.0 cm in front of a converging lens with focal length 12.0 cm. To the right of the converging lens is a second converging lens, 30.0 cm from the first lens, of focal length 10.0 cm. Find the location of the final image by ray tracing and verify by using the lens equations.

5. Verify the locations and sizes of the images formed by the two lenses in Fig. 24.1 using the lens equation and the following data: $f_1 = +4.00$ cm, $f_2 = -2.00$ cm, $s = 8.00$ cm, $p_1 = +6.00$ cm, and $h = 2.00$ mm. (Note that the vertical scale is different from the horizontal scale.)

◆ 6. Show that if two thin lenses are close together (s is negligibly small), the two lenses can be replaced by a single equivalent lens with focal length f_{eq}. Find the value of f_{eq} in terms of f_1 and f_2.

24.2 Cameras

7. Esperanza uses a 35-mm camera with a standard lens of focal length 50.0 mm to take a photo of her son Carlos, who is 1.2 m tall and standing 3.0 m away. (a) What must be the distance between the lens and the film to get a properly focused picture? (b) What is the magnification of the image? (c) What is the height of the image of Carlos on the film?

8. A person on a safari wants to take a photograph of a hippopotamus from a distance of 75.0 m. The animal is 4.00 m long and its image is to be 1.20 cm long on the film. (a) What focal length lens should be used? (b) What would be the size of the image if a lens of 50.0-mm focal length were used? (c) How close to the hippo would the person have to be to capture a 1.20-cm-long image using a 50.0-mm lens?

9. Jim plans to take a picture of McGraw Tower with a 35-mm camera that has a 50.0-mm focal length lens. A roll of 35-mm film is 35 mm wide; each frame is 24 mm by 36 mm (Fig. 24.24). The tower has a height of 52 m and Jim wants a detailed close-up picture. How close to the tower should Jim be to capture the largest possible image of the entire tower on his film?

10. A photographer wishes to take a photograph of the Eiffel Tower (300 m tall) from across the Seine River, a distance of 300 m from the tower. What focal length lens should she use to get an image that is 20 mm high on the film?

Figure 24.24   A strip of 35-mm film (Problems 9, 11, 12, 38, and 48)

11. If a slide of width 36 mm (see Fig. 24.24) is to be projected onto a screen of 1.50 m width located 12.0 m from the projector, what focal length lens is required to fill the width of the screen?

12. A slide projector has a lens of focal length 12 cm. Each slide is 24 mm by 36 mm (see Fig. 24.24). The projector is used in a room where the screen is 5.0 m from the projector. How large must the screen be?

24.3 The Eye

Unless the problem states otherwise, assume that the distance from the cornea-lens system to the retina is 2.0 cm and the normal near point is 25 cm.

13. If the distance from the lens system (cornea + lens) to the retina is 2.00 cm, show that the focal length of the lens system must vary between 1.85 cm and 2.00 cm to see objects from 25.0 cm to infinity.

14. Suppose that the lens system (cornea + lens) in a particular eye has a focal length that can vary between 1.85 and 2.00 cm, but the distance from the lens system to the retina is only 1.90 cm. (a) Is this eye nearsighted or farsighted? Explain. (b) What range of distances can the eye see clearly without corrective lenses?

15. If Michaela needs to wear reading glasses with refractive power of +3.0 D, what is her uncorrected near point? Neglect the distance between the glasses and the eye.

16. The uncorrected far point of Charlotte's eye is 2.0 m. What refractive power contact lens enables her to clearly distinguish objects at large distances?

17. The distance from the lens system (cornea + lens) of a particular eye to the retina is 1.75 cm. What is the focal length of the lens system when the eye produces a clear image of an object 25.0 cm away?

✦18. A nearsighted woman cannot clearly see objects more than 2.0 m away. The distance from the lens of the eye to the retina is 2.0 cm, and the eye's power of accommodation is 4.0 D (the focal length of the cornea-lens system increases by a maximum of 4.0 D over its relaxed focal length when accommodating for nearby objects). (a) As an amateur optometrist, what corrective eyeglass lenses would you prescribe to enable her to clearly see distant objects? Assume the corrective lenses are 2.0 cm from the eyes. (b) Find the nearest object she can see clearly with and without her glasses.

24.4 The Simple Magnifier

19. (a) What is the focal length of a magnifying glass that gives an angular magnification of 8.0 when the image is at infinity? (b) How far must the object be from the lens? Assume the lens is held close to the eye.

20. Keesha is looking at a beetle with a magnifying glass. She wants to see an upright, enlarged image at a distance of 25 cm. The focal length of the magnifying glass is +5.0 cm. Assume that Keesha's eye is close to the magnifying glass. (a) What should be the distance between the magnifying glass and the beetle? (b) What is the angular magnification?

21. Callum is examining a stamp that is 3.00 cm square in size with a magnifying glass of refractive power +40.0 D. The magnifier forms an image of the stamp at a distance of 25.0 cm. Assume that Callum's eye is close to the magnifying glass. (a) What is the distance between the stamp and the magnifier? (b) What is the angular magnification? (c) How large is the image formed by the magnifier?

22. An insect that is 5.00 mm long is placed 10.0 cm from a converging lens with a focal length of 12.0 cm. (a) What is the position of the image? (b) What is the size of the image? (c) Is the image upright or inverted? (d) Is the image real or virtual? (e) What is the angular magnification if the lens is close to the eye?

✦23. A simple magnifier gives the *maximum* angular magnification when it forms a virtual image at the near point of the eye instead of at infinity. For simplicity, assume that the magnifier is right up against the eye, so that distances from the magnifier are approximately the same as distances from the eye. (a) For a magnifier with focal length f, find the object distance p such that the image is formed at the near point, a distance N from the lens. (b) Show that the angular size of this image as seen by the eye is

$$\theta = \frac{h(N+f)}{Nf}$$

where h is the height of the object. [*Hint:* Refer to Fig. 24.15.] (c) Now find the angular magnification and compare it to the angular magnification when the virtual image is at infinity.

24.5 Compound Microscopes

24. Figure 24.25 shows a schematic diagram of a microscope. For the object and image locations shown, which of the points (A, B. C, or D) represents a focal point of the eyepiece? Draw a ray diagram.

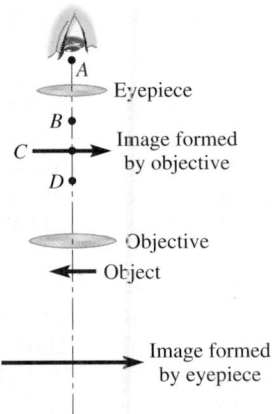

Eyepiece

Image formed by objective

Objective

Object

Image formed by eyepiece

Figure 24.25 Problem 24

25. The wing of an insect is 1.0 mm long. When viewed through a microscope, the image is 1.0 m long and is located 5.0 m away. Determine the angular magnification.

26. A microscope has an eyepiece that gives an angular magnification of 5.00 for a final image at infinity and an objective lens of focal length 15.0 mm. The tube length of the microscope is 16.0 cm. (a) What is the transverse magnification due to the objective lens alone? (b) What is the angular magnification due to the microscope? (c) How far from the objective should the object be placed?

27. Repeat Problem 26(c) using a different eyepiece that gives an angular magnification of 5.00 for a final image at the viewer's near point (25.0 cm) instead of at infinity.

28. A microscope has an objective lens of focal length 5.00 mm. The objective forms an image 16.5 cm from the lens. The focal length of the eyepiece is 2.80 cm. (a) What is the distance between the lenses? (b) What is the angular magnification? The near point is 25.0 cm. (c) How far from the objective should the object be placed?

29. Repeat Problem 28 if the eyepiece location is adjusted slightly so that the final image is at the viewer's near point (25.0 cm) instead of at infinity.

30. Use the thin lens equation to show that the transverse magnification due to the objective of a microscope is $m_o = -L/f_o$. [*Hints:* The object is *near* the focal point of the objective; do not assume it is *at* the focal point. Eliminate p_o to find the magnification in terms of q_o and f_o. How is L related to q_o and f_o?]

24.6 Telescopes

31. A telescope mirror has a radius of curvature of 10.0 m. It is used to take a picture of the Moon (radius 1740 km, distance from Earth 385,000 km). What is the diameter of the image of the Moon produced by this mirror?

32. The objective lens of an astronomical telescope forms an image of a distant object at the focal point of the eyepiece, which has a focal length of 5.0 cm. If the two lenses are 45.0 cm apart, what is the angular magnification?

33. A refracting telescope is used to view the Moon (diameter 3480 km, distance from Earth 385,000 km). The focal lengths of the objective and eyepiece are +2.40 m and +16.0 cm, respectively. (a) What should be the distance between the lenses? (b) What is the diameter of the image produced by the objective? (c) What is the angular magnification?

34. The eyepiece of a Galilean telescope is a diverging lens. The focal points F_o and F_e' coincide (Fig. 24.26). In one such telescope, the lenses are a distance $d = 32$ cm apart and the focal length of the objective is 36 cm. A rhinoceros is viewed from a large distance. (a) What is the focal length of the eyepiece? (b) At what distance from the eyepiece is the final image? (c) Is the final image formed by the eyepiece real or virtual? Upright or inverted? (d) What is the angular magnification? [*Hint:* The angular magnification is β/α.]

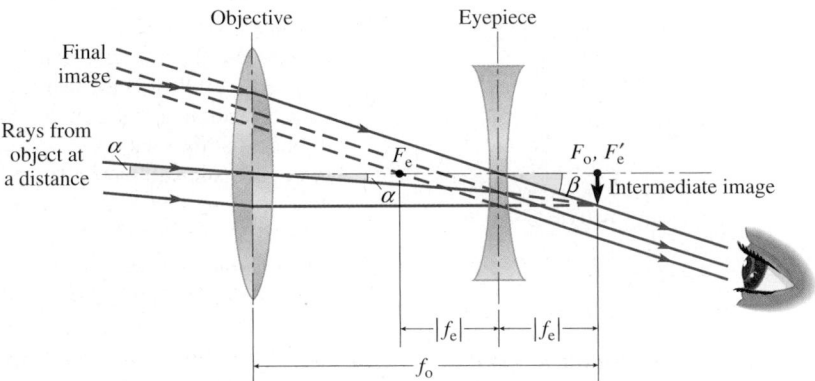

Figure 24.26 Problem 34

COMPREHENSIVE PROBLEMS

35. Two converging lenses, separated by a distance of 50.0 cm, are used in combination. The first lens, located to the left, has a focal length of 15.0 cm. The second lens, located to the right, has a focal length of 12.0 cm. An object, 3.00 cm high, is placed at a distance of 20.0 cm in front of the first lens. (a) Find the intermediate and final image distances relative to the corresponding lenses. (b) What is the total magnification? (c) What is the height of the final image?

36. A camera has a telephoto lens of 240-mm focal length. The lens can be moved in and out a distance of 16 mm from the film plane by rotating the lens barrel. If the lens can focus objects at infinity, what is the closest object distance that can be focused?

© 37. You have two lenses of focal length 25.0 cm (lens 1) and 5.0 cm (lens 2). (a) To build an astronomical telescope that gives an angular magnification of 5.0, how should you use the lenses (which for objective and which for eyepiece)? Explain. (b) How far apart should they be?

38. The Ortiz family is viewing slides from their summer vacation trip to the Grand Canyon. Their slide projector has a projection lens of 10.0-cm focal length and the screen is located 2.5 m from the projector. (a) What is the distance between the slide and the projection lens? (b) What is the magnification of the image? (c) How wide is the image of a slide of width 36 mm on the screen? (See Fig. 24.24.)

✦ 39. Veronique is nearsighted; she cannot see clearly anything more than 6.00 m away without her contacts. One day she doesn't wear her contacts; rather, she wears an old pair of glasses prescribed when she could see clearly up to 8.00 m away. Assume the glasses are 2.0 cm from her eyes. What is the greatest distance an object can be placed so that she can see it clearly with these glasses?

40. An object is placed 20.0 cm from a converging lens with focal length 15.0 cm (Fig. 24.27, not drawn to scale). A concave mirror with focal length 10.0 cm is located 75.0 cm to the right of the lens. (a) Describe the final image—is it real or virtual? Upright or inverted? (b) What is the location of the final image? (c) What is the total transverse magnification?

Figure 24.27   Problem 40

41. Two lenses, of focal lengths 3.0 cm and 30.0 cm, are used to build a small telescope. (a) Which lens should be the objective? (b) What is the angular magnification? (c) How far apart are the two lenses in the telescope?

42. (a) If Harry has a near point of 1.5 m, what focal length contact lenses does he require? (b) What is the power of these lenses in diopters?

43. An astronomical telescope provides an angular magnification of 12. The two converging lenses are 66 cm apart. Find the focal length of each of the lenses.

44. Two lenses, separated by a distance of 21.0 cm, are used in combination. The first lens has a focal length of +30.0 cm; the second has a focal length of –15.0 cm. An object, 2.0 mm long, is placed 1.8 cm before the first lens. (a) What are the intermediate and final image distances relative to the corresponding lenses? (b) What is the total magnification? (c) What is the height of the final image?

© 45. A camera lens has a fixed focal length of magnitude 50.0 mm. The camera is focused on a 1.0-m-tall child who is standing 3.0 m from the lens. (a) Should the image formed be real or virtual? Why? (b) Is the lens converging or diverging? Why? (c) What is the distance from the lens to the film? (d) How tall is the image on the film? (e) To focus the camera, the lens is moved away from or closer to the film. What is the total distance the lens must be able to move if the camera can take clear pictures of objects at distances anywhere from 1.00 m to infinity?

46. A camera with a 50.0-mm lens can focus on objects located from 1.5 m to an infinite distance away by adjusting the distance between the lens and the film. When the focus is changed from that for a distant mountain range to that for a flower bed at 1.5 m, how far does the lens move with respect to the film?

© 47. Anne is farsighted; the nearest object she can see clearly without corrective lenses is 2.0 m away. It is 1.8 cm from the lens of her eye to the retina. (a) Sketch a ray diagram to show (qualitatively) what happens when she tries to look at something closer than 2.0 m, without corrective lenses. (b) What should the focal length of her contact lenses be so that she can see clearly objects as close as 20.0 cm from her eye?

48. The area occupied by one frame on 35-mm film is 24 mm by 36 mm—see Fig. 24.24. The focal length of the camera lens is 50.0 mm. A picture is taken of a person 182 cm tall. What is the minimum distance from the camera for the person to stand so that the image fits on the film? Give two answers; one for each orientation of the camera.

© 49. A statue is 6.6 m from the opening of a pinhole camera, and the screen is 2.8 m from the pinhole. (a) Is the image erect or inverted? (b) What is the magnification of the image? (c) To get a brighter image, we enlarge the pinhole to let more light through, but then the image looks blurry. Why? (d) To admit more light and still have a sharp image, we replace the pinhole with a lens. Should it be a converging or diverging lens? Why? (e) What should the focal length of the lens be?

50. A dissecting microscope is designed to have a large distance between the object and the objective lens. Suppose the focal length of the objective of a dissecting microscope is 5.0 cm, the focal length of the eyepiece is 4.0 cm, and the distance between the lenses is 32.0 cm. (a) What is the distance between the object and the objective lens? (b) What is the angular magnification?

51. A cub scout makes a simple microscope by placing two converging lenses of +18 D at opposite ends of a 28-cm-long tube. (a) What is the tube length of the microscope? (b) What is the angular magnification? (c) How far should an object be placed from the objective lens?

◆◆52. A microscope has an eyepiece of focal length 2.00 cm and an objective of focal length 3.00 cm. The eyepiece produces a virtual image at the viewer's near point (25.0 cm from the eye). (a) How far from the eyepiece is the image formed by the objective? (b) If the lenses are 20.0 cm apart, what is the distance from the objective lens to the object being viewed? (c) What is the angular magnification?

53. A convex lens of power +12 D is used as a magnifier to examine a wildflower. What is the angular magnification if the final image is at (a) infinity or (b) the near point of 25 cm?

54. A refracting telescope has an objective lens with a focal length of 2.20 m and an eyepiece with a focal length of 1.5 cm. If you look through this telescope the wrong way, that is, with your eye placed at the objective lens, by what factor is the angular size of an observed object reduced?

55. Suppose the distance from the lens system of the eye (cornea + lens) to the retina is 18 mm. (a) What must the power of the lens be when looking at distant objects? (b) What must the power of the lens be when looking at an object 20.0 cm from the eye? (c) Suppose that the eye is farsighted; the person cannot see clearly objects that are closer than 1.0 m. Find the power of the contact lens you would prescribe so that objects as close as 20.0 cm can be seen clearly.

◆56. A man requires reading glasses with +2.0 D power to read a book held 40.0 cm away *with a relaxed eye*. Assume the glasses are 2.0 cm from his eyes. (a) What is his uncorrected far point? (b) What refractive power lenses should he use for distance vision? (c) His uncorrected near point is 1.0 m. What should the refractive powers of the two lenses in his bifocals be to give him clear vision from 25 cm to infinity?

ANSWERS TO PRACTICE PROBLEMS

24.1 The object can be either real or virtual. If the real image forms before the second lens, it becomes a real object; if the second lens interrupts the light rays before they form the real image, it becomes a virtual object.

24.2 $q_1 = +16.7$ cm; $q_2 = 4.3$ cm; $m = -0.43$

24.3 51.7 mm

24.4 13.3 cm

24.5 49.9 cm

24.6 (a) 3.08; (b) 2.08

24.7 (a) 18.0 cm; (b) 1.9 cm

24.8 −780

Interference and Diffraction

When we look at plants and animals, most of the colors we see—brown eyes, green leaves, yellow sunflowers—are due to the selective absorption of light by pigments. In the leaves and stems of green plants, the chief pigment that absorbs some wavelengths and reflects the wavelengths we perceive as green is chlorophyll.

In some animals, color is produced in a different way. The shimmering, intense blue color of the wing of many species of the *Morpho* butterfly of Central and South America makes colors produced by pigments look flat. When the wing or the viewer moves, the color of the wing changes slightly, causing the shimmering quality we call *iridescence*. Iridescent colors are found in the wings or feathers of the Oregon swallowtail butterflies, ruby-throated hummingbirds, and many other species of butterflies and birds. Iridescent colors also appear in some beetles, in the scales of fish, and in the skins of snakes. How are these iridescent colors produced?

25.1 CONSTRUCTIVE AND DESTRUCTIVE INTERFERENCE

Chapters 23 and 24 dealt with topics in *geometric optics:* reflection, refraction, and image formation. For the most part we were able to trace light rays propagating in straight-line paths; the rays changed direction only due to reflection or refraction at boundaries. Geometric optics is a useful approximation when objects and apertures are large compared to the wavelength of the light.

The present chapter is concerned with *physical optics,* in which the wave nature of light is made manifest. In physical optics we consider what happens when light propagates around obstacles or through apertures that are *not* large compared to the wavelength. In such situations we encounter interference and diffraction.

The distinction between interference and diffraction is not always clear-cut. Generally, **interference** refers to situations where waves from a small number of sources travel different paths and arrive at an observer with different phases. **Diffraction** is the bending of waves when they travel around obstacles or through apertures. According to Huygens's construction, *every point* on a wavefront is a source of wavelets. The superposition of light from all these point sources must account for the phase differences due to the different paths traveled. Thus, instead of a small number of sources, in diffraction we have the superposition of waves from an *infinite* number of sources.

Any kind of wave can exhibit interference and diffraction because they are just manifestations of the principle of superposition, which says that the net wave disturbance at any point due to two or more waves is the sum of the disturbances due to each wave individually. Superposition is not a new principle for light. We used it earlier in our study of sound and other mechanical waves. We also used it to find the electric and magnetic fields due to more than one source; the resultant electric and magnetic fields are the vector sums of the fields due to each source individually. Now we apply the principle of superposition to EM waves.

Coherent and Incoherent Sources

Why do we not commonly see interference effects with visible light? With light from a source such as the Sun, an incandescent bulb, or a fluorescent bulb, we do not see regions of constructive and destructive interference; rather, the *intensity* at any point is the sum of the intensities due to the individual waves. Light from any one of these sources is, at the atomic level, emitted by a vast number of *independent* sources. Waves from independent sources are **incoherent**; they do not maintain a fixed phase relationship with each other. We cannot accurately predict the phase (for instance, whether the wave is at a maximum or at a zero) at one point given the phase at another point. Incoherent waves have *rapidly fluctuating* phase relationships. The result is an averaging out of interference effects, so that the total intensity (or power per unit area) is just the sum of the intensities of the individual waves.

Only the superposition of **coherent** waves produces interference. Coherent waves must be locked in with a fixed phase relationship. *Coherent* and *incoherent* waves are idealized extremes; all real waves fall somewhere between the extremes. The light emitted by a laser can be highly coherent—the waves can remain coherent for as far as several kilometers. Light from a distant point source (such as a star other than the Sun) has some degree of coherence.

Thomas Young (1773–1829) performed the first visible-light interference experiments using a clever technique to obtain two coherent light sources from a single source (Fig. 25.1). When a single narrow slit is illuminated, the light wave that passes through the slit diffracts or spreads out. The single slit acts as a single coherent source to illuminate two other slits. These two other slits then act as sources of coherent light for interference.

Interference of Two Waves

If two waves are in step with each other, with the crest of one falling at the same point as the crest of the other, they are said to be *in phase*. The phase difference between the two

In the superposition of *incoherent* waves, the resultant intensity is the sum of the intensities due to each wave individually.

Figure 25.1 Young's technique for illuminating two slits with coherent light

waves that are in phase is an integral multiple of 2π rad. The superposition of two waves that are in phase has an amplitude equal to the sum of the amplitudes of the two waves. For instance, in Fig. 25.2 two sinusoidal waves are in phase. The amplitudes of the two are $2A_0$ and $5A_0$. When the two waves are added together, the resulting wave has an amplitude of $2A_0 + 5A_0 = 7A_0$. The superposition of two waves that are *in phase* is called **constructive interference**.

For constructive interference, the intensity of the resulting wave (I) is *greater than* the sum of the intensities of the two waves individually ($I_1 + I_2$). Since intensity is proportional to amplitude squared, let $I = CA^2$, $I_1 = CA_1^2$, and $I_2 = CA_2^2$, where C is a constant. Since $A = A_1 + A_2$,

$$A^2 = (A_1 + A_2)^2 = A_1^2 + A_2^2 + 2A_1A_2$$

Therefore,

$$I = I_1 + I_2 + 2\sqrt{I_1I_2}$$

Since intensity is power per unit area, where does the extra energy come from? Don't worry; energy is still conserved. If in some places $I > I_1 + I_2$, then in other places $I < I_1 + I_2$. Constructive interference does not manufacture energy; the energy is just redistributed. To summarize:

Constructive interference of two waves	
Phase difference $\Delta\phi = 2m\pi$ rad ($m = 0, \pm1, \pm2, \dots$)	(25-1)
Amplitude $A = A_1 + A_2$	(25-2)
Intensity $I = I_1 + I_2 + 2\sqrt{I_1I_2}$	(25-3)

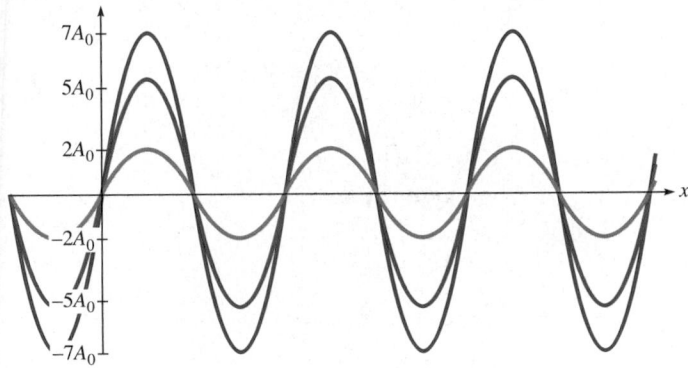

Figure 25.2 Two coherent waves with amplitudes $2A_0$ and $5A_0$. Since they are in phase, they interfere constructively. The superposition of the two (red graph) has an amplitude of $7A_0$.

Two waves that are 180° out of phase are a half cycle apart; where one is at a crest the other is at a trough (Fig. 25.3). The superposition of two waves that are 180° out of phase is called **destructive interference**. The phase difference for destructive interference is 180° plus any integral multiple of 360° (or π rad plus any integral multiple of 2π rad). The destructive interference of two waves with amplitudes $2A_0$ and $5A_0$ gives a resulting amplitude of $3A_0$. If the two waves had the same amplitude, there would be complete cancellation—the resultant wave would have an amplitude of zero. Two waves can have any phase relationship between the two extremes of in phase or 180° out of phase. To summarize:

Destructive interference of two waves	
Phase difference $\quad \Delta\phi = (m + \frac{1}{2})2\pi$ rad $\quad (m = 0, \pm1, \pm2, \ldots)$	(25-4)
Amplitude $\quad A = |A_1 - A_2|$	(25-5)
Intensity $\quad I = I_1 + I_2 - 2\sqrt{I_1 I_2}$	(25-6)

Phase Difference due to Different Paths

In interference, two or more coherent waves travel different paths to a point where we observe the superposition of the two. The paths may have different lengths, or pass through different media, or both. The difference in path lengths introduces a phase difference—it changes the phase relationship between the waves.

Suppose two waves start in phase but travel different paths in the same medium to a point where they interfere (Fig. 25.4). If the difference in path lengths Δl is an integral number of wavelengths,

$$\Delta l = m\lambda \quad (m = 0, \pm1, \pm2, \ldots) \tag{25-7}$$

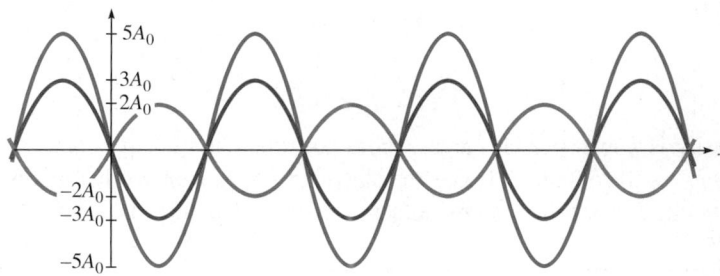

Figure 25.3 Destructive interference of waves with amplitudes $2A_0$ and $5A_0$. The superposition of the two (red graph) has amplitude $3A_0$.

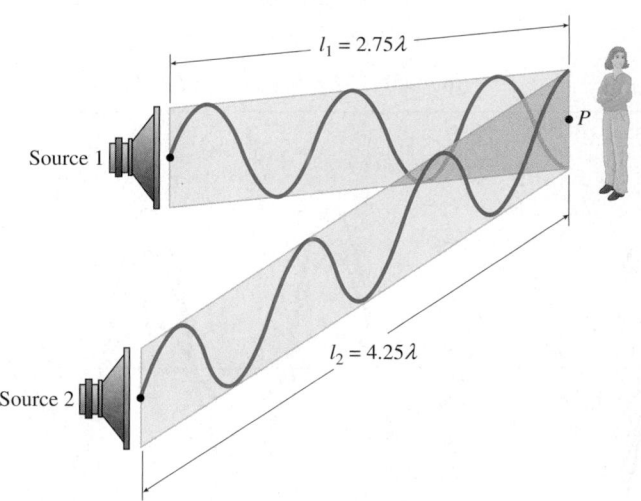

Figure 25.4 Two loudspeakers are fed the same electrical signal. The sound waves travel different distances to reach the observer. The phase difference between the two waves depends on the difference in the distances traveled. In this case, $l_2 - l_1 = 1.50\lambda$, so the waves arrive at the observer 180° out of phase.

then one wave is simply going through a whole number of extra cycles, which leaves them in phase—they interfere constructively. Path lengths that are integral multiples of λ can be ignored since they do not change the relative phase between the two waves.

On the other hand, suppose two waves start in phase but the difference in path lengths is an odd number of *half* wavelengths:

$$\Delta l = \pm \tfrac{1}{2}\lambda, \pm \tfrac{3}{2}\lambda, \pm \tfrac{5}{2}\lambda, \ldots = (m + \tfrac{1}{2})\lambda \quad (m = 0, \pm 1, \pm 2, \ldots) \qquad \textbf{(25-8)}$$

One wave travels a half cycle farther than the other (plus a whole number of cycles, which can be ignored). Now the waves are 180° out of phase; they interfere destructively.

In cases where the two paths are not completely in the same medium, we have to keep track of the number of cycles in each medium separately (since the wavelength changes as a wave passes from one medium into another).

A path difference equal to an integral number of wavelengths does not change the phase relationship between two waves.

Example 25.1

Interference of Microwave Beams

A microwave transmitter (T) and receiver (R) are set up side by side (Fig. 25.5a). Two flat metal plates (M) that are good reflectors for microwaves face the transmitter and receiver, several meters away. The beam from the transmitter is broad enough to reflect from both metal plates. As the lower plate is slowly moved to the right, the microwave power measured at the receiver is observed to oscillate between minimum and maximum values (Fig. 25.5b). Approximately what is the wavelength of the microwaves?

Strategy   Maximum power is detected when the waves reflected from the two plates interfere *constructively* at the receiver. Thus, the positions of the mirror that give maximum power must occur when the path difference is an integral number of wavelengths.

Solution   When the lower plate is farther from the transmitter and receiver, the wave reflected from it travels some extra distance before reaching the receiver. If the metal plates are far enough from the transmitter and receiver, then the microwaves approach the plates and return almost along the same line. Then the extra distance traveled is approximately $2x$.

Constructive interference occurs when the path lengths differ by an integral number of wavelengths:

$$\Delta l = 2x = m\lambda \quad (m = 0, \pm 1, \pm 2, \ldots)$$

From one position of constructive interference to an adjacent one, the path length difference must change by one wavelength:

$$2\Delta x = \lambda$$

The maxima are at $x = 3.9$, 5.2, and 6.5 cm, so $\Delta x = 1.3$ cm. Then

$$\lambda = 2.6 \text{ cm}$$

Discussion   Note that the distance the lower plate is moved between maxima is *half* a wavelength, since the wave makes a round trip.

Practice Problem 25.1   Path difference for destructive interference

Verify that at positions where minimum power is detected, the difference in path lengths is a half-integral number of wavelengths $[\Delta l = (m + \tfrac{1}{2})\lambda]$.

Figure 25.5
(a) Microwave transmitter and receiver and reflecting plates;
(b) microwave power detected as a function of x.

Power at R

3.5   4.0   4.5   5.0   5.5   6.0   6.5

x (cm) ⟶

(b)

(a)

Figure 25.6 (a) Cross-sectional view of a CD. A laser beam passes through the polycarbonate plastic and reflects from the aluminum layer. (b) The pits are arranged in a spiral track. Surrounding the pits, the flat aluminum surface is called *land*. When the laser reflects from a pit, it also reflects from the land on either side. (c) A motor spins the CD at between 200 and 500 rpm, keeping the track speed constant. Light from a laser is reflected by a semitransparent mirror toward the CD; light reflected by the CD is transmitted through this same mirror to the detector. The detector produces an electrical signal proportional to the variations in the intensity of reflected light.

How a CD Is Read

In Example 25.1, EM waves from a single source are reflected from metal surfaces at two different distances from the source; the two reflected waves interfere at the detector. A similar system is used in reading an audio CD or a CD-ROM.

To manufacture a CD, a disk of polycarbonate plastic 1.2 mm thick is impressed with a series of *pits* arranged in a single spiral track (Fig. 25.6). The pits are 0.5 μm wide and at least 0.83 μm long. The disk is coated with a thin layer of aluminum and then with acrylic to protect the aluminum. To read the CD, a laser beam illuminates the aluminum layer from below; the reflected beam enters a detector. The laser beam is wide enough that when it reflects from a pit, part of it also reflects off the *land* (the flat part of the aluminum layer) on either side of the track. The height *h* of the pits is chosen so that light reflected from the land interferes destructively with light reflected from the pit (see Problem 48). Thus, a pit causes a minimum intensity to be detected. On the other hand, when the laser reflects from the land between pits, the intensity at the detector is a maximum. Changes between the two intensity levels represent the binary digits (the 0's and the 1's).

Making The Connection:
reading a compact disc (CD)

25.2 THE MICHELSON INTERFEROMETER

Making The Connection:
Michelson interferometer

Albert Michelson (1852–1931) invented the interferometer to determine whether the Earth's motion has any effect on the speed of light as measured by an observer on Earth. The concept behind the Michelson interferometer (Fig. 25.7) is not complicated, yet it

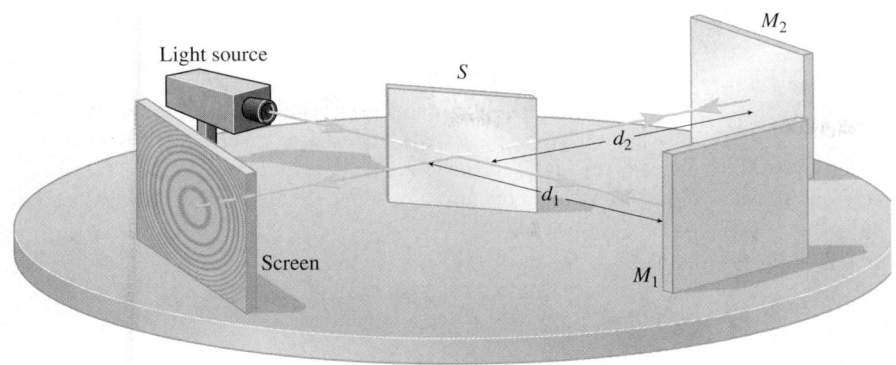

Figure 25.7   A Michelson interferometer

is an extremely precise tool. A beam of coherent light is incident on a beam splitter S (a half-silvered mirror) that reflects only half of the incident light, while transmitting the rest. Thus, a single beam of coherent light from the source is separated into two beams, which travel different paths down the *arms* of the interferometer and are reflected back by fully silvered mirrors (M_1, M_2). At the half-silvered mirror, again half of each beam is reflected and half transmitted. Light sent back toward the source leaves the interferometer. The remainder combines into a single beam and is observed on a screen. A phase difference between the two beams may arise because the arms have different lengths or because the beams travel through different media in the two arms. If the two beams arrive at the screen in phase, they interfere constructively to produce maximum intensity (a *bright fringe*) at the screen; if they arrive 180° out of phase, they interfere destructively to produce a minimum intensity (a *dark fringe*).

Example 25.2

Measuring the Index of Refraction of Air

Suppose a transparent vessel 30.0 cm long is placed in one arm of a Michelson interferometer. The vessel initially contains air at 0°C and 1 atm. With light of vacuum wavelength 633 nm, the mirrors are arranged so that a bright spot appears at the center of the screen. As air is gradually pumped out of the vessel, the central region of the screen changes from bright to dark and back to bright 274 times—that is, 274 bright fringes are counted (not including the initial bright fringe). Calculate the index of refraction of air.

Strategy   As air is pumped out, the path lengths traveled in each of the two arms do not change, but the *number of wavelengths traveled* does change, since the index of refraction inside the vessel begins at some initial value n and decreases gradually to 1. Each new bright fringe means that the number of wavelengths traveled has changed by one more wavelength.

Solution   Let the index of refraction of air at 0°C and 1 atm be n. If the *vacuum* wavelength is $\lambda_0 = 633$ nm, then the wavelength in air is $\lambda = \lambda_0/n$. Initially, the number of wavelengths traveled during a round-trip through the air in the vessel is

$$\text{initial number of wavelengths} = \frac{\text{round-trip distance}}{\text{wavelength in air}} = \frac{2d}{\lambda} = \frac{2d}{\lambda_0/n}$$

where $d = 30.0$ cm is the length of the vessel. As air is removed, the number of wavelengths decreases since, as n

decreases, the wavelength gets longer. Assuming that the vessel is completely evacuated in the end (or nearly so), the final number of wavelengths is

$$\text{final number of wavelengths} = \frac{\text{round-trip distance}}{\text{wavelength in vacuum}} = \frac{2d}{\lambda_0}$$

The change in the number of wavelengths traveled, N, is equal to the number of bright fringes observed:

$$N = \frac{2d}{\lambda_0/n} - \frac{2d}{\lambda_0} = \frac{2d}{\lambda_0}(n-1)$$

Since $N = 274$, we can solve for n.

$$n = \frac{N\lambda_0}{2d} + 1$$

$$= \frac{274 \times 6.33 \times 10^{-7}\ \text{m}}{2 \times 0.300\ \text{m}} + 1$$

$$= 1.000289$$

Discussion   The measured value for the index of refraction of air is slightly less than that given in Table 23.1 ($n = 1.000\,293$). One reason is that the vacuum in the vessel is surely not perfect. If we had been able to reduce the pressure in the vessel all the way to zero, we would have counted a larger number of fringes, resulting in a larger value for n.

continued on next page

Example 25.2 *continued*

Conceptual Practice Problem 25.2 A possible alternative method

Instead of counting the fringes, another way to measure the index of refraction of air might be to move one of the mirrors as the air is slowly pumped out of the vessel, maintaining a

bright fringe at the screen. The distance the mirror moves could be measured and used to calculate *n*. If the mirror moved is the one in the arm that does *not* contain the vessel, should it be moved in or out? In other words, should that arm be made longer or shorter?

The Interference Microscope

Making The Connection:
interference microscope

An *interference microscope* enhances contrast in the image when viewing objects that are transparent or nearly so. A cell in a water solution is difficult to see with an ordinary microscope. The cell reflects only a small fraction of the light incident on it, so it transmits almost the same intensity as the water does and there is little contrast between the cell and the surrounding water. However, if the cell's index of refraction is different from that of water, light transmitted through the cell is phase-shifted compared to the light that passes through water. The interference microscope exploits this phase difference. As with the Michelson interferometer, a single beam of light is split into two and then recombined. The light in *one* arm of the interferometer passes through the sample. When the beams are recombined, interference translates the phase differences that are invisible in an ordinary microscope into intensity differences that are easily seen.

25.3 THIN FILMS

The rainbow-like colors seen in soap bubbles and oil slicks are produced by interference. Suppose a wire frame is dipped into soapy water and then held vertically aloft with a thin film of soapy water clinging to the frame (Fig. 25.8a). Due to gravity pulling downward, the film at the top of the wire frame is very thin—just a few molecules thick. The film gets thicker and thicker toward the bottom, so the cross section is wedge-shaped (Fig. 25.8b). The film is illuminated with white light from behind the camera; the photo shows light *reflected* from the film.

Figure 25.9 shows a light ray incident on a portion of a thin film. At each boundary, some light is reflected while most is transmitted. When looking at the light reflected from the film, we see the superposition of all the reflected rays (of which only the first three are shown). The interference of these rays determines what color we see. In most cases, we can consider the interference of the first two reflected rays and ignore the rest. Unless the indices of refraction on either side of a boundary are nearly the same, the amplitude of a reflected wave is a small fraction of the amplitude of the incident wave. Rays 1 and 2 each reflect only once; their amplitudes are nearly the same. Ray 3 reflects three times, so its amplitude is much smaller. Other reflected rays are even weaker.

Interference effects are much less pronounced in the transmitted light. Ray A is strong since it does not suffer a reflection. Ray B suffers two reflections, so it is much

Figure 25.8 (a) Viewing a film of soapy water by reflected light. (The background is dark so that only reflected light is shown in the photo; the camera and the light source are both on the same side of the film.) (b) A wedge-shaped cross section of the film. The thickness *t* gradually increases from the top of the frame toward the bottom.

Making The Connection:
colors in soap films, oil slicks

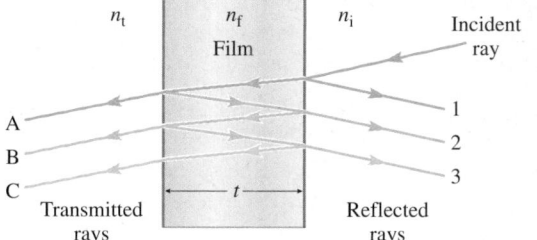

Figure 25.9 Rays reflected and transmitted by a thin film

Figure 25.10   (a) A wave pulse on a string heads for a boundary with a slower medium (greater mass per unit length). The reflected pulse is inverted. (b) A pulse reflected from a faster medium is not inverted.

Lighter string   Heavier string
Incident pulse

Heavier string   Lighter string
Incident pulse

Reflected pulse *is* inverted
Transmitted pulse

Reflected pulse is *not* inverted   Transmitted pulse

(a)

(b)

weaker than A. Ray C is even weaker since it goes through four reflections. Thus, the amplitude of the transmitted light for constructive interference is not much larger than the amplitude for destructive interference. Nevertheless, interference in the transmitted light must occur for energy to be conserved: if more of the energy of a particular wavelength is reflected, less is transmitted. In Problem 17 you can show that if a certain wavelength interferes constructively in reflected light, then it interferes destructively in transmitted light, and *vice versa*.

Phase Shifts due to Reflection

In Section 11.8, we saw that reflected waves are sometimes inverted, which is to say they are phase-shifted 180° with respect to the incident wave. Whenever a wave hits a boundary where the wave speed suddenly changes, reflection occurs. The reflected wave is inverted (Fig. 25.10a) if it reflects off a slower medium (a medium in which the wave travels more slowly); it is *not* inverted if it reflects off a faster medium (Fig. 25.10b). The transmitted wave is never inverted.

The same thing happens to EM waves:

Higher *n*   Lower *n*

Inverted
(180° phase change)

Not inverted
(no phase change)

Figure 25.11   Phase change due to reflection

> When light reflects from a boundary with a slower medium (*higher* index of refraction), it is inverted (180° phase change); when light reflects from a faster medium (*lower* index of refraction), it is *not* inverted (no phase change). (See Fig. 25.11.)

To determine whether rays 1 and 2 in Fig. 25.9 interfere constructively or destructively, we must consider both the relative phase change due to reflection and the extra path length traveled by ray 2 in the film. Depending on the indices of refraction of the three media (the film and the media on either side), it may be that *neither* of the rays is inverted upon reflection, or that *both* are, or that one of the two is. If the index of refraction of the film n_f is *between* the other two indices (n_i and n_t), there is no *relative* phase difference due to reflection; either both are inverted or neither is. If the index of the film is the largest of the three or the smallest of the three, then one of the two rays is inverted; in either case there is a relative phase difference of 180°.

Problem-Solving Strategy for Thin Films

- Sketch a ray picture to show the first two reflected rays. If the problem concerns normal incidence, draw the incident ray with a *nonzero* angle of incidence to separate the various rays. Label the indices of refraction.
- Decide whether there is a relative phase difference of 180° between the rays due to reflection.
- If there is no relative phase difference due to reflection, then an extra path length of $m\lambda$ keeps the two rays in phase, resulting in constructive interference. An extra path length of $(m + \frac{1}{2})\lambda$ causes destructive interference. Remember that λ is the wavelength *in the film*, since that is the medium in which ray 2 travels the extra distance.
- If there is a 180° relative phase difference due to reflection, then an extra path length of $m\lambda$ preserves the 180° phase difference and leads to *destructive* interference. An extra path length of $(m + \frac{1}{2})\lambda$ causes *constructive* interference.
- Remember that ray 2 makes a round-trip in the film. For normal incidence, the extra path length is $2t$.

Example 25.3

Appearance of a Film of Soapy Water

A film of soapy water in air is held vertically and viewed in reflected light (as in Fig. 25.8a). The film has index of refraction $n = 1.36$. (a) Explain why the film appears black at the top. (b) The light reflected perpendicular to the film at a certain point is missing the wavelengths 504 nm and 630.0 nm. No wavelengths between these two are missing. What is the thickness of the film at that point? (c) What other visible wavelengths are missing, if any?

Strategy   First we sketch the first two reflected rays, labeling the indices of refraction and the thickness t of the film (Fig. 25.12). The sketch helps determine whether there is a relative phase difference of 180° due to reflection. Since the top of the film appears black, there must be destructive interference for all visible wavelengths. Farther down the film, the wavelengths missing in reflected light are those that interfere destructively; we consider phase shifts both due to reflection and due to the extra path ray 2 travels in the film. We must remember to use the wavelength *in the film,* not the wavelength in vacuum, because ray 2 travels its extra distance within the film.

Figure 25.12
The first two rays reflected by the soap film

Solution   (a) The speed of light in the film is slower than in air. Therefore ray 1, which reflects from a slower medium (the film), is inverted; ray 2, which reflects from a faster medium (air), is not inverted. There is a relative phase difference of 180° between the two *regardless of wavelength.* Due to gravity, the film is thinnest at the top and thickest at the bottom. Ray 2 has a phase shift compared to ray 1 due to the extra distance traveled in the film. The only way to preserve destructive interference for *all* wavelengths is if the top of the film is thin compared to the wavelengths of visible light; then the phase change of ray 2 due to the extra path traveled is negligibly small.

(b) For light reflected perpendicular to the film (normal incidence), reflected ray 2 travels an extra distance $2t$ compared to ray 1, which introduces a phase difference between them. Since there is already a relative phase difference of 180° due to reflection, the path difference $2t$ must be an *integral* number of wavelengths to preserve destructive interference:

$$2t = m\lambda = m\frac{\lambda_0}{n}$$

Suppose $\lambda_{0,m} = 630.0$ nm is the vacuum wavelength for which the path difference is $m\lambda$ for a certain value of m. Since

there are no missing wavelengths between the two, $\lambda_{0,(m+1)} = 504$ nm must be the vacuum wavelength for which the path difference is $m + 1$ times the wavelength in the film. Why not $m - 1$? 504 nm is smaller than 630.0 nm, so a *larger* number of wavelengths fits in the path difference $2t$.

$$2nt = m\lambda_{0,m} = (m + 1)\lambda_{0,(m+1)}$$

We can solve for m:

$$m \times 630.0 \text{ nm} = (m + 1) \times 504 \text{ nm} = m \times 504 \text{ nm} + 504 \text{ nm}$$

$$m \times 126 \text{ nm} = 504 \text{ nm}$$

$$m = 4.00$$

Then the thickness is

$$t = \frac{m\lambda_0}{2n} = \frac{4.00 \times 630.0 \text{ nm}}{2 \times 1.36} = 926.47 \text{ nm} = 926 \text{ nm}$$

(c) We already know the missing wavelengths for $m = 4$ and $m = 5$. Let's check other values of m.

$$2nt = 2 \times 1.36 \times 926.47 \text{ nm} = 2520 \text{ nm}$$

For $m = 3$,

$$\lambda_0 = \frac{2nt}{m} = \frac{2520 \text{ nm}}{3} = 840 \text{ nm}$$

which is IR rather than visible. There is no need to check $m = 1$ or 2 since they give wavelengths even larger than 840 nm—wavelengths even farther from the visible range. Therefore, we try $m = 6$:

$$\lambda_0 = \frac{2nt}{m} = \frac{2520 \text{ nm}}{6} = 420 \text{ nm}$$

This wavelength is generally considered to be visible. What about $m = 7$?

$$\lambda_0 = \frac{2nt}{m} = \frac{2520 \text{ nm}}{7} = 360 \text{ nm}$$

360 nm is UV. Thus, the only other missing visible wavelength is 420 nm.

Discussion   As a check, we can verify directly that the three missing wavelengths in vacuum travel an integral number of wavelengths in the film:

λ_0	$\lambda = \dfrac{\lambda_0}{1.36}$	$m\lambda$
420 nm	308.8 nm	$6 \times 308.8 \text{ nm} = 1853 \text{ nm}$
504 nm	370.6 nm	$5 \times 370.6 \text{ nm} = 1853 \text{ nm}$
630 nm	463.2 nm	$4 \times 463.2 \text{ nm} = 1853 \text{ nm}$

Since the path difference is $2t = 2 \times 926.47 \text{ nm} = 1853 \text{ nm}$, the extra path is an integral number of wavelengths for all three.

Practice Problem 25.3   Constructive interference in reflected light

What visible wavelengths interfere *constructively* in the reflected light where $t = 926$ nm?

(a)   (b)

Figure 25.13 (a) Two glass slides with a narrow air gap between them. When illuminated with white light, interference fringes form in the reflected light. (b) Pressing on the glass changes the thickness of the air gap and distorts the interference fringes.

Thin Films of Air

A thin air gap between two solids can produce interference effects. Figure 25.13a is a photograph of two glass slides separated by an air gap. The thickness of the air gap varies because the glass surfaces are not perfectly flat. The photo shows colored fringes. Each fringe of a given color traces out a curve along which the thickness of the air film is constant. In Fig. 25.13b, an instrument pushes gently down on the top slide. The resulting distortion in the surface of the top slide causes the fringes to move.

If a glass lens with a convex spherical surface is placed on a flat plate of glass, the air gap between the two increases in thickness as we move out from the contact point (Fig. 25.14a). Assuming a perfect spherical shape, we expect to see alternating bright and dark circular fringes in reflected light (Fig. 25.14b). The fringes are called Newton's rings. Well past Newton's day, it was a puzzle that the center was a *dark* spot. Thomas Young figured out that the center is dark because of the phase shift on reflection. Young did an experiment producing Newton's rings with a lens made of crown glass ($n = 1.5$) on top of a flat plate made of flint glass ($n = 1.7$). When the gap between the two was filled with air, the center was dark in reflected light. Then he immersed the experiment in sassafras oil (which has an index of refraction between 1.5 and 1.7). Now the center spot was bright, since there was no longer a relative phase difference of 180° due to reflection.

Source of light

(a)   (b)

Figure 25.14 (a) The air gap between a convex, spherical glass surface and an optically flat plate. (b) A pattern of circular interference fringes, known as Newton's rings, is seen in reflected light.

Newton's rings can be used to check a lens to see if its surface is spherical. A perfectly spherical surface gives circular interference fringes that occur at predictable radii (see Problem 16).

Antireflective Coatings

Making The Connection:
antireflective
coatings

A common application of thin film interference is the antireflective coatings on lenses (Fig. 25.15). The importance of these coatings increases as the number of lenses in an instrument increases—if even a small percentage of the incident light intensity is reflected at each surface, reflections at each surface of each lens can add up to a large fraction of the incident intensity being reflected.

The most common material used as an antireflective coating is magnesium fluoride (MgF_2). It has an index of refraction $n = 1.38$, between that of air ($n = 1$) and glass ($n \approx 1.5$ or 1.6). Ideally, we would like destructive interference in reflected light for *all* visible wavelengths, but that is not possible with a single film. We choose the thickness of the film so that destructive interference occurs in the middle of the visible spectrum.

Figure 25.15 The left side of this lens has an antireflective coating; the right side does not.

Making The Connection:
iridescent colors in
butterfly wings

Butterfly Wings

Interference from light reflected by step structures or partially overlapping scales produces the iridescent colors seen in many butterflies, moths, birds, and fish. A stunning example is the shimmering blue of the *Morpho* butterfly. Figure 25.16a shows the *Morpho* wing as viewed under an electron microscope. The tree-like structures that project up from the top surface of the wing are made of a transparent material. Light is thus reflected from a series of steps. Let us focus on two rays reflected from the tops of successive steps of thickness t_1 with spacing t_2 between the steps (Fig. 25.16b). Both are inverted on reflection, so there is no relative phase difference due to reflection. At normal incidence, the path difference is $2(t_1 + t_2)$. However, the ray passes through a thickness t_1 of the step where the index of refraction is $n = 1.5$. We cannot find the wavelength for constructive interference simply by setting the path difference equal to a whole number of wavelengths: which wavelength would we use?

To solve this sort of problem, we think of path differences in terms of numbers of wavelengths. The number of wavelengths traveled by ray 2 in a distance $2t_1$ (round-trip) through a thickness t_1 of the wing structure is

$$\frac{2t_1}{\lambda} = \frac{2t_1}{\lambda_0/n}$$

where λ_0 is the wavelength in vacuum and $\lambda = \lambda_0/n$ is the wavelength in the medium with index of refraction n. The number of wavelengths traveled in a distance $2t_2$ in air is

$$\frac{2t_2}{\lambda} = \frac{2t_2}{\lambda_0}$$

For constructive interference, the number of extra wavelengths traveled by the lower ray, relative to the first ray, must be an integer:

$$\frac{2t_1}{\lambda_0/n} + \frac{2t_2}{\lambda_0} = m$$

We can solve this equation for λ_0 to find the wavelengths that interfere constructively:

$$\lambda_0 = \frac{2}{m}(nt_1 + t_2)$$

For $m = 1$,

$$\lambda_0 = 2(1.5 \times 64 \text{ nm} + 127 \text{ nm}) = 2 \times 223 \text{ nm} = 446 \text{ nm}$$

Figure 25.16 (a) *Morpho* wing as viewed under an electron microscope. (b) Light rays reflected from two successive steps interfere. Constructive interference produces the shimmering blue color of the wing. For clarity, the rays shown are not at normal incidence. (c) Two other pairs of rays that interfere.

This is the wavelength of the blue color we see when looking at the butterfly wing at normal incidence. Constructive interference at higher values of m are outside the visible spectrum (in the UV).

We only considered reflections from two adjacent steps, but if those interfere constructively, so do all the other reflections from the tops of the steps. Since there are typically seven or eight steps, the interference maxima are much narrower than the broad ones of Example 25.1 (where there are only two steps). The more steps there are, the narrower are the interference maxima.

Since the path length traveled by ray 2 depends on the angle of incidence, the wavelength of light that interferes constructively depends on the angle of view (see Conceptual Question 16). Thus, the color of the wing changes as the viewing angle changes, which gives the wing its shimmering iridescence.

So far we have ignored reflections from the bottoms of the steps. Rays reflected from the bottoms of two successive steps interfere constructively at the same wavelength of 446 nm, since the path difference is the same. The interference of two other pairs of rays (Fig. 25.16c) gives constructive interference only in UV since the path length difference is so small.

25.4 YOUNG'S DOUBLE-SLIT EXPERIMENT

In 1801, Thomas Young performed a double-slit interference experiment that not only demonstrated the wave nature of light, but also allowed the first measurement of the wavelength of light. Figure 25.17 shows the setup for Young's experiment. Coherent light of wavelength λ illuminates a mask in which two parallel slits have been cut. Each

(a)

(b)

Figure 25.17 Young's double-slit interference experiment. (a) The slit geometry. (b) Cylindrical wavefronts emerge from the slits and interfere to form a pattern of fringes on the screen.

slit has width a, which is comparable to the wavelength λ, and length $L \gg a$; the centers of the slits are separated by a distance d. When light from the slits is observed on a screen at a great distance D from the slits, what pattern do we see—how does the intensity I of light falling on the screen depend on the angle θ?

Light from a *single* narrow slit spreads out primarily in directions perpendicular to the slit, since the wavefronts coming from it are cylindrical. Thus, the light from one narrow slit forms a band of light on the screen. The light does *not* spread out significantly in the direction *parallel* to the slit since the slit length L is *large* compared to the wavelength.

With *two* narrow slits, the two bands of light on the screen interfere with one another. The light from the slits starts out in phase, but travels different paths to reach the screen. We expect constructive interference at the center of the screen ($\theta = 0$) since the waves travel the same distance and so are in phase when they reach the screen. Constructive interference also occurs wherever the path difference is an integral multiple of λ. Destructive interference occurs when the path difference is an odd number of half wavelengths. A gradual transition between constructive and destructive interference occurs since the path difference increases continuously as θ increases. This leads to the characteristic alternation of bright and dark bands (fringes) that are shown in Fig. 25.18a, a photograph of the screen from a double-slit experiment. Figure 25.18b and c are a graph of the intensity on the screen and a Huygens construction for the same interference pattern, respectively.

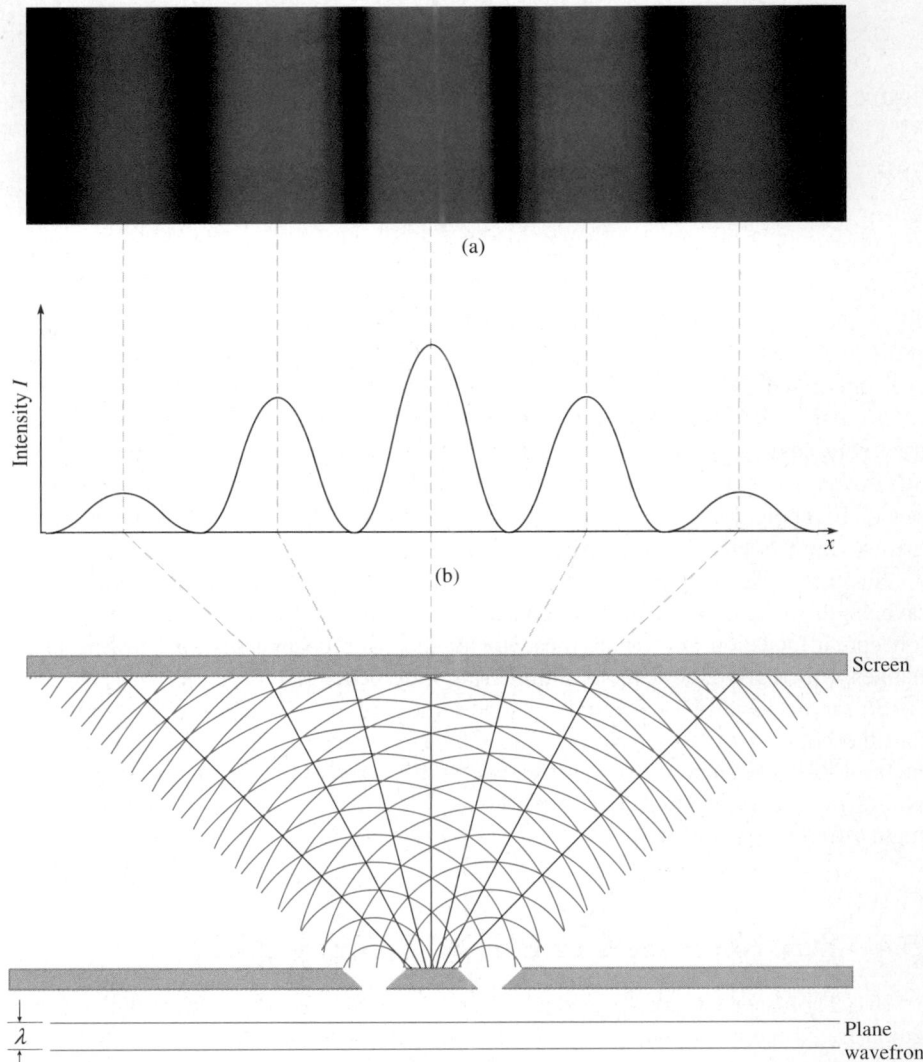

(a)

(b)

Screen

(c)

λ

Plane wavefronts

Figure 25.18 Two-slit interference. (a) Photo of the screen. (b) The intensity as a function of position on the screen. (c) A Huygens construction for the two-slit experiment.

To find where constructive or destructive interference occurs, we need to calculate the path difference. Figure 25.19a shows two rays going from the slits to a *nearby* screen. If the screen is moved farther from the slits, the angle α gets smaller. When the screen is far away, α is small and the rays are nearly parallel. In Fig. 25.19b, the rays are drawn as parallel for a distant screen. The distance that the rays travel from points A and B to the screen are equal; the path difference is the distance from the right slit to point B:

$$\Delta l = d \sin \theta \qquad (25\text{-}9)$$

Maximum intensity at the screen is produced by constructive interference; for constructive interference, the path difference is an integral multiple of the wavelength:

Double-slit maxima

$$d \sin \theta = m\lambda \quad (m = 0, \pm 1, \pm 2, \ldots) \qquad (25\text{-}10)$$

The absolute value of m is often called the **order**. Thus, the third-order maxima are those for which $d \sin \theta = \pm 3\lambda$. Minimum (zero) intensity at the screen is produced by destructive interference; for destructive interference, the path difference is an odd number of half wavelengths:

Double-slit minima

$$d \sin \theta = (m + \tfrac{1}{2})\lambda \quad (m = 0, \pm 1, \pm 2, \ldots) \qquad (25\text{-}11)$$

In Fig. 25.18 the bright and dark fringes appear to be equally spaced. In Problem 18, you can show that the interference fringes *are* equally spaced near the center of the interference pattern, where θ is a small angle.

Figure 25.20 shows the interference of *water waves* in a ripple tank. Surface waves are generated in the water by two point sources that vibrate up and down at the same frequency and in phase with each other, so they are coherent sources. The pattern of interference of the water waves far from the two sources is similar to the two-slit interference pattern for light. If d represents the distance between the sources, Eqs. (25-10) and (25-11) give the correct angles θ for constructive and destructive interference of *water* waves at a large distance. The advantage of the ripple tank is that it lets us see what the wavefronts look like. Notice the similarity between Fig. 25.20 and Fig. 25.18c.

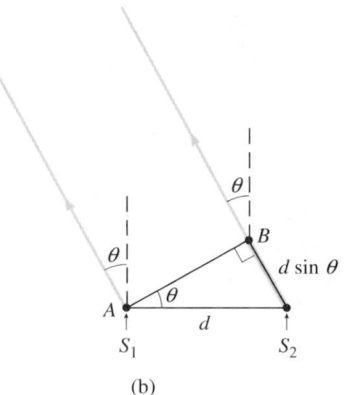

Figure 25.19 (a) Rays from two slits to a nearby screen. As the screen is moved farther away, α decreases—the rays become more nearly parallel. (b) In the limit of a distant screen, the two rays are parallel (but still meet at the same point on the screen). The difference in path lengths is $d \sin \theta$.

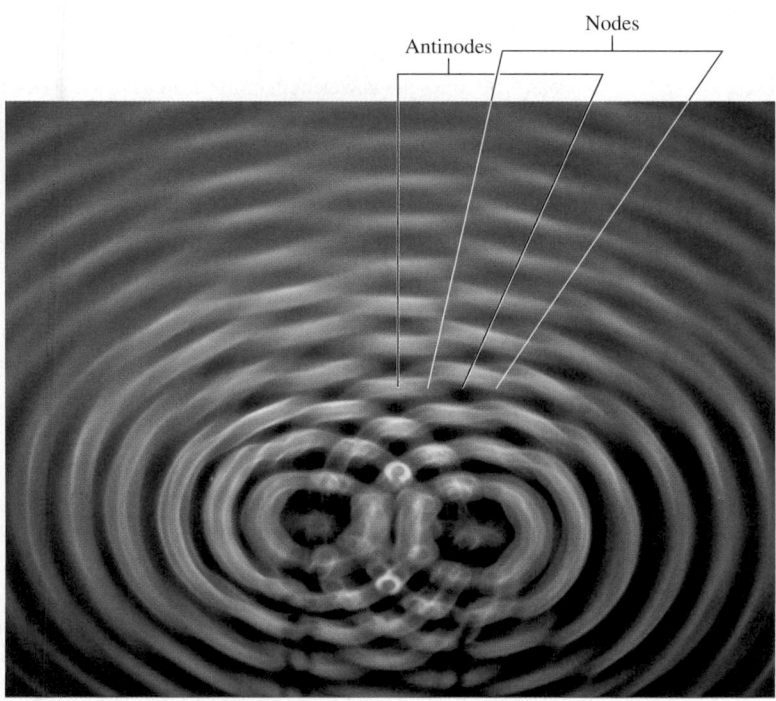

Figure 25.20 Water waves in a ripple tank exhibit two-source interference. Lines of antinodes correspond to the directions of maximum intensity in two-slit interference for light; lines of nodes correspond to the minima.

Points where the interference is constructive are called **antinodes**. Just as for standing waves, here the superposition of two coherent waves causes some points—the antinodes—to have maximum amplitudes. There are also **nodes**—points of complete destructive interference. In a one-dimensional standing wave on a string, nodes and antinodes were single points. For the two-dimensional water waves in a ripple tank, the nodes and antinodes are *curves*. For three-dimensional light waves (or three-dimensional sound waves), the nodes and antinodes are *surfaces*.

Example 25.4

Interference from Two Parallel Slits

A laser ($\lambda = 690.0$ nm) is used to illuminate two parallel slits. On a screen that is 3.30 m away from the slits, interference fringes are observed. The distance between adjacent bright fringes in the center of the pattern is 1.80 cm. What is the distance between the slits?

Strategy   Bright fringes occur at angles θ given by $d \sin \theta = m\lambda$. The distance between the $m = 0$ and $m = 1$ maxima is $x = 1.80$ cm. A sketch helps us see the relationship between the angle θ and the distances given in the problem.

Solution   The central bright fringe ($m = 0$) is at $\theta_0 = 0$. The next bright fringe ($m = 1$) is at an angle given by

$$d \sin \theta_1 = \lambda$$

Figure 25.21 is a sketch of the geometry of the situation. The angle between the lines going to the $m = 0$ and $m = 1$ maxima is θ_1. The distance between these two maxima on the screen is

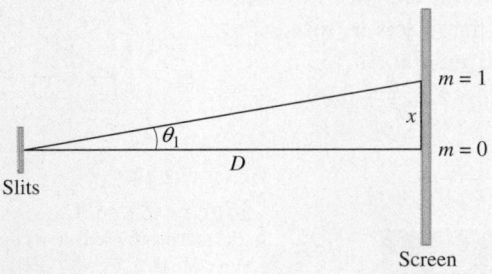

Figure 25.21
Sketch of the two-slit experiment for Example 25.4

x and the distance from the slits to the screen is D. We can find θ_1 from x and D using trigonometry:

$$\tan \theta_1 = \frac{x}{D} = \frac{0.0180 \text{ m}}{3.30 \text{ m}} = 0.005455$$

$$\theta_1 = \tan^{-1} 0.005455 = 0.3125°$$

Now we substitute θ_1 into the condition for the $m = 1$ maximum.

$$d = \frac{\lambda}{\sin \theta_1} = \frac{690.0 \text{ nm}}{\sin 0.3125°} = \frac{690.0 \text{ nm}}{0.005454} = 0.127 \text{ mm}$$

Discussion   We might have noticed that since $x \ll D$, θ_1 is a small angle—that's why the sine and the tangent are the same to three significant figures. Using the small angle approximation ($\sin \theta = \tan \theta = \theta$ in radians) from the start gives

$$d\theta_1 = \lambda$$

and

$$\theta_1 = \frac{x}{D}$$

so

$$d = \frac{\lambda D}{x} = \frac{690.0 \text{ nm} \times 3.30 \text{ m}}{0.0180 \text{ m}} = 0.127 \text{ mm}$$

Practice Problem 25.4   Fringe spacing when wavelength is changed

In a particular two-slit experiment, the distance between the slits is 50 times the wavelength of the light. (a) Find the angles in radians at which the $m = 0$, 1, and 2 maxima occur. (b) Find the angles at which the first two minima occur. (c) What is the distance between two maxima at the center of the pattern on a screen 2.0 m away?

Conceptual Example 25.5

Changing the Slit Separation

A laser is used to illuminate two narrow parallel slits. The interference pattern is observed on a distant screen. What happens to the pattern observed if the distance between the slits is slowly decreased?

Solution and Discussion   When the slits are closer together, the path difference $d \sin \theta$ for a given angle gets smaller. Larger angles are required to produce a path difference that is

a given multiple of the wavelength. The interference pattern therefore spreads out, with each maximum (other than $m = 0$) and minimum moving out to larger and larger angles.

Conceptual Practice Problem 25.5   Interference pattern for $d < \lambda$

If the distance between two slits in a double-slit experiment is less than the wavelength of light, what would you see at a distant screen?

25.5   GRATINGS

Instead of having two parallel slits, a **grating** consists of a large number of parallel, narrow, evenly spaced slits. Typical gratings have hundreds or thousands of slits. The slit separation of a grating is commonly characterized by the number of slits per cm (or the number per any other unit of distance), which is the reciprocal of the slit separation d:

$$\text{slits per cm} = \frac{1}{\text{cm per slit}} = \frac{1}{d}$$

Gratings are made with up to about 50,000 slits/cm, so slit separations are as small as 200 nm. The smaller the slit separation, the more widely different wavelengths of light are separated by the grating.

Figure 25.22 shows light rays traveling from the slits of a grating to a distant screen. Suppose light from the first two slits is in phase at the screen because the path difference $d \sin \theta$ is a whole number of wavelengths $m\lambda$. Then, since the slits are evenly spaced, the light from *all* the slits arrives at the screen in phase. The path difference between any pair of slits is an integral multiple of $d \sin \theta$ and therefore an integral multiple of λ. Therefore, the angles for constructive interference for a grating are the same as for two slits with the same separation:

Maxima for a grating

$$d \sin \theta = m\lambda \quad (m = 0, \pm 1, \pm 2, \ldots) \tag{25-10}$$

For two slits, there is a gradual change in intensity from maximum to minimum and back to maximum. By contrast, for a grating with a large number of slits, the maxima are narrow and the intensity everywhere else is negligibly small. How does the presence of many slits make the maxima so narrow?

Suppose we have a grating with $N = 100$ slits, numbered 0 to 99. The first-order maximum occurs at angle θ such that the path length difference between slits 0 and 1 is $d \sin \theta = \lambda$. Now suppose we look at a *slightly* greater angle $\theta + \Delta\theta$ such that $d \sin (\theta + \Delta\theta) = 1.01\lambda$. The rays from slits 0 and 1 are almost in phase; if there were only two slits, the intensity would be almost as large as the maximum. With 100 slits, each ray is 1.01λ longer than the previous ray. If the length of ray 0 is l_0, then the length of ray 1 is $l_0 + 1.01\lambda$, the length of ray 2 is $l_0 + 2.02\lambda$, and so forth. The length of ray 50 is $l_0 + 50.50\lambda$; thus, rays 0 and 50 interfere destructively since the path difference is an odd number of half wavelengths. Likewise, slits 1 and 51 interfere destructively ($51.51\lambda - 1.01\lambda = 50.50\lambda$); slits 2 and 52 interfere destructively; and so on. Since the light from every slit interferes destructively with the light from some other slit, the intensity at the screen is *zero*. The intensity goes from maximum at θ to zero at $\theta + \Delta\theta$.

The angle $\Delta\theta$ is called the *half width* of the maximum since it is the angle from the center of the maximum to one edge of the maximum (rather than from one edge to the other). By generalizing the argument, we find that the widths of the maxima are inversely proportional to the number of slits ($\Delta\theta \propto 1/N$). The larger the number of slits, the narrower the maxima. Increasing N also makes the maxima *brighter*. More slits let more light pass through and bunch the light energy into narrower maxima. Since light from N slits interferes constructively, the amplitudes of the maxima are proportional to N and the intensities are proportional to N^2. The maxima for a grating are narrow and occur at different angles for different wavelengths. Therefore:

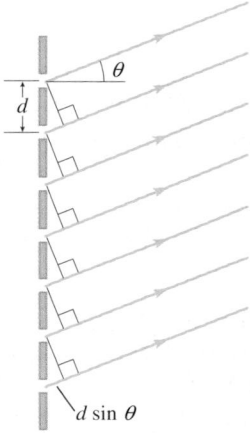

Figure 25.22 Light rays traveling from the slits of a grating to a point on a screen. Since the screen is far away, the rays are nearly parallel to each other; they all leave the grating at (nearly) the same angle θ. Since the distance between any two adjacent slits is d, the path difference between two adjacent rays is $d \sin \theta$.

Maxima for a grating are at the same angles as maxima for two slits with the same d.

A grating separates light with a mixture of wavelengths into its constituent wavelengths.

Example 25.6

Slit Spacing for a Grating

Bright white light shines on a grating. A cylindrical strip of color film is exposed by light emerging at all angles (–90° to +90°) from the grating (Fig. 25.23a). Figure 25.23b shows the resulting photograph. Estimate the number of slits per cm in the grating.

Strategy The grating separates white light into the colors of the visible spectrum. Each color forms a maximum at angles given by $d \sin \theta = m\lambda$. From Fig. 25.23b, we see that more than just first-order maxima are present. If we can estimate the wavelength of the light that exposed the edge of the photo—light that left the grating at ±90°—and if we know what order maximum that is, we can find the slit separation.

Solution The central ($m = 0$) maximum appears white due to constructive interference for *all* wavelengths. On either side of the central maximum lie the first-order maxima. The first-order violet (shortest wavelength) comes first (at the smallest angle), and red is last. Next comes a gap where there

are no maxima. Then the second-order maxima begin with violet. The colors do not progress through the pure spectral colors as before because the third-order maxima start to appear before the second order is finished. The third-order spectrum is not complete; the last color we see at either extreme ($\theta = \pm90°$) is blue-green. Thus, the third-order maximum for blue-green light occurs at ±90°.

Wavelengths that appear blue-green are around 500 nm (Section 22.4). Using $\lambda = 500$ nm and $m = 3$ for the third-order maximum, we can solve for the slit separation.

$$d \sin \theta = m\lambda$$

$$d = \frac{m\lambda}{\sin \theta} = \frac{3 \times 500 \text{ nm}}{\sin 90°} = 1500 \text{ nm}$$

Then the number of slits per cm is

$$\frac{1}{d} = \frac{1}{1500 \times 10^{-9} \text{ m}} = 670{,}000 \text{ slits/m} = 6700 \text{ slits/cm}$$

Discussion The final answer is reasonable for the number of slits per cm in a grating. We would suspect an error if it came out to be 67 million slits/cm, or 67 slits/cm.

⚠️ For a maximum occurring at 90°, we *cannot* use the small angle approximation! We often look at maxima formed by gratings that occur at large angles for which the small angle approximation is not valid.

Practice Problem 25.6 Slit spacing for a full third order

How many slits per cm would a grating have if it just barely produced the full third-order spectrum? Would any of the fourth-order spectrum be produced by such a grating?

Grating

Film

(a)

Figure 25.23
(a) White light incident on a grating. (b) The developed film

(b)

CD Tracking

Data on a CD is encoded as bumps arranged in a spiral track 0.5 μm wide (Fig. 25.24). A 1.6-μm width of flat aluminum on either side separates two adjacent tracks. One of the hardest jobs of a CD player (or CD-ROM drive) is to keep the laser beam centered on the spiral track. One method to keep the laser on track uses a grating to split the laser beam into three beams. The central ($m = 0$) maximum is centered on the track. The first-order ($m = \pm1$) maxima are tracking beams. They reflect from the flat aluminum surfaces (called *land*) on either side of the track onto detectors. Normally the reflected intensity is constant. If one of the tracking beams encounters the bumps in an adjacent track, the changes in reflected intensity signal that the position of the laser must be corrected.

Some fast CD-ROM drives use a grating to split a laser beam into seven beams—zeroth- through third-order maxima. Each beam is used to read a different track; thus, the drive reads seven tracks simultaneously.

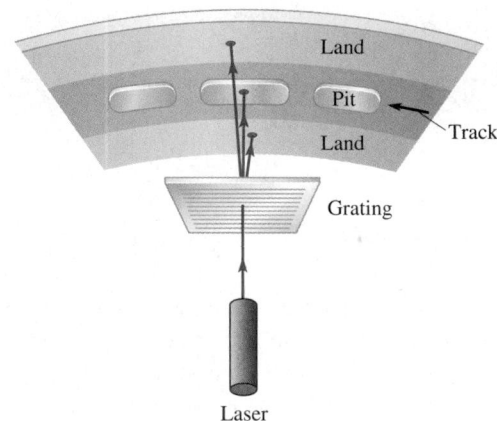

Figure 25.24 A three-beam tracking system

Spectroscopy

The **grating spectroscope** is a precision instrument to measure wavelengths of visible light (Fig. 25.25). *Spectroscope* means (roughly) *spectrum-viewing*. Light from the source first passes through a narrow, vertical slit, which is at the focal point of the collimating lens. Thus, the rays emerging from the lens are parallel to each other. The grating rests on a platform that is adjusted so that the incident rays strike the grating at normal incidence. The telescope can be moved in a circle around the grating to observe the maxima and to measure the angle θ at which each one occurs. The angles are then used along with the spacing of slits in the grating to determine the wavelength(s) present in the light source. The maxima are often called *spectral lines*—they appear as thin lines because they have the shape of the entry slit of the collimator.

Although thermal radiation (such as sunlight and incandescent light) contains a continuous spectrum of wavelengths, other sources of light contain a discrete spectrum composed of only a few narrow bands of wavelengths. A discrete spectrum is also called a line spectrum due to its appearance as a set of lines when viewed through a spectroscope. For example, fluorescent lights and gas discharge tubes produce discrete spectra. In a gas discharge tube, a glass tube is filled with a single gas at low pressure and an electrical current is passed through the gas. The light that is emitted is a discrete spectrum that is characteristic of the gas. Some older streetlights are sodium discharge tubes; they have a characteristic yellow color.

Figure 25.26 shows the spectrum of a sodium discharge tube, which includes two yellow lines at wavelengths of 589.0 nm and 589.6 nm (the *sodium doublet*). Imagine using a grating with fewer slits. The maxima would be wider; if they were too wide, the two yellow lines would overlap and appear as a single line. So a large number of slits is an advantage if we need to *resolve* (distinguish) wavelengths that are close together.

Making The Connection:
grating spectroscope

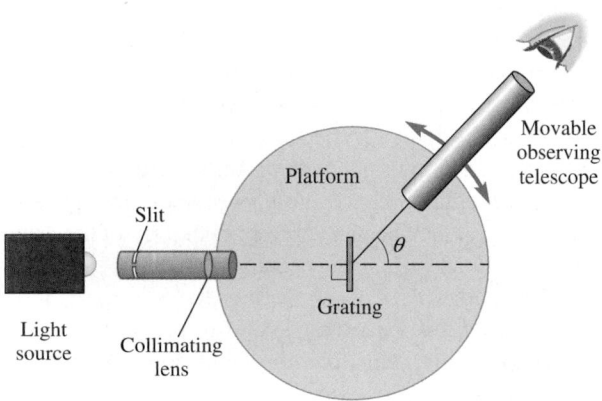

Figure 25.25 Overhead view of a grating spectroscope

Figure 25.26 Emission spectrum of sodium

Reflection Gratings

In the **transmission gratings** we have been discussing, the light viewed is that transmitted by the transparent slits of the grating. Another common kind of grating is the **reflection grating**. Instead of slits, a reflection grating has a large number of parallel, thin reflecting surfaces separated by absorbing surfaces. Using Huygens's principle, the analysis of the reflection grating is exactly the same as for the transmission grating, except that the direction of travel of the wavelets is reversed. Reflection gratings are used in high resolution spectroscopy of astronomical x-ray sources. The spectra enable scientists to identify elements such as iron, oxygen, silicon, and magnesium in the corona of a star or in the remnants of a supernova.

Physics at Home

A compact disc can be used as a reflection grating, since it has a large number of equally spaced reflective tracks. Hold a CD at an angle so that the side without the label reflects light from the Sun or another light source. Tilt the CD back and forth slightly and look for the rainbow of colors that results from the interference of light reflecting from the narrowly spaced grooves of the CD. Next place the CD, label side down, on the floor directly below a ceiling light. Look down at the CD as you slowly walk from it. The first-order maxima form a band of colors (violet to red). Once you are a meter or so away, gradually lower your head to the floor, watching the CD the whole time. You have now observed from $\theta = 0$ to $\theta = 90°$. Count how many orders of maxima you see for the different colors. Now estimate the spacing between tracks on the CD.

25.6 DIFFRACTION AND HUYGENS'S PRINCIPLE

Suppose a plane wave approaches an obstacle. Using geometric optics, we would expect the rays not blocked by the obstacle to continue straight ahead, forming a sharp, well-defined shadow on a screen beyond the obstacle. If the obstacle is large compared to the wavelength, then geometric optics gives a good *approximation* to what actually happens. If the obstacle is *not* large compared to the wavelength, we must return to Huygens's principle to show how a wave bends (diffracts).

In Fig. 25.27a, a wavefront just reaches a barrier with an opening in it. Every point on that wavefront acts as a source of *spherical* wavelets. Points on the wavefront that are behind the barrier have their wavelets absorbed or reflected. Therefore, the propagation of the wave is determined by the wavelets generated by the unobstructed part of the wavefront. The Huygens constructions in Figs. 25.27b–d show that the wave bends around the edges of the barrier, something that would not be expected in geometric optics.

Figure 25.28 shows water waves in a ripple tank that pass through three openings of different widths. For the opening that is much wider than the wavelength (Fig. 25.28a), the bending of the wavefronts is a small effect. Essentially, the part of the wavefront that is not obstructed just travels straight ahead, producing a sharp shadow. As the opening gets narrower (Fig. 25.28b), the bending of the wavefronts becomes more pronounced. Diffraction is appreciable when the size of the opening approaches the size of the wavelength or is even smaller. In the case of Fig. 25.28c, where the opening is about the size of the wavelength, the opening acts almost like a point source of circular waves.

For the openings of intermediate size, a careful look at the waves shows that the amplitude is larger in some directions than in others (Fig. 25.28b). The source of this structure, due to the interference of wavelets from different points, is examined in Section 25.7.

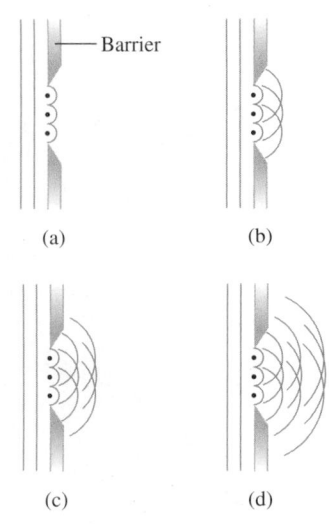

(a)   (b)

(c)   (d)

Figure 25.27 (a) A plane wave reaches a barrier. Points along the wavefront act as sources of spherical wavelets. (b) to (d) At later times, the initial wavelets are propagating outward as new ones form; the wavefront bends around the edges of the barrier.

(a)

(b)

(c)

Figure 25.28 Water waves in a ripple tank pass through openings of different widths.

Since EM waves are three-dimensional, we must be careful when interpreting two-dimensional sketches of Huygens wavelets. Figure 25.29a might represent light incident on a small circular hole or a long thin slit. If it represents a hole, the light bends in all directions, yielding spherically shaped wavefronts (Fig. 25.29b). If the opening represents a slit, we can think of the two directions in turn. The more narrowly restricted the wavefront, the more it spreads out. In the direction of the length of the slit, we get essentially a geometric shadow. In the direction of the width, the wavefront is restricted to a short distance, so the wave spreads out in that direction. The wavefronts past the slit are cylindrically shaped (Fig. 25.29c).

(a)

(b)

(c)

Figure 25.29 (a) Sketch of wavefronts that could represent either a small circular hole or a slit. (b) For a small circular hole, the emerging wavefronts are spherical. (c) For a slit, the emerging wavefronts are cylindrical.

Conceptual Example 25.7

Diffraction and Photolithography

 The CPU chip in a computer contains about 10^8 transistors, numerous other circuit elements, and the electrical connections between them, all in a very small package. One process used to fabricate such a chip is called photolithography. In photolithography, a silicon wafer is coated with a photosensitive material. The chip is then exposed to ultraviolet radiation through a mask that contains the desired pattern of material to be removed. The wafer is then etched. The areas of the wafer not exposed to UV are not susceptible to etching. In areas that were exposed to UV, the photosensitive material and part of the silicon underneath are removed. Why does this process work better with UV than it would with visible light? Why are researchers trying to develop x-ray lithography to replace UV lithography?

Strategy   Without knowing details of the chemical processes involved, we think about the implications of different wavelengths. X-ray wavelengths are shorter than UV wavelengths, which are in turn shorter than visible wavelengths.

Solution and Discussion   The photolithography process depends on the formation of a *sharp shadow* of the mask. To make smaller chips contain more and more circuit elements, the lines in the mask must be made as thin as possible. The danger is that if the lines are too thin, diffraction will spread out the light going through the mask. To minimize diffraction effects, the wavelength should be small compared to the openings in the mask. UV has smaller wavelengths than visible light, so the openings in the mask can be made smaller. X-ray lithography would permit even smaller openings.

Conceptual Practice Problem 25.7   Sunlight through a window

Sunlight streams through a rectangular window, illuminating a bright area on the floor. The edges of the illuminated area are fuzzy rather than sharp. Is the fuzziness due to diffraction? Explain. If not diffraction, what does blur the boundaries of the illuminated area?

(a)

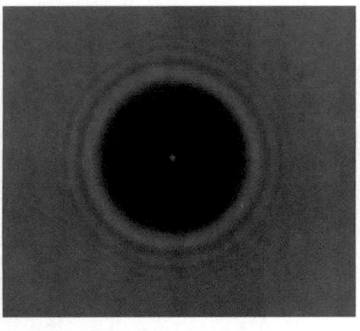

(b)

Figure 25.30   (a) Diffraction pattern formed when a razor blade is illuminated with laser light. (b) Diffraction pattern formed by a small sphere. Note the bright Poisson spot at the center.

Physics at Home

Figure 25.30a shows the shadow of a razor blade illuminated by coherent laser light. Bright and dark fringes are formed by the diffraction of light. One of the counterintuitive predictions of the wave theory of light is that the shadow of a circular or spherical object in coherent light should have a *bright spot at the center* due to diffraction (Fig. 25.30b). Fresnel's prediction of this bright spot was considered by some eminent scientists of the nineteenth century (such as Poisson) to be ridiculous—until it was shown experimentally to exist. You can see the Poisson spot yourself. Use superglue to attach a ball bearing or small opaque marble to a glass microscope slide. View the shadow of the ball bearing by holding it near your eye so that it blocks a distant source of light. The source must be distant enough to act like a point source; there should be a minimal amount of light from other sources. Try a bright streetlight at night or a single lightbulb in an otherwise dark room. Remember that it is never safe to stare directly into a bright light.

Physics at Home

Find a finely woven piece of cloth with a regular mesh pattern, such as a piece of silk, a nylon curtain, an umbrella, or a piece of lingerie. Look through the cloth at a distant, bright light source in an otherwise darkened room—or at a streetlight outside at night. Can you explain the origin of the pattern you see? Could it be simply a geometric shadow of the threads in the cloth? Observe the pattern as you rotate the cloth. Also try stretching the cloth slightly in one direction.

25.7   DIFFRACTION BY A SINGLE SLIT

In a more detailed treatment of diffraction, we must consider the *phases* of all the Huygens wavelets and apply the principle of superposition. Interference between the wavelets causes structure in the diffracted light. In the ripple tanks of Fig. 25.28, we saw structure in the diffraction pattern. In some directions, the wave amplitude was large; in other directions it was small. Figure 25.31 shows the diffraction pattern formed by light passing through a single slit. A wide central maximum contains most of the light energy. (*Central maximum* is the usual way to refer to the entire bright band in the center of the pattern, although the actual *maximum* is just at $\theta = 0$. A more accurate name is *central bright fringe*.) The intensity is brightest right at the center and falls off gradually until the first minimum on either side, where the screen is dark (intensity is zero). Continuing away from the center, maxima and minima alternate, with the intensity changing gradually between them. The lateral maxima are quite weak compared to the central maximum and they are not as wide.

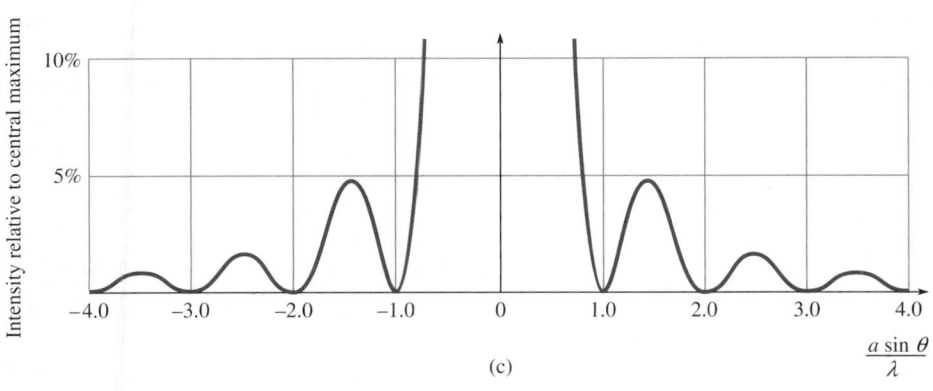

Figure 25.31 Single-slit diffraction. (a) Photo of the diffraction pattern as viewed on a screen. (b) Intensity (as a percentage of the intensity of the central maximum) as a function of the number of wavelengths difference in the path length from top and bottom rays [$(a \sin \theta)/\lambda$]. Minima occur at angles where $(a \sin \theta)/\lambda$ is an integer other than zero. (c) Close-up of the same graph. Intensities of the first four lateral maxima (as percentages) are 4.72%, 1.65%, 0.834%, and 0.503%. The first four lateral maxima occur when $a \sin \theta = 1.43\lambda$, 2.46λ, 3.47λ, and 4.48λ.

(a)

(b)

Figure 25.32 (a) Every point along a slit serves as a source of Huygens wavelets; (b) the ray from the center of the slit travels a greater distance to reach the screen than the ray from the top of the slit; the extra distance is $\frac{1}{2}a \sin \theta$.

According to Huygens's principle, the diffraction of the light is explained by considering every point along the slit as a source of wavelets (Fig. 25.32a). The light intensity at any point beyond the slit is the *superposition* of these wavelets. The wavelets start out in phase, but travel different distances to reach a given point on the screen. The structure in the diffraction pattern is a result of *the interference of the wavelets*. This is a much more complicated interference problem than any we have considered, because an *infinite* number of waves interfere—*every point* along the slit is a source of wavelets. Despite this complication, a clever insight—similar to one we used with the grating—lets us find out where the minima are without the need to resort to complicated math.

Figure 25.32b shows two rays that represent the propagation of two wavelets: one from the top edge of the slit and one from exactly halfway down. The rays are going off at the same angle θ to reach the same point on a *distant* screen. The lower one travels an extra distance $\frac{1}{2}a \sin \theta$ to reach the screen. If this extra distance is equal to $\frac{1}{2}\lambda$, then *these two wavelets* interfere destructively. Now let's look at two other wavelets, shifted down a distance Δx so that they are still separated by half the slit width ($\frac{1}{2}a$). The path difference between these two must also be $\frac{1}{2}\lambda$, so these two interfere destructively. *All the wavelets* can be paired off; since each pair interferes destructively, no light reaches the screen at that angle. Therefore, the first diffraction minimum occurs where

$$\tfrac{1}{2}a \sin \theta = \tfrac{1}{2}\lambda$$

or

$$a \sin \theta = \lambda$$

The other minima are found in a similar way, by pairing off wavelets separated by a distance of $\dfrac{1}{4}a, \dfrac{1}{6}a, \dfrac{1}{8}a, \ldots, \dfrac{1}{2m}a$, where m is any integer *other than zero*. The diffraction minima are given by

$$\frac{1}{2m}a \sin \theta = \frac{1}{2}\lambda \quad (m = \pm1, \pm2, \pm3, \ldots)$$

Simplifying algebraically yields:

Single-slit diffraction *minima*

$$a \sin \theta = m\lambda \quad (m = \pm1, \pm2, \pm3, \ldots) \tag{25-12}$$

Be careful: Eq. (25-12) looks a lot like Eq. (25-10) for the interference maxima due to N slits, but it gives the locations of the diffraction *minima*. Also, $m = 0$ is excluded in Eq. (25-12); a maximum, not a minimum, occurs at $\theta = 0$.

What happens if the slit is made narrower? As a gets smaller, the angles θ for the minima get larger—the diffraction pattern spreads out. If the slit is made wider, then the diffraction pattern shrinks as the angles for the minima get smaller.

The angles at which the lateral maxima occur are much harder to find than the angles of the minima; there is no comparable simplification we can use. The central maximum is at $\theta = 0$, since the wavelets all travel the same distance to the screen and arrive in phase. The other maxima are *approximately* (not exactly) halfway between adjacent minima (see Fig. 25.31c).

Example 25.8

Single-Slit Diffraction

The diffraction pattern from a single slit of width 0.020 mm is viewed on a screen. If the screen is 1.20 m from the slit and light of wavelength 430 nm is used, what is the width of the central maximum?

Strategy The central maximum extends from the $m = -1$ minimum to the $m = +1$ minimum. Since the pattern is symmetric, the width is twice the distance from the center to the $m = +1$ minimum. A sketch helps relate the angles and distances in the problem.

Solution The $m = 1$ minimum occurs at an angle θ satisfying

$$a \sin \theta = \lambda$$

We draw a sketch (Fig. 25.33) showing the angle θ for the $m = 1$ minimum, the distance x from the center of the diffraction pattern to the first minimum, and the distance D from the slit to the screen. The width of the central maximum is $2x$. From Fig. 25.33,

$$\tan \theta = \frac{x}{D}$$

Assuming that $x \ll D$, θ is a small angle. Therefore, $\sin \theta \approx \tan \theta$:

$$\frac{x}{D} = \frac{\lambda}{a}$$

$$x = \frac{\lambda D}{a} = \frac{430 \times 10^{-9}\ \text{m} \times 1.20\ \text{m}}{0.020 \times 10^{-3}\ \text{m}} = 0.026\ \text{m}$$

Comparing the values of x and D, our assumption that $x \ll D$ is justified. The width of the central maximum is $2x = 5.2$ cm.

Discussion The width of the central maximum depends upon the angle θ for the first minimum and the distance D between the slit and the screen. The angle θ, in turn, depends on the wavelength of light and the slit width. For larger values of θ, which means either a longer wavelength or a smaller slit width, the diffraction pattern is more spread out on the screen. For a given wavelength, narrowing the slit increases the diffraction. For a given slit width, the diffraction pattern is wider for longer wavelengths so the pattern for red light ($\lambda = 690$ nm) is more spread out than that for violet light ($\lambda = 410$ nm).

Practice Problem 25.8   Location of first lateral maximum

Approximately how far from the center of the diffraction pattern is the first lateral maximum?

Figure 25.33
A diffraction pattern is formed on a distant screen by light of wavelength λ from a single slit of width a at a distance D from the screen.

Intensities of the Maxima in Two-Slit Interference

In a two-slit interference experiment, the bright fringes are equally spaced but are not equal in intensity (Fig. 25.18). Light diffracts from each slit; the light reaching the screen from either slit forms a diffraction pattern (Fig. 25.31). The two diffraction patterns have the same amplitude at any point on the screen, but different phases. Where the interference is constructive, the amplitude is twice what it would be at that point for a single slit (and therefore four times the intensity).

Figure 25.18 shows only interference maxima within the central diffraction maximum of each slit. If the light incident on the slits is bright enough, interference maxima beyond the first diffraction minimum can be observed (Fig. 25.34).

Figure 25.34 A graph of the intensity for two-slit interference where the spacing d between the two slits is five times the slit width a (that is, $d = 5a$). The first *diffraction* minimum occurs where $a \sin \theta = \lambda$; at that same angle, $5a \sin \theta = d \sin \theta = 5\lambda$. The fifth-order interference maximum is missing because it falls at the first diffraction minimum, where no light reaches the screen.

25.8 DIFFRACTION AND THE RESOLUTION OF OPTICAL INSTRUMENTS

Cameras, telescopes, binoculars, microscopes—practically all optical instruments, including the human eye—admit light through circular apertures. Thus, the diffraction of light through a circular aperture is of great importance. If an instrument is to resolve (distinguish) two objects as being separate entities, it must form separate images of the two. If diffraction spreads out the image of each object enough that they overlap, the instrument cannot resolve them. Figure 25.35 shows photos of a car at night from three different distances. When the car is far away (Fig. 25.35a), the angular separation of the headlights is too small for the camera to resolve the images; the photo shows a single bright spot instead of two.

When light passes through a circular aperture of diameter a, the light is restricted (the wavefronts are blocked) in *all* directions rather than being restricted primarily in a single direction (as for a slit). Thus, for a circular opening, light spreads out in all directions. The

(a)   (b)   (c)

Figure 25.35 (a) When the car is far from the camera, the images of the two headlights overlap so much that there seems to be a single headlight. (b) As the car moves closer, the images begin to separate. (c) Closer yet, the headlights can be distinguished as two separate images.

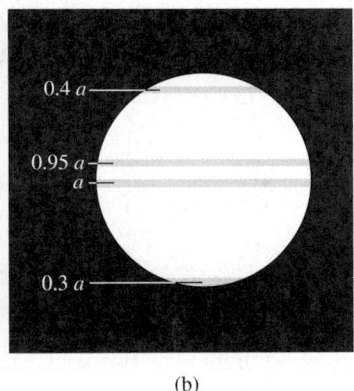

(a)   (b)

Figure 25.36 (a) Diffraction pattern from a circular aperture on a distant screen. (b) A circular aperture with diameter a can be thought of as a large number of thin parallel slits of varying lengths, decreasing from width a at the center to zero at an outer edge.

diffraction pattern due to a circular aperture (Fig. 25.36a) reflects the circular symmetry of the aperture. The diffraction pattern has many similarities to that of a slit. It has a wide, bright central maximum, beyond which minima and weaker maxima alternate; but now the pattern consists of concentric circles reflecting the circular shape of the aperture.

Calculating the angles for the minima and maxima is a difficult problem. Of greatest interest to us is the location of the *first* minimum, which is given by

$$a \sin \theta = 1.22\lambda \qquad (25\text{-}13)$$

Deriving the factor of 1.22 is difficult, but we can make some sense of it. The width of the aperture along any particular direction varies between 0 and the diameter a (Fig. 25.36b); we can think of $a/1.22$ as an effective average aperture width. Then Eq. (25-13) becomes $a_{\text{eff}} \sin \theta = \lambda$.

The reason that the first minimum is of particular interest is that it tells us the diameter of the central maximum, which contains 84% of the intensity of the diffracted light. The size of the central maximum is what limits the resolution of an optical instrument.

When we look at a distant star through a telescope, the star is far enough to be considered a point source, but since the light passes through the circular aperture of the telescope, it spreads out into a circular diffraction pattern like Fig. 25.36a. What if we look at two or more stars that appear close to each other? With the unaided eye, people with good vision can see two separate stars, Mizar and Alcor, in the handle of the Big Dipper. With a telescope, one can see that Mizar is actually *two* stars, called Mizar A and Mizar B; the eye cannot resolve (separate) the images of these two stars, but a telescope with its much wider aperture can. Spectroscopic observations reveal periodic Doppler shifts in the light coming from Mizar A and Mizar B, showing that each is a *binary star*—a pair of stars so close together that they rotate about their center of mass. The companion stars to Mizar A and Mizar B cannot be seen with even the best telescopes available. When light rays from these five stars pass through a circular aperture, diffraction spreads out the images, so that we see only three stars through a telescope or two stars when viewed directly.

Rayleigh's Criterion

Light from a single star (or other point source) forms a circular diffraction pattern after passing through a circular aperture. Two stars with a small angular separation form two overlapping diffraction patterns. Since the stars are incoherent sources, their diffraction patterns overlap without interfering with each other (Fig. 25.37). How far apart must the diffraction patterns be in order to resolve the stars?

A somewhat arbitrary but conventional criterion is due to Rayleigh, who said that the images must be separated by at least half the width of each of the diffraction patterns. In other words, **Rayleigh's criterion** says that two sources can just barely be resolved if the center of one diffraction pattern falls at the first minimum of the other

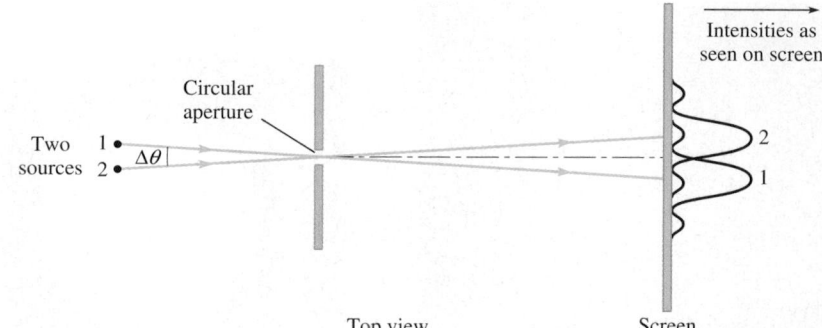

Figure 25.37 Two point sources with an angular separation $\Delta\theta$ form overlapping diffraction patterns when the light passes through a circular aperture.

one. Suppose light from two sources travels through vacuum (or air) and enters a circular aperture of diameter a. If $\Delta\theta$ is the angular separation of the two sources as measured from the aperture and λ_0 is the wavelength of the light in vacuum (or air), then the sources can be resolved if

---

Rayleigh's criterion

$$a \sin \Delta\theta \geq 1.22\lambda_0 \tag{25-14}$$

Example 25.9

Resolution with a Laser Printer

 A laser printer puts tiny dots of ink (toner) on the page. The dots should be sufficiently close together (and therefore small enough) that we don't see individual dots; rather, we see letters or graphics. Approximately how many dots per inch (dpi) ensure that you don't see individual dots when viewing a page 0.40 m from the eye in bright light? Use a pupil diameter of 2.5 mm.

Strategy If the angular separation of the dots exceeds Rayleigh's criterion, then you are able to resolve individual dots. Therefore, the angular separation of the dots must be *smaller* than that given by Rayleigh's criterion—we do *not* want to be able to resolve individual dots.

Solution Call the distance between the centers of two adjacent dots Δx, the diameter of the pupil a, and the angular separation of the dots $\Delta\theta$ (Fig. 25.38). The page is held a distance

$D = 0.40$ m from the eye. Then, since $\Delta x \ll D$, the angular separation of the dots is

$$\Delta\theta \approx \frac{\Delta x}{D}$$

In order for the dots to merge, the angular separation $\Delta\theta$ must be *smaller* than the angle given by the Rayleigh criterion for resolution. The minimum $\Delta\theta$ for resolution is given by

$$a \sin \Delta\theta \approx a\,\Delta\theta = 1.22\lambda_0$$

Since we do *not* want the dots to be resolved, we want

$$a\,\Delta\theta < 1.22\lambda_0$$

Substituting for $\Delta\theta$,

$$a\frac{\Delta x}{D} < 1.22\lambda_0$$

To guarantee that Δx is small enough so that the dots blend together for *all* visible wavelengths, we take $\lambda_0 = 400.0$ nm, the smallest wavelength in the visible range. Now we solve for the distance between dots Δx:

$$\Delta x < \frac{1.22\lambda_0 D}{a} = \frac{1.22 \times 400.0 \text{ nm} \times 0.40 \text{ m}}{0.0025 \text{ m}} = 7.81 \times 10^{-5} \text{ m}$$

$$= 0.0781 \text{ mm}$$

Figure 25.38
Angular separation $\Delta\theta$ of two adjacent dots

continued on next page

Example 25.9 *continued*

To find the minimum number of dots per *inch*, first convert the dot separation Δx to inches.

$$\Delta x = 0.0781 \text{ mm} \div 25.4 \frac{\text{mm}}{\text{in}} = 0.00307 \text{ in}$$

$$\text{dots per inch} = \frac{1}{\text{inches per dot}}$$

$$\frac{1}{0.00307 \text{ inches/dot}} = 330 \text{ dpi}$$

Discussion   Based on this estimate, we expect the printout from a 300-dpi printer to be slightly grainy, since we can just barely resolve individual dots. Output from a 600-dpi printer should look smooth.

You might wonder whether Eq. (25-14) applies to diffraction that occurs within the eye since it uses the wavelength in vacuum (λ_0). The wavelength in the vitreous fluid of the eye is $\lambda = \lambda_0/n$, where $n \approx 1.36$ is the index of refraction of the vitreous fluid. Equation (25-14) *does* apply in this situation because the factor of n in the wavelength is canceled by a factor of n due to refraction (see Problem 45).

Practice Problem 25.9   Effect of wavelength on resolution

The Postimpressionist painter Georges Seurat perfected a technique known as *pointillism,* in which paintings are composed of closely spaced dots of different colors, each about 2 mm in diameter (Fig. 25.39). A close-up view reveals the individual dots; from farther away the dots blend together. Estimate the minimum distance away a viewer should be in order to see the dots blend into a smooth variation of colors. Assume a pupil diameter of 2.2 mm.

(a)

(b)

Figure 25.39
(a) Seurat's *La Parade de Cirque* (Circus Sideshow). (b) A close-up view of the same painting.

Resolution of the Human Eye

In bright light, the pupil of the eye narrows to about 2 mm; diffraction caused by this small aperture limits the resolution of the human eye. In dim light, the pupil is much wider. Now the limit on the eye's resolution in dim light is not diffraction, but the spacing of the photoreceptor cells on the fovea (where they are most densely packed). For an *average* pupil diameter, the spacing of the cones is optimal (see Problem 42). If the cones were less densely packed, resolution would be lost; if they were more densely packed, there would be no gain in resolution due to diffraction.

Making The Connection:
resolution of the human eye

25.9   X-RAY DIFFRACTION

The interference and diffraction examples discussed so far have dealt mostly with visible light. However, the same effects occur for wavelengths longer and shorter than those visible to our eyes. Is it possible to do an experiment that shows interference or diffraction effects with x-rays? X-ray radiation has wavelengths much shorter than those of visible light, so to do such an experiment, the size and spacing of the slits in a grating (for example) would have to be much smaller than in a visible-light grating.

Making The Connection:
x-ray diffraction

Figure 25.40 (a) Crystal structure of aluminum. The dots represent the positions of the aluminum atoms. (b) The x-ray diffraction pattern formed by polycrystalline aluminum (a large number of randomly-oriented aluminum crystals). The central spot, which is formed by x-rays that are not scattered by the sample, has been mostly blocked from reaching the film. Rings form at angles for which the scattered x-rays interfere constructively.

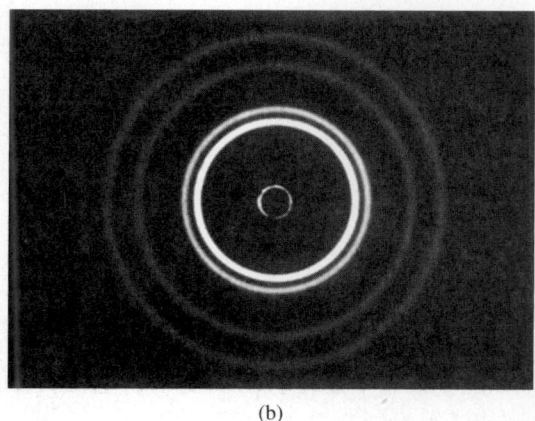

(a)

(b)

Typical x-ray wavelengths range from about 10 nm to about 0.01 nm. There is no way to make a parallel-slit grating small enough to work for x-rays: the diameter of an atom is typically around 0.2 nm, so the slit spacing would be about the size of a single atom.

In 1912, Max von Laue (1879–1960) realized that the regular arrangement of atoms in a crystal makes a perfect grating for x-rays. The regular arrangement and spacing of the atoms is analogous to the regular spacing of the slits in a conventional grating, but a crystal is a *three-dimensional* grating (as opposed to the two-dimensional gratings we use for visible light).

Figure 25.40a shows the atomic structure of aluminum. When a beam of x-rays passes through the crystal, the x-rays are scattered in all directions by the atoms. The x-rays scattered in a particular direction from different atoms interfere with each other. In certain directions they interfere constructively, giving maximum intensity in those directions. Photographic film records those directions as a collection of spots for a single crystal, or as a series of rings for a sample consisting of many randomly-oriented crystals (Fig. 25.40b).

Determining the directions for constructive interference is a difficult problem due to the three-dimensional structure of the grating. W. L. Bragg discovered a great simplification. He showed that we can think of the x-rays *as if they reflect from planes of atoms* (Fig. 25.41a). Constructive interference occurs if the path difference between x-rays reflecting from an adjacent pair of planes is an integral multiple of the wavelength. Figure 25.41b shows that the path difference is $2d \sin \theta$, where d is the distance between the planes and θ is the angle that the incident and reflected beams make with the plane (*not* with the normal). Then, constructive interference occurs at angles given by **Bragg's law**:

X-ray diffraction maxima

$$2d \sin \theta = m\lambda \quad (m = 1, 2, 3, \ldots) \qquad (25\text{-}15)$$

Although Bragg's law is a great simplification, x-ray diffraction is still complicated because there are many sets of parallel planes in a crystal, each with its own plane spacing. In practice, the largest plane spacings contain the largest number of scattering centers (atoms) per unit area, so they produce the strongest maxima.

Figure 25.41 (a) Incident x-rays behave as if they reflect from parallel planes of atoms. (b) Geometry for finding the path difference for rays reflecting from two adjacent planes.

(a)

(b)

X-ray diffraction has many uses:

- Just as a grating separates white light into the colors of the spectrum, a crystal is used to extract an x-ray beam with a narrow range of wavelengths from a beam with a continuous x-ray spectrum.

- If the structure of the crystal is known, then the angle of the emerging beam is used to determine the wavelength of the x-rays.

- The x-ray diffraction pattern can be used to determine the structure of a crystal. By measuring the angles θ at which strong beams emerge from the crystal, the plane spacings d are found and from them the crystal structure.

- X-ray diffraction patterns are used to determine the molecular structures of biological molecules such as proteins. X-ray diffraction studies by Rosalind Franklin were a key clue to James Watson and Francis Crick in their 1953 discovery of the double helix structure of DNA (Fig. 25.42). Intense beams of x-rays radiated by electrons in synchrotrons have even been used to study the structure of viruses.

Figure 25.42 This x-ray diffraction pattern of DNA (deoxyribonucleic acid) was obtained by Rosalind Franklin in 1953. Some aspects of the structure of DNA can be deduced from the pattern of spots and bands. Franklin's data convinced James Watson and Francis Crick of DNA's helical structure, which is revealed by the cross of bands in the diffraction pattern.

25.10 HOLOGRAPHY

An ordinary photograph is a record of the intensity of light that falls on the film at each point. For incoherent light, the phases vary randomly, so it would not be useful to record phase information. A hologram is made by illuminating the subject with *coherent* light; the hologram is a record of the intensity *and the phase* of the light incident on the film. Holography was invented in 1948 by Dennis Gabor, but holograms were difficult to make until lasers became available in the 1960s.

Imagine using a laser, a beam splitter, and some mirrors to produce two coherent plane waves of light that overlap but travel in different directions (Fig. 25.43). Let the waves fall on a photographic plate. The exposure of the plate at any point depends on the intensity of the light falling on it. Since the two waves are coherent, a series of parallel fringes of constructive and destructive interference occur. The spacing between fringes depends on the angle θ_0 between the two waves; a smaller angle makes the spacing between fringes larger. In Problem 63 the spacing between fringes is found to be

$$d = \frac{\lambda}{\sin \theta_0}$$

When the plate is developed, the equally spaced fringes make a grating. If the plate is illuminated at normal incidence with coherent light at the same wavelength λ, the central ($m = 0$) maximum is straight ahead, while the $m = 1$ maximum is at an angle given by

$$\sin \theta = \frac{\lambda}{d} = \sin \theta_0$$

Thus, the $m = 0$ and $m = 1$ maxima re-create the original two waves.

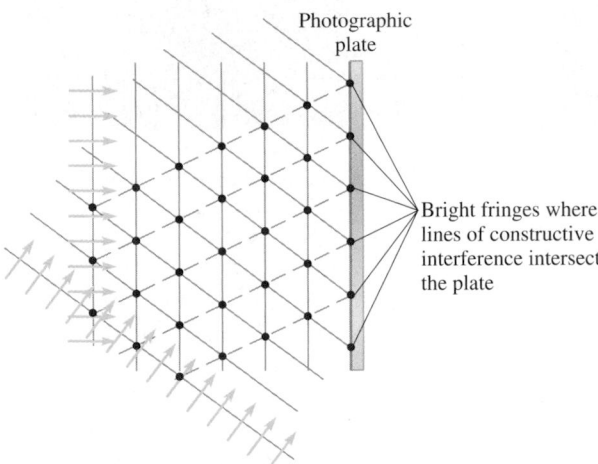

Photographic plate

Bright fringes where lines of constructive interference intersect the plate

Figure 25.43 Two coherent plane waves traveling in different directions expose a photographic plate. An interference pattern is formed on the plate.

Now imagine a plane wave with a point object (Fig. 25.44). The point object scatters light, producing spherical waves just as a point source does. The interference of the original plane wave with the scattered spherical wave gives a series of circular fringes. When this plate is developed and illuminated with laser light, both the plane and spherical waves are re-created. The spherical waves appear to come from a point behind the plate, which is a virtual image of the point object. The plate is a hologram of the point object.

With a more complicated object, each point on the surface of the object is a source of spherical waves. When the hologram is illuminated with coherent light, a virtual image of the object is created. This image can be seen from different perspectives (Fig. 25.45) since the hologram *re-creates wavefronts just as if they were coming from the object.*

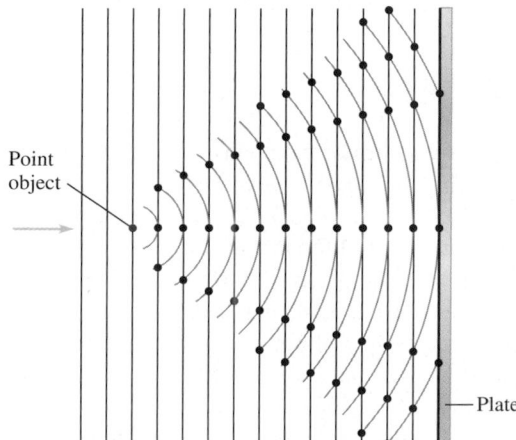

Figure 25.44 Coherent plane waves are scattered by a point object. The spherical waves scattered by the object interfere with the plane wave to form a set of circular interference fringes on a photographic plate.

Figure 25.45 A holographic image of a dragon behind a lens. Notice that the part of the dragon that is magnified by the lens in the hologram depends on the viewing angle.

MASTER THE CONCEPTS

Summary

- When two coherent waves are in phase, their superposition results in constructive interference:

  Phase difference $\Delta\phi = 2m\pi$ rad $(m = 0, \pm1, \pm2, \ldots)$ (25-1)

  Amplitude $A = A_1 + A_2$ (25-2)

  Intensity $I = I_1 + I_2 + 2\sqrt{I_1 I_2}$ (25-3)

- When two coherent waves are 180° out of phase, their superposition results in destructive interference:

  Phase difference $\Delta\phi = (m + \frac{1}{2}) 2\pi$ rad
  $(m = 0, \pm1, \pm2, \ldots)$ (25-4)

  Amplitude $A = |A_1 - A_2|$ (25-5)

  Intensity $I = I_1 + I_2 - 2\sqrt{I_1 I_2}$ (25-6)

- A path length difference equal to λ causes a phase shift of 2π (360°). A path length difference of $\frac{1}{2}\lambda$ causes a phase shift of π (180°).

- When light reflects from a boundary with a slower medium (higher index of refraction), it is inverted (180° phase change); when light reflects from a faster medium (lower index of refraction), it is not inverted (no phase change).

- The angles at which the maxima and minima occur in a double-slit interference experiment are

  Maxima: $d \sin\theta = m\lambda$ $(m = 0, \pm1, \pm2, \ldots)$ (25-10)

  Minima: $d \sin\theta = (m + \frac{1}{2})\lambda$ $(m = 0, \pm1, \pm2, \ldots)$ (25-11)

  The distance between the slits is d. The absolute value of m is called the order.

- A grating with N slits produces maxima that are narrow (width $\propto 1/N$) and bright (intensity $\propto N^2$). The maxima occur at the same angles as for two slits.

- The minima in a single-slit diffraction pattern occur at angles given by

  $$a \sin\theta = m\lambda \quad (m = \pm1, \pm2, \pm3, \ldots) \quad (25\text{-}12)$$

  A wide central maximum contains most of the light energy. The other maxima are approximately (not exactly) halfway between adjacent minima.

- The first minimum in the diffraction pattern due to a circular aperture is given by

  $$a \sin\theta = 1.22\lambda \quad (25\text{-}13)$$

- Rayleigh's criterion says that two sources can just barely be resolved if the center of one diffraction pattern falls at the first minimum of the other one. If $\Delta\theta$ is the angular separation of the two sources, then the sources can be resolved if

  $$a \sin\Delta\theta \geq 1.22\lambda_0 \quad (25\text{-}14)$$

- The regular arrangement of atoms in a crystal makes a grating for x-rays. We can think of the x-rays as if they reflect from planes of atoms. Constructive interference occurs if the path difference between x-rays reflecting from an adjacent pair of planes is an integral multiple of the wavelength.

- A hologram is made by illuminating the subject with coherent light; the hologram is a record of the intensity and the phase of the light incident on the film. The hologram recreates wavefronts just as if they were coming from the object.

Highlighted Figures and Tables

F25.2 Constructive interference of waves with amplitudes $2A_0$ and $5A_0$ (p. 911)

F25.3 Destructive interference of waves with amplitudes $2A_0$ and $5A_0$ (p. 912)

F25.4 Sound waves from two loudspeakers travel different distances to reach the observer (p. 912)

F25.6 Cross-sectional view of a CD (p. 914)

F25.11 Phase change due to reflection (p. 917)

F25.16 Interference caused by the *Morpho* wing (p. 921)

F25.18 Two-slit interference (p. 922)

F25.19 Rays from two slits to a nearby screen; in the limit of a distant screen, the two rays are parallel and the difference in path lengths is $d \sin\theta$ (p. 923)

F25.25 A spectroscope (p. 927)

F25.31 Single-slit diffraction (p. 931)

F25.33 Geometry for finding the first minimum in single-slit diffraction (p. 933)

F25.37 Two point sources with an angular separation $\Delta\theta$ form overlapping diffraction patterns when the light passes through a circular aperture (p. 936)

CONCEPTUAL QUESTIONS

1. Explain why two waves of significantly different frequencies cannot be coherent.

2. Why do eyeglasses, cameras lenses, and binoculars with antireflective coatings often look faintly purple?

3. Telescopes used in astronomy have large lenses (or mirrors). One reason is to let a lot of light in—important for seeing faint astronomical bodies. Can you think of another reason why it is an advantage to make these telescopes so large?

4. The Hubble telescope uses a mirror of radius 1.2 m. Is its resolution better when detecting visible light or UV? Explain.

5. Why can you easily hear sound around a corner due to diffraction, while you cannot see around the same corner?

6. Stereo speakers should be wired with the same polarity. If by mistake they are wired with opposite polarities, the bass (low frequencies) sound much weaker than if they are wired correctly. Why? Why is the bass (low frequencies) weakened more than the treble (high frequencies)?

7. Two antennas driven by the same electrical signal emit coherent radio waves. Is it possible for two antennas driven by *independent* signals to emit radio waves that are coherent with each other? If so, how? If not, why not?

8. A radio station wants to ensure good reception of its signal everywhere inside a city. Would it be a good idea to place several broadcasting antennas at roughly equal intervals around the perimeter of the city? Explain.

9. The size of an atom is about 0.1 nm. Can a light microscope make an image of an atom? Explain.

10. What are some of the advantages of a UV microscope over a visible light microscope? What are some of the disadvantages?

11. The *f-stop* of a camera lens is defined as the ratio of the focal length of lens to the diameter of the aperture. A large f-stop therefore means a small aperture. If diffraction is the only consideration, would you use the largest or the smallest f-stop to get the sharpest image?

12. In Section 25.3 we studied interference due to thin films. Why must the film be *thin*? Why don't we see interference effects when looking through a window or at a poster covered by a plate of glass—even if the glass is optically flat?

13. Describe what happens to a single-slit diffraction pattern as the width of the slit is slowly decreased.

14. Explain, using Huygens's principle, why the Poisson spot is expected.

15. What effect places a lower limit on the size of an object that can be clearly seen with the best optical microscope?

16. Make a sketch (similar to Fig. 25.16b) of the reflected rays from two adjacent steps of the *Morpho* butterfly wing for a large angle of incidence (around 45°). Refer to your sketch to explain why the wavelength at which constructive interference occurs depends on the viewing angle.

17. A lens ($n = 1.51$) has an antireflective coating of MgF_2 ($n = 1.38$). Which of the first two reflected rays has a phase shift of 180°? Suppose a different antireflective coating on a similar lens had $n = 1.62$. Now which of the first two reflected rays has a phase shift of 180°?

18. In the microwave experiment of Example 25.1 and in the Michelson interferometer, we ignored phase changes due to reflection from a metal surface. Microwaves and light *are* inverted when they reflect from metal. Why were we able to ignore the 180° phase shifts?

19. Why does a crystal act as a three-dimensional grating for x-rays but not for visible light?

MULTIPLE CHOICE QUESTIONS

1. If Fig. 25.46 shows the wavefronts for a two-slit interference experiment with light, at which of the labeled points is the intensity zero?

    (a) *A* only   (b) *B* only   (c) *C* only   (d) *A* and *B*
    (e) *B* and *C*   (f) *A* and *C*   (g) *A, B,* and *C*

2. If Fig. 25.46 shows the surface water waves in a ripple tank with two coherent sources, at which of the labeled points would a bit of floating cork bob up and down with the greatest amplitude? (Same answer choices as Question 1.)

3. In a two-slit experiment, light rays from the two slits that reach the second maximum on one side of the central maximum travel distances that differ by

    (a) 2λ (b) λ (c) $\lambda/2$ (d) $\lambda/4$

4. A Michelson interferometer is set up for microwaves. Initially the reflectors are placed so that the detector reads a maximum. When one of the reflectors is moved 12 cm, the needle swings to minimum and back to maximum six times. What is the wavelength of the microwaves?

    (a) 0.5 cm   (b) 1 cm   (c) 2 cm   (d) 4 cm
    (e) Cannot be determined from the information given.

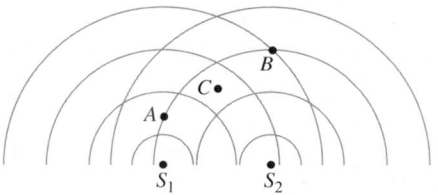

Figure 25.46 Multiple Choice Questions 1 and 2. The wavefronts represent wave *crests* only (not crests and troughs).

5. In a two-slit experiment with coherent light, the intensity of the light reaching the center of the screen from one slit alone is I_0 and the intensity of the light reaching the center from the other slit alone is $9I_0$. When both slits are open, what is the intensity of the light at the interference *minima* nearest the center? The slits are very narrow.

    (a) 0   (b) I_0 (c) $2I_0$
    (d) $3I_0$ (e) $4I_0$ (f) $8I_0$

6. Which of these actions will improve the resolution of a microscope?

(a) increase the wavelength of the light
(b) decrease the wavelength of the light
(c) increase the diameter of the lenses
(d) decrease the diameter of the lenses
(e) both (b) and (c)   (f) both (b) and (d)
(g) both (a) and (c)   (h) both (a) and (d)

7. Coherent light of a single frequency passes through a double slit, with slit separation d, to produce a pattern of maxima and minima on a screen a distance D from the slits. What would cause the separation between adjacent minima on the screen to decrease?

(a) decrease the frequency of the incident light
(b) increase of the screen distance D
(c) decrease the separation d between the slits
(d) increase the index of refraction of the medium in which the setup is immersed

8. Two narrow slits, of width a, separated by a distance, d, are illuminated by light with a wavelength of 660 nm. The resulting interference pattern is labeled (1) in Fig. 25.47. The lighter areas represent regions of maximum intensity. The same light source is then used to illuminate another group of slits and produces pattern (2). The second slit arrangement is

(a) many slits, spaced d apart.
(b) many slits, spaced $2d$ apart.
(c) two slits, each of width $2a$, spaced d apart.
(d) two slits, each of width $a/2$, spaced d apart.

Figure 25.47
Multiple Choice
Question 8

9. Figure 25.48 shows the interference pattern obtained in a double-slit experiment. Which letter indicates a third-order maximum?

Central bright fringe

(a)(b)   (c) (d) (e)

Figure 25.48
Multiple Choice
Question 9

10. The intensity pattern in Fig. 25.49 is due to
(a) two slits.   (b) a single slit.
(c) a grating.   (d) a circular aperture.

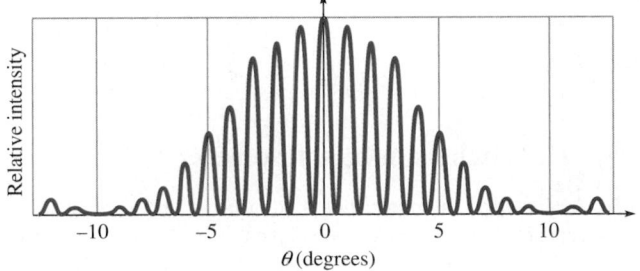

Relative intensity

θ (degrees)

Figure 25.49   Multiple Choice Question 10

PROBLEMS

Note: **C** indicates a combination conceptual/quantitative problem. Gold diamonds ✦, ✦✦ are used to indicate the increasing level of difficulty of each problem. Problem numbers appearing in blue, 9., denote problems that have a detailed solution available in the Student Solutions Manual. Some problems are *paired* by concept; their numbers are connected by a ruled box.

25.1 Constructive and Destructive Interference

1. The light falling on the center of a desk surface from two different lamps has intensities I_0 and $4I_0$, respectively. What is the intensity of light falling on the desk when both lamps are on? If there is more than one possibility, give the range of possibilities.

2. Coherent light from a laser is split into two beams with intensities I_0 and $4I_0$, respectively. What is the intensity of the light when the beams are recombined? If there is more than one possibility, give the range of possibilities.

✦ 3. A *simplified model* of the step structure of the wing of the *Morpho* butterfly is shown in Fig. 25.50. Assume that the step height is $h = 223$ nm, which gives constructive interference for $\lambda = 446$ nm *at normal incidence*. Using this model, find the wavelength of reflected visible light that interferes constructively if the wing is viewed at an angle θ to the normal. Evaluate numerically for $\theta = 0, 10.0°$, and $30.0°$. [*Hint:* The extra path length for reflection from each adjacent step is greater than 446 nm.]

✦ 4. The feathers of the ruby-throated hummingbird have an iridescent green color due to interference. A simplified model of the

Figure 25.50   Problems 3 and 4

step structure of the feather is shown in Fig. 25.50. (a) If the strongest reflection for normal incidence is at λ = 520 nm, what is the step height h? Assume h has the smallest possible value. (b) At what wavelength is the strongest reflection for incidence at θ = 45.0°?

5. An experiment similar to Example 25.1 is performed; the power at the receiver as a function of x is shown in Fig. 25.51. (a) Approximately what is the wavelength of the microwaves? (b) What is the ratio of the *amplitudes* of the microwaves entering the detector for the two maxima shown?

Figure 25.51 Problem 5

25.2 The Michelson Interferometer

6. A Michelson interferometer is adjusted so that a bright fringe appears on the screen. As one of the mirrors is moved 25.8 μm, 92 bright fringes are counted on the screen. What is the wavelength of the light used in the interferometer?

7. In Practice Problem 25.2, calculate the distance that the mirror would be moved. Assume that, outside of the vessel, the light travels through vacuum.

© 8. A Michelson interferometer is set up using white light. The arms are adjusted so that a bright white spot appears on the screen (constructive interference for all wavelengths). A slab of glass (n = 1.46) is inserted into one of the arms. To return to the white spot, the mirror in the other arm is moved 6.73 cm. (a) Is the mirror moved in or out? Explain. (b) What is the thickness of the slab of glass?

25.3 Thin Films

9. A thin film of oil (n = 1.50) is spread over a puddle of water (n = 1.33). In a region where the film looks red from directly above (λ = 630 nm), what is the minimum possible thickness of the film?

10. A thin film of oil (n = 1.50) of thickness 0.40 μm is spread over a puddle of water (n = 1.33). For which wavelength *in the visible spectrum* do you expect constructive interference for reflection at normal incidence?

11. A transparent film (n = 1.3) is deposited on a glass lens (n = 1.5) to form a nonreflective coating. What is the minimum thickness that would minimize reflection of light with wavelength 500.0 nm in air?

12. A camera lens (n = 1.50) is coated with a thin film of magnesium fluoride (n = 1.38) of thickness 90.0 nm. What wavelength in the visible spectrum is most strongly transmitted through the film?

13. A soap film has an index of refraction n = 1.50. The film is viewed in reflected light. (a) At a spot where the film thickness is 910.0 nm, which wavelengths are missing in the reflected light? (b) Which wavelengths are strongest in reflected light?

14. A soap film has an index of refraction n = 1.50. The film is viewed in *transmitted* light. (a) At a spot where the film thickness is 910.0 nm, which wavelengths are missing in the transmitted light? (b) Which wavelengths are strongest in transmitted light?

15. Two optically flat plates of glass are separated at one end by a wire of diameter 0.200 mm; at the other end they touch (Fig. 25.13a). Thus, the air gap between the plates has a thickness ranging from 0 to 0.200 mm. The plates are 15.0 cm long and are illuminated from above with light of wavelength 600.0 nm. How many bright fringes are seen in the reflected light?

◆ 16. A lens is placed on a flat plate of glass to test whether its surface is spherical (Fig. 25.52). (a) Show that the radius r_m of the mth dark ring should be

$$r_m = \sqrt{m \lambda R}$$

where R is the radius of curvature of the lens surface facing the plate and the wavelength of the light used is λ. Assume that $r_m \ll R$. [*Hint:* Start by finding the thickness t of the air gap at a radius $r = R \sin \theta \approx R\theta$. Use small angle approximations.] (b) Are the dark fringes equally spaced? If not, do they get closer together or farther apart as you move out from the center?

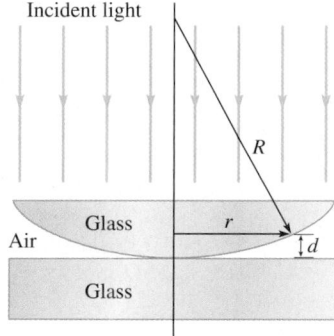

Figure 25.52 Problem 16

© ◆ 17. A thin film is viewed both in reflected and transmitted light at normal incidence. Figure 25.53 shows the strongest two rays for each. Show that if rays 1 and 2 interfere constructively, then rays 3 and 4 must interfere destructively, and if rays 1 and 2 interfere destructively, then rays 3 and 4 interfere constructively. Consider all of the following possibilities: n is the largest of the three indices, n is the smallest of the three indices, or n is in between the other two indices.

Figure 25.53 Problem 17

25.4 Young's Double-Slit Experiment

18. Show that the interference fringes in a two-slit experiment are equally spaced on a distant screen near the center of the interference pattern. [*Hint:* Use the small angle approximation for θ.]

Ⓒ 19. Use a compass to make an accurate drawing of the wavefronts in a two-slit interference experiment similar to Fig. 25.18c. Place the slits 2.0 cm apart and let the wavelength of the incident wave be 1.0 cm. Using a straightedge, draw lines of constructive interference (antinodes) and use them to find the locations of the $m = \pm 1$ maxima on a screen 12 cm from the slits. Measure the angles of the maxima with a protractor; do they agree with those given by Eq. (25.10)? Explain any discrepancy.

20. In a double-slit interference experiment, the wavelength is 546 nm, the slit separation is 0.10 mm, and the screen is 30.0 cm away from the slits. What is the linear distance between adjacent maxima on the screen?

21. Light incident on a pair of slits produces an interference pattern on a screen 2.50 m from the slits. If the slit separation is 0.0150 cm and the distance between the bright fringes in the pattern is 0.760 cm, what is the wavelength of the light?

22. Light from a helium-neon laser (630 nm) is incident on a pair of slits. In the interference pattern on a screen 1.5 m from the slits, the bright fringes are separated by 1.35 cm. What is the slit separation?

23. Light of wavelength 589 nm incident on a pair of slits produces an interference pattern on a distant screen in which the separation between the bright fringes is 0.530 cm. A second light source when incident on the same slits produces an interference pattern with a separation of 0.640 cm between bright fringes. What is the wavelength of the second source? [*Hint:* Use a small-angle approximation.]

24. A double slit is illuminated with monochromatic light of wavelength 600.0 nm. The $m = 0$ and $m = 1$ bright fringes are separated by 3.0 mm on a screen 40.0 cm away from the slits. What is the separation between the slits?

25.5 Gratings

25. A grating has exactly 8000 lines uniformly spaced over 2.54 cm and is illuminated by light from a mercury vapor discharge lamp. What is the expected angle for the third-order maximum of the green line ($\lambda = 546$ nm)?

26. A red line (wavelength 630 nm) in the third order overlaps with a blue line in the fourth order for a particular grating. What is the wavelength of the blue line?

27. Red light of 650 nm can be seen in three orders in a particular grating. About how many rulings per cm does this grating have?

28. A grating has 5000.0 lines per cm. How many orders of violet light of wavelength 412 nm can be observed with this grating?

29. A grating is made of exactly 8000 slits; the slit spacing is 1.50 μm. Light of wavelength 0.600 μm is incident normally on the grating. (a) How many maxima are seen in the pattern on the screen? (b) Sketch the pattern that would appear on a screen 3.0 m from the grating. Label distances from the central maximum to the other maxima.

Ⓒ 30. A reflection grating spectrometer is used to view the spectrum of light from a helium discharge tube. The three brightest spectral lines seen are red, yellow, and blue in color. These lines appear at the positions labeled *A*, *B*, and *C* in Fig. 25.54, though not necessarily in that order of color. In this spectrometer, the distance between the grating and slit is 30.0 cm and the line spacing in the grating is 1870 nm. (a) Which is the red line? Which is the yellow line? Which is the blue line? (b) Calculate the wavelength (in nm) of spectral line *C*. (c) What is the highest order of spectral line *C* that it is possible to see using this grating?

Figure 25.54
Problem 30

Ⓒ ✦ 31. A spectrometer is used to analyze a light source. The screen-to-grating distance is 50.0 cm and the grating has 5000.0 slits/cm. Spectral lines are observed at the following angles: 12.98°, 19.0°, 26.7°, 40.6°, 42.4°, 63.9°, and 77.6°. (a) How many different wavelengths are present in the spectrum of this light source? Find each of the wavelengths. (b) If a different grating with 2000.0 slits/cm were used, how many spectral lines would be seen on the screen on one side of the central maximum? Explain.

✦ 32. White light containing wavelengths from 400 nm to 700 nm is shone through a grating. Assuming that at least part of the third-order spectrum is present, show that the second- and third-order spectra always overlap, regardless of the slit separation of the grating.

Ⓒ 33. A grating 1.600 cm wide has exactly 12,000 slits. The grating is used to resolve two nearly equal wavelengths in a light source: $\lambda_a = 440.000$ nm and $\lambda_b = 440.936$ nm. (a) How many orders of the lines can be seen with the grating? (b) What is the angular separation $\theta_b - \theta_a$ between the lines in each order? (c) Which order best resolves the two lines? Explain.

✦ 34. A grating spectrometer is used to resolve wavelengths 660.0 nm and 661.4 nm in second order. (a) How many slits/cm must the grating have to produce both wavelengths in second order? (The answer is either a maximum or a minimum number of slits per cm.) (b) The minimum number of slits required to resolve two closely spaced lines is $N = \lambda/(m\,\Delta\lambda)$, where λ is the average of the two wavelengths, $\Delta\lambda$ is the difference between the two wavelengths, and m is the order. What minimum number of slits must this grating have to resolve the lines in second order?

25.7 Diffraction by a Single Slit

35. The central maximum in a single-slit diffraction pattern is 1.0 cm wide on a distant screen. How wide are the other maxima?

36. The first two dark fringes on one side of the central maximum in a single-slit diffraction pattern are 1.0 mm apart. The wavelength of the light is 610 nm and the screen is 1.0 m from the slit. What is the slit width?

37. Light of wavelength 630 nm is incident upon a single slit with width 0.40 mm. Fig. 25.55 shows the pattern observed on a screen positioned 2.0 m from the slit. Determine the distance *s*.

$\longleftarrow s \longrightarrow$

Figure 25.55
Problem 37

38. Light from a red laser passes through a single slit to form a diffraction pattern on a distant screen. If the width of the slit is increased by a factor of two, what happens to the width of the central maximum on the screen?

39. The diffraction pattern from a single slit is viewed on a screen. Using blue light, the width of the central maximum is 2.0 cm. (a) Would the central maximum be narrower or wider if red light is used instead? (b) If the blue light has wavelength 0.43 μm and the red light has wavelength 0.70 μm, what is the width of the central maximum when red light is used?

40. Light of wavelength 490 nm is incident on a narrow slit. The diffraction pattern is viewed on a screen 3.20 m from the slit. The distance on the screen between the central maximum and the third minimum is 2.5 cm. What is the width of the slit?

25.8 Diffraction and the Resolution of Optical Instruments

41. A beam of yellow laser light (590 nm) passes through a circular aperture of diameter 7.0 mm. What is the angular width of the central diffraction maximum formed on a screen?

42. The photosensitive cells (rods and cones) in the retina are most densely packed in the fovea—the part of the retina used to see straight ahead. In the fovea the cells are all cones spaced about 1 μm apart. Would our vision have much better resolution if they were closer together? To answer this question, assume two light sources are just far enough apart to be resolvable according to Rayleigh's criterion. Assume an average pupil diameter of 5 mm and an eyeball diameter of 25 mm. Also assume that the index of refraction of the vitreous fluid in the eye is 1; in other words, treat the pupil as a circular aperture with air on both sides. What is the spacing of the cones if the centers of the diffraction maxima fall on two nonadjacent cones with a single intervening cone? (There must be an intervening dark cone in order to resolve the two sources; if two *adjacent* cones are stimulated the brain assumes a single source.)

43. A pinhole camera doesn't have a lens; a small circular hole lets light into the camera, which then exposes the film. For the sharpest image, light from a distant point source makes as small a spot on the film as possible. What is the optimum size of the hole for a camera in which the film is 16 cm from the pinhole? A hole smaller than the optimum makes a larger spot since it diffracts the light more. A larger hole also makes a larger spot because the spot cannot be smaller than the hole itself (think in terms of geometrical optics). Let the wavelength be 560 nm.

44. The radio telescope at Arecibo, Puerto Rico, has a reflecting spherical bowl of 305 m (1000 ft) diameter. Radio signals can be received and emitted at various frequencies with appropriate antennae at the focal point of the reflecting bowl. At a frequency of 300 MHz, what is the angle between two stars that can barely be resolved?

45. To understand Rayleigh's criterion as applied to the pupil of the eye, notice that rays do *not* pass straight through the center of the lens system (cornea + lens) of the eye except at normal incidence because the indices of refraction on the two sides of the lens system are different. In a simplified model, suppose light from two point sources travels through air and passes through the pupil (diameter *a*). On the other side of the pupil, light travels through the vitreous fluid (index of refraction *n*). Figure 25.56 shows two rays, one from each source, that pass through the center of the pupil. (a) What is the relationship between $\Delta\theta$, the angular separation of the two *sources*, and β, the angular separation of the two *images*? [*Hint:* Use Snell's law.] (b) The first diffraction minimum for light from source 1 occurs at angle ϕ, where $a \sin \phi = 1.22\lambda$ [Eq. (25-13)]. Here, λ is the wavelength *in the vitreous fluid*. According to Rayleigh's criterion, the sources can be resolved if the center of image 2 occurs no closer than the first diffraction minimum for image 1; that is, if $\beta \geq \phi$ or, equivalently, $\sin \beta \geq \sin \phi$. Show that this is equivalent to Eq. (25-14), where λ_0 is the wavelength *in air*.

Figure 25.56 Problem 45

COMPREHENSIVE PROBLEMS

46. A thin layer of an oil ($n = 1.60$) floats on top of water ($n = 1.33$). One portion of this film appears green ($\lambda = 510$ nm) in reflected light. How thick is this portion of the film? Give the three smallest possibilities.

47. If diffraction were the only limitation, what would be the maximum distance at which the headlights of a car could be resolved (seen as two separate sources) by the naked human eye? The diameter of the pupil of the eye is about 7 mm when dark-adapted. Make reasonable estimates for the distance between the headlights and for the wavelength.

48. Find the height h of the pits on a CD (Fig. 25.6a). When the laser beam reflects partly from a pit and partly from land (the flat aluminum surface) on either side of the pit, the two reflected beams interfere destructively; h is chosen to be the smallest possible height that causes destructive interference. The wavelength of the laser is 790 nm and the index of refraction of the polycarbonate plastic is $n = 1.5$.

49. The Very Large Array (VLA) is a set of 30 dish radio antennas located near Socorro, New Mexico. The dishes are spaced 1.0 km apart and form a Y-shaped pattern, as in Fig. 25.57. Radio pulses from a distant pulsar (a rapidly rotating neutron star) are detected by the dishes; the arrival time of each pulse is recorded using atomic clocks. If the pulsar is located 60.0° above the horizontal direction parallel to the right branch of the Y, how much time elapses between the arrival of the pulses at adjacent dishes in that branch of the VLA?

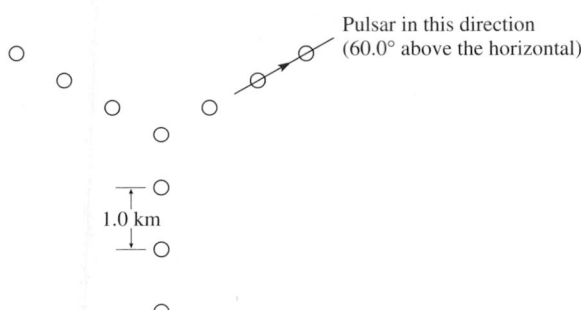

Figure 25.57 Problem 49

Problems 50 and 51: Two radio towers are a distance d apart as shown in the overhead view of Fig. 25.58. Each antenna *by itself* would radiate equally in all directions in a horizontal plane. The radio waves have the same frequency and start out in phase. A detector is moved in a circle around the towers at a distance of 100 km.

50. The power radiated in a horizontal plane by both antennas together is measured by the detector and is found to vary with angle. (a) Is the power detected at $\theta = 0$ a maximum or a minimum? Explain. (b) Sketch a graph of P versus θ to show qualitatively how the power varies with angle θ (from $-180°$ to $+180°$) if $d = \lambda$. Label your graph with values of θ at which the power is maximum or minimum. (c) Make a qualitative graph of how the power varies with angle for the case $d = \lambda/2$. Label your graph with values of θ at which the power is maximum or minimum.

51. The waves have frequency 3.0 MHz and the distance between antennas is $d = 0.30$ km. (a) What is the difference

Figure 25.58 Problems 50 and 51

in the path lengths traveled by the waves that arrive at the detector at $\theta = 0°$? (b) What is the difference in the path lengths traveled by the waves that arrive at the detector at $\theta = 90°$? (c) At how many angles ($0 \le \theta < 360°$) would you expect to detect a maximum intensity? Explain. (d) Find the angles (θ) of the maxima in the first quadrant ($0 \le \theta \le 90°$). (e) Which (if any) of your answers to parts (a) to (d) would change if the detector were instead only 1 km from the towers? Explain. (Don't calculate new values for the answers.)

52. Two narrow slits with a center-to-center distance of 0.48 mm are illuminated with coherent light at normal incidence. The intensity of the light falling on a screen 5.0 m away is shown in Fig. 25.59, where x is the distance from the central maximum on the screen. (a) What would be the intensity of the light falling on the screen if only one slit were open? (b) Find the wavelength of the light.

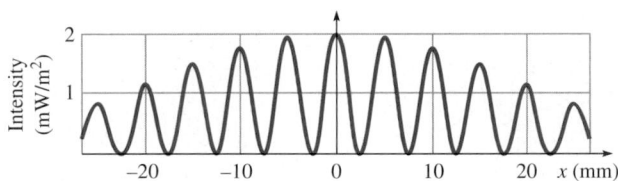

Figure 25.59 Problem 52

53. A lens ($n = 1.52$) is coated with a magnesium fluoride film ($n = 1.38$). (a) If the coating is to cause destructive interference in reflected light for $\lambda = 560$ nm (the peak of the solar spectrum), what should its minimum thickness be? (b) At what two wavelengths closest to 560 nm does the coating cause *constructive* interference in reflected light? (c) Is any visible light reflected? Explain.

54. A thin soap film ($n = 1.35$) is suspended in air. The spectrum of light reflected from the film is missing two visible wavelengths of 500.0 nm and 600.0 nm, with no missing wavelengths between the two. (a) What is the thickness of the soap film? (b) Are there any other visible wavelengths missing from the reflected light? If so, what are they? (c) What wavelengths of light are strongest in the *transmitted* light?

55. Instead of an antireflective coating, suppose you wanted to coat a glass surface to *enhance* the reflection of visible light. Assuming that $1 < n_{coating} < n_{glass}$, what should the minimum thickness of the coating be to maximize the reflected intensity for wavelength λ?

56. If you shine a laser (wavelength 0.60 μm) with a small aperture at the Moon, diffraction makes the beam spread out and the spot on the Moon is large. Making the aperture smaller only makes the spot on the Moon *larger*. On the other hand,

shining a wide searchlight at the Moon can't make a tiny spot—the spot on the Moon is at least as wide as the searchlight. What is the radius of the *smallest* possible spot you can make on the Moon by shining a light from Earth? Assume the light is perfectly parallel before passing through a circular aperture.

57. A mica sheet 1.00 μm thick is suspended in air. In reflected light, there are gaps in the visible spectrum at 450, 525, and 630 nm. Calculate the index of refraction of the mica sheet.

58. In bright light, the pupils of the eyes of a cat narrow to a vertical *slit* 0.30 mm across. Suppose that a cat is looking at two mice 18 m away. What is the smallest distance between the mice for which the cat can tell that there are two mice rather than one using light of 560 nm? Assume the resolution is limited by diffraction only.

59. Parallel light of wavelength λ strikes a slit of width a at normal incidence. The light is viewed on a screen that is 1.0 m past the slits. In each case that follows, sketch the intensity on the screen as a function of x, the distance from the center of the screen, for $0 \leq x \leq 10$ cm. (a) $\lambda = 10a$. (b) $10\lambda = a$. (c) $30\lambda = a$.

60. About how close to each other are two objects on the Moon that can just barely be resolved by the 5.08 m (200 in.) diameter Mount Palomar reflecting telescope? (Use a wavelength of 520 nm.)

61. A grating in a spectrometer is illuminated with red light ($\lambda = 690$ nm) and blue light ($\lambda = 460$ nm) simultaneously. The grating has 10,000.0 slits per cm. Sketch the pattern that would be seen on a screen 2.0 m from the grating. Label distances from the central maximum. Label which lines are red and which are blue.

62. Two slits separated by 20.0 μm are illuminated by light of wavelength 0.50 μm. If the screen is 8.0 m from the slits, what is the distance between the $m = 0$ and $m = 1$ bright fringes?

✦63. Two coherent plane waves travel at angle θ_0 toward a photographic plate. Show that the distance between fringes of constructive interference on the plate is given by $d = \lambda/\sin \theta_0$. See Fig. 25.43.

64. In a double-slit experiment, what is the linear distance on the screen between adjacent maxima if the wavelength is 546 nm, the slit separation is 0.100 mm, and the slit-screen separation is 20.0 cm?

65. The radio telescope at Arecibo, Puerto Rico, has a reflecting spherical bowl of 305 m (1000 ft) diameter. Radio signals can be received and emitted at various frequencies with the appropriate antennae at the focal point. If two Moon craters 499 km apart are to be resolved, what wavelength radiowaves must be used?

ANSWERS TO PRACTICE PROBLEMS

25.2 The mirror should be moved in (shorter path length). Since the number of wavelengths traveled in the arm with the vessel decreases, we must decrease the number of wavelengths traveled in the other arm.

25.3 560 nm and 458 nm

25.4 (a) 0, 0.02 rad, 0.04 rad; (b) 0.01 rad, 0.03 rad; (c) 4 cm

25.5 The intensity is maximum at the center ($\theta = 0$) and gradually decreases to either side but never reaches zero.

25.6 4760 slits/cm; fourth-order maxima are present for wavelengths up to 525 nm.

25.7 No, the window is large compared to the wavelength of light, so we expect diffraction to be negligible. The Sun is not distant enough to treat it as a point source; rays from different points on the Sun's surface travel in slightly different directions as they pass through the window.

25.8 3.9 cm

25.9 9 m

Chapter
26

Relativity

The centers of some galaxies are much brighter than the rest of the galaxy. These active galactic nuclei, which may be only about as big as our solar system, can give off 20 billion times as much light as the Sun. The core of the galaxy known as NGC 6251 emits a narrow, extremely energetic jet of charged particles in a direction roughly toward Earth. The photo shows the jet as imaged by the Very Large Array of radiotelescopes in New Mexico; the galactic core is at the extreme left.

When scientists first measured the speed of the tip of this jet, they used two radiotelescope images, taken on two successive days. They measured how far the tip of the jet moved, divided by the time elapsed between the two images, and came up with a speed greater than the speed of light! Is it possible for the charged particles in the jet to move faster than light? If not, what was the scientists' mistake?

PART FIVE   Quantum and Particle Physics and Relativity

26.1   POSTULATES OF RELATIVITY

Reference Frames

The idea of *relativity* is not something entirely new; it goes all the way back to Galileo. Aristotle had previously said that a body continues to move only if a force continues to propel it; take away the force and the body comes to rest. The authority of Aristotle's opinion prevailed for many centuries. Galileo turned this thinking totally around by saying that a body maintains a constant velocity (which can be zero or nonzero) in the absence of any external forces acting on it; this concept is the basis for the law of inertia as stated by Newton.

All motion must be measured in some particular reference frame, which we usually represent as a set of coordinate axes. Suppose two people walk hand-in-hand on a moving sidewalk in an airport. They might walk at 1.3 m/s with respect to a reference frame attached to the moving sidewalk and at 2.4 m/s with respect to a reference frame attached to the building. The two reference frames are *equally valid*.

An **inertial reference frame** is one in which no accelerations are observed in the absence of external forces. In noninertial reference frames, bodies have accelerations in the absence of applied forces because *the reference frame itself is accelerating* with respect to an inertial frame. For example, suppose two people sit across from each other on a rapidly rotating merry-go-round. When one tosses a ball to the other, the ball is deflected sideways as viewed by observers on the merry-go-round. The sideways acceleration is not caused by any force acting on the ball; the rotating merry-go-round is a *noninertial* reference frame (Fig. 26.1). The law of inertia does not hold in noninertial frames.

For many purposes, the Earth's surface can be considered to be an inertial reference frame, even though strictly speaking it is not. The Earth's rotation causes phenomena such as the rotary motion of hurricanes and trade winds, which, in a reference frame attached to Earth's surface, are accelerations not caused by applied forces.

Any reference frame that moves with constant velocity with respect to an inertial frame is itself inertial; if the acceleration of a body in one inertial frame is zero, its acceleration in any of the other inertial frames is also zero. In our earlier example, if a reference frame fixed to the airport terminal is inertial and the moving sidewalk moves at constant velocity with respect to the terminal, then a reference frame fixed to the moving sidewalk is also inertial.

(a)   (b)

Figure 26.1 A ball tossed across a merry-go-round. (a) Trajectory of the ball as viewed in the noninertial frame fixed to the platform. The ball is thrown straight toward the catcher but then is deflected sideways. In this frame, the platform is at rest and the tree is moving. (b) Trajectory of the ball as viewed in the inertial frame fixed to the ground. In this frame, the catcher rotates away from the straight-line path of the ball.

Principle of Relativity

Ever since Galileo and Newton, scientists have been careful to formulate the laws of physics so that the *same laws* hold in any inertial reference frame. Particular quantities (velocity, momentum, kinetic energy) have different values in different inertial reference frames, but the **principle of relativity** requires that the *laws* of physics (such as Newton's laws of motion and the conservation of momentum and energy) be the *same* in all inertial frames.

Principle of Relativity
The *laws* of physics are the *same* in all inertial frames.

The laws and equations in this chapter—just like those of all other chapters in this book—are only valid in inertial frames. The laws of physics must be modified if they are to apply in noninertial (accelerated) reference frames.

Apparent Contradictions with the Principle of Relativity

In the nineteenth century, James Clerk Maxwell used the four basic laws that describe electromagnetic fields (*Maxwell's equations*, Section 22.2) to show that electromagnetic waves travel through vacuum at a speed of $c = 3.00 \times 10^8$ m/s. In fact, Maxwell's equations show that EM waves travel at the *same speed in every inertial reference frame*, regardless of the motion of the source or of the observer.

This conclusion, that the speed of light is the same in any inertial reference frame, contradicts the Galilean laws of relative velocity (Section 4.7). Suppose a car travels at velocity $\vec{\mathbf{v}}_{CG}$ with respect to the ground (Fig. 26.2). Light coming from the car's headlights travels at velocity $\vec{\mathbf{v}}_{LC}$ with respect to the car. Galilean velocity addition says that the speed of the light beam with respect to the ground is

$$\vec{\mathbf{v}}_{LG} = \vec{\mathbf{v}}_{LC} + \vec{\mathbf{v}}_{CG} \tag{4-9}$$

Thus, the speed of light would have two *different* values (v_{LG} and v_{LC}) in two different inertial reference frames.

A possible resolution to the contradiction would be if Maxwell's equations give the speed of light with respect to the medium in which light travels. Nineteenth-century scientists believed that light was a vibration in an invisible, elusive medium called the *ether*. If the speed of light c derived by Maxwell is the speed *with respect to the ether*,

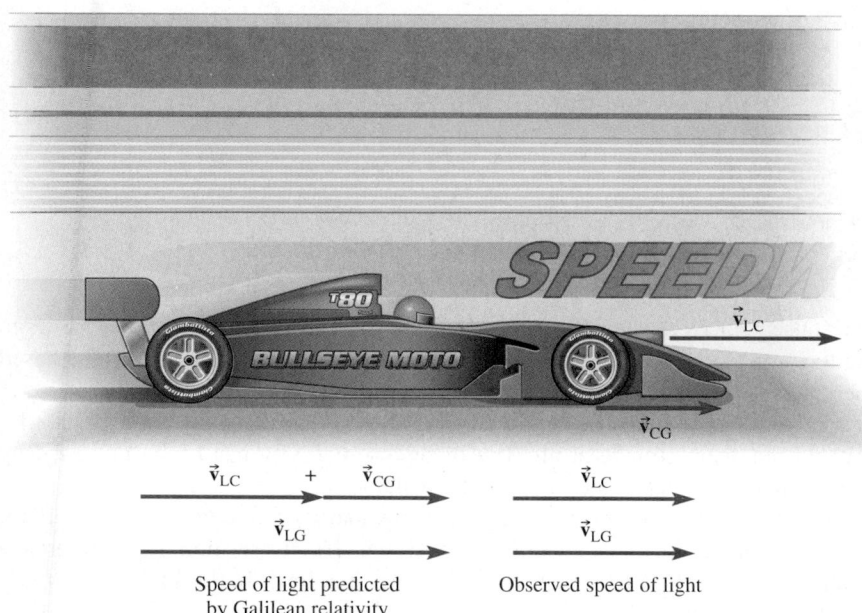

Speed of light predicted   Observed speed of light
by Galilean relativity

Figure 26.2 Measurement of light speed according to Galilean relativity

Figure 26.3 Simplified version of the Michelson-Morley experiment. Assume the apparatus moves to the right at speed v with respect to the ether. With respect to the lab, the light beam moves at speed $c - v$ to the right in arm 1 and at speed $c + v$ to the left. The round-trip time in arm 1 is then $\Delta t_1 = L_1/(c - v) + L_1/(c + v) \neq (2L_1)/c$. The number of cycles of the wave in arm 1 is $\Delta t_1/T = f\,\Delta t_1$, where f is the frequency of the light. Thus, the number of cycles in arm 1 depends on the speed of the apparatus with respect to the ether. As the entire apparatus is rotated in a horizontal plane, the interference pattern viewed through the telescope should change as the difference in the number of cycles in arms 1 and 2 changes. No change in the interference pattern was observed by Michelson and Morley.

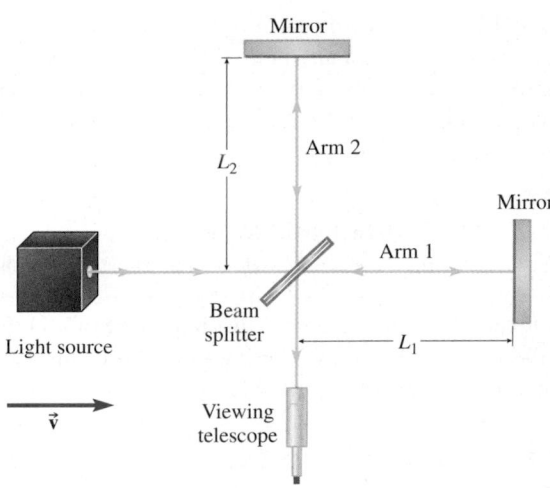

then in any inertial frame moving with respect to the ether, the speed of light should differ as predicted by the Galilean relative velocity laws.

Does the speed of light as measured on Earth really depend on the motion of the Earth through the ether? In 1881, Albert Michelson designed a sensitive instrument, now called the Michelson interferometer (Section 25.2), to find out. In a later, more sensitive version of the experiment, Michelson was joined by E. W. Morley. The Michelson-Morley experiment showed no change in the speed of light due to the motion of the Earth relative to the ether (Fig. 26.3). This led to the conclusion that there is no ether.

Einstein's Postulates

Albert Einstein (1879–1955) resolved these contradictions in his theory of special relativity (1905), now recognized as one of the cornerstones of modern physics. Einstein started with two postulates. The first is identical to Galileo's principle of relativity: the laws of physics are the same in any inertial reference frame. The second is that light travels at the same speed through vacuum in any inertial reference frame, regardless of the motion of the source or of the observer.

Einstein's Postulates of Special Relativity:

(I) The laws of physics are the same in all inertial reference frames.
(II) The speed of light in vacuum is the same in all inertial reference frames, regardless of the motion of the source or of the observer.

The consequences Einstein derived from these two postulates deliver fatal blows to our intuitive notions of space and time. Our intuition about the physical world is based on experience, which is limited to things moving much slower than light. If moving at speeds approaching the speed of light were part of our everyday experience, then relativity would not seem strange at all. The theory of relativity has been confirmed by many experiments—which is the true test of any theory.

Einstein's theory of *special* relativity concerns inertial reference frames. In 1915, Einstein published his theory of general relativity, which concerns noninertial reference frames and the effect of gravity on intervals of space and time. In this chapter we study inertial reference frames only.

26.2 SIMULTANEITY AND IDEAL OBSERVERS

The postulate that the speed of light is the same in all inertial reference frames leads to a startling conclusion: observers in different inertial reference frames disagree about whether two events are simultaneous if the events occur at different places. In Newtonian physics, time is absolute. Observers in different reference frames can use the same clock to measure time; they all agree on whether two events are simultaneous or not. Einstein's relativity does away with the notion of absolute time.

The idea of an event is crucial in relativity. The location of an event can be specified by three spatial coordinates (x, y, z); the time at which the event occurs is specified by t. Einstein's relativity treats space and time as four-dimensional *space-time* in which an event has four space-time coordinates (x, y, z, t).

Imagine two spaceships piloted by astronauts named Abe and Bea (Fig. 26.4). Each ship has zero acceleration because external forces are negligible and they are not firing

(a) Abe's reference frame

(b) Bea's reference frame

Figure 26.4 Two space probes emit flashes of light. (a) The sequence of events as viewed in Abe's reference frame. (b) The sequence of events as viewed in Bea's reference frame.

their engines. Then Abe and Bea are observers in inertial reference frames. They are not at rest with respect to each other, though. According to Abe, Bea moves past him at speed v; according to Bea, Abe moves past her at speed v.

Abe and Bea observe two events: two space probes each emit a flash of light. Abe sees the two flashes of light simultaneously. From the long measuring sticks in front of and behind his ship, which record the positions at which events occur, he finds that the flashes were emitted at *equal distances* from his ship. In Abe's frame of reference (Fig. 26.4a), the flashes travel the same distance at the same speed (c) and arrive at the same time, so they must have been emitted *simultaneously*. Bea's ship happened to be along-side Abe's at the instant the flashes were emitted, but the flash from the probe on our right reaches Bea before the flash from the probe on our left. In Abe's reference frame, that happens because Bea is moving toward one probe and away from the other; the flashes do not travel equal distances to reach Bea.

In Bea's reference frame (Fig. 26.4b), the right flash arrives before the left flash. Bea has measuring sticks similar to those of Abe; when she consults them, she finds that the flashes were emitted at equal distances from her ship. Since the flashes from each probe travel equal distances at the same speed (c), the right flash must arrive first because it was emitted first. In Bea's reference frame, *the flashes are not emitted simul-taneously*. Bea's explanation for why the flashes reach Abe at the same time is that Abe is moving away from the first flash and toward the second at just the right speed. In Bea's reference frame, the flashes do not travel equal distances to reach Abe.

According to Einstein's postulates, the two reference frames are equally valid and light travels at the same speed in each. The inescapable conclusion is that the events *are* simultaneous in one frame and *are not* in the other.

Ideal Observers

Making The Connection:
**observing active
galactic nuclei**

Since the high-speed jet of charged particles from the core of NGC 6251 moves toward Earth, the time it takes light to travel from the tip of the jet to Earth is continually decreasing. In Fig. 26.5a, light from the tip of the jet travels a distance x in elapsed time T. In Fig. 26.5b, light from the tip of the jet leaves at a time Δt later than that in part (a), travels a distance $x - \Delta x$, and arrives 1 day later. The elapsed travel time for light to travel a distance $x - \Delta x$ is $T + 1$ day $- \Delta t$, because the *jet* had already moved a distance Δx in time Δt. Since light travels at speed c,

$$x = cT$$
$$x - \Delta x = c(T + 1 \text{ day} - \Delta t)$$

We can solve these two equations for Δt:

$$\Delta t = 1 \text{ day} + \frac{\Delta x}{c}$$

The speed at which the tip moves is $\Delta x / \Delta t$. The scientists' mistake was to assume that $\Delta t = 1$ day. The jet is fast, but not as fast as light.

In Abe and Bea's disagreement about simultaneity, we were careful not to make a similar mistake. Each of them sees two flashes that travel *equal distances* to reach them.

Figure 26.5 Calculating the jet speed

To avoid the confusion of light signals traveling different distances, we can imagine an *ideal observer* who has placed sensors with synchronized clocks *at every point in space*; these sensors record the time at which any event occurs at their location. Even if Abe and Bea were ideal observers, the data recorded by their sensors would still show that they reach different conclusions about the time sequence of the two flashes.

Cause and Effect

Continuing with the same reasoning, an observer moving to the left with respect to Abe would say that the *left* flash occurs first. Thus, the time order of the two events is different in different reference frames. How can there possibly be any cause-effect relationships if the time order of events depends on the observer? What would it mean if, in some reference frames, the effect occurs *before* the cause?

In order for event 1 to cause event 2, some sort of signal—some information—must travel from event 1 to event 2. One conclusion of Einstein's postulates is that no signal can travel faster than c. If there is enough time, in some reference frame, for a signal at light speed to travel from event 1 to event 2, then it can be shown—through a more advanced analysis than we can do here—that a signal can travel from event 1 to event 2 in *all* inertial reference frames. The cause comes before the effect for all observers. On the other hand, if there is *not* enough time for a signal at light speed to travel from event 1 to event 2, then the two cannot have a causal relationship in *any* reference frame. For such events, some observers say that event 1 happens first, some say that event 2 happens first, and one particular observer says the events are simultaneous.

26.3   TIME DILATION

Since inertial observers in relative motion disagree about simultaneity, can two such observers agree about the time kept by clocks in relative motion? Two ideal clocks, both at rest, keep the same time by ticking simultaneously. However, if the clocks are in relative motion, the ticks are events that occur at different spatial locations, so two different inertial observers may disagree on whether the ticks are simultaneous, or about which clock ticks first.

The situation is easiest to analyze by imagining a conceptually simple kind of clock— a light clock (Fig. 26.6a). A light clock is a tube of length L with mirrors at each end. A light pulse bounces back and forth between the two mirrors. One tick of the clock is one round-trip of the light pulse. The time interval between ticks for a stationary clock is $\Delta t_0 = 2L/c$.

Imagine now that Abe and Bea have two identical light clocks. Bea holds the clock vertically as she flies past Abe in her spaceship at speed $v = 0.8c$. What is the time interval between ticks of Bea's clock, as measured by Abe?

The light pulse in Bea's clock is not traveling in the y-direction in Abe's reference frame (Fig. 26.6b). The pulse must have an x-component of velocity if it is to meet up with the mirrors, which move to the right at speed v. During one tick, the light pulse moves along the *diagonal paths* shown.

Let us analyze one tick of Bea's clock as observed in Abe's reference frame. Suppose the time interval for one tick of Bea's clock as measured by Abe is Δt. Then the distance traveled by the light pulse during one tick is $c\,\Delta t$. During this same time interval the clock moves horizontally a distance $v\,\Delta t$. By the Pythagorean theorem (see Fig. 26.6b):

$$L^2 + \left(\frac{v\,\Delta t}{2}\right)^2 = \left(\frac{c\,\Delta t}{2}\right)^2$$

In Bea's reference frame, the clock is at rest. The distance traveled by the light pulse during one tick is $2L$, so the time interval for one tick as measured in Bea's reference frame is $\Delta t_0 = 2L/c$. We can therefore make the substitution

$$L = \left(\frac{c\,\Delta t_0}{2}\right)$$

Figure 26.6 (a) A light pulse bounces up and down between the mirrors in a light clock. The time interval for one tick of the clock is the round-trip time for the light pulse. (b) In Abe's reference frame, Bea's light clock moves to the right at speed $v = 0.8c$. The path of the light pulse in Bea's clock is along the diagonals shown.

into the Pythagorean equation:

$$\left(\frac{c\,\Delta t_0}{2}\right)^2 + \left(\frac{v\,\Delta t}{2}\right)^2 = \left(\frac{c\,\Delta t}{2}\right)^2$$

Solving for Δt yields (Problem 10)

$$\Delta t = \frac{1}{\sqrt{1 - v^2/c^2}} \times \Delta t_0 \qquad (26\text{-}1)$$

The factor multiplying Δt_0 in Eq. (26-1) occurs in many relativity equations, so we assign it a symbol (γ, the Greek letter gamma) and a name (the **Lorentz factor**, after physicist Hendrik Lorentz).

$$\gamma = \frac{1}{\sqrt{1 - v^2/c^2}} \qquad (26\text{-}2)$$

See Fig. 26.7 for a graph of γ as a function of v/c. Using γ, Eq. (26-1) becomes

Time dilation

$$\Delta t = \gamma\,\Delta t_0 \qquad (26\text{-}3)$$

Notice that when $v \ll c$, $\gamma \approx 1$. Thus, for an object moving at nonrelativistic speeds—speeds small compared to the speed of light—$\Delta t = \gamma \Delta t_0 \approx \Delta t_0$.

Since $\gamma > 1$ for any $v \neq 0$, the time interval between ticks as measured in the reference frame in which the clock is moving, Δt, is *longer* than the time interval Δt_0 as measured in the clock's **rest frame**—the frame in which the clock is at rest. In a short phrase, *moving clocks run slow*. This effect is called **time dilation**; the time between ticks of the moving clock is dilated or expanded.

Abe's and Bea's reference frames are equally valid. Wouldn't Bea say that it is *Abe's* clock that runs slow? Yes, and they are *both* correct. Imagine that both of the clocks tick just as Abe and Bea pass each other—when they're at the same place. They agree that the clocks tick simultaneously. To see which clock runs slow, we compare the time of the *next* tick of the two clocks. Since the clocks are then at different places, the two observers disagree about the sequence of the ticks. Abe observes his clock ticking first, while Bea observes hers ticking first. They are both correct: there is no absolute or preferred reference frame from which to measure the time intervals.

The time interval Δt_0 measured in the rest frame of the clock is called the **proper** time interval; in that frame, the clock is at the same position for both ticks. When using the time dilation relation $\Delta t = \gamma \Delta t_0$, Δt_0 always represents the proper time interval—the time interval between two events measured in an inertial reference frame in which the events *occur in the same place*. The proper time interval is always shorter than the time interval Δt measured in any other inertial frame. The time dilation equation does *not* apply to a time interval between events that occur at different locations for *all* inertial observers.

Although we have analyzed time dilation using light clocks, *any* clock must show the same effect; otherwise there would be a preferred reference frame—the one in which light clocks behave the same as other clocks. Furthermore, a *clock* can be anything that measures a time interval. Biological processes such as the beating of a heart or the aging process are subject to time dilation. The nature of space and time, not the workings of a particular kind of device, is responsible for time dilation.

Time dilation may seem strange, but it has been verified in many experiments. One straightforward one was done in 1971 by J. C. Hafele and R. E. Keating using extremely precise cesium beam atomic clocks. The clocks were loaded onto airplanes and flown around for nearly two days. When the clocks were compared to reference clocks at the U.S. Naval Observatory, the clocks that had been in the air were behind those on the ground by an amount consistent with relativity.

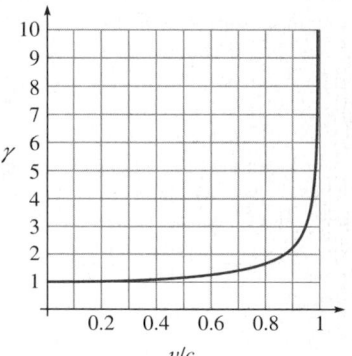

Figure 26.7 A graph of the Lorentz factor γ as a function of v/c. For low speeds, $\gamma \approx 1$. For speeds approaching that of light, γ increases without bound.

Problem-Solving Strategy: Time Dilation

- Identify the two events that mark the beginning and end of the time interval in question. The "clock" is whatever measures this time interval.
- Identify the reference frame in which the clock is at rest. In that frame, the clock measures the proper time interval Δt_0.
- In any other reference frame, the time interval is *longer* by a factor of $\gamma = (1 - v^2/c^2)^{-1/2}$, where v is the speed of one frame relative to the other.

Speed and Distance Units Commonly Used in Relativity

- Speeds are usually written as a fraction times the speed of light (for example, $0.13c$).
- Distances are often measured in **light-years** (symbol ly). A light-year is the distance that light travels in 1 yr. Calculations involving light-years are simplified by writing the speed of light as 1 ly/yr.

Example 26.1

Slowing the Aging Process

 A 20.0-yr-old astronaut named Ashlin leaves Earth in a spacecraft moving at 0.80c. How old is Ashlin when he returns from a trip to a star 30.0 light-years from Earth, assuming that he moves at 0.80c relative to Earth during the entire trip?

Strategy and Solution According to an earthbound observer, the trip takes (60.0 ly) ÷ (0.80 ly/yr) = 75 yr to complete. Since the astronaut is moving at high speed relative to Earth, all clocks on board—including biological processes such as aging—run slow as observed by Earth observers. Therefore, when the astronaut returns he is *less than* 95 yr old. Maybe he'll have time for another trip!

The two events that measure the time interval are the departure and return of the astronaut. Let the "clock" be the astronaut's aging process. This "clock" measures a 75-yr time interval *according to Earth observers*, for whom the clock is *moving*. The proper time interval is that measured by the astronaut himself. Thus, $\Delta t = 75$ yr and we want to find Δt_0. The Lorentz factor is calculated using the relative speed of the two reference frames, which is 0.80c.

$$\gamma = (1 - 0.80^2)^{-1/2} = \tfrac{5}{3}$$

From the time dilation relation $\Delta t = \gamma \, \Delta t_0$,

$$\Delta t_0 = \frac{1}{\gamma} \, \Delta t = \frac{3}{5} \times 75 \text{ yr} = 45 \text{ yr}$$

If the astronaut ages 45 yr during the trip, he is 65 yr old on his return.

Discussion If the astronaut could travel 60.0 ly in 45 yr, then he would be traveling faster than light; his speed would be 60.0 ly/45 yr = 1.3c. As we see in Section 26.4, the astronaut has traveled *less than 60.0 ly* in his reference frame. Just as time intervals are different in different reference frames, so are distances.

Suppose that Ashlin has a twin brother, Earnest, who stays behind on Earth. When Ashlin returns, he is 65 yr old while Earnest is 95 yr old (Fig. 26.8). Why can't Ashlin say that, in *his* reference frame, *Earnest* is the one who was moving at 0.80c, and therefore *Earnest's* biological clock should run slow so that Earnest is *younger* rather than older? This question is sometimes called the *twin paradox*.

We analyzed the situation in the reference frame of Earnest, which is assumed to be inertial. If Ashlin's reference frame were also inertial, he would have to move at *constant velocity* with respect to Earnest. If he did, he could never return to Earth! Ashlin must turn around; his reference frame is *accelerated*. We cannot analyze the situation from Ashlin's point of view using special relativity. Analysis of the trip *in Ashlin's noninertial reference frame* requires general relativity, which confirms that Ashlin returns to Earth younger than Earnest.

Practice Problem 26.1   Journey to newly formed stars near Earth

In 1998, using the Keck telescope, scientists discovered some previously undetected, young stars that are only 150 ly from Earth. Suppose a space probe is flown to one of these stars at a speed of 0.98c. The battery that powers the communications systems can run for 40 yr. Will the battery still be good when the space probe reaches the star?

0.80c

(a)   (b)

Figure 26.8
(a) Both Ashlin and Earnest are 20.0 yr old at the beginning of the voyage. (b) On Ashlin's return, he is 65 yr old while Earnest is 95 yr old.

26.4   LENGTH CONTRACTION

Suppose Abe has two identical metersticks, which he has verified to have exactly the same length. He gives one to Bea, the astronaut. As Bea flies past Abe with speed $v = 0.8c$, holding her meterstick in the direction of motion, they compare the lengths of the two metersticks (Fig. 26.9). Are they still equal?

Figure 26.9 Comparing the lengths of two metersticks in relative motion

No; Abe finds that Bea's meterstick is *less than* 1 m in length. To measure the length of Bea's moving meterstick, Abe might start a timer when the front end of her meterstick passes a reference point and stop the timer when the other end passes. The length L that Abe measures for Bea's stick is the measured time interval Δt_0 multiplied by the speed at which she is moving:

$$L = v \, \Delta t_0 \quad \text{(Abe; moving stick)}$$

Since Abe measures the time interval between two events that occur at the same place—at his reference point—Δt_0 is the *proper* time interval between the events.

Bea can measure the length of her own stick (L_0) the same way—by recording the time interval Δt between when Abe's reference point passes the two ends of her meterstick.

$$L_0 = v \, \Delta t \quad \text{(Bea; stick at rest)}$$

The time interval Bea measures is dilated; it is longer than the proper time interval by a factor of γ:

$$\Delta t = \gamma \, \Delta t_0$$

Therefore, the length of Bea's meterstick as measured by Abe (L) is *shorter* than the length measured by Bea ($L_0 = 1$ m):

$$\frac{L}{L_0} = \frac{\Delta t_0}{\Delta t} = \frac{1}{\gamma}$$

Length contraction

$$L = \frac{L_0}{\gamma} \qquad (26\text{-}4)$$

In the length contraction relation $L = L_0/\gamma$, L_0 represents the **proper** length or rest length—the length of an object in its rest frame. L is the length measured by an observer for whom the object is moving.

Meanwhile, Bea can also measure Abe's meterstick. Bea would say that *Abe's* meterstick is the one that is shorter. How can they both be right? Which meterstick *really* is shorter?

To resolve the issue once and for all, they might want to hold the metersticks together, but they cannot: the metersticks are in relative motion. To compare the lengths, they could wait until the left ends of the two metersticks coincide (Fig. 26.10). They must compare the positions of the right ends of the metersticks *at the same instant*. Since Abe and Bea disagree about simultaneity, they disagree about which meterstick is shorter, but they are *both* correct. Just as an observer always finds that a moving clock runs slow compared to a stationary clock, an observer always finds that a moving object is contracted (shortened) *along the direction of its motion*. Lengths perpendicular to the direction of motion are not contracted.

Stationary meterstick

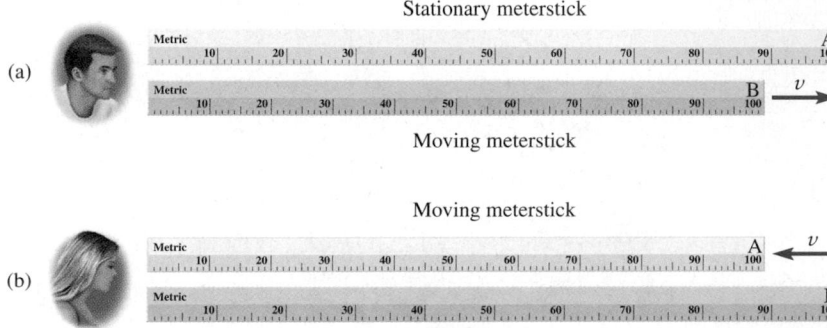

(a)

Moving meterstick

Moving meterstick

(b)

Stationary meterstick

Figure 26.10 Comparing two metersticks by lining up the left ends. (a) As seen in Abe's rest frame. (b) As seen in Bea's rest frame.

Problem-Solving Strategy: Length Contraction

- Identify the object whose length is to be measured in two different frames. The length is contracted only in the direction of the object's motion. If the length in question is a distance rather than the length of an actual object, it often helps to imagine the presence of a long measuring stick.

- Identify the reference frame in which the object is at rest. The length in that frame is the proper length L_0.

- In any other reference frame, the length L is *contracted*: $L = L_0/\gamma$. $\gamma = (1 - v^2/c^2)^{-1/2}$, where v is the speed of one frame relative to the other.

Example 26.2

Muon Survival

Cosmic rays are energetic particles—mostly protons—that enter the Earth's upper atmosphere from space. The particles collide with atoms or molecules in the upper atmosphere and produce showers of particles. One of the particles produced is the muon, which is something like a heavy electron. The muon is not stable. Half of the muons present at any particular instant of time still exist 1.5 μs later; the other half decay into an electron plus two other particles. In a shower of muons streaming toward Earth's surface, some decay before reaching the ground. If 1 million muons are moving toward the ground at speed $0.995c$ at an altitude of 4500 m above sea level, how many survive to reach sea level?

Strategy Imagine a measuring stick extending from the upper atmosphere to sea level (Fig. 26.11). In the reference frame of the Earth, the measuring stick is at rest; its proper length is $L_0 = 4500$ m. In the reference frame of the muons, the muons are at rest and the measuring stick moves past them at speed $0.995c$. In the muon frame, the measuring stick is contracted. In the muon frame, sea level is *not* 4500 m away when the upper end of the measuring stick passes by; the distance is shorter due to length contraction. Once we find the contracted distance L, the time the measuring stick takes to move past them at speed v is $\Delta t = L/v$. From the elapsed time, we can determine how many muons decay and how many are left.

Muon

4.5 km   0.995c

Earth's view

(a)

450 m   0.995c

Muon's view

(b)

Figure 26.11
The muons' trip as viewed by (a) an Earth observer and (b) the muons.

continued on next page

Example 26.2 *continued*

Solution   The contracted distance is

$$L = \frac{L_0}{\gamma}$$

The Lorentz factor is

$$\gamma = (1 - 0.995^2)^{-1/2} = 10$$

Therefore, the contracted distance is $L = \frac{1}{10} \times 4500$ m = 450 m. The elapsed time is

$$\Delta t = L/v = (450 \text{ m})/(0.995 \times 3 \times 10^8 \text{ m/s}) = 1.5 \text{ μs}$$

During 1.5 μs, half of the muons decay, so 500,000 muons reach sea level.

Discussion   If there were no length contraction, the elapsed time would be

$$\Delta t = \frac{4500 \text{ m}}{0.995 \times 3 \times 10^8 \text{ m/s}} = 15 \text{ μs}$$

This time interval is equal to ten successive intervals of 1.5 μs. During each of those intervals, half of the muons present at the start of the interval decay. Therefore, the number that survive to reach sea level would be only

$$1{,}000{,}000 \times \left(\tfrac{1}{2}\right)^{10} \approx 980 \text{ muons}$$

The relative numbers of muons at sea level compared to the numbers at higher elevations has been studied experimentally; the results are consistent with relativity. Problem 57 looks at muon decay in the reference frame of an Earth observer.

Practice Problem 26.2   Rocket velocity

An astronaut in a rocket passes a meterstick moving parallel to its long dimension. The astronaut measures the meterstick to be 0.80 m long. How fast is the rocket moving with respect to the meterstick?

Figure 26.12 Space probe launched with velocity v_{PB} as measured in the frame of the moving spaceship

26.5   VELOCITIES IN DIFFERENT REFERENCE FRAMES

Figure 26.12 shows Abe and Bea in their spaceships; in Abe's reference frame, Bea moves at velocity v_{BA}. Bea launches a space probe, which, in *her* reference frame, moves at velocity v_{PB}. What is the velocity of the probe in Abe's reference frame (v_{PA})? (Since we consider only velocities along a straight line—in this case, along a horizontal line—we write the velocities as *components* along that line.)

If v_{PB} and v_{BA} are small compared to c, time dilation and length contraction are negligible; then v_{PA} is given by the Galilean velocity addition formula:

$$v_{\text{PA}} = v_{\text{PB}} + v_{\text{BA}} \tag{4-9}$$

However, Eq. (4-9) cannot be correct in general, since the probe cannot move faster than light in *any* inertial reference frame. (If $v_{\text{PB}} = +0.6c$ and $v_{\text{BA}} = +0.7c$, the Galilean formula gives $v_{\text{PA}} = 1.3c$.) The relativistic equation that replaces Eq. (4-9) takes time dilation and length contraction into account and predicts $|v_{\text{PA}}| < c$ for *any* values of v_{PB} and v_{BA} whose magnitudes are less than c.

The relativistic velocity transformation formula is

$$v_{\text{PA}} = \frac{v_{\text{PB}} + v_{\text{BA}}}{1 + v_{\text{PB}} v_{\text{BA}}/c^2} \tag{26-5}$$

The denominator in Eq. (26-5) can be thought of as a correction factor to account for both time dilation and length contraction. When v_{PB} and v_{BA} are small compared to c, the denominator is approximately 1; then Eq. (26-5) reduces to the Galilean approximation

[Eq. (4-9)]. For example, if $v_{PB} = v_{BA} = +3$ km/s (fast by ordinary standards, but small compared to the speed of light), the denominator is

$$1 + \frac{v_{PB}v_{BA}}{c^2} = 1 + \frac{(3 \times 10^3 \text{ m/s})^2}{(3 \times 10^8 \text{ m/s})^2} = 1 + 10^{-10}$$

In this case the Galilean velocity addition formula is off by only 0.000 000 01%.

Next, we verify that Einstein's second postulate holds—that light has the same speed in any inertial reference frame. Suppose that instead of launching a space probe, Bea turns on her headlights. The velocity of the light beam in Bea's frame is $v_{LB} = +c$. The speed of the light beam in Abe's frame is

$$v_{LA} = \frac{v_{LB} + v_{BA}}{1 + v_{LB}v_{BA}/c^2} = \frac{c + v_{BA}}{1 + cv_{BA}/c^2} = \frac{c(1 + v_{BA}/c)}{1 + v_{BA}/c} = c$$

Thus, even if the jet of charged particles from the active nucleus of a galaxy moves toward Earth at a speed close to that of light in an Earth observer's reference frame, the light it emits travels at the speed of light in the Earth frame. In the calculation of the speed of the jet outlined in Section 26.2, it is correct to use c for the speed of light emitted by the jet.

Problem-Solving Strategy: Relative Velocity

- Sketch the situation as seen in two different reference frames. Label the velocities with subscripts to help keep them straight. The subscripts in v_{BA} mean *the velocity of B as measured in A's reference frame*.

- The velocity transformation formula [Eq. (26-5)] is written in terms of the *components* of the three velocities along a straight line. The components are positive for one direction (your choice) and negative for the other.

- If A moves to the right in B's frame, then B moves to the left in A's frame:

$$v_{BA} = -v_{AB}.$$

- To get Eq. (26-5) right, make sure that the inner subscripts on the right side are the same and "cancel" to leave the left-side subscripts *in order*:

$$v_{LA} = \frac{v_{LB} + v_{BA}}{1 + v_{LB}v_{BA}/c^2}$$

Example 26.3

Observation in Space

Two spaceships travel at high speed in the same direction along the same straight line. As measured by an observer on a nearby planet, ship 1 is behind ship 2 and moves at speed $0.90c$; ship 2 moves at speed $0.70c$. According to an observer aboard ship 1, how fast and in what direction is ship 2 moving?

Strategy   The two reference frames of interest are that of the planet and that of ship 1. Then a sketch of the planet and the two ships as seen in each of the reference frames helps us assign subscripts to the velocities. After choosing a positive direction, we carefully assign the correct algebraic signs to each velocity. Then we are ready to apply the velocity transformation formula.

Solution   First we draw Fig. 26.13a showing the two ships moving to the right in the reference frame of the planet. Let the right be the positive direction. In this frame, the velocities

of the ships are $v_{1P} = +0.90c$ and $v_{2P} = +0.70c$. We want to find the velocity of ship 2 as seen by observers on ship 1 (v_{21}), so we draw Fig. 26.13b in the reference frame of ship 1. Since ship 1 moves to the right in the planet frame, the planet moves to the left in ship 1's frame: $v_{P1} = -v_{1P} = -0.90c$.

Now we apply Eq. (26-5). Since we want to calculate v_{21}, it goes on the left side of the equation. The two velocities on the right side are v_{2P} and v_{P1} so that the P's "cancel" to leave v_{21}.

$$v_{21} = \frac{v_{2P} + v_{P1}}{1 + v_{2P}v_{P1}/c^2}$$

Substituting $v_{2P} = +0.70c$ and $v_{P1} = -0.90c$ yields

$$v_{21} = \frac{0.70c + (-0.90c)}{1 + [0.70c \times (-0.90c)]/c^2} = \frac{-0.20c}{1 - 0.63} = -0.54c$$

So according to observers on spaceship 1, spaceship 2 is moving to the left at speed $0.54c$.

continued on next page

Example 26.3 *continued*

Discussion Note how important it is to get the signs correct. For instance, if we had made an error by writing $v_{P1} = +0.90c$, then we would have calculated $v_{21} = +0.98c$. This answer has spaceship 2 moving to the *right* relative to spaceship 1, which doesn't make sense. In the planet frame, ship 1 is catching up to ship 2, so in ship 1's frame, ship 2 must move toward ship 1. In one dimension, the velocity is always in the same *direction* that you would expect in Galilean velocity addition; only the speed is different.

Practice Problem 26.3 Relative velocity of approaching rocket

According to an observer on a space station, two rocket ships are moving toward each other in opposite directions along the x-axis, ship A with velocity $0.40c$ to the right and ship B with velocity $0.80c$ to the left. With what speed does an observer on ship B observe ship A to be moving?

(a)

(b)

Figure 26.13
(a) An observer on the planet measures the velocities of the two spaceships. (b) The same events seen by an observer on spaceship 1.

26.6 RELATIVISTIC MOMENTUM

When a particle's speed is not small compared to the speed of light, the nonrelativistic expressions for momentum and kinetic energy are not valid. If we try to use them for particles moving at high speeds, it appears that the momentum and energy conservation laws are violated. We must redefine momentum and kinetic energy so that the conservation laws hold for *any* speed. The nonrelativistic expressions $\vec{\mathbf{p}} = m\vec{\mathbf{v}}$ and $K = \frac{1}{2}mv^2$ are *good approximations* as long as $v \ll c$. The relativistic expressions are more general; they give the correct momentum and kinetic energy for *any* speed. Thus, the relativistic expressions must give the same momentum and kinetic energy as the nonrelativistic expressions when $v \ll c$.

The relativistically correct expression for the momentum of a particle with mass m and speed v is

> **Momentum**
> $$\vec{\mathbf{p}} = \gamma m\vec{\mathbf{v}} \tag{26-6}$$

For speeds small compared to c, $\gamma \approx 1$ and $\vec{\mathbf{p}} \approx m\vec{\mathbf{v}}$. For example, consider an airplane traveling at 300 m/s (670 mi/h), which is just under the speed of sound in air. Compared to the speed of light, 300 m/s is quite slow; it's just one one-millionth of the speed of light. When $v = 300$ m/s, the Lorentz factor is $\gamma = 1.000\,000\,000\,000\,5$. In this case, using the nonrelativistic expression $\vec{\mathbf{p}} = m\vec{\mathbf{v}}$ to find the momentum of the plane is a *very* good approximation! But for a proton ejected from the Sun at nine-tenths the speed of light, $\gamma = 2.3$. The proton's momentum is more than twice as large as we would expect from the nonrelativistic expression $\vec{\mathbf{p}} = m\vec{\mathbf{v}}$. Therefore, $\vec{\mathbf{p}} = m\vec{\mathbf{v}}$ should not be used for a proton traveling at $0.9c$.

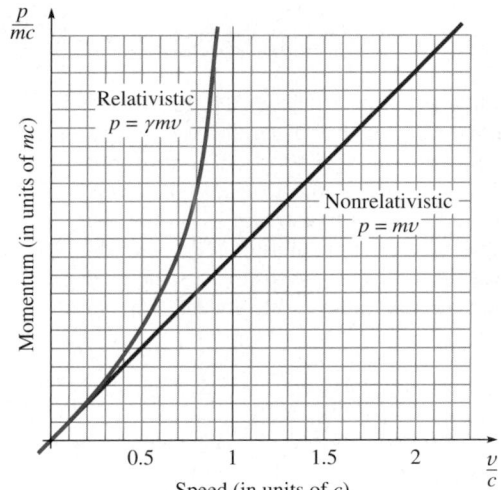

Figure 26.14 A graph of momentum versus v showing the nonrelativistic and relativistic expressions. At low speeds the two expressions are in agreement.

 What is the cutoff speed below which the nonrelativistic formula can be used? There's no clear boundary. As a rule of thumb, the nonrelativistic formula is less than 1% off as long as $\gamma < 1.01$. Setting $\gamma = 1.01$ and solving for v, we get $v = 0.14c \approx 4 \times 10^7$ m/s. As long as the speed of the particle is less than about $\frac{1}{7}$ the speed of light, the nonrelativistic momentum expression is correct to within 1%.

The relativistic expression for momentum has some dramatic consequences. Consider the momentum of a particle as v gets close to the speed of light (Fig. 26.14). As v approaches the speed of light, the momentum increases *without bound*. The momentum can get as large as you want *without the speed ever reaching the speed of light*. Or, in other words, it is impossible to accelerate something to the speed of light. You can give something as much momentum as you like, but you can never get the speed up to c.

With relativistic momentum, it is still true that the impulse delivered equals the change in momentum ($\vec{\mathbf{F}}_{net} \, \Delta t = \Delta\vec{\mathbf{p}}$), but $\vec{\mathbf{F}}_{net} = m\vec{\mathbf{a}}$ is *not* true: the acceleration due to a constant net force gets smaller and smaller as the particle's speed approaches c. The longer the force is applied, the larger the momentum, but the speed never reaches the speed of light. This fact is verified in the daily operation of particle accelerators, which are used in high-energy physics research. Particles such as electrons and protons are "accelerated" to larger and larger momenta (and kinetic energies) as their speeds get closer and closer to—but never exceed—the speed of light.

Example 26.4

Collision in the Upper Atmosphere

 Cosmic rays collide with atoms or molecules in the upper atmosphere (Fig. 26.15). If a proton moving at $0.70c$ makes a head-on collision with a nitrogen atom, initially at rest, and the proton recoils at $0.63c$, what is the speed of the nitrogen atom after the collision? (The mass of a nitrogen atom is about 14 times the mass of a proton.)

Strategy We apply the principle of momentum conservation to solve this collision problem. The only change from the way we have analyzed collisions previously is that we must use the relativistic momentum expression for the proton. It remains to be seen whether the nitrogen atom is moving at relativistic speed after the collision. If it is not, we can simplify the calculations by using the nonrelativistic momentum expression. It is perfectly fine to "mix" the two; they aren't different kinds of momentum. The nonrelativistic expression is just an approximation—when it is a good approximation, we use it.

Figure 26.15
A proton moving with speed v_i collides head on with a nitrogen atom at rest in the upper atmosphere. The nitrogen atom moves with speed v_N after the collision, while the proton rebounds with speed v_f.

continued on next page

Example 26.4 *continued*

Solution Choose the direction of the proton's initial velocity as the +x-direction. The initial momentum of the proton is

$$p_{ix} = \gamma m_p v_{ix}$$

where $v_{ix} = +0.70c$ and

$$\gamma = (1 - 0.70^2)^{-1/2} = 1.4003$$

Therefore, the x-component of the initial momentum is

$$p_{ix} = 1.4003 m_p \times 0.70c = +0.9802 m_p c$$

After the collision, the proton's momentum is

$$p_{fx} = \gamma m_p v_{fx}$$

where $v_f = -0.63c$ since it moves in the –x-direction and

$$\gamma = (1 - 0.63^2)^{-1/2} = 1.288$$

The final momentum of the proton is

$$p_{fx} = -0.8114 m_p c$$

The change in the proton's x-component of momentum is

$$\Delta p_x = -0.8114 m_p c - 0.9802 m_p c = -1.7916 m_p c$$

To conserve momentum, the nitrogen atom's final momentum is $P_x = +1.7916 m_p c$. To find the velocity of the atom, we set $P_x = \gamma M v_{Nx}$.

Since the mass of a nitrogen atom is about fourteen times that of a proton,

$$1.7916 m_p c = \gamma \times 14 m_p \times v_{Nx}$$

Canceling m_p from both sides and simplifying, we have

$$0.1280c = \gamma v_{Nx} = [1 - (v_{Nx}/c)^2]^{-1/2} \times v_{Nx}$$

This equation can be solved for v_{Nx} with some very messy algebra, but it's better to realize that an approximation is appropriate. Since γ is never less than 1, v_{Nx} cannot be greater than $0.1280c$. Therefore, v_{Nx} is small enough to use the nonrelativistic momentum expression $P_x = Mv_{Nx}$—or, in other words, to set $\gamma = 1$. Then $v_{Nx} = 0.1280c$. Rounding to two significant figures, the speed of the nitrogen atom is $0.13c$.

Discussion Using the nonrelativistic momentum *throughout*, for the proton as well as the atom, would have given:

$$0.70 m_p c = -0.63 m_p c + 14 m_p v_{Nx}$$

$$v_{Nx} = \frac{1.33c}{14} = 0.095c$$

which is 26% smaller than the correct value. On the other hand, you can verify (by doing a lot of algebraic manipulation) that using the relativistic expression for the nitrogen without approximating would have given $0.13c$—the same answer (to within two significant figures). All that extra algebra would have yielded nothing. It pays to decide whether it is necessary to use relativistic expressions or whether the nonrelativistic ones are perfectly adequate.

Practice Problem 26.4 A change in momentum

A chunk of space debris with mass 1.0 kg is moving with a speed of $0.707c$. A constant force of magnitude 1.0×10^8 N, in the direction opposite to the chunk's motion, acts on it. How long must this force act to bring the space debris to rest? [*Hint*: The impulse delivered is equal to the change in momentum.]

26.7 MASS AND ENERGY

A particle at rest has no kinetic energy, but that doesn't mean it has no energy. Relativity tells us that mass is a measure of energy content. The **rest energy** E_0 of a particle is its energy as measured in its rest frame. Thus, rest energy does *not* include kinetic energy. The relationship between rest energy and mass is:

Rest energy

$$E_0 = mc^2 \qquad (26\text{-}7)$$

The interpretation of mass as a measure of rest energy is confirmed by observations of radioactive decay, in which particles at rest decay into products of smaller total mass; the products carry off kinetic energy equal to the decrease in total mass.

A kilogram of coal has a rest energy of $(1 \text{ kg}) \times (3 \times 10^8 \text{ m/s})^2 = 9 \times 10^{16}$ J. If the entire rest energy of the coal were converted into electrical energy, it would be enough to supply the electricity needs of a typical American household for *millions of years*. When coal is burned, only a tiny fraction (about one part in a billion) of the coal's rest energy is released. The change in mass—the difference between the mass of the coal and the total mass of all the products—is immeasurably small. In chemical reactions it *seems* as if mass is conserved.

On the other hand, in nuclear reactions and radioactive decays, a much larger fraction of the mass of a nucleus is transformed into the kinetic energy of the reaction products. The total mass of the **daughter** particles (particles present after the reaction) is not the same as the total mass of the **parent** particles (particles present before the reaction).

Mass is *not* conserved, but total energy (the sum of rest energy and kinetic energy) is conserved. If there is a decrease in mass, then energy is released by the reaction. Total energy is still conserved; it has just been changed from one form to another—from rest energy to kinetic energy and/or radiation. If there is an increase in rest energy (that is, if the daughter particles have more total mass than the parent particles), the reaction does not occur spontaneously. The reaction can occur only if the energy deficit is supplied by the initial kinetic energies of the parent particles.

The Electron-Volt

A unit of energy commonly used in atomic and nuclear physics is the electron-volt (symbol eV). One electron-volt is equal to the kinetic energy that a particle with charge $\pm e$ (such as an electron or a proton) gains when it is accelerated through a potential difference of magnitude 1 V. Since 1 V = 1 J/C and $e = 1.60 \times 10^{-19}$ C, we can find the conversion between electron-volts and joules:

$$1 \text{ eV} = e \times 1 \text{ V} = 1.60 \times 10^{-19} \text{ C} \times 1 \text{ J/C} = 1.60 \times 10^{-19} \text{ J}$$

For larger amounts of energy, keV represents kilo-electron-volts (10^3 eV) and MeV represents mega-electron-volts (10^6 eV).

Example 26.5

Energy Released in Radioactive Decay

Carbon dating is based on the radioactive decay of a carbon-14 nucleus (a nucleus with 6 protons and 8 neutrons) into a nitrogen-14 nucleus (with 7 protons and 7 neutrons). In the process, an electron (e⁻) and a particle called an antineutrino ($\bar{\nu}$) are created. The reaction is written as

$$^{14}\text{C} \rightarrow {}^{14}\text{N} + e^- + \bar{\nu}$$

Find the energy released by this reaction. The masses of the nuclei are 13.999 950 u for ^{14}C and 13.999 234 u for ^{14}N. [The *atomic mass unit* (u) is commonly used in atomic and nuclear physics; 1 u = 1.66×10^{-27} kg.] The antineutrino's mass is negligibly small.

Strategy We compare the total masses of the particles before and after the decay. A decrease in total mass means that rest energy has been transformed into other forms: the kinetic energies of the nitrogen atom and electron and the energy of the antineutrino. The energy released by the reaction is equal to the change in the amount of rest energy.

Solution Before decay, the total mass is 13.999 950 u. After decay, the total mass is

$$13.999\ 234 \text{ u} + \frac{9.11 \times 10^{-31} \text{ kg}}{1.66 \times 10^{-27} \text{ kg/u}} = 13.999\ 783 \text{ u}$$

The change in mass is

$$\Delta m = -0.000\ 167 \text{ u}$$

Let Q be the quantity of energy released. Since total energy is conserved:

rest energy before decay = rest energy after decay + energy released

$$m_i c^2 = m_f c^2 + Q$$

$$Q = m_i c^2 - m_f c^2$$

$$= -\Delta m \times c^2 = 0.000\ 167 \text{ u} \times 1.66 \times 10^{-27} \text{ kg/u} \times (3.00 \times 10^8 \text{ m/s})^2$$

$$= 2.495 \times 10^{-14} \text{ kg·m}^2/\text{s}^2 = 2.50 \times 10^{-14} \text{ J}$$

In units of keV,

$$Q = \frac{2.495 \times 10^{-14} \text{ J}}{1.60 \times 10^{-19} \text{ J/eV}} \times 10^{-3} \frac{\text{keV}}{\text{eV}} = 156 \text{ keV}$$

Discussion The fraction of the original rest energy of the ^{14}C nucleus that is released is 0.000 167 u/13.999 950 u = 0.0012%. This may not seem like a huge fraction, but it is about 10^4 times larger than the fractional decrease in mass that occurs when carbon is burned. In nuclear fusion, the fractional mass change approaches 1%.

Practice Problem 26.5   How fast is the Sun losing mass?

The Sun radiates energy at a rate of 4×10^{26} J/s. At what rate is the mass of the Sun decreasing?

Invariance

So far we have seen two quantities that are *invariant*—that have the same value as measured in all inertial frames. One is the speed of light; the other is mass. Distances and time intervals are not the same in different frames of reference, so they are not invariant.

Let us emphasize the difference between a conserved quantity and an invariant quantity. A conserved quantity maintains the same value *in a given reference frame*; the value may differ from one frame to another, but in any given frame its value is constant. An invariant is a quantity that, at a given moment in history, has the same numerical value in *all inertial frames*. Thus, momentum is conserved but is not invariant. Mass is invariant but is not conserved; as in Example 26.5, the total mass can change in a radioactive decay or other nuclear reaction. The total energy *is* conserved in such a reaction; but total energy is not invariant, since particles have different kinetic energies in different frames.

26.8 RELATIVISTIC KINETIC ENERGY

A more general, relativistic expression for momentum is required in order to preserve the principle of momentum conservation for particles moving at relativistic speeds. We need to do the same for kinetic energy.

With the relation between force and momentum ($\vec{F} = \Delta\vec{p}/\Delta t$) and the concept of work as the product of force and distance, we can deduce a formula for the kinetic energy of a particle. As in the nonrelativistic case, the kinetic energy of a body is equal to the work done to accelerate it from rest to its present velocity. The result is

---

Kinetic energy

$$K = (\gamma - 1)mc^2 \tag{26-8}$$

Kinetic energy is energy of motion—the *additional* energy that a moving object has, compared to the energy of the same body when at rest. Einstein proposed identifying the kinetic energy expression above as the difference of two terms. The first term in the above expression, γmc^2, is the **total energy** E of the particle, which includes both kinetic energy and rest energy. The second term, mc^2, is the rest energy E_0—the energy of the particle when at rest. Therefore, we can rearrange Eq. (26-8) as

$$E = K + mc^2 = K + E_0 = \gamma mc^2 \tag{26-9}$$

With total energy and kinetic energy defined in this way, we find that if any reaction conserves total energy in one inertial reference frame, the total energy is automatically conserved in all other inertial frames. In other words, energy conservation is restored to the status of a universal law of physics.

Recall that as v approaches c, γ increases without bound. Then from Eq. (26-9), we can conclude that no object with mass can travel at the speed of light since it would need to have an *infinite total energy* to do so.

At first it may look as if kinetic energy, $K = (\gamma - 1)mc^2$, doesn't depend on speed, but remember that γ is a function of speed. As the speed of a particle increases, γ increases, and therefore so does the kinetic energy. There is no limit to the kinetic energy of a particle. As is true with momentum, the kinetic energy gets large without bound as the speed gets closer to c.

It's not at all obvious that the relativistic expression for K approaches the nonrelativistic expression $\frac{1}{2}mv^2$ for objects moving much slower than the speed of light, but it does. To show that, we make use of the binomial approximation $(1 - x)^n \approx 1 - nx$ for $x \ll 1$ (see Appendix A.5). If we let $x = v^2/c^2$ and $n = -\frac{1}{2}$ in the binomial approximation, γ becomes

$$\gamma = \left(1 - \frac{v^2}{c^2}\right)^{-1/2} \approx 1 + \frac{1}{2}\left(\frac{v^2}{c^2}\right)$$

No object with mass can travel at the speed of light.

The kinetic energy is then

$$K = (\gamma - 1)mc^2 \approx \left[1 + \frac{1}{2}\left(\frac{v^2}{c^2}\right) - 1\right]mc^2 = \frac{1}{2}\left(\frac{v^2}{c^2}\right)mc^2 = \frac{1}{2}mv^2$$

The relativistic expression for K is valid for both relativistic and nonrelativistic motion. The nonrelativistic expression $\frac{1}{2}mv^2$ is an approximation that is only valid for speeds much less than c. If $K \ll mc^2$, then γ is very close to 1; the particle is not moving at a relativistic speed, so nonrelativistic approximations can be used.

Example 26.6

An Energetic Electron

The radioactive decay of carbon-14 into nitrogen-14 ($^{14}C \rightarrow {}^{14}N + e^- + \bar{\nu}$) releases 156 keV of energy (Fig. 26.16). If all of the energy released appears as the kinetic energy of the electron, how fast is the electron moving?

Strategy   We are given the kinetic energy of the electron and need to find its speed. The electron-volt (eV) and its multiples, keV or MeV, are commonly used energy units for atomic, nuclear, and high-energy particle physics. We could do a straightforward calculation but how do we know whether it is appropriate to use the nonrelativistic expression for kinetic energy? We compare the kinetic energy of the electron (156 keV) to its rest energy (mc^2).

Solution   The rest energy of the electron is

$$E_0 = mc^2 = 9.109 \times 10^{-31} \text{ kg} \times (2.998 \times 10^8 \text{ m/s})^2 = 8.187 \times 10^{-14} \text{ J}$$

Since we know K in keV, let us convert E_0 to keV.

$$E_0 = 8.187 \times 10^{-14} \text{ J} \times \frac{1 \text{ eV}}{1.602 \times 10^{-19} \text{ J}} \times \frac{1 \text{ keV}}{1000 \text{ eV}} = 511 \text{ keV}$$

Before   After

^{14}C ^{14}N

$\bar{\nu}$

e^-

Figure 26.16
Radioactive decay of carbon-14

Thus, K is of the same order of magnitude as E_0. Since the electron is moving at a relativistic speed, the relativistic equations must be used.

The Lorentz factor is

$$\gamma = 1 + \frac{K}{mc^2} = 1 + \frac{156 \text{ keV}}{511 \text{ keV}} = 1.305$$

From γ, we determine the speed. First we square γ:

$$\gamma^2 = \frac{1}{1 - v^2/c^2}$$

Now we solve for v/c to find the speed of the electron as a fraction of c.

$$1 - \frac{v^2}{c^2} = \frac{1}{\gamma^2}$$

$$v/c = \sqrt{1 - 1/\gamma^2} = 0.6425$$

$$v = 0.6425c$$

Discussion   The Lorentz factor is *not* very close to 1, which is another indication that we cannot use $K = \frac{1}{2}mv^2$ for this electron. Doing so would give

$$v = (2K/m)^{1/2} = 2.344 \times 10^8 \text{ m/s} = 0.782c$$

A result this close to c should cause concern about using a nonrelativistic approximation.

The speed $0.6425c$ is an upper limit on the kinetic energy of the electron. The electron cannot carry off all of the energy released in the reaction; the kinetic energy must be divided among the three particles in such a way as to conserve momentum.

Practice Problem 26.6   Accelerating a proton

How much work must be done to accelerate a proton from rest to $0.75c$? Express the answer in MeV.

Momentum-Energy Relationships

Tips

Momentum is usually written in units of eV/c (or multiples such as MeV/c) to avoid lots of unit conversions. Masses are commonly written in units of eV/c^2, keV/c^2, MeV/c^2, or GeV/c^2.

In Newtonian physics the relation between kinetic energy and momentum is $K = p^2/2m$ (Problem 43), but that relation no longer holds for particles moving at relativistic speeds. From the relativistic definitions of $\vec{p}$, E, and K, you can derive these useful relations (try it—see Problems 44 and 45):

$$E^2 = E_0^2 + (pc)^2 \tag{26-10}$$

$$(pc)^2 = K^2 + 2KE_0 \tag{26-11}$$

Equations (26-10) and (26-11) are valuable for calculating total energy or kinetic energy from momentum or *vice versa* without going through the intermediate step of calculating the speed of the particle. Since $E = \gamma mc^2$ and $\vec{\mathbf{p}} = \gamma m\vec{\mathbf{v}}$, another useful relationship is

$$\frac{\vec{\mathbf{v}}}{c} = \frac{\vec{\mathbf{p}}c}{E} \qquad (26\text{-}12)$$

Equation (26-12) makes it easier to calculate the velocity or momentum or total energy when any two of these three quantities are known. It also shows that pc can never exceed the total energy, but approaches E as $v \rightarrow c$.

Deciding Whether to Use Relativistic Equations

There are several ways of deciding whether relativistic calculations are called for, depending on the information given in a particular problem. Section 26.6 suggested that for $v = 0.14c$, the nonrelativistic expression for momentum differs from the relativistic by about 1%. We may not need that degree of accuracy. Even for $v = 0.2c$, the differences between the nonrelativistic and relativistic expressions for momentum and kinetic energy are only about 2% and 3%, respectively. For speeds higher than about $0.3c$, γ rises rapidly and the differences between nonrelativistic and relativistic physics become appreciable.

Comparing a particle's kinetic energy to its rest energy is another way to decide whether to use relativistic or nonrelativistic expressions. If $K \ll mc^2$, then γ is very close to 1 and the particle's speed is nonrelativistic.

A particle is *nonrelativistic* if any of the following equivalent conditions are true:	
$v \ll c$	(26-13)
$\gamma - 1 \ll 1$	(26-14)
$K \ll mc^2$	(26-15)
$p \ll mc$	(26-16)

Example 26.7

Speed and Momentum of an Electron

An electron has a kinetic energy of 1.0 MeV. Find the electron's speed and its momentum.

Strategy   Use energy-momentum relations and momentum units of MeV/c to simplify the calculation.

Solution   In Example 26.6, we found that the rest energy of an electron is $E_0 = 0.511$ MeV. Since the kinetic energy is almost twice the rest energy, we definitely must do relativistic calculations. The total energy of the electron is $E = K + E_0 = 1.511$ MeV. We can immediately find the momentum:

$$E^2 = E_0^2 + (pc)^2$$

$$(pc)^2 = E^2 - E_0^2$$

$$pc = \sqrt{(1.511 \text{ MeV})^2 - (0.511 \text{ MeV})^2} = 1.422 \text{ MeV}$$

Dividing both sides of this equation by c gives the momentum in units of MeV/c:

$$p = 1.4 \text{ MeV}/c$$

Now that we know the momentum, we can find the speed:

$$\frac{v}{c} = \frac{pc}{E} = \frac{1.422}{1.511} = 0.9411$$

$$v = 0.94c$$

Discussion   A good check is to use the speed to calculate the kinetic energy. First we find the Lorentz factor:

$$\gamma = \frac{1}{\sqrt{(1 - 0.9411^2)}} = 2.957$$

Then the kinetic energy is

$$K = (\gamma - 1)mc^2 = 1.957 \times 0.511 \text{ MeV} = 1.0 \text{ MeV}$$

The momentum in SI units can be obtained as

$$p = 1.422 \frac{\text{MeV}}{c} \times \frac{10^6 \text{ eV}}{\text{MeV}} \times \frac{1.60 \times 10^{-19} \text{ J}}{\text{eV}} \times \frac{c}{3.00 \times 10^8 \text{ m/s}}$$

$$= 7.6 \times 10^{-22} \text{ kg·m/s}$$

continued on next page

Example 26.7 *continued*

Practice Problem 26.7   Protons and antiprotons at Fermilab

The Tevatron at the Fermi National Accelerator Laboratory accelerates protons and antiprotons to kinetic energies of 0.980 TeV (tera-electron-volts). Antiprotons have the same mass as protons (938.3 MeV/c^2) but charge $-e$ instead of $+e$. (a) What is the magnitude of the momentum of the protons and antiprotons in units of TeV/c? (b) At what speed are the protons and antiprotons moving relative to the lab? [*Hint*: Since $\gamma \gg 1$, use the binomial approximation: $v/c = \sqrt{1 - 1/\gamma^2} \approx 1 - 1/(2\gamma^2)$.]

MASTER THE CONCEPTS

Summary

- The two postulates of relativity are:
  (I) The laws of physics are the same in all inertial frames.
  (II) The speed of light in vacuum is the same in all inertial frames.

- The speed of light in vacuum in any inertial reference frame is
  $$c = 3.00 \times 10^8 \text{ m/s}$$

- Observers in different reference frames disagree about the time order of two events (including whether the events are simultaneous) if there is *not* enough time for a signal at light speed to travel from one event to the other.

- The Lorentz factor occurs in many relativity equations.
  $$\gamma = \frac{1}{\sqrt{1 - v^2/c^2}} \qquad (26\text{-}2)$$

- In time dilation problems, identify the two events that mark the beginning and end of the time interval in question. The "clock" is whatever measures this time interval. Identify the reference frame in which the clock is at rest. In that frame, the clock measures the proper time interval Δt_0. In any other reference frame, the time interval is *longer*:
  $$\Delta t = \gamma \, \Delta t_0 \qquad (26\text{-}3)$$

- In length contraction problems, identify the object whose length is to be measured in two different frames. The length is contracted only in the direction of the object's motion. If the length in question is a distance rather than the length of an actual object, it often helps to imagine the presence of a long measuring stick. Identify the reference frame in which the object is at rest. The length in that frame is the proper length L_0. In any other reference frame, the length L is *contracted*:
  $$L = \frac{L_0}{\gamma} \qquad (26\text{-}4)$$

- Velocities in different reference frames are related by
  $$v_{PA} = \frac{v_{PB} + v_{BA}}{1 + v_{PB}v_{BA}/c^2} \qquad (26\text{-}5)$$

  The subscripts in v_{BA} mean *the velocity of B as measured in A's reference frame*. Equation (26-5) is written in terms of the *components* of the three velocities along a straight line. The components are positive for one direction (your choice) and negative for the other. If A moves to the right in B's frame, then B moves to the left in A's frame: $v_{BA} = -v_{AB}$.

- The relativistic expression for momentum is
  $$\vec{\mathbf{p}} = \gamma m \vec{\mathbf{v}} \qquad (26\text{-}6)$$

  With relativistic momentum, it is still true that the impulse delivered equals the change in momentum ($\vec{\mathbf{F}}_{net} \, \Delta t = \Delta \vec{\mathbf{p}}$), but $\vec{\mathbf{F}}_{net} = m\vec{\mathbf{a}}$ is *not* true: the acceleration due to a constant net force gets smaller and smaller as the particle's speed approaches c. Thus, it is impossible to accelerate something to the speed of light.

- The rest energy E_0 of a particle is its energy as measured in its rest frame. The relationship between rest energy and mass is:
  $$E_0 = mc^2 \qquad (26\text{-}7)$$

  Kinetic energy is
  $$K = (\gamma - 1)mc^2 \qquad (26\text{-}8)$$

  Total energy is rest energy plus kinetic energy:
  $$E = \gamma mc^2 = K + E_0 \qquad (26\text{-}9)$$

- Useful relations between momentum and energy:
  $$E^2 = E_0^2 + (pc)^2 \qquad (26\text{-}10)$$
  $$(pc)^2 = K^2 + 2KE_0 \qquad (26\text{-}11)$$
  $$\frac{\vec{\mathbf{v}}}{c} = \frac{\vec{\mathbf{p}}c}{E} \qquad (26\text{-}12)$$

Highlighted Figures and Tables

F26.1 A ball tossed across a merry-go-round: trajectory of the ball as viewed in the noninertial frame fixed to the platform and as viewed in the intertial frame fixed to the ground (p. 952)

F26.3 Simplified version of the Michelson-Morley experiment (p. 954)

F26.4 Two space probes emit flashes of light; observers disagree about whether the flashes are emitted simultaneously. (p. 955)

F26.6 Time dilation illustrated with a pair of light clocks. (p. 958)

F26.9 Comparing the lengths of two metersticks in relative motion (p. 961)

F26.14 A graph of momentum versus v showing the nonrelativistic and relativistic expressions (p. 966)

CONCEPTUAL QUESTIONS

1. An *invariant* is a quantity that has the same value in all inertial reference frames. (a) According to Galilean relativity, which of these quantities are invariant: position, displacement, length, time interval, velocity, acceleration, force, momentum, mass, kinetic energy, the speed of light in vacuum? (b) Which of them are invariant according to Einstein's special relativity?

2. A friend argues with you that relativity is absurd: "It's obvious that moving clocks don't run slow and that moving objects aren't shorter than when they're at rest." How would you reply?

3. An electron is moving at nearly light speed. A constant force of magnitude F is acting on the electron in the direction of its motion. Is the acceleration of the electron less than, equal to, or greater in magnitude than F/m? Explain.

4. A sprinter crosses the start line (event 1) and runs at constant velocity until she crosses the finish line (event 2). In what reference frame would an observer measure the proper time interval between these two events? In what reference frame would an observer measure the proper length of the track from start line to finish line?

5. As you talk on a cell phone, does the mass of the phone's battery change at all? If so, does it increase or decrease?

6. A particle with nonzero mass m can never move faster than the speed of light. Is there also a maximum momentum that the particle can have? A maximum kinetic energy? Explain.

7. An astronaut in top physical condition has an average resting pulse on Earth of about 52 beats per minute. Suppose the astronaut is in a spaceship traveling at $0.87c$ ($\gamma = 2$) with respect to Earth when he takes his own resting pulse. Does he measure about 52 beats per minute, about 26 beats per minute, or about 104 beats per minute? Explain.

8. A constant force is applied to a particle initially at rest. Sketch qualitative graphs of the particle's speed, momentum, and acceleration as functions of time. Assume that the force acts long enough so the particle achieves relativistic speeds.

9. You are in a special compartment on a train that admits no light, sound, or vibration. Is there any way you can tell whether the train is at rest or moving at constant nonzero velocity? Explain.

10. Harry and Sally are on opposite sides of the room at a wedding reception. They simultaneously (in the frame of the room) take flash pictures of the bride and groom cutting the cake in the center of the room. What would an observer moving at constant velocity from Harry to Sally say about the time order of the two flashes?

11. In an Earth laboratory, an astronaut measures the length of a rod to be 1.00 m. The astronaut takes the rod aboard a spaceship and flies away from Earth at speed $0.5c$. Is the length of the rod measured by an observer on Earth greater than, less than, or equal to 1.00 m as measured by the astronaut in the spaceship? Explain. Does the answer depend on the orientation of the rod?

12. In Section 26.2, suppose that Celia moves in a spaceship to the left with respect to Abe (see Fig. 26.4). What would Celia conclude about the time order of the two flashes?

13. Explain why it is impossible for a particle with mass to move faster than the speed of light.

14. Does a stretched spring have the same mass as when it is relaxed? Explain.

MULTIPLE CHOICE QUESTIONS

1. Which of these statements are *postulates* of Einstein's special relativity?

   (1) The speed of light is the same in all inertial reference frames.
   (2) Moving clocks run slow.
   (3) Moving objects are contracted along the direction of motion.
   (4) The laws of nature are the same in all inertial reference frames.
   (5) $E_0 = mc^2$

   (a) 1 only   (b) 2 and 3 only   (c) all 5
   (d) 4 only   (e) 1 and 4 only   (f) 4 and 5 only

2. Which of these statements correctly defines an inertial frame?

   (a) An inertial frame is a frame in which there are no forces.
   (b) An inertial frame is one in which Newton's second and third laws hold, but not his first.

   (c) An inertial frame is a frame of reference in which Newtonian mechanics holds true, but relativistic mechanics does not.
   (d) An inertial frame is a frame where there are no accelerations without applied forces.
   (e) An inertial frame is a frame of reference in which relativistic mechanics holds true, but Newtonian mechanics does not.

3. An astronaut in a rocket moving with a speed $v = 0.6c$ relative to Earth performs a collision experiment with two small steel balls and concludes that both momentum and energy are conserved in his reference frame. What would an Earth observer conclude?

   (a) Momentum and energy are conserved.
   (b) Momentum is conserved, but energy is not.
   (c) Energy is conserved, but momentum is not.
   (d) The collision never takes place because the two balls are never at the same place at the same time.
   (e) Neither energy nor momentum is conserved.

4. A spaceship moves away from Earth at constant velocity 0.60c, according to Earth observers. In the reference frame of the spaceship,

   (a) Earth moves away from the spaceship at 0.60c.
   (b) Earth moves away from the spaceship at a speed less than 0.60c.
   (c) Earth moves away from the spaceship at a speed greater than 0.60c.
   (d) The speed of Earth cannot be accurately measured because the reference frame is moving.
   (e) The speed of Earth is not constant.

5. A clock ticks once each second and is 10 cm long when at rest. If the clock is moving at 0.80c parallel to its length with respect to an observer, the observer measures the time between ticks to be _____ and the length of the clock to be _____.

   (a) more than 1 s; more than 10 cm
   (b) less than 1 s; more than 10 cm
   (c) more than 1 s; less than 10 cm
   (d) less than 1 s; less than 10 cm
   (e) equal to 1 s; equal to 10 cm

6. Which best describes the *proper time interval* between two events?

   (a) the time interval measured in a reference frame in which the two events occur at the same place
   (b) the time interval measured in a reference frame in which the two events are simultaneous
   (c) the time interval measured in a reference frame in which the two events occur a maximum distance away from each other
   (d) the longest time interval measured by any inertial observer

PROBLEMS

Note: **C** indicates a combination conceptual/quantitative problem. Gold diamonds ✦, ✦✦ are used to indicate the increasing level of difficulty of each problem. Problem numbers appearing in blue, 9., denote problems that have a detailed solution available in the Student Solutions Manual. Some problems are *paired* by concept; their numbers are connected by a ruled box.

26.1 Postulates of Relativity

1. An engineer in a train moving toward the station with a velocity $v = 0.60c$ lights a signal flare as he reaches a marker 1.0 km from the station (according to a scale laid out on the ground). By how much time, on the stationmaster's clock, does the arrival of the optical signal precede the arrival of the train?

2. The light-second is a unit of distance; 1 light-second is the distance that light travels in one second. (a) Find the conversion between light-seconds and meters: 1 light-second = ? m. (b) What is the speed of light in units of light-seconds per second?

3. A spaceship traveling at speed 0.13c away from Earth sends a radio transmission to Earth. (a) According to Galilean relativity, at what speed would the transmission travel relative to Earth? (b) Using Einstein's postulates, at what speed does the transmission travel relative to Earth?

26.3 Time Dilation

4. An astronaut wears a new Rolex watch on a journey at a speed of 2.0×10^8 m/s with respect to Earth. According to mission control in Houston, the trip lasts 12.0 h. How long is the trip as measured on the Rolex?

5. An unstable particle called the *pion* has a mean lifetime of 25 ns in its own rest frame. A beam of pions travels through the laboratory at a speed of 0.60c. (a) What is the mean lifetime of the pions as measured in the laboratory frame? (b) How far does a pion travel (as measured by laboratory observers) during this time?

6. A spaceship is traveling away from Earth at 0.87c. The astronauts report home by radio every 12 h (by their own clocks). At what interval are the reports *sent* to Earth, according to Earth clocks?

7. A spaceship travels at constant velocity from Earth to a point 710 ly away as measured in Earth's rest frame. The ship's speed relative to Earth is 0.9999c. A passenger is 20 yr old when departing from Earth. (a) How old is the passenger when the ship reaches its destination, as measured by the ship's clock? (b) If the spaceship sends a radio signal back to Earth as soon as it reaches its destination, in what year, by Earth's calendar, does the signal reach Earth? The spaceship left Earth in the year 2000.

✦ 8. A clock moves at a constant velocity of 8.0 km/s with respect to Earth. If the clock ticks at intervals of one second in its rest frame, how much more than a second elapses between ticks of the clock as measured by an observer at rest on Earth? [*Hint*: Use the binomial approximation $(1 - v^2/c^2)^{-1/2} \approx 1 + v^2/(2c^2)$.]

✦ 9. A plane trip lasts 8.0 h; the average speed of the plane during the flight relative to Earth is 220 m/s. What is the time difference between an atomic clock on board the plane and one on the ground, assuming they were synchronized before the flight? (Neglect General Relativistic complications due to gravity and the acceleration of the plane.)

10. Fill in the missing algebraic steps in the derivation of the time dilation equation [Eq. (26-1)].

26.4 Length Contraction

11. A UFO flies directly over a football field, from one goal line to the other, at a speed of $0.50c$. (a) If the length of the field between goal lines in the Earth frame is 91.5 m (100 yards), what length is observed by the pilot of the UFO? (b) How long does it take the UFO to go from one goal line to the other according to Earth observers? (c) How long does it take according to the pilot?

12. A laboratory measurement of the coordinates of the ends of a moving meterstick, taken at the same time in the laboratory, gives the result that one end of the stick is 0.992 m due north of the other end. If the stick is moving due north, what is its speed with respect to the lab?

13. Two spaceships are moving directly toward each other with a relative velocity of $0.90c$. If an astronaut measures the length of his own spaceship to be 30.0 m, how long is the spaceship as measured by an astronaut in the other ship?

14. A spaceship is moving at a constant velocity of $0.70c$ relative to an Earth observer. The Earth observer measures the length of the spaceship to be 40.0 m. How long is the spaceship as measured by its pilot?

15. A spaceship moves at a constant velocity of $0.40c$ relative to an Earth observer. The pilot of the spaceship is holding a rod, which he measures to be 1.0 m long. (a) The rod is held perpendicular to the direction of motion of the spaceship. How long is the rod according to the Earth observer? (b) After the pilot rotates the rod and holds it parallel to the direction of motion of the spaceship, how long is it according to the Earth observer?

16. A rectangular plate of glass, measured at rest, has sides 30.0 cm and 60.0 cm. (a) As measured in a reference frame moving parallel to the 60.0-cm edge at speed $0.25c$ with respect to the glass, what are the lengths of the sides? (b) How fast would a reference frame have to move in the same direction so that the plate of glass viewed in that frame is square?

17. A futuristic train moving in a straight line with a uniform speed of $0.80c$ passes a series of communications towers. The spacing between the towers, according to an observer on the ground, is 3.0 km. A passenger on the train uses an accurate stopwatch to see how often a tower passes him. (a) What is the time interval the passenger measures between the passing of one tower and the next? (b) What is the time interval an observer on the ground measures for the train to pass from one tower to the next?

18. An astronaut in a rocket moving at $0.50c$ toward the Sun finds himself halfway between Earth and the Sun. According to the astronaut, how far is he from Earth? In the frame of the Sun, the distance from Earth to the Sun is 1.50×10^{11} m.

19. The mean (average) lifetime of a muon in its rest frame is 2.2 μs. A beam of muons is moving through the lab with speed $0.994c$. How far on average does a muon travel through the lab before it decays?

26.5 Velocities in Different Reference Frames

20. Particle A is moving with a constant velocity $v_{AE} = +0.90c$ relative to an Earth observer. Particle B moves in the opposite direction with a constant velocity $v_{BE} = -0.90c$ relative to the same Earth observer. What is the velocity of particle B as seen by particle A?

21. A man on the Moon observes two spaceships coming toward him from opposite directions at speeds of $0.60c$ and $0.80c$. What is the relative speed of the two ships as measured by a passenger on either one of the spaceships?

22. Rocket ship *Able* travels at $0.400c$ relative to an Earth observer. According to the same observer, rocket ship *Able* overtakes a slower moving rocket ship *Baker* that moves in the same direction. The captain of *Baker* sees *Able* pass her ship at $0.114c$. Determine the speed of *Baker* relative to the Earth observer.

23. Relative to the laboratory, a proton moves to the right with a speed of $\frac{4}{5}c$, while relative to the proton, an electron moves to the left with a speed of $\frac{5}{7}c$. What is the speed of the electron relative to the lab?

24. As observed from Earth, rocket *Alpha* moves with speed $0.90c$ and rocket *Bravo* travels with a speed of $0.60c$. They are moving along the same line toward a head-on collision. What is the speed of rocket *Alpha* as measured from rocket *Bravo*?

25. Electron A is moving west with speed $\frac{3}{5}c$ relative to the lab. Electron B is also moving west with speed $\frac{4}{5}c$ relative to the lab. What is the speed of electron B in a frame of reference in which electron A is at rest?

26. Captain Kirk is measuring the speed of light in an evacuated chamber aboard a spaceship traveling with a constant velocity of $0.60c$ with respect to Earth. The light is moving in the direction of motion of the spaceship. Mr. Spock is on Earth watching the experiment. With what speed does the light in the vacuum chamber travel, according to Spock's observations?

26.6 Relativistic Momentum

27. A proton moves at $0.90c$. What is its momentum in SI units?

28. An electron has momentum of magnitude 2.4×10^{-22} kg·m/s. What is the electron's speed?

29. A body has a mass of 12.6 kg and a speed of $0.87c$. (a) What is the magnitude of its momentum? (b) If a constant force of 16.4×10^8 N acts in the direction opposite to the body's motion, how long must the force act to bring the body to rest?

30. A constant force, acting for 3.6×10^4 s (10 h), brings a spaceship of mass 2200 kg from rest to speed $0.70c$. (a) What is the magnitude of the force? (b) What is the *initial* acceleration of the spaceship? Comment on the magnitude of the answer.

26.7 Mass and Energy

31. How much energy is released by a nuclear reactor if the total mass of the fuel decreases by 1.0 g?

32. Two lumps of putty are moving in opposite directions, each one having a speed of 30.0 m/s. They collide and stick together. After the collision the combined lumps are at rest. If the mass of each lump was 1.00 kg before the collision, and no energy is lost to the environment, what is the change in mass of the system due to the collision?

33. A lambda hyperon Λ^0 (mass = 1115 MeV/c^2) at rest decays into a neutron n (mass = 940 MeV/c^2) and a pion (mass = 135 MeV/c^2):
$$\Lambda^0 \rightarrow n + \pi^0$$
What is the total kinetic energy of the neutron and pion?

34. Radon decays as follows: ^{222}Rn $\rightarrow$ ^{218}Po + α. The mass of the radon-222 nucleus is 221.970 39 u, the mass of the polonium-218 nucleus is 217.962 89 u, and the mass of the alpha particle is 4.001 51 u. How much energy is released in the decay? (1 u = 931.5 MeV/c^2.)

26.8 Relativistic Kinetic Energy

35. A muon with rest energy 106 MeV is created at an altitude of 4500 m and travels downward toward Earth's surface. An observer on Earth measures its speed as 0.980c. (a) What is the muon's total energy in the Earth observer's frame of reference? (b) What is the muon's total energy in the muon's frame of reference?

36. An object of mass 0.12 kg is moving at 1.80×10^8 m/s. What is its kinetic energy in joules?

37. When an electron travels at 0.60c, what is its total energy in MeV?

38. Find the conversion between the momentum unit MeV/c and the SI unit of momentum.

39. Find the conversion between the mass unit MeV/c^2 and the SI unit of mass.

40. An observer in the laboratory finds that an electron's total energy is 5.0mc^2. What is the magnitude of the electron's momentum (as a multiple of mc), as observed in the laboratory?

41. The rest energy of an electron is 0.511 MeV. What momentum (in MeV/c) must an electron have in order that its total energy be 3.00 times its rest energy?

42. An electron has a total energy of 6.5 MeV. What is its momentum (in MeV/c)?

43. For a *nonrelativistic* particle of mass m, show that $K = p^2/2m$. [*Hint*: Start with the nonrelativistic expressions for kinetic energy K and momentum p.]

♦ 44. Derive the energy-momentum relation
$$E^2 = E_0^2 + (pc)^2 \qquad (26\text{-}10)$$
Start by squaring the definition of total energy ($E = K + E_0$) and then use the relativistic expressions for momentum and kinetic energy [Eqs. (26-6) and (26-8)].

♦ 45. Starting with the energy-momentum relation $E^2 = E_0^2 + (pc)^2$ and the definition of total energy, show that $(pc)^2 = K^2 + 2KE_0$ [Eq. (26-11)].

♦ 46. Show that Eq. (26-11) reduces to the nonrelativistic relationship between momentum and kinetic energy, $K \approx p^2/(2m)$, for $K \ll E_0$.

♦ 47. Show that each of these statements implies that $v \ll c$, which means that v can be considered a nonrelativistic speed: (a) $\gamma - 1 \ll 1$ [Eq. (26-14)]; (b) $K \ll mc^2$ [Eq. (26-15)]; (c) $p \ll mc$ [Eq. (26-16)]; (d) $K \approx p^2/(2m)$.

© 48. In a beam of electrons used in a diffraction experiment, each electron is accelerated to a kinetic energy of 150 keV. (a) Are the electrons relativistic? Explain. (b) How fast are the electrons moving?

COMPREHENSIVE PROBLEMS

♦♦ 49. Two atomic clocks are synchronized. One is put aboard a spaceship that leaves Earth at $t = 0$ at a speed of 0.750c. (a) When the spaceship has been traveling for 48.0 h (according to the atomic clock on board), it sends a radio signal back to Earth. When would the signal be received on Earth, according to the atomic clock on Earth? (b) When the Earth clock says that the spaceship has been gone for 48.0 h, it sends a radio signal to the spaceship. At what time (according to the spaceship's clock) does the spaceship receive the signal?

♦ 50. A particle decays in flight into two pions, each having a rest energy of 140.0 MeV. The pions travel at right angles to each other with equal speeds 0.900c. Find (a) the momentum of the original particle, (b) its kinetic energy, and (c) its mass in units of MeV/c^2.

51. Consider the following decay process: $\pi^+ \rightarrow \mu^+ + \nu$. The mass of the pion ($\pi^+$) is 139.6 MeV/$c^2$, the mass of the muon ($\mu^+$) is 105.7 MeV/$c^2$, and the mass of the neutrino (ν) is negligible. If the pion is initially at rest, what is the total kinetic energy of the decay products?

52. A neutron (mass 1.008 66 u) disintegrates into a proton (mass 1.007 28 u), an electron (mass 0.000 55 u), and an antineutrino (mass 0). What is the sum of the kinetic energies of the particles produced, if the neutron is at rest? (1 u = 931.5 MeV/c^2.)

♦ 53. A spaceship is moving away from Earth with a constant velocity of 0.80c with respect to Earth. The spaceship and an Earth station synchronize their clocks, setting both to zero, at an instant when the ship is near Earth. By prearrangement, when the clock on Earth reaches a reading of 1.0×10^4 s, the Earth station sends out a light signal to the spaceship. (a) In the frame of reference of the Earth station, how far must the signal travel to reach the spaceship? (b) According to an Earth observer, what is the reading of the clock on Earth when the signal is received?

54. The starship *Enterprise* takes 3.0 days to travel between two distant space stations according to its own clocks. Instruments on one of the space stations indicate that the trip took 4.0 days. How fast did the *Enterprise* travel, relative to that space station?

55. Two spaceships are observed from Earth to be approaching each other along a straight line. Ship A moves at $0.40c$ relative to the Earth observer, while ship B moves at $0.50c$ relative to the same observer. What speed does the captain of ship A report for the speed of ship B?

56. A neutron, with rest energy 939.6 MeV, has momentum 935 MeV/c downward. What is its total energy?

C ✦57. Refer to Example 26.2. One million muons are moving toward the ground at speed $0.9950c$ from an altitude of 4500 m. *In the frame of reference of an observer on the ground*, what are (a) the distance traveled by the muons; (b) the time of flight of the muons; (c) the time interval during which half of the muons decay; and (d) the number of muons that survive to reach sea level. [*Hint:* The answers to (a) to (c) are *not* the same as the corresponding quantities in the muons' reference frame. Is the answer to (d) the same?]

58. Suppose that as you travel away from Earth in a spaceship, you observe another ship pass you heading in the same direction and measure its speed to be $0.50c$. As you look back at Earth, you measure Earth's speed relative to you to be $0.90c$. What is the speed of the ship that passed you according to Earth observers?

59. Verify that the collision between the proton and the nitrogen nucleus in Example 26.4 is elastic.

60. Muons are created by cosmic-ray collisions at an elevation h (as measured in Earth's frame of reference) above Earth's surface and travel downward with a constant speed of $0.990c$. During a time interval of 1.5 μs in the rest frame of the muons, half of the muons present at the beginning of the interval decay. If one-fourth of the original muons reach Earth before decaying, about how big is the height h?

C 61. Refer to Example 26.1. Ashlin travels at speed $0.800c$ to a star 30.0 ly from Earth. (a) Find the distance between Earth and the star in the astronaut's frame of reference. (b) How long (as measured by the astronaut) does it take to travel this distance at a speed of $0.80c$? Compare your answer to the result of Example 26.1 and explain any discrepancy.

✦62. A charged particle is observed to have a total energy of 0.638 MeV when it is moving at $0.600c$. If this particle enters a linear accelerator and its speed is increased to $0.980c$, what is the new value of the particle's total energy?

✦63. A cosmic-ray proton entering the atmosphere from space has a kinetic energy of 2.0×10^{20} eV. (a) What is its kinetic energy in joules? (b) If all of the kinetic energy of the proton could be

harnessed to lift an object of mass 1.0 kg near Earth's surface, how far could the object be lifted? (c) What is the speed of the proton? [*Hint:* v is so close to c that most calculators do not keep enough significant figures to do the required calculation, so make use of the binomial approximation:

$$\text{if } \gamma \gg 1, \quad \sqrt{1 - \frac{1}{\gamma^2}} \approx 1 - \frac{1}{2\gamma^2}$$

✦64. A spaceship is traveling away from Earth at $0.87c$. The astronauts report home by radio every 12 h (by their own clocks). (a) At what interval are the reports *sent* to Earth, according to Earth clocks? (b) At what interval are the reports *received* by Earth observers, according to their own clocks?

65. An extremely relativistic particle is one whose kinetic energy is much larger than its rest energy. Show that for an extremely relativistic particle $E \approx pc$.

✦✦66. *Derivation of the Doppler formula for light.* A source and receiver of EM waves move relative to each other at velocity v. Let v be positive if the receiver and source are moving apart from each other. The source emits an EM wave at frequency f_s (in the source frame). The time between wavefronts as measured by the source is $T_s = 1/f_s$. (a) In the receiver's frame, how much time elapses between the *emission* of wavefronts by the source? Call this T'_r. (b) T'_r is *not* the time that the receiver measures between the *arrival* of successive wavefronts because the wavefronts travel different distances. Say that, according to the receiver, one wavefront is emitted at $t = 0$ and the next at $t = T'_0$. When the first wavefront is emitted, the distance between source and receiver is d_r. When the second wavefront is emitted, the distance between source and receiver is $d_r + vT'_r$. Each wavefront travels at speed c. Calculate the time T_r between the arrival of these two wavefronts as measured by the receiver. (c) The frequency detected by the receiver is $f_r = 1/T_r$. Show that f_r is given by

$$f_r = f_s \sqrt{\frac{1 - v/c}{1 + v/c}}$$

✦✦67. A lambda hyperon Λ^0 (mass = 1115.7 MeV/c^2) at rest in the lab frame decays into a neutron n (mass = 939.6 MeV/c^2) and a pion (mass = 135.0 MeV/c^2):

$$\Lambda^0 \rightarrow n + \pi^0$$

What are the kinetic energies (in the lab frame) of the neutron and pion after the decay? [*Hint:* Use Eq. (26-11) to find momentum.]

✦✦68. Radon decays as $^{222}\text{Rn} \rightarrow {}^{218}\text{Po} + \alpha$. The mass of the radon-222 nucleus is 221.970 39 u, the mass of the polonium-218 nucleus is 217.962 89 u, and the mass of the alpha particle is 4.001 51 u. (1 u = 931.5 MeV/c^2.) If the radon nucleus is initially at rest in the lab frame, at what speeds (in the lab frame) do the (a) polonium-218 nucleus and (b) alpha particle move?

ANSWERS TO PRACTICE PROBLEMS

26.1 Yes. The trip takes 30 yr as measured in the rest frame of the battery.

26.2 $0.60c$

26.3 $0.909c$

26.4 3.0 s

26.5 4×10^9 kg/s

26.6 480 MeV

26.7 (a) 0.981 TeV/c; (b) 0.999 999 54c

Early Quantum Physics and the Photon

The police respond to a call and find an empty motel room that looks clean in normal light. A detective sprays a colorless liquid on the walls and floor, closes the blinds, and turns off the lights. In the dark she sees spots on the floor that give off a blue glow. The detective tells a uniformed officer to cordon off the room as a possible crime scene. What causes the blue glow? Why does the detective suspect that a crime took place in the room?

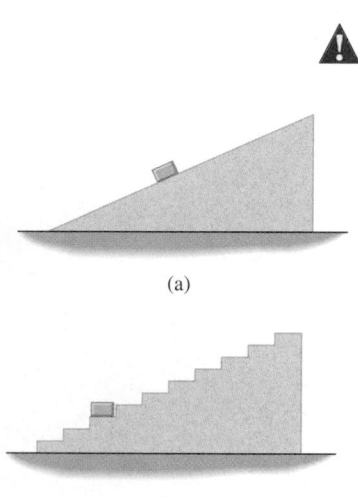

(a)

(b)

Figure 27.1 (a) A crate resting on a ramp; gravitational potential energy is continuous. (b) A crate resting on a staircase. Gravitational potential energy is quantized; it can have only one of a set of discrete values.

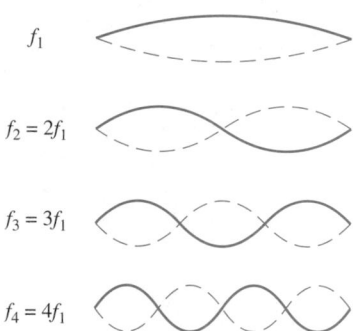

f_1

$f_2 = 2f_1$

$f_3 = 3f_1$

$f_4 = 4f_1$

Figure 27.2 Standing waves on a string can only exist at one of a discrete set of frequencies.

27.1   QUANTIZATION

As the nineteenth century ended, much progress had been made in physics—so much that some physicists feared that everything had been discovered. Newton had laid the foundations of mechanics in his *Principia*, the laws of thermodynamics were well established, and Maxwell had explained electromagnetism. Nevertheless, as scientists developed new experimental techniques and new kinds of equipment, questions arose that could not be explained by the set of physical laws that had seemed nearly complete until then—the laws now known as *classical physics*. The new laws that were developed in the first decades of the twentieth century were the foundation of what we now call *quantum physics*.

In classical physics, most quantities are continuous: they can take any value in a continuous range. As an analogy, Fig. 27.1a shows a crate resting on a ramp. The gravitational potential energy of the crate is continuous—it can have *any* value between the minimum and maximum. By contrast, a crate resting on a staircase can only have certain allowed values (Fig. 27.1b). A quantity is **quantized** when its possible values are limited to a discrete set. A salient feature of quantum physics is the quantization of quantities that were thought to be continuous in classical physics.

The staircase is an imperfect analogy of quantization. While a crate is being moved from one step to another, the gravitational potential energy passes through all the intermediate values. By contrast, something that is truly quantized does *not* pass through intermediate values; it changes suddenly from one value to another.

Standing waves provide an example of quantization in classical physics. The frequency of a standing wave on a string fixed at both ends is quantized (Fig. 27.2). The allowed frequencies are integral multiples of the fundamental frequency ($f_n = nf_1$). In classical physics, the *energy* of the standing wave is continuous; in quantum physics both the frequency *and* the energy are quantized.

This chapter considers several experiments whose results were difficult or impossible to explain with the laws of classical physics, but relatively easy to explain once electromagnetic waves are assumed to be quantized.

27.2   BLACKBODY RADIATION

A major problem vexing late nineteenth-century physics was blackbody radiation (see Section 14.8). An ideal blackbody absorbs all the radiant energy that falls upon it; the radiation emitted by a blackbody is a continuous spectrum characteristic only of the temperature of the body. Figure 27.3 shows experimental blackbody radiation curves—graphs of the relative intensity of the EM radiation as a function of the frequency—at three temperatures. As the temperature increases, the peak of the radiation curve shifts to higher frequencies. At 2000 K, almost all of the power is radiated in the infrared. At 2500 K, the object is red hot—it radiates significantly in the red and orange parts of the visible spectrum. An object at 3000 K, such as the filament of an incandescent lightbulb, radiates light that we perceive as white, but most of the radiation is still infrared. The total area under the curve, for any temperature, represents the total radiated power per unit surface area; the total power is proportional to T^4.

From the classical theory of electromagnetism, physicists understood why objects emit EM radiation: the vibrating atoms and molecules have charged electrons and protons, so the acceleration of these charges produces EM radiation. However, the classical theory predicted that the blackbody radiation curve should continue to increase with increasing frequency (into the ultraviolet and beyond), instead of peaking and then decreasing to zero (Fig. 27.3). As a result, classical theory predicted that a blackbody should radiate an *infinite amount of energy*, an impossibility dubbed the *ultraviolet catastrophe*.

In 1900, Max Planck (1858–1947) found a mathematical expression that fit the experimental radiation curves. He then sought a physical model to be the basis for his

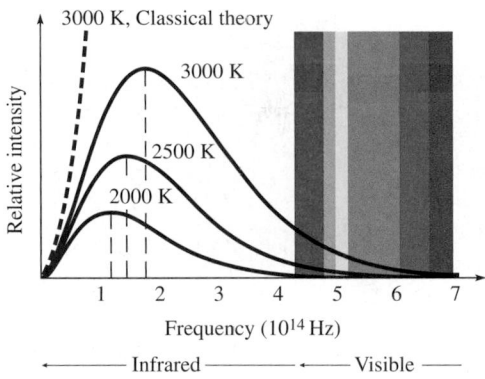

Figure 27.3 Three blackbody curves showing the relative intensity of blackbody radiation as a function of frequency for three different temperatures: 2000 K, 2500 K, and 3000 K. Also, the blackbody curve as predicted by theory before Max Planck's proposal.

mathematical expression. He proposed something revolutionary: that the energy emitted and absorbed by oscillating charges must occur only in discrete amounts called **quanta** (singular, **quantum**). He associated a fundamental amount of energy E_0 with each oscillator; the oscillator could emit E_0, or $2E_0$, or any integral multiple of E_0, but nothing in between. As an analogy, imagine that the economy of the oscillator is limited to $10 bills; an oscillator can have in its bank $10, $20, $30, but no intermediate amounts such as $15 or $4. When it spends its capital, it can only give away multiples of $10.

Planck found that the theoretical expression based on quantization matched the experimental radiation curves if E_0 is directly proportional to the frequency f of the oscillator:

$$E_0 = hf \qquad (27\text{-}1)$$

where the constant of proportionality has the unique value $h = 6.626 \times 10^{-34}$ J·s.

Planck's assumption of quantization was a bold break with the fundamental ideas of classical physics. No one knew it at the time, but Planck had launched a half century of exciting developments in physics. He chose the value of h so that his theory would match the experimental data; now h is called **Planck's constant** and is included among the fundamental physical constants such as the speed of light c and the elementary charge e.

> The value of Planck's constant causes the step size in Fig. 27.1b to be so small that the staircase (quantized energies) looks like a ramp (continuous energies) unless we examine it on the atomic or molecular scale.

27.3   THE PHOTOELECTRIC EFFECT

In 1886 and 1887, Heinrich Hertz (1857–1894) did experiments that confirmed Maxwell's classical theory of electromagnetic waves. In the course of these experiments, Hertz discovered the effect that Einstein later used to introduce the *quantum* theory of EM waves. Hertz produced sparks between two metal knobs by applying a large potential difference. He noticed that when the knobs were exposed to ultraviolet light, the sparks became stronger. He had discovered the **photoelectric effect** in which EM radiation incident on a metal surface causes electrons to be ejected from a metal.

Figure 27.4 shows an apparatus to study the photoelectric effect. A photocell is made by enclosing a metal plate and a collecting wire in an evacuated glass tube. EM radiation (visible light or UV) falls on the metal plate; some of the emitted electrons make their way to the collecting wire, which completes the circuit. An ammeter measures the current in the circuit and thus the number of electrons per second that move from the plate to the collecting wire.

An applied potential difference holds the collecting wire at a lower potential than the plate so that electrons lose kinetic energy as they move from the plate to the wire. The larger the potential difference, the smaller the number of electrons with enough kinetic energy to complete the circuit. The **stopping potential** V_s is the magnitude of the potential difference that stops even the most energetic electrons. Therefore, the maximum kinetic energy of the electrons is equal to the increase in potential energy for an electron moving through a potential difference $-V_s$:

$$K_{max} = q\,\Delta V = (-e) \times (-V_s) = eV_s \qquad (27\text{-}2)$$

> Photoelectric effect: EM radiation supplies the energy to remove electrons from a metal.

Figure 27.4 Apparatus used to study the photoelectric effect

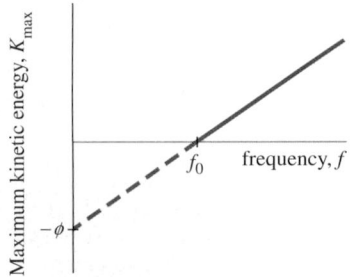

Figure 27.5 Maximum kinetic energy of the electrons ejected from a metal as a function of the frequency f of the incident light

Albert Einstein in 1910. In 1921, Einstein was awarded the Nobel prize in physics "for his service to Theoretical Physics, and especially for his discovery of the law of the photoelectric effect," according to the citation from the Nobel committee.

Experimental Results

The photoelectric effect itself seems reasonable according to classical physics: the EM wave supplies the energy needed by the electrons to break free from the metal. However, several *details* of the photoelectric effect were puzzling.

1. *Brighter* light causes an increase in current (more electrons ejected) but does *not* cause the individual electrons to gain higher energies. In other words, the maximum kinetic energy of the electrons is independent of the intensity of the light. Classically, more intense light has larger amplitude EM fields and thus delivers more energy. That should not only enable a larger number of electrons to escape from the metal; it should also enable the electrons emitted to have more kinetic energy.

2. The maximum kinetic energy of the emitted electrons *does* depend on the *frequency* of the incident radiation (Fig. 27.5). Thus, if the incident light is very dim (low intensity) but high in frequency, electrons with large kinetic energies are released. Classically, there is no explanation for a frequency dependence.

3. For a given metal, there is a **threshold frequency** f_0. If the frequency of the incident light is below the threshold, *no electrons are emitted*—no matter what the intensity of the incident light. Again, classical physics has no explanation for the frequency dependence.

4. When EM radiation falls on the metal, electrons are emitted virtually instantaneously; the time delay observed experimentally is about 10^{-9} s, regardless of the light intensity. If the EM radiation behaves as a classical wave, its energy is evenly distributed across the wavefronts. If the intensity of the light is low, it should take some time for enough energy to accumulate on a particular spot to liberate an electron. Experiments have used intensities so low that, classically, there ought to be a time delay of hours before the first electrons escape the metal. Instead, electrons are detected almost immediately!

The Photon

Planck's explanation of blackbody radiation said that the possible energies of the oscillating charges in matter are quantized; the energy of an oscillator at frequency f can only have the values $E = nhf$, where n is an integer and

$$h = 6.626 \times 10^{-34} \text{ J·s} \tag{27-3}$$

In 1905, the same year that he published his special theory of relativity, Einstein explained the photoelectric effect and correctly predicted the results of some experiments that had not yet been performed. Einstein said that *EM radiation itself* is

Figure 27.6 In the photoelectric effect, a photon is absorbed. If the energy of the photon is sufficient, an electron can be ejected from the metal.

quantized. The quantum of EM radiation—that is, the smallest indivisible unit—is now called the **photon**. The energy of a photon of EM radiation with frequency f is

$$E = hf \qquad (27-4)$$

| Energy of a photon

According to Einstein, the reason a blackbody can only emit or absorb energy in integral multiples of hf is because the EM radiation emitted or absorbed by a blackbody is itself quantized. A blackbody can emit or absorb only a whole number of photons.

The key to understanding the photoelectric effect is that the electron has to absorb a whole photon (Fig. 27.6); it cannot absorb a fraction of a photon's energy. The energy of a photon is proportional to frequency; thus, the photon theory explains the frequency dependence in the photoelectric effect that had mystified scientists.

Example 27.1

Energies of Visible and X-Ray Photons

Find the energy of a photon of visible red light of wavelength 670 nm and compare it with the energy of an x-ray photon with frequency 1.0×10^{19} Hz.

Strategy The product of Planck's constant with each frequency gives the corresponding photon energy. For the 670-nm photon, the frequency and wavelength are related by $c = f\lambda$.

Solution The frequency of the red light is

$$f = \frac{c}{\lambda}$$

To find the energy we multiply the frequency by Planck's constant.

$$E = hf = \frac{hc}{\lambda}$$

$$= \frac{6.626 \times 10^{-34} \text{ J·s} \times 3.00 \times 10^8 \text{ m/s}}{670 \times 10^{-9} \text{ m}} = 3.0 \times 10^{-19} \text{ J}$$

For the x-ray photon,

$$E = hf = 6.626 \times 10^{-34} \text{ J·s} \times 1.0 \times 10^{19} \text{ Hz} = 6.6 \times 10^{-15} \text{ J}$$

The energy of the x-ray photon is more than 20,000 times the energy of a photon of red light.

Discussion $E = 2.97 \times 10^{-19}$ J is the smallest amount of energy for red light of wavelength 670 nm that can be absorbed or emitted in any process. Similarly, 6.6×10^{-15} J is the energy of one quantum–one photon–of x-rays at the given frequency.

The much larger energy of an x-ray photon is the reason that x-rays can be far more damaging to the human body and that exposure to x-rays should be minimized (Fig. 27.7).

Practice Problem 27.1 Energy of a photon of blue light

Find the energy of one photon of visible blue light of frequency 6.3×10^{14} Hz.

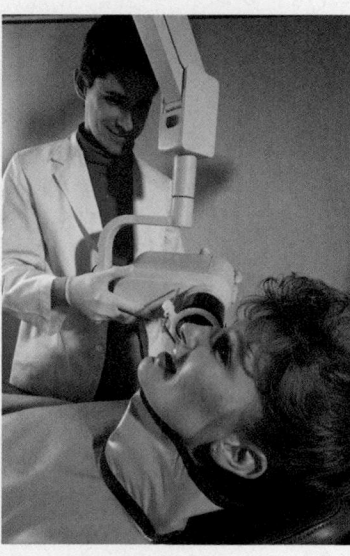

Figure 27.7
The body of a person having an x-ray done for dental purposes is protected by a lead apron. Lead is a good absorber of x-rays, so the apron minimizes the exposure of the rest of the body to x-rays.

Example 27.2

Photons Emitted by a Laser

A laser produces a beam of light 2.0 mm in diameter. The wavelength is 532 nm and the output power is 20.0 mW. How many photons does the laser emit per second?

Strategy   The photons all have the same energy since the beam has a single wavelength. The output power is the energy output per unit time. Then the energy output per second is the energy of each photon times the number of photons emitted per second.

Solution

energy per second = energy per photon × photons per second

Since $\lambda f = c$, the energy of a photon of wavelength λ is

$$E = hf = h \times \frac{c}{\lambda} = \frac{hc}{\lambda}$$

The energy of each photon emitted by the laser is

$$E = \frac{6.626 \times 10^{-34} \text{ J·s} \times 3.00 \times 10^8 \text{ m/s}}{532 \times 10^{-9} \text{ m}} = 3.736 \times 10^{-19} \text{ J}$$

The number of photons emitted per second yields

$$\text{photons per second} = \frac{\text{energy per second}}{\text{energy per photon}}$$

$$= \frac{0.0200 \text{ J/s}}{3.736 \times 10^{-19} \text{ J/photon}} = 5.35 \times 10^{16} \text{ photons/s}$$

Discussion   Notice that the diameter of the beam is irrelevant to the solution. If the power output were the same but the diameter of the beam were larger, the same number of photons per second would be emitted; they would just be spread across a wider beam. If the problem had stated the *intensity* (power per unit area) of the beam rather than the total power output, the diameter of the beam would have been relevant.

The quantization of light is not noticed in many situations due to the extremely large numbers of photons. An ordinary 100-W incandescent lightbulb or a 23-W compact fluorescent bulb both emit about 10 W of power as visible light. Thus the number of photons per second in the visible range emitted by an ordinary lightbulb is around 3×10^{19}.

Practice Problem 27.2   Radio wave photons

A radio station broadcasts at 90.9 MHz. The power output of the transmitter is 50.0 kW. How many radio wave photons per second are emitted by the transmitter?

Units of Energy

The energies of the photons in Examples 27.1 and 27.2 are small compared to energies of macroscopic bodies, so it is often convenient to express them in *electron-volts* (symbol eV) rather than in joules. One electron-volt is equal to the kinetic energy that a particle with charge $\pm e$ (such as an electron or a proton) gains when it is accelerated through a potential difference of magnitude 1 V. Since 1 V = 1 J/C and $e = 1.60 \times 10^{-19}$ C, the conversion between electron-volts and joules is

$$1 \text{ eV} = e \times 1 \text{ V} = 1.60 \times 10^{-19} \text{ C} \times 1 \text{ J/C} = 1.60 \times 10^{-19} \text{ J} \qquad (27\text{-}5)$$

For larger amounts of energy, keV represents kilo-electron-volts (10^3 eV) and MeV represents mega-electron-volts (10^6 eV). The photon of red light in Example 27.1 has energy 1.9 eV; the x-ray photon has energy 41 keV.

When finding the energy of a photon given its wavelength (or *vice versa*) using $E = hc/\lambda$, the energy of a photon is often expressed in electron-volts (eV) and wavelengths are often stated in nanometers (nm). For this reason, it is useful to express the constant hc in units of eV·nm:

$$h = \frac{6.626 \times 10^{-34} \text{ J·s}}{1.602 \times 10^{-19} \text{ J/eV}} = 4.136 \times 10^{-15} \text{ eV·s}$$

$$c = 2.998 \times 10^8 \text{ m/s} \times 10^9 \text{ nm/m} = 2.998 \times 10^{17} \text{ nm/s}$$

$$hc = 4.136 \times 10^{-15} \text{ eV·s} \times 2.998 \times 10^{17} \text{ nm/s} = 1240 \text{ eV·nm} \qquad (27\text{-}6)$$

The Photon Theory Explains the Photoelectric Effect

The amount of energy that must be supplied to break the bond between a metal and one of its electrons is called the **work function** (ϕ). Each metal has its own characteristic work function. According to Einstein, if the photon energy (hf) is at least equal to the work function, then absorption of a photon can eject an electron. If the photon energy is greater than the work function, some or all of the extra energy can appear as the ejected electron's kinetic energy. The maximum kinetic energy of an electron is the difference between the photon energy and the work function.

Einstein's photoelectric equation

$$K_{max} = hf - \phi \qquad (27\text{-}7)$$

The photoelectric equation is based on conservation of energy.

According to Eq. (27-7), a graph of K_{max} versus f is a straight line with a slope of h and a "y"-intercept of $-\phi$; this equation fits the experimental relationship (Fig. 27.5). The intercept on the frequency axis is the threshold frequency f_0. Setting

$$K_{max} = hf_0 - \phi = 0$$

predicts a threshold frequency of

$$f_0 = \frac{\phi}{h} \qquad (27\text{-}8)$$

Threshold frequency (photoelectric effect)

The four puzzling results of photoelectric effect experiments are easily explained using the photon concept:

1. Light of greater intensity (but constant frequency) delivers more photons per unit time to the metal surface, so the *number* of electrons ejected per second increases as the intensity of the light increases. However, the energy of each photon remains the same. The maximum kinetic energy of the emitted electrons does not depend on the number of photons striking the metal per second because each emitted electron is the result of the absorption of *one photon*.

2. Higher-frequency light has larger energy photons. As the frequency of the light increases, the photons have more excess energy that can potentially become the electron's kinetic energy. Thus, K_{max} increases with increasing frequency.

3. Below the threshold frequency, a photon does not have enough energy to free an electron from the metal, so no electrons are emitted.

4. At low intensities, the number of photons per second is small, but the energy is still delivered in discrete packets. Just after the light is turned on, some photons hit the surface; some of them are absorbed and eject electrons from the metal. There is no time delay because the electrons cannot gradually accumulate energy; they either absorb a photon or they don't.

Example 27.3

A Photoelectric Effect Experiment

Cesium has a work function of 1.8 eV. When cesium is illuminated with light of a certain wavelength, the electrons ejected from the surface have kinetic energies ranging from 0 to 2.2 eV. What is the maximum wavelength of the light?

Strategy The work function and the maximum kinetic energy (2.2 eV) are given. To eject an electron, the photon must supply 1.8 eV of energy. Some or all of the remainder of the photon's energy ($hf - \phi$) gives the electron its kinetic energy. The maximum kinetic energy occurs when all of the remainder goes to the electron's kinetic energy.

Solution The energy of a photon is hf. The maximum kinetic energy of the photoelectrons is

$$K_{max} = hf - \phi = 2.2 \text{ eV}$$

The problem asks for the maximum wavelength, so we substitute $f = c/\lambda$ and solve for λ:

$$K_{max} = \frac{hc}{\lambda} - \phi$$

$$\lambda = \frac{hc}{K_{max} + \phi}$$

Substituting $hc = 1240$ eV·nm yields

$$\lambda = \frac{1240 \text{ eV·nm}}{2.2 \text{ eV} + 1.8 \text{ eV}} = 310 \text{ nm}$$

Discussion The photon has energy 2.2 eV + 1.8 eV = 4.0 eV. 1.8 eV raises the potential energy of the electron so that it is free to leave the metal. The remaining 2.2 eV does

continued on next page

Example 27.3 *continued*

not necessarily all become the electron's kinetic energy; some of it can be absorbed by the metal. Therefore, 2.2 eV is the *maximum* kinetic energy of a photoelectron. Since the wavelength is less than 400 nm, the photon is in the ultraviolet part of the spectrum.

Practice Problem 27.3   Wavelength of incident light

A metal with a work function of 2.40 eV is illuminated with monochromatic light. If the stopping potential that prevents electrons from being ejected is 0.82 V, what is the wavelength of the light? [*Hint:* What is the maximum kinetic energy of the electrons ejected from the surface in electron-volts?]

Applications of the Photoelectric Effect

While our principal interest in the photoelectric effect is how clearly it illustrates the concept of the photon, many practical applications exist. Most of them exploit the dependence of the photoelectric current on the intensity of the light. A movie sound track is a strip along one side of the film that encodes sound as variations in the transparency of the film (Fig. 27.8). Light shines through the sound track and then onto a photocell. The photocell generates a current proportional to the intensity of light incident on it. This current is amplified and fed to the speakers.

Devices such as garage door openers, burglar alarms, and smoke detectors often use a light beam and a photocell as a switch. When the light beam is interrupted, the current through the photocell stops. A child walking underneath a garage door that is being closed interrupts a light beam; when the current stops, a switch stops the motion of the door. In some smoke alarms, particles of smoke in the air reduce the intensity of the light that reaches a photocell; when the current drops below a certain level, the alarm is activated.

Making The Connection:
photocells in sound tracks, burglar alarms, garage door openers

27.4   X-RAY PRODUCTION

Another confirmation of the quantization of EM radiation is found in the production of x-rays. Figure 27.9a shows an x-ray tube; it looks something like a photocell operated in reverse. In the photoelectric effect, EM radiation incident on a target causes the emission of electrons; in an x-ray tube, electrons incident on a target cause the emission of EM radiation. Electrons move through a large potential difference V to give them large kinetic energies $K = eV$. In the target, they are deflected as they pass by atomic nuclei (Fig. 27.9b). Sometimes an x-ray photon is emitted; the energy of the photon comes from the electron's kinetic energy, so the electron slows down. This process for creating x-rays is called *bremsstrahlung*, from the German for *braking radiation*, since the x-rays are emitted as electrons slow down.

Charges radiate only when they are accelerating, not when they move at constant velocity.

Figure 27.8   An optical sound track

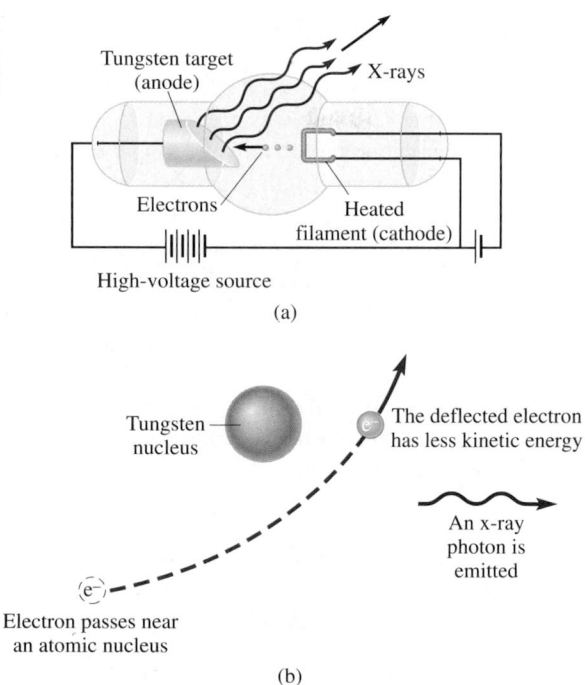

Tungsten target
(anode)

X-rays

Electrons

Heated
filament (cathode)

High-voltage source

(a)

Tungsten
nucleus

The deflected electron
has less kinetic energy

An x-ray
photon is
emitted

Electron passes near
an atomic nucleus

(b)

Figure 27.9   (a) An x-ray tube. An electric current heats the filament to "boil off" electrons. The electrons are accelerated through a large potential difference between the filament and the target. When electrons strike the target, x-rays are emitted as the electrons lose kinetic energy. (b) An electron is deflected by an atomic nucleus. An x-ray photon is emitted, carrying away some of the electron's kinetic energy.

The x-rays produced in this way do not all have the same frequency; there is a continuous spectrum of frequencies up to a maximum, called the **cutoff frequency**. Typically an electron emits many photons as it slows down; each of the photons takes away *part of* the electron's kinetic energy. The maximum frequency occurs when all of the electron's kinetic energy is carried away by *a single photon*:

$$hf_{max} = K \qquad\qquad (27-9) \qquad | \quad \text{Cutoff frequency (x-ray production)}$$

Example 27.4

Diagnostic X-Rays in Medicine

A potential difference of 87.0 kV is applied between the filament and target in the x-ray tube used at the local clinic to look for broken bones. What are the shortest wavelength x-rays produced by this tube?

Strategy   The shortest wavelength corresponds to the highest frequency. The highest frequency is produced when all of the electron kinetic energy is given up in the emission of a single photon.

The accelerating potential of 87.0 kV supplies the electrons with kinetic energy before they hit the target. We do not need to use the numerical value of e to find the kinetic energy. An electron traveling through 1 V of potential difference gains an energy of 1 eV, so an electron traveling through a potential difference of 87.0 kV gains 87.0 keV of kinetic energy. The constants h and c can be looked up individually, but it is easier to use the combination $hc = 1240$ eV·nm.

Solution   The maximum frequency occurs when the energy of the photon is equal to the electrons' kinetic energy:

$$hf_{max} = K = 87.0 \text{ keV}$$

$$f_{max} = \frac{K}{h}$$

The minimum wavelength is

$$\lambda_{min} = \frac{c}{f_{max}}$$

Substituting for f_{max} yields

$$\lambda_{min} = \frac{hc}{K} = \frac{1240 \text{ eV·nm}}{87.0 \times 10^3 \text{ eV}} = 0.0143 \text{ nm} = 14.3 \text{ pm}$$

Discussion   Notice how much simpler the calculation is made by using the electron-volt for energy. The electron-volt saves physicists from having to constantly multiply and divide by the numerical value of the elementary charge e.

Practice Problem 27.4   Potential difference across an x-ray tube

If the shortest wavelength detected for x-rays from an x-ray tube is 0.124 nm, what is the potential difference applied to the tube?

27.5 COMPTON SCATTERING

In 1922, A. H. Compton (1892–1962) noticed that when x-rays of a single wavelength impinged on matter, some of the radiation was scattered in various directions. Further study showed that some of the scattered radiation had longer wavelengths than the incident radiation. The increase in the wavelength depended only on the angle between the incident radiation and the scattered radiation. According to classical theory, the incident radiation should induce vibration of the electrons in the target material *at the same frequency* as the incident wave. A scattered wave results when some of the incident energy is absorbed and re-emitted in a different direction. Thus, according to classical EM theory, the scattered radiation should have the same frequency and wavelength as the incident radiation.

In the photon picture, **Compton scattering** is viewed as a collision between a photon and an electron (Fig. 27.10). The scattered photon must have less energy than the incident photon since some energy is given to the recoiling electron. Thus, conservation of energy requires that

$$E = K_e + E'$$

or

$$\frac{hc}{\lambda} = K_e + \frac{hc}{\lambda'} \qquad (27\text{-}10)$$

where E is the energy of the incident photon, K_e is the kinetic energy given to the recoiling electron, and E' is the energy of the scattered photon. Since the scattered photon has less energy, its wavelength is longer. Although the scattered photon has less energy, it does *not* move any more slowly than the incident photons. All photons move at the same speed c.

Energy conservation alone does not explain why the wavelength of the photons scattered in a particular direction (at angle θ with respect to the incident photons) is *always the same* for a given incident wavelength λ. If energy conservation were the only restriction, photons of *any* energy $E' < E$ could be scattered at *any* angle θ. Just as in other collisions, we must consider conservation of *momentum*.

According to classical electromagnetic theory, EM waves carry momentum of magnitude E/c, where E is the energy of the wave and c is the speed of light. In the photon picture, each photon carries a little bit of that momentum in proportion to the amount of energy it carries. The **momentum of a photon** is

$$p = \frac{\text{photon energy}}{c} = \frac{hf}{c} = \frac{h}{\lambda} \qquad (27\text{-}11)$$

The direction of the photon's momentum is in its direction of propagation.

In most cases the initial energy and momentum of the electron are negligible compared to the energy and momentum imparted by the collision. The energy of an x-ray photon is large compared to the work function of the target material, so we can neglect the work function and treat the electron as free. Compton's explanation ignores the initial energy and momentum of the electron and the work function; the scattering process is viewed as a collision between a photon and a *free* electron *initially at rest*.

Conservation of momentum requires:

$$\vec{\mathbf{p}} = \vec{\mathbf{p}}_e + \vec{\mathbf{p}}'$$

Figure 27.10 In Compton scattering, momentum and energy are transferred to the electron. Since momentum and energy are both conserved, the scattered photon has less energy—and therefore a longer wavelength—than the incident photon. The interaction can be analyzed as an elastic collision.

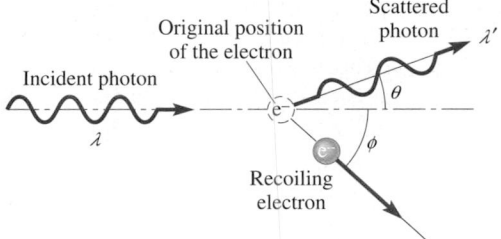

Using the incident photon's direction as the x-axis, we can separate this into two component equations:

$$\frac{h}{\lambda} = p_e \cos \phi + \frac{h}{\lambda'} \cos \theta \quad \text{(x-component)} \qquad (27\text{-}12)$$

and

$$0 = -p_e \sin \phi + \frac{h}{\lambda'} \sin \theta \quad \text{(y-component)} \qquad (27\text{-}13)$$

From the equations for conservation of energy and momentum [Eqs. (27-10), (27-12), and (27-13)], Compton derived this relationship:

Compton shift

$$\lambda' - \lambda = \frac{h}{m_e c} (1 - \cos \theta) \qquad (27\text{-}14)$$

Compton shift: difference in wavelength between the scattered and incident photons

In Eq. (27-14), the incident photon has wavelength λ, the scattered photon has wavelength λ', m_e is the mass of the electron, and θ is called the scattering angle. Equation (27-14) correctly predicts the wavelength shifts observed in experiment.

The quantity $h/(m_e c)$ is known as the **Compton wavelength** because it has the *dimensions* of a wavelength.

Compton wavelength

$$\frac{h}{m_e c} = \frac{6.626 \times 10^{-34} \text{ J·s}}{9.109 \times 10^{-31} \text{ kg} \times 2.998 \times 10^{8} \text{ m/s}}$$

$$= 2.43 \times 10^{-3} \text{ nm} = 2.43 \text{ pm} \qquad (27\text{-}15)$$

Since $\cos \theta$ can vary between +1 and −1, the quantity $(1 - \cos \theta)$ varies from 0 to 2 and the wavelength change varies from 0 to twice the Compton wavelength (4.85 pm).

Calculation of the shift in the photon wavelength requires putting together just two principles: conservation of energy and conservation of momentum. In many cases, the electron's recoil speed is fast enough that we *cannot* use the nonrelativistic equations $K_e = \frac{1}{2}mv^2$ and $p_e = mv$ for the kinetic energy and momentum of the electron. Compton used the relativistic equations for the momentum [Eq. (26-6)] and kinetic energy [Eq. (26-8)] of the electron in his derivation, so Eq. (27-14) is valid for any recoil speed.

Example 27.5

Energy of Recoiling Electron

An x-ray photon of wavelength 10.0 pm is scattered through 110.0° by an electron. What is the kinetic energy of the recoiling electron?

Strategy   Since we know the scattering angle, we can find the Compton shift [Eq. (27-14)]. The Compton shift and the wavelength of the incident photon enable us to find the wavelength of the scattered photon; from the wavelength we can find the energy of the scattered photon. By energy conservation, the kinetic energy of the electron plus the energy of the scattered photon is equal to the energy of the incident photon.

Solution   The Compton shift formula is

$$\Delta \lambda = \lambda' - \lambda = \frac{h}{m_e c} (1 - \cos \theta)$$

where $h/m_e c = 2.43$ pm. With $\lambda = 10.0$ pm and $\theta = 110.0°$,

$$\Delta \lambda = \lambda' - \lambda = 2.43 \text{ pm} \times (1 - \cos 110.0°) = 3.26 \text{ pm}$$

Then the scattered photon has wavelength

$$\lambda' = \lambda + \Delta \lambda = 10.0 \text{ pm} + 3.26 \text{ pm} = 13.26 \text{ pm}$$

The kinetic energy of the electron is

$$K_e = E - E' = \frac{hc}{\lambda} - \frac{hc}{\lambda'}$$

$$= 1240 \text{ eV·nm} \times \left(\frac{1}{0.0100 \text{ nm}} - \frac{1}{0.01326 \text{ nm}} \right)$$

$$= 30.5 \text{ keV}$$

 Discussion   Avoid the common algebraic mistake of substituting $hc/\Delta \lambda$ for $hc/\lambda - hc/\lambda'$. That error would have given an answer of 380 keV for the kinetic energy of the electron—wrong by more than a factor of 12.

Practice Problem 27.5   Change in wavelength

In a Compton scattering experiment, x-rays scattered through an angle of 37.0° with respect to the incident x-rays have a wavelength of 4.20 pm. What is the wavelength of the incident x-rays?

27.6 SPECTROSCOPY AND EARLY MODELS OF THE ATOM

In 1853, A. J. Ångström (1814–1874) used spectroscopy to study the light emitted by various low-pressure gases in a *discharge tube* (Fig. 27.11). The gas is kept at low pressure so that the atoms are far apart from one another; thus, the light is emitted by a collection of essentially independent atoms. Electrons are injected into the gas, either by heating the electrodes or by applying a large potential difference between them. The electrons collide with gas atoms in the tube. Electric current flows between the electrodes; electrons move in one direction and positive gas ions in the other. As long as the current is maintained, the tube emits light. A neon sign (Fig. 27.11b) is a familiar example of a discharge tube. A fluorescent lamp is a mercury discharge tube with a coating on the inside of the glass.

In the spectroscopic analysis of the light emitted by a discharge tube, the light first passes through a thin slit. Then it passes through either a prism or a grating so that light of different wavelengths emerges at different angles. While the light emitted by a hot solid object forms a *continuous* spectrum, Ångström discovered that the light from a gas discharge tube forms a *discrete* spectrum (Fig. 27.12a–d). A discrete spectrum is also called a *line* spectrum because each discrete wavelength forms an image of the slit; the spectrum appears as a set of narrow, parallel lines of different colors with dark space between the lines.

Making The Connection:
neon signs

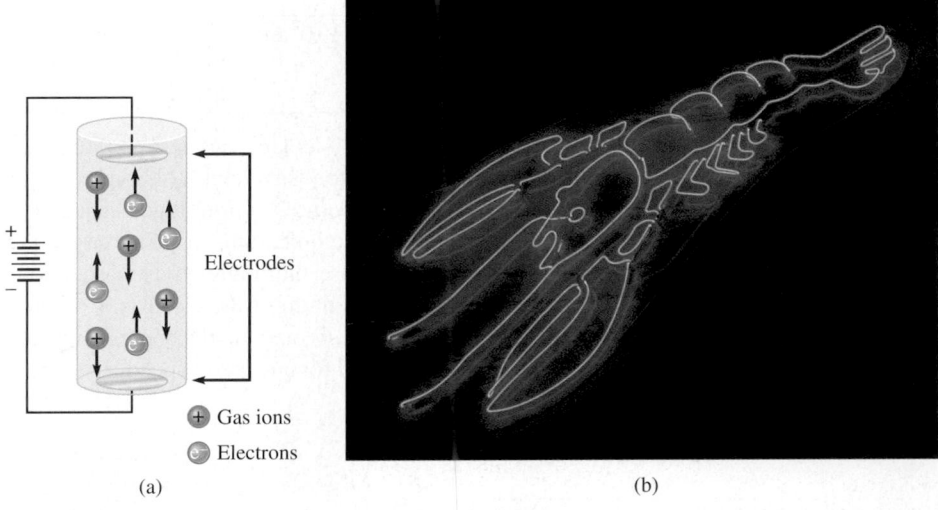

Figure 27.11 (a) A gas discharge tube. (b) A neon sign is a discharge tube containing neon gas.

(+) Gas ions
(e⁻) Electrons

(a)

(b)

Figure 27.12 From top to bottom, emission spectra of barium, calcium, hydrogen, and sodium, and a solar absorption spectrum.

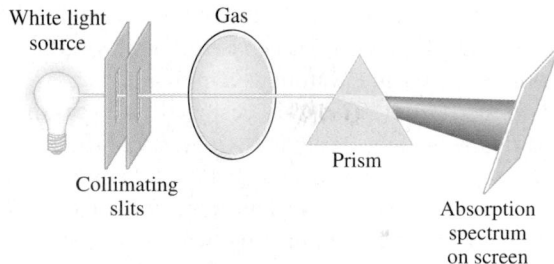

Figure 27.13 Setup for obtaining an absorption spectrum. When white light passes through a gas, some wavelengths are absorbed. The missing wavelengths cause dark lines to appear in the otherwise continuous spectrum on the screen.

In addition to examining the light emitted by a gas, scientists also studied the light absorbed by a gas. A beam of white light is sent through the gas and the transmitted light is analyzed with a spectrometer (Fig. 27.13). The resulting absorption spectrum is the continuous spectrum expected for white light except for some dark lines. Most wavelengths are transmitted through the gas, but the dark lines show that a few discrete wavelengths are absorbed (Fig. 27.12e). The wavelengths absorbed are a subset of the wavelengths emitted by the same gas when in a discharge tube.

Each element has its own characteristic emission spectrum. For instance, the characteristic red color of a neon sign is caused by the emission spectrum of neon. Scientists soon began to use spectroscopy to identify the elements present in substances. Many previously unknown elements were discovered through spectroscopy. Cesium was named for its bright blue spectral lines (in Latin, *cesius* = "sky blue"); rubidium was named for its prominent red lines (in Latin, *rubidius* = "dark red"). Turning their spectroscopes toward the Sun and stars, scientists identified elements such as helium, which had not yet been discovered on Earth. (The Greek word for the Sun is *helios*.)

The spectra of most elements show no obvious pattern, but hydrogen—the simplest atom—does show a striking pattern. Figure 27.14 shows an emission spectrum for hydrogen that includes lines in the ultraviolet, visible, and near infrared. In 1885, J. J. Balmer (1825–1898) found a simple formula for the four wavelengths of the hydrogen emission lines in the *visible* range:

$$\frac{1}{\lambda} = R\left(\frac{1}{n_f^2} - \frac{1}{n_i^2}\right) \qquad (27\text{-}16)$$

where $n_f = 2$ and $n_i = 3, 4, 5,$ or 6. The experimentally measured quantity $R = 1.097 \times 10^7 \text{ m}^{-1}$ is called the Rydberg constant.

Subsequently, it was found that Eq. (27-16) gives the wavelengths of *all* of the hydrogen lines, not just the four visible lines. Each value of n_f gives rise to a series of lines; each line in a series has a unique value of $n_i > n_f$.

The experimental observation that individual atoms in a gas absorb and emit EM radiation only at discrete wavelengths was impossible to explain using early models of atomic structure.

Discovery of the Atomic Nucleus

At the beginning of the twentieth century, the most common model of the atom was the *plum pudding model*. The positive charge and most of the mass of the atom were thought to be spread evenly throughout the volume of the atom, with negatively

Making The Connection:
spectroscopic analysis of the elements

91 nm
122 nm
365 nm
656 nm
820 nm
1875 nm

Lyman series   Balmer series   Paschen series
$n_f = 1$ $n_f = 2$ $n_f = 3$

Figure 27.14 Emission spectrum of atomic hydrogen

Figure 27.15 Thomson's plum pudding model of the atom

charged electrons sprinkled here and there like plums in a pudding (Fig. 27.15). J. J. Thomson, who discovered the electron, accepted the uniform distribution of positive charge but said that the electrons in the atom were moving.

In 1911, Ernst Rutherford (1871–1937) designed an experiment in which a thin gold foil was bombarded with alpha particles. (*Alpha particles* are emitted in the decay of some radioactive elements. They have charge $+2e$ and approximately four times the mass of a hydrogen atom. We now know that an alpha particle consists of 2 protons and 2 neutrons.) After striking the foil, the alpha particles were detected by observing the flashes of light produced when they hit a fluorescent screen (Fig. 27.16a).

In the plum pudding model of the atom, charge and mass are distributed evenly, with no points at which mass or charge are concentrated. Based on this model, Rutherford expected the alpha particles to pass through the atoms of the foil barely deflected at all. He was surprised to find alpha particles that were deflected through large angles—sometimes more than 90°, so they bounced back from the foil instead of passing through it. Rutherford expressed his surprise by saying, "it was almost as incredible as if you fired a fifteen-inch [artillery] shell at a piece of tissue paper and it came back and hit you."

The alpha particles deflected through large angles must have collided with something massive; the massive object must be tiny since most of the alpha particles are deflected through much smaller angles. Based on the results of scattering experiments, Rutherford proposed a new model of the atom in which a central dense nucleus with a radius of about 10^{-15} m contains all of the positive charge and most of the mass of the atom (Fig. 27.16b). The positively charged nucleus repels the positively charged alpha

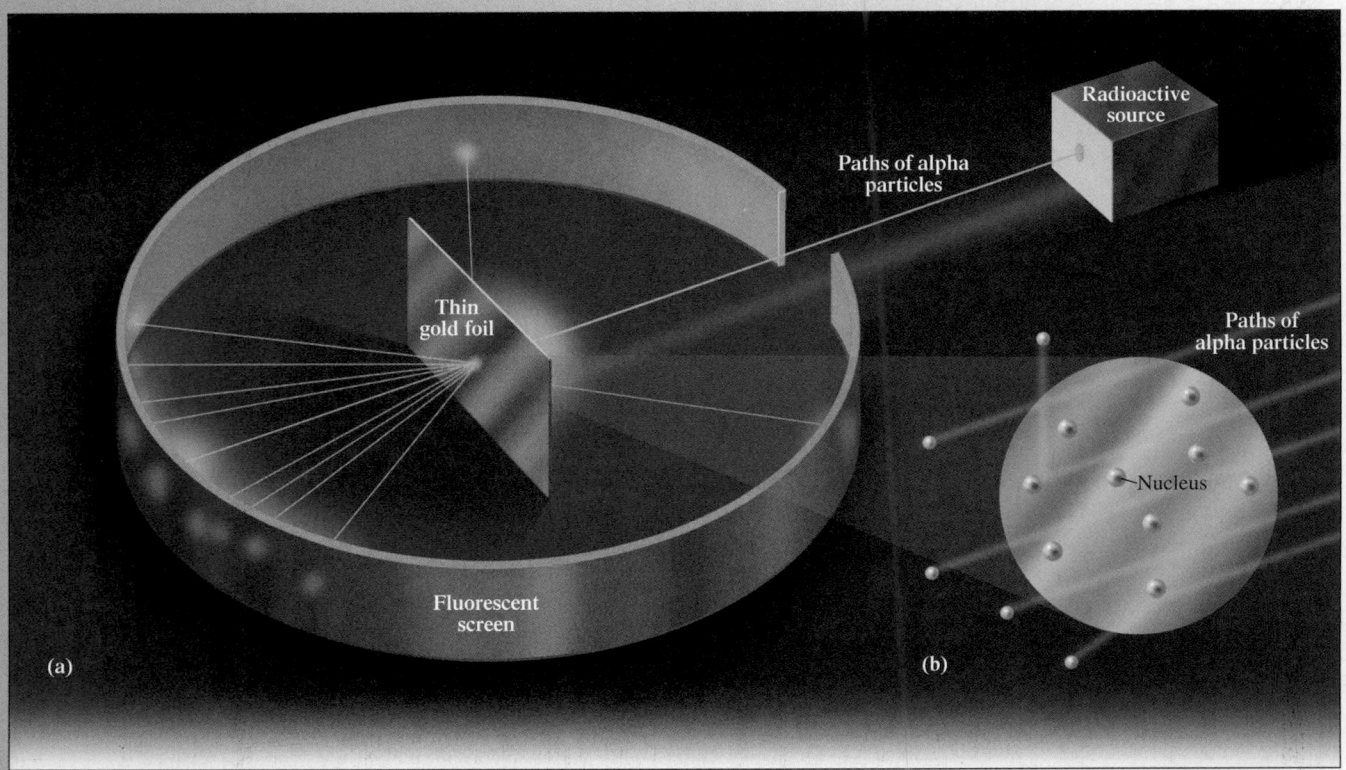

Figure 27.16 Rutherford scattering experiment. (a) Alpha particles from a radioactive source are aimed at a thin gold foil. The scattered particles are detected by light emitted when they hit a fluorescent screen. (b) Alpha particles that get close to a gold nucleus are deflected through large angles; alpha particles that do not pass near a nucleus are barely deflected at all.

particles that come near it. The radius of the nucleus is only one hundred-thousandth (10^{-5}) times the radius of the atom; thus, most of the alpha particles pass right through the gold foil without significant deflection. The few alpha particles that pass near to the nucleus feel a large repulsive force and are deflected through large angles.

After the discovery of the nucleus, the planetary model of the atom replaced the plum pudding model: electrons were thought to revolve around the nucleus like a small solar system, with the electric force on the electrons due to the nucleus playing the role that gravity plays in the solar system.

Two serious questions bothered scientists. First, in classical electromagnetic theory, an accelerating electric charge gives off EM radiation. An electron orbiting the nucleus has an acceleration—the direction of its velocity is always changing—so it ought to be continually radiating. As the radiation carries off energy, the electron's energy should decrease, causing the electron to spiral into the nucleus. Thus, atoms ought to radiate for a short while—only about 0.01 μs—until they collapse; atoms could not be stable according to classical electromagnetism. The second question: when atoms do radiate, as in a discharge tube, why only at certain wavelengths? In other words, why are emission spectra from atoms seen as line spectra rather than as continuous spectra?

27.7 THE BOHR MODEL OF THE HYDROGEN ATOM; ATOMIC ENERGY LEVELS

In 1913, the Danish physicist Niels Bohr (1885–1962) constructed the first atomic model that addressed these questions. Bohr's model is of the hydrogen atom—the simplest atom, with one electron and a single proton as the nucleus.

Assumptions of the Bohr Model

1. *The electron can exist without radiating energy only in certain circular orbits* (Fig. 27.17). Bohr asserts that, since the accelerating electron does not radiate, some aspects of classical electromagnetic theory *do not apply* to an electron orbiting the nucleus at certain discrete radii. The electron is allowed to be in only one of a discrete set of orbits called **stationary states**. (The *electron* is not stationary; it orbits the nucleus. The *state* of the electron is stationary because the electron orbits at a fixed radius without radiating.) Each stationary state has a definite energy associated with it; the set of energies of the states are called **energy levels**. Thus, Bohr extends quantum theory to the *structure of the atom* itself: both the radii and energies of the orbits are quantized.

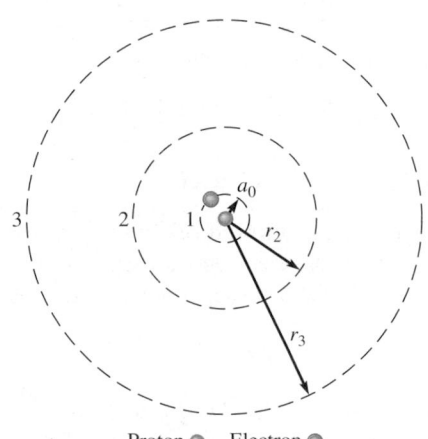

Proton ● Electron ●

Figure 27.17 In the Bohr model of the hydrogen atom, the electron orbits the nucleus in a circle. The radius of the orbit must be one of a discrete set of radii.

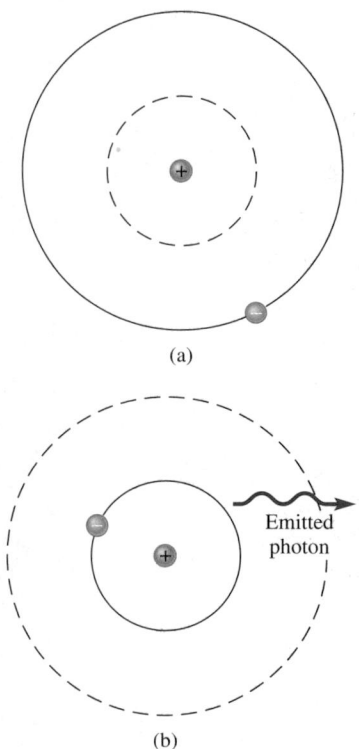

Figure 27.18 (a) A hydrogen atom in an allowed orbit. (b) The atom emits a photon and the electron drops down into a different allowed orbit with lower energy.

2. *The laws of Newtonian mechanics apply to the motion of the electron in any of the stationary states.* The force on the electron due to the nucleus is given by Coulomb's law. Newton's second law ($\Sigma \vec{\mathbf{F}} = m\vec{\mathbf{a}}$) relates the Coulomb force to the centripetal acceleration of the electron in its circular orbit. The energy of the orbit is the electron's kinetic energy plus the electric potential energy of the interaction between the electron and the nucleus.

3. *The electron can make a transition between stationary states through the emission or absorption of a single photon* (Fig. 27.18). The energy of the photon is equal to the difference between the energies of the two stationary states:

$$\Delta E = hf$$

Since the electron energy levels have only certain discrete values, emission and absorption spectra are made up of photons of discrete energies—they are line spectra. Bohr made no attempt to explain *how* an electron "jumps" from one orbit to another.

4. *The stationary states are those circular orbits in which the electron's angular momentum is quantized in integral multiples of $h/(2\pi)$.*

$$L_n = n\frac{h}{2\pi} = n\hbar \quad (n = 1, 2, 3, \ldots) \tag{27-17}$$

The combination of constants $h/(2\pi)$ is commonly abbreviated as $\hbar$ ("h-bar"). Bohr chose these values of angular momentum because they gave agreement with the experimental data on the hydrogen emission spectrum.

Radii of the Bohr Orbits

A particle of mass m in a circular orbit of radius r at speed v has rotational inertia $I = mr^2$ [Eq. (8-2)] and angular momentum

$$L = I\omega = mr^2\omega = mvr \tag{8-14}$$

since $\omega = v/r$. Then the Bohr condition on angular momentum becomes

$$mvr_n = n\hbar \quad (n = 1, 2, 3, \ldots) \tag{27-18}$$

where r_n is the radius of the orbit with angular momentum $n\hbar$. Using Newton's second law ($\Sigma \vec{\mathbf{F}} = m\vec{\mathbf{a}}$) applied to an electron held in circular orbit by the Coulomb force (see Problem 32), Bohr showed that the only orbital radii that satisfy Eq. (27-18) are

$$r_n = \frac{n^2\hbar^2}{mke^2} \quad (n = 1, 2, 3, \ldots) \tag{27-19}$$

The smallest radius, which occurs for $n = 1$, is known as the **Bohr radius**:

$$a_0 = \frac{\hbar^2}{mke^2} = 5.29 \times 10^{-11} \text{ m} = 52.9 \text{ pm} \tag{27-20}$$

The allowed orbital radii of the electron are then

$$r_n = n^2 a_0 \quad (n = 1, 2, 3, \ldots) \tag{27-21}$$

Energy Levels of the Bohr Orbits

Now that we know the radii of the allowed orbits, we can calculate the energies. The energy of an orbit is the sum of the electron's kinetic energy and the electric potential energy when the electron and nucleus are separated by a distance r:

$$E = K + U = \frac{1}{2}mv^2 - \frac{ke^2}{r}$$

The potential energy U is negative because we assume that the potential energy is zero at infinite separation; the potential energy decreases as the distance between the oppositely charged electron and proton decreases. As you can show in Problem 33, the energy E for the electron in the nth Bohr orbit is

$$E_n = -\frac{mk^2e^4}{2n^2\hbar^2} \qquad (27\text{-}22)$$

The energy E_n is negative because the energy of the atom with the electron bound to the nucleus is less than the energy of the ionized atom. In the ionized atom, the electron is at rest infinitely far from the nucleus, so $E = 0$ (both the kinetic and potential energies are zero). An electron in one of the bound states must be supplied with energy for it to escape from the nucleus and cause the atom to become ionized.

For the state $n = 1$, called the **ground state**, the orbit has the smallest possible radius and the lowest possible energy. The ground state energy is

$$E_1 = -\frac{mk^2e^4}{2\hbar^2} = -2.18 \times 10^{-18}\,\text{J} = -13.6\,\text{eV} \qquad (27\text{-}23)$$

The energies of the **excited states** ($n > 1$) are given by

$$E_n = \frac{E_1}{n^2} \qquad (27\text{-}24)$$

Figure 27.19 is an energy level diagram for hydrogen. Each horizontal line represents an energy level. The vertical arrows show transitions between levels, accompanied

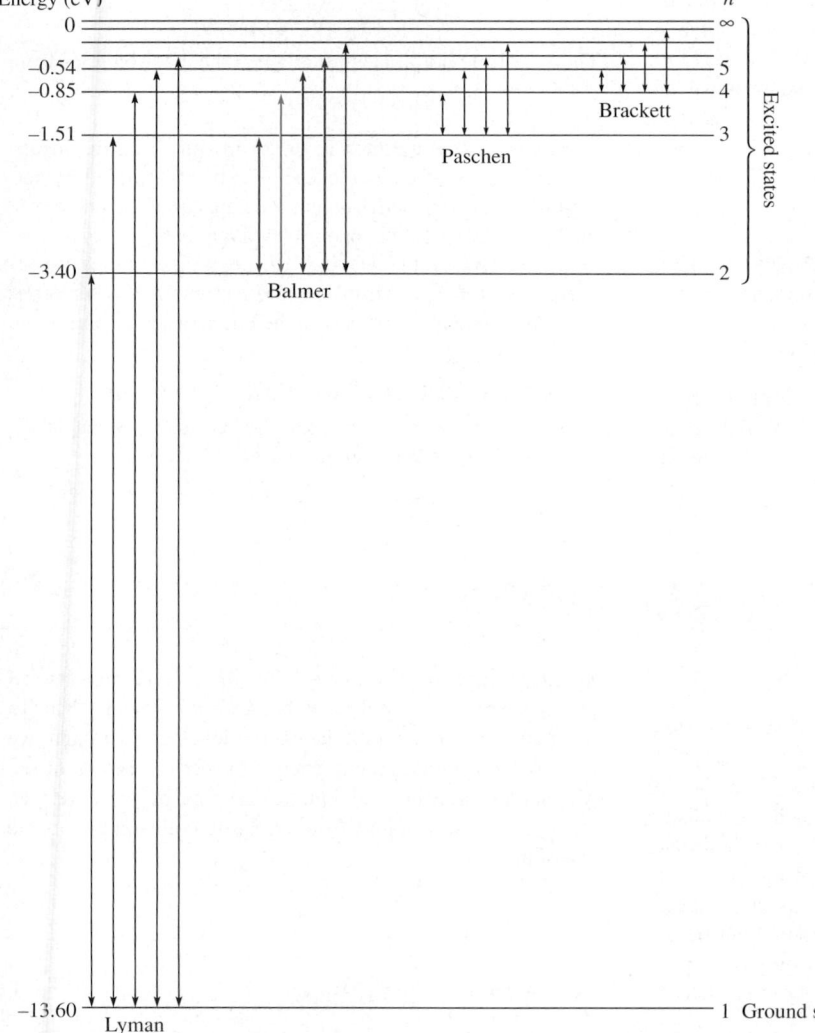

Figure 27.19 Energy level diagram for the hydrogen atom. Energy $E = 0$ at level $n = \infty$ corresponds to the ionized atom. Arrows represent transitions between energy levels. The length of an arrow represents the energy of the photon emitted or absorbed.

by either the emission or absorption of a photon of the appropriate energy. The energy of the photon emitted when the electron goes from state n_i to state n_f is

$$E = \frac{hc}{\lambda} = E_i - E_f = E_1 \left(\frac{1}{n_i^2} - \frac{1}{n_f^2} \right)$$

If we take the general form of the Balmer formula [Eq. (27-16)] and multiply both sides by hc, we get

$$\frac{hc}{\lambda} = hcR \left(\frac{1}{n_f^2} - \frac{1}{n_i^2} \right) = -hcR \left(\frac{1}{n_i^2} - \frac{1}{n_f^2} \right)$$

where R is the Rydberg constant. Thus, Bohr's theory is in perfect agreement with the spectroscopic data as long as $E_1 = -hcR$. When Bohr did the calculation, he found the two in agreement to within 1%.

Example 27.6

Identifying Initial and Final States

One wavelength in the infrared part of the hydrogen emission spectrum has wavelength 1.28 μm. What are the initial and final states of the transition that results in this wavelength being emitted?

Strategy   The energy of the 1.28-μm photon must be the difference in two energy levels. Rather than trying to solve an equation with two unknowns (the initial and final values of n), we can use the energy level diagram to narrow down the choices first.

Solution   The energy of the photon emitted is

$$E = \frac{hc}{\lambda} = \frac{1240 \text{ eV·nm}}{1280 \text{ nm}} = 0.969 \text{ eV}$$

Looking at the energy level diagram (Fig. 27.19), the photon must be in the Paschen series. The smallest photon energy in the Balmer series is

$$(-1.51 \text{ eV}) - (-3.40 \text{ eV}) = 1.89 \text{ eV}$$

The photons in the Lyman series have even larger energies. The *largest* energy photon in the Brackett series has energy 0.85 eV. Only the Paschen series can include a photon around

1 eV. Therefore, the final state is $n = 3$. Now we solve for the initial state n.

energy of photon emitted =
initial energy level – final energy level

$$0.969 \text{ eV} = \frac{-13.6 \text{ eV}}{n^2} - (-1.51 \text{ eV})$$

$$n = \sqrt{\frac{13.6 \text{ eV}}{1.51 \text{ eV} - 0.969 \text{ eV}}} = 5$$

The 1.28-μm photon is emitted when the electron goes from $n = 5$ to $n = 3$.

Discussion   For a photon in the hydrogen spectrum, identifying the lower of the two energy levels is simplified by noting that the various series do not overlap. All of the photons in the Lyman series (lower energy level $n = 1$) have larger energies than any of the photons in the Balmer series (lower energy level $n = 2$); all of the photons in the Balmer series have larger energies than any in the Paschen series; and so on.

Practice Problem 27.6   Fifth Balmer line

The first four Balmer lines are easily visible. What is the wavelength of the fifth Balmer line?

Example 27.7

Thermal Excitation

Absorbing or emitting a photon is not the *only* way an atom can make a transition between energy levels. One of the other ways is called thermal excitation. If their kinetic energies are sufficiently large, two atoms can undergo an inelastic collision in which one of them makes a transition into an excited state, leaving the atoms with less total translational kinetic energy after the collision than before. (a) What is the average translational kinetic energy of an atom in a gas at room temperature (300 K)? (b) Explain why, in atomic hydrogen gas at room temperature, almost all of the atoms are in the ground state.

Strategy   In Section 13.6, we found the average translational kinetic energy of an ideal gas to be $\langle K \rangle = \frac{3}{2}kT$ [Eq. (13-20)]. To facilitate comparison with the energy levels in hydrogen, we convert the average kinetic energy to eV. The key is to see whether the translational kinetic energies of the hydrogen atoms are large enough that an inelastic collision can excite one of the atoms.

continued on next page

Example 27.7 *continued*

Solution and Discussion (a) At $T = 300$ K,

$$\langle K \rangle = \frac{3}{2}kT \tag{13-20}$$

$$= \frac{3}{2} \times 1.38 \times 10^{-23} \text{ J/K} \times 300 \text{ K} \times \frac{1 \text{ eV}}{1.60 \times 10^{-19} \text{ J}}$$

$$= 0.04 \text{ eV}$$

(b) Suppose two atoms, both in the ground state, collide. To excite one of them into $n = 2$ (the transition from the ground state that requires the smallest energy) requires

$$\Delta E = E_2 - E_1 = (-3.40 \text{ eV}) - (-13.6 \text{ eV}) = 10.2 \text{ eV}$$

This is 260 times the average kinetic energy. At any given instant, some atoms have more than the average and some have less; very few have much more than average. The number of atoms with kinetic energies *hundreds of times* the average kinetic energy is extremely small. (See the Maxwell-Boltzmann distribution curves in Fig. 13.13.) The tiny number of atoms excited in this way quickly decay back to the ground state by emitting a photon. Thus, at any given instant, a negligibly small fraction of the atoms are excited; for all practical purposes they are all in the ground state.

Conceptual Practice Problem 27.7 Absorption spectrum of hydrogen atoms

At room temperature, the *absorption* spectrum of atomic hydrogen has no black lines in the visible part of the spectrum. At high temperatures, the absorption spectrum of atomic hydrogen has four dark lines in the visible—at the same wavelengths as the four visible lines in the Balmer series of the emission spectrum. Explain.

Successes of the Bohr Model

The Bohr model has been replaced by the quantum mechanical version of the atom (Chapter 28). Despite serious deficiencies, the Bohr model was an important step in the development of quantum physics. Some important ideas that carry over from the Bohr atom to the quantum mechanical atom include:

- The electron can be in one of a discrete set of stationary states with quantized energy levels.
- The atom can make a transition between energy levels by emitting or absorbing a photon.
- Angular momentum is quantized.
- Stationary states can be described by quantum numbers (n is now called the *principal quantum number*).
- The *quantum jump*—the discontinuous transition between energy levels.

Bohr's model gives the correct numerical values—even if for the wrong reasons—of the energy levels in the hydrogen atom. It also correctly predicts the size of the H atom: the Bohr radius a_0 is now understood as the *most likely* distance between the electron and the nucleus when the H atom is in the ground state.

Problems with the Bohr Model

- The whole idea of the electron orbiting the nucleus—indeed, of the electron having any kind of trajectory—is incorrect. Newtonian mechanics does *not* apply to the motion of the electron. The electron must be described quantum mechanically; quantum mechanics gives the *probabilities* of the electron being located at various distances from the nucleus.
- Angular momentum is quantized, but *not* in integral multiples of $\hbar$.
- The Bohr model gives no way to calculate the probabilities of an electron absorbing or emitting a photon.

Applications of Bohr Model to Other One-Electron Atoms

The Bohr model can be applied to ions that have *a single electron* such as ionized helium, He^+, and doubly ionized lithium, Li^{2+}. Instead of having nuclear charge $+e$, these ions have a nuclear charge of $+Ze$, where Z is the atomic number (the number of protons in the nucleus). Every time e^2 appears in equations for the hydrogen atom, one factor of e came from the electron's charge and one from the charge of the nucleus. For a nucleus with charge Ze, we replace each factor of e^2 with Ze^2. Then the orbital radii are smaller by a factor of Z:

$$r_n = \frac{n^2\hbar^2}{mkZe^2} = \frac{n^2}{Z}a_0 \quad (n = 1, 2, 3, \ldots) \tag{27-25}$$

and the energy levels are larger by a factor of Z^2:

$$E_n = -\frac{mk^2Z^2e^4}{2n^2\hbar^2} = -\frac{Z^2}{n^2} \times 13.6 \text{ eV} \quad (n = 1, 2, 3, \ldots) \tag{27-26}$$

Example 27.8

Electron Speed in He⁺

Compare the speed of the electron in the ground state of He⁺ with that of the electron in the first excited state of Li²⁺.

Strategy Singly ionized helium has one electron orbiting a nucleus with an atomic number $Z = 2$. Doubly ionized lithium has one electron orbiting a nucleus with atomic number $Z = 3$. The speed of the electron is related to the orbital radius through Bohr's assumption of angular momentum quantization: $mvr_n = n\hbar$. We can find the radii in terms of the Bohr radius, then set up a ratio and solve for one speed in terms of the other.

Solution From Eq. (27-25) for the orbital radii,

$$r_n = \frac{n^2}{Z}a_0$$

For the ground state of He⁺, $n = 1$ and $Z = 2$.

$$r_{\text{He}^+} = \frac{1}{2}a_0$$

For the first excited state of Li²⁺, $n = 2$ and $Z = 3$.

$$r_{\text{Li}^{2+}} = \frac{4}{3}a_0$$

We solve the Bohr condition of quantized angular momentum for speed:

$$v = \frac{n\hbar}{mr_n}$$

The electron speeds are

$$v_{\text{He}^+} = \frac{\hbar}{m\left(\frac{1}{2}a_0\right)} = \frac{2\hbar}{ma_0}$$

and

$$v_{\text{Li}^{2+}} = \frac{2\hbar}{m\left(\frac{4}{3}a_0\right)} = \frac{3\hbar}{2ma_0}$$

Then the ratio of the electron speeds is

$$\frac{v_{\text{He}^+}}{v_{\text{Li}^{2+}}} = \frac{2}{\frac{3}{2}} = \frac{4}{3}$$

The speed of the electron in the ground state of He⁺ is $\frac{4}{3}$ times that of the electron in the first excited state of Li²⁺.

Discussion The different masses of the nuclei do not enter into the calculation; the nucleus is so much more massive than the electron that we assume the nucleus to be stationary. Thus, in all of the Bohr equations, m is the mass of the *electron*.

Practice Problem 27.8 Radius of an orbiting electron

What is the radius of the orbit for the second excited state ($n = 3$) of He⁺?

Example 27.9

He⁺ Energy Levels

Calculate the first few energy levels for singly ionized helium. Draw an energy level diagram for singly ionized helium and compare it with that for hydrogen.

Strategy Helium has an atomic number $Z = 2$. We use the ground state energy for hydrogen along with Z and the various values of n to find the energy levels.

Solution and Discussion The ground state energy for *hydrogen* is −13.6 eV. A one-electron atom in which the nucleus has charge $+Ze$ has energy levels

$$E_n = -\frac{mk^2Z^2e^4}{2n^2\hbar^2} = -\frac{Z^2}{n^2} \times 13.6 \text{ eV} \quad (n = 1, 2, 3, \ldots)$$

For He⁺, $Z = 2$:

$$E_n = -\frac{4}{n^2} \times 13.6 \text{ eV} = -\frac{1}{n^2} \times 54.4 \text{ eV} \quad (n = 1, 2, 3, \ldots)$$

The first six energy levels for He⁺ are

$$E_1 = -54.4 \text{ eV}; E_2 = -13.6 \text{ eV}; E_3 = -6.04 \text{ eV}; E_4 = -3.40 \text{ eV};$$
$$E_5 = -2.18 \text{ eV}; E_6 = -1.51 \text{ eV}$$

Now we draw an energy level diagram (not to scale) for He⁺ next to one for hydrogen (Fig. 27.20). Due to the factor of

continued on next page

Example 27.9 *continued*

Z^2, each energy level in He$^+$ is four times the energy level for the same value of n in hydrogen. The first excited state ($n = 2$) for He$^+$ has the same energy as the ground state of hydrogen; the third excited state ($n = 4$) for He$^+$ has the same energy as the first excited state ($n = 2$) of hydrogen. In general, the energy of state $2n$ in He$^+$ is the same as the energy of state n in hydrogen.

Conceptual Practice Problem 27.9   Ionization energy

The *ionization energy* is the energy that must be supplied to an atom in its ground state to separate the electron from the nucleus. (a) What is the ionization energy for H? (b) What is the ionization energy for He$^+$? (c) Give a qualitative explanation for why He$^+$ has a larger ionization energy than H.

Figure 27.20
Energy levels for H and He$^+$

Fluorescence, Phosphorescence, and Chemiluminescence

Suppose atomic hydrogen gas is illuminated with ultraviolet radiation of wavelength 103 nm. Some of the atoms absorb a photon and are excited into the $n = 3$ level. When one of the excited atoms decays back to the ground state, it does not necessarily emit a 103-nm photon. It can decay first to $n = 2$ (by emitting a 656-nm photon) and then to $n = 1$ (by emitting a 122-nm photon). *The presence of intermediate energy levels enables the atom to absorb a photon of one wavelength and emit photons of longer wavelengths.*

Fluorescent materials absorb ultraviolet radiation and decay in a series of steps; at least one of the steps involves the emission of a photon of visible light. In a molecule or solid, not all of the transitions involve the emission of a photon. Some of the transitions increase the vibrational or rotational energy of the molecule of the solid; this energy is ultimately dissipated into the surroundings.

A fluorescent lamp is a mercury discharge tube whose interior is coated with a mixture of fluorescent materials called phosphors. The mercury spectrum does include visible light, although much of it is in the ultraviolet. The phosphors absorb this ultraviolet radiation and in turn emit visible light. A "black light"—a source of ultraviolet radiation—makes fluorescent dyes glow brightly in the dark.

"Day-glo" or "neon" colored paints are fluorescent. Fluorescent dyes are added to laundry detergents; the dyes absorb UV and emit blue light to "make whites whiter" (Fig. 27.21). The emission of blue light compensates for the yellowing of a fabric as it ages.

Phosphorescence is similar to fluorescence but involves a time delay. Most excited states of atoms and molecules decay quickly (typically within a few nanoseconds), but certain *metastable* excited states last for several seconds or even longer before a transition occurs. Watch dials, wall switch plates, and toys that glow in the dark absorb photons when illuminated and get stuck in a metastable state so that the emission of light occurs much later.

In Rutherford's scattering experiment, alpha particles were detected by a phosphor screen. The phosphors were excited by a collision with an alpha particle rather than by absorbing a photon. The phosphor dots on a TV screen are excited by a beam of high-speed electrons; the decay back to the ground state involves emitting a visible photon. The screen uses three different phosphors to produce blue, green, and red. The phosphorescence must persist long enough so the images from successive scans of the electron gun blend together, but not *too* much longer.

Making The Connection:
fluorescent lamps

Making The Connection:
fluorescent dyes in
laundry detergent

(a)   (b)

Figure 27.21   A freshly laundered blouse and the soap powder used viewed in (a) natural light and (b) ultraviolet light.

Physics at Home

Watch TV for awhile in a dark room. Turn off the TV but continue to look at it. What do you see? Does the screen go dark immediately?

The blue glow from the floor of the motel room described at the beginning of this chapter is caused by *chemiluminescence.* The colorless liquid solution contains luminol (3-aminophthalhydrazide) and hydrogen peroxide. Traces of hemoglobin (which is found in blood) catalyze an oxidation reaction between the luminol and the hydrogen peroxide. The reaction leaves one of the products in an excited state, which then decays to the ground state by emitting a photon. The luminol test is effective even on clothing or surfaces that have been carefully washed. Thus, the blue glow reveals the location of possible bloodstains. Fireflies light up due to a similar chemical reaction. The reaction is controlled by enzymes (biological catalysts), allowing the firefly to turn the light on and off.

27.8   PAIR ANNIHILATION AND PAIR PRODUCTION

The Positron

In 1929, P. A. M. Dirac (1902–1984) made a theoretical prediction of the existence of a particle with the same mass as the electron but opposite charge ($q = +e$). Experiments later verified the existence of this particle, now called the *positron.* Some radioactive elements emit a positron spontaneously as they decay.

The positron was the first *antiparticle* discovered. Each of the particles that make up ordinary matter (electron, proton, and neutron) has an antiparticle (positron, antiproton, and antineutron). Cosmologists struggle with the question of why there is more matter than antimatter in the universe. We introduce the positron here because two processes, the production of and the annihilation of an electron-positron pair, provide some of the clearest and most direct evidence for the photon model of EM radiation.

Pair Production

An energetic photon can *create* a positron and an electron where no such particles existed before. The photon is totally absorbed in the process. Energy must be conserved in any process, so in order for **pair production** to occur,

$$E_{photon} = E_{electron} + E_{positron}$$

The total energy of a particle with mass is the sum of its kinetic energy and its rest energy (the energy of the particle when at rest). A particle of mass m has *rest energy*

$$E_0 = mc^2 \qquad (26\text{-}7)$$

(see Section 26.7). Thus, a photon must have an energy of at least $2m_ec^2$ in order to create an electron-positron pair. If the photon's energy is greater than $2m_ec^2$, the excess energy appears as kinetic energy of the electron and positron. A photon is massless and thus, has no rest energy; the total energy of a photon is $E = hf = hc/\lambda$.

Momentum must also be conserved. For the photon, $p = E/c$. For an electron or a positron,

$$p = \frac{1}{c}\sqrt{E^2 - (m_ec^2)^2} < \frac{E}{c}$$

An electron or positron with total energy E has a momentum less than E/c—that is, less than the momentum of a photon with the same energy. Even if the electron and positron move in the same direction, their total momentum cannot be as large as the momentum of the photon. Therefore, it is impossible for both the pair's total momentum and total energy to be equal to the photon's momentum and total energy. Another particle must take part in the reaction: pair production can only occur when the photon passes near a massive particle such as an atomic nucleus (Fig. 27.22). The recoil of the massive particle satisfies momentum conservation without carrying off a significant amount of energy, so our assumption that all of the energy of the photon goes into the electron-positron pair is a good approximation.

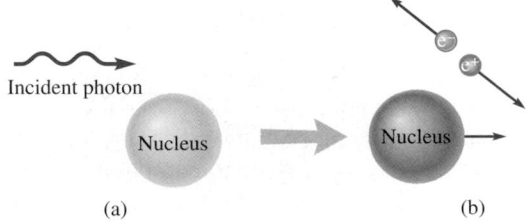

Incident photon

Nucleus

Nucleus

(a)   (b)

Figure 27.22   Pair production. (a) A photon passes near an atomic nucleus. (b) The photon vanishes by creating an electron-positron pair. The nucleus recoils with an insignificant kinetic energy but, due to its large mass, with a significant momentum.

Pair Annihilation

Since ordinary matter contains plenty of electrons, sooner or later a positron gets near an electron. For a short while, the pair forms something like an atom; then—poof!—both particles disappear by creating *two* photons (Fig. 27.23). Pair annihilation cannot create just one photon; two photons are required to conserve both energy and momentum. The total energy of the two photons must be equal to the total energy of the electron-positron pair. Ordinarily the kinetic energies of the electron and positron are negligible compared to their rest energies, so for simplicity we assume they are at

(a)   (b)

Figure 27.23   Pair annihilation. (a) An electron and positron vanish, creating (b) a pair of photons.

rest; then their total energy is just their rest energy, $2m_ec^2$, and their total momentum is zero. Annihilation of the pair then produces two photons, each with energy $E = hf = m_ec^2 = 511$ keV, traveling in opposite directions. Annihilation is the ultimate fate of positrons; the characteristic 511-keV photons are the sign that pair annihilation has taken place.

Besides confirming the photon model of EM radiation, pair annihilation and pair production clearly illustrate Einstein's ideas about mass and rest energy.

Example 27.10

Threshold Wavelength for Pair Production

Find the threshold wavelength for a photon to produce an electron-positron pair.

Strategy The photon must have at least enough energy to create the electron and positron, each of which has a rest energy of $m_ec^2 = 511$ keV. From the minimum photon energy, we find the threshold wavelength—the *maximum* wavelength, since larger wavelengths correspond to smaller photon energies.

Solution The minimum photon energy to create an electron-positron pair is

$$E = 2m_ec^2 = 1.022 \text{ MeV}$$

Now we find the wavelength of a photon with this energy.

$$E = hf = \frac{hc}{\lambda}$$

Solving for the wavelength,

$$\lambda = \frac{hc}{E} = \frac{1240 \text{ eV·nm}}{1.022 \times 10^6 \text{ eV}} = 0.00121 \text{ nm} = 1.21 \text{ pm}$$

Discussion Quick check: a visible photon has a wavelength of about 500 nm and an energy of about 2 eV. Here, the photon energy is half a million times that of a visible photon, so the wavelength is about 500 nm/500,000 = 0.001 nm = 1 pm.

Practice Problem 27.10   Muon-antimuon pair production

What is the longest wavelength that a photon can have if it is to supply enough energy to create a muon and an antimuon? The masses of the muon and the antimuon are 106 MeV/c^2 (their rest energies are 106 MeV). The pair production occurs as a result of the interaction between a photon and a nucleus.

Positron Emission Tomography

Positron emission tomography (PET) is a medical imaging technique based on pair annihilation that is used to diagnose diseases of the brain and heart as well as certain types of cancer. A tracer is first injected into the body. The tracer is a compound—commonly glucose, water, or ammonia—that incorporates a radioactive atom. When the radioactive atom emits a positron in the body, the positron annihilates with an electron, producing two 511-keV gamma rays traveling in opposite directions. The two photons are detected by a ring of detectors around the body (Fig. 27.24a); then the atom that emitted the positron lies along the line between the two detectors. A computer analyzes the directions of many gamma rays and locates the regions of highest concentration of the tracer. Then the computer constructs an image of that slice of the body (Fig. 27.24b).

While other imaging techniques such as x-rays, CAT scans, and MRIs show the structure of body tissues, PET scans show the biochemical activity of an organ or tissue. For example, a PET scan of the heart can differentiate normal heart tissue from nonfunctioning heart tissue, which helps the cardiologist determine whether the patient can benefit from bypass surgery or from angioplasty.

Because rapidly growing cancer cells gobble up a glucose tracer faster than healthy cells, PET scans can accurately distinguish malignant tumors from benign tumors. They help oncologists to determine the best treatment for a patient with cancer as well as to monitor the efficacy of a course of treatment. A brain tumor can be precisely located without cutting into the patient's skull for a biopsy. PET is used to evaluate diseases of the brain such as Alzheimer's, Huntington's, and Parkinson's diseases; epilepsy; and stroke.

Figure 27.24 (a) A PET scan detects the gamma rays emitted when a positron and an electron annihilate within the body. (b) A PET scan of the brain. Color is used to distinguish regions with differing levels of positron emission.

MASTER THE CONCEPTS

Summary

- A quantity is quantized when its possible values are limited to a discrete set.

- Max Planck found an equation to match experimental results for blackbody radiation. The equation led him to postulate that the energy of an oscillator must be quantized in integral multiples of hf, where f is the frequency of the oscillator. Planck's constant is now recognized as one of the fundamental constants in physics:

$$h = 6.626 \times 10^{-34} \text{ J·s} \qquad (27\text{-}3)$$

- In the photoelectric effect, EM radiation incident on a metal surface causes electrons to be ejected from the metal. To explain the photoelectric effect, Einstein said that *EM radiation itself* is quantized. The quantum of EM radiation—that is, the smallest indivisible unit—is now called the photon. The energy of a photon with frequency f is

$$E = hf \qquad (27\text{-}4)$$

The maximum kinetic energy of an electron is the difference between the photon energy and the work function ϕ,

which is the amount of energy that must be supplied to break the bond between an electron and the metal.

$$K_{max} = hf - \phi \qquad (27\text{-}7)$$

- One electron-volt is equal to the kinetic energy that a particle with charge $\pm e$ (such as an electron or a proton) gains when it is accelerated through a potential difference of magnitude 1 V.

$$1 \text{ eV} = 1.60 \times 10^{-19} \text{ J} \qquad (27\text{-}5)$$

- In an x-ray tube, electrons are accelerated to kinetic energy K and then strike a target. The maximum frequency of the x-ray radiation emitted occurs when all of the electron's kinetic energy is carried away by *a single photon*:

$$hf_{max} = K \qquad (27\text{-}9)$$

- In Compton scattering, x-rays scattered from a target have longer wavelengths than the incident x-rays; the wavelength shift depends on the scattering angle θ:

$$\lambda' - \lambda = \frac{h}{m_e c}(1 - \cos\theta) \qquad (27\text{-}14)$$

MASTER THE CONCEPTS *continued*

Compton scattering can be viewed as a collision between a photon and a free electron at rest. The momentum and kinetic energy of the incident photon must equal the total momentum and kinetic energy of the scattered photon and recoiling electron.

- Emission and absorption by individual atoms form line spectra. Each element has its own characteristic spectrum. Spectroscopy provided clues to the structure of the atom.

- Assumptions of the Bohr model of the hydrogen atom:
  1. The electron can exist without radiating energy only in certain circular orbits.
  2. The laws of Newtonian mechanics apply to the motion of the electron in any of the stationary states.
  3. The electron can make a transition between stationary states through the emission or absorption of a single photon.
  4. The stationary states are those circular orbits in which the electron's angular momentum is quantized in integral multiples of $h/(2\pi)$.

- Fluorescent materials absorb ultraviolet radiation and decay in a series of steps; one or more of the steps involve the emission of a photon of visible light.

- In pair production, an energetic photon passing by a massive particle creates an electron-positron pair. In pair annihilation,

an electron-positron pair is annihilated and two photons are created.

Highlighted Figures and Tables

F27.2 Standing waves on a string can only exist at one of a discrete set of frequencies (p. 980)

F27.3 Three blackbody curves showing the relative intensity of blackbody radiation as a function of frequency for three different temperatures (p. 981)

F27.4 A photoelectric effect apparatus (p. 982)

F27.5 A graph of K_{max} versus f for the photoelectric effect (p. 982)

F27.9 X-ray production (p. 987)

F27.10 Compton scattering (p. 988)

F27.12 Emission spectra and a solar absorption spectrum (p. 990)

F27.14 The emission spectrum of atomic hydrogen (p. 991)

F27.16 Rutherford scattering experiment (p. 992)

F27.19 Energy level diagram for the hydrogen atom (p. 995)

F27.24 A PET scan (p. 1003)

CONCEPTUAL QUESTIONS

1. An incandescent lightbulb is connected to a dimmer switch. When the bulb operates at full power, it appears white, but as it is dimmed it looks more and more red. Explain.

2. Describe the photoelectric effect and four aspects of the experimental results that were puzzling to nineteenth-century physicists. How does the photon model of light explain the experimental results in each case?

3. Use the photon model to explain why ultraviolet radiation is harmful to your skin while visible light is not.

4. An experiment shines visible light on a target and measures the wavelengths of light scattered at different angles. Would the experiment show that the scattered photons are Compton shifted? Explain.

5. Some stars are reddish in color, others bluish, and others yellowish-white (like the Sun). How is the color related to the surface temperature of the star? What color are the hottest stars? What color are the coolest?

6. How does the observation of the sharp lines seen in the hydrogen emission spectrum verify the notion that all electrons have the same charge?

7. In the photoelectric effect, what is the relationship between the maximum kinetic energy of ejected electrons and the frequency of the light incident upon the surface?

8. In the photoelectric effect, why are no electrons emitted from the metal when the incident light is below the threshold frequency?

9. Describe the process by which a continuous spectrum of x-rays is produced. Does the spectrum have a maximum wavelength or a minimum wavelength? Explain.

10. A darkroom used for developing black-and-white film can be dimly lit by a red lightbulb without ruining the film. Why is a red lightbulb used rather than white or blue or some other color?

11. List the assumptions of the Bohr theory of the hydrogen atom.

12. If green light causes the ejection of electrons from a metal in a photoelectric effect experiment and yellow light does not, what would you expect to happen if red light were used to illuminate the same metal? Do you expect more intense yellow light to eject electrons? What about very faint violet light?

13. In both Compton scattering and the photoelectric effect, an electron gains energy from an incident photon. What is the essential difference between the two processes?

14. Why does a photon that has been scattered from an electron, initially at rest, have a longer wavelength than the incident photon? Why is the wavelength change more noticeable for an incident x-ray photon than for a photon of visible light?

15. What process becomes especially important for photons with energies in excess of 1.02 MeV?

16. Explain how Rutherford's experiment, in which alpha particles are incident on a thin gold foil, refutes the plum pudding model of the atom.

17. In a photoelectric effect experiment, how is the stopping potential determined? What does the stopping potential tell us about the electrons emitted from the metal surface?

18. A fluorescent substance absorbs EM radiation of one wavelength and then emits EM radiation of a different wavelength. Which wavelength is longer? Explain.

19. Explain why every line in the absorption spectrum of hydrogen is present in the emission spectrum, but not every line in the emission spectrum is present in the absorption spectrum. [*Hint:* The excited states are very short-lived.]

20. A solar cell is used to generate electricity when sunlight falls on it. How would you expect the current produced by a solar cell to depend on the intensity of the incident light? How would you expect the current to depend on the wavelength of the incident light?

21. The photoresponse of the retina of the human eye at low light levels depends on individual photosensitive molecules in rod cells being excited by the incident light. When excited, these molecules change shape, leading to other changes in the cell that trigger a nerve impulse to the brain. How does the photon model of light do a better job than the wave model in explaining how these changes can happen even at low light levels?

22. Explain why the annihilation of an electron and a positron creates a *pair* of photons rather than a single photon.

23. In a photoelectric effect experiment, two different metals (1 and 2) are subjected to EM radiation. Metal 1 produces photoelectrons for both red and blue light; metal 2 produces photoelectrons for blue light but not for red. Which metal produces photoelectrons for ultraviolet radiation? Which *might* produce photoelectrons for infrared radiation? Which has the larger work function?

MULTIPLE CHOICE QUESTIONS

1. How many emission lines are possible for atomic hydrogen gas with atoms excited to the $n = 4$ state?
   (a) 1 (b) 2 (c) 4 (d) 5 (e) 6

2. An electron, passing close to a target nucleus, decelerates and radiates away some of its energy. What is this process called?
   (a) Compton effect (b) photoelectric effect
   (c) bremsstrahlung (d) blackbody radiation
   (e) stimulated emission

3. In the Compton effect a photon of wavelength λ and frequency f is scattered from an electron, initially at rest. In this process,
   (a) the electron gains energy from the photon so that the scattered photon's wavelength is less than λ.
   (b) the electron gives energy to the scattered photon so that the photon's frequency is greater than f.
   (c) momentum is not conserved, but energy is conserved.
   (d) the photon loses energy so that the scattered photon has a frequency less than f.

4. The number of electrons per second ejected from a metal in the photoelectric effect
   (a) is proportional to the intensity of the incident light.
   (b) is proportional to the frequency of the incident light.
   (c) is proportional to the wavelength of the incident light.
   (d) is proportional to the threshold frequency of the metal.

5. Two lasers emit equal numbers of photons per second. If the first laser emits blue light and the second emits red light, the power radiated by the first is
   (a) greater than that emitted by the second.
   (b) less than that emitted by the second.
   (c) equal to that emitted by the second.
   (d) impossible to determine without knowing the time interval during which emission occurs.

6. If a photoelectric material has a work function ϕ, the threshold wavelength for the material is given by
   (a) $\frac{\phi}{hc}$ (b) hf (c) $\frac{hc}{\phi}$ (d) $\frac{\phi}{e}$ (e) $\frac{\phi}{hf}$

7. In analyzing data from a spectroscopic experiment, the inverse of each experimentally determined wavelength of the Balmer series is plotted versus $1/(n_i^2)$, where n_i is the initial energy level from which a transition to the $n = 2$ level takes place. The slope of the line is
   (a) the shortest wavelength of the Balmer series.
   (b) $-h$, where h is Planck's constant.
   (c) one divided by the longest wavelength in the Balmer series.
   (d) $-hc$, where h is Planck's constant.
   (e) $-R$, where R is the Rydberg constant.

8. Electrons are accelerated through a potential difference V and then strike a dense target. In the x-rays that are produced,
   (a) there is a maximum wavelength.
   (b) there is a minimum wavelength.
   (c) there is a single wavelength.
   (d) there is neither a maximum nor a minimum wavelength.
   (e) there is both a maximum and a minimum wavelength.

9. In a photoelectric effect experiment, light of a single wavelength is incident on the metal surface. As the intensity of the incident light is increased,
   (a) the stopping potential increases.
   (b) the stopping potential decreases.
   (c) the work function increases.
   (d) the work function decreases.
   (e) none of the above.

10. In a photoelectric effect experiment, the stopping potential is determined by
    (a) the work function of the metal.
    (b) the wavelength of the incident light.
    (c) the intensity of the incident light.
    (d) all three (a), (b), and (c).
    (e) (b) and (c).
    (f) (a) and (b).
    (g) (a) and (c).

PROBLEMS

Note: $\mathbf{C}$ indicates a combination conceptual/quantitative problem. Gold diamonds ✦, ✦✦ are used to indicate the increasing level of difficulty of each problem. Problem numbers appearing in blue, 9., denote problems that have a detailed solution available in the Student Solutions Manual. Some problems are *paired* by concept; their numbers are connected by a ruled box.

27.3 The Photoelectric Effect

1. What is the energy of a photon of light of wavelength 0.70 μm?

2. Find the (a) wavelength and (b) frequency of a 3.1-eV photon.

3. The photoelectric threshold frequency of silver is 1.04×10^{15} Hz. What is the minimum energy required to remove an electron from silver?

4. A rubidium surface has a work function of 2.16 eV. (a) What is the maximum kinetic energy of ejected electrons if the incident radiation is of wavelength 413 nm? (b) What is the threshold wavelength for this surface?

5. A clean iron surface is illuminated by ultraviolet light. No photoelectrons are ejected until the wavelength of the incident UV light falls below 288 nm. (a) What is the work function (in eV) of the metal? (b) What is the maximum kinetic energy for electrons ejected by incident light of wavelength 140 nm?

6. The minimum energy required to remove an electron from a metal is 2.60 eV. What is the longest wavelength photon that can eject an electron from this metal?

7. Ultraviolet light of wavelength 220 nm illuminates a tungsten surface and electrons are ejected. A stopping potential of 1.1 V is able to just prevent any of the ejected electrons from reaching the opposite electrode. What is the work function for tungsten?

$\mathbf{C}$ 8. Two different monochromatic light sources, one yellow (580 nm) and one violet (425 nm), are used in a photoelectric effect experiment. The metal surface has a photoelectric threshold frequency of 6.20×10^{14} Hz. (a) Are both sources able to eject photoelectrons from the metal? Explain. (b) How much energy is required to eject an electron from the metal? (Use $h = 4.136 \times 10^{-15}$ eV·s.)

9. Calculate the value of (a) Planck's constant and (b) the work function of the metal from the data of Fig. 27.25 obtained by Robert A. Millikan in 1916. Millikan was attempting to disprove the Einstein photoelectric equation; instead he found that his data supported Einstein's prediction.

Frequency (Hz × 10^{13})

43.9×10^{13} Hz

Figure 27.25
Problem 9

27.4 X-Ray Production

10. If the shortest wavelength produced by an x-ray tube is 0.46 nm, what is the voltage applied to the tube?

11. What is the minimum potential difference applied to an x-ray tube if x-rays of wavelength 0.250 nm are produced?

12. What is the cutoff frequency for an x-ray tube operating at 46 kV?

13. The potential difference in an x-ray tube is 40.0 kV. What is the minimum wavelength of the continuous x-ray spectrum emitted from the tube?

14. In a color TV tube, electrons are accelerated through a potential difference of 20.0 kV. Some of the electrons strike the metal mask (instead of the phosphor dots behind holes in the mask), causing x-rays to be emitted. What is the smallest wavelength of the x-rays emitted?

15. Show that the cutoff frequency for an x-ray tube is proportional to the potential difference through which the electrons are accelerated.

27.5 Compton Scattering

16. X-rays of wavelength 10.0 pm are incident on a target. Find the wavelengths of the x-rays scattered at (a) 45.0° and (b) 90.0°.

17. An x-ray photon of wavelength 0.150 nm collides with an electron initially at rest. The scattered photon moves off at an angle of 80.0° from the direction of the incident photon. Find (a) the Compton shift in wavelength and (b) the wavelength of the scattered photon.

18. An incident beam of photons is scattered through 100.0°; the wavelength of the scattered photons is 124.65 pm. What is the wavelength of the incident photons?

19. An x-ray photon of initial frequency 3.0×10^{19} Hz collides with a free electron at rest; the scattered photon moves off at 90°. What is the frequency of the scattered photon?

20. A photon of energy 240.0 keV is scattered by a free electron. If the recoil electron has a kinetic energy of 190.0 keV, what is the wavelength of the scattered photon?

✦ 21. An incident photon of wavelength 0.0100 nm is Compton scattered; the scattered photon has a wavelength of 0.0124 nm. What is the change in kinetic energy of the electron that scattered the photon?

✦ 22. A Compton scattering experiment is performed using an aluminum target. The incident photons have wavelength λ. The scattered photons have wavelengths λ' and energies E that depend on the scattering angle θ. (a) At what angle θ are scattered photons with the smallest energy detected? (b) At this same scattering angle θ, what is the ratio λ'/λ for $\lambda = 10.0$ pm?

27.6 Spectroscopy and Early Models of the Atom; 27.7 The Bohr Model of the Hydrogen Atom; Atomic Energy Levels

23. What is the orbital radius of the electron in the $n = 3$ state of hydrogen?

24. Find the energy for a hydrogen atom in the stationary state $n = 4$.

25. How much energy must be supplied to a hydrogen atom to cause a transition from the ground state to the $n = 4$ state?

26. A hydrogen atom in its ground state absorbs a photon of energy 12.1 eV. To what energy level is the atom excited?

27. The Bohr theory of the hydrogen atom neglects gravitational forces between the electron and the proton. Make a calculation to justify this omission. [*Hint:* Find the ratio of the gravitational and electrostatic forces acting on the electron due to the proton.]

28. Use the Bohr theory to find the energy necessary to remove the electron from a hydrogen atom initially in its ground state.

29. How much energy is required to ionize a hydrogen atom initially in the $n = 2$ state?

30. What is the smallest energy photon that can be absorbed by a hydrogen atom in its ground state?

31. Find the wavelength of the radiation emitted when a hydrogen atom makes a transition from the $n = 6$ to the $n = 3$ state.

✦ 32. An electron orbits a proton at constant speed in a circle of radius r. (a) Using Coulomb's law, write an expression for the magnitude of the electric force on the electron in terms of r, the elementary charge e, and the Coulomb constant k. (b) Apply Newton's second law to the electron and use it to show that the electron's speed is

$$v = \sqrt{\frac{ke^2}{m_e r}}$$

[*Hint:* The electron is in uniform circular motion.] (c) Use the Bohr assumption about the electron's angular momentum, Eq. (27-17), to show that the radius of the nth Bohr orbit is

$$r_n = \frac{n^2 \hbar^2}{mke^2} \qquad (27\text{-}19)$$

✦ 33. An electron orbits a proton at constant speed in a circle of radius r. (a) What is the electron's kinetic energy in terms of k, e, and r? Use the expression for the electron's speed found in Problem 32. (b) What is the electron's electric potential energy in terms of k, e, and r? (Assume $U = 0$ when $r = \infty$.) (c) Show that the electron's mechanical energy ($K + U$) is $E = -ke^2/(2r)$. (d) Use Eq. (27-19) to show that the energy of the nth Bohr orbit is

$$E_n = -\frac{mk^2 e^4}{2n^2 \hbar^2} \qquad (27\text{-}22)$$

34. By directly substituting the values of the fundamental constants, show that the Bohr radius $a_0 = \hbar^2/(mke^2)$ has the numerical value 5.29×10^{-11} m.

35. By directly substituting the values of the fundamental constants, show that the ground state energy for hydrogen in the Bohr model $E_1 = -mk^2 e^4/(2\hbar^2)$ has numerical value -13.6 eV.

36. Calculate, according to the Bohr model, the speed of the electron in the ground state of the hydrogen atom.

37. Find the speed of the electron in the ground state of He⁺.

38. One line in the helium spectrum is bright yellow and has the wavelength 587.6 nm. What is the difference in energy (in eV) between two helium levels that produce this line?

✦ 39. A particle collides with a hydrogen atom in the $n = 2$ state, transferring 15.0 eV of energy to the atom. As a result, the electron breaks away from the hydrogen nucleus. What is the kinetic energy of the electron when it is far from the nucleus?

40. A hydrogen atom has an electron in the $n = 5$ level. (a) If the electron returns to the ground state by emitting radiation, what is the *minimum* number of photons that can be emitted? (b) What is the *maximum* number that might be emitted?

41. The *Paschen series* in the hydrogen emission spectrum is formed by electron transitions from $n_i > 3$ to $n_f = 3$. (a) What is the longest wavelength in the Paschen series? (b) What is the wavelength of the series limit (the lower bound of the wavelengths in the series)? (c) In what part or parts of the EM spectrum is the Paschen series found (IR, visible, UV, etc.)?

42. A fluorescent solid absorbs a photon of ultraviolet light of wavelength 320 nm. If the solid dissipates 0.500 eV of the energy and emits the rest in a single photon, what is the wavelength of the emitted light?

27.8 Pair Annihilation and Pair Production

43. What is the maximum wavelength of a photon that can create an electron-positron pair?

44. A detector shows evidence of the creation of an electron-positron pair. If the tracks of the particles indicate that each one has an energy of 0.22 MeV, what is the energy of the photon that created the two particles?

45. A positron and an electron that were at rest suddenly vanish and two photons of identical frequency appear. What is the wavelength of each of these photons?

© 46. In gamma-ray astronomy, the existence of positrons (e⁺) can be inferred by a characteristic gamma ray that is emitted when a positron and an electron (e⁻) annihilate. For simplicity, assume that the electron and positron are at rest with respect to an Earth observer when they annihilate and that nothing else is in the vicinity. (a) Consider the reactions e⁻ + e⁺ → γ, where the annihilation of the two particles at rest produces one photon, and e⁻ + e⁺ → 2γ, where the annihilation produces two photons. Explain why the first reaction does not occur, while the second does. (b) Suppose the reaction e⁻ + e⁺ → 2γ occurs and one of the photons travels toward Earth. What is the energy of the photon?

COMPREHENSIVE PROBLEMS

47. A surgeon is attempting to correct a detached retina by using a pulsed laser. (a) If the pulses last for 20.0 ms and if the output power of the laser is 0.500 W, how much energy is in each pulse? (b) If the wavelength of the laser light is 643 nm, how many photons are present in each pulse?

48. These data are obtained for photoelectric stopping potentials using light of four different wavelengths. (a) Plot a graph of the stopping potential versus the reciprocal of the wavelength. (b) Read the values of the work function and threshold wavelength for the metal used directly from the graph. (c) What is the slope of the graph? Compare the slope to the expected value (calculated from fundamental constants).

Color	Wavelength (nm)	Stopping Potential (V)
Yellow	578	0.40
Green	546	0.60
Blue	436	1.10
Ultraviolet	366	1.60

49. An FM radio station broadcasts at a frequency of 89.3 MHz. The power radiated from the antenna is 50.0 kW. (a) What is the energy in eV of each photon radiated by the antenna? (b) How many photons per second does the antenna emit?

50. An owl has good night vision because its eyes can detect a light intensity as small as 5.0×10^{-13} W/m^2. What is the minimum number of photons per second that an owl eye can detect if its pupil has a diameter of 8.5 mm and the light has a wavelength of 510 nm?

51. What is the shortest wavelength x-ray produced by a 0.20-MV x-ray machine?

52. How much energy is required to remove an electron from a hydrogen atom in the $n = 4$ state?

✦53. During a Compton scattering experiment, an electron that was initially at rest recoils in the direction of motion of the incident x-ray photon. If the recoil electron has a kinetic energy of 0.20 keV, what is the wavelength of the incident x-ray? What is the wavelength of the scattered x-ray?

✦54. Compare the orbital radii of the He$^+$ and H atoms for levels of *equal energy* (not the same value of n). Can you draw a general conclusion from your results?

55. According to the Bohr model, the speed of the electron in the ground state of singly ionized helium (He$^+$, with $Z = 2$) is 4.4×10^6 m/s. Use this information to find the speed of an electron in the first excited state of triply ionized beryllium (Be^{3+} with $Z = 4$).

56. The output power of a laser pointer is about 1 milliwatt. (a) What are the energy and momentum of one laser photon if the laser wavelength is 670 nm? (b) How many photons per second are emitted by the laser? (c) What is the average force on the laser due to the momentum carried away by these photons?

57. In a photoelectric experiment using sodium, when incident light of wavelength 570 nm and intensity 1.0 W/m^2 is used, the measured stopping potential is 0.28 V. (a) What would the stopping potential be for incident light of wavelength 400.0 nm and intensity 1.0 W/m^2? (b) What would the stopping potential be for incident light of wavelength 570 nm and intensity 2.0 W/m^2? (c) What is the work function of sodium?

C✦58. A hydrogen atom in its ground state is immersed in a continuous spectrum of ultraviolet light with wavelengths ranging from 96 nm to 110 nm. After absorbing a photon, the atom emits one or more photons to return to the ground state. (a) What wavelength(s) can be absorbed by the H atom? (b) For each of the possibilities in (a), if the atom is at rest before absorbing the UV photon, what is its recoil speed after absorption (but before emitting any photons)? (c) For each of the possibilities in (a), how many different ways are there for the atom to return to the ground state?

C59. A 100-W lightbulb radiates *visible* light at a rate of about 10 W; the rest of the EM radiation is mostly infrared. Assume that the lightbulb radiates uniformly in all directions. Under ideal conditions, the eye can see the lightbulb if at least 20 visible photons per second enter a dark-adapted eye with a pupil diameter of 7 mm. (a) Estimate how far from the source the lightbulb can be seen under these rather extreme conditions. Assume an average wavelength of 600 nm. (b) Why do we not normally see lightbulbs at anywhere near this distance?

C60. A thin aluminum target is illuminated with photons of wavelength λ. A detector is placed at 90.0° to the direction of the incident photons. The scattered photons detected are found to have half the energy of the incident photons. (a) Find λ. (b) What is the wavelength of backscattered photons (detector at 180°)? (c) What (if anything) would change if a copper target were used instead of an aluminum one?

61. What potential difference must be applied to an x-ray tube to produce x-rays with a minimum wavelength of 45.0 pm?

62. Nuclei in a radium-226 radioactive source emit photons whose energy is 186 keV. These photons are scattered by the electrons in a metal target; a detector measures the energy of the scattered photons as a function of the angle θ through which they are scattered. Find the energy of the scattered γ-rays scattered through $\theta = 90.0°$ and 180.0°.

63. What is the ground state energy, according to Bohr theory, for (a) He$^+$, (b) Li^{2+}, (c) deuterium (an isotope of hydrogen whose nucleus contains a neutron as well as a proton)?

C64. Follow the steps outlined in this problem to estimate the time lag (predicted classically but *not* observed experimentally) in the photoelectric effect. Let the intensity of the incident radiation be 0.01 W/m^2. (a) If the area of the atom is $(0.1 \text{ nm})^2$, find the energy per second falling on the atom. (b) If the work function is 2.0 eV, how long would it take (classically) for enough energy to fall on this area to liberate one photoelectron? (c) Explain briefly, using the photon model, why this time lag is not observed.

C✦65. A hydrogen atom in its ground state absorbs a 97-nm ultraviolet photon. It then emits one or more photons to return to the ground state. (a) If the atom is at rest before absorbing the UV photon, what is its recoil speed after absorption? (b) There are several different possible ways for the atom to return to the ground state. How many? (c) For each of the ⓒ possibilities in (b), give the wavelength of each photon emitted, and classify it as visible, UV, IR, x-ray, etc.

ⓒ 66. The photoelectric effect is studied using a tungsten target. The work function of tungsten is 4.5 eV. The incident photons have energy 4.8 eV. (a) What is the threshold frequency? (b) What is the stopping potential? (c) Explain why, in classical physics, no threshold frequency is expected.

67. An x-ray photon with wavelength 6.00 pm collides with a free electron initially at rest. What is the maximum possible kinetic energy acquired by the electron?

68. A photoelectric effect experiment is performed with tungsten. The work function for tungsten is 4.5 eV. (a) If ultraviolet light of wavelength 0.20 μm is incident on the tungsten, calculate the stopping potential. (b) If the stopping potential is turned off (i.e., the cathode and anode are at the same voltage), the 0.20-μm incident light produces a photocurrent of 3.7 μA. What is the photocurrent if the incident light has wavelength 400 nm and the same intensity as before?

69. In a television tube, electrons of kinetic energy 2.0 keV strike the screen. No EM radiation is emitted below a certain wavelength. Calculate this wavelength.

70. The *Lyman series* in the hydrogen emission spectrum is formed by electron transitions from an excited state to the ground state. Calculate the longest three wavelengths in the Lyman series.

✦ 71. Photons of energy $E = 4.000$ keV undergo Compton scattering. What is the largest possible change in photon energy, measured as a fraction of the incident photon's energy $(E - E')/E$?

72. Consider the emission spectrum of singly ionized helium (He^+). Find the longest three wavelengths for the series in which the electron makes a transition from a higher excited state to the first excited state (*not* the ground state).

ⓒ ✦ 73. Suppose that you have a glass tube filled with atomic hydrogen gas (H, not H_2). Assume that the atoms start out in their ground states. You illuminate the gas with monochromatic light of various wavelengths, ranging through the entire IR, visible, and UV parts of the spectrum. At some wavelengths, visible light is emitted from the H atoms. (a) If there are two and only two visible wavelengths in the emitted light, what is the wavelength of the incident radiation? (b) What is the largest wavelength of incident radiation that causes the H atoms to emit visible light? What wavelength(s) is/are emitted for incident radiation at that wavelength? (c) For what wavelengths of incident light are hydrogen ions (H^+) formed?

✦ 74. Figure 27.26 shows the lowest six energy levels of the outer electron in sodium. In the ground state, the electron is in the "3s" level. (a) What is the ionization energy of sodium? (b) What is the wavelength of the radiation emitted in the transition from the 3d to the 3p level? (c) What is the transition that gives rise to the characteristic yellow light of sodium at 589 nm?

Energy (eV)
——————— 0

5s ——————— −1.1
4p ——————— −1.4
3d ——————— −1.6
4s ——————— −1.9

3p ——————— −3.0

3s ——————— −5.1

Figure 27.26 Problem 74

ANSWERS TO PRACTICE PROBLEMS

27.1 4.2×10^{-19} J

27.2 8.30×10^{29} photons per second

27.3 385 nm ($K_{max} = 0.82$ eV)

27.4 10.0 kV

27.5 3.71 pm

27.6 397 nm—difficult to see for most people

27.7 At room temperature, the atoms are almost all in the ground state. The absorption spectrum shows only transitions that start from the ground state—the Lyman series; all of them are in the ultraviolet. At high temperatures, some of the atoms are excited into the $n = 2$ energy level by collisions. These atoms can absorb photons in the Balmer series, causing a transition from $n = 2$ to a higher energy level.

27.8 $4.5a_0$

27.9 (a) 13.6 eV; (b) 54.4 eV; (c) In He^+, the electron is more tightly bound since the nucleus has twice the charge.

27.10 5.85 fm

Quantum Physics

Biologists and medical researchers commonly use electron microscopes instead of light microscopes when very fine detail is desired. This photo shows a colorized transmission electron micrograph of *Clostridium butyricum* bacteria, magnified approximately 50,000 times. Spores (orange) and granules (blue) are visible, surrounded by cytoplasm (yellow) within the cell. What enables an electron microscope to achieve a greater resolution than a light microscope? Are there any limits to the resolution of an electron microscope?

28.1 THE WAVE-PARTICLE DUALITY

Classical physics maintains a sharp distinction between particles and waves; quantum physics blurs the distinction. Interference and diffraction experiments (Chapter 25) demonstrate the wave nature of light. On the other hand, in the photoelectric effect, Compton effect, and pair production and annihilation (Chapter 27), EM radiation behaves as if it is composed of *particles* called photons. So which is it—wave or particle?

In quantum physics the two descriptions, particle and wave, are complementary. In some circumstances, light behaves more like a wave and less like a particle; in other circumstances, more like a particle and less like a wave. *Wave* and *particle* are really two idealized extremes.

Imagine a double-slit interference experiment in which the screen is replaced by a set of photomultipliers—devices that can count individual photons. Each photomultiplier records the number of photons that arrive during a set time interval. Since the intensity is proportional to the number of photons counted, we can then plot the intensity as a function of position along the "screen." The intensity pattern is just like that recorded by photographic film; the photomultiplier records alternating maxima and minima with smooth transitions between them (Fig. 28.1).

Now suppose the intensity of the incident light is reduced until only *one photon at a time* leaves the source. In the wave picture, the interference pattern arises from the superposition of EM waves from each of the slits. When only one photon at a time leaves the source, we might not expect an interference pattern—common sense suggests that each photon reaching the detector must have gone through either one slit or the other, but not both.

What do we observe in this experiment? At first, photons seem to appear at random places (Fig. 28.2a); we cannot predict where the next photon will be detected. If we continue the experiment, the photons are clearly more numerous in some places than in others (Fig. 28.2b). We still cannot predict where the next photon will land, but we see that the *probability* of detecting a photon is higher in some places than in others. If the experiment is allowed to run for a long time, the photons form distinct interference fringes (Fig. 28.2c). After a very long time, the intensity pattern is exactly that of Fig. 28.1—the double-slit interference pattern—*even though only one photon at a time passes through the slits.* Nevertheless, even after a clear interference pattern forms, we still cannot predict where the *next* photon will be detected.

Suppose we add to the experiment a detector that observes which slit each photon passes through. Such a detector would show that each photon goes through one slit or the other, but never both. However, when the detector is in place, *no interference fringes are formed.* By making an observation that pins down the path of each photon to one slit or the other, we make the photon behave more like a particle *and less like a wave.*

If this wave-particle duality seems strange, rest assured that even the greatest physicists have felt the same way. Niels Bohr said: "Anyone who has not been shocked by quantum mechanics has not understood it." Common sense is formed from observations in which quantum effects are not noticeable. While studying quantum mechanics, don't be discouraged when it seems confusing; quantum mechanics never seems obvious to anyone, but that's partly what makes it fascinating. Richard P. Feynman (1918–1988) put it this way: "I am going to tell you what nature behaves like. If you

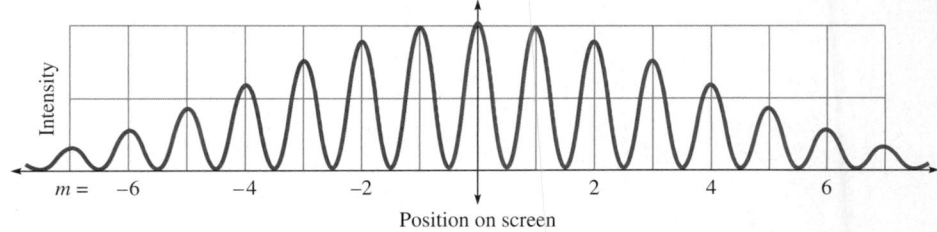

Figure 28.1 Two-slit interference pattern: the intensity as a function of position on the screen.

will simply admit that maybe she does behave like this, you will find her a delightful, entrancing thing. Do not keep saying to yourself, if you can possibly avoid it, 'But how can it be like that?' . . . Nobody knows how it can be like that."

Probability

In the double-slit experiment, we can never predict where any one photon will end up, but we can calculate the *probability* that it will fall in a given location. Two photons that are initially *identical* can end up at different places on the screen. The intensity pattern calculated by treating light as a wave is a statistical average that assumes a large number of photons.

The intensity of an EM wave is the energy flow per unit time per unit cross-sectional area.

$$I = \frac{\text{energy}}{\text{time} \cdot \text{area}}$$

In the wave picture, the intensity is proportional to the square of the electric field amplitude:

$$I \propto E^2$$

In the photon picture, each photon carries a definite quantity of energy, so

$$I = \frac{\text{number of photons}}{\text{time} \cdot \text{area}} \times \text{energy of one photon}$$

The number of photons that cross a given area is proportional to the *probability* that a photon crosses the area:

$$I = \frac{\text{number of photons}}{\text{time} \cdot \text{area}} \propto \frac{\text{probability of finding a photon}}{\text{time} \cdot \text{area}}$$

Therefore, the probability of finding a photon is proportional to the square of the electric field amplitude. The electric field as a function of position and time can be regarded as the wave function—the mathematical function that describes the wave—so the probability of finding a photon in some region of space is proportional to the square of the wave function in that region.

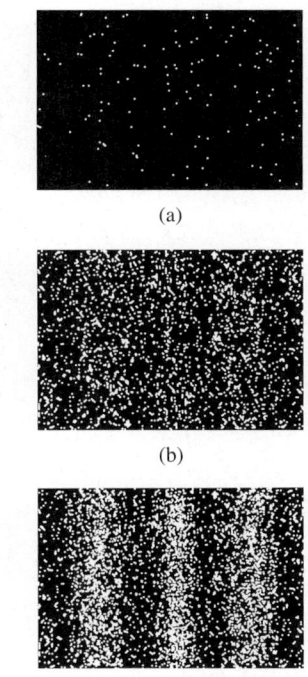

Figure 28.2 A two-slit experiment in which only one photon at a time passes through the slits. The experiment replicates the usual two-slit interference pattern once a large number of photons are recorded.

Probability ∝ square of the wave function

28.2   MATTER WAVES

We have seen that light has particle characteristics as well as wave characteristics. In 1923, French physicist Louis de Broglie (pronounced roughly *lwee duh broy*) suggested that this wave-particle duality may also pertain to particles such as electrons and protons. If light, which was so successfully established as a wave by Maxwell, could also have particle properties, why couldn't an electron have wave properties? But what would the wavelength of an electron be? De Broglie proposed that the relationship between the momentum and wavelength of any particle is the same as that for a photon [Eq. (27-11)]. It was not long before overwhelming experimental evidence confirmed de Broglie's hypothesis of the wave nature of electrons and other particles. The wavelength of the matter wave describing the behavior of a particle is now called its **de Broglie wavelength**.

---

de Broglie wavelength

$$\lambda = \frac{h}{p} \qquad (28\text{-}1)$$

Electron Diffraction

How can we see wave characteristics of particles such as electrons? The hallmarks of a wave are interference and diffraction. In 1925, C. Davisson and L. H. Germer shone a low-energy electron beam at a crystalline nickel target and observed the number of electrons scattered as a function of the scattering angle ϕ (Fig. 28.3). The maximum number of

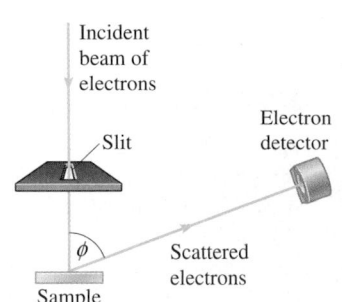

Figure 28.3 The Davisson-Germer experimental setup

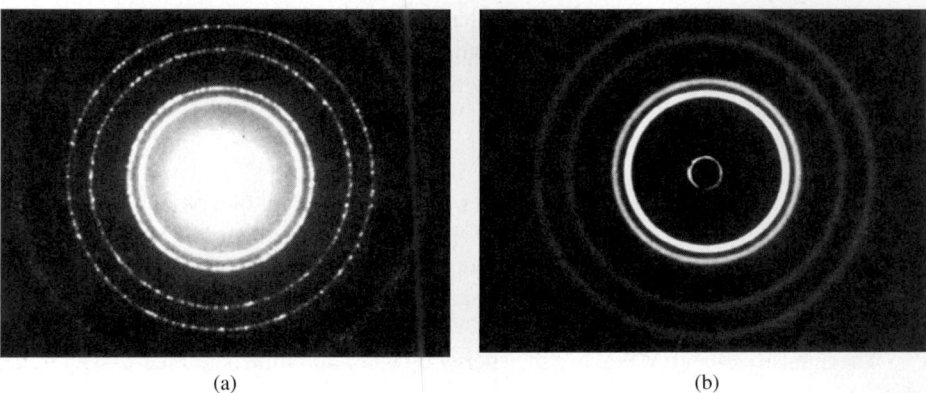

(a)   (b)

Figure 28.4 (a) Electron diffraction pattern from a polycrystalline aluminum sample. (b) X-ray diffraction pattern from the same sample. Since the electron energy in (a) was chosen so that the de Broglie wavelength of the electrons was the same as the wavelength of the x-rays in (b), the bright fringes occur at the same angles.

electrons was detected at a scattering angle $\phi = 130°$. What could make the number of scattered electrons maximum at one particular angle? Could the maximum be due to interference or diffraction? If so, then electrons must have wave-like properties.

Later analysis showed that the maximum occurred at the angle predicted by Bragg's law for x-ray diffraction [Eq. (25-15)] if the wavelength of the electrons is given by de Broglie's relation (see Problem 52). The scattered electrons interfere just as scattered x-rays interfere, giving a maximum intensity at angles where the path difference is an integral number of wavelengths.

Davisson and Germer observed a broad maximum. The low-energy electrons they used did not penetrate very far into the crystal, so the electrons were scattered from a relatively small number of planes. Just as a large number of slits in a grating makes the maxima narrow, if the electrons scatter from all of the planes in the crystal, the electron-diffraction maxima become sharp. In 1927, G. P. Thomson performed an electron diffraction experiment using higher-energy electrons. Instead of a single crystal, his sample was polycrystalline—many small crystals with random orientations. In x-ray diffraction, a polycrystalline sample produces maxima in a series of bright concentric rings due to constructive interference. Thomson saw a ring pattern for electron diffraction that was identical to an x-ray diffraction pattern when the x-rays had the same wavelength as the electrons (Fig. 28.4). These experiments showed that de Broglie's hypothesis was correct; electrons with a wavelength $\lambda = h/p$ diffract just as do x-rays of the same wavelength.

An interesting historical aside: J. J. Thomson is credited with the discovery of the electron in the late 1890s due to his measurement of the electron's charge-to-mass ratio. His son, G. P. Thomson, performed groundbreaking experiments in electron diffraction. The experiments of the father showed that electrons are *particles* while those of his son demonstrated the wave nature of electrons.

Example 28.1

Electron Diffraction Experiment

An electron diffraction experiment is performed using electrons that have been accelerated through a potential difference of 8.0 kV. (a) Find the de Broglie wavelength of the electrons. (b) Find the wavelength and energy of x-ray photons that would give the same diffraction pattern on the same sample.

⚠ **Strategy** The relationship between wavelength and momentum is the same for both electrons and photons,

but the relationship between wavelength and energy is not the same. The Bragg condition [Eq. (25-15)] for diffraction maxima in x-ray diffraction requires the path difference between x-rays reflecting off adjacent planes to be an integral multiple of the wavelength. The conditions for interference and diffraction maxima and minima always relate path differences to wavelengths. So to give the same diffraction pattern, the x-rays must

continued on next page

Example 28.1 *continued*

have the same wavelength as the electrons. We expect the energy of the x-ray photons to be different from the kinetic energy of the electrons—the relationship between momentum and energy is not the same for a photon as for a particle with mass.

Solution (a) If electrons are accelerated through a potential difference of magnitude 8.0 kV, they have a kinetic energy of 8.0 keV. We need the kinetic energy in SI units to find the momentum in SI units:

$$K = 8000 \text{ eV} \times 1.6 \times 10^{-19} \text{ J/eV} = 1.28 \times 10^{-15} \text{ J}$$

The electron's kinetic energy (8.0 keV) is small compared to its rest energy (511 keV), so the electron is nonrelativistic—we can use $p = mv$ and $K = \frac{1}{2}mv^2$. Solving for p in terms of K by eliminating the speed v, the momentum is

$$p = \sqrt{2mK} = \sqrt{2 \times 9.11 \times 10^{-31} \text{ kg} \times 1.28 \times 10^{-15} \text{ J}}$$

$$= 4.83 \times 10^{-23} \text{ kg·m/s}$$

The wavelength is then

$$\lambda = \frac{h}{p} = \frac{6.63 \times 10^{-34} \text{ J·s}}{4.83 \times 10^{-23} \text{ kg·m/s}} = 1.37 \times 10^{-11} \text{ m} = 13.7 \text{ pm}$$

(b) The x-rays would need to have the same wavelength, 13.7 pm. The energy of a photon with this wavelength is

$$E = hf = \frac{hc}{\lambda} = \frac{1.24 \text{ keV·nm}}{0.0137 \text{ nm}} = 90.5 \text{ keV}$$

Discussion As a check, we use an alternative solution to part (a) that does not require conversion to SI units. Multiplying both sides of $p = \sqrt{2mK}$ by c yields $pc = \sqrt{2mc^2K}$. For an electron, $mc^2 = 511$ keV. Then

$$\lambda = \frac{h}{p} = \frac{hc}{pc} = \frac{hc}{\sqrt{2mc^2K}} = \frac{1.24 \text{ keV·nm}}{\sqrt{2 \times 511 \text{ keV} \times 8.0 \text{ keV}}}$$

$$= 0.0137 \text{ nm} = 13.7 \text{ pm}$$

Practice Problem 28.1 A neutron's de Broglie wavelength

Find the kinetic energy of a neutron with the same de Broglie wavelength as a 22 keV photon.

Conceptual Example 28.2

Size of Diffraction Pattern and Electron Energy

An electron diffraction experiment is performed on a polycrystalline aluminum sample. The electrons produce a ring pattern as in Fig. 28.4a. If we increase the accelerating potential of the electrons, what happens to the radius of the rings? See Fig. 28.5, which shows the formation of one of the rings.

Strategy A ring is formed by constructive interference; the path difference between electrons reflecting from two successive planes is an integral number of wavelengths. As the accelerating potential is increased, we must find out what happens to the wavelength. Then we determine how ϕ must change to keep the extra path length equal to a fixed number of wavelengths.

Solution and Discussion A larger accelerating potential gives the electrons a larger kinetic energy and a larger momen-

tum. With a larger momentum, the de Broglie wavelength is smaller. For a smaller wavelength, it takes a smaller path difference to produce constructive interference since the path difference must remain equal to a fixed integer times a smaller wavelength. From Fig. 28.5b, a smaller path difference is produced by a smaller ϕ; from Fig. 28.5a, a smaller ϕ makes the radius of the ring smaller. Thus, the radius of each of the bright rings gets smaller as the electron energy is increased.

Conceptual Practice Problem 28.2 Double-slit pattern

In a double-slit experiment using a beam of monoenergetic electrons (electrons that all have the same kinetic energy) instead of light, the same interference pattern is obtained as for light. The interference maxima are found at angles satisfying $d \sin \theta = m\lambda$ [Eq. (25-10)], where d is the slit separation and λ is the de Broglie wavelength of the electron beam. What happens to the interference pattern as the accelerating potential is increased?

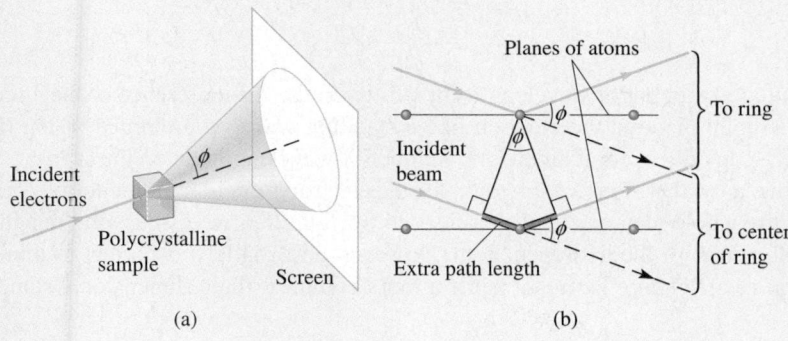

Planes of atoms

Incident electrons

Polycrystalline sample

Screen

Incident beam

Extra path length

To ring

To center of ring

(a) (b)

Figure 28.5

(a) One ring in the diffraction pattern is formed by electrons scattered at an angle ϕ from the incident beam. (b) Rays of electrons reflected from two successive planes of atoms (not shown), showing the path length difference.

Later, *neutron* diffraction experiments were performed on crystals; again, the results confirmed de Broglie's hypothesis that $\lambda = h/p$. Today, x-ray, electron, and neutron diffraction are commonly used tools for probing microscopic structures. There are some differences among them. Electrons do not penetrate as well as x-rays, so electrons are better for studying microscopic structures of surfaces. X-rays primarily interact with the atomic electrons. If a sample is made primarily of lighter elements, which have few electrons, x-ray diffraction studies are not as effective. In these cases, neutron diffraction is often used. Neutrons interact with the nuclei in the sample; since they are electrically neutral they hardly interact with electrons at all. Neutron diffraction is especially useful in determining the position of hydrogen atoms within the structure of a protein or other biological macromolecule.

In the last decade, interference and diffraction experiments have been performed using beams of atoms or molecules. Even these beams of complicated particles, with sizes much greater than the size of the neutron (not to mention that of an electron), interfere and diffract according to quantum theory.

Matter Waves and Probability

Consider a double-slit interference experiment using an electron beam rather than light. Just as for light, the interference pattern emerges even if we send only one electron at a time toward the slits. Each electron hits the screen as a localized particle and makes a small spot, just as a photon does. After many electrons have hit the screen, the interference pattern becomes evident—just as for photons (see Fig. 28.2). The interference of the matter waves emerging from the two slits determines the probability that an electron lands at a particular spot on the screen. Where the matter wave interferes constructively, the probability is high; where it interferes destructively, the probability is low.

Thus, we must conclude that the electron wave propagates through *both slits*. As for photons, we can add a detector to record which slit each electron passes through. Such a detector always finds that an electron goes through one slit or the other but never both. Just as for photons, when this detector is in place, the interference pattern disappears. The particle and wave natures of a beam of electrons are complementary.

Electron Standing Waves in the Bohr Model

In his theory of the hydrogen atom, Bohr quantized the electron energies by assuming that the electron's angular momentum must be an integral multiple of $\hbar$:

$$L_n = mvr_n = n\hbar \qquad \text{(27-17, 27-18)}$$

This quantization was later given an alternative interpretation by de Broglie. In a circular orbit, the magnitude of the angular momentum is the momentum times the radius; then Bohr's assumption says

$$mvr = pr = n\hbar = \frac{nh}{2\pi}$$

With some algebraic rearrangement, we get

$$2\pi r = n\frac{h}{p} = n\lambda$$

Thus, an integral number of wavelengths fit exactly in the circumference of the circular orbit. De Broglie pictured the electron like a standing wave on a circular string (Fig. 28.6), which can only occur if an integral number of wavelengths fit on the string.

We now know that Bohr's expression for the electron's angular momentum is incorrect, as is the whole idea of circular orbits—in Section 28.6 we discuss the quantum-mechanical picture of the hydrogen atom. However, de Broglie was correct to think of the electron as a *standing wave*—it's just a more complex, three-dimensional standing wave.

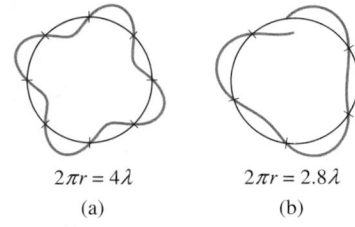

$2\pi r = 4\lambda$ $2\pi r = 2.8\lambda$

(a)   (b)

Figure 28.6 (a) A circular standing wave requires an integral number of wavelengths to fit on the circle. (b) If $2\pi r \neq n\lambda$, the ends do not join together.

28.3   ELECTRON MICROSCOPES

As we saw in Section 25.8, the resolution of a light microscope is limited by diffraction. Rayleigh's criterion says that for a microscope with aperture diameter a, the smallest angle θ that we can hope to resolve is given by $\sin \theta = 1.22\lambda/a$ [Eq. (25-13)]. Using the small angle approximation $\sin \theta \approx \theta$ (Appendix A.7), the smallest distance we can resolve is (see Fig. 28.7)

$$\Delta x \approx \theta D \approx 1.22 \frac{D}{a} \lambda$$

It is not practical to place the sample closer to the microscope than the size of the aperture, so $D > a$ and

$$\Delta x > \lambda$$

Thus, the best we can hope to do with a light microscope is to resolve distances about the size of the wavelength of visible light, which is about 400–700 nm. That's a large distance on the scale of atoms and molecules; the distance between atoms in a solid is typically only about 0.2 nm.

To get better resolution, we must use a shorter wavelength. One possibility is to use an ultraviolet microscope. These microscopes exist, with a wavelength down to about 200 nm. For wavelengths shorter than that, making effective lenses becomes too difficult.

A beam of electrons can easily be made with a wavelength around 0.2 nm or smaller. To make electrons with a wavelength of 0.2 nm, we would need to accelerate them through a potential difference of only 37.4 V. Typically the electrons used in an electron microscope are more energetic than that, and so have shorter wavelengths. However, the resolution of an electron microscope is also limited by lens aberrations—imperfections in the electromagnetic "lenses" used to focus the electron beam and form the image.

The workings of an electron microscope can be explained *without* talking explicitly about the wave nature of the electrons. We described light microscopes using geometric optics by tracing light rays (which we can think of as trajectories of photons). Similarly, we can follow the trajectories of electrons as they are bent by magnetic lenses and scattered by the sample being studied. The advantage of the electron microscope over the light microscope is the smaller wavelength of the electrons, which extends geometric "optics" to much smaller objects. A disadvantage is that the electron microscope requires a vacuum.

Electron microscopes come in several forms. The one closest to the familiar light microscope is called the transmission electron microscope or TEM (Fig. 28.8a,b). When a beam of parallel electrons passes through the sample, electrons that are scattered by a point within the sample are focused back to a point on a screen by magnetic lenses, forming a real image of the sample on the screen. The electrons must pass through the sample without being slowed down appreciably, so the TEM works only for thin samples—about 100 nm of thickness maximum. The TEM can resolve details as small as 0.2 nm—about 1000 times better than an ultraviolet microscope with wavelength 200 nm.

Another kind of electron microscope, the scanning electron microscope (SEM), uses a magnetic lens to focus a beam of electrons onto one point on the sample at a time (Fig. 28.8c,d). These primary electrons knock *secondary* electrons out of the sample; an electron collector detects the number of secondary electrons produced. The electrical signal from the collector is fed to a monitor to be viewed directly or to a computer. The focused electron beam is swept across the sample by a beam deflector. The number of secondary electrons emitted at each spot on the sample is measured and the result displayed on a screen to give an image of the specimen. The resolution of the SEM is not as good as the TEM—about 10 nm at best. But the SEM doesn't require thin samples and, since it is sensitive to the surface contour of the specimen, it is much better at imaging three-dimensional structures.

Making The Connection:
electron microscopes

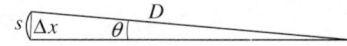

Figure 28.7   The separation between two resolvable points is approximately equal to the arc length s of a circle of radius D subtended by angle θ; $\Delta x \approx s = \theta D$. (The curvature of the arc is exaggerated in the drawing for clarity.)

Figure 28.8 Two types of electron microscope. In both types, electrons emitted from a heated filament are accelerated by the electric field between the cathode and the anode. (a) In a TEM, a condensing lens forms a parallel beam and an aperture restricts its diameter. After the beam passes through the specimen, the objective lens forms a real image. One or more projection lenses magnify the image and project it onto film, a fluorescent screen, or a CCD camera (similar to a video camera). (b) Transmission electron micrograph of a parenchyma cell from the voodoo lily plant (*Sauromatum guttatum*). The image shows the nucleolus (purple), nucleus (light green), DNA (blue), mitochondria (red), cell wall (dark green), and starch grains (light yellow). (c) In an SEM, the condensing lens forms a narrow beam. The beam deflector is a series of coils that sweep the beam across the sample. The objective lens focuses the electron beam into a small spot on the specimen. Secondary electrons knocked out of the specimen at that spot are detected by the electron collector and the electrical signal is fed to a monitor or computer. (d) Colorized scanning electron micrograph of the fruit fly claw and pulvillar pad.

The scanning transmission electron microscope (STEM) scans the sample point by point, like the SEM, but it detects the electrons transmitted through the sample. Another kind of electron microscope, the scanning tunneling microscope (STM), is discussed in Section 28.10.

28.4   THE UNCERTAINTY PRINCIPLE

In the latter part of the nineteenth century, Newtonian mechanics, Maxwell's equations of electromagnetism, and thermodynamics were thought to be so highly developed—

and so well confirmed by experiment—that some scientists thought no new basic laws were left to discover. Some people even became complete determinists. Their rationale was that the state of the universe at one instant of time (the position and velocity of every particle) determined the state at all later times. In principle, the future position and velocity of every particle could be calculated using Newton's laws.

Quantum mechanics does not allow complete determinism. In the double-slit experiments described in Sections 28.1 and 28.2, it is impossible to predict, even in principle, where any one photon or electron will appear on the screen. In 1927, Werner Heisenberg formulated the **uncertainty principle**, which describes the nature of this indeterminacy. Suppose we try to determine simultaneously the position and momentum of a particle. The uncertainty principle says there are limits to how precisely we can determine them, even in an ideal experiment. If Δx is the uncertainty in the x-coordinate of position and Δp_x is the uncertainty in the x-component of the momentum, then

(a)

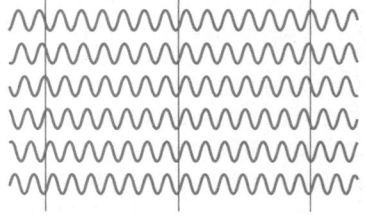

(b)

Position-momentum uncertainty principle

$$\Delta x \, \Delta p_x \geq \tfrac{1}{2}\hbar \qquad (28\text{-}2)$$

To apply the uncertainty principle in a rigorous way, we need precise definitions of the uncertainty in x and in p_x. Those definitions are beyond the level of this text. Instead, we apply the uncertainty principle in a rough, order-of-magnitude way, which means we can get by with rough estimates of the uncertainties.

Why should the precise determination of position and momentum be incompatible? It is a result of the wave-particle duality. In quantum physics a localized particle is represented as a *wave packet*—a wave with a finite extent in space (Fig. 28.9a). The momentum of a particle is related to the wavelength. To make a localized wave packet, we need to add waves *with different wavelengths* (Fig. 28.9b). These waves cancel each other everywhere except in the wave packet. The shorter the length of the wave packet, the larger the range of wavelengths that must go into the mix (Fig. 28.9c). Equivalently, the smaller the uncertainty in the particle's position, the larger the uncertainty in the momentum. The superposition of waves with a smaller range of wavelengths produces a longer wave packet—since the wavelengths are close together, they stay in phase with each other over a longer distance. Therefore, the smaller the uncertainty in momentum, the larger is the uncertainty in the position.

In Newtonian mechanics, the forces acting on a particle determine the object's motion. There is no fundamental limit to how precisely a particle's trajectory can be calculated or measured. By contrast, the uncertainty principle places a fundamental limit on the precision with which the position and momentum can simultaneously be known. The more precisely we know the position of a particle at time t, the less precisely its momentum at the same instant can be known. Uncertainty in the momentum at time t means that we cannot predict precisely where the particle will be at time $t + \Delta t$. Thus, it is not possible, even in principle, to track the motion of a particle as a function of time. A major problem with the Bohr theory of the hydrogen atom is that the electron cannot move along any well-defined trajectory such as a circular orbit without violating the uncertainty principle.

(c)

Figure 28.9 (a) A wave packet representing a localized particle. The uncertainty in the particle's position is the width of the wave packet. (b) These six waves have slightly different wavelengths; they are all in phase at the center. Moving away from the center, phase differences accumulate due to the differing wavelengths. The sum of these six waves is the wave packet in (a). (Adding just these six actually produces a recurring packet like a beat pattern. To get a true localized wave packet that does not repeat, we need to add an infinite number of waves over a small range of wavelengths.) (c) A larger range of wavelengths—around the same *average* wavelength—is needed to form a narrower wave packet like this one. A particle with a smaller uncertainty in its position is represented as a wave packet with a larger range of wavelengths and therefore a larger uncertainty in its momentum.

Example 28.3

Uncertainty in a Single-Slit Experiment

An electron diffraction experiment is performed using a single horizontal slit of width a (Fig. 28.10a). Let the center of the slit be at $y = 0$. The y-coordinates of the electrons that pass through the slit are between $y = -a/2$ and $y = +a/2$. Thus, y is within $\pm a/2$ of the average position ($y = 0$), so an estimate of the uncertainty in the y-coordinate (Δy) is $a/2$. (a) What is the y-component of the momentum of an electron that leaves the slit at angle θ? Write the answer in terms of p and θ. (b) Most of the electrons fall within the central diffraction maximum.

continued on next page

Example 28.3 *continued*

Use this fact to estimate the uncertainty Δp_y of the electrons as they pass through the slit. (c) Find the product $\Delta y \, \Delta p_y$. How does it compare to the limiting value given by the uncertainty principle?

Strategy For a wide slit ($a \gg \lambda$), we expect little diffraction; a large uncertainty in y allows for a small uncertainty in p_y, and the electrons travel straight ahead to form a geometric shadow. For a narrow slit, the electrons form a diffraction pattern on the screen. The electrons spread out into the diffraction pattern because their y-components of momentum vary as the electrons pass through the slit. The wider the diffraction pattern, the greater is Δp_y as they pass through the slit.

Solution (a) Figure 28.10b shows the momentum vector of an electron moving toward the screen at angle θ. The y-component is

$$p_y = p \sin \theta$$

(b) The angle of the first diffraction minimum is

$$\sin \theta = \frac{\lambda}{a} \qquad (25\text{-}12)$$

Thus, the range of momentum y-components for electrons that land in the central maximum is

$$-\frac{p\lambda}{a} < p_y < \frac{p\lambda}{a}$$

The uncertainty in the y-component of momentum is approximately

$$\Delta p_y = \frac{p\lambda}{a}$$

(c) The product of the uncertainties is

$$\Delta y \, \Delta p_y = \frac{a}{2} \times \frac{p\lambda}{a} = \frac{p\lambda}{2}$$

Since $\lambda = h/p$,

$$\Delta y \, \Delta p_y = \frac{ph}{2p} = \frac{1}{2} h$$

This estimate of $\Delta y \, \Delta p_y$ is a factor of 2π larger than the minimum value required by the uncertainty principle ($\Delta y \, \Delta p_y \geq \frac{1}{2}\hbar$).

Discussion This rough calculation shows that the product $\Delta y \, \Delta p_y$ is on the order of Planck's constant h, *regardless of the width of the slit or the wavelength of the electrons*. In accordance with the uncertainty principle, the two uncertainties are inversely related. A wide slit (Δy large) produces little diffraction (Δp_y small); a narrow slit (Δy small) produces a large diffraction pattern (Δp_y large).

Practice Problem 28.3 Confined electron

An electron is confined to a quantum wire of length 15 nm. What is the minimum uncertainty in the electron's momentum component along the length of the wire? What is the minimum uncertainty in the electron's velocity component along the length of the wire?

Figure 28.10
(a) A single-slit electron diffraction experiment. (b) An electron heading off at an angle θ has a momentum vector $\vec{\mathbf{p}}$ as shown. Components of $\vec{\mathbf{p}}$ are found using a right triangle.

Energy-Time Uncertainty Principle

Another uncertainty principle has to do with energy. If a system (such as an atom) is in a quantum state for a time interval Δt, then the uncertainty in the energy of that state is related to the lifetime of that state (Δt) by

$$\Delta E \, \Delta t \geq \tfrac{1}{2}\hbar \qquad (28\text{-}3)$$

28.5 WAVE FUNCTIONS FOR A CONFINED PARTICLE

A free particle can have *any* momentum and energy. In an electron diffraction experiment or electron microscope, there is no theoretical restriction on the de Broglie wavelength of the electrons used. By contrast, electrons in atoms have only certain discrete or quantized energy levels available to them. The difference is due to the *confinement* of the electron. *A confined particle has quantized energy levels.*

A good analogy is that of a transverse wave on a string. Any wavelength is possible for a traveling wave on a long string. However, for a standing wave, in which the wave is confined to a length L of the string, only certain wavelengths are possible (see Section 11.10). If the string is fixed at both ends, the allowed wavelengths are

$$\lambda_n = \frac{2L}{n} \quad (n = 1, 2, 3, \ldots) \tag{11-12}$$

For the longest wavelength ($\lambda = 2L$), the string vibrates at its lowest possible frequency (the fundamental). The standing wave is a classical example of quantization.

The same thing is true for particles such as electrons. If they are not confined, there is no restriction on their de Broglie wavelengths or energies. When they are confined, then only certain allowed values of the wavelength and energy are possible.

The wave function $y(x, t)$ for a string wave is the displacement y as a function of position along the string (x) and time (t). For the quantum mechanical wave function of a particle in one dimension we write $\psi(x, t)$, where ψ is the Greek letter psi. The interpretation of the wave function for a transverse wave on a string is easy: it tells how far a certain point on the string is displaced from its equilibrium position. For now we defer the question of what ψ stands for.

The simplest model of a confined particle is a particle that can only move in one dimension and is confined by absolutely impenetrable "walls" to a length L. We assume that the particle is free in the region between $x = 0$ and $x = L$, but it cannot leave that region, no matter how much energy we give it. This model is called the **particle in a box** (but remember that the "box" is one-dimensional).

The wave function of the particle confined in this way is exactly analogous to a transverse wave on a string fixed at both ends, so we obtain the same result for the possible wavelengths:

$$\lambda_n = \frac{2L}{n} \quad (n = 1, 2, 3, \ldots) \tag{28-4}$$

Wavelengths for a particle in a box

The de Broglie wavelength of the particle is related to its momentum:

$$p_n = \frac{h}{\lambda_n} = \frac{nh}{2L} \tag{28-5}$$

Figure 28.11 shows the wave functions for the ground state (the quantum state of lowest energy) and the first three excited states—that is, for $n = 1, 2, 3,$ and 4.

What is the energy of the confined particle? The energy is the sum of the potential and kinetic energies. Since the particle is free inside the box, the potential energy is constant inside the box; for simplicity we choose $U = 0$ inside the box. The kinetic energy can be found from the momentum:

$$K = \frac{1}{2} mv^2 = \frac{(mv)^2}{2m} = \frac{p^2}{2m} \tag{28-6}$$

$$E = K + U = \frac{p^2}{2m} + 0 = \frac{n^2h^2}{8mL^2} \tag{28-7}$$

Just as the string wave has a fundamental mode with the lowest possible frequency, the confined particle has a *minimum possible energy* in its ground state ($n = 1$). The energy of the ground state is

$$E_1 = \frac{h^2}{8mL^2} \tag{28-8}$$

The existence of a nonzero minimum energy has important ramifications. A confined particle *cannot* have zero kinetic energy. A particle confined to a smaller box has a *larger* ground-state energy. This conclusion is supported by the uncertainty principle: a smaller box means a smaller uncertainty in position ($\Delta x \approx L/2$) and therefore a greater uncertainty in momentum. Although the *magnitude* of the momentum is well defined for the particle in a box ($p = h/\lambda$), the momentum

Figure 28.11 Wave functions for a particle in a box ($n = 1, 2, 3,$ and 4)

x-component can be either $+p$ or $-p$. Thus, $\Delta p_x \approx p = h/2L$ for the ground state. The product of the uncertainties is

$$\Delta x\, \Delta p_x \approx \frac{1}{2}L \times \frac{h}{2L} = \frac{1}{4}h = \frac{\pi}{2}\hbar$$

If the uncertainty principle is used to estimate the ground state energy for the particle in a box, using *estimates* for the two uncertainties, the result is only off by a factor of π.

The energies of the excited states are

$$E_n = n^2 E_1 \tag{28-9}$$

Just as for the H atom, the particle in a box can make a transition from an excited state n to a lower energy state m by radiating a photon with energy

$$E = E_n - E_m \tag{28-10}$$

Note that the energy levels get farther apart as n increases. In contrast, the energy levels of the H atom get closer together as n increases. Why the difference? The particle in a box is confined to the same length L, no matter how much energy it has. The potential energy that confines the electron in the H atom changes gradually (Fig. 28.12). For electrons with higher energies, the box is longer.

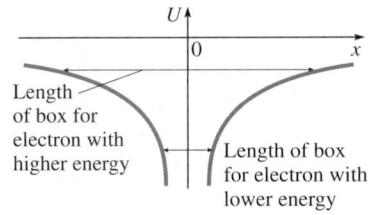

Figure 28.12 Potential energy of the electron in a hydrogen atom as a function of x; we assume for simplicity that the electron is somewhere on the *x*-axis. The nucleus is at $x = 0$.

Finite Box

A slightly more realistic model of a particle confined in one dimension is the *particle in a finite box*. In this model, the particle is still free inside the box, but the walls are not impenetrable; as shown in Fig. 28.13, the potential energy outside the box ($U = U_0$) is higher than that inside the box ($U = 0$). For a particle in a finite box, the energies are still quantized for *bound states* ($E < U_0$), but the number of bound states is finite. If the particle has an energy E greater than U_0, then it is no longer confined to the box. For these free states, since the particle is no longer confined, a continuum of wavelengths and energies is possible.

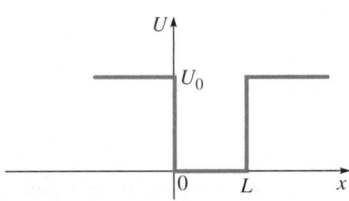

Figure 28.13 Potential energy for a particle in a finite box

In a *finite* box, the wave functions for bound states do not have to be zero at the walls and everywhere outside; instead, they extend past the walls a bit, decaying exponentially as the distance from the wall increases (Fig. 28.14). According to classical physics, a particle with $E < U_0$ can *never* be in the region outside the box since that would make the kinetic energy negative. Many experiments have verified that the wave function of a confined particle *does* extend outside the box, in accordance with the predictions of quantum mechanics.

Quantum Corral

In 1993, researchers at IBM fabricated a quantum corral (Fig. 28.15)—a two-dimensional finite box for electrons. The corral is a circular ring of radius 7.13 nm composed of 48 iron atoms on the surface of a copper crystal. The ring of iron atoms traps some electrons inside. The ripples reflect the circular standing wave formed by the electrons' wave functions; thus, the quantum corral provides a unique opportunity to see what a wave function looks like. A scanning tunneling microscope (Section 28.10) was used first to move the iron atoms into place one at a time and then to form an image of the corral.

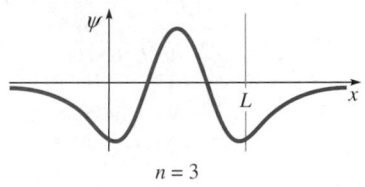

Figure 28.14 Wave functions for a particle in a finite box ($n = 1$, 2, and 3)

Interpretation of the Wave Function

In 1925, Erwin Schrödinger (1887–1961) obtained de Broglie's thesis concerning the wavelike nature of particles. Within a few weeks, Schrödinger formulated a fundamental equation of quantum mechanics. Quantum-mechanical wave functions are solutions of the Schrödinger equation.

The statistical interpretation of the wave function is due to Max Born (1882–1970): we cannot observe the wave function directly, but the probability of finding a particle in

Figure 28.15 Scanning tunneling microscope picture of a quantum corral. Clearly visible in this false-color image is an electron standing wave. The electrons are confined by 48 iron atoms that form the "fence" of the corral on a copper surface. The radius of the corral is 7.13 nm.

a certain location is proportional to the *square* of the magnitude of the wave function: $P \propto |\psi|^2$. In Section 28.1, we saw that the probability that a photon in an interference experiment hits the screen at a certain location is proportional to the intensity of light at that location. The intensity, in turn, is proportional to the square of the electric field amplitude. Classically, the field serves as the wave function for an EM wave, so the probability of finding a photon at a certain location is proportional to the square of the magnitude of the wave function.

To be more precise, we can't ever expect to find a particle exactly at a single mathematical point; rather, we want to know the probability of finding a particle in a small region of space. In one dimension, $|\psi(x)|^2 \, \Delta x$ is the probability of finding the particle between x and $x + \Delta x$.

Quantum physics is probabilistic in a way that classical physics is not. A particle's future is *not* completely determined by its present. Two particles, even if identical and in identical environments, may not behave in the same way. Two hydrogen atoms in the same excited state may not return to the ground state in the same way or at the same time. One may stay in the excited state longer than the other, and they may take different intermediate steps. The best we can do is to find the probability per unit time that photons of various energies are radiated.

Probability is central in nuclear physics (Chapter 29). A collection of identical radioactive nuclei, for instance, decay at different times and possibly by different processes. We can predict and measure the half-life—the time interval during which half of the nuclei decay—but there is no way to know *which* nuclei will decay *when*, or by which process.

28.6 THE HYDROGEN ATOM: WAVE FUNCTIONS AND QUANTUM NUMBERS

The quantum picture of the hydrogen atom is quite different from Bohr's model. The electron doesn't orbit the proton in a circular orbit—or any other kind of orbit. The best we can expect is to know the probability of finding the electron in a given place.

You may have seen the electron depicted as an electron cloud similar to Fig. 28.16. The electron cloud is one way to represent the electron's probability distribution. But the electron is *not* spread out into a fuzzy cloud; any measurement to locate the electron would find a point particle. (If the electron is not a point particle, experiments have shown that its size is less than 10^{-17} m, which is $\frac{1}{100}$ the size of the proton and less than 10^{-7} times the size of an atom.) Although the electron does not follow an orbit, it does have kinetic energy and can have angular momentum associated with its motion.

Since an electron bound to a nucleus is confined in space to the region surrounding the nucleus, its energies are quantized. As we have seen, a confined particle in a stationary

Figure 28.16 Electron cloud representation of the ground state of the hydrogen atom state of definite energy is a standing wave. The wave function for the electron is a three-dimensional standing wave (not the one-dimensional standing wave around an orbit, as in de Broglie's interpretation of Bohr's model).

The potential energy of the electron at a distance r from the proton is

$$U = -\frac{ke^2}{r}$$

The electron in the ground state has energy $E_1 = -13.6$ eV, just as in the Bohr model. As you can show in Problem 35, the electric potential energy is equal to E_1 at a distance $2a_0$ from the nucleus (Fig. 28.17). Since $E = K + U$, the kinetic energy at $r = 2a_0$ is zero. According to classical physics, the electron could never be found at distances $r > 2a_0$; but the wave function of the electron extends into the region $r > 2a_0$, just as the wave function extends past the walls of a finite box.

Since the potential energy is not constant, the wave function does not have a single, constant wavelength. The wave function $\psi(r)$ for the ground state ($n = 1$) is shown in Fig. 28.18a. Although the wave function has its maximum value at $r = 0$, the distance from the nucleus at which the electron is most likely to be found is not 0 but a_0 (see Figs. 28.18b,c).

Quantum Numbers

It turns out that the quantum state of the electron is not determined by n alone. Specifying the quantum state requires four **quantum numbers**. The integer n is called the **principal quantum number**. The energy levels are the same as the Bohr energies:

$$E_n = \frac{E_1}{n^2}, \quad E_1 = -\frac{mk^2e^4}{2\hbar^2} = -13.6 \text{ eV} \tag{28-11}$$

For a given principal quantum number n, the electron can have n different quantized amounts of orbital angular momentum L.

$$L = \sqrt{l(l+1)}\,\hbar, \quad l = 0, 1, 2, \ldots, n-1 \tag{28-12}$$

For a given n, the **orbital angular momentum quantum number** l can be any integer from 0 to $n - 1$. In the ground state ($n = 1$), $l = 0$ is the only possible value; the angular momentum in the ground state must be $L = 0$. For higher n, there are states both with nonzero and zero L. Note that L is called the *orbital* angular momentum because it is associated with the *motion* of the electron, but remember that the electron does not follow a well-defined orbit.

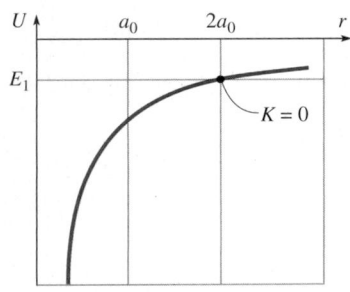

Recall that $a_0 = \hbar^2/(mke^2) = 52.9$ pm is the Bohr radius for the hydrogen atom and $k = 1/(4\pi\epsilon_0) = 8.99 \times 10^9$ N·m²/C² is the Coulomb constant.

Figure 28.17 The graph shows the electric potential energy of the electron as a function of distance r from the nucleus ($U = -ke^2/r$). E_1 is the ground-state energy. Since $E = K + U$ the kinetic energy at any distance r is the difference between the horizontal line representing E_1 and the curve representing $U(r)$.

(a)

(b)

(c)

Figure 28.18 (a) Ground-state wave function of the electron in the H atom. (b) Graphs of $|\psi|^2$ and $4\pi r^2$, the two competing factors that determine the probability of finding the electron at a given *distance* from the proton. $|\psi|^2$ is the probability *per unit volume*. The volume of space at distances between r and $r + \Delta r$ from the nucleus is the area of a thin spherical shell ($4\pi r^2$) times its thickness Δr. (c) Graph of $4\pi r^2 |\psi|^2$, which is proportional to the probability of finding the electron at distances between r and $r + \Delta r$ from the nucleus. The probability is maximum at $r = a_0$.

The orbital angular momentum quantum number l determines only the magnitude L of the orbital angular momentum; what about the direction? The direction also turns out to be quantized. For a given n and l, the component of L along some direction that we'll call the z-axis can have one of $2l + 1$ quantized values:

$n = 2, l = m_l = 0$

$$L_z = m_l \hbar, \quad m_l = -l, -l + 1, \ldots, -1, 0, +1, \ldots, l - 1, l \qquad \text{(28-13)}$$

The **orbital magnetic quantum number** m_l can be any integer from $-l$ to $+l$.

Figure 28.19 shows the wave functions of several states of the H atom. Notice that the states with zero orbital angular momentum ($l = 0$) are spherically symmetric, while $l \neq 0$ states are not.

$n = 3, l = m_l = 0$

In addition to the angular momentum associated with its motion, every electron has an intrinsic angular momentum whose magnitude is $S = (\sqrt{3}/2)\hbar$. Originally, it was thought that the electron was spinning about an axis—we still call S the *spin angular momentum*—but it can't be. The electron is, as far as we know, a point particle; to generate this angular momentum by spinning, the electron would have to be large and would have to violate relativity. The spin angular momentum is an intrinsic property of the electron, like its charge or mass.

$n = 2, l = 1, m_l = 0$

Electrons always have the same *magnitude* spin angular momentum, but the z-component of $\vec{S}$ has two possible values:

$$S_z = m_s \hbar, \quad m_s = \pm \tfrac{1}{2} \qquad \text{(28-14)}$$

The two values of the **spin magnetic quantum number** m_s are often referred to as *spin up* and *spin down*. (The quantum numbers m_l and m_s are called *magnetic* because the energy of a state depends on their values when the atom is in an external magnetic field.)

$n = 3, l = 1, m_l = 0$

The state of the electron in a hydrogen atom is completely determined by the values of the four quantum numbers n, l, m_l, and m_s.

$n = 3, l = 2, m_l = 2$

28.7 THE EXCLUSION PRINCIPLE; ELECTRON CONFIGURATIONS FOR ATOMS OTHER THAN HYDROGEN

Figure 28.19 Hydrogen atom wave functions

According to the **Pauli exclusion principle**—named after Wolfgang Pauli (1900–1958)—no two electrons in an atom can be in the same quantum state. The quantum state of an electron in any atom is specified by the same four quantum numbers we used for hydrogen: n, l, m_l, and m_s (see Table 28.1). However, the electron energy levels are not the same as those of hydrogen. In atoms with more than one electron, the interaction between the electrons has to be taken into account. In addition, the nuclear charge varies from one element to another. Thus, the same set of four quantum numbers do not correspond to the same energy level from one species of atom to another.

Table 28.1

Quantum Numbers for Electron States in an Atom

Symbol	Quantum Number	Possible Values
n	principal	$1, 2, 3, \ldots$
l	orbital angular momentum	$0, 1, 2, 3, \ldots, (n-1)$
m_l	orbital magnetic	$-l, -l + 1, \ldots, -1, 0, 1, \ldots, l - 1, l$
m_s	spin magnetic	$-\tfrac{1}{2}, +\tfrac{1}{2}$

Table 28.2

Electron Subshells Summarized

$l =$	0	1	2	3	4	5
Spectroscopic notation	s	p	d	f	g	h
Number of states in subshell	2	6	10	14	18	22

Making The Connection:
electron orbitals,
electronic
configuration of
atoms

All the electron states with the same value of n are called a **shell**. Each shell is composed of one or more **subshells**. A subshell is a unique combination of n and l. Subshells are often represented by the numerical value of n followed by a lowercase letter representing the value of l. The letters s, p, d, f, g, and h stand for $l = 0, 1, 2, 3, 4$, and 5, respectively (see Table 28.2). For example, $3p$ is the subshell with $n = 3$ and $l = 1$. The letters s, p, and d came from the appearance of the associated spectral lines long before the advent of quantum theory. The dominant or **p**rincipal spectral lines came from the $l = 1$ subshell; the spectral lines from the $l = 0$ subshell were especially **s**harp in appearance; and those from the $l = 2$ subshell looked more **d**iffuse than the others.

Since the orbital angular momentum quantum number l can be any integer from 0 to $n - 1$, with n possible values, there are n subshells in a given shell. Thus, there are three subshells in the $n = 3$ shell: $3s, 3p$, and $3d$. A superscript following the subshell label indicates how many electrons are present in that subshell. This compact notation represents the configuration of electrons in an atom. For example, the ground state of the nitrogen atom is $1s^2 2s^2 2p^3$; it has two electrons in the $1s$ subshell, two in the $2s$ subshell, and three in the $2p$ subshell.

Each subshell, in turn, consists of one or more **orbitals**, which are specified by n, l, and m_l. Since m_l can be any integer from $-l$ to $+l$, there are $2l + 1$ orbitals in a subshell. Therefore, s subshells have only one orbital, p subshells have three orbitals, d subshells have five orbitals, and so forth. Each orbital can accommodate two electrons: one spin up $\left(m_s = +\frac{1}{2}\right)$ and one spin down $\left(m_s = -\frac{1}{2}\right)$. It can be shown (Problem 67) that

> the number of electron states in a subshell is $4l + 2$, (28-15)
> and the number of states in a shell is $2n^2$

The ground-state (lowest energy) electronic configuration of an atom is found by filling up electron states, starting with the lowest energy, until all the electrons have been placed. According to the exclusion principle, there can only be one electron in each state. Generally, the subshells in order of increasing energy are

> $1s, 2s, 2p, 3s, 3p, 4s, 3d, 4p, 5s, 4d, 5p, 6s, 4f, 5d, 6p, 7s$ (28-16)

However, there are some exceptions. The energies of the subshells are not the same in different atoms; different nuclear charges and the interaction of the electrons make the energy levels differ from one atom to another. So, for example, the ground state of chromium (Cr, atomic number 24) is $1s^2 2s^2 2p^6 3s^2 3p^6 4s^1 3d^5$ instead of $1s^2 2s^2 2p^6 3s^2 3p^6 4s^2 3d^4$. Similarly, the ground state of copper (Cu, atomic number 29) is $1s^2 2s^2 2p^6 3s^2 3p^6 4s^1 3d^{10}$ instead of $1s^2 2s^2 2p^6 3s^2 3p^6 4s^2 3d^9$. There are only 8 elements among the first 56 that are exceptions to the subshell order in Eq. (28-16):

Cr, Cu, Nb, Mo, Ru, Rh, Pd, Ag

Many more exceptions are found in the electron configurations of elements with atomic numbers greater than 56.

Example 28.4

Electron Configuration of Arsenic

What is the ground-state electron configuration of arsenic (atomic number 33)?

Strategy   Arsenic has atomic number 33, so there are 33 electrons in the neutral atom. Arsenic is not one of the above-mentioned exceptions for atomic numbers ≤ 56, so we can fill subshells in the order listed in Eq. (28-16) until the total number of electrons reaches 33. A subshell can hold up to $4l + 2$ electrons. Each s subshell ($l = 0$) holds a maximum of $4 \times 0 + 2 = 2$ electrons, each p ($l = 1$) subshell holds $4 \times 1 + 2 = 6$, and each d subshell ($l = 2$) holds $4 \times 2 + 2 = 10$.

Solution   We fill up subshells and keep track of the total number of electrons: $1s^2 2s^2 2p^6 3s^2 3p^6 4s^2 3d^{10}$ has $2 + 2 + 6 + 2 + 6 + 2 + 10 = 30$ electrons. Then the remaining 3 go into the

subshell with the next highest energy—$4p$. The ground-state configuration of arsenic is therefore

$$1s^2 2s^2 2p^6 3s^2 3p^6 4s^2 3d^{10} 4p^3$$

Tips   **Discussion**   To double-check an electron configuration for an element that is not one of the exceptions:
- Add up the total number of electrons.
- Check that the subshells go in the order of Eq. (28-16).
- Make sure that all subshells except the last are full (s^2, p^6, d^{10}).

If the configuration passes those three tests, it is correct.

Practice Problem 28.4   Electron configuration of phosphorus

What is the electron configuration of phosphorus (atomic number 15)?

If a subshell is not full, how are the electrons distributed among that subshell's orbitals? Recall that a subshell contains $2l + 1$ orbitals and each orbital contains two electron states. As a rule, electrons do not double up in an orbital until each orbital has one electron in it. The two electrons in an orbital have the same spatial distribution—the same electron cloud. Thus, the two electrons in a single orbital are closer together, on average, than are two electrons in different orbitals. Due to the electrical repulsion, the energy is lower if the electrons are in different orbitals, since they are farther apart. For example, the three $4p$ electrons in arsenic (Example 28.4) are in different orbitals in the ground state: one has $m_l = 0$, one has $m_l = +1$, and one has $m_l = -1$.

Understanding the Periodic Table

The elements in the Periodic Table (see inside back cover) are arranged in order of increasing atomic number Z. The nucleus of an element has charge $+Ze$ and the neutral atom has Z electrons. Furthermore, the elements are arranged in columns according to the configuration of their electrons (see Table 28.3). Elements with similar electronic configurations tend to have similar chemical properties.

Making The Connection:
Periodic Table

Table 28.3

The Periodic Table Organizes the Elements According to Electronic Configuration

1A	2A	3B–8B, 1B, 2B	3A	4A	5A	6A	7A	8A
Alkali Metals	Alkaline Earths	Transition Elements, Lanthanides, and Actinides					Halogens	Noble Gases
s^1	s^2	$d^n s^2$, $d^n s^1$, or $f^m d^n s^2$	$s^2 p^1$	$s^2 p^2$	$s^2 p^3$	$s^2 p^4$	$s^2 p^5$	$s^2 p^6$ (except He)

The Periodic Table of the elements is arranged in columns by electronic configuration. Elements with similar electronic configurations tend to have similar chemical properties. The table lists only the subshells beyond the configuration of the previously occurring noble gas.

Figure 28.20 Energy level diagram for atomic subshells. The energies of the subshells differ from one atom to another, but this diagram gives a general idea of the relative spacing of the energies. The subshells are filled from the bottom (lowest energy) up.

Although the energy level of a subshell differs from one atom to another, Fig. 28.20 gives a *general* idea of the energies of the various atomic subshells. Note the larger than usual spacing between each *s*-subshell and the subshell below it. The *s*-subshell is the lowest energy subshell in a given shell. When starting a new shell (with a higher value of n), the electrons are farther from the nucleus and more weakly bound. The most stable electronic configurations—those that are difficult to ionize and are chemically nonreactive—are those that have all the subshells below an *s*-subshell full. Elements with this stable configuration are called the *noble gases* (Group 8A). Helium has configuration $1s^2$—the only subshell below $2s$ is full. The rest of the noble gases have a full *p*-subshell as their highest energy subshell: neon (all subshells below $3s$ full), argon (full below $4s$), krypton (full below $5s$), xenon (full below $6s$), and radon (full below $7s$).

The energy required to excite a helium atom into its first excited state ($1s^1 2s^1$) is quite large—about 20 eV—due to the large energy gap between the $1s$ and $2s$ subshells. The energy required to excite a lithium atom into its first excited state is much smaller (about 2 eV). Lithium and the other *alkali metals* (Group 1A) have one electron beyond a noble gas configuration. As a shorthand, we often write the spectroscopic notation only for the electrons in an atom that are beyond the configuration of a noble gas, since only those electrons participate in chemical reactions; thus, lithium's configuration is $[He]2s^1$, sodium's is $[Ne]3s^1$, and so on. The single electron in the *s*-subshell is quite weakly bound, so it can easily be removed from the atom, making the alkali metals highly reactive. They can easily give up their "extra" electron to achieve a noble gas configuration as an ion with charge $+e$ (alkali metals have valence +1).

The alkali metals form ionic bonds with the highly reactive *halogens* (Group 7A), which are one electron shy of a noble gas configuration. For instance, chlorine (Cl, $[Ne]3s^2 3p^5$) needs to gain only one electron to have the electron configuration of the noble gas argon (Ar, $[Ne]3s^2 3p^6$). Thus, the halogens have valence −1. Sodium can give its weakly bound electron to chlorine, leaving both ions (Na^+ and Cl^-) in stable noble gas configurations. The electrostatic attraction between the two forms an ionic bond: NaCl.

The *alkaline earths* (Group 2A) all have a full *s*-subshell (s^2) beyond a noble gas configuration. They are not as reactive as the alkali metals, since the full *s*-subshell lends some stability, but they can give up both *s* electrons to achieve a noble-gas configuration, so alkaline earths usually act with valence +2.

Toward the middle of the periodic table, the chemical properties of elements are more subtle. Covalent bonds tend to form when two or more elements have unpaired electrons in orbitals that they can share. Carbon is particularly interesting. Its ground state is $1s^2 2s^2 2p^2$. The two $2p$ electrons are in different orbitals; then there are two unpaired electrons and carbon in the ground state has a valence of 2. However, it takes only a small amount of energy to raise a carbon atom into the state $1s^2 2s^1 2p^3$. Now there are four unpaired electrons (the $2s$ orbital and the three $2p$ orbitals each have one electron). Thus, carbon can have a valence of 4 as well.

In the Groups numbered 1A, 2A, . . . , 7A, the numeral before the "A" represents the number of electrons beyond a noble gas configuration. In the *transition elements*, a *d*-subshell is being filled; their electronic configurations are usually [noble gas]$d^n s^2$ but sometimes [noble gas]$d^n s^1$, where $0 \leq n \leq 10$. In the *lanthanides* and the *actinides*, an *f*-subshell is being filled; their electronic configurations are [noble gas]$f^m d^n s^2$, where $0 \leq m \leq 14$ and $0 \leq n \leq 10$. The *d*- and *f*-subshells participate less in chemical reactions than the *s*- and *p*-subshells, so the chemical properties of the transition elements, lanthanides, and actinides are chiefly based on their outermost *s*-subshell.

28.8   ELECTRON ENERGY LEVELS IN A SOLID

An isolated atom radiates a discrete set of photon energies that reflect the quantized electron energy levels in the atom. Although a gas discharge tube contains a large number of gas atoms (or molecules), the pressure is low enough that the atoms are, on average, quite far apart. As long as the wave functions of electrons in different atoms do not overlap appreciably, each atom radiates photons of the same energies as would a single isolated atom.

On the other hand, solids radiate a continuous spectrum rather than a line spectrum. What has happened to the quantization of electron energies? The energy levels are still quantized, but they are so close together that in many circumstances we can think of them as continuous.

Splitting of Electron Energy Levels

Imagine two atoms that are initially far apart. The two isolated atoms have the same energy levels. As the atoms come closer together, the electron wave functions begin to overlap. This overlap changes the wave functions and their energies—just as the energy levels in multielectron atoms differ from the energy levels in a hydrogen atom due in part to the overlap of the electron wave functions. Each energy level splits into two energy levels. With significant overlap of the wave functions, we can no longer tell which electron "belongs" to which nucleus. Instead of two identical atomic energy levels, we have two energy levels for the system of two atoms. The first levels to split are the higher energy levels, since their wave functions extend farther from the nucleus and thus begin to overlap first as the atoms are brought closer together. As the atoms get closer and closer together, the lower energy levels start to split in order of decreasing energy (Fig. 28.21).

Energy level splitting is the basis of the formation of covalent chemical bonds. Imagine two hydrogen atoms in their ground states, initially far apart. Each has an electron in its $1s$ state. As the atoms are brought closer together, the electron wave functions overlap each other. The interaction modifies the wave functions and the energies—the $1s$ energy levels split. The electron wave function is no longer spherically symmetric around a single nucleus; now the electrons are attracted by two nuclei. In the lower of the two energy levels that result from the split of the $1s$ atomic level, the electron probability is high in the region between the two protons, where its electric potential energy is low. In

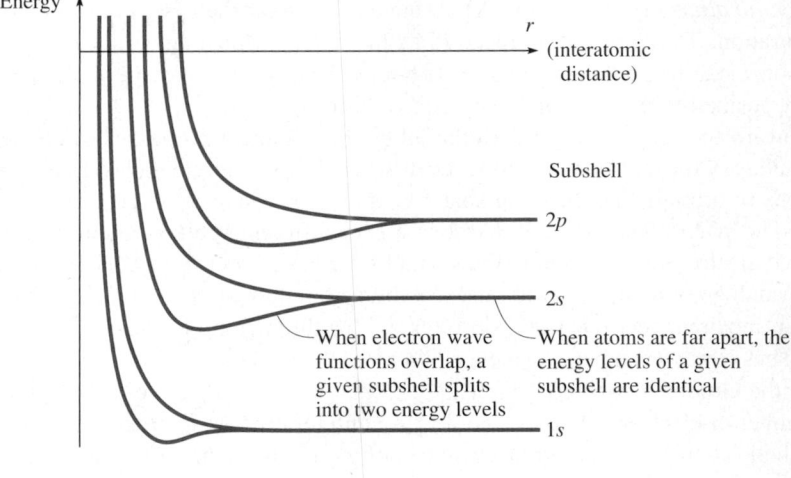

Figure 28.21 Graph showing the energy levels for two lithium atoms as a function of their separation. The ground-state configuration of a single Li atom is $1s^2 2s^1$. When the atoms are far apart, each atom has one electron in the $2s$ subshell; the two have the same energy. The energy level starts to split when the atoms are sufficiently close together that the wave functions begin to overlap. Both $2s$ electrons are in the lower of the two energy levels (one spin up, the other spin down).

the higher of the two energy levels, the electron is less likely to be between the two protons, so it is less tightly bound—its energy is higher. In the ground state, both electrons go into the lower of the two levels, one with spin up and one with spin down. The equilibrium separation of the protons in the H_2 molecule is that which makes the energy minimum. If the protons are pushed closer together than that, the energy of the molecule rises again, reflecting in part the Coulomb repulsion between the two protons.

Bands of Energy Levels

A solid is an assembly of a huge number N of atoms instead of the two atoms in the H_2 molecule. Instead of each atomic energy level splitting into two levels, now they split into N levels—one for each atom. N might typically be of the order of Avogadro's number, $\approx 10^{24}$. Each atomic energy level becomes a **band** of N closely spaced energy levels. Overlap between the bands associated with adjacent atomic energy levels is common, but there are also **band gaps**: ranges of energy in which there are no electron energy levels (Fig. 28.22).

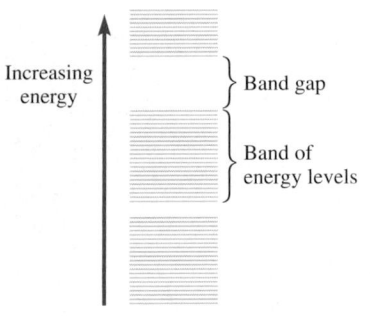

Figure 28.22 Electron energies in a solid form bands of closely spaced energy levels. Band gaps are ranges of energy in which there are no electron energy levels.

Constructing the electronic ground state of a solid is similar to constructing the ground state of an atom: the electron states are filled up in order of increasing energy starting from the lowest energy states, according to the exclusion principle. A solid at room temperature isn't in the ground state, but its electron configuration is not very different from the ground state; the extra thermal energy available promotes a small fraction of the electrons (still a large number, though) into higher energy levels, leaving some lower energy states vacant. The energy range of electron states that are thermally excited is small—of the order of $k_B T$, where k_B is the Boltzmann constant ($k_B = 1.38 \times 10^{-23}$ J/K).

Conductors, Semiconductors, and Insulators

The ground-state electron configuration (that is, the configuration at absolute zero) of a solid determines its electrical conductivity. If the highest-energy electron state filled at $T = 0$ is in the middle of a band, so that this band is only partially filled, the solid is a conductor (Fig. 28.23a). In order for current to flow, an electric field (perhaps due to a battery connected to the conductor) must be able to change the momentum and energy of the conduction electrons; this can only happen if there are vacant electron states nearby into which the conduction electrons can make transitions. Since the band is only partly full, there are plenty of available electron states at energies only slightly higher than the highest occupied.

On the other hand, if the ground-state configuration fills up the electron states right to the top of a band, then the solid is a semiconductor or an insulator. The difference between the two depends on how the size of the band gap E_g above the completely occupied band (the *valence band*) compares to the available thermal energy ($\approx k_B T$) and thus depends on the temperature of the solid (Fig. 28.23b,c).

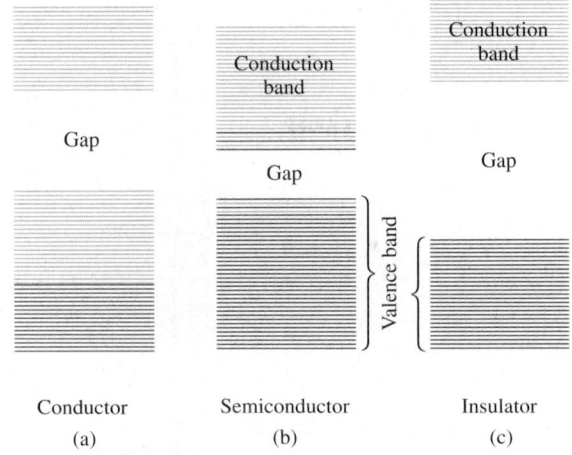

Figure 28.23 Electron energy bands in (a) a conductor, (b) a semiconductor, and (c) an insulator. Horizontal lines indicate electron energy levels; the darker lines are those levels that are occupied by electrons. In a semiconductor at room temperature (b), the valence band is mostly full but a relatively small number of electrons are thermally excited into the conduction band, leaving some vacancies near the top of the valence band.

Most materials considered semiconductors at room temperature have band gaps between about 0.1 eV and 2.2 eV. The technologically most important semiconductor, silicon, has a gap of 1.1 eV, which is about 40 times the available thermal energy at room temperature (≈ 0.025 eV). The number of electrons excited to higher energy states is much smaller than in a conductor, since there are no available energy levels nearby in the same band. The only electrons that can carry current are those promoted to the mostly empty band above the gap (the *conduction band*).

Because a relatively small number of electrons move into the conduction band, an equal number of vacant electron states exist near the top of the valence band. Electrons in nearby states can easily "fall" into these **holes**, filling one vacancy and creating another. The holes act like particles of charge $+e$ that, in response to an external electric field, move in a direction opposite that of the conduction electrons. The electric current in a semiconductor has two components: the electron current and the hole current.

The crystalline form of carbon known as diamond has a larger band gap (about 5.5 eV) than the semiconductors. It is considered to be an insulator at room temperature. The number of electrons in the conduction band is approximately proportional to $e^{-E_g/(k_BT)}$, so a gap five times larger than that of silicon means that diamond has only $e^{-5} \approx 0.7\%$ as many conduction electrons as silicon and a correspondingly smaller number of holes.

As discussed in Section 18.4, the dependence of resistivity on temperature is opposite for conductors and semiconductors. In a conductor, the effect of higher temperature is largely to make collisions between electrons and ions more likely, since the ions are vibrating with larger amplitudes. Thus, the resistivity of a conductor increases with increasing temperature. In a semiconductor, the main effect of a higher temperature is to excite a larger number of electrons into the conduction band. The greater number of conduction electrons translates into a *decrease* of resistivity with increasing temperature.

28.9   LASERS

Making The Connection:
lasers

A laser produces an intense, parallel beam of coherent, monochromatic light. The word **laser** is an acronym for *light amplification by stimulated emission of radiation*.

Stimulated Emission

When a photon has energy $\Delta E = E' - E$, where E' is a vacant energy level in an atom and E is a lower energy level that is filled, the photon can be absorbed, kicking the electron up into the higher energy level (Fig. 28.24a). If the higher energy level is filled and the lower one is vacant, the electron can drop into the lower energy level by spontaneously emitting a photon of energy ΔE (Fig. 28.24b).

Incoming photon   Electron transition   Outgoing photon(s)

(a) Absorption

(b) Spontaneous emission

(c) Stimulated emission

Figure 28.24 Absorption, spontaneous emission, and stimulated emission of a photon by an atom. All of the photons have energy $E' - E$ (the difference between the two energy levels). For a photon to be emitted, either spontaneously or when stimulated by an incoming photon, the electron must initially be in the higher energy level E'. In stimulated emission, an incident photon of energy $E' - E$ stimulates the atom to emit a photon. The two photons are identical in energy, phase, and direction.

In addition to absorption and spontaneous emission, a third interaction between an atom and a photon was first proposed by Einstein in 1917. Called **stimulated emission** (Fig. 28.24c), this process is a kind of resonance. If the electron is in the higher energy level and the lower level is vacant, an incident photon of energy ΔE can stimulate the emission of a photon as the electron drops to the lower energy level. The photon emitted by the atom is *identical* to the incident photon that stimulates the emission: they have the same energy and wavelength, move in the same direction, and are in phase with each other.

If a cascade of stimulated emissions occur, the number of identical photons increases—the *light amplification* in the acronym *laser.* The beam is *coherent* because the photons are all in phase; the beam is *monochromatic* because the photons all have the same wavelength; and the beam is *parallel* because the photons all move in the same direction.

Metastable States

How can a cascade of stimulated emissions occur when most of the atoms are in their ground states, with the electrons populating the lowest energy levels? An atom in an excited state returns to the ground state quickly by spontaneous emission of a photon. In such circumstances, the probability that a photon of energy ΔE stimulates emission is extremely small, because almost none of the atoms are in the excited state; the photon is overwhelmingly more likely to be absorbed by an atom in the ground state.

In order to produce a cascade of identical photons, we need to make stimulated emission *more likely than absorption*: more of the atoms must be in the higher-energy state than are in the lower-energy state. Since this is the reverse of the usual case, it is called a **population inversion**. A population inversion is difficult to achieve if the higher-energy state is short lived—that is, if the atom quickly emits a photon. However, some excited states—called **metastable states**—last for a relatively long time before spontaneous emission occurs. If atoms can be *pumped* up into a metastable state fast enough, a population inversion can occur.

The Ruby Laser

One way to achieve a population inversion is called **optical pumping**. Incident light of the correct wavelength is absorbed, causing the atoms to make transitions into a short-lived excited state, from which the atoms spontaneously decay to the metastable state. The ruby laser (Fig. 28.25a), developed in 1960, uses optical pumping. Ruby is an aluminum oxide crystal (sapphire) in which some of the aluminum atoms are replaced by

Figure 28.25 (a) A ruby laser. (b) Energy level diagram for a ruby laser. Optical pumping occurs when incident 2.25-eV photons are absorbed by the chromium ion, leaving it in the excited state E^*. The ion quickly decays to the metastable state E_m. While the ion is in the metastable state, a 1.79-eV photon passing by can stimulate emission of an identical 1.79-eV photon.

chromium atoms. The energy levels of the chromium ion Cr^{3+} are shown in Fig. 28.25b. The state labeled E_m is a metastable state of energy 1.79 eV above the ground state E_0. At an energy of about 2.25 eV above the ground state, a band of closely spaced energy levels E^* exists; we may think of the band as a single level for our purposes. An atom excited to the E^* energy level quickly decays to the metastable E_m state; the atom remains in the metastable state for a relatively long time.

To make a laser, a ruby rod has its ends polished and silvered to become mirrors. One end is partially transparent. A high-intensity flash lamp coils around the rod. The lamp produces a series of rapid, high-intensity bursts of light. Absorption of light with wavelength 550 nm (photon energy 2.25 eV) pumps Cr^{3+} ions to the E^* state, from which they spontaneously decay to the metastable state E_m. Strong optical pumping results in a population inversion. Eventually a few ions decay from the metastable state to the ground state by spontaneously emitting photons of wavelength 694 nm (energy 1.79 eV, in the red part of the spectrum). These photons then cause stimulated emission in other chromium ions in the metastable state. Only photons emitted parallel to the axis of the rod are reflected back and forth many times by the mirrors at the ends to continue stimulating emissions. Some of the photons escape out through the end of the rod that is partially silvered to form a narrow, intense, coherent beam of light.

Other Lasers

Similar to the ruby laser, the Nd:YAG laser consists of an optically pumped rod. Nd:YAG is yttrium aluminum garnate (YAG), a colorless crystal once used to make imitation diamonds, into which some neodymium atoms (Nd) have been introduced as impurities. The Nd ions have a metastable state suitable for lasing. Unlike ruby, which can only operate as a pulsed laser, Nd:YAG can operate either as a continuous beam or as a pulsed beam (see Conceptual Question 11). The Nd:YAG laser can produce a high-power beam at wavelength 1064 nm (in the infrared); it is commonly used in industry and in medicine.

Helium-neon lasers are commonly used in school laboratories and in older bar-code readers. A gas discharge tube contains a low pressure mixture of helium and neon. The He-Ne laser is *electrically* pumped: the electrical discharge excites helium atoms into a metastable state with energy 20.61 eV above the ground state (Fig. 28.26). Neon has a metastable state 20.66 eV above its ground state—0.05 eV higher than the energy of the metastable state of helium. An excited helium atom can make an inelastic collision with a neon atom in the ground state, leaving the neon atom in its metastable state and returning the helium atom to its ground state; the extra 0.05 eV of energy comes from the kinetic energies of the atoms. Stimulated emission leaves the atom in an excited state of energy 18.70 eV; spontaneous transitions quickly take it back to the ground state.

The carbon dioxide laser, which produces an infrared beam (10.6 μm wavelength), is similar in operation to the He-Ne laser. A gas discharge tube contains a low pressure mixture of CO_2 and N_2. The N_2 molecule is excited by the electrical discharge; the CO_2 molecule is excited into a metastable state by colliding with an excited N_2 molecule. The most powerful continuous wave lasers in common use are carbon dioxide lasers; the power of a single beam can exceed 10 kW. An almost perfectly parallel beam can be focused onto a very small spot, allowing CO_2 lasers to cut, drill, weld, and machine the hardest metal with ease. CO_2 lasers are also widely used in medicine.

Semiconductor lasers are small, inexpensive, efficient, and reliable. They are used in CD and DVD players, barcode readers, laser printers, and laser pointers. A semiconductor laser is electrically pumped: an electric current pumps electrons from the valence band to the conduction band. A photon is emitted when an electron jumps back from the conduction band to the valence band. Thus, the wavelength of the laser light depends on the band gap in the semiconductor.

Lasers in Medicine

Making The Connection:
laser surgery

Lasers are widely used in surgery to destroy tumors, to cauterize blood vessels, and to pulverize kidney stones and gallstones (Fig. 28.27). A detached retina can be "welded" back into place by a laser beam shone through the pupil of the eye. Laser surgery is used to reshape the cornea of the eye to correct nearsightedness. The laser beam can be guided by an optical fiber (Section 23.4) in an endoscope to the site of a tumor; an optical fiber can also guide a laser beam into an artery to remove plaque from the artery walls. In photodynamic cancer therapy, a photosensitizing drug is injected into the bloodstream; the drug accumulates selectively in cancerous tissues. Laser light of the

Figure 28.26 Simplified energy level diagram for the He-Ne laser.

(a)

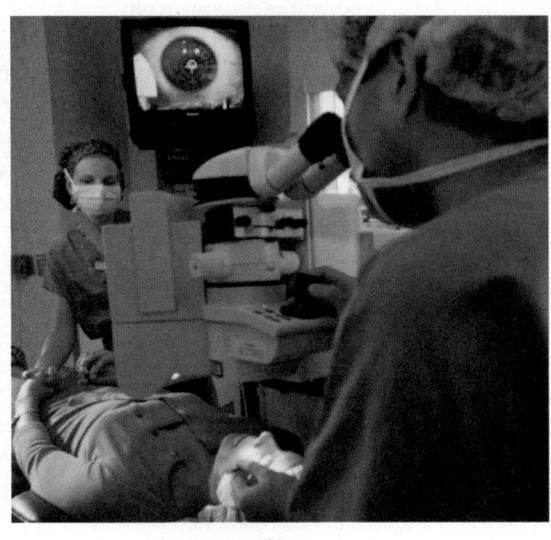

(b)

Figure 28.27 (a) An optical fiber guides a laser beam into the patient's esophagus in an experimental treatment for tumors at the Knoxville Cancer Center. (b) A patient undergoes laser optical surgery to correct her vision. The procedure is called LASIK (laser-assisted in situ keratomileusis).

correct frequency is delivered to the tumor site by an endoscope. The light causes a chemical reaction that activates the drug; it becomes toxic, destroying tumor cells and the blood vessels that supply oxygen to the tumor.

Conceptual Example 28.5

Photocoagulation

An argon ion laser is used to repair vascular abnormalities and fissures in the retina of the eye in a process known as photocoagulation. Laser light *absorbed* by the tissue raises its temperature until proteins become coagulated, forming the scar tissue that repairs the split. The principal wavelengths emitted by the argon laser are 514 nm and 488 nm. (a) What are the photon energies for these wavelengths? (b) What are the colors associated with these two wavelengths? (See Fig. 22.6.) Are both wavelengths effective on blood vessels?

Strategy The energy of a photon is related to its wavelength by $E = hc/\lambda$. Section 22.4 lists the colors of the visible spectrum and the associated wavelengths. A wavelength is effective if it is strongly absorbed.

Solution and Discussion (a) The photon energies are

$$E = \frac{hc}{\lambda} = \frac{1240 \text{ eV·nm}}{514 \text{ nm}} = 2.41 \text{ eV}$$

and

$$E = \frac{hc}{\lambda} = \frac{1240 \text{ eV·nm}}{488 \text{ nm}} = 2.54 \text{ eV}$$

(b) The color associated with 514 nm is green and with 488 nm is blue. Both wavelengths are effective on blood vessels because red blood vessels reflect red and absorb radiation of other colors.

Conceptual Practice Problem 28.5   Ruby laser and blood

Would red light from a ruby laser be effective on blood and thus useful in the treatment of vascular abnormalities?

28.10 TUNNELING

In Section 28.5 we saw that the wave function of a particle in a finite box extends into regions where, according to classical physics, the particle can never go because it has insufficient energy. In these *classically forbidden regions*, the wave function decays exponentially. If the classically forbidden region is of finite length, a curious but significant phenomenon called **tunneling** can occur.

Figure 28.28a shows a situation in which tunneling is possible. A particle is initially confined to a one-dimensional box. On the right side, the barrier is of finite thickness a. According to classical physics, if $E < U_0$, the particle can *never* get out of the box; it doesn't have enough energy.

However, if $E > U_1$, the classical prediction is not what we observe; instead, there is a nonzero probability of finding the particle *outside the box* at a later time. The wave function of the particle decays exponentially only from $x = 0$ to $x = a$; for $x > a$ it becomes sinusoidal again, although with a reduced amplitude due to the exponential decay in the barrier (Fig. 28.28b). The amplitude of the wave function for $x > a$ determines the probability per second that the particle is found outside the box.

Since the wave function decays exponentially in the barrier, the tunneling probability decreases dramatically as the barrier thickness increases. For a relatively wide barrier, the tunneling probability decreases exponentially with barrier thickness:

$$P \propto e^{-2\kappa a} \tag{28-17}$$

In Eq. (28-17), P is the probability per unit time that tunneling occurs, a is the barrier thickness, and κ is a measure of the barrier height:

$$\kappa = \sqrt{\frac{2m}{\hbar^2}(U_0 - E)} \tag{28-18}$$

Equation (28-17) is an approximation valid when $e^{-2\kappa a} \ll 1$. The tunneling probability's dependence on barrier thickness is more complicated for an extremely thin barrier.

A particle does not have to be confined in order to tunnel. The particle can be free on both sides of the barrier (which can be envisioned by letting $L \to \infty$ in Fig. 28.28). It is also possible for a free particle to tunnel *into* a box (which means the particle is initially to the right of the barrier in Fig. 28.28).

(a)

(b)

Figure 28.28 (a) A particle of energy $E < U_0$ is confined to a finite box of length L. The potential energy inside the box is taken to be zero; the potential energy on either side of the box is U_0. To the right of the barrier, the potential energy is U_1. For $U_1 < E < U_0$, the particle can tunnel out of the box. (b) Sketch of the wave function for a particle that can tunnel out of the box.

The Scanning Tunneling Microscope

Making The Connection:
scanning tunneling
microscope

The scanning tunneling microscope (STM) exploits the exponential dependence of tunneling probability on barrier thickness (Fig. 28.29) to produce highly magnified images of surfaces. In an STM, a very fine metal tip is placed very close to a surface of interest. The tip must be much finer than an ordinary needle—it ideally should have a single atom at the tip. The distance between the tip and the sample is typically only a few nanometers. The apparatus must be isolated from vibrations, which under ordinary circumstances have amplitudes of 1000 nm or more. The sample and tip are in an evacuated chamber.

Figure 28.29 (a) Simplified model of the potential energy of an electron that tunnels from the tip of an STM to a sample a distance a away. An applied potential difference ΔV causes a potential energy difference of magnitude $e\,\Delta V$ between tip and sample. Normally, an electron must be supplied with an energy equal to the work function of the metal—a few electron-volts—to break free of the metal. Here, because the tip and sample are only a few nanometers apart, an electron can tunnel through the barrier presented by the work functions of the metals. (b) Schematic of an STM. (c) Scanning tunneling micrograph of a section of a DNA molecule. The average distance between the coils of the helix (seen as yellow peaks) is 3.5 nm.

A small potential difference $\Delta V \approx 10$ mV is applied between the tip and the sample. Electrons now tunnel between the tip and the sample. The barrier they tunnel through is due to the work functions of the tip and the sample; an electron bound to the metal has a lower energy than one that is free outside of the metal.

As the tip is scanned over the surface, its distance from the sample is adjusted to keep the tunneling current constant. Since the current depends exponentially on the distance a, the tip is moved to keep a constant. Thus, the movements of the tip accurately reflect the surface beneath. An STM is easily able to image individual atoms on a surface.

Example 28.6

Change in Tunneling Current

Suppose that an STM scans a surface at a distance of $a = 1.000$ nm. Take the height of the potential energy barrier to be $U_0 - E = 2.00$ eV. If the distance between the surface and the STM tip decreases by 1.0% (= 0.010 nm, which is about $\frac{1}{5}$ the radius of the smallest atom), estimate the percentage change in the tunneling current.

Strategy   The tunneling current is proportional to the number of electrons that tunnel per second, which is in turn proportional to the tunneling probability per second [P in Eq. (28-17)]. Thus, the ratio of the probabilities per second is equal to the ratio of the tunneling currents.

Solution   The tunneling probability per unit time is

$$P \propto e^{-2\kappa a} \tag{28-17}$$

where

$$\kappa = \sqrt{\frac{2m}{\hbar^2}(U_0 - E)} \tag{28-18}$$

$$= \sqrt{\frac{2 \times 9.109 \times 10^{-31}\ \text{kg}}{[6.626 \times 10^{-34}\ \text{J·s}/(2\pi)]^2} \times (2.00\ \text{eV} \times 1.602 \times 10^{-19}\ \text{J/eV})}$$

$$= 7.245 \times 10^9\ \text{m}^{-1}$$

Since the tip moves 0.010 nm closer to the surface, the distance changes from $a = 1.000$ nm to $a' = 0.990$ nm. The ratio of the tunneling probabilities is

$$\frac{P_{a'}}{P_a} = \frac{e^{-2\kappa a'}}{e^{-2\kappa a}} = e^{-2\kappa(a'-a)}$$

The quantity in the exponent is

$$2\kappa(a' - a) = 2 \times 7.245 \times 10^9\ \text{m}^{-1} \times (-0.010 \times 10^{-9}\ \text{m}) = -0.1449$$

The ratio of the probabilities per unit time is

$$\frac{P_{a'}}{P_a} = e^{0.1449} = 1.16$$

Then the ratio of the currents is also 1.16. A 1.0% decrease in the distance between tip and sample causes a 16% increase in the tunneling current.

Discussion   A decrease in distance means an increase in tunneling current, as expected. The large change in current for a small change in distance is due to the exponential falloff of the wavelength in the forbidden region; it makes the STM a very sensitive instrument.

Let us check the units in the calculation for κ:

$$\sqrt{\frac{\text{kg}}{\text{J}^2 \cdot \text{s}^2} \times \text{J}} = \sqrt{\frac{\text{kg}}{\text{s}^2} \times \frac{1}{\text{J}}} = \sqrt{\frac{\text{kg}}{\text{s}^2} \times \frac{\text{s}^2}{\text{kg} \cdot \text{m}^2}} = \text{m}^{-1}$$

Practice Problem 28.6   Change in tunneling current when tip moves away

Estimate the percentage change in tunneling current if the tip moves *away* by 1.00% (from 1.0000 nm to 1.0100 nm).

An Atomic Clock Based on Tunneling

Tunneling in the ammonia molecule (NH_3) was exploited to make the first atomic clocks. The three-dimensional structure of the molecule has the three hydrogen atoms in an equilateral triangle. The nitrogen atom is equidistant from the three hydrogen atoms (Fig. 28.30a). The nitrogen atom has two possible equilibrium positions: it can be on either side of the plane of the H atoms.

The potential energy of the nitrogen atom is shown in Fig. 28.30b. The equilibrium positions are the two minima in $U(z)$. The barrier between the two is due to the Coulomb repulsion between the atoms. In the ground state of the NH_3 molecule, the N atom does not have enough energy to move back and forth along the z-axis between the two equilibrium positions. However, it *does* oscillate back and forth between the two positions: the N atom *tunnels* back and forth through the potential energy barrier. The tunneling probability determines the frequency of oscillation, which is 2.4×10^{10} Hz. Since the oscillation depends on tunneling, this frequency is much lower than a typical molecular vibration frequency, making it easier to use as a time standard for the first atomic clocks.

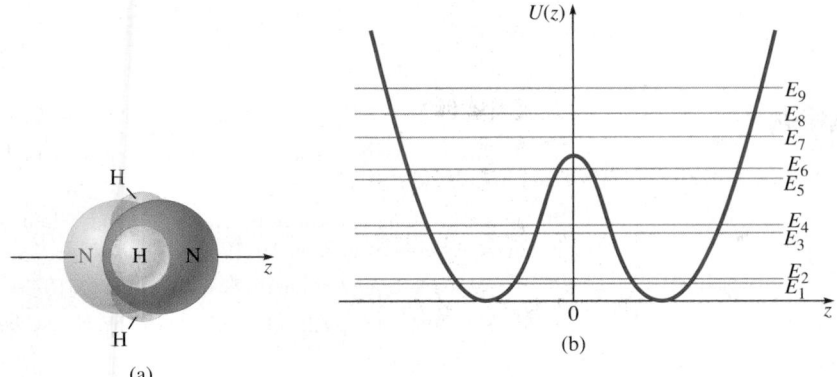

Figure 28.30 (a) Three-dimensional structure of the NH₃ molecule. The alternate equilibrium position of the N atom is shown in light blue. (b) Potential energy of the nitrogen atom as a function of its distance from the center of the three H atoms. For the lowest six vibrational energy levels, the nitrogen atom tunnels from one side to the other.

MASTER THE CONCEPTS

Summary

- In quantum physics the two descriptions, particle and wave, are complementary. The wavelength of a particle is called its de Broglie wavelength:

$$\lambda = \frac{h}{p} \qquad (28\text{-}1)$$

- The uncertainty principle sets limits on how precisely we can simultaneously determine the position and momentum of a particle:

$$\Delta x \, \Delta p_x \geq \tfrac{1}{2}\hbar \qquad (28\text{-}2)$$

- If a system is in a quantum state for a time interval Δt, then the uncertainty in the energy of that state is related to the lifetime of that state by the energy-time uncertainty principle:

$$\Delta E \, \Delta t \geq \tfrac{1}{2}\hbar \qquad (28\text{-}3)$$

- Free particles have wave functions that are traveling waves; confined particles have wave functions that are standing waves. Confinement leads to the quantization of de Broglie wavelengths and energies.

- A particle in a one-dimensional box has wavelengths analogous to those of a standing wave on a string:

$$\lambda_n = \frac{2L}{n} \quad (n = 1, 2, 3, \ldots) \qquad (28\text{-}4)$$

- The square of the magnitude of the wave function is proportional to the probability of locating the particle in a given region of space.

- The quantum state of the electron in an atom can be described by four quantum numbers:

principal quantum number $n = 1, 2, 3, \ldots$

orbital angular momentum quantum number
$l = 0, 1, 2, 3, \ldots, n - 1$

magnetic quantum number
$m_l = -l, -l + 1, \ldots, -1, 0, 1, \ldots, l - 1, l$

spin magnetic quantum number $m_s = -\tfrac{1}{2}, +\tfrac{1}{2}$

- According to the exclusion principle, no two electrons in an atom can be in the same quantum state.

- All the electron states with the same value of n are called a shell. A subshell is a unique combination of n and l. Spectroscopic notation for a subshell is the numerical value of n followed by a letter representing the value of l.

- In a solid, the electron states form bands of closely spaced energy levels. Band gaps are ranges of energy in which there are no electron energy levels. Conductors, semiconductors, and insulators are distinguished by their band structure.

- If an electron is in a higher energy level and a lower level is vacant, an incident photon of energy ΔE can stimulate the emission of a photon. The photon emitted by the atom is *identical* to the incident photon.

- Lasers are based on stimulated emission. In order for stimulated emission to occur more often than absorption, a population inversion must exist (the state of higher energy must be more populated than the state of lower energy).

- The wave function of a confined particle extends into regions where, according to classical physics, the particle can never go because it has insufficient energy. If the classically forbidden region is of finite length, tunneling can occur.

Highlighted Figures and Tables

F28.4 Electron diffraction pattern from a polycrystalline aluminum sample and the x-ray diffraction pattern from the same sample (p. 1014)

F28.8 Two types of electron microscope (p. 1018)

F28.9 A wave packet representing a localized particle (p. 1019)

F28.11 Wave functions for a particle in an infinite box ($n = 1, 2, 3,$ and 4) (p. 1021)

F28.14 Wave functions for a particle in a finite box ($n = 1, 2,$ and 3) (p. 1022)

CONCEPTUAL QUESTIONS

1. An electron diffraction experiment gives the same pattern as an x-ray diffraction experiment with the same sample. How do we know the wavelengths of the electrons and x-rays are the same? Would they give the same pattern if their *energies* were the same?

2. In the Bohr model, the electron in the ground state of the hydrogen atom is in a circular orbit of radius 0.0529 nm. Explain how the Bohr model is incompatible with the uncertainty principle. How does the quantum mechanical picture of the H atom differ from the Bohr model? In what ways are the two similar?

3. It is sometimes said that, at absolute zero, all molecular motion, vibration, and rotation would cease. Do you agree? Explain.

4. The uncertainty principle does not allow us to think of the electron in an atom as following a well-defined trajectory. Why, then, are we able to define trajectories for golf balls, comets, and the like? [*Hint:* How are the uncertainties in momentum and velocity related?]

5. We often refer to the state of the hydrogen atom as "the $n = 3$ state," for example. Under what circumstances do we only need to specify one of the four quantum numbers? Under what circumstances would we have to be more specific?

6. Why does a particle confined to a finite box have only a finite number of bound states?

7. How should we interpret electron cloud representations of electron states in atoms?

8. Describe some differences between the beam of light from a flashlight and from a laser.

9. In an optically pumped laser, the light that causes optical pumping is always shorter in wavelength than the laser beam. Explain.

10. Explain why a population inversion is necessary in a laser.

11. The Nd:YAG laser operates in a four-state cycle (Fig. 28.31), while the ruby laser operates in a three-state cycle (compare Fig. 28.25b). In which laser is it easier to maintain a population inversion? Why? Explain why the Nd:YAG laser can produce a continuous beam while the ruby laser can produce only brief pulses of laser light.

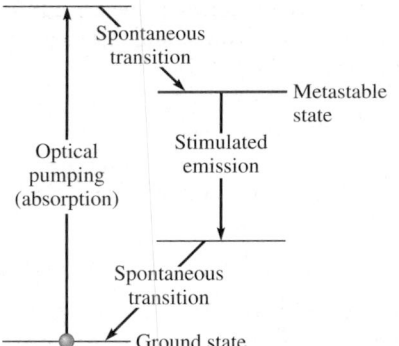

Figure 28.31
Conceptual Question 11

12. What do the ground-state electron configurations of the noble gases have in common? Why are the noble gases chemically nonreactive?

13. Central to the operation of a photocopy machine (see Section 16.2) is a drum coated with a photoconductor—a semiconductor that is a good insulator in the dark but allows charge to flow freely when illuminated with light. How does light allow charge to flow freely through the semiconducting material? How large should the band gap be for a good photoconductor? If the drum gets hot, is the contrast between light and dark areas on the image improved or degraded?

14. Why does a confined particle have quantized energy levels?

15. How can we demonstrate the existence of matter waves?

16. When a particle's kinetic energy increases, what happens to its de Broglie wavelength?

17. Explain why the electrical resistivity of a semiconductor decreases with increasing temperature.

18. When aluminum is exposed to oxygen, a *very thin* layer of aluminum oxide forms on the outside. Aluminum oxide is a good insulator. Nevertheless, if two aluminum wires are twisted together, electric current can flow from one to the other, even if the oxide layer has not been cleaned off. How is this possible?

MULTIPLE CHOICE QUESTIONS

1. What happens to the energy level spacings for a particle in a box when the box is made longer?

   (a) The spacings decrease.

   (b) The spacings increase.

   (c) The spacings stay the same.

   (d) Insufficient information to answer this question.

2. Which of the lettered transitions on the energy level diagram of Fig. 28.32 would be the best candidate for light amplification by stimulated emission?

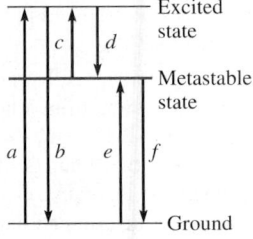

Figure 28.32 Multiple Choice Question 2

3. A bullet is fired from a rifle. The end of the barrel is a circular aperture. Is diffraction a measurable effect?

   (a) No, because only charged particles have de Broglie wavelengths.

   (b) No, because a circular aperture never causes diffraction.

   (c) No, because the de Broglie wavelength of the bullet is too large.

   (d) No, because the de Broglie wavelength of the bullet is too small.

   (e) Yes.

4. An electron and a neutron have the same de Broglie wavelength. Which is true?

   (a) The electron has more kinetic energy and a higher speed.

   (b) The electron has less kinetic energy but a higher speed.

   (c) The electron has less kinetic energy and a lower speed.

   (d) The electron and neutron have the same kinetic energy but the electron has the higher speed.

   (e) The neutron has more kinetic energy but the two have the same speed.

5. Which one of these statements is true?

   (a) The atomic spacing in crystals is too fine to produce observable diffraction effects with matter waves.

   (b) Only charged particles have matter waves associated with them.

   (c) Identical diffraction patterns are obtained when either electrons or neutrons of the same kinetic energy are incident on a single crystal.

   (d) Electrons, neutrons, and x-rays of appropriate energies can all produce similar diffraction patterns when incident on single crystals.

   (e) Wave phenomena are not observed for macroscopic objects such as baseballs because the de Broglie wavelength associated with such macroscopic objects is too long.

6. A particle is incident from the left on a slit of width w and the particle passes through the slit opening (Fig. 28.33). The uncertainty principle restricts which of these quantities?

   (a) the product of the width w and the minimum possible uncertainty in the y-component of the particle's momentum

   (b) the product of the width w and the minimum possible uncertainty in the x-component of the particle's momentum

   (c) the product of the width w and the minimum possible uncertainty in the z-component of the particle's momentum

   (d) the product of the width w and the minimum possible de Broglie wavelength of the particle

   (e) the maximum possible width w

Figure 28.33 Multiple Choice Question 6

7. The exclusion principle:

   (a) Implies that in an atom no two electrons can have identical sets of quantum numbers.

   (b) Says that no two electrons in an atom can have the same orbit.

   (c) Excludes electrons from atomic nuclei.

   (d) Excludes protons from electron orbits.

8. Which one of these statements is true?

   (a) The principal quantum number of the electron in a hydrogen atom does not affect its energy.

   (b) The principal quantum number of an electron in its ground state is zero.

   (c) The orbital angular momentum quantum number of an electron state is always less than the principal quantum number of that state.

   (d) The electron spin quantum number can take on any one of four different values.

9. If a particle is confined to a three-dimensional, cubical region of length L on each side:

   (a) The particle has an uncertainty in each component of momentum of about $h/(\pi L)$.

   (b) The particle cannot have a wavelength less than $2L$.

   (c) The components of the particle's momentum in the y- and z-directions can be determined exactly as long as there is a finite uncertainty in the x-component of momentum.

   (d) The particle's kinetic energy has an upper limit but no lower limit.

10. Which of these statements about electron energy levels in hydrogen atoms is true?

   (a) An electron in the hydrogen atom is best represented as a traveling wave.

   (b) An electron with positive total energy is a bound electron.

   (c) An electron in a stable energy level radiates electromag-

netic waves because the electron is accelerating as it moves around the nucleus.

   (d) The orbital angular momentum of an electron in the ground state is zero.

   (e) An electron in state n can make transitions only to the state $n + 1$ or $n - 1$.

PROBLEMS

Note: **C** indicates a combination conceptual/quantitative problem. Gold diamonds ✦, ✦✦ are used to indicate the increasing level of difficulty of each problem. Problem numbers appearing in blue, 9., denote problems that have a detailed solution available in the Student Solutions Manual. Some problems are *paired* by concept; their numbers are connected by a ruled box.

28.2 Matter Waves

1. What is the de Broglie wavelength of a basketball of mass 0.50 kg when it is moving at 10 m/s? Why don't we see diffraction effects when a basketball passes through the circular aperture of the hoop?

2. A fly with a mass of 1.0×10^{-4} kg crawls across a table at a speed of 2 mm/s. Compute the de Broglie wavelength of the fly and compare it to the size of a proton (about 1 fm).

3. What are the de Broglie wavelengths of electrons with the following values of kinetic energy? (a) 1.0 eV; (b) 1.0 keV.

4. What is the ratio of the wavelength of a 0.100-keV photon to the wavelength of a 0.100-keV electron?

5. What is the momentum of an electron with a de Broglie wavelength of 0.40 nm?

6. What is the de Broglie wavelength of an electron moving at speed $\frac{3}{5}c$?

7. The distance between atoms in a crystal of NaCl is 0.28 nm. The crystal is being studied in a neutron diffraction experiment. At what speed must the neutrons be moving so that their de Broglie wavelength is 0.28 nm?

8. An x-ray diffraction experiment using 16-keV x-rays is repeated using electrons instead of x-rays. What should the kinetic energy of the electrons be in order to produce the same diffraction pattern as the x-rays (using the same crystal)?

✦ 9. Neutron diffraction by a crystal can be used to make a velocity selector for neutrons. Suppose the spacing between the relevant planes in the crystal is $d = 0.20$ nm. A beam of neutrons is incident at an angle $\theta = 10.0°$ with respect to the planes. The incident neutrons have speeds ranging from 0 to 2.0×10^4 m/s. (a) What wavelength(s) are strongly reflected from these planes? [*Hint:* Bragg's law, Eq. (25-15), applies to neutron diffraction as well as to x-ray diffraction.] (b) For each of the wavelength(s), at what angle with respect to the incident beam do those neutrons emerge from the crystal?

10. A nickel crystal is used as a diffraction grating for x-rays. Then the same crystal is used to diffract electrons. If the two diffraction patterns are identical, and the energy of each x-ray photon is $E = 20.0$ keV, what is the kinetic energy of each electron?

28.3 Electron Microscopes

11. If diffraction were the only limitation on resolution, what would be the smallest structure that could be resolved in an electron microscope using 10-keV electrons?

12. A scanning electron microscope is used to look at cell structure with 10-nm resolution. A beam of electrons from a hot filament is accelerated with a voltage of 12 kV and then focused to a small spot on the specimen. (a) What is the wavelength in nanometers of the beam of incoming electrons? (b) If the size of the focal spot were determined only by diffraction, and if the diameter of the electron lens is $\frac{1}{5}$ the distance from the lens to the specimen, what would be the minimum separation resolvable on the specimen? (In practice, the resolution is limited much more by aberrations in the magnetic lenses and other factors.)

C 13. An image of a biological sample is to have a resolution of 5 nm. (a) What is the kinetic energy of a beam of electrons with a de Broglie wavelength of 5.0 nm? (b) Through what potential difference should the electrons be accelerated to have this wavelength? (c) Why not just use a light microscope with a wavelength of 5 nm to image the sample?

14. The phenomenon of Brownian motion is the random motion of microscopically small particles as they are buffeted by the still smaller molecules of a fluid in which they are suspended. For a particle of mass 1.0×10^{-16} kg, the fluctuations in velocity are of the order of 0.010 m/s. For comparison, how large is the change in this particle's velocity when the particle absorbs a photon of light with a wavelength of 660 nm, such as might be used in observing its motion under a microscope?

28.4 The Uncertainty Principle

15. If the momentum of the basketball in Problem 1 has a fractional uncertainty of $\Delta p/p = 10^{-6}$, what is the uncertainty in its position?

16. An electron passes through a slit of width 1.0×10^{-8} m. What is the uncertainty in the electron's momentum component in the direction perpendicular to the slit but in the plane containing the slit?

✦ 17. A beam of electrons passes through a single slit 40.0 nm wide. The width of the central fringe of a diffraction pattern formed on a screen 1.0 m away is 6.2 cm. What is the kinetic energy of the electrons passing through the slit?

✦ 18. Electrons are accelerated through a potential difference of 38.0 V. The beam of electrons then passes through a single slit. The width of the central fringe of a diffraction pattern formed on a screen 1.00 m away is 1.13 mm. What is the width of the slit?

19. A hydrogen atom has a radius of about 0.05 nm. (a) Estimate the uncertainty in any component of the momentum of an electron confined to a region of this size. (b) From your answer to (a), estimate the electron's kinetic energy. (c) Does the estimate have the correct order of magnitude? (The ground-state kinetic energy predicted by the Bohr model is 13.6 eV.)

C 20. A bullet with mass 10.000 g has a speed of 300.00 m/s; the speed is accurate to within 0.04%. (a) Estimate the minimum uncertainty in the position of the bullet, according to the uncertainty principle. (b) An electron has a speed of 300.00 m/s, accurate to 0.04%. Estimate the minimum uncertainty in the position of the electron. (c) What can you conclude from these results?

✦ 21. The omega particle (Ω) decays on average about 0.1 ns after it is created. Its rest energy is 1672 MeV. Estimate the fractional uncertainty in the Ω's rest energy ($\Delta E_0/E_0$). [Hint: Use the energy-time uncertainty principle, Eq. (28-3).]

✦ 22. Nuclei have energy levels just as atoms do. An excited nucleus can make a transition to a lower energy level by emitting a gamma-ray photon. The lifetime of a typical nuclear excited state is about 1 ps. What is the uncertainty in the energy of the gamma-rays emitted by a typical nuclear excited state? [Hint: Use the energy-time uncertainty principle, Eq. (28-3).]

23. A radar pulse has an average wavelength of 1.0 cm and lasts for 0.10 μs. (a) What is the average energy of the photons? (b) Approximately what is the least possible uncertainty in the energy of the photons?

28.5 Wave Functions for a Confined Particle

24. What is the minimum kinetic energy of an electron confined to a region the size of an atomic nucleus (1.0 fm)?

25. An electron is confined to a box of length 1.0 nm. What is the magnitude of its momentum in the $n = 4$ state?

26. A marble of mass 10 g is confined to a box 10 cm long and moves at a speed of 2 cm/s. (a) What is the marble's quantum number n? (b) Why can we not observe the quantization of the marble's energy? [Hint: Calculate the energy difference between the nth and $n + 1$st states. How much does the marble's speed change?]

27. An electron confined to a one-dimensional box has a ground-state energy of 40.0 eV. (a) If the electron makes a transition from its first excited state to the ground state, what is the wavelength of the emitted photon? (b) If the box were somehow made twice as long, how would the photon's energy change for the same transition (first excited state to ground state)?

✦ 28. An electron is confined to a one-dimensional box. When the electron makes a transition from its first excited state to the ground state, it emits a photon of energy 1.2 eV. (a) What is the ground-state energy (in eV) of the electron? (b) List all energies (in eV) of photons that could be emitted when the electron starts in its second excited state and makes transi-

tions downward to the ground state either directly or through intervening states. Show all these transitions on an energy level diagram. (c) What is the length of the box (in nm)?

29. The particle in a box model is often used to make rough estimates of energy level spacings. For a metal wire 10 cm long, treat a conduction electron as a free particle confined to a one-dimensional box of length 10 cm. (a) Sketch the wave function ψ as a function of position for the electron in this box for the ground state and each of the first three excited states. (b) Estimate the spacing between energy levels of the conduction electrons by finding the energy *spacing* between the ground state and the first excited state.

30. The particle in a box model is often used to make rough estimates of ground-state energies. Suppose that you have a *neutron* confined to a one-dimensional box of length equal to a nuclear diameter (say 10^{-14} m). What is the ground-state energy of the confined neutron?

✦ 31. (a) Make a qualitative sketch of the wave function for the $n = 5$ state of an electron in a *finite* box [$U(x) = 0$ for $0 < x < L$ and $U(x) = U_0 > 0$ elsewhere]. (b) If $L = 1.0$ nm and $U_0 = 1.0$ keV, *estimate* the number of bound states that exist.

C ✦ 32. An electron is confined to a one-dimensional box of length L. (a) Sketch the wave function for the third excited state. (b) What is the energy of the third excited state? (c) The potential energy can't really be infinite outside of the box. Suppose that $U(x) = +U_0$ outside the box, where U_0 is large but finite. Sketch the wave function for the third excited state of the electron in the finite box. (d) Is the energy of the third excited state for the finite box less than, greater than, or equal to the value calculated in part (b)? *Explain your reasoning.* [Hint: Compare the wavelengths inside the box.] (e) Give a rough estimate of the number of bound states for the electron in the *finite* box in terms of L and U_0.

28.6 The Hydrogen Atom: Wave Functions and Quantum Numbers; 28.7 The Exclusion Principle: Electron Configurations for Atoms Other than Hydrogen

33. How many electron states of the H atom have the quantum numbers $n = 3$ and $l = 1$? Identify each state by listing its quantum numbers.

34. What are the possible values of L_z (the component of angular momentum along the z-axis) for the electron in the second excited state ($n = 3$) of the hydrogen atom?

C ✦ 35. (a) Show that the ground-state energy of the hydrogen atom can be written $E_1 = -ke^2/(2a_0)$, where a_0 is the Bohr radius. (b) Explain why, according to classical physics, an electron with energy E_1 could never be found at a distance greater than $2a_0$ from the nucleus.

36. What is the largest number of electrons with the same pair of values for n and l that an atom can have?

37. List the number of electron states in each of the subshells in the $n = 7$ shell. What is the total number of electron states in this shell?

38. What is the ground-state electron configuration of bromine (Br, atomic number 35)?

39. What is the ground-state electron configuration of nickel (Ni, atomic number 28)?

© 40. (a) What are the electron configurations of the ground states of lithium ($Z = 3$), sodium ($Z = 11$), and potassium ($Z = 19$)? (b) Why are these elements placed in the same column of the Periodic Table?

© 41. (a) What are the electron configurations of the ground states of fluorine ($Z = 9$) and chlorine ($Z = 17$)? (b) Why are these elements placed in the same column of the Periodic Table?

42. What is the electronic configuration of the ground state of the carbon atom? Write it in the following ways: (a) using spectroscopic notation ($1s^2$ . . .); (b) listing the four quantum numbers for each of the electrons. Note that there may be more than one possibility in (b).

28.8 Electron Energy Levels in a Solid

43. A light-emitting diode (LED) has the property that electrons can be excited into the conduction band by the electrical energy from a battery; a photon is emitted when the electron drops back to the valence band. (a) If the band gap for this diode is 2.36 eV, what is the wavelength of the light emitted by the LED? (b) What color is the light emitted? (c) What is the minimum battery voltage required in the electrical circuit containing the diode?

44. A photoconductor (see Conceptual Question 13) allows charge to flow freely when photons of wavelength 640 nm or less are incident on it. What is the band gap for this photoconductor?

28.9 Lasers

45. What is the wavelength of the light usually emitted by a helium-neon laser? (See Fig. 28.26.)

46. Many lasers, including the helium-neon, can produce beams at more than one wavelength. Photons can stimulate emission and cause transitions between the 20.66-eV metastable state and several different states of lower energy. One such state is 18.38 eV above the ground state. What is the wavelength for this transition? If only these photons leave the laser to form the beam, what color is the beam?

47. In a ruby laser, laser light of wavelength 694.3 nm is emitted. The ruby crystal is 6.00 cm long, and the index of refraction of ruby is 1.75. Think of the light in the ruby crystal as a standing wave along the length of the crystal. How many wavelengths fit in the crystal? (Standing waves in the crystal help to reduce the range of wavelengths in the beam.)

48. The beam emerging from a ruby laser passes through a circular aperture 5.0 mm in diameter. (a) If the spread of the beam is limited only by diffraction, what is the angular spread of the beam? (b) If the beam is aimed at the Moon, how large a spot would be illuminated on the Moon's surface?

28.10 Tunneling

♦ 49. A proton and a deuteron (which has the same charge as the proton but 2.0 times the mass) are incident on a barrier of thickness 10.0 fm and "height" 10.0 MeV. Each particle has a kinetic energy of 3.0 MeV. (a) Which particle has the higher probability of tunneling through the barrier? (b) Find the ratio of the tunneling probabilities.

♦ 50. Refer to Example 28.6. Estimate the percentage change in the tunneling current if the distance between the sample surface and the STM tip increases 2.0%.

COMPREHENSIVE PROBLEMS

© 51. A bullet leaves the barrel of a rifle with a speed of 300.0 m/s. The mass of the bullet is 10.0 g. (a) What is the de Broglie wavelength of the bullet? (b) Compare λ to the diameter of a proton (about 1 fm). (c) Is it possible to observe wave properties of the bullet, such as diffraction? Explain.

♦ 52. In the Davisson-Germer experiment (Section 28.2), the electrons were accelerated through a 54.0-V potential difference before striking the target. (a) Find the de Broglie wavelength of the electrons. (b) Bragg plane spacings for nickel were known at the time; they had been determined through x-ray diffraction studies. The largest plane spacing (which gives the largest intensity diffraction maxima) in nickel is 0.091 nm. Using Bragg's law [Eq. (25-15)], find the Bragg angle for the first-order maximum using the de Broglie wavelength of the electrons. (c) Does this agree with the observed maximum at a scattering angle of 130°? [*Hint:* The scattering angle and the Bragg angle are not the same. Make a sketch to show the relationship between the two angles.]

♦ 53. A beam of neutrons has the same de Broglie wavelength as a beam of photons. Is it possible that the energy of each photon is equal to the kinetic energy of each neutron? If so, at what de Broglie wavelength(s) does this occur? [*Hint:* For the neutron, use the relativistic energy-momentum relation $E^2 = E_0^2 + (pc)^2$.]

54. A beam of neutrons is used to study molecular structure through a series of diffraction experiments. A beam of neutrons with a wide range of de Broglie wavelengths comes from the core of a nuclear reactor. In a time-of-flight technique, used to select neutrons with a small range of de Broglie wavelengths, a pulse of neutrons is allowed to escape from the reactor by opening a shutter very briefly. At a distance of 16.4 m downstream, a second shutter is opened very briefly 13.0 ms after the first shutter. (a) What is the speed of the neutrons selected? (b) What is the de Broglie wavelength of the neutrons? (c) If each shutter is open for 0.45 ms, estimate the *range* of de Broglie wavelengths selected.

55. In Fig. 28.4b, the x-rays had a frequency of 1.0×10^{19} Hz. Through what potential difference were the electrons in Fig. 28.4a accelerated?

56. The neutrons produced in fission reactors have a wide range of kinetic energies. After the neutrons make several collisions with atoms, they give up their excess kinetic energy

and are left with the same *average* kinetic energy as the atoms, which is $\frac{3}{2}k_BT$. If the temperature of the reactor core is $T = 400.0$ K, find (a) the average kinetic energy of the thermal neutrons, and (b) the de Broglie wavelength of a neutron with this kinetic energy.

57. A double-slit interference experiment is performed with 2.0-eV photons. The same pair of slits is then used for an experiment with electrons. What is the kinetic energy of the electrons if the interference pattern is the same as for the photons (that is, the spacing between maxima is the same)?

58. The particle in a box model is often used to make rough estimates of energy level spacings. Suppose that you have a proton confined to a one-dimensional box of length equal to a nuclear diameter (about 10^{-14} m). (a) What is the energy difference between the first excited state and the ground state of this proton in the box? (b) If this energy is emitted as a photon as the excited proton falls back to the ground state, what is the wavelength and frequency of the electromagnetic wave emitted? In what part of the spectrum does it lie? (c) Sketch the wave function ψ as a function of position for the proton in this box for the ground state and each of the first three excited states.

C 59. An electron is confined to a one-dimensional box of length L. When the electron makes a transition from its first excited state to the ground state, it emits a photon of energy 0.20 eV. (a) What is the ground-state energy (in eV) of the electron in this box? (b) What are the energies (in eV) of the photons that can be emitted when the electron starts in its third excited state and makes transitions downwards to the ground state (either directly or through intervening states)? (c) Sketch the wave function of the electron in the third excited state. (d) If the box were somehow made longer, how would the electron's new energy level spacings compare with its old ones? (Would they be greater, smaller, or the same? Or is more information needed to answer this question? Explain.)

◆ 60. An electron in a one-dimensional box has ground-state energy 0.010 eV. (a) What is the length of the box? (b) Sketch the wave functions for the lowest three energy states of the electron. (c) What is the wavelength of the electron in its second excited state ($n = 3$)? (d) The electron is in its ground state when it absorbs a photon of wavelength 15.5 μm. Find the wavelengths of the photon(s) that could be emitted by the electron subsequently.

61. An electron is confined in a one-dimensional box of length L. Another electron is confined in a box of length $2L$. Both are in the ground state. What is the ratio of their energies E_{2L}/E_L?

C 62. Before the discovery of the neutron, one theory of the nucleus proposed that the nucleus contains protons and electrons. For example, the helium-4 nucleus would contain 4 protons and 2 electrons instead of—as we now know to be true—2 protons and 2 neutrons. (a) *Assuming that the electron moves at nonrelativistic speeds*, find the ground-state

energy in MeV of an electron confined to a one-dimensional box of length 5.0 fm (the approximate diameter of the ^{4}He nucleus). (The electron actually does move at relativistic speeds, but we are only interested in a rough estimate.) (b) What can you conclude about the electron-proton model of the nucleus? The binding energy of the ^{4}He nucleus—the energy that would have to be supplied to break the nucleus into its constituent particles—is about 28 MeV. (c) Repeat (a) for a neutron confined to the nucleus (instead of an electron). Compare your result to (a) and comment on the viability of the proton-neutron theory compared to the electron-proton theory.

C ◆ 63. The kinetic energy of the electron found in Problem 62 part (a) is much greater than the electron's rest energy (0.511 MeV). (a) Use relativistically correct equations to calculate the electron's kinetic energy. (b) Is the conclusion in Problem 62 part (b) still true? Explain.

64. What is the ground-state electron configuration of tellurium (Te, atomic number 52)?

65. A free neutron (that is, a neutron on its own rather than in a nucleus) is not a stable particle. Its average lifetime is 15 min, after which it decays into a proton, an electron, and an antineutrino. Use the energy-time uncertainty principle [Eq. (28-3)] and the relationship between mass and rest energy to estimate the inherent uncertainty in the mass of a free neutron. Compare to the average neutron mass of 1.67×10^{-27} kg. (While the uncertainty in the neutron's mass is far too small to be measured, unstable particles with extremely short lifetimes have marked variation in their measured masses.)

C ◆ 66. A particle is confined to a *finite* box of length L. In the nth state, the wave function has $n - 1$ nodes. The wave function must make a smooth transition from sinusoidal inside the box to a decaying exponential outside—there can't be a kink at the wall. (a) Make some sketches to show that the wavelength λ_n inside the box must fall in the range $2L/n < \lambda_n < 2L/(n-1)$. (b) Show that the energy levels E_n in the finite box satisfy $(n-1)^2E_1 < E_n < n^2E_1$, where $E_1 = h^2/(8mL^2)$ is the ground-state energy for an *infinite* box of length L.

◆ 67. (a) Show that the number of electron states in a subshell is $4l + 2$. [*Hint:* First, how many states are in each orbital? Second, how many orbitals are in each subshell?] (b) By summing the number of states in each of the subshells, show that the number of states in a shell is $2n^2$. [*Hint:* Use the following theorem: the sum of the first n odd integers, from 1 to $2n - 1$, is n^2. That comes from regrouping the sum in pairs, starting by adding the largest to the smallest:

$$1 + 3 + 5 + \cdots + (2n-5) + (2n-3) + (2n-1)$$
$$= [1 + (2n-1)] + [3 + (2n-3)] + [5 + (2n-5)] + \cdots$$
$$= 2n + 2n + 2n + \cdots = 2n \times \frac{n}{2} = n^2$$

ANSWERS TO PRACTICE PROBLEMS

28.1 0.26 eV

28.2 Increasing energy ⇒ decreasing wavelength; decreasing the wavelength decreases θ for a given fringe, so the spacing between fringes decreases (the pattern contracts).

28.3 $\Delta p_x \approx 10^{-26}$ kg·m/s; $\Delta v_x = \Delta p_x/m \approx 10$ km/s

28.4 $1s^2 2s^2 2p^6 3s^2 3p^3$

28.5 A ruby laser would be ineffective. Blood appears red because it *reflects* red light; the red light emitted by a ruby laser would be largely reflected rather than absorbed.

28.6 −13.5% (a decrease)

Nuclear Physics

After more than 300 years, Rembrandt's 1653 painting *Aristotle with a Bust of Homer* needed to be cleaned. Aristotle's black apron showed signs of damage; it was unclear whether any of the original paint had survived underneath the apron. Conservators at the Metropolitan Museum of Art (New York) needed to know as much as possible about the damaged area before undertaking the painting's restoration and cleaning. Art historians wanted to know whether Rembrandt altered the composition as he worked on the painting. To help provide such information, the painting was taken to a nuclear reactor at the Brookhaven National Laboratory. How can a nuclear reactor help conservators and art historians learn about a painting?

29.1 NUCLEAR STRUCTURE

In an atom, electrons are bound electrically to a positively charged nucleus. In Chapters 27 and 28, we generally treated the nucleus as a point charge so massive that it is not affected by electrical forces on it due to the electrons. In reality, the atomic nucleus is several thousand times more massive than the electrons in an atom and occupies only a tiny fraction of the atom's volume (about 1 part in 10^{12} or less). The finite mass and volume of the nucleus have subtle effects on the electronic configuration and thus on the chemical properties of atoms. However, the nucleus has a complex structure of its own that manifests itself in radioactive decay and nuclear reactions.

The nucleus is a bound collection of protons and neutrons. Together, protons and neutrons are referred to as **nucleons** (particles found inside the nucleus). The **atomic number** Z is the number of protons in the nucleus. Each proton has a charge of $+e$ and the neutron is uncharged, so the electrical charge of a nucleus is $+Ze$. The number of electrons in a neutral atom is also equal to Z. The value of Z alone determines to which element, or chemical species, an atom belongs.

Once it was thought that all atoms of a given element were identical. However, we now know that there exist different **isotopes** of a given element. The isotopes of an element all have the same number of protons in the nucleus, but they have different masses because the number of neutrons (N) differs. The total number of nucleons therefore also differs from one isotope to another. The **nucleon number** A is the total number of protons and neutrons:

$$A = Z + N \tag{29-1}$$

Any particular species of nucleus, called a **nuclide**, is characterized by the values of A and Z. The nucleon number A is also called the **atomic mass number**. Since almost all of the mass of an atom is found in the nucleus, and since protons and neutrons have *approximately* the same mass, the mass of an atom is roughly proportional to the number of nucleons.

Since their masses differ, the isotopes of an element can be separated using a mass spectrometer (Section 19.3). Sometimes the differing masses of isotopes have an effect on chemical reaction rates, but on the whole, the chemical properties of different isotopes are virtually identical. On the other hand, different nuclides have *very* different nuclear properties. The number of neutrons present affects how strongly the nucleus is held together, so that some are stable while others are unstable (**radioactive**). Nuclear energy levels, radioactive half-lives, and radioactive decay modes are all particular to a specific nuclide; they are very different for two isotopes of the same element.

Several notations are used to distinguish nuclides. The chemical symbol O stands for the element oxygen. To specify a particular isotope of oxygen, the mass number must also be specified. Oxygen-18, O-18, O^{18}, and ^{18}O all stand for the isotope of oxygen with $A = 18$. Sometimes it is helpful to include the atomic number as well, even though it is redundant; oxygen by definition has $Z = 8$. When including the atomic number, the preferred form is $^{18}_{8}O$, although $_{8}O^{18}$ is found in some older texts.

Example 29.1

Finding the Number of Neutrons

How many neutrons are present in an ^{18}O nucleus?

Strategy The superscript gives the number of nucleons (A). We consult the Periodic Table to find the atomic number (Z) for oxygen. The number of neutrons is $N = A - Z$.

Solution An ^{18}O nucleus has 18 nucleons. Oxygen has atomic number 8, so there are 8 protons in the nucleus. That leaves $18 - 8 = 10$ neutrons.

Discussion Different isotopes of oxygen have different numbers of neutrons but the same number of protons.

Practice Problem 29.1   Identifying the element

Write the symbol (in the form $^{A}_{Z}X$) for the nuclide with 44 protons and 60 neutrons and identify the element.

It is usually more convenient to write the mass of a nucleus in **atomic mass units** instead of kg. The modern symbol for the atomic mass unit is "u"; in older literature it is often written "amu." The atomic mass unit is defined as exactly $\frac{1}{12}$ the mass of a neutral ^{12}C atom. The conversion factor between u and kg is

$$1 \text{ u} = 1.660\ 539 \times 10^{-27} \text{ kg} \qquad (29\text{-}2)$$

Nucleons have masses of *approximately* 1 u, while the electron is much less massive (Table 29.1). Therefore, the mass of a nucleus (or an atom) is *approximately* A atomic mass units—which is why A is called the mass number.

Table 29.1

Masses of Proton, Neutron, and Electron in Atomic Mass Units

Particle	Mass (u)
Proton	1.007 276 5
Neutron	1.008 664 9
Electron	0.000 548 6

Example 29.2

Estimating Mass

Estimate the mass in kg of one mole of ^{14}C.

Strategy   We can estimate 1 u of mass for each nucleon and ignore the relatively small mass of the electrons. One mole contains Avogadro's number of atoms. Then we convert u to kg.

Solution   A ^{14}C nucleus has 14 nucleons, so the mass of the ^{14}C atom is roughly 14 u. One mole contains Avogadro's number of atoms; therefore the mass of one mole is roughly

$$M = N_A m = 6.02 \times 10^{23} \times 14 \text{ u} = 8.4 \times 10^{24} \text{ u}$$

Now we convert to kg:

$$8.4 \times 10^{24} \text{ u} \times 1.66 \times 10^{-27} \text{ kg/u} = 0.014 \text{ kg}$$

Discussion   Note that the mass of one mole of an isotope with mass number 14 is approximately 14 g. The atomic mass unit is defined so that the mass of one atom in u is numerically equal to the mass of one mole of atoms in grams.

The mass of a nucleus is not exactly equal to A atomic mass units for two reasons. The masses of the proton and neutron are not exactly 1 u. Even if they were, as we see in Section 29.2, the mass of a nucleus is *less than* the total mass of its individual protons and neutrons.

Practice Problem 29.2   Estimating the mass of a nucleus in u

Approximately what is the mass in u of an oxygen nucleus that has nine neutrons?

The atomic mass of an element given in the Periodic Table is an average over the isotopes of that element in their natural relative abundances on Earth. In nuclear physics we must consult a table of nuclides (Appendix B) for the mass of a specific nuclide.

Sizes of Nuclei

How do we know the *sizes* of nuclei? The first experimental evidence came from the Rutherford scattering of alpha particles from gold nuclei (Section 27.6). From analysis of the number of alpha particles observed at different scattering angles, we can estimate the size of the gold nucleus. Similar experiments were performed on other nuclei. More recently, electron diffraction has been used to probe the structure of the nucleus. Using electrons of very short wavelength, we can determine not only the size of the nucleus but learn about its internal structure as well.

These and other experiments show that the mass density of all nuclei is approximately the same—the volume of a nucleus is proportional to its mass. Imagine a nucleus to be like a spherical container full of marbles (Fig. 29.1); each marble represents a nucleon. The nucleons are tightly packed together, as if touching each other. Both the mass and volume of the nucleus are proportional to the number of nucleons, so the mass per unit volume (density ρ) is approximately independent of the number of nucleons. If M is the mass of a nucleus and V is its volume, then

$$M \propto A \quad \text{and} \quad V \propto A$$
$$\Rightarrow \rho = \frac{M}{V} \text{ is independent of } A$$

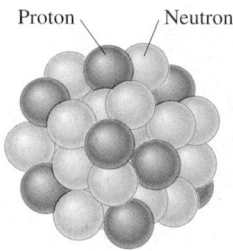

Proton   Neutron

Figure 29.1   Model of a nucleus as a set of hard spheres (representing the nucleons) packed together into a sphere.

Most nuclei are approximately spherical in shape, so

$$V = \tfrac{4}{3}\pi r^3 \propto A$$

$$\Rightarrow r^3 \propto A \quad \text{or} \quad r \propto A^{1/3} \tag{29-3}$$

The radius of a nucleus is proportional to the cube root of its mass number. Experiment shows that the constant of proportionality is approximately 1.2×10^{-15} m:

Radius of a nucleus	
$r = r_0 A^{1/3}$	(29-4)
$r_0 = 1.2 \times 10^{-15} \text{ m} = 1.2 \text{ fm}$	(29-5)

The SI prefix "f-" stands for *femto*; the fm is properly called a *femtometer* but is also called a *fermi*, after Enrico Fermi (1901–1954). The nuclear radius ranges from 1.2 fm (for ordinary hydrogen) to 7.7 fm (for $A \approx 260$).

 Although nuclei all have about the same mass density, *atoms* do not. More massive atoms are generally more dense than lighter atoms. The increase in volume of an atom does not keep pace with the increase in mass. Although larger atoms have more electrons, these electrons are on average more tightly bound, due to the increased charge of the nucleus. Thus, some solids and liquids (in which the atoms are tightly packed) are more dense than others.

Example 29.3

Radius and Volume of Barium Nucleus

What are the radius and volume of the barium-138 nucleus?

Strategy To find the radius of a nucleus, all we need to know is the mass number A, which in this case is 138. To find the volume, we assume that the nucleus is approximately spherical.

Solution To find the radius we apply Eq. (29-4), substituting $A = 138$:

$$r = r_0 A^{1/3}$$

$$r = 1.2 \text{ fm} \times 138^{1/3} = 6.2 \text{ fm}$$

The approximate volume of the nucleus is

$$V = \tfrac{4}{3}\pi r^3$$

Cubing Eq. (29-4) yields

$$r^3 = r_0^3 A$$

Therefore, the volume of a nucleus is approximately

$$V = \tfrac{4}{3}\pi r_0^3 A$$

Now we substitute numerical values.

$$V = \tfrac{4}{3}\pi \times (1.2 \times 10^{-15} \text{ m})^3 \times 138 = 1.0 \times 10^{-42} \text{ m}^3$$

Discussion The radius (6.2 fm) is within the expected range of 1.2 fm to 7.7 fm. The equation $V = \tfrac{4}{3}\pi r_0^3 A$ says the volume of a nucleus is proportional to the number of nucleons (A), as expected; each nucleon occupies a volume of $\tfrac{4}{3}\pi r_0^3$.

Practice Problem 29.3   Volume of a radium nucleus

What is the volume of a radium-226 nucleus?

29.2   BINDING ENERGY

What holds the nucleons together in a nucleus? Gravity is far too weak to do it; electrical forces push protons *away* from each other. The nucleons are held together by the **strong nuclear force**, one of the three fundamental interactions (along with gravity and the electroweak interaction) discussed in Section 2.2. The strong nuclear force makes little distinction between protons and neutrons.

Unlike gravity and the electromagnetic forces, the strong nuclear force is extremely short range. The ranges of the gravitational and electromagnetic forces are infinite, with the magnitude of the force between point objects falling off with distance as $1/r^2$. By contrast, the strong nuclear force between two nucleons is significant only at distances of about 3.0 fm or less. Because the strong force is so short range, a nucleon is attracted only to its *nearest neighbors* in the nucleus. On the other hand, since electrical repulsion is long range, every proton in the nucleus repels *all* the other protons. These two competing forces determine which nuclei are stable.

The Liquid Drop Model

In 1936, Niels Bohr proposed the **liquid drop model** of the nucleus. This model exploits several similarities between the nucleons in a nucleus and the molecules in a liquid such as water. Water is nearly incompressible, so the density of a drop of water is the same regardless of the size of the drop—just as the density of the nucleus is independent of size.

The forces that act between water molecules are similar in many ways to the forces between nucleons in the nucleus. In a drop of water, a water molecule exerts an attractive force on each of its nearest neighbors, but essentially no force on molecules that are not "touching." Energy must be supplied to remove a water molecule from the drop, since work must be done to overcome the attractive forces between the molecule and its neighbors. If two water molecules get *too* close together, then the intermolecular force pushes them apart. The intermolecular forces are electromagnetic in origin, but they are much weaker than the electromagnetic forces that hold the three atoms of the water molecule together.

The intermolecular forces in a drop of water give rise to surface tension (see Section 9.11). The molecules at the surface have fewer neighbors than do molecules in the interior. The cohesive forces that act in all directions on a molecule inside the drop produce a net inward pull on surface molecules. The surface molecules are less tightly bound than are the molecules in the interior. Thus, to minimize its energy, the surface tension pulls the drop into a spherical shape (in the absence of gravity).

Forces between nucleons in a nucleus share many of these features. A nucleon exerts an attractive force on its nearest neighbors only; this force turns repulsive if the nucleons get too close to one another. The nucleus is a bound system; energy must be supplied to remove a nucleon, since it is bound to its neighbors.

The nucleus also experiences something analogous to surface tension. A nucleon in the interior of the nucleus is completely surrounded by other nucleons and is attracted to all of them. A nucleon on the surface of the nucleus does not have as many nearest neighbors, so it is not as tightly bound as the nucleons in the interior. Because nucleons at the surface are pulled inward, the nucleus forms a spherical shape. However, if disturbed, the shape of the nucleus can be deformed from its normal shape (see Section 29.7). The nucleons are not locked into place like the atoms in a solid; they move around, more like the molecules in a liquid.

Binding Energy and Mass Defect

The **binding energy** E_B of a nucleus is the energy that must be supplied to separate a nucleus into individual protons and neutrons. Since the nucleus is a bound system, its total energy is *less* than the energy of Z protons and N neutrons that are far apart and at rest.

Binding energy

E_B = (total energy of Z protons and N neutrons) − (total energy of nucleus)   (29-6)

The concept of binding energy gives a way of looking at how the nucleus is held together in terms of energy instead of forces.

The concept of binding energy applies to systems other than nuclei. The total energy of a proton and an electron far from each other is 13.6 eV higher than the energy when the two are bound together in a hydrogen atom (in its ground state). Thus, the binding energy of the hydrogen *atom* is 13.6 eV. In atoms with more than one electron, the binding energy is not the same as the ionization energy. The ionization energy is the energy required to remove *one* electron; the binding energy is the energy required to remove *all of the electrons*.

The mass of a particle is a measure of its *rest energy*—its total energy in a reference frame in which it's at rest (see Section 26.7):

$$E_0 = mc^2 \qquad (26\text{-}7)$$

Since the rest energy of a nucleus is *less than* the total rest energy of Z protons and N neutrons, the mass of the nucleus is less than the total mass of the protons and neutrons. The **mass defect** Δm comes about because we would have to *add* energy to a nucleus to break it up into Z individual protons and N individual neutrons. The mass defect is related to the binding energy via Eq. (26-7).

Mass defect and binding energy

$$\Delta m = (\text{mass of } Z \text{ protons and } N \text{ neutrons}) - (\text{mass of nucleus}) \qquad (29\text{-}7)$$

$$E_B = \Delta m c^2 \qquad (29\text{-}8)$$

The energy unit most commonly used in nuclear physics is the MeV, which stands for mega-electron-volt and is read "em-ee-vee." In order to use MeV for energy and atomic mass units for mass in Eq. (29-8), it is convenient to know the value of c^2 in units of MeV/u. It can be shown (Problem 14) that

---

$$c^2 = 931.494 \text{ MeV/u} \qquad (29\text{-}9)$$

Mass tables such as Appendix B give the masses of *neutral atoms*, which include the masses of the electrons as well as the mass of the nucleus. To find the mass of a nucleus with atomic number Z, subtract the mass of Z electrons from the mass of the neutral atom. The binding energy of the electrons to the nucleus is much smaller than the rest energy of the electrons and can be ignored.

Example 29.4

Binding Energy of a Nitrogen-14 Nucleus

Find the binding energy of the ^{14}N nucleus.

Strategy From Appendix B, the mass of the ^{14}N *atom* is 14.003 074 u. The mass of the N atom includes the mass of seven electrons. Subtracting $7m_e$ from the mass of the atom gives the mass of the nucleus. Then we can find the mass defect and the binding energy.

Solution

$$\text{mass of } {}^{14}\text{N nucleus} = 14.003\,074 \text{ u} - 7m_e$$
$$= 14.003\,074 \text{ u} - 7 \times 0.000\,548\,6 \text{ u}$$
$$= 13.999\,234 \text{ u}$$

The ^{14}N nucleus has seven protons and seven neutrons. The mass defect is

$$\Delta m = (\text{mass of 7 protons and 7 neutrons}) - (\text{mass of nucleus})$$
$$= 7 \times 1.007\,276\,5 \text{ u} + 7 \times 1.008\,664\,9 \text{ u} - 13.999\,234 \text{ u}$$
$$= 0.112\,356 \text{ u}$$

The binding energy is therefore,

$$E_B = \Delta m c^2 = 0.112\,356 \text{ u} \times 931.494 \text{ MeV/u}$$
$$= 104.659 \text{ MeV}$$

Discussion Since the binding energy of the electrons in an atom is so small, we can assume that the mass of an atom is equal to the mass of its nucleus plus the mass of the electrons. As a shortcut, we can calculate the mass defect using the mass of the nitrogen *atom* instead of the nitrogen nucleus and the mass of the *hydrogen atom* instead of the proton. Since each term contains the extra mass of seven electrons, the masses of the electrons subtract out:

$$\Delta m = (\text{mass of 7 } {}^{1}\text{H atoms and 7 neutrons}) - (\text{mass of } {}^{14}\text{N atom})$$
$$= 7 \times 1.007\,825 \text{ u} + 7 \times 1.008\,664\,9 \text{ u} - 14.003\,074 \text{ u}$$
$$= 0.112\,355 \text{ u}$$

Practice Problem 29.4   Binding energy of nitrogen-15

Calculate the binding energy of the ^{15}N nucleus. The mass of the ^{15}N *nucleus* is 14.996 269 u. [*Hint:* This time you have been given the mass of the *nucleus*, not the mass of the atom.]

Nuclear Energy Levels

Neutrons and protons obey the exclusion principle: no two identical nucleons can be in the same quantum state. As for atomic energy levels, a group of closely spaced nuclear energy levels is called a *shell*. We can describe the quantum state of the nucleus by specifying how the proton and neutron states are occupied, much as we did for electron states in the atom (Fig. 29.2). Two protons can occupy each proton energy level (one spin up, one spin down) and two neutrons can occupy each neutron energy level. The energy levels for the proton and neutron are similar; as far as the nuclear force is concerned, protons and neutrons are pretty much the same. The main difference is that the protons are affected by the Coulomb repulsion in addition to the strong nuclear force.

In Problem 18, an order of magnitude calculation shows that the energy level spacings in nuclei are expected to be in the MeV range. The structure of the nucleus is complex; energy level spacings range from tens of keV to several MeV. A nucleus in an excited state can return to the ground state by emitting one or more *gamma-ray* photons. [The distinction between gamma rays and x-rays is based more on the source than the energy. A photon emitted by an excited nucleus or in pair annihilation (Section 27.8) is called a gamma ray; a high-energy photon emitted by an excited *atom,* by an electron slowing down on striking a target (Section 27.4), or by a circulating charged particle in a synchrotron is usually called an x-ray.] Just as the energy levels of atoms can be deduced by measuring the wavelengths of photons radiated by excited atoms, measurement of the gamma-ray energies emitted by excited nuclei enable us to deduce the nuclear energy levels. Each nuclide has its own characteristic gamma-ray spectrum, which can be used to identify it. A gamma-ray spectrum usually identifies the energy of the photons, in contrast to a visible spectrum where the wavelength is usually specified. In both cases, the quantity used is the one that is easiest to measure.

Energy level diagrams help explain why, in stable light nuclides, the number of neutrons and protons tends to be approximately equal. Figure 29.2 shows energy level diagrams for four different nuclides, each of which has 12 nucleons. The energy levels are *not* quantitatively correct, but serve to illustrate the general idea. A maximum of two protons can be in any proton energy level and a maximum of two neutrons can be in any neutron energy level. The proton and neutron energy levels are similar; the proton levels are slightly higher in energy than the neutron levels to account for the Coulomb repulsion between the protons. The energy is lower with 6 protons and 6 neutrons than is possible with 5 of one and 7 of the other.

The story is more complicated for heavier nuclides. The Coulomb repulsion between protons favors more neutrons ($N > Z$) since the neutrons are immune to the Coulomb repulsion. For larger nuclides, the Coulomb repulsion becomes more and

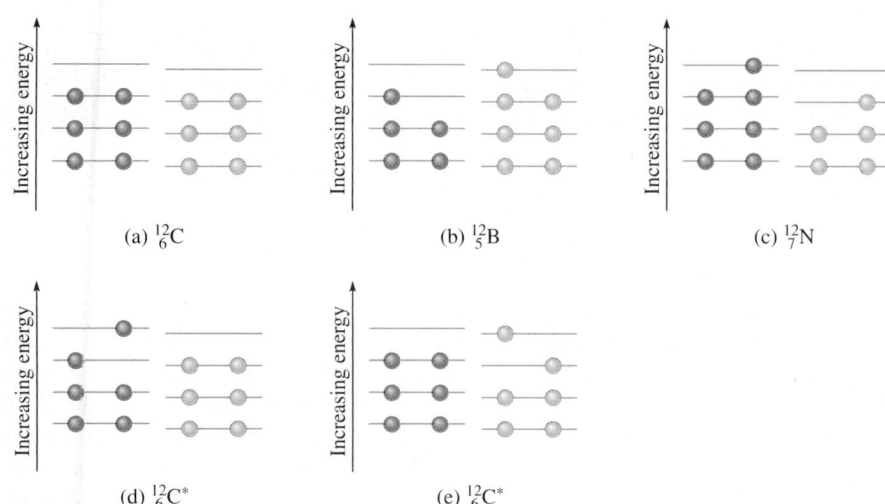

(a) $^{12}_{6}$C

(b) $^{12}_{5}$B

(c) $^{12}_{7}$N

(d) $^{12}_{6}$C*

(e) $^{12}_{6}$C*

Figure 29.2 Qualitative energy level diagrams for some nuclides with $A = 12$. Red spheres represent protons and gray spheres represent neutrons. $^{12}_{6}$C is stable, while $^{12}_{5}$B and $^{12}_{7}$N are unstable. The asterisks in (d) and (e) indicate that $^{12}_{6}$C is in an excited state.

Figure 29.3 Chart of the known nuclides. Stable nuclides are shown as green points. The stability curve is a best-fit curve that shows the general trend of increasing N/Z ratio for stable nuclides with respect to the reference line $N = Z$.

more important since it is long range: each proton repels *every other proton* in the nucleus. The proton energy levels get higher and higher with respect to the neutron energy levels as the electric potential energy of all those repelling protons adds up. Thus, large nuclides tend to have an excess of neutrons ($N > Z$). On the other hand, there is a limit to the neutron excess: neutrons are slightly more massive than protons, so if there is too much of a neutron excess, the mass (and therefore the energy) of the nucleus is higher than it would be if one or more neutrons were changed into protons.

Figure 29.3 shows the number of protons (Z) and number of neutrons (N) for the stable nuclides (represented as points in green). For the smallest nuclides, $N \approx Z$. As the total number of nucleons ($A = Z + N$) increases, the number of neutrons increases faster than the number of protons. The largest stable nuclides have about 1.5 times as many neutrons as protons.

In Chapter 28 we found that certain elements, called the noble gases, are chemically nonreactive due to the unusual stability of their electronic configurations. The numbers of electrons in the noble gas configurations are 2, 10, 18, 36, 54, or 86. These numbers arise when a group of electron energy levels is full and there is a larger than usual energy gap to the next higher energy level.

For nuclei, the so-called **magic numbers** are 2, 8, 20, 28, 50, 82, 126. Nuclides with a magic number of protons or a magic number of neutrons are particularly stable; they are more tightly bound than would otherwise be expected. If N and Z are *both* magic, then the nuclide is extremely stable. Only five doubly magic nuclei exist: $^4_2$He, $^{16}_8$O, $^{40}_{20}$Ca, $^{48}_{20}$Ca, and $^{208}_{82}$Pb. Calcium appears twice in this list with two different mass numbers, 40 and 48. These nuclides are relatively abundant in the universe due to their unusual stability.

The magic numbers are no more "magic" than the numbers of electrons in the noble gases; they just tell us how many nucleons fill up a shell (a closely spaced group of nuclear energy levels). Since there are two kinds of nucleons, a nuclide can have a full shell of neutrons, a full shell of protons, or both. A more general observation is that even numbers of protons or neutrons are more stable than odd numbers due to the exclusion principle. Protons pair up in energy levels—one with spin up and one with spin down—as do neutrons (see Fig. 29.2). Only four stable nuclides have both N and Z odd: ^{2}H, ^{6}Li, ^{10}B, and ^{14}N. By contrast, more than half of the stable nuclei have both N and Z even.

Figure 29.4 Binding energy per nucleon (E_B/A) as a function of nucleon number A. Only the most stable nuclide for each value of A is shown.

Binding Energy Curve

Figure 29.4 shows a graph of the binding energy *per nucleon* as a function of mass number. The smallest nuclides have such a small number of nucleons that the strong nuclear force is not saturated. Recall that the strong force binds nucleons only to their nearest neighbors. In small nuclides there are not enough nucleons for all to fully bind since the average number of nearest neighbors is small. Increasing the number of nucleons leads to a larger binding energy per nucleon, up to a point, because the average number of nearest neighbors is increasing. Thus, we see a steep increase in the binding energy per nucleon as A increases from 1.

Once nuclei reach a certain size, all nucleons except those on the surface have as many nearest neighbors as possible. Adding more nucleons doesn't increase the average binding energy due to the strong force very much, but the Coulomb repulsion keeps adding up since it is long range. Thus, above $A \approx 60$, adding more nucleons *decreases* the average binding energy per nucleon. The decrease is relatively gentle, compared to the steep increase for small A, since the Coulomb repulsion is weak compared to the strong nuclear force.

The binding energy per nucleon is within the range 7–9 MeV for all but the smallest nuclides. For example, in Example 29.4 we found that the binding energy of ^{14}N is 104.659 MeV. The binding energy *per nucleon* for ^{14}N is

$$\frac{104.659 \text{ MeV}}{14 \text{ nucleons}} = 7.475\,64 \text{ MeV/nucleon}$$

The most tightly bound nuclides are around $A \approx 60$, where the binding energy is about 8.7 MeV/nucleon. The first two doubly magic nuclides, $^{4}_{2}$He and $^{16}_{8}$O, stick way up above the general trend, showing their higher than usual binding energies.

29.3 RADIOACTIVITY

Henri Becquerel (1852–1908) discovered radioactivity in 1896 when, quite by accident, he found that a uranium salt spontaneously emitted radiation in the absence of an external source of energy, such as sunlight. The radiation exposed a photographic plate even though the plate was wrapped in black paper to keep light out.

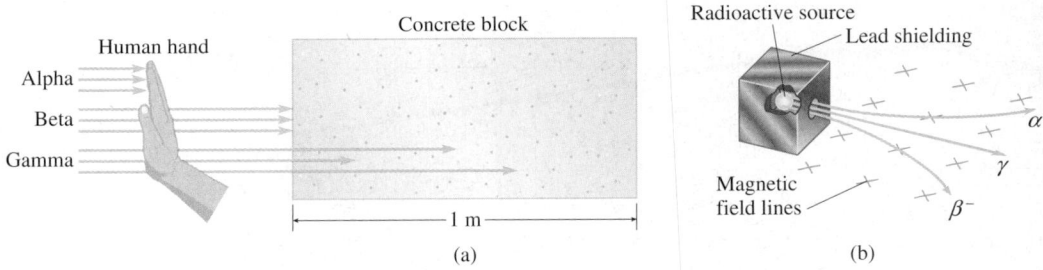

Figure 29.5 Alpha, beta, and gamma rays differ (a) in their abilities to penetrate matter as well as (b) in their electrical charges.

Nuclides can be divided into two broad categories. Some are stable; others are unstable or **radioactive**. An unstable nuclide **decays**—takes part in a spontaneous nuclear reaction—by emitting radiation. (The radiation may include but is not limited to *electromagnetic* radiation.) Depending on the kind of radiation emitted, the reaction may change the nucleus into a different nuclide, with a different charge or nucleon number or both.

Scientists studying radioactivity soon identified three different kinds of radiation emitted by radioactive nuclei; they were named alpha rays, beta rays, and gamma rays after the first three letters of the Greek alphabet. The initial distinction between the three was their differing abilities to penetrate matter (Fig. 29.5). Alpha rays are the least penetrating; they can only make it through a few centimeters of air and are completely blocked by human skin, thin paper, and other solids. Beta rays can travel farther in air—about a meter typically—and can penetrate a hand or a thin metal foil. Gamma rays are much more penetrating than either alpha or beta rays. Later, the electrical charge and mass were determined and used to distinguish the three types of radiation—and ultimately to identify them.

Of the approximately 1500 known nuclides, only about 20% are stable. All of the largest nuclides (those with $Z > 83$) are radioactive. As far as we know, stable nuclei last forever without decaying spontaneously. Each radioactive nuclide decays with an average lifetime characteristic of that nuclide. The lifetimes cover an enormous range, from about 10^{-22} s (roughly the time it takes light to travel a distance equal to the diameter of a nucleus) to 10^{+28} s (10^{10} times the age of the universe).

Conservation Laws in Radioactive Decay

In a nuclear reaction, whether spontaneous or not, the total electric charge is conserved. Another conservation law says that the number of nucleons must stay the same. We *balance* a nuclear reaction by applying these two conservation laws. It is helpful to write symbols for electrons, positrons, and neutrons as if they were nuclei, with a superscript for the number of nucleons and a subscript for the electrical charge in units of e (see Table 29.2). Then the reaction is balanced with regard to nucleon number if the sum of the superscripts is the same on both sides; it is balanced with regard to charge if the sum of the subscripts is the same on both sides.

Another conservation law is important in radioactive decay: all nuclear reactions also conserve energy. How can a nucleus with little or no kinetic energy decay, leaving products with significant kinetic energies? Where did this energy come from? In a spontaneous nuclear reaction, some of the rest energy of the radioactive nucleus is converted into kinetic energy of the products. The amount of rest energy that is converted into other forms of energy is called the **disintegration energy**. In order for kinetic energy to increase, there must be a corresponding decrease in rest energy. The total mass of the products must be less than the mass of the original radioactive nucleus in order for that nucleus to decay spontaneously. In other words, the products must be more tightly bound than is the original nucleus; the disintegration energy is the difference between the binding energy of the radioactive nucleus and the total binding energy of the products.

Disintegration energy = binding energy original nucleus – total binding energy products

Table 29.2

Particles Commonly Involved in Radioactive Decay and Other Nuclear Reactions

Particle Name	Symbols	Charge (in units of e)	Nucleon Number
Electron	$e^-, \beta^-, {}_{-1}^{0}e$	−1	0
Positron	$e^+, \beta^+, {}_{1}^{0}e$	+1	0
Proton	$p, {}_{1}^{1}p, {}_{1}^{1}H$	+1	1
Neutron	$n, {}_{0}^{1}n$	0	1
Alpha particle	$\alpha, {}_{2}^{4}\alpha, {}_{2}^{4}He$	+2	2
Photon	γ	0	0
Neutrino	ν	0	0
Antineutrino	$\bar{\nu}$	0	0

Alpha Decay

Alpha rays are now known to be ^{4}He nuclei; the helium nucleus is a group of two protons and two neutrons that is very tightly bound, since it is doubly magic. The mass of an alpha particle is 4.001 506 u and its charge is $+2e$.

In alpha decay, the original (*parent*) nuclide is converted to a *daughter* by the emission of an alpha particle. (Historically, the product nuclide is called a daughter nuclide because, in a chain of radioactive decays, only a daughter can "give birth" to new nuclides.) Balancing the reaction shows that the daughter nuclide has a nucleon number reduced by four and a charge reduced by two. Using P for the parent nuclide and D for the daughter nuclide, the spontaneous reaction in which an alpha particle is emitted is

Alpha decay

$$_{Z}^{A}P \rightarrow {}_{Z-2}^{A-4}D + {}_{2}^{4}\alpha \qquad (29\text{-}10)$$

Example 29.5

An Alpha Decay

Polonium-210 decays via alpha decay. Identify the daughter nuclide.

Strategy First we look up the atomic number of polonium. Next we write the nuclear reaction with an unknown nuclide and an alpha particle as the products. Balancing the reaction gives us the values of Z and A of the daughter nucleus.

Solution Polonium is atomic number 84. Then the reaction is

$$_{84}^{210}Po \rightarrow {}_{Z}^{A}(?) + {}_{2}^{4}\alpha$$

where A and Z are the nucleon number and atomic number of the daughter nucleus. To conserve charge,

$$84 = Z + 2$$

Thus, $Z = 82$. To conserve nucleon number,

$$210 = A + 4$$

and $A = 206$. Looking up atomic number 82 in the Periodic Table reveals that the element is lead. Thus, the daughter nucleus is lead-206 ($_{82}^{206}$Pb).

Discussion Writing out the reaction makes it easy to check that the total number of nucleons and the total electric charge are both conserved by the reaction:

$$_{84}^{210}Po \rightarrow {}_{82}^{206}Pb + {}_{2}^{4}\alpha$$

Practice Problem 29.5 Finding the parent given the daughter

Radon-222, the radioactive gas that poses a significant health risk in some areas, is itself produced by the alpha decay of another nuclide. Identify its parent nuclide.

In alpha decay, the disintegration energy released is shared between the daughter nucleus and the alpha particle. Momentum conservation determines exactly how the energy is shared. Therefore, the alpha particles released in a particular radioactive decay have a characteristic energy (assuming that the initial kinetic energy of the parent is insignificant and can be taken to be zero).

Example 29.6

Alpha Decay of Uranium-238

The ^{238}U nuclide can decay by emitting an alpha particle:

$$^{238}\text{U} \rightarrow {}^{234}\text{Th} + \alpha$$

The *atomic* masses of ^{238}U, ^{234}Th, and $^4_2$He are 238.050 78 u, 234.043 60 u, and 4.002 60 u, respectively. (a) Find the disintegration energy. (b) Find the kinetic energy of the alpha particle, assuming the parent ^{238}U nucleus was initially at rest.

Strategy The calculations can be performed using *atomic* masses. The mass of the $^{238}_{92}$U atom includes 92 electrons; the combined masses of the $^{234}_{90}$Th and $^4_2$He atoms also include 90 + 2 = 92 electrons.

We expect *most* of the kinetic energy to go to the alpha particle, since its mass is much smaller than that of the thorium nucleus. Momentum conservation determines exactly how the kinetic energy splits between the two particles.

Solution (a) The total mass of the products is

$$234.043\,60 \text{ u} + 4.002\,60 \text{ u} = 238.046\,20 \text{ u}$$

which is less than the mass of the parent nucleus. The change in mass is

$$\Delta m = 238.046\,20 \text{ u} - 238.050\,78 \text{ u} = -0.004\,58 \text{ u}$$

Δm stands for the *change* in mass: final mass minus initial mass. (When we write the mass defect of a nucleus as Δm, we imagine a reaction that separates the nucleus into its constituent protons and neutrons.) The decrease in mass for this reaction means that the rest energy decreases. According to Einstein's mass-energy relation, the change in rest energy is

$$E = \Delta mc^2 = -0.000\,458 \text{ u} \times 931.5 \text{ MeV/u} = -4.27 \text{ MeV}$$

By conservation of energy, the kinetic energy of the products is 4.27 MeV more than the kinetic energy of the parent. The disintegration energy is 4.27 MeV.

(b) Assuming for the moment that the daughter nucleus and the alpha particle can be treated nonrelativistically, their kinetic energies are related to their momenta by

$$K = \frac{p^2}{2m}$$

Momentum conservation says that their momenta must be equal in magnitude and opposite in direction. Therefore, the ratio of the kinetic energies is

$$\frac{K_\alpha}{K_\text{d}} = \frac{p^2/(2m_\alpha)}{p^2/(2m_\text{d})} = \frac{m_\text{d}}{m_\alpha} = \frac{234.043\,60 \text{ u}}{4.002\,60 \text{ u}} = 58.5$$

where the subscript "d" is used for the daughter. The two kinetic energies must add up to 4.27 MeV. The alpha gets 58.5

parts of the total to one part for the thorium nucleus. Then the alpha gets 58.5 parts out of the total of 59.5; the fraction of the kinetic energy that goes to the alpha particle is 58.5/59.5.

$$K_\alpha = \frac{58.5}{59.5} \times 4.27 \text{ MeV} = 4.20 \text{ MeV}$$

We must now check the speed of the alpha to make sure it is not close to the speed of light, to verify that assumption. Using the nonrelativistic expression for kinetic energy, we calculate the speed of the alpha particle. If the speed comes out $\leq 0.3c$, then the nonrelativistic expression $K = \frac{1}{2}mv^2$ is a reasonable approximation (see Section 26.8). Solving for v^2,

$$v^2 = \frac{2K}{m} = \frac{2 \times 4.20 \text{ MeV}}{4.00 \text{ u}} = 2.10 \text{ MeV/u}$$

Dividing by $c^2 = 931.5$ MeV/u yields

$$\frac{v^2}{c^2} = \frac{2.10 \text{ MeV/u}}{931.5 \text{ MeV/u}} = 0.002\,25$$

Taking the square root,

$$v/c = \sqrt{0.002\,25} \approx 0.05$$

The alpha does not need to be treated relativistically.

Discussion The change in mass is *negative*: the total mass after the decay is less than the mass before. Some of the mass (or, more accurately, rest energy) of the U nucleus is converted into the kinetic energy of the products. The disintegration energy is positive because it is the quantity of energy *released*.

A tacit approximation was used in applying conservation of momentum. We really should subtract the masses of the atomic electrons and use the masses of the bare nuclei. The electron masses are small enough that this doesn't make much difference. Repeating the calculation of K_d/K_α using the *nuclear* masses gives the same answer to three significant figures:

$$\frac{K_\text{d}}{K_\alpha} = \frac{234.043\,60 - 90 \times 0.000\,548\,6}{4.002\,60 - 2 \times 0.000\,548\,6} = 58.5$$

Practice Problem 29.6   Alpha energy in the decay of polonium-210

Find the kinetic energy of the alpha particle emitted by the decay of ^{210}Po:

$$^{210}_{84}\text{Po} \rightarrow {}^{206}_{82}\text{Pb} + \alpha$$

Emission of an alpha particle is the most common type of radioactive decay for large nuclides ($Z > 83$). Since no nuclide with $Z > 83$ is stable, emitting an alpha particle moves toward the line of stability most directly by decreasing both Z and N by 2. Emission of an alpha decay increases the ratio of neutrons to protons. For example, $^{238}_{92}U$ has a neutron-to-proton ratio of $(238 - 92)/92 = 1.587$. By emitting an alpha particle, $^{238}_{92}U$ becomes $^{234}_{90}Th$ with a higher neutron-to-proton ratio: $(234 - 90)/90 = 1.600$. Thus, large nuclides with a smaller neutron-to-proton ratio are more likely to alpha decay than are similar nuclides with a greater neutron-to-proton ratio. There is also a cluster of alpha-emitters in the range $Z = 62$ to 72 (see Fig. 29.3); these fall below the line of stability and thus are neutron-deficient.

Beta Decay

Beta rays are electrons or positrons (sometimes still called beta-minus and beta-plus particles). The **positron** is the antiparticle of the electron (see Section 27.8); it has the same mass as the electron but a positive charge of $+e$. In beta-minus decay, an electron is emitted and a neutron in the nucleus is converted to a proton. Thus, the mass number does not change, but the charge of the nucleus increases by one:

Beta-minus decay

$$^{A}_{Z}P \rightarrow ^{A}_{Z+1}D + e^- + \bar{\nu} \tag{29-11}$$

The symbol $\bar{\nu}$ represents an antineutrino, an uncharged particle with negligible mass. In beta-plus decay, a positron is emitted and a proton in the nucleus is converted into a neutron. This time the charge of the nucleus decreases by one:

Beta-plus decay

$$^{A}_{Z}P \rightarrow ^{A}_{Z-1}D + e^+ + \nu \tag{29-12}$$

The symbol e^+ represents the emitted positron and ν is a neutrino with no charge and negligible mass.

Unlike alpha decay, beta decay of a radionuclide does not change the number of nucleons. In essence, beta decay changes a neutron into a proton or vice versa. Since the mass of the neutron is greater than the combined mass of a proton plus an electron, free neutrons decay spontaneously by beta-minus emission. The half-life of this process is 10.2 min. A free proton cannot spontaneously decay into a neutron plus a positron; that would violate energy conservation. But within a nucleus, a proton can change into a neutron by emitting a positron; the energy required to make this happen comes from the change in the binding energy of the nucleus. Thus, the basic beta decay reactions that take place inside the nucleus are

$$\beta^-: \quad n \rightarrow p + e^- + \bar{\nu}$$

$$\beta^+: \quad p \rightarrow n + e^+ + \nu$$

Beta decay does not change the mass number, but it does change the ratio of neutrons to protons. A nuclide that has too many neutrons to be stable is likely to decay via β^-. By emitting an electron, a neutron is changed into a proton inside the nucleus. A nuclide that has too few neutrons is likely to decay by β^+, emitting a positron and turning a proton into a neutron. In either case, total electric charge is conserved.

Figure 29.3 shows all of the known nuclides, whether the nuclide is stable or radioactive, and if radioactive, the principal decay mode. Nuclides above the line of stability—those with too many neutrons to be stable—tend to be β^- emitters. Nuclides below the line of stability—those with too few neutrons to be stable—tend to be β^+ emitters.

Figure 29.6 Typical energy spectrum of electrons emitted in beta decay from a particular nuclide

Beta decay was a puzzle at first because a *continuous spectrum* of electron (or positron) energies was observed. In alpha decay, the definite kinetic energy of the alpha particles emitted in a given decay reaction is understood to come from conservation of both energy and linear momentum. For the same reasons, scientists thought that beta particles emitted in a given decay reaction should also be monoenergetic. However, when the kinetic energies were measured, the emitted beta particles had a continuous range of kinetic energies up to a maximum value (Fig. 29.6). The *maximum* kinetic energy was consistent with what scientists thought the beta particle's kinetic energy should have been.

Why did many of the beta particles have lower energies than expected? Had scientists found an exception to one of the conservation laws (energy or momentum)? Although some quite respectable scientists—including Niels Bohr—started to think that energy conservation had been violated, Wolfgang Pauli finally suggested another explanation, which turned out to be correct. Pauli speculated that not one, but two particles were being emitted, the beta particle and another, as yet undetected, particle. If a nucleus emits two particles instead of one, then they can conserve both energy and momentum while splitting up the kinetic energy in every possible way. Two momentum vectors that add to zero must be equal in magnitude and opposite in direction, but *three* momentum vectors can add to zero in an infinite number of ways and still share the same total kinetic energy.

Enrico Fermi named this hypothetical particle the **neutrino**. The symbol for the neutrino is the Greek letter "nu" (ν). An **antineutrino** is written with a bar over it ($\bar{\nu}$). For reasons that we study in Chapter 30, an antineutrino ($\bar{\nu}$) is emitted in β^- decay, while a neutrino (ν) is emitted in β^+ decay. Neutrinos are famously hard to detect because they do not interact via the electromagnetic or strong interactions. It took 25 years after Pauli predicted their existence before one was actually observed. A neutrino can pass through the Earth with only about a 1 in 10^{12} chance of interacting. Enormous numbers of neutrinos, streaming toward us from the Sun, pass through your body every second but cause no ill effects.

Example 29.7

Beta Decay of Nitrogen-13

The isotope of nitrogen with mass number 13 (^{13}N) is unstable to beta decay. (a) ^{14}N and ^{15}N are the stable isotopes of nitrogen. Do you expect ^{13}N to decay via β^- or β^+? Explain. (b) Write the decay reaction. (c) Calculate the maximum kinetic energy of the emitted beta particle.

Strategy The key in deciding between β^- and β^+ is whether the nucleus has too many or too few neutrons to be stable.

Solution (a) The stable isotopes of nitrogen have more neutrons than ^{13}N, so ^{13}N has too few neutrons to be stable. The beta decay should convert a proton into a neutron to increase the neutron-to-proton ratio. That means the charge of the nucleus decreases by e, so a positron (charge = $+e$) must be created to conserve charge. We expect the isotope ^{13}N to undergo β^+ decay.

continued on next page

Example 29.7 *continued*

(b) Since a positron is emitted, it must be accompanied by a neutrino (not an antineutrino). Z decreases by 1, from 7 (for nitrogen) to 6 (which is carbon). A is unchanged. The reaction is

$$^{13}_{7}N \rightarrow\ ^{13}_{6}C + ^{0}_{+1}e + \nu$$

Both charge and nucleon number are conserved: $13 = 13 + 0$ and $7 = 6 + 1$.

(c) From Appendix B, the atomic masses of $^{13}_{7}N$ and $^{13}_{6}C$ are 13.005 739 u and 13.003 355 u. To get the masses of the nuclei, we subtract Zm_e from each. The mass of the positron is the same as that of the electron: $m_e = 0.000\,548\,6$ u. The neutrino mass is negligibly small. If M_N and M_C represent atomic masses, then

$$\Delta m = [(M_C - 6m_e) + m_e] - (M_N - 7m_e)$$
$$= M_C - M_N + 2m_e$$
$$= 13.003\,355\ u - 13.005\,739\ u + 2 \times 0.000\,548\,6\ u$$
$$= 13.003\,355\ u - 13.005\,739\ u + 0.001\,097\,2 = -0.001\,286\,8\ u$$

The mass decreases, as it must for a spontaneous decay. The disintegration energy is

$$E = |\Delta m|c^2 = 0.001\,286\,8\ u \times 931.494\ \text{MeV/u} = 1.199\ \text{MeV}$$

This *is* the maximum kinetic energy of the positron, since it can get virtually all of the energy and leave the neutrino and daughter nucleus with a negligibly small amount.

Discussion   From Fig. 29.3, we see that ^{13}N lies below the line of stability, confirming that it has too few neutrons to be stable. Emission of a positron moves the nucleus toward the line of stability by converting a proton into a neutron.

It is *usually* possible to determine whether a radioactive nuclide decays via β^+ or β^-, but there are exceptions. For example, $^{40}_{19}K$ can decay by *either* β^+ or β^-. The only way to be sure is to compare the masses of the products to the mass of the radionuclide to see whether the spontaneous decay is energetically possible.

⚠ Note that in β^+ decay the electron masses (which are included in the atomic masses) do not automatically "cancel out" as they did for alpha decay.

Practice Problem 29.7   Maximum electron energy in the decay of potassium-40

Find the maximum energy of the electron emitted in the β^- decay of $^{40}_{19}K$. The atomic masses are $^{40}_{19}K$ = 39.963 999 u and $^{40}_{20}Ca$ = 39.962 591 u.

Electron Capture

A nuclide that decays via β^+ can also decay by **electron capture**. Both processes convert a proton into a neutron. In electron capture, instead of emitting a positron, the nucleus absorbs an electron. The basic reaction is

$$e^- + p \rightarrow n + \nu \tag{29-13}$$

When a nucleus captures an electron, the only reaction products are the daughter nucleus and the neutrino. With only two particles, conservation of momentum and energy determine what fraction of the energy released is taken by each particle. The neutrino, with its tiny mass, takes almost all of the kinetic energy, leaving the daughter to recoil with only a few eV of kinetic energy.

Gamma Decay

Gamma rays are high-energy photons. Emission of a gamma ray does not change the nucleus into a different nuclide, since neither the charge nor the number of nucleons is changed. A gamma-ray photon is emitted when a nucleus in an excited state makes a transition to a lower energy state, just as photons are emitted when electrons in atoms make transitions between energy levels.

Figure 29.7 shows some of the energy levels of the thallium-208 ($^{208}_{81}Tl$) nucleus. The nucleus in an excited state can radiate a photon to jump to a state of lower energy. For instance, the third arrow from the right, from 492 keV to 40 keV, shows a transition that results in the emission of a 452-keV photon. (Although the nucleus must recoil in order to conserve momentum, in Problem 67 you can show that the recoil kinetic energy of the nucleus is negligibly small compared to the photon energy.)

Figure 29.7   An energy level diagram for $^{208}_{81}Tl$. Downward arrows show the allowed transitions for gamma decay.

To emphasize that a nucleus is in an excited state, we put an asterisk as a super-script after the symbol: $^{208}_{81}\text{Tl}^*$. The gamma decay of an excited Tl-208 nucleus by emitting one photon is written as

$$^{208}_{81}\text{Tl}^* \to {}^{208}_{81}\text{Tl} + \gamma$$

Alpha and beta decay do not always leave the daughter nucleus in its ground state. Sometimes the daughter nucleus is left in an excited state that then emits one or more gamma-ray photons until it reaches the ground state. In alpha decay, therefore, there may be different possible kinetic energies of the alpha particles emitted, depending on which excited state of the daughter nucleus is produced by the decay. For example, $^{212}_{83}\text{Bi}$ can alpha decay to form any of the five energy states of $^{208}_{81}\text{Tl}$ shown in Fig. 29.7 (the ground state and four excited states). The alpha-particle spectrum in the decay of $^{212}_{83}\text{Bi}$ is still discrete, but there are five discrete values instead of one (see Problem 57). In beta decay, if the daughter nucleus can be left in an excited state, then the amount of kinetic energy shared by the electron (or positron), the antineutrino (or neutrino), and the daughter nucleus is smaller. The spectrum of electron (or positron) kinetic energies is still continuous.

29.4 RADIOACTIVE DECAY RATES AND HALF-LIVES

What determines when an unstable nucleus decays? Radioactive decay is a quantum-mechanical process that can only be described in terms of *probability*. Given a collection of identical nuclides, they do not all decay at the same time, and there is no way to determine which one decays when. The decay probability for one nucleus is independent of its past history and of the other nuclei. Each radioactive nuclide has a certain decay probability per unit time, written λ (no relation to wavelength). The decay probability per unit time is also called the **decay constant**. Since probability is a pure number, the decay constant has SI units s^{-1} (probability *per second*).

$$\text{decay constant } \lambda = \frac{\text{probability of decay}}{\text{unit time}} \quad (29\text{-}14)$$

The probability that a nucleus decays during a short time interval Δt is $\lambda \Delta t$.

In a collection of a large number N of identical radioactive nuclei, each one has the same decay probability per unit time. The nuclei are independent—the decay of one has nothing to do with the decay of another. Since the decays are independent, the average number that decay during a short time interval Δt is just N times the probability that any one decays:

$$\Delta N = -N\lambda \Delta t \quad (29\text{-}15)$$

The minus sign is necessary because as nuclei decay, the number of nuclei that *remain* is *decreasing*, so the change in N is *negative*. Equation (29-15) gives the *average* number that are expected to decay during Δt. Since radioactive decay is a statistical process, we may not observe exactly that number of decays. If N is sufficiently large, then we expect Eq. (29-15) to be very close to what we observe; for small N, however, deviations from the expected number can be significant. Statistical fluctuations in the number of decays $|\Delta N|$ actually measured are of order $\sqrt{|\Delta N|}$; that is, if the average number of decays expected is 10,000, the actual number of decays that occur in a particular experiment can vary by about $\sqrt{10{,}000} = 100$ above or below the average number.

Equation (29-15) is only valid for a *short* time interval $\Delta t \ll 1/\lambda$ because it assumes that the number of nuclei is a constant N. If the time interval is long enough that N changes significantly, then what N should we use? The initial number? The final number? Some sort of average? As long as $\Delta t \ll 1/\lambda$, then we are sure that $|\Delta N| \ll N$, which means that N has not changed significantly.

The number of radioactive decays from a sample per unit time is called the decay rate or **activity** (symbol R). The SI unit of activity is the becquerel (Bq), which is one decay per second. These three ways of writing the SI unit of activity are equivalent:

$$1 \text{ Bq} = 1\,\frac{\text{decay}}{\text{s}} = 1 \text{ s}^{-1} \quad (29\text{-}16)$$

Another unit of activity in common use is the curie (Ci):

$$1 \text{ curie} = 1 \text{ Ci} = 3.7 \times 10^{10} \text{ Bq} \qquad (29\text{-}17)$$

If the number of decays during a short interval Δt is $|\Delta N|$, then the activity is

$$R = \frac{\text{number of decays}}{\text{unit time}} = \frac{-\Delta N}{\Delta t} = \lambda N \qquad (29\text{-}18)$$

In Eq. (29-18), the rate of change of N ($\Delta N/\Delta t$) is a negative constant ($-\lambda$) times N. We have seen this sort of mathematical relationship before. For example, in a discharging RC circuit, current and charge on the capacitor are both decaying exponential functions (see Section 18.10).

$$I = I_0 e^{-t/\tau} \quad \text{and} \quad q = q_0 e^{-t/\tau}$$

Therefore, the current is proportional to the charge at any instant:

$$I = \frac{I_0}{q_0} q$$

The initial current I_0 and initial charge q_0 are both related to the initial voltage V_0:

$$I_0 = \frac{V_0}{R} \quad \text{and} \quad q_0 = CV_0$$

Then

$$I = \frac{I_0}{q_0} q = \frac{V_0/R}{CV_0} q = \frac{1}{RC} q = \frac{1}{\tau} q$$

where $\tau = RC$ is the time constant. The current is the rate of flow of charge, but here a positive current corresponds to a decrease in charge on the capacitor, so

$$I = \frac{-\Delta q}{\Delta t} = \frac{1}{\tau} q \qquad (29\text{-}19)$$

In Eq. (29-19), the rate of change of q is a negative constant ($-1/\tau$) times q. Equations (29-18) and (29-19) are mathematically identical: the rate of change of a quantity is a negative constant times the quantity. Therefore, the number of remaining nuclei N in radioactive decay (the number that have *not* decayed) must be an exponential function (Fig. 29.8), just as the charge on the capacitor is an exponential function:

$$N(t) = N_0 e^{-t/\tau} \qquad (29\text{-}20)$$

For radioactive decay, the time constant is

$$\tau = \frac{1}{\lambda} \qquad (29\text{-}21)$$

and N_0 is the number of nuclei at $t = 0$. The time constant is also called the **mean lifetime** since it is the *average* time that a nucleus survives before decaying. However, it would be a misconception to think that nuclei "get old." A uranium-238 nucleus that has been sitting in rock for millions of years has the *same* probability per second of decay as one that

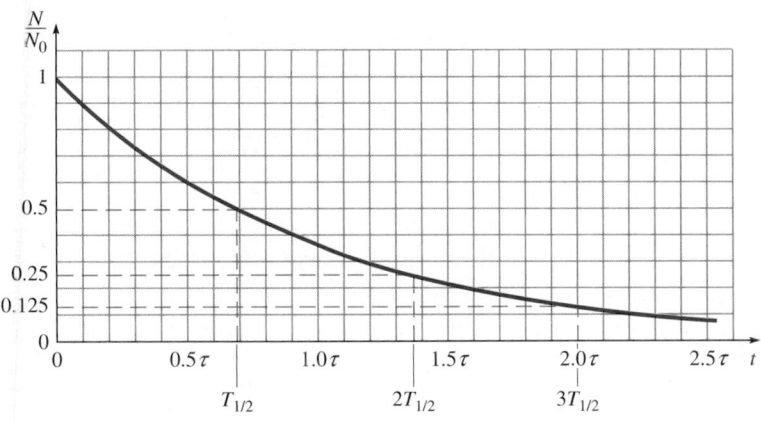

Figure 29.8 Fraction of radioactive nuclei remaining (N/N_0) as a function of time.

has just been created seconds ago in a nuclear reaction; no more, no less. Equations such as (29-18) and (29-20) tell us *how many* nuclei are expected to decay, but not *which ones.*

Since the decay rate is proportional to the number of nuclei, the rate also decays exponentially:

$$R(t) = R_0 e^{-t/\tau} \tag{29-22}$$

As for exponential decay in any other context, the time constant τ is the time for the quantity to decrease to $1/e \approx 36.8\%$ of its initial value. During a time interval τ, 63.2% of the nuclei decay, leaving 36.8%. After a time interval of 2τ, $1/e^2 \approx 13.5\%$ of the nuclei still have not decayed, while $1 - 1/e^2 \approx 86.5\%$ have decayed.

Radioactive decay is often described in terms of the **half-life** $T_{1/2}$ instead of the time constant τ. The half-life is the time during which half of the nuclei decay. After two half-lives, $\frac{1}{4}$ of the nuclei remain; after m half-lives, $\left(\frac{1}{2}\right)^m$ remain. It can be shown (Problem 38) that

$$T_{1/2} = \tau \ln 2 \approx 0.693\,\tau \tag{29-23}$$

where ln 2 is the natural (base-e) logarithm of 2. Then

$$N(t) = N_0 \times (2^{-t/T_{1/2}}) = N_0 \times \left(\tfrac{1}{2}\right)^{t/T_{1/2}} \tag{29-24}$$

Example 29.8

Radioactive Decay of Nitrogen-13

The half-life of ^{13}N is 10.0 min. (a) If a sample contains 3.20×10^{12} ^{13}N atoms at $t = 0$, how many ^{13}N nuclei are present 40.0 min later? (b) What is the ^{13}N activity at $t = 0$ and at $t = 40.0$ min? Express the activities in Bq. (c) What is the probability that any one ^{13}N nucleus decays during a one-second time interval?

Strategy (a, b) The number of nuclei at $t = 0$ is $N_0 = 3.20 \times 10^{12}$ and the half-life is $T_{1/2} = 10.0$ min. The problem asks for N at $t = 40.0$ min and for R at both $t = 0$ and at $t = 40.0$ min. Since the time interval of 40.0 min is four times the half-life, we can simplify the solution: both N and R are multiplied by $\frac{1}{2}$ during each 10.0 min interval.

(c) The probability of decay during a 1-s interval is λ *only if* 1 s can be considered a short time interval. Since the half-life is 10 min = 600 s, 1 s is a tiny fraction of the half-life and therefore *can* be considered a short time interval.

Solution (a) Half of the nuclei are left after one half-life, $\frac{1}{2} \times \frac{1}{2} = \left(\frac{1}{2}\right)^2$ after two half-lives, and $\left(\frac{1}{2}\right)^4$ after four half-lives. Therefore, the number remaining after 40.0 min is

$$N = \left(\tfrac{1}{2}\right)^4 \times 3.20 \times 10^{12} = 2.00 \times 10^{11}$$

Using Eq. (29-24) gives the same result:

$$N(t) = N_0 \times \left(\tfrac{1}{2}\right)^{t/T_{1/2}} = N_0 \times \left(\tfrac{1}{2}\right)^{40.0/10.0} = \tfrac{1}{16} N_0 = 2.00 \times 10^{11}$$

(b) The activity and number of nuclei are related by Eq. (29-18):

$$R = \lambda N = \frac{N}{\tau}$$

The time constant τ is related to the half-life by Eq (29-23):

$$\tau = \frac{T_{1/2}}{\ln 2} = \frac{6.0 \times 10^2 \text{ s}}{0.693\,15} = 865.6 \text{ s}$$

Next we substitute the number of nuclei N at $t = 0$ and at $t = 40.0$ min to determine the rate of decay at those two times. The time constant does not change.

At $t = 0$,

$$R_0 = \frac{N_0}{\tau} = \frac{3.20 \times 10^{12}}{865.6 \text{ s}} = 3.70 \times 10^9 \text{ Bq}$$

At $t = 40$ min,

$$R = \frac{N}{\tau} = \frac{2.00 \times 10^{11}}{865.6 \text{ s}} = 2.31 \times 10^8 \text{ Bq}$$

(c) The probability per second is

$$\lambda = \frac{1}{\tau} = 1.16 \times 10^{-3} \text{ s}^{-1}$$

A nucleus has a 0.001 16 probability of decaying during a 1 s interval.

Discussion As a check, we note that R after four half-lives should be $\frac{1}{16}$ of R_0:

$$\tfrac{1}{16} \times 3.70 \times 10^9 \text{ Bq} = 2.31 \times 10^8 \text{ Bq}$$

The probability of decay in one second would *not* be equal to λ if the half-life were not large compared to 1 s. For a longer time interval, we find the decay probability as follows:

$$\text{probability of decay} = \frac{\text{number expected to decay}}{\text{original number}}$$

$$= \frac{|\Delta N|}{N_0} = \frac{N_0 - N}{N_0} = 1 - e^{-t/\tau}$$

Practice Problem 29.8 Number remaining after one half of a half-life

How many ^{13}N atoms are present at $t = 5.0$ min?

Physics at Home

You can simulate radioactive decay by tossing pennies. Assemble a collection of as many pennies (or other coins) as you can; around 100 would be ideal. The pennies serve as models of radioactive nuclei. Record the original number of pennies, then toss them. Let each penny that comes up "tails" represent a nucleus that has decayed. Remove all the tails and count them. Then toss the remaining pennies. Continue until only 1–3 are left. Make a table like this:

t (Number of Tosses)	ΔN (Number of Tails)	N (Number Remaining)
0	—	100
1	54	46 (= 100 – 54)
2	22	24 (= 46 – 22)
3	. . .	. . .

One toss serves as a unit of time. The half-life of the pennies is expected to be 1.0 toss. Plot your data points N and ΔN as functions of t on the same graph. Also plot the functions $N(t) = N_0 2^{-t/T_{1/2}}$ and $|\Delta N|(t) = (\Delta N)_0 2^{-t/T_{1/2}}$ as smooth curves; $-t/T_{1/2}$ is the number of tosses. (Use semilog graph paper if you have some and know how to use it.) Are the deviations of your data points for ΔN from the corresponding curve reasonable, given that the expected size of statistical fluctuations is approximately $\sqrt{|\Delta N|}$? The fluctuations are relatively large here because of the small number of "nuclei."

Radiocarbon Dating

Making The Connection:
radiocarbon dating

The immensely useful radiocarbon dating technique (or carbon dating as it is frequently called) is based on the radioactive decay of a rare isotope of carbon. Almost all of the naturally occurring carbon on Earth is one of the two stable isotopes—98.9% is ^{12}C and 1.1% is ^{13}C. However, there is also a trace amount of ^{14}C—about one in every 10^{12} carbon atoms. The carbon-14 isotope, ^{14}C, has a relatively short half-life of 5730 yr. Since the Earth is about 4.5 billion (4.5×10^9) yr old, we would expect to find no carbon-14 at all if it were not continually being replenished.

The production of carbon-14 occurs because the Earth's atmosphere is bombarded by cosmic rays. Cosmic rays are extremely high-energy charged particles—mostly protons—from space. When one of these particles hits an atom in Earth's upper atmosphere, a shower of secondary particles is created, which includes a large number of neutrons. Typically about 1 million neutrons are produced by each cosmic ray particle. Some of these neutrons then react with ^{14}N nuclei in the atmosphere to form ^{14}C:

$$n + {}^{14}N \rightarrow {}^{14}C + p \qquad (29\text{-}25)$$

The ^{14}C forms CO_2 molecules and drifts down to the surface, where it is absorbed from the air by plants and incorporated into carbonate minerals. Animals take in the ^{14}C by eating plants and other animals. The ^{14}C in an organism or mineral decays via beta decay:

$$^{14}C \rightarrow {}^{14}N + e^- + \bar{\nu} \qquad (29\text{-}26)$$

Balance between the rate at which ^{14}C is continually being created by cosmic rays and the rate at which the ^{14}C decays results in an equilibrium ratio of ^{14}C to ^{12}C atoms in the atmosphere equal to 1.3×10^{-12}. While an organism is alive, carbon is exchanged with the environment, so the organism maintains the same relative abundance of ^{14}C as the environment. The carbon-14 activity in the atmosphere or in a living organism is 0.25 Bq per gram of carbon (see Problem 34). When an organism dies, or when ^{14}C is incorporated into a mineral, carbon exchange with the environment stops. As the ^{14}C present in the organism decays, the ratio of ^{14}C to ^{12}C decreases. The ratio of ^{14}C to ^{12}C in a sample can be measured and used to determine the age of the sample. One way to do this is to measure the carbon-14 activity per gram of carbon.

Example 29.9

Dating a Charcoal Sample

A piece of charcoal (essentially 100% carbon) from an archaeological site in Egypt is subjected to radiocarbon dating. The sample has a mass of 3.82 g and a ^{14}C activity of 0.64 Bq. What is the age of the charcoal sample?

Strategy   While a tree is alive, it maintains the same relative abundance of ^{14}C as the environment. After a tree is cut down to make charcoal, the relative abundance of ^{14}C decreases since ^{14}C is no longer being replaced from the environment. As the number of ^{14}C nuclei decreases, so does the ^{14}C activity. The activity decreases exponentially from its initial value with a *half-life* of 5730 yr. We assume the relative abundance in the environment in ancient Egypt was similar to today, so the initial activity is 0.25 Bq per gram of carbon.

Solution   The activity of ^{14}C decreases exponentially:

$$R = R_0\, e^{-t/\tau}$$

The initial activity is

$$R_0 = 0.25 \text{ Bq/g} \times 3.82 \text{ g} = 0.955 \text{ Bq}$$

The present activity is $R = 0.64$ Bq. Now we solve for t from the values of R and R_0.

$$\frac{R}{R_0} = e^{-t/\tau}$$

Taking the natural logarithm of each side gets t out of the exponent:

$$\ln \frac{R}{R_0} = \ln e^{-t/\tau} = -\frac{t}{\tau}$$

$$t = -\tau \ln \frac{R}{R_0} = -\frac{T_{1/2}}{\ln 2} \ln \frac{R}{R_0}$$

$$= -\frac{5730 \text{ yr}}{\ln 2} \times \ln \frac{0.64 \text{ Bq}}{0.955 \text{ Bq}} = 3300 \text{ yr}$$

The charcoal is 3300 yr old.

Discussion   As a check, we can test to see whether

$$R_0(2^{-t/T_{1/2}}) = R$$

$$R_0(2^{-t/T_{1/2}}) = 0.955 \text{ Bq} \times 2^{-3300 \text{ yr}/5730 \text{ yr}} = 0.955 \text{ Bq} \times 0.671$$

$$= 0.64 \text{ Bq} = R$$

Practice Problem 29.9   The age of Ötzi

In 1991, a hiker found the frozen, naturally mummified remains of a man protruding from a glacier in the Italian Alps. The man was called Ötzi by researchers and became popularly known as the Iceman. The ^{14}C activity of the Iceman's remains was measured to be 0.131 Bq per g of carbon. How long ago did the Iceman die?

Example 29.10

Yearly Decrease in Carbon-14 Activity of a Nonliving Sample

By what percentage does the ^{14}C activity of a nonliving sample decrease in one year?

Strategy   We are given neither the activity at the beginning nor at the end of the one year period, but we only want to find the change *expressed as a percentage of the initial activity*. The percentage change is a way to express the fractional change (the change in activity as a fraction of the initial activity). Let the initial activity be R_0 and the activity one year later be R. The quantity to be determined is

$$\frac{\Delta R}{R_0} = \frac{R - R_0}{R_0}$$

expressed as a percentage.

Solution   The activities R_0 and R are related by

$$R(t) = R_0(2^{-t/T_{1/2}})$$

We choose this form rather than the exponential form $R = R_0 e^{-t/\tau}$ because we are given the half-life rather than the time

constant. We don't know R_0 or R, but we can find the ratio of the two.

$$\frac{R}{R_0} = 2^{-t/T_{1/2}} = 2^{-1/5730} = 0.999\,879$$

Now we find the fractional change during 1 yr.

$$\frac{\Delta R}{R_0} = \frac{R - R_0}{R_0} = \frac{R}{R_0} - 1 = 0.999\,879 - 1 = -0.000\,121$$

The carbon-14 activity decreases 0.012% in a year.

Discussion   The tiny change in activity illustrates one reason why we do not expect carbon-14 dating to give dates precise to a specific year.

Practice Problem 29.10   Dating precision

If the ^{14}C activity of a shard of pottery can be determined to a precision of ±0.1%, to what precision can we expect to date the shard (assuming no other sources of imprecision)? [*Hint:* In what time interval does the activity change by 0.1%?]

Carbon dating can be used for specimens up to about 60,000 yr old, which is about 10 half-lives of ^{14}C. The older a specimen is, the smaller its ^{14}C activity; for very old samples, it is difficult to measure the ^{14}C activity accurately. The half-life also imposes constraints on the precision with which a sample can be dated. Even if all our assumptions were true, we cannot expect carbon dating to give ages down to the exact year. One year is only a small fraction of the half-life, so the activity changes very little during a year's time (as shown in Example 29.10).

A major assumption of the simplest kind of carbon dating presented here is that the equilibrium ratio of ^{14}C to ^{12}C in Earth's atmosphere has been the same for the past 60,000 years. Is that a good assumption? How can we test it? One way to test it for relatively short times is by taking core samples from very old trees—or from the remains of ancient trees—and measuring ^{14}C activities from various times. The tree rings give an independent way to determine the age of different parts of the sample.

At present, scientists believe that the relative abundance of ^{14}C in the atmosphere hasn't changed much in the past 1000 years (until the beginning of the twentieth century) although it has varied considerably during the past 60,000 years, reaching peaks as much as 40% higher than at present. Fortunately, radiocarbon dating can be adjusted for the changes in the relative abundance of ^{14}C in the atmosphere. Tree rings allow such adjustment going back about 11,000 years. In Japan's Lake Suigetsu, layers of dead algae sink to the bottom annually and are covered by a layer of clay sediment before the next algae layer. The alternating layers of light-colored algae and dark clay can be read like tree rings, allowing radiocarbon data to be adjusted for the varying abundance of ^{14}C in the atmosphere going back about 43,000 yr.

The relative abundance of ^{14}C in the atmosphere began changing rapidly in the twentieth century due to human activity. An enormous increase in the burning of fossil fuels introduced large quantities of old carbon—that is, carbon with a low abundance of ^{14}C—into the atmosphere. Beginning about 1940, open-air nuclear testing, nuclear bombs, and nuclear reactors have increased the relative abundance of ^{14}C in the atmosphere. In the distant future it will be difficult to use radiocarbon dating for artifacts from the twentieth century.

Other Isotopes Used in Radioactive Dating

Besides ^{14}C, other radioactive nuclides are also used for radioactive dating. Isotopes commonly used to date geologic formations (with their half-lives in billions of years) include uranium-235 (0.7), potassium-40 (1.25), uranium-238 (4.5), thorium-232 (14), and rubidium-87 (49). One direct way to calculate the Earth's age is based on the abundances of various lead isotopes in terrestrial samples and in meteorites. Pb-206 and Pb-207 are the final products of long chains of radioactive decays that begin with U-238 and U-235, respectively.

Making The Connection:
radioactive dating of
geologic formations

Lead-210, with a half-life of only 22.3 yr, is used for geologic dating over the last 100 to 150 yr. It forms in rocks containing uranium-238 as a decay product of radon gas. After forming from radon in the atmosphere, the lead isotope falls to Earth, where it collects on the surface and is stored in the soil, or in the sediment of lakes and oceans, or in glacial ice. The age of a sediment layer can be determined by measuring the amount of lead-210 present.

Quantum-Mechanical Tunneling Explains Radioactive Half-Lives for Alpha Decay

An early triumph of quantum mechanics was its explanation of the correlation between the half-life of a particular alpha decay and the kinetic energy of the alpha particle. The kinetic energies vary over a narrow range (4–9 MeV) but the half-lives range from 10^{-5} s to 10^{25} s (10^{17} yr). Despite this discrepancy in ranges, the two quantities are closely correlated (Fig. 29.9a); higher alpha particle energies consistently go with shorter half-lives.

Figure 29.9 (a) Correlation between half-life and the energy (E) of the alpha particle. Z is the atomic number of the parent nuclide. Note that the vertical scale is logarithmic and that increasing numbers on the horizontal axis represent *decreasing* energies. (b) Simplified model of the potential energy U of an alpha particle as a function of its distance r from the center of the nucleus. The energy of the alpha particle is E.

The correlation arises because the alpha particle must tunnel its way out of the nucleus. Think of an alpha particle in a nucleus as facing the simplified potential energy graph of Fig. 29.9b. Inside the nucleus, the potential energy of the alpha particle is roughly constant. Beyond the edge of the nucleus, where the strong attractive force no longer pulls the alpha particle toward the nucleus, the alpha particle feels only a Coulomb repulsion from the nucleus [which has charge $+(Z-2)e$ since it has lost two protons]. The potential energy barrier is higher than the energy E of the alpha particle. Since the barrier tapers off gradually, decreasing with distance as $1/r$, lower energy alpha particles are not only farther below the top of the barrier; they face a much *wider* barrier as well. Higher energy alpha particles have much higher tunneling probabilities and therefore much shorter half-lives.

29.5 BIOLOGICAL EFFECTS OF RADIATION

Making The Connection:
biological effects
of radioactivity

We are all continually exposed to radiation. The biological effects of radiation depend on what kind of radiation it is, how much of it is absorbed by the body, and the duration of the exposure. *Ionizing* radiation is radiation with enough energy to ionize an atom or molecule—typically between 1 eV and a few tens of eV. An alpha particle, beta particle, or gamma ray with a typical energy of about 1 MeV can potentially ionize tens of thousands of molecules. Molecules in living cells that are ionized due to radiation become chemically active and can interfere with the normal operation and reproduction of the cell.

The **absorbed dose** is the amount of radiation energy absorbed per unit mass of tissue. The SI unit of absorbed dose is the gray (Gy):

$$1 \text{ Gy} = 1 \text{ J/kg} \tag{29-27}$$

Another common unit for absorbed dose is the rad (rd):

$$1 \text{ rad or } 1 \text{ rd} = 0.01 \text{ Gy} \tag{29-28}$$

The name "rad" stands for <u>r</u>adiation <u>a</u>bsorbed <u>d</u>ose.

Different kinds of radiation cause different amounts of biological damage, even if the absorbed dose is the same. The health effects also depend on what kind of tissue is exposed. To account for these factors, a quantity called the quality factor (QF)—sometimes called the relative biological effectiveness (RBE)—is assigned to each type of radiation. The QF is a relative measure of the biological damage caused by different kinds of radiation compared to 200-keV x-rays (which are assigned QF = 1). The QF varies depending on the kind of radiation (alpha, beta, gamma), the energy of the radiation, the kind of tissue exposed, and the biological effect under consideration. Table 29.3 gives some typical QF values.

Table 29.3

Typical Quality Factors (QFs) for Various Forms of Radiation

Gamma rays	0.5–1
Beta particles	1
Protons, neutrons	2–10
Alpha particles	10–20

To measure the biological damage caused by exposure to radiation, we calculate the **biologically equivalent dose** in rem:

$$\text{biologically equivalent dose (in rem)} = \text{absorbed dose (in rad)} \times \text{QF} \quad (29\text{-}29)$$

Thus, one rem of biologically equivalent dose is the amount of biological damage caused by 1 rad of exposure to 200-keV x-rays.

Example 29.11

Biologically Equivalent Dose in a Brain Scan

A 60.0-kg patient about to have a brain scan is injected with 20.0 mCi of the radionuclide technetium-99, $^{99}\text{Tc}^m$. (The superscript "m" stands for metastable. The metastable state $^{99}\text{Tc}^m$ decays to the ground state with a half-life of 6.0 h.) The $^{99}\text{Tc}^m$ nucleus decays by emitting a 143-keV photon. Assuming that half of these photons escape the body without interacting, what biologically equivalent dose does the patient receive? The QF for these photons is 0.97. Assume that all of the $^{99}\text{Tc}^m$ decays while in the body.

Strategy The activity (20.0 mCi) together with the half-life (6.0 h) enable us to calculate the number of $^{99}\text{Tc}^m$ nuclei. Then we can determine how many photons are absorbed in the body; multiplying the number of photons absorbed by the energy of each photon (143 keV) gives the total radiation energy absorbed. The absorbed dose is the radiation energy absorbed per unit mass of tissue. The biologically equivalent dose in rem is the absorbed dose in rad times the quality factor.

Solution The activity of the injected material in Bq is

$$R_0 = 20.0 \times 10^{-3} \text{ Ci} \times 3.7 \times 10^{10} \text{ Bq/Ci} = 7.4 \times 10^8 \text{ Bq}$$

The activity is related to the number of nuclei N by

$$R_0 = \lambda N_0 = \frac{N_0}{\tau}$$

Then the number of nuclei injected is

$$N_0 = \tau R_0 = \frac{T_{1/2}}{\ln 2} R_0 = \frac{6.0 \text{ h} \times 3600 \text{ s/h}}{\ln 2} \times 7.4 \times 10^8 \text{ s}^{-1} = 2.306 \times 10^{13}$$

Each of these nuclei emits a photon and half of the photons are absorbed by the body. The energy of each photon is 143 keV. Therefore, the total energy absorbed in joules is

$$E = \tfrac{1}{2} \times (2.306 \times 10^{13} \text{ photons}) \times 143{,}000 \text{ eV/photon} \times (1.60 \times 10^{-19} \text{ J/eV})$$

$$= 0.264 \text{ J}$$

The absorbed dose is

$$\frac{0.264 \text{ J}}{60.0 \text{ kg}} = 0.004\ 4 \text{ Gy} = 0.44 \text{ rad}$$

The biologically equivalent dose in rem is the absorbed dose in rad times the quality factor:

$$0.44 \text{ rad} \times 0.97 = 0.43 \text{ rem}$$

Discussion A quantity of radioactive material is often specified by its activity ("20.0 mCi of $^{99}\text{Tc}^m$") rather than by mass, number of moles, or number of nuclei. As already shown, the number of radioactive nuclei can be calculated from the activity and the half-life.

Practice Problem 29.11   Determining mass from activity

What is the mass of 5.0 mCi of $^{60}_{27}\text{Co}$?

The average radiation dose received by a person in 1 yr due to natural sources is about 0.3 rem. Of this, an average of 0.2 rem is due to inhaled radon-222 gas and its decay products. Radon-222 is constantly produced by the alpha decay of radium-226 present in soil and rocks. Radon gas enters houses through cracks in the foundation, and tends to collect there since it is more dense than air. The ^{222}Rn nuclei are inhaled, where they can give a significant dose of alpha-particle radiation to the lungs. The amount of radon gas that enters a building varies greatly from one place to another. In some localities, radon is not much of a problem. In other places, with large amounts of radium in the soil and geological formations that make it easy for radon gas to find its way into a basement, it is a major cause of lung cancer (second only to smoking). Fortunately, an inexpensive test can be used to determine the concentration of radon gas in the air. Where radon is a problem, sealing cracks in the basement and adding ventilation are generally all that is needed.

About 0.07 rem/yr is due to radioactive nuclides that enter the body in food and water (such as ^{14}C and ^{40}K) or are present in the soil and in building materials (such as polonium, radium, thorium, and uranium). Another 0.03 rem/yr is due to cosmic rays. The cosmic ray dose is significantly higher for people living at high altitudes or who spend a lot of time in airplanes. In a commercial jet at 35,000 ft, the dose received is about 0.0007 rem per hour, so 40 h of flight doubles the average person's cosmic ray dose; 90 h of flight is equivalent to the average annual dose due to medical and dental exposure.

Human activities have added an average dose of 0.06 rem/yr to the 0.3 rem/yr from natural sources. Most of this additional radiation dose comes from medical and dental diagnosis and treatment (principally diagnostic x-rays). The average dose due to fallout from testing of nuclear weapons and due to nuclear reactors is about 0.001 rem/yr, but is much higher in some places (for instance in Ukraine, due to the Chernobyl disaster).

A single large dose of radiation causes radiation sickness. Symptoms can include nausea, diarrhea, vomiting, and hair loss. Radiation sickness can be fatal if the dose is large enough. A single dose of 50 rem or less causes few, if any, short-term symptoms. A single dose of about 400–500 rem is fatal about half of the time. Long-term effects of exposure to radiation include increased risk of cancer and genetic mutations.

Penetration of Radiation

Different kinds of radiation have different abilities to penetrate biological tissue (or other materials). The range of an alpha particle in human tissue is about 0.03 mm to 0.3 mm, depending on the energy of the particle. Alpha particles are stopped by a few centimeters of air or by an aluminum foil only 0.02 mm thick. Alpha particles are potentially the most damaging form of radiation, since each can ionize large numbers of molecules. On the other hand, they cannot penetrate the skin, so alpha-emitters outside the body are not so dangerous. Radon gas is dangerous because the alpha decay occurs *inside* the body, exposing the lung tissue directly. Similarly, if alpha emitters are present in food, they can deliver a significant dose of radiation to the digestive tract, and those with longer half-lives may then be incorporated into other body tissue (for example, radioactive iodine collects in the thyroid and radioactive iron collects in the blood).

Beta-minus particles (electrons) are more penetrating than alphas. Their range in human tissue can be as much as a few centimeters (again, depending on energy). They can penetrate several meters of air; it takes an aluminum plate about 1 cm thick to stop them. High-speed electrons not only ionize molecules, but also emit x-rays through bremsstrahlung (Section 27.4); the x-rays are much more penetrating than are the electrons themselves. β^+ particles (positrons) have a very limited range—they quickly annihilate with an electron, producing two photons.

Beta emitters are more dangerous when found inside the body, though the difference is not as striking as for alpha-emitters. Atmospheric tests of nuclear weapons in the 1950s produced many dangerous radioactive nuclides. One of them, radioactive strontium-90, is produced by the fission of ^{235}U. Strontium is chemically similar to calcium. Both are alkali metals; Sr is directly below Ca in the Periodic Table. The strontium-90 produced by atmospheric tests entered the human food supply and was incorporated into the bones and teeth of growing children. Strontium-90 undergoes beta decay with a half-life of 28 yr, but since calcium (and strontium) stays in the body for a long time, the presence of this radionuclide in the bones ends up delivering a significant radiation dose and probably increases the incidence of leukemia and other cancers. Fortunately, atmospheric tests are now banned internationally, and the incidence of strontium-90 and other artificially produced radionuclides is smaller than it once was.

Both alphas and electrons have a fairly definite range for a given material and energy. They lose their energy through a large number of collisions with molecules. By contrast, a photon can lose a large proportion or even all of its energy in a *single* interaction (via the photoelectric effect, Compton scattering, or pair production). The probability of one of these interactions occurring can be calculated using quantum mechanics. For photons of a certain energy, we can only predict the *average* distance traveled in a given material. For example, half of 5-MeV photons can penetrate 23 cm or more into the body. Half of 5-MeV photons can penetrate 1.5 cm or more in lead. The penetrating ability of photons is measured as a *half-value layer*, which is the thickness of material that half of the photons can penetrate.

Medical Applications of Radiation

There are many medical applications of radioactive materials and of radiation. **Radioactive tracers** are important diagnostic tools. One example was mentioned in Example 29.11. The metastable state of technetium-99 is the product of the beta decay of molybdenum-99. Most nuclear excited states decay to the ground state in very short times, ranging from about 10^{-15} s to 10^{-8} s. The metastable state of technetium-99 has an unusually long half-life of 6.0 h, perfect for use as a radioactive tracer. If the half-life were much shorter, much of the $^{99}\text{Tc}^{\text{m}}$ would decay before it reached the tumor cells. If the half-life were much longer, then the activity would be small and only a small fraction of the gamma rays could be detected within a reasonable length of time.

Making The Connection:
radioactive tracers in medical diagnosis

The blood-brain barrier prevents the technetium (which is injected in the form of technetium oxide and attaches to red blood cells) from diffusing into normal brain cells, but the abnormal cells in a tumor do not have such a barrier. Therefore, the tumor can be located and imaged by observation of the gamma rays that are emitted from the brain.

One way to do the imaging is to use an **Anger camera** (pronounced ahn-zhay; see Fig. 29.10). A lead collimating plate has parallel holes drilled in it. The lead absorbs gamma rays, so only photons emitted parallel to one of the holes can get through the plate. Behind the plate is a scintillation crystal; when a gamma photon hits this crystal, a pulse of light is produced. Photomultiplier tubes, one for each hole in the collimator, detect these light pulses. By moving the Anger camera around at different angles, we can "triangulate" back and figure out where the tumor is.

Similarly, TlCl (thallium chloride) tends to collect at the site of a blood clot. Thallium-201 has a half-life of 73 h. When thallium-201 undergoes beta decay in the body, gamma rays are also emitted as the daughter nucleus drops down into its ground state. An Anger camera can then be used to locate the clot.

In positron emission tomography (PET) positron-emitters (radioisotopes whose decay mode is β^+) are injected into the body. Examples of positron emitters include carbon-11, nitrogen-13, and oxygen-15. A positron emitted in the body quickly annihilates with an electron to produce two gamma rays traveling in opposite directions. The photons are detected by a ring of detectors around the body (Fig. 27.24).

Making The Connection:
positron emission tomography (PET) scans

Radioactive tracers are used in research as well as in clinical diagnosis. For example, radioactive iron-59 was used to determine that iron, unlike most other elements, is not constantly being eliminated from the body and then replaced. Rather, once an iron atom is incorporated into a hemoglobin molecule, it stays there for the entire life of the red blood cell. Even when a red blood cell dies, the iron is recycled for use in another cell.

Radiation therapy is used in cancer treatment. Cancer cells are more vulnerable to the destructive effects of radiation, in part because they are rapidly dividing. Thus, the idea of radiation therapy is to supply enough radiation to destroy the malignant cells without causing too much damage to normal cells. The radiation can be administered

Making The Connection:
radiation therapy

(a)

Photomultiplier tubes

Scintillation crystal
Lead collimating plate

Gamma rays

Tumor site

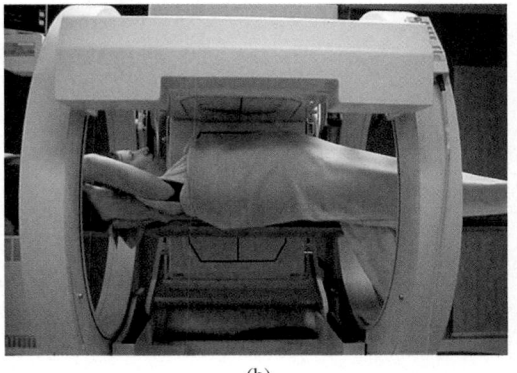

(b)

Figure 29.10 (a) Simplified diagram of an Anger camera. (b) This photo shows an Anger camera with two detector heads, one above the patient's chest and the other to her left. The lead plates, scintillation crystals, and photomultiplier tubes are hidden behind the grids marking the detector heads.

Gamma rays

Lead helmet

Tumor

(a)

(b)

Figure 29.11   (a) Diagram of the lead "helmet" used in gamma knife radiosurgery. (b) The patient is carefully positioned in the helmet to ensure that the gamma rays converge at the desired point in the brain. A lead apron protects the body from exposure to radiation.

Making The Connection:
**gamma knife
radiosurgery**

internally or externally. Internally applied radiation treatment uses radionuclides that are either injected into the tumor, or which collect at the tumor site (much as tracers do). In a promising new technique for targeting cancer cells with radiation, a single radioactive atom is enclosed in a microscopic cage made of carbon and nitrogen atoms. Attached to the cage is a protein that locks onto a specific protein on the surface of a cancerous cell, after which the cage moves inside the cell. The alpha particles emitted in a series of radioactive decays then kill the cell.

Externally applied radiation can be x-rays produced by bremsstrahlung or by other means. Cobalt-60 emits gamma rays that are also used for radiation therapy. The cobalt-60 is kept in a lead box with a small hole so that the gamma rays can be limited to the site of the tumor.

An advanced form of cobalt-60 therapy is called **gamma knife radiosurgery**. In this technique, a spherical lead "helmet" with hundreds of holes (Fig. 29.11) enables the gamma rays to converge at a small region in the brain. In this way, the radiation dose to the tumor, where all the gamma rays converge, can be much larger than the dose to the surrounding tissue.

Some hospitals have cyclotrons (Section 19.3) or other particle accelerators on site. Their purpose is twofold. The accelerator can be used to manufacture radionuclides that have short half-lives. Radionuclides with longer half-lives can be manufactured offsite at either an accelerator or at a nuclear reactor. Second, beams of accelerated charged particles are used in radiation therapy.

29.6   INDUCED NUCLEAR REACTIONS

In radioactivity, an unstable nucleus decays in a *spontaneous* nuclear reaction, releasing energy in the process. An *induced* nuclear reaction is one that does not occur spontaneously; it is caused by a collision between a nucleus and something else. The other reactant can be another nucleus, a proton, a neutron, an alpha particle, or a photon.

We have already seen an example of an induced nuclear reaction; carbon-14 is formed in a nuclear reaction induced when an energetic neutron collides with a nitrogen-14 nucleus:

$$n + {}^{14}N \rightarrow {}^{14}C + p \tag{29-25}$$

Equation (29-25) is an example of **neutron activation**, in which a stable nucleus is transformed into a radioactive one by absorbing a neutron. Nearly all of the elements have one or more isotopes that can be activated (made radioactive) by absorbing a neutron.

A spontaneous nuclear reaction always releases energy, so that the total mass of the products is always less than the total mass of the reactants. By contrast, an induced reaction can convert some of the kinetic energy of the reactants into rest energy. Thus, the total mass of the products can be greater than, less than, or equal to the total mass of the reactants. A nucleus involved in such a reaction does not have to be radioactive; a stable nucleus can participate in a reaction when struck by some other particle. The first such reaction ever observed, by Rutherford in 1919, was

$$\alpha + {}^{14}N \rightarrow {}^{17}O + p \qquad (29\text{-}30)$$

A reaction takes place when the target nucleus absorbs the incident particle, forming an intermediate compound nucleus. In the reaction of Eq. (29-30), the compound nucleus is ^{18}F:

$$_2^4He + {}_7^{14}N \rightarrow {}_9^{18}F \rightarrow {}_8^{17}O + {}_1^1H$$

Example 29.12

A Neutron Activation

Consider the reaction

$$n + {}^{24}Mg \rightarrow p + ?$$

(a) Determine the product nucleus and the intermediate compound nucleus. (b) Is this reaction exoergic or endoergic? That is, does it release energy, or does it require the input of energy to occur? Calculate either the energy released (if exoergic) or the energy absorbed (if endoergic).

Strategy The product nucleus and compound nucleus can be identified by balancing the reaction: the total charge and total number of nucleons must remain the same. The energetics are determined by whether the total mass of the products is greater or less than the total mass of the reactants.

Solution (a) Magnesium is atomic number 12. The reaction, written out more fully, is

$$_0^1n + {}_{12}^{24}Mg \rightarrow {}_{12}^{25}(?)^* \rightarrow {}_{11}^{24}(?) + {}_1^1p$$

where we have made sure that the total electrical charge and the total number of nucleons remain unchanged. From the Periodic Table, atomic number 11 is Na and we already know that atomic number 12 is Mg. Therefore, the product nucleus is $_{11}^{24}Na$ and the intermediate nucleus is $_{12}^{25}Mg$.

(b) We compare the total mass of the reactants to the total mass of the products. From Table 29.1 and Appendix B, the atomic masses are

$$\text{mass of } {}^{24}Mg = 23.985\,042 \text{ u}$$
$$\text{mass of } {}^{24}Na = 23.990\,963 \text{ u}$$
$$\text{mass of } {}^1H = \ 1.007\,825 \text{ u}$$
$$\text{mass of } n = \ 1.008\,665 \text{ u}$$

Using atomic masses is fine, since both sides include the extra mass of the same number of electrons (12). Then the total mass of the reactants is

$$1.008\,665 \text{ u} + 23.985\,042 \text{ u} = 24.993\,707 \text{ u}$$

and the total mass of the products is

$$1.007\,825 \text{ u} + 23.990\,963 \text{ u} = 24.998\,788 \text{ u}$$

Thus, the total mass *increases* when this reaction takes place:

$$\Delta m = 24.998\,788 \text{ u} - 24.993\,707 \text{ u} = +0.005\,081 \text{ u}$$

Since the mass of the products is greater than the mass of the reactants, the reaction is endoergic; there is less kinetic energy after the reaction than there was before. The energy absorbed is

$$E = \Delta mc^2 = 0.005\,081 \text{ u} \times 931.5 \text{ MeV/u}$$
$$= 4.733 \text{ MeV}$$

Discussion We expect this reaction to be *possible* only if the total kinetic energy of the reactants is at least 4.733 MeV more than the total kinetic energy of the products. It is not necessarily the *most likely* outcome. Other reactions compete, such as the emission of one or more photons:

$$_0^1n + {}_{12}^{24}Mg \rightarrow {}_{12}^{25}Mg^* \rightarrow {}_{12}^{25}Mg + \gamma$$

In other cases, the competing reaction might include alpha decay, beta decay, or fission.

Practice Problem 29.12 The reaction that produces carbon-14

Determine whether the reaction $n + {}^{14}N \rightarrow {}^{14}C + p$ is exoergic or endoergic. How much energy is released or absorbed?

Neutron Activation Analysis

Neutron activation analysis (NAA) is a technique used to study precious works of art, rare archaeological specimens, geological objects, and the like. It is used to determine which elements are present in the sample being studied—even those present in only trace amounts. The great advantage of NAA over other methods of analysis is that it is

Making The Connection:
neutron activation analysis of geological objects

nondestructive. An entire painting can be analyzed without the need to scrape off some of the paint, as would have to be done to use a mass spectrometer. Art historians know that different paint pigments were used in different historical periods. Determination of the pigments used can help establish the date of a painting; it can also detect forgeries, repairs, and canvasses that have been painted over.

Suppose the ground (the paint first applied to bare canvas) contains the manganese-containing pigment called umber. When the painting is irradiated with neutrons from a fission reactor, some of the neutrons are captured by ^{55}Mn nuclei:

$$n + {}^{55}Mn \rightarrow {}^{56}Mn^* \rightarrow {}^{56}Mn + \gamma$$

The manganese nucleus has been activated—it has been changed from the stable ^{55}Mn nucleus to the radioactive ^{56}Mn nucleus. Initially the ^{56}Mn* is in an excited state, but it quickly decays to the ground state ^{56}Mn by emitting one or more photons while the sample is still being irradiated.

The painting is now taken out of the neutron beam for analysis. The ^{56}Mn nuclei in the sample decay via beta decay:

$$^{56}Mn \rightarrow {}^{56}Fe^* + e^- + \bar{\nu}$$

The half-life for this decay is 2.58 h. The asterisk in the symbol ^{56}Fe* emphasizes that the ^{56}Fe nucleus is initially in an excited state; it quickly (in 10^{-12} s or less) decays to the ground state of ^{56}Fe by emitting one or more photons. A gamma-ray spectrometer detects individual photons and measures their energies.

The elements present in a sample are identified by the characteristic gamma-ray energies emitted by the activated nuclei when they decay. By taking gamma-ray spectra at different times, the half-lives can also be used for identification purposes. Quantitative analysis of the gamma-ray spectrum yields the concentrations of various elements in the samples being studied. Neutron activation analysis of this type has been used to study lunar samples from the Apollo missions, bullets and gunshot residue swabs used as forensic evidence in criminal investigations, oceanographic fossils and sediments, textiles, and artifacts from archaeological excavations, just to name a few examples.

Most of the gamma-ray photons emitted immediately after activation come from activated nuclides that beta decay with short half-lives. Later, the short-lived nuclides have mostly decayed, so most of the photons emitted are from the nuclides with longer half-lives. For example, if a sample contains both ^{27}Al and ^{31}P, gamma rays emitted shortly after activation are mostly from ^{27}Al:

$$n + {}^{27}Al \rightarrow {}^{28}Al \rightarrow {}^{28}Si^* + e^- + \bar{\nu} \quad \text{(half-life} \approx 2.3 \text{ min)}$$

Several days later, the ^{28}Al nuclei have essentially all decayed, so the gamma rays are now mostly from ^{31}P:

$$n + {}^{31}P \rightarrow {}^{32}P \rightarrow {}^{32}S^* + e^- + \bar{\nu} \quad \text{(half-life} \approx 14.3 \text{ days)}$$

NAA enables the art historian to determine which pigments have been used on which parts of the painting, even in the underlying layers, without damaging the painting. First a canvas is exposed to neutrons, activating some of the nuclei. Then the painting is covered with photographic film, which is exposed by the electrons emitted (the gamma rays pass through the film without interacting). The exposed film is called an *autoradiograph* (auto- because the radiation comes from within the sample rather than from an external source). Autoradiographs are taken at several different times after activation. Earlier films show where nuclides with shorter half-lives are found, while later films show where the longer-lived nuclides are found.

In *Aristotle with a Bust of Homer*, the sixth autoradiograph (Fig. 29.12a) shows the extent of the damage to the apron and to the hat. The lightest areas, in which the weave of the canvas can be seen, show where the damage extends to and includes the first layer of ground. In the adjacent areas that are slightly darker, the first layer of ground is intact but the second layer of ground is damaged. By studying the different layers of paint as revealed by the nine autoradiographs, art historians drew some con-

(a)   (b)

Figure 29.12   (a) Sixth autoradiograph of *Aristotle with a Bust of Homer*. Exposure of the film began 48 h after activation and ended 96 h after activation. (b) Eighth autoradiograph. Exposure of the film began 8 days after activation and ended 20 days after activation.

clusions about how Rembrandt's composition was altered as he worked. For example, comparing the sixth and eighth (Fig. 29.12b) autoradiographs to the completed painting reveals changes in Aristotle's costume, changes in the positions of the arms and shoulders, a change in the position of the medal, and a change in the height of the bust of Homer. Historians knew that the canvas had lost 14 inches of its original height; the early position of the bust helped them conclude that most of the missing canvas was at the bottom.

29.7   FISSION

As shown in Fig. 29.4, very large nuclei have a smaller binding energy per nucleon than do nuclides of intermediate mass. The binding energy of large nuclides is reduced by the long-range Coulomb repulsion of the protons. Each proton in the nucleus repels every other proton. The strong nuclear force, which holds the nucleons together in a nucleus, is short range. Each nucleon is bound only to its nearest neighbors. In large nuclides the average number of nearest neighbors is approximately constant, so the strong nuclear force does not increase the binding energy per nucleon to compensate for the Coulomb repulsion, which decreases the binding energy per nucleon.

A large nucleus can therefore release energy by splitting into two smaller, more tightly bound nuclei in the process called **fission**. The term is borrowed from biology; a cell fissions when it splits into two. Nuclear fission was first observed by Otto Hahn and Fritz Strassman in 1938.

Some very large nuclei can fission spontaneously. Radioactive uranium-238, for instance, can break apart into two fission products, though it is much more likely to decay by emitting an alpha particle. Fission can also be induced by an incident neutron, proton, deuteron (a ^{2}H nucleus), alpha particle, or photon. Fission due to the capture of a slow neutron allows the possibility of a chain reaction. Uranium-235 is the only naturally occurring nuclide that can be induced to fission by slow neutrons.

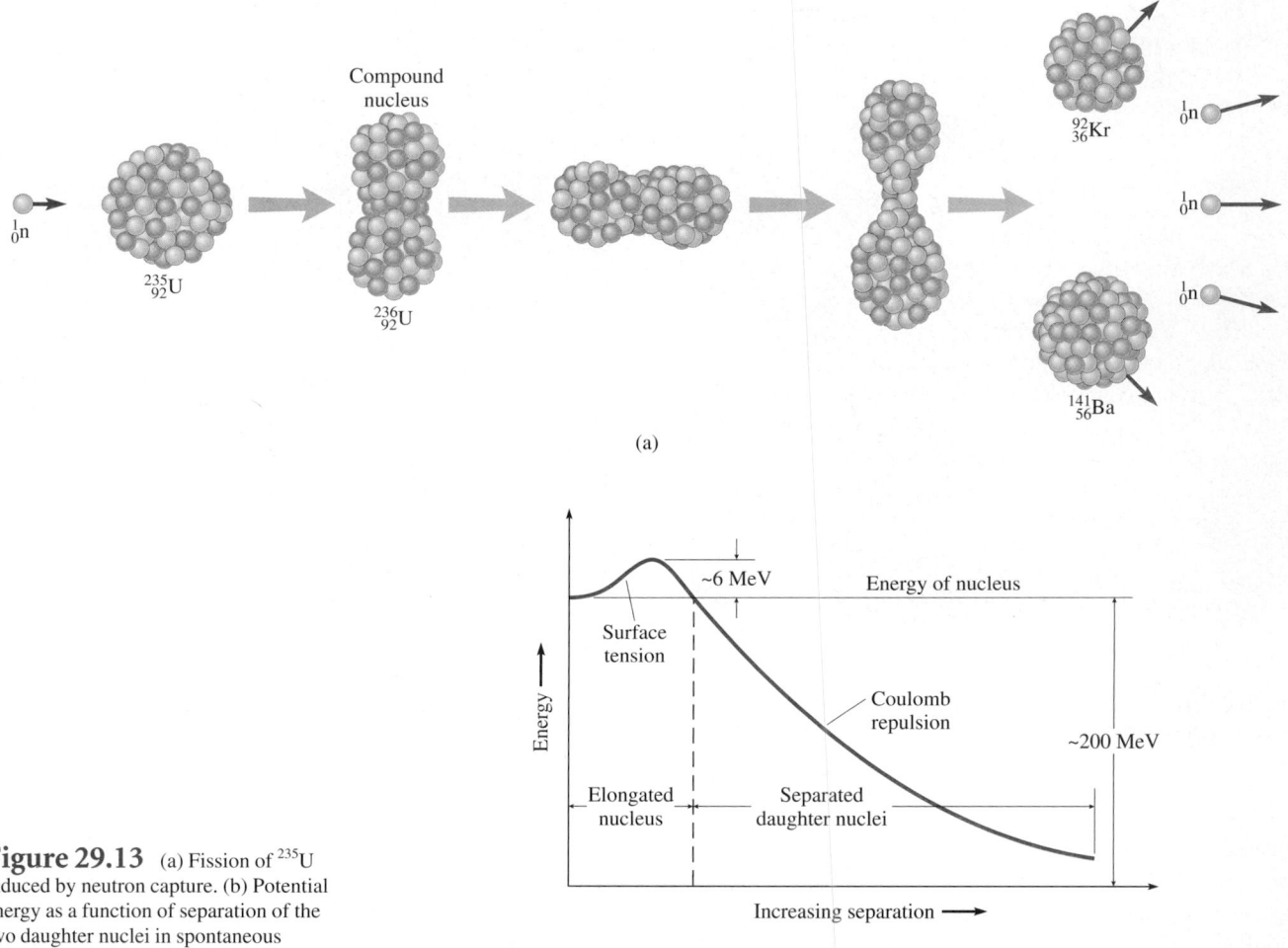

(a)

(b)

Figure 29.13 (a) Fission of ^{235}U induced by neutron capture. (b) Potential energy as a function of separation of the two daughter nuclei in spontaneous fission.

The liquid drop model illustrates how induced fission occurs. Suppose that a slow neutron is captured by a ^{235}U nucleus. The compound nucleus formed, ^{236}U, is in an excited state since the neutron gives up energy when it becomes bound to the nucleus (Problem 48). The excited nucleus is no longer spherical; now it oscillates (Fig. 29.13a). When the shape of the nucleus is elongated, surface tension tends to pull it back into a sphere, while the Coulomb repulsion between protons tends to push the ends apart. If the excitation energy is sufficient, the surface tension is not strong enough to hold the compound nucleus together. A neck forms, and the nucleus splits into two parts. The Coulomb repulsion then pushes the two fragments apart so they do not recombine into a single nucleus.

Spontaneous fission occurs by quantum-mechanical tunneling (Section 28.10). With no incident particle to supply the extra energy, the nucleus cannot form an elongated shape. To form a shape with a larger surface area, work must be done against the surface tension, which pulls surface nucleons inward and tends to minimize surface area. Figure 29.13b shows the potential energy as a nucleus elongates and splits into two. Since the final potential energy after fission is about 200 MeV lower than the potential energy of a single nucleus, spontaneous fission can occur—but only by tunneling through the higher-energy barrier (of about 6 MeV) caused by surface tension. The tunneling probability is much lower than the probability of alpha decay. If U-238 decayed only by spontaneous fission, its half-life would be about 10^{16} yr (instead of 4×10^9 yr).

Many different fission reactions are possible for a given parent nuclide. The fission products are not always the same, and there is no way to predict which of the many possibilities actually occur. Here are two examples of the induced fission of ^{235}U after it captures a slow neutron:

$$\frac{1}{0}n + ^{235}_{92}U \rightarrow ^{236}_{92}U^* \rightarrow ^{141}_{56}Ba + ^{92}_{36}Kr + 3^1_0n \tag{29-31a}$$

$$\frac{1}{0}n + ^{235}_{92}U \rightarrow ^{236}_{92}U^* \rightarrow ^{139}_{54}Xe + ^{95}_{38}Sr + 2^1_0n \tag{29-31b}$$

Notice that the masses of the two daughter nuclei differ significantly in these two examples. The ratio of the masses of the two U-235 fission fragments varies from 1 (equal masses) to a little over 2 (one slightly more than twice as massive as the other). The most likely split is a mass ratio of approximately 1.4–1.5 (Fig. 29.14).

Besides the daughter nuclei, neutrons are released in a fission reaction. Large nuclei are more neutron-rich than are smaller nuclei; a few excess neutrons are released when a large nucleus fissions. As many as five neutrons can be released in the fission of uranium-235; the average number released in a large number of fission reactions is about 2.5. The fission fragments themselves are often still too neutron-rich. The unstable fragments undergo beta decay one or more times, stopping when a stable nuclide is formed. In a fission chain reaction (Fig. 29.15), hundreds of different radioactive nuclides—most of which do not occur naturally—are produced.

Figure 29.14 Mass distribution of fission fragments from ^{235}U. Note that the vertical scale is logarithmic.

Figure 29.15 A fission chain reaction. Neutrons released when fission occurs can go on to induce fission in other nuclei.

Example 29.13 shows that the energy released in a fission reaction is enormous—typically around 200 MeV for the split of a single nucleus.

Example 29.13

Energy Produced in a Fission Reaction

Estimate the energy released in the fission reaction of Eq. (29-31a). Use Fig. 29.4 to estimate the binding energies per nucleon for $^{235}_{92}U$, $^{141}_{56}Ba$, and $^{92}_{36}Kr$.

Strategy   The energy released is equal to the increase in binding energy. The binding energies are estimated by reading the binding energy per nucleon from Fig. 29.4 and multiplying by the number of nucleons.

Solution   From Fig. 29.4, the binding energies per nucleon of $^{235}_{92}U$, $^{141}_{56}Ba$, and $^{92}_{36}Kr$ are approximately 7.6 MeV, 8.25 MeV, and 8.75 MeV, respectively. We find the total binding energies by multiplying by the number of nucleons. Binding energy:

$$\text{for } ^{235}_{92}U \approx 235 \times 7.6 \text{ MeV} = 1786 \text{ MeV}$$

$$\text{for } ^{141}_{56}Ba \approx 141 \times 8.25 \text{ MeV} = 1163 \text{ MeV}$$

$$\text{for } ^{92}_{36}Kr \approx 92 \times 8.75 \text{ MeV} = 805 \text{ MeV}$$

The increase in binding energy is

$$1163 \text{ MeV} + 805 \text{ MeV} - 1786 \text{ MeV} = 182 \text{ MeV}$$

The energy released by the fission reaction is about 180 MeV.

Discussion   The energy released doesn't vary much from one fission reaction to another. A nuclide with $A \approx 240$ has a binding energy of about 7.6 MeV/nucleon. The fission products have an average binding energy of about 8.5 MeV/nucleon. Thus, we expect the energy released to be a little less than 1 MeV per nucleon.

To refine this estimate, we can do a precise calculation of the energy released in the reaction using the masses of the parent and daughter nuclides (Problem 50).

Conceptual Practice Problem 29.13   Can smaller nuclides fission?

Suppose that a $^{54}_{24}Cr$ nucleus captures a slow neutron:

$$^{1}_{0}n + ^{54}_{24}Cr \rightarrow ^{55}_{24}Cr^{*}$$

Explain why fission does not occur. What might happen instead?

To get a *macroscopically* significant amount of energy from fission, a large number of nuclei must split. A neutron can induce fission in ^{235}U, and each fission produces an average of 2.5 neutrons, each of which can go on to induce other nuclei to fission, resulting in a chain reaction. An uncontrolled chain reaction is the basis of the fission bomb. To make constructive use of the energy released by fission, the chain reaction must be controlled.

Fission Reactors

Making The Connection:
nuclear fission
reactors

Most modern fission reactors (Fig. 29.16) use *enriched uranium* as fuel. Only U-235 carries on the chain reaction; U-238 can capture neutrons without splitting. Naturally occurring uranium is 99.3% U-238 and only 0.7% U-235; with so much U-238 absorbing neutrons, it would be difficult to maintain a chain reaction. In enriched uranium, the U-235 content is increased to a few percent. The neutrons produced in a fission reaction have large energies. These fast neutrons are equally likely to be captured by U-238 nuclei or U-235 nuclei. But if the neutrons are slowed down, then they are much more likely to be captured by U-235 and induce fission. For this reason, a substance called a *moderator* is included in the fuel core. Moderators include hydrogen (in water or zirconium hydride), deuterium (^2H, in molecules of heavy water), beryllium, or carbon (as graphite). The moderator's function is to slow down the neutrons by colliding with them

Figure 29.16 A pressurized water fission reactor. Water under high pressure is the primary coolant that carries heat from the core into a heat exchanger in a closed loop. (In some other reactors, liquid sodium is used as the primary coolant.) The heat exchanger extracts heat from the primary coolant to make steam; the steam drives a turbine connected to an electric generator. In essence, the fission reactions in the core act like a furnace to supply the heat needed to run a heat engine. As in all heat engines, waste heat must be exhausted to the environment.

without capturing too many. Light nuclei are most effective at slowing down the neutrons, since the fractional loss in kinetic energy decreases with increasing target mass.

To control the chain reaction, a substance that readily absorbs neutrons, such as cadmium or boron, is formed into *control rods*. The control rods are lowered into the fuel core to absorb more neutrons, or retracted to absorb fewer neutrons. In normal operation, the reactor is *critical*: on average one neutron from each fission goes on to initiate another fission. A critical reactor produces a steady power output. If the reactor is *subcritical*, on average less than one of the neutrons produced by a fission reaction goes on to cause another fission. As fewer and fewer fission reactions occur, the chain reaction dies out. A reactor is shut down by lowering the control rods to make the reactor subcritical. If the reactor is *supercritical*, then on average more than one neutron from each fission causes another fission. Thus, the number of fission reactions per second is increasing in a supercritical reactor. A reactor must be allowed to be supercritical for a *brief* time while it is starting up.

Fission reactors have purposes other than power generation. They also provide the neutron sources for neutron activation analysis and neutron diffraction experiments. The neutrons from a reactor are also used to produce artificial radioisotopes for medical use. A by-product of the fission reactions in a *breeder reactor* is to produce more fissionable

Figure 29.17 Aerial view of the exploded fourth reactor of the Chernobyl nuclear power plant.

material (plutonium-239) from uranium-238 in its fuel core than is consumed. The Pu-239 can be left in the core, to fission and generate power, or it can be extracted and used to make bombs. Thus, breeder reactors are a concern for those working to halt the proliferation of nuclear weapons.

Problems with Fission Reactors

Fission reactors must be carefully designed to prevent the release of harmful radioactive materials into the environment. By far the worst accident at a nuclear reactor occurred in 1986 at the Chernobyl reactor in Ukraine, then part of the Soviet Union (Fig. 29.17). Poor reactor design and a series of mistakes by the operators of the reactor resulted in two explosions, releasing radioactive fission products into the atmosphere and allowing the graphite moderator in the core to burst into flame. The graphite fire continued for 9 days. The estimated amount of radiation that escaped is of the order of 10^{19} Bq; winds dispersed radioactive material over Ukraine, Belarus, Russia, Poland, Scandinavia, and eastern Europe.

Another major problem is the safe storage of radioactive waste. Spent fuel rods, which are removed from the reactor core when the fissionable material is depleted, contain highly radioactive fission products that must be stored for thousands of years. Other parts of the reactor become radioactive by being exposed to neutrons. After about 30 yr of operation, the structural materials of the reactor have been weakened by radiation and the reactor must be decommissioned.

In 1988, the U.S. government chose Yucca Mountain in Nevada as the nation's permanent repository for approximately 77,000 tons of high-level radioactive waste from fission reactors. Subsequent studies showed that the desert site might not be as geologically stable as was originally thought. The Nevada state government and some other groups oppose the planned facility; as this book goes to press, the legal battles continue. The earliest the Yucca Mountain facility can receive high-level waste is about 2010; the repository would be expected to operate for 50 to 300 yr. Until then, high-level nuclear waste continues to be stored on site at the reactors.

29.8   FUSION

The energy radiated by the Sun and other stars is produced by nuclear **fusion**. Fusion is essentially the opposite of fission. Instead of a large nucleus splitting into two smaller pieces, fusion combines two small nuclei to form a larger nucleus. Both fission and fusion release energy, since they move toward larger binding energies per nucleon (see Fig. 29.4). Due to the steep slope of the binding energy per nucleon curve at low mass numbers, fusion can be expected to release significantly more energy per nucleon than fission.

Here is an example of a fusion reaction:

$$^{2}\text{H} + {}^{3}\text{H} \rightarrow {}^{4}\text{He} + \text{n} \tag{29-32}$$

Two hydrogen nuclei fuse to form a helium nucleus. The reaction releases 17.6 MeV of energy, which is 3.52 MeV per nucleon—much more than the 0.75–1 MeV per nucleon typical of fission reactions. Although this reaction produces a tremendous amount of energy, it cannot occur at room temperature. The deuterium (^{2}H) and tritium (^{3}H) nuclei must get close enough to react. At room temperature, the two positively charged nuclei have kinetic energies much too small to overcome their mutual Coulomb repulsion. However, in the Sun's interior the temperature is about 2×10^{7} K and the average kinetic energy of the nuclei is $\frac{3}{2}k_{\text{B}}T \approx 2.52$ keV. This *average* kinetic energy is still far too small to allow a fusion reaction (see Example 29.14), but some of the more energetic nuclei do have enough kinetic energy to overcome the Coulomb repulsion. Fusion reactions are also called thermonuclear reactions because they depend on the large kinetic energies available at high temperatures.

Two fusion cycles to explain energy production in stars were proposed by Hans Bethe (1906–). One cycle is called the **proton-proton cycle**:

$$p + p \rightarrow {}^{2}H + e^{+} + \nu \tag{29-33a}$$

$$p + {}^{2}H \rightarrow {}^{3}He + \gamma \tag{29-33b}$$

$$^{3}He + {}^{3}He \rightarrow {}^{4}He + 2p \tag{29-33c}$$

The net effect of the proton-proton cycle is to fuse four protons into a ${}^{4}He$ nucleus. (The first two reactions must each take place twice to form the two ${}^{3}He$ nuclei needed for the third reaction.) The three steps can be summarized as

$$4p \rightarrow {}^{4}He + 2e^{+} + 2\nu + 2\gamma$$

Each positron annihilates with an electron

$$e^{+} + e^{-} \rightarrow 2\gamma$$

so the overall reaction due to the proton-proton cycle is

$$4p + 2e^{-} \rightarrow {}^{4}He + 2\nu + 6\gamma \tag{29-34}$$

The other cycle of fusion reactions that occurs in some stars is the **carbon cycle**:

$$p + {}^{12}C \rightarrow {}^{13}N + \gamma \tag{29-35a}$$

$$^{13}N \rightarrow {}^{13}C + e^{+} + \nu \tag{29-35b}$$

$$p + {}^{13}C \rightarrow {}^{14}N + \gamma \tag{29-35c}$$

$$p + {}^{14}N \rightarrow {}^{15}O + \gamma \tag{29-35d}$$

$$^{15}O \rightarrow {}^{15}N + e^{+} + \nu \tag{29-35e}$$

$$p + {}^{15}N \rightarrow {}^{12}C + {}^{4}He \tag{29-35f}$$

Here the carbon-12 nucleus acts as a catalyst; it is present in the beginning and at the end. After the annihilation of the two positrons, the net effect is the same as the proton-proton cycle except for the number of photons produced:

$$4p + 2e^{-} \rightarrow {}^{4}He + 2\nu + 7\gamma$$

The *total* energy released by the carbon cycle is the same as the total energy released by the proton-proton cycle. By "total energy released" we mean the total energy of all the photons and neutrinos produced plus the kinetic energy of the ${}^{4}He$ nucleus minus the initial kinetic energies of the protons and electrons.

Example 29.14

First Step of the Carbon Cycle

(a) Calculate the energy released in the first step of the carbon cycle. (b) Estimate the minimum kinetic energy of the proton and ${}^{12}C$ nucleus required for this reaction to take place.

Strategy To calculate the energy released, we must determine the mass difference between reactants and products. For the minimum initial kinetic energy, we know that the two positively charged particles repel each other. We can find the distance between the two when they just "touch," and find the electric potential energy in that position.

Solution (a) The reaction in question is

$$p + {}^{12}C \rightarrow {}^{13}N + \gamma$$

We use atomic masses in the calculations since the extra mass of seven electrons is equally present in the atomic masses of the reactants and products. The initial mass is then

$$1.007\,825\text{ u} + 12.000\,000\text{ u} = 13.007\,825$$

The mass change is

$$\Delta m = 13.005\,739\text{ u} - 13.007\,825 = -0.002\,086\text{ u}$$

The energy released is

$$E = 0.002\,086\text{ u} \times 931.5\text{ MeV/u} = 1.943\text{ MeV}$$

continued on next page

Example 29.14 *continued*

(b) The radii of the proton and ^{12}C nucleus are 1.2 fm and

$$1.2 \text{ fm} \times 12^{1/3} = 2.75 \text{ fm}$$

For an *estimate* of the electric potential energy when the proton and ^{12}C nucleus are just "touching," we find the electric potential energy of two *point charges*, $+e$ and $+6e$, at a separation of 3.95 fm.

$$U_E = \frac{6ke^2}{r} = \frac{6 \times (9 \times 10^9 \text{ N·m}^2/\text{C}^2) \times (1.60 \times 10^{-19} \text{ C})^2}{3.95 \times 10^{-15} \text{ m}}$$

$$= 3.50 \times 10^{-13} \text{ J} = 2 \text{ MeV}$$

The minimum total kinetic energy of the proton and ^{12}C nucleus that allows the reaction to take place is about 2 MeV.

Discussion The energy released, 1.943 MeV, includes both the increase in kinetic energy and the energy of the photon.

Practice Problem 29.14   Second step of carbon cycle

Calculate the energy released in the second step in the carbon cycle.

Making The Connection:
**nuclear fusion
in stars**

Stars act as factories to form heavier nuclides from lighter nuclides. In a star like our Sun, most of the fusion reactions produce helium from hydrogen. At higher core temperatures, heavier nuclides can take part in fusion reactions. Nuclides all the way up to the peak of the binding energy curve (Fig. 29.4), around $A = 60$, are formed by fusion reactions in the interiors of stars. Once a star core is rich in iron and nickel, elements near the top of the binding energy curve, fusion reactions die out. Heavier nuclides are *less* tightly bound than iron and nickel, so fusion reactions no longer release energy. Eventually the large star implodes under its own gravity; the implosion provides the energy for the fusion of heavier nuclides. Ultimately the star may explode, an event called a supernova. Additional fusion and neutron-capture reactions occur in the shock waves caused by the explosion, forming the heaviest nuclides. The nuclides formed in the supernova, plus the ones that had already been formed in the star's core, are dispersed by the explosion into space. The atoms that make up all of us and our surroundings were distributed into space by one or more supernovae before making their way to Earth. Other than hydrogen (and a small fraction of some of the other light elements), all of the elements found on Earth were either made in the core of a star or in a supernova (or are radioactive decay products of these elements).

Controlled Fusion

In a thermonuclear bomb (or hydrogen bomb), a fission bomb creates the high temperatures that enable an uncontrolled fusion reaction to take place. For decades, researchers have been trying to make possible a sustained, *controlled* fusion reaction. Fusion as an energy source would have several advantages over fission. The fuel for fusion is more easily obtained than the fuel for fission. The most promising reactions for controlled fusion are deuteron-deuteron fusion (^{2}H + ^{2}H) or deuteron-triton fusion [^{2}H + ^{3}H, as in Eq. (29-32)]. Deuterium is readily available in seawater; about 0.015% of the water molecules in the ocean contain a deuterium atom. Tritium is not as abundant in seawater but is not difficult to produce from lithium.

Making The Connection:
fusion reactors

One of the biggest problems associated with *fission* reactors is the radioactive waste that must be safely stored for thousands of years. A fusion reactor would produce less radioactive waste and the waste would not have to be stored for as long.

However, a sustained, controlled fusion reaction has not yet been achieved. The main problem is containing the fuel at the extremely high temperatures (estimated to be about 10^8 K, which is higher than the temperature of the Sun's interior) needed for fusion to take place while maintaining a high density of nuclei so that they collide into each other. An ordinary container cannot be used; the nuclei would lose too much kinetic energy when they collide with the walls of such a container and the container would be vaporized by the high temperatures. Two principal confinement schemes are being tried. One is *magnetic confinement* (see Fig. 29.18). The other is *inertial confinement*, in which a small fuel pellet is heated rapidly by intense laser beams from all sides, causing the fuel pellet to implode and the fusion reactions to take place before the pellet is vaporized.

Figure 29.18 The tokamak is one of the most promising methods for containing a controlled fusion reaction. At such high temperatures, the atoms in the fuel break apart to form a *plasma*—a mixture of electrons and positively charged nuclei. Magnetic fields confine the charged nuclei to the interior of a doughnut-shaped, evacuated chamber. The nuclei spiral around magnetic field lines and are confined without colliding with the walls of the vacuum chamber.

Labels on figure: Toroidal magnets; Protective shielding; Access port; Plasma

MASTER THE CONCEPTS

Summary

- A particular nuclide is characterized by its atomic number Z (the number of protons) and its nucleon number A (the total number of protons and neutrons). The isotopes of an element have the same atomic number but different numbers of neutrons.

- The mass density of all nuclei is approximately the same. The radius of a nucleus is

$$r = r_0 A^{1/3} \qquad (29\text{-}4)$$

where

$$r_0 = 1.2 \times 10^{-15} \text{ m} = 1.2 \text{ fm} \qquad (29\text{-}5)$$

- The **binding energy** E_B of a nucleus is the energy that must be supplied to separate a nucleus into individual protons and neutrons. Since the nucleus is a bound system, its total energy is *less* than the energy of Z protons and N neutrons that are far apart and at rest.

$$E_B = \Delta m c^2 \qquad (29\text{-}8)$$

- In any nuclear reaction, the total electric charge and the total number of nucleons are conserved.

- An unstable or radioactive nuclide decays by emitting radiation.

Alpha decay $\quad {}^{A}_{Z}P \rightarrow {}^{A-4}_{Z-2}D + {}^{4}_{2}\alpha \qquad (29\text{-}10)$

Beta-minus decay $\quad {}^{A}_{Z}P \rightarrow {}^{A}_{Z+1}D + e^{-} + \overline{\nu} \qquad (29\text{-}11)$

Beta-plus decay $\quad {}^{A}_{Z}P \rightarrow {}^{A}_{Z-1}D + e^{+} + \nu \qquad (29\text{-}12)$

Gamma decay $\quad P^{*} \rightarrow P + \gamma$

- Each radioactive nuclide has a characteristic decay probability per unit time λ. The activity R of a sample with N nuclei is

$$R = \frac{\text{number of decays}}{\text{unit time}} = \frac{-\Delta N}{\Delta t} = \lambda N \qquad (29\text{-}18)$$

Activity is commonly measured in becquerels (1 Bq = 1 decay per second) or curies (1 Ci = 3.7×10^{10} Bq).

- The number of remaining nuclei N in radioactive decay (the number that have *not* decayed) is an exponential function:

$$N(t) = N_0 e^{-t/\tau} \qquad (29\text{-}20)$$

where the time constant is $\tau = 1/\lambda$. The half-life is the time during which half of the nuclei decay:

$$T_{1/2} = \tau \ln 2 \approx 0.693\,\tau \qquad (29\text{-}23)$$

- The absorbed dose is the amount of radiation energy absorbed per unit mass of tissue, measured in grays (1 Gy = 1 J/kg) or rads (1 rad = 0.01 Gy).

- The quality factor (QF) is a relative measure of the biological damage caused by different kinds of radiation. The biologically equivalent dose in rem is

biologically equivalent dose (in rem) =
absorbed dose (in rad) × QF   (29-29)

- A large nucleus can release energy by splitting into two smaller, more tightly bound nuclei in the process called fission. The energy released in a fission reaction is enormous—typically around 200 MeV for the split of a single nucleus.

MASTER THE CONCEPTS *continued*

- Nuclear fusion combines two small nuclei to form a larger nucleus. Fusion typically releases significantly more energy per nucleon than fission.

Highlighted Figures and Tables

T29.1 Masses of proton, neutron, and electron in atomic mass units (p. 1049)

F29.3 Chart of the known nuclides, showing N versus Z (p. 1054)

F29.4 Binding energy per nucleon (E_B/A) as a function of nucleon number A (p. 1055)

F29.5 Distinguishing alpha, beta, and gamma radiation (p. 1056)

T29.2 Particles commonly involved in radioactive decay and other nuclear reactions (p. 1057)

F29.8 Fraction of radioactive nuclei remaining (N/N_0) as a function of time (p. 1063)

T29.3 Typical quality factors (QFs) for various forms of radiation (p. 1068)

F29.13 Fission of ^{235}U induced by neutron capture (p. 1076)

F29.15 A fission chain reaction (p. 1077)

F29.16 A pressurized water fission reactor (p. 1079)

F29.18 The tokamak is one of the most promising methods for containing a controlled fusion reaction (p. 1083)

CONCEPTUAL QUESTIONS

1. How could Henri Becquerel and other scientists determine that there were three *different* kinds of radiation *before* having determined the electrical charges or masses of the alpha, beta, and gamma rays?

2. What technique could Becquerel and others have used to determine that alpha rays are positively charged, beta rays negatively charged, and gamma rays uncharged? Explain how they could find that alpha rays have a charge-to-mass ratio half that of the H$^+$ ion, while beta rays have the same charge-to-mass ratio as "cathode rays" (electrons). (See Chapter 19 for some ideas.)

3. Why is a slow neutron more likely to induce a nuclear reaction (as in neutron activation and induced fission) than a proton with the same kinetic energy?

4. Explain why neutron-activated nuclides tend to decay by β^- rather than β^+.

5. Why can we ignore the binding energies of the atomic *electrons* in calculations such as Example 29.4? Isn't there a mass defect due to the binding energy of the electrons?

6. Why would we expect atmospheric testing of nuclear weapons to increase the relative abundance of C-14 in the atmosphere? Why would we expect the widespread burning of fossil fuels to *decrease* the relative abundance of C-14 in the atmosphere?

7. Isolated atoms (or atoms in a dilute gas) radiate photons at discrete energies characteristic of that atom. In dense matter, the spectrum radiated is quasi-continuous. Why doesn't the same thing happen with nuclear spectra: why do the gamma rays have the same characteristic energies even when emitted from a solid?

8. Section 29.8 states that the total energy released by the proton-proton cycle is the same as that released by the carbon cycle. Why must the total energy released be the same?

9. Iodine is eliminated from the body through *biological* processes with an effective half-life of about 140 days. The radioactive half-life of iodine-131 is 8 days. Suppose some radioactive ^{131}I nuclei are present in the body. Assuming that no new ^{131}I nuclei are introduced into the body, how much time must pass until only half as much ^{131}I is left in the body: less than 8 days, between 8 and 140 days, or more than 140 days? Explain your reasoning.

10. Why does a fission reaction tend to release one or more neutrons? Why is the release of neutrons necessary in order to sustain a chain reaction?

11. Radioactive alpha-emitters are relatively harmless outside the body, but can be dangerous if ingested or inhaled. Explain.

12. Fission reactors and cyclotrons tend to produce different kinds of isotopes. A reactor produces isotopes primarily through neutron activation; thus, the isotopes tend to be neutron-rich (high neutron-to-proton ratio). A cyclotron can only accelerate charged particles such as protons or deuterons. When stable nuclei are bombarded with protons or deuterons, the resulting radioisotopes are neutron-deficient (low neutron-to-proton ratio). (a) Explain why a cyclotron cannot accelerate neutrons. (b) Suppose a hospital needs a supply of radioisotopes to use in positron-emission tomography (PET). Would the radioisotopes more likely come from a reactor or a cyclotron? Explain.

13. Why would a fusion reactor produce less radioactive waste than a fission reactor? [*Hint:* Compare the products of a fission reaction to the products of a fusion reaction.]

14. Radon-222 is created in a series of radioactive decays starting with $^{238}_{92}$U and ending with $^{206}_{82}$Pb. The half-life of Rn-222 is 3.8 days. (a) If the half-life is so short, why hasn't all the Rn-222 gas decayed by now? (b) If the half-life of Rn-222 were much shorter, say a few seconds, would it be more dangerous to us or less dangerous? What if the half-life were much longer, say thousands of years?

MULTIPLE CHOICE QUESTIONS

1. Radioactive $^{215}_{83}$Bi decays into $^{215}_{84}$Po. Which of these particles is released in the decay?

   (a) a proton   (b) an electron   (c) a positron
   (d) an alpha particle   (e) a neutron   (f) none of the above

2. For all stable nuclei

   (a) there are equal numbers of protons and neutrons.
   (b) there are more protons than neutrons.
   (c) there are more neutrons than protons.
   (d) none of the above have to be true.

3. For all stable nuclei

   (a) the mass of the nucleus is less than $Zm_p + (A - Z)m_n$.
   (b) the mass of the nucleus is greater than $Zm_p + (A - Z)m_n$.
   (c) the mass of the nucleus is equal to $Zm_p + (A - Z)m_n$.
   (d) none of the above have to be true.

4. Of the *hypothetical* nuclear reactions below, which would violate conservation of charge?

   (a) $^{10}_{5}B + ^{4}_{2}He \rightarrow ^{13}_{7}N + ^{1}_{1}H$
   (b) $^{10}_{5}B + ^{1}_{0}n \rightarrow ^{11}_{5}B + \beta^- + \bar{\nu}$
   (c) $^{23}_{11}Na + ^{1}_{1}H \rightarrow ^{20}_{10}Ne + ^{4}_{2}He$
   (d) $^{14}_{7}N + ^{1}_{1}H \rightarrow ^{13}_{6}C + \beta^+ + \nu$
   (e) none of them   (f) all of them
   (g) all but (c)   (h) (a) and (d)

5. Of the *hypothetical* nuclear reactions listed in Multiple Choice Question 4, which would violate conservation of nucleon number?

6. In a fusion reaction, two deuterons produce a helium-3 nucleus. What is the other product of the reaction?

   (a) an electron   (b) a proton   (c) a neutron
   (d) an alpha particle   (e) a positron   (f) a neutrino

7. The activity of a radioactive sample (with a single radioactive nuclide) decreases to $\frac{1}{8}$ its initial value in a time interval of 96 days. What is the half-life of the radioactive nuclide present?

   (a) 6 days   (b) 8 days   (c) 12 days
   (d) 16 days   (e) 24 days   (f) 32 days

8. Solid lead has more than four times the mass density of solid aluminum. What is the main reason that lead is so much more dense?

   (a) The Pb atom is smaller than the Al atom.
   (b) The Pb nucleus is smaller than the Al nucleus.
   (c) The Pb nucleus is more massive than the Al nucleus.
   (d) The Pb nucleus is more dense than the Al nucleus.
   (e) The Pb atom has many more electrons than the Al atom.

9. Which of these are appropriate units for the decay constant λ of a radioactive nuclide?

   (a) s   (b) Ci   (c) rd   (d) s^{-1} (e) rem   (f) MeV

10. Which of the units listed in Multiple Choice Question 9 are appropriate for the biologically equivalent dose that results when a person is exposed to radiation?

PROBLEMS

Note: **C** indicates a combination conceptual/quantitative problem. Gold diamonds ◆, ◆◆ are used to indicate the increasing level of difficulty of each problem. Problem numbers appearing in blue, 9., denote problems that have a detailed solution available in the Student Solutions Manual. Some problems are *paired* by concept; their numbers are connected by a ruled box.

29.1 Nuclear Structure

1. Calculate the mass density of nuclear matter.

2. A neutron star is a star that has collapsed into a collection of tightly packed neutrons. Thus, it is something like a giant nucleus; but since it is electrically neutral, there is no Coulomb repulsion to break it up. The force holding it together is gravity. Suppose the Sun were to collapse into a neutron star. What would its radius be? Assume that the density is about the same as for a nucleus. Express your answer in km.

3. Write the symbol (in the form $^A_Z X$) for the nuclide with 38 protons and 50 neutrons and identify the element.

4. Write the symbol (in the form $^A_Z X$) for the isotope of potassium with 21 neutrons.

5. How many neutrons are found in a ^{35}Cl nucleus?

6. How many protons are found in a ^{136}Xe nucleus?

7. Write the symbol (in the form $^A_Z X$) for the nuclide that has 78 neutrons and 53 protons.

8. Find the radius and volume of the $^{107}_{43}$Tc nucleus.

29.2 Binding Energy

9. What is the binding energy of an alpha particle (a ^{4}He nucleus)? The mass of an alpha particle is 4.001 51 u.

10. Find the binding energy of a deuteron (a ^{2}H nucleus). The mass of a deuteron (*not* the deuterium atom) is 2.013 553 u.

◆ 11. Using a mass spectrometer, the mass of the $^{238}_{92}U^+$ *ion* is found to be 238.050 24 u. (a) Use this result to calculate the mass of the $^{238}_{92}U$ *nucleus*. (b) Now find the binding energy of the $^{238}_{92}U$ nucleus.

12. (a) Find the binding energy of the ^{16}O nucleus. (b) What is the average binding energy per nucleon? Compare your answer with Fig. 29.4.

13. Calculate the binding energy per nucleon of the $^{31}_{15}P$ nucleus.

14. Show that $c^2 = 931.494$ MeV/u. [*Hint:* Start with the conversion factors to SI units for MeV and u.]

15. What is the mass defect of the ^{14}N nucleus?

16. What is the mass of an ^{16}O atom in units of MeV/c^2? (1 MeV/c^2 is the mass of a particle with rest energy 1 MeV.)

17. (a) What is the mass defect of the ^{1}H *atom* due to the binding energy of the *electron* (in the ground state)? (b) Should we worry about this mass defect when we calculate the mass of

the ^{1}H nucleus by subtracting the mass of one electron from the mass of the ^{1}H atom?

✦ 18. To make an order-of-magnitude estimate of the energy level spacings in the nucleus, assume that a nucleon is confined to a one-dimensional box of width 10 fm (a typical nuclear diameter). Calculate the energy of the ground state.

29.3 Radioactivity

19. Identify the daughter nuclide when $^{40}_{19}$K decays via β^- decay.

20. $^{232}_{90}$Th decays via α decay. Identify the daughter nuclide.

21. Identify the daughter nuclide when $^{22}_{11}$Na decays by electron capture.

22. Identify the daughter nuclide when $^{22}_{11}$Na decays by emitting a positron.

23. Radium-226 decays as $^{226}_{88}$Ra $\rightarrow$ $^{222}_{86}$Rn $+$ $^4_2$He. If the $^{226}_{88}$Ra nucleus is at rest before the decay and the $^{222}_{86}$Rn nucleus is in its ground state, *estimate* the kinetic energy of the alpha particle. (Assume that the $^{222}_{86}$Rn nucleus takes away an insignificant fraction of the kinetic energy.)

✦ 24. Calculate the kinetic energy of the alpha particle in Problem 23. This time, do *not* assume that the $^{222}_{86}$Rn nucleus is at rest after the reaction. Start by figuring out the ratio of the kinetic energies of the alpha particle and the $^{222}_{86}$Rn nucleus.

25. Which decay mode would you expect for radioactive $^{31}_{14}$Si: α, β^-, or β^+? Explain. [*Hint:* Look at the neutron-to-proton ratio.]

26. Show that the spontaneous alpha decay of ^{19}O is not possible.

27. Calculate the maximum kinetic energy of the beta particle when $^{40}_{19}$K decays via β^- decay.

28. Calculate the maximum kinetic energy of the beta particle when $^{90}_{38}$Sr decays via β^- decay.

29.4 Radioactive Decay Rates and Half-Lives

29. A certain radioactive nuclide has a half-life of 200.0 s. A sample containing just this one radioactive nuclide has an initial activity of 80,000.0 s^{-1}. (a) What is the activity 600.0 s later? (b) How many nuclei were there initially? (c) What is the probability per second that any one of the nuclei decays?

30. Calculate the activity of 1.0 g of radium-226 in Ci.

31. What is the activity in Bq of 1.0 kg of ^{238}U?

32. The half-life of I-131 is 8.0 days. A sample containing I-131 has an activity of 6.4×10^8 Bq. How many days later will the sample have an activity of 2.5×10^6 Bq?

33. Some bones discovered in a crypt in Guatemala are carbon dated. The ^{14}C activity of the bones is measured to be 0.242 Bq per gram of carbon. Approximately how old are the bones?

34. In this problem, you will verify the statement (in Section 29.4) that the ^{14}C activity in a living sample is 0.25 Bq per gram of carbon. (a) What is the decay constant λ for ^{14}C? (b) How many ^{14}C atoms are in 1.00 g of carbon? One mole of carbon atoms has a mass of 12.011 g, and the relative abundance of ^{14}C is 1.3×10^{-12}. (c) Using your results from parts (a) and (b), calculate the ^{14}C activity per gram of carbon in a living sample.

35. Carbon-14 dating is used to date a bone found at an archaeological excavation. If the ratio of C-14 to C-12 atoms is 3.25×10^{-13}, how old is the bone? [*Hint:* Note that this ratio is $\frac{1}{4}$ the ratio of 1.3×10^{-12} that is found in a living sample.]

36. A sample of radioactive $^{214}_{83}$Bi, which has a half-life of 19.8 min., has an activity of 0.058 Ci. What is its activity 1.0 h later?

37. The activity of a sample containing radioactive ^{108}Ag is 6.4×10^4 Bq. Exactly 12 min later, the activity is 2.0×10^3 Bq. Calculate the half-life of ^{108}Ag.

✦ 38. Show mathematically that $2^{-t/T_{1/2}} = \left(\frac{1}{2}\right)^{t/T_{1/2}} = e^{-t/\tau}$ if and only if $T_{1/2} = \tau \ln 2$. [*Hint:* Take the natural logarithm of each side.]

✦✦ 39. The Physics at Home in Section 29.4 suggests tossing coins as a model of radioactive decay. An improved version is to toss a large number of dice instead of coins: each die that comes up a "one" represents a nucleus that has decayed. Suppose that N dice are tossed. (a) What is the average number of dice you expect to decay on one toss? (b) What is the average number of dice you expect to remain undecayed after three tosses? (c) What is the average number of dice you expect to remain undecayed after four tosses? (d) What is the half-life in numbers of tosses?

29.5 Biological Effects of Radiation

40. An alpha particle produced in radioactive alpha decay has a kinetic energy of typically about 6 MeV. When an alpha particle passes through matter (such as biological tissue), it makes ionizing collisions with molecules, giving up some of its kinetic energy to supply the binding energy of the electron that is removed. If a typical ionization energy for a molecule in the body is around 20 eV, roughly how many molecules can the alpha particle ionize before coming to rest?

41. If meat is irradiated with 2000.0 Gy of x-rays, most of the bacteria are killed and the shelf life of the meat is greatly increased. (a) How many 100.0-keV photons must be absorbed by a 0.30-kg steak so that the absorbed dose is 2000.0 Gy? (b) Assuming steak has the same specific heat as water, what temperature increase is caused by a 2000.0-Gy absorbed dose?

✦ 42. Make an order-of-magnitude estimate of the amount of radon-222 gas, measured in curies, found in the lungs of an average person. Assume that 0.1 rem/yr is due to the alpha particles emitted by radon-222. The half-life is 3.8 days. You will need to calculate the energy of the alpha particles emitted.

29.6 Induced Nuclear Reactions

43. A certain nuclide absorbs a neutron. It then emits an electron, and then breaks up into two alpha particles. (a) Identify the original nuclide and the two intermediate nuclides (after absorbing the neutron and after emitting the electron). (b) Would any (anti)neutrino(s) be emitted? Explain.

44. A neutron-activated sample emits gamma rays at energies that are consistent with the decay of mercury-198 nuclei from an excited state to the ground state. If the reaction that takes place is n $+$ (?) $\rightarrow$ ^{198}Hg* $+$ e$^-$ $+$ $\bar{\nu}$, what is the

nuclide "(?)" that was present in the sample before neutron activation?

45. A small pottery fragment is studied using neutron activation analysis. After exposure to neutrons, gamma-ray photons are counted during two successive time periods, each of which lasts for 300.0 s.

Photon Energy (keV)	Number of Photons Detected ($t = 0$ to $t = 300.0$ s)	Number of Photons Detected ($t = 332.0$ s to $t = 632.0$ s)
166	15,730	15,028
662	87,103	20,102
834	15,258	13,318
1,345	6,907	1,594

(a) Based on the numbers of counts, could two of these gamma-ray energies be emitted by the same nuclide? [*Hint:* Find the half-lives or time constants based on the change in the number of counts.] Explain. (b) Suppose nuclide A emits the 166-keV photons and nuclide B emits the 662-keV photons. Assuming that the detector is equally effective at counting gamma rays of different energies, what is the ratio of the initial number of A nuclides to the initial number of B nuclides? [*Hint:* The number of nuclides that decay during the time interval between $t = 0$ and a later time t is $|\Delta N| = N_0 - N(t)$, where $N(t)$ is given by Eqs. (29-20) and (29-24)].

46. In a neutron-activation experiment, gamma-ray photons are counted during two successive time periods, each of which lasts for 1.0 min. The second time period starts 80.0 min after the beginning of the first time period.

Photon Energy (keV)	Number of Photons Detected ($t = 0$ to $t = 1.0$ min)	Number of Photons Detected ($t = 80.0$ min to $t = 81.0$ min)
845	50,000	49,000
1,435	32,000	2,000
1,780	5,000	2,500
1,810	10,000	9,800
2,115	4,000	3,920

(a) What is the minimum number of different radioactive nuclides (different *species* of radioactive nucleus) in this sample that would account for the data shown? [*Hint:* Find the half-lives or time constants based on the change in the number of counts.] Briefly give your reasoning. (b) What is

the half-life of the nuclide that emits the 1435-keV gamma ray? (c) Suppose nuclide A emits the 845-keV photons and nuclide B emits the 1435-keV photons. Assuming that the detector is equally effective at counting gamma rays of different energies, what is the ratio of the initial number of A nuclides to the initial number of B nuclides? [*Hint:* The number of nuclides that decay during the time interval between $t = 0$ and a later time t is $|\Delta N| = N_0 - N(t)$, where $N(t)$ is given by Eqs. (29-20) and (29-24)].

29.7 Fission

47. One possible fission reaction for ^{235}U is ^{235}U + n → ^{141}Cs + ^{93}Rb + ?n, where "?n" represents one or more neutrons. (a) How many neutrons? (b) From the graph in Fig. 29.4, you can read the approximate binding energies per nucleon for the three nuclides involved. Use that information to *estimate* the total energy released by this fission reaction. (c) Do a precise calculation of the energy released. (d) What fraction of the rest energy of the U-235 nucleus is released by this reaction?

48. A ^{235}U nucleus captures a low-energy neutron to form the compound nucleus ^{236}U*. Find the excitation energy of the compound nucleus. Ignore the small initial kinetic energy of the captured neutron.

49. *Estimate* the energy released in the fission reaction of Eq. (29-31b). Look up the binding energy per nucleon of the nuclides in Fig. 29.4.

50. Calculate the energy released in the fission reaction of Eq. (29-31a). The atomic masses of $^{141}_{56}$Ba and $^{92}_{36}$Kr are 140.914 u and 91.926 u, respectively.

29.8 Fusion

51. Consider the fusion reaction of a proton and a deuteron: $^1_1$H + $^2_1$H → X. (a) Identify the reaction product X. (b) The binding energy of the deuteron is about 1.1 MeV *per nucleon* and the binding energy of "X" is about 2.6 MeV *per nucleon*. Approximately how much energy (in MeV) is released in this fusion reaction? (c) Why is this reaction unlikely to occur in a room temperature setting?

52. What is the total energy released by the proton-proton cycle [Eq. (29-34)]? (The total energy released is the total energy of the neutrinos and gamma rays plus the kinetic energy of the ^{4}He nucleus minus the initial kinetic energies of the protons and electrons.)

53. Estimate the minimum total kinetic energy of the ^{2}H and ^{3}H nuclei necessary to allow the fusion reaction of Eq. (29-32) to take place.

COMPREHENSIVE PROBLEMS

54. Which of these unidentified nuclides are isotopes of each other? $^{175}_{71}(?)$, $^{71}_{32}(?)$, $^{175}_{74}(?)$, $^{167}_{71}(?)$, $^{71}_{30}(?)$, and $^{180}_{74}(?)$.

55. An alpha particle with a kinetic energy of 1.0 MeV is headed straight toward a gold nucleus. (a) Find the distance of closest approach between the centers of the alpha particle and gold nucleus. (Assume the gold nucleus remains stationary. Since its mass is much larger than that of the alpha particle, this assumption is a fairly good approximation.) (b) Will the two get close enough to "touch"? (c) What is the minimum initial kinetic energy of an alpha particle that will make contact with the gold nucleus?

56. Figure 29.7 is an energy level diagram for ^{208}Tl. What are the energies of the photons emitted for the six transitions shown?

♦57. $^{212}_{83}$Bi can alpha-decay to the ground state of $^{208}_{81}$Tl, or to any of the four excited states of $^{208}_{81}$Tl shown in Fig. 29.7. The maximum kinetic energy of the alpha particles emitted by $^{212}_{83}$Bi is 6.090 MeV. What other alpha-particle kinetic energies are possible? [*Hint:* Estimate the atomic mass of $^{208}_{81}$Tl.]

58. Approximately what is the total energy of the neutrino emitted when $^{22}_{11}$Na decays by electron capture?

♦59. Suppose that a radioactive sample contains equal numbers of two radioactive nuclides A and B at $t = 0$. A has a half-life of 3.0 h, while B has a half-life of 12.0 h. Find the ratio of the decay rates or activities R_A/R_B at (a) $t = 0$, (b) $t = 12.0$ h, and (c) $t = 24.0$ h.

♦60. The power supply for a pacemaker is a small amount of radioactive ^{238}Pu. This nuclide decays by alpha decay with a half-life of 86 yr. The pacemaker is typically replaced every 10.0 yr. (a) By what percentage does the activity of the ^{238}Pu source decrease in 10 yr? (b) The energy of the alpha particles emitted is 5.6 MeV. Assume an efficiency of 100%—all of the alpha-particle energy is used to run the pacemaker. If the pacemaker starts with 1.0 mg of ^{238}Pu, what is the power output initially and after 10.0 yr?

♦61. The first nuclear reaction ever observed (in 1919 by Ernest Rutherford) was $\alpha + {}^{14}_{7}$N $\rightarrow$ p + X. (a) Show that the reaction product "X" must be $^{17}_{8}$O. (b) For this reaction to take place, the alpha particle must come in contact with the nitrogen nucleus. Calculate the distance d between their centers when they just make contact. (c) If the alpha particle and the nitrogen nucleus are initially far apart, what is the minimum value of their kinetic energy necessary to bring the two into contact? Express your answer in terms of the elementary charge e, the contact distance d, and whatever else you need. (d) Is the total kinetic energy of the reaction products more or less than the initial kinetic energy in part (c)? Why? Calculate this kinetic energy difference.

♦62. The last step in the carbon cycle that takes place inside stars is p + ^{15}N $\rightarrow$ ^{12}C + (?). This step releases 5.00 MeV of energy. (a) Show that the reaction product "(?)" must be an alpha particle. (b) Calculate the atomic mass of helium-4 from the information given. (c) In order for this reaction to occur, the proton must come into contact with the nitrogen nucleus. Calculate the distance d between their centers when they just "touch."

(d) If the proton and nitrogen nucleus are initially far apart, what is the minimum value of their total kinetic energy necessary to bring the two into contact?

63. Radon gas (Rn) is produced by the α decay of radium $^{226}_{88}$Ra. (a) How many neutrons and how many protons are present in the nucleus of the isotope of Rn produced by this decay? (b) In the air in an average size room in a student basement apartment in Ithaca, NY, there are about 10^7 Rn nuclei. The Rn nucleus itself is radioactive; it too decays by emitting an α particle. The half-life of Rn is 3.8 days. How many α particles per second are emitted by decaying Rn nuclei in the room?

🅒 64. (a) What fraction of the ^{238}U atoms present at the formation of the Earth still exist? Take the age of the Earth to be 4.5 × 10^9 yr. (b) Answer the same question for ^{235}U. Could this explain why there are more than 100 times as many ^{238}U atoms as ^{235}U atoms in the Earth today?

🅒 65. The radioactive decay of ^{238}U produces alpha particles with an energy of 4.17 MeV. (a) At what speed do these alpha particles move? (b) Put yourself in the place of Rutherford and Geiger. You know that alpha particles are positively charged (from the way they are deflected in a magnetic field). You want to measure the speed of the alpha particles using a velocity selector. If your magnet produces a magnetic field of 0.30 T, what strength electric field would allow the alpha particles to pass through undeflected? (c) Now that you know the speed of the alpha particles, you measure the radius of their trajectory in the same magnetic field (without the electric field) to determine their charge-to-mass ratio. Using the charge and mass of the alpha particle, what would the radius be in a 0.30-T field? (d) Why can you determine only the ratio q/m by this experiment, but not the individual values of q and m?

🅒 66. Once Rutherford and Geiger determined the charge-to-mass ratio of the alpha particle (Problem 65), they performed another experiment to determine its charge. An alpha source was placed in an evacuated chamber with a fluorescent screen. Through a glass window in the chamber, they could see a flash on the screen every time an alpha particle hit it. They used a magnetic field to deflect beta particles away from the screen so they were sure that every flash represented an alpha particle. (a) Why is the deflection of a beta particle in a magnetic field much larger than the deflection of an alpha particle moving at the same speed? (b) By counting the flashes, they could determine the number of alphas per second striking the screen (R). Then they replaced the screen with a metal plate connected to an electroscope and measured the charge Q accumulated in a time Δt. What is the alpha-particle charge in terms of R, Q, and Δt?

♦67. A $^{208}_{81}$Tl* nucleus (mass 208.0 u) emits a 452-keV photon to jump to a state of lower energy. Assuming the nucleus is initially at rest, calculate the kinetic energy of the nucleus after the photon has been emitted. [*Hint:* Assume the nucleus can be treated nonrelativistically.]

ANSWERS TO PRACTICE PROBLEMS

29.1 $^{104}_{44}$Ru (ruthenium)

29.2 17 u

29.3 1.6×10^{-42} m^3

29.4 115.492 MeV

29.5 $^{226}_{88}$Ra (radium-226)

29.6 5.31 MeV

29.7 1.31 MeV

29.8 2.26×10^{12}

29.9 5300 yr ago

29.10 ±8 yr

29.11 4.4 µg

29.12 exoergic; 0.626 MeV released

29.13 From Fig. 29.4, nuclides around $A \approx 60$ are the most tightly bound; they have the highest binding energies per nucleon. Fission cannot occur because the total mass of the daughter nuclides and any neutrons released would be *greater* than the mass of the $^{55}_{24}$Cr* compound nucleus. More likely, $^{55}_{24}$Cr* would emit an electron and one or more gamma rays, leaving a stable $^{55}_{25}$Mn nucleus as the final product.

29.14 1.199 MeV

Particle Physics

Currently, the particle accelerator that produces the highest energy collisions is the Tevatron at the Fermi National Accelerator Laboratory (Fermilab) near Batavia, Illinois. In the Tevatron, protons and antiprotons with kinetic energies of 1 TeV (= 1000 GeV) travel in opposite directions around a ring of radius 1.0 km (6.3 km circumference) before being smashed together, so that the resulting collisions have energies of 2 TeV. The Large Hadron Collider (LHC), currently under construction at the European Center for Nuclear Research (CERN) near Geneva, Switzerland, will achieve collisions between particles with kinetic energies up to 7 TeV so that collisions at an energy of 14 TeV can be examined. What is the goal of studying particle collisions with higher and higher energies?

30.1   FUNDAMENTAL PARTICLES

One of the overarching goals of physics is to find the fundamental building blocks of the universe. In the fifth century B.C., the Greek philosopher Democritus speculated that all matter was composed of fundamental units so tiny they could not be seen; the English word *atom* comes from the Greek word *atomos*, meaning indivisible. However, atoms are *not* indivisible: they consist of one or more electrons bound to a nucleus. The nucleus, in turn, is a bound collection of protons and neutrons. Are electrons, protons, and neutrons the fundamental building blocks of matter?

Quarks

We now know that protons and neutrons have internal structure and thus are *not* fundamental particles. Each proton or neutron is composed of three **quarks**. Quarks are fundamental particles whose existence was proposed independently in 1963 by Murray Gell-Mann and George Zweig. Gell-Mann took the name *quark* from a line in *Finnegans Wake* by James Joyce: "Three quarks for Muster Mark." Although three quarks were originally proposed, subsequent experiments have shown that there are six altogether (see Table 30.1). The quark masses are expressed in GeV/c^2, a mass unit commonly used in high-energy physics. Since $c^2 = 0.931\,494$ GeV/u [see Eq. (29-9)], $1\ u = 0.931\,494\ GeV/c^2$.

For each of the six quarks, there is a corresponding *antiquark* with the same mass and opposite electric charge. In Section 27.8, we saw that the electron and its antiparticle, the positron, can *annihilate*, producing two photons to carry away the energy and momentum. Electron-positron pairs can also be *created*. Similarly, other particle-antiparticle pairs can be created or annihilated. Annihilation does not always produce a pair of photons; it can, for instance, produce a different particle-antiparticle pair. The antiquarks are written with a bar over the symbol; for example, the antiparticle of the u quark is written $\bar{u}$ (read *u-bar*).

Quarks were first detected in a scattering experiment similar to the way the nucleus was discovered in Rutherford's experiment (Section 27.6). In 1968–1969, scientists at the Stanford Linear Accelerator (SLAC) studied the effects of scattering high-energy electrons from protons and neutrons. The experiment showed that the electrons scattered from point-like objects inside each proton or neutron.

Although many experiments have looked for them, an isolated quark has never been observed. A bound quark-antiquark pair is called a **meson**; a bound triplet of quarks or antiquarks is called a **baryon** (Fig. 30.1). Collectively, the mesons and baryons are called **hadrons**. The proton is a baryon made up of two up quarks and one down quark (uud); the neutron is a baryon made up of one up quark and two down quarks (udd).

Table 30.1					
The Six Quarks					
Name	**Symbol**	**Antiparticle**	**Charge**	**Mass (GeV/c^2)**	
Up	u	$\bar{u}$	$\pm\frac{2}{3}e$	0.001–0.005	**I**
Down	d	$\bar{d}$	$\mp\frac{1}{3}e$	0.005–0.009	
Strange	s	$\bar{s}$	$\mp\frac{1}{3}e$	0.08–0.16	**II**
Charm	c	$\bar{c}$	$\pm\frac{2}{3}e$	1.0–1.4	
Bottom	b	$\bar{b}$	$\mp\frac{1}{3}e$	4.0–4.5	**III**
Top	t	$\bar{t}$	$\pm\frac{2}{3}e$	170–180	

The upper symbol in $\pm$ or $\mp$ is for the particle and the lower is for the antiparticle.

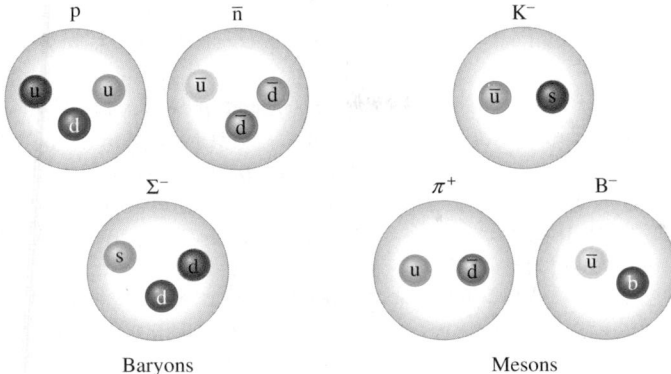

Figure 30.1 Some hadrons

Hundreds of hadrons have been observed. Other than the proton and neutron, all of them have short half-lives—less than 0.1 μs for the longest-lived ones. A neutron inside a nucleus is stable, but a *free* neutron decays with a half-life of 10.2 min into a proton, an electron, and an antineutrino ($n \rightarrow p + e^- + \bar{\nu}_e$). The proton appears to be stable; experiments have shown that if it is unstable, its half-life is at least 10^{25} yr—roughly 10^{15} times the age of the universe.

Table 30.1 organizes the quarks into three generations (indicated by Roman numerals). Each generation has two quarks, one with charge $+\frac{2}{3}e$ and one with charge $-\frac{1}{3}e$. The quarks in each successive generation are more massive than those in the previous generation. Ordinary matter contains only quarks from the first generation (u and d).

Leptons

While the proton and neutron are composed of quarks, the electron seems to be truly fundamental; no experiment has suggested that the electron has any internal structure. The electron belongs to another group of fundamental particles called the **leptons** (Table 30.2).

The six leptons (and their antiparticles) are grouped into three generations like the quarks. Each generation has one particle with charge $-e$ and an uncharged neutrino. The masses again increase from one generation to the next. As is true for the quarks, ordinary matter contains only first generation leptons. Electrons are a basic building block of atoms. The positron (e^+) is the antiparticle of the electron and is emitted in the β^+ decay of a radioactive nucleus, along with an electron neutrino ν_e (Section 29.3). In β^- decay, an electron antineutrino ($\bar{\nu}_e$) is emitted in addition to the electron. Electron

Table 30.2

The Six Leptons

Name	Symbol	Antiparticle	Charge	Mass (GeV/c^2)	
Electron	e^-	e^+	$\mp e$	0.000 511	**I**
Electron neutrino	ν_e	$\bar{\nu}_e$	0	*	
Muon	μ^-	μ^+	$\mp e$	0.106	**II**
Muon neutrino	ν_μ	$\bar{\nu}_\mu$	0	*	
Tau	τ^-	τ^+	$\mp e$	1.777	**III**
Tau neutrino	ν_τ	$\bar{\nu}_\tau$	0	*	

*See text discussion of neutrino masses.

The upper symbol in $\mp$ is for the particle and the lower is for the antiparticle. Note that the antiparticles of the negatively charged leptons are written with plus signs to indicate their positive charges but without a bar over them.

A muon created in a cosmic ray collision is detected by the Super-Kamiokande detector, located 1 km under Mt. Ikenoyama in Japan. The 11,200 photo-tubes record the light emitted by the muon as it moves through a cylinder containing 50,000 tons of ultrapure water. The muon entered through the bottom of the cylinder of ultra-pure water and exited through the side wall near the center of the image.

neutrinos and antineutrinos are also released in nuclear fusion (Section 29.8); Earth is bathed in a steady stream of neutrinos from the fusion reactions taking place in the Sun's interior.

Neutrinos are difficult to observe because they can pass through matter with only a small probability of interacting with anything. For a long time the neutrinos were thought to be massless, but experiments have recently shown that neutrinos do have mass. There are more neutrinos in the universe than all of the other leptons and quarks combined; therefore, neutrinos might make up a significant fraction of the total mass and energy of the universe.

Muons were the first second-generation particles to be observed. Cosmic rays—streams of energetic particles, mostly protons, traveling from outer space—continually bombard the Earth's upper atmosphere. The particles are capable of extremely high-energy interactions: most have energies in the GeV range, but some have been observed with energies over 10^{11} GeV, much higher than the energies that can be achieved by particle accelerators. When cosmic rays collide with atoms high in the Earth's atmosphere, the resulting shower of secondary particles—including electrons, muons, and gamma rays—can be detected at Earth's surface. Muons rain down on us at the rate of about one per cm^2 per minute.

Neither the muon nor the tau is stable; they are considered to be fundamental or *elementary* particles because they do not appear to have any substructure. A neutrino can transform from one type of neutrino to another. This effect, called **neutrino oscillation**, explains why the number of electron neutrinos reaching Earth from the Sun is smaller than had been predicted—some of the electron neutrinos are transformed into muon or tau neutrinos before they reach Earth. Because of neutrino oscillation, we cannot assign a definite mass value to the three types of neutrino.

30.2   FUNDAMENTAL INTERACTIONS

Quarks and leptons are not the whole story; what about the interactions between them? In Section 2.2 we described the four fundamental interactions in the universe: strong, electromagnetic, weak, and gravitational. The interactions are sometimes called *forces* but in a sense much broader than in Newtonian physics (where force is the rate of change of momentum). The fundamental "forces" do much more than push or pull; they include every change that occurs between particles: annihilation and creation of particle-antiparticle pairs, decay of unstable particles, binding of quarks into hadrons, and all kinds of reactions.

Each interaction can be understood as the exchange of a particle called a **mediator** or **exchange particle** (Table 30.3). The exchange particle is emitted by one particle and absorbed by another; it can transfer momentum and energy from one particle to another. The photon mediates the electromagnetic interaction. The weak interaction is mediated by one of three particles (W^+, W^-, and Z^0) whose existence was predicted in the 1960s by Steven Weinberg, Sheldon Glashow, and Abdus Salam. A team of scientists led by Carlo Rubbia first observed the three particles in 1982–1983. The strong interaction is mediated by the eight *gluons*; gravity is believed to be mediated by a particle called the *graviton*, which has not yet been observed. Like the photon, the gluons and graviton have no electric charge and are massless. Like the quarks and leptons, the exchange particles are considered to be fundamental; they apparently have no substructure.

Physics at Home

To get a feel for a particle that mediates a force, collect a friend and a heavy medicine ball. Then toss the heavy ball back and forth. The ball is the mediating or exchange particle that carries momentum and energy from one of you to the other. This is a repulsive force. It is even more dramatic if you are both standing on skateboards or wearing ice skates. Unfortunately, this analogy can only illustrate a repulsive force.

Table 30.3

The Four Fundamental Interactions and Their Exchange Particles

Interaction	Relative Strength	Range (m)	Affects Which Fundamental Particles?	Exchange Particles	Masses of Exchange Particles (GeV/c^2)
Strong	1	10^{-15}	Quarks and gluons	Gluons (g)	0
Electromagnetic	10^{-2}	∞	Electrically charged	Photon (γ)	0
Weak	10^{-6}	10^{-17}	Quarks and leptons	W^+, W^-, Z^0	80.4, 80.4, 91.2
Gravitational	10^{-43}	∞	All	Graviton[†]	0

[†]Not yet observed.

The relative strengths are for a pair of u quarks a distance of 0.03 fm apart.

The Strong Interaction

The strong interaction holds quarks together to form hadrons. Quarks interact via the strong interaction but leptons do not. Quarks carry *strong charge* (or *color charge*) that determines their strong interactions, just as a particle's electrical charge determines its electromagnetic interactions. While electrical charge comes in only one kind (positive) and its opposite (negative), strong charge comes in *three* kinds (called red, blue, and green), each of which has an opposite (called antired, antiblue, and antigreen). Color charge has nothing to do with the colors of light that we perceive visually. Rather, they are based on an analogy: just as the red, blue, and green dots produced by different phosphors on a TV screen combine to make white, a red quark, blue quark, and green quark combine to form a colorless (white) combination.

A baryon always contains one quark of each color, an antibaryon always contains one quark of each anticolor, and a meson always contains a quark of one color and an antiquark of the corresponding anticolor, such as a red quark and an antired antiquark (see Fig. 30.1). In each case, the strong force holds quarks together in *colorless* combinations, similar to the way the electromagnetic force holds negative and positive charges together to form a neutral atom with zero net electrical charge.

While it is possible to pull an electron from an atom, leaving an ion with a net electrical charge, the theory of quark confinement says that the strong force is too strong to pull a quark out of a colorless group—which is why isolated quarks are not observed. Just as two ions exert a much greater electromagnetic force on each other than do two neutral atoms, pulling a quark out of a colorless group would leave two groups of quarks with colors other than white; the force between the two groups would be extremely strong and, unlike the electromagnetic force, the strong force grows *stronger* with increasing distance within its short range.

Gluons, the mediators of the strong force, are the "glue" that holds the quarks together. Quarks continually emit and absorb gluons; the gluons carry strong charges so that emission or absorption of a gluon changes the color of the quark. Gluons themselves can emit and absorb gluons. If a quark is pulled out of a colorless combination, the gluons exchanged between the quarks have to travel a greater distance; this greater distance allows more time for the gluons to emit additional gluons, so the force gets stronger as the distance gets bigger. If there is enough energy to pull the quarks apart, some of the energy is used to create a quark-antiquark pair. The newly created quark goes off with one group and the newly created antiquark goes with the other group in such a way that both groups are colorless. Thus, even in high-energy collisions where hadrons are created and decay into other particles, quarks always end up in colorless combinations.

When we say that a proton consists of three quarks (uud), we really mean that its *net* quantum numbers match that picture. Quarks are surrounded by clouds of gluons being emitted and absorbed; from these gluons quark-antiquark pairs are continually created and annihilated, all within the volume of the proton. The energy of the clouds of gluons and the quark-antiquark pairs are part of the rest energy of the proton, which is why the proton mass (0.938 GeV/c^2) is much larger than the combined mass of two up quarks and one down quark (less than 0.02 GeV/c^2). The same fundamental interaction that holds three quarks together to form a nucleon also binds nucleons together to form a nucleus. However, the force between quarks is much stronger than the force between the colorless nucleons, just as the electromagnetic force between two ions is much stronger than the electromagnetic force between two neutral atoms.

The Weak Interaction

The weak interaction proceeds by the exchange of one of three particles (W^+, W^-, Z^0), two of which are electrically charged. All three of these particles have mass, which effectively limits the range of the weak interaction. While leptons do not take part in the strong interaction because they have no color charge, both leptons and quarks have weak charge and thus can take part in weak interactions.

Among other things, the weak interaction allows one quark *flavor* (u, d, s, c, b, t) to change into another. In general, a quark is most likely to decay to the next most massive quark. Since isolated quarks cannot be observed, the transformation of one quark flavor into another occurs within a hadron.

For example, the β^- decay of a radioactive nucleus was described as the transformation of a neutron into a proton within the nucleus (see Section 29.3):

$$n \rightarrow p + e^- + \bar{\nu}_e$$

Since a neutron is udd and a proton is uud, at a more fundamental level a d quark within the neutron is transformed into a u quark by emitting a W^-:

$$d \rightarrow u + W^-$$

The W^- then quickly decays into an electron and an electron antineutrino.

The Standard Model

The successful quantum mechanical description of the strong, weak, and electromagnetic interactions and the three generations of quarks and leptons is called the **standard model** of particle physics. The standard model, equipped with experimentally measured quantities (such as the particle masses and force charges), correctly predicts the results of decades of experiments in particle physics. While a remarkable achievement, the standard model is at best incomplete; it generates as many questions as it does answers.

30.3   UNIFICATION

One of the main goals of physics is to understand how the world works at its most basic and fundamental level. Part of that goal is to describe the immense variety of forces in the universe in terms of the fewest number of *fundamental* interactions.

Newton's law of gravity is an early example of unification. Before Newton, scientists did not understand that the same force that makes an apple fall to the ground from a tree also keeps the planets in orbit around the Sun. In the nineteenth century, Maxwell showed that the electric and magnetic forces are aspects of the same fundamental electromagnetic interaction.

A more recent success of unification is the electroweak theory. In ordinary matter, the electromagnetic and weak interactions have entirely different ranges, strengths, and effects. Glashow, Salam, and Weinberg showed that at energies of about 1 TeV or higher, the differences between the two fade until they are indistinguishable; they merge into a single **electroweak interaction**.

The ultimate goal is to describe all forces in terms of a single interaction. Many physicists believe that there was only one fundamental interaction immediately after the **Big Bang**—the explosion that gave birth to the universe. As the universe cooled and expanded, first gravity split off; then the strong force split, leaving three fundamental interactions (gravity, strong, and electroweak). Finally, the electroweak split into the weak and electromagnetic interactions. The splitting apart of the interactions all took place in about the first 10^{-10} s after the Big Bang. Higher energy accelerators will tell us whether the electroweak and strong interactions are unified into a single interaction.

Gravity remains a stumbling block, even though it has been a goal of physics since Einstein to unify gravity with other forces. The standard model does not include gravity; attempts to develop a quantum theory of gravity have so far been unsuccessful. General relativity, Einstein's theory of gravity, works well for large-scale phenomena—astronomically large distances and long times—but not for the microcosmos of quantum mechanics. Physicists working on the small scale can use quantum mechanics without worrying about general relativity, since gravitational effects are negligible. Thus, the standard model has successfully explained experiments in particle physics even though it omits gravity.

Making The Connection:
the Big Bang
and the history
of the universe

Higher Dimensions

Theorists have found that including extra dimensions in the universe enables them to reconcile gravity with quantum mechanics and to unify gravity with the other fundamental forces. Higher dimensions also provide a theoretical basis for quantities such as the masses of the quarks and leptons that, within the standard model, can only be measured experimentally. String theory and brane-world theory represent radical changes in our ideas about space and time and what a particle is.

According to *string theory* and *M-theory*, the various leptons and quarks are not fundamental entities; they are different vibrational patterns of a one-dimensional entity called a *string* that exists in a universe with 10 or 11 dimensions. The extra 6 or 7 dimensions are so small that we cannot observe them directly. As a visual aid, imagine the surface of a thin wire: it is a two-dimensional surface, but one of the dimensions is very small. In a similar way, the extra dimensions proposed by string theory are "curled up" over a length scale of about 10^{-35} m. To probe distances so small requires accelerator energies of about 10^{16} TeV—not possible in the foreseeable future. Experiments must look for indirect tests of string theory.

Other theories propose that the particles we can observe live in a four-dimensional membrane (the familiar three space dimensions and one time dimension) within a six- or seven-dimensional universe. In *brane-world theory*, the additional dimensions don't have to be as small as in string theory. They could be as large as a fraction of a millimeter, while the brane in which we're trapped extends only $\frac{1}{1000}$ the radius of a proton into the additional dimensions. If there are two extra dimensions, the force of gravity is predicted to be proportional to $1/r^4$ at distances less than a millimeter; scientists are now trying to measure the strength of gravity over small distances to see if this prediction is correct. Even if there were seven extra dimensions, as in the leading variant of string theory, it would cause a measurable effect on the strength of gravity over distances about the size of a nucleus. Experiments at the Tevatron are looking for indirect evidence of the existence of these extra dimensions; similar experiments are being planned for the LHC.

How can we know if there really are higher dimensions in the universe? Even the most elegant theory in physics would be abandoned if it cannot describe—or predict—the results of experiments. In brane-world theories, if a particle is given enough energy, it may be able to pop out of the membrane and move through the extra dimensions. Another possibility is that a particle moving into the higher dimensions will look like a new particle from our point of view; the discovery of new particles predicted by theory would be evidence for the extra dimensions.

30.4 "WHO ORDERED THAT?"

In the 1930s, it seemed as if all the fundamental particles had been discovered. Protons, neutrons, electrons, positrons, and (electron) neutrinos seemed to be all the particles necessary to explain everything. When the muon was discovered in the secondary particles created by cosmic ray collisions, physicist Isidor Isaac Rabi reacted rather unenthusiastically by asking, "Who ordered that?"

Today we might ask who ordered the tau, the two additional flavors of neutrinos, and the four quark flavors not found in ordinary matter. After all, one goal of physics is to understand the vast complexity of phenomena in terms of the *fewest* number of laws and fundamental building blocks. These other particles must be included to explain the variety of objects and phenomena observed in cosmic rays, in particle accelerators, and in other parts of the universe. Our "ordinary matter" is only a small subset of the states of matter and energy found elsewhere in the universe. The universe contains many kinds of stars, black holes, supernovae, quasars, and many other exotic objects—including some postulated and some not even dreamed of yet. Not only are we at a particular *place* in the universe, but we live at a particular *time* (see Fig. 30.2).

The standard model may seem complicated, but it is actually a great simplification in the quest to understand the universe's fundamental building blocks. In the past, the existence of a hundred or so different atoms led scientists to question whether atoms were built out of a smaller number of more fundamental particles. Our understanding that all the atoms are built from electrons, protons, and neutrons is a great simplification. Similarly, before the standard model was established, hundreds of different hadrons were identified in particle accelerator experiments, each with its own mass, charge, and other attributes. Scientists questioned whether all of these hadrons could really be *fundamental* building blocks. Now we know that all of the hadrons are built from six flavors of quark and their antiparticles.

Are quarks and leptons fundamental, or are they made up of smaller, more fundamental particles? So far, quarks and leptons seem to be indivisible; no evidence exists that they have finite size or any kind of internal structure. Some theories propose that quarks and leptons are not fundamental; ultimately experiments will establish whether these theories are correct.

Accelerators

Making The Connection:
high-energy particle accelerators

A *synchrotron* (Fig. 19.19) is a ring-shaped particle accelerator with many separate accelerating tubes. Each time a bunch of particles passes through an accelerating tube, an electric field gives the particles a little boost in kinetic energy. The machine is ring-shaped so that the particles can pass through the accelerating tubes many times. The particles also pass between the poles of strong magnets placed all around the ring to bend the paths of the particles approximately into a circle. Because charged particles radiate when they are accelerated, the particles lose energy each time a magnet changes their direction of motion. This lost kinetic energy must be replenished by the accelerating tubes. Once the particles reach the desired energy, they continue to circulate in a *storage ring* until they are made to collide inside a detector. A storage ring is similar to a synchrotron; it has magnets to bend the paths of the particles and accelerating tubes to replenish the lost kinetic energy. In some cases, the same machines act both as synchrotron and storage ring.

In a *linear accelerator*, the charged particles move in a straight line rather than around a circular ring. Therefore, much less energy is lost due to radiation because there is no centripetal acceleration. SLAC is currently the only large linear accelerator in operation; it accelerates electrons and positrons to kinetic energies of 50 GeV. Smaller linear accelerators are used to feed particles into a synchrotron.

After increasing the kinetic energies of charged particles, particle accelerators then slam them together. The resulting cascade of decays proceeds until particles are formed that live long enough to be detected. By creating the high energies that existed in the early moments of the universe, unstable particles that once existed are created in the laboratory. Going to higher energies allows us to probe matter on shorter and shorter length scales; recall that a particle's de Broglie wavelength is inversely proportional to

Figure 30.2 The History of the Universe

- For the first 10^{-43} s after the Big Bang, the smooth fabric of spacetime does not yet exist; the universe is a frothing, bubbling "quantum foam." There is just one fundamental interaction: a single superforce.
- At about 10^{-43} s, gravity splits off from the other interactions, leaving two fundamental interactions (gravity and strong-electroweak); spacetime comes into existence. Free quarks, leptons, and exchange particles exist. Particle-antiparticle pairs are created and annihilated. Radiation dominates the universe: the total energy of the photons is much greater than the total energy of matter.
- At 10^{-34} s, the strong force splits off from the electroweak.
- At $10^{-1\cdot}$ s, the weak and electromagnetic interactions split, so there are now four fundamental interactions.

- At 10^{-5} s, quark confinement begins as hadrons are formed. As the universe continues to cool, the heavy hadrons annihilate or decay, leaving light hadrons (such as protons, neutrons, and pions), leptons, and photons.
- Nuclei begin to form at 10 s, but there are few if any atoms due to the large numbers of photons with more than enough energy to ionize an atom.
- At 3×10^{5} yr, the temperature of the universe has cooled to about 3000 K. The energies of the photons in equilibrium at this temperature are mostly too small to ionize atoms, so atoms begin to form and the universe is for the first time transparent to photons. The cosmic microwave background radiation that we observe today is left over from this era, but the photon energies are much smaller now since the universe has cooled to only 2.7 K.

its momentum [$\lambda = h/p$, Eq. (28-1)]. Currently, the highest energy collisions (2 TeV) are produced at Fermilab's Tevatron.

The Large Hadron Collider is scheduled for completion in 2005 (Fig. 30.3). It will share a circular tunnel of circumference 27 km with the Large Electron Positron Collider (LEP), which is currently in operation. The LHC will achieve proton-proton collisions with energies up to 14 TeV. Five thousand superconducting magnets located along the circumference of the ring will steer and focus the two particle beams, traveling in opposite directions, through the 27-km-long, high-vacuum tube. Large detectors will be located at four crossing points of the beams. In addition to proton-proton collisions, the LHC will also collide beams of heavy ions, such as lead, sulfur, calcium, and oxygen. A head-on collision between two lead ions will result in a total collision energy of 1148 TeV. Scientists expect these collisions to produce a mixture of quarks and gluons similar to one that existed one-millionth of a second after the Big Bang, before the quarks and gluons coalesced into hadrons.

30.5   TWENTY-FIRST-CENTURY PARTICLE PHYSICS

Particle physics is at the brink of a revolution, according to many physicists. The standard model is extremely successful so far, but it is incomplete. It cannot predict the masses of the quarks and leptons and it cannot predict what happens at the energies that will soon be available at the LHC. It cannot unify gravity with the other forces.

Some of the many questions that particle physicists are trying to answer include:

Is there a pattern to the masses of quarks and leptons? Scientists searched for the top quark believing that the third generation should include two quarks like the other generations, but there was no way to predict the mass of the top quark. Why is there such a large range of masses of fundamental particles? The top quark has a mass about 175 times that of the proton. Can something with so much mass really be fundamental?

Are there exactly three generations of quarks and leptons, or are there more yet to be discovered? Why are there three generations (or however many there turn out to be)? Are there fundamental particles or forces we don't know about? Some recent experiments have suggested that new fundamental particles exist at energies just beyond the reach of current accelerators. Will the Higgs particle be observed? Existence of the Higgs particle would explain how the weak and electromagnetic forces unify into a single electroweak interaction. The Higgs particle and the force associated with it resists the motion of particles and effectively gives them mass. Without the Higgs particle, quarks and leptons would be massless and the weak force would be much stronger.

Are quarks and leptons truly fundamental?

Is the proton truly stable?

Why is there a vast range of strengths of forces? Why is gravity so weak?

What makes up the dark matter in the universe? There is apparently more gravitational attraction between nearby galaxies and between inner and outer parts of individual galaxies than can be accounted for by the mass of the visible objects making up the galaxies. What is the invisible (hence, *dark*) matter that supplies this extra gravitational force? Do the masses of neutrinos account for a significant fraction of this missing mass? Or is the dark matter chiefly made up of particles not yet observed?

What happened to the antimatter? If there is a symmetry between matter and antimatter, why do we observe almost no antimatter in the universe? If the Big Bang created equal amounts of matter and antimatter, what happened to the antimatter? To help answer this question, experiments planned for the LHC will look for differences in the behavior of particles and antiparticles.

Can gravity be unified with the other fundamental interactions?

Why is the universe four-dimensional (three spatial dimensions and one time dimension)? Or does it actually have more than four dimensions; if so, why does it *appear* to have four?

With all these open questions, particle physics in the twenty-first century promises to be even more exciting than it was in the twentieth.

Figure 30.3   The LEP/LHC tunnel is about 100 m below ground; it straddles the border of France and Switzerland near Geneva. The LHC will have four detectors. ATLAS, about the size of a five-story building, and CMS, which weighs over 12,000 tons, are general-purpose detectors. ALICE will concentrate on the products of heavy ion collisions. LHCb will focus on proton-proton collisions that produce the short-lived B meson.

MASTER THE CONCEPTS

Summary

- Protons and neutrons are not fundamental particles; they are made up of quarks.

- According to the standard model, the fundamental particles are the six quarks (up, down, strange, charm, bottom, and top), the six leptons (electron, muon, tau, and the three kinds of neutrinos), the antiparticles of the quarks and leptons, and the exchange particles for the strong, weak, and electromagnetic interactions.

- Isolated quarks are not observed; quarks are always confined by the strong force to colorless groups. Color charge plays a role in the strong interaction similar to, but more complicated than, that of electric charge in the electromagnetic interaction.

- Only the first generation of quarks and leptons (up, down, electron, and electron neutrino) are found in ordinary matter.

- Just after the Big Bang, there was only a single interaction. First gravity split off, then the strong interaction; finally the weak and electromagnetic interactions split, giving the four fundamental interactions we now recognize.

- New particle accelerators at higher energies will put the standard model, as well as theories competing to be its successor, to the test. Several new theories propose that the universe has more than the familiar four dimensions (three space dimensions + one time dimension).

Highlighted Figures and Tables

T30.1 The six quarks (p. 1092)

T30.2 The six leptons (p. 1093)

T30.3 The four fundamental interactions and their exchange particles (p. 1095)

F30.2 The history of the universe (p. 1099)

F30.3 The planned LHC (p. 1101)

CONCEPTUAL QUESTIONS

1. What *fundamental* particles make up an atom?
2. What tool enabled scientists to observe hundreds of different hadrons in the latter half of the twentieth century?
3. Why is the number of electron neutrinos reaching Earth from the Sun smaller than had originally been predicted?
4. How many different hadrons are stable (as far as we know)?
5. What particles are in the lepton family?
6. Why is the muon sometimes called a "heavy electron"?
7. Is the charge on the electron the smallest fundamental unit of charge? Explain.
8. Explain the use of "color" as a quantum number for the quarks.
9. Why do we not notice the effects of billions of neutrinos passing through each square centimeter of our body surface every second?
10. Why are there no observed particles composed of two quarks (qq), two antiquarks ($\overline{q}\overline{q}$), or four quarks (qqqq)? [*Hint:* Consider the color charges of the quarks.]

11. In a synchrotron, charged particles are accelerated as they travel around in circles; in a linear accelerator they move in a straight line. What are some of the advantages and disadvantages of each?

12. In a fixed-target experiment, high-energy charged particles from an accelerator are smashed into a stationary target. By contrast, in a colliding beam experiment, two beams of particles are accelerated to high energies; particles moving in opposite directions suffer head-on collisions when the two beams are steered together. Describe one advantage of each type of experiment over the other. [*Hint:* For an advantage of the colliding beam experiment, consider not only the total kinetic energy of the particles involved in the collision, but also how much of that energy is available to create new particles. Remember that momentum must be conserved in the collision.]

MULTIPLE CHOICE QUESTIONS

1. Baryons are always composed of
   (a) three gluons.   (b) any odd number of quarks.
   (c) three leptons.
   (d) three quarks with three different colors.
   (e) three quarks of matching color.
   (f) two quarks of complementary colors.
   (g) a colorless quark-antiquark pair.

2. Mesons are always composed of
   (a) three gluons.   (b) any odd number of quarks.
   (c) three leptons.
   (d) three quarks with three different colors.
   (e) three quarks of matching color.
   (f) a colorless quark-antiquark pair.

3. Quark flavors include
   (a) up, down.   (b) red, green.   (c) muon, pion.
   (d) cyan, magenta.   (e) lepton, gluon.

4. Hadrons that contain one or more strange quarks are called strange particles. The particles were originally called *strange*— before quark theory had been formulated—due to their anomalously long lifetimes of 10^{-10} to 10^{-7} s (compared to about 10^{-23} to 10^{-20} s for the other hadrons known at the time). When a strange hadron decays into particles that are not strange, the decay is a manifestation of the
   (a) strong interaction.   (b) weak interaction.
   (c) electromagnetic interaction.
   (d) gravitational interaction.

5. The weak interaction is moderated by
   (a) leptons.   (b) photons.   (c) gluons.
   (d) W^+, W^-, Z^0.   (e) mesons.

COMPREHENSIVE PROBLEMS

Note: **ⓒ** indicates a combination conceptual/quantitative problem. Gold diamonds ◆, ◆◆ are used to indicate the increasing level of difficulty of each problem. Problem numbers appearing in blue, 9., denote problems that have a detailed solution available in the Student Solutions Manual. Some problems are *paired* by concept; their numbers are connected by a ruled box.

Note: a particle is extremely relativistic when its rest energy is negligible compared to its kinetic energy. Then $E = K + E_0 \gg E_0$ and $E = \sqrt{(pc)^2 + E_0^2} \approx pc$.

1. A pion (mass 0.140 GeV/c^2) at rest decays by the weak interaction into a muon of mass 0.106 GeV/c^2 and a muon antineutrino: $\pi^- \rightarrow \mu^- + \bar{\nu}_\mu$. Neglecting the mass of the antineutrino, how much energy is released by this decay?

◆ 2. In Problem 1, what is the kinetic energy of the muon? [*Hint:* The muon is nonrelativistic, so its kinetic energy-momentum relationship is $K = p^2/(2m)$. The antineutrino is extremely relativistic.]

3. A muon decay is described by $\mu^- \rightarrow e^- + \nu_\mu + \bar{\nu}_e$. What is the maximum kinetic energy of the electron, if the muon was at rest? Assume that the electron is extremely relativistic and ignore the small masses of the neutrinos.

4. A neutral pion (mass 0.135 GeV/c^2) decays via the electromagnetic interaction into two photons: $\pi^0 \rightarrow \gamma + \gamma$. What is the energy of each photon, assuming the pion was at rest?

5. A proton of mass 0.938 GeV/c^2 and an antiproton, at rest relative to an observer, annihilate each other as described by $p + \bar{p} \rightarrow \pi^- + \pi^+$. What are the kinetic energies of the two pions, each of which has mass 0.14 GeV/c^2?

6. In an accelerator, two protons with equal kinetic energies collide head-on. The following reaction takes place: $p + p \rightarrow p + p + p + \bar{p}$. What is the minimum possible kinetic energy of each of the incident proton beams?

7. A charged pion can decay either into a muon or an electron. The two decay modes of a π^- are: $\pi^- \rightarrow \mu^- + \bar{\nu}_\mu$ and $\pi^- \rightarrow e^- + \bar{\nu}_e$. Write the two decay modes for the π^+. [*Hint:* π^+ is the antiparticle of π^-. Replace each particle in the decay reaction with its corresponding antiparticle.]

◆ 8. When a proton and an antiproton annihilate, the annihilation products are usually pions. (a) Suppose three pions are produced. What combination(s) of π^+, π^-, and π^0 are possible? The mass of a charged pion is 0.140 GeV/c^2; the mass of a neutral pion is 0.135 GeV/c^2. (b) Suppose five pions are produced. What combination(s) of π^+, π^-, and π^0 are possible? (c) What is the maximum number of pions that could be produced if the kinetic energy of the proton and antiproton are negligibly small?

9. What are the quarks composing an antiproton? [*Hint:* Replace each of the three quarks that compose a proton with its corresponding antiquark.]

10. The energy at which the fundamental forces are expected to unify is about 10^{19} GeV. Find the mass (in kg) of a particle with rest energy 10^{19} GeV.

11. What is the de Broglie wavelength of a proton with kinetic energy 1.0 TeV?

12. What is the de Broglie wavelength of a proton with kinetic energy 7.0 TeV?

13. In the Cornell Electron Storage Ring, electrons and positrons circulate in opposite directions with kinetic energies of 6.0 GeV each. When an electron collides with a positron and the two annihilate, one possible (though unlikely) outcome is the production of one or more proton-antiproton pairs. What is the *maximum possible* number of proton-antiproton *pairs* that could be formed?

14. A proton in the Tevatron is accelerated through a potential difference of 2.5 MV during each revolution around the ring of radius 1.0 km. In order to reach an energy of 1 TeV, how many revolutions must the proton make? How far has it traveled?

15. Estimate the magnetic field strength required at the LHC to make 7.0-TeV protons travel in a circle of circumference 27 km. Start by deriving an expression, using Newton's second law, for the field strength B in terms of the particle's momentum p, its charge q, and the radius r. Even though derived using classical physics, the expression is relativistically correct. (The estimate will come out lower than the actual value of 8.33 T. In the LHC, the protons do not travel in a constant magnetic field; they move in straight-line segments between magnets.)

APPENDIX A
Mathematics Review

A.1 ALGEBRA

There are two basic kinds of algebraic manipulations.

- The same operation can always be performed on both sides of an equation.
- Substitution is always permissible (if $a = b$, then any occurrence of a in any equation can be replaced with b).

Products distribute over sums

$$a(b + c) = ab + ac \tag{A-1}$$

The reverse—replacing $ab + ac$ with $a(b + c)$—is called *factoring*. Since dividing by c is the same as multiplying by $1/c$,

$$\frac{a + b}{c} = \frac{a}{c} + \frac{b}{c} \tag{A-2}$$

Equation (A-2) is the basis of the procedure for adding fractions. To add fractions, they must be expressed with a *common denominator*.

$$\frac{a}{b} + \frac{c}{d} = \frac{a}{b} \times \frac{d}{d} + \frac{c}{d} \times \frac{b}{b} = \frac{ad}{bd} + \frac{bc}{bd}$$

Now applying Eq. (A-2), we end up with

$$\frac{a}{b} + \frac{c}{d} = \frac{ad + bc}{bd} \tag{A-3}$$

To divide fractions, remember that dividing by c/d is the same as multiplying by d/c:

$$\frac{a}{b} \div \frac{c}{d} = \frac{\dfrac{a}{b}}{\dfrac{c}{d}} = \frac{a}{b} \times \frac{d}{c} = \frac{ad}{bc}$$

A product in a square root can be separated:

$$\sqrt{ab} = \sqrt{a} \times \sqrt{b} \tag{A-4}$$

Pitfalls to Avoid

These are some of the most common *incorrect* algebraic substitutions. Don't fall into any of these traps!

$$\sqrt{a + b} \neq \sqrt{a} + \sqrt{b}$$

$$\frac{a}{b + c} \neq \frac{a}{b} + \frac{a}{c}$$

$$\frac{a}{b} + \frac{c}{d} \neq \frac{a + c}{b + d}$$

$$(a + b)^2 \neq a^2 + b^2$$

In the last one, the cross term is missing: $(a + b)^2 = a^2 + 2ab + b^2$.

Graphs of Linear Functions

If the graph of y as a function of x is a straight line, then y is a *linear function* of x. The relationship can be written in the standard form

$$y = mx + b \tag{A-5}$$

where m is the *slope* and b is the *y-intercept*. The slope measures how steep the line is. It tells how much y changes for a given change in x:

$$m = \frac{\Delta y}{\Delta x} = \frac{y_2 - y_1}{x_2 - x_1} \tag{A-6}$$

The y-intercept is the value of y when $x = 0$. On the graph, the line crosses the y-axis at $y = b$.

Example

What is the equation of the line graphed in Fig. A.1?

Solution   The y-intercept is –2. To find the slope, note that when x has values from 0 to 6, y values range from –2 to 0.

$$m = \frac{y_2 - y_1}{x_2 - x_1} = \frac{0 - (-2)}{6 - 0} = \frac{1}{3}$$

Then $y = mx + b = \frac{1}{3}x - 2$.

Figure A.1

A.2   SOLVING EQUATIONS

Solving an equation means using algebraic operations to isolate one variable. Many students tend to substitute numerical values into an equation as soon as possible. In many cases, that's a mistake. Although at first it may seem easier to manipulate numerical quantities than to manipulate algebraic symbols, there are several advantages to working with symbols:

*   Symbolic algebra is much easier to follow than a series of numerical calculations. Plugging in numbers tends to obscure the logic behind your solution. If you need to trace back through your work (to find an error or review for an exam), it'll be much clearer if you have worked through the problem symbolically. It will also help your instructor when grading your homework papers or exams. When your work is clear, your instructor is better able to help you understand your mistakes. You may also get more partial credit on exams!

*   Symbolic algebra lets you draw conclusions about how one quantity depends on another. For instance, working symbolically you might see that the horizontal range of a projectile is proportional to the *square* of the initial speed. If you had substituted the numerical value of the initial speed, you wouldn't notice that. In particular, when an algebraic symbol cancels out of the equation, you know that the answer doesn't depend on that quantity.

*   On the most practical level, it's easy to make arithmetic or calculation errors. The later on in your solution that numbers are substituted, the fewer number of steps you have to check for such errors.

When solving equations that contain square roots, be careful not to assume that a square root is positive. The equation $x^2 = a$ has *two* solutions for x: $x = \pm\sqrt{a}$. (The symbol $\pm$ means *either + or –*.)

Solving Quadratic Equations

An equation is quadratic in x if it contains terms with no powers of x other than a squared term (x^2), a linear term (x^1), and a constant (x^0). Any quadratic equation can be put into the standard form:

$$ax^2 + bx + c = 0 \tag{A-7}$$

The quadratic formula gives the solutions to any quadratic equation written in standard form:

$$x = \frac{-b \pm \sqrt{b^2 - 4ac}}{2a} \qquad \text{(A-8)}$$

The symbol "$\pm$" (read "plus or minus") indicates that in general there are two solutions to a quadratic equation; that is, two values of x will satisfy the equation. One solution is found by taking the + sign and the other by taking the − sign in the quadratic formula. If $b^2 - 4ac = 0$, then there is only one solution (or, technically, the two solutions happen to be the same). If $b^2 - 4ac < 0$, then there is no solution to the equation (for x among the real numbers).

The quadratic formula still works if $b = 0$ and/or $c = 0$, although in such cases the equation can easily be solved without recourse to the quadratic formula.

Example

Solve the equation $5x(3 - x) = 6$.

Solution   First put the equation in standard quadratic form:

$$15x - 5x^2 = 6$$
$$-5x^2 + 15x - 6 = 0$$

We identify $a = -5$, $b = 15$, $c = -6$. Then

$$x = \frac{-b \pm \sqrt{b^2 - 4ac}}{2a}$$

$$= \frac{-15 \pm \sqrt{15^2 - 4 \times (-5) \times (-6)}}{-10}$$

$$\simeq \frac{-15 \pm 10.25}{-10} = 0.475 \text{ or } 2.525$$

Solving Simultaneous Equations

Simultaneous equations are a set of N equations with N unknown quantities. We wish to solve these equations *simultaneously* to find the values of all of the unknowns. We *must* have at least as many equations as unknowns. It pays to keep track of the number of unknown quantities and the number of equations in solving more challenging problems. If there are more unknowns than equations, then look for some other relationship between the quantities—perhaps some information given in the problem that has not been used.

One way to solve simultaneous equations is by *successive substitution*. Solve one of the equations for one unknown (in terms of the other unknowns). Substitute this expression into each of the other equations. That leaves $N - 1$ equations and $N - 1$ unknowns. Repeat until there is only one equation left with one unknown. Find the value of that unknown quantity, and then work backward to find all the others.

Example

Solve the equations $2x - 4y = 3$ and $x + 3y = -5$ for x and y.

Solution   First solve the second equation for x in terms of y:

$$x = -5 - 3y$$

Substitute $-5 - 3y$ for x in the first equation:

$$2 \times (-5 - 3y) - 4y = 3$$

This can be solved for y:

$$-10 - 10y = 3$$
$$-10y = 13$$
$$y = \frac{13}{-10} = -1.3$$

Now that y is known, use it to find x:

$$x = -5 - 3y = -5 - 3 \times (-1.3) = -1.1$$

It's a good idea to check the results by substituting into the original equations.

A.3   EXPONENTS AND LOGARITHMS

These identities show how to manipulate exponents.

$$a^{-x} = \frac{1}{a^x} \tag{A-9}$$

$$(a^x) \times (a^y) = a^{x+y} \tag{A-10}$$

$$\frac{a^x}{a^y} = (a^x) \times (a^{-y}) = a^{x-y} \tag{A-11}$$

$$(a^x) \times (b^x) = (ab)^x \tag{A-12}$$

$$(a^x)^y = a^{xy} \tag{A-13}$$

$$a^{1/n} = \sqrt[n]{a} \tag{A-14}$$

$$a^0 = 1 \quad \text{(for any } a \neq 0) \tag{A-15}$$

$$0^a = 0 \quad \text{(for any } a \neq 0) \tag{A-16}$$

A common mistake to avoid: $(a^x) \times (a^y) \neq a^{xy}$ [see Eq. (A-10)].

Logarithms

Taking a logarithm is the inverse of exponentiation:

$$x = \log_b y \quad \text{means that} \quad y = b^x \tag{A-17}$$

Thus, one undoes the other:

$$\log_b b^x = x \tag{A-18}$$

$$b^{\log_b x} = x \tag{A-19}$$

The two commonly used bases b are 10 (the *common* logarithm) and $e = 2.71828\ldots$ (the *natural* logarithm). The common logarithm is written "$\log_{10}$," or sometimes just "log" if base 10 is understood. The natural logarithm is usually written "ln" rather than "$\log_e$."

These identities are true for any base logarithm.

$$\log xy = \log x + \log y \tag{A-20}$$

$$\log \frac{x}{y} = \log x - \log y \tag{A-21}$$

$$\log x^a = a \log x \tag{A-22}$$

Here are some common mistakes to avoid:

$$\log (x + y) \neq \log x + \log y$$

$$\log (x + y) \neq \log x \times \log y$$

$$\log xy \neq \log x \times \log y$$

$$\log x^a \neq (\log x)^a$$

Semilog Graphs

A semilog graph uses a logarithmic scale on the vertical axis and a linear scale on the horizontal axis. Semilog graphs are useful when the data plotted is thought to be an exponential function. If

$$y = y_0 e^{ax}$$

then

$$\ln y = ax + \ln y_0$$

so a graph of $\ln y$ versus x will be a straight line with slope a and y-intercept $\ln y_0$.

Rather than calculating $\ln y$ for each data point and plotting on regular graph paper, it is convenient to use special semilog paper. The vertical axis is marked so that the values of y can be plotted directly, but the markings are spaced proportional to the log of y. (If you are using a plotting calculator or a computer to make the graph, log scale should be chosen for one axis from the menu of options.) The slope a on a semilog graph is *not* $\Delta y/\Delta x$ since the logarithm is actually being plotted. The correct way to find the slope is

$$a = \frac{\Delta(\ln y)}{\Delta x} = \frac{\ln y_2 - \ln y_1}{x_2 - x_1}$$

Note that there cannot be a zero on a logarithmic scale.

The two graphs of Figs. A.2 and A.3 are linear and semilog plots, respectively, of the function $y = 3e^{-2x}$.

Figure A.2

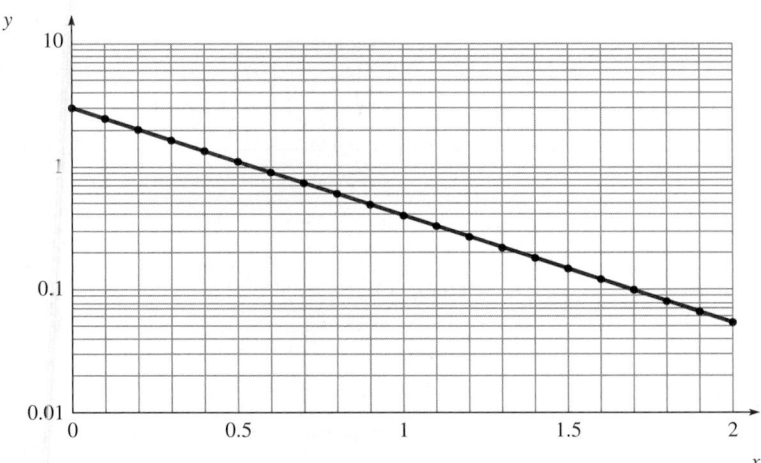

Figure A.3

Log-Log Graphs

A log-log graph uses logarithmic scales for both axes. Log-log graphs are useful when the data plotted is thought to be a power function

$$y = Ax^n$$

For such a function,

$$\log y = n \log x + \log A$$

so a graph of $\log y$ versus $\log x$ will be a straight line with slope n and y-intercept $\log A$. The slope (n) on a log-log graph is found as

$$n = \frac{\Delta(\log y)}{\Delta(\log x)} = \frac{\log y_2 - \log y_1}{\log x_2 - \log x_1}$$

The graphs of Figs. A.4 and A.5 are linear and log-log plots, respectively, of the function $y = 130x^{3/2}$.

Figure A.4

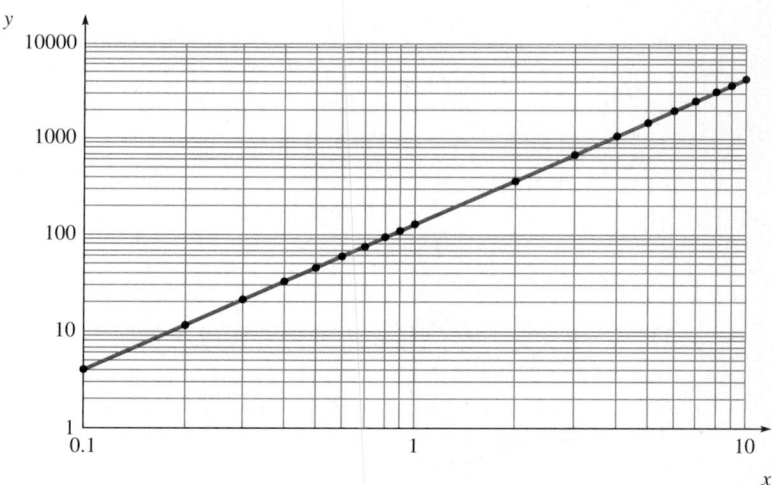

Figure A.5

A.4   PROPORTIONS AND RATIOS

The notation

$$y \propto x$$

means that y is directly proportional to x. A proportionality can be written as an equation

$$y = kx$$

if the constant of proportionality k is written explicitly. Be careful: an equation can *look like* a proportionality without being one. For example, $V = IR$ means that $V \propto I$ *if and only if* R is constant. If R depends on I, then V is *not* proportional to I.

The notation

$$y \propto \frac{1}{x}$$

means that y is inversely proportional to x. The notation

$$y \propto x^n$$

means that y is proportional to the nth power of x.

Writing out proportions as ratios usually simplifies solutions when some common items in an equation are unknown but we do know the values of all but one of the proportional quantities. For example if $y \propto x^n$, we can write

$$\frac{y_1}{y_2} = \left(\frac{x_1}{x_2}\right)^n$$

Percentages

Percentages require careful attention. Look at these four examples:

"B is 30% of A" means $B = 0.30A$

"B is 30% larger than A" means $B = (1 + 0.30)A = 1.30A$

"B is 30% smaller than A" means $B = (1 - 0.30)A = 0.70A$

"A increases by 30%" means $\Delta A = +0.30A$

Example

If $P \propto T^4$, and T increases by 10.0%, by what percentage does P increase?

Solution

$$\Delta T = +0.100T_i$$
$$T_f = T_i + \Delta T = 1.100T_i$$
$$\frac{P_f}{P_i} = \left(\frac{T_f}{T_i}\right)^4 = 1.100^4 \approx 1.464$$

Therefore, P increases by about 46.4%.

A.5   APPROXIMATIONS

Binomial Approximations

A binomial is the sum of two terms. The general rule for the nth power of an algebraic sum is given by the binomial expansion:

$$(a + b)^n = a^n + na^{n-1}b + \frac{n(n-1)}{1 \times 2}a^{n-2}b^2 + \frac{n(n-1)(n-2)}{1 \times 2 \times 3}a^{n-3}b^3 + \cdots$$

The binomial approximations are used when a binomial in which one term is much smaller than the other is raised to a power n. Only the first two terms of the binomial expansion are of significant value; the other terms are dropped. A common case for physics problems is that in which $a = 1$, or can be made equal to one by factoring. The basic approximation forms are then given by

$$(1 + x)^n \approx 1 + nx \quad \text{when } x \ll 1 \tag{A-23}$$

$$(1 - x)^n \approx 1 - nx \quad \text{when } x \ll 1 \tag{A-24}$$

The power n can be any real number, including negative as well as positive numbers. It does not have to be an integer. An *estimate* of the error—the difference between the approximation and the exact expression—is given by

$$\text{error} \approx \tfrac{1}{2}n(n-1)x^2 \tag{A-25}$$

Of course, the larger term in a binomial is not necessarily 1, but the larger term can be factored out and then Eq. (A-23) or Eq. (A-24) applied. For instance, if $A \gg b$, then

$$(A + b)^n = \left[A \times \left(1 + \frac{b}{A}\right)\right]^n = A^n\left(1 + \frac{b}{A}\right)^n$$

Another common expansion is

$$e^x = 1 + x + \frac{x^2}{2!} + \frac{x^3}{3!} + \cdots$$

where, for any integer n, $n! = n \times (n-1) \times (n-2) \times \cdots \times [n - (n-1)]$; for example, $3! = 3 \times 2 \times 1 = 6$.

Small-Angle Approximations

Approximations for small angles appear in Section A.7 on trigonometry.

A.6 GEOMETRY

Geometric Shapes

Table A.1 lists the geometric shapes that commonly appear in physics problems. It is often necessary to determine the area or volume of one of these simple shapes to complete the solution of a problem. The formulae for the properties associated with each geometric form are listed in the column to the right.

Angular Measure

An angle between two straight lines that meet at a single point is specified in degrees. If the two lines are perpendicular to each other, as shown in Fig. A.6a, the angular separation is said to be a right angle or 90°. In Fig. A.6b two such 90° angles are placed side by side; they add to 180°, so a straight line represents an angular separation of 180°. When four right angles are grouped as shown in Fig. A.6c, the angles add to 360° and a full circle contains 360° as shown in Fig. A.6d. An angle that is less than 90° is called an **acute** angle; one greater than 90° is called an **obtuse** angle.

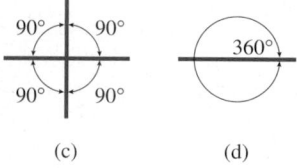

Figure A.6 (a) A right angle; (b) two adjacent right angles, or a straight line; (c) four adjacent right angles; (d) a full circle

Table A.1

Common Geometric Shapes Met in Physics Problems

Geometric Shape	Properties of Common Shapes	Geometric Shape	Properties of Common Shapes
Circle	Diameter $d = 2r$ Circumference $C = \pi d = 2\pi r$ Area $A = \pi r^2$	Sphere	Surface area $A = 4\pi r^2$ Volume $V = \frac{4}{3}\pi r^3$
Rectangle	Perimeter $P = 2b + 2h$ Area $A = bh$	Parallelepiped	Surface area $A = 2(ab + bc + ac)$ Volume $V = abc$
Right triangle	Perimeter $P = a + b + c$ Area $A = \frac{1}{2}\text{base} \times \text{height} = \frac{1}{2}ba$ Pythagorean theorem $c^2 = a^2 + b^2$ Hypotenuse $c = \sqrt{a^2 + b^2}$	Right circular cylinder	Surface area $A = 2\pi r^2 + 2\pi rh$ $= 2\pi r(r + h)$ Volume $V = \pi r^2 h$
Triangle	Area $A = \frac{1}{2}bh$		

When two lines meet, as shown in Fig. A.7, there are two possible angles that might be specified; one is the acute angle α and the other is the obtuse angle β. The symbol used to indicate an angle is $\angle$. When two angles placed adjacent to each other form a straight line, they are called supplementary angles; angles α and β in Fig. A.7 are supplementary angles. When two angles placed adjacent to each other form a right angle, they are called complementary angles.

Various triangles are shown in Fig. A.8. The sum of the interior angles of any triangle is 180°. An isosceles triangle has two sides of equal length; the angles opposite to the equal sides are equal angles. An equilateral triangle has all three sides of equal length; it is also equiangular. Right triangles have one right angle, 90°, and the sum of the other two angles is 90°, so those angles are acute angles. Commonly used right triangles have sides in the ratio of 3:4:5 and 5:12:13.

Triangles are similar when all three angles of one are equal to the three angles of the other. If two angles of one triangle are equal to two angles of the other, the third angles are necessarily equal and the triangles are similar. The ratio of corresponding sides of similar triangles are equal, as shown in Fig. A.9. Similar triangles of the same size are called congruent triangles.

Figure A.10 shows other useful relations among angles between intersecting lines. When two angles add to 180°, as $\angle\alpha + \angle\beta$ in three of the small figures, the angles are supplementary. Another small figure shows two angles, $\angle\alpha + \angle\beta$ adding to 90°, so in that case the angles are complementary.

For many physics problems it is convenient to use angles measured in radians rather than in degrees; the abbreviation for radians is rad. The arc length s measured along a circle is proportional to the angle between the two radii that define the arc, as shown in Fig. A.11. One radian is defined as the angle subtended when the arc length is equal to the length of the radius.

For θ measured in radians,

$$s = r\theta$$

When the angular displacement is all the way around the circle, 360°, the arc length is equal to the circumference of the circle.

$$s = 2\pi r = r\theta$$

The equivalent to 360° measured in radians is thus $\theta = 2\pi$ radians and the equivalence between radians and degrees is

$$1 \text{ rad} = \frac{360°}{2\pi} = 57.3° \quad \text{or} \quad 1° = \frac{2\pi}{360°} = 0.01745 \text{ rad}$$

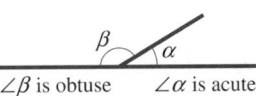

$\angle\beta$ is obtuse $\angle\alpha$ is acute

Figure A.7   Acute and obtuse angles

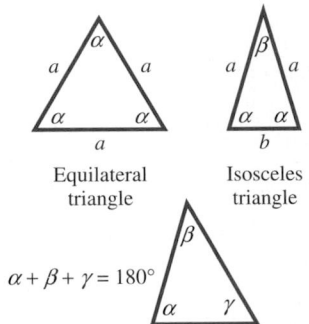

Equilateral triangle   Isosceles triangle

$$\alpha + \beta + \gamma = 180°$$

Figure A.8   Triangles

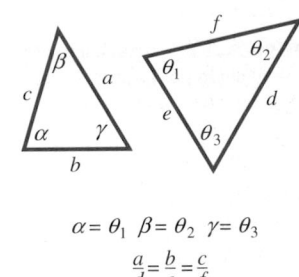

$$\alpha = \theta_1 \quad \beta = \theta_2 \quad \gamma = \theta_3$$
$$\frac{a}{d} = \frac{b}{e} = \frac{c}{f}$$

Figure A.9   Similar triangles

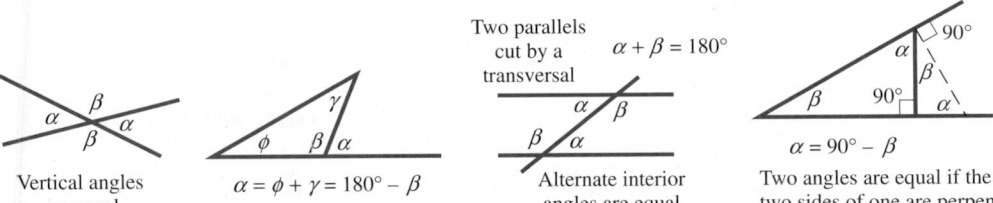

Vertical angles are equal

$$\alpha = \phi + \gamma = 180° - \beta$$

Two parallels cut by a transversal $\alpha + \beta = 180°$

Alternate interior angles are equal

$$\alpha = 90° - \beta$$

Two angles are equal if the two sides of one are perpendicular to the two sides of the other, left side to left side and right side to right side as seen from the vertex

Figure A.10   Angles formed by intersecting lines

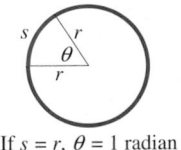

If $s = r$, $\theta = 1$ radian

Figure A.11   Radian measure

Note that the radian has no physical dimensions; it is a ratio of two lengths so it is a pure number. We use the term *rad* to remind us of the angular units being used.

A.7   TRIGONOMETRY

The basic trigonometric functions used in physics are shown in Fig. A.12. Note that in determining the function values, the units of length cancel, so the sine, cosine, and tangent functions are dimensionless.

Since the side opposite and the side adjacent to either of the angles in the right triangle are of lesser length than the hypotenuse according to the Pythagorean theorem, the values of the sine and cosine cannot exceed unity. The value of the tangent does exceed unity whenever the side opposite exceeds the side adjacent to the angle considered.

Figure A.13 shows the signs (positive or negative) associated with the trigonometric functions for an angle θ located in each of the four quadrants. The hypotenuse r is positive, so the sign for the sine or cosine is determined by the signs of x or y as measured along the positive or negative x- and y-axes. The sign of the tangent then depends on the signs of the sine and cosine. The angle θ is measured in a counterclockwise direction starting from the positive x-axis, which represents $0°$. Angles measured from the x-axis going in a clockwise direction (below the x-axis) are negative angles; an angle of $-60°$, which is located in the fourth quadrant, is the same as an angle of $+300°$. There are some useful trigonometric identities given in Table A.2.

Figure A.12 Trigonometric functions used in physics problems; angles θ and ϕ are complementary angles.

Right triangle

$\phi = 90° - \theta$

$$\sin\theta = \frac{\text{side opposite }\angle\theta}{\text{hypotenuse}} = \frac{b}{c} = \cos\phi$$

$$\cos\theta = \frac{\text{side adjacent }\angle\theta}{\text{hypotenuse}} = \frac{a}{c} = \sin\phi$$

$$\tan\theta = \frac{\text{side opposite }\angle\theta}{\text{side adjacent }\angle\theta} = \frac{b}{a} = \frac{\sin\theta}{\cos\theta} = \frac{1}{\tan\phi}$$

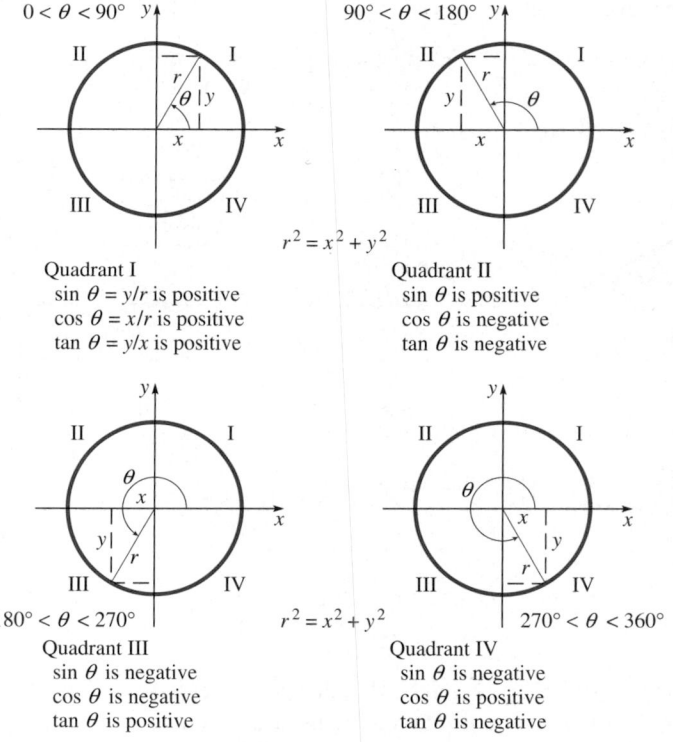

Figure A.13 Signs of trigonometric functions in various quadrants

$0 < \theta < 90°$

Quadrant I
$\sin\theta = y/r$ is positive
$\cos\theta = x/r$ is positive
$\tan\theta = y/x$ is positive

$90° < \theta < 180°$

$r^2 = x^2 + y^2$

Quadrant II
$\sin\theta$ is positive
$\cos\theta$ is negative
$\tan\theta$ is negative

$180° < \theta < 270°$

Quadrant III
$\sin\theta$ is negative
$\cos\theta$ is negative
$\tan\theta$ is positive

$r^2 = x^2 + y^2$

$270° < \theta < 360°$

Quadrant IV
$\sin\theta$ is negative
$\cos\theta$ is positive
$\tan\theta$ is negative

Table A.2

Useful Trigonometric Identities

$\sin^2 \theta + \cos^2 \theta = 1$

$\sin 2\theta = 2 \sin \theta \cos \theta$

$\cos 2\theta = \cos^2 \theta - \sin^2 \theta = 2 \cos^2 \theta - 1 = 1 - 2 \sin^2 \theta$

$\tan 2\theta = \dfrac{2 \tan \theta}{1 - \tan^2 \theta}$

$\sin (\alpha \pm \beta) = \sin \alpha \cos \beta \pm \cos \alpha \sin \beta$

   For $\alpha = 90°$, $\sin (90° \pm \beta) = \cos \beta$

   For $\alpha = \beta$, $\sin 2\beta = 2 \sin \beta \cos \beta$

$\cos (\alpha \pm \beta) = \cos \alpha \cos \beta \mp \sin \alpha \sin \beta$

   For $\alpha = 90°$, $\cos (90° \pm \beta) = \mp \sin \beta$

   For $\alpha = \beta$, $\cos 2\beta = \cos^2 \beta - \sin^2 \beta = 1 - 2 \sin^2 \beta$

$\tan (\alpha \pm \beta) = \dfrac{\tan \alpha \pm \tan \beta}{1 \mp \tan \alpha \tan \beta}$

$\sin (-\theta) = -\sin \theta$

$\cos (-\theta) = \cos \theta$

$\tan (-\theta) = -\tan \theta$

$\sin \alpha + \sin \beta = 2 \sin \left[\frac{1}{2}(\alpha + \beta)\right] \cos \left[\frac{1}{2}(\alpha - \beta)\right]$

$\sin \alpha - \sin \beta = 2 \cos \left[\frac{1}{2}(\alpha + \beta)\right] \sin \left[\frac{1}{2}(\alpha - \beta)\right]$

Two laws that apply to any triangle labeled as shown in Fig. A.14:

Law of sines: $\dfrac{\sin \alpha}{a} = \dfrac{\sin \beta}{b} = \dfrac{\sin \gamma}{c}$

Law of cosines: $c^2 = a^2 + b^2 - 2ab \cos \gamma$ (where γ is the interior angle formed by the intersection of sides a and b)

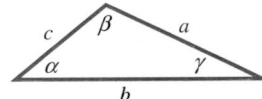

Figure A.14 A general triangle

Table A.3

Two Common Right Triangles

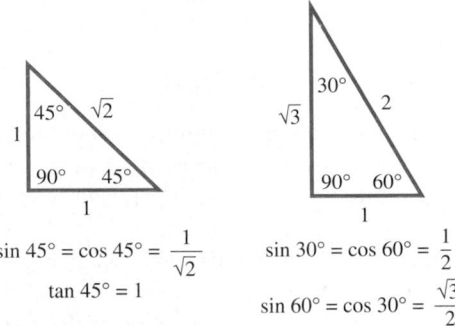

$\sin 45° = \cos 45° = \dfrac{1}{\sqrt{2}}$

$\tan 45° = 1$

$\sin 30° = \cos 60° = \dfrac{1}{2}$

$\sin 60° = \cos 30° = \dfrac{\sqrt{3}}{2}$

Small-Angle Approximations

These approximations are written for θ *in radians* and are valid when $\theta \ll 1$ rad.

$$\sin \theta \approx \theta \tag{A-26}$$

$$\cos \theta \approx 1 - \tfrac{1}{2}\theta^2 \tag{A-27}$$

$$\tan \theta \approx \theta \tag{A-28}$$

Figure A.15

The sizes of the errors involved in using these approximations are roughly $\frac{1}{6}\theta^3$, $\frac{1}{24}\theta^4$, and $\frac{2}{3}\theta^3$ respectively. In *some* circumstances it may be all right to ignore the $\frac{1}{2}\theta^2$ term and write

$$\cos\theta \approx 1 \qquad\qquad \text{(A-29)}$$

The origin of these approximations can be illustrated using a right triangle of hypotenuse 1 with one very small angle θ (Fig. A.15). If θ is very small, then the adjacent side ($\cos\theta$) will be nearly the same length as the hypotenuse (1). Then we can think of those two sides as radii of a circle that subtend an angle θ. The relationship between the arc length s and the angle subtended is

$$s = \theta r$$

Since $\sin\theta \approx s$ and $r = 1$, we have $\sin\theta \approx \theta$. To find an approximate form for $\cos\theta$ (but one more accurate than $\cos\theta \approx 1$), we can use the Pythagorean theorem:

$$\sin^2\theta + \cos^2\theta = 1$$
$$\cos\theta = \sqrt{1 - \sin^2\theta} \approx \sqrt{1 - \theta^2}$$

Now, using a binomial approximation,

$$\cos\theta \approx (1 - \theta^2)^{1/2} \approx 1 - \tfrac{1}{2}\theta^2$$

A.8   VECTORS

The distinction between vectors and scalars is discussed in Chapter 2. Scalars have magnitude while vectors have magnitude and direction. A vector is represented by an arrow of length proportional to the magnitude of the vector and aligned in a direction that corresponds to the vector direction.

In print, a vector is sometimes written in bold font, or in roman font with an arrow over it, or in bold font with an arrow over it (as done in this book). When writing by hand, a vector is designated by drawing an arrow over the symbol: $\vec{A}$. When we write just plain A, that stands for the *magnitude* of the vector. We also use absolute value bars to stand for the magnitude of a vector, so $A = |\vec{A}|$.

Addition and Subtraction of Vectors

When vectors are added or subtracted, the magnitudes and direction must be taken into account. When adding two collinear vectors $\vec{A}$ and $\vec{B}$, as shown in Fig. A.16a, the second vector ($\vec{B}$) is placed with its initial point, or origin, at the terminal point, or terminus, of the first vector ($\vec{A}$). The resultant vector ($\vec{C}$) is drawn from the origin of the first vector to the terminus of the second vector (tip to tail), as shown in Fig. A.16a. If collinear vectors are to be subtracted (Fig. A.16b), the procedure is the same, but the negative of one vector is used, which effectively reverses the direction of that vector. The resultant vector sum ($\vec{D}$) is again the magnitude of the vector formed by drawing an arrow from the tail of the first ($\vec{A}$) to the tip of the second ($\vec{B}$).

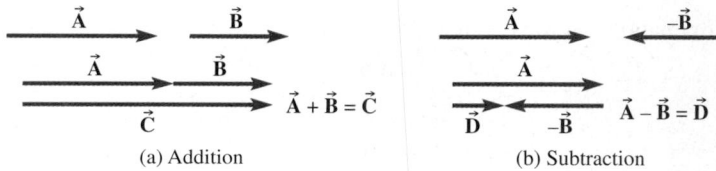

Figure A.16 Addition and subtraction of collinear vectors

(a) Addition   (b) Subtraction

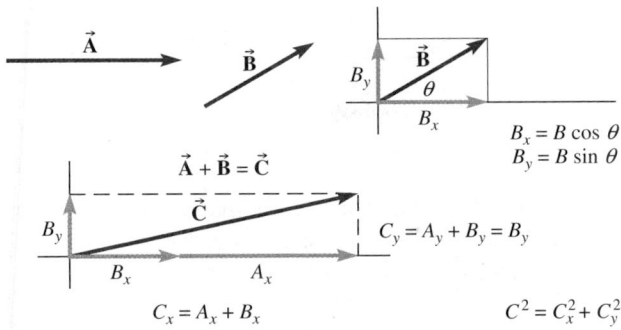

$$B_x = B \cos \theta$$
$$B_y = B \sin \theta$$

$$C_y = A_y + B_y = B_y$$

$$C_x = A_x + B_x \qquad\qquad C^2 = C_x^2 + C_y^2$$

Figure A.17 Addition of two vectors by the component method

When the vectors are at an arbitrary angle with respect to each other, orthogonal components of each vector may be added or subtracted; the resultant is then found by applying the Pythagorean theorem to the newly formed orthogonal components.

In Fig. A.17, vector $\vec{A}$ is along the x-axis and vector $\vec{B}$ is at an angle θ with respect to the x-axis. Components of vector $\vec{B}$ are taken along the x- and y-axes. Then the x-components are added and separately the y-components are added. Since the y-component of vector $\vec{A}$ is zero, the y-component of the resultant vector $\vec{C}$ is the same as the y-component of vector $\vec{B}$. The sum of the x-components of vectors $\vec{A}$ and $\vec{B}$ is the x-component of the resultant vector $\vec{C}$.

Another method for adding vectors at an arbitrary angle involves placing the vectors tip to tail and then drawing from the tail of the first to the tip of the second, as shown in Fig. A.18. This is called the method of triangles, parallelogram method, or more simply, the geometric or graphical method.

Figure A.19 shows both methods of addition applied to two arbitrary vectors. One method requires graph paper and careful geometric construction of the vectors; the other method involves finding components on a set of common axes. The vectors specified in problems have their directions given with respect to a set of axes, so the component method is commonly used for finding the magnitude and direction of a resultant vector. The same result is found by either method. The direction of the resultant vector can be found from the resultant components.

$$\tan \phi = \frac{C_y}{C_x}$$

and

$$\phi = \tan^{-1} \frac{C_y}{C_x} = \arctan \frac{C_y}{C_x}$$

$$\vec{A} + \vec{B} = \vec{C}$$

$$\vec{A} - \vec{B} = \vec{D}$$

Figure A.18 Addition and subtraction of two vectors by the method of triangles

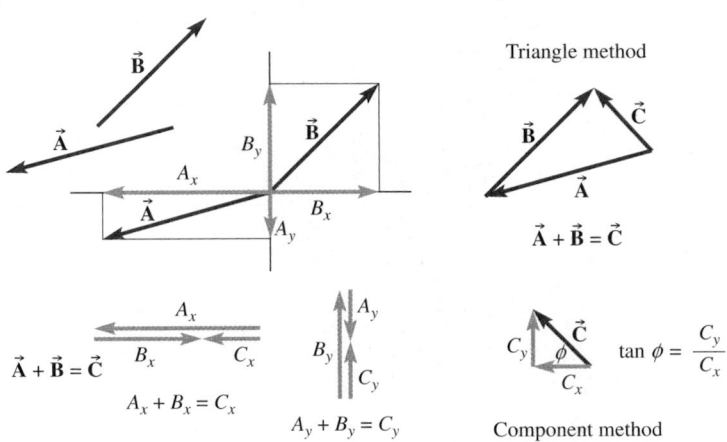

Triangle method

$$\vec{A} + \vec{B} = \vec{C}$$

$$\vec{A} + \vec{B} = \vec{C}$$

$$A_x + B_x = C_x$$

$$A_y + B_y = C_y$$

$$\tan \phi = \frac{C_y}{C_x}$$

Component method

Figure A.19 Adding two arbitrary vectors by two different methods

Figure A.20 Multiplication of a
vector by a scalar

Multiplication of Vectors: Product of a Vector and a Scalar

When a vector is multiplied by a scalar, the vector magnitude changes by the scalar factor, as shown in Fig. A.20. The direction of the vector does not change unless the scalar factor is negative, in which case the direction is reversed.

Cross Product of Two Vectors

One type of product between two vectors is the *cross product*, which is introduced in Chapter 19. It is denoted by

$$\vec{A} \times \vec{B} = \vec{C}$$

The product is a vector quantity; it has magnitude and direction. $\vec{A} \times \vec{B}$ is read as "$\vec{A}$ cross $\vec{B}$."

For two vectors, $\vec{A}$ and $\vec{B}$, separated by an angle θ (with θ chosen to be the *smaller* angle between the two), the magnitude of the cross product $\vec{C}$ is

$$|\vec{C}| = |\vec{A} \times \vec{B}| = AB \sin \theta$$

See Chapter 19 for the right-hand rule for finding the direction of a cross product.

The cross product depends on the order of the multiplication.

$$\vec{A} \times \vec{B} = -\vec{B} \times \vec{A}$$

The magnitude is $AB \sin \theta$ in both cases, but the direction of one cross product is opposite to the direction of the other.

APPENDIX B

Table of Selected Nuclides

Atomic Number Z	Element	Symbol	Mass Number A	Atomic Mass, neutral atom (u)	Percentage Abundance (or Decay Mode)	Half-life (if Unstable)
1	Hydrogen	H	1	1.007 825 0	99.989	
	(Deuterium)	(D)	2	2.014 101 8	0.0115	
	(Tritium)	(T)	3	3.016 049 3	β^-	12.33 yr
2	Helium	He	3	3.016 029 3	0.000 14	
			4	4.002 603 2	99.999 86	
3	Lithium	Li	6	6.015 122 3	7.6	
			7	7.016 004 0	92.4	
4	Beryllium	Be	7	7.016 929 2	EC	53.29 d
			8	8.005 305 1	2α	6.8×10^{-17} s
			9	9.012 182 1	100	
5	Boron	B	10	10.012 937 0	19.9	
			11	11.009 305 5	80.1	
6	Carbon	C	11	11.011 433 8	EC, β^+	20.39 min
			12	12.000 000 0	98.89	
			13	13.003 354 8	1.1	
			14	14.003 242 0	β^-	5730 yr
			15	15.010 599 3	β^-	2.449 s
7	Nitrogen	N	13	13.005 738 6	EC, β^+	9.965 min
			14	14.003 074 0	99.634	
			15	15.000 108 9	0.366	
8	Oxygen	O	15	15.003 065 4	EC, β^+	122.24 s
			16	15.994 914 6	99.757	
			17	16.999 131 5	0.038	
			18	17.999 160 4	0.205	
			19	19.003 578 7	β^-	26.91 s
9	Fluorine	F	19	18.998 403 2	100	
10	Neon	Ne	20	19.992 440 2	90.48	
			22	21.991 385 5	9.25	
11	Sodium	Na	22	21.994 436 8	EC, β^+	2.6019 yr
			23	22.989 769 7	100	
			24	23.990 963 3	β^-	14.9590 h
12	Magnesium	Mg	24	23.985 041 9	78.99	
15	Phosphorus	P	31	30.973 761 5	100	
			32	31.973 907 2	β^-	14.262 d
16	Sulfur	S	32	31.972 070 7	94.93	
19	Potassium	K	39	38.963 706 9	93.2581	
			40	39.963 998 7	0.0117; β^-	1.28×10^9 yr
20	Calcium	Ca	40	39.962 591 2	96.941	
27	Cobalt	Co	59	58.933 200 2	100	
			60	59.933 822 2	β^-	5.27 yr
36	Krypton	Kr	84	83.911 506 6	57.0	
			86	85.910 610 3	17.3	
			92	91.926 152 8	β^-	1.840 s

Atomic Number Z	**Element**	**Symbol**	**Mass Number A**	**Atomic Mass, *neutral atom* (u)**	**Percentage Abundance (or Decay Mode)**	**Half-life (if Unstable)**
					continued	
37	Rubidium	Rb	85	84.911 789 3	72.165	
			93	92.922 032 8	β^-	5.84 s
38	Strontium	Sr	88	87.905 614 3	82.58	
			90	89.907 737 6	β^-	28.79 yr
39	Yttrium	Y	89	88.905 847 9	100	
			90	89.907 151 4	β^-	64.00 h
55	Cesium	Cs	133	132.905 446 9	100	
			141	140.920 044 0	β^-	24.94 s
56	Barium	Ba	138	137.905 241 3	71.70	
			141	140.914 406 4	β^-	18.27 min
82	Lead	Pb	206	205.974 449 0	24.1	
83	Bismuth	Bi	214	213.998 698 7	β^-	19.8 min
84	Polonium	Po	210	209.982 857 4	α	138.376 d
			218	218.008 965 8	α	3.10 min
86	Radon	Rn	222	222.017 570 5	α	3.8235 d
88	Radium	Ra	226	226.025 402 6	α	1600 yr
90	Thorium	Th	232	232.038 050 4	100; α	1.41×10^{10} yr
			234	234.043 595 5	β^-	24.10 d
92	Uranium	U	235	235.043 923 1	0.720; α	7.038×10^8 yr
			236	236.045 561 9	α	2.34×10^7 yr
			238	238.050 782 6	99.27; α	4.468×10^9 yr
			239	239.054 287 8	β^-	23.45 min

EC = electron capture.

Answers to Selected Questions and Problems

Chapter 1

Multiple Choice Questions
1. (b) **2.** (a) **3.** (b) **4.** (c) **5.** (d) **6.** (b) **7.** (d) **8.** (b)
9. (d) **10.** (c)

Problems
1. 7.7% **3.** 4 **5.** 434 m/s **7.** (a) 3; 5.74×10^{-3} kg (b) 1; 2 m
(c) 3; 4.50×10^{-3} m (d) 3; 4.50×10^{1} kg (e) 4; 1.009×10^{5} s
(f) 4; 9.500×10^{3} mL **9.** (a) 1.29×10^{8} (b) 1.3×10^{8}
11. (a) -3.63×10^{7} (b) 2.268×10^{-178} **13.** 0.278 m/s
15. (a) 220 markers (b) 221 markers **17.** 0.14 W/cm^{2}
19. (a) 3.3×10^{-8} m (b) 3.3×10^{-2} μm (c) 1.3×10^{-6} in

21. 13.6 g/cm^{3} **23.** 6.0×10^{-6} m^{3} **25.** (a) $\sqrt{\dfrac{hG}{c^{5}}}$

(b) 1.3×10^{-43} s **27.** (a) 8.8×10^{-2} m (b) 3.4 in

29. (a) $a = K\dfrac{v^{2}}{r}$, where K is a dimensionless constant.

(b) 21.0% **31.** 0.46 s^{-1} **33.** (a) 1.004 (1.00399)
(b) 1.02 (1.0198) (c) 1.2 (1.183) (d) 1.33 (1.288)
35. (a) 30° (b) 10° **37.** 100 m **39.** 10^{11} gal
41.

(a) 101.8°F (b) 0.9°F/h (c) no; the patient would die before 12 hours passed and the temperature reached 113°F.
43. (a) a (b) $+v_0$
45. (a)

(b) about 100 g
(c) ; 0.30 s^{-1}

47. 56% **49.** 42 km **51.** $\dfrac{\text{kg·m}}{\text{s}^{2}}$ **53.** \$90,000,000,000
55. (a) 4 (b) 21.6% **57.** 104.5°F
59. (a) 2.4×10^{5} km/h (b) 10 min

Chapter 2

Multiple Choice Questions
1. (e) **2.** (d) **3.** (d) **4.** (c) **5.** (c) **6.** (b) **7.** (c) **8.** (c)
9. (d) **10.** (c)

Problems
1. 3200 N/cm **3.** (a) 6.0×10^{10} N/m (b) 8.0 nm **5.** 778 N
7. (a) the upward force on the Earth due to the skydiver; the downward force on the parachute due to the skydiver; the downward force on the air due to the skydiver (b) the upward force on the skydiver due to the parachute; the downward force on the air due to the parachute; the upward force on the Earth due to the parachute (c) dynamic equilibrium
9. p = system of plant, soil, pot; c = cord; e = Earth; h = hook; C = ceiling; s = system of plant, soil, pot, cord, hook

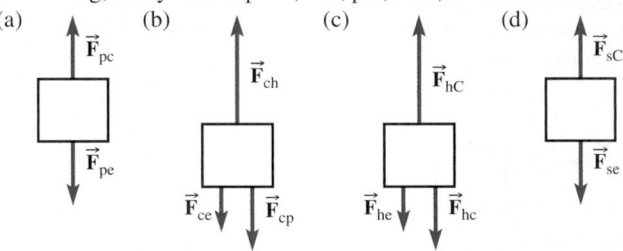

11. s = sailboat; e = Earth; w = wind; l = lake; m = mooring line

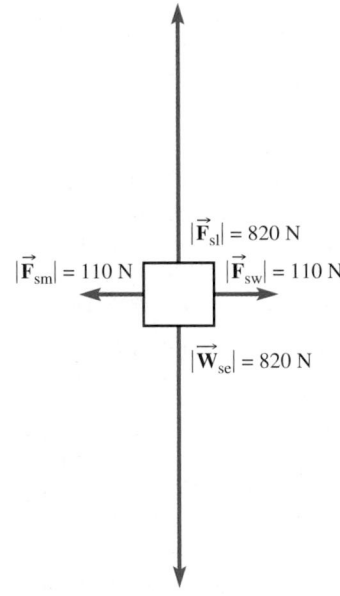

13. (a) 30 N to the right (b) 0 (c) 18 N downward

15. (a) 50.0 N upward (total for both feet)
(b) 650.0 N upward
(c) s = woman and chair system; e = Earth; f = floor

17. the upwardly directed drag due to the air; the downwardly directed force due to gravity
19. (a) 392 N (b) 88.2 lb **21.** (a) 1.9 N/cm (b) 2.4 kg
23. (a) 1432 km (b) 4664 km **25.** (a) 5.6 N up (b) 17.6 N
27. (a) 250 N (b) 3100 N (c) 250 N toward the satellite
29. (a) 1.98×10^{20} N (b) the same
31. b = book; d = desk; e = Earth; h = hand
(a)   (b)   (c)

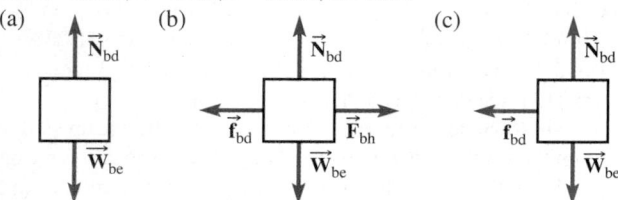

33. 12 N **35.** (a) 0.41 (b) 40 N
37. (a) $\mu_s > 0.48$ (b) 0.60 (c) 0.48
39. t = table; e = Earth; 1 = block 1; 2 = block 2; 3 = block 3; 4 = block 4; h = horizontal force; 1234 = system of blocks
(The blocks are numbered from left to right.)
(a)

(b)

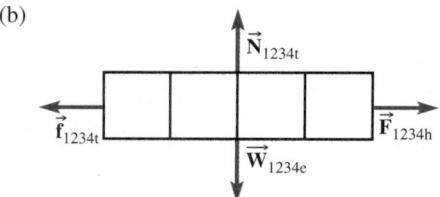

41. 1) the force of gravity; 2) the vertical force of the water opposing gravity and the forces of the water's currents (if any); 3) the force of the wind; 4) the force of the line tied to the mooring
43. Scale A reads 120 N; scale B reads 240 N.
45. Scale B reads 120 N; scale A reads 120 N. **47.** (a) $\dfrac{k}{2}$ (b) $2k$
49. i = ice; e = Earth; s = stone; o = opponent's stone
(a)   (b)   (c)

51. (a) 1) the gravitational forces between the magnet and the Earth; 2) the contact forces, normal and frictional, between the magnet and the photo; 3) the magnetic forces between the magnet and the refrigerator

(b) m = magnet; p = photo; e = Earth; r = refrigerator

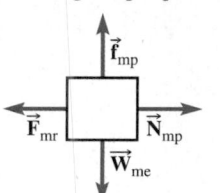

(c) The long-range forces are gravity and magnetism. The contact forces are friction and the normal force.
(d) $W_{me} = 0.14$ N; $F_{mr} = 2.10$ N; $f_{mp} = W_{me} = 0.14$ N; $N_{mp} = F_{mr} = 2.10$ N

53. (a) In both cases the two ropes pull on the scale with forces of 550 N in opposite directions, so the scales give the same reading.
(b) 550 N **55.** (a) 6×10^{-6} N (b) $F \propto r$ (c) The surface tension would only double, since the feet would only be twice as wide, yet they would have to support eight times the weight.
57. (a) 15 N (b) 20 N
59.

(a) 0.15 N/cm
(b) 12 cm

61. $\dfrac{k_1 k_2}{k_1 + k_2}$ **63.** 0.027 m/s² **65.** 98 N

Chapter 3

Multiple Choice Questions

1. (d) **2.** (c) **3.** (a) **4.** (b) **5.** (a) **6.** (d) **7.** (a) **8.** (b)
9. (a) **10.** (a) **11.** (c) **12.** (a) **13.** (a) **14.** (d) **15.** (c)
16. (a) **17.** (b) **18.** (a) **19.** (d) **20.** (c) **21.** (b)

Problems

1. 32 s **3.** 8 m **5.** 160 m **7.** (a) 1.5 m/s (b) 1.2 m/s
9. (a) 170 cm to the left (b) 28 cm/s (c) 9.4 cm/s to the left
11. 27 m/s west **13.** 13 s **15.** (a) –10 m/s² (b) 0 (c) 5.0 m
17. 2.5 m/s² **19.** 1.0×10^5 N in the direction of motion **21.** 3.5 N
23. 17 kN **25.** (a) 3.0 kN upward (b) 3.3 m/s² downward
27. 23 N downward **29.** (a) 86.4 m (b) 14.4 m/s
31. (a) 4.0 m/s (b) 14.0 m/s **33.** 80 m **35.** 1.5 m/s² northeast
39. No; it takes 236 m for the train to stop. **41.** 85 m/s down
43. 5.0 m/s **45.** 30.0 m/s **47.** 13 m **49.** (a) 5 m (b) 1 km
51. (a) 72 N up (b) 41 N up (c) 0 (d) 4.3 m/s² down
53. (a) 567 N (b) 629 N **55.** 620 N **57.** 4×10^{10} m
59. (a) 330 m/s up (b) 16 m/s² up **61.** $2v_0$ **63.** (a) 0.14 mi
(b) $2.2 \dfrac{\text{mi/h}}{\text{s}} = 0.98$ m/s² in the direction of motion
65. (a) 46 m (b) 30.0 m/s² up **67.** 0.81 s **69.** 1.8 m/s² down
71. (a) t_3 and t_4 (b) $t_0, t_2, t_5,$ and t_7 (c) t_1 and t_6
(d) $t_0, t_3,$ and t_7 (e) t_6 **73.** 68 s **75.** (a) 25.0 km (b) 150 s
(c) 76 km (d) 1200 m/s downward **77.** (a) $1.10mg$
(b) $1.10mg$ **79.** $\Delta x = 160$ m **81.** 3260 ft; 25.5 s
83. (a) 1 mm/s (b) 20 ms (c) 100 m/s

Chapter 4

Multiple Choice Questions

1. (c) **2.** (a) **3.** (d) **4.** (d) **5.** (a) **6.** (a) **7.** (a) **8.** (c)
9. (d) **10.** (e)

Problems

1. (a)

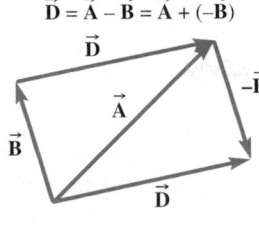

$$\vec{D} = \vec{A} - \vec{B} = \vec{A} + (-\vec{B})$$

$$\vec{C} = \vec{A} + \vec{B}$$

(b)

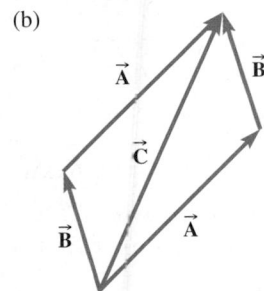

$$\vec{C} = \vec{A} + \vec{B} = \vec{B} + \vec{A}$$

3. $\Delta\vec{r} = \vec{r}_n - \vec{r}_0$ **5.** x-comp $= -17.3$ m; y-comp $= 10.0$ m
7. (a) 2.0 units at 30° CCW from the $+y$-axis
(b) 2.0 units at 30° CW from the $+y$-axis
(c) x-comp $= -1.0$ unit; y-comp $= -\sqrt{3.0}$ units
9. 120 N   **11.** (a) 9.4 cm, 32° CCW from the $+y$-axis
(b) 130 N, 27° CW from the $+x$-axis
(c) 16.3 m/s, 33° CCW from the $-x$-axis
(d) 2.3 m/s², 1.6° CCW from the $+x$-axis
13. 2.0 km, at 70° south of east   **15.** 3.15 nm; 0.85 nm at 18° CCW
from the horizontal   **19.** 1810 N; 5 times the force with which
Yoojin pulls; from the oak tree   **21.** (a) 34 N   (b) 39 N

23. $\dfrac{mg}{\cos\theta}$

25. (a) 75.2 N   (b) 27.4 N   (c) 0.364   (d) 80.0 N upward

27. (a) $W\tan 12°$ (b) $\dfrac{W}{\cos 12°}$ **29.** 23 km/h at 37° north of east

31. (a) 9.4 m/s at 45° north of east   (b) 15 m/s² at 45° south of east
(c) Changing the direction of the velocity requires an acceleration.
33. (a)

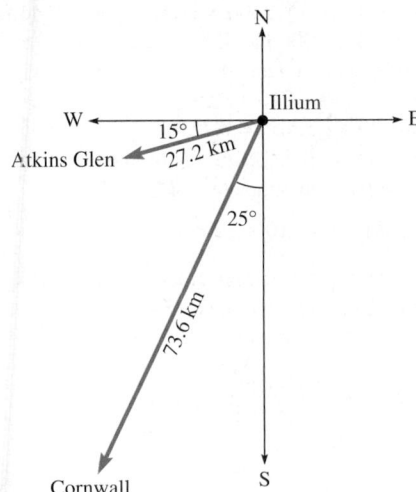

(b) 59.9 km at 85° north of east   (c) 80 km/h at 85° north of east

35. (a)

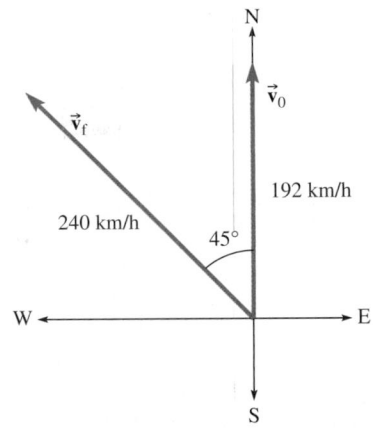

(b) 170 km/h at 7° south of west   (c) 57 km/h² at 7° south of west
37. (a) $R = \dfrac{2v_0^2 \sin\theta \cos\theta}{g}$ (b) 221 m   (c) 4 m   **41.** (a) 2.0 s
(b) 2.0 s   (c) 1.5 s   **43.** (a) 3.50 s   (b) 2.01 s   (c) 80.8 m
45. (a)

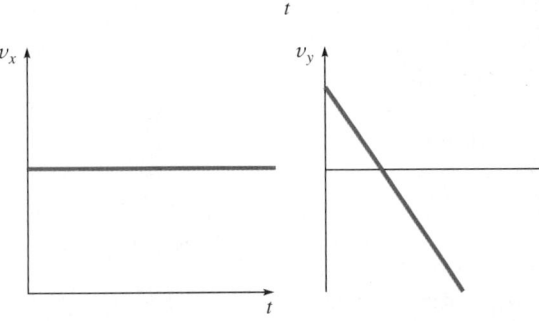

(b) 27.6 m/s at 25.0° above the horizontal   (c) 37 m   (d) 44 m
above the ground   **49.** 0.8 s   **51.** (a) 69 N directed down the
incline   (b) 6.9 m/s² directed down the incline   (c) 1.4 s
53. (a) $mg\tan\theta$ (b) $mg\tan\theta$ (c) $mg\tan\theta + \dfrac{ma}{\cos\theta}$
55. (a) 6.00×10^2 N directed along the 8.00×10^2-N vector
(b) 0.0414 m/s² in the same direction as the force   **57.** (a) 88 N
(b) 2 s   (c) 70 N   (d) 10 kg   **59.** 130 km/h north
61. 0.42 km/h   **63.** 27° upstream   **65.** (a) 76.37° N of E
(b) 2.717 h   **67.** 39 s   **69.** (a) 30.0° N of W   (b) 9.1 min
71. 13 nautical miles at 23° south of east   **73.** (a) 375.0 m
(b) 257.4 m at 10.5° N of W
75.   ; 400 N

```
           T sin 30.0°
             ↑
  ←─────── ┌──┐ ──────→
T cos 30.0°│  │  F_sw
           └──┘
             ↓
          W_sign
```

77. (a) 7.67 m   (b) 13.9 m/s   (c) 12.5 m/s   **79.** 23 m

81. (a) v (m/s)

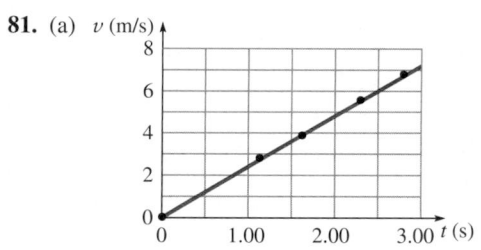

(b) yes; 2.43

m/s^2 (c) 14°   (d) larger

83. (a) $v_x = v_0 \cos \theta$ and $v_y = 0$.   (b) $\dfrac{v_0 \sin \theta}{g}$

85. (a) 28.6 cm   (b) smaller   (c) larger
(d) $H = 21.3$ cm; $R = 85.1$ cm   **87.** 46 m

89. (a) a = air; w = water; s = sailboat

Case (1):   Case (2):   Case (3):

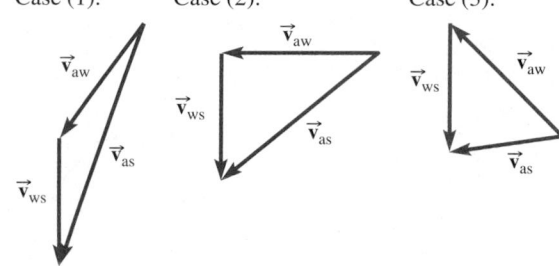

(b) 1 and 2   (c) all three
91. (a) 98 N   (b) 60° above the horizontal
93. (a) in front of   (b) 21.4 m   (c) 3.6 m   (d) earlier

Chapter 5

Multiple Choice Questions

1. (b)   **2.** (a)   **3.** (f)   **4.** (b)   **5.** (a)   **6.** (b)   **7.** (b)   **8.** (a)
9. (e)   **10.** (c)

Problems

1. 17 m   **3.** 0.105 rad/s   **5.** 26 rad/s   **7.** (a) 3.49 rad/s
(b) 0.45 m/s   **9.** 3800 ft   **11.** (a) 31 m/s   (b) 31 rad/s
13. 3.37 cm/s^2 **15.** (a) 1.72×10^{-2} rad
(b) 514 m/s perpendicular to the average velocity
(c) 0.00595 m/s^2 perpendicular to the average velocity
(d) $\vec{a}_c = 0.00595$ m/s^2 perpendicular to the velocity, which is the
same as $\vec{a}_{av}$ within 3 significant figures.

17. (a) $\dfrac{mg}{\cos \phi}$ (b) $2\pi \sqrt{\dfrac{L \cos \phi}{g}}$ **19.** 7.9 m/s   **21.** 59°

23. (a) 2300 N   (b) 19 m/s   **25.** $\tan^{-1} \dfrac{v^2}{rg}$ **27.** 130 h

29. 3.6×10^{22} N   **31.** 16 h   **33.** (a) 13 N
(b) The bob has an upward acceleration, so the net F_y must be
upward and greater than the weight of the bob.   **35.** $g \sin \theta$
37. 4.0 rad/s^2 **39.** (a) 1.7 rad/s^2 (b) 0.56 rev
41. (a) 1.0×10^5 m/s   (b) 0.080 m/s^2 (c) 5.0×10^{10} m/s^2
43. 0.045 Hz   **45.** (a) 518.5 N   (b) 521.5 N   (c) 45 m
47. 0.0257 m/s^2 **49.** 464 m/s   **51.** 1.80×10^6 degrees
53. (a) 7.3×10^{-5} rad/s   (b) 0.02 rad (arm-length $\approx$ 1 m and
finger-width $\approx$ 2 cm)   (c) 5 min   **55.** 200 km/s
57. smallest; 4.1 s   **59.** 0.40ω **61.** (a) 0.60 m/s
(b) The dolls do not stay on the record.   **63.** 1.0 rad/s

65. 110 μm/s   **67.** (a) 90°   (b) $T = \dfrac{2\pi m}{k}$ (c) $r = \dfrac{m}{k}v$

Chapter 6

Multiple Choice Questions

1. (e)   **2.** (b)   **3.** (b)   **4.** (a)   **5.** (c)   **6.** (c)   **7.** (c)   **8.** (c)
9. (b)   **10.** (b)   **11.** (f)

Problems

1. 75 J   **3.** No work is done.   **5.** 210 kJ   **7.** 720 kJ
9. (a) 0.70 J   (b) 0.37 m/s   **11.** 0   **13.** 5.8 MJ (meteor);
0.46 MJ (car); the meteor has more than 12 times the kinetic energy
of the car.   **15.** 1.6 J   **17.** 0   **19.** 5.2 J   **21.** (a) 0
(b) 3.4 kJ   (c) dissipated as heat   **23.** (a) 0   (b) 2.9 J
25. 2   **27.** 53 kJ   **29.** $v_1 = 25$ m/s; $v_2 = 18$ m/s; $v_3 = 21$ m/s
31. (a) −1.8 kJ   (b) 7.0 m/s   (c) 1100 N   (d) 5.7 m/s
33. 1.9 m   **35.** (a) $\sqrt{v^2 + 2gh}$ (b) The final speed is
independent of the angle.   **37.** 22.4 km/s   **39.** 0.33 m

41. (a) $\sqrt{5g(L - d)}$ (b) $\cos^{-1}\left(\dfrac{5d}{2L} - \dfrac{3}{2}\right)$ **43.** 13 m

45. (a) $d\sqrt{\dfrac{k}{m}}$ (b) d **47.** 2.5 kJ   **51.** 150 W   **53.** 60 kW

55. 6.2 g; the other 90% of the energy is lost as heat.
57. (a) 20 N   (b) 6.7 m/s   **59.** (a) 10 kW   (b) 5.8°
61. −52 kJ   **63.** (a) −500 J   (b) 3 GW
(c) 300,000 households   **65.** 60.0 km/s   **67.** 11.2 km/s
69. 43.5 km/s   **71.** (a) 2.62 kW   (b) 7.86 kW   **73.** 6.1 m
75. (a) 2200 kcal/day   (b) more than 0.51 lb   **77.** 27 N
79. 5×10^{24} J (equivalent to 50 million thermonuclear bombs)
81. 1.3 cm; 32 J   **83.** (a) 500 m^3 (b) 600 kg   (c) 30 kJ
(d) 10 kW   (e) It decreases to $\frac{1}{8}$ of its initial value; the power
production of wind turbines is inconsistent, since modest changes
in wind speed produce large changes in power output.
85. approximately 3%–5%

Chapter 7

Multiple Choice Questions

1. (c)   **2.** (d)   **3.** (c)   **4.** (b)   **5.** (d)   **6.** (b)   **7.** (f)   **8.** (d)
9. (a)   **10.** (e)   **11.** (d)   **12.** (b)

Problems

1. 2.0 kg·m/s to the right   **3.** (a) 11 m/s   (b) 1300 N
5. 3 kg·m/s north   **7.** 20 kg·m/s in the −x-direction
9. 1.0×10^2 kg·m/s downward   **11.** 320 s   **13.** 6.0×10^3 N
opposite the car's direction of motion   **15.** (a) 750 kg·m/s upward
(b) 990 N·s downward   (c) 2500 N downward   **17.** 1.8 m/s
19. 2.6×10^5 m/s   **21.** 1500 kg   **23.** (4.2 cm, 0)
25. (1.9 m, 1.4 m)   **27.** (6 m/s, −4 m/s)   **31.** 3.0 m/s east
33. 0.20 kg   **35.** 43 m/s   **37.** 4.8 m/s   **39.** 0.066 m/s
45. (a) $\Delta p_{1x} = -1.00m_1v_0$; $\Delta p_{1y} = 0.751m_1v_0$
(b) $\Delta p_{2x} = m_1v_0$; $\Delta p_{2y} = -0.751m_1v_0$; the momentum changes
for each mass are equal and opposite.   **47.** $1.73v_{1f}$

49. $\left(\dfrac{v_0}{2}, -\dfrac{v_0}{2\sqrt{3}}\right)$ **51.** 5.0×10^9 kg·m/s   **53.** 34 N   **55.** 410 N

57. (a) 0.01 kg·km/h opposite the car's motion   (b) 0.01 kg·km/h
along the car's velocity   (c) 10^5 flies   **59.** 2.8 m/s
61. 0.83 m/s   **63.** (a) 2.5 m   (b) 4.0 m   **65.** 10^{-18} N

67. $\dfrac{1}{9}h$ **69.** 10 m/s   **71.** (a) $\dfrac{111}{2}$ (b) 1   (c) $\dfrac{111}{2}$

Chapter 8

Multiple Choice Questions

1. (b) **2.** (d) **3.** (a) **4.** (c) **5.** (e) **6.** (b) **7.** (a) **8.** (f)
9. (e) **10.** (c)

Problems

3. $0.98\,\text{kg·m}^2$ **5.** $0.0512\,\text{J}$ **7.** 0.019 **9.** $570\,\text{J}$ **11.** $4.5\,\text{N·m}$
13. $25\,\text{N·m}$ **15.** $0.30\,\text{N·m}$ **17.** $5.83\,\text{m}$ **19.** $(0.42s, 0.58s)$
21. $150\,\text{J}$ **23.** (a) $53.0\,\text{kJ}$ (b) $1.51\,\text{MN·m}$ **25.** (a) $540\,\text{N}$
(b) $390\,\text{N}$ **27.** $200\,\text{N}$ **29.** $180\,\text{N}$ toward the wall
31. $(\frac{1}{2}mg + W)/\tan\theta$; For $\theta = 0$, $T \to \infty$, and for $\theta = 90°$, $T \to 0$.
33. palms: $390\,\text{N}$; feet: $270\,\text{N}$ **35.** $7\,\text{kN}$—too much for a human,
but maybe not for a cyborg **37.** $130\,\text{N}$ **41.** $0.0012\,\text{N·m}$
43. $15\,\text{N·m}$ **45.** (a) $48\,\text{N·m}$ (b) $19\,\text{N}$ **47.** $0.09\,\text{N·m}$
49. solid sphere: $K = \frac{7}{10}mv^2$; solid cylinder: $K = \frac{3}{4}mv^2$; hollow
cylinder: $K = mv^2$ **51.** (a) The drilled cylinder takes more time
because its rotational inertia is larger. (b) 4% longer **53.** $3r$
55. $0.0864\,\text{kg·m}^2/\text{s}$ **57.** $1.4 \times 10^7\,\text{kg·m}^2/\text{s}$ **59.** $1.60\,\text{s}$
61. $1.5\,\text{rev/s}$ **63.** $3.15\,\text{rad/s}$ **65.** $2.10 \times 10^6\,\text{N·m}$ **67.** 1.14
69. (a) $16\,\text{kg·m}^2$ (b) $8.0 \times 10^7\,\text{J}$ (c) 320 (d) $120\,\text{km}$
73. $98\,\text{N·m}$ **75.** $5.4\,\text{rad/s}$ **77.** (a) $98\,\text{N}$ (b) This does not
help the person trying to lift the ladder, since the torque problem
is not alleviated by exerting a force at the point of rotation.
79. $\dfrac{m_2 g}{m_1 + m_2 + \dfrac{I}{R^2}}$ **81.** (a) $6.28\,\text{rad/s}$ (b) $0.955\,\text{kg·m}^2/\text{s}$
(c) friction (d) $0.300\,\text{N}$ **83.** (b) $mr^2\omega$ (c) $\frac{1}{2}r^2\omega\Delta t$
85. (a) $\dfrac{I_0\omega_0}{I_0 + mR^2}$ (b) before: $K_{\text{rot}} = \frac{1}{2}I_0\omega_0^2$ and $L = I_0\omega_0$;
after: $K_{\text{rot}} = \frac{1}{2}\dfrac{I_0^2\omega_0^2}{I_0 + mR^2}$ and $L = I_0\omega_0$ **87.** (a) $735.0\,\text{N}$
(b) $0.88\,\text{m}$ (c) $0.55h$ **89.** $23\,\text{N}$ **91.** $T_1 = 67\,\text{kN}$;
$T_2 = 250\,\text{kN}$; $\vec{\mathbf{F}}_p = 380\,\text{kN}$ at $51°$ with the horizontal
93. (a) $1.3\,\text{rev/s}$ (b) stays the same; no net torque (The falling
dumbbells are no longer part of the system.)
95. (a) $9.6\,\text{m/s}$ (b) $3.1\,\text{m/s}$ (c) $21\,\text{m/s}$

Chapter 9

Multiple Choice Questions

1. (b) **2.** (b) **3.** (d) **4.** (a) **5.** (a) **6.** (b) **7.** (a) **8.** (a)
9. (d) **10.** (c)

Problems

1. $50\,\text{atm}$ **3.** $22\,\text{kPa}$ **5.** $4.0\,\text{kN}$ southward **7.** (a) $625\,\text{N}$
(b) $6.25\,\text{mm}$ (c) 16.0 **9.** (a) $30\,\text{N}$ (b) $5.8\,\text{N·m}$ **11.** $2\,\text{atm}$
13. (a) $343\,\text{kPa}$ (b) $410\,\text{Pa}$ **15.** $2.9\,\text{N}$ **17.** $4.65 \times 10^{-5}\,\text{m}^3$;
$\dfrac{V_{\text{Pt}}}{V_{\text{Al}}} = 0.126$ **19.** (a) $2.2 \times 10^5\,\text{Pa}$ (b) $1700\,\text{torr}$ (c) $2.2\,\text{atm}$
21. $114.0\,\text{cm Hg}$ **23.** (a) $5.6\,\text{cm}$ (b) $0.37\,\text{cm}$ **25.** $250\,\text{kg/m}^3$
27. (a) $8.8\,\text{N}$ upward (b) $9.6\,\text{N}$ upward **29.** $0.80g$ downward
31. (a) 0.910 (b) $1.28\,\text{cm}$ (c) $0.13\,\text{cm}$ **33.** $0.17\,\text{cm}^3$
35. $50\,\text{m/s}$ **37.** (a) $39.1\,\text{cm/s}$ (b) $78.5\,\text{cm}^3/\text{s}$ (c) $78.5\,\text{g/s}$
39. $1.9 \times 10^5\,\text{N}$ **41.** (a) $78\,\text{W}$ (b) $390\,\text{kPa}$ (c) at the bottom
45. (a) $6850\,\text{Pa}$ (b) $0.685\,\text{N}$ **47.** $0.040\,\text{m}^3/\text{s}$ **49.** (a) $50\,\text{Pa}$
(b) $1100\,\text{Pa}$ (c) approximately $10\,\text{kPa}$ **51.** (a) $1.3 \times 10^{-10}\,\text{N}$
(b) $2.6 \times 10^{-14}\,\text{W}$ **53.** $0.4\,\text{Pa·s}$ **55.** $2.9\,\text{cm/s}$
57. Since m/v_t is constant, the drag force is primarily viscous.
59. $5\,\text{Pa}$ **61.** (a) $\gamma L\Delta s$ (b) $\Delta E = \gamma\Delta A$ **63.** (a) $1.1 \times 10^8\,\text{Pa}$
(b) $1.1 \times 10^8\,\text{N}$ **65.** 20% **67.** (a) $1.4\,\text{N}$ (b) $0.43\,\text{N}$ upward

(c) $6.8\,\text{m/s}^2$ downward **69.** (a) $10.0\,\text{cm}$ (b) 0.814 (c) 0.545
71. $0.4\,\text{mm/s}$ **73.** $270\,\text{m/s}$ **75.** $76\,\text{Pa}$ **77.** (a) $2.2\,\text{m/s}$ up
(b) $21\,\text{kPa/s}$ **79.** (a) force is weight of column above the area,
definition of pressure (b) $8.0\,\text{km}$ (c) lower limit
81. (a) $10\,\text{m}$ (b) The pump pushes the water up from below.

Chapter 10

Multiple Choice Questions

1. (c) **2.** (b) **3.** (b) **4.** (a) **5.** (a) **6.** (a) **7.** (c) **8.** (b)
9. (c) **10.** (a) **11.** (f) **12.** (e) **13.** (e) **14.** (f) **15.** (f)
16. (e) **17.** (c) **18.** (c) **19.** (j) **20.** (k)

Problems

1. $0.1\,\text{mm}$ **3.** $2.2\,\text{cm}$ **5.** (a) $1.2 \times 10^{-4}\,\text{W}$ (b) $5 \times 10^{-6}\,\text{W}$; no
(c) $3.7 \times 10^{-7}\,\text{J}$ (d) $3.7 \times 10^{-4}\,\text{W}$; yes **7.** $0.8\,\text{mm}$
9. tension: $1.5 \times 10^{10}\,\text{N/m}^2$; compression: $9.0 \times 10^9\,\text{N/m}^2$
11. $8.7 \times 10^{-5}\,\text{m}$ **13.** $630\,\text{N}$ **15.** (a) $2.8 \times 10^7\,\text{Pa}$
(b) 4.7×10^{-4} (c) $9.3 \times 10^{-4}\,\text{m}$ (d) $5.0 \times 10^5\,\text{N}$
17. human: $3\,\text{cm}^2$; horse: $7\,\text{cm}^2$ **19.** volume: 7.7×10^{-4};
radius: 2.6×10^{-4} **21.** $7.5 \times 10^5\,\text{N}$ **23.** $0.30\,\text{N}$ **25.** $0.63\,\text{m/s}$
27. $7.0\,\text{cm/s}$ **31.** $5.0\,\text{rad/s}$ **35.** $0.157\,\text{m/s}$; $24.7\,\text{m/s}^2$
37. (a) g (b) $0.78\,\text{m}$ **39.** $-0.031\,\text{J}$ **41.** $0.250\,\text{Hz}$
43. (a)
(b)
(c)
45.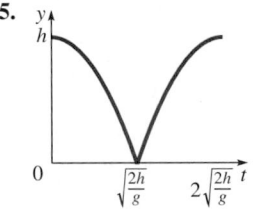
not a sine or cosine function
47. $4.0\,\text{s}$ **49.** $1.5\,\text{s}$
51. (a) $v_x = \omega A\cos\omega t$ (b) $\frac{1}{2}m\omega^2 A^2$
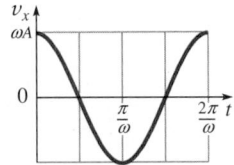
53. 1.11 **55.** 1st method: $3.14\,\text{cm/s}$; 2nd method: $3.14\,\text{cm/s}$
57. 5 cycles **59.** $2.5\,\text{s}$ **61.** (a) $1.64\,\text{s}$ (b) $1.53\,\text{s}$ (c) $1.94\,\text{s}$

63. 91 Hz **65.** (a) $2\pi\sqrt{\dfrac{L\left(\dfrac{m_1}{3}+m_2\right)}{g\left(\dfrac{m_1}{2}+m_2\right)}}=2\pi\sqrt{\dfrac{2L(m_1+3m_2)}{3g(m_1+2m_2)}}$

(b) For $m_1 \gg m_2$, $T=2\pi\sqrt{\dfrac{2L}{3g}}$, and for $m_1 \ll m_2$, $T=2\pi\sqrt{\dfrac{L}{g}}$.

67. 2.1 m/s; 370 m/s² **69.** (b) $k=\dfrac{mg}{L}$

71. $y=(1.6\text{ cm})\cos[(25\text{ rad/s})t]$ **73.** 2.0° **75.** $2\pi\sqrt{\dfrac{R}{g}}$

77. (a) 8×10^{-4} (b) 8.0 kN (c) 5×10^{-5} m² (d) no
79. (a) 3.6 cm² (b) 7.8×10^{6} Pa (c) 3.3×10^{-4} m
81. (a) 5.1 N (b) 7.7×10^{-2} J

Chapter 11

Multiple Choice Questions
1. (b) **2.** (c) **3.** (d) **4.** (f) **5.** (a) **6.** (b) **7.** (d) **8.** (a)
9. (b) **10.** (d)

Problems
1. 52 W/m² **3.** 1.7 MW **5.** (a) 1.5 m/s (b) 21 cm/s
7. 168 m/s **9.** 16 ms **11.** 0.375 m
13. (a) 340 Hz (b) 3.0×10^{8} Hz **15.** 0.33 Hz
17. (a) 3.5 cm (b) 6.0 cm **19.** (a) 4.0 mm (b) 1.0 m
(c) 0.010 s (d) 100 m/s (e) in the $+x$-direction (to the right)
21. (a)

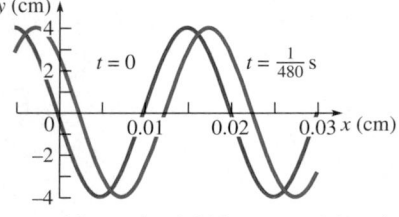

$y_{max}=4.0$ cm; $\lambda=0.020$ cm; $v=1.2$ cm/s

(b)

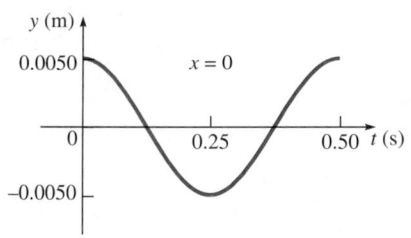

$\dfrac{\pi}{189}\text{ s}\approx16.6$ ms

23. $v_m=0.063$ m/s; $a_m=0.79$ m/s²

25.

27. 96.0°

29. (a) ; 6.9 cm

(b) 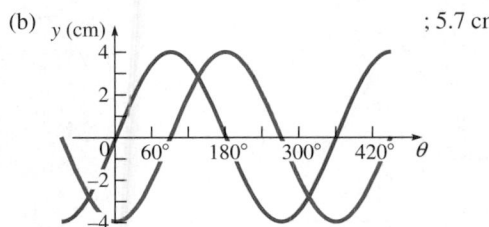 ; 5.7 cm

31. 4.8 m/s   **33.** 1.7 s   **35.** (a) 0°; 9.0 cm   (b) 180°; 3.0 cm
(c) 9:1   **37.** 80 µW/m²   **39.** 7.8%
43. (a) 1350 m/s   (b) 45.6 N   (c) 0.76 m and 450.0 Hz
45. 4.5×10^{-4} kg/m   **47.** 190 m   **49.** 3 m
51. (a) $y(x, t) = (0.020$ m$) \sin[(1.6$ rad/s$)t + (0.0016$ rad/m$)x]$
(b) 0.031 m/s   (c) 1.0 km/s   **53.** $v \propto \sqrt{\lambda g}$ **55.** 80 km
57. 3.64 cm; 7.07 cm; 10.32 cm
59.

61. 1.4 kHz

Chapter 12

Multiple Choice Questions

1. (c)   **2.** (a)   **3.** (b)   **4.** (c)   **5.** (b)   **6.** (c)   **7.** (b)   **8.** (c)
9. (b)   **10.** (d)

Problems

1. 3.4 mm   **3.** 173 ms   **7.** 3.5 km/s   **9.** (a) 338 m/s
(b) 2.8 km   **11.** 0.099 W   **13.** 26%   **15.** (a) 125.0%
(b) 3.522 dB   **17.** (a) 65.6 cm   (b) 252.4 Hz   **19.** 43.3 cm
21. (a) There is a displacement node (pressure antinode) at the
center of the rod and displacement antinodes (pressure nodes) at
the ends.   (b) 5100 m/s   (c) 13.1 cm   (d) The ends move in
opposite directions, and thus, they are out of phase.
23. (a) 1.20 m   (b) 90.0 cm   (c) 338 m/s   **25.** (a) 290.0 Hz
(b) 1.4%   **27.** 2 Hz   **29.** (a) 1.5 kHz   (b) 500 Hz
33. 49 m/s   **35.** (a) Mach 2.67   (b) 860 m/s   **37.** 5.42 km
39. 34 cm   **41.** (a) $\dfrac{2fv}{c - v}$ (b) $f_{\text{beat}} = \dfrac{2fv}{c}$
43. 17.9 Hz, 53.6 Hz, 89.3 Hz, 125 Hz
45. (a) 10 m   (b) 2 ms   (c) No, the bat will still be emitting
the first chirp.   **47.** 3 kHz   **49.** 15 m   **51.** 0.0955 s
53. 534 m/s   **55.** (a) 138 dB   (b) 88.0 dB

Chapter 13

Multiple Choice Questions

1. (e)   **2.** (d)   **3.** (b)   **4.** (c)   **5.** (b)   **6.** (d)   **7.** (a)   **8.** (c)
9. (c)   **10.** (e)

Problems

1. (a) 29°C   (b) 302 K   **3.** (a) −6.0 K   (b) −11°F
5. (a) 3.6 mm   (b) 10.8 mm   **7.** 3.8×10^{-4} mm²
9. (b) 2.4×10^{-3} **11.** (a) 44°C   (b) −21°C   **13.** 1.3 m
17. 150°C   **21.** 7.31×10^{-26} kg   **23.** 1.7×10^{27}
25. 10^{18} atoms   **27.** 2.65×10^{25} atoms   **29.** 8.9985 mol
31. 2.5×10^{19} molecules   **35.** 0.0136   **37.** 1.55   **39.** 3.1 cm³
41. 3.3×10^5 Pa   **43.** (a) 90 ft³   (b) 5.3 h   **45.** 4×10^{-17} Pa
47. 0.38 m³   **49.** 1550 K   **51.** (a) 1.52×10^5 J/m³

(b) 4.559×10^7 J/m³   **53.** 50%   **55.** $\dfrac{1}{\sqrt{2}}$

57. (a) 493 m/s   (b) 461 m/s   (c) 393 m/s   **59.** yes
61. 1550 K   **65.** 0.14°C   **67.** (a) 100 nm   (b) 200 nm
(c) 1 m   **69.** 2.5×10^4 s   **71.** 4.8×10^7 N/m²   **73.** 3.05 mm
75. (b) 0.7 m   **77.** (a) 52 cm   (b) 12 m   **81.** HNO₃
83. (a) 6.42×10^{-21} J   (b) 0.25%   **85.** 1.9×10^{14} molecules
87. 25 m/s

Chapter 14

Multiple Choice Questions

1. (a)   **2.** (b)   **3.** (d)   **4.** (d)   **5.** (c)   **6.** (b)   **7.** (c)   **8.** (d)
9. (d)   **10.** (b)   **11.** (c)   **12.** (c)

Problems

1. (a) 34 J   (b) Yes; the increase in internal energy causes a
slight temperature increase.   **3.** 4.90 kJ   **5.** (a) 250 J
(b) all three   **7.** 5.4 J   **9.** 1.16×10^{-3} kWh   **11.** 0.153 cal/K
13. 0.50 MJ   **15.** 700 m   **17.** (a) 581 kcal/K   (b) 860 kcal/K
19. 177 kcal   **21.** 0.030 kcal/(kg·K)   **23.** 0.090 J
25. 27.5 kJ   **27.** (a) 180°C   (b) 20.9°C
29. 4.83×10^{21} molecules   **31.** 36 g   **33.** 539 cal/g
35. 730 kcal

37. (a)

Segment	Pressure	Temperature	Phase Change
AB	Decreases	Constant	Liquid to solid
BC	Constant	Increases	Solid to liquid
CD	Decreases	Constant	Liquid to vapor
DE	Constant	Decreases	Vapor to solid

(b) a is the critical point; b is the triple point.
39. 10 g   **41.** 46.3 g   **43.** 2 g   **45.** 250 W
47. (a) 0.12 K/W   (b) 2.7×10^{-4} K/W   (c) 5.0×10^{-5} K/W
49. 6.67 W/m²   **51.** (a) 0.32 W   (b) 800 K/m   (c) 0.16 W
(d) 0.64 W   (e) 64°C   **53.** −37°C
55. (a) the skier with the down jacket   (b) The person with the
down jacket can stay outside 6.4 times longer.   **57.** 105 W
59. 2.24 kW   **61.** 4.8×10^{-5} m²   **63.** (a) 8.9×10^{-3}°C/s
(b) 9.4 min   **65.** 1.70°C/s   **67.** (a) 103.81°C   (b) 0.565 g/s
69. 0.58 L/h   **71.** 0.189 kcal/(kg·K)   **73.** 4.0 g
75. (a) 2.4 kcal   (b) 350 g   **77.** 6.6×10^{-3}°C
79. reduced to 75% of the original

81.

Animal	(a) BMR/kg	(b) BMR/m²
Mouse	210	1200
Dog	51	1000
Human	32	1000
Pig	18	1000
Horse	11	960

(a) It is true.   (c) Radiative loss depends upon surface area.

83. 0.23 kg **85.** 7.86 g **87.** 140 m **89.** 22.6°C
91. (a) 7.00 times higher (b) 35.7°C; the dog is a much better regulator of temperature and, as a result, has more endurance.

Chapter 15

Multiple Choice Questions

1. (d) **2.** (d) **3.** (c) **4.** (c) **5.** (a) **6.** (c) **7.** (a) **8.** (c)
9. (c) **10.** (d) **11.** (e) **12.** (b) **13.** (d)

Problems

1. −290 J **3.** 95 J **5.** 101.3 J **7.** (a) 1372 J
(b) $\Delta U = 1216$ J; $Q = 2588$ J **9.** −5.00 kJ; the heat flows out of the gas and into the reservoir. **11.** (a) 3.00 kJ (b) 2.00 kJ
13. (a) 1.2×10^{17} J (b) 1.4×10^{13} kg **15.** 0.182
17. 3.00 kJ **19.** 6.0 kcal **21.** 14 W **23.** The coal-fired plant and the nuclear plant exhaust 0.43 MJ and 0.60 MJ of heat, respectively. **25.** (a) 433 K (b) 233 J **27.** 4.2%
29. 12 kJ **31.** 0.0174 **33.** 110 kJ **35.** 62.7 kW
37. (c), (a), (b) **39.** +1.4 kcal/K **41.** 9.81 cal/K
43. 56.6 cal/K **45.** (a) 100 W (b) 0.3 W/K
47. $1.79k_B$ **49.** (a) 3 heads and 3 tails (b) 6 heads or 6 tails
(c) $\dfrac{20}{64} = \dfrac{5}{16} = 0.3125$ **51.** $k \ln 2$ **53.** 31 kg **55.** 4.84 kJ
57. decreasing the low temperature reservoir
59. (c), (a), (b), (d) **61.** 0.079 J **63.** 24°C

65.

W (J)	e (%)
9.27×10^{-4}	40.0
5.91×10^{-4}	26.5
9.06×10^{-5}	4.67

67. 350 J/K

69. (a) 0.22 cal/K (b) −0.68 cal/K **71.** 15 min
73. (a) 7.61 J/K (b) −5.57 J/K (c) 2.04 J/K

Chapter 16

Multiple Choice Questions

1. (j) **2.** (d) **3.** (e) **4.** (a) **5.** (c) **6.** (b) **7.** (d) **8.** (c)
9. (b) **10.** (b)

Problems

1. 9.6×10^5 C **3.** (a) added (b) 3.7×10^9
5. (a) negative charge (b) an equal magnitude of positive charge **7.** -5.0×10^{-7} C **9.** 2.268×10^{39} **11.** 1.2 N at 28° below the negative x-axis **13.** 0.020 nC
15. $\dfrac{kq^2}{2d^2}$ to the left **17.** $16F$ **19.** $-\dfrac{3\sqrt{3}}{4}q \approx -1.3q$
21. (a) into the cell (b) 1.6×10^{-12} N

23.

$25.$ 3.2×10^{12} m/s² up
$27.$ $\dfrac{kq}{2d^2}$ in the +x-direction
$29.$ 400 N/C
$31.$ −0.43q

33.

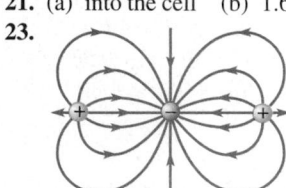

35. (a) The electric field is directed from the top plate to the bottom plate. (b) 1.77×10^{-7} C/m² (c) 0.44 mm
37. (a) vertically downward (b) 2600 N/C (c) 4.3×10^{-17} m
39. −5 μC **41.** (a) 7 μC (b) −11 μC **43.** (a) $4\pi r^2 E$
45. 0.866EA **47.** (a) 0 (c) No—applying the superposition

principle to two parallel sheets of charge gives the same result.

51. (a)

(b) $E(r \le a) = 0$;
$E(a < r < b) = \dfrac{2k\lambda}{r}$;
$E(r \ge b) = 0$

53. (a) $E(r_1 < r < r_2) = \dfrac{\sigma_1}{\epsilon_0}\left(\dfrac{r_1}{r}\right)$

(b)

55. 1.3 m

57.

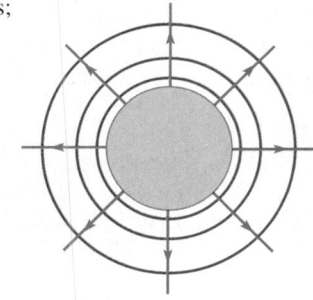

59. 3.20×10^{-14} N upward
61. 2.2×10^6 m/s
63. (a) $|Q_S| = 1.712 \times 10^{20}$ C and $|Q_E| = 5.148 \times 10^{14}$ C
(b) No, the force would be repulsive.

65. (a) $\dfrac{kqd}{x^3}$ (b) negative y-direction for all x **67.** −1.5 nC
69. 0.80 **71.** (a) 2 mN (b) Coulomb's law is only valid for point charges or when the sizes of the charge distributions are much smaller then their separation. (c) smaller
73. (a) $\sqrt{\dfrac{9\eta v_t}{2(\rho_{oil} - \rho_{air})g}}$ (b) $\dfrac{4\pi R^3(\rho_{oil} - \rho_{air})g}{3E}$

Chapter 17

Multiple Choice Questions

1. (a) **2.** (f) **3.** (c) **4.** (e) **5.** (d) **6.** (b) **7.** (f) **8.** (e)
9. (b) **10.** (b) **11.** (b) **12.** (d)

Problems

1. −18 mJ **3.** 2.3×10^{-13} J **5.** 2.8 mJ
7. $\vec{E} = 0$; $V = 2.3 \times 10^7$ V **9.** (a) −1.5 kV (b) −900 V
(c) 600 V; increase (d) -6.0×10^{-7} J; decrease
(e) 6.0×10^{-7} J **11.** (a) positive (b) 10.0 cm
13. (a) $\sqrt{3}R_0$ (b) $3R_0$ **15.** (a) $V_a = 270$ V; $V_b = -160$ V
(b) −430 V (c) -6.5×10^{-7} J **17.** (a) $V_a = 1350$ V;
$V_b = -899$ V; $V_c = 0$ (b) $W_{a\text{ to }b} = -4.49$ μJ; $W_{b\text{ to }c} = 1.80$ μJ
19. (a) 1.0 μN to the right (b) 0.25 μJ (c) 60 V **21.** 1.0 cm
23. cylinders;

25. (a)

(b)

27. 1.6×10^{-19} C $\approx e$ **29.** 150 V   **31.** 0

33. (a) upward   (b) $\dfrac{v_y m d}{e \, \Delta V}$ (c) decreases

35. 612 µC   **37.** (a) stays the same   (b) increases

39. 1500 km   **41.** 300 m   **43.** 83 pF   **45.** 89 nF

47. (a) 0.18 µF   (b) 8.9×10^8 J   **49.** (a) 18 nC   (b) 1.3 µJ

51. (a) 630 V   (b) 0.063 C   **53.** 0.27 mJ   **55.** (a) 0.14 C

(b) 0.30 MW   **57.** (a) 10.0 GJ   (b) 443 kg   (c) 0.694 mo

59. (a) $U_a = -6.3$ µJ; $U_b = U_c = 0$ (b) -6.3 µJ   **61.** 450 kV

63. (a)

(b)

65.

67. (a) 7.0×10^6 m/s upward
     (b) 7.0 mm
69. (a) 6 nC
     (b) 3×10^{10} ions
71. (a) 1.25 µm   (b) 1600 m²
73. 8.0 V/m
75. (a) 2.7 kV   (b) 6.8 µJ

77. The energy is reduced by 20.0%.   **79.** 1.44×10^{-20} J
81. (a) $U_i = 20.0$ mJ; $U_f = 3.3$ mJ
(b) positive work to remove the dielectric

Chapter 18

Multiple Choice Questions

1. (a)   **2.** (d)   **3.** (f)   **4.** (d)   **5.** (c)   **6.** (b)   **7.** (b)   **8.** (d)
9. (d)   **10.** (b)

Problems

1. (a) 5.00 C   (b) 3.12×10^{19} electrons   **3.** 9.6 µA
5. 8.0 mA   **7.** scooter: 48 kJ; automobile: 360 kJ
9. (a) 6.480×10^5 C   (b) 7.78 MJ   (c) 54.5 h   **11.** (a) 260 J
(b) 8.1 mW   **13.** 81 µm   **15.** 50 h   **17.** 0.25 mV
19. (a) 15 Ω   (b) 10 Ω   **21.** 2.5 mm   **23.** 3.8×10^{-3}°C^{-1}
25. 5 A   **27.** (a) 1.1   (b) 0.48   (c) aluminum   **29.** (a) 30 Ω
(b) 30 Ω   (c) 80°C   (d) The electric field stays the same, the
resistivity decreases, and the drift speed increases.   **31.** 7.0 V
33. 23 µF   **35.** 5 Ω   **37.** 3 Ω   **39.** 3 Ω   **41.** 8.0 µF

43.

Branch	I (mA)	Direction
Top	85	right to left
Middle	56	left to right
Bottom	29	left to right

45. The current through the 5.00-Ω resistor flows upward and is
0.195 A. The current through the 4.00-Ω resistor flows right to left
and is 0.245 A. The emf is 0.13 V.

47. (a)

Branch	I (A)	Direction
alternator	22.1	up
battery	7.9	down
R	14.3	down

; 12.1 V; charging

(b) $R < 0.51$ Ω

49. 20 W   **51.** 360 Ω   **53.** 1.13 W   **57.** $P_4 = 4.82$ W;
$P_5 = 1.36$ W   **59.** (a) $\mathscr{E} - Ir$ (b) $P = \mathscr{E}I - I^2 r$; $\mathscr{E}I$ is the rate
that an ideal emf supplies power, and $I^2 r$ is the rate of energy
dissipation by the internal resistance.   (c) $\mathscr{E}I$ (d) $\mathscr{E}I + I^2 r$
61. 833 kΩ   **63.** (a) 25 kΩ   (b) 250 kΩ   **65.** (a) 6.27 mA
(b) 5.37 mA   **67.** (a) 140 µF; 90 Ω; 5.8 mJ   (b) 4.4 ms
(c)

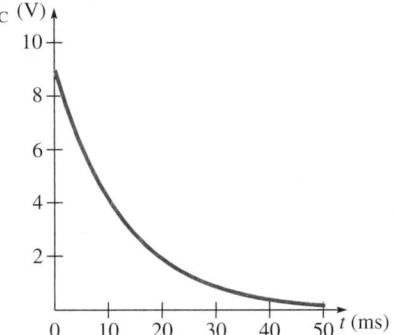

69. (a) 632 V   (b) 63.2 mC   (c) 6.7 Ω
71. (a) 140 μF; 18 Ω; 2.4×10^{-4} C   (b) 1.7 ms
73. (a)

(b) 5.0 min

75. (a) 4.6τ   (b) 6.9τ   (c) 4.6τ   **77.** (a) 50 mA
(b) 7.4 mA   **79.** (a) *D*   (b) The devices cannot all be operated at the same time since the total current would be 25 A, which is greater than the rated 20.0 A.   **81.** 6.5 kJ   **83.** 3 A
85. (a) 1.91 A   (b) 0.907 A   **87.** (a) 175 μJ
(b) $Q_2 = 16.0$ μC; $Q_3 = 24.0$ μC; 160 μJ   (c) heat loss
89. $P_1 = 6$ mW; $P_2 = 16$ mW   **91.** (a) (a), (e), and (f)
(b) Bulb 3 is the brightest, and bulb 1 and bulb 2 are the same.
(c) $I_3 = 0.25$ A; $I_1 = I_2 = 0.13$ A
93. (a)

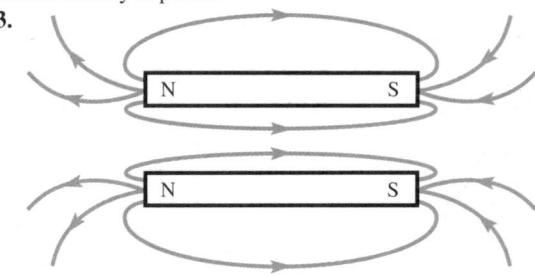

(b) No, because the 1.5-V non-rechargeable battery is being recharged.

95. $9R_0$ **97.** $v_{Au} = 3v_{Al}$ **99.** 20 W   **101.** (a) $R_{60} = 240$ Ω; $R_{100} = 140$ Ω   (b) 60.0-W bulb; $P = I^2R$ and $I_{60} = I_{100}$, and since $R_{60} > R_{100}$, $P_{60} > P_{100}$.   (c) 100.0-W bulb; $P = \dfrac{V^2}{R}$ and $V_{60} = V_{100}$, and since $R_{60} > R_{100}$, $P_{100} > P_{60}$.   **103.** (a) 2×10^5 W
(b) Cu: 1.2 cm; Al: 1.5 cm   (c) Cu: 1.0 kg/m; Al: 0.48 kg/m
105. (a) 2 mA   (b) 3 mA   (c) 3 mW
(d) increase by a factor greater than 2   **107.** (b) $R = \dfrac{8\eta L}{\pi r^4}$
(c) yes   (d) yes   **109.** (a) 8.00 V
(b) No current passes through the source.   **111.** (a) $C\mathscr{E}$
(b) $\dfrac{1}{2}C\mathscr{E}^2$ (c) $C\mathscr{E}^2$ (d) Energy is dissipated by the resistor.

Chapter 19

Multiple Choice Questions

1. (g)   **2.** (f)   **3.** (e)   **4.** (e)   **5.** (c)   **6.** (b)   **7.** (c)   **8.** (g)
9. (b)   **10.** (b)   **11.** (d)   **12.** (d)

Problems

1. (a) *F*   (b) *A*; highest density of field lines at point *A* and lowest density at point *F*
3.

5.

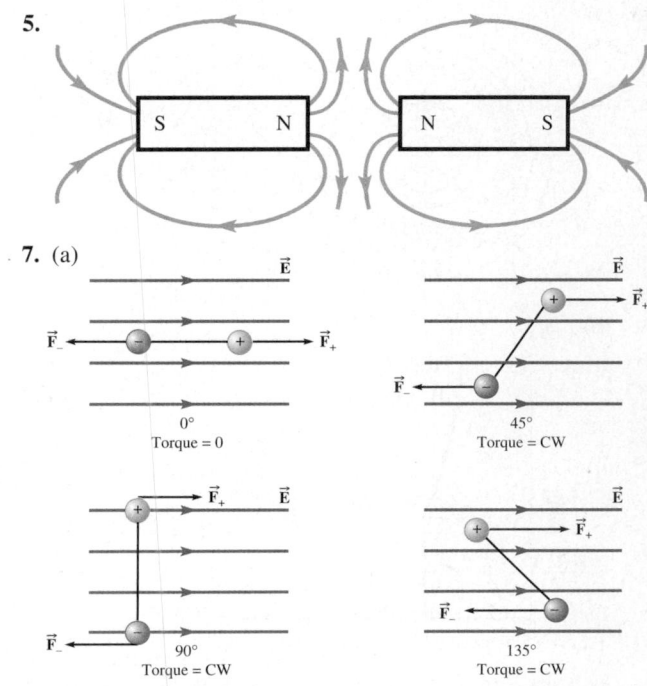

7. (a)

(b) parallel to the electric field lines
9. 7.4×10^{-12} N east   **11.** (a) 3.0×10^7 m/s   (b) 4.2×10^{18} m/s^2
(c) 3.0×10^7 m/s   (d) 2.4×10^7 V/m   (e) Since the force due to the magnetic field is always perpendicular to the velocity of the electrons, it does not increase the electrons' speed but only changes their direction.   **13.** 12 cm into the page   **15.** There are two possibilities: 56° N of W and 56° N of E.   **17.** 8.0×10^{-13} N
19. 0.21 m   **21.** 13.0 u   **23.** (a) 14 u   (b) nitrogen
25. (a) 29 cm   (b) $f_{12}/f_{14} = 1.17$ (c) 27 cm; $f_{12}/f_{14} = 1.17$
27. $\dfrac{E^2}{2B^2\Delta V}$ **29.** 0.48 mm/s   **31.** (a) No; the probe measures the perpendicular component of the field.   (b) $V_H = 0$
33. (a) 0.35 m/s   (b) 4.4×10^{-6} m^3/s   (c) the east lead
35. (a) 2.2 N   (b) Only the maximum possible force can be calculated since only the magnitudes, and not the directions, of $\vec{B}$ and $\vec{L}$ are given.   **37.** 0.072 N north   **39.** (a) $\vec{F}_{top} = 0.75$ N in the $-y$-direction; $\vec{F}_{bottom} = 0.75$ N in the $+y$-direction; $\vec{F}_{left} = 0.50$ N in the $+x$-direction; $\vec{F}_{right} = 0.50$ N in the $-x$-direction   (b) 0
41. (a) 18° below the horizontal with the horizontal component due south   (b) 42 A   **43.** 0.0013 N·m
45. (a) 0.0014 N·m   (b) 0
49.

51. (a) 9×10^{-8} T out of the page
(b) 1.6×10^{-5} T out of the page
53. 3.2×10^{-16} N parallel to the current
55. 750 A into the page
57. (a) $5I$ out of the page
(b) $2I$ into the page

61. (c) S and (d) N   **63.** (a) 1.1 mA   (b) 9.3×10^{-24} A·m^2
(c) The orbital and intrinsic dipole moments are the same.
65. (a) (a)   (b) (c)   **67.** 21 cm   **69.** $\dfrac{2\pi m}{qB}$ **71.** (a) 1.1 mm/s
(b) 17 μV   (c) left side   **73.** (a) positive   (b) west
(c) north   (d) $\Delta V_1 = 1.6$ kV; $\Delta V_2 = 320$ V   (e) $\Delta V_1 = 1.4$ kV; $\Delta V_2 = 280$ V   **75.** (a) $\vec{F}_{top} = 0.65$ N into the page; $\vec{F}_{bottom} = 0.65$ N out of the page; $\vec{F}_{right} = 0.25$ N out of the page; $\vec{F}_{left} = 0.25$ N into

the page   (b) 0   **77.** 5 A   **79.** $-y$-direction   **81.** 6.4×10^{-14} N

at 86° below west   **83.** (a) $\dfrac{\epsilon_0 A m v^2}{2de}$; the west plate is positive

and the east plate is negative.   (b) vB north

(c)

Ion trajectory $\vec{E}$ $\vec{B} \odot$

(d) $\dfrac{eBD}{2v}$

85. 52.4 cm

Chapter 20

Multiple Choice Questions

1. (c)   **2.** (c)   **3.** (d)   **4.** (b)   **5.** (d)   **6.** (e)   **7.** (e)   **8.** (b)
9. (b)   **10.** (c)

Problems

1. (a) $\dfrac{vBL}{R}$ (b) CCW   (c) left   (d) $\dfrac{vB^2L^2}{R}$ **3.** (a) $\dfrac{vB^2L^2}{R}$

(b) $\dfrac{v^2B^2L^2}{R}$ (c) $\dfrac{v^2B^2L^2}{R}$ (d) Energy is conserved since the
rate that the external force does work is equal to the power
dissipated in the resistor.   **5.** (a) toward the left (front) end

(b) up the incline   (c) $\dfrac{ma_0R}{L^2B^2}$ **7.** (a) $\mathscr{E} = 0$; $I = 0$; no magnetic

force   (b) 32 nV; 400 pA CW; 2.9×10^{-17} N to the left
(c) 1.3×10^{-17} W   **9.** 5.2×10^{-4} Wb   **11.** (a) CW
(b) away from the long straight wire   (c) 2.0 Wb/s
13. (a) 0.070 A   (b) CCW   **15.** (a) CW   (b) 8.0 V
(c) yes; upward   (d) R decreases; I increases; F_M increases
17. CCW as viewed from the left   **19.** 110 V

21. (a) $\dfrac{1}{20}$ (b) 1000   **23.** 2.00   **25.** (a) 14   (b) 29 mA

27.

29. 8.1 mN/C tangent to the loop   **31.** 8.5×10^{-4} H
33. (a) 3.1×10^{-5} V   (b) 14 mV   **39.** $L_{eq} = L_1 + L_2$
41. (a) 38 mJ   (b) −7.5 W   (c) −38 mW   (d) 69 ms
43. (a) $I_1 = 1.7$ mA; $I_2 = 0$; $V_{3.0} = 0$; $V_{27} = 45$ V; $P = 75$ mW;
$V_L = 45$ V   (b) $I_1 = 1.7$ mA; $I_2 = 15$ mA; $V_{3.0} = V_{27} = 45$ V;
$P = 0.75$ W; $V_L = 0$ **45.** (a) 180 mA   (b) 2.5 mJ   (c) 1.1 W
(d) 0   **47.** (a) 50 A   (b) The attempt to suddenly stop the
current would induce a huge emf in the windings of the
electromagnet, possibly damaging it. It is likely that sparks would
complete the circuit across the open switch. The shunt resistor is
used to reduce the rate of change of the current, and allow the
electromagnet to be safely shut off.   (c) 50 kW   (d) 2.3 s

(e) yes; $\tau \propto \dfrac{1}{R_L + R_{shunt}}$ **49.** (a) $[L] = $V·s/A; $[R] = $V/A;

$[\mathscr{E}] = $V   **51.** (a) $2\pi r^2 NB$ (b) $\dfrac{2\pi r^2 NB}{\Delta t}$ (c) $\dfrac{2\pi r^2 NB}{R\,\Delta t}$

(d) $\dfrac{2\pi r^2 NB}{R}$ **53.**

55. $\dfrac{\mu_0 N^2 a^2}{2\pi R}$ **57.** (a) $\dfrac{\mu_0 N_1 N_2 \pi r^2}{L}$ (b) $\dfrac{\mu_0 N_1 N_2 \pi r^2 I_m \sin \omega t}{L}$

(c) $\dfrac{\mu_0 N_1 N_2 \pi r^2 \omega I_m \cos \omega t}{L}$ **59.** 250 rad/s   **61.** 68 nWb

63. (a) 3.1 V   (b) the northernmost wing   **65.** (a) 0.6 GJ/m^3

(b) 1×10^{10} V/m   **67.** 85   **69.** (a) $\dfrac{\mu_0 N_1 N_2 \pi r^2}{\ell}$

(b) $\dfrac{\mu_0 N_1 N_2 \pi r^2}{\ell} \dfrac{\Delta I_1}{\Delta t}$ **71.** (a) No; for DC the magnetic field is

constant so no emf is induced.   (b) No; the currents flow in
opposite directions.

Chapter 21

Multiple Choice Questions

1. (i)   **2.** (b)   **3.** (a)   **4.** (d)   **5.** (a)   **6.** (d)   **7.** (c)   **8.** (c)
9. (c)   **10.** (b)

Problems

1. 120 times per second   **3.** 18 A   **5.** 4.0 kW   **7.** (a) 0
(b) 6.25 A^2 (c) 3.54 A   **9.** (a) 12 Ω   (b) 10 A   (c) 2.4 kW
11. (a) 119.5 V; 59.50 W   (b) $V_{bulb} = 111.0$; $P_{bulb} = 51.34$ W;
$V_{A \text{ to ground}} = 4.5$ V   (c) An appliance that draws a large startup
current causes a larger-than-usual voltage drop across the wiring.
(d) Normally, the ground wire carries no current and therefore is
all at the same potential. The neutral wire does carry current, so
there is a potential difference between its ends.   **13.** (a) 60.0 Hz

(b) 13.3 kΩ   **15.** 53 nF   **17.** (a) $\dfrac{2\omega CV}{\pi}$ (b) $\dfrac{\omega CV}{\sqrt{2}}$

(c) The rms current is the square root of the average of the *square*
of the AC current. Squaring tends to emphasize higher values of
current, so they contribute more to the resulting average than
lower current values.

19.

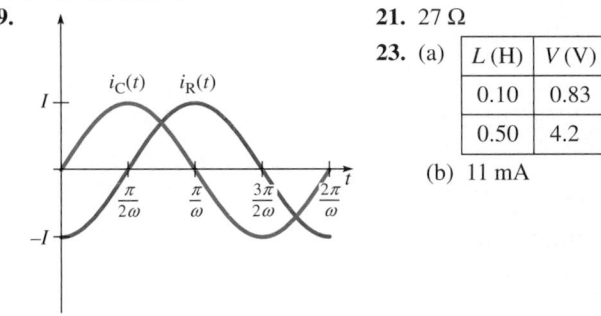

25. (a) $\omega LI \cos \omega t$

21. 27 Ω

23. (a)

L (H)	V (V)
0.10	0.83
0.50	4.2

(b) 11 mA

(c) $v_L(t)$ leads $i(t)$ by $\frac{\pi}{2}$.

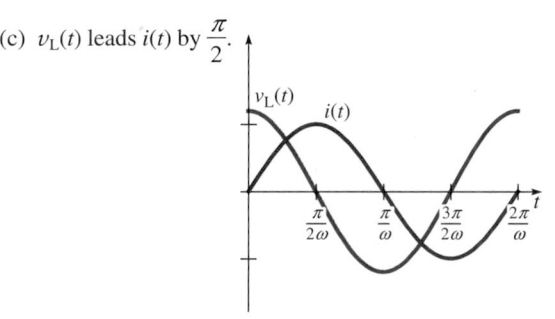

27. (a) negative   (b) $-\underline{}\mathrm{mmm}\,+$ (d) positive; $+\mathrm{mmm}\,-$

29. 71.2 mH   31. (a) $V_L = 919$ V; $V_R = 771$ V
(b) no; $\mathcal{E} = \sqrt{V_C^2 + V_R^2}$
(c)

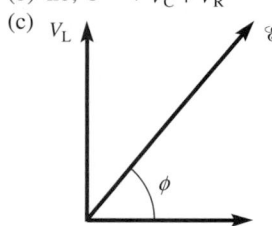

33. 500 Ω
35. (a) 32 Hz
(b) no; $\dfrac{V_R}{\mathcal{E}} = \dfrac{V_L}{\mathcal{E}} = \dfrac{1}{\sqrt{2}}$
(c) $\mathcal{E}$ leads I by $\dfrac{\pi}{4}$ rad $= 45°$.
(d) 210 Ω
37. (a) 0.71   (b) 44°
39. 8.0×10^{-7} F; 310 Ω

41. (a) $V_2 \sin \phi_2 = V \sin \phi$; the result indicates that the y-component of V_2 is equal to the y-component of V.
(b) $V_1 + V_2 \cos \phi_2 = V \cos \phi$; the result indicates that the sum of the x-components of V_1 and V_2 is equal to the x-component of V.
43. (a) 3.18×10^5 rad/s or 50.6 kHz   (b) 2.4 kV   45. 500.0 Ω
47. (a) 10 kHz   (b) 840 Hz   (c) Low-pass filter; for low frequencies, $X_C \gg R$, so V_C (= output voltage) is approximately equal to the input voltage.   49. 650 μF   51. 120 times per second   53. (a) 1.1×10^{10} W   (b) Aluminum is cheaper and less dense.   55. (a) 4.53 kΩ (b) 24 mA   57. 1.0 mA   59.
(a) $X_{C1} = 2.65$ kΩ; $X_{C2} = 21.2$ Ω (b) $Z_1 = 3.32$ kΩ; $Z_2 = 2.00$ kΩ (c) 1.00 mA   (d) The current leads the voltage; 1.50 kHz: 0.608°; 12.0 Hz: 53.0°.   61. 11 kHz   63. (a) 33.8 Ω
(b) $V_{peak} = 286$ V; $V_{rms} = 202$ V   (c) 8.46 A   (d) 1.07 kW
(e) $i(t) = (8.46\ \text{A}) \sin [(360\ \text{rad/s})t - 0.480\ \text{rad}]$ 65. 390 pF
67. (a) 0.95   (b) 470 Ω (c) 4.2 A   (d) 4.0 kW
69. (a) 6.4 MW   (b) 0.12 MW; 15 kV: 53%; 110 kV: 1.0%
71. (a) 53 kV·A   (b) no   (c) The load factor may be less than 1 since the transformer may have to supply a reactive load. Even though the load draws less power than a purely resistive load, the transformer must supply the same current and has the same heating in its windings and core.

Chapter 22

Multiple Choice Questions

1. (d)   2. (f)   3. (b)   4. (f)   5. (d)   6. (b)   7. (a)   8. (c)
9. (d)   10. (d)

Problems

1. $\dfrac{\mu_0 I}{2\pi r}$ 3. $\dfrac{\mu_0 I r}{2\pi R^2}$ 5. east-west   9. 3.3 m
11. (a) 5.00×10^6 m   (b) The radius of the Earth is 6.4×10^6 m, which is close in value to the wavelength.   (c) radio waves
13. (a) about one octave   (b) approximately 3 octaves
15. 1.62   17. 1.67 ns   19. (a) 455 nm   (b) 4.34×10^{14} Hz
23. (a) 7.5 mV/m; 3.0 MHz   (b) 4.5 mV/m; in the $+x$-direction
25. (a) $+y$-direction   (b) $B_x = \dfrac{E_m}{c} \sin (ky - \omega t + \pi/6)$, $B_y = B_z = 0$
27. (a) 4.7×10^{-6} J/m^3 (b) 730 V/m; 2.4×10^{-6} T

29. 9×10^{26} W   33. (a) 4.2 kW   (b) 840 W
(c) Excess energy produced at midday should be stored for use at night; a supplemental energy source is necessary.   35. 0.25
37. (a) $I_1 = I_0 \cos^2 90.0° = 0$ (b) $0.250 I_0$ (c) $0.531 I_0$
39. (a) $0.125 I_0$ (b) 0   41. The star is moving at 680 km/s away from the Earth.   43. 5×10^7 m/s   45. 7.5 cm
47. (a) infrared   (b) 6.00 mm   (c) 4.51 mm   (d) 5660
(e) 1.20 pJ   51. 2.2×10^4 rad/s, 4.4×10^4 rad/s, 6.6×10^5 rad/s
53. (a) 8.0×10^5 W/m^2 (b) 1.8×10^{-9} W/m^2

Chapter 23

Multiple Choice Questions

1. (b)   2. (d)   3. (f)   4. (c)   5. (d)   6. (c)   7. (b)   8. (d)
9. (a)   10. (d)

Problems

1.

3.

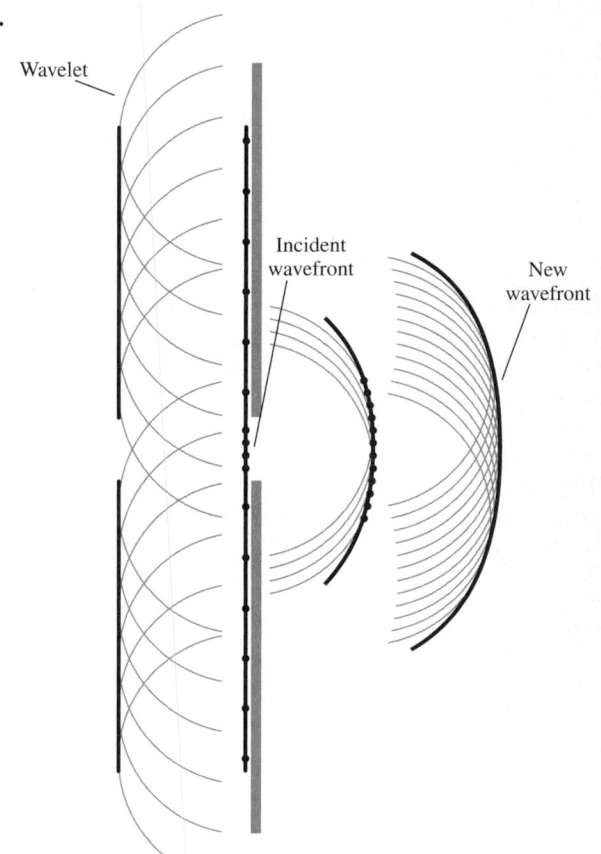

incident side: two planar waves;
transmitted side: one hemispherical wave

5.

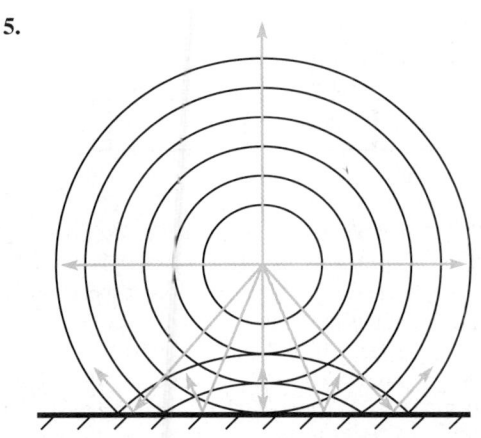

7. 100° **11.** 63.1° **13.** 44.6° **15.** 1.83° **17.** $\delta = \beta(n-1)$
19. $44.1° \le \theta \le 45.9°$ **21.** (a) 24.42° (b) 33.44°
(c) Under water, the larger critical angle means that fewer light rays are totally reflected at bottom surfaces of the diamond. Thus, less light is reflected back toward the viewer. **23.** Since the angle of incidence (45°) is greater than the critical angle (39°) no light exits the back of the prism. The light is totally reflected downward, and then passes through the bottom surface ($\theta_i < \theta_c$), with a small amount reflected back into the prism. **25.** no
27. (a) 36.88° (b) perpendicular to the plane of incidence
(c) yes; 53.12° **29.** (a) The critical angle is always greater than Brewster's angle, regardless of n_1 and n_2. (b) no critical angle; $\theta_B > 45°$ **31.** 10.8 mm **33.** 0.82 m **35.** 2.25 m
37. He sees 7 images total.
41. 6.67 cm in front of the mirror

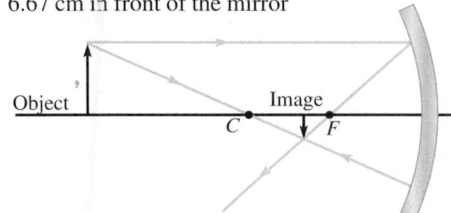

43. 12.5 cm in front of the mirror

49.

51.

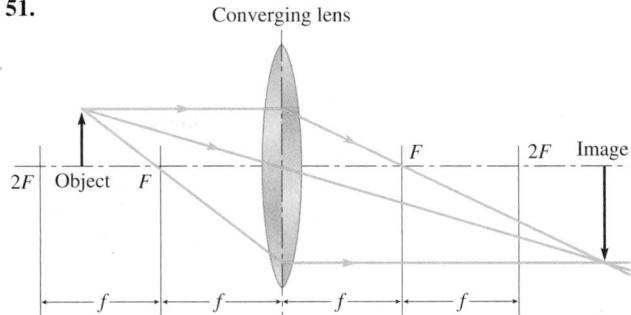

53. 18 cm

55. (a)

p (cm)	q (cm)
5.00	−3.08
8.00	−4.00
14.0	−5.09
16.0	−5.33
20.0	−5.71

; All images are virtual, upright, and diminished in size.

(b) 5.00 cm: 2.46 m; 20.0 cm: 1.14 cm

57.

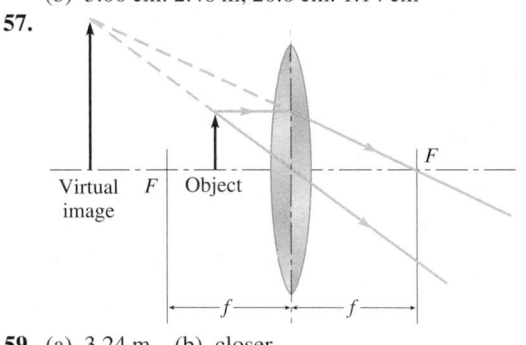

59. (a) 3.24 m (b) closer
61.

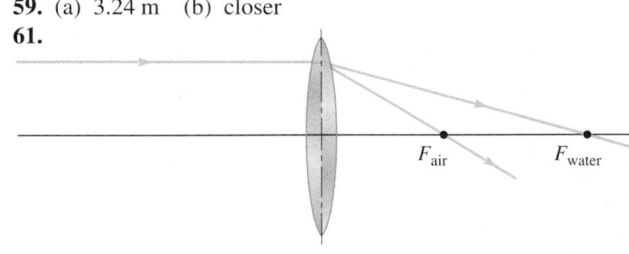

63. 19.2° **65.** no; blue **67.** (a) total internal reflection at 75°
(b) reflection at 25° and transmission at 34° **69.** 8.0 km/h
71. pin-image distance = 40.0 cm; pin-mirror distance = 10.0 cm;

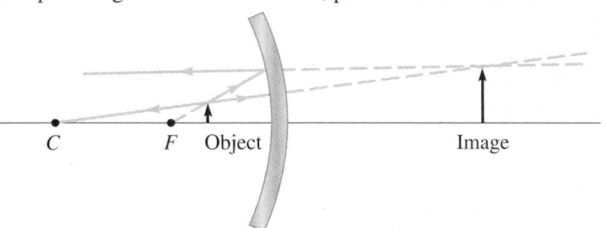

73. 15.0 cm; 60.0 cm behind the lens **75.** 15.0 cm; converging
81. red and yellow.
83. 32 cm in front of the mirror

85.

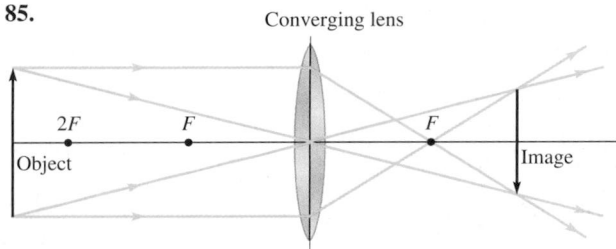

87. 50 cm

Chapter 24

Multiple Choice Questions

1. (d)   **2.** (b)   **3.** (b)   **4.** (a)   **5.** (f)   **6.** (b)   **7.** (b)   **8.** (b)
9. (d)   **10.** (c)

Problems

1. (a) 2.5 cm past the 4.0-cm lens; real   (b) −0.27   **3.** 15.6 cm
to the left of the diverging lens   **5.** $q_1 = 12.0$ cm; $q_2 = -4.0$ cm;
$h'_1 = -4.00$ mm; $h'_2 = 4.0$ mm   **7.** (a) 50.8 mm   (b) −0.0169
(c) 20.3 mm   **9.** 72 m   **11.** 280 mm   **15.** 1.0 m   **17.** 1.64 cm
19. (a) 3.1 cm   (b) 3.1 cm   **21.** (a) 2.27 cm   (b) 11
(c) 33 cm   **23.** (a) $\dfrac{Nf}{N+f}$ (c) $M = \dfrac{N}{f} + 1 = M_\infty + 1$ **25.** 50
27. 1.63 cm   **29.** (a) 19.0 cm   (b) −318   (c) 5.16 mm
31. 4.52 cm   **33.** (a) 2.56 m   (b) 2.17 cm   (c) −15
35. (a) $q_1 = 60.0$ cm; $q_2 = 5.45$ cm   (b) −1.64   (c) 4.91 cm
37. (a) Lens 1 is the objective and lens 2 is the eyepiece.
(b) 30.0 cm   **39.** 23.9 m   **41.** (a) the lens with the 30.0-cm
focal length   (b) −10   (c) 33.0 cm   **43.** $f_e = 5.1$ cm; $f_o = 61$ cm
45. (a) real   (b) converging   (c) 51 mm   (d) 17 mm
(e) 2.6 mm
47. (a)   (b) 22.2 cm

49. (a) inverted   (b) −0.42   (c) The cone of light rays,
diverging from each point on the object, is larger, so the eye can
detect that the rays do not converge to a single point.
(d) converging   (e) 2.0 m   **51.** (a) 17 cm   (b) −14
(c) 7.4 cm   **53.** (a) 3.0   (b) 4.0   **55.** (a) 56 D   (b) 61 D
(c) 4.0 D

Chapter 25

Multiple Choice Questions

1. (a)   **2.** (e)   **3.** (a)   **4.** (d)   **5.** (e)   **6.** (e)   **7.** (d)   **8.** (a)
9. (e)   **10.** (a)

Problems

1. $5I_0$ **3.** 0°: 446 nm; 10.0°: 453 nm; 30.0°: 515 nm
5. (a) 3.2 cm   (b) 1.1   **7.** 86.7 μm   **9.** 105 nm   **11.** 96 nm
13. (a) 683 nm, 546 nm, and 455 nm   (b) 607 nm, 496 nm,
and 420 nm   **15.** 667   **19.** maxima: ±6.9 cm from the central
maximum; ±30°; yes, they agree.   **21.** 456 nm   **23.** 711 nm
25. 31.1°   **27.** 5000 lines/cm
29. (a) five

(b)

31. (a) two; 449.2 nm and 651 nm   (b) 18 lines   **33.** (a) three
(b) $\theta_{b1} - \theta_{a1} = 0.04°$, $\theta_{b2} - \theta_{a2} = 0.11°$, $\theta_{b3} - \theta_{a3} = 0.91°$
(c) The two lines are best resolved by the third-order spectra
since the angular separation is the largest.   **35.** 0.50 cm
37. 6.3 mm   **39.** (a) wider   (b) 3.3 cm   **41.** 0.012°
43. 0.33 mm   **45.** (a) $\sin \Delta\theta = n \sin \beta$ **47.** 20 km   **49.** 1.7 μs
51. (a) 0   (b) 0.30 km   (c) There will be 12 maxima total.
(d) 0; 19°; 42°; 90°   (e) The answers in parts (a), (b), and (c)
will be unchanged. The angles calculated in part (d) for θ_0 and θ_3
will be unchanged, but θ_1 and θ_2 will be different than before.
53. (a) 100 nm   (b) 280 nm; 140 nm   (c) Yes; although
perfectly constructive interference does not occur for any visible
wavelength, some visible light is reflected at all visible
wavelengths except 560 nm (the only wavelength with perfectly
destructive interference).   **55.** $t = \dfrac{\lambda}{2n_{coating}}$ **57.** 1.6
59. (a)

(b)

(c)

61.

65. shorter than 32.4 cm

Chapter 26

Multiple Choice Questions

1. (e) 2. (d) 3. (a) 4. (a) 5. (c) 6. (a)

Problems

1. 22 µs 3. (a) 0.87c (b) c 5. (a) 31 ns (b) 5.6 m
7. (a) 30 years old (b) 3420 9. 7.7 ns 11. (a) 79 m
(b) 610 ns (c) 530 ns 13. 13 m 15. (a) 1.0 m (b) 0.92 m
17. (a) 7.5 µs (b) 13 µs 19. 6.0 km 21. 0.946c
23. $\frac{1}{5}c$ 25. $\frac{5}{13}c$ 27. 1.04×10^{-18} kg·m/s

29. (a) 6.7×10^9 kg·m/s (b) 4.1 s 31. 9.0×10^{13} J
33. 40 MeV 35. (a) 530 MeV (b) 106 MeV 37. 0.64 MeV
39. 1 MeV/$c^2 = 1.782 \times 10^{-30}$ kg 41. 1.45 MeV/c
49. (a) 127 h (b) 127 h 51. 33.9 MeV
53. (a) 1.2×10^{13} m (b) 5.0×10^4 s 55. 0.75c
57. (a) 4500 m (b) 15 µs (c) 15 µs (d) 500,000
61. (a) 18.0 ly (b) 22.5 yr 63. (a) 32 J (b) 3.3 m
(c) $\approx (1 - 1.1 \times 10^{-23})c = 0.999\,999\,999\,999\,999\,999\,999\,989c$
67. neutron: 5.7 MeV; pion: 35.4 MeV

Chapter 27

Multiple Choice Questions

1. (e) 2. (c) 3. (d) 4. (a) 5. (a) 6. (c) 7. (e) 8. (b)
9. (e) 10. (f)

Problems

1. 1.8 eV 3. 4.30 eV 5. (a) 4.31 eV (b) 4.6 eV
7. 4.5 eV 9. (a) 6.66×10^{-34} J·s (b) 1.82 eV 11. 4.96 kV
13. 31 pm 17. (a) 2.01 pm (b) 152 pm 19. 2.4×10^{19} Hz
21. 2.4×10^4 eV 23. 0.476 nm 25. 12.8 eV
27. $F_G/F_e = 4.41 \times 10^{-40}$ 29. 3.40 eV 31. 1.09 µm
33. (a) $\frac{ke^2}{2r}$ (b) $-\frac{ke^2}{r}$ 37. 4.38×10^6 m/s 39. 11.6 eV
41. (a) 1876 nm (b) 820.6 nm (c) IR 43. 1.21 pm
45. 2.43 pm 47. (a) 10.0 mJ (b) 3.23×10^{16}
49. (a) 3.69×10^{-7} eV (b) 8.45×10^{29} photons/s 51. 6.2 pm
53. 171 pm; 176 pm 55. 4.4×10^6 m/s 57. (a) 1.2 V
(b) 0.28 V (c) 1.9 eV 59. (a) 2000 km
(b) Light is scattered by the atmosphere, and other light sources overwhelm the light from a single bulb. 61. 27.6 kV
63. (a) −54.4 eV (b) −122 eV (c) −13.6 eV
65. (a) 4.1 m/s (b) 4 ways emitting 6 different photons

(c)

Transition	λ (nm)	Class
$4 \to 3$	1875	IR
$4 \to 2$	486	visible
$4 \to 1$	97	UV
$3 \to 2$	656	visible
$3 \to 1$	103	UV
$2 \to 1$	122	UV

67. 92.4 keV
69. 0.62 nm
71. 0.01544
73. (a) 97.3 nm
(b) 102.6 nm; 656.3 nm
(c) $\lambda \le 91.2$ nm

Chapter 28

Multiple Choice Questions

1. (a) 2. (f) 3. (d) 4. (a) 5. (d) 6. (a) 7. (a) 8. (c)
9. (a) 10. (d)

Problems

1. 1×10^{-34} m; the wavelength is much smaller than the diameter

of the hoop—a factor of 10^{-34} smaller! 3. (a) 1.2 nm
(b) 39 pm 5. 1.7×10^{-24} kg·m/s 7. 1.4 km/s 9. (a) 69 pm,
35 pm, and 23 pm (b) 20.0° for all three wavelengths
11. 0.01 nm 13. (a) 0.060 eV (b) 0.060 V (c) 5 nm is an
x-ray wavelength. 15. 1×10^{-29} m 17. 0.98 eV
19. (a) 1×10^{-24} kg·m/s (b) 4 eV (c) yes 21. 2×10^{-15}
23. (a) 1.2×10^{-4} eV (b) 3×10^{-9} eV 25. 1.3×10^{-24} kg·m/s
27. (a) 10.3 nm (b) one fourth as much
29. (a)

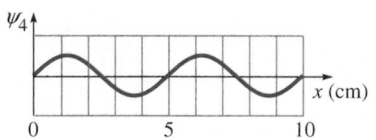

(b) 1×10^{-16} eV

31. (a)

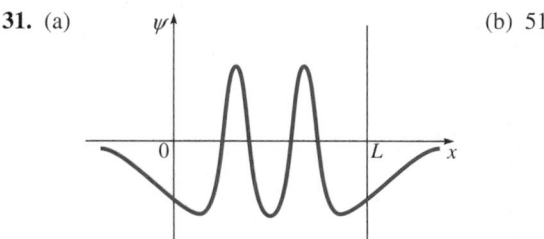

(b) 51

33. 6;

n	3	3	3	3	3	3
l	1	1	1	1	1	1
m_l	−1	−1	0	0	1	1
m_s	$-\frac{1}{2}$	$+\frac{1}{2}$	$-\frac{1}{2}$	$+\frac{1}{2}$	$-\frac{1}{2}$	$+\frac{1}{2}$

35. (b) The kinetic energy of an electron at $r = 2a_0$ is zero, so
there is no available energy to move the electron past this point.
37. For $l = 0, 1, 2, 3, 4, 5$, and 6, there are 2, 6, 10, 14, 18, 22,
and 26 electron states, respectively; the total is 98.
39. $1s^2 2s^2 2p^6 3s^2 3p^6 4s^2 3s^8$ 41. (a) F: $1s^2 2s^2 2p^5$;
Cl: $1s^2 2s^2 2p^6 3s^2 3p^5$ (b) p^5 subshell 43. (a) 525 nm
(b) green (c) 2.36 V 45. 633 nm 47. 151,000 wavelengths
49. (a) proton (b) $P_p/P_d = 120$ 51. (a) 2.21×10^{-34} m
(b) about 10^{-19} smaller (c) No; the wavelength is so much
smaller than any aperture that diffraction is negligible.
53. No; $E_{\text{photon}} = K_{\text{neutron}} = 0$ is the only possibility.

55. 1.7 kV   **57.** 3.9×10^{-6} eV   **59.** (a) 0.070 eV
(b) 0.21 eV, 0.35 eV, 0.49 eV, 0.56 eV, 0.84 eV, and 1.05 eV
(c)   (d) smaller since $E \propto 1/L^2$

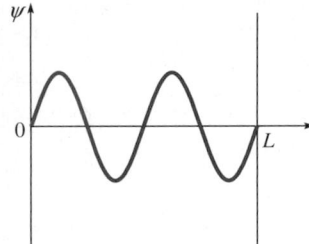

61. $\dfrac{1}{4}$ **63.** (a) 0.12 GeV   (b) Yes, since 0.12 GeV $\gg$ 28 MeV.
65. $\Delta m = 6.5 \times 10^{-55}$ kg; $\Delta m/m = 3.9 \times 10^{-28}$

Chapter 29

Multiple Choice Questions
1. (b)   **2.** (d)   **3.** (a)   **4.** (g)   **5.** (d)   **6.** (c)   **7.** (f)   **8.** (c)
9. (d)   **10.** (e)

Problems
1. 2.3×10^{17} kg/m^3 **3.** $^{88}_{38}$Sr   **5.** 18   **7.** $^{131}_{53}$I   **9.** 28.29 MeV
11. (a) 238.00032 u   (b) 1.80169 GeV
13. 8.48114 MeV/nucleon   **15.** -0.112355 u
17. (a) 1.46×10^{-8} u   (b) no   **19.** $^{40}_{20}$Ca   **21.** $^{22}_{10}$Ne
23. 4.871 MeV   **25.** β^-; n/p = 1.2   **27.** 1.312 MeV
29. (a) 10,000 s^{-1} (b) 2.308×10^7 (c) 3.466×10^{-3} s^{-1}
31. 1.2×10^7 Bq   **33.** 270 yr   **35.** 11,500 yr   **37.** 2.4 min

39. (a) $\dfrac{N}{6}$ (b) $\left(\dfrac{5}{6}\right)^3 N$ (c) $\left(\dfrac{5}{6}\right)^4 N$ (d) about 3.8 tosses
41. (a) 3.8×10^{16} photons   (b) 0.48°C   **43.** (a) $^7_3$Li; $^8_3$Li; $^8_4$Be
(b) Yes; the emission of an electron (beta-minus decay) is
accompanied by the emission of one antineutrino.
45. (a) 662 keV and 1345 keV   (b) $\dfrac{N_{0A}}{N_{0B}} = 3.28$
47. (a) 2   (b) 200 MeV   (c) 179.9 MeV   (d) 0.000822
49. 200 MeV   **51.** (a) $^3_2$He   (b) 5.6 MeV   (c) At room
temperature, the kinetic energies of the proton and the deuteron
are much too small to overcome their Coulomb repulsion.
53. 0.44 MeV   **55.** (a) 230 fm   (b) no   (c) 26 MeV
57. 6.050 MeV, 5.763 MeV, 5.618 MeV, 5.598 MeV
59. (a) 4.0   (b) 0.50   (c) 0.063   **61.** (b) 4.8 fm   (c) $\dfrac{14ke^2}{d}$
(d) 1.192 MeV   **63.** (a) 136 neutrons; 86 protons
(b) 20 alpha particles per second   **65.** (a) 1.42×10^7 m/s
(b) 4.3×10^6 V/m   (c) 98 cm   (d) Both m and q affect the
radius of the trajectory.   **67.** 0.527 eV

Chapter 30

Multiple Choice Questions
1. (d)   **2.** (f)   **3.** (a)   **4.** (b)   **5.** (d)

Problems
1. 34 MeV   **3.** 53 MeV   **5.** 0.80 GeV
7. $\pi^+ \to \mu^+ + \nu_\mu$ and $\pi^+ \to e^+ + \nu_e$ **9.** $\overline{u}\overline{u}\overline{d}$
11. 1.2×10^{-18} m   **13.** 6 pairs   **15.** 5.4 T

CREDITS

About the Authors
Photo courtesy of Phil Krasicky, Department of Physics, Cornell University.

Chapter 1
Opener: © Lucasfilm, Ltd & ™ All Rights Reserved; p. 4: © Corbis R-F Website; 1.1(1): © Science, Visuals Unlimited; 1.1(2): © Hans Gelderblom/Getty Images; 1.1(3): Photo by Jennifer Merlis; 1.1(4): © Jonathan Blair/CORBIS; 1.1(5): © Vol. 34/PhotoDisc; 1.1(6): © Vol. 56/Corbis; 1.1(7): © Vol. 56/Corbis; 1.2: © Vol. 86/PhotoDisc; 1.4: © Yorgos Nikas/Getty Images.

Chapter 2
Opener: Bernard Photo Productions/Animals/Animals; p. 27 left: AP/Wide World Photos; 2.1: © Eric Meola/Getty Images; p. 29: © Doug Pensinger/Getty Images; 2.5: © David T. Roberts/Photo Researchers. Inc.; p. 33 top: © Vol. 85/Corbis; 2.8: © Steven Walker, Peter Arnold; p. 42: Photo by author R. C. Richardson; 2.56: © Mike Hewitt/Getty Images.

Chapter 3
Opener: Courtesy Al Macdonald; p. 64: © Cosmo Condina/Getty Images; 3.25: © T. Bannor/Custom Medical Stock Photo; p. 89: © Michael S. Yamashita/ CORBIS.

Chapter 4
Opener: © John Bova/Photo Researchers; p. 104: © Vol. 35/Corbis; p. 117: © EyeWire R-F Website; 4.20: © Loren M. Winters/Visuals Unlimited; 4.49: Stouffer Productions/Animals/Animals.

Chapter 5
Opener: © Robert Landau/CORBIS; p. 149: © Vol. 29 /PhotoDisc; p. 153: © Vol. 7/Corbis; p. 159: © Chris Sattlberger/Getty Images; p. 165: © Robin Smith/Getty Images; p. 169: © Corbis R-F Website; 5.17: Supplied by Globe Photos; 5.21: © Richard T. Nowitz; 5.22: © Matthew Stockman /Allsport; p. 176: © Angelo Hornak/CORBIS.

Chapter 6
Opener: © Mark Newman/PictureQuest; 6.1: © Galen Rowell/CORBIS; p. 189: Gerard.Sioen/FRANCEDIAS.COM; 6.8a: USA Archery; p. 195: NASA; 6.16: © Stephen Dalton/Animals/Animals; p. 199: NASA; 6.19: © Craig Jones/Allsport; 6.25: © John Sohlden/Visuals Unlimited; 6.38: © James L. Amos/CORBIS; 6.41: © Corbis R-F Website.

Chapter 7
Opener: Edward William Cooke (1811–80) Private Collection/Bridgeman Art Library; 7.3: © Erica Lansner/Globe Photos; 7.8: © Vol. 56/Corbis; 7.11b: © Loren M. Winters/Visuals Unlimited; 7.11c: © Michael Steele/Allsport; 7.17: © Loren M. Winters/Visuals Unlimited; p. 243 left: © Jeffrey L. Rotman/CORBIS; 7.22: © Doug Armand/Getty Images.

Chapter 8
Opener: © R.W. Jones/CORBIS; 8.5: AP/Wide World Photos; 8.20: © The Frank Lloyd Wright Foundation; 8.27: © Mike Powell/Allsport; 8.33: © Michael Newman/PhotoEdit; p. 276: © Kevin Fleming/CORBIS; 8.37a: Photography by Leah; 8.37b: Photography by Leah; 8.50: © Doug Pensinger/Allsport; 8.74a: © Tony Duffy/Allsport; 8.74b: © Jonathan Daniel/Allsport; 8.77: © Michel Hans, Agence Vandystadt, Paris—France/Allsport.

Chapter 9
Opener: © Corbis R-F Website; p. 309: © Tom Pantages; 9.10: © Susan Van Etten/PhotoEdit; 9.13: © PhotoDisc R-F Website; p. 316: © Corbis R-F Website; 9.14: © Takeshi Takahara/Photo Researchers, Inc.; 9.17: © David Wrobel/Visuals Unlimited; 9.27: © Courtesy NASA (adaptation); 9.28: © Larry Stepanowicz/Visuals Unlimited.

Chapter 10
Opener: © Ladybird Books; 10.1: © 1996 Amoz Eckerson/Visuals Unlimited; 10.5: Courtesy of Tim Zellner, Judd Wire Inc.; 10.6: © Natalie Forbes/CORBIS; p. 347: © John Springer/CORBIS; 10.9a: © Doug Pensinger/Getty Images; 10.9b: © SIU/Visuals Unlimited; 10.15: © Vol. 54/PhotoDisc; 10.28: © Bettmann/CORBIS; 10.29: © Ray Malace Photo; 10.30: © Wolfgang Kaehler/CORBIS.

Chapter 11
Opener: AP/Wide World Photo; p. 376: © David Young-Wolff/PhotoEdit; 11.6a: © Loren M. Winters/Visuals Unlimited; 11.6b: © Loren M. Winters/Visuals Unlimited; p. 387: U.S. Navy photo by Photographer's Mate 1st Class Spike Call.

Chapter 12
Opener: Bernard Benoit//Science Photo Library/Photo Researchers; p. 406: Photo by Jennifer Merlis; p. 411: © Steve Kaufman/CORBIS; p. 413: U.S. Navy photo by Photographer's Mate Airman Lawrence Braxton; 12.4: Image courtesy of Berghaus Organ Company; 12.6: © The Picture Source/Terry Oakley; p. 429 top: © Loren M. Winters/Visuals Unlimited; p. 429 bottom: U.S. Navy photo by Ensign John Gay; p. 430: © Merlin D. Tuttle, Bat Conservation International; 12.18: AP/Wide World Photos.

Chapter 13
Opener: © John Eastcott/YVA Momatiuk; 13.6: © John Sohlden/Visuals Unlimited; p. 454: © Vol. 112/Corbis; 13.15a: © Joe McDonald/CORBIS; 13.15b: © Craig Lovell/CORBIS; 13.15c: © Ken Kaminesky/CORBIS.

Chapter 14
Opener: AP/Wide World Photos; 14.6: Photo by author R. C. Richardson; 14.18: © Montrose/Custom Medical Stock.

Chapter 15
Opener: © David Davis/Index Stock Imagery/PictureQuest.

Chapter 16
Opener: Image courtesy Massashi Kawasaki, University of Virginia; 16.1: © Dean Conger/CORBIS; 16.3: © Tom Pantages; 16.4: © Tony Freeman/Photo Edit; 16.27: © John Tinning; Frank Lane Picture Agency/CORBIS.

1137

Chapter 17

Opener: © MEDNET/Phototake; 17.11: © Roger Ressmeyer/CORBIS; 17.13a: © Melanie Brown/Photo Edit; 17.14a: © Richard T. Nowitz/Photo Researchers; 17.20: © Loren M. Winters/Visuals Unlimited; 17.24: © Tom Pantages; 17.30: © Dan Robinson, WVlightning.com; 17.34: © Yoav Levy/Phototake.

Chapter 18

Opener: © Chuck Swartzell/Visuals Unlimited; 18.5: © Tom McHugh/Photo Researchers, Inc.; 18.7: © Richard Megna/Fundamental Photographs, NYC; 18.8a: © Loren Winters/Visuals Unlimited; 18.12: Photo courtesy of IBM Corporation; 18.13: © Tony Freeman/ Photo Edit, Inc.; 18.39: © Tom Pantages.

Chapter 19

Opener: Image courtesy Hojatollah Vali, McGill University; 19.1a: © Martyn F. Chillmaid/Photo Researchers, Inc.; 19.2: © Tom Pantages; 19.15a: © CERN Geneva; 19.18: IBA Proton Therapy Cyclotron; 19.30a: © Richard Megna/ Fundamental Photographs, NYC; 19.30c: © Loren Winters/Visuals Unlimited; 19.33a: © The Image Works Archives; 19.38: © Tom Pantages.

Chapter 20

Opener: © The Picture Source/Terry Oakley; p. 728: U.S. Army Corps of Engineers; 20.14: Dr. Scriba/SPL/Custom Medical Stock; 20.21: © Mark Antman/ The Image Works; 20.22: © Tom Pantages; 20.29: © The Image Works Archives.

Chapter 21

Opener: Alan Giambattista; 21.2: © Loren Winters/Visuals Unlimited; 21.14: © Loren Winters/Visuals Unlimited; p. 784: © The Image Works Archives.

Chapter 22

Opener: © George D. Lepp/CORBIS; 22.8a: © Dan McCoy/DoctorStock.com; 22.8b: © Richard Lowenberg/Science Source/Photo Researchers, Inc.; 22.9 a,b:

© Charles Mazel; 22.13 a,b: © Anglo-Australian Observatory. Photograph by David Malin; 22.15: © D. Parker/Photo Researchers, Inc.; 22.17a: © Inga Spence/Visuals Unlimited; 22.25 a,b: © Tom Pantages; 22.28 a,b: © Heather Angel; 22.29: Digital Image © 1996 CORBIS; Original image courtesy of NASA/CORBIS; 22.32a: Image courtesy Lewis Ling, Carleton University.

Chapter 23

Opener: © Bettmann/CORBIS; 23.1: © 1998 Jeff J. Daly/Fundamental Photographs, NYC; 23.3: © Anthony Cooper Ecoscene/CORBIS; 23.12 a,b: © The Picture Source/Terry Oakley; 23.13a: © Peter Turner/Getty Images; 23.14a: © Pekka Parviainen/Polar Image; 23.21b: © Yoav Levy/Phototake; 23.22b: © Collection CNRI/Phototake; 23.23: Bell, Alexander Graham. "Alexander Graham Bell's design sketch of the photophone, ca. 1880." Ca 1880. Alexander Graham Bell Family Papers, Manuscript Division, Library of Congress; 23.30: © Tom Pantages; 23.31b: © Tom Uhlmany/Visuals Unlimited; 23.37a: © Felicia Martinez/PhotoEdit, Inc.; p. 863: © RDF/Visuals Unlimited.

Chapter 24

Opener: NASA; 24.5c: © Bettmann/ CORBIS; 24.18: © Roger Ressmeyer/CORBIS; 24.20 (all): NASA, H. Ford (JHU), G. Illingworth (USCS/LO), M. Clampin (STScI), G. Hartig (STScI), the ACS Science Team, and ESA; 24.21: Courtesy of NAIC, Cornell University.

Chapter 25

Opener: © Kevin Schafer/Getty Images; 25.8a, 25.13 a,b: © The Picture Source/Terry Oakley; 25.14b: Tom Pantages; 25.15, 25.17a: © The Picture Source/Terry Oakley; 25.20: © 1986 Richard Megna—Fundamental Photographs, NYC; 25.23b: Alan Giambattista; 25.26: © Tom Pantages; 25.28 a-c: © The Picture Source/Terry Oakley; 25.30a: © 1987 Ken Kay/ Fundamental Photographs, NYC;

25.30b: © The Picture Source/Terry Oakley; 25.31a: © Tom Pantages; 25.35 a–c: © The Image Finders; 25.36a: © Tom Pantages; 25.39 a,b: © Frances G. Mayer/CORBIS; 25.40b: © Education Development Center, Inc.; 25.42: Science Source/Photo Researchers, Inc.; 25.45 (both): © Holographics North Inc.

Chapter 26

Opener: Map of the core and jet of the radio galaxy NGC6251 at 1.4 GHz made by P. N. Werner, M. Birkinshaw & D. M. Worrall using the NRAO Very Large Array.

Chapter 27

Opener: © Spencer Grant/Photo Edit, Inc.; p. 982: © Hulton-Deutsch Collection/CORBIS; 27.7: © Richard T. Nowitz/CORBIS; 27.11b: © W. Cody/ CORBIS; 27.12 (a–e): © 1994 Wabash Instrument Corp./Fundamental Photographs, NYC; 27.21 a,b: © Charles Mazel.

Chapter 28

Opener: Collection CNRI/Phototake; 28.4 a,b: Education Development Center, Inc.; 28.8 b,d: © Dennis Kunkel/ Phototake; 28.15: Courtesy: IBM Research, Almaden Research Center. Unauthorized use not permitted; 28.27a: © Fritz Hoffman/The Image Works; 28.27b: © Yoav Levy/Phototake; 28.29c: © Lawrence Berkeley Lab/Photo Researchers, Inc.

Chapter 29

Opener: © Geoffrey Clements/CORBIS; 29.10b: Vesper V. Grantham & Lindsay Huckabaa, University of Oklahoma Health Sciences Center; 29.11b: Image courtesy Elekta Instruments, Inc.; 29.12 a,b: Metropolitan Museum of Art; 29.17: REUTERS/Getty Images.

Chapter 30

Opener: © Michael S. Yamashita/ CORBIS; p. 1094: © Tomasz Barszczak/ Super-Kamiokande Collaboration/Photo Researchers, Inc.; 30.3: © CERN Geneva.

Text Credits

Some examples and problems adapted from Alan H. Cromer, *Physics for the Life Sciences*. Copyright © 1977 by the McGraw-Hill Companies, Inc., New York. All rights reserved. Reprinted by permission of Alan H. Cromer.

Some examples and problems adapted from Alan H. Cromer, Study Guide for *Physics for the Life Sciences*. Copyright © 1977 by

the McGraw-Hill Companies, Inc., New York. All rights reserved. Reprinted by permission of Alan H. Cromer.

Some examples and problems adapted from Alan H. Cromer, *Physics for the Life Sciences*, 2d. ed. Copyright © 1994 by the McGraw-Hill Companies, Inc. Primis Custom Publishing, Dubuque, Iowa. All rights reserved. Reprinted by permission of Alan H. Cromer.

Chapter 28: Quote by Richard Feynman from Richard Phillips Feynman, *The Character of Physical Law;* introduction by James Gleick. Copyright © 1994 by Modern Library, New York. All rights reserved. Reprinted by permission of MIT Press.

INDEX

UNIT CONVERSIONS

Length
1 in = 2.540 cm
1 cm = 0.3937 in
1 ft = 30.48 cm
1 m = 39.37 in = 3.281 ft
1 mi = 5280 ft = 1.609 km
1 km = 0.6214 mi
1 light-year (ly) = 9.461×10^{15} m

Time
1 y = 365.24 d = 3.156×10^7 s
1 d = 24 h = 1440 min = 8.64×10^4 s

Speed
1 mi/h = 1.467 ft/s
$\quad$ = 1.609 km/h = 0.4470 m/s
1 km/h = 0.2778 m/s
$\quad$ = 0.6214 mi/h = 0.9113 ft/s
1 ft/s = 0.3048 m/s = 0.6818 mi/h
1 m/s = 3.281 ft/s = 3.600 km/h

Volume
1 liter (L) = 1000 cm^3 = 10^{-3} m^3
1 cm^3 = 0.0610 in^3 = 1 mL
1 m^3 = 1×10^6 cm^3

Mass
1 kg = 1000 g
1 mass unit (u) = 1.6605×10^{-27} kg
$\quad$ [1 kg weighs 2.20 lb where g = 9.80 m/s^2]

Force
1 N = 0.2248 lb
1 lb = 4.448 N

Energy
1 J = 0.7376 ft·lb = 0.2389 cal
1 ft·lb = 1.36 J = 3.24×10^{-4} kcal
1 kcal = 4.186 kJ
1 Btu = 1055 J
1 kWh = 3.600 MJ = 860.0 kcal
1 eV = 1.602×10^{-19} J

Mass-Energy Equivalents
1 u = 931.5 MeV/c^2
$\quad$ = 1.492×10^{-10} J/c^2

Power
1 W = 1 J/s = 1.341×10^{-3} hp
1 hp = 550 ft·lb/s = 745.7 W

Pressure
1 Pa = 1 N/m^2 = 1.450×10^{-4} lb/in^2
1 atm = 0.1013 MPa = 14.7 lb/in^2
1 lb/in^2 = 6.90×10^3 N/m^2
1 mm Hg = 1.333×10^2 Pa
1 in Hg = 3.386×10^3 Pa

Angle
1 rad = 57.30°
1° = 0.01745 rad
360° = 2π rad
1 rad/s = 9.55 rpm
1 rpm = 0.1047 rad/s

SI PREFIXES

Power	Prefix	Symbol
10^{12}	tera	T
10^9	giga	G
10^6	mega	M
10^3	kilo	k
10^{-1}	deci	d
10^{-2}	centi	c
10^{-3}	milli	m
10^{-6}	micro	μ
10^{-9}	nano	n
10^{-12}	pico	p
10^{-15}	femto	f

SI DERIVED UNITS

Quantity	Units		Equivalents	
Force	newton	N	J/m	kg·m/s^2
Energy	joule	J	N·m	kg·m^2/s^2
Power	watt	W	J/s	kg·m^2/s^3
Pressure	pascal	Pa	N/m^2	kg/(m·s^2)
Frequency	hertz	Hz	cycle/s	s^{-1}
Electric charge	coulomb	C		A·s
Electric potential	volt	V	J/C	kg·m^2/(A·s^3)
Electric resistance	ohm	Ω	V/A	kg·m^2/(A^2·s^3)
Capacitance	farad	F	C/V	A^2·s^4/(kg·m^2)
Magnetic field	tesla	T	N·s/(C·m)	kg/(A·s^2)
Magnetic flux	weber	Wb	T·m^2	kg·m^2/(A·s^2)
Inductance	henry	H	V·s/A	kg·m^2/(A^2·s^2)